A TEXTBOOK OF

ELECTRONIC CIRCUIT

For the B.E. Students of Anna University, VTU, Mumbai University & useful for Other Indian Technical Universities & also for B.Sc. Courses of Indian Universities

Dr. R.S. SEDHA

Ph.D. (U.K.); M. Tech. (IIT Delhi); Life Fellow IETE (Delhi), MIEEE

School of Engineering

Republic Polytechnic

SINGAPORE

S. CHAND
PUBLISHING

S. Chand And Company Limited

(ISO 9001 Certi ied Company)

S. Chand And Company Limited
(ISO 9001 Certified Company)
Head Office: D-92, Sector–2, Noida – 201301, U.P. (India), Ph. 91-120-4682700
Registered Office: A-27, 2nd Floor, Mohan Co-operative Industrial Estate, New Delhi – 110 044, Phone: 011-49731800
www.**schandpublishing.com**; e-mail: **info@schandpublishing.com**

Marketing Offices:

Chennai	:	Ph: 23632120; chennai@schandpublishing.com
Guwahati	:	Ph: 2738811, 2735640; guwahati@schandpublishing.com
Hyderabad	:	Ph: 40186018; hyderabad@schandpublishing.com
Jalandhar	:	Ph: 4645630; jalandhar@schandpublishing.com
Kolkata	:	Ph: 23357458, 23353914; kolkata@schandpublishing.com
Lucknow	:	Ph: 4003633; lucknow@schandpublishing.com
Mumbai	:	Ph: 25000297; mumbai@schandpublishing.com
Patna	:	Ph: 2260011; patna@schandpublishing.com

First Edition 2007
Second Revised Edition 2009; Reprint 2010
Third Revised Edition 2012; Reprints 2014, 2016, 2018, 2020

Reprint 2023

ISBN: 978-81-219-2803-8 **Product Code:** H3ELC53ETRN10ENAC12R

PRINTED IN INDIA

By Vikas Publishing House Private Limited, Plot 20/4, Site-IV, Industrial Area Sahibabad, Ghaziabad – 201 010 and Published by S. Chand And Company Limited, A-27, 2nd Floor, Mohan Co-operative Industrial Estate, New Delhi – 110 044.

PREFACE TO THE THIRD REVISED EDITION

I take an opportunity to present this standard treatise entitled ***A Textbook of Electronic Circuits*** to especially the students of Anna University (Chennai), Bharathidasan University (Trichi), VTU, Mumbai University (Mumbai), as well as B.E./ B.Sc. of all the other Indian Universities in general.

The textbook covers the syllabus of Electronics Engineering subject. 5 New chapters have been added, as per the latest syllabus :

- Communication Systems
- Number Systems & Codes
- Logic gates
- Boolean Algeber & Logic Simplification
- Arithmetic Operations & Circuits

The objective of this book is to present the subject matter in a most concise, compact, to the point and lucid manner. While writing this book, I have constantly kept in mind the requirements of all the students, regarding the latest as well as changing trend of their examinations. To make it really useful for the students, latest examination questions of various Indian and Foreign universities as well as other important examining bodies have been included. The book has been written in an easy style with full details and illustration. All along the approach to the subject matter, every care has been taken to arrange the matter from simpler to harder, known to unknown, in a self-study style. In-short, it is expected that the book will meet the crying need of the students for whom it has been produced.

The Textbook has been divided into 36 Chapters. This material can accommodate a variety of scheduling and program requirements.

Dr. R.S. SEDHA

Singapore

e-mail: **ranjit_singh@rp.sg**

PREFACE TO THE FIRST EDITION

I take an opportunity to present this standard treatise entitled ***A Textbook of Electronic Circuits*** to especially the students of Anna University, Chennai, Bharathidasan University, Trichi, VTU, Mumbai University, Mumbai, as well as B.E./ B.Sc. of all the Other Indian Universities in general.

The Book covers topics from the following modules of Anna University's Electronics & Communication Engineering, Electrical & Electronics Engineering and Computer Science and Engineering degree programmes.

EC234-Electronic Circuits I

EC241-Electronic Circuits II

EC250 and EC254-Electronic Circuits

The objective of this book is to present the subject matter in a most concise, compact, to the point and lucid manner. While writing this book, I have constantly kept in mind the requirements of all the students, regarding the latest as well as changing trend of their examinations. To make it really useful for the students, latest examination questions of various Indian and foreign universities as well as other important examining bodies have been included. The book has been written in an easy style with full details and illustration. All along the approach to the subject matter, every care has been taken to arrange the matter from simpler to harder, known to unknown, in a self-study style. In-short, it is expected that the book will meet the crying need of the students for whom it has been produced.

The textbook has been divided into 34 Chapters. This material can accommodate a variety of scheduling and program requirements.

Dr. R.S. SEDHA

Singapore

e-mail: **ranjit_singh@rp.sg**

TO THE STUDENT

Anything worthwhile requires hard work, and electronic is no exception. The best way to learn the new material is by reading, thinking and doing. This text is designed to help you along the way by providing an outline and objectives for each chapter, numerous worked-out examples and tutorial questions.

Don't expect everything to be crystal clear after a single reading. Read each article of the text carefully and think about what you have read. Work through the example problems step by step before trying the tutorial problems. Sometimes more than one reading of the article will be necessary. Review the chapter summary. Take the multiple choice questions and work the problems at the end of the chapter.

Although every care has been taken to check mistakes and misprints, yet it is very difficult to claim perfection. Any errors, omissions and suggestions for the improvement of this book, brought to my notice will be thankfully acknowledged and incorporated in the next edition.

Dr. R.S. SEDHA

Singapore

e-mail: **ranjit_singh@rp.sg**

SYLLABUS

ANNA UNIVERSITY

EC 9151 **ELECTRON DEVICES** **3003**

Unit I **Semiconductor Diode** **9**
PN junction, current equations, Diffusion and drift current densities, V–I characteristics, Forward and Reverse characteristics, Switching Times.

Unit II **Bipolar junction Transistor** **9**
NPN-PNP-Junctions-Early effect-Current equations-Input and Output characteristics of CE, CB CC-Hybrid pi-model h-parameter model—Ebers Moll Model-Power BJT Gummel poon-model.

Unit III **Field Effect Transistors**
JFETs–Drain and Transfer characteristics, current equations-pinch off voltage and its significance, MOSFET-characteristic DMOSFET, EMOSFET, current equation-model-parameters, threshold voltage modifications by ion implantation-channel length modulation-power MOSFET.

Unit IV **Special Semicoductor devices** **9**
Metal-Semiconductor, Junction, Schottky barrier diode, Zener diode, Varacter diode, Tunnel diode, Gallium Arsenic device, LASER diode, LDR, and MOSFETs

Unit V **Power Devices and Display Devices** **9**
UJT, SCR, Diac, Triac, DMOS, VMOS, FINFET, DUALGATE, MOSFET, LED, LCD, Photo transistor, Opto Coupler, Solar cell, CCD, MULTI EMITTER Transistor.

EC 9202 **ELECTRONIC CIRCUITS – I** **3003**

Unit I **Biasing of Discrete BJT and MOSFET** **9**
DC Load line, operating point, Various biasing methods for BJT-Design-Stability-Bias Compensation, Thermal stability, Design of biasing for MOSFET and JEET.

Unit II **BJT Amplifiers** **9**
Small signal Analysis of Common Emitter-AC Loadline, Voltage swing limitations, Common Collector and Common Base amplifiers – JFET Amplifiers – Differential Amplifiers – CMRR, Darlington Amplifier–Bootstrap technique - Cascaded stages - Cascode Amplifier

Unit III **MOSFET Amplifiers** **9**
Small Signal Analysis of Common source, Source follower and Common Gate Amplifiers-CMOS Inverters–DC Analysis of CMOS Inverters – Voltage Transfer Curve – BiMOS Cascade - Design of NMOS inverter using resistive load – Noise Margin – VTC.

Unit IV **IC MOSFET Amplifers** **9**
Single Stage IC MOS Amplifiers – Active Loads – Depletion MOS, Enhancement, MOS, MOS in Triode region, NMOS Current Source and PMOS Current Source, their equivalent circuits and load line on the VI characteristics– Current steering circuit using MOSFET – CMOS common source amplifier and CMOS Common source follower – CMOS differential amplifier – CMRR.

Unit V **High Frequency Analysis and Large Signal Amplifiers** **9**
Short circuit current gain, cut off frequency – f_α and f_β unity gain and Bandwidth – Miller effect-frequency, Analysis of CS and CE Amplifiers-Determinations of BW of single stage and Multistage Amplifier–Analysis of class *A*, class *B*, *AB* with darlington output stage and with output stage utilizing MOSFETs – Class *C*, Class *D*, Class *E* power amplifiers.

EC 9252 ELECTRONIC CIRCUITS - II 3104

Unit I Feedback Amplifiers and Stability **9**

Basic Feedback Concepts–Properties of Negative Feedback – Four feedback topologies with amplifier circuit. Examples–Analysis of series–shunt feedback amplifiers–stability problem–Frequency compensation.

Unit II Oscillators **9**

Barkhausen criteria for oscillator–Analysis of RC oscillators–phase shift Wein bridge oscillators–LC oscillators–Colpitt, Hartley, Clapp, Crystal, Armstrong, Franklin and Ring Oscillators

Unit III Tuned Amplifiers **9**

Basic principles— inductor losses – use of tranformers – single tuned amplifier frequency analysis - Amplifier with multiple tuned circuits – Cascade – Synchronous tuning – Stagger tuning – Stability of tuned amplifiers using Neutralization techniques.

Unit IV Multivibrators and Time Base generators **9**

Switching characteristics of transistors – Bistable, Monostable and Astable operation – Collector coupled and Emitter coupled circuits – Schmitt trigger – Voltage sweep generators – Current sweep generators.

Unit V Rectifiers and Power Supplies **9**

Halfwave and fullwave rectifiers with filters – Ripple factor – Series Voltage Regulator analysis and design – lGBT – working and characteristics – AC voltage control using thyristors – SMPS – DC/DC convertors – Buck, Boost, Buck-Boost analysis and design.

Total 45 = Periods

EC 9211 ELECTRONIC DEVICES AND CIRCUITS 3 LTPC 3104

AIM To Study the characteristics and applications of electronic devices.

OBJECTIVES

To acquaint the students with construction, theory and characteristics of the following electronic devices :

P-N junction diode, Bipolar transistor, Field Effect transistor, LED, LCD and other photo electronic devices, Power control/regulator devices, Feedback amplifiers and oscillators.

Unit I PN Junction Devices **9+3**

PN junction diode – structure, operation and V–I characteristic - Current equation of drift current density and diffusion current density-diffusion and transient capacitance – display devices – LED, LASER diodes, Zener, breakdown, Zener reverse characteristic, Zener as regulator.

Unit II Bipolar Junction Transistors **9+3**

Structure, operation and V-I characteristic- MOSFET– structure, operation and VI characteristics – types of MOSFET – JFET –structure, operation and V-I characteristics

Unit III Amplifiers **9+3**

BJT small signal model –biasing –analysis of CE, CB, CC amplifiers- Gain and frequency response–MOSFET small signal model – Biasing – analysis of CS and Source follower – gain and Frequency response.

Unit IV Multistage Amplifiers and Differential Amplifier **9+3**

BIMOS cascade amplifier, differential amplifier – Common mode and Difference mode analysis – FET input stages – Tuned amplifiers – single tuned amplifiers – Gain and Frequency response – Neutralization methods.

Unit V Feedback Amplifiers and Oscillators **9+3**

Advantages of negative feedback – voltage / current, series, shunt feedback – positive feedback – condition for oscillations, phase shift – Wien bridge, Hartley, colpitts and Crystal oscillators.

Lecture-45 + Tutorils-15 = Total : 45 Periods

SYLLABUS

VTU (Karnataka)

FIRST SEMESTER B.E. (Common to All Branches)

BASIC ELECTRONICS

Sub Code : 10 ELN 15/10ELN 25

PART–A

Unit – I Semiconductor Diodes and Applications :

p–n junction diode, Characteristics and Parameters, Diode approximations, DC load line, Temperature dependence of p-n characteristics, AC equivalent circuits, Zener Other full-wave circuits, Shunt capacitor – Approximate analysis of capacitor filters, Power supply performance, zener diode voltage regulators, Numerical examples as applicable.

Unit II Transistors:

Bipolar Junction transistor, Transistor Voltages and currents, amplification, Common Base, Common Emitter and Common Collector Characteristics, DC Load line and Bias Point.

Unit III Biasing Methods :

Base Bias, Collector to Base Bias, Voltage divider Bias, Comparison of basic bias circuits, Bias circuit design, Thermal Stability of Bias Circuits (Qualitative discussions only).

Unit IV Other Devices:

Silicon Controlled Rectifier (SCR), SCR Control Circuits, More SCR applications; Unijunction transistor, UJT applications, Junction Field Effect Transistors (Exclude Fabrication and Packaging), JFET Characteristics, FET Amplifications, Numerical examples as applicable.

PART–B

Unit V Amplifiers and Oscillators :

Decibels and Half power points, Single Stage CE Amplifier and Capacitor coupled two stage CE Amplifier and Capacitor coupled two stage CE amplifier (Qualitative discussious only), The Barkhausen Criterion for Oscillations, BJT RC phase shift oscillator, Hartley, Colpitts and crystal oscillator (Qualitative discussions only) Numerical Problems as applicable.

Unit VI Introduction to operational Amplifiers :

Ideal OPAMP, Saturable property of an OP AMP inverting and non inverting OPAMP

ANALOG ELECTRONIC CIRCUITS

(Common to EC/TC/EE/IT/BM/ML)

Sub Code : 10ES32

PART–A

Unit I Diode Circuits :

Diode Resistance, Diode equivalent circuits, Transition and Diffusion capacitance, Reverse recovery time, Load line analysis, Rectifiers, Clippers and Clampers.

Unit II Transistor Biasing:

Operating point, Fixed bias circuits, Emitter stabilized biased circuits, Voltage divider biased, DC bias with voltage feedback, Miscellaneous bias configurations, Design operations, Transistor switching networks, PNP transistors, Bias stabilization.

Unit II Transistor Biasing:

Operating point, Fixed bias circuits, Emitter stabilized biased circuits, Voltage divider biased, DC bias with voltage Feedback, Miscellaneous bias configurations, Design operations, Transistor switching networks, PNP transistors, Bias stabilization.

Unit III Transistor at Low Frequencies:

BJT transistor modeling, Hybrid equivalent model, CE Fixed bias configuration, Voltage divider bias, Emitter follower, CB configuration, Collector feedback configuration, Hybrid equivalent model. Design of simple amplifers.

Unit IV Transistor Frequency Response:

General frequency considerations, low frequency response, Miller effect capacitance, High frequency response, multistage frequency effects.

PART–B

Unit V

(*a*) **General Amplifiers:** Cascaded connections, Cascode connections, Darlington connections.

(*b*) **Feedback Amplifier:** Feedback concept, Feedback connections type, Practical feedback circuits, Design procedures for the feedback amplifiers.

Unit VI Power Amplifiers:

Definitions and amplifier, Transformer coupled Class A amplifiers, Class B amplifier operations, Class B amplifier circuits, Amplifier distortions. Designing of Power amplifiers.

Unit VII Oscillators:

Oscillator operation, Phase shift Oscillator, Wienbridge Oscillator, Tuned Oscillator circuits, Crystal Oscillator. (BJT Version Only) Simples design methods of Oscillators.

Unit VIII FET Amplifiers:

FET small signal model, Biasing of FET, Common drain and common gate configurations, MOSFETs, FET amplifier networks.

CONTENTS

1

CHAPTER

Introduction

OUTLINE

Objectives

After completing this chapter, you should be able to:

- Known the definition of electronics,
- Descirbe applications of electronics in the field of communication, defence, industries and medicines.
- Identify different electronic design automation (EDA) tools for designing and simulating electronic circuits.

1-1. Definition

Let us first begin our study about what is engineering is the science of transforming knowledge in physics, chemistry, and mathamatics into products and system that meet the needs and wants of everyday life.

A physicist may have a knowledge of the contribution cycle, but it is the engineer who builds the car. A chemist may have discovered the properties of silicon, but it is the engineer who produces the micro-processor chip. A Mathematical may have invented Fast Fourier Transform. but again it is the engineer who designed the DVD player.

Electronics is the field of science and technology that deals with the behaviour of electric currents in electronic devices, components, system or equipment. The electronic device, component, system or equipment could be carry thing raging from lasers, photo cells, transistor, micro-processors, microcontrollers, robots, CD-player, DVD, mobile telephones, fax machines, the Internet, computers, high definition plasma television, disk drives digital camera, camcorder, MP3 player, mobile phones, *etc*.

These days, the field of electronics has become the most important branch of engineering in our society. Several thousand developments have taken place in the last country especially in the past five decodes. You name of the item and it has an electronic circuit inside. The electronic devices and gadgets are being used in almost all industries for manufacturing inspection, quality check, control and automation of industrial process, fixed mobile communication, computer networks, internet and so on. Because of the tremendous growth of electronics and ever growing applications of electronics in almost all fields, the students of all engineering disciplines (computer engineering, information technology, electrical engineering, mechanical production engineering, machatronics) have to be taught electronics at undergraduate level.

1–2. Beginning and Development of Electronics

The era of electronics is believed to have dawned with the experiments involving electric currents in glass vacuum tubes. Such experiments were performed around 1850 by a German scientist named Geissler. He found that if the air is removed from a glass tube, it glows when an electric current passes through it. Around 1878, British scientist, Sir William Crookes found that the current in vacuum tubes seemed to consist of particles.

As a matter of fact, the electron was discovered in the 1890's. A French physicist, Perrin demonstrated that current in a vacuum tube consists of the movement of negatively charged particles in a given direction. Some of the properties of these particles were measured by a British physicist Thomson. These negatively charged particles were later on, named as electrons. An American physicist, Milikan, in 1909 measured the charge on an electron. As a result of these discoveries, the movement of electrons could be controlled, and thus the electron era started.

Fleming, in 1904, invented a vacuum tube that allowed electrical current only in one direction. This tube was named as Fleming valve or diode vacuum tube and used to detect electromagnetic waves. In 1907, Lee deForest, an American scientist invented a tube, which could amplify weak electrical a.c. signals. This tube was named as triode vacuum tube.

Around 1915, few great improvements were made in the triode vacuum tube. It was used in oscillator circuit. It was also used in telephone system, which made possible the communication between the different continents. The tetrode was invented in 1916 by a German engineer and a pentode was invented in 1926 by a Dutch engineer. The tetrode, alongwith the pentode, greatly improved the triode. The first television picture tube, called the kinescope, was developed in 1920's by an American researcher.

Several types of microwave tubes were developed during the World War II. Such tubes made the modern RADAR and other communication systems possible. The magnetron was invented in Britain in 1939. In the same year, klystron microwave tube was developed by two Americans. The travelling wave tube (TWT) was invented in 1943 by an American scientist.

The era of solid state electronics began with the invention of transistor in 1947 at Bell laboratory. The inventors were Walter Britain, John Bardeen and William Shockley. The transistors were produced commercially in 1951. The companies like RCA, Raythesn, General Electric, Westinghouse and Western Electric were the first to fabricate transistors.

In 1958, it was thought that germanium and silicon can be used to build an entire circuit, called monolithic circuit. Resistors were formed with the bulk semiconductor or by diffusing one semiconductor into another. The capacitors were formed by using a metallic layer and the semiconductor for the plates and an oxide layer for the dielectric. In 1959, Noyce gave an idea for making multiple devices on a single piece of silicon, in order to make an interconnection between devices as part of manufacturing process. It could reduce the size, weight and cost per device. The monolithic-circuit was named an integrated circuit (IC). The IC's were produced, commercially in 1961 by Fairchild and Texas instruments.

1-3. Modern Trends in Eletronics

The development of an integrated circuit (IC) in 1961 was a great achievement in the field of electronics. But, as there is no end to the man's desire, the growth accelerated further every year. Rapid developments have been made in the integrated circuit technology. Thus starting from the *small-scale integration (SSI) in 1961, then **medium-scale integration (MSI) in 1966, and †large-

* It indicates that the number of components fabricated on a single-chip was less than 100.

** It indicates that the number of components fabricated on a single-chip was 100 to 1000.

† It indicates that number of components fabricated on a single-chip is between 1000 to 10,000.

scale integration (LSI) in 1970. In 2004 we have a very large-scale integration (VLSI) Ultra large-scale integration (ULSI and Giga-scale integration (GSI). It means that now we have a technology, which can fabricate several million components on a single chip.

It will be interesting to know that the use of vacuum tube has become almost obsolete in 1960's. It was replaced by the solid state devices (bipolar transistors and field-effect transistors). Similarly, the developments in integrated circuit technology, has made the use of individual transistors as unnecessary. These days, we see all electronic systems made of IC's. The developments which, have taken place in last 60 years, may be understood from the fact that today's microcomputer is several thousand times faster, has a huge memory, million times more reliable, consumes negligible power, occupies 1/100,000 the volume and costs 1/50,000 as compared to the first large electronic computer.

The microelectronics (term used for the fabrication of VLSI/ULSI/GSI-circuits) has made a great impact on communication control and computer industries. The applications, which were difficult to realize earlier, have now become possible. In last decade, silicon based technology has dominated the sphere of electronics and will continue to play the role for the next decade as well. However, after that gallium arsenide (Gas) might play a significant role in electronics.

1-4. Applications of Electronics

These days, we find that the sphere of electronics has spread so vast, that it plays an important role in almost every activity of life. It has penetrated into our homes, our places of work and our means of communication from one place to another. It is believed that by the end of 2025, the electronics will be like the electric motor of today (*i.e.*, its presence will go unnoticed). Although, electronics has a large number of applications, yet the following are important from the subject point of view:

(*i*) Communication and entertainment applications.

(*ii*) Instrumentation and control applications.

(*iii*) Defence applications.

(*iv*) Applications in medicine.

(*v*) Computer-aided design of electronic circuits.

Now we shall discuss all these applications, in brief, in the following pages.

1-5. Communication and Entertainment Applications

In the beginning of 20th century, the elctronics was used only in the field of telephony and telegraphy. It required a pair of conducting wires. But these days we have cordless telephones, mobile phones which send or receive information without the use of any wire. Such a communication is known as wireless communication. It is possible to have live video conference from anywhere on the earth or in fact millions of kilometers in space. Several space agencies in the world like NASA and ISRO have been using wireless communication to control satallites position controlling digital cameras to take picture and send them back to the earth.

These days, the radio and television provides a means of communication, education and entertainment. The electronics gadgets like HiFi stereos, CD-player, DVD-player, MP3-player, Video games are widely used for entertainment. The use of ICs (integrated circuits) have reduced the size and cost, improved the reliability and performance of the electronic equipment.

We have the digital cameras and digital camcorders where the picture can be stored on a memory stick (flash memory). The picucture can be downloaded to a computer to keep a record in albums for sharing with the relatives and friends through the Internet. The MP3-player can be used to download the music from the Internet and listen at your convenience.

1-6. Instrumentation and Control Applications

The instrumentation plays an important role in our industry. It helps in precise measurement and testing of a number of quantities of interest. The electronics is also playing an increasingly important role in instrumentation. The cathode ray oscilloscopes (CRO's) frequency counters, signal generators, spectrum analysers, logic analizers, automatic test equipment, vision systems, security surveilance systems digital multimeters (DMM's) power supplies, *etc.* are some of the instruments, which are found in every industry and research organisations.

The control industry has improved drastically by the introduction of semiconductor electronics. Devices (discrete as well as integrated circuits, in some traditional areas, such as motor speed control and power rectifiers and inverters, there has been tremendous advancement. Instead of gas filled tubes (thyratrons) we have silicon controlled rectifiers (SCRs), power semiconductor diodes, power bipolar transistors, power MOSFETs, SITs and IGBT'S.

Most of the control systems, which are available today, are possible only because of the advances in electronics. Besides analog control systems, we have digital control systems, analog neural network control systems and fuzzy logic control systems. All these systems are in use in digital camcorders, industrial and game robots, factory automation and many more.

The introduction of microprocessors, microcomputers, and digital integrated circuits and devices, has led to "smart" instruments and continuously increasing variety of digital control systems. The computers have become integral component of digital control systems.

1-7. Defence Applications

The role of electronics in defence started with the introduction of RADAR during the World War II. The RADAR is an abbreviation used for Radio Detection and Ranging. A radar is capable of detecting and finding exact location of enemy aircraft. These days, we have aeroplanes, ships, anti-aircraft guns and guided missiles completely controlled by electronics.

1-8. Applications in Medicines

These days, big hospitals and drug research organisations are utilising an increasing role of electronic instrumentation. In many developed countries, the computers are used to hold patients records and assist in arriving at correct diagnosis. A large number of electronic instruments are also used in the diagnosis and treatment of various diseases. Some of the important ones are given below:

(*i*) *Electrocardiograph* (*ECG*). It is used to find the condition of the heart of a patient.

(*ii*) *X-ray equipment*. It is used for taking pictures of internal bone structures. It is also used for the treatment of some diseases.

(*iii*) *Ultrasound scanner*. It is used to take pictures and examine the functions of a brain, kidney etc. It is also used for the treatment of some diseases.

(*iv*) *Electrical enceplograph* (*EEG*). It is used for neurological investigations.

(*v*) *Cathode ray oscilloscope*. It is used for studying muscle actions within the body.

(*vi*) *Magnetic Resonance Imaging* (*MRI*). It is used for investigating tumours or other defects within the body.

1-9. Computer-Aided Design of Electronic Circuits

The circuit designers, of previous generations, used to build a circuit on printed boards, and then study its performance. These days, it is possible to use a computer not only as a means of simulating a particular system design, but also to model the effects of statistical variations in the manufacturing process. The computer is also used to optimise parameters in a given system design. It is also used to meet some set of design objectives better and to explore various alternative designs. It is generally done to determine the relative trade-off between the various competing objectives. These days, the VLSI (very large-scale integrated) circuits are designed by means of a computer. Even for a product design, the engineer can design the product in the computer, carryout the studies on various parameters by simulation on the computer. Rapid prototyping is also done by using computers.

1-10. Careers in the field of Electronics

The field of electronics is very diverse and career opportunities are available in many areas. Since electronics is currently found in so many different applications ranging from toys to spacecrafts and new technology is being developed at a very fast rate, its future appears limitless. There is hardly any area of our lives that is not enhanced to some degree by electronics technology. Those who acquire a sound, basic knowledge of electrical and electronic principles and are willing to continue learning will always be in demand.

The importance of obtaining a through understanding of the basic principles contained in this text cannot be overemphasized. Most employers prefer to hire people who have both a through grounding in the basics and the ability and eagerness to grasp new concepts and techniques. If you have a good training in basics, an employer will train you in the specifics of the job to which you are assigned.

There are many types of job classifications for which a person with training in electronics technology may qualify. A few of the most common job functions are discussed briefly in the following paragraphs.

Scientists and Engineers- Personnel in this category are involved in designing, prototyping and testing electronic systems in research and development laboratories during the development phase of the product.

Industrial Manufacturing Engineers- Manufacturing personnel are involved in testing of electronic products at the assembly-line level or in the maintenance and troubleshooting of electronic and electromechanical system used in testing and manufacturing of products. Virtually every type of manufacturing plant, regardless of its product, uses automated equipment that is electronically controlled.

Technical Sales and Marketing- Technically trained people are in demand as sales and marketing representatives for high technology products. The ability both to understand technical concepts and to communicate the technical aspects of the product to a potential customer is very valuable. In this area, as in technical writing, competency in expressing yourself orally, and in writing is essential. Actually, being able to communicate well is very important in any technical job category because you must be able to record data clearly and explain procedures, conclusions, and actions taken so that others can readily understand what you are doing.

Technical writers- These personal compile technical information and then use the information to write and produce manuals and audiovisual materials. A broad knowledge of a particular system and the ability to clearly explain its principles and operation are essential.

Besides this, the diploma engineering people can find job as a customer support officers, laboratory technician, field service technician, service-shop technician, engineering assistant and many more.

SUMMARY

In this chapter you have learned that:

1. Electronics is that field of science and engineering that deals with the behaviour of electric currents in electronic devices, components, systems or equipment.
2. Electronics has numerous applications in the area of mobile and fixed communication, entertainment, automation and control, defence, surgery, medicine, computer-aided design.

GLOSSARY

Engineering is the science of transforming knowledge in physics, chemistry and mathematics into products and systems that meet to needs and wants of everyday life.

Electronics is field of science and technology that deals with the behaviour of electric currents in electronic devices, compounds, system or equipment.

Electronic communication is a field of electronics that deals with sending/receiving information using electronics devices computers, system or equipment.

Instrumentation is a field of electronics that deals with precise measurement and testing of current, voltage, power, *etc*.

❑ ❑ ❑

2

CHAPTER

Structure of Solids

OUTLINE

Objectives

After completing this chapter, you should be able to :

- Know the different types of crystalline structure.
- Define, atomic number and atomic mass.
- Understand the distribution of electrons in different orbits or shells.
- Know the concept of bonding in solids.
- Identify three different types of bonds- ionic, covalent and metallic.
- Describe energy band diagrams of conductors, insulators and semiconductors.

2-1. Introduction

It has been established since long that the operation of electronic devices depends upon the *motion of charged particles* within them. Therefore, we must have an understanding of the forces, which control the motion of these particles. But as a matter of fact, it is the physical structure of a solid, which exerts this control. Therefore, we shall study the structure and arrangement of atoms in solids. Though this topic is very large, yet we shall only study the structure of solids with a brief introduction to crystal structure.

2-2. Structure of Solids

A detailed study of X-ray scattering and many others have revealed that most of the solids are crystalline in structure. It means that a solid consists of *atoms* or *molecules* (technically speaking, ions) which are arranged in a periodic fashion.

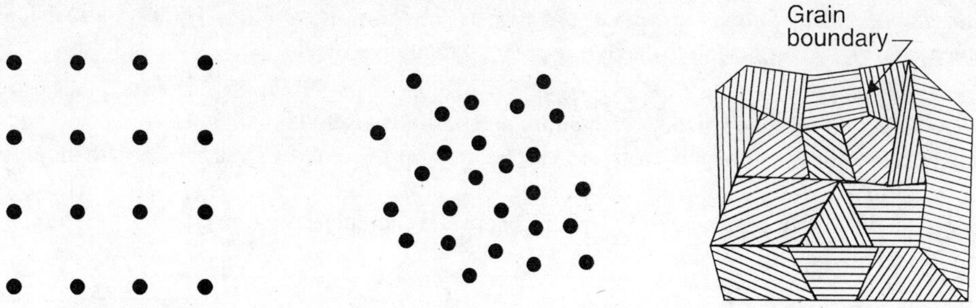

(*a*) Crystalline (*b*) Amorphous (*c*) Polycrystalline

Fig. 2-1. *Structure of solids.*

There is always some basic *arrangement* of atoms, which is repeated throughout the entire solid material. Such an arrangement of atoms, within a solid, is called *crystal lattice*. Such solids are called *crystalline solids*. However, there are some other solid materials, which do not have crystalline structure. Such solid materials are called *non-crystalline* or *amorphous solids*. All metals and elements like silicon and germanium are crystalline materials while wood, plastic, paper, glass *etc*. are amorphous materials.

A crystalline material may be in the form of single crystal or an aggregate of many crystals known as *polycrystalline*. A polycrystalline material is always separated by well defined boundaries called grain boundaries. Fig. 2-1 (*a*), (*b*) and (*c*) distinguishes among crystalline, amorphous and polycrystalline structures.

In a crystalline structure, since there are many different ways of placing atoms in a volume, the distances and orientation between atoms can take many forms. However, in every case, the *lattice* contains a volume called a *unit cell*. The unit cell is a representative of the entire lattice, and is regularly repeated throughout the crystal. Fig. 2-2 shows a two-dimensional arrangement of atoms within a unit cell *OABC*. This cell has an atom at each corner, which is shared with the adjacent cells. The importance of the unit cell is that we can analyse the crystal by determining its volume. For example, from the unit cell, we can find out the distance between one atom and the next nearest atom. This can be used to calculate the forces holding the crystal lattice together. However, the most important point is that the properties of crystal lattice determine the allowed energies of the electrons, which participate in the conduction process.

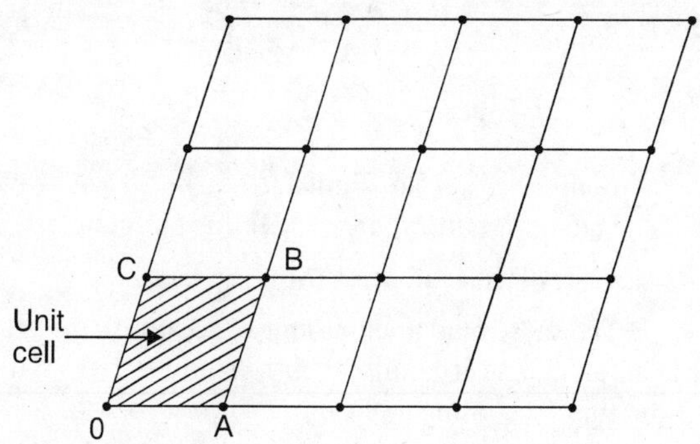

Fig. 2-2. *Two dimensional crystal lattice showing a unit cell.*

2-3. Types of Crystalline Structures

Although there are many types of crystalline structures, yet the following four are important from the subject point of view :

1. *Simple cubic* (*SC*). The unit cell of this structure contains *atoms at all the corners of a cube* as shown in Fig. 2-3 (*a*). The compound cesium chloride (CsCl) possess this type of structure.

2. *Body centred cubic* (*BCC*). The unit cell of this structure is a *simple cube, with an additional atom at the body centre of the cube* as shown in Fig. 2-3 (*b*). The crystals such as tungsten, vanadium, chromium and alkali metals (*i.e*., sodium, potassium *etc*.) have a body centred cubic structure.

3. *Face centred cubic* (*FCC*). The unit cell of this structure is a *simple cube with an additional atom in the centre of each six faces of the cube* as shown in Fig. 2-3 (*c*). The crystals such as gold, silver, copper, nickel, lead, *etc.* have a face centred cubic structure.

4. *Diamond or tetrahedral.* This is one of the most important crystal structure, because the *semiconductors* such as silicon, germanium and gallium arsenide (GaAs) have this type of crystal structure. The unit cell of this structure can be thought of as an FCC structure with an extra atom located at $a/4$, + $b/4$, + $c/4$ from each of the FCC atoms as shown in Fig. 2-1 (*d*). The *a*, *b* and *c* represents he lattice constants (*i.e.*, spacing between the adjacent atoms, along the *X*-axis, *Y*-axis and *Z*-axis respectively.

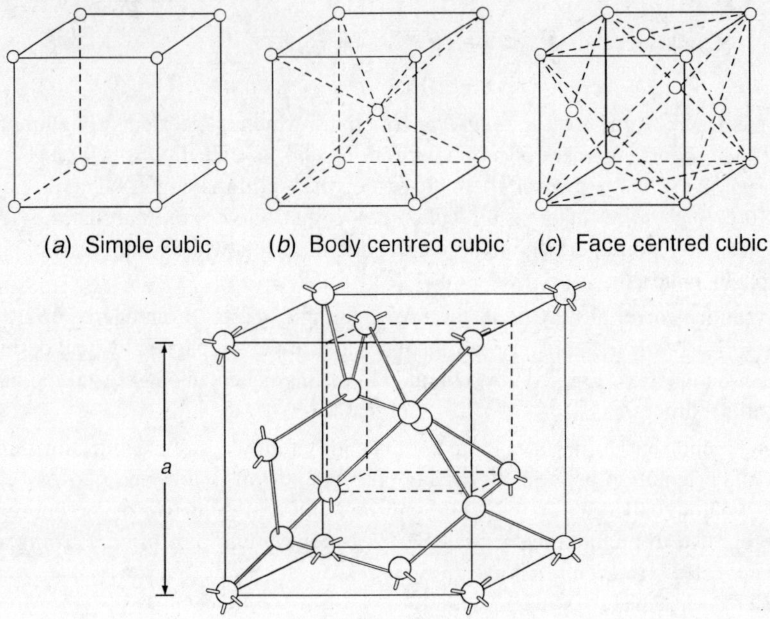

(*a*) Simple cubic (*b*) Body centred cubic (*c*) Face centred cubic

(*d*) Diamond

Fig. 2-3. *Crystalline Structures*

If all the atoms are similar, (then $a = b = c$), we call this structure as diamond lattice. However, if the atoms are different, we call this structure as a Zinc blends lattice.

2-4. Structure of an Atom

The most fundamental unit of a solid is an atom, and it is important to study its structure. According to the present concept, the structure of an atom is mainly composed of three fundamental particles, *i.e.*, electrons, protons and neutrons. An electron has a negative charge equal to 1.6×10^{-19} coulomb. A neutron has no charge at all. The protons and neutrons are located in a small nucleus, at the centre of an atom. The electrons revolve around the nucleus in fixed circular as well as elliptical orbits called energy levels or main shells as shown in Fig. 2-4.

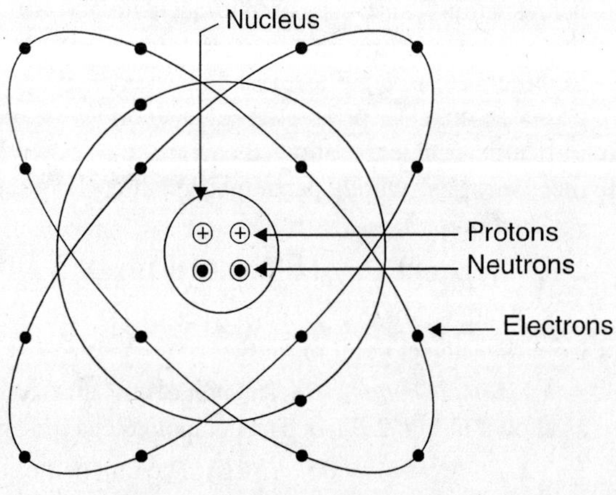

Fig. 2-4. *Structure of an atom.*

The electrons moving in the various energy levels, occupy a vast empty space around the nucleus. The aggregate of all the electrons forms an electron cloud around the nucleus, by the electrostatic force of attraction between the nucleus and the electrons. In a normal atom, the number of protons is equal to the number of electrons. Thus the entire atom is electrically neutral. The mass of an electron is 9.1×10^{-31} kg. However, the mass of a proton or a neutron is 1840 times that of an electron. Thus, the mass of an atom is mainly due to protons and neutrons only.

2-5. Atomic Number

The atomic number (Z) of an atom is equal to the number of protons (or electrons) contained in an atom. For example, hydrogen (H) atom has one proton in its nucleus (or one electron around its nucleus). Therefore atomic number of hydrogen atom is 1. The atomic numbers of some important atoms are: $Z = 5$ for boron, $Z = 6$ for carbon, $Z = 13$ for aluminum, $Z = 14$ for silicon, $Z = 15$ for phosphorus and $Z = 32$ for germanium.

2-6. Atomic Mass

The atomic mass (also called atomic weight) of an atom is equal to the total number of protons and neutrons contained in its nucleus. For example, a silicon atom has 14 protons and 14 neutrons in its nucleus. Therefore, its atomic mass is 28. Similarly, a germanium atom has 32 protons and 41 neutrons. Therefore, its atomic mass is 73.

2-7. Electron Orbits

We have already discussed in Article 2-4 that an electron revolves around the nucleus in fixed orbits, called energy levels or shells. These orbits are represented by the letters *K, L, M, N, etc.* counted from the nucleus to outwards. Thus the *K*-shell is closest to the nucleus. The next is *L*-shell and so on. Sometimes *K, L, M, N, etc.* are also designated as 1, 2, 3, 4, *etc.*

The maximum number of electrons, in any shell, is given by the relation $2n^2$ where n is an integer and is called **principal quantum number**. Thus, for *K*-shell, $n = 1$. Therefore, it can have a maximum number of $2(1)^2 = 2$ electrons. Similarly, *L*-shell, with $n = 2$ can have $2\,(2)^2 = 8$ electrons. The *M*-shell with the $n = 3$ can accommodate a maximum of $2(3)^2 = 18$ electrons. The *N*-shell with $n = 4$ can have a maximum of $2(4)^2 = 32$ electrons. However, the distribution of electrons in an atom is subject to the following two rules :

(1) The outermost shell of an atom cannot accommodate more than 8 electrons, even if it has a capacity to accommodate more electrons.

(2) The shell of an atom just prior to the outermost shell, cannot accommodate more than 8 electrons.

The above mentioned points help us in understanding the distribution of electrons in the atom as discussed below:

2-8. Distribution of Electrons in Atoms

Fig. 2-5 shows the distribution of electrons in different orbits of boron (B) silicon (Si) and germanium (Ge) atoms. The distribution of electrons in these atoms is explained as below.

We know that the atomic number of boron is 5. It means that it has 5 protons and 5 electrons. Two electrons occupy the *K*-shell, which is then said to be completely filled. The other 3 electrons occupy the *L*-shell as shown in Fig. 2-5 (*a*). Similarly, the atomic number of silicon (Si) is 14. It has 14 protons and 14 electrons. The 14 electrons are distributed as 2 in the *K*-shell, 8 in *L*-shell and 4 in *M*-shell as shown in Fig. 2-5 (*b*). The atomic number of germanium (Ge) is 32. It has 32 protons and 32 electrons. These electrons are distributed as 2 in the *K*-shell, 8 in *L*-shell, 18 in *M*-shell and 4 in *N*-shell as shown Fig. 2-5 (*c*).

Example 2-1. *Phosphorus atom has Z = 15. Give its electronic distribution, indicating which orbits are full and which are not.*

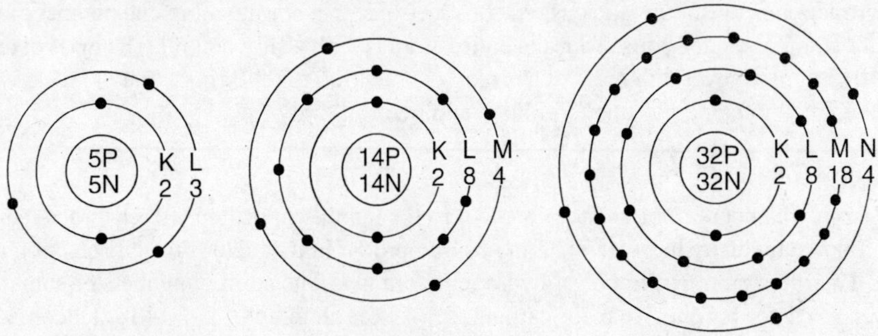

(*a*) Boron (B) (*b*) Silicon (Si) (*c*) Germanium (Ge)

Fig. 2-5. *Distribution of electrons in different atoms.*

Solution. The phosphorus atom has 15 electrons. These are distributed in the various shells as given below:

K-shell	2 electrons	Full
L-shell	8 electrons	Full
M-shell	5 electrons	Incomplete

2-9. Electron Sub-orbits or Sub-shells

It has been found that each electron orbit (or main-shell) in an atom consists of number of sub-orbits or sub-shells. The number of sub-orbits is equal to the number of its principal quantum number (n). For example, for *K*-shell (n = 1), there is only one orbit, which is its own sub-orbit. However, for *L*-shell (n = 2), there are two sub-orbits. For *M*-shell (n = 3), there are three sub-orbits and for *N*-shell (n = 4), there are four sub-orbits.

These sub-orbits are designed by the letters *s*, *p*, *d*, *f*, etc. The total number of electrons (which equals $2n^2$) are distributed in the various sub-orbits as shown in Table 2.1.

Table 2-1: *Distribution of electrons in sub-orbits*

Orbit	*Total No. of electrons*	*No. of the sub-orbits*	*Name of the sub-orbits*	*Distribution of electrons in sub-orbits*
K (n = 1)	2	1	1*s*	2
L (n = 2)	8	2	2*s*, 2*p*	2, 6
M (n = 2)	18	3	3*s*, 3*p*, 3*d*	2, 6, 10
N (n = 4)	32	4	4*s*, 4*p*, 4*d*, 4*f*	2, 6, 10, 14

It is evident from the above table that *K*-orbit has one sub-orbit, which is also the main orbit. This sub-orbit is called ls sub-orbit and contains only two electrons. The *L*-orbit has two sub-orbits called 2*s* and 2*p* sub-orbits. The 2*s* sub-orbit contains 2 electrons and 2*p* sub-orbit contains 6 electrons, thus making up a total of 8 electrons. The *M*-orbit has three sub-orbits called 3*s*, 3*p* and 3*d*, which contains 2, 6 and 10 electrons respectively, thus making a total of 18 electrons. The *N*-orbit has four sub-orbits called 4s, 4*p*, 4*d* and 4*f* which contains 2, 6, 10 and 14 electrons respectively. Thus making a total of 32 electrons. The arrangement of these sub-orbits, for first three orbits is as known in Fig. 2-6 (*a*).

Fig. 2-6 (*b*) shows a carbon atom having a total of 6 electrons. Two of these electrons fill up the *K*-orbit (n = 1) which is also its own sub-orbit. The other 4 electrons fill up the *L*-orbit in such a way that 2 electrons are in the 2*s* sub-orbit and other 2 in the 2*p* sub-orbit. It may be noted that 2*p* sub-orbit can contain four more electrons. Therefore, the *L*-orbit of the carbon atom is said to be partially filled.

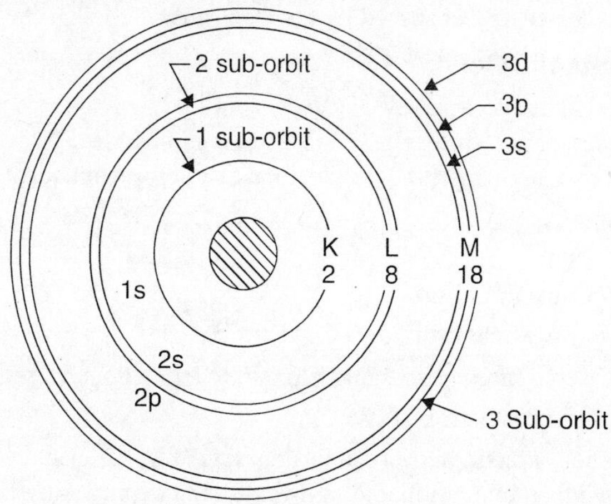

(a) Sub-orbits for first three orbits.

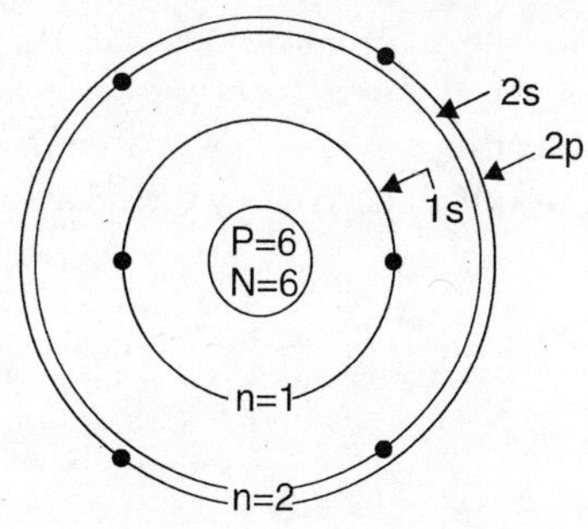

(b) Sub-orbits for carbon atom.

Fig. 2-6.

2-10. Valence Electrons

The electrons, which exists in the *outermost shell* of an atom, are called valence electrons. For example, Fig. 2-5 shows that boron has 3 valence electrons, while silicon and germanium have 4 valence electrons. It will be interesting to know that it is the number of valence electrons, which determine the action of an another element to form compounds. If the outermost shell of an electron is completely filled, the element is said to be stable. It means that this atom is not capable of combining with the atoms of other element to form a compound. However, if the outermost shell of an atom is not complete, the atom is susceptible of forming compound. The general nature of these formations is that an atom will accept valence electrons, from another atom, so that the combined electrons fill up the outer shell completely. In turn, an atom may contribute its valence electrons to those of another atom to form a compound. The general rule for the acceptance or contribution of valence electrons is that if an atom contains more than 4 valence electrons, it will usually receive electrons from another atom. However, if an atom contains less than 4 valence electrons, it will usually contribute (*i.e.*, give away) its electrons to those of the other atom.

2-11. Ionized Atoms

The atoms have an important property that electric charge on the nucleus is always equal to the negative charge as determined by the number of electrons. Thus for hydrogen, the nucleus has a positive charge of one unit as its atom contains only one electron. Similarly, silicon atom has a positive charge of 14 units and germanium has a charge of 32 units. It means that every atom, itself, is electrically neutral. However, if an atom loses or gains an electron, it will be in a charged condition. The atom in a charged condition is said to be an ionized atom or an ion. If an atom loses its electron (or electrons), it becomes a positive ion. On the other hand, if an atom gains an electron (or electrons), it becomes a negative ion.

2-12. Application of Energy to an Atom

There are two different ways of applying (*i.e.*, supplying or giving) energy to an atom. It may be by the application of heat or direct light. The light represents visible portion of the entire electromagnetic spectrum. Thus the light consists of the same radiations as radio waves. But the frequency is higher for the visible region. Whenever a body receives the electromagnetic radiations, the radiations absorbed are converted into energy. As a result of this, the energy of its electrons is raised and the electron is said to be excited. The energy is absorbed by the electrons in quantas only. The amount of energy in a quanta ($E = h.f$) depends upon the frequency (f) of radiations. Thus, greater the frequency of incident radiations, higher is the energy absorbed by the electrons.

2-13. Energy Possessed by an Electron in the Orbit

In his Atomic Model Theory, Bohr suggested that some energy is always emitted by an electron, when it moves from a higher orbit to a lower orbit. Similarly, some energy is absorbed by an electron, when it moves from a lower orbit to a higher orbit. The amount of energy emitted (or absorbed) is given by the relation:

$$W_2 - W_1 = h.f \text{ (in joules)}$$

where W_1 = Energy of the initial orbit,

W_2 = Energy of the final orbit,

h = Planck's constant whose value is 6.626×10^{-34} in joules sec,

and f = Frequency of radiation.

It has been found that the total energy (both kinetic and potential) possessed by an electron, when it revolves in the nth orbit of an atom, with atomic number Z, is given by the expression:

$$W_n = -21.76 \times 10^{-19} \frac{Z^2}{n^2} \text{ (in joules)}$$

A joule is a unit of energy in S.I. system. But in the study of electronics, the energies involved are much smaller. Therefore, we use a much smaller unit of energy called electron volt (eV) which is defined as:

$$1 \text{ eV} = 1.602 \times 10^{-19}$$

The energy of an electron (in electron volts) is given by the expression,

$$E_n = -\frac{21.76 \times 10^{-19}}{1.6 \times 10^{-19}} \times \frac{Z^2}{n^2} = -13.6 \times \frac{Z^2}{n^2} \text{ (in electron volts)}$$

The above equation is a general expression and can be used to determine the energy possessed by the electron in any orbit. It may be noted that energy in joules is designated by W_n and energy in electron volt (eV) is designed by E_n.

2-14. Energy Level Diagram

In the last article, we have obtained an expression for the total energy of an electron, when it is revolving in some orbit around its nucleus. Now, let us obtain the energies E_1, E_2 E_3*....... *etc*., by using the relation :

$$E = -13.6 \frac{Z^2}{n^2}$$

If we plot the energies corresponding to $K, L, M,$...... *etc.* shells along the positive side of the vertical axis by horizontal lines, we shall obtain a diagram a shown in Fig. 2-7. This diagram is popularly known as energy level diagram.

It will be interesting to know that when the electron is present in the first orbit ($n = 1$), it is said to be in its normal or ground state. And if it is present in the higher orbits, it is said to be in the excited state.

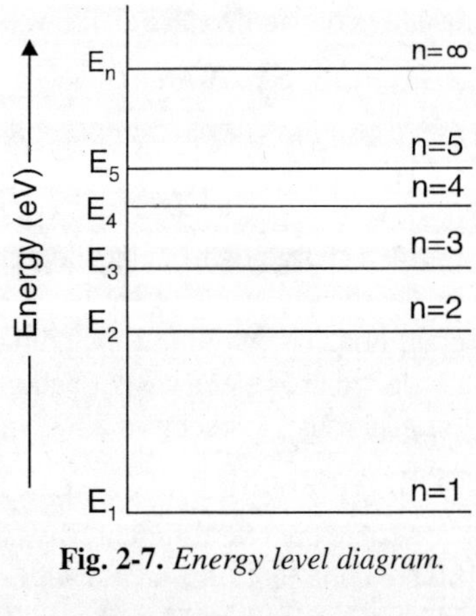

Fig. 2-7. *Energy level diagram.*

* Actually, the values of energy levels are negative and hence should be plotted on the negative side of the vertical axis by horizontal lines. But in actual practice, it is more convenient to indicate energy levels on the positive side of the vertical axis.

2-15. Bonding in Solids

The rigidity of a solid arises from the fact that the atoms in the solid are held together by interatomic forces, called interatomic bonds. The spatial arrangement of atoms, in a solid, is strongly influenced by the nature of these bonds, which in turn is influenced by the electronic structure of the atoms. The interatomic bonds, in solids, have the following two important characteristics :

1. ***Bond energy*** : The energy corresponding to the position of the atoms, at which the bonds are formed, is called bond energy and is designated by W_0. It is expressed in kilojoules per mole of bonds (kJ/mole). The bonds, which possess higher bond energy, are called stronger bonds and those which possess smaller bond energy are called weaker bonds.
2. ***Bond length*** : It may be defined as centre-to-centre distance of the bonding atoms. The stronger bonds have smaller bond length, while weaker bonds have larger bond length.

As mentioned above, the bond energy is used to classify the various types of bonds in solids in two broad categories, *i.e.*, stronger or primary bonds and weak or secondary bonds. Now we shall discuss the formation of only primary bonds in solids. There are three types of primary bonds, *i.e.*, ionic, covalent and metallic. Since, the formation of each such bond is different, therefore, we shall discuss the formation of each of the above mentioned primary bonds.

2-16. Ionic Bond

The bond formed between two oppositively charged ions, which are produced by the transfer of electrons from one atom to another, is called **ionic bond** or **electrovalent bond**. In the formation of an ionic bond, the atoms acquire electronic arrangement or configuration of *noble gases (or inert gases). We know that an atom, which loses the electron, acquires a positive charge and becomes a positive ion. On the other hand, an atom which gains the electron, acquires a negative charge and becomes a negative ion. In this way, both the atoms become oppositely charged ions. The electrostatic force of attraction between the oppositely charged ions forms an ionic bond.

It has been found that an ionic bond is readily formed between the atoms of alkali metals and halogen atoms. The ionic bond is also formed between the atoms of metals like magnesium, aluminum, calcium *etc*., and the atoms of non-metals like oxygen, chlorine, bromine *etc*. The formation of ionic bond in the case of sodium chloride is illustrated below:

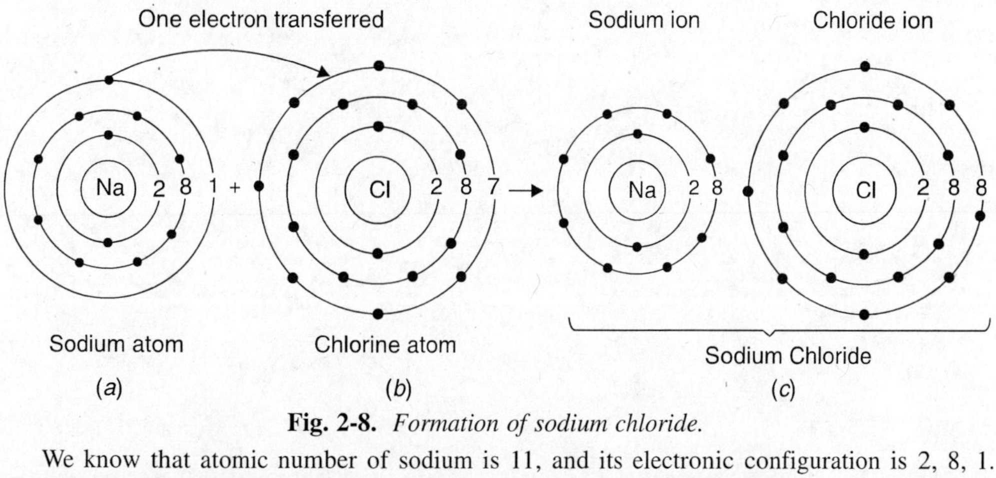

Fig. 2-8. *Formation of sodium chloride.*

We know that atomic number of sodium is 11, and its electronic configuration is 2, 8, 1. Similarly, atomic number of chlorine is 17 and its electronic configuration is 2, 8, 7. The atomic structure of both these atoms is as shown in Fig. 2-8 (*a*) and (*b*). The sodium atom has 1 electron in

* The elements like, helium, neon, argon, krypton etc. are known as noble gases or inert gases. Such elements do not form bonds. The arrangement of electrons in these elements is such that their outermost orbits are completely filled and hence are said to be in stable arrangement.

its outermost energy level, whereas the chlorine atom has 7 electrons. We know that sodium atom has a tendency to lose its electron in the outermost energy level to have a stable configuration of neon inert gas. After losing its outermost electron, sodium atom becomes a sodium ion (with configuration 2, 8). Moreover, chlorine atom has a tendency to gain or accept an electron (lost by the sodium atom) in its outer most energy level to have a stable configuration of argon inert gas (with configuration 2, 8, 8). After gaining an electron in its outermost energy level, the chlorine atom becomes chlorine ion. The binding forces between sodium ion and chloride ion give rise to the formation of sodium chloride as shown in Fig 2-8 (*c*).

The compounds containing ionic bonds are called *ionic solids*. The ionic solids consist of oppositively charged ions. As a result of this, the ions are tightly bound to one another by the electrostatic force of attraction. Therefore, a large amount of energy is required to break away an ionic bond. The ionic solids have some interesting properties which are mentioned as below:

- These are rigid, unidirectional and crystalline in nature.
- These are bad conductors of electricity in their solid state but good conductors of electricity in their molten state.
- They have high melting and boiling temperatures.

2-17. Covalent Bond

The bond, which is formed by the sharing of electrons between two atoms, is called covalent bond. The sharing of electrons takes place in such a way that each atom, in the resulting molecule, gets the stable electron arrangement of an inert or a noble gas. A covalent bond is formed when both the reacting atoms need electrons to achieve the inert gas electron configuration. This happens in non-metals, *i.e.*, the elements which have 5, 6 and 7 electrons in the outermost energy levels of their atoms. In addition to this, covalent bonds are also formed in elements containing 4 electrons in their outermost energy level. The example of such elements are carbon, silicon and germanium, *etc*. The formation of a covalent bond in silicon is as illustrated below:

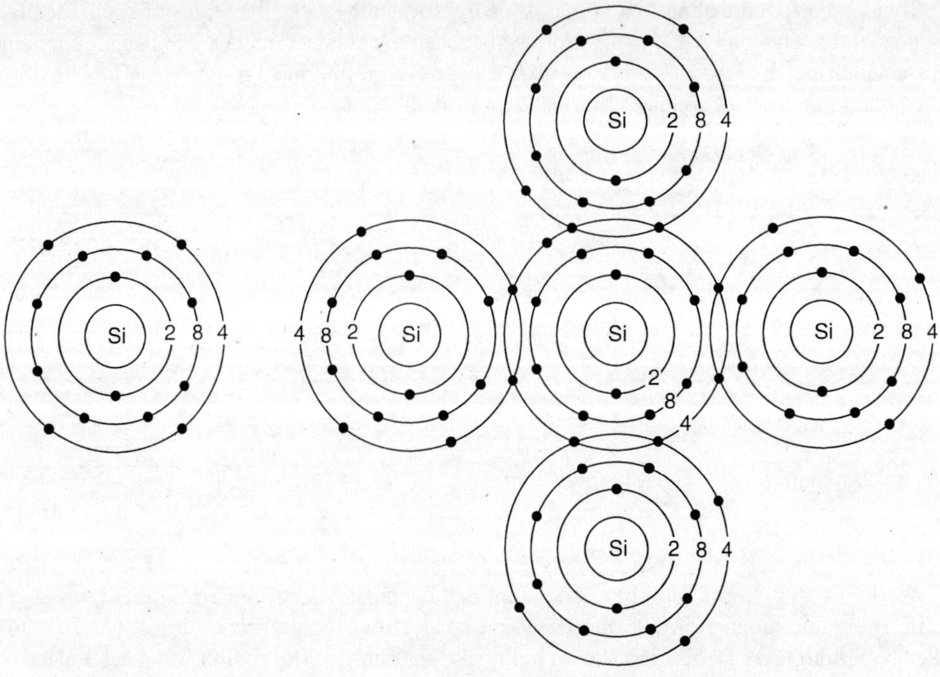

(*a*) Silicon atom. (*b*) Formation of covalent bond in silicon crystal.

Fig. 2-9.

We know that atomic number of silicon atom is 14 and its electronic configuration is 2, 8, 4. The arrangement of these 14 electrons is as shown in Fig. 2-9 (*a*). The silicon atom has 4 electrons in its outermost shell and, it needs 4 more electrons to complete its outermost shell to become stable. It gets these electrons by sharing four electrons, one each from its neighbouring silicon atoms, as shown in Fig. 2-9 (*b*). Similarly, each neighbouring silicon atom shares one electron each with the other neighbouring silicon atoms. It means that a pair of electrons is shared between any two silicon atoms.

It is convenient to represent the covalent bond in silicon structure as shown in Fig. 2-10. In this figure, + 4 indicates, a nucleus and the inner (10) electrons of silicon atom. A part of closed (or broken) lines represent a covalent bond between two silicon atoms, while two dark circles within dashed lines represent a pair of shared electrons.

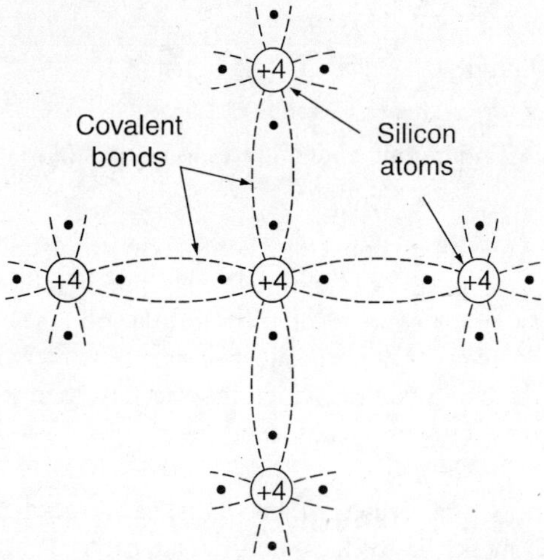

Fig. 2-10. *Covalent bond in silicon structure.*

A covalent bond formed by the sharing of one pair of electrons between the atoms (as in silicon) is called a *single covalent bond.* There are some molecules or compounds, which are formed by the sharing two or three pairs of electrons. A covalent bond, formed by the sharing of two pair of electrons between the atoms, is called a *double covalent bond.* Similarly, a covalent bond formed by the sharing of three pairs of electrons between the atom is known as a *triple covalent bond.*

The compounds, containing covalents bonds, are called *covalent compounds.* They are known as compounds, because they can exist as solids, liquids and gases under normal conditions of temperature and pressure. The covalent compounds are not as strong as the ionic solids. Some of the interesting properties of covalent solids are as follows :

- They are rigid, directional and crystalline in nature.
- Usually, they are bad conductors of electricity. However, some of the solids such as silicon and germanium can be made to conduct electricity by adding certain impurities in the controlled amount.
- They have low melting and boiling temperatures as compared to ionic solids.

2-18. Metallic Bond

It has been observed that in all the metal atoms, the electrons in their outermost shells are *loosely held* by their nucleus. As a result of this, the electrons in the outermost shell require a very small amount of energy to detach themselves from their nuclei. At room temperature, all the metal atoms lose their outermost shell electrons, which form an electron cloud or common pool of electrons. These electrons have a freedom to move anywhere within the crystal. The atoms, after losing their outermost shell electrons, acquire positive charges and become positive ions. Thus *a metal may be considered as an array of positive ions surrounded by a large number of free electrons, forming an electron gas or electron cloud* as shown in Fig. 2-11. The electrostatic force of attraction between the electron cloud and positive ions forms a bond, which is known as metallic bond (as it takes place in metals only).

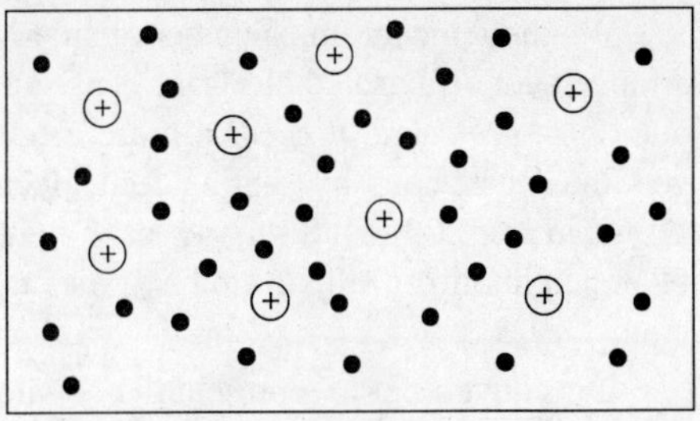

Fig. 2-11. *Metallic bond.*

It may be noted that in metallic bonding, the electrons are not bound to any atom or between two atoms as in the case of covalent bond. This freedom makes metallic bonds *non-directional.* The solids containing metallic bonds, are called metallic solids. Some of the interesting properties of metallic solids are as follows:

- They are rigid, non-directional and crystalline in nature.
- They are good conductors of heat as well as electricity.
- They have low melting and boiling temperatures as compared to the ionic and covalent solids.

2-19. Energy Bands

It has been found that when atoms come together to form a molecule, they exert an influence on one another. As a result of this, the allowed energy levels 'split' into closely spaced levels. A similar situation exists, when many atoms come together to form a solid. The splitted energy levels form continuous bands of allowed energy, which the electrons may occupy. Since, the motion of electrons in a solid is controlled by the energy levels, which they occupy, therefore the nature of these bands is of primary interest in determining the electrical properties of the solid. This has been explained in the following example.

Consider an isolated atom of a carbon. We know that each carbon atom ($Z = 6$) has 6 electrons, two of which occupy the *K*-shell whereas the other four are in the *L*-shell. The energy level diagram of a carbon atom, alongwith its state of electron occupancy, is shown in Fig. 2-12. (*a*).

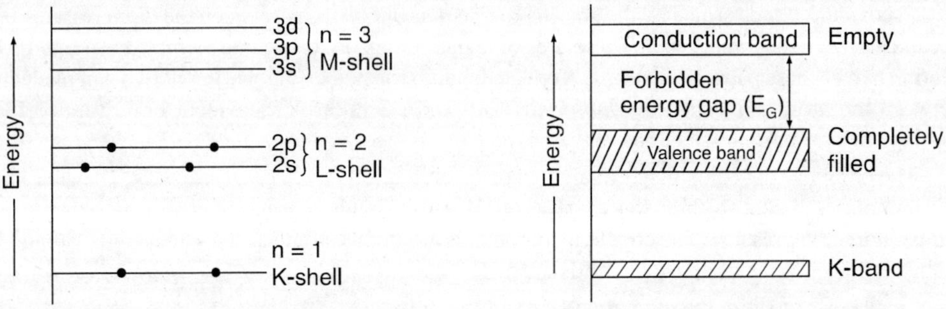

(*a*) Energy level diagram of an isolated carbon atom.

(*b*) Energy level diagram of a diamond crystal.

Fig. 2-12.

Now consider imaginary formation of a diamond crystal from isolated carbon atoms. When the carbon atoms are brought closer to each other, then every atom exerts some force on its neighbouring atoms. Because of this, a coupling will exist between the electrons of one atom and those of its neighbours. This coupling is strongest for the electrons in the higher lying levels. These levels correspond to the outermost shells of the atoms. As a result of this coupling, the splitting of the energy levels take place. The splitting is greatest for the outermost shells and negligible for the innermost shells. After splitting, the separation between the energy levels is small. But since, the number of atoms in a crystal is very large ($\approx 10^{23}$ atoms/cm^3), the total spread between the maximum and minimum energy may be several electronvolts (eV). This large number of discrete, but closely spaced energy levels, are called energy bands. The *K*-shell of an atom forms one energy band, called *K*-band (or 1*s*-band). Similarly, the bands are formed by 2*s*, 2*p* ; 3*s*, 3*p*, 3*d* and high lying sub-orbits of the atoms. For the diamond crystal, the bands formed by 2*s* and 2*p*, after overlapping splits into two bands separated by a forbidden energy gap (E_G) as shown in the Fig. 2-12 (*b*). The lower band contains all the 4 valence electrons of the *L*-shell of the isolated atom and is called valence band. This band is completely filled with the electrons. The upper band is an empty band, *i.e.*, it contains no electrons. The empty band corresponds to the first excited state, where the valence electrons will move under the influence of an external field. The electrons are free to move in this band. This empty band, which lies just above the valence band, is called *conduction band*. The conduction band and valence band are separated by a forbidden energy gap (E_G) as shown in the diagram.

Notes:1. The number of energy bands in a solid is equal to the number of energy levels in any isolated atom of the same substance.

2. Each band consists of a large number of very closely spaced discrete energy levels. The number of such levels in each band is the same as that of number of atoms in the solids.

3. In considering the electrical properties, we are only interested in the two upper bands. These two bands are the conduction band and the valence band. The electrons, having energies in the conduction band, are those which take part in the conduction of current.

2-20. Simplified Representation of Energy Level Diagram of a Solid

Fig. 2-13 shows a simplified representation of an energy level diagram of a solid. In this diagram, only the conduction and the valence bands are shown. The inner (or lower) bands, which are completely filled, are not shown in the diagram. This has been done due to the fact that the inner bands are not important in determining the electrical as well as electronic properties of a solid. It may be noted that at absolute zero (0 K) temperature, the electrons will occupy the lowest energy levels available to them. Thus at 0 K, the valence band is filled, while the conduction band is completely empty. This arrangement of filled and empty energy bands has an important effect on the electrical as well as electronic properties of a solid. If an electron in the valence band is to move throughout the crystal, it must be elevated first from valence band into the conduction band. This can be done only, if a certain amount of energy equal to the energy gap is supplied or added to the material. When an electron in the valance band acquires enough energy to move into the conduction band, it is called a free electron. The free electron can move throughout the crystal. *This provides* an electric current, when a voltage source is connected across a material.

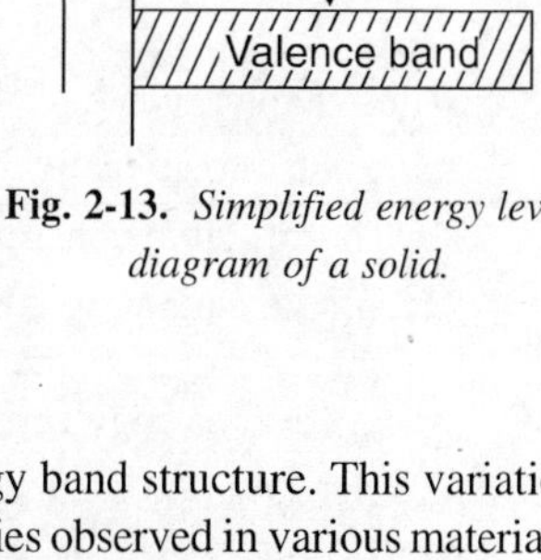

Fig. 2-13. *Simplified energy level diagram of a solid.*

2-21. Insulators, Conductors and Semiconductors

As a matter of fact, every solid has its own characteristic energy band structure. This variation in band structure is responsible for the wide range of electrical properties observed in various materials. Now we shall discuss the electrical properties of insulators, conductors and semiconductors.

(i) Insulators : The materials, in which the conduction and valence bands are separated by a wide energy gap ($\approx$ 15 eV) as shown in Fig. 2-14 (*a*) are called insulators (also referred to as dielectric materials). A wide energy gap means that a large amount of energy is required, to free the electrons, by moving them from the valence band into the conduction band. Since at room temperature, the valence electrons of an insulator do not have enough energy to jump into the conduction band, therefore insulators do not have an ability to conduct current. Thus insulators have very high resistivity (or extremely low conductivity) at room temperatures. However, if the temperature is raised, some of the valence electrons may acquire energy and jump into the conduction band. It causes the resistivity of insulators to decrease. Therefore, the insulators have negative-temperature coefficient of resistance.

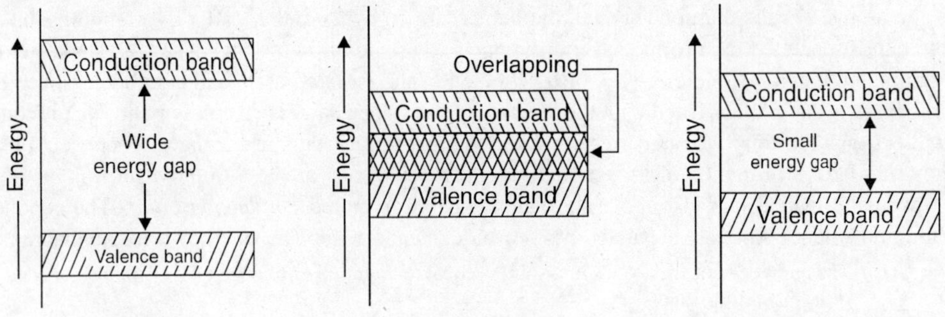

(*a*) Insulators (*b*) Conductors (*c*) Semiconductors

Fig. 2-14.

(ii) Conductors **:** The materials, in which conduction and valence bands overlap as shown in Fig. 2-14 (*b*) are called conductors. The overlapping indicates a large number of electrons available for conduction. Hence, the application of a small amount of voltage results in a large amount of current.

(iii) Semiconductors **:** The materials, in which the conduction and valence bands are separated by a small energy gap (1 eV) as shown in Fig. 2-14 (*c*) are called semiconductors. Silicon and germanium are the commonly used semiconductors. A small energy gap means that a small amount of energy is required to free the electrons by moving them from the valence band into the conduction band. The semiconductors behave like insulators at 0 K, because no electrons are available in the conduction band. However, at room temperature, a significant number of electrons are available in the conduction band. If the temperature is further increased (above room temperature), more valence electrons will acquire energy to jump into the conduction band. Thus like insulators, semiconductors also have negative temperature coefficient of resistance. It means that conductivity of semiconductors increases with the increase in temperature.

SUMMARY

In this chapter, you have learned that :

1. Solids are crystalline in structure.

2. The structure consists of atoms or molecules.

3. Simple Cubic (SC), body-centred cubic (BCC), face-centred Cubic (FCC) and diamond or tetrahedral are the four major types of crystalline structure of solids used in the field of electronics engineering.

4. The structure of an atom is composed of electons, protons and neutrons.

5. The electrons in an atom revolve around the nucleus in fixed orbits, called energy levels or shells. The maximum number of electrons in any shell is given by $2n^2$ where n is an integer and is called principal quantum number.

6. Energy is emitted by an electron when it falls from a higher orbit to a lower orbit. Similarly energy is absorbed by an electron when it jumps from a lower orbit to a higher orbit.

GLOSSARY

Amorphous structure : Solid materials which do not have crystalline structure.

Crystalline structure : Solid materials that have ordered arrangement of atoms repeated throughout the structure.

Body-centred cubic : A crystalline structure with a unit cell in the form of a cube. It contains atoms at all the corners of a cube with an additional atom at the body centre of a cube.

Energy Bond : A group of energy levels corresponding to a particular orbit for different atoms in a solid.

Energy Level Diagram : A plot indicating energy levels of different orbits starting from inner most orbit at the lower level.

Semiconductors : The materials in which conductor and valence bands are separated by a small energy gap ($\approx$ 1eV).

Face-centred cubic : A crystalline structure with a unit cell in the form of a cube. It contains atom at all the corners of a cube with an additional atom in the centre of each six faces of the cube.

Simple cubic : A crystalline structure with a unit cell in the form of a cube. It contains atoms only at all the corners of a cube.

DESCRIPTIVE QUESTIONS

1. What are the various types of crystal structures?

2. Define the terms atomic number and atomic mass.

3. What are valence electrons? Do they help in bonding the solids?

4. Explain the procedure for bonding in a solid.

5. Differentiate between an ionic bond and a covalent bond.

6. Describe the insulators, conductors and semiconductors using energy band diagram.

7. Explain the energy band diagram for conductors, insulators and semiconductors.
(*Electron Devices, Anna University, April/May 2008*)

8. Derive expression for kinetic energy of electron.
(*Electron Devices, Anna University, April/May 2008*)

9. What is Fermi-Dirac energy distribution function?
(*Electorn Devices, Anna University, April/May 2008*)

MULTIPLE CHOICE QUESTIONS

1. The atomic number of boron is
(*a*) 2 (*b*) 5 (*c*) 6 (*d*) 14

2. An atom is said to be ionized when any of its orbiting electron
(*a*) jumps from one orbit to another (*b*) is raised to a higher orbit
(*c*) comes to the ground state (*d*) is completely removed.

3. The maximum number of electrons which the M-shell of an atom can contain is,
(*a*) 32 (*b*) 8 (*c*) 18 (*d*) 50

4. Electronic distribution of Si atom is
(*a*) 2, 10, 2 (*b*) 2, 8, 4 (*c*) 2, 7, 5 (*d*) 2, 4, 8

5. Which of the following bond is formed by transferring electrons from one atom to another?
   (*a*) ionic (*b*) covalent (*c*) metallic (*d*) either '*a*' or '*b*'
6. Which of the following solids have least electrical conductivity?
   (*a*) ionic (*b*) covalent (*c*) metallic (*d*) all of them
7. In a semiconductor material, the energy gap is
   (*a*) large (*b*) small (*c*) zero (*d*) infinity

ANSWERS

1. (*b*) **2.** (*d*) **3.** (*c*) **4.** (*b*) **5.** (*a*)
6. (*a*) **7.** (*b*)

❑ ❑ ❑

3
CHAPTER
Semiconductors

OUTLINE

Objectives

After completing this chapter, you should be able to :

- Know the difference between intrinsic and extrinsic semiconductors.
- Describe the generation of electrons and holes in an intrinsic semiconductor.
- Draw energy band diagram for an intrinsic and extrinsic semiconductor.
- Explain N-type and P-type semiconductors using the atomic structure and energy level diagrams.
- Understand the meaning of majority and minority charge carriers.
- Know the effect of immobile ions in semiconductors.

- Explain the mass-action law.
- Understand the effect of temperature on intrinsic concentration, mobility conductivity and energy-gap of a semiconductor.
- Calculate the diffusion current density.
- Describe the diffusion current density.
- Describe the Hall-effect and its applications.
- Calculate the charge carrier concentration or mobility using the Hall equation,

$$\mu_n = \sigma/n \cdot q = \sigma \cdot R_H \text{ or } \mu_p = \sigma/p \cdot q = \sigma \cdot R_H$$

3-1. Introduction

The materials, whose electrical properties lie between those of conductors and insulators, are known as *semiconductors*. The examples of such materials are germanium (Ge), silicon (Si), gallium arsenide (GaAs), cadmium sulfide (CdS), lead telluride *etc*.

In terms of energy bands, the semiconductor may be defined as those materials, which at room temperature, have

1. Partially filled conduction band.
2. Partially filled valence band.
3. A very narrow energy gap, E_G (≈ 1 eV) between conduction and valence bands.

We know that at absolute zero temperature (*i.e.*, 0 K) there are no electrons in the conduction band of semiconductors and the valence band is completely filled. Thus, the semiconductors behave like perfect insulators at 0 K. However, as the temperature is increased, the width of energy gap reduces. Consequently, some of the electrons jump into conduction band and semiconductors show some conductivity. It is thus obvious, that the conductivity of semiconductors increases with the increase in temperature. The electrical conductivity of semiconductors is in the range of 10^{-3} to 10^{-6} per ohm per cm.

3-2. Types of Semiconductors

The semiconductors are of the following two types :

(1) Intrinsic semiconductors (2) Extrinsic semiconductors

3-3. Intrinsic Semiconductors

A semiconductor, which is in its extremely pure form, is known as an *intrinsic* (or pure) semiconductor. The nature of semiconductors is such that even a small amount of certain impurities can change their electrical properties drastically. It is due to this fact, that a semiconductor would not be called truly intrinsic, unless the impurity level is very small. For germanium, the impurity level is less than 1 part in 10^8 parts. And for silicon, it is less than 1 part in 10^{12} parts. In actual practice, however, a semiconductor material, with somewhat larger impurity concentrations, than those mentioned above, is still called as an intrinsic.

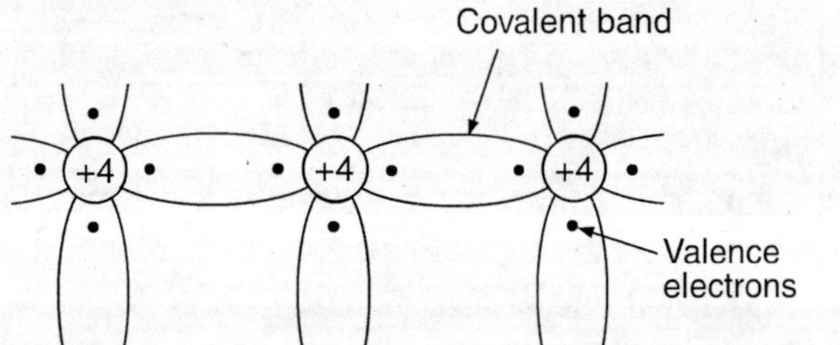

Fig. 3-1. *Two dimensional representation of a silicon or a germinium crystal.*

The silicon and germanium are the two most widely used intrinsic semiconductors. The crystal structure of these materials consists of a regular repetition in three dimensions of a unit cell having the form of tetrahedron, with one atom at each vertex. The two dimensional representation of such a structure is shown in Fig. 3-1. For example, the silicon has a total of 14 electrons in its atomic structure. Each atom in a silicon crystal contributes 4 valence electrons, so that the atom is tetravalent. The four valence electrons are shown by the dots in the figure. The circle, with + 4 designation, represents the central portion of the atom, including the nucleus and the electrons except those in the outermost orbit. This portion is often known as core of the atom. These atoms are bonded together by a force, which results from the sharing of the neighbouring atoms. This bonding force is called a covalent bond and is shown by a pair of curved lines in the figure.

3-4. Generation of Electrons and Holes

At very low temperatures (*i.e.*, in the range of – 273°C or 0 K), an intrinsic semiconductor will have exactly the ideal structure as shown in Fig. 3-1. In that structure, all the valence electrons are tightly held by the parent atoms and through covalent bonds by other atoms. Since, the electrons are not free to move, through the crystal structure, therefore they cannot conduct electricity. Thus an intrinsic semiconductor, at absolute zero temperature (*i.e.*, 0 K), behaves like an insulator.

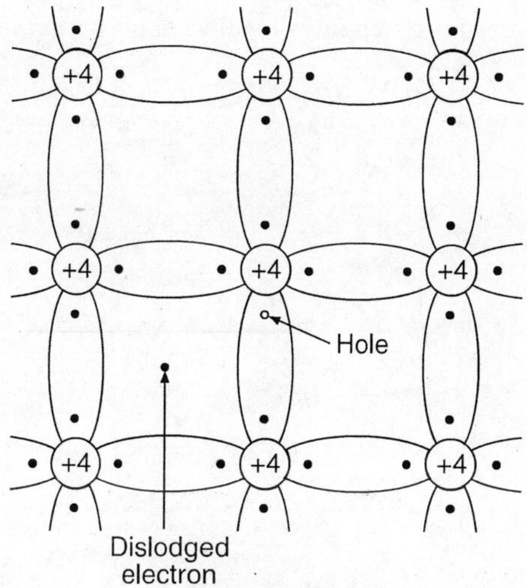

Fig. 3-2. *Intrinsic semiconductor with a broken covalent bond.*

Now, let us assume that the temperature of the semiconductor is increased to room temperature (*i.e.*, 27°C or 300 K). At this temperature, some of the covalent bonds are broken. This situation is as shown in Fig. 3-2. Here an electron, which was a part of covalent bond earlier, is now shown as a dislodged electron. This electron is free to move anywhere within the crystal. The energy required to break such a covalent bond is equal to the band gap energy (E_G). The value of E_G at room temperature is about 0.72 eV for germanium and 1.1 eV for silicon. The vacancy of an incomplete covalent bond, left behind the dislodged electron (represented by a small circle in the figure), is called a hole. A hole can serve as a carrier of electricity like the free electron. A combination of such a free electron and a hole is known as electron-hole pair.

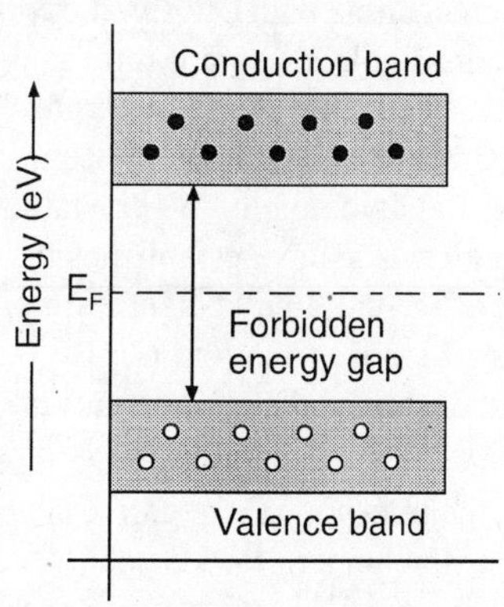

Fig. 3-3. *Energy band diagram for an intrinsic semiconductor.*

It may be noted that if the temperature of an intrinsic semiconductor is increased, beyond room temperature, a large number of electron-hole pairs are generated. Since, the electron and holes are generated in pairs, the free electron concentration 'n' (*i.e.*, number of electrons/cm^3) is equal to the concentration of holes 'p' (*i.e.*, number of holes/cm^3). Each of these intrinsic carrier concentration is commonly denoted by the letter n_i.

Thus for an intrinsic semiconductor, concentration of free electrons or concentration of holes,

$$n = \text{Intrinsic carrier concentration}$$
$$= p = n_i$$

Fig. 3-3 shows the energy band diagram for an intrinsic semiconductor at room temperature. Here, the horizontal line marked indicates the *Fermi level. It may be noted that the Fermi level lies at the middle of the forbidden energy gap.

3-5. Mechanism of Holes Contributing to Conductivity

Consider the structure of an intrinsic semiconductor as shown in Fig. 3-4. Now, let us imagine that a hole is created by an electron breaking away from the covalent bond. At this instant, it is possible that an electron from the neighbouring atom leaves its covalent bond to fill up the hole. However, this electron generates a new hole in its initial position. When this occurs, it appears that a hole has moved in a direction opposite to that of the electron. This hole, in its new position, may now be filled up by an electron from another covalent bond. And the hole will move, correspondingly, one more step in a direction opposite to the motion of the electron. This process is illustrated in Fig. 3-4.

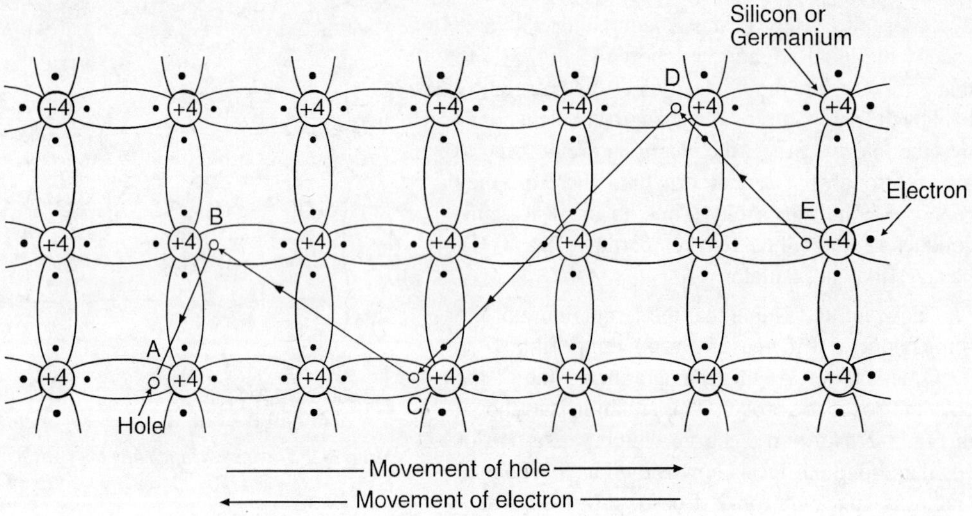

Fig. 3-4. *Mechanism of holes contributing to conductivity.*

Here a circle with a dot in it, represents an electron present in a covalent bond. An empty circle represents a hole, and a dot represents an electron broken away from the covalent bond. Suppose a hole exists at point *A* in the figure. The electron which had broken away from a covalent bond, creating, this hole has moved elsewhere in the crystal lattice and is not shown in the figure. Due to the thermal agitation, in the crystal lattice, an electron at *B* may find it more favourable to jump to the position *A*. This re-establishes the covalent bond at *A*. But this creates a new hole at *B*. Thus, we can say that the hole at *A* has simply moved to the new position at *B*. Similarly, the hole may continue to move from *B* to *C*, *C* to *D* and finally from *D* to *E*. The net displacement of the hole from *A* to *E* constitutes the electric current.

It is evident from the above discussion, that the motion of a hole, in one direction, actually means the transport of a negative charge by an equal distance in the opposite direction. Thus, we can consider the hole, in fact, a positive charge equal in magnitude to the electronic charge.

3-6. Free Charge Carriers in Semiconductors

We know that the conduction properties of a semiconductor depend upon the number of valence electrons. When a sufficient energy (thermal or electrical) is given to valence electrons, they break

* The Fermi level (E_F) is the maximum energy level, which is occupied by an electron at absolute zero (0K) temperature. At higher temperatures, some electrons are excited into higher energy states. Then Fermi level corresponds to the centre of gravity of conduction electrons and holes weighed according to their energies.

away from their parent atoms, and take part in conduction through the crystal. Such valence electrons are called *conduction band electrons*. These electrons leave behind the holes in the valence band. The electrons and holes, generated in such a way, are called free charge carriers. They are free to move anywhere within the crystal. Hence, such electrons and holes (called charge carriers) are known as free charge carriers. It will be interesting to know that the number of free charge carriers is an important factor, because it determines the property of a semiconductor.

3-7. Generation and Recombination of Carriers

We have already discussed in Art. 3-4 that in a pure semiconductor, the number of holes is equal to the number of electrons. However, thermal agitation continues to produce new hole-electron pairs, while other hole-electron pairs disappear as a result of recombination. Although the recombination does not occur very frequently, but when it occurs, the free electrons fall into empty covalent bonds. This results into a loss of a pair of free electron and a hole. The recombination also results in the release of energy equivalent to energy gap (E_G) as a photon of light or thermal energy. In some materials, the energy appears in the visible spectrum also.

On an average, a hole (or an electron) may exist for a very small time before recombination. This time of existence is known as the mean life time. These life times are, usually, in the range of 1 to 1000 μs ($1 \mu s = 1 \times 10^{-6}$ sec). This life time is a very important parameter in semiconductor devices.

3-8. Extrinsic Semiconductors

We have already discussed in Art. 3-4 that an intrinsic or a pure semiconductor has a negligible conductivity at room temperature. Therefore, it is not of any practical signification. In order to make the intrinsic semiconductor useful, their characteristics have to be changed, by adding a certain amount of desired impurity atoms. The resulting semiconductors are called ***impure*** or ***extrinsic semiconductors***.

The process of adding impurity atoms, to the intrinsic semiconductor is called ***doping***. Generally, the impurities are added at the rate of only one atom of impurity per 10^6 to 10^{10} semi- conductor atoms. As a result of this, the impurity atoms are separated in the crystal by hundreds or thousands of semiconductor atoms in all directions. These impurity atoms are so scarce, that they do not exert any influence on the manner in which the crystal grows. It may be noted that an impurity atom enters the crystal by substituting for one of germanium or silicon atoms.

The purpose of adding impurity is to increase either the number of free electrons or holes in a semiconductor. Generally, two types of impurity atoms are added to the semiconductor namely the impurity atoms containing 5 valence electrons (called pentavalent impurity atoms) and the impurity atoms containing 3 valence electrons (called trivalent impurity atoms). Depending upon the type of impurity atoms added to the semiconductor, the resulting semiconductor (*i.e.*, extrinsic semiconductor) may be of the following two types :

(*i*) *N-type semiconductors.*

(*ii*) *P-type semiconductors.*

3-9. N-Type Semiconductors

The semiconductors, which are obtained by introducing pentavalent impurity atoms, (*i.e.*, atoms containing 5 valence electrons) are known as ***N-type semiconductors***. The pentavalent impurity is an element from Group V of the Periodic Table. The elements in this group contain 5 valence electrons. The typical examples of such elements are phosphorus (P), antimony (Sb), arsenic (As) and bismuth (Bi). (refer to Table 3.1). These elements donate excess (negative) electron carriers. Therefore, such elements are known as donor or N-type impurities.

Table 3-1. *Pentavalent impurities*

No	*Element*	*Symbol*
1	Phosphorus	P
2	Arsenic	As
3	Antimony	Sb
4	Bismuth	Bi

It has been observed that when a pentavalent impurity is added to a pure semiconductor, it displaces some of its atoms. Fig. 3-5. shows the structure of a silicon crystal lattice containing as antimony atom at the central position. It may be noted that out of 5 valence electrons, 4 electrons will form covalent bonds by sharing one electron each with the electrons of the neighbouring atoms. The 5th electron is an extra electron, and is loosely bound with the antimony atom. This extra electron, if detached from the antimony atom, will be available as a carrier of the current. The energy required to detach this electron is of the order of 0.05 eV for silicon and 0.01 eV for germanium.

It will be interesting to know that if a pure semiconductor is doped with N-type impurities, the number of electrons in the conduction band increases above a level that would be available in a pure semiconductor. However, the number of holes in the valence band decreases below a level, that would be available in a pure semiconductor. It is due to the fact that a large number of electrons present in the semiconductor increases the rate of recombination of electrons with holes. Because of the greater number of electrons in conduction band than that of holes in the valence band. The Fermi level shifts upward towards the bottom of the conduction band as shown in Fig. 3-6.

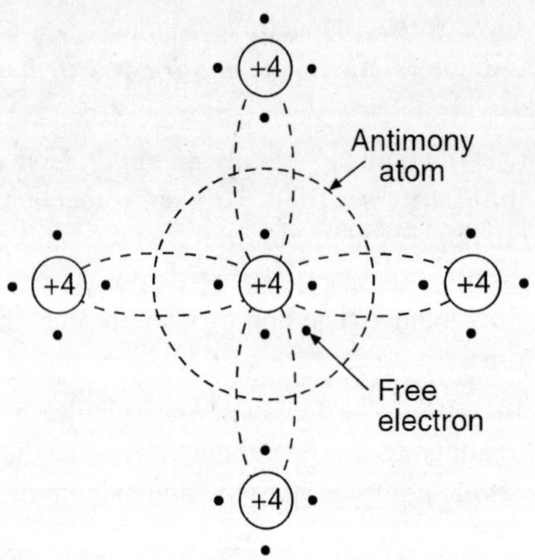

Fig. 3-5. *A silicon atom replaced by stationary atom in a crystal lattice.*

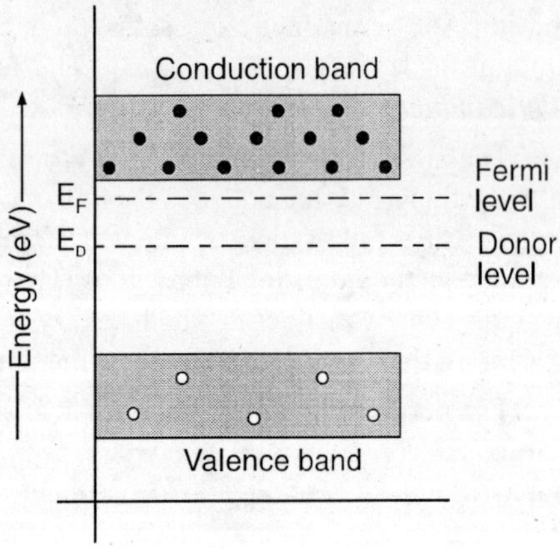

Fig. 3-6. *Energy band diagram of an N-type semiconductor.*

The addition of a pentavalent impurity atoms gives away or donates its 5th electron as discussed above. After the donation, the impurity atom becomes a positively charged ion and is known as a *donor ion*. This ion is bounded by four covalent bonds in the crystal lattice and therefore can not be made to move within the lattice. As a result of this, donor ions do not contribute to the conduction of current in N-type semiconductor. Thus in an N-type semiconductor, the current flows due to the movement of electrons and holes. But a major part of the current flows due to the movement of electrons. Therefore the electrons in an N-type semiconductor are known as majority carriers and holes as minority carriers.

3-10. P-Type Semiconductors

The semiconductors, which are obtained by introducing a trivalent impurity atoms (*i.e.*, atoms containing 3 valence electrons) are known as P-type semiconductors. The trivalent impurity is an element of group III of the Periodic Table. The elements in this group contain 3 valence electrons. Typical examples of such elements are gallium (Ga), indium (In), aluminium (Al), Boron (B) *etc.* (refer to Table 3-2). These elements make available positive charge carriers because they create holes, which can accept electrons. Therefore, such elements are known as acceptor or P-type impurities.

Table 3-2. *Trivalent Impurities*

No.	*Element*	*Symbol*
1	Aluminium	Al
2	Gallium	Ga
3	Boron	B
4	Indium	In

It has been observed that when a trivalent impurity is added to a pure semiconductor, it displaces some of its atoms. Fig. 3-7 shows the structure of a silicon crystal containing an indium atom at the central position. The 3 valence electrons, of an indium atom, forms 3 covalent bonds by sharing one electron with the electrons of neighbouring atoms. However, the fourth covalent band is incomplete. A vacancy, which exists in the incomplete covalent bond constitutes a hole. Now, the indium atom seeks its surrounding atoms so as to acquire the 4th electron, to complete the covalent bond. It does so by taking an advantage of thermal motion, which brings one electron from the surrounding atoms. Thus, an electron, which is in a favourable position, is captured by an indium atom. After doing this, the indium atom becomes an immobile ion. The energy, involved in capturing the electron to 0.05 eV for silicon and 0.01 eV for germanium.

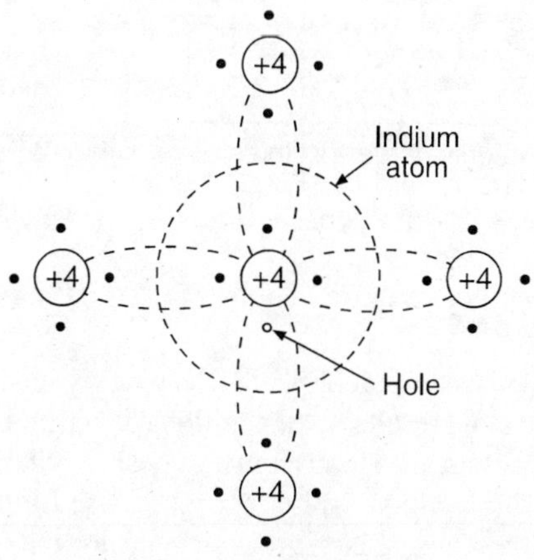

Fig. 3-7. *A silicon atom replaced by indium atom in a crystal lattice.*

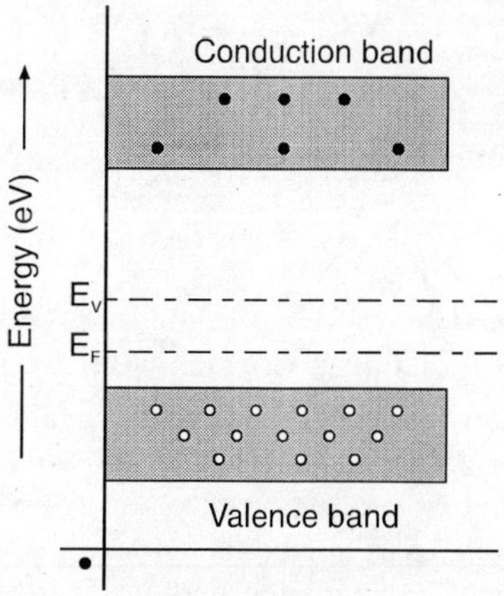

Fig. 3-8. *Energy band diagram of a P-type semiconductor.*

It will be interesting to know that if a pure semiconductor is doped with P-type impurities, the number of holes in the valence band increases above a level, which would be available in a pure semiconductor. However, the number of electrons in the conduction band decreases below a level, which would be available in a pure semiconductor. It is due to the fact, that a large number of holes present in the semiconductor increases the rate of recombination of holes with electrons. Because of the greater number of holes in the valence band than that of electrons in the conduction band, the Fermi level shifts downwards towards the top of the valence band as shown in Fig. 3-8.

The addition of a trivalent impurity atom takes or accepts its fourth electron as discussed above. After accepting, the impurity atom becomes a negatively charged ion and is known as an acceptor ion. This ion is bounded by four covalent bonds in the crystal lattice and therefore cannot be made to move within the lattice. As a result of this, the acceptor ions do not contribute to the conduction of current in a P-type semiconductor. Thus in a P-type semiconductor, the current flows due to the movement of holes and electrons. But a major part of the current flows due to the movement of holes. Therefore, holes in a P-type semiconductor are known as majority carriers and electrons as minority carriers.

3-11. Majority and Minority Charge Carriers

We have already discussed that a pure (or intrinsic) semiconductor has no free charge carriers at absolute zero (0 K) temperature. However, as the temperature of the semiconductor is raised to room temperature, some of the covalent bonds are broken, due to which few electron-hole pairs are created. These electron-hole pairs are called *thermally generated charge carriers*. The number of these charge carriers is, usually, very small. However, an extrinsic semiconductor (either N- or P-type) has a larger number of charge carriers. For instance, an N-type semiconductor has a large number of electrons. Besides this, N-type semiconductor also contains fewer number of thermally generated holes. Since, the number of electrons, are much larger than the number of holes in a N-type semiconductor, therefore ***electrons are called majority carriers*** and ***holes a minority carriers***. This situation is as shown in Fig. 3-9 (*a*).

Similarly, in a ***P-type semiconductor***, the number of holes are much larger than the number of electrons. Therefore in a P-type semiconductor, ***holes are called majority carriers*** and ***electrons as minority carriers***. This situation is as shown in Fig. 3-9 (*b*).

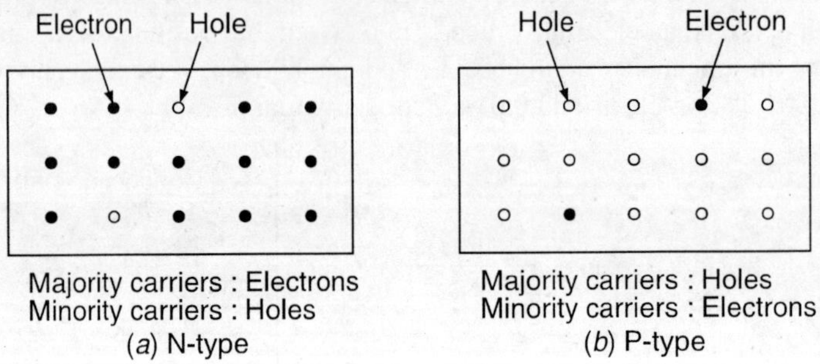

Fig. 3-9.

3-12. Mobile Charge Carriers and Immobile Ions

We know that an N-type semiconductor is formed by the addition of pentavalent atoms such as phosphorus, antimony, bismuth *etc*. The addition of such atom results in an additional electron in a crystal lattice. As the electron moves away from the parent atom, the atom acquires a positive charge. The positively charged atom is known as a donor ion. The position of a donor ion is fixed in the crystal lattice. Therefore, it cannot take part in conduction. Thus an N-type semiconductor may be regarded as a neutral material studded with donor ions (*i.e.*, immobile ions) and electrons (*i.e.*, mobile charge carriers) as shown in Fig. 3-10 (*a*).

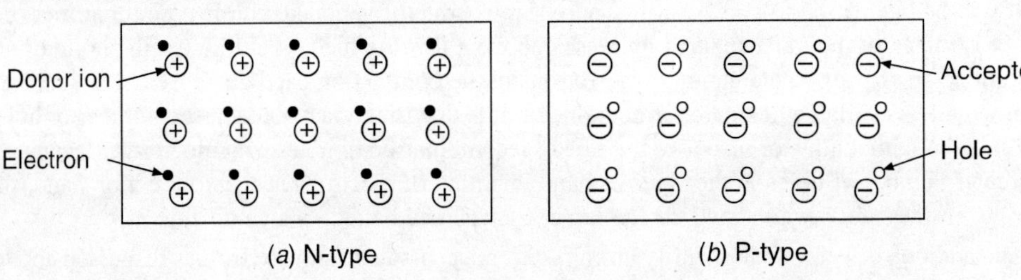

Fig. 3-10. *Mobile charge carriers and immobile ions.*

Similarly, a P-type semiconductor is formed by the addition of trivalent atoms such as gallium, indium, aluminium, *etc*. The addition of each such atom results in an additional hole in a crystal lattice. As a hole moves away from the parent atom, the atom acquires a negative charge. This negatively charged atom is known as an *acceptor ion*. The position of an acceptor ion is fixed in the crystal lattice. In other words, it is immobile and therefore cannot take part in conduction. Thus, a P-type material may be regarded as a neutral material studded with acceptor ions (*i.e.*, immobile ions) and holes (mobile charge carriers) as shown in Fig. 3-10 (*b*).

3-13. Conduction in Metals

In order to study the characterstics of a conductor in semiconductor, it is important to discuss first, the conduction in metals. We shall derive few relations for drift velocity, current density and conductivity for metals. Later on, we shall use these relations to develop similar relations for semiconductors.

We know that valence electrons in a metal move continuously at random. During their movement, the valence electrons collide with heavy (stationery ions). After each collision, the direction of its movement is changed. The average distance between collisions is known as mean free path. Since, the motion of electrons is at random, therefore there will be as many electrons passing through a unit area in the metal in any direction as in the opposite direction in a given time. Hence, the average value of the current is zero.

Now, if an electric field is applied to a metal, than due to the electrostatic force, the electrons would be accelerated and the velocity would increase indefinitely, if there would have been no collision with the ions. However, at each collision with the ion, the electron loses some energy. A steady-state condition is reached, when a finite value of drift velocity is obtained. The drift velocity is in a direction opposite to that of the applied electric field. Now let us obtain an expression for the drift velocity in a metal.

Let $\boldsymbol{E}$ = Strength of the applied electric field,

$\boldsymbol{F}$ = Force experienced by an electron due to the applied electric field,

$\boldsymbol{a}$ = Acceleration, which the electron obtains due to the force experienced by it,

$\boldsymbol{t}$ = Time between the collisions,

$\boldsymbol{v}$ = Drift velocity,

$\boldsymbol{q}$ = Charge of an electron, and

$\boldsymbol{m}$ = Mass of an electron.

We know that force experienced by the electron due to applied electric field.

$$F = q \cdot E$$

and acceleration,
$$a = \frac{\text{Force}}{\text{Mass}} = \frac{F}{m} = \frac{q \cdot E}{m}$$

∴ Drift velocity,
$$v = a \cdot t = \frac{q \cdot E}{m} \cdot t = \mu \cdot E$$

where μ is equal to $\frac{q \cdot t}{m}$ and is known as *mobility of an electron*. The mobility is expressed in m^2/(volt-sec) and its value is different for different metals.

3-14. Electric Current in a Conductor

Consider a piece of a metal or conductor in which electrons are uniformly distributed as shown in Fig. 3-11.

Let $\boldsymbol{N}$ = No. of free electrons distributed in the conductor

$\boldsymbol{L}$ = Length of the conductor

$\boldsymbol{A}$ = Cross-sectional area of the conductor

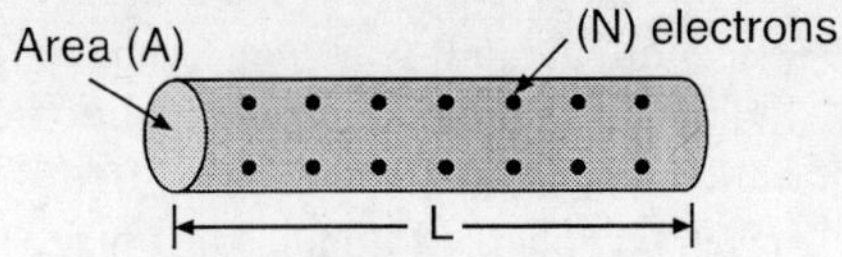

Fig. 3-11. *Metal or conductor.*

We know that when an electric field (E) is applied to the conductor, the electrons start moving in the conductor. Therefore, avarage velocity of the electrons

$$= \frac{L}{T}$$

where, T is the time taken by the electrons to travel through a distance.

The number of electrons passing through any area per second,

$$= \frac{N}{T}$$

∴ Total charge passing through any area per second (called current),

= Charge of an electron × Number of electrons per second passing through any area

$$= q \times \frac{N}{T}$$

Multiplying and dividing the right hand side of the above equation by L,

$$= q \times \frac{N}{T} \times \frac{L}{L} = \frac{q \cdot N \cdot v}{L} \quad \ldots\left(\because v = \frac{L}{T}\right)$$

Thus, the current in a conductor,

$$I = \frac{q \cdot N \cdot v}{L}$$

and the current per unit area (called current density),

$$J = \frac{I}{A} = \frac{q \cdot N \cdot v}{L \cdot A}$$

Since, the term '$L \cdot A$' is the volume of the conductor containing N electrons, therefore concentration of electrons per unit volume,

$$n = \frac{N}{L \cdot A}$$

and $$J = q \cdot n \cdot v$$

3-15. Conductivity of a Conductor

We have already discussed in the last article that electric current density (*i.e.*, electric current per unit area),

$$J = q \cdot n \cdot v$$

Substituting the value of from Art. 10.13 in the above equation,

$$J = q \cdot n \cdot (\mu \cdot E) = q \cdot n \cdot \mu \cdot E = \sigma \cdot E$$

where, σ (equal to $q \cdot n \cdot \mu$) is called *conductivity of a metal*. The above relation indicates that electric current density (J) within a metal is directly proportional to the applied electric field (E). Now we know that resistivity of a metal is reciprocal of conductivity, *i.e.*,

$$\rho = \frac{1}{\sigma} = \frac{1}{q \cdot n \cdot \mu}$$

and $$\sigma = \frac{1}{\rho} = \frac{1}{q \cdot n \cdot \mu}$$

Example 3-1. *Find the length of a round copper wire of diameter 1 mm which will have a resistance of 1kΩ and conductivity 5.8 × 10^7 S/m. If the wire is subjected to an electric field of 10 mV/m, what is the current density.* *(Pune Univ. 1994)*

Solution. Given : Resistance of a copper wire (R) = 1kΩ = 1000Ω; Conductivity (σ) = 5.8×10^7 S/m; diameter (d) = 1mm = 0.001m.

Length of Copper wire

Let l = length of the copper wire (m).

We know that resistance of a wire (R),

$$1000 = \rho \cdot \frac{l}{A} = \frac{1}{\sigma} \cdot \frac{l}{A} = \frac{1}{\sigma} \cdot \frac{l}{\pi\,(d^2/4)}$$

$$= \frac{1}{5.8 \times 10^7} \times \frac{l}{\pi \times (0.001)^{2/4}}$$

$$= \frac{4l}{182.12}$$

$$l = \frac{1000 \times 182.12}{4} = 45.5 \times 10^3 \text{ m} = 45.5 \text{ km} \textbf{ Ans.}$$

Current density in copper wire.

Given $E = 10 \text{ mV/m} = 10 \times 10^{-3} \text{ V/m}$

We know that current density,

$$J = \sigma E$$

$$= (5.8 \times 10^7) \times (10 \times 10^{-3})$$

$$= 5.8 \times 10^5 \text{ A/m}^2 \textbf{ Ans.}$$

Example 3-2. *A copper wire of 2 mm diameter with conductivity of 5.8 × 10^7 Siemens/m and electron mobility of 0.0032 m^2/V-s is subjected to an electric field of 20 m V/m. Find **(a)** the charge density of free electrons, **(b)** the current density, **(c)** the current flowing in the wire, and **(d)** the electron drift velocity.* *(UPSC Engg Services 2002)*

Solution. Given : Diameter of copper wire (d) =2 mm = 2×10^{-3} m; Conductivity, (σ) = 5.8×10^7 S/m; electron mobility, (μ) = 0.0032 m^2/V-s; Electric field (E) = 20 mV/m.

***(a)** Charge density of free electrons*

Let n = number of free electrons per unit volume of the copper wire (*i.e.*, electron density) per m^3.

We know that conductivity (σ),

$$5.8 \times 10^7 = q \cdot n \cdot \mu = 1.6 \times 10^{-19} \times n \times 0.0032$$

$$\therefore \quad n = \frac{5.8 \times 10^7}{1.6 \times 10^{-19} \times 0.0032} = 1.132 \times 10^{29}/\text{m}^3 \textbf{ Ans.}$$

***(b)** Current density*

We know that current density,

$$J = \sigma E = 5.8 \times 10^7 \times 20 \times 10^{-3} = 1.16 \times 10^6 \text{ A/m}^2 \textbf{ Ans.}$$

***(c)** Current flowing in the wire*

Let I = the value of current flowing in the wire (A)

We know that area of cross-section of wire,

$$A = \frac{\pi\, d^2}{4} = \frac{\pi \times (2 \times 10^{-3})^2}{4} = 3.16 \times 10^{-6} \text{ m}^2$$

We know that current density (J),

$$1.16 \times 10^6 = \frac{I}{3.16 \times 10^{-6}}$$

or $$I = 1.16 \times 3.16 = 3.67 \text{ A} \textbf{ Ans.}$$

(*d*) *Electron Drift velocity*

We know that drift velocity,

$$v = \mu E = 0.0032 \times (20 \times 10^{-3}) = 64 \text{ m/s} \quad \textbf{Ans.}$$

Example 3-3. *A conduction wire has a resistivity of 1.54 × 10^{-8} Ω-m at room temperature. There are 5.8 × 10^{28} conduction electrons per m^3. Calculate the mobility and relaxation time of electrons.* *(UPSC Engg Services 2000)*

Solution. Given : $\rho = 1.54 \times 10^{-8}$ Ω-m and $n = 5.8 \times 10^{28}$.

Let μ = mobility and

t = relaxation time (*i.e.*, time between the collisions)

Mobility

We know that resistivity (ρ),

$$1.54 \times 10^{-8} = \frac{1}{\sigma}$$

$$\therefore \quad \sigma = \frac{1}{1.54 \times 10^{-8}} = 0.649 \times 10^{8} \text{ S/m}$$

We also know that conductivity (σ)

$$0.649 \times 10^{8} = q \cdot n \cdot \mu.$$

$$= 1.60 \times 10^{-19} \times 5.8 \times 10^{28} \times \mu$$

$$= 9.28 \times 10^{9} \mu$$

$$\therefore \quad \mu = \frac{0.649 \times 10^{8}}{9.28 \times 10^{9}} = 0.699 \text{ m}^2/\text{V-s} \quad \textbf{Ans.}$$

Relaxation time

We know that mobility (μ),

$$0.699 = \frac{q \cdot t}{m} = \frac{1.6 \times 10^{-19} \times t}{9.1 \times 10^{-31}}$$

$$= 0.176 \times 10^{12}\, t$$

$$\therefore \quad t = \frac{0.699}{0.176 \times 10^{12}} = 3.976 \times 10^{-12} \text{ S} = 3.976 \text{ ps} \quad \textbf{Ans.}$$

3-16. Conduction in Intrinsic Semiconductors

We have already discussed in Art. 3-4 that in an intrinsic semiconductor at room temperature (*i.e.*, 300 K), the thermal energy is sufficient to produce a large number of electron hole pairs. When an electric field is applied, the current flows through the semiconductor. The flow of current, in a semiconductor, is due to the movement of free electrons in one direction and holes in the opposite direction. We know that electric current in a metal (which is due to the movement of free electrons) is given by the relation,

$$\boldsymbol{J = q \cdot n \cdot \mu \cdot E}$$

Using the above relation, we can write an equation for the current density in an intrinsic semiconductor. Thus current density, which results due to the movement of electrons only, is given by the relation,

$$J_n = q \cdot n \cdot \mu_n \cdot E$$

where, q = Charge of an electron,

n = Number of electrons per unit volume also called (electron concentration) in an intrinsic semiconductor,

μ_n = Mobility of electrons in a semiconductor, and
E = Applied electric field strength.

A similar expression can be written for current density due to holes in an intrinsic semiconductor and is given by the relation,

$$J_p = q \cdot p \cdot \mu_p \cdot E$$

where,
q = Charge of a hole and is equivalent to that of an electron,
p = Number of holes per unit volume also called the concentration in an intrinsic semiconductor,
μ_p = Mobility of holes in a semiconductor, and
E = Applied electric field strength.

The total current density with in an intrinsic semiconductor,

$$J = J_n + J_p$$
$$= q \cdot n \cdot \mu_n \cdot E + q \cdot p \cdot \mu_p \cdot E \quad \text{... }(i)$$
$$= q\,(n \cdot \mu_n + p \cdot \mu_p)\,E \quad \text{... }(ii)$$
$$= \sigma \cdot E$$

where, σ equal to $q\,(n \cdot \mu_n + p \cdot \mu_p)$ is called conductivity of a semiconductor. It is evident from equation (*ii*) that current density within a semiconductor is directly proportional to applied electric field (E).

We know that for an intrinsic semiconductor, the number of holes is equal to the number of electrons, *i.e.*, $p = n = n_i$ where n_i is called intrinsic carrier concentration. Substituting $n = p = n_i$ in equation (*ii*),

$$\boldsymbol{J = q \cdot n_i\,(\mu_n + \mu_p)\,E}$$

and conductivity of an intrinsic semiconductor,

$$\boldsymbol{\sigma_i = q \cdot n_i\,(\mu_n + \mu_p)}$$

It is evident from the above relation that conductivity of an intrinsic semiconductor depends upon its intrinsic concentration (n_i) and the mobility of electrons (μ_n) and holes (μ_p).

Note. The value of q (*i.e.*, charge of electron or hole) is taken as 1.6×10^{-19} coulomb.

Example 3-4. *The mobilities of free electrons and holes in pure germanium are 0.38 and 0.18 m²/V-s. The corresponding values for pure silicon are 0.13 and 0.05 m²/V-s respectively. Find the values of intrinsic conductivity for both the materials. Assume* $n_i = 2.5 \times 10^{19}/m^3$ *for Ge and* $n_i = 1.5 \times 10^{16}/m^3$ *for at room temperature.* *(Anna University Electronic Devices, May/Jun 2007)*

Solution. Given : For germanium, $\mu_n = 0.38$ m²/V-s; $\mu_p = 0.18$ m²/V-s; $n_i = n = p = 2.5 \times 10^{19}/\text{m}^3$ and for silicon, $\mu_n = 0.13$ m²/V-s; $\mu_p = 0.05$ m²/V-s and $n_i = n = p = 1.5 \times 10^{16}/\text{m}^3$.

(*a*) *Intrinsic conductivity for germanium*

We know that the intrinsic conductivity for germanium,

$$\sigma_i = q\,.\,n_i\,(\mu_n + \mu_p) = (1.6 \times 10^{-19}) \times (2.5 \times 10^{19}) \times (0.38 + 0.18)\ (\Omega\text{-m})^{-1}$$
$$= 2.24\ (\Omega\text{-m})^{-1}\ \textbf{Ans.}$$

(*b*) *Intrinsic conductivity for silicon*

We also know that intrinsic conductivity for silicon,

$$\sigma_i = q\,.\,n_i\,(\mu_n + \mu_p)$$
$$= (1.6 \times 10^{-19}) \times (1.5 \times 10^{16}) \times (0.13 + 0.05)\ (\Omega\text{-m})^{-1}$$
$$= 4.32 \times 10^{-4}\ (\Omega\text{-m})^{-1}\ \textbf{Ans.}$$

Example 3-5. *Calculate the intrinsic conductivity of silicon at room temperature if $n_i = 1.41 \times 10^{16}\ m^{-3}$, $\mu_n = 0.145\ m^2/V\text{-}s$, $\mu_p = 0.05\ m^2/V\text{-}s$ and $q = 1.6 \times 10^{-19}$ C. What are the individual contributions made by electrons and holes?* (*Electronics Engg. Nagpur Univ. 1991*)

Solution. Given : $n = 1.41 \times 10^{16}\ \text{m}^{-3}$, $\mu_n = 0.145\ \text{m}^2/\text{V-s}$, $\mu_p = 0.05\ \text{m}^2/\text{V-s}$, and $q = 1.6 \times 10^{-19}$ C.

We know that conductivity of an intrinsic semiconductor,

$$\sigma_i = q \cdot n_i\,(\mu_n + \mu_p)$$

$$= \underbrace{q\,n_i\,\mu_n}_{\text{Contribution by electrons}} + \underbrace{q\,n_i\,\mu_p}_{\text{Contribution by holes}}\ (\Omega\text{-}m)^{-1}$$

$$= (1.6 \times 10^{-19}) \times (1.41 \times 10^{16}) \times 0.145 + (1.6 \times 10^{-19}) \times (1.41 \times 10^{16}) \times 0.05$$

$$= 0.325 \times 10^{-3} + 0.112 \times 10^{-3}\ \Omega\text{-}m^{-1}$$

∴ Contribution by electrons $= 0.325 \times 10^{-3}\ (\Omega\text{-}m)^{-1}$ **Ans.**

and, Contribution by holes $= 0.112 \times 10^{-3}\ (\Omega\text{-}m)^{-1}$ **Ans.**

Example 3-6. *A specimen of silicon is 0.2 mm long and has a cross-section of 0.2 × 0.2 mm. One volt impressed across the bar results in a current of 8 mA. Assuming that the current is due to electrons, calculate **(a)** concentration of free electrons, and **(b)** the drift velocity. Given at 300 K, = 1300 cm²/V-s and q = 1.60 × 10¹⁹ C.*

Solution. Given : $l = 0.2$ mm $= 0.2 \times 10^{-3}$ m; $A = 0.2 \times 0.2 = 0.04\ \text{mm}^2 = 0.04 \times 10^{-6}\ \text{m}^2$; $V = 1$ volt; $I = 8\ \text{mA} = 8 \times 10^{-3}$ A; $\mu_n = 1300\ \text{cm}^2/\text{V-s} = 0.13\ \text{m}^2/\text{V-s}$.

(*a*) *Concentration of free electrons*

Let n = Concentration of free electrons, and

ρ = Resistivity of the specimen.

We know that resistance of a specimen,

$$R = \frac{V}{I} = \frac{1}{8 \times 10^{-3}} = 125\ \Omega$$

and, resistance of a specimen (R),

$$125 = \rho \times \frac{l}{A} = \rho \times \frac{0.2 \times 10^{-3}}{0.04 \times 10^{-6}} = 5000\ \rho$$

$$\rho = 125/5000 = 0.025\ \Omega\text{-m}$$

and, conductivity, $$\sigma = \frac{1}{\rho} = \frac{1}{0.025} = 40$$

We know that the conductivity of a silicon specimen (σ),

$$40 = q \cdot n \cdot \mu_n = (1.60 \times 10^{-19}) \times n \times 0.13 = 0.208 \times 10^{-19}\ n$$

$$n = \frac{40}{(0.208 \times 10^{-19})} = 1.92 \times 10^{21}\ /\text{m}^3 \ \textbf{Ans.}$$

(*b*) *Drift velocity*

Let v = Drift velocity of free electrons in silicon.

We know that current density,

$$J = \frac{I}{A} = \frac{8 \times 10^{-3}}{0.04 \times 10^{-6}} = 2 \times 10^5\ \text{A/m}^2$$

and drift velocity,

$$v = \frac{J}{n \cdot q} = \frac{2 \times 10^5}{1.92 \times 10^{21} \times 1.60 \times 10^{-19}} = 651\ \text{m/sec}\ \textbf{Ans.}$$

Example 3-7. *Find the intrinsic carrier concentration of germanium, if its intrinsic resistivity at 300 K is 0.47 Ω-m. It is given that the electronic charge is 1.6×10^{-19} coulombs and that electron and hole mobilities at 300 K are 0.39 and 0.19 m²/volt-sec respectively.*

Solution. Given : $n_i = 0.47$ Ω-m; $q = 1.6 \times 10^{-19}$ C; $\mu_n = 0.39$ m²/V-sec and $\mu_p = 0.19$ m²/V-sec.

Let n_i = Intrinsic carrier concentration of germanium.

We know that conductivity of a semiconductor,

$$\sigma = \frac{1}{\rho} = \frac{1}{0.47} = 2.13\ (\Omega\text{-m})^{-1}$$

We also know that conductivity of intrinsic semiconductor (σ),

$$2.13 = q \cdot n\ (\mu_n + \mu_p) = 1.6 \times 10^{-19} \times n_i\ (0.39 + 0.19)$$
$$= 0.93 \times 10^{-19}\ n_i$$

$$n_i = \frac{2.13}{(0.93 \times 10^{-19})} = 2.3 \times 10^{19}/\text{m}^3 \ \textbf{Ans.}$$

3-17. Conductivity of N- and P-type Semiconductors

We know that conductivity of an intrinsic semiconductor,

$$\sigma_i = q \cdot n_i\ (\mu_n + \mu_p) = q\ (n_i \cdot \mu_n + n_i \cdot \mu_p)$$

Now substituting $n_i = n = p$ in the above relation,

$$\sigma_i = q\ (n \cdot \mu_n + p \cdot \mu_p)$$

We also know that for N-type semiconductor ($n >> p$), therefore conductivity,

$$\sigma = q \cdot n \cdot \mu_n$$

and for a P-type semiconductor ($p >> n$), therefore conductivity,

$$\sigma = q \cdot p \cdot \mu_p$$

3-18. Mass-Action Law

We have already discussed in Art. 3-9 that the addition of N-type impurities decreases the number of holes below a level, which would have been available in a pure semiconductor. Similarly, the addition of P-type impurities decreases the number of electrons below a level, which would have been available in a pure semiconductor. It has been experimentally found that under thermal equilibrium, the product of the number of holes and the number of electrons is constant and is independent of the amount of donor and acceptor impurity doping. This relation is known as mass action law and is given by the relation,

$$\boldsymbol{n \cdot p = n_i^2}$$

where, n = Free electron concentration (*i.e.*, number of free electrons per m³),

p = Hole concentration (*i.e.*, number of holes per m³), and

n_i = Intrinsic concentration.

3-19. Charge Densities in N- and P-type Semiconductors

We have already discussed in the last article the relationship between free electron concentration (n) and hole concentration (p), which is known as mass-action law. These concentrations are further related by the law of electrical neutrality as discussed below :

Let N_D = Concentration of donor atoms (*i.e.*, the number of positive charges/m³ contributed by donor ions),

p = Hole concentration (*i.e.*, number of holes/m³),

N_A = Concentration of acceptor atoms (*i.e.*, the number of negative charges/m³ contributed by acceptor ions), and

n = Free electron concentration (*i.e.*, number of free electrons/m^3).

∴ Total number of positive charges/m^3 (called total positive charge density) in a semiconductor

$$= N_D + p$$

and the total number of negative charges/m^3 (called total negative charge density) in a semiconductor

$$= N_A + n$$

Since, the semiconductor, as a whole, is electrically neutral, therefore magnitude of total positive charge density is equal to total negative charge density, *i.e.*,

$$N_D + p = N_A + n \qquad \text{... } (i)$$

Now consider an N-type semiconductor. We know that in N-type semiconductor, the concentration of acceptor atoms is zero (*i.e.*, $N_A = 0$), and also the number of free electrons is much larger than the number of holes, (*i.e.*, $n >> p$). Therefore for an N-type semiconductor, the equation (*i*) may be written as :

$$N_D = n$$

The above relation shows that in an N-type semiconductor, the free electron concentration (n) is equal to the number of donor atoms/m^3. The concentration of holes in an N-type semiconductor may be obtained by using mass-action law (*i.e.*, $n \cdot p = n_i^2$) and its value is given by the relation,

$$p = \frac{n_i^2}{n} = \frac{n_i^2}{N_D} \qquad \text{... } (ii)$$

Similarly, in a P-type semiconductor, the concentration of donor atoms is zero (*i.e.*, $N_D = 0$) and the number of holes is much larger than the number of free electrons (*i.e.*, $p >> n$). Therefore, for a P-type semiconductor, the equation (*i*) may be written as :

$$p = N_A$$

The above relation shows that in a P-type semiconductor, the hole concentration (p) is equal to the number of acceptor atoms/m^3. The concentration of free electrons may be obtained by using mass-action law (*i.e.*, $n \cdot p = n_i^2$) and its value is given by the relation,

$$n = \frac{n_i^2}{p} = \frac{n_i^2}{N_A} \qquad \text{... } (iii)$$

In order to avoid unnecessary confusion regarding the type of semiconductor under consideration, it is better to add subscript 'N' for a N-type semiconductor and subscript 'P' for a P-type semiconductor. Thus, if

n_N = Free electron concentration in a N-type semiconductor,

p_N = Hole concentration in a N-type semiconductor,

n_P = Free electron concentration in a P-type semiconductor, and

p_P = Hole concentration in P-type semiconductor.

Then the equations (*ii*) and (*iii*) may be rewritten as :

$$p_N = \frac{n_i^2}{n_N} = \frac{n_i^2}{N_D}$$

and $$n_N = \frac{n_i^2}{p_P} = \frac{n_i^2}{N_A}$$

It will be interesting to know that it is possible to add donor atoms into a P-type semiconductor. Similarly, it is also possible to add acceptor atoms into an N-type semiconductor. If an equal concentration of donor atoms and acceptor atoms is added into an intrinsic semiconductor, it remains intrinsic semiconductor. It is due to the fact that holes supplied by the acceptor atoms combine with the free electrons supplied by the donor atoms to give no additional carriers. The above statement indicates that if the concentration of donor atoms added to a P-type semiconductor exceeds the concentration of acceptor ions, then the semiconductor is changed from a P-type to N-type. This concept is used in the preparation of P-N junction, which is an essential part of all electronic devices.

Example 3-8. *A sample of silicon is doped with phosphorus to a density of $10^{21}/m^3$ as well as with boron to a density of $5 \times 10^{20}/m^3$. What will be conductivity of the silicon sample. The electron mobility in silicon is 0.18 m^2/V-s.*

Solution. Given : $N_D = 10^{21}/m^3$ and $N_A = 5 \times 10^{20}/m^3$

Since, phosphorous atoms are donor and boron atoms are acceptor, therefore net donor density,

$$= N_D' = N_D - N_A = 10^{21} - 5 \times 10^{20} = 5 \times 10^{20}/m^3$$

Number of free electrons, $n = N_D' = 5 \times 10^{20}/m^3$

We know that conductivity of silicon,

$$\sigma = q \cdot n \cdot \mu_n = (1.6 \times 10^{-19}) \times (5 \times 10^{20}) \times 0.18\ (\Omega\text{-m})^{-1}$$

$$= 14.4\ (\Omega\text{-m})^{-1}$$ **Ans.**

Example 3-9. *Calculate the donor concentration in N-type germanium having resistivity of 100 Ω-m. Take $q = 1.6 \times 10^{-19}$ C and $\mu_n = 0.36\ m^2\ V^{-1}\ s^{-1}$.*

Soluction. Given : Resistivity of N-type germanium (ρ) = 100 Ω-m,

$q = 1.6 \times 10^{-19}$ and $\mu_n = 0.36\ m^2\ v^{-1} s^{-1}$

Let N_D = Donor concentration in N–type germanium

We know that conductivity of N-type germanium,

$$\sigma = \frac{1}{p} = \frac{1}{100} = 0.01\ (\Omega - \text{m})^{-1}$$

We also know that conductivity of N-type germanium (σ),

$$0.01 = q\, n\, \mu_n$$

$$= 1.6 \times 10^{-19} \times n \times 0.36$$

$$\therefore \quad n = \frac{0.01}{(1.6 \times 10^{-19}) \times 0.36} = 1.74 \times 10^{17}\ \text{atoms/m}^3$$

and $N_D \approx n = 1.74 \times 10^{17}$ atoms/m^3 **Ans.**

Example 3-10. *Determine the concentration of the free electrons and holes in a sample of Ge at 300 K which has a concentration of donor atoms equal to 2×10^{14} atoms/cm^3 and a concentration of acceptor ions equal to 3×10^{14} atoms/cm^3.* *(Nagpur Univ. Winter 2003)*

Solution. Given : $N_D = 2 \times 10^{14}$ atoms/cm^3; $N_A = 3 \times 10^{14}$ atoms/cm^3,

Let n = number of free electrons

p = number of holes.

We know that number of free electrons,

$$n = \frac{n_i^2}{N_A} \qquad ...(i)$$

where n_i = intrinsic concentration = 2.3×10^{19}

Substituting the volue of n_i (= 2.3×10^{19}) and N_A (= 3×10^{14}) in equation (*i*), we get,

$$n = \frac{\left(2.3 \times 10^{19}\right)^2}{3 \times 10^{14}} = 1.76 \times 10^{24}\ \text{electrons/cm}^3$$ **Ans.**

We know that number of holes,

$$p = \frac{n_i^2}{N_D} = \frac{(2.3 \times 10^{19})^2}{2 \times 10^{14}} = 2.65 \times 10^{24}\ \text{holes/cm}^3$$ **Ans.**

Example 3-11. *A germanium PN junction at 300 K has the following parameters : $N_D = 5 \times 10^{18}/cm^3$, $N_A = 6 \times 10^{16}/cm^3$, $n_i = 1.5 \times 10^{10}/cm^3$. Calculate the minority electron density in the P-region and the minority hole density in the N-region.* (*Banglore Univ. July 2003*)

Soluction. Given : $N_D = 5 \times 10^{18}/cm^3$, $N_A = 6 \times 10^{16}/cm^3$, $n_i = 1.5 \times 10^{10}/cm^3$.

Let n = number of electrons (or electron density) in the P-region and

p = number of holes (or hole density) in the N-region,

We know that number of electrons, in the P-region,

$$n = \frac{n_i^2}{N_A} = \frac{(1.5 \times 10^{10})^2}{6 \times 10^{16}} = 0.375 \times 10^4 = 3750 \textbf{ Ans.}$$

and the number of hole in the N-region,

$$p = \frac{n_i^2}{N_D} = \frac{(1.5 \times 10^{10})^2}{5 \times 10^8} = 0.45 \times 10^{12} \textbf{ Ans.}$$

Example 3-12. *A cylindrical piece of silicon having a diameter of 1mm is doped with $10^{20}/m^3$ atoms of phosphorus which are fully ionized. What length of this silicon would be required to give a resistance of 1 kΩ if electron mobility in silicon is 0.1 m^2/V-s.* (*Pune Univ. 1994*)

Solution. Given : Diameter of silicon piece (d) = 1 mm = 0.001 m; Number of phosphorus atom (N_D) = $10^{20}/m^3$; Resistance (R) = 1kΩ = 1000Ω; electron mobility (μ_n) = 0.1 m^2/V-s.

Let l = length of silicon piece (m),

We know that number of free electrons,

$$n = N_D = 10^{20}/m^3$$

We also know that conductivity,

$$\sigma = q \cdot n \cdot \mu_n$$

$$= 1.6 \times 10^{-19} \times 10^{20} \times 0.1$$

$$= 1.6 \text{ S/m}$$

and the resistance of silicon piece (R),

$$1000 = \frac{1}{\sigma} \cdot \frac{l}{A} = \frac{1}{1.6} \times \frac{l}{\pi\left(\frac{d^2}{4}\right)}$$

$$= \frac{1}{1.6} \times \frac{l}{\pi \times \frac{(0.001)^2}{4}}$$

$$= 7.96 \times 10^{+5}\, l$$

$$\therefore \quad l = \frac{1000}{7.96 \times 10^5} = 1.26 \times 10^{-3} \text{m} = 1.26 \text{ mm } \textbf{Ans.}$$

Example 3-13. *Find the concentration of holes and electrons in a P-type germanium at 300 K, if the conductivity is 100 per ohm per cm. Also find these values for N-type silicon, if the conductivity is 0.1 per ohm per cm. Given for germanium, $n_i = 2.5 \times 10^{13}/cm^3$; μ_n = 3800 cm^2/V-sec; μ_p = 1800 cm^2/V-sec and for silicon $n_i = 1.5 \times 10^{10}/cm^3$; μ_n = 1300 cm^2/V-sec and μ_p = 500 cm^2/V-sec.*

(*Banglore Univ. 2004*)

Solution. Given : For germanium σ = 100 $(\Omega\text{-cm})^{-1}$;

For silicon, ρ = 0.1 $(\Omega\text{-cm})^{-1}$; $n_i = 1.5 \times 10^{10}/cm^3$;

μ_n = 1300 cm^2/V-sec, and μ_p = 500 cm^2/V-sec

$n_i = 2.5 \times 10^{13}/cm^3$; μ_n = 3800 cm^2/V-sec; μ_p = 1800 cm^2/V-sec and

Let p = Concentration of holes, and

n = Concentration of electrons.

Concentration of P-type germanium

We know that conductivity of a P-type germanium (σ),

$$100 = q \cdot p \cdot \mu_p = 1.602 \times 10^{-19} \times p \times 1800 = 2.88 \times 10^{-16} \times p$$

$$\therefore \quad p = \frac{100}{2.88 \times 10^{-16}} = 3.47 \times 10^{17}/\text{cm}^3 \textbf{ Ans.}$$

and the concentration of electrons,

$$n = \frac{n_i^2}{p} = \frac{(2.5 \times 10^{13})^2}{3.47 \times 10^{17}} = 1.8 \times 10^9/\text{cm}^3 \textbf{ Ans.}$$

Concentration in N-type silicon

We also know that conductivity of an N-type silicon, (σ),

$$0.1 = q \cdot n \cdot \mu_n = 1.602 \times 10^{-19} \times n \times 1300 = 2.08 \times 10^{-16} \times n$$

$$\therefore \quad n = \frac{0.1}{2.08 \times 10^{-16}} = 4.8 \times 10^{14}/\text{cm}^3 \textbf{ Ans.}$$

and the number of holes,

$$p = \frac{n_i^2}{n} = \frac{(1.5 \times 10^{10})^2}{4.8 \times 10^{14}} = 4.7 \times 10^5/\text{cm}^3 \textbf{ Ans.}$$

Example 3-14. *In a germanium sample, a donor type impurity is added to the extent of 1 atom per 10^8 germanium atoms. Show that resistivity of the germanium sample drops to 3.7 ohm-cm. Given : $\mu_n = 3800$ cm²/V-sec; $\mu_p = 1800$ cm²/V-sec; $n_i = 2.5 \times 10^{13}$/cm³ ; $N_{ge} = 4.41 \times 10^{22}$ /cm³ and $q = 1.602 \times 10^{-19}$ coulombs.*

Solution. Given : $\mu_n = 3800$ cm²/V-sec; $\mu_p = 1800$ cm²/V-sec; $n_i = 2.5 \times 10^{13}$/cm³; $N_{Ge} = 4.41 \times 10^{22}$/cm³ and $q = 1.602 \times 10^{-19}$ coulombs.

We know that the number of donor atoms,

$$N_D = \frac{N_{Ge}}{10^8} = \frac{4.41 \times 10^{22}}{10^8} = 4.41 \times 10^{14}/\text{cm}^3$$

Since, the number of free electrons (n) is approximately equal to the donor atoms (N_D), therefore $n \approx 4.41 \times 10^{14}$ /cm³. We know that number of holes,

$$p = \frac{n_i^2}{N_D} = \frac{(2.5 \times 10^{13})^2}{4.41 \times 10^{14}} = 1.417 \times 10^{12}/\text{cm}^3$$

It may be noted that the number of electrons (4.41×10^{14}) is very large as compared to the number of holes (1.417×10^{12}) in a doped germanium. Therefore, conductivity of an N-type germanium,

$$\sigma = q \cdot n \cdot \mu_n = (1.602 \times 10^{-19}) \times (1.41 \times 10^{14}) \times 3800 = 0.268\ (\Omega\text{-cm})^{-1}$$

$\therefore$ Resistivity of a doped germanium,

$$\rho = \frac{1}{\sigma} = \frac{1}{0.268} = 3.73\ \Omega\text{-cm} \textbf{ Ans.}$$

Example 3-15. *Find the resistivity of intrinsic silicon. What will this change to, when the silicon is doped with a pentavalent impurity to the extent of one impurity atom for each 50 million silicon atoms ? Following data may be used.*

Number of silicon atoms = 4.96×10^{22}/cm³

Intrinsic carrier concentration = 1.52×10^{10} /cm²

Electron charge = 1.6×10^{-19} *coulomb*

Electron mobility = 0.135 m²/Volt-sec

Hole mobility = 0.048 m²/Volt-sec

Solution. Given : $N_{Si} = 4.96 \times 10^{22}$ /cm^3; $n_i = 1.52 \times 10^{10}$/cm^2; $q = 1.6 \times 10^{-19}$ C; $\mu_n = 0.135$ m^2/V-sec = 1350 cm^2/V-sec and $\mu_p = 0.048$ m^2/V-sec = 480 cm^2/V-sec.

Resistivity of intrinsic silicon

We know that conductivity of an intrinsic silicon,

$$\sigma = q \cdot n_i (\mu_n + \mu_p) = (1.6 \times 10^{-19}) \times (1.52 \times 10^{10}) \times (1350 + 480)\ (\Omega\text{-cm})^{-1}$$
$$= 4.5 \times 10^{-16}\ (\Omega\text{-cm})^{-1}$$

$\therefore$ Resistivity of intrinsic silicon,

$$\rho = \frac{1}{\sigma} = \frac{1}{4.5 \times 10^{-6}} = 2.22 \times 10^5\ \Omega\text{-cm}\ \textbf{Ans.}$$

Resistivity of doped silicon

We know that if the silicon is doped with a pentavalent impurity to the extent of one impurity atom for each 50 million (50×10^6) silicon atoms, then the number of donor atoms,

$$N_D = \frac{N_{si}}{50 \times 10^6} = \frac{4.96 \times 10^{22}}{50 \times 10^6} = 9.92 \times 10^{14}/\text{cm}^3$$

$\therefore$ Number of free electrons in a semiconductor,

$$n = N_D = 9.92 \times 10^{14}/\text{cm}^3$$

We also know that number of holes in a doped silicon,

$$p = \frac{n_i^2}{n} = \frac{(1.52 \times 10^{10})^2}{9.92 \times 10^{14}} = 2.33 \times 10^5/\text{cm}^3$$

Since, the number of electrons (equal to 9.92×10^{14}) in a doped silicon is much larger than the number of holes (equal to 2.33×10^5), therefore, conductivity of a doped silicon,

$$\sigma = q \cdot n \cdot \mu_n = (1.6 \times 10^{-19}) \times (9.92 \times 10^{14}) \times 1350 = 0.21\ (\Omega\text{-cm})^{-1}$$

$\therefore$ Resistivity of a doped silicon,

$$\rho = \frac{1}{\sigma} = \frac{1}{0.21} = 4.67\ \Omega\text{-cm}\ \textbf{Ans.}$$

3-20. Variation in Semiconductor Parameters

There are many parameters of a semiconductor but the following four are important from the subject point of view

1. Intrinsic concentration (n_i)
2. Mobility (μ)
3. Conductivity (σ)
4. Energy gap (E_G)

As a matter of fact, most of the semiconductors operate on a wide range of temperature. Therefore variations of these parameters, with temperature, are very important. The typical values of these parameters for silicon and germanium are as shown in Table 3-3.

Table 3-3: *Properties of intrinsic silicon and germanium*

S. No.	*Property*	*Silicon*	*Germanium*
1.	Atomic Number	14	32
2.	Atomic weight	28.1	72.6
3.	Density (in gm/cm^3)	2.33	5.32
4.	Number of atoms/cm^3	5.0×10^{22}	4.4×10^{22}
5.	Energy gap (E_G) at 0 K (in eV)	1.21	0.785
6.	Energy gap (E_G) at 300 K (in eV)	1.1	0.72
7.	Resistivity at 300 K (in Ω-cm)	2.3×10^5	45
8.	Electron mobility (μ_n) at 300 K (in cm^2/Volt-sec)	1300	3800
9.	Hole mobility (μ_p) at 300 K (in cm^2/Volt-sec)	500	1800
10.	Electron diffusion constant (D_n) at 300 K (in cm^2/sec)	34	99
11.	Hole diffusion constant (D_p) at 300 K (in cm^2/sec)	13	47

3-21. Intrinsic Concentration

In an intrinsic semiconductor, the concentration (n_i) increases with the increase in temperature. The dependence of intrinsic concentration on the temperature is given by the relation,

$$n_i^2 = A_0 \cdot T^3 \cdot e^{-E_{G0} \cdot k \cdot T}$$

where,

A_0 = A constant independent of the temperature,

T = Temperature (in K),

E_{G_0} = Energy gap at 0 K (in eV), and

k = Boltzmann's constant (in eV/K).

The increase in n_i^2 with the temperature has an effect on the charge densities in extrinsic semiconductor. For example in *N*-type semiconductor, the number of free electrons (n) do not change appreciably with the increase in temperature, but number of holes (p) increases. Similarly, for a *P*-type semiconductor, the number of free electrons (n) increases with the increase in temperature, but the number of holes (p) remains constant.

3-22. Mobility

The mobility (μ) of an intrinsic semiconductor varies as T^{-m} over a temperature range (T) of 100 to 400 K. For silicon, the value of $m = 2.5$ for electrons and 2.7 for holes. Similarly, for germanium, the value of $m = 1.66$ for electrons and 2.33 for holes. The mobility of an intrinsic semiconductor decreases with the increase in temperature because at higher temperatures, the number of carriers are more and they are more energetic also. This causes an increased number of collisions with the atoms and thus the mobility decreases.

3-23. Conductivity

The conductivity (σ) of an intrinsic semiconductor depends upon the number of hole-electron pairs and mobility. The number of hole-electron pairs increases with the increase in temperature, while its mobility decreases. However, the increase in hole-electron pairs is greater than the decrease in their mobilities. Therefore, the conductivity of an intrinsic semiconductor increases with the increase in temperature. The conductivity at any temperature (T) is given by the relation,

$$\sigma = \sigma_0 (1 + \alpha \cdot \Delta T)$$

α = Temperature coefficient, *i.e.*, the rate of which conductivity increases per degree Kelvin, and

ΔT = Rise in temperature = $(T - T_0)$

The conductivity of extrinsic semiconductors decreases with the increase in temperature. It is because of the fact, that with the increase in temperature, the number of majority carriers is nearly constant, but mobility decreases. This causes the conductivity to decrease.

Example 3-16. *An intrinsic semiconductor (silicon) has 5×10^{28} atoms/m^3 at 20°C room temperature. At this temperature, there are 1.5×10^{16} electron-hole pairs. Find the conductivity of silicon (σ) at 20°C. If the above material is doped with indium atoms at the rate of 1 atom per 10^7 silicon atoms, find the conductivity of the doped material at room temperature. If the conductivity increases at the rate of 5% per °C then find the conductivity of silicon at 34°C also. Take*

Note: *Hole mobility,* $\mu_p = 0.048$ *m^2/Volt-sec*

Electron mobility, $\mu_n = 0.135$ *m^2/Volt-sec*

Charge of an electron, $q = 1.602 \times 10^{-19}$ *coulombs*

Solution. Given : $\mu_p = 0.048$ m²/Volt-sec; $\mu_n = 0.135$ m²/Volt-sec and $q = 1.602 \times 10^{-19}$ C.

(*a*) *Conductivity of intrinsic silicon*

Given : $Ns_i = 5 \times 10^{28}/\text{m}^3$ and $n_i = 1.5 \times 10^{16}$

We know that conductivity of intrinsic silicon,

$$\sigma = q \cdot n_i (\mu_n + \mu_p) = (1.602 \times 10^{-19}) \times (1.5 \times 10^{16}) \times (0.135 + 0.048)$$

$$= 0.44 \times 10^{-3} \ (\Omega\text{-m})^{-1} \ \textbf{Ans.}$$

(*b*) *Conductivity of doped silicon*

We know that if intrinsic silicon is doped with indium atoms at the rate of 1 atom in silicon atoms, then number of indium atoms,

$$N_A = \frac{Ns_i}{10^7} = \frac{5 \times 10^{28}}{10^7} = 5 \times 10^{21}/\text{m}^3$$

Number of holes, $p = N_A = 5 \times 10^{21}/\text{m}^3$

We also know that number of free electrons,

$$n = \frac{n_i^2}{p} = \frac{(1.5 \times 10^{16})^2}{5 \times 10^{21}} = 4.5 \times 10^{10}/\text{m}^3$$

Since, the number of holes (equal to 5) in a doped silicon is much larger than the number of free electrons (equal to 4.5 × 10^{10}), therefore the doped material is a P-type semiconductor. We know that conductivity of a P-type silicon,

$$\sigma = q \cdot p \cdot \mu_p = (1.602 \times 10^{-19}) \times (5 \times 10^{21}) \times 0.048 \ (\Omega\text{-m})^{-1}$$

$$= 38.45 \ (\Omega\text{-m})^{-1} \ \textbf{Ans.}$$

(*c*) *Conductivity of silicon at 34°C*

Given : $\alpha = 0.05/°\text{C}$; $\Delta T = 34 - 20 = 14°\text{C}$; and $\sigma_{20} = 0.44 \times 10^{-3} \ (\Omega\text{-m})^{-1}$

We also know that conductivity at 34°C,

$$\sigma_{34} = \sigma_{20} (1 + \alpha \cdot \Delta T) = (0.44 \times 10^{-3}) \times (1 + 0.05 \times 14) \ (\Omega\text{-m})^{-1}$$

$$= 0.75 \times 10^{-3} \ (\Omega\text{-m})^{-1} \ \textbf{Ans.}$$

3-24. Energy Gap

The energy required to break a covalent bond in a semiconductor is known as energy gap. It is equal to the difference of energy levels between the conduction band and valence band of the semiconductor crystal structure. It has been found that energy gap decreases with the increase in temperature. The energy gap at any temperature (T) is expressed by the relation,

$$E_G = E_{G_0} - \beta \cdot T \quad \text{... } (i)$$

where, E_{G_0} = Energy gap at 0 K. It value for silicon is equal to 1.21 eV and for germanium 0.785eV, and

β = A constant, whose value depends upon the nature of the material. Its value for silicon is equal to 3.60×10^{-4} and for germanium, 2.23×10^{-4} .

Substituting values of E_{G_0} (equal to 1.21) and β (equal to 3.60×10^{-4}) for silicon in equation (*i*),

$$E_G = 1.21 - 3.60 \times 10^{-4}\,T$$

∴ Energy gap for silicon at room temperature (*i.e.*, 300 K),

$$E_G = 1.21 - 3.60 \times 10^{-4} \times 300 = 1.1 \text{ eV}$$

Similarly, substituting the values of E_{G_0} (equal to 0.785) and β (equal to 2.23×10^{-4}) for germanium in equation (*i*),

$$E_G = 0.785 - 2.23 \times 10^{-4}\,T$$

∴ Energy gap for germanium at room temperature (*i.e.*, 300 K),

$$E_G = 0.785 - (2.23 \times 10^{-4} \times 300) = 0.72 \text{ eV}$$

3-25. Diffusion Current in a Semiconductor

We know that if some voltage is applied to a semiconductor bar, the holes move towards the negative terminal of the battery and electrons towards the positive terminal. This combined effect of the movement of holes and electrons constitute electric current, which is known as *drift current.*

It is possible for an electric current to flow in a semiconductor even in the absence of an applied voltage, provided a concentration gradient exists in the material. A concentration gradient exists, if the number of either electrons or holes is greater in one region of a semiconductor as compared to the rest of the region. It has been observed that when the concentration gradient of carriers (either electrons or holes) exists in a material, the carriers tend to move from the region of higher concentration to the region of lower concentration. This process is called *diffusion* and the electric current, produced due to this process, is known as *diffusion current*.

Consider a bar of P-type semiconductor having a non-uniform concentration of holes as shown in Fig. 3-12. Now consider an imaginary surface *Y-Y'* (shown by a dotted line in the figure). The density of holes on left side of Y-Y is larger than that of the right side. This means that there exists a concentration gradient (*dp*/*dx*) whose value is given by the relation :

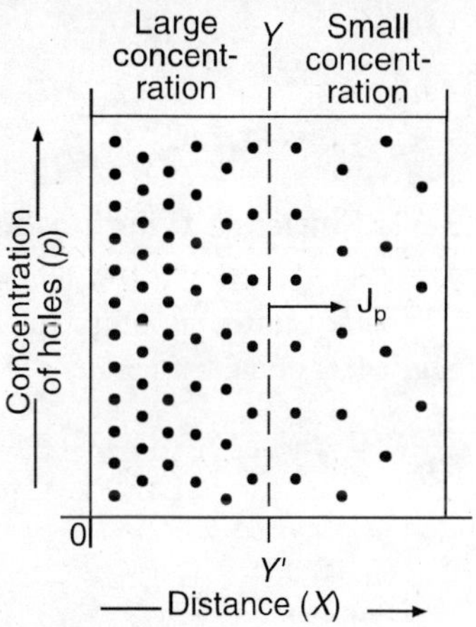

Fig. 3-12. *Illustrating diffusion holes in a P-type semiconductor.*

$$\frac{dp}{dx} = -\frac{p_0 - p_x}{x}$$

where, p_0 = Concentration of holes at left end of the semiconductor baa and may be referred to as an origin plane,

p_x = Concentration of holes at the imaginary surface *Y-Y'*, and

x = Distance of the imaginary surface from the origin plane.

The negative sign indicates that the hole concentration decreases with an increase in the value of *x*. Because of the concentration gradient (*dp*/*dx*), the holes from left of the *Y-Y'* surface diffuse towards the right side. This produces an electric current, called diffusion current. The magnitude of diffusion current per unit area (called diffusion density) is given by the relation,

$$J = -q D_p \cdot \frac{dp}{dx}$$

where, q = Charge of a hole in coulombs,

D_p = Diffusion constant for holes in m^2/sec, and

$\frac{dp}{dx}$ = Concentration gradient.

The negative sign indicates that concentration gradient is negative with the increasing value of x. Similarly, it can be shown that for an N-type semiconductor, the diffusion current density is given by the relation,

$$J_n = -q \cdot D_n - \frac{dn}{dx}$$

where, D_n is the diffusion constant for electrons in m^2/sec.

3-26. Total Current in a Semiconductor

Sometimes, it is possible for the electric field concentration gradient to exist simultaneously within a semiconductor. In that case, the total current is the sum of drift current and diffusion current. Thus for a P-type semiconductor, the total current per unit area (*i.e.*, total current density),

$$J_p = q \cdot \mu_p \cdot p \cdot E + \left(-q \cdot D_p \cdot \frac{dp}{dx} \right)$$

$$= q \cdot \mu_p \cdot p \cdot E - q \cdot D_p \cdot \frac{dp}{dx}$$

Similarly, the total current density for an N-type semiconductor,

$$J_n = q \cdot \mu_n \cdot n \cdot E + \left(-q \cdot D_n \cdot \frac{dn}{dx} \right)$$

$$= q \cdot \mu_n \cdot n \cdot E - q \cdot D_n \cdot \frac{dn}{dx}$$

3-27. Einstein's Relationship for a Semiconductor

According to the Einstein's relationship for a semiconductor, the ratio of diffusion constant (D_p or D_n) to the mobility (μ_p or μ_n) of the charge carriers is constant and is equal to the volt equivalent of the temperature (V_T). Mathematically, the Einstein relationship is expressed as :

$$\boxed{\frac{D_p}{\mu_p} = \frac{D_n}{\mu_n} = V_T}$$

where, D_p = Diffusion constant for holes in m^2/sec,

D_n = Diffusion constant for electrons in m^2/sec,

μ_p = Hole mobility in m^2/V-sec,

μ_n = Electron mobility in m^2/V-sec, and

V_T = Volt equivalent of temperature whose value is equal to $T/11600$

where, T is the temperature in K. The value of V_T is 26 mV at room temperature (*i.e.*, 300 K).

The importance of Einstein's relationship is that it can be used to determine the value of D_p (or D_n), if the mobility of holes (or electrons) is measured experimentally. The values of D_p and D_n for an intrinsic silicon and germanium are as given below :

D_p = 34 cm^2/sec (= 0.0034 m^2/sec)

and D_p = 13 cm^2/sec (= 0.0013 m^2/sec)

Similarly, for germanium the corresponding values are :

D_p = 47 cm^2/sec (= 0.0047 m^2/sec)

and D_n = 97 cm^2/sec (= 0.0097 m^3/sec)

Example 3-17. *Find the diffusion coefficients of holes and electrons for germanium at 300 K. The carrier mobilities in cm²/volt-sec at 300 K for electrons and holes are respectively 3600 and 1700. Density of carriers is 2.5 × 10²³. Boltzmann constant. k = 1.38 × 10²³. (UPSC Engg. Services 1998)*

Solution. Given : μ_n = 3600 m²/V-sec, μ_p = 1700 m²/V-sec.

We know that the diffusion coefficient of holes,

$$D_p = \mu_p \cdot V_T$$

$$= 1700 \times \frac{T}{11600}$$

$$= 1700 \times \frac{300}{11,600} = 4.36 \times 10^{-3} \text{ m}^2/\text{s} \ \textbf{Ans.}$$

and the diffusion coefficient of electrons,

$$D_n = \mu_n V_T$$

$$= 3600 \times \frac{T}{11600}$$

$$= 3600 \times \frac{300}{11,600} = 9.2 \times 10^{-3} \text{ m}^2/\text{sec} \ \textbf{Ans.}$$

3-28. Effect of Temperature on Semiconductors

The product of the majority and minority carrier densities in an intrinsic or extrinsic semiconductor, is fixed at a constant temperature. The ratio of majority to minority carrier densities is in the range of 10^8 to 10^{24} at 300 K. This indicates that in an extrinsic semiconductor, the contribution of minority carriers to the current flow is insignificant. However, we have also discussed in Art. 3-21, that the intrinsic carrier density (n_i) increases with an increase in temperature. Therefore, in an extrinsic semiconductor, both the majority and minority carrier densities increase with the increase in temperature. Since, thermal generation produces electron-hole pairs, therefore the ratio of majority to minority carrier densities reduces as the temperature increases. This ratio continues to decrease until a temperature called *transition-temperature* is reached. At this temperature, the density of thermally generated carriers is the same as the density of the carriers due to doping. Thus, for N-type, $n_i = N_D$ and for P-type, $n_i = N_A$.

However, at and above the transition-temperature, the number of donor electrons in a N-type semiconductor is not equal to the number of donor atoms. Similarly, in a P-type semiconductor, the number of acceptor electrons is not equal to the number of acceptor atoms. This happens due to the significant proportion of the thermally generated carriers. Therefore, an extrinsic semiconductor loses its properties. It acquires properties those of a intrinsic semiconductor as the electron and hole densities are of the same order.

The generation of electron-hole pairs, at elevated temperature, is important in certain semiconductor devices, because it increases the current flow resulting form the rise in carrier densities. This causes the power dissipation of the device to increase. These in turn, increase the temperature of the semiconductor device due to which the carrier densities further increase. Beyond a certain limit, it results in the failure of the device. This effect is called *thermal run away*. It will be taken up again in the subsequent chapters.

3-29. The Hall Effect

It was discovered in 1874 by E.H. Hall, which effect is used to determine the nature of a semiconductor material (*i.e.*, whether it is P-type or N-type), mobile carrier concentration and its mobility. Hall discovered that if a specimen of semiconductor carrying a current is placed in a transverse magnetic field (B) then an electric field, is induced in a direction perpendicular to both I and B. This phenomenon is known as *Hall effect*. It is also observed in metals.

Consider a semiconductor (P-type or N-type) in which the current (I) and magnetic field (B) are applied in the directions as shown in Fig. 3-13. Because of this, a force (F) acts on the charge carriers pushing them towards the back as shown in the figure. The current (I) may be due to the free electrons moving in a direction opposite to that of the current or due to holes moving in the same direction as the current. Thus whether the carriers are electrons or holes, they will be pushed towards the back. This means that the force on the carriers is independent of the type of carriers. But it is depended on the current.

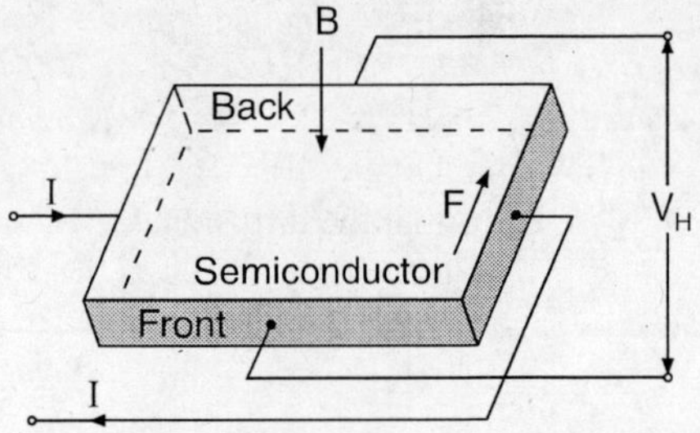

Fig. 3-13. *Hall effect.*

It will be interesting to know that when the mobile carriers (*i.e.*, electrons or holes) are pushed towards the back, the front becomes depleted and the semiconductor loses its neutrality. Now there is an excess of mobile carriers at the back and an excess of opposite charge due to the ionized impurity atoms at the front. If the semiconductor is N-type material, the electrons will be in excess at the back surface and this surface become negatively charged with respect to the front surface. This gives rise to a potential difference called the Hall Voltage (V_H) between the front and back surfaces as shown in the figure. On the other hand, if the semiconductor is P-type material, then holes will be in excess at the back and this surface becomes positively charged with respect to the front surface producing thereby a Hall Voltage (V_H). Thus for a P-type semiconductor, the polarity of V_H is in the reverse direction.

3-30. Experimental Determination of Carrier Concentration and Mobility

Consider an N-type semiconductor carrying a current (I) and placed in a transverse magnetic field (B) in a direction as shown in Fig. 3-14. Then due to the forces exerted on the electrons by the electric and magnetic field, a Hall Voltage (V_H) is induced across the semiconductor as shown in the figure.

In order to maintain a steady flow of current through the semiconductor (called equilibrium state) the force exerted on the electrons due to the electric field and magnetic field must balance each other, *i.e.*,

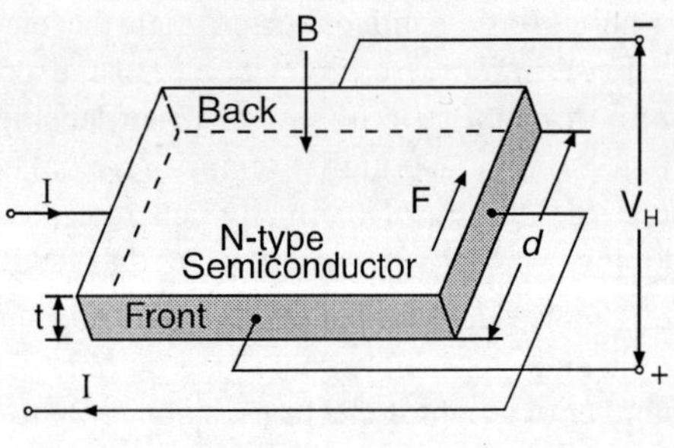

Fig. 3-14. *Determination of carrier concentration.*

$$F_{(electric)} + F_{(magnetic)} = 0$$

$$n \cdot q \cdot E + B \cdot J = 0$$

$$E = -\frac{B \cdot J}{n \cdot q} \qquad \dots (i)$$

where, J = Current density,

n = Electron concentration, and

q = Electronic charge.

It is customary to introduce a term called Hall coefficient in equation (i), which is defined as follows : The Hall coefficient,

$$R_H = -\frac{1}{n \cdot q}$$

Substituting the above value of in equation (i),

$$E = B \cdot J \cdot R_H \qquad \dots (ii)$$

We know that the electric field,

$$E = V_H/d$$

where, V_H = Hall voltage, and

d = Distance between the front and back surface.

We also know that current density is defined as the current per unit area,

$$J = \frac{I}{A} = \frac{I}{d \cdot t}$$

where, t is the thickness of the semiconductor in the direction of the magnetic field. Now substituting the values of E and J in equation (ii),

$$\frac{V_H}{d} = B \times \frac{I}{d \cdot t} \times \frac{1}{n \cdot q}$$

$$V_H = \frac{B \cdot I}{n \cdot q \cdot t}$$

or

$$n = \frac{B \cdot I}{V_H \cdot q \cdot t} \qquad ...(iii)$$

It means that by measuring the values of B, I, V_H and t, the concentration of free electrons can be determined in an N-type semiconductor. The equation (iii) is also valid to determine free electrons in a metal. A similar procedure may be sued to determine the concentration of holes (p) in a P-type semiconductor, which is given by the relation :

$$p = \frac{B \cdot I}{V_H \cdot q \cdot t}$$

The Hall coefficient for a P-type semiconductor is given by the relation :

$$R_H = +\frac{1}{p \cdot q}$$

The Hall coefficient for a P-type semiconductor is positive, whereas it is negative for a N-type semiconductor. It is true because the Hall Voltage in a P-type semiconductor is of opposite polarity to that in a N-type semiconductor. We know that electrical conductivity for an extrinsic semiconductor is given by the relation :

$$\sigma = n \cdot q \cdot \mu_n \qquad \text{... (For N-type semiconductor)}$$

$$= p \cdot q \cdot \mu_p \qquad \text{... (For P-type semiconductor)}$$

where, μ_n = Electron mobility, and

μ_p = Hole mobility

Thus, if the conductivity of a semiconductor is also measured alongwith R_H, then mobility can be determined from the following relations :

$$\mu_n = \frac{\sigma}{n \cdot q} = \sigma \cdot R_H \qquad \text{... (For N-type semiconductor)}$$

and

$$\mu_p = \frac{\sigma}{p \cdot q} = \sigma \cdot R_H \qquad \text{... (For P-type semiconductor)}$$

Example 3-18. *A sample of N-type semiconductor has a Hall coefficient of 160 cm³/coulomb. If its resistivity is 0.16 ohm-cm estimate the electron mobility in the sample.*

Solution. Given : $R_H = 160$ cm^3/coulomb; $\rho = 0.16$ ohm-cm.

We know that electron mobility,

$$\mu_n = \sigma \cdot R_H = \frac{1}{\rho} \cdot R_H = \frac{1}{0.16} \times 160 = 1000 \text{ cm}^2\text{/volt-sec} \textbf{ Ans.}$$

Example 3-19. *A current of 50 A is passed through a metal strip, which is subjected to a magnetic flux of density 1.2 Wb/m². The magnetic field is directed at right angles to the current direction and the thickness of the strip in the direction of the magnetic field is 0.5 mm. The Hall voltage is found to be 100 V. Calculate the number of conduction electrons per cubic metre in the metal.*

Solution. Given : $I = 50$ A; $B = 1.2$ Wb/m²; $t = 0.5$ mm $= 0.5 \times 10^{-3}$ m and $V_H = 100$ volts.

We know that number of conduction electrons,

$$n = \frac{B \cdot I}{V_H \cdot q \cdot t} = \frac{1.2 \times 50}{100 \times 1.6 \times 10^{-19} \times 0.5 \times 10^{-3}}$$

$$= 7.51 \times 10^{21}/\text{m}^3 \textbf{ Ans.}$$

Example 3-20. *An N-type semiconductor has a resistivity of 20×10^{-2} ohm-m. The mobility of the electrons through a separate experiment was found to be 100×10^{-4} m² V⁻¹ s⁻¹. Find the number of electron carriers per m³.*

Solution. Given : $\rho = 20 \times 10^{-2}$ ohm-m; $\mu = 100 \times 10^{-4}$ m² V⁻¹ s⁻¹.

Let n = Number of electron carriers per m³.

We know that mobility (μ)

$$100 \times 10^{-4} = \frac{\sigma}{n \cdot q} = \frac{1}{\rho \cdot n \cdot q} = \frac{1}{20 \times 10^{-2} \times n \times 1.6 \times 10^{-19}} = \frac{1}{32 \times 10^{-21} \times n}$$

$$n = \frac{1}{(100 \times 10^{-4}) \times (32 \times 10^{-21})} = 3.1 \times 10^{21}/\text{m}^3 \textbf{ Ans.}$$

Example 3-21. *The Hall-coefficient of a specimen of doped silicon is found to be 3.66×10^{-4} m³C ; the resistivity of the specimen is 8.93×10^{-3} Ω-m. Find the mobility and density of charge carriers assuming single carriers conduction.* (*UPSC Engg. Services 2002*)

Solution. Given : $R_H = 3.66 \times 10^{-4}$ m³C; $\rho = 8.93 \times 10^{-3}$ Ω-m.

Let μ = mobility and

n = density of charge carriers.

We know that mobility of charge carriers,

$$\mu = \sigma R_H = \frac{1}{\rho} \cdot R_H = \frac{1}{8.93 \times 10^{-3}} \times 3.66 \times 10^{-4}$$

$$= 0.041 \text{ m}^2/\text{V-s}.$$

and the Hall coefficient, (R_H),

$$3.66 \times 10^{-4} = \frac{1}{n \cdot q} = \frac{1}{n \times 1.6 \times 10^{-19}}$$

$$n = \frac{1}{3.66 \times 10^{-4} \times 1.6 \times 10^{-19}} = 1.7 \times 10^{22}/\text{m}^3 \textbf{ Ans.}$$

Example 3-22. *Resistivity of a sample semiconductor is 9 mΩ-m. Its holes have mobility of 0.03 m²/V-s. Calculate the value of Hall-coefficient of the sample. State assumptions made.* (*UPSC Engg. Services 2002*)

Solution. Given : $\rho = 9$ mΩ-m $= 9 \times 10^{-3}$ Ω-m ; $\mu_p = 0.03$ m²/V-s.

We know that conductivity,

$$\sigma = \frac{1}{\rho} = \frac{1}{9 \times 10^{-3}} = 111.1 \text{ S/m}.$$

We also know that mobility (μ_p),

$$0.03 = \sigma \cdot R_H = 111.1 \times R_H$$

$$\therefore \quad R_H = \frac{0.03}{111.1} = 2.7 \times 10^{-4} \text{ m}^3\text{C} \textbf{ Ans.}$$

The assumption is that conductivity in a P-type semiconductor is due to majority carriers (holes) only. In other words, the conductivity due to minority carriers (electrons) is neglected.

3-31. Hall Effect Applications

The Hall effect is used to determine whether the semiconductor is N-type or P-type. It is also used to measure semiconductor parameters like electron or hole concentration and mobility. The Hall effect is used in a magnetic field meter to sense the angular position of static magnetic fields. It is also used in an instrument called Hall effect multiplier, which gives the output proportional to the product of two input signals. Hall devices for such applications are made from N-type Indium antimonide (InSb), because it has a very high electron mobility as well as high and high sensitivity.

SUMMARY

In this chapter you have learned that :

1. Semiconductors are the materials with electrical properties in between that of conductors and insulators.
2. Intrinsic semiconductors are semiconductors is their most pure form.
3. An electron-hole pair is formed by the breakage of a covalent bond when ever the electron acquires more energy.
4. Intrinsic semiconductor, when added with desired amount of impurity atoms is called an extrinsic semiconductor.
5. The process of adding injurity atom to a semiconductor is called doping.
6. Based on the type of impurity atom, the semiconductor is N-type and P-type.
7. Under thermal equilibrium, the product of the number of holes and the number of electrons is constant and is independent of the amount of donor and acceptor impurity doping. This is called mass action law.

GLOSSARY

Acceptor atoms : Atoms which can accept electron to complete covalent bond.

Diffusion current : A small current produced by the diffusion of carriers due to concentration gradient in a PN junction.

Donor atoms : Atoms that have a tendency to donate its free electrons.

Extrinsic semiconductor : Semiconductors obtained by adding certain desired amount of impurity atoms.

Hall effect : Whenever a semiconductor, carrying a current is placed in a transverse magnetic field (B), then an electric field is induced in a direction perpendicular to both I and B.

Impurity : Substances added to an intrinsic semiconductor to change its electrical properties in some desired manner.

Intrinsic concentration (n_i) : Number of charge carriers (*i.e.*, electron or holes) per unit volume of the intrinsic semiconductor.

Intrinsic semiconductor : Semiconductors, in their extremely pure form.

Majority charge carriers : The charge carriers (*i.e.*, electrons or holes) having a higher concentration than that of the other in a semiconductor.

Minority charge : The charge carriers (*i.e.*, electrons or holes) having a lower concentration than that of the other in a semiconductor.

Mobility : The ratio of drift velocity to the electric field (*i.e.*, v/E) and is expressed in $m^2/(V\text{-sec})$.

N-type semiconductor : Semiconductor obtained by introducing pentavalent impurity atoms *e.g.* antimony (Sb) bismuth (Bi) *etc*.

P-type semiconductor : Semiconductor obtained by introducing trivalent impurity atoms, *e.g.*, gallium (Ga), indium (In) *etc*.

Recombination : A process through which an electron-hole pair disappear in a semiconductor.

Semiconductor : The materials, whose electrical properties lie between those of conductors and insulators.

NUMERICAL PROBLEMS

1. Determine the conductivity and resistivity of an intrinsic sample of silicon at normal room temperature (*i.e.*, 300 K). Assume electron mobility, μ_n = 1350 cm²/(V-sec); hole mobility, μ_p = 480 cm²/(V/sec), intrinsic concentration at 300 K = 1.52×10^{10} and charge of an electron or hole = 1.6×10^{-19} coulomb. (**Ans.** 4.45×10^{-6} ; 2.25×10^5)

2. Density of free electrons in pure or intrinsic germanium at 300 K is about 2.4×10^{19} electron per m³. Determine the resistance of a bar of germanium 1 mm × 2 mm and 10 mm long. Given : $\mu_n = 0.39$; $\mu_p = 0.19$ and $q = 1.6 \times 10^{-19}$ coulomb. (**Ans.** 2250 Ω)

3. A sample of germanium is doped to the extent of 10^{14} donor atoms/cm³ and 7×10^{13} acceptor atoms/cm³. At the temperature of 300K, the resistivity of pure germanium is 60Ω-cm. If the applied electric field is 2V/cm, find the total conduction current density. (**Ans.** 2250 Ω)

4. Given that pure silicon has 4.99×10^{22} atoms/cm³ and the intrinsic concentration $n_i = 1.5 \times 10^{10}$ /cm³ at 300 K. Calculate its resistivity and number of holes and free electrons. If there is a donor atom per 10^{12} atoms of silicon, calculate the number of holes and electrons and resistivity of the sample. Other constants are: $q = 1.6 \times 10^{-19}$ coulomb; μ_n = 1300 cm²/V-sec and μ_p = 500 cm²/V-sec).
Ans. ($\sigma_1 = 2.3 \times 10^5$ Ω-cm; $n_i = n = p = 1.5 \times 10^{10}$/cm³; $\rho = 9.3 \times 10^4$; $n = 4.99 \times 10^{10}$/cm³ and $p = 0.45 \times 10^{10}$/cm³)

5. Silicon has a conductivity of 5×10^{-4} ohm^{-1} m^{-1} when pure. How many aluminum atom/m³ are required so that a saturation conductivity of 200 ohm^{-1} m^{-1} could be achieved in silicon using aluminium as an impurity? Given mobility of holes in silicon is 0.05 m² V^{-1} s^{-1} and of electrons is 0.13 m² V^{-1} s^{-1}. (**Ans.** 2.5×10^{22} m³)

6. A current of 10 A is passed through a metal strip which is subjected to a magnetic flux density of 1 Wb/m². The magnetic field is directed at right angles to the current direction and the thickness of the strip in the direction of the magnetic field is 1 mm. The Hall voltage is found to be 10 V. Calculate the number of conduction electrons per cubic metre in the metal. (**Ans.** 6.25×10^{15})

7. The Hall effect is used to determine the mobility of holes in a P-type silicon bar. Assume the bar resistivity as 200000 Ω-cm, the magnetic field β = 0.1 Wb/m² and $d = t = 3$ mm. The measured value of current and Hall voltages are 10 μA and 50 mV respectively. Find the hole mobility (μ_p). (**Ans.** 7.5 m/sec)

DESCRIPTIVE QUESTIONS

1. What is meant by intrinsic semiconductor ?

2. (*a*) Explain what a hole is. How do they move in intrinsic semiconductor ?
(*b*) What is the role of doping of impurities in pure silicon or germanium ?

3. Name three acceptor and three donor materials for doping of a semiconductor.

4. Draw roughly, to scale, the energy band diagrams for a metal, an N-type semiconductor and an insulator at room temperature.

5. Explain the diffusion and drift current for a semiconductor.

6. Give a simple diagram of intrinsic silicon crystal structure and explain briefly the generation and movements of electrons and holes. How does doping modify the conductivity of semiconductors? What are P-type and N-type of semiconductors ?

7. Discuss the meaning of the term covalent bond and its significance in relation to the conduction of electricity in intrinsic P-type and N-type semiconductors. Draw the energy level diagrams in each case. Develope an expression for the resistivity of an intrinsic semiconductor from the concept of free electron and mobility of charge carriers ?

8. Describe the Hall effect and explain its relation to mechanical force exerted by magnetic field on a conduction.

9. What is Hall effect? (*Electron Devices, Anna University, Nov/Dec 2007*)

10. Explain thermal generation of carriers in intrinsic semiconductors.
(*Electron Devices, Anna University, April/May 2008*)

11. State and explain Hall effect. Give the application of Hall Effect.
(*Electron Devices, Anna University, April/May 2008*)

12. What is meant by N-type impurities in a semiconductor?
(*Electorn Devices, Anna University, April/May 2008*)

13. Draw and explain the energy band diagrams of metal semiconductor junction.
(*Electron Devices, Anna University, April/May 2008*)

14. How do you increase the conductivity of intrinsic semiconductor?
(*Electron Devices, Anna University, Nov/Dec 2007*)

15. Derive the continuity equation for a semiconductor.
(*Electron Devices, Anna University, Nov/Dec 2007*)

MULTIPLE CHOICE QUESTIONS

1. The bonding forces in compound semiconductors such as GaAs arise from
(*a*) ionic bounding (*b*) metallic bonding
(*c*) Covalent bonding (*d*) Combination of ionic and covalent bonding
(*UPSC Engg. Services* 2002)

2. For silicon, the energy gap at 300 K is
(*a*) 1.1W (*b*) 1.1 J (*c*) 1.1eV (*d*) all of these

3. The forbidden gap for germanium is,
(*a*) 0.12 eV (*b*) 0.72 eV (*c*) 7.2 eV (*d*) none of these

4. The most commonly used semiconducting material in manufacturing of electronic devices is
(*a*) germanium (*b*) silicon
(*c*) gallium arsenide (*d*) indium antimonide

5. The electric charge on a hole is
(*a*) 1.6×10^{-19} C (*b*) 1.6×10^{-8} C (*c*) 1.6×10^{8} C (*d*) 1.6 C

6. Electron-hole pairs are produced by
(*a*) recombination (*b*) thermal energy (*c*) ionization (*d*) doping

7. Recombination takes place when
(*a*) an electron falls into a hole
(*b*) a positive and a negative ion bond together
(*c*) a valence electron becomes a conduction
(*d*) a crystal is formed

8. The process of adding impurities to a pure semiconductor is called
(*a*) mixing (*b*) doping (*c*) diffusing (*d*) refining

9. An intrinsic semiconductor at the absolute zero temperature
(*a*) behaves like an insulator
(*b*) has a large number of holes
(*c*) has few holes and same number of electrons
(*d*) behaves like a metallic conductor

10. With a rise in temperature around 300 K, the conductivity of an intrinsic semiconductor,

(*a*) remains constant (*b*) increases

(*c*) decreases (*d*) becomes zero

11. The donor type impurities

(*a*) create excess holes

(*b*) can be added to germanium but not to silicon

(*c*) must have only three valence electrons

(*d*) must have only five valence electrons

12. The pentavalent impurities like antimony and phosphorus, added to intrinsic semiconductors are called

(*a*) acceptor or P-types impurities (*b*) donor or P-types impurities

(*c*) acceptor or N-type impurities (*d*) donor or N-type impurities

13. Impurities like boron, gallium or indium are added to intrinsic semicondutor to form

(*a*) N-type doped semiconductor

(*b*) P-type doped semiconductor

(*c*) A junction diode

(*d*) all of these

14. In a N-type semiconductor, the position of Fermi-level

(*a*) is lower than the centre of energy gap

(*b*) is at the centre of energy gap

(*c*) is higher than the centre of energy gap

(*d*) can be any where

15. Which of the following doping will produce a P-type semiconductor,

(*a*) Germanium with phosphorus

(*b*) Silicon with phosphorus

(*c*) Germanium with antimony

(*d*) Silicon with indium

16. The mobility of electrons in a material is expressed in units of :

(*a*) V/s (*b*) m^2/V-s (*c*) m^2/s (*d*) J/K

17. If for intrinsic silicon at 27°C, the charge concentration and mobilities of free electrons and holes are 1.5×10^{16} per m^3, 0.13 m^2/(Vs) and 0.05 m^2 (Vs) respectively the conductivity will be :

(*a*) 2.4×10^{-3} $(\Omega\text{-m})^{-1}$ (*b*) 3.15×10^{-3} $(\Omega\text{-m})^{-1}$

(*c*) 5×10^{-4} $(\Omega\text{-m})^{-1}$ (*d*) 4.32×10^{-4} $(\Omega\text{-m})^{-1}$

18. In a metal

(*a*) the electrical conduction is by electrons and holes

(*b*) the conductivity decreases with the rise in temperature

(*c*) the conduction band is empty

(*d*) None of the above

19. The conductivity of a conductor may be increased by

(*a*) increasing its temperature.

(*b*) decreasing its temperature

(*c*) increasing its vibration

(*d*) changing its shape

20. The energy gap in a semiconductor

(*a*) increases with temperature

(*b*) does not change with temperature
(*c*) decreases with temperature
(*d*) is zero.

21. The Hall-effect occurs only in
(*a*) metals
(*b*) N-type semiconductors
(*c*) intrinsic semiconductors
(*d*) all of the above.

22. If an intrinsic semiconductor is doped with a very small amount of boron, then in the extrinsic semiconductor so formed, the number of electrons and holes will,
(*a*) decrease
(*b*) increase and decrease respectively
(*c*) increase
(*d*) decrease and increase respectively.

23. Two initially identical samples A and B of pure germanium are doped with donors to concentrations of 1×10^{20} and 3×10^{20} respectively. If the hole concentration in A is 9×10^{12}, then the hole concentration in B at the same temperature will be
(*a*) $3 \times 10^{12}\ m^{-3}$
(*b*) $7 \times 10^{12}\ m^{-3}$
(*c*) $11 \times 10^{12}\ m^{-3}$
(*d*) $27 \times 10^{12}\ m^{-3}$

24. Hall effect is observed in a specimen when it (metal or a semiconductor) is carrying current and is placed in a magnetic field. The resultant electric field inside the specimen will be in
(*a*) a direction normal to both current and magnetic field
(*b*) the direction of current
(*c*) a direction antiparallel to the magnetic field
(*d*) an arbitrary direction depending upon the conductivity of the specimen.

25. The unit of Hall-coefficient is :
(*a*) $Vm^3\ A^{-1}\ Wb^{-1}$
(*b*) $Vm^2\ A\ Wb^{-1}$
(*c*) $Vm^3\ A\ Wb^{-1}$
(*d*) $Vm^2\ A^{-2}\ Wb$ *(Chatrapati Shahuji Maharaj Univ. 2004)*

ANSWERS

1. (*a*)	**2.** (*c*)	**3.** (*b*)	**4.** (*b*)	**5.** (*a*)
6. (*b*)	**7.** (*a*)	**8.** (*b*)	**9.** (*a*)	**10.** (*b*)
11. (*a*)	**12.** (*d*)	**13.** (*b*)	**14.** (*c*)	**15.** (*d*)
16. (*b*)	**17.** (*d*)	**18.** (*b*)	**19.** (*b*)	**20.** (*c*)
21. (*b*)	**22.** (*d*)	**23.** (*a*)	**24.** (*a*)	**25.** (*a*)

❑ ❑ ❑

4

CHAPTER

PN Junction

OUTLINE

Objectives

After completing this chapter, you should be able to:

- Know what is a PN junction diode.
- Explain the formation of depletion layer in a PN junction.
- Understand junction or potential barrier.
- Define an unbiased PN junction.
- Explain the behaviour of forward biased PN junction.
- Draw the circuit diagram of a forward biased PN junction.
- Explain the behaviour of reverse biased PN junction.
- Draw the circuit diagram of a reverse biased PN junction.
- Understand the presence of leakage current in a PN junction.
- Know the difference between Zener breakdown and Avalenche breakdown.
- Calculate the depletion layer capacitance and diffusion (or storage) capacitance.
- Describe the different techniques for the fabrication of PN junction.

4-1. Introduction

We have already discussed about the P-type and N-type semiconductors in the last chapter. As a matter of fact, the P-type and N-type semiconductors, taken separately, are of little use in actual practice. If we join a piece of P-type semiconductor to a piece of N-type semiconductor such that the crystal structure remains continuous at the boundary as shown in Fig. 4-1, a PN junction is formed.

Such a PN junction forms a very useful device and is called a semiconductor diode, PN junction diode or simply a crystal diode.

It will be interesting to know that a PN junction cannot be formed by simply joining or welding the two pieces together, because it would produce a discontinuous crystal structure. Special fabrication techniques are used to prepare PN junctions.

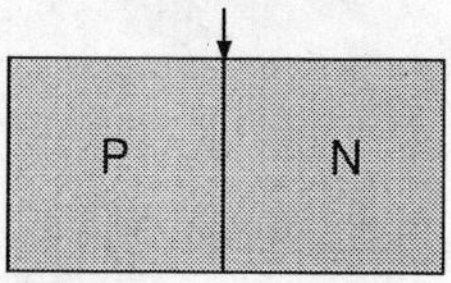

Fig. 4-1. *PN junction*

The PN junction, itself, is an important device. Besides this, all semiconductor devices contain at least one PN junction. Therefore it is very important to understand the behavior of a PN junction, when connected in an electric circuit.

4-2. Formation of Depletion Layer in a PN Junction

Fig. 4-2 (*a*) shows a semiconductor consisting of a PN junction, which has been just formed. It may be noted that this entire sample is a single crystal. Its left half is a P-type and the right half is N-type. The P-region has holes (as majority carriers) and negatively charged impurity atoms, called negative ions (or acceptor ions). The N-region has free electrons (as majority carriers) and positively charged impurity atoms, called positive ions (or donor ions). The minority carriers in the P-and N-regions are not shown just for simplicity.

We know that holes and electrons are mobile charges, and therefore are known as mobile charge carriers. On the other hand, positive and negative ions are immobile charges and therefore do not take part in the conduction. The sample, as a whole, is electrically neutral and so are the P- and *N*-regions, considered separately. Thus in the *P*-region, the total positive charge on the holes is equal to the total negative charges on free electrons and immobile ions. Similarly, in the *N*-region, the total negative charge on free electrons is equal to the total positive charge on holes and immobile ions.

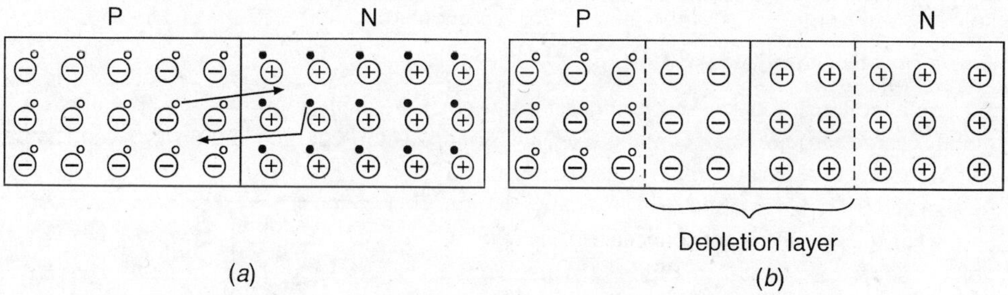

Fig. 4-2. *Formation of deplection layer in a PN junction.*

It may be noted that no external voltage has been connected to the PN junction of Fig. 4-2. As soon as the junction is formed, the conduction and valence bands of P- and N-type materials overlap. As a result of this the following processes take place:

(*i*) The holes, from the P-region diffuse to the N-region, where they combine with the free electrons.

(*ii*) The free electrons, from the N-region diffuse to the P-region, where they combine with holes.

(*iii*) The diffusion of holes (from P-region to N-region) and free electrons (from N-region to P-region) takes place due to the reason that there is a difference of concentrations in the two regions, *i.e.,* the P-region has more number of holes, whereas N-region has more number of free electrons. This difference in concentration creates a concentration gradient across the junction. It results in the diffusion of mobile charge carries across the junction. Moreover, the electrons and holes move at random in all directions due to thermal energy. In doing so, some carriers manage to crossover the junction.

(*iv*) The diffusion of holes and free electrons across the junction takes place for a short time. After a few recombination of holes and free elections, in the vicinity of the junction, a restraining force is automatically set up. This force is produced due to depletion region, which exists on either side of the junction. As a result of this, further diffusion of holes and free electrons from one region to the other is stopped by this depletion layer. The formation of this depletion is explained below:

We know that as soon as the PN junction is formed, some of the holes in P-region and the free electrons in the N-region diffuse in each other and disappear due to recombination. In this process, the negative acceptor ions in the P-region and positive donor ions in the N-region are left uncovered (or uncompensated) in the vicinity of junction as shown in Fig. 4-2 (*b*). The additional holes, trying to diffuse to the N-region, are repelled by the uncovered positive charge of the donor ions. Similarly, the electrons, trying to diffuse into the P-region, are repelled by the uncovered negative charge of the acceptor ions. As a result of this, the further diffusion of free electrons and holes across the junction is stopped.

The region containing the uncovered accepter and donor ions, in the vicinity of the junction, is called *depletion region*. Since, this region has immobile (or fixed) ions, which are electrically charged, therefore the depletion region is also known as *space-charge region*. Moreover, as the uncovered charges within the depletion region exists in the form of parallel rows or plates of opposite charges, therefore, it is known as *depletion layer*. The depletion layer behaves like an insulator. Because of the presence of rows of fixed charges, the depletion layer possesses capacitance. It will be discussed in detail in Art. 4-12.

It will be interesting to know that the width of depletion layer depends upon the doping level of the impurity in *N*-type or *P*-type semiconductor. The higher the doping level, the thinner will be the depletion layer and vice-versa. It is due to the fact that a highly doped PN junction contains a large number of electrons and holes. Because of this, a diffusing charge carrier (either free electron or hole) has not to travel across the junction for recombination.

4-3. Junction or Barrier Voltage

We have already discussed in the last article that the depletion layer of a PN junction has no mobile charge carriers. But it contains fixed rows of oppositely charged ions on its two sides. Because of this charge separation, an electric potential (designated as V_B) is established across the junction, even when the junction is not connected to any external voltage source as shown in Fig. 4-3. This electric potential is called *junction* or *potential barrier.*

It will be interesting to know that this potential barrier exerts a repelling force on the mobile charge carriers, trying to crossover the junction. This force stops the mobile charge carriers to crossover the junction, unless the energy is supplied from an external source. At room temperature (*i.e.,* at 300 K), the value of V_B

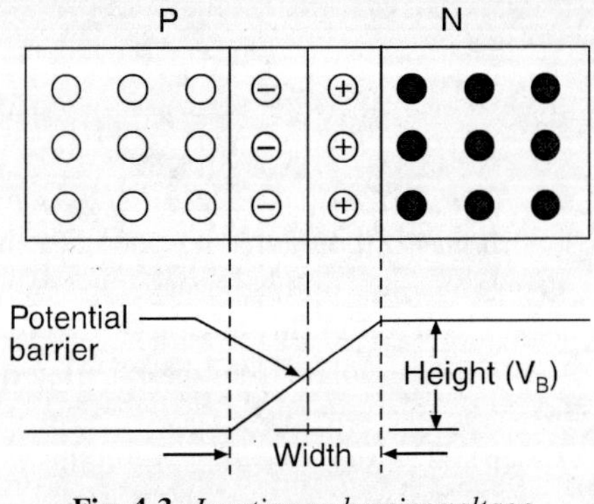

Fig. 4-3. *Junction or barrier voltage.*

$$V_B = \mathbf{0.7\ V} \text{ for silicon and } \mathbf{0.3\ V} \text{ for germanium}$$

4-4. Effect of Temperature on Barrier Voltage

The barrier voltage of a PN junction depends upon three factors namely density, electronic charge and temperature. For a given PN junction, the first two factors are constant, thus making the value of V_B dependent only on temperature. It has been observed that for both germanium and silicon the value of V_B decrease by 2 mV/°C. Mathematically, the decreases in barrier voltage,

$$\Delta V_B = -\ 0.002 \times \Delta t$$

where, Δt is the increase in temperature in °C.

4-5. Currents in an Unbiased PN Junction

A PN junction, across which *no external voltage* source is connected, is known as *unbiased* PN *junction.* Now, consider such an unbiased PN junction as shown in Fig. 4-3. We know that immediately after the formation of a PN junction, a depletion layer is formed in the vicinity of junction. Because of this, there exists a barrier potential (V_B), which stops the further diffusion (or crossover) of carriers across the junction.

The significance of V_B is that, a hole in the P-region requires an energy (equal to $q \cdot V_B$) in order to crossover the junction. Similarly, an electron in the N-region requires the same amount of energy (equal to $q \cdot V_B$) in order to crossover the junction. If the junction is at room temperature, thermal energy is added continuously. As a result of this, few holes and electrons will acquire enough energy to get over the potential barrier, and diffuse the junction. Since, the diffusion of electrons and holes is *opposite* in direction, therefore there is a single current across the junction. This component of current is called a current due to majority carriers or simply *majority carrier current.*

There is another component of current, which flows through the junction. This current is due to the diffusion of minority carriers across the junction. We know that thermal energy causes electron-hole pairs to be generated within the semiconductor materials. Such electron-hole pairs are assisted in diffusing across the junction by barrier potential (V_B). The current produced due to the diffusion of minority carriers, across the junction, is called *minority carrier current.* The minority carrier current flows in a direction opposite to that of the majority carrier current. In an unbiased PN junction, the majority carrier current and minority carrier current are equal in magnitude and flow in opposite directions. It is thus obvious, that there is no net flow of current across the junction.

4-6. Biasing the PN Junction

We have already discussed in the last article that there is no net flow of current across an unbiased PN junction. As a result, the unbiased PN junction is of no use in actual practice. A PN junction, connected to an external voltage source is called a *biased* PN *junction.* Such a PN junction finds a variety of applications in semiconductor devices.

By applying an external voltage across a PN junction, we are able to control the width of the depletion layer. This allows us to control the resistance of the PN junction and also the amount of current that can pass through the device.

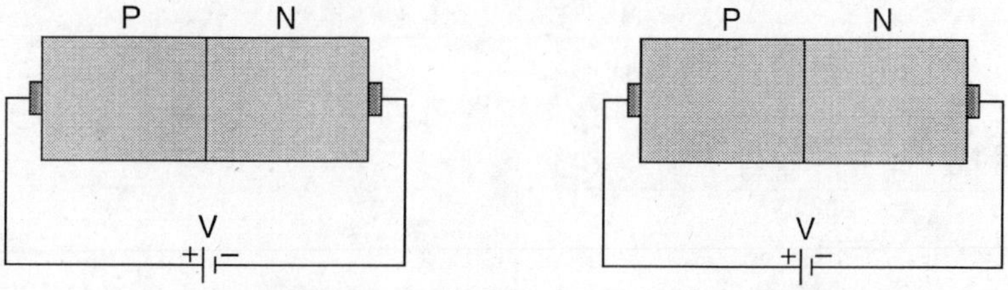

(*a*) Forward bias (*b*) Reverse bias

Fig. 4-4. *Biasing the PN junction.*

There are two ways of connecting voltage source to a PN junction as discussed below:

(*i*) ***Forward bias*** **:** In this case, positive terminal of the voltage source is connected to the P-side and negative terminal to the N-side as shown in Fig. 4-4(*a*). A large amount of current flows through the junction under this condition.

(*ii*) ***Reverse bias*** **:** In this case, positive terminal of the voltage source is connected to the N-side and negative terminal to the P-side as shown in Fig. 4-4(*b*). Practically, no current flows through the junction under this condition.

The application of forward bias across a PN junction shown in Fig. 4-4(*a*) can also be shown as in Fig. 4-5(*a*). Another way of applying the forward bias is as shown in Fig. 4-5(*b*).

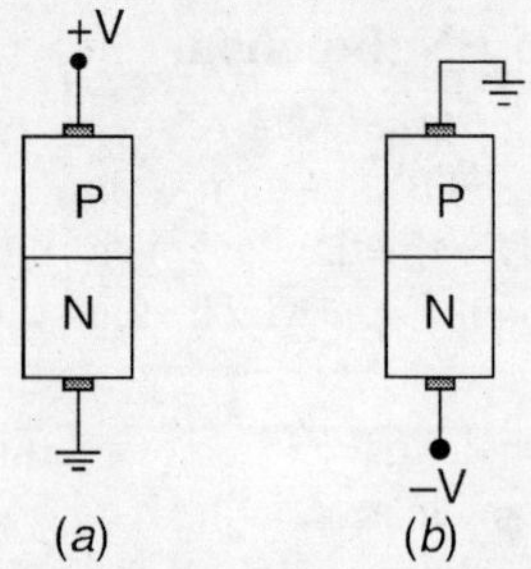

Fig. 4-5. *Forward biasing a PN junction.*

Similarly, the application of reverse bias across a PN junction shown in Fig. 4-4(*b*) can also be shown as in Fig. 4-6(*a*). Another way of applying the reverse bias is as shown in Fig. 4-6(*b*).

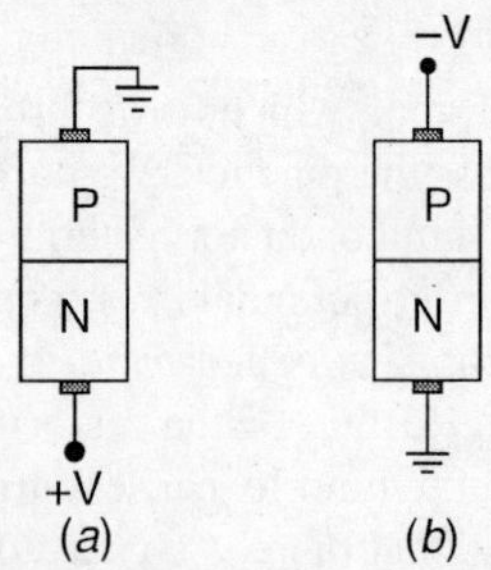

Fig. 4-6. *Reverse biasing a PN junction.*

Now, we shall discuss the forward and reverse-biased PN junctions in greater detail in the following pages.

4-7. Forward Biased PN Junction

We have already discussed in the last article that if we connect voltage source to the PN junction, such that the positive terminal is connected to the P-region and negative terminal to the N-region, the PN junction is said to be forward biased as shown in Fig. 4-7.

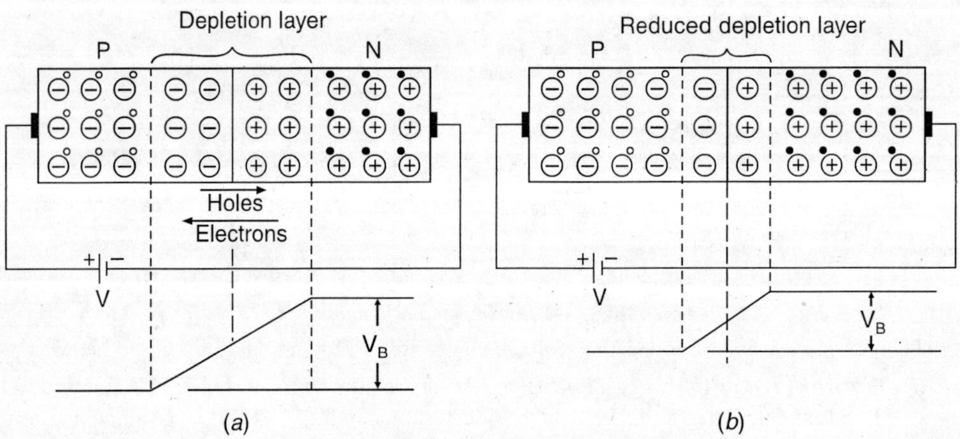

Fig. 4-7. *PN junction with forward bias.*

When a PN junction is forward biased as shown in Fig. 4-7(*a*) the holes are repelled by the positive terminal of the voltage source and are forced to move towards the junction. Similarly, the electrons are repelled by the negative terminal of the voltage source and move towards the junction. Because of their acquired energy (from the voltage source), some of the holes and electrons enter the depletion layer and recombine themselves. *This reduces the width as well as height of the potential barriers* (V_B) as shown in Fig. 4-7 (*b*). In other words, the width of depletion layer and the barrier potential reduces with the forward bias. As a result of this, more majority carriers diffuse across the junction. Therefore, it causes a large current to flow through the PN junction.

It may be noted that for each recombination of free electron and hole, which occurs, an electron from the negative terminal of the voltage source enters the N-type region. Then it moves towards the junction. Similarly, in the P-type region near the positive terminal of the voltage source, an electron breaks a covalent bond in the crystal and enters the positive terminal of the voltage source. Thus, for each electron, which breaks its bond, a hole is created. This hole drifts towards the junction. The current through the external circuit, is due to the movement of electrons only. On the other hand, the current within the PN junction is the sum of electron current (in the N-region) and hole current (in the P-region).

The current in the external circuit continues to flow as flow as long as the voltage source is present in the circuit. The current increases with the increase in applied voltage and is of the order of several milliamperes. The maximum value of current depends upon the actual resistance, called bulk resistance of the semiconductor material.

4-8. Effect of Barrier Potential on the Forward Biased *PN* Junction

The barrier potential of the depletion layer can be considered as a small battery, which opposes the external d.c. voltage as shown in Fig. 4-8. The resistance R_p and R_n represent the bulk resistances of the P-type and N-type semiconductors.

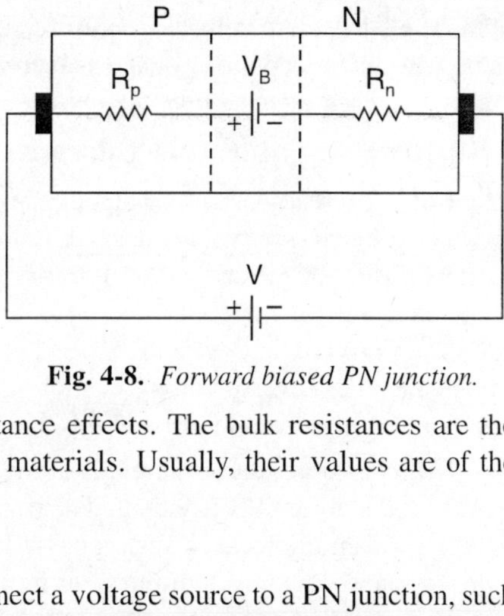

Fig. 4-8. *Forward biased PN junction.*

The PN junction does not permit the current to flow, until the external bias voltage overcomes the barrier potential, (*i.e.*, $V > V_B$). For example, the *silicon* PN junction, does not conduct as long as the external applied voltage is below 0.7 V. Similarly, the germanium PN junction does not conduct as long as the external voltage is below 0.3 V.

It may be noted, that once the current starts flowing through the junction, the voltage drop across it remains equal to the barrier potential. It changes very little with the change in current, except for bulk resistance effects. The bulk resistances are the ohmic resistances of P-type and N-type semiconductor materials. Usually, their values are of the order of few ohms.

4-9. Reverse Biased *PN* Junction

We have already discussed in Art. 4-6 that if we connect a voltage source to a PN junction, such that positive terminal of the voltage source is connected to the N-region and negative to the P-region, then the PN junction is said to be reverse biased. Fig. 4-9 shows a reverse biased PN junction.

When a PN junction is reverse biased, as shown in Fig. 4-9(*a*), the holes in the P-region are attracted towards the negative terminal of the voltage source. And the electrons in the N-region are attracted to the positive terminal of the voltage source. Thus the majority carriers are drawn away from the junction. *This widens the depletion layer and increases the barrier potential* as shown in Fig. 4-9(*b*).

The increased barrier potential makes it very difficult for the majority carriers to diffuse across the junction. Thus, there is no current due to majority carriers in a reverse biased PN junction. In other words, the junction offers very high resistance under reverse biased condition. However, the barrier potential helps the minority carriers in crossing the junction. As a matter of fact, as soon as

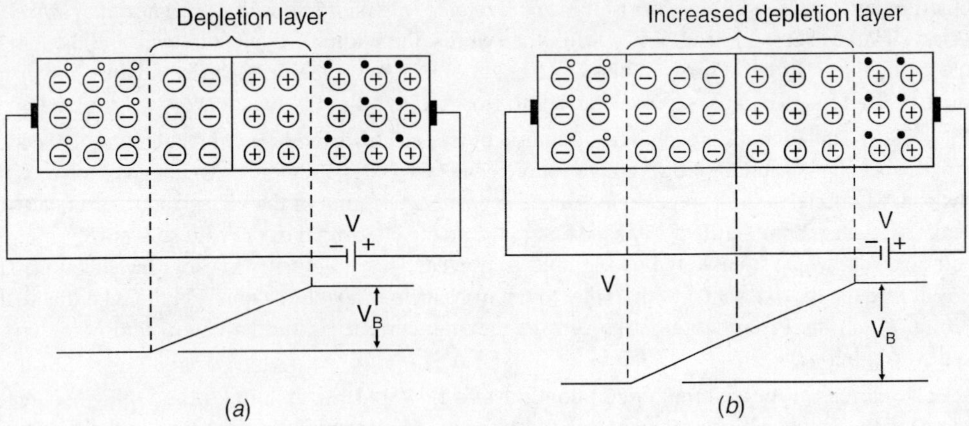

Fig. 4-9. *PN junction with reverse bias.*

a minority carrier is generated, it is swept (*i.e.*, drifted) across the junction because of the barrier potential. Hence, a small amount of current does flow through the reverse biased PN junction. The amount of this current depends upon the number of minority carriers diffusing across the junction. This, in turn, depends upon the generation of minority carriers within the P-region and N-region. It may be noted that generation of minority carriers is dependent upon the temperature and is independent of the applied reverse voltage. Therefore, the current due to the flow of minority carriers, remains the same whether the applied voltage is increased or decreased. Because of this reason, the current is known as *reverse saturation current.*

4-10. Reverse Saturation Current

We have already discussed that in a reverse biased PN junction, practically no current flows due to majority carriers. However, a small amount of current does flow due to the diffusion of minority carriers across the junction. The minority carriers are the hole-electron pairs generated throughout the semiconductor material as a result of thermal energy. The current so produced is known as *reverse saturation current* (or *leakage current*) and is designated by I_0 (or I_s) The word saturation implies that the reverse current cannot be increased by increasing the reverse bias across the PN junction. But it can be increased with the increase in temperature.

The reverse saturation current is of the order of nanoamperes (nA) for silicon and microamperes (μA) for germanium PN junctions. The reverse saturation current, for silicon devices, is smaller than that of germanium because silicon has a large energy gap (≈ 1.1 eV) between the conduction and valence bands. As a result of this, less minority carriers are thermally generated at any temperature in silicon than that of germanium.

It has been experimentally found that the value of reverse saturation current (I_0) is about double in magnitude for every 10°C rise in temperature. For example, if for silicon I_0 is 5 *n*A at 25°C, then it is approximately 10 *n*A at 35°C, 20 *n*A at 45°C, 40 *n*A at 55°C and approximately 1000 *n*A (or 1 μA) at 100°C and so on. However, for germanium, I_0 is in the order of 1-2 μA at 25°C, but its value increases to 100 μA at 100°C.

It is evident from the above discussion that typical values of I_0 for silicon are much lower than that of germanium for similar power and current levels. It has been found that even at high temperatures, the level of I_0 for silicon devices do not reach the same high levels obtained for germanium. It is because of this reason that silicon devices enjoy a significantly higher level of development and utilization in design.

4-11. Reverse Breakdown

We have already discussed that a PN junction allows a very small amount of current (called reverse saturation current), when it is reverse biased. This current is due to the movement of minority carriers across the junction and is independent of the applied reverse voltage. If the reverse bias is increased to a large value, the current through the junction increase abruptly. The voltage, at which this action (*i.e.,* abrupt increase) occurs is known as breakdown voltage. At this voltage, the crystal structure breaks down. It may be noted, that the junction break down is avoided in normal applications. If the excess reverse bias voltage is removed, the crystal structure will return to normal, provided that overheating has not permanently damaged the crystal. The following two processes cause junction breakdown due to the increase in reverse voltage:

1. ***Zener breakdown :*** In this case the breakdown occurs in junctions, which are heavily doped. The heavily doped junctions have a narrow depletion layer. When the reverse voltage is increased, the electric field at the junction also increases. A strong electric field causes a covalent bond to break from the crystal structure. As a result of this, a large number of minority carriers are generated and a large current flows through the junction.
2. ***Avalanche breakdown :*** In this case, the increased reverse voltage increases the amount of energy imparted to minority carriers, as they diffuse across the junction. As the reverse voltage is increased, further, the minority carriers acquire a large amount of energy (or momentum). When these carriers collide with silicon (or germanium) atoms, within the crystal structure, they impart sufficient energy to break a covalent bond and generate additional carriers (*i.e.,* electron-hole pairs). These additional carriers pick up energy from the applied voltage and generate still more carriers. As a result of this, the reverse current increases rapidly. This cummulative process of carrier generation (or multiplication) is known as avalanche breakdown or avalanche multiplication.

4-12. Depletion Layer Capacitance

We have already discussed that when a PN junction is formed, a layer of positive and negative impurity ions, called depletion layer, is formed on either side of the junction. It is also known as depletion-region, space-charge region or transition region. The depletion-layer acts as a dielectric (*i.e.,* non-conductive) medium between P-region and N-region. We know that the P-region and N-region on either side of the junction, has a low resistance. Therefore, these regions act as two plates of a capacitor, separated by a dielectric (*i.e.,* depletion layer) as shown in Fig. 4-10.

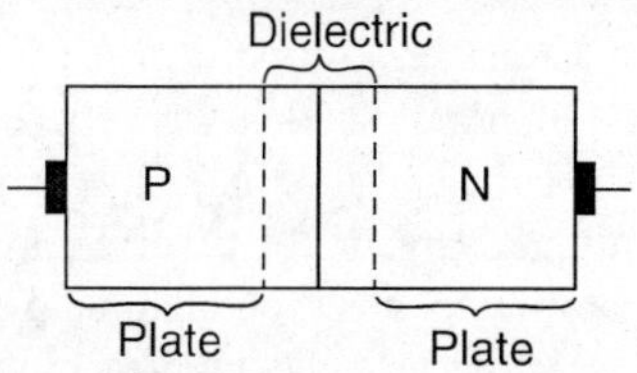

Fig. 4-10. *Depletion layer capacitance.*

The capacitance formed in a junction area is called *depletion layer capacitance.* It is also called *depletion region-capacitance, space charge capacitance, transition region capacitance* or *simply junction capacitance*. The capacitance of a parallel plate capacitor is given by the equation, $C = \varepsilon \cdot A/d$, where ε is the permittivity of the dielectric (insulator) medium between the plates of cross-sectional area A separated by a distance d. Since, the depletion layer width (d) increases with the increase in reverse bias voltage, the resulting depletion layer capacitance will decrease with the increased reverse bias.

The depletion layer capacitance depends upon the nature of a PN junction, semiconductor material and magnitude of the applied reverse voltage. It is given by the relation,

$$\boxed{C_T = \frac{K}{(V_B - V)^n}}$$

where, K = A constant, depending upon the nature of semiconductor material,

V_B = Barrier voltage. It is 0.7 V for silicon and 0.3 V for germanium,

V = Applied reverses voltage, and

n = A constant depending upon the nature of the junction.

It is evident from the above relation that the value of depletion layer capacitance (C_T) can be controlled by varying the applied reverse voltage. This property of variable capacitance, possessed by a reverse biased PN junction, is used in the construction of a device called *varicap* or *varacter*.

Notes: 1. The value of constant K is given by the relation:

$$K = A \times \frac{\varepsilon \cdot q}{2}\left(\frac{N_A \cdot N_D}{N_A + N_D}\right)$$

2. The value of n is taken as 1/2 for step or abrupt junction, 1/3 for linearly graded junction.

4-13. Diffusion or Storage Capacitance

We have already discussed in the last artical that a PN junction behaves like a capacitor. The capacitance, which exists in a forward-biased junction, is called a *diffusion* or *storage capacitance.* It is different from the depletion layer capacitance, which exists in a reverse-biased junction. The diffusion capacitance arises due to the arrangement of minority carrier density. And its value is much larger than the depletion layer capacitance. Its value for an abrupt junction is given by the relation:

$$\boxed{C_D = \frac{\tau \cdot I}{\eta \cdot V_T}}$$

where, τ = Mean life time of the carriers,

I = Value of forward current,

η = A constant (whose value is 1 for germanium and 2 for silicon),

and V_T = Volt equivalent of temperature.

It is evident from the above relation, that diffusion capacitance is directly proportional to the forward current (I). The effect of diffusion capacitance may be understood from the following discussion:

Consider a forward-biased PN junction carrying a current of (I) amperes through it. Suppose the junction is suddenly reverse biased. It causes the forward current to reduce quickly to zero. But it leaves a number of majority charge carriers stored within the depletion region. This charge represents a stored charge in the reverse biased condition and must be removed out of the space charge region. The removal of stored charge takes a finite time. It represents an effect similar to the discharging of a capacitor. Thus the quantity of the stored charge represents the magnitude of diffusion capacitance.

The typical value of diffusion capacitance (C_D) is about 0.02 μF, which is 5000 times the depletion layer capacitance. The effect of diffusion capacitance is negligible for a reverse-biased PN junction.

4-14. Effect of Depletion Layer Capacitance and Diffusion Capacitance

Fig. 4-11 shows a depletion layer and diffusion capacitance versus applied bias for a silicon diode. From the figure, we find that the value of C_D increases with the increase in applied bias while value of C_T decreases with the increase in reverse bias.

It is a well established fact that the electronic devices are inherently sensitive to very high frequencies. Most shunt capacitive effects can be ignored at low frequencies, because the reactance, $X_C = 1/2\pi f C$ is very large (equivalent to open circuit). However, at very high frequencies, the value of X_C becomes sufficiently small due to the high value of f. This introduces a low reactance (or shorting) path for the current. Thus, the depletion layer capacitance and diffusion capacitance do not effect much the performance of PN junction device at low frequencies. Hence, we can ignore their presence in the PN junction at low frequencies. However, their effect is significant at high frequencies.

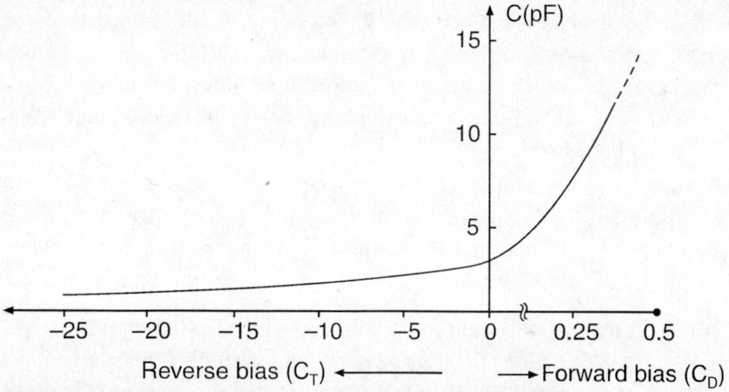

Fig. 4-11. *Depletion layer and diffusion capacitance versus applied bias for a silicon diode.*

4-15. Fabrication of a PN Junction

There are a number of techniques for the fabrication of PN junctions :

(*i*) Grown junction (*ii*) Alloy junction (*iii*) Diffused junction
(*iv*) Epitaxial growth (*v*) Point contract junction

The first process in the manufacture of any semiconductor device is to obtain the semiconductor material in an extremely pure form. In actual practice, the accepted impurity level is less than one part of impurity in ten billion parts of the semiconductor material. To begin with, the material is heated to a temperature above its melting point in an inert gas atmosphere. This gives us a material with uncertain impurity and in polycrystalline form. The material is further purified by a process called *zone refining process*. This process is based on the fact that when a molten semiconductor solidifies prograssively, most of the impurities tend to remain in the liquid phase and will be swept to the end that solidifies in the last. The next process is to obtain single crystals of silicon or germanium, which are required for the fabrication of devices. This is obtained by using a process called Czochralski. These crystals are then cut into wafers. Let us now discuss the basic processes commonly used in the manufacturing of a PN junction.

4-16. Grown Junction

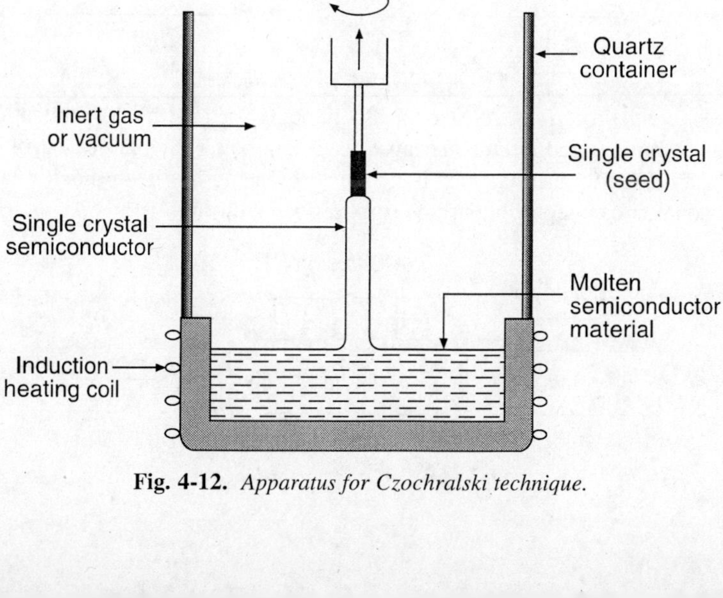

Fig. 4-12. *Apparatus for Czochralski technique.*

This junction is prepared by using either the Czochralski or floating zone technique. The apparatus used for Czochralski technique is as shown in Fig. 4-12. It consists of a graphite crucible, a quartz container, a rotating pulling rod and the induction heating coils placed around the graphite crucible. The graphite crucible contains the molten semiconductor material (*i.e.,* silicon or germanium). A single crystal semiconductor (called seed) of the desired impurity level is placed at the lower end of the rotating rod.

First of all, a single semiconductor seed is immersed in the molten semiconductor material. Then it is gradually withdrawn, while the rod holding the seed is slowly rotating. The PN junctions are grown by first adding P-type impurities to the melt and then changing it to N-type.

4-17. Alloy Junction

The alloy junction technique produces PN junction which have high (PIV) peak-inverse-voltage and current ratings. Such junctions have large capacitance, due to their large junction area. In the alloy junction technique, a small dot of indium is placed on the N-type silicon wafer as shown in Fig. 4-13(*a*).

It is then heated to a temperature of about 150°C. At this temperature, the indium melts and dissolves some of the silicon. Then its temperature is lowered and silicon refreezes to form a single crystal having a PN junction as shown in Fig. 4-13(*b*). The junction location obtained by this method is difficult to control precisely.

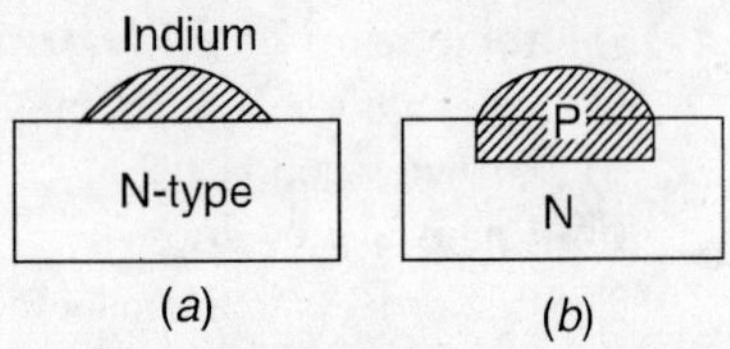

Fig. 4-13. *Alloy junction.*

4-18. Diffused Junction

This technique gives us a precise control of impurity concentration for manufacturing *PN* junctions. The diffused junctions may be obtained by the following three methods:

1. ***Diffused mesa junction :*** In this method, an N-type silicon wafer called substrate (or base) is exposed to a gaseous impurity of P-type as shown in Fig. 4-14(*a*). Then the wafer is heated to a sufficiently high temperature at which the impurities diffuse slowly into the surface of the wafer. After the diffusion, portions of the surface are protected and the rest are etched out as shown in Fig. 4-14(*b*).

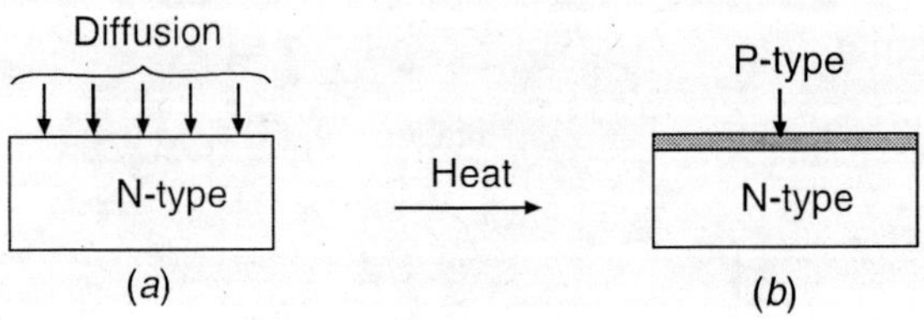

Fig. 4-14. *Diffused mesa junction.*

2. ***Diffused planar junction :*** It is the most commonly used method of fabricating the semiconductor devices and integrated circuits, because it gives a better degree of control over the lateral geometry of the diffused junction. This is achieved by using a silicon dioxide layer, which can prevent most donor and acceptor impurities from diffusing through it.

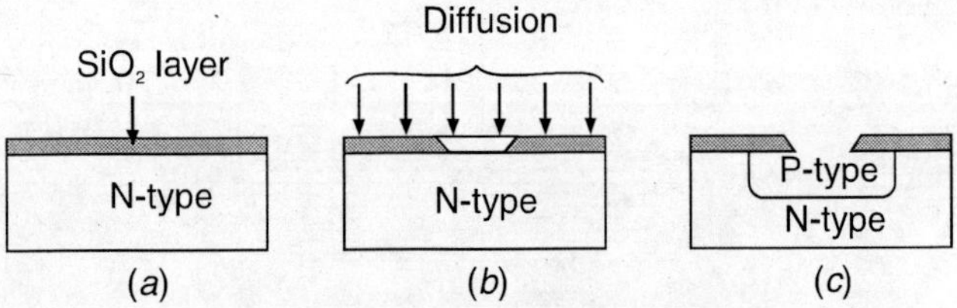

Fig. 4-15. *Diffused planar junction.*

A very thin layer of silicon dioxide (approximately 1 μm) is thermally grown on N-type silicon wafer as shown in Fig. 4-15(*a*). Then with the use of lithographic techniques, portions of the oxide layer can be removed and the windows (or patterns) are cut in the oxide, as shown in Fig. 4-15(*b*) for introducing P-type impurity. This P-type impurity will diffuse only through the exposed silicon surface, and PN junction will be formed as shown in Fig. 4-15(*c*). This process of forming the *PN* junction is called a planar process.

3. *Ion implantation :* It is a process of introducing energetic charged atomic particles into a substrate for changing the electrical, metallurgical or chemical properties of the substrate. The ion implantation method gives the most precise control of an impurity concentration at present.

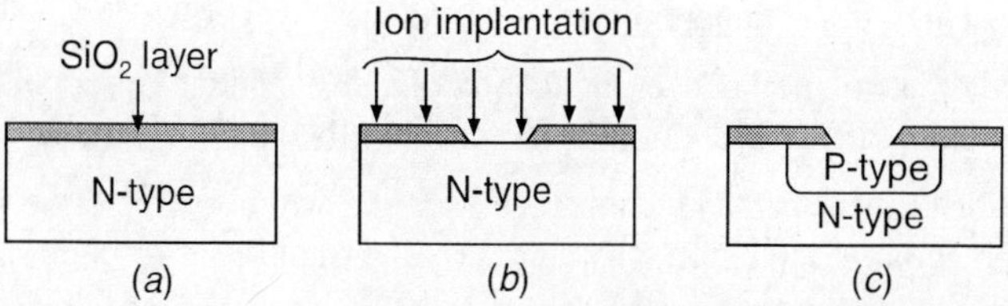

Fig. 4-16. *Ion implantation.*

The formation of a PN junction proceeds by growing a silicon dioxide (SiO_2) layer on the substrate as shown in Fig. 4-16(*a*). The layer is then removed by lithographic techniques and the windows are cut in the oxide as shown in Fig. 4-16(*b*). The ions of P-type impurity are implanted through the exposed silicon surface and the PN junction is formed as shown in Fig. 4-16(*c*).

4-19. Epitaxial Junction

This junction differs from the diffused junction only in one aspect that the junction is fabricated not on the substrate, but on the epitaxial layer grown above the substrate. The epitaxial junctions have an advantage of low resistance.

4-20. Point Contact Junction

It consists of an N-type semiconductor (silicon or germanium) wafer, whose one face is soldered to a metallic base and the other face has a phosphor bronze (or tungsten) spring (called Cat's whisker) present against it as shown in Fig. 4-17(*a*). The whole assembly is encapsulated in a ceramic or glass envelop to give it mechanical strength.

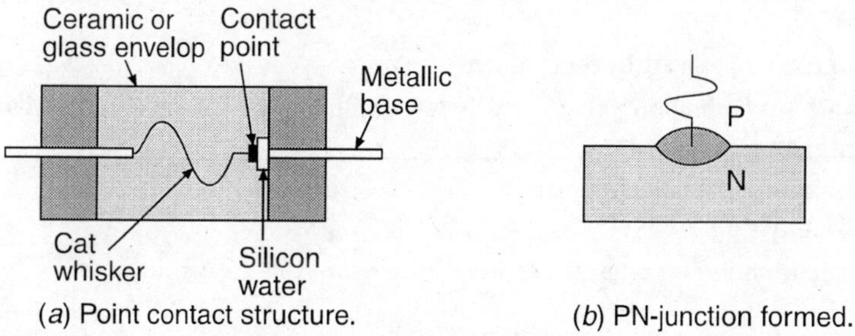

(*a*) Point contact structure.

(*b*) PN-junction formed.

Fig. 4-17.

The PN junction is formed by passing a large current (about 200 mA) for 1 to 100 millisecond duration. The junction is formed at the contact point because of the metling of silicon surface and diffusion of the whisker material into the surface at that point as shown in Fig. 4-17(*b*).

The point contact junctions have very low value of the capacitance. Because of this, such junctions are very useful for operation at the frequencies as high as 10 GHz.

SUMMARY

In this chapter you have learned that :

1. A depletion layer is formed at a junction whenever a P- and N-regions of a semiconductor are joined together.
2. Due to charge separation in the depletion layer, a potential barrier is developed across the junction, Its value is equal to 0.7 V for silicon and 0.3 V for germanium.
3. Forward biasing of a PN junction reduces the width of depletion layer and value of barrier potential. As a result of this a large current flows through the junction.
4. There are two types of junction breakdown: zener breakdown and avalanche breakdown.
5. There are variety of methods for fabrication of a PN junction. This includes grown junction, alloy junction, diffused junction, epitaxial growth and point-contact junction.
6. Reverse biasing of a PN junction increases the width of depletion layer and value of barrier potential. As a result, only a negative current flows through the junction.
7. If the reverse voltage in increased beyond a certain vlaue (V_{BR}), the junction breakdown occurs. The breakdown means the current increases rapidly to a very large value.

GLOSSARY

Avalanche breakdown : It occurs due to breakage of covalent bonds within a PN junction when a reverse voltage is increased beyond a certain limit. Due to avalanche breakdown, the reverse saturation current in a PN junction increases abruptly.

Barrier Voltage : The electric potential estabilished across the PN junction due to charge separation.

Biased PN junction : A PN junction connected to an external voltage source or a battery.

Depletion layer : A small region on both sides of the PN junction which contains fixed ions but no mobile charge carriers.

Depletion layer capacitance : The capacitance offered by the depletion layer of a PN junction. It is also called space-charge region or transition region capacitance.

Diffusion (or storage) : The capacitance offered by the depletion layer of a forward-biased PN junction.

Junction Voltage : Refer to Barrier voltage.

PN junction : A combination of P- and N-type semiconductor in which crystalline structure remains continuous.

Reverse breakdown : A phenomenon in which the current through a reverse biased PN junction increases abruptly.

Reverse saturation current : A small value of current that flows through a PN junction when reverse biased.

Zener breakdown : It occurs due to the breakage of covalent bonds within a PN junction when a strong electric field is applied across the junction.

DESCRIPTIVE QUESTIONS

1. What is a PN junction? Explain the formation of depletion layer in a PN junction.
2. Explain the formation of the depletion region in an open circuit PN junction.
3. Briefly explain (without derivations) the behaviour of junction in forward and reverse bias mode.
4. Discuss how a barrier potential is developed at the PN junction. Using this concept, discuss the behaviour of a PN junction when it is (*i*) forward biased, and (*ii*) reverse biased.
5. Explain how the process of avalanche breakdown occurs in a PN junction. How it is different from a zener breakdown?
6. Explain why a PN junction possesses capacitance.
7. Describe in your own words how diffusion and depletion layer capacitance differ.
8. Give the theory of current components of PN junction diode. (*Electron Devices, Anna University, Nov/Dec. 2007*)
9. Derive an expression for the diffusion capacitance of a PN junction diode. (*Electron Devices, Anna University, Nov/Dec. 2007*)
10. How do the transition region width and contact potential acrosses a PN junction vary with the applied bias voltage? (*Electron Devices, Anna University, Nov/Dec 2007*)
11. Mention the two mechanisms of breadkdown in a PN junction. (*Electron Devices, Anna University, Nov/Dec 2007*)
12. Descirbe the operation of PN junction diode with neat diagram and derive for current. (*Electron Devices, Anna University, April/May 2008*)
13. What is avalanche breakdown? (*Electron Devices, Anna University, April/May 2008*)
14. Define Diffusion Current. (*Electron Devices, Anna University, April/May 2008*)
15. Capacitance of diodes (*Electronic Devices, Anna University, Nov/Dec 2007*)
16. What is meant by depletion region? (*Electronic Devices, Anna University, Nov/Dec 2007*)
17. Write a note on temperature dependence of breakdown voltages. (*Electron Devices, Anna University, Nov/Dec 2007*).
18. Define the transition capacitance of a diode. (*Electronic Devices, Anna University, Nov/Dec 2007*)
19. Explain junction capacitance with reference to a PN diode. (*VIC, Dec 08/Jan 09*)

MULTIPLE CHOICE QUESTIONS

1. When a PN junction is formed, diffusion current causes
   (*a*) mixing of current carriers (*b*) foward bias
   (*c*) reverse bias (*d*) barrier potential
2. The depletion region of a PN junction is one, that is depleted of
   (*a*) Atoms (*b*) Mobiles charges
   (*c*) Immobile charges (*d*) Velocity of the carriers
3. The depletion region within a PN junction is reduced when the junction has:
   (*a*) Zero bias (*b*) Forward bias (*c*) Reverse bias (*d*) all of these
4. A silicon PN junction in forward conduction has a voltage drop closer to
   (*a*) 0.1 V (*b*) 0.7 V (*c*) 1.7 V (*d*) 2.1 V

5. For a reverse biased PN junction, the current through the junction increases abruptly at

(*a*) breakdown voltage (*b*) 0V

(*c*) 0.2 eV (*d*) 7.2 eV

6. The reverse saturation current of a PN junction varies with temperature (*T*) as

(*a*) T (*b*) $1/T$

(*c*) Independent of T (*d*) T^2

7. The transition capacitance of a reverse biased PN junction having uniform doping on both sides, varies with junction voltage (V_B) as

(*a*) $1/V_B$ (*b*) V_B (*c*) $V_B^{-1/2}$ (*d*) V_B^2

8. The leakage current of a PN junction is caused by

(*a*) heat energy (*b*) chemical energy

(*c*) barrier potential (*d*) majority carriers

9. The junction capacitance of a linearly graded junction varies with the applied reverse bias, V_R as

(*a*) V_R^{-1} (*b*) $V_R^{-1/2}$ (*c*) $V_R^{-1/3}$ (*d*) $V_R^{1/2}$

(*UPSC Engg. Services* 2002)

10. The diffusion capacitance of a forward biased P^+N junction diode with a steady current I depends on

(*a*) width of the depleted region (*b*) mean life-time of the holes

(*c*) mean life-time of the electrons (*d*) junction area.

(A P^+ N junction diode is a diode with very heavily doped *P*-region)

(*UPSC Engg. Services* 2002)

ANSWERS

1. (*a*)	**2.** (*b*)	**3.** (*b*)	**4.** (*b*)	**5.** (*a*)
6. (*a*)	**7.** (*c*)	**8.** (*a*)	**9.** (*c*)	**10.** (*c*)

❑ ❑ ❑

5

CHAPTER

PN Junction Diode

OUTLINE

Objectives

After completing this chapter, you should be able to :

- Identify the two terminals (anode and cathode) of commercially available PN junction diodes.
- Draw the V-I characteristics (formed and reverse characteristics) of a typical PN junction diode.
- Define knee voltage and break down voltage.
- Calculate current using the diode current equation.
- Describe the static and dynamic resistance of a diode.
- Compare the real diode with an ideal diode.
- Draw the equivalent circuit of a real diode.
- List the diode applications.
- Read the specifications of a diode from the manufacturer data sheet.
- Determine the suitability of a given diode for a given application using diode specifications sheets and /or selector guides.

5-1. Introduction

A PN junction diode consists of a PN junction, formed either in *germanium* or *silicon* crystal. The diode has two terminals namely *anode* and *cathode*. The anode refers to the P-type region and cathode refers to the N-type region as shown in Fig. 5-1(*a*). Its circuit symbol is as shown in Fig. 5-1(*b*). The arrow head, shown in the circuit symbol, points the direction of current flow, when it is forward biased. It may be noted that it is the same direction in which the movement of holes takes place.

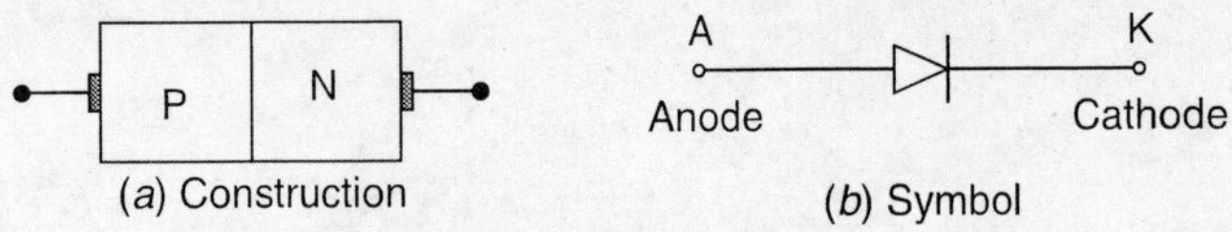

Fig. 5-1. *PN junction diode.*

The commercially available diodes, usually have some notations to identify the P and N terminals (or leads). The standard notation consists of type numbers preceded by IN, such as IN 240 and IN 1250. Here, 240 and 1250 correspond to colour bands. In some diodes, the schematic symbol of a diode is painted or the colour dots are marked on the bodies.

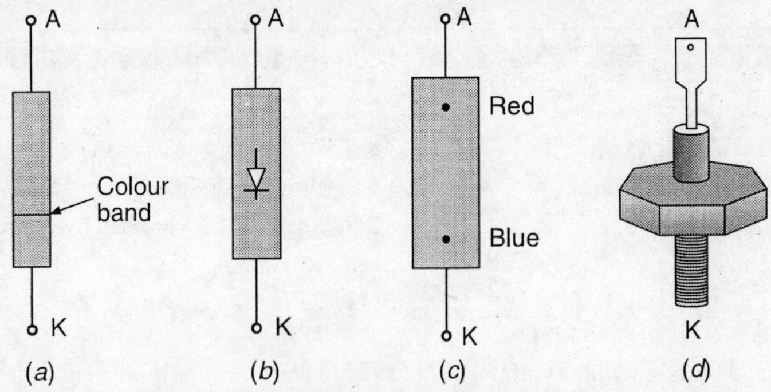

Fig. 5-2. *Low, medium and high current diodes.*

The diode shown in Fig. 5-2(*a*) has a colour band located near one of the ends. The end, which is near the colour band, is identified as cathode (K). And the other end, obviously, is the anode (A). The diode of Fig. 5-2(*b*), has a schematic symbol actually painted as its cathode (K) and the other end as anode. The diodes of Fig. 5-2(*a*) and (*b*), can pass a forward current of 100 mA and are known as *low current diodes.*

The diode shown in Fig. 5-2(*c*) has a colour dots marked on its body. The end lying rear the blue dot is a cathode, while the other end is anode. Sometimes, the diode of Fig. 5-2(*c*) is shown bigger in size than that of diodes shown in Fig. 5-2(*a*) and (*b*). The diodes of this size can pass a forward current of 500 mA and are known as medium current diodes. Fig. 5-2(*d*) shows a diode, which can pass a forward current of several amperes. Therefore, it is known as a *power diode* or a *high current diode*.

5-2. V-I Characteristic of a PN Junction Diode

It is very important to know how a device responds (or behaves) when it is connected in an electrical circuit. This information is obtained by means of a graph known as its volt-ampere (or V-I) characteristics or simply characteristics.

It is a graph between the voltage applied across the terminals of a device and the current that flows through it. Fig 5-3 shows the V-I characteristics of a typical PN junction diode with respect to break down voltage (V_{BR}). It may be noted that the complete graph can be divided into two parts namely forward characteristic and reverse characteristic.

Now we shall discuss the forward characteristics and reverse characteristics of a PN junction diode along with the circuits for obtaining them in the laboratory.

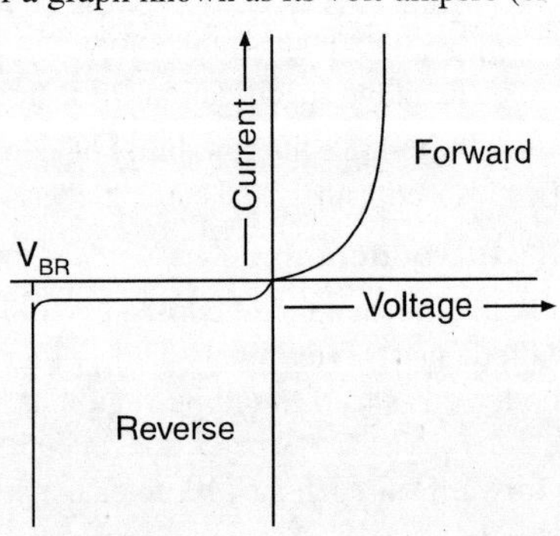

Fig. 5-3. *VI characteristic of PN-junction diode.*

5-3. Forward Characteristic

Figure 5-4(*a*) shows the circuit arrangement for obtaining the forward characteristic of a diode. In this circuit, the diode is connected to a d.c. (V_{AA}) through a potentiometer (*P*) and a resistance (*R*). The potentiometer helps in varying the voltage applied across the diode. The resistance (*R*) is included in the circuit, so as to limit the current through the diode.

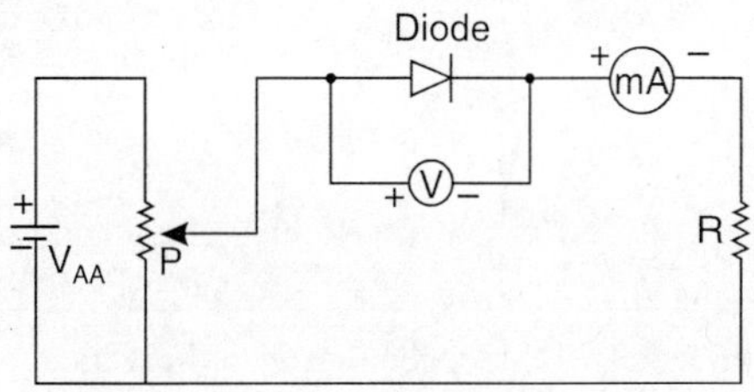

(*a*) Circuit arrangement.

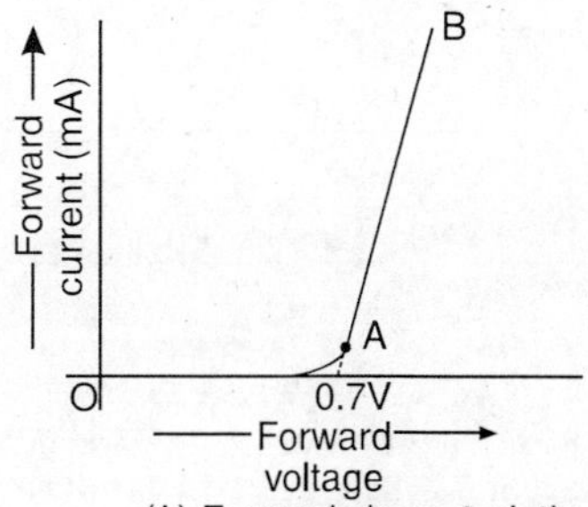

(*b*) Forward characteristic.

Fig. 5-4.

It will be interesting to know that if excessive current is permitted to flow through the diode, it may get permanently damaged. A voltmeter is connected across the diode to measure the voltage, whereas a milliammeter measures current in the circuit.

The positive terminal of the voltage source connected to the anode of a diode and negative terminal to the cathode. Hence, the diode is forward-biased. Let us gradually increase the voltage in small steps of about 0.1 V and record the corresponding values of diode current. Now, if we plot a graph with voltage across the diode, along the horizontal axis and diode current, along the vertical axis, we shall obtain a curve *OAB* as shown in Fig. 5-4(*b*). The curve *OAB* is called the forward characteristic of a silicon PN junction diode.

A careful study of the forward characteristic indicates that there is no diode current till the point *A* is reached. It is because of the fact, that the external applied voltage is being opposed by the junction voltage, whose value is 0.7 V for silicon and 0.3 V for germanium. However, as the voltage is increased above that of the point A, the diode current increases rapidly. It has been observed that a voltage of about 1 volt produces a forward current of about 20 to 50 mA. The applied voltage should not be increased beyond a certain safe limit, otherwise the diode is likely to burn out.

The voltage, at which the diode starts conducting, is called a *knee voltage*, barrier voltage cut-in voltage or threshold voltage. The knee voltage is designated either by V_K V_B or V_γ. Its value is equal to 0.7 V for silicon and 0.3 V for germanium. The knee voltage may be obtained from the forward characteristic by extending the curve *AB* backwards, till it meets the horizontal axis. The value on the horizontal axis is equal to the knee voltage.

5-4. Reverse Characteristic

The circuit arrangement for obtaining the reverse characteristic of a diode is as shown in Fig. 5-5(*a*). This circuit is similar to that shown in Fig. 5-4, except two changes namely the diode terminals are reversed and the milliammeter is replaced by a microammeter.

It may be noted that negative terminal of the voltage source is connected to the anode of a diode and positive terminal to the cathode. Hence, the diode is reverse biased. The applied reverse voltage is gradually increased above zero in suitable steps and the values of diode current are recorded at each step. Now, if we plot a graph with reverse voltage along the horizontal axis and the diode current along the vertical axis, we shall obtain a curve marked *OCD* as shown in Fig. 5-5(*b*). The curve *OCD* is called *reverse characteristic of the diode*.

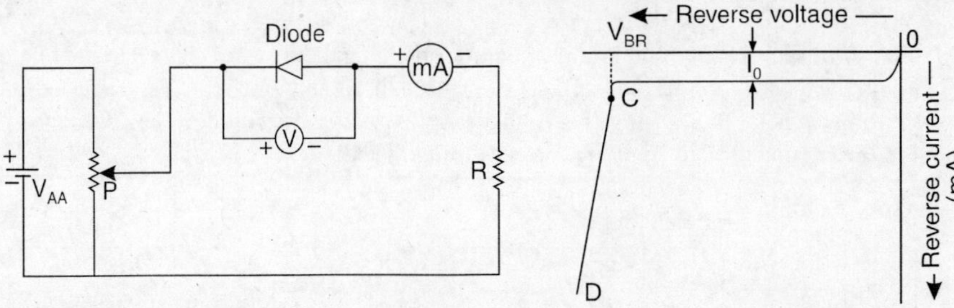

(*a*) Circuit arrangement. (*b*) Reverse characteristic.

Fig. 5-5.

A careful study of the reverse characteristic indicates that when the applied reverse voltage is below the breakdown voltage (V_{BR}), the diode current is small and remains constant. This value of current is called *reverse saturation current* (I_0). It is of the order of nanoamperes (1 nA = 10^{-9}A) for silicon diode and microamperes (1μA = 10^{-6} A) for germanium diode. When the reverse voltage is increased to a sufficiently large value, the diode reverse current increases as rapidly as shown by the curve *CD* in the figure. The applied reverse voltage, at which this happens, is known as break down voltage (V_{BR}) of a diode.

5-5. Diode Current Equation

The mathematical equation, which describes the forward and reverse characteristic of a semi-conductor diode is called the diode current equation.

Let I = Forward (or reverse) diode current,

I_0 = Reverse saturation current,

V = External voltage. It is positive for forward bias and negative for reverse bias.

η = A constant, whose value is equal to 1 for germanium diodes and 2 for silicon diodes for relatively low value of diode current (*i.e.*, at or below the knee of the curve) and = η = 1 for germanium and silicon for higher levels of diode current (*i.e.*, in the rapidly increasing section of the curve)

V_T = Volt-equivalent of temperature. Its value is given by the relation, $\frac{T}{11\,600}$, where T is the absolute temperature. At room temperature (*i.e.*, 300 K), V_T is equal to 26 mV.

The current equation for a forward-biased diode is given by the relation,

$$\boxed{I = I_0\,(e^{V/\eta . V_T} - 1)} \quad ...(i)$$

We know that at room temperature, V_T is equal to 26 mV. Substituting the value of V_T (equal to 26 mV or 0.026 V) in equation (*i*),

$$I = I_0\,(e^{40\,V/\eta} - 1)$$

∴ Diode current at or below the knee of the curve for germanium

$$= I_0\,(e^{40\,V} - 1) \quad ...(\because \eta = 1)$$

and for silicon

$$= I_0\,(e^{20\,V} - 1) \quad ...(\because \eta = 2)$$

If the value of applied voltage is greater than unity (*i.e.*, for the diode current in the rapidly increasing section of the curve), then the equation of diode current for germanium or silicon

$$I = e \cdot {}^{20\,V}$$

The current equation for a reverse biased diode may be obtained from equation (*i*) by changing the sign of the applied voltage (V). Thus the diode current for reversed bias,

$$I = I_0 \, (e^{-V/\eta . V_T} - 1)$$

If the value of $V >> V_T$, then the term $e^{-V/\eta . V_T} << 1$. Therefore, $I = I_0$. Thus, the diode current under reverse-bias is equal to the reverse saturation current as long as the external voltage is below its breakdown value.

Example 5-1. *The current flowing in a certain PN junction diode at room temperature is 2×10^{-7} A, when the large reverse voltage is applied. Calculate the current flowing, when 0.1 V forward bias is applied at room temperature.*

Solution. Given : $I_0 = 2 \times 10^{-7}$ A and $V_F = 0.1$ volt.

We know that the current flowing through the diode under forward bias,

$$I = I_0 \, (e^{40 \, V_F} - 1) = 2 \times 10^{-7} \, (e^{40 \times 0.1} - 1)$$

$$= 2 \times 10^{-7} \, (e^4 - 1) = 1.07 \times 10^{-5} \text{ A}$$

$$= 10.7 \mu \text{A} \textbf{ Ans.}$$

Example 5-2. *Determine the germanium PN junction diode current for the forward bias voltage of 0.22 V at room temperature 25°C with reverse saturation current equal to 1 mA. Take $\eta = 1$.*

Solution. Given: $V_F = 0.22$ volt; $T = 25°\text{C} = 298$ K; $I_0 = 1$ mA $= 1 \times 10^{-3}$ A and $\eta = 1$.

We know that volt equivalent of the temperature (T),

$$V_T = \frac{T}{11\,600} = \frac{298}{11\,600} = 25.7 \, mV = 25.7 \times 10^{-3} \text{ V}$$

$\therefore$ The diode current,

$$I = I_0 \, (e^{V_F / \eta \cdot V_T} - 1)$$

$$= (1 \times 10^{-3}) \, (e^{0.22/(1 \times 25.7 \times 10^{-3})} - 1) \quad ... (\because \eta = 1 \text{ for germanium})$$

$$= (1 \times 10^{-3}) \, (e^{8.56} - 1) = 5.22 \text{ A} \textbf{ Ans.}$$

Example 5-3. *The current voltage characteristic of a PN junction diode is given by the relation:*

$$I = I_0 \, (e^{qV/\eta \cdot k \cdot T} - 1)$$

The diode current is 0.5 mA at V = 340 mV and 15 mA at V = 440 mV. Determine the value of η. Assume kT/q = 25 mV.

Solution. Given: $I_1 = 0.5$ mA $= 0.5 \times 10^{-3}$ A; $V_1 = 340$ mV; $I_2 = 15$ mA $= 15 \times 10^{-3}$ A; $V_2 = 440$ mV and $k.T/q = 25$ mV.

We know that the diode current (I_1)

$$0.5 \times 10^{-3} = I_0 \, (e^{qV_1/\eta \cdot k \cdot T} - 1) = I_0 \cdot e^{qV_1/\eta \cdot k \cdot T}$$

$$= I_0 \cdot e^{340/25\eta}$$

$$= I_0 \cdot e^{13.6/\eta} \qquad ...(i)$$

Similarly, the diode current (I_2)

$$15 \times 10^{-3} = I_0 \cdot e^{440/25\eta} = I_0 \cdot e^{17.6/\eta} \qquad ...(ii)$$

Dividing equation (*ii*) by (*i*),

$$30 = I_0 \cdot e^{(17.6 - 13.6)/\eta} = e^{4/\eta}$$

Taking natural logarithms on both sides,

$$\log_e 30 = \log_e (e^{4/\eta}) = \frac{4}{\eta} \cdot \log_e = \frac{4}{\eta} \qquad .....(\because \log_e = 1)$$

$$3.4 = 4/\eta \text{ or } \eta = 4/3.4 = 1.18 \textbf{ Ans.}$$

5-6. Effect of Temperature on Diode Characteristic

We have already discussed in the last article that the diode current is a function of temperature. And the temperature appears in the denominator of the exponent term of the diode current equation (*i.e.*, $V/\eta \cdot V_T$, where V_T is equal to $T/11\,600$). *It is thus obvious that with the increase in temperature, the exponent will reduce and hence the diode current should also decrease.* However, it has been found that the variation of saturation current (I_0) is much greater than the exponential term. The above fact may be expressed in the form of a mathematical relation as given below :

Let I_{01} = Saturation current at some temperature (T_1) for either germanium or silicon diode,

and I_{02} = Saturation current at some other temperature (T_2).

Then $I_{02} = I_{01} \cdot 2^{(T2-T1)}/10$ ... (For either germanium or silicon)

As discussed in last chapter (Art. 11.10), the reverse saturation current (IO) will be just about double in magnitude for every 10°C increase in temperature. For example, a germanium diode with an I_0 in the order of 1 or 2μ A at 25°C has a leakage current of 100 μA (= 0.1 mA) at a temperature of 100°C. Current levels of this magnitude in the reverse bias region would certainly question our desired open-circuit condition in the reverse bias region. However, typical values of I_0 for silicon diode are much lower than that of germanium for similar power and current levels.

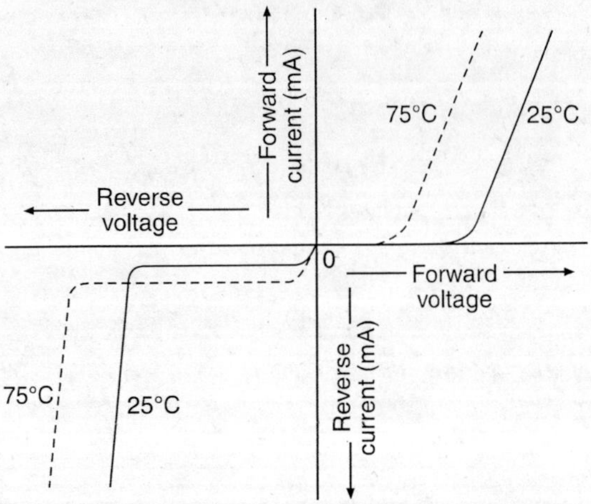

Fig. 5-6. *Effect of temperature on the diode characteristic.*

The increasing level of I_0 with temperature account for the lower threshold voltage as shown in Fig. 5- 6. Because of this reason, the forward characteristic at 75°C is shown to the left to that of the characteristic at 25°C.

As the temperature increases, the forward characteristics shifts more and more to the left of the characteristic at 25°C (*i.e.*, become more and more "ideal"). However, temperatures beyond the normal operating range can have a very detrimental effect on the diode's maximum power and current levels. We see in Fig. 5-6, that in the reverse bias region, the breakdown voltage is increasing with the increase in temperature.

Example 5-4. *A germanium diode has a saturation current of 10 μA at room temperature (i.e., 300 K). Find the saturation current at 400 K.*

Solution. Given: $I_{300} = 10\mu\text{A} = 10 \times 10^{-6}\,\text{A}$; $T_1 = 300$ K and $T_2 = 400$ K.

We know that saturation current at 400 K,

$$I_{400} = I_{300} \times 2^{(T_2 - T_1/10)} = 10 \times 10^{-6} \times 2^{(400-300)/10}$$
$$= 10 \times 10^{-6} \times 2^{10} = 10.2\text{ mA} \textbf{ Ans.}$$

5-7. The Ideal Diode

We know that a PN junction diode conducts well, when it is forward biased and poorly (practically no current) when reverse biased. It will behave as an ideal, if the diode acts as a perfect conductor (with no voltage drop across it) when forward biased and as a perfect insulator (with no current through it) when reverse biased. The V-I characteristic of such an ideal diode is shown in Fig. 5-7(*a*). An ideal diode acts like an automatic switch. The switch is closed when the diode is forward biased and is open when it is reverse biased as shown in Fig. 5-7 (*b*).

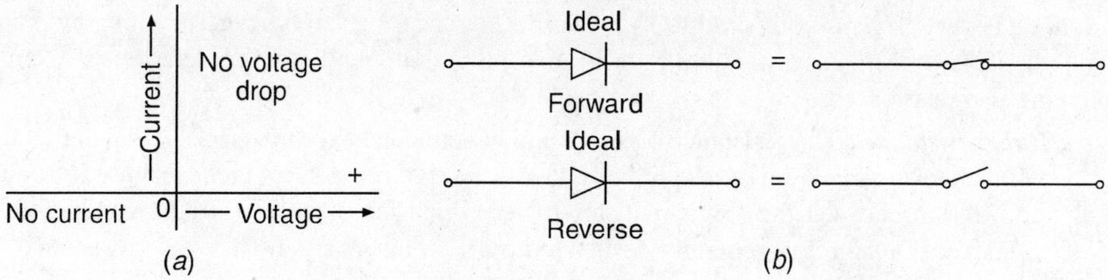

Fig. 5-7. *Ideal diode.*

In other words, an ideal diode is a device, which conducts with zero resistance when forward biased and appears as an infinite resistance when reverse biased. As a matter of fact, an ideal diode cannot be manufactured in actual practice. It is only a theoretical approximation of a real diode. However, in a well-designed electric circuit, a real diode behaves almost like an ideal diode because the forward voltage across the diode is small as compared to the input and output voltages.

5-8. Static and Dynamic Resistance of a Diode

We know that a real diode does not behave as a perfect conductor, when it is forward biased. Similarly, it does not behave as a perfect insulator, when it is reverse biased. In other words, it does not offer zero resistance when forward biased and an infinite resistance when reverse biased. It means that a diode has a definite value of resistance when forward biased. This resistance is known as the d.c or static forward resistance of the diode. It is given by the ratio of the d.c. voltage across the diode to the d.c. current flowing through it. Mathematically, the static forward resistance (in ohms)

$$R_F = V_F / I_F$$

The static forward resistance may be obtained graphically from the diode forward characteristic as shown in Fig. 5-8.

For the operating point *P*, the static forward resistance,

$$R_F = \frac{0.8}{16} = 0.05\ \Omega$$

As seen, the value of R_F is quite low, as it should be. Because of the non-linear shape of the

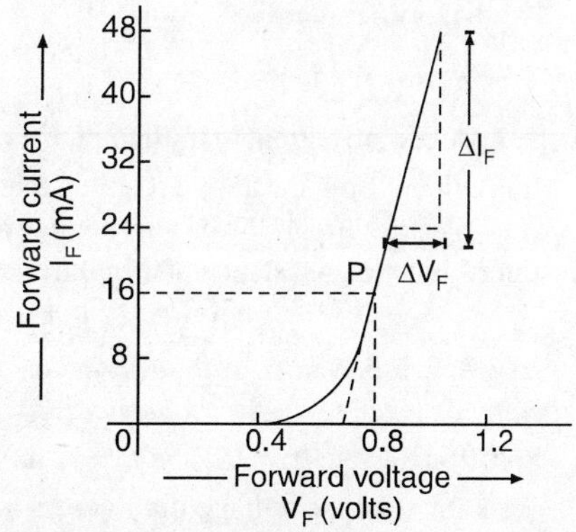

Fig. 5-8. *Static and dynamic forward resistances of a diode from the characteristic curve.*

diode characteristic, the value of R_F depends upon the exact location of the point on the curve. In practice, we don't use the static forward resistance. Instead, we use the dynamic or a.c. resistance. The resistance offered by the diode to an a.c. signal is called its dynamic or a.c. resistance. The a.c. resistance of a diode, at a particular d.c. voltage, is equal to the reciprocal of the slope of the characteristic at that point, *i.e.*, the a.c resistance,

$$r_{ac} = \frac{1}{\Delta I_F / \Delta V_F} = \frac{\Delta V_F}{\Delta I_F} = \frac{\text{Change in voltage}}{\text{Resulting change in current}}$$

Because of the non-linear shape of the forward characteristic, the value of a.c. resistance of a diode is in range of 1 to 25 Ω. Usually, it is smaller than the d.c. resistance of a diode.

In addition to the forward resistance, the diode also possesses another resistance, when it is reverse biased. Such a resistance is known as *reverse resistance*. It can be either d.c. or a.c. depending upon whether the reverse bias is direct or alternating voltage. It may be noted that, ideally, the reverse resistance of a diode is infinite. However, in actual practice, the reverse resistance is never infinite. It is due to the existance of leakage current in a reverse biased diode. The reverse resistance of a diode is very large as compared to its forward resistance. Its value for germanium and silicon diodes is of several meghoms. The a.c. resistance of a diode may also be determined from the following two resistances:

1. *Bulk resistance :* The resistance of the P- and N-semiconductor materials of which the diode is made of, is known as bulk resistance or body resistance. It also includes the resistance introduced by the connection between the semiconductor material and the external metallic conductor also called *contact resistance*. It is generally designated by r_B. Mathematically, the bulk resistance,

$$\boldsymbol{r_B = r_P + r_N}$$

where, r_P = Ohmic resistance of the P-type semiconductor, and

r_N = Ohmic resistance of the N-type semiconductor.

The bulk resistance can range from typically 0.1Ω for high power devices to 2 Ω for some low-power general purpose diodes. The diode current passing through the bulk resistance will develop a voltage equal to ($I_F\ r_B$). Then total voltage drop across the diode is, $V_F = V_B + I_F\ r_B$

$= 0.7 + I_F \cdot r_B$ ... (For silicon diode)

$= 0.3 + I_F \cdot r_B$ ... (For germanium diode)

The above equation indicates that total voltage drop across a diode is directly proportional to I_F.

2. *Junction resistance :* Its value for a forward-biased PN junction depends upon the value of forward d.c. current. It is given by the relation,

$$r_j = \frac{26}{I_F}$$

where, I_F is forward current in milliamperes.

It is evident from the above relation, that the junction resistance is a variable resistance. Higher the value of d.c. current, lower will be the value of r_j and vice versa. The a.c. resistance is equal to the sum of junction resistance and bulk resistance. Mathematically, the a.c. resistance,

$$\boldsymbol{r_{ac} = r_j + r_B}$$

Example 5-5. *A silicon diode has a bulk resistance of 2 Ω and a forward current of 12 mA. What is the actual value of V_F for the device.*

Solution. Given: $r_B = 2\ \Omega$ and $I_F = 12\text{ mA} = 12 \times 10^{-3}$ A.

We know that the voltage drop across a silicon diode,

$$V_F = 0.7 + I_F\ r_B = 0.7 + (12 \times 10^{-3}) \times 2 = 0.724\text{ V}$$ **Ans.**

Example 5-6. *An ideal germanium diode at a temperature of 125°C has a reverse saturation current of 30 μ A. At a temperature of 125°C, find the dynamic resistance for a 0.2 V bias in **(a)** the forward direction, and **(b)** the reverse direction.*

Solution. Given: $T = 125°C = 398$ K; $I_0 = 30\ \mu A = 30 \times 10^{-6}$ A and $V = 0.2$ volt.

(*a*) *Dynamic resistance in forward direction*

We know that volt equivalent of the temperature,

$$V_T = \frac{T}{11\,600} = \frac{398}{11\,600} = 0.0343$$

∴ Diode current,

$$I = I_0 \cdot (e^{V/\eta \cdot V_T} - 1) = I_0 \cdot (e^{V/V_T} - 1) \qquad (\because \eta = 1 \text{ for germanium})$$

Differentiating the above expression with respect to V,

$$\frac{dI}{dV} = \left(\frac{I_0}{V_T}\right) \times e^{V/V_T}$$

or

$$\frac{1}{r_{ac}} = \left(\frac{I_0}{V_T}\right) \times e^{V/V_T} \qquad \ldots \because \frac{\Delta V}{\Delta I} = \frac{dV}{dI} = \frac{r}{ac}$$

∴

$$r_{ac} = \frac{V_T}{I_0} \cdot e^{-V/V_T} \qquad \ldots (i)$$

$$r_{ac} = \frac{0.0343}{30 \times 10^{-6}} \times e^{-0.2/0.0343} = 1143.3 \times e^{-5.83}$$

$$= 1143.3 \times (2.938 \times 10^{-3})$$

$$= 3.36\ \Omega \quad \textbf{Ans.}$$

(*b*) *Dynamic resistance in reverse direction*

Now substituting the given values in equation (*i*) with *V* equal to – 0.2 (due to reverse direction), we find that the dynamic resistance in reverse direction,

$$= \frac{0.0343}{30 \times 10^{-6}} \times e^{(-0.2)/0.0343} = 1143.3 \times e^{5.83}$$

$$= 1143.3 \times 340.35 = 0.389 \times 10^6\ \Omega$$

$$= 0.389\ \text{M}\Omega \quad \textbf{Ans.}$$

Example 5-7. *Show that the junction resistance for a silicon or germanium diode is given by the equation :*

$$r_j = \frac{26}{I_F}$$

where, I_F is the diode current in milliamperes.

Solution. We know that the equation for diode current,

$$I = I_0\ (e^{V/\eta V_T} - 1)$$

and the equation for the diode current in the rapidly increasing section of the forward characteristic,

$$I_F = I_0\ (e^{V/V_T} - 1) \qquad \ldots(\because \eta = 1)$$

$$= I_0\ (e^{11600\,V/T} - 1) \qquad \ldots(\because = V_T\ T/11600)$$

Differentiating the above equation with respect to the applied bias,

$$\frac{dI_F}{dV} = \frac{d}{dV}\left[I_0\ (e^{11600\,V/T} - 1)\right] = \frac{11600}{T}\ (I_F + I_0)$$

In general, $I_F >> I_0$ in the vertical slope section of the characteristics. Therefore,

$$\frac{dI_F}{dV} = \frac{11600}{T}\ I_F$$

And at room temperature, taking the value of T as 25°C or 298 K,

$$\therefore \quad \frac{dI_F}{dV} = \frac{11600}{298} I_F$$

$$\text{or} \quad \frac{dV}{dI_F} = r_j = \frac{298}{11600\, I_F} = \frac{0.026}{I_F}$$

$$\therefore \quad r_j \approx \frac{0.026}{I_F} = \frac{26 \text{ mV}}{I_F}$$

Now, if I_F is expressed in milliamperes, then the junction resistance,

$$r_j = \frac{26 \text{ mV}}{I_F \text{ mA}} = \frac{26}{I_F}$$

5-9. The Real/Ideal Diode

We have already discussed in Art. 5-7 that a diode can be approximated to an automatic switch, which is closed when a diode is forward biased and open when it is reverse biased. But in actual practice, we have to consider two more factors, in addition to a simple switch, for a diode operation as discussed below:

When a diode is forward biased, it does not pass any current until the applied voltage exceeds the knee voltage or the cut-in voltage. Therefore, a diode may be represented by an automatic switch (or an ideal diode) in series with a battery of e.m.f. equal to the knee voltage. The knee voltage has no effect on the diode operation, when it is reverse biased. Therefore, the equivalent circuit of a real diode may be represented as shown in Fig. 5-9(a) and its V-I characteristic as shown in Fig. 5-9(b).

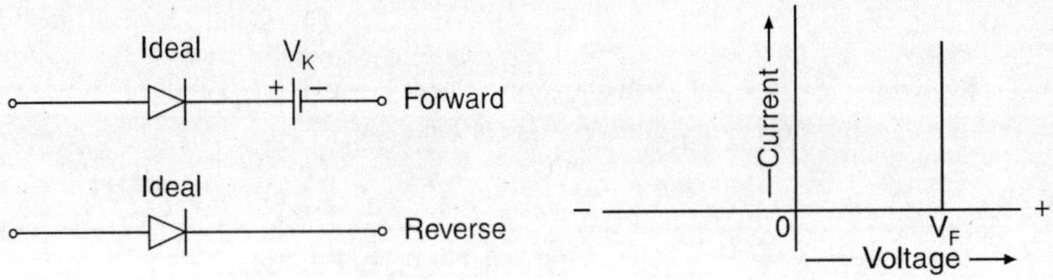

(a) Equivalent circuit. (b) V-I characteristic.

Fig. 5-9.

In addition to a knee voltage (V_K), a diode possesses a.c. or dynamic resistance, which is equal to the sum of bulk resistance and the junction resistance. Thus, when a diode is forward biased, it does not conduct, until the applied voltage exceeds the knee voltage. As the applied voltage exceeds further, the current through a diode increases and produces a voltage drop across the a.c. resistance. Since, the a.c. resistance is linear, therefore the voltage drop across a real diode is given by the relation,

$$V_F = V_K + I_F \cdot r_{ac}$$

where, V_K = Knee voltage whose value is 0.7 V for silicon and 0.3 V for germanium,

I_F = Forward current, and

r_{ac} = A.C resistance.

When a diode is reverse biased, it does not behave like an open switch. In other words, the reverse current is not zero, though it has a small value. It is known as *reverse saturation current*. This means that a reverse biased diode is equivalent to an ideal diode in series with a high resistance (r_r) which is of the order of MΩ. Therefore, the equivalent circuit of a real diode may be represented as shown in Fig. 5-10(a) and its V-I characteristic as shown in Fig. 5-10(b).

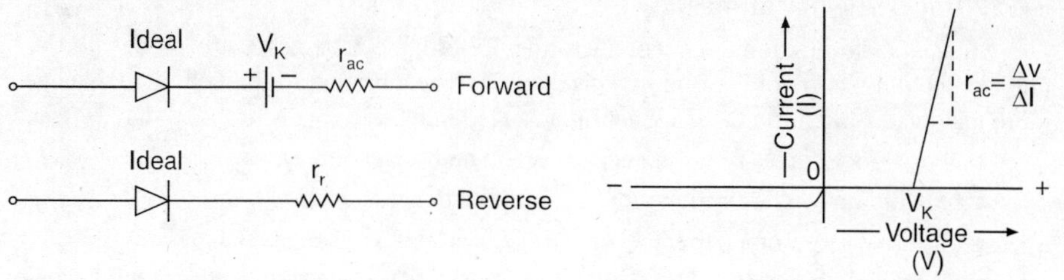

(*a*) Equivalent circuit. (*b*) V.I. characteristic.

Fig. 5.10. *The real diode*

5-10. Diode Applications

A PN junction diode has an important characteristic that it conducts well in forward direction and poorly in reverse direction. This characteristic makes a diode very useful in a number of applications given below:

- As rectifiers or power diodes in d.c. power supplies.
- As signal diodes in communication circuits.
- As zener diodes in voltage stabilizing circuits.
- As varactor diodes in radio and TV receivers.
- As a switch in logic circuits used in computers.

5-11. Power and Current Ratings of a Diode

We know that every diode carrying certain current dissipates some power, whose value is given by the product of voltage and the current through it. Mathematically, the power dissipation for a forward biased diode,

$$P_D = V_F \times I_F$$

where, V_F = Forward voltage drop, and

I_F = Forward current.

Similarly, power dissipation for a reverse biased diode,

$$P_D = V_R \times I_R$$

where, V_R = Reverse voltage drop, and

I_R = Reverse current.

The maximum value of power, which a diode can dissipate without failure, is called ***power rating***. Thus, the power dissipation should not exceed power rating in any case. If it happens, the diode gets destroyed. The manufactures specify power rating of a diode on data sheets. For example, the diode 1 N 914 has a power rating of 250 mW. It means that the diode can dissipate a maximum power of 250 mW.

It will be interesting to know, that the diode manufactures more oftenly list the maximum current, which a device can handle, (called current rating), rather than power rating. It is because of the fact, that it is easy to measure current rating than power rating. For example, the data sheet of 1N 4003 does not indicate power rating. But it does list a maximum d.c. current of 1 ampere. This means that if we pass a d.c. current of magnitude more than 1 ampere, through 1N 4003, diode it may be destroyed.

Broadly speaking, there are two categories of diodes namely small signal diodes and rectifiers or power diodes. The small signal diodes are those which have a power rating below 0.5 W whereas the rectifier diodes have a power rating above 0.5 W. Thus, 1N 914 is a small signal diode because its power rating is 0.25 W, whereas 1N 4003 is a rectifier diode because its power rating is 1 W.

5-12. Diode Specification Sheets

All manufacturers of the semiconductor device provide data on specific diodes for the users to make proper utilization of the device. This data could be a brief description limited to a one page or more than that. It includes the information arranged in tables, graphs, *etc*. The data is usually for:

- Forward voltage, V_F (at a specific current and temperature).
- Maximum forward current, I_F (at a specified temperature).
- Reverse saturation current I_R or I_O (at a specified voltage and temperature).
- Reverse voltage rating [*PIV*, *PRV*, *VRRM* or *V*(*BR*)], where *BR* implies breakdown at a specified temperature.
- Maximum power dissipation level at a particular temperature.
- Capacitance values.
- Reverse recovery time, $t_{rr.}$
- Operating temperature range.

Besides this, depending on the type of diode being considered, more data may also be provided such as frequency range, noise level, switching time, thermal resistance level and peak repetitive values. For the application in mind, the significance of the data will usually be self-apparent. If the maximum power or dissipation rating is also provided, it is understood to be equal to the produce,

$$P_{D_{max}} = V_D I_D$$

where, I_D and V_D are the diode current and voltage at a particular point of operation.

Fig 5-11(*a*) and (*b*) [on page 80, 81 and 82] shows the data sheet for 1N 4001 through 1N 4007 silicon diodes the information important from the subject point of view is as follows :

(*i*) From the "maximum ratings" table, we find that a peak repetitive voltage or reverse voltage rating for 1N 4001 is 50 V and for 1N 4007 is 1000 V. The average value of forward current for all the diodes in this series (*i.e.*, 1N 4001 through 1N 4007) is 1.0 A.

(*ii*) From the "Electrical characteristics" table we find that average value of forward voltage drop for all diodes is 0.8v. the maximum value of reverse current at room temperature (25°c) is 10 μA and it can increase to 50 mA at 100°C.

(*iii*) From the tables on page 81C, we find that for handling different values of current, we have different packages for the diode. For example, if the forward current is in the range of 1.0 to 6.0 A, the small plastic packaging is used. However, if the forward current is in the range of 12 to 50 A, the metal packaging is more suitable. The student is advised to carefully observe the shapes of the packages.

Example 5-8. *A diode has maximum power dissipation of 0.5 W.* ***(a)*** *What is the maximum d.c. current allowed in the forward direction if we allow an approximate forward voltage drop of one volt?* ***(b)*** *What is the approximate breakdown current that burns out the diode, if it has a breakdown voltage of 150 V ?*

Solution. Given: $P_{D_{max}} = 0.5$ W, $V_F = 1$ volt and $V_{BR} = 150$ volts.

(***a***) *Maximum forward d.c. current*

Let $I_{F(max)}$ = Maximum forward d.c. current.

We know that maximum power dissipation, $P_{D(max)}$

$$0.5 = V_F \times I_{F(max)} = 1 \times I_{F(max)}$$

$\therefore$ $I_{F(max)} = 0.5/1 = 0.5$ A **Ans.**

MOTOROLA SEMICONDUCTORS

BOX 20912 PHOENIX ARIZONA 85036

1N4001 thru 1N4007

Designers ▲Data Sheat

"SURMETIC"▲ RECTIFIERS

. . . subminiature size, axial lead mounted rectifiers for general-purpose low-power applications.

Designers Dat for "Wirst Case" Conditions

The Designers Data Sheets permit the design of most circuits entirely from the information presented. Limit curves – represented boundaries on device characteristics – are guveb ti facukutate "worst case" design.

LEAD MOUNTED SILICON RECTIFIERS

50-1000 VOLTS DIFFUSED JUNCTION

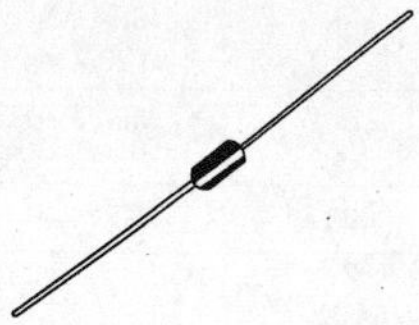

***MAXIMUM RATING**

Rating	Symbol	1N4001	1N4002	1N4003	1N4004	1N4005	1N4006	1N4007	Unit
Peak Repetitive Reverse Voltage Working Peak Reverse Voltage DC Blocking Voltage	V_{RRM} V_{RWM} V_R	50	100	200	400	600	800	1000	Volts
Non-Repetitive Peak Reverse Voltage (halfwave, single phase, 60 Hz)	V_{RWM}	60	120	240	480	720	1000	1200	Volts
RMS Reverse Voltage	$V_{R(RMS)}$	35	70	140	280	420	560	700	Volts
Average Rectified Forward Current (single phase, resistive load, 60 Hz, see Figure 8, T_A = 75°C)	I_O	1.0							Amp
Non-Repetitive Peak Surge Current (surge applied at rated load conditions, see Figure 2)	I_{FMS}	30 (for 1 cycle)							Amp
Operating and Storage Junction Temperature Range	T_J, T_{stg}	– 65 to + 175							°C

***ELECTRICAL CHARACTERISTICS**

Characteristic and Conditions	Symbol	Typ	Max	Unit
Maximum Instantaneous Forward Voltage Drop (i_F = 1.0 Amp, T_J = 25°C) Figure 1	v_F	0.93	1.1	Volts
Maximum Full-Cycle Average Forward Voltage Drop (I_O = 1.0 Amp, T_L = 75°C, 1 inch leads)	$V_{F(AV)}$	-	0.8	Volts
Maximum Reverse Current (rated dc voltage) T_J = 25°C T_J = 100°C	I_R	 0.05 1.0	 10 50	μA
Maximum Full-Cycle Average Reverse Current (I_O = 1.0 Amp, T_L = 75°C, 1 inch leads)	$I_{R(AV)}$	-	30	μA

*Indicates JEDEC Registered Data.

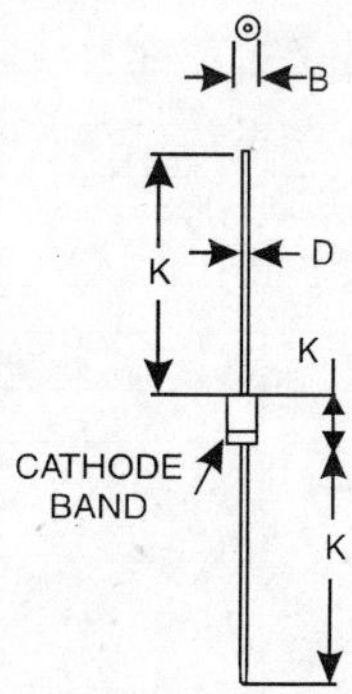

DIM	MILLIMETERS MIN	MILLIMETERS MAX	INCHES MIN	INCHES MAX
A	5.97	6.60	0.235	0.260
B	2.97	3.05	0.110	0.120
D	0.76	0.86	0.030	0.034
K	27.94	—	1.100	—

CASE 59-04
Does Not Conform to Do-41 Outline

MECHANICAL CHARACTERISTICS

CASE: Void free, Transfer Molded

MAXIMUM LEAD TEMPERATURE FOR SOLDERING PURPOSES: 350°C, 3/8" from case for 10 seconds at 5 lbs. tension

FINISH: All external surfaces are corrosion-resistant, leads are readily solderable

POLARITY: Cathode indicated by color band

WEIGHT: 0.40 Grams (approximately)

▲Trademark of Motorola Inc.

MOTOROLA INC. 1975 DS 6015 R3

Fig. 5-11(*a*) *Courtesy of Motorola Inc.*

V_{RRM} (Volts)	I_O, AVERAGE RECTIFIED FORWARD CURRENT (Amperes)							
	12	20	24	25	30		40	50
	245A-02 (DO-203AA) Metal		339-02 Plastic Note 1	193-04 Plastic Note 2	43-02 (DO-21) Metal		42A-01 (DO-203AB) Metal	43-04 Metal

50	MR1120 1N1199, A, B	MR2000	MR2400	MR2500	1N3491	1N3659	1N1183A	MR5005
100	MR1121 1N1200, A, B	MR2001	MR2401	MR2501	1N3492	1N3660	1N1184A	MR5010
200	MR1122 1N1202, A, B	MR2002	MR2402	MR2502	1N3493	1N3661	1N1186A	MR5020
400	MR1124 1N1204, A, B	MR2004	MR2404	MR2504	1N3495	1N3663	1N1188A	MR5040
600	MR1126 1N1206, A, B	MR2006	MR2406	MR2506		Note 3	1N1190A	Note 3
800	MR1128	MR2008		MR2508		Note 3	Note 3	Note 3
1000	MR1130	MR2010		MR2510		Note 3	Note 3	Note 3
I_{FMS} (Amps)	300	400	400	400	300	400	800	600
T_A @ Rated I_O (°C)								
T_C @ Rated I_O (°C)	150	150	125	150	130	100	150	150
T_J (Max) (°C)	190	175	175	175	175	175	190	195

Note 1. Meets mounting configuration of TO-220 outline
Note 2. Request Data Sheet for Mounting Information.
Note 3. Available on special order.

V_{RRM} (Volts)	I_O, AVERAGE RECTIFIED FORWARD CURRENT (Amperes)					
	1.0	1.5		3.0		6.0
	59-03 (DO-41) Plastic	59-04 Plastic	60-01 Metal	267-03 Plastic	267-02 Plastic	194-04 Plastic
				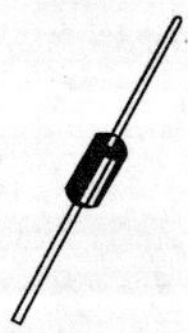		
50	†1N4001	**1N5391	1N4719	**MR500	1N5400	MR750
100	†1N4002	**1N5392	1N4720	**MR501	1N5401	MR751
200	†1N4003	1N5393 *MR5059	1N4721	**MR502	1N5402	MR752
400	†1N4004	1N5395 *MR5060	1N4722	**MR504	1N5404	MR754
600	†1N4005	1N5397 *MR5061	1N4723	**MR506	1N5406	MR756
800	†1N4006	1N5398	1N4724	MR508		MR758
1000	†1N4007	1N5399	1N4725	MR510		MR760

(Continued)

I_{FMS} (Amps)	30	50	300	100	200	400
T_A @ Rated I_O (°C)						
T_C @ Rated I_O (°C)	150	150	125	150	130	100
T_J (Max) (°C)	190	175	175	175	175	175

† Meets mounting configuration of TO-220 outline
* Request Data Sheet for Mounting Information.
** Available on special order.

Fig. 5-11(*b*) *Courtesy of Motorola Inc.*

(***b***) *Breakdown current*

Let $\quad I_R$ = Breakdown current that burns out the diode.

We also know that maximum power dissipation $P_{D(max)}$

$$0.5 = V_{BR} \times I_R = 150 \times I_R$$

$$\therefore \quad I_R = 0.5/150 = 0.003\,33\text{ A} = 3.33\text{ mA} \textbf{ Ans.}$$

5-13. Testing of Diode with an Ohmmeter

The internal battery in certain ohmmeters will forward-bias or reverse bias a diode. This permits a simple and quick check for proper functioning. This works for ohmmeters as such VOMS (volt-ohm meter).

In order to check the diode, in the forward direction, the positive meter lead is connected to the anode and the negative lead to the cathode as shown in Fig. 5-12(*a*). When a diode is forward biased, its internal resistance is low. Typically, it is around 100 ohms. But it can be more or less also.

When the meter leads are reversed as shown in Fig. 5-12(*b*), the internal ohmmeter battery reverse biases the diode. The meter indicates a very large resistance (ideally infinite). The PN junction is short-circuited, if a low resistance is indicated in both forward and reverse bias conditions. It is open, if a high resistance is indicated for both the forward and reverse biasing.

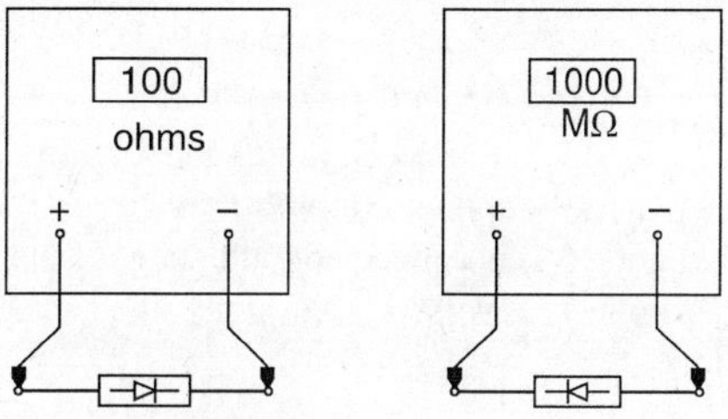

(*a*) Forward-biased diode check. (*b*) Reverse-biased diode check.

Fig. 5-12.

5-14. Diode Circuits with a DC Voltage Source

Let us consider an ideal diode connected to a voltage source and a resistor as shown in Fig. 5-13(*a*). Our aim is to determine the various voltage and current through the circuit. The first thing to be noted carefully is that the cathode terminal of the diode is connected to the positive terminal of the source through a resistor (Alternatively diode arrow points towards the positive terminal of the source.) This indicates that the diode is reverse biased. Because of this, there will be no current through the circuit (*i.e.,* I = 0), therefore full applied voltage is dropped across the diode (D). Further since, there is no current through, R, there is no voltage drop across this component, *i.e.,* $V_R = 0$.

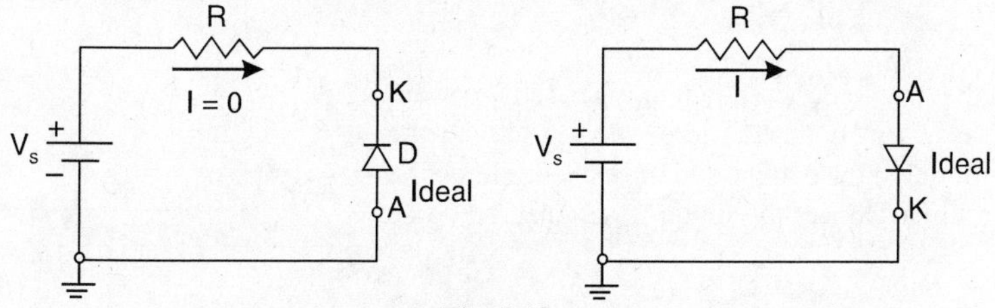

Fig. 5.13.

Now consider the circuit shown in Fig 5-13(*b*). Notice that the diode arrow points toward the negative terminal of the source. This indicates that the diode is forward-biased. As a result of this the diode conducts current through it, hence there is no voltage drop across it *i.e.*,

$$V_D = 0$$

Further the voltage drop across the resistor is equal to the dc source voltage, *i.e.*,

$$V_R = V_S$$

and the current through the circuit,

$$I = \frac{V_R}{R} = \frac{V_S}{R}$$

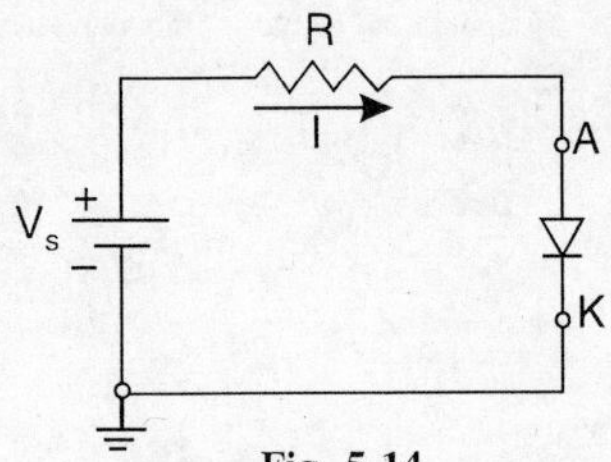

Fig. 5-14.

Next consider the circuit shown in Fig 5-14. Notice that this circuit is similar to the one shown in Fig 5-13 (*b*) except that the diode is a real 'one'. In such a circuit, there will be a voltage drop across the diode whose value for a silicon diode is 0.7 V. The circuit is given by the equation,

$$I = \frac{V_S - 0.7}{R} \ (A)$$

Example 5-9. *Determine the values of ia voltage drop across* **(i)** *the diode,* **(ii)** *the resistance, and* **(iii)** *the current for the circuit shown in Fig 5-15. Assume the diode to be an ideal.*

Solution. Given: D = An ideal diode,

$$R = 330\,\Omega$$

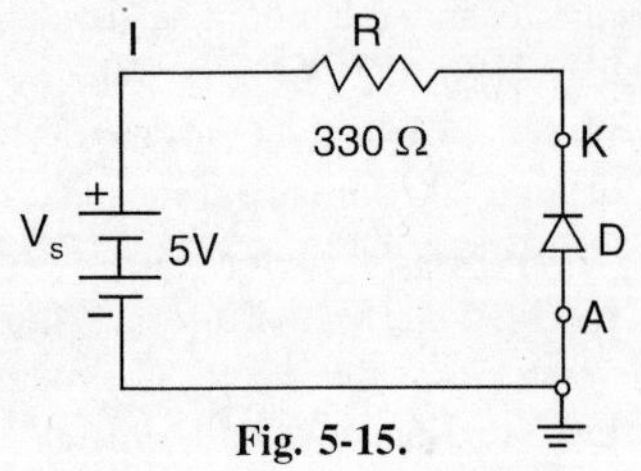

Fig. 5-15.

(*i*) *Voltage drop across the diode*

Notice that in the circuit of Fig 5-15, the cathode (K) of the ideal diode is connected to possitive terminal of the battery and the anode (A) is connected to the negative terminal. Therefore, the diode is reverse biased hence the voltage drop across the diode.

$$V_D = V_S = 5\,V \text{ Ans.}$$

(*ii*) *Voltage drop across the resistance*

Since the diode is reverse biased, therefore, there is no current through the diode or resistance, hence, no voltage drop across resistance, *ie.*,

$$V_R = 0 \text{ Ans.}$$

(*iii*) *Current through the circuit*

Since the diode is reverse biased, therefore the circuit current is zero, *ie.*,

$$I = 0 \text{ Ans.}$$

Example 5-10. *A series circuit consist of a 12-V source, a 470-Ω resistor (R) and a diode (D) as shown in Fig 5-16. Determine the value of V_D, V_R, and I.*

Solution. Given : V_S = 12 V; R = 470 Ω and D = An ideal diode

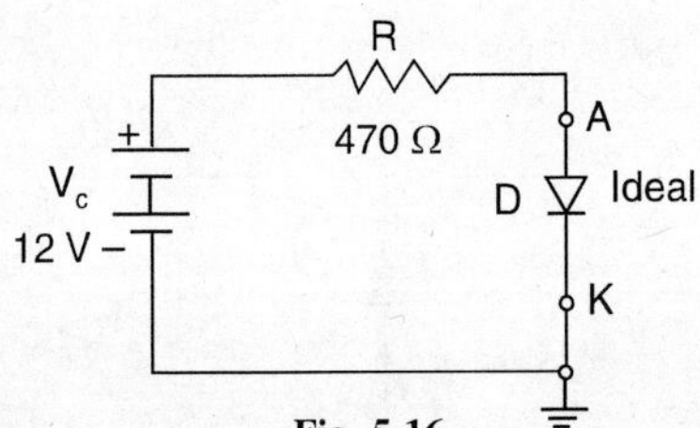

Fig. 5-16.

Value of V_D

Noice that the diode, D shown in Fig 5-15 is forward biased because its anode (A) is connected (K) to the positive terminal of the battery and the cathode (K) to the negative terminal. Since, the diode is assumed to be an ideal.

$\therefore \quad V_D = 0$ **Ans.**

Value of V_R

Since, there is no voltage drop across an ideal diode, the voltage drop across, the resistor.

$$V_R = V_S = 12 \text{ V } \textbf{Ans.}$$

Current through the circuit

The current through the circuit,

$$I = \frac{V_S}{R} = \frac{12}{470} = 2.553 \times 10^{-2} \text{ A}$$
$$= 25.53 \text{ mA } \textbf{Ans.}$$

Example 5-11. *A series-circuit consist of a forward biased diode, and two resistors 330 Ω and 470 Ω connected as shown in Fig. 5-17. Find the value of current through the circuit. Assume the diode to be real diode.*

Solution. Given: $V_S = 6 \text{ V}; R_1 = 330\,\Omega; R_2 = 470\,\Omega; V_D = 0.7 \text{ V}.$

We known that total resistance in the circuit shown in Fig. 5-17,

$R_T = 330 + 470 = 800\Omega$ and the current through the diode,

$$I = \frac{V_S - 0.7}{R_T} = \frac{6 - 0.7}{800} = 0.00662 \text{ A}$$
$$= 6.62 \times 10^{-3} \text{ A}$$
$$= 6.62 \text{ mA } \textbf{Ans.}$$

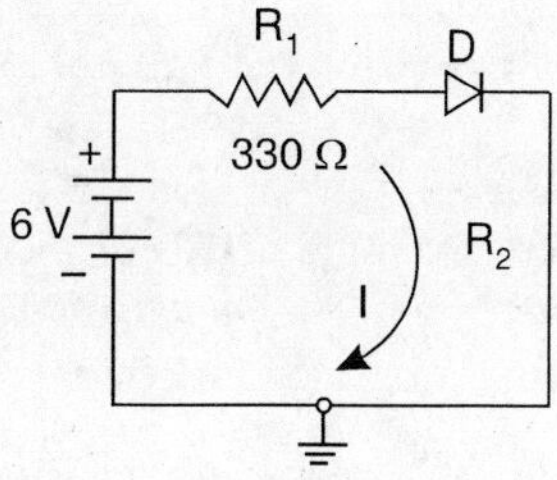

Fig. 5-17.

Note. The resistor R_1, and R_2 in Fig 5-16 are connected in series. To understand more about series circuit. never to Appendix B: DC resistore circuits.

Example 5-12. *Fig 5-18 shows a series circuit consisting of a voltage source, a resistor and a real diode. Determine the voltage across the resistor and the circuit current.*

Solution. Given: $V_S = 5\text{V}; R = 510\,\Omega; V_F = 0.7 \text{ V}$

(*a*) **Voltage across R**

We know that the voltage drop across R,

$$V_R = V_S - 0.7 = 5 - 0.7 = 4.3 \text{ V } \textbf{Ans.}$$

(*b*) **Circuit current**

We know that current through the given series circuit,

$$I = \frac{V_R}{R} = \frac{4.3}{510} = 0.00843 \text{ A}$$
$$= 8.43 \times 10^{-3} \text{ A}$$
$$= 8.43 \text{ mA } \textbf{Ans.}$$

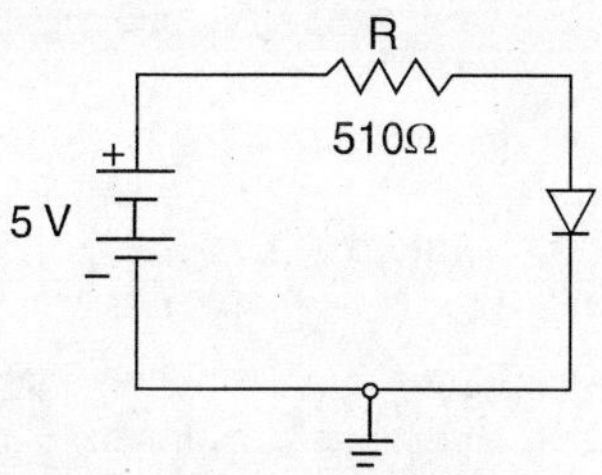

Fig. 5-18.

Example 5-13. *Determine the value of the total current through the series circuit shown in Fig. 5-19. Assume both the diodes D_1 and D_2 to be real.*

Solution. Given, $V_S = 6\text{V}$; $V_{D_1} = V_{D_2} = 0.7\text{V}$; $V_R = 1.5$ k$\Omega = 1.5 \times 10^3\,\Omega$

We know that the total series current through the given circuit,

$$I = \frac{V_S - V_{D_1} - V_{D_2}}{R}$$

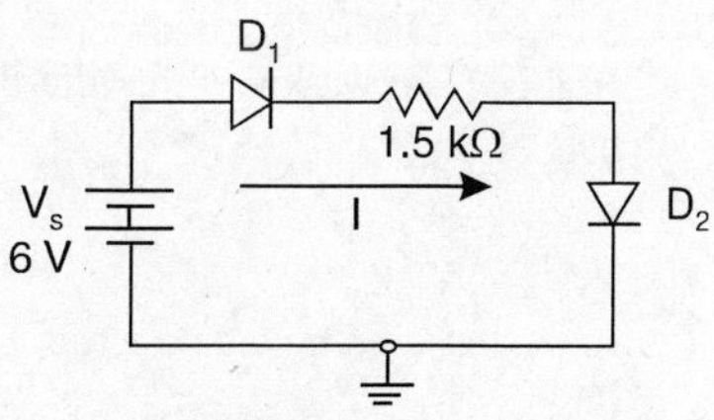

Fig. 5-19.

$$= \frac{6 - 0.7 - 0.7}{1.5 \times 10^3}$$
$$= 3.067 \times 10^{-3}\text{ A}$$
$$= 3.067\text{ mA } \textbf{Ans.}$$

Example 5-14. *A series circuit consists of two forward biased real diodes, and two resistors connected as shown in Fig 5-20. Determine the value of total current through the circuit.*

Solution. Given : $V_S = 12$ V; $R_1 = 1.5$ kΩ $= 1.5 \times 10^3$ Ω; $R_2 = 1.8$ kΩ $= 1.8 \times 10^3$ Ω; $V_{D_1} = V_{D_2} = 0.7$V.

We know that current through the given series circuit,

$$I = \frac{V_S - 0.7 - 0.7}{R_T} = \frac{12 - 0.7 - 0.7}{(1.5 \times 10^3) + (1.8 \times 10^3)}$$
$$= \frac{10{\cdot}6}{3{\cdot}3 \times 10^3} = 3.212 \times 10^{-3}\text{ A}$$
$$= 3.212\text{ mA } \textbf{Ans.}$$

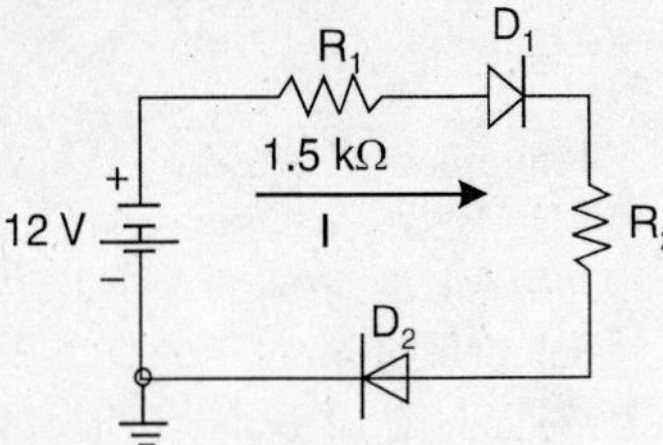

Fig. 5-20.

Example 5-15. *For the circuit. shown in Fig. 5-21, determine the output voltage when*

(*a*) $V_1 = V_2 = 0V$ (*b*) $V_1 = 0, V_2 = 5V$
(*c*) $V_1 = 5$ V, $V_2 = 0$ V (*d*) $V_1 = V_2 = 5$ V

(*Banglore Univ. B.Sc. Electronics, May* 2001)

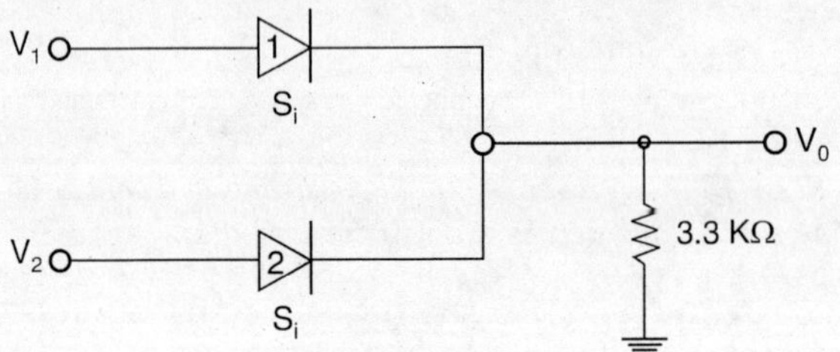

Fig. 5-21.

Solution.

(*a*) $V_1 = V_2 = 0$

In this case, we know that $V_1 = V_2 = 0$, both the Si diodes are reverse biased. Therefore, the output voltage $V_0 = 0$. **Ans.**

(*b*) $V_1 = 0$, $V_2 = 5$ V

In this case we know that diode '1' is reverse biased while diode '2' is formed biased. Therefore, the diode '1' is 'off' while diode '2' is ON. Therefore the output voltage,

$$V_0 = V_2 - 0.7V = 5 - 0.7 = 4.3\text{ V } \textbf{Ans.}$$

(*c*) $V_1 = 5V$, $V_2 = 0V$

In this case, we know that the diode '1' is ON while diode '2' is OFF. Therefore, the output voltage,

$$V_0 = V_1 - 0.7 = 5 - 0.7 = 4.3\text{ V } \textbf{Ans.}$$

(*d*) $V_1 = V_2 = 5V$

In this case both we know that both diodes are ON, therefore, the output voltage,

$$V_0 = V_1 - 0.7\text{ V or } V_2 - 0.7$$
$$= 5 - 0.7\text{ V}$$
$$= 4.3\text{ V } \textbf{Ans.}$$

5-15. Diode Circuit with D.C. and A.C. Voltage Sources

The electronic circuits used in radios and televisions contain both d.c. and a.c. sources. Such circuits can be easily analysed by using *superposition theorem. According to this theorem, first we draw a d.c. equivalent circuit (neglecting a.c. sources) and find the required current and voltage values. Next we draw the a.c. equivalent circuit (neglecting the d.c. sources) and again find the current and voltage values. Finally, we superimpose the two sets of values to obtain the result. While drawing the equivalent circuits, the following two points must be taken care of :

(1) Direct current cannot flow through a capacitor. Therefore, all capacitors look like an open to a d.c. source.

(2) The capacitors used in most of the circuits are large enough to look like a short to an a.c. source. It is particularly true for circuits operating at high frequency. Similarly, d.c. sources would also act as short circuit, unless they have very high internal resistances.

Example 5-16. *The silicon diode shown in Fig. 5-22 has a bulk resistance of 1Ω . The frequency of the 10 mV (peak) signal is so high that the reactance of the coupling capacitor may be taken as zero.*

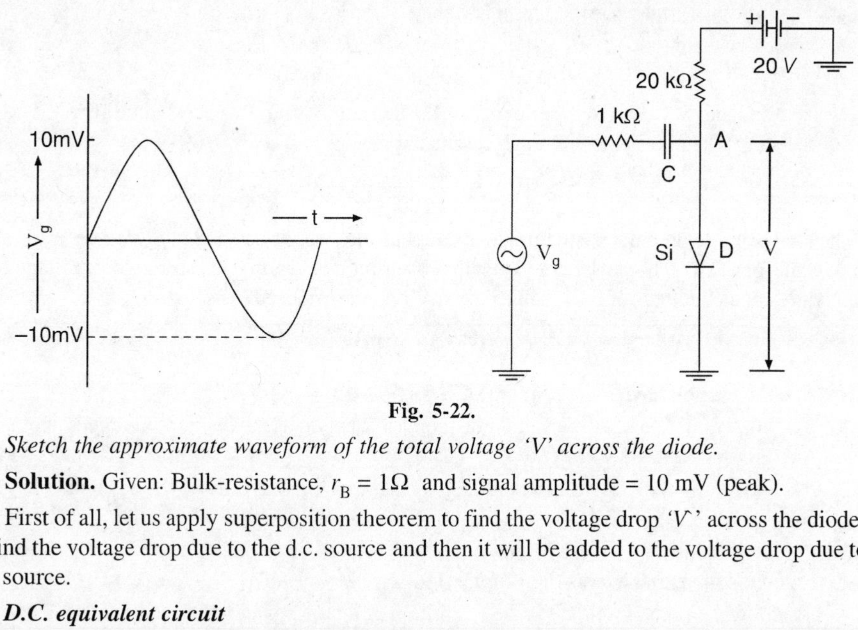

Fig. 5-22.

Sketch the approximate waveform of the total voltage 'V' across the diode.

Solution. Given: Bulk-resistance, $r_B = 1\Omega$ and signal amplitude = 10 mV (peak).

First of all, let us apply superposition theorem to find the voltage drop '*V*' across the diode. Let us find the voltage drop due to the d.c. source and then it will be added to the voltage drop due to the a.c. source.

(*a*) *D.C. equivalent circuit*

It may be noted from Fig. 5-21 that the circuit to the left of the point *A* is 'open' to d.c. source, because of the capacitor 'C'. Thus the d.c. equivalent circuit is as shown in Fig. 5-23 (*a*). In this circuit, the diode is forward biased by the d.c. source. Under this condition, the voltage drop across the diode is equal to 0.7 V d.c. and current through it,

$$I = \frac{20 - 0.7}{20} \approx 1 \text{ mA} \quad \textbf{Ans.}$$

(*b*) *A.C. equivalent circuit*

Here, the capacitor 'C' and the 20 V d.c. source are treated as shorts. This gives us the a.c. equivalent circuit as shown in Fig. 5-23(*b*). Now, for a given silicon diode, the junction resistance,

* To study more about superposition theorem, refer to Appendix C: Kirchoff's Laws and Network Theorems.

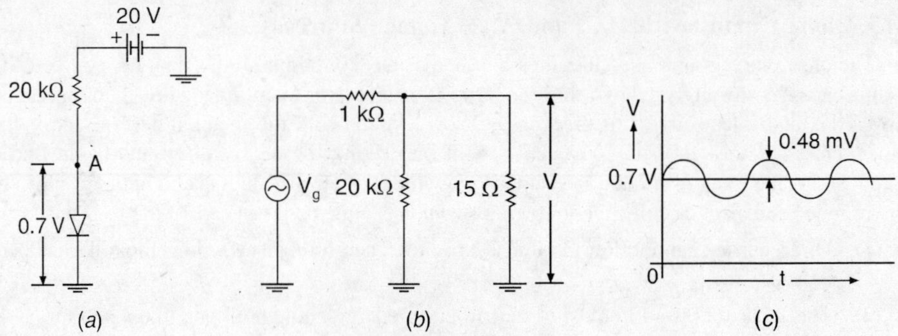

Fig. 5-23.

$$r_j = \frac{50}{1} = 50\ \Omega$$

and the a.c. resistance, $r_e' = r_B + r_j = 1 + 50 = 51\ \Omega$

It may be noted from Fig. 5-23(*b*), that for an a.c. signal source, 20 k Ω and 51 Ω are in *parallel. The equivalent resistance of this parallel combination

$$= 20\ k\Omega \,||\, 51\ \Omega = \frac{20{,}000 \times 51}{20{,}000 + 51} \approx 51\ \Omega$$

Hence, 1 k Ω and 51Ω are put in series across the a.c. signal source of peak value 10 mV. The peak value of the a.c. voltage drop across 51Ω resistance

$$= 10 \times \frac{51}{51 + 1000} = 0.48\text{ mV}$$

The total voltage drop across the diode is equal to the sum of d.c. and a.c. voltage drops and is as shown in Fig. 5-23(*c*). It may be noted that the resulting waveform consists of a d.c. voltage of 0.7 V over which an a.c. voltage of peak value 0.48 mV is superimposed. **Ans.**

SUMMARY

In this chapter, you have learned that :

1. The current in a forward biased diode increase as soon as the applied voltage is above the the junction voltage (*i.e.*, > 0.7V).
2. The current in a reverse biased diode increases abruptly as soon as the applied voltage is increased to a value equal to (or greater than) the breakdown voltage of a diode.
3. A real diode when forward biased can be considered as an equivalent of an ideal diode, a voltage source and a resistor connected in series.
4. The power rating of a diode is the maximum value of power which a diode can dissipate without failure.

GLOSSARY

Anode : P-type region of a PN junction diode.

Bulk resistance : The resistance of the *P*-and N-semiconductor materials of which the diode is made of.

Cathode : N-type region of a PN junction diode.

Data sheet : Document supplied by the manufacturer with brief descriptions of a maximum ratings of electrical and mechanical characteristies of the device.

* To study more about parallel circuits, refer to Appendix B: DC Resistive Circuits.

Forward voltage : Voltage drop across PN Junction diode when it is forward biased.

Ideal diode : A diode, which conducts with zero resistance when forward biased and appears as an infinite resistance when reverse biased

Junction diode : Diode, consisting a PN junction, formed either in germanium or silicon.

Junction resistance : The resistance of a PN junction, which varies with current.

Real diode : The diode in actual practice, which has a voltage drop across its terminals when forward biased and has small amount of leakage current when reverse biased.

Reverse saturation current : The small value of current (usaually in μA) that flows through the diode when it is reverse biased.

Reverse voltage : The maximum reverse biased voltage arcoss the terminals of a diode above which reverse breakdown occurs.

NUMERICAL PROBLEMS

1. A silicon diode has a saturation current of 0.01 μA at room temperature (300 K). Find the saturation current at 400 K. (**Ans.** 10.24 μA)

2. For silicon diode with $I_0 = 2.5$ μA at 300 K, find the forward voltage at a forward current of 10 mA. (**Ans.** 0.4 V)

3. Calculate the voltage which has to be applied across a PN semiconductor junction to cause a forward current of 5 A/cm^2, assuming that the saturation current density at 300 K is 25 μA/cm^2. (**Ans.** 0.316 V)

4. For what voltage will the reverse current in a PN junction germanium diode attains a value of 90% of its saturation value at room temperature? (**Ans.** 0.06 V)

5. A junction diode has a maximum power dissipation of 250 mW and the reverse saturation current of 100 mA. Calculate the value of the maximum forward current at a temperature of 27°C. (**Ans.** 2.85 A)

Hint. Forward voltage drop across the diode

$$V_F = 250 \times 10^{-3} / 100 \times 10^{-3} = 2.5 \text{ V}$$

Use this value in the relation

$$I = I_0 \cdot (e^{V_F / V_T^1})$$

6. Determine the voltage across a silicon diode for value for value of $I_F = 5$ mA. Assume that bulk resistance of the diode is 5Ω . (**Ans.** 0.625 V)

7. Determine the applied voltage to achieve a forward current of 0.45 μA in a PN junction silicon diode at T = 300 K, if the reverse saturation current is 1.0 nA. (*b*) If the reverse saturation current in a germanium diode is 10 μA, what current would result if the voltage in part (*a*) is applied in the forward direction? (**Ans.** 0.3 V and 1.63 A)

8. Using approximate Boltzmann's diode equation, find the change in forward bias for doubling the forward current of a germanium semiconductor at 290 K.

(*Basic Electronics, Osmania Uni.* 1993) (**Ans.** 17.3 mV)

9. A certain PN junction diode has a leakage current of 10^{-14} A at room temperature of 27°C and 10^{-9} A at 125 °C. The diode is forward biased with a constant current source of 1 mA at room temperature. If the current is assumed to remain constant, calculate the junction barrier voltage at room temperature and at 125° C.

(**Ans.** At 25°C ⇒ 0.66 V; at 125° C ⇒ 0.5V)

10. A silicon diode has a forward voltage drop of 1.2 V for a forward dc current of 100 mA. It has a reverse current of 1μA for a reverse voltage of 10 V. Calculate:

(*a*) bulk and reverse registance of a diode

(*b*) ac resistance of forward dc current of (*i*) 2.5 mA and (*ii*) 25 mA.

Ans. (*a*) $r_B = 5\Omega$; $R_R = 10M\Omega$ (*b*) r_{ac} at 2.5 mA = 15 Ω and at 25 mA = 6Ω

11. Using analytical expression for diode current, calculate the dynamic slope resistance of a germanium diode at 290 K, when forward biased at current of (*i*) 10 μA and (*ii*) 5 mA. (**Ans.** (*i*) $r_{ac} = 2500\ \Omega$, (*ii*) $r_{ac} = 5\ \Omega$)

12. A germanium diode draws 40 mA with a forward bias of 0.25 V. the junction is at room temperature of 293 K. Calculate the reverse saturation current of the diode.(**Ans.** 1.82 μA)

13. Calculate forward current in Germanium diode at 20°C when forward voltage is 0.3 V. Compare this value with that after a temperature rise of 50°C. Assume that reverse saturation current doubled for every 10°C rise in temperature.

(**Ans.** $I_{20°C} = 162,755\, I_O$; $I_{70°C} = 704,800\, I_O$; This indicates that $I_{70°C}$; $= 4.3\, I_{20°C}$ *i.e.*, for a rise in temperature of 50°C, I_O has increased 32 times whereas the forward current has increased only 4.3 times for the same rise in temperature.)

14. Determine the value of V_D, V_R and I_T for the circuits shown in Fig. 5-24(*a*) and respectively Assume the diode to be an ideal.

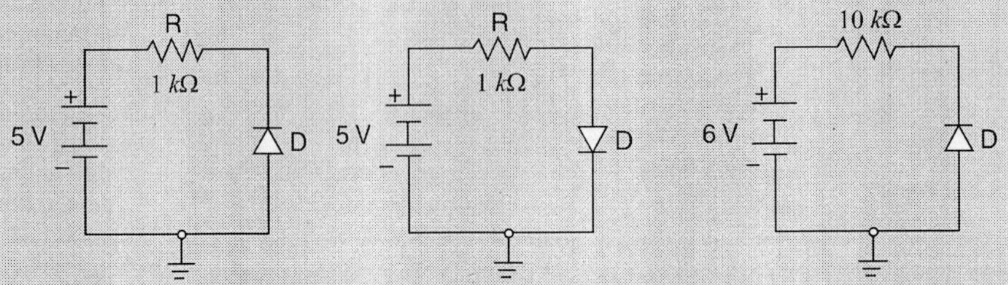

Fig. 5-24.

(**Ans.** (*a*) $V_D = 5V$, $V_R = 0$, $I_T = 0A$ (*b*) $V_D = 0$; $V_R = 5V$; $I_T = 5$ mA; (*c*) $V_D = 0$; $V_R = 6$ V; $I_T = .6$ mA.)

15. Determine the current through resistor, (R) in the circuits shown in Fig. 5-25 (*a*), (*b*) and (*c*) respectively. Assume the diode to be real. (**Ans.** (*a*) 0.53 mA; (*b*) 1.26 mA; (*c*) 2.61 mA)

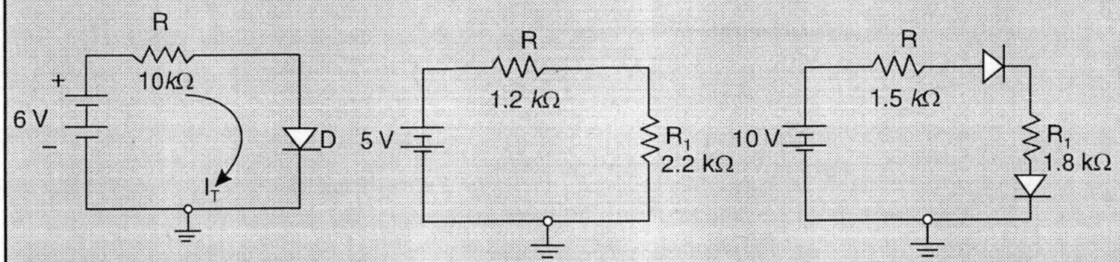

Fig. 5-25.

DESCRIPTIVE QUESTIONS

1. What is a PN junction diode? How its terminals are identified?

2. Draw the V-I characteristics of a junction diode when it is (*a*) forward biased, and (*b*) reverse biased.

3. For silicon and germanium diode, state approximately, cut in voltages. Also compare their reverse saturation currents.
4. Briefly explain (without derivations) the behaviour of junction in forward bias and reverse bias mode and draw its volt-ampere characteristics.
5. Explain the effect of temperature on the volt-ampere characteristic of a diode.
6. What is an ideal diode and real diode?
7. Explain the following terms: ***(a)*** static resistance ***(b)*** bulk resistance ***(c)*** junction resistance, ***(d)*** a.c. or dynamic resistance and ***(e)*** reverse resistance of a diode.
8. What are the important applications of a diode?
9. With the volt–ampere characteristics, explain the working principle of the diode and also explain the static dynamic resistance of the diodes?

   (*Electronic Devices, Anna University, Nov/Dec 2007*)
10. Define Cut in Voltage of a diode.(*Electorn Devices, Anna University, April/May 2008*)
11. Application of diodes. (*Electron Devices, Anna University, Nov/Dec 2007*)
12. Explain the different diode equivalent circuits with necessary approximation if any.

    (*VTU, Dec 08/Jan 09*)
13. Explain with the help of lod line the effect of variation of V_{cc} I_B on Q - point of a transistor. (*VIU, Dec 08/Jan 09*)

MULTIPLE CHOICE QUESTIONS

1. The schematic symbol for a PN junction diode is:

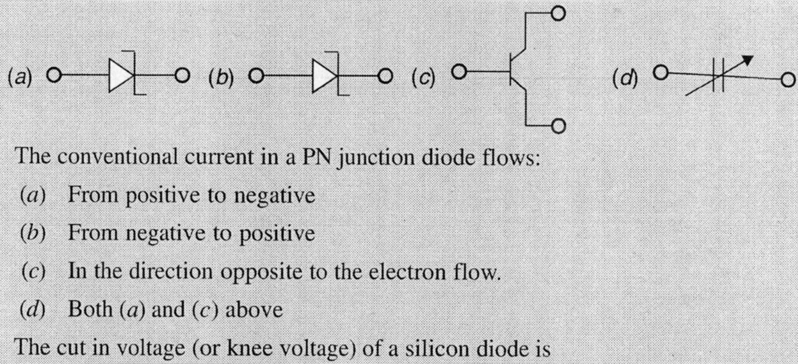

2. The conventional current in a PN junction diode flows:
   (*a*) From positive to negative
   (*b*) From negative to positive
   (*c*) In the direction opposite to the electron flow.
   (*d*) Both (*a*) and (*c*) above
3. The cut in voltage (or knee voltage) of a silicon diode is
   (*a*) 0.2 V (*b*) 0.7 V (*c*) 0.8 V (*d*) 1.0 V
4. When a diode is reverse biased, it is equivalent to:
   (*a*) An OFF switch (*b*) An ON switch
   (*c*) A high resistance (*d*) None of the above
5. The resistance of a diode is equal to:
   (*a*) Ohmic resistance of the P- and N- semiconductors
   (*b*) Junction resistance
   (*c*) Reverse resistance
   (*d*) Algebraic sum of (*a*) and (*b*) above

6. For a silicon diode, the value of the forward - bias voltage typically,
(*a*) must be grater than 0.3V
(*b*) must be greater than 0.7V
(*c*) depends on the width of the depletion region
(*d*) depends on the concentration of majority carriers

7. When forward biased, a diode
(*a*) blocks current (*b*) conducts current
(*c*) has a high resistance (*d*) drops a large voltage.

8. A PN junction diode's dynamic conductance is directly proportional to
(*a*) the applied voltage (*b*) the temperature
(*c*) the current (*d*) the thermal voltage

Hint: Conductance is reciprocal of the resistance.

9. The forward region of a semiconductor diode characteristic curve is where diode appears as
(*a*) constant current source (*b*) a capacitor
(*c*) on OFF switch (*d*) an ON switch

10. At room temperature of 25 °C, the barrier potential for silicon is 0.7 V. Its value at 125° C is
(*a*) 0.5V (*b*) 0.3 V (*c*) 0.9 V (*d*) 0.7 V

11. Junction breakdown of a PN junction occurs
(*a*) with forward bias (*b*) with reverse bias
(*c*) because of manufacturing defect (*d*) None of these

12. Reverse saturation current in a silicon PN junction diode nearly doubles for every
(*a*) 2° C rise in temperature (*b*) 5° C rise in temperature
(*c*) 6° C rise in temperature (*d*) 10° C rise in temperature

13. The approximate value of v_0 across the diode shown in Fig 5-26 is,

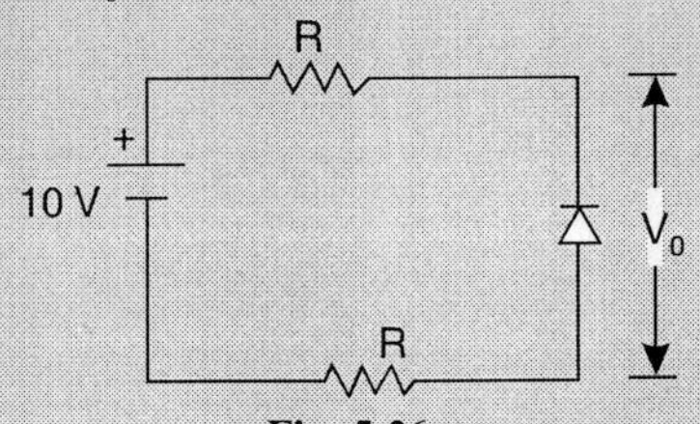

Fig. 5-26.

(*a*) zero (*b*) 10V
(*c*) 5V (*d*) dependent on the value of R

14. The diode D is an ideal in the circuit shown in Fig 5-27. The circuit, I will be:

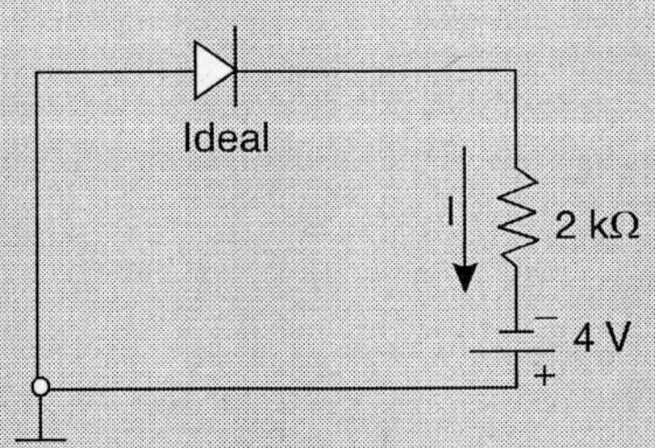

Fig. 5-27.

(*a*) – 2nA (*b*) zero (*c*) 2 mA (*d*) 4 mA

15. The voltage at V_1 and V_2 of the arragement shown in Fig. 5-27 will be respectively:

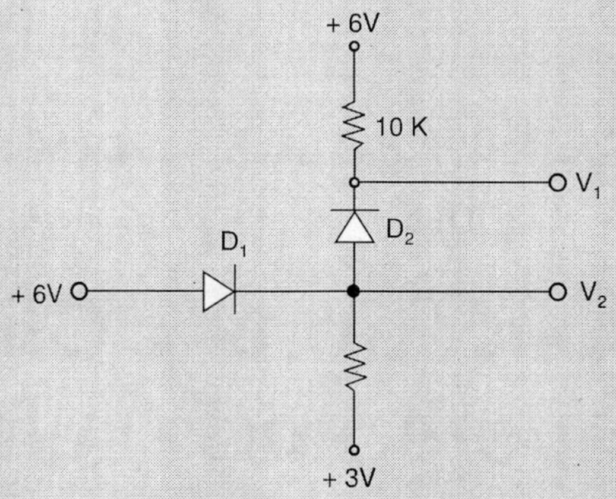

Fig. 5-28.

(*a*) 6 V and 5·4 V (*b*) 5·4 V and 6 V

(*c*) 3 V and 5·4 V (*d*) 6 V and 5 V

16. The transition capacitance of a diode is 1nF and it can withstand a reverse potential of 400 V. A capacitance of 2nF which can withstand a reverse potential of 1 kV is obtained by connecting

(*a*) two 1 nF diodes in series

(*b*) six parallel branches with each branches comparising there1 nF diodes in series

(*c*) two 1 nF diodes in series

(*d*) three parallel branches with each branch comprising 1nF diodes in series

ANSWERS

1. (*b*)	**2.** (*d*)	**3.** (*b*)	**4.** (*a*)	**5.** (*d*)
6. (*b*)	**7.** (*b*)	**8.** (*c*)	**9.** (*d*)	**10.** (*a*)
11. (*b*)	**12.** (*d*)	**13.** (*a*)	**14.** (*c*)	**15.** (*d*)
16. (*b*)				

❑❑❑

6

CHAPTER

PN Junction Diode Applications : Clippers, Clampers and Multipliers

OUTLINE

Objectives

After completing this chapter, you should be able to :

- Understand clipping circuits or clippers.
- Draw the circuit for four types of clippers namely, positive clipper, negative clipper, biased clipper and combination clipper.
- Explain the clipping action of different clippers.
- Understand the clamping circuits or clampers.
- Draw the circuit for different types of clampers namely, positive clamper, negative clamper and biased clamper.
- Explain the clamping action of different clampers.
- Understand the practical clamper.
- Know the application of clamping circuits namely, voltage multiplier, voltage doubler, voltage trippler and voltage quadrupler.

6-1. Introduction

The PN junction diode is an extremely useful device in non-linear wave shaping circuits. Such circuits are used to perform functions of amplitude limiting, clipping and clamping. Now we shall discuss these circuits under the following two heads :

(*i*) Clipping circuits, and (*ii*) Clamping circuits.

6-2. Clipping Circuits

The circuit, with which the waveform is shaped by removing (or clipping) a certain portion of the input signal voltage above or below a certain level, is called a clipping circuit or simply clipper. Such circuits are used in digital computers, radars, radio and television receivers *etc*. A clipping circuit may also be used to limit the amplitude of the input signal voltage as required in several application. In that case, a clipping circuit is known as amplitude limiter or simply limiter. The clipping circuits are of the following four types :

(*i*) Positive Clipper (*ii*) Negative Clipper

(*iii*) Biased Clipper (*iv*) Combination Clipper

6-3. Positive Clipper

Fig. 6-1(*a*) shows the circuit of a positive clipper. It consists of a diode (*D*) and a resistor (*R*) with output taken across the resistor. The diode acts as an ideal switch between the source and the load. It acts as a closed switch, when the input voltage is negative (*i.e.*, $v_i < 0$) and as an open switch when the input voltage is positive or zero (*i.e.*, $v_i \geq 0$). The purpose of resistor (*R*) is to limit the current through diode, when it acts as closed switch. The operation of the circuit may be explained as follows :

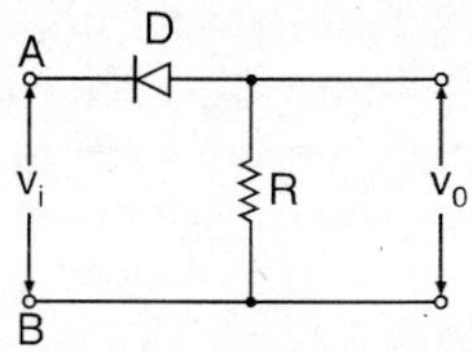

(*a*) Positive clipper.

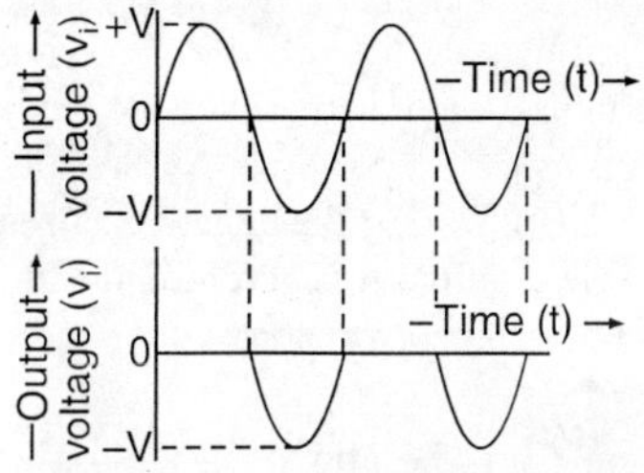

(*b*) Input and output waveforms.

Fig. 6-1.

Fig. 6-1(*b*) shows that waveform of an input voltage. During the positive half cycle of the input voltage, the terminal *A* is positive with respect to *B*. This reverse biases the diode and it acts as an open switch. Therefore, all the applied voltage drops across the diode and none across the resistor. As a result of this, there is no output voltage during the positive half cycle of the input voltage as shown in the figure. During the negative half cycle of the input voltage, the terminal *B* is positive with respect to *A*. Therefore, it forward biases the diode and it acts as a closed switch. Thus, there is no voltage drop across diode during the negative half cycle of the input voltage. All the input voltage is dropped across the resistor as shown in the output waveform.

It is evident from the above discussion as well as the output waveform that the positive clipper has allowed to pass the negative half cycle of the input voltage and clipped the positive half cycle completely. Hence, the circuit of Fig. 6-1(*a*) is called a *positive clipper*. It may be noted in Fig. 6-1(*a*) that the diode acts as a series switch between the source and the load. Due to this reason, the circuit is called *series positive clipper*. It is also possible to connect a diode as a shunt switch between the source and the load as shown in Fig. 6-2. Such a circuit is known as shunt positive clipper. In this circuit, the diode acts as a closed switch when the input voltage is positive (*i.e.*, $v_i \geq 0$) and as an open switch when the input voltage is negative (*i.e.*, $v_i < 0$). The output waveform, is the same as that of a series positive clipper.

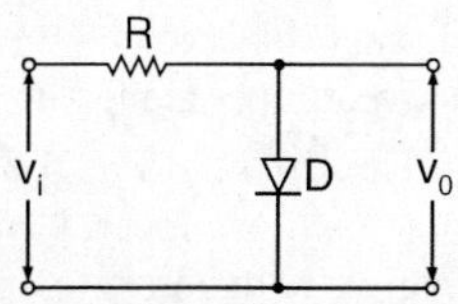

Fig. 6-2. *Shunt positive clipper.*

6-4. Negative Clipper

Fig. 6-3(*a*) shows the circuit of a negative clipper. Here the diode is shown connected in a direction opposite to that of a positive clipper. The function of the resistor (*R*) is to limit the current when the diode is forward biased (*i.e.*, when it acts as a closed switch). The diode acts as a closed switch for a positive input voltage (*i.e.*, $v_i \geq 0$) and as an open switch for a negative input voltage (*i.e.*, $v_i < 0$). The operation of the circuit may be explained in the same way as discussed for a positive clipper and is as given below.

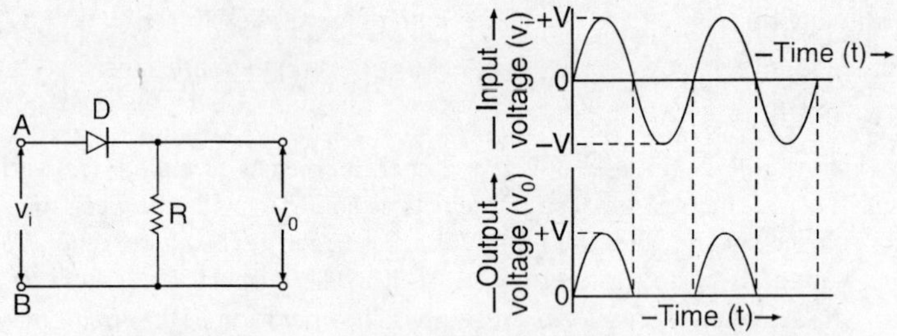

(*a*) Negative clipper. (*b*) Input and output waveforms.

Fig. 6-3.

Fig. 6-3(*b*) shows the waveform of the input voltage. During the positive half cycle of the voltage, the terminal *A* is positive with respect to the terminal *B*. Therefore, the diode is forward biased and it acts as a closed switch. As a result, all the input voltage appears across the resistor as shown in Fig. 6-3 (*b*). During the negative half cycle of the input voltage, the terminal *B* is positive with respect to the terminal *A*. Therefore, the diode is reverse biased and it acts as an open switch. Thus, there is no voltage drop across the resistor during the negative half cycle as shown in the output waveform.

It is evident from the above discussion that the negative clipper has allowed to pass the positive half cycle of the input voltage and clipped the negative half cycle completely. The negative clipper circuit of Fig. 6-3(*a*) is called *series negative clipper*, because the diode acts as a series switch between the source and the load. It is also possible to connect a diode as a shunt switch between the source and the load as shown in Fig. 6-4. Such a circuit is known as shunt negative clipper. In this circuit, the diode acts as a closed switch for a negative input voltage (*i.e.*, $v_i < 0$) and as an open switch for a positive input voltage (*i.e.*, $v_i \geq 0$). The output waveform of the circuit is the same as that of series negative clipper.

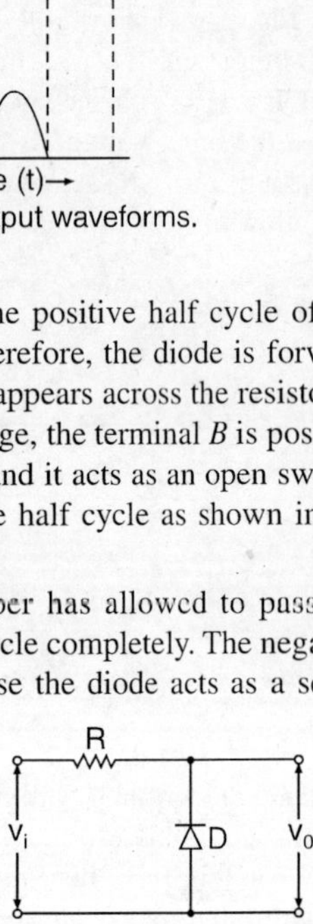

Fig. 6-4. *Shunt negative clipper.*

6-5. Biased Clipper

We have already discussed in the last articles that a positive clipper clips off the entire positive portions of the input signal. Similarly, a negative clipper clipps off the entire negative portions off the input signal. In such circuits, there is no provision for adjustment of the clipping level.

A clipping circuit, which has a provision for the adjustment of a clipping level is called a *biased clipper*. The name '*bias*' is designated because the adjustment of the clipping level is achieved by additing a bias voltage in series with the diode or resistor. Fig. 6-5 shows the circuit of a biased positive clipper alongwith the input and output voltage waveforms.

The circuit of Fig. 6-5(*a*) is a biased series positive clipper, while that of Fig. 6-5(*b*) is a biased shunt positive clipper. It may be noted that the clipping takes place during the positive cycle when the input voltage is greater than battery voltage (*i.e.*, $v_i \geq V_B$). The clipping level can be shifted up or down by varying the bias voltage (V_B). It will be interesting to know that if the polarity of the

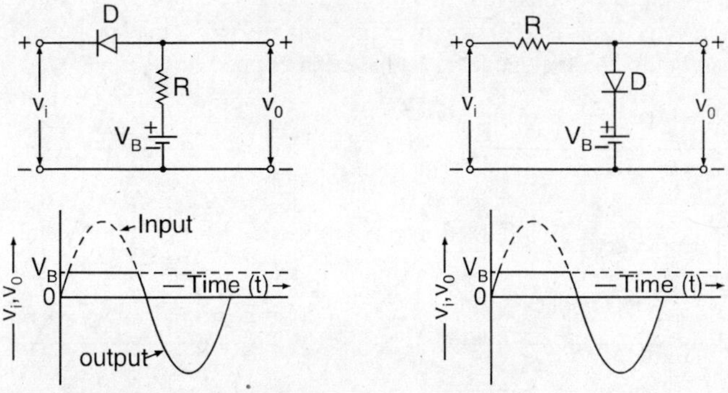

(*a*) Biased series positive clipper. (*b*) Biased shunt positive clipper.

Fig. 6-5. *Biased clipper.*

bias voltage of the circuits will be as shown in Fig. 6-5 is reversed, then the resulting circuits will be as shown in Fig. 6-6(*a*) and (*b*). Here, the input signal lying above the voltage $-V_B$ level is clipped. The resulting waveforms of the output voltage are shown in figures.

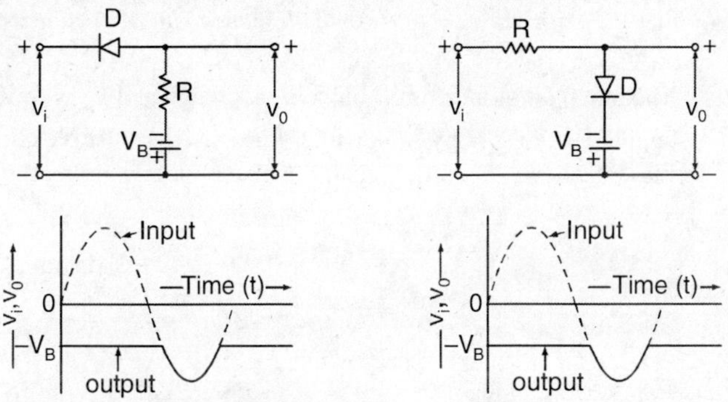

(*a*) Biased series positive clipper. (*b*) Biased shunt positive clipper.

Fig. 6-6. *Biased clipper with reversed polarity of battery V_B.*

Fig. 6-7 shows the circuit of a biased negative clipper. The circuit of Fig 6-7(*a*) corresponds to the biased series negative clipper, while that of Fig. 6-7(*b*) a biased shunt negative clipper. In these circuits, clipping takes place during the negative half cycle only when the input voltage $v_i \leq -V_B$. The clipping level can be shifted up or down by varying the bias-voltage ($-V_B$).

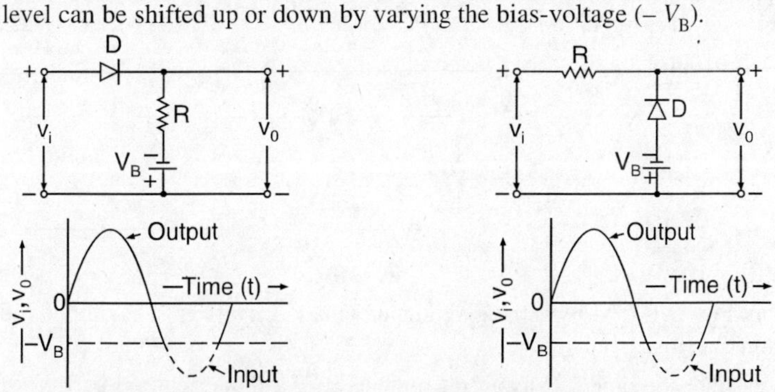

(*a*) Biased series negative clipper. (*b*) Biased shunt negative clipper.

Fig. 6-7.

Now suppose, the battery connections in the biased negative clipper are reversed. The resulting circuit alongwith the waveforms is as shown in Fig. 6-8(*a*) and (*b*). It may be noted from this circuit that the entire signal below the voltage level V_B has been clipped off.

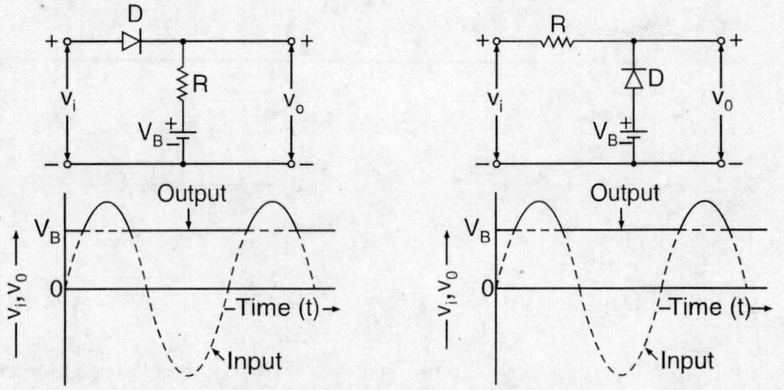

(*a*) Biased series negative clipper. (*b*) Biased shunt negative clipper.

Fig. 6-8.

Example 6-1. *Draw the circuit diagram of a diode clipper to clip a 5 V sine wave input signal at + 2 V.*

Solution. Given : Maximum amplitude of the input signal = 5 V and V_B = 2 V.

We can obtain a clipping level + 2 V by using either biased series positive clipper or a biased shunt positive clipper. Fig. 6-9 shows the biased shunt positive clipper clipping the signal at +2V.

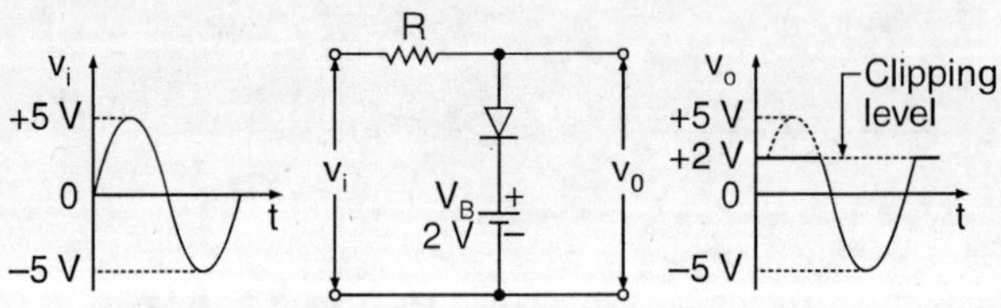

Fig. 6-9.

It is evident from the output waveform shown above that the circuit has clipped off the input signal lying above the level of + 2 V.

Example. 6-2. *Sinusoidal waveform of 10 V peak to peak is applied to the circuit shown in Fig. 6-10. Sketch the waveform at the circuit output.* (*AMIE, Summer* 1996)

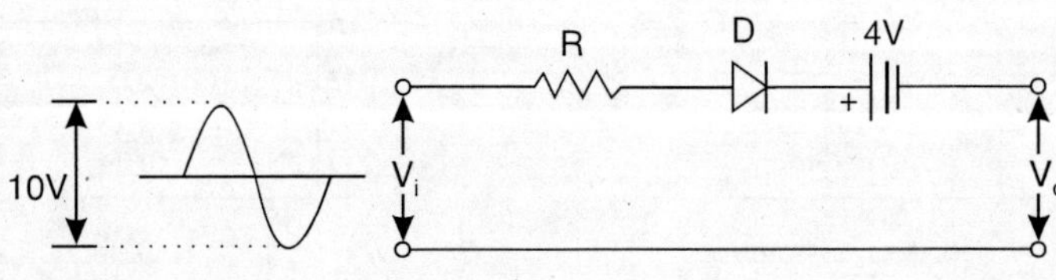

Fig. 6-10.

Solution. Fig. 6-10(*a*) shows the given input voltage waveform. The output voltage waveform can be obtained as discussed below.

Let us assume the diode, *D* as an ideal diode (*i.e.*, a diode with no barrier voltage and zero forward resistance).

The diode, *D*, acts like an open circuit for an input voltage below (+ 4 V). Therefore, the output voltage for an input greater than or equal to 4 V can be calculated as $v_0 = (v_i - 4)$ V. The output remains at zero for input voltages less than or equal to 4 V. Therefore, the output voltage waveform can be sketched as shown in Fig. 6-11.

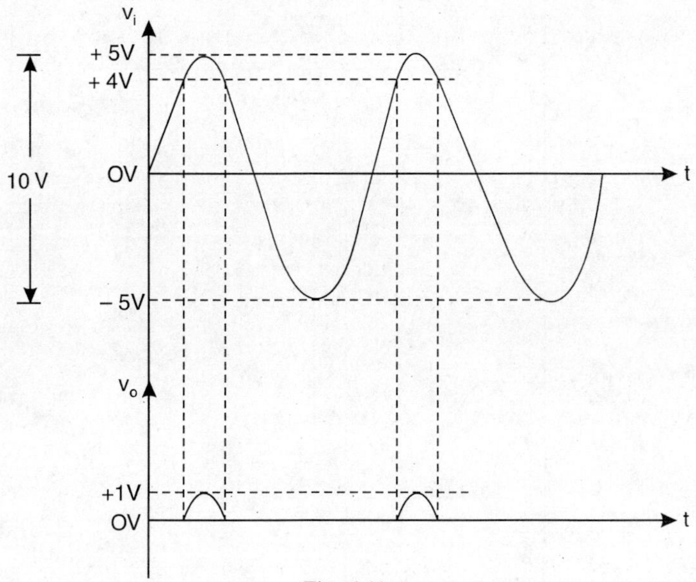

Fig. 6-11.

Note. If we assume the diode to be real *i.e.*, with forward voltage drop of 0.7 V. Then it will turn on only when the input is greater than 4.7 V.

Example. 6-3. *Sinusoidal waveform of 10 V peak-to-peak is applied to the circuit shown in Fig. 6-12. Sketch the waveform at the circuit output.*

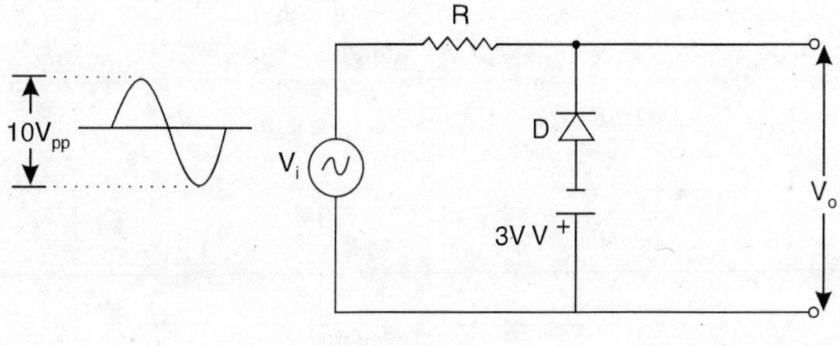

Fig. 6-12.

(*AMIE, Summer* 1996)

Solution. Given circuit shown in Fig. 6-12. It resembles that of a biased negative shunt chipping circuit.

When the input signal is below – 3 V, it is clear from the circuit, that the diode, *D*, turns 'ON' and starts conducting. So, the output is clipped at –3 V for any input lower than this value. Therefore the output waveform is as shown in Fig. 6-13.

Note. If the diode is assumed to be a 'real' or practical; the output will be clipped at – 3.7 V.

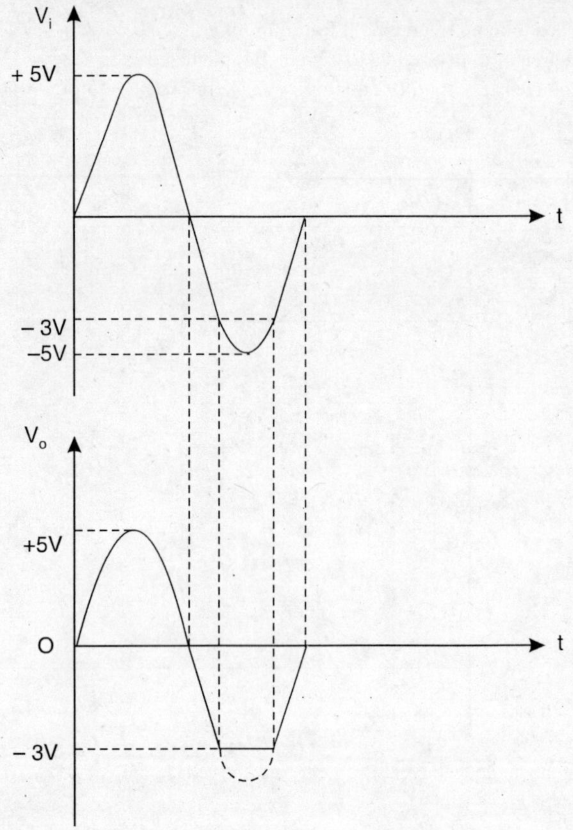

Fig. 6-13.

6-6. Some Other Biased Clippers

We have already discussed the circuit for biased positive and negative clippers. In all these circuits, the battery for the bias voltage (V_B) is shown in shunt between the source and the load. It is also possible to use the battery (V_B) in series between the source and the load for both the positive and negative clippers as discussed above :

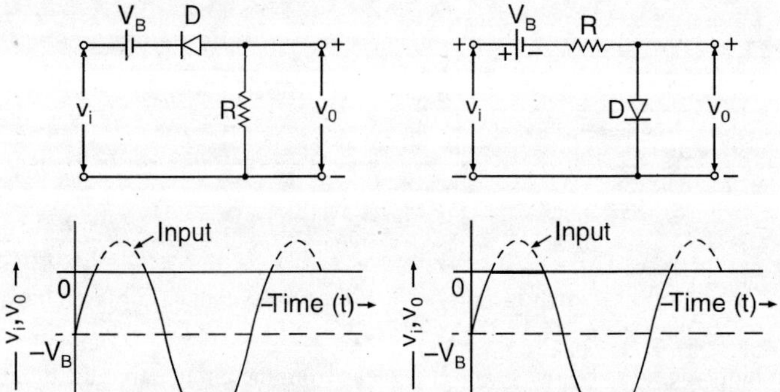

(*a*) Biased series positive clipper.

(*b*) Biased shunt positive clipper.

Fig. 6-14.

Fig. 6-14(*a*) shows the circuit of a biased series positive clipper alongwith its output waveform. Here, the addition of bias voltage V_B (with polarity of the battery as shown in the figure) shifts the entire input signal below the zero-voltage axis by an amount equal to the bias voltage (V_B).Then the circuit clipps off the part of the positive half cycle lying above the zero-voltage axis. Fig. 6.14(*b*) shows the circuit of a biased shunt positive clipper. It produces an output similar to that of a series positive clipper as shown in the figure.

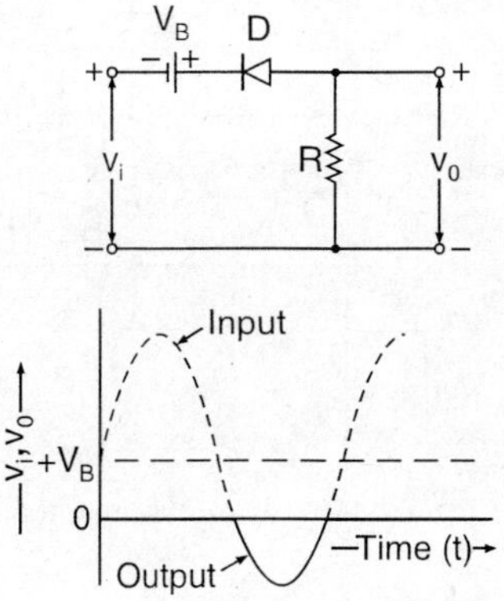

(*a*) Biased series positive clipper.

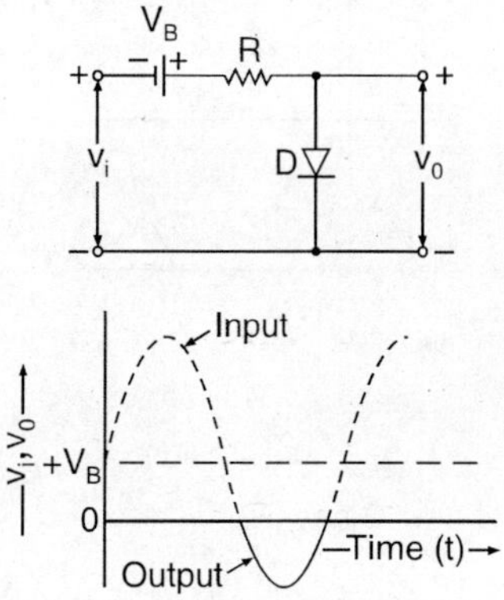

(*b*) Biased shunt positive clipper.

Fig. 6-15.

Now suppose the connections of the battery V_B in a biased positive clipper shown in Fig. 6-14 (*a*) and (*b*) are reversed. Then the resulting circuits for the series and shunt-type positive clipper are as shown in Fig. 6-15(*a*) and (*b*) respectively. It may be noted that the addition of bias voltage V_B (with polarity of the battery as shown in the figure) shifts the entire input signal above the zero voltage axis by an amount equal to the bias voltage (*i.e.*, V_B). Then the circuit clipps off the part of the negative cycle lying below the zero voltage axis as shown in the figure.

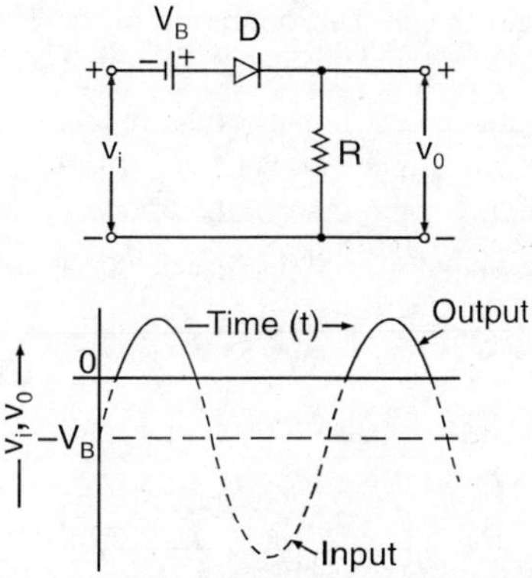

(*a*) Biased series negative clipper.

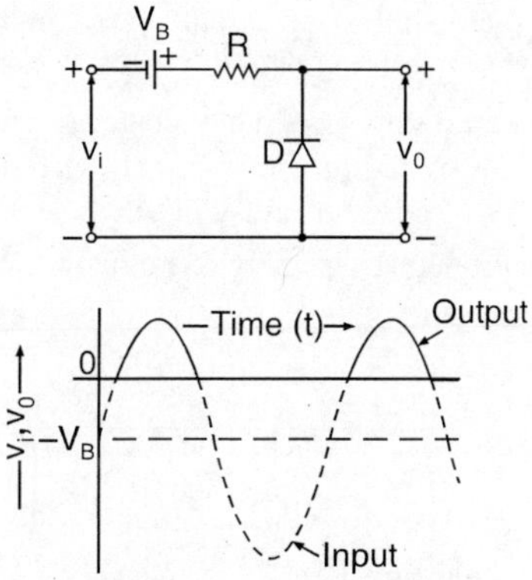

(*b*) Biased shunt negative clipper.

Fig. 6-16.

Fig. 6-16(*a*) shows the circuit of a biased series negative clipper alongwith its output waveform. Here, the addition of bias voltage V_B (with the polarity of the battery as shown in the figure) shifts the entire input signal below the zero-voltage axis by an amount equal to the bias voltage (V_B). Then the circuit, clipps off the part of the negative half cycle lying below the zero voltage axis. Fig. 6-16 (*b*), shows the circuit of a biased shunt negative clipper. It produces an output similar to that of a series negative clipper as shown in the figure.

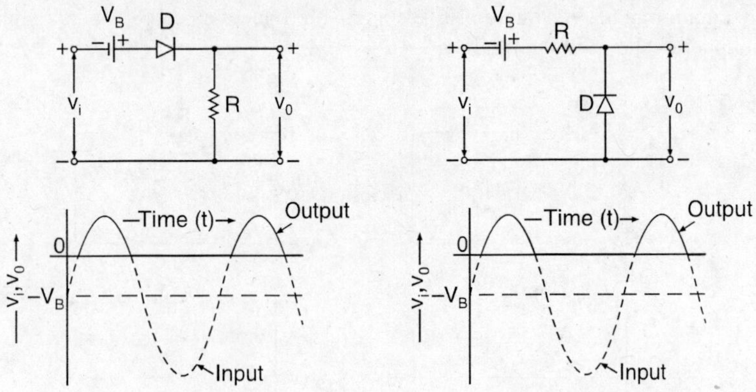

(*a*) Biased series negative clipper.

(*b*) Biased shunt negative clipper.

Fig. 6-17.

Now suppose the connections of the battery V_B in a biased negative clipper [Fig. 6.16(*a*) and (*b*)] are reversed. Then the resulting circuits for the series and shunt-type negative clipper are as shown in Fig. 6-17(*a*) and (*b*) respectively. It may be noted that the addition of a bias voltage (with polarity of the battery as shown in the figure) shifts the entire input signal above the zero voltage axis by an amount equal to the bias voltage (V_B). Then the circuit clipps off the part of the negative half cycle lying below the zero voltage axis as shown in the figure.

6-7. Combination Clipper

The combination of a biased positive clipper and a biased negative clipper is called combination clipper. Such a clipper circuit can clip at two-independent levels depending upon the bias voltages. Fig. 6-18 (*a*) shows the circuit of a combination clipper. The operation of the circuit may be explained as discussed below.

Suppose a sinusoidal a.c. voltage is applied at the input terminals of the circuit. Then during the positive half-cycle, the diode D_1 is forward biased, while diode D_2 is reverse biased. Therefore, the diode D_1 will conduct and will act as short-circuit. On the other hand, diode D_2 will act as an open-circuit. However, the value of output voltage cannot exceed the voltage level of V_{B_1} as shown in Fig. 6-18 (*b*).

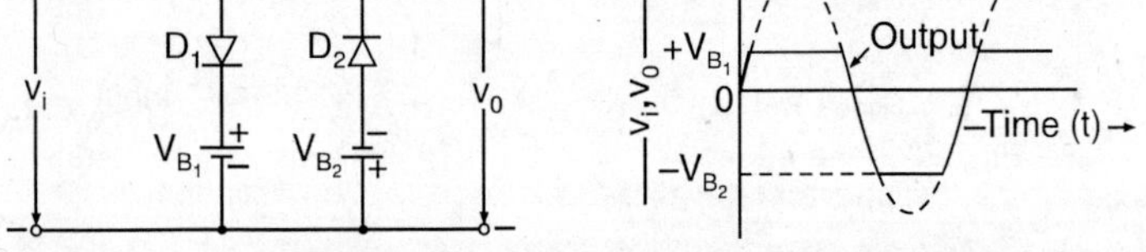

(*a*) Combination clipper.

(*b*) Input and output wave forms.

Fig. 6-18.

Similarly, during the negative input half-cycle, the diode D_2 acts as short circuit, while the diode D_1 as an open circuit. However, the value of output voltage cannot exceed the voltage level of V_{B_2}. It may be noted that the clipping levels of the circuit can be varied by changing the values of V_{B_1} and V_{B_2}. If the values of V_{B_1} and V_{B_2} are equal, the circuit will clip both the positive and negative half cycles at the same voltage level. Such a circuit is known as a symmetrical clipper.

Example 6-4. *Draw the circuit diagram of a clipper to clip a 5 V sine wave input signal at +3 V and – 2 V.*

Solution. Fig. 6-19 shows the circuit of a combination clipper alongwith the input and output waveforms.

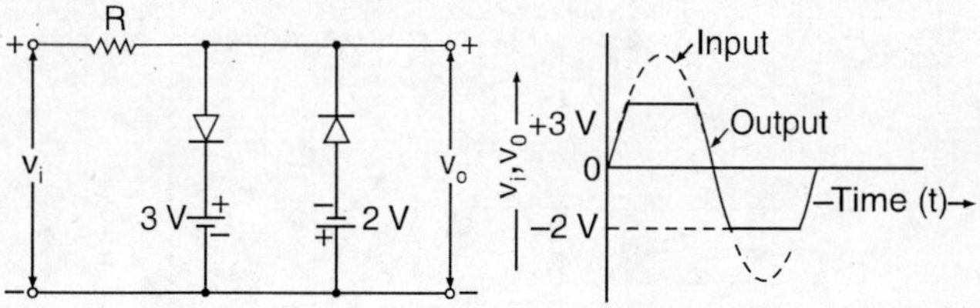

Fig. 6-19.

It is evident from the waveforms, that the positive half cycle of the input voltage is clipped at + 3 V and the negative cycle is clipped at – 2 V.

Example 6-5. *Sinusoidal waveform of 10 V peak-to-peak is applied to the circuit shown in Fig. 6-20. Sketch the waveform at the circuit output.*

(*Electronic Circuits II, Anna University May/June 2007*)

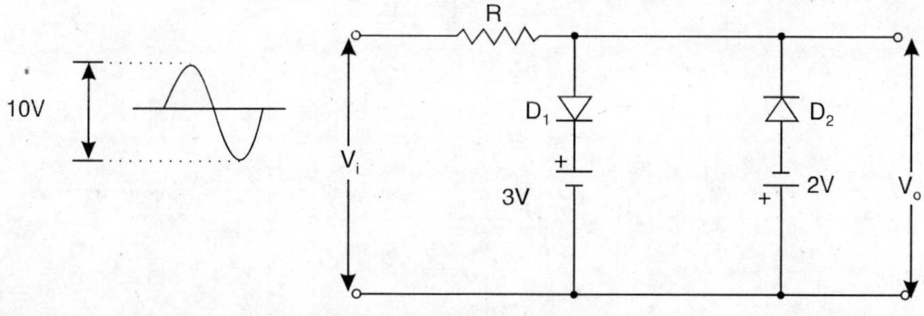

Fig. 6-20.

Solution. Given the circuit in Fig. 6-20. The input signal is $10V_{P\text{-}P}$ sinusoidal wave. That means it varies between +5 V and – 5 V sinusoidally.

By analyzing the circuit, it can be seen that, the section containing diode D_1, and 3 V source acts as a biased positive clipper that clips the output at 3 V for any input above this value.

Similarly, the section containing the diode, D_2, and 2 V source cuts as a biased negative clipper that clips the output at –2V for any input below this value.

Therefore, the output waveform will be as shown in Fig. 6-21.

Example 6-6. *A biased shunt clipper circuit as shown in Fig. 6-22 is to be designed to protect a circuit which cannot accept voltage axceeding, v_0 = ± 2.7 V. The input to the clipper is a ± 8 V square wave and output current is to be a maximum of 1 mA.*

Calculate the value of R_1 and specify the diode features required. (*AMIE, Summer* 2002)

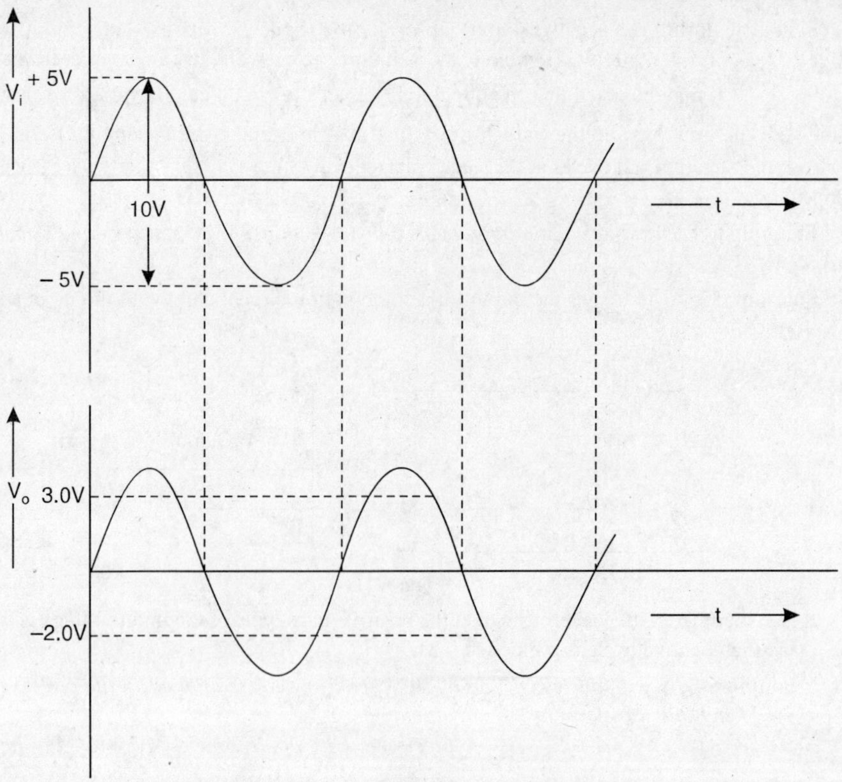

Fig. 6-21.

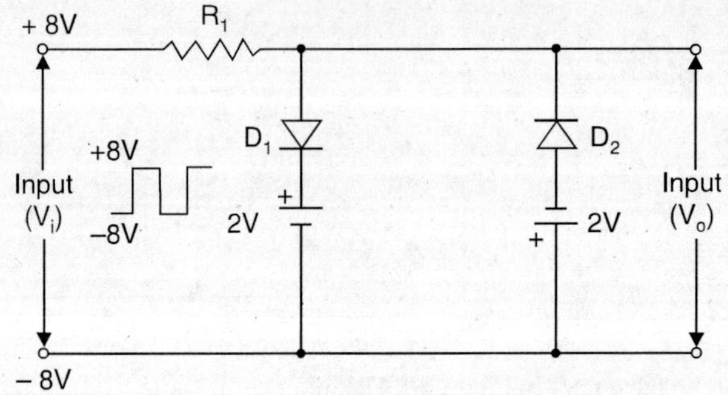

Fig. 6-22.

Solution. Given the circuit diagram of Fig. 6.22. We know that diode D_1 acts like an open-circuit for an input voltage below + 2.7 V voltage below an input voltage above. Therefore, the similar diode D_2 acts like-an open-circuit for – 2.7 V sketched as shown in Fig. 6-23.

We know that the maximum value of voltage drop across R_1,

$$= 8 - 2.7 = 5.3 \text{ V}$$

$\therefore$ Value of R_1, so that the current does not exceed 1 mA

$$R_{1(min)} = \frac{5.3}{1 \times 10^{-3}} = 5.3 \times 10^3\,\Omega \text{ or } 5.3 \text{ k}\Omega \textbf{ Ans.}$$

The diodes D_1 and D_2 has to be silicon diodes.

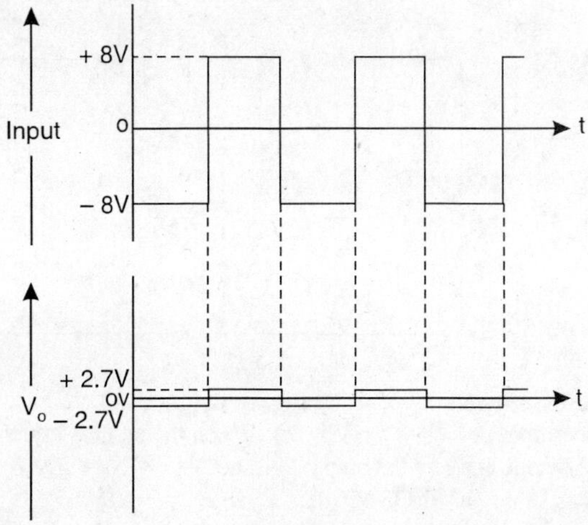

Fig. 6-23.

6-8. Two Level Slicer

Fig. 6-24(*a*) shows the circuit of a two level slicer. This is similar to the combination clipper but with the diode connections circuit reversed.

Fig. 6-24(*b*) shows the input and output waveforms for a two level slicer.

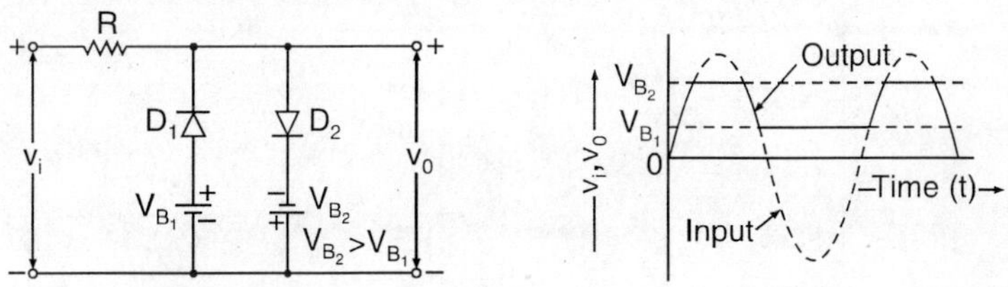

(*a*) Two level slicer. (*b*) Input and output waveforms.

Fig. 6.24.

It is evident from this figure that the circuit has a slice cut from positive half cycle of the input signal.

Notes. 1. In all the clipper circuits discussed above, the resistor (R) is used to limit the current through the diode when it is forward biased.

2. It is assumed that the value of resistor (R) is much larger than the forward resistance of the diode (R_f), and much smaller than the reverse resistance (R_r). This condition is fulfilled only if the value of resistor (R) is selected as the geometric mean of the resistances R_f and R_r *i.e.*,

$$R = \sqrt{R_f \cdot R_r}$$

If this condition is not fulfilled, then the output waveform of the clipping circuit will be rounded near the clipping level (*i.e.*, it will not remain flat as indicated in the figures).

Example 6-7. *A sinusoidal waveform of 10 V peak- to-peak. is applied to the circuit shown in Fig. 6-25. Sketch the waveform at the circuit output.*

Solution. Given a sinusoidal waveform of 10 V peap-to-peak applied to the circuit shown in Fig. 6-25.

In order to determine the output waveform, let us assume both the diodes D_1 and D_2 to be an ideal diodes. The input voltage waveform is sketched in Fig 6-26(*a*). Notice that the diode D_1 is off as long as the input voltage is below 3 V and D_2 is off as long as the input voltage is below 2 V.

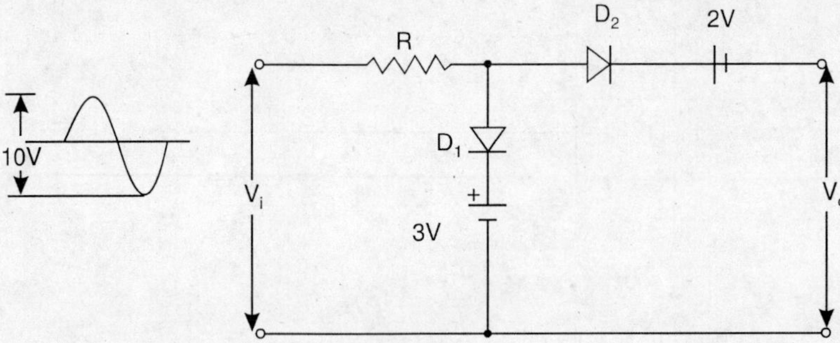

Fig. 6-25.

For an input voltage below 2 V, the output voltage is 0. For an input voltage above 2V but below 3 V, D_2 is ON, therefore the output voltage $V_0 = V_i - 2$ V. When the input is equal to or greater than, 3 V, both the diodes D_1 and D_2 one ON and the output voltage. $V_0 = -2$ V + 3 V = 1 V. The output voltage waveform is sketched as shown in Fig. 6-17(*b*).

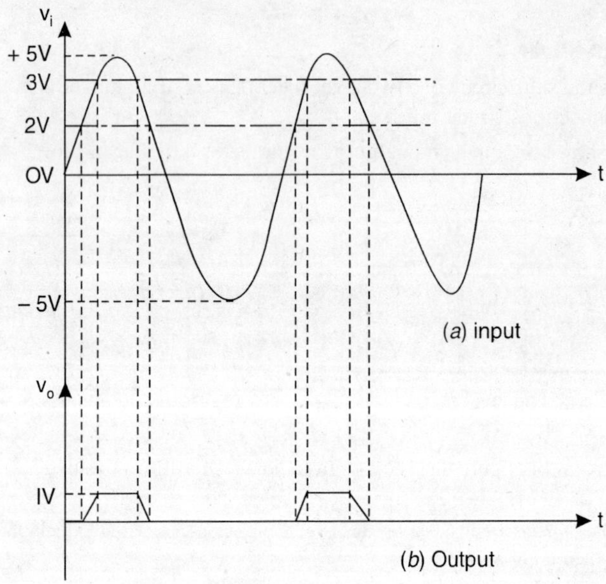

Fig. 6-26. *Input and output waveforms of circuit shown in this Figure.*

Example 6-8. *Sketch the output of the clipper circuit shown Fig. 6-27 for the given input. Assume diodes are made of silicon and practical.* *(Dip. IETE, Dec. 2000)*

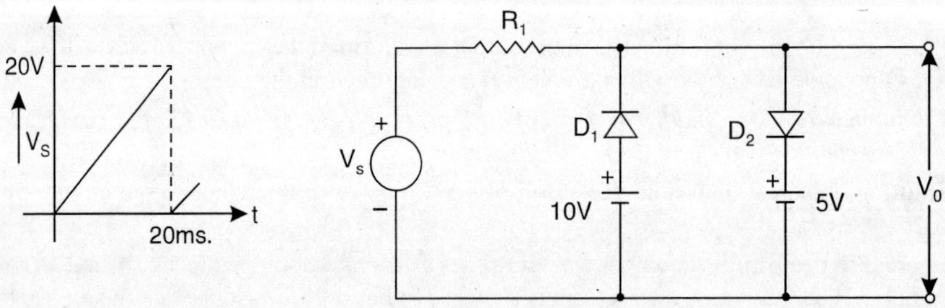

Fig. 6-27.

Solution. Given, The input voltage waveform and a clipper circuit shown in Fig. 6-27. The diodes D_1 and D_2 are silicon and practical. It means, when the diode D_1 or D_2 is forward biased (> 0.7 V), it acts like a closed switch and the voltage drop across it is 0.7V. However, when the diode D_1 or D_2 is reversed biased, it acts like an open switch with no voltage drop across it.

From the circuit we can see that, the diode, D_1, is reverse biased initially and the diode, D_2, is forward biased, due to the biasing sources connected in series with them. Therefore, the output can be calculated as,

$$v_0 = 5 \text{ V} - \text{voltage drop across the diode } D_2$$
$$= 5 - 0.7 = \mathbf{4.3\ V}$$

The output remains at 4.3 V. Until the input is increased above 4.3V. At this point, the diode D_2 become reverse biased and as a result the output follows the input.

When the input voltage, v_s, reaches 10.7 V, the diode D_1 become forward biased, since it can overcome the barrier potential. When the input is increased further up, the diode, D_1, conducts maintaining the output at 10.7 V for the remaining part of the input pulse.

Therefore, the output waveform is as shown in Fig. 6-28.

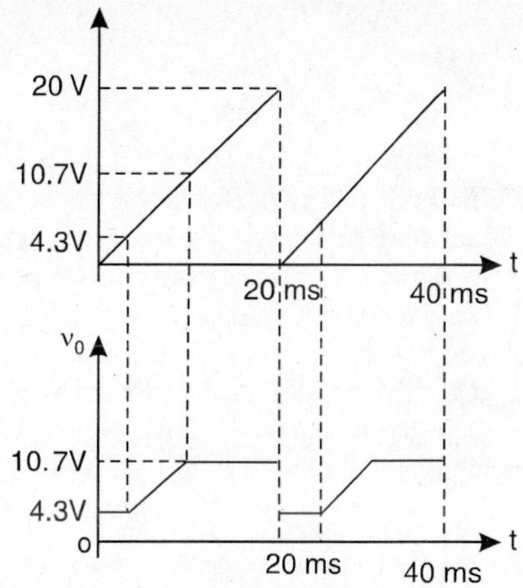

Fig. 6-28.

6-9. Clamping Circuits

The circuits, with which the waveform can be shifted in such a way so that a particular part of it (say positive or negative peak) is maintained at a specified voltage level, is called a *clamping circuit* (or simply a clamper). As a matter of fact, a clamping circuit introduces (or restores) a d.c. level to an a.c. signal. Thus, a clamping circuit is also known as d.c. restorer. Such circuits are used in television receivers to restore the original. *d.c. reference signal to the video signal.

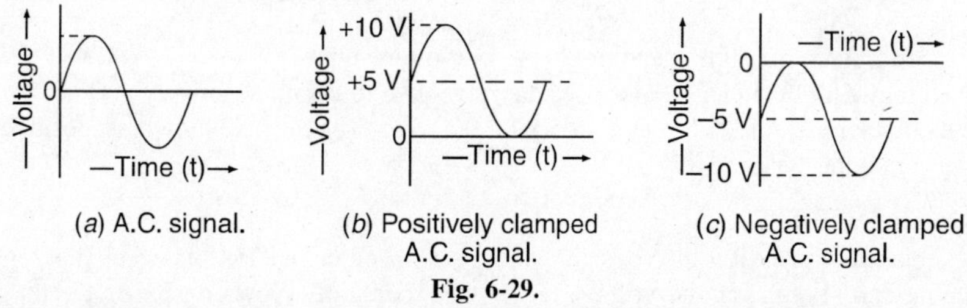

(*a*) A.C. signal. (*b*) Positively clamped A.C. signal. (*c*) Negatively clamped A.C. signal.

Fig. 6-29.

* In television systems, the d.c. reference level corresponding to the brightness level of the picture is not transmitted with video signal.

Now, consider an a.c. signal waveform as shown in Fig. 6-29(*a*). It is a sine wave with equal positive and negative swings of 5 V above and below the zero level. It has no d.c. level. On the other hand, the signal waveform shown in Fig. 6-29(*b*) is lifted up so as to just touch the horizontal axis. It is said to have a d.c. level of + 5 V and is known in Fig. 6-29(*c*) has a d.c. level of – 5 V and is known as waveform negatively clamped at 0 V.

It is possible to clamp a waveform positively at 0V by connecting a d.c. battery in series with the source, whose e.m.f is equal to the peak value of the sine wave ($v_i = V_m \sin \omega t$) as shown in Fig. 6-30. However, it is a very crude way of achieving the required clamping action and therefore is not used in actual practice.

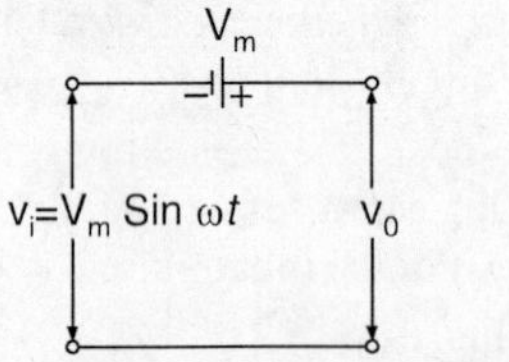

(*a*) Battery connected in series with the source.

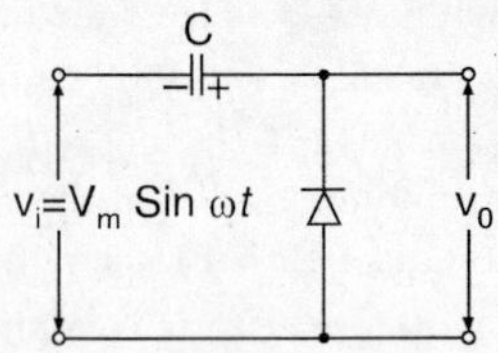

(*b*) Diode clamper.

Fig. 6-30.

A correct way of clamping the input signal, positively at 0 V, is to use circuit as shown in Fig. 6-30(*b*). This circuit consists of a diode and capacitor and is popularly known as a diode clamper. The clamping action of the circuit may be understood from the fact that during the negative half cycle of the input voltage (v_i), the diode is forward biased and the current flows through the circuit. As a result of this, the capacitor (*C*) is charged to a voltage equal to the negative peak value (*i.e.*, $-V_m$). Once the capacitor is fully charged to $-V_m$, it cannot discharge because the diode cannot conduct in the reverse direction. It means that this capacitor acts as a battery with an e.m.f, equal to the $-V_m$. The polarity of this voltage is such that it adds to the input signal. Therefore, the output voltage is equal to the sum of the a.c. input signal and the capacitor voltage V_m,

i.e., $$v_o = v_i + V_m = V_m \cdot \sin \omega t + V_m$$

It is evident from the above relation, that a d.c. voltage equal to V_m is added to input signal. This causes the waveform to clamp positively at 0 V.

Now suppose it is required to clamp the input signal waveform ($v_i = V_m \sin \omega t$) negatively at 0V. This can be done by reversing the diode connections as shown in Fig. 6-31. Here the diode is forward biased during the positive half-cycle of the input signal. Therefore, the capacitor is charged to a voltage equal to the positive peak of the input signal (*i.e.*, V_m), the polarity of this voltage is such that it is subtracted from the input signal. Hence, the output voltage,

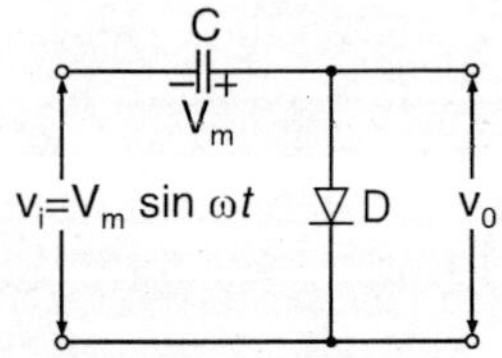

Fig. 6-31.

$$v_o = v_i - V_m = V_m \sin \omega t - V_m$$

It is evident from the above relation that a d.c. voltage equal to $-V_m$ is added to the input signal. This causes the waveform to clamp negatively at 0 V.

The diode clamper circuits shown in Fig. 6-30(*b*) and 6-31 are also known as positive clamper and negative clamper respectively.

6-10. Biased Clampers

Like a biased clipper, it is also possible to have a biased clamper. A biased clamper means that the clamping can be done at any voltage level other than zero. Now consider a circuit shown in Fig. 6-32(*a*). Here a battery of 5 V is added in such a way that the clamping takes place positively at – 5 V as shown in Fig. 6-32(*b*).

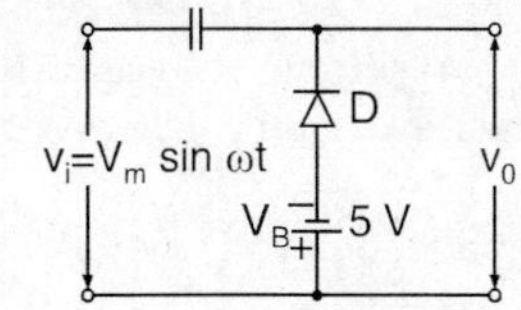

(*a*) Biased clamper.

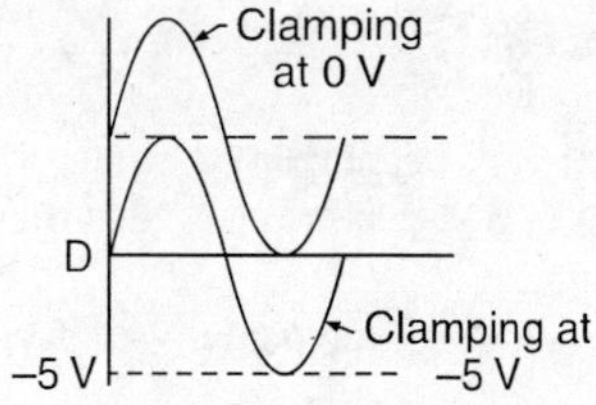

(*b*) Output waveforms.

Fig. 6-32.

Similarly, it is also possible to clamp the input waveform positively at + 5 V by reversing the battery connections.

6-11. Practical Clamper Circuit

The clamper circuits used in last articles work properly, only when the amplitude of the input signal does not change with time. However, in actual practice, the amplitude of the input signal changes continuously, with time, as it happens in a televisions signal. Therefore, such circuits are not suitable in actual practice. Fig. 6-33 shows a more practical clamper circuit. Here a resistor (*R*) having a very large value of resistance is connected in parallel with the diode. The purpose of this resistor is to provide a discharge path for the capacitor. The large value for this resistor is required to produce a time constant ($\tau = R \cdot C$) which is large enough than the period of the input signal waveform. For good clamping action, the value of time constant should be at least 10 times the period of the input signal waveform.

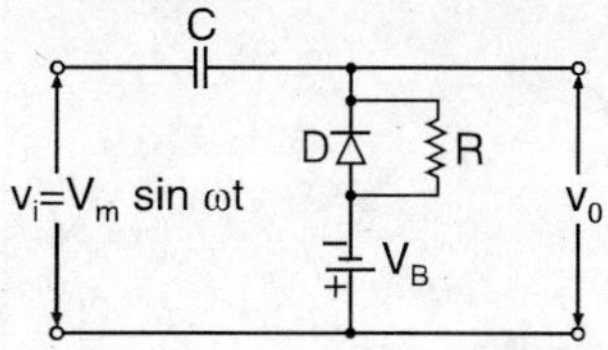

Fig. 6-33. *Practical clamper circuit.*

Example 6-9. *The input signal with waveform shown in Fig. 6-34(a) is applied to the clamper circuit shown in Fig. 6-34(b).*

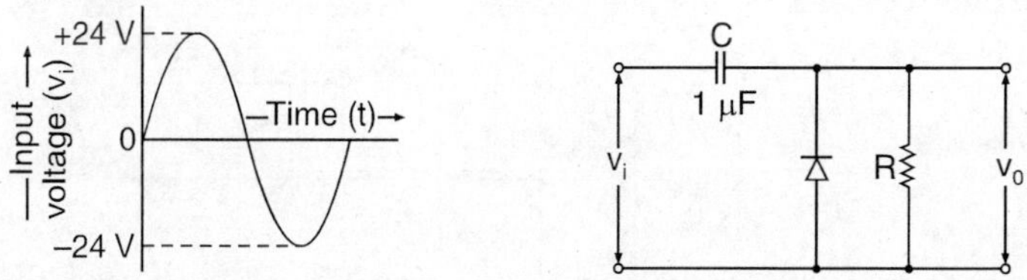

(*a*) (*b*)

Fig. 6.34.

Sketch the waveform of the output voltage.

Solution. From the figure, we find the diode is forward biased during the negative half cycle of the input signal. Therefore, it acts as short circuit. As a result of this, capacitor is charged to a voltage equal to the negatives peak of the input signal (*i.e.*, 24 V). The polarity of this voltage is such that it gets added to the input signal. Therefore the output voltage v_o = 24 sin ω*t* + 24. This causes the voltage to clamp positively at zero volt as shown in Fig. 6-35.

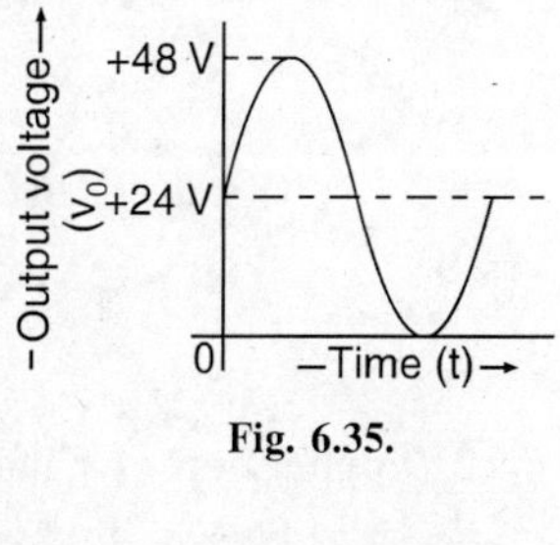

Fig. 6.35.

Example 6-10. *A ± 10 V square wave is applied (a) to the input terminals of negative voltage clamping circuit and (b) to the input of a negative shunt clipper. Sketch the output waveform that result in each case.* (*AMIE, Summer* 1999)

Solution. (*a*) The output of a negative voltage (refer to Fig. 6-31., page. 107) clamping circuit remains at 0 V, during positive half cycle of the input wave. During negative half cycle, the output is given as 2 V_{in}.

Therefore, the resulting output waveform is as shown in Fig. 6-36.

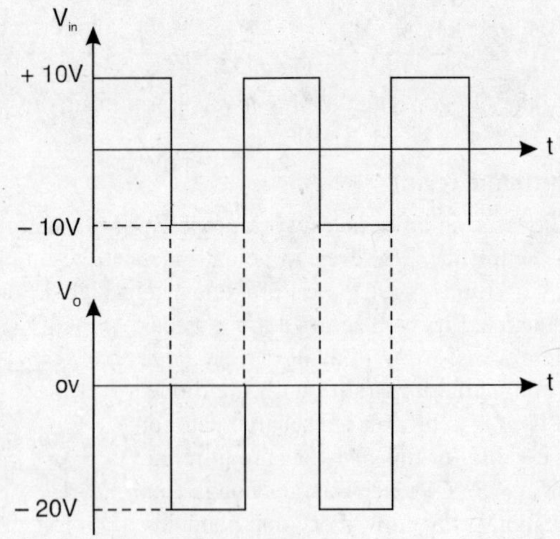

Fig. 6-36.

(*b*) The output of a negative shunt clipper (refer to Fig. 6-17. page 101) remains at 0V, during negative half cycle of the input wave. During positive half cycle, the output follows the input. Therefore, the resulting output waveform is as shown in Fig. 6-37.

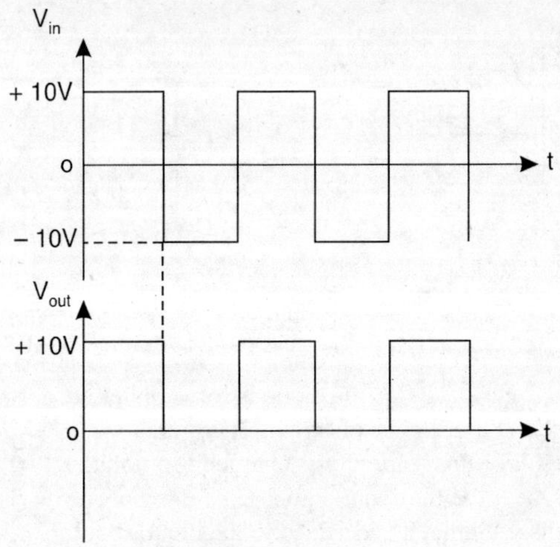

Fig. 6-37.

6-12. Application of Clamping Circuits

We have already discussed that the clamping circuits are used to shift any part of the input signal waveform and can be maintained at a specified voltage level. Such circuits are used in television receivers to restore the original d.c. reference signal (corresponding to the brightness level of the picture) to the video signal. The clamping circuits are also used to produce a d.c. voltage whose value is a multiple of peak a.c. input voltage (*i.e.*, $2V_m$, $3V_m$, $4V_m$ *etc.*). Such circuits are known as *voltage multipliers*. These circuits are used to supply power to high voltage/low current devices like cathode ray tubes used in television receivers, oscilloscopes and computer displays.

6-13. Voltage Multipliers

We have already discussed in the last article that a voltage multiplier is a circuit, which produces an output d.c. voltage whose value is multiple of peak a.c. input voltage (*i.e.*, $2V_m$, $3V_m$, $4V_m$ and so on). Such circuits are used as a power supply for high voltage/low current device like cathode ray tubes.

6-14. Voltage Doubler

A voltage multiplier, whose output d.c. voltage is double the peak a.c. input voltage, is called a voltage doubler. Fig. 6-38 shows the circuit, of a half-wave voltage doubler. In this circuit, each section of a diode and capacitor (say D_1 and C_1 or D_2 and C_2) is called a *peak rectifier*. The operation of the circuit may be explained as given below :

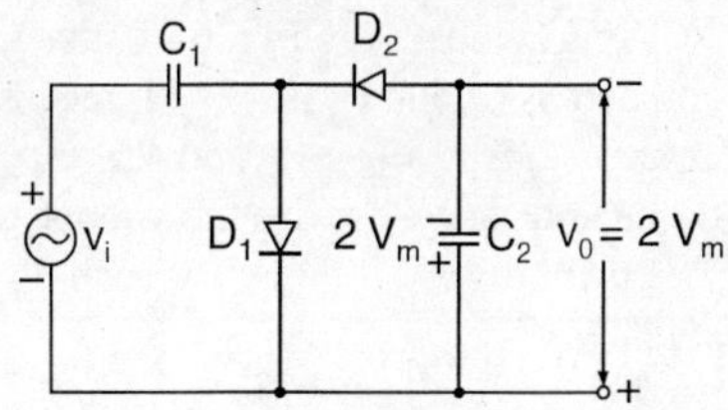

Fig. 6-38.

During the positive half-cycle of the input signal, the diode D_1 conducts (and diode D_2 is cut-off), charging the capacitor C_1 up to the peak rectified voltage (*i.e.*, V_m). During the negative half-cycle, diode, D_1 is cut-off and diode D_2 conducts charging capacitor C_2. It may be noted that during negative half-cycle, the voltage across capacitor C_1 is in series with the input voltage. Therefore, the total voltage presented to capacitor C_2 is equal to 2 V_m. As a result of this, the capacitor C_2 is charged to a voltage of 2 V_m during the negative half-cycle. On the next positive half-cycle, the diode D_2 is non-conducting and the capacitor will discharge through the load. If no load is connected across capacitor C_2, both capacitors stay charged at their full values (*i.e.*, C_1 to V_m and C_2 to 2 V_m). It may be noted that both the diodes D_1 and D_2 have a peak inverse voltage (PIV) of 2 V_m each.

Fig. 6-39 shows another voltage doubler circuit known as full-wave voltage doubler. In this circuit during, the positive half-cycle of a.c. input voltage, diode D_1 conducts charging capacitor C_1 to a peak voltage V_m with polarity as shown in the Fig. 6-39. The diode D_2 is cut-off at this time. During the negative half-cycle the diode D_2 conducts (while D_1 is at cut-off) charging capacitor C_2 to V_m. If there is no load connected across the output, then the output voltage is equal to $2V_m$. However, if the load is connected, then the voltage would be less then $2V_m$.

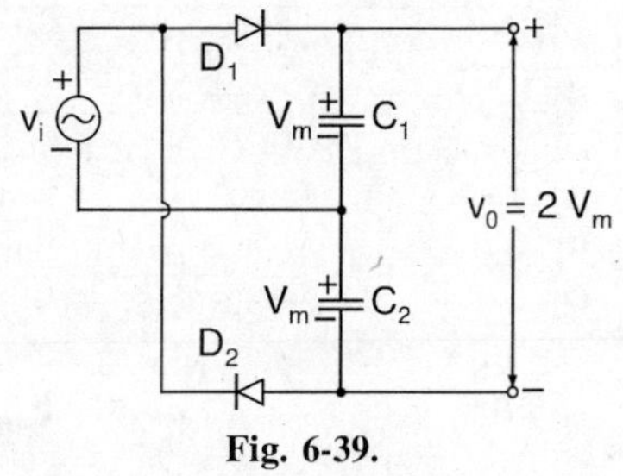

Fig. 6-39.

The peak inverse voltage (PIV) across each diode, in a full-wave voltage doubler, is equal to $2V_m$. The full-wave voltage doubler has an advantage over a conventional centre-tapped full-wave rectifier that it does not require any centre-tapped transformer.

6-15. Voltage Trippler and Quadrupler

Fig. 6-40 shows the extension of a half-wave voltage doubler. This circuit develops an output voltage equal to 3 and 4 times the peak input voltage. It is obvious from the pattern of the circuit connection as to how additional diodes and capacitors may be connected for obtaining higher multiplications of the basic peak voltage. Theoretically, there is no upper limit for the amount of voltage multiplication. But it has been observed that if we increase the voltage multiplication beyond four, the regulation is very poor. It is because of this fact, that voltage multipliers are not used in low-voltage supplies.

The operation of the voltage trippler and quadrupler circuit may be understood from the fact, that during the first positive half-cycle, the capacitor C_1 charges through diode D_1 to a peak voltage V_m. During the negative half-cycle, capacitor C_2 charges through diode D_2 to twice the peak voltage $2V_m$ developed by the sum of the voltages across capacitor C_1 and the input signal.

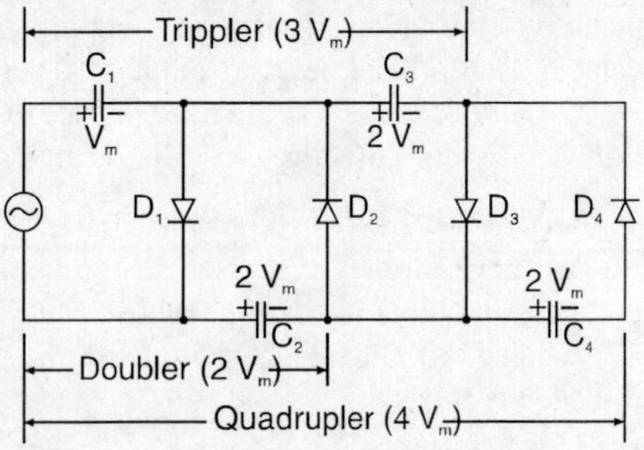

Fig. 6-40.

During the second positive half-cycle, the diode D_3 conducts and the voltage across capacitor C_2 charges the capacitor C_3 to the same 2 V_m peak voltage. And during the negative half-cycle, diodes $D_2$and D_4 conduct allowing capacitor C_3 to charge capacitor C_4 to the same 2 V_m peak voltage.

It is evident from Fig. 6-40, that the voltage across capacitor C_2 is 2 V_m, across C_1 and C_3 is 3 V_m and across C_2 and C_4 is 4 V_m. It means that the voltage quadrupler circuit can provide three different voltages, *i.e.*, 2 V_m, 3 V_m and 4 V_m. It will be interesting to know that if we use additional sections of diode and capacitor with the voltage quadrupler circuit, then each capacitor will be charged to 2 V_m. The peak inverse voltage for each diode in the circuit is equal to 2 V_m.

SUMMARY

In this chapter you have learned that :

1. PN junction diodes can be used in clipping and clamping circuits.
2. Clipping circuits are used for shaping a waveform by removing a certain portion of it.
3. In a positive clipper, positive portion of the waveform is removed.
4. In a negative clipper, negative portion of the waveform is removed.
5. A biased clipper can be used to remove the portions of a waveform with reference to a voltage other then zero.
6. Clamping circuits are used for shifting the level of a waveform.
7. Clamping circuits have applications such as voltage multiplier, voltage doubler, voltage trippler and voltage quadrupler.

GLOSSARY

Bias voltage (V_B) : The voltage applied in series with the diode or resistor in a biased clipper circuit.

Biased clamper : A Clamper with which clamping can be done at any voltage level other than zero.

Biased clipper : A clipper with which clipping can be done at any voltage level other than zero.

Clamper : The circuit, with which the waveform can be shifted in such a way that a particular part of it is maintained at a specified voltage level.

Clipper : The circuit, with which the waveform is shaped by removing a certain portion of the combination of a biased positive and a biased negative clipper.

Combination clippes : Clipper formed by the combination of a biased positive and a biased negative clipper.

D.C. restorer : refer to clamper

Negative clamper : Clamper circuit that shifts the wave form such that positive peak voltage reaches to zero level.

Negative clipper : Clipper circuit that removes the negative portion of the waveform.

Positive clamper : Clamper circuit that shifts the waveform such that negative peak voltage reaches to zero level.

Positive clipper : Clipper circuit that removes the positive portion of the waveform.

NUMERICAL PROBLEMS

1. Find the output voltage shape for the clipping circuit shown in Fig. 6-41(*a*) for the inputs shown in Fig. 6-41 (*b*).

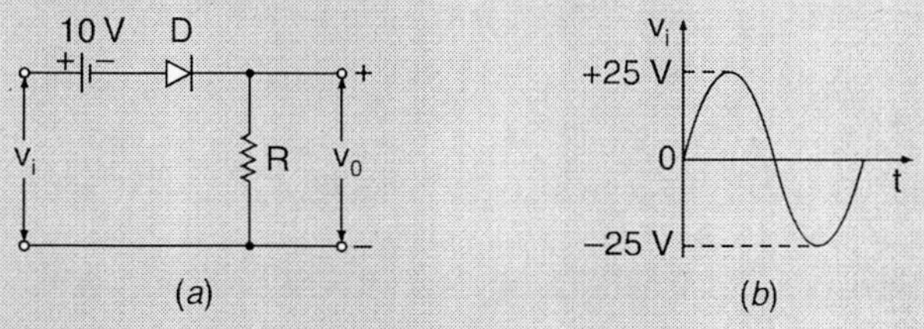

Fig. 6-41.

2. Sketch the output waveform for a diode clamper circuit shown in Fig. 6-42(*a*) for a sine wave input with a peak value of 30 V.

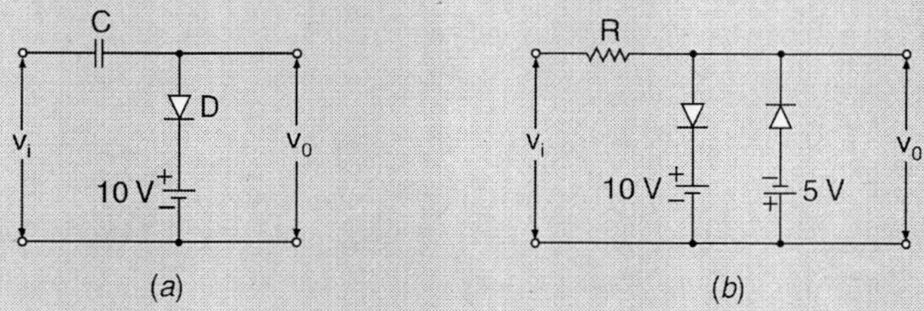

Fig. 6-42.

3. Draw neatly the output waveform for a diode clamper circuit shown in Figure 6-42 (*b*) for a sine wave input with a peak voltage of 20 V.
4. Draw the output waveform for the given input waveform. Also draw the output waveform when the polarity of the bias voltage is reversed. (*Banglore University*, 2001)

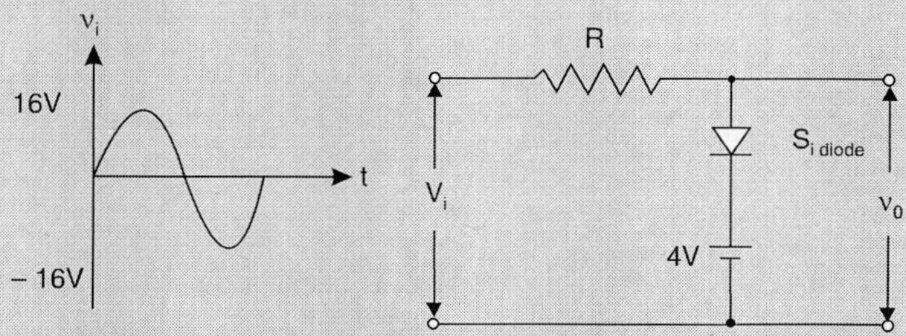

Fig. 6-43.

DESCRIPTIVE QUESTIONS

1. What is the difference between positive and negative clippers? Explain with the help of output waveform.
2. With the help of circuit diagram, describe the operation of clipper that can clip at two independent levels. Choose a suitable waveform for clipper description. Explain how the two clipping levels can be controlled independently.
3. Explain with suitable diagrams how a diode can be used as a peak clipper and a base clipper as series element and shunt element. Draw a circuit for a slicer.

4. Sketch the circuit of a positive series clipper showing the input and output waveforms. Briefly explain its operation.
5. Explain the operation of clipping at two independent levels if a sinusoidal input is fed to a two-diode clipper circuit. Can it be called a clamping circuit. Explain with justification.
6. Draw the circuit for double ended clipper using avalanche diodes and explain briefly its operation.
7. Distinguish between clipping and clamping circuits with suitable circuits with suitable diagram and supporting input and output waveforms.
8. Sketch the circuit of a positive biased clamper circuit. Show the input and output waveforms. Briefly explain how it functions. (*AMIE Summer* 2002)
9. Distinguish between clamper and clipping circuit showing patterns of outputs generated in each case. (*AMIE Summer* 2002)
10. (*a*) Describe the function of a double diode clipper.
    (*b*) A square wave from a source resistance, R_S is applied to the input of a diode clamping circuit. Find analytically the expressions for output transition (*AMIE Winter* 2001)
11. Define the difference between clipping and clamping circuits. (*AMIE., Summer* 1999)
12. What is the purpose of a clamping circuit? Explain the working of a diode clamper. How clamping to a dc level is achieved.
13. Explain the operation of clipping at two independent levels if a sinusoidal input is fed to a two-diode clipper circuit. Can it be called a clamping circuit? Explain with justification. (*AMIE., Summer* 1995)
14. Draw the circuit for double ended clipper using avalanche diodes and explain briefly its operation. (*AMIE., Summer* 1995)
15. How does an ideal diode behave as am amplitude limiter? (*AMIE., Summer* 1993)
16. Draw the circuit of a diode clamper where the output is clamped to a maximum of zero volt. State and prove the clamping circuit theorem and apply the same to the circuit already drawn to obtain the steady state output for any typical pulse train input. (*AMIE., Winter* 1993)
17. Briefly explain how a diode-capicitor circuit serves as a dc restorer. What are the modifications in the circuit to accommodate amplitude changes in either directions of the input waveform? (*AMIE., Summer* 1994)
18. Explain the working of a diode clamping circuit taking source and diode resistance into account.
19. Draw neatly the circuit diagram of a voltage quadrupler. Is it possible to obtain an output of 3 times the peak voltage?
20. Draw the schematic of a voltage doubler circuit and explain how the output voltage obtained is twice of the input voltage. (*AMIE., Summer* 2001)
21. What is the purpose of a clamping circuit? Explain the working of a diode clamper. How clamping to a d.c level is achieved. (*AMIE., Summer* 1996)
22. Explain the working of a transistor clipper in base clipping and peak clipping modes. (*AMIE., Summer* 1996)
23. Explain the working of a transistor clipper in base clipping and peak clipping modes. (*AMIE., Summer* 1996)
24. Draw and explain the working of positive clamper (*Eelctronic Circuits, VTU June/July 2008*)
25. Draw the circuit of negative clipper with input and output waveforms. (*Electronic Circuit-II, Anna University, May/June 2007*)

MULTIPLE CHOICE QUESTIONS

1. The main function of a clipping circuit is to

(*a*) remove a certain portion of the input signal above or below a certain level

(*b*) restore d.c. level to the signal

(*c*) suppress amplitude variations in the input signal voltage

(*d*) both (*a*) and (*c*) above

2. A positive clipping circuit,

(*a*) removes positive portion of the waveform

(*b*) removes negative portion of the waveform

(*c*) passes the entire waveform

(*d*) none of these (*AMIE., Summer* 2001)

3. In a clamping circuit,

(*a*) there is only resistor and capacitor

(*b*) there is a resistor, a capacitor and diode

(*c*) there is a diode and resistor

(*d*) there is a diode and capacitor (*AMIE., Summer* 2001)

4. A cascade voltage doubler can be used to provide any degree voltage multiplication by

(*a*) using diodes with large PIV rating (*b*) adding a step-up transformer

(*c*) cascading diodes and capacitors (*d*) increasing the value of capacitors

5. The clamper circuit is used to (*AMIE Summer* 2002)

(*a*) introduce a d.c. level to a.c. signal

(*b*) suppress variations in amplitude of the input signal

(*c*) obtain an output which is integral of the input signal

(*d*) none of the above

ANSWERS

1. (*a*) **2.** (*a*) **3.** (*d*) **4.** (*c*) **5.** (*a*)

❑❑❑

7

CHAPTER

Special Purpose Diodes and Opto-Electronic Devices

OUTLINE

Objectives

After completing this chapter, you should be able to :

- Discuss the basic operating principles of the zener diode.
- Identify the circuit symbol of zener diode and determine the direction of current through the device.
- List the main zener diode parameters and explain how each limits the use of component.
- Determine the suitability of a given zener diode for a specific application using zener diode specification data sheet.
- Describe the operation of tunnel diode.
- List and explain the main tunnel diode parameters.
- Discuss the purpose served by the varactor.
- List, define and explain the significance of the commonly used varactor parameters and specifications.
- Describe the construction and operation of schottky diode.
- Discuss the basic operating principles of the light-emitting diode (LED).
- Calculate the value of the current limiting resistor that is needed for LED in a given circuit.
- Describe the construction of seven-segment displays and liquid crystal displays.
- Explain the operation of photoconductive cell and solar cell.
- Describe the construction and operation of laser diode.
- List the applications of laser diode.
- Identify the circuit symbol of phototransistor and photo darlington.

7-1. Introduction

We have already discussed about rectifier or general purpose diodes in the last chapter. Now we shall discuss several other types of diodes, which are used for specific applications. Such diodes are zener, varactor, schottky, tunnel, step recovery, photo and light emitting diodes. These diodes are also called *special purpose diodes* or simply special diodes. The devices that produce light or responds to the light (opto-electronic devices) are also discussed in this chapter.

7-2. Zener Diode

A zener diode is also called a voltage-reference, voltage regulator or breakdown diode. Like a rectifier diode, it is also important in many power applications. Fig. 7-1(a) shows a schematic symbol for a zener diode. It will be interesting to know that the line at the end of the arrow looks like the letter Z.

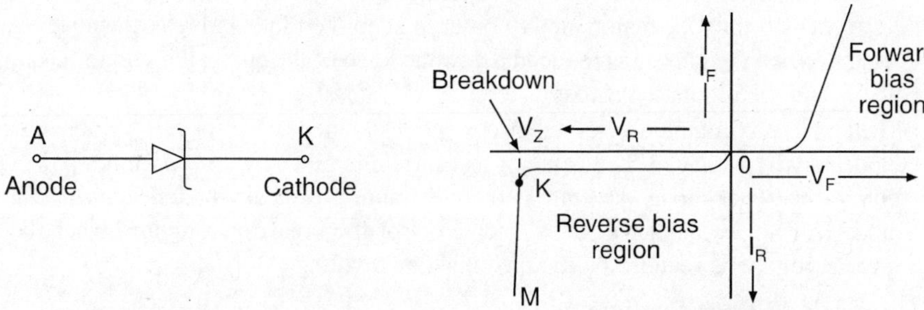

(a) Symbol (b) V.I. characteristic

Fig. 7-1. *Zener diode.*

The zener diode is a silicon PN junction device, which differs from a rectifier diode, in the sense, that it is operated in the reverse break down region. The breakdown voltage of a zener diode is set by carefully controlling the *doping level* during manufacture. As discussed in the last chapter, when a reverse voltage across a diode is increased, a critical voltage called breakdown voltage is reached at which the reverse current increases sharply as shown by the curve KM in Fig. 7-1(b).

The reverse breakdown of a PN junction may occur either due to *avalanche or zener effect.* As discussed in chapter on PN-junctions the avalanche breakdown occurs, when the accelerated free electrons acquire sufficient energy to ionize a lattice atom by bombardment. The additional free electrons, created in this way, are accelerated by the reverse field causing more and more ionization. The multiplication of the number of free carriers causes the reverse current to increase rapidly.

The zener breakdown occurs when the electric field across the junction, produced due to reverse voltage, is sufficiently high. This electric field exerts a force on the electrons in the outer most shell. This force is so high that the electrons are pulled away from their parent nuclei and become free carriers. This ionization, which occurs due to the electrostatic force of attraction, is known as *zener effect.* It causes an increase in the number of free carriers and hence, an increase in the reverse current.

The zener diodes, with breakdown voltages of less than 6 V, operate predominantly in zener breakdown. Those with breakdown, voltages greater than 6 V, operate predominantly in avalanche breakdown. Strictly speaking, the first one should be called a zener diode and the second an avalanche diode. But in actual practice, both the types are called zener diodes.

7-3. Reverse Characteristics of Zener Diode

We have just discussed that a zener diode is operated only in the reverse-bias region. That is why we shall examine its characteristic in this region only. Fig. 7-2 shows the reverse portion of the V-I characteristic of the zener diode. It may be noted from this figure that as the reverse voltage (generally written as V_R) is increased, the reverse current (usually called zener current, I_Z) remains negligibly small up to the 'knee' of the curve (point K in Fig. 7-2). At this point, the effect of breakdown process begins. From the bottom of the knee, the breakdown voltage (also called zener breakdown voltage or simply zener voltage V_Z) remains essentially constant. This ability of a diode is called regulating ability and is an important feature of a zener diode. It maintains, an essentially a constant voltage across its terminals over a specified range of zener current values. Following two points are important from the characteristics of a zener diode:

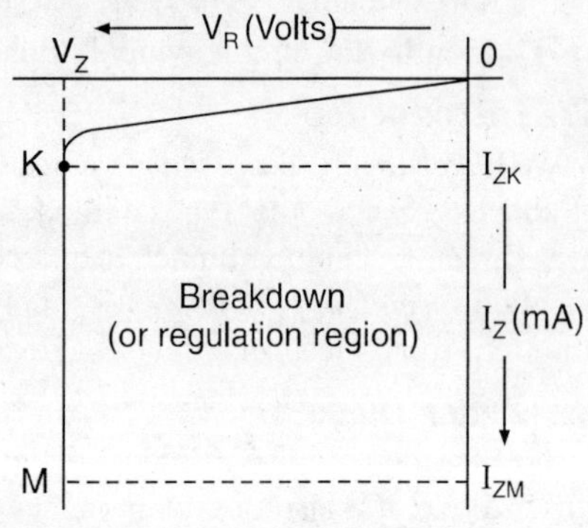

Fig. 7-2. *Reverse characteristic of a zener diode.*

(1) There is a minimum value of zener current called breakover current designated as I_{ZK} or $I_{Z\,(min)}$ which must be maintained in order to keep the diode in breakdown (or regulation) region. When the current is reduced below the knee of the curve, the voltage changes drastically and the regulation is lost.

(2) There is a maximum value of zener current designated as I_{ZM} or $I_{Z\,(max)}$ above which the diode may be damaged. The value of this current is given by the maximum power dissipation of the zener diode. As long as the maximum power dissipation is not exceeded, the diode will not be damaged. It will come out of the breakdown region, when the applied reverse voltage is reduced below the breakdown voltage.

7-4. Zener Diode Specifications

The zener diodes are generally specified in terms of four factors namely zener voltage (V_Z) maximum power dissipation [P_{ZM}, $P_{D\,(max)}$ or $P_{Z(max)}$], breakover current (I_{ZK}) and zener resistance (r_Z). They are available with breakdown voltages ranging from 1.8 V to 2000 V.

The power dissipation of a zener diode is the product of breakdown voltage (V_Z) and reverse current (I_Z). Mathematically the power dissipation,

$$P_Z = V_Z \times I_Z$$

The maximum value of power dissipation, which a zener can dissipate, without failure, is called power rating and is designated by P_{ZM}. The zener diodes are available with power rating from 150 mW to 50 W. Generally, the data sheets of zener diodes include the maximum current, a zener can handle without exceeding its power rating. The value of this maximum current is given by the relation,

$$I_{ZM} = \frac{P_{ZM}}{V_Z}$$

where, P_{ZM} = Power rating of a zener diode, and

V_Z = Breakdown voltage.

Fig 7-3 (a) and (b) shows data sheet of silicon zener diodes. Notice that it shows zener diodes with three different number series, (1) 1N 746 through 1N 759, (2) 1N 957 A through 1N 986 A and (3) 1N 4370 through 4372. All these motorola zener diode have a maximum power rating of 500 mW as seen from the table marked "Maximum Ratings". The devices can handle this amount of power until 50ºC. If the device is used beyond 50ºC temperature, power handling capability reduces by 3.33 mW/ºC. From the table marked "Electrical Characteristics", we can find the nominal zener voltage (V_Z), maximum value of zener current (I_Z), corresponding to any zener diode given in the list. For example, 1N 4370 has V_Z = 2.4 V and I_{ZM} = 190 mA, 1N 4372 has V_Z = 3.0 V and I_{ZM} = 150 mA. Similarly, 1N 746 has V_Z = 3.3 V and I_{ZM} = 135 mA, 1N 759 has V_Z = 12 V and I_{ZM} = 35 mA. The zener diode 1N 957 A has V_Z = 6.8 V and I_{ZM} = 61 mA and 1N 986 A has V_Z = 110 V and I_{ZM} = 4.1 mA. The data sheet can help the circuit designer to select the proper device to handle the required zener voltage (V_Z) and the required zener current.

Usually, the zener diodes with maximum power dissipation below 1W have glass packaging while those with 1W uses metal case and those with power dissipation below 5W uses Sumetic packaging. It may be carefully noted that zener diodes are available with leads as well as without leads.

Example 7-1. *A 1N 754 zener diode has a dc power dissipation rating of 500 mW and a zener voltage rating of 6.8 V. What is the value of I_{ZM} for the device?*

Solution. Given: P_{ZM} = 500 mW = 500 10^{-3} W and V_Z = 6.8 V.

We know that maximum value of zener current,

$$I_{ZM} = \frac{P_{ZM}}{V_Z} = \frac{500 \times 10^{-3}}{6.8} = 73.5 \times 10^{-3} \text{ A} = 73.5 \text{ mA}$$

Note: While designing circuits using zener diodes, we must make sure that the current through the zener does not exceed the I_{ZM} value. For example, in the above situation, we find I_{ZM} = 73.5 mA. If the current through the zener exceeds this value, we shall have to replace the diode.

Example 7-2. *A Motorola's 1N 746 zener diode is being operated at 75°C. What is the maximum power dissipation for the device? Assume that the d.c. power dissipation rating of the device is 500 mW and the derating factor is 3.33 mW/ºC above 50°C.*

Solution. Given: P_{ZM} = 500 mW and derating factor = 3.33 mW.

We know that so long the zener diode is being operated at or below 50°C, its maximum power dissipation is 500 mW. However, if it is to operated at 75°C, its maximum power dissipated will be below 500 mW, because of the excessive heat generated inside.

We know that total derating value

$$= \text{Derating factor} \times (75 - 50)$$

$$= (3.33) \times (75 - 50) = 83.25 \text{ mW}$$

∴ Maximum power dissipation for the device,

$$P_Z = 500 - 83.25 = 416.75 \text{ mW} \quad \textbf{Ans.}$$

MOTOROLA
SEMICONDUCTOR
TECHNICAL DATA

1.5KE.8, A thru 1.5KE250,A
See Page 4- 59

Designers Data

500-MILLIWATT HERMATICALLY SEALED GLASS SILICON ZENER DIODES

- Compete Voltage Range – 2.4 to 110 Volts
- Do-35 Package–Smaller than Conventional D0-7 package
- Double slug Type Construction
- Metallurgically Bonded Construction
- Oxide Passivated Die

Designers' Data for 'Worst Case"Conditions

The Designer's Data sheets permit the design of must circuits entirely from the information presented. Limit curves – representing boundaries on device characteristics – are given to facillitate "worst case" design.

1N1746 thru 1N759
1N957A thru 1N986A
1N4370 thru 1N4372

GLASS ZENER DIODES
500 MILLIWATTS
2.4-110 VOLTS

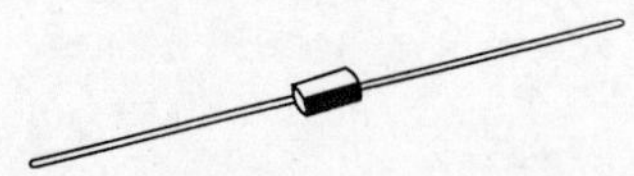

MAXIMUM RATING

Rating	Symbo	Value	Unit
DC Power Dissipation @ $T_L < 50°c$.	P_D		
•JEDEC Registration		400	mW
•Derate above $T_L = 50°c$		3.2	mW/°c
Motorola Device Rating		500	mW
Derate above $T_L = 50°c$		3.33	mW/°c
Operating and Storage Junction Temperature Range	T_J T_{stg}		°c
•JEDEC Registration		-65 to + 175	
Motorola Device Rating		-65 to + 200	

•Indicates JEDEC Registered Date.

MECHANICAL CHARACTERISTICS

MAXIMUM LEAD TEMPERATURE FOR SOLDERING PURPOSES:230 °C, 1/16" from case for 10 seconds
FINISH: All external surface are corrosion resistant with readily solderable leads.
POLARITY: Cathode indicated by colour band. When operated in zener mode, cathode
MOUNTING POSITION: Any will be possitive with respect to anode

STEADY STATE POWER DERATING

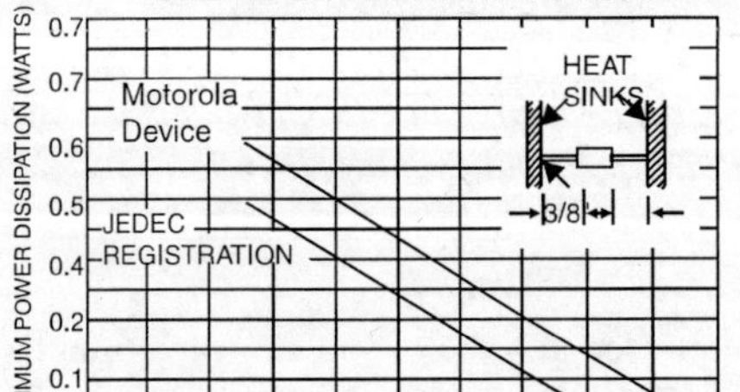

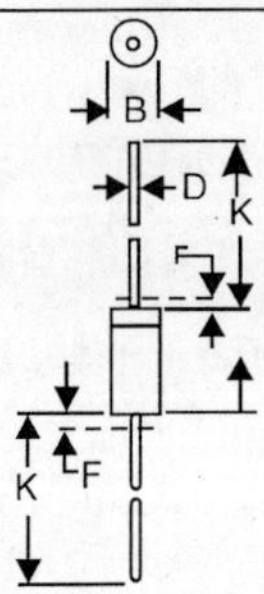

NOTES

1. PACKAGE CONTOUR OPTIONAL WITHIN A AND B. HEAT SLUGS,IF ANY. SHALL BE NOT SUBJECT TO THE MININIUM LIMIT OF B.
2. LEAD DIAMETER NOT CONTROLLED IN ZONE TO ALLOW FOR FLASH,LEAD FINISH BUILDUP AND MINOR IRRUGU-LARITIES OTHER THAN HEAT SLUGS.
3. POLARITY DENDTED BY CATHODE BAND. DIMENSIONNING AND TOLERANCINGER
4. ANSL Y14.5, 1973.L

DIM	MILLIMETERS		INCHES	
A	3.05	5.08	0.102	0.200
D	1.52	2.29	0.060	0.090
B	0.46	0.56	0.018	0.022
F	—		—	0.050
K	25.40	38.10	1.000	1.500

All JEDEC dimensions and notes apply.

CASE 299-02
DO-204AH
GLASS

Fig. 7-3 (*a*).

Electrical Characteristics (T_A = 25°C, V_f = 1.5 V max at 200 mA for all types)

Type Number (Note1)	Nominal zoner voltage V_z @I_{ZT} (Note2) volts	Test Current I_{ZT} mA	Maximum zonner impedence Z_{ZT} @ I_{ZT} (Note3) Ohms	Maximum DC zenner current Z_{ZM} mA		Maximum Reverse Leakage current T_A= 25°c I_R @ V_R 1 V μA	T_A= 250°c I_R @ V_R 1 V μA
1N4370	2.4	20	30	150	190	100	200
1N4371	2.7	20	30	135	165	75	150
1N4372	3.0	20	29	120	150	50	100
1N746	3.3	20	28	110	135	10	30
1N747	3.6	20	24	100	125	10	30
1N748	3.9	20	23	95	115	10	30
1N749	4.3	20	22	85	105	2	30
1N750	4.7	20	19	75	95	2	30
1N751	5.1	20	17	70	85	1	20
1N752	5.6	20	11	65	80	1	20
1N753	6.2	20	7	60	70	0.1	20
1N754	6.8	20 20	5	55	65	0.1	20
1N755	7.5		6	50	60	0.1	20
1N756	8.2	20	8	45	55	0.1	20
1N757	9.1	20	10	40	50	0.1	20
1N758	10	20	17	35	45	0.1	20
1N759	12	20	30	30	35	0.1	20

Type Number (Note1)	Nominal zoner voltage V_z (Note2) volts	Test Current I_{ZT} mA	Maximum zonner impedence (Note3) Z_{ZT} @ I_{ZT} Ohms	Z_{ZT} @ I_{ZT} Ohms	I_{ZK} mA	Maximum DC zonner currenter I_{ZM} Ohms μA		Maximum Reverce current I_R Maximum μA	Test voltage Vdc V_R 5%	10%
1N957A	6.8	18.5	4.5	700	1.0	47	61	150	5.2	4.9
1N958A	7.5	16.5	5.5	700	0.5	42	55	75	5.7	5.4
1N959A	8.2	15	6.5	700	0.5	38	50	50	6.2	5.9
1N960A	9.1	14	7.5	700	0.5	35	45	25	6.9	6.6
1N961A	10	12.5	8.5	700	0.25	32	41	10	7.6	7.2
1N962A	11	11.5	9.5	700	0.25	28	47	5	8.4	8.0
1N963A	12	10.5	11.5	700	0.25	26	34	5	9.1	8.6
1N964A	13	8.5	13	700	0.25	24	32	5	9.9	9.4
1N965A	15	9.5	16	700	0.25	21	27	5	11.4	10.8
1N966A	16	7.8	17	750	0.25	29	37	5	12.2	11.5
1N967A	18	7.0	21	750	0.25	17	23	5	13.7	13.0
1N968A	20	6.2	25	750	0.25	15	20	5	15.2	14.4
1N969A	22	5.6	29	750	0.25	14	18	5	16.7	15.8
1N950A	24	5.2	33	750	0.25	13	17	5	18.2	17.3
1N951A	27	4.6	41	750	0.25	11	15	5	20.8	19.4
1N952A	30	3.2	49	1000	0.25	10	13	5	22.8	21.6
1N953A	33	3.8	58	1000	0.25	9.2	12	5	25.1	23.8
1N954A	36	3.4	70	1000	0.25	8.5	11	5	27.4	25.9
1N955A	39 43	3.2	80	1000	0.25	7.8	10	5	29.7	28.1
1N956A		3.0	93	1500	0.25	7.0	9.6	5	32.7	31.0
1N957A	47	2.7	105	1500	0.25	6.4	8.8	5	35.8	33.8
1N958A	51	2.5	125	1500	0.25	5.9	8.1	5	38.8	36.7
1N959A	56	2.2	150	2000	0.25	5.4	7.4	5	42.6	40.3
1N950A	62	2.0	185	2000	0.25	4.9	6.7	5	47.1	44.6
1N951A	68	1.8	230	2000	0.25	4.5	6.1	5	51.7	49.0
1N952A	75	1.7	270	2000	0.25	1.0	5.5	5	56.0	54.0
1N953A	82	1.5	300	3000	0.25	3.7	5.0	5	62.2	59.2
1N956A	91	1.4	400	3000	0.25	3.3	4.5	5	69.2	65.2
1N954A	100	1.3	500	3000	0.25	3.0	4.5	5	76	72
1N955A	110	1.1	750	4000	0.25	2.7	4.1	5	83.6	79.2

Fig. 7-3 (*b*). *Courtesy Motorola Inc.*

7-5. Zener Diode Equivalent Circuit

Fig. 7-4(*a*) shows an ideal approximation of a zener diode in reverse breakdown. It shows that a zener diode is simply equivalent to a battery having a voltage equal to the zener voltage (V_Z) Fig. 7-4(*b*) shows a practical equivalent circuit of a zener diode. This circuit shows that a zener diode is equivalent to a battery with voltage, (V_Z) in series with a resistance (r_Z) called *zener resistance*. The zener resistance is also known as *dynamic resistance* or *a.c. resistance.*

The presence of resistance (r_Z) indicates that the reserve V-I characteristic of a zener diode is not ideally vertical but it is slightly tilted as shown in Fig. 7-4. The value of zener resistance may be obtained from the V-I characteristic as discussed below:

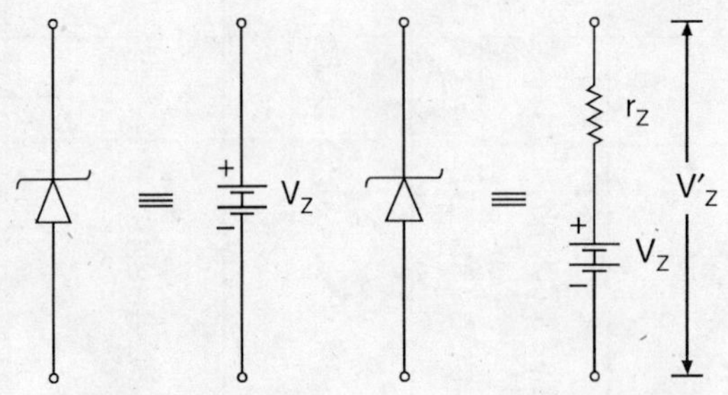

(*a*) Ideal zener equivalent diode circuit.

(*b*) Practical zener diode equivalent circuit.

Fig. 7-4.

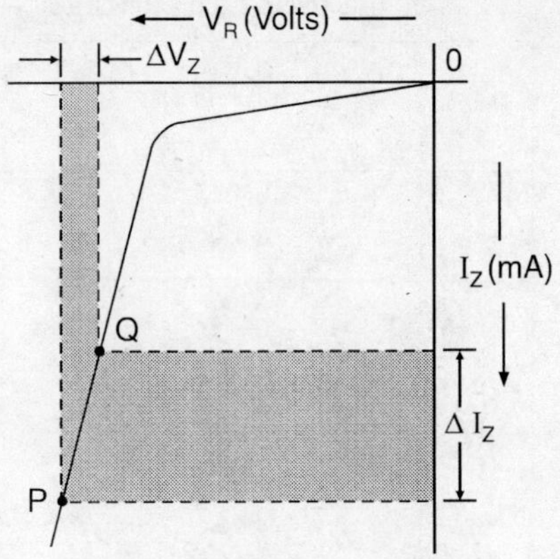

Reverse V-I characteristic curve illustrating zener resistance.

Fig. 7-5.

Consider any two point *P* and *Q* on the reverse V-I curve as shown in Fig. 7-5.

Let ΔV_Z = Change in the values of zener voltage between the points *P* and *Q*, and

ΔI_Z = Change in the values of zener current at the corresponding points.

We know that zener resistance is given by the relation,

$$r_Z = \frac{\Delta V_Z}{\Delta I_Z}$$

For a perfectly vertical breakdown curve, the value of zener resistance is zero. But in actual practice, its value may vary from few ohms to several hundred ohms. The variation in its value depends upon the particular zener voltage and the value of operating current. It may be noted that if the zener current flows through the zener diode in the direction as shown in Fig. 7-6, then voltage drop (V_Z) across the zener resistance has the same polarity as that of a battery. In this case, the voltage across the terminals *A* and *B*,

$$V_Z' = I_Z \cdot r_Z + V_Z$$

If the value of r_Z is very small, then the voltage across the terminals *A* and *B*,

$$V_Z' = V_Z$$

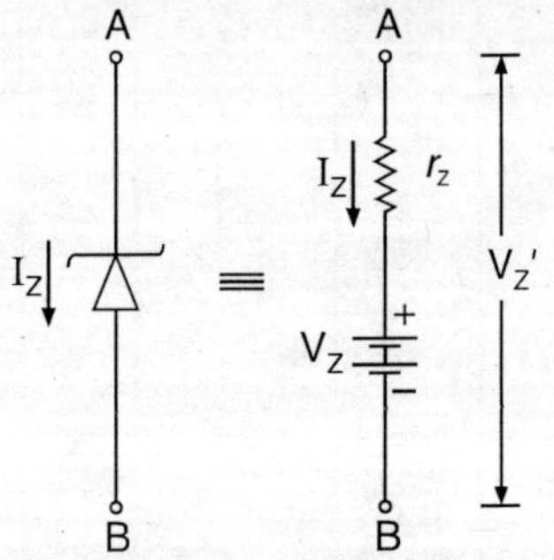

Fig. 7-6. *Terminal voltage of a zener diode.*

Example 7-3. *When the reverse current in a particular zener diode increases from 20 mA to 30 mA, the zener voltage changes from 5.6 V to 5.65 V. What is the resistance of device?*

Solution. Given: $\Delta I_Z = 30 - 20 = 10$ mA = 10×10^{-3}A and $\Delta V_Z = 5.65 - 5.6 = 0.05$ V.

We know that resistance of the zener diode,

$$r_Z = \frac{\Delta V_Z}{\Delta I_Z} = \frac{0.05}{10 \times 10^{-3}} = 5\ \Omega$$ **Ans.**

Example 7-4. *A 4.7 V zener has a resistance of 15 Ω . What is the terminal voltage, when the current is 20 mA?*

Solution. Given $V_Z = 4.7$ volts; $r_Z = 15\ \Omega$ and $I_Z = 20$ mA = 20×10^{-3} A.

We know that the terminal voltages of a zener diode,

$$V_Z' = (V_Z + I_Z \cdot r_Z) = 4.7 + (20 \times 10^{-3}) \times 15 = 5 \text{ V } \textbf{Ans.}$$

7-6. Zener Diode Applications

Though the zener diodes have a number of applications, yet the following are important from the subject points of view :

(*i*) As a voltage regulator.
(*ii*) As a fixed reference voltage in transistor biasing circuits.
(*iii*) As peak clippers or limiters in wave shaping circuits.
(*iv*) For meter protection against damage from accidental applications.

7-7. Tunnel Diode

It has been observed that if the concentration of impurity atoms is greatly increased in a normal PN junction (by 1000 times or more), its characteristics are completely changed. This gives rise to a new type of diode known as *tunnel diode.* It was invented in 1958 by Dr. Leo Esaki. That is why, a tunnel diode is also known as *Esaki diode*.

In a normally doped PN junction, the impurity concentration (*i.e.,* doping level) is of very small value (*i.e.,* about one part in 10^8 atoms). With this amount of doping, the width of depletion layer is of the order of one micron (which is equal to 10^{-6} m). This constitutes a potential barrier at the junction, which controls the flow of charge carriers (*i.e.,* electrons and holes) across the junction. The charge carriers cannot cross-over the potential barrier, unless they acquire sufficient energy to overcome it. However, when impurity concentration is increased (say about one part in 10^3 atoms), the width of depletion layer reduces to about 10 nanometer (where, one nanometer $=10^{-9}$m). Under such conditions, the charge carriers will penetrate through the junction at the speed of light, even though they do not have enough energy to overcome the potential barrier. As a result of this, a large forward current is produced, even if the applied forward voltage is much less than 0.3 V.

The phenomenon of penetrating the charge carriers, directly through the potential barrier, instead of climbing over it, is called *tunneling*. That is why, highly doped PN junction devices are called tunnel diodes. These diodes are usually made of germanium (Ge) or gallium-arsenide (GaAs).

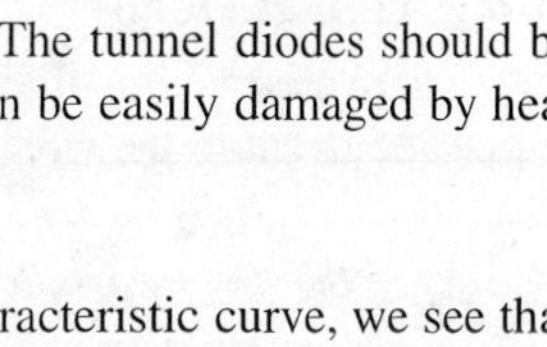

Fig. 7-7. *Symbols for tunnel diodes.*

Fig. 7-7 shows two commonly used symbols for a tunnel diode. The tunnel diodes should be handled with a great care, because they are low power devices and can be easily damaged by heat and static electricity.

7-8. V-I Characteristics of a Tunnel Diode

Fig. 7-8 shows V-I characteristics of a tunnel diode. From the characteristic curve, we see that as the applied forward voltage is increased from zero, the current increases very rapidly, till it reaches its maximum value known as peak current (I_P) as indicated by the point *A*. The corresponding value of the forward voltage is indicated by peak voltage (V_P). The value of this voltage is typically 65 mV for germanium and 160 mV for gallium arsenide tunnel diode.

It will be interesting to know that if the forward voltage is further increased (*i.e.,* beyond V_P), the current decreases, till it reaches its minimum value known as *valley current* (I_V) as indicated by the point *B*. As the voltage is further increased, the current increases in a usual manner as in a normal PN junction diode. It has been observed that the current again reaches its peak value (*i.e.,* I_P) as indicated by the point *C*. The corresponding value of the voltage is indicated by V_F as shown in the figure. For larger values of voltages, the current increases beyond this values.

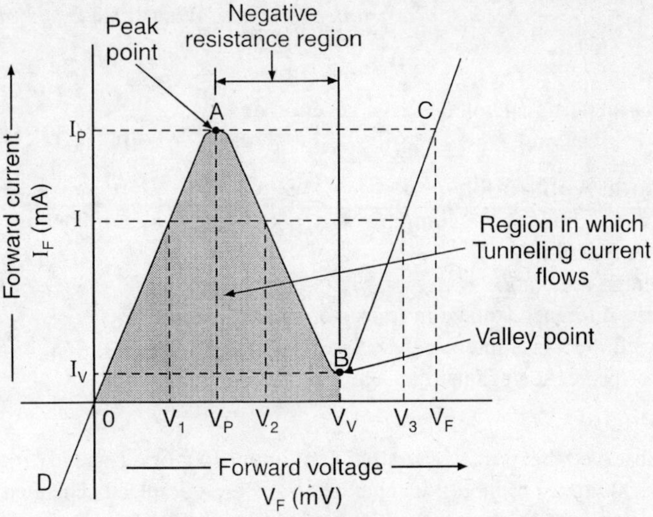

Fig. 7-8. *V-I characteristic of tunnel diode.*

If the tunnel diode is reversed biased (*i.e.,* P side of junction is made negative with respects to its N-side), it acts like an excellent conductor, *i.e.,* the reverse current increases with the increase in reverse voltage. It is indicated by the curve *OD* in the figure. Following are some of the important points of the characteristics of a tunnel diode.

1. Between the peak point *A* and valley point *B*, the current decreases with the increase in voltage. Therefore, the tunnel diode possess a negative resistance in this region as indicated in the figure. This feature makes the tunnel diode useful in high frequency oscillators.
2. For currents, whose values are between I_V and I_P, the curve is triple valued. It means that each current can be obtained at three different applied voltages. It is indicated in the figure by the voltages V_1, V_2 and V_3 for the current *I*. This multivalued feature makes the tunnel diode useful in pulse and digital circuits.
3. The portion *BC* of the characteristic is similar to that of a forward-characteristic of a normal PN junction.
4. The shaded region in the figure indicates the region in which the tunneling current flows through the device.

7-9. Tunnel Diode Parameters

There are two important parameters of a tunnel diode namely negative resistance and the ratio of peak current to the valley current (*i.e.,* I_P/I_V). These parameters are discussed in more detail as below :

(1) ***Negative resistance :*** It is the resistance of a tunnel diode which it offers when operated in a negative resistance region. The negative resistance is given by the relation,

$$\boldsymbol{R_n = -\frac{\Delta V_F}{\Delta I_F}}$$

where, ΔV_F = Change in forward voltage between any two points lying within the negative resistance region of the V-I characteristic, and

ΔI_F = Corresponding change in forward current.

The negative resistance region may also be expressed in terms of slope of the characteristic in the negative resistance region. Its value depends upon the semiconductor material used for manufacturing tunnel diodes and the value may range from 10 Ω to 200 Ω.

(2) ***Ratio of peak current to valley current*** (I_P / I_V) **:** This parameter is important in high speed switching circuits, which are used in computers. The ratio I_P/I_V for germanium type tunnel diode is 6 and for gallium arsenide type is 10. The silicon type tunnel diode has a low I_P/I_V ratio which is about 3. It is because of this fact that silicon type tunnel diodes are not manufactured in actual practice.

7-10. Tunnel Diode Equivalent Circuit

Fig. 7-9 shows a small signal equivalent circuit for a tunnel diode. It consists of a series resistance (R_S), inductance (L_S), the junction capacitance (C) and a negative resistance ($-R_N$). The series resistance is the resistance due to leads, contacts and semiconductor material. Its typical value may range from 1 to 5 ohms. The series inductance is due to lead lengths. Its typical value may range from 0.1 to 4 *n*H.

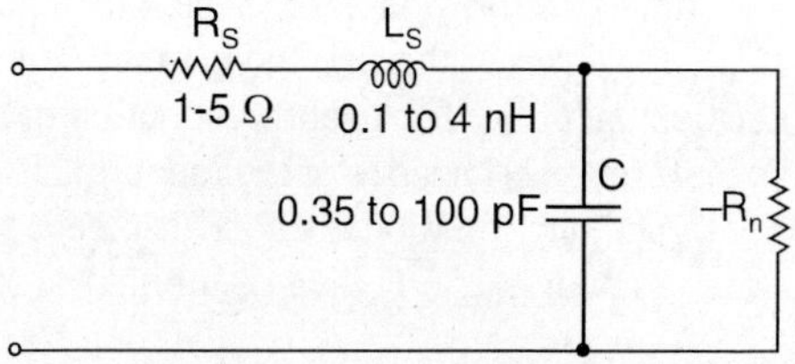

Fig. 7-9. *Equivalent circuit of a tunnel diode.*

The junction capacitance is due to diffusion capacitance and the applied voltage. Its typical value may range from 0.35 to 100 pF as shown in Fig. 7-9.

7-11. Tunnel Diode Applications

The tunnel diodes exhibit a specific characteristic called negative resistance. They have extremely low values of inductance and capacitance. These features make them useful in a number of applications as discussed below:

(*a*) As an ultra high-speed switching device–it is possible due to the tunneling mechanism, which takes place at the speed of light. The switching time is of the order of nano-seconds (10^{-9} sec).

(*b*) As a logic memory storage device–it is possible due to the triple valued of the curve for current between I_P and I_V.

(*c*) As a microwave oscillator at frequencies in the order of 10 GHz–it is possible due to extremely low values of inductance and capacitance of the device.

(*d*) In relaxation oscillator–it is possible due to negative resistance of the device.

7-12. Varactor Diode

A varacter diode is, basically, a reverse biased PN junction, which utilizes the inherent capacitance of the depletion layer. It is also known as varicap, voltcap or tunning diode. It is used as a voltage-variable capacitor.

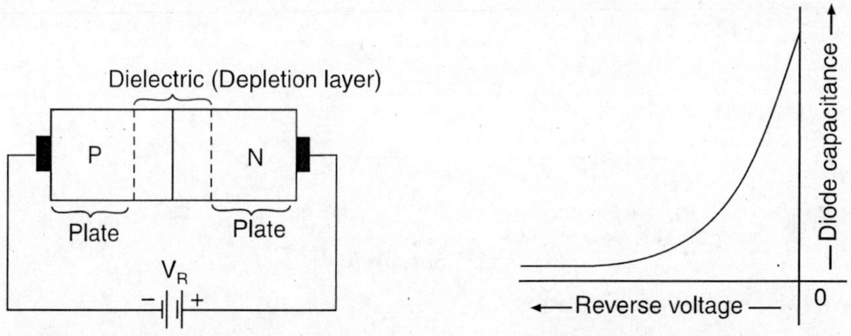

(*a*) Reverse biased PN junction. (*b*) Capacitance versus reverse voltage.

Fig. 7-10.

We have already discussed in chapter on PN junction, that the depletion layer created by the reverse bias acts as a capacitor dielectric, whereas the P- and N-regions act as the capacitor plates as shown in the Fig. 7-10(*a*).

When the reverse bias voltage increases, the depletion layer widens. This increases the dielectric thickness, which in turn, reduces the capacitance. When the reverse bias voltage decreases, the depletion layer narrows down. This decreases the dielectric thickness, which in turn increases the capacitance. Fig. 7-10(*b*) shows the variation of capacitance with the reverse voltage. This indicates that the variation of capacitance is maximum, when the reverse voltage is equal to zero. It reduces in a non-linear manner, as the value of reverse voltage is increased.

In a varactor diode, the capacitance parameters are controlled by the method of doping in the depletion layer, or the size and geometry of the diode construction. For example, Fig. 7-11 (*a*) shows the doping profile for an abrupt junction diode (*i.e.,* the normal PN junction). In this type, the doping is uniform on both sides of the junction. The range of capacitance variation (called tuning range) of an abrupt junction diode is 4 : 1. It means that if its maximum transition capacitance is 100 pF and the minimum is 25 pF, then its tuning range is 4 : 1.

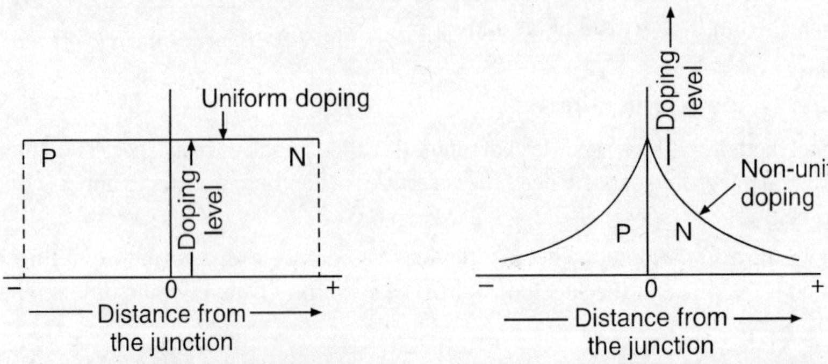

(*a*) Abrupt doping profile. (*b*) Hyper abrupt profile.

Fig. 7-11.

The higher tuning range is obtained for varactors, which have hyper abrupt junction. The doping profile of such a junction is as shown in Fig. 7-11(*b*). In this type, the doping level increases as we approach the junction. The heavy doping at the junction results in a narrower depletion layer and hence, a larger capacitance. Moreover, any change in reverse voltage produces a larger variation in capacitance than that of abrupt junction. A hyper abrupt varactor diode has a tuning range of 10 : 1. This range is enough to tune a broadcast receiver, over its frequency range of medium wave band (*i.e.*, frequencies from 550 kHz to 1650 kHz).

Fig. 7-12(*a*) shows a schematic symbol for a varactor diode and Fig. 7-12(*b*) shows its equivalent circuit. Here, R_S is a reverse series resistance and C_T is a variable capacitance. The value of C_T ranges typically from 2 pF to 100 pF.

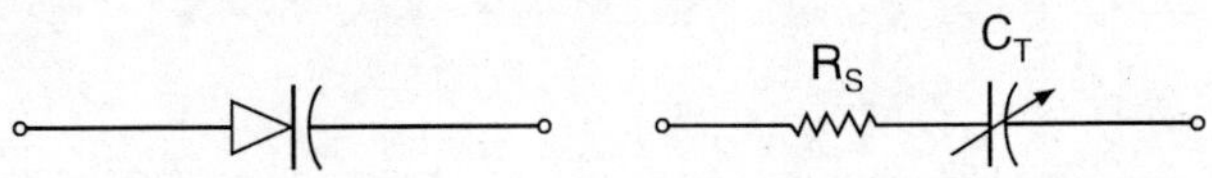

(*a*) Symbol (*b*) Equivalent circuit

Fig. 7-12. *Varactor diode.*

7-13. Varactor Diode Specifications

The specification sheet of any varactor diode will provide us with the information we need for any circuit analysis. Fig. 7-13 shows the specification (or data) sheet and operating curves for the

Motorola MV 209 series varactor diodes. We will use this specification sheet for our discussion on the commonly used varactor diode parameter and characteristics.

SILICON EPICAP DIODE

... designed for gernal frequency control and tuning applications; providing solid-state reliability in replacement of mechinical tuning methods.

- High Q with Guranteed minimum Values at VHF Frequencies
- Controlled and Uniform Tuning Ratio
- Available in Surface Mount Package

MMBV109L
MV209

CASE 182-02 STYLE 1
(TO-226 AC)

2 Cathode — 1 Anode

CASE 318-03 STYLE 8
SOT-23(TO236 AB)

2 Cathode — 1 Anode

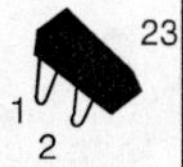

26-32 pF
VOLTAGE VARIABLE
CAPACITANCE DIODES

MAXIMUM RATINGS

Rating	Symbol	Value MV209	Value MMBV209,L	Unit
Reverse Voltage	V_R	30		Volts
Forward Current	I_F	200		mA
Forward Power Dissipation @ $T_A = 25°c$ Derate above 25°	P_D	200 2.8	200 2.0	mW mW/°c
Junction Temperature		+125		°c
Storage Temperature Range	T_{stg}	– 55 to + 150		°c

DEVICE MARKING

MMBV109 = 4A

ELECTRICAL CHARACTERISTICS (T_A = 25°Unless Otherwise noted.)

Characteristic	Symbol	Min	Typ	Max	Unit
Reverse Breakdown Voltage (I_R = 10 μAdc)	$V_{(BR)R}$	30	—	—	Vdc
Reverse Voltage Leakage Current (V_R = 25 Vdc)	I_R	—	—	o.1	μAdc
Diode Capacitance Temperature Coefficient	TC_C	—	300	—	ppm/°C

Device	C_t. Diode. Capacitance V_R = 3.0 Vdc, f = 1.0 MHz pF			Q. Figure of Merit V_R = 3.0 Vdc t = 50 MHz (Note 1)	C_R, Capacitance Ratio $C_3/C25$ f = 1.0 MHz (Note 2)	
	Min	Nom	Max	Min	Min	Max
MMBV109L, MV209	26	29	32	200	5.0	6.5

FIGURE 1 – DIODE CAPACITANCE

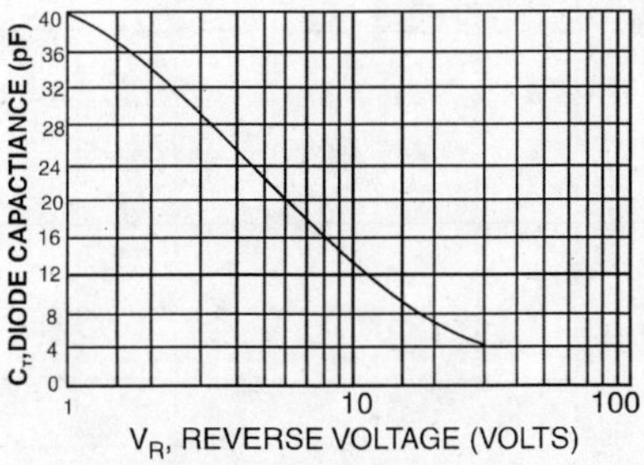

Fig. 7-13. *Varactor diode specification (or data) sheet.*

(Courtery of Motorola Inc)

The maximum ratings for the varactor diode are the same as those that are used for the PN junction diode and the zener diode. The same holds true for the reverse breakdown voltage and reverse leakage current ratings that appear in the "Electrical characteristics" section of the specification sheet. We have already discussed these ratings in the last chapter. However, some of the additional ratings are given below:

FIGURE 2 – FIGURE OF MERIT

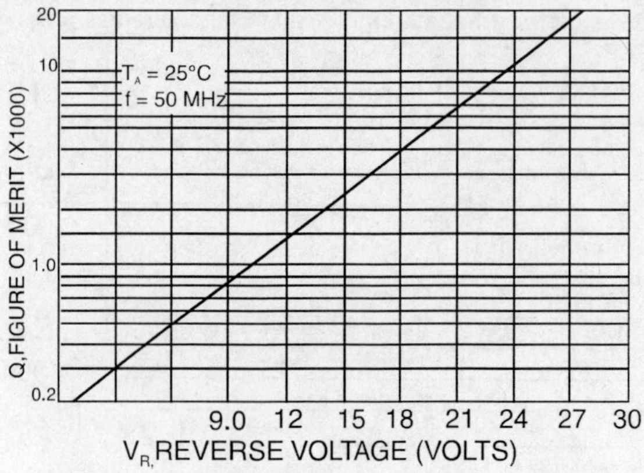

FIGURE 3 – LEAKAGE CURRENT

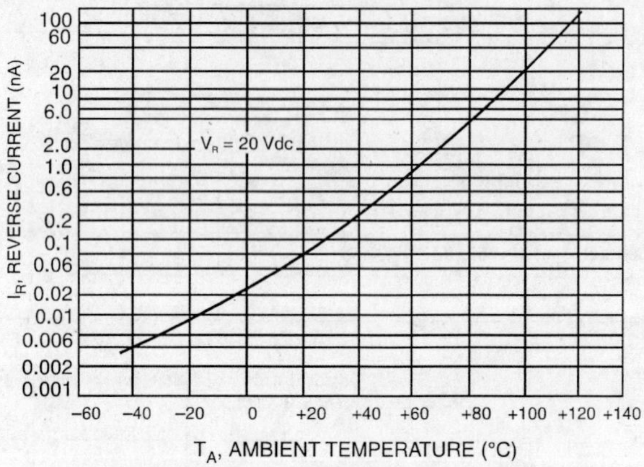

FIGURE 4 – DIODE CAPACITANCE

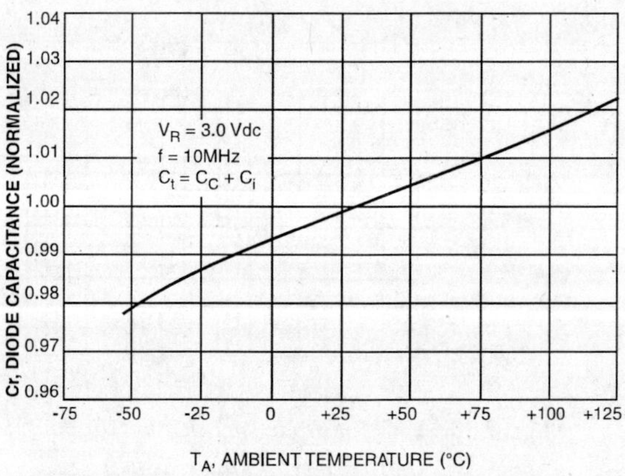

NOTES ON TESTING AND SPECIFICATIONS

1. Q is calculated by taking the G and C readings of an admittance bridge, such as Boonton Electronic Model 33AS8, at the specified frequency and Substituting in the following equation :

$$Q = \frac{2\pi fC}{G}$$

2. C_R is the ratio of C_t measured at 3.0 Vdc divided by C_t measured at 25 Vdc.

Fig. 7-13. *Vector diode specification (or data) sheet continued. (Courtesy of Motorola Inc)*

(1) ***The diode capacitance temperature coefficient (TC_C) rating :*** This rating of a varactor diode gives the charge in the component's capacitance for each 1°C rise in temperature above 25°C. The TC_C rating for the MV 209 series varactor diodes is 300 ppm/°C. This means that varactor diode capacitance will increase by 300 parts per million for each 1°C rise in temperature above 25°C. For example, the normal capacitance rating of the MV 209 is typically 29 pF. Therefore, increase in capacitance for per 1°C rise in temperature would be,

$$\Delta C = (0.0003) \; 29 \text{ pF} = 0.0087 \text{ pF}$$

Thus, the MV 209 varactor diode would have to experience a temperature increase of nearly 115°C in order to experience an increase in capacitance of 1 pF. This leads us to conclude that the capacitance of a varacter diode is relatively independent of operating temperature.

(2) ***Diode capacitance rating (C_t) :*** This rating for MV 209 series varactor diode is 26 to 32 pF when V_R is 3 V d.c. We know that the nominal value (29 pF) is the rated value of the component. In other words, MV 209 would be rated as 29 pF even though the actual value could fall anywhere within the specified range of V_R = 3 V dc.

(3) ***Capacitance ratio (C_R) :*** This rating gives us that variation of the junction capacitance over the given range of voltages. For the MV 209 series, the C_R rating ranges from 5.0 to 6.5 for V_R = 3 to 25 V dc.

(4) ***Figure of merit (Q) :*** It is the quality factor of a junction capacitance. We know that 'Q' of a capacitor is the ratio of energy stored in the capacitor to the energy lost through leakage current. In other words, it is the ratio of power returned to the circuit by the capacitor to the power dissipated by the capacitor. For the MV 209, Q rating is 200 (minimum).

7-14. Varactor Diode Applications

A major application of varactor diode is in *tuning circuits. It is used in electronic tuners in radio, television and other commercial receivers. When a varactor is used in a tuning circuit, it allows a resonant frequency to be adjusted by a variable voltage level as shown in Fig. 7-14. In this circuit, two varactor diodes D_1 and D_2 provide the total variable capacitances in a parallel resonant circuit and V_C is a variable d.c. voltage, which controls the reverse bias and the capacitance of the diodes. The resonant frequency of the tank circuit is given by the relation,

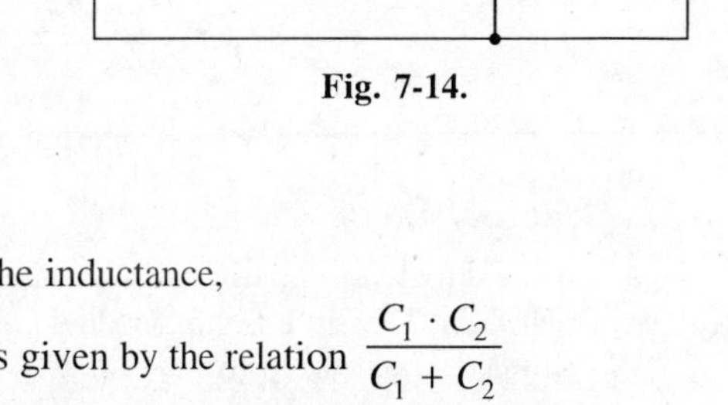

Fig. 7-14.

$$f_0 = \frac{1}{2\pi\sqrt{LC}}$$

where, L = The value of the inductance,

C = The value of total capacitance, which is given by the relation $\dfrac{C_1 \cdot C_2}{C_1 + C_2}$

where, C_1 and C_2 are the maximum and minimum values of the capacitances of diode.

The varactor diode is also used in automatic frequency control device, adjustable band-pass filter and parametric amplifier.

* It is a circuit in which the resonant frequency is altered to accept the desired channel frequency.

Example 7-5. *The capacitance of certain varactor diode can be varied from 5 pF to 50 pF. The diode is used in a tuned circuit of a radio receiver as shown in Fig. 7-15.*

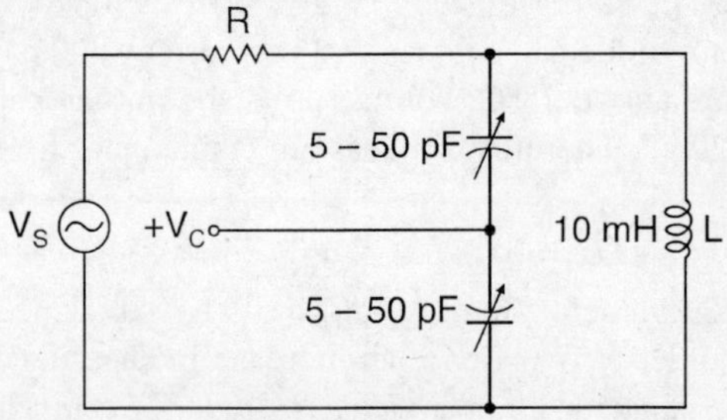

Fig. 7-15.

Determine the tuning range for circuit, if the inductance is 10 mH.

Solution. Given: C_1 (*i.e.*, C_{min}) = 5 pF; C_2 (*i.e.*, C_{max}) = 50 pF and L = 10 mH = 10×10^{-3} H.

Maximum resonant frequency

We know that the maximum resonant frequency is obtained, when the total capacitance of the varactor diode is minimum and its value,

$$C_{T\ (max)} = \frac{C_1 \cdot C_2}{C_1 + C_2} = \frac{5 \times 5}{5 + 5} = 2.5 \text{ pF} = 2.5 \times 10^{-12} \text{ F}$$

∴ Maximum resonant frequency,

$$f_{o\ (max)} = \frac{1}{2\pi\sqrt{L.C_{T(\min)}}} = \frac{1}{2\pi\sqrt{(10 \times 10^{-3}) \times (2.5 \times 10^{-12})}} \text{Hz}$$

$$= 1 \times 10^6 \text{ H}_Z = 1 \text{ MHz } \textbf{Ans.}$$

Minimum resonant frequency

We also know that the minimum resonant frequency is obtained, when the capacitance of the varactor diode is maximum and its value

$$C_{T\ (max)} = \frac{C_1 \cdot C_2}{C_1 + C_2} = \frac{50 \times 50}{50 + 50} = 25 \text{ pF} = 25 \times 10^{-12} \text{ F}$$

∴ Minimum resonant frequency,

$$f_{o\ (min)} = \frac{1}{2\ \pi\sqrt{L \cdot C_{T\ (\max)}}} = \frac{1}{2\ \pi\sqrt{(10 \times 10^{-3}) \times (2.5 \times 10^{-12})}} \text{ Hz}$$

$$= 318 \times 10^3 \text{ Hz} = 318 \text{ kHz } \textbf{Ans.}$$

7-15. Schottky Doide

It is formed by joining a doped semiconductor region (usually N-type) with a metal such as gold, silver or platinum. Thus in a Schottky diode, there is a metal-to-semiconductor junction rather than a simple PN junction as shown in Fig. 7-16(*a*). A Schottky diode is also called a hot carrier diode or a Schottky barrier diode. Fig. 7-16(*b*) shows the schematic symbol of a Schottky diode. It will be interesting to know that in the symbol, that the line at the end of a arrow looks like the letter *S*.

Fig. 7-17 shows the characteristic curve of a Schottky diode. As shown in the figure, the Schottky diode has been lower forward voltage drop (V_F) and reverse breakdown voltage (V_{BR}) than those of PN junction diode. Typically, the Schottky diode will have a V_F of approximately 0.3 V and a V_{BR} of about –50 V. These are much lower than the typical PN junction ratings of V_F = 0.7 V and V_{BR} = –150V.

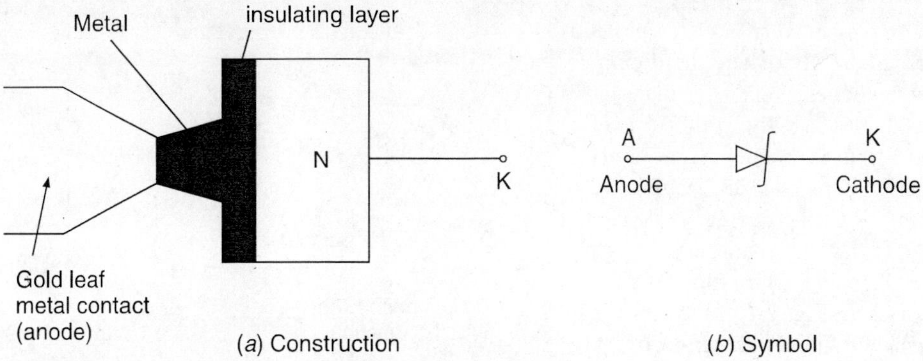

Fig. 7-16. *Schottky diode.*

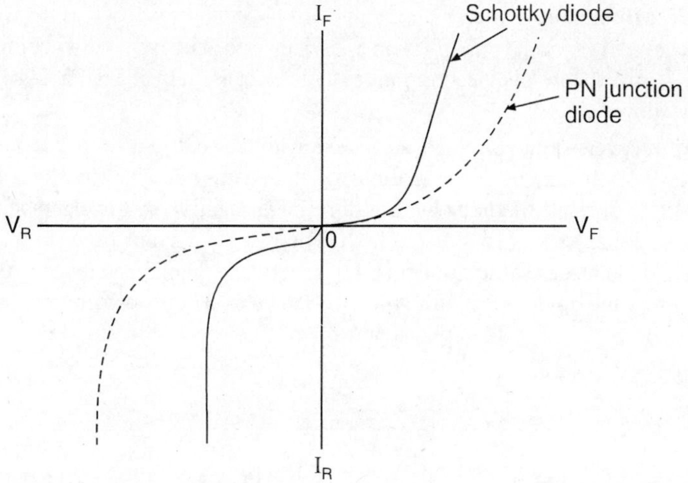

Fig. 7-17. *Schottky diode.*

The Schottky diode has a very little junction capacitance. Because of this, it can be operated at much higher frequencies (of 20 GHz or more) than the typical PN junction diode. The reduced junction capacitance also results in a much *faster switching time*. Because of this reason, Schottky devices are being used more and more in digital switching applications.

7-16. Schottky Diode Applications

We know that Schottky diodes have two important features namely it operates only with majority carriers, and there is no charge storage when the device is switched from its 'ON' to 'OFF' state. These features make it useful for a variety of applications as discussed below:

- To rectify very high frequency (above 300 MHz) signals.
- As a switching device in digital computers. The speed of computers depends on how fast their diodes and transistors can turn 'ON' and 'OFF'.
- In clipping and clamping circuits.
- In low power Schottky transistor-transistor logic (Schottky TTL) circuits.
- In mixing and detecting circuits used in communication systems.
- In low voltage power supply circuits. The Schottky diode has a threshold voltage of 0.25V only. It is an advantage in low voltage rectifiers because we have to subtract only 0.25V (instead of 0.7 V) for each diode from the low input voltage to get an output.

7-17. Step-Recovery Diode

A step-recovery diode employs a graded doping, where the doping level of the semiconductor material is reduced as the junction is approached. Such a doping profile is as shown in Fig. 7-18(*a*). This profile indicates that the doping level (*i.e.,* the density) of charge carriers decreases as the junction is approached. Fig. 7-18(*b*) shows the schematic symbol of a step-recovery diode.

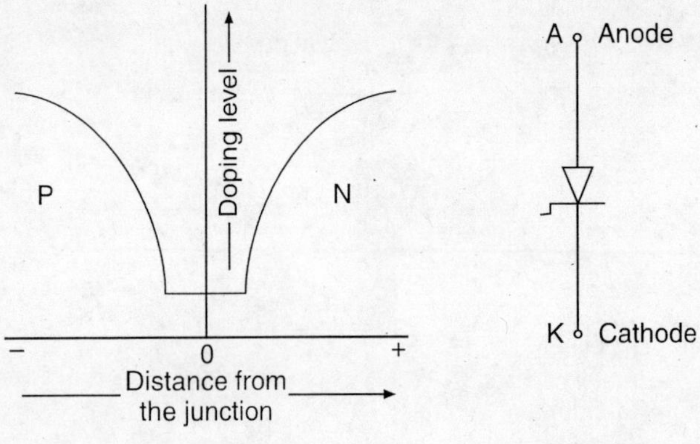

(*a*) Doping profile. (*b*) Symbol.

Fig. 7-18. *Step-recovery diode*

The unusual doping of the step-recovery diode affects the time required for the device to switch from off to on and vice versa. The typical switching time for a step-recovery diode is in the low-picosecond range. This makes them ideal for ultra-high frequency switching applications.

When a step-recovery diode is connected to a sinusiodal a.c. input source, it produces an output as shown in Fig. 7-19. It may be noted that during the positive half cycle of the input signal, the diode conducts like a normal silicon diode. But during the negative half cycle, reverse current exists for a short time and then suddenly drops to zero. The reverse current flows because of the stored charge in a diode, which takes some finite time for its removal. Since, the reverse recovery period is very sharp, therefore the diode is called *step-recovery diode.* The step-recovery diode may also be thought of as if the diode is suddenly snapped open. Therefore, the step-recovery diode is also known as a *snap-diode.*

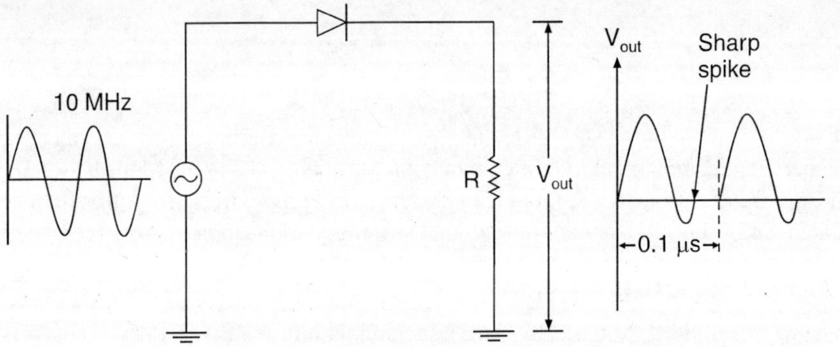

Fig. 7-19. *Output showing a sharp spike.*

A fourier analysis of such a transition (*i.e.,* sharp reverse recovery) shows that the output consists of a number of frequency components called *harmonics*. These harmonics have frequencies of f, $2f$, $3f$, ... nf. The frequency f (10 MHz in the present case) is called fundamental frequency whereas $2f$, $3f$, ..., nf are called second, third ..., harmonics. The output waveform can be filtered to produce a sine wave of higher frequency. Because of this fact, the step-recovery diodes are useful in frequency multipliers (*i.e.,* circuits whose output frequency is a multiple of the input frequency).

7-18. Harmonic Generator (or Frequency Multiplier)

In the last article, we have discussed the step-recovery diode. As a matter of fact, the step-recovery diode has a specific feature that when it is switched from ON to OFF position, it gives a spike at its output. This feature makes it useful in harmonic generator or frequency multiplier circuit. The output frequency of such a circuit is a multiple of the input frequency signal.

Fig. 7-20 shows a harmonic generator circuit. When a sinewave input signal of a known high-frequency (say 10 MHz) is applied at the input, the step recovery diode generates spikes as indicated in the figure. A fourier analysis of such spikes shows that the output consists of a number of frequency components called harmonics. These harmonics have frequencies of f, $2f$, $3f$ ..... These spikes act as an input to a tank-circuit (or a resonant-circuit). The resonant circuit generates a desired harmonic multiple of the input frequency signal. The output of the resonant circuit may be filtered to get a pure multiple of the input frequency. The generation of 3rd hormonic (*i.e.,* 30 MHz) is indicated in the figure.

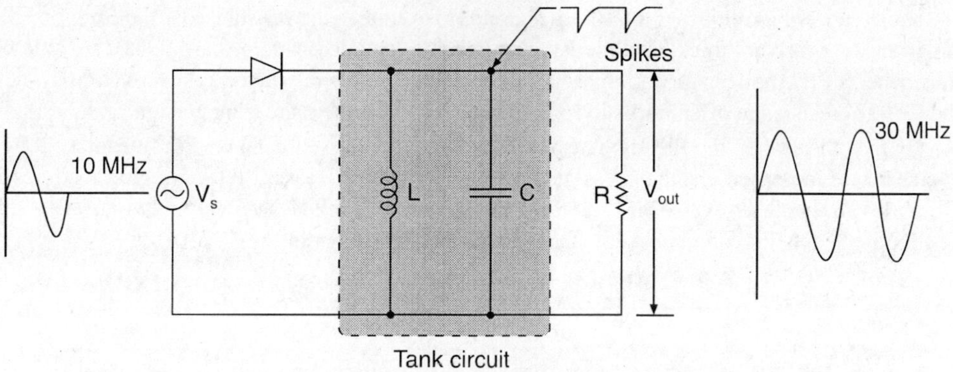

Fig. 7-20. *Harmonic generator.*

Example 7-6. *A step recovery diode is used in a frequency multiplier. The period of the rectified output signal is 0.04 μ sec. What is the frequency of 5th harmonic?*

Solution. Given: T = 0.04 μ sec = 0.04×10^{-6} sec.

We know that the frequency of the output signal,

$$f = \frac{1}{T} = \frac{1}{0.04 \times 10^{-6}} = 25 \times 10^6 \text{ Hz} = 25 \text{ MHz}$$

∴ Frequency of the 5th harmonic

$$= 5f = 5 \times 25 = 125 \text{ MHz} \quad \textbf{Ans.}$$

7-19. PIN Diode

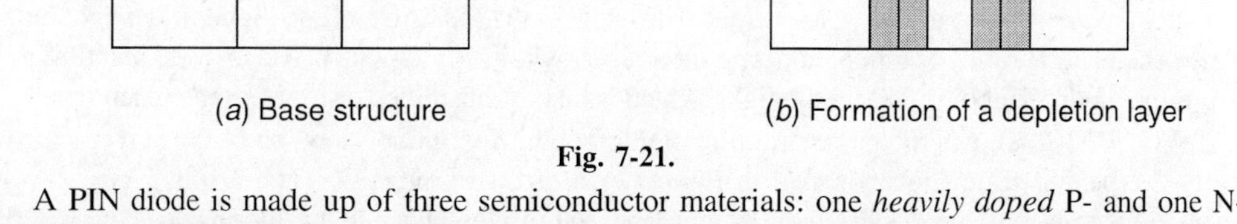

Fig. 7-21.

A PIN diode is made up of three semiconductor materials: one *heavily doped* P- and one N-type material separated by an *intrinsic* (*i.e.,* undoped) semiconductor (I) as shown in Fig. 7-21(*a*). The intrinsic region offers high resistance to the current flowing through it. The PIN diode has the following two advantages over the normal PN junction diode :

(*i*) The capacitance between the P and N region *decreases* because of the increased separation between the P and N regions. This advantage allows the PIN diode to have fast response time. Hence, these diodes are useful at very-high-frequencies (*i.e.,* above 300 MHz).

(*ii*) There is a greater electron-hole pair generation because of the increased electric field between the P and N regions. This advantage allows the PIN diode to process even weak signals.

We know that when a PIN diode is unbiased (*i.e.,* no voltage is applied across the diode), there is a diffusion of electron and holes across the junction due to the different concentration of atoms in the P, I and N regions. The diffusion of electrons and holes produce a depletion layer across the PI and IN junctions as shown in Fig. 7-21(*b*). The depletion layers penetrate to a little distance in the P-type and N-type semiconductor regions but to a larger distance in the I-region. Under such condition, the device has a high value of resistance.

When the PIN diode is forward biased, the width of depletion layers decreases. As a result of this, more carriers are injected into the I-region. This reduces the resistance of the I-region. If the depletion layers are not thick, then the I-region becomes flooded with the carriers at a suitable bias. Thus, when a PIN diode is forward biased, it acts like a variable resistance as shown in Fig. 7-22(*a*). The forward resistance of an intrinsic region decreases with the increasing current.

On the other, when the PIN diode is reverse biased, the depletion layers become thicker, As the reverse bias is increased, the thickness of the depletion layer increases till the I-region becomes free of mobile carriers. The reverse bias, at which this happens, is called *swept out voltage*. At this stage, the PIN diode acts like an almost constant capacitance as shown in Fig. 7-22(*b*).

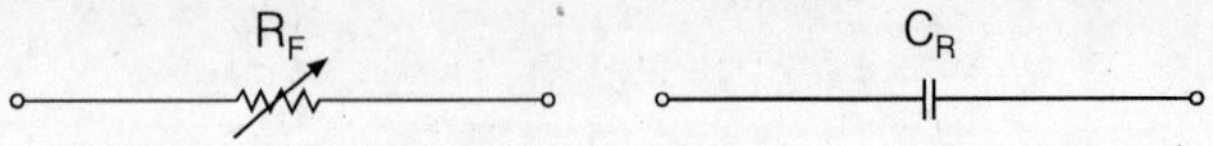

(*a*) Forward biased. (*b*) Reverse biased.

Fig. 7-22. *Equivalent circuit of a PIN diode.*

7-20. PIN Diode Applications

We know that when a PIN diode is forward biased, it acts like a variable resistance which decreases with the increasing current. And when it is reverse biased, the PIN diode acts like an almost constant capacitance. These features make the PIN diode useful for a variety of applications as discussed below :

- As a d.c. controlled microwave switch operated by rapid changes in bias or as a modulating device. This application takes advantage of the variable forward-resistance characteristic. Modulators are used to combine two signals of different frequencies into a single signal.
- In attenuater applications because its resistance can be controlled by the current.

7-21. Light Emitting Diode (LED)

A PN junction diode, which emits light when forward biased, is known as a *light emitting diode* (abbreviated as LED). The emitted light may be visible or invisible. The amount of light output is directly proportional to the forward current. Thus, higher the forward current, higher is the light output. The schematic symbol of a light emitting diode is shown in Fig. 7-23(*a*). The arrows, pointing away from the diode symbol represent the light, which is being transmitted away from the junction.

Fig. 7-23(*b*) shows the basic structure of a light emitting diode. Here, an N-type layer is grown on a P-type substrate (not indicated in figure) by a diffusion process. Then a thin P-type layer is grown on the N-type layer. The metal connections to both the layers make anode and cathode terminals as indicated. The light energy is released at the junction, when the recombination of electrons with holes take place. After passing through the P-region, the light is emitted through the window provided at the top of the surface.

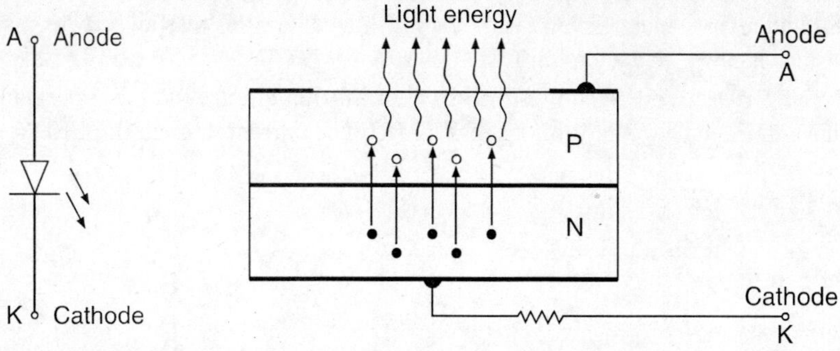

(*a*) Symbol (*b*) Basic structure

Fig. 7-23. *Light emitting diode.*

It will be interesting to know that when the LED is forward biased, the electrons and holes move towards the junction and the recombination takes place. After recombination, the electrons, lying in the conduction bands of N-region, fall into the holes lying in the valence band of a P-region. The difference of energy between the conduction band and valence band is radiated in the form of light energy. In ordinary diodes, this energy is radiated in the form of heat.

The semiconducting materials used for manufacturing light emitting diodes are gallium arsenide, gallium arsenide phosphide. The silicon and germanium is not used for manufacturing light emitting diodes because these are heat producing materials. Moreover, these materials are very poor in emitting light radiations.

The LED's radiate light in different colours such as red, green, yellow, blue, orange *etc*. Some of the LED's emit infrared (*i.e.,* invisible) light also. The colour, of the emitted light, depends upon the type of the semiconductor used. Thus gallium arsenide emits infrared radiations, gallium arsenide phosphide produces either red or yellow light, gallium phosphide emits red or green light and gallium nitrite produces blue light.

Since, LED's have clear (or semiclear) cases, there is normally no label on the cased to identify the leads. The two leads of a LED are identified using one of the several schemes as discussed below.

(*i*) The leads may have different lengths as shown in Fig. 7-24(*a*), when the schemes is used, the longer of the two leads is usually the anode.

(*ii*) One of the leads may be flattened as shown in Fig. 7-24(*b*). The flattened lead is usually the anode

(*iii*) One side of the case is flattened as shown in Fig. 7-24(*c*). The lead closest to the flattened side is usually the anode.

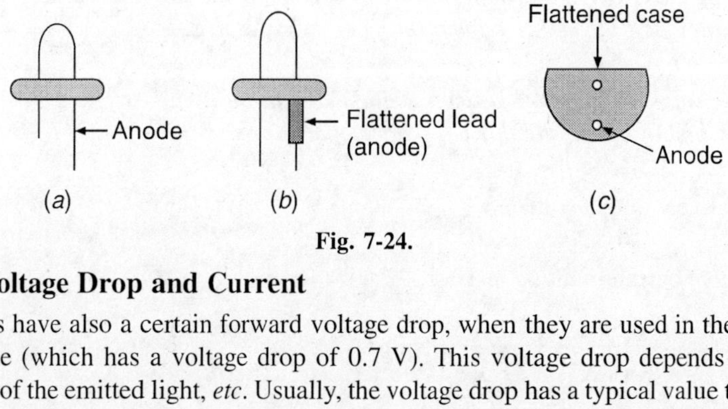

(*a*) (*b*) (*c*)

Fig. 7-24.

7-22. LED Voltage Drop and Current

The LED's have also a certain forward voltage drop, when they are used in the circuit like an ordinarily diode (which has a voltage drop of 0.7 V). This voltage drop depends upon the LED current, colour of the emitted light, *etc*. Usually, the voltage drop has a typical value that varies from

1.5 to 2.5 V for currents between 10 and 50 mA. It will be interesting to know that LED's are not able to withstand reverse bias of even every small voltages. For this reason, it is necessary to assure that reverse bias is never applied to an LED.

Consider a LED circuit as shown in Fig. 7-25. In this circuit, the LED is connected to the supply voltage through a current limiting resistor (R). The current through the LED is given by the relation.

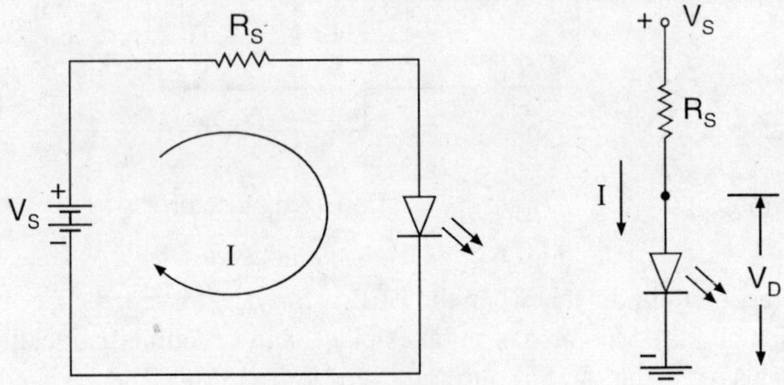

Fig. 7-25. *LED Circuit.*

$$I = \frac{V_S - V_D}{R_S}$$

where, V_S = Supply voltage,

V_D = Voltage drop across LED, and

R_S = Current limiting resistor.

The value of current limiting resistor is given by the relation,

$$R_S = \frac{V_S - V_D}{L_{F(\max)}}$$

where, V_S = Peak output voltage

V_D = LED voltage drop, and

$I_{F(\max)}$ = Maximum value of LED current drop.

Since, the voltage drop, across the diode, depends upon the value of current through it, therefore it controls the amount of light output (or brightness) of the LED. In order to avoid fluctuations in the brightness, the current through the LED must be kept constant. In actual practice, it is done by selecting a large supply voltage and a large value of current limiting resistor.

Example 7-7. *The driving circuit shown in Fig. 7-26 has a peak output voltage of 8 V. The LED has a rating of V_F = 1.8 to 2.0 V and $I_{F\,(max)}$ = 16 mA.*

What is the smallest standard resistor value that can be used as R_S. (Note. A listing of the standard resistor values appears in Appendix)

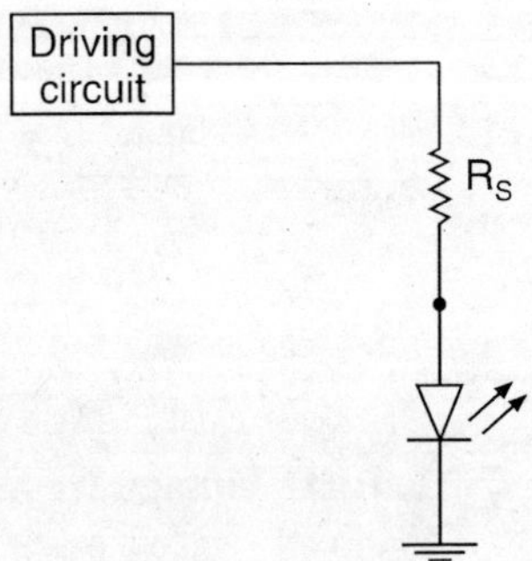

Fig. 7-26.

Solution. Given: output voltage $V_{out\,(pk)} = V_S = 8$ V; $V_F = V_D$ 1.8 to 2.0 V and $I_{F\,(max)} = 16$ mA $= 16 \times 10^{-3}$ A.

We know that current limiting resistor,

$$R_S = \frac{V_S - V_D}{I_{F(\max)}} = \frac{8 - 1.8}{16 \times 10^{-3}} = 387.5\ \Omega$$

and if V_D or $V_F = 2.0$ V then $R_S = 375\ \Omega$. In either case, the smallest standard-value resistor that has a value greater than 387.5 Ω or 375 Ω is the 390 Ω resistor. This is the component which we have to use in the circuit of Fig. 7-26.

Example 7-8. *Fig. 7-8 shows an LED circuit. The LED has a minimum drop of 1.5 V and a maximum drop of 2.3 V. If the supply voltage is 10 V and $R_l = 470\ \Omega$. What are the minimum and maximum values of LED current?*

Solution. Given : $V_{D\ (min)} = 1.5$ volts; $V_{D\ (max)} = 2.3$ volts; $V_S = 10$ volts and $R_1 = 470\ \Omega$.

We know that the maximum value of current,

$$I_{max} = \frac{V_S - V_{D\ (min)}}{R_1} = \frac{10 - 1.5}{470}\ \text{A}$$

$$= 18.1 \times 10^{-3}\ \text{A} = 18.1\ \text{mA}\ \textbf{Ans.}$$

and minimum value of current,

$$I_{min} = \frac{V_S - V_{D\ (max)}}{R_1} = \frac{10 - 2.3}{470}\ \text{A}$$

$$= 16.4 \times 10^{-3}\ \text{A} = 16.4\ \text{mA}\ \textbf{Ans.}$$

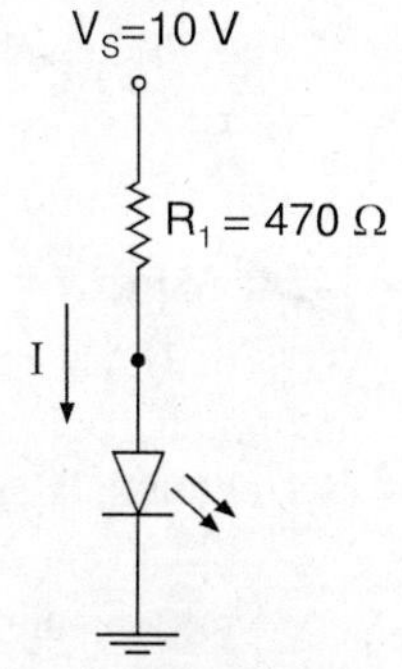

Fig. 7-27.

Example 7-9. *A light emitting diode with minimum and maximum voltage drops of 1.8 V and 3 V respectively is connected alternately with 24 volt supply (with 820 resistor), and 5 volt supply (with 120 resistor). Indicate which supply voltage will keep the brightness of the diode as constant.*

Solution. Given: $V_{D\ (min)} = 1.1$ volts and $V_{D\ (max)} = 3$ volts.

We know that minimum and maximum values of the current through the LED with 24 volt supply (V_S) and 820 Ω resistor (R_S).

$$I_{min} = \frac{V_S - V_{D\ (max)}}{R_1} = \frac{5 - 3}{120} = 16.7 \times 10^{-3}\ \text{A} = 16.7\ \text{mA} \quad ...(i)$$

and
$$I_{max} = V_S - \frac{V_{D\ (min)}}{R_S} = \frac{24 - 1.8}{820} = 27.1 \times 10^{-3}\ \text{A} = 27.1\ \text{mA} \quad ...(ii)$$

We also know that minimum and maximum values of the current through the LED with 5 volt supply (V_S) and 120 Ω resistor (R_S).

$$I_{min} = \frac{V_S - V_{D\ (max)}}{R_S} = \frac{5 - 3}{120} = 16.7 \times 10^{-3}\ \text{A} = 16.7\ \text{mA} \quad ...(iii)$$

and
$$I_{max} = \frac{V_S - V_{D\ (min)}}{R_S} = \frac{5 - 1.8}{120} = 26.7 \times 10^{-3}\ \text{A} = 26.7\ \text{mA} \quad ...(iv)$$

It may be noted that the variation in current for the first case (27.1 – 25.6 = 1.5 mA) is small whereas in the second case it is (26.7 – 16.7 = 10.0 mA) large. Therefore, the brightness in the first case will remain constant whereas in the second case it will be changing. Hence, in order to get an approximately constant brightness, we use as large a supply voltage as possible.

7-23. LED Applications

The LED's operate at low voltages, *i.e.*, from 1.5 V to 2.5 V. They have a long life of about 10000 hours and can be switched 'ON' and 'OFF' at a very fast speed (≈ 1 n sec). These features make LED's very important electronic device. Following are the important applications of the LED's:

(*i*) In 7-segment, 16-segment and dot matrix displays. Such displays are used to indicate alphanumeric characters and symbols in various systems such as digital clocks, microwave-ovens, stereo tuners, calculators, electronic d.c. power supplies, *etc.*

(*ii*) For indicating power ON/OFF conditions, power level indicators in stereo amplifiers.

(*iii*) In optical switching applications.

(*iv*) For solid state video displays, which are rapidly replacing cathode ray tubes (CRT's).

(*v*) In the field of optical communication, where LED's are used to transfer (or couple) energy from one circuit to another. They are also used to send light energy to fibre optical cable, which transmits energy by means of total internal reflection. The fibre optical cable is of light weight, flexible, often transparent and as small a 0.043 mm in diameter.

(*vi*) For image sensing circuits in picturephone.

(*vii*) In burglar alarm systems. In such applications, LED's radiating infrared light are preferred.

7-24. Multicolour LEDs

The LEDs, which emit one colour of light when forward biased and another when reverse biased, are called multicoloured LEDs. One commonly used symbol for a multicolour LED is as shown in Fig. 7-28.

Multicolour LEDs actually contain two PN junction that are connected in reverse-parallel, *i.e.*, they are in parallel, with the anode of one being connected to the cathode of the other. Multicolour LEDs are typically red when biased in one direction and green when biased in the other direction. Incidently if a multicolour LED is switched fast, the LED will produced a third colour. For example, a red/green LED will produce a yellow light when rapidly switched back and forth between biasing polarities.

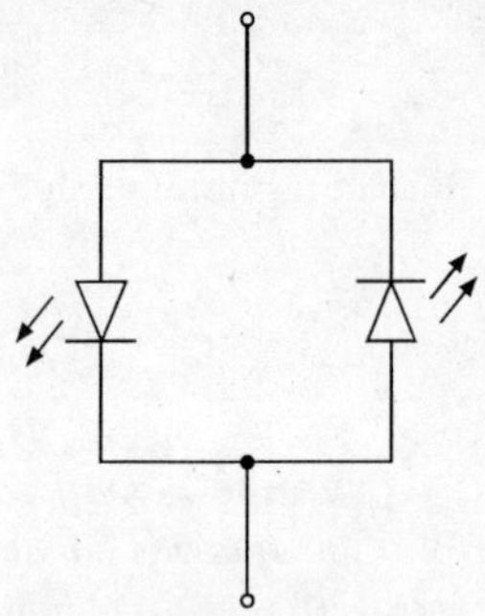

Fig. 7-28. *Schematic symbol for multicolour-LED.*

7-25. Seven-segment Displays

Fig. 7-29(*a*) shows a seven segment display. It is used to display alphanumeric characters. It consist of 7 rectangular light emitting diodes designated by the letters *a*, *b*, *c*, *d*, *e*, *f* and *g*. Each LED is called a segment, because it forms a part of the character being displayed.

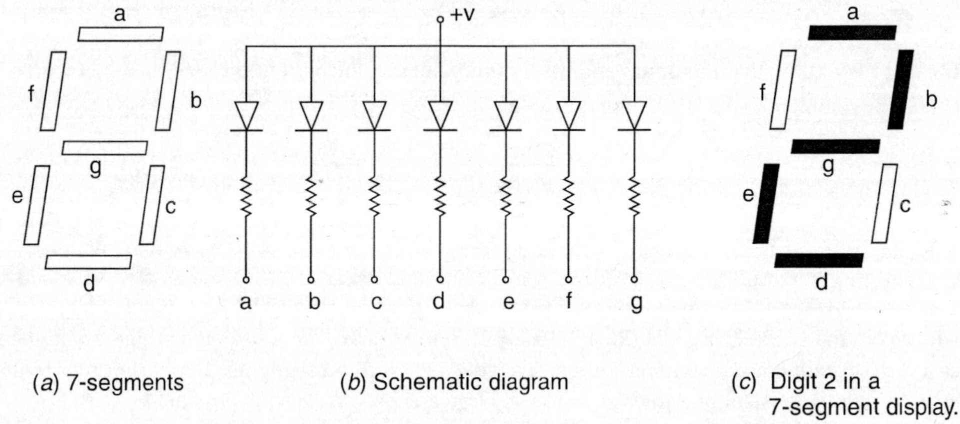

Fig. 7-29. *Seven-segment display.*

Fig. 7-29(*b*) shows a schematic diagram of a seven-segment display. In this circuit, the anodes of all the diodes are connected together to the positive terminal of the dc voltage source. The cathodes are connected to the external resistors. The external resistor is necessary to limit the current through the LED. By grounding the external resistors, we can form any decimal digit from 0 to 9. For example, by grounding *a*, *b*, *g*, *e* and *d*, we can form the digit '2' as shown in Fig. 7-29 (*c*). Similarly, by grounding *f*, *g* and *c*, we can form the digit '4' and so on. The digit 0 can be formed by grounding all the terminals except g. A seven segment display can also display the capital letters, *A*, *C*, *E* and *F*. Besides this, it can also display the lower-case letters *b* and *d*.

An average LED requires a current of about 20 mA for its operation. If we have a two-digit (*i.e.*, two seven-segment displays connected together) display in a circuit, then there are 14 LED's. If all the LEDs are glowing, the total current drawn from the supply voltage is 280 mA. With this much current, required to operate the display, alongwith the current, which is needed to operate the remainder of the circuit, the battery life could be very short. In battery powered applications, it is a common practice to instal an electronic switch, which is used to activate the display for short periods of time. The battery life can be increased by turning off the display, when it is not needed. In an equipment, where power consumption is not problem, the LED's make an excellent display. These days, the seven-segment displays are used in digital clocks, calculators microwave ovens, stereo tuners, digital multimeters, microprocessor trainer kits, *etc*.

The seven-segment displays are available in a single package as an integrated circuit. In such a package, the emitted light is conducted through the pipes to make it more useful. Thus, TIL 312, 313 and FND 500 seven-segment displays are available in a single package. Usually, these displays are referred to as common-anode or common-cathode type of displays depending upon whether their anodes or cathodes are connected together. Both these types are useful in actual practice. The choice depends upon the type of application where it is to be used.

7-26. Use of LEDs in Facsimile Machines

Fig. 7-30 shows a simplified schematic diagram of a facsimile (or fax) machine. As seen, the light from the LED array is focussed on the document paper. The light reflected at the paper is focussed on a charge-coupled device (CCD) by a combination. The electrical information is then sent through the data-processing unit to its destination *via* telephone line.

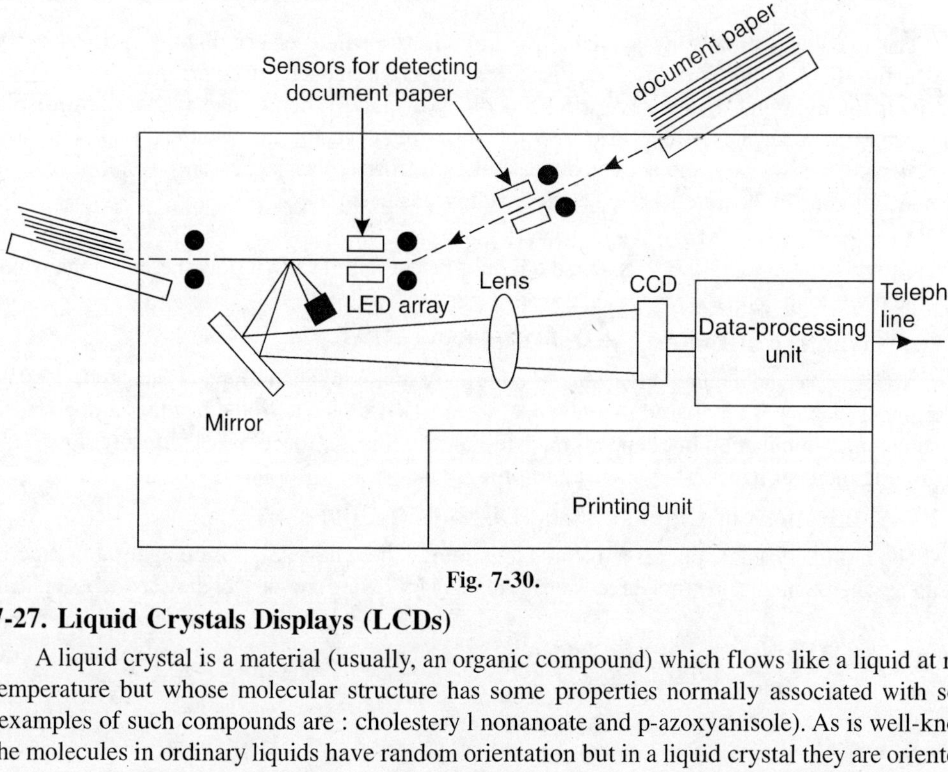

Fig. 7-30.

7-27. Liquid Crystals Displays (LCDs)

A liquid crystal is a material (usually, an organic compound) which flows like a liquid at room temperature but whose molecular structure has some properties normally associated with solids (examples of such compounds are : cholestery l nonanoate and p-azoxyanisole). As is well-known, the molecules in ordinary liquids have random orientation but in a liquid crystal they are oriented in a definite crystal pattern. Normally, a thin layer of liquid crystal is transparent to incident light but when an electric field is applied across it, its molecular arrangement is disturbed causing changes in falls on an activated layer of a liquid crystal, it is either absorbed or else is scattered by the disoriented molecules.

As shown in Fig. 7-31(*a*), a liquid crystal 'cell' consists of a thin layer about 10 μm of a liquid crystal sandwiched between two glass sheets with transparent electrodes deposited on their inside faces. With both glass sheets transparent, the cell is known as *transmittive type cell*. When one glass is transparent and the other has a reflective coating, the cell is called *reflective type*. The LCD does not produce any illumination of its own. It, in fact, depends entirely on illumination falling on it from an external source for its visual effect.

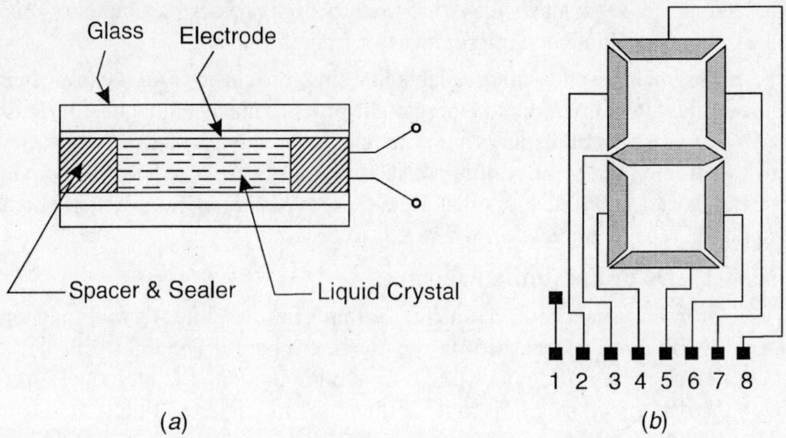

Fig. 7-31.

The two types of display available are known as (*i*) **field-effect display** and (*ii*) **dynamic scattering display.** When field-effect display is energized, the energized areas of the LCD absorb the incident light and, hence give localized black display. When dynamic scattering display is energized, the molecules of energized area of the display become turbulent and scatter light in all directions. Consequently, the activated areas take on frosted glass appearance resulting in a silver display. Of course, the un-energized areas remain translucent.

As shown in Fig. 7-31(*b*), a digit on an LCD has a segment appearance. For example, if number 5 is required, the terminals 8, 2, 3, 6 and 5 would be energized so that only these regions would be activated while the other areas would remain clear.

7-28. Advantage of Liquid Crystal Displays (LCDs)

An LCD has the distinct advantage of extremely low power requirement (about 10-15 μW per 7-segment display as compared to a few mW for a LED). It is due to the fact that it does not itself generate any illumination but depends on external illumination for its visual effect (colour depending on the incident light). They have a life-time of about, 50,000 hours.

7-29. Applications of Liquid Crystal Displays (LCDs)

The liquid crystal display is a matured technology these days. So it is being used in practically all areas where sunlight does not directly on the display following are some of the important applications of LCDs :

(*i*) Field-effect LCDs are normally used in watches and portable instruments such as digital thermometers, blood pressure, blood sugar, pressure monitoring instruments, *etc.* where source of energy is a prime consideration.

(*ii*) Recent desk top LCD monitors.

(*iii*) Notebook computer display.

(*iv*) Cellular phone display,

(*v*) To display data on personal digital assistant (PDAs) such as Palm V_x and pocket PCs, *etc*.

The liquid crystal display (LCDs) commonly used on notebook computers and handheld PADs are also appearing on desktop. These flat panel displays promise great clarity at increasingly high resolutions and are available in screen sizes upto 15 inches. The LCD monitor offers benefits and drawbacks. The first benefit is size. Because of the need to house the tube itself, cathode-ray tube (CRT) monitors are big and heavy. LCD monitors are expensive than CRTs at present. Another problem is the viewing angle. The optimal viewing angle of an LCD is from straight in front and as you move further to the side the screen becomes harder to read, much more so than with a CRT. Moreover, screen resolutions generally reach only as high as 1,024 × 768, which is insufficient for some applications.

7-30. Photodiodes

It is a two-terminal junction device which is operated by first reverse-biasing the junction and then illuminating it. A reverse-biased *P-N* junction has a small amount of reverse saturation current I_s (or I_0) due to thermally-generated electron-hole pairs. In silicon, I_s is the range of nanoamperes. The number of these minority carriers depends on the intensity of light incident on the junction. When the diode is in glass package, light can reach the junction and thus change the reverse current.

The basic biasing arrangement, construction and symbols of a photodiode are shown in Fig. 7-32. As seen, a lines has been used in the cap of the unit to focus maximum light on the reverse-biased junction. The active diameter of these devices is about 2.5 mm but they are mounted in standard TO-5 packages with a window to allow maximum incident light.

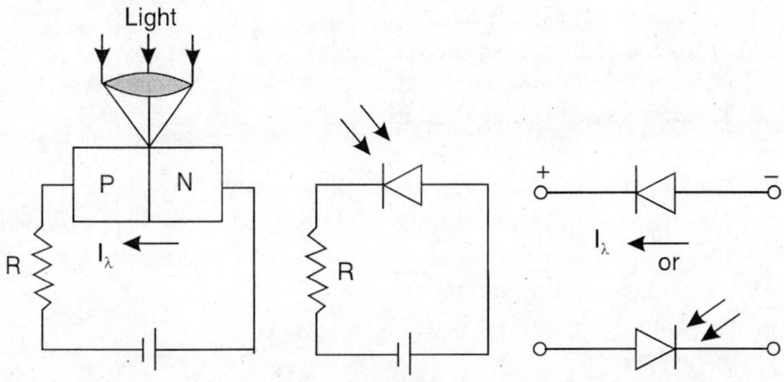

Fig. 7-32.

The characteristics of Fig. 7-33 shown that for a given reverse voltage, I_λ (or I_s) increases with increase in the level of illumination. The dark current refers to the current that flows when no light is incident. By changing the illumination level, reverse current can be changed. In this way, reverse resistance of the diode can be changed by a factor of nearly 20.

7-31. Photodiode Applications

A photodiode can turn its current ON and OFF in nanoseconds. Hence, it is one of the fastest photodetectors. It is used where it is required to switch light ON and OFF at a maximum rate. Applications of a photodiode include

(*i*) detection, both visible and invisible;

(*ii*) demodulation;

(*iii*) switching ;

(*iv*) logic circuit that require stability and high speed;

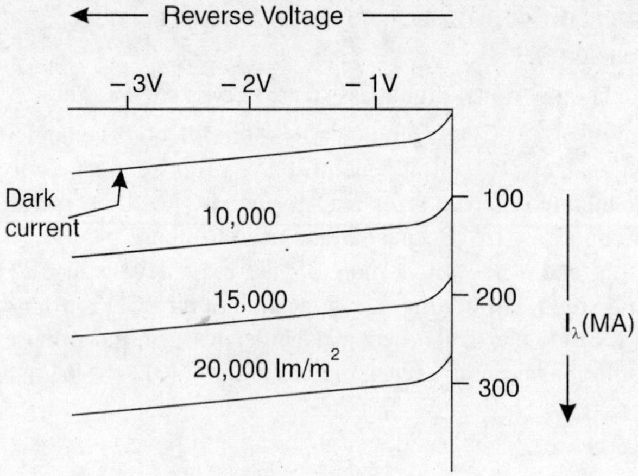

Fig. 7-33.

(*v*) character recognition ;

(*vi*) optical communication equipment;

(*vii*) encoders, *etc.*

7-32. Dust Sensor

Fig. 7-34 shows a combination of an LED and a photodiode used as a dust sensor. As seen, the light emitted from the LED gets reflected by the dust particles. The reflected light is collected by the dust particles. The reflected light is collected by the photodiode and is converted into an electrical signal. The dust sensor is employed in cleaners.

The combination of an LED and a photodiode is also used as : **(1)** a paper sensor in facsimile machines, **(2)** as a tape-end sensor in videotape recorders/players, and **(3)** as a dirt detector for rinsing in washing machines.

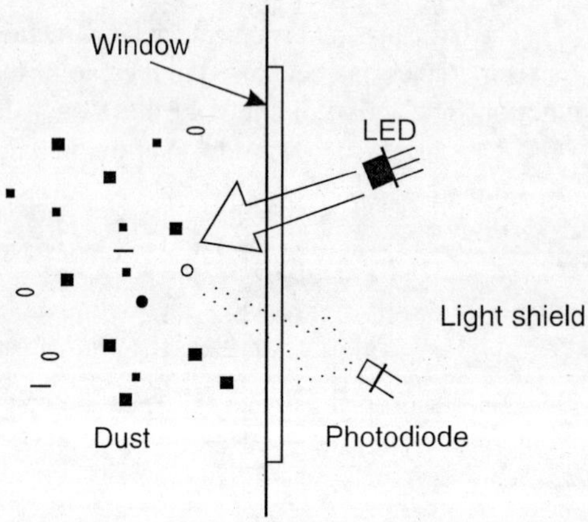

Fig. 7-34.

7-33. Photoconductive Cell

It is a semiconductor device whose resistance varies inversely with the intensity of light falls upon it. It is also known as *photoresistive* cell or *photoresistor* because it operates on the principle of photoresistivity.

The resistivity (and hence, resistance) of a semiconductor depends on the number of free charge carriers available in it. When the semiconductor is not illuminated, the number of charge carriers is small and hence, resistivity is high. But when light in the form of photons strikes the semiconductor, each photon delivers energy to it. If the photon energy is greater than the energy band gap of the semiconductor, free mobile charge carriers are liberated and, as a result, resistivity of the semiconductor is decreased.

Photoconductive cells are generally made of cadmium compounds such as cadmium sulphide (CdS) and cadmium selenide (CdSe). Spectral response of CdS cell is similar to the human eye, hence such cells are often used to simulate the human eye. That is why they find use in light metering circuits in photographic cameras.

The construction of a typical photoconductive cell and its two alternative circuit symbols are shown in Fig. 7-35 (*a*) and (*b*) respectively. As seen, a thin layer of photosensitive semiconductor material is deposited in the form of a long strip zig-zagged across a disc-shaped ceramic base with protective sides. For added protection, a glass lens or plastic cover is used. The two ends of the strip are bought out to connecting pins below the base.

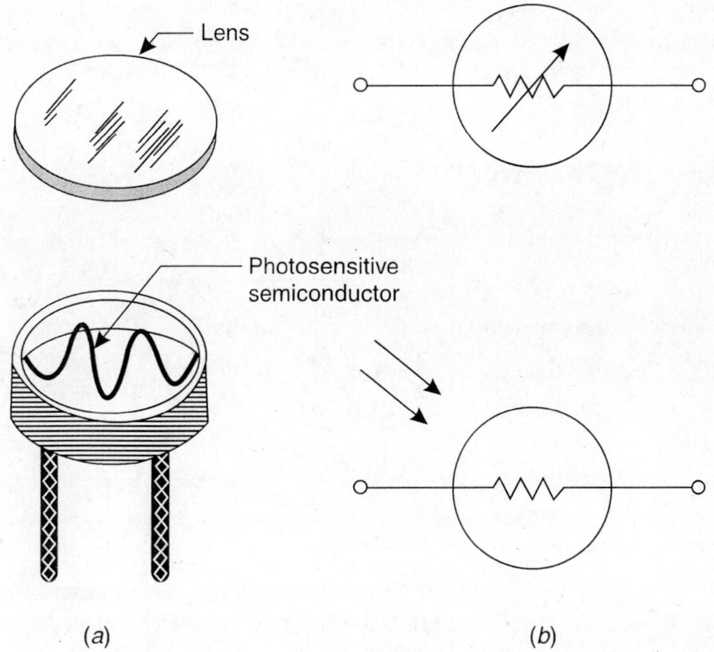

Fig. 7-35.

The terminal characteristic of a photoconductive cells is shown in Fig. 7-36. It depicts how the resistance of the cell varies with light intensity. Typically, the dark resistance of the cells is 1 MΩ or larger. Under illumination, the cell resistance drops to a value between 1 and 100 kΩ depending on surface illumination.

A photoconductive cell is an inexpensive and simple detector which is widely used in OFF/ON circuits, light-measurement and light-detecting circuits.

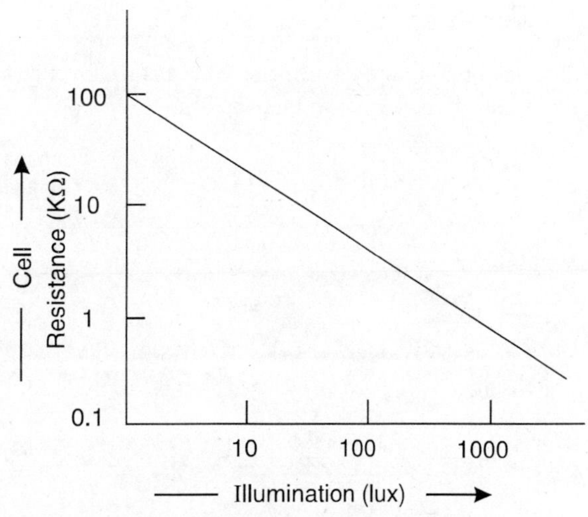

Fig. 7-36.

Example 7-10. *A relay is controlled by a photo-conductive cell which has resistance of 100 k Ω when illuminated and 1 kΩ when in the dark. The realy is supplied with 10 mA from a 30-V supply when cell is in illuminated and is required to be de-energized when the cell is in the dark Sketch a suitable circuit and calculate the required series resistance and value of dark current.*

(*Optoelectronic Devices, Gujarat Univ.* 1993)

Solution. The circuit is as shown in Fig. 7-37 where R is a current-limiting resistor.

The current through the circuit,

$$I = \frac{30}{R+r}$$

where, R is the series resistance and r is the cell resistance Reassranging the equation (i),

$$R = \frac{30}{I} - r$$

when illuminated, from equation (ii),

$$R = \frac{30}{10\times10^{-3}} - (1\times10^{3}) = 2\times10^{3}\Omega = 2k\Omega$$

Ans.

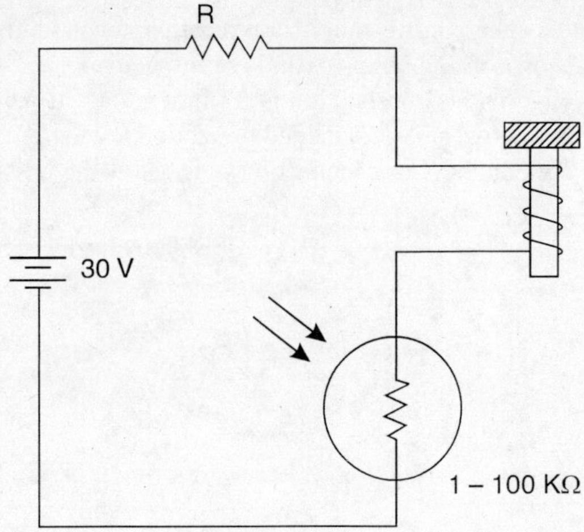

Fig. 7-37.

From equation (i), the dark current,

$$I_d = \frac{30}{10\times10^{-3}} - (1\times10^{3}) = 2\times10^{3}\Omega = 2k\Omega \textbf{ Ans.}$$

$$I_d = \frac{30}{(2\times10^{3})+(100\times10^{3})} = 0.3\times10^{-3}\text{A or } 0.3\text{mA } \textbf{Ans.}$$

7-34. Photo Voltaic or Solar Cell

Such cells operate on the principle of photovoltaic action, *i.e.* conversion of light energy into electrical energy. This action occurs in all semiconductors which are constructed to absorb energy.

As shown in Fig. 7-38(a), a basic solar cell consists of P-type and N-type semiconductor material (usually, silicon or selenium) forming a P-N junction. The bottom surface of the cell (which is always away from light) covered with a continuous conductive contact to which a wire lead is attached. The upper surface has a maximum area exposed to light with a small contact either along the edge or around the perimeter. The surface layer of P-type material is extremely thin (0.5 mm) so that light can penetrate to the junction.

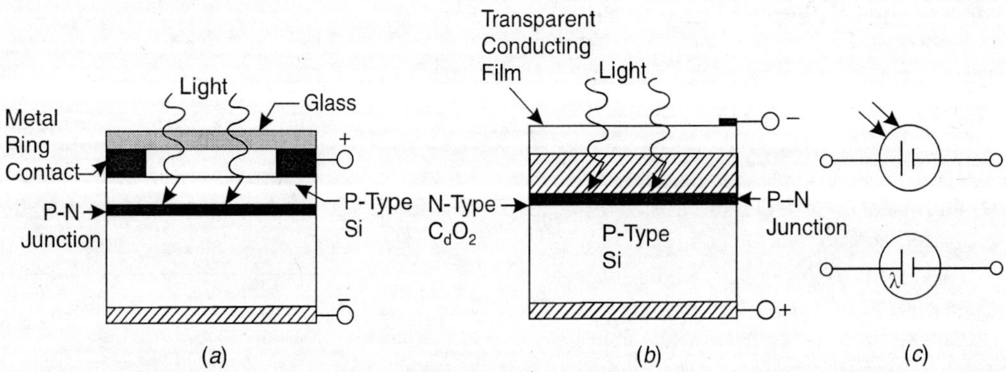

Fig. 7-38.

Although silicon is commonly used for fabricating solar cells, another construction consists of *P*-type selenium covered with a layer of *N*-type cadmium oxide *P-N* junction as shown in Fig.7-38(*b*). Two alternative circuit symbols are shown in Fig. 7-38(*c*). Power solar cells are also fabricated in flat strips to form efficient coverage of available surface area. Incidentally, the maximum efficiency of a solar cell in converting sunlight into electrical energy is nearly 15 percent at the present.

When the *P-N* junction of a solar cell is illuminated, electron-hole pairs are generated in much the same way, as in photovoltaic cell. An electric field is established near the *P-N* junction by the positive and negative ions created due to the production of electron-hole pairs which leads to the development of potential across the junction. Since, the number of electron-hole pairs far exceeds the number exceeds for thermal equilibrium, many of the electrons are pulled across the junction by the force of the electric field. Those that cross the junction contribute to the current in the cell and through the external load. The terminal voltage of the cell is directly proportional to the intensity of the incident light. The voltage may be as high as 0.6 V depending on the external load. Usually a large number of cells are arranged in an array in order to obtained higher voltages and currents as shown in Fig. 7-39(*a*).

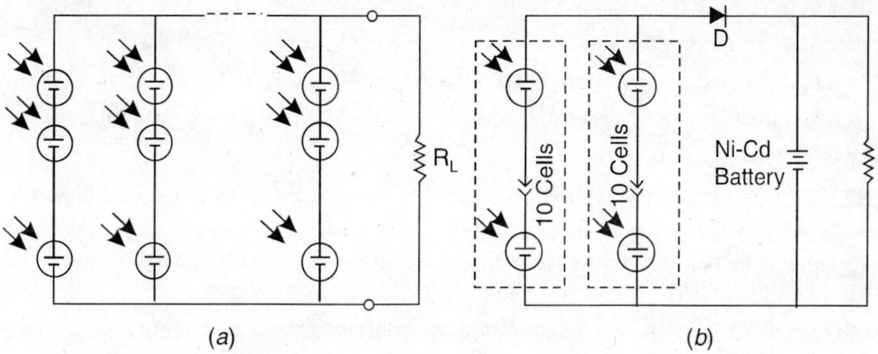

Fig. 7-39.

Solar cells act like a battery when connected in series or parallel. Fig. 7-39(*b*) show two groups of 10 series cells connected in parallel with each other. If each cell provides 0.5 V at 150 mA, the overall value of the solar bank is 5 V at 150 mA. The two parallel solar banks provides 5 V at 300 mA. This solar power source supplies the load and also charges the Ni-Cd battery. The battery provides power in the absence of light. A blocking diode *D* is used to isolate the solar cells from Ni-Cd battery otherwise in the absence of light, the battery will discharge through the cells thereby damaging them.

A solar cell operates with fair efficiency, has unlimited life can be easily mass-produced and has a high power capacity per weight. It is because of these qualities that it has becomes an important source of power for earth satellites.

Example 7-11. *An earth satellite has on board 12- V battery which supplies a continuous current of 0.5 A. Solar cells are used to keep the battery charged. The solar cells are illuminated by the sun for 12 hours in every 24 hours. If during exposure, each cell gives 0.5 V at 50 mA, determine the number of cells required.* (*Optoelectronics Devices, Gujarat Univ.* 1994)

Solution. The solar cell battery-charging circuit is shown in Fig. 7-40. The cells must be connected in series to provided the necessary voltage and such groups must be connected in parallel to provide the necessary current. The charging voltage has to be greater than the battery voltage of 12 V. Allowing for different drops, let the solar bank voltage be 13.5 V.

Number of series connected solar cells = 13.5/0.5 = 27

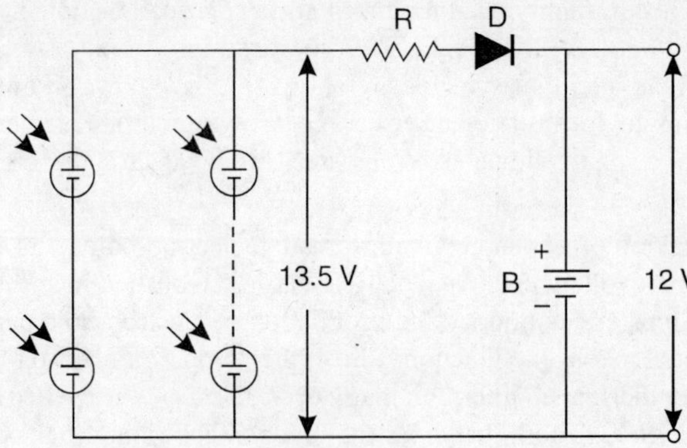

Fig. 7-40.

The charge given out by batteries during a 24 hour period = 12 × 0.5 = 6 Ah. Hence, solar cells must supply this much charge over the same period. However, solar cells deliver current only when they illuminated, *i.e.,* for 12 hours in every 24 hours. Necessary charging current required from the solar cells = 6 Ah/ 12 h = 0.5 A.

Total number of groups of solar cells required to be connected in parallel is

$= \text{output current / cell current} = 0.5/\ 50 \times 10^{-3} = 10$

∴ total number of solar cells required for the earth satellite = 27 × 10 = **270**

7-35. LASER Diode

Like LEDs, laser diodes are typical *PN* junction devices used under a forward-bias. The word LASER in an acronym for *Light Amplification* by *Stimulated Emission of Radiation.* The use of laser is (becoming increasing common) in medical equipment used in surgery and in consumer products like compact disk (CD) players, laser printers, hologram scanners, *etc*.

Broadly speaking, the laser diode structure can be divided into two categories :

- ***Surface-emitting LASER diodes :*** These laser diodes emit light in a direction **perpendicular** to the PN junction plane.
- ***Edge-emitting LASER diodes* :** These laser diodes emit light in a direction **parallel** to the *PN* junction plane.

Fig. 7-41(*a*) shows the structure of an edge-emitting laser diode. This type of structure is called Febry-Perot type laser. As seen from the figure, a *P-N* junction is formed by two layers of doped gallium arsenide (GaAs). The length of the *PN* junction bears a precise relationship with the wavelength of the light to be emitted. As seen, there is a highly reflective surface at one end of the junction and a partially reflective surface at the other end. External leads provide the anode and cathode connections.

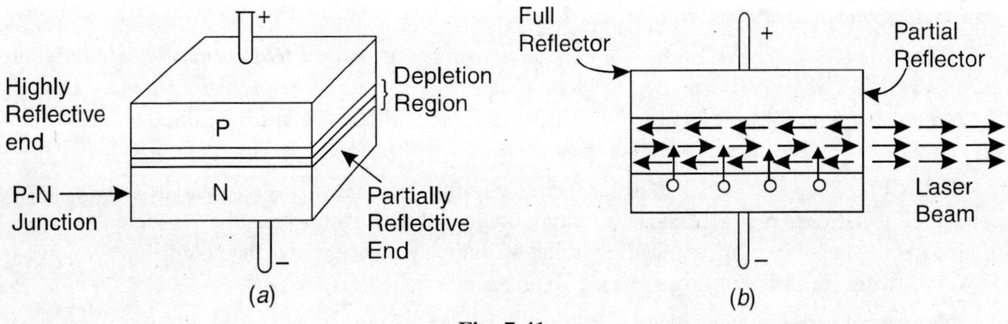

Fig. 7-41.

When the *P-N* junction is forward-biased by an external voltage source, electrons move across the junction and usual recombination occurs in the depletion region which results in the production of photons. As forward current is increased, more photons are produced which drift at random in the depletion region. Some of these photons strike the reflective surface perpendicularly. These reflected photons enter the depletion region, strike other atoms and release more photons. All these photons move back and forth between the two reflective surfaces [Fig. 7-41(*b*)]. The photon activity becomes so intense that at some point, a strong beam of laser light comes out of the partially reflective surface of the diode.

Unique Characteristics of LASER Light

The beam of laser light produced by the diode has the following unique characteristics:

- It is coherent, *i.e.,* there is no path difference between the waves comprising the beam;
- It is monochromatic, *i.e.,* it consists of one wavelength and hence, one colour only;
- It is collimated, *i.e.,* emitted light waves travel parallel to each other.

Laser diodes have a threshold level of current above which the laser action occurs but below which the laser diode behaves like a LED emitting incoherent light. The schematic symbol of a laser diode is similar to that of LED. Incidentally, a filter or lens is necessary to view the laser beam.

7-36. LASER Diode Applications

Laser diodes used in variety of applications ranging from medical equipment used in surgery to consumer products like optical disk equipment, laser printers, hologram scanners *etc*. Laser diodes emitting visible light are used as pointers. Those emitting visible and infrared light are used to measure range (or distance). The laser diodes are also widely used in parallel processing of information and in parallel interconnections between computers. Some of these applications are discussed in the following articles.

7-37. Optical Discs

The major application field for laser diodes is in optical disc equipment is used for reading or recording information and can be broadly divided into two groups :

1. Reading-only **2.** Recording-and -reading type.

The optical disc equipment of either type make use of a laser diode, lenses and photodiodes. During recording, it changes electrical information into optical information and then records the information on the optical disc. During reading (or playback), the head optically reads the recorded information and changes the optical information into electrical information. Fig. 7-42 shows the different types of optical discs used in practice. The commercial system make use of discs that are 90, 120, 130 and 300 mm in diameter. A mini disk, 64 mm in diameter is also used for digital audio.

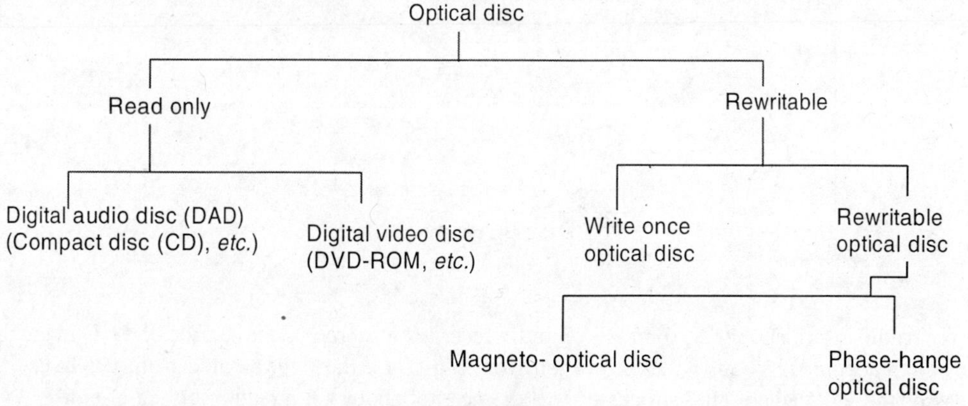

Fig. 7-42.

The optical discs have several advantages over semiconductor memories. Some of these include their larger data storage capacity, shorter access time and smaller size. Therefore, they are used in terminal equipment of computers as well as in audio visual equipment.

7-38. Read-only Optical Discs Equipment

Fig. 7-43 shows an optical equipment for reading data from digital audio (compact) discs. Compact discs (CDs) which are 120 mm in diameter are typical digital audio discs. Compact discs usually means digital audio compact disc, but it also includes the read-only memory (CD-ROM) for data memory and interactive compact disc (CD-I) for multimedia use.

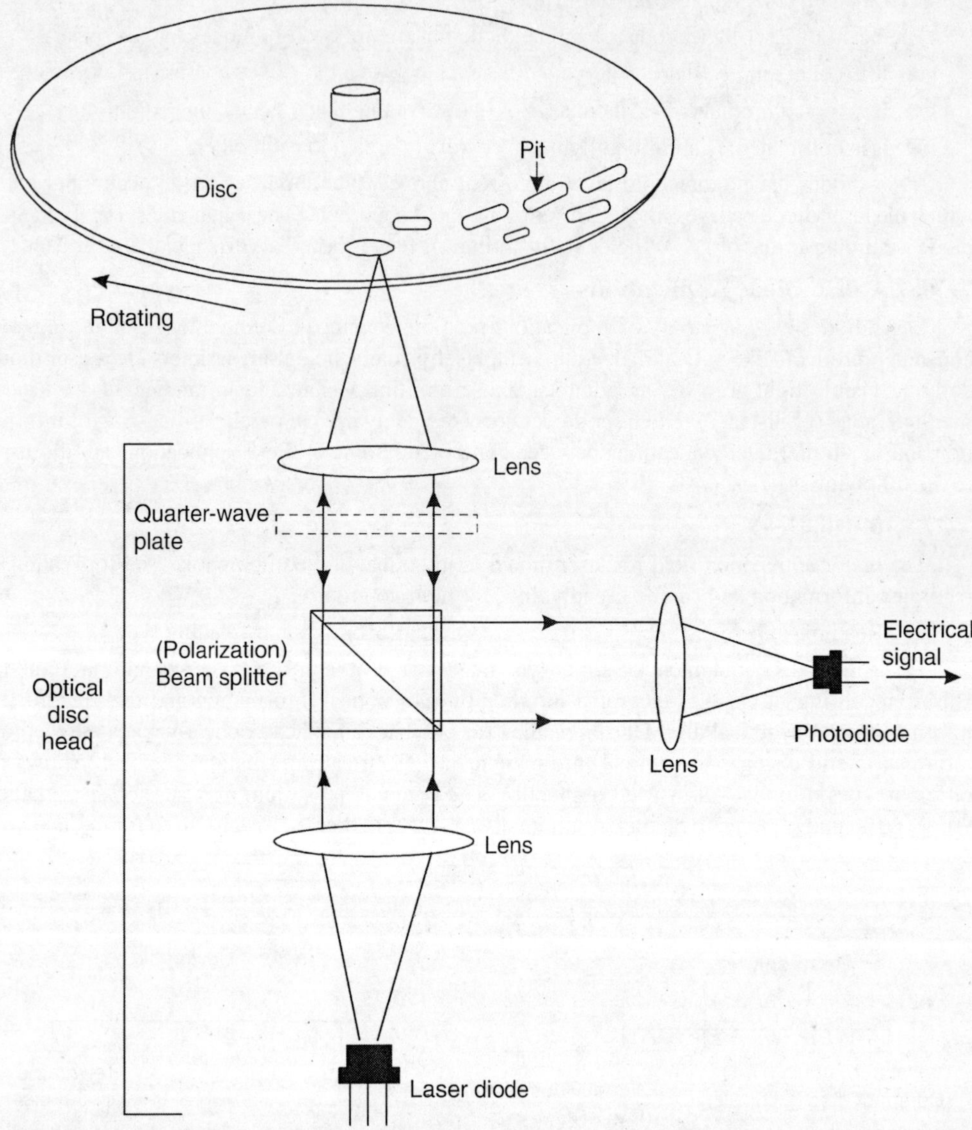

Fig. 7-43.

Audio information (*i.e.*, sound) is digitally recorded in stereo on the surface of a CD in the form of microscopic "pits" and "flats". As seen from Fig. 7-43, the light emitted from the laser diode passes through the lens and is focussed to a diameter of about 1 μm on the surface of a disc. As the

CD rotates, the lens and beam follow the track under control of a servo motor. The laser light which is altered by the pits and flats along the recorded track is reflected back from the track through the lens and optical system to infrared photodiodes. The signal from the photodiodes in then used to reproduce the digitally recorded sound.

7-39. Printers using LASER Diodes

These are two types of optical source usually used in printers; (1) laser diodes and (2) LED arrays. The printers using laser diodes are called laser beam printers (or simply laser printers). These are one of the most attractive type of equipment in office automation in today's world. Words and figures can be printed rapidly and clearly more easily by a laser printer than by other type of printers.

Fig. 7-44 shows a simplified diagram of a laser printer. As seen the laser diode is driven by modulated signals from the computer. The optical beam after passing through the lens is reflected by

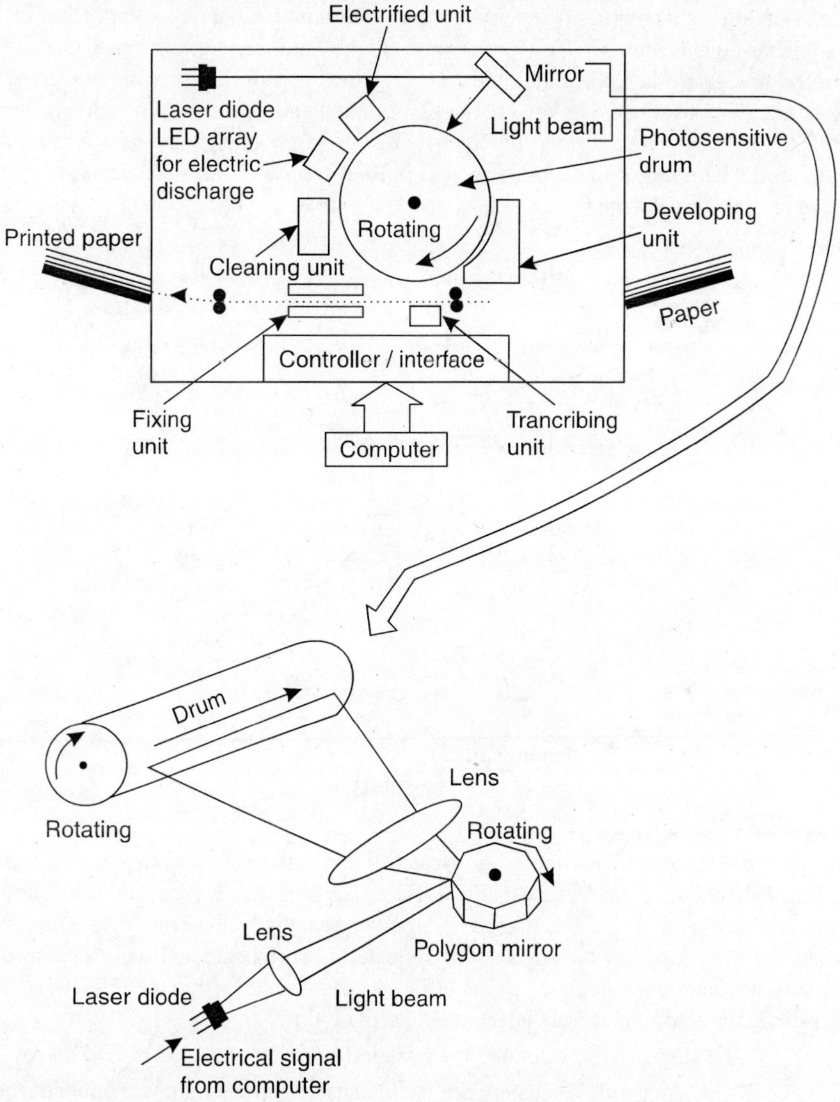

Fig. 7-44. *Courtesy optical semiconductor devices by M.Fukuda published by John Weliy & Sons Inc.*

the rotating polygon mirror and scanned on the photosensitive drum. The drum is homogeneously charged when it passes through the charging unit consisting of an LED array. The homogeneous electrification is partially erased in accordance with the scanned optical beam. This is because of the fact that the electrical resistance at the light-irradiated part decreases and the electric charge is released. This causes the signals (*i.e.*, data) from the computer to be written on to the drum. At the developing unit, an electrically charged powder (called toner) is electrostatically attached to the written parts. At the transcribing unit, the powder is transferred to the paper. Next, the transferred pattern is fixed by heating and pressing at the fixing unit. The data from the computer is thus printed on the paper.

7-40. Hologram Scanners

The hologram scanner is widely used in various equipment and is ordinarily used in bar-code readers in point-of-sale systems (such as super marked checkout counters). It is also used in laser printers for scanning the laser beam on the drum precisely.

Fig. 7-45 shows a simplified schematic of a hologram scanner. As seen, the optical beam for reading the bar-code is focussed by a lens through the hologram disk and scanned on the bar-code by rotating the hologram disk. Gratings with coaxial circles are formed on the hologram disk. This causes the incident laser beam to bend at the grating by an amount determined by the grating pitch. The reflected light modulated according to the bar-code is reflected by the mirror and monitored by the photodiode. The monitored optical signal is then translated into an electrical signal is then translated into electrical signal.

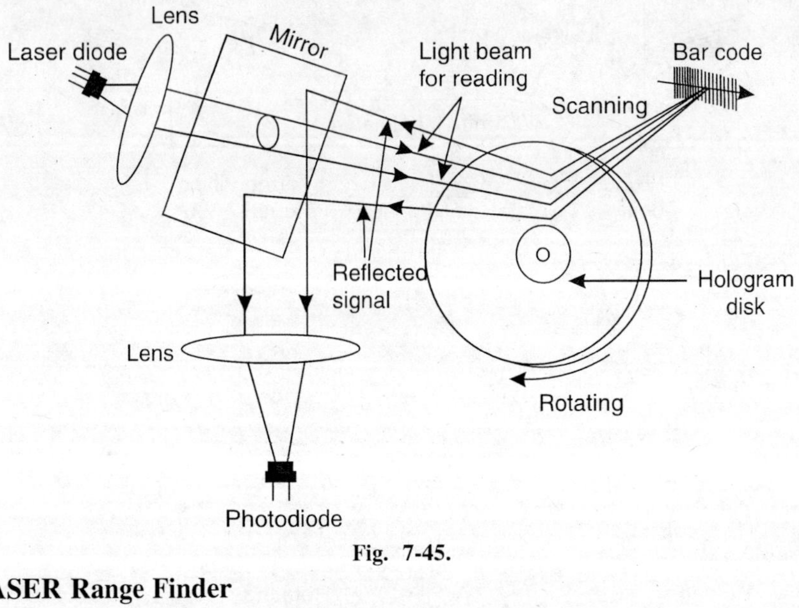

Fig. 7-45.

7-41. LASER Range Finder

The laser diodes along with photodiodes can be used to measure the range (*i.e.*, a distance) of an object. Fig. 7-46 shows a simple schematic of a laser range finder. As seen, the laser diode is modulated with high current pulses. The pulsed high-power beam is emitted in the direction of an object. The beam is reflected from the object. The reflected beam is detected with a photo detector (or photodiode). The range is calculated as the difference between the time the light was emitted from the laser diode and the time it was detected by the photodiode.

Let D = distance between the laser range finder and the object.

ΔT = Time difference between the instance when the light was emitted from the laser diode and the instance when it was detected by the photodiode

Then $$D = \frac{1}{2} \times \text{speed of light} \times \Delta T$$

A 2-D array of laser diodes and photodetectors can be constructed. Such a system is used to obtain 3-D images of an object.

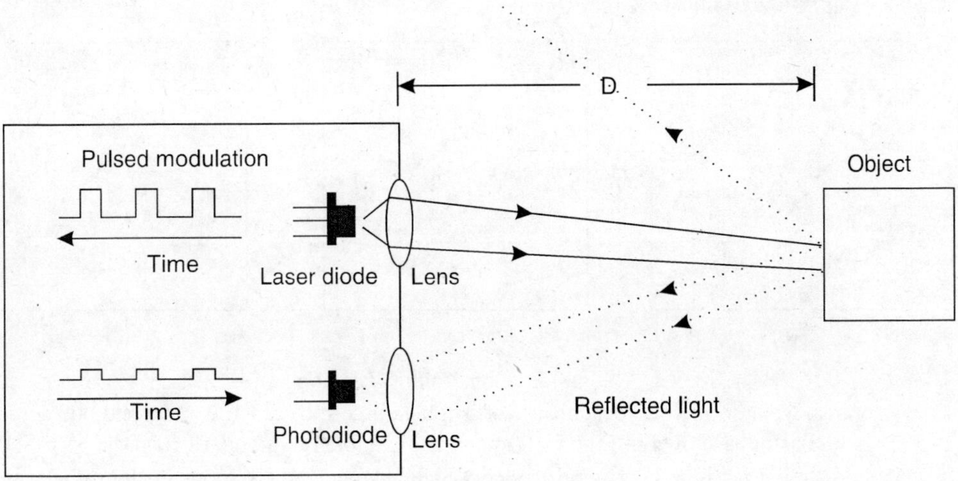

Fig. 7-46.

7-42. Light-Activated SCR (LASCR)

The operation of a LASCR is essentially similar to that of a conventional SCR (refer to chapter 12 Art 12-3) except that it is light-triggered (Fig. 7-47). Moreover it has a window and lens which focuses light on the gate junction area. The LASCR operates like a latch. It can be triggered ON by a light input on the gate area but does not turn OFF when light source is removed. It can be turned OFF only be reducing the current through it below its holding current . Depending on its size, a LASCR is capable of handling larger amount of current than be handled by photodiode or a photo-transistor.

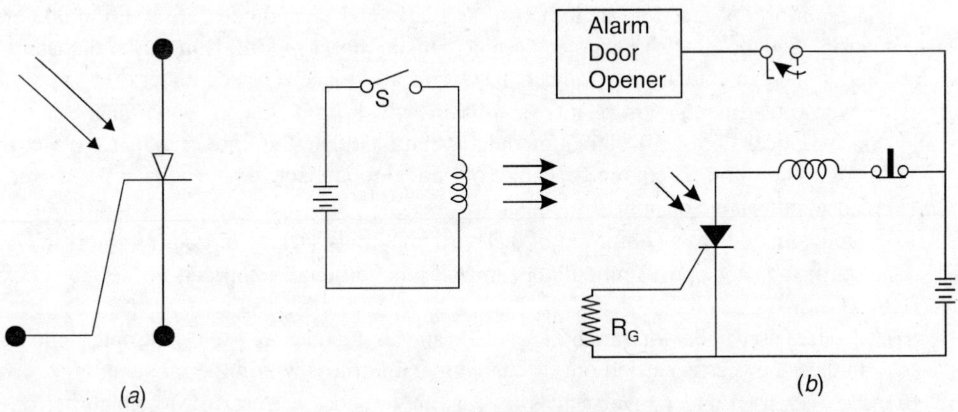

Fig. 7-47. *LASCR*

Fig. 7-47(*b*) shows how a LASCR can be used for energizing a latching relay. The input dc source turns on the electric lamp and the resulting incident light triggers the LASCR into conduction. The anode current energizes the relay and closes the contact. It is seen that the input dc source is electrically isolated from the rest of the circuit.

7-43. Optical Isolators

Optical isolators are designed to electrically isolate one circuit from another allowing one circuit to control the other. The usual purpose of isolation is to provide protection from high-voltage transients, surge voltages and low-level electrical noise that could possible result in an erroneous output or damage to the device. Such isolators allow interfacing of circuits with different voltage levels and different grounds, *etc*.

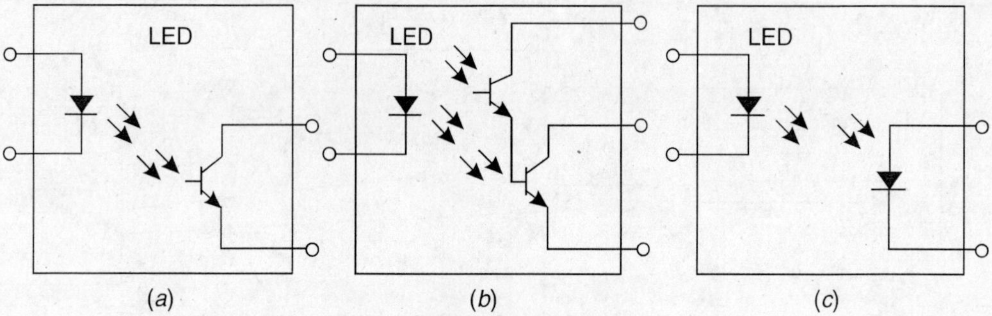

Fig. 7-48.

An optical isolator (or coupler) consists of a light source such as LED and a photodetector such as a photo transistor as shown in Fig. 7-48(*a*) and is available in a standard *IC* package.

A photo transistor is a 3 terminal device which responds to its input in the same way as a photodiode. In addition it produces larger output.

When LED is forward-biased, the light produced by it is transferred to the photo transistor which is turned ON thereby producing current through the external load.

Fig. 7-48(*b*) shows a Darlington transistor coupler which is used when increased output current capability is needed beyond that provided by the phototransistor output. The LASCR output coupler of Fig. 7-48(*c*) can be used in applications where a low-level input voltage is required to latch a high voltage relay for activating some king of electro-mechanical device.

7-44. Optical Modulators

Light emitting *PN* junction devices such as LEDs and laser diodes are easily modulated by superimposing signals no to the injected current. This is **direct modulation**. Laser diodes in high-bit rate and long-span optical communication systems are frequently used under direct modulation.

However, direct modulation results in *chirping* which limits transmission quality because of *dispersion* in optical fibres. An optical modulator can modulate the light output from laser diodes with little or no chirping. There are two types of optical modulators :

(1) The semiconductor optical modulators

(2) Optical modulators composed of dielectric materials such as *lithium nitrate* ($LiNO_3$)

The semiconductor optical modulators are *PN* junction diodes and can further be subdivided into two types :

(*i*) Devices used under forward bias (as LEDs and laser diodes are used). The optical modulation in these devices is carried out by changing gain or loss within the modulators.

(*ii*) Devices used under reverse bias (*i.e.,* as photodiodes are used). Most high-performance semiconductor optical modulators are used under reverse bias. The reverse bias is needed to generate strong electric field. Optical modulation is basically performed by modulating the *refractive index* or *optical absorption coefficient* of the modulators. The devices which make use of refractive index phenomenon of modulation are called phase modulation type devices while those that use optical absorption coefficient phenomenon are called intensity modulation type devices.

There are several different of optical modulators available today. But the waveguide type optical modulator is more common in use. Further there are several different waveguide type optical modulator structures possible. Fig. 7-49 shows a mesa type optical modulator structure.

It may be noted that although we have shown the structure making use of a simple *N*-and *P*-layer but in reality each layer (*N*-type or *P*-type) is made up of several different semiconductors.

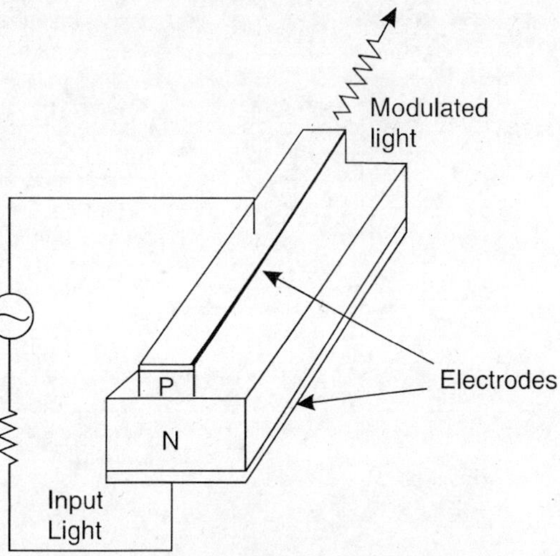

Fig. 7-49.

7-45. Optical Fibre Communication Systems

The optical fibre communication systems (such as public communication networks and data links) are the basic infrastructure of the information hungry society. There are several advantages of the optical fibre system over metallic transmission systems as listed below :

(*i*) Data can be transmitted at a very high-frequency over longer distances without much loss.

(*ii*) Electromagnetic induction (EMI) noise is never induced during transmission through optical fibre cables.

(*iii*) Optical fibre cable is light, flexible and economical.

Fig. 7-50 shows the public optical fibre communication system broadly divided into two groups : **(1)** Submarine systems, and **(2)** Land systems. Submarine systems have already been used to connect countries all over the world. The submarine systems help people to talk overseas without any time delay.

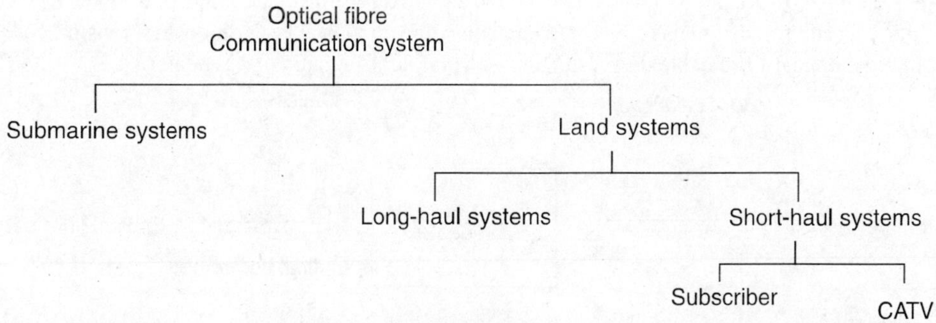

Fig. 7-50.

In land systems, long-haul systems have been connected between large cities. The land systems also include systems such as subscriber systems and CA TV (*i.e.*, community or common antenna television, cable and telecommunication television systems, or cable television system).

Fig. 7-51 shows an application of LEDs, laser diodes and photodiodes in a simplified optical fibre communication systems. The LEDs and laser diodes emit light modulated with a signal. The optical signal is then transmitted through the optical fibre and is received with photodiodes on the destination side. In this type of system LEDs or laser diodes emit the light directly through the optical fibre and therefore is referred to as direct modulation type systems. But in more recent systems **optical modulators** modulate the light emitted from the laser diodes and then the modulated light is transmitted through the optical fibre [refer to Fig. 7-52].

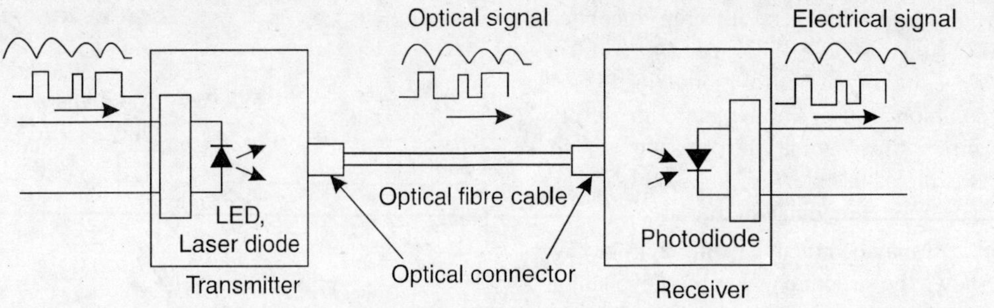

Fig. 7-51.

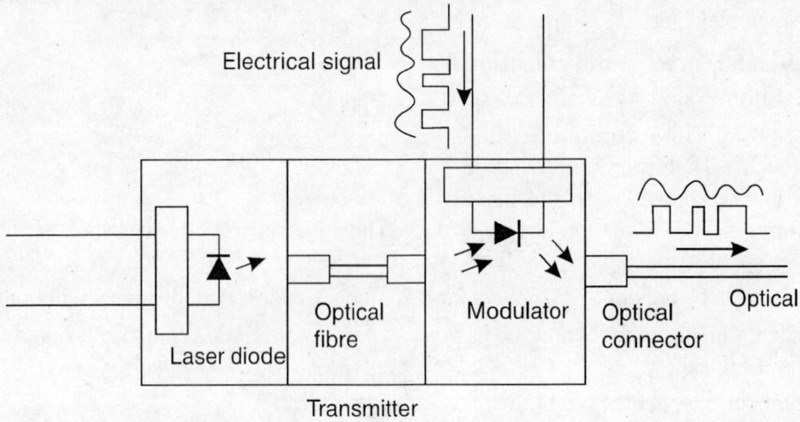

Fig. 7-52.

In long-haul systems, **repeaters** (which include photodiode and laser diodes and electronic circuits) are inserted. In the repeater, the weak optical signal being transmitted through the optical fibre is detected by the photodiode. The detected signal is reformed and amplified by the electronic circuits. The amplified signal is converted again into an signal by a laser diode and transmitted again through the optical fibre cable. Fig. 7-53 shows a simple schematic of a repeater.

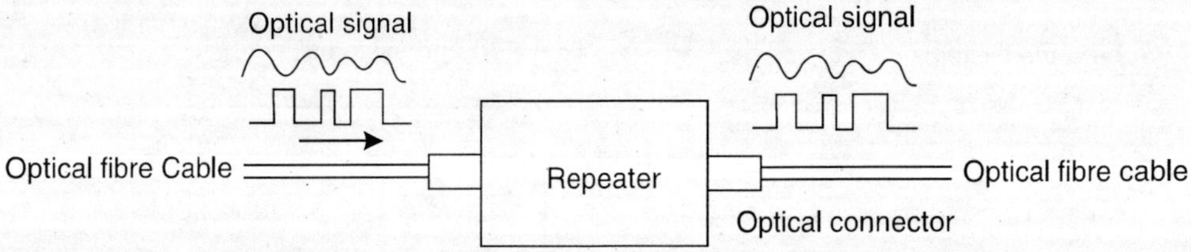

Fig. 7-53.

From the modulation point of view, the optical fibre communication systems can be divided into *digital systems* and *analog systems.* Most long-haul and large capacity optical fibre communication systems are digital systems. The analog systems are used for transmitting information over a short distance.

7-46. Optical Fibre Data Links

The use of optical fibre data links has wide spread in the past few decades. Its application ranges from local area networks (LANs) to the computer, digital audio and mobile fields. Several different types of LEDs and laser diodes emitting light at wavelengths ranging from visible to the infrared are used as optical sources. The transmission data rate is a function of transmission distance and varies from application. For computer links where the distance varies from 1 m to 100 m, the data transmission rate varies from 1 M bits/s to 100 M bits/s. For local-area -networks used in factory, office and building automation, the data transmission rate varies from 10 K bits/s to 5 M bits/s. In digital audio field, where the distance is below 1 m the data transmission rate varies from 500 bits/s. to over 10 M bits/s. Similarly, in mobile fields (such as chip, aircraft, train and automotive applications) where the distance could vary from 1 m to 100 m, the data transmission rate varies from 1 K bits/s to 1 M bits/s.

(1) *Optical fibre local area networks :* The optical fibre local area networks (LANs) are similar to the public communication systems. Some of their advantages over the systems using metallic cables are : **(1)** high transmission capacity and bit rate, and **(2)** longer transmission distance. However, the range of LANs is restricted. They are more commonly used within factories, offices, buildings *etc*. Computers, printers, facsimile machines and other office equipment are connected with each other by optical fibre cables as shown in Fig. 7-54.

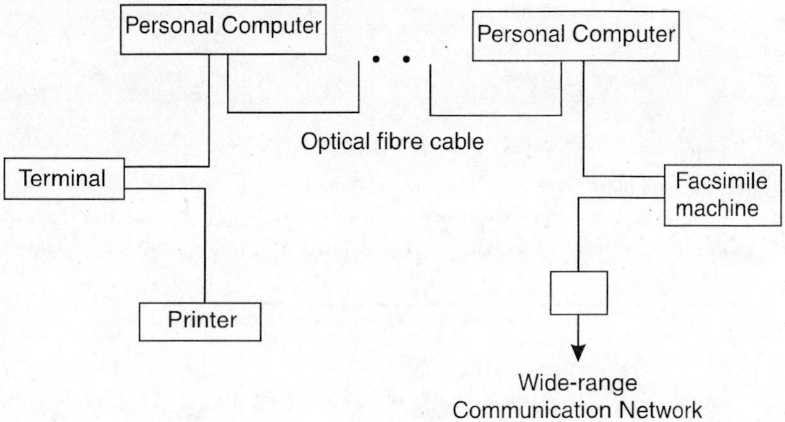

Fig. 7-54.

The LEDs and laser diodes are used to transmit data through the optical fibre cable and photodiodes are used to receive data. The different types of equipment connected in the LAN could be one of the following two types : **(1)** An optical ethernet having a radial-shape network as shown in Fig. 7-55(*a*) or **(2)** a fibre-distributed data interface (FDDI) having a ring-shape network as shown in Fig. 7-55(*b*).

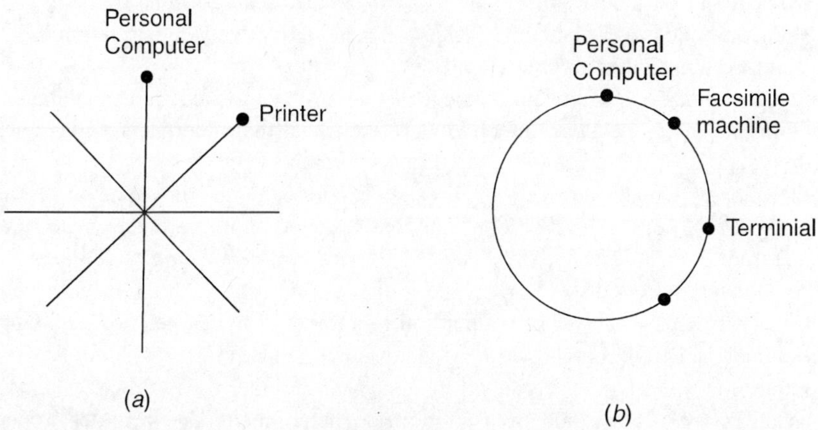

Fig. 7-55.

(2) *Digital audio field :* Fig.7-56 shows an example of a data link in digital audio field. As seen, the optical fibre cable is used to connect compact disc (CD) player, laser disc (LD) player, digital audio type (DAT) and tuner with the amplifier and speaker. The connection between the amplifier and everything except DAT is unidirectional. The audio digital signals from CD, LD player, DAT and tuner are converted into optical signals by LEDs or laser diodes at one end of the fibre optic cable and then transmitted through the cable to the opposite end. At the opposite end, the signals are received by photodiodes and converted into an electrical signal for amplification and finally speaker for reproduction to a sound.

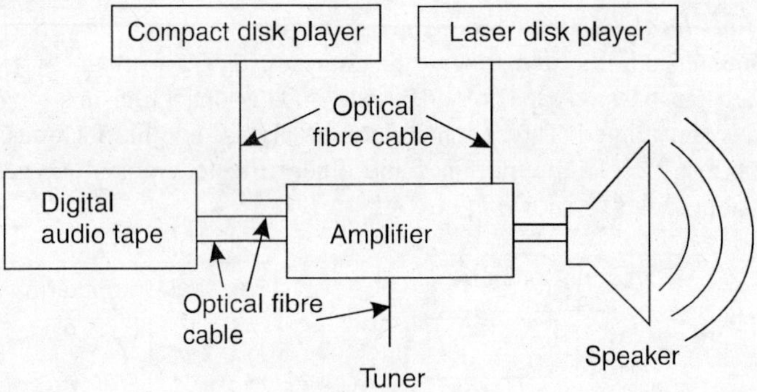

Fig. 7-56.

(3) ***Mobile fields :*** The optical data links are very suitable in mobile fields such as ship, aircraft, train, automotive *etc*. The reason is that optical data links are very compact and light in weight than metalic data links. In addition to this, the optical data links are not subjected to noise induced by electromagnetic induction.

SUMMARY

In the chapter, you have learned that,

1. There are several different types of special purpose diodes. This includes zener, light-emitting, tunnel, varactor, schottky, PIN, photodiode, and laser diode.

2. The zener diode is designed to work in the reverse breakdown region of the diode characteristic curve.

3. There are two types of zener diode breakdown : *avalanche breakdown* and *zener breakdown.*

4. Tunnel diode is a highly doped PN junction device. The Gallium Arseride (GaAs) is used as a material for fabricating tunnel diodes. Tunnel diode is used in oscillator circuits. Operating at frequencies of the order of 10 GHz (*i.e.,* microwave) frequency range.

5. Varactor diode is a PN junction device which has a relatively high junction capacitance. It is most often used as a voltage-controlled capacitor.

6. Schottky diode is a PN junction device that has a very little junction capacitance. Because of this, it can be operated at much higher frequencies than a normal PN junction. Schottky diodes are used more commonly in digital switching circuits.

7. PIN diodes is a device made of three semiconductor materials. The centre material is made of pure (or intrinsic) silicon. The P and N-type regions are very heavily doped and therefore have low resistance. The PIN diodes are used as a modulator in microwave communication circuits.

8. LEDs are diodes that will provide a limited amount of light when properly biased.

9. Multicolour LEDs contain two PN junctions that are connected is parallel.

10. Laser diodes are special PN junction devices used under forward bias. There diodes are usually made of GaAs and are used in medical equipment and consumer produces.

11. Phototransistor is a light-sensitive device and is basically used for some applications as a photo diode.

GLOSSARY

Breakover current : The minimum value of zener current which must be maintained in order to keep the diode in breakdown (or regulation) region.

Esaki diode : Refer to tunnel diode.

Intrinsic semiconductor : Semiconductor in their extremely pure form.

Light emitting diode : A PN junction diode. which emits light when forward biased.

Photo diode : A PN junction device, which shows an increase in current through it when the intensity of light falling on to the junction increases, when reverse biased.

PIN diode : A semiconductor device in which heavily doped *P* and *N*-type materials are separated by an intrinsic semiconductor. It acts as a variable resistor when forward biased and as a constant capacitor when reverse biased.

Schottky diode : Semiconductor device forward by joining a doped semiconductor region (usually *N*-type) with a metal such as gold, silver or platinum. It has a very fast switching time.

Step-recovery diode : Semiconductor diode that has lesser doping level towards the junction they are used for frequency multiplication purpose.

Swept-out voltage : The reverse bias voltage of a PIN diode junction at which I-region becomes free of charge carriers.

Tunnel diode : A PN junction diode which is doped with a high concentration of inpurity atoms (by 1000 times or more) than that of normal diodes.

Tunneling : The phenomenon of penetration of charge carriers directly through the potential barrier due to reduced width of depletion layer.

Varactor diode : A reverse biased PN junction that utiliges the inherent capacitance of the depletion layer.

Zener breakdown : The reverse breakdown of a PN junction due to the electric field across the junction produced by the sufficiently large reverse voltage.

Zener diode : A PN junction diode that is operated in the reverse biased condition, possessing voltage regulator capabilities

Zener effect : The ionization, which occurs due to the electrostatic force of attraction, when the reverse bias voltage across a zener diode is sufficiently high.

Zener voltage (V_z) : The reverse biased voltage at which breakdown occurs in a zener diode.

NUMERICAL PROBLEMS

1. A zener diode has a dc power dissipation of 1 W and a zener voltage ratings of 27 V. What is the value of I_{ZM} for the device ? (**Ans.** 37 mA)
2. Determine the power dissipating rating of Motorola's IN 746 zener diode at 125°C. Assume that the dc power dissipation rating of the device is 500 mW and the derating factor is 3.33 mW/°C above 50°C. (**Ans.** 250.25 mW)
3. A 6.2 V zener diode has a resistance of 20 Ω. What is the terminal voltage when the current in 25 mA? (**Ans.** 6.7 V)
4. A step recovery diode is used in a frequency multiplier. The period of the rectified output signal is 0.05 sec. What is the frequency of the 5th harmonic ? (**Ans.** 100 MHz)
5. A light remitting diode (LED) is connected across a voltage source of + 10 volts through a series resistance of 820 Ω. Calculate the LED current. Assume the voltage drop across an LED of 1.5 volt. (**Ans.** 10.4 mA)

DESCRIPTIVE QUESTIONS

1. What is a zener diode? Draw the equivalent circuit of an ideal zener in the breakdown region.
2. Explain the working of a tunnel diode. Give its two applications.
3. Draw the V-I characteristics of a tunnel diode and indicate the useful region in the curve.
4. What is a varactor diode? Explain briefly, how it can be used in tuning circuits.
5. Briefly explain the PIN diode operation. Give its applications.
6. In what respect is an LED different from an ordinary PN junction diode? State applications of LEDs.
7. Explain V–I characteristics of zener diode

   (*Electron Devices, Anna University, APril/May 2008.*)
8. Write short notes no

   (*i*) LED

   (*ii*) Photo Diode (*Electron Devices, Anna University, April/May 2008*)
9. With structural diagram explain the V-I characteristics of the schottky diode.

   (*Electron Devices, Anna University, April/May 2008*)
10. Describe the construction, working, and applications of photo diode.

    (*Electron Devices, Anna University, Nov/Dec 2007*)
11. What is zener effect? Explain the function of a zener diode and draw its characteristics.

    (*Electron Devices, Anna University, Nov/Dec 2007*)
12. Characteristic of the Photocoupler.

    (*Linear Integrated Circuits, Anna University, April/May 2008*)
13. What is a backward diode? (*Electron Devices, Anna University, Nov/Dec 2007*)
14. What is LDR? (*Electronic Devices, Anna University, Nov/Dec 2007*)

MULTIPLE CHOICE QUESTIONS

1. A zener diode

   (*a*) has a high forward-voltage rating

   (*b*) has a sharp breakdown at low reverse voltage

   (*c*) is useful as an amplifier

   (*d*) has a negative resistance
2. A 5 V reference is drawn from the circuit shown in Fig. 7.57 if the zener diode is of 5 mA, then R_z will be

   (*a*) 50 Ω (*b*) 500 Ω

   (*c*) 5000 Ω (*d*) 50,000 Ω
3. A tunnel diode is

   (*a*) a very heavily-doped PN junction diode

   (*b*) a high resistivity PN junction diode

   (*c*) a slow switching device

   (*d*) used with reverse bias
4. A PIN diode is frequently used as a

   (*a*) voltage regulator

   (*b*) peak detector

   (*c*) harmonic generator

   (*d*) switching diode for very high frequencies

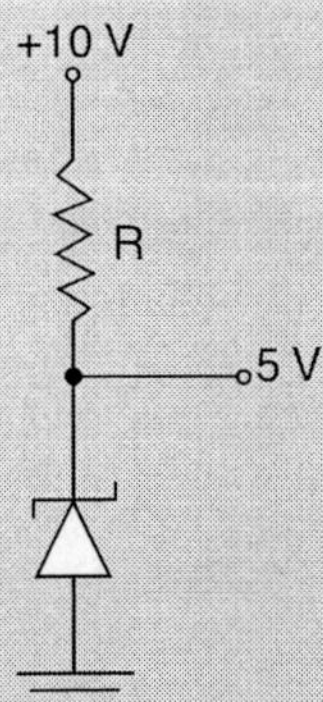

Fig. 7.57

5. A step recovery diode
(*a*) has an extremely short recovery twice
(*b*) is an ideal rectifier of high frequency signals
(*c*) is mainly used as a harmonic generator
(*d*) conducts equally well in both director

6. The light-emitting diode (LED),
(*a*) is usually made from silicon
(*b*) uses a reverse-biased junction
(*c*) gives a light output which increases with the increase in temperature
(*d*) depends on the recombination of holes and electrons

7. LED's do not require
(*a*) heating (*b*) warm-up time
(*c*) both (*a*) and (*b*) above (*d*) non of above

8. The sensitivity of a photodiode depends upon
(*a*) light intensity and depletion region width
(*b*) depletion region width and excess carrier life time
(*c*) excess carrier life time and forward bias current
(*d*) forward bias current and light intensity

9. LEDs are commonly fabricated from gallium compounds like gallium arsenide and gallium phosphide because they
(*a*) are cheap (*b*) are easily available
(*c*) emit more heat (*d*) emit more light.

10. A LED is basically a ....P-N junction.
(*a*) forward-biased (*b*) reverse-biased
(*c*) lightly-doped (*d*) heavily-doped

11. As compared to a LED display, the distinct advantage of an LCD display is that it requires
(*a*) no illumination (*b*) extremely-bias
(*c*) no forward-bias (*d*) a solid crystal

12. Before illuminating a P-N junction photodiode, it has to be
(*a*) reverse-biased (*b*) forward-biased
(*c*) switched ON (*d*) switched OFF.

13. In a photoconductive cell, the resistance of the semiconductor material varies .....with the intensity of incident light.
(*a*) directly (*b*) inversely
(*c*) exponentially (*d*) logarithmically

14. A photoconductive cell is known as ....cell.
(*a*) phototransistor (*b*) photoresistor
(*c*) photovoltaic (*d*) both (*a*)and (*b*).

15. A phototransistor excels a photodiode in the matter of
(*a*) faster switching (*b*) greater sensitivity
(*c*) higher current capacity (*d*) both (*a*) and (*b*)
(*e*) both (*b*) and (*c*)

16. A photodarlington comprises of
(*a*) a phototransistor (*b*) a transistor
(*c*) a photodiode (*d*) both (*a*) and (*b*)

17. Solar cell operates on the principle of
(*a*) diffusion (*b*) recombination
(*c*) photo voltaic action (*d*) carrier flow.

18. Solar cells are used as source of power in earth satellites because they have
(*a*) very high efficiency (*b*) unlimited life
(*c*) higher power capacity per weight (*d*) both (*b*) and (*c*)
(*e*) both (*a*) and (*b*).

19. The device possessing the highest sensitivity is a
(*a*) photo conductive cell (*b*) photovoltaic cell
(*c*) photodiode (*d*) phototransistor

20. The unique characteristics of LASER light are that it is
(*a*) coherent (*b*) monochromatic
(*c*) collimated (*d*) all of the above

21. The LASCR operates like a
(*a*) latch (*b*) LED (*c*) photodiode (*d*) phototransistor.

22. Optical couplers are designed to .....one circuit from another.
(*a*) control (*b*) isolate (*c*) disconnect (*d*) protect.

23. The main purpose of using optical isolators is to provide protection to devices from
(*a*) high-voltage transients (*b*) surge voltages
(*c*) low-level noise (*d*) all of the above.

24. A LED emits visible light when its........
(*a*) P-N junction is reverse-biased (*b*) depletion region widens
(*c*) holes and electrons recombine (*d*) P-N junction becomes hot.

25. In LED, light is emitted because
(*a*) recombination of charge carriers takes place
(*b*) diode gets heated up
(*c*) light falling on the diode gets amplified
(*d*) light gets reflected due to lens action.

26. GaAs, LEDs emit radiation in the
(*a*) ultraviolet region
(*b*) violet-blue green range of the visible region
(*c*) visible region
(*d*) infra-red region

27. Phototransistors respond much like a conventional transistor except that, in their case, light energy is used to...
(*a*) alter leakage current (*b*) change base voltage
(*c*) switch it ON (*d*) alter emitter current.

ANSWERS

1. (*b*)	**2.** (*c*)	**3.** (*a*)	**4.** (*d*)	**5.** (*c*)
6. (*d*)	**7.** (*c*)	**8.** (*a*)	**9.** (*d*)	**10.** (*a*)
11. (*b*)	**12.** (*a*)	**13.** (*b*)	**14.** (*d*)	**15.** (*e*)
16. (*d*)	**17.** (*c*)	**18.** (*d*)	**19.** (*d*)	**20.** (*d*)
21. (*a*)	**22.** (*b*)	**23.** (*d*)	**24.** (*c*)	**25.** (*a*)
26. (*d*)	**27.** (*c*)			

❑❑❑

8

CHAPTER

PN Junction Diode Applications : Rectifiers & Filters

OUTLINE

Objectives

After completing this chapter, you should be able to :

- Understand the function of a half-wave rectifier circuit.
- Calculate d.c. output voltage, d.c. load current, power delivered to the load and peak inverse voltage of a diode in a single-phase half-wave rectifier.
- Understand the function of a single-phase full-wave rectifier circuit.
- Differentiate between centre-tapped full-wave rectifier and bridge full-wave rectifier.
- Calculate d.c. output voltage, d.c. load current, and peak inverse voltage of a diode in a Single-phase full-wave rectifier.
- Understand the ripple factor of single-phase half-wave and full-wave rectifiers.
- Know the efficiency of a single-phase half-wave and full-wave rectifiers.
- Explain the filtering action of the important filters namely, induction filter, capacitor filter, L-C filter and π-filter.
- Describe the circuit operation of three-phase half-wave rectifier, full-wave rectifier, six-phase half-wave rectifier, and three-phase bridge rectifier.
- Read the specifications data sheet of a rectifier diode and bridge rectifier to select the suitable device for circuit design.

8-1. Introduction

The rectifiers and filters are the circuits, which are used in the initial stages of a d.c. power supply. The d.c. power supply is essential for the operation of many electronic devices and circuits. A rectifier is a circuit, which uses one or more diodes to convert a.c. voltage into pulsating d.c. voltage. It may be broadly categorized in the following two types :

(1) Half-wave rectifier, and **(2)** Full-wave rectifier.

Since, the rectifier need a.c. mains supply as an imput, it is more appropriate to distinguish the input rectifiers whether they use single-phase $(1 - \phi)$ or three-phase $(3 - \phi)$ a.c. supply. Accordingly we will consider the following rectifier circuits :

- Single-phase half-wave rectifier
- Single-phase full-wave rectifier
- Full-wave bridge circuit
- Three-phase half-wave rectifier
- Three-phase full-wave rectifier
- Six-phase half-wave rectifier
- Three-phase bridge circuit
- Voltage multiplier circuits.

Now we shall discuss all these rectifiers one by one in the following pages.

8-2. Single-Phase Half-Wave Rectifier

Fig 8-1(*b*) shows a half-wave rectifier circuit. It consists of a single diode in series with a load resistor. The input to the half-wave rectifier is supplied from the 50 Hz a.c. supply, whose waveform is shown in Fig. 8-1(*a*). The working of a half-wave rectifier circuit may be studied by considering separately the positive and negative half cycles of the a.c. input voltage.

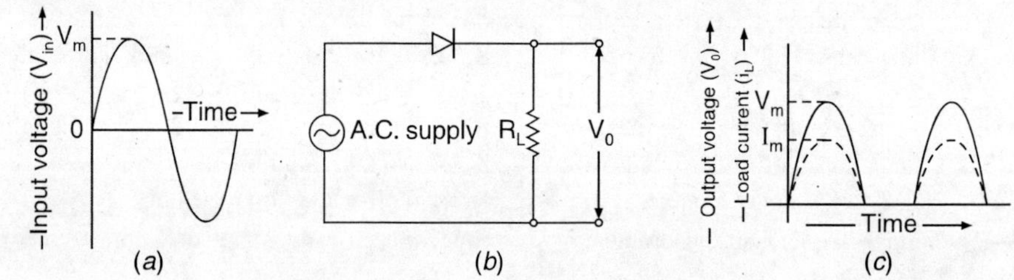

Fig. 8-1. *Half-wave rectifier.*

During the positive half cycle of the a.c. input voltage, the diode is forward biased and conducts for all instantaneous voltages greater than the offset voltage (0.7 V for silicon and 0.3 V for germanium diodes). However, for all practical purposes, we assume that the diode is forward biased, whenever the a.c. input voltage goes above zero. While conducting, the diode acts as a short-circuit, so that the circuit current flows and produces a voltage across the load resistor (R_L). The voltage produced across the load resistor has the same shape as that of the positive input half cycle of a.c. input voltage as shown in Fig. 8-1(*c*). The waveform of diode current (which is equal to the load current) is also shown in Fig. 8-1(*c*).

During the negative half-cycle, the diode is reverse biased and hence, it does not conduct. Thus, there is no current flow or voltage drop across load resistor (R_L), *i.e.*, $i_D = 0$ and $v_0 = 0$. The net result is that only the positive half cycle of the a.c. input voltage appears across R_L. It means that only the positive half cycle of the a.c. input voltage is utilized for delivering a.c. power. It is evident that output voltage (v_0) is not a steady d.c. But it is a pulsating d.c. wave having a ripple frequency equal to the input voltage frequency. It will be interesting to know that we can see the waveform of output voltage on an oscilloscope connected across R_L. If we measure the output voltage (or output current) by a d.c. meter, it will indicate some average positive value for both voltage and current.

It is evident from the above discussion, that *as the circuit uses only one-half cycle of the a.c. input voltage, therefore it is popularly known as a half-wave rectifier.*

8-3. Single-Phase Half-Wave Rectifier with Transformer Coupled Input

These days, most of the electronic circuits use a transformer to couple a.c. input voltage from source to the rectifier circuit as shown in Fig. 8-2. The transformer coupling provides the following two advantages :

(*i*) It allows the source voltage to be stepped up or stepped down.

(*ii*) The a.c. power source is electrically isolated from the rectifier circuit. It reduces the chances of getting a shock.

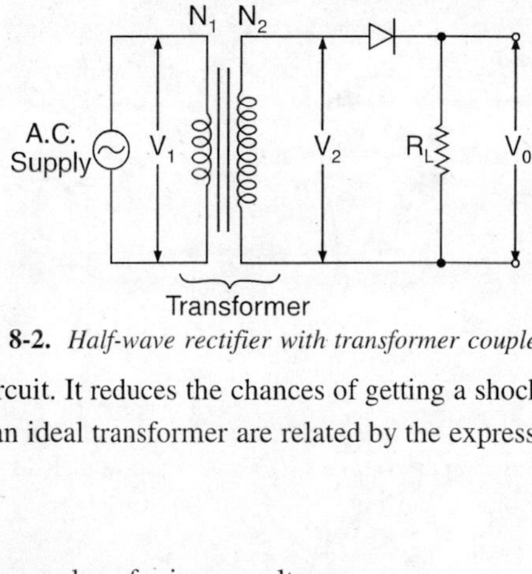

Fig. 8-2. *Half-wave rectifier with transformer coupled input.*

The primary and secondary voltages of an ideal transformer are related by the expression,

$$\boxed{\frac{V_2}{V_1} = \frac{N_2}{N_1}}$$

where, V_1 = The r.m.s. value of primary voltage.

V_2 = The r.m.s. value of secondary voltage.

N_1 = Number of turns in primary winding, and

N_2 = Number of turns in secondary winding.

8-4. Average Values of Output Voltage and Load Current in a Single-phase Half-wave Rectifier

Consider a half-wave rectifier circuit with a sinusoidal a.c. input voltage. We know that instantaneous value of the sinusoidal a.c. input voltage is given by the relation,

$$v = V_m \sin \omega t = V_m \sin \theta \qquad \ldots (\because \theta = \omega t)$$

where, V_m = Maximum value of the a.c. input voltage

It may be noted that in case of a transformer coupling, the equation of a.c. input voltage would remain the same as given above. However, the maximum value of a.c. input voltage (V_m) will represent the maximum voltage across the secondary winding of the transformer.

Now, let I_m = Maximum value of diode current or load current. It is the current through the resistor R_L.

V_{dc} = Average or d.c. value of output (or load) voltage across the load resistor, and

I_{dc} = Average or d.c. value of load current.

Fig. 8-3(*a*) shows the waveform of an output voltage of a half-wave rectifier. It indicates that there is an output voltage (varying sinusoidally) for a period from 0 to π and nothing for a period from π to 2π. The average or d.c. value of the output voltage is given by the relation,

$$V_{dc} = \frac{\text{Area under the curve over the full cycle}}{\text{Base}}$$

$$= \frac{\int_0^{\pi} v \cdot d\theta}{2\pi} = \frac{1}{2\pi}\int_0^{\pi} V_m \cdot \sin\theta \cdot d\theta$$

$$= \frac{V_m}{2\,\pi}\left|(-\cos\theta)\right|_0^{\pi} = \frac{V_m}{2\pi}(+1-(-1))$$

$$= \frac{V_m}{\pi} = 0.318\,V_m$$

The above expression indicates that the average or d.c. value of the output voltage is 31.8% of the maximum a.c. input voltage. This value is indicated by a broken horizontal line in Fig. 8-3 (*a*).

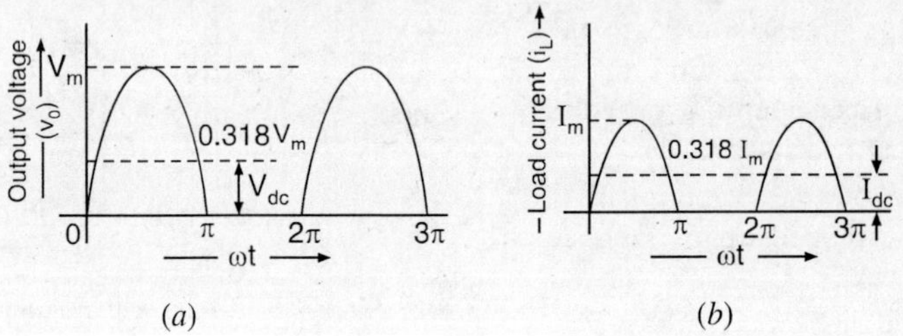

Fig. 8-3. *Average values of output voltage and load current in a half-wave rectifier.*

Let us now find the average or d.c. value of load current. It may be found by dividing the d.c. value of output voltage (V_{dc}) by the value of load resistor (R_L). Mathematically, the average or d.c. value of load current,

$$I_{dc} = \frac{V_{dc}}{R_L} = \frac{V_m}{\pi \cdot R_L} = \frac{I_m}{\pi} \qquad \ldots\ldots\left(\because I_m = \frac{V_m}{R_L}\right)$$

$$= 0.318\,I_m$$

The above expression indicates that the average or d.c. value of load current is 31.8% of the maximum load current. This value is indicated by a broken horizontal line in Fig. 8-3(*b*).

Notes: 1. In deriving the average values of output voltage and load current, we have assumed that the diode used in the rectifier circuit, is an ideal one, *i.e.*, it has zero forward resistance. However, In actual practice, it is not so. If R_f is the diode forward resistance, then the average output voltage across the load,

$$V_{dc} = \frac{V_m}{\pi} - I_{dc} \cdot R_f$$

and the average value of load current,

$$I_{dc} = \frac{V_m}{\pi(R_f + R_L)}$$

2. If the a.c. input voltage is applied through the transformer, then it is necessary to include the secondary winding resistance (R_s) in the relation of average values of output voltage and current. In this case, average value of output voltage,

$$V_{dc} = \frac{V_m}{\pi} - I_{dc}(R_f + R_s)$$

and the average value of load current,

$$I_{dc} = \frac{V_m}{\pi(R_f + R_s + R_L)}$$

8-5. Peak Inverse Voltage of a Diode in a Single-phase Half-wave Rectifier

We have already discussed in Art. 8-2 that during negative half-cycle of the a.c. input voltage, the diode in a half-wave rectifier of Fig. 8-1 is reverse biased. And it does not conduct. As a result of this, there is no current in the circuit and hence no voltage drop across the load resistance R_L. Now, if we apply Kirchoff's Voltage Law to the circuit, then we find that whatever be the negative voltage at the input, the same amount appears as a reverse voltage across the diode. The maximum value of reverse voltage occurs at the peak of the negative alternation of the input cycle, which is equal to V_m. This maximum reverse voltage is called *peak-inverse voltage* (abbreviated as *PIV*). Thus, peak inverse voltage of a diode in a half-wave rectifier,

$$\textbf{PIV} = V_m$$

Example 8-1. *An a.c. supply of 200 V is applied to a half-wave rectifier circuit through a transformer of turns ratio 10 : 1 as shown in Fig. 8-4.*

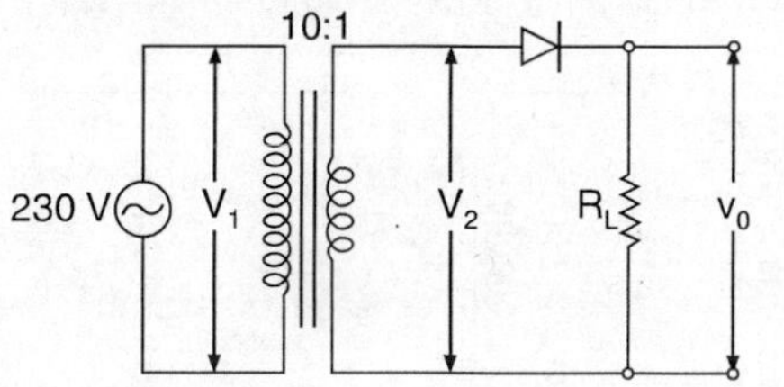

Fig. 8-4.

*Find **(a)** d.c. output voltage, and **(b)** PIV of a diode. Assume the diode to be an ideal one.*

Solution. Given: V_1 = 230 volts; N_2/N_1 = 1/10

(*a*) *D.C. output voltage*

We know that the secondary voltage,

$$V_2 = V_1 \times \frac{N_2}{N_1} = 230 \times \frac{1}{10} = 23 \text{ V}$$

and maximum value of secondary voltage,

$$V_m = \sqrt{2} \times V_2 = \sqrt{2} \times 23 = 32.5 \text{ V}$$

∴ D.C. output voltage,

$$V_{dc} = 0.318\, V_m = 0.318 \times 32.5 = 10.3 \text{ V } \textbf{Ans.}$$

(*b*) *PIV of a diode*

We also know that peak-inverse voltage of a diode,

$$PIV = V_m = 32.5 \text{ V } \textbf{Ans.}$$

Example 8-2. *Fig. 8-5 shows a half-wave rectifier circuit with a transformer coupled input. Determine the value of dc load current.*

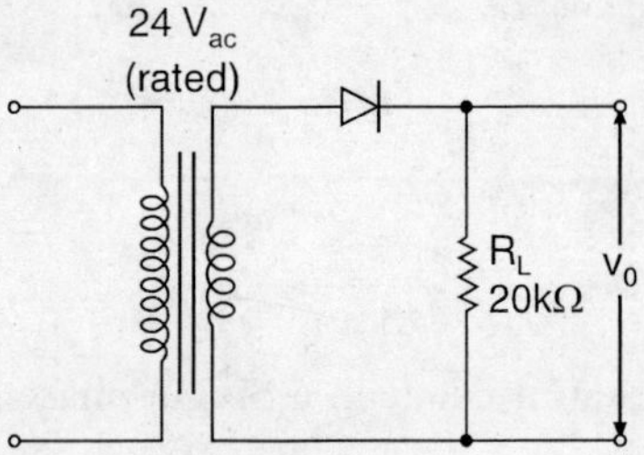

Fig. 8-5.

Solution. Given: $R_L = 20\ k\Omega = 20000\ \Omega$.

We know that maximum value of secondary voltage,

$$V_m = \sqrt{2} \times V_2 = \sqrt{2} \times 24 = 33.94\ V$$

∴ Maximum value of load current,

$$I_m = \frac{V_m}{R_L} = \frac{33.94}{20000} = 1.697 \times 10^{-3}\ A = 1.697\ mA$$

and d.c. or average value of load current,

$$I_{dc} = 0.318\ I_m = 0.318 \times (1.697 \times 10^{-3}) = 0.54 \times 10^{-3}\ A \ \textbf{Ans.}$$

Example 8-3. *Fig. 8-6 shows a half-wave rectifier circuit with a transformer coupled input.*

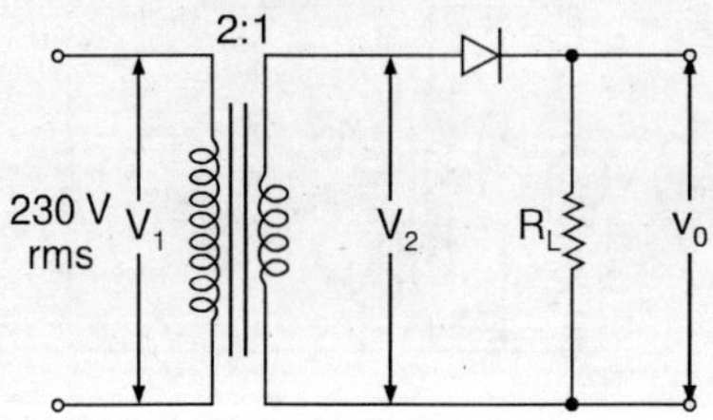

Fig. 8-6.

Determine the maximum and average values of power delivered to the load. Take R_L equal to 200 Ω.

Solution. Given: $V_1 = 230$ volts; $N_2/N_1 = 1/2 = 0.5$ and $R_L = 200\ \Omega$.

Maximum value of load power

We know that r.m.s. value of the secondary voltage.

$$V_2 = V_1 \times \frac{N_2}{N_1} = 230 \times \frac{1}{2} = 115\ V$$

and maximum value of secondary voltage,

$$V_m = \sqrt{2} \times V_2 = \sqrt{2} \times 115 = 162.6\ V$$

We also know that maximum value of load current,

$$I_m = \frac{V_m}{R_L} = \frac{162.6}{200} = 0.813\ A$$

and maximum value of load power,

$$P_m = I_m^2 \times R_L = (0.813)^2 \times 200 = 132.3\ W \ \textbf{Ans.}$$

Average value of load power

We know that average value of output voltage,

$$V_{dc} = 0.318\ V_m = 0.318 \times 162.6 = 51.7\ V$$

and the average value of load current,

$$I_{dc} = \frac{V_{dc}}{R_L} = \frac{51.7}{200} = 0.26 \text{ A}$$

∴ Average value of load power,

$$P_{dc} = I^2_{dc} \times R_L = (0.26)^2 \times 200 = 13.4 \text{ W } \textbf{Ans.}$$

Example 8-4. *A half-wave rectifier is used to supply 30 V d.c. to a resistive load of 600 Ω.. The diode has a forward resistance of 25 Ω. Find the maximum value of the a.c. voltage required at the input.*

Solution. Given: $V_{dc} = 30$ volt; $R_L = 600\ \Omega$ and $R_f = 25\ \Omega$

We know that the average value of load current,

$$I_{dc} = \frac{V_{dc}}{R_L} = \frac{30}{600} = 0.05 \text{ A}$$

and maximum value of load current,

$$I_m = \pi \times I_{dc} = \pi\ 0.05 = 0.157 \text{ A}$$

∴ Maximum a.c. voltage required at the input,

$$v_{in\ (max)} = I_m \times (R_f + R_L) = 0.157 \times (25 + 600) \text{ V} = 98.125 \text{ V } \textbf{Ans.}$$

8-6. Single-phase Full-wave Rectifier

A full-wave rectifier is a circuit, which allows a unidirectional current to flow through the load during the entire input cycle as shown in Fig. 8-7. The result of full-wave rectification is a d.c. output voltage that pulsates every half-cycle of the input. On the other hand, a half-wave rectifier allows the current to flow through the load during positive half-cycle only.

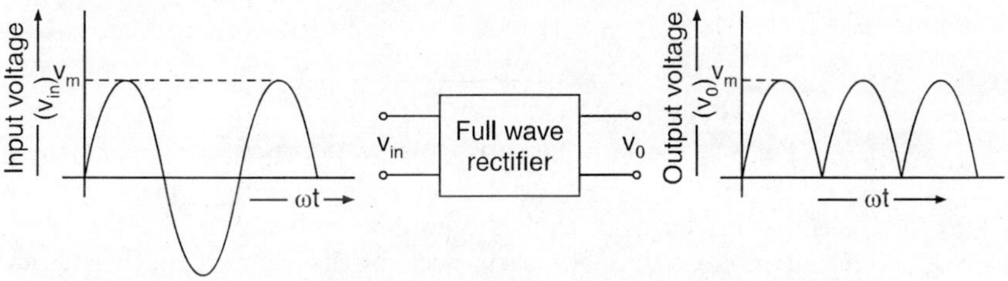

Fig. 8-7. *Full wave rectifier, with input and output voltage waveforms.*

There are two types of full-wave rectifiers namely centre-tapped and bridge rectifier. Now we shall discuss both these types of rectifiers in the following pages.

8-7. Centre-tapped Single-phase Full-wave Rectifier

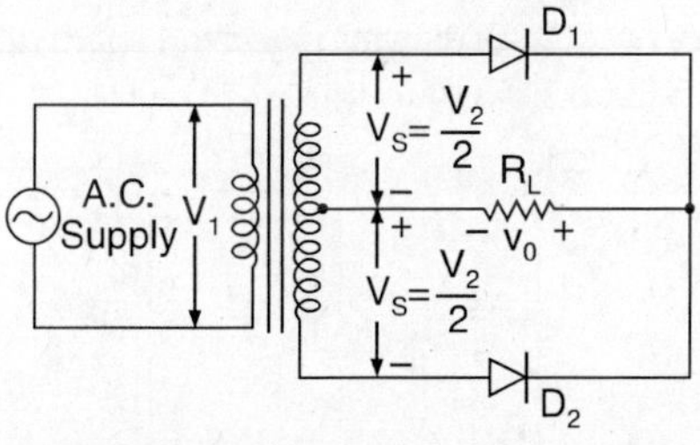

Fig.8-8. *Centre-tapped full-wave rectifier.*

Fig. 8-8 shows the circuit of a centre-tapped full-wave rectifier. The circuit uses two diodes, which are connected to the centre-tapped secondary winding of the transformer. The input signal is applied to the primary winding of the transformer. The centre-tap on the secondary winding of a transformer is, usually, taken as the ground or zero voltage reference point. It may be noted that the voltage between the centre-tap and either end of the secondary winding is half of the secondary voltage, *i.e.*,

$$V_S = \frac{V_2}{2}$$

The operation of a centre-tapped full-wave rectifier circuit may be studied as follows:

During the positive input half-cycle, the polarities of the secondary voltage are as shown in Fig. 8-9(*a*). This forward biases the diode D_1 and reverse-biases the diode D_2. As a result of this, the diode D_1 conducts some current whereas the diode, D_2 is OFF. The current through load R_L is as indicated in the figure.

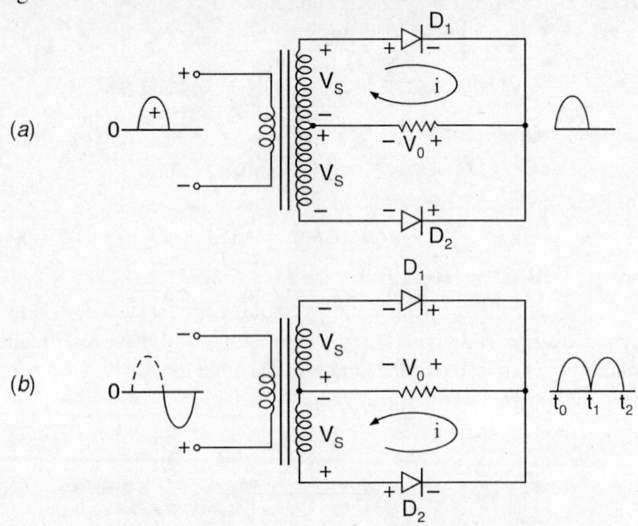

Fig. 8-9. *Operation of centre-tapped full-wave rectifier.*

During the negative input half-cycle, the polarities of the secondary voltage are as shown in Fig. 8-9(*b*). This reverse-biases the diode D_1 and forward-biases the diode, D_2. As a result of this, the diode D_1 is OFF and the diode D_2 conducts some current. The current through the load R_L is as indicated in the figure.

It may be noted that current through the load flows in the same direction, during both the positive and negative portions of the input cycle. Therefore, the output voltage developed across load R_L is full-wave rectified d.c. voltage as shown in Fig. 8-9(*b*).

8-8. Ways of Drawing Single-phase Full-wave Rectifier Circuit

As a matter of fact, there are two ways of drawing the centre-tapped full-wave rectifier circuit as shown in Fig. 8-10(*a*) and (*b*). In Fig. 8-10(*a*), the load resistor (R_L) is connected to the center-tap as indicated,

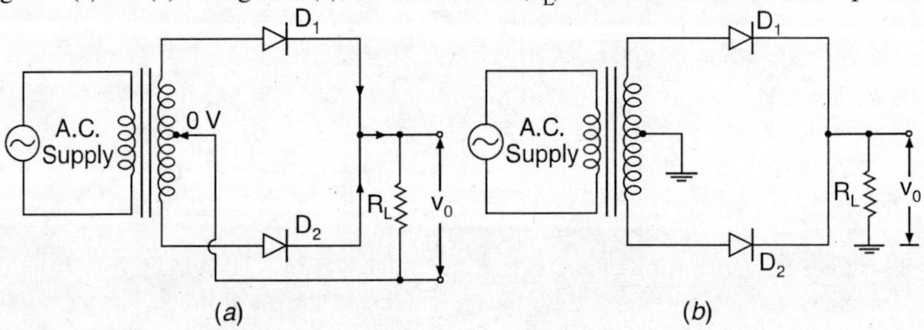

Fig. 8-10. *Ways of drawing full-wave rectifier circuit.*

Fig. 8-10(*b*) shows another way of drawing the full-wave rectifier circuit. In this circuit, the load resistor (R_L) is connected to the centre-tap *via* the ground.

8-9. Effect of Turns Ratio on Single-phase Full-wave Rectifier Output Voltage

We have already discussed in Art. 8-7, that in a centre-tapped full-wave rectifier, the voltage between center tap and either end of secondary winding is half of the secondary voltage. However, the value of secondary voltage depends upon the turns ratio (*i.e.*, N_2/N_1) of the transformer. Thus, if the transformer turns ratio is '1' (*i.e.*, $N_2 = N_1 = 1$), the primary and secondary voltages are equal. Therefore, output voltage is equal to half of the primary voltage (or secondary voltage).

It is possible to obtain an output voltage equal to the input, by using a step-up transformer with a turns ratio of 2 : 1 (*i.e.*, $N_2 = 2$ and $N_1 = 1$). In this case, the total secondary voltage (V_2) is twice the primary voltage. Therefore, the voltage between centre-tap and either end of the secondary winding is V_1 (which is equal to $V_2/2$). Mathematically,

$$V_S = V_1 = \frac{V_2}{2}$$

8-10. Average Values of Output Voltage and Load Current in a Single-phase Full-wave Rectifier

Consider a centre-tapped full-wave rectifier with a sinusoidal a.c. input voltage.

Let V_m = Maximum value of the voltage across each half of the secondary winding.

V_S = The r.m.s. value of the voltage across each half of the secondary winding,

I_m = Maximum value of the load current,

V_{dc} = Average or d.c. value of the output voltage across the load resistor, and

I_{dc} = Average or d.c. value of the current through the load resistor (R_L).

We know that the equation for the voltage across each half of the secondary winding,

$$v_S = V_m \sin \omega t = V_m \sin \theta \qquad \text{... } (\because \theta = \omega t)$$

where, v_S = Instantaneous value of the voltage across each half of the secondary winding.

We also know that the average or d.c. value of the output voltage,

$$V_{dc} = \frac{\text{Area under the curve over a * half-cycle}}{\text{Base}}$$

$$= \frac{\int_0^{\pi} v_S \cdot d\theta}{\pi} = \frac{\int_0^{\pi} V_m \sin \theta \, d\theta}{\pi}$$

$$V_{dc} = \frac{1}{\pi} \int_0^{\pi} V_m \cdot \sin \theta \, d\theta$$

$$V_{dc} = \frac{V_m}{\pi} \left| (-\cos \theta) \right|_0^{\pi} = \frac{V_m}{\pi} [+1 - (-1)]$$

$$= \frac{2\,V_m}{\pi} = 0.636\, V_m.$$

It is evident from the above relation that the average value of a full-wave rectifier is 0.636 V_m and is as shown in Fig. 8-11(*a*). It may be noted that this value is twice that of a half-wave rectifier.

We also know that the average or d.c. value of load current,

$$I_{dc} = \frac{V_{dc}}{R_L} = \frac{2V_m}{\pi \cdot R_L} = \frac{2I_m}{\pi} = 0.636\, I_m$$

---

* Here, we have taken into account only the half cycle (and not full cycle) of the output voltage.

It is evident from the above relation that average value of load current is 0.636 I_m and is as shown in Fig. 8-11(*b*). It may be noted that this value is twice that of a half-wave rectifier.

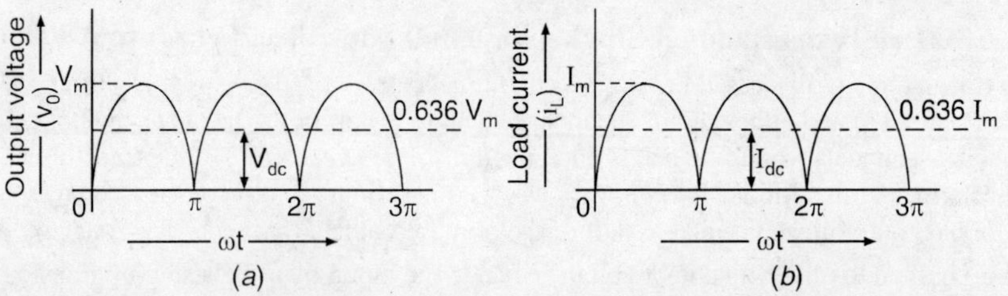

Fig. 8-11. *Average values of output voltage and load current in a full-wave rectifier.*

Note. In deriving the average values of output, voltages and load current, we have assumed that the diodes and the transformer used in the rectifier circuit is an ideal one. However, in actual practice it is not so. Thus, if R_f is the diode forward resistance and R_S is the secondary winding resistance, then average value of output voltage,

$$V_{dc} = \frac{2\,V_m}{\pi} - I_{dc}\,(R_f + R_S)$$

and the average value of load current,

$$I_{dc} = \frac{2\,I_m}{\pi} = \frac{2\,V_m}{\pi\,(R_f + R_S + R_L)}$$

Example 8-5. *Fig. 8-12 shows the circuit of a centre-tapped full-wave rectifier circuit.*

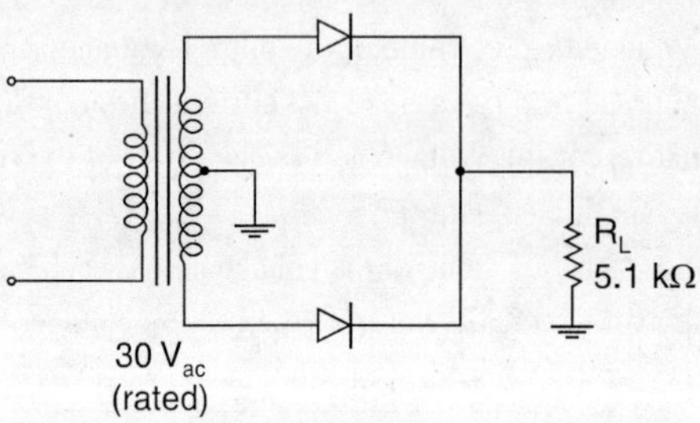

Fig. 8-12.

Determine the d.c. output voltage and the average (or d.c.) load current.

Solution. Given: $V_2 = 30\ V_{ac}$ and $R_L = 5.1\ \text{k}\ \Omega = 5100\ \Omega$.

DC output Voltage

We know that voltage between the centre-tap and either end of secondary winding of a transformer,

$$V_S = \frac{V_2}{2} = \frac{30}{2} = 15\ V_{ac}.$$

and maximum value of this voltage,

$$V_m = \sqrt{2} \times V_S = \sqrt{2} \times 15 = 21.2\ \text{V}$$

∴ The d.c. output voltage,

$$V_{dc} = 0.636\ V_m = 0.636 \times 21.2 = 13.5\ \text{V}\ \textbf{Ans.}$$

Average or dc load current

We also know that the average of d.c. load current,

$$V_{dc} = \frac{V_{dc}}{R_L} = \frac{13.5}{5100} = 2.647 \times 10^{-3}\ \text{A} = 2.647\ \text{mA}\ \textbf{Ans.}$$

8-11. Peak Inverse Voltage of a Diode in Centre-tapped Single-phase Full-wave Rectifier

We know that each diode in a full-wave rectifier, is alternately forward biased and reversed biased. The maximum value of reverse voltage, which each diode must withstand, is equal to the maximum secondary voltage (*i.e.*, $2V_m$) as discussed.

Consider the circuit of a centre-tapped full-wave rectifier as shown in Fig. 8-13. It may be noted that when the secondary voltage (v_S) has the polarity as shown in the figure, the anode of the diode D_1 is $+V_m$ (where, V_m is the maximum half secondary voltage) and the anode of D_2 is $-V_m$. Since, D_1 is forward biased, its cathode is at the same voltage as its anode (neglecting barrier potential), *i.e.*, $+V_m$. This is also the voltage on the cathode of the diode D_2. Therefore, the total reverse voltage across the diode D_2

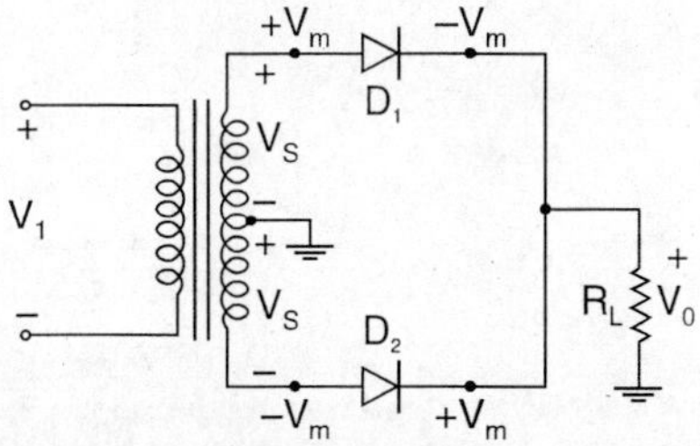

Fig.8-13. *Peak-inverse voltage of a diode in centre-tapped full-wave rectifier.*

$$= V_m - (-V_m) = 2V_m$$

∴ Peak-inverse voltage of each diode in a centre-tapped full-wave rectifier,

$$PIV = 2\,V_m$$

8-12. Advantages and Disadvantages of Centre-tapped Single-phase Full-wave Rectifier

A centre-tapped full-wave rectifier has the following advantages and disadvantages:

Advantages

- The d.c. output voltage and load current values are twice than those of a half-wave rectifier.
- The *ripple factor is much less (0.482) than that of half-wave rectifier (1.21).
- The **efficiency is twice that of half-wave rectifier. For a full-wave rectifier, the maximum possible value of efficiency is 81.2%, while that of half-wave rectifier is 40.6%.

Disadvantages

- The output voltage is half of the secondary voltage.
- The peak-inverse voltage (PIV) of a diode is twice that of the diode used in the half-wave rectifier.
- It is expensive to manufacture a centre-tapped transformer, which produces equal voltages on each half of the secondary winding.

8-13. Single-phase Full-wave Bridge Rectifier

It uses four diodes connected across the main supply, as shown in Fig. 8-14(*a*). The operation of the circuit may be studied as follows :

When the input voltage is positive as shown in Fig. 8-14(*b*), the diodes D_1 and D_2 are forward biased and conduct some current in the direction as indicated in the figure. A voltage is developed across the resistance R_L due to the current flow through it. The voltage looks like the positive half of the input cycle. At this time the diodes D_3 and D_4 are reverse biased.

When the input voltage is negative as shown in Fig. 8-14(*c*), the diodes D_3 and D_4 are forward biased and conduct some current in the same direction through R_L as during the positive half-cycle. During this time, the diodes D_1 and D_2 are reverse biased. As a result of this action, a full-wave rectified output voltage is developed across the resistance R_L.

* Please see Article 8-19.

** Please see Article 8-22.

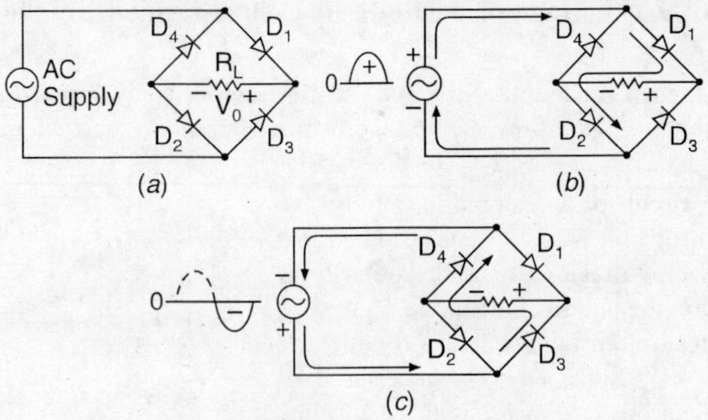

Fig. 8-14. *Bridge rectifier operation.*

The bridge rectifier shown in Fig. 8-14 (*a*) is directly connected across the mains supply. But in actual practice, the mains supply is applied through a transformer as shown in Fig. 8-15.

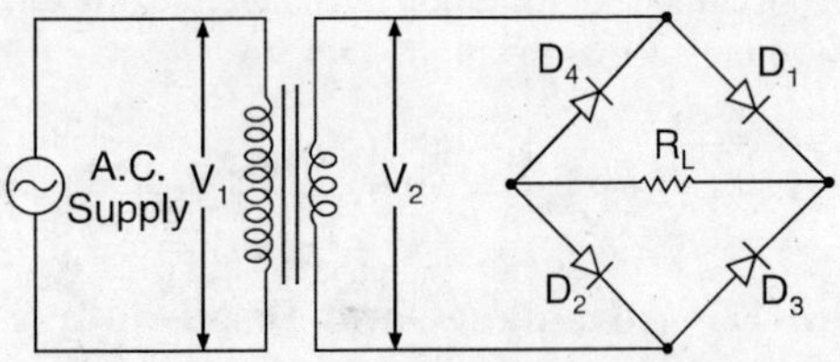

Fig. 8-15. *Bridge rectifier with transformer.*

8-14. Ways of Drawings a Single-phase Bridge Rectifier Circuit

As a matter of fact, there are two ways of drawing a bridge rectifier circuit as shown in Fig. 8-16 (*a*) and (*b*). In Fig. 8-16(*a*), the load resistor (R_L) is connected to the points *B* and *C* of the diode bridge as indicated.

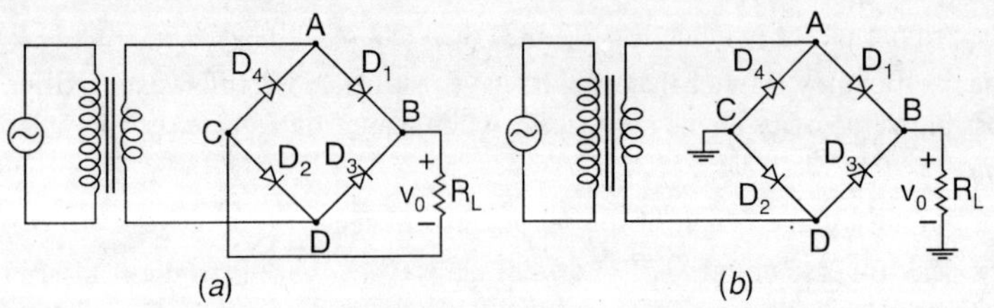

Fig. 8-16. *Ways of drawing bridge rectifier circuit.*

Fig. 8-16(*b*) shows another way of drawing the bridge rectifier circuit. In this circuit, the load resistor (R_L) is connected to the point C of the diode bridge *via* the ground.

8-15. Average Values of Output Voltage and Load Current in a Single-phase Bridge Rectifier

We have already discussed these values in Art. 8-10. The average values of output voltage and load current are the same as for a centre-tapped full wave rectifier. Thus these values in a bridge rectifier may also be obtained in the same way as in Art. 8-10. The average values of output voltage and load current for bridge rectifier are:

$$V_{dc} = \frac{2\,V_m}{\pi} \quad \text{and} \quad I_{dc} = \frac{2\,I_m}{\pi} = \frac{2\,V_m}{\pi \cdot R_L}$$

where, V_m = Maximum value of secondary voltage.

Note: If the values of diode forward resistance (R_F) and the secondary winding resistance (R_S) are also taken into account, then the average value of output voltage.

$$V_{dc} = \frac{2V_m}{\pi} - I_{dc}(R_S + R_F)$$

and the average value of load current,

$$I_{dc} = \frac{2I_m}{\pi} = \frac{2V_m}{\pi(R_S + R_F + R_L)}$$

8-16. Peak Inverse Voltage of a Diode in a Single-phase Bride Rectifier

Consider a bridge rectifier circuit as shown in Fig. 8-17. The polarities of primary and secondary windings correspond to the positive half cycle of the input. Under this condition, the diodes D_1 and D_2 are forward biased. Therefore, these diodes are shown as shorted points in the figure.

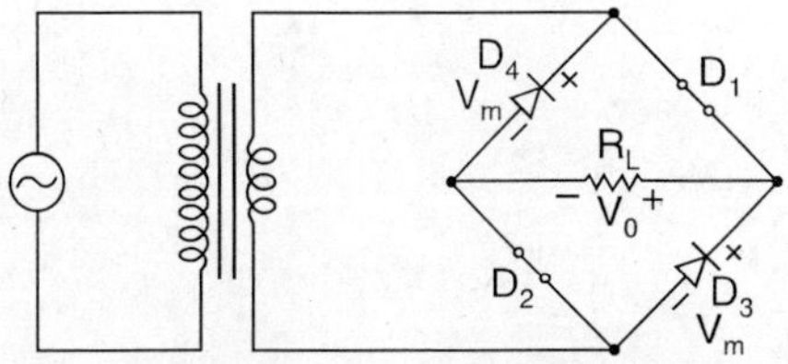

Fig. 8-17. *Peak-inverse voltage of a diode in bridge a rectifier.*

On the other hand, the diodes D_3 and D_4 are reverse biased and have a maximum reverse voltage equal to the maximum secondary voltage (V_m). Thus, peak-inverse voltage of a diode in a bridge rectifier,

$$\textbf{PIV} = V_m$$

8-17. Advantages and Disadvantages of Single-phase Bridge Rectifier

A bridge rectifier has the following advantages and disadvantages:

Advantages

- It can be used, with advantage, in applications allowing floating output terminals, *i.e.*, no output terminal is grounded.
- The transformer is less costly as it is required to provide only half the voltage of an equivalent centre-tapped transformer used in a full wave rectifier circuit.
- No centre-tap is required on the transformer.

Disadvantages

It has only one disadvantage that it uses four diodes as compared to two diodes for centre-tapped full-wave rectifier. But ready availability of low cost silicon diodes have made it more economical despite its requirement of four diodes. Besides this, PIV rating required of the diodes, in a bridge rectifier, is only half of that for a centre-tapped full-wave rectifier. This is a great advantage, which offsets the disadvantage of using four diode instead of two in a full-wave rectifier circuit.

Note. These days, the bridge rectifiers are so common that manufacturers are packing them as a single unit with bakelite or some other plastic encapsulation with external connections brought out.

8-18. Frequency of Output in a Single-phase Half-wave and Full-wave Rectifiers

Fig 8-18(*a*) shows the waveform of an a.c. input voltage while Fig. 8-18(*b*) and (*c*) shows the waveforms of a half-wave and full-wave rectifiers respectively.

It may be noted that frequency of half-wave rectifier output is the same as of its input, *i.e.*, $f_{out} = f_{in}$.

However, the frequency of full-wave rectifier is twice that of input, *i.e.*, $f_{out} = 2f_{in}$.

Thus in power supply applications, the a.c. power line voltage is of 50 Hz frequency. Therefore, d.c. output voltage of a half-wave rectifier is 50 Hz. But d.c. output voltage of a full-wave rectifier is 100 Hz.

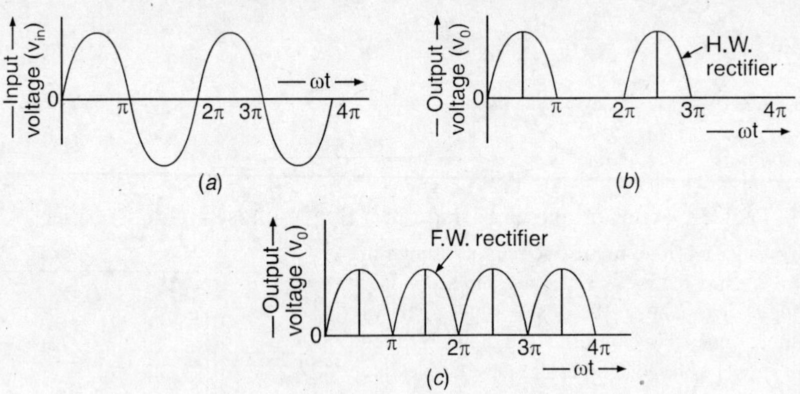

Fig. 8-18. *Frequency of output in half-wave and full-wave rectifiers.*

Example 8-6. *Fig. 8-19 shows a bridge rectifier circuit.*

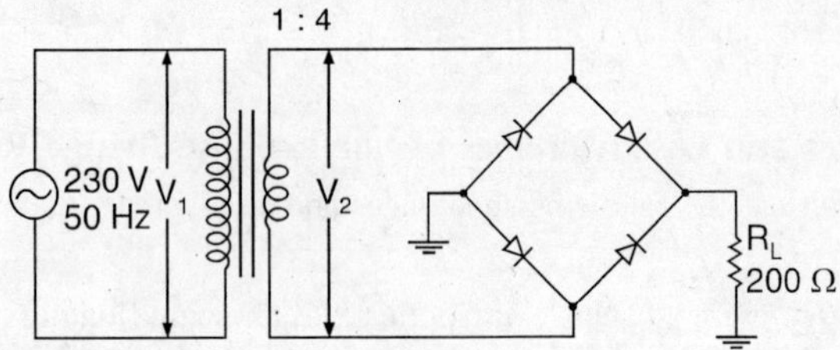

Fig. 8-19.

Assuming the diodes to be ideal, determine the quantities ***(a)*** *d.c. output voltage,* ***(b)*** *peak-inverse voltage of a diode, and* ***(c)*** *output frequency.*

Solution. Given: $V_1 = 230$ volt; $N_2/N_1 = 1/4 = 0.25$; $R_L = 200\ \Omega$ and $f_{in} = 50$ Hz.

(***a***) *D.C. output voltage*

We know that r.m.s. value of secondary voltage,

$$V_2 = V_1 \times \frac{N_1}{N_2} = 230 \times 0.25 = 57.5 \text{ V}$$

and maximum value of the secondary voltage,

$$V_m = \sqrt{2} \times V_2 = \sqrt{2} \times 57.6 = 81.3$$

∴ D.C. output voltage,

$$V_{dc} = 0.636\ V_m = 0.636 \times 81.3 = 51.7 \text{ V } \textbf{Ans.}$$

(***b***) *Peak-inverse voltage of a diode*

We know that peak-inverse voltage of a diode,

$$PIV = V_m = 81.3 \text{ V } \textbf{Ans.}$$

(***c***) *Output frequency*

We also know that output frequency,

$$f_{out} = 2 f_{in} = 2 \times 50 = 100 \text{ Hz } \textbf{Ans.}$$

8-19. Ripple Factor

The output voltage (or load current) of a rectifier consists of two components namely d.c. component and a.c. component. The a.c. component present in the output is called a ***ripple***. As a matter of fact, the ripple is undesirable and accounts for pulsations in the rectifier output. The effectiveness of a rectifier depends upon the magnitude of ripple in the output. Smaller the ripple, more effective will be the rectifier. Mathematically, the ripple factor,

$$\gamma = \frac{\text{The r.m.s value of a.c. component of output voltage}}{\text{The d.c. component of output voltage}}$$

$$= \frac{V_{r(rms)}}{V_{dc}} = \frac{I_{r(rms)}}{I_{dc}}$$

where, $V_{r(rms)}$ = The r.m.s. value of the a.c. component of the output voltage.

V_{dc} = The average or d.c. value of the output voltage.

$I_{r(rms)}$ = The r.m.s. value of the a.c. component of current, and

I_{dc} = The average or d.c. value of the load current.

We know that the r.m.s. value of the rectified load current,

$$I_{rms} = \sqrt{I_{dc}^2 + I_{r(rms)^2}}$$

Dividing the above equation on both sides by I_{dc}.

$$\frac{I_{rms}}{I_{dc}} = \frac{\sqrt{I_{dc}^2 + I_{r(rms)^2}}}{I_{dc}} = \sqrt{1 + \left(\frac{I_{r(rms)}}{I_{dc}}\right)^2}$$

Squaring and rearranging the above expression,

$$\frac{I_{r(rms)}}{I_{dc}} = \sqrt{\frac{I_{rms}}{I_{dc}}^2 - 1}$$

or
$$\gamma = \sqrt{\frac{I_{rms}}{I_{dc}}^2 - 1}$$

Now, we shall use the above relation to find out the ripple factor for half-wave and full-wave rectifier circuits.

8-20. Ripple Factor of a Single-phase Half-wave Rectifier

We have already discussed in Art. 8-4 that the average value of load current in a half-wave rectifier,

$$I_{dc} = \frac{I_m}{\pi} \qquad \text{... } (i)$$

where, I_m is the maximum value of load current. The r.m.s. value of the load current for a half-wave rectifier may be found by using *calculus and its value,

$$I_{rms} = \frac{I_m}{2} \qquad \text{... } (ii)$$

---

* We know that expression for the load current,

$$iL = I_m \cdot \sin\theta$$

$$\therefore \quad I_{rms} = \sqrt{\frac{1}{2\pi}\int_0^{2\pi} i_L^2 \cdot d\theta} = \sqrt{\frac{1}{2\pi}\left[\int_0^{\pi} I_m^2 \sin^2\theta \cdot d\theta + \int_{\pi}^{2\pi}(0 \cdot d\theta)\right]}$$

$$\sqrt{\frac{I_m^2}{2\pi}\int_0^{\pi}\left(\frac{1-\cos 2\theta}{2}\right)d\theta} = \sqrt{\frac{I_m^2}{4\pi}\left(\theta - \frac{\sin 2\theta}{2}\right)_0^{\pi}} = \frac{I_m}{2}$$

Substituting these values of I_{dc} and I_{rms} in the expression for ripple factor (Art. 8-9),

$$\gamma = \sqrt{\left(\frac{I_{rms}}{I_{dc}}\right)^2 - 1} = \sqrt{\left(\frac{I_m / 2}{I_m / \pi}\right)^2 - 1} = 1.21$$

If we express the ripple factor in terms of a percentage, its value is 121%. This indicates that the amount of a.c. component present in the output of a half-wave rectifier is 121% of d.c. output voltage. Hence, the half-wave rectifier is not very successful in converting the current from a.c. to d.c.

8-21. Ripple Factor of a Single-phase Full-wave Rectifier

We have already discussed in Art. 8-10 that the average value of load current in a full-wave rectifier (either centre-tapped or bridge rectifier) is given by the relation, $I_{dc} = 2\, I_m/\pi$ ... (*i*)

The r.m.s. value of the load current may be found by using the *calculus and its value is found to be,

$$I_{rms} = \frac{I_m}{\sqrt{2}} \qquad \text{... (ii)}$$

Substituting these values of I_{dc} and I_{rms} in the expression for ripple factor (Art. 8-9),

$$\gamma = \sqrt{\left(\frac{I_{rms}}{I_{dc}}\right)^2 - 1} = \sqrt{\left(\frac{I_m / \sqrt{2}}{2I_m / \pi}\right) - 1} = 0.482$$

It is evident from the above result that ripple factor of a full-wave rectifier is 0.482 and is much smaller than that of a half-wave rectifier. Because of this reason, full-wave rectifier is used more commonly in actual practice.

8-22. Efficiency of a Single-phase Rectifier

It may be defined as the ratio of d.c. power delivered to the load to the a.c. input power from the secondary winding of the transformer. Mathematically, the rectifier efficiency,

$$\eta = \frac{\text{D.C. power delivered to the load}}{\text{A.C. input power from the transformer secondary}}$$

$$= \frac{P_{dc}}{P_{ac}} \qquad \text{... (i)}$$

We know that d.c. power delivered to the load,

$$P_{dc} = I_{dc}^{\,2} \times R_L$$

and a.c. input power from the transformer secondary,

$$P_{ac} = I_{rms}^{\,2}\,(R_f + R_L)$$

where, R_L = Load resistance, and

R_f = Forward resistance of a diode.

Substituting the values of P_{dc} and P_{ac} in equation (*i*),

$$\eta = \frac{I_{dc}^2 \times R_L}{I_{rms}^2\,(R_f + R_L)}$$

---

* We know that expression for load current is given by the relation,

$$i_L = I_m \cdot \sin\theta$$

$$\therefore \quad I_{rms} = \sqrt{\frac{1}{2\pi}\int_0^{\pi} i_L^2 \cdot d\theta} = \sqrt{\frac{1}{\pi} I_m^2 \sin^2\theta \cdot d\theta}$$

$$= \sqrt{\frac{I_m^2}{\pi}\int_0^{\pi}\left(\frac{1-\cos 2\theta}{2}\right) d\theta} = \sqrt{\frac{I_m^2}{2\,\pi}\left(\theta - \frac{\sin 2\theta}{2}\right)_0^{\pi}} = \sqrt{\frac{I_m^2}{\pi} \times \frac{\pi}{2}} = \frac{I_m}{\sqrt{2}}$$

8-23. Efficiency of a Single-phase Half-wave Rectifier

We have already discussed in the last article that efficiency of a rectifier is given by the expression,

$$\eta = \frac{I_{dc}^2 \times R_L}{I_{rms}^2 (R_f + R_L)} \qquad ...(i)$$

We also know that for a half-wave rectifier,

$$I_{dc} = \frac{I_m}{\pi} \text{ and } I_{rms} = \frac{I_m}{2}$$

Substituting the values of I_{dc} and I_{rms} in equation (*i*),

$$\eta = \frac{(I_m/\pi)^2 \times R_L}{(I_m/2)^2 \times (R_f + R_L)} = \frac{4}{\pi^2} \times \frac{R_L}{R_f + R_L} = \frac{0.406}{1 + \frac{R_f}{R_L}}$$

Now efficiency will be maximum, if $R_L >> R_f$. Thus

$$\eta_{max} = 0.406 \text{ or } 40.6\%.$$

It shows that efficiency of a half-wave rectifier is 40.6% under the condition that the value of load resistance is very large as compared to the forward resistance of a diode (*i.e.*, $R_L >> R_f$). However, in actual practice, the efficiency is always less then 40.6%.

8-24. Efficiency of a Single-phase Full-wave Rectifier

We have already discussed in Art. 8-22 that rectifier efficiency is given by the relation,

$$\eta = \frac{I_{dc}^2 \times R_L}{I_{rms}^2 (R_f + R_L)} \qquad ...(i)$$

We know that for a full-wave rectifier,

$$I_{dc} = 2\frac{I_m}{\pi} \text{ and } I_{rms} = \frac{I_m}{\sqrt{2}}$$

Substituting the values of I_{dc} and I_{rms} in equation (*i*),

$$\eta = \frac{\left(\frac{2I_m}{\pi}\right)^2 \times R_L}{\left(\frac{I_m}{\sqrt{2}}\right)^2 \times (R_f + R_L)} = \frac{8}{\pi^2} \times \frac{R_L}{R_f + R_L} = \frac{0.812}{1 + \frac{R_f}{R_L}}$$

Now, efficiency will be maximum if $R_L >> R_f$. Thus, $\eta_{max} = 0.812$ or 81.2%

It is evident from the above result, that maximum efficiency of a full-wave rectifier is twice that of half-wave rectifier. It means that a full-wave rectifier is twice as effective as a half-wave rectifier.

Example 8-7. *Fig. 8-20 shows a centre-tapped full-wave rectifier circuit:*

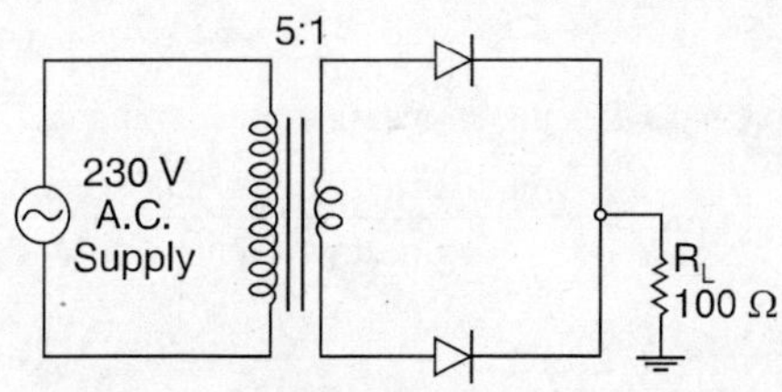

Fig. 8-20.

*Assuming both the diodes to be ideal, determine **(a)** d.c. output voltage; **(b)** peak-inverse voltage, and **(c)** rectification efficiency.*

Solution. Given: $V_1 = 230$ volts; $N_2/N_1 = 1/5 = 0.2$ and $R_L = 100\ \Omega$.

(*a*) *D.C. output voltage*

We know that the r.m.s. value of secondary voltage,

$$V_2 = V_1 \times \frac{N_2}{N_1} = 230 \times 0.2 = 46 \text{ V}$$

and half of the secondary voltage,

$$V_S = \frac{V_2}{2} = \frac{46}{2} = 23 \text{ V}$$

∴ Maximum value of half of the secondary voltage,

$$V_m = \sqrt{2} \times V_S = \sqrt{2} \times 23 = 32.5 \text{ V}$$

and d.c. output voltage,

$$V_{dc} = \frac{2\,V_m}{\pi} = \frac{2 \times 32.5}{\pi} = 20.7 \text{ V } \textbf{Ans.}$$

(*b*) *Peak-inverse voltage of a diode*

We know that peak-inverse voltage of a diode,

$$\text{PIV} = 2\,V_m = 2 \times 32.5 = 65 \text{ V } \textbf{Ans.}$$

(*c*) *Rectification efficiency*

We also know that rectification efficiency of a full-wave rectifier,

$$\eta = 0.812 \text{ or } 81.2\% \textbf{ Ans.}$$

Example 8-8. *A full-wave rectifier, using a centre-tap transformer of 200-0-200 V, has two rectifying diodes. Each one of them is rated at I_{max} = 700 mA and I_{avg} 250 mA. Assuming ideal rectifying diodes, determine:*

(***a***) *The value of load resistor that can be connected to give largest d.c. power output without exceeding any diode rating.*

(***b***) *The d.c. load voltage and current for the condition of part (a) above.*

(***c***) *The peak-inverse voltage of each diode that can be safely used in this rectifier.*

Solution. Given: $V_S = 200$ volts; $I_{max} = 700$ mA and $I_{avg} = 250$ mA.

(*a*) *Value of load resistor*

We know that maximum value of current through the diode for normal operation should not exceed 80% of its current rating.

$$\therefore \quad I_{max} = 0.8 \times 700 = 560 \text{ mA} = 560 \times 10^{-3}$$

We also know that the maximum value of the half secondary voltage,

$$V_m = \sqrt{2} \times R.M.S. \text{ value of half-secondary voltage}$$

$$= \sqrt{2} \times V_S = \sqrt{2} \times 200 = 282.8 \text{ V}$$

∴ Value of load resistor that gives the largest d.c. power output,

$$R_L = \frac{\text{Maximum value of half-secondary voltage}}{\text{Maximum current through the diode}}$$

$$= \frac{V_m}{I_{max}} = \frac{282.8}{560 \times 10^{-3}} = 505\ \Omega \textbf{ Ans.}$$

(*b*) *D.C. load voltage and current*

We know that d.c. load voltage (or output voltage),

$$V_{dc} = \frac{2V_m}{\pi} = \frac{2 \times 282.8}{\pi} = 180 \text{ V} \quad \textbf{Ans.}$$

and the d.c. load current,

$$I_{dc} = \frac{V_{dc}}{R_L} = \frac{180}{505} = 0.36 \text{ A} \quad \textbf{Ans.}$$

(*c*) *Peak-inverse voltage of a diode*

We also know that peak-inverse voltage across each diode,

$$\text{PIV} = 2\,V_m = 2 \times 282.8 = 565.6 \quad \textbf{Ans.}$$

8-25. Transformer Utilization Factor

It may be defined as ***the ratio of d.c. power delivered to the load and the a.c. rating of the transformer secondary***. Mathematically, transformer utilization factor,

$$\text{T.U.F.} = \frac{\text{D.C. power delivered to the load}}{\text{A.C. rating of the transformer secondary}} = \frac{P_{dc}}{P_{ac(rated)}}$$

For a half-wave rectifier, the value of transformer utilization factor is found to be equal to 0.287. For a centre-tapped full-wave rectifier, its value is equal to 0.693 and for a bridge rectifier, its value is equal to 0.812. The transformer utilization factor is very useful in determining the rating of a transformer to be used with rectifier circuit.

8-26. Comparison of Single-phase Rectifiers

The characteristics of the three rectifiers, which we have discussed, in the previous articles are summarized in table 8-1.

Table 8-1. *Comparison of rectifiers*

S. No.	*Item*	*Half-wave*	*Full-wave*	
			Centre-tapped	*Bridge*
1.	Number of diodes	1	2	4
2.	Peak-inverse voltage of diode	V_m	$2\,V_m$	V_m
3.	D.C. Output voltage	$0.318\,V_m$	$0.636\,V_m$	$0.636\,V_m$
4.	Ripple factor	1.21	0.482	0.482
5.	Ripple frequency	$2f_{in}$	$2f_{in}$	$2f_{in}$
6.	Transformer utilization factor	0.287	0.693	0.812

8-27. Filters

We have already discussed in Art. 8-19 that the output of a rectifier contains d.c. component as well as an a.c. component. The presence of an a.c. component is most undesirable and therefore must be removed from the rectifier output. It is achieved by a circuit called **filter**.

Fig. 8-21 shows the concept of a filter, where the full-wave rectified output voltage is applied at its input. The output of a filter is not exactly a constant d.c. level. But it contains a small amount of a.c. component as well. Thus, a filter circuit may be defined as a *circuit which removes (or minimizes) the unwanted a.c. component of the rectifier output and allows only the d.c. component to reach the load.*

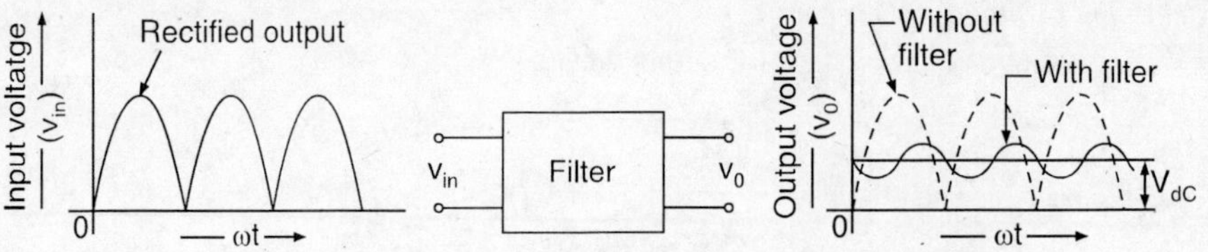

Fig. 8-21. *Filter.*

A filter circuit consists of passive circuit elements, *i.e.*, inductors, capacitors, resistors and their combination. The filtering action depends upon the basic electrical properties of passive circuit elements. For example, an inductor allows the direct current to pass through it. But it blocks the alternating current. On the other hand, a capacitor allows the alternating current to pass through it. But it blocks the direct current. Some of the important filters used in the field of electronics are as given below:

(*i*) Inductor filter (*ii*) Capacitor filter

(*iii*) Inductor-capacitor or LC filter (*iv*) π-filter

8-28. Inductor Filter

It is also called choke filter. It consists of an inductor (L), which is suitably inserted between the rectifier and the load resistance (R_L) as shown in Fig. 8-22. the rectifier output contains a.c. components as well as d.c. components. When the output passes through an inductor, it offers a high resistance to the a.c. components and no resistance to d.c. components. Therefore, a.c. component of the rectified output is blocked and only d.c. component reaches at the load. Theoretically, the output of inductor filter should consist of only d.c. voltage. But in actual practice, it contains a small a.c. components as well.

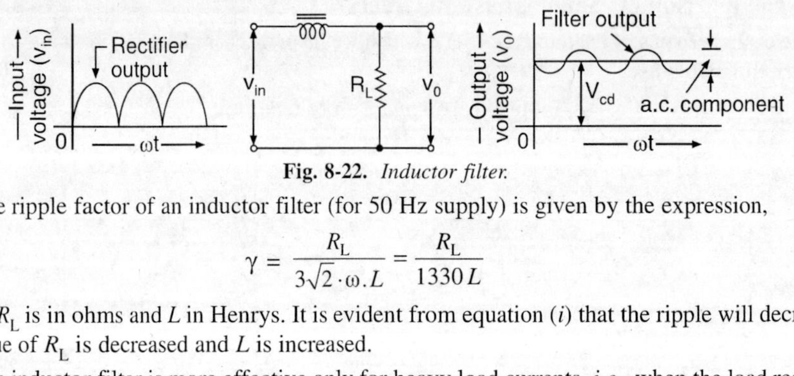

Fig. 8-22. *Inductor filter.*

The ripple factor of an inductor filter (for 50 Hz supply) is given by the expression,

$$\gamma = \frac{R_L}{3\sqrt{2}\cdot\omega.L} = \frac{R_L}{1330\,L} \qquad \text{... }(i)$$

where, R_L is in ohms and L in Henrys. It is evident from equation (*i*) that the ripple will decrease as the value of R_L is decreased and L is increased.

The inductor filter is more effective only for heavy load currents, *i.e.*, when the load resistance (R_L) is small. The a.c. component, present at the output of a filter, can be reduced if we use an inductor of a large value. But it may be noted that an inductor of a large value will also have a higher d.c. resistance, which will result in a lower d.c. output voltage.

Example 8-9. (***a***). *What should be the value of inductance to use in an inductor filter connected to a full-wave rectifier operating at 50 Hz, if the ripple is not to exceed 5% for a 100 Ω load.*

(***b***) *Repeat the above problem for the standard aircraft power frequency of 400 Hz.*

Solution. (***a***) Given: f = 50 Hz; γ = 5% = 0.05 and R_L = 100 Ω.

We know that ripple factor (γ),

$$0.05 = \frac{R_L}{3\sqrt{2}\cdot\omega.L} = \frac{100}{3\sqrt{2}\times(2\pi\times 50)\times L} = \frac{0.075}{L}$$

∴ $L = 0.075/0.05 = 1.5$ H **Ans.**

(*b*) Given: $f = 400$ Hz; $\gamma = 0.05$ and $R_L = 100\ \Omega$.

We also know that ripple factor (γ),

$$0.05 = \frac{R_L}{3\sqrt{2}\cdot\omega\cdot L} = \frac{100}{3\sqrt{2}\times(2\pi\times 400)\times L} = \frac{1}{106.6L}$$

∴ $L = 1/(106.6 \times 0.05) = 0.188$ H **Ans.**

8-29. Capacitor Filter

We have already discussed in the last article that an inductor filter is suitable only for heavy loads (*i.e.*, a load with smaller values of resistance). An inexpensive filter for light loads (*i.e.*, a load with larger value of resistance) is available in the form of a capacitor filter. The operation of a capacitor filter depends upon the property of a capacitor to oppose any change in voltage, when it is connected across a pulsating d.c. voltage. The action of a capacitor filter is to smoothed out (or filter out) the voltage ripples or pulsations.

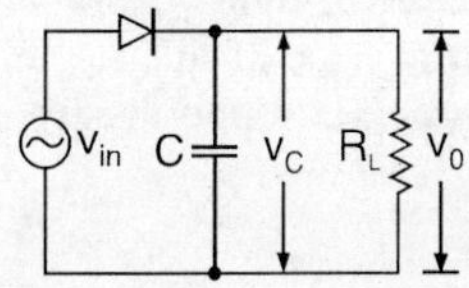

Fig. 8-23. *Half-wave rectifier with capacitor filter*

Fig. 8-23. shows a half-wave rectifier with a capacitor filter. The operation of this circuit may be explained in terms of the waveforms as shown in Fig. 8-24(*a*), (*b*) and (*c*) respectively. Fig. 8-24(*a*) shows the waveform of the a.c. input voltage (indicated by the broken curve) and the half-wave rectifier output. Fig. 8-24(*b*) shows the waveform of a voltage across a capacitor and Fig. 8-24(*c*) shows the waveform of an output voltage. These waveforms may be better understood from the following discussion.

During the positive half-cycle of the a.c. input voltage, the diode is forward biased. It allows the capacitor to charge quickly to its maximum value of voltage (V_m) as indicated by the curve *oa* in Fig. 8-24(*b*). Since, there is no resistance in the charging path, except diode forward resistance (which is negligible), the charging time is almost zero. Therefore, the capacitor follows the charging voltage. When the a.c. input voltage begins to decrease below its maximum value, the capacitor returns its charge. We know that during the negative half-cycle of the input, the capacitor tries to discharge through the load resistance as shown by the curve *ab* in Fig. 8-24(*b*). It may be noted that the capacitor cannot discharge through the diode, because it is reverse biased. The discharging time constant may be determined by the product of the capacitance value of the capacitor (C) and the value of load resistance (R_L). Mathematically, time constant,

$$\tau = C \cdot R_L$$

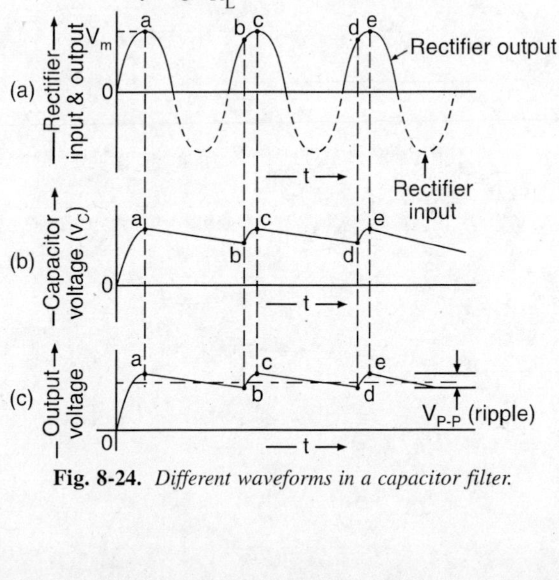

Fig. 8-24. *Different waveforms in a capacitor filter.*

It will be interesting to know that larger the time-constant, lesser will be the capacitor discharge. In actual practice, the discharging time constant is kept about 100 times larger than that of the charging time. Therefore, the capacitor does not have enough time to discharge appreciably. As a result of this, the capacitor maintains a large voltage across the load resistance (R_L).

We also know that during the next half cycle, when the a.c. input voltage exceeds the capacitor voltage, as indicated by point '*b*' in Fig. 8-24(*a*), the diode is again forward biased and the capacitor charges to its maximum value of the voltage (*i.e.*, V_m). It is indicated by the curve *dc* in Fig. 8-24(*b*). During the negative half-cycle, the diode is reverse biased and the capacitor tries to discharge through the load resistance. The process continues for the other cycles and we get an output voltages as shown in Fig. 8-24(*c*).

It may be noted from Fig. 8-24(*c*) that due to charging and discharging of a capacitor, there is a ripple (*i.e.*, a.c. component) present in the output voltage of a capacitor filter. As a matter of fact, smaller the value of this ripple, better will be the filtering action. The ripple factor of a capacitor filter (for 50 Hz supply) is given by the relation,

$$\gamma = \frac{1}{4\sqrt{3}\cdot f\cdot C\cdot R_L} = \frac{2890}{C\cdot R_L}$$

where, C is in μF and R_L in ohms.

Example 8-10. *A full-wave rectifier, using a capacitor filter, has to supply 30 V d.c. to a load resistance of 1 kΩ.. Assuming the diode and transformer winding resistance to be negligible, estimate the value of capacitor filter for a ripple factor of 0.01.*

Solution. Given: V_{dc} = 30 volts; $R_L = 1\text{ k}\Omega = 1\times10^3\ \Omega$ and $\gamma = 0.01$.

Let C = Value of capacitor in μF

We know that ripple factor for a 50 Hz supply (γ)

$$0.01 = \frac{2890}{C\cdot R_L} = \frac{2890}{C\times(1\times10^3)} = \frac{2.89}{C}$$

∴ $C = 2.89/0.01 = 289\ \mu\text{F}$ **Ans.**

Example 8-11. *Find the ripple factor for a FWR with capacitor filter with the output waveform as shown in the Fig. 16 – 25 . Assume R_L = 100 Ω with capacitor C = 1000 μF.*

(*Electronic Circuits - 1, Anna University, April/May 2008*)

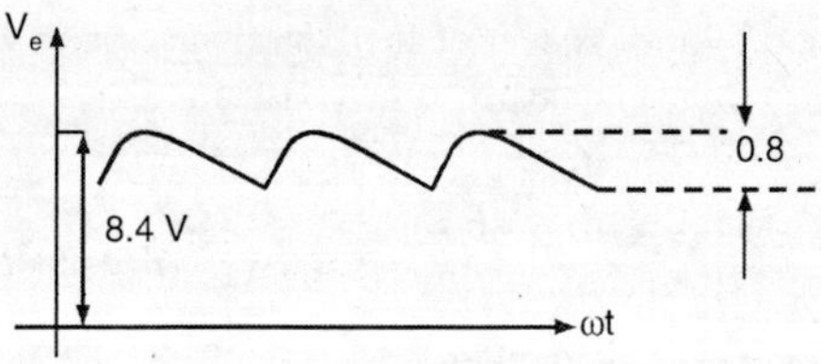

Fig. 8-25

Solution . Given: R_L = 100 Ω, C = 1000 μF, V_r (peak-to-peak) = 0.8 V, E_m = 8.4 V

Assuming this is to be triangular,

We know that,

$$V_{r\,(r.m.s)} = \frac{V_{r\,\text{(peak-to-peak)}}}{2\sqrt{3}} \quad \frac{-0.8}{2\sqrt{3}} = 0.231\text{ V}$$

$$V_{r\,\text{(peak-to-peak)}} = \frac{E_x}{z\,f\,c\,R_L} \qquad ...(i)$$

$$I_{DC} = \frac{E_x}{R_L} \qquad ...(ii)$$

$$V_{r\,(\text{peak-to-peak})} = \frac{I_{DC}}{z f C}$$

$$0.8 = \frac{I_{DC}}{2 \times 50 \times 1000 \times 10^{-6}}$$

$$\therefore \quad I_{DC} = 0.08 \text{ A}$$

and $$E_{DC} = E_m - \frac{I_{DC}}{4 f C} = 8V$$

We also know that the value of ripple factor,

$$\text{Ripple factor,} = \frac{V_{r(r.m.s)}}{E_{DC}} = \frac{0.2309}{8}$$

$$= 0.02886 = 2.886\% \textbf{ Ans.}$$

8-30. LC Filter

We have already discussed that in an inductor filter, the ripple factor is directly proportional to the load resistance. On the other hand in a capacitor filter, it is varying inversely with the load resistance. Hence, if we combine the inductor (*L*) filter and the capacitor (*C*), the ripple factor will become almost independent of the load resistance. The combination of *L* and *C* filter is called *LC* filter and is as shown in Fig. 8-25. The *LC* filter is also known as inductor input filter, choke input filter, *L* input filter or *LC*-section.

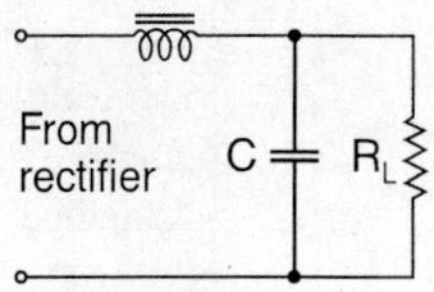

Fig. 8-26. *LC filter.*

The ripple factor for 50 Hz supply of a *LC* filter is given by the relation,

$$\gamma = \frac{\sqrt{2}}{12\,\omega^2 \cdot LC} = \frac{1.195}{LC}$$

where, *L* is in Henrys and *C* in μF.

Example 8-12. *A full-wave rectifier circuit, with LC filter supplies of 12 V d.c. at 100 mA with a maximum ripple constant of 1%. If the value of L = 1 H, estimate the size of the capacitor.*

Solution. Given: $V_{dc} = 12$ volts; $I_{dc} = 100$ mA; $\gamma = 1\% = 0.01$ and $L = 1$ H

Let C = Size or value of the capacitor in μF.

We know that ripple factor (γ)

$$0.01 = \frac{1.195}{LC} = \frac{1.195}{1 \times C} = \frac{1.195}{LC}$$

$$\therefore \quad C = 1.195/0.01 = 119\text{-}5 \ \mu\text{F} \textbf{ Ans.}$$

8-31. Bleeder Resistor

It is a common practice to use a resistor (called bleeder resistor) across the filter output as shown in Fig. 8-26. Such a resistor provides the following three advantages :

- It maintains the minimum current necessary for optimum operation of the inductor.
- It improves voltage regulation of the supply by acting as preload on the supply.
- It provides safety to the persons handling the equipment, by acting as a discharging path for capacitors.

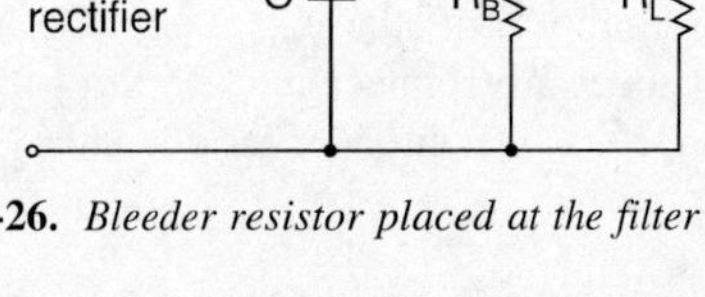

Fig. 8-26. *Bleeder resistor placed at the filter output.*

8-32. π-Filter

It is also called capacitor input filter, or *CLC* filter. The π-filter is used wherever a low output current and a high d.c. output voltage is required. It consists of two capacitors C_1 and C_2 and an inductor (L) connected in the form of a greek letter π (pie) as shown in Fig. 8-27. The pulsating output from the rectifier is applied at the input terminals of the π-filter. The filtering action of each component may be studied as follows:

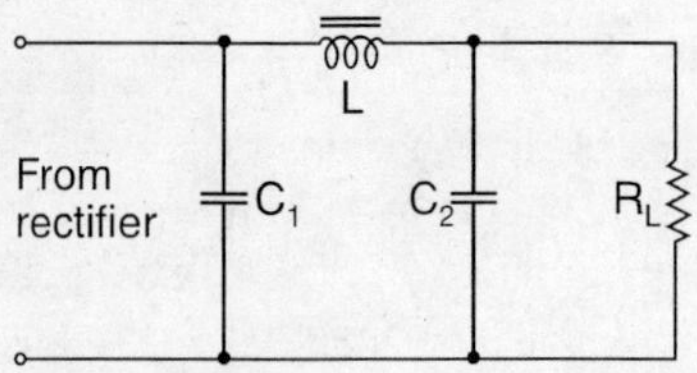

Fig. 8-27. *π-filter*

(*i*) ***Capacitor C_1* :** It offers a low reactance to a.c. component of rectifier output. But it offers infinite resistance to the d.c. component. Therefore, the capacitor C_1 bypasses an appreciable amount of a.c. component to the ground, while d.c. component moves towards *L*.

(*ii*) ***Inductor L* :** It offers a high reactance to the a.c. component of rectifier output but zero resistance to the d.c. component. Therefore, it allows the d.c. component to pass through it, and blocks the a.c. component, which could not be bypassed by the capacitor C_1.

(*iii*) ***Capacitor C_2* :** Its behaviour is similar to the capacitor C_1. It bypasses the a.c. component of rectifier output, which could not be blocked by an inductor *L*. As a result, only the d.c. component is available at the output.

It has been found that the ripple factor in a π-filter is dependent only on the product of the capacitance values of the capacitors C_1 and C_2 (and not on the individual values of capacitors C_1 and C_2). But the d.c. output voltage will be high only if the value of capacitor C_1 is large. Hence, for a given ripple factor, the -filter is designed with a large value of capacitor C_1. The ripple factor (for 50 Hz supply) of a -filter is given by the relation,

$$\gamma = \frac{1}{4\sqrt{2}\cdot\omega^3\cdot C_1\cdot C_2 L\cdot R_L} = \frac{5700}{C_1\cdot C_2\cdot L\cdot R_L}$$

where, C_1 and C_2 are in μF, *L* in Henrys and R_L in ohms.

Example 8-12. *A full-wave rectifier supplies 0.2 A current at 30 V d.c. Find the ripple factor to be expected when two 100 μF, capacitors and a 5 H inductor are used in a π-filter with a 50 Hz supply.*

Solution. Given: I_{dc} = 0.2 A; V_{dc} = 30 volt; $C_1 = C_2$ = 100 μF; *L* = 5 H and *f* = 50 Hz

We know that the load resistance,

$$R_L = \frac{V_{dc}}{I_{dc}} = \frac{30}{0.2} = 150\ \Omega$$

and the ripple factor for 50 Hz supply,

$$\gamma = \frac{5700}{C_1\cdot C_2\cdot R_L} = \frac{5700}{100\times 100\times 5\times 150}$$

$$= 7.6\times 10^{-4} = 0.076\%\ \textbf{Ans.}$$

8-33. Three-phase Half-wave Rectifier

Rectification of a three-phase supply with the help of diodes is shown in Fig. 8-28(*a*) along with a smoothing circuit. The three diodes are connected to the three phases of star-connected secondary of a three-phase transformer. Neutral point *N* of the secondary is the negative terminal for the rectified output and is earthed as shown.

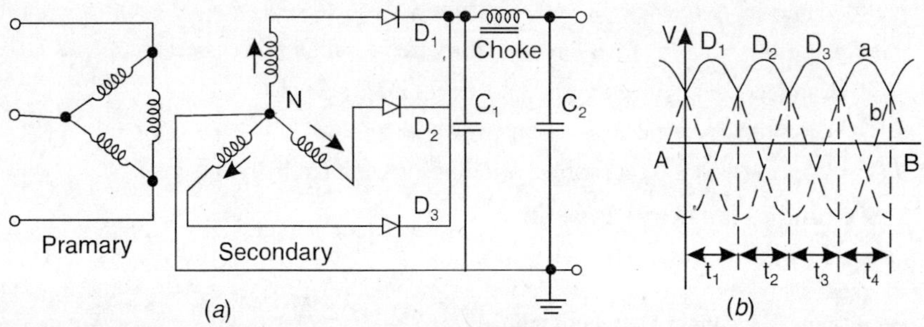

Fig. 8-28.

The shape of the output is shown in Fig. 8-28(*b*). The horizontal line *AB* represents the potential of the negative d.c. terminal output and the sine waves 1, 2 and 3 each represent the anode potentials of the three diodes.

During one-third of the cycle, *i.e.*, during time t_1, only diode D_1 will conduct. It will cease conducting at t_2 and then D_2 will conduct current upto t_2 after which D_2 will take over and will supply anode current till t_3. When one diode conducts, the other two remain inactive because then their anodes are more negative than their cathodes. This process repeats itself during each ensuing cycle, with the conducting period of each diode being as indicated. The output is given by the vertical distance 'ab' between the upper envelope and the line *AB*. Obviously, the output fluctuates between the maximum and minimum values thrice in each cycle. The variations of output lie between V_{sm} and 0.5 V_{sm} (neglecting voltage drop in diodes) and has a mean value of V_{dc} = 0.83 V_{sm} or 1.17 V_S where V_S is the r.m.s. value of the secondary phase voltage. Its maximum conversion η = 96.5% and γ = 0.17. But it should be noted that the magnitude of these fluctuations or pulsations is lesser than for a single-phase, full-wave rectified output since the current never touches zero. It is further smoothened up by a *C-L-C* filter circuit as shown in Fig. 8-28.

It is seen that direct current of each diode appears in the secondary phase winding and so causes transformer saturation resulting in large primary current. It can be avoided by using zig-zag secondary.

8-34. Full-wave Rectification of three-phase Currents

As shown in Fig. 8-29, a three-phase full-wave rectifier requires a transformer with six secondary winding connected to give two separate three-phase supplies 180° out of phase with each other. The centre taps are connected by an interphase transformer which enables the two rectifier units to operate independently of each other. Obviously, each diode conducts for one-third cycle. However, the addition of two outputs cancels the lowest frequency component of the ripple.

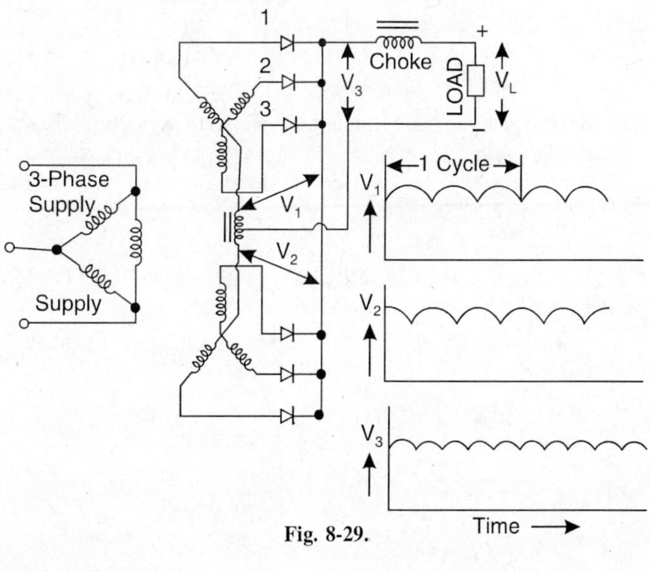

Fig. 8-29.

The output voltage has a mean value of 0.83 V_{sm} (less the diode voltage drop) and a ripple having a fundamental frequency six times the supply frequency.

Three-phase full-wave rectifier circuit is preferred for high power because

(*i*) Each secondary carries current for one-third of a cycle;

(*ii*) Each primary carries current for two-thirds of a cycle;

(*iii*) Copper loss in the transformer windings is comparatively lower.

8-35. Six-phase Half-wave Reactifier

Such a rectifier can be operated from a three-phase supply using a transformer with three centre-tapped secondary windings with all the centre taps connected together as shown in Fig. 8-30. Obviously, each diode conducts for one-sixth of a cycle. The output voltage has a mean value of 0.955 V_{sm} (less the voltage drop in the diode). Ripples has a very small value and a fundamental frequency six times the supply frequency.

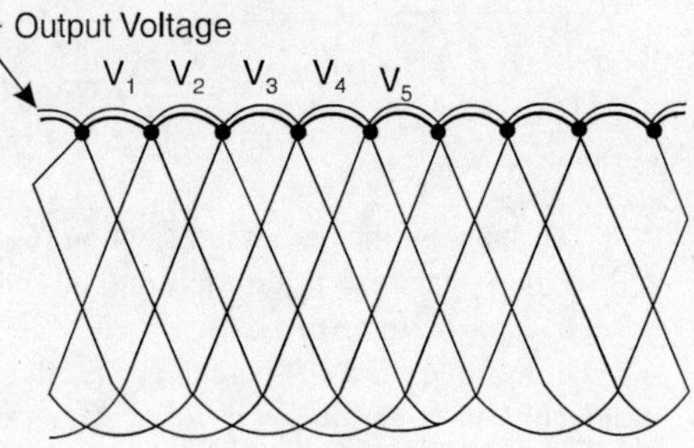

Fig. 8-30.

8-36. Three-phase Bridge Rectifier Circuit

Fig. 8-31 shows a three-phase bridge rectifier circuit. This circuit is very frequently used because apart from being simple, it does not require centre-tap transformer. Only that diode supplies the load whose phase voltage is more positive than the others. For example, when D_1 is forward-biased, D_2 and D_3 are reverse-biased. The current returns to the supply *via* D_5 and D_6.

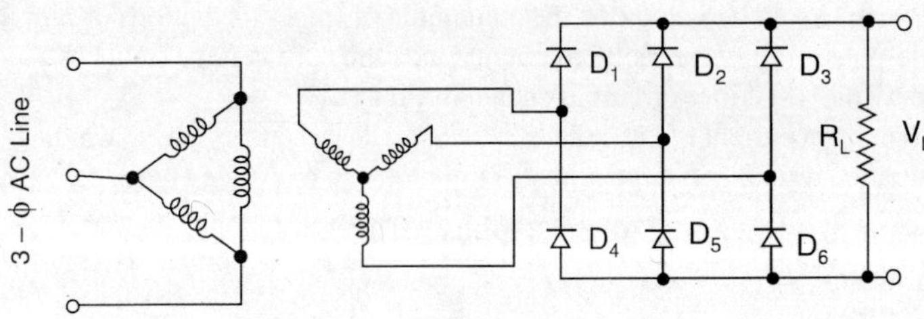

Fig. 8-31.

In this circuit, I_{dc} is 0.955 times the peak current through each diode and only one-third of it flows through each diode (rather than one-sixth as in Fig.). Similarly, V_{dc} is twice of the V_{in} 3-ϕ, half-wave rectifier or $2 \times 1.17 = 2.34$ times the r.m.s. a.c. voltage across each secondry leg. Accordingly, r.m.s. voltage across each secondary leg need be only 1/2.34 = 0.428 times the desired d.c. output voltage.

Example 8-13. *The r.m.s. value of transformer secondary voltage per leg (V_S) in a full-wave, D/Y six-phase rectifier is 150 V. If average value of load current is 2 A, find* **(i)** V_{dc} **(ii)** *peak and average current through each diode, and* **(iii)** *average power delivered to the load, i.e.* P_{dc}.

Solution. Given : r.m.s. value of secondary voltage, V_S = 150 V, average value of load current, I_{dc} = 2 A.

(*i*) *Value of* V_{dc}

We know that,
$$V_{dc} = 2.34 \times V_S$$
$$= 2.34 \times 150$$
$$= \mathbf{351\ V\ Ans.}$$

(*ii*) *Peak and average current through each diode*

We know that peak current per diode, $= \dfrac{1}{0.955} \times I_{dc} = 1.05 \times 2 = 2.1$ A **Ans.**

and average current per diode, $= \dfrac{2}{3} = 0.667$ A **Ans.**

(In D/Y half-wave rectifier its value is $1/6^{th}$ rather than $1/3^{rd}$)

(*iii*) *Average power delivered to the load*

We know that average power delivered to the load,

$$P_{dc} = V_{dc} \cdot I_{dc}$$
$$= 351 \times 2 = 702 \text{ W } \textbf{Ans.}$$

8-37. Calculations with Resistive Load

Let us suppose that no filter is connected across the rectifier but only a load of resistance R. Further, we will neglect transformer resistance and leakage and internal diode drops. In that case,

$$V_{dc} = 0.827\, V_{sm} \; ; I_{dc} = 0.827\, V_{sm}/R$$
$$P_{dc} = V_{dc}\, I_{dc} = (0.827\, V_{sm})\; 2/R \; ; P_{ac} = P_{in} = 0.706\, V_{sm}^{\;2}/R$$

$$\therefore \qquad \text{Rectifier } \eta = \frac{P_{dc}}{P_{ac}} = \frac{(0.827\, V_{sm})^2/R}{0.706\, V_{sm}^2/R} = 0.965 = 96.5\%$$

$$\gamma = \sqrt{1/dc^2 - 1} = \sqrt{(1.014)^2 - 1} = 0.17 \text{ or } 17\%$$

Thus, we can conclude that here, average d.c. current is 0.827 times the peak current as compared to 0.318 times for 1-ϕ, half-wave circuit and 0.636 times for single-phase, full-wave circuit. Moreover, V_{dc} is also correspondingly high, it is 51.17 times the r.m.s. voltage of each secondary leg ($\because$ 1.414 $\times$ 0.827 = 1.17). Conversely, r.m.s. voltage (V_S) across each leg of the secondary need only be 1/1.17 = 0.855 times the average desired d.c. output voltage across the load.

8-38. Voltage Multipliers

A voltage multiplier is a circuit which produces a greater d.c. output voltage than a.c. input voltage to the rectifiers. Multipliers are required in many circuit applications where it is necessary to have high voltages with low currents as for electron accelerating purposes in a cathode-ray tube (*CRT*).

We will consider the following circuit :

(*i*) half-wave voltage doubler, (*ii*) full-wave voltage doubler.

(*iii*) voltage tripler, (*iv*) voltage quadrupler.

8-39. Half-Wave Voltage Doubler

It is also known as *cascade voltage doubler*. The circuit is shown in Fig. 8-32. The operation of the circuit may be explained as follows :

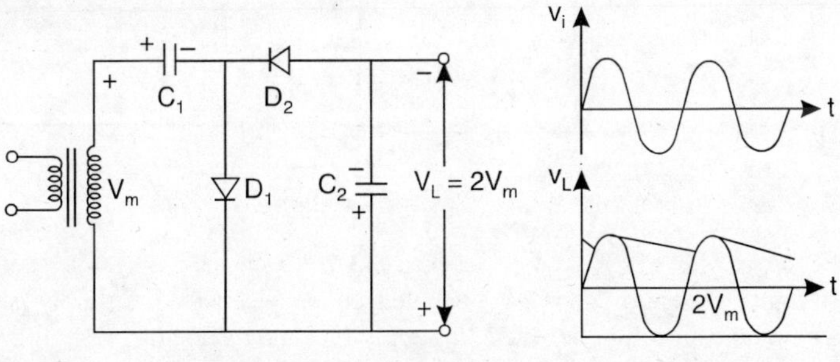

(*a*) (*b*)

Fig. 8-32.

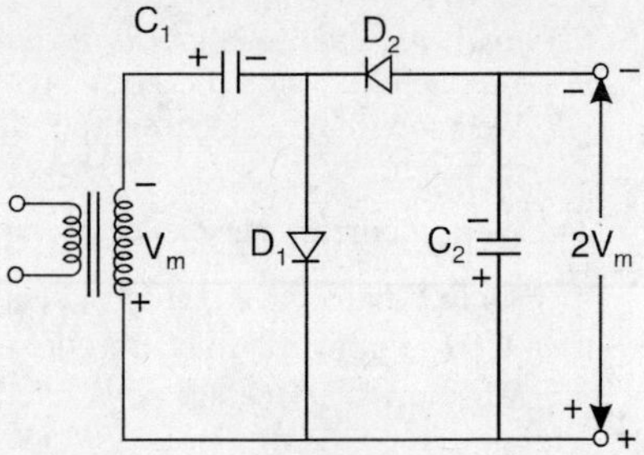

Fig. 8.33.

During the positive half-cycle of the input voltage, D_1 conducts (not D_2) and charges C_1 to peak value of secondary voltage (V_m) with the polarity as shown in Fig. 8-32 (*a*). During the negative half-cycle, D_2 conducts (not D_1) and charges C_2. The voltage across C_2 is the sum of peak supply voltage and the voltage across C_1 (Fig. 8-32). It can be proved by applying *KVL* to the outer loop. Starting from the bottom of the transformer secondary in Fig. 8-33 and going clockwise, we get

$$-V_m - V_m + V_{c2} = 0$$

or $\quad V_{c2} = 2\,V_m = 2 \times \text{peak input voltage}$

During the next positive half-cycle, D_2 is open and C_2 will discharge through the load if it is connected. If no load is connected across C_2, then both capacitors stay charged, *i.e.*, C_1 to V_m and C_2 to 2 V_m. If there is a load connected across C_2, it will discharge a little bit and, as a result of it, voltage across it will drop slightly. But it will get recharged in the next half-cycle.

The output wageform shown in Fig. 8-32(*b*) is that of a half-wave rectifier filtered by a shunt capacitor. *PIV* across each diodes is 2 V_m. Ripple frequency is equal to the supply frequency.

This circuit has very poor regulation and its ripple content is also high. This circuit has a common connection between the line and load (which a full-wave doubler does not have).

8-40. Full-Wave Voltage Doubler

This circuit is shown in Fig. 8-34(*a*). During the positive half-cycle of the point voltage, D_1 conducts (but not D_2) and charges capacitor C_1 to the peak voltage V_m with the polarity as shown.

During the negative half-cycle, D_2 conducts (but not D_1) charging C_2 to V_m. As far as the load is concerned, voltages across C_1 and C_2 are in series-aiding. If there is no load connected across the output, then load voltage $V_L = 2\,V_m$ as shown in Fig. 8-34(*a*). For example, if 220-V, 50-Hz is the supply, then $V_{dc} = 2\,V_m = 2 \times \sqrt{2} = 620$ V. Of course, if a load is connected across the output terminals, then V_L would be less than 2 V_m.

The waveform of the output voltage is shown in Fig. 8-34(*b*).

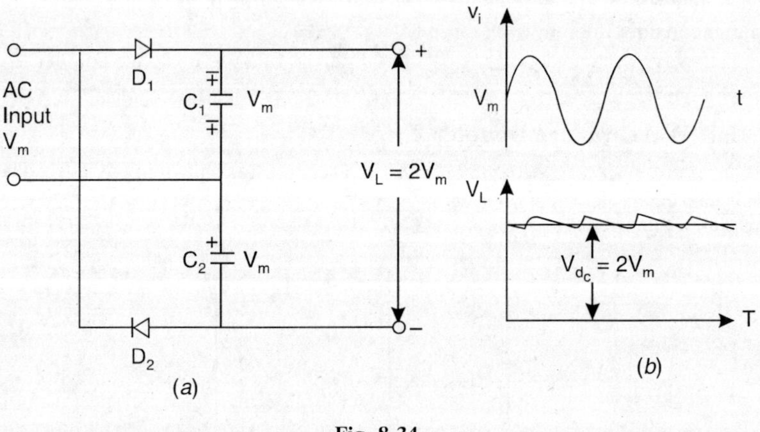

Fig. 8-34.

The *PIV* rating of each diode is 2 V_m. Ripple frequency is twice the supply frequency. As seen, there is no common connection between the supply line and the load.

It is worth noting that where the expense of a line transformer is justified, it is preferable to use the superior conventional full-wave rectifier.

8-41. Voltage Tripler and Quadrupler Circuits

The half-wave voltage doubler circuit discussed in Art. 8-39 (Fig. 8-32) can be extended to obtain any multiple of the peak input voltage (V_m), *i.e.*, $3V_m$, $4V_m$, $5V_m$ *etc*. Theoretically speaking, there is no upper limit to the amount of voltage multiplication that can be obtained. Though voltage triplers and quadruplers are commonly used, practical consideration limit additional multiplications. The main handicap is that total amount of capacitance becomes unduly large to maintain the desired d.c. output voltage for any thing except extremely light loads.

The circuit for different multipliers is shown in Fig. 8-35. It should be obvious from the repetitive pattern of the circuit connections how additional diodes and capacitors may be connected to the doubler circuit for obtaining higher multiplications of the peak output voltage V_m.

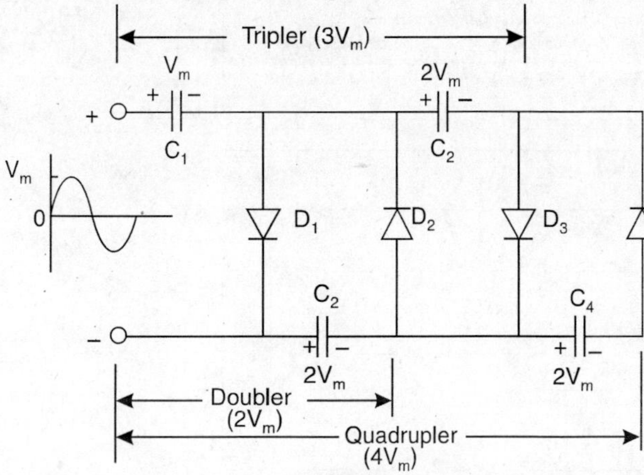

Fig. 8-35.

The operation of the circuit may be explained as follows :

During the first positive half-cycle, C_1 charges to V_m as diode D_1 conducts. During negative half-cycle, C_2 is charged through D_2 to $3V_m$, *i.e.*, to the sum of voltage across C_1 and peak input voltage V_m.

During the second positive half-cycle, D_3 conduct and voltage across C_2* charges C_3 to same voltage $2V_m$. During the negative half-cycle, diodes D_2 and D_4 conduct allowing C_3 to charge C_4 to the same peak voltage 2 V_m.

If is seen from Fig. 8-35 that voltage across C_2 is $2V_m$, across C_1 and C_3 is $3V_m$ and across C_2 and C_4 is $4V_m$.

If additional diodes and capacitors are used, each capacitor would be charged to a peak voltage of $2V_m$.

The *PIV* rating of each diode is $2V_m$ and ripple frequency is twice the line frequency. Generally, these circuits are used where both the supply voltage and load are maintained constant.

8-42. Specifications Sheet of Rectifier diodes and Bridge Rectifier

Fig. 8-36 shows the specifications datasheet of IN5391 through IN5399 rectifier diodes for general purpose low power applications. As seen from the maximum ratings table, the maximum value of reverse voltage which these diodes can withstand is in the range of 50 V to 1000 V. The average value of forward current is 1.5 A.

Fig. 8-37. shows the specifications data sheet for another set of rectifier diodes *MUR* 5005, 5010, 5015 and 5020. These diodes have the maximum value of reverse voltage in the range of 50 V to 200 V. However, the average value o f current which all these diodes can handle is 50 A. Therefore, these diodes are more suitable for rectifiers with high current applications.

Fig. 8-38. shows the specifications datasheet of MDA 2500 series single-phase full-wave bridge rectifiers. As seen from the maximum ratings table, the maximum value of reverse voltage for MDA 2500 series devices ranges from 50 V to 1000 V whereas the d.c. output voltage ranges from 30 V to 620 volts. The average value of rectified current for all the devices in MDA 2500 series in 25 A.

* C_1 can not charge C_3 because it is shorted by D_1.

MOTOROLA
SEMICONDUCTOR
TECHNICAL DATA

1N5391 thru 1N5399

Designers Data Sheet

"SURMETIC" RECTIFIERS

...subminiature size, axial lead-mounted rectifiers for general-purpose, low-power applications.

Designers Data for "Worst Case" Conditions

The Designers Data Sheets permit the design of most circuits entirely from the information presented. Limits curves–representing boundaries on device characteristics–are given to facilitate "worst-case" design.

LEAD-MOUNTED SILICON RECTIFIERS

50-1000 VOLTS DIFFUSED JUNCTION

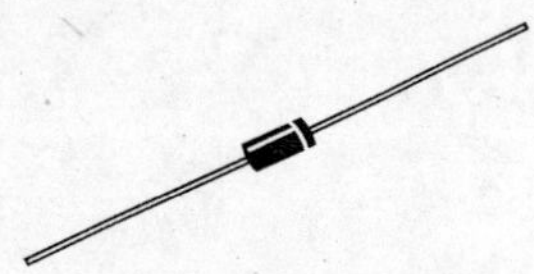

***MAXIMUM RATINGS**

Rating	Symbol	1N5391	1N5392	1N5393	1N5395	1N5397	1N5398	1N5399	Unit
Peak Repetitive Reverse Voltage Working Peak Reverse Voltage DC Blocking Voltage	V_{RRM} V_{RWM} V_R	50	100	200	400	600	800	1000	Volts
Nonrepetitive Peak Reverse Voltage (Halfwave, Single Phase, 60 Hz)	V_{RSM}	100	200	300	525	800	1000	1200	Volts
RMS Reverse Voltage	$V_{R(RMS)}$	35	70	140	280	420	560	700	Volts
Average Rectified Forward Current (Single Phase, Resistive Load, 60 Hz, T_L = 70°C, 1/2" From Body)	I_O	← 1.5 →							Amp
Nonrepetitive Peal Surge Current (Surge Applied at Rated Load Conditions, See Figure 2)	I_{FSM}	← 50 (for 1 cycle) →							Amp
Storage Temperature Range	T_{stg}	← – 65 to + 175 →							°c
Operating Temperature Tange	T_L	← – 65 to + 170 →							°c
DC Blocking Voltage Temperature	T_L	← 150 →							°c

***ELECTRICAL CHARACTERISTICS**

Characteristic and Conditions	Symbol	Typ	Max	Unit
Maximum Instantaneous Forward Voltage Drop (i_F = 4.7 Amp Peak, T_L = 170°C, 1/2 Inch Leads)	v_F	–	1.4	Volts
Maximum Reverse Current (Rated dc Voltage) (T_L = 150°C)	I_R	250	300	μA
Maximum Full-Cycle Average Reverse Current (1) (I_O = 1.5 Amp, T_L = 70°C, 1/2 Inch Leads)	$I_{R(AV)}$	–	300	μA

*Indicates JEDEC Registered Data.

NOTE 1: Measured in a single-phase, halfwave circuit such as shown in Figure 6.25 of EIA RS-282, November 1963. Operated at rated load conditions I_O = 1.5 A, V_r = V_{RWM}, T_L = 70°C.

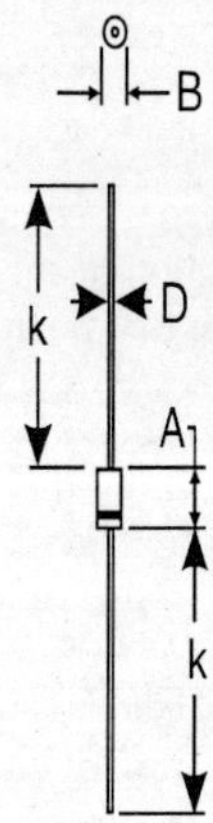

NOTES:
1. ALL RULES AND NOTES ASSOCIATED WITH JEDEC DO-41 OUTLINE SHALL APPLY.
2. POLARITY DENOTED BY CATHODE BAND.
3. LEAD DIAMETER NOT CONTROLLED WITHIN "F" DIMENSION.

	MILLIMETERS		INCHES	
DIM	MIN	MAX	MIN	MAX
A	5.97	6.60	0.235	0.260
B	2.79	3.05	0.110	0.120
D	0.76	0.86	0.030	0.034
K	27.94	–	1.100	–

CASE 59-04 PLASTIC

MECHANICAL CHARACTERISTICS

CASE: Transfer molded plastic

MAXIMUM LEAD TEMPERATURE FOR SOLDERING PURPOSES: 240°C, 1/8" form case for 10 seconds at 5 lbs. tension

FINISH: All external surfaces are corrosion-resistant, leads are readily solderable

POLARITY: Cathode indicated by color band

WEIGHT: 0.40 grams (approximately)

Fig. 8-36. *(Courtesy of Motorola Inc.)*

MOTOROLA
SEMICONDUCTOR
TECHNICAL DATA

MUR5005
MUR5010
MUR5015
MUR5020

ULTRAFAST RECTIFIERS
50 AMPERES
50 to 200 VOLTS

SWITCHMODE POWER RECTIFIERS

...designed for use in switching power supplies, inverters and as free wheeling diodes, these state-of-the-art devices have the following features:

- Ultrafast 50 Nanosecond Recovery Time
- Low Forward Voltage Drop
- Hermetically Sealed Metal DO-203AB Package

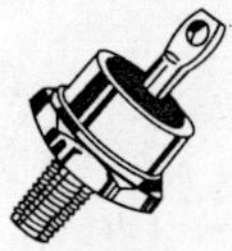

MAXIMUM RATINGS

Rating	Symbol	MUR 5005	5010	5015	5020	Unit
Peak Repetitive Reverse Voltage Working Peak Reverse Voltage DC Blocking Voltage	V_{RRM} V_{RWM} V_R	50	100	150	200	Volts
Nonrepetitive Peak Reverse Voltage	V_{RSM}	55	110	165	220	Volts
Average Forward Current TC = 150°C	$I_{F(AV)}$	50				Amps
Nonrepetitive Peak Surge Forward Current (half cycle, 60 Hz, Sinusoidal Waveform)	I_{FSM}	600				Amps
Operating Junction and Storage Temperature	T_J, T_{stg}	– 55 to + 175				°C

THERMAL CHARACTERISTICS

Tating	Symbol	All Devices	Unit
Thermal Resistance, Junction to Case	$R_{\theta JC}$	1.0	°C/W

ELECTRICAL CHARACTERISTICS

Maximum Instantaneous Forward Voltage Drop (i_F = 50 Amp, T_J = 25°C) (i_F = 50 Amp, T_J = 125°C) (i_F = 100 Amp, T_J = 125°C)	v_F	 1.15 0.95 1.10	Volts
Maximum Reverse Current @ DC Voltage (T_J = 25°C) (T_J = 125°C)	I_F	 10 1.0	 μA mA
Maximum Reverse Recovery Time (I_F = 1.0 Amp, di/dt = 50 Amp/ms, V_R = 30 V, T_J = 25°C)	t_{rr}	50	ns

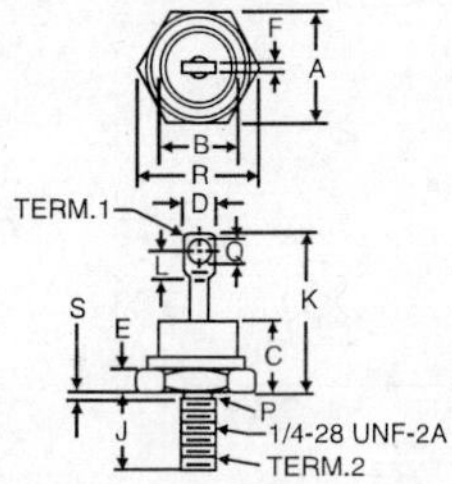

NOTES:
1. DIM "P" IS DIA.
2. CHAMFER OR UNDERCUT ON ONE OR BOTH ENDS OF HEXAGONAL BASE IS OPTIONAL
3. ANGULAR ORIENTATION AND CONTOUR OF TERMINAL ONE OPTIONAL.
4. THREADS ARE PLATED.
5. DIMENSIONING AND TOLERANCING PER ANSI Y14.5, 1973.

DIM	MILLIMETERS MIN	MAX	INCHES MIN	MAX
A	16.94	17.45	0.669	0.687
B	—	16.94	—	0.667
C	—	11.43	—	0.450
D	—	9.53	—	0.375
E	2.92	5.08	0.115	0.200
F	—	2.03	—	0.080
J	10.72	11.51	0.422	0.453
K	—	25.40	—	1.000
L	3.86	—	0.156	—
P	5.59	6.32	0.220	0.249
Q	3.56	4.45	0.140	0.175
R	—	20.16	—	0.794
S	—	2.26	—	0.089

CASE 257-01
DO-203AB
METAL

MECHANICAL CHARACTERISTICS

CASE: Welded, hermetically sealed

FINISH: All external surface correction resistant and terminal leads are readily solderable

POLARITY: Cathode to Case

MOUNTING POSITIONS: Any

MOUNTING TORQUE: 25 in-lb max

Fig. 8-37. *(Courtesy of Motorola, Inc.)*

MOTOROLA
SEMICONDUCTOR
TECHNICAL DATA

MDA2500 Series

RECTIFIER ASSEMBLY

...utilizing individual viod-free molded rectifiers, interconnected and mounted on an electrically isolated aluminum heat sink by a high thermal-conductive epoxy resin.

- 400 Ampere Surge Capability
- Electrically Isolated Base
- UL Recognized
- 1800 Volt Heat Sink Isolation

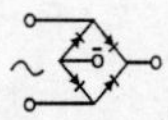

SINGLE-PHASE
FULL-WAVE BRIDGE

25 AMPERES
50-600 VOLTS

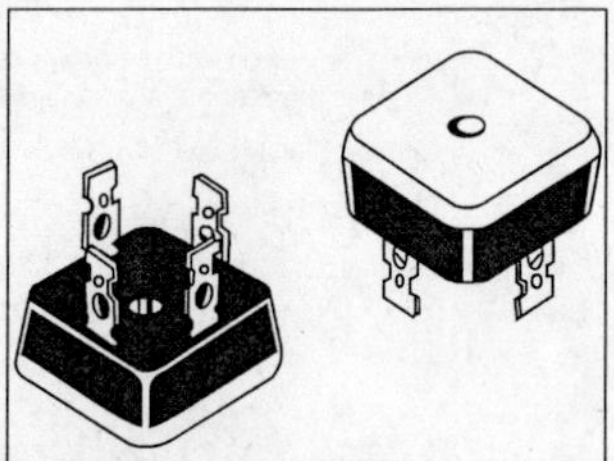

MAXIMUM RATINGS

Rating (Per Diode)	Symbol	MDA 2500	2501	2502	2504	2506	2508	2510	Unit
Peak Repetitive Reverse Voltage Working Peak Reverse Voltage DC Blocking Voltage	V_{RRM} V_{RWM} V_R	50	100	200	400	600	800	1000	Volts
DC Output Voltage	Vdc								Volts
Resistive Load		30	62	124	250	380	500	620	
Capacitive Load		50	100	200	400	600	600	1000	
Sine Wave RMS Input Voltage	$V_{R(RMS)}$	35	70	140	280	420	560	700	Volts
Average Rectified Forward Current (Single phase bridge resistive load, 60 Hz, T_C = 55°C)	I_O	25							Amp
Nonrepetitive Peak Surge Current (Surge applied at rated load conditions)	I_{FSM}	400							Amp
Operating and Storage Junction Temperature Range	T_J, T_{stg}	– 65 to + 175							°C

THERMAL CHARACTERISTICS

Characteristic	Symbol	Typ	Max	Unit
Thermal Resistance, Junction to Case	$R_{\theta JC}$			°C/W
Each Die		4.5	6.0	
Total Bridge		2.0	2.8	

ELECTRICAL CHARACTERISTICS (TC = 25°C unless otherwise noted)

Characteristic	Symbol	Min	Typ	Max	Unit
Instantaneous Forward Voltage (Per Diode) (i_F = 40 A)*	v_F	–	0.95	1.05	Volts
Reverse Current (Per Diode) (Rated V_R)	I_F	–	–	10	μA

MECHANICAL CHARACTERISTICS

CASE: Plastic case with an electrically isolated aluminum base.

POLARITY: Terminal designation embossed on case:
+ DC output
– DC output
AC not marked

MOUNTING POSITION: Bolt down. Highest heat transfer efficiency accomplished through the surface opposite the terminals. Use silicone heat sink compound on mounting surgace for maximum heat transfer.

WEIGHT: 25 grams (approx.)

TERMINALS: Suitables for fast-on connections. Readily solderable, corrosion resistant. Soldering recommended for applications greater than 15 amperes.

MOUNTING TORQUE: 20 in-lb max

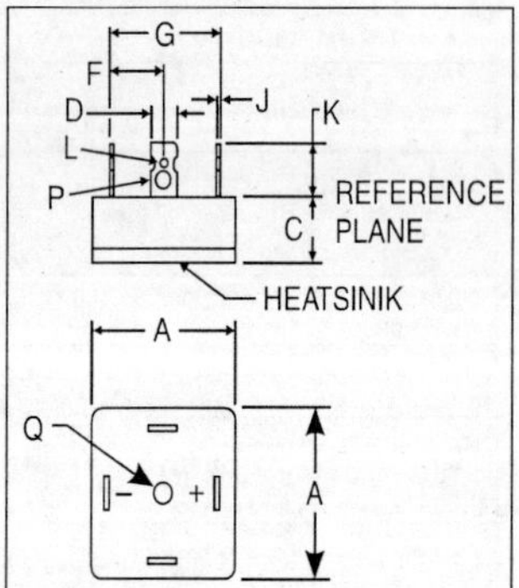

NOTES:
1. DIMENSION "Q" SHALL BE MEASURED ON HEATSINK SIDE OF PACKAGE.
2. DIMENSIONS "F" AND "G" SHALL BE MEASURED AT THE REFERENCE PLANE.

DIM	MILLIMETERS MIN	MAX	INCHES MIN	MAX
A	25.65	26.16	1.010	1.030
C	12.44	13.97	0.490	0.550
D	6.10	6.60	0.240	0.260
F	10.01	10.49	0.394	0.413
G	19.99	21.01	0.787	0.827
J	0.71	0.86	0.028	0.034
K	9.52	11.43	0.375	0.450
L	1.52	2.06	0.060	0.081
P	2.79	2.92	0.110	0.115
Q	4.42	4.67	0.174	0.184

CASE 309A-03

* Pulse Width = 100 ms, Duty Cycle ≤ 2%.

Fig. 8-38. *(Courtesy of Motorola, Inc.)*

SUMMARY

In this chapter, you have learned that

1. The average (or dc) value of output voltage for a single-phase half-wave rectifier, $V_{dc} = V_m/\pi$ (or $0.318\ V_m$), where V_m is the maximum value of ac input voltage. The average or dc value of current, $I_{dc} = 0.318\ I_m$, where I_m is the maximum value of diode current (or load current).
2. The peak inverse voltage of a diode in single-phase half-wave rectifier, $PIV = V_m$.
3. The average (or dc) value of output voltage for a centre-tapped single-phase full-value rectifier, $V_{dc} = 0.636\ V_m$. The average (or dc) value of load current, Idc = $0.636\ I_m$. The peak inverse voltage of each diode $PIV = 2\ V_m$.
4. The average (or dc) value of output voltage and load current for a single-phase bridge rectifier are the same as for a centre-tapped single-phase full-wave rectifier. However, the peak inverse voltage of each diode, $PIV = V_m$.
5. The ripple-factor is the ac component present in the output voltage (or load current) of a rectifier. It is given by the equation,
$$\gamma = \sqrt{\left(\frac{I_{rms}}{I_{dc}}\right)^2 - 1}$$
6. The ripple factor of a single-phase half-wave rectifier is 1.21 and for a full-wave rectifier, 0.482.
7. The efficiency of a rectifier is the ratio of dc power delivered to the load to the ac input power from the secondary winding of the transformer. It is given by the equation,
$$\eta = \frac{P_{dc}}{P_{ac}} = \frac{I_{dc}^2 \times R_L}{I_{rms}^2 (R_f + R_L)}$$
8. The efficiency of a single-phase half-wave rectifier is 40.6% while that of a full-wave rectifier, 81.2%.
9. Filter are used to remove ripples from the rectifier output.
10. Inductors, capacitors and their combination can be used as a filter circuit.
11. The inductor filter is more suitable for heavy load currents.
12. The capacitor filter is more suitable for light load currents.
13. The ripple factor of an inductor filter is given by,
$$\gamma = \frac{R_L}{1330 L}$$
where, R_L is in ohms and L is in Henrys.
14. The ripple factor of a capacitor filter is given by,
$$\gamma = \frac{2890}{C R_L}$$
where, R_L is in ohms and C in μF.
15. The ripple factor of an LC filter is given by,
$$\gamma = \frac{1.195}{LC}$$
where, L is in Henrys and C in μF.
16. The π-filter is more suitable for circuits where a low output current and a high dc output voltage is required. The ripple factor of a π-filter,
$$\gamma = \frac{5700}{C_1 . C_2 . C . R_L}$$
where, C_1 and C_2 are in μF, L in Henrys and R_L in Ohms.
17. The ripple factor a three-phase half-wave rectifier, $\gamma = 0.17$ and the efficiency, $\gamma = 96.5\%$.
18. A voltage multiplier can be used to produce dc output voltage greater than ac input voltage.

GLOSSARY

Bleeder Resistor : A resistor connected at the output of a filter to (*i*) maintain minimum current required circuit operation, (*ii*) improve voltage regulation, and (*iii*) provide safety to the persons handling the circuit.

Bridge Rectifier : A full-wave rectifier circuit that makes use of four diodes connected in a Wheat Stone Bridge configuration across the main supply.

Filter : A circuit that removes the unwanted a.c. component of the rectifier output and allows only d.c. component to reach the output. Usually it consists of passive circuit devices like inductors and capacitors.

Full-wave Rectifier : A rectifier circuit that allows a unidirectional current to flow through the load during the entire cycle of the a.c. input voltage.

Half-wave Rectifier : A rectifier circuit that allows the current to flow through the load during only one-half cycle of the a.c. input voltage.

Peak Inverse voltage (PIV) : The maximum reverse voltage that appears across the diode in a rectifier circuit.

π-filter : A filter circuit made by arranging two capacitors, and one inductors in the form of a symbol (π). It is also called CLC filter.

Rectifier efficiency : A ratio of the dc power delivered to the load by the rectifier to the a.c. input power from the secondary winding of the transformer.

Ripple factor : A ratio of the r.m.s. value of ac component of output voltage to the d.c. component of output voltage.

Transformer : A electrical device to step-up or step-down the voltage. It is also used to isolate the two electrical/electronic circuits.

Transformer Utilization Factor (TUF) : A ratio of dc power delivered to the load to the a.c. rating of the transformer secondary. For a half-wave rectifier, TUF = 0.287, for a bridge rectifier, TUF = 0.812.

Voltage multiplier : A circuit to produce a dc output voltage greater than ac input voltage.

NUMERICAL PROBLEMS

1. A half-wave rectifier circuit supplies 100 mA d.c. to a 250 Ω load. Find the d.c. output voltage, the *PIV* rating of a diode and the r.m.s. voltage for the transformer supplying the rectifier. (**Ans.** 25 V; 35.4 V and 17.7 V)

2. An a.c. supply of 230 V is applied to a half-wave rectifier circuit through a transformer of turns ratio 20 : 1. If the load resistance is 1kΩ, find (***a***) the d.c. output voltage, (***b***) maximum value of the output voltage, (***c***) d.c. load current, and (***d***) maximum value of load current. (**Ans.** 5.17 V; 16.3 V; 5.17 mA; 16.3 mA)

3. A half-wave rectifier uses a diode having a forward resistance of 2 Ω . If the transformer secondary voltage is 10 V (r.m.s.) and has a winding resistance of 5 Ω. Find (***a***) the d.c. output voltage at a load of 100 mA, and (***b***) *PIV* of a diode. (**Ans.** 3.8 V and 14.1 V)

4. A full-wave rectifier with a centre-tapped transformer (10 – 0 – 10 V) supplies a load current of 100 mA. Neglecting the diode forward resistance and secondary winding resistance, find (***a***) the d.c. output voltage, (***b***) *PIV* of each diode, and (***c***) ripple frequency. (**Ans.** 9 V; 28.3 V and 100 Hz)

5. A full-wave rectifier is fed by a 24 V_{ac} (rated) centre-tapped transformer. What is the d.c. output voltage of the circuit. If the rectifier is connected to a load resistance of 2.2 k Ω load, what is the d.c. value of load current. (**Ans.** 10.8 V; 4.9 mA)

6. A full-wave rectifier uses two diodes, the forward resistance of each diode is 20 Ω. The transformer r.m.s. secondary voltage from centre-tapped to each end of the secondary is 50 V and the load resistance is 980 Ω. Determine (*a*) the average value of load current, and (*b*) the r.m.s. value of load current. (**Ans.** 45 mA and 25 mA)

7. A transformer with 20 V r.m.s., 50 Hz secondary, supplies a bridge rectifier having a load of 200 Ω. Determine the d.c. output voltage, d.c. load current and *PIV* rating of the diode. (**Ans.** 18 V; 0.09 A and 28.3 V)

8. A 50 Hz transformer having 50 V r.m.s. on each side of the centre-tap supplies a full-wave rectifier circuit. The circuit load is 150 Ω with a shunt capacitor filter of 1000 μF. Find the ripple factor. (**Ans.** 1.9%)

9. A π-filter in a full-wave rectifier uses $C_1 = C_2 = 100$ μF and $L = 10$ H. The load current is 0.5 A at 100 V d.c. Find the ripple factor? (**Ans.** 0.0285%)

DESCRIPTIVE QUESTIONS

1. Draw the circuit diagram of half-wave rectifier. Explain its working. What is the frequency of ripple in its output?

2. Draw the circuit diagram of a full-wave rectifiers (*a*) with centre-tap connection, and (*b*) bridge connection. Explain their working. What is the peak-inverse voltage of a diode in each case?

3. How does a half-wave rectifier differs from the full-wave circuit in the following aspects?

(*a*) Fundamental ripple frequency in the output,

(*b*) Rectified d.c. voltage,

(*c*) Ripple factor,

(*d*) Average d.c. load current, and

(*e*) Peak-inverse voltage.

4. What are the advantages of a bridge rectifier as compared to a full-wave centre-tapped rectifier?

5. What is efficiency of rectification? Compare half-wave and full-wave rectifiers from the point of view of rectification efficiency.

6. Define (*i*) a ripple factor, (*ii*) rectification efficiency, and (*iii*) transformer utilization factor for a rectifier.

7. Why do we need filters in a power supply? Under what condition we shall prefer a capacitor filter?

8. What is the function of a bleeder resistor?

9. Describe the terms capacitor-input filter and inductor input filter.

10. Explain the operation of full wave voltage doubler circuit.
(*Electronic Circuits, VTU June/July 2008*)

11. What is a voltage multiplier? (*Electronic Circuits*-I, *Anna University, April/May 2008*)

12. Derive the ripple factor for *FWR* with capacitor filter.

13. Derive the expression for ripple factor of *FWR* with *TT* section filter.
(*Electronic Circuits* - I, *Anna University, April/May 2008*)

14. Define Transformer Utilization Factor.

(*Electronic Circuits - I, Anna University, April/May 2008*)

(*Electronic Circuits - I, Anna University, Nov/Dec 2007*)

15. Compare the performance of half wave rectifier and full wave rectifier.

(*Electronic Circuits - I, Anna University, Nov/Dec 2007*)

16. Draw the circuit diagram of a FWR with capacitor input filter. With suitable waveform explain its working. Derive the expression for ripple factor.

(*Electronic Circuits - I, Anna University, Nov/Dec 2007*)

17. Draw and explain a half-wave rectifier circuit using SCR.

(*Electronic Devices, Anna University, Nov/Dec 2007*)

18. What is a switched capacitor filter?

(*Linear Integrated Circuits - Anna University, April/May 2008*)

MULTIPLE CHOICE QUESTIONS

1. In a half-wave rectifier, the load current flows for
(*a*) complete cycle of the input signal,
(*b*) less than half-cycle of the input signal,
(*c*) more than half-cycle but less than complete cycle of the input signal.
(*d*) only for the positive half-cycle of the input signal.

2. In a full-wave rectifier, the current in each of the diodes flows for
(*a*) complete cycle of the input signal.
(*b*) half cycle of the input signal.
(*c*) less than half of the input signal.
(*d*) none of the above.

3. The ripple factor of a bridge rectifier is
(*a*) 0.482 (*b*) 0.812
(*c*) 1.11 (*d*) 1.21

4. A bridge rectifier is preferable to an ordinary two-diode full-wave rectifier because
(*a*) it needs much smaller transformer for the same output.
(*b*) it uses four diodes.
(*c*) its transformer does not require centre-tap.
(*d*) none of the above.

5. The basic purpose of a filter is to:
(*a*) minimise variations in a.c. input signal.
(*b*) suppress harmonics in rectified output.
(*c*) remove ripples from the rectified output.
(*d*) stabilise d.c. output voltage.

6. The use of a capacitor filter in a rectifier circuit gives satisfactory performance only when the load
(*a*) current is high (*b*) current is low
(*c*) voltage is high (*d*) voltage is low

(*UPSC Engg. Services* 2001)

7. A half-wave rectifier is equivalent to
(*a*) a clamper circuit (*b*) a clipper circuit
(*c*) a clamper circuit with negative bias (*d*) a clamper circuit with positive bias

8. Bridge rectifiers are prefered because

(*a*) they require small transformer

(*b*) they have less peak-inverse voltage

(*c*) they need small transformer and also have less peak-inverse voltage

(*d*) they have low ripple factor.

9. If V_m is the peak value of an applied voltage in a half-wave rectifier with a large capacitor across the load, then the peak-inverse voltage will be :

(*a*) $V_m/2$ (*b*) V_m

(*c*) $\sqrt{2}\, V_m$ (*d*) $2\, V_m$

10. A voltage of 200 cos 100 *t* is applied to a half-wave rectifier with a load resistance of 5 kΩ. The rectifier is represent by an ideal diode in series with a resistance of 1 kW. The maximum value of current, d.c. component of current and r.m.s. value of current will be respectively,

(*a*) 33.33 mA, 10.61 mA and 16.67 mA (*b*) 22.22 mA, 8.61 mA and 12.38 mA

(*c*) 28.33 mA, 14.61 mA and 13.33 mA (*d*) 40 mA, 20 mA and 25 mA

[**Hint:** Refer to Fig. 8-39]

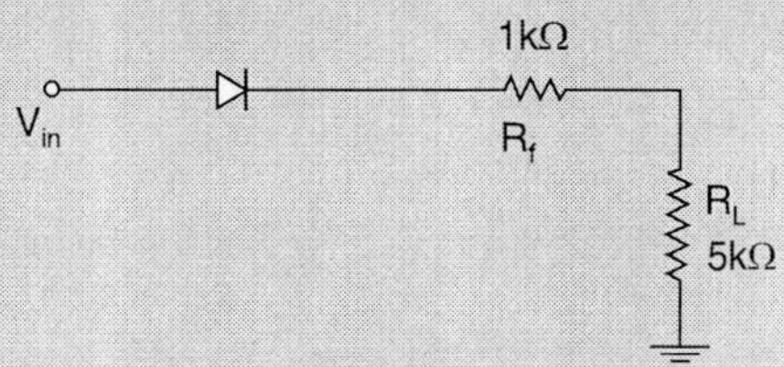

Fig. 8-39

$$I_m = \frac{V_m}{R_f + R_L} = \frac{200}{1k + 5k} = 33.33 \text{ mA } \textbf{Ans.}$$

$$I_{dc} = \frac{I_m}{\pi} = \frac{33.33 \text{ mA}}{\pi} = 10.61 \text{ mA } \textbf{Ans.}$$

$$I_{rms} = \sqrt{I_{dc}^2 + I_{L(ac)}^2}$$

where, $I_{L(ac)} = 0.385\, I_m$

Substituting the values, we get, $I_{rms} = 16.67$ mA.

11. The basic reason why a full-wave rectifier has a twice the efficiency of a half-wave rectifier is that

(*a*) it makes use of transformer

(*b*) its ripple factor is much less

(*c*) it utilizes both half-cycle of the input

(*d*) its output frequency is double the line frequency.

12. The output of a half-wave rectifier is suitbale only for

(*a*) running car radios (*b*) running a.c. motors

(*d*) charging batteries (*d*) running tape-recorders.

13. The ripple factor of a bridge rectifier is

(*a*) 0.406 (*b*) 0.812

(*c*) 1.21 (*d*) 1.11.

14. The ripple factor of a power supply is given by (symbols have the ususal meaning).

(*a*) $\frac{P_{dc}}{P_{ac}}$ (*b*) $\sqrt{\left(\frac{I_{rms}}{I_{dc}}\right)^2 - 1}$

(c) $\sqrt{\left(\frac{I_{dc}}{I_{rms}}\right)^2 - 1}$ (d) $\frac{I_{dc}}{I_{rms}}$

15. The *PIV* of a half-wave rectifier circuit with a shunt capacitor filter is
(a) $2\,V_{sm}$ (b) V_{sm}
(c) $V_{sm}/2$ (d) $3\,V_{sm}$.

16. The primary function of a rectifier filter is to
(a) minimise a.c. input variations
(b) suppress odd harmonics in the rectifier output
(c) stabilise d.c. level of the output voltage
(d) remove ripples from the rectified output

17. In a rectifier, larger the value of shunt capacitor filter
(a) larger the *peak-to-peak* value of ripple voltage
(b) larger the peak current in the rectifying diode
(c) longer the time that current pulse flows through the diode
(d) smaller the d.c. voltage across the load.

18. In a LC filter, the ripple factor,
(a) increases with the load current (b) increases with the load resistance
(c) remains constant with the load current (d) has the lowest value.

19. The main reason why a bleeder resistor is used in a d.c. power supply is that it
(a) keeps the supply ON (b) improves voltage regulation
(c) improves filtering action (d) both (b) and (c).

20. Which rectifier requires four diodes ?
(a) half-wave voltage doubler (b) full-wave voltage doubler
(c) full-wave bridge circuit (d) voltage quadrupler.

21. If, by mistake, a.c. source in a bridge rectifier, is connected across the d.c. terminals, it will burn out and hence short .... diodes.
(a) one (b) two
(c) three (d) four

22. The d.c. output polarity from a half-wave rectifier can be reversed by reversing
(a) the diode (b) transformer primary
(c) transformer secondary (d) both (b) and (c).

23. In a half-wave rectifier if a resistance equal to load resistance is connected in parallel with the diode, then
(a) output voltage would be halved (b) output voltage would be doubled
(c) circuit will stop rectifying. (d) output voltage will remain unchanged.

24. If the input supply frequency is 50 Hz, the output ripple frequency of a bridge rectifier is .... Hz.
(a) 100 (b) 75
(c) 50 (d) 25.

ANSWERS

1. (d)	**2.** (b)	**3.** (a)	**4.** (a)	**5.** (c)
6. (b)	**7.** (b)	**8.** (c)	**9.** (b)	**10.** (a)
11. (c)	**12.** (b)	**13.** (c)	**14.** (b)	**15.** (a)
16. (d)	**17.** (b)	**18.** (c)	**19.** (d)	**20.** (b)
21. (d)	**22.** (a)	**23.** (c)	**24.** (a)	

❑❑❑

9

CHAPTER

Bipolar Junction Transistors

OUTLINE

Objectives

After completing this chapter, you should be able to :

- Know bipolar junction transistor.
- Understand the three terminals of a bipolar junction transistor (*i.e.*, emitter, base and collector).
- Know the four different techniques for the manufacturing of bipolar junction transistor (*i.e.*, grown-junction, alloy to fused-junction, diffused junction and epitaxial).
- Differentiate the symbols of NPN and PNP transistors.
- Understand the formation of depletion layer in an unbiased bipolar junction transistor.
- Know the three modes of bipolar junction transistor operation (*i.e.*, forward-active, saturation and cut-off).
- Describe the operation of a NPN and PNP bipolar junction transistors.

- Find bipolar junction transistor currents (*i.e.*, collector current I_C, base current I_B and emitter current I_E).
- Understand the three different configurations (*i.e.*, common-base, common-emitter and common-collector) of a bipolar junction transistor
- Find the current gain of a bipolar junction transistor in different configurations.
- Know base spreading resistance.

9-1. Introduction

The Bipolar junction transistor or simply a BJT, was developed by Dr. Shockley alongwith Bell Laboratories team in 1951. For the first time, this component was used in any type of commercial venture in October 1952, when the Bell System employed transistor circuits in the telephone switching circuits in Englewood, N.J. Since, then it has revolutionized the field of electronics.

The BJT is a main building block of all modern electronic systems. It is three-terminal device whose output current, voltage, and/or power are controlled by its input current. In communication systems, it is urdely used as the primary component in the amplifier. *An amplifier is a circuit that is used to increase the strength of an ac signal.* In digital computer electronics, the BJT is used as a high-speed electronic switch that is capable of switching between two operating states (open and closed) at a rate of several billions of times per second.

There is another types of transistors namely *field-effect transistors* (abbreviated as FET). The field-effect transistors will be discussed in chapter-11. The bipolar junction transistor is more commonly known as junction transistor, bipolar transistor or simply a transistor.

A bipolar junction transistor has a very important property that it can raise the strength of a weak signal. This property is called *amplification*. Because of this property, the transistor is one of the most widely used semiconductor devices. At present, the transistors are used in digital computers, satellites, mobile phones and other communication systems, control systems, *etc*.

A bipolar junction transistor consists of two PN junctions. The junctions are formed by sandwitching either P-type or N-type semiconductor layers between a pair of opposite types as shown in Fig. 9-1(*a*) and (*b*).

Fig. 9-1. *Transistor.*

9-2. Bipolar Junction Transistor Construction

A bipolar junction transistor has, essentially, three regions known as emitter, base and collector. All these three regions are provided with terminals, which are labelled as E (for emitter), B (for base) and C (for collector) respectively. A brief description of the above regions is discussed below :

(*i*) ***Emitter :*** It is a region situated in one side of transistor, which supplies charge carries (*i.e.*, electrons or holes) to the other two regions. The emitter is a heavily doped region.

(*ii*) ***Base :*** It is the middle region that forms two P-N junctions in the transistor. The base of transistor is thin, as compared to the emitter and is a lightly doped region.

(*iii*) ***Collector :*** It is a region situated in the other side of transistor (*i.e.*, the side opposite to the emitter), which collects charge carriers (*i.e.*, electrons or holes). The collector of a transistor is always larger than the emitter and base of a transistor. The doping level of the collector is intermediate between the heavy doping of emitter and the light doping of the base.

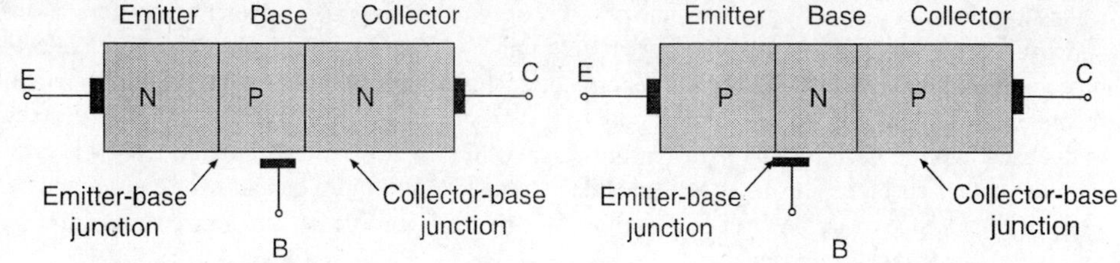

(*a*) NPN Transistor (*a*) PNP Transistor

Fig. 9-2. *Bipolar Junction Transistor construction.*

As a matter of fact, the transistor has two PN junctions J_C and J_E as shown in Fig. 9-2(*a*) and (*b*). The junction J_E is a junction between emitter and base regions. Thus, it is known as emitter-base junction. Similarly, the junction J_C is a junction between collector and base regions. Thus, it is known as collector-base junction.

9-3. Bipolar Junction Transistor Manufacturing Techniques

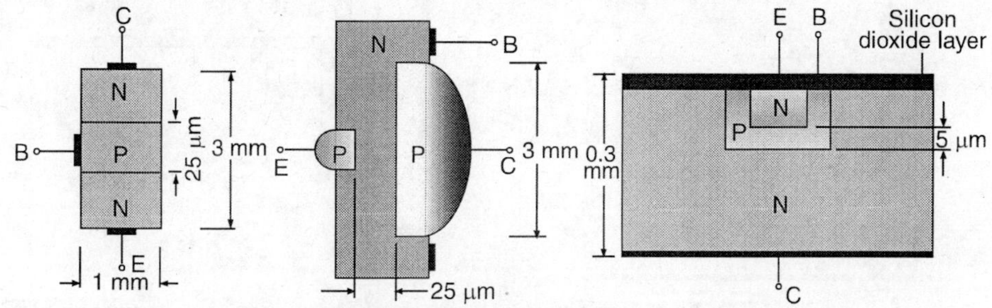

(*a*) Grown-junction type (*b*) Alloy-junction type (*c*) Diffusion type

Fig. 9-3. *Transistor manufacturing techniques.*

The techniques used for manufacturing the bipolar junction transistor (BJT) are the same as discussed for the PN junctions. Following are the basic techniques used for the BJT manufacturing :

(*i*) Grown-junctions
(*ii*) Alloy of fused-junctions
(*iii*) Diffused-junctions
(*iv*) Epitaxial

Depending upon the manufacturing technique, a transistor may be called grown-junction type transistor, alloy-junction type transistor, diffused-junction transistor and epitaxial type transistor. Each type of the transistor construction is discussed as below :

(1) *Grown-junction type :* In this type, a single crystal is drawn from the molten silicon or germanium. The impurity concentration of single crystal is changed during the crystal drawing operation by adding N-type or P-type impurities as required. Then the grown crystal is cut into small area devices and the contacts are made to each region as shown in Fig. 9-3(*a*).

(2) *Alloy-type :* In this type, two small dots of indium (a P-type impurity) are placed on opposite sides of a thin wafer of N-type material. The whole structure is heated for a short time to a temperature high enough to melt indium into the base material. The alloying of indium, with the original wafer, produces the PN junctions. The leads may be connected to the structure as shown in Fig. 9-3(*b*). It may be noted, that collector region is made larger than the emitter region. This has on advantage that the collector collects the whole emitter current and thus, prevents the current from diffusing into the base region.

(3) *Diffused junction type :* In this type, a semiconductor wafer is subjected to gaseous diffusion of both N and P-type impurities to form both the emitter-base and collector-base junctions. This process is called planar process and will be discussed in detail in chapter on integrated circuits. The transistor, manufactured by using planar process, are known as silicon planer transistors and are shown in Fig. 9-3(*c*). The manufacturing process starts with a N-type silicon wafer. The collector-base junction area is determined by diffusion mask. This area is photoetched by photolithographic techniques. Then the P-type base is applied by a gaseous diffusion process. Next the N-type emitter is diffused into the base through a different mask. A thin layer of silicon dioxide (SiO_2) is grown over the entires surface. This layer is photoetched to make aluminium contacts for emitter and base leads. The presence of oxide layer avoids most of the surface problems, and hence results in a very low *transistor leakage current*.

Sometimes, the planar transistor is formed in such a way that the collector area is reduced by the photoetching process. The resulting structure of a transistor is as shown in Fig. 9-4(*a*). This structure has a flat topped called mesa, and is called mesa transistor. This structure is very rugged and has large power dissipation capability. It can operate at very high frequencies.

(4) *Epitaxial type* : In this type, a very thin single-crystal layer of silicon (or germanium) is grown on a heavily doped substrate of the same material. This layer forms a collector on which base and emitter regions may be diffused. Fig. 9-4(*b*) shows an epitaxial type of transistor.

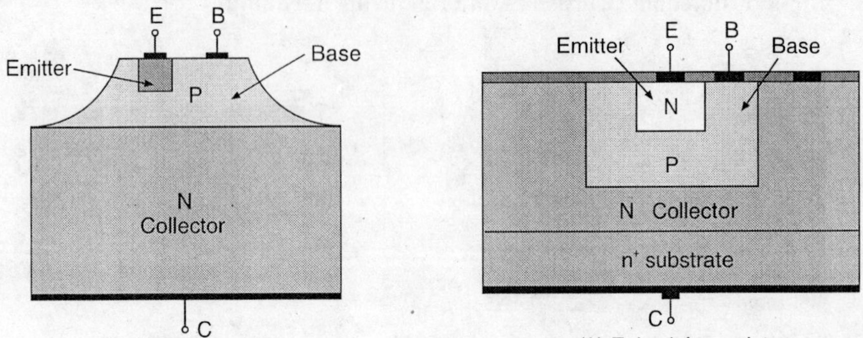

(*a*) Mesa transistor (*b*) Epitaxial transistor

Fig. 9-4.

The techniques discussed above may be combined to form a large number of methods for constructing transistors. For example, there are diffused-alloy types, grown-diffused types, alloy-emitter epitaxial-base transistors, *etc*.

9-4. Bipolar Junction Transistor Packaging and Terminal Identification

We have already discussed in the last article the different techniques of transistor manufacturing. After the transistor is manufactured, by using any one of the techniques, leads of gold, aluminium or nickel are then attached and the entire structure is packed or encapsulated in a container as shown in Fig. 9-5(*a*). The transistors with small can (top hat) or plastic body are low-to medium-power devices, while those with studs and heat sinks are high power devices.

(*a*) Package types.

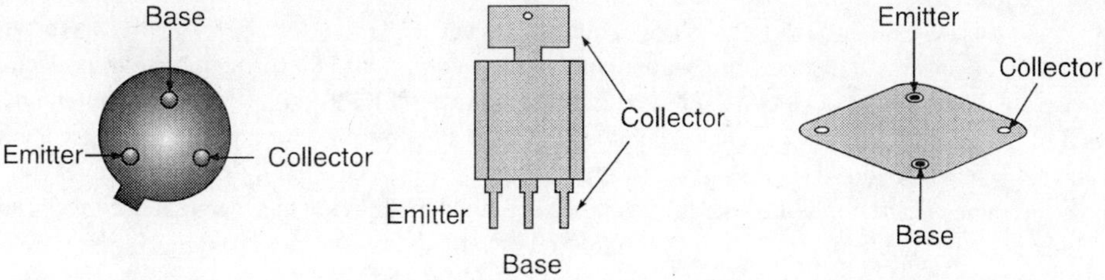

(*b*) Terminal identification methods.

Fig. 9-5.

Most of the transistor packages have some marking to identify its terminals (*i.e.*, to identify which leads are connected to the emitter, base or collector of a transistor). Some of the commonly used methods are as shown in Fig. 9-5 (*b*). In the metal 'top hat' cases (*i.e.*, packages), the tab is closest to emitter lead and the collector is often connected to the case. Generally, the heat sink mounting or stud on the power transistors is the collector terminal.

9-5. Bipolar Junction Transistor Symbols

We have already mentioned in the Art. 9-2 that there are two types of transistors namely NPN and PNP. When the transistor is used as circuit element, it is difficult to represent it with Fig. 9-2(*a*) and (*b*). In order to make it convenient, both the transistors are represented by their circuit symbols as shown in Fig. 9-6(*a*) and (*b*) respectively. The transistor symbol carries an arrowhead in the emitter pointing from the P-region towards the N-region. The arrowhead indicates the direction of a conventional current flow in a transistor.

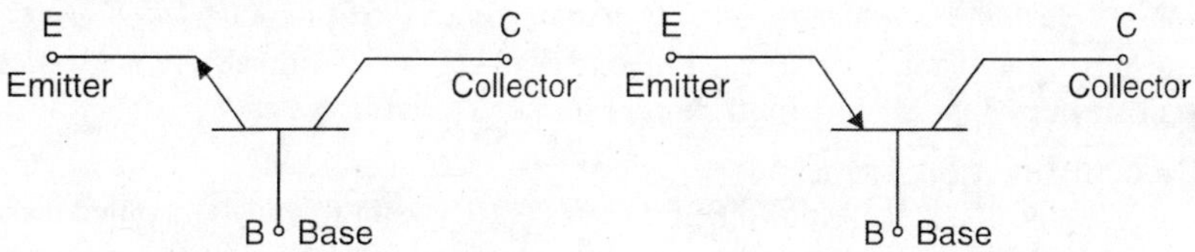

(*a*) Symbol for NPN transistors. (*b*) Symbol for PNP transistors.

Fig. 9-6. *Bipolar Junction Transistor symbols.*

It may be noted that the direction of arrowheads at the emitter in NPN and PNP transistors is opposite to each other. Sometimes, it is more convenient to represent the symbols for NPN and PNP transistors as shown in Fig. 9-7(*a*) and (*b*).

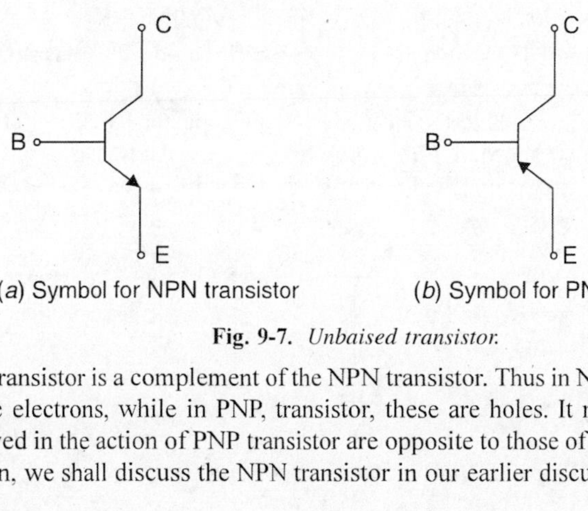

(*a*) Symbol for NPN transistor (*b*) Symbol for PNP transistor

Fig. 9-7. *Unbaised transistor.*

The PNP transistor is a complement of the NPN transistor. Thus in NPN transistor, the majority carries are free electrons, while in PNP, transistor, these are holes. It means that the current and voltages involved in the action of PNP transistor are opposite to those of NPN transistor. In order to avoid confusion, we shall discuss the NPN transistor in our earlier discussion.

9-6. Unbiased Bipolar Junction Transistor

A transistor, with three terminals (*i.e.*, Emitter, Base and Collector) left open, is called an unbiased transistor or an open-circuited transistor. Under these conditions, the diffusion of free electrons across the junction produces two depletion layers as shown in Fig. 9-8 (*a*). The barrier potential, for each of these layers at 25°C, is approximately 0.7 V for silicon transistor and 0.3 V for germanium transistor. Since, the three regions have different doping levels, therefore the depletion layers do not have the same width. It may be noted that a more heavily doped region has the greater concentration of ions near the junction.

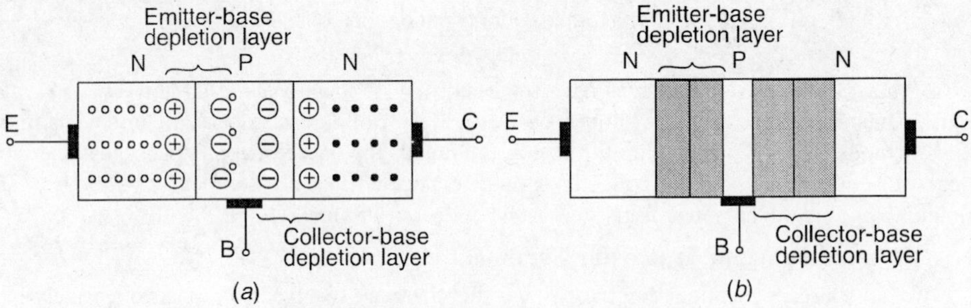

Fig. 9-8. *Unbaised Bipolar Junction transistor.*

It has been observed that an emitter-base depletion layer penetrates slightly into the emitter, as it is a heavily doped region, whereas it penetrates deeply into the base as its a lightly doped region. Similarly, the collector-base depletion layer penetrates more into the base region and less into the collector region. Both these depletion layers are as shown in Fig. 9-8(*b*). It may be noted that *emitter-base depletion layer width is smaller than that of collector base depletion layer.*

An unbiased transistor is never used in actual practice. Its terminals are always connected, suitably, to the dc voltage sources (or a battery) for proper transistor action.

9-7. Bipolar Junction Transistor Biasing

The application of a suitable dc voltages, across the transistor terminals, is called biasing. Each junction of a transistor may be forward biased or reverse-biased independently. There are following three different ways of biasing a transistor, which are also known as modes of transistor operation.

(1) ***Forward-active :*** In this mode, the emitter-base junction of a transistor is forward biased and the collector base junction is reverse biased as shown in Fig. 9-9(*a*). In a forward active biasing the negative terminal of a battery is connected to N-side and positive terminal to P-side. The reverse biasing requires all the connections to be opposite to those for forward biasing.

(2) ***Saturation :*** In this mode, both the emitter-base and collector-base junctions of a transistors are forward-biased as shown in Fig. 9-9(*b*). In this mode, the transistor has a very large value of current. The transistor is operated in this mode, when it is used as a closed switch.

(3) ***Cut-off*** **:** In this mode, both the emitter-base and collector-base junctions of a transistor are reverse biased as shown in Fig. 9-9(*c*). In this mode, the transistor has practically zero current. The transistor is operated in this mode, when it is used as an open switch.

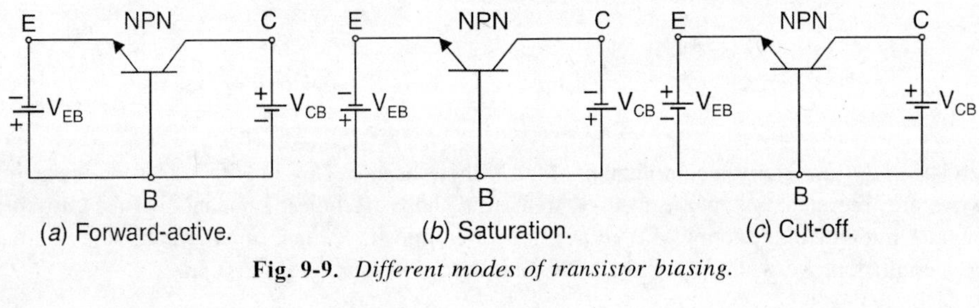

Fig. 9-9. *Different modes of transistor biasing.*

The three different modes are summarized in Table 9-1. It must be remembered that for amplifier, a transistor is used in forward-active mode and for switching applications (*i.e.*, ON and OFF applications) a transistor is used alternatively is saturation and cut-off modes.

Table 9-1. *Modes of transistor operation*

S. No.	*Mode*	*Junction-bias condition*	
		Emitter Base	***Collector Base***
1.	Forward active	Forward	Reverse
2.	Saturation	Forward	Forward
3.	Cut-off	Reverse	Reverse

Note. The modes of operation are the same for both NPN and PNP transistors. However, in PNP transistor biasing, the polarity of battery voltages are opposite to that of NPN transistor.

9-8. Operation of an NPN Bipolar Junction Transistor

Fig. 9-10(*a*) shows an NPN transistor biased in forward-active mode *i.e.*, the emitter-base of a transistor is forward-biased and collector-base junction is reverse-biased. The emitter-base junction is forward biased only if V_{EB} is greater than barrier potential which is 0.7 V for silicon and 0.3 V for germanium transistors. The forward bias on the emitter-base junction causes the free electrons in the N-type emitter to flow towards the base region. This constitutes the emitter current (I_E). It may be noted that the direction of conventional current (I_C) is opposite to the flow of electrons. Therefore, electrons after reaching the base region as shown in Fig. 9-10(*b*), tend to combine with the holes. If these free electrons combine with the holes in the base, they constitute base current (I_B). However, most of the free electrons do not combine with the holes in the base. This is because of the fact that the base width is made extremely small and electrons do not get sufficient holes for recombination. Thus most of the electrons will diffuse to the collector region as shown in Fig. 9-10(*c*) and constitutes collector current (I_C). This collector current is also called injected current because this current is produced due to electrons injected from the emitter region.

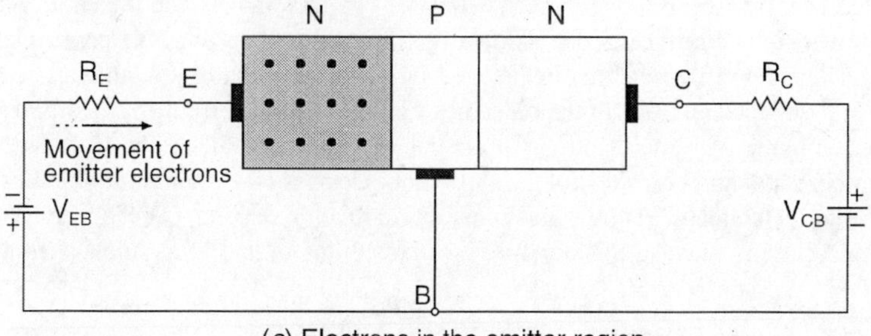

(*a*) Electrons in the emitter region.

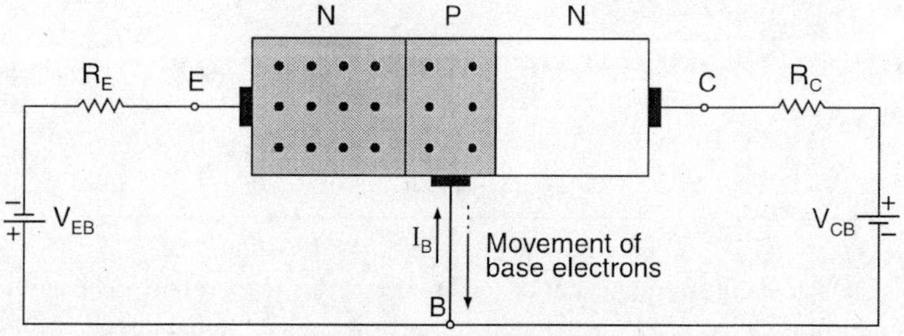

(*b*) Electrons in the base region.

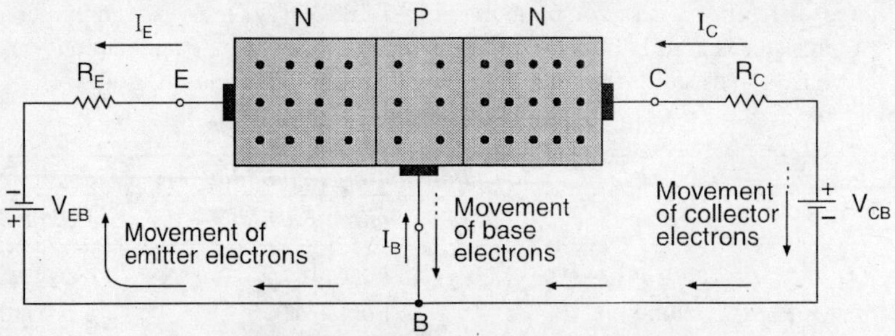

(*c*) Electrons entered in the collector region.

Fig. 9-10.

It will be interesting to know that there is another component of collector current due to the thermally generated carriers. This current component is called reverse saturation current and is quite small.

Notes. 1. The emitter current of a transistor consists of two components namely base current and collector current. The base current is about 2% of the emitter current, while collector current is about 98% of the emitter current.

2. The collector current is mainly due to the electrons injected from the emitter. However, there is another small component of collector current due to thermally generated carriers. This small component of collector current is called reverse saturation current.

3. The resistances R_E and R_C shown in the circuit of Fig. 14.10 are added to limit the magnitude of current in the transistor.

9-9. Operation of a PNP Bipolar Junction Transistor

The operation of a PNP transistor is similar to that of an NPN transistor. However, the current within a PNP transistor is due to the movement of holes, whereas in an NPN transistor it is due to the movement of free electrons. Fig. 9-11 shows a PNP transistor with its emitter-base junction as forward-biased and collector-base junction as reverse biased. The forward bias on the emitter-base junction causes the holes in the emitter region to flow towards the base region. This constitutes the emitter current (I_E). The holes, after reaching in the base region, tend to combine with the electrons. If these holes combine with the electrons in the base, they constitute base current (I_B). However, most of the holes do not combine with the electrons in the base region. It is due to the fact that base width is made extremely small and holes do not get sufficient electrons for recombination. Thus, most of the holes diffuse to the collector regions and constitute collector current (I_C). This collector current is called injected current because this current is produced due to the holes injected from the emitter region.

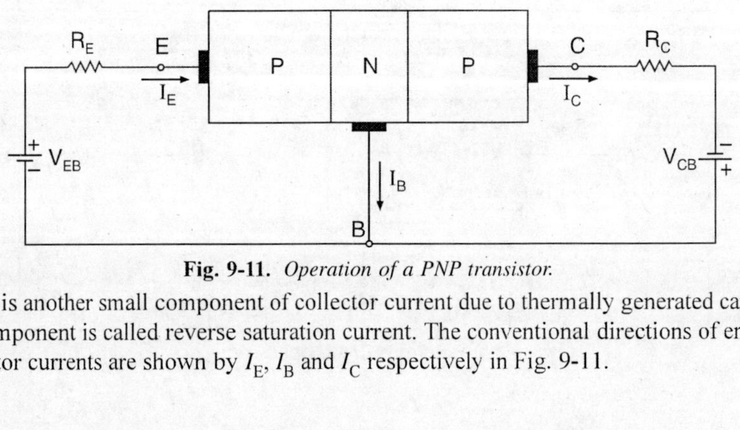

Fig. 9-11. *Operation of a PNP transistor.*

There is another small component of collector current due to thermally generated carriers. This current component is called reverse saturation current. The conventional directions of emitter, base and collector currents are shown by I_E, I_B and I_C respectively in Fig. 9-11.

9-10. Bipolar Junction Transistor Currents

We know that the direction of a conventional current is always opposite to the electron current in any electronic device. Therefore the conventional currents in a NPN transistor are as shown in Fig. 9-12 (*a*). However, the direction of a conventional current is the same as that of a hole current in a PNP transistor and is as shown in Fig. 9-12 (*b*). In this figure, I_E denotes the emitter current, I_B the base current and I_C the collector current respectively.

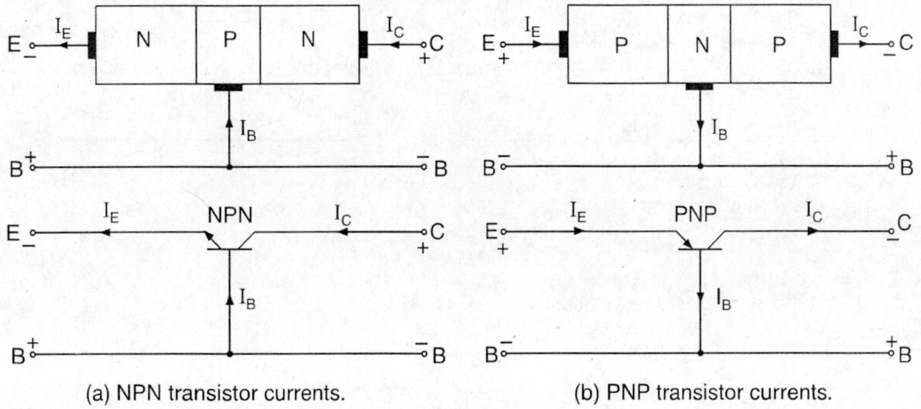

Fig. 9-12.

It is evident from these diagrams that the *emitter current is the sum of the collector and base currents*. Mathematically, the emitter current,

$$I_E = I_B + I_C$$

Since, the base current is very small, therefore

$$I_E \approx I_C$$

Example 9-1. *In a common-base connection, a certain transistor has an emitter current of 10 mA and a collector current of 9.8 mA. Calculate the value of the base current.*

Solution. Given: $I_E = 10$ mA and $I_C = 9.8$ mA

We know that emitter current (I_E),

$$10 = I_B + I_C = I_B + 9.8$$

$$\therefore \quad I_B = 10 - 9.8 = 0.2 \text{ mA } \textbf{Ans.}$$

9-11. Bipolar Junction Transistor Circuit Configurations

We know that a transistor has three terminals or leads namely emitter (*E*), base (*B*) and collector (*C*). However, when a transistor is connected in a circuit, we require four terminals, *i.e.*, two terminals for input and two for output. This difficulty is overcome by using one of the three terminals as a common terminal to the input and output terminals. Depending upon the terminals, which are used as a common terminals, the transistors can be connected in the following three different connections or configurations :

(1) ***Common-Base (CB) connection :*** In this configuration, the transistor is connected with the base as a common terminal as shown in Fig. 9-13(*a*). The input is applied between the emitter and base terminals. The output is taken between the collector and base terminals. This type of configuration is used to explain the operation of NPN and PNP transistors.

(2) ***Common-Emitter (CE) connection :*** In this configuration, the transistor is connected with the emitter as a common terminal as shown in Fig. 9-13(*b*). The input is applied between the base and collector terminals. The output is taken between collector and emitter terminals. This is one of the most commonly used connections of a transistor.

3. ***Common-Collector (CC) connection :*** In this configuration, the transistor is connected with collector as a common terminal as shown in Fig. 9-13 (*c*). The input is applied between the base and collector terminals. The output is taken between emitter and collector terminals.

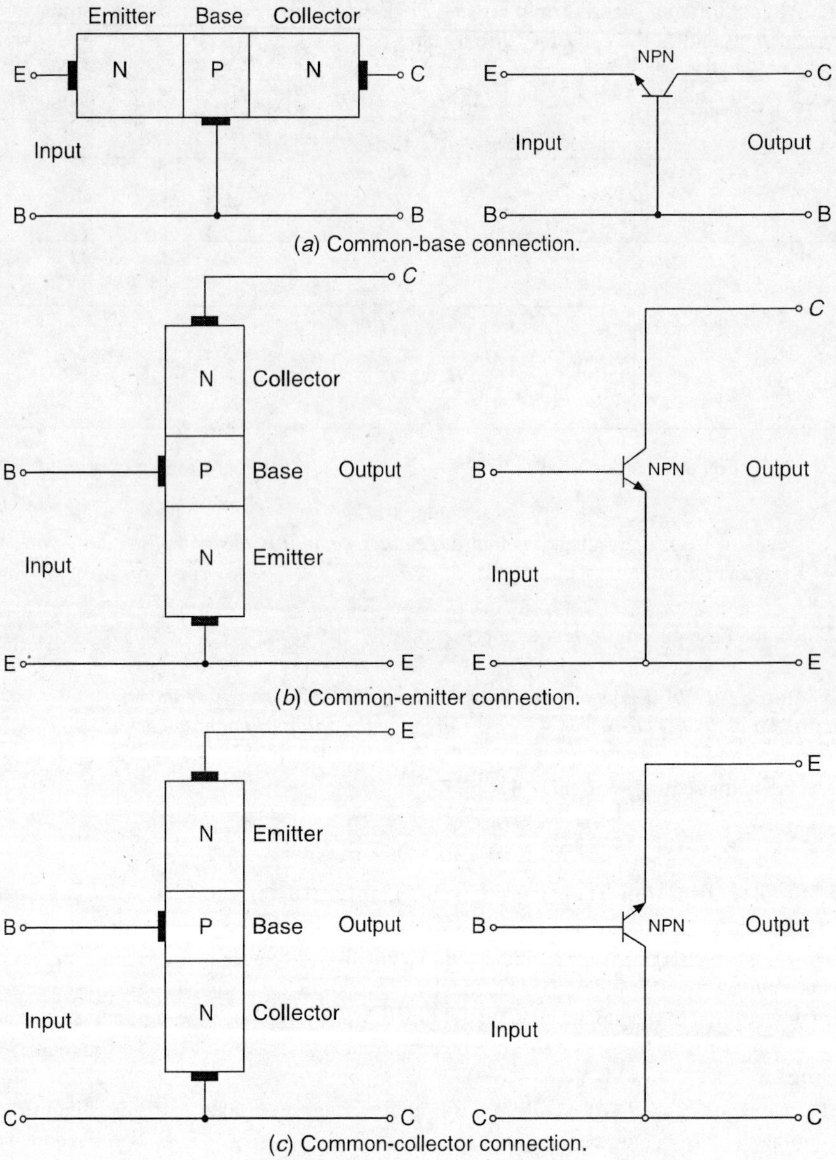

Fig. 9-13. *Transistor circuit connections or configurations.*

It may be noted that each circuit connection has its own advantages and disadvantages. We shall discuss these in Chapter-10. However, it may be noted that regardless of the connection type, the emitter-base junction of a transistor is forward-biased and collector-base junction is reverse-biased.

9-12. Current Gain of a Bipolar Junction Transistor in Common-Base Configuration

Consider a transistor (either NPN or PNP) in a common-base configuration as shown in Fig. 9-14(*a*) or (*b*). Here the emitter current is the input current and collector current is the output current. *The ratio of the transistor output current to the input current is called current gain of a transistor*. Since, the input current and output current may be either direct current or alternating current, therefore we define two types of current gains namely d.c. current gain and a.c. current gain.

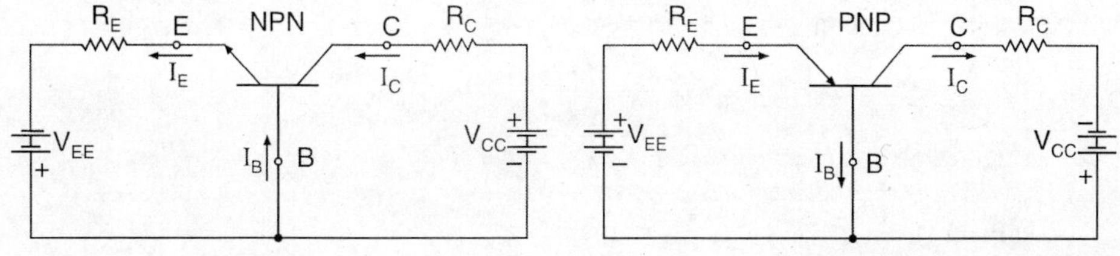

(*a*) Common-base NPN transistor circuit. (*b*) Common-base PNP transistor circuit.

Fig. 9-14.

(1) *Common-Base d.c. current gain* (α) : It is defined as the ratio of collector current (I_C) to emitter current (I_E) and is usually designated by α, α_{DC} or h_{FB}. Mathematically, the *common-base d.c. current gain,

$$\alpha = I_C/I_E \qquad ...(i)$$

We know that in a transistor, the collector current is always less than the emitter current. Therefore current-gain of a transistor in common-base configuration is always less than unity. For example, if $I_C = 9.8$ mA and $I_E = 10$ mA, then common-base d.c. current gain,

$$\alpha = 9.8/10 = 0.98$$

The above value of indicates that the collector current (I_C) is 98% of the emitter current (I_E). Therefore, the base current is just 2% of the emitter current. The value of d.c. current gain (α) is made closer to unity by making the width and doping level of base region as small as possible. This causes the majority of electrons in the NPN transistor (or holes in a PNP transistor) to reach the collector. Hence, the collector current will be higher. The actual value of α ranges from 0.95 to 0.998. From equation (*i*), we find that the collector current,

$$I_C = \alpha \cdot I_E$$

We also know that the emitter current,

$$I_E = I_B + I_C$$

$$\therefore \quad I_B = I_E - I_C = I_E - \alpha \cdot I_E = (1 - \alpha)I_E$$

(2) *Common-Base a.c. current gain* (α_0) : It is defined as the ratio of small change in collector current (ΔI_C) to a small change in emitter current (ΔI_E) for a constant collector-to-base voltage (V_{CB}). It is designated by α_0, α_{ac} or h_{fb}. Mathematically, the common-base a.c. current gain,

$$\alpha_0 = \Delta I_C/\Delta_E$$

The term α_0 is also called common-base short-circuit current gain or small signal current gain. The difference between d.c. current gain (h_{FB}) and a.c. current gain (h_{fb}) should be carefully noted. Here the upper case subscript (*i.e.*, FB) designates the d.c. value and lower-case subscript (*i.e.*, *fb*) designates the a.c. value. The values of α_0 is also less than unity and is approximately the same as .α For all practical purposes, d.c. current gain is considered equal to the a.c. current gain, *i.e.*,

$$\alpha = \alpha_0$$

* It is also referred to as large-signal common base d.c. current gain.

It may be noted that current gain of a transistor in a common-base configuration (α) is less than unity. But still it is called current gain. It is due to the fact that the output resistance of a common-base transistor is much higher than the input resistance. This produces a large voltage gain and hence the large power gain.

Example 9-2. *In a common-base connection, the emitter current is 6.28 mA and the collector current is 6.20 mA. Determine the common-base d.c. current gain.*

Solution. Given: $I_E = 6.28$ mA and $I_C = 6.20$ mA.

We know that common-base d.c. current gain,

$$a = \frac{I_C}{I_E} = \frac{6.20}{6.28} = 0.987 \textbf{ Ans.}$$

Example 9-3. *The common-base d.c. current gain of a transistor is 0.967. If the emitter current is 10 mA, what is the value of base current?*

Solution. Given: α = 0.967 and $I_E = 10$ mA.

We know that common-base d.c. current gain (α),

$$0.967 = \frac{I_C}{I_E} = \frac{I_C}{10}$$

$$\therefore \quad I_C = 0.967 \times 10 = 9.67 \text{ mA}$$

We also know that emitter current (I_E),

$$10 = I_B + I_C = I_B + 9.67$$

$$\therefore \quad I_B = 10 - 9.67 = 0.33 \text{ mA } \textbf{Ans.}$$

Example 9-4. *The transistor shown in Fig. 9-15 has values of $I_E = 10$ mA and α = 0.987. Determine the value of I_C and I_B for device.*

Solution. Given: $I_E = 10$ mA and α = 0.987

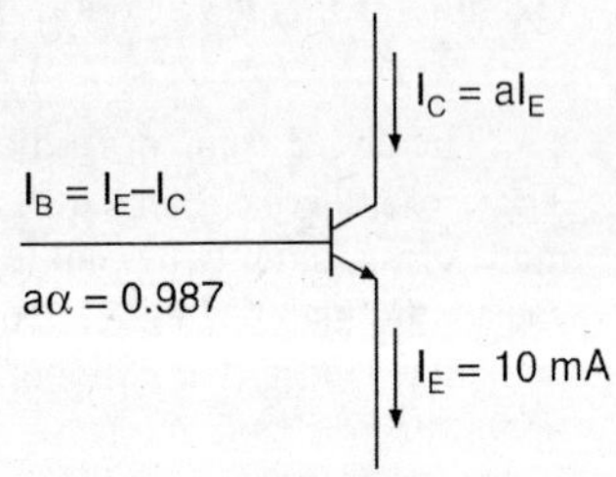

Fig. 9-15.

Value of I_C

We know that common-base d.c. current gain, (α),

$$0.987 = \frac{I_C}{I_E} = \frac{I_C}{10}$$

$$\therefore \quad I_C = 0.987 \times 10 = 9.87 \text{ mA } \textbf{Ans.}$$

We also know that emitter current (I_E),

$$10 = I_B + I_C = I_B + 9.87$$

$$\therefore \quad I_B = 10 - 9.87 = 0.13 \text{ mA } \textbf{Ans.}$$

9-13. Current Gain of Bipolar Junction Transistor in Common-Emitter configuration

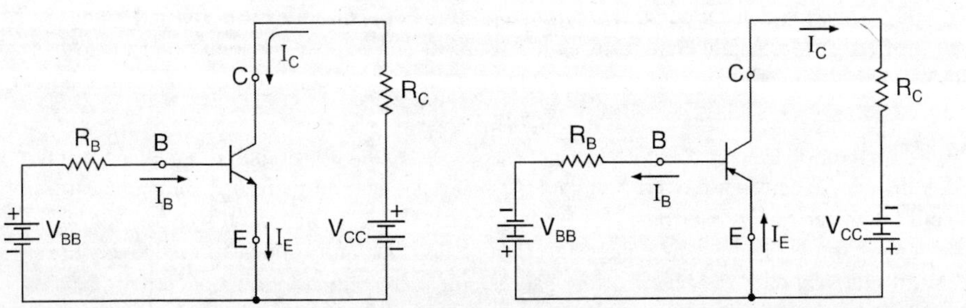

(*a*) Common-emitter NPN transistor circuit. (*b*) Common-emitter PNP transistor circuit.

Fig. 9-16.

Consider a transistor (either NPN or PNP) in a common-emitter configuration as shown in Figure 9-16(*a*) and (*b*). Here, the base current is the input current and the collector current is the output current. The current gain of a transistor in the common-emitter configuration is the ratio of collector current to the base current. Since, the base current and collector current may be direct or alternating current, therefore we define two types of current gains namely d.c. current gain and a.c. current gains.

(*a*) *Common-Emitter d.c. current gain* (β) : It is defined as the ratio of collector current (I_C) to base current (I_B) and is designated by β, β_{dc} or h_{FE}. Mathematically, the common-emitter d.c. current gain,

$$\beta = I_C/I_B \qquad ...(i)$$

Now we know that collector current of a transistor is much larger than the base current. Therefore, the value of β is much greater than unity. For example, if I_C = 5 mA and I_B = 0.05 mA, then common-emitter d.c. current gain,

$$\beta = 5/0.05 = 100$$

The above value of β indicates that collector current is 100 times that of base current. Typical values of β may range from 20 to 250. Sometimes β of a transistor is also known as large-signal common-emitter current gain.

(*b*) *Common-Emitter a.c. current gain* (β_0) : It is defined as the ratio of small change in collector current (ΔI_C) to the small change in base current (ΔI_B) for a constant collector-to-emitter voltage (V_{CE}). It is designated by β_0. β_{ac} or h_{FE}. Mathematically common-emitter a.c. current gain,

$$\beta_0 = \Delta I_C/\Delta I_B \qquad ...(ii)$$

The term β_0 is also called common-emitter short circuit current gain or small-signal common-emitter current gain. It will be interesting to know that β_0 or h_{fe} is a parameter used in the analysis of small-signal transistor amplifier circuits. β_0 have a range of values similar to that mentioned for β. For all practical purposes, d.c. current gain and a.c. current gain are considered to be equal. *i.e.*,

$$\beta = \beta_0$$

9-14. Relation Between Current Gain α and β

We know that emitter current (I_E) of a transistor is the sum of its base current (I_B) and collector current (I_C), *i.e.*,

$$I_E = I_B + I_C$$

Dividing the above equation on both sides by I_C,

$$\frac{I_E}{I_C} = \frac{I_B}{I_C} + 1$$

Since, $I_C/I_E = \alpha$ and $I_C/I_B = \beta$, therefore

$$\frac{1}{\alpha} = \frac{1}{\beta} + 1 = \frac{1+\beta}{\beta}$$

$$\therefore \qquad \boldsymbol{\alpha = \frac{\beta}{\beta + 1}}$$

The above expression may be written as

$$\alpha(\beta + 1) = \beta$$

$$\alpha \cdot \beta + \alpha = \beta$$

$$\alpha = \beta - \alpha \cdot \beta = \beta(1 - \alpha)$$

$$\therefore \qquad \boldsymbol{\beta = \frac{\alpha}{1 - \alpha}}$$

Example 9-5. *In a Common emitter configuration, find the value of **(a)** β, when a = 0.975, and **(b)** α, when β = 200.*

Solution. Given: α = 0.975 and β = 200

Value of β when α is 0.975

We know that

$$\beta = \frac{\alpha}{1-\alpha} = \frac{0.975}{1-0.975} = 39 \textbf{ Ans.}$$

Value of β when is 200

We also know that

$$\alpha = \frac{\beta}{\beta+1} = \frac{200}{200+1} = 0.995 \textbf{ Ans.}$$

Example 9-6. *A transistor has a typical β of 100. If the collector current is 40 mA, what is the value of emitter current?*

Solution. Given: β = 100 and I_C = 40 mA.

We know that (β), $$100 = \frac{I_C}{I_B} = \frac{40}{I_B}$$

$$I_B = 40/100 = 0.4 \text{ mA}$$

and the emitter current,

$$I_E = I_B + I_C = 0.4 + 40 = 40.4 \text{ mA } \textbf{Ans.}$$

Example 9-7. *A transistor has β = 150. Calculate the approximate collector and base currents, if the emitter current is 10 mA.*

Solution. Given: β = 150 and I_E = 10 mA.

Collector current

Let I_C = Collector current.

We know that common-base current gain,

$$\alpha = \frac{\beta}{\beta+1} = \frac{150}{150+1} = 0.993$$

and common-base current gain (α),

$$0.993 = \frac{I_C}{I_E} = \frac{I_C}{10}$$

$$\therefore \quad I_C = 0.993 \times 10 = 9.93 \text{ mA } \textbf{Ans.}$$

Base current

Let I_B = Base current.

We also know that emitter current (I_E),

$$10 = I_B + I_C = I_B + 9.93$$

$$\therefore \quad I_B = 10 - 9.93 = 0.07 \text{ mA } \textbf{Ans.}$$

Example 9-8. *Determine the value of I_B and I_E for the circuit shown in Fig. 9-17.*

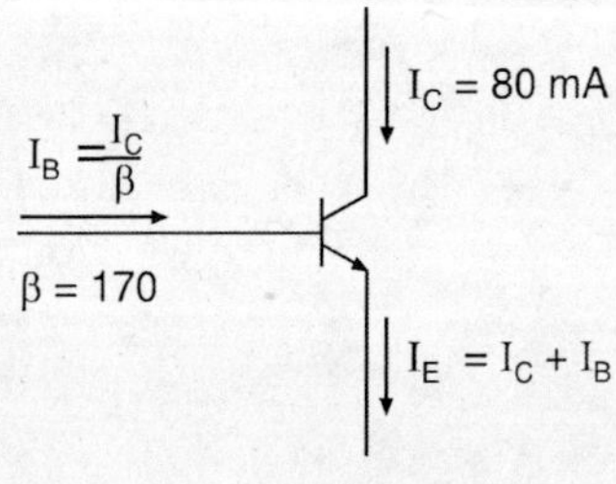

Fig. 9-17.

Solution. Given: β = 170 and I_C = 80 mA.

We know that (β),

$$170 = \frac{I_C}{I_B} = \frac{80}{I_B}$$

$$\therefore \qquad I_B = \frac{80}{170} = 0.47 \text{ mA } \textbf{Ans.}$$

and emitter current,

$$I_E = I_B + I_C = 0.47 + 80 = 80.47 \text{ mA } \textbf{Ans.}$$

Example 9-9. *Determine the value of I_C and I_E for the circuit shown in Figure 9-18.*

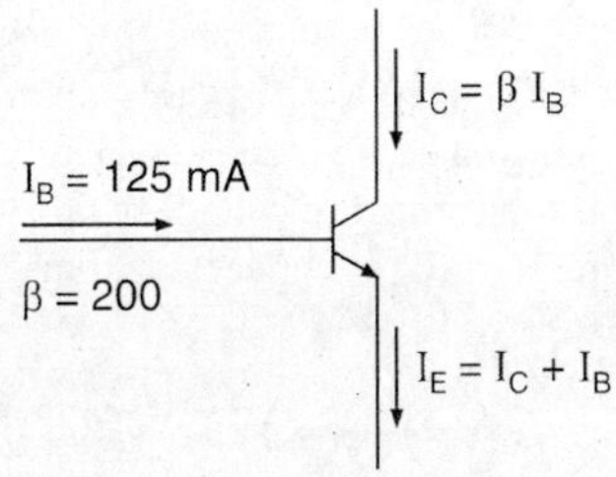

Fig. 9-18.

Solution. Given: $I_B = 125\ \mu\text{A} = 0.125$ mA and $\beta = 200$.

We know that (β),

$$200 = \frac{I_C}{I_B} = \frac{I_C}{0.125}$$

$$\therefore \qquad I_C = 200 \times 0.125 = 25 \text{ mA } \textbf{Ans.}$$

and emitter current,

$$I_E = I_B + I_C = 0.125 + 25 = 25.125 \text{ mA. } \textbf{Ans.}$$

Example 9-10. *Determine the value of I_C and I_B for the circuit shown in Fig. 9-19.*

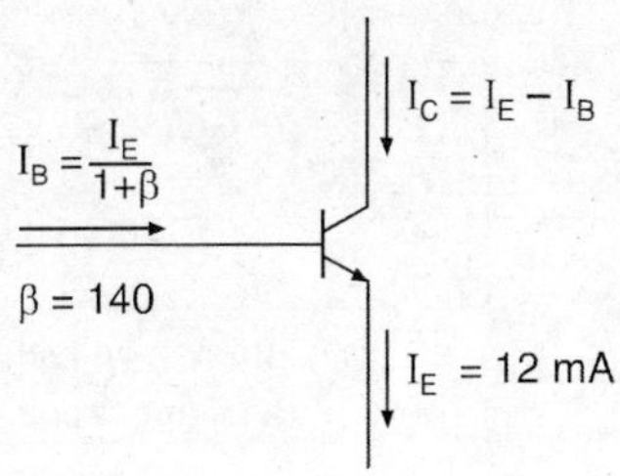

Fig. 9-19.

Solution. Given: $I_E = 12$ mA and $\beta = 140$.

We know that base current,

$$I_B = \frac{I_E}{1+\beta} = \frac{12}{1+140} = 0.085 \text{ mA } \textbf{Ans.}$$

and collector current,

$$I_C = I_E - I_B = 12 - 0.085 = 11.915 \text{ mA } \textbf{Ans.}$$

Example 9-11. *A transistor has $I_B = 105\ \mu A$ and $I_C = 2.05$ mA. Find:* ***(a)*** *β of the transistor;* ***(b)*** *of the transistor,* ***(c)*** *emitter current I_E,* ***(d)*** *Now, if I_B changes by + 27 μA and I_C changes by + 0.65 mA, find the new value of β.*

Solution. Given: $I_B = 105\ \mu\text{A} = 105 \times 10^{-6}$ A and $I_C = 2.05 \text{ mA} = 2.05 \times 10^{-3}$ A.

(*a*) *β of the transistor*

We know that β of the transistor,

$$\beta = \frac{I_C}{I_B} = \frac{2.05 \times 10^{-3}}{105 \times 10^{-6}} = 19.5 \textbf{ Ans.}$$

(*b*) *α of the transistor*

We know that of the transistor,

$$\alpha = \frac{\beta}{1+\beta} = \frac{19.5}{1+19.5} = 0.95 \textbf{ Ans.}$$

(*c*) *Emitter current, I_E*

We know that emitter current,

$$I_E = I_B + I_C = 105\ \ 10^{-6} + 2.05 \times 10^{-3} \text{ A}$$
$$= (0.105 + 2.05) \times 10^{-3} = 2.155 \times 10^{-3} \text{ A}$$
$$= 2.155 \text{ mA } \textbf{Ans.}$$

(*d*) *New value of β when*

$$\Delta I_B = 27\ \mu\text{A and } \Delta I_C = 0.65 \text{ mA.}$$

We know that new base current,

$$I_B = 105 + 27 = 132\ \mu\text{A} = 132 \times 10^{-6} \text{ A}$$

and new $\qquad I_C = 2.05 + 0.65 = 2.7 \text{ mA} = 2.7 \times 10^{-3}$ A

∴ New value of β of the transistor,

$$\beta = \frac{I_C}{I_B} = \frac{2.7 \times 10^{-3}}{132 \times 10^{-6}} = 20.5 \text{ Ans.}$$

9-15. Current Gain of a Bipolar Junction Transistor in Common-Collector Configuration

Figure 9-20(*a*) and (*b*) show a NPN transistor connected in a common-collector configuration. The voltage V_{BB} forward-biases the emitter-base junction and V_{CC} reverse-biases the collector-base junction of the transistor. Here, the collector is not at d.c. ground. But it is at a.c. ground. It is due to the fact, that the voltage of d.c. supply source (V_{CC}) has zero resistance to an a.c. signal. An external resistor R_E is connected from the emitter to ground. The output of the circuit is taken across the resistor R_E.

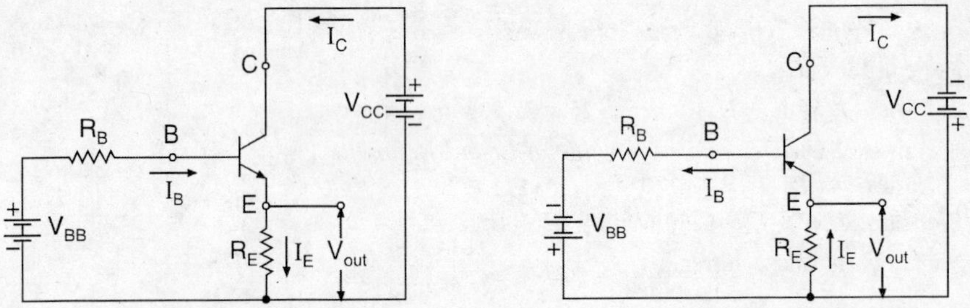

(*a*) Common-collector NPN transistor circuit. (*b*) Common-collector PNP transistor circuit.

Fig. 9-20.

It may be noted that in a common-collector transistor circuit, the input current is the base current (I_B) and the output current is the emitter current (I_E). Therefore, common-collector current gain is given by the relation,

$$= \frac{I_E}{I_B} = \frac{I_E}{I_C} \times \frac{I_C}{I_B} = \frac{1}{\alpha} \times \beta$$

Substituting the value of $\alpha = \dfrac{\beta}{1+\beta}$ in the above equation,common-collector current gain

$$= \frac{\beta}{\beta/(1+\beta)} = 1 + \beta.$$

It means that the output current of a common-collector transistor circuit is $(1 + \beta)$ times that of the input current. We know that $\beta >> 1$, therefore common-collector current gain ≈ β.

9-16. Leakage Current in a Common-Base Bipolar Junction Transistor

Consider a common base (CB) PNP transistor circuit as shown in Figure 9-21. When the switch is in its closed position, the forward-bias voltage (V_{EE}) injects the holes in the base region. The reverse-bias voltage (V_{CC}) on the collector-base junction attracts the majority of holes, from the base region, and constitute collector current. A small number of holes combine with the electrons in the base region and constitute base current. It means that the total emitter current splits up into two components namely the base current (I_B) and the collector current (I_C).

Now, let the switch '*S*' be opened. It disconnects the emitter from the base and hence, the emitter-base junction of the transistor is open-circuited. Thus, there is no emitter current and therefore no base current (I_B) or collector current (I_C). However, it may be noted that the collector-base junction of the transistor is reverse-biased due to the holes injected from the emitter. But this junction is forward-biased due to the thermally generated minority carriers (*i.e.*, electrons in the P-type collector

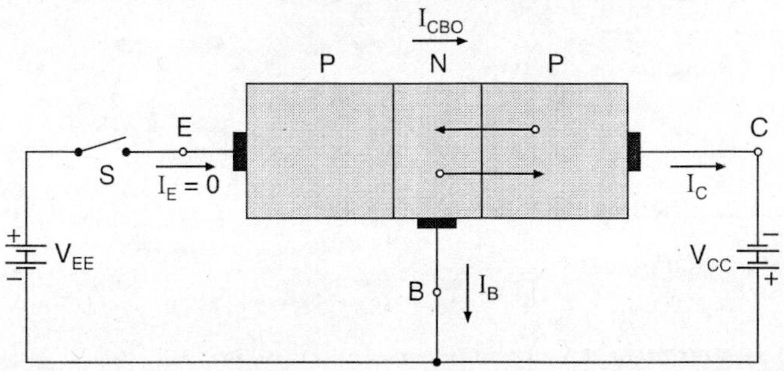

Fig. 9-21. *Leakage current in a common-base transistor.*

region and holes in the N-type base region). The minority carriers diffuse across the collector-base junction and hence, produce a certain value of current known as *leakage current*. This current is called the leakage current from collector to base with emitter open and is designated by I_{CBO}. It is also known as the *reverse saturation current* or *collector cut-off current* (I_{CO}). Thus, I_{CBO} or I_{CO} is similar to the reverse saturation current in a PN junction. The direction of I_{CBO} or I_{CO} is the same as that of a injected current. It is evident from the above discussion that the total collector current, in a transistor, consists of the following two components :

(1) The current produced by normal transistor action. This component of a current is called injected current. Its value is equal to $\alpha \cdot I_E$ and is due to majority carriers.

(2) The reverse saturation current (I_{CO}) produced by the thermally generated carriers or minority carriers.

Hence, the total collector current,

$$I_C = \alpha I_E + I_{CO}$$

The above relation gives us a more accurate definition of common base current gain (α) of a transistor.

$$\therefore \qquad \alpha = \frac{I_C - I_{CO}}{I_E}$$

Substituting the value of $I_E = I_B + I_C$ in this equation,

$$\alpha = \frac{I_C - I_{CO}}{I_B + I_C}$$

$$\therefore \qquad I_C = \alpha (I_B + I_C) + I_{CO} = \alpha \cdot I_B + \alpha \cdot I_C + I_{CO}$$

$$(1 - \alpha) I_C = \alpha \cdot I_B + I_{CO}$$

$$\therefore \qquad I_C = \frac{\alpha \cdot I_B}{1 - \alpha} + \frac{I_{CO}}{1 - \alpha}$$

It will be interesting to know that at room temperature, for silicon transistor, the value of I_{CO} is usually of the order of nanoamperes and thus is negligible. However, for germanium transistors the value of I_{CO} is of the order of microamperes and thus cannot be neglected. The value of I_{CO} is strongly temperature dependent for both silicon and germanium. It doubles for every 10°C increase in temperature. However, because of the lower magnitude of I_{CO} in silicon, the transistors can be used up to 200°C. On the other hand, germanium transistors are limited to about 100°C only.

Example 9-12. *A certain transistor has* $\alpha = 0.98$, $I_{CO} = 5\ \mu A$ *and* $I_B = 100\ \mu A$. *Find the values of collector and emitter currents.*

Solution. Given: $\alpha = 0.98$, $I_{CO} = 5\ \mu A$ and $I_B = 100\ \mu A$

Collector current

We know that the collector current,

$$I_C = \frac{\alpha \cdot I_B}{1-\alpha} + \frac{I_{CO}}{1-\alpha} = \frac{0.98 \times 100}{1-0.98} + \frac{5}{1-0.98}\ \mu A$$

$$= 4900 + 250 = 5150\ \mu A = 5.15\ mA\ \textbf{Ans.}$$

Emitter current

We also know that the emitter current,

$$I_E = I_B + I_C = 100 + 5150 = 5250\ \mu A = 5.25\ mA\ \textbf{Ans.}$$

9-17. Leakage Current in a Common-Emitter Bipolar Junction Transistor

Consider a common-emitter (CE) PNP transistor circuit as shown in Fig. 9-22. When the switch is in its closed position (*i.e.*, 'ON' position), the current flows in the same way as discussed in the last article. However, if the switch is in open position (*i.e.*, 'OFF' position), then the base is disconnected from the emitter. Hence, the base-emitter junction of the transistor is opened and therefore base current is zero. Under this condition, a leakage current flows from emitter to collector terminal. This current is designated by I_{CEO}. This leakage current is not just due to the thermally generated carriers across the collector-base junction, but also due to the movement of electrons across the base-emitter junction. This flow of electrons slightly forward biases the base-emitter junction. It appears as if a base current (equal to I_{CO}) is being supplied to the transistor. This produces an additional collector current (equal to $\beta \cdot I_{CO}$). Thus, the total leakage current flowing through the transistor with base open,

$$\boldsymbol{I_{CEO} = I_{CO} + \beta \cdot I_{CO} = (1+\beta)\, I_{CO}}$$

The total collector current in a transistor consists of the injected current ($\beta \cdot I_B$) and the leakage current (I_{CEO}). Thus the total collector current,

$$I_C = \beta \cdot I_B + \beta_{CEO} = \beta \cdot I_B + (1+\beta)\, I_{CO}$$

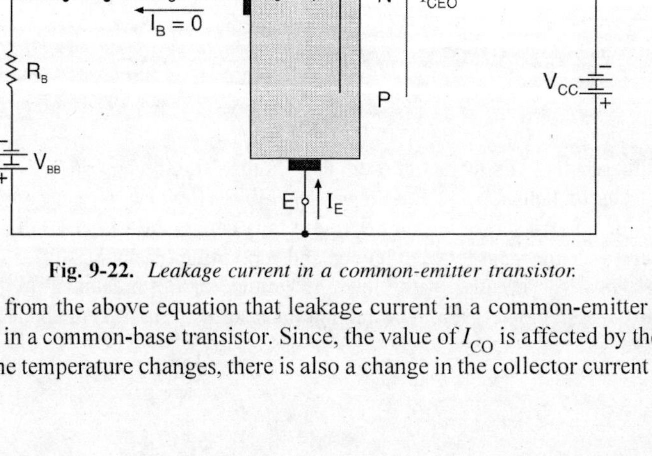

Fig. 9-22. *Leakage current in a common-emitter transistor.*

It is evident from the above equation that leakage current in a common-emitter transistor is β times larger than in a common-base transistor. Since, the value of I_{CO} is affected by the temperature, therefore when the temperature changes, there is also a change in the collector current of a common-emitter circuit. For example, let a silicon transistor has I_{CO} at 25°C equal to 0.025 μA and at 150°C, I_{CO} equal to 15 A. The transistor has a β of 100. Thus, if this transistor is used in a common-emitter circuit, then its leakage current (I_{CEO}) would be 25 μA at 15°C and 1500 μA at 150°C. This may lead to instability in the common-emitter transistor circuit. Thus, we conclude from the above discussion that common-base transistor circuit is much less temperature dependent than the common-emitter transistor circuit.

Example 9-13. *A germanium transistor used in a complementary symmetry amplifier has a collector cut-off current I_{CBO} = 10 μA at a temperature of 27°C and h_{FE} = 50.*

(a) *What is the collector current, when the base current is 0.25 mA?*

(b) *Assuming h_{FE} does not increase with temperature, what would be the value of new collector current, if the transistor's temperature rises to 50°C?*

Solution. Given: $I_{CBO} = 10\ \mu A = 10 \times 10^{-6}$ A and h_{FE} (= β) = 50

(a) *Value of collector current when I_B = 0.25 mA (0.25 × 10^{-3} A)*

We know that collector current,

$$I_C = \beta I_B + (1 + \beta)\, I_{CBO}$$
$$= 50 \times (0.25 \times 10^{-3}) + (1 + 50) \times (10 \times 10^{-6})\text{A}$$
$$= 13.01 \times 10^{-3}\ \text{A} = 13.01\ \text{mA}\ \textbf{Ans.}$$

(b) *Value of new collector current if temperature rises to 50°C*

We also know that I_{CBO} doubles for every 10°C rise in temperature. Thus, I_{CBO} at 50°C.

$$I_{CBO50} = I_{CBO} \times 2^{(T_2 - T_1)/10} = 10 \times 2^{(50-27)/10}\ \mu A$$
$$= 10 \times 2^{2.3} = 49.2\ \mu A = 49.2 \times 10^{-6}\ \text{A}$$

∴ Collector current at 50°C,

$$I_C = \beta \cdot I_B + (1+\beta)\, I'_{CBO}$$
$$= 50 \times (0.25 \times 10^{-3}) + (1 + 50) \times 49.2 \times 10^{-6}\text{A}$$
$$= 15.01 \times 10^{-3}\ \text{A} = 15.01\ \text{mA}\ \textbf{Ans.}$$

9-18. Base Spreading Resistance

We know that the width of the base region of a transistor is extremely small. Therefore, the current, which enters the base region, across the emitter-base junction, has a narrow path to reach the base terminal. The resistance, offered by this narrow path region, is called *base spreading resistance* and is designated by r_b'. The value of this resistance can be increased by increasing the reverse-bias voltage (V_{CB}) across the collector-base junction. It is because of the fact that by increasing the value of V_{CB}, the width of collector-base depletion layer increases. Since, the depletion layer penetrates more into the base-region than that of collector, therefore the cross-sectional area the base-region is reduced. The smaller cross-sectional means a large base-spreading resistance. It may be noted that the resistances of emitter and collector regions are very small, as these regions are much larger in size as compared to the base.

Usually, the value of base spreading resistance (r_b') is about 50 to 150 ohms. But in some cases, it may be as high as 1000 ohms. When a current passes through base spreading resistance, it produces a voltage drop across it. It has been observed that the effect of base spreading resistance is significant in high frequency circuits. At low frequencies, it has practically no effect.

9-19. Current Gain or β-Curve

We have already discussed in Art. 9-13 that common-emitter current gain (β) is very important transistor parameter. Therefore, its study is very essential for every electronic engineer and technician. It has been experimentally found that the value of β varies with the collector current (I_C) and the junction temperature (T_j).

Figure 9-23 shows a graph of β versus I_C at different junction temperatures (*i.e.*, – 55°C, 25°C and 125°C). Such a graph is called current gain graph, (or β curves). It may be noted from the graph that for a constant junction temperature, β has a maximum value at certain value of I_C. At a fixed

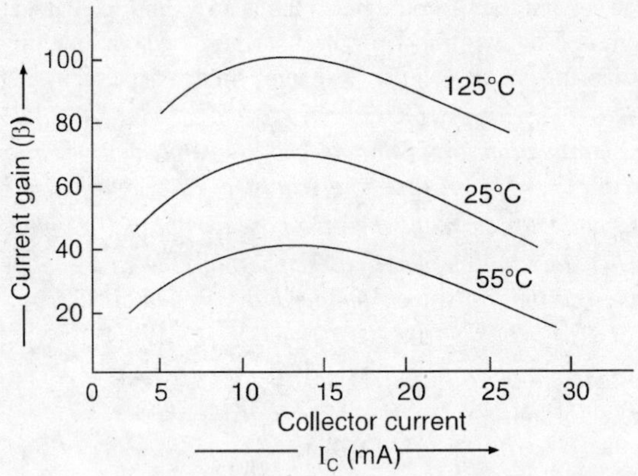

Fig. 9-23. *Current gain β curves.*

value of I_C, the value of β increases with the increase in junction temperature. The variation in β over a large temperature range may be as high as 3 : 1. The manufacturer always specifies the value of β for the fixed values of I_C and junction temperature. It will be interesting to know that for a fixed value of I_C and junction temperature, the value of β varies from device to device for a given transistor.

SUMMARY

In this chapter you have learned that :

1. Bipolar junction transistor consist of two PN junctions namely emitter base junction and collector-base junction.
2. Emitter, base and collector are the three terminals at a bipolar junction transistor.
3. NPN and PNP are the two types of bipolar junction transistor.
4. Forward active, saturation and cut-off are the three modes of operation of a junction transistor.
5. Common-base, common-emitter and common-collector are the bipolar junction transistor circuit connections.

GLOSSARY

Base (B) : The middle region of a bipolar junction transistor that forms two P-N junction which is lightly doped.

Base spreading resistance ($r_{b'}$) : The resistance offered by the base of a bipolar junction transistor to the current entering across the emitter-base junction.

Bipolar junction transistor (BJT) : A three terminal device formed by sandwitching either P-type or N-type semiconductor layers between a pair of opposite types.

Collector (C) : The region situated in one end of the bipolar junction transistor which is larger than the other regions and moderately doped.

Common-Base a.c. current gain (α_0) : The ratio of small change in collector current to a small change in emitter current for a constant collector-to-base configuration.

Common-Base d.c. current gain (α) : The ratio of collector current to emitter current of a bipolar junction transistor in common-base configuration.

Common-Emitter a.c. current again (β_0) : The ratio of small change in collector current to the small change in base current for constant collector-to-emitter voltage of a bipolar junction transistor in common-emitter configuration.

Common-Emitter d.c. current gain (β) : Ratio of collector current to base current of a bipolar junction transistor in common-emitter configuration.

Current gain : The ratio of output current to the input current.

Cut-off mode : The operation mode of a bipolar junction transistor in which its emitter-base and collector-base junctions are reverse biased.

Emitter (E) : The region situated in one end of the bipolar junction transistor which is heavily doped.

Forward-active mode : The operation mode of a bipolar junction transistor in which emitter-base junction is forward biased and collector-base junction is reverse biased.

NPN transistor : Bipolar junction transistor formed by sandwitching P-type semiconductor layer in between two N-type semiconductor layers.

PNP transistor : Bypolar juction transistor formed by sandwiching N-type semiconductor type in between two types of semiconductor layers.

Saturation mode : The operation mode of a bipolar junction transistor in which its emitter -base and collector-base junction are forward-biased. In this mode a very large of current flows through it.

NUMERICAL PROBLEMS

1. A certain transistor has of 0.98 and a collector leakage current I_{CO} of 1 μA. Calculate the collector and the base currents, when I_E = 1 mA. (**Ans.** 0.981 mA and 19 μA)

2. In a CB connection, the emitter current is 5.80 mA and the collector current is 5.6 mA. Find the value of α . (**Ans.** 0.966)

3. The beta of a transistor is 100. If the collector current is 50 mA, find the value of base current. (**Ans.** 0.5 mA)

4. A germanium transistor has I_{CBO} = 10 μA, = 0.98 and I_C = 1 mA. Determine the emitter current and β of the transistor. (**Ans.** 1.01 mA; 98)

5. The transistor shown in Figure 9-24 has values of I_B = 50 μA and β = 400. Determine the value of I_C and I_E for the device. (**Ans.** 20 mA, 20.05 mA)

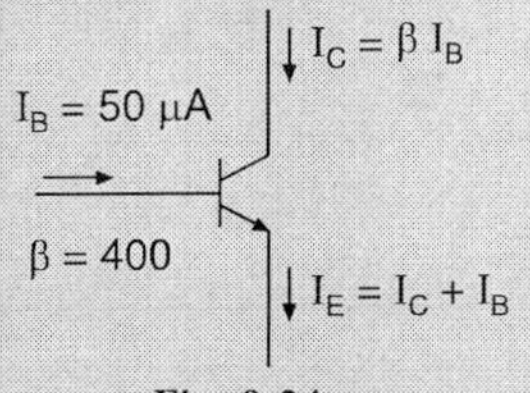

Fig. 9-24.

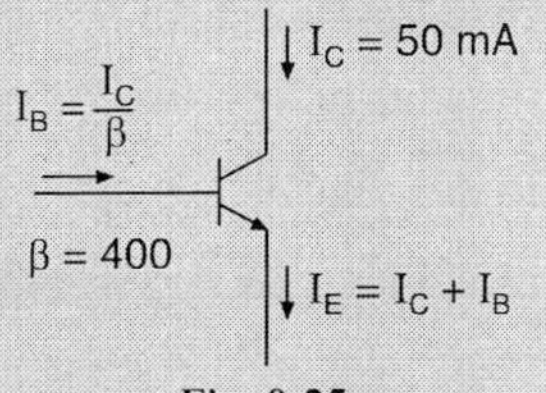

Fig. 9-25.

6. Determine the value of I_B and I_E for the circuit shown in Figure 9-25. (**Ans.** 0.125 mA, 50.125 mA)

7. A germanium transistor's collector current is 50 mA, when the base current is 0.49 mA. If h_{FE} = 100, what is the transistor's collector cut-off current I_{CEO}. (**Ans.** 1mA)

8. **(*a*)** Find the value of d.c. beta of a transistor if its collector current is 3 mA and base current is 25 μA.

(*b*) Determine the value of d.c. alpha using the value of d.c. beta obtained in part (*a*).

(*c*) Calculate the value of I_{CBO} by using the value of d.c. beta obtained in part (*a*) if I_{CBO} is 600 μA. (**Ans.** (*a*) 120 (*b*) 0.992 and (*c*) 4.96 μA)

DESCRIPTIVE QUESTIONS

1. Explain, briefly, what do you mean by a transistor.
2. Write a short note on transistor manufacturing techniques.
3. Draw the current symbols for NPN and PNP transistors. Also indicate the conventional directions for three currents.
4. What are the different ways of biasing a transistor? Which method of biasing is used for operating transistor as an amplifier?
5. Describe the operation of a NPN transistor.
6. Define the current gain and current gain β of a transistor.
7. Derive the relation between alpha and beta of a transistor.
8. What is the order of magnitude of the reverse collector saturation current I_{CBO} for a silicon transistor? How does I_{CBO} vary with the temperature?
9. Define h_{FE} (*i.e.*, beta) of a transistor and sketch a graph of h_{FE} versus I_C.
10. Write a short note on "Transistor construction"
    (*Electronic Devices, Anna University, Nov/Dec 2007*)
11. Show that $I_C = \alpha I_E + I_{CBO}$ and discuss the parameters given in the equation.
    (*Electron Devices, Anna University, Nov/Dec 2007*)
12. Draw the power flow diagram of the transistor.
    (*Electronic Circuits - I, Anna University, April/May 2008*)

MULTIPLE CHOICE QUESTIONS

1. In a PNP transistor with normal bias
   (*a*) Only holes cross the collector junction
   (*b*) Only majority carriers cross the collector junction
   (*c*) The collector junction has a low resistance
   (*d*) The emitter junction is forward-biased and collector junction is reverse biased
2. For transistor action
   (*a*) The collector must be more heavily doped than the base region
   (*b*) The base region must be N-type material
   (*c*) The base region must be very narrow
   (*d*) The collector-base junction must be forward biased
3. In a PNP transistor, the electrons flow
   (*a*) Into the transistor at the emitter and base leads
   (*b*) Out of the transistor at the emitter and base leads
   (*c*) Into the transistor at the collector and base leads
   (*d*) Out of the transistor at the collector and base leads
4. The leakage current I_{CBO} flows in
   (*a*) The emitter, base and collector leads (*b*) The emitter and base leads
   (*c*) The emitter and collector leads (*d*) The base and collector leads
5. A small increase in collector-base reverse bias will cause
   (*a*) A large increase in emitter current
   (*b*) A large increase in collector current
   (*c*) Very little change in collector reverse saturation current
   (*d*) A large decrease in collector current

6. Early effect in BJT refers to
(*a*) avalanche breakdown (*b*) thermal breakdown
(*c*) base narrowing (*d*) Zener breakdown
(*UPSC Engg. Services* 2002)

7. A PNP transistor has
(*a*) only donor ions (*b*) only acceptor ions
(*c*) two P-regions an one N-region (*d*) three PN junctions

8. The emitter of a transistor is generally doped the heaviest because it
(*a*) has to dissipate maximum power (*b*) has to supply the charge carriers
(*c*) is the first region of the transistor (*d*) must possess low resistance.

9. For current working of an *NPN* bipolar junction transistor, the different electrodes should have the following polarities with respect to emitter.
(*a*) collector + *ve*, base – *ve* (*b*) collector – *ve*, base + *ve*
(*c*) collector – *ve*, base – *ve* (*d*) collector + *ve*, base + *ve*

10. Select the CORRECT alternative. In a bipolar transistor
(*a*) emitter region is of low/high resistivity matterial which is lightly/heavily-doped.
(*b*) collector region is of lower/higher conductivity than emitter region
(*c*) base region is of high/low resistivity material which is only lightly/heavily-doped.

11. In a properly-biased *NPN* transistor, most of the electrons from the emitter
(*a*) recombine with holes in the base (*b*) recombine in the emitter itself
(*c*) pass through the base to the collector (*d*) are stopped by the junction barrier.

12. The following relationship between α and β are correct EXCEPT
(*a*) $\beta = \dfrac{\alpha}{1-\alpha}$ (*b*) $\alpha = \dfrac{\beta}{1-\beta}$
(*c*) $\alpha = \dfrac{\beta}{1+\beta}$ (*d*) $1-\alpha = \dfrac{1}{1+\beta}$.

13. The value of total collector current in a *CB* circuit is
(*a*) $I_C = \alpha I_E$ (*b*) $I_C = \alpha I_E + I_{CO}$
(*c*) $I_C = \alpha I_E - I_{CO}$ (*d*) $I_C = \beta I_E$.

14. In a junction transistor, the collector cut off current I_{CBO} reduces considerably by doping the
(*a*) emitter with high level of impurity (*b*) emitter with low level of impurity
(*c*) collector with high level of impurity (*d*) collector with low level of impurity

15. In a transistor amplifier, the reverse saturation current I_{CO}
(*a*) doubles for every 10°C rise in temperature
(*b*) doubles for every 1°C rise in temperature
(*c*) increase linearly with the temperature
(*d*) doubles for every 5°C rise in temperature

16. In the case of a bipolar transistor, α is
(*a*) positive and > 1 (*b*) positive and < 1
(*c*) negative and > 1 (*d*) negative and < 1.

17. The emitter-base of a given transistor is forward-biased and its collector-base reverse-biased. If the base current is increased, then its

(*a*) I_C will decrease (*b*) V_{CE} will increase
(*c*) I_C will increase (*d*) V_{CC} will increase.

18. The collector characteristics of a common-emitter connected transistor may be used to find its
(*a*) input resistance (*b*) base current
(*c*) output resistance (*d*) voltage gain.

19. Which of the following approximations is often used in electronic circuits ?
(*a*) $I_C \cong I_E$ (*b*) $I_B \cong I_C$
(*c*) $I_B \cong I_E$ (*d*) $I_E \cong I_B + I_C$.

20. If in a bibolar junction transistor, I_B = 100 μA and I_C = 10 mA, in what range does the value of its beta lie ?
(*a*) 0.1 to 1.0 (*b*) 1.01 to 10
(*c*) 10.1 to 100 (*d*) 100.1 to 1000

21. In a BJT, largest current flow occurs
(*a*) in the emether (*b*) in the collector
(*c*) in the base (*d*) through CB junction.

22. Which of the following transistor configuration circuits is much less temperature dependent ?
(*a*) common-base (*b*) common-emitter
(*c*) common-collector (*d*) none of the above

(Bharthiar Univ. April 2004)

ANSWERS

1. (*d*)	**2.** (*c*)	**3.** (*c*)	**4.** (*d*)	**5.** (*d*)
6. (*c*)	**7.** (*c*)	**8.** (*b*)	**9.** (*d*)	**10.** (*a*)
11. (*c*)	**12.** (*b*)	**13.** (*b*)	**14.** (*d*)	**15.** (*a*)
16. (*b*)	**17.** (*a*)	**18.** (*c*)	**19.** (*a*)	**20.** (*a*)
21. (*b*)	**22.** (*c*)			

❑❑❑

10

CHAPTER

BJT Characteristics

OUTLINE

Objectives

After completing this chapter, you should be able to :

- Know the characteristic curves of a transistors (*i.e.*, input characteristics and output characteristics).
- Understand base width modulation or early effect.
- Differentiate the characteristics between common-base and common-emitter configurations of a transistor.
- The ratings (*i.e.*, maximum collector current, maximum power dissipation and maximum output voltage) of a transistor.
- Understand thermal runaway.
- Know integrated, high-voltage and high-current transistors.

10-1. Introduction

We have already discussed in the last chapter about the construction of a transistor and its current gains α and β. As a matter of fact, the values of α and β do not completely describe the behavior of a transistor. Many more details can be studied with the help of curves, which relate the transistor currents and voltages. These curves are known as characteristic curves. Following are the two sets of characteristic curves, which are important from the subject point of view :

(1) ***Input characteristics :*** These curves give the relationship between the input current and input voltage for a given output voltage.

(2) ***Output characteristics :*** These curves give the relationship between the output current and the output voltage for a given input current.

These two sets of characteristics curves completely describe the static operation of a transistor. Depending upon the configuration, the transistor characteristics may be studied under the following two heads :

(*i*) Characteristics of a transistor in a common-base configuration.

(*ii*) Characteristics of a transistor in a common-collector configuration.

The characteristics of a transistor, in a common-collector configuration, are not needed as they can be treated as a special case of common-emitter configuration (with feedback applied). That is why such characteristics are not discussed in this chapter.

10-2. Characteristics of a Transistor in a Common-Base Configuration

Following are two important characteristics of a transistor in a common-base (CB) configuration :

(*i*) ***Input characteristics :*** These curves give the relationship between the emitter current (I_E) and the emitter-to-base voltage (V_{EB}) for a constant collector-to-base voltage (V_{CB}). These curves are also known as *base curves* of a transistor.

(*ii*) ***Output characteristics :*** These curves give the relationship between the collector current (I_C) and the collector-to-base voltage (V_{CB}) for a constant emitter current (I_E). These curves are also known as *collector curves* of a transistor.

These characteristics may be obtained by means of a circuit arrangement as shown in Fig. 10-1. In this circuit, the NPN transistor is connected in a common-base configuration. The collector-to-base voltage can be varied by changing the position of potentiometer R_2, while the emitter-to-base voltage can be varied by changing the position of potentiometer R_1. The d.c. milliammeters and d.c. voltmeters are connected in the emitter and collector circuits of a transistor to measure the voltages and currents.

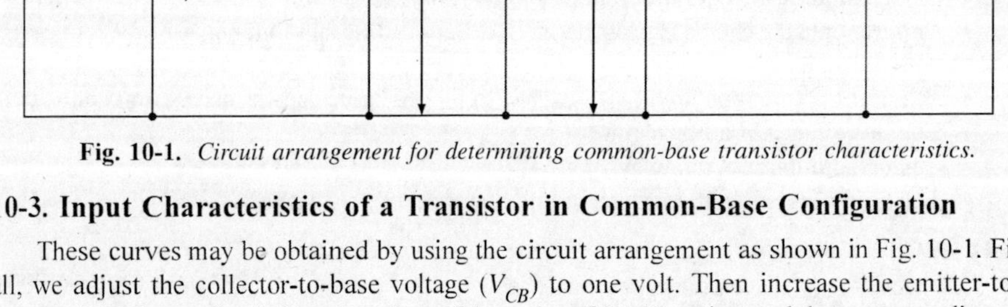

Fig. 10-1. *Circuit arrangement for determining common-base transistor characteristics.*

10-3. Input Characteristics of a Transistor in Common-Base Configuration

These curves may be obtained by using the circuit arrangement as shown in Fig. 10-1. First of all, we adjust the collector-to-base voltage (V_{CB}) to one volt. Then increase the emitter-to-base voltage (V_{EB}) in small suitable steps (*i.e.*, of the order of 0.1 V) and record the corresponding values of emitter current (I_E) at each step. Now, if we plot a graph with emitter-to-base voltage (V_{EB}) along the horizontal axis and the emitter current (I_E) along the vertical axis, we shall obtain a curve marked V_{CB} = 1 V as shown in Fig. 10-2. A similar procedure may be used to obtain curves at different collector-to-base voltage 5 V and 10 V as shown in the Fig. 10-2.

It will be interesting to know that the input characteristics give us the information about the following important points :

(i) There exists a threshold voltage (also called offset voltage, cut-in voltage or knee voltage) as indicated by the region *OA*, below which the emitter current is negligibly small. The knee-or threshold voltage is designated by V_K (orV_γ). The value of knee voltage (V_K) is approximately 0.5 V for silicon and 0.1 V for germanium transistors.

(ii) Beyond the point *A*, for a fixed collector-to-base voltage the emitter current (I_E) increases rapidly with a small increase in emitter-to-base voltage (V_{EB}). It means that input resistance (R_i) of a transistor in common-base configuration is very small.

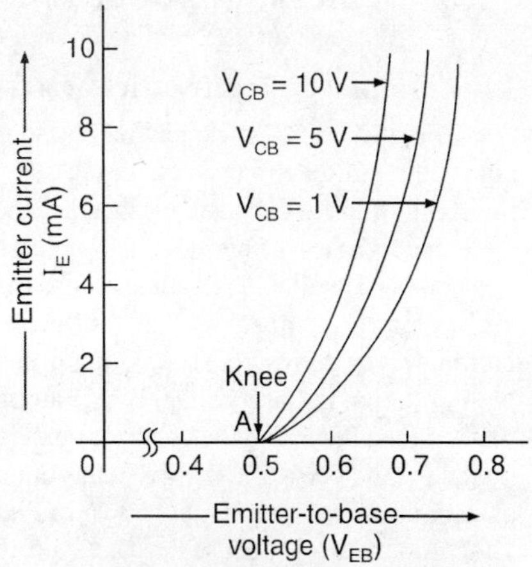

Fig. 10-2. *Input characteristics of a common-base transistor.*

(iii) As the collector-to base voltage (V_{CB}) is increased above one volt, the curves shift upwards. It occurs due to a phenomenon called base-width modulation or early effect. It will be explained in the next article.

(iv) The input characteristic may be used to determine the value of *a.c. input resistance. Its value at any point on the curve is given by the ratio of a change in emitter-to-base voltage (ΔV_{EB}) to the resulting change in emitter current (ΔI_E) for a constant collector-to-base voltage (V_{CB}). Mathematically the a.c. input resistance,

$$R_i = \frac{\Delta V_{EB}}{\Delta I_E}$$

It may be noted that input characteristic is linear in the upper region, but non-liner (*i.e.*, curved) in the lower region. Therefore, *the a.c. input resistance depends upon the location of the operating point selected along the curve*. Its value, in the linear region of the curve, is about 50 ohms.

10-4. Base Width Modulation or Early Effect

We have already discussed in the last article that the emitter-base diode characteristic shifts upwards by increasing the value of collector-to-base voltage (V_{CB}). This occurs due to the phenomenon called base width modulation or early effect, which may be explained as follows :

We know that in a PN junction, the width of a depletion region increases as the reverse-bias voltage increase. In a transistor, since the emitter-base junction is forward-biased, therefore this has no effect on the width of depletion region. However, the collector-base junction is reverse biased. Therefore, as the reverse-bias voltage across the collector-base junction increases, the width of depletion region also increases. Since, the base is lightly doped, as compared to the collector, therefore depletion region penetrates deeper into the base region. This reduces the effective width of the base region. This variation or modulation of the effective base width, by the collector voltage, is known as *base width modulation* or *early effect*. The decrease in base width by the collector voltage, has the following three effects:

- It reduces the chances of recombination of electrons with the holes in the base region. Hence, the common base current gain (α) increases with the increase of collector-to-base voltage (V_{CB}).
- The concentration gradient of minority carriers within the base increases. This, in turn, increases the emitter current.

* The d.c. input resistance of a transistor is not of any use in actual practice.

- For extremely large collector voltage, the effective base width may be reduced to zero, causing voltage breakdown of a transistor. This phenomenon is known as *punch through*.

10-5. Output Characteristics of a Transistor in Common-Base Configuration

These characteristics may be obtained by using the circuit shown in Fig. 10-1. First of all, we adjust, the emitter-to-base voltage (V_{EB}) to get a suitable value of emitter current (say 2 mA). Keeping the emitter current constant, we increase the collector-to-base voltage (V_{CB}) from zero in a number of suitable steps and record the corresponding values of the collector current (I_C) at each step. If we plot a graph with collector-to-base voltage (V_{CB}) along the horizontal axis and the collector current (I_C) along the vertical axis, we shall obtain a curve marked I_E = 2 mA as shown in Fig. 10-3. A similar procedure may be used to the obtain the characteristics at different values of emitter current, *i.e.*, I_E = 4, 6, and 8 mA. The output characteristic gives us the information about the following important points :

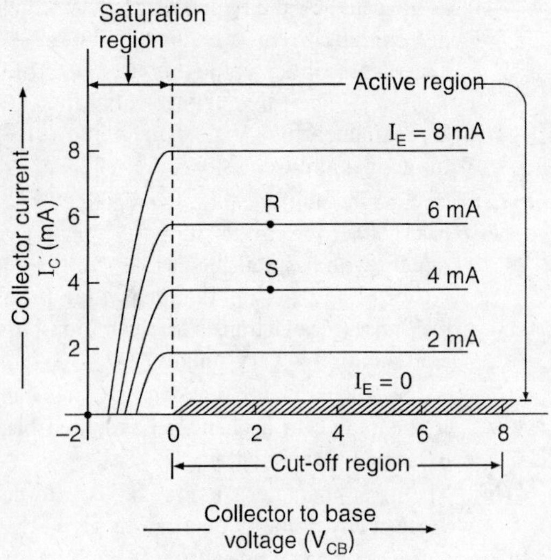

Fig. 10-3. *Output characteristics of a common-base transistor.*

(1) The curve may be divided into three important regions namely saturation region, active region and cut-off region. The saturation region is the region to the left of the vertical dashed line. It maybe noted that in this region, collector-to-base voltage (V_{CB}) is negative for a NPN transistor. It means that collector-base junction of a transistor is also forward biased in the saturation region. In this region, a small change in V_{CB} results in a large value of current. The active region is the region between the vertical dashed line and the horizontal axis. In the active region, the collector current is constant and is equal to the emitter current. The cut-off region is the region along the horizontal axis as shown by a shaded region in the Fig. 10-3. It corresponds to the curve marked I_E = 0. In the cut-off region, both the junctions of a transistor are reverse-biased.

(2) The collector current flows even when the collector-to-base voltage (V_{CB}) is zero. This is due to the fact that electrons are injected into the base under the action of a forward-biased emitter-base junction. These electrons are collected by the collector due to the internal junction voltage (*i.e.*, the barrier potential) at the collector-base junction.

(3) A small collector current flows even when emitter current (I_E) is zero. This current is called collector leakage current and is designated as I_{CBO}.

(4) The collector current is practically independent of collector-to-base voltage (V_{CB}) in the active region. However, if V_{CB} is increased beyond a certain large value, the collector current increases rapidly due to avalanche breakdown (not shown in the figure) and the transistor action is lost.

(5) The characteristic may be used to determine the common-base transistor a.c. output resistance. Its value at any point is given by the ratio of a change in collector-to-base voltage (ΔV_{CB}) to the resulting change in collector current (ΔI_C) for a constant emitter current (I_E). Mathematically, the a.c. output resistance,

$$R_0 = \frac{\Delta V_{CB}}{\Delta I_C}$$

It may be noted that the characteristic curves of common-base transistor are almost horizontal. This indicates that the value of output resistance (R_0) is very high. Its typical value is about 500 kΩ.

(6) The characteristic may be used to determine small-signal common-base current gain or a.c. alpha (α_0) of transistor. This can be done by selecting two point R and S on the characteristics and note down the corresponding values of ΔI_C and I_C. Thus, if $\Delta I_C = 6 - 4 = 2$ mA and $I_E = 6 - 4 = 2$ mA, then small signal common-base current gain.

$$\alpha_0 = \frac{\Delta I_C}{\Delta I_E} = \frac{2}{2} = 1$$

10-6. Characteristics of a Transistor in Common-Emitter Configuration

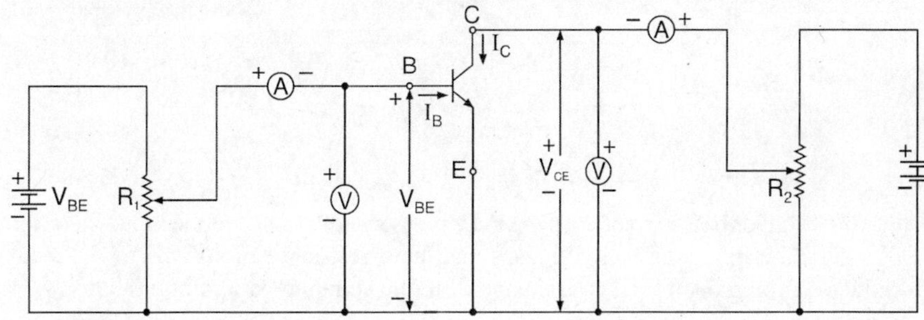

Fig. 10-4. *Circuit arrangement for determining common-emitter transistor characteristics.*

A transistor, in a common-emitter configuration, has also two important characteristics namely input characteristics and output characteristics as discussed below :

(1) ***Input characteristics :*** These curves give the relationship between the base current (I_B) and the base-to-emitter voltage (V_{BE}) for a constant collector-to-emitter voltage (V_{CE}).

(2) ***Output characteristics :*** These curves give the relationship between the collector current (I_C) and collect-to-emitter voltage (V_{CE}) for a constant base current (I_B).

These characteristics may be obtained by means of a circuit arrangement as shown in Fig. 10-4. In this circuit, an NPN transistor is connected in a common-emitter configuration. The collector-to-emitter voltage (V_{CE}) can be varied by changing the position of potentiometer , while the base-to-emitter voltage (V_{BE}) can be varied by changing the position of potentiometer R_1. The dc milliammeters and dc voltmeters are connected in the base and the collector circuit of a transistor to measure the voltages and currents.

10-7. Input Characteristics of a Transistor in Common-Emitter Configuration

These characteristics may be obtained by using the circuit arrangement shown in Fig. 10-4. First of all, we adjust the collector-to-emitter voltage (V_{CE}) to 1 volt. Then we increase to base-to-emitter voltage (V_{BE}) in small suitable steps and record the corresponding values of base current (I_B) at each step. If we plot a graph with base-to-emitter voltage (V_{BE}) along the horizontal axis and the base current (I_B) along the vertical axis, we shall obtain a curve marked V_{BE} = 1 V as shown in Fig. 10-5. A similar procedure may be used to obtain characteristics at different values of collector-to-emitter voltage, *i.e.*, V_{CE} = 2, 10 and 20V. The input characteristic gives us the information about the following important points :

(*i*) There exists a threshold or knee voltage (V_K) below which the base current is very small. The value of knee voltage is 0.5 V for silicon and 0.1 V for germanium transistors.

(*ii*) Beyond the knee, the base current (I_B) increases with the increase in base-to-emitter voltage (V_{BE}) for a constant collector-to-emitter voltage (V_{CE}). However, it may be noted

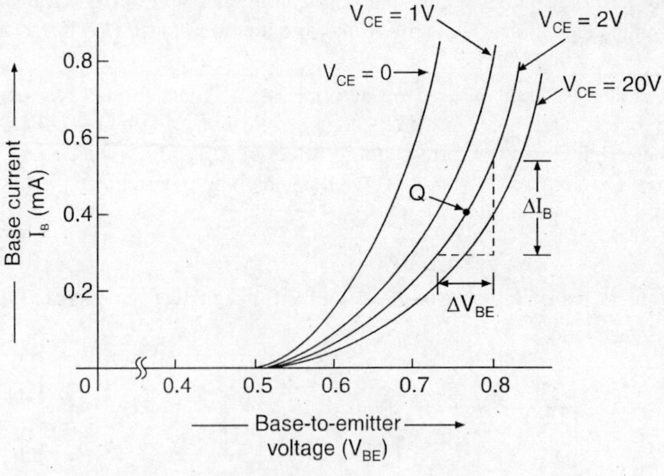

Fig. 10-5. *Input characteristics of a common-emitter transistor.*

that the value of base current does not increase as rapidly as that of the input characteristic of a common-base transistor. It means that input resistance of a transistor in common-emitter configuration is higher as compared to the common base configuration.

(iii) As the collector-to-emitter voltage (V_{CE}) is increased above 1 V, the curves shift downwards. It occurs because of the fact, that as V_{CE} is increased, the depletion width in the base-region increases. The reduces the effective base width, which in turn reduces the base current.

(iv) The input characteristic may be used to determine the value of common-emitter transistor a.c. input resistance (R_i). Its value is given by the ratio of change in base-to-emitter voltage (ΔV_{BE}) to the resulting change in base current (I_B) at a constant collector-to-emitter voltage (V_{CE}). Mathematically, the a.c. input resistance,

$$R_i = \frac{\Delta V_{BE}}{\Delta I_B}$$

It may be noted that the input characteristic is not linear in the lower region of the curve. Therefore, the input resistance varies with the location of the operating point. The value of a.c. input resistance ranges from 600 Ω to 4000 Ω.

10-8. Output Characteristics of a Transistor in Common-Emitter Configuration

These characteristics may be obtained by using the circuit shown in Fig. 10-4. To begin with, adjust the base current (I_B) to 40 μA value. Then increase the collector-to-emitter voltage (V_{CE}) in a number of steps and record the corresponding values of collector current (I_C) at each step. If we plot a graph with collector-to-emitter voltage (V_{CE}) along the horizontal axis and collector current (I_C) along the vertical axis, we shall obtain a curve marked I_B = 40 μA as shown in Fig. 10-6. A similar procedure may be used to obtain characteristics at I_B = 80 μA , 120 μA, 160 μA and so on. The output characteristics give us the information about the following important points :

(i) The output characteristics may be divided into three important regions namely saturation region, active region and cut-off region. The saturation and cut-off regions are shown by the shaded areas, while the active region is the region between the saturation and cut-off region.

(ii) As the collector-to-emitter voltage (V_{CE}) is increased above zero, the collector current (I_C) increases rapidly to a saturation value, depending upon the value of base current. It may be noted that collector current (I_C) reaches to a saturation value when V_{CE} is about 1 V.

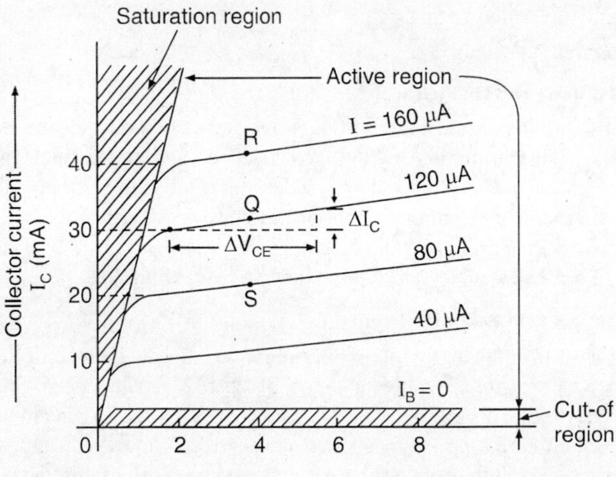

Fig. 10-6. *Output characteristics of a common-emitter transistor.*

(iii) When collector-to-emitter voltage (V_{CE}) is increased further, the collector current slightly increases. This increase in collector current is due to the fact that increased value of collector-to-emitter voltage (V_{CE}) reduces the base current and hence, the collector current increases. This phenomenon is called an *early effect.*

(iv) When the base current is zero, a small collector current exists. This is called leakage current. However, for all practical purposes, the collector current (I_C) is zero, when the base current (I_B) is zero. Under this condition the transistor is said to be cut-off.

(v) The characteristic may be used to determine the common-emitter transistor a.c. output resistance. Its value at any given operating point Q is given by the ratio of a change in collector-to-emitter voltage (ΔV_{CE}) to the resulting change in collector current (ΔI_C) for a constant base current. Mathematically, the a.c. output resistance,

$$R_0 = \frac{\Delta V_{CE}}{\Delta I_C}$$

The common-emitter output resistance of a transistor range from 10 kΩ to 50 kΩ.

(vi) The characteristic may be used to determine the small-signal common-emitter current gain or a.c. beta (β_0) of a transistor. This can be done by selecting two points R and S on the characteristics and note down the corresponding value of ΔI_C and ΔI_B. Thus, if $\Delta I_C = 40 - 20 = 20$ mA $= 20 \times 10^{-3}$ A and $\Delta I_B = 160 - 80 = 80$ μA $= 80 \times 10^{-6}$ A, then small signal common emitter current again,

$$\beta_0 = \frac{\Delta I_C}{\Delta I_B} = \frac{20 \times 10^{-3}}{80 \times 10^{-6}} = 250$$

10-9. Transistor Ratings

A transistor, like any other electronic device, has limitations on its operation. These limitations are stated in terms of maximum ratings. If these ratings are exceeded, it may cause either permanent damage to transistors or temporarily change their operating characteristics. The maximum ratings are normally specified on the manufacturer's data sheet. Following are some of the important maximum ratings of a transistor :

(*i*) Maximum collector current.

(*ii*) Maximum power dissipation.

(*iii*) Maximum output voltage.

10-10. Maximum Collector Current

This is the absolute maximum value of the current, which a collector can handle safely and is designated by $I_{C(max)}$. This limitation is due to the localized hot spots, which can occur within the transistor due to high current densities at unavoidable current path restrictions or high resistance points within the transistor. A common source of difficulty is the junction area and the wire bonds that connect the transistor terminals to the external leads. For example, for a general purpose NPN silicon transistor, 2 N 2222 A, $I_{C(max)}$ is 800 mA and for 2 N 3904, $I_{C(max)}$ is 200 mA.

10-11. Maximum Power Dissipation

It is a rating used to indicate the limits on the power handling capability of a transistor. The power dissipation in a transistor is mainly the power dissipated in its junctions. Under d.c. conditions, the junction power is found from the product of junction voltage and junction current. In normal operation, two junctions carry approximately the same current. But the collector-base junction has a much higher voltage across it than emitter-base junction. As a result of this, the power dissipation of a collector-base junction is of more importance. Mathematically, the maximum power dissipation of a transistor is given by the relation,

$$P_{D(max)} = V_{CE} \times I_C(max) = V_{CE(max)} \times I_C$$

It may be noted that in the equation, both collector-to-emitter voltage (V_{CE}) and collector current (I_C) cannot be maximum at the same time. For a given transistor, a maximum power dissipation curve can be plotted on the *CE* output characteristics as shown in Fig. 10-7. The curve can be obtained by joining a number of points obtained by assuming different values of and finding the corresponding values of from the relation,

$$V_{CE} = \frac{P_{D\,(max)}}{I_C}$$

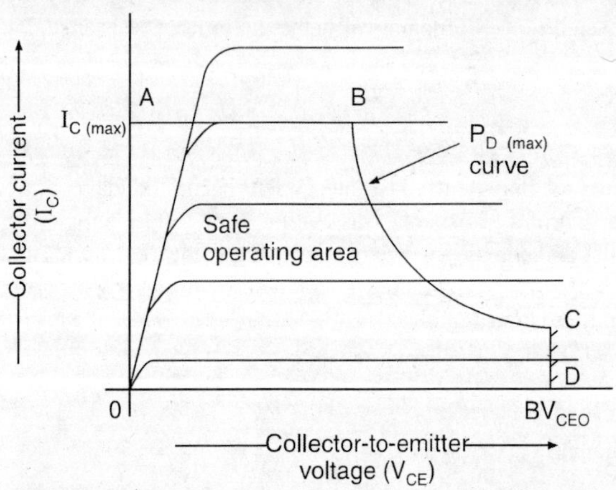

Fig. 10-7. *Maximum power dissipation curve for a transistor.*

It may also be noted from the power dissipation curve that a transistor cannot be operated in the shaded portion of the graph. Thus, the maximum collector current $I_{C(max)}$ is the limiting rating between the points *A* and *B*, is the limiting rating between the points *B* and *C* and breakdown voltage from collector-to-emitter with base open (BV_{CEO}) is the limiting rating between the points *C* and *D*. The region inside the curve *ABCD* (or $P_{D(max)}$ curve) is called a *safe operating area*.

The value of maximum power dissipation $P_{D(max)}$ is, usually, specified at an ambient temperature of 25°C. For example, $P_{D(max)}$ of 2 N 2222 A is 0.5 W at 25°C. For higher temperatures, its value is less. The data sheets often give derating factors for determining the value of $P_{D(max)}$ at any temperature above 25°C. For example, a derating factor of 2 mW/°C means that the maximum power dissipation is reduced 2 mW for each °C increase in temperature. The maximum power dissipation at any temperature (T) may be obtained from the relation,

$$P_{D(max)}(T) = P_{D(max)} - DF \times (T - 25)$$

where, $P_{D(max)}$ = Maximum power dissipation at 25°C,

DF = Derating factor, and

T = Any temperature in °C.

Example 10-1. *A 2 N 2222 A transistor has a $P_{D(max)}$ of 500 mW at 25°C. The derating factor is 2.28 mW/°C. What is the $P_{D(max)}$ at a temperature of 70°C ?*

Solution. Given: $P_{D(max)}$ = 500 mW; DF = 2.28 mW/°C and T = 70°C.

We know that maximum power dissipation at a temperature of 70°C,

$$P_{D(max)}(70) = P_{D(max)} - DF\,(T - 25) = 500 - 2.28 \times (70 - 25)$$
$$= 500 - 102.6 = 397 \text{ mW} = 0.397 \text{ W } \textbf{Ans.}$$

10-12. Maximum Output Voltage

In most of the applications, the collector-base junction is subjected to the highest voltages. Since, there is a possibility of voltage breakdown at high voltages, therefore it is necessary to specify the maximum allowable collector-base voltage or collector-emitter voltage or both. The voltage breakdown, which can occur in a transistors is of two types namely avalanche breakdown and punch-through (also called *reach-through*).

(*i*) ***Avalanche breakdown :*** It is a form of voltage breakdown that arises, when the electrons and holes in the semiconductor are accelerated enough by the applied voltage. The accelerated electrons and holes collide with the bound electrons and produce more free electrons. This causes even more collisions due to which current increases to a large value.

(*ii*) ***Punch-through :*** It is a form of voltage breakdown that arises due to the increased width of collector-base junction depletion region with increased collector-base junction voltage (*i.e.*, early effect). The depletion region at a junction is the region of fixed ions on both sides of the junction. As the applied voltage across the junction increases, the depletion region penetrates deeper into the base. Since, the base is very thin, therefore it is possible that at moderate voltages, the depletion region will have spread completely across the base to reach the emitter-base junction. As a result of this, the current increases to a very large value. The punch-through differs from the avalanche breakdown, in the sense, that it takes place at fixed voltage between the collector and base. It is not dependent on the circuit configuration.

The avalanche breakdown in a transistor may take place because of avalanche multiplication of the reverse saturation current (I_{CO}) that crosses the collector junction. The manufacturer, usually, specifies two voltage ratings namely the maximum reverse voltage, (BV_{CEO}) which may be applied between the collector and emitter terminals under the condition that the base lead is open (*i.e.*, when base current, I_B = 0) and the maximum reverse voltage (V_{CBO}) which may be applied between the collector and base terminals under the condition that the emitter lead is open (*i.e.*, when emitter current (I_E = 0). Usually, the value of BV_{CBO} is double than that of the value of BV_{CEO}. The BV_{CEO} collector-to-emitter breakdown voltage is affected by the condition of emitter-base circuit. Consequently, additional ratings are given either for a specified resistance connected from base to emitter (designated by BV_{CER}) or for the base shorted to emitter (designated by B_{CEB}). If the base is connected to the emitter through a resistance R, the breakdown voltage BV_{CER} will lie between BV_{CED} and BV_{CBO}.

Figure 10-8. shows the *CE* output characteristics extended into the breakdown region. It may be noted that the breakdown voltage decreases for the larger collector currents. The data sheets list the minimum value of BV_{CEO}, which is approximately 40 V as shown in the figure. Transistors are available with breakdown voltage ranging from 20 V to 500 V. In actual practice, the transistor is never operated near the breakdown voltage. Therefore, the collector-to-emitter voltage (V_{CE}) of a transistor is kept well below the breakdown voltage.

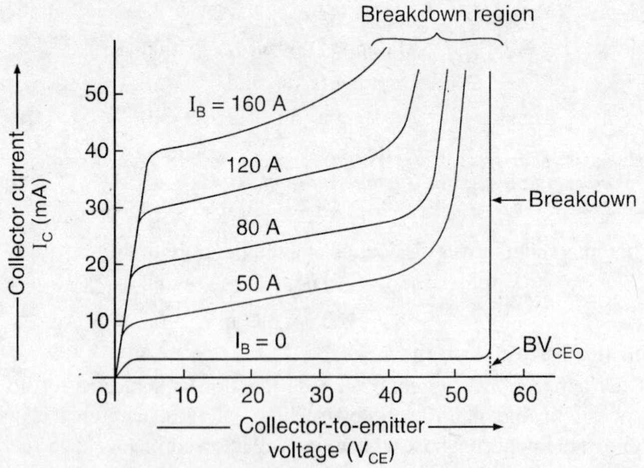

Fig. 10-8. *CE output characteristics extended into the breakdown region.*

10-13. Bipolar Junction Transistor Specification Sheet

As a matter of fact the transistor specification sheet (or datasheet) contains a wide variety of *dc* and *ac* operating characteristics. But in this article we will focus only on some of the commonly used dc maximum ratings and electrical characteristics.

Fig. 10-9 this shows the datasheet for the Motorola 2N3903 and 2N3904 transistors. As seen from this datasheet, we find that the right top corner box shows the transistor numbers, *i.e.*, 2N3903 and 2N3903 and 2N3904. A box below that shows that there transistors are NPN silicon type devices. Both the transistor can be used in switching and amplifier circuits. The box below that shows the transistor package. Besides the three boxes, the specification sheet also shows four lists namely *Maximum Ratings, Thermal Characteristics, OFF Characteristics and ON Characteristics.*

(i) ***Maximum Ratings :*** This list gives us the information about the maximum values of the junction voltage which the transistor can handle safely. From this list we find that, 2N3903/2N3904 has maximum *collector-base voltage* (V_{CB}) of 60 V, the maximum *collector-emitter voltage with base open* (V_{CEO}) of 40 V and maximum *emitter-base voltage* (V_{EB}) of 6.0 V.

(ii) ***Thermal Characteristics :*** These are used in circuit development applications and will not be discussed here.

(iii) ***OFF Characteristics :*** These characteristics define the operation of the transistor when it is being operated in cut-off. The cut-off means that the collector current through the transistor, I_C is nearly zero. The first three ratings in this list, *i.e.*, BV_{CBO}, BV_{CEO} and BV_{CBO} are the same ratings that appear in the **Maximum Ratings** list. They are repeated here for the convenience of engineer who must rapidly find the ratings.

The collector cut-off current (I_{CEX}) rating indicates the maximum value of I_C when the device is cut-off. From the list, we find that for 2N3903/2N3904 the maximum value of I_{CEX} is 50 nA.

2N3903
2N3904

NPN SILICON SWITCHING & AMPLIFIER TRANSISTORS

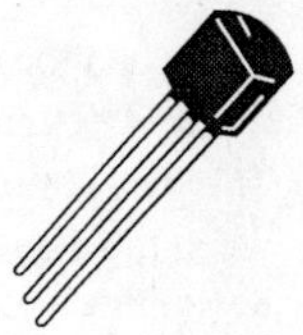

MAXIMUM RATINGS

Rating	Symbol	Value	Unit
*Collector-Base Voltage	V_{CB}	60	Vdc
*Collector-Emiter Voltage	V_{CEO}	40	Vdc
*Emitter-Base Voltage	V_{EV}	6.0	Vdc
*Collector Current	I_C	200	mAdc
Total Power Dissipation @ $T_A = 60°C$	P_D	250	mW
**Total Power Dissipation @ $T_A = 25°C$	P_D	350	mW
Derate above 25°C		2.8	mW/°C
**Total Power Dissipation @ $T_C = 25°C$ Derate above 25°C	P_D	1.0 8.0	Watts mW/°C
**Junction Operating Temperature	T_J	150	°C
**Storage Temperature Range	T_{stg}	– 55 to + 150	°C

THERMAL CHARACTERISTICS

Characteristics	Symbol	Max	Unit
Thermal Resistance, Junction to Ambient	$R_{\theta JA}$	357	°C/W
Thermal Resistance, Junction to Case	$R_{\theta JC}$	125	°C/W

OFF CHARACTERISTICS ($T_A = 25°C$ unless otherwise noted)

Characteristic	Fig. No.	Symbol	Min	Max	Unit
Collector-Base Breakdown Voltage ($I_C = 10\ \mu Adc$, $I_E = 0$)		BV_{CBO}	60	-	Vdc
Collector-Emitter Breakdown Voltage (1) ($I_C = 10$ mAdc, $I_B = 0$)		BV_{CEO}	40	-	Vdc
Emitter-Base Breakdown Voltage ($I_E = 10\ \mu Adc$, $I_C = 0$)		BV_{EBO}	6.0	-	Vdc
Collector-Cut-off Current ($V_{CE} = 30$ Vdc, $V_{EB(off)} = 0.3$ Vdc)		I_{CEX}	-	50	nAdc
Base-Cut-off Current ($V_{CE} = 30$ Vdc, $V_{EB(off)} = 0.3$ Vdc)		I_{BL}	-	50	nAdc

ON CHARACTERISTICS

Characteristic	Fig. No.	Symbol	Min	Max	Unit
DC Current Gain (1)	15	h_{FE}			-
($I_C = 0.1$ mAdc, $V_{CE} = 1.0$ Vdc) 2N3903			20	-	
2N3904			40	-	
($I_C = 1.0$ mAdc, $V_{CE} = 1.0$ Vdc) 2N3923			35	-	
2N3904			70	-	
($I_C = 10$ mAdc, $V_{CE} = 1.0$ Vdc) 2N3903			50	150	
2N3904			100	300	
($I_C = 50$ mAdc, $V_{CE} = 1.0$ Vdc) 2N3903			30	-	
2N3904			60	-	
($I_C = 100$ mAdc, $V_{CE} = 1.0$ Vdc) 2N3903			15	-	
2N3904			30	-	
Collector-Emitter Saturation Voltage (1)	16, 17	$V_{CE(sat)}$			Vdc
($I_C = 10$ mAdc, $I_B = 1.0$ mAdc)			-	0.2	
($I_C = 50$ mAdc, $I_B = 5.0$ mAdc)			-	0.3	
Base-Emitter Saturation Voltage (1)	17	$V_{CE(sat)}$			Vdc
($I_C = 10$ mAdc, $I_B = 1.0$ mAdc)			0.65	0.85	
($I_C = 50$ mAdc, $I_B = 5.0$ mAdc)			-	0.95	

Fig. 10-9. *Bipolar Junction Transistor specification sheet (Courtesy, Motorola Inc.)*

The base cut-off current (I_{BL}) ratings, indicates the maximum amount of base current that will be present when the emitter-base junction is in cut-off. From the list, we find that for 2N3903/2N3904 the maximum value of I_{BL} is 50 nA.

From the I_{BL} and I_{CEX} ratings, we also find that thermal currents of the transistor at cut-off are extremely low. For 2N3903/2N3904 I_B and I_C will be no greater than 50 nA. The value of I_E will be no greater than the sum of the base and collect currents (*i.e.*, 100 nA).

(*iv*) ON Characteristics: These characteristics describe the dc operating characteristics for both the active and saturation regions of transistor operation.

The dc current gain (h_{FE}) rating of the transistor is the value of dc beta. Note that the label (h_{FE}) is normally used to represent dc beta, so we will be using this label in all our discussions from now onwards. As seen from the list, the value of h_{FE} is measured at different values of I_C. this goes along with the discussion we had in the last article on current gain or beta curve.

The collector-to-emitter saturation voltage [$V_{CE(sat)}$] rating indicates the maximum value of V_{CE} when the device is in saturation. Recall that saturation means that I_C is maximum. For the 2N3904, the value of [$V_{CE(sat)}$] is 0.2 V when I_C is 10 mA and 0.3 V when I_C is 50 mA.

The base-emitter saturation voltage [$V_{BE(sat)}$] ratings indicates the maximum value of V_{BE} when the device is in saturation. As seen from the list, we find that this value is 0.85 or 0.95 depending upon the rated value of I_B.

Note: Specification of another bipolar junction transistor 2N4123 is given in Appendix F.

10-14. Transistor Testing

You may recall from the discussion in the last chapter that diode can be tested with an ohmmeter. The transistor can be checked in the same basic manner. However, there are many more measurements required to verify whether the transistor under test is good or bad.

Fig 10-10 shows the process for testing the transistor. The emitter-base junction is tested for a low forward resistance and a high reverse resistance. When connected to the ohmmeter as shown in Fig. 10-10(*a*), the meter should read a low resistance, typically 1 kΩ or less. When connected as shown in Fig 10-10(*b*), the resistance, should be high, typically 100 kΩ or higher. If both of these tests show the emitter-base junction to be good, the collector-base junction should be tested in the same manner. These tests illustrated in Fig 10-10 (*c*) and (*d*) should yield the same results. The forward resistance of the collector-base junction should be 1 kΩ and the reverse resistance should be 100 kΩ or higher.

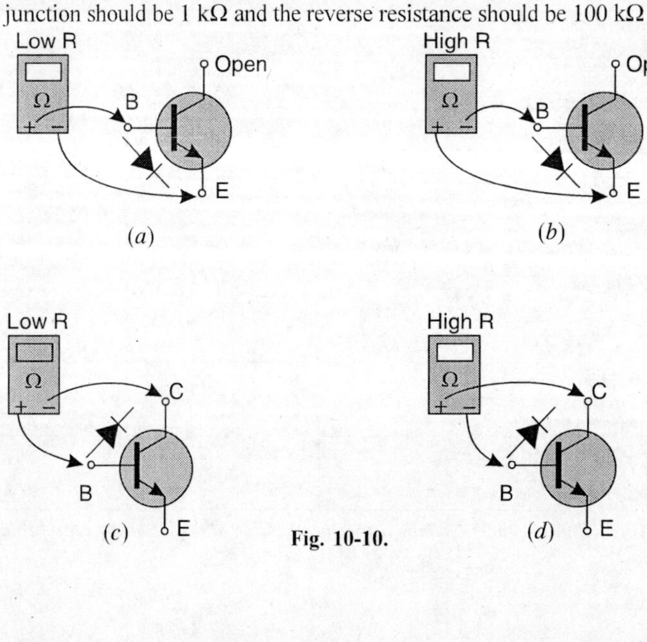

Fig. 10-10.

The last test involves checking the collector-emitter terminals. These terminals should have extremely high resistance regardless of polarity of the meter. If the transistor passes all the resistance checks, it is a good transistor. If it fails any one of the checks mentioned above, transistor if faulty and must be replaced.

Note: The polarities shown in Fig. 10-10 are for an NPN transistor. When testing a PNP transistor, the ohmmeter leads must be reversed.

10-15. Integrated Transistors

Integrated transistors are single devices that contain more than one transistor in integrated form. Such devices contain more than one device in a single casing. Fig. 10-11(*a*) shows a typical IC transistor casing. It may be noted from the pin listing shows in Fig 10-11(*a*) or the internal connection shows in Fig. 10-11 (*b*) that the package contains four individual transistors.

Strictly speaking, each of the transistors in an IC like the one shown has the same type of maximum ratings and electrical characteristics as individual (or discrete) transistors. The only real difference is the small amount of space that is taken up by the four transistors in the IC package.

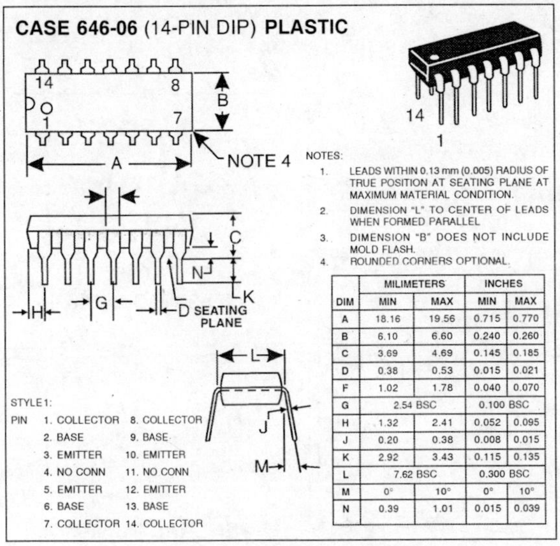

DIM	MILIMETERS		INCHES	
	MIN	MAX	MIN	MAX
A	18.16	19.56	0.715	0.770
B	6.10	6.60	0.240	0.260
C	3.69	4.69	0.145	0.185
D	0.38	0.53	0.015	0.021
F	1.02	1.78	0.040	0.070
G	2.54 BSC		0.100 BSC	
H	1.32	2.41	0.052	0.095
J	0.20	0.38	0.008	0.015
K	2.92	3.43	0.115	0.135
L	7.62 BSC		0.300 BSC	
M	0°	10°	0°	10°
N	0.39	1.01	0.015	0.039

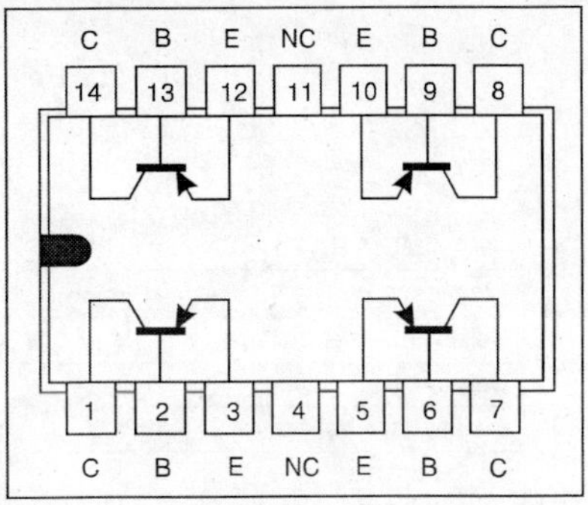

NC - No internal connection

Fig. 10-11. *Integrated transistor (Courtesy : Motorola Inc).*

10-16. High Voltage Transistors

High-voltage transistors have unusually high reverse breakdown voltage ratings. For example, the maximum ratings for the Motorola BFW 43 transistor are shown in Fig. 10-12 As seen from the "maximum ratings" list, the V_{CEO} and V_{CBO} are much higher than those for the 2N3904 transistor. High-voltage transistors are used in circuits that are extremely high supply voltages, such as television CRT

10-17. High Current Transistors

High-current transistors have very high maximum current ratings. Fig 10-13 shows the maximum ratings of the Motorola 2N4237 transistor. As seen from the list, this device has an I_C rating of 1A. This is much higher as compared to the maximum ratings of 2N3904 (200 mA only). These devices are used in applications that have high current demands such as current regulator circuits, motor drives, *etc*.

MAXIMUM RATINGS

Rating	Symbol	Value	Unit
Collector-Emitter Voltage	V_{CEO}	150	Vdc
Collector-Base Voltage	V_{CBO}	150	Vdc
Emitter-Base Voltage	V_{EBO}	6.0	Vdc
Collector Current – Continuous	I_C	0.1	Adc
Total Device Dissipation @ $T_A = 25°C$ Derate above 25°C	P_D	0.4 2.66	Watt mW
Total Device Dissipation @ $T_C = 25°C$ Derate above 25°C	P_D	1.4 8.0	Watts mW/°C
Operating and Storage Junction Temperature Range	T_J, T_{stg}	– 65 to + 200	°C

BFW43

CASE 22-03, STYLE 1
TO -18 (TO-206AA)

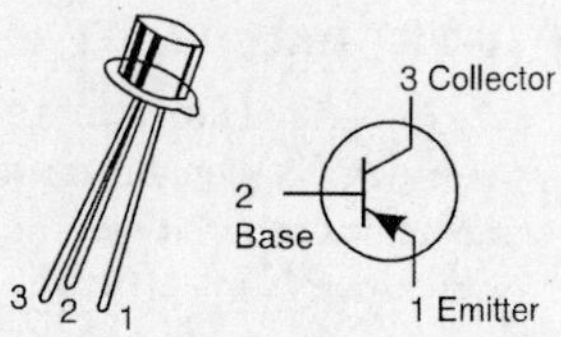

HIGH VOLTAGE TRANSISTOR

PNP SILICON

Fig. 10-12. *High-Voltage Transistors (Courtesy: Motorola Inc.)*

MAXIMUM RATINGS

Rating	Symbol	2N4237	2N4238	2N4239	Unit
Collector-Emitter Voltage	V_{CEO}	40	60	80	Vdc
Collector-Base Voltage	V_{CBO}	50	80	100	Vdc
Emitter-Base Voltage	V_{EBO}		6.0		Vdc
Base Current	I_B		500		mA
Collector Current – Continuous	I_C		1.0		Adc
Total Device Dissipation @ $T_A = 25°C$ Derate above 25°C	P_D		1.0 5.3		Watt mW/°C
Total Device Dissipation @ $T_C = 25°C$ Derate above 25°C	P_D		6.0 34		Watts mW/°C
Operating and Storage Junction Temperature Range	T_J, T_{stg}		– 65 to + 200		°C

2N4237
thru
2N4239

CASE 79-04STYLE 1
TO-39 (TO-250AD)

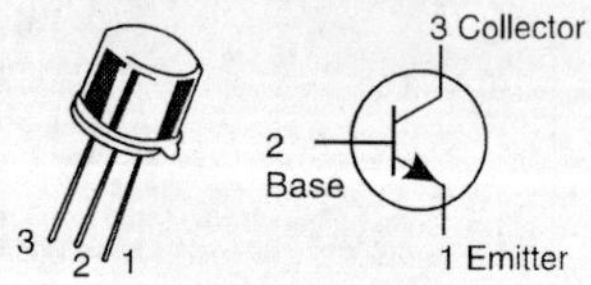

GENERAL PURPOSE TRANSISTOR

NPN SILICON

Fig. 10-13. *High Current Transistors (Courtesy: Motorola Inc.)*

10-18. High-Power Transistors

These transistors are designed for use in high-power circuits, such as regulated and switch-mode power supplies. The high-power transistors are capable of dissipating large amount of power. Fig.10-14 shows the specifications sheets of Motorola 2N3902. As seen from the list, this transistor can dissipate 100 W of power Also notice from the picture that the casing of power transistors is different from the low and medium power transistors. This device has V_{CEO} of 400 V, V_{CBO} of 700 V, V_{EB} of 5.0 V and I_C of 3.5 A.

The high-power transistors are used in regulated power supply applications. The regulated power supply circuits are discussed in Chapter-31.

MOTOROLA

SEMICONDUCTOR

TECHNICAL DATA

NPN
2N3902

HIGH VOLTAGE NPN SILICON TRANSISTORS

. . . designed for use in high-voltage inverters, converters, switching regulators and line operated amplifiers.

- High Collector-Emitter Voltage – V_{CEX} = 700 Vdc
- Excellent DC CURRENT GAIN – h_{FE} = 10 (Min) @ I_C = 2.5 Adc
- Low Collector-Emitter Saturation Voltage – $V_{CE(sat)}$ = 0.8 Vdc (Max). @ I_C = 1.0 Adc

3.5 AMPERE
POWER TRANSISTORS
NPN SILICON
400/VOLTS
100 WATTS

***MAXIMUM RATINGS**

Rating	Symbol	2N3902	Unit
Collector-Emitter Voltage	V_{CEO}	400	Vdc
Collector-Emitter Voltage	V_{CEX}	700	Vdc
Emitter-Base Voltage	V_{EB}	5.0	Vdc
Collector Current – Continuous	I_C	3.5	Adc
Base Current	I_B	2.0	Adc
Total Device Dissipation @ T_C = 75°C Derate above 75°C	P_D	100 1.33	Watts W/°C
Operating Junction Temperature Range	T_J	– 65 to + 150	°C
Storage Temperature Range	T_{stg}	– 65 to + 200	°C

THERMAL CHARACTERISTICS

Characteristic	Symbol	Max	Unit
Thermal Resistance, Junction to Case	q_{JC}	0.75	°C/W

*Indicates JEDEC Registered Data

FIGURE 1 – POWER DERATING

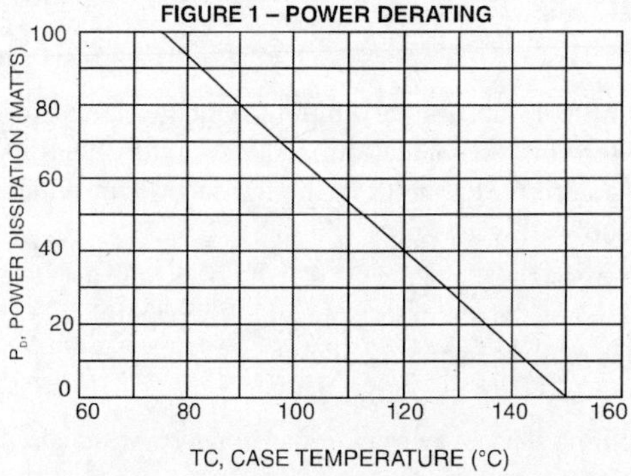

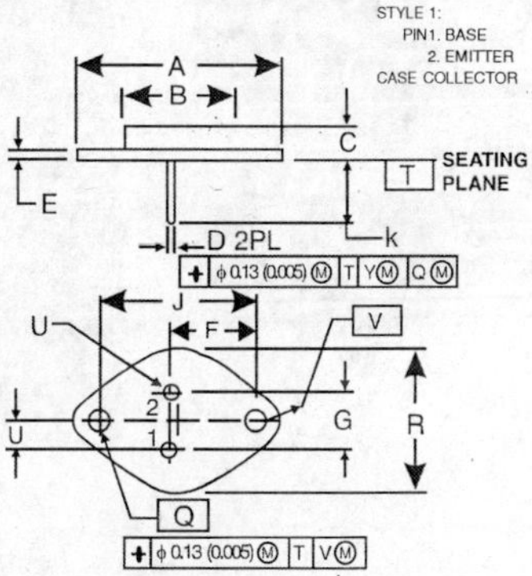

NOTES:
1. DIMENSIONING AND TOLERANCING PER ANSI Y14.5M, 1982.
2. CONTROLLING DIMENSION: INCH.
3. ALL RULES AND NOTES ASSOCIATED WITH REFERENCED TO-204AA OUTLINE SHALL APPLY.

DIM	MILIMETERS MIN	MILIMETERS MAX	INCHES MIN	INCHES MAX
A	–	39.37	–	1.550
B	–	21.08	–	0.830
C	6.35	8.25	0.250	0.325
D	0.97	1.09	0.038	0.043
E	1.40	1.77	0.055	0.070
F	30.15 BSC		1.187 BSC	
G	10.92 BSC		0.430 BSC	
H	5.46 BSC		0.215 BSC	
J	16.89 BSC		0.665 BSC	
K	11.18	12.19	0.440	0.480
Q	3.84	4.19	0.151	0.165
R	–	26.67	–	1.050
U	4.83	5.33	0.190	0.210
V	3.84	4.19	0.151	0.165

CASE 1-06
TO-204AA
(TO-3)

Fig. 10-14. *Specification sheet of high power transistor 2N3902 (Courtesy, Motorola Inc.)*

10-19. Thermal Runaway and Heat Sink

We have already discussed that the reverse saturation current in semiconductor devices, changes with temperature. Moreover, the reverse saturation current approximately doubles for every 10°C rise in temperature. This fact may cause considerable practical difficulty in using a transistor as an amplifier. It is because of the fact, that if the temperature of the collector-base junction increases, the leakage current of the transistor increases. And the collector current also increases. The increase in collector current produces an increase in the power dissipation at the collector-base junction. This, in turn, further increases the temperature of the collector-base junction causing the collector current to further increase. This process may become cumulative and it is possible that the ratings of the transistor are exceeded. If it happens, the device gets burnt out. This process is described as the thermal runaway of the transistor. In actual practice, a thermal runaway is avoided by using a stabilization circuitry or a heat sink with the transistor.

A heat sink is a metallic heat-conducting device placed in close contact with a transistor to increase the power dissipation capability of the transistor. The heat sink is, usually, not required for the transistors handling small signals. However, for power transistors, its use is essential. For low-power transistors, a fin-type heat sink is used as shown in Fig. 10-15 (left). On the other hand, for large power transistors, a rectangular heat sink is used as shown in Fig. 10-15 (right).

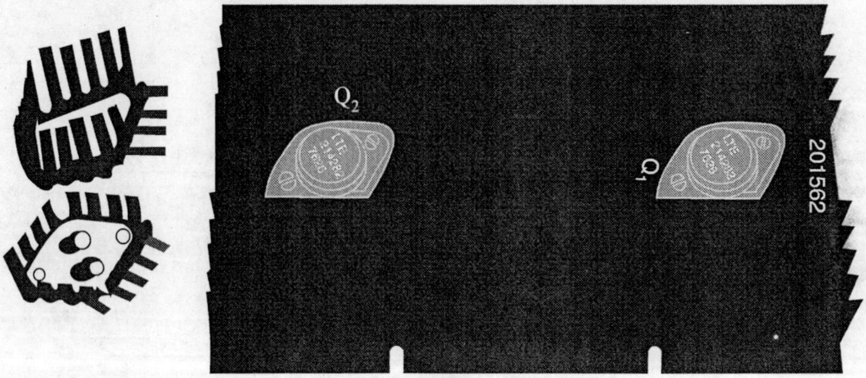

Fig. 10-15. *Heat Sink.*

It may be noted that connecting a heat sink to the transistor, increases the surface area from which the heat is to be dissipated to the atmosphere. The heat moves from the transistor to the heat sink by conduction and then it is removed from the sink to the atmosphere by convection and radiation. In order to radiate maximum heat, the heat sink is generally painted black and is mounted at such a position where the free air can flow around it.

10-20 Phototransistor

It is light-sensitive transistor and is similar to an ordinary bipolar junction transistor (BJT) except that it has no connection to the base terminal. Its operation is based on the photodiode that exists at the *CB* junction. Instead of the base current, the input to the transistor is provided in the form of light as shown in the schematic symbol of Fig.10-16(*a*).

Silicon NPNs are mostly used as phototransistor. The device is usually packed in a TO-type can with a lens on top although it is sometimes encapsulated in clear plastic. When there is no incident light on the *CB* junction, there is a small thermally-generated collector-to-emitter leakage current I_{CEO} which, in this case, is called *dark current* and is in the *nA* range.

When light is incident on the *CB* junction, a base current I_λ is produced which is directly proportional to the light intensity. Hence, collector current $I_C = \beta I_\lambda$.

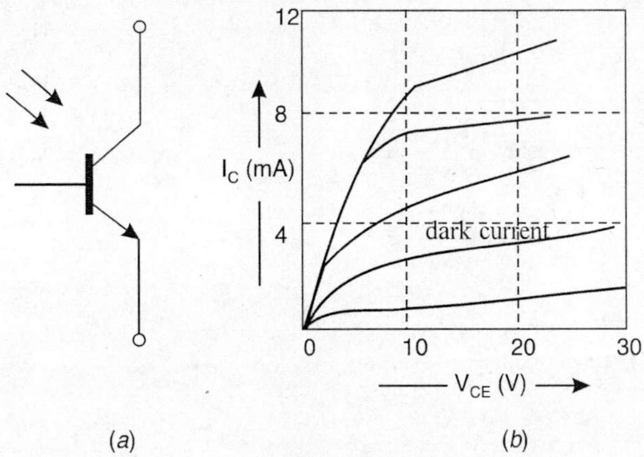

Fig. 10-16.

Typical collector characteristic curves of a phototransistor are shown in Fig. 10-16 (*b*). Each individual curve corresponds to a certain value of light intensity expressed in mW/cm^2. As seen, I_C increases with light intensity.

The phototransistor has applications similar to those of a photodiode. Their main differences are in the current and response time. The photo-transistor has the advantages of greater sensitivity and current capacity than photodiodes. However, photodiodes are faster of the two, switching in less than a nanosecond.

10-21. Photodarlington

As shown in Fig. 10-17, a photodarlington consists of a phototransistor in a Darlington arrangement with a common transistor. It has a much greater sensitivity to incident radiant energy than a phototransistor because of higher current gain. However, its switching time of 50 μs is much longer than the phototransistor (2μs) or the photodiode (1ns). Its circuit symbol is shown in Fig. 10-17.

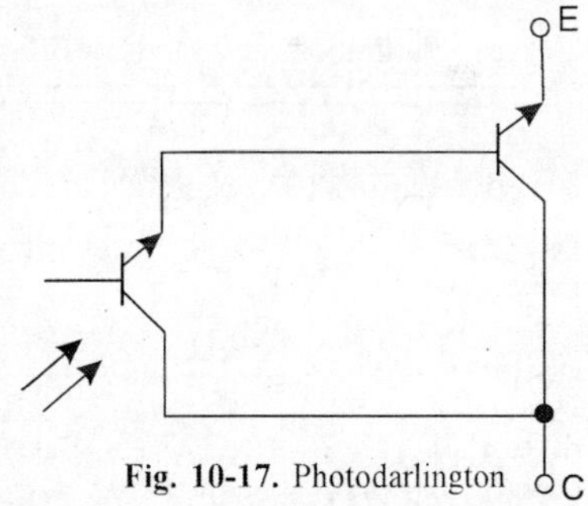

Fig. 10-17. Photodarlington

Photodarlington are used in a variety of applications some of which are given below.

A light-operated relay is shown in Fig. 10-18(*a*). The phototransistor Q_1 drives the bipolar transistor Q_2. When sufficient light falls on Q_1. It is driven into saturation so that I_C is increased manifold. This collector current while passing through the realy coil energizes the relay.

Fig. 10-18 (*b*) shows a dark-operated relay circuit, *i.e.,* one in which relay is deenergized when light falls on the phototransistor. Here, Q_1 and R form a potential divider across V_{CC}. With insufficient light incident on Q_1, transistor Q_2 is biased ON thereby keeping the relay energized. However, when there is sufficient light, Q_1 turns ON and pulls the base of Q_2 low thereby turning Q_2 OFF and hence, deenergizing the relay.

Such relays are used in many applications such as ***(i)*** automatic door activators, ***(ii)*** process counters, and ***(iii)*** various alarm systems for smoke or intrusion detection.

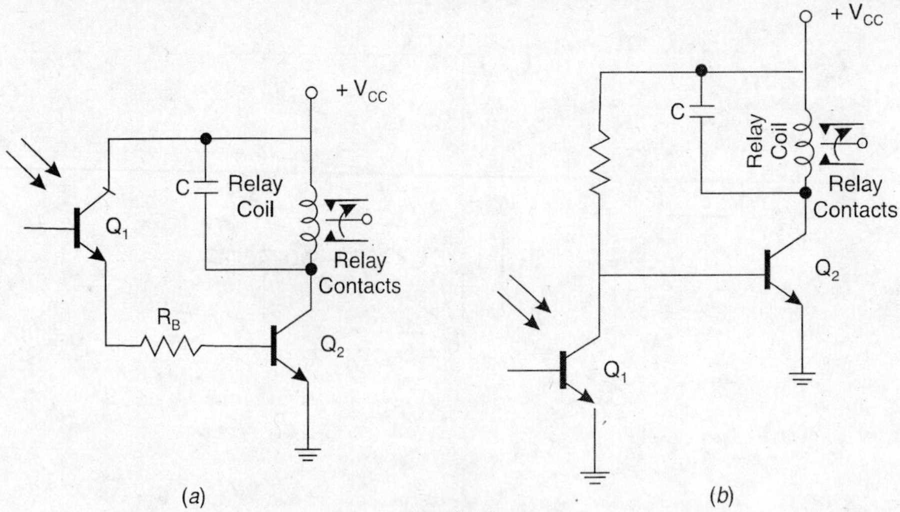

Fig. 10-18.

SUMMARY

In this chapter you have learned that :

1. The input characteristics of a transistor is non-linear in the lower region of the graph and become linear in the upper region in the case of both common-base and common-emitter configuration.
2. The value of a.c. input resistance depends on the operating point of the transistor.
3. Saturation active and cut-off are the three regions of the output characteristic of a transistor.

GLOSSARY

Active regions : The region in the output characteristic of a transistor where the collector current remains almost constant for a given emitter current in the case of common-base configuration, for a given base current in the case of a common-collector configuration. Transistors are normally operated in this region.

Cut-off region : The region in the output characteristics of a transistor where the output current is zero.

High-current transistor : Transistors with very high value of maximum collector current ratings.

High-voltage transistor : Transistors with high reverse breakdown voltage ratings.

Input characteristic : Curves that give the relationship between the input current and input voltage for a given output voltage.

Integrated transistor : Single devices that have more than one transistor in integrated form.

Maximum collector current : Absolute maximum value of the current, that a collector of a transistor can handle safety.

Maximum Power dissipation : The value of maximum power that the transistor junction can dissipate.

Output characteristics : Curves that gives the relationship between the output current and the output voltage for a given input current of a transistor.

Punch through : Voltage breakdown of a transistor caused by the reduction of effective base width to zero due to extremely large collector voltage.

Saturation region : The region in the output characteristics of a transistor where a small charge in input voltage can change the output current abruptly.

Thermal runaway : Failure of a transistor, caused by the cumulative increase in collector current due to the increase in junction temperature.

DESCRIPTIVE QUESTIONS

1. Sketch a family of common-base (CB) transistor input characteristics. Label all variables and explain how will you calculate the a.c. input resistance of a transistor from these curves.
2. Describe the circuit arrangement for obtaining the output characteristics of common-base transistor.
3. Sketch a family of output characteristics of a common-base transistor. Indicate the active, cut-off and saturation regions. Explain the shape of the curves quantitatively.
4. What do you mean by saturation of a transistor?
5. Explain the term base-width modulation or early effect.
6. Give the circuit arrangement for obtaining the input and output characteristics of transistor in common-emitter configuration.
7. Sketch the V-I characteristics of an NPN transistors in common emitter operation and indicate the different regions of importance.
8. Explain the terms **(*a*)** Maximum collector current **(*b*)** Maximum power dissipation, and **(*c*)** Maximum output voltage.
9. Draw a derating curve (*i.e.*, power dissipation curve) for a power transistor.
10. Discuss thermal runaway. How it can be prevented in a high power transistor?
11. What is a heat-sink ?
12. What is a heat-sink compound ?
13. Define Heat-sink. (*Electronic circuits I, Anna University, April/May 2008*)
14. For common-emitter-bipolar junction transistor configuration, analyze the input and output characteristics. Also, give the infernces and necessary analytical expression for the same. (*Electronic Devices, Anna University, Nov/Dec 2007*)
15. Draw the input and output characteristics of a transistor in CE configuration and mark the cutoff, saturation and active regions. (*Electron Devices, Anna University, April/May 2008*)
16. Draw and explain the input and output characteristics of a common base configuration. (*Electron Devices, Anna University, Nov/Dec 2007*)
17. Which of the BJT configuration is suitable for impedance matching applications? Why? (*Electronic Devices, Anna University, Nov/Dec 2007*)
18. What are Power transistors ? (*Electronic Devices, Anna University, Nov/Dec 2007*)

MULTIPLE CHOICE QUESTIONS

1. The "cut-in" voltage of a silicon small-signal transistor is
   (*a*) 0 V (*b*) 0.2 V (*c*) 0.5 V (*d*) 0.8 V
2. When the collector junction in transistors is biased in the reverse direction and the emitter junction in the forward direction, the transistor is said to be in the
   (*a*) active region (*b*) cut-off region (*c*) saturation region (*d*) none of them.
3. The transistor is said to be in saturation region when
   (*a*) both collector and emitter junctions are forward biased
   (*b*) both collector and emitter junctions are reversed biased
   (*c*) emitter junction is forward biased, but the collector junction is reverse biased
   (*d*) emitter junction is reverse biased, but the collector junction is forward biased

4. For a silicon transistor in the common emitter configuration the cut-off condition is achieved by applying a minimum reverse voltage across the emitter junction of the order of

(*a*) 0 V (*b*) 0.7 V (*c*) 1.5 V (*d*) 5 V

5. A transistor connected in common-base configuration has

(*a*) a high input resistance and a low output resistance.

(*b*) a low input resistance and high output resistance

(*c*) a low input resistance and a low output resistance

(*d*) a high input resistance and a high output resistance

ANSWERS

1. (*c*) **2.** (*a*) **3.** (*a*) **4.** (*a*) **5.** (*b*)

❑❑❑

11

CHAPTER

BJT Low and High Frequency Models

OUTLINE

Objectives

After completing this chapter, you should be able to :

- Know the standard notations used for BJT currents and voltages in electronic circuit analysis.
- Understand the Ebers Moll model of a BJT.
- Understand the low and high frequency small signal model of a BJT.
- Know the low and high frequency hybrid-π model of a BJT.

11-1. Introduction

A model is a combination of a circuit elements, properly chosen, that best approximates the actual behaviour of a semiconductor device under specific operating conditions.

The *biopolar junction transistor (BJT) models* are useful for analysing amplifier circuits. The models are more easy to use than graphical procedures for the analysis purpose. When the signal is small, a transistor behaves as a linear device. However, when the signal is large, the transistor behaves as a non-linear device. Therefore, separate models are required for transistors in small-signal operation and large-signal operation. The large-signal model is also known as a ***d.c.-model.***

On the other hand, a small signal model is known as an ***a.c.-model.*** Following are the two different approaches of developing transistor models :

(1) In terms of current gain and circuit components.

(2) In terms of hybrid parameters.

In this chapter, we shall discuss transistor models in terms of the first approach. The transistor models, in terms of hybrid parameters will be discussed in the chapter on hybrid parameters of a transistor.

11-2. Notations for Transistor Currents and Voltages

It is important to introduce a standard terminology for the symbols representing specific current and voltage components to avoid confusion between a.c. and d.c. components. This terminology is taken from the list recommended by the Institution of Electrical and Electronic Engineers (IEEE) of U.S.A. The system underlying the notation is as follows :

(1) Lower case letters (such as v and i) designate instantaneous voltage and current respectively.

(2) Upper case letters (such as V and I) designate the r.m.s. or peak values of a.c. components or average (d.c.) values of total quantities.

(3) Lower case subscripts are used only with time varying components of voltage and current. For example, the instantaneous value of collector voltage is represented by v_c and r.m.s. value by V_c.

(4) Upper case subscripts refer to total or d.c quantities. For example, the instantaneous value of the total collector voltage is denoted by v_C. Similarly, the average value of the total collector voltage is denoted by V_C.

(5) Repeated upper case subscripts refer to the magnitudes of supply quantities. For example, collector voltage supply is designated by V_{CC} and base voltage supply by V_{BB}.

(6) The currents are taken as positive, when entering a terminal from the external circuit. Fig. 11-1(*a*) and (*b*) shows the positive directions for the currents in a NPN and PNP transistors respectively. It may be noted that the actual direction of the emitter current in a NPN transistor is opposite to that indicated by the arrow (I_E) in Fig. 11-1(*a*). Similarly, for a PNP transistor, the actual directions of base current and collector current are opposite to that indicated by the arrows I_B and I_C in Fig.11-1(*b*) respectively.

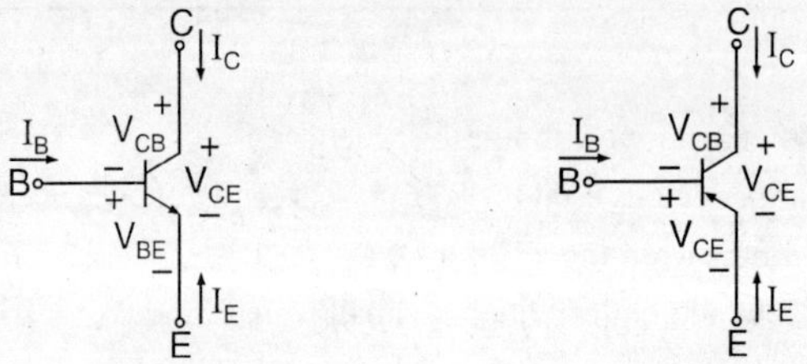

(*a*) NPN transistor (*b*) PNP transistor

Fig. 11-1. *Standard terminology for voltages and currents in a transistor.*

(7) The voltages are taken as positive, when measured with respect to a reference node (usually ground). For voltages measured between a pair of terminal (not the reference) an additional subscript is used as shown in Fig. 11-1(*a*) and (*b*). For example, the instantaneous collector-to-emitter voltage is indicated by v_{CE}.

(8) The additional subscript Q is added to circuit variables to indicate quiescent or operating point values. For example, the quiescent value of collector current is denoted by I_{CQ}.

Table 11-1 summarize the notations for transistor voltage and currents. It may be noted that total quantity (voltage or current) is the sum of d.c. and a.c. components. For example, the total value of collector voltage may be expressed as,

$$v_C = \underset{d.c.\ value}{V_C} + \underset{a.c.\ value}{v_c}$$

where, V_C = D.C. or the quiescent collector voltage, and

v_c = Instantaneous value of a.c. collector voltage.

Table 11-1. *Notations for transistor voltages and currents*

S. No.	*Item*	*Supply*	*Quiescent (static)*	*A. C. or time varying component*		*Total d.c. + a.c.*	
				Instantaneous	*r.m.s.*	*Instantaneous*	*Avg.*
1.	Collector voltage	V_{CC}	V_{CQ}	v_c	V_c	v_C	V_C
2.	Collector current	I_{CC}	I_{CQ}	i_c	I_c	i_C	I_C
3.	Base voltage	V_{BB}	V_{BQ}	v_b	V_b	v_B	V_B
4.	Base current	I_{BB}	I_{BQ}	i_b	I_b	i_B	I_B
5.	Emitter voltage	V_{EE}	V_{EQ}	v_e	V_e	v_E	V_E
6.	Emitter current	I_{EE}	I_{EQ}	i_e	I_e	i_E	I_E

11-3. Ebers-Moll Model of a Transistor

We know that a transistor consists of two coupled PN junctions. The base region is common to both junctions and forms a coupling between them. Since the transistors are constructed with very narrow base regions, therefore a significant electrical interaction exists between the junctions. This interaction is called *transistor action*. Fig. 11-2(*a*) shows the current components of a PNP transistor, which constitute the terminals current, *i.e.*, emitter current (I_E) and collector current (I_C). The voltages V_{EB} and V_{CB} are the voltage drops from emitter-to-base and collector-to-base respectively. Assuming that there is no voltage drop across all the semiconductors forming the emitter, base and collector regions (*i.e.*, $r_{ee'} = r_{bb'} = r_{cc'} = 0$), the voltages V_{EB} and V_{CB} appear across the respective junctions. It may be noted in Fig. 11-2(*a*) that emitter current has two components namely the current associated with the emitter-base diode indicated by I_{ED} and the current associated with the collector-base diode indicated by I_{CD}. The component $\alpha_R \cdot I_{CD}$ is the portion of collector-base diode current, which is coupled through the base to emitter. Similarly, $\alpha_F \cdot I_{ED}$ is the portion of emitter-base diode current, which is coupled through the base to the collector.

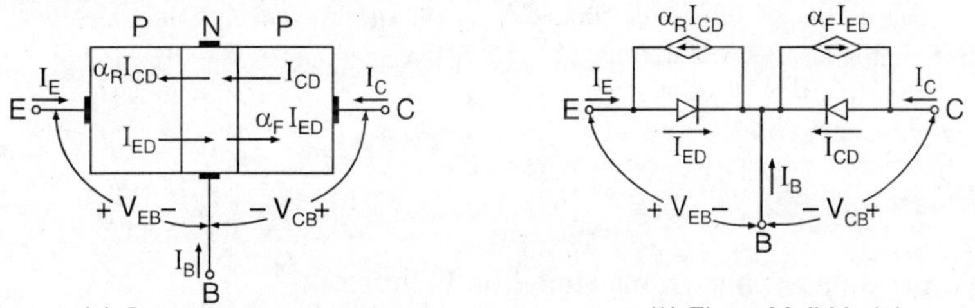

(*a*) Current components. (*b*) Ebers-Moll Model.

Fig. 11-2. *PNP transistor.*

On the basis of above discussion, we can construct Ebers Moll model, as shown in Fig. 11-2(*b*). From this figure, the following important points may be carefully noted:

(1) The model consist of two diodes connected back-to-back (*i.e.*, the cathodes are connected together). These diodes represent the emitter-base junction and collector-base junction of the transistor.

(2) There are two controlled current sources indicated by $\alpha_R \cdot I_{CD}$ and $\alpha_F \cdot I_{CD}$. These sources indicate coupling between the two junctions of the transistor.

(3) The currents I_{ED} and I_{CD} are related to V_{EB} and V_{CB} by the following equations :

$$I_E = I_{ED} - \alpha_R \cdot I_{CD}$$

$$= I_{ES}\,(e^{V_{EB}/V_T} - 1) - \alpha_R \cdot I_{CS}\,(e^{V_{CB}/V_T} - 1) \qquad ...(i)$$

and

$$I_C = -\,\alpha_F \cdot I_{ED} + I_{CD}$$

$$= -\,\alpha_F \cdot I_{ES}\,(e^{V_{EB}/V_T} - 1) + I_{CS}\,(e^{V_{CB}/V_T} - 1) \qquad ...(ii)$$

where, I_{ES} = Reverse saturation current of the emitter base junction,

I_{CS} = Reverse saturation current of the collector-base junction.

α_F = Forward transmission from emitter to collector (*i.e.*, I_C/I_B). It is also called common-base current gain (α) of transistor. Its value is in the range of 0.98 to 0.998.

α_R = Reverse transmission from collector to emitter (*i.e.*, I_E/I_C). It is also called inverse alpha (α_1) of a transistor. Its value is in the range of 0.4 to 0.8.

The equations (*i*) and (*ii*) are known as *Ebers-Moll equations* and give us the values of emitter current and collector current respectively. The four quantities I_{ES}, I_{CS}, α_E and α_R are functions of doping densities and transistor geometry. These quantities are related by the equation,

$$\alpha_F \cdot I_{ES} = \alpha_R \cdot I_{CS} \qquad ...(iii)$$

The above equation is, sometimes, called reciprocity condition for the transistor. Fig. 11-3(*a*) shows the current components in an NPN transistor and Fig. 11-3(*b*) shows its Ebers-Moll model. It may be noted that for an NPN transistor, the direction of all current components and junction voltages are reversed from those of a PNP transistor.

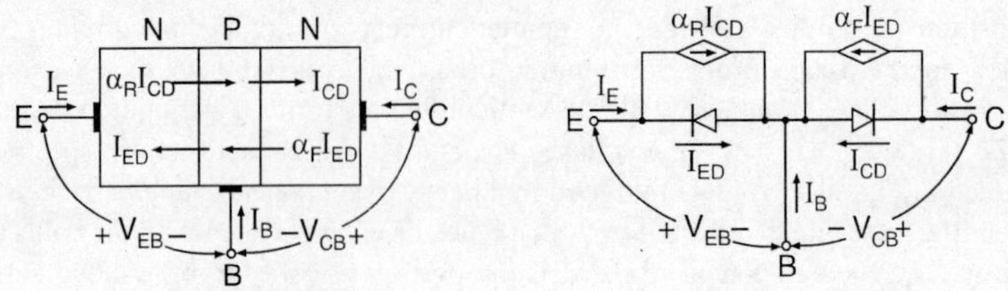

(*a*) Current components. (*b*) Ebers-Moll Model.

Fig. 11-3. *NPN transistor.*

The Ebers-Moll equations, *i.e.*, equations (*i*) and (*ii*) are valid for an NPN transistor, provided a negative sign is inserted before the voltages V_{EB}, V_{CB} and each current component. Thus Ebers-Moll equations for an NPN transistor are :

$$I_E = -I_{ES}\,(e^{-V_{EB}/V_T} - 1) + \alpha_R \cdot I_{CS}\,(e^{-V_{CB}/V_T} - 1)$$

and

$$I_C = \alpha_F \cdot I_{ES} \cdot (e^{-V_{EB}/V_T} - 1) - I_{CS}\,(e^{-V_{CB}/V_T} - 1)$$

11-4. Approximate Ebers-Moll Model of Transistors

The Ebers-Moll model of a transistor can be simplified, if the junction diodes are replaced by an ideal ones. An ideal diode is one, which acts as an open circuit when it is reverse biased; and as a short circuit when it is forward biased. Now, we shall discuss the behaviour of this approximate model in the saturation, active and cut-off regions of a PNP transistor.

(*i*) ***Saturation region*** **:** In this region, both the emitter-base and collector-base junctions are forward biased. Therefore, both the diodes act as short circuits. Under this condition, the Ebers-Moll model in saturation region is as shown in Fig. 11-4(*a*). It may be noted that in saturation region the transistor acts as a short-circuit.

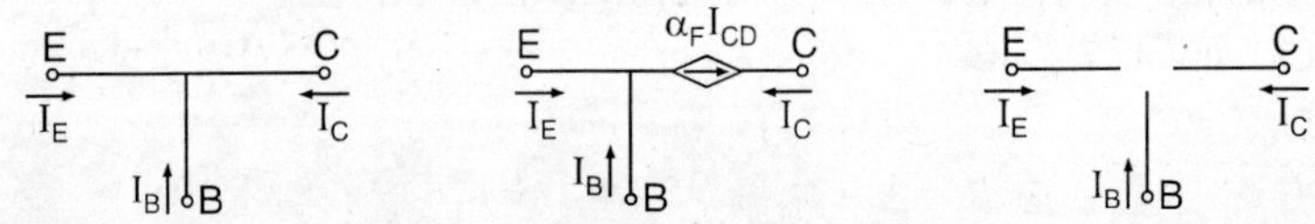

(*a*) In saturation region. (*b*) In active region. (*c*) In cut-off region.

Fig. 11-4. *Approximate Ebers-Moll model.*

(*ii*) ***Active region :*** In this region, the emitter-base junction is forward biased and collector-base junction is reverse biased. Therefore the emitter diode acts as a short-circuit and collector diode as an open circuit. Under this condition, the Ebers-Moll model in active region is as shown in Fig. 11-4(*b*).

(*iii*) ***Cut-off region :*** In this region, both the emitter-base and collector-base junctions are reverse biased. Therefore, both the diodes and current sources act as open circuits. Under this condition, the Ebers-Moll model in cut-off region is as shown in Fig. 11-4(*c*). It may be noted that in the cut-off region, the transistor acts as an open-circuit.

11-5. More Accurate Ebers-Moll Model of Transistor

As a matter of fact, the Ebers-Moll models discussed in the last article are very simple. Although these models provide accurate results in most of the times, yet in some other times we require complex models to get more accurate results. We shall now discuss as to how these complex models can be obtained in the three regions.

(1) ***More accurate saturation region model :*** The model shown in Fig. 11-4(*a*) indicates that voltages V_{EB} and V_{CB} are zero in the saturation region. But in actual practice, it is not the case. As a matter of fact, there is always some voltage drop between the transistors terminals. The values of voltages V_{EB} and V_{CB} are essentially those of forward biased diodes. Above the cut-in or knee voltage (V_K), the diode current increases very rapidly with the applied voltage. Thus, if the external circuitry limits the current as it often does the forward biased PN junction voltage will remain essentially constant. Therefore, saturation models which are more accurate, are shown in Fig. 11-5(*a*) and (*b*) for PNP and an NPN transistors respectively. The values of battery voltages $V_{EB\ (sat)}$ and $V_{CB\ (sat)}$ are 0.8 V and 0.6 V respectively for both the transistors.

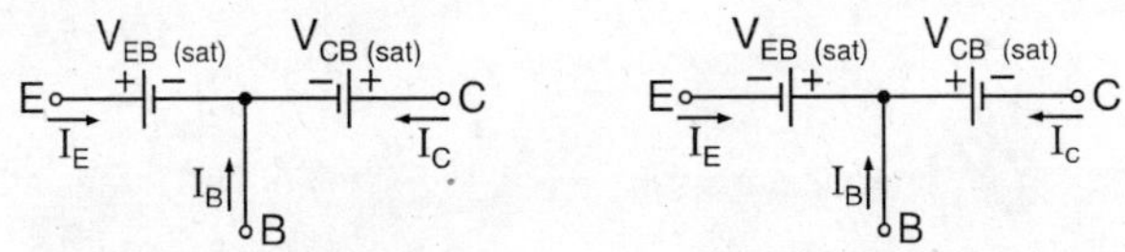

(*a*) For PNP transistors. (*b*) For NPN transistors.

Fig. 11-5. *More accurate saturation region model.*

(2) ***More accurate active region model :*** The active region model shown in Fig. 11-4(*b*) can be made, somewhat, more accurate by adding a battery voltage $V_{EB\ (active)}$. The value of this battery voltage is 0.7 V.

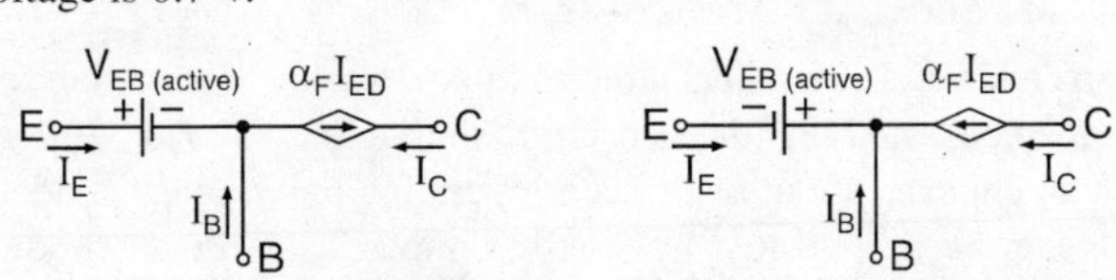

(*a*) For PNP transistors. (*b*) For NPN transistors.

Fig. 11-6. *More accurate active region model.*

Fig. 11-6(*a*) and (*b*) shows, somewhat, more accurate model for a PNP and NPN transistors respectively.

(3) ***More accurate cut-off region model :*** Generally, the model shown in Fig. 11-4(*c*) is sufficient in the cut-off region, *i.e.*, the transistor in cut-off region can be considered to be an open-circuited. But sometimes we are interested in few additional details of the transistor in this region. Consider the model shown in Fig. 11-2(*b*). Since, the diodes are not ideal, therefore they have an effective cut-in or knee voltage (V_K). For instance, if in Fig. 11-2(*b*) the voltage V_{EB} is less than V_K, then the emitter diode will be cut-off and the emitter current will be zero. If the collector diode is cut-off, then the transistor will also be cut-off.

However, if the external voltages are such that the collector diode is cut-off, then the transistor will be cut off only if the emitter-base voltage is not of the proper polarity and its magnitude is less that V_K. This defines the edges of cut-off region.

11-6. Ebers-Moll Model of Transistor in Common Emitter Configuration

The Ebers-Moll model of a transistor discussed in the last articles is that of a common-base configuration. However, the transistors are used more commonly in a common-emitter configuration. Therefore, it is important to discuss the Ebers-Moll Model also called d.c. model of a transistor in common emitter configuration.

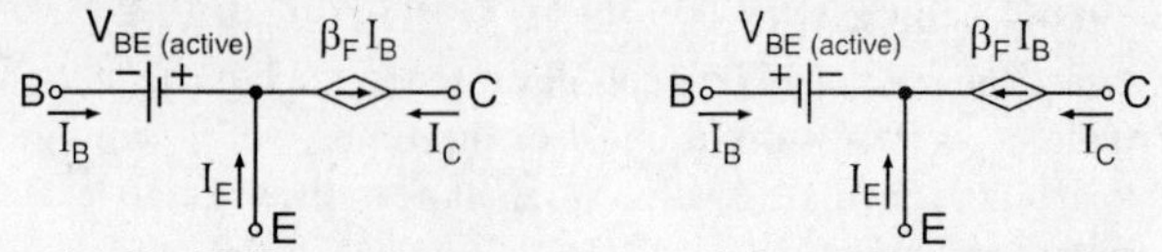

(*a*) For PNP transistors. (*b*) For NPN transistors.

Fig. 11-7. *D.C. models for a transistor in common emitter configuration.*

Fig. 11-7 shows d.c. model of a transistor in the active region. Here, β_F (equal to β) is a common emitter current gain. It is the ratio of collector current to base current. The term $V_{BE\,(active)}$ is the barrier potential across the base-emitter diode and its value is 0.7 V for silicon transistor.

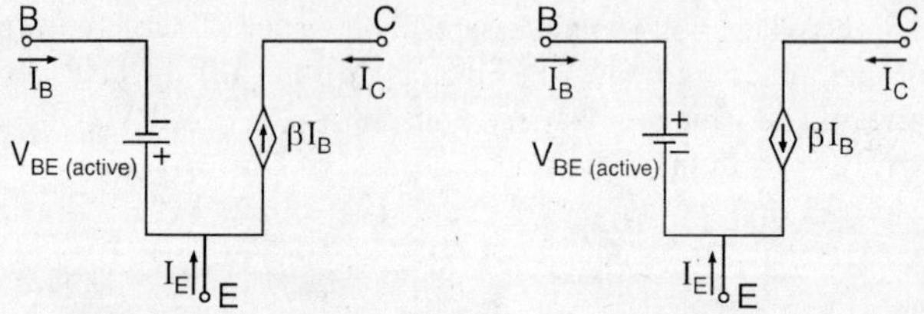

(*a*) For PNP transistor. (*b*) For NPN transistor.

Fig. 11-8.

Sometimes, it is convenient to represent the d.c. model of a transistor in a slightly different way as shown in Figure 11-8(*a*) and (*b*) for PNP and an NPN transistor respectively. In this figure, we have replaced the current gain β_F by β.

11-7. Small Signal Models of Transistor

These are also called a.c. equivalent circuits of a transistor. The models, which we shall discuss here are linear. It may be noted that the transistor is not a linear device. But when we are interested in determining small changes about a quiescent operating point, linear models can be accurate. We know that signals can be represented by small changes. Thus linear models can be used in the analysis of small signal circuits for a variety of amplifier applications. It has been observed that the behaviour of transistor is different for low-frequencies and high-frequencies. Therefore, we shall discuss the small signal models of a transistor under the following two heads :

- Low-frequency small signal model.
- High-frequency small signal model.

11-8. Low-Frequency Small Signal Model of Transistor

A low frequency small signal model can be obtained from the Ebers-Moll model of a transistor as shown in Fig. 11-2(*a*). Since, in the normal operation of a transistor, the emitter-base junction is forward biased and collector-base junction is reverse biased, therefore the Ebers-Moll model shown in Fig. 11-2(*a*), can be reduced to a circuit as shown in Fig. 11-9(*a*).

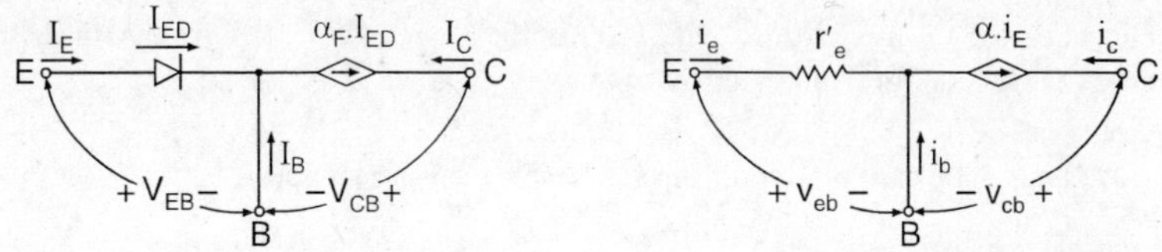

(*a*) Ebers-Moll model. (b) Small signal model.

Fig. 11-9. *PNP transistor.*

From this circuit, we can obtain a small signal model of a transistor as shown in Fig. 11-9(*b*). This model is for a PNP transistor in a common base configuration. Following are some of the importants points, which may be carefully noted from the Fig. 11-9(*b*):

(*i*) The emitter diode is replaced by an a.c. resistance r'_e. The resistance r'_e represents the a.c. resistance (or dynamic resistance) of a forward-biased emitter base junction. Its value is given by the relation,

$$r'_e = \frac{25}{I_E}$$

where, I_E is the d.c. or quiescent operating current in milliamperes.

(*ii*) All currents and voltages of a transistor are indicated by lower case letter *i* and *v* respectively.

(*iii*) The diode current is represented by i_e instead of i_{ed}, as it does not make any difference. Similarly, α_F is shown as α.

The small-signal model of a PNP transistor in common-base configuration can be represented in a more convenient way as shown in Fig. 11-10(*a*).

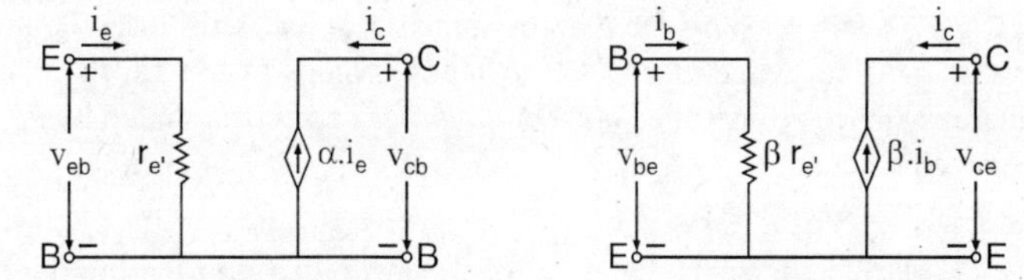

(*a*) For a PNP transistor in common base configuration. (*b*) For a PNP transistor in common emitter configuration.

Fig. 11-10. *Low-frequency small signal model.*

The small signal model of a PNP transistor in common emitter configuration may by obtained in the same way as discussed above and is as shown in Fig. 11-10(*b*).

It may be noted that in a common emitter model, the a.c. resistance of a base-emitter junction is represented by $\beta \cdot r'_e$. The current of a controlled source is indicated by $\beta \cdot i_b$. The small signal models for an NPN transistor in a common base and common emitter configurations are shown in Fig. 11-11(*a*) and (*b*) respectively.

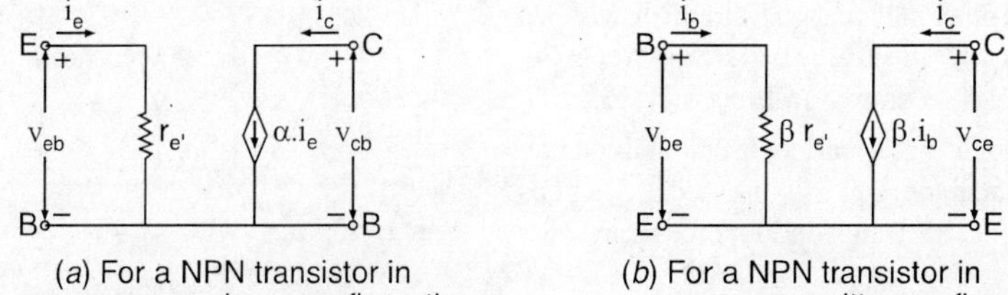

(*a*) For a NPN transistor in common base configuration. (*b*) For a NPN transistor in common emitter configuration.

Fig. 11-11. *Low frequency small signal model.*

Note. The a.c. resistance of the emitter-base junction in common emitter configuration is equal to $\beta \cdot r_e'$ (and not equal to r_e'). It is because of the fact that the a.c. resistance, looking into the base in common emitter configuration, is given by the relation,

$$r_{ac} = 25/I_B$$

where, I_B is the base current in milliamperes. Since, the base current

$$I_B = I_C/\beta = I_E/\beta$$

$$\therefore \quad r_{ac} = \frac{25}{I_B} = \frac{25}{I_E/\beta} = \beta\frac{25}{I_E} = \beta \cdot r_e'$$

11-9. Low Frequency Hybrid-π Model of Transistor

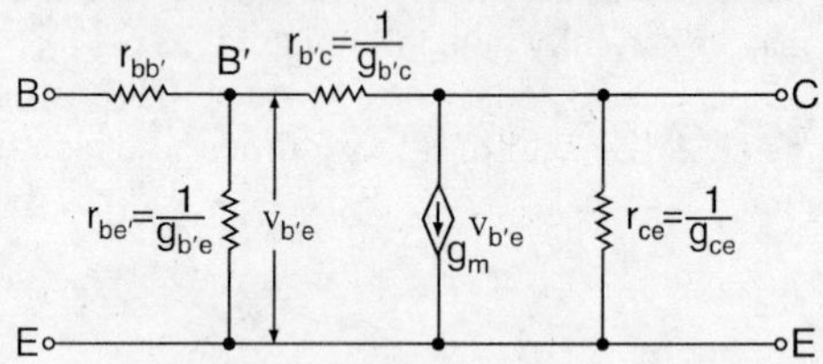

Fig. 11-12. *Low frequency hybrid-π model of a transistor in a common emitter configuration.*

Figure 11-12 shows the low frequency hybrid-π model of a transistor with common emitter configuration. It is more accurate than that of the Ebers-Moll model discussed in the last article. Following are the components of a hybrid-π model :

(*i*) ***Resistance*** $(r_{b'b})$: It represents the base spreading resistance and accounts for the voltage drop in the path between the base contact and the active base region. Its value decreases with the increase in current value. The value of resistance $r_{b'b}$ is in the range of 40 Ω to 400 Ω.

(*ii*) ***Resistance*** $(r_{b'e})$: It represents the a.c. resistance of a forward-biased emitter-base junction. Its value is in the range of about few hundred ohms to several thousand ohms. The reciprocal of resistance $r_{b'e}$ is called *a.c. conductance* and is designated by $g_{b'e}$. Mathematically, the a.c. conductance

$$g_{b'e} = 1/r_{b'e}$$

(*iii*) ***Resistance*** $(r_{b'c})$: It represents the effect of feedback between the emitter-base junction and collector-base junction due to an early effect. Its value is of the order of several megaohms. Because of its extremely high value, it is neglected in computation.

(*iv*) ***Controlled current source*** $(g_m \cdot v_{b'e})$: It represents the coupling between the junctions. Its value (equal to $g_m \cdot v_{b'e}$) is proportional to the base current (i_b).

(*v*) ***Resistance*** (r_{ce}) : It is called *output resistance*. Its typical value is in the range of about few kiloohms to hundreds of kiloohms. The reciprocal of resistance r_{ce} is called *output conductance* and is designated by g_{ce}. Mathematically, the output conductance, $g_{ce} = 1/r_{ce}$. *Sometimes, the resistance r_{ce} is also represented by a symbol r_o.

The hybrid-π model may be simplified, if we omit $r_{bc'}$ from the model. The simplified low frequency model is as shown in Figure 11-13.

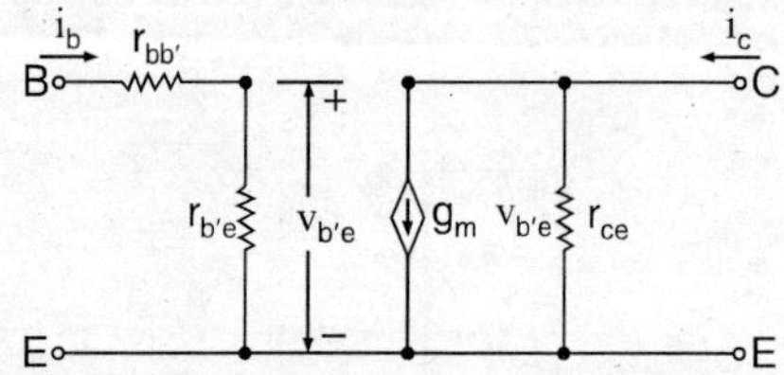

Fig. 11-13. *Simplified hybrid-π low frequency model of a common-emitter transistor.*

The parameter g_m is called transconductance (or mutual conductance) and is expressed in Siemens (*S*) or ohms (Ω). The transconductance represents

* According to this phenomenon, the width of the depletion region of collector-base junction increases as the magnitude of the reverse bias voltage (across the collector base junction) increases.

the small changes in collector current about the operating point produced by the small changes in the base-emitter voltage. Its value is given by the relation,

$$g_m = \frac{I_c}{25}$$

where, I_c is the collector current in milliamperes.

It may be noted from Fig. 11-13 that the voltage drop across resistance $r_{be'}$ is given by the relation,

$$v_{b'e} = i_b \cdot r_{b'e}.$$

where, i_b is the base current.

It may also be noted that if the collector-to-emitter voltage (v_{ce}) is equal to zero, then there is no current through resistance r_{ce}. In this case, the whole current will pass through the controlled current generator. The value of collector current,

$$i_c = g_m \cdot v_{b'e}.$$

Substituting the value of $v_{b'e'}$ (equal to $i_{b'} \cdot r_{b'e'}$) is the above equation, the collector current,

$$i_c = g_m \cdot i_b \cdot r_{b'e'}$$

or $$i_c/i_b = g_m \cdot r_{b'e'}$$

The ratio i_c/i_b is defined as *a.c. current gain* (β_{ac}) of a common emitter transistor. It is also called small signal short circuit current gain, and is designated by h_{fe}. Thus a.c. current gain,

$$h_{fe} = i_c/i_b = g_m \cdot r_{b'e'}$$

$$\therefore \quad r_{b'e} = h_{fe}/g_m \text{ and } g_{b'e} = \frac{1}{r_{b'e'}} = gm/h_{fe}$$

The parameter h_{fe} is supplied by the manufacturer. Knowing the value of quiescent collector current, the value of transconductance may be determined by using the relation $g_m = I_c/25$. And then may be calculated by using the relation $r_{b'e} = h_{fe}/g_m$. The value of base spreading resistance may be obtained by using the relation,

$$r_{bb'} = h_{ie} - r_{b'e'}$$

where, h_{ie} is the input resistance of a common emitter transistor and is usually specified by the manufacturer. The value of output conductance may be obtained by the relation,

$$g_{ce} = h_{oe} - (1 + h_{fe})\, g_{be}$$

where, h_{oe} = Output conductance with input as short-circulated. Its value is usually supplied by the manufacturer.

$g_{b'c}$ = Feedback conductance. Its value is given by the relation $h_{re} \cdot g_{b'e,}$. The parameter h_{re} is called reverse voltage gain and is usually specified by the manufacturer.

In actual practice, the value of h_{fe} >>1. In that case, the output conductance,

$$g_{ce} = h_{oe} - h_{fe} \cdot g_{b'e'}$$

Substituting the value of $g_{b'c'}$. (equal to $h_{re} \cdot g_{b'c'}$) in the above expression, the value of output conductance,

$$g_{ce} = h_{oe} - h_{fe} \cdot h_{re}\, g_{b'c'}$$

$$= h_{oe} - h_{re} \cdot g_m \qquad ...(\because g_{b'e'} = g_m/h_{fe})$$

$\therefore$ Value of output resistance,

$$r_{ce} = \frac{1}{g_{ce}} = \frac{1}{h_{oe} - (1 + h_{fe})\, g_{b'c}}$$

$$= 1/(h_{oe} - h_{re}\, g_m)$$

Example 11-1. *A given transistor with I_C = 10 mA, V_{CE} = 10 V and at room temperature has the following set of low frequency parameters:*

$$h_{ie} = 500\ \Omega\ h_{oe} = 10^{-5}\ A/V$$

$$h_{fe} = 100\ h_{re} = 10^{-4}$$

Find the values of all the hybrid - π parameters of a low frequency model.

Solution. Given: I_C = 10 mA; V_{CE} = 10 volt; h_{ie} = 500 Ω; h_{oe} = 10^{-5} A/V; h_{fe} = 100 and h_{re} = 10^{-4}

Resistance $r_{b'e}$

We know that the value of transconductance,

$$g_m = \frac{I_C}{25} = \frac{10}{25} = 0.4\ \Omega$$

and the value of resistance, $r_{b'e}$.

$$= \frac{h_{fe}}{g_m} = \frac{100}{0.4} = 250\ \Omega\ \textbf{Ans.}$$

Base spreading resistance

We know that the value of base spreading resistance,

$$r_{bb'} = h_{ie} - r_{b'e'} = 500 - 250 = 250\ \textbf{Ans.}$$

Output resistance

We know that the value of feedback conductance,

$$g_{b'c} = \frac{h_{re}}{r_{b'e}} = \frac{10^{-4}}{250} = 4 \times 10^{-7}$$

and the value of output resistance,

$$r_{ce} = \frac{1}{h_{oe} - (1 + h_{fe})\ g_{b'c}} = \frac{1}{10^{-5} - (1 + 100) \times (4 \times 10^{-7})}$$

$$= 3.3 \times 10^4\ \Omega = 33\ k\Omega\ \textbf{Ans.}$$

11-10. High Frequency Hybrid-π Model of a Transistor

Figure 11-14 shows a high frequency hybrid-π model of a transistor in common emitter configuration. This model is similar to the low frequency hybrid-π model, except the addition of two capacitances, c_e and c_c respectively. The values of resistance $r_{bb'}$, $r_{b'c}$, $r_{b'e}$ and r_{ce} may be obtained in the same way as we discussed for the low-frequency hybrid -π model.

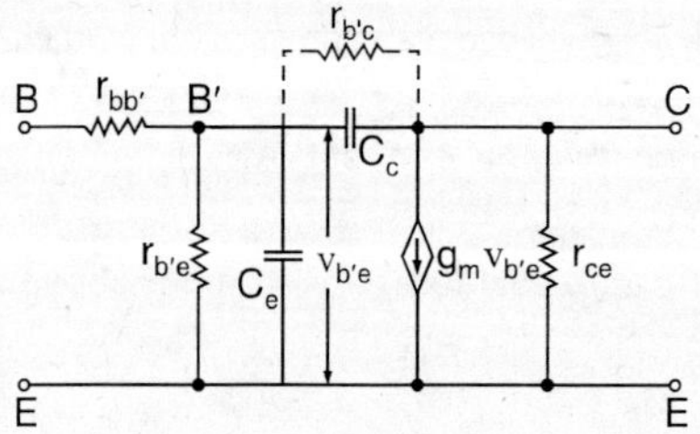

Fig. 11-14. *High frequency hybrid-π model of a transistor in a common emitter configuration.*

The capacitance c_e is the sum of the emitter diffusion capacitance (C_{D_e}) and the emitter junction capacitance (C_{T_e}). Mathematically, the capacitance,

$$C_e = C_{D_e} + C_{T_e}$$

For a forward biased emitter-base junction, the emitter diffusion capacitance (C_{D_e}) is, usually much larger than the emitter junction capacitance (C_{T_e}). Therefore, the capacitance

$$C_e = C_{D_e}$$

It has been observed that the capacitance C_{D_e} is proportional to the emitter bias current, I_E and is almost independent of temperature. The value of C_c may be determined experimentally from measurement of frequency (f_T) at which the common emitter short-circuit current gain drops to unity. Its value is given by the relation,

$$C_e = \frac{g_m}{2f_T}$$

where, g_m is the transconductance.

The capacitance C_e is the depletion region capacitance of the reverse biased collector-base junction. It varies as $(V_{CE})^{-n}$, where n is 1/2 or 1/3 for an abrupt or graded junction.

SUMMARY

In this chapter you have learned that :

1. The large signal model of a tansistor is known as d.c. model.
2. The small signal model of a transistor is known as a.c. model.
3. Low frequency small signal and high frequency small signal are the two small signal models of transistor.

GLOSSARY

The BJT model : A set of mathematical equations for the calculation of terminal voltages and currents.

Ebers-Moll model : A BJT model used in analysis of small-grids amplifiers.

Hybrid-π model : A BJT model used for the same purpose as Ebers-Mall model. But Hybrid-π model is accurate as compared to the Ebers-Moll model.

DESCRIPTIVE QUESTIONS

1. For a PNP transistor in the active region, what is the sign (positive or negative) of emitter current, collector current, base current, collector-to-base voltage and emitter-to-base voltage?
2. Draw the Ebers-Moll model for an NPN transistor. Also give the equations for emitter current and collector current.
3. Draw an approximate d.c. model of a PNP transistor in each of the three regions of operation.
4. Draw the low-frequency hybrid-π model and explain the meaning of each component of a model.
5. (*a*) Define the transconductance g_m.
   (*b*) write an equation, which relates g_m to current gain β (or h_{fe}).
6. Draw a neat diagram of a high-frequency hybrid-π transistor equivalent circuit. Explain the meaning of each component.
7. Draw the small signal high frequency common emitter model of a transistor. (*A.M.I.E., Winter,* 1987)
8. Explain the high frequency model of a bipolar transistor. (*Grad. I.E.T.E., Dec.* 1987)
9. Derive an Ebers-Moll model of a transistor. (*Electron Devices, Anna Univ., April/May 2008*)

MULTIPLE CHOICE QUESTIONS

1. Which of the following is not a time varying quantity

(*a*) v_{ce} (*b*) V_{CE} (*c*) v_{CE} (*d*) V_{ce}

2. In the Ebers-Model of a bipolar transistor, the parameter is the

(*a*) Forward transmission from emitter to collector

(*b*) Reverse transmission from collector to emitter

(*c*) Common base current gain

(*d*) Both (*a*) and (*c*) above

3. The value of transconductance of a bipolar transistor for a collector current of 1.5 mA is

(*a*) 0.05 Ω (*b*) 0.05×10^{-3} Ω

(*c*) 37.5 Ω (*d*) None of the above

4. The resistance $r_{b'z}$ in the low frequency hybrid-π model of a bipolar transistor represents

(*a*) Base spreading resistance

(*b*) A.C. resistance for forward biased emitter-base junction

(*c*) The effect of feedback between the emitter-base junction and collector-base junction due to Early effect

(*d*) None of the above

5. The capacitance c_e in the high frequency hybrid-π model of a bipolar transistor represents the

(*a*) Depletion region capacitance (*b*) Emitter diffusion capacitance

(*c*) Emitter-base junction capacitance (*d*) Sum of the (*b*) and (*c*) above

ANSWERS

1. (*b*) **2.** (*d*) **3.** (*a*) **4.** (*c*) **5.** (*d*)

❑❑❑

12

CHAPTER

BJT Biasing and Stabilisation

OUTLINE

Objectives

After completing this chapter, you should be able to :

- Know transistor biasing.
- Understand d.c. load line of a transistor.
- Know Q-point of a transistor.
- Find the stability of common base and common emitter transistor circuits.
- Understand different methods of transistor biasing namely, base bias, base bias with emitter feed back, base bias with collector feedback, voltage divider bias and emitter bias.
- Know bias compensation circuits.

12-1. Introduction

It is a well known fact these days, that transistors are used in a large variety of applications and in many different ways. To use a transistor in any application, it is necessary first to bias the transistor. The usual reason for this biasing is to turn the device ON and to place it in a region where it operates linearly and provides a constant amount of voltage gain.

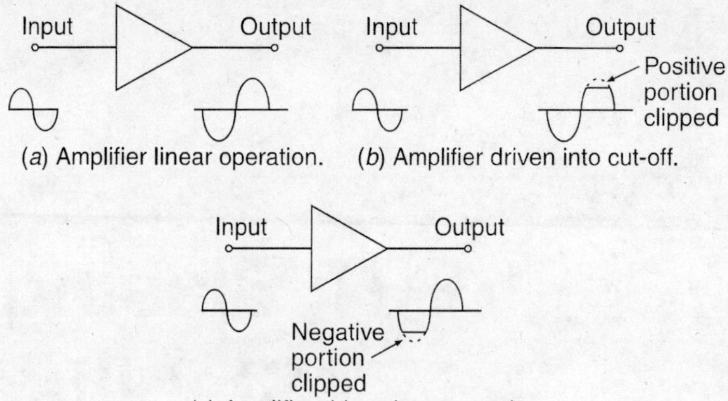

(a) Amplifier linear operation. (b) Amplifier driven into cut-off.

(c) Amplifier driven into saturation.

Fig. 12-1.

The biasing deals with setting a fixed level of the current, which should flow through the transistor with a desired fixed voltage drop across the transistor junctions. Usually, the collector current, base current, collector-to-emitter voltage and the base-to-emitter voltage are the currents and voltages, which are required to be set by the biasing circuit. The proper values of these currents and voltages allow a transistor to amplify the weak signals faithfully. It means that the transistor must increase the magnitude of the signal without changing its shape.

Fig. 12-1 (*a*) shows a properly biased transistor amplifier. Here, the signal variations at the input of a transistor are accurately reproduced at the input. However, Fig. 12-1(*b*) and (*c*) shows an improperly biased transistor amplifiers. It may be noted that in Fig. 12-1(*b*), the positive portion of the output signal is clipped. It is because of the fact, that the transistor is biased near its cut-off point. Similarly, in Fig. 12-1 (*c*), the negative portion of the output signal is clipped. In this case, the transistor is biased near its saturation-point.

12-2. The D.C. Operating Point and Load Line

We have already discussed in the last article that of proper operation of a transistor, in any application, we set a fixed level of certain currents and voltages in a transistor. These values of currents and voltages define the point, at which the transistor operates. This point is called ***operating point***. It is also known as ***quiescent point*** or simply ***Q-point***. Since, the level of currents and voltages are fixed, therefore the operating point is also called ***d.c. operating point***.

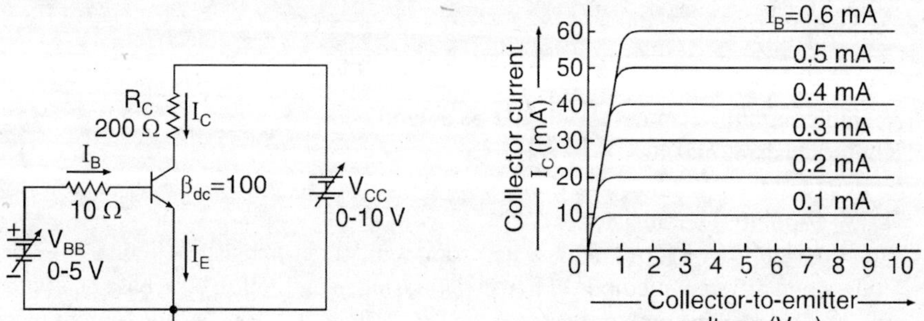

(a) Transistor circuit. (b) Collector characteristic curves showing Q points.

Fig. 12-2.

Now let us consider a transistor circuit as shown in Fig. 12-2(*a*). Here the transistor is biased with two d.c. supplies (or batteries) namely V_{CC} and V_{BB} to obtain the fixed values of collector current (I_C), base current (I_B) and collector-to-emitter voltage (V_{CB}).

Fig. 12-2(*b*) shows the collector characteristic curves for this particular transistor. We shall use these characteristics to graphically illustrate the effects of d.c. biasing. In order to draw these curves, we set various values of base current (I_B) say from 0.1 mA to 0.6 mA, by varying the V_{BB} supply and then observe the changes in collector current and collector-to-emitter voltage.

First of all, let us adjust the V_{BB} supply to produce a base current of 0.3 mA. The corresponding value of collector current,

$$I_C = \beta \cdot I_B = 100 \times 0.3 = 30 \text{ mA} = 30 \times 10^{-3} \text{ A}$$

and the collector-to-emitter voltage,

$$V_{CE} = V_{CC} - I_C \cdot R_C = 10 - (30 \times 10^{-3}) \times 200 = 10 - 6 = 4 \text{ V}$$

These values of collector current and collector-to-emitter voltage define a *Q*-point marked as Q_1 on the characteristic curves of Fig. 12-3(*a*). Now the V_{BB} supply is increased to produce a base current of 0.4 mA. This gives a collector current (I_C) of 40 mA and collector-to-emitter voltage (V_{CE}) of 2 V. These values of collector current and collector-to-emitter voltage define another *Q*-point marked as Q_2 on the characteristic curves of Fig. 12-3(*a*). Finally, the V_{BB} supply is reduced to produce a base current of 0.2 mA. This gives a collector current of 20 mA and collector-to-emitter voltage of 6 V. These values of collector current and collector-to-emitter voltage define *Q*-point marked Q_3 on the characteristic curves as shown in Fig. 12-3(*a*).

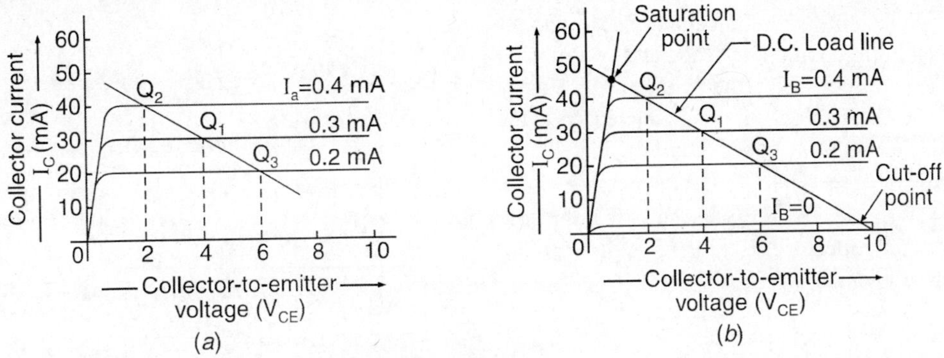

Fig. 12-3. *D.C. load line.*

It is evident from the above discussion that if the value of base current increases, the collector current also increases, while the collector-to-emitter voltage decreases. On the other hand, if the value of base current decreases, the collector current also decreases while the collector-to-emitter voltage increases. It means that as the V_{BB} supply is adjusted up or down, the d.c. operating point of a transistor moves up and down along a sloping straight line connecting the *Q*-points as shown in Fig. 12-3(*a*).

If we extend the line joining the operating points Q_1, Q_2 and Q_3 on both sides, we shall obtain a straight line as shown in Fig. 12-3(*b*). This line is called a ***d.c. load line***. The values of base current, collector current and collector-to-emitter voltage can be read at any point along this line.

The d.c. load line gives us the information about the following two important points :

(*i*) The load line intersects the horizontal axis at a point marked V_{CC}, which is equal to 10 V. This point is called *transistor cut-off point* or *lower end of the load line*. At this point, the values of base current and collector current are *zero* (ideally). But actually, there is a small value of reverse saturation current (I_{CBO}) at cut off point as indicated in the Fig.12-3. As a result of this, the collector-to-emitter voltage (V_{CE}) is slightly less than the value of V_{CC}.

(***ii***) The load line intersects the vertical axis at a point marked I_C equal to 50 mA. This point is called transistor *saturation point* or *upper end of the load line*. At this point, the collector current is maximum and the collector-to-emitter voltage is very small equal to $V_{CE\,(sat)}$.

It may be noted that if the upper and lower ends of the load line are known, the load line can be obtained by joining the upper and lower ends.

12-3. Simpler Way of Drawing a D.C. Load Line

We have already discussed in the last article about the d.c. load line and the way it is obtained from the collector characteristic curves. A much simpler way of drawing a d.c. load line is possible, if we know the values of V_{CC} supply and the collector resistance (R_C). For example, consider again the transistor circuit shown in Fig. 12-2(a). For this circuit, we know that the value of collector current is given by the relation,

$$I_C = \frac{V_{CC} - V_{CE}}{R_C} \quad ...(i)$$

where, V_{CC} = Value of d.c. supply voltage in the collector circuit,

V_{CE} = The value of collector-to-emitter, and

R_C = The value of collector resistance.

We know that the value of collector-to-emitter voltage (V_{CE}), at saturation point, is very small as compared to V_{CC} supply. Therefore, the collector current at saturation point (neglecting V_{CE}),

$$I_C = \frac{V_{CC}}{R_C} \quad ...(ii)$$

It means that the value of collector current at saturation point designated as $I_{C\,(sat)}$ may be obtained by dividing the value of V_{CC} supply by the value of collector resistance (R_C). This value gives us the upper end of the load line as shown in Fig. 12-4.

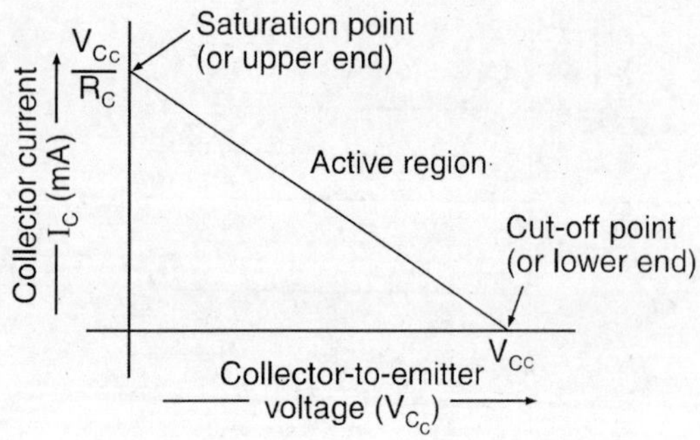

Fig.18.4. *D.C. load line.*

We also know that at cut-off point, the value of collector current is zero. Substituting $I_C = 0$ in equation (i),

$$0 = \frac{V_{CC} - V_{CE}}{R_C}$$

or $V_{CE} = V_{CC} = V_{CE\,(cut\text{-}off)}$

It means that the value of collector-to- emitter voltage at cut-off point designated as $V_{CE\,(cut\text{-}off)}$ is equal to the value of V_{CC} supply. This value gives us the lower end of the load line as shown in the figure.

It may be noted that region lying in between the saturation and cut-off points of the load line is called *active region* (or linear region), of the transistor's operation. As long as the transistor is operated in this active region, the output voltage is a *linear reproduction* of the input. It will be interesting to know that for normal operation, the Q-point of a transistor is selected *midway* between the saturation and cut-off point, *i.e.*, where the collector-to-emitter voltage (V_{CE}) is equal to half of the V_{CC} supply.

Example 12-1. *A transistor circuit of a pulse amplifier is shown Fig. 12-5. What is the value of saturation current and cut-off voltage? Also draw the d.c. load line.*

Solution. Given : V_{BB} = 10 V; R_B = 47 kΩ; V_{CC} = 20 V; R_C = 10 kΩ = 10 × 10^3 Ω and β = 100.

Value of saturation current

We know that the value of saturation current,

$$I_{C\,(sat)} = \frac{V_{CC}}{R_C} = \frac{20}{10 \times 10^3} = 0.002 \text{ A or 2 mA } \textbf{Ans.}$$

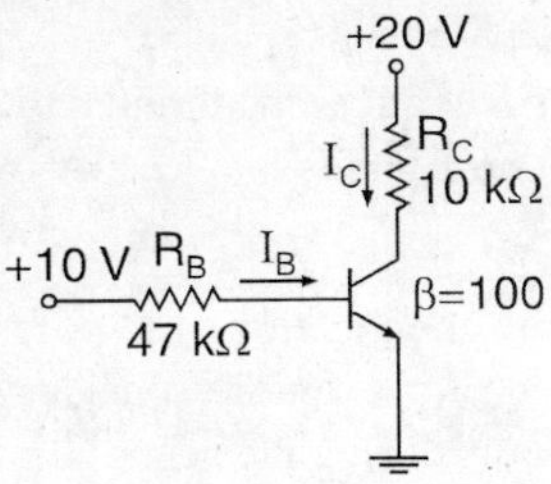

Fig. 12-5. *D.C. load line.*

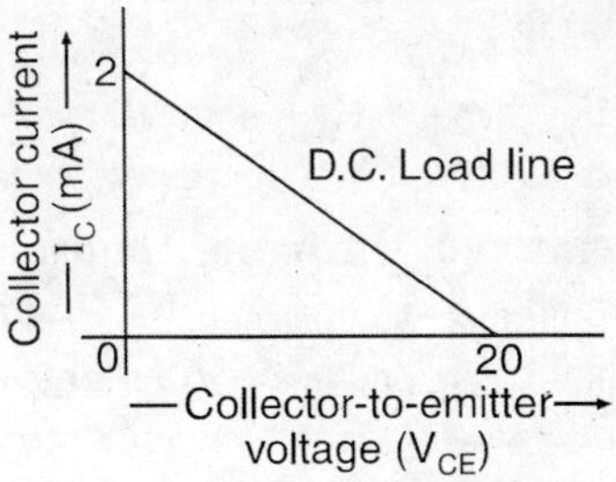

Fig. 12-6.

Value of cut-off voltage

We also know that the value of cut-off voltage,

$$V_{CE\,(cut\text{-}off)} = V_{CC} = 20 \text{ V } \textbf{Ans.}$$

D.C. load line

The values of $I_{C\,(sat)}$ and $V_{CC\,(cut\text{-}off)}$ gives the upper and lower ends of d.c. load line. By using these two points, we get a straight line (called d.c. load line) as shown in Fig. 12-6.

Example 12-2. *Find the upper and lower ends of the d.c. load line for the circuit shown in Fig. 12-7. Also find the Q-point and locate it on the d.c. load line.*

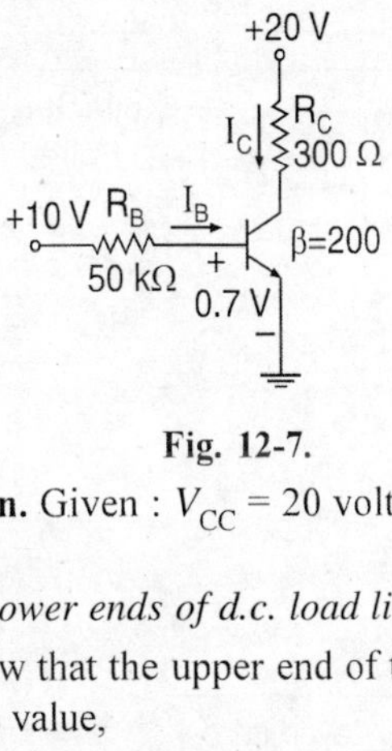

Fig. 12-7.

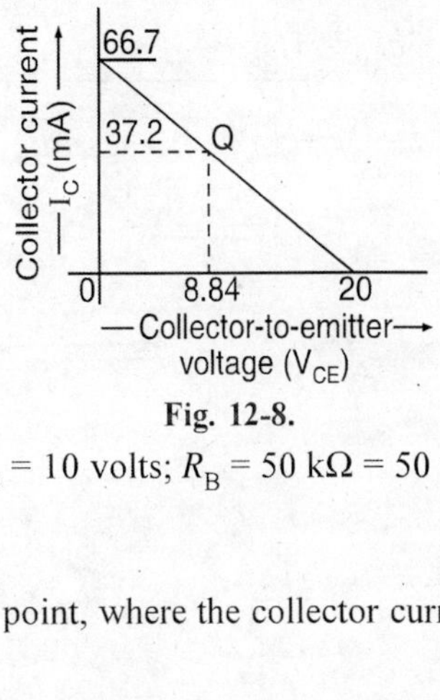

Fig. 12-8.

Solution. Given : V_{CC} = 20 volts; R_C = 300 Ω; V_{BB} = 10 volts; R_B = 50 kΩ = 50 × 10^3 Ω and β = 200.

Upper and lower ends of d.c. load line

We know that the upper end of the load line is at a point, where the collector current is maximum and its value,

$$I_{C\,(sat)} = \frac{V_{CC}}{R_C} = \frac{20}{300} = 0.0667 \text{ A} = 66.7 \text{ mA}$$

Similarly, the lower end of the load line is at a point, where the collector-to-emitter voltages maximum and its value,

$$V_{CE\,(cut\text{-}off)} = V_{CC} = 20 \text{ V } \textbf{Ans.}$$

Now if we join the upper and lower ends, we get a d.c. load line as shown in Fig. 12-8.

Q-point

Applying Kirchoff's Voltage Law to the base loop of the given circuit,

$$I_B \cdot R_B + 0.7 = 10$$

or $$I_B \times (50 \times 10^3) + 0.7 = 10$$

∴ $$I_B = \frac{10 - 0.7}{50 \times 10^3} = 0.186 \times 10^{-3} \text{ A}$$

and collector current, $$I_C = \beta \cdot I_B = 200 \times 0.186 \times 10^{-3} \text{ A} = 37.2 \times 10^{-3} \text{ A}$$

∴ Collector-to-emitter voltage,

$$V_{CE} = V_{CC} - I_C \cdot R_C = 20 - (27.2 \times 10^{-3}) \times 300 = 20 - 11.16 = 8.84 \text{ V}$$

Now let us locate the Q-point on the d.c. load line by drawing a vertical line through the 8.84 V as shown in Fig. 12-8. From the figure, we find that the Q-point corresponds to 37.2 mA **Ans.**

12-4. *Q*-Point and Maximum Undistorted Output

It has been observed that under certain input signal conditions, the location of Q-point, on the load line, may cause one peak of the output signal to be clipped. For example consider Q-point located near the saturation point as shown in Fig. 12-9(a). In this case, during the negative half cycle of the input, the transistor is driven into saturation. As a result of this, the negative peak of the input signal is clipped at the output.

Now consider a Q-point located near the cut-off point as shown in Fig. 12-9 (b). In this case, the transistor is driven into cut-off. Therefore, the positive peak of the input signal is clipped at the output. It may be noted that if the Q-point is located at the centre of the load line as shown in Fig. 12-9 (c), we get an undistorted signal at the output.

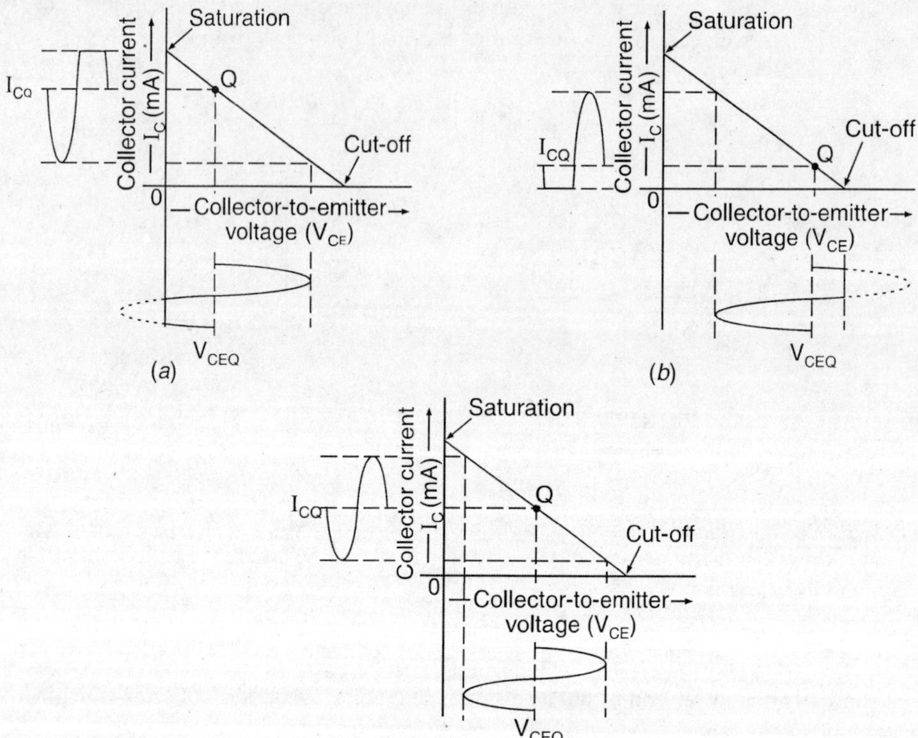

Fig. 12-9. *Q-point and maximum undistorted output.*

12-5. Factors Affecting Stability of *Q*-Point

It will be interesting to know that even after the suitable selection of Q-point, it tends to shift from its position. It happens because of the following two factors :

(1) ***Inherent variations of transistor parameters :*** We know that the collector current for a common-emitter transistor amplifier is given by the relation,

$$I_C = \beta \cdot I_B + (I + \beta) \cdot I_{CO} \qquad ...(i)$$

where, β = Common-emitter current gain,

I_B = Base-current and

I_{CO} = Reverse saturation current.

The above equation contains three variables namely, β_2, I_B and I_{CO}. All the three variables are found to be strongly dependent upon temperature. As the temperature, increases, all the three variables also increase. It results in the increase of collector current. This causes the operating point to *shift* towards the saturation point, resulting in clipping and bad distortion of the output. In some cases, a transistor may even *burn* out because of the excessive collector current.

(2) ***Variation in parameter values of transistors of the same type :*** It has been observed that inspite of the tremendous advancement, made in the semiconductor device manufacturing technology, some transistors of a particular type with a large variation in parameter values are manufactured. This variability is provided by the manufacturer in the data sheet, where minimum, maximum and typical values of parameters are mentioned. This gives rise to an instability of the Q-point.

We know that if the temperature increases, the reverse saturation current also increases. This causes the collector current to increase by a factor of $(1 + \beta)$ as seen from equation (*i*). However if by some means, the base current is made to decrease with the increase in temperature, then the collector current will decrease by a factor of β. This will compensate for the increase in $(1 + \beta)I_{CO}$ and hence, the collector current will remain nearly constant. This principle is used to provide stabilization of collector current by designing it as biasing circuit of different types and later providing compensation techniques.

12-6. Stability Factor

The stability factor may be defined as ***the rate of change of collector current with respect to the reverse saturation current keeping the common-emitter current gain (β) and base current (I_B) as constant.*** Mathematically, the stability factor,

$$S = \frac{dI_C}{dI_{CO}}$$

The stability factor is a measure of *bias stability* of a transistor circuit. It will be interesting to know that a *higher value of stability factor indicates poor stability, whereas a lower value indicates good stability*. In this sense, the stability factor should be called as an instability factor.

It may be noted that if the rate of change of collector current (dI_C) is equal to the rate of change of reverse saturation current (dI_{CO}), then the value of stability factor is unity. The unity is the lowest value of stability factor. It may be noted that if the value of stability factor is closer to the unity, the variation of collector current with the temperature will be less. Therefore, there will be less variation in the Q-point.

The stability factor may also be expressed alternatively by using the relationship between base current (I_B), collector current (I_C) and reverse saturation current (I_{CO}). We know that the value of collector current in a transistor is given by the relation,

$$I_C = \beta \cdot I_B + (1 + \beta)\, I_{CO}$$

Differentiating the above expression with respect to I_C,

$$1 = \frac{d\,(\beta \cdot I_B)}{dI_C} + \frac{d\,(1 + \beta)\, I_{CO}}{dI_C}$$

$$= \beta \times \frac{dI_B}{dI_C} + (1 + \beta)\,\frac{dI_{CO}}{dI_C} \qquad \text{... (Assuming } \beta \text{ as constant)}$$

$$= \beta \times \frac{dI_B}{dI_C} + (1 + \beta)\,\frac{1}{S} \qquad \text{...}\left(\because S = \frac{dI_C}{dI_{CO}}\right)$$

$$\therefore \qquad S = \frac{1+\beta}{1-\beta\left(\dfrac{dI_B}{dI_C}\right)} \qquad ...(i)$$

The above expression is a general expression, which is very useful for determining the stability factor (S) of any biasing circuit. It can be done first by finding the relationship between the base current and collector current and then using the above equation to determine the value of stability factor.

12-7. Stability Factor of Common Base Circuit

We have already discussed in the previous chapter on transistors that the collector current of a transistor, in common base configuration, is given by the relation,

$$I_C = \alpha \cdot I_E + I_{CO}$$

Differentiating this equation with respect to I_{CO},

$$\frac{dI_C}{dI_{CO}} = 0 + 1$$

$$\therefore \qquad S = 1$$

The value of stability factor indicates that the *common base circuits are highly stable*. Thus, there is no need of bias stabilization in these circuits.

12-8. Stability Factor of Common Emitter Circuit

We have already discussed in previous chapter on transistors that the collector current of a transistor in common emitter configuration is given by the relation,

$$I_C = \beta \cdot I_B + (1 + \beta)\, I_{CO}$$

Differentiating this equation with respect to I_{CO}.

$$\frac{dI_C}{dI_{CO}} = 0 + (1 + \beta)$$

$$\therefore \qquad S = 1 + \beta$$

It will be interesting to know that if $\beta = 100$, then the value of $S = 1 + 100 = 101$. It means that the collector current changes 101 times the change in reverse saturation current. Thus in a common emitter circuit, the collector current is highly dependent upon the reverse saturation current and hence upon the temperature. Thus, *there is a strong need to provide bias stabilization in common emitter circuits to improve the stabilization factor.*

12-9. Conditions for Proper Biasing of a Transistor

We have already discussed that for faithful amplification of an input signal, a transistor must be biased properly. Following are some of the important conditions for biasing the transistor properly :

- Proper d.c. value of the collector current.
- Proper value of V_{BE} (0.7 V for silicon and 0.3 V for germanium) at any instant.
- Proper value of V_{CE} (1 V for silicon and 0.5 V for germanium) at any instant.

The conditions 1 and 2 make sure that base-emitter junction of a transistor shall remain forward biased, during all parts of the signal. The Third condition makes sure that collector-base junction of input a transistor shall remain reverse biased at all instants.

12-10. Methods of Transistor Biasing

In the transistor amplifier circuits, drawn so far, the biasing is done with two power supplies V_{BB} and V_{CC}. The V_{BB} supply is used for biasing of the emitter-base junction and V_{CC} supply for biasing the collector base junction. However, in actual practice, one power supply is used for biasing both the junctions of a transistor. Following are the most commonly used methods for biasing the transistors :

- Base bias (also called *fixed bias*).
- Base bias with emitter feedback (also called *emitter feedback bias*).
- Base bias with collector feedback (also called *collector feedback bias*).
- Voltage divider bias (also called *self bias*).
- Emitter bias.

Now we shall discuss the above mentioned methods of biasing in the following pages.

12-11. Base Bias

Fig. 12-10(*a*) shows a base bias circuit for a NPN transistor. The base bias circuit is also known as *fixed bias circuit*. In this case, the circuit uses two d.c. supplies namely V_{BB} and V_{CC}. But a more practical method is to use a single d.c. supply (V_{CC}) as shown in Fig. 12-10 (*b*). In this case, both the base and collector resistance are connected to the positive side of the V_{CC} supply. To simplify the circuit diagram, we may omit the battery symbol and replace it by a line termination with a voltage indicated as shown in Fig. 12-10(*c*).

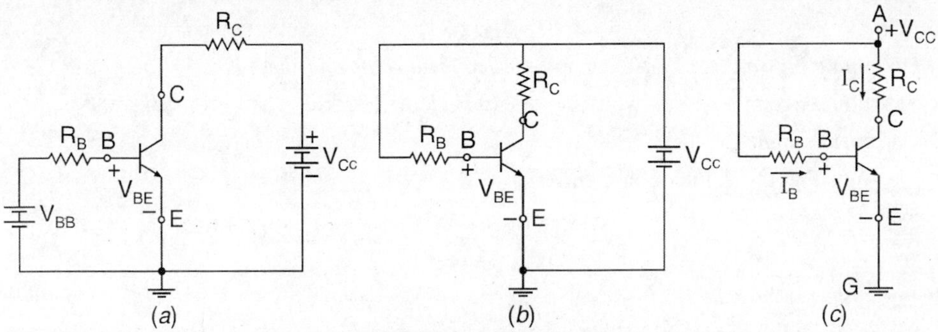

Fig. 12-10. *Base bias circuit.*

Now, we shall determine the d.c. bias currents and voltages in the base and collector of the transistor. First of all, consider the base-emitter circuit loop (AFBEGA) in the base bias circuit. Applying Kirchoff's Voltage Law for the given loop,

$$I_B \cdot R_B + V_{BE} - V_{CC} = 0$$

or the base current,
$$I_B = \frac{V_{CC} - V_{BE}}{R_B} \qquad ...(i)$$

Since, the supply voltage V_{CC} and the base-emitter voltage (V_{BE}) have fixed values of voltage, the selection of base bias resistor (R_B) fixes the value of the base current. The equation (*i*) may be simplified, if we neglect the value of base-emitter voltage, because it is usually very small as compared to the value of V_{CC}. Thus, the approximate value of base current,

$$I_B = \frac{V_{CC}}{R_B} \qquad ...(\because V_{CC} >> V_{BE})$$

Now, consider the collector-emitter circuit loop (AFCEGA) in the base bias circuit. Applying Kirchoff's Voltage Law for this loop,

$$I_C \cdot R_C + V_{CE} = V_{CC}$$

$$\therefore \quad V_{CE} = V_{CC} - I_C \cdot R_C \qquad ...(ii)$$

The above equation gives the voltage drop across the collector-emitter terminals of the transistor. The value of collector current is given by the relation,

$$I_C = \beta \cdot \frac{V_{CC}}{R_B} = \frac{V_{CC}}{R_B / \beta} \qquad ...(iii)$$

The above relation shows that the collector current is β times greater than the base current and is not at all dependent on the resistance of the collector circuit.

It may be noted from the equations (*ii*) and (*iii*) that the values of collector current (I_C) and collector-to-emitter voltage (V_{CE}) are dependent on β. But β is strongly dependent upon temperature. It means that collector current and collector-to-emitter voltage of a base bias circuit (which sets the *Q*-points of a transistor) will vary with the change in value of β due to variation in temperature. It means that it is impossible to obtain a stable *Q*-point in a base-bias circuit. Because of this fact, the base bias is never used in amplifier circuits. However, it is used in digital circuits, where the transistor is used as a switch between saturation and cut-off-regions.

Example 12-3. *Fig. 12-11 shows a fixed bias circuit using an NPN transistor circuit.*

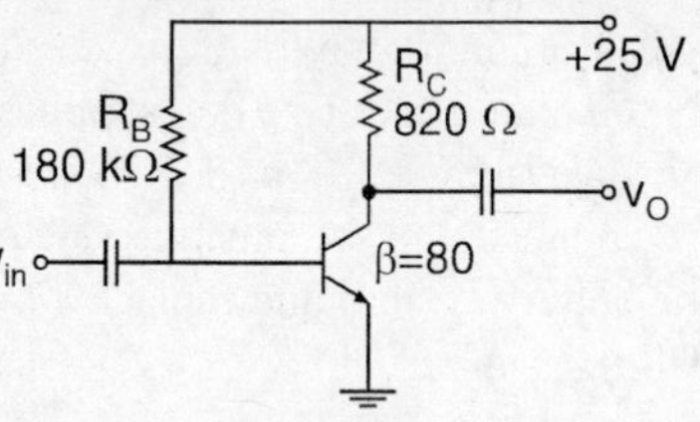

Fig. 12-11.

Determine the values of base current, collector current and the collector-to-emitter voltage.

Solution. Given : $V_{CC} = 25$ volts; $R_C = 820\ \Omega$; $R_B = 180\ k\Omega = 180 \times 10^3\ \Omega$; $\beta = 80$.

Value of base current

We know that the value of base current,

$$I_B = \frac{V_{CC}}{R_B} = \frac{25}{180 \times 10^3}\ A = 0.14 \times 10^{-3}\ A = 0.14\ mA\ \textbf{Ans.}$$

Value of collector current

We know that the value of collector current,

$$I_C = \beta \cdot I_B = 80 \times 0.14 \times 10^{-3} = 11.2 \times 10^{-3}\ A$$
$$= 11.2\ mA\ \textbf{Ans.}$$

Value of collector-to-emitter voltage

We know that the value of collector-to-emitter voltage,

$$V_{CE} = V_{CC} - I_C \cdot R_C = 25 - (11.2 \times 10^{-3}) \times 820\ V$$
$$= 25 - 9.2 = 15.8\ V\ \textbf{Ans.}$$

12-12. Stability Factor of Base-Bias Circuit

We know that the basic circuit used in the base bias (or fixed bias) current is that of common-emitter configuration. We have already discussed in Art. 12-8 that stability factor, is equal to (1+ β). Therefore, for base-bias circuit, the stability factor.

$$S = 1 + \beta$$

Example 12-4. *Determine the value of bias resistor R_B and the stability factor for the fixed bias circuit shown in Fig. 12-12. Also determine the voltage between the collector and ground.*

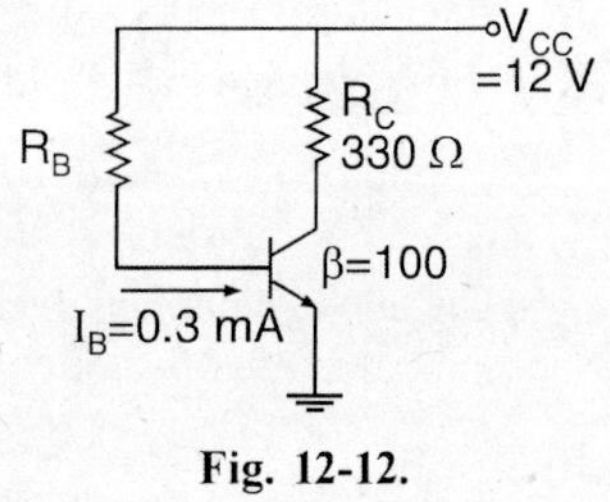

Fig. 12-12.

Solution. Given : $V_{CC} = 12$ volts; $R_C = 330\ \Omega$; $I_B = 0.3$ mA = 0.3×10^{-3} A and $\beta = 100$.

Value of bias resistor

Let R_B = Value of base resistor.

We know that the bias current (I_B),

$$0.3 \times 10^{-3} = \frac{V_{CC}}{R_B} = \frac{12}{R_B}$$

$$R_B = 12/(0.3 \times 10^{-3}) = 40 \text{ k}\Omega \textbf{ Ans.}$$

Stability factor

We know that the stability factor for the given circuit,

$$S = 1 + \beta = 1 + 100 = 101 \textbf{ Ans.}$$

Voltage between collector and ground

We know that the collector current,

$$I_C = \beta \cdot I_B = 100 \times 0.3 \times 10^{-3} = 30 \times 10^{-3} \text{ A}$$

and the voltage between collector and ground (or emitter),

$$V_{CE} = V_{CC} - I_C \cdot R_C = 12 - (30 \times 10^{-3}) \times 330 = 2.1 \text{ V} \textbf{ Ans.}$$

12-13. Base Bias with Emitter Feedback

Fig 12-13 shows the circuit of a base bias with emitter resistor. The emitter resistor provides better bias stability than the base bias circuit, as discussed in Art. 12-12.

When the current gain (β) increases (say due to temperature), the collector currents in the circuit also increases. It produces more voltage across the emitter resistor (R_E), which reduces the voltage drop across the base resistor. This causes the base current to decrease. As a result of this, the collector current also decreases. Thus, the increase in collector current, due to the increase in value of current gain, is compensated.

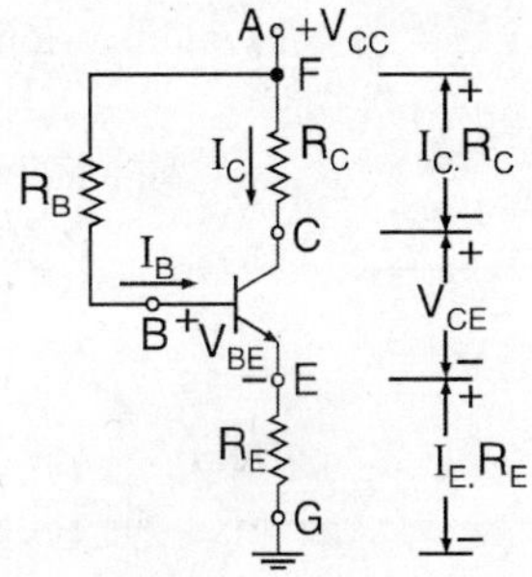

Fig. 12-13. *Base bias with emitter feedback.*

It is evident from the above discussion that the emitter resistor provides compensation for the variation in the value of current gain (β) by reducing the base current. Thus, the emitter resistor is called emitter feedback resistor. The emitter feedback circuit is effective only if the emitter resistor is as large as possible. But in actual practice, it is not possible because the emitter resistor is kept relatively small to avoid the transistor to operate in saturation.

Let us now determine the d.c. bias currents and voltages in the base, collector and emitter of the transistor. First of all, consider the base-emitter circuit loop (AFBEGA) in the d.c. bias circuit. Applying Kirchoff's Voltage Law to the base-emitter circuit loop,

$$V_{BE} + I_E \cdot R_E - V_{CC} + I_B \cdot R_B = 0$$

Substituting I_E equal to $(\beta + 1)\, I_B$ in the above equation, and solving it for the base current,

$$I_B = \frac{V_{CC} - V_{BE}}{R_B + (\beta + 1)R_E} = \frac{V_{CC} - V_{BE}}{R_B + \beta \cdot R_E} \quad \text{...(Taking } \beta + 1 \approx \beta)$$

∴ Collector current,

$$I_C = \beta \cdot I_B = \frac{V_{CC} - V_{BE}}{R_E + R_B/\beta} = \frac{V_{CC}}{R_E + R_B/\beta} \quad \text{...}(\because V_{CC} >> V_{BE})$$

It is evident from the above equation that if the value of emitter resistor (R_E) is made very large as compared to the value of (R_B/β), the bias circuit is independent of variations in current gain (β). But in actual practice, it is not possible to make the emitter resistor large enough to compensate for the effects of current gain (β) without saturating the transistor. Thus, some compromise will have to be made between the bias stability and the performance of the circuit. Now consider the collector-emitter circuit loop (AFCEGA) in the d.c. bias circuit. Applying Kirchoff's Voltage Law to this loop.

$$V_{CE} + I_E \cdot R_E - V_{CC} + I_C \cdot R_C = 0$$

Replacing I_E with I_C in the above equation and solving it for the voltage across collector-emitter of a transistor,

$$V_{CE} = V_{CC} - I_C (R_C + R_E) \quad ...(i)$$

The voltage measured from emitter to ground,

$$V_E = I_E \cdot R_E = R_C \cdot R_C \quad ...(\because I_E \approx I_C)$$

The voltage measured from collector to ground,

$$V_C = V_{CC} - I_C \cdot R_C$$

Notes: 1. The voltage, at which the transistor is biased, may also be calculated by the relation,

$$V_{CE} = V_C - V_E$$

2. The saturation value of collector current may be obtained by using the equation (*i*) and its value is given by the relation,

$$I_{C\,(sat)} = \frac{V_{CC} - V_{CE}}{R_C + R_E} = \frac{V_{CC}}{R_C + R_E} \quad ...(\text{if } V_{CC} >> V_{CE})$$

Example 12-5. *Calculate all the d.c. bias voltages and currents in circuit shown in Fig. 12-13, if $R_B = 400$ kΩ, $R_C = 2$ kΩ; $R_E = 1$ kΩ, $V_{CC} = 20$ V and β = 100. Neglect V_{BE} of the transistor.*

Solution. Given : $R_B = 400$ KΩ = 400 × 10^3 Ω; $R_C = 2$ kΩ = 2 × 10^3 Ω; $R_E = 1$ kΩ = 1 × 10^3 Ω; $V_{CC} = 20$ volts and β = 100.

Base current

We know that the base current,

$$I_B = \frac{V_{CC}}{R_B + \beta \cdot R_E} = \frac{20}{(400 \times 10^3) + [100 \times (1 \times 10^3)]} \text{ A}$$

$$= 0.04 \times 10^{-3} \text{ A} = 0.04 \text{ mA } \textbf{Ans.}$$

Collector current

We know that the collector current,

$$I_C = \beta \cdot I_B = 100 \times 0.04 = 4 \text{ mA } \textbf{Ans.}$$

Collector-to-emitter voltage

We also know that the collector-to-emitter voltage,

$$V_{CE} = V_{CC} - I_C \cdot (R_C + R_E)$$

$$= 20 - (4 \times 10^{-3}) [(2 \times 10^3) + (1 \times 10^3)] \text{ V}$$

$$= 20 - 12 = 8 \text{ V } \textbf{Ans.}$$

Example 12-6. *In the circuit shown in Fig. 12-14, a silicon transistor with β = 50 is used. Find I_C and V_{CE}. Draw the D.C. load line on output characteristics and indicate the Q-point. Take $V_{BE} = 0.7$ V.*

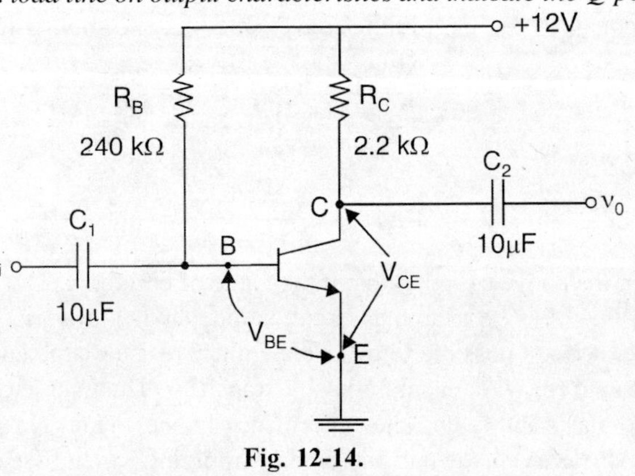

Fig. 12-14.

Solution. Given : $V_{CC} = 12$ V, $R_C = 2.2$ kΩ; $R_B = 240$ kΩ; β = 50 and $V_{BE} = 0.7$ V.

We know that the value of collector current,

$$I_C = \frac{V_{CC} - V_{BE}}{R_E + \frac{R_\beta}{\beta}}$$

$$= \frac{12 - 0.7}{0 + \left(\frac{240 \times 10^3}{50}\right)} = 2.35 \text{ mA}$$

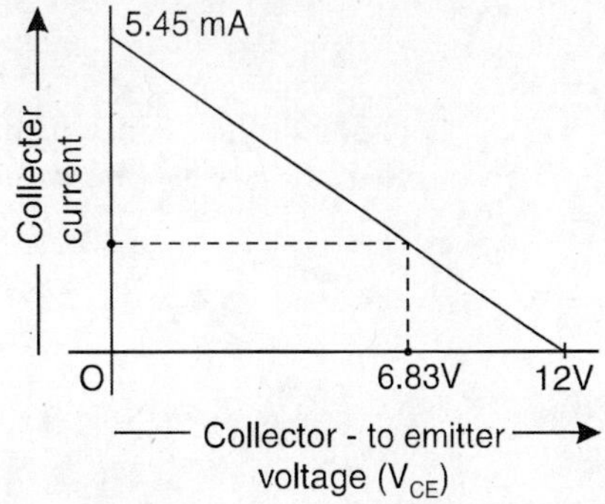

Fig. 12-15.

and collector-to-emitter voltage,

$$V_{CE} = V_{CC} - I_C \cdot R_C = 12 - (2.35 \times 10^{-3}) \times (2.2 \times 10^3) \text{ V}$$

$$= 12 - 5.17 \text{ V}$$

$$= 6.83 \text{ V}$$

We also know that,

$$I_{C(sat)} = \frac{V_{CC}}{R_C} = \frac{12V}{2.2 \times 10^3} = 5.45 \text{ mA}$$

and the cut-off voltage,

$$V_{CE(cut\text{-}off)} = V_{CC} = 12 \text{ V}$$

Using the values of $I_{C(sat)}$, I_C, V_{CE} and V_{CC} the d.c. load line can be obtained as shown in Fig. 12-15.

Example 12-7. *For the common-emitter circuit shown in Fig. 12-16, draw the d.c. load line and mark the d.c. operating point on it. Assume, β = 100 and neglect V_{BE}.*

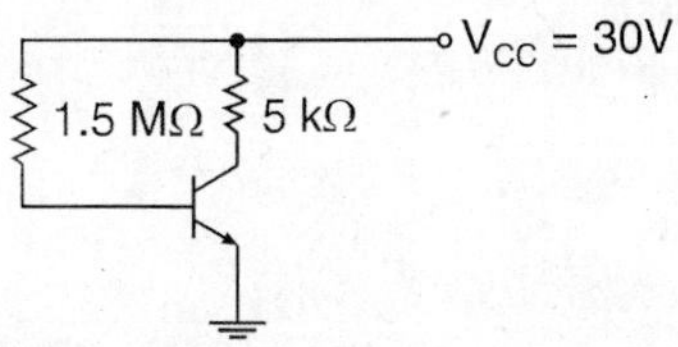

Fig. 12-16.

(*Banglore Univ. July* 2003)

Solution. Given : $V_{CC} = 30$ V, $R_C = 5$ kΩ $= 5 \times 10^3$ Ω; $R_B = 1.5$ MΩ $= 1.5 \times 10^6$ Ω; β = 100.

We know that collector current at saturation,

$$I_{C(sat)} = \frac{V_{CC}}{R_C} = \frac{30}{5 \times 10^3} = 6 \text{ mA}$$

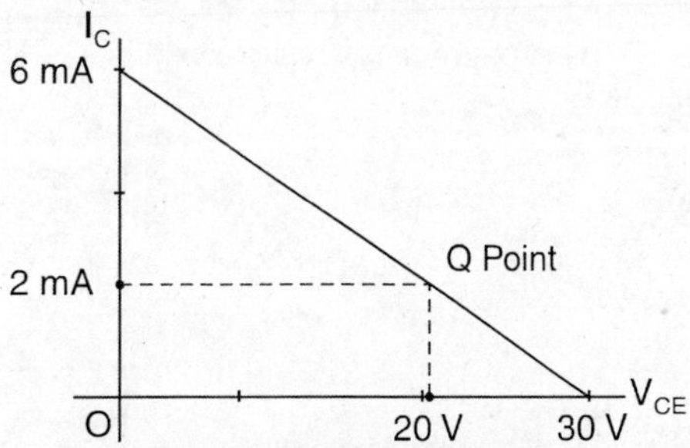

Fig. 12-17.

and the value of cut-off voltage,

$$V_{CE(cut\text{-}off)} = V_{CC} = 30 \text{ V}$$

We also know that base current,

$$I_B = \frac{V_{CC}}{R_B} = \frac{30}{1.5 \times 10^6} = 20 \times 10^{-6} \text{ A} = 20 \text{ μA}$$

and the collector current,

$$I_C = \beta I_B = 100 \times 20 \times 10^{-6}$$

$$= 20 \times 10^{-4} \text{ A} = 2 \text{ mA}$$

The collector-to-emitter voltage,

$$V_{CE} = V_{CC} - I_C R_C = 30 - (2 \times 10^{-3}) \times (5 \times 10^3)$$
$$= 30 - 10 \text{ V}$$
$$= 20 \text{ V}$$

∴ The Q-point is located on the d.c. load line as shown in Fig. 12-17.

Example 12-8. *For the circuit shown in the Fig. 12-18 determine the operation point with* $\beta = 100$. (*Electronic Circuits - I, Anna University, April/May 2008*)

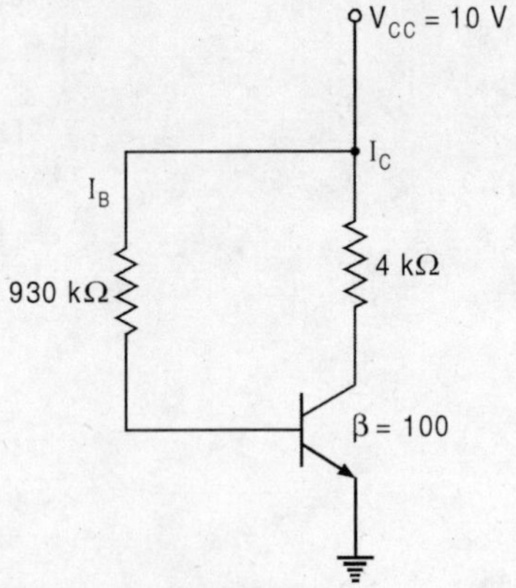

Fig. 12-18.

Solution. Given: $V_{CC} = 10\ V$, $R_C = 4\ k\Omega = 4 \times 10^3\ \Omega$; $R_B = 930 \times 10^3$; $\beta = 100$.

We know that collector current at saturation,

$$I_{C\,(sat)} = \frac{V_{CC}}{R_C} = \frac{10}{4 \times 10^3} = 2.5 \text{ mA}$$

and the value of cut-off voltage,

$$V_{CE} \text{ (cut-off)} = V_{CC} = 30 \text{ V}$$

We also know that the base current,

$$I_B = \frac{V_{CC}}{R_B} = \frac{10}{930 \times 10^3} = 10.7\ \mu\text{A}\ .$$

$$I_C = \beta I_B = 100 \times (10.7 \times 10^{-6})$$
$$= 10.7 \times 10^3 A = 1.07 \text{ mA}.$$

The collector-to-emitter voltage,

$$V_{CC} = V_{CC} - I_C\, R_C = 10 - (1.07 \times 10^{-3}) \times (4 \times 10^3)$$
$$= 5.72 \text{ V}.$$

The Q-point is located on the d.c. load line as shown in Fig. 12-19.

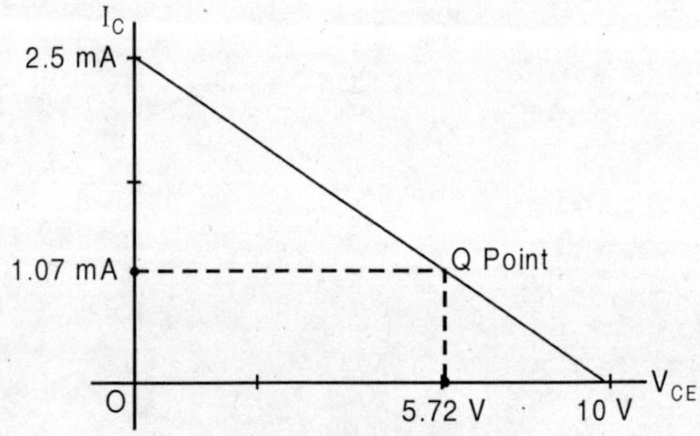

Fig. 12-19.

Example 12-8. *For the circuit shown in Fig. 12-20. What will be the state of LED when*

(a) $V_{BB} = 0.2$ V

(b) $V_{BB} = 3$ V

Given reasons. (*Banglore Univ. B.Sc. Electronics, May* 2001)

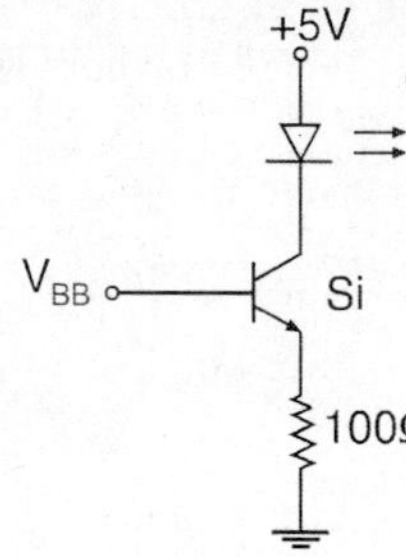

Fig. 12-20.

Solution. Given : $V_{CC} = 5$ V, $R_E = 100\ \Omega$

We know that the state of LED will be determined by the fact whether the transistor is ON or OFF.

(*a*) $V_{BB} = 0.2$ V

Since, the transistor in the given circuit is Si, it needs 0.7 V between the base-emitter junction to turn ON. Therefore, the LED is OFF.

(*b*) $V_{BB} = 3$ V

Since, $V_{BB} = 3$ V is more that 0.7 V, the transistor will turn ON and therefore the LED will be ON.

12-14. Stability Factor of Base Bias with Emitter Feedback

We have already discussed in the last article that if we apply Kirchoff's Voltage Law to the base-emitter circuit loop shown in Fig. 12-13, then we get

$$V_{BE} + I_E \cdot R_E - V_{CC} + I_B \cdot R_B = 0$$

Replacing I_E with $I_B + I_C$ in the equation and solving it for base current,

$$I_B = \frac{V_{CC} - V_{BE} - I_C \cdot R_E}{R_E + R_B}$$

Differentiating the above expression with respect to I_C,

$$\frac{dI_B}{dI_C} = \frac{0 - 0 - 1 \cdot R_E}{R_E + R_B} = -\frac{R_E}{R_E + R_B}$$

Substituting the value of dI_B/dI_C in the general expression for stability factor,

$$S = \frac{1+\beta}{1-\beta\left(\dfrac{dI_B}{dI_C}\right)} = \frac{1+\beta}{1-\beta\left(-\dfrac{R_E}{R_E+R_B}\right)} = \frac{1+\beta}{1+\beta \cdot \dfrac{R_E}{R_E+R_B}}$$

Example 12-9. *Fig. 12-21 shows the base-bias with emitter feedback resistor circuit. The transistor has a current gain (β) of 80. Determine the value of collector current, collector-to-emitter voltage and stability factor. Assume $V_{BE} = 0.7$ V.*

Solution. Given : $\beta = 80$; $V_{BE} = 0.7$ volt; $V_{CC} = 25$ volts; $R_C = 820\ \Omega$; $R_B = 180\ \text{k}\Omega = 180 \times 10^3\ \Omega$ and $R_E = 200\ \Omega$.

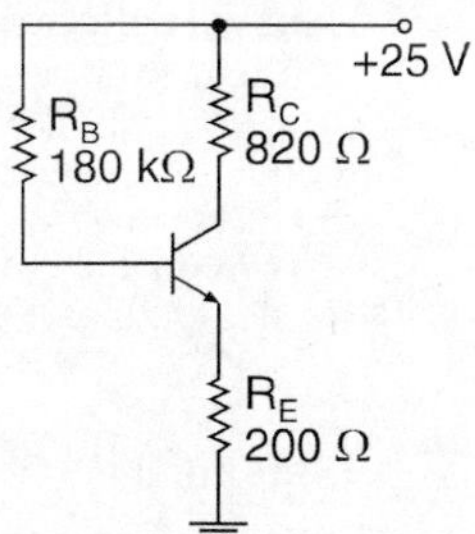

Fig. 12-21

Value of collector current

We know that the value of collector current,

$$I_C = \frac{V_{CC} - V_{BE}}{R_E + \dfrac{R_B}{\beta}} = \frac{25 - 0.7}{200 + \dfrac{180 \times 10^3}{80}}\ \text{A}$$

$$= 9.9 \times 10^{-3}\ \text{A or } 9.9\ \text{mA}\ \textbf{Ans.}$$

Value of collector-to-emitter voltage

We also know that the value of collector-to-emitter (or ground) voltage,

$$V_{CE} = V_{CC} - I_C \cdot R_C = 25 - (9.9 \times 10^{-3}) \times 820 \text{ V}$$
$$= 25 - 8.12 = 16.88 \text{ V } \textbf{Ans.}$$

Value of Stability factor

We know that the value of stability factor,

$$S = \frac{1+\beta}{1+\beta \cdot \dfrac{R_E}{R_E + R_B}} = \frac{1+80}{1+80 \times \dfrac{200}{200+(180\times 10^3)}}$$

$$= \frac{81}{1.089} = 74.394 \textbf{ Ans.}$$

12-15. Base Bias with Collector Feedback

Fig. 12-22 shows a d.c. bias circuit in which the base resistor (R_B) is connected to the collector terminal of a transistor rather than to the V_{CC} supply as in a base bias circuit. Here, the collector voltage provides the bias-emitter junction. The resistor R_B acts as a feedback resistor. It provides a very stable Q-point by reducing the effect of variations in current gain (β).

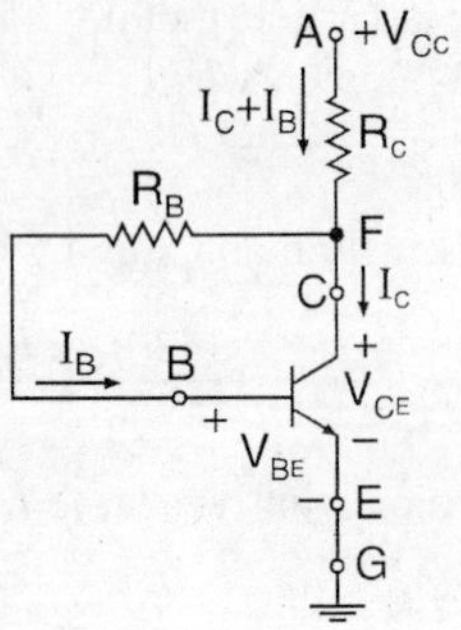

Fig. 12-22. *Base bias with collector feedback.*

Let us now, determine the d.c. bias currents and voltages in the base and collector of the transistor. First of all, consider the base-emitter circuit loop (AFBEGA) in the d.c. bias circuit. Applying Kirchoff's Voltage Law to the base-emitter circuit loop,

$$V_{BE} - V_{CC} + (I_B + I_C)\, R_C + I_B \cdot R_B = 0$$

Substituting I_C equal to β in the above equation and solving for the base current,

$$I_B = \frac{V_{CC} - V_{BE}}{R_B + (\beta+1)R_C} = \frac{V_{CC} - V_{BE}}{R_B + \beta \cdot R_C} \quad \text{...(Taking } \beta + 1 = \beta\text{)}$$

and the collector current,

$$I_C = \beta \cdot I_B = \frac{V_{CC} - V_{BE}}{R_C + R_B/\beta} = \frac{V_{CC}}{R_C + R_B/\beta} \quad \text{...}(\because V_{CC} >> V_{BE})$$

Now consider the collector-emitter circuit loop (*i.e.*, AFCEGA) in d.c. bias circuit. Applying Kirchoff's Voltage Law to this loop,

$$V_{CE} - V_{CC} + (I_C + I_B)\, R_C = 0$$

Substituting $I_C + I_B$ equal to I_C in the above equation and solving it for collector-to-emitter voltage,

$$V_{CE} = V_{CC} - I_C \cdot R_C$$

Example 12-10. *Calculate the Q-point values, i.e., the values of collector current and collector-to-emitter voltage for the d.c. bias circuit shown in Fig. 12-23.*

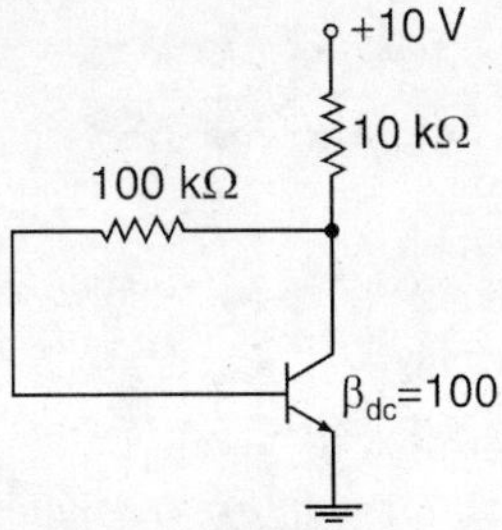

Fig. 12-23.

Also draw the load line and locate Q-point on it. Assume V_{BE} = 0.7 V.

Solution. Given : V_{CC} = 10 volts; $R_C = 10\text{ k}\Omega = 10 \times 10^{-3}\ \Omega$; $R_B = 100\text{ k}\Omega = 100 \times 10^3\ \Omega$; β_{dc} = 100 and V_{BE} = 0.7 volt.

Q-point values

We know that the value of collector current,

$$I_C = \frac{V_{CC} - V_{BE}}{R_C + \dfrac{R_B}{\beta}} = \frac{10 - 0.7}{10 \times 10^3 + \dfrac{100 \times 10^3}{100}}$$

$$= 0.845 \times 10^{-3}\text{ A} = 0.845\text{ mA } \textbf{Ans.}$$

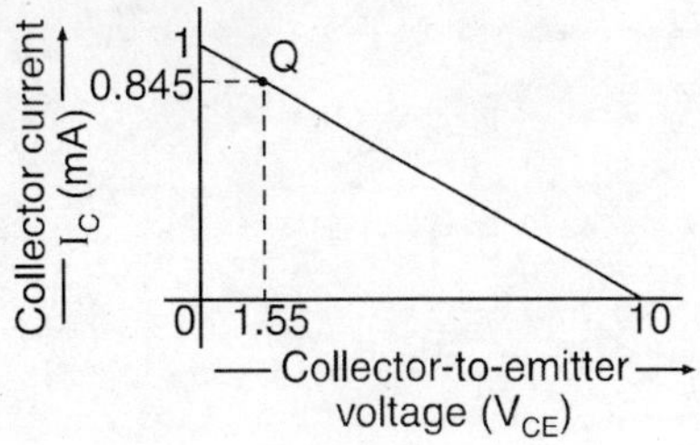

Fig. 12-24.

and the collector-to-emitter voltage,

$$V_{CE} = V_{CC} - I_C \cdot R_C = 10 - [(0.845 \times 10^{-3}) \times (10 \times 10^3)]$$

$$= 10 - 8.45 = 1.55\text{ V } \textbf{Ans.}$$

D.C. load line

We know that upper end of the load line

$$I_{C(sat)} = \frac{V_{CC}}{R_C} = \frac{10}{10 \times 10^3} = 1 \times 10^{-3}\text{ A} = 1\text{ mA}$$

and lower end of the load line, $$V_{CE\,(cut\text{-}off)} = V_{CC} = 10\text{ V}$$

Thus the load line may be obtained by joining the upper and lower ends as shown in Fig. 12-24. The *Q*-point corresponding to the collector current of 0.845 mA and collector-to-emitter voltage of 1.55 V is indicated on the load line.

Example 12-11. *For the circuit shown in Fig. 12-25, Calculate I_B, I_C and I_E if Si transistor is used with β = 50 and stability factor 5.* (*Nagpur Univ. Winter* 2003)

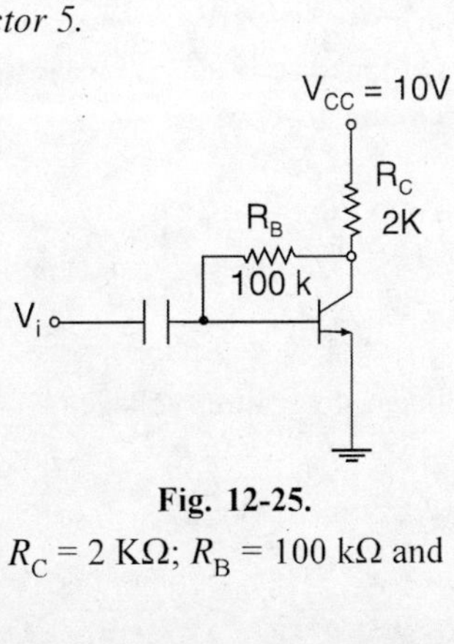

Fig. 12-25.

Solution. Given V_{CC} = 10 V; R_C = 2 KΩ; R_B = 100 kΩ and β = 50.

We know that base current,

$$I_B = \frac{V_{CC} - V_{BE}}{R_B + \beta \cdot R_C} = \frac{10 - 0.7}{100\text{k} + (50 \times 2\text{k})}$$

...(For *Si* transistor $V_{BE} = 0.7$ V)

$$= 0.0465 \text{ mA } \textbf{Ans.}$$

and the collector current,

$$I_C = \beta \cdot I_B = 50 \times 0.0465 \ \mu\text{A} = 2.325 \text{ mA } \textbf{Ans.}$$

We also know that emitter current,

$$I_E \approx I_C = 2.325 \text{ mA } \textbf{Ans.}$$

Example 12-12. *For the circuit shown in Fig. 12-26, what are the possible causes if **(a)** $V_C = 0$; **(b)** $V_C = 9$ V.* *(Banglore Univ. B.Sc. Electronics, May 2001)*

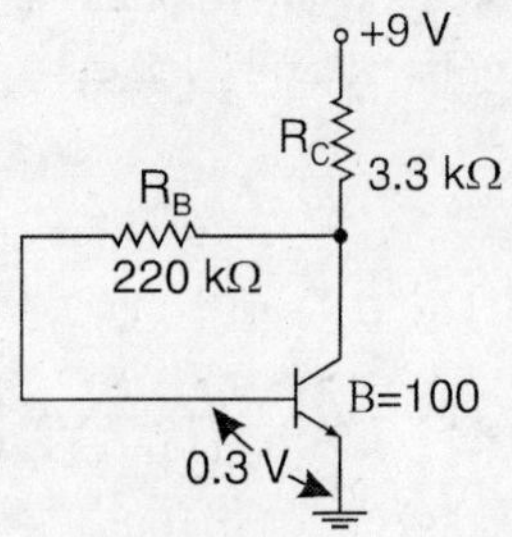

Fig. 12-26.

Solution. Given : $V_{CC} = q_v$; $R_B = 220 \text{ k}\Omega = 220 \times 10^3 \ \Omega$; $R_C = 3.3 \text{ k}\Omega = 3.3 \times 10^3 \ \Omega$; $V_{BE} = 0.3$ V and $\beta = 100$.

(a) *If* $V_C = 0$

We know that

$$I_B = \frac{V_{CC} - V_{BE}}{R_B + \beta \cdot R_C} = \frac{9 - 0.3}{(220 \times 10^3) + (100 \times 3.3 \times 10^3)}$$

$$= 0.016 \times 10^{-3} \ \Omega$$

$$= 16 \ \mu\text{A}$$

and the collector current,

$$I_C = \beta \cdot I_B = 100 \times 16 \ \mu\text{A} = 1600 \ \mu\text{A} = 1.6 \ \mu\text{A}.$$

Also $\quad V_{CE} = V_{CC} - I_C \cdot R_C$

Since, $\quad V_C = V_{CE} = 0$; from equation (*i*) we know that

$$I_C R_C = V_{CC} = 9 \text{ V}.$$

This means the entire V_{CC} is dropped across 3.3 kΩ resistor. In other words the collector-emitter junction of a transistor is short-circuited.

(b) *If* $V_C = 9V$

We know that the base current,

$$I_B = 16 \ \mu\text{A}$$

and the collector current,

$$I_C = \beta \cdot I_B = 1.6 \text{ mA}$$

But also notice that, the collector-to-emitter voltage (V_{CE}),

$$9 = V_{CC} - I_C \cdot R_C = 9 - I_C \cdot R_C$$

$$\therefore \quad I_C \cdot R_C = 0$$

Since, $I_C \neq 0$, therefore $R_C = 0$.

This means the collector resistance (R_C) is short-circuited.

Example 12-13. *If* $\alpha = 0.98$, *and* $V_{BE} = 0.7$ *V, find* R_1 *in the circuit shown in Fig. 12-27 for an emitter current* $I_E = 2nA$. *Neglect reverse saturation current.* (*Nagpur Univ. Winter* 2003)

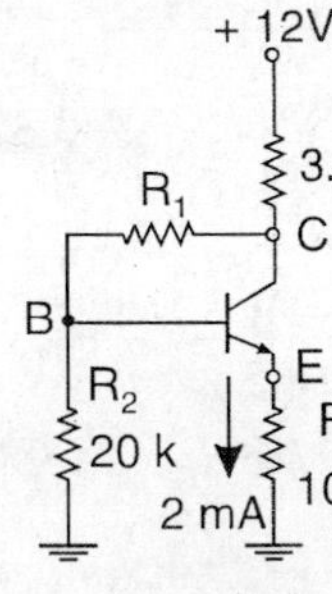

Fig. 12-27.

Solution. Given : $V_{CC} = 12$ V; $R_C = 3.3$ kΩ; $R_E = 100$ Ω; $I_E = 2$ mA; $V_{BE} = 0.7$ V, $\alpha = 0.98$.

We know that collector current,

$$I_C = \alpha \cdot I_E = 0.98 \times 2 \text{ mA} = 1.96 \text{ mA}$$

We know that the base voltage, (measured with respect to ground)

$$V_B = V_{BE} + I_E \cdot R_E$$
$$= 0.7 \text{ V} + 2 \text{ mA} \times 100 \ \Omega$$
$$= 0.9 \text{ V}$$

and the collector voltage (measured with respect to ground),

$$V_C = V_{CC} - I_C \cdot R_C = 12 - (1.96 \text{ mA}) \times (3.3 \text{ k}\Omega)$$
$$= 12 - 6.488$$
$$= 5.532 \text{ V}$$

We also know that current through R_2,

$$I_{R_2} = \frac{V_B}{R_2} = \frac{5.532}{20 \text{ k}} = 0.277 \text{ mA}$$

Also the base current,

$$I_B = I_E - I_C$$
$$= 2 \text{ mA} - 1.96 \text{ mA}$$
$$= 0.04 \text{ mA}$$

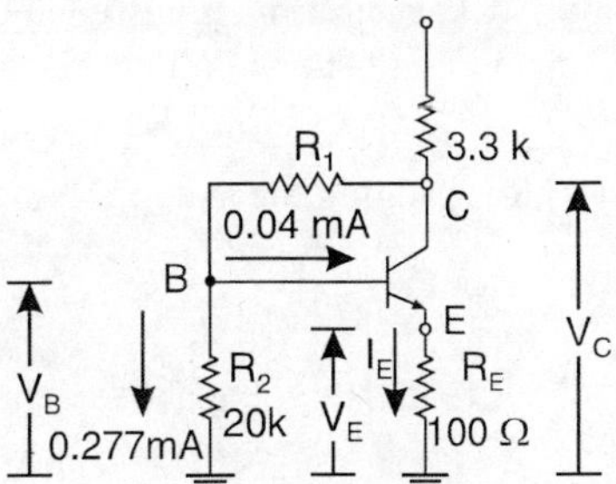

Fig. 12-28.

∴ The current through R, (refer to Fig. 12-28),

$$I_{R_1} = I_{R_2} + I_B = 0.277 \text{ mA} + 0.04 \text{ mA} = 0.317 \text{ mA}$$

and the value of the resistance,

$$R_1 = \frac{V_C - V_B}{I_{R_1}} = \frac{5.532 - 0.9}{0.317 \text{ mA}} = 14.6 \text{ k}\Omega \ \textbf{Ans.}$$

Example 12-14. *In a circuit shown in Fig. 12-29,* $V_{CC} = 24$ *V,* $R_C = 10$ *k and* $R_E = 270$ Ω. *If a Si transistor is used with* $\beta = 45$ *and if* $V_{CE} = 5$ V. Find R_B, neglect reverse saturation current.

(*Nagpur Univ. Winter* 2003)

Solution. Given $V_{CC} = 24$ V; $R_C = 10$ k Ω; $R_E = 270$ Ω; For a transistor, $V_{BE} = 0.7$ V; $\beta = 45$ and $V_{CE} = 5$ V.

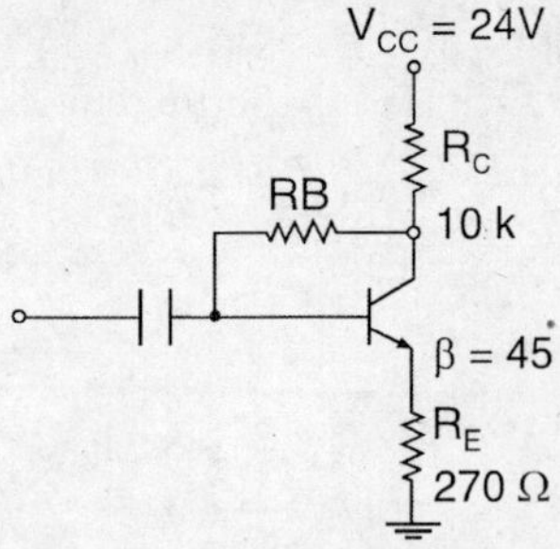

Fig. 12-29.

We know that collector-to emitter voltagte, (V_{CE})

$$5 = V_{CC} \cdot I_C \cdot R_C = 24 - I_C$$

$$\therefore \quad I_C = \frac{24-5}{10 \times 10^3} = 1.9 \text{ mA}$$

We also knwo that collector current (I_C)

$$1.9 \times 10^{-3} = \frac{V_{CC} - V_{BE}}{R_C + \dfrac{R_B}{\beta}} = \frac{24 - 0.7}{(10 \times 10^3) + \dfrac{R_B}{45}}$$

$$\therefore \quad \frac{R_B}{45} = \frac{23.93}{1.9 \times 10^{-3}} - (10 \times 10^3) = (12.6 \times 10^3) - (10 \times 10^3)$$

$$= 2.6 \times 10^3$$

or $\quad R_B = (2.6 \times 10^3) \times 45 = 117 \text{ k}\Omega$ **Ans.**

12-16. Stability Factor of Base Bias with Collector Feedback

We have already discussed in the last article that if we apply Kirchoff's Voltage Law to the base-emitter circuit shown in Fig. 12-15, then we get

$$V_{BE} - V_{CC} + (I_B + I_C) R_C + I_B \cdot R_B = 0$$

Rearranging the above equation and solving for base current,

$$I_B = \frac{V_{CC} - V_{BE} - I_C \cdot R_C}{R_C + R_B}$$

Differentiating the above expression with respect to I_C,

$$\frac{dI_B}{dI_C} = \frac{0 - 0 - 1 \cdot R_C}{R_C + R_B} = -\frac{R_C}{R_C + R_B}$$

Substituting the value of dI_B/dI_C in the general expression for stability factor,

$$S = \frac{1+\beta}{1 - \beta\left(\dfrac{dI_B}{dI_C}\right)} = \frac{1+\beta}{1 - \beta\left(-\dfrac{R_C}{R_C + R_B}\right)} = \frac{1+\beta}{1 + \beta \times \dfrac{R_C}{R_C + R_B}}$$

It is evident from the above expression that the stability factor (S) of a collector feedback bias is smaller than $(1 + \beta)$. Therefore, the biasing arrangement of collector feedback resistor is certainly an improvement over the fixed bias circuit.

Example 12-15. *Determine the d.c. bias current and voltage for the d.c. bias circuit shown in Fig. 12-30.*

Also determine the stability factor of the bias circuit. Assume V_{BE} = 0.7 V.

Solution. Given : $V_{CC} = 3$ volts; $R_C = 1.8$ k$\Omega = 1.8 \times 10^3\ \Omega$; $R_B = 33$ k$\Omega = 33 \times 10^3\ \Omega$; $\beta = 90$ and $V_{BE} = 0.7$ volt

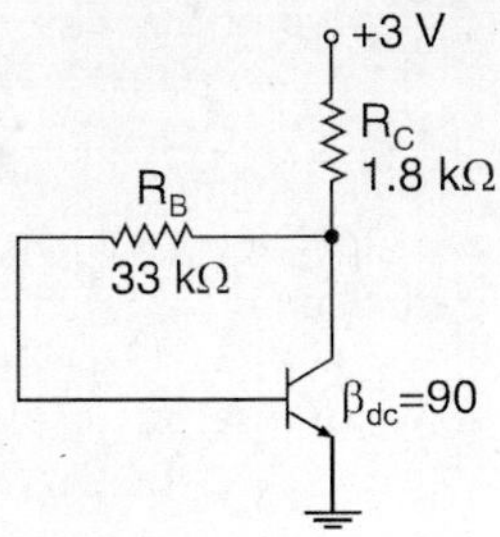

Fig. 12-30.

D.C bias current and voltage

We know that the base current,

$$I_B = \frac{V_{CC} - V_{BE}}{R_B + \beta \cdot R_C} = \frac{3 - 0.7}{(33 \times 10^3) + 90\ (1.8 \times 10^3)}\ \text{A}$$

$$= 0.0118 \times 10^{-3}\ \text{A} = 0.0118\ \text{mA}$$

and the collector current,

$$I_C = \beta \cdot I_B = 90 \times 0.0118 = 1.06\ \text{mA}\ \textbf{Ans.}$$

∴ Collector-to-emitter voltage,

$$V_{CE} = V_{CC} - I_C \cdot R_C = 3 - (1.06 \times 10^{-3}) \times (1.8 \times 10^3)\ \text{V}$$

$$= 3 - 1.9 = 1.1\ \text{V}\ \textbf{Ans.}$$

Stability factor

We also know that stability factor,

$$S = \frac{1 + \beta}{1 + \beta \times \dfrac{R_C}{R_C + R_B}} = \frac{1 + 90}{1 + 90 \times \dfrac{1.8 \times 10^3}{(1.8 \times 10^3) + (33 \times 10^3)}}$$

$$= \frac{91}{5.65} = 16.1\ \textbf{Ans.}$$

Example 12-16. *Determine the bias resistor R_B for fixed bias and collector to base bias and compare the stability factor S for both of them. Given V_{CC} = 12V, R_L = 330 Ω, I_B = 0.3 mA, β = 100 and V_{cEQ} = 6V.* (*Electronic Circuits- I, Anna University, April/May 2008*)

Solution. Given: $V_{CC} = 12$ V, $R_L = 330\ \Omega$, $I_B = 0.3$ mA, $\beta = 100$, $V_{CEQ} = 6$ V.

Fig. 12-31 Shows fixed bias circuit,

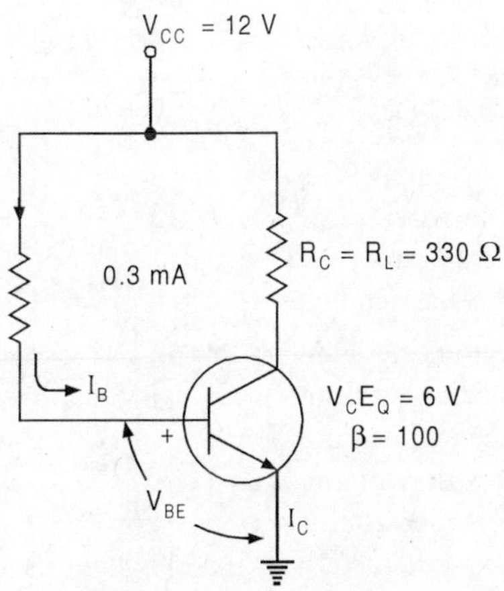

Fig. 12-31

For a fixed bias circuit, we know that the base resistor,

$$R_B = \frac{V_{CC} - V_{BE}}{I_B} = \frac{12 - 07}{0.3 \times 10^{-3}} = 37.67\ k\Omega$$

Value of Stability factor,

$$S = 1 + \beta = 100$$

Fig. 12-32 shows the collector to base bias circuit,

Applying kirchhoffs Voltage Law to the base-emitter circuit loop,

Value of Stability factor,

$$S = \frac{1+\beta}{1+\beta\left(\dfrac{R_C}{R_C+R_B}\right)}$$

$$= \frac{1+100}{1+100\left(\dfrac{330}{300+4337}\right)} = 12.443 \textbf{ Ans.}$$

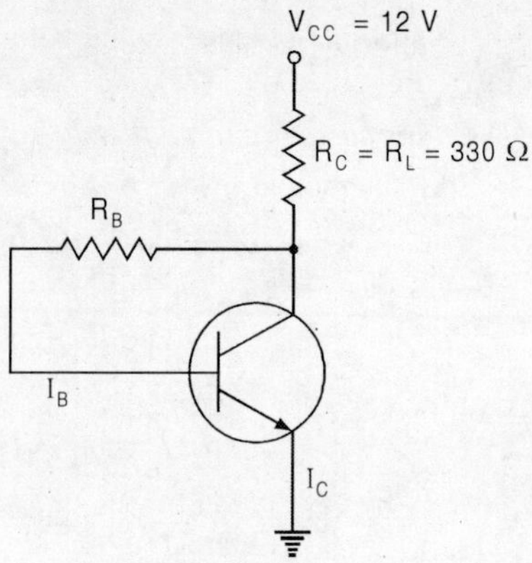

Fig. 12-32

$$V_{CC} - (I_B + I_C)\,R_L - I_B \,.\, R_B - V_{BE} = 0$$

$$12 - [(0.3 \times 10^{-3}) + (0.3 \times 10^{-3}) \times 100] \times 330 - (0.3 \times 10^{3}) \times R_B - 0.7 = 0$$

$$R_B = \frac{12 - 9.999 - 0.7}{0.3 \times 10^{-3}} = \frac{1.301}{0.1 \times 10^{-3}}$$

$$= 4.337 \text{ k}\Omega \textbf{ Ans.}$$

12-17. Voltage Divider Bias

In all the d.c. bias circuits, discussed in the previous articles, we have found that the values of d.c. bias current and voltage of the collector depends upon the current gain (β) of the transistor. But we know that the value of current gain (β) is temperature sensitive especially for silicon transistors. Since, the nominal value of current gain (β) is not well defined, therefore it would be desirable to provide a d.c. bias circuit which is independent of the transistor current gain (β). The d.c. bias circuit shown in Fig. 12-33 (*a*) meets this condition and is thus a very popular bias circuit. It is commonly known as *voltage divider bias* or *self bias circuit.*

The name voltage divider is derived from the fact that resistors R_1 and R_2 form a potential divider across the V_{CC} supply. The voltage drop across resistor R_2 forward biases the base-emitter junction of a transistor. The emitter resistor (R_E) provides the d.c. stability.

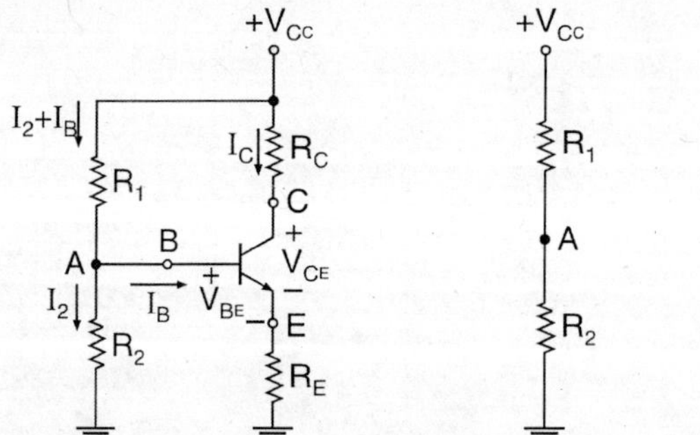

(*a*) Voltage divide bias circuit. (*b*) Voltage divider.

Fig. 12-33.

Now let us analyse the base-emitter loop of the voltage divider bias circuit. A basic assumption is that the resistance looking into the base is much larger than that of the resistor R_2. If this is so, then the current through resistor R_1 flows almost completely into resistor R_2 and the two resistors may be considered effectively *in series* as shown in Fig. 12-33(*b*). The voltage at the junction of the resistor (*i.e.*, point *A*), which is also the voltage at the base of the transistor, is then determined simple by the voltage divider network of R_1 and R_2 and the supply voltage. If the current through resistors R_1 and R_2 is of the order of milliamperes and that through the base is of the order of the microamperes, then the base current component can be neglected.

It is evident from Fig. 12-33(*b*), that the voltage at the transistor base (due to the voltage divider network of resistors R_1 and R_2),

$$V_B = V_{CC} \times \frac{R_2}{R_1 + R_2}$$

Since, the voltage drop across the base-emitter junction (V_{BE}) when forward biased is very small, as compared to the voltage at the base (V_B), therefore the voltage at the emitter is almost equal to the voltage at the base, *i.e.*,

$$V_E = V_B \qquad \text{... (Neglecting } V_{BE}\text{)}$$

∴ Value of emitter current,

$$I_E = \frac{V_E}{R_E}$$

and the value of collector current,

$$I_C \approx I_E$$

The voltage drop across the collector resistor,

$$V_{R_C} = I_C \cdot R_C$$

and the voltage at the collector (measured with respect to the ground)

$$V_C = V_{CC} - V_{R_C} = V_{CC} - I_C \cdot R_C$$

∴ The voltage from collector-to-emitter,

$$V_{CE} = V_C - V_E = V_{CC} - I_C \cdot R_C - I_E \cdot R_E$$
$$= V_{CC} - I_E (R_C + R_E) \qquad \text{...}(\because I_C = I_E)$$

Note. The values of d.c. bias voltages and currents obtained above are only approximate. The more accurate values of these quantities will be obtained in the next article.

Example 12-16. *Determine the value of collector current and collector-to-emitter voltage for the voltage divider bias circuit shown in Fig. 12-34.*

Assume V_{BE} *= 0.7 V and β = 100.*

(Electronic Circuits-I, Anna Univ. Nov./Dec. 2007)

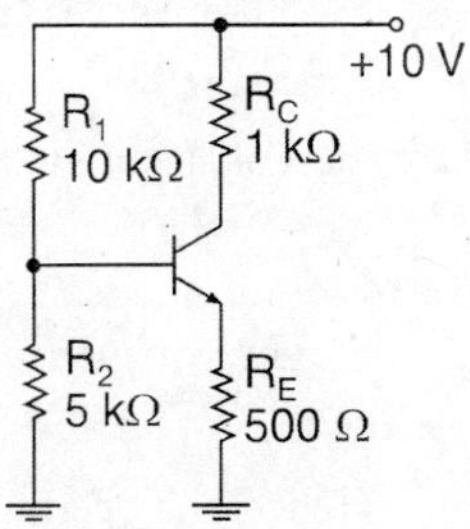

Fig. 12-34.

Solution. Given : V_{CC} = 10 volts; R_C = 1 kΩ = 1 × 10^3 Ω; R_1 = 10 kΩ; R_2 = 5 kΩ; R_E = 500 Ω; V_{BE} = 0.7 volt and β = 100.

Collector current

We know that voltage at the base,

$$V_B = V_{CC}\left(\frac{R_2}{R_1 + R_2}\right) = 10\left(\frac{5}{10+4}\right) \text{V}$$
$$= 3.33 \text{ V}$$

and voltage at the emitter,

$$V_E = V_B - V_{BE} = 3.33 - 0.7 = 2.63 \text{ V}$$

∴ Emitter current,

$$I_E = \frac{V_E}{R_E} = \frac{2.63}{500} = 5.26 \times 10^{-3} \text{ A} = 5.26 \text{ mA}$$

and collector current,

$$I_C = I_E = 5.26 \text{ mA } \textbf{Ans.}$$

Collector-to-emitter voltage

We also know that the collector-to-voltage,

$$V_{CE} = V_{CC} - (R_C + R_E) = 10 - 5.26 \times 10^{-3} (1 \times 10^3 + 500) \text{ V}$$
$$= 10 - 7.89 = 2.11 \text{ V } \textbf{Ans.}$$

Example 12-17. *A Si transistor is biased using voltage divider method as shown in Fig. 12-35. Determine the operating point.* (*Banglore Univ. B.Sc. Electronics, May* 2001)

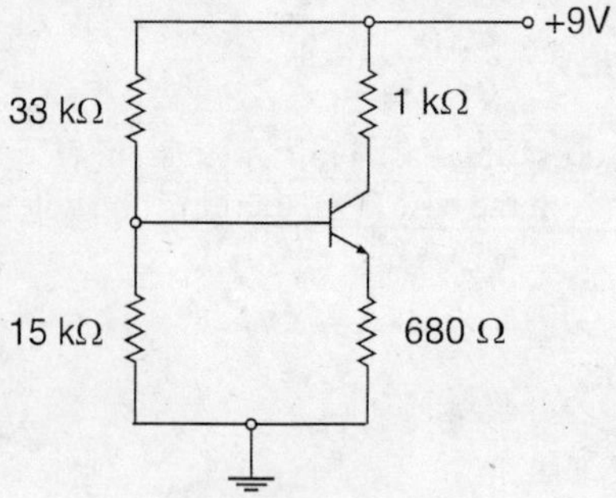

Fig. 12-35.

Solution. Given : $V_{CC} = 9$ V; $R_C = 1$ kΩ $= 1 \times 10^3$ Ω; $R_E = 680$ Ω; $R_1 = 33$ kΩ and $R_2 = 15$ kΩ.

We know that the base voltage,

$$V_B = V_{CC} \times \frac{R_2}{R_1 + R_2} = 9 \times \frac{15\text{ k}}{33\text{ k} + 15\text{ k}} = 2.81\text{ V}$$

Since, the voltage drop across the base emitter junction for a Si transistor is 0.7 V, therefore,

$$V_E = V_B - 0.7\text{ V} = 2.81 - 0.7 = 2.11\text{ V}$$

and the value of emitter current,

$$I_E = \frac{V_E}{R_E} = \frac{2.11}{680} = 3.1 \times 10^{-3}\text{ A} = 3.1\text{ mA}$$

∴ Collector current,

$$I_C \approx I_E = 3.1\text{ mA}$$

We know that voltage across the collector resistor,

$$VR_C = I_C \cdot R_C = 3.1\text{ mA} \times 1\text{ k}\Omega = 3.1\text{ V}$$

and the voltage at the collector (measured with respect to the ground),

$$V_C = V_{CC} - V_{R_C} = 9 - 3.1 = 5.9\text{ V}$$

∴ The voltage from collector-to-emitter,

$$V_{CE} = V_C - V_E = 5.9 - 2.11 = 3.79\text{ V.}$$

Hence, the operating point values are,

$$I_C = 3.1\text{ mA and } V_{CE} = 3.79\text{ V } \textbf{Ans.}$$

Example 12-18. *The BJT in amplifier of Fig. 12-36, has $h_{fe} = 100$, $V_{BE} = 0.7$ V, $I_{CO} = 0$. Calculate the value of R_1 and R_C such that its $I_C = 1$ mA and $V_{CE} = 2.5$ V.*

(*UPSC Engg. Services* 2001)

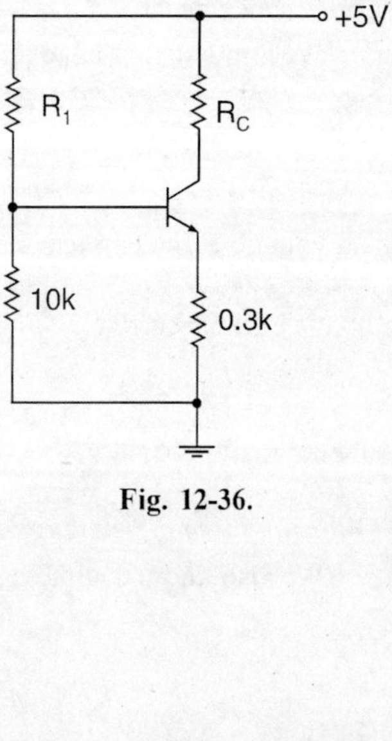

Fig. 12-36.

Solution. Given : $V_{CC} = 5$ V; $R_E = 0.3$ kΩ $= 300$ Ω; $I_C = 1$ mA; $V_{CE} = 2.5$ V; $h_{fe} = \beta = 100$; $V_{BE} = 0.7$ V; $I_{CO} = 0$.

We know that the emitter current,

$$I_E \approx I_C = 1\text{ mA}$$

and the voltage from collector-to-emitter, (V_{CE})

$$2.5 = V_{CC} - I_E (R_C + R_E) = 5 - (1 \times 10^{-3}) (R_C + 300)$$

or $$R_C = \frac{5 - 2.5}{1 \times 10^{-3}} - 300 = 2200\ \Omega \textbf{ Ans.}$$

We also know that the emitter voltage,

$$V_E = I_E \cdot R_E = (1 \times 10^{-3}) \times 300\ \Omega = 0.3\ V$$

and the base voltage,

$$V_B = V_E + V_{BE} = 0.3\ V + 0.7\ V = 1.0\ V$$

For a voltage-divider bias, (V_B),

$$1V = V_{CC} \times \frac{R_2}{R_1 + R_2} = 5\ V \times \frac{10\ k}{R_1 + 10\ k}$$

$\therefore$ $$R_1 = 50\ k - 10\ k = 40\ k\Omega \textbf{ Ans.}$$

Example 12-19. *Calculate the value of emitter current in the voltage divider circuit shown in Fig. 12-37, Also find the value of V_{CE} and collector potential V_C.* (*Dip. IETE., June* 2001)

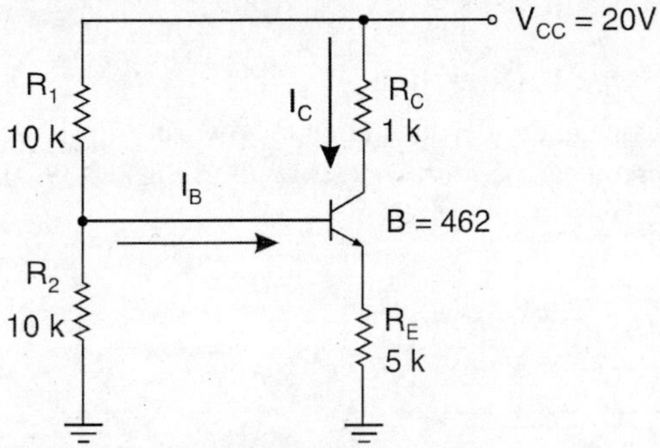

Fig. 12-37.

Solution. Given : $V_{CC} = 20$ V; $R_C = 1\ k\Omega$; $R_E = 5\ k\Omega$; $R_1 = 10\ k\Omega$; $R_2 = 10\ k\Omega$. $\beta = 462$; assume $V_{BE} = 0.7$ V.

We know that voltage at the base,

$$V_B = V_{CC} \times \frac{R_2}{R_1 + R_2} = 20 \times \frac{10\ k}{10\ k + 10\ k} = 10\ V$$

and voltage at the emitter,

$$V_E = V_B - V_{BE} = 10 - 0.7 = 9.3\ V$$

$\therefore$ Emitter current,

$$I_E = \frac{V_E}{R_E} = \frac{9.3}{5\ k} = 1.86\ mA \textbf{ Ans.}$$

and collector current,

$$I_C \approx I_E = 1.86\ mA$$

We also know that collector-to-emitter voltage,

$$V_{CE} = V_{CC} - I_C \cdot R_C = 20 - (1.86\ mA) \times (1\ k\Omega)$$
$$= 18.14\ V \textbf{ Ans.}$$

Example 12-20. *A transistor is connected in common-emitter (CE) configuration in which collector supply is 8V and the voltage drop across resistance R_C connected in the collector circuit is 0.5 V. The value of R_C is 800 Ω. If $\alpha = 0.96$ determine **(a)** Collector-emitter voltage (V_{CE}), and **(b)** base current.* (*Dip, IETE,. June* 2001)

Solution. Given : $V_{CC} = 8$ V; $V_{R_C} = 0.5$ V; $R_C = 800\ \Omega$; $\alpha = 0.96$.

Fig. 12.38 shows the arrangement.

We know that the collector-to-emitter voltage,

$$V_{CE} = V_{CC} - I_C \cdot R_C$$
$$= 8 - 0.5$$
$$= 7.5 \text{ V } \textbf{Ans.}$$

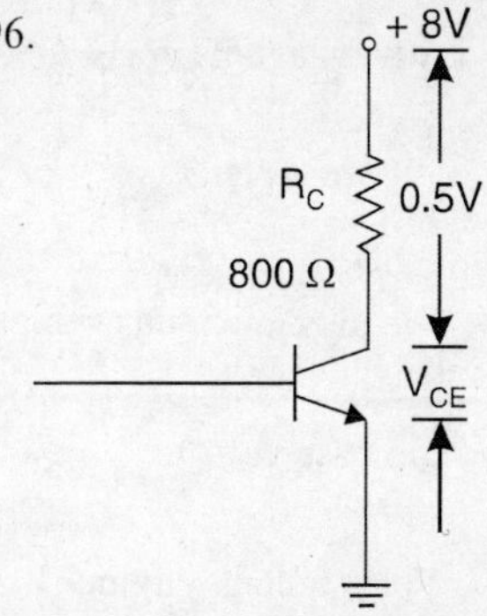

Fig. 12-38.

and the collector current,

$$I_C = \frac{V_{R_C}}{R_C} = \frac{0.5}{800} = 625 \times 10^{-4} \text{ A}$$
$$= 0.625 \text{ mA}$$

We also know that emitter current,

$$I_E = \frac{I_C}{\alpha} = \frac{0.625}{0.96} = 0.651 \text{ mA}$$

∴ The base current,

$$I_B = I_E - I_C = 0.651 - 0.625 \text{ mA}$$
$$= 0.026 \text{ mA or } 26\ \mu\text{A } \textbf{Ans.}$$

12-18. Stability of Voltage Divider Bias

Consider again the voltage divider bias circuit shown in Fig. 12-33(*a*). Looking out from the base terminal, the bias circuit can be redrawn as shown in Fig. 12-39(*a*).

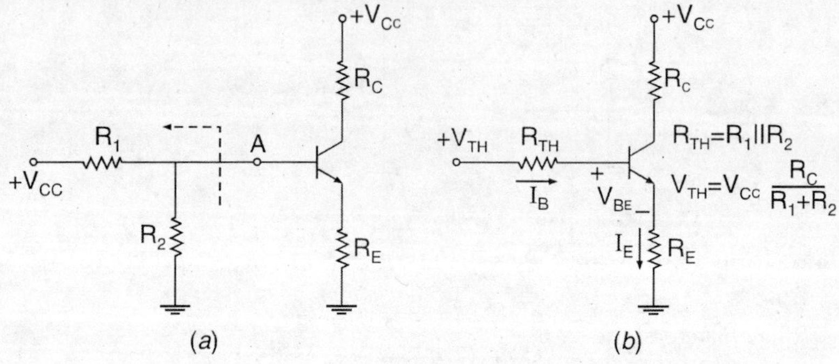

Fig. 12-39. *Thevenizing the voltage divider bias circuit.*

Now applying the Thevenin's Theorem for detailed procedure refer to Appendix C to the circuit lying on the left of the point *A*, we get the Thevenin's Voltage,

$$V_{TH} = \left(\frac{R_2}{R_1 + R_2}\right) V_{CC}$$

and the Thevenin's Equivalent Resistance

$$R_{TH} = R_1 \parallel R_2 = \frac{R_1 \cdot R_2}{R_1 + R_2}$$

We know that the Thevenin's Voltage (V_{TH}) is a voltage drop across resistor R_2. As a matter of fact, it is the voltage at the base (V_B) measured with respect to the ground. Fig. 12-39(*b*) shows the voltage divider bias circuit in which the series combination of resistors R_1 and R_2 is replaced by a Thevenin's equivalent circuit. Applying, Kirchoff's Voltage Law to the base-emitter loop of this circuit,

$$I_B \cdot R_{TH} + V_{BE} + I_E \cdot R_E - V_{TH} = 0$$

Replacing I_B with I_E/β (equal to I_C/β) in the above relation and solving for emitter current,

$$I_E = \frac{V_{TH} - V_{BE}}{R_E + \frac{R_{TH}}{\beta}} \quad ...(i)$$

$$= \frac{V_{TH} - V_{BE}}{R_E} \quad ...(R_E >> R_{TH}/\beta)$$

It is evident from the above relation that *the emitter current is independent of current gain* (β). In actual practice, it is achieved by making the emitter resistor 10 times greater than R_{TH}/β. Mathematically, emitter resistance,

$$R_E \geq 10\, R_{TH}/\beta$$

or

$$R_{TH} \leq 0.1\, \beta \cdot R_E \quad ...(ii)$$

Usually, the value of resistor R_2 is smaller than the value of resistor R_1. Therefore the Thevenin's equivalent resistance R_{TH} (equal to $R_1 \parallel R_2$) is approximately equal to R_2. In that case, the equation (*ii*) may be written as

$$R_2 \leq 0.1\, \beta \cdot R_E \quad ...(iii)$$

Example 12-21. *Fig. 12-40 shows that d.c. bias circuit of a common emitter transistor amplifier.*

Find the percentage change in collector current, if the transistor with $h_{FE} = 50$ is replaced by another transistor with $h_{FE} = 150$. It is given that the base emitter drop $V_{BE} = 0.6$ V.

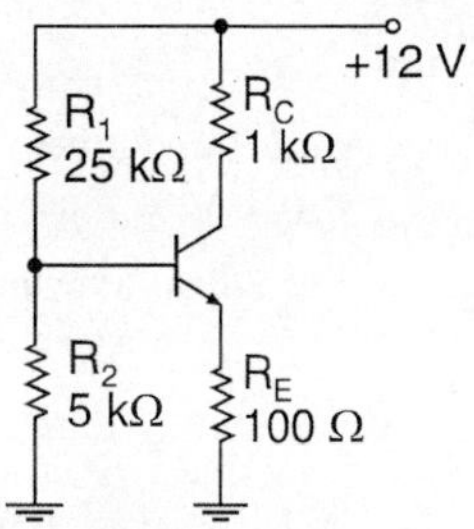

Fig. 12-40.

Solution. Given : V_{CC} = 12 volts; R_C = 1 kΩ; R_E = 100 Ω; R_1 = 25 kΩ; R_2 = 5 kΩ; h_{FE} (=β) = 50 and 150 and V_{BE} = 0.6 volt.

We know that the Thevenin's Voltage,

$$V_{TH} = V_{CC}\left(\frac{R_2}{R_1 + R_2}\right) = 12\left(\frac{5}{25+5}\right) = 2 \text{ V}$$

and the Thevenin's Equivalent Resistance,

$$R_{TH} = R_1 \parallel R_2 = \frac{R_1 \cdot R_2}{R_1 + R_2} = \frac{5 \times 25}{5 + 25} \text{ k}\Omega = 4.17 \text{ K}\Omega = 4170\ \Omega$$

We also know that the emitter current,

$$I_E = \frac{V_{TH} - V_{BE}}{R_E + \frac{R_{TH}}{\beta}}$$

For β = 50, the emitter current,

$$I_E = \frac{2 - 0.6}{100 + \frac{4170}{50}} = 7.63 \times 10^{-3} \text{ A} = 7.63 \text{ mA}$$

and for β = 150, the emitter current,

$$I_E = \frac{2 - 0.6}{100 + \frac{4170}{150}} = 10.95 \times 10^{-3} \text{ A} = 10.95 \text{ mA}$$

∴ Change in the collector current, when the transistor with h_{FE} equal to 50 is changed with another transistor with h_{FE} equal to 150,

$$\frac{10.95 - 7.63}{7.63} = 0.435 = 43.5\% \textbf{ Ans.}$$

Example 12-22. *A silicon transistor uses voltage divider biasing V_{CC} = 12 V, R_1 = 10 kΩ, R_2 = 5 kΩ, R_L = 1 kΩ and R_E = 3 kΩ.* Determine the operating point using Thevenin's theorem.

(*Electronic Circuits - I, Anna University, April/May 2008*)

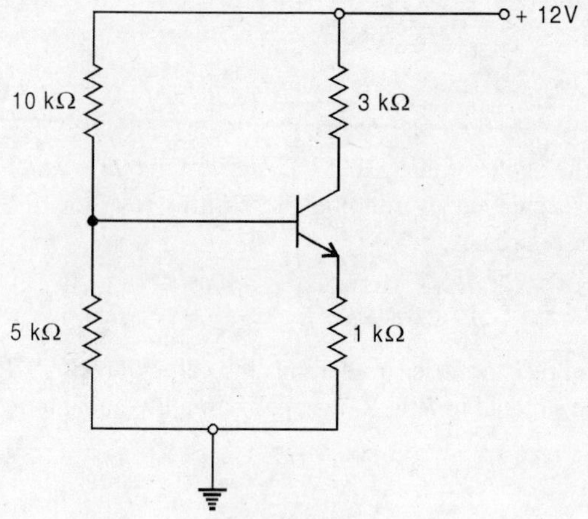

Fig. 12-41.

Solution. Given: V_{CC} = 12 V, R_1 = 10 kΩ, R_2 = 5 kΩ, R_L = 1 kΩ and R_E = 3 kΩ, assume V_{BE} = 0.7 V.

We know that the Thevenin's Voltage,

$$V_{TH} = V_{CC}\left(\frac{R_2}{R_1 + R_2}\right) = 12 \times \left(\frac{5k}{10k + 5k}\right) = 4 \text{ V}$$

and the Thevenin's Equivalent Resistance

$$R_{TH} = R_1 \parallel R_2 = \frac{R_1 \,.\, R_2}{R_1 + R_2}$$

$$= \frac{10k + 5k}{10k + 5k} = 3.333 \text{ k}\Omega = 3333\Omega$$

We also know that the emitter current,

$$I_E = \frac{V_{TH} - V_{BE}}{R_E} = \frac{4 - 0.7}{3333}$$

$$= 0.9 \times 10^{-3} \text{ A.}$$

∴ Collector Current,

$$I_C \approx I_E = 0.9 \text{ mA}$$

We know that voltage across the collector resistor,

$$V_{R_C} = I_C \,.\, R_C = 0.9 \text{ mA} \times 3 \text{ k}\Omega = 2.7 \text{ V}$$

and the voltage at the collector (measured with respect to the ground),

$$V_C = V_{CC} - V_{R_C} = 12 - 2.7 = 9.3 \text{ V}$$

Hence, the operating-point values are,

$$I_C = 0.9 \text{ mA and } V_{CE} = 9.3 \text{ V}$$ **Ans.**

12-19. Stability Factor of Voltage Divider Bias

We have already discussed in the last article that if we apply Kirchoff's Voltage Law to the base-emitter loop of the voltage divider bias circuit shown in Fig. 12-33 (*b*), we get,

$$I_B \cdot R_{TH} + V_{BE} + I_E \cdot R_E - V_{TH} = 0 \qquad ...(i)$$

Substituting the value of I_E (equal to $I_B + I_C$) in the above equation and solving it for base current,

$$I_B = \frac{V_{TH} - V_{BE} - I_C \cdot R_E}{R_E + R_{TH}}$$

Differentiating the above expression with respect to I_C,

$$\frac{dI_B}{dI_C} = \frac{0 - 0 - 1 \cdot R_E}{R_E + R_{TH}} = -\frac{R_E}{R_E + R_{TH}}$$

Substituting the above value of dI_B/dI_C in the general expression for stability factor,

$$S = \frac{1+\beta}{1 - \beta\left(\dfrac{dI_B}{dI_C}\right)} = \frac{1+\beta}{1 - \beta\left(-\dfrac{R_E}{R_E - R_{TH}}\right)} = \frac{1+\beta}{1 + \beta\left(\dfrac{R_E}{R_E + R_{TH}}\right)}$$

Rearranging the above expression,

$$S = \frac{(1+\beta)\left(1 + \dfrac{R_{TH}}{R_E}\right)}{1 + \beta + \dfrac{R_{TH}}{R_E}}$$

Substituting the value of R_{TH} equal to $\dfrac{R_1 \cdot R_2}{(R_1 + R_2)}$ in the above expression and rearranging,

$$S = \frac{R_E\,(R_1 + R_2) + R_1 \cdot R_2}{R_E\,(R_1 + R_2) + R_1 \cdot R_2\,(1 - \alpha)}$$

Example 12-22. *Show that the stability factor* $\left(S = \dfrac{dI_C}{dI_{CO}}\right)$ *for the d.c. bias circuit shown in Fig. 12-42 is given by the expression :*

$$S = (1+\beta) \times \frac{1 + \dfrac{R_B}{R_E}}{1 + \beta + \dfrac{R_B}{R_E}}$$

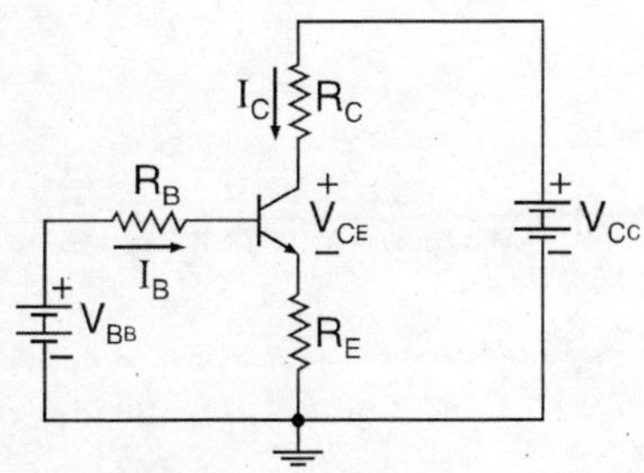

Fig.12-42.

Solution. The given circuit is similar to the voltage divider bias circuit and may by compared with the Thevenin's Equivalent Circuit of a voltage divider bias circuit. By comparing, we find that the V_{BB} supply is the same as the Thevenin's Voltage (V_{TH}) and R_B is the same as the Thevenin's Equivalent Resistance (R_{TH}).

We know that the stability factor,

$$S = \frac{(1+\beta)\left(1 + \dfrac{R_{TH}}{R_E}\right)}{1 + \beta + \dfrac{R_{TH}}{R_E}}$$

Replacing R_{TH} with R_B in the above expression,

$$S = \frac{(1+\beta)\left(1+\frac{R_B}{R_E}\right)}{1+\beta+\frac{R_B}{R_E}} \quad \textbf{Ans.}$$

Example 12-23. *Show that in a voltage divider single battery bias type common emitter voltage amplifier using NPN transistor, the stability factor is given by the expression,*

$$S = \frac{\delta I_C}{\delta I_{CBO}} = \frac{R_E\,(R_1+R_2)+R_1 \cdot R_2}{R_E\,(R_1+R_2)+R_1 \cdot R_2\,(1-\alpha)}$$

where, R_1, R_2, R_E *and* α *have their usual meanings.*

Solution. We know that the stability factor for a voltage divider bias circuit,

$$S = \frac{(1+\beta)}{1+\beta\left(\frac{R_E}{R_E+R_{TH}}\right)}$$

Substituting the value of R_{TH} (equal to $R_1 \cdot R_2/(R_1+R_2)$ in the above expression,

$$S = \frac{(1+\beta)}{1+\beta \cdot \frac{R_E}{R_E+\frac{R_1 \cdot R_2}{R_1+R_2}}} = \frac{(1+\beta)\left(R_E+\frac{R_1 \cdot R_2}{R_1+R_2}\right)}{R_E+\frac{R_1 \cdot R_2}{R_1+R_2}+\beta \cdot R_E}$$

$$= \frac{(1+\beta)\,[R_E\,(R_1+R_2)+R_1 \cdot R_2]}{R_E\,(R_1+R_2)+R_1 \cdot R_2+\beta \cdot R_E\,(R_1+R_2)}$$

$$= \frac{(1+\beta)\,[R_E\,(R_1+R_2)+R_1 \cdot R_2]}{(1+\beta)\,R_E\,(R_1+R_2)+R_1 \cdot R_2}$$

$$= \frac{(1+\beta)\,[R_E\,(R_1+R_2)+R_1 \cdot R_2]}{(1+\beta)\left[R_E\,(R_1+R_2)+\frac{R_1 \cdot R_2}{1+\beta}\right]}$$

$$= \frac{R_E\,(R_1+R_2)+R_1 \cdot R_2}{R_E\,(R_1+R_2)+\frac{R_1 \cdot R_2}{1+\beta}} \quad ...(i)$$

We also know that the current gain,

$$\beta = \frac{\alpha}{1-\alpha}$$

Adding unity on both sides of the above expression,

$$1+\beta = 1+\frac{\alpha}{1-\alpha} = \frac{1-\alpha+\alpha}{1-\alpha} = \frac{1}{1-\alpha}$$

Replacing $1+\beta$ with $1/(1-\alpha)$ in equation (*i*),

$$S = \frac{R_E\,(R_1+R_2)+R_1 \cdot R_2}{R_E\,(R_1+R_2)+R_1 \cdot R_2\,(1-\alpha)} \quad \textbf{Ans.}$$

Example 12-24. *A transistor with* $\beta = 50$; $V_{BE} = 0.7$ V, $V_{CC} = 22.5$ V *and* $R_C = 5.6$ kΩ *is used in a biasing circuit as shown in Fig. 12-43. It is designed to establish the quiescent point at* $V_{CE} = 12$ V, $I_C = 1.5$ mA *and stability factor* $S \leq 3$.

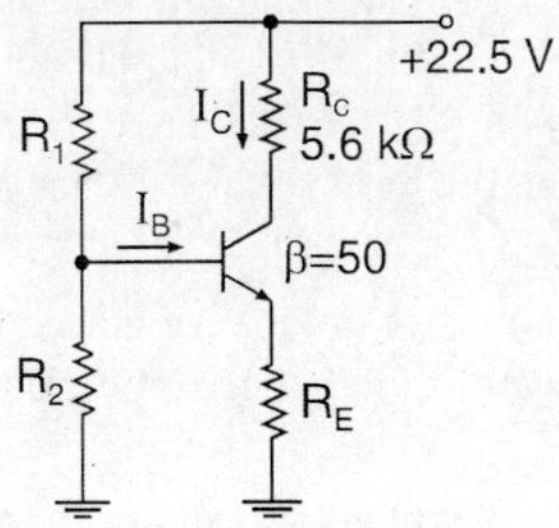

Fig. 12-43.

Find the values of resistor R_E, R_1 and R_2, where the symbols have their there usual meanings.

Solution. Given : $\beta = 50$; $V_{BE} = 0.7$ volt; $V_{CC} = 22.5$ volts; $R_C = 5.6\ k\Omega = 5.6 \times 10^3\ \Omega$, $V_{CE} = 12$ volts; $I_C = 1.5$ mA $= 1.5 \times 10^{-3}$ A and $S \leq 3$.

Emitter resistance (R_E)

We know that the value of collector-to-emitter voltage (V_{CE}),

$$12 = V_{CC} - I_C \cdot (R_C + R_E)$$
$$= 22.5 - (1.5 \times 10^{-3})\ [(5.6 \times 10^3) + R_E]$$
$$= 14.1 - 1.5 \times 10^{-3}\ R_E$$
$$\therefore \quad R_E = 1.4 \times 10^3\ \Omega = 1.4\ k\Omega \ \textbf{Ans.}$$

Resistances R_1 and R_2

We know that the stability factor (S)

$$3 = \frac{\beta + 1}{1 + \beta \times \dfrac{R_E}{R_{TH} + R_E}} = \frac{50 + 1}{1 + 50 \times \dfrac{1.4 \times 10^3}{R_{TH} + 1.4 \times 10^3}}$$

$$= \frac{51}{1 + \dfrac{70 \times 10^3}{R_{TH} + 1.4 \times 10^3}}$$

$$3\left[\frac{(R_{TH} + 1.4 \times 10^3) + 70 \times 10^3}{R_{TH} + 1.4 \times 10^3}\right] = 51$$

$$3\ [(R_{TH} + (1.4 \times 10^3)] + (70 \times 10^3)] = 51\ [R_{TH} + (1.4 \times 10^3)]$$
$$48\ [R_{TH} + (1.4 \times 10^3)] = 210 \times 10^3$$

or
$$R_{TH} + (1.4 \times 10^3) = (210 \times 10^3)/48 = 4375$$

$$\therefore \quad R_{TH} = 4375 - (1.4 \times 10^3) = 2975\ \Omega = 2.98\ k\Omega$$

We also know that for a good voltage divider, the value of resistor

$$R_2 = 0.1\ \beta \cdot R_E = 0.1 \times 50 \times (1.4 \times 10^3)\ \Omega = 7 \times 10^3\ \Omega = 7\ k\Omega \ \textbf{Ans.}$$

and the Thevenin's Resistance (R_{TH}),

$$2.98 = R_1 \parallel R_2 = \frac{R_1 \cdot R_2}{R_1 + R_2} = \frac{7 R_1}{R_1 + 7}$$

or
$$2.98\ (R_1 + 7) = 7\ R_1$$
$$4.02\ R_1 = 20.86$$
$$\therefore \quad R_1 = 20.86/4.02 = 5.2\ k\Omega \ \textbf{Ans.}$$

12-20. Emitter Bias

Fig. 12-44 shows the emitter bias circuit. It uses both the positive (*i.e.*, V_{CC} supply) and negative (*i.e.*, $-V_{EE}$ supply) voltages. If the base resistance (R_B) is small enough in the emitter bias circuit, then the base voltage (V_B) is approximately zero.

The base-emitter junction of a transistor is forward biased by the $-V_{EE}$ supply. The d.c. bias currents and voltages may be obtained as given below :

The base voltage (V_B) is approximately zero, because the value of resistance R_B is usually very high. Thus, $V_B = 0$. And the voltage at the emitter (measured with respect to ground),

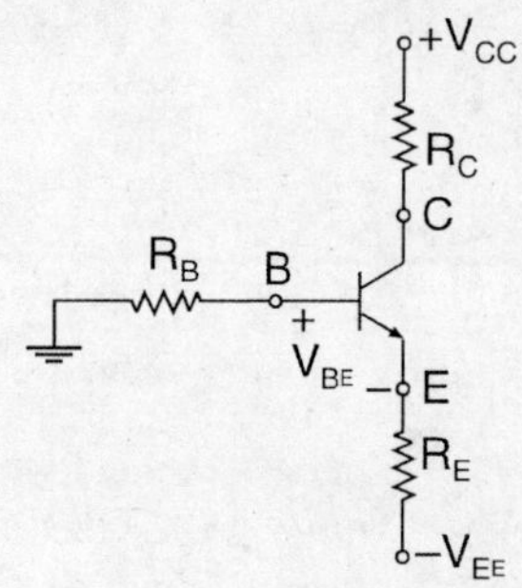

Fig. 12-44.

$$V_E = -V_{BE}$$

∴ Value of emitter current,

$$I_E = \frac{V_E - V_{EE}}{R_E} = \frac{-V_{BE} - V_{EE}}{R_E} \qquad ...(\because V_E = -V_{BE})$$

Since, V_{EE} supply is negative, therefore, the emitter current,

$$I_E = \frac{-V_{BE} - (-V_{EE})}{R_E} = \frac{V_{EE} - V_{BE}}{R_E}$$

We know that the value of collector current is approximately the same as that of emitter current, *i.e.*, $I_C = I_E$.

∴ Collector voltage (measured with respect to ground),

$$V_C = V_{CC} - I_C \cdot R_C$$

and the collector-to-emitter voltage,

$$V_{CE} = V_C - V_E$$

Example 12-25. *Find the values of emitter current, collector current, and collector-to-emitter voltage for the emitter bias circuit shown in Fig. 12-45. Assume* $V_{BE} = 0.7$ *V.*

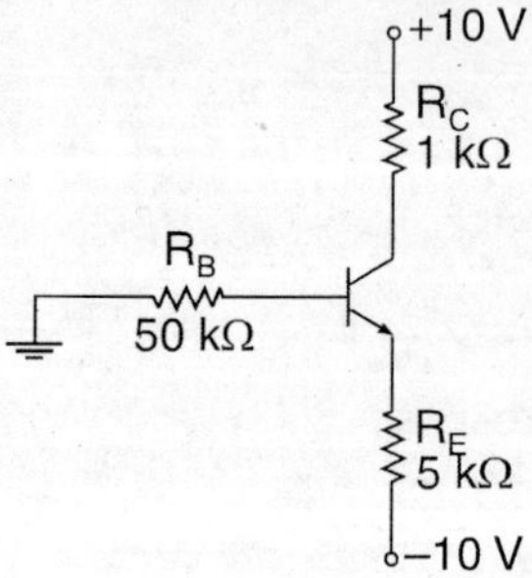

Fig. 12-45.

Solution. Given : $V_{CC} = 10$ volts; $V_{EE} = 10$ volts; $R_C = 1\ \text{k}\Omega = 1 \times 10^3\ \Omega$; $R_E = 5\ \text{k}\Omega = 5 \times 10^3\ \Omega$ and $R_B = 50\ \text{k}\Omega$.

Emitter current

We know that the emitter voltage,

$$V_E = -V_{BE} = -0.7\ \text{V}$$

and the value of emitter current,

$$I_E = \frac{V_{EE} - V_{BE}}{R_E} = \frac{10 - 0.7}{5 \times 10^3}\ \text{A}$$

$$= 1.86 \times 10^{-3}\ \text{A} = 1.86\ \text{mA}\ \textbf{Ans.}$$

Collector current

We know that the value of collector current,

$$I_C = I_E = 1.86 \times 10^{-3} = 1.86\ \text{mA}\ \textbf{Ans.}$$

Collector-to-emitter voltage

We also know that the collector voltage,

$$V_C = V_{CC} - I_C \cdot R_C = 10 - (1.86 \times 10^{-3}) \times (1 \times 10^3) = 8.14\ \text{volts}$$

and the collector-to-emitter voltage,

$$V_{CE} = V_C - V_E = 8.14 - (-0.7) = 8.84\ \text{volts}\ \textbf{Ans.}$$

12-21. Stability of Emitter Bias

Consider again the emitter bias circuit as shown in Fig. 12-44. Applying Kirchoff's Voltage Law to the base-emitter loop of this circuit,

$$I_B \cdot R_B + V_{BE} + I_E \cdot R_E - V_{EE} = 0$$

Substituting the value of base current I_B (equal to $I_C/\beta = I_E/\beta$) in the above equation,

$$I_E\left(\frac{R_B}{\beta} + R_E\right) = V_{EE} - V_{BE}$$

$$\therefore \qquad I_E = \frac{V_{EE} - V_{BE}}{\dfrac{R_B}{\beta} + R_E}$$

$$= \frac{V_{EE} - V_{BE}}{R_E} \qquad \text{...(If } R_E >> R_B/\beta\text{)}$$

$$= V_{EE}/R_E \qquad \text{...(If } V_{EE} >> V_{BE}\text{)}$$

The above relation indicates that the emitter current is independent of current gain (β) and base-to-emitter voltage (V_{BE}) as long as the value of resistance R_E and V_{EE} supply are very large. In actual practice, the value of R_E is selected in such a way that it is 100 times greater than R_B/β. Mathematically, it is equivalent to,

$$R_E \geq 100\ R_B/\beta$$

or

$$R_B \leq 0.01\ \beta \cdot R_E$$

Note. Since, the emitter current is independent of β and V_{BE}, therefore the Q-point is not affected appreciably by the variations in these parameters. Thus, the emitter bias provides a reasonably stable bias point.

12-22. Stability Factor of Emitter Bias

We have already discussed in the last article that if we apply Kirchoff's Voltage Law to the base-emitter loop of the emitter bias circuit shown in Fig. 12-44, we get

$$I_B \cdot R_B + V_{BE} + I_E \cdot R_E - V_{EE} = 0$$

Substituting the value of I_B (equal to $I_B + I_C$) in the above equation and solving it for base current,

$$I_B = \frac{V_{EE} - V_{BE} + I_C \cdot R_C}{R_E + R_B}$$

Differentiating the above expression with respect to I_C

$$\frac{dI_B}{dI_C} = \frac{0 - 0 - 1 \cdot R_E}{R_E + R_B} = -\frac{R_E}{R_E + R_B}$$

Substituting the above value of dI_B/dI_C in the general expression for stability factor,

$$S = \frac{1+\beta}{1-\beta\left(\dfrac{dI_B}{dI_C}\right)} = \frac{1+\beta}{1-\beta\left(-\dfrac{R_E}{R_E+R_B}\right)} = \frac{1+\beta}{1+\beta \times \dfrac{R_E}{R_E+R_B}}$$

Example 12-26. *Fig. 12-46 shows the emitter bias current using NPN transistor.*

Determine how much the Q-point will change over a temperature range, where β increases from 50 to 100 and V_{BE} decreases from 0.7 V to 0.6 V.

Solution. Given : $V_{CC} = 20$ volts; $V_{EE} = 20$ volts; $R_C = 5\ k\Omega = 5 \times 10^3\ \Omega$; $R_E = 10\ k\Omega = 10 \times 10^3\ \Omega$; $R_B = 10\ k\Omega = 10 \times 10^3\ \Omega$; $\beta = 50$ and 100; $V_{BE} = 0.7$ volt and 0.6 volt.

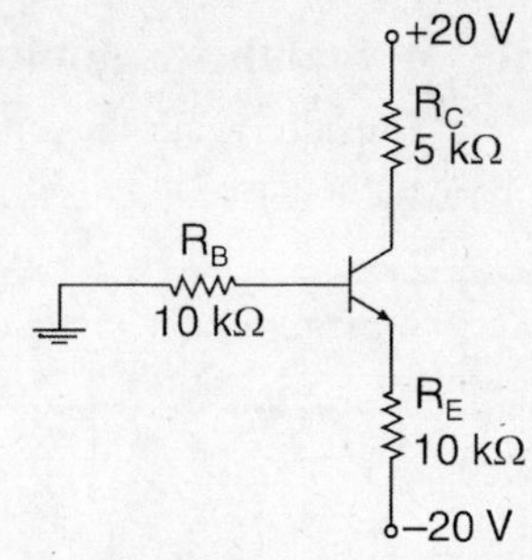

Fig. 12-46.

We know that the *Q*-point (*i.e.*, operating point) is defined by the values of collector current (I_C) and collector-to-emitter voltage (V_{CE}). Therefore, in order to determine the change in *Q*-point, we shall find the variation in collector current and the collector-to-emitter voltage in both the cases.

We know that the value of emitter current, when $\beta = 50$ and $V_{BE} = 0.7$ V,

$$I_E = \frac{V_{EE} - V_{BE}}{R_E + \dfrac{R_B}{\beta}} = \frac{20 - 0.7}{10 \times 10^3 + \dfrac{10 \times 10^3}{50}}$$

$$= 1.892 \times 10^{-3} \text{ A} = 1.892 \text{ mA}$$

and the collector current,

$$I_C = I_E = 1.892 \times 10^{-3} \text{ A} = 1.892 \text{ mA} \quad ...(i)$$

We also know that the collector voltage,

$$V_C = V_{CC} - I_C \cdot R_C = 20 - (1.892 \times 10^{-3}) \times (5 \times 10^3)$$

$$= 20 - 9.46 = 10.54 \text{ volts}$$

and the collector-to-emitter voltage,

$$V_{CE} = V_C - V_E = 10.54 - (-0.7) = 11.24 \text{ V} \quad ...(ii)$$

Similarly, value of the emitter current, when $\beta = 100$ and $V_{BE} = 0.6$,

$$I_E = \frac{V_{EE} - V_{BE}}{R_E + \dfrac{R_B}{\beta}} = \frac{20 - 0.6}{10 \times 10^3 + \dfrac{10 \times 10^3}{100}}$$

$$= 1.921 \times 10^{-3} \text{ A} = 1.921 \text{ mA}$$

and the collector current,

$$I_C = I_E = 1.921 \times 10^{-3} = 1.921 \text{ mA}$$

We also know that the collector voltage,

$$V_C = V_{CC} - I_C \cdot R_C = 20 - (1.921 \times 10^{-3}) \times (5 \times 10^3)$$

$$= 20 - 9.605 = 10.395 \text{ V}$$

and the collector to emitter voltage,

$$V_{CE} = V_C - V_E = 10.395 - (-0.6) = 10.995 \text{ V}$$

∴ Change in collector current,

$$\Delta I_C = \frac{1.921 - 1.892}{1.892} = 0.0153 = 1.53\% \textbf{ Ans.}$$

and change in collector to emitter voltage,

$$\Delta V_{CE} = \frac{11.24 - 10.995}{11.24} = 0.0218 = 2.18\% \textbf{ Ans.}$$

12-23. Additional Stability Factors

We have already discussed a number of d.c. biasing circuits, which provide stabilization of collector current (I_C) against the variation in reverse saturation current (I_{CO}). However, there are two other factors, which affect the stability of the d.c. biasing circuit, namely *variation in the base current* (I_B) *and the current gain* (β) *with temperature*. The stabilities, due to these two quantities, are measured by the stability factors S' and S'' respectively.

The stability factor S' may be defined as ***the rate of change of collector current (I_C) with respect to the base current (I_B) keeping the common-emitter current gain (β) and reverse saturation current (I_{CO}) as constant.*** Mathematically, the stability factor,

$$S' = \frac{dI_C}{dI_B}$$

In actual practice, the variation in base current occurs because of the change in base-to-emitter voltage (V_{BE}) with the change in temperature. We know that the base-to-emitter voltage (V_{BE}) decreases at a rate of 2.5 m V/°C for a constant collector current. Therefore, it is more convenient to define the stability factor S' as ***the rate of change of collector current with respect to the base-to-emitter voltage i.e.,***

$$S' = \frac{dI_C}{dV_{BE}}$$

For a voltage divider bias, the stability factor,

$$S' = -\frac{S}{R_{TH} + R_E} \times \frac{\beta}{\beta + 1} \qquad ...(i)$$

where, R_{TH} = Thevenin's Equivalent Resistance
$= (R_1 \cdot R_2)/(R_1 + R_2)$
R_E = Emitter resistance, and
S = Stability factor (equal to dI_C/dI_{CO}).

Similarly, the stability factor, S'' may be defined as ***the rate of change of collector current (I_C) with respect to current gain (β) keeping the base current (I_B) and reverse saturation current (I_{CO}) as constant.*** Mathematically, the stability factor,

$$S'' = \frac{dI_C}{d\beta}$$

For a voltage divider bias, the stability factor,

$$S'' = \frac{I_C \cdot S}{\beta_1 (\beta_2 + 1)} \qquad ... (ii)$$

where, S = Stability factor (equal to dI_C/dI_{CO}),
β_1 = Current gain at temperature T_1, and
β_2 = Current gain at temperature T_2.

From the expressions (*i*) and (*ii*) above, we find that the stability factors S' and S'' increase as the value of stability factor S is increased. The total change in collector current as a function of all the three stability factors,

$$\Delta I_C = S \cdot \Delta I_{CO} + S' \cdot \Delta V_{BE} + S'' \cdot \Delta\beta$$

This variation may be caused either due to temperature variation or due to transistor replacement. It will be interesting to know that for a silicon transistor, the collector current (I_C) is least dependent on reverse saturation current (I_{CO}). Therefore, the manufacturer specifies the average range of operation for a silicon transistor as – 55°C to + 175°C. However, for a germanium transistor, the effect of variation in the reverse saturation current on collector current is large. Therefore, the manufacturer specifies the average range of operation for a germanium transistor as –55°C to + 75°C.

12-24. Biasing Circuits for PNP Transistors

A PNP transistor is considered as a complement of an NPN transistor. It means that all voltages and currents in a PNP transistor are opposite to those of the NPN transistor. Thus, the d.c. biasing circuit for a PNP transistor may be obtained by the following two actions :

(*i*) Replace the NPN transistor by a PNP transistor.

(*ii*) Reverse all voltages and currents.

A positive V_{CC} supply in a PNP transistor biasing circuit will become the negative supply with positive terminal of the supply grounded. It is true because to forward bias the base-emitter junction of a PNP transistor, the base-emitter voltage (V_{BE}) has a negative value. Similarly, to reverse bias the collector-base junction, the collector-to-base voltage (V_{CB}) has a negative value. Fig. 12-47(*a*), (*b*) and (*c*) shows the d.c. biasing circuits for PNP transistor. Fig. 12-47(*a*) is a collector feedback bias, (*b*) is a emitter feedback bias and (*c*) is a voltage divider bias.

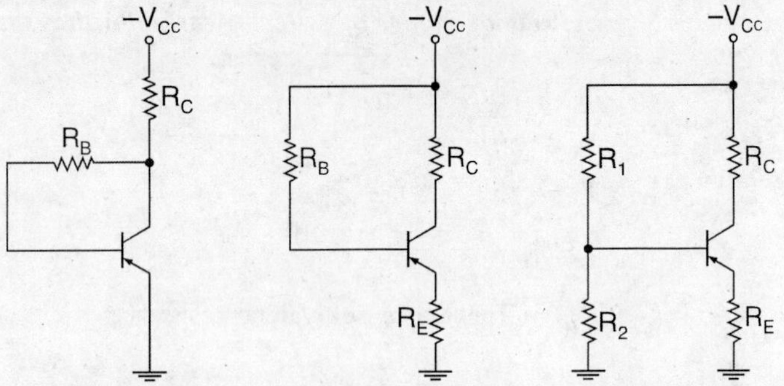

(*a*) Collector feed-back bias. (*b*) Emitter feed-back bias. (*c*) Voltage divider bias.

Fig. 12-47. *Biasing circuits for PNP transistors.*

Sometimes, the power supply to be used is positive. In such a case, the PNP transistors are, usually, drawn upside down as shown in Fig. 12-48(*a*), (*b*) and (*c*). It may be noted that there is no difference in the working of a transistor, if it is oriented in any direction in the space. It is just for convenience that we use this kind of convention.

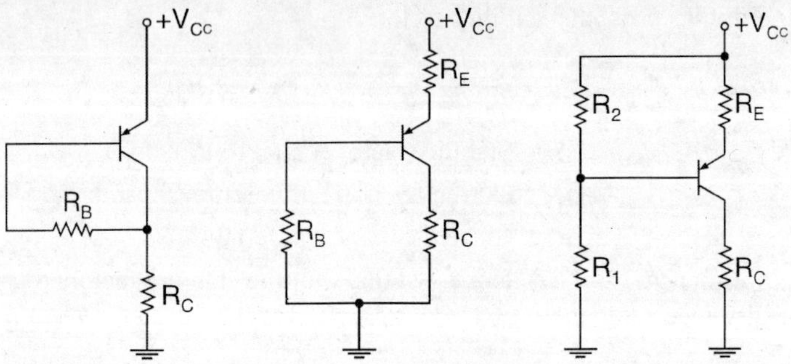

(*a*) Collector feed-back bias. (*b*) Emitter feed-back bias. (*c*) Voltage divider bias.

Fig. 12-48. *Upside down conversion for PNP transistor*

The current and voltages for any PNP transistor biasing circuit may be determined in the same way as for NPN transistor biasing circuits.

Now consider a voltage divider bias circuit using PNP transistor. We know that the voltage at the base of a PNP transistor is given by the relation,

$$V_B = V_{EE}\left(\frac{R_1}{R_1 + R_2}\right)$$

and the voltage at the emitter,

$$V_E = V_B - V_{BE}$$

∴ Value of emitter current,

$$I_E = \frac{V_{EE} - V_E}{R_E}$$

and the value of collector current,

$$I_C = I_E$$

∴ Voltage at the collector,

$$V_C = I_C \cdot R_C$$

and the collector-to-emitter voltage,

$$V_{CE} = V_C - V_E = I_C \cdot R_C - (V_B - V_{BE})$$
$$= -[(V_B - V_{BE}) - I_C \cdot R_C]$$

Note. If a d.c. biasing circuit for PNP transistor is shown using a V_{CC} supply, the expressions for calculating the d.c. biasing currents and voltages are in the same as that of a NPN transistor.

Example 12-27. *Calculate the d.c. bias voltages and currents for a voltage divider bias circuit using PNP transistor. The circuit has V_{CC} = 12 V; R_C = 2 kΩ; R_E = 1 kΩ; R_1 = 100 kΩ; and R_2 = 20 kΩ. The transistor has β = 100 and V_{BE} = – 0.2 V.*

Solution. Given : V_{CC} = 12 volts; R_C = 2 kΩ = 2 × 10³ Ω; R_E = 1 kΩ = 1 × 10³ Ω; R_1 = 100 kΩ; R_2 = 20 kΩ; β = 100 and V_{BE} = – 0.2 volt.

Base voltage

We know that the base voltage,

$$V_B = -V_{CC} \times \frac{R_1}{R_1 + R_2} = -12 \times \frac{20}{100 + 20} = -2 \text{ V } \textbf{Ans.}$$

Emitter voltage

We know that the emitter voltage,

$$V_E = V_B - V_{BE} = -2 - (-0.2) = -1.8 \text{ V } \textbf{Ans.}$$

Collector current

We know that the emitter current,

$$V_E = \frac{V_E}{R_E} = \frac{1.8}{1 \times 10^3} = 1.8 \times 10^{-3} \text{ A} = 1.8 \text{ mA}$$

and the collector current,

$$I_C = I_E = 1.8 \times 10^{-3} \text{ A} = 1.8 \text{ mA } \textbf{Ans.}$$

Collector voltage

We know that the collector voltage,

$$V_C = -[V_{CC} - I_C \cdot R_C] = -[12 - (1.8 \times 10^{-3}) \times (2 \times 10^3) \text{ V}]$$
$$= -(12 - 3.6) = -8.4 \text{ V } \textbf{Ans.}$$

Collector-to-emitter voltage

We know that the collector-to-emitter voltage,

$$V_{CE} = V_C - V_E = -8.4 - (-1.8) = -6.6 \text{ V } \textbf{Ans.}$$

Example 12-28. *A P-N-P Ge transistor is used in the self-biasing arrangement shown in Fig. 12-49. The circuit component values are V_{CC} = 4.5 V; R_C = 1.5 kΩ; R_E = 0.27 kΩ; R_2 = 2.7 kΩ and R_1 = 27 kΩ. If β = 44, find the quiescent points.* (*Nagpur Univ. Winter* 2003)

Solution. Given : V_{CC} = 4.5 V; R_C = 1.5 kΩ; R_E = 0.27 kΩ; R_2 = 2.7 kΩ; R_1 = 27 kΩ; β = 44; V_{BE} for a Ge transistor = 0.3 V.

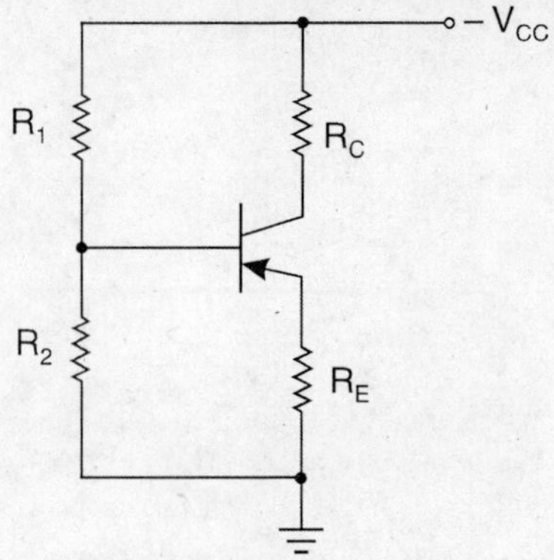

Fig. 12-49.

Quiescent Point

We need to find out the values of collector current (I_C) and collector-emitter voltage (V_{CE}).

We know that base voltage,

$$V_B = -V_{CE} \times \frac{R_2}{R_1 + R_2} = 4.5 \times \frac{2.7\text{ k}}{27\text{ k} + 2.7\text{ k}} = -0.409\text{ V}$$

and the voltage at the emitter

$$V_E = V_B - V_{BE} = -0.409 - (-0.3) = -0.109\text{ V}$$

∴ Value of emitter current,

$$I_E = \frac{V_E}{R_E} = \frac{0.109}{0.27\text{ k}\Omega} = 0.404\text{ mA}$$

and the collector current,

$$I_C \approx I_E = 0.404\text{ mA}$$

We also know that voltage across the collector resistor,

$$V_{R_C} = I_C \cdot R_C = 0.404\text{ mA} \times 1.5\text{ k} = 0.606\text{ V}$$

and the voltatge at the collector (measured with respect to the ground),

$$V_C = -[V_{CC} - V_{R_C}] = -[4.5 - 0.606] = -3.894\text{ V}$$

∴ The voltage from collector to emitter,

$$V_{CE} = -V_C - (-V_E) = -3.894 - (-0.109) = -3.785\text{ V.}$$

∴ The operating points values are,

$$I_C = 0.404 \text{ and } V_{CE} = -3.785\text{ V} \quad \textbf{Ans.}$$

12-25. Bias Compensation

The d.c. biasing circuits, which we have discussed in the previous articles, use a *negative feedback* to stabilize the operating point. The negative feedback means that a fraction of the collector voltage (or collector current) is supplied back at the input of the transistor. We know that the use of negative feedback reduces the gain of amplifier. However, in some cases, it is not tolerable that the gain of an amplifier should reduce. In such cases, we use *compensation techniques* to stabilize the *Q*-point instead of d.c. biasing circuits. The compensation technique is the use of temperature sensitive devices such as diodes, thermistors, transistors, *etc*. Sometimes, a combination of the biasing circuits and temperatures sensitive devices is also used.

Fig. 12-50 (*a*) shows a diode of the same material as that of the transistor for providing bias compensation. The diode is reverse biased by the base-emitter junction voltage. It allows the reverse saturation current or leakage current (I_0) to flow through it. When the temperature increases, the reverse saturation current (I_{CO}) through the transistor also increases. This, in turn, increases the leakage current through the diode, which will reduce the base current by keeping the current (I) constant. This action is required to keep constant value of collector current. It may be noted that this method of compensation does not need a change in collector current to effect the change in base current. The collector current is dependent only upon the temperature.

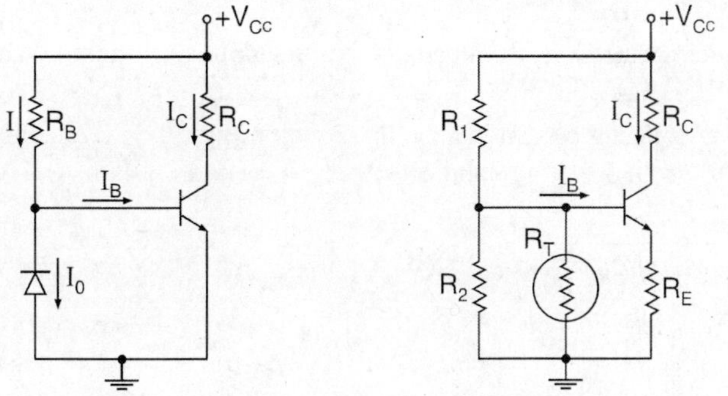

(*a*) Use of a diode. (*b*) Use of a thermistor.

Fig. 12-50. *Bias compensation.*

Fig. 12-50 (*b*) shows a thermistor (R_T) for providing bias compensation. The thermistor has a *negative temperature coefficient, i.e.*, its resistance decreases with the increase in temperature. It is placed in parallel with the resistance R_2. The resistance values are selected in such a way so that a correct bias is applied at the normal temperature. As the temperature increases, the drop across thermistor decreases. This, in turn, reduces the forward bias across the base emitter junction, which reduces the base current and hence the collector current.

NUMERICAL PROBLEMS

1. Find the collector current and collector-to-emitter voltage for the circuit shown in Fig. 12-50. (**Ans.** 1.5 mA and 6 V)

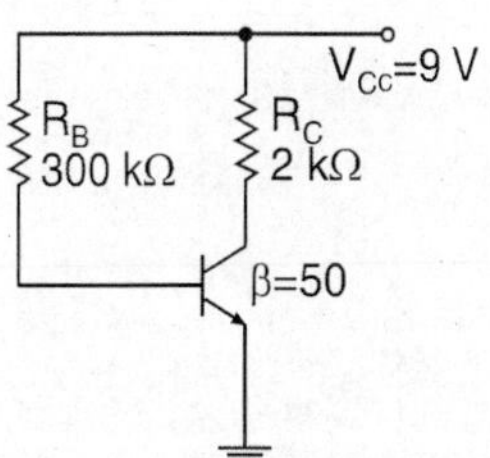

Fig. 12-51.

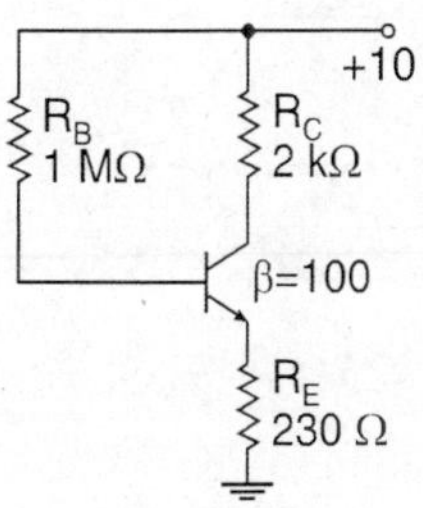

Fig. 12-52.

2. Find the d.c. load line for problem 1 and also locate the Q-point on the d.c load line. (**Ans.** Upper end: 4.5 mA; Lower end: 9 V)
3. Calculate the values of emitter, base and collector currents for the circuit shown in Fig. 12-52. (**Ans.** 0.909 mA; 9.09 μA; 0.909 mA)
4. For the collector feed back bias circuit shown in Fig. 12-53, find $I_{C\,(sat)}$, V_{CE} and stability factor (S). Take $\beta = 100$ and neglect V_{BE}. (**Ans.** 1 mA 5 V, 34.1)

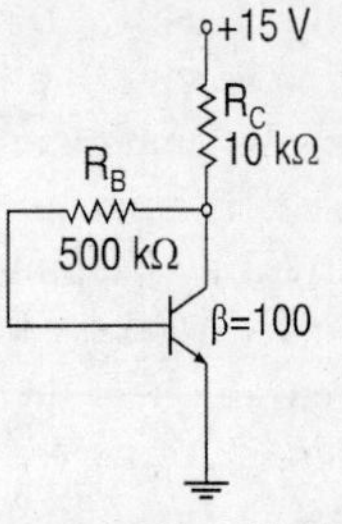

Fig. 12-53.

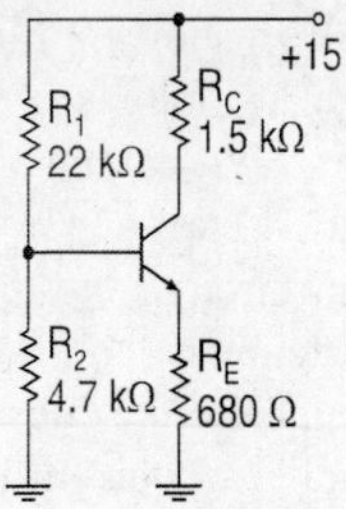

Fig. 12-54.

5. Find collector current and collector-to-emitter voltage for the voltage divider bias circuit shown in Fig. 12-54. (**Ans.** 2.75 mA and 9 V)

6. What is the collector current in a circuit shown in Fig. 12-55 ? Also determine the following voltages to ground : V_B , V_E and V_C. (**Ans.** 1.43 mA; 5 V; 4.3 V and 17.14 V)

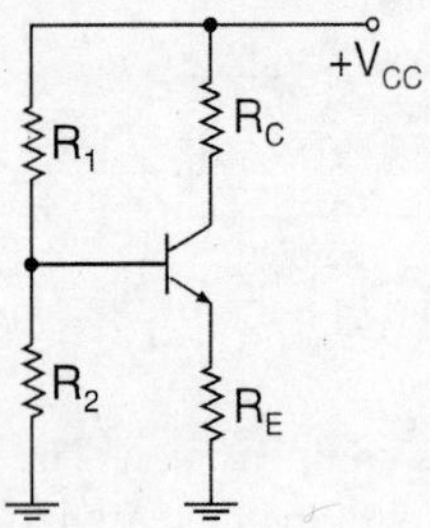

Fig. 12-55.

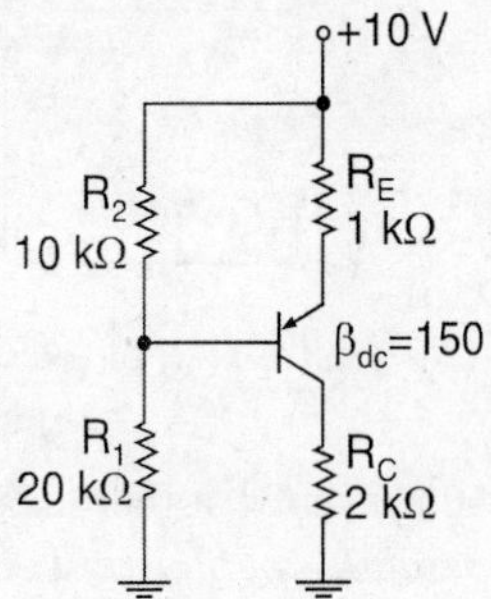

Fig. 12-56.

7. Consider a common emitter NPN circuit with voltage divider bias as shown in Fig. 12-56. Compute the percentage change in the collector current (I_C), if h_{FE} of the transistor varies from 50 and 200 in this circuit. Neglect V_{BE} and I_{CBO}. (**Ans.** 5.7%)

8. Determine the values of resistances for the circuit shown in Fig. 12-57 such that $I_C = 5$ mA; $V_{CE} = 6$ V; $V_C = 8$ V; $S = 10$; $h_{FE} = 200$ and $V_{CC} = 20$ volts. (**Ans.** $R_C = 2.4$ kΩ; $R_E = 400$ Ω; $R_1 = 4.2$ kΩ and $R_2 = 2$ kΩ)

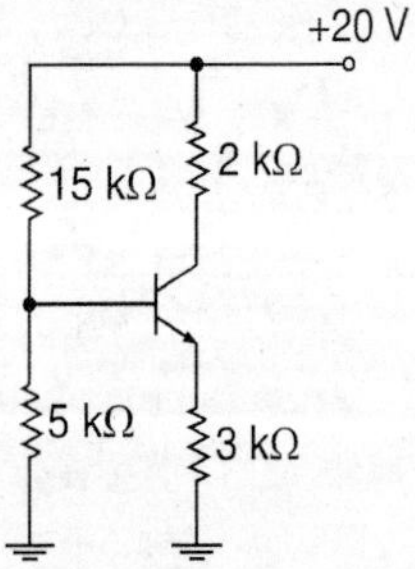

Fig. 12-57.

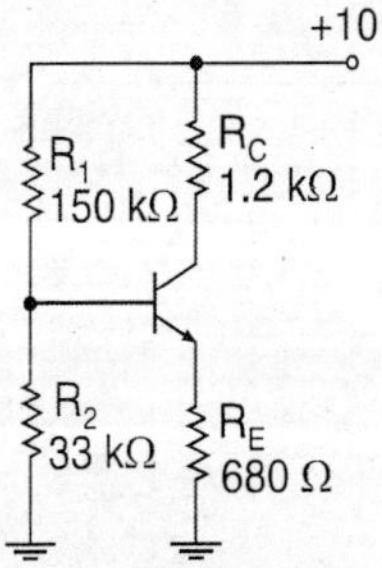

Fig. 12-58.

9. Find the Q-point values (*i.e.*, the values of collector current and collector-to-emitter voltage) for the voltage divider bias circuit shown in Fig. 12-58. Also draw a load line and locate Q-point on it. (**Ans.** $I_C = 2.63$ mA and $V_{CE} = -2.11$ V)

10. Calculate the d.c. bias currents and voltages for a voltage divider bias circuits using *PNP* transistor. The circuit has $R_C = 1.5$ kΩ; $R_E = 2$ kΩ; $R_1 = 40$ kΩ; $R_2 = 10$ kΩ and $V_{EE} = 10$ V. The transistor has $\beta = 100$ and $V_{BE} = 0.7$ V.

(**Ans.** $V_B = 8$ V; $V_E = 8.7$ V; $I_C = 0.65$ mA; $V_C = 0.975$ V; $V_{CE} = -7.725$ V)

SUMMARY

In this chapter you have learned that :

1. Biasing deals with setting a fixed level of the collector current, which should flow through the transistor with a desired voltage drop across the transistor junctions.
2. D.C. load line for a transistor can be drawn, if the supply voltage and collector resistance is known.
3. The dipping of the output signal can be eliminated by choosing the operating point to be at the middle of the d.c. load line.
4. The stability factor of common base circuits is 1 and therefore it requires no bias stabilization.
5. The stability factor of common emitter circuits depend upon the current gain, β of the transistor and is given by S = (1 + β).
6. There are several methods of transistor biasing. This includes base bias, emitter feedback bias collector feedback bias, voltage divider bias and emitter bias.
7. Bias compensation is achieved by using temperature sensitive devices such as diodes, thermistors, transistors etc.

GLOSSARY

Active region : The region lying between the saturation and cut-off points of the load line.
Base bias with collector feed back : A transistor biasing circuit in which, base resistor R_B is connected between base and collector to act as a feedback resistor.
Base bias with emitter feedback : A transistor biasing circuit in which, emitter resistor is used to control base current acting as a feedback.
Base biasing : A transistor biasing circuit, in which the base is biased through base resistor by V_{CC} or V_{BB}.
D.C. load line : A straight line obtained by joining two operating points of a transistor for a given supply voltage and collector resistance.
Emitter bias : A transistor biasing circuit in which emitter resistor is connected to negative supply voltage while base resistor is grounded.
Operating point : The fixed values of certain currents and voltages in a transistor that defines a point at which it operates.
Quiescent point (Q-point) : Refer operating point.
Saturation point : The upper end of the d.c. load line where collector-emitter voltage is very small.
Stability factor (S) : Rate of change of collector current with respect to the reverse saturation current keeping the common-emitter current gain and base current as constant.
Stability factor S′ : The rate of change of collector current with respect to the base current keeping the common-emitter current gain and reverse saturation current as constant.
Stability factor S″ : The rate of change of collector current with respect to current gain keeping the base current and reverse saturation current as constant.
Transistor biasing : Setting a fixed level of current, which should flow through the transistor with a desired fixed voltage drop across the transistor junctions.
Transistor cut-off point : The lower end of the d.c. load line, where collector-emitter voltage is equal to supply voltage.
Voltage divider bias : A transistor biasing circuit, in which the biasing voltage for base is supplied by a voltage divider resistor circuit.

DESCRIPTIVE QUESTIONS

1. Briefly explain the reasons for keeping the operating point of a transistor as fixed.
2. Explain why common emitter amplifier require a form of d.c. stabilization, whereas common base amplifiers are usually unstabilized.
3. Can the value of stability factor be less than unity ? Explain it briefly.
4. Name different types of biasing circuits and give three circuit configurations.
5. Compare the relative stability of **(*a*)** emitter bias and fixed bias circuit, **(*b*)** emitter bias and voltage divider bias circuits.
6. Derive an expression for the stability factor of a fixed bias current.
7. Draw a voltage divider bias circuit and derive an expression for its stability factor.
8. List the three sources of instability of collector current and hence define the three stability factors.
9. How would you provide temperature compensation for the variations of V_{BE} and stabilization of the operating point ?
10. Discuss the cause for bias instability in a transistor. (*Electronic Circuits, VTU June/July 2008*)
11. With neat circuit diagram, explain how compensation for V_{BE} can be obtained using diode in emitter circuit of a transistor. (*Electronic Circuits, VTU June/July 2008*)
12. Discuss the biasing techniques used for linear interated circuits. (*Electronic Circuits, VTU June/July 2008*)
13. Define three stability factors.(*Electronic Circuits - I, Anna University, April/May 2008*)
14. Define quiescent point. (*Electronic Circuits - I, Anna University, April/May 2008*)
15. What is bias compensation using thermistor? (*Electronic Circuits - I, Anna University, April/May 2008*)
16. Define stability factor. Derive and explain the condition to avoid thermal runaway. (*Electronic Circuits - I, Anna University, April/May 2008*)
17. Write short notes on temperature compensation methods. (*Electronic Circuits - I, Anna University, April/May 2008*)
18. Why biasing is necessary n BJT amplifier and explain the concept of DC load line with neat diagram. (*Electronic Circuits - I, Anna University, Nov/Dec 2007*)

MULTIPLE CHOICE QUESTIONS

1. The d.c. load line of a transistor circuit
   (*a*) is a graph between I_C and V_{CE} (*b*) is a graph between I_C and I_B
   (*c*) does not contain the locating point (*d*) is a curved line
2. The positive part of the output signal in a transistor circuit starts clipping, if *Q*-point of the circuit moves
   (*a*) towards the saturation point (*b*) towards the cut-off point
   (*c*) towards the centre of the load line (*d*) none of the above
3. The negative part of the output signal in a transistor circuit starts clipping, if *Q*-point of the circuit moves
   (*a*) towards the saturation point (*b*) towards the cut-off point
   (*c*) towards the centre of the load line (*d*) none of the above

4. The voltage divider biasing circuit is used in amplifiers quite often because it

(*a*) limits the a.c. signal going to the base

(*b*) makes the operating point almost independent of β

(*c*) reduces the d.c. base current (*d*) reduces the cost of the circuit

5. Which of the following components are used for bias compensation in transistor circuits

(*a*) resistors (*b*) rectifier diodes

(*c*) thermistors (*d*) both (*b*) and (*c*) above

6. To avoid thermal run away in the design of analog circuit, the operating point of the BJT should be such that it satisfies the condition

(*a*) $V_{CE} = 1/2\ V_{CE}$ (*b*) $V_{CE} = \leq 1/2\ V_{CC}$

(*c*) $V_{CE} > 1/2\ V_{CC}$ (*d*) $V_{CE} \leq 0.78\ V_{CC}$

7. In the case of a BJT amplifier, bias stability is achieved by

(*a*) keeping the base current constant

(*b*) changing the base current in order to keep the I_C and V_{CB} constant

(*c*) keeping the temperature constant

(*d*) keeping the temperature and the base current constant.

8. For a transistor amplifier with self-biasing network, the following components are used : $R_1 = 4\ k\Omega$, $R_2 = 4\ k\Omega$ and $R_E = 1\ k\Omega$ the approximate value of the stability factor 'S' will be

(*a*) 4 (*b*) 3 (*c*) 2 (*d*) 1.5

9. A transistor circuit employing base bias with collector feedback has greater stability than the one without feedback because

(*a*) I_C decreasc in magnitude (*b*) V_{BE} is decreased

(*c*) of negative feedback effect (*d*) I_C becomes independent of β.

10. The universal bias stabilization circuit is most popular because

(*a*) I_C does not depend on transistor characteristics

(*b*) its b-sensitive is high

(*c*) voltage divider is heavily loaded by transistor base

(*d*) I_C equals to I_E.

11. Improper biasing of a transistor circuit leads to

(*a*) excessive heat production at collector terrminal

(*b*) distortion in output signal

(*c*) faulty location of load line

(*d*) heavy loading of emitter terminal.

12. The negative output swing in a transistor circuit starts clipping first when *Q*-point

(*a*) has optimum value (*b*) is near saturation point

(*c*) is near cut-off point (*d*) is in the active region of the load line.

13. When a *BJT* is employed as an amplifier, it operates

(*a*) in cut-off (*b*) in saturation

(*c*) well into saturation (*d*) over the active region.

14. Which of the following method used for biasing a *BJT* in integrated circuits is considered independent of transistor beta ?

(*a*) fixed biasing (*b*) voltage divider bias

(*c*) collector feedback bias (*d*) base bias with collector feedback.

15. The voltage V_0 of the circuit shown in Fig. 12-59 is,

(*a*) 5.1 V (*b*) 3.1 V

(*c*) 2.5 V (*d*) zero

16. The collector voltage V_C of the circuit

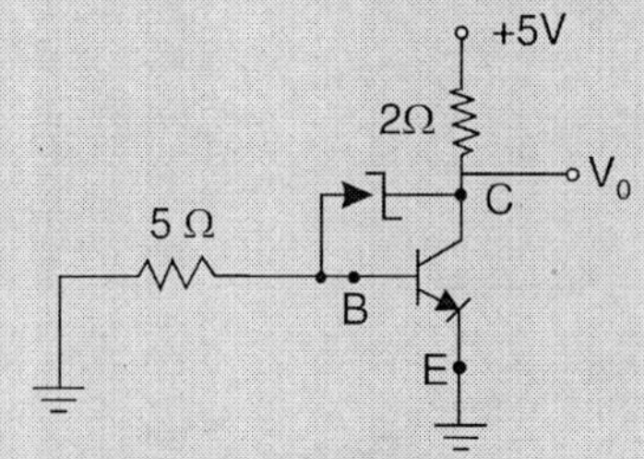

Fig. 12-59.

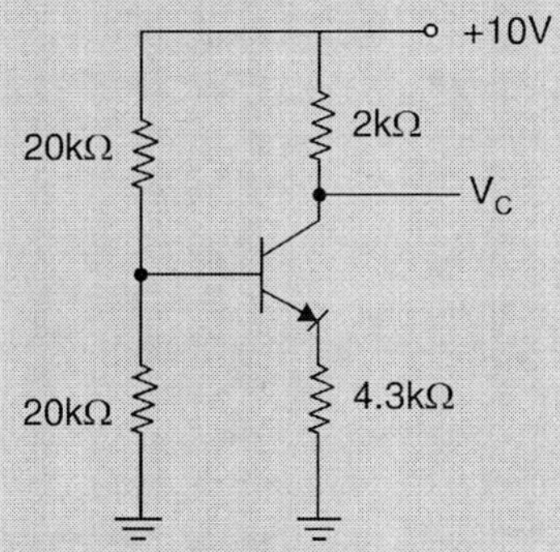

Fig. 12-60.

shown in Fig. 12-60, is approximately

(*a*) 2 V (*b*) 4.6 V

(*c*) 8 V (*d*) 8.6 V

17. In the circuit shown in Fig. 12-61. The transistor is biased at :

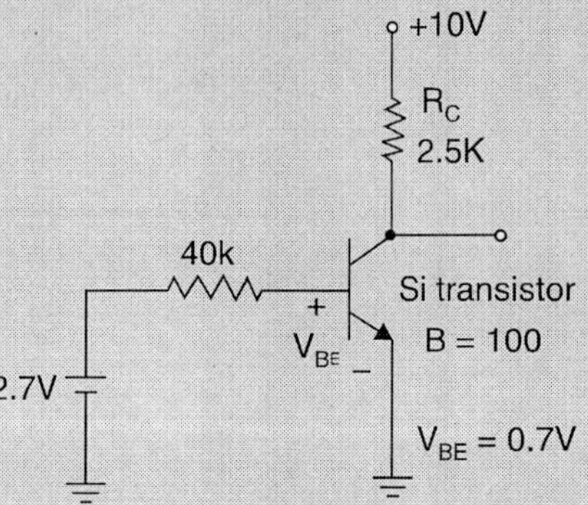

Fig. 12-61.

(*a*) 0 mA (*b*) 5 mA (*c*) 3.9 mA (*d*) ∞

(*UPSC Engg. Services* 2002)

Hint. $[\,I_B = \dfrac{2.7 - 0.7}{40\text{ k}} = 0.05\text{ mA}\,;\; I_C = \beta I_B = 100 \times 0.05 = 5\text{ mA}\,]$

18. Assume $V_{BE} = 0.7$ V and $\beta = 50$ for the transistor shown in Fig. 12-62.

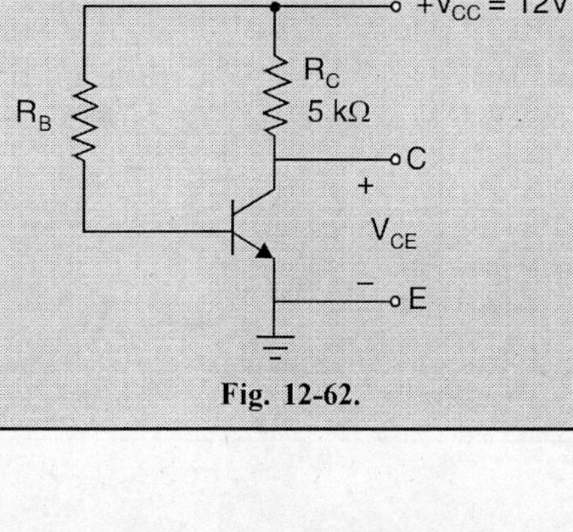

Fig. 12-62.

For $V_{CE} = 2$ V, the value of R_B is :

(*a*) 200 kΩ (*a*) 242 kΩ

(*c*) 283 kΩ (*d*) 300 kΩ

Hint. [$V_{CE} = V_{CC} - I_C \cdot R_C \Rightarrow 2\text{ V} = 12 - I_C = 2$ mA

$\therefore \quad I_B = I_C/\beta = 2\text{ mA}/50 = 0.04$ mA;

$$R_B = \frac{12 - 0.7}{0.04\text{ mA}} = 282.5\text{ k}\Omega \approx 283\text{ k}\Omega\ ]$$

19. The collector voltage V_C of the circuit shown in Fig. 12-63, is approximately,

(*a*) 2 V (*b*) 4.6 V (*c*) 8 V (*d*) 8.6 V

(*UPSC Engg. Services* 1999)

Hint. [$V_B = 10 \times \dfrac{20\text{ k}}{20\text{ k} + 20\text{ k}} = 5\text{ V}; V_E = V_B - 0.7 = 5 - 0.7 = 4.3$ V

$$I_E = \frac{V_E}{R_E} = \frac{4.3}{4.3\text{ k}} = 1\text{ mA } I_C \approx I_E = 1\text{ mA};$$

$V_C = V_{CC} - I_C \cdot R_C = 10 - (1\text{mA} \times 2\text{k}) = 8$ V]

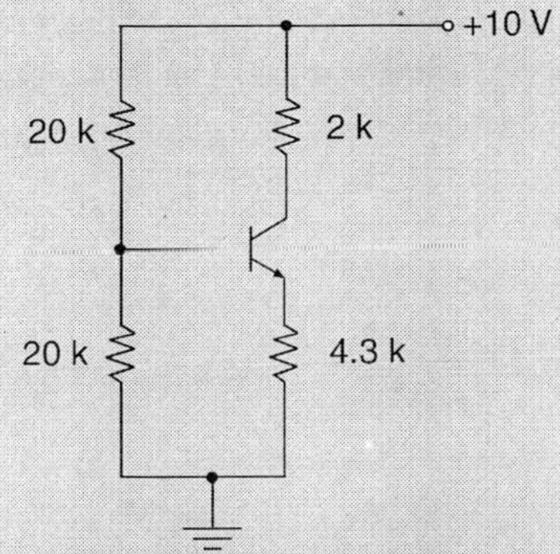

Fig. 12-63.

20. In the circuit shown in Fig. 12-64, the approximate voltages at the transistor,

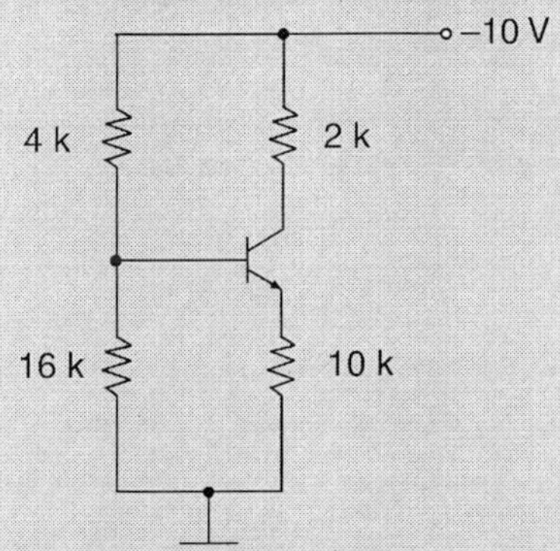

Fig. 12-64.

(*a*) base and emitter are – 8 V and – 7.3 V

(*b*) base and collector respectively are – 8 V and – 7.3 V
(*c*) collector and emitter respectively are – 5 V and – 7.3 V
(*d*) base, emitter and collector respectively are – 8 V, – 7.3 V and – 5 V.

(*UPSC Engg. Services* 2000)

Hint. [$V_B = -10 \times \dfrac{16\text{ k}}{4\text{ k} + 16\text{ k}} = -8\text{V}; V_E = V_B - 0.7 = -7.3\text{V}$

$I_E = \dfrac{V_E}{R_E} = \dfrac{7.3}{10\text{ k}} = 0.73\text{ mA } I_C \approx I_E = 0.73\text{ mA};$

$\therefore V_C = V_{CC} - I_C \cdot R_C = -10 - (0.73\text{ mA} \times 2\text{k}) = -8.54\text{ V}$]

21. A circuit using BJT is shown in Fig. 12-65. The value of β is

(*a*) 120 (*b*) 150 (*c*) 165 (*d*) 186

Hint. [$I_E = \dfrac{-V_{BE} - V_{EE}}{R_E} = \dfrac{-0.7-(-10)}{5\text{ k}} = 1.86\text{ mA}, I_C \approx I_B = 1.86\text{ mA}$

$\beta = I_C/I_B = 1.86/0.01 = 1.86$ **Ans.**]

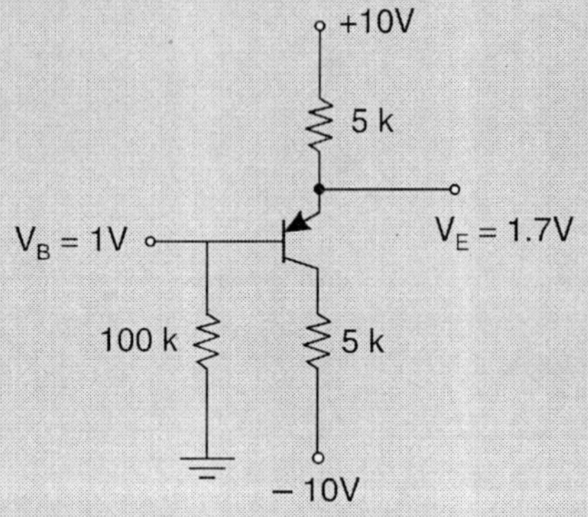

Fig. 12-65.

22. Thermal runway will take place if the quiescent point is such that

(*a*) $V_{CE} > \dfrac{1}{2} V_{CC}$ (*b*) $V_{CE} < V_{CC}$

(*c*) $V_{CE} < 2\,V_{CC}$ (*d*) $V_{CE} < \dfrac{1}{2} V_{CC}$

(*UPSC Engg. Services* 1999)

ANSWERS

1. (*a*)	**2.** (*b*)	**3.** (*a*)	**4.** (*b*)	**5.** (*d*)
6. (*c*)	**7.** (*a*)	**8.** (*b*)	**9.** (*b*)	**10.** (*d*)
11. (*b*)	**12.** (*b*)	**13.** (*d*)	**14.** (*b*)	**15.** (*d*)
16. (*a*)	**17.** (*b*)	**18.** (*c*)	**19.** (*c*)	**20.** (*a*)
21. (*d*)	**22.** (*d*)			

❑❑❑

13

CHAPTER

Field Effect Transistors

OUTLINE

Objectives

After completing this chapter, you should be able to :

- Understand the constrution and operation of junction field-effect transistors (JFETs).
- Know the characteristics of JFET namely drain and transfer characteristics.
- Describe the various parameters of JFET.
- Distinguish between field-effect transistor and bipolar junction transistor.
- Understand the two types of MOSFETs : depletion-type and enhancement-type.
- Sketch the characteristics of depletion-type MOSFET and enhancement-type MOSFET.
- Understand the advantages of N-channel MOSFETs over P-channel MOSFETs.

13-1. Introduction

The field-effect transistor (FET) is a three terminal device used for a variety of applications that match, to a large extent, those of the BJT described in Chapter-8. Although there are lots of differences between the two types of devices, there are also many similarities that will be pointed out later.

The primary difference between the two types of transistors is highlighted in Fig.13-1. It may be carefully noted that the BJT transistor is a *current-controlled* device Fig. 13-1(*a*), while the JFET transistor is a *voltage-controlled* device Fig. 13-1(*b*). In other words, the current I_C in Fig. 13-1(*a*) is a direct function of the level of I_B. For the FET, the current *I* will be a function of the voltage VGS applied to the input circuit as shown in Fig. 13-1(*b*). In each case the current of the output circuit is being controlled by a parameter of the input circuit-in one case a current level and in the other an applied voltage.

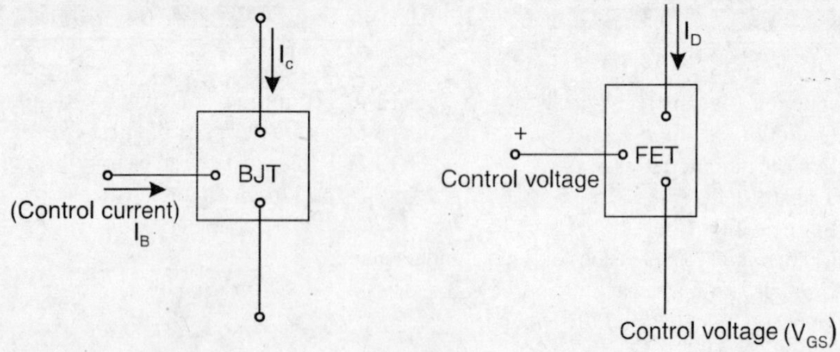

Fig. 13-1.

Field effect transistors offer many attractive features for applications in analog switching, high input impedance amplifiers, microwave amplifications and integrated circuits, The FETs have considerably higher input impedance than bipolar transistors which allows the input of a FET to be more readily matched to the standard microwave system. The FET has a negative temperature at high current lives; that is the current decreases as temperature increases. This characteristic leads to a more uniform temperature distribution over the device area and prevents the FET from thermal runaway or second breakdown which occurs in the bipolar transistor. The device is thermally stable, even when the active area is large or when many devices are connected in parallel.

Because FETs are unipolar devices, they do not suffer from minority carrier storage effects and consequently have higher switching speeds and higher cut-off frequencies. In addition the devices are basically square-law or linear devices; inter-modulation and cross-modulation products are much smaller.

13-2. Types of Field-Effect Transistor

Broadly speaking there are two main types of field-effect transistors :

(*i*) Junction field-effect transistor (JFET).

(*ii*) Metal oxide semiconductor field-effect transistor (MOSFET)

Both of these types of FET's can be further subdivided as shown in Fig. 13-2.

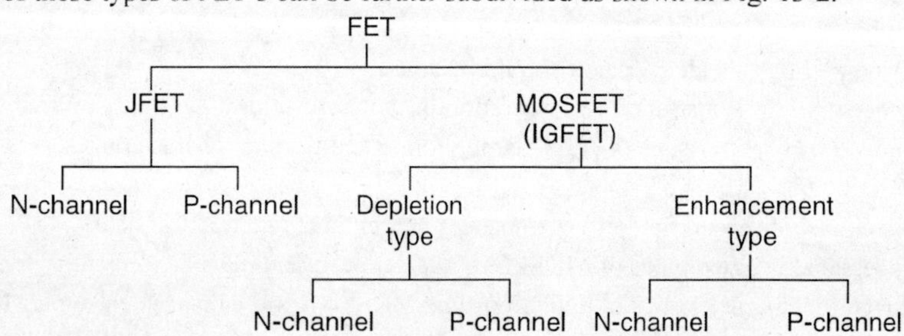

Fig. 13-2.

Now we shall discuss the construction, operation and characteristic of all the above mentioned field-effect transistors in the following pages.

13-3. Junction Field-Effect Transistor

The junction field-effect transistors (JFET's) can be divided depending upon their structure into the following two categories :

(*i*) N-channel JFET and **(*ii*)** P-channel JFET.

The basic construction of an N-channel JFET is as shown in Fig. 13-3(*a*). It consists of an N-type semiconductor bar with two P-type heavily doped regions diffused on opposite sides of its middle part. The P-type regions form two PN junctions. The space between the junctions (*i.e.*, N-type region) is called a *channel*. Both the P-type regions are connected internally and a single wire is taken out in the form of a terminal called the *gate* (*G*). The electrical connections (called ohmic contacts) are made to both ends of the N-type semiconductor and are taken out in the form of two terminals called *drain* (*D*) and *source* (*S*). The drain (*D*) is a terminal through which electrons leave the semiconductor bar and source (*S*) is a terminal through which the electrons enter the semiconductor.

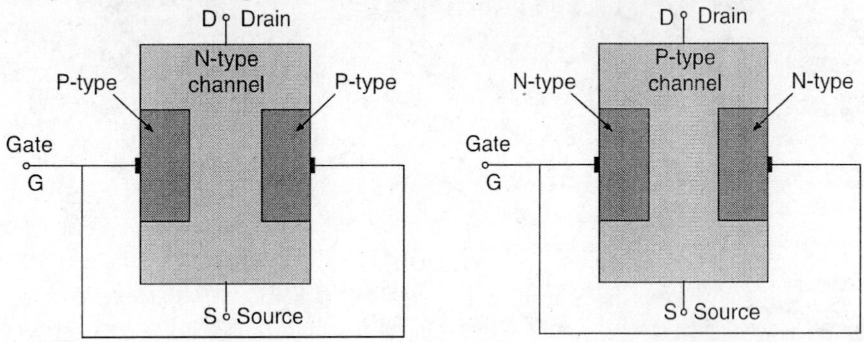

(*a*) N-channel. (*b*) P-channel.

Fig. 13-3. *JFET's.*

Whenever a voltage is applied across the drain and source terminals, a current flows through the N-channel. The current consists of only one type of carriers (*i.e.*, electrons) Therefore, the field-effect transistor (FET) is called a *unipolar device*. This distinguishes a FET from a BJT (*i.e.*, a bipolar junction transistor) where the current consists of the flow of both the electrons and holes.

A P-channel JFET is shown in Fig. 13-4(*b*). Its construction is similar to that of N-channel JFET, except that it consists of a P-channel and N-type regions. The current carriers in JFET are the holes, which flow through the P-type channel. Fig. 13-4(*a*) shows the schematic symbol for a N-channel JFET. Similarly, Fig. 13-4(*b*) shows the schematic symbol for a P-channel JFET. In an N-channel JFET, the arrow points towards the vertical line. The vertical line represents the N-channel. On the other hand, in a P-channel symbol, the arrow points away from the vertical line. Here, the vertical line represents the P-channel.

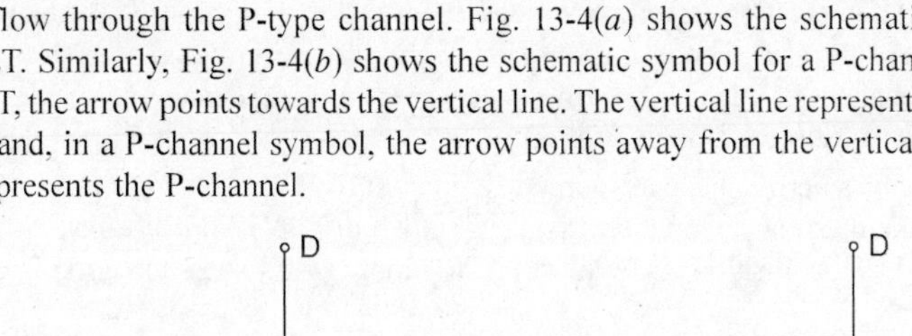

(*a*) N-Channel. (*a*) P-Channel.

Fig. 13-4. *Symbols for JFET's.*

Now we shall discuss the operation and characteristics of JFET's with reference to N-channel JFET. The operation of a P-channel JFET is similar to N-channel except that all voltages and currents are reversed.

13-4. Formation of Depletion Region in JFET

Consider an N-channel JFET as shown in Fig. 13-5(*a*). Here the P-type gate and N-type channel constitute PN junction. This PN junction is always reverse biased in JFET operation. The reverse bias is applied by a voltage V_{GG} connected between the gate and source terminals as shown in Fig. 13-5(*b*). It may be noted that positive terminal of the voltage (V_{GG}) is connected to source (*S*) and negative terminal to gate (*G*).

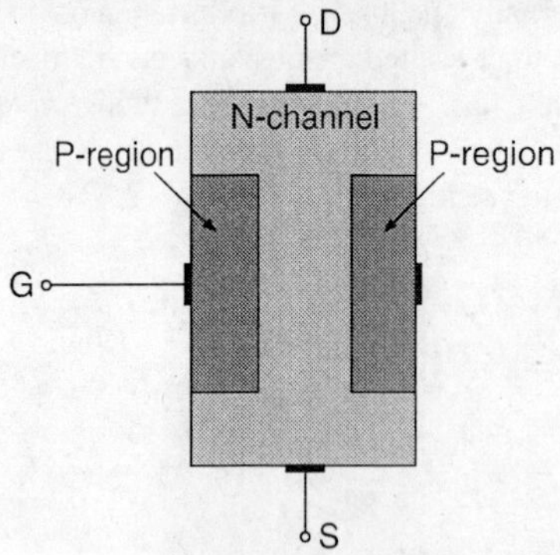

(*a*) N-channel JFET.

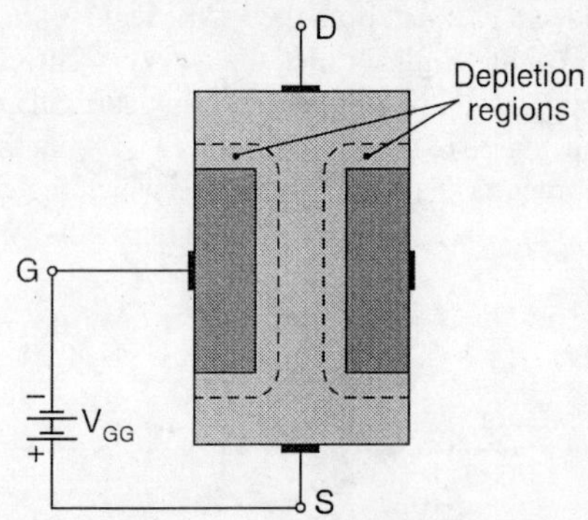

(*b*) Depletion regions in JFET.

Fig. 13-5.

We know that whenever a PN junction is reverse biased, the electrons and holes diffuse across the junction and leave behind the positive ions on N-side and negative ions on P-side. The region, containing these immobile ions, is known as depletion region. As the reverse bias voltage across the junction is increased, thickness of the depletion region is also increased. If both the P-side and N-side of the junctions are equally doped, the depletion region will extend equally in both the regions. However, if one side of the junction is heavily doped, as compared to the other side, the depletion region extends more into the region of lower doping.

A similar action takes place in JFET. Since, the *P-region of a N-channel JFET is heavily doped, as compared to the N-channel, the depletion region extends less into the P-region and deeper into the N-channel* as shown in Fig. 13-5(*b*). Incidentally, when there is no applied voltage between the gate and source, the depletion region is symmetrical around the junction. The conductivity of depletion region is zero because there are no mobile charge carries in this region. Hence, the effective width of the N-channel is reduced. It further reduces with the increased reverse-bias voltage applied across the gate and source terminals of the JFET.

The reverse-bias across the gate source junction of a JFET may also be achieved by applying a voltage across the drain and source terminals as shown in Fig. 13-6(*a*). It may be noted that drain (*D*) is connected to the position terminal of the dc supply (V_{DD}) and source (*S*) is connected to the negative terminal.

Now we shall discuss as to how the drain-to-source voltage produces a reverse-bias across the gate-source junction in the same way as the gate-to-source voltage. In order to do so, we represent the channel resistances r_a and r_b as shown in the Fig. 13-6(*a*). These resistances are shown variable, as their values depends upon the magnitude of the drain-to-source voltage (V_{DS}) and gate-to-source voltage (V_{GS}). In the presence of positive supply voltage V_{DD} (with gate open), the electrons flow from source to drain through the N-channel and constitutes a current known as *drain current* (I_D). The conventional direction of drain current is indicated from drain-to-source through the device.

The drain current causes a voltage drop across the resistance r_b, which has the effect of reverse biasing the gate-to-source junction. Thus, even if the gate is open, the gate-to-source junction is reverse biased by the drain-to-source voltage. It creates depletion regions within the channel as shown in Fig. 13-6(*b*).

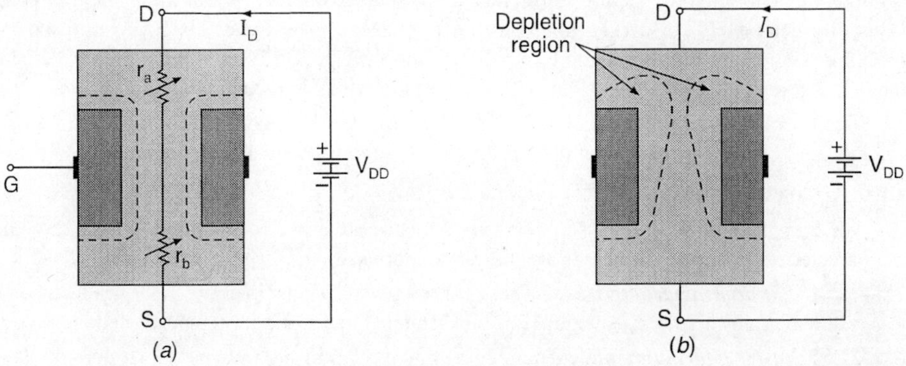

Fig. 13-6. *Effect of drain-to-source supply.*

It may be noted from Fig. 13-6(*b*) that depletion region is not symmetrical around the gate-to-source junction. As a matter of fact, it is extended deeper into the channel near the drain terminal and less on the source terminal. It is because of the fact, *that voltage drop across r_a is greater than that across r_b. Due to this, reverse bias voltage is higher near the drain end of the channel as compared to the source end.*

13-5. Operation of JFET

We have already discussed in the last article the application of negative gate voltage or positive drain voltage, with respect to source, reverse biases the gate-source junction of an N-channel JFET. The effect of reverse bias voltage is to form depletion regions within the channel. When a voltage is applied between the drain and source with a d.c. supply voltage (V_{DD},) the electrons flows from source to drain through the narrow channel existing between the depletion regions. This constitutes the drain current (I_D) and its conventional direction is indicated from drain-to-source. The value of drain current is maximum, when no external voltage is applied between the gate and source and is designated by the symbol I_{DSS}.

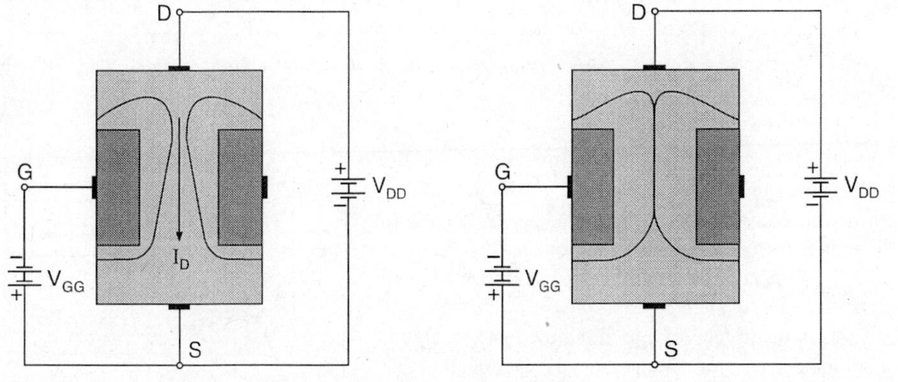

Fig. 13-7. *Operation of JFET.*

When the gate-to-source voltage (V_{GS}), if applied by a d.c. supply, (V_{GG})and increased above zero as shown in Fig. 13-7(*a*), the reverse-bias voltage across the gate-source junction is increased.

As a result of this, the depletion regions are widened. This reduces the effective width of the channel and therefore controls the flow of drain current through the channel. When the gate-to-source voltage (V_{GS}) is increased further, a stage is reached at which two depletion regions touch each other as shown in Fig. 13-7 (*b*). At this gate-to-source voltage, the channel is completely blocked or pinched off and drain current is reduced to zero. The gate-to-source voltage (V_{GS}) at which the drain current is zero (or completely cut-off) is called pinch off voltage. It is designated by the symbol V_P or V_{GS} (off). The value of pinch off voltage V_P is negative for N-channel JFET's. It depends on (*i*) doping of the N and P regions of the device, and (*ii*) width of the original channel structure.

The operation of P-channel JFET is exactly similar to N-channel JFET, except that current carriers are holes and polarities of the d.c. supply, voltage V_{GG} and V_{DD} are reversed.

13-6. Characteristics of JFET

We know that a family (or a set) of curves which relate device current and voltages are known as characteristic curves. Followings are the two important characteristics of a JFET:

(1) ***V-I or drain characteristics :*** These curves give relationship between the drain current (I_D) and drain-to-source voltage (V_{DS}) for different values of gate-to-source voltage (V_{GS}).

(2) ***Transfer characteristics :*** These curves give relationship between drain current (I_D) and gate-to-source voltage (V_{GS}) for different values of drain-to-source (V_{DS}) voltage.

The drain and transfer characteristics of JFET may be obtained by using an N-channel JFET connected in the common source mode as shown in Fig. 13-8. Here the potentiometers R_1 and R_2 are used to vary the voltages V_{GS} and V_{DS} respectively. The voltages V_{DS} and V_{GS} may be measured by the voltmeters connected across the JFET terminals. The drain current (I_D) can be measured by the milliammeter connected in series with the JFET and the supply voltage V_{DD}.

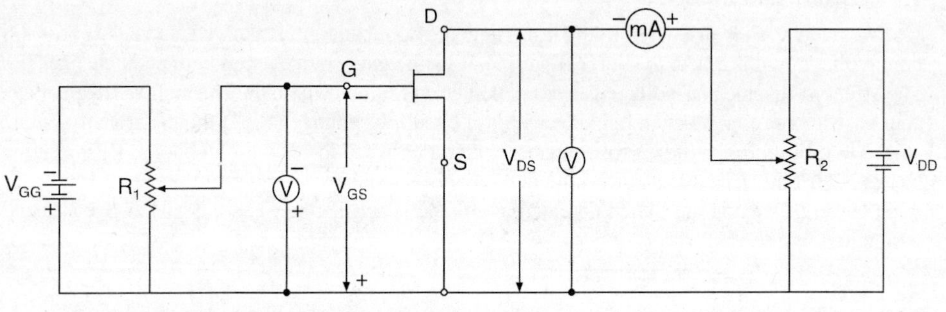

Fig. 13-8. *Circuit arrangement for plotting JFET characteristics.*

13-7. Drain Characteristicss

These curves may be obtained by using the circuit arrangement shown in Fig. 13-8. First of all, we adjust the gate-to-source voltage (V_{GS}) to zero volt. Then increase the drain-to-source voltage (V_{DS}) in small suitable steps and record the corresponding values of drain current (I_D) at each step. Now if we plot a graph with drain-to-source voltage along the horizontal axis and drain current along the vertical axis, we shall obtain a curve marked $V_{GS} = 0$ as shown in Fig. 13-9. A similar procedure may be used to obtain curves for different values of gate-to-source voltage, *i.e.*, V_{GS} = 1, 2, 3 and 4 V.

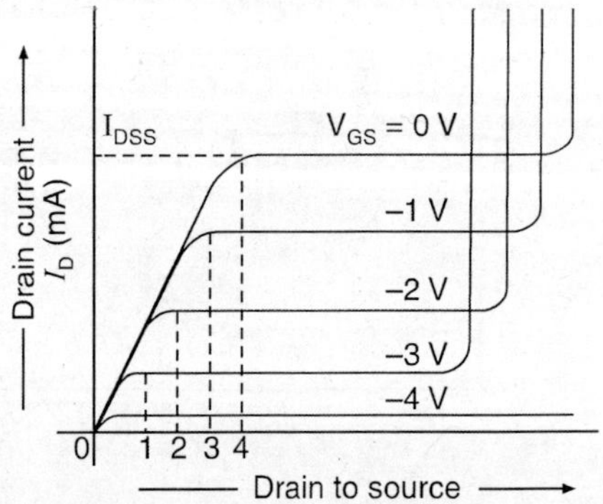

Fig. 13-9 *Drain characteristics of JFET.*

In order to explain the typical shape of drain characteristics, let us select the curve with $V_{GS} = 0$ volt as shown in Fig. 13-10. The curve may be sub-divided into the following regions:

(*i*) ***Ohmic region :*** This region is shown as a curve *OA* in the Fig. 13-10. In this region, the drain current increases linearly with the increase in drain-to-source voltage, obeying Ohm's Law. The linear increase in drain current is due to the fact that N-type semiconductor bar acts like a simple resistor.

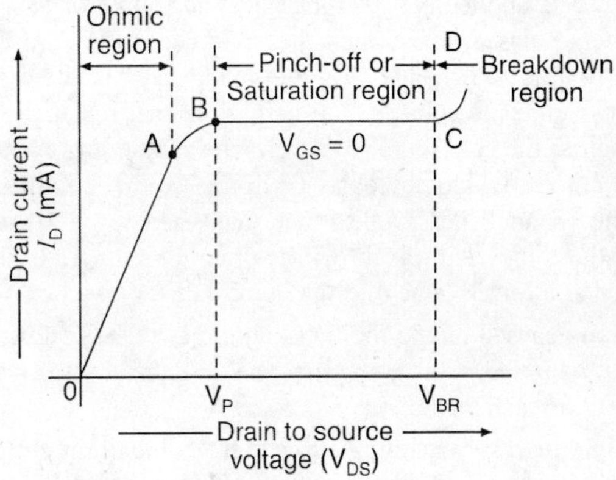

Fig. 13-10. *Drain characteristics with $V_{GS} = 0$ volt.*

(*ii*) ***Curve AB :*** In this region, the drain current increase at the reverse square law rate with the increase in drain-to-source voltage. It means that drain current increases slowly as compared to that in ohmic region. It is because of the fact, that with the increase in drain-to-source voltage, the drain current increases. This in turn increases the reverse bias voltage across the gate-source junction. As a result of this, the depletion region grows in size, thereby reducing the effective width of channel. At the drain-to-source voltage, corresponding to point *B*, the channel width is reduced to a minimum value and is known as pinch off. The drain-to-source voltage, at which the channel pinch-off occurs, is known as *pinch-off* voltage (V_P).

(*iii*) ***Pinch-off region :*** This region is shown by the curve *BC*. It is also called saturation region or constant current region. In this region, the drain current remains constant at its maximum value (*i.e.*, I_{DSS}). The drain, current in the pinch off region, depends upon the gate-to-source voltage and is given by the relation,

$$I_D = I_{DSS}\left(1 - \frac{V_{GS}}{V_P}\right)^2$$

The above relation is known as *Shockly's equation.* The pinch off region is the normal operating region of JFET, when used as an amplifier.

(*iv*) ***Breakdown region :*** This region is shown by the curve *CD*. In this region, the drain current increases rapidly as the drain-to-source voltage is also increased. It happens because of the breakdown of gate-to-source junction due to avalanche effect. The drain-to-source voltage corresponding to point *C* is called *breakdown voltage*.

13-8. Effect of Gate-to-Source Voltage on Drain Characteristics

We have already discussed in the last article about the drain characteristics of JFET. The effect of increasing gate-to-source voltage on the drain characteristics may be noted from Fig. 13-9. As the gate-to-source voltage (V_{GS}) is increased above zero to the 1, 2, 3 and 4 volts, following two effects take place:

- The value of pinch off voltage is reached at a smaller value of drain current as compared to that when $V_{GS} = 0$, and
- The value of drain-to-source voltage (V_{DS}) is decreased as compared to that when $V_{GS} = 0$.

The first effect may be understood from the fact that reverse bias voltage across the gate-source junction of JFET is the sum of gate-to-source voltage (V_{GS}) and drain-to-source voltage (V_{DS}). Thus for fixed V_{DS}, the increased value of V_{GS} will increase the reverse bias across the junction. For example, if V_{GS} of – 1V is applied between the gate and source of JFET, then the gate-to-source junction becomes reverse biased, even when the drain current is zero. This causes the depletion regions to be formed within the channel. As the voltage V_{DS} is increased above zero, the drain current increases. This produces voltage drop across channel resistance, which in turn produces reverse bias voltage across the gate-source junction. The amount of reverse bias voltage required to be produced by drain current would obviously be decreased by one volt. In other words, a smaller voltage drop along the channel (*i.e.*, smaller than that when $V_{GS} = 0$) will increase the depletion regions to the point where they will pinch-off the drain current. As a result of this, the pinch-off voltage (V_P) is reached at a lower value of drain current than that when $V_{GS} = 0$.

The second effect is caused due to the fact that the increased value of gate-to-source voltage (V_{GS}) keeps adding to the reverse bias at the junction produced by the drain current. Therefore, its lower value is required to reach the breakdown voltage.

It may be noted that with zero gate-to-source voltage, the drain current saturates at I_{DSS} value and the characteristic indicates the pinch off voltage (V_P) equal to 4 V. However, if V_{GS} of – 1 V, is applied, the channel still requires – 4 V to achieve pinch off condition. It means that now the channel requires only 3 V drop (instead of previously 4 V) from the drain-to-source voltage. This 3 V drop may be achieved with the lower value of drain current. Similarly, when V_{GS} is – 2 and –3 V, pinch off is achieved with 2 V and 1 V drops respectively along the channel. These voltage drops are obtained with further reduced values of the drain current. It may be noted that when V_{GS} is – 4 V (*i.e.*, equal to pinch off voltage), no channel drop is required and hence the drain current is zero.

13-9. Transfer Characteristics

These are also called transconductance curves, which give us the relationship between drain current (I_D) and gate-to-source voltage (V_{GS}) for a constant value of drain-to-source voltage (V_{DS}). The transfer characteristics may be obtained by using the circuit arrangement shown in Fig. 13-8. First of all, we adjust the drain-to-source voltage to some suitable value and increase the gate-to-source voltage in small suitable steps. Now record the corresponding values of drain current at each step. If we plot a graph with gate-to-source voltage (V_{GS}) along the horizontal axis and the drain current (I_D) along the vertical axis, we shall obtain a curve as shown in Fig. 13-11. A similar procedure may be used to obtain curves at different values of gate-to-source voltage (V_{GS}).

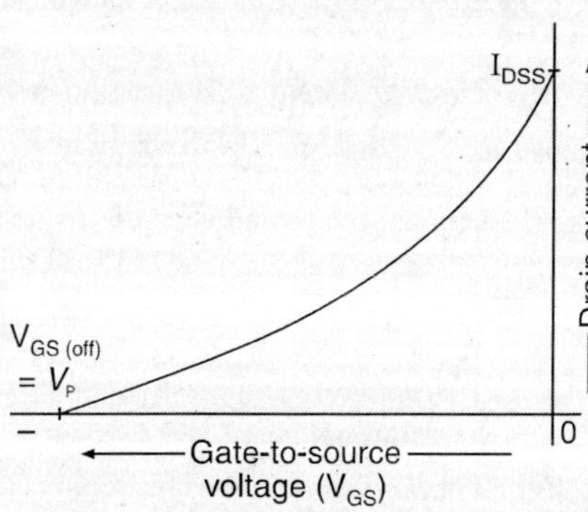

Fig. 13-11. *Transfer characteristics of N-channel.*

The upper end of the curve is shown by the drain current value equal to I_{DSS}, while the lower end is indicated by a voltage equal to $V_{GS(off)}$ or V_P. It may be noted that the curve is part of a parabola and therefore may be expressed by the equation,

$$I_D = I_{DSS}\left(1 - \frac{1 - V_{GS}}{V_{GS(off)}}\right)^2 = \mathbf{I_{DSS}}\left(\frac{\mathbf{1 - V_{GS}}}{\mathbf{V_P}}\right)^2$$

There is one important point which must be mentioned. Most JFET specifications sheet will list more than one value of $V_{GS(off)}$ and I_{DSS}. For example, the specifications sheet for the JFET 2N5457 lists the following:

$$V_{GS(off)\,(minimum)} = -0.5\ V \text{ and } I_{DSS(min)} = 1 \text{ mA}$$
$$V_{GS(off)\,(maximum)} = -6\ V \text{ and } I_{DSS(max)} = 5 \text{ mA}$$

In such a situation, we must use the two minimum values to plot one curve and the two maximum values to plot another curve on the same graph.

Example 13-1. *The data sheet of a certain JFET indicates that I_{DSS} equal to 15 mA and V_{GS} (off) equal to –5V. Determine the drain current for V_{GS} equal to 0 V, – 1 V and – 4 V.*

Solution. Given: I_{DSS} = 15 mA and $V_{GS\,(off)}$ = – 5 V

We know that drain current,

$$I_D = I_{DSS}\left(1 - \frac{V_{GS}}{V_{GS(off)}}\right)^2$$

∴ Drain current when (V_{GS} = 0)

$$I_D = 15\left(1 - \frac{0}{-5}\right)^2 = 15 \text{ mA} \quad \textbf{Ans.}$$

Similarly, drain current when V_{GS} = – 1 V,

$$I_D = 15\left(1 - \frac{-1}{-5}\right)^2 = 9.6 \text{ mA} \quad \textbf{Ans.}$$

and drain current when V_{GS} = – 4 V,

$$I_D = 15\left(1 - \frac{-4}{-5}\right)^2 = 0.6 \text{ mA} \quad \textbf{Ans.}$$

Example 13-2. *A JFET has parameters of $V_{GS\,(off)}$ equal to – 20 V and I_{DSS} equal to 12 mA. Plot the transconductance curve for the device using V_{GS} values of* 0 V, – 5 V, –10 V, – 15 V and – 20 V.

Solution. Given: $V_{GS\,(off)}$ = – 20V and I_{DSS} = 12 mA

First of all, let us find the value drain currents for the given values of V_{GS} (0 V,– 5 V, – 10 V, – 15 V and – 20 V) as discussed below:

We know that drain current,

$$I_D = I_{DSS}\left(1 - \frac{V_{GS}}{V_{GS(off)}}\right)^2$$

∴ Drain current when V_{GS} = – 5 V,

$$I_D = 12\left(1 - \frac{-5}{-20}\right)^2 = 6.75 \text{ mA}$$

Similarly, drain current when V_{GS} = – 10 V,

$$I_D = 12\left(1 - \frac{-10}{-20}\right)^2 = 3 \text{ mA}$$

and drain current when V_{GS} = – 15 V,

$$I_D = 12\left(1 - \frac{-15}{-20}\right)^2 = 0.75 \text{ mA}$$

The value of V_{GS} and the corresponding values of I_D obtained above are shown in Table 13-1.

Table 13-1:

$V_{GS}(V)$	I_D (mA)
– 20	0
– 15	0.75
– 10	3
– 5	6.75
– 0	12

These points are now plotted and joined together to form a curve shown in Fig. 13-12.

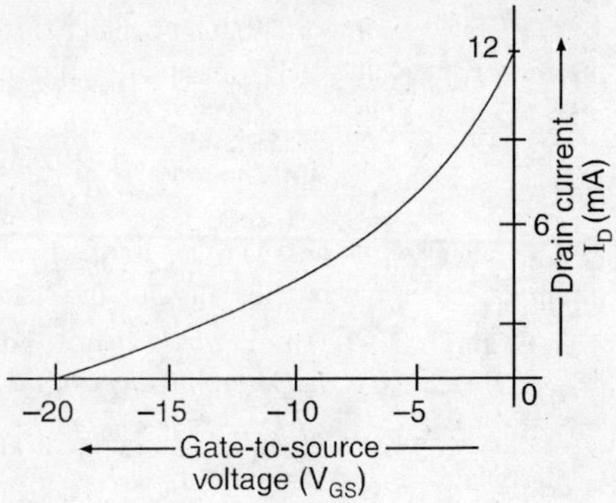

Fig. 13-12.

Example 13-3. *A2N5486 JFET has values of $V_{GS(off)}$ equal to – 2 V to – 6 V and I_{DSS} equal to 8 mA to 20 mA. Plot the minimum and maximum transconductance curves for the device.*

Solution: Given: $V_{GS\ (off)}$ = –2 V to – 6 V and I_{DSS} = 8 mA to 20 mA.

First fall, let us find the values of drain currents for the given values of V_{GS}.

We know that drain current,

$$I_D = I_{DSS}\left(1 - \frac{V_{GS}}{V_{GS(off)}}\right)^2 \quad ...(i)$$

Substituting the values of V_{GS} = – 0.5, – 1.0 and – 1.5 with $V_{GS\ (off)}$ = – 2.0 V, we get the corresponding values of I_D = 4.5 mA, 2 mA and 0.5 mA. With these values, we have the set of V_{GS} versus I_D value, shown in Table 13-2

Table 13-2:

For minimum transconductance curve

$V_{GS}(V)$	I_D (mA)
– 2.0	0
– 1.5	0.5
– 1.0	2.0
– 0.5	4.5
0	8

Table 13-3:

For maximum transconductance curve

$V_{GS}(V)$	I_D (mA)
– 6	0
– 4	2.2
– 2	8.9
0	20

Similarly, using the maximum values of $V_{GS(off)}$ = – 6 V and I_{DSS} = 20 mA, we can use equation (*i*) again for V_{GS} = – 4 V and – 2 V to obtain the values of I_D in between the end points. The end points of the transconductance curve are (– 6 V, 0 mA) and (0 V, 20 mA). The values of I_D for in between points are 2.2 mA and 8.9 mA. The values obtained are shown in Table 13-3.

Now if we plot the values corresponding to the sets shown in Tables 13-2 and 13-3 and joining them, we get the minimum and maximum transconductance curve as shown in Fig. 13-13.

13-10. Specifications Sheet of JFET

Although the general content of specification sheets may very from the absolute minimum to an extensive display of graph and charts, there are a few fundamental parameters that will be provided by all manufactures. The specifications sheet for 2N5484 through 2N5486 JFETs as provided by the Motorola as shown in Fig. 13-13.

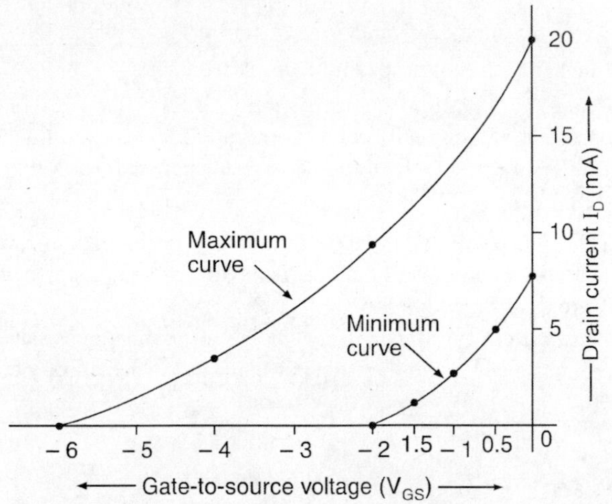

Fig. 13-13.

Maximum Ratings: Most of the maximum ratings for these devices are self-explanatory at this point. In our discussion for the BJT specifications sheet, we covered all the commonly used maximum voltage, current and power ratings. The JFET maximum ratings are no different. As seen from the table for maximum ratings, we find that for 2N5484, the maximum ratings for V_{DG} and V_{GSR} is 25 V.

The next maximum ratings in the list are for the drain current which is indicated as 30 mA dc. Then it is the maximum value of forward gate current. This rating is indicated as 10 mAdc. This rating may produce a though in your mind that the FET could be operated in the forward region. But the answer to such a thought is NO. Recall that the control of drain current in FET depends upon the amount of *reverse* bias on the JFET. The maximum drain current (I_{DSS}) is reached when the reverse bias reaches 0 V. Any increase in V_{GS} above 0V will not cause any increase in drain current. However, should the gate accidentally becomes forward biased; I_G must be greater than 10 mAdc, for the device to be destroyed. As long as I_G is below 10 mAdc, the device will be safe.

The maximum power dissipation which the device under consideration can safely handle at 25°C is 310 mW. As the temperature exceeds 25°C temperature, the maximum power dissipation reduces at the rate of 2.82 mW for each 1°C increase in temperature above 25°C.

OFF Characteristics : In this list three values are indicated :

(i) Gate-source breakdown voltage, $V_{(BR)GSS}$,

(ii) Gate reverse current, I_{GSS}, and

(iii) Gate-source cut-off voltage, $V_{GS(off)}$. The gate-source breakdown voltage defines the limit on V_{GS}. If V_{GS} is allowed to exceed this value (–25 V for 2N5484), the JFET will have to be replaced.

The gate reverse current rating indicates the maximum value of gate current that will occur when the gate-source junction is reverse biased. For the 2N5484 JFET, this rating is 1 nAdc when the ambient temperature (T_A) is 25 °C. Note that the negative current value is used to indicate the direction of the gate current. I_{GSS} is a temperature dependent rating. Notice that the value of I_{GSS} increases form 1 nAdc to 1μAdc as the ambient temperature increases from 25° C to 100° C.

Since, we have already disscussed $V_{GS(OFF)}$ and I_{DSS} (listed under "**ON Characteristics**") in detail, we will not discuss them further here. It may be carefully noted that there are wide ranges of $V_{GS(OFF)}$ and I_{DSS} Values within the 2N 5484-5486 series. For example, for 2N5484, the values of V_{GS} (off) vary from –0.3 to –3V and I_{DSS} from 1mA to 5 mA. Similarly for 2N5485, the values of

V_{GS} (off) vary from –0.2 V to 4V and I_{DSS} from 4 mA to 10 mA. And for 2N5486, the Value of V_{GS} (off) vary from –2V to – 6V and I_{DSS} from 8 mA to 20 mA.

Small signal characteristics: These will be discussed later.

Case Construction and Terminal Identification : The 2N5484-6 JFETs have the appearance as shown of right top corner of the specification sheet (page 227). The terminal identification is also provided there. The JFETs are available in top-hat containers as shown in Fig. 13-14.

13-11. JFET Parameters

The electrical behaviour of JFET may be described in terms of certain parameters called JFET parameters. Such parameters may be obtained from the characteristic curves. The important parameters of JFET are as discussed below :

(i) D.C drain resistance (R_{DS}) : It is also called the static or ohmic resistance of the channel and is given by the ratio of voltage (V_{DS}) to the drain current (I_D). Mathematically d.c. drain resistance,

$$R_{DS} = \frac{V_{DS}}{I_D}$$

(ii) A.C. drain resistance (r_d) : It is also called dynamic drain resistance and is the a.c. resistance between the drain and source terminal when the JFET is operating in the pinch-off or saturation region. It is given by the ratio of small change in drain-to-source voltage (ΔV_{DS}) to the corresponding change in drain current (ΔI_D) for a constant gate-to-source voltage (V_{GS}). Mathematically, the a.c. drain resistance,

$$r_d = \frac{\Delta V_{DS}}{\Delta I_D}$$

It is sometimes written as r_{ds}, which indicates that it is the resistance from drain-to-source terminals. Since, r_d is the output resistance of JFET, it may also be expressed as an output admittance y_{DS}.(*i.e.*, equal to $1/r_d$).

(iii) Transconductance (g_m) : It is also called forward transconductance or forward transmittance. It is given by the ratio of small change in drain current (ΔI_D) to the corresponding change in gate-to-source voltage (ΔV_{GS}) for a constant drain-to-source voltage (V_{DS}). Mathematically, the transconductance,

$$g_m = \frac{\Delta I_D}{\Delta V_{GS}}$$

The value of transconductance (g_m) is expressed in siemens (S) or mhos. We know that transfer characteristic of a JFET is a part of a parabola, *i.e.*, it is non-linear. Therefore, the value of g_m varies depending upon the location on the curve as set by the gate-to-source voltage (V_{GS}). It has a greater value near the top of the curve than near the bottom.

(iv) Amplification factor (μ) : It is given by the ratio of small change in drain-to-source voltage (ΔV_{DS}) to the corresponding change in gate-to-source voltage (ΔV_{GS}) for a constant drain current (I_D). Mathematically, the amplification factor,

$$\mu = \frac{\Delta V_{DS}}{\Delta V_{GS}}$$

The amplification factor (μ) may also be expressed as a product of transconductance (g_m) and a.c. drain resistance (r_d).

Thus, $$\mu = \frac{\Delta V_{DS}}{\Delta V_{GS}} = \frac{\Delta V_{DS}}{\Delta I_D} \times \frac{\Delta I_D}{\Delta V_{GS}} = r_d \times g_m$$

(v) Input Resistance (R_i) : We know that a JFET operation with its gate-source junction reverse biased. As the reverse current of a junction is, usually, very small. Therefore gate-reverse current of a JFET is also very small. Since the gate-source junction of a JFET is used as an input,

2N5484 thru 2N5486

CASE 29-04, STYLE 5
TO-92 (TO-226AA)

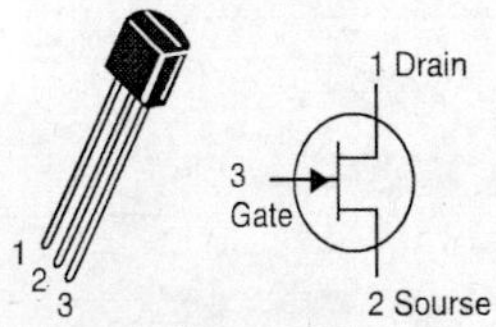

JFET
VHF/UHF AMPLIFIERS

N-CHANNEL– DEPLETION

Refer to 2N4416 for graphs.

MAXIMUM RATINGS

Rating	Sympol	Value	Unit
Drain-Gate Voltage	V_{DG}	25	Vdc
Reverse Gate-Source Voltage	V_{GSR}	25	Vdc
Drain Current	I_D	30	mAdc
Forward Gate Current	$I_{G(f)}$	10	mAdc
Total Device Dissipation @ $T_C = 25°C$	P_D	310	mW
Derate above 25°C		2.82	mW/°C
Operating and Storage Junction Temperature Range	T_J, T_{stg}	– 65 to + 150	°C

ELECTRICAL CHARACTERISTICS ($T_A = 25°C$ unless otherwise noted.)

Characteristic		Symbol	Min	Typ	Max	Unit
OFF CHARACTERISTICS						
Gate-Source Breakdown Voltage ($I_G = -1.0\ \mu$Adc, $V_{DS} = 0$)		$V_{(BR)GSS}$	– 25	—	—	Vdc
Gate Reverse Current		I_{GSS}				
($V_{GS} = -20$ Vdc, $V_{DS} = 0$)			—	—	– 1.0	nAdc
($V_{GS} = -20$ Vdc, $V_{DS} = 0$, $T_A = 100°C$)			—	—	– 0.2	μAdc
Gate Source Cutoff Voltage		$V_{GS(off)}$				Vdc
($V_{DS} = 15$ Vdc, $I_D = 10$ nAdc)	2N5484		– 0.3	—	– 3.0	
	2N5485		– 0.5	—	— 4.0	
	2N5486		– 2.0	—	– 6.0	
ON CHARACTERISTICS						
Zero-Gate-Voltage Drain Current		I_{DSS}				mAdc
($V_{DS} = 15$ Vdc, $V_{GS} = 0$)	2N5484		1.0	—	5.0	
	2N5485		4.0	—	10	
	2N5486		8.0	—	20	
SMALL-SIGNAL CHARACTERISTICS						
Forward Transfer Admittance		$\lvert y_{fs} \rvert$				μmhos
($V_{DS} = 15$ Vdc, $V_{GS} = 0$, f = 1.0 kHz)	2N5484		3000	—	6000	
	2N5485		3500	—	7000	
	2N5486		4000	—	8000	
Input admittance		$Re(y_{is})$				μmhos
($V_{DS} = 15$ Vdc, $V_{GS} = 0$, f = 100 MHz)	2N5484		—	—	100	
($V_{DS} = 15$ Vdc, $V_{GS} = 0$, f = 400 MHz)	2N5485, 2N5486		—	—	1000	
Output Admittance		$\lvert y_{os} \rvert$				μmhos
($V_{DS} = 15$ Vdc, $V_{GS} = 0$, f = 1.0 kHz)	2N5484		—	—	50	
	2N5485		—	—	60	
	2N5486		—	—	75	
Output Conductance		$Re(y_{os})$				μmhos
($V_{DS} = 15$ Vdc, $V_{GS} = 0$, f = 100 MHz)	2N5484		—	—	75	
($V_{DS} = 15$ Vdc, $V_{GS} = 0$, f = 400 MHz)	2N5485, 2N5486		—	—	100	
Forward Transconductance		$Re(y_{fs})$				μmhos
($V_{DS} = 15$ Vdc, $V_{GS} = 0$, f = 100 MHz)	2N5484		2500	—	—	
($V_{DS} = 15$ Vdc, $V_{GS} = 0$, f = 400 MHz)	2N5485		3000	—	—	
	2N5486		3500	—	—	

Fig. 13-14. *(Courtesy : Motorola Inc.)* **(Continued)**

ELECTRICAL CHARACTERISTICS (TA = 25°C unless otherwise noted)

Substrate connected to source.

Characteristic	Fig. No.	Symbol	Min	Max	Unit
OFF CHARACTERISTICS					
Drain-Source Breakdown Voltage (I_D = 10 μAdc, V_{GS} = 0)	–	$V_{(BR)DSS}$	25	–	Vdc
*Gate Leakage Current	–	I_{GSS}			pAdc
(V_{GS} = – 35 Vdc, V_{DS} = 0)			–	10	
(V_{GS} = – 35 Vdc, V_{DS} = 0, T_A = 125°C			–	100	
*Zero-Gate-Voltage Drain Current	–	I_{DSS}			nAde
(V_{DS} = 10 Vdc, V_{GS} = 0)			–	10	
(V_{DS} = 10 Vdc, V_{GS} = 0, T_A = 125°C)			–	1.0	μAdc
*ON CHARACTERISTICS					
Gate-Source Threshold Voltage (V_{DS} = 10 Vdc, I_D = 10 μAdc) 3N169	–	$V_{GS(th)}$	0.5	1.5	Vdc
3N170			1.0	2.0	
3N171			1.5	3.0	
"ON" Drain Current (V_{GS} = 10 Vdc, V_{DS} = 10 Vdc)	3	$I_{D(on)}$	10	–	mAdc
Drain-Source "ON" Voltage (I_D = 10 mAdc, V_{GS} = 10 Vdc)	–	$V_{DS(on)}$	–	2.0	Vdc
SMALL SIGNAL CHARACTERISTICS					
*Drain-Source Resistance (V_{GS} = 10 Vdc, I_D = 0, f = 1.0 kHz)	4	$r_{ds(on)}$	–	200	Ohms
Forward Transfer Admittance (V_{DS} = 10 Vdc, I_D = 2.0 mAdc, f = 1.0 kHz)	1	$\|y_{fs}\|$	1000	–	μmhos
*Reverse Transfer Capacitance (V_{DS} = 0, V_{GS} = 0, f = 1.0 MHz)	2	C_{rss}	–	1.3	pF
*Input Capacitance (V_{DS} = 10 Vdc, V_{GS} = 0, f = 1.0 MHz)	2	C_{iss}	–	5.0	pF
*Drain-Substrate Capacitance ($V_{D(SUB)}$ = 10 Vdc, f = 1.0 MHz	–	$C_{d(sub)}$	–	5.0	pF
* SWITCHING CHARACTERISTICS					
Turn-On Delay Time (V_{DD} = 10 Vdc, $I_{D(on)}$ = 10 mAdc, $V_{GS(on)}$ = 10 Vdc, $V_{GS(off)}$ = 0, R_G = 50 Ohms)	6,10	$t_{d(on)}$	–	3.0	ns
Rise Time	7,10	t_r	–	10	ns
Turn-Off Delay Time	8,10	$t_{d(off)}$	–	3.0	ns
Fall Time	9,10	t_f	–	15	ns

*Indicates JEDEC Registered Data.

FIGURE 1 FORWARD TRANSFER ADMITTANCE

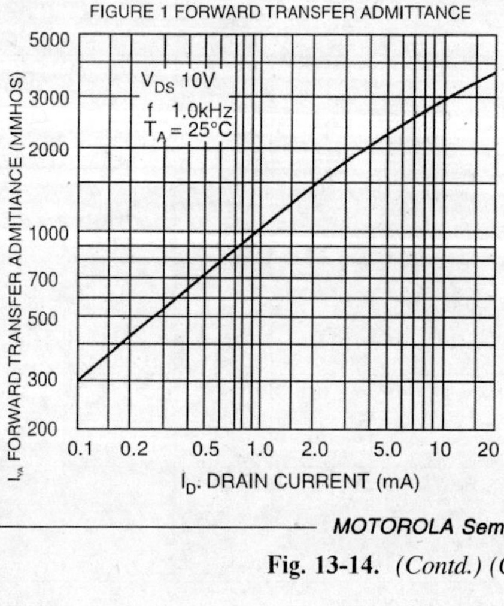

FIGURE 2 – CAPPACITANCE

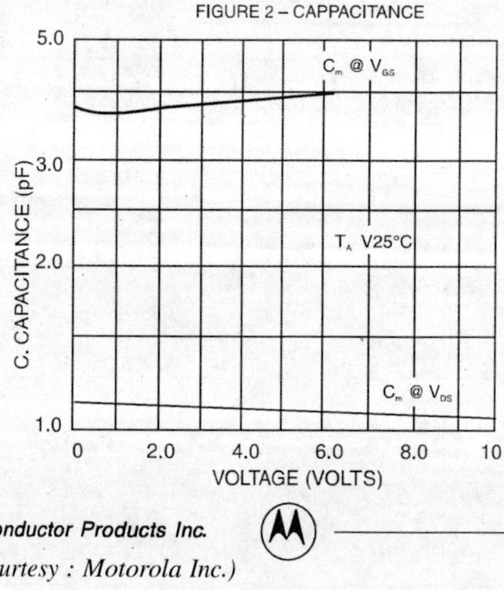

MOTOROLA Semiconductor Products Inc.

Fig. 13-14. *(Contd.) (Courtesy : Motorola Inc.)*

therefore the input resistance of a JFET is very high. This high input resistance is an advantage of JFET over the bipolar transistor. Mathematically, the input resistance,

$$R_i = \frac{V_{GS}}{I_{GSS}}$$

where, V_{GS} = Gate-to-source voltage and

I_{GSS} = Gate reverse current. It is of the order of nano-amperes.

13-12. Mathematical Expression for Transconductance

We have already discussed in Art 13-6 that the drain current is given by the relation,

$$I_D = I_{DSS}\left(1 - \frac{V_{GS}}{V_P}\right)^2$$

Differentiating on both sides with respect to V_{GS},

$$\frac{dI_D}{dV_{GS}} = 2\,I_{DSS}\left(1 - \frac{V_{GS}}{V_P}\right) \times \left(\frac{0-1}{V_P}\right)$$

$$g_m = -\frac{2\,I_{DSS}}{V_P}\left(1 - \frac{V_{GS}}{V_P}\right)$$

Substituting V_{GS} = 0 in the above expression,

$$g_m = -\frac{2\,I_{DSS}}{V_P} = g_{mo}$$

The value of g_m at V_{GS} = 0 is the maximum and is equal to $-2I_{DSS}/V_P$. This value of g_m is designated as g_{mo}. The equation (*i*) may now be rewritten as:

$$\boldsymbol{g_m = g_{mo}\left(1 - \frac{V_{GS}}{V_P}\right)}$$

Example 13-4. *The following information is included on the data sheet for an N-channel JFET:*

$$I_{DSS} = 20\text{ mA},\ V_P = -8\text{ V and } g_{mo} = 5000\ \mu s.$$

Determine the values of drain current and transconductance at V_{GS} = – 4 volts.

Solution. Given: I_{DSS} = 20 mA, V_P = – 8 volts; g_{mo} = 5000 μs and V_{GS} = – 4 volts.

Value of drain current

We know that drain current,

$$I_D = I_{DSS}\left(1 - \frac{V_{GS}}{V_P}\right)^2 = 20\left(1 - \frac{-4}{-8}\right)^2 = 5\text{ mA} \quad \textbf{Ans.}$$

Value of transconductance

We also know that transconductance,

$$g_m = g_{mo}\left(1 - \frac{V_{GS}}{V_P}\right) = 5000\left(1 - \frac{-4}{-8}\right)^2 = 2500\ \mu s \quad \textbf{Ans.}$$

13-13. Comparison Between FET and BJT

Table 13–4 shows some of the important points of comparison between the field-effect transistor and bipolar junction transistor.

Table 13-4. *Comparison between FET and BJT*

S.No.	***Field-Effect Transistor (FET)***	***Bipolar junction transistor (BJT)***
1.	It is a unipolar device. *i.e.*, current in the device is carried either by electrons or holes.	It is a bipolar device. *i.e.* current in the device is carried by both electrons and holes.
2.	It is a voltage-controlled device *i.e.*, voltage at the gate (or drain) terminal controls the amount of current flowing through the device.	It is a current-controlled device. *i.e.* the base current controls the amount of collector current.
3.	Its input resistance is very high and is of the order of several megaohms.	Its input resistance is very low as compared to FET and is of the order of few kilohms.
4.	It has a negative temperature coefficient at high current levels. It means that current decreases as the temperature increases. This characteristic prevents the FET from thermal breakdown.	It has a positive temperature coefficient at high current level. It means that collector current increases with the increase in temperature. This characteristic leads the BJT to thermal breakdown.
5.	It does not suffer from minority-carrier storage effects and therefore has higher switching speeds and cut-off frequences.	It suffers from minority carrier storage effects and therefore has lower switching speed and cut-off frequencies than that of FET's.
6.	It is less noisy than a BJT or vacuum tube and is thus more suitable as an input amplifier for low-level signals. It is used extensively in high fidelity frequency modulated receivers.	It is comparatively more noisy than a field-effect transistor.
7.	It is much simpler to fabricate as a integrated circuit (IC) and occupies a less space on IC chip than that of BJT.	It is comparatively difficult to fabricate as an integrated circuit and occupies more space on IC chip than that of FET.

13-14. MOSFETs

The MOSFET is an abbreviation for metal-oxide semiconductor field-effect transistor. Like JFET, the MOSFET has a source, gate and drain. However, unlike JFET, the gate of a MOSFET is insulated form the channel. Because of this, the MOSFET is sometimes known as IGFET which stands for insulated-gate field-effect transistor. Other names used for MOSFET are MISFET (metal-insulator-semiconductor field-effect transistor) and MOST (metal-oxide-semiconductor transistor).

The metal-oxide-semiconductor field-effect transistor (MOSFET) is the most important device for the very-large-scale integrated (VLSI) circuits such as microprocessors and semiconductor memories. MOSFET is also becoming an important power device.

The JFET has a main drawback that its gate must be reverse biased for its proper operation. Recall that the reverse bias of the gate is adjusted to deplete the channel of free carriers, thus controlling the effective size of the channel. This type of operation is know as *depletion-mode* operation. A depletion-type device would be the one which uses an input voltage to reduce the channel size from its *zero-bias* size. On the other hand, *MOSFET* can be operated in the *enhancement mode*. This means that the input signal can be used to increase the size of the channel. This also means that the device is not restricted to operating with its gate reverse biased. This is an improvement over the standard FET.

13-15. Types of MOSFETs

There are two basic types of MOSFETs; depletion-type MOSFETs and enhancement-type MOSFETs. The depletion type MOSFETs are also called D-type MOSFETs and enhancement-type

MOSFETs. as E-type MOSFETs. The primary difference between the two types of MOSFETs is the difference between the constructions. Both the D-type and E-type MOSFETs are disscussed one by one in the following pages.

13-16. Depletion-Type MOSFET

Fig. 13-15 shows the basic structure of an N-channel depletion type MOSFET. It consists of a conducting bar of N-type material with an insulated gate on the left and P-region on the right. Free electrons can flow from source to drain through the N-type material. The P-region is called substrate (or body). It physically reduces the conducting path to a narrow channel. A thin layer of silicon dioxide is deposited on the left side of the channel. This layer insulates the gate from the channel. Because of this, a negligible gate current flows even when the gate voltage is positive. It will be interesting to know that a PN junction, which exists in a JFET, has been eliminated in the MOSFET.

The basic construction of a depletion-type P-channel MOSFET is similar to that of N-channel except that the conducting bar is of P-type material and the substrate is of N-type material.

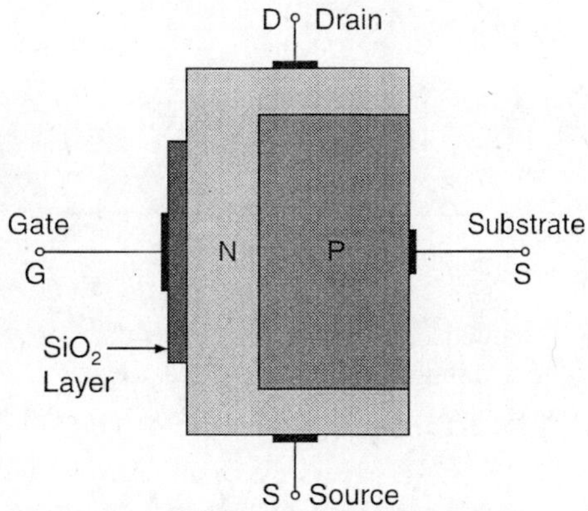

Fig. 13-15. *Basic structure of an N-channel depletion-type MOSFET.*

13-17. Working of a Depletion-Type MOSFET

The depletion-type MOSFET can be operated in two different modes as given below :

(1) ***Depletion mode :*** The device operates in this mode, when the gate voltage in negative.

(2) ***Enhancement mode :*** The device operates in this mode, when the gate voltage is positive.

Since, the depletion-type MOSFET can be operated in either depletion or enhancement mode, therefore this device is commonly known as depletion-enhancement (DE) type MOSFET. The working of a MOSFET may be explained easily, if we visualize the entire structure of the device as a parallel plate capacitor. One of the plates is formed by the gate and the other by the semiconductor channel. The plates are separated by a dielectric (SiO_2 layer). We know that if one plate of a capacitor is made negative, it induces a positive charge on the opposite plate and vice versa. This principle is used below in explaining the working of MOSFET's in the depletion and enhancement modes.

(i) Depletion mode : Fig. 13-16(*a*) shows a MOSFET with a negative gate-to-source voltage. The negative voltage, on the gate, induces a positive charge in the channel. Because of this, free electrons in the vicinity of positive charge are repelled away in the channel. As a result of this, the channel is depleted of free electrons. This reduces the number of free electrons (which constitute the drain current) passing through the channel. Thus as the value of negative gate-to-source voltage is increased, the value of drain voltage, called $V_{GS(off)}$, the channel is totally depleted of free electrons and therefore the drain current reduces to zero. Thus with the negative gate voltage, the operation of MOSFET is similar to that of a JFET.

It is evident from the above discussion that negative gate voltage depletes the channel of free electrons. It is due to this fact that the working of a MOSFET, with a negative gate voltage, is called *depletion mode*.

(ii) Enhancement mode : Fig. 13-16(*b*) shows a MOSFET with a positive gate-to-source voltage. The positive gate voltage increases the number of free electrons passing through the channel. The greater the gate voltage, greater is the number of free electrons passing through the channel. This increases, *i.e.*, enhances the conducting of the channel. Because of this fact, positive gate operation is called *enhancement mode.*

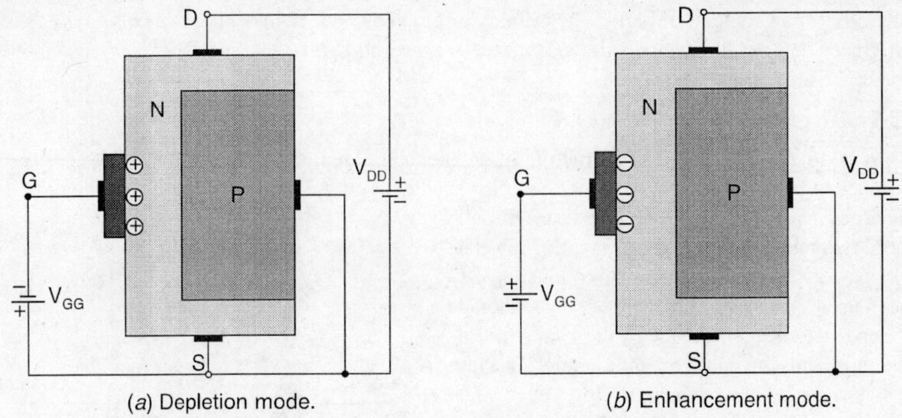

Fig. 13-16. *Working of a depletion-type MOSFET.*

It will be interesting to know that depletion-type MOSFET can conduct even if the gate-to-source voltage (V_{GS}) is zero. Because of this, it is commonly known as Normally-ON MOSFET.

13-18. Drain Characteristic of Depletion-type MOSFET

Fig. 13-17 shows the drain characteristic for the N-channel depletion-type MOSFET in the common source configuration. These curves are plotted for both negative and positive values of gate-to-source voltage (V_{GS}). The curves shown above the curve for V_{GS} = 0 have a positive zero whereas those below it have a negative value of V_{GS}. When V_{GS} is zero and negative, the MOSFET operates in the depletion-mode. On the other hand, if VGS is zero and positive, the MOSFET operates in the enhancement-mode.

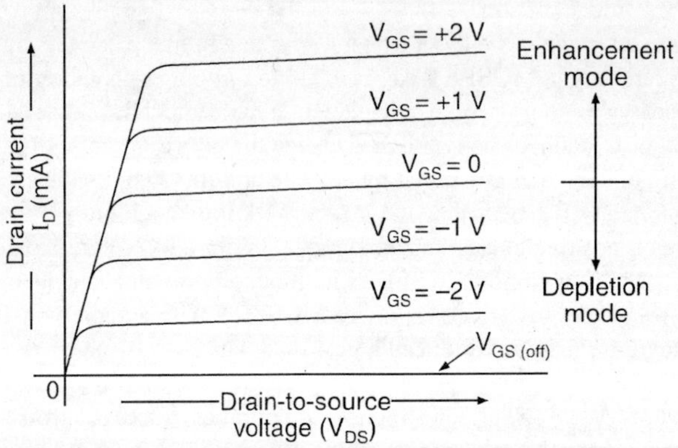

Fig. 13-17. *Drain characteristics of N-channel depletion-type MOSFET.*

It may be noted that the drain characteristics of depletion-type MOSFET's are similar to that of JFET. The only difference is that JFET does not operate for positive values of gate-to-source voltage (V_{GS}).

13-19. Transfer Characteristic of Depletion-type MOSFET

Fig. 13-18 shows the transfer characteristic (also called transconductance curve) for an N-channel depletion-type MOSFET. It may be noted from this curve that the region AB of the characteristic is similar to that of JFET. But here, this curve extends for the positive values of gate-to-source voltage (V_{GS}) also. The value I_{DSS} represents the current from drain-to-source with V_{GS} = 0. The drain current at any point along the transfer characteristic (*i.e.*, the curve *ABC*) is given by the relation,

$$I_D = I_{DSS}\left(1 - \frac{V_{GS}}{V_{GS(off)}}\right)^2$$

It may be noted that even if $V_{GS} = 0$, the device has a drain current equal to I_{DSS}. Due to this fact, it is called normally-**ON** MOSFET.

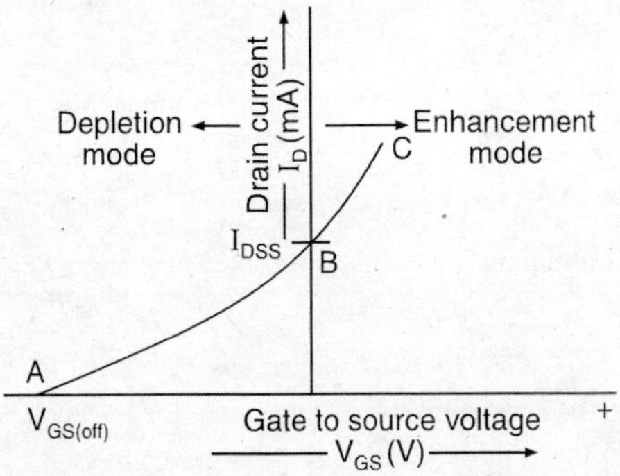

Fig. 13-18. *Transfer characteristics of a N-channel depletion-type MOSFET.*

13-20. Circuit Symbol for Depletion-type MOSFET

Fig. 13-19 (*a*) and (*b*) shows the circuit symbols for the N-channel depletion-type MOSFET. In these Figs, a thin vertical line (just right to the gate) represents the channel. The drain and source terminals are connected to the top and bottom of the channel as shown. The arrow, on the P-type substrate, points towards the channel. This indicates that the channel is N-type. In some MOSFET's, a connection from the substrate is also taken out. Such MOSFET's have 4-terminals as indicated in the Fig. 13-19 (*a*). But in most of the MOSFET's the substrate is internally connected to the source. This results in a three terminal device, whose circuit symbol is as shown in Fig. 13-19 (*b*).

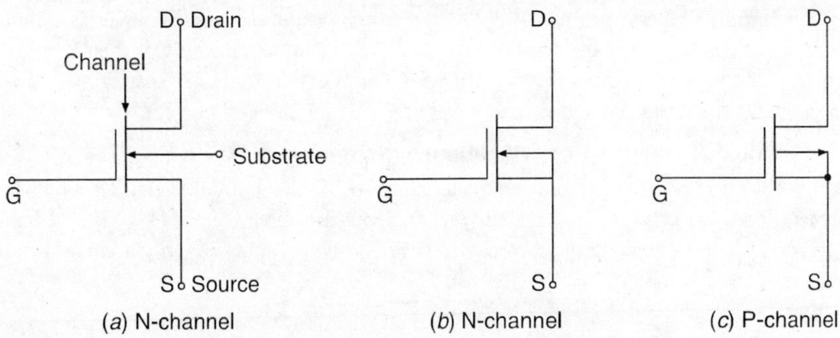

Fig. 11-19. *Circuit symbols for depletion type MOSFET's.*

Fig. 13-19(*c*) shows the circuit symbol for a P-channel depletion type MOSFET. It may be noted that the symbol is similar to that of N-channel, except the direction of the arrow on the substrate. Its direction is away from the channel, which indicates that the channel is of P-type material.

13-21. Enhancement-type MOSFET

The enhancement-type MOSFET has no depletion mode and it operates only in enhancement mode. It differs in construction from the depletion-type MOSFET in the sense that it has no physical channel. Fig. 13-20 (*a*) shows the basic structure of the N-channel enhancement-type MOSFET. It may be noted, that the P-type substrate extends the silicon dioxide layer completely.

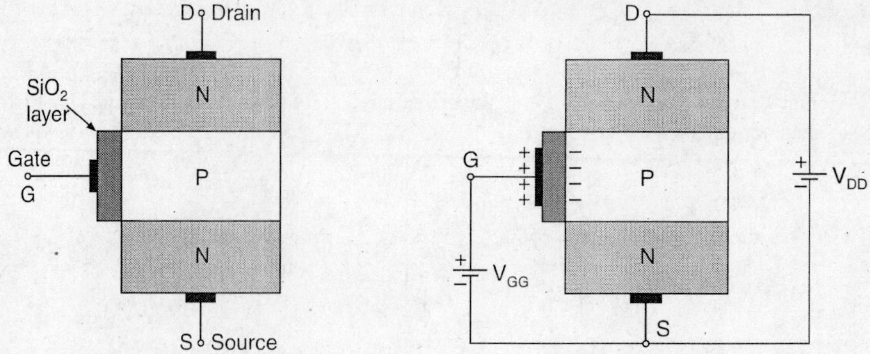

(*a*) Basic construction. (*b*) Formation of inversion layer.

Fig. 13-20. *Enhancement type MOSFET.*

Fig. 13-20 (*b*) shows the normal biasing polarities for the N-channel enhancement-type MOSFET. It must be remembered that this MOSFET is always operated with the positive gate-to-source voltage (V_{GS}). When the gate-to-source voltage is zero, the V_{DD} supply tries to force free electrons from source-to-drain. But the presence of P-region does not permit the electrons to pass through it. Thus, there is no drain current for $V_{GS} = 0$. Due to this fact, the enhancement type MOSFET is also called normally-**OFF** MOSFET.

Now, if some positive voltage is applied at the gate, it induces a negative charge in the P-type substrate just adjacent to the silicon dioxide layer. The induced negative charge is produced by attracting the free electrons from the source. When the gate is positive enough, it can attract a number of free electrons. This forms a thin layer of electrons, which stretches from source to drain. This effect is equivalent to producing a thin layer of N-type channel in the P-type substrate. This layer of free electrons is called N-type inversion layer.

The minimum gate-to-source voltage (V_{GS}), which produces inversion layer, is called threshold voltage and is designated by the symbol $V_{GS(th)}$. When the voltage V_{GS} is less than $V_{GS(th)}$, no current flows from drain to source. However, when the voltage V_{GS} is greater than $V_{GS(th)}$, the inversion layer connects the drain and source and we get significant value of current.

13-22. Drain Characteristics for Enhancement-type MOSFET

Fig. 13-21(*a*) shows the drain characteristics for N-channel enhancement-type MOSFET. It may be noted from this Fig. 13-21, that when the gate-to-source voltage (V_{GS}) is less than threshold voltage, $V_{GS(th)}$, there is no drain current. However, in actual practice, an extremely small value of

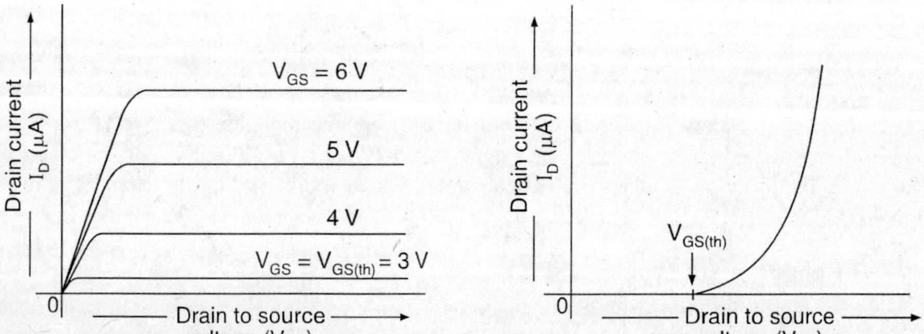

(*a*) Drain characteristic. (*b*) Transfer characteristic.

Fig. 13-21. *Characteristics for N-channel enhancement-type MOSFET.*

drain current does flow through the MOSFET. This current flow is due to the presence of thermally generated electrons in the P-type substrate. When the value of V_{GS} is kept above $V_{GS(th)}$, a significant drain current flows.

The values of drain current increases with the increase in gate-to-source voltage. It is because of the fact that the width of inversion layer widens for increased values of V_{GS} and therefore, allows more number of free electrons to pass through it. The drain current reaches its saturation value above a certain value of drain-to-source voltage (V_{DS}).

13-23. Transfer Characteristic of Enhancement-type MOSFET

Fig. 13-21(*b*) shows the transfer characteristic for N-channel enhancement type MOSFET. It may be noted from this Figure that there is no drain current when the gate-to-source voltage, $V_{GS} = 0$. However, if V_{GS} is increased above the threshold voltage, $V_{GS(th)}$, the drain current increases rapidly as shown in Fig. 13-21. The drain current at any point along the curve is given by the relation,

$$I_D = K\left[V_{GS} - V_{GS(th)}\right]^2$$

where, K is a constant, whose value depends on the type of MOSFET. Its value can be determined from the data sheet by taking specified value of drain current called $I_{D(ON)}$ at the given value of V_{GS} and then substituting these values in the above equation. Incidently, it may be noted that enhancement-type MOSFET does not have an I_{DSS} parameter like JFET and depletion-type MOSFET.

13-24. Circuit Symbol of Enhancement-type MOSFET

Fig. 13-22(*a*) shows the circuit symbol for *N*-channel enhancement-type MOSFET. In this Figure the broken line indicates that there is no conducting channel between drain and source, when $V_{GS} = 0$. Due to this fact, this device is also known as "Normally-OFF MOSFET". The drain and source terminals are shown at the top and bottom end of the broken line. The substrate is internally connected to the source as shown. The arrow points in the direction of channel (or inversion layer), which is created when V_{GS} is increased above $V_{GS(th)}$.

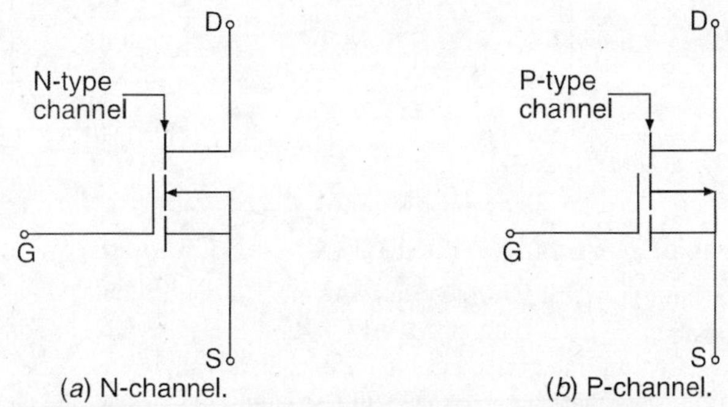

Fig. 13-22. *Circuit symbols for enhancement-type MOSFET.*

Fig. 13-22(*b*) shows the circuit symbol for the P-channel enhancement-type MOSFET. This symbol is similar to N-channel except that the arrow points outwards. This indicates that the channel (or inversion layer) created in the substrate is of P-type. In actual practice, the N-channel and P-channel enhancement-type MOSFET's are known as NMOS and PMOS field-effect transistors respectively.

Example 13-5. *The data sheet for a certain enhancement-type MOSFET reveals that* $I_{D(on)} = 10$ *mA at* $V_{GS} = -12$ *V and* $V_{GS(th)} = -3$ *V. Is this device P-channel or N-channel? Find the value of* I_D, *when* $V_{GS} = -6$ *V.*

Solution. Given: $I_{D(on)} = 10$ mA; $V_{GS} = -12$ volts and $V_{GS(th)} = -3$ volts.

Since, the value of V_{GS} is negative for the enhancement-type MOSFET, this indicated that the device is P-channel. We know that the drain current,

$$I_D = K\,[\,V_{GS} - V_{GS(th)}]^2 \qquad ...(i)$$

or $$I_{D(on)} = K\,[(V_{GS} - V_{GS(th)})]^2$$

$$10 = K\,[-12-(-3)]^2 = 81\ \text{K}$$

$\therefore$ $$K = 10/81 = 0.12\ \text{mA/V}$$

Substituting this value of K and V_{GS} (equal to – 6) in the equation (*i*),

$$I_D = 0.12\,[-6-(-3)]^2 = 1.08\ \text{mA}\ \textbf{Ans.}$$

13-25. The MOSFET as a Resistor

The MOSFET's have an important property that they can be used as a resistor, capacitor, amplifier and a switch. This makes the design of electronic circuits very simple, because the entire circuit consists of only MOSFETs and no other component. The examples of such circuits are microprocessors and memory circuits.

The depletion type MOSFET may be connected as a resistor, if its gate is connected to the source so that the gate-to-source voltage (V_{GS}) is zero, as shown in Fig. 13-23(*a*).

The enhancement-type MOSFET may be connected as a resistor, if its gate is connected to drain so that $V_{GS} = V_{DS}$ as shown in Fig. 13-23(*b*).

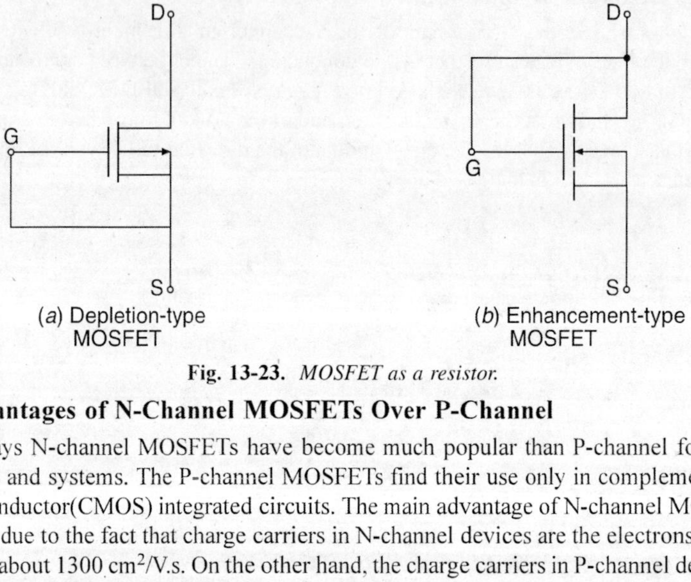

Fig. 13-23. *MOSFET as a resistor.*

13-26. Advantages of N-Channel MOSFETs Over P-Channel

These days N-channel MOSFETs have become much popular than P-channel for fabricating MOS circuits and systems. The P-channel MOSFETs find their use only in complementary metal-oxide semiconductor(CMOS) integrated circuits. The main advantage of N-channel MOSFET, over P-channel, is due to the fact that charge carriers in N-channel devices are the electrons, which have a mobility of about 1300 cm^2/V.s. On the other hand, the charge carriers in P-channel devices are the holes which have a mobility of about 500 cm^2/V.s. Since, the current in a semiconductor is directly proportional to the mobility, therefore the current in a N-channel MOSFET is more than two times that of P-channel for the same dimension.

The ON resistance of the N-channel MOSFET is one-third of that for P-channel. It means that in order to achieve the same value of current and ON resistance, the P-channel MOSFET requires three times the area of an equivalent N-channel MOSFET. Thus, electronic currents using N-channel MOSFETs are much smaller in size than those of P-channel. This results in higher packing density.

13-27. Complementary MOSFETs (CMOS)

MOSFETs are used extensively in the field of digital electrons to manufacture logic gate and many other synchronous and asynchronous circuits. The logic gates include AND, OR and NOT gates and their various combinations. The synchronous circuits include flip-flops, counters, memories, A/D and D/A converters the asynchronous circuits, include combination of logic gates such as decoders, decoders, encoders muntiplexers and demultiplexers.

A CMOS logic gate consists of a combination of NMOS and PMOS devices. The detailed discussion of CMOS devices is beyond the scope of this book.

13-28. Dual-gate MOSFETs

As a matter of fact, the operation of MOSFETs is limited at high frequencies because of their high gate to-channel capacitance. The cause of this high capacitance can be seen in Fig. 13-24. The metal plate that is used for the gate is a conductor. The silicon dioxide between the gate and the channel in an insulator. The channel itself can be viewed as being a conductor to the silicon dioxide layer. Thus, the combination of the three forms a parallel-plate capacitor.

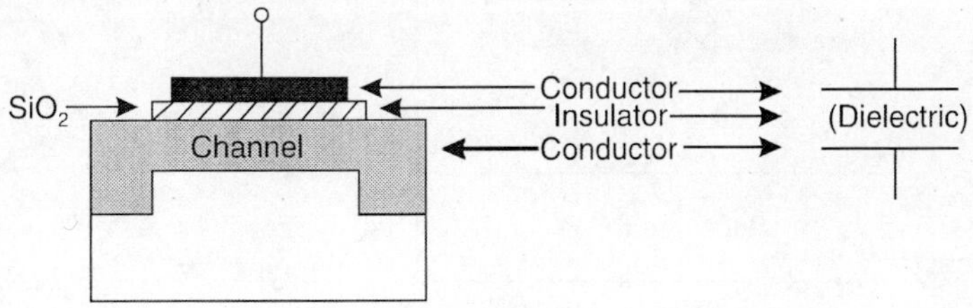

Fig. 13-24.

The capacitance offered by this parallel plate capacitor can be reduced by employing a dual gate MOSFET structure. The physical construction and schematic symbols for the dual-gate MOSFET are shown in Fig. 13-25.

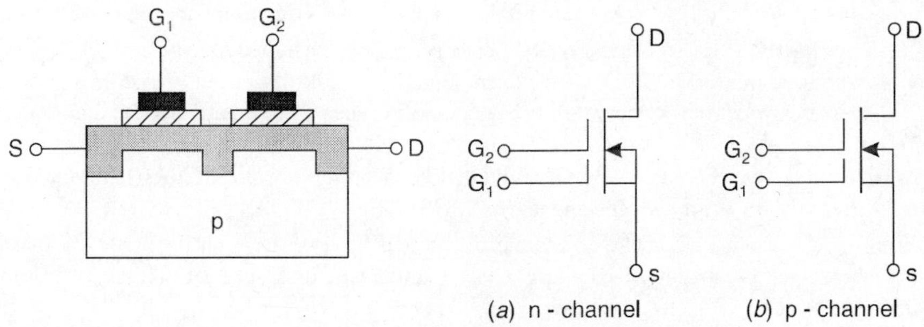

(*a*) n - channel (*b*) p - channel

Fig. 13-25.

The reduced capacitance of the dual-gate MOSFET is the result of the way in which the component is used. Normally the component is used in such a way as to act as two series-connected MOSFETs. When the dual gate MOSFET is used as series MOSFETs, the effect is similar to connecting two capacitors in series Recall that the total capacitance in a sries connection is lower than either individual component value. Thus by connecting the two gates in a series configuration, the overall capacitance is reduced. This allows the dual-gate MOSFET to have a better high-frequency response as compared to a normal MOSFET.

13-29. Power MOSFETs

With the advancement of technology, the engineers have produced a wide variety of MOSFETs that are designed specifically for high current, high voltage and high power applications. Some applications for these components are found in cascode amplifiers, RF amplifier, power MOSFET derivers. The detail of these applications is beyond the scope of this book.

Following are some of the power MOSFETs which are important from the subject point of view.

(*i*) Vertical MOSFET (VMOS) (*ii*) Lateral Double-Diffused MOSFET(LDMOS)

Both there power MOSFETs are discussed below one by one in the following pages.

13-30. VMOS (or V-Groove MOSFET)

One of the major disavantages of a typical MOSFET is the reduced power handling level as compared to BJT transistors. The power handling level of a typical MOSFET is less than 1W. This drawback of the MOSFET can be overcome by changing the construction mode from one of the planar nature to one with a vertical structure as shown in Fig. 13-26.

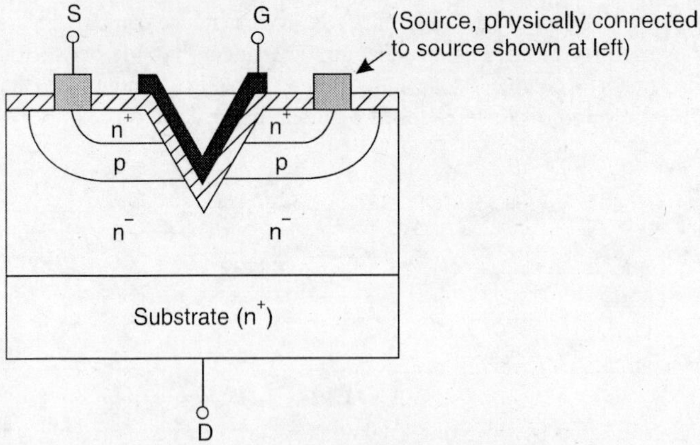

Fig. 13-26. *Illustrating the construction of VMOS.*

As seen from this figure, all the elements of the planar MOSFET are present in the vertical metal-oxide silicon FET (or simply VMOS) the metallic surface connection to the terminals of the device.

The *vertical*-MOSFET (or simply VMOS) is a component disigned to handle much larger drain currents than the standard MOSFET. The current handling capability of the VMOS is a result of its physical consturction which is illustrated in Fig. 13-26. As seen from this figure, the component has materials that are labelled as P, N^+ and N^-. The N- material labels indicate the differences in doping levels. Also notice that there is no physical channel connecting the source (at top) and the drain (at bottom). Thus VMOS is an *enchancement type* MOSFET.

With the V-shaped gate, a larger channel is formed by a positive gate voltage. With a large channel, the device is capable of handling a large amount of drain current. The operation of the VMOS is illustrated in Fig. 13-27 below.

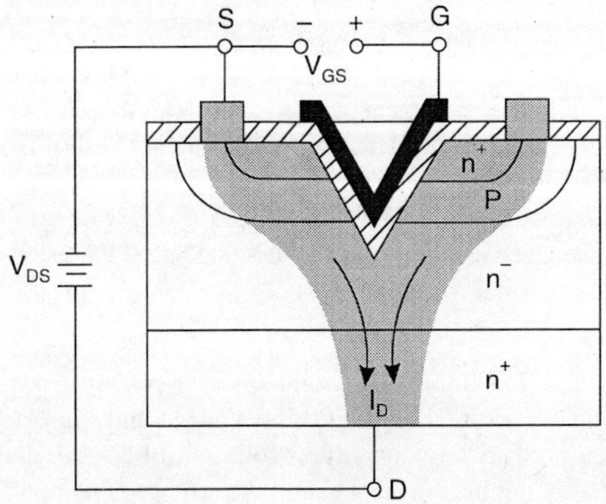

Fig. 13-27. *Illustrating VMOS Operator.*

VMOS Operation : When a positive gate voltage is applied to the device, an N- type channel forms in the P-type region. This effective channel connects the source to the drain. As seen from Fig. 13-27 the shape of the gate causes a wider channel to form than is created in the standard MOSFET. Because of this, the amount of drain current is much higher for this component. Moreover, the VMOS can exhibit a higher transconductance and a lower turn-on resistance than the conventional planer MOSFET.

Another advantage of using VMOS is the fact that it is not susceptible to thermal runaway. The VMOS has a *positive* temperature coefficient. This means that the resistance of the component *increases* when temperature increases. Thus an increases in temperature will cause a *decrease* in drain current.

The VMOS device can be fabricated with more than one V-groove to increase amount of drain current and some other performance characteristics.

13-31. LDMOS

The LDMOS (*i.e.*, lateral double-diffused MOSFET) is another type of power MOSFET. This MOSFET uses a very small channel region and a heavily doped N-type region (N^+) to obtain a high drain current and low channel resistance [$r_{d(ON)}$].

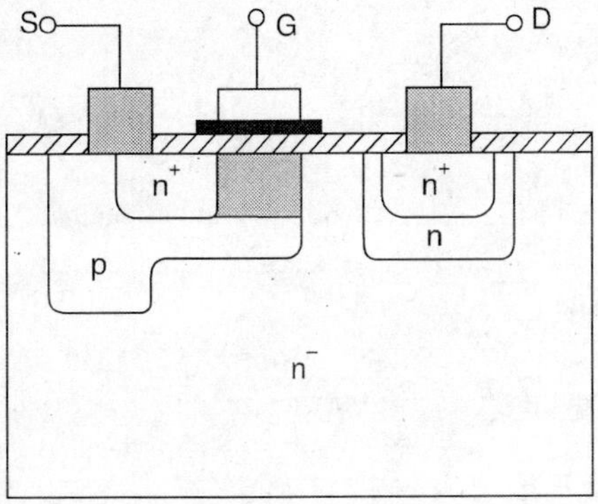

Fig. 13-28. *LDMOS*

Fig. 13-28 shows the basic construction of LDMOS. As seen from this figure, the narrow channel (shaded region) is made up of the P-type material that lies between the N-substrate (lightly doped) and the N^+ (heavily doped) source region. Since, only the N-type material lies between the channel and the drain, the effective length of the channel is externally short. This coupled with the N-type material in the channel-to-drain path provides an extermely low value of $r_{d(ON)}$. With a low channel resistance, the LDMOS device can handle very high amounts of current without generating any damaging amount of heat.

The LDMOS has typical values of $r_{d(ON)}$ that are in the range of 2 Ω or less. With this low channel resistance, it is typically capable of handling currents as high as 20 A.

13-32. Multichannel FETs

To improve the FET's power handling capability, Multichannel devices have been studied. The multichannel FET is basically many JFETs connected in parallel. The current-voltage characteristic of such devices is similar to a single-channel JFET except the drain current is substantially higher.

13-33. Handling Precautions for MOSFETs

It has been observed that the MOSFETs get damaged easily if they are not properly handled. The damage occurs due to the static electricity, which ruptures the silicon dioxide layer between the gate and channel of a MOSFET. The static electricity may result from the sliding of a device in the plastic bag. It may also result from the person, who picks up the device by its case and touches the gate lead against some grounded object.

In order to protect a MOSFET from damaging, it is kept with its terminals shorted by a ring. The ring is removed at the time of soldering MOSFET in a circuit. The MOSFET's can also be protected by keeping them in a conducting foam. The MOSFET's should never be inserted or removed from the circuits with power ON.

These days, some of the MOSFETs have a built-in protection against static electricity and high voltages. This is done by connecting a pair of zener diodes back-to-back between the gate and source of a MOSFET as shown in Fig. 13-29. These diodes are designed in such a way that if the gate-to-source voltage exceeds say by + 10 V (or – 10 V) typically, one of the zener diodes will conduct and the other will breakdown. This provides a path for the excessive charge from gate-to-source. But this protection provides a minor disadvantage that it lowers the input resistance of the MOSFET.

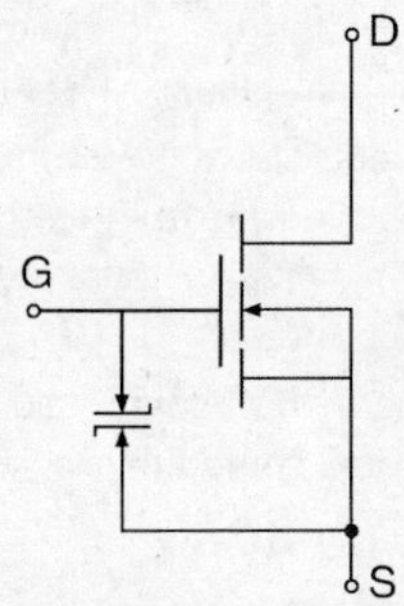

Fig. 13-29. *Pair of zener diodes as a protection against static electricity.*

SUMMARY

In this chapter you have learned that :

1. The two main types of FETs are junction field-effect transistor (JFET) and metal oxide semiconductor field-effect transistor (MOSFET).
2. Output charcteristic of FET are controlled by input voltage.
3. Acording to the type of smiconductor bar used for construction, JFET is divided into two types: N-channel and P-channel JFET.
4. MOSFETs are divided into two types : depletion-type and enhancement-type MOSFET
5. A depletion-type MOSFET can be operated in depletion mode or enchancement mode.
6. MOSFETs can be used as a resistor, capacitor amplifier and a switch.
7. The current in a N-channel MOSFET is two times more than that of P-channel MOSFET of the surce dimension.
8. The frequency responce of MOSFETs can be enhanced by modifying its back structure to that of a dual-gate MOSFET.
9. The MOSFETs can be designed to handle large current/voltage by modifying its basic structure to VMOS or LDMOS.
10. The MOSFETs need to be handled carefully.

GLOSSARY

A.C. drain resistance : The a.c. resistance between the drain and source terminal when the JFET is operating in the pinch-off or satruation region.

Amplification factor : The ratio of small change in drain-to-source voltage to the corresponding change in gate-to-source voltage.

D.C. drain resistance : The ratio of drain-to-source voltage to the drain current.

Deplecation-type MOSFET : A semiconductor device consisting of a conducting bar of N-type or P-type material with gate insulated and a substrate of opposite material sitting apposite to the gate.

Drain (D) : The terminal of FET through which the current carrier leaves the semiconductor

Drain characteristic : Curves that gives the relationship between the drain current and drain-to-source voltage for different gate-to-source voltage.

Enhancement-type MOSFET : MOSFET that has no physical channel and operates only in the enchancement mode.

Gate (G) : The terminal of a FET, using which the current through the channel is controlled

N-channel JFET : A field-effect transistor, consisting of a N-type semiconductor bar with two P-type heavily doped regions diffused on opposite sides of its middle part.

P-channel JFET : A field-effect transistor, consisting of a P-type semiconductor bar with two N-type heavily doped regions diffused on apposite sides of its middle part.

Pinch-off voltage : The gate-to-source voltage of a JEFT, at which the channel is completely blocked or pinched off and drain current is reduced to zero.

Source (S) : The terminal of a FET through which the current carriers enter the semiconductor.

Transconductance : Ratio of small change in drain current to the corresponding change in gate-to -source voltage, for a constant drain-to-source voltage.

Transfer characteristic : Curves that give the relationship between drain current and gate-to-source voltage for different drain-to-source voltage.

Unipolar device : A semiconductor device in which current carriers are either electrons or holes but not both.

NUMERICAL PROBLEMS

1. A JFET with parameters of I_{DSS} = 3 mA and $V_{GS(off)}$ = –6V is used in a circuit shown in Fig. 13-30, that provides a V_{GS} = –3V. Determine the value of I_D in the circuit.(**Ans.** 1.33 mA)

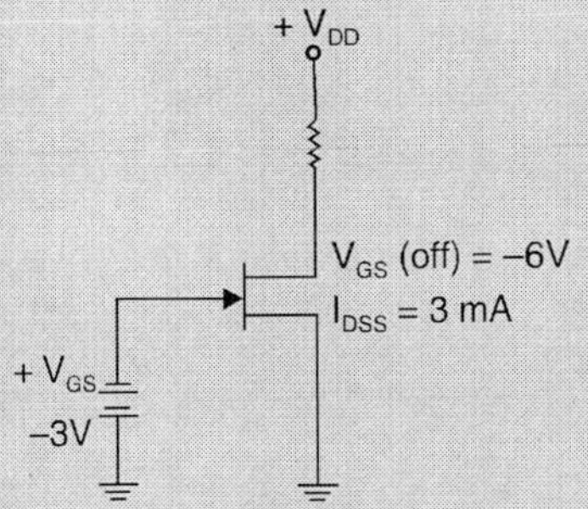

Fig. 13-30

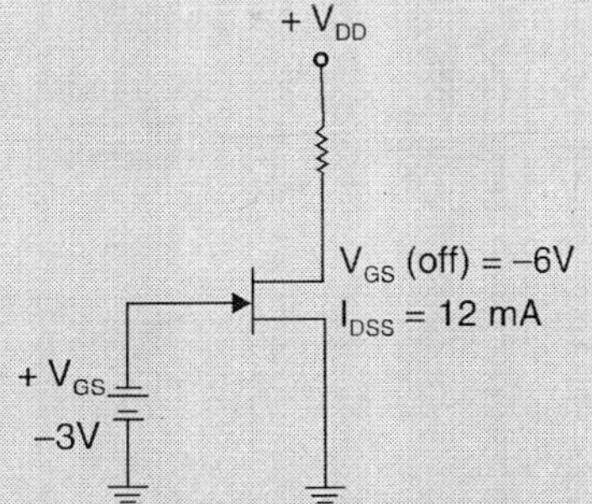

Fig. 13-31

2. Determine the value of drain current for the circuit shown in Fig. 13-31. (**Ans.** 5.32 mA)

3. A JFET has parameters of $V_{GS\,(off)}$ = –6V and I_{DSS} = 3 mA. Plot the transconductance curve for the device using V_{GS} values of 0V, –1V, –3V, –5V and –6V.

4. Plot the transconductance curve for a JFET $V_{GS(off)}$ = –20 V and I_{DSS} is 12 mA. Using V_{GS} values of 0V, –5V, –10V, –15V and –20V.

DESCRIPTIVE QUESTIONS

1. (*a*) Explain the working of a JFET (junction field-effect transistor).

(*b*) Draw the V-I characteristics of an N-channel JFET.

(*c*) Define dynamic drain resistance (r_d) and amplification factor (μ) for a JFET.

2. Define the following terms for a JFET:
   (*a*) The pinch-off voltage
   (*b*) Channel ohmic region
   (*c*) Drain resistance
   (*d*) Transconductance
   (*e*) I_{DSS}
   (*f*) I_{GSS}
3. Explain with the help of neat diagrams, the structure of a N-channel FET, and its volt-ampere characteristic. In what ways it is different from a bipolar junction transistor?
4. List the advantages and disadvantages of FET over bipolar transistors.
5. Bring out a neat comparison between a JFET and a MOSFET.
6. Describe the construction and explain the operation of depletion enhancement IGFET. Also draw the static characteristics.
7. Discuss with the aid of diagrams, the essential differences between enhancement mode FETs and depletion mode FETs.
8. Define the pinch-off voltage.
   (*Electronic Devices, Anna University, Nov/Dec 2007*)
9. Give the drain current equation of JFET.
   (*Electronic Devices, Anna University, Nov/Dec 2007*)
10. State the advantages of FET over of BJT.
    (*Electron Devices, Anna University, April/May 2008*)
11. Describe the operation of *N* Channel JFET with diagrams and characteristic curves. Explain the important parameters of JFET.
    (*Electron Devices, Anna University, April/May 2008*)
12. Mention the applications of FETs.
    (*Electron Devices, Anna University, Nov/Dec 2007*)
13. With neat diagram, Explain the operation, V-I and transfer characteristics of N-channel depletion type MOSFET. (*Electron Devices, Anna University, Nov/Dec 2007*)
14. Compare depletion and enhancement type of MOSFETs.
    (*Electron Devices, Anna University, Nov/Dec 2007*)

MULTIPLE CHOICE QUESTIONS

1. A field-effect transistor (FET) operates on
   (*a*) Majority carriers only
   (*b*) Minority carriers only
   (*c*) Positively charged ions only
2. In JFET operating above pinch-off voltage, the
   (*a*) Drain current remains practically constant
   (*b*) Drain current starts decreasing
   (*c*) Drain current increases rapidly
   (*d*) Depletion region becomes smaller
3. The JFET is oftenly called square-law device because its
   (*a*) Transconductance curve is parabolic
   (*b*) A.C. resistance from drain-to-source varies inversely as square of the drain current
   (*c*) Drain current varies as square of drain voltage for a fixed gate-to-source voltage
   (*d*) Reverse gate leakage current varies as a square of the reverse gate voltage

4. For the operation of depletion-type MOSFET, the gate voltage has to be
(*a*) Low positive (*b*) High positive (*c*) High negative (*d*) Zero

5. The N-channel MOSFET devices are preferred more than P-channel's because
(*a*) N-channel devices are faster than P-channel devices
(*b*) N-channel devices consumes less power then P-channel devices
(*c*) N-channel devices have higher packing density than P-channel devices
(*d*) Both (*a*) and (*c*) above

6. As compared to N-channel MOS switch, the P-channel MOS switch has
(*a*) Same ON resistance (*b*) Less ON resistance
(*c*) More ON resistance (*d*) either '*a*' or '*b*'

7. Thermal runway is not possible in FET because as the temperature of the FET increases:
(*a*) the mobility decreases (*b*) the transconductance increases
(*c*) the drain current increases (*c*) the mobility increases
(*UPSE Engg Service* 2001)

8. The trans conductance '*g*' of a JFET is equal to
(*a*) $\frac{-2\,I_{DSS}}{V_P}$ (*b*) $\frac{2}{|V_P|}\sqrt{I_{DSS}\cdot I_D}$
(*c*) $-\frac{2\,I_{DSS}}{V_P}\left\{1-\frac{V_{GS}}{V_P}\right\}$ (*d*) $\frac{I_{DSS}}{V_P}\left\{1-\frac{V_{GS}}{V_P}\right\}$

9. An FET is a better chopper than a BJT because it has :
(*a*) lower offset voltage (*b*) higher series ON resistance
(*c*) lower input current (*d*) higher input impedance.
(*e*) both (*c*) and (*d*) (*UPSC Engg Service* 2000)

10. In a biased JFET, the shape of the channel is as shown in Fig 13-32.

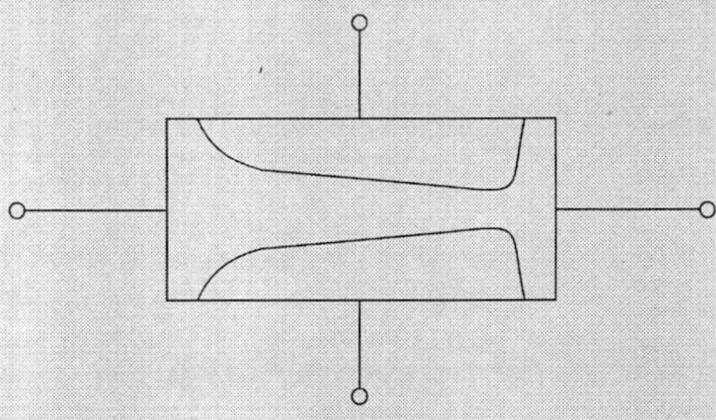

Fig. 13-32.

(*a*) it is the property of the material used.
(*b*) the drain end is more reverse biased than source end.
(*c*) the drain end is more forward biased than source end.
(*d*) the impurity profile varies with the distance from the source.
(*UPSC Engg. Services* 1999)

11. Which of the following statement is not true in case of FET.
(*a*) It has high input impedance (*b*) It is less noisy than bipolar transistor.
(*c*) It has a large gain × band width (*d*) all of the above.

12. The JFET is a
(*a*) current controlled device with high input resistance
(*b*) voltage controlled device with high input resistance
(*c*) voltage controlled device with low input resistance
(*d*) current controlled device with low input reisistance

13. The input impedance of a JFET is in the range of
(*a*) above 2 MΩ (*b*) 200 to 400 kΩ
(*c*) 20 to 40 kΩ (*d*) below 2 kΩ

(Bharthiar Univ., 2004)

14. FET is
(*a*) current controlled device (*b*) voltage controlled device
(*c*) resistance controlled device (*d*) reactance controled device

(Chatrapati Shiva Maharaj Univ., 2004)

15. In a FET, 10 volts reverse voltage is applied. If gate current is 0.1 μA, the input resistance is,
(*a*) 100 MΩ (*b*) 10 MΩ
(*c*) 100 MΩ (*d*) none of these

(Chatrapati Shival Maharaj Univ., 2004)

ANSWERS

1. (*a*)	**2.** (*a*)	**3.** (*a*)	**4.** (*c*)	**5.** (*d*)
6. (*c*)	**7.** (*b*)	**8.** (*a*)	**9.** (*e*)	**10.** (*e*)
11. (*c*)	**12.** (*b*)	**13.** (*a*)	**14.** (*b*)	**15.** (*c*)

❑❑❑

14

CHAPTER

Thyristors

OUTLINE

Objectives

After completing this chapter, you should be able to :

- Describe the various types of thyristors.
- Recognise the different terminals of the SCR.
- Explain the operation of SCR.
- Understand how the SCR is triggered ON and OFF.
- Know the SCR ratings.
- Describe the basic construction of UJT.
- Sketch an equivalent circuit of UJT.
- Explain the UJT operation.
- Sketch the V-I characterstics of all the thyristor devices.

14-1. Introduction

A thyristor is a four-layer PNPN device, which has three PN junctions. It has two stable switching states namely the ON or conducting state and the OFF or non-conducting state. There is no other state in between these two, as is in bipolar and field-effect transistors. The thyristors are used specifically for high power switching applications such as control of a.c. power to the load, motor speed control, light dimmers, *etc*. Unlike BJT's and operational amplifiers (OP-AMPS), which can also be used as switches, most thyristors are not designed to be used as linear amplifying devices.

14-2. Types of Thyristors

Following are two main types of thyristors :

(*i*) ***Unidirectional :*** The thyristors, which conduct in forward direction only, are known as unidirectional thyristors. The examples of such thyristors are Silicon Controlled Rectifiers (SCR's), Light Activated Silicon Controlled Rectifiers (LASCR's), and Silicon Controlled Switch (SCS).

(*ii*) ***Bidirectional :*** The thyristors, which can conduct in forward as well as in reverse direction, are known as bidrectional thyristors. The example of such thyristors is a Triode A.C. Switch (TRIAC).

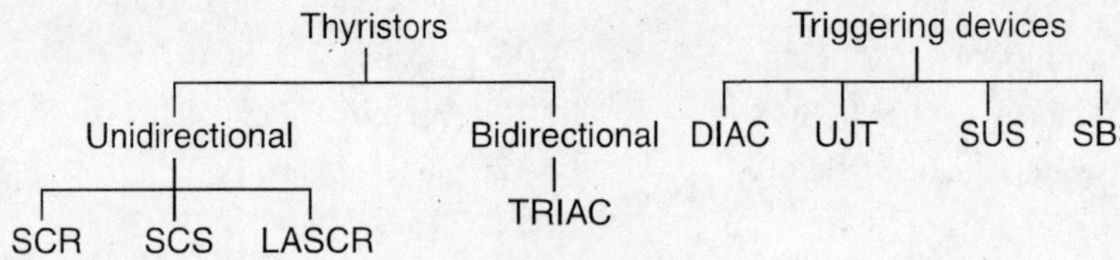

Fig. 14-1.

The thyristors require a control signal to switch from the non-conducting to the conducting state. The devices, which generate such signals, are called triggering devices.

Fig. 14-1 (*b*) shows the various types of triggering devices, which are available in these days. It includes Unijunction Transistor (UJT), Diode A.C. Switch (DIAC). Silicon Unilateral Switch (SUS) and Silicon Bilateral Switch (SBS).

14-3. Silicon Controlled Rectifier (SCR)

These days, the Silicon Controlled Rectifier briefly written as (SCR) is one of the most important semiconductor device in the industrial or power electronics field. It is used as a controlled switch to perform a variety of functions such as rectification, d.c. to a.c. inversion and power control. It is an important element in the control of electrical motor speed, electrical furnace heat, lighting and many other uses.

A Silicon Controlled Rectifier (SCR) consists of four semiconductor layers forming a PNPN structure as shown in Fig. 14-2 (*a*). It has three PN junctions namely J_1, J_2 and J_3. There are three terminals called anode (*A*), cathode (*K*) and the gate (*G*) taken out as shown in the Figure. The anode (*A*) terminal is taken out from the P_1-layer. The cathode (*K*) terminal is taken out from the N_2-layer, and the gate (*G*) terminal from the P_2-layer. The schematic symbol of SCR is shown in Fig. 14-2(*b*). It may be noted that the SCR symbol is quite similar to that of a diode. In fact, the SCR resembles the diode electrically, since it conducts the current in one direction only, when forward biased. However, the SCR's different from a diode because it has an additional gate terminal. This gate is used to turn 'ON' the device.

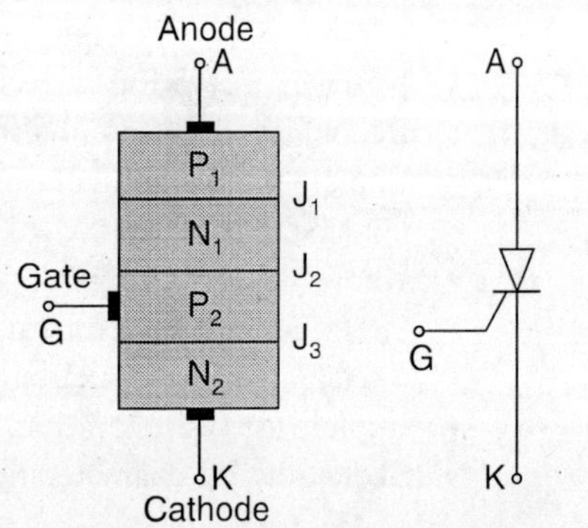

(*a*) Construction. (*b*) Symbol.

Fig. 14-2. *Silicon controlled rectifier.*

14-4. SCR Biasing

The SCR can be biased in two modes depending upon the polarity of the applied voltage across anode and cathode terminals. When the anode is positive, with respect to the cathode, the SCR is said to be forward-biased as shown in Fig. 14-3(*a*).

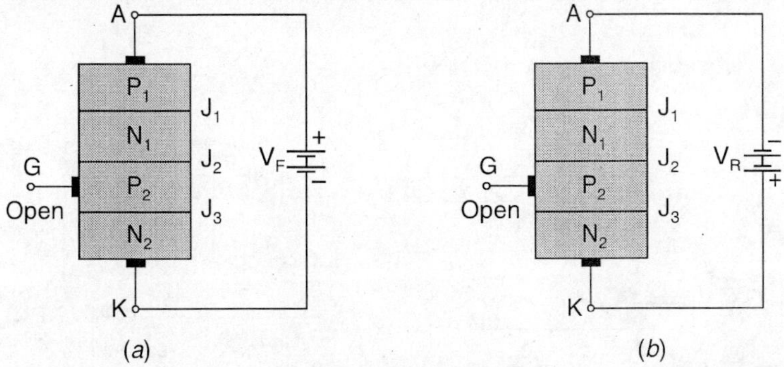

Fig. 14-3. *SCR biasing.*

In this mode, the junctions J_1 and J_3 are forward biased and the junction J_2 is reverse biased. There is no current (except leakage current) through the SCR. Therefore, the SCR is in 'OFF' (non-conducting) state. Under this condition, the device offers a very high resistance. The value of this resistance is several megaohms.

On the other hand, if the anode is negative, with respect to cathode, the SCR is said to be reverse biased as shown in Fig. 14-3(*b*). In this mode, the junctions J_1 and J_3 are reverse biased and the junction J_2 is forward biased. Again there is no flow of current (except leakage current) through the SCR. Therefore, the SCR is in 'OFF' (non-conducting) state.

14-5. SCR Operation

The SCR operation may be understood with reference in to Fig. 14-3(*a*) and (*b*). We know that when SCR is forward-biased with a small voltage, it is in 'OFF' position and no current flows through the device. However, if the applied forward voltage is increased, a certain critical value called forward breakover voltage (V_{BO}) is reached at which the junction J_2 breaks down. This causes the SCR to quickly switch to its 'ON' (*i.e.*, conducting) position. Under this condition, the SCR offers very small resistance (of about 0.1 to 1.0 Ω) and the voltage very across it, drops to a low value. The value of this voltage is about one volt. In the ON state, the current through the SCR is very large and is controlled by the applied voltage and the external resistance.

Now, if the battery connections of the applied voltage are reversed, the SCR is reverse biased. Under this condition, the junction J_1 and J_3 are reverse biased and junction J_2 is forward biased. However, it has been found that most of the voltage will drop across junction J_1 only. When the applied reverse voltage is small, the SCR is OFF and therefore no current (except leakage current) flows through the device. If the reverse voltage is increased to reverse breakdown voltage, the junction J_1 will breakdown due to avalanche effect. This causes a large current to flow through the SCR, due to which it may get damaged in the same way as the reverse biased PN-junction diode.

It is evident from the above discussion that SCR can be used to conduct in one forward direction only, like a rectifier diode. Therefore, SCR is unidirectional semiconductor device, which remains OFF so long the applied anode voltage is below the breakover voltage and turns ON when it exceeds the breakover voltage.

It will be interesting to know that SCR is never operated with the anode to cathode voltage (*i.e.*, supply voltage across SCR) equal to the forward breakover voltage. In fact, it is operated at a supply voltage much smaller than the forward breakover voltage. In that case, SCR is turned ON by the gate voltage and gate current.

14-6. SCR Equivalent Circuit

The basic structure of SCR can be divided into 3-layer structure as shown in Fig. 14-4(*a*). It may noted that the upper 3-layer structure is a PNP transistor, whereas the lower one is an NPN transistor. Thus, SCR can be represented by two transistors Q_1 (*i.e.*, PNP) and Q_2 (*i.e.*, NPN) interconnected as shown in Fig. 14-4(*b*). Sometimes, this circuit is also called as the *two transistor analogy* or *an ideal latch of a SCR.*

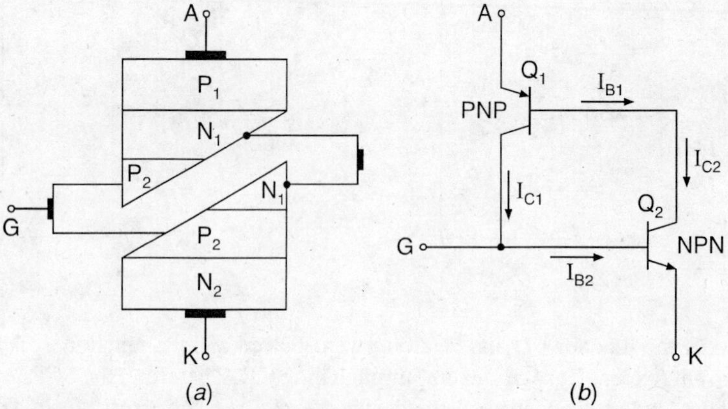

Fig. 14-4. *SCR equivalent circuit.*

It is evident from the Fig. 14-4(b) that the collector of each transistor is connected to the base of other transistor. Therefore, the collector current of transistor Q_1, is the base current of transistor Q_2, and the base current of transistor Q_1 is the collector current of transistor Q_2. For this circuit, we have an action of positive feedback or regeneration. It means that if there is a change in current, at any point in the loop (formed by transistors Q_1 and Q_2), it is amplified and returned to the starting point with the same phase. For example, if the base current of transistor Q_2 increases, the collector current of transistor. Q_2 will also *increase. It causes more base current through transistors Q_1 due to which the collector current of transistor Q_1 increases. This action will continue till both the transistors are driven into saturation. In this case, the SCR acts like a ON switch and it will pass the current from anode to cathode.

On the other hand, if the base current of transistor Q_2 decreases, the collector current of transistor Q_2 will also decrease. It causes the reduced base current through transistor Q_1 due to which the collector current of transistor Q_1 decreases. This action will continue till both the transistors are driven into cut-off. In this case, the SCR acts like a OFF switch and hence, it will block the current from anode to cathode.

14-7. Turning ON (or Triggering) SCR

The SCR can be turned ON, from its OFF position, by several methods as discussed below :

(1) ***Forward breakover voltage :*** If the voltage across the SCR exceeds the rated forward breakover voltage, the SCR will start conducting due to avalanche breakdown.

(2) ***Gate triggering :*** This is the most commonly used method to trigger the SCR. In this method, the SCR is operated with an anode voltage slightly less than the rated forward breakover voltage and is triggered into conduction by a low-power gate pulse. It may be noted that once the SCR is switched ON, the gate has no further control on the device current. The gate pulse signals can be supplied either from a d.c. source or an a.c. source.

* This will happen due to d.c. current gain (β) of the transistor. If the base current I_B increases, then the collector current become $I_C = \beta \cdot I_B$.

Fig. 14-5 (*a*) shows an SCR connected to the d.c. source through a load. In this case, the gate signal is generated by a push button switch (*S*). When the switch (*S*) is pressed, momentarily, a positive voltage is applied at the gate. As a result of this, the SCR is turned ON, and the current flows through the load. The SCR will remain in its ON position, until the supply voltage is removed or reversed.

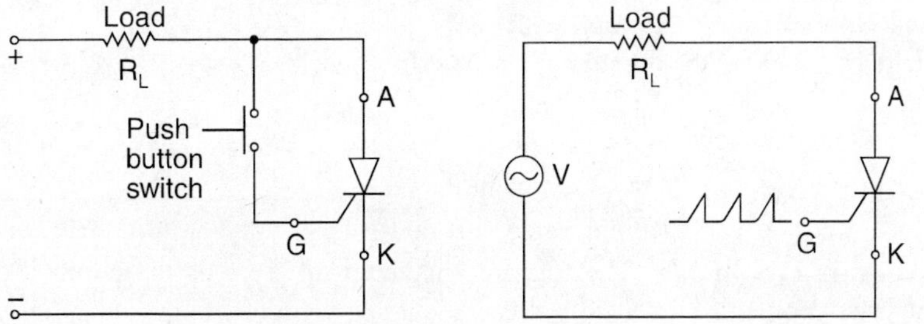

(*a*) Triggering SCR from d.c. source. (*b*) Triggering SCR from pulses.

Fig. 14-5.

Fig. 14-5 (*b*) shows a SCR connected to the a.c. source through a load. In this circuit, the gate signal is provided by the timing pulses. Such pulses can be generated by a number of devices, called triggering devices. Such devices are unijunction transistor, diac, silicon unilateral switch *etc*.

(3) ***Rate-effect or dV/dt triggering*** **:** In this method, the SCR is turned ON by rapidly increasing the anode-to-cathode voltage. The rapidly increasing anode-to-cathode voltage produces a charging current, which triggers the SCR to conduction.

(4) ***Light triggering*** **:** In this method, the SCR is triggered by irradiating it with light. When the light falls on the middle junction (J_2) of the SCR, the device turns ON. Such a device is called Light-Activated Silicon Controlled Rectifier (LASCR).

14-8. Turning OFF SCR

We have already discussed in the last article, that once the SCR turns ON (*i.e.*, starts conducting), it continues to conduct even when the gate signal is removed. This ability of the SCR to remain conducting, even when the gate signal is removed, is known as latching. It means that SCR cannot be turned OFF simply by removing the gate signal. A number of methods are used for turning OFF the SCR as mentioned below :

(*i*) Reversing polarity of anode-to-cathode voltage.

(*ii*) Interrupting anode current by means of momentarily series or parallel switching arrangement. This method is known as *anode current interruption*.

(*iii*) Reducing the current through SCR below the holding current. This method is known as *forced commutation*.

14-9. V-I Characteristics of SCR

It gives the relationship between the anode current and the anode-to-cathode voltage of SCR for different values of gate current. The SCR has two types of V-I characteristics namely forward characteristic and reverse characteristics.

The forward characteristic may be obtained by using the SCR circuit having adjustable d.c. supplies V_{AA} and V_{GG} as shown in Fig. 14-6. The reverse characteristic may be obtained by reversing the connections of both the d.c. supplies.

An ammeter is used to measure the value of anode current, while the voltmeter is used to measure the anode-to-cathode voltage.

14-10. Forward Characteristic

The forward characteristic of SCR may be obtained by using the circuit shown in Fig. 14-6. First of all, adjust the gate current to zero value by keeping the switch (*S*) open. Then increase the applied voltage across the SCR in small suitable steps. At each step, record the value of anode current. Now, if we plot a graph with anode-to-cathode voltage (V_{AK}) along the horizontal axis and the anode current (I_A) along the vertical axis, we shall obtain a curve marked *OABC* as shown in Fig. 14-7 (*a*).

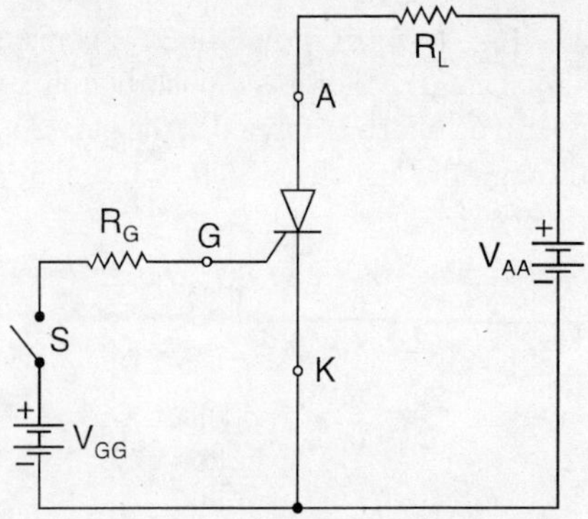

Fig. 14-6. *SCR circuit.*

A similar procedure may be used to obtain curves marked I_{G_1}, I_{G_2} and I_{G_3} for different values of gate-current as shown in Fig. 14-7(*b*). The forward characteristic gives us the information about the following important points:

(*i*) As the applied anode-to-cathode voltage (V_{AK}) is increased above zero, a very little anode current flows through the device. Under this condition, the SCR is OFF and it offers very high resistance (ideally open). It continues till the voltage V_{AK} reaches the forward breakover voltage marked by point *A* in Fig. 14-7(*a*).

(*ii*) As the anode-to-cathode voltage exceeds the breakover voltage the SCR turns ON and the anode-to-cathode voltage decrease quickly to a value marked by point *B*. The value of breakover voltage is about one volt. At this stage, the current through the SCR increases rapidly to a large value, which is determined by the supply voltage and the value of load resistance in the circuit.

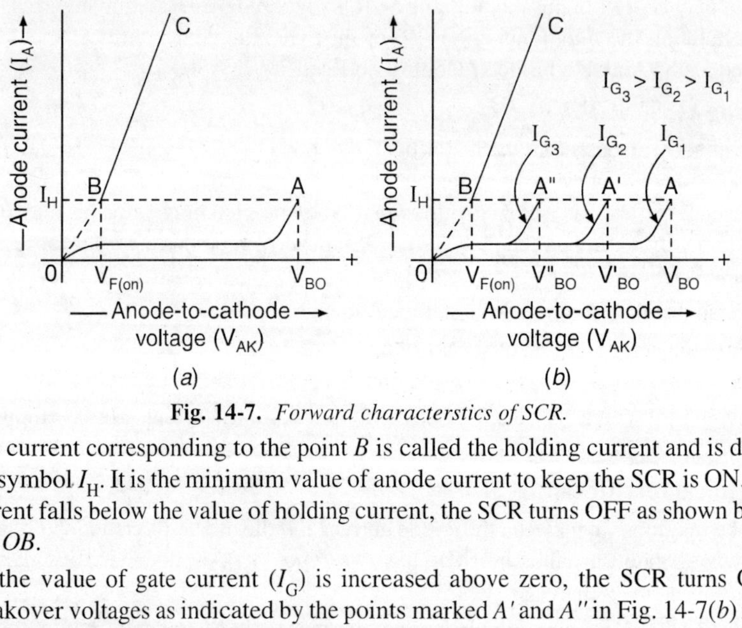

Fig. 14-7. *Forward characterstics of SCR.*

(*iii*) The current corresponding to the point *B* is called the holding current and is designated by the symbol I_H. It is the minimum value of anode current to keep the SCR is ON. If the anode current falls below the value of holding current, the SCR turns OFF as shown by the broken line *OB*.

(*iv*) As the value of gate current (I_G) is increased above zero, the SCR turns ON at lower breakover voltages as indicated by the points marked *A'* and *A"* in Fig. 14-7(*b*) respectively.

(*v*) The region lying between the points *O* and *A* is called forward blocking region. In this region, the SCR is OFF and blocks the forward anode-to-cathode voltage.

(*vi*) The region lying between the points *B* and *C* is called forward conduction region. In this region, the SCR is ON and conducts current.

14-11. Reverse Characteristics of SCR

The reverse characteristic of SCR may be obtained by using the circuit as shown in Fig. 14-6 and reversing the connections of the d.c. supplies V_{AA} and V_{GG}. First of all, adjust the gate current to any suitable value. Then increase the reverse applied voltage in suitable steps and record the value of reverse anode current. Now, if we plot a graph with applied reverse anode-to-cathode voltage (V_R) along the horizontal axis and anode current (I_R) along the vertical axis, we shall obtain a curve marked *ODE* as shown in Fig. 14-8. The reverse characteristic of SCR gives us the information about the following important points :

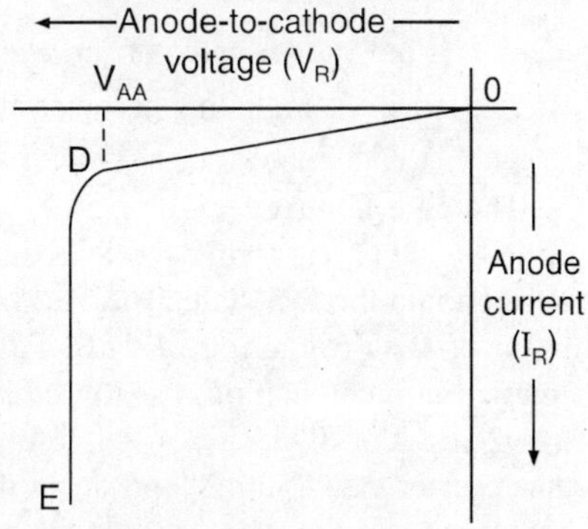

Fig. 14-8. *Reverse characterstic of SCR.*

(1) As the applied reverse voltage (V_R) is increased above zero, a very little current (called leakage current) flows through the SCR. Under this condition, the SCR is OFF and it offers a very high resistance. It continues till the applied reverse voltage reaches breakdown voltage (V_{BR}).

(2) As the applied reverse voltage is increased, above the breakdown voltage, the reverse current increase more rapidly as shown by the curve *DE*. This rapid increase is because of the breakdown of SCR (due to avalanche effect) and may damage the device if the current exceeds the rated value.

(3) The region lying between the points *O* and *D* is called reverse blocking region. In this region, the SCR is OFF. Therefore, it blocks the reverse anode-to-cathode voltage.

(4) The region lying between the points *D* and *E* is called reverse avalanche region. In this region, a large value of reverse current flows through the device.

14-12. SCR Ratings

Following are some of the ratings of SCR, which are important from the subject point of view :

- ***Forward breakover voltage :*** It is the voltage at which the SCR is switched from its OFF position to ON position. Its value is maximum for zero gate current. Commercially available SCR's have a forward breakover voltage ranging from 50 to 1200 V.
- ***Holding current :*** It is the value of anode current below which the SCR switches from its ON position to OFF position under the given conditions. In other words, holding current is the minimum value of current required to maintain conduction.
- ***Gate trigger current :*** It is the value of anode current necessary to switch SCR from its OFF position to ON position under specified conditions.
- ***Average forward currents :*** It is the maximum continuous value of anode current, which the SCR can handle in its ON position under specified conditions. It is also called current rating. Commercially available SCR's have a current rating ranging from 1 to 1800 A.
- ***Reverse breakdown voltage :*** It is the value of reverse voltage from cathode-to-anode at which the device breaks into avalanche region and starts conducting heavily.

Other parameters, usually, included on the specification sheet of an SCR are the turn-on-time, turn-off-time, junction temperature and case temperature, circuit fusing (I^2t) rating, gate non-trigger voltage (V_{GD}) rating and critical rise (dv/dt) rating.

The circuit fusing (I^2t) rating indicates the maximum forward surge current capability of the SCR device. For example, the I^2t rating of Motorola's 2N 682-92 series SCR's is 93 A^2s. *If the product of the square of the surge current times the duration (time) of the surge exceeds 93 A^2s, the SCR is destroyed by excessive power dissipation.*

When the circuit fusing rating of an SCR is known, we can determine the maximum allowable duration of the surge with a known current value as follows :

$$t_{max} = \frac{I^2 t \,(\text{rated})}{I_s^2}$$

where, I_s is the known value of surge current.

The circuit fusing rating becomes critical in applications where SCR is used as a surge protection device.

The gate nontrigger voltage (V_{GD}) rating indicates the maximum gate voltage that can be applied without triggering the SCR into conduction. If V_G exceeds this rating, the SCR will be triggered into the ON state. The V_{GD} rating is important because it points out one of the potential reason of false triggering of SCR. False triggering is a situation where the SCR is accidentally triggered into conduction, usually by some type of electrical noise. For example, Motorola's 2N 682-92 series of SCRs has a V_{GD} rating of 250 mV. This means that if a noise signal with a peak value greater than 250 mV appears at the gate, the device may be triggered into conduction.

Another common reason of false triggering is a rise in anode voltage that exceeds the critical rise (dv/dt) rating of the SCR. This rating indicates the maximum rate of increase in anode-to-cathode voltage that the SCR can handle without false triggering occurring. The dv/dt rating of Motorola's 2N 682-92 series SCRs is 30 V/μs. This means that a noise signal in V_{AK} with a rate of rise equal to 30 V/μs may cause false triggering.

Example 14-1. *The 2N 682 SCR is subjected to 40 A surge that lasts for 15 ms. Determine whether or not this surge will destroy the device. The circuit fusing rating of the device is 93 A^2s.*

Solution. Given: $I = 40$ A and $t = 15$ ms $= 15 \times 10^{-3}$ sec.

We know that I^2t value for the SCR,

$$= (40^2) \times (15 \times 10^{-3}) = 24 \text{ A}^2\text{s}.$$

Since, this value is well below the maximum rating of 93 A^2s, therefore the device will not be destroyed.

Example 14-2. *The C35 series SCR have a circuit fusing rating of 75 A^2s. Determine the maximum allowable duration of a 100 A surge that passes through one of these devices.*

Solution. Given: $I^2t = 75$ A^2s and $I_s = 100$ A

We know that maximum allowable duration,

$$t_{max} = \frac{I^2 t \,(\text{rated})}{I_s^2} = \frac{75}{(100)^2} = 7.5 \times 10^{-3} \text{ s} = 7.5 \text{ ms} \quad \textbf{Ans.}$$

14-13. Applications of SCR

The SCR has a number of applications, yet the following industrial applications are important from the subject point of view:

1. Motor speed control **2.** Light-dimming control **3.** Heater control **4.** Phase control **5.** Battery charger **6.** Inverters **7.** Static switches **8.** Rectifier power supplies **9.** Relay control

14-14. Triac

A triac is a three terminal device, which can conduct in either direction, when triggered either by a positive or a negative pulse irrespective of the polarity of the voltage across its main terminals. The triac behaves like two SCR's connected in parallel but in opposite directions, with a common gate terminal, *i.e.*, the anode of one SCR is connected to the cathode of the other and their gates are connected together. Thus, anode and gate voltage applied in either direction will trigger the triac. It is due to the fact that the applied voltage will trigger at least one of the SCR's connected in opposite direction. The triacs are available with current ratings up to 25 A and voltage ratings up to 500 volts.

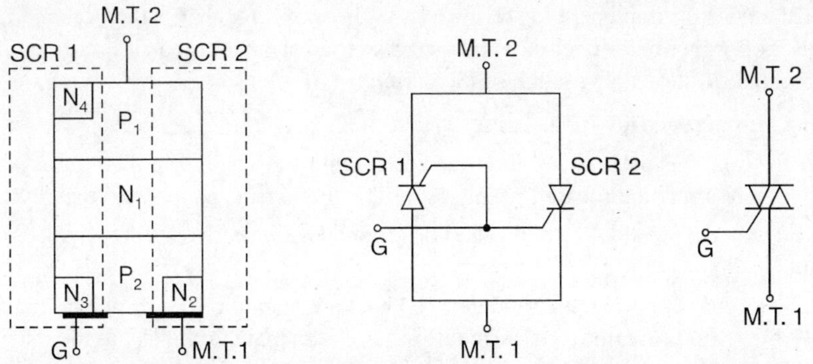

(*a*) Construction (*b*) Equivalent circuit (*c*) Symbol

Fig. 14-9. *Triac.*

The basic construction of a triac is shown in Fig. 14-9 (*a*). It may be noted from this figure that the triac consists of two four layer switches in parallel. These switches are $P_1N_1P_2N_2$ and $P_2N_1P_1N_4$ as shown by the broken lines in the figure. Since, the triac conducts in both directions, therefore the terminals are designated by numbers instead of anode and cathode as in SCR. The triac has two main terminals namely main terminal 1 [*i.e.*, (M.T. 1)] and main terminal 2 (*i.e.*, M.T.2) and one gate terminal (*G*). Fig. 14-9 (*b*) shows the equivalent circuit of a triac, which consists of two SCR's connected in parallel but in opposite directions with a common gate terminal. Fig. 14-9 (*c*) shows the schematic symbol of a triac.

14-15. Triac Operation

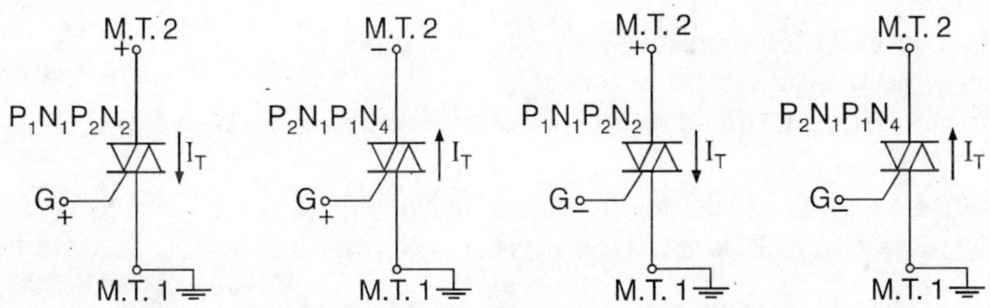

(*a*) Mode 1 (*b*) Mode 2 (*c*) Mode 3 (*d*) Mode 4

Fig. 14-10. *Triac operation.*

There are four modes of triac operation, depending upon the polarity of voltage across its main terminals and gate terminal. These modes are described as below :

(*a*) ***M.T 2 is positive and G is positive :*** In this mode, the operation of the triac is identical to SCR. The current flows through the switch $P_1N_1P_2N_2$ from M.T. 2 to M.T. 1 as shown in Fig. 14-10 (*a*).

(*b*) ***M.T. 2 is negative and G is positive :*** In this mode, the current flows through the switch $P_2N_1P_1N_4$ from M.T. 1 to M.T. 2 as shown in Fig. 14-10 (*b*). This is an inefficient mode and must be avoided.

(*c*) ***M.T. 2 is positive and G is negative :*** In this mode, the current flows through the switch $P_1N_1P_2N_2$ from M.T. 2 to M.T 1 as shown in Fig. 14-10 (*c*). This mode is less efficient than mode 1 but not as poor as mode 2.

(*d*) ***M.T. 2 is negative and G is negative :*** In this mode, the current flows through the switch $P_2N_1P_1N_4$ from M.T. 1 to M.T. 2 as shown in Fig. 14-10 (*d*). This mode is slightly less efficient than mode 1.

The mode 1 and mode 4 are efficient modes in triac operation. Therefore, these two modes are called normal modes of triac operation. However, as discussed for SCR's the triac is triggered with the gate pulses of proper polarity of d.c. voltages. The triac can be turned OFF only by reducing the device current below the holding value of the current.

14-16. V-I Characteristics of a Triac

It gives the relationship between the triac current and the voltage applied across its two main terminals. We have already discussed in the last article that a triac is operated usually in two ways

(*i*) Main terminal 2 (*i.e.*, M.T. 2) and gate (*G*) both are positive with respect to main terminal 1 (*i.e.*, M.T. 1), and

(*ii*) Main terminal 2 (*i.e.*, M.T. 2) and gate (*G*) both are negative with respect to main terminal 1 (*i.e.*, M.T. 1). In the first case, the triac current flows from the main terminal 2 to the main terminal 1. And in the second case, the triac current flows from the main terminal 1 to the main terminal 2 (*i.e.*, in a direction opposite to that of the first case).

When the triac is operated, with its main terminal 2 and gate both positive with respect to main terminal 1, the V-I characteristic obtained is as shown in Fig. 14-11 by the curve marked *OABC*. Similarly, when the Triac is operated with its main terminal 2 and gate both negative with respect to main terminal 1, the V-I characteristic obtained is shown in Fig. 14-11 by the curve marked *ODEF*.

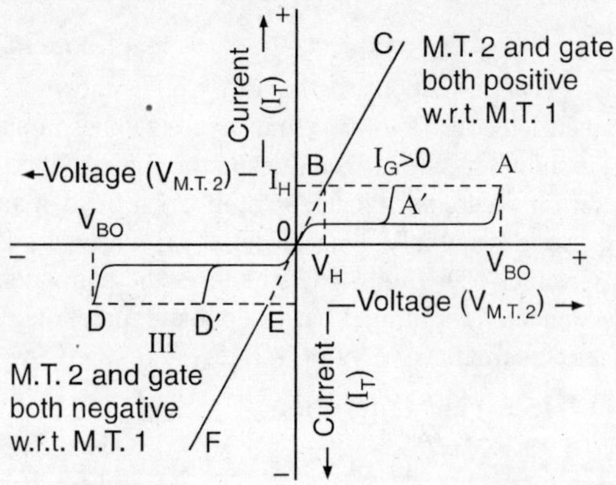

Fig. 14-11. *V-I characterstics of a Triac.*

The V-I characteristic of a triac gives us the information about the following important points :

(*i*) The curves OABC and ODEF are symmetrical and identical to the forward characteristic of silicon controlled rectifier (SCR).

(*ii*) The Triac is OFF until the applied voltage of either polarity (*i.e.*, whether M.T 2 is positive with respect to M.T. 1 or M.T. 2 is negative with respect to M.T. 1) exceeds the breakover voltage.

(*iii*) As the applied voltage of either polarity exceeds the breakover voltage, the triac turns ON and the voltage drop across the triac decreases to a low value (indicated by V_H). The triac current increases to a value determined by the supply voltage and load resistance.

(*iv*) As the value of gate current (I_G) is increased above zero, the breakover voltage is lowered, (indicated by points *A′* and *D′*). Like SCR, the triac is never operated with the zero gate current. When the gate current of a suitable value is applied (usually in the form of pulses), the triac turns ON at much lower breakover voltage.

14-17. Applications of Triac

The triac has an important property that it can conduct current in either forward or reverse direction, depending upon the polarity of the voltage across its terminals. This property makes the triac very useful in a large number of industrial applications. Some of the triac application are given below:

1. Phase control **2.** Motor speed control **3.** Heater control **4.** Light-dimming control **5.** Static switch to turn a.c. power ON and OFF

14-18. Difference Between SCR and Triac

Table 14-1 gives some of the important points of difference between SCR and a triac.

Table 14-1. *Difference between SCR and Triac*

S.No.	*SCR*	*Triac*
1.	It is unidirectional, *i.e.*, it can conduct current in forward direction only.	It is bidirectional, *i.e.*, it can conduct in forward as well as reverse direction.
2.	It is triggered by a narrow positive pulse applied at the gate terminal.	It is triggered by a narrow pulse of either positive or negative polarity applied at the gate terminal.
3.	SCR's are available with a large current rating.	Triacs are available for lower current rating as compared to that of SCR's.
4.	It has a fast turn-off and therefore can be used to switch a.c. supply frequencies up to few kHz.	It has a comparatively longer turn-off time than SCR. Therefore its use is limited to a.c. supply frequencies up to 400 Hz.

14-19. Unijunction Transistor (UJT)

A unijunction transistor (abbreviated as UJT) is a three terminal silicon semiconductor device. As the name indicates, the UJT has only one PN junction like an ordinary diode. However, it is different from the ordinary diode in the sense that it has three terminals. It will be interesting to know that the behaviour of unijunction transistor is quite different from the other transistors like bipolar junction transistor (BJT) and the field-effect transistor (FET).

14-20. Basic Construction of UJT

The basic construction of a unijunction transistor (UJT) is as shown in Fig. 14-12 (*a*). It consists of an N-type silicon semiconductor bar and a P-type silicon region. The N-type bar is called a base and the P-type region as the emitter. Thus a PN junction is formed between the emitter and base regions. The emitter region is heavily doped, while the base region is lightly doped. Due to this reason, the resistivity of the base material is very high. Three terminals are taken out of the whole structure : one from the emitter region and two from the ends of the base region. These terminals are labelled as emitter (E), base 1 (B_1) and base 2 (B_2). It may be noted that the emitter region is shown closer to base 2 terminal than base 1. Fig. 14-12 (*b*) shows the schematic symbol of UJT.

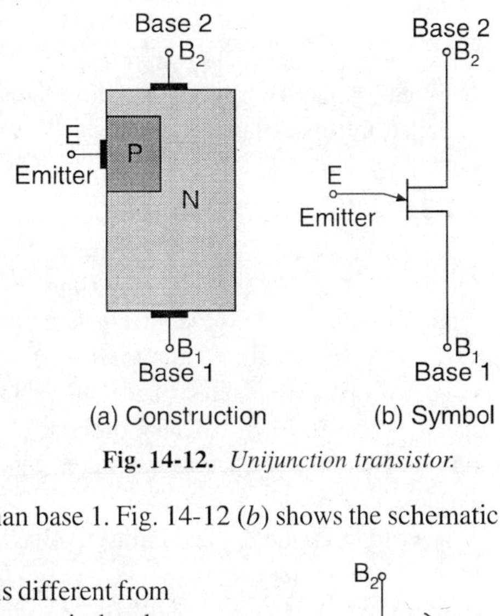

Fig. 14-12. *Unijunction transistor.*

It may be noted that the schematic symbol of UJT is different from that of JFET (junction field-effect transistor). The difference is that the arrow is at some angle in the schematic symbol of UJT.

14-21. Equivalent Circuit of UJT

Figure 14-13 shows the equivalent circuit of a unijunction transistor (UJT). It consists of a diode and a resistance. The diode (D) represents the PN junction, while the resistance (r_{BB}) is the internal bulk resistance of the silicon bar from one end to the other. In other words, the resistance r_{BB} represents the total resistance between the

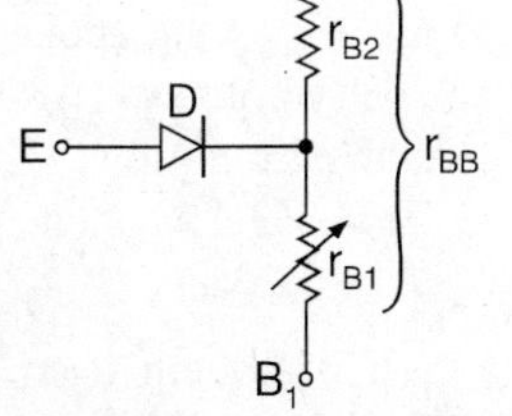

Fig. 14-13. *UJT equivalent circuit.*

base terminals and is called the *interbase resistance*. The resistance r_{BB} is represented by the sum of two separate resistances r_{B1} and r_{B2} in the equivalent circuit of UJT. The resistances r_{B1} represents the bulk resistance between the emitter (E) and base 1 (B_1), whereas resistance r_{B2} is the bulk resistance between the emitter (E) and base 2 (B_2). Mathematically, the resistance

$$r_{BB} = r_{B1} + r_{B2}$$

When there is no voltage applied to the UJT, the value of resistance, r_{BB} is typically 5 to 10kΩ. The resistance r_{B1} is shown as a variable resistance in the UJT equivalent circuit. It is because of the fact that the value of resistance r_{B1} varies inversely with the emitter current (I_E). Depending upon the value of emitter current, the value of resistance r_{B1} can vary typically from 4 kΩ to 40Ω.

14-22. Intrinsic Stand-off Ratio

Consider the equivalent circuit of a unijunction transistor (UJT) with a battery voltage V_{BB} applied across its base terminals B_1 and B_2 as shown in Fig. 14-14. As the emitter is open, the applied voltage V_{BB}, divides itself across resistance r_{B1} and r_{B2}. The voltage across the resistance r_{B_1},

$$V_1 = \frac{r_{B1}}{r_{B1} + r_{B2}} \times V_{BB} = \frac{r_{B1}}{r_{BB}} \times V_{BB}$$

The resistance ratio $\frac{r_{B1}}{r_{BB}}$ is an important characteristic of unijunction transistor. It is known as intrinsic stand-off ratio and is designated by η. Mathematically, the intrinsic stand-off ratio,

$$\eta = \frac{r_{B1}}{r_{BB}} = \frac{r_{B1}}{r_{B1} + r_{B2}}$$

Generally, the value of intrinsic stand-off ratio is between 0.5 and 0.8. The voltage drop across the resistance is called intrinsic stand off voltage. It reverse biases the emitter diode.

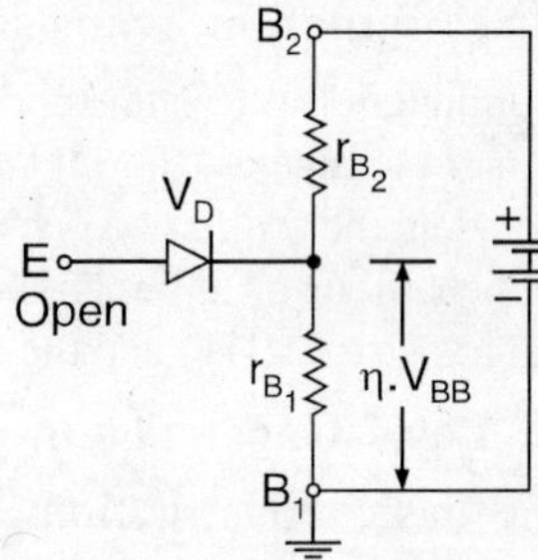

Fig. 14-14. *UJT equivalent circuit with battery V_{BB}.*

14-23. UJT Operation

Consider the equivalent circuit of a unijunction transistor (UJT) with the voltage source V_{EE} (connected across the emitter and base 1 terminal) and V_{BB} (connected across the base terminals B_1 and B_2) as shown in Fig. 14-15. Hence, the emitter diode (D) is reverse biased by a voltage drop across the resistance r_{B_1} (whose value is equal to $\eta \cdot V_{BB}$) and its own barrier potential (V_D). Thus, the total reverse bias voltage across a diode is equal to the sum of $\eta \cdot V_{BB}$ and V_D.

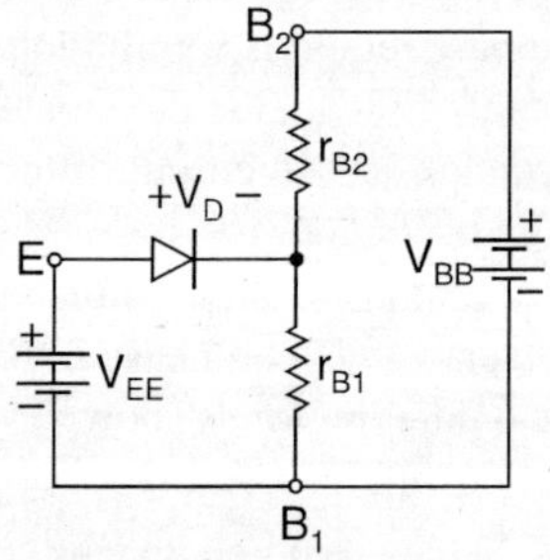

Fig. 14-15. *UJT equivalent circuit with V_{EE} and V_{BB} supplies.*

As long as the applied emitter voltage is below the total reverse bias voltage (*i.e.*, $\eta \cdot V_{BB} + V_D$) across the diode, it remains reverse biased. And there is no emitter current. However, as the applied emitter voltage reaches (or exceeds) the value equal to ($\eta \cdot V_{BB} + V_D$), the diode conducts and the emitter current flows. The value of emitter voltage, which causes the diode to conduct, is called peak-point voltage. Mathematically, the peak-point voltage,

$$V_P = \eta \cdot V_{BB} + V_D$$

It is evident from the above discussion that as the emitter voltage reaches the peak-point voltage, the diode conducts (*i.e.*, it becomes forward biased) and the emitter current begins to flow. Under this condition, the unijunction transistor is said to be fired, triggered or turned ON. At this instant, the holes from the P-type emitter region are injected into the base region and are swept by the electric field towards the base terminals B_1. The presence of excess holes, slightly reduces the resistance r_{B1} which in turn reduces the intrinsic stand-off voltage ($\eta \cdot V_{BB}$). This action is called conductivity modulation because the conductivity of the material between the emitter and base of terminals increases as the holes are injected into it. It is, obviously, a regenerative process because a smaller value of intrinsic stand-off voltage results in a stronger forward bias across the diode. Due to this, more holes are injected and the intrinsic stand-off voltage ($\eta \cdot V_{BB}$) is further reduced. As a result of this, the emitter current increases, while the voltage at the emitter (V_E equal to $\eta \cdot V_{BB} + V_D$) decreases. It produces a negative resistance region in the V-I characteristic of UJT. And the UJT switches from its OFF position to ON position.

Notes: 1. When the diode is forward biased, the current entering from the positive terminal of the supply voltage V_{BB} in the base 2 terminal also increases. Therefore, it is necessary to place an external resistor (called current limiting resistor) to the UJT in series with base 2 terminal. The value of current limiting resistor for the UJT's 2N 2646 and 2N 2647 is given by the relation,

$$R = \frac{1000}{\eta \cdot V_{BB}}$$

The value of resistor (R) ranges from 100 to 200 Ω.

2. Sometimes, it is required to obtain the output voltage from base of terminal of UJT. This can be done by placing a resistor of small value (ranging from 39 Ω to 47 Ω) in the base 1 lead.

3. A UJT can be switched ON from its OFF position by applying a positive trigger at its emitter terminal. It can be brought back to its OFF position by applying a negative trigger at the emitter terminal.

4. The output of UJT can be taken at anyone of the three terminals *i.e.*, emitter, base 1 or base 2.

Example 14-3. *Determine the peak-point voltage (V_P) value for the 2N 4870 UJT shown in Fig. 14-16. Take V_D = 0.7 V.*

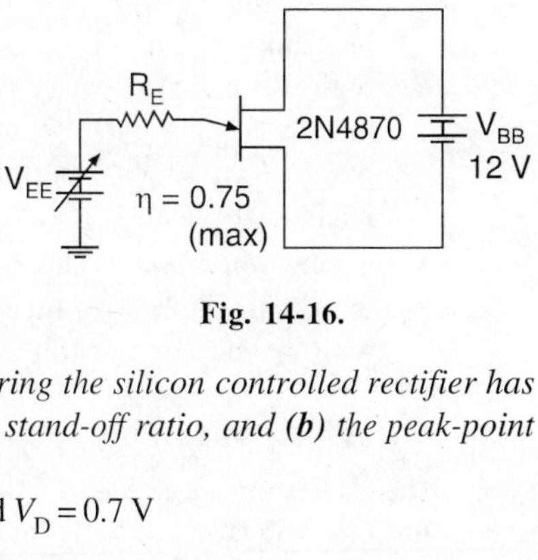

Fig. 14-16.

Solution. Given: V_D = 0.7 V; η = 0.75 (max) and $V_{BB} = 12$ V

We know that peak-point voltage of the circuit,

$$V_P = \eta \cdot V_{BB} + V_D$$
$$= (0.75 \times 12) + 0.7 = 9.7 \text{ V } \textbf{Ans.}$$

Example 14-4. *A unijunction transistor used for triggering the silicon controlled rectifier has r_{B1} = 4 kΩ and r_{B2} = 2.5 kΩ. Find **(a)** the value of intrinsic stand-off ratio, and **(b)** the peak-point voltage, if V_{BB} = 15 V. Take the barrier potential as 0.7 V.*

Solution. Given: $r_{B1} = 4$ kΩ $r_{B2} = 2.5$ kΩ $V_{BB} = 15$ V and $V_D = 0.7$ V

(*a*) *Intrinsic stand-off ratio*

We know that intrinsic stand-off ratio,

$$\eta = \frac{r_{B1}}{r_{B1} + r_{B2}} = \frac{4}{4 + 2.5} = 0.62 \textbf{ Ans.}$$

(*b*) *Peak-point voltage*

We also know that the peak-point voltage,

$$V_P = \eta \cdot V_{BB} + V_D = (0.62 \times 15) + 0.7 \text{ V}$$
$$= 10 \text{ V } \textbf{Ans.}$$

Example 14-5. *The intrinsic stand-off ratio for a UJT is determined to be 0.60. A measurement of its interbase resistance indicates 7 kΩ. What are the UJT's static values of r_{B1} and r_{B2}?*

Solution. Given: η = 0.60 and $r_{BB} = 7$ kΩ.

We know that intrinsic stand-off ratio (η)

$$0.60 = \frac{r_{B1}}{r_{BB}} = \frac{r_{B1}}{7}$$

$$\therefore \quad r_{B1} = 0.60 \times 7 = 4.2 \text{ k}\Omega \textbf{ Ans.}$$

and the total interbase resistance (r_{BB}),

$$\therefore \quad 7 = r_{B1} + r_{B2} = 4.2 + r_{B2}$$

$$r_{B2} = 7 - 4.2 = 2.8 \text{ k}\Omega \textbf{ Ans.}$$

14-24. V-I Characteristics of UJT

Fig. 14-17 shows the V-I characteristic of UJT. There are two important points on the characteristic curve namely the peak-point and the valley-point. These points divide the curve into three important regions, *i.e.*, cut-off region, negative resistance region and saturation region. These regions are explained below :

(1) ***Cut-off region :*** The region, to the left of peak-point, is called cut-off region. In this region, the emitter voltage is below the peak-point voltage (V_P) and the emitter current is approximately zero. The UJT is in its OFF position in this region.

(2) ***Negative resistance region :*** The region, between the peak-point and the valley-point, is called negative-resistance region. In this region, the emitter voltage decreases from V_P to V_V and the emitter current increases from I_P to I_V. The increase in emitter current is due to the decrease in resistance r_{B1}. It is because of this fact that this region is called negative-resistance region. It is the most important region from the application point of view. For example, when the UJT is operated as an oscillator, it works in the negative-resistance region.

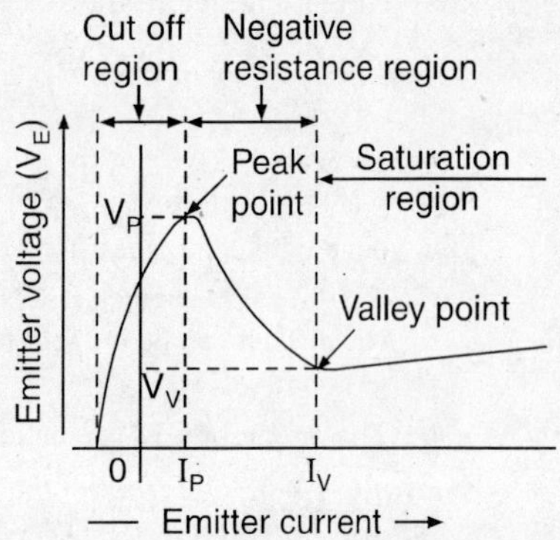

Fig. 14-17. *V-I characteristics of UJT.*

(3) ***Saturation region :*** The region, beyond the valley point, is called saturation region. In this region, the device is in its ON position. The emitter voltage (V_E) remains almost constant with the increasing emitter current.

14-25. UJT Applications

The UJT is not classified as a thyristor but rather as a trigger device. This device is mainly used to control the triggering of SCRs and triacs. The UJT applications may be summerized as below.

(*i*) Trigger device for SCR's and TRIACs

(*ii*) Non-sinusiodal oscillators.

(*iii*) Saw-tooth generators

(*iv*) Timing circuits

14-26. Sweep Circuit Using Unijunction Transistor Switch

Fig. 14-18 shows a sweep circuit using a unijunction transistor (abbreviated as UJT) switch. The circuit is commonly known as UJT relaxation oscillator.

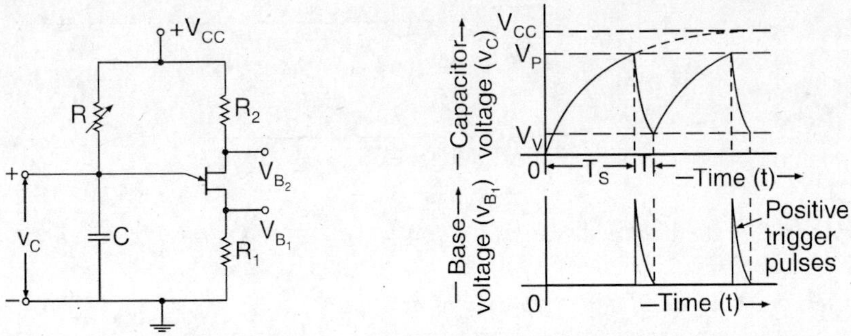

(*a*) UJT relaxation oscillator. (*b*) Associated waveforms.

Fig. 14-18.

The operation of the circuit may be understood form the condition that when the supply voltage (V_{CC}) is switched ON, the capacitor (*C*) charges through resistor (*R*), till the capacitor voltage reaches the voltage level (V_P) which is called peak-point voltage. At this voltage, the unijunction transistor turns ON. As a result of this, the capacitor (*C*) discharges rapidly through resistor (R_1). When that capacitor voltage drops to level V_v (called valley-point voltage) the unijunction transistor switches OFF allowing the capacitor (*C*) to charge again.

Fig. 14-18 (*b*) shows the collector voltage (v_c) and the base 1 voltage waveforms during the capacitor charging and discharging interval. It may be noted that the voltage developed at base 1 terminal is in the form of narrow pulses commonly known as trigger pulses. The similar pulses (but with opposite polarity) are also available at the base terminal of the unijunction transistor.

It will be interesting to know that the sweep period, also called period of oscillation, depends upon time constant ($R \cdot C$) and the type of unijunction transistor used in the circuit. However, it is independent of supply voltage (V_{CC}) and the temperature. The sweep period is given by the relation,

$$T = R \cdot C \cdot \log_e \frac{1}{1-\eta}$$

$$= 2.3\,\mathrm{R \cdot C} \cdot \log_{10} \frac{1}{1-\eta}$$

where η is a intrinsic stand-off ratio. Its value is approximately equal to the ratio of peak-point voltage to the supply voltage *i.e.*, $\eta = V_P/V_{CC}$. The sweep frequency (or the frequency of oscillation) is the reciprocal of sweep period *i.e.*,

$$f = \frac{1}{T} = \frac{1}{2.3\, R \cdot C \cdot \log_{10} \dfrac{1}{1-\eta}}$$

The sweep frequency can be varied by changing values of either resistance (*R*) or capacitor (*C*). However, it is more convenient to change the value of (*R*) than (*C*). It is due to this fact that the resistor (*R*) is shown as a variable resistor in Fig. 14-18 (*a*).

Example 14-6. *In a sample UJT sweep circuit, the resistance and capacitance are 100 kΩ and 0.4 μF. The ratio of the peak-point voltage to the supply voltage is 0.57. Find the frequency of sweep.* *(Electronic Circuits-II, Anna Univ., May/June 2007)*

Solution. Given: $R = 100\ \text{k}\Omega = 100 \times 10^3\ \Omega$; $C = 0.4\ \mu\text{F} = 0.4 \times 10^{-6}$ F and $\eta = V_P/V_{CC} = 0.57$.

We know that the frequency of sweep,

$$f = \frac{1}{2.3\, R \cdot C \cdot \log_{10} \dfrac{1}{1-\eta}}$$

$$= \frac{1}{2.3 \times (100 \times 10^3) \times (0.4 \times 10^{-6}) \log_{10} \frac{1}{1 - 0.57}}$$

$$= \frac{1}{0.92 \times \log_{10} 2.33} = \frac{1}{0.092 \times 0.367} \text{ Hz}$$

$$= 29.6 \text{ Hz } \textbf{Ans.}$$

Example 14-7. *A unijunction transistor with η = 0.62 is used in a relaxation oscillator circuit with R = 5 kΩ and C = 0.05 μF.*

(*a*) *Determine the period and frequency of oscillation.*

(*b*) *Determine the new value of R, which must be changed in order to obtain a frequency of oscillation of 50 Hz.*

(*c*) *If C is increased to 0.5 μF, (i.e., by a factor of 10) what should be the value of R, if the frequency is to be 50 Hz ?* (*Electrinic Circuits-II, Anna Univ., Nov/Dec 2007*)

Solution. Given: $\eta = 0.62$; $R = 5 \text{ k}\Omega = 5 \times 10^3 \ \Omega$ and $C = 0.05 \ \mu\text{F} = 0.05 \times 10^{-6}$ F.

(*a*) *Period and frequency of oscillation*

We know that period of oscillation,

$$T = 2.3 \ R \cdot C \log_{10} \frac{1}{1 - \eta}$$

$$= 2.3 \times (5 \times 10^3) \times (0.05 \times 10^{-6}) \log_{10} \frac{1}{1 - 0.62}$$

$$= (5.75 \times 10^{-4}) \times \log_{10} 2.63$$

$$= 2.4 \times 10^{-4} \text{ sec} = 0.24 \text{ msec } \textbf{Ans.}$$

and the frequency of oscillation,

$$f = \frac{1}{T} = \frac{1}{2.4 \times 10^{-4}} = 4140 \text{ Hz } \textbf{Ans.}$$

(*b*) *New value of R*

Here $f = 50 \text{ Hz}$

Let R = Value of resistance which gives 50 Hz frequency.

We know that period of oscillation,

$$= \frac{1}{f} = \frac{1}{50} = 0.02 \text{ sec}$$

and the period (T),

$$0.02 = 2.3 \ R \cdot C \log_{10} \frac{1}{1 - \eta}$$

$$= 2.3 \times R \times (0.05 \times 10^{-6}) \times \log_{10} \frac{1}{1 - 0.62}$$

$$= 2.3 \times R \times (0.05 \times 10^{-6}) \times \log_{10} 2.36$$

$$= 2.3 \times R \times (0.05 \times 10^{-6}) \times 0.42$$

$$= 4.83 \times 10^{-8} \ R$$

$$\therefore \quad R = 0.02/(4.83 \times 10^{-8}) = 4.14 \times 10^3 \ \Omega$$

$$= 414 \text{ k}\Omega \ \textbf{Ans.}$$

(*c*) *Value of R with C = 0.5 μF*

Here $f = 50 \text{ Hz}$ and $C = 0.5 \ \mu\text{F} = 0.5 \times 10^{-6}$ F

Let R = Value of resistance which gives 50 Hz frequency.

We know that period of oscillation,

$$T = \frac{1}{f} = \frac{1}{50} = 0.02 \text{ sec}$$

and the period of oscillation (T),

$$0.02 = 2.3\, R \cdot C \log_{10} \frac{1}{1-\eta}$$

$$= 2.3 \times R \times (0.5 \times 10^{-5}) \times \log_{10} \frac{1}{1-0.62}$$

$$= 2.3 \times R \times (0.5 \times 10^{-6}) \times 0.42 = 4.83 \times 10^{-7}\ R$$

$$\therefore \quad R = 0.02/(4.83 \times 10^{-7}) = 4.14 \times 10^{4}\ \Omega$$

$$= 41.4\ \text{k}\Omega\ \textbf{Ans.}$$

The value of R can also be determined straight forward by looking at the expression for the period, when the capacitance is increased by a factor of 10 (*i.e.*, from 0.05 μF to 0.5 μF) then to keep frequency of 50 Hz as constant. The value of R will have to reduced by the same factor, *i.e.*, from 414 kΩ to 41.4 kΩ. **Ans.**

14-27. Programmable Unijunction Transistor (PUT)

(*Electronic Devices, Anna University, Nov/Dec 2007*)

The programmable unijunction transistor (or in short PUT) is a four-layer PNPN device with a gate connected directly to the sandwiched N - type layer. The symbol for the device and the basic biasing arrangement is as shown in Fig. 14-19 and 14-20 respectively. As seen from the symbol, it is essentially an SCR with a control mechanism that permits a duplication of the characteristics of the typical SCT. The term *programmable* is applied because the bulk resistance (RBB), and peak-point voltage (V_P) as defined for the UJT can be controlled through the resistors RB_1, RB_2, and the supply voltage VBB.

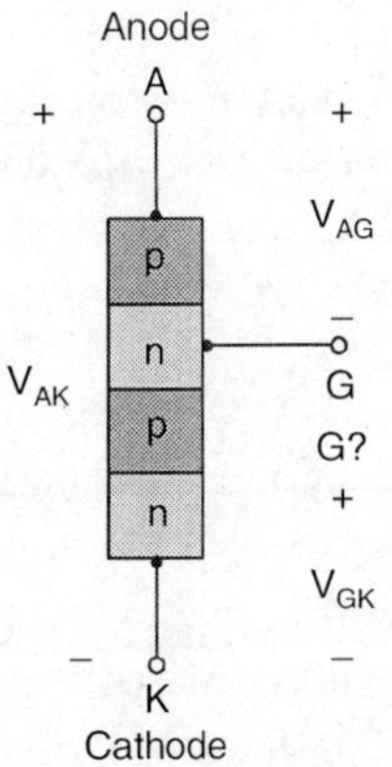

Fig. 14-19. *Programmable UJT (PUT).*

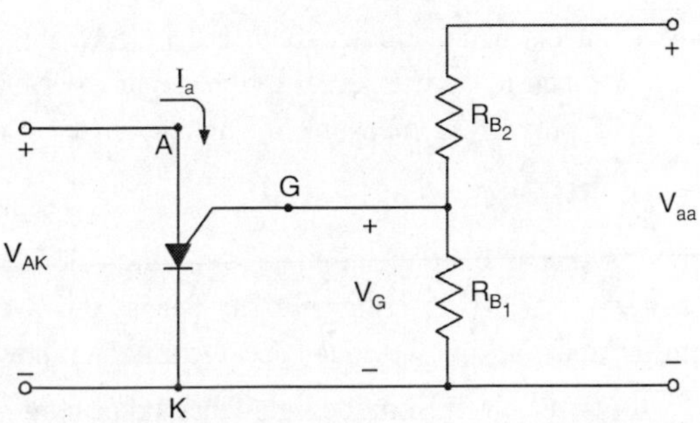

Fig. 14-20. *Basic biasing arrangement for the PUT*

It may be noted from Fig. 14-18 that throught an application of the voltage-divider rule, when $I_G = 0$, the gate voltage is given by the equation,

$$V_G = \frac{R_{B_1}}{R_{B_1} + R_{B_2}} V_{BB} = \eta V_{BB}$$

The above equation can be rewritten as,

$$\eta = \frac{R_{B_1}}{R_{B_1} + R_{B_2}}$$

This equation shows that that the intrinsic stand-off ration for PUT is the same as defined for the UJT.

Fig. 14-21 shows the characteristics of the PUT device. As seen form this diagram, the "OFF" state (I low, V between 0 and V_P) and the "ON" state ($I \geq Iv$, $V \geq Vv$) are separated by the unstable region as occurred for the UJT. That is, the device cannot stay in the unstable state–it will simply shift to either the "OFF" or "ON" stable states.

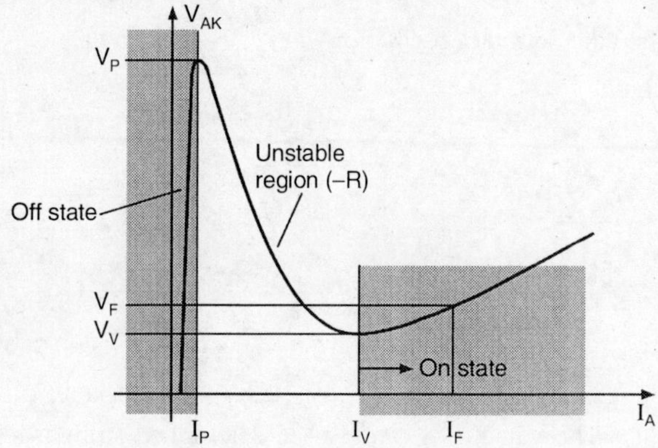

Fig. 14-21

The firing potential (V_P) or voltage necessary to "Fire" the device is given by the equation.

$$V_P = \eta V_{BB} + V_D$$

Notice that this equation is the same as defined for the UJT. However, V_P represents the voltage drop V_{AK} in Fig. 14-20 (the forward voltage drop across the conducting diode). For silicon devices, V_D is typically 0.7 V. Therefore,

$$V_{AK} = V_{AG} + V_{GK}$$

$$V_P = V_D + V_G$$

$$V_P = \eta V_{BB} + 0.7\ V$$

Substituting, $V_G = \eta .\ V_{BB}$ in the above equation, we get,

$$V_P = V_G + 0.7$$

It may be noted that for the UJT both R_{B1} and R_{B2} represent the bulk resistance and ohmic base contacts of the device–both are inaccessible to the application engineers. However, in the PUT, R_{B_1} and R_{B_2} are external to the device, permitting an ajustment of V_G above. In other words, the PUT provides a measure of control on the level of V_P required to turn ON the device.

14-28. DIAC

The diac is a two terminal device, which can pass current in either direction when the breakover voltage is reached in either polarity across the two terminals. The basic construction of a diac is similar to a triac, but without a gate terminal as shown in Fig. 14-22(*a*).

It may be noted from the figure that a diac consists of two 4-layer diodes connected in parallel but in opposite directions. The four layer diodes are $P_1N_1P_2N_2$ and $P_2N_1P_1N_3$ as shown by the broken lines in the figure. Since, the diac conducts in both directions, therefore the terminals are designated by numbers instead of anode and cathode. The diac has two main terminals namely M.T. 1 and M.T. 2. Fig. 14-22(*b*) shows the equivalent circuit of a diac, which consists of two four-layer diodes connected in parallel but in opposite directions. Fig. 14-22(*c*) shows the schematic symbol of the diac.

The diac can pass current in either direction depending upon the polarity of voltage across its main terminals. It can be turned ON only when the applied voltage across its main terminals reaches

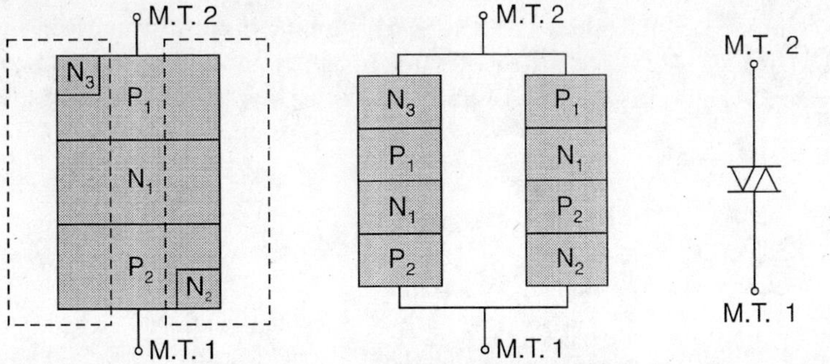

(*a*) Basic construction. (*b*) Equivalent circuit. (*c*) Symbol.

Fig. 14-22. DIAC.

the breakover voltage. When the applied voltage makes the M.T. 2 positive with respect to M.T. 1, the diac passes current through the diode $P_1N_1P_2N_2$ from M.T. 2 to M.T. 1 as shown in Fig. 14-23(*a*).

However, if the applied voltage makes M.T. 2 negative with respect to the M.T. 1, the diac passes current through the diode $P_2N_1P_1N_3$ from M.T. 1 to M.T. 2 as shown in Fig. 14-23(*b*). The diac turns OFF when the current drops below the holding value. A diac is used as a triggering device for triac in phase control circuits such as light dimming, heat control and motor speed control, *etc*.

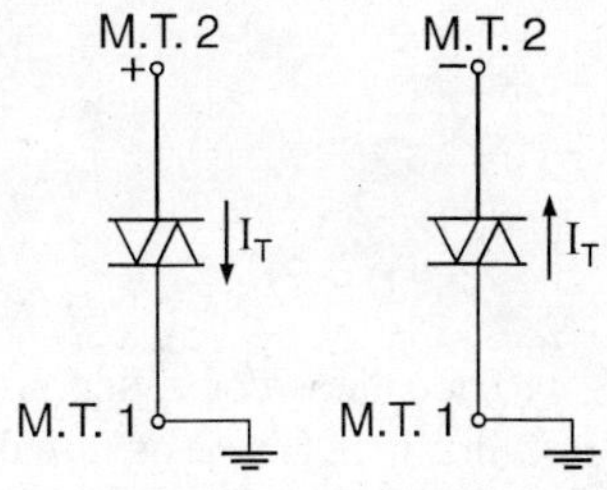

(*a*) (*b*)

Fig. 14-23. *Diac current flow.*

14-29. V-I Characteristics of a Diac

We have already discussed in the last article that a diac can conduct in either direction, depending upon the polarity of the applied voltage across its main terminals. When a diac is operated with M.T. 2 positive with respect to M.T 1, the V-I characteristic obtained is as shown in Fig. 14-24 by the curve marked OAB. Similarly, when the diac is operated with its M.T. 2 negative with respect to M.T. 1, the V-I characteristic obtained is as shown in Fig. 14-24 by the curve marked OCD.

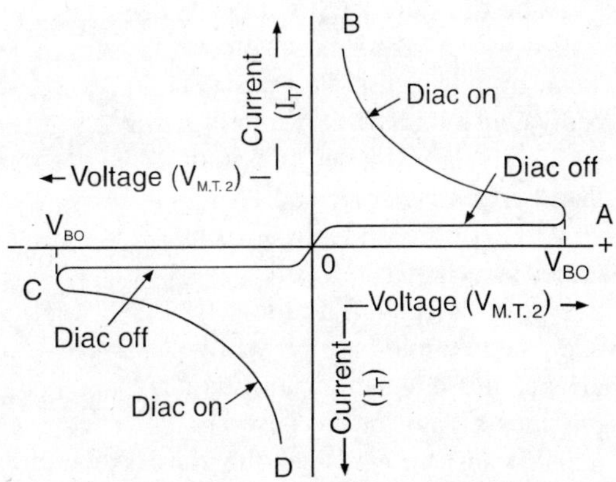

Fig. 14-24. *V-I characterstics of a diac.*

The V-I characteristic of a diac gives us an important information about the following points:

(*i*) The curves OAB and OCD are symmetrical and identical. It is true because the operation of a diac is identical in both directions of conduction.

(*ii*) The diac does not pass any current, until the applied voltage of either polarity reaches the breakover voltage.

(*iii*) At the breakover voltage, the diac turns ON and the current through the diac increases rapidly as shown in the figure.

(*iv*) The operating voltage and currents are the same in either direction. Therefore, the diac is a symmetrical bilateral diode.

The diacs are available with a breakdown voltage ranging from 28 to 36 V and maximum pulse current of 2 A.

14-30. Silicon Controlled Switch (SCS)

A silicon controlled switch (abbreviated as SCS) is similar in construction to a silicon controlled rectifier as shown in Fig. 14-25(*a*). It has four terminals namely anode, cathode and two gate terminals called anode gate and cathode gate. Fig. 14-25(*b*) shows the schematic symbol of silicon controlled switch.

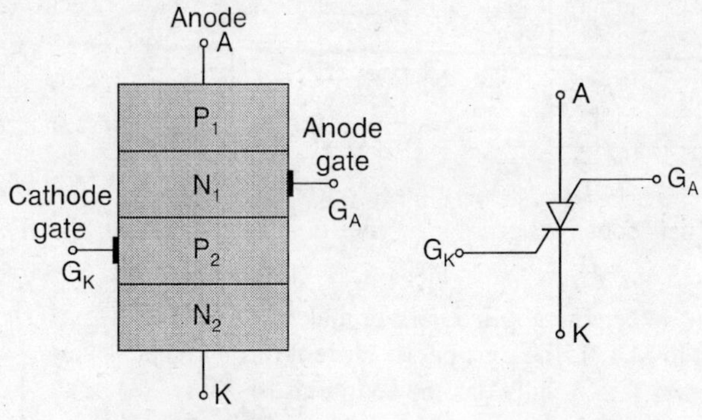

(*a*) Construction. (*b*) Symbol

Fig. 14-25. *Silicon controlled switch.*

The silicon controlled switch has an advantages that it can be turned ON and OFF using either gate terminal. On the other hand, an SCR can be turned ON only using its gate terminal. But it cannot be turned off using the same gate terminal.

14-31. SCS Operation

The operation of silicon controlled switch (SCS) may be understood by referring to its transistor equivalent circuit as shown in Fig. 14-26. To start with, let us assume that the transistors Q_1 and Q_2 are OFF. Therefore, the silicon controlled switch is also OFF. If a positive pulse is applied on the cathode gate, the transistor Q_2 will be driven into conduction and thus provides a path for the base current of transistor Q_1. When the transistor Q_1 turns ON, its collector current provides the base drive to the transistor Q_2. This action drives both the transistors Q_1 and Q_2 into saturation and hence the silicon controlled switch turns ON. The silicon controlled switch can also be turned ON by applying a negative pulse at the anode gate terminal as shown in the figure.

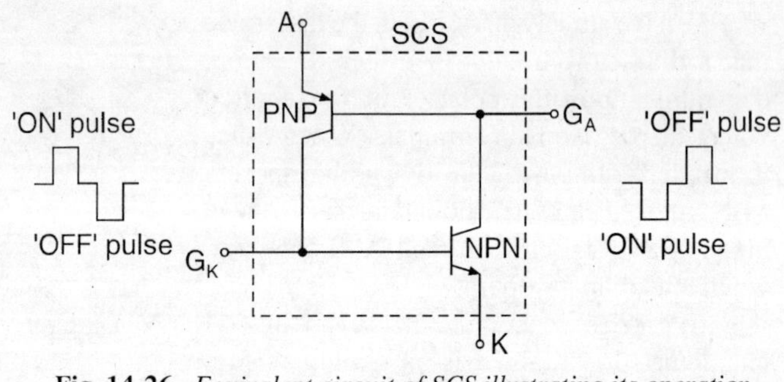

Fig. 14-26. *Equivalent circuit of SCS illustrating its operation.*

The silicon controlled switch can be turned OFF by applying a positive pulse at the anode gate or a negative pulse at the cathode gate. The applied anode gate voltage pulse reverse biases the base-emitter junction of the transistor Q_1 and turns it OFF. This, in turn, cuts-off the transistor Q_2 due to which the silicon controlled switch is also turned OFF. It will be interesting to know that the silicon controlled switch can also be turned OFF by reducing the anode current below the holding current value.

14-32.SCS Applications

The silicon controlled switch (SCS) has two major advantages over SCR, namely it has a fast turn OFF and it can be turned OFF with pulses on either gate terminal. But the SCS has one disadvantage that it is available only in lower power ratings. Because of this fact, SCS is used only in low power circuits. Some of the important applications of SCS are given below:

- In counters, registers and timing circuits.
- In pulse generators.
- In voltage sensors.
- In oscillators.

14-33. Silicon Unilateral Switch (SUS)

A silicon Unilateral Switch (abbreviated as SUS) is used for triggering SCR's and triacs. Basically, a silicon unilateral switch is a four-layer diode that resembles the SCR in which the contacts are made only to anode and cathode.

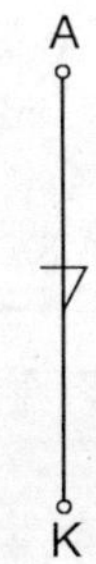

Fig. 14-27. *SUS symbol.*

Fig. 14-27 shows the schematic symbol of silicon unilateral switch. The operation of a silicon unilateral switch is similar to that of SCR. A typical silicon unilateral switch is in its OFF position until the applied voltage is below the switching voltage (about 6 to 10 V). The voltage drop across, a silicon unilateral switch when it is ON, is about 1.5 V. The maximum forward current capability is about 200 mA. The holding current ranges from about 0.5 mA to 1.5 mA.

A silicon unilateral switch can be turned OFF by reversing the anode voltage or disconnecting the supply. The maximum reverse voltage rating of a silicon unilateral switch is about 30 V.

14-34. Silicon Bilateral Switch (SBS)

The silicon bilateral switch (abbreviated as SBS) is used for triggering SCR's and triacs. Basically, it consist of two silicon unilateral switches, connected in parallel but in opposite directions as shown in Fig. 14-28(a). The schematic symbol of silicon bilateral switch is shown in Fig. 14-28(b).

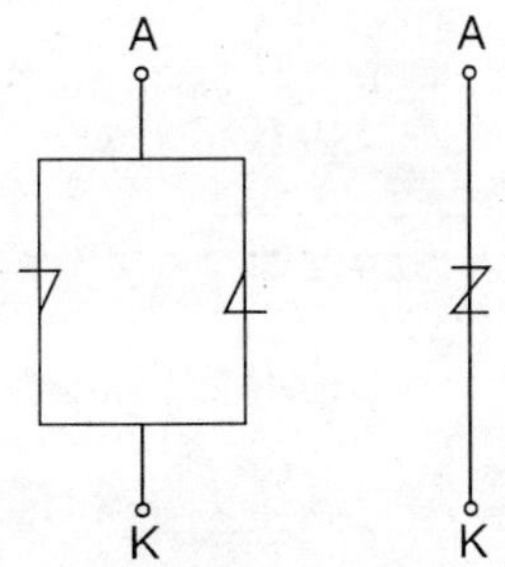

(a) Construction (b) Symbol

Fig. 14-28. *Silicon bilateral switch.*

The operation of a silicon bilateral switch is similar to that of a triac. It is in its OFF position so long as the applied voltage is below the switching voltage (about 6 to 10 volts). It turns ON as soon as the applied voltage is increased above the switching voltage. When the silicon bilateral switch is ON, the voltage drop across it, is about 1.7 volts.

Since, the silicon bilateral switch conducts in either direction, therefore it does not have any reverse voltage rating like silicon unilateral switch. The silicon bilateral switch may be used in phase control of SCR's and triacs.

SUMMARY

In this chapter, you have learned that :

1. A thyristor is a four-layer PNPN device having three junctions. A thyristor has two stable states : ON and OFF.
2. Thyristors could be unidirectional or bidirectional. The SCR, SCS and LASCR are the unidirectional devices whereas TRIAC is a bidirectional thyristor.
3. The SCR is operated with an anode voltage slightly less than rated forward break-over voltage.
4. The SCR is triggered by applying a low power pulse at the gate-input. This method of triggering is called gate triggering.

5. The SCR can be turned off by using any one of the following methods : (1) by reversing the polarity of anode-to-cathode voltage, (2) by interrupting anode current by means of momentarily series or parallel switching arrangement, or (3) by reducing the current through SCR below the holding value.
6. TRIAC is a bidirectional type of thyristor, *i.e.*, it can conduct current in either forward or reverse direction depending upon the polarity of the voltage across its terminals.
7. A UJT can be switched ON by applying a positive trigger at its emitter terminal. It can be switched OFF by applying a negative trigger at the emitter.
8. A diac can pass current in either direction when the breakover voltage is reached in either polarity across its two terminals.
9. A silicon controlled switch (SCS) can be turned ON/OFF at a much faster speed as compared to SCR but is suitable only for lower power ratings.

NUMERICAL PROBLEMS

1. In a given circuit, 2N 6394 SCR is subjected to 50 A current surge, that lasts for 10 ms. Determine whether or not, this surge will destroy the device. Given the circuit fusing rating of 2N6394 as 40 A^2s. **(Ans.** No)
2. The 2N1843 A SCR has a circuit fusing rating of 60 A^2s. The device is being used in a circuit where it could be subjected to a 100-A surge. Determine the limit on the duration of such a surge. (**Ans.** 6 ms)
3. The 2N5431 UJT has a rating of $\eta = 0.8$ (max.) Determine the maximum value of V_P for the device when it is being used in the circuit with $V_{BB} = +18$ V. Take $V_D = 0.7$ V. (**Ans.** 144.7 V)

DESCRIPTIVE QUESTIONS

1. Discuss briefly, the construction, working, characteristics and applications of silicon controlled rectifier.
2. Give a schematic diagrams of the SCR and derive its equivalent representation in terms of transistors.
3. Discuss two transistor analogy of a SCR.
4. Explain the working of SCR with the help of its two transistor equivalent circuits. Draw its forward and reverse characteristics.
5. Explain the construction and working of a Triac.
6. Elaborate the difference between SCR and Triac.
7. What is a Unijunction Transistor (UJT)? How does it differ from a JFET?
8. Explain the V-I characteristics of UJT. Why it is called a current controlled negative resistance device?
9. Explain the construction and working of a diac.
10. What is SCS ? How does it differ from SCR ?
11. Explain briefly about the difference between SUS and SBS.
12. Draw the transistor equivalent circuit of SCR. (*Electronic Devices, Anna University, Nov/Dec 2007*)
13. With neat diagram explain the construction, working characteristics of Unijunction transistor. Give its equivalent circuit. (*Electronic Devices, Anna University, Nov/Dec 2007*)

14. Give the characteristic features of devices belonging to the thyristors family. *(Electronic Devices, Anna University, Nov/Dec 2007)*

15. Write a note on PUT. *(Electronic Devices, Anna University, Nov/Dec 2007)*

16. Draw the two transistor model of SCR. *(Electron Devices, Anna University, April/May 2008)*

17. Define intrinsic stand off ratio of UJT. *(Electron Devices, Anna University, April/May 2008)*

18. Explain the operation of DIAC. *(Electron Devices, Anna University, April/May 2008)*

19. Describe the operation of UJT with its equivalent circuit and its emitter characteristic. *(Electron Devices, Anna University, April/May 2008)*

20. How the forced turn-off of an SCR is different from natural turn-off ? *(Electron Devices, Anna University, Nov/Dec 2007)*

21. Explain the terms peak voltage and valley voltage with respect to UJT. *(Electron Devices, Anna University, Nov/Dec 2007)*

22. Diac is a bidirectional device. Explain. *(Electron Devices, Anna University, Nov/Dec 2007)*

23. Mention the applications of UJT, Diac, and Triac. *(Electron Devices, Anna University, Nov/Dec 2007)*

24. Describe the working of UJT relaxation oscillotor. *(Electronics Circuit-II, Anna University, May/June 2007)*

MULTIPLE CHOICE QUESTIONS

1. An SCR is a semiconductor device which consists of

(*a*) four PN junctions (*b*) three PN junctions

(*c*) two PN junctions (*d*) one PN junction.

2. After triggering an SCR, the gate pulse is removed. The current in the SCR will

(*a*) immediately fall to zero (*b*) rise up

(*c*) remain the same (*d*) immediately rise a little and then falls to zero

3. An SCR may be turned OFF by

(*a*) reversing polarity of the anode-to-cathode voltage

(*b*) interrupting its anode current

(*c*) low-current drop out

(*d*) all of the above.

4. A triac can be triggered into conduction by applying

(*a*) only positive voltage at the gate

(*b*) positive or negative voltage at the gate

(*c*) positive or negative voltage at either main terminal

(*d*) both (*a*) and (*c*) above.

5. Which of the following acts like a diode and two resistors

(*a*) SCR (*b*) triac (*c*) diac (*d*) UJT

6. A four layer diode is sometimes called

(*a*) UJT (*b*) PNPN diode (*c*) DIAC (*d*) Switch

7. Which of the following PNPN device has two gates.

(*a*) SCR (*b*) DIAC (*c*) TRIACn (*d*) TRANSISTOR

8. When a SCR is conducting, it has

(*a*) infinitely large resistance (*b*) resistance of few megohms

(*c*) low resistance (*d*) zero resistance

ANSWERS

1. (*b*)	**2.** (*c*)	**3.** (*d*)	**4.** (*d*)	**5.** (*d*)
6. (*c*)	**7.** (*c*)	**8.** (*c*)		

❑❑❑

15

CHAPTER

Controlled Rectifiers

OUTLINE

Objectives

After completing this chapter, you should be able to :

- Know the different types of controlled rectifier.
- Sketch the output waveforms across the controlled rectifier for different firing angles for the sine wave input.
- Calculate the values of output voltage and current for different values of firing angle in controlled rectifier.
- Explain the operation of 90° and 180° variable half-wave rectifier.
- Describe the operation of SCR full-wave rectifier.
- Know how a UJT can be used as a trigger device in SCR Phase Control.
- Explain the operation of triac and triac-diac phase control circuit.
- Describe the operation of a chopper, inverter and push-pull inverter circuits.

15-1. Introduction

We have already discussed in chapter on rectifiers and filters about diodes, half-wave and full-wave rectifiers. In such rectifier circuits, the current flows through the diode, whenever the instantaneous value of the a.c. input voltage is greater then the voltage across the load at that instant. This point is made clear in the half-wave diode rectifier circuit as shown in Fig. 15-1. In such case, the load current flows during all times of the positive half cycles of the input voltage.

We shall now introduce a new class of rectifier called *controlled rectifier*. In such a rectifier, special semiconductor device such as *silicon controlled rectifier* abbreviated as SCR or a *triac* is used instead of an ordinary PN junction diode. Such a device has three terminals, namely anode,

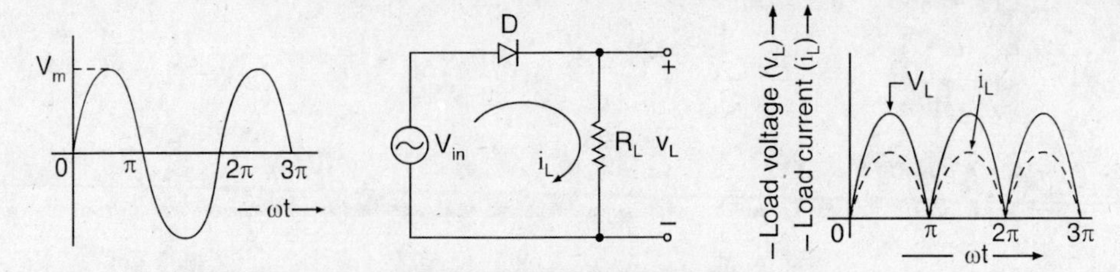

Fig. 15-1. *Diode half-wave rectifier.*

cathode and gate. The load current, in a controlled rectifier, flows only when a control signal is applied at the gate terminal to turn 'ON' the rectifier.

Fig. 15-2(*a*) shows the half-wave controlled rectifier circuit and Fig. 15-2(*b*) shows the waveform for input voltage across the controlled rectifier and load current. It may be noted that the control signal is applied to turn on the rectifier in such a way that the conduction starts at a specific point marked A in Fig. 15-2(*b*). The point A (called *firing point*) corresponds to the angle θ, so that the conduction is delayed by this much period. It is because of this delay that the amount of load current has reduced in a controlled rectifier as compared to an ordinary rectifier.

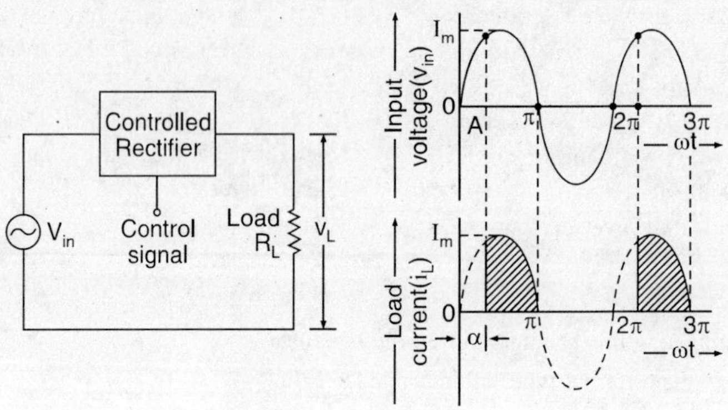

(*a*) (*b*)

Fig. 15-2. *Half-wave controlled rectifier.*

It will be interesting to know that once the rectifier is turned 'ON' it remains in conduction for the rest of the positive half-cycle, *i.e.*, up to 180° (or π-radians). Thus, the point *A* (at which the conduction starts) is determined by the angle of delay in applying control signal to the rectifier circuit. As the angle θ increases, conduction is further delayed and the load current is also reduced.

It is evident from the above discussion, that load current in a controlled rectifier may be controlled easily by selecting the firing point. This feature (*i.e.*, control over the load current) is of great advantage in controlled rectifiers.

Let the equation of the input voltage be,

$$v_i = V_m \sin \omega t$$

$$= V_m \sin \theta$$

For a half-wave rectifier, the average value is found by taking the average over the whole cycle (0 to 2π) even though the conduction takes place only from 0 to π in the entire cycle.

$$\therefore \quad V_L = \frac{1}{2\pi} \int_\alpha^\pi v_{in} \cdot d\theta$$

$$= \frac{1}{2\pi} {}_\alpha^\pi V_m \sin \theta \cdot d\theta$$

$$= \frac{V_m}{2\pi}\int_{\alpha}^{\pi} \sin\theta \cdot d\theta$$

$$= \frac{V_m}{2\pi}\left| -\cos\theta \right|_{\alpha}^{\pi}$$

$$= \frac{V_m}{2\pi}\left|(-\cos\pi) - (-\cos\alpha)\right|$$

$$= \frac{V_m}{2\pi}(1 + \cos\alpha)$$

and dc output voltage,

$$V_{dc} = V_L = \frac{V_m}{2\pi}(1 + \cos\alpha)$$

For a resistive load,

$$I_{dc} = I_L = \frac{V_m}{2\pi R_L}(1 + \cos\alpha)$$

Notice that if $\alpha = 0$, then

$$V_{dc} = \frac{V_m}{\pi} \quad \text{and} \quad I_{dc} = \frac{V_m}{\pi R_L}$$

Recall that these are the same values as obtained for a uncontrolled half-wave rectifier. (Refer to Chapter-8, Article 8-4)

15-2. Output Waveforms for Different Firing Angles

Fig. 15-3(*a*), (*b*), (*c*) and (*d*) shows the output waveforms for different firing angles. (0°, 45°, 90° and 135°) along with the waveforms for input voltage and voltage drop across the controlled rectifier. For all these waveforms, we have assumed that a.c. supply voltage is sinusoidal ($V_{in} = V_m \sin\theta$). Also a resistive load is assumed. Further, the diode is assumed to be an ideal, *i.e.*, when it conducts, the voltage drop across it is 0. When the diode does not conduct, it is 'open' circuit.

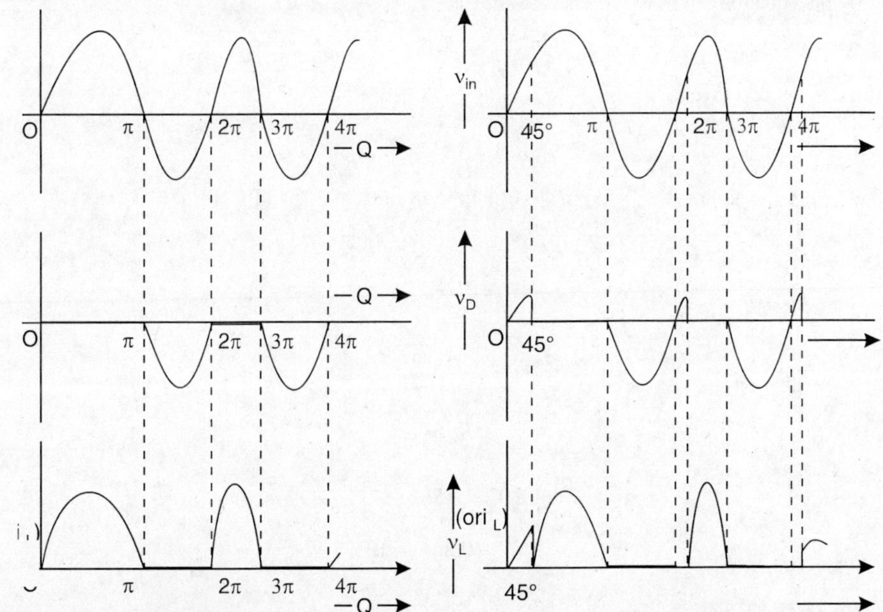

(*a*) α = 0° (*b*) α = 45°

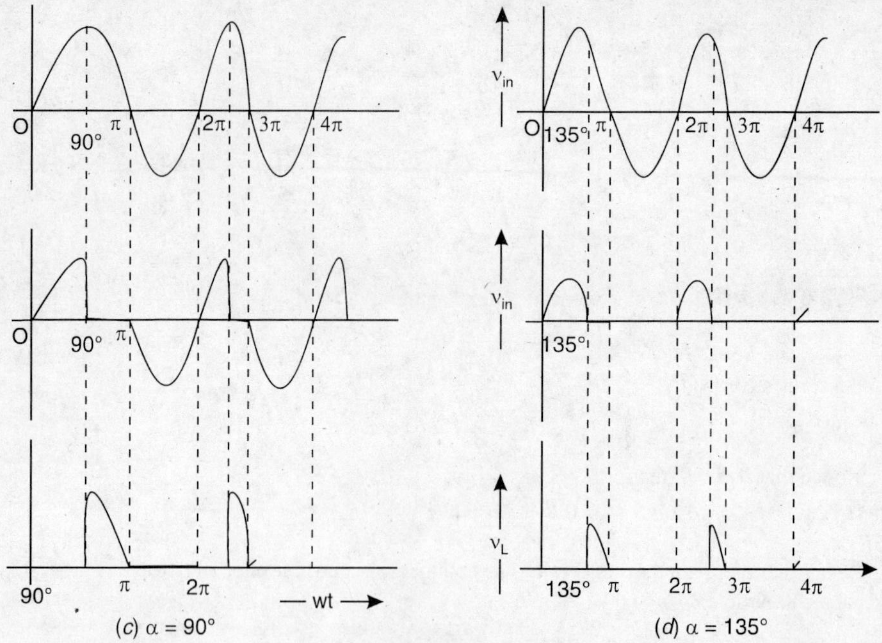

Fig. 15-3.

It is evident from the waveforms that as the firing angle 'α' is increased from 0 through 135°, the output voltage keeps decreasing till for α = 180°, the output is zero.

15-3. Values of Output Voltage and Current in Controlled Rectifiers

We have already mentioned in Art. 15-1. that the d.c. output voltage for a half-wave rectifier is given by the equation,

$$V_{dc} = \frac{V_m}{2\pi}(1 + \cos\alpha) \quad ...(i)$$

and the d.c. output current,

$$I_{dc} = \frac{V_m}{2\pi R_L}(1 + \cos\alpha) = \frac{V_{dc}}{R_L} \quad ...(ii)$$

Let us use the equations (*i*) and (*ii*) to obtain the values of V_{dc} and I_{dc} for α = 0°, 30°, 45°, 60°, 90°, 120° and 135°.

(***a***) When α = 0,

$$\cos\alpha = \cos 0° = 1$$

$$\therefore \quad V_{dc} = \frac{V_m}{2\pi}(1 + 1) = \frac{V_m}{\pi} = 0.318\, V_m$$

$$\text{and} \quad I_{dc} = \frac{V_{dc}}{R_L} = \frac{V_m}{\pi R_L} = 0.318\, \frac{V_m}{R_L}$$

(***b***) When α = 30°

$$\cos\alpha = \cos 30° = 0.866$$

$$\therefore \quad V_{dc} = \frac{V_m}{2\pi}(1 + 0.866) = 0.297\, V_m$$

$$\text{and} \quad I_{dc} = \frac{V_{dc}}{R_L} = 0.297\, \frac{V_m}{R_L}$$

(*c*) When $\alpha = 45°$

$$\cos\alpha = \cos 45° = 0.707$$

$$\therefore \quad V_{dc} = \frac{V_m}{2\pi}(1 + 0.707) = 0.27\, V_m$$

and $$I_{dc} = \frac{V_{dc}}{R_L} = 0.27\,\frac{V_m}{R_L}$$

(*d*) When $\alpha = 60°$

$$\cos\alpha = \cos 60° = 0.5$$

$$\therefore \quad V_{dc} = \frac{V_m}{2\pi}(1 + 0.5) = 0.239\, V_m$$

and $$I_{dc} = \frac{V_{dc}}{R_L} = 0.239\,\frac{V_m}{R_L}$$

(*e*) When $\alpha = 90°$

$$\cos\alpha = \cos 90° = 0$$

$$\therefore \quad V_{dc} = \frac{V_m}{2\pi}(1 + 0) = \frac{V_m}{2\pi} = 0.159\, V_m$$

and $$I_{dc} = \frac{V_{dc}}{R_L} = \frac{V_m}{2\pi R_L} = 0.159\,\frac{V_m}{R_L}$$

(*f*) When $\alpha = 120°$

$$\cos\alpha = \cos 120° = -0.5$$

$$\therefore \quad V_{dc} = \frac{V_m}{2\pi}(1 - 0.5) = 0.08\, V_m$$

and $$I_{dc} = \frac{V_{dc}}{R_L} = 0.08\,\frac{V_m}{R_L}$$

(*g*) When $\alpha = 135°$

$$\cos\alpha = \cos 135° = -0.707$$

$$\therefore \quad V_{dc} = \frac{V_m}{2\pi}(1 - 0.707) = 0.0466\, V_m$$

and $$I_{dc} = \frac{V_{dc}}{R_L} = 0.0466\,\frac{V_m}{R_L}$$

Example 15-1. *A 100 Ω load is connected to a peak supply of 300 V through a controlled half-wave diode rectifier. The load power is to varied from 25 W to 80 W. What is the angular firing control required ? Neglect forward voltage drop of the diode.* (*Pune Univ.* 1990)

Solution. Given : $R_L = 100\ \Omega$; and $V_m = 300$ V.

We know that load power,

$$P = V_{dc} \cdot I_{dc}$$

$$= \left[\frac{V_m}{2\pi}(1 + \cos\alpha)\right] \times \left[\frac{V_m}{2\pi R_L}(1 + \cos\alpha)\right]$$

$$= \left(\frac{V_m}{2\pi}\right)^2 \cdot \frac{1}{R_L} \cdot (1 + \cos\alpha)^2$$

$$= \left(\frac{300}{2\pi}\right)^2 \cdot \frac{1}{100} \cdot (1 + \cos\alpha)^2$$

$$= 22.8\ (1 + \cos\alpha)^2$$

Angular firing control when Load power P = 25 W,

$$25 = 22.8\ (1 + \cos\alpha)^2$$

or $$1 + \cos\alpha = \sqrt{\left(\frac{25}{22.8}\right)} = 1.096$$

$\therefore$ $$\cos\alpha = 0.096 \text{ or } \alpha = 89.4°$$

Angular firing control when Load power P = 80 W,

$$80 = 22.8\ (1 + \cos\alpha)^2$$

or $$1 + \cos\alpha = \sqrt{\frac{80}{22.8}} = 1.873$$

$\therefore$ $$\cos\alpha = 0.873 \text{ or } \alpha = 29.2°$$

This shows that when the load power is varied from 25 W to 80 W, the firing angle is reduced from 89.4° to 29.2°. **Ans.**

Example 15-2. *In a controlled half-wave rectifier, peak supply voltage is 200 V and the value of load resistor is 1 kΩ. Calculate the power delivered to the load circuit for firing angles of* **(i)** *0°,* **(ii)** *45°,* **(iii)** *90°, and* **(iv)** *135°.* *(BHU, 1988)*

Solution. Given : $V_m = 200$, and $R_L = 1$ kΩ

(*i*) $\alpha = 0°$

We know that d.c. output voltage,

$$V_{dc} = 0.318\ V_m = 0.318 \times 200 = 63.6 \text{ V}$$

and the d.c. output current,

$$I_{dc} = \frac{V_{dc}}{R_L} = \frac{63.6}{1000} = 63.6 \text{ mA}$$

$\therefore$ Load power, $$P = V_{dc} \cdot I_{dc} = 63.6 \text{ V} \times 63.6 \text{ mA}$$
$$= 4045 \text{ mW or } 4.045 \text{ W}$$ **Ans.**

(*ii*) $\alpha = 45°$

We know that d.c. output voltage,

$$V_{dc} = 0.27\ V_m = 0.27 \times 200 = 54 \text{ V}$$

and the d.c. output current,

$$I_{dc} = \frac{V_{dc}}{R_L} = \frac{54}{1000} = 54 \text{ mA}$$

$\therefore$ Load power,

$$P = V_{dc} \cdot I_{dc}$$
$$= 54 \text{ V} \times 54 \text{ mA}$$
$$= 2916 \text{ mW or } 2.916 \text{ W}$$ **Ans.**

(*iii*) $\alpha = 90°$

We know that d.c. output voltage,

$$V_{dc} = 0.159\ V_m = 0.159 \times 200 = 31.8 \text{ V}$$

and $$I_{dc} = \frac{V_{dc}}{R_L} = \frac{31.8}{1000} = 31.8 \text{ mA}$$

$\therefore$ $$P = V_{dc} \cdot I_{dc} = 31.8 \text{ V} \times 31.8 \text{ mA} = 1011 \text{ mW or } 1.011 \text{ W}$$ **Ans.**

(iv) $\alpha = 135°$

We know that d.c. output voltage,

$$V_{dc} = 0.0466\ V_m = 0.0466 \times 200 = 9.3 \text{ V}$$

and
$$I_{dc} = \frac{V_{dc}}{R_L} = \frac{9.3}{1000} = 9.3 \text{ mA}$$

∴
$$P_{dc} = V_{dc} \cdot I_{dc} = 9.3 \text{ V} \times 9.3 \text{ mA} = 86.49 \text{ mW} \textbf{ Ans.}$$

15-4. Average values for Full-wave controlled Rectifier

Fig. 15-4. shows the full-wave rectified waveform to be controlled by the rectifier for $\theta = 60°$. As seen from this diagram, the average values will be doubled because there are two half sinusoids to be arranged.

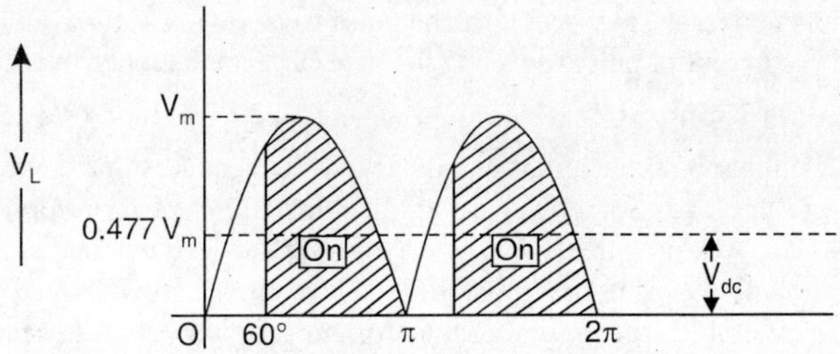

Fig. 15-4.

Thus the d.c. output voltage,

$$V_{dc} = 2 \times \left[\frac{V_m}{2\pi}(1 + \cos\theta_1)\right] = \frac{V_m}{\pi}(1 + \cos\theta_1)$$

In the present case, $\theta = 60°$,

∴
$$V_{dc} = \frac{V_m}{\pi}(1 + \cos 60°) = \frac{V_m}{\pi}(1 + 0.5) = 0.477\ V_m$$

and the d.c. output current,

$$I_{dc} = \frac{V_{dc}}{R_L}$$

15-5. Silicon Controlled Rectifier (SCR)

It is a three junction, four-layer and three terminal semiconductor device. The three terminals are anode, cathode and gate. The basic construction and circuit symbol for the silicon controlled rectifier (SCR) is as shown in Fig. 15-5(*a*) and (*b*).

When the anode is made positive, with respect to cathode, the junctions J_1 and J_3 are forward biased, while the junction J_2 is reverse biased. As a result of this, the SCR does not conduct. However, if a positive voltage of suitable amplitude is applied at the gate, with respect to the cathode the junction J_2 is forward biased and the SCR is triggered, *i.e.*, it starts conducting like an ordinary diode.

It will be interesting to know that once the SCR turns 'ON' the gate has no control over the SCR. The conduction can only be terminated by removing the applied voltage from the anode.

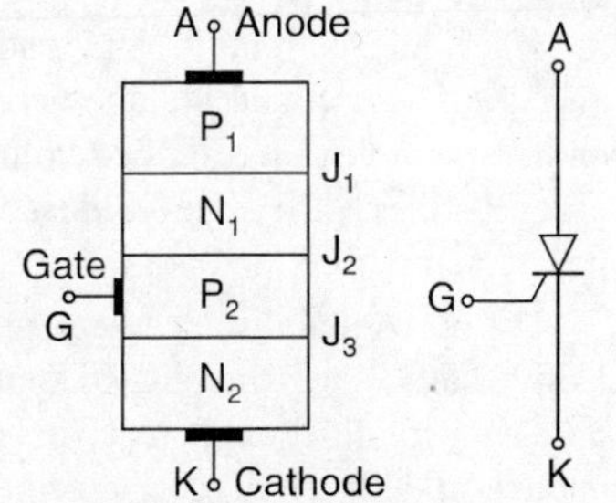

(*a*) Basic construction. (*b*) Symbol.

Fig. 15-5. *Silicon Controlled rectifier.*

As mentioned above, the SCR needs positive voltage at the gate terminal to turn 'ON'. This voltage is provided by a control circuit as shown in Fig. 15-6. Although there are a number of methods of turning SCR in its 'ON' position, yet the following two are important from the subject point of view :

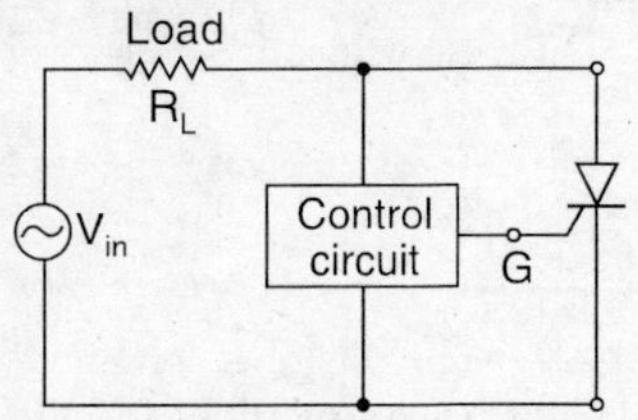

Fig. 15-6. *SCR with control circuit.*

(*i*) ***Amplitude triggering :*** In this method, the control circuit produces a gate voltage of suitable level or amplitude by using ordinary diodes and resistors. Here, the gate current flows for the entire period of SCR conduction.

(*ii*) ***Pulse triggering :*** In this method, the control circuit produces pulses of narrow width (1 μsec to 1000 μsec) and of suitable amplitude. The pulses of desired width may be generated by using unijunction transistor oscillator. The pulse triggering has an advantage over amplitude triggering that in case of pulse triggering the gate current is supplied only for a short duration.

15-6. Power Control Using SCR

There are large number of applications in industry, which require some control over the amount of current supplied to the load. Some of these applications are *electric welding, lighting control, motor speed control etc.* We know that a simple way to vary the load current is either to control the transformer secondary voltage or insert a resistor in the load circuit. But in actual practice, neither of these two methods is desirable. The control of transformer secondary voltage requires an expensive equipment, whereas the insertion of resistor in the load circuit results in a loss of power. Therefore, the use of silicon controlled rectifier (SCR)* is an efficient way of controlling the power. When an a.c. voltage is applied across an SCR, it will turn 'ON' during each positive half cycle and 'OFF' during each negative half cycle. The average load current can be varied over a wide limit by controlling the point in each positive half-cycle at which the SCR is turned ON. The point, at which the SCR turns ON, may be controlled either by amplitude triggering or pulse triggering.

15-7. SCR Half-wave Rectifier

Fig. 15-7(*a*) shows the circuit of a silicon controlled rectifier (SCR) half-wave rectifier. In this circuit, the a.c. supply is connected to the SCR through a load resistance R_L. The gate current is obtained from the a.c. supply through the resistances R and R_1 and diode D connected in series.

The diode D is connected in such a way, so that it blocks the reverse voltage on the gate during the negative half-cycle.

Fig. 15-7(*b*), (*c*) and (*d*) shows the waveforms of the voltage across SCR (V_{SCR}), the load current (I_L) and voltage across the load (V_L) respectively. The various curves in the waveforms may be explained as follows : During the positive half-cycle, the SCR remains 'OFF' till the input voltage reaches the gate trigger voltage, *i.e.*, point A in Fig. 15-7(*b*). As a result, there is no current through R_L. Hence, load current (I_L) and load voltage (V_L) are zero. At point *A*, SCR is fired into conduction. It acts like a short and voltage across it drops to zero, *i.e.*, curve *AB* in Fig. 15-7(*b*). Under this condition, whole of the applied voltage drops across the load resistance. During the negative input-half cycle, the SCR is reverse biased and hence does not conduct. As a result, all the applied voltage appears across SCR and none across the load resistance (R_L).

The horizontal distance between the points *O* and *A* in Fig. 15-7(*b*) represents the time in the positive half-cycle, when SCR is not conducting. This distance in degrees is called the firing angle, phase angle or delay angle (α). This angle gives us an idea about the position at which the SCR starts conduction with respect to the origin. It may be noted from the figure that the SCR is conducting between the points *B* and *C*. This angle is called conduction angle (ϕ) and its value is equal to ($\pi - \alpha$).

* The gate trigger voltage must be of suitable amplitude, which can provide the necessary gate current for SCR triggering.

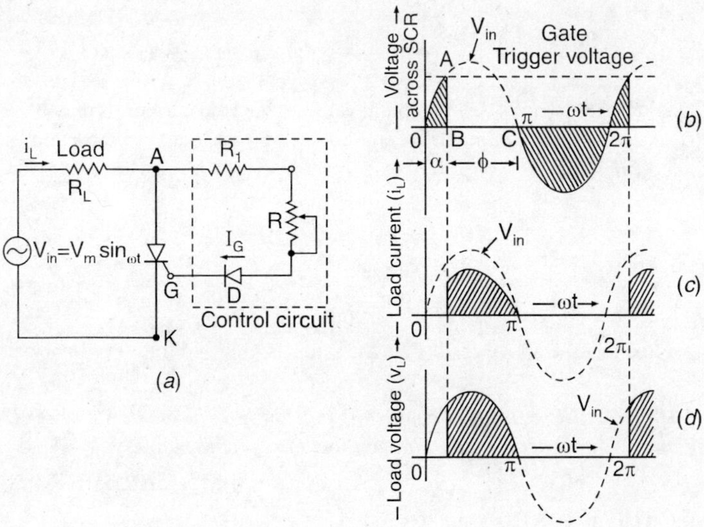

Fig. 15-7. *SCR half-wave rectifier and its waveforms.*

15-8. Average Values of Load Voltage and Current for SCR Half-wave Rectifier

We have already discussed in the last article, about the SCR half-wave rectifier. Now we shall determine its average values of load voltage (V_L) and load current (i_L). We know that the equation of a.c. supply voltage is

$$v_{in} = V_m \cdot \sin \omega t = V_m \cdot \sin \theta \qquad ...(\because \theta = \omega t)$$

We also know that for a half-wave rectifier, the average value of a load voltage is determined by the relation,

$$V_{dc} = \frac{\text{Area under the curve over a complete cycle}}{\text{Base}}$$

$$= \frac{\int_0^{2\pi} v_{in} \cdot d\theta}{2\pi} = \frac{1}{2\pi} \int_0^{2\pi} V_m \cdot \sin \theta \, d\theta \qquad ...(i)$$

Now the load voltage is developed only during the period SCR* conducts. This period lies between the angles α and π. Therefore, we must take the average over the limits α to π instead of 0 to in equation (*i*).

$$\therefore \quad V_{dc} = \frac{1}{2\pi} \int_\alpha^\pi V_m \cdot \sin \theta \, d\theta$$

$$= \frac{V_m}{2\pi} \int_\alpha^\pi \sin \theta \cdot d\theta$$

$$= \frac{V_m}{2\pi} \left| -\cos \theta \right|_\alpha^\pi$$

$$= \frac{V_m}{2\pi} \left| (-\cos \pi) - (-\cos \alpha) \right|$$

$$= \frac{V_m}{2\pi} (1 + \cos \alpha) \qquad ...(ii)$$

* Here we have neglected the forward voltage drop across the device. Its value is about 1 volt.

We know that load current, $i_L = v_L/R_L$, therefore the average value of the load current,

$$I_{dc} = \frac{V_{dc}}{R_L} = \frac{V_m}{2\pi R_L}(1 + \cos\alpha) = \frac{I_m}{2\pi}(1 + \cos\alpha) \qquad ...(iii)$$

where, I_m is equal to V_m/R_L and is the maximum value of the load current. It may be noted that if the firing angle (α) is equal to zero, then the average values of load voltage and current is obtained from equations (*ii*) and (*iii*)

$$\therefore \qquad V_{dc} = \frac{V_m}{\pi}$$

and

$$I_{dc} = \frac{I_m}{\pi} = \frac{V_m}{\pi \cdot R_L}$$

It may be noted that the above results are the same as obtained for an ordinary diode half-wave rectifier.

Example 15-3. *An SCR is used for converting a.c. to d.c. . The anode supply is 220 volts, 50 Hz and the firing angle is adjusted at 60°. Find out the d.c. output voltage.*

(*U.P.S.C., Engg. Services,* 1985)

Solution. Given : V_{rms} = 220 V and α = 60°.

We know that maximum value of the applied voltage, *i.e.*, anode supply,

$$V_m = \sqrt{2} \times V_{rms} = \sqrt{2} \times 220 = 311 \text{ V}$$

and the d.c. output voltage,

$$V_{dc} = \frac{V_m}{2\pi}(1 + \cos\alpha) = \frac{311}{2\pi}(1 + \cos 60°) = 74.2 \text{ V } \textbf{Ans.}$$

Example 15-4. *Find the value of resistance to limit the average current to 0.5 A, when 100 V r.m.s. is applied to an SCR with a firing angle of 45°.*

Solution. Given : I_{dc} = 0.5 A; V_{rms} = 100 volts and α = 45°

Let R_L = Value of resistance, which limits the anode current to 0.5 A

We know that maximum value of applied voltage,

$$V_m = \sqrt{2} \times V_{rms} = \sqrt{2} \times 100 = 141.4 \text{ V}$$

and the average current (I_{dc})

$$0.5 = \frac{V_m}{2\pi \cdot R_L}(1 + \cos\alpha) = \frac{141.4}{2\pi \cdot R_L}(1 + \cos 45°)$$

$$= \frac{141.4}{2\pi \cdot R_L}(1 + 0.707) = \frac{38.42}{R_L}$$

$$\therefore \qquad R_L = 38.42/0.5 = 76.84 \ \Omega \ \textbf{Ans.}$$

15-9. 90° Variable Half-wave Rectifier

The SCR half-wave rectifier circuit shown in Fig. 15-5 (*a*) is known as 90° variable half-wave rectifier (or 90° variable phase control circuit). The term 90° variable implies that we can make SCR to conduct anywhere from the origin to the maximum value of the positive half-cycle, *i.e.*, the firing angle can be varied from 0° to 90°. The firing angle depends upon the value of variable resistance (*R*). If the resistance *R* is set at a low value, the SCR will start conducting near the origin. On the other hand, if *R* is set at a large value, the SCR may not conduct until the maximum value of positive half-cycle.

15-10. 180° Variable Half-wave Rectifier

Fig. 15-8 (*a*) shows that circuit of a 180° variable half-wave rectifier. It is also called 180° variable phase control circuit. We can see the difference between this circuit and that of a 90° variable phase control circuit. In this circuit, a capacitor (*C*) has been added between the variable resistance (*R*) and the cathode of SCR. The combination of resistance (*R*) and capacitor (*C*) forms a phase-shift network. This network delays the point at which SCR will trigger on the positive half of the applied voltage. The delay is provided by charging of the capacitor through the resistance *R*. This allows the firing angle to be controlled from 0° up to a maximum of 180°.

Fig. 15-8(*b*) shows the waveforms of input voltage (V_{in}), capacitor voltage (V_C) and load voltage (V_L). It may be noted that capacitor voltage lags behind the input voltage. Their phase relationship is given by the relation,

$$V_C = \frac{V_{in}}{1 + \omega \cdot C \cdot R}$$

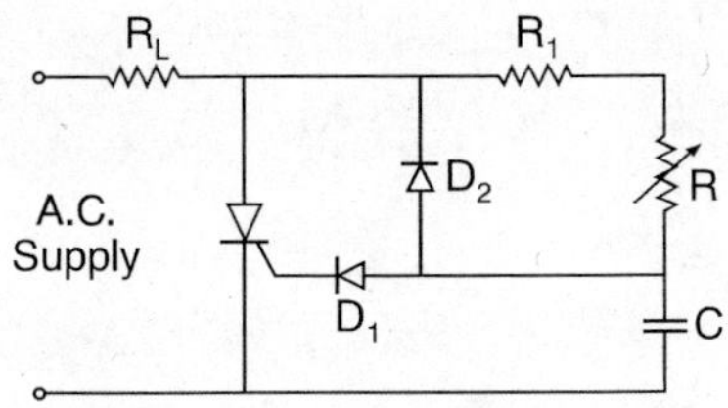

(*a*) 180° Variable half-wave rectifier.

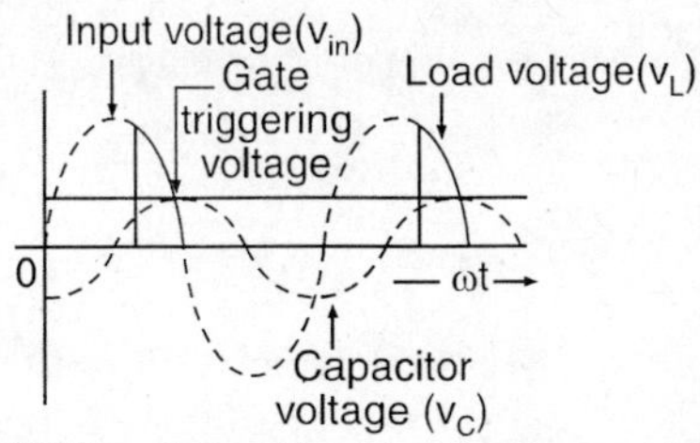

(*b*) Waveform of input voltage, capacitor voltage and load voltage.

Fig. 15-8.

It may be noted from the above expression that if the value of $\omega \cdot C \cdot R$ is much smaller than unity, then the capacitor voltage (V_C) and the input voltage (V_{in}) are in phase with each other. As a result of this, the firing will occur almost immediately after the input voltage is positive (*i.e.*, $\alpha = 0°$, $\phi = 180°$). On the other hand, if the value of $\omega \cdot C \cdot R$ is much larger than unity, then capacitor voltage is delayed by practically 90° from the input voltage and is much reduced in amplitude. This allows the firing to be delayed to almost the end of half a cycle (*i.e.*, $\alpha = 180°$, $\phi = 0°$).

The diode D_2 has been added to reset capacitor (*C*) by discharging it during the negative half cycle. This allows the firing to occur at the same point in each cycle.

15-11. SCR Full-wave Rectifier

Fig. 15-9(*a*) shows one of the circuits used for SCR full-wave rectifier. It consists of a diode bridge in series with SCR. The gate control input is left open, because a number of different firing circuits can be used.

The working of the SCR full-wave rectifier circuit may be understood from the following discussion. When the terminal *A* is positive, the current will flow through the diode D_1, through SCR, then through load resistance and then finally through diode D_2 it passes to terminal *B*. However, if the terminal *B* is positive, the current will flow through diode D_3, SCR, load resistance, diode D_4 and then finally to terminal *A*. Thus, the four diodes forming a bridge provide full-wave rectification, while the SCR controls the amount of voltage applied to the load.

Fig. 15-9(*b*) shows the full-wave rectified and load voltage waveforms. The amount of voltage applied to the load may be controlled by adjusting the firing angle of the SCR. This can be achieved by adjusting the gate current supplied by the control circuit.

The average value of load voltage and load current may be obtained from the equation of the applied voltage, *i.e.*,

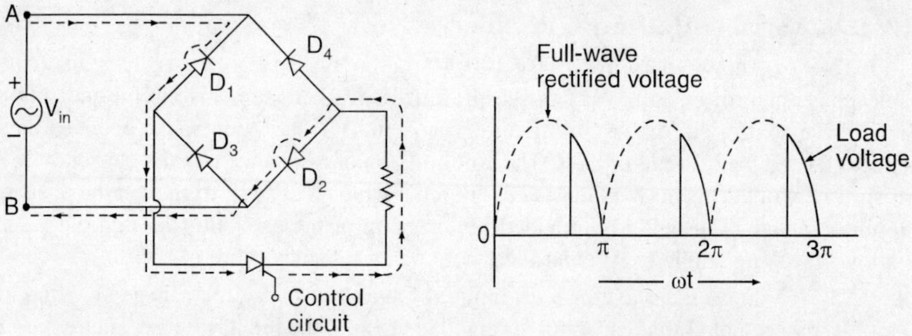

(*a*) SCR full-wave rectifier.

(*b*) The full-wave rectified and load voltage waveforms.

Fig. 15-9.

$$V_{in} = V_m \sin \omega t = V_m \sin \theta$$

Then average value of load voltage,

$$V_{dc} = \frac{1}{\pi}\int_{\alpha}^{\pi} V_m \sin\theta \cdot d\theta = \frac{V_m}{\pi}\int_{\alpha}^{\pi} \sin\theta \, d\theta$$

$$= \frac{V_m}{\pi}\left| -\cos\theta \right|_{\alpha}^{\pi} = \frac{V_m}{\pi}[(-\cos\pi) - (-\cos\alpha)]$$

$$= \frac{V_m}{\pi}(1+\cos\alpha) \qquad ...(i)$$

and the average value of load current,

$$I_{dc} = \frac{V_{dc}}{R_L} = \frac{V_m}{\pi R_L}(1+\cos\alpha) \qquad ...(ii)$$

It may be noted that if $\alpha = 0$ in equation (*i*) and (*ii*) then average value of voltage and current,

$$V_{dc} = \frac{2V_m}{\pi} \quad \text{and} \quad I_{dc} = \frac{2V_m}{\pi R_L}$$

These values are the same as in an ordinary diode full-wave rectifier.

15-12. UJT Triggered SCR Phase Control

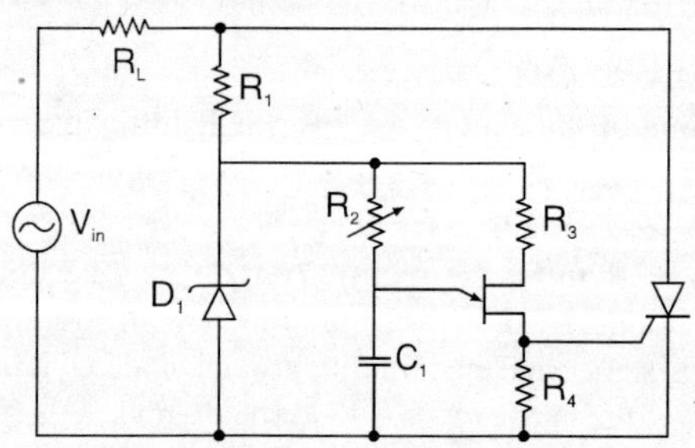

Fig. 15-10. *UJT triggered SCR phase control.*

We have already discussed in previous articles that SCR's are used in power control circuits. In these circuits, a control circuit is needed to vary the firing or the phase angle at which the SCR is triggered. The value of phase angle controls the power delivered to the load. If we want to deliver the maximum power to the load, then the SCR must be triggered as soon as the a.c. voltage across it goes positive, *i.e.*, the firing angle must be zero. Therefore, the SCR will conduct for 180° after triggering. If the power requirement is below the maximum value, the firing angle must be between 0° and 180°. In order to adjust the firing angle between 0° and 180°, we need a control circuit. We have already seen that a circuit shown in Fig. 15-8 (*a*) can be used to accomplish the control. But this circuit is affected by loading and supply voltage variation. Therefore, this circuit does not give a better phase control. However, a better control can be achieved by using a unijunction transistor (UJT) relaxation oscillator circuit.

Fig. 15-10 shows a UJT oscillator controlling a SCR circuit. The SCR is in it 'OFF' state until the UJT fires. When the UJT fires, the capacitor will discharge quickly through the resistor R_4. This positive pulse will turn 'ON' the SCR. The SCR will remain 'ON' until the SCR line voltage approaches zero. At this point, the SCR will turn 'OFF' throughout the entire negative circle. A similar circuit can also be made for triac phase control.

15-13. Triac Power Control

The triacs, like SCR's are also used to control average power delivered to the load by the method of phase control. The triac can be triggered such that the a.c. power is supplied to the load for a controlled position of each half-cycle.

The triac is in its 'OFF' position during each positive half-cycle* of the a.c. for a certain interval called the delay angle or firing angle (measured in degrees). Then triac is turned 'ON' and passes current through the load for the remaining portion of the positive half-cycle, except that the current passes in the opposite direction through the load. The action during both half-cycles is illustrated in Fig. 15-11(*a*). The resulting waveform of load current is as shown Fig. 15-11(*b*).

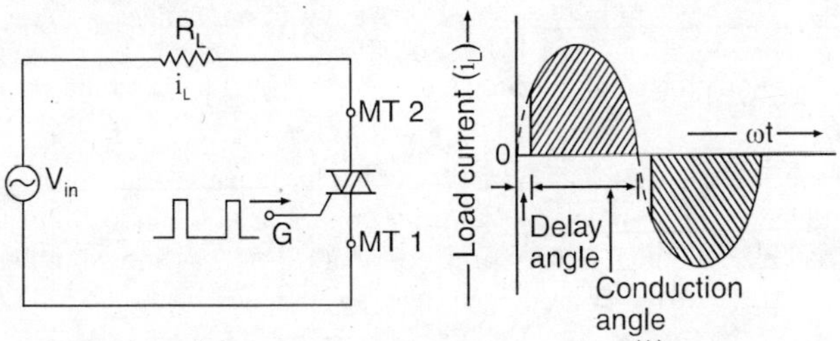

Fig. 15-11. *Basic triac phase control.*

A simple example of phase control using a triac is shown in Fig. 15-12(*a*). The diodes are used to provide trigger pulses to the gate of the triac. The diode D_1 conducts during the positive half-cycle. The value of resistance R_1 sets the point on the positive half-cycle at which the triac triggers. It may be noted that during this portion of the a.c. cycle, M.T. 2 and *G* are positive with respect to M.T. 1.

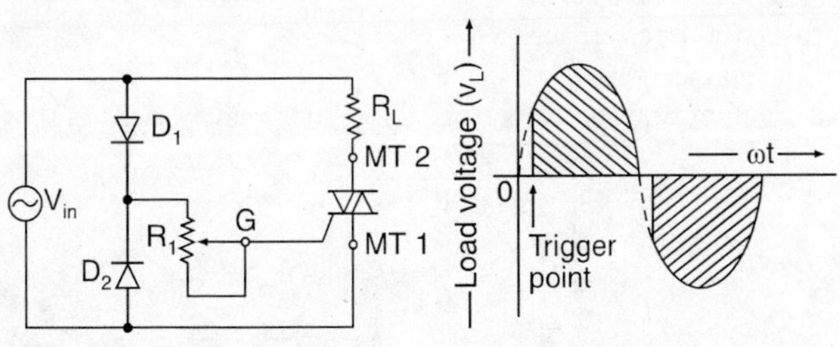

Fig.15-12. *Triac phase control circuit.*

* For construction and working of triacs, please see chapter-14 on Thyristors.

The diode D_2 conducts during the negative half-cycle and the resistance R_1 sets the trigger point. It may be noted that during this portion of the a.c. cycle, M..T 2 and G are negative with respect to M.T. 1. The resulting waveform of the voltage across the load resistor is shown in Fig. 15-12 (*b*). In the control circuit applications, it is necessary that the triac turn 'OFF' occurs at the end of each positive and negative alteration of the input voltage. But the triac turn 'OFF' occurs, usually, before the alteration of (positive or negative) when the load current falls below the holding current value.

15-14. Diac-Traic Phase Control Circuit

Fig. 15-13(*a*) shows a basic diac-traic phase control circuit, which may be used for light dimming, heat control or universal motor speed control. The operation of the circuit may be explained as follows: As the input voltage increases, either positively or negatively, the capacitor C_2 charges at the rate determined by setting of resistance R_2. When the voltage across capacitor C_2 exceeds the breakover voltage of diac, it turns 'ON'. As a result of this, the capacitor C_2 discharges through the conducting diac into the gate of triac. Hence, the triac turns ON and it conducts current through the load. The waveform of voltage across the load as shown in Fig. 15-13 (*b*).

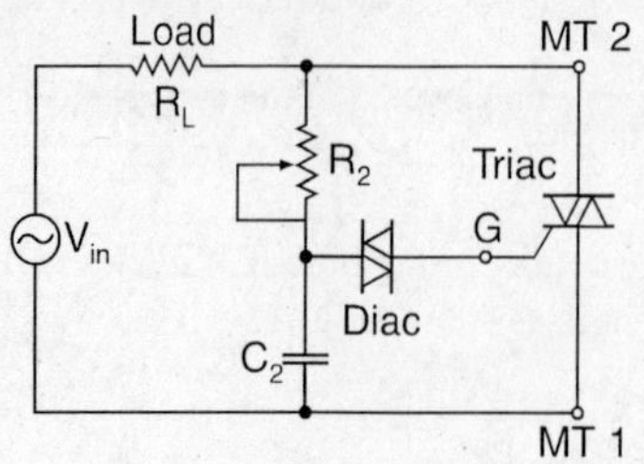

(*a*) Diac-triac phase control circuit.

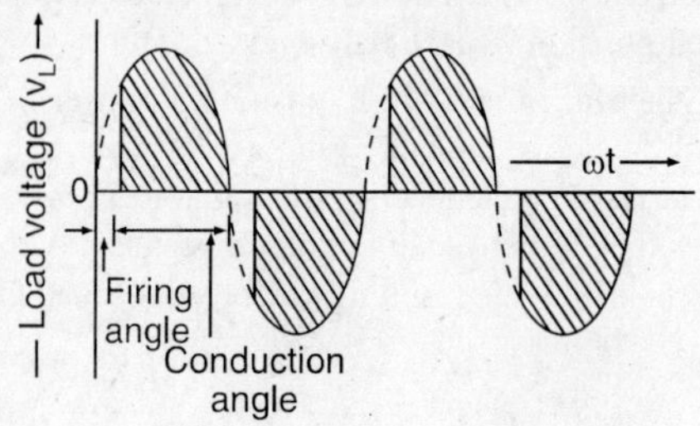

(*b*) Load voltage waveform.

Fig. 15-13.

The circuit of Fig. 15-13(*a*) has a drawback that it can cause triac to trigger before the desired turn-ON time. Such early turn-ON gives a prior control at low conduction angles. This drawback can be minimized by including another time constant ($R_3 \cdot C_3$) as shown in Fig. 15-14. This circuit can be adjusted to trigger triac over a wide range, *i.e.*, practically zero to full ON. The waveform of load voltage shown in Fig. 15-14(*b*) indicates that the current rises quickly (in about 1 or 2 μsec) each time the triac conducts. This rapid rise produces noise or radio frequency interference (RFI), which may be radiated and can cause direct interference equipment. The following methods may be used to reduce radio frequency interference, depending upon the application :

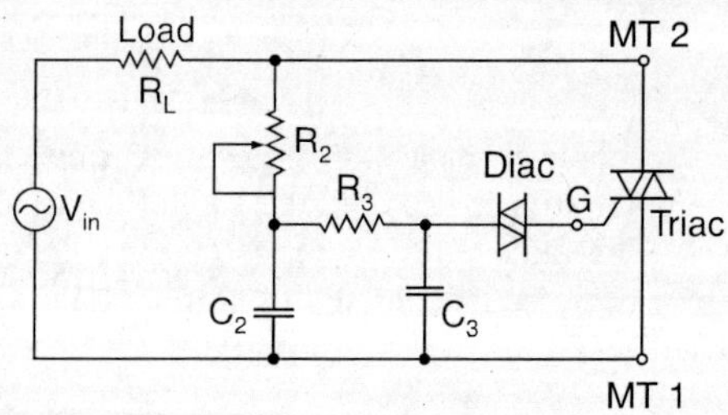

Fig. 15-14. *Phase control circuit using double time constant.*

(*i*) In light dimming and speed controls, a RFI filter may be used as shown in Fig. 15-15.

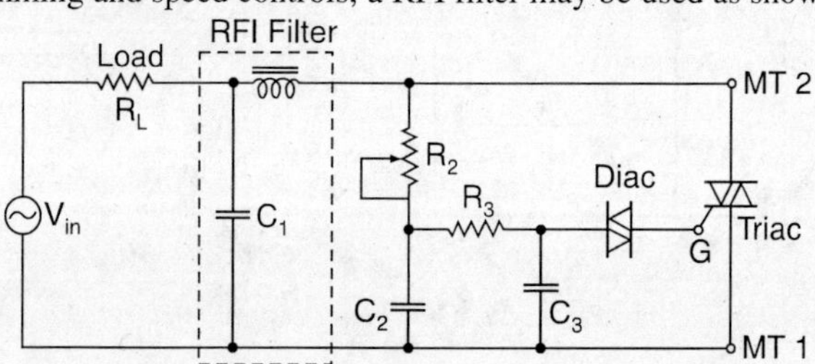

Fig. 15-15. *Phase control circuit using RFI filter.*

(*ii*) In heat controls, the current through an electrical heater may be turned ON for a few cycles and then turned OFF for a few cycles to maintain a desired temperature. If the turn ON is only allowed to take place when the voltage is zero, no rapid increase in current takes place as it does if the switching takes place during the cycle.

Several manufactures provide IC's that accomplish the above mentioned type of switching called zero-voltage switching (ZVS). One such circuit is the CA 3059 manufactured by RCA. This device is a silicon monolithic integrated circuit, which is designed to act as a triggering circuit for SCR's and triacs. This circuit provides an output pulse (typically 100 μsec duration) each time the applied line voltage crosses zero on each half cycle. The pulses are suitable for frequencies ranging from 50 Hz to 400 Hz. This circuit may also be used with light dimming circuits, if the a.c. frequency is 400 Hz as is found in aircraft applications. But it is not very useful in controlling the speed of motors.

15-15. Chopper

A chopper is a *dc-to-dc converter.* It converts a given constant d.c. voltage into a variable average d.c. voltage across a load by placing a high-speed static switch between the d.c. source and the load. This high-speed static switch is called a chopper because it chops off the d.c. supply into ON and OFF periods of flow.

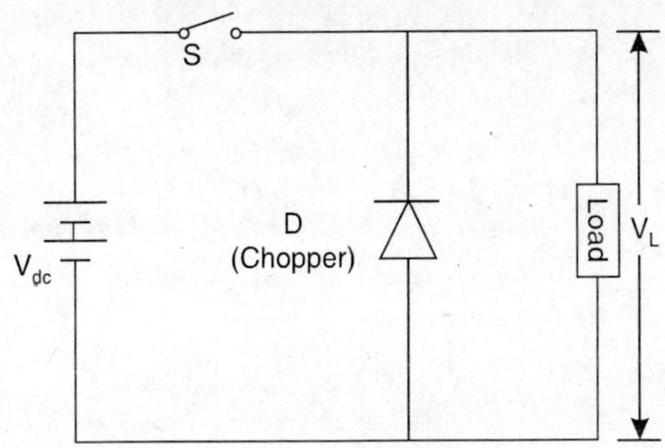

Fig. 15-16.

A basic chopper circuit is shown in Fig. 15-16. When switch *S* is closed, the d.c. supply voltage V_{dc} is applied across the load and when it is open, the load is disconnected from the supply. By varying the ratio of the switch-closed time (T_{ON}) to the switch-open time (T_{OFF}) at a fixed frequency, the value of the average output d.c. voltage can be controlled.

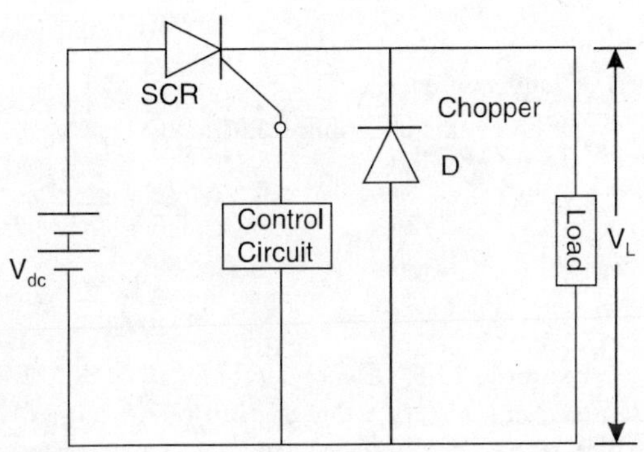

Fig. 15-17.

The switch *S* in Fig. 15-16 could be either a transistor or an *SCR* depending on the amount of power involved. An *SCR* is used in high-power applications whereas transistors are used when power involved is low. A d.c. chopper circuit using an *SCR* is shown in Fig. 15-17. The *SCR* acts like a static switch and has two states of ON and OFF. The duration of ON and OFF states can be varied with the help of triggering and commutating circuits (not shown) respectively. By changing the values of ON and OFF periods, average d.c. load voltage can be changed.

The operation of the circuit may understood as explained below. When *SCR* is triggered into conduction with the help of control circuitry, full d.c. voltage V_{dc} is applied across the load for a period of T_{ON}. When the *SCR* is switched OFF by the control circuitary, there is no voltage across the load. The output load voltage is in the form of a square wave as shown in Fig. 15-18. The waveshapes of the load current and chopper current for an inductive load are as shown. The free-wheeling diode *D* provides path for the stored inductive energy to flow.

If T_{ON} is the ON time and T_{OFF} time of the chopper, the duty cycle of the chopper is given by

$$\text{duty cycle} = \frac{T_{ON}}{T} = \frac{T_{ON}}{T_{ON} + T_{OFF}}$$

The load voltage V_L is given by

$$V_L = V_{dc}\left(\frac{T_{ON}}{T}\right) = V_{dc} \times \text{duty cycle}$$

$$= f\,V_{dc}\,T_{ON}$$

where f (= 1/T) is the switching frequency of the chopper.

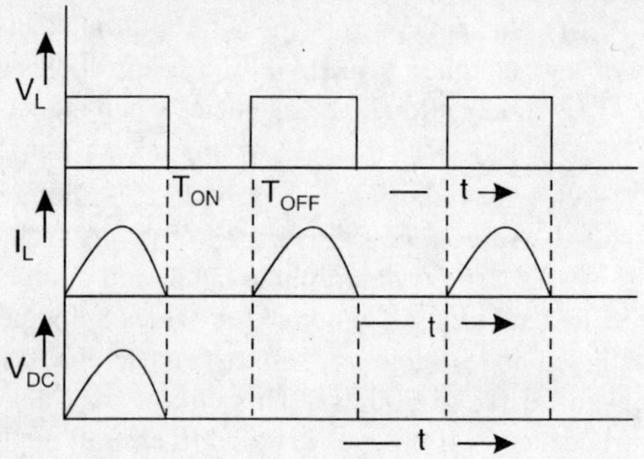

Fig. 15-18.

It is evident from the above discussion that V_L depends on the duty cycle since V_{dc} is constant. Hence, following different methods of controlling V_L are available.

- Keeping T_{OFF} constant and varying T_{ON}.
- Keeping T_{ON} constant and varying T_{OFF}.
- Varying the ratio T_{ON}/T_{OFF}.
- Any combination of the above.

Example 15-5. *A dc chopper has ON time of 30 μs and OFF time of 10 μs. Calculate* ***(i)*** *chopper duty cycle, and* ***(ii)*** *chopping frequency.*

Solution. Given : T_{ON} = 30 μs, and T_{OFF} = 10 μs.

(*i*) *Chopper duty cycle*

We know that chopper duty cycle,

$$= \frac{T_{ON}}{T} = \frac{T_{ON}}{T_{ON} + T_{OFF}}$$

$$= \frac{30\ \mu s}{30\ \mu s + 10\ \mu s} = \frac{30}{40} = 0.75 \text{ or } 75\% \textbf{ Ans.}$$

(*ii*) *Chopping frequency*

We know that chopping frequency,

$$f = \frac{1}{T} = \frac{1}{T_{ON} + T_{OFF}} = \frac{1}{30\ \mu s + 10\ \mu s}$$

$$= \frac{1}{40\ \mu s} = 0.025 \times 10^6 \text{ Hz or 25 kHz } \textbf{Ans.}$$

Example 15-6. *A chopper supplied by a 200 V d.c. has ON time of 30 ms and OFF time of 10 ms. Determine the value of average d.c. output voltage.* (*Mysore Univ.* 1998)

Solution. Given : T_{ON} = 30 ms and T_{OFF} = 10 ms.

We know that the duty cycle,

$$= \frac{T_{ON}}{T_{ON} + T_{OFF}} = \frac{30\ ms}{30\ ms + 10\ ms} = 0.75$$

∴ Average value of d.c. output voltage,

$$V_L = V_{dc} \times \text{duty cycle}$$

$$= 200 \times 0.75 = 150 \text{ V } \textbf{Ans.}$$

15-16. Inverters

An inverter is a device that *changes d.c. power into a.c. power* (just the opposite of converters). The inversion process can be achieved with the help of transistors, *SCR*s and tunnel diodes, *etc*. For low and medium outputs, transistorised inverters are suitable but for high power outputs, SCR inverters are essential. For very low voltage and high current requirements, tunnel diode invertors are used.

For inverter applications, transistors have definite advantages over *SCR*s regarding the switching speed, simplicity of control circuitry, high efficiency and greater reliability. It is mainly due to this fact that *SCR* inverters require complicated circuitry for triggering and commutation.

The basic working principle of an inverter may be explained with the help of circuit shown in Fig. 15-19. It is called voltage-driven inverter because a d.c. voltage source is connected through semiconductor switches directly to the primary of a transformer.

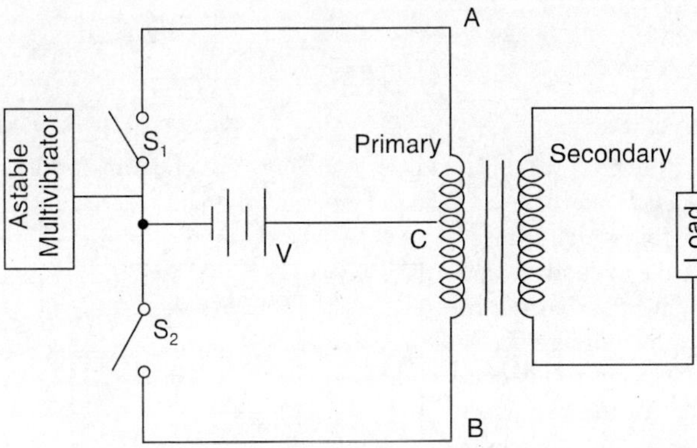

Fig. 15-19.

In Fig. 15-19, S_1 and S_2 are switching devices (transistors or *SCR*s) which open and close alternately at regular intervals of time. The two switching devices are generally driven by an astable multi-vibrator operating at the desired frequency. When S_1 is closed, the entire d.c. source voltage V is applied across points A and B of the transformer primary. S_1 remains closed for a certain period of time after which it is cut off and S_2 closes. It also remains closed for the same period of time during which the source voltage V is impressed across points B and C of the primary. S_2 then opens out and S_1 closes. In this way, an alternating voltage is applied across the primary which induces an a.c. voltage in the secondary. Since, d.c. supply voltage is connected directly across the primary, the output waveform of the secondary voltage is a square wave (Fig. 15-20.) irrespective of the type of load and load power factor. However, the waveforms of both the primary and secondary currents depend on the type of load whether resistive, inductive or capacitive.

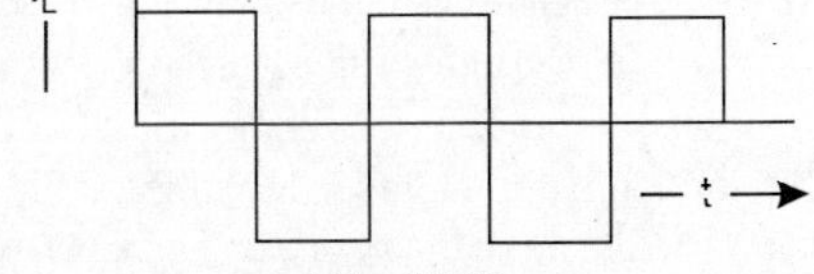

Fig. 15-20.

15-17. Single-phase Inverter

Fig. 15-21(*a*) shows a single-phase inverter with a load resistor using 4 *SCR*s working in pairs. The triggering and commutating circuitry of the *SCR*s has not been shown in the figure. The two thyristors SCR_1 and SCR_4 are triggered simultaneously so that load current passes through R_L from left to right. Exactly when these two *SCR*s are switched off by the commutating circuitry, thyristors

SCR_2 and SCR_3 are triggered into conduction thereby sending current through R_L from right to left. Hence, an a.c. voltage is developed across the load whose waveform is as shown in Fig. 15-21.

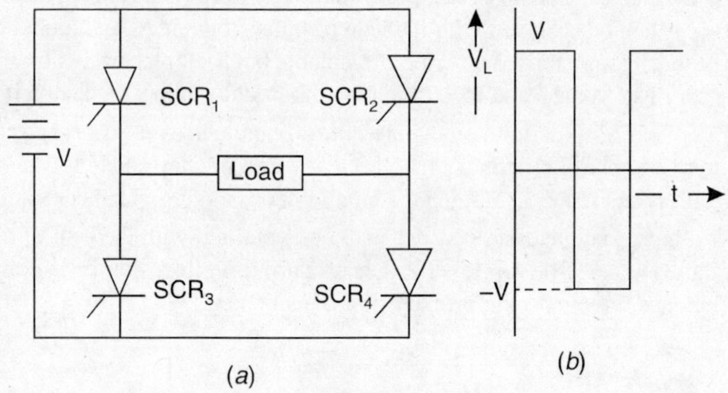

Fig. 15-21.

15-18. Push-Pull Inverter

Fig. 15-22 shows an inverter which employees two *SCR*s and one transformer. These two *SCR*s are triggered into conduction alternately for the same period of time. As a result, current through the primary becomes alternating which induces an a.c. voltages across the secondary and hence the load. As explained earlier in Art. 15-17, the secondary a.c. voltage has a square waveform. The capacitor *C* is connected across the anodes of the two *SCR*s and provides commutation, *i.e.*, switching off of the *SCR*s. The capacitor charges to double the supply voltage as a result of transformer action between the two halves of the primary winding. This large voltage is sufficient to reverse-bias the *SCR*s and drive the holding current below its rated value.

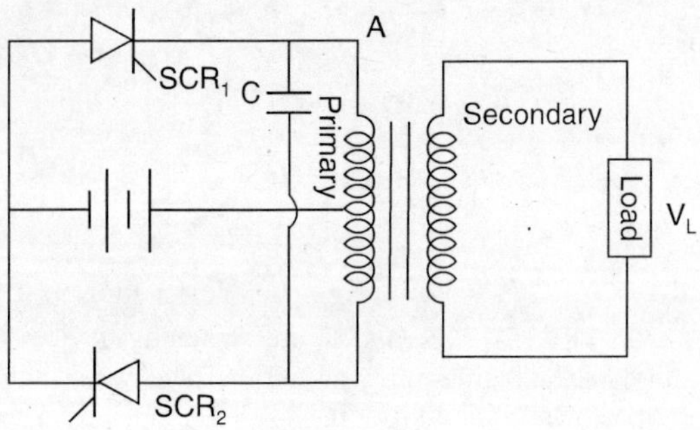

Fig. 15-22.

SUMMARY

In this chapter you have learned that :

1. A controlled rectifier have three terminals instead of two as in a general purpose PN junction diode.
2. The load current in a controlled rectifier flows only when a control signal is applied at the gate terminal to turn 'ON' the rectifier.
3. The *SCR* has three terminals: anode, cathode and gate. Once the positive voltage of suitable amplitude is applied at the gate input, the SCR is triggered ; *i.e.*, it starts conducting like an ordinary diode. Once the *SCR* turns ON, the gate has no control over the *SCR*. The conduction can only be terminated by removing the applied voltage from the anode.
4. For a *SCR* half-wave rectifier, the d.c. output voltage,

$$V_{dc} = \frac{V_m}{2\pi}(1 + \cos\alpha)$$

where α is the conduction angle in degrees.

For a resistive load, the d.c. output current,

$$I_{dc} = \frac{V_m}{2\pi R_L}(1 + \cos\alpha)$$

5. *UJT* can be used to provide a trigger signal to *SCR* and a *TRIAC*
6. a chopper converts a given constant d.c. voltage into a variable average d.c. voltage across a load by placing a high-speed static switch between the d.c. source and the load.
7. an inverter is a device that changes d.c. power into a.c. power.

GLOSSARY

Chopper : A device circuit that converts d.c. to a.c.

Controlled rectifier : A device which allows the flow of current through the load only when it turns 'ON' for the desired duration.

Inverter : A device circuit that changes d.c. power into a.c. power.

Silicon Controlled Rectifier : A semiconductor device with four layer, three junctions and three terminals: anode, cathode and gate.

NUMERICAL PROBLEMS

1. An SCR is rated at 50 A peak and 10 A average. What is the greatest possible delay in the trigger angle, if the direct current is at the rated value ? (**Ans.** 75°)
2. Find the value of load resistance, which can limit the average current to 0.8 A, if 220 V r.m.s. is applied to a SCR with a firing angle 60°. (**Ans.** 92.8Ω)
3. A silicon controlled rectifier with anode supply of 220 V r.m.s. 50 Hz, applied through a 100 Ω resistor is fired at 50° of the anode cycle. Assuming the internal drop to be zero; find
   (*a*) the average current and
   (*b*) average power delivered to the load. (**Ans.** 0.74 A and 54.9 W)

DESCRIPTIVE QUESTIONS

1. What is a controlled rectifier ? How it is different from an ordinary diode rectifier ?
2. Write a short note on the use of SCR for power control. (*Grad. I.E.T.E., June* 1985)
3. Explain how a silicon controlled rectifier is used for controlled rectification of a.c. supply voltage. (*A.M.I.E., Winter* 1986)
4. Draw a 90° variable phase control circuit using SCR and explain its working.
5. Explain the circuit of a light dimmer using a triac and employing phase control.

MULTIPLE CHOICE QUESTIONS

1. A silicon controlled rectifier is a semiconductor device which is made up of
   (*a*) Two layers and one junction (*b*) Three layers and two junctions
   (*c*) Four layers and three junctions (*d*) None of the above

2. Which of the following is preferred for controlling the amount of power supplied to the load

(*a*) Controlling the transformer secondary voltage

(*b*) Inserting a resistor in the load circuit

(*c*) Use of a silicon controlled rectifier

(*d*) None of the above

3. The value of d.c. load voltage for a SCR full-wave rectifier circuit is

(*a*) $\frac{V_m}{\pi}(1 + \cos\alpha)$ (*b*) $\frac{V_m}{2\pi}(1 + \cos\alpha)$

(*c*) $\frac{V_m}{\pi}(1 - \cos\alpha)$ (*d*) $\frac{V_m}{2\pi}(1 - \cos\alpha)$

4. The power delivered to the load in maximum when the SCR firing angle is

(*a*) zero (*b*) 90°

(*c*) 180° (*d*) Between 90° and 180°

5. The triac is a device which conducts during

(*a*) Positive half cycle of the input voltage

(*b*) Negative half cycle of the input voltage

(*c*) Both positive and negative half cycles of the input voltage

(*d*) None of the above

6. For a half wave controlled rectifier, the average value of output d.c. voltage is given by

(*a*) $V_{dc} = \frac{V_m}{2\pi}(1 - \cos\theta_1)$ (*b*) $V_{dc} = \frac{2V_m}{\pi}(1 + \cos\theta_1)$

(*c*) $V_{dc} = \frac{V_m}{\pi}(\cos\theta_1 - 1)$ (*d*) $V_{dc} = \frac{V_m}{2\pi}(\cos\theta_1 + 1)$

ANSWERS

1. (*c*) **2.** (*c*) **3.** (*a*) **4.** (*a*) **5.** (*c*)

6. (*c*)

❑❑❑

16

CHAPTER

Single-Stage BJT Amplifiers

OUTLINE

Objectives

After completing this chapter, you should be able to :

- Know amplifier and their classification.
- Understand the amplifying action of a transistor.
- Describe the operation of common-emitter transistor amplifier.
- Do the analysis of common-emitter amplifier.
- Know the operation of swamped amplifier.
- Describe the operation of common-base amplifier.
- Describe the operation of common collector amplifier.
- Do the analysis of common-collector amplifier.
- Compare common-emitter, common-base and common collector amplifiers characteristics.
- Explain the effects of non-linear distortion in amplifiers.

16-1. Introduction

When we cannot hear a voice on the mobile phone, we increase the volume, when the picture on our television or the computer is too dark, we increase the brightness control. In both of these cases, we are taking a relatively weak signal and making its stronger (*i.e.*, increasing of its power). The process of increasing the power of an a.c. signal is called ***amplification***. The circuits used to performs this function are called ***amplifiers***.

An amplifier may also be defined as ***a device which produces a larger electrical output of similar characteristics than that of the input parameters.*** It increases the magnitude of the current or voltage input by means of energy drawn from an external source. The input signal may be obtained from a phonograph cartridge, tape head, or a transducer such as thermocouple, pressure gauge, *etc*. The output signal may be supplied to a loudspeaker in an audio amplifier, a motor in a servo amplifier, a relay in a control application, *etc*.

As mentioned above, an amplifier needs a source of energy for amplification. This may be provided by a battery (usually a dry battery) or a d.c. source resulting from a rectifier and filter combination. The amplifier also contains atleast one active device. This may be an electron tube, bipolar transistor or a field-effect transistor, which can provide a control function. Basically, the active device converts the energy from the d.c. source into energy at the output of the amplifier, which is proportional to the input signal. The a.c. input signal merely supplies a means of controlling the d.c. to a.c. convertion, which takes place in the tube or transistor. The control is, usually, accomplished with comparatively little input signal power.

The amplifier, in which the instantaneous output signal is directly proportional to the corresponding input signal, is called a ***linear amplifier***. On the other hand, an amplifier, in which the output is not directly proportional to the input, is called a ***non-linear amplifier***.

16-2. Classifications of Amplifiers

The linear amplifiers may be classified according to their mode of operation, *i.e.*, the way they operate on the predetermined set of values. Their descriptions are based on the following factors :

(1) *As based on the input*

(*a*) Small-signal amplifier (*b*) Large-signal amplifier

(2) *As based on the output*

(*a*) Voltage amplifier (*b*) Power amplifier

(3) *As based on transistor configuration*

(*a*) Common-emitter (CE) amplifier (*b*) Common-base (CB) amplifier

(*c*) Common-collector (CC) amplifier

(4) *As based on biasing conditions*
   (*a*) Class-A (*b*) Class-B (*c*) Class-AB (*d*) Class-C
   (*e*) Class -D
(5) *As based on nature of load resistance*
   (*a*) Untuned amplifiers (Wide-band amplifier)
   (*b*) Tuned amplifier (Narrow-band amplifier)
(6) *As based on frequency response*
   (*a*) Direct coupled (DC) amplifier
   (*b*) Audio frequency (AF) amplifier
   (*c*) Radio frequency (RF) amplifier
   (*d*) Ultra high frequency (UHF) and microwave frequency amplifier
(7) *As based on number of stages*
   (*a*) Single-stage amplifier (*b*) Multistage amplifier
(8) *As based on the method of coupling between the stages*
   (*a*) DC (direct coupled) amplifier (*b*) R-C coupled amplifier
   (*c*) Transformer coupled amplifier

Now we shall discuss the small signal, single-stage common-emitter, common base and common collector amplifiers in the following pages.

16-3. Transistor as an Amplifier

Fig. 16-1 shows a basic circuit of an amplifier. In this circuit, an NPN transistor is used in the common emitter (CE) configuration. Therefore, the basic circuit is known as a basic common emitter amplifier. Here, V_{BB} supply forward-biases the emitter-base junction and V_{CC} supply reverse-biases the collector-base junction. This biases the transistor to operate in the active region V_S is a sinusoidal a.c. input signal source. It has a source resistance R_S. The magnitude of signal source voltage is such that it always forward-biased the emitter-base junction regardless of the polarity of the signal. A brief description of the operation of amplifier circuit is given below :

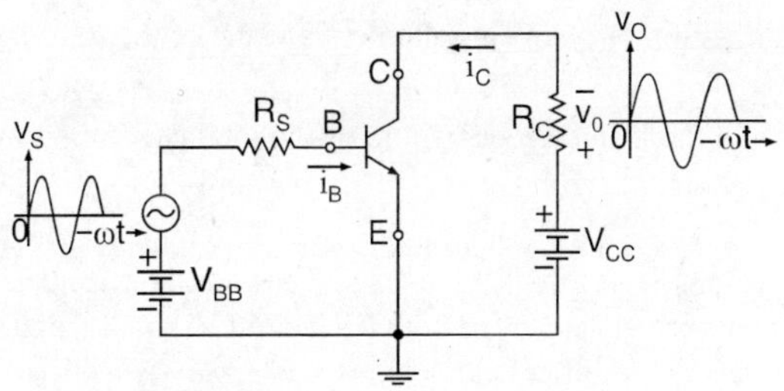

Fig. 16-1. *Basic common emitter amplifier.*

First of all, let us assume that there is no a.c. signal source. Under this condition, a d.c collector current (I_C) flows through the collector load (R_C). This is called zero signal current or quiescent operating current. Now let an a.c. signal be applied between the emitter-base junction. During the positive half-cycle of the input signal, the forward-bias across the emitter-base junction is increased. As a result of this, more electrons are injected into the base and reach the collector, which increase the collector current. The increased collector current produces greater voltage drop across the resistance R_C. However, during the negative half-cycle, the forward-bias across the emitter-base junction is decreased. Due to this, the collector-current decreases. The decreased collector current produces smaller voltage drop across resistance R_C.

It is evident from the above discussion, that a small a.c. signal at the input produces a large a.c. signal at the output or load resistance. Thus, the transistor acts as an amplifier. The ratio of output voltage to input voltage is called voltage gain of the amplifier.

16-4. Common-Emitter Transistor Amplifier

Fig. 16-2 shows a circuit of a common-emitter amplifier using NPN transistor. It may be noted from this circuit that the input a.c. signal is applied across the base-emitter terminals and the output signals is taken across the collector and emitter terminals. The emitter-base junction of a transistor is forward-biased by the V_{BB} supply. The collector-base junction is reverse-biased by the V_{CC} supply. The Q-point is determined by the V_{CC} supply alongwith the resistance R_B and R_C. The capacitors C_1 and C_2 are called blocking capacitors (or coupling capacitors). Each capacitor acts like a switch, which is open to a direct current but shorted to an alternating current. Because of this, a blocking capacitor blocks the direct current, but passes the alternating current. This action isolates d.c. bias from an a.c. signal in the circuit. The function of capacitor C_1 is to connect the a.c. signal source to the input circuit (or base-emitter circuit) of the amplifier. The function of capacitor C_2 is to connect the amplifier to its load resistance or to the next stage of the amplifier.

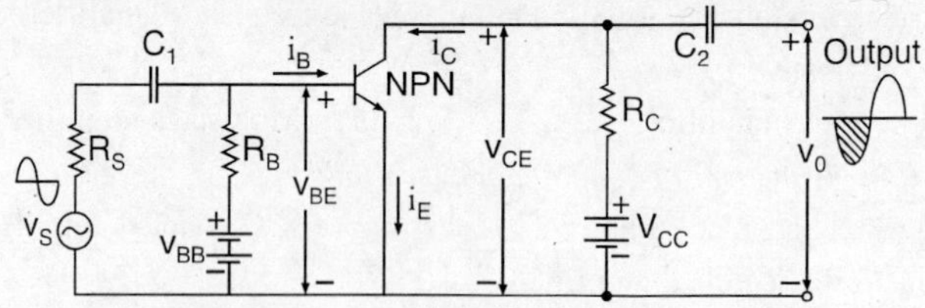

Fig. 16-2. *Common-emitter transistor amplifier.*

All the voltages and currents of a transistor are indicated by the instantaneous total values (*i.e.*, d.c + a.c.). Thus the instantaneous value of the total base current is given by the relation,

$$i_B = I_B + i_b$$

and the instantaneous value of the total collector current,

$$i_C = I_C + i_c$$

Similarly, the instantaneous value of the total base-to-emitter voltage,

$$v_{BE} = V_{BE} + v_{be}$$

and the instantaneous value of the total collector-to-emitter voltage,

$$v_{CE} = V_{CE} + v_{ce}$$

Here, the upper case letters I and V indicate the d.c. (or quiescent-operating) current and voltage values respectively. The lower case letters i and v represent the a.c. current and voltage values respectively. It may be noted that emitter terminal of a transistor is shown grounded, therefore common emitter amplifier is also known as *grounded emitter amplifier*.

16-5. Different Ways of Drawing Common Emitter Amplifier Circuit

In most of the commonly used amplifier circuits, the polarity of bias voltage required for the base, is the same as that required for a collector. Consequently, the circuit shown in Fig. 16-2 may be simplified by eliminating the V_{BB} supply and simply connecting it to the V_{CC} supply as shown in Fig. 16-3(*a*).

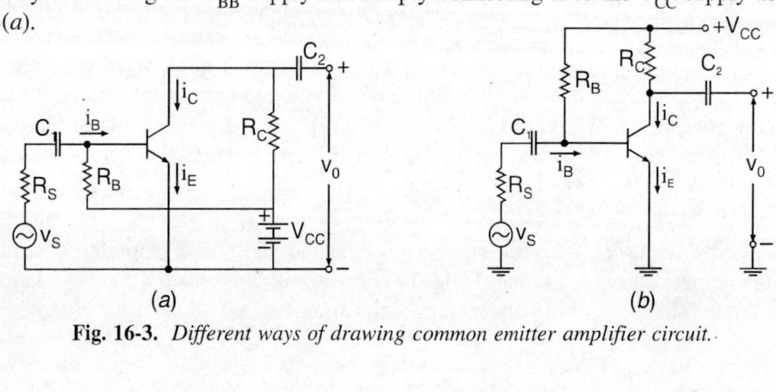

(*a*) (*b*)

Fig. 16-3. *Different ways of drawing common emitter amplifier circuit.*

Fig. 16-3(*b*) shows a circuit drawn in another form. It is more practical way of drawing an amplifier circuit. It may be noted that the battery symbol is not shown in this circuit. It is always understood to be connected from the terminal marked + V_{CC} to the ground terminal, even though it may not be actually shown. Moreover, the grounded terminals of the transistor and the signal source are shown separately for the sake of convenience.

The common emitter amplifier circuit shown in Fig. 16-3 has a base bias or fixed bias circuit. The biasing circuit can be a voltage-divider circuit, voltage-divider with feedback emitter resistor, collector feedback resistor, *etc*.

16-6. A.C. Load

Fig. 16-4 (*a*) shown a common-emitter amplifier with an external load resistance (R_L). Under this condition, the effective load as seen by the transistor is not equal to the resistance R_C. But it is a parallel combination of resistance R_C and R_L. A brief description of this is given as below:

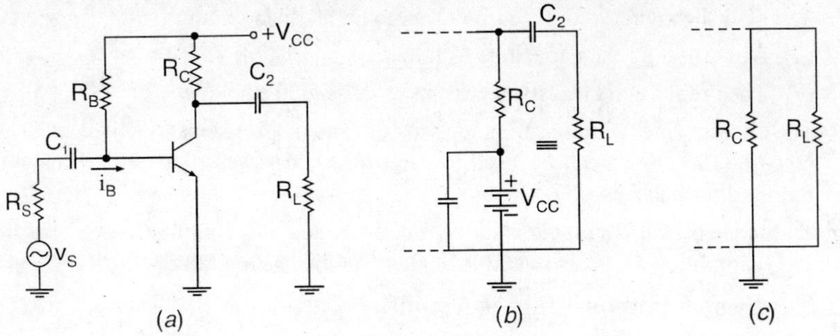

Fig. 16-4. *A.C. load.*

The Fig. 16-4 (*b*) shows an equivalent circuit of an amplifier output circuit. This circuit shows a capacitor across V_{CC} supply. This capacitor exists within the electronic power supplies. Because of this capacitor, the d.c. supply, *i.e.*, (V_{CC}) supply presents a very low resistance to the a.c. signal currents. Thus, an a.c. signal current flowing through resistance R_C and into the supply, produces a very little a.c. voltage drop within the power supply. In other words, the entire a.c. voltage is dropped across the resistance R_C. Hence, the resistance R_C may be considered as connected to ground as for as a.c. signal is concerned. It is for this reason that the equivalent circuit shown in Fig. 16-4 (*c*) does not show power supply.

Now consider the effect of coupling the capacitor C_2. We know that for a.c. current, the capacitor C_2 acts as short-circuit. Therefore, the a.c. voltage drop across the resistance R_L must be equal to that across the resistance R_C. It means that the two resistances R_C and R_L may be considered in parallel as far as a.c. currents are concerned. It is illustrated in the circuit shown in Fig. 16-4 (*c*).

It is evident from the above discussion, that the actual signal load, on the amplifier is a parallel combination of resistances R_C and R_L. This load is called a.c. load or effective load and is designated by r_L. Thus, the a.c. load

$$r_L = R_C \parallel R_L = \frac{R_C \times R_L}{R_C + R_L}$$

If $R_L = \infty$ or $R_L >> R_C$, then the a.c. load is equal to the collector resistance, *i.e.*, $r_L = R_C$.

Sometimes, it is very convenient to call resistance R_C as the d.c. load and the parallel combination of resistance R_C and R_L (designated as r_L) as a.c. load on an amplifier. It is the a.c. load, which determines the voltage gain of an amplifier.

16-7. Analysis of Transistor Amplifier

We know that in a transistor amplifier, the d.c. sources set up d.c. biasing currents and voltages, which are known as quiescent operating currents and voltages. The a.c. input signal source produces fluctuations in these currents and voltages. As a result of this, we get an amplifier signal at the output. We have to analyse the amplifier circuit in order to determine the a.c. and d.c. currents and voltage gain, input resistance and output resistance of an amplifier. The simplest way to analyse the circuit is to split the analysis into the following parts :

(1) ***D.C. analysis :*** The following procedure is adopted for the d.c. analysis :

(***a***) First of all, reduce the a.c. sources to zero. It means that a voltage source is replaced by a short-circuit and a current source by an open-circuit.

(***b***) Now open all the capacitors, as they block direct current.

The circuit, which remains after applying the above two steps, is called a d.c. equivalent circuit. This circuit can be used to determine the d.c. currents and voltages, which are required for the analysis.

(2) ***A.C. analysis :*** The following procedure is adopted for the a.c. analysis :

(***a***) First of all, reduce the d.c. sources to zero. It means that a voltage source is replaced by a short-circuit and a current source by an open-circuit.

(***b***) Now short all the capacitors as they pass alternating current.

(***c***) Then replace the transistor with its small-signal equivalent circuit.

The circuit, which remains after applying the above three steps, is called an **a.c. equivalent circuit.** The circuit can be used to determine the a.c currents and voltages, which are required for analysis.

This technique is valid for single-stage common-emitter (CE) common-base (CB) and common-collector (CC) amplifiers. It is also valid to analyse multistage amplifiers.

16-8. Analysis of Common Emitter Amplifier

Fig. 16-5 (*a*) shows the circuit of a simple common emitter amplifier with a base-bias (or fixed-bias) arrangement. The analysis of this amplifier can be done by splitting the circuit into two parts, *i.e.*, d.c. equivalent circuit and a.c. equivalent circuit.

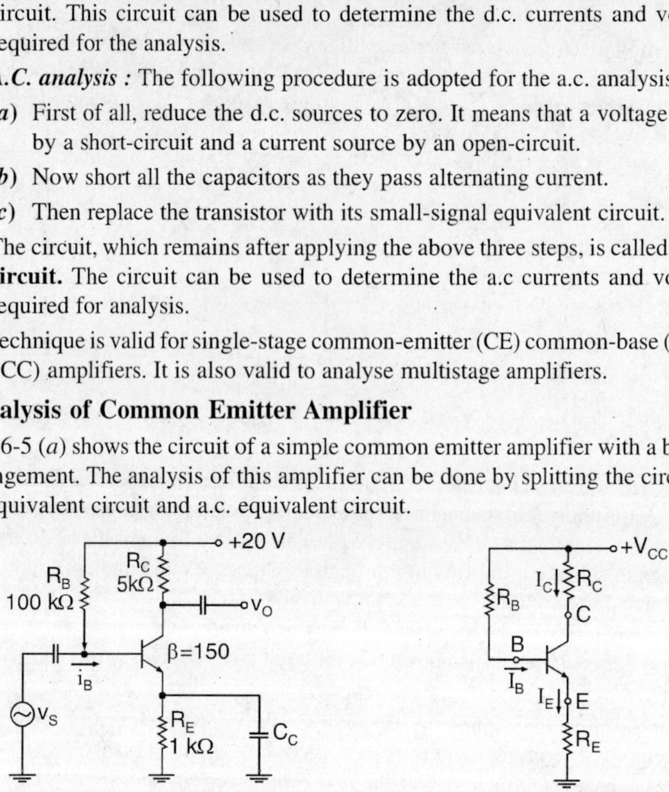

(*a*) Common emitter amplifier with emitter bypass capacitor.

(*b*) D.C. equivalent circuit of common emitter amplifier.

Fig. 16-5.

(*a*) ***D.C. equivalent circuit :*** It may be obtained from the amplifier circuit by opening (or disconnecting) the capacitors. The d.c. equivalent circuit is shown in Fig. 16-5 (*b*). We have already discussed in chapter on transistor biasing that base current in a base-bias circuit is given by the relation,

$$I_B = \frac{V_{CC} - V_{BE}}{R_B} = \frac{V_{CC}}{R_B} \qquad ...(\because V_{CC} >> V_{BE})$$

and the collector current,

$$I_C = \beta \cdot I\beta$$

$\therefore$ Emitter current,

$$I_E \approx I_C \approx \beta \cdot I_B$$

(b) *A.C. equivalent circuit :* It may be obtained from the amplifier circuit by replacing the d.c. supply (V_{CC}) by a ground and short-circuiting the capacitors. The a.c. equivalent circuit is shown in Fig. 16-6.

We shall use d.c. and a.c. equivalent circuits to determine the amplifier parameters in the next article.

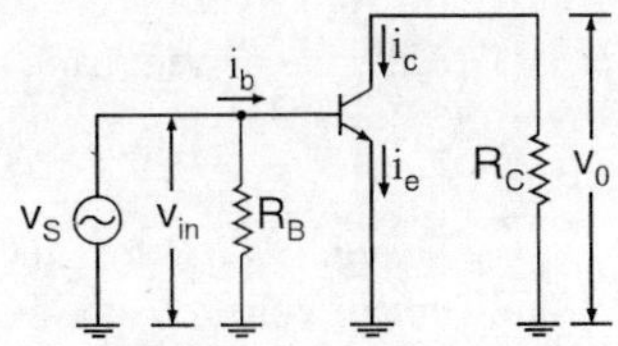

Fig. 19.6. *A.C. equivalent circuit.*

16-9. Common Emitter Amplifier Parameters

We have already discussed in the last article a simple a.c. equivalent circuit of a common emitter amplifier with base-bias. A more detailed a.c. equivalent circuit may be obtained by replacing NPN transistor with a small-signal model as discussed in the last chapter. Fig. 16-7 shows such an a.c. equivalent circuit. This circuit is used to determine amplifier parameters such as input resistance (R_i), output resistance (R_o), current gain (A_i), voltage gain (A_v) and power gain (A_p). It may be noted that we have not shown the signal source resistance R_s in the Figure. Therefore, we shall take resistance R_s as zero. The effect of signal source resistance shall be discussed later in this chapter.

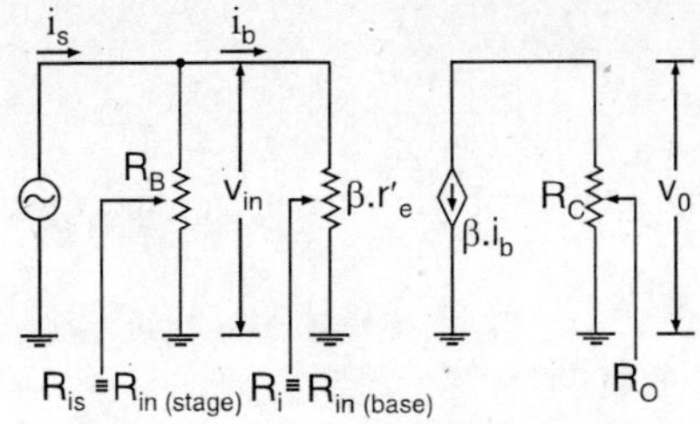

Fig. 19.7. *Detailed a.c. equivalent circuit of common emitter.*

It may be noted from Fig. 16-7 that $\beta \cdot r'_e$ represents the a.c. resistance of the emitter-base junction as seen by the input signal. It is known as input resistance, looking directly into the base and is designated by R_i or R_{in} (base). Here, β is common emitter d.c. current gain (or h_{FE}) of a transistor. The input resistance, (R_i) does not include the effect of external biasing resistors connected to the base. The input resistance, which does include this is called an input resistance of amplifier stage. It is designated by R_{is} or R_{in} (stage). It may be noted that in the absence of signal source resistance (R_s), all the source voltage appears as the input voltage across the base-emitter junction. Another important point is that since $\beta \cdot r'_e$ is much smaller than resistance R_B, therefore a major part of source current (i_s) passes through $\beta r'_e$ and a negligible part through the resistance R_B. Following parameters of a common emitter amplifier are important from the subject point of view :

(1) ***Input resistance :*** It is the resistance looking directly into the base and is given by the ratio of base voltage to base current. Thus, input resistance

$$R_i = \frac{v_b}{i_b} = \frac{v_{in}}{i_b}$$

Since, $\quad v_{in} = \beta \cdot r'_e \cdot i_b,\quad$ therefore, input resistance,

$$R_i = \frac{\beta \cdot r'_e \cdot i_b}{i_b} = \beta \cdot r'_e$$

and the input resistance of the amplifier stage,

$$R_{is} = R_B \parallel (\beta \cdot r'_e)$$

Since, $\quad R_B >> \beta \cdot r'_e \quad$ therefore,

$$R_{is} = \beta \cdot r'_e = R_i$$

(2) ***Output resistance :*** It is the resistance, looking into collector, and is approximately equal to the collector resistance (R_C). Mathematically the output resistance,

$$R_0 = R_C$$

(3) ***Current gain :*** It is the ratio of collector current (i_c) to the base current (i_b). Mathematically, the current gain,

$$A_i = \frac{i_c}{i_b} = \beta$$

(4) ***Voltage gain :*** It is the ratio of output voltage (v_0) to the input voltage (v_{in}). Since, the output voltage is the same as collector voltage and input voltage is the same as base voltage, therefore it is also known as voltage gain from base to collector. Mathematically the voltage gain,

$$A_v = \frac{v_0}{v_{in}}$$

We also know that the input voltage,

$$v_{in} = i_b \cdot \beta \cdot r'_e$$

and the output voltage,

$$v_0 = i_e \cdot R_C = \beta \cdot i_b \cdot R_C \qquad \ldots (\because (i_e = \beta \cdot i_b))$$

$\therefore$ Voltage gain,

$$A_v = \frac{\beta \cdot i_b \cdot R_C}{i_b \cdot \beta\, r'_e} = \frac{R_C}{r'_e}$$

(5) ***Power gain :*** It is the product of current gain (A_i) and voltage gain (A_v). Mathematically, the power gain,

$$A_p = A_i \cdot A_v = \beta \cdot R_C / r'_e$$

and the power gain in decibels,

$$G_p = 10 \log_{10} A_p$$

Example 16-1. *Fig. 16-8 shows a common emitter amplifier with a base-bias arrangement.*

*Determine the values of **(a)** input resistance looking into the base; **(b)** input resistance of the stage; **(c)** output resistance; and **(d)** voltage gain.*

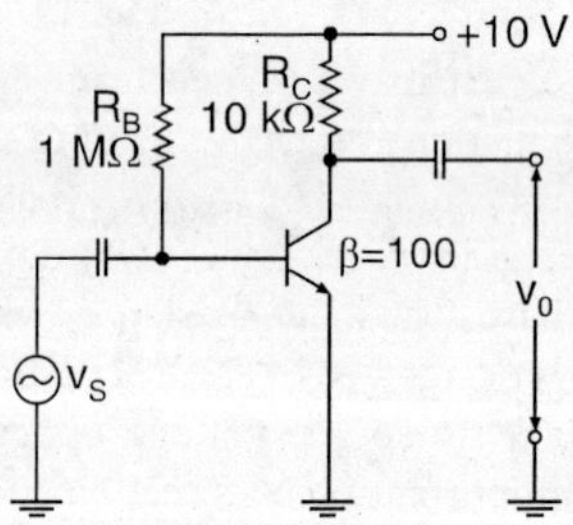

Fig. 16-8.

Solution. Given: V_{CC} = 10 volts; R_C = 10 kΩ = 10 × 10^3 Ω; R_B = 1 MΩ = 1 × 10^6 Ω and β = 100.

Fig. 16-9 shows the d.c. equivalent circuit for the common emitter amplifier.

(*a*) *Input resistance looking into the base*

We know that the base current,

$$I_B = \frac{V_{CC} - V_{BE}}{R_B} = \frac{10 - 0.7}{1 \times 10^6} = 9.3 \times 10^{-6} \text{ A}$$

$$= 9.3 \ \mu\text{A}$$

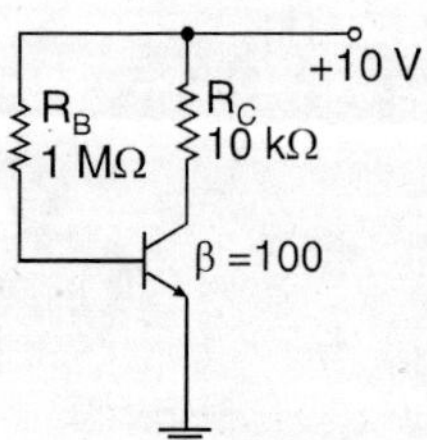

Fig. 16-9.

and the collector current,

$$I_C = \beta \cdot I_B = 100 \times (9.3 \times 10^{-6})$$

$$= 0.93 \times 10^{-3} \text{ A} = 0.93 \text{ mA}$$

$\therefore$ Emitter current,

$$I_E \approx I_C = 0.93 \text{ mA}$$

and a.c. resistance of emitter diode,

$$r'_e = \frac{25}{I_E} = \frac{25}{0.93} = 26.9 \ \Omega$$

∴ Input resistance looking directly into the base,

$$R_i = \beta \cdot r'_e = 100 \times 26.9 = 2690\ \Omega = 2.69\ \text{k}\Omega\ \textbf{Ans.}$$

(*b*) *Stage input resistance*

We know that stage input resistance,

$$R_{is} = R_B \parallel (\beta \cdot r'_e) = (1 \times 10^6) \parallel (2.69 \times 10^3)$$

$$= \frac{(1\times 10^6)\times(2.69\times 10^3)}{(1\times 10^6 + 2.69\times 10^3)} = 2.69\ \text{k}\Omega\ \textbf{Ans.}$$

(*c*) *Output resistance*

We know that the output resistance,

$$R_0 = R_C = 10\ \text{k}\Omega\ \textbf{Ans.}$$

(*d*) *Voltage gain*

We also know that the voltage gain,

$$A_v = \frac{R_C}{r'_e} = \frac{10\times 10^3}{26.9} = 372\ \textbf{Ans.}$$

Example 16-2. *A common emitter amplifier has an input resistance = 2.5 kΩ and a voltage gain of 200. If the input signal voltage is 5 mV, find* **(a)** *the base current* **(b)** *the collector current* **(c)** *the power gain, and* **(d)** *dB power gain. Take β = 50.*

Solution. Given: $R_i = 2.5\ \text{k}\Omega = 2.5 \times 10^3\ \Omega$; $A_v = 200$; $V_s = 5\ \text{mV} = 5 \times 10^{-3}$ V and $\beta = 50$.

(*a*) *Base current*

Let i_b = Base current.

We know that the input resistance (R_i),

$$2.5 \times 10^3 = \frac{V_s}{i_b} = \frac{5\times 10^{-3}}{i_b}$$

$$\therefore \quad i_b = 2 \times 10^{-6}\ \text{A} = 2\ \mu\text{A}\ \textbf{Ans.}$$

(*b*) Collector current

We know that the collector current,

$$i_c = \beta \cdot i_b = 50 \times 2 = 100\ \mu\text{A}\ \textbf{Ans.}$$

(*c*) *Power gain*

We know that the current gain,

$$A_i = \beta = 50$$

and power gain,

$$A_p = A_i \cdot A_v = 50 \times 200 = 10{,}000\ \textbf{Ans.}$$

(*d*) *dB Power gain*

We also know that *dB* power gain,

$$G_p = 10 \log_{10} A_p = 10 \log_{10} 10000$$

$$= 10 \log 10^3 = 40\ \text{dB}.\ \textbf{Ans.}$$

16-10. Effect of Emitter Feedback Resistor on Common Emitter Amplifier Parameters

Fig. 16-10 (*a*) shows a common emitter amplifier employing base bias with emitter feedback resistor. The d.c. equivalent circuit for the given amplifier circuit is shown in Fig. 16-10 (*b*). It is obtained from the amplifier circuit by disconnecting the capacitors.

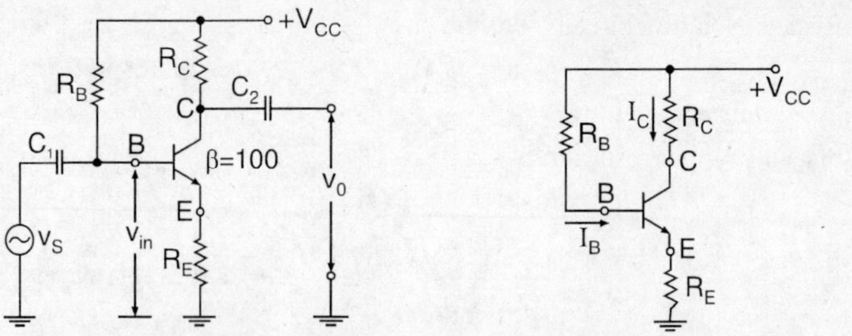

(*a*) Common emitter amplifier. (*b*) D.C. equivalent circuit.

Fig. 16-10.

We have already discussed in chapter on transistor biasing that for such a biasing circuit, the collector current is given by the relation,

$$I_C = \frac{V_{CC} - V_{BE}}{R_E + \dfrac{R_B}{\beta}} = \frac{V_{CC}}{R_E + \dfrac{R_E}{\beta}} \qquad ...(\because V_{CC} >> V_{EE})$$

The a.c. equivalent circuit for the amplifier circuit is shown in Fig. 16-11. This circuit is obtained from the amplifier circuit by short circuiting the capacitors and replacing the d.c. supply by a ground.

It may be noted that the emitter resistance, when referred to base circuit, becomes $\beta \cdot R_E$. Therefore, the total resistance in the base circuit is $\beta\,(r'_e + R_E)$

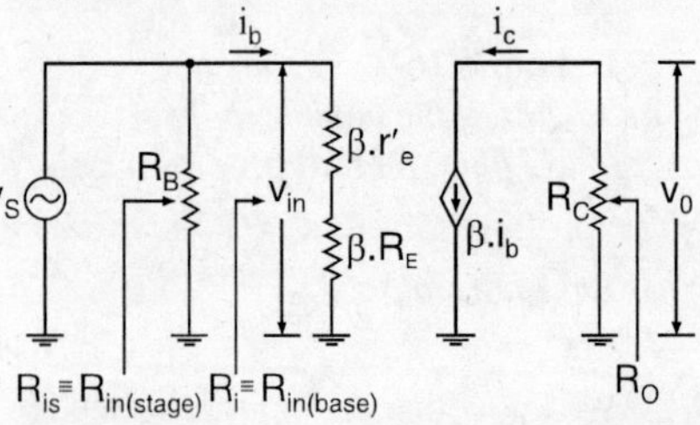

Fig. 16-11. *A.C. equivalent circuit.*

The input resistance looking into the base,

$$R_i = \beta\,(r'_e + R_E)$$

If $\qquad R_E >> r'_e$, then $R_i \approx \beta \cdot R_E$

The input resistance of amplifier stage (*i.e.*, the total input resistance as seen by the source),

$$R_{is} = R_B \parallel (\beta \cdot R_E) = R_B \parallel R_i$$

This indicates that input resistance of an amplifier increases with the addition of emitter resistance. Now we shall see the effect of emitter resistance on the voltage gain. As soon from Fig. 16-11, the input voltage,

$$v_{in} = i_b \cdot \beta\,(r'_e + R_E) = i_b \cdot \beta \cdot R_E$$

and the output voltage,

$$v_0 = i_c \cdot R_C = \beta \cdot i_b \cdot R_C$$

∴ Voltage gain,

$$A_v = \frac{v_o}{v_{in}} = \frac{\beta \cdot i_b \cdot R_C}{i_b \cdot \beta \cdot (r'_e + R_E)} = \frac{R_C}{r'_e + R_E}$$

$$= \frac{R_E}{R_B} \qquad ...(\because R_E >> r'_e)$$

We know that voltage gain of a common emitter amplifier without emitter feedback resistor is given by the relation,

$$A_v = \frac{R_C}{r'_e}$$

Since, the resistance R_E is much larger than a.c resistance r'_e, therefore the addition of emitter resistance in the common emitter amplifier reduces the voltage gain. Incidently, it may be noted that the emitter feedback resistor has no effect on the output resistance.

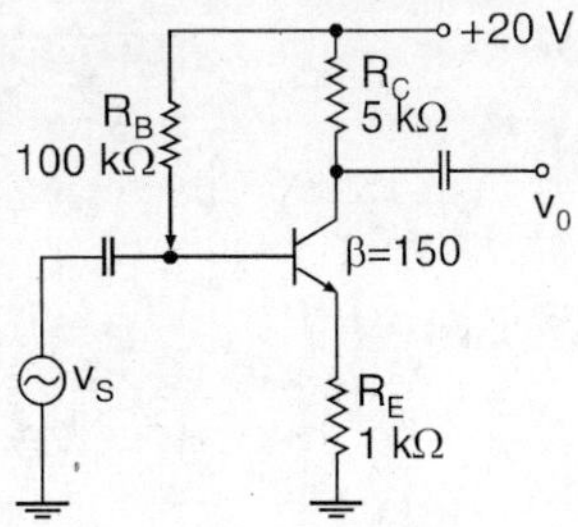

Fig. 16-12.

Example 16-3. *A common emitter amplifier is shown in Fig. 16-12.*

Determine the values of ***(a)*** *Input resistance looking directly into the base:* ***(b)*** *Input resistance of the stage;* ***(c)*** *Voltage gain; and* ***(d)*** *dB voltage gain.*

Solution. Given: $V_{CC} = 20$ volts; $R_C = 5\ k\Omega = 5 \times 10^3\ \Omega$; $R_E = 1\ k\Omega, = 1 \times 10^3\ \Omega$; $R_B = 100\ \Omega = 100 \times 10^3\ \Omega$ and $\beta = 150$.

(*a*) *Input resistance looking directly into the base*

We know that the collector current,

$$I_C = \frac{V_{CC}}{R_E + \frac{R_B}{\beta}} = \frac{20}{1 \times 10^3 + \frac{100 \times 10^3}{150}} = 12\ \text{mA}$$

and emitter current,

$$I_E \approx I_C = 12\ \text{mA}$$

$$\therefore \qquad r'_e = \frac{25}{I_E} = \frac{25}{12} = 2.1\ \Omega$$

and input resistance looking directly into the case,

$$R_i = \beta\,(r'_e + R_E) = 150\,[2.1 + (1 \times 10^3)]\ \Omega = 150 \times 10^3\ \Omega$$
$$= 150\ k\Omega\ \textbf{Ans.}$$

(*b*) *Input resistance of the stage*

We know that input resistance of the stage,

$$R_{is} = R_B \parallel R_i = \frac{(R_B \cdot R_i)}{(R_B + R_i)} = \frac{(100 \times 150)}{(100 + 150)} = 60\ k\Omega\ \textbf{Ans.}$$

(*c*) *Voltage gain*

We know that voltage gain,

$$A_v = \frac{R_C}{R_E} = \frac{5}{1} = 5\ \textbf{Ans.}$$

(*d*) *dB Voltage gain*

We also know that dB voltage gain,

$$G_p = 10 \log_{10} A_v = 10 \log_{10} 5 = 7\ dB\ \textbf{Ans.}$$

16-11. Effect of Emitter Bypass Capacitor on Common Emitter Amplifier Parameters

Fig. 16-13 (*a*) shows a common emitter amplifier with a capacitor connected across the resistance R_E. This capacitor is called a **bypass capacitor*, because it connects the emitter terminal to ground for a.c. signal current. The emitter bypass capacitor acts as an open switch to the direct current. Therefore, it does not affect the d.c. bias of the circuit. The d.c. equivalent circuit for the given amplifier is as shown in Fig. 16-13 (*b*). It may be noted that it is the same circuit as shown in Fig. 16-13 (*b*).

The a.c. equivalent circuit for the given amplifier is shown in Fig. 16-14. It may be noted, in this circuit, that input is directly looking into the base is $\beta \cdot r'_e$ only because resistance R_E is assumed to be shorted by the capacitor for a.c. signal current. Thus,

$$R_i = \beta \cdot r'_e \qquad ...(i)$$

* The action of a bypass capacitor depends upon its value of capacitance (*c*). Usually, a capacitor is selected with a large value. The advantage of this is that the reactance offered by the capacitor to a.c. current is much less than resistance R_E. Therefore resistance R_E effectively appears as a short for a.c. currents.

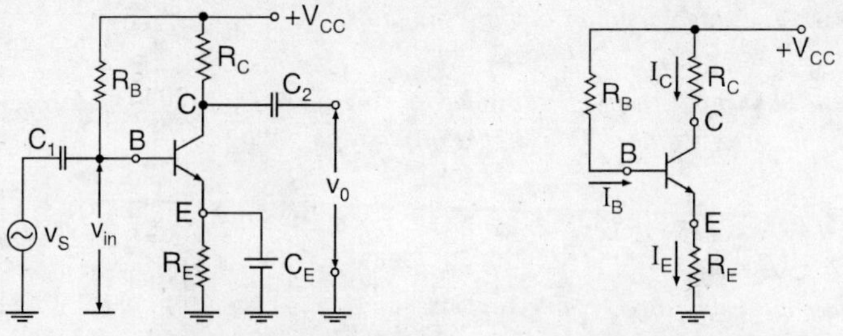

(*a*) Common emitter amplifier with emitter feedback.

(*b*) D.C. equivalent circuit.

Fig. 16-13.

and $R_{is} = R_B \parallel (\beta \cdot r'_e)$ ...(*ii*)

It is evident from equations (*i*) and (*ii*) that the input resistance is lower than that obtained without emitter bypass capacitor in the last article. It means that emitter bypass capacitor reduces the input resistance of a common emitter amplifier than that of an amplifier without emitter capacitor. Now let us study the effect of emitter bypass capacitor on the voltage gain. We know that the input voltage,

$$v_{in} = i_b \cdot \beta \cdot r'_e$$

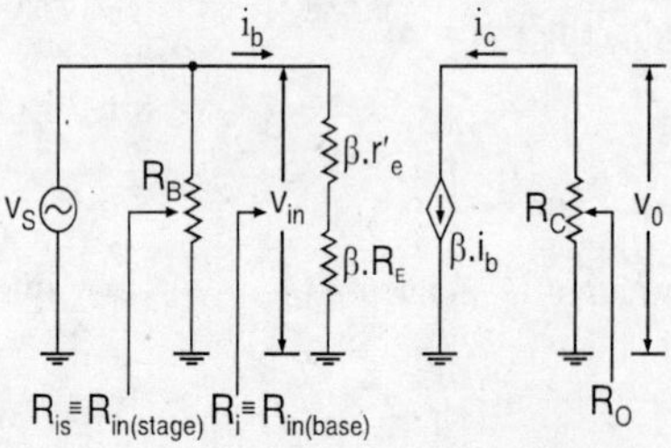

Fig. 16-14. *A.C. equivalent circuit of common emitter amplifier with emitter bypass capacitor.*

and the output voltage,

$$v_0 = i_c \cdot R_C = \beta \cdot i_b \cdot R_C \quad \text{... } (\because i_e = \beta \cdot i_b)$$

∴ Voltage gain,

$$A_v = \frac{v_0}{v_{in}} = \frac{\beta \cdot i_b \cdot R_C}{i_b \cdot \beta \cdot r'_e} = \frac{R_C}{r'_e}$$

The above value of voltage gain is the same as obtained for a common emitter amplifier without feedback resistor. However, the voltage gain with feedback resistor has been found to be equal to R_C/R_E. As the value of the resistance (R_E) is very large as compared to that of r'_e, therefore, we say that a common emitter amplifier with emitter bypass capacitor provides a higher voltage gain than that of an amplifier without emitter bypass capacitor.

Example 16-4. *A transistor used in a single-stage common emitter amplifier shown in Fig. 16-15 has β = 50. Determine **(a)** the input resistance looking directly into the base; **(b)** the total input resistance; **(c)** the output resistance; and **(d)** the voltage gain. What will be the voltage gain, if the emitter bypass capacitor is removed ?*

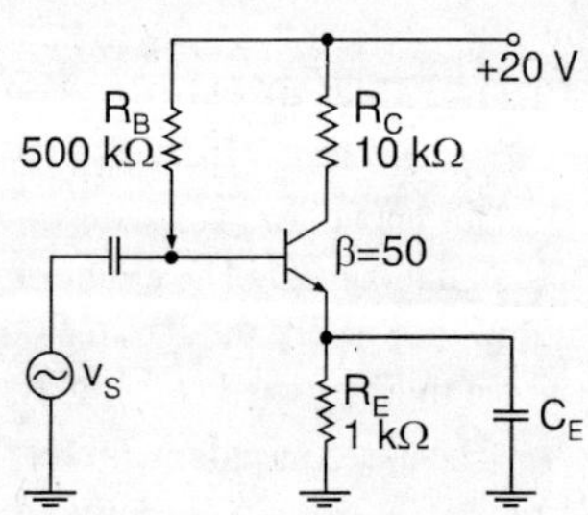

Fig. 16-15.

Solution. Given: V_{CC} = 12 volts; R_C = 10 kΩ = 10 × 10^3 Ω; R_E = 1 kΩ = 1 × 10^3 Ω; R_B = 500 kΩ = 500 × 10^3 Ω and β = 50.

(*a*) *Input resistance looking directly into the base*

We know that collector current,

$$I_C = \frac{V_{CC}}{R_E + \dfrac{R_B}{\beta}} = \frac{12}{1 \times 10^3 + \dfrac{500 \times 10^3}{50}} \text{ A}$$

$$= 1.09 \times 10^{-3}\ \text{A} = 1.09\ \text{mA}$$

$$\therefore \qquad I_E \approx I_C = 1.09\ \text{mA}$$

and $$r'_e = \frac{25}{I_E} = \frac{25}{1.09} = 22.9\ \Omega$$

$\therefore$ Input resistance looking directly into the base,

$$R_i = \beta \cdot r'_e = 50 \times 22.9 = 1146\ \Omega\ \textbf{Ans.}$$

(*b*) *Total input resistance*

We know that total input resistance,

$$R_{is} = R_B \parallel R_i = (500 \times 10^3) \parallel 1146 = 1146\ \Omega$$

(*c*) *Output resistance*

We know that output resistance,

$$R_0 = R_C = 10\ \text{k}\Omega\ \textbf{Ans.}$$

(*d*) *Voltage gain*

We know that the voltage gain,

$$A_v = \frac{R_C}{r'_e} = \frac{(10 \times 10^3)}{22.9} = 436.7\ \textbf{Ans.}$$

Voltage gain if the emitter bypass capacitor is removed

We also know that voltage gain, if the emitter bypass capacitor is removed,

$$A_v = \frac{R_C}{R_E} = \frac{10}{1} = 10\ \textbf{Ans.}$$

16-12. Effect of A.C. Load on Common Emitter Amplifier Parameters

Fig. 16-16 (*a*) shows a common emitter amplifier with a voltage divider-bias and connected to an external load resistance (R_L). The external load may be a loud speaker, an electric motor or the next stage of an amplifier. The voltage divider bias has been selected to illustrate change in a.c. equivalent circuit from base-bias of the amplifier. The d.c. equivalent circuit for the given amplifier is as shown in Fig. 16-16 (*b*). This has been obtained by opening (or disconnecting) all the capacitors.

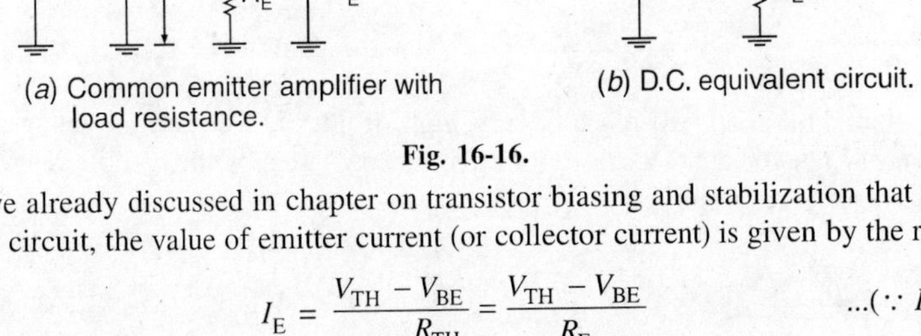

(*a*) Common emitter amplifier with load resistance.

(*b*) D.C. equivalent circuit.

Fig. 16-16.

We have already discussed in chapter on transistor biasing and stabilization that for a voltage divider bias circuit, the value of emitter current (or collector current) is given by the relation,

$$I_E = \frac{V_{TH} - V_{BE}}{R_E + \dfrac{R_{TH}}{\beta}} = \frac{V_{TH} - V_{BE}}{R_E} \qquad ...(\because R_E >> R_{TH}/\beta)$$

where, V_{TH} = Thevenin's Voltage and its value is equal to $V_{CC}\left(\frac{R_2}{R_1 + R_2}\right)$ and

R_{TH} = Thevenin's Equivalent Resistance and its value is equal to $\frac{(R_1 \cdot R_2)}{(R_1 + R_2)}$

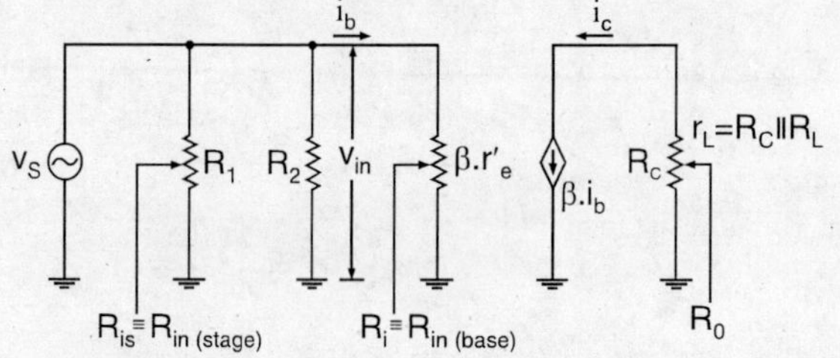

Fig. 16-17. *A.C. equivalent circuit of a common emitter amplifier with load resistance.*

The a.c. equivalent circuit of the given amplifier is shown in Fig. 16-17. It has been obtained by replacing the d.c. supply by ground and capacitor by a short-circuit. It may be noted that the input resistance directly looking into the base,

$$R_i = \beta \cdot r'_e$$

The total input resistance as seen by the source (*i.e.*, the input resistance of the amplifier stage),

$$R_{is} = (R_1 \parallel R_2) \parallel (\beta \cdot r'_e)$$

Now let us discuss the effect of an a.c. load on the voltage gain. We have already discussed in Art. 16-5 that the a.c. load (or effective load) as seen by the amplifier, at its output, is a parallel combination of R_C and R_L. Thus, a.c. load

$$r_L = R_C \parallel R_L$$

and input voltage,

$$v_{in} = i_b \cdot \beta \cdot r'_e$$

Output voltage,

$$v_0 = i_c \cdot r_L = \beta \cdot i_b \cdot r_L$$

and voltage gain,

$$A_v = \frac{v_0}{v_{in}} = \frac{\beta \cdot i_b \cdot r_L}{i_b \cdot \beta \cdot r'_c} = \frac{r_L}{r'_e}$$

Since, the value of a.c. load is smaller than the value of collector resistance, therefore the addition of load resistance reduces the voltage gain of common emitter amplifier. However, if $R_L >> R_C$, then the a.c. load,

$$r_L \approx R_C$$

and the voltage gain,

$$A_v = \frac{R_C}{r'_e}$$

It means that if the load resistance is very high, it does not effect the voltage gain of the amplifier. It may be noted that the a.c. load (or effective load) seen by the amplifier is r_L. Therefore, the output resistance of an amplifier,

$$R_0 = r_L$$

Example 16-5. *Draw the a.c. equivalent circuit for the common emitter amplifier shown in Fig. 16-18. Given $\beta = 200$, $V_{BE} = 0.7$ V and a.c. resistance of the emitter diode is of the form,*

$$r'_e = \frac{25}{\text{D.C emitter current in mA}} \quad (\Omega)$$

Also find out the a.c. output voltage and input impedance for this circuit.

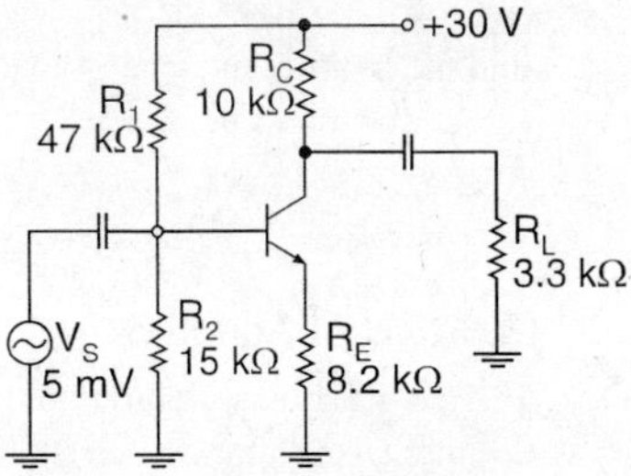

Fig. 16-18.

Solution. Given: $\beta = 200$; $V_{BE} = 0.7$ volts; $V_{CC} = 30$ volts; $R_C = 10\ k\Omega$; $R_L = 3.3\ k\Omega$; $R_1 = 47\ k\Omega$; $R_2 = 15\ k\Omega$; $R_E = 8.2\ k\Omega$ and $V_S = 5$ mV.

A.C. equivalent circuit

We know that the Thevenin's Voltage,

$$V_{TH} = V_{CC}\left(\frac{R_2}{R_1 + R_2}\right) = 30 \times \left(\frac{15}{47 + 15}\right) = 7.26 \text{ V}$$

and the equivalent resistance,

$$R_{TH} = R_1 \parallel R_2 = \frac{R_1 \cdot R_2}{R_1 + R_2} = \frac{47 \times 15}{47 + 15} = 11.4\ k\Omega = 11.4 \times 10^3\ \Omega$$

∴ Value of the emitter current,

$$I_E = \frac{V_{TH} - V_{BE}}{R_E + \frac{R_{TH}}{\beta}} = \frac{7.26 - 0.7}{8.2 \times 10^3 + \frac{11.4 \times 10^3}{200}}$$

$$= 0.794 \text{ mA} = 0.8 \text{ mA}$$

and the value of a.c. resistance of the emitter diode,

$$r'_e = \frac{25}{(I_E)(\text{in mA})} = \frac{25}{0.794} = 31.5\ \Omega$$

We also know that the a.c. load seen by the amplifier,

$$r_L = R_C \parallel R_L = \frac{R_C \cdot R_L}{R_C + R_L} = \frac{10 \times 3.3}{10 + 3.3} = 2.5\ k\Omega$$

Now let us draw the a.c. equivalent circuit for the given common emitter amplifier as shown in Fig. 16-19. **Ans.**

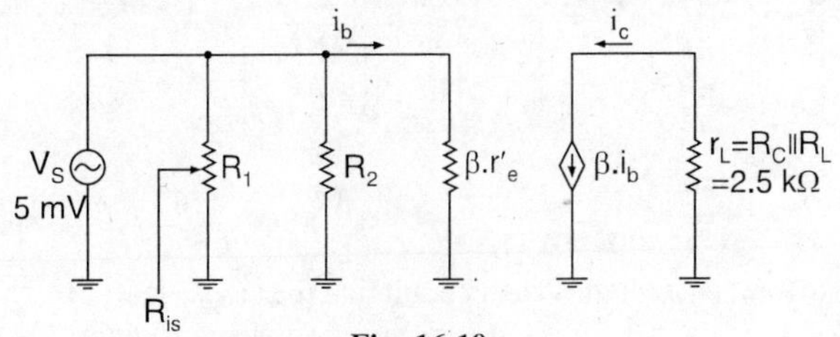

Fig. 16-19.

Output voltage

Let v_0 = Output voltage,

We know that the voltage gain,

$$A_v = \frac{r_L}{r'_e} = \frac{2.5 \times 10^3}{31.5} \approx 79 \text{ Ans.}$$

and the voltage gain (A_v)

$$79 = \frac{v_0}{v_{in}} = \frac{v_0}{5}$$

∴ $v_0 = 79 \times 5 = 395$ mV **Ans.**

Input impedance (or input resistance)

We know that the input resistance looking directly into the base,

$$R_i = \beta \cdot r'_e = 200 \times 31.5 = 6.3 \times 10^3\ \Omega = 6.3\ \text{k}\Omega$$

and the input resistance of the stage (*i.e.*, total input resistance) of the circuit,

$$R_{is} = (R_1 \parallel R_2) \parallel R_i = R_{TH} \parallel R_i$$
$$= 11.4 \parallel 6.3 = (11.4 \times 6.3) / (11.4 + 6.3)$$
$$= 4\ \text{k}\Omega\ \textbf{Ans.}$$

16-13. Effect of Signal Source Resistance on Common Emitter Amplifier Parameters

Fig. 16-20 shows a common emitter amplifier connected to a signal source. The signal source has an internal resistance designated as R_S. If the internal resistance of the source is zero, then all the input signal voltage appears at the base terminal. However, if the a.c. source has a non-zero internal resistance, then the actual signal voltage at the base will be less by a factor equal to the voltage drop across resistance R_S. Thus, the actual signal voltage,

$$v_{in} = v_S = \text{Drop across the resistance } R_S$$

It is evident from the above relation, that greater the internal resistance of an a.c. signal source, greater the voltage drop across the resistance R_S and hence lesser the input voltage.

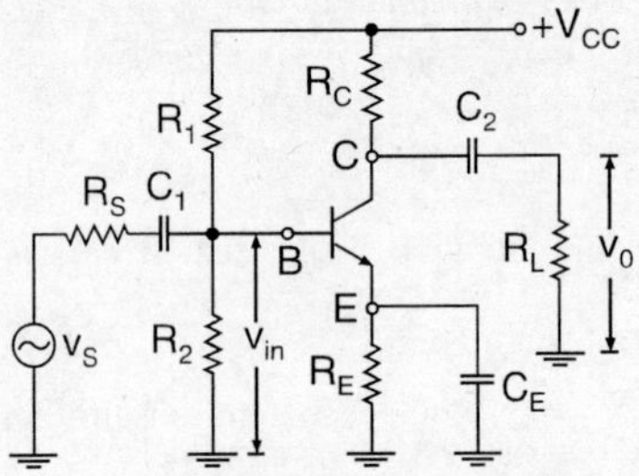

Fig. 16-20. *Common emitter amplifier with signal source resistance.*

The a.c. equivalent circuit of the given amplifier is shown in Fig. 16-21. The total input resistance of the amplifier as seen by the signal source,

$$R_{is} = (R_1 \parallel R_2) \parallel (\beta \cdot r'_e)$$

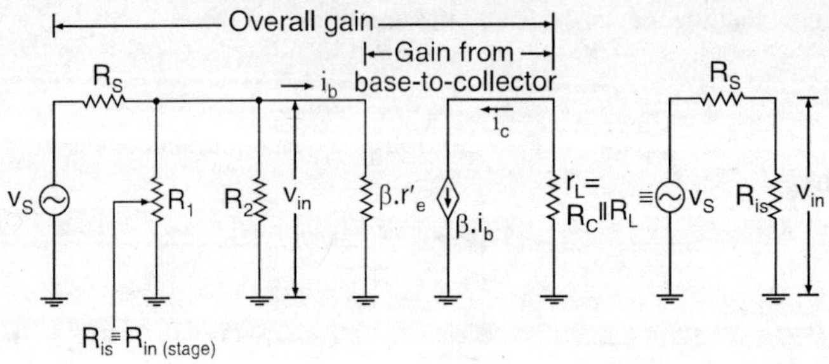

Fig. 16-21. *A.C. equivalent circuit.*

It may be noted from the a.c. equivalent circuit that the signal source resistance (R_S) is in series with the total input resistance. Therefore, the a.c. source voltage is divided between resistances R_S and R_{is}. The voltage at the base is given by the relation,

$$v_{in} = \left(\frac{R_{is}}{R_S + R_{is}}\right) v_S$$

If $R_S << R_{is}$, then the voltage at the base,

$$v_{in} \approx v_S$$

We know that voltage gain from base to collector (or output) is given the relation,

$$A_v = \frac{r_L}{r'_e} = \frac{R_C \parallel R_L}{r'_e}$$

The overall voltage gain (called stage gain) of an amplifier is given by the relation,

$$A_{vs} = A_v \cdot \frac{v_{in}}{v_S}$$

Example 16-6. *For the amplifier shown in Fig. 16-22, determine (**a**) output voltage, and (**b**) the overall voltage gain,*

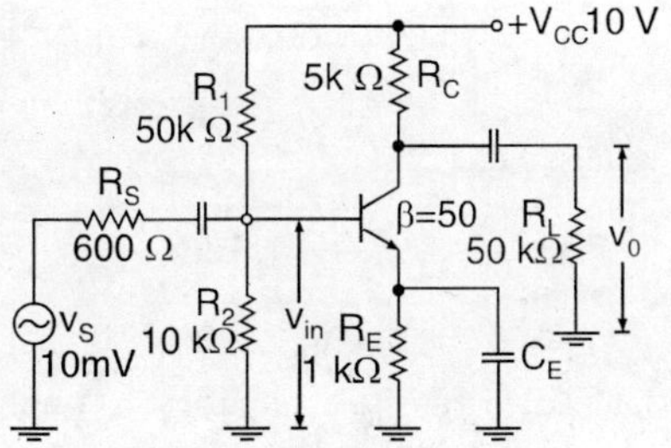

Fig. 16-22.

Solution. Given: V_{CC} = 10 volts; R_C = 5 kΩ; R_E = 1 kΩ; R_1 = 50 kΩ; R_2 = 10 kΩ; R_S = 600 Ω; R_L = 50 kΩ; β = 50 and *V_S = 10 mV.

(**a**) *Output voltage*

We know that Thevenin's Voltage,

$$V_{TH} = V_{CC}\left(\frac{R_2}{R_1 + R_2}\right) = 10 \times \left(\frac{10}{50+10}\right) = 1.67 \text{ V}$$

and the Thevenin's Resistance,

$$R_{TH} = R_1 \parallel R_2 = (R_1 \cdot R_2)/(R_1 + R_2)$$
$$= (50 \times 10)/(50 + 10) = 8.3 \text{ k}\Omega = 8.3 \times 10^3 \ \Omega$$

∴ Value of emitter current,

$$I_E = \frac{V_{TH} - V_{BE}}{R_E + \dfrac{R_{TH}}{\beta}} = \frac{1.67 - 0.7}{1000 + \dfrac{8.3 \times 10^3}{50}} \text{ A}$$
$$= 0.83 \times 10^{-3} \text{ A} = 0.83 \text{ mA}$$

and the value of a.c. resistance,

$$r'_e = \frac{25}{I_E} = \frac{25}{0.83} = 30 \ \Omega$$

We also know that the total input resistance,

$$R_{is} = (R_1 \parallel R_2) \parallel (\beta \cdot r'_e) = R_{TH} \parallel (\beta \cdot r'_e)$$
$$= 8.3 \times 10^3 \parallel (50 \times 30) = 1270 \ \Omega$$

and the value of a.c. load resistance,

$$r_L = R_C \parallel R_L = 5 \parallel 50 = 4.545 \text{ k}\Omega$$
$$= 4.545 \times 10^3 \ \Omega$$

∴ Voltage gain,

* Here the r.m.s. value of the voltage is given. We shall use this value to calculate the r.m.s. value of the output voltage. That is why there is change in voltage symbol from v to V.

$$A_v = \frac{r_L}{r'_e} = \frac{4.545 \times 10^3}{30} = 151.1$$

and the input voltage,

$$V_{in} = V_S\left(\frac{R_{is}}{R_{is} + R_s}\right) = 10\left(\frac{1270}{1270 + 600}\right)$$

$$= 6.79 \text{ mV}$$

Now the output voltage,

$$V_O = A_v \times V_{in} = 151.5 \times 6.79 = 1029 \text{ mV}$$

$$= 1.029 \text{ V } \textbf{Ans.}$$

(*b*) *Overall voltage gain*

We know that overall voltage gain,

$$A_{vs} = A_v \times \frac{v_{in}}{v_s} = 151.5 \times \frac{6.79}{10}$$

$$= 102.9 \textbf{ Ans.}$$

Example 16-7. For a single-style transistor amplifier shown in Fig. 18-23, find the voltage *Av*, input impedance Z_{in} and V_{CE}. Take $\beta = 200$ and $r_e = 30\ mV/I_E$.

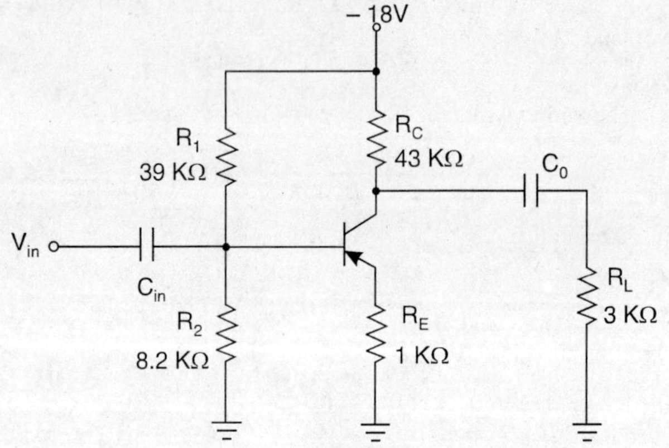

Fig. 16-23.

(*U.P.S.C. Engg. Services* 2002)

Solution. Given : $V_{CC} = -18\ V$, $R_C = 4.3\ k\ \Omega$; $R_E = 1\ k\ \Omega$; $R_1 = 39\ k\Omega$, $R_2 = 8.2k\Omega$; $R_L = 3\ k\Omega$; $\beta = 200$.

We know that Thevenin's voltage,

$$V_{TH} = V_{CC}\left(\frac{R_2}{R_1 + R_2}\right) = -18 \times \frac{8.2\ k}{39\ k + 8.2\ k}$$

and the Thevenin's resistance,

$$R_{TH} = R_1 \| R_2 = (R_1 \cdot R_2) / R_1 + R_2$$

$$= (39\ \text{k} \times 8.2\ \text{k}) / (39\ \text{k} + 8.2\ \text{k}) = 6.78\text{k}\ \Omega$$

∴ Value of emitter current,

$$I_E = \frac{V_{TH} - V_{BE}}{R_E + \dfrac{R_{TH}}{\beta}} = \frac{-3.13 - (-0.7)}{1000 + \dfrac{6780}{200}}$$

$$= -2.35 \text{ mA}$$

and value of a.c. resistance,

$$r'_e = \frac{30\ mV}{I_E} = \frac{30\ mV}{2.35\ mA} = 12.77\ \Omega$$

We also know that the total input resistance,

$$R_{is} = (R_1 \| R_2) \| (\beta \cdot r'_e) = R_{TH} \| (\beta . r'_e)$$
$$= (6.78\ \text{k}) \| (200 \times 12.77\Omega)$$
$$= 6.78\ \text{k} \| 2.55\ \text{k}$$
$$= 1.85\ \text{k}\Omega$$

and the value of a.c. load resistance,

$$r_e = R_C \| R_2 = 4.3\ \text{k} \| = 1.77\ \text{k}\ \Omega$$
$$= 1.77 \times 10^3\ \Omega$$

∴ Voltage gain,

$$A_v = \frac{r_L}{r_e'} = \frac{1.77 \times 10^3}{12.77} = 138.6 \textbf{ Ans.}$$

Example 16-8. *For a self-biased circuit show in Fig. 18-24. Calculate A_V, R_i and R_o. Also calculate A_{VS}.*

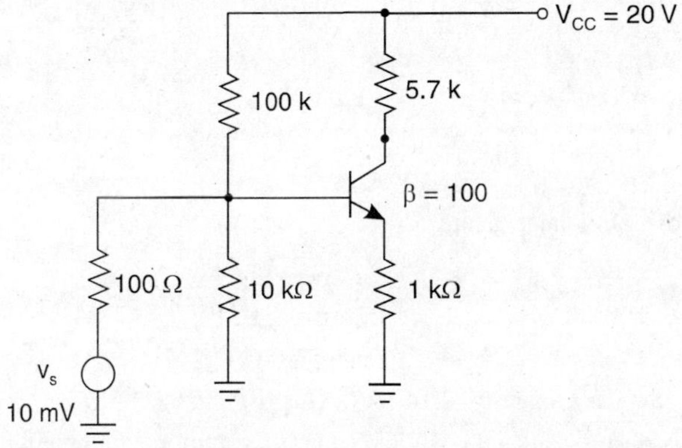

Fig. 16-24.

(*Nagpur Univ. Winter* 2003)

Solution. Given : $V_{CC} = 20$ V; $R_c = 5.7$ kΩ; $R_E = 1$ k Ω; $R_1 = 100$ kΩ, $R_2 = 10$ kΩ; $R_S = 100$ Ω; β = 100.

We know that Thevenin's voltage,

$$V_{TH} = V_{CC}\left(\frac{R_2}{R_1 + R_2}\right) = 20 \times \frac{10\ k}{100\ k + 10\ k} = 1.82\ V$$

and the Thevenin's resistance,

$$R_{TH} = R_1 \| R_2 = (R_1 \cdot R_2) / (R_1 + R_2)$$
$$= (100\ \text{k} \times 10\ \text{k}) / (100\ \text{k} + 10\ \text{k}) = 9.09\ \text{k}\Omega$$

∴ Voltage of emitter current,

$$I_E = \frac{V_{TH} - V_{BE}}{R_E + \frac{R_{TH}}{\beta}} = \frac{1.82 - 0.7}{1000 + \frac{9.09 \times 10^3}{100}} = 1.027\ m\text{A}$$

and the value of a.c. resistance,

$$r'_e = \frac{25}{I_E} = \frac{25}{1{\cdot}027} = 24.34\ \Omega$$

We also known that the total input resistance,

$$R_{is} = (R_1 \parallel R_2) \parallel (\beta.r'_e) = R_{TH} \parallel (\beta\, r'_e)$$
$$= (9.09\ \text{k}) \parallel (100 \times 24.34)$$
$$= (9.09\ \text{k}) \parallel (2.434\ \text{k})$$
$$= 1.92\ \text{k}\Omega$$
$$= 1920\ \Omega$$

and the value of ac load resistance,

$$r_L = R_C = 5.7\ \text{k} = 5700\ \Omega$$

∴ Voltage gain,

$$A_v = \frac{r_L}{r'_e} = \frac{5700}{2434} = 234.18\ \textbf{Ans.}$$

and the input voltage,

$$v_{in} = v_s \left(\frac{R_{is}}{R_{is} + R_S} \right) = 10\ mV \left(\frac{1920}{1920 + 100} \right)$$
$$= 9.50\ \text{mV}$$

Now the output voltage,

$$V_o = A_V \times V_{in} = 234.18 \times 9.50\ \text{mV} = 2224.71\ \text{mV}$$
$$= 2.22\ \text{V}\ \textbf{Ans.}$$

We know that overall voltage gain,

$$A_{VS} = A_v \times \frac{V_{in}}{V_S} = 234.18 \times \frac{9.50\ mV}{10\ mV}$$
$$= 222{\cdot}47\ \textbf{Ans.}$$

16-14. Gain Stability in Common Emitter Amplifier

We have already discussed that the voltage gain of a common emitter amplifier is given by the ratio of r_L/r'_e , where r_L is the a.c. load resistance and r'_e is the emitter diode a.c. resistance. It indicates that the voltage given depends upon to value of a.c resistance r'_e. We have also discussed that the a.c. resistance depends upon the temperature. Thus, as the temperature increases, the a.c. resistance also increases and the voltage gain decreases. The gain variations may be acceptable in some applications, like a transistor radio receiver where we can adjust the volume control. But there are many applications where we need as stable a voltage gain as possible. This can be achieved by inserting a resistor (r_E) in series with the emitter as shown in Fig. 16-25. Now the emitter is no longer at the a.c. ground. Because of this, a.c. current flows through resistor (r_E) and produces voltage at the emitter. If the value of resistor (r_E) is much greater than a.c. resistance (r'_e), almost all the a.c. input signal appears at the emitter. Therefore, the emitter is said to be bootstrapped for alternating current as well as direct current. Because of this, the d.c. emitter voltage is within 0.7 V of the d.c. base voltage, and the a.c. emitter voltage is approximately equal to the a.c. base voltage.

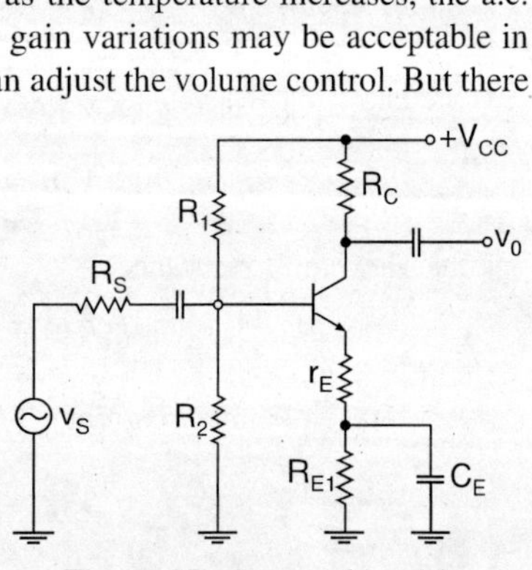

Fig. 16-25. *Swamped amplifier.*

The voltage gain for this amplifier circuit is given by the relation,

$$A_v = \frac{R_C}{r'_e + r_E}$$

If $\qquad r_E >> r'_e,$

Then, $\qquad A_v = R_C/r_E$

It means that if the value of resistor r_E is made much larger than that of a.c. resistance (r'_e), the voltage gain is independent of r'_e. Therefore, any variation in r'_e due to temperature change will have, practically, no effect on the voltage gain and hence, the gain is stable. In other words, the resistor has swamped the variations in r'_e. Because of this, the circuit shown in Fig. 16-25 is called swamped amplifier.

16-15. A.C. Equivalent Circuit of a Swamped Amplifier

Fig. 16-26 shows an a.c. equivalent circuit of a swamped amplifier. The input resistance of a swamped amplifier, when looking directly into the base, is given by the relation,

$$R_i = \beta\,(r_E + r'_e)$$

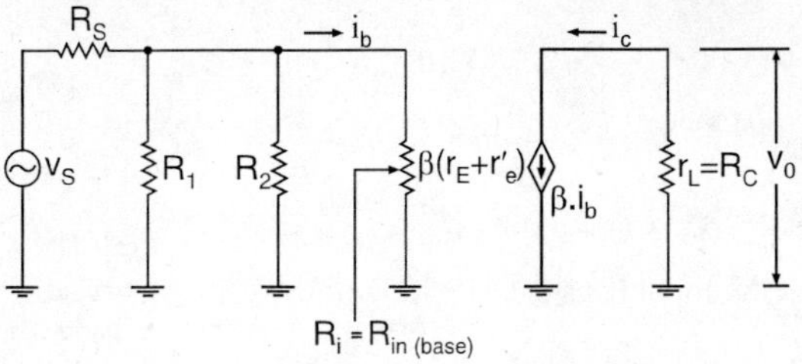

Fig. 16-26. *A.C. equivalent circuit of a swamped amplifier.*

It is evident from the above relation, that the addition of resistor (r_E) increases the input resistance of the amplifier as compared to the input resistance of the unswamped amplifier (whose value is equal to $\beta \cdot r'_e$).

16-16. Advantages of a Swamped Amplifier

Following are some of the important advantages of a swamped amplifier over an unswamped common emitter amplifier:

(*i*) Stable voltage gain.

(*ii*) Higher input resistance.

(*iii*) Produces less distortion in the output signal.

Example 16-9. *Fig. 16-27 shows the circuit of a swamped amplifier.*

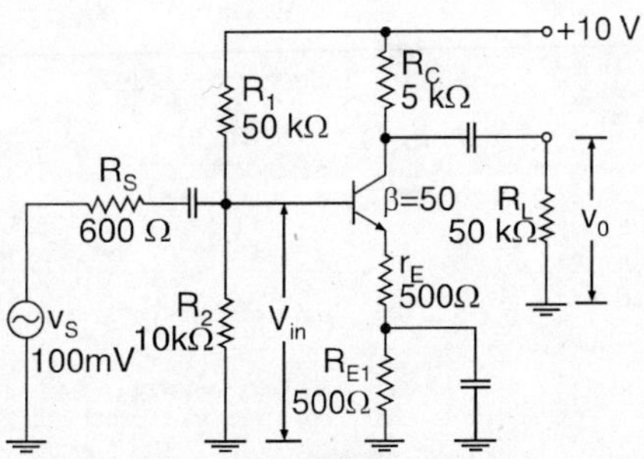

Fig. 16-27.

Determine the values of **(a)** *input resistance looking directly into the base;* **(b)** *input resistance of the stage;* **(c)** *voltage gain;* **(d)** *overall voltage gain; and* **(e)** *the output voltage.*

Solution. Given: V_{CC} = 10 volt; R_C = 5 kΩ; R_L = 50 kΩ; β = 50; R_1 = 50 kΩ; R_2 = 10 kΩ; r_E = 500 Ω; R_{E1} = 500 Ω; R_S = 600 Ω and V_S = 100 mV.

(*a*) *Input resistance looking directly into the base*

We know that the Thevenin's Voltage,

$$V_{TH} = V_{CC}\left(\frac{R_2}{R_1 + R_2}\right) = 10 \times \left(\frac{10}{50 + 10}\right) \text{V} = 1.67 \text{ V}$$

and the value of Thevenin's Equivalent Resistance,

$$R_{TH} = \text{R}_1 \parallel \text{R}_2 = 50 \parallel 50 = 8.3 \text{ k}\Omega = 8.3 \times 10^3 \ \Omega$$

We also know that emitter resistance,

$$R_E = R_{E1} + r_E = 500 + 500 = 1000 \ \Omega$$

∴ Value of emitter current,

$$I_E = \frac{V_{TH} - V_{BE}}{R_E + \dfrac{R_{TH}}{\beta}} = \frac{1.67 - 0.7}{1000 + \dfrac{8.3 \times 10^3}{50}}$$

$$= 0.83 \times 10^{-3} \text{ A} = 0.83 \text{ mA}$$

and the value of a.c. resistance,

$$r'_e = \frac{25}{I_E} = \frac{25}{0.83} = 30 \ \Omega$$

We also know that the input resistance looking directly into the base,

$$R_i = \beta\,(r_E + r'_e) = 50\,(500 + 30) = 26.5 \times 10^3 \ \Omega. = 26.5 \text{ k}\Omega \ \textbf{Ans.}$$

(*b*) *Input resistance of the stage*

We know that input resistance of the stage,

$$R_{is} = (R_1 \parallel R_{2)} \parallel [\beta\,(r_E + r'_e)]$$

$$= 8.3 \parallel 26.5 = 6.32 \text{ k}\Omega = 6320 \ \Omega \ \textbf{Ans.}$$

(*c*) *Voltage gain*

We know that the value of a.c. load resistance,

$$r_L = R_C \parallel R_L = 5 \parallel 50 = 4.55 \text{ k}\Omega = 4.55 \times 10^3 \ \Omega$$

∴ Voltage gain

$$A_v = \frac{r_L}{r_E + r'_e} = \frac{4.55 \times 10^3}{500 + 30} = 8.58 \ \textbf{Ans.}$$

(*d*) *Overall voltage gain*

We know that the ratio of base voltage to the source voltage,

$$\frac{v_{in}}{v_s} = \frac{R_{is}}{R_s + R_{is}} = \frac{6320}{600 + 6400} = 0.913$$

∴ Overall voltage gain,

$$A_{vs} = A_v \times \frac{v_{in}}{v_s} = 8.58 \times 0.913 = 7.83 \ \textbf{Ans.}$$

(*e*) *Output voltage*

We also know that the value of output voltage,

$$V_o = A_{vs} \times V_s = 7.83 \times 100 = 7.83 \text{ mV} \ \textbf{Ans.}$$

16-17. Characteristics and uses of Common Emitter Amplifier

A common emitter amplifier has the following important characteristics :

- Its input resistance is in the range of 1 kΩ to 2 kΩ, which is considered to be moderately low.
- Its output resistance is about 50 kΩ and is considered to be moderately large.
- Its current gain (β) is very high and is in the range of 50 to 300.
- Its voltage gain is of the order of 1500.
- It produces very large power gain and is of the order of 10,000 (equal to 40 dB) or so.
- It produces phase reversal of the input signal, *i.e.*, output is 180° out of phase with respect to the input signal.

The common emitter amplifier is the most widely used amplifier in actual practice because of its large voltage and power gains. In addition to this, its input and output resistances are suitable for most of the applications.

16-18. Common Base Amplifier

Fig. 16-28 shows the circuits of a common-base (CB) amplifier using NPN transistor. Since, the base is grounded, therefore this circuit is also called a grounded-base amplifier. In this circuit, the input is applied across the emitter-base terminals and the output is taken across collector-base terminals. The emitter-base junction is forward biased by V_{EE} supply and the collector-base junction is reverse biased by V_{CC} supply. The capacitors C_1 and C_2 are called blocking capacitors (or coupling capacitors). These capacitors block the direct current, but pass the alternating current. Therefore, they are used to isolate d.c. biase from a.c. signal current in the circuit.

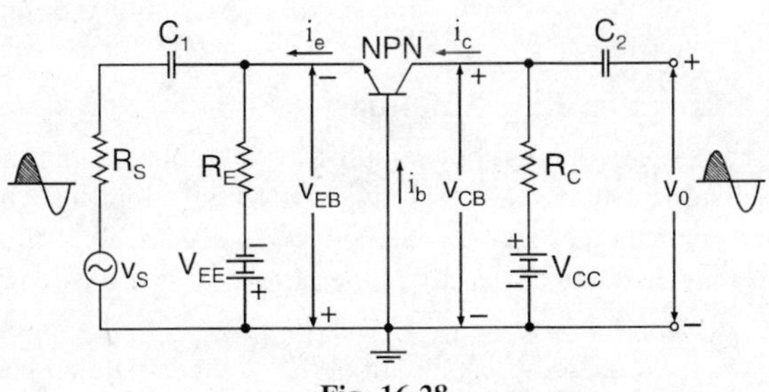

Fig. 16-28.

The Q-point or d.c. working conditions in the amplifier circuit are obtained by the d.c. batteries (V_{EE} and V_{CC}) alongwith the resistances R_E and R_C. The d.c. collector-to-base-voltage is given by the equation,

$$V_{CB} = V_{CC} - I_C \cdot R_C \quad ...(i)$$

It will be interesting to know that when there is no a.c. signal source (*i.e.*, no v_s), the current flowing through the load resistor (R_L) is the quiescent operating current. Now supposes an a.c. signal is applied across the emitter-base terminals. Then during positive half-cycle of the input signal, the forward-bias across the emitter-base junction decreases because the voltage V_{EB} is negative with respect to the ground. As a result of this, a small number of electrons are injected into the base and reach the collector. Therefore, the base and collector currents decrease. Now the decrease in collector current produces smaller voltage drop across the resistor R_C. Therefore, the voltage V_{CB} is increased as seen from the equation (i) During the negative half cycle, the forward-bias across emitter-base junction increases. As a result of this, more number of emitter electrons are injected into the base, due to which collector current increases. The increased collector current, produces a larger voltage drop across the resistance R_C and hence, the voltage V_{CB} decreases.

It is evident from the above discussion, that the positive-going input signal produces a positive going output signal. Therefore, there is no phase reversal between the input and output signals. It may be noted that the collector current is less than emitter current. Therefore, current gain of the common base amplifier (α) is less than unity. But still we get a voltage gain from the amplifier circuit. It is because of the fact that the output resistance of the common-base amplifier is very large as compared to the input resistance. Thus, even though the changes in input and output currents are same, yet the voltage drop across output resistance is very large. Therefore, the changes in output voltage are much larger than the input a.c. signal.

16-19. Different Ways of Drawing Common Base Amplifier Circuit

Fig. 16-29 (*a*) and (*b*) shows a common base a amplifier circuit drawn in a different form. In Fig. 16-29 (*a*), the battery symbol is not indicated. It is always understood to be connected from the terminal marked + or – to the ground terminal, even though it may not be actually shown.

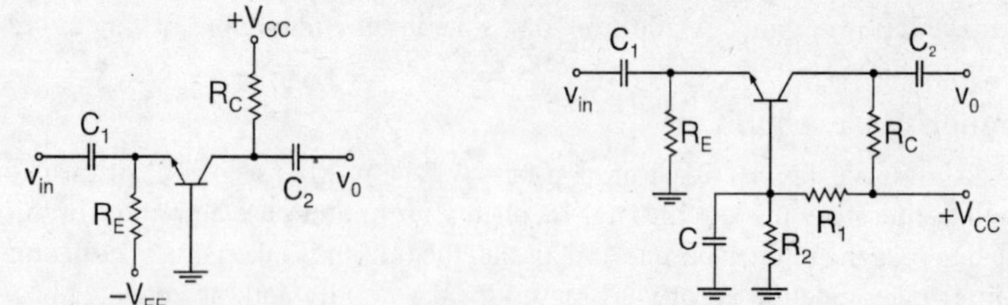

(*a*) (*b*)

Fig. 16-29. *Different ways of drawing common base amplifier circuit.*

This circuit has a emitter-bias arrangement. A common base amplifier can have a voltage-divider bias arrangement as shown in Fig. 16-27 (*b*). This circuit uses a single d.c. supply. Here the base is at a.c. ground through a capacitor (*C*).

16-20. Analysis of a Common Base Amplifier

Fig. 16-30 shows a simple common base amplifier with a emitter-bias arrangement and connected to an external load resistance (R_L). The analysis of this amplifier circuit can be done by splitting it into two parts namely d.c. equivalent circuit and a.c. equivalent circuit. Both the d.c. and a.c. equivalent circuits may be obtained exactly in the same way as discussed for a common emitter amplifier.

Fig. 16-30 (*b*) shows the d.c. equivalent circuit of the common base amplifier. We have already discussed in chapter on transistor bias that such a circuit is called emitter-bias arrangement.

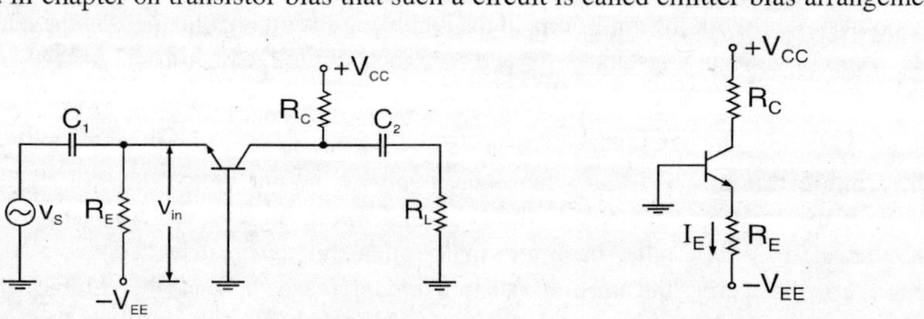

(*a*) Common base amplifier. (*b*) D.C. equivalent circuit.

Fig. 16-30.

The emitter current in such a circuit is given by the relation,

$$I_E = \frac{V_{EE} - V_{BE}}{R_E}$$

and the collector current,

$$I_C = I_E$$

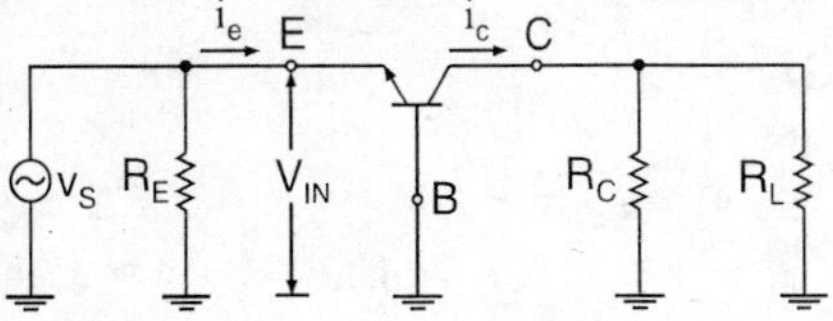

Fig. 16-31. *A simple a.c. equivalent circuit for CB amplifier.*

Fig. 16-31 shows a simple a.c. equivalent circuit for a common base amplifier. The input signal drives the emitter terminal and the output is taken from the collector.

16-21. Common Base Amplifier Parameters

Fig. 16-32 shows a more detailed a.c. equivalent circuit of a common-base amplifier as compared to that discussed in the last article. In this circuit, the NPN transistor has been replaced by its small signal equivalent circuit. This circuit is used to determine the amplifier parameters.

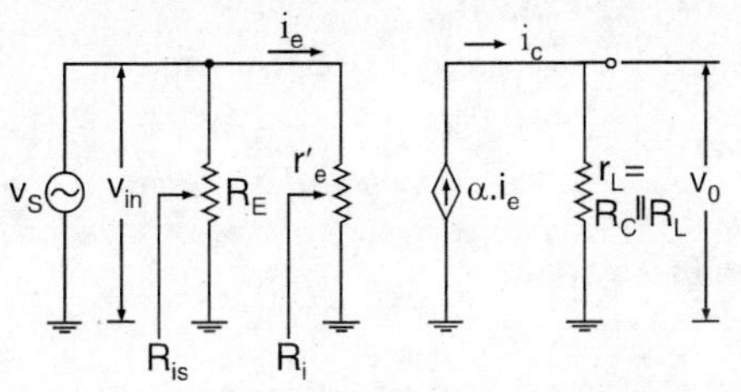

Fig. 16-32. *Detailed a.c. equivalent circuit of a common base amplifier.*

It may be noted from the Fig. 16-32, that the signal source has no resistance, *i.e.*, R_s is equal to zero. Therefore, all of the signal voltage appears as the input voltage (*i.e.*, $v_{in} = v_s$). Now we shall determine the voltage of amplifier parameters in terms of circuit components.

(*i*) ***Input resistance :*** The input resistance, looking directly into the emitter, is given by the relation,

$$R_i = r'_e$$

and the input resistance of the stage,

$$R_{is} = R_E \| r'_e = r'_e \qquad \text{...(if } R_E >> r'_e\text{)}$$

where, r'_e is the a.c. resistance of the emitter diode (*i.e.*, emitter-base junction). It means that input resistance of a common base amplifier is just equal to r'_e. Since, the value of r'_e is usually very small, therefore the input resistance is also very small.

(*ii*) ***Output resistance :*** It is given by the relation,

$$R_0 = r_L$$

where, r_L is the a.c. load resistance and its value is equal to the parallel combination of the resistances R_C and R_L (*i.e.*, $r_L = R_C \| R_L$). However, if R_L is not connected

$$R_0 = R_C$$

(*iii*) ***Current gain :*** It is given by the ratio of output current to the input current or the ratio of collector current to the emitter current. Mathematically the current gain,

$$A_i = \frac{i_c}{i_e} = \alpha$$

(*iv*) ***Voltage gain :*** It is given by the ratio of output voltage to the input voltage. Mathematically, the voltage gain.

$$A_v = \frac{v_0}{v_{in}} \quad \text{.. } (i)$$

We know that the input voltage,

$$v_{in} = i_e \cdot r'_e$$

and the output voltage,

$$v_0 = i_c \cdot r_L = i_e \cdot r_L \quad \text{... } (\because i_c = i_e)$$

Substituting the values of v_{in} and v_0 in equation (*i*) the voltage gain,

$$A_v = \frac{i_e \cdot e_L}{i_e \cdot r'_e} = \frac{r_L}{r'_e}$$

It indicates that voltage gain of a common-base amplifier is the ratio of a.c. load resistance to the a.c. resistance of emitter diode. It is also called the voltage gain from emitter to collector.

(*v*) ***Power gain*** **:** It is given by the product of voltage gain and current gain. Mathematically, the power gain,

$$A_p = A_v \cdot A_i$$

The power gain, in decibels, is given by the relation,

$$G_p = 10 \log_{10} A_p$$

Example 16-10. *For the single-stage common base amplifier shown in Fig. 16-33, find the values of* ***(a)*** *input resistance,* ***(b)*** *current gain,* ***(c)*** *voltage gain and* ***(d)*** *power gain.*

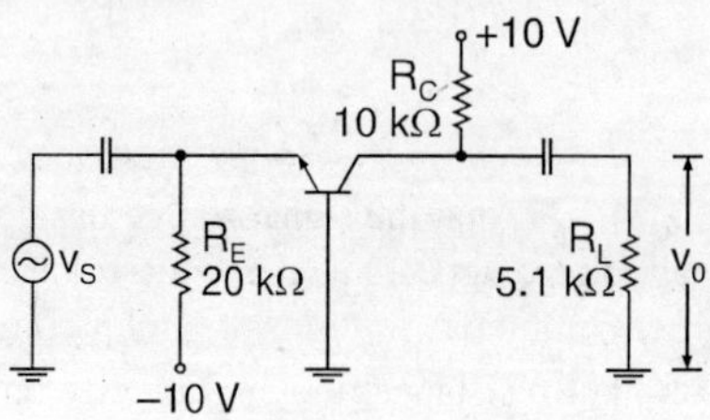

Fig. 16-33.

What would be the r.m.s. value of the signal voltage across the load, if the input signal voltage is 10 mV? Assume the transistor to be made of silicon with current gain = 0.98.

Solution. Given: $V_S = 10$ mV; $\alpha = 0.98$; V_{BE} (for silicon transistor) = 0.7 volt; $V_{CC} = 10$ volts; $R_C = 10$ kΩ; $R_L = 5.1$ kΩ; $R_E = 20$ kΩ $= 20 \times 10^3$ Ω and $V_{EE} = 10$ volts.

Input resistance

We know that the value of emitter current,

$$I_E = \frac{V_{EE} - V_{BE}}{R_E} = \frac{10 - 0.7}{20 \times 10^3} = 0.465 \times 10^3 \text{ A} = 0.465 \text{ mA}$$

and the value of a.c. emitter resistance,

$$r'_e = \frac{25}{I_E} = \frac{25}{0.465} = 53.8\ \Omega$$

∴ Value of input resistance, looking directly into the emitter,

$$R_i = r'_e = 53.8\ \Omega$$

and the input resistance of the stage,

$$R_{is} = R_E \parallel r'_e = (20 \times 10^3) \parallel 53.8 = 53.8\ \Omega \textbf{ Ans.}$$

Current gain

We know that the value of current gain,

$$A_i = \alpha = 0.98 \textbf{ Ans.}$$

Voltage gain

We know that the value of a.c. load resistance,

$$r_L = R_C \parallel R_L = 10 \parallel 5.1 = 3.38 \text{ k}\Omega = 3.38 \times 10^3 \ \Omega$$

∴ Voltage gain,

$$A_v = \frac{r_L}{r'_e} = \frac{3.38 \times 10^3}{53.8} = 62.8 \textbf{ Ans.}$$

Power gain

We know that the value of power gain,

$$A_p = A_v \cdot A_i = 62.8 \times 0.98 = 61.5 \textbf{ Ans.}$$

and the power gain in decibels,

$$G_P = 10 \log_{10} A_P = 10 \log_{10} 61.5 = 17.9 \ dB \textbf{ Ans.}$$

Output voltage

We know that the value of input voltage,

$$v_{in} = v_s = 10 \text{ mV} \qquad \text{... } (R_s = 0)$$

∴ Value of output voltage,

$$V_0 = A_v \times v_{in} = 62.8 \times 10 = 628 \text{ mV} \textbf{ Ans.}$$

16-22. Effect of Signal Source Resistance on Common Base Amplifier Parameters

Fig. 16-34 shows the circuit of a common base amplifier connected to an a.c. signal source of internal resistance (R_S). It has been found that greater value of internal resistance of a.c. signal source, greater will be the voltage drop across internal resistance, and lesser will be the input voltage. Consequently, the voltage gain will be lower with a large value of internal resistance.

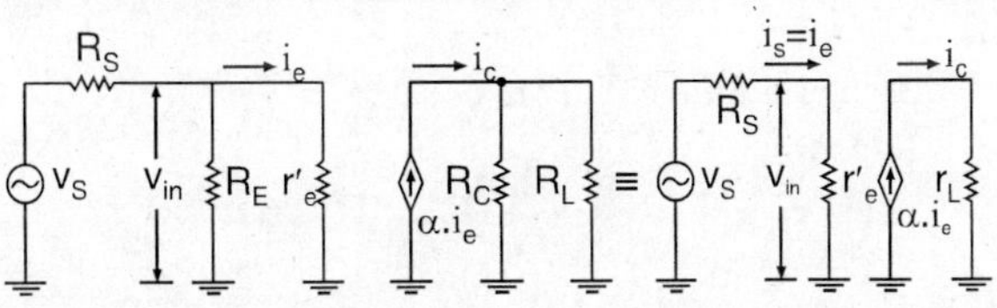

Fig. 16-34. *A.C. equivalent circuit of a common base amplifier.*

The a.c. equivalent circuit of the given amplifier is shown in Fig. 16-34. The total resistance of the amplifier as seen by the signal source,

$$R_{is} = R_E \parallel r'_e$$

$$= r'_e \qquad \text{...(If } R_E >> r'_e)$$

It may be noted from the a.c. equivalent circuit, that the internal resistance of a signal source (R_S) is in series with the total input resistance. Therefore, the a.c. source voltage is divided between the internal resistance (R_S) and the total input resistance (R_{is}). The voltage at the base is given by the relation,

$$v_{in} = v_s \left(\frac{R_{is}}{R_s + R_{is}} \right)$$

$$= v_s \left(\frac{r'_e}{R_s + r'_e} \right) \qquad \text{...(If } R_{is} = r'_e)$$

$$= v_s \left(\frac{1}{1 + R_s / r'_e} \right)$$

It is evident from the above relation that as the value of R_s increase, in comparison to r'_e, the ratio R_s/r'_e also increases. It decreases the value of voltage at the base (v_{in}). Now as seen from the source, the value of input resistance,

$$= R_s + r'_e$$

∴ Voltage gain of the amplifier stage,

$$A_{vs} = \frac{v_0}{v_{in}} = \frac{r_L}{R_s + r'_e} \approx \frac{r_L}{R_s} \qquad ...(\text{If } R_S >> r'_e)$$

The above relation indicates that if the value of R_s is much larger than r'_e the voltage gain of the amplifier stage becomes independent of r'_e. In other words, the voltage gain is independent of the transistor to be used in the amplifier circuit. This technique is called swamping out the emitter diode resistance of a transistor. It may be noted that the voltage gain of the stage is different from the voltage gain from emitter to collector, whose value is calculated from the ratio r_L/r'_e.

Example 16-11. *Consider the common base amplifier circuit as shown in Fig. 16-33 of Example 16-10. Assuming, the signal sources resistance (R_S) = 50 Ω, find **(a)** the voltage gain from source to output, **(b)** the voltage gain from emitter to output, and **(c)** approximate value of v_{in}.*

Solution. Given: $R_S = 50\ \Omega$.

We have already obtained the following in Example 16-10.

$I_E = 0.465$ mA; $r'_e = 53.8\ \Omega$; $R_i = 53.8\ \Omega$; $R_{is} = 52.4\ \Omega$ and $R_L = 3.38$ kΩ. Now we shall use the above values in this example.

(*a*) *Voltage gain from source to output*

We know that the voltage gain from source to output (called stage gain) is given by the relation,

$$A_{vs} = \frac{r_L}{R_s + r'_e} = \frac{3.38 \times 10^3}{50 + 53.8} = 32.6 \textbf{ Ans.}$$

(*b*) *Voltage gain from emitter to output*

We know that voltage gain from emitter to output,

$$A_v = \frac{r_L}{r'_e} = \frac{3.38 \times 10^3}{53.8} = 62.8 \textbf{ Ans.}$$

(*c*) *Value of v_{in}*

We know that the voltage gain from source to output, (A_{vs}),

$$32.6 = \frac{v_0}{v_s} = \frac{v_0}{10}$$

$$\therefore \quad v_0 = 32.6 \times 10 = 326 \text{ mV}$$

We also know that the voltage gain from emitter to collector (A_v),

$$62.8 = \frac{v_0}{v_{in}} = \frac{326}{v_{in}}$$

$$\therefore \quad v_{in} = 326/62.8 = 5.2 \text{ mV } \textbf{Ans.}$$

16-23. Characteristics and Uses of Common Base Amplifier

A common base amplifier has the following important characteristics :

- It has very low input resistance and is of the order of 30 to 150 ohms.
- It has very high output resistance and is of the order of 500 kΩ.

- Its current gain (α) is less than unity.
- Its voltage gain is of the order of 1500.
- Its power gain is of the order of 30 dB.
- Its input and output are in phase with each other, *i.e.*, there is no phase reversal between the input and output signals.

A common base amplifier has a very important application as it is used for matching circuits with low output resistance to those with a high input resistance.

16-24. Common-Collector Amplifier

Fig. 16-35 shows a circuit of common-collector (CC) amplifier using *NPN* transistor. Since, the collector resistance (R_C) is zero, therefore the collector is at a.c. ground. That is why this circuit is also known as *grounded collector amplifier*. In this circuit, the input is applied between the base and collector terminals and the output is taken across the emitter and collector terminals. The capacitors C_1 and C_2 block the direct current; but pass the a.c. signal current in the circuit.

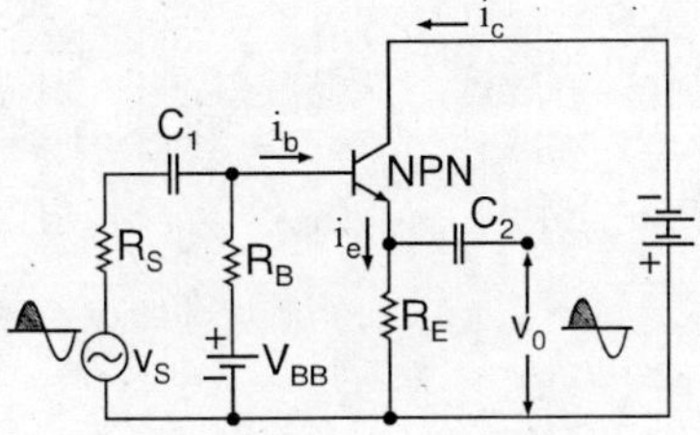

Fig. 16-35. *Common-collector transistor amplifier.*

The *Q*-point in the circuit is obtained by the d.c. batteries V_{BB} and V_{CC} alongwith the resistors R_E and R_B. The d.c. collector to emitter voltage is given by the equation,

$$V_{CE} = V_{CC} - I_E \cdot R_E = V_{CC} - V_0 \qquad \text{...(where, } V_0 \text{ is the output voltage)}$$

It will be interesting to know that when there is no a.c. signal source (*i.e.*, v_s is equal to zero), the current flowing through the emitter (or load) resistor is the quiescent operating current. Now let an a.c. signal be applied across the input terminals. Then during the positive half cycle of the input a.c. signal, the forward bias is increased because the voltage V_B is positive with respect to the collector (*i.e.*, ground). This causes the base current to increase, due to which the emitter current is also increased. As a result of this, the voltage drop across resistor R_E is increased and hence the output voltage (v_0) is also increased. Consequently, we get positive half cycle of the output. It means that a positive going input signal results in a positive going output signal. Thus, the input and output signals are in phase with each other as shown in Fig. 16-35. In other words, the emitter voltage follows the base voltage. That is why the common collector amplifier is known as *emitter follower*. The voltage gain of an emitter follower is less than unity. But it has a large current gain and large power gain.

16-25. Different Ways of Drawing Common Collector Amplifier Circuit

Fig. 16-36 (*a*) and (*b*) indicates a common collector amplifier drawn in different forms. In Fig. 16-36 (*a*) the battery is indicated with the emitter terminal. This circuit has a base bias arrangement. A common collector amplifier may have a voltage divider bias arrangement as shown in Fig. 16-36 (*b*). This circuit uses a single d.c. supply. It may be noted that in all the circuits, collector is at a.c. ground.

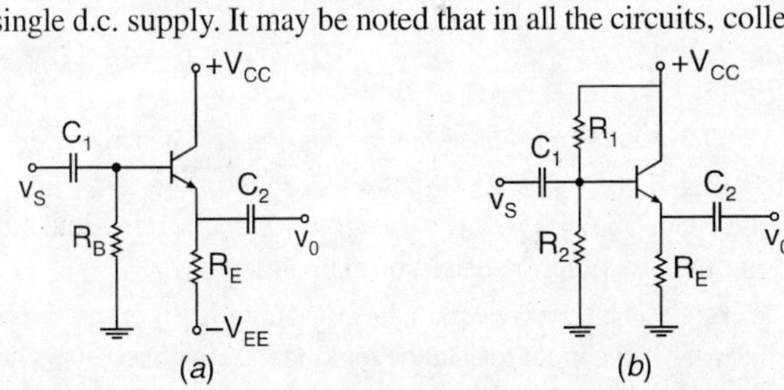

Fig. 16-36. *Different ways of drawing common collector amplifier circuit.*

16-26. Analysis of Common Collector Amplifier

Fig. 16-37(*a*) shows a simple common collector amplifier with a base-bias arrangement. The amplifier is connected to a load resistance (R_L). The d.c. equivalent circuit for the given amplifier is as shown in Fig. 16-37 (*b*). The value of emitter current for such a circuit is given by the relation,

$$I_E = \frac{V_{EE} - V_{BE}}{R_E - R_B/\beta} \approx \frac{V_{EE}}{R_E + R_B/\beta}$$

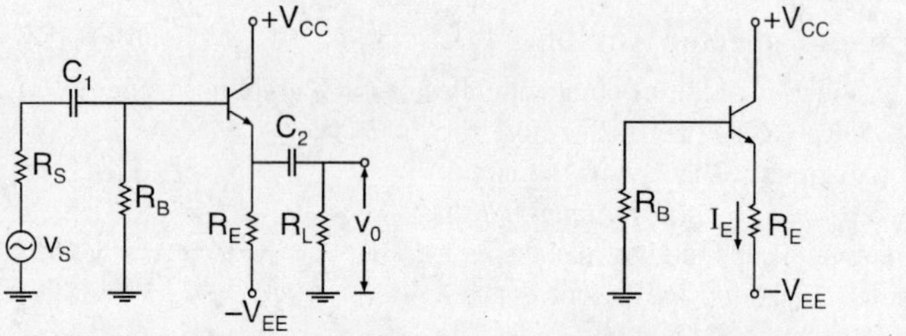

(*a*) Common collector amplifier. (*b*) D.C. equivalent circuit.

Fig. 16-37.

Fig. 16-38 shows a simple a.c. equivalent circuit for a common collector amplifier. The input signal drives the base terminal and the output is taken from the emitter.

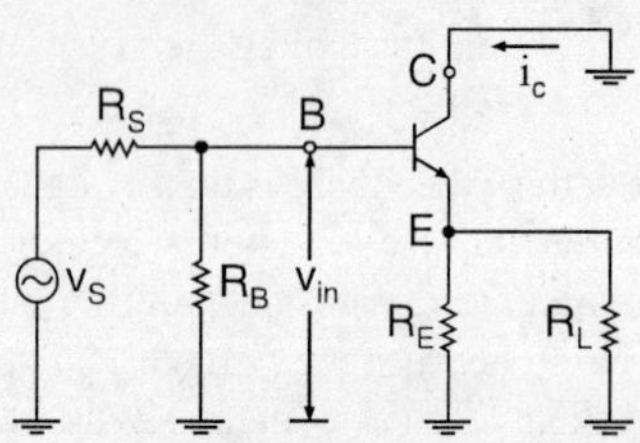

Fig. 16-38.

16-27. Common Collector Amplifier Parameters

Fig. 16-39 shows a more detailed a.c. equivalent circuit of a common collector amplifier as compared to that of Fig. 16-38. In this circuit, an NPN transistor is replaced by its small-signals equivalent circuit. This circuit is used to determine input and output resistances, current gain, voltage gain and power gain of common collector amplifier.

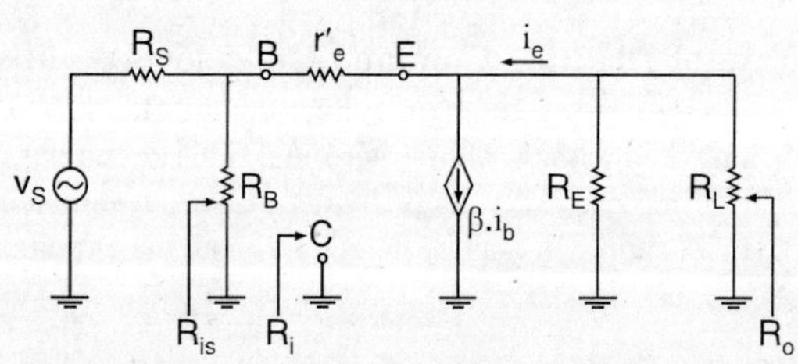

Fig. 16-39. *Detailed a.c. equivalent circuit of a common collector amplifier.*

It may be noted that if the value of signal source resistance (R_S) is zero, then all the signal voltage appears as the input voltage (*i.e.*, $v_{in} = v_S$). Now we shall determine the values of common-collector amplifier parameters in terms of circuit components.

(*i*) ***Input resistance*** **:** The input resistance of the stage is given by the parallel combination of R_E and R_i. Now R_i is the input resistance looking into the base. Its value is equal to $\beta\,(r'_e + r_L) \approx \beta \cdot r_L$ · *i.e.*, β times the a.c. load resistance seen by the emitter.

$$\therefore \quad R_{is} = R_B \parallel \beta\,(r'_e + r_L)$$
$$\approx \beta\,(r'_e + r_L) \qquad \text{...[If } R_B >> \beta\,(r'_e + r_L)]$$
$$\approx \beta \cdot r_L \qquad \text{...(If } r_L >> r'_e)$$
$$= \beta \cdot R_E \qquad \text{(If } R_L \text{ is equal to zero)}$$

(*ii*) ***Output resistance :*** Its value is given by the relation,

$$R_O = r_L \parallel \left(r'_e \frac{R_B \parallel R_S}{\beta} \right)$$
$$= r'_e + \frac{R_B \parallel R_S}{\beta} \qquad \text{... (If } r_L >> r'_e)$$

(*iii*) ***Current gain :*** It is given by the ratio of output current to input current, *i.e.*, the ratio of emitter current to base current. Thus current gain,

$$A_i = \frac{i_c}{i_b} \approx \frac{i_e + i_b}{i_b} = 1 + \frac{i_c}{i_b} = 1 + \beta \approx \beta$$

(*iv*) ***Voltage gain :*** It is given by the ratio of output voltage to the input voltage. Thus, voltage gain

$$A_v = v_0 / v_{in}$$

We know that input voltage,

$$v_{in} = i_b \cdot R_i = i_b \cdot \beta\,(r'_e + r_L)$$

and output voltage,

$$v_0 = i_e \cdot r_L = \beta \cdot i_b \cdot r_L$$

$$\therefore \quad A_v = \frac{\beta \cdot i_b \cdot r_L}{i_b \cdot \beta\,(r'_e + r_L)} = \frac{r_L}{r'_e + r_L} \approx \frac{r_L}{r_L} = 1 \qquad \text{(If } r'_e << r_L)$$

It means that output signal from the emitter has the same magnitude as the input signal at the base.

(*v*) ***Power gain :*** It is given by the product of voltage and current gain. Mathematically, the power gain.

$$A_p = A_v \times A_i = 1 \times \beta = \beta$$

and the power gain in decibels,

$$G_p = 10 \log_{10} A_p \text{ dB}$$

It is evident from the above discussion that the voltage gain of emitter follower is unity and current gain is β. However, input resistance is β times greater than the output resistance. It means that emitter follower steps up the resistance level of the output to a much higher value at its input. It is the most useful characteristic of an emitter follower (or common collector amplifier). This characteristic makes the emitter follower very useful as a resistance matching device, *i.e.*, it is used to match circuits with high output resistance to those with low input resistance.

Example 16-12. *Fig. 16-40 shows the circuit of an emitter follower. The transistor used in the circuit has current gain (β) of 50 and V_{BE} = 0.7 V*

Determine the values of input resistance, and the exact voltage gain.

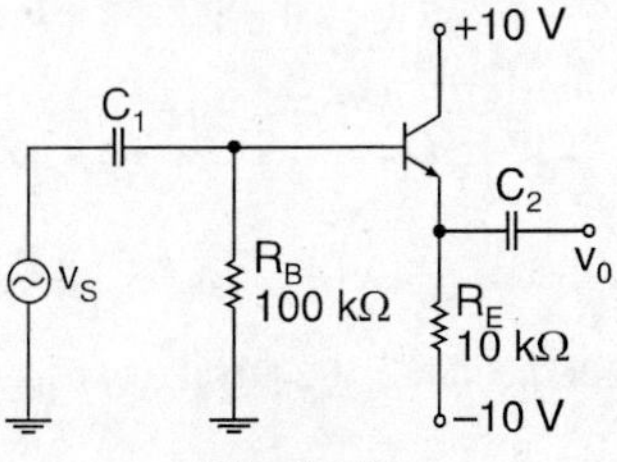

Fig. 16-40.

Solution. Given: V_{EE} = 10 volts ; $R_E = 10\text{ k}\Omega = 10 \times 10^3\ \Omega$; $R_B = 100\text{ k}\Omega = 100 \times 10^3\ \Omega$; β = 50 and V_{BE} = 0.7 volt.

Input resistance

We know that the value of emitter current,

$$I_E = \frac{V_{EE} - V_{BE}}{R_E + \dfrac{R_B}{\beta}} = \frac{10 - 0.7}{(10 \times 10^3) + \dfrac{100 \times 10^3}{50}}$$

$$= 0.77 \times 10^{-3} \text{ A} = 0.775 \text{ mA}$$

and a.c. resistance of emitter diode,

$$r'_e = \frac{25}{I_E} = \frac{25}{0.775} = 32.3\ \Omega$$

∴ Input resistance looking directly into the base,

$$R_i = \beta\,(R_E + r'_e) = 50\,[(10 \times 10^3) + 32.3] = 501.6\ \text{k}\Omega \textbf{ Ans.}$$

and input resistance of the stage,

$$R_{is} = R_B \parallel R_i = 100 \parallel 501.6 = 83.4\ \text{k}\Omega \textbf{ Ans.}$$

Output resistance

We know that the output resistance,

$$R_0 = r'_e + \frac{R_B \parallel R_S}{\beta} = r'_e = 32.3\ \Omega \textbf{ Ans.} \qquad ...(\because R_s = 0)$$

Voltage gain

We also know that the voltage gain,

$$A_v = \frac{R_E}{r'_e + R_E} = \frac{10 \times 10^3}{32.3 + (10 \times 10^3)} = 0.997 \textbf{ Ans.}$$

Example 16-13. *The emitter follower circuit shown in Fig. 16-41 uses a voltage divider bias arrangement. The transistor has a current gain (β) of 80 and V_{BE} = 0.7 V.*

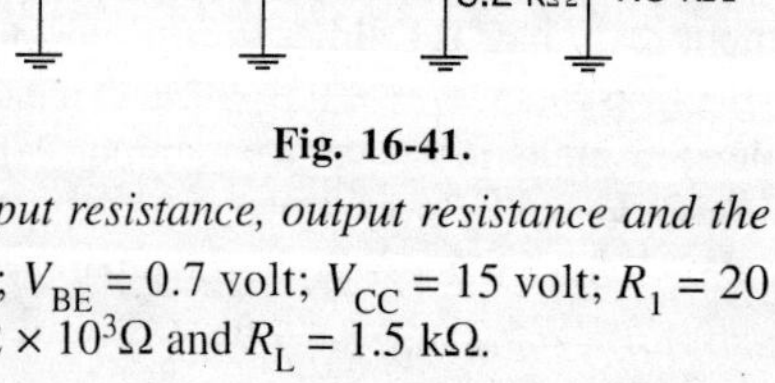

Fig. 16-41.

Calculate the value of input resistance, output resistance and the a.c. output voltage.

Solution. Given: $\beta = 80$; $V_{BE} = 0.7$ volt; $V_{CC} = 15$ volt; $R_1 = 20$ kΩ; $R_2 = 20$ *k*Ω $R_S = 2$ kΩ; $V_S = 5$ mV; $R_E = 8.2$ kΩ $= 8.2 \times 10^3 \Omega$ and $R_L = 1.5$ kΩ.

Input resistance

We know that Thevenin's voltage

$$V_{TH} = V_{CC} \times \frac{R_2}{R_1 + R_2} = 15 \times \frac{20}{20 + 20} = 7.5 \text{ V}$$

and the value of equivalent resistance,

$$R_{TH} = R_1 \parallel R_2 = 20 \parallel 20 = 10\text{k}\Omega = 10 \times 10^3\ \text{k}\Omega$$

∴ Value of emitter current,

$$I_E = \frac{V_{TH} - V_{BE}}{R_E + \frac{R_{TH}}{\beta}} = \frac{7.5 - 0.7}{8.2 \times 10^3 + \frac{(10 \times 10^3)}{80}}$$

$$= 0.817 \times 10^{-3} \text{ A} = 0.817 \text{ mA}$$

and a.c. resistance of emitter diode,

$$r'_e = \frac{25}{I_E} = \frac{25}{0.817} = 30.6\ \Omega$$

We also know that the a.c. load resistance,

$$r_L = R_E \parallel R_L = 8.2 \parallel 1.5 = 1.27\ \text{k}\Omega = 1.27 \times 10^3\ \Omega$$

∴ Value of input resistance, looking directly into the base,

$$R_i = \beta\,(r_L + r'_e) = 80\,(1.27 \times 10^3 + 30.6) = 104 \times 10^3\ \Omega = 104\ \text{k}\Omega$$

and input resistance of the stage,

$$R_{is} = R_{TH} \parallel R_i = 10 \parallel 104 = 9.1\ \text{k}\Omega\ \textbf{Ans.}$$

Output resistance of the stage,

We know that the output resistance of the stage

$$R_0 = r'_e + \frac{R_S \parallel R_{TH}}{\beta} = 30.6 + \frac{(2 \times 10^3 \parallel 10 \times 10^3)}{80} = 51.4\ \Omega\ \textbf{Ans.}$$

A.C. output voltage

We know that the input voltage,

$$V_{in} = V_s \times \left(\frac{R_{is}}{R_s + R_{is}}\right) = 5 \times \left(\frac{9.1}{2 + 9.1}\right) \text{mV} = 4.1 \text{ mV}$$

and the output voltage,

$$v_0 = v_{in} = 4.1 \text{ mV}\ \textbf{Ans.}$$

16-28. Characteristics and Uses of Common Collector Amplifier

A common collector (or emitter follower) has the following important characteristics :

- It has a high input resistance, which is of the order of 20 to 500 kΩ.
- It has a low input resistance, which is of the order of 50 to 1000Ω.
- It has a high current gain. Its value depends upon the current gain (β) of the transistor, which is in the range of 50 to 500.
- Its voltage gain is less than unity.
- Its power gain is much smaller than that of common emitter and common base amplifiers.
- Its output signal is in phase with the input signal.

The main application of common collector amplifier is for resistance matching. It is used to transfer energy from a high output resistance circuit to that of a low input resistance circuit. The common collector amplifier can also be used as a two-way amplifier, because it can pass a signal in either direction.

16-29. Comparison of Three Amplifier Configurations

Table 16-1 show the comparison of important characteristics of three amplifier configurations, namely common emitter, common base and common collector.

Table 16-1. *Comparison of three amplifier configurations.*

S. No.	*Charactristics*	*Types of configuration*		
		Common Emitter	*Common Base*	*Common Collector*
1.	Input resistance (R_i)	$\beta \cdot r'_e$ (Moderate)	r'_e (Lowest)	$\beta \cdot R_E$ (Highest)
2.	Output resistance (R_0)	R_c (Moderate)	R_c (Highest)	$\approx r'_e$ (Lowest)
3.	Current gain (A_i)	β (High)	≈ 1 (Lowest)	$1 + \beta$ (Highest)
4.	Voltage gain (A_v)	r_L/r'_e (Highest)	r_L/r'_e (High)	≈ 1
5.	Power gain (A_p)	Highest	Moderate	Lowest
6.	Phase reversal	Yes	No	No

16-30. Nonlinear Distortion

Any unwanted change in the shape of an a.c. signal is called ***distortion***. One problem that commonly occurs in common-emitter amplifiers is called *nonlinear distortion*. The output from an waveform from an amplifier experiencing nonlinear distortion is shown in Fig. 16-42 (*a*).

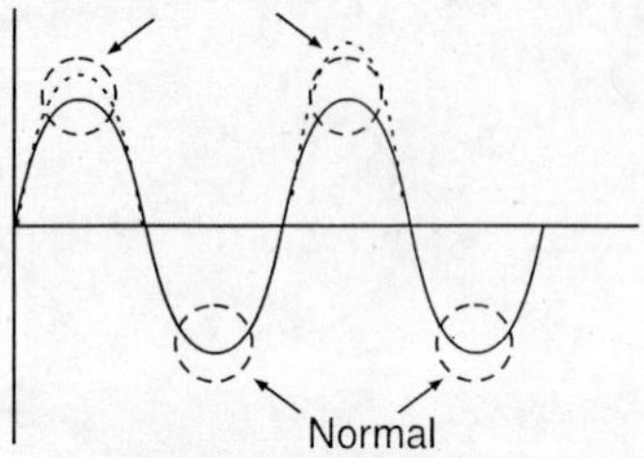

(*a*) Output waveform illustrating nonlinear distortion.

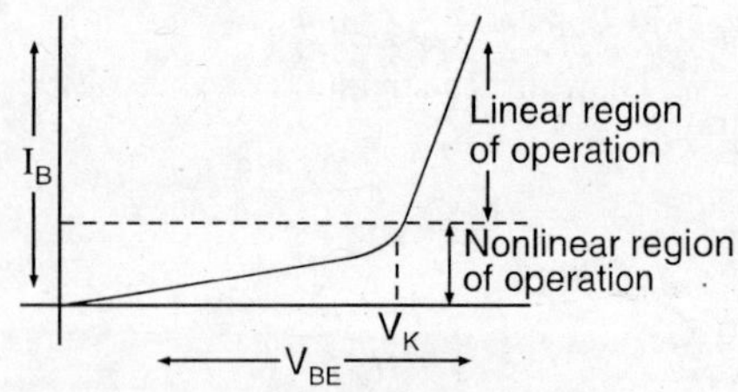

(*b*) Base characteristic curve.

Fig. 16-42.

Notice the difference between the shape of the negative alteration of the signal (normal) and that of its positive alteration (distorted). Nonlinear distortion is caused by driving the base-emitter junction of the transistor into its nonlinear operating region. This point is illustrated with the help of Fig. 16-42 (*b*) which shows the basic characteristic curve of a transistor.

Normally a transistor is operated so that the base-emitter junction stays in the linear region of operation. In this region, a change in V_{BE} causes a linear (*i.e.*, constant rate) change in I_B. If the transistor is operated in the nonlinear operating region, a change in V_{BE} will cause a change in I_B that is not linear. It is evident if we compare the slope of the curve at the points immediately above and below V_K. Since, the slope of the curve changes, the rate of change in I_B also changes.

The base-emitter junction of a transistor may be driven into its non linear operating region in one of two sets of circumstances :

(*i*) A transistor is normally biased so that I_B will have a value well within the linear region of operation. If an amplifier is poorly biased so that the *Q*-point value of I_B is near the nonlinear region of operation, a relatively small a.c. input signal will drive the amplifier into the nonlinear region of operation, and nonlinear distortion will result.

(*ii*) The amplifier input signal may be sufficient to drive even a well-biased amplifier into nonlinear operation, causing nonlinear distortion.

The solution to the first problem is to redesign the amplifier for a higher value of I_B. While for the second problem, the solution is to reduce the amplitude of the amplifier input signal.

Nonlinear distortion can cause a variety of problems in communications electronics. In stereos, it may cause the audio to sound "grainy" or may even cause the tones of various musical instruments to change. In televisions, crossover distortion in the video circuitary can cause distortion in the picture on the CRT. To avoid these types of problems, amplifier design normally emphasizes avoiding *crossover distortion*. Since, amplifiers are normally designed to avoid nonlinear distortion problems, the most common cause of nonlinear distortion is *overdriving* an amplifier.

SUMMARY

In this chapter you have learned that ;

1. Based on the transistor configuration, amplifiers are classified into common-emitter, common-base, and common-collector amplifiers.
2. The a.c. load, which is a parallel combination of collector resistance and load resistance, determines the voltage gain of a common-emitter amplifier.
3. The addition of an emitter resistor in common-emitter amplifier reduces its voltage gain but increases the input resistance
4. The addition of an emitter by-pass capacitor parallel to emitter resistor, in common-emitter amplifier increases the voltage gain.
5. A swamped amplifier have stable voltage gain, higher input resistance, and it produces less distortion in the output signal as compared to a normal common-emitter amplifier.
6. Common-base amplifier is used for matching circuits with low output resistance to those with high input resistance.
7. Common-collector amplifier is used for matching circuits with high output resistance to those with low input resistance
8. Nonlinear distortion in common-emitter amplifier can be reduced by redesigning the circuit with Q-point-value of I_B, away from the nonlinear region or by reducing the amplifier of the input signal.

GLOSSARY

A.C load : The actual signal load on the amplifier, which is a parallel combination of collector resistance, R_e and load resistance, R_L.

Amplification : The process of increasing the power of an a.c. signal.

Amplifiers : The circuits used to perform amplification.

Common-base amplifier : A transistor amplifier in common-base configuration.

Common-collector amplifier : A transistor amplifier in common-collector configuration.

Common-emitter amplifier : A transistor amplifier in common-emitter configuration.

Distortion : Any unwanted change in the shape of an a.c. signal.

Ground emitter amplifier : Refer to common-emitter amplifier.

Linear amplifier : The amplifier, in which. the instantaneous output-signal is directly proportional to the corresponding input signal.

Non linear amplifier : The amplifier, in which , instantaneous output signal is not directly proportional to the corresponding input signal.

Nonlinear distortion : Distortion caused by the operation of transistor in its nonlinear region.

Swamped amplifier : A common-emitter amplifier with a resistor (r_E) connected in series with emitter so that it is not shorted to a.c. ground with this the amplifier achieves gain stability.

NUMERICAL PROBLEMS

1. For the single-stage common emitter amplifier shown in Fig. 16-43, find the values of input resistance looking directly into the base, output resistance, voltage and dB voltage gain. **[Ans.** 415 Ω; 5 kΩ; 602.4 and 55.6 dB]
2. If for the common emitter amplifier shown in Fig. 16-43, the input voltage v_s = 5 mV, find the value of output voltage and the overall voltage gain. **[Ans.** 681 mV and 136]

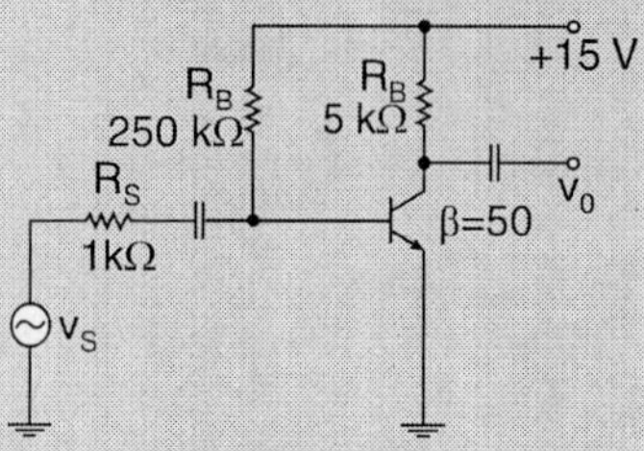

Fig. 16-43.

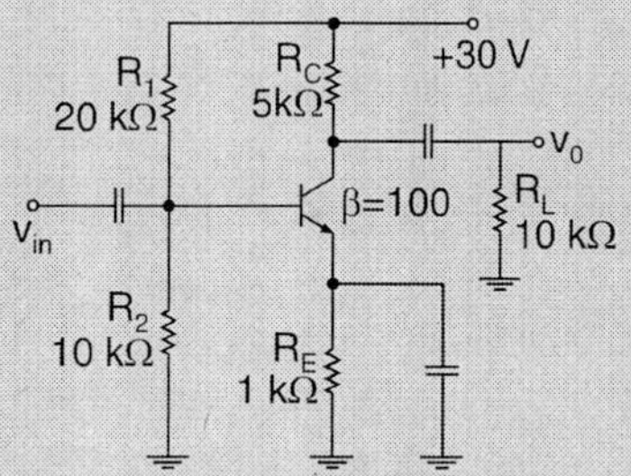

Fig. 16-44.

3. Draw the a.c. equivalent circuit for the single-stage common emitter amplifier shown in Fig. 16-44. Also find the values of R_i, R_{is}, R_0 and A_v.**[Ans.** 270 Ω; 270 Ω; 3.3 kΩ, 1222]
4. Fig. 16-45 shows the circuit of a single stage common base amplifier. Find the values of (*a*) stage input resistance (*b*) stage output resistance, (*c*) current gain and (*d*) stage voltage gain. **[Ans.** 25 Ω; 5 kΩ; 1 and 200]

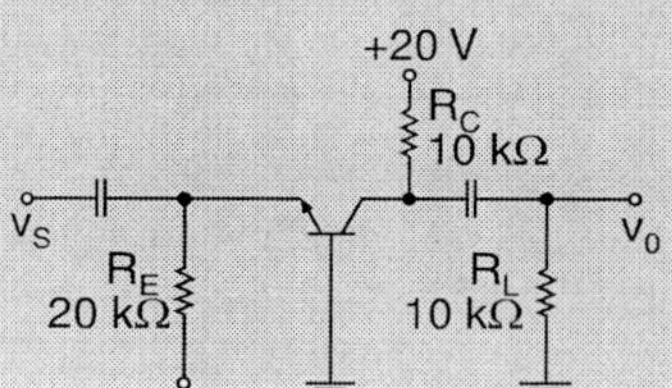

Fig. 16-45.

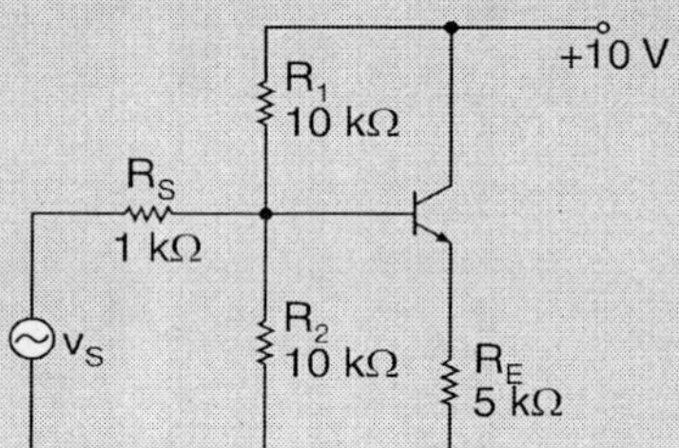

Fig. 16-46.

5. For the single-stage common collector amplifier circuit shown in Fig. 16-46, calculate the values of R_i, R_0, A_i and A_v. Take transistor β = 100 and $r'_e = 25/I_E$ (mA). **[Ans.** 500 kΩ; 29.4 Ω; 100 and 0.99]

DESCRIPTIVE QUESTIONS

1. With a neat circuit diagram, explain the working of a transistor amplifier.
2. Draw the circuit diagram of a common emitter amplifier with emitter and voltage divider biasing circuit. Why normally an emitter bypass capacitor is used?
3. Derive the expressions for input resistance, output resistance and voltage gain of a common emitter amplifier.
4. With a neat circuit diagram, explain the working of a transistor amplifier in which phase inversion of the input signal does not take place. Obtain the expressions for input resistance for such an amplifier.
5. Discuss the merits and limitations of common base amplifier.

6. Draw the circuit diagram of common base amplifier and derive expressions for current gain and voltage gain.

7. With suitable derivations, bring out the salient features of an emitter follower and indicate its applications.

8. Derive the expressions for input resistance, output resistance and voltage gain of an emitter follower circuit.

9. Give neat circuit diagrams showing complete biasing arrangements, including bypass and coupling capacitors for each of the common emitter, common base and common collector transistor amplifiers.

10. Give a comparison of the voltage and current gains, input and output resistances of common emitter, common base and common collector transistor amplifiers. Also give one important application of each transistor amplifier.

MULTIPLE CHOICE QUESTIONS

1. For a common emitter amplifier having a small unbypassed emitter resistance (R_E) the input resistance is approximately equal to

(*a*) R_E (*b*) h_{fe} (*c*) $h_{fe} \cdot R_E$ (*d*) R_E/h_{fe}

2. The voltage gain of a common base amplifier is

(*a*) zero (*b*) less than unity

(*c*) unity (*d*) greater than unity

3. For a common base transistor amplifier having input resistance (R_i) and output resistance (R_0), which of the following statements holds good

(*a*) R_i is low, R_0 is high (*b*) R_i is high, R_0 is low

(*c*) R_i and R_0 are both medium

4. The current gain of an emitter follower is

(*a*) zero (*b*) greater than unity

(*c*) less than unity (*d*) all of them

5. Which of the following transistor amplifiers has the highest voltage gain

(*a*) common-base (*b*) common-collector

(*c*) common-emitter (*d*) none of them

6. In an ac amplifier, larger the internal resistance of the ac signal source

(*a*) greater the overall voltage gain (*b*) greater the input impedance

(*c*) smaller the current gain (*d*) smaller the circuit voltage gain.

7. The main use of an emitter follower is as

(*a*) power amplifier (*b*) impedance matching device

(*c*) low-input impedance circuit (*d*) follower of base signal.

8. An ideal amplifier is one which

(*a*) has infinite voltage gain

(*b*) responds only to signal at its input terminals

(*c*) has positive feedback

(*d*) gives uniform frequency response.

9. The voltage gain of a single-stage amplifier is increased when

(*a*) its ac load is decreased (*b*) resistance of signal source is increased

(*c*) emitter resistance R_E is increased. (*d*) as load resistance is increased.

10. When emitter bypass capacitor in a common-emitter amplifier is removed, its....................is considerably reduced.

(*a*) input resistance (*b*) output load resistance

(*c*) emitter current (*d*) voltage gain

11. Unique features of a *CC* amplifier circuit is that it

(*a*) steps up the impedance level

(*b*) does not increases signal voltage

(*c*) acts as an impedance matching device

(*d*) all of the above.

12. An amplifier circuit is shown in Fig 16-47. The voltage gain (V_o/V_s) is

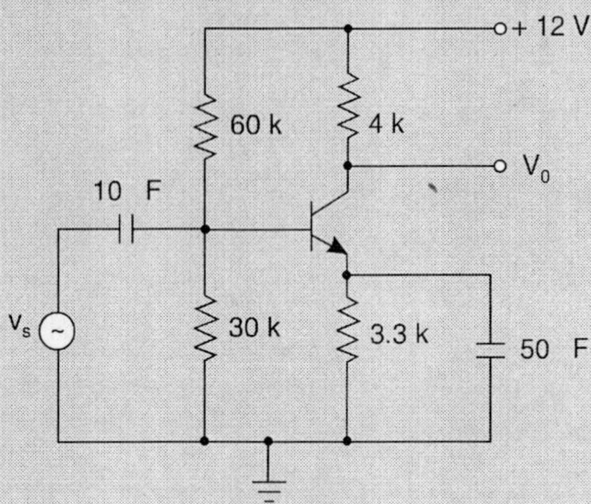

Fig. 16-47.

(*a*) 4/3.33 (*b*) 100

(*c*) 150 (*d*) 160

(*UPSC Engg. Services* 2000)

Hint. $V_B = 12 \times \dfrac{30}{30k + 160\,k} = 4V;\ I_E \doteq \dfrac{V_B - V_{BE}}{R_E} = \dfrac{4 - 0{\cdot}7V}{3{\cdot}3k}$

$\therefore \qquad r_e' = \dfrac{25mV}{I_E} = \dfrac{25\ mV}{1mA} = 25\ \Omega$

And gain $\qquad A_V = \dfrac{R_C}{r_e'} = \dfrac{4k}{25\ \Omega} = 160\ \Omega$

ANSWERS

1. (*c*)	**2.** (*d*)	**3.** (*a*)	**4.** (*b*)	**5.** (*c*)
6. (*d*)	**7.** (*b*)	**8.** (*b*)	**9.** (*d*)	**10.** (*d*)
11. (*d*)	**12.** (*d*)			

❑❑❑

17

CHAPTER

Hybrid Parameters

OUTLINE

Objectives

After completing this chapter you should be able to :

- Define h-parameters of a circuit is terms of voltages and currents.
- Calculate the h-parameters of a circuit.
- Sketch hybrid equivalent circuit for CE, CB and CC configuration.
- Derive the amplifier parameters in terms of h-parameters of a transistor for CE, CB and CC amplifiers.
- Determine h-parameters of a transistor from its imput and output characteristics.

17-1. Introduction

The hybrid parameters are commonly known as *h-parameters*. These are generally used to determine amplifier characteristic parameters such as voltage gain, input and output resistances, *etc.* We have already discussed in the last chapter that amplifier characteristic parameters may be determined by using current gain (β) and values of other circuit components. This approach had the following advantages :

(*i*) Value of circuit components are easily available, and

(*ii*) The procedure followed is quite simple and easy to understand.

However, this method has a drawback that the results obtained are not very accurate. It is mainly because of the fact that base and collector circuits of a transistor are not completely independent of each other. For example, we know that collector current should be β-times of the base current. Moreover, it should remain constant at this value. However, in actual practice, it is not so, as it is affected by the value of load resistance. Similarly, changes in the output voltage also cause changes in input voltage. These drawbacks are removed by the use of hybrid parameters. The hybrid parameters give very accurate results in transistor amplifier circuit analysis.

17-2. The *h*-parameters of a Linear Circuit

Fig. 17-1(*a*) shows a model of any linear device or a circuit. This device or a circuit is represented by a box and has four terminals, *i.e.*, two-terminals 1-1′ for input and two-terminals 2-2′ for output. The behaviour of this circuit is specified by two voltages and two currents. The two voltages are the input voltage (v_1) and the output voltage (v_2). The two currents are the input current (i_1) and output current (i_2). The polarity of voltages and direction of currents may be carefully noted. The polarity of voltage is such that the upper terminal is positive and the lower terminal is negative. Both the input an output currents are assumed to flow into the box. It is a standard convention for representing currents and voltages and do not correspond with the actual polarity of voltages and direction of currents. Thus, if in any circuit the voltages are of opposite polarity and currents are flowing out of the box; then these voltages and currents are taken as negative quantities.

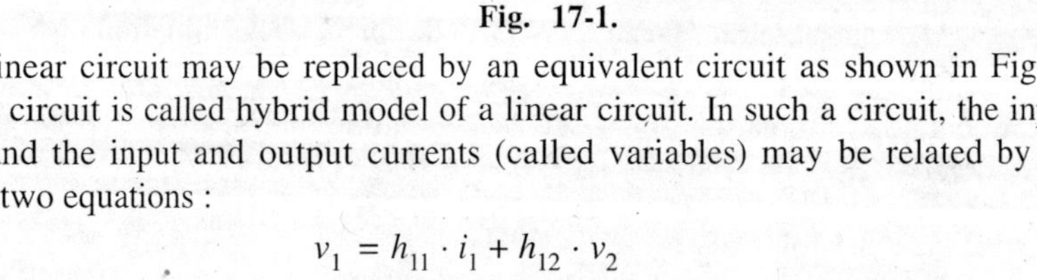

(*a*) Linear circuit. (*b*) Hybrid model for a linear circuit.

Fig. 17-1.

The linear circuit may be replaced by an equivalent circuit as shown in Fig. 17-1(*b*). The equivalent circuit is called hybrid model of a linear circuit. In such a circuit, the input and output voltages and the input and output currents (called variables) may be related by the set of the following two equations :

$$v_1 = h_{11} \cdot i_1 + h_{12} \cdot v_2 \quad \text{... } (i)$$

$$i_2 = h_{21} \cdot i_1 + h_{22} \cdot v_2 \quad \text{... } (ii)$$

where, v_1 = Input voltage, v_2 = Output voltage,

i_1 = Input current, i_2 = Output current, and

h_{11}, h_{12}, h_{21} and h_{22} = The hybrid or *h*-parameters.

The equations (*i*) and (*ii*) have been obtained by applying Kirchoff's Voltage Law to the input and output circuits of the hybrid model. Thus, a linear circuit has a set of four parameters namely h_{11}, h_{12}, h_{21} and h_{22}. These parameters completely describe the behaviour of the circuit and are constant for a given circuit. They are expressed in different units. The parameter h_{11} is expressed in ohms (Ω) h_{12}, and h_{21} are dimensionless and h_{22} is expressed in ohms (Ω) or (S). Since, these parameters have mixed units, therefore these are called hybrid (means mixed) or h-parameters.

17-3. Determination and Meaning of *h*-parameters

We have already discussed in the last article that every linear circuit may be represented by a set of four h-parameters namely h_{11}, h_{12}, h_{21} and h_{22} . The parameters h_{11} and h_{21} may be determined by short-circuiting the output terminals of a given circuit. On the other hand, h_{12} and h_{22} may be determined by open-circuiting the input terminals of the given circuits.

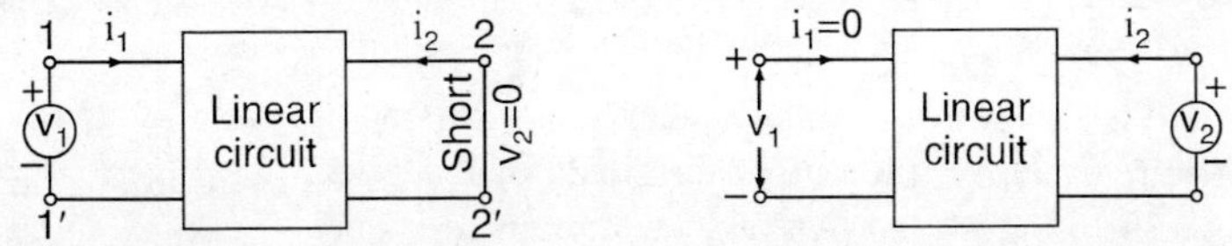

(*a*) Output short-circuited. (*b*) Input open-circuited.

Fig. 17-2.

(a) Determination of h_{11} and h_{21} : These are determined by short-circuiting the output terminals of a given circuit as shown in Fig. 17-2 (*a*). A short-circuit at the output terminals makes the voltage v_2 equal to zero. We know that the input voltage is given by the relation,

$$v_1 = h_{11} \cdot i_1 + h_{12} \cdot v_2$$

Substituting the value of v_2 (equal to zero) in the above equation, the input voltage,

$$v_1 = h_{11} \cdot i_1 \text{ or } h_{11} = v_1/i_1$$

Thus h_{11} may be determined from the ratio v_1/i_1. The value of i_1 is obtained by applying a voltage at the input and then measuring the value of input current (i_1). Since, h_{11} is the ratio of voltage to current, therefore it has the units of ohms, *i.e.*, the same unit as that of a resistance. Because of this fact, h_{11} is also called input resistance of the circuit with output short-circuited. Similarly, we know that the output current is given by the relation,

$$i_1 = h_{21} \cdot i_1 \text{ or } h_{22} \cdot v_2$$

Again substituting the value of v_2 (equal to zero) in the above equation, the output current,

$$i_1 = h_{21} \cdot i_1 \text{ or } h_{21} = i_2/i_1$$

Thus, h_{21} may be determined from the ratio i_2/i_1. The values of i_1 and i_2 may be obtained by applying a voltage at the input and then measuring the input current (i_1) and output current (i_2). Since, h_{21} is the ratio of currents, therefore it has no units. The parameter h_{21} is called the forward current gain of the circuit with output short-circuited.

(b) Determination of h_{12} and h_{22} : These are determined by open circuiting the input terminals of a given circuit as shown in Fig. 17-2 (*b*). An open circuit, at the input terminals, makes the current (i_1) equal to zero. We also know that the input voltage is given by the relation,

$$v_1 = h_{11} \cdot i_1 + h_{12} \cdot v_2$$

Substituting the value of i_1 (equal to zero) in the above equation, the input voltage,

$$v_1 = h_{12} \cdot v_2 \text{ or } h_{12} = v_1/v_2$$

Thus, h_{12} may be determined from the ratiov_1/v_2 . The value of v_1 may be obtained by applying a voltage v_2 at the output and then measuring the input voltage (v_1). Since, h_{12} is a ratio of voltages, therefore it has no units. As h_{12} is the ratio of input voltage (v_1) to the output voltage (v_2), therefore, its value is known as the reverse voltage gain in order to distinguish it from to forward voltage gain, whose value is equal to v_2/v_1 .

Similarly, we know that the output current is given by the relation,

$$i_2 = h_{21} \cdot i_1 + h_{22} \cdot v_2$$

Again substituting the value of (equal to zero) in the above equation, the output current,

$$i_2 = h_{22} \cdot v_2 \text{ or } h_{22} = i_2/v_2$$

Thus h_{22} may be determined from the ratio i_2/v_2. The value of current (i_2) may be obtained by applying a voltage at the output (v_2) and then measuring the output current with input open. Since, h_{22} is the ratio of current to voltage, therefore it has the units of ohms (Ω) or Siemens (*S*). The parameter h_{22} is also called output conductance with input open.

Example 17-1. *Determine the value of h-parameters for a circuit shown in Fig. 17-3.*

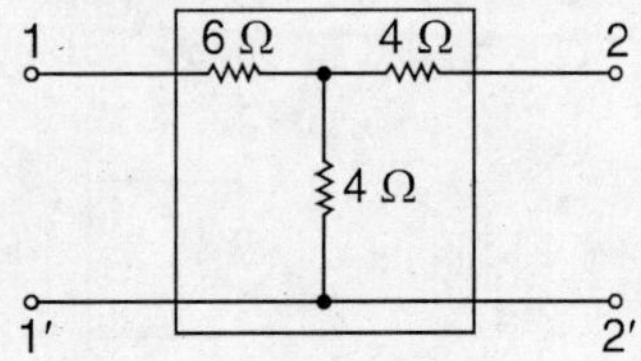

Fig. 17-3.

Solution. ***(a)*** *Parameters h_{11} and h_{21}*

We know that the values of h_{11} and h_{21} can be determined by putting a short-circuit across the output terminals of the given circuit as shown in Fig. 17-4 (*a*). And h_{11} (*i.e.*, input resistance looking from the input terminals 1-1′) is given by the relation,

$$h_{11} = 6 + (4 \,||\, 4) = 6 + \frac{4 \times 4}{4 + 4} = 8\ \Omega \textbf{ Ans.}$$

It may be noted from Fig. 17-4 (*a*), that the input current is divided into two equal parts at point N. A current $i_1/2$ of flows outwards and $i_1/2$ flows outwards and $i_1/2$ flows downward. The output current,

$$i_2 = -i_1/2$$

The negative sign indicates that the current is actually flowing out of the box.

$$\therefore \qquad h_{21} = \frac{i_2}{i_1} = \frac{-i_1/2}{i_1} = -\frac{1}{2} = -0.5 \textbf{ Ans.}$$

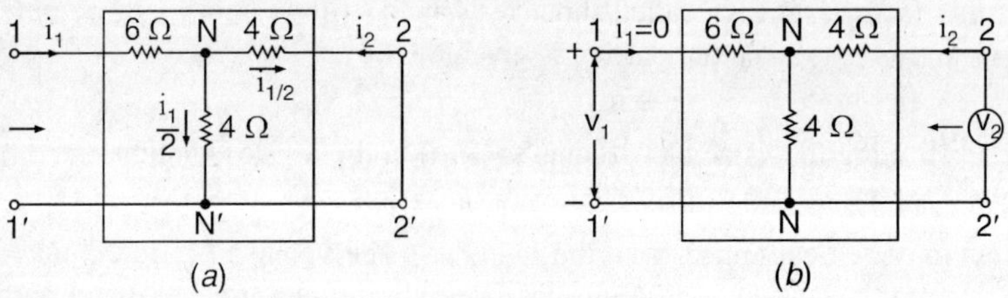

Fig. 17-4.

(b) *Parameters h_{12} and h_{22}*

We know that the values of h_{12} and h_{22} may be determined by keeping the input terminals open and applying a voltage source v_2 across the output terminals as shown in Fig. 17-4 (*b*). Due to the open input, i_1 = 0. The current flowing through 6 Ω resistor is zero, and hence there is no voltage drop across it. The voltage (v_2) produces a current (i_1) in the circuit. This current will produce equal voltage drops across the two 4 Ω resistors. The voltage drop across the 4 Ω resistor connected between *N* and *N′* appears as the voltage v_1 across the input terminals 1 : 1′. Thus $v_1 = v_2/2$.

$$\therefore \qquad h_{12} = \frac{v_1}{v_2} = \frac{v_2/2}{2} = 0.5 \textbf{ Ans.}$$

The output resistance (*i.e.*, the resistance looking from the output terminals 2 : 2′ with input terminals open)

$$= 4 + 4 = 8\ \Omega$$

∴ Output conductance,

$$h_{22} = 1/8 = 0.125 \text{ ohms } \textbf{Ans.}$$

17-4. Another Representation for *h*-parameters

Table 17-1 summaries the meaning of each *h*-parameter and the required condition.

Table 17-1. *h-parameters*

No.	*Parameter*	*Meaning*	*Condition*
1.	h_{11}	Input resistance	Output shorted
2.	h_{12}	Reverse voltage gain	Input open
3.	$h_{2,}$	Forward current gain	Output shorted
4.	h_{22}	Output conductance	Input open

It will be interesting to know that the parameters h_{11} and h_{21} are called *forward parameters* sh_{12} and h_{22} are called *reverse parameters*. The forward and reverse parameters indicate the way they represent themselves. For example, h_{12} is a reverse parameter because it is a voltage gain of v_1/v_2 (and not a forward gain v_2/v_1). Similarly, h_{22} is a reverse parameter because it is an output conductance. Sometimes, it is more convenient to represent the parameters h_{11}, h_{12} and h_{22} as h_i, h_r, h_f and h_o respectively. In other words,

$h_i = h_{11}$ = Input resistance with output shorted,

$h_r = h_{12}$ = Reverse voltage gain with input open,

$h_f = h_{21}$ = Forward current gain with output shorted, and

$h_o = h_{22}$ = Output conductance with input open.

These notations for *h*-parameters may be easy to remember by knowing the meaning of *i*, *r*, *f* and 0, which are given as below:

i = Input $\qquad$ r = Reverse

f = Forward $\qquad$ o = Output

17-5. The *h*-parameter Notations for Transistors

The *h*-parameters of a transistor depend upon the types of the connection (or configuration) used, *i.e.*, common-emitter (CE), common-collector (CC) or common-base (CB). Because of this, each of the four *h*-parameters carries a second subscript letter *e*, *b* or *c*. The letter '*e*' is used to designate common-emitter; '*c*' for common collector and '*b*' for common base configuration. Table 17-2 shows the notations commonly used for transistor *h*-parameters.

Table 17-2.

S. No.	*General Parameter*	*Transistor configurations*		
		Common emitter	*Common base*	*Common collector*
1.	h_{11}	h_{ie}	h_{ib}	h_{ic}
2.	h_{12}	h_{re}	h_{rb}	h_{rc}
3.	h_{21}	h_{fe}	h_{fb}	h_{fc}
4.	h_{22}	h_{oe}	h_{ob}	h_{oc}

17-6. *h*-parameters of a Transistor

It has been observed that for small signal operation, the behaviour of a transistor is like a linear device (*i.e.*, a device whose output is directly proportional to the input and varies continuously with the input signal). Thus for small signal operation, every transistor has a set of four *h*-parameters. The values of these parameters depend upon the following factors :

- Transistor type.
- Transistor configuration.

- Operating point.
- Frequency.
- Temperature.

The values of h-parameters may be obtained experimentally or graphically from the transistor characteristics. The procedure for determining h-parameters will be discussed later in this chapter.

17-7. Hybrid Equivalent Circuit of a Transistor

Fig. 17-5(a) shows the transistor with its input and output terminals. The amplifier can be formed by connecting a signal source at its input terminals and a load resistance at its output terminals. Here the transistor represents in any one of the three possible configurations. Fig. 17-5(b) shows the general form of a h-parameter equivalent circuit of such a transistor. The equivalent circuit is known as small-signal low-frequency hybrid model of a transistor.

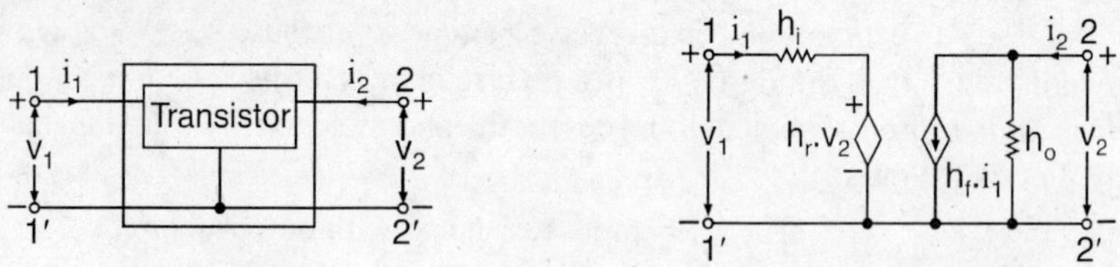

(a) Transistor. (b) Hybrid equivalent circuit of a transistor.

Fig. 17-5.

It may be noted in Fig. 17-5 (b) that the input resistance of a transistor (h_i) appears in series at the input. The reverse voltage gain (h_r) is multiplied by the output voltage to produce an equivalent voltage source (equal to $h_r \cdot v_2$) in series with the input. The equivalent voltage source indicates that in a practical transistor, some of the output voltage is feedback to the input. This represents an unavoidable interaction between the input and output circuits. The forward current gain (h_f) is multiplied by the input current and appears as an equivalent current source ($h_f \cdot i_1$) in the output. The output resistance ($i.e.$, reverse of output conductance, which is equal to $1/h_o$) appears across the output terminals. Since, the transistor can be connected in any one of the three different ways namely common emitter, common base and common collector, therefore there are three hybrid equivalent circuits, which will be discussed in the following pages.

17-8. Hybrid Equivalent Circuit for Common Emitter Transistor

Fig. 17-6 (a) shows the transistor connected in common-emitter (CE) configuration and Fig. 17-6 (b) shows the hybrid equivalent circuit of such a transistor.

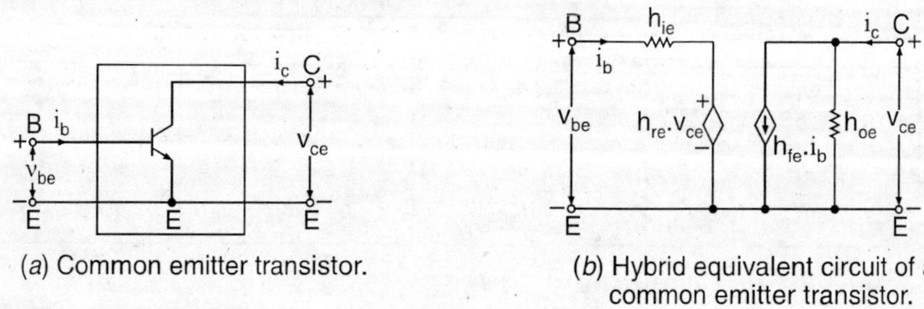

(a) Common emitter transistor. (b) Hybrid equivalent circuit of a common emitter transistor.

Fig. 17-6.

In a common emitter transistor configuration, the input signal is applied between the base and emitter terminals of a transistor and output appears between the collector and emitter terminals. The input voltage (v_{eb}) and the output current (i_c) are given by the following equation :

$$v_{be} = h_{ie} \cdot i_b + h_{re} \cdot v_{ce}$$
$$i_c = h_{fe} \cdot i_b + v_{ce}$$

17-9. Hybrid Equivalent Circuit for Common Base Transistor

Fig. 17-7 (*a*) shows the transistor connected in a common-base (CB) configuration and Fig. 17-7 (*b*) shows its hybrid equivalent circuit. In a common-base configuration, the input signal is applied between emitter and base terminals and output appears between collector and base terminals.

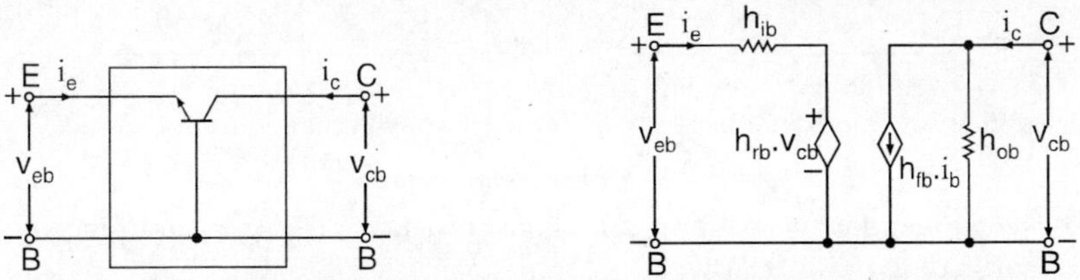

(*a*) Common emitter transistor.

(*b*) Hybrid equivalent circuit of a common emitter transistor.

Fig. 17-7.

The input voltage (v_{eb}) and the output current (i_c) are given by the following equations :

$$v_{eb} = h_{ib} \cdot i_e + h_{rb} \cdot v_{cb}$$
$$i_e = h_{fb} \cdot i_c + h_{ob} \cdot v_{cb}$$

17-10. Hybrid Equivalent Circuit for Common Collector Transistor

Fig. 17-8 (*a*) shows the transistor connected in common-collector (CC) configuration and Fig. 17-8 (*b*) shows its hybrid equivalent circuit. In a common-collector configuration, the input signal is applied between base and collector terminals and the output appears between emitter and collector terminals. The input voltage (v_{bc}) and the output current (i_e) are given by the following equations :

$$v_{bc} = h_{ic} \cdot i_b + h_{rc} \cdot v_{ec}$$
$$i_e = h_{fc} \cdot i_b + h_{oc} \cdot v_{ec}$$

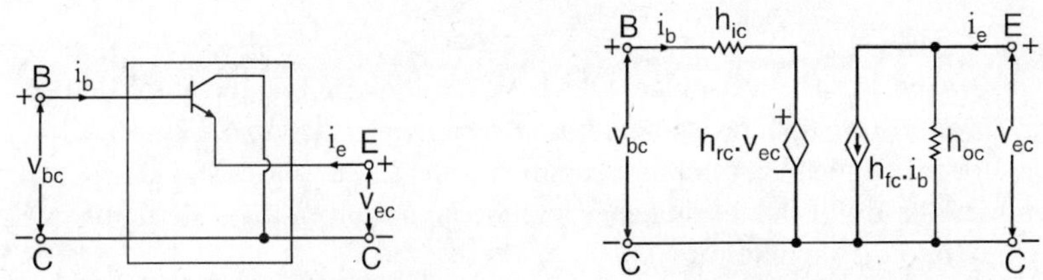

(*a*) Common collector transistor.

(*b*) Hybrid equivalent circuit of a common collector transistor.

Fig. 17-8.

17-11. Amplifier Expressions

Fig. 17-9 (*a*) shows a general amplifier circuit. In this circuit, a transistor is connected in any one of the three possible configurations (*i.e.*, common emitter, common base or common collector) to a voltage source (v_s) and load resistance (r_L). The voltage source has an internal resistance (R_s) as shown in the figure. The load resistance (r_L) is the effective or a.c. load resistance seen by the transistor at its output. Fig. 17-9(*b*) shows the hybrid equivalent circuit of a general transistor amplifier circuit. Now we shall derive the expressions (called hybrid formulas) for input resistance, output resistance, current gain and voltage gain of a transistor.

The expressions for transistor amplifier input and output resistances, voltage and current gains may be obtained from the relations :

$$v_1 = h_i \cdot i_1 + h_r \cdot v_2 \qquad ...(i)$$
$$i_2 = h_f \cdot i_1 + h_0 \cdot v_2 \qquad ...(ii)$$

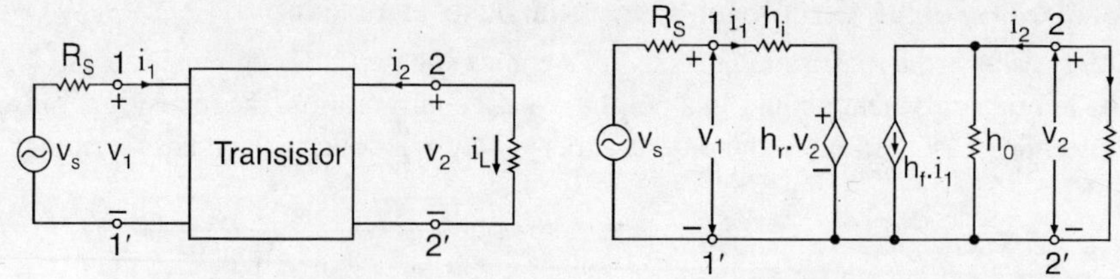

(*a*) General amplifier circuit. (*b*) Hybrid equivalent circuit.

Fig. 17-9.

It may be noted that the voltage drop across load resistance (r_L) is equal to the voltage across the output terminals of a transistor. Thus,

$$v_2 = i_L \cdot r_L = -i_2 \cdot r_L \qquad \text{... } (\because\ i_2 = -i_L)$$

(*a*) *Current gain :* It is the ratio of output current (i_L) to input currents (i_1). Mathematically, the current gain,

$$A_i = \frac{i_L}{i_1} = -\frac{i_2}{i_1}$$

Substituting the value of v_2 (equal to $-i_2 \cdot r_L$) in equation (*ii*),

$$i_2 = h_f \cdot i_1 + h_0(-i_2 \cdot r_L) = h_f \cdot i_1 - h_0 \cdot i_2\, r_L$$

$$i_2 + h_0 \cdot i_2 \cdot r_L = h_f \cdot i_1$$

$$i_2(1 + h_0 \cdot r_L) = h_f \cdot i_1$$

$$\frac{i_2}{i_1} = \frac{h_f}{1 + h_0 \cdot r_L}$$

$$\therefore \qquad A_i = -\frac{i_2}{i_1} = -\frac{h_f}{1 + h_0 \cdot r_L} \qquad \text{... } (iii)$$

(*b*) *Input resistance :* The resistance, which we see looking into the amplifier input terminals (1,1') is called the amplifier input resistance. It is also known as the input resistance looking into the base of a transistor for common emitter and common collector amplifier and is designated by R_i. Its value is given by the ratio of input voltage (v_1) to the input current (i_1). Mathematically, the input resistance looking into the input terminals,

$$R_i = \frac{v_1}{i_1}$$

Again substituting the value of v_2 (equal to $-i_2 \cdot r_L$) in equation (*i*)

$$v_1 = h_i \cdot i_1 + h_r(-i_2 \cdot r_L)$$

$$= h_i \cdot i_1 - h_r \cdot i_2 \cdot r_L$$

Dividing the above equation by i_1 on both sides,

$$\frac{v_1}{i_1} = \frac{h_i \cdot i_1 - h_r \cdot i_2 \cdot r_L}{i_1} = h_i - h_r \left(\frac{i_2}{i_1}\right) r_L$$

Replacing v_1/i_1 with R_i and i_2/i_1 by $-A_i$ in the above expression,

$$R_i = h_i - h_r(-A_i) \cdot r_L = h_i + h_r \cdot A_i \cdot r_L \qquad \text{... } (iv)$$

Substituting the value of A_i equal to $-h_f/(1 + h_0 \cdot r_L)$ from equation (*iii*) in the above expression,

$$R_i = h_i + h_r \left(-\frac{h_f}{1 + h_0 \cdot r_L}\right) r_L = h_i - \frac{h_r \cdot h_f}{h_0 + \dfrac{1}{r_L}} \qquad \text{... } (v)$$

(*c*) ***Voltage gain :*** It is defined as the ratio of output voltage (v_2) to the input voltage (v_1). Mathematically, the voltage gain,

$$A_v = \frac{v_2}{v_1} = -\frac{i_2 \cdot r_L}{v_1} \qquad \ldots (\because\ v_2 = -i_2 \cdot r_L)$$

Substituting the value of i_2 (equal to $-A_i \cdot i_1$) from equation (*iii*) the above expression,

$$A_v = \frac{A_i \cdot i_1 \cdot r_L}{v_1} = A_i \cdot r_L \left(\frac{i_1}{v_1}\right)$$

Replacing i_1/v_1 by R_i in the above expression,

$$A_v = \frac{A_i \cdot r_L}{R_i}$$

Substituting the value of A_i (equal to $-h_f/(1 + h_0 \cdot r_L)$) and R_i (equal to $\dfrac{h_i - h_r \cdot h_f}{(h_0 + 1/r_L)}$) in the above expression and rearranging,

$$A_v = -\frac{h_f \cdot r_L}{h_i + (h_i \cdot h_0 - h_r \cdot h_f)\, r_L}$$

If we replace($h_i \cdot h_0 - h_r . h_f = \Delta h$) in the above expression, then the voltage gain,

$$A_v = -\frac{h_f \cdot r_L}{h_i + \Delta h \cdot r_L}$$

(*d*) ***Output resistance :*** It is obtained by setting the source voltage (V_S) to zero and the load resistance (r_L) to infinity and by driving the output terminals from a generators v_2 as shown in Fig.17-10.

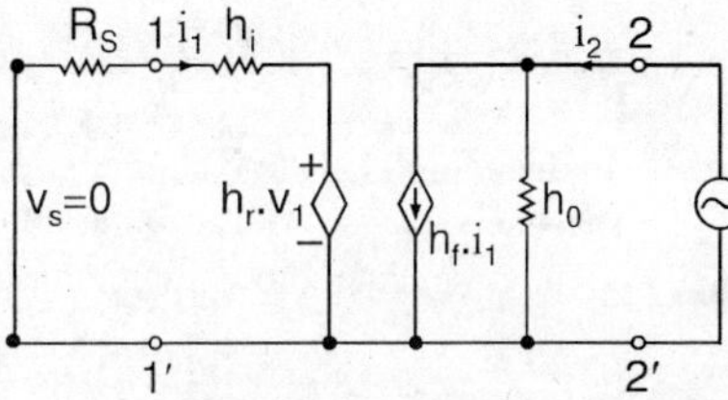

Fig. 17-10. *Hybrid equivalent circuit of a general transistor amplifier with voltage source connected across output terminals.*

Then output resistance is the ratio of voltage (v_2) to the current drawn from the voltage source (i_2). Mathematically, the output resistance,

$$R_0 = \frac{v_2}{i_2} = \frac{v_2}{h_f \cdot i_1 + h_0 \cdot v_2} \qquad \ldots(vi)$$

The value of current (i_1) may be obtained by applying Kirchoff's Voltage Law to the input side of transistor amplifier circuit. Thus

$$R_s = i_1 + h_r \cdot v_2 = 0$$

$$\therefore \qquad i_1 = -\frac{h_r \cdot v_2}{R_s + h_i}$$

Substituting this value of i_1 in equation (*vi*),

$$R_0 = \frac{v_2}{h_f\left(-\dfrac{h_r \cdot v_2}{R_s + h_i}\right) + h_0 \cdot v_2}$$

$$= \frac{v_2}{-\dfrac{h_f \cdot h_i \cdot v_2}{R_s + h_i} + h_0 \cdot v_2}$$

Rearranging the above equation,

$$R_0 = \frac{R_s + h_i}{(R_s + h_i)\, h_0 - h_f \cdot h_i} = \frac{R_s + h_i}{R_s \cdot h_0 + (h_i \cdot h_0 - h_f \cdot h_r)}$$

$$= \frac{R_s + h_i}{R_s \cdot h_0 + \Delta h}$$

It may be noted from the above relation that if source resistance (R_s) is zero, then output resistance,

$$R_0 = \frac{h_i}{\Delta h}$$

(*e*) ***Power gain :*** It is defined as the product of voltage gain (A_v) and the current gain (A_i). Mathematically, the power gain,

$$\Delta P = A_v \cdot A_i$$

Notes: 1. The input resistance (R_i) of a transistor is the value of resistance looking directly into the transistor from the input terminals. It does not include the values of resistor connected externally for the biasing purpose. The input resistance, which takes into account the biasing resistors, is called resistance of the stage (or stage input resistance) and is designated by R_{is}. For a fixed bias current, the value of stage input resistance,

$$R_{is} = R_i \| R_B$$

and for voltage divider bias circuit,

$$R_{is} = R_i \| (R_1 \| R_2)$$

2. The output resistance (R_o) of a transistor is the resistance looking directly into the transistor from the output terminals. It does not include the value of collector resistance (R_C) and load resistance (R_L). The output resistance, which includes the effect of R_C and , $R_L$is called output resistance of a stage (or stage output resistance) and is designated by R_{os}. Its value is given by the relation,

$$R_{os} = R_o \| (R_C \| R_L) = R_O \| r_L \qquad \text{... } (\because r_L = R_C \| R_L)$$

3. In the expressions for output resistance, current and voltage gains, we require the value of a.c load resistance (*i.e.*,r_L). The value of depends upon the type of connection used for transistor. For common emitter and common base connections, the a.c. load resistance,

$$r_L = R_C \qquad \text{... (If } R_L \text{ is not connected)}$$

$$= R_C \| R_L \qquad \text{... (If } R_L \text{ is connected)}$$

However, for a common collector (or emitter follower connection) the value of a.c. load resistance,

$$r_L = R_E \qquad \text{... (If } R_L \text{ is not connected)}$$

$$= R_E \| R_L \qquad \text{... (If } R_L \text{ is connected)}$$

17-12. Effect of Source Resistance on Voltage and Current Gains

We have already discussed in the last article the current and voltage gains of a transistor amplifier. The expressions for current and voltage gains derived there are from the amplifier input terminals (1, 1′) to the output terminal (2, 2′). In other words, these values do not take into consideration the source resistance (R_S). However, it is important to take into account the source resistance (R_S), while calculating the voltage and current gains of the transistor amplifier. In this case, the voltage and current gains are known as overall gains or stage gains.

The overall gains of the transistor amplifier is the ratio of output voltage (v_2) to the source voltage (v_s). Mathematically, the overall voltage gain,

$$A_{vs} = \frac{v_2}{v_s}$$

Multiplying and dividing the right hand side of the above equation by v_1,

$$A_{vs} = \frac{v_2}{v_s} \times \frac{v_1}{v_1} = \frac{v_2}{v_1} \times \frac{v_1}{v_s} = A_v \times \frac{v_1}{v_s}$$

where, A_v (equal to v_2/v_1) is the voltage gain of amplifier from the input terminals (1, 1′) to the output terminals (2, 2′). In order to determine the voltage ratio (v_1/v_s), consider the circuit shown in Fig. 17-11(*a*).

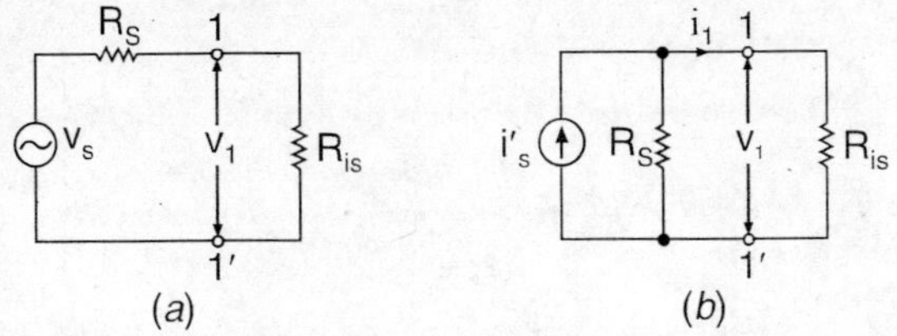

Fig. 17-11. *Effect of source resistance on amplifier voltage and current gains.*

In this figure, the resistance R_{is} represents the amplifier input resistance. It is evident from this figure that the voltage drop across input resistance,

$$v_1 = v_s \times \frac{R_{is}}{R_s + R_{is}}$$

$$\therefore \quad \frac{v_1}{v_s} = \frac{R_{is}}{R_s + R_{is}}$$

$$\text{and} \quad A_{vs} = A_v \times \frac{R_{is}}{R_s + R_{is}}$$

Similarly, the overall current gain of the transistor amplifier is the ratio of output current (i_L) to the current delivered by the source (i_s). Mathematically, the overall current gain,

$$A_{is} = \frac{i_L}{i_s} = -\frac{i_2}{i_s} \qquad \ldots (\because\ i_L = -i_2)$$

Multiplying and dividing the right hand side of the above equation by i,

$$A_{is} = -\frac{i_2}{i_s} \times \frac{i_1}{i_1} = -\frac{i_2}{i_1} \times \frac{i_1}{i_s} = -A_i \times \frac{i_1}{i_s}$$

where, A_i (equal to) (i_2 / i_1) is the current gain of the amplifier from the input terminals (1, 1′) to the output terminals (2,2′). In order to determine the current ratio, consider the circuit shown in Fig. 17-11 (*b*). Here, the voltage source has been replaced by its equivalent current source. It may be noted that the current through the amplifier input resistance,

$$i_1 = i_s \times \frac{R_s}{R_s \times R_{is}}$$

$$\therefore \quad \frac{i_1}{i_s} = \frac{R_s}{R_s + R_{is}}$$

$$\text{and} \quad A_{is} = A_i \times \frac{i_1}{i_s} = \frac{R_s}{R_s + R_{is}}$$

17-13. Hybrid Formulae for Common Emitter Amplifier

These formulae may be obtained from the general hybrid formulas derived in Art. 17-11 by adding a second subscript letter '*e*' (which stands for common emitter) with the *h*-parameters and are as discussed below:

(*a*) *Current gain :* It is given by the relation,

$$A_i = -\frac{h_{fe}}{1 + h_{oe} \cdot r_L}$$

where r_L is the a.c. load resistance. Its value is equal to the parallel combination of resistances R_C and R_L. Since h_{fe} of a transistor is a positive number, therefore A_i of a common emitter amplifier is negative.

(*b*) ***Input resistance :*** The resistance looking into the amplifier input terminals (*i.e.*, base of a transistor) is given by the relation,

$$R_i = h_{ie} + h_{re} \cdot A_i \cdot r_L$$

$$= h_{ie} - \frac{h_{re} \cdot h_{fe}}{h_{oe} + \dfrac{1}{r_L}}$$

The input resistance of the amplifier stage (called stage input resistance R_{is}) depends upon the biasing arrangement. For a fixed bias circuit, the stage input resistance,

$$R_{is} = R_i \parallel R_B$$

If the circuit has no biasing resistances, then $R_{is} = R_i$.

(*c*) ***Voltage gain :*** It is given by the relation,

$$A_v = \frac{A_i \cdot r_L}{R_i}$$

Since, the current gain (A_i) of a common emitter amplifier is negative, therefore the voltage gain (A_v) is also negative. It means that there is a phase difference of 180° between the input and output. In other words, the input signal is inverted at the output of a common emitter amplifier. The voltage gain, in terms of h-parameters, is given by the relation,

$$A_v = -\frac{h_{fe} \cdot r_L}{h_{ie} + \Delta h \cdot r_L}$$

where, $\Delta h = h_{ie} \cdot h_{oe} - h_{re} \cdot h_{fe}$

(*d*) ***Output resistance :*** The resistance looking into the amplifier output terminals is given by the relation,

$$R_O = \frac{R_s + h_{ie}}{R_s \cdot h_{oe} + \Delta h}$$

where, R_s = Resistance of the source, and

$$\Delta h = h_{ie} \cdot h_{oe} - h_{re} \cdot h_{fe}$$

The output resistance of the stage,

$$R_{os} = R_O \parallel r_L$$

(*e*) ***Overall voltage gain :*** It is given by the relation,

$$A_v = \frac{A_v \cdot R_s}{R_s + R_{is}}$$

(*f*) ***Overall current gain :*** It is given by relation,

$$A_{is} = \frac{A_i \cdot R_s}{R_s + R_{is}}$$

Example 17-2. *The h-parameters of a transistor used in a common emitter circuit are h_{ie} = 1.0 kΩ, h_{re} = 10, × 10^{-4} , h_{fe} = 50 and h_{oe} = 100 μ mhos. The load resistor for the transistor is 1 kΩ in the collector circuit. The transistor is supplied from a signal source of resistance 1000 Ω. Determine the value of input and output impedances (or resistances), voltage and current gains in the amplifier stage.*

Solution. Given: h_{ie} = 1.0 kΩ = 1000 Ω; h_{re} 1.0 × 10^{-4} ; h_{oe} = 100 μ mhos = 100 × 10^{-6} mhos; R_C = 1 kΩ = 1000 Ω and R_s = 1000 Ω.

Fig. 17-12. (*a*) shows the common emitter transistor amplifier circuit and (*b*) shows its hybrid equivalent circuit.

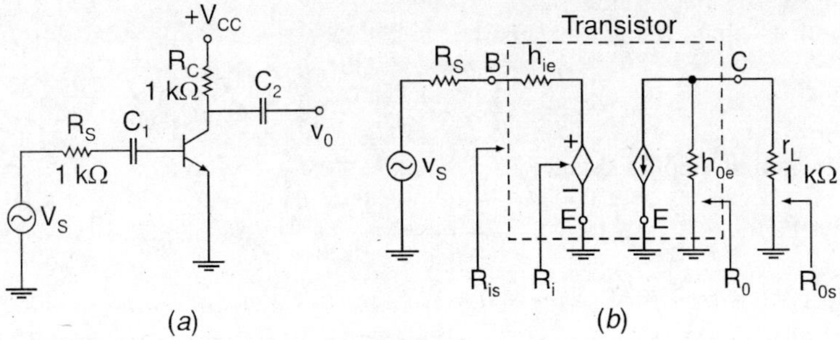

Fig. 17-12.

(a) *Input resistance of the amplifier stage*

We know that there is no load connected at the output of the amplifier (*i.e.*, $R_L = 0$), therefore the value of a.c. load resistance,

$$r_L = R_C = 1000\ \Omega$$

We also know that current gain of a transistor,

$$A_i = \frac{h_{fe}}{1 + h_{oe} \cdot r_L} = -\frac{50}{1 + (100 \times 10^{-6}) \times 1000} = -45.5$$

and the input resistance of a transistor, looking directly into the base,

$$R_i = h_{ie} + h_{re} \cdot A_i \cdot r_L$$
$$= 1000 + [(1.0 \times 10^{-4}) \times (-45.5) \times 1000]\ \Omega$$
$$= 995\ \Omega$$

∴ Input resistance of the amplifier stage,

$$R_{is} = R_i = 995\ \Omega \textbf{ Ans.}$$

(b) *Output resistance of amplifier stage*

We know that

$$\Delta h = h_{ie} \cdot h_{oe} - h_{re} \cdot h_{fe}$$
$$= [1000 \times (100 \times 10^{-6})] - (1.0 \times 10^{-4}) \times 50$$
$$= 95 \times 10^{-3} = 0.095$$

∴ Output resistance of the transistor looking directly into collector,

$$R_0 = \frac{R_s + h_{is}}{R_s \cdot h_{oe} + \Delta h}$$
$$= \frac{1000 + 1000}{[1000 \times (100 \times 10^{-6})] + 0.095}\ \Omega$$
$$= \frac{2000}{0.1 + 0.095} = 10.256\ \Omega$$

and output resistance of the amplifier stage,

$$R_{os} = R_o \parallel r_L = 10256 \parallel 1000 = 911\ \Omega \textbf{ Ans.}$$

(c) *Current gain of amplifier stage*

We know that the current gain of amplifier stage,

$$A_{is} = \frac{A_i \cdot R_s}{R_s + R_{is}} = \frac{(-45.5) \times (1000)}{1000 + 995} = -22.8 \textbf{ Ans.}$$

(*d*) *Voltage gain of amplifier stage*

We know that voltage gain of a transistor,

$$A_v = \frac{A_i \cdot r_L}{R_i} = \frac{(-45.5) \times (1000)}{995} = -45.7$$

∴ Voltage gain of amplifier stage,

$$A_{vs} = \frac{A_v \cdot R_{is}}{R_s + R_{is}} = \frac{(-45.7) \times 995}{1000 + 995} = -22.8 \textbf{ Ans.}$$

Example 17-3. *A transistor connected in CE configuration has the following h-parameters.* $h_{ie} = 1{\cdot}1\ k\Omega$; $h_{re} = 2{\cdot}5 \times 10^{-4}$; $h_{fe} = 50$, $h_{oe} = 25\ \mu$ siemens and $r_S = r_L = 1\ k\Omega$
Calculate current gain, input impedance and voltage gain.

(*Banglore University, BSc. April/May* 2001)

Solution. Given: $h_{ie} = 1{\cdot}1\ k\Omega = 1{\cdot}1 \times 10^3\Omega$; $h_{re} = 2{\cdot}5 \times 10^{-4}$; $h_{fe} = 50$, $h_{oe} = 25\ \mu S = 25 \times 10^{-6}$ S; and $r_S = r_L = 1\ k\Omega = 1 \times 10^3\ \Omega$

Current Gain

We know that current gain,

$$A_i = \frac{h_{fe}}{1 + h_{oe} \cdot r_L} = \frac{50}{1 + (25 \times 10^{-6}) \times (1 \times 10^3)}$$

$$= 48.78 \textbf{ Ans.}$$

Input Impedance

We know that input impedance,

$$R_i = h_{ie} + h_{re} \cdot A_i \cdot r_L$$

$$= (1{\cdot}1 \times 10^3) + (2.5 \times 10^{-4}) \times (48{\cdot}78) \times (1 \times 10^3)$$

$$= 1112.2\ \Omega \textbf{ Ans.}$$

Voltage Gain

We know that voltage gain,

$$A_v = \frac{A_i \cdot r_L}{R_i} = \frac{48{\cdot}78 \times (1 \times 10^3)}{1112.2} = 43{\cdot}86 \textbf{ Ans.}$$

17-14. Hybrid Formulae for Common Emitter Amplifier with Collector Feedback Bias

Consider a common emitter amplifier with collector feedback bias as shown in Fig. 17-13 (*a*). Such a circuit can be analysed by using Miller's Theorem. According to this theorem, the collector feedback bias resistance (R_B) can be disconnected from its position and can be shown as an equivalent of two resistances R_{B_1} and R_{B_2} as shown in Fig. 17-13 (*b*).

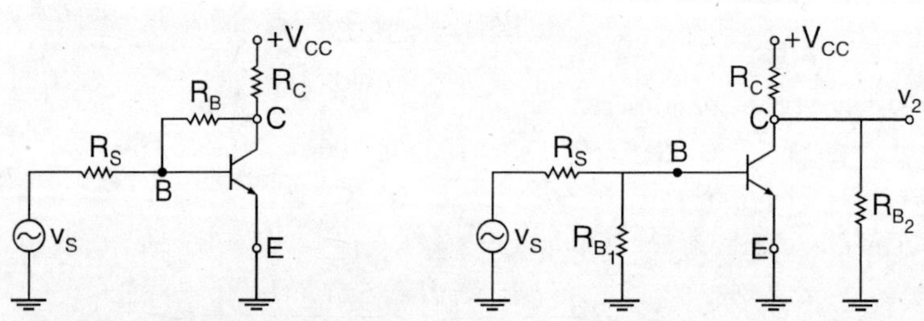

(*a*) Common emitter amplifier with a collector feedback bias.

(*b*) Equivalent circuit.

Fig. 17-13.

The resistance R_{B_2} is shown to be connected between the base and ground and its value is given by the relation,

$$R_{B_1} = \frac{R_B}{1 - A_v}$$

where, A_v = Voltage gain

Since, the value of voltage gain of common emitter amplifier is very large, as compared to unity (*i.e.*, $A_v >> 1$), therefore the value of resistance R_{B_1} is much smaller than that of resistance R_B. The resistance R_{B_2} is shown to be connected between the collector and ground and its value is given by the relation,

$$R_{B_2} = \frac{R_B}{1 - \frac{1}{A_v}}$$

Since, the voltage gain is very large, as compared to unity (*i.e.*, $A_v >> 1$), therefore value of resistance R_{B_2} is approximately equal to resistance R_B. The circuit shown in Fig. 17-13 (*b*) can now be analysed easily as discussed earlier in Art. 17-13. Thus, a.c. equivalent load resistance of the circuit,

$$r_L = R_C \,||\, R_{B_2} = \frac{R_C \times R_{B_2}}{R_C + R_{B_2}}$$

(*a*) ***Current gain :*** It is given by the relation,

$$A_i = \frac{h_{fe}}{1 + h_{oe} \cdot r_L}$$

(*b*) ***Input resistance :*** The resistance looking into the amplifier input terminal (*i.e.*, base of the transistor) is given by the relation,

$$R_i = h_{ie} + h_{re} \cdot A_i \cdot r_L$$

and input resistance of the amplifier stage,

$$R_{is} = R_i \,||\, R_{B_1}$$

(*c*) ***Voltage gain :*** It is given by the relation,

$$A_v = \frac{A_i \cdot r_L}{R_i}$$

(*d*) ***Output resistance :*** The resistance looking into the amplifier output terminals is given by the relation,

$$R_0 = \frac{R_s + h_{ie}}{R_s \cdot h_{oe} + \Delta h}$$

where, R_s = Resistance of the source, and

$$\Delta h = h_{ie} \cdot h_{0e} - h_{re} \cdot h_{fe}$$

The output resistance of the amplifier stage,

$$R_{0s} = R_0 \,||\, r_L$$

(*e*) ***Overall voltage gain :*** It is given by the relation,

$$A_{vs} = \frac{A_v \cdot R_{is}}{R_s + R_{is}}$$

(*f*) ***Overall current gain :*** It is given by the relation,

$$A_{is} = \frac{A_i \cdot R_s}{R_s + R_{is}}$$

Example 17-4. *Consider a common emitter transistor amplifier with collector load resistance R_C = 4 kΩ, collector-to-base feedback resistance R_B = 40 kΩ and a source resistance R_S = 10 kΩ. The transistor parameters are = 1100 Ω; h_{fe} = 50; h_{re} = h_e = 0. Calculate the voltage gain, the input resistance looking from the source terminals and the output resistance.*

(*Grad. I.E.T.E., June* 1987)

Solution. Given: $R_C = 4$ kΩ; $R_B = 40$ kΩ; $R_S = 10$ k Ω $= 10 \times 10^3$ Ω; h_{ie} = 1100 Ω; h_{fe} = 50 and $h_{re} = h_{oe} = 0$.

Fig. 17-14 (*a*) shows the circuit with the given resistance value, while Fig. 17-14 (*b*) shows its equivalent circuit.

Input resistance

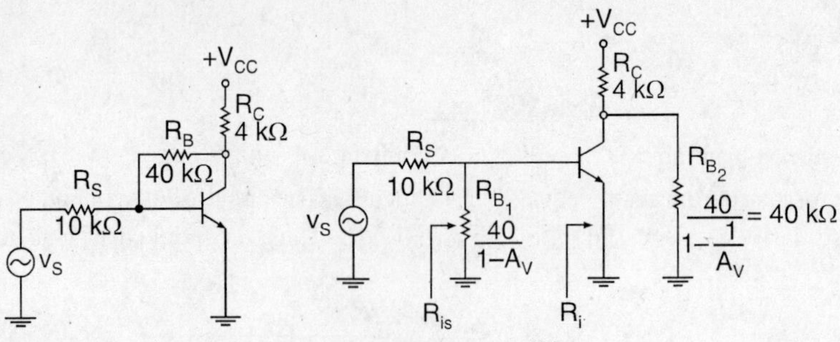

(*a*) (*b*)

Fig. 17-14.

We know that the resistance,

$$R_{B_2} = \frac{40}{1 - \frac{1}{A_v}} = 40 \text{ k}\Omega \qquad \text{....}(\because A_v >> 1)$$

∴ A.C. load resistance,

$$r_L = R_C \parallel R_{B_2} = 4 \parallel 40 \text{ k}\Omega = 3.64 \text{ k}\Omega = 3640\ \Omega$$

We also know that current gain,

$$A_i = \frac{h_{fe}}{1 + h_{oe} \cdot r_L} = -h_{fe} = -50 \qquad \text{...} (\because h_{oe} = 0)$$

and the input resistance of the amplifier looking into the base,

$$R_i = h_{ie} + h_{re} \cdot A_i \cdot r_L = h_{ie} = 1100\ \Omega \qquad \text{...} (\because h_{re} = 0)$$

We also know that the value of voltage gain,

$$A_v = \frac{A_i \cdot r_L}{R_i} = \frac{(-50) \times 3640}{1100} = -166$$

and the value of resistance,

$$R_{B_1} = \frac{40}{1 - A_v} = \frac{40}{1 - (-166)} \text{ k}\Omega$$

$$= 0.240 \text{ k}\Omega = 240\ \Omega$$

∴ Input resistance (*i.e.*, resistance looking from the source terminals),

$$R_{is} = R_i \parallel R_{B_1} = 1100 \parallel 240 = 197\ \Omega \ \textbf{Ans.}$$

Output resistance

We know that output resistance,

$$R_o = \frac{R_s + h_{ie}}{R_s \cdot h_{oe} + \Delta h} = \infty \qquad \text{... } (\because h_{oe} = \Delta h = 0)$$

$\therefore$ Output resistance of the stage,

$$R_{os} = R_0 \parallel r_L = r_L = 3640\ \Omega \textbf{ Ans.} \qquad \text{... } (\because R_0 = \infty)$$

Voltage gain of the stage

We know that voltage gain of the stage (*i.e.*, voltage gain of the amplifier looking from the source terminals called stage gain),

$$A_{vs} = \frac{R_v \cdot R_{is}}{R_s + R_{is}} = \frac{(-166) \times 197}{(10 \times 10^3) + 197} = -3.2 \textbf{ Ans.}$$

17-15. Hybrid Formulae for Common Base Amplifier

These formulae may be obtained from the general hybrid formulas derived in Art. 17-11 by adding a second subscript letter '*b*' (which stands for common base) with *h*-parameters and are as discussed below :

(*a*) ***Current gain :*** It is given by the relation,

$$A_i = -\frac{h_{fb}}{1 + h_{ob} \cdot r_L}$$

where, r_L is the a.c. load resistance. Its value is equal to the parallel combination of resistances R_C and R_L. Since, h_{fb} of a transistor is a negative number, therefore the value of current gain (A_i) of a common base amplifier is positive.

(*b*) ***Input resistance :*** The resistance looking into the input terminals (*i.e.*, emitter of a transistor) is given by the relation,

$$R_i = h_{ib} + h_{rb} \cdot A_i \cdot r_L$$

$$= h_{ib} - \frac{h_{rb} \cdot h_{fb}}{h_{ob} + \dfrac{1}{r_L}}$$

The input resistance of the amplifier stage (called stage input resistance) R_{is} depends upon the biasing arrangement. If the amplifier circuits has no biasing resistance, then $R_{is} = R_i$

(*c*) ***Voltage gain :*** It is given by the relation,

$$A_v = \frac{A_i \cdot r_L}{R_i}$$

Since, current gain (A_i) of a common base amplifier is positive, therefore the voltage gain (A_v) is also positive. It means that there is no phase difference between the input and output signals of a common base amplifier. The voltage gain, in terms of *h*-parameters, is given by the relation,

$$A_v = -\frac{h_{fb} \cdot r_L}{h_{ib} + \Delta h \cdot r_L}$$

where, $\Delta h = h_{ib} \cdot h_{ob} - h_{rb} \cdot h_{fb}$

(*d*) ***Output resistance :*** The resistance looking into the amplifier output terminals is given by the relation,

$$R_0 = \frac{R_s + h_{ib}}{R_s \cdot h_{ob} + \Delta h}$$

where, R_s = Resistance of the source, and

$$\Delta h = h_{ib} \cdot h_{ob} - h_{rb} \cdot h_{fb}$$

The output resistance of the amplifier stage,

$$R_{os} = R_o \parallel r_L$$

(*e*) ***Overall voltage gain :*** It is given by the relation,

$$A_{vs} = \frac{A_v \cdot R_s}{R_s + R_{is}}$$

(*f*) ***Overall current gain :*** It is given by the relation,

$$A_{is} = \frac{A_i \cdot R_s}{R_s + R_{is}}$$

Example 17-5. *A transistor used in a common base amplifier has the following values of h-parameters:*

$$h_{ib} = 28\ \Omega;\ h_{fb} = -0.98;\ h_{rb} = 5 \times 10^{-4};\ \text{and}\ h_{ob} = 0.34 \times 10^{-6}\ S$$

Calculate the values of input resistance, output resistance, current gain and voltage gain, if the load resistance is 1.2 kΩ. Assume source resistance as zero.

Solution. Given: $h_{ib} = 28\ \Omega$; $h_{fb} = -0.98$; $h_{rb} = 5 \times 10^{-4}$; $h_{ob} = 0.34 \times 10^{-6}$ S;
$r_L = 1.2\ \text{k}\Omega = 1200\ \Omega$ and $R_s = 0$.

Current gain

We know that current gain,

$$A_i = -\frac{h_{fb}}{1 + h_{ob} \cdot r_L}$$

$$= -\frac{(-0.98)}{1 + [(0.34 \times 10^{-6}) \times 1200]}$$

$$= 0.98\ \textbf{Ans.}$$

Input resistance

We know that input resistance,

$$R_i = h_{ib} + h_{rb} \cdot A_i \cdot r_L$$

$$= 28 + [(5 \times 10^{-4}) \times 0.98 \times 1200]\ \Omega$$

$$= 28 + 0.6 = 28.6\ \Omega\ \textbf{Ans.}$$

Output resistance

We know that

$$\Delta h = h_{ib} \cdot h_{ob} - h_{rb} \cdot h_{fb}$$

$$= [28 \times (0.34 \times 10^{-6})] - [(5 \times 10^{-4}) \times (-0.98)]$$

$$= 5 \times 10^{-4}$$

∴ Output resistance,

$$R_o = \frac{R_s + h_{ib}}{R_s \cdot h_{ib} + \Delta h} = \frac{h_{ib}}{\Delta h} \quad \ldots (\because R_s = 0)$$

$$= \frac{28}{5 \times 10^{-2}} = 5.6 \times 10^4\ \Omega = 56\ \text{k}\Omega\ \textbf{Ans.}$$

Voltage gain

We know that voltage gain,

$$A_v = \frac{A_i \cdot r_L}{R_i} = \frac{0.98 \times 1200}{28.6} = 41\ \textbf{Ans.}$$

17-16. Hybrid Formulae for Common Collector Amplifier

These formulas may be obtained from the general hybrid formulas derived in Art. 17-11 by adding a second subscript letter '*c*' (which stands for common collector) with *h*-parameters and are as discussed below :

(*a*) ***Current gain :*** It is given by the relation,

$$A_i = -\frac{h_{fc}}{1 + h_{oe} \cdot r_L}$$

where, r_L is the a.c. load resistance. Its value is equal to the parallel combination of resistancs R_E and R_L. Since, h_{fc} of a transistor is a negative number, therefore current gain A_i is positive.

(*b*) ***Input resistance :*** The resistance looking into the amplifier input terminals (*i.e.*, base of a transistor) is given by the relation,

$$R_i = h_{ic} + h_{rc} \cdot A_i \cdot r_L$$

$$= h_{ic} - \frac{h_{rc} \cdot h_{fc}}{h_{oe} + \frac{1}{r_L}}$$

The input resistance of the amplifier stage R_{is} (called stage input resistance) depends upon the biasing arrangement, if the amplifier circuit has no biasing resistance, then $R_{is} = R_s$.

(*c*) ***Voltage gain :*** It is given by the relation,

$$A_v = \frac{A_i \cdot r_L}{R_i}$$

Since, current gain (A_i) of a common collector amplifier is positive, therefore the voltage gain (A_v) is also positive. It means that there is no phase difference between the input and output signals of a common collector amplifier. The voltage gain, in terms of *h*-parameters, is given by the relation,

$$A_i = -\frac{h_{fc} \cdot r_L}{h_{ic} + \Delta h \cdot r_L}$$

where, $\Delta h = h_{ic} \cdot h_{oc} - h_{rc} \cdot h_{fc}$

(*d*) ***Output resistance :*** The resistance looking into the amplifier output terminals is given by the relation,

$$R_0 = \frac{R_s + h_{ic}}{R_s \cdot h_{oc} + \Delta h}$$

where, R_s = Resistance of the source, and

$$\Delta h = h_{ic} \cdot h_{oc} - h_{rc} \cdot h_{fc}$$

(*e*) ***Overall voltage gain :*** It is given by the relation,

$$A_{vs} = \frac{A_v \cdot R_{is}}{R_s + R_{is}}$$

(*f*) ***Overall current gain :*** It is given by the relation,

$$A_{is} = \frac{A_i + R_s}{R_i + R_{is}}$$

Example 17-6. *A transistor used in a common collector (or emitter follower) circuit as shown in Fig. 17-15 has the following set of h-parameters:*

$$h_{ic} = 2k\Omega;\ h_{fc} = -51; = 1 \text{ and } h_{oc} = 25 \times 10^{-6} \text{ ohms}$$

Find the values of input and output resistances, current and voltage gains of the amplifier stage. Use the approximate or exact hybrid formulae.

Solution. Given: $h_{ic} = 2\ k\Omega = 2000\ \Omega$; $h_{fc} = -51$; $h_{rc} = 1$; $h_{oc} = 25 \times 10^{-6}$ ohms; $r_L = R_E = 5\ k\Omega = 5 \times 10^3\ \Omega$; $R_s = 1\ k\Omega = 1000\ \Omega$ and $R_1 = R_2 = 10\ k\Omega$

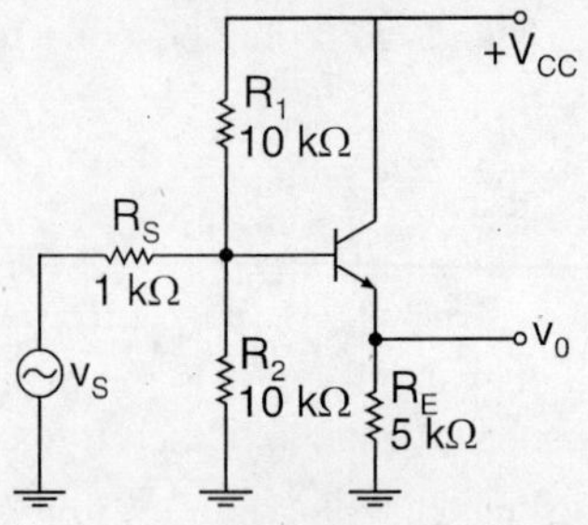

Fig. 17-15.

Fig. 17-16 shows the hybrid equivalent circuit of the given common collector amplifier.

Input resistance of the amplifier stage

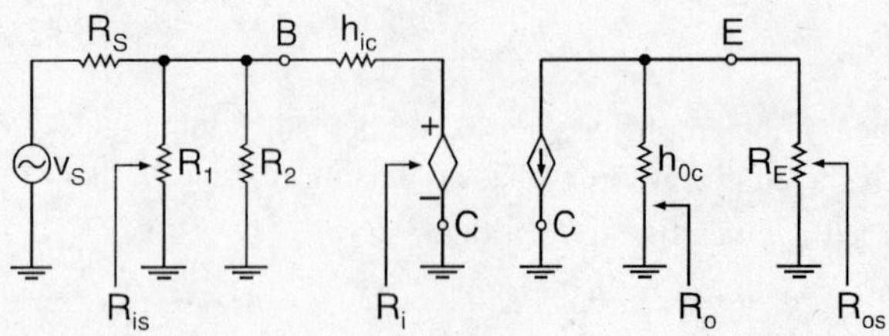

Fig. 17-16.

We know that current gain,

$$A_i = \frac{h_{fc}}{1 + h_{oe} \cdot r_L} = \frac{(-51)}{1 + [(25 \times 10^{-6}) \times (5 \times 10^3)]}$$

$$= 45.3$$

and input resistance,

$$R_i = h_{ic} + h_{rc} \cdot A_i \cdot r_L$$
$$= 2000 + (1 \times 45.3 \times 5 \times 10^3)\ \Omega$$
$$= 228 \times 10^3\ \Omega = 228\ k\Omega$$

$\therefore$ Input resistance of the amplifier stage,

$$R_{is} = R_i \parallel (R_1 \parallel R_2) = 228 \parallel (10 \parallel 10) = 4.9\ k\Omega$$
$$= 4900\ \Omega\ \textbf{Ans.}$$

Output resistance of the amplifier stage

We know that output resistance,

$$R_o = -\frac{R_s + h_{ic}}{h_{fc}} = -\frac{1000 + 2000}{(-51)} = 58.8\ \Omega$$

and output resistance of the amplifier stage,

$$R_{os} = R_o \parallel R_E = 58.8 \parallel (5 \times 10^3) = 58.1\Omega\ \textbf{Ans.}$$

Current gain of the amplifier stage

We know that current gain of the amplifier stage,

$$A_{is} = \frac{A_i \cdot R_s}{R_s + R_{is}} = \frac{45.3 \times 1000}{1000 + 4900} = 7.7\ \textbf{Ans.}$$

Voltage gain of the amplifier stage

We know that voltage gain,

$$A_v = \frac{A_i \cdot r_L}{R_i} = \frac{45.3 \times (5 \times 10^3)}{228 \times 10^3} = 1$$

and voltage gain of the amplifier stage,

$$A_{vs} = \frac{A_v \cdot R_{is}}{R_s + R_{is}} = \frac{1 \times 4900}{1000 + 4900} = 0.83 \textbf{ Ans.}$$

17-17. Approximate Hybrid Formulae for Transistor Amplifier

The hybrid formulae derived in the previous articles for common emitter, common base and common collector amplifiers are called *exact formulae*. These formulas are used to calculate the exact input resistance (R_i), output resistance (R_o), current gain (A_i) and voltage gain (A_v). As a matter of fact, the calculation of exact values is time consuming and liable to some calculation error. Therefore, we can use approximate values of R_i, R_o, A_i and A_v in actual practice. The approximate values of R_i, R_o, A_i and A_v may be obtained from their exact values by assuming that the a.c. load resistance (r_L) is small enough, so that the product $h_{oe} \cdot r_L$ is less than unity. With this assumption, the approximate values of R_i, R_o, A_i and A_v for a common emitter amplifier may be obtained from the following hybrid formulae :

$$R_i = h_{ic} \qquad R_o = 1/h_{oe}$$

$$A_i = h_{fe} \qquad A_v = h_{fe} \cdot r_L / h_{ie}$$

A similar hybrid formulae can also be obtained for common base and common collector (or emitter follower) amplifier. A summary of such hybrid formulae for all the three types of transistor amplifiers is given in Table 17-3.

Table 17-3. *Approximate hybrid formulas for common emitter, common base and common collector amplifiers*

S. No.	*Item*	*Symbol*	*Common emitter*	*Common base*	*Common collector*
1.	Input resistance	R_i	h_{ie}	h_{ib}	$h_{ic} + h_{fc} \cdot r_L$
2.	Output resistance	R_o	$\frac{1}{h_{oe}}$	$\frac{1}{h_{ob}}$	$\frac{R_s + h_{ic}}{h_{fc}}$
3.	Current gain	A_i	$-h_{fe}$	$h_{fb} = 1$	h_{fc}
4.	Voltage gain	A_v	$\frac{h_{fe} \cdot r_L}{h_{ie}}$	$\frac{h_{fb} \cdot r_L}{h_{ib}}$	1

Example 17-7. *In a single stage common emitter amplifier as shown in Fig. 17-17, the transistor h-parameters are:*

$h_{ie} = 1500\ \Omega$; $h_{fe} = 50$; $h_{re} = 50 \times 10^{-4}$ *and* $h_{oe} = 20 \times 10^{-6}$

Calculate the values of input and output resistances, current and voltage gains of the stage. Use approximate hybrid formulas for calculations.

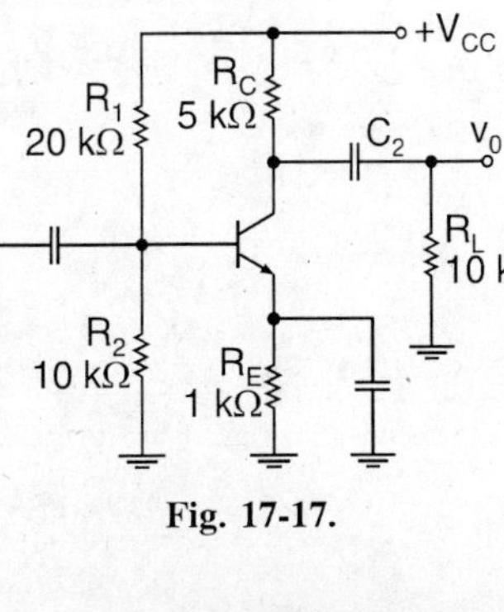

Fig. 17-17.

Solution. Given $h_{ie} = 1500\ \Omega$; $h_{fe} = 50$; $h_{re} = 50 \times 10^{-4}$ and $h_{oe} = 20 \times 10^{-6}$.

Input resistance of the voltage

We know that the value of a.c. load resistance,

$$r_L = R_C \parallel R_L = 5 \parallel 10 = 3.3\ \text{k}\Omega$$

$$= 3.3 \times 10^3\ \Omega$$

and current gain, $A_i = -h_{fe} = -50$

We also know that the input resistance,

$$R_i = h_{ie} = 1500\ \Omega = 1.5\ \text{k}\Omega$$

∴ Input resistance of the stage,

$$R_{is} = R_i \parallel (R_1 \parallel R_2) = 1.5 \parallel (20 \parallel 10)\ \text{k}\Omega\ 1.22\ \text{k}\Omega\ \textbf{Ans.}$$

Output resistance of the stage

We know that output resistance,

$$R_o = \frac{1}{h_{oe}} = \frac{1}{20 \times 10^{-6}} = 0.05 \times 10^6\ \Omega = 3.1\ \text{k}\Omega$$

and output resistance of the stage,

$$R_{os} = R_o \parallel R_L = 50 \parallel 3.3 = 3.1\ \text{k}\Omega\ \textbf{Ans.}$$

Voltage gain of the stage

We know that the voltage gain,

$$A_{vs} = \frac{A_i \cdot r_L}{R_i} = \frac{(-50) \times (3.3 \times 10^3)}{1500} = -110$$

and voltage gain of the stage,

$$A_{vs} = \frac{A_v \cdot R_{is}}{R_s + R_{is}} = A_s = -110\ \textbf{Ans.} \qquad \dots (\because R_s = 0)$$

Current gain of the stage

We know that current gain of the stage,

$$A_{is} = \frac{A_i \cdot R_s}{R_s + R_{is}} = A_v = -50\ \textbf{Ans.} \qquad \dots (\because R_s = 0)$$

Example 17-8. *A common-emitter amplifier with voltage divider bias has values of R_C = 12 kΩ, R_L = 4.7 kΩ, R_1 = 33 kΩ, R_2 = 4.7 kΩ and I_C = 1 mA. At 1 mA, the transistor has h-parameter values of [1] h_{ie} = 1 kΩ to 5 kΩ and h_{fe} = 70 to 350. Determine the value of **(a)** input impedance, and **(b)** the overall voltage gain for the circuit.*

Solution. Given: R_C = 12 kΩ, R_L = 4.7 kΩ, R_1 = 33 kΩ, R_2 = 4.7 kΩ, I_C = 1 mA; h_{ie} = 1 kΩ to 5 kΩ and h_{fe} = 70 to 350.

With these values, the circuit for the common-emitter amplifier is as shown in Fig. 17-18.

We know that, $^*h_{ie} = \sqrt{h_{ie(min)} \times h_{ie(max)}}$

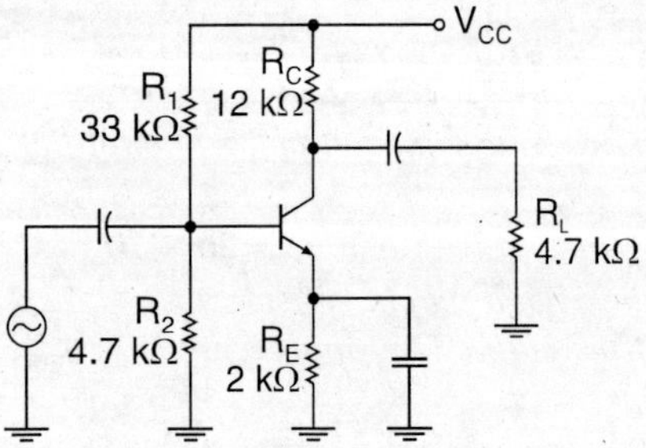

Fig. 17-18.

$$= \sqrt{1\ \text{k}\Omega \times 5\text{k}\Omega}$$

* In commercially available transistors, the value of h_{ie} and h_{fe} varies over a certain range, $h_{ie\,(min)}$ to $h_{ie\,(max)}$, and $h_{fe\,(man)}$ to $h_{fe\,(min)}$ respectively. In such a case, the value of $h_{ie} = \sqrt{h_{ie\,(min)} \times h_{ie\,(max)}}$ and $h_{fe} = \sqrt{h_{fe\,(min)} \times h_{fe\,(max)}}$

$$= 2.24\ k\Omega$$

and $$h_{fe} = \sqrt{h_{fe(min)} \times h_{fe(max)}} = \sqrt{70 \times 350}$$

$$= 104.67$$

(*a*) *Input resistance (or impedance) of the amplifier*

We know that input resistance of a transistor is,

$$R_i = h_{ie} = 2.24\ k\Omega$$

and input resistance of the amplifier,

$$R_{is} = R_i \parallel R_1 \parallel R_2 = 2.24k\ \Omega \parallel 33\ k\Omega \parallel 4.7k\Omega \approx 1.45\ k\Omega \textbf{ Ans.}$$

(*b*) *Overall voltage gain*

We know that current gain of a transistor,

$$A_i = h_{fe} = 104.67$$

We also know a.c. load resistance,

$$r_c = R_C \parallel R_L = 12\ k\Omega \parallel 4.7\ k\Omega = 3.38\ k\Omega,$$

and the overall voltage gain,

$$A_{vs} = A_v = \frac{A_i \cdot r_L}{R_i} = \frac{104.67 \times 3.38\ k\Omega}{2.24\ k\Omega} = 157.94 \textbf{ Ans.} \quad (\because R_S = 0)$$

Example 17-9. *Fig. 17-19 shows the circuit of a common-emitter amplifier with fixed bias. Determine the value of input resistance, output resistance, current gain and voltage gain of the amplifier circuit.*

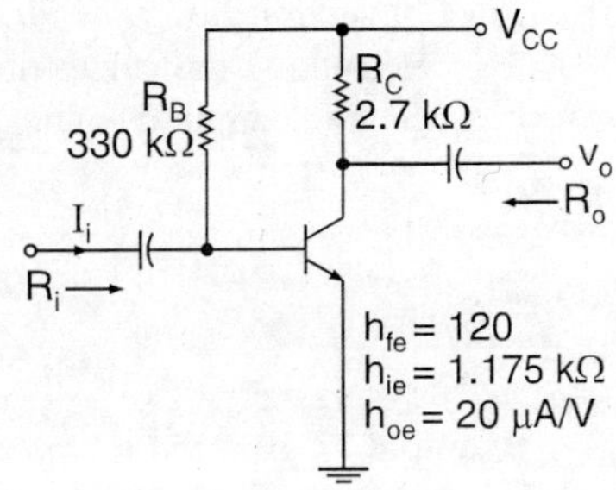

Fig. 17-19.

Solution. Given: $R_B = 330\ k\Omega$; $R_C = 2.7\ k\Omega$; $h_{fe} = 120$; $h_{ie} = 1.175\ k\Omega$ and $h_{oe} = 20\ \mu A/V$.

Input resistance of the amplifier

We know that input resistance of the transistor,

$$R_i = h_{ie} = 1.175\ k\Omega$$

and input resistance of the circuit,

$$R_{is} = h_{ie} \parallel R_B = 1.175\ k\Omega \parallel 330\ k\Omega = 1.17\ k\Omega \textbf{ Ans.}$$

Output resistance at the amplifier

We know that output resistance, of the transistor,

$$R_o = \frac{1}{h_{oe}} = \frac{1}{20\ \mu\ A/V} = \frac{1}{20 \times 10^{-6}} = 5 \times 10^4\ \Omega = 50\ k\Omega$$

and output resistance of the circuit,

$$R_{os} = R_o \parallel r_L$$

$$= R_o \parallel R_C \quad (\because r_L = R_C)$$

$$= 50\ k\Omega \parallel 2.7\ k\Omega.$$

$$= 2.56\ k\Omega \textbf{ Ans.}$$

Current gain of the amplifier

We know that current gain of the circuit,

$$A_i = h_{fe} = 120 \textbf{ Ans.}$$

Voltage gain of the amplifier

We know that voltage gain of the amplifier,

$$A_{vs} = \frac{A_i \cdot r_L}{R_i} = \frac{h_{fe} \times R_C}{R_i} = \frac{120 \times 2.7\ k\Omega}{1.175\ k\Omega} = 275.7 \textbf{ Ans.}$$

17-18. Approximate Conversion Formulae for *h*-parameters

These days, transistors are used in most of the circuits in common emitter configuration. Therefore, the manufacturers list only the common emitter *h*-parameters in their data sheets. However, if the transistor is to be used in a configuration, other than the common emitter, then we must use the specified conversion formulae for determining its *h*-parameters. The conversion formulas for common base and common collector configuration are as given below.

(1) ***Converstion formulae for common base configuration :*** The convertion formulas for determining the values of *h*-parameters of a transistor in common base configuration, from the common emitter *h*-parameter values, are as given below :

$$h_{ib} = \frac{h_{ie}}{1 + h_{fe}}$$

$$h_{rb} = \frac{h_{ie} \cdot h_{oe}}{1 + h_{fe}} = -h_{re}$$

$$h_{fb} = -\frac{h_{fe}}{1 + h_{fe}}$$

$$h_{ob} = \frac{h_{oe}}{1 + h_{fe}}$$

(2) ***Conversion formulae for common collector configuration :*** The conversion formulae for determining the values of *h*-parameters of a transistor in common collector configuration from the common emitter *h*-parameter value, are as given below:

$$h_{ic} = h_{ie}$$

$$h_{rc} = 1 - h_{re} \approx 1 \qquad \ldots (\because h_{rc} << 1)$$

$$h_{fc} = -(1 + h_{fe})$$

$$h_{oe} = h_{oe}$$

Example 17-10. *Find the values of h_{fb} and h_{fc}, if the value of h_{fe} of a transistor is 50.*

Solution. Given: $h_{fe} = 50$

We know that $$h_{fb} = -\frac{h_{fe}}{1 + h_{fe}} = -\frac{50}{1 + 50} = -0.98 \textbf{ Ans.}$$

and $$h_{fc} = -(1 + h_{fe}) = -(1 + 50) = -51 \textbf{ Ans.}$$

Example 17-11. *A transistor is connected in common collector configuration and its h-parameters are:*

$$h_{ie} = 1100\ \Omega;\ h_{re} = 2.5 \times 10^{-4}\ h_{fe} = 50;\ h_{oe} = 24\ \mu A/V$$

The circuit uses $R_L = 10$ kΩ and $R_S = 1$ kΩ. Calculate gain A_i, the input resistance R_i and the voltage gain, A_v of this amplifier. (*A.M.I.E., Winter,* 1988)

Solution. Given: $h_{ie} = 1100\ \Omega$; $h_{re} = 2.5 \times 10^{-4}$; $h_{fe} = 50$; $h_{oe} = 24\ \mu$ A/V $= 24 \times 10^{-6}$ A/V; r_L (R_L) $= 10$ kΩ $= 10 \times 10^3\ \Omega$ and $R_S = 1$ kΩ.

Current gain

First of all, let us find the values of *h*-parameters for a transistor in common collector configuration because the given *h*-parameter values are for a transistor in common emitter configuration. We know that

$$h_{ic} = h_{ie} = 1100\ \Omega$$

and $$h_{rc} = 1 - h_{re} = 1 - (2.5 \times 10^{-4}) \approx 1$$

Similarly we know that

$$h_{fc} = -(1 + h_{fe}) = -(1 + 50) = -51$$

and
$$h_{oe} = h_{oe} = 24 \times 10^{-6} \ A/V$$

∴ Current gain,

$$A_i = -\frac{h_{fc}}{1 + h_{oe} \cdot r_L} = \frac{(-51)}{1 + (24 \times 10^{-6})(10 \times 10^3)} = 41.1 \textbf{ Ans.}$$

Input resistance

We know that input resistance,

$$R_i = h_{ic} + h_{rc} \cdot A_i \times r_L = 1100 + [1 \times 41..1 \times (10 \times 10^3)]$$
$$= 1100 + (41.1 \times 10^4) = 412.1 \times 10^3 \ \Omega$$
$$= 412.1 \text{ k}\Omega \textbf{ Ans.}$$

Voltage gain

We know that voltage gain,

$$A_v = \frac{A_i \cdot r_L}{R_i} = \frac{41.1 \times (10 \times 10^3)}{412.1 \times 10^3} = 0.997 \textbf{ Ans.}$$

17-19. Determination of Transistor *h*-parameters

The transistor *h*-parameters may be determined by any one of the following two methods :

(*i*) Graphical determination of the *h*-parameters; and

(*ii*) Experimental determination of the *h*-parameters.

Now we shall discuss both these methods one by one in the following pages.

17-20. Graphical Determination of *h*-parameters

We have already discussed in Art. 17-2 that the input and output voltages and currents of any linear circuit are related by the following two equations :

$$v_1 = h_i \cdot i_1 + h_r \cdot v_2 \qquad ...(i)$$

and
$$i_2 = h_f \cdot i_1 + h_o \cdot v_2 \qquad ...(ii)$$

where, v_1 = Input voltage; v_2 = Output voltage

i_1 = Input current; i_2 = Output current and

h_f, h_i, h_r, h_o, = Hybrid or *h*-parameters

These equations may also be used to relate input and output voltages and currents for a transistor. For example, in a common emitter transistor, the input voltage (v_1) is the base-to-emitter voltage (v_{BE}) and the output voltage (v_2) is the collector-to-emitter voltage (v_{CE}). Similarly, the input current (i_1) is the base current (i_B) and the output current (i_2) is the collector current (i_C). Thus, for a common emitter transistor, the equations (*i*) and (*ii*) may be written as :

$$v_{BE} = h_{ie} \cdot i_B + h_{re} \cdot v_{CE}$$
$$i_C = h_{fe} \cdot i_B + h_{oe} \cdot v_{CE}$$

With the help of partial derivatives for the above two equations, it can be shown that the *h*-parameters for the common emitter transistor can be found by using the following equations:

$$h_{ie} = \frac{\delta v_{BE}}{\delta i_B} = \frac{\Delta v_{BE}}{\Delta i_B} \qquad \text{... (for a constant } v_{CE})$$

$$h_{re} = \frac{\delta v_{BE}}{\delta v_C} = \frac{\Delta v_{BE}}{\Delta v_{CE}} \qquad \text{... (for a constant } i_B)$$

$$h_{fe} = \frac{\delta i_C}{\delta i_B} = \frac{\Delta i_C}{\Delta i_B} \quad \text{... (for a constant } v_{CE})$$

$$h_{oe} = \frac{\delta i_C}{\delta v_{CE}} = \frac{\Delta i_C}{\Delta v_{CE}} \quad \text{... (for a constant } i_B)$$

It is evident from equation (*i*) that parameter h_{ie} is determined by the ratio of a small change in base-to-emitter voltage (Δv_{BE}) to the corresponding change in base current (Δi_B) for a constant value of collector-to-emitter voltage (v_{CE}). Similarly, parameter h_{re} is given by the ratio of a small change in base-to-emitter voltage (Δv_{BE}) to the corresponding change in collector-to-emitter voltage (Δv_{CE}) for a constant base current (i_B). On the other hand, the parameter h_{fe} is determined by the ratio of a small change in collector current (Δi_C) to the corresponding small change in base current (i_B) for a constant collector to emitter voltage (v_{CE}). Similarly, the parameter h_{oe} is given by the ratio of small change in collector current (Δi_C) to the corresponding small change in collector-to-emitter valtage (ΔV_{CE}) for a constant base current (i_B). It will be interesting to know that the values of h_{ie} and h_{re} parameters are determined graphically from the input or base characteristics, while the values of h_{fe} and h_{oe} parameters are determined from the output or collector characteristics.

17-21. Graphical Determination of h_{ie} and h_{re} Parameters

We have already discussed in the last article that the value of h_{ie} and h_{re} parameters are determined from the input or base characteristics. In order to determine the values of these parameters, the first step is to locate a point on the input or base characteristics as shown in Fig. 17-20. Now to find the value of h_{ie}, we draw a line which is tangent to the curve marked v_{CE_2} and passes through the *Q*-point.

A small change in base-to-emitter voltage (Δv_{BE}) is then chosen and the corresponding change in base current (i_B) is measured. Substituting these values in the relation $\Delta v_{BE}/\Delta i_B$, we get the value of h_{ie}.

The parameter h_{re} can be found by selecting another point *P* on the curve marked v_{CE_1} in such a way so that the point *P* and *Q* have the same value of base current. Then measure the values of changes in collector-to-emitter voltage Δv_{CE} (*i.e.*, $v_{CE_2} - v_{CE_1}$). and the resulting change in base-to-emitter valtage $\Delta\ V_{BE.}$ (equal to $v_{BE_2} - v_{BE_1}$). Substituting these values in the relation $\Delta v_{BE}/\Delta v_{CE}$, we get the value of h_{re} .

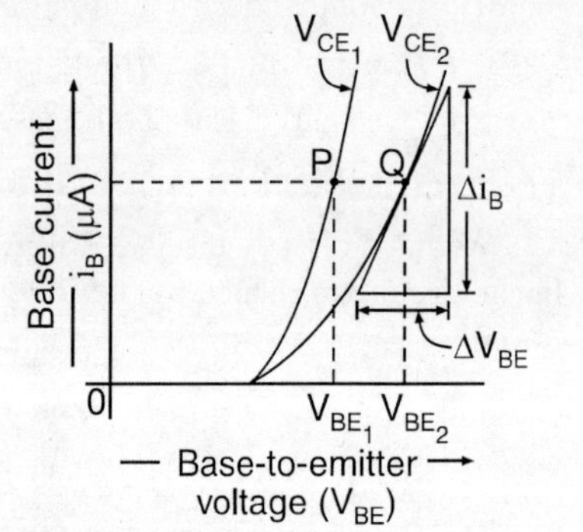

Fig. 17-20. *Determination of h_{ie} and h_{re}, parameters.*

17-22. Graphical Determination of h_{fe} and h_{oe} Parameters

We have already discussed in Art. 17-20 that the values of h_{fe} and h_{oe} parameters are determined from the output or collector characteristics as shown in Fig. 17-21. Now in order to find the value of h_{fe}, we select two points *R* and *S* (above and below the *Q*-point) in such a way that these points have the same collector-to-emitter voltage (*i.e.*, v_{CE_2}). Then measure the values of selected change in base current i_B (equal to $(i_{B_2} - i_{B_1})$ and the corresponding change in collector current Δi_C (equal to $i_{C_2} - i_{C_1}$). Substituting these values in relation $\Delta i_C / \Delta i_B$, we get the value of h_{fe}.

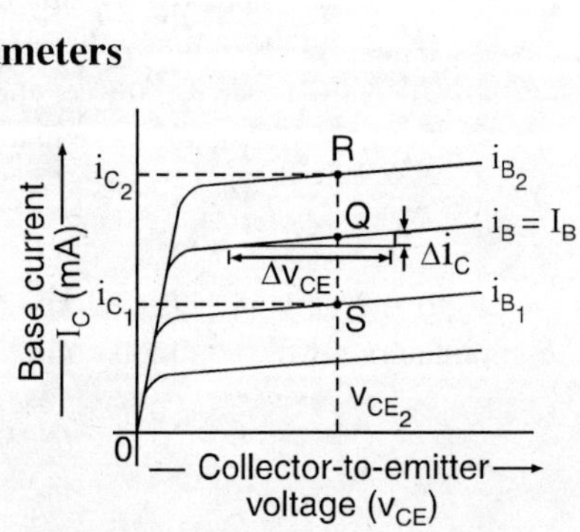

Fig. 17-21. *Determination of h_{fe} and h_{oe} parameter.*

The parameter h_{oe} can be found by drawing a line, which is tangent to the curve marked $i_B = i_B$ and passes through the Q-point as shown in the figure. A small change in collector-to-emitter voltage (Δv_{CE}) is then chosen and the corresponding change in collector current (Δi_C) is measured. Substituting these values in the relation Di_C / v$_{CE}$, we get the value of h_{oe}.

17-23. Experimental Determination of *h*-parameters

The *h*-parameters may be determined experimentally by constructing a circuit, which will satisfy the conditions specified in the definition of *h*-parameters. Although there are four parameters, which are to be determined, yet only two circuits are needed for the parameters. The circuit shown in Fig. 17-22 (*a*) may be used to determine h_{ie} and h_{fe}, while the circuit shown in Fig. 17-22 (*b*) may be used to determine h_{re} and h_{oe}.

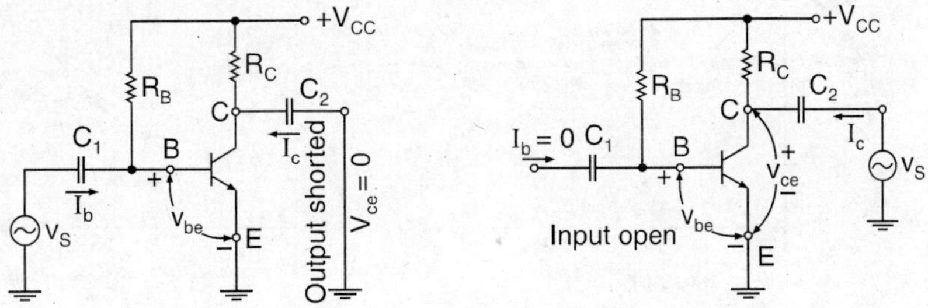

Fig. 17-22. *Experimental determination of h-parameters.*

The transistor in the circuits shown above is connected in a common emitter configuration. The capacitors C_1 and C_2 are the coupling capacitors, *i.e.*, they act as a short circuit for the a.c. signal. But they prevent any d.c. level associated with the signal from upsetting the biasing conditions set up by the V_{CC} supply and the resistors R_B and R_C. It may be noted that all voltages and currents shown in the figures are the r.m.s. values,

The values of h_{ie} and h_{fe} may be obtained by measuring the values of base-to-emitter voltage (V_{be}),base current (I_b) and collector current (I_c) from the Fig. 17-22 (*a*). Then,

$$h_{ie} = \frac{V_{be}}{I_b} \quad \text{and} \quad h_{fe} = \frac{I_c}{I_b}$$

Similarly, the values of h_{re} and h_{oe} may be obtained by measuring the values of collector-to-emitter voltage (V_{ce}), base-to-emitter voltage (V_{be}) and collector current (I_c) from Fig. 17-22 (*b*). Then,

$$h_{re} = \frac{V_{be}}{V_{ce}} \quad \text{and} \quad h_{oe} = \frac{I_c}{V_{ce}}$$

17-24. Variation of Transistor *h*-parameters

We have already discussed the procedures for determining transistor *h*-parameters from the characteristic curves. The parameters can be obtained at the operating point from the slopes and spacing between the curves at that point. Since, the characteristic curves are not in general straight lines and not equally spaced for equal changes in base current (I_B) or collector-to-emitter voltages (V_{CE}) therefore the values of *h*-parameters varies depending upon the position of the operating point of the curves. The variation of *h*-parameters for a typical transistor depends upon the following factors:

- Collector current (I_C).
- Junction temperature (T_j).
- Collector-emitter voltage (V_{CE}).

However, the variation of *h*-parameters due to the collector current is most important from subject point of view and is discussed as below :

(*a*) ***Variation of*** h_{ie} : Fig. 17-23 (*a*) shows the variation of h_{ie} with the operating value of collector current (I_C). it may be noted that the value of h_{ie} decreases as the operating value of collector current increases.

(*b*) ***Variation of*** h_{re} : Fig. 17-23 (*b*) shows the variation of h_{re} with the collector current. It may be noted that first h_{re} decreases with the increase in collector current. It reaches to a minimum value at a certain collector current. Beyond that, the parameter h_{re} increases with the increase in collector current.

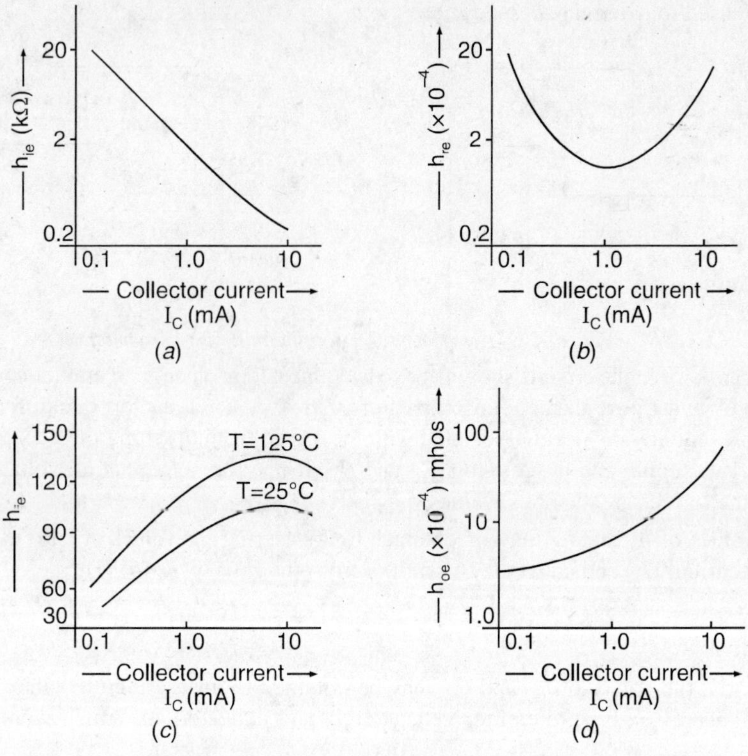

Fig. 17-23. *Variation of transistor h-parameters.*

(*c*) ***Variation of*** h_{fe} : Fig. 17-23 (*c*) shows the variation of h_{fe} (a.c. beta) with the operating value of collector current. It may be noted that h_{fe} increases with the increase in collector current and reaches a maximum at certain collector current. Beyond that value, h_{fe} decreases with the increase in collector current. The parameter h_{fe} is very sensitive to temperature. Its value increases with the increase in temperature as shown in figure by the curves market T = 25°C and T = 125°C.

(*d*) ***Variation of*** h_{oe} : Fig. 17-23 (*d*) shows the variation of h_{oe} with the operating value of collector current. It may be noted that h_{oe} increases with the increase of collector current.

It is evident from the above discussion that *h*-parameters vary over a wide range with the change in collector current. Therefore, the data sheet of a transistor, usually, specifies the values of *h*-parameters at a certain value of collector current. For example, for a transistor 2 N 3905, the data sheet specifies the following values of *h*-parameters at a collector current of I_C = 1 mA :

$$h_{ie} = 3.5\ \text{k}\Omega;\ h_{re} = 1.3 \times 10^{-4};\ h_{fe} = 120 \text{ and } h_{oe} = 8.5\ \mu \text{ ohms.}$$

SUMMARY

In this chapter you have learned that :

1. *h*-parameters are used to determine amplifier characteristic parameters are used determine amplifier characteristic parameters such as voltage gain, input and output resistances, *etc*.
2. There are four general parameter of a transistor.

h_i = imput resistance with output shorted

h_r = reverse voltage gain with input open

h_f = forward current gain with output shorted and

h_o = output conductance with input open

3. The current gain of a CE transistor amplifier is given by,

$$A_i = -\frac{h_{fe}}{1 + h_{oe} \cdot r_L}$$

4. The input resistance of a CE transistor amplifier is given by,

$$R_i = h_{ie} - \frac{h_{re} \cdot h_{fe}}{h_{oe} + \dfrac{1}{r_L}}$$

where, r_L = a.c. load resistance.

5. The voltage gain of a CE transistor amplifier is given by

$$A_v = \frac{A_i \cdot r_L}{R_i}$$

6. The overall voltage gain of a CE transistor amplifier is given by,

$$A_{v_s} = \frac{A_v \cdot R_{is}}{R_s + R_{is}}$$

where, $R_{is} = R_i \parallel R_B$

7. The *h*-parameters of a transistor depends upon (*i*) the collecter current, (*ii*) junction temperature (T_j) and (*iii*) collector-emitter voltage (V_{CE})

GLOSSARY

Current Gain : Ratio of output current (i_L) to input current (i_1).

***h*-parameter :** Acronym for hybrid parameters.

Power gain : Product of voltage gain (A_v) and current gain (A_i).

Voltage gain : Ratio of output voltage (V_o) to the input voltage (V_i)

NUMERICAL PROBLEMS

1. The hybrid parameters for a transistor used in common emitter configuration are:

h_{ie} = 1000 Ω; h_{fe} = 150

$h_{re} = 1.2 \times 10^{-4}$, $h_{oe} = 25 \times 10^{-6}$ mhos

The transistor has a load resistance of 10 kΩ in the collector and is supplied from a signal source of resistance 5 kΩ. Compute the values of input resistance, output resistance, current gain and voltage gain of the stage. (**Ans.** 8.56 Ω; 8.2 kΩ; 102.5 and 205)

2. A single common emitter amplifier circuit is shown in Fig. 17-24
Calculate the values of R_i, R_o, A_i, A_v, A_{vs}, A_{is}. The transistor has a set of following parameters :

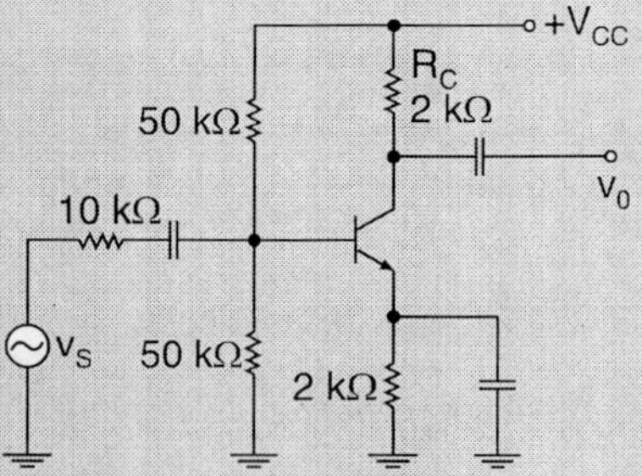

Fig. 17-24.

$h_{fe} = 50;$ $h_{ie} = 1.1$ kΩ

$h_{re} = 0;$ $h_{oe} = 0$ (**Ans.** 1100 Ω; ∞; 50; 90.9; 9 and 45)

3. Find the values of voltage gain, current gain, input resistance and power gain for a common emitter transistor amplifier, with $r_L = 1600$ Ω and $R_S = 1000$ Ω. The transistor has a following set of *h*-parameters: $h_{ie} = 1100$ Ω; $h_{re} = 2.5 \times 10^{-4}$; $h_{fe} = 25$ μ A/V.

(*U.P.S.C. Engg. Service,* 1987) (**Ans.** 69.8, 48, 1084 Ω; 50 kΩ and 3350)

4. Fig. 17-25 shows the circuit of a common emitter amplifier. Determine the value of input resistance and overall voltage gain for the circuit. The values of *h*-parameter for the transistor 2N 4400 are as below: $h_{ie} = 0.5$ kΩ to 7.5 kΩ and $h_{fe} = 20$ to 250.

(**Ans.** 1.35 kΩ and 244)

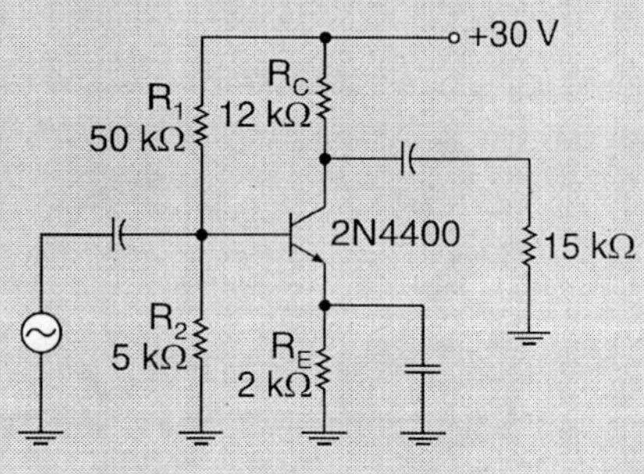

Fig. 17-25.

5. A transistor used in a common base circuit has a following set of parameters: $h_{ib} = 20$ Ω; $h_{fb} = 0.98$; $h_{rb} = 3 \times 10^{-4}$ and $h_{ob} = 0.5 \times 10^{-6}$. Find the values of R_i, k_O, A_i, A_v and A_{vs}; if R_s = 600 W, and r_L = 1.5k. (**Ans.** 20.4 Ω; 1 MΩ; 0.98; 72 and 2.3)

6. Design an emitter follower circuit having $R_i = 500$ kΩ, $R_o = 20$Ω , which uses a transistor with $h_{fe} = 50$; $h_{ie} = 1$ kΩ and $h_{oe} = 25$μ A/V. Find the values of *Ai* and A_v for this circuit.

(*A.M.I.E., Summer, 1984*) (**Ans.** $R_E = 10$ kΩ; $A_i = 51$ and $A_v = 0.998$)

7. A transistor used in common emitter configuration as shown in Fig. 17-26 has the following set of *h*-parameters $h_{ie} = 1$ kΩ; $h_{fe} = 50$; $h_{re} = 5 \times 10^{-4}$ and $h_{oe} = 2 \times 10^{-5}$ mhos; Determine R_i, R_{is}, R_o. A_i and A_v. Use the approximate hybrid formulae.

(**Ans.** 1 kΩ; 996 Ω; 50 kΩ; 50 and 250)

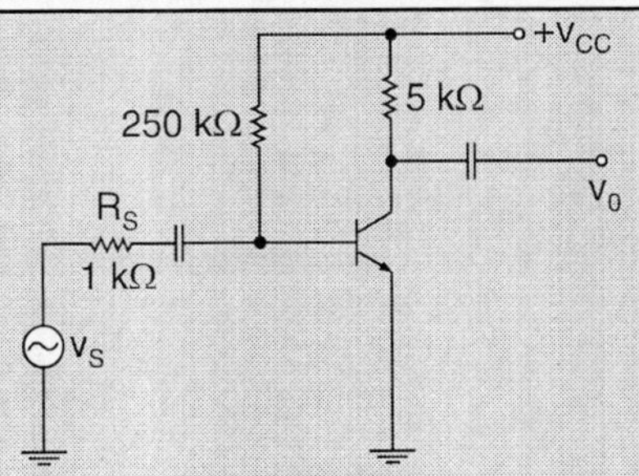

Fig. 17-26.

8. For $I_E = 1$ mA, give the order of magnitudes (including the sign) of (*a*) h_{ib} (*b*) h_{ie} (*c*) h_{re} (*d*) h_{rc} (*e*) h_{fb} (*f*) h_{fe} (*g*) h_{fc} (*h*) h_{oe} (*i*) h_{ob} and (*j*) h_{oc}.

(**Ans.** (*a*) 29.4 (*b*) 4 kΩ (*c*) 2×10^{-4} Ω (*d*) 1 (*e*) – 0.99 (*f*) 120 (*g*) – 120 (*h*) 50×10^{-6}Ω (*i*) 0.5×10^{-6} Ω and (*j*) 50×10^{-6}Ω)

DESCRIPTIVE QUESTIONS

1. Draw the equivalent circuit of a common emitter amplifier in terms of *h*-parameter and obtain the expression for voltage gain.

2. Draw an approximate h-parameter model of common emitter transistor configuration driven by a voltage source with zero source resistance (R_S).

3. Develop a low-frequency equivalent circuit for a basic common collector amplifier and drive the relations for the circuit gain, voltage gain, input resistance in terms of *h*-parameters. Make suitable assumptions and simplify the final results. Also justify the name 'emitter follower' for this type of amplifier.

4. Draw the hybrid model for a common emitter transistor and write the input and output equations. Find the common collector (CC) parameters in terms of common emitter *h*-parameters.

5. Explain, how will you determine *h*-parameters of a transistor experimentally?

6. Draw the hybrid model of CE amplifier and obtain its gain, input impendance and output impedance. Compare the performance of this CE amplifier with CC and CB configuration. (*Electronic Circuits - I, Anna University, April/May 2008*)

7. Define the four *h*-parameters. (*Electronic Circuits - I, Anna University, April/May 2008*)

8. Determine voltage gain and current gain expression for CB configuration using transistor hybrid model. (Figures are necessary)

(*Electronic Devices, Anna University, Nov/Dec 2007*)

9. With the neat sketch explain hybrid - pi (π) common Emitter transistor model.

(*Electronic Circuits - I, Anna University, Nov/Dec 2007*)

10. Derive the expression for transistor conductance (gm) for hybrid π common emitter transistor model. (*Electronic Circuits - I, Anna University, Nov/Dec 2007*)

11. Sketch the small signal high frequency ciruit of a common source amplifier and derive the expression for a voltage gain.

(*Electronic Circuits - I, Anna University, Nov/Dec 2007*)

12. What specific capacitance has the greatest effect on the high frequency response of a cascade of FET amplifier? Explain.

(*Electronic Circuits - I, Anna University, Nov/Dec 2007*)

MULTIPLE CHOICE QUESTIONS

1. The h-parameters are called hybrid parameters because they
(*a*) are different from y- and z-parameters.
(*b*) are mixed with other parameters
(*c*) apply to circuits contained in a box
(*d*) are defined by using both open-circuit and short-circuit terminations

2. Which of the following statement is not correct regarding the h-parameters of a transistor
(*a*) the values of h-parameters can be obtained from transistor characteristics.
(*b*) their values depends upon the transistor configuration
(*c*) their values depend on operating point
(*d*) they are four in number

3. Which of the following four h-parameters of a transistor has a greatest value?
(*a*) h_i (*b*) h_r
(*c*) h_o (*d*) h_f

4. Which of the following four h-parameters of a transistor has a smallest value?
(*a*) h_i (*b*) h_r
(*c*) h_o (*d*) h_f

5. The typical value h_{ic} of is
(*a*) 1 kΩ (*b*) 40 kΩ
(*c*) 100 kΩ (*d*) 2 mΩ

6. The smallest of the four h-parameters of a transistor is
(*a*) h_i (*b*) h_r
(*c*) h_o (*d*) h_f

7. The input impedance h_{11} of a network with output shorted is given by the ratio
(*a*) v_1/i_1 (*b*) v_1/v_2
(*c*) i_2/i_1 (*d*) i_2/v_2

8. The h-parameters of a transistor depend on its
(*a*) configuration (*b*) operating point
(*c*) temperature (*d*) all of the above

9. The output admittance h_0 of an ideal transistor connected is common-base configuration is .... siemens
(*a*) 0 (*b*) $1/r$
(*c*) $1/\beta\, r_e$ (*d*) – 1

10. A transistor has $h_{fe} = 100$, $h_{ie} = 5.2$ kΩ, and $r_{bb} = 0$.
At room temperature, $V_T = 26$ mV. The collector current, I_C will be
(*a*) 10 mA (*b*) 5 mA
(*c*) 1 mA (*d*) 0.5 mA

ANSWERS

1. (*d*)	**2.** (*c*)	**3.** (*a*)	**4.** (*c*)	**5.** (*a*)
6. (*c*)	**7.** (*a*)	**8.** (*a*)	**9.** (*a*)	**10.** (*a*).

❑❑❑

18

CHAPTER

Multistage BJT Amplifiers

OUTLINE

Objectives

After completing this chapter, you should be able to :

- Know multistage amplifiers.
- Calculate the voltage gain of a multistage amplifiers.
- Understand different coupling schemes, namely; resistance-capacitance coupling, impedance coupling, transformer coupling and direct-coupling.
- Know the frequency response of RC coupled amplifier.
- Understand the frequency response of a transformer coupled amplifier.
- Know the darlington amplifier.

18-1. Introduction

It has been observed that the voltage (or power) gain, obtained from a single stage small signal amplifier, is limited. Moreover, it is not sufficient for all practical applications. Therefore, in order to achieve greater voltage and power gain, we have to use more than one stage of amplification. Such an amplifier is called a *multistage amplifier*. It will be interesting to know that when we use multistage amplifier, then we feed the output of one stage to the input of the next as shown in Fig. 18-1. Such a connection of amplifiers is called cascading. The amplifier used in radio and television receivers is, usually, a multistage amplifier.

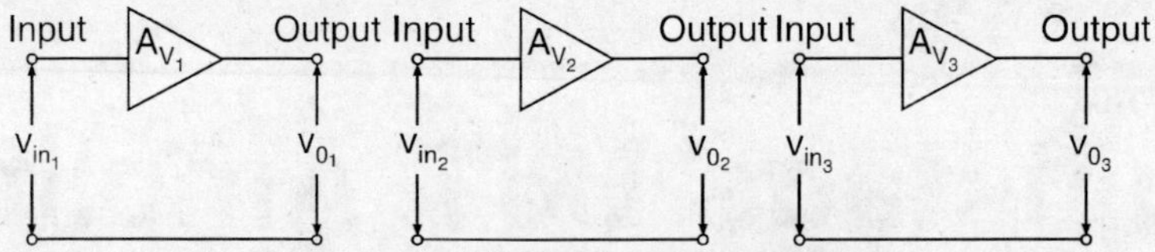

Fig. 18-1. *A general multistage amplifier.*

A multistage amplifier, using two or more single stage common emitter amplifiers, is called a *cascade amplifier*. The term cascade denotes the type of connection used for coupling two amplifier stages. On the other hand, a multistage amplifier with common emitter amplifier as the first stage and common base (or common collector) amplifier as the second stage, is called a *cascode amplifier*. It will be interesting to know that cascade and cascode connections are not only used in transistor amplifiers but also in field-effect transistors.

18-2. Gain of a Multistage Amplifier

We have already discussed in the last article that in a multistage amplifier the output voltage of the first stage acts as an input voltage to the second stage. And the output voltage of second stage as an input to the third stage and so on. *The voltage gain of a multistage amplifier is equal to the product of the gains of the individual stages*. Thus, if A_{v_1}, A_{v_2} and A_{v_3} are the individual stage gains, then overall voltage gain,

$$A_v = A_{v_1} \times A_{v_2} \times A_{v_3}$$

The amplifier voltage gain may also be expressed in decibels (dB). However, in that case the overall decibel voltage gain is the sum of the decibel gains of the individual stages. Thus, if G_{v1}, G_{v2} and G_{v3} are the individual stage decibel gains, then overall decibel voltage gain,

$$G_v = G_{v_1} + G_{v_2} + G_{v_3}$$

$$= 20 \log_{10} A_{v_1} + 20 \log_{10} A_{v_2} + 20 \log_{10} A_{v_3}$$

$$= 20 \log_{10} A_v \text{ (dB)}$$

Similarly, the overall current amplification,

$$A_i = A_{i_1} \times A_{i_2} \times A_{i_3}$$

Then the overall power gain,

$$A_p = A_v \cdot A_i$$

and the overall decibel power gain,

$$G_p = 10 \log_{10} A_p$$

Example 18-1. *A given amplifier arrangement has the following voltage gains. A_{v1} = 10, A_{v2} = 20 and A_{v3} = 40. What is the overall voltage gain? Also express each gain in dB and determine the total dB voltage gain.*

Solution. Given: $A_{v_1} = 10$; $A_{v_2} = 20$ and $A_{v_3} = 40$

Overall voltage gain

We know that the overall voltage gain,

$$A_v = A_{v_1} \times A_{v_2} \times A_{v_3} = 10 \times 20 \times 40 = 8000$$ **Ans.**

Total dB voltage gain

We know that dB voltage gain of the first stage

$$G_{v_1} = 20 \log_{10} A_{v_1} = 20 \log_{10} 10 = 20 \text{ dB}$$

Similarly, $$G_{v_2} = 20 \log_{10} A_{v_2} = 20 \log_{10} 20 = 26 \text{ dB}$$

and $$G_{v_3} = 20 \log_{10} A_{v_3} = 20 \log_{10} 40 = 32 \text{ dB}$$

∴ Total dB voltage gain,

$$G_v = G_{v_1} \times G_{v_2} \times G_{v_3} = 20 \times 26 \times 32 \text{ dB} = 78 \text{ dB}$$ **Ans.**

Note. Check $$G_v = 20 \log_{10} A_v = 20 \log_{10} 8000 = 78 \text{ dB}.$$

Example 18-2. *Three amplifier stages are working in cascade with 0.05 V peak-to-peak input providing 150 V peak-to-peak output. If the voltage gain of the first stage is 20 and input to the third stage is 15 V peak-to-peak, determine,*

(*i*) *the overall voltage gain*

(*ii*) *voltage gain of 2nd and 3rd stages*

(*iii*) *input voltage of the 2nd stage.* (*Banglore Univ. Jan.* 2004)

Solution. Given : No. of amplifier stages = 3. Input to the first stage, $V_{in_1} = 0.05\ v_{pk-pk}$, Output of the final stage $(v_{out})_3 = 150\ V_{pk-pk}$. Voltage gain of the first stage $A_{v_1} = 20$; Input to the third stage $V_{in_3} = 15\ V_{pk-pk}$. Refer to Fig. 18-2 for clear explanation.

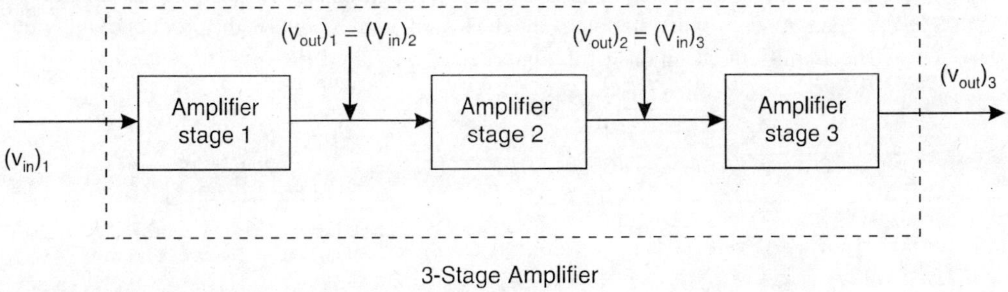

Fig. 18-2.

(*i*) *Overall Voltage gain*

We know that overall voltage gain,

$$A_v = \frac{(v_{out})_3}{(v_{in})_1} = \frac{150}{0.05} = 3000 \textbf{ Ans.}$$

(*ii*) *Voltage gain of 2nd and 3rd stage*

We know that voltage gain of the 3rd stage,

$$A_{v_3} = \frac{(v_{out})_3}{(v_{in})_3} = \frac{150}{15} = 10 \textbf{ Ans.}$$

Let A_{v_2} = Gain of the 2nd stage,

We know that the overall voltage gain (A_v),

$$3000 = A_{v_1} \cdot A_{v_2} \cdot A_{v_3} = 20 \times A_{v_2} \times 10 = 200\ A_{v_2}$$

$$\therefore \quad A_{v_2} = \frac{3000}{200} = 15 \textbf{ Ans.}$$

(*iii*) *Input voltage of the second stage*

We know that voltage gain of the 2nd stage (A_{v_2}),

$$15 = \frac{(v_{out})_2}{(v_{in})_2} = \frac{(v_{in})_3}{(v_{in})_2} = \frac{15}{(v_{in})_2}$$

$$\therefore \qquad (v_{in})_2 = \frac{15}{15} = 1\ V_{kp-pk} \ \textbf{Ans.}$$

18-3. Different Coupling Schemes Used in Amplifiers

As a matter of fact, all amplifiers, need some kind of *coupling network*. Even a single stage amplifier, needs coupling to the input source and output load. The multistage amplifiers need coupling between their individual stages. This type of coupling is called interstage coupling. It serves the following two purposes :

(*i*) It transfers a.c. output of one stage to the input of the next stage.

(*ii*) It isolates the d.c. conditions of one stage to the next. It is necessary to prevent the shifting of *Q*-points.

The coupling network (or coupling device) must ensure that both the above purposes are fulfilled, when an a.c. signal is to be amplified. Following are the four coupling schemes used in amplifiers.

(1) ***Resistance-capacitance (RC) coupling :*** It is the most important method of coupling the signal from one stage to the next and is shown in Fig. 18-3 (*a*). In this method, the signal developed across the collector resistor of each stage is coupled through capacitor into the base of the next stage. The cascaded stages amplify the signal and the overall gain is equal to the product of individual stage gains. The amplifiers, using this coupling scheme, are called *RC-coupled amplifiers*.

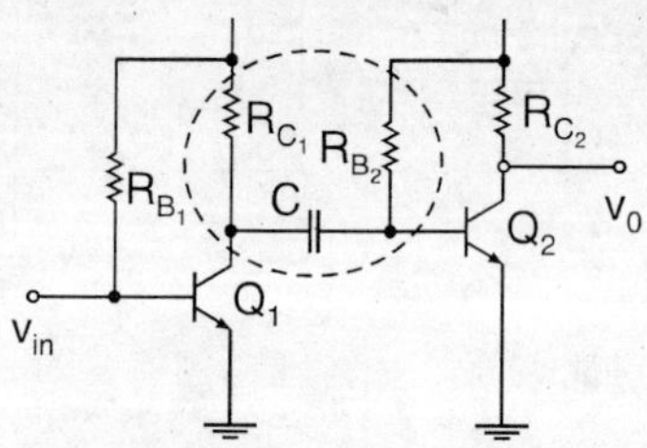

(*a*) Resistance-capacitance coupling.

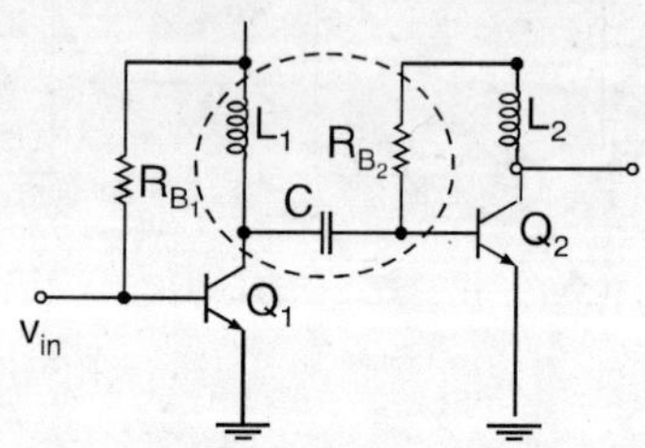

(*b*) Impedance coupling.

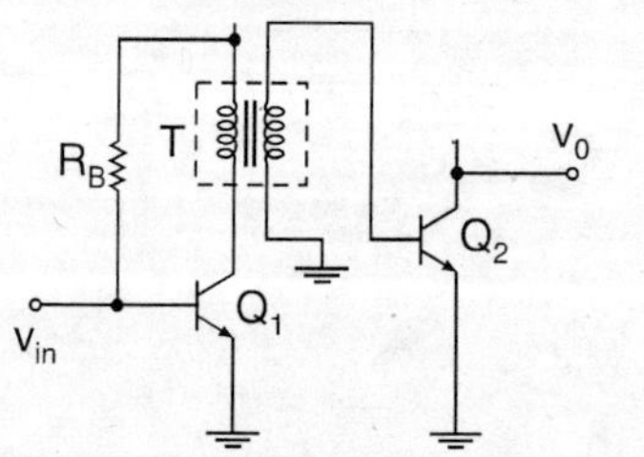

(*c*) Transformer coupling.

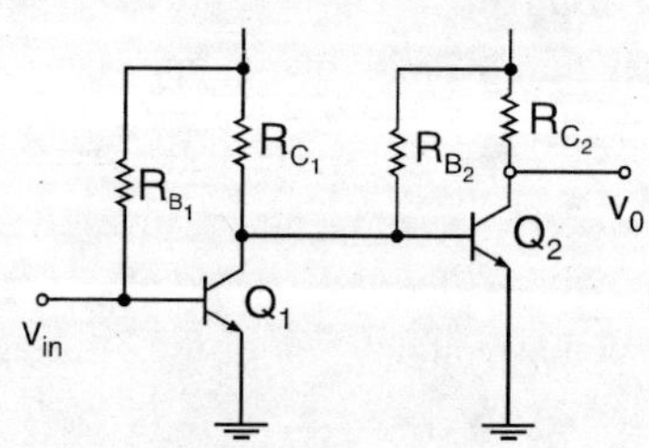

(*d*) Direct coupling.

Fig. 18-3. *Coupling schemes in amplifiers.*

(2) ***Impedance coupling :*** This coupling is shown in Fig. 18-3 (*b*). It may be noted that the collector resistor (R_C) is replaced by an inductor (L). As the frequency increases, inductive reactance X_L (equal to $\omega \cdot L$) approaches infinity and each inductor appears open. In other words, the inductors pass direct current but block alternating current. The amplifiers, using this coupling scheme, are called *impedance-coupling amplifiers*.

(3) *Transformer coupling :* This coupling is shown in Fig. 18-3 (*c*). In this method, primary winding of the transistor acts as a collector load and the secondary winding conveys the a.c. output signal directly to the base of the next stage. It may be noted that there is no need of coupling capacitor in the transformer coupling. The amplifiers, using this coupling scheme, are called *transformer-coupled amplifiers.*

(4) *Direct-coupling :* This coupling is shown in Fig. 18-3 (*d*). In this method, the a.c. output signal is fed directly to the next stage. This type of coupling is used where low frequency signals are to be amplified. The coupling devices such as capacitors, inductors and transformers cannot be used at low frequencies because their size becomes very large. The amplifiers, using this coupling scheme, are called *direct coupled amplifiers or d.c. amplifiers.*

21.4. RC Coupled Amplifier

Fig. 18-4 shows a two-stage RC coupled transistor amplifier. The circuit consists of two single-stage common emitter transistor amplifiers. The resistors R_C, R_B and capacitor C_C form the coupling network. The capacitor C_1 is used to couple the input signal to the base of Q_1, while the capacitor C_2 is used to couple the output signal from the collector of Q_2 to the load. The capacitor C_E, connected at the emitters of Q_1 and Q_2, are needed because they *bypass* the emitter to the ground. Without these capacitors, the voltage gain of each stage will be lost.

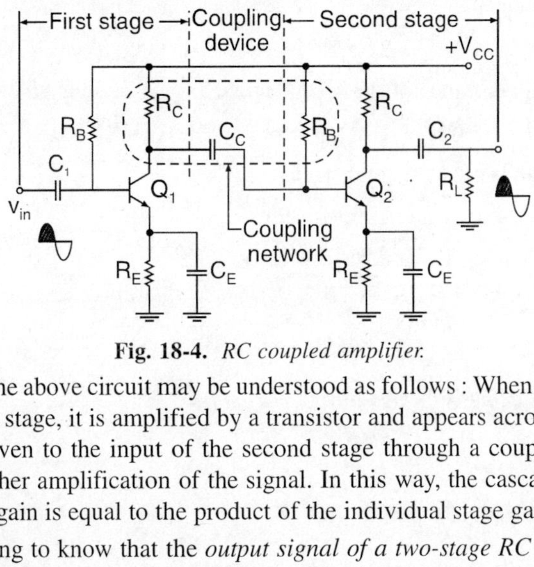

Fig. 18-4. *RC coupled amplifier.*

The operation of the above circuit may be understood as follows : When an a.c. signal is applied to the input of the first stage, it is amplified by a transistor and appears across the collector resistor (R_C). This signal is given to the input of the second stage through a coupling capacitor C_C. The second stage does further amplification of the signal. In this way, the cascaded stages amplify the signal and the overall gain is equal to the product of the individual stage gains.

It will be interesting to know that the *output signal of a two-stage RC coupled amplifier is in phase with the input signal.* It is because of the fact that its phase has been reversed twice by the amplifier.

18-5. Calculation of Voltage Gain for RC Coupled Amplifier

Consider a two stage RC coupled transistor amplifier circuit as shown in Fig. 18-4. The a.c. equivalent circuit for each amplifying stage of that circuit is shown in Fig. 18-5 (*a*) and (*b*) respectively. Here we have not shown the complete small signal a.c. equivalent circuit of the transistor. Rather, it is represented by an a.c. emitter diode resistance ($\beta \cdot r'_e$) and the three points marked *E* (for emitter), *B* (for base) and *C* (for collector). It has been done because a.c. equivalent circuit of a transistor is slightly different at low and high-frequencies.

The parameters R_{i_1} and R_{o_1} in the equivalent circuits shown below, represent the input resistance and output resistance of the first stage respectively. Similarly, the parameters R_{i_2} and R_{o_2}, represent the input resistance and output resistance of the second stage.

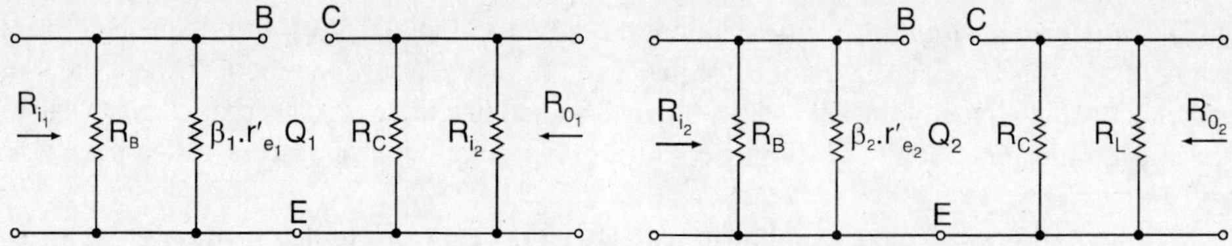

(*a*) A.C. equivalent circuit for first stage. (*b*) A.C. equivalent circuit for second stage.

Fig. 18-5. *A.C. equivalent circuit of a two-stage RC coupled amplifier.*

The parameters r'_{e_1} and r'_{e_2} represent the a.c. emitter diode resistances of the transistors Q_1 and Q_2 respectively. The values of r'_{e_1} and r'_{e_2} may be obtained from the relations,

$$r'_{e_1} = \frac{25}{I_{E_1}} \text{ and } r'_{e_2} = \frac{25}{I_{E_2}}$$

where, I_{E_1} and I_{E_1} are the values of emitter current in milliamperes. Now the input resistance of the first stage as seen from the Fig. 18-5 (*a*),

$$R_{i_2} = R_B \parallel (\beta_1 \cdot r'_{e_1}) = \beta_1 \cdot r'_{e_1} \qquad \text{...(If } R_B >> \beta_1 \cdot r'_{e_1})$$

and the output resistance,

$$R_{o_1} = R_C \parallel R_{i2}$$

where, R_{i2} is the input resistance of the second stage and its value is given by the relation,

$$R_{i_1} = R_B \parallel (\beta_2 \cdot r'_{e_2}) = \beta_2 \cdot r'_{e_2} \qquad \text{...(If } R_B >> \beta_2 \cdot r'_{e_2})$$

The output resistance of the second stage,

$$R_{o_2} = R_C \parallel R_L$$

Now the voltage gain of the first stage,

$$A_{v_1} = \beta_1 \times \frac{R_{o_1}}{R_{i_1}} = \beta_1 \times \frac{R'_{o_1}}{\beta_1 \cdot r'_{e_1}} = \frac{R_{o_1}}{r'_{e_1}}$$

Similarly, voltage gain of the second stage,

$$A_{v_2} = \frac{R_{o_2}}{r'_{e_2}}$$

∴ Overall voltage gain of the amplifier,

$$A_v = A_{v_1} \cdot A_{v_2} = \frac{R_{o_1}}{r'_{e_1}} \times \frac{R_{o_2}}{r'_{e_2}}$$

If the transistors used in both the stages are identical, then current gain of the transistors Q_1 and Q_2 (*i.e.*, β_1 and β_2) will be equal. In that case, the emitter currents will also be equal. Thus, $r'_{e_1} = r'_{e_2} = r'_{e_1}$. Then overall voltage gain,

$$A_v = R_{o_1} \cdot R_{o_2} / r'^2_e$$

Example 18-3. *Fig. 18-5 shows the circuit of a two stage RC coupled amplifier.*

Determine the values of input resistance, output resistance, current gain, voltage gain for the first and second stages. Also determine the overall voltage gain and overall dB voltage gain.

Solution. Given: V_{CC} = 10 volts; R_C = 5 kΩ = 5000 Ω ; R_B = 1 MΩ = 1 × 10^6 Ω; R_E = 1 kΩ = 1 × 10^3 Ω; R_L = 10 kΩ = 10 000 Ω and $\beta_1 = \beta_2$ = 100.

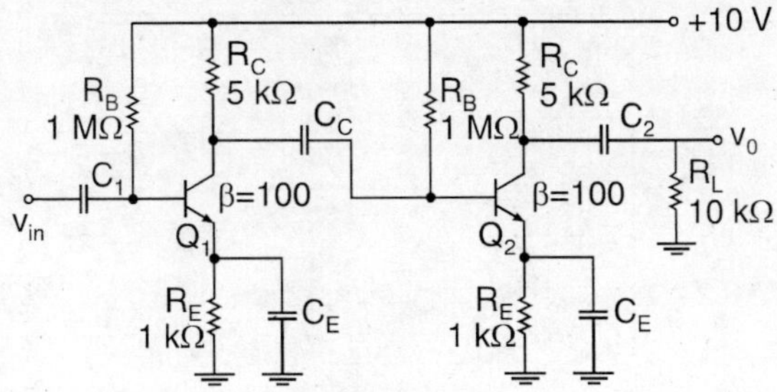

Fig. 18-6.

Input resistance of the first and second stage

We know that the value of emitter current for the first as well as second stage,

$$I_E = \frac{V_{CC}}{R_E + \dfrac{R_B}{\beta}} = \frac{10}{1 \times 10^3 + \dfrac{1 \times 10^4}{100}}$$

$$= 0.91 \times 10^{-3} \text{ A} = 0.91 \text{ mA}$$

and the value of a.c. emitter diode resistance,

$$r'_e = \frac{25}{I_E} = \frac{25}{0.91} = 27.5\ \Omega$$

∴ Input resistance of the first stage,

$$R_{i_1} = \beta \cdot r'_e = 100 \times 27.5 = 2750\ \Omega \textbf{ Ans.}$$

and input resistance of the second stage,

$$R_{i_2} = \beta \cdot r'_e = 100 \times 27.5 = 2750\ \Omega \textbf{ Ans.}$$

Output resistance of the first and second stage

We know that output resistance of the first stage,

$$R_{o_1} = R_C \parallel R_{i_2} = 5000 \parallel 2750 = 1774\ \Omega \textbf{ Ans.}$$

and output resistance of the second stage,

$$R_{o_2} = R_C \parallel R_L = 5000 \parallel 10000 = 3333\ \Omega \textbf{ Ans.}$$

Voltage gain of the first and second stage,

We know that the voltage gain of the first stage,

$$A_{v_1} = \frac{R_{o_1}}{r'_e} = \frac{1774}{27.5} = 64.5 \textbf{ Ans.}$$

and voltage gain of the second stage,

$$A_{v_2} = \frac{R_{o_2}}{r'_e} = \frac{3333}{27.5} = 121.2 \textbf{ Ans.}$$

Overall voltage gain and dB voltage gain

We also know that the overall voltage gain,

$$A_v = A_{v_1} \cdot A_{v_2} = 64.5 \times 121.2 = 7817.4 \textbf{ Ans.}$$

and the overall dB voltage gain,

$$G_v = 20 \log_{10} A_v = 20 \log_{10} 7817.4 = 77.9 \text{ dB} \textbf{ Ans.}$$

Example 18-4. *A radio receiver uses a two stage RC coupled amplifier as shown in Fig.18-7. Determine the values of voltage gain of each stage and overall voltage gain. Also express the overall voltage gain in decibels.*

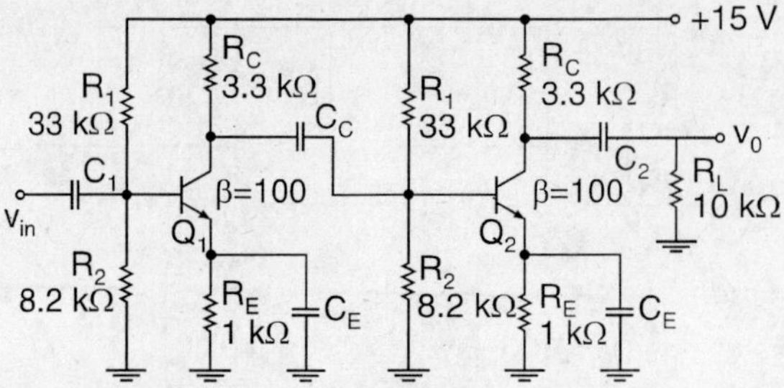

Fig. 18-7.

Solution. Given: V_{CC} = 15 volts; R_C = 3.3 kΩ = 3300 Ω; R_E = 1 kΩ = 1000 Ω; R_1 = 33 kΩ; R_2 = 8.2 kΩ; R_L = 10 kΩ = 10 000 Ω and $\beta_1 = \beta_2 = 100$.

Voltage gain of each stage

We know that voltage drop across the resistor R_2,

$$V_{TH} = V_{CC}\left(\frac{R_2}{R_1 + R_2}\right) = 15\left(\frac{8.2}{33 + 8.2}\right) \text{ V} = 2.985 \text{ V}$$

and the parallel combination of resistances R_1 and R_2,

$$R_{TH} = R_1 \parallel R_2 = 33 \parallel 8.2 = 6.57 \text{ k}\Omega = 6570\ \Omega$$

∴ Emitter current,

$$I_E = \frac{V_{TH} - V_{BE}}{R_E + \dfrac{R_{TH}}{\beta}} = \frac{2.985 - 0.7}{1000 + \dfrac{6570}{100}} \text{ A}$$

$$= 2.14 \times 10^{-3} \text{ A} = 2.14 \text{ mA}$$

and the a.c. emitter diode resistance,

$$r'_e = \frac{25}{I_E} = \frac{25}{2.14} = 11.7\ \Omega$$

We also know that input resistance of the second stage,

$$R_{i_2} = b \cdot r'_e = 100 \times 11.7 = 1170\ \Omega$$

and the output resistance of the first stage,

$$R_{o_1} = R_C \parallel R_{i_2} = 3300 \parallel 1170 = 864\ \Omega$$

Similarly, output resistance of the second stage,

$$R_{o_2} = R_C \parallel R_L = 3300 \parallel 10\,000 = 2481\ \Omega$$

∴ Voltage gain of the first stage,

$$A_{v_1} = \frac{R_{o_1}}{r'_e} = \frac{864}{11.7} = 73.8 \textbf{ Ans.}$$

and voltage gain of the second stage,

$$A_{v_2} = \frac{R_{o_2}}{r'_e} = \frac{2481}{11.7} = 212 \textbf{ Ans.}$$

Overall voltage gain

We know that the overall voltage gain,

$$A_v = A_{v_1} \cdot A_{v_2} = 73.8 \times 212 = 15\,646 \textbf{ Ans.}$$

Overall voltage in decibels

We also know that the overall voltage gain in decibels (called the dB voltage gain),

$$G_v = 20 \log_{10} A_v = 20 \log_{10} 15\,646 = 83.9 \text{ dB } \textbf{Ans.}$$

18-6. Frequency Response of RC Coupled Amplifier

The frequency response of an amplifier is a graph, which indicates the relationship between the voltage gain as a function of frequency. Usually, the voltage gain (in decibels) is plotted along the vertical axis and the frequency (in hertz or kilohertz) along the horizontal axis of the frequency response graph. Fig. 18-7 shows the frequency response of RC coupled amplifier.

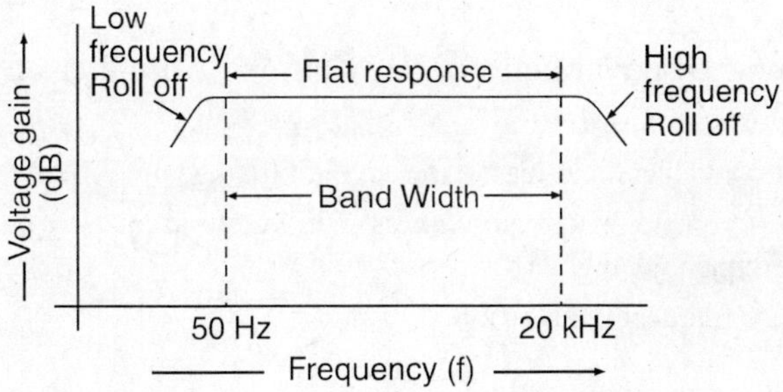

Fig. 18-8. *Frequency response of RC coupled amplifier.*

It is evident from the graph that the voltage gain *drops off* (or rolls off) at low frequencies (*i.e.*, frequencies below 50 Hz) and at high frequencies (*i.e.*, frequencies above 20 kHz), while it remains constant in the mid-frequency range called band width. The behaviour is discussed in more detail as follows:

(i) At low frequencies (i.e., below 50 Hz) : We know that the capacitive reactance (X_C) is inversely proportional to the frequency. Thus, at low frequencies, the reactance of the capacitor C_C is quite large. Therefore, it will allow only a small part of the signal to pass from one stage to the next stage. In addition to this, the emitter bypass capacitor (C_E) cannot shunt the emitter resistor effectively, because of its large reactance at low frequencies. As a result of these two factors, the voltage gain rolls off at low frequencies.

(ii) At high frequencies (i.e., above 20 kHz) : In this frequency range, the reactance of capacitor C_C becomes quite small, therefore it behaves like a short-circuit. As a result of this, the loading effect of the next stage increases, which reduces the voltage gain. In addition to this, the capacitance of the emitter diode plays an important role at high frequencies. It increases the base current of the transistor due to which the *current gain (β) reduces*. Hence, the voltage gain rolls off at high frequencies.

(iii) At mid-frequencies (i.e., 50 Hz to 20 kHz) : The effect of coupling capacitor, in this frequency range, is such that it maintains a constant voltage gain. Thus as the frequency increases, the reactance of capacitor C_C decreases, which tends to increase the gain. However, at the same time, the lower capacitive reactance increases the loading effect of the next stage due to which the gain reduces. These two factors almost cancel each other. Thus, a constant gain is maintained throughout this frequency range.

Example 18-5. *An amplifier has a band width of 500 kHz. If the lower cut-off frequency is 25 Hz, what is the upper cut-off frequency? Also find the voltage gain at the lower cut-off frequency if the midband gain is 120.* (*Banglore Univ. July* 2003)

Solution. Given: Bandwidth (BW) = 500 kHz $A_{v(max)}$ = 120, The lower cut-off frequency, (f_1) = 25 Hz, Refer to Fig. 18-8.

Let f_2 = Upper cut-off frequency

We know that bandindth (BW),

$$500 \times 10^3 = f_2 - f_1$$
$$= f_2 - 25$$
$$\therefore \quad f_2 = 500 \times 10^3 + 25$$
$$= 500.025 \text{ kHz } \textbf{Ans.}$$

We know that the voltage gain at the lower cut-off frequency,

$$A_v = \frac{A_{v(\max)}}{\sqrt{2}} = \frac{120}{\sqrt{2}} = 118.6 \textbf{ Ans.}$$

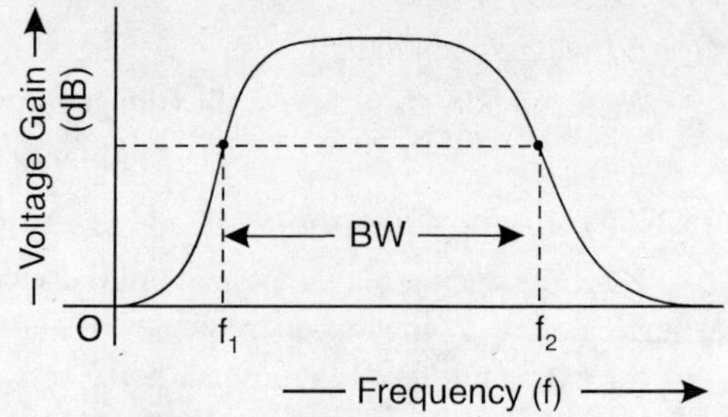

Fig. 18-9.

18-7. Advantages and Disadvantages of RC Coupled Amplifier

Advantages

- It is the most convenient and least expensive multistage amplifier.
- It has a wide frequency response, *i.e.*, its gain versus frequency curve remains constant over a wide frequency range.
- It provides less frequency distortion.

Disadvantages

- The overall gain of the amplifier is comparatively small because of the loading effect of successive stages.
- It has a tendency to become noisy with age, especially in moist climates.
- It provides poor resistance (or impedance) matching between the stages.

18-8. Applications of RC Coupled Amplifiers

We have already discussed in Art. 18-6 that an RC coupled amplifier has an excellent frequency response from 50 Hz to 20 kHz. This property makes it very useful in the initial stages of all the *public address systems*. However, it may be noted that a coupled amplifier cannot be used as a final stage of the amplifier because of its poor resistance matching characteristics. The resistance mismatching is because of the fact that the output resistance of the amplifier is several hundred ohms, whereas that of the load (such as speaker) is only few ohms.

18-9. Impedance Coupled Amplifier

Fig. 18-10 shows a two stage impedance coupled amplifier. The circuit consists of two single-stage common emitter transistor amplifiers. The inductor (L), capacitor (C_C) and resistor (R_B) forms the coupling network. The function of C_1, C_2 and C_E are the same as explained for RC coupled amplifier in Art. 18-4.

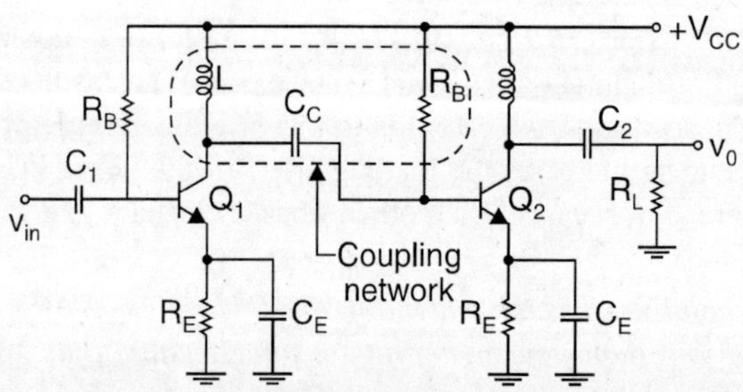

Fig. 18-10. *Two-stage impedance coupled amplifier.*

The expression for overall voltage gain and individual stage gains may be obtained in the same way as obtained for RC coupled amplifier. The only change to be made there is that R_C should be replaced by inductive reactance X_L (= ωL). Thus, voltage gain of the first stage,

$$A_{v_1} = \frac{R_{o_1}}{r'_{e_1}} = \frac{X_L \parallel R_{i_2}}{r'_{e_1}} = \frac{\omega L \parallel R_{i_2}}{r'_{e_1}}$$

$$\approx \frac{R_{i_2}}{r'_{e_1}} \qquad \text{...(If } \omega_L >> R_{i2}\text{)}$$

where, R_{i_2} is the input resistance of the second stage and is given by the expression,

$$R_{i_2} = R_B \parallel (\beta_2 \cdot r'_{e_2}) = \beta_2 \cdot r'_{e_2} \qquad \text{...(If } R_B >> \beta_2 \cdot r'_{e2}\text{)}$$

and voltage gain of the second stage,

$$A_{v_2} = R_{o_2} / r'_{e_2}$$

∴ Overall voltage gain,

$$A_v = A_{v_1} \times A_{v_2}$$

The main advantage of an impedance coupled amplifier is that it *can be operated at low collector supply voltage*, as there is hardly any drop across inductor (L). The disadvantage of impedance coupled amplifier is that it *is relatively expensive because of the use of inductors*. Moreover, the impedance of the inductor drops off at lower frequencies. Therefore as a rule, *an impedance coupled amplifier is suitable only at frequency above 20 kHz (called radio frequencies).*

18-10. Transformer Coupled Amplifier

Fig. 18-11 shows a two-stage transformer coupled amplifier. The circuit consists of two single-stage common emitter transistor amplifiers. The function of transformer (T_1) is to couple the a.c. output signal from the output of the first stage to the input of the second stage, while transformer (T_2) couples the output signal to the load. The input coupling capacitor is C_1, while the emitter bypass capacitor is C_E,

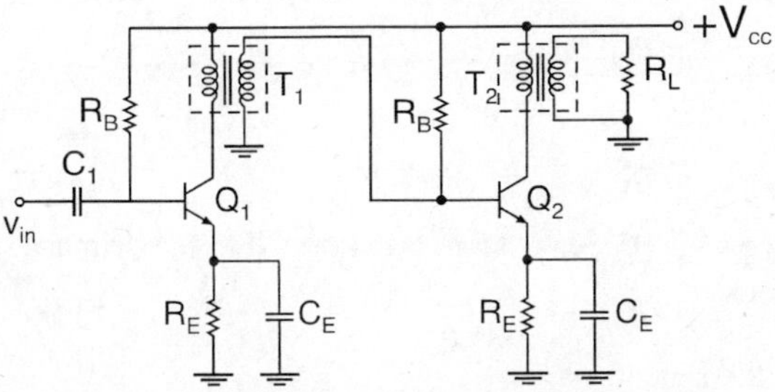

Fig. 18-11. *Two stage transformer coupled amplifier.*

The operation of the above circuit may be understood from the condition that when an a.c. input signal is applied to the base transistor Q_1, it appears in the amplified form across primary winding of the transformer (T_1). The voltage developed across the primary winding is then transferred to the input of the next stage by the secondary winding of the transformer (T_1). The second stage does amplification in an exactly similar manner.

In actual practice, a bypass capacitor is used on the bottom of each primary winding to produce an a.c. ground. This avoids the load inductance of the connecting line that returns to the d.c. supply point. Similarly, a bypass capacitor is used on the bottom of each secondary winding to get an a.c. ground. This prevents the signal power loss in the biasing resistors.

18-11. Calculations of Voltage Gain for Transformer Coupled Amplifier

Consider the circuit of a two-stage transformer coupled amplifier shown in Fig. 18-10. The a.c. equivalent circuit for the amplifier is shown in Fig. 18-11 (a) and (b). In these figures, the a.c. equivalent circuit of the transistor is shown by the a.c. emitter diode resistance and three points marked *E*, *B* and *C*. Assuming the transformer to be an ideal (*i.e.*, neglecting its leakage current, winding resistance, *etc.*), whatever be the load on its secondary winding side, it transforms the same on the primary side. The value of the transformed load,

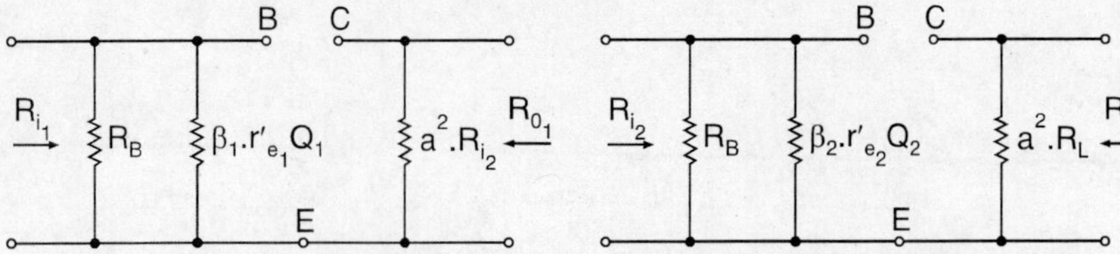

(*a*) For first stage. (*b*) For second stage.

Fig. 18-12. *A.C. equivalent circuit for two stage transformer coupled amplifier.*

$$R'_L = \left(\frac{N_1}{N_2}\right)^2 R_L = a^2 \cdot R_L \qquad \ldots\left(\text{where a} = \frac{N_1}{N_2}\right)$$

where, N_1 = Number of primary turns, and

N_2 = Number of secondary turns.

Similarly, if R_{i_2} is the input resistance of the second stage, then its value transformed to the primary side,

$$R'_{i_2} = a^2 \cdot R_{i_2}$$

We know that the input resistance of the first stage is given by the relation,

$$R_{i_1} = R_B \parallel (\beta_1 \cdot r'_e) \qquad \ldots(\text{If } R_B >> \beta_1 \cdot r'_{e_1})$$

and the input resistance of the second stage, transformed to the primary side,

$$R'_{i_2} = a^2 \cdot R_{i_2} = a^2 \cdot R_B \parallel (\beta_2 \cdot r'_{e_2})$$

$$= a^2 (\beta_2 \cdot r'_{e_2}) \qquad \ldots(\text{If } R_B >> \beta_2 \cdot r'_{e_2})$$

Similarly, output resistance of the first stage,

$$R_{o_1} = R'_{i_2} = a^2 \cdot R_{i_2}$$

and the output resistance of the second stage, transformed to the primary side,

$$R_{o_2} = a^2 \cdot R_L$$

∴ Voltage gain of the second stage,

$$A_{v_1} = R_{o_1} / r'_{e_1}$$

and voltage gain of the second stage,

$$A_{v_2} = \frac{R'_{o_2}}{r'_{e_2}} = \frac{a^2 \cdot R_L}{r'_{e_2}}$$

∴ Overall voltage gain,

$$A_v = A_{v_1} \cdot A_{v_2} = \frac{R_{o_1}}{r'_{e_1}} \times \frac{a^2 \cdot R_L}{r'_{e_2}}$$

Example 18-6. *Fig. 18-13 shows the circuit diagram of a two-stage transformer coupled transistor amplifier. Calculate the values of individual stage gains and the overall voltage gain in decibels. Neglect* V_{BE} *and take* $r'_e = 25\ mV/I_E\ (mA)$.

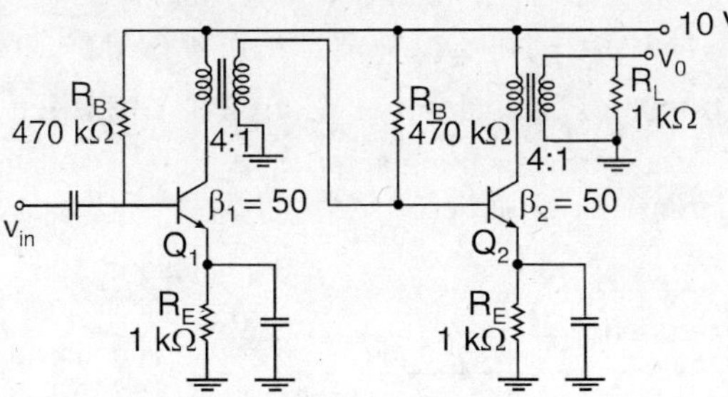

Fig. 18-13.

Solution. Given: $V_{CC} = 10$ volts; $R_E = 1\ k\Omega = 1 \times 10^3\ \Omega$; $R_B = 470\ k\Omega = 470 \times 10^3\ \Omega$; $a = 4$; $\beta_1 = \beta_2 = 50$ and $R_L = 1\ k\Omega = 1000\ \Omega$.

Individual stage gains

We know that the value of emitter current,

$$I_E = \frac{V_{CC}}{R_E + \frac{R_B}{\beta}} = \frac{10}{(1 \times 10^3) + \frac{470 \times 10^3}{50}}\ A$$

$$= 0.96 \times 10^{-3}\ A = 0.96\ mA$$

and the value of a.c. emitter diode resistance,

$$r'_e = \frac{25\ mV}{I_E\ (mA)} = \frac{25}{0.95} = 26\ \Omega$$

We also know that input resistance of the first stage,

$$R_{i_1} = R_B \parallel (\beta_1 \cdot r'_{e_1}) = (470 \times 10^3) \parallel (50 \times 26)\ \Omega$$

$$= 1296\ \Omega$$

and input resistance of the second stage,

$$R_{i_2} = R_B \parallel (\beta_2 \cdot r'_{e_2}) = (470 \times 10^3) \parallel (50 \times 26)$$

$$= 1296\ \Omega$$

The input resistance of the second stage, transformed to the primary side,

$$R'_{i_2} = a^2 \cdot R_{i_2} = (4)^2 \times 1296 = 20736\ \Omega$$

Similarly, output resistance of the first stage,

$$R_{o_1} = R'_{i_2} = 20736\ \Omega$$

and output resistance of the second stage, transformed to the primary side,

$$R'_{i_2} = a^2 \cdot R_L = (4)^2 \times 1000 = 16\ 000\ \Omega$$

∴ Voltage gain of the first stage,

$$A_{v_1} = \frac{R_{o_1}}{r'_{e_1}} = \frac{20\ 736}{26} = 797.5 \textbf{ Ans.}$$

and voltage gain of the second stage,

$$A_{v_2} = \frac{R'_{o_2}}{r'_{e_2}} = \frac{16\ 000}{26} = 615.4 \textbf{ Ans.}$$

Overall voltage gain in decibels

We know that the overall voltage gain,

$$A_v = A_{v_1} \cdot A_{v_2} = 797.5 \times 615.4 = 490\ 782 \textbf{ Ans.}$$

and the overall voltage gain in decibels (called dB voltage gain),

$$G_v = 20 \log_{10} A_v = 20 \log_{10} 490\ 782 = 114 \text{ dB } \textbf{Ans.}$$

18-12. Frequency Response of Transformer Coupled Amplifier

Fig. 18-14 shows the frequency response graph (*i.e.*, a graph of voltage gain *versus* frequency) for a transformer coupled transistor amplifier.

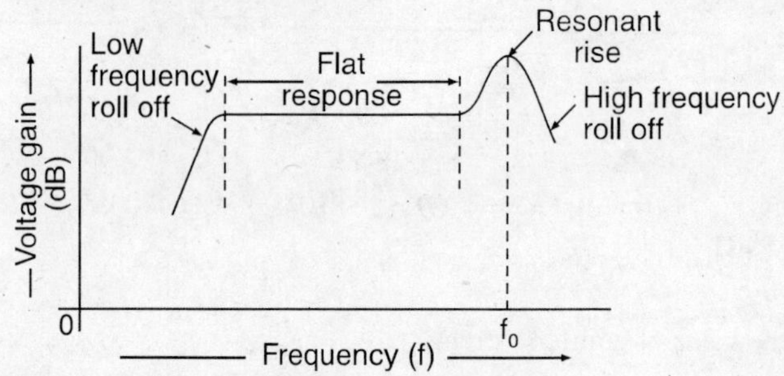

Fig. 18-14. *Frequency response of a transformer coupled amplifier.*

It is evident from this Fig. 18-14, that the voltage gain drops off (or rolls off) at low as well as at high frequencies , whereas it remains constant in mid-frequency range. Another noticeable feature is that *at one particular frequency (f_0) the voltage gain increases and then rolls off continuously.* This typical behaviour may be explained as follows :

We know that output voltage of a transformer coupled amplifier is equal to the product of collector current and the reactance of the primary winding of coupling transformer. At low frequencies, the reactance of primary winding ($X_L = \omega \cdot L$) begins to decrease and hence, the voltage gain reduces. At high frequencies, the effect of leakage inductance and distributed capacitance (*i.e.*, the capacitance between the turns of the winding) becomes significant and hence, the voltage gain reduces. *The peak gain results due to the resonance (or turning) effect of inductance and distributed capacitance, which forms a resonant circuit.* ***The frequency, at which the peak occurs, is called resonant frequency (f_0).***

It has been found that flat part of frequency response curve of transformer coupled amplifiers is small as compared to that of RC coupled amplifiers. As a result of this, *these amplifiers can not be used over a wide range of frequencies.* Moreover, if *they are used, they produce frequency distortion*, which means that all frequency components in a complete input signal (such as music, speech signal) are not equally amplified. However, the transformers can be suitably designed to provide a fairly flat response curve and excellent fidelity over the entire audio frequency range (*i.e.*, 20 Hz to 20 kHz).

18-13. Advantages and Disadvantages of Transformer Coupled Amplifier

Advantages

- No signal power is lost in the collector or base resistors, because of the low winding resistance of the transformer.
- It provides a higher voltage gain than the RC coupled amplifier.
- It provides an excellent resistance (or impedance) matching between the stages. The resistance matching is desirable for maximum power transfer.

Disadvantages

- The coupling transformer is expensive and bulky, particularly when operated at audio frequencies.
- At radio frequencies, the winding inductance and distributed capacitance produces reverse frequency distortion.
- It tends to produce '*hum*' in the circuit.

18-14. Applications of Transformer Coupled Amplifier

We have already discussed in the last article that transformer coupled amplifiers provide excellent impedance matching between the individual stages. This ability makes it very useful in a multistage amplifier, where it is used as a final stage. *It is used to transfer power to the low impedance load (such as speaker).* The impedance of a speaker varies from 4 Ω to 16 Ω, whereas the output resistance of a transistor amplifier is several hundred ohms. In order to match the load impedance, with that of the amplifier output, a step-down transformer of proper turns ratio is used. The resistance of the secondary winding of the transformer is made equal to the speaker impedance, while that of the primary winding is made equal to the output resistance of the amplifier.

18-15. Direct Coupled Amplifier

It is also called as *DC amplifier* and is used to *amplify very low frequency* (*i.e.*, below 10 Hz) signals including direct current or zero frequency. It may be noted that the capacitors, inductors and transformers can not be used as a coupling network at very low frequencies because the electrical size of these devices, at low frequencies, becomes very large.

Fig. 18-15 shows a two-stage direct coupled transistor amplifier. It may be noted that the output of the first stage is directly connected to the base of the next transistor. Moreover, there are no input or output coupling capacitors. The operation of this circuit is discussed below :

The signal to be amplified is applied directly to the input of the first stage. Due to the transistor action, it appears in the amplified form across the collector resistor or transistor Q_1. This voltage then drives the base of the second transistor Q_2 and the amplified output is obtained across the collector resistor of transistor Q_2.

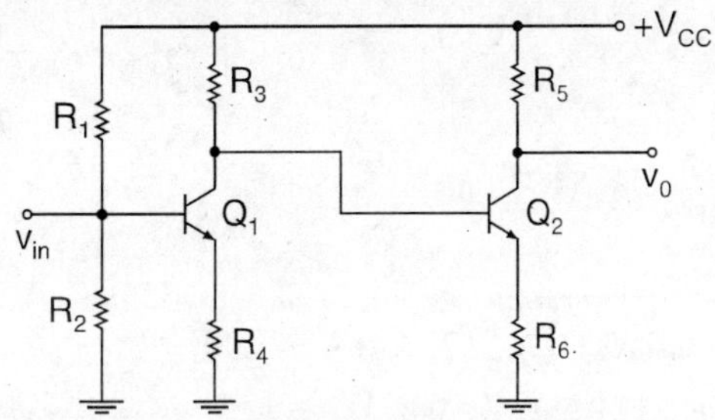

Fig. 18-15. *Two stage direct coupled amplifier.*

18-16. Calculation of Voltage Gain of Direct Coupled Amplifier

Consider the circuit diagram of a direct coupled amplifier shown in Fig. 18-15. The a.c. equivalent circuit for such an amplifier is shown in Fig. 18-15.

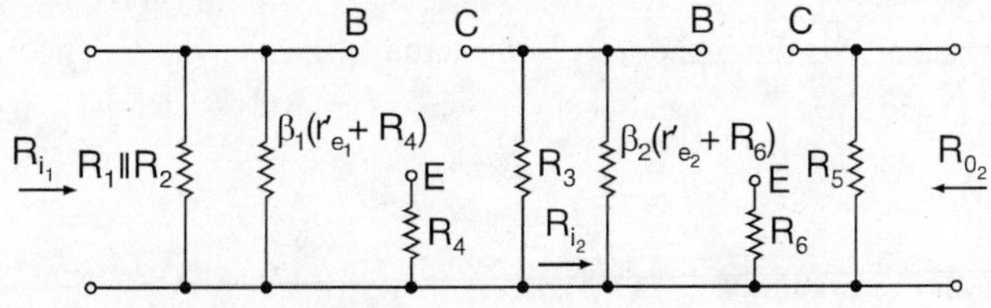

Fig. 18-16. *A.C. equivalent circuit of a two stage direct coupled amplifier.*

It is evident from the figure shown above that the input resistance of the first stage,

$$R_{i_1} = (R_1 \parallel R_2) \parallel \beta_1 (r'_{e_1} + R_4)$$

and input resistance of the second stage,

$$R_{i_2} = \beta_2 (r'_{e_2} + R_6)$$

Similarly, the output resistance of the first stage,

$$R_{o_1} = R_3 \parallel R_{i_2}$$

and output resistance of the second stage,

$$R_{o_2} = R_5$$

∴ Voltage gain of the first stage,

$$A_{v_1} = \beta \times \frac{R_{o_1}}{R_{i_1}} = \frac{R_{o_1}}{r'_{e1} + R_4}$$

and voltage gain of the second stage,

$$A_{v_2} = \beta_2 \times \frac{R_{o_2}}{R_{i_3}} = \frac{R_{o_2}}{r'_{e_2} + R_4}$$

Now the overall voltage gain is given by the relation,

$$A_v = A_{v_1} \cdot A_{v_2}$$

Example 18-7. *Fig. 18-17 shows a direct coupled two-stage amplifier. The d.c. bias of the first stage sets the d.c. bias of the second.*

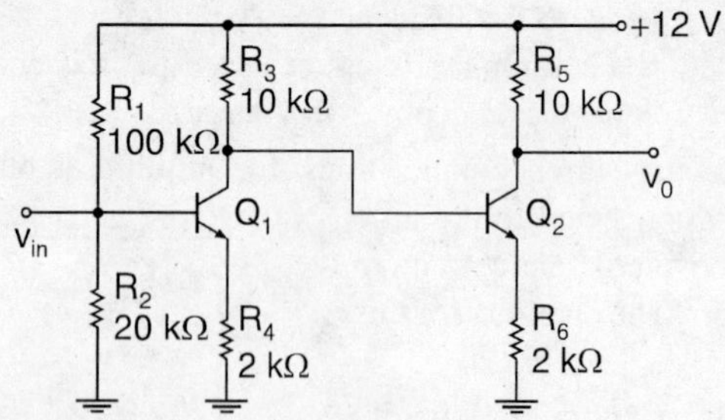

Fig. 18-17.

Determine the individual stages a.c. voltage gain and the overall voltage gain. Neglect V_{BE} *and use* $r'_e = 25/I_E$ *in mA. Take* $\beta_1 = \beta_2 = 100$.

Solution. Given: $V_{CC} = 12$ volts; $R_1 = 100$ kΩ; $R_2 = 20$ kΩ; $R_3 = 10$ kΩ = 10 000 Ω; $R_4 = 2$ kΩ = 2000 Ω; $R_5 = 10$ kΩ = 10 000 Ω; $R_6 = 2$ kΩ = 2000 Ω; $r'_e = 25/I_E$ and $\beta_1 = \beta_2 = 100$.

Individual stages a.c. voltage gain

We know that voltage drop across resistor R_2,

$$V_{TH} = V_{CC}\left(\frac{R_2}{R_1 + R_2}\right) = 12\left(\frac{20}{100 + 20}\right) = 2 \text{ V}$$

As the voltage drop across the resistor R_4 is the same as that across R_2. Therefore emitter current of Q_1 transistor,

$$I_{E_1} = \frac{V_{TH}}{R_4} = \frac{2}{2000} = 0.001 \text{ A} = 1 \text{ mA}$$

and the a.c. emitter diode resistance,

$$r'_{e_1} = \frac{25}{I_{E_1}} = \frac{25}{1} = 25 \ \Omega$$

Again neglecting V_{BE}, the voltage drop across resistor, R_e is the same as the collector voltage of Q_1 transistor and its value,

$$V_{R_6} = V_{C_1} = V_{CC} - I_{E_1} \cdot R_3$$

$$= 12 - (0.001 \times 10\,000) = 12 - 10 = 2 \text{ V}$$

∴ Emitter current of transistor,

$$I_{E_2} = \frac{V_{R_6}}{R_6} = \frac{2}{2000} = 0.001 \text{ A} = 1 \text{ mA}$$

and the a.c. emitter diode resistance,

$$r'_{e_2} = \frac{25}{I_{E_2}} = \frac{25}{1} = 25\ \Omega$$

We also know that input resistance of the second stage

$$R_{i_2} = \beta_2\,(r'_{e_2} + R_6) = 100\,(25 + 2000) = 202\ 500$$

and output resistance of the first stage,

$$R_{o_1} = R_3 \parallel R_{i_2} = 10\ 000 \parallel 202\ 500 = 10\ 000\ \Omega$$

Similarly, output resistance of the second stage,

$$R_{o_2} = R_5 = 10\ 000\ \Omega$$

∴ Voltage gain of the first stage

$$A_{v1} = \frac{R_{o_1}}{r'_{e_1} + R_4} = \frac{10\ 000}{25 + 2000} = 4.9 \textbf{ Ans.}$$

and voltage gain of second stage,

$$A_{v_2} = \frac{R_{o_2}}{r'_{e_2} + R_6} = \frac{10000}{25 + 2000} = 4.9 \textbf{ Ans.}$$

Overall voltage gain

We know that overall voltage gain,

$$A_v = A_{v_1} \cdot A_{v_2} = 4.9 \times 4.9 = 24 \textbf{ Ans.}$$

18-17. Direct Coupled Amplifier Using Complementary Transistor

Fig. 18-18 shows a two-stage direct coupled amplifier using *complementary transistors*. The term '*complementary*' means that one stage uses *NPN* transistor, while the second stage uses *PNP* transistor. The operation of the circuit may be understood from the condition that when the positive going input signal is applied to the base of transistor Q_1, its collector current increases. This increases the voltage drop across resistor R_3 due to which the collector voltage of transistor Q_1 decreases. Consequently, the base of transistor Q_2 becomes less positive (or more negative). It means that a negative going signal is applied to the base of Q_2 transistor. It causes the forward bias across emitter-base junction of Q_2 (*i.e.*, PNP transistor) to increase, which in turn increases the collector current. As a result of this, the voltage drop across resistor R_6 increases and we get a positive going amplified output. Similarly, during the positive going input signal, we get a negative going amplified output across the resistor R_6.

Fig. 18-18. *Direct coupled amplifier using complementary.*

It may be noted that if we take an output from the emitter terminal of Q_2 transistor, its magnitude and polarity will be the same as that of its input (*i.e.*, at the base of Q_2 transistor).

Example 18-8. *For the circuit shown in Fig. 18-19, determine the d.c. collector currents and collector-to-emitter voltages of transistors Q_1 and Q_2.*

Also determine the approximate overall a.c. voltage gain of the cascaded amplifier. Assume h_{FE} (i.e., β) of both the transistors to be 100 and C_1 and C_2 to be very large. Assume V_{BE} = 0.7 V.

Solution. Given: V_{CC} = 10 volts; R_1 = 800 Ω; R_2 = 200 Ω; R_3 = 600 Ω; R_4 = 200 Ω; R_5 = 100 Ω; R_6 = 1 kΩ = 1 × 10^3 Ω, $\beta_1 = \beta_2$ = 100 and V_{BE} = 0.7 volt.

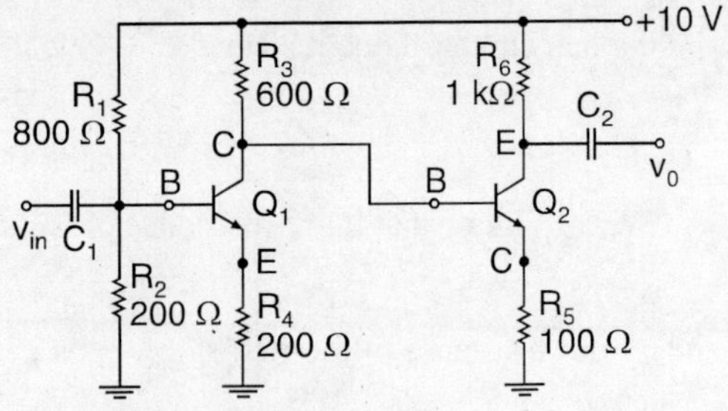

Fig. 18-19.

D.C. collector currents and collector-to-emitter voltages

We know that the voltage drop across resistor R_2,

$$V_{R_2} = V_{CC}\left(\frac{R_2}{R_1 + R_2}\right)$$

$$= 10\left(\frac{200}{800 + 200}\right) = 2 \text{ V}$$

and the emitter current of Q_1 transistor,

$$I_{E_1} = \frac{V_{R_2} - V_{BE_1}}{R_E} = \frac{2 - 0.7}{200} = 6.5 \times 10^{-3} \text{ A}$$

∴ Collector current of Q_1, transistor,

$$I_{C_1} = I_{E_1} = 6.5 \times 10^{-3} \text{ A} = 6.5 \text{ mA } \textbf{Ans.}$$

We also know that the voltage at the collector of Q_1 transistor,

$$V_{C_1} = V_{CC} - I_{C_1} \cdot R_3 = 10 - (6.5 \times 10^{-3}) \times 600 = 6.1 \text{ V}$$

and the voltage at the emitter of Q_1 transistor,

$$V_{E_1} = I_{E_1} \cdot R_4 = (6.5 \times 10^{-3}) \times 200 = 1.3 \text{ V}$$

∴ Collector-to-emitter voltage of Q_1 transistor,

$$V_{CE_1} = V_{C_1} - V_{E_1} = 6.1 - 1.3 = 4.8 \text{ V } \textbf{Ans.}$$

We also know that the voltage at the emitter of Q_2 transistor,

$$V_{E_2} = V_{C_1} - (-V_{BE}) \qquad (\because V_{BE} \text{ for a PNP Transistor is negative})$$

$$= 6.1 - (-0.7) = 6.8 \text{ V}$$

and the emitter current of Q_2 transistor,

$$I_{E_2} = \frac{V_{CC} - V_{E_2}}{R_6} = \frac{10 - 6.8}{1 \times 10^3} = 3.2 \times 10^{-3} \text{ A}$$

∴ Collector current of Q_2 transistor,

$$I_{C_2} = I_{E_2}$$

and the voltage at the collector of Q_2 transistor,

$$V_{C_2} = I_{C_2} \cdot R_5 = (3.2 \times 10^{-3}) \times 100 = 0.32 \text{ V}$$

∴ Collector-to-emitter voltage of Q_2 transistor,

$$V_{CE_2} = V_{C_2} - V_{E_2} = 0.32 - 6.8 = -6.48 \text{ V } \textbf{Ans.}$$

Overall voltage gain

We know that a.c. emitter diode resistance of Q_1 transistor,

$$r'_{e_1} = \frac{25}{I_{E_1}} = \frac{25}{6.5} = 3.85\ \Omega$$

and a.c. emitter diode resistance of Q_2 transistor,

$$r'_{e_2} = \frac{25}{I_{E_2}} = \frac{25}{3.2} = 7.8\ \Omega$$

We also know that the input resistance of the second stage,

$$R_{i_2} = \beta_2(r'_{e_2} + R_6) = 100\ (7.8 + 1000) = 100.8 \times 10^3\ \Omega$$

and output resistance of the first stage,

$$R_{o_1} = R_3 \parallel R_{i_2} = 600 \parallel (100.8 \times 10^3) = 596.3\ \Omega$$

∴ Voltage gain of the first stage,

$$A_{v_1} = \frac{R_{o_1}}{r'_{e_1} + R_4} = \frac{596.3}{3.85 + 200} = 2.93$$

and voltage gain of the second stage,

$$A_{v_2} \approx 1 \qquad \text{...(}\because \text{ It is an emitter follower stage)}$$

∴ Overall voltage gain,

$$A_v = A_{v_1} \cdot A_{v_2} = 2.93 \times 1 = 2.93 \textbf{ Ans.}$$

18-18. Frequency Response of Direct Coupled Amplifier

Fig. 18-20 shows the frequency response (*i.e.*, a graph of dB voltage gain versus frequency) of a direct coupled amplifier. It is evident from this Fig., the *gain is uniform up to a certain frequency denoted by f_2. Beyond this frequency, the gain rolls off slowly*. The gain rolls off at high frequencies due to the increased emitter diode capacitance and stray wiring capacitance.

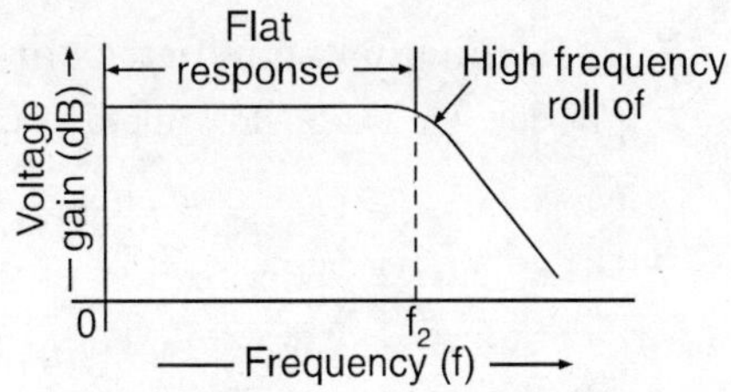

Fig. 18-20. *Frequency response of direct coupled amplifier.*

18-19. Advantages and Disadvantages of Direct Coupled Amplifier

Advantages

- The circuit arrangement is very simple because it uses a minimum number of resistors.
- The circuit cost is low because of the absence of expensive coupling devices.
- It can amplify very low frequency signals down to zero frequency.

Disadvantages

- It cannot amplify high frequency signals.
- It has a poor temperature stability. Because of this, its, *Q*-point shifts. In a multistage direct coupled amplifier, the *Q*-point shifts are amplified in succeeding stages. Thus a small d.c. shift, in the first stage can cause the final stage to be either saturated or cut-off. All integrated circuit amplifiers are direct coupled because of the difficulty of fabricating large integrated capacitors. It leads to special problems in their design.

18-20. Applications of Direct Coupled Amplifiers

The direct coupled amplifiers are used in many electronic systems that handle signals, which change very slowly with time. Some of the important applications are as given below :

(*i*) Analog computation.

(*ii*) Power supply regulators.

(*iii*) Bioelectric measurements.

(*iv*) Linear integrated circuits.

18-21. Darlington Amplifier

The Darlington Amplifier consists of two cascaded emitter followers as shown in Fig. 18-21. The configuration results in a set of improved amplifier characteristics.

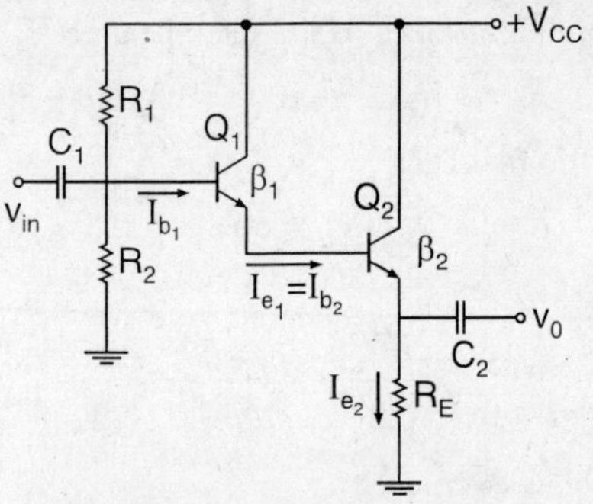

Fig. 18-21. *Darlington amplifier.*

The Darlington amplifier has a high input resistance, low output resistance and high current gain. These characteristics make it very useful as a current amplifier. The voltage gain of a Darlington amplifier is less than unity.

18-22. Darlington Amplifier Characteristics

Consider the Darlington amplifier circuit shown in Fig. 18-21.

Let I_{b_1} = Base current of Q_1 transistor,

I_{e_1} = Emitter current of Q_1 transistor,

β_1 = Current gain of Q_1 transistor,

I_{b_2} = Base current of Q_2 transistor. Its value is equal to the emitter current of Q_1 transistor,

I_{e_2} = Emitter current of Q_2 transistor, and

β_2 = Current gain of Q_2 transistor.

(*i*) ***Current gain :*** We know that the emitter current of Q_2 transistor,

$$I_{e_1} = \beta_1 \cdot I_{b_1}$$

and emitter current Q_2 of transistor,

$$I_{e_2} = \beta_2 \cdot I_{b_2} = \beta_2 \cdot I_{e_1} \qquad ...(\because I_{b_2} = I_{e_1})$$

$$= \beta_2 (\beta_1 \cdot I_{b_1}) \qquad ...(\because I_{e_1} = \beta_1 \cdot I_{b_1})$$

$$= \beta_1 \cdot \beta_2 \cdot I_{b_1} = \beta^2 \cdot I_{e_1}$$

Now let both the transistors used in the amplifier be identical. Then current gain β_1 and $\beta_2$will be equal. In that case, emitter current of Q_2 transistor,

$$I_{e_2} = b_2 \cdot I_{b_1} \qquad ...(\because \beta_1 \cdot \beta_2 = \beta^2)$$

$\therefore$ Overall current gain,

$$A_i = \frac{I_{e_2}}{I_{b_1}} = \beta^2$$

(*ii*) ***Input resistance :*** We know that input resistance of the second stage is given by the relation,

$$R_{i_2} = \beta_2 (r'_{e_2} + R_E)$$

$$= \beta_2 \cdot R_E \qquad ...(\text{If } R_E \; r_{e_2})$$

It is the value of resistance seen by the emitter of Q_1 transistor and is given by the relation,

$$R'_{i_1} = \beta_1 (r'_{e_1} + R_{i_2})$$

$$= \beta_1 \cdot R_{i_2} \quad ...(\text{If } R_{i_2} >> r'_{e_1})$$

$$= \beta_1 \cdot \beta_2 \cdot R_E \quad ...(\because R_{i_2} = \beta_2 \cdot R_E)$$

For identical transistor, we know that current gain β_1 and β_2 are also equal. Therefore, input resistance

$$R'_{i_1} = \beta^2 \cdot R_E$$

and input resistance of the amplifier stage,

$$R_{i_1} = (R_1 \parallel R_2) \parallel R'_{i_1} = (R_1 \parallel R_2) \parallel \beta^2 \cdot R_E$$

$$= R_1 \parallel R_2 \quad ...(\because (R_1 \parallel R_2) << \beta^2 \cdot R_E)$$

(iii) Output resistance : The output resistance of the first stage is given by the relation,

$$R_{o_1} = r'_{e_1} + \frac{R_1 \parallel R_2}{\beta_1}$$

and output resistance of the second stage,

$$R_{o_2} = r'_{e_2} + \frac{R_{o_1}}{\beta_2}$$

$$= r'_{e_2} + \frac{r'_{e_1} + \frac{R_1 \parallel R_2}{\beta_1}}{\beta_2}$$

$$= r'_{e_2} + \frac{r'_{e_1}}{\beta_2} + \frac{R_1 \parallel R_2}{\beta_1 \cdot \beta_2}$$

$$= r'_{e_2} \quad ...\left(\frac{r'_{e_1}}{\beta_2} \text{ and } \frac{(R_1 \parallel R_2)}{\beta_1 \cdot \beta_2} \text{ are negligible}\right)$$

(iv) Voltage gain : The voltage gain of Darlington amplifier is given by the relation,

$$A_v = \frac{v_o}{v_{in}} \quad ...(i)$$

The value of output voltage of a Darlington amplifier,

$$v_o = I_{e_2} \cdot R_E$$

and the input voltage,

$$v_{in} = I_{e_1} \cdot r'_{e_1} + (r'_{e_2} + R_E) \quad ...(ii)$$

The above expression is obtained by applying Kirchoff's Voltage Law to the input circuit of the Darlington amplifier. Substituting the value of I_{e_1} (equal to I_{e_2}/β_2) in equation (*ii*)

$$v_{in} = \frac{I_{e_2}}{\beta_2} r'_{e_1} + I_{e_2}(r'_{e_2} + R_E)$$

∴ Voltage gain,

$$A_v = \frac{I_{e_2} \cdot R_E}{\frac{I_{e_2}}{\beta_2} \cdot r'_{e_1} + I_{e_2}(r'_{e_2} + R_E)} = \frac{R_E}{\frac{r'_{e_1}}{\beta_2} + (r'_{e_2} + R_E)}$$

$$\approx 1 \quad ...(\because R_E >> r'_{e_2} \text{ or } r'_{e_1}/\beta_2)$$

18-23. Darlington Pair

These days, the semiconductor device manufacturers are supplying a pair of identical transistors, connected suitably, in a single package. It is known as a *Darlington pair*. The two transistors are connected in such a way that the emitter of one transistor is connected to the base of the other as shown in Fig. 18-22 (*a*) and (*b*). The Darlington pairs shown in Fig. 18-22 (*a*) and (*b*) are called NPN Darlington pair and PNP Darlington pair respectively

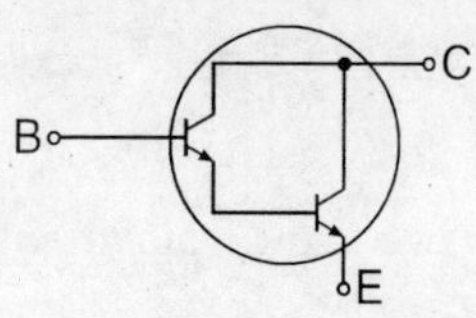

(*a*) NPN transistor Darlington pair.

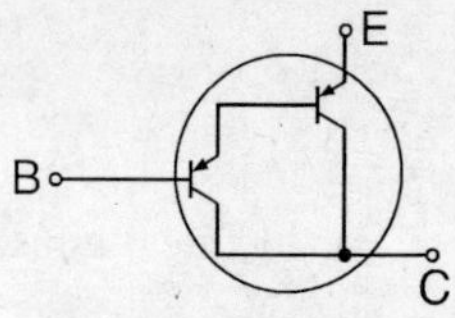

(*b*) PNP transistor Darlington pair.

Fig. 18-22.

A Darlington pair is a three terminal device namely base (*B*) emitter (*E*) and collector (*C*). It acts like a transistor, with an extremely high current gain (β). The Darlington pairs are available for both NPN and PNP transistors. Even photo-transistor Darlington pairs are also available. Such a device is β times more sensitive than that of a photo-transistor.

Example 18-9. *Fig. 18-23 shows the circuit diagram of a Darlington amplifier.*

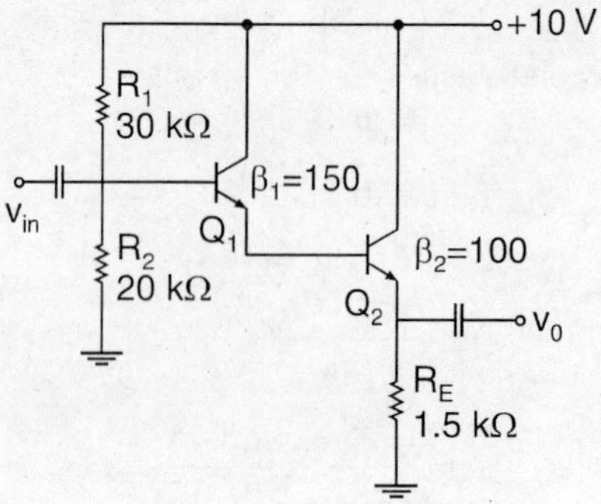

Fig. 18-23.

Determine **(a)** *the overall gain;* **(b)** *the a.c. emitter diode resistance for each transistor;* **(c)** *total input resistance; and* **(d)** *the overall voltage gain.*

Solution. Given $V_{CC} = 10$ volt; $R_1 = 30$ kΩ; $R_2 = 20$ kΩ; $R_E = 1.5$ kΩ $= 1.5 \times 10^3$ Ω; $\beta_1 = 150$ and $\beta_2 = 100$.

(*a*) *Overall current gain*

We know that overall current gain,

$$A_i = \beta_1 \cdot \beta_2 = 150 \times 100 = 15\ 000 \textbf{ Ans.}$$

(*b*) *A.C. emitter diode resistance for each transistor*

We know that voltage drop across resistor R_2

$$V_{R_2} = V_{CC} \times \frac{R_2}{R_1 + R_2} = 10\left(\frac{20}{30 + 20}\right) = 4 \text{ V}$$

and the voltage at the base of Q_2 (or emitter of Q_1),

$$V_{B_2} = V_{R_2} - V_{BE_1} = 4 - 0.7 = 3.3 \text{ V}$$

Similarly, voltage at the emitter of Q_2,

$$V_{E_2} = V_{B_2} - V_{BE_2} = 3.3 - 0.7 = 2.6 \text{ V}$$

and the value of emitter current of Q_2,

$$I_{E_2} = \frac{V_{E_2}}{R_E} = \frac{2.6}{1.5 \times 10^3} = 1.73 \times 10^{-3}\ A = 1.73 \text{ mA}$$

∴ A.C. emitter diode resistance of Q_2 transistor

$$r'_{e_2} = \frac{25}{I_{E_2} \text{ (in } mA)} = \frac{25}{1.73} = 14.5\ \Omega \textbf{ Ans.}$$

Now the base current of Q_2,

$$I_{b_2} = \frac{I_{E_2}}{\beta_2} = \frac{1.73}{100} = 0.0173 \text{ mA}$$

and the emitter current of Q_1,

$$I_{E_1} = I_{b_2} = 0.0173 \text{ mA}$$

$\therefore$ A.C. emitter diode resistance of Q_1,

$$r'_{e_1} = \frac{25}{I_{E_1} (\text{in } mA)} = \frac{25}{0.0173} = 1445\ \Omega \textbf{ Ans.}$$

(*c*) *Total input resistance*

We know that total input resistance (*i.e.*, input resistance of the amplifier stage),

$$R_{i_1} = R_1 \parallel R_2 = 30 \parallel 20 = 12 \text{ k}\Omega \textbf{ Ans.}$$

(*d*) *Overall voltage gain*

We also know that overall voltage gain,

$$A_v = \frac{R_E}{\frac{r'_{e_1}}{\beta_2} + (r'_{e_2} + R_E)} = \frac{1.5 \times 10^3}{\frac{1445}{100} + [14.4 + (1.5 \times 10^3)]}$$

$$= 0.98 \textbf{ Ans.}$$

SUMMARY

In this chapter you have learned that :

1. In multistage amplifiers, output of one stage is connected to the input of the next.
2. The voltage gain of a multistage amplifier is equal to the product of the gains of the individual stages.
3. Interstage coupling helps to transfer the a.c. output of one stage to the next, while isolating d.c. conditions between the two stages.
4. R.C. coupled amplifiers have a good frequency response from 50 Hz to 20 kHz range, which is in the audible region.
5. The transformer coupled amplifier is used to transfer power to the low impedance load.
6. Direct coupled amplifiers are used for amplification of D.C. or very low frequency singals.
7. Darlington amplifiers are used for current amplification.

GLOSSARY

Cascade amplifier : A multistage amplifier, using two or more single stage common-emitter amplifiers.

Cascading : Connecting the output of one single stage amplifier to the input of another single stage amplifier.

Cascade amplifier : A multistage amplifier with common-emitter amplifier as the first stage and common-base (or common-collector) amplifier as the second stage.

Darlington pair : A three terminal device of two identical transistors, connected in such a way that emitter of one transistor is fed to the base of the other.

Direct-coupling : An interstage coupling scheme, in which output of one stage is directly fed to the input of the next without using any coupling device such as capacitors, inductors or transformers.

Frequency response : A graph (of an amplifier), in which voltage gain (in dB) is plotted against frequency.

Impedance coupling : An interstage coupling scheme, in which collector resistance of the transistor circuit is replaced with an inductor and a.c. voltage drop across this inductor is fed to the next stage.

Interstage coupling : The coupling between individual stages of a multistage amplifier.

Multistage amplifier : Amplifiers with more than one single stage amplifier is cascaded.

Resistance-capacitance coupling : An interstage coupling scheme, involving two resistors, namely, collector resistor of first stage and base resistor of second stage, and a capacitor in between them.

Transformer coupling : An interstage coupling scheme, in which the collector resistor of the first stage is replaced with the primary of a transformer and secondary is used to convey the a.c. output signal to the second stage.

NUMERICAL PROBLEMS

1. For the two-stage RC coupled amplifier shown in Fig. 18-24, determine the individual stage gains, input resistance and the overall voltage gain,

(**Ans.** $A_{v_1} = 66$; $A_{v_2} = 193$; $R_i = 1.98$ kΩ; $A_v = 82.1$ dB)

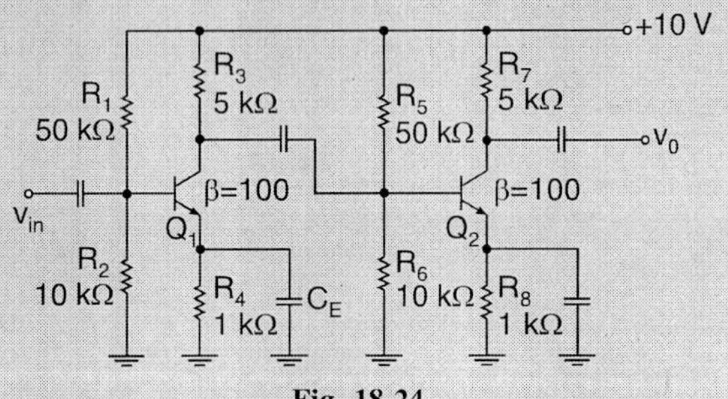

Fig. 18-24.

2. Fig. 18-25 shows a two-stage RC coupled amplifier consisting of common emitter stage followed by an emitter follower. Determine A_{v1}, A_{v2}, A_v, R_{i1} and the a.c. output voltage.

(**Ans.** $A_{v_1} = 159$; $A_{v_2} = 1$; $A_v = 159$; $R_{i_1} = 3.6$ kΩ and $V_0 = 100$ mV)

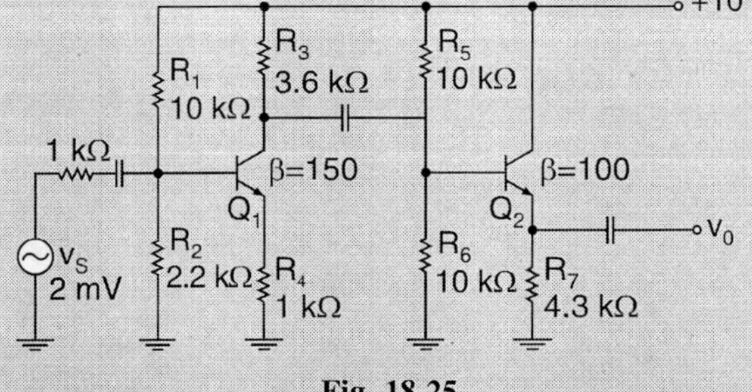

Fig. 18-25.

3. Fig. 18-26 shows a direct coupled two-stage amplifier. The d.c. bias of the first stage sets the d.c. base for the second stage. Find the individual stage voltage gain and the overall voltage gain. Neglect V_{BE} and use $r'_e = 25$ mV/I_E. (mA).Take $\beta_1 = \beta_2 = 100$.

(**Ans.** $A_{v_1} = 259$, $A_{v_2} = 1$; $A_v = 259$)

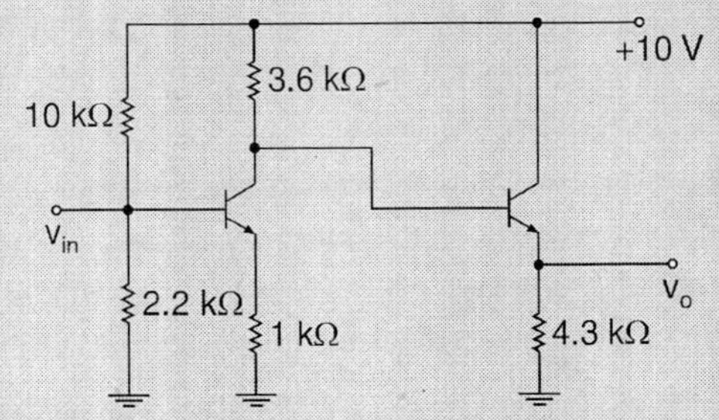

Fig. 18-26.

DESCRIPTIVE QUESTIONS

1. What do you mean by a multistage amplifier? Explain it briefly.
2. State the various methods of cascading transistor amplifiers.
3. Give a circuit diagram of a two-stage transistorized RC coupled amplifier. Also draw the frequency response of the amplifier.
4. With the help of a suitable circuit diagram, explain the working of a RC coupled amplifier. Derive the expression for voltage gain of the amplifier.
5. Draw the circuit diagram of a RC coupled amplifier using *PNP* transistor.
6. Explain with suitable circuit diagram, the operation of transformer coupled transistorized amplifier.
7. Explain the essential difference between the RC coupled and direct coupled amplifier.
8. Draw the circuit diagram of a Darlington emitter follower and derive the expressions for its voltage gain and input resistance.
9. Give the equation of overall upper and lower cut off frequencies of multistage amplifier.
   (*Electronic Circuits - I, Anna University, April/May 2008*)
10. Derive for the upper and lower cut off frequencies of RC coupled BJT amplifier.
    (*Electronic Circuits - I, Anna University, April/May 2008*)
11. Discuss the frequency response characteristics of *RC* coupled amplifiers. Derive the general expressions for gain at low, middle and high frequencies. Draw Bode plots for low and high frequency regions.
    (*Electronic Circuits - I, Anna University, April/May 2008*)
12. Explain how the characteristics are modified with transformer coupling.
    (*Electronic Circuits - I, Anna University, Nov/Dec 2007*)
13. Why capacitive coupling used to connect a signal source to an amplifier?
    (*Electronic Circuits - I, Anna University, Nov/Dec 2007*)

MULTIPLE CHOICE QUESTIONS

1. The gain of a cascaded amplifier is equal to the
   (*a*) product of individual gains (*b*) sum of individual gains
   (*c*) ratio of stage gains (*d*) none of these
2. Two stages of a multistage amplifier have a gains of 50 and 20. The dB voltage gain is
   (*a*) 3 (*b*) 30
   (*c*) 300 (*d*) 1000

3. The resistance-capacitance coupling is popular in audio amplifiers because
(*a*) it provides an output signal in phase with the input signal
(*b*) it needs low voltage collector supply
(*c*) it has better audio frequency response
(*d*) none of these

4. The major advantage of a direct coupled amplifier is that
(*a*) it uses less number of components
(*b*) it has a very good temperature stability
(*c*) it does not use frequency sensitive components
(*d*) it can amplify direct current and low frequency signals

5. In a Darlington pair,
(*a*) the two transistors are connected in parallel
(*b*) the emitter of the first transistor feeds the base of the second
(*c*) the collector of the first transistor feeds the base of the second

6. Cascading two amplifiers will result in
(*a*) reduction in overall gain and increase in overall bandwidth
(*b*) reduction in overall gain and reduction in overall bandwidth
(*c*) increase in overall gain and increase in overall bandwidth
(*d*) increase in overall gain and reduction in overall bandwidth

7. The overall bandwidth of two identical voltage amplifiers connected in cascade will
(*a*) remain the same as that of a single stage
(*b*) be worse than that of a single stage
(*c*) be better than that of a single stage
(*d*) be better if stage gain is low and worse if stage gain is high.

8. RC coupling is popular in low-level audio amplifiers because it
(*a*) has better low frequency response
(*b*) is inexpensive and needs no adjustments
(*c*) provides an output signal in phase with the input signal
(*d*) needs low voltage battery for collector supply

9. Frequency response characteristic at a single-stage RC coupled amplifier is shown in Fig. 18-27. The fall in gain at both ends of the characteristic is due to

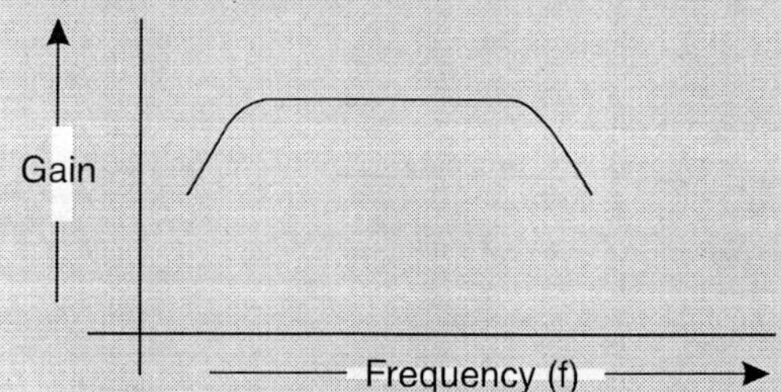

Fig. 18-27.

(*a*) transistor shunt capacitances
(*b*) bypass and coupling capacitances of the circuit
(*c*) transistor shunt capacitances of the lower end and bypass and coupling capacitances at the higher end
(*d*) transistor shunt capacitances at the higher end and bypass and coupling capacitances at the lower end.

10. The most desireable feature of a transformer coupling is
(*a*) higher voltage gain
(*b*) wide frequency range
(*c*) ability to provide impedance matching between the stages
(*d*) ability to eliminate hum from the output

11. A transformer coupled amplifier would give
(*a*) maximum voltage gain (*b*) impedance metching
(*c*) maximum current gain (*d*) larger bandwidth

12. In multistage amplifiers, direct coupling is especially suited for amplifying
(*a*) high frequency a.c. singals (*b*) changes in d.c. voltages
(*c*) high-level voltages (*d*) sinusoidal singals.

13. The outstanding characteristic of a direct coupled amplifier is its
(*a*) utmost economy
(*b*) temperature stability
(*c*) avoidance of frequency-sensitive components
(*d*) ability to amplify direct current and low frequency signals.

14. The coupling that produces minimum interferance with frequency response is
(*a*) direct coupling (*b*) impedance coupling
(*c*) RC coupling (*d*) transformer coupling

15. A signal has frequency components lying in the range of 0.001 Hz to 10 Hz. Which one of the following types of couplings should be chosen in a multistage amplifier designed to amplify this signal ?
(*a*) RC coupling (*b*) transformer coupling
(*c*) direct coupling (*d*) double-tuned transformer

16. The Darlington pair consists of the two stages :
(*a*) CE & CC (*b*) both CE
(*c*) both CC (*d*) CE & CB

17. The Darlington pair is mainly used for
(*a*) impedance matching (*b*) wide band voltage amplification
(*c*) power amplification (*d*) reducing distortion.

ANSWERS

1. (*a*)	**2.** (*d*)	**3.** (*c*)	**4.** (*d*)	**5.** (*b*)
6. (*c*)	**7.** (*c*)	**8.** (*b*)	**9.** (*d*)	**10.** (*c*)
11. (*b*)	**12.** (*b*)	**13.** (*d*)	**14.** (*a*)	**15.** (*c*)
16. (*c*)	**17.** (*c*)			

❑❑❑

19

CHAPTER

FET Amplifiers

OUTLINE

Objectives

After completing this chapter, you should be able to :

- Describe different methods of biasing a JFET.
- Calculate the operating point for a self-baised JFET.
- Understand the different biasing methods of MOSFET's.
- Explain small-signal FET models.
- Analyse common source, common drain and common gate amplifiers.
- Understand the effect of a.c. load on common source amplifier parameters.
- Know the effect of external source resistance on the voltage gain of common source amplifier.

19-1. Introduction

We have already discussed the various types of amplifiers using bipolar junction transistors. Now we shall discuss some amplifiers, using field-effect transistors (abbreviated as FET's). The field-effect transistor is a three terminal unipolar device in which the current is controlled by an electric-field. A field-effect transistor has a number of advantages over a bipolar junction transistor namely *extremely high input resistance, less noisy, better thermal stability, etc.* Its only disadvantage is that it has a relatively *small gain-bandwidth product.* However, this drawback is being improved considerably by using advanced manufacturing techniques. Because of the above mentioned advantages, the field-effect transistor amplifier has become more popular.

19-2. Biasing the FET

To use a FET in any application, it is necessary first to bias the device. The usual reason for this biasing is to turn the device ON and to place it in a region where it operates properly and provide a constant amount of voltage gain. The biasing deals with setting a fixed level of current which should flow through the FET with a desired fixed voltage drop across the FET junctions.

19-3. Biasing the JFET

There are several methods of biasing the JFET. These are (*a*) Gate bias, (*b*) Self-bias, (*c*) Voltage divider bias, and (*d*) Source bias. Whatever be the biasing method, the gate-source junction of a JFET will always be reverse biased. Even when the gate and source terminals are shorted, the drain current (I_D) will cause the junction to be reverse biased. As a matter of fact, the gate-source junction of a JFET must never be allowed to become forward biased. It is because of the reason that the gate material is not constructed to handle any significant amount of current. If the junction is allowed to become forward biased, current will start to flow in the gate circuit. Under most circumstances, this current will destroy the JFET.

Now we shall discuss the JFET biasing methods one by one in the following pages.

19-4. Gate Bias

The gate bias for JFET is similar to the base bias of bipolar transistor Fig. 19-1(*a*) shows the gate bias circuit for N-channel, JFET (junction field-effect transistor). In this circuit, the gate supply voltage ($-V_{GG}$) is used to ensure that the gate-source junction is reverse biased. Since, there is no gate current, there is no voltage drop across R_G.

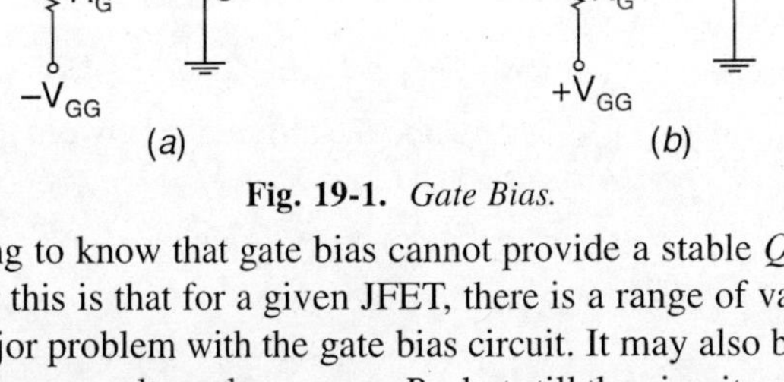

Fig. 19-1. *Gate Bias.*

It will be interesting to know that gate bias cannot provide a stable *Q*-point from one JFET to another. The reason for this is that for a given JFET, there is a range of values for $V_{GS(off)}$ and I_{DSS} which can lead to a major problem with the gate bias circuit. It may also be noted that even though there is no gate current or any voltage drop across R_G, but still the circuit contains R_G. Actually, R_G is used for a.c. operation purpose. Thus R_G resistor has little to do with the d.c. operation of the amplifier, but it is a vital part of the a.c. operation of the circuit.

Fig. 19-1 (*b*) shows the gate bias circuit for P-channel JFET. This circuit is similar to the one for a N-channel JFET, except that the polarity of V_{DD} and V_{GG} supplies are reversed.

19-5. Self-Bias

Fig. 19-2 (*a*) shows the self-bias circuit for an N-channel junction field-effect transistor (JFET). In this circuit, there is only one drain supply and no gate supply. The gate terminal is connected through a resistor (R_G) to the ground. The source terminal is connected through a resistor (R_S) to the ground. When the drain voltage is applied, a drain current flows even if there is no gate current (V_G). The drain current produces a voltage drop across resistor R_S (equal to $I_D . R_S$). This voltage drop produces the gate-to-source reverse voltage required for an FET operation. The resistor R_S is called a feedback resistor. Its function is to prevent any variation in the FET drain current, which may be understood from the following discussion.

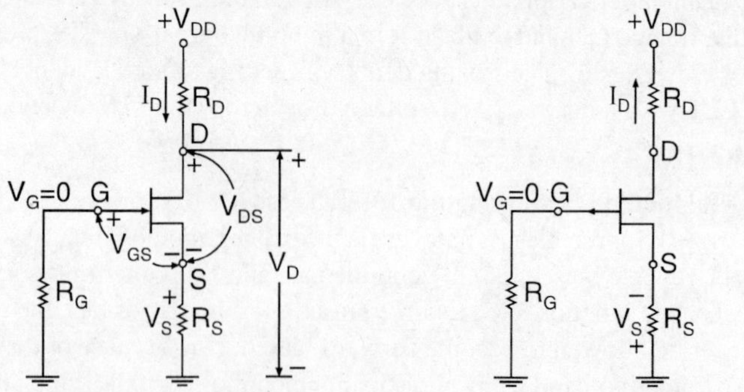

(*a*) For N-channel JFET. (*b*) For P-channel JFET.

Fig. 19-2. *Gate Bias.*

First of all, consider an increase in drain current, which will increase the voltage drop across the resistor (R_S). The increased voltage drop increases the reverse gate-to-source voltage, which decreases the effective width of the channel. It reduces the value of drain current. Now, if the drain current decreases, then the reverse action taken place, *i.e.*, the reduced drain current decreases the gate-to-source voltage, which in turn increases the effective width of channel, thereby increasing the value of drain current.

Fig. 19-2 (*b*) shows the self-bias circuit for P-channel JFET. It is similar to N-channel JFET circuit, except that the polarity of drain supply voltage (*i.e.*, V_{DD}) is reversed. We know that in a JFET, no current flows through the gate terminal to the ground because the gate-source junction is reverse biased. In actual practice, a small current called reverse leakage current (I_{GSS}) does flow through the gate terminal. Usually, its value is very small and can be neglected. Therefore, we shall assume that the gate terminal is at zero voltage. In other words, the gate voltage with respect to ground, *i.e.*, $V_G = 0$. We know that the source voltage with respect to ground,

$$V_S = I_D \cdot R_S$$

and the drain voltage,

$$V_D = V_{DD} - I_D \cdot R_D$$

The drain-to-source voltage (V_{DS}) is equal to the difference between the drain voltage (V_D) and the source voltage (V_S),

i.e.,
$$V_{DS} = V_D - V_S = (V_{DD} - I_D \cdot R_D) - I_D \cdot R_S$$
$$= V_{DD} - I_D (R_D + R_S)$$

The gate-to-source voltage (V_{GS}) is equal to the difference between the gate voltage (V_G) and the source voltage (V_S),

i.e.,
$$V_{GS} = V_G - V_S$$

$$= 0 - I_D \cdot R_S$$

$$= - I_D \cdot R_S \qquad ...(i)$$

It is evident from the above relation that the gate-to-source voltage is equal to the negative of the voltage across the source resistor. It is thus obvious that greater the value of drain current, more negative will be the gate-to-source voltage. Rearranging equation (*i*),

$$I_D = -\frac{V_{GS}}{R_S} \qquad ...(ii)$$

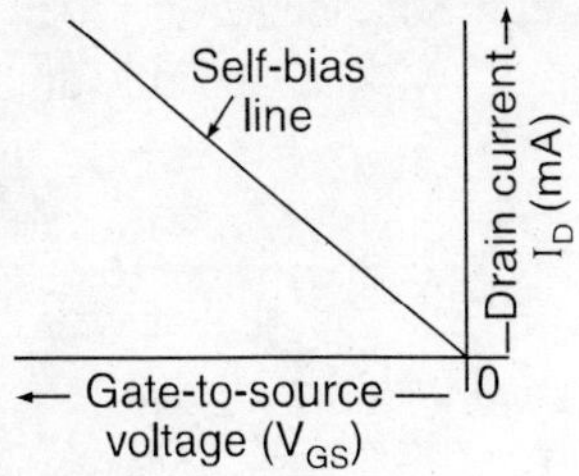

Fig. 19-3. *Self bias line.*

Now it is plot a graph using the value of drain current obtained from equation (*ii*) for a constant value of resistor (R_S) against the gate-to-source voltage we get a straightline as shown in Fig. 19-3. This straight-line is called a ***self-bias line.***

Example 19-1. *Find the values of drain-to-source voltage (V_{DS}) and the gate-to-source voltage (V_{GS}) for the circuit shown in Fig. 19-4. Given I_D = 5 mA.*

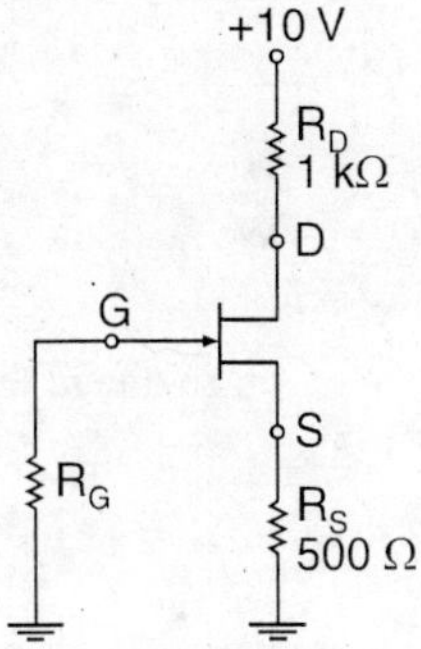

Fig. 19-4.

Solution. *Given:* $I_D = 5\ mA = 5 \times 10^{-3}$ A; V_{DD} = 10 volts; $R_D = 1\ k\Omega = 1 \times 10^3\ \Omega$ and $R_S = 500\ \Omega$

Drain-to-source voltage

We know that the source voltage,

$$V_S = I_D \cdot R_S = (5 \times 10^{-3}) \times 500 = 2.5 \text{ V}$$

and the drain voltage,

$$V_D = V_{DD} - I_D \cdot R_D = 10 - (5 \times 10^{-3}) \times (1 \times 10^3) \text{ V}$$

$$= 10 - 5 = 5 \text{ V}$$

∴ Drain-to-source voltage,

$$V_{DS} = V_D - V_S = 5 - 2.5 \text{ V} = 2.5 \text{ V } \textbf{Ans.}$$

Gate-to-source voltage

We know that gate-to-source voltage,

$$V_{GS} = - V_S = - 2.5 \text{ V } \textbf{Ans.}$$

Example 19-2. *The N-channel FET shown is Fig. 19-5 has I_{DSS} = 1.5 mA, V_P = – 1.5 V. If the quiscient drain to ground voltage is 10 volts, find R_1.* (*Nagpur Univ. Winter* 2003)

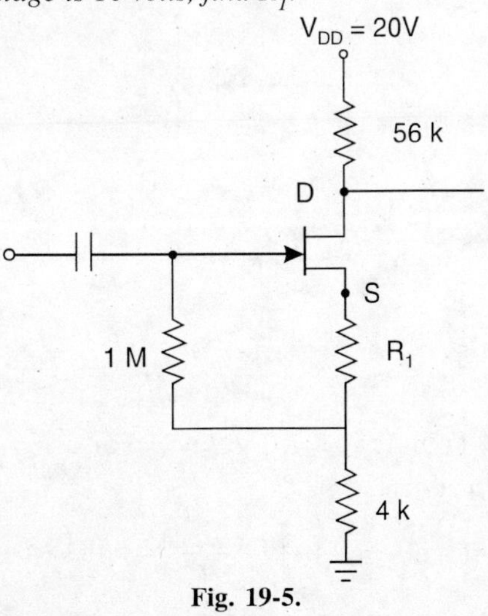

Fig. 19-5.

Solution. Given : $R_D = 56$ kW; $R_G = 1$ MΩ; $I_{DSS} = 1.5$ mA; $V_P = -1.5$ V. $V_D = 10$ V and $V_{DD} = 20$ V

We know that the drain voltage (measured with respect to ground), (V_D)

$$10 = V_{DD} - I_D \cdot R_D = 20 - I_D \times 56 \text{ k}.$$

$$\therefore \quad I_D = \frac{10}{56\ k} = 0.179 \text{ mA}$$

and the drain current,

$$0.179 \times 10^{-3} = I_{DSS}\left[1 - \frac{V_{GS}}{V_P}\right]^2 = (1.5 \times 10^{-3}) \times \left[1 - \frac{V_{GS}}{1.5}\right]^2$$

or
$$\left[1 - \frac{V_{GS}}{1.5}\right]^2 = \frac{0.179 \times 10^{-3}}{1.5 \times 10^{-3}} = 0.119$$

$$\therefore \quad 1 - \frac{V_{GS}}{1.5} = \sqrt{0.119} = 0.345$$

or
$$V_{GS} = -1.5 \times (1 - 0.345) = -0.98 \text{ V}$$

We also know that the source voltage,

$$V_S = V_{GS} = -0.98 \text{ V}$$

and voltage drop across the source resistor (V_S),

$$0.98 = I_D \cdot (R_1 + 4\ k) = 0.179 \times 10^{-3}\ (R_1 + (4 \times 10^3))$$

$$R_1 + (4 \times 10^3) = \frac{0.98}{0.179 \times 10^{-3}} = 5.475 \times 10^3 \approx 5.5 \times 10^3\ \Omega$$

$$\therefore \quad R_1 = (5.5 - 4) \times 10^3 = 1.5 \text{ k}\Omega$$ **Ans.**

Example 19-3. *In a self-bias N-channel JFET shown in Fig. 19-6, the operating point is to be set at $I_D = 1.5$ mA and $V_{DS} = 10$ V. The JFET parameters are $I_{DSS} = 5$ mA and $V_P = -2$ V. Find the values of R_S and R_D. Given that $V_{DD} = 20$ V.* (*Dip IETE., June* 2001)

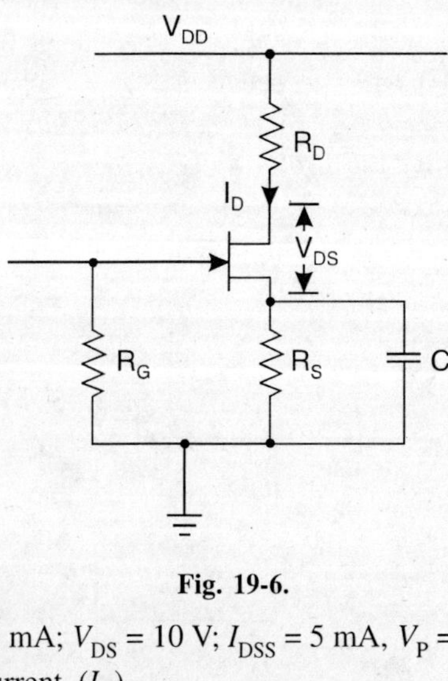

Fig. 19-6.

Solution. Given : $I_D = 1.5$ mA; $V_{DS} = 10$ V; $I_{DSS} = 5$ mA, $V_P = -2$ V and $V_{DD} = 20$ V.

We know that the drain current, (I_D),

$$1.5 \text{ mA} = I_{DSS}\left(1 - \frac{V_{GS}}{V_P}\right) = 5 \text{ mA}\left(1 - \frac{V_{GS}}{-2}\right)$$

$$\therefore \quad 1 + \frac{V_{GS}}{2} = \frac{1.5}{5} \Rightarrow \frac{V_{GS}}{2} = \frac{1.5}{5} - 1 = -0.7$$

or $$V_{GS} = -0.7 \times 2 = -1.4 \text{ V}$$

We also know that the d.c. voltage from gate-to-source,

$$V_{GS} = V_G - V_S = -V_S$$

$$\therefore \quad V_S = +1.4 \text{ V.}$$

The drain current (I_D),

$$1.5 \text{ mA} = \frac{1.4 \text{ } V}{R_S}; \; R_S = \frac{1.4 \text{ } V}{1.5 \text{ mA}} = 0.933 \text{ k}\Omega = 933 \; \Omega \text{ \textbf{Ans.}}$$

We also know that drain-to-source voltage (V_{DS}),

$$10 = V_{DD} - I_D (R_D + R_S)$$

$$= 20 - 1.5 \times 10^{-3} (R_D + 933)$$

$$\therefore \quad R_D = \left(\frac{20 - 10}{1.5 \times 10^{-3}} \right) - 933$$

$$= 5733.1 \; \Omega = 5.7 \text{ k}\Omega \text{ \textbf{Ans.}}$$

19-6. Setting a *Q*-point

We know that the *Q*-point (*quiescent operating point* or bias point) for a self-biased JFET is established by determining the value of drain current (I_D) for a desired value of gate-to-source voltage (V_{GS}) or vice versa. However, if the data sheet of a JFET includes a transfer characteristic curve (also called *transconductance curve*), then we can find the *Q*-point by using the procedure as given below :

(*i*) First of all, select a convenient value of drain current. Usually, its value is taken half of the maximum possible value of drain current (*i.e.*, I_{DSS}). Then find the voltage drop across source resistor (R_S) by the relation,

$$V_S = I_D \cdot R_S$$

and the gate-to-source voltage from the relation,

$$V_{GS} = -V_S$$

(*ii*) Now plot the assumed value of drain current (I_D) and the corresponding gate-to-source voltage (V_{GS}) on the transfer characteristic curve.

(*iii*) Draw a line through the plotted point and the origin. The point of intersection of the line and the curve gives the desired *Q*-point. Read the coordinates of *Q*-point.

As a matter of fact, it is desirable to set the *Q*-point near the mid-point of the transfer characteristic curve of a JFET. Under signal conditions, the mid-point bias allows a maximum amount of drain current swing between the values of I_{DSS} and *O*.

Now if we want to design a self-bias circuit, then any one of the following two methods is used for this purpose :

(1) *Analytical method*

In this method, we note down the values of maximum drain current (I_{DSS}) and the gate-to-source cut-off voltage ($V_{GS(off)}$) from the data sheets of JFET supplied by the manufacturer. Now the value of drain current is obtained by the relation,

$$I_D = I_{DSS} \left[1 - \frac{V_{GS}}{V_{GS(off)}} \right]^2$$

For example, if we select the gate to source voltage (V_{GS}) equal to $V_{CS(off)}/4$, then the value of drain current,

$$I_D = I_{DSS}\,[1 - 0.25]^2 = I_{DSS}\,(0.75)^2 = 0.56\,I_{DSS}$$

It indicates that the drain current is slightly more than one-half of I_{DSS}. But it will bias the JFET close to the mid-point of the curve. The value of drain resistor (R_D) is selected in such a way that the drain voltage (V_D) is equal to half the drain supply voltage (V_{DD}). The value of gate resistor (R_G)is chosen arbitrarily large, so that it prevents loading on the driving stages.

(2) *Graphical method*

In this method, we draw a self-bias line in such a way that it intersects the transfer characteristic curve near its mid-point. The point of intersection of the self-bias line and the transfer characteristic curve gives us the required *Q*-point. Then read the coordinaties of the *Q*-point. The value of source resistance (R_S) is given by the ratio of gate-to-source voltage (V_{GS}) to the drain current (I_D). Mathematically, the source resistance,

$$R_S = \frac{V_{GS}}{I_D}$$

However, a more accurate method is to draw a self-bias line through the coordinates of I_{DSS} and $V_{GS(off)}$as shown in Fig. 19-7. Then the point of intersection of self-bias line and the transfer characteristic curve locates the *Q*-point. The value of source resistor is given by the relation,

$$R_S = \frac{V_{GS(off)}}{I_{DSS}}$$

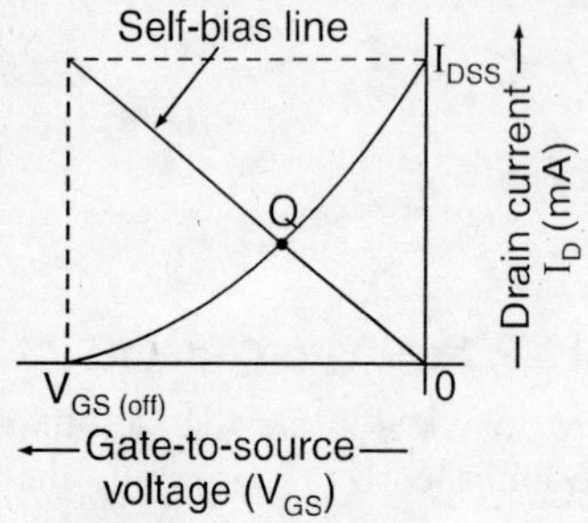

Fig. 19-7. *Setting a point*

The value of drain resistor (R_D) and gate resistor (R_G) are selected in the same way as discussed above for the analytical method.

Example 19-4. *Fig. 19-8 (a) shows that circuit of a self-biased JFET amplifier and Fig. 19-8 (b) indicates the transfer characteristic curve of the JFET.*

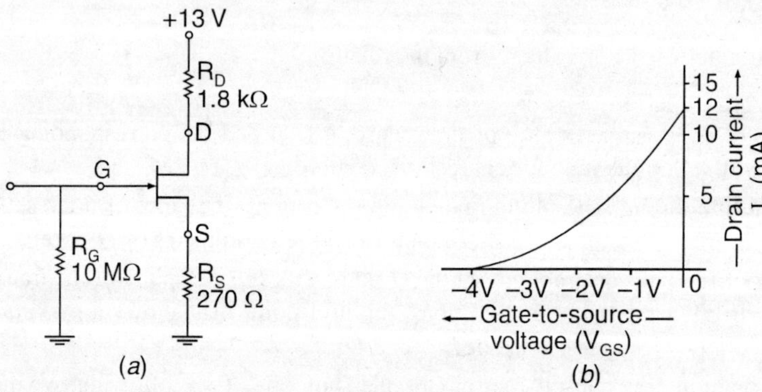

Fig. 19-8.

Find the quiescent values of I_D and V_{GS}. Calculate the values of d.c. voltage between the drain and ground.

Solution. Given: $R_D = 1.8\text{ k}\Omega = 1.8 \times 10^3\ \Omega$; $R_S = 270\ \Omega$; $R_G = 10\text{ M}\Omega$ and I_{DSS} (from transfer characteristic curve) = 12 mA.

(*a*) *Quiescent values of I_D and V_{GS}*

(*i*) First of all, assume the drain current (I_D) equal to the half of I_{DSS} (*i.e.*, 6 mA or 6×10^3 A). Using this value of drain current, the voltage drop across resistor (R_S),

$$V_S = I_D \cdot R_S = (6 \times 10^{-3}) \times 270 = 1.62\text{ V}$$

and the gate-to-source voltage,

$$V_{GS} = -V_S = -1.62 \text{ V}$$

(*ii*) Now plot the values of drain current (I_D = 6 mA) and the corresponding value of voltage V_{GS} (*i.e.*, – 1.62 V) on the graph of transfer characteristic curve as shown in Fig. 19-9.

(*iii*) Draw a line through the plotted point (P) and the origin as shown in the figure and read the coordinates of Q-point.

It is evident from this graph that the coordinates of Q-point are

$$I_D = 5 \text{ mA}$$

$$V_{GS} = -1.4 \text{ V} \textbf{ Ans.}$$

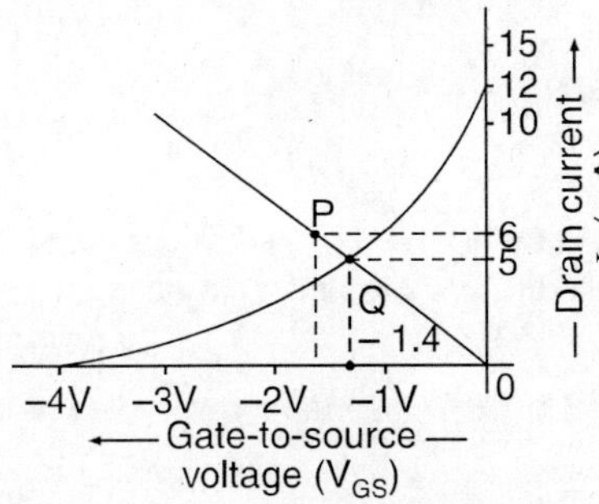

Fig. 19-9.

(**b**) *D.C. voltage between drain and ground*

We know that the d.c. voltage between drain and ground,

$$V_D = V_{DD} - I_D \cdot R_D$$
$$= 15 - (5 \times 10^{-3}) \times (1.8 \times 10^3) \text{ V}$$
$$= 15 - 9 = 6 \text{ V } \textbf{Ans.}$$

Example 19-5. *A self-biased P-channel JFET has a pinch off voltage* (V_P) = 5 V *and* I_{DSS} = 12 mA. *The supply voltage available is 12 V. Determine the values of resistors* R_D *and* R_S, *so that* I_D = 5 mA *and* V_{DS} = 6 V.

Solution. Given: V_P *i.e.*, $V_{GS(off)}$ = 5 volt; I_{DSS} = 12 mA = 12 × 10^{-3}A; V_{DD} = 12 volt, I_D = 4 mA = 4 × 10^{-3} A and V_{DS} = 6 volt.

We know that the drain current (I_D),

$$4 \times 10^{-3} = I_{DSS}\left[1 - \frac{V_{GS}}{V_{GS(off)}}\right]^2 = 12 \times 10^{-3}\left[1 - \frac{V_{GS}}{5}\right]^2$$

or $$\frac{1}{3} = 1 - \frac{V_{GS}}{5}^{\;2}$$

or $$\frac{1}{\sqrt{3}} = 1 - \frac{V_{GS}}{5} \qquad \text{...(Taking square root)}$$

$\therefore$ $$V_{GS} = 1 - \frac{1}{\sqrt{3}} \times 5 = (1 - 0.58) \times 5 \text{ V} = 2.1 \text{ V}$$

We also know that the source voltage,

$$V_S = V_{GS} = 2.1 \text{ V}$$

and voltage drop across the source resistor (V_S),

$$2.1 = I_D \cdot R_S = (4 \times 10^{-3})\, R_S$$

$\therefore$ $$R_S = \frac{2.1}{(4 \times 10^{-3})} = 525\ \Omega \textbf{ Ans.}$$

and drain-to-source voltage (V_{DS}),

$$6 = V_{DS} - I_D \cdot R_D = 12 - (4 \times 10^{-3}) \times R_D$$

$$R_D = 1500\ \Omega = 1.5 \text{ k}\Omega \textbf{ Ans.}$$

19-7. Setting a *Q*-point Using D.C. Load Line

The Q-point for the self-biased JFET circuit may also be obtained graphically by using d.c. load line. The d.c. load line may be drawn easily on the drain characteristics as shown in Fig. 19-10 (*a*).

The values of drain current (I_D) and drain-to-source voltage (V_{DS}) at the upper and lower ends of the d.c. load line may be obtained by using the relation,

$$V_{DS} = V_{DD} - I_D \cdot (R_D + R_S) \quad ...(i)$$

At the upper end (*i.e.*, point *A*), the value of drain-to-source voltage (V_{DS}) is equal to zero. Therefore, substituting $V_{DS} = 0$ in equation (*i*),

$$0 = V_{DD} - I_D \cdot (R_D + R_S)$$

$$\therefore \quad I_D = \frac{V_{DD}}{(R_D + R_S)}$$

Similarly, at the lower end (*i.e.*, point *B*) the value of drain current (I_D) is equal to zero. Therefore, substituting, $I_D = 0$ in equation (*i*),

$$V_{DS} = V_{DD} - 0\ (R_D + R_S) = V_{DD}$$

Thus the upper and lower ends of the d.c. load line are given by the points,

$$I_D = V_{DD}/(R_D + R_S) \text{ and } V_{DS} = 0 \quad ...\text{(Upper end)}$$

$$I_D = V_{DS} = V_{DD} \quad ...\text{(Lower end)}$$

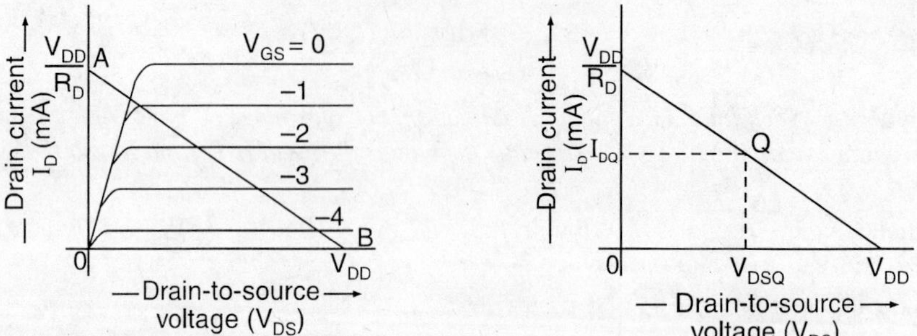

(*a*) Drawing d.c. load line. (*b*) Setting a Q-point.

Fig. 19-10.

The *Q*-point is set at the mid-point of the d.c. load line as shown in Fig. 19-10 (*b*). In that case, the value of drain current at *Q*-point,

$$I_{DQ} = \frac{V_{DD}}{2(R_D + R_S)}$$

and the drain-to-source voltage at the *Q*-point,

$$V_{DSQ} = \frac{V_{DD}}{2}$$

Example 19-6. *Calculate the self-bias operation point for the FET circuit shown in Fig. 19-11. Also calculate the values of resistors R_D and R_S to obtain this bias condition. Given the maximum value of drain current as 10 mA and $V_{GS} = -2.2$ V at $I_D = 5$ mA.*

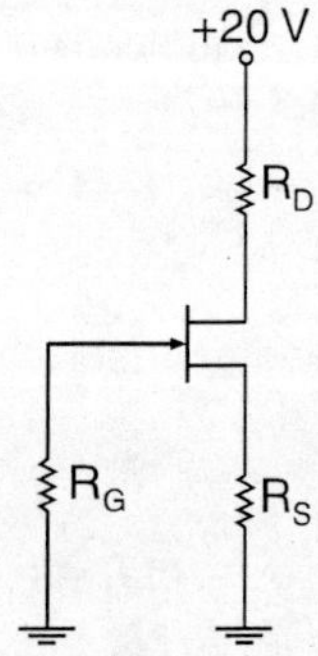

Fig. 19-11.

Solution. Given: $I_{DSS} = 10$ mA and $V_{DD} = 20$ volt.

(*a*) *Operating point*

We know that the value of drain current at *Q*-point,

$$\therefore \quad I_{DQ} = \frac{I_{DSS}}{2} = \frac{10}{2} = 5 \text{ mA}$$

and the value of drain-to-source voltage,

$$V_{DSQ} = \frac{V_{DD}}{2} = \frac{20}{2} = 10 \text{ V}$$

∴ Operating point,

$$I_D = 5 \text{ mA and } V_{DS} = 10 \text{ V } \textbf{Ans.}$$

(*b*) *Values of R_D and R_S*

Here $V_{GS} = -2.2$ volt and $I_D = 5 \text{ mA} = 5 \times 10^{-3}$ A

We know that the drain-to-source voltage (V_{DS}),

$$10 = V_{DD} - I_D \cdot R_S = 20 - (5 \times 10^{-3})\, R_D$$

$$\therefore \quad R_D = \frac{20 - 10}{5 \times 10^{-3}} = 2 \times 10^3 \ \Omega = 2 \text{ k}\Omega \ \textbf{Ans.}$$

We also know that the source voltage,

$$V_S = -V_{GS} = -(-2.2) = 2.2 \text{ V}$$

and the value of voltage drop across the source resistor (V_S),

$$2.2 = I_D \cdot R_S = (5 \times 10^{-3}) \times R_S$$

$$\therefore \quad R_S = 2.2/(5 \times 10^{-3})\ 440\ \Omega \ \textbf{Ans.}$$

19-8. Biasing Against Device Parameter Variation

We have already discussed that for a self-biased JFET, the *Q*-point is set at the mid-point of the transfer characteristic curve. However, it has been found that the JFET parameters *like maximum drain current (I_{DSS}) and the gate-to-source cut-off ($V_{GS(off)}$) vary with the temperature. In that case, the Q-point will also vary with the temperature and hence will not remain stable.* This degrades the performance of the FET amplifier circuit.

In order to provide compensation for the device parameter variation, the manufacturers provide two transistor characteristic curves as shown in Fig. 19-12. The first curve corresponds to the maximum values of I_{DSS} and $V_{GS(off)}$. In such a case, we draw a bias line through both the curves, which is near the mid-point of either curve. Then read the values of gate-to-source voltage (V_{GS}) and the drain current (I_D) for either *Q*-point. The optimum value of source resistor (R_S) is given by the ratio of gate-to-source voltage (V_{GS}) and drain current (I_D).

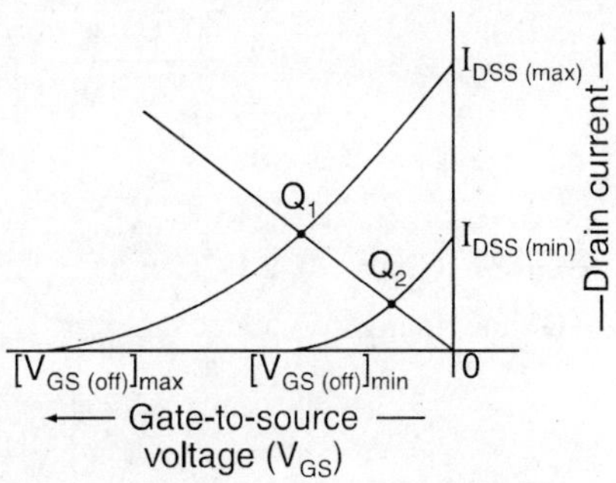

Fig. 19-12. *Use of two transfer characteristic curves for setting a Q-point.*

Sometimes, the manufacturers do not supply the transfer characteristic curves, but provide the maximum and minimum values of the I_{DSS} and $V_{GS(off)}$. In that case, we calculate the maximum and minimum values of source resistor (R_S) and then take the average of these two values. This average value will provide a *Q*-point near the mid-point of the either transfer characteristic curve.

19-9. Voltage Divider Bias

Fig. 19-13 shows the voltage divider bias circuit. The name 'voltage divider' is derived from the fact that resistor R_1 and R_2 are connected on the gate side (between the V_{DD} supply and ground) form a voltage divider. Assuming the current to be zero, the voltage is given by the relation,

$$V_G = \left(\frac{R_2}{R_1 + R_2}\right) V_{DD}$$

and the voltage from source to ground,

$$V_S = V_G - V_{GS}$$

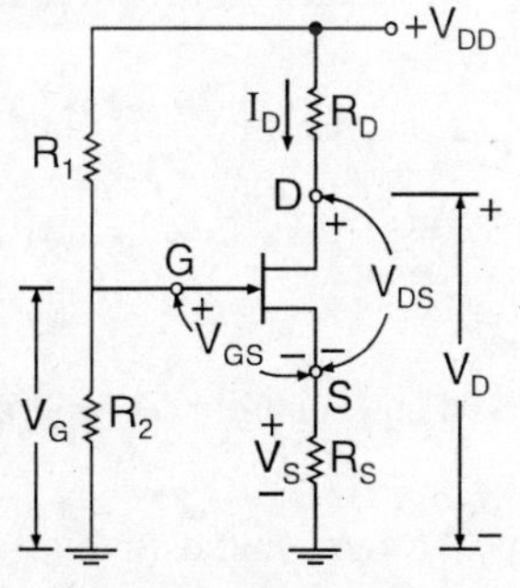

Fig. 19-13. *Voltage divider bias.*

∴ Value of drain current,

$$I_D = \frac{V_S}{R_S} = \frac{V_G - V_{GS}}{R_S} \quad ...(i)$$

and the d.c. voltage from drain to ground,

$$V_D = V_{DD} - I_D \cdot R_D$$

Now, if the gate voltage is very large, as compared to gate-to-source voltage, then the drain current is approximately constant for any JFET. It has been found that in actual practice the voltage divider bias is less effective with JFET's than bipolar transistors. It is because of the fact that in bipolar transistors the base-to-emitter voltage (V_{BE}) is approximately 0.7 V with only minor variations from one transistor to next. However, in a JFET, the gate-to-source voltage (V_{GS}) can vary several volts from one JFET to another. As a result of this, it is difficult to make gate voltage (V_G) large enough than the gate-to-source voltage (V_{GS}).

Example 19-7. *Fig. 19-14 shows an FET voltage divider bias circuit. If I_D = 4 mA, determine V_{GS} and V_{DS} respectively.*

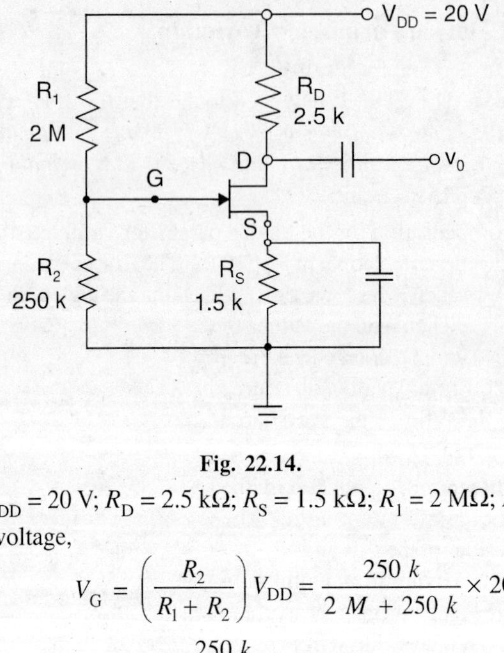

Fig. 22.14.

Solution. Given : V_{DD} = 20 V; R_D = 2.5 kΩ; R_S = 1.5 kΩ; R_1 = 2 MΩ; R_2 = 250 kΩ; I_D = 4 mA.

We know that gate voltage,

$$V_G = \left(\frac{R_2}{R_1 + R_2}\right) V_{DD} = \frac{250\ k}{2\ M + 250\ k} \times 20$$

$$= \frac{250\ k}{2000\ k + 250\ k} \times 20 = 2.22 \text{ V}$$

and the voltage from source to ground,

$$V_S = V_G - V_{GS} \quad ...(i)$$

The value of V_{GS} is obtained as follows. We know that drain current, (I_D)

$$4 \text{ mA} = \frac{V_S}{R_S} = \frac{V_S}{1.5\ k}$$

$$\therefore \quad V_S = 4 \text{ mA} \times 1.5 \text{ k} = 6.0 \text{ V}$$

Using equation (*i*), we find that

$$V_{GS} = V_G - V_S = 2.22 - 6.0 = -3.8 \text{ V } \textbf{Ans.}$$

We know that d.c. voltage from drain to ground,

$$V_D = V_{DD} - I_D \cdot R_D = 20 - (4 \text{ mA} \times 2.5 \text{ k}) = 10 \text{ V}$$

∴ Drain-to-source voltage,

$$V_{DS} = V_D - V_S = 10 - 6.0 = 4.0 \text{ V } \textbf{Ans.}$$

19-10. Source Bias

Fig. 19-15 shows the circuit of a source bias for JFET. It may be noted that this circuit is similar to the emitter bias circuit used for biasing bipolar transistors.

The value of drain current is given by the relation,

$$I_D = \frac{V_{SS} - V_{GS}}{R_S}$$

Since, $V_{SS} >> V_{GS}$, therefore drain current,

$$I_D \approx \frac{V_{SS}}{R_S}$$

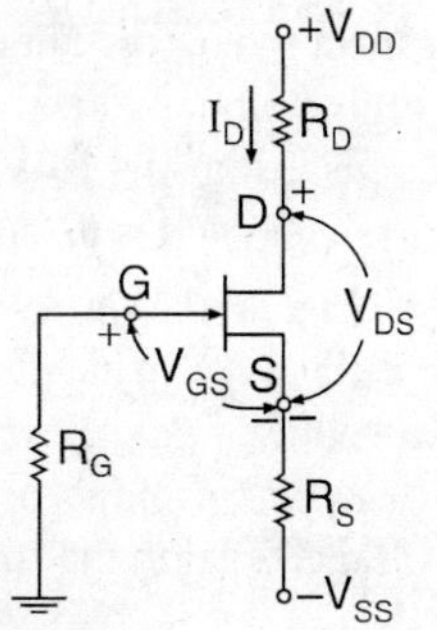

Fig. 19-15. *Source bias.*

19-11. Current Source Bias

Fig. 19-16 shows the current-source bias circuit for a JFET. The circuit of Fig. 19-16 (*a*) uses two supplies (positive and negative), while that of Fig. 19-16 (*b*) employs only one (*i.e.*, positive) supply.

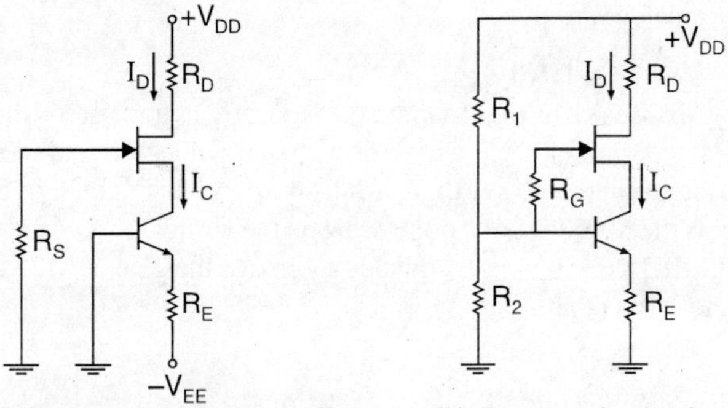

(*a*) Two supply circuits. (*b*) One supply circuits.

Fig. 19-16. *Current-source bias circuit.*

First of all, consider the two supply current-source bias circuit. Here, the bipolar transistor is emitter biased. Therefore, its collector current is given by the relation,

$$I_C = \frac{V_{EE} - V_{BE}}{R_E} = \frac{V_{EE}}{R_E} \quad \text{...(If } V_{EE} >> V_{BE})$$

The drain current of the JFET is equal to the collector current because the bipolar transistor acts like a d.c. current-source, *i.e.*,

$$I_D = I_C$$

Now consider the one supply current-source bias circuit. Here, the bipolar transistor is voltage divider biased. Its collector current is given by the relation,

$$I_C = \frac{V_B - V_{BE}}{R_E}$$

and the drain current,

$$I_D = I_C$$

The current-source bias circuit provides a solid *Q*-point with JFET. Hence, this circuit is very useful in actual practice.

19-12. Biasing the Enhancement Type MOSFET's

We have already discussed in the chapter on field-effect transistors that the operation of an enhancement types MOSFET requires a gate-to-source voltage greater than its threshold value (*i.e.*, $V_{GS} > V_{GS(th)}$). This eliminates the use of self-bias because it makes the gate voltage negative with respect to source. There are two methods for biasing the enhancement type MOSFET's namely the voltage divider bias and the drain feedback bias. The voltage divider bias has been already discussed in Art. 19-9. The drain feedback bias is similar to the collector feedback bias of bipolar transistors and is discussed as below :

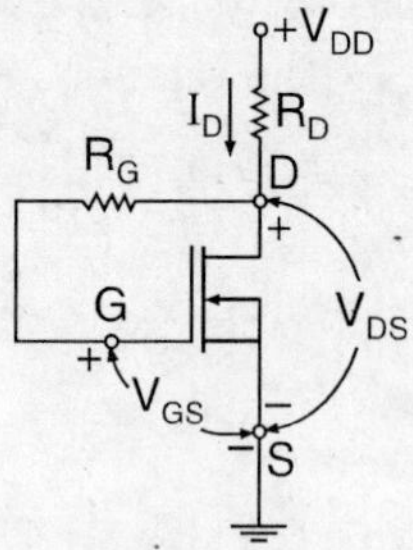

Fig. 19-17.
Drain feedback bias

Fig. 19-17 shows the circuit of a drain feedback bias. It may be noted that this type of bias is used only for enhancement type MOSFET's. Since, the value of gate current in FET's is approximately zero, therefore there is no voltage drop across gate resistor (R_G). This makes the gate-to-source voltage equal to the drain-to-source voltage, *i.e.*,

$$V_{GS} = V_{DS}$$

The value of drain current is given by relation,

$$I_D = \frac{V_{DD} - V_{DS}}{R_D}$$

The drain feedback bias provides a stable *Q*-point, because it tends to compensate for any changes in device parameter variation.

19-13. Biasing the Depletion Type MOSFET's

We have already discussed in the chapter on field-effect transistors that the depletion type MOSFET can be operated with either positive or negative values of gate-to-source voltage (V_{GS}). Therefore, the simplest method is to set the gate-to-source voltage equal to zero as shown in Fig. 19-18 (*a*). This will bias the MOSFET at a point lying on the vertical axis of the transfer characteristic curve as shown in Fig. 19-18 (*b*). If an a.c. signal is applied at the gate, it will vary the gate-to-source voltage above and below the *Q*-point.

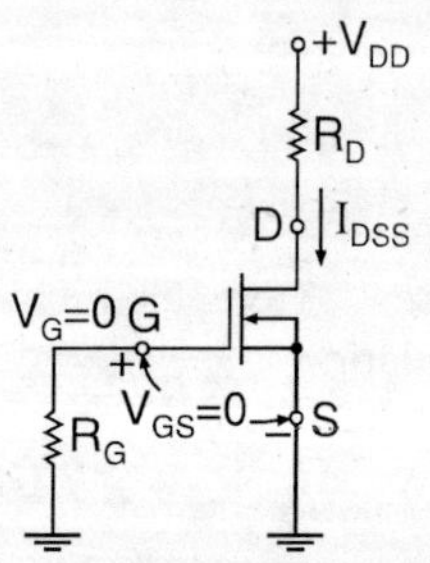

(*a*) Bias circuit for depletion type MOSFET.

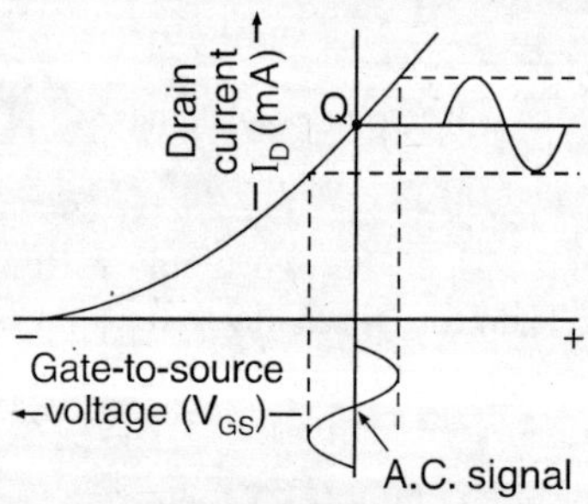

(*b*) Location of Q-point on the transfer characteristic curve.

Fig. 19-18.

It may be noted that the biasing circuit of a depletion type MOSFET has no applied gate or source voltage. Therefore, the gate-to-source voltage is equal to zero and the drain current is equal to its maximum value, (*i.e.*, $I_D = I_{DSS}$). The drain-to-source voltage is given by the relation,

$$V_{DS} = V_{DD} - I_D \cdot R_D = V_{DD} - I_{DSS} \cdot R_D$$

It may be noted that the zero bias is possible only with depletion type MOSFET. It cannot be used for biasing the JFET or bipolar transistor. However, all the biasing methods, discussed for the JFET, can be used for depletion type MOSFET in the depletion mode but not in enhancement mode.

19-14. Small-Signal FET Models

The small-signal FET *model (valid both for JFET and MOSFET) is used to relate small changes in FET current and voltages about the quiescent operating point. The model is different at low and high-frequencies. Therefore, we shall study the small-signal models separately as the low-frequency FET model and high-frequency model. In both these models, the FET will be considered in common source configuration.

19-15. Small-Signal Low-Frequency FET Model

Fig. 19-19 shows the small-signal low-frequency model of a field-effect transistor. In this model, the gate-to-source junction is represented by an open circuit and no current is drawn by the input terminal of the field-effect transistor. It is because of the fact, that the input resistance (*i.e.*, the resistance of gate-source junction) is very large. Its value at d.c. or zero frequency is typically 10^8 to 10^{10} Ω for JFET's and 10^{10} to 10^{14} for MOSFET's. For all practical purposes, these values can be considered to be so large that the input of FET can be considered as an open circuit.

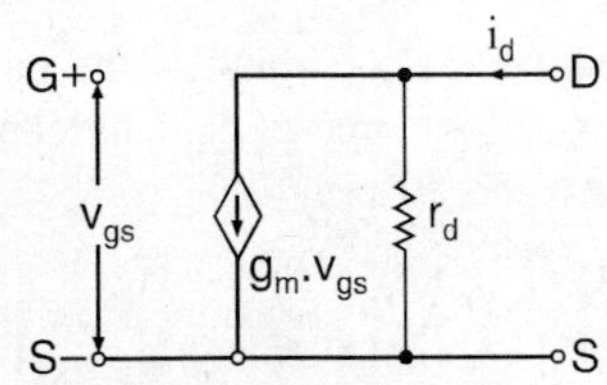

Fig. 19-19. *Low-frequency FET model.*

It will be interesting to know that although the gate-source junction appears as an open circuit, yet the gate-to-source voltage affects the value of drain current. It is indicated by a voltage-controlled current source ($g_m \cdot v_{gs}$) whose value is proportional to the gate-to-source voltage. The FET transconductance (g_m) is measured in milliamperes per volt (mA/V), milli siemens (mS) or milli mhos (mΩ). Typical values of transconductance are from 0.5 mA/V to 10 mA/V for JFET's and 0.5 mA/V to 20 mA/V for MOSFET's.

The FET drain resistance (also called FET output resistance) is represented by the resistance (r_d). Typical values of drain resistance are from 100 kΩ to 1 MΩ for JFET's and 1 kΩ to 50 kΩ for MOSFET's.

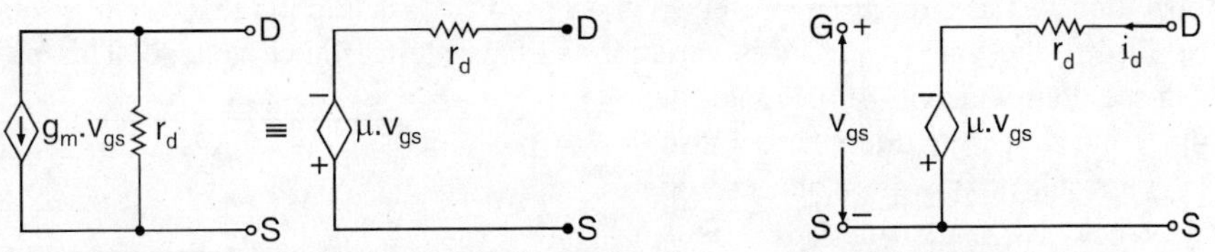

(*a*) Voltage-controlled current source with its equivalent voltage-controlled voltage source.

(*b*) Another form of low-frequency FET model.

Fig. 19-20.

It is possible to convert the voltage-controlled current source ($g_m \cdot v_{gs}$) into its equivalent voltage-controlled voltages source as shown in Fig. 19-20 (*a*).

The equivalent circuit consists of a voltage-controlled voltage source ($\mu \cdot v_{gs}$) in series with the drain resistance (r_d). The parameter (μ) is the amplification factor of the field effect transistor and its value is equal to the product of transconductance (g_m) and the drain resistance (r_d). Mathematically, the amplification factor,

$$\mu = g_m \cdot r_d$$

Fig. 19-20 (*b*) shows another form of the low frequency model of a field effect transistor using the voltage source representation. Sometimes this representation is also very useful in analysing the FET amplifiers.

---

* It is also called an equivalent circuit.

19-16. Small-Signal High-Frequency FET Model

Fig. 19-21 shows the small-signal high-frequency model of a field-effect transistor. It is identical to the low-frequency model, except the addition at capacitances between each pair of terminals. The capacitor, (c_{gs}) represents the barrier capacitance between the gate and source. Its typical value is from 1 pF to 10 pF for both JFET's and MOSFET's. The capacitor (C_{gd}) represents the barrier capacitance between the gate and drain. Its typical value is also from 1 pF to 10 pF for both JFET's and MOSFET's. Similarly, the capacitor (C_{ds}) represents the drain-to-source capacitance. The typical value of C_{ds} is from 0.1 pF to 1 pF.

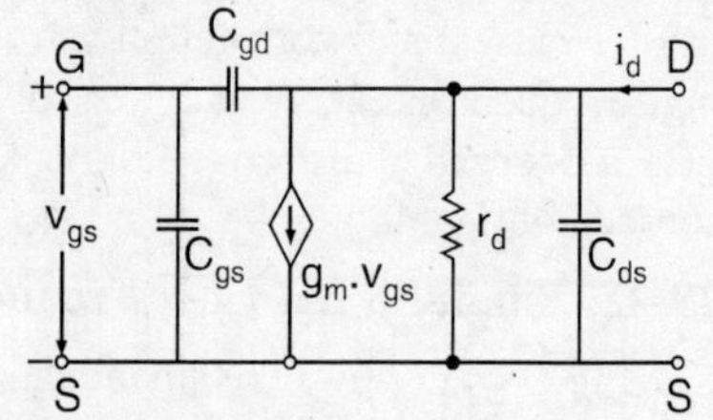

Fig. 19-21. *High-frequency FET model.*

19-17. The Field-Effect Transistor Amplifier

The field-effect transistor (FET) has a capability to amplify a.c. signals like a bipolar transistor. Depending open the configuration, the FET amplifiers may be studied under the following three heads :

(*a*) Common source amplifier.
(*b*) Common drain amplifier, and
(*c*) Common gate amplifier.

19-18. Common Source Amplifier

Fig. 19-22 shows the circuit of a common source N-channel JFET amplifier. It is similar to a common emitter amplifier. Here, the resistors R_1 and R_2 (called a voltage divider) are used to bias the field-effected transistor. The capacitor (C_1) and (C_2) are used to couple the a.c. input voltage source and the output voltage respectively, these are known as *coupling capacitors*. The capacitor (C_S) keeps the source of the FET effectively at a.c. ground and is known as bypass capacitor.

The operation of the circuit may be understood from the assumption that when a small a.c. signal is applied to the gate, it produces variations in the gate-to-source voltage. This produces variations in the drain current. As the gate-to-source voltage increases, the drain current also increases. As a result of this, the voltage drop across the resistor (R_D) also increases. This causes the drain voltage to decrease. It means that the positive half cycle of the input voltage produces the negative half cycle of the output voltage. In other words, the output voltage (at the drain) is 180° out-of-phase with the input voltage (at the gate). This phenomenon of *phase inversion* is similar to that exhibited by a common emitter bipolar transistor amplifier.

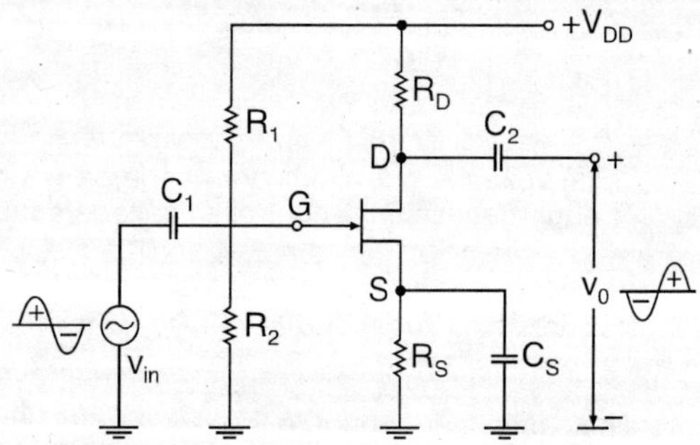

Fig. 19-22. *Common source amplifier.*

19-19. Analysis of Common Source Amplifier

Fig. 19-23 shows the a.c. equivalent circuit of a common source amplifier. This circuit has been obtained from the amplifier circuit shown in Fig. 19-22 by short-circuiting the capacitors and the d.c. voltage supplies. The field-effect transistor is also replaced by its low-frequency model (or equivalent circuit). Now we shall use this circuit to find the expressions for amplifier voltage gain, *input resistance and output resistance.

* In some books, the input and output resistances are referred as the input and output impedance respectively. But as we are studying the amplifier performance at low frequencies, at which the effect of capacitances is negligible, therefore it is more appropriate to refer the impedances as resistances.

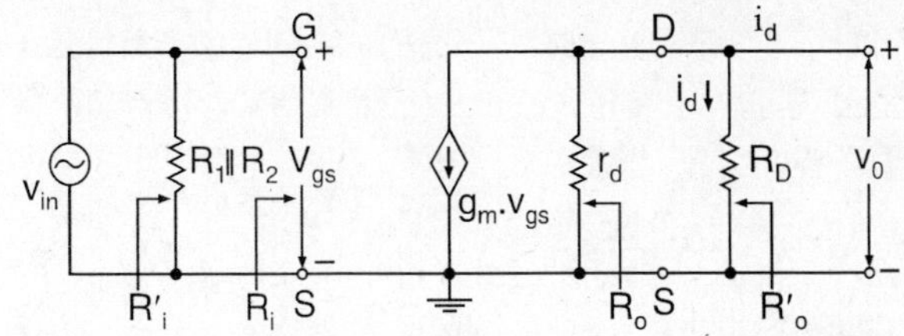

Fig. 19-23. *A.C. equivalent circuit of a common source amplifier.*

(1) ***Voltage gain :*** It is the ratio of the output voltage (v_o) to the input voltage (v_{in}). Mathematically the voltage gain,

$$A_v = \frac{v_o}{v_{in}}$$

It may be noted that the current from the current source splits between the resistors r_d and R_D. The current through resistor R_D (as per current divider rule) is given by the relation,

$$i_d = \frac{r_d}{R_D + r_d}(g_m \cdot v_{gs})$$

and the output voltage,

$$v_o = -i_d \cdot R_D = -\left(\frac{r_d}{R_D + r_d}\right)(g_m \cdot v_{gs})\, R_D$$

$$= -g_m\left(\frac{r_d \cdot R_D}{R_D + r_d}\right) v_{gs}$$

$$= -g_m (r_d \| R_D) \cdot v_{gs}$$

$$= -g_m \cdot r_L \cdot v_{gs} \qquad \ldots(\because r_L = r_d \| R_D)$$

$$= -g_m \cdot r_L \cdot v_{in} \qquad \ldots(\because v_{in} = v_{gs})$$

The minus sign indicates that the output voltage is 180° out-of-phase with the input voltage.

∴ Voltage

$$A_v = \frac{v_o}{v_{in}} = -g_m \cdot r_L$$

It may be noted that if the drain resistance (r_d) is sufficiently greater than resistor R_D, (*i.e.*, $r_d \geq 10\, R_D$) then the equivalent resistance of the resistors r_d and R_D in parallel,

$$r_L = R_D$$

and the voltage gain,

$$A_v = g_m \cdot R_D$$

Note. If the resistance R_D is much larger than r_d (*i.e.*, $R_D >> r_d$) then a.c. load resistance (r_L) is approximately equal to r_d. In that case, the amplifier gain is equal to the FFT amplification factor (μ), which is the theoretical maximum voltage gain of the FET. But in actual practice, we cannot select R_D to be much larger than r_d because it makes the d.c. bias current in the range of microamperes. The FET transconductances (*i.e.*, g_m) is near zero at this low bias current value.

(2) ***Input resistance :*** It is the ratio of the input voltage (v_{in}) to the input current (i_{in}). Mathematically, the input resistance,

$$R_i = \frac{v_{in}}{i_{in}}$$

We know that input resistance (R_i) of a field-effect transistor is very high and hence, can be considered to be infinity (*i.e.*, open-circuit). However, the input resistance of the amplifier stage (R'_i) is equal to the parallel combination of resistors R_1 and R_2 and the FET input resistance (R_i). Thus,

$$R'_i = (R_1 \parallel R_2) \parallel R_i$$

$$= R_1 \parallel R_2 \qquad \text{...(When } R_i \text{ is infinite)}$$

It may be noted that if we use a self-bias circuit (instead of a voltage divider bias), then the input resistance of the amplifier stage,

$$R'_i = R_G$$

(3) ***Output resistance :*** It is the ratio of the output voltage (v_o) to the output current (i_d). Mathematically, the output resistance,

$$R'_0 = \frac{v_o}{i_d}$$

We know that the output resistance (R_O) of a field-effect transistor is r_d. But for the amplifier stage, the output resistance is the parallel combination of resistors R_D and r_d, *i.e.*,

$$R'_0 = R_D \parallel r_d$$

However, if $r_d >> R_D$, then the output resistance of amplifier stage,

$$R_0 = R_D$$

Example 19-8. *A certain JFET has a g_m of 4 mS. With an external drain resistance of 1.5 kΩ, find the value of ideal voltage gain.*

Solution. Given: $g_m = 4$ mS $= 4 \times 10^{-3}$ S and $R_D = 1.5$ kΩ $= 1.5 \times 10^3$ Ω

We know that the voltage gain,

$$A_v = -g_m \cdot R_D = -(4 \times 10^{-3}) \times (1.5 \times 10^3) = -6 \textbf{ Ans.}$$

Example 19-9. *A JFET amplifier has g_m = 2.5 mA/V and r_d = 500 kΩ. The load resistance is 10 kΩ.* Find the value of voltage gain.

Solution. Given: $g_m = 2.5$ mA/V $= 2.5 \times 10^{-3}$ A/V; $r_d = 500$ kΩ and $R_D = 10$ kΩ

We know that the a.c. equivalent resistance,

$$r_L = \frac{R_D \times r_d}{R_D + r_d} = \frac{10 \times 500}{10 + 500} \text{ k}\Omega = 9.8 \text{ k}\Omega = 9.8 \times 10^3 \ \Omega$$

∴ Voltage gain,

$$A_v = -g_m \cdot r_L = -(2.5 \times 10^{-3}) \times (9.8 \times 10^3)$$

$$= -24.5 \textbf{ Ans.}$$

Example 19-10. *The input and output resistances of the FET amplifier are shown in Fig. 19-24. Calculate the value of voltage gain. The FET amplifier has g_m = 2 mA/V and r_d = 40 kΩ.*

Solution. Given: $g_m = 2$ mA/V $= 2 \times 10^{-3}$ A/V; $r_d = 40$ kΩ; $R_D = 20$ kΩ and $R_G = 100$ MΩ.

Voltage gain

We know that the a.c. equivalent resistance,

$$r_L = \frac{R_D \times r_d}{R_D + r_d} = \frac{20 \times 40}{20 + 40} = 13.3 \text{ k}\Omega$$

$$= 13.3 \times 10^3 \ \Omega$$

∴ Voltage gain

$$A_v = -g_m \cdot r_L = -(2 \times 10^{-3}) \times (13.3 \times 10^3) = -26.7$$

Ans.

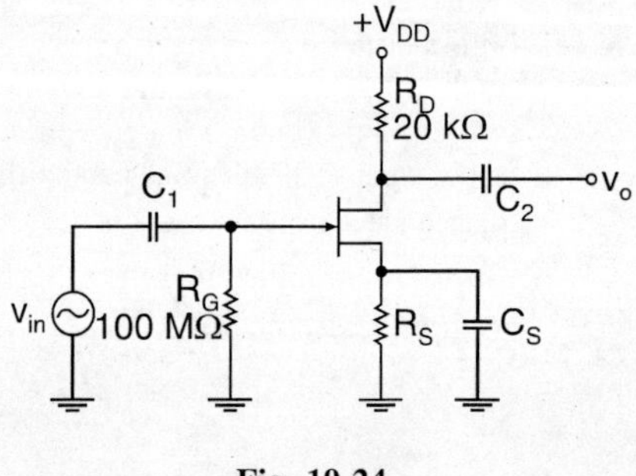

Fig. 19-24.

Input resistance

We know that the input resistance,

$$R'_i = R_G = 100 \text{ M}\Omega \textbf{ Ans.}$$

Output resistance

We also know that the output resistance,

$$R'_0 = r_L = 13.3 \text{ k}\Omega \textbf{ Ans.}$$

Example 19-11. *Calculate the voltage gain, $A_v = \dfrac{V_o}{V_i}$ at 1 kHz for the circuit in Fig. 19-25. The FET parameters are g_m = 2 mA/V and r_d = 10 k. Neglect capacitances.*

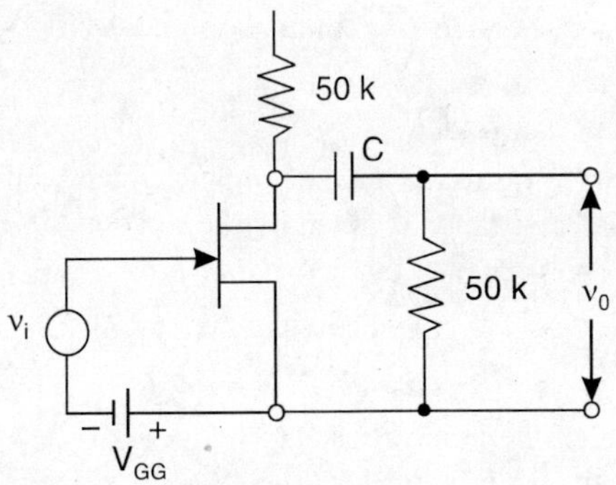

Fig. 19-25.

(*Nagpur Univ. Winter* 2003)

Solution. Given : $g_m = 2$ mA/V; $r_d = 10$ kΩ; $R_D = 50$ kΩ.

We know that a.c. load resistance,

$$r_L = r_d \parallel R_D = 10 \text{ k} \parallel 50 \text{ k}$$

$$= \frac{10\, k \times 50\, k}{10\, k + 50\, k} = 8.33 \text{ k}\Omega$$

and voltage gain of the FET,

$$A_v = -g_m \cdot r_L = -(2 \text{ mA/V}) \times (8{\cdot}33 \text{ k}\Omega)$$

$$= -16{\cdot}67 \textbf{ Ans.}$$

Example 19-12. *A MOSFET has a drain resistance R_D of 100 kΩ and operates at 20 kHz. Calculate the voltage gain of this device as a single-stage amplifier. The MOSFET parameters are :*

$$g_m = 1{\cdot}6 \text{ mA/V},\ r_d = 44 \text{ k}\Omega,\ C_{gs} = 3{\cdot}0 \text{ pF}$$

$$C_{ds} = 1{\cdot}0 \text{ pF and } C_{gd} = 2{\cdot}8 \text{ pF}$$

(*UPSC Engg Services* 1996)

Solution. Given : $R_D = 100$ kΩ; $g_m = 1{\cdot}6$ mA/V, $r_d = 44$ kΩ, $C_{gs} = 3{\cdot}0$ pF; $C_{ds} = 1{\cdot}0$ pF and $C_{gd} = 2{\cdot}8$ pF.

Since, the operating frequency is 20 kHz (a low frequency), interelectrode capacitances, C_{gs}, C_{ds} and C_{gd} are not be considered.

We know that the a.c. load resistance,

$$r_L = R_D \parallel r_d = (100 \text{ k}\Omega) \parallel (44 \text{ k}\Omega)$$

$$= \frac{100\, k \times 44\, k}{100\, k + 44\, k} = 30.56 \text{ k}\Omega$$

and the voltage gain,

$$A_v = -g_m \cdot r_L$$

$= 1.6 \text{ mA/V} \times 30.56 \text{ k}\Omega$

$= 48.9$ **Ans.**

19-20. Effect of A.C. Load on Amplifier Parameters

Consider a common source amplifier shown in Fig. 19-26. Here a load resistance (R_L) is connected to the amplifier's output through a coupling capacitor. In this case, the total or effective drain resistance is the parallel combination of resistors R_D and R_L *i.e.*,

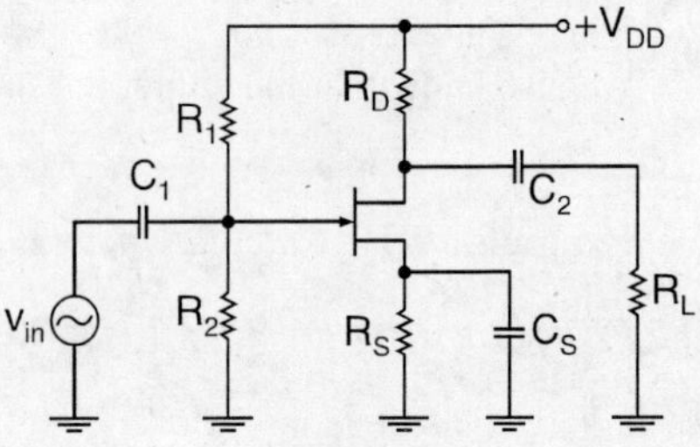

Fig. 19-26.

$$r_L = R_D \parallel R_L$$

It may be noted that in the above expression, if we also consider the FET drain resistance (r_d), then the effective drain resistance,

$$r_L = (R_D \parallel R_L) \parallel r_d$$

The effect of load resistor (R_L) is to reduce the voltage gain as well as the output resistance. The value of voltage gain with load resistance is given by the relation,

$$A_v = -g_m \cdot r_L$$

$$= -g_m (R_D \parallel R_L) \parallel r_d$$

$$= -g_m (R_D \parallel R_L) \qquad \text{...(If } r_d >> (R_D \parallel R_L))$$

and the output resistance, of the amplifier stage

$$R'_0 = (R_D \parallel R_L) \parallel r_d$$

$$= R_D \parallel R_L \qquad \text{...(If } r_d >> (R_D \parallel R_L))$$

Example 19-13. *Fig. 19-27 shows the circuit of a common source FET amplifier.*

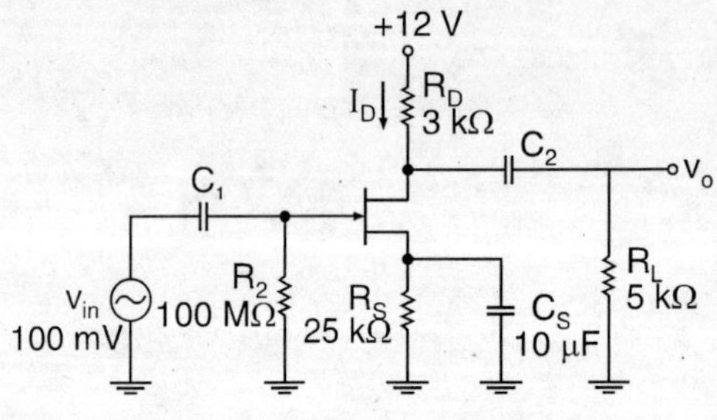

Fig. 19-27.

Find the value of r.m.s. output voltage. The g_m is 4500 μS and I_D is 2 mA. Neglect the FET drain resistance (r_d).

Solution. Given: $g_m = 4500\ \mu S = 4500 \times 10^{-6}\ S$; $R_D = 3\ \text{k}\Omega$, $R_L = 5\ \text{k}\Omega$; $v_{in} = 100\ \text{mV} = 100 \times 10^{-3}$ V and $I_D = 2$ mA.

We know that the effective value of a.c. drain resistance,

$$r_L = R_D \parallel R_L = \frac{R_D \times R_L}{R_D + R_L} = \frac{3 \times 5}{3 + 5} \text{ k}\Omega = 1.875 \text{ k}\Omega = 1875\ \Omega$$

$\therefore$ Output voltage,

$$v_o = -g_m \cdot r_L \cdot v_{in} = -(4500 \times 10^{-6}) \times (1875) \times (100 \times 10^{-3})$$

$= 0.844$ V **Ans.**

19-21. Effect of External Source Resistance on Voltage Gain

We have already discussed a field-effect transistor amplifier in which the source resistor is completely bypassed for a.c. signals. It means that for a.c. signals, the source is grounded. Now consider the amplifier circuit in which we include an external resistor (V_S) to the source resistance (R_S) as shown in Fig. 19-28. The source is no longer at a.c. ground. The drain current through resistor (r_s) produces an a.c. voltage between the source and ground.

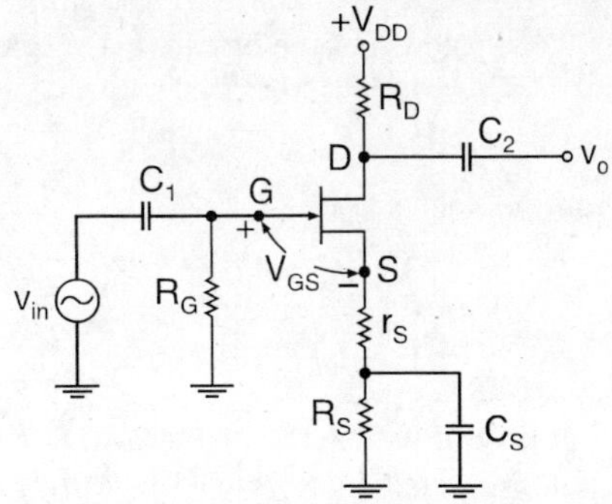

Fig. 19-28. *JFET amplifier with external source resistance.*

It may be noted that the total input voltage between the gate and ground,

$$v_{in} = v_{gs} + i_d \cdot r_s$$

and the output voltage,

$$v_o = -i_d \cdot r_L$$

where, r_L is the parallel combination of resistor R_D and the FET output resistance (r_d). Therefore, voltage gain

$$A_v = \frac{v_o}{v_{in}} = -\frac{i_d \cdot r_L}{v_{gs} + i_d \cdot r_s}$$

Substituting the value of i_d (equal to $g_m \cdot v_{gs}$) in the above expression,

$$A_v = -\frac{g_m \cdot v_{gs} \cdot r_L}{v_{gs} + g_m \cdot v_{gs} \cdot r_s} = -\frac{g_m \cdot v_{gs} \cdot r_L}{v_{gs}(1 + g_m \cdot r_s)}$$

$$= -\frac{g_m \cdot r_L}{1 + g_m \cdot r_s} = \frac{r_L}{r_e + \frac{1}{g_m}}$$

It is evident from the above expression that the external source resistance (r_s) reduces the voltage gain. If the value of resistor is made sufficiently large as compared to $1/g_m$, then the voltage gain,

$$A_v = -\frac{r_L}{r_s}$$

It means that the voltage gain is independent of the changes in the transconductance value. It is known as the *swamping effect*. Because of this fact the external source resistance (r_s) is referred to as the swamping resistor and the FET amplifier as the swamped amplifier.

Example 19-14. *Fig. 19-29 shows the circuit of a swamped FET amplifier.*

Determine the voltage gain **(a)** *when $R_L = 0$, and* **(b)** *when R_L is 100 kΩ. Neglect the FET output resistance (r_d) Take g_m = 4 mS.*

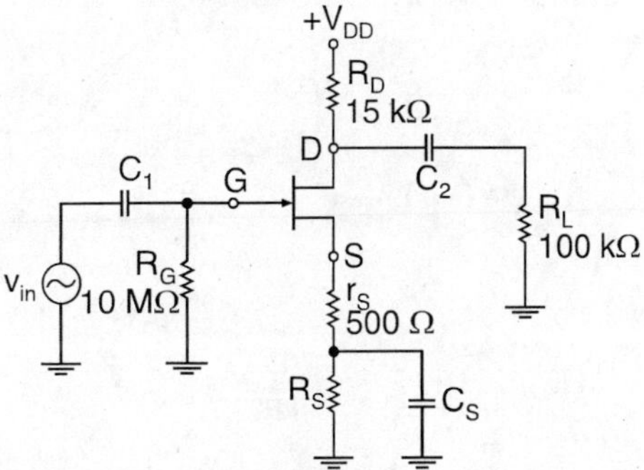

Fig. 19-29.

Solution. Given: g_m = 4 mS = 4 × 10^{-3} *S*; R_D = 1.5 kΩ; R_G = 10 MΩ and r_s = 500Ω .

(*a*) *Voltage gain when R_L is zero*

We known that the effective a.c. drain resistance,

$$r_L = R_D = 1.5 \times 10^3\ \Omega$$

and the voltage gain,

$$A_v = -\frac{g_m \cdot r_L}{1 + g_m \cdot r_s}$$

$$= -\frac{(4 \times 10^{-3}) \times (1.5 \times 10^3)}{1 + (4 \times 10^{-3}) \times 500}$$

$$= -2 \textbf{ Ans.}$$

(*b*) *Voltage gain when R_L is 100 kΩ*

We know that the effective a.c. drain resistance,

$$r_L = \frac{R_D \times R_L}{R_D + R_L} = \frac{1.5 \times 100}{1.5 + 100} = 1.48 \text{ k}\Omega = 1.48 \times 10^3 \ \Omega$$

and the voltage gain,

$$A_v = -\frac{g_m \cdot r_L}{1 + g_m \cdot r_s} = \frac{(4 \times 10^3) \times (1.48 \times 10^3)}{1 + (4 \times 10^{-3}) \times 500}$$

$$= -1.97 \textbf{ Ans.}$$

Example. 19-15. *Determine the voltage gain of the amplifier shown in Fig. 19-30, with (i) unbypassed R_S , and (ii) with bypassed R_S. Assume the output persistance (r_d) of the FET is imfinity. The JFET parameters :*

I_{DSS} = 10 mA and V_p = – 3·5 V. The Q-point
I_{DQ} = 2·3 mA and V_{GSQ} = – 1·8 V. (*Dip. IETE., Dec.* 2000)

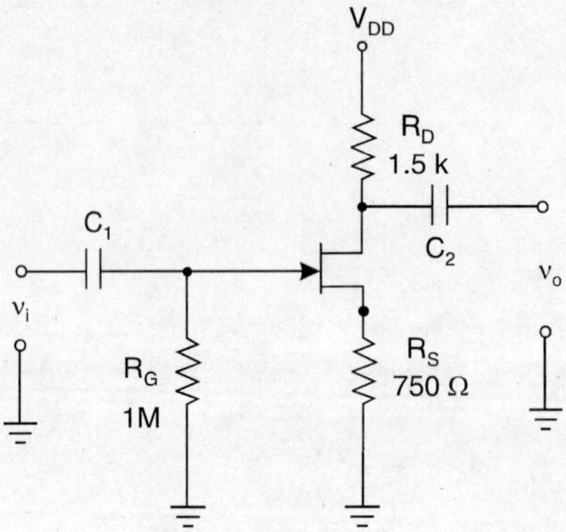

Fig. 19-30.

Solution. Given : $R_D = 1{\cdot}5 \text{ k}\Omega$; $R_S = 750 \ \Omega$; $R_G = 1 \text{ M}\Omega$, $I_{DSS} = 10$ mA; $V_p = -3{\cdot}5$ V. The Q-point $I_{DQ} = 2{\cdot}3$ mA and $V_{GSQ} = -1{\cdot}8$ V.

We know that,

$$g_{mo} = -\frac{2 \ I_{DSS}}{V_p} = -\frac{2 \times 10 \text{ mA}}{-3{\cdot}5 \text{ V}} = 5{\cdot}71 \text{ mA/V}$$

At the Q-point,

$$g_m = g_{mo}\left(1 - \frac{V_{GS}}{V_p}\right) = 5{\cdot}71\left(1 - \frac{-1{\cdot}8}{-3{\cdot}5}\right)$$

$$= 2{\cdot}77 \text{ mA/V}$$

The a.c. load resistance,

$$r_L = r_d \parallel R_D = R_D$$

(*i*) *Unbypassed R_S*

We know that the Voltage gain with unbypassed R_S

$$A_v = -\frac{g_m \cdot r_L}{1 + g_m R_S} = -\frac{2{\cdot}77 \text{ mA/V} \times 1{\cdot}5 \ k}{1 + 2{\cdot}77 \text{ mA/V} \times 0{\cdot}750 \ k}$$

$$= -\frac{4{\cdot}155}{1 + 2{\cdot}08} = -1{\cdot}35 \textbf{ Ans}$$

(*ii*) *Bypassed* R_S

We know that the voltage gain with bypassed R_S,

$$A_v = -g_m \cdot r_L = -2{\cdot}77 \text{ mA/V} \times 1{\cdot}5 \text{ k} = -4{\cdot}155 \textbf{ Ans.}$$

19-22. Common Drain Amplifier

Fig. 19-31 shows the circuit of a common drain amplifier. It is similar to common collector (or emitter follower) amplifier. Self-biasing is used in the circuit. The input signal is applied to the gate through a coupling capacitor (C_1). And the output is taken from the source terminal through the coupling capacitor (C_2).

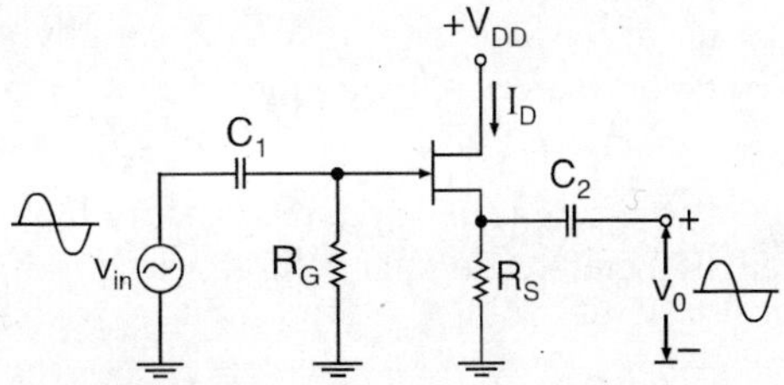

Fig. 19-31. *Common drain amplifier.*

The operation of the amplifier may be understood by assuming that when a small a.c. signal is applied to the gate, it produces variations in the gate-to-source voltage. This further produces the variations in drain current (I_D). As the gate-to-source voltage increases, the drain current also increases. As a result of this, the voltage drop across the source resistor (R_S) also increases. Thus, the output voltage (v_o) increases. It may be noted that the output voltage of a common drain amplifier is approximately equal to and in phase with the input voltage. Because of this fact, the circuit is known as *source follower*.

A common drain amplifier (or a source follower) has a very high value of input resistance. It is because of this property that a common drain amplifier is used at the front end of measuring instruments like electronic voltmeters and cathode ray oscilloscopes.

19-23. Analysis of a Common Drain Amplifier

Fig. 19-32 shows the a.c. equivalent circuit of a common drain amplifier. This circuit has been obtained from the amplifier circuit shown in Fig. 19-32 by short circuiting the capacitors and d.c. voltage supply. The field-effect transistor is also replaced by its low-frequency model. Now we shall use this circuit to find the expressions for voltage gain, input resistance and output resistance.

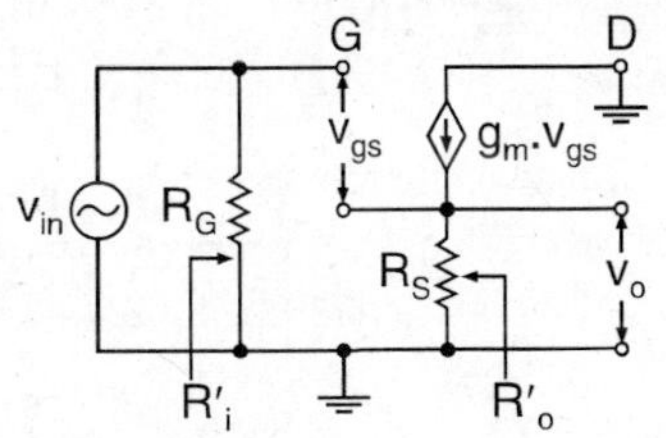

Fig. 19-32. *A.C. equivalent circuit of a common drain amplifier.*

(*i*) *Voltage gain :* It is the ratio of output voltage (v_o) to the input voltage (v_{in}). Mathematically, the voltage gain,

$$A_v = \frac{v_o}{v_{in}}$$

For the common drain amplifier, the input voltage,

$$v_{in} = v_{gs} + i_d \cdot R_s$$

and the output voltage,

$$v_o = i_d \cdot R_S$$

$\therefore$ Voltage gain,

$$A_v = \frac{v_o}{v_{in}} = \frac{i_d \cdot R_s}{v_{gs} + i_d \cdot R_s}$$

Substituting the value of i_d ($= g_m \cdot v_{gs}$) in the above expression,

$$A_v = \frac{(g_m \cdot v_{gs})\, R_s}{v_{gs} + g_m \cdot v_{gs} \cdot R_s}$$

$$= \frac{g_m \cdot v_{gs} \cdot R_s}{v_{gs}\,(1 + g_m \cdot R_s)}$$

$$= \frac{g_m \cdot R_s}{1 + g_m \cdot R_s} = \frac{R_s}{R_s + \dfrac{1}{g_m}}$$

It may be noted that when R_s is much greater than $1/g_m$, the value of voltage gain approaches unity.

(ii) Input resistance : We know that the input signal is applied to the gate of the amplifier. As a result of this, the input resistance seen by the input signal source is extremely high just as in the common source configuration. The total input resistance (or the input resistance of the amplifier),

$$R'_i = R_G \parallel R_i$$

where, R_i is the input resistance of the field effect transistor. Since, the value of R_i is extremely high, therefore the total input resistance,

$$R'_i = R_G$$

(iii) Output resistance : We know that voltage gain of the common drain amplifier,

$$A_v = \frac{v_o}{v_{in}} = \frac{R_s}{R_s + \dfrac{1}{g_m}}$$

∴ Output voltage,

$$v_o = \left(\frac{R_s}{R_s + \dfrac{1}{g_m}} \right) \times v_{in}$$

The above expression is a voltage-divider equation. It implies that the input voltage (v_{in}) drives two resistors R_s and $1/g_m$ with output voltage taken across resistor (R_s). It means that the output side of the amplifier appears as shown in Fig. 19-33 (*a*). It is evident from this figure, that the source resistor is driven by an a.c. source with an output resistance, *i.e.*,

$$R_D = 1/g_m$$

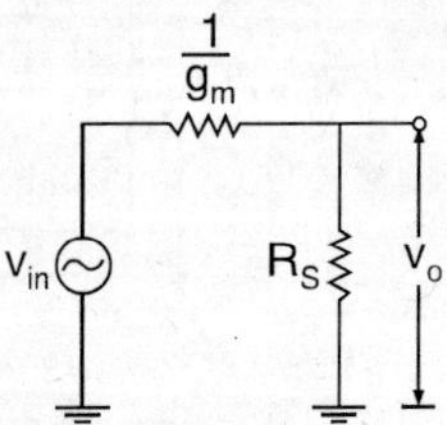

(*a*) A.C. equivalent circuit of output side of the amplifier.

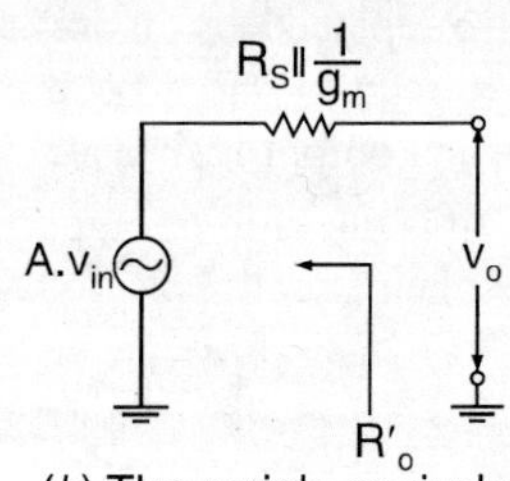

(*b*) Thevenin's equivalent circuit of the amplifier.

Fig. 19-33.

It may be noted that output resistance (R_o) is the value of resistance looking back into the source terminal of the JFET. Now let us Thevenize the output circuit. The resulting circuit is as shown in Fig. 19-33 (*b*). It is evident from this figure, that the resistance R_S is in parallel with $1/g_m$ and the output resistance of the amplifier stage,

$$R'_o = R_s \parallel \frac{1}{g_m}$$

If the value of source resistance (R_s) is much larger than $1/g_m$, then the output resistance of the amplifier stage,

$$R'_o = \frac{1}{g_m}$$

Example 19-16. *Fig. 19-34 shows the circuit of a source follower. Determine the voltage gain of the amplifier.*

Also determine the input and output resistance of the amplifier. Assume g_m = 8000 μS, infinite input resistance and neglect FET output resistance.

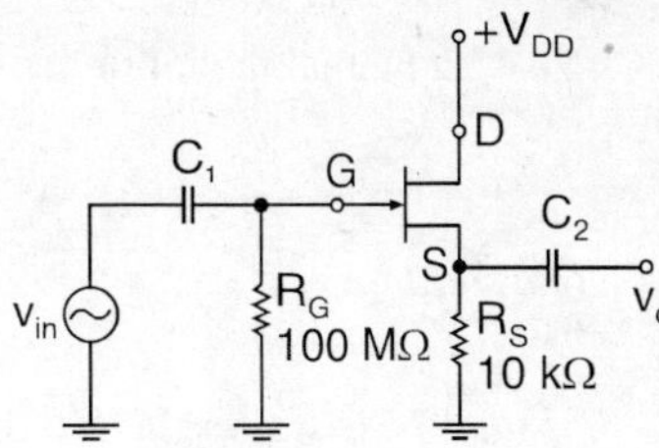

Fig. 19-34.

Solution. Given: $g_m = 8000\ \mu S = 8000 \times 10^{-6} S$; $R_G = 10\ k\Omega = 10 \times 10^3\ \Omega$ and $R_G = 100 M\Omega$.

Voltage gain

We know that the reciprocal of transconductance,

$$\frac{1}{g_m} = \frac{1}{8000 \times 10^{-6}} = 125\ \Omega$$

and the voltage gain,

$$A_v = \frac{R_s}{R_s + \frac{1}{gm}} = \frac{10 \times 10^3}{(10 \times 10^3) + 125} = 0.988\ \textbf{Ans.}$$

Input resistance

We know that the input resistance

$$R'_i = R_G = 100\ M\Omega\ \textbf{Ans.}$$

Output resistance

We also know that the output resistance,

$$R'_o = \frac{1}{g_m} = \frac{1}{8000 \times 10^{-6}} = 125\ \Omega\ \textbf{Ans.}$$

Example 19-17. *Find the voltage gain for the source follower shown in Fig. 19-35.*

Also find the input and output resistances. If the input voltage is 2 mV, find the value of the output voltage. Assume g_m = 5500 μS.

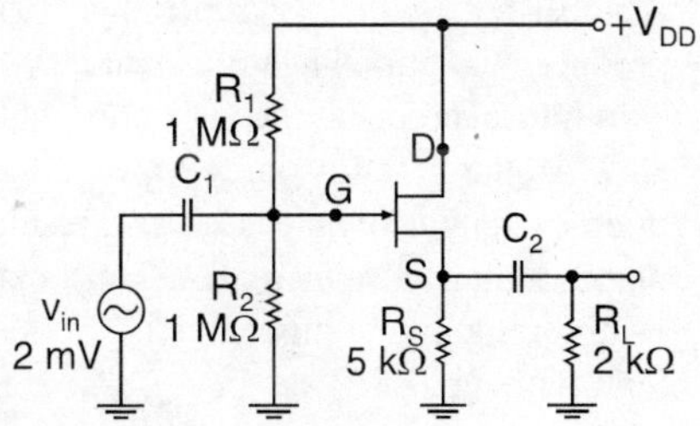

Fig. 19-35

Solution. Given: $v_{in} = 2$ mA; $g_m = 5500\ \mu S = 5500 \times 10^{-6}$ S; $R_1 = R_2 = 1\ M\Omega$; $R_S = 5\ k\Omega = 5000\ \Omega$ and $R_L = 2\ k\Omega = 2000\ \Omega$.

Voltage gain

We know that the reciprocal of transconductance,

$$\frac{1}{g_m} = \frac{1}{5500 \times 10^{-6}} = 181.8\ \Omega$$

and the voltage gain,

$$A_v = \frac{R_s}{R_s + \frac{1}{g_m}} = \frac{5000}{5000 + 181.8} = 0.965\ \textbf{Ans.}$$

Input resistance

We know that input resistance,

$$R'_i = R_1 \parallel R_2 \frac{R_1 \cdot R_2}{R_1 + R_2} = \frac{1 \times 1}{1+1} M\Omega = 0.5\ M\Omega\ \textbf{Ans.}$$

Output resistance

We know that the input resistance,

$$R'_0 = R_s \parallel \frac{1}{g_m} = 5000 \parallel 181.8 = 175.4\ \Omega \quad \textbf{Ans.}$$

Output voltage

Let v_o = Value of the output voltage.

Fig. 19-36 shows the a.c. equivalent circuit of the output side of the source follower. It is evident from this figure, that an a.c. source of voltage $A_v \cdot v_{in}$ (equal to 0.965 × 2 or 1.93 V) is in series with an output resistance of 175.4 Ω.

We know that the a.c. voltage across the load resistor,

$$v_o = \left(\frac{R_L}{R_L + R'_o}\right) \times A_v \cdot v_{in}$$

$$= \left(\frac{2000}{2000 + 175.4}\right) \times 1.93 \text{ mV}$$

$$= 1.77 \text{ mV} \quad \textbf{Ans.}$$

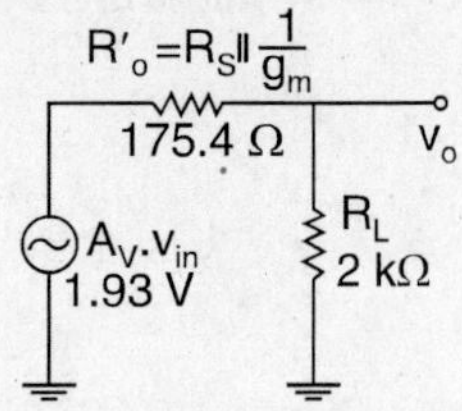

Fig. 19-36.

19-24. Common Gate Amplifier

Fig. 19-37 shows the circuit of a common gate amplifier. It is similar to a common base amplifier. The input signal is applied at the source through a coupling capacitor (C_1) and the output is taken from the drain through the coupling capacitor (C_2). The gate is effectively at a.c. ground because of the capacitor, (C_G).

The operation of the common gate amplifier may be understood from the assumption that when a small a.c. signal is applied to the source, it produces variation in gate-to-source voltage (V_{GS}). This inturn, produces the variations in drain current (I_D). As the gate-to-source voltage increases, the drain current also increases. As a result of this, the output voltage also increases. Thus, output voltage of a common gate amplifier is in phase with the input voltage.

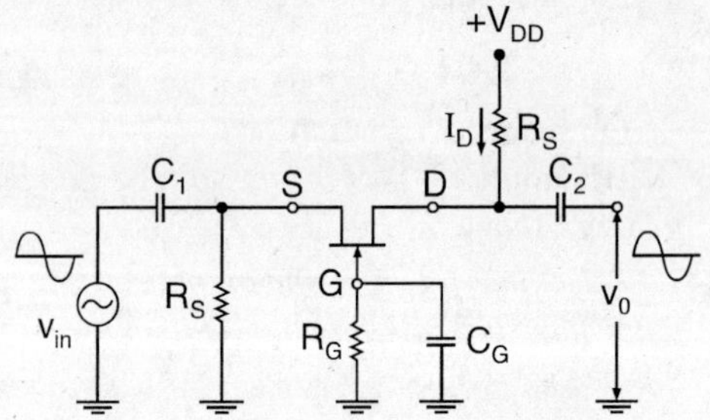

Fig. 19-37. *Common gate amplifier.*

A common gate amplifier has a low input resistance, high output resistance, high voltage gain and no phase reversal.

19-25. Analysis of a Common Gate Amplifier

Fig. 19-38 shows an a.c. equivalent circuit of a common gate amplifier. This circuit has been obtained from the amplifier circuit shown in Fig. 19-37 by short-circuiting the capacitors and the d.c. supply. The field-effect-transistor is also replaced by its low-frequency model.

Now we shall use this circuit to find the expressions for amplifier voltage gain, input resistance and output resistance.

(i) Voltage gain : It is the ratio of a.c. output voltage (v_o) to the a.c. input voltage (v_{in}). Mathematically, the voltage gain,

$$A_v = \frac{v_o}{v_{in}}$$

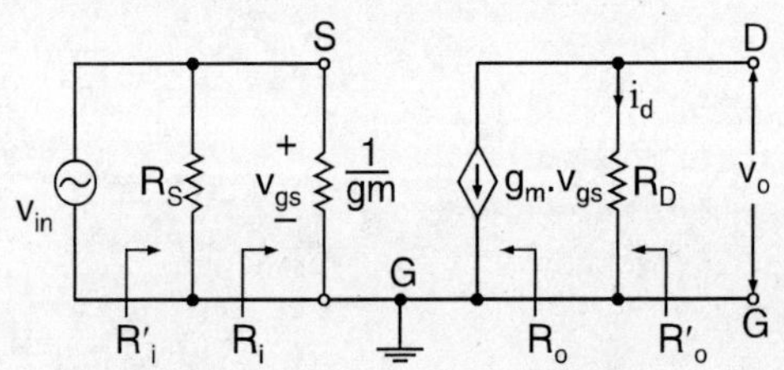

Fig. 19-38. *Common gate amplifier.*

We know that the a.c. input voltage,

$$v_{in} = v_{gs}$$

and the a.c. output voltage,

$$v_o = i_d \cdot R_D$$

∴ Voltage gain, $$A_v = \frac{i_d \cdot R_D}{v_{gs}}$$

Substituting the value of i_d (equal to $g_m \cdot v_{gs}$) in the above expression,

$$A_v = \frac{(g_m \cdot v_{gs}) \cdot R_D}{v_{gs}} = g_m \cdot R_D$$

It may be noted that the above expression for voltage gain is true only when there is no load resistor. If there is a load resistor (R_L) connected to the amplifier output, then the voltage gain.

$$A_v = g_m (R_D \parallel R_L)$$

(ii) Input resistance : It is the ratio of input voltage (v_{in}) to the input current (i_{in}). Mathematically, the input resistance,

$$R_i = \frac{v_{in}}{i_{in}}$$

For a common gate amplifier, the input voltage (v_{in}) is equal to the gate-to-source voltage (v_{gs}) and the input current (i_{in}) is equal to the drain current (i_d). Therefore the input resistance,

$$R_i = \frac{v_{gs}}{i_d} = \frac{v_{gs}}{g_m \cdot v_{gs}} \qquad ...(\because i_d = g_m \cdot v_{gs})$$

$$= \frac{1}{g_m}$$

It may be noted that the resistance (R_i) is the value of input resistance looking directly into the source terminal of the JFET. The input resistance of the amplifier,

$$R'_i = R_s \parallel \frac{1}{g_m}$$

However, if R_s is much larger than the reciprocal of transconductance ($1/g_m$) then the input resistance of the amplifier,

$$R'_i = R_s$$

It is evident from the above relation that a common gate amplifier has a low input resistance (equal to the value of external source resistance only). It is in contrast to the common source and common drain amplifier, which have extremely high input resistances.

(iii) Output resistance : It is the ratio of a.c. output voltage (v_o) to the a.c. output current (i_o). Mathematically, the output resistance of the amplifier,

$$R'_o = \frac{v_o}{i_o} = \frac{i_d \cdot R_D}{i_d} \qquad ...(\because v_o = i_d \cdot R_D \text{ and } i_o = i_d)$$

$$= R_D$$

It means that the output resistance of the common gate amplifier is equal to the external drain resistance (R_D). If there is a load resistor (R_L) connected to the amplifier output, then the output resistance of the amplifier,

$$R'_o = R_D \parallel R_L$$

Example 19-18. *Fig. 19-39 shows the circuit of a common gate JEET amplifier. The JFET has transconductance g_m = 2500 μS. Determine the amplifier voltage gain and input resistance.*

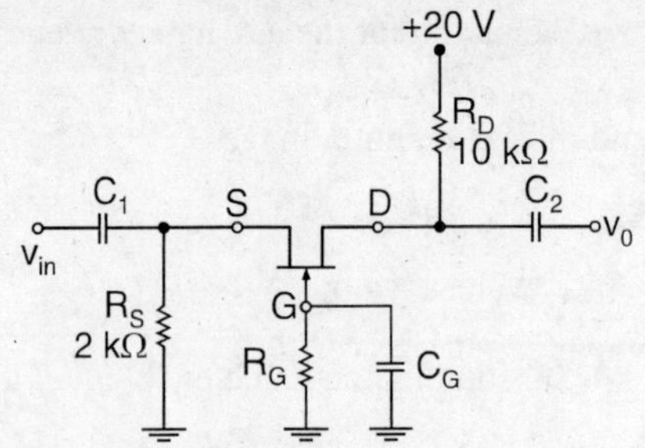

Fig. 19-39.

Solution. Given: $g_m = 2500\ \mu S = 2500 \times 10^{-6} S$; $R_D = 2\ k\Omega = 2000\Omega$ and $R_D = 10\ k\Omega = 10{,}000\ \Omega$.

Voltage gain

We know that voltage gain,

$$A_v = g_m \cdot R_D = (2500 \times 10^{-6}) \times 10{,}000 = 25 \textbf{ Ans.}$$

Input resistance

We know that reciprocal of transconductance,

$$\frac{1}{g_m} = \frac{1}{2500 \times 10^{-6}} = 400\ \Omega$$

and the amplifier input resistance,

$$R'_i = R_i \parallel \frac{1}{g_m} = 2000 \parallel 400\ \Omega$$

$$= \frac{2000 \times 400}{2000 + 400} = 333\ \Omega \textbf{ Ans.}$$

Example 19-19. *A FET source follower has g_m of 2m℧ and r_d of 50 kΩ. If the source resistance R_S is 1 kΩ, find the output resistance of the amplifier.*

Solution. Given : $g_m = 2$ m℧, $r_d = 50$ kΩ, and $R_S = 1$ kΩ.

We know that for a source follower (*i.e.*, common drain amplifier), the output resistance,

$$R_o' = R_S \parallel \frac{1}{g_m} = 1\text{ k} \parallel \left(\frac{1}{2\text{ m℧}}\right)$$

$$= 1000 \parallel \left(\frac{1}{2 \times 10^{-3}\ ℧}\right) = 1000 \parallel 500\ \Omega$$

$$= 333{\cdot}3\ \Omega \textbf{ Ans.}$$

Example 19-20. *For a JFET, the transconductance at zero gate-to-source voltage, g_{mo} = 5 mS. Determine the amplifier input resistance (R'_i) and a.c. voltage gain (v_o / v_{in}) for the circuit shown in Fig. 19-40. Assume the capacitors to be short circuit for a.c.*

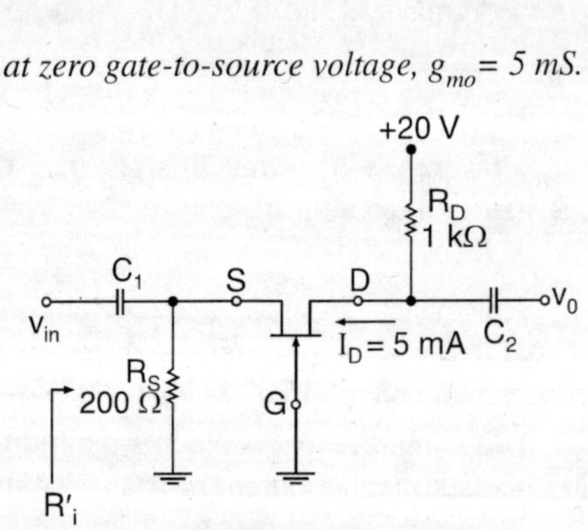

Fig. 19-40.

Solution. Given: $g_{mo} = 5$ mS $= 5 \times 10^{-3}$ *S*; $R_D = 1$ kΩ $= 1 \times 10^3\ \Omega$; $R_S = 200\ \Omega$ and $I_D = 5$ mA $= 5 \times 10^{-3}$ A.

Input resistance

We know that reciprocal of transconductance

$$\frac{1}{g_{mo}} = \frac{1}{5 \times 10^{-3}} = 200\ \Omega$$

and the input resistance of the amplifier,

$$R'_i = R_s \parallel \frac{1}{g_{mo}} = 200 \parallel 200\ \Omega = 100\ \Omega \textbf{ Ans.}$$

Voltage gain

We know that the d.c. source voltage,

$$V_S = I_D \cdot R_S = (5 \times 10^{-3}) \times 200 = 1\text{ V}$$

and the gate-to-source voltage,

$$V_{GS} = V_S = 1\text{V}$$

We also know that the *maximum value of drain current,

$$I_{DSS} = 2\,I_D = 2 \times (5 \times 10^{-3}) \times 10 \times 10^{-3}\text{ A}$$

and gate-to-source cut-off voltage

$$V_{GS(off)} = -\frac{2\,I_{DSS}}{I_D} = -\frac{2 \times (10 \times 10^{-3})}{5 \times 10^{-3}} = -4\text{ V}$$

and the transconductance (g_m)

$$= g_{mo}\ 1 - \frac{V_{GS}}{V_{GS(off)}}$$

$$= (5 \times 10^{-3}) \left[1 - \frac{1}{4}\right]$$

$$= 3.75 \times 10^{-3}\text{ S}$$

∴ The voltage gain,

$$A_v = g_m \cdot R_D = (3.75 \times 10^{-3}) \times (1 \times 10^3) = 3.75$$ **Ans.**

SUMMARY

In This chapter you have learned that :

1. Biasing an FET deals with setting a fixed level of current which should follow through the device with a fixed voltage drop across the FET junctions.
2. The different methods of biasing a JFET are gate bias, self-bias, voltage divider bias and source bias.
3. The two methods for designing a self-biased circuit are analytical and graphical.
4. The enternal source resistance reduces the voltage gain of a common source amplifier.
5. A common drain amplifier has very high value of input resistance.
6. A common gate amplifier has a low input resistance, high output resistance, high voltage gain and no phase reversal.

GLOSSARY

Common drain amplifier : A JFET amplifier configuration in which drain of the device is directly connected to V_{DD} supply.

Common gate amplifier : A JFET amplifier configuration in which gate of the device is effectively connected to the a.c. ground.

Common source amplifier : A JFET amplifier configuration in which source of the device is effectively connected to the a.c. ground.

Field effect transistor : Three terminal unipolar device in which the current is controlled by an electric-field.

Source follower : Refer to common drain amplifier.

---

* In actual practice, the amplifiers are biased in such a way that the Q-point is located at the mid-point of the transconductance curve. In that case, the maximum value of drain current () is twice the value of drain current.

NUMERICAL PROBLEMS

1. Find the values of drain-to-source voltage (V_{DS}) and the gate-to-source voltage (V_{GS}) for the circuit shown in Fig. 19-41. (**Ans.** 7.5 V and – 1.5 V)

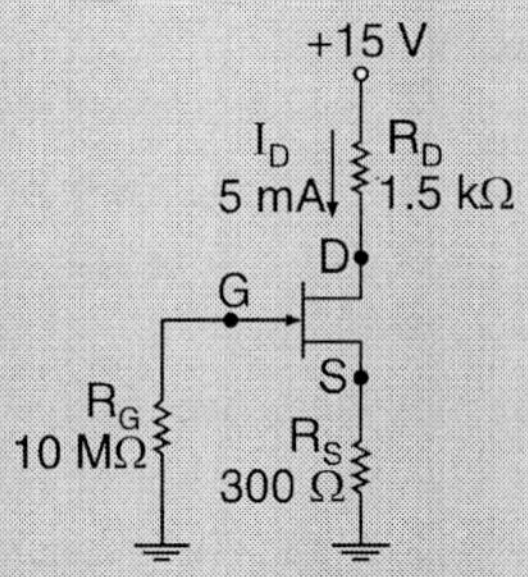

Fig. 19-41.

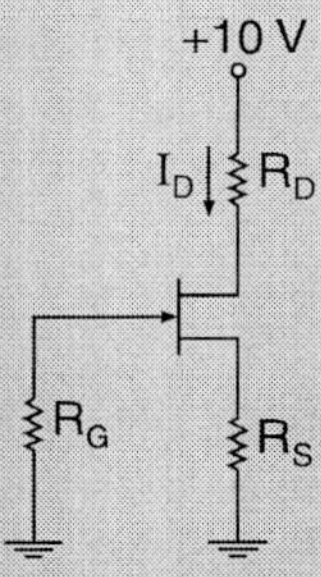

Fig. 19-42.

2. Calculate the self-bias operating point for the FET shown in Fig. 19-42. Also calculate the values of resistors R_D and R_S to obtain this bias condition. Given the maximum value of drain current as 5 mA and V_{GS} = 0.6 V or I_D = 2.5 mA.

(**Ans.** I_{DQ} = 2.5 mA : V_{DSQ} = 5 V: R_D = 2 kΩ and R_S = 240 Ω)

3. Fig. 19-43 (*a*), shows the circuit of a self-biased JFET amplifier and Fig. 19-43 (*b*) indicates the transfer characteristic curve of the JFET.

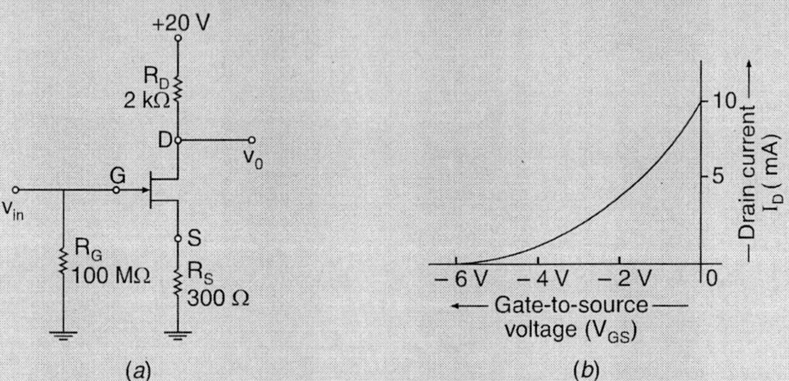

(*a*) (*b*)

Fig. 19-43.

Find the quiescent values of I_D and V_{GS}. Also find the value of d.c. voltage between the drain and ground. (**Ans.** I_{DQ} = 5.5 mA and V_{GSQ} = – 1.6 V: V_D = 9 V)

4. A certain JFET has a transconductance (g_m) of 2500 μS. With an external drain resistance of 2 kΩ, find the value of ideal voltage gain. (**Ans.** 5)

5. The gain of a certain JFET amplifier with a source resistance of zero is 25. Determine the value of drain resistance, if the transconductance is 5500 μS. (**Ans.** 4.55 kΩ)

6. A certain JFET amplifier has g_m of 4 mS, r_d = 10 kΩ and R_D = 5 kΩ. What is the voltage gain? Assume the source resistance to be zero. (**Ans.** 13.3)

7. Calculate the value of voltage gain, input resistance and output resistance of the amplifier shown in Fig. 19-44.

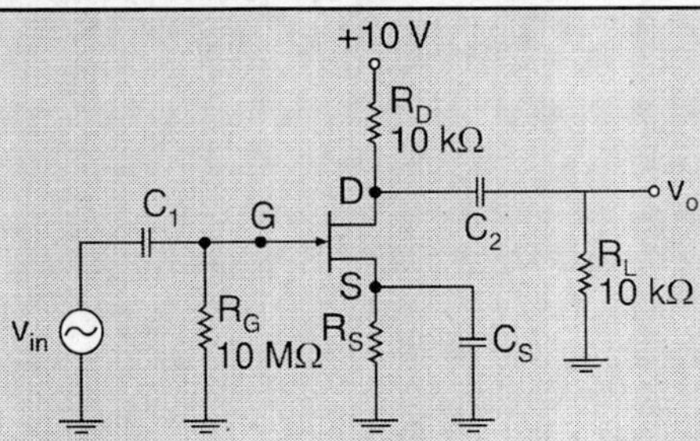

Fig. 19-44.

The FET has a transconductance of 4.2 mA/V and r_d = 400 kΩ.

(**Ans.** 18.6 : 10 MΩ and 4.4 kΩ)

8. Fig. 19-45 shows the circuit of a common source amplifier. Find its voltage gain and output resistance. Assume g_m = 4500 µS. (**Ans.** 0.845 and 220 Ω)

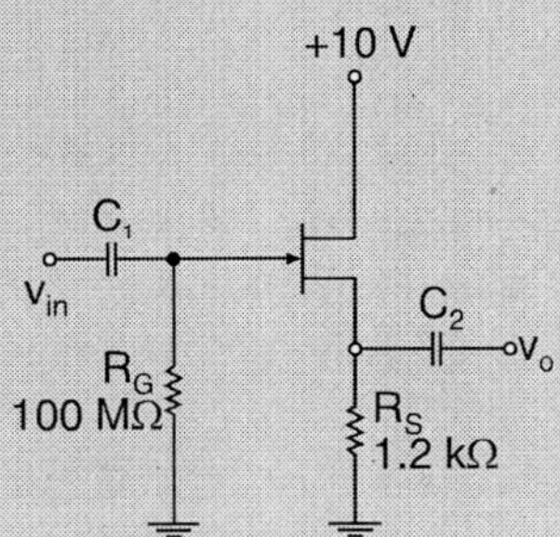

Fig. 19-45

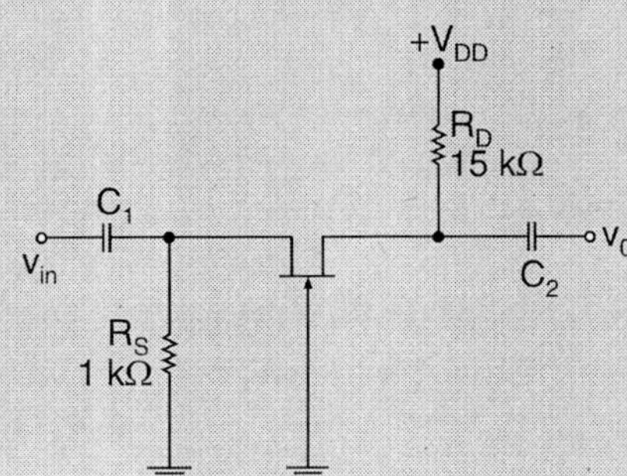

Fig. 19-46

9. What is that value of voltage gain and input resistance of a common gain amplifier shown in Fig. 19-46 ? Assume g_m = 4000 µS. (**Ans.** 6 and 200Ω)

10. For the circuit shown in Fig. 19-47, obtain V_{o_1} and V_{o_2}. (*UPSC Engg. Services* 2002)

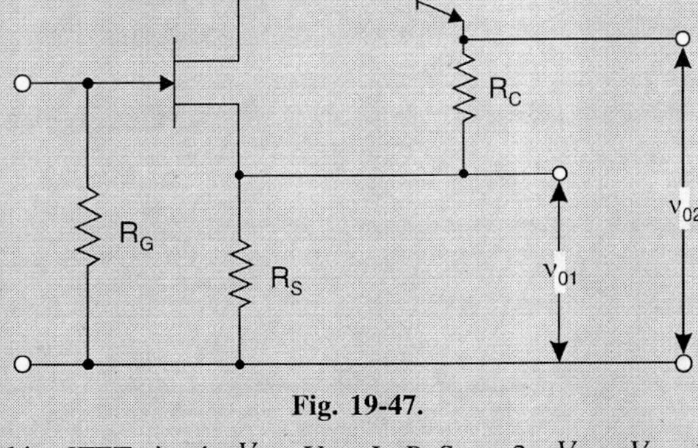

Fig. 19-47.

[**Hint :** State 1 : Self-bias JFET circuit; $V_{o_1} = V_S = I_D R_S$ Stage 2 : $V_{o_2} = V_{o_1} - I_C R_C = I_D R_S - I_C R_C$]

DESCRIPTIVE QUESTIONS

1. What are the biasing schemes available to achieve the required bias in a junction field-effect transistor (JFET)? Explain any one of the biasing schemes.
2. How do you set a *Q*-point in a self-biased JFET?
3. Draw a small signal low-frequency model of a field-effect transistor and explain its various elements. Also give the approximate range of the elements.
4. How a small-signal high frequency model is different from a low-frequency model? Explain it briefly.
5. Draw neatly the circuit diagram of a common source JFET amplifier and explain its working.
6. Derive the expression for voltage gain and output resistance for a common source JFET amplifier.
7. What is the effect of external source resistance on the voltage gain of a common source amplifier? Explain with necessary derivation.
8. Draw the circuit diagram of a common gate amplifier. Derive the expression for its voltage gain and input resistance.
9. Draw and explain voltage divider bias using *FET* and derive for its stability factor. (*Electronic Circuits - I, Anna University, April/May 2008*)
10. Draw the high frequency equivalent circuit for a *FET* amplifier and derive the values of all the parameters. (*Electronic Circuits - I, Anna University, April/May 2008*)
11. With the help of a neat diagram explain the voltage divider biasing method for JFET (*Electronic Circuits - I, Anna University, April/May 2008*)
12. How is a *JFET* used as a voltage variable resistance? Explain. (*Electronic Circuits - I, Anna University, Nov/Dec 2007*)
13. Write the construction, operation and characteristics behavior of *JFET* under various biasing conditions. Give the necessary figures. (*Electronic Devices, Anna University, Nov/Dec 2007*)
14. Draw a MOS differential pair and discuss its small-signal operations. (*Electronic Devices. Anna University, June 2009*)

MULTIPLE CHOICE QUESTIONS

1. The best location for setting a *Q*-point on d.c. load line of an FET amplifier is at
   (*a*) saturation point (*b*) cut-off point
   (*c*) mid-point (*d*) none of these
2. Which of the following bias methods provides a solid *Q*-point in JFET, amplifiers
   (*a*) Gate bias (*b*) Self-bias
   (*c*) Voltage divider bias (*d*) Current source bias
3. Which of the following technique is used for biasing the enhancement type MOSFET's?
   (*a*) Voltage divider bias (*b*) Collector feedback bias
   (*c*) Current source bias (*d*) Self-bias
4. The voltage gain of a common source JFET amplifier depends upon its
   (*a*) transconductance (g_m) (*b*) amplification factor (μ)
   (*c*) external load resistance (R_D) (*d*) both (*a*) and (*c*) above

5. A common gate amplifier has
(*a*) high input resistance and high output resistance
(*b*) low input resistance and high output resistance
(*c*) low input resistance and low output resistance
(*d*) high input resistance and low output resistance

6. The transconductance 'g_m' of a JFET is equal to
(*a*) $-\frac{2I_{DSS}}{V_P}$ (*b*) $\frac{2}{|V_P|}\sqrt{I_{DSS}\,I_D}$
(*c*) $-\frac{2I_{DSS}}{V_P}\left(1-\frac{V_{GS}}{V_P}\right)$ (*d*) $\frac{I_{DSS}}{V_P}\left(1-\frac{V_{GS}}{V_P}\right)$

7. A transconductance amplifier has
(*a*) high input impedance and low output impedance
(*b*) low input impedance and high output impedance
(*c*) high input and output impedances
(*d*) low input and output impedances

8. The threshold voltage of an-n-channel enhancement mode MOSFET is 0.5 V, when the device is biased at a gate voltage of 3 V, pinch-of would occur at a drain voltage of
(*a*) 1.5 V (*b*) 2.5 V
(*c*) 3.5 V (*d*) 4.5 V

9. The zero gate bias channel resistance of a junction field-effect transistor is 750 and the pinch-off voltage is 3 V. For a gate bias of 1.5 V and very low drain voltage, the device would behave as a resistance of
(*a*) 320 Ω (*b*) 816 Ω
(*c*) 1000 Ω (*d*) 1270 Ω

10. If properly biased, JFET with act as a
(*a*) current controlled current source (*b*) voltage controlled voltage source
(*c*) voltage controlled current source (*d*) current controlled voltage source

ANSWERS

1. (*c*)	**2.** (*d*)	**3.** (*b*)	**4.** (*d*)	**5.** (*b*)
6. (*c*)	**7.** (*a*)	**8.** (*c*)	**9.** (*b*)	**10.** (*b*)

❑❑❑

20

CHAPTER

Amplifiers with Compound Configurations

OUTLINE

Objectives

After completion of this chapter, you should be able to :

- Know the different types of compound configurations.
- Calculate the voltage gain, input impedance and output impedance of a cascade amplifier.
- Explain the operation of Darlington transistor and feedback pair.
- Describe the operation of CMOS circuit.
- Explain how a BJT and FET can be connected to operate as a constant current source.

- Analyse the current mirror circuits.
- Understand the d.c. and a.c. operation of a differential amplifier circuit.
- Calculate a.c. voltage gain of a single ended and double ended differential amplifier.
- Calculate the a.c. voltage gain of a differential amplifier in a common-mode configuration.

20-1. Introduction

We have already discussed amplifiers with standard circuit connections. The example of such amplifiers are common-emitter, common-base, common-collector, common-source, common-gate and common-drain type amplifiers. In this chapter, we will study some non-standard circuit connections which are used quite frequently by the electronics industry. Although there is a wide variety of non-standard circuit connections but the following ones are important from the subject point of view:

- Cascade connection
- Cascode connection
- Darlington connection
- Feedback pair
- CMOS circuit
- Current source circuits
- Current mirror circuit
- Differential amplifier cirucit

All these non-standard circuit connections for the BJT and FET based amplifiers are discussed one by one in the following pages.

20-2. Cascade Connection

A Connection Scheme in which the output of one amplifier stage is connected to the input of the second stage is called cascade connection (Refer to Fig. 20-1 below).

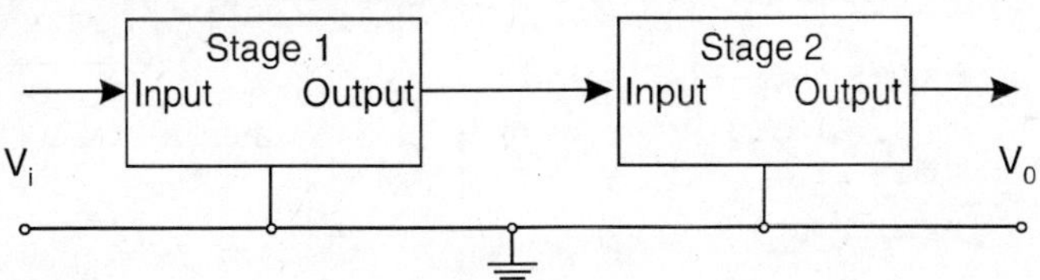

Fig. 20-1.

Now we shall discussed the cascade connection for an RC coupled amplifier using FET and BJT respectively.

20-3. FET Based Cascade Connection

We have already mentioned in the last article that this type of connection, the output of one amplifier stage is applied as an input to the second stage. The FET based RC coupled amplifier shown in Fig. 18-4 is an example of a cascade connection. For calculation of input impedance, output impedance, voltage gain, *etc.*, refer to Art. 18-5.

Fig. 20-2 shows a cascade connection of two FET amplifier stages. This connection provides a multiplication of the gain of each stage for a larger overall gain.

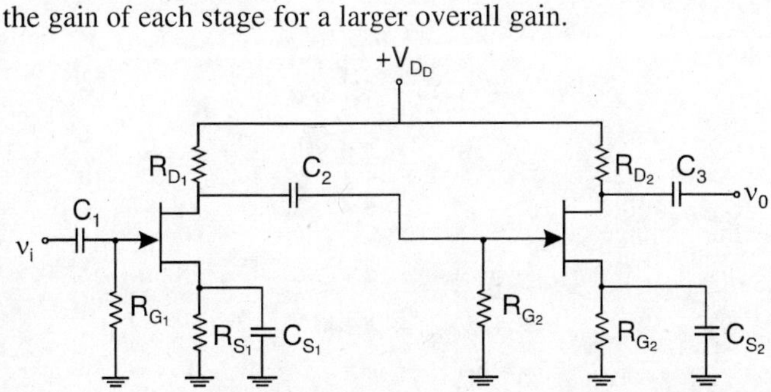

Fig. 20-2. *Cascaded FET amplifier.*

It may be noted that the voltage gain to the overall cascade amplifier is the product of stage gains, thus,

$$A_v = A_{v_1} \times A_{v_2}$$

where, A_{v_1} = voltage gain of stage 1 and

A_{v_2} = voltage gain of stage 2.

Since, the voltage gain of a common-source JFET amplifier is $- g_m R_D$. Therefore, the overall voltage gain,

$$A_v = A_{v_1} \times A_{v_2} = (- g_{m_1} R_{D_1}) \times (- g_{m_2} R_{D_2})$$

The input impedance of the cascade amplifier is the same as the input impedance of stage 1. Thus,

$$R_i = R_{G_1}$$

and the output impedance is the same as that of stage 2. Thus,

$$R_o = R_{G_2}$$

Example 20-1. *Fig. 20-3 shows the circuit of a JFET cascade amplifier. The JFET used in this circuit has I_{DSS} = 8 mA and V_p = – 4·5 V.*

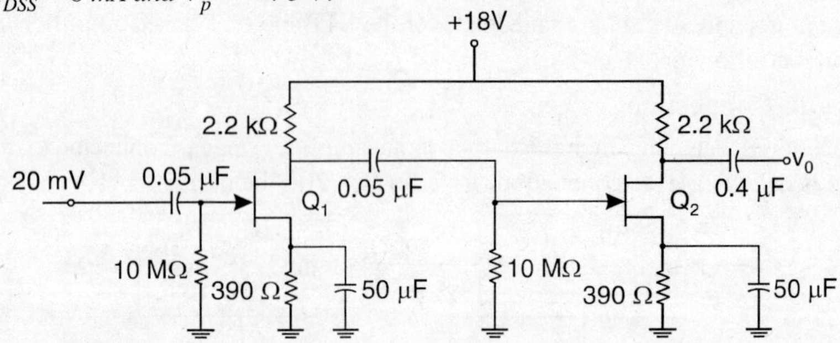

Fig. 20-3.

Calculate the d.c. bias, voltage gain, input impedance, output impedance and the resulting output for the cascade amplifier. Also calculate the load voltage if a 10 kΩ load is connected across the output.

Solution. Since, both the stages are identical, therefore they have the same d.c. bias. Using the d.c. bias techniques, given in chapter 18, *i.e.*, choosing $I_D = I_{DQ}$ = 4 mA, (≈ ½ of I_{DSS}) we get,

$$V_{GSQ} = - I_D R_S = - 4 \text{ mA} \times 390 = - 1.560 \text{ V}$$

and

$$g_{mo} \approx \frac{2 I_{DSS}}{|V_P|} \approx \frac{2 \text{ (8 mA)}}{4.5 \, V} = 3.55 \text{ mA/V or } 3.55 \text{ mS}$$

At the d.c. point (or *Q*-point), we know that,

$$g_m = g_{mo} \quad 1 - \frac{V_{GSQ}}{V_P}$$

$$= 3.55 \quad 1 - \frac{-1.56}{-4.5 \, V} = 2.32 \text{ mS}$$

We also know that voltage gain of each stage,

$$A_{v_1} = A_{v_2} = - g_m R_D$$

$$= - (2.32) \times 2.2 \text{ k}\Omega$$

$$= - 5.1$$

∴ The voltage gain of the cascade amplifier,

$$A_v = A_{v_1} \cdot A_{v_2}$$
$$= (-5.1) \times (-5.1)$$
$$= 26.01 \textbf{ Ans.}$$

We know that the voltage gain of the cascade amplifier, (A_v),

$$26.01 = \frac{v_o}{v_i}$$
$$= \frac{v_o}{20 \text{ mV}}$$

∴ $v_o = 26.01 \times 20 \text{ mV} = 520.2 \text{ mV}$ **Ans.**

We know that the input impedance of the cascaded amplifier,

$$Z_i = R_G = 10 \text{ M}\Omega \textbf{ Ans.}$$

We also know that the output impedance of the cascaded amplifier, (assuming $r_d = \infty$) is,

$$Z_o = R_D = 2.2 \text{ k}\Omega \textbf{ Ans.}$$

The output voltage if a 10-kΩ load is connected across the output,

$$V_L = \frac{R_L}{Z_o + R_L} V_o$$
$$= \frac{10\ k\Omega}{2.2\ k\Omega + 10\ k\Omega} \times (520.2 \text{ mV})$$
$$= 426.4 \text{ V} \textbf{ Ans.}$$

20-4. BJT Based RC Coupled Amplifier

Fig. 20-4 shows a BJT based RC coupled cascade amplifier. The advantage of cascading is the large overall voltage gain. DC bias calculations can be carried out in the same way as discussed in the chapter on Single-Stage Small-Signal BJT Amplifier, (Chapter-16 Article 16-12).

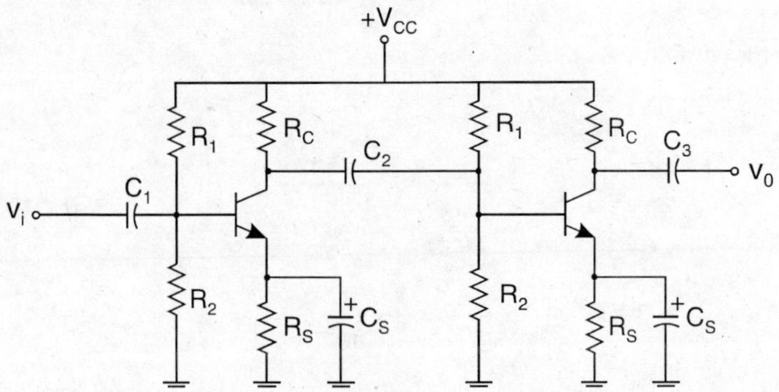

Fig. 20-4.

The voltage gain of each stage is given by the equation,

$$A_v = -\frac{R_C \parallel R_L}{r'_e}$$

The overall amplifier input or resistance is the same as that of stage 1, *i.e.*,

$$Z_i = R_1 \parallel R_2 \parallel \beta\, r'_e$$

and the overall output or resistance is the same as that of stage 2, *i.e.*,

$$Z_o = R_C \parallel r_o$$

Example 20-2. *For the BJT cascade amplifier shown in Fig. 20-5, calculate the d.c. bias voltages and collector current for each stage.*

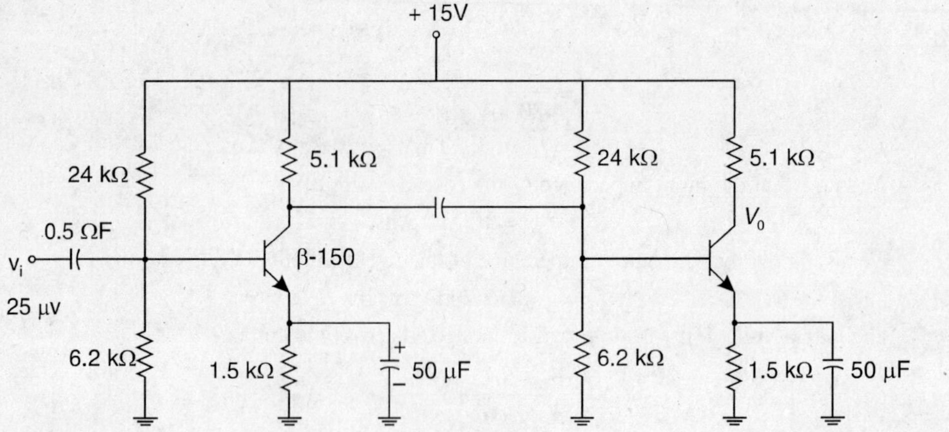

Fig. 20-5.

Solution.

We know that the base voltage,

$$V_B = V_{CC} \cdot \frac{R_2}{R_1 + R_2} = 15. \frac{6.2\,k}{6.2\,k + 24\,k} = 3.08\text{ V}$$

and
$$V_E = V_B - 0.7\text{ V} = 3.08 - 0.7 = 2.38\text{ V}$$

The emitter current,

$$I_E = \frac{V_E}{R_E} = \frac{2.38}{1.5\,k} = 1.587\text{ mA}$$

and the collector current,

$$I_C \approx I_E = 1.587\text{ mA.}$$

The a.c. resistance, at the bias point,

$$r'_e = \frac{26\text{ mV}}{I_E} = \frac{26\text{ mV}}{1.587\text{ mA}} = 16.4\ \Omega$$

We also know that voltage gain of stage 1,

$$A_{v_1} = \frac{R_{0_1}}{r'e_1} = \frac{R_e \parallel R_{ir}}{r'r_1}$$

$$= \frac{i_C \parallel \{(R_1 \parallel R_2) \parallel (\beta_1 r'e_1)\}}{r'_{e_1}} - \frac{R_C \parallel (R_1 \parallel R_2 \parallel \beta r_e)}{r_e}$$

$$= -\frac{5.1\,k \parallel \{(24\,k \parallel 6.2\,k \parallel (150 \times 16.4)\ \Omega)\}}{16.4\ \Omega}$$

$$= -\frac{5.1\,k \parallel (24\,k \parallel 6.2\,k \parallel 2.46\,k)}{16.4\ \Omega}$$

$$= -\frac{5.1\,k \parallel (1.76\,k)}{16.4\ \Omega}$$

$$= -\frac{1.31\,k}{16.4\,\Omega} = -\,79.9 \textbf{ Ans}$$

and the voltage gain of stage 2,

$$A_{v_2} = -\frac{R_C}{r_e'} = -\frac{5.1\,k}{16.4} = -\,311$$

∴ The overall voltage gain,

$$A_v = A_{v_1} \cdot A_{v_2} = (-\,79.9) \times (-\,311) = 24849$$

The output voltage,

$$v_o = A_v \,.\, v_i = 24849 \times 25\ \mu\text{V} = 621225\ \mu\text{V}$$
$$= 0.62\ \text{V} \textbf{ Ans}$$

The amplifier input resistance,

$$R_i = R_1 \parallel R_2 \parallel \beta\, r_e'$$
$$= 24\ \text{k} \parallel 6.2\ \text{k} \parallel (150 \times 16.4)$$
$$= 1.76\ \text{k}\Omega \textbf{ Ans}$$

The amplifier output resistance,

$$R_o = R_C = 5.1\ \text{k}\Omega \textbf{ Ans.}$$

If a 10 kΩ load is connected to the amplifier output, the resulting voltage across the load,

$$v_L = v_o \times \frac{R_L}{R_0 + R_L} = 0.62 \times \frac{10\,k}{1.76\,k + 10\,k} = 0.527\ V \textbf{ Ans.}$$

20-5. Cascode Connection

A cascode connection has one bipolar junction transistor on top of (in series with) another. Fig. 20-6 shows a cascode connection with a common-emitter (CE) amplifier stage feeding a common-base (CB) amplifier stage. The connection provides a high input resistance with low voltage gain to ensure that the input Miller effect capacitance (discussed in chapter-21) is at a minimum with a common-base stage providing good high-frequency operation.

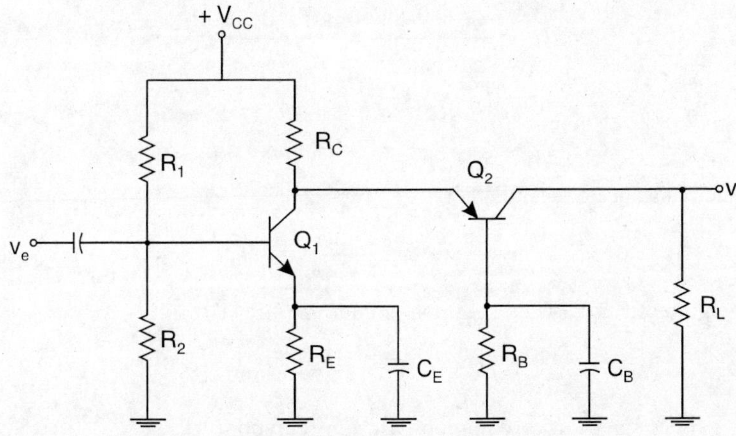

Fig. 20-6. *Illustrating a cascode connection for a CE amplifier feeding a CB stage.*

Example 20-3. *Fig. 20-7 shows a circuit of a BJT cascode amplifier. Calculate the overall voltage gain of the amplifier.*

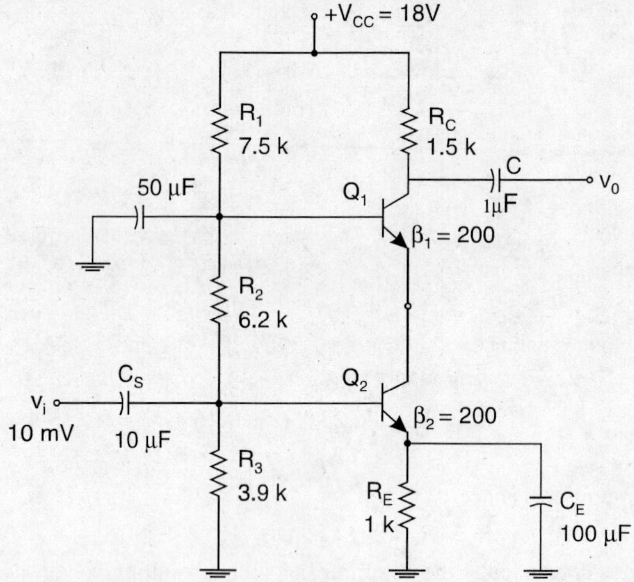

Fig. 20-7.

Solution. Given $V_{CC} = 18V$; $R_1 = 7.5\ k\Omega$; $R_2 = 6.2\ k\Omega$; $R_3 = 3.9\ k\Omega$; $R_C = 1.5\ k\Omega$; $\beta_1 = \beta_2 = 200$; $C_E = 100\ \mu F$.

We know that the base voltage,

$$V_{B_1} = V_{CC} \times \frac{R_2 + R_3}{R_1 + R_2 + R_3} = 18 \times \frac{6.2\ k + 3.9\ k}{7.5\ k + 6.2\ k + 3.9\ k} = 10.3\ V$$

and

$$V_{B_2} = V_{CC} \times \frac{R_3}{R_1 + R_2 + R_3} = 18 \times \frac{3.9\ k}{7.5\ k + 6.2\ k + 3.9\ k} = 4.0\ V$$

We also know that

$$I_{E_2} = \frac{V_{B_2} - 0.7\ V}{R_E} = \frac{4.0 - 0.7\ V}{1\ k} = 3.3\ mA$$

and $I_{C_2} \approx I_{E_2} = 3.3\ mA$

Further $I_{E_1} \approx I_{C_2}$ and $I_{C_1} \approx I_{E_1} \approx 3.3\ mA.$

The dynamic resistance of each BJT is,

$$r_e' = \frac{26\ mV}{I_E} = \frac{26\ mV}{3.3\ mA} = 7.9\ \Omega$$

The voltage gain of stage 1 (*i.e.*, common-emitter amplifier),

$$A_{v_1} = -\frac{R_C}{r_e'} = -\frac{r_e'}{r_e'} = -1$$

and the voltage gain of stage 2 (*i.e.*, common-base configuration),

$$A_{v_2} = -\frac{R_C}{r_e'} = \frac{1.5\ k}{7.9\ \Omega} = 189.9$$

∴ The overall voltage gain, of the cascode amplifier,

$$A_v = A_{v_1} \times A_{v_2} = (-1)(189.9) = -189.9$$ **Ans.**

20-6. Darlington Connection of BJTs

This connection is a very popular connection of two bipolar junction transistors for operation as one "Superbeta" transistor. Fig. 20-8 shows a Darlington connection of two NPN transistors. The main feature of Darlington connection is that the composite transistor acts as a single unit with a current gain that is equal to the product of the current gains of the individual transistors.

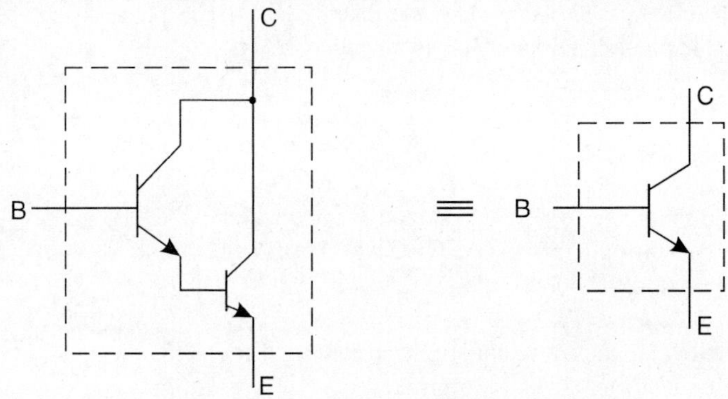

Fig. 20-8. *Darlington connection*

Let β_1 = Current gain of transistor, Q_1 and

β_2 = Current gain of transistor, Q_2.

Then, current gain of a darlington transistor (or a composite transistor),

$$\beta = \beta_1 \cdot \beta_2$$

If the two transistors are fabrication in such a way they are "well matched", then $\beta_1 = \beta_2 = \beta$. In such a case, the current gain of a composite transistor,

$$\beta_D = \beta \cdot \beta = \beta^2$$

Example 20-4. *A certain IC has a Darlington Connection of two transistors each having a current gain, 160 find the current gain of the composite transistor.*

Solution. Given : $\beta_1 = \beta_2 = 160$

We know that current gain of a composite transistor with $\beta_1 = \beta_2$ is,

$$\beta_D = \beta^2 = 160^2 = 25600 \textbf{ Ans.}$$

Example 20-5. *Calculate the value of current gain provided by each Darlington connection of two indentical transistors. Given : overall β_D = 6000.*

Solution. Given β_D of two identical transistors = 6000

Let $\beta_1 = \beta_2 = \beta$ = current gain each transistor.

We know that current gain of a Darlington (or composite) transistor, (β_D),

$$6000 = \beta_1 \times \beta_2 = \beta^2$$

$$\therefore \quad \beta = \sqrt{6000} = 77.46 \textbf{ Ans.}$$

20-7. Packaged Darlington Transistor

Several Darlington transistors are available commercially. Fig. 20-9 shows a specification data sheet of a typical Darlington transistor (2N999). As seen from the fig. 20-9, the current gain of the Darlington transistor (β_D or h_{FE}) ranges from 4000 to 70, 000. This value is very high as compared to other typical single transistors.

Parameter	Test conditions	Min	Max
V_{BE}	I_C = 100 mA		1.8 V
h_{FE} (or β_D)	I_C = 10 mA	4000	
	I_C = 100 mA		70, 000

Fig. 20-9. *Information from the Specification Data Sheet of a Darlington transistor (2N999).*

20-8. DC Bias Operation of Darlington Circuit

Fig. 20-10 a basic Darlington circuit with a current gain of β_D. The base current for this circuit may be obtained in the same way as discussed in Art 12-13 Chapter-12. It is given by the equation :

$$I_B = \frac{V_{CC} - V_{BE}}{R_B + \beta_D R_E} \quad \text{...(i)}$$

Notice that equation (*i*) is the same as for a regular transistor. But the value of β_D is much higher and V_{BE} is much larger. The emitter current,

$$I_E = (\beta_D + 1)\, I_B \approx \beta_D\, I_B$$

and the collector current,

$$I_C \approx I_E$$

In order to find the values of V_E, V_B, V_C and V_{CE}, consider the collector-emitter loop circuit (AFCEGA). Applying Kirchoff's Voltage Law to this loop,

$$V_{CE} + I_E \cdot R_E - V_{CC} + I_C \cdot R_C = 0$$

Replacing I_C with I_E in the above equation and solving if for the voltage across collector-emitter of a transistor,

$$V_{CE} = V_{CC} - I_C\,(R_C + R_E),$$
$$V_E = I_E R_E \approx I_C \cdot R_E$$
$$V_C = V_{CC} - I_C \cdot R_C$$
$$V_{CE} = V_C - V_E$$

and $$V_B = V_E + V_{BE}$$

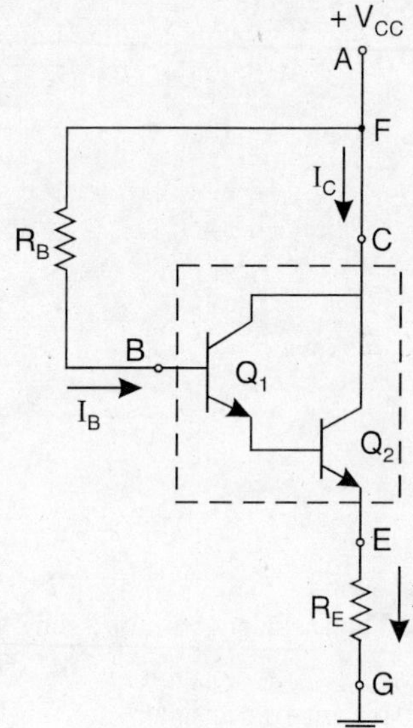

Fig. 20-10.

Example 20-6. *In a Darlington amplifier R_E = 100. If β_1, = 100 and β_2, = 200, Calculate the overall input impedance.*
(*Electronic Circuits - I, Anna University, April/May 2008*)

Solution. Given, $\beta_1 = 100$, $\beta_2 = 200$, $R_E = 100$

We know that over all input impedance,

$$\text{Over all input impedance} = \beta_1 \cdot \beta_2 \cdot R_E$$
$$= 100 \times 100 \times 200 = 2{,}00{,}00{,}00\ \Omega$$
$$= 2000\ \text{k}\Omega$$

Ans.

Example 20-7. *Fig. 20-11 shows a circuit of a Darlington connection. Given : β_D = 6000 and V_{BE} = 10.6 V.*

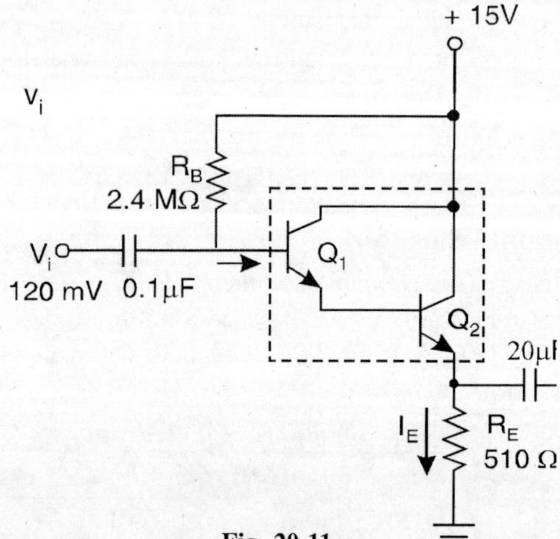

Fig. 20-11.

Calculate the dc bias voltage, V_{E_2}, emitter current I_{E_2}, and the amplifier voltage gain.

Solution. Given : $V_{CC} = 15\text{V}$ $R_B = 2.4\text{ M}\Omega = 2.4 \times 10^6\ \Omega$; $\beta_D = 6000$; $R_E = 510\ \Omega$; $V_i = 120\text{ mV}$

We know that the base current,

$$I_B = \frac{V_{CC} - V_{BE}}{R_B + \beta_D R_E} = \frac{15 - 1.6}{2.4 \times 10^{+6} + (6000 \times 510)} = 2.45 \times 10^{-6}\text{ A} = 2.45\ \mu\text{A}$$

∴ The emitter current,

$$I_E \approx \beta_D \cdot I_B = 6000 \times (2.45 \times 10^{-6}) = 14700 \times 10^{-6}\text{ A} = 14.7\text{ mA}$$

$$I_{E_2} = I_E = 14.7\text{ mA. } \textbf{Ans}$$

We also know that,

$$V_{E_2} = I_{E_2} \cdot R_E = 14.7\text{ mA} \times 510\ \Omega = 7.5\text{ V } \textbf{Ans.}$$

20-9. AC Analysis of Darlington Pair

Fig. 20-12 shows a Darlington emitter-follower circuit. As seen, the a.c. input signal is applied to the base of the Darlington circuit through capacitor C_1. While the a.c. output, v_o is obtained from the emitter through capacitor C_2.

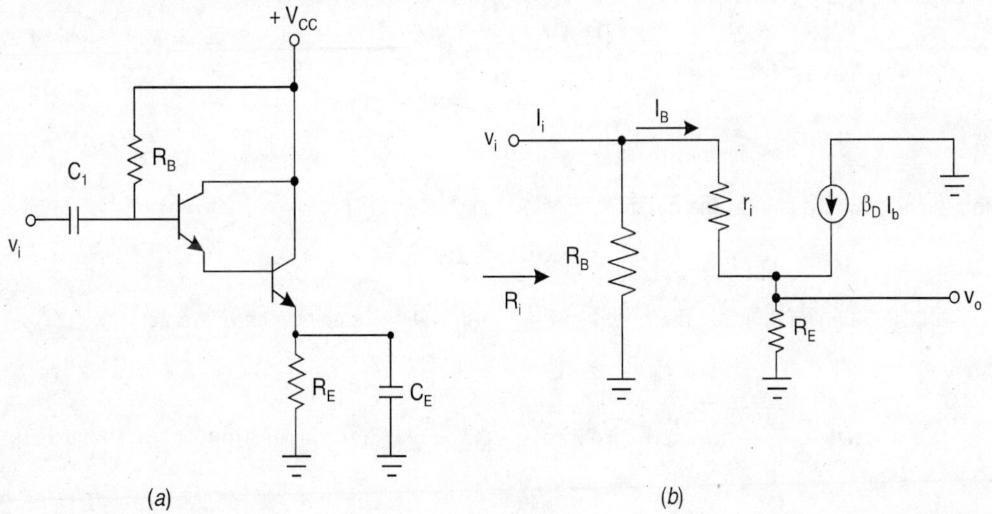

Fig. 20-12.

The a.c. equivalent circuit for a Darlington emitter-follower circuit may be obtained by replacing the Darlington by a circuit composed of an input resistance r_i, and an output current source $\beta_D\ I_b$. Refer to the circuit shown is Fig. 20-12(b). We shall use this circuit to find the values of a.c. input impedance, a.c. current gain, a.c. output impedance and a.c. voltage gain.

AC Input Impedance

We know that the a.c. base current through the input resistance r_i is given by the equation.

$$i_b = \frac{v_i - v_o}{r_i} \qquad ...(i)$$

and the output voltage,

$$v_o = (I_b + \beta_D\ I_b)\ R_E \qquad ...(ii)$$

Rerranging the equation (*i*), we get,

$$i_b \,.\, r_i = v_i - v_o$$

or $$v_i = i_b \,.\, r_i + v_o \qquad ...(iii)$$

Substituting the value of v_o from equation (*ii*) to equation (*iii*), we get,

$$v_i = i_b \,.\, r_i + (i_b + \beta_D\, i_b)\, R_E$$

or $$v_i = i_b\,[r_i + (1 + \beta_D)\, R_E]$$

Now the a.c. input impedance looking into the transistor base,

$$\frac{v_i}{i_b} = r_i + (1 + \beta_D)\, R_E$$

$$\approx r_i + \beta_D\, R_E \qquad ...(\because \beta_D >> 1)$$

And the a.c. input impedance looking into the circuit,

$$R_i = R_B \parallel \frac{V_i}{i_b}$$

or $$R_i = R_B \parallel (r_i + \beta_D\, R_E)$$

AC Current Gain

We know that a.c. output current through R_E (refer to fig. 20-12 (*b*)) is given by the equation,

$$i_o = i_b = \beta_D\, i_b$$

$$= (1 + \beta_D)\, i_b$$

$$= \beta_D\, i_b \qquad ...(\beta_D >> 1)$$

We also that transistor gain, is

$$\frac{i_o}{i_b} = \beta_D \qquad ...(i)$$

and the a.c. current gain of the circuit,

$$A_i = \frac{i_o}{i_i}$$

Multiplying and dividing the right hand side of the above equation by i_b, we get,

$$A_i = \frac{i_o}{i_b} \cdot \frac{i_b}{i_i} \qquad ...(ii)$$

Now the ratio i_b/i_i can be determined by using the current division rule from circuit shown in fig. 20-12 (*b*)

$$i_b = \frac{R_B}{(r_i + \beta_D\, R_E) + R_B}\, i_i$$

$$= \frac{R_B}{\beta_D\, R_E + R_B}\, i_i \qquad ...(\because r_i << \beta_D\, R_E \text{ or } R_B)$$

or $$\frac{i_b}{i_i} = \frac{R_B}{\beta_D\, R_E + R_B} \qquad ...(iii)$$

Substituting the values of i_o/i_b (= β_D) and i_b/i_i (= $\beta_D/(\beta_D\, R_E + R_B)$) in equation (*ii*), the ac circuit current gain,

$$A_i = \beta_D \cdot \frac{R_B}{\beta_D\, R_E + R_B} = \frac{\beta_D\, R_B}{R_B + \beta_D\, R_E}$$

AC Output Impedance

The a.c. output impedance of a Darlington emitter follower circuit shown in Fig. 20-12. Can be determined by using the a.c. circuit shown in Fig. 20-13 (*a*) and (*b*).

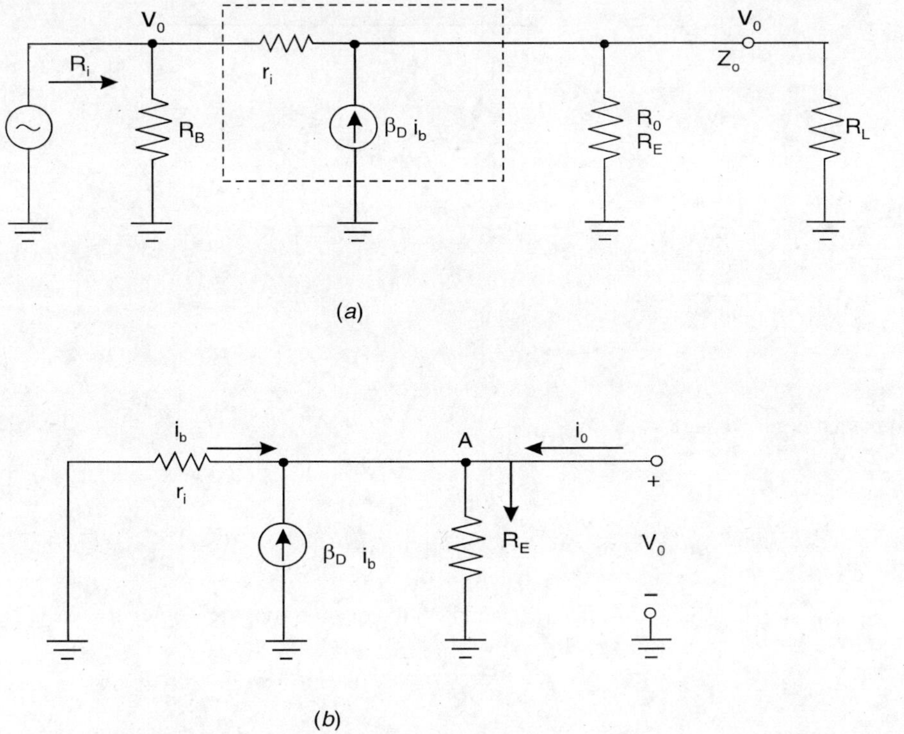

Fig. 20-13. *AC, equivalent circuit for calculation of a.c. output impedance*

We know that as seen by load R_L is determined by applying a voltage, v_o and measuring the current i_o (with input v_S set to zero).

Applying Kirchoff's Current Law at node A,

$$I_o = \frac{v_o}{R_E} + \frac{v_o}{r_i} - \beta_D \cdot i_b$$

Substituting for i_b ($= v_o/r_i$) in the above equation,

$$i_o = \frac{v_o}{R_E} + \frac{v_o}{r_i} - \beta_D \quad - \frac{v_o}{r_i}$$

$$= v_o \quad \frac{1}{R_E} + \frac{1}{r_i} + \frac{\beta_D}{r_i}$$

Therefore, the ac output impedance,

$$R_o = \frac{v_o}{i_o} = \frac{1}{\frac{1}{R_E} + \frac{1}{r_i} + \frac{\beta_D}{r_i}}$$

$$= R_E \parallel r_i \parallel \left(\frac{r_i}{\beta_D}\right)$$

$$\approx \frac{r_i}{\beta_D}$$

AC Voltage Gain

The a.c. voltage gain of a Darlington emitter follower circuit shown in Fig. 20-14 can be determined by using the a.c. equivalent circuit shown in Fig. 20-14.

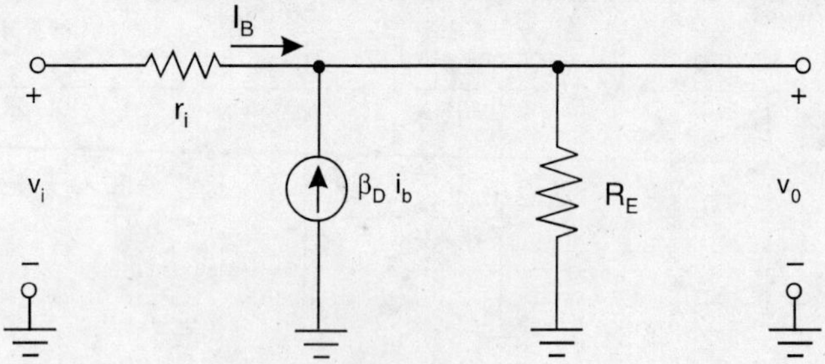

Fig. 20-14.

As seen from this circuit, the a.c. output voltage,

$$v_o = (i_b + \beta_D \cdot i_b)\, R_E$$

$$= i_b\,(R_E + \beta_D\, R_E) \qquad \text{...}(iv)$$

and the a.c. input voltage,

$$v_i = i_b\, r_i + (i_b + \beta_D\, i_b)\, R_E$$

$$= i_b\,(r_i + R_E + \beta_D\, R_E)$$

or

$$i_b = \frac{v_i}{r_i + R_E + \beta_D\, R_E} \qquad \text{...}(v)$$

Substituting the value of Ib from equation (*v*) to equation (*iv*), the ac output voltage,

$$v_o = \frac{v_i}{r_i + R_E + \beta_D\, R_E}\,(R_E + \beta_D\, R_E)$$

or

$$A_v = \frac{v_o}{v_i} = \frac{R_E + \beta_D\, R_E}{r_i + R_E + \beta_D\, R_E} \approx 1$$

Example 20-7. *For the Darlington amplifier shown in Fig. 20-15, obtain its Z-matrix for indentical Q_1 and Q_2 having $h_{fe} = 100$, $R_E = h_{ie} = 1\ k\Omega$,* (*UPSC Engg. Services.* 2001)

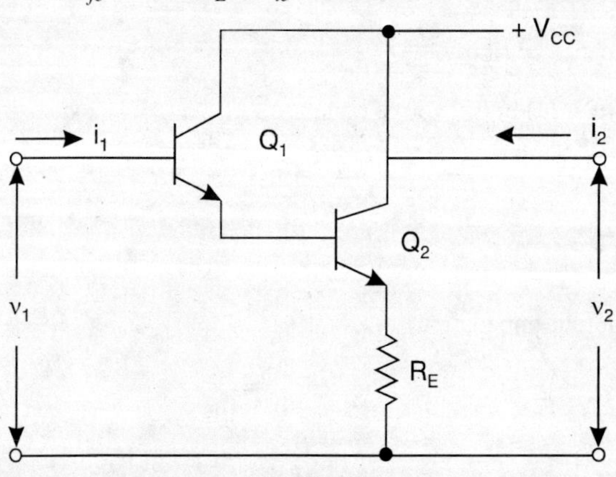

Fig. 20-15.

Solution. Given : $h_{fe} = \beta = 100$; $R_E = 1\ k\Omega$ and $h_{ie} = r_i = 1\ k\Omega$

We know that input resistance of the circuit,

$$R_i = R_B \,||\, (r_i + \beta_D\, R_E)$$

Since, there is no R_B (or rather $R_B = \infty$), therefore input resistance,

$$R_i = r_i + \beta_D R_E$$
$$= 1000 + (100 \times 100) \times 1000$$
$$\approx 10 \times 10^6\ \Omega \text{ or } 10\ \text{M}\Omega \textbf{ Ans.}$$

The output resistance of the circuit,

$$R_o = \frac{r_i}{\beta_D} = \frac{1000}{(100)^2} = 0.1\ \Omega \textbf{ Ans.}$$

20-10. Feedback Pair

Fig. 20-16 shows a feedback pair connection, As seen, this connection is a two transistor circuit that operater like a Darlington circuit. Notice that the feedback pair uses a PNP transistor driving an NPN transistor. Overall, the feedback pair acts like one PNP transistor which provides very high current gain.

The feedback pair is used along with a Darlington connection to provide a complementary transistor operation in Power Amplifiers. This is covered later in Chapter 22, Art 22-21.

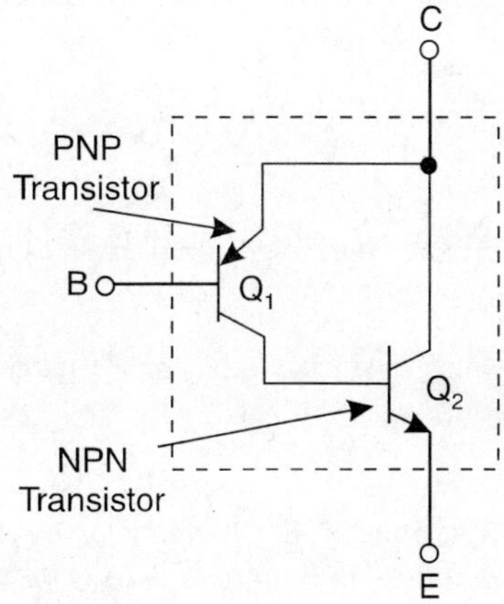

Fig. 20-16. *Illustrating a feedback pair connection*

20-11. DC Bias Operation of Feedback Pair

Fig. 20-17 shows a basic circuit of a feedback pair. The circuit currents and voltages may be obtained in the same way as discussed in chapter-19.

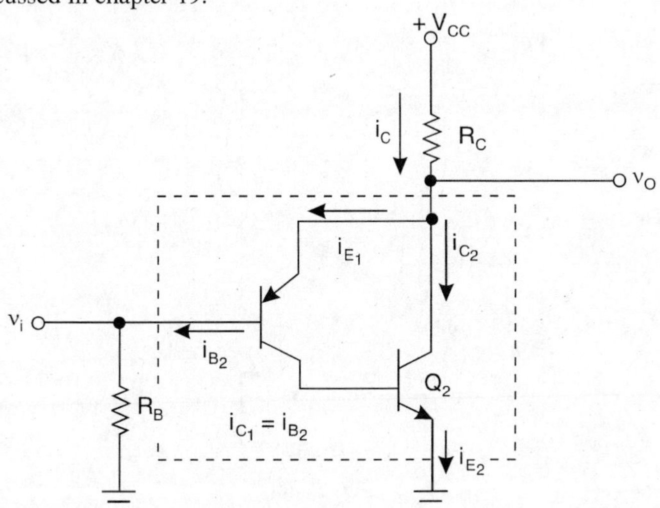

Fig. 20-17. *Illustrating the operation of a feedback pair*

From the base-emitter loop of transistor, Q_1, we obtain,

$$V_{CC} - I_C R_C - V_{BE_1} - I_{B_1} . R_B = 0$$

Substituting, $I_C = \beta_1 . \beta_2 . I_B$ in the above equation.

$$V_{CC} - \beta_1 . \beta_2 . I_B . R_C - V_{BE_1} - I_{B_1} . R_B = 0$$

Solving for I_{B_1}, we get,

$$I_{B_1} = \frac{V_{CC} - V_{BE_1}}{R_B + \beta_1 . \beta_2 . R_C}$$

The collector current of transistor, Q_1,

$$I_{C_1} = \beta_1 . I_{B_1}$$

This is also the base current of transistor, Q_2

$$\therefore \qquad I_{C_1} = I_{B_2}$$

Next, the collector current of transistor, Q_2,

$$I_{C_2} = \beta_2 \, I_{B_2} \approx I_{E_2}$$

The current through R_C,

$$I_C = I_{E_1} + I_{C_2}$$

Since, $I_{E_1} \approx I_{C_1}$, therefore,

$$I_C \approx I_{C_1} + I_{C_2}$$

Further since, $I_{C_1} = I_{B_2}$ (usually a small value because it is a base current and β_1 is high), therefore,

$$I_C \approx I_{C_2}.$$

Example 20-8. *Calculate the d.c. bias currents and voltages for the circuit shown in Fig. 20-18 to provide V_o at one-third the supply voltage.*

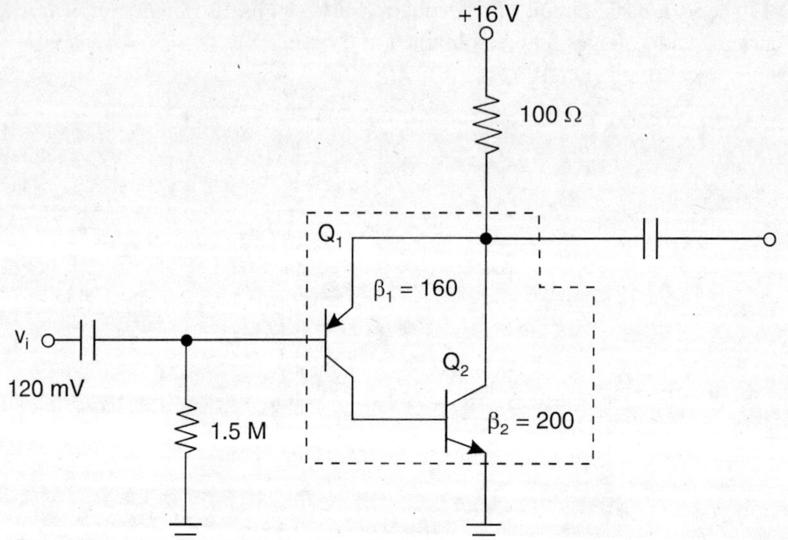

Fig. 20-18.

Solution. Given : V_{CC} = 16 V; β_1 = 160, β_2 = 200 ; R_B = 1.5 MΩ; V_i = 120 mV

We know that the base current,

$$I_{B_1} = \frac{V_{CC} - V_{EB_1}}{R_B + \beta_1 \, \beta_2 \, R_C} = \frac{16 - 0.7}{(1.5 \times 10^6) + (160 \times 200 \times 100)} \text{ A}$$

$$= 3.255 \times 10^{-6} \text{ A or } 3.255 \ \mu\text{A}$$

and the base current of Q_2,

$$I_{B_2} = I_{C_1} = \beta_1 \, I_{B_1} = 160 \times (3.255 \times 10^{-6})$$

$$= 0.520 \times 10^{-3} \text{ or } 0.520 \text{ mA}$$

We also know that collector current of Q_2,

$$I_{C_2} = \beta_2 I_{B_2} = 200 \times (0.520 \times 10^{-3})\ \text{A}$$
$$= 0.104\ \text{A or } 104\ \text{mA}$$

and the current through R_C,

$$I_C = I_{E_1} + I_{C_2} = 0.520 + 104 \approx 104.5\ \text{mA}$$

The d.c. voltage at the output,

$$V_{o(dC)} = V_{CC} - I_C R_C$$
$$= 16 - (104.5 \times 10^{-3}) \times 100\ \text{V}$$
$$= 5.55\ \text{V}$$

and the d.c. voltage at the input,

$$V_{i(dc)} = V_{o(dc)} - V_{BE} = 5.55 - 0.7 = 4.85\ \text{V}\ \textbf{Ans.}$$

Notes : Voltage at the base of Q_1, $V_{B_1} = V_{i(dc)} = 4.85$ V; the

Voltage at the collector of Q_2, $V_{C_2} = V_{o(dc)} = 5.55$ V.

20-12. AC Analysis of Feedback Pair

Fig. 20-19 shows a feedback pair circuit. Its a.c. equivalent circuit is as shown in Fig. 20-20. It may be carefully noted that Fig. 20-20 has been drawn to show clearly each transistor and its base and collector resistors. Then the a.c. equivalent circuit of Fig. 20-21 is redrawn. We will use this circuit to carryout a.c. analysis.

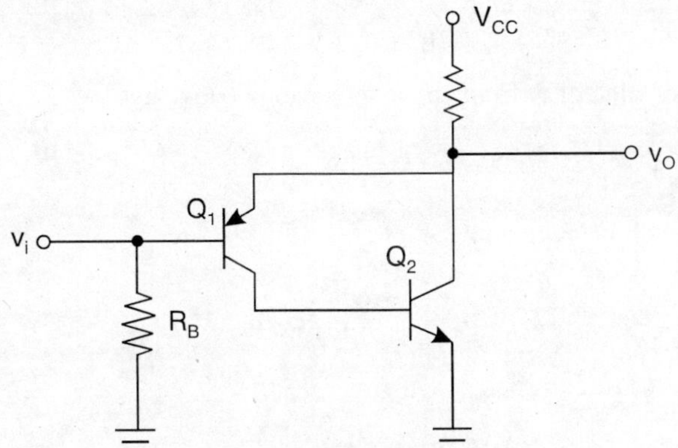

Fig. 20-19.

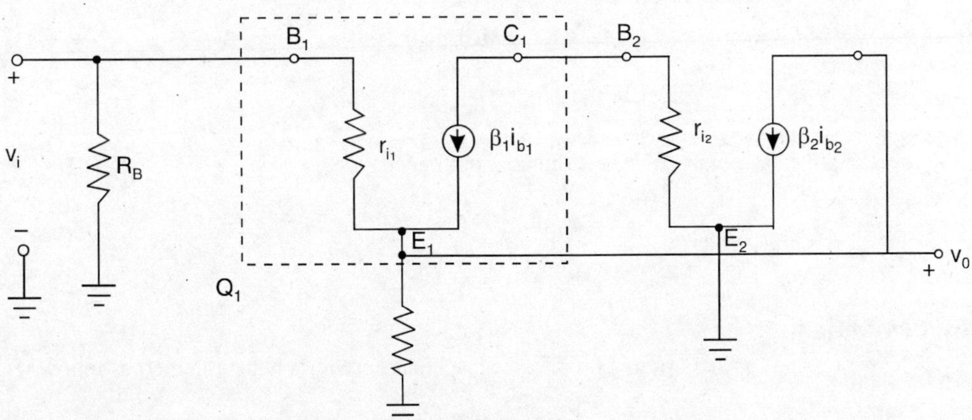

Fig. 20-20.

AC Input resistance (R_i)

Refer to circuit shown in Fig. 20-21. The a.c. input resistance seen looking into the base of transistor Q_1 is obtained as follows.

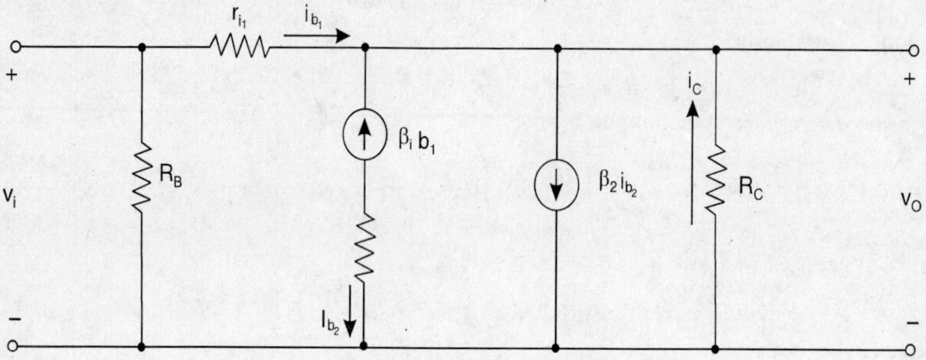

Fig. 20-21.

The base current of Q_1,

$$i_{b_1} = \frac{v_i - v_o}{r_{i_1}} \qquad \text{...}(i)$$

where,

$$v_o = -i_C R_C$$
$$\approx \left(-\beta_1 i_{b_1} + \beta_2 i_{b_2}\right) R_C$$
$$\approx \left(\beta_2 i_{b_2}\right) R_C$$

Substituting the value of v_o from above to equation (*i*) we get,

$$i_{b_1} = \frac{v_i - \left(\beta_2 i_{b_2}\right) R_C}{r_{i_1}}$$

Rearranging the above equation,

$$i_{b_1} . r_{i_1} = v_i - \beta_2 i_{b_2} R_C$$

or

$$v_i = i_{b_1} r_{i_1} + \beta_2 i_{b_2} R_C$$

Substituting for $i_{b_2} \left(= i_{C_1} = \beta_1 i_{b_1}\right)$, we get,

$$V_i = i_{b_1} r_{i_1} + \beta_2 (\beta_1 i_{b_1}) R_C$$
$$= i_{b_1} r_{i_1} + \beta_1 \beta_2 i_{b_1} R_C$$
$$= i_{b_1} (r_{i_1} + \beta_1 \beta_2 R_C)$$

$$\therefore \qquad \frac{v_i}{i_{b_1}} = r_{i_1} + \beta_1 \beta_2 R_C$$

Including the base bias resistance, the ac input impedance,

$$R_i = R_B \parallel \frac{v_i}{i_{b_1}}$$
$$= R_B \parallel (r_{i_1} + \beta_1 \beta_2 R_C)$$

AC Current Gain (A_i)

The expression for ac current gain of a feedback pair circuit may be obtained as follows :

$$i_o = \beta_2 i_{b_2} - \beta_1 i_{b_1} - i_{b_1}$$

$$= \beta_2 (\beta_1 i_{b_1}) - (1 + \beta_1) i_{b_1}$$

$$\approx \beta_1 \beta_2 i_{b_1}$$

$$\therefore \quad \frac{i_o}{i_{b_1}} = \beta_1 \beta_2$$

Including R_B, the current gain is given by,

$$A_i = \frac{i_o}{i_i} = \frac{i_o}{i_{b_1}} \cdot \frac{i_{b_1}}{i_i}$$

$$= \beta_1 \beta_2 \cdot \frac{R_B}{R_B + R_i}$$

AC Output resistance (R_0)

The a.c. output impedance can be obtained by applying v_o with v_i set to zero. Following a similar approach as discussed earlier for Darlington circuit, the output impedance,

$$R_o = \frac{v_o}{i_o} = R_C \parallel r_{i_1} \parallel \frac{r_{i_1}}{\beta_1} \parallel \frac{r_{i_1}}{\beta_1 \beta_2}$$

$$\approx \frac{r_{i_1}}{\beta_1 \beta_2}$$

This results is a low output impedance.

AC Voltage Gain (A_v)

The output voltage (refer to circuit 20-21)

$$v_o = - i_C R_C \approx \beta_1 \beta_2 i_{b_1} R_C$$

Since, $$i_{b_1} = \frac{v_i - v_o}{r_{i_1}}$$

$$\therefore \quad v_o = v_i - i_{b_1} \cdot r_{i_1}$$

$$= v_i - \frac{v_o}{\beta_1 \beta_2 R_C} r_{i_1}$$

$\therefore$ The voltage gain,

$$A_v = \frac{v_o}{v_i} = \frac{1}{1 + \frac{r_{i_1}}{\beta_1 \beta_2 R_C}}$$

$$= \frac{\beta_1 \beta_2 R_C}{\beta_1 \beta_2 R_C + r_{i_1}}$$

20-13. CMOS Circuit

The CMOS (called complementary metal-oxide semiconductor) circuit connection is very popular in digital electronic circuitary. The connection uses both *N*-channel and *P*-channel, enhancement MOSFETs. Fig. 20-22 shows the circuit of a CMOS INVERTER, *i.e.*, a circuit that inverts the input voltage. This CMOS circuit uses complemantary (or opposite)-type transistors. Notice that the input (v_i) is applied to both gates and the output is taken from the connected drains.

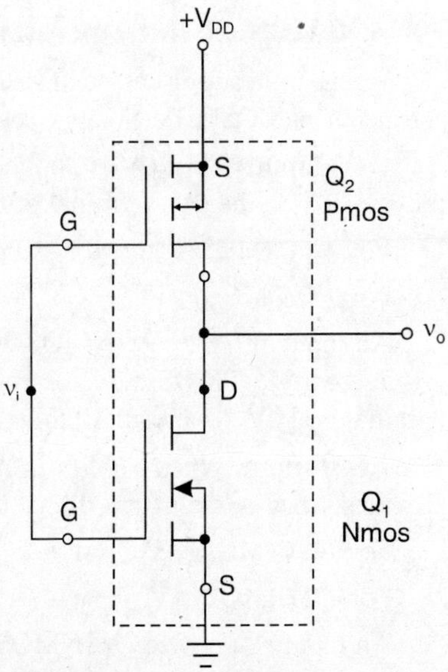

Fig. 20-22. *A CMOS Inverter*

Before we discuss the operation of a CMOS circuit, let us review the operation of the *N*-channel and *P*-channel enhancement type MOSFETs.

NMOS ON/OFF Operation : Fig. 20-23(*a*) shows the drain characteristic of an *N*-channel enhancement MOSFET. With $V_{GS} = 0$ V applied to gate-source junction, there is no drain current. There is no current through the device until $V_{GS} = V_{TH}$. When $V_{GS} = +5$ V, the device turns on and a current I_D flows through the device. Thus to summarize, an input of *OV* leaves the NMOS device "OFF" while an input of + 5 V turns the device "ON".

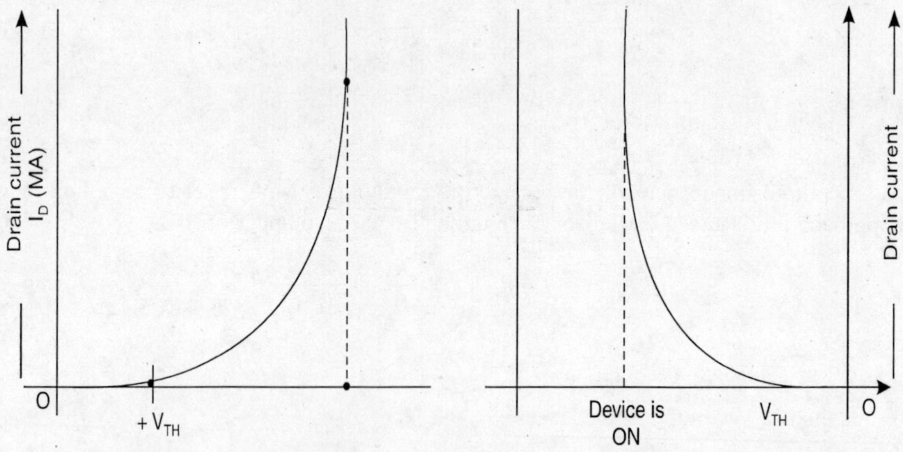

(a) Nmos ON / OFF Operation (b) Pmos ON / OFF Operation

Fig. 20-23.

PMOS ON/OFF Operation. Fig. 20-23 (*b*) shows the drain characteristic of a *P*-channel enhancement MOSFET. With $V_{GS} = 0$ V, there is no drain current. While for an input of – 5 V, the device is ON with drain current (I_D) present. Thus to summerize, an input of $V_{GS} = 0$ V leaves PMOS device OFF while $V_{GS} = -5$ V turns the device ON.

20-14. CMOS Circuit Operation

We have already discussed in the last article that a CMOS circuit used both *N*-channel and *P*-channel enhancement MOSFETs. Now we shall study the operation of a CMOS circuit for $v_i = 0$ V and + 5 V.

O-V Input. When 0V is applied to the input of the CMOS circuit (refer to Fig. 20-24 (*a*)), it provides 0V to the gate terminal of both NMOS and PMOS devices. In such a case.

For Q_1 (NMOS) : $V_{GS} = v_i - 0 = 0 - 0 = 0$ V

For Q_2 (PMOS) : $V_{GS} = v_i - (+5) = 0 - 5 = -5$ V

An input of 0V to Q_1 (NMOS device) leaves it off. However, the same 0V input produces $V_{GS} = -5$ V for Q_2 (PMOS device). This is due to the fact that source of Q_2 is at + 5 V and its gate terminal is at 0V. As a result of this Q_2 device turns ON. This produces an output voltage $V_o = +5$ V.

+ 5 V-Input. When a + 5 V is applied to the input of CMOS circuit (refer to Fig. 20-24 (*b*)), it provides + 5 V to the gate terminal of both NMOS and PMOS devices. In such a case :

For Q_1 (NMOS) : $V_{GS} = v_i - 0 = 5 - 0 = +5$ V

For Q_2 (PMOS) : $V_{GS} = v_i - (+5) = +5 - 5 = 0$ V

An input of + 5 V to Q_1 (NMOS device) turns it ON while Q_2 (PMOS device) turns OFF. This produces an output voltage, $v_o = 0$ V.

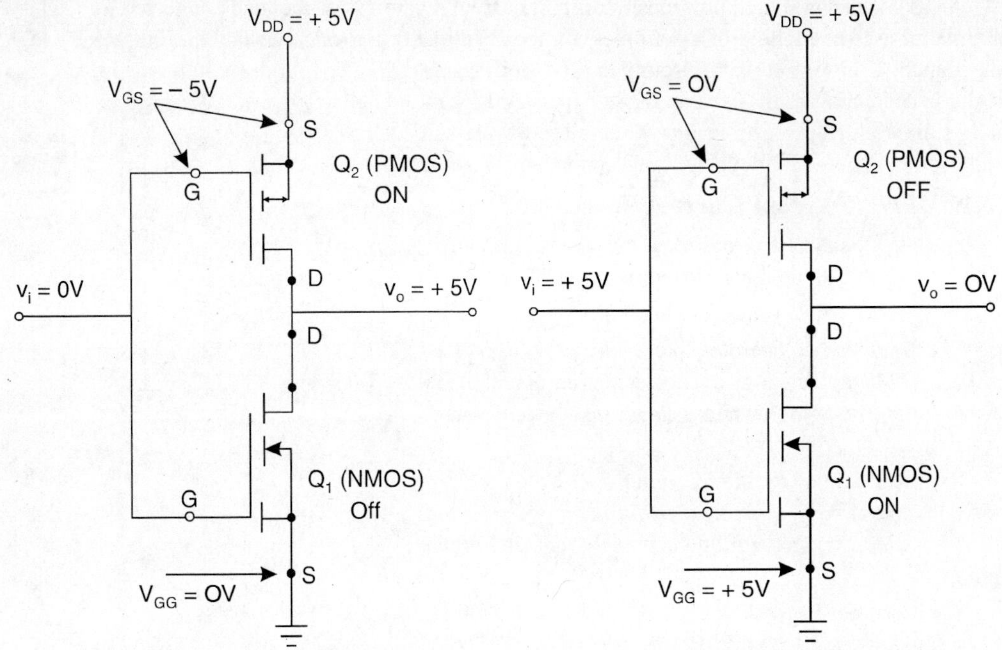

(*a*) $V_i = 0,\ V_0 = +5V$ (*b*) $V_i = +5V,\ V_0 = 0V$

Fig. 20-24. *CMOS Circuit operation*

Thus to summerize, the CMOS connection of Fig 'A', provides operation as a logic 'INVERTER', *i.e.* the circuit inverts its input. In other words, if V_i = O V, the circuit output (V_o) is + 5 V and if V_i = + 5 V, the circuit output (V_o) is OV. This is reproduced in Fig. 20-25 for your convenience.

v_i	Q_1	Q_2	v_o
0V	OFF	ON	+ 5 V
+ 5 V	ON	OFF	0V

Fig. 20-25. *CMOS Circuit operation*

20-15. Current Source Circuits

We have discussed about voltage and current sources in detail in Appendix E. There we defined a practical voltage source (shown in Fig. 20-26 (*a*)) is a voltage supply in series with a resistance. An ideal voltage source has $R = 0$ while a practical source includes some small value of resistance (refer to Fig. 20-26 (*b*)). On the other hand, a practical current source (shown in Fig. 20-26 (*c*)) is a current supply in parallel with a large value of resistance. An ideal current source has $R = \infty$ (refer to Fig. 20-26 (*d*)).

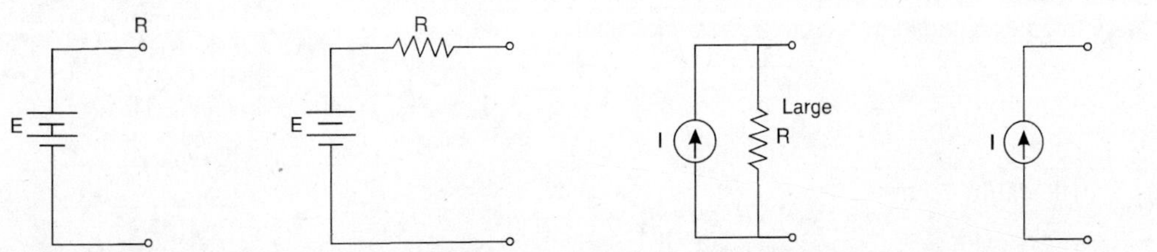

(*a*) an ideal voltage source (*b*) A practical current source (*c*) Practical current source (*d*) An ideal current source

Fig. 20-26.

An ideal current source provides a constant current regardless of the load connected to it. There are several uses in the field of electronics, for a circuit that provides a constant current at very high impedance. Constant current circuits can be built using FET, BJT and a combination of these devices. There are circuits used in discrete from and others more suitable for operation in integrated circuits. We will study some forms of both types in this chapter.

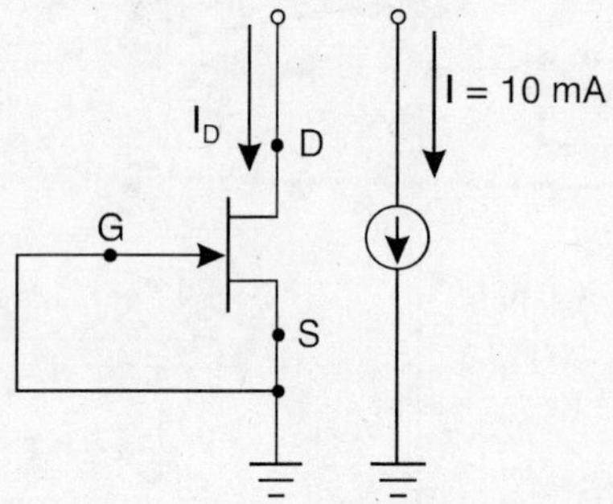

Fig. 20-27. *JFET constant current source*

20-16. JFET Constant Current Source

Fig. 20-27 shows a circuit of a JFET current source. With V_{GS} set to 0V, the drain current is fixed at

$$I_D = I_{DSS} \text{ (say 10 mA)}$$

The JFET device operates like a current source of 10 mA value. It may be carefully noted that an actual JFET has an output resistance where as an ideal current source would be a 10 mA supply.

Example 20-9. *Fig. 20-28 shows a JFET connected as a constant current source. Assuming I_{DSS} = 6 mA and V_P = – 3 V, determine the load current, I_D and the output voltage (V_o).*

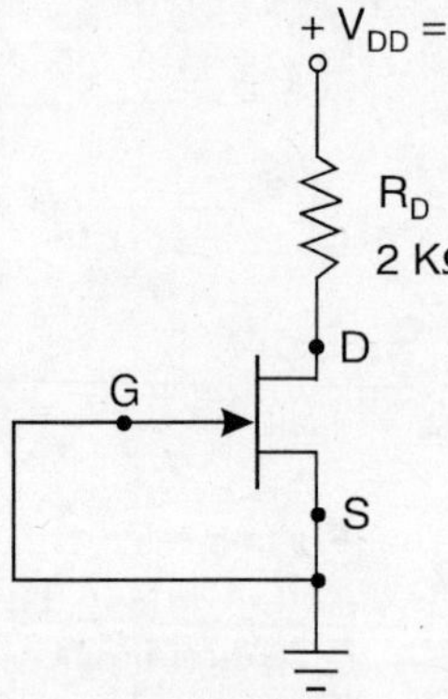

Fig. 20-28.

Solution. Given : V_{DD} = 18 V, R_D = 2 kΩ = 2000 Ω, I_{DSS} = 6 mA = 6 × 10^{-3} A and V_P = – 3 V.

Value of I_D

Notice that the gate terminal of the JFET is connected to ground. This means, $V_{GS} = 0$

We know that if V_{GS} = 0 V, the drain current,

$$I_D = I_{DSS} = 6 \text{ mA } \textbf{Ans.}$$

Output voltage (V_o)

We know that the output voltage,

$$V_o = V_{DD} - I_D R_D = 18 - (6 \times 10^{-3}) \times 2000 \text{ V}$$
$$= 6 \text{ V } \textbf{Ans.}$$

20-17. BJT Constant Current Source

Bipolar junction transistors (BJT) can be connected in a circuit which acts as a constant current source in a variety of ways.

Fig. 20-29 shows a circuit of an NPN bipolar transistor operating as a constant current source. The current through the emitter resistor can be determined as follows :

Assuming that the base input resistance of the transistor is very large as compared to R_1 and R_2, we know that,

$$V_B = (-V_{EE}) \cdot \left(\frac{R_1}{R_1 + R_2}\right)$$

and
$$V_E = V_B - 0.7 \text{ V}$$

∴ The emitter current,

$$I_E = \frac{V_E - (-V_{EE})}{R_E} \approx I_C$$

where, I_C is the constant current provided by the circuit of Fig. 20-29.

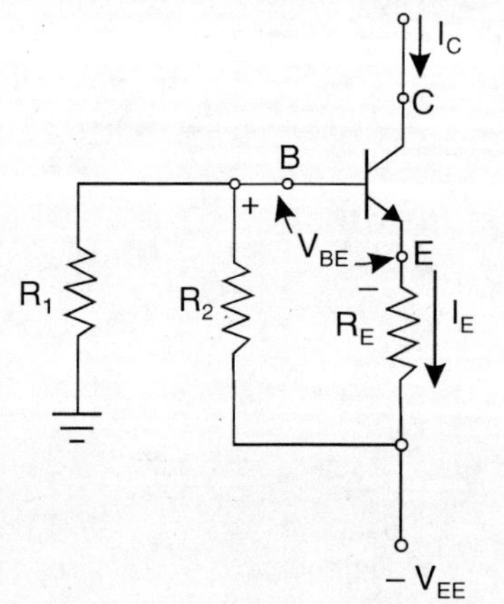

Fig. 20-29. *BJT constant current source*

Example 20-10. *Fig. 20-30 shows a circuit of a bipolar junction transistor connected as a constant current source.*

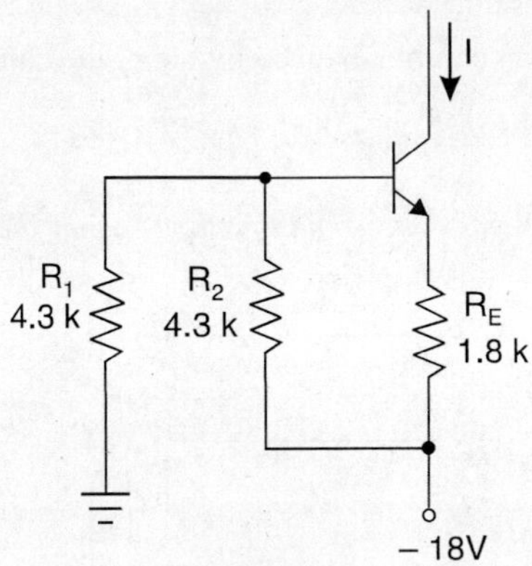

Fig. 20-30.

Assuming β of a transistor = 100, calculate the value of constant current (I_C) in the circuit.

Solution. Given : $V_{EE} = -18$ V, $R_1 = R_2 = 4.3$ kΩ $= 4.3 \times 10^3$ Ω; $R_E = 1.8$ k $= 1.8 \times 10^3$ Ω and $\beta = 100$.

We know that the base voltage,

$$V_B = (-V_{EE}) \cdot \frac{R_2}{R_1 + R_2} = -18 \times \frac{4.3}{4.3 + 4.3} = -9 \text{ V}$$

and the emitter voltage,

$$V_E = V_B - 0.7 \text{ V} = -9 - 0.7 = -9.7 \text{ V}$$

We also know that the emitter current,

$$I_E = \frac{V_E - (-V_{EE})}{R_E} = \frac{-9.7 - (-18)}{1.8 \times 10^3} = 4.61 \times 10^{-3} \text{ A}$$

$$= 4.61 \text{ mA} \textbf{ Ans.}$$

20-18. BJT-Zener Diode Constant Current Source

The BJT constant-current source circuit of Fig. 20-29 can be improved by replacing the resistor R_2 by a zener diode. The resulting circuit is as shown in Fig. 20-31.

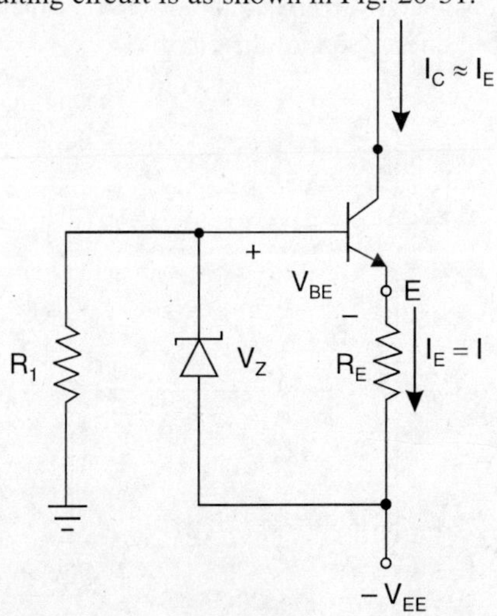

Fig. 20-31.

The constant current depends on the zener diode voltage and the emitter resistor (R_E). The voltage supply (V_{EE}) has no effect on the value of I. The advantage of this circuit is that the zener diode voltage remains quite constant.

The value of constant-current can be obtained by using the equation,

$$I \approx I_E = \frac{V_Z - V_{BE}}{R_E}$$

Example 20-11. *Fig. 20-32 shows a circuit of a BJT/Zener diode constant-current source.*

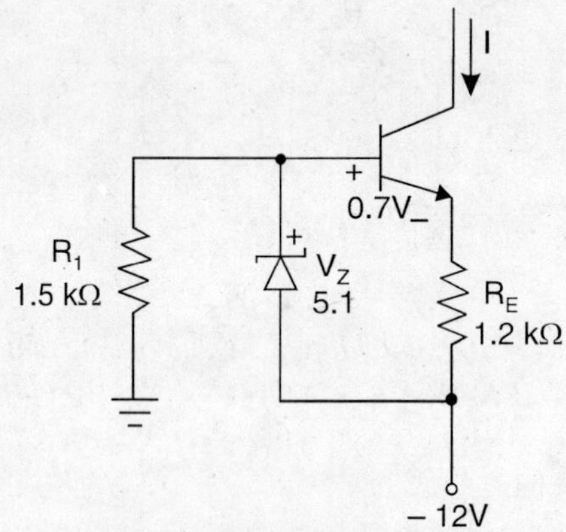

Fig. 20-32.

Assuming β of a transistor = 200, determine the value of constant current (I) in the circuit.

Solution. Given : $V_Z = 5.1$ V; $V_{BE} = 0.7$ V; $R_E = 1.2$ kΩ $= 1.2 \times 10^3$ Ω and $\beta = 200$

We know that the value of constant current,

$$I = \frac{V_Z - V_{BE}}{R_E} = \frac{5.1 - 0.7}{1.2 \times 10^3} = 3.67 \times 10^{-3} \text{ A}$$

$$= 3.67 \text{ mA } \textbf{Ans.}$$

20-19. Current Mirror Circuits

Fig. 20-33 shows a current mirror circuit. Notice that the collector of transistor Q_1 is shorted to its base. Also the base terminals of both the transistors Q_1, Q_2 are connected together. This circuit provides a constant current is obtained from an output current (I) which is the reflection or mirror of a constant current (I_x) developed on one side of the circuit.

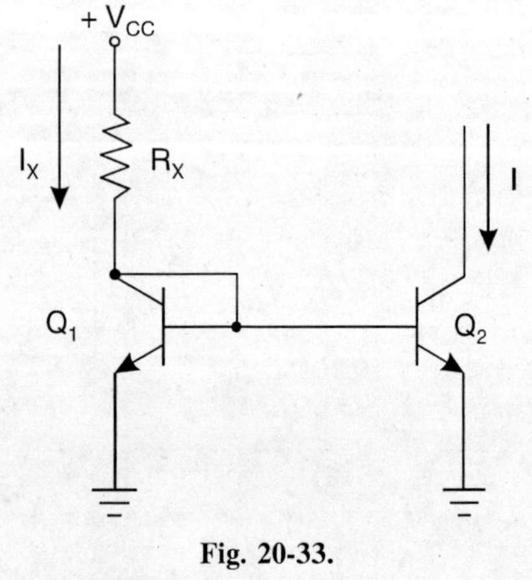

Fig. 20-33.

As mentioned earlier, the current mirror circuit is particularly suited to linear *IC* manufacture. This is due to the fact that the *IC* circuit requires that the transistors used have identical base-emitter voltage drops and identical values of 'β'. Moreover for best results transistors must be fabricated at the same time in *IC* manufacture. Notice from fig. 20-33, that the current, I_x set by transistor, Q_1 and resistor R_x is mirrored in the current, I through transistor, Q_2.

Fig. 20-34 shows the circuit currents for the current mirror circuit. Notice that both the transistors carry the same amount of emitter current (I_E). Since, 'β' of both the transistors are the same, then their base currents are also equal, *i.e.*,

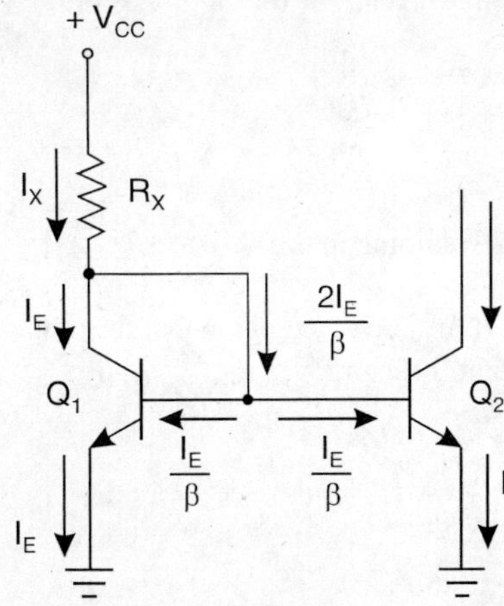

Fig. 20-34.

$$I_B = \frac{I_E}{\beta + 1} \approx \frac{I_E}{\beta}$$

Then the collector current of each transistor,

$$I_C \approx I_E$$

Finally, the collector current through R_x resistor,

$$I_x = I_E + \frac{2\,I_E}{\beta} = \frac{\beta\,I_E}{\beta} + \frac{2\,I_E}{\beta} = \frac{\beta + 2}{\beta}\,I_E$$

$$\approx I_E \qquad ...(\because \beta >> 2)$$

Thus, the constant-current provided at the collector of Q_2 mirrors that of Q_1 because

$$I_x = \frac{V_{CC} - V_{BE}}{R_x}$$

The current I_x set by V_{CC} and R_x is mirrored in the current into the collector of Q_2.

Example 20-12. *Fig. 20-35 shows a current mirror circuit. Determine the value of mirrored current I. Assume V_{BE} = 0.7 V.*

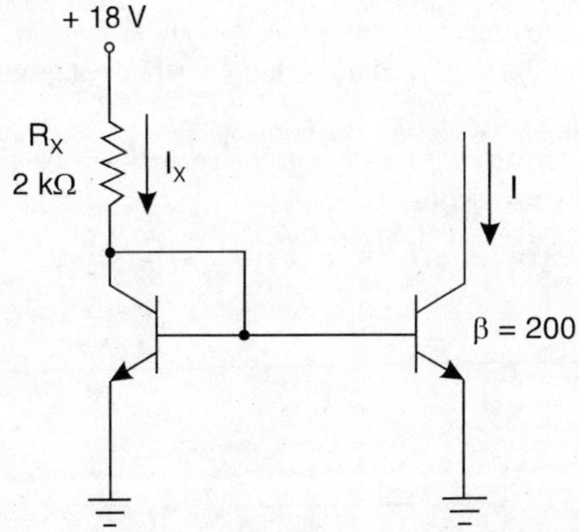

Fig. 20-35.

Solution. Given : V_{CC} = 18 V; R_x = 2 kΩ (= 2 × 10³ Ω) and V_{BE} = 0.7 V

We know that the value of current,

$$I_x = \frac{V_{CC} - V_{BE}}{R_x} = \frac{18 - 0.7}{2 \times 10^3} = 8.65 \times 10^{-3}\text{ A}$$

$$= 8.65\text{ mA}$$

∴ The valued of mirrored current,

$$I = I_x = 8.65\text{ mA } \textbf{Ans.}$$

20-20. Current Mirror Circuit with Higher output Resistance

Fig. 20-36 shows a current mirror circuit that provides higher output resistance than that of fig. 20-33. The current through resistor, R_x is given by

$$I_x = \frac{V_{CC} - 2V_{BE}}{R_x}$$

$$\approx I_E + \frac{I_E}{\beta} = \frac{\beta + 1}{\beta} I_E \approx I_E$$

Assuming that transistors Q_1 and Q_2 are well matched, the output current, I is held constant at,

$$I \approx I_E = I_x$$

It may be noted again that the output current, I is a mirrored value of the current set by the fixed current through R_x.

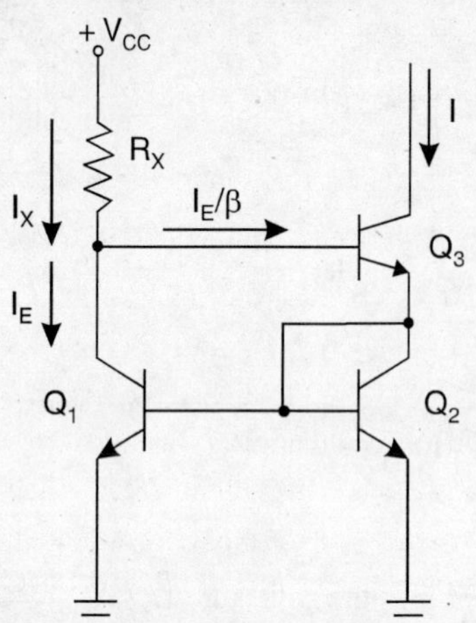

Fig. 20-36. *Current mirror circuit with higher output resistance.*

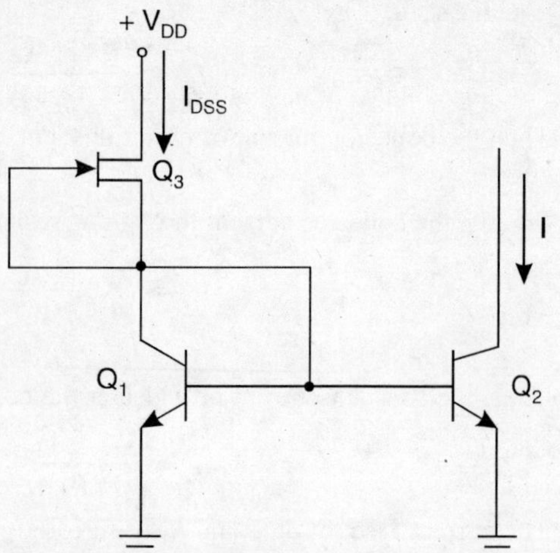

Fig. 20-37. *Current mirror connection with JFET*

Fig. 20-37 shows still another form of current mirror. Here, the JFET provides a constant-current set at the value of I_{DSS}. This current is mirrored resulting in a current through Q_2 of the same value, *i.e.*

$$I = I_{DSS}$$

Example 20-13. *In the circuit shown in Fig. 20-38, all the transistors are matched silicon transistors. Symbols carry their usual meaning.*

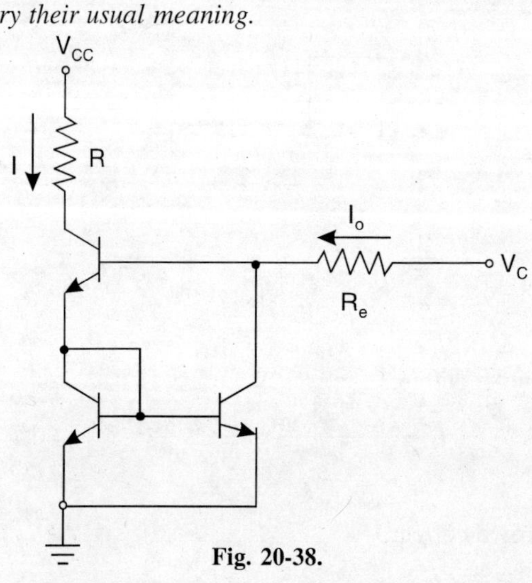

Fig. 20-38.

(**a**) *Find the expression of current I in terms of* I_o.

(**b**) *Calculate the value of I if* $V_c = 5$ *V,* $R_e = 2$ *k*Ω*,* $V_{CC} = 6$ *V and R = 2.2 k*Ω*. Take* $V_{BE} = 0.7$ *V and* $\beta = 100$.

(**c**) *What can be the maximum value of I ? If* R_e *is varied between 1 k*Ω *and 4 k*Ω *give the range of variation of I.* (*UPSC Engg Services* 2001)

Solution.

(**a**) *Expression for current I*

$$I = \frac{V_{CC} - 2V_{BE}}{R_e} \approx I_o + \frac{I_o}{\beta} \approx I_o \quad \textbf{Ans.}$$

(**b**) Given : $V_C = 5$ V; $R_e = 2$ kΩ; $V_{CC} = 6$ V, $R = 2.2$ kΩ; $V_{BE} = 0.7$ V and $\beta = 100$.

We know that,

$$I = \frac{V_{CC} - 2V_{BE}}{R_e} = \frac{6 - 2 \times 0.7}{2\,k} = 2.3 \text{ mA} \quad \textbf{Ans.}$$

(**c**) When $R_e = 1$ kΩ,

$$I = \frac{V_{CC} - 2V_{BE}}{R_e} = \frac{6 - 2 \times 0.7}{1\,k} = 4.6 \text{ mA}$$

and when $R_e = 4$ kΩ,

$$I = \frac{V_{CC} - 2V_{BE}}{R_e} = \frac{6 - 2 \times 0.7}{4\,k} = 1.15 \text{ mA.}$$

Thus, when R_e is varied between 1 kΩ and 4 kΩ, the current *I* will vary between 4.6 mA to 1.15 mA **Ans.**

20-21. Differential Amplifier Circuit

It is an extremely popular connection in linear integrated circuits. The connection can be described by considering the basic differential amplifier circuit shown in Fig. 20-39.

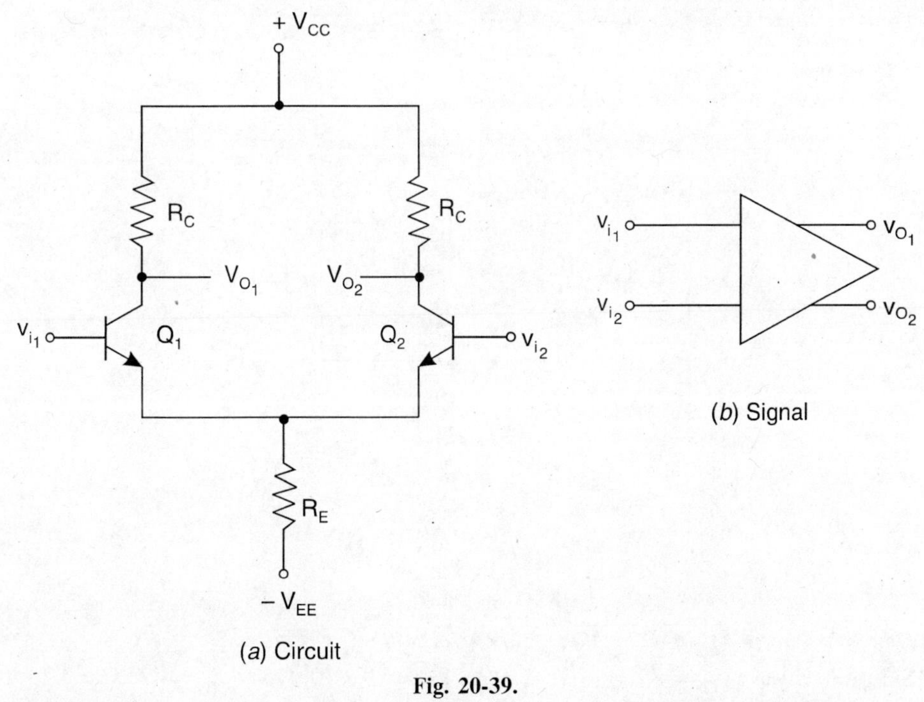

(*a*) Circuit

(*b*) Signal

Fig. 20-39.

As seen from Fig. 20-39 (*a*), the differential amplifier circuit has two separate inputs (v_{i_1} and v_{i_2}), two separate outputs (v_{o_1} and v_{o_2}) and that emitters of Q_1 and Q_2 are connected together. While most differential amplifier circuits use two separate voltage supplies, the circuit can also operate using a single supply. Fig. 20-39 (*b*) shows a circuit symbol of a differential amplifier.

The differential amplifier circuit can have a number of input signal combinations :

(1) The input signal is applied to either input and the other input is connected to ground. In this configuration, differential amplifier circuit is called single-ended differential amplifier.

(2) Two input signals with opposite polarity are applied to the two inputs. In this configuration, the differential amplifier circuit is called double-ended differential amplifier.

It may be carefully noted that although in a single-ended operation a single input in applied, but due to common-emitter connection, the input signal operates both the transistors, As a result, the output is produced on both collectors. In a double-ended operation, outputs from both collectors is due to the difference of the signals applied to both inputs.

In common-mode operation, the common input signal results in opposite signals at each collector. Ideally speaking these collector signals should cancel each other so that the resulting output signal is zero. However, in actual practice, the opposite signals do not completely cancel and a small-signal is produced at the collector outputs.

The major advantage of the differential amplifier is the very large gain when opposite signals are applied to its inputs as compared to the very small gain resulting from common inputs. The ratio of this differential gain to the common gain is called common-mode rejection ratio. These concepts are discussed later in Chapter-30 on OP-AMPS.

Example 20-14. *Fig. 20-40 shows the circuit of a differential amplifier commonly used in integrated circuits.*

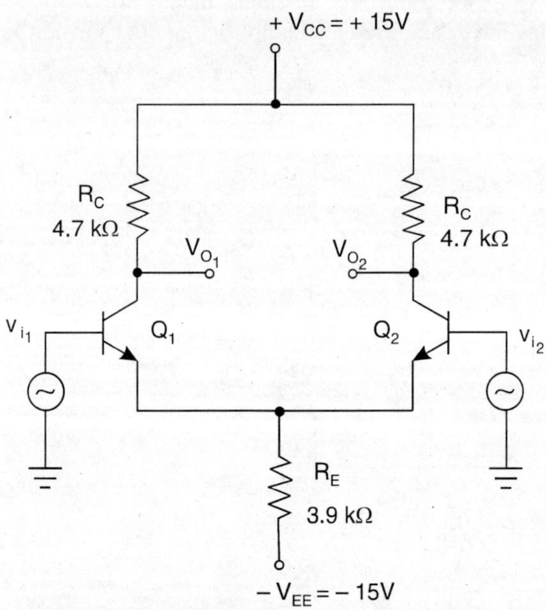

Fig. 20-40.

Determine the values of d.c. voltage and currents in this circuit.

Solution. Given : $V_{CC} = V_{EE} = 15$ V; $R_E = 3.9$ kΩ $= 3.9 \times 10^3$ Ω and $R_C = 4.7$ kΩ $= 4.7 \times 10^3$ Ω

Notice that the two base inputs are connected to a.c. voltage sources. However, the base voltage at the inputs of Q_1 and Q_2 is 0V.

We know that emitter current through resistor R_E,

$$I_E = \frac{V_{EE} - 0.7}{R_E} = \frac{15 - 0.7}{3.9 \times 10^3} = 3.67 \times 10^{-3} \text{ A}$$

$$= 3.67 \text{ mA}$$

and the value of collector current,

$$I_C = \frac{I_E}{2} = \frac{3.67}{2} = 1.83 \text{ mA } \textbf{Ans.}$$

We also know that the collector voltage,

$$V_C = V_{CC} - I_C R_C = 15 - (1.83 \times 10^{-3}) \times (4.7 \times 10^3)$$

$$= 15 - 8.601 \approx 6.4 \text{ V } \textbf{Ans.}$$

20-22. DC Bias Operation of Differential Amplifier

In order to study the d.c. bias operation, consider the differential amplifier circuit shown in Fig. 20-41. As seen, the d.c. voltage at each input is essentialy connected to 0V. With each base voltage at 0V, the common-emitter d.c. bias voltage is,

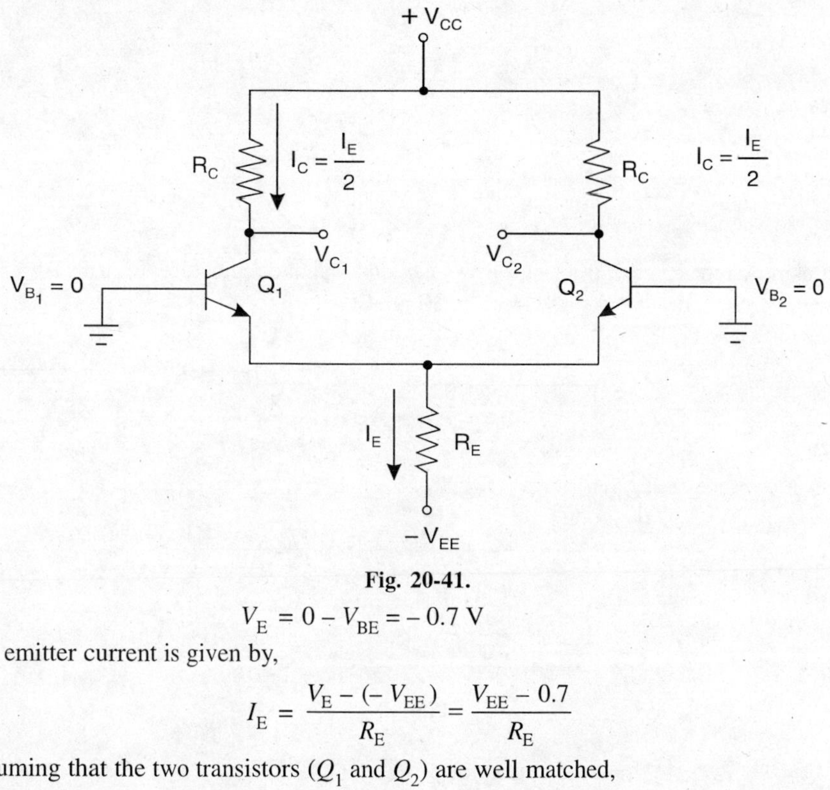

Fig. 20-41.

$$V_E = 0 - V_{BE} = -0.7 \text{ V}$$

The emitter current is given by,

$$I_E = \frac{V_E - (-V_{EE})}{R_E} = \frac{V_{EE} - 0.7}{R_E}$$

Assuming that the two transistors (Q_1 and Q_2) are well matched,

$$I_{C_1} = I_{C_2} = \frac{I_E}{2}$$

The value of collector voltage,

$$V_{C_1} = V_{C_2} = V_{CC} - I_C \cdot R_C = V_{CC} - \left(\frac{I_E}{2}\right) \cdot R_C$$

20-23. AC Operation of Differential Amplifier

In order to understand the a.c. operation, consider the differential amplifier circuit shown in Fig. 20-42. As seen, separate input signals are applied as v_{i_1} and v_{i_2} producing separate outputs, v_{o_1} and v_{o_2} .

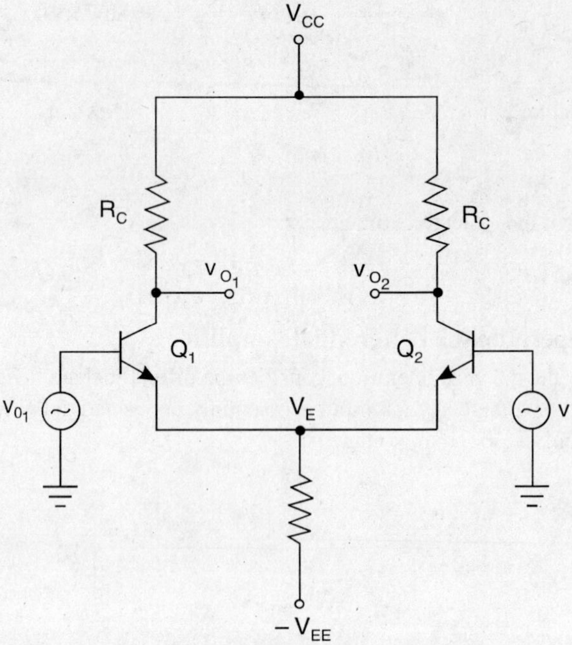

Fig. 20-42.

Let us now replace each transistor by its a.c. equivalent circuit (refer to Art. 11-8 Chapter-11). The circuit can be redrawn as shown in Fig. 20-43.

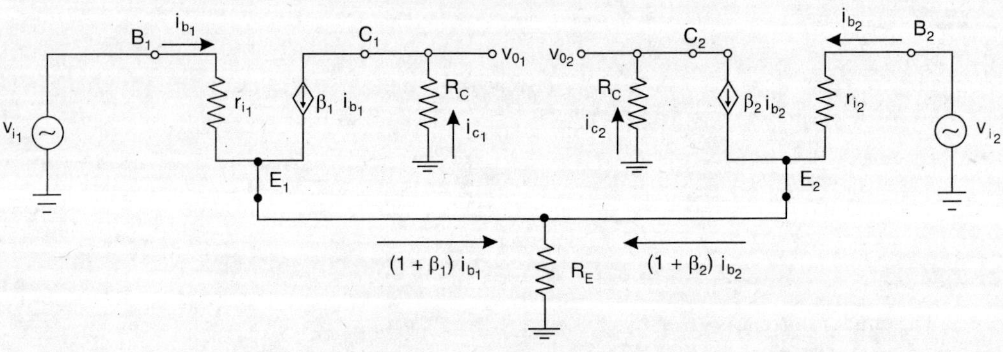

Fig. 20-43.

where, i_{b_1} = base current of Q_1; i_{b_2} = base current of Q_2

i_{c_1} = collect current of Q_1; i_{c_2} = collector current of Q_2

β_1 = current gain of Q_1; β_2 = current gain of Q_2

r_{i_1} = input resistance of Q_1 $(\beta_1\ r'_{e_1})$ and

r_{i_2} = input resistance of Q_2 $(\beta_2\ r'_{e_2})$

20-24. Single-ended AC Voltage Gain of Differential Amplifier

In order to calculate the single-ended a.c. voltage gain (v_o/v_i) of a differential amplifier, apply signal to one input with the other connected to the ground as shown in Fig. 20-44.

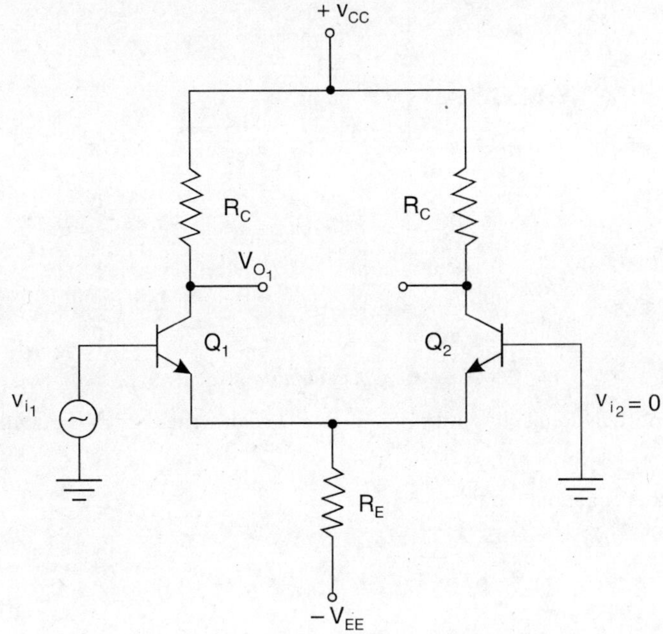

Fig. 20-44. *Connections of a differential amplifier for the calculation of single-ended a.c. voltage gain.*

Replacing the two transistors Q_1 and Q_2 by their a.c. equivalent circuit, Fig. 20-44 can be redrawn as shown in Fig. 20-45.

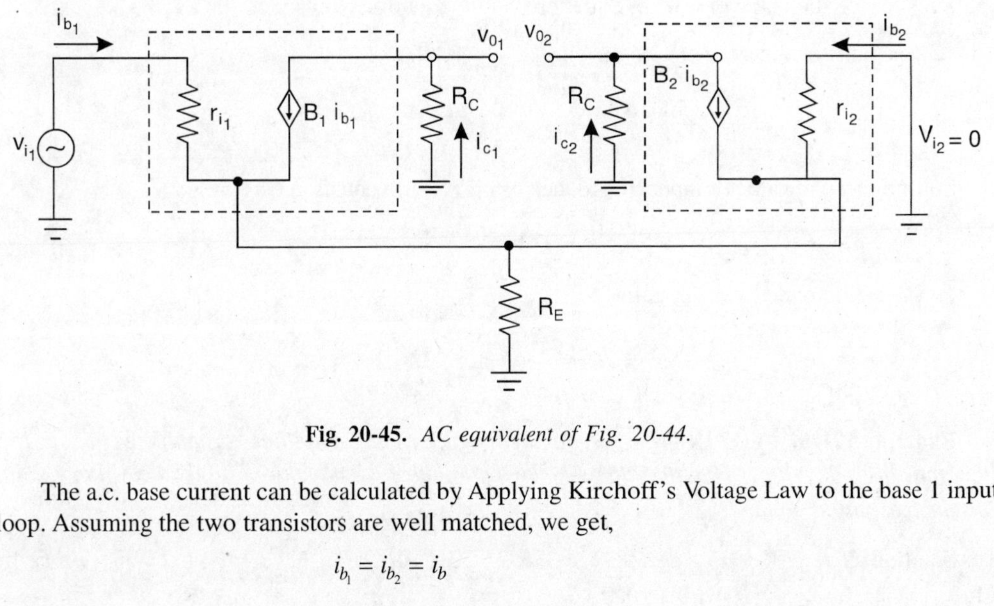

Fig. 20-45. *AC equivalent of Fig. 20-44.*

The a.c. base current can be calculated by Applying Kirchoff's Voltage Law to the base 1 input loop. Assuming the two transistors are well matched, we get,

$$i_{b_1} = i_{b_2} = i_b$$

and $$r_{i_1} = r_{i_2} = r_i$$

With R_E very large (ideally infinite), the circuit for obtaining the KVL equation for the base current (i_b) simplifies to the one shown in Fig. 20-46.

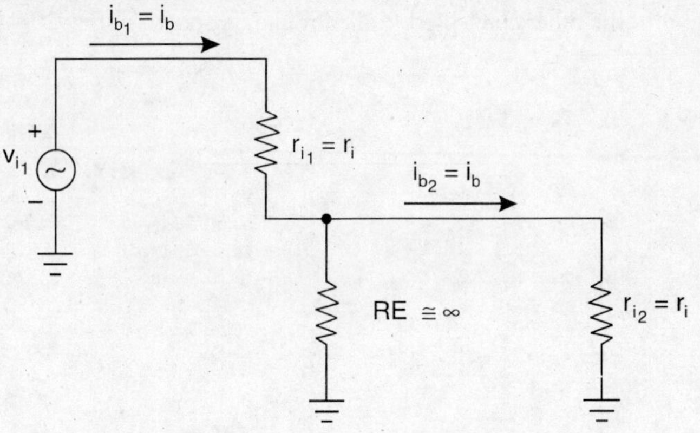

Fig. 20-46. *A simplified circuit to calculate i_b.*

From the simplified circuit shown in Fig. 20-46; we can write the KVL equation as :

$$v_{i_1} - i_b\, r_i - i_b\, r_i = 0$$

Rearranging this equation, we get,

$$i_b = \frac{v_{i_1}}{2\, r_i}$$

Assuming $\beta_1 = \beta_2 = \beta$, we get,

$$i_c = \beta\, i_b = \beta \cdot \frac{v_{i_1}}{2\, r_i} \qquad ...(i)$$

The value of output voltage at either collector is given by,

$$v_o = i_c \cdot R_c \qquad ...(ii)$$

Substituting the value of i_c from equation (*i*) into equation (*ii*) we get,

$$v_o = \beta \cdot \frac{v_{i_1}}{2\, r_i} \cdot R_C$$

$$= \frac{\beta\, R_C}{2\, r_i} \cdot v_{i_1} \qquad ...(iii)$$

Substituting the value of input impedance $r_i = \beta\, r'_e$, into equation (*iii*) we get,

$$v_o = \frac{\beta\, R_C}{2\, \beta\, r'_e} \cdot v_{i_1}$$

∴ The single-ended voltage gain,

$$A_v = \frac{v_o}{v_{i_1}} = \frac{\beta\, R_C}{2\, \beta\, r'_e} = \frac{R_C}{2 r'_e}$$

Example 20-15. *Fig. 20-47 shows a circuit of a differential amplifier. Assuming $\beta_1 = \beta_2 = 150$ determine the d.c. values of emitter current, collection current, collector voltage, a.c. voltage gain, and the a.c. output voltage v_{o_1}.*

Solution. Given : $V_{CC} = V_{EE}$; 12 V; $R_E = 33\ k\Omega = 33 \times 10^3\ \Omega$; $R_{C_1} = R_{C_2} = 36\ k\Omega = 36 \times 10^3\ \Omega$; $\beta_1 = \beta_2 = 150$ and $v_{i_1} = 2\ mV = 2 \times 10^{-3}$ V.

Value of Emitter Current

We know that the emitter current,

$$I_C = \frac{V_{EE} - 0.7\text{ V}}{R_E} = \frac{12 - 0.7}{33 \times 10^3} = 0.342 \times 10^{-3}\text{ A}$$

$$= 0.342\text{ mA } \textbf{Ans}$$

Value of Collector Current

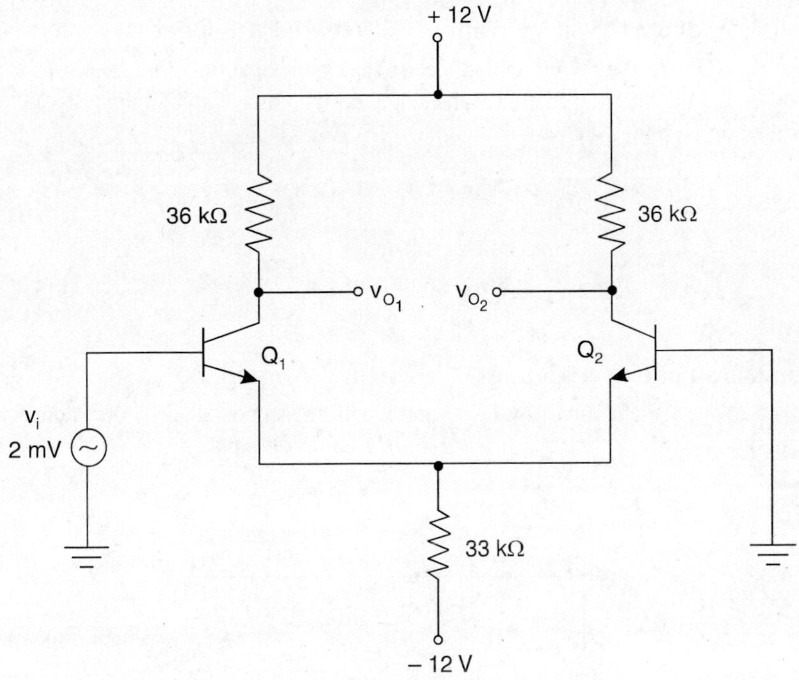

Fig. 20-47.

We know that the collector d.c. current,

$$I_C = \frac{I_E}{2} = \frac{0.342}{2} = 0.171\text{ mA } \textbf{Ans.}$$

Value of Collector d.c. output voltage

We know that the collector output voltage,

$$V_C = V_{CC} - I_C R_C = 12 - (171 \times 10^{-6}) \times (36 \times 10^3)$$

$$= 12 - 6.16$$

$$= 6.14\text{ V}$$

Value of a.c. voltage gain

We know that the value of r'_e is,

$$r'_e = \frac{25}{0.342} = 73.1\ \Omega$$

and the a.c. voltage gain,

$$A_v = \frac{v_{o_1}}{v_{i_1}} = \frac{R_C}{2\,r'_e} = \frac{36 \times 10^3}{2 \times 73.1} = 246.2\ \textbf{Ans.}$$

AC output voltage (v_{o_1})

We also know that a.c. voltage gain (A_v)

$$246.2 = \frac{v_{o_1}}{v_{i_1}} = \frac{v_{o_1}}{2 \times 10^{-3}}$$

$$\therefore \quad v_{o_1} = 246.2 \times (2 \times 10^{-3}) = 492.4 \times 10^{-3} \text{ V}$$

$$= 0.492 \text{ V } \textbf{Ans.}$$

20-25. Double-ended AC Voltage Gain of Differential Amplifier

In order to calculate the double-ended a.c. voltage gain of differential amplifier we can do a similar analysis like the one carried out for the single-ended voltage gain in the last article. It can be shown that the double-ended a.c. voltage gain,

$$A_d = \frac{v_o}{v_d} = \frac{\beta R_C}{2 r_i}$$

where, v_o = output voltage,

v_d = differential input voltage

$= v_{i_1} - v_{i_2}$

20-26. Common-mode operation of a Differential Amplifier

Fig. 20-48 shows a differential amplifier circuit with the same voltage source connected at both the inputs. In other words, both the transistors have a common input.

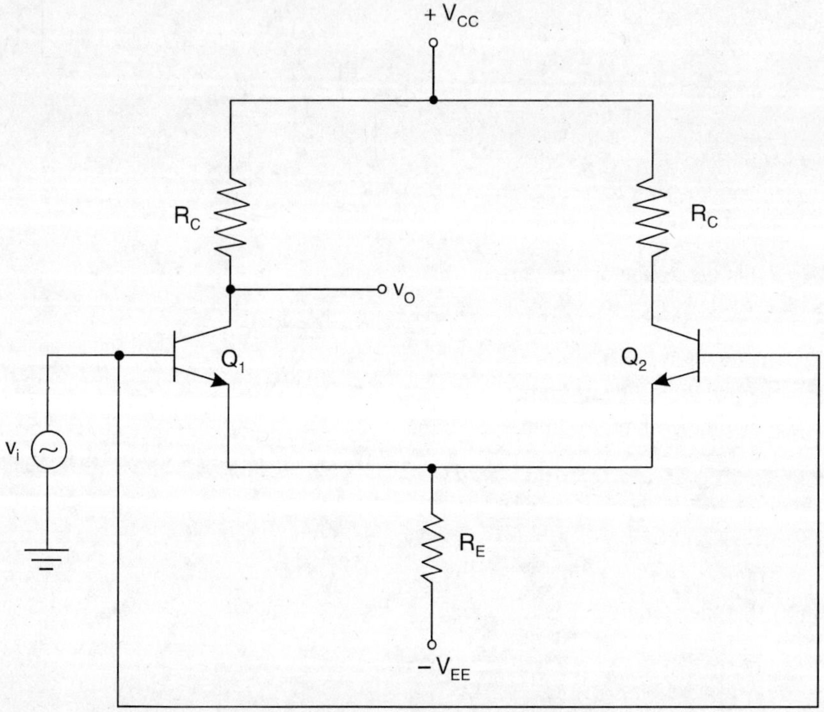

Fig. 20-48. *Common-mode connection in a differential amplifier.*

In order to calculate the value of common-mode gain, we will have to obtain the a.c. equivalent circuit of Fig. This can be obtained from Fig. 20-48 by replacing the transistors Q_1 and Q_2 by their a.c. equivalent circuits. The resulting circuit is as shown in Fig. 20-49.

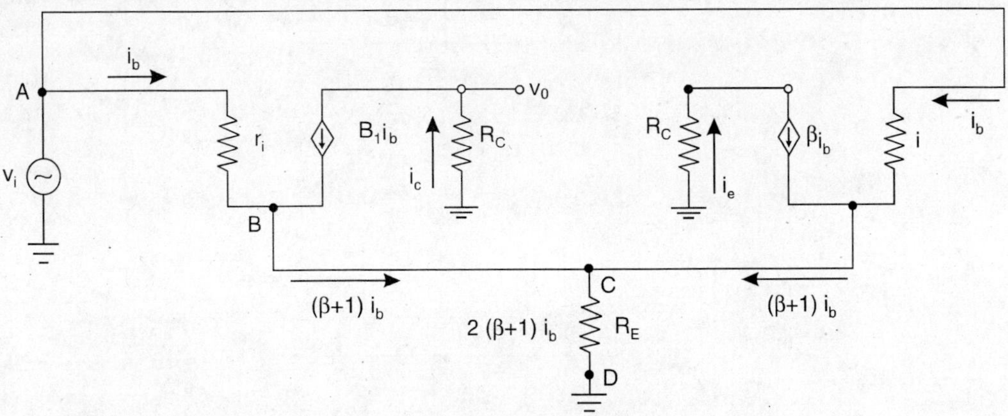

Fig. 20-49.

Applying Kirchoff's Voltage Law to the loop ABCDA in the a.c. equivalent circuit of Fig. 20-49. we get,

$$i_b\, r_i + 2\,(\beta + 1)\, i_b\, R_E - v_i = 0$$

$$\therefore \qquad i_b = \frac{v_i}{r_i + 2\,(\beta + 1)\, R_E} \qquad ...(i)$$

The value of output voltage,

$$v_o = i_c \,.\, R_C$$

Substituting the value of i_c (= $\beta\, i_b$) in equation (*ii*) we get,

$$v_o = \beta\, i_b \,.\, R_C$$

Substituting the value of i_b from equation (*i*), we get,

$$v_o = \frac{\beta\, v_i\, R_C}{r_i + 2\,(\beta + 1)\, R_E}$$

$\therefore$ The common-mode a.c. voltage gain,

$$A_C = \frac{V_o}{V_i} = \frac{\beta\, R_C}{r_i + 2\,(\beta + 1)\, R_E}$$

Note : 1. A differential amplifier provides large amplification of a difference signal applied to both inputs.

2. A differential amplifier provides a small amplification of the signal common to both inputs.

Example 23.16. *Determine the value of common-mode voltage gain for the differential amplifier circuit shown in Fig. 20-50. Assume* $\beta_1 = \beta_2 = 200;\ r_{i_1} = r_{i_2} = 20\ k\Omega$

Solution. Given : $\beta_1 = \beta_2 = 200$, $r_{i_1} = r_{i_2} = 20\ k\Omega = 20 \times 10^3\ \Omega$; $R_C = 47\ k\Omega = 47 \times 10^3\ \Omega$; $R_E = 43\ k\Omega = 43 \times 10^3\ \Omega$.

We know that the common-mode voltage gain,

$$A_C = \frac{v_o}{v_i} = \frac{\beta\, R_C}{r_i + 2\,(\beta + 1)\, R_E}$$

$$= \frac{200 \times (47 \times 10^3)}{(20 \times 10^3) + 2\,(200 + 1) \times (43 \times 10^3)}$$

$$= 0.543$$ **Ans.**

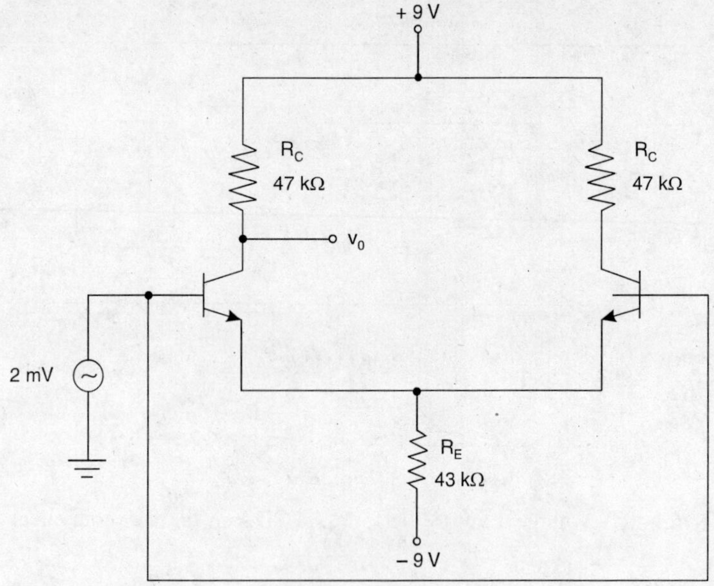

Fig. 20-50.

20-27. Use of Constant-Current Source in Differential Amplifier

We have already mentioned in the last article that a good differential amplifier has a very large differential gain (A_d) and a very small common-mode gain (A_c). The common-mode rejection ability of the differential amplifier can be considerably improved by making the common-mode gain as small as possible (ideally, speaking, 0) Recall from Art, 20-26 that the common-mode gain is given by the equation,

$$A_C = \frac{v_o}{v_i} = \frac{\beta R_C}{r_i + 2\,(\beta + 1)\,R_E}$$

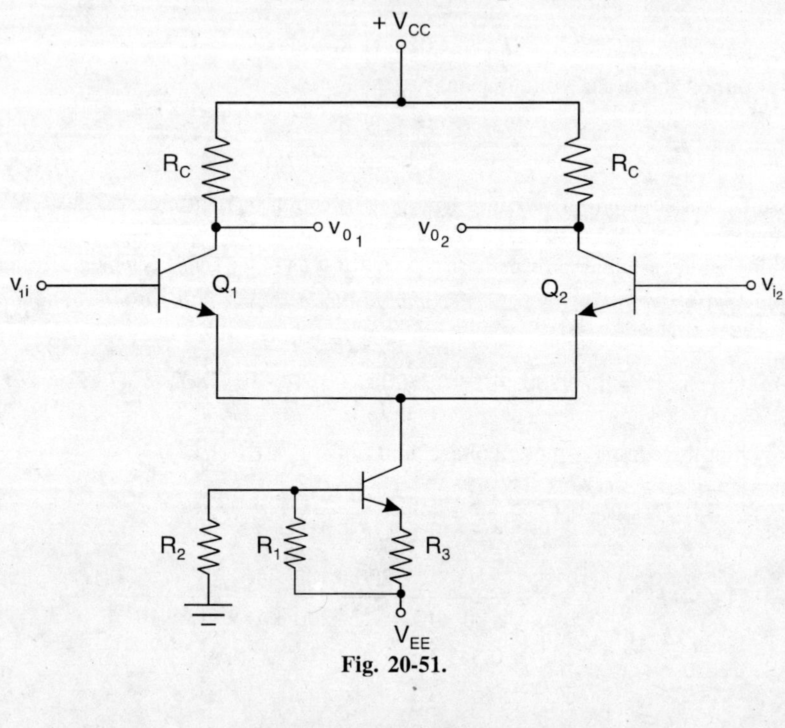

Fig. 20-51.

It is evident from this equation that larger the value of R_E, the smaller will be the value of A_C. One popular method for increasing the a.c. value of R_E is using a constant-current source.

Fig. 20-51 shows a differential amplifier circuit with constant-current source to provide a large value of resistance from common emitter to a.c. ground. The major advantage of this circuit over that shown in fig. 20-42 is the much larger a.c. resistance for R_E obtained using the constant current source.

Fig. 20-52 shows an a.c. equivalent circuit for the differential amplifier circuit shown in Fig. 20-51. Notice that a practical constant current source is shown as a high resistance (R_o) in parallel with an ideal constant-current source (I).

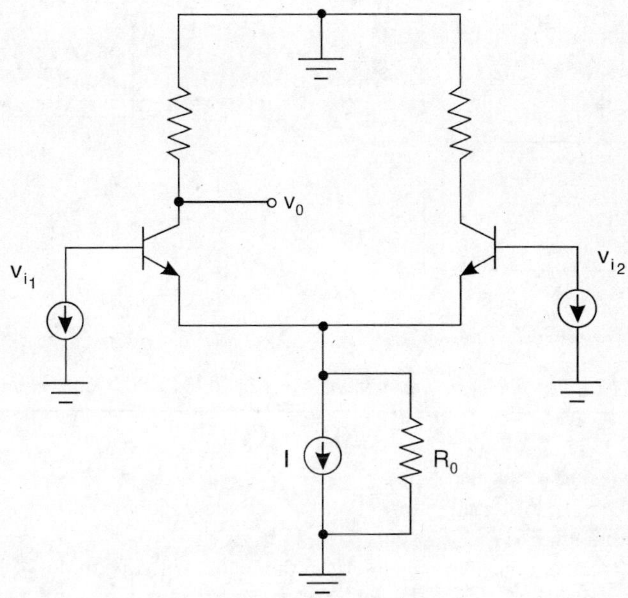

Fig. 20-52.

20-28. BiFET, BiMOS and CMOS Differential Amplifier Circuits

The differential amplifier circuits discussed so far uses bipolar transistors. But the circuits which are commercially available also use JFETs, MOSFET devices, and a combination of JFETs/MOSFETs and bipolar transistors. The integrated circuit (IC) containing a differential amplifier built using both bipolar (Bi) and JFETs is referred to as BiFET circuit. Similarly an IC containing a differential amplifier built using both bipolar (Bi) and MOSFET is called BiMOS circuit. Finally a circuit built using opposite (or complementary) type MOSFETs is called a CMOS circuit. Now we will show the differential amplifier circuits using various multidevices. However, it may be carefully noted that these circuits are only symbolic and not the actual circuits. The actual circuits are much more complex than the symbolic ones.

Fig. 20-53 shows a BiFET circuit with JFET at its inputs and BJTS to provide the current source. The current mirror ensures that each JFET is operated at the same bias current. For a.c. operation, the JFET provides a high input impedance.

Fig. 20-54 shows a circuit using MOSFET and BJT for the current sources. The BiMOS circuit provides even higher input impedance than the BiFET due to the use of MOSFET transistors.

A differential amplifier circuit can also be built using complementary MOSFETs. Such a circuit is more suitable for battery operated circuits.

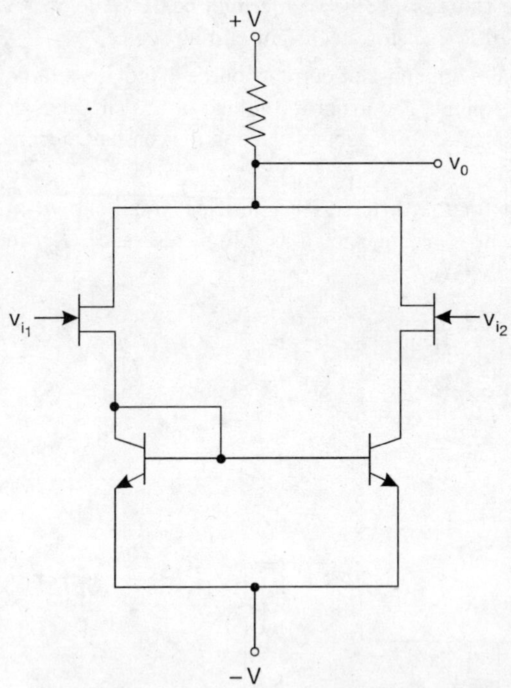

Fig. 20-53. *BiFET differential amplifier circuit.*

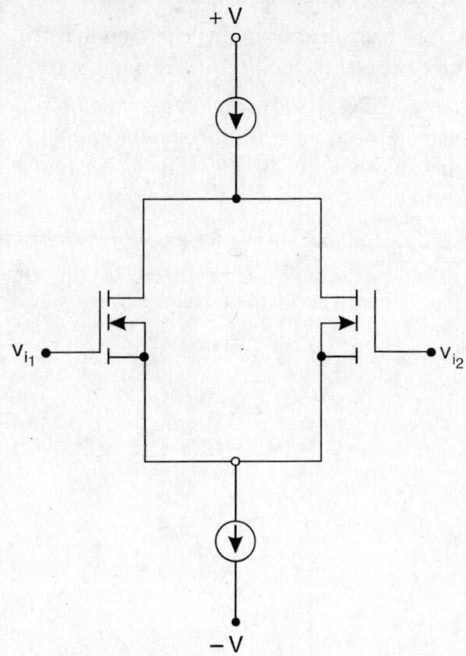

Fig. 20-54. *BiMOS differential amplifier circuit.*

SUMMARY

In this chapter you have learned that :

1. There are number of circuit connections which are widely used in either discrete or in integrated circuits.
2. A cascade connection is basically a series connection with the output of one stage applied as input to the second stage. The gain of the overall cascade amplifier is the product of individual stage gains, *e.g.*, for a two-stage cascade amplifier,

$$A_v = A_{v_1} \times A_{v_2}$$

3. A cascode connection is one where one transistor sits on top of another one. This connection provides a high input resistance with low voltage gain.
4. A Darlington transistor is one where the collectors of two identical transistors are connected together and emitter of the first transistor feeds to the base of the second transistor. This connection provides a composite transistor with very large value of current gain.
5. The feedback pair connection is a two transistor circuit that operates like Darlington circuit. However it uses a PNP transistor driving an NPN transistor.
6. A differential amplifier circuit can be operated as a single-ended or double-ended circuit. The major advantage of differential amplifier is that it produces a very large gain when oposite signals are applied to its inputs as compared to a very small gain resulting from common inputs. The ratio of this difference gain to the common gain is called *common-mode rejection ratio* (CMRR).

GLOSSARY

Current Source : a circuit that can deliver constant current

Current Mirror : a circuit that provides a constant current and is used in integrated circuits.

Differential Amplifier : an amplifier that provides a very large gain to the signals of opposite polarity applied to its inputs.

NUMERICAL PROBLEMS

1. Calculate the d.c. bias, voltage gain, input resistance output resistance and output voltage for the cascade amplifier shown in fig. 20-55. What is the load voltage if a 10 kΩ load is connected across its output. Assume I_{DSS} = 10 mA, V_p = – 4 V.

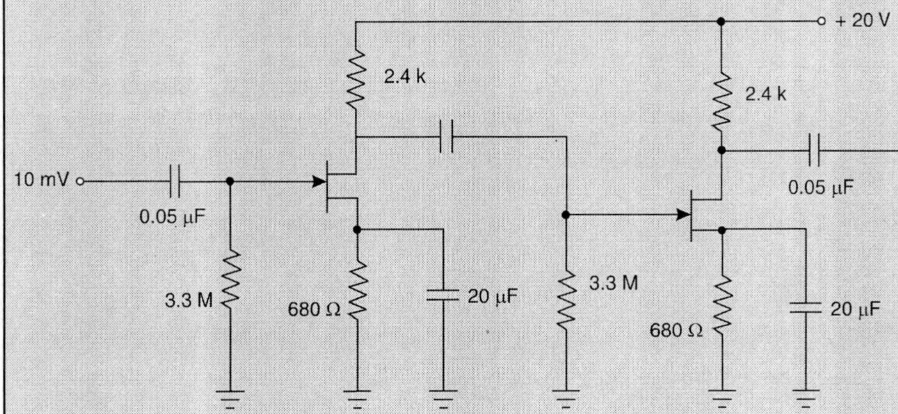

Fig. 20-55.

(**Ans.** V_{GSQ} = – 1.9 V; I_{DQ} = 2.8 mA, g_m = 2.6 mS, $A_{v_1} = A_{v_2}$ = – 6.2, R_i = 3.3 MΩ; R_o = 2.4 kΩ and V_L = 310 mV)

2. Calculate the voltage gain, output voltage, input resistance and output resistance for the cascade BJT amplifier of Fig. 20-56. What is the output voltage if 10 kΩ load is connected to the output. Assume β = 200.

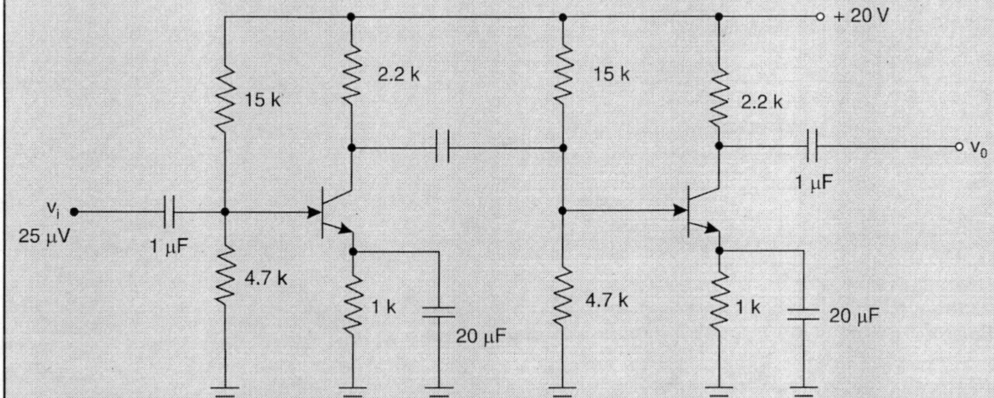

Fig. 20-56.

(**Ans.** V_B = 4.8 V, V_E = 4.1 V, V_C = 11 V, I_C = 4.1 mA, A_v = 36292, V_o = 0.9 V, R_i = 932 Ω; R_o = 2.2 kΩ; V_L = 0.7 V)

3. Calculate the value of current gain provided by a Darlington connection of two identical transistors each having a current gain of $\beta = 200$. (**Ans.** 40, 000)

4. Calculate the d.c. bias voltages and currents for the Darlington circuit shown in Fig. 20-57. Assume $\beta_D = 8000$ and $V_{BE} = 1.6$ V.
(**Ans.** $I_B = 2.56\ \mu$A; $I_C = 20.48$ mA, $V_E = 8$ V, $V_B = 9.6$ V; $V_C = 18$ V)

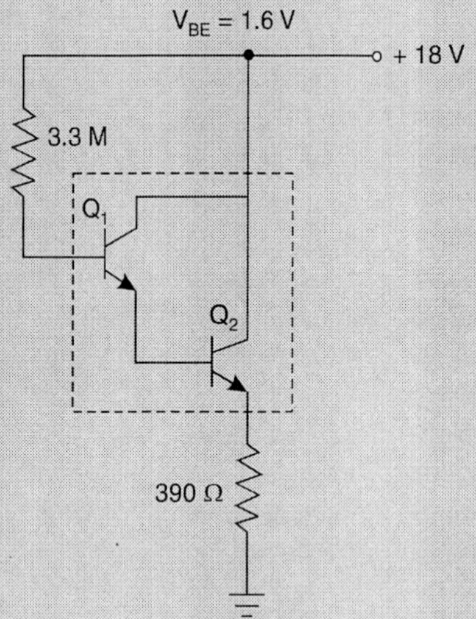

Fig. 20-57.

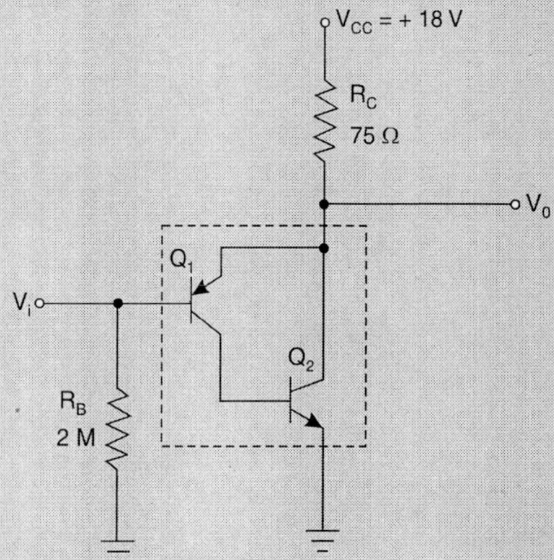

Fig. 20-58.

5. Calculate the input resistance, current gain, output resistance and a.c. voltage gain of the circuit shown in Fig. 20-57. (**Ans.** $R_i = 1.6$ MΩ; $A_i = 4112$; $R_o = 625$ Ω; $A_v = 0.998$)

6. Calculate the d.c. bias currents and voltages for the feedback pair circuit shown in Fig. 20-58, to provide V_o at one-half the supply voltage. Assume $\beta_1 = 140$ and $\beta_2 = 180$.
(**Ans.** $I_{B_1} = 4.45\ \mu$A; $I_{C_1} = 0.623$ mA, $I_{C_2} = 112.1$ mA; V_o (dc) = 9.6 V; V_i (dc) = 8.9 V)

7. Calculate the values of input resistance, output resistance, current gain and voltage gain of the feedback pair circuit shown in Fig. 20-58.
(**Ans.** $R_i = 974$ kΩ; $R_o = 0.12$ Ω; $A_i = 3.7 \times 10^6$; $A_v = 0.998$)

8. Calculate the value of constant current (I_C) in the BJT circuit shown in Fig. 20-59.
(**Ans.** 4.56 mA)

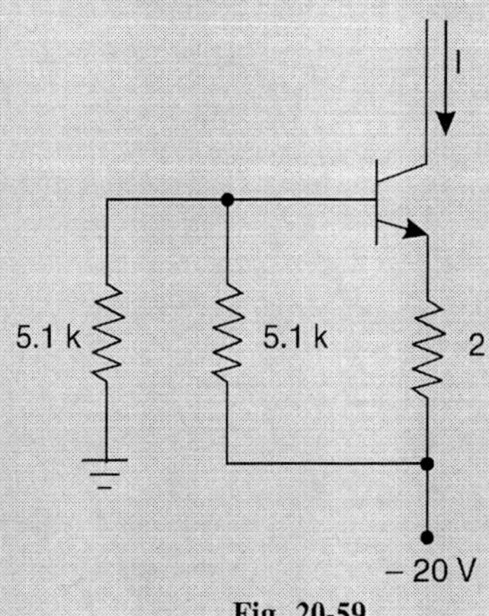

Fig. 20-59.

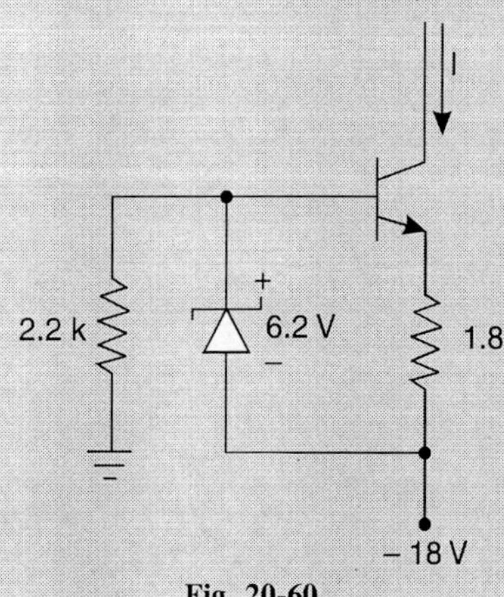

Fig. 20-60.

9. Calculate the constant current (I) in the BJT/Zener circuit shown in Fig. 20-60. (**Ans.** 3 mA)

10. Calculate the value of mirror current (I) in the circuit shown in Fig. 20-61. (**Ans.** 10.27 mA)

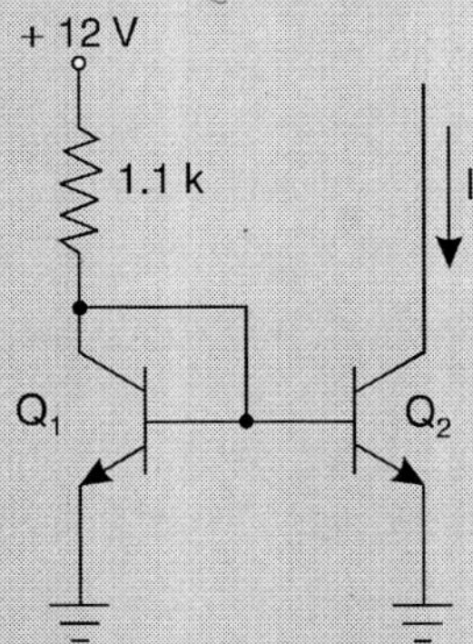

Fig. 20-61.

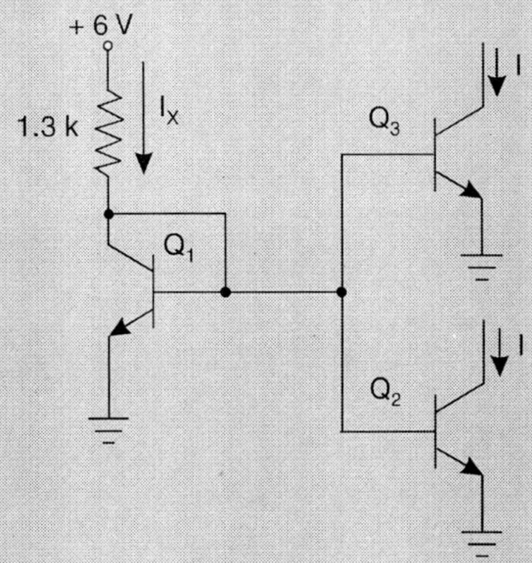

Fig. 20-62.

11. Calculate the values of currents through each of the transistors Q_2 and Q_3 in the current mirror circuit shown in Fig. 20-62. (**Ans.** 4.08 mA)

12. Calculate the d.c. voltages and currents in the differential amplifier circuits shown in Fig. 20-62. (**Ans.** I_E = 2.5 mA, I_C = 1.25 mA, V_C = 4.1 V)

13. Calculate the single-ended output voltage for the differential amplifier circuit shown in Fig. 20-63. Given : $r_{i_1} = r_{i_2}$ = 20 kΩ; $\beta_1 = \beta_2 = 75$. (**Ans.** 0.175 V)

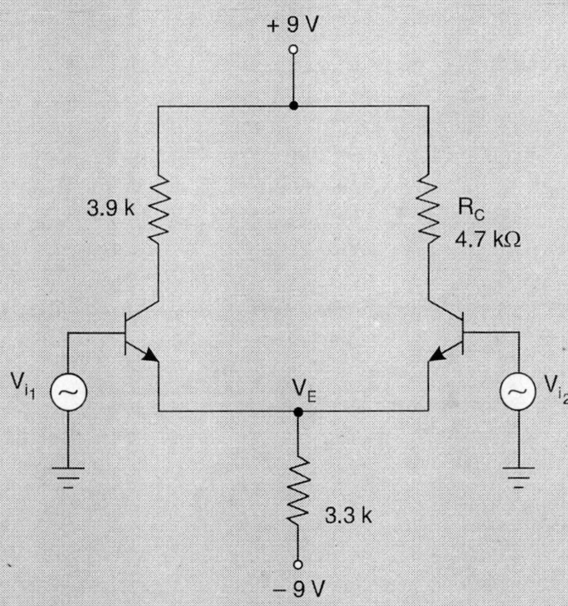

Fig. 20-63.

14. What is the value of common-mode gain for the differential amplifier circuit shown in Fig. 20-64. (**Ans.** 0.54)

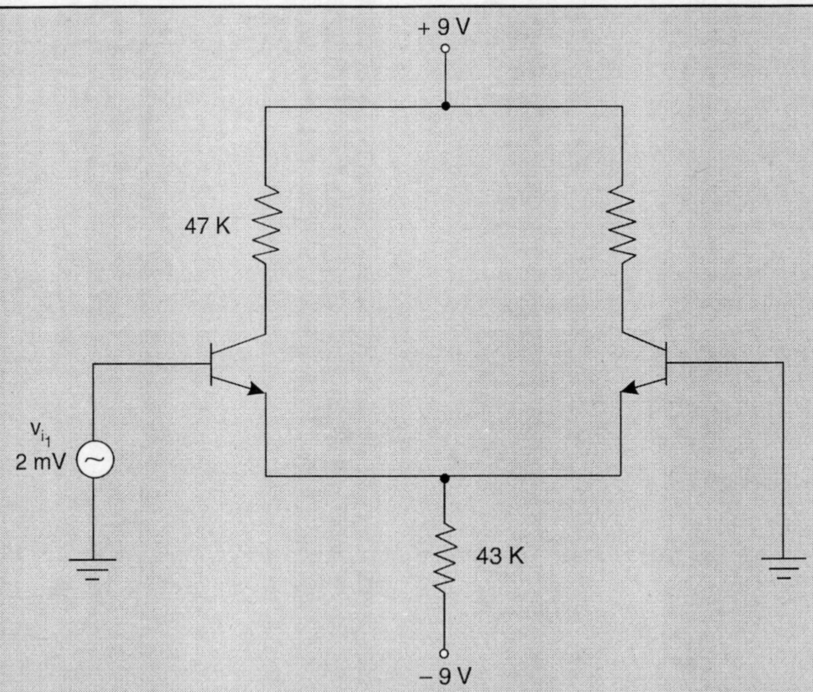

Fig. 20-64.

DESCRIPTIVE QUESTIONS

1. What is the difference between a cascade and a cascode connection ? Explain briefly with the help of sketches.
2. How a Darlington circuit is different from a feedback pair circuit ? Explain with reasons.
3. Explain the operation of a CMOS cricuit.
4. Sketch the circuit of a current mirror used in integrated circuits. Explain the circuit operation.
5. What is a differential amplifier? Explain the d.c. and a.c. operation of the circuit for single-ended and double-ended a.c. voltage gain.
6. How does input impedance increase due to Darlington connection? *(Electronic Circuits - I, Anna University, April/May 2008)*
7. Draw the circuit diagram for a differential amplifier using BJTs. Describe common mode and differential modes of working. Describe common mode and differential modes of working. *(Electronic Circuits - I, Anna University, April/May 2008)*
8. What is cascode amplifier? Exaplain its significance. *(Electronic Circuits - I, Anna University, April/May 2008)*
9. What is current mirror circuit? Discuss in detail with Widlar current source. *(Linear Integrated Circuits - I, Anna University, April/May 2008)*
10. Explain the emitter coupled difference amplifier with neat diagram. *(Electronic Circuits - I, Anna University, Nov/Dec 2007)*
11. Explain the operation of basic emitter coupled differential amplifier and derive its *CMRR*. *(Electronic Circuits - I, Anna University, April/May 2008)*
12. How does the constant current source increase the gain and hence CMRR in a differential amplifier. *(Electronic Circuits - I, Anna University, April/May 2008)*
13. Explain the operation of emitter coupled BJT differential amplifier circuits with active load *(Electronic Circuits - I, Anna University, June 2009)*

MULTIPLE CHOICE QUESTIONS

1. When some input signal is applied to both the inputs of an ideal differential amplifier, the output
   (*a*) is zero (*b*) depends on CMRR
   (*c*) depends on its voltage gain (*d*) is determined by its summetry
2. The common-mode rejection ratio of an ideal differential amplifier is
   (*a*) zero (*b*) infinity
   (*c*) less than unity (*d*) greater than unity
3. Which of the following mode of operation is possible with a differential amplifier?
   (*a*) single-ended input (*b*) differential input
   (*c*) common-mode input (*d*) all of the above
4. In a single-stage differential amplifier, the output offset voltage is basically dependent on the mismatch of
   (*a*) V_{BE}, I_B and β (*b*) V_{BE} and I_B
   (*c*) I_B and β (*d*) V_{BE} and β
   (*UPSC Engg. Services* 1999)
5. Increasing the value of R_E in a differential amplifier.
   (*a*) increase the differential gain, A_d (*b*) increases the common-mode gain, A_c
   (*c*) decreases CMRR (*d*) increases CMRR

ANSWERS

1. (*a*) **2.** (*b*) **3.** (*d*) **4.** (*d*) **5.** (*d*)

❑❑❑

21

CHAPTER

Frequency Response of BJT and JFET Amplifiers

OUTLINE

Objectives

After studying this chapter, you should be able to :

- List the advantages of using dB gain values.
- Calculate the amplifier gain in dB.
- Define dBm reference and discuss the difference between dBm values and standard dB values.
- Compare and contrast the Bode plot with the frequency response curve.
- Calculate the values of f_{L_S}, f_{L_C} and f_{L_E} for a given BJT amplifier.
- Calculate the values of f_{L_G}, f_{L_C} and f_{L_S} for a given JFET amplifier.
- Explain Miller's Effect and the purpose it serves.
- Explain why BJT or FET internal capacitances are not considered in low-frequency analysis.
- Calculate the values of f_{H_i} and f_{H_o} for a given BJT amplifier and FET amplifier.
- Discuss and analyse the frequency response of multistage amplifier.

21-1. Introduction

We have already studied the small-signal analysis of BJT and FET amplifier in chapter-16 and 19 respectively. Some discription was also done in a chapter on Multi-stage Amplifier. But the discussion throughout was limited to the mid-frequency range. In this chapter, we shall study the low-and high-frequency response of both BJT and FET amplifiers. This will include the frequency effects introduced by larger capacitive elements of the circuit at the low-frequency end and the smaller capacitive elements of the active device at the high frequencies.

21-2. Bel

Let P_i = power level at the input and

p_o = power level at the output,

Then the change (increase/decrease) in power level is given by the equation,

$$G = \log_{10} \frac{P_o}{P_i} \text{ bel} \quad ...(i)$$

The term 'bel' was derived from the family name of Alexander Grahm Bell.

However, the engineers found that the bel was too large a unit of measurement for all practical purpose. So a small unit called decibel (dB) was defined such that,

10 decibels = 1 bel

Therefore,

$$G_{dB} = 10 \log_{10} \frac{P_o}{P_i} \text{ dB} \quad ...(ii)$$

As a matter of fact, the terminal rating (or maximum power rating) of electronic communication equipment (like microphones, emplifiers, *etc.*) is commonly expressed in decibels. It may be noted from equation (*ii*) that decibel rating is a measure of the difference in magnitude between the power levels.

For a specified terminal (or output) power, P_o, there must be a reference power level (P_i). Generally the reference level (P_i) is accepted to be 1 mW. The resistance to be associated with the 1 mW power level is 600 Ω. The value 600 Ω is chosen because it is the characteristic impedance of the audio transmission line. When the 1 mW level is employed as the reference level, the decibel symbol frequently appears as dbm. Mathematically,

$$G_{dbm} = 10 \log_{10} \frac{P_o}{1 \text{ mW}} \text{ dBm} \quad ...(iii)$$

The equation (*ii*) can also be expressed in terms of voltage ratings rather than power ratings. Thus,

$$G_{dB} = 10 \log_{10} \frac{P_o}{P_i}$$

$$= 10 \log_{10} \frac{V_o^2/R_i}{V_i^2/R_i}$$

$$= 10 \log_{10} \left(\frac{v_o}{v_i}\right)^2$$

$$= 20 \log_{10} \left(\frac{v_o}{v_i}\right) \text{ dB} \quad ...(iv)$$

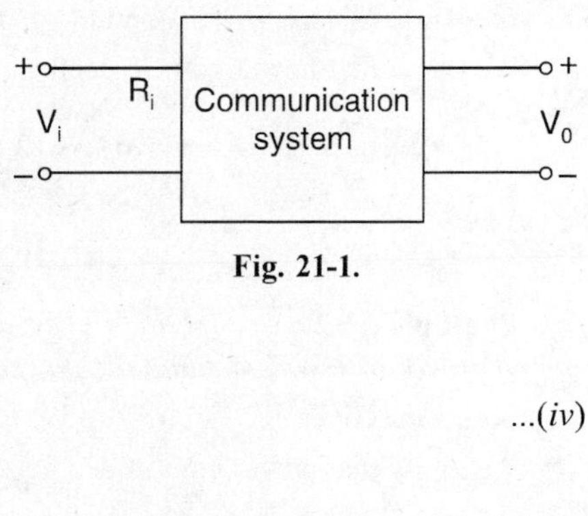

Fig. 21-1.

where, v_i = voltage level at the input,

v_o = voltage level at the output,

R_i = Input resistance of the communication system. (Refer to Fig. 21-1).

Now let us consider, a cascaded system with '*n*' stages and each stage having a voltage gain expressed as $A_{V_1}, A_{V_2}, \ldots\ldots\ldots A_{V_n}$.

Then the magnitude of overall voltage gain,

$$|A_{V_T}| = |A_{V_1}| \cdot |A_{V_2}| \cdot |A_{V_3}| \ldots\ldots\ldots |A_{V_n}|$$

This can be rewritten as,

$$G_v = 20 \log_{10} = 20 \log_{10} \left| A_{V_1} \right| + 20 \log_{10} \left| A_{V_2} \right|$$

$$+ 20 \log_{10} \left| A_{V_3} \right| + \ldots\ldots\ldots + 20 \log_{10} \left| A_{V_n} \right| \text{ dB} \qquad ...(v)$$

or
$$G_v = G_{v_1} + G_{v_2} + G_{v_3} + \ldots\ldots\ldots + G_{v_n} \text{ dB} \qquad ...(iv)$$

Table 21-1 shows the dB level for the various values of voltage gain varying from 0.5 to 10,000. It may be noted from this table that a voltage gain of 2 results in a dB level of +6 dB while a drop of voltage to 1/2 results in a –6 dB level. A change in v_o/v_i from 1 to 10, 10 to 100 or 100 to 1000 results in the same 20-dB change in level.

Table 21-1. *Illustration of the relation between voltage gain and dB level.*

Voltage gain (v_o/v_i)	*dB level*
0.5	– 6
0.707	– 3
1	0
2	6
10	20
40	32
100	40
1000	60
10000	80

Notice from table 21-1; that when $v_o/v_i = 1$ (*i.e.*, $v_o = v_i$), the dB level is 0. At a very high gain of 1000, the dB level is 60. While a much higher gain of 10,000, the dB level is 80 (*i.e.*, an increase of only 20dB). It will be interesting to know that voltage gains of 50 dB or higher are usually recognized as quite high.

Example 21-1. *An audio power-amplifier has an input power of 5 W and produces an output power of 100 W. Calculate the power gain in decibels.*

Solution. Given : P_i = 5W and P_o = 100 W.

We know that power gain in decibels,

$$G = 10 \log_{10} \frac{P_2}{P_i} \text{ dB}$$

$$= 10 \log_{10} \frac{100}{5} = 10 \log_{10} 20 = 13 \text{ dB } \textbf{Ans.}$$

Example 21-2. *A cellphone power-amplifier has an input power of 5 mW and produces an output power of 1 W. Calculate the power gain in decibels.*

Solution. Given : $P_i = 5 \text{ mW} = 5 \times 10^{-3}$ W and P_o = 1 W

We know that power gain in decibels,

$$G = 10 \log_{10} \frac{P_o}{P_i}$$

$$= 10 \log_{10} \frac{1}{5 \times 10^{-3}} = 10 \log_{10} 200 = 130 \text{ dB } \textbf{Ans.}$$

Example 21-3. *An audio amplifier in a portable radio receiver has an input power of 20 μW and an output power of 100 μW. Calculate the power gain in decibels.*

Solution. Given : $P_i = 20\ \mu W$ and $P_o = 100\ \mu W$.

We know that power gain in decibels,

$$G = 10 \log_{10} \frac{P_2}{P_1} \text{ dB}$$

$$= 10 \log_{10} \frac{100\ \mu W}{20\ \mu W} = 10 \log_{10} \frac{100}{20} = 10 \log_{10} 5 \text{ dB}$$

$$= 6.99 \text{ dB } \textbf{Ans.}$$

Example 21-4. *A certain communication system has an output power level of 25 W. Determine the power gain G_{dBm}.*

Solution. Given : $P_o = 25$ W

We know that power gain in dBm,

$$G_{dBm} = 10 \log_{10} \frac{P_o}{1 \text{ mW}}$$

$$= 10 \log_{10} \frac{25 \text{ W}}{1 \text{ mW}} = 10 \log_{10} \frac{25 \text{ W}}{1 \times 10^{-3} \text{ W}}\ 10 \log_{10} 25000$$

$$= 10 \times 4.398 \text{ dBm} = 43.98 \text{ dBm } \textbf{Ans.}$$

Example 21-5. *Two voltage measurements were made across the some resistance are $V_1 = 25$ and $V_2 = 100$ (refer to Fig. 21-2 below):*

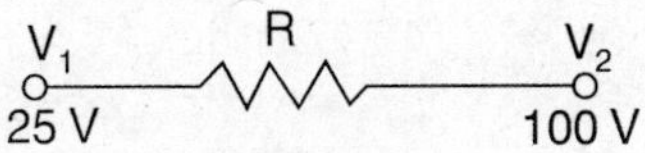

Fig. 21-2.

Calculate the power gain in decibels of the second reading over the first reading.

Solution. Given : $V_1 = 25$ V and $V_2 = 100$ V.

We know that power gain in decibels,

$$G_{dB} = 10 \log_{10} \frac{V_o}{V_i} \text{ dB}$$

$$= 10 \log_{10} \frac{V_2}{V_1} = 10 \log_{10} \frac{100}{25} = 10 \log_{10} 4 \text{ dB}$$

$$= 6.02 \text{ dB } \textbf{Ans.}$$

Example 21-6. *The total decibel gain of a 3-stage system is 120 dB. Determine the decibel gain of each stage if the second stage has twice the decibel gain of the first and the third has 2.7 times the decibel gain of the first. Also determine the voltage gain of each stage.*

Solution. Given : Total decibel gain $G_{TdB} = 120$ dB.

Power Gain of each stage

Let G_{dB_1} = power gain of stage 1, dB

G_{dB_2} = the power gain of stage 2, dB

$= 2\ G_{v_1}$ dB

G_{dB_3} = the power gain of stage 3, dB

$= 2.7\ G_{v_1}$ dB

We know that the total decibel gain $(G_{T_{dB}},)$

$$120 = G_{dB_1} + G_{dB_2} + G_{dB_3}$$

$$= G_{dB_1} + 2G_{dB_1} + 2.7G_{dB_1}$$

$$= 5.7G_{dB_1}$$

$\therefore$ $G_{dB_1} = 120/5.7 = 21.05$ dB

Thus, power gain of the first stage,

$$G_{dB_1} = 21.05 \text{ dB } \textbf{Ans.}$$

power gain of the second stage,

$$G_{dB_2} = 2\ G_{dB_1} = 2 \times 21.05 = 42.1 \text{ dB } \textbf{Ans.}$$

and power gain of the third stage,

$$G_{dB_3} = 2.7\ G_{dB_1} = 2.7 \times 21.05 = 56.8 \text{ dB } \textbf{Ans.}$$

Voltage gain of each stage

Let, A_{v_1} = voltage gain of the first stage,

A_{v_2} = voltage gain of the second stage, and

A_{v_3} = voltage gain of the third stage,

We know that gain of the first stage (G_{dB_1}),

$$21.05 = 20 \log_{10} \left| A_{v_1} \right|$$

$$\therefore \quad \log_{10} \left| A_{v_1} \right| = \frac{21.05}{20} = 1.0525 \text{ or } \left| A_{v_1} \right| = 10^{1.0525} = 11.285 \textbf{ Ans.}$$

Similarly, gain of the second stage (G_{dB_2}),

$$42.10 = 20 \log_{10} |Av_2|$$

$$\therefore \log_{10} | Av_2 | = \frac{42.1}{20} = 2.105 \text{ or } |Av_2| = 10^{2.105} = 127.35 \textbf{ Ans.}$$

and gain of the third stage, (G_{dB_3}),

$$56.8 = 20 \log_{10} |Av_3|$$

$$\log_{10} |Av_3| = \frac{56.8}{20} = 2.84 \quad \text{or} \quad | Av_3 | = 10^{2.84} = 691.43 \textbf{ Ans.}$$

Example 21-7. *An audio amplifier rated at 40 W is connected to a 10 Ω speaker.* ***(a)*** *Calculate the input power required for full-power output if the power gain is 25 dB.* ***(b)*** *Calculate the input voltage for rated output if the amplifier voltage gain is 40 dB.*

Solution. Given rated output power P_o = 40 W, speaker resistance = 10 Ω.

(*a*) *Input power required*

Let P_i = input power required for full-power output for a 25 dB power gain

We know that the power gain, (G_{dB_1})

$$25 = 10 \log_{10} \frac{P_o}{P_i} = 10 \log_{10} \frac{40 \text{ W}}{P_i}$$

$$\text{or} \quad \log_{10} \frac{40 \text{ W}}{P_i} = \frac{25}{10} = 2.5$$

$$\text{or} \quad \frac{40}{P_i} = 10^{2.5} = 316.23$$

$$\therefore \quad P_i = \frac{40}{316.23} = 0.1265 \text{ W or } 126.5 \text{ mW } \textbf{Ans.}$$

(*b*) *Input voltage for rated output*

Let v_i = Input voltage for rated output if amplifier voltage gain 40 dB.

v_o = output voltage

We know that the voltage gain in decibels (G_{dB_1})

$$40 = 20\log_{10}\frac{v_o}{v_i} = 20\log_{10}\frac{v_o}{v_i}$$

or $$\log_{10}\frac{v_o}{v_i} = \frac{40}{20} = 2 \Rightarrow \frac{v_o}{v_i} = 10^2 = 100 \quad ...(i)$$

We also know that output power (p_o),

$$40\text{ W} = \frac{v_o^2}{R} = \frac{v_o^2}{10} \Rightarrow v_o = \sqrt{40\times 10} = 20\,.$$

From equation (i),

$$\frac{20}{v_i} = 100 \text{ or } v_i = \frac{20}{100} = 0.2\text{ V or }200\text{ mV } \textbf{Ans.}$$

21-3. General Frequency Considerations

Strictly speaking, the frequency of the applied signal can have a significant effect on the frequency response of single or multistage amplifier. We have described earlier in chapter-18 the mid-frequency response of amplifiers.

In this chapter, we will study that at low frequencies, the coupling and bypass capacitors can no longer be replaced by the short-circuit approximation. This is because of the increase in reactance of capacitive elements. On the other hand the high-frequency response of the system is limited by the frequency dependent parameters of the small-signal equivalent circuits and the stray capacitive elements associated with the active device (BJT or FET) and the circuit . An increase in the number of stages of a cascaded system will also limit the high-and low frequency response.

Fig. 21-3(a), and direct-coupled shows the gain-frequency plot of an RC-coupled, amplifier. Similarly Fig. 21-3 (b) and (c) shows gain frequency plot of transformer-coupled and direct-coupleed amplifier respectively. Notice that in all the three diagrams, the horizontal scale is a logarithmic scale to permit a plot extending from the low to the high-frequency regions. For each plot, a low-high- and mid-frequency region has been defined. In addition, the primary reasons for the drop in gain at low and high frequencies have also been indicated in the parenthesis.

For the R-C coupled amplifier, the drop at low frequencies is due to the increasing reactance of capacitors $(C_C, C_S$ and $C_E)$. While the upper frequency limit of RC coupled amplifier is determined by either the parasitic capacitive elements of the circuit and frequency dependance of the gain of the active device.

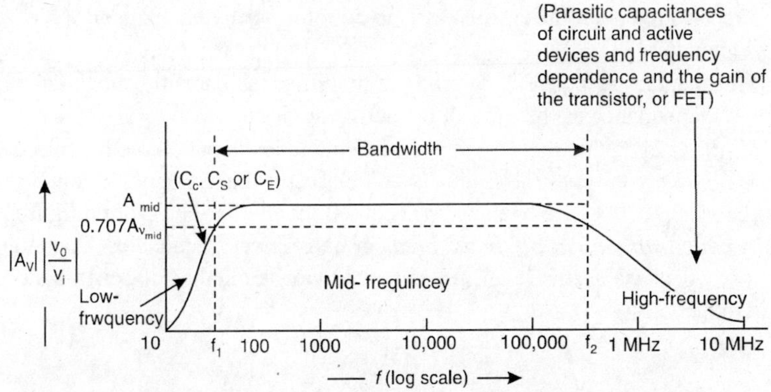

(a) Gain-frequency plot of a RC-coupled amplipier

* Recall that inductive reactance, $X_L = 2\pi fL$. Thus at low frequencies, $X_L \approx 0$

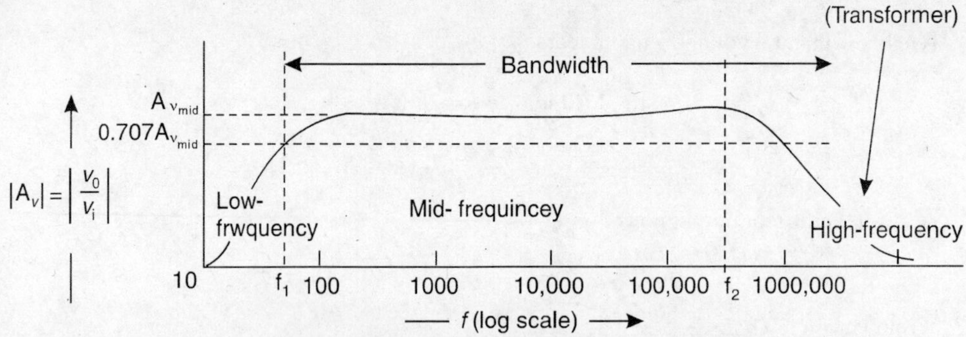

(*b*) Gain frequency plot of a RC-coupled amplifier

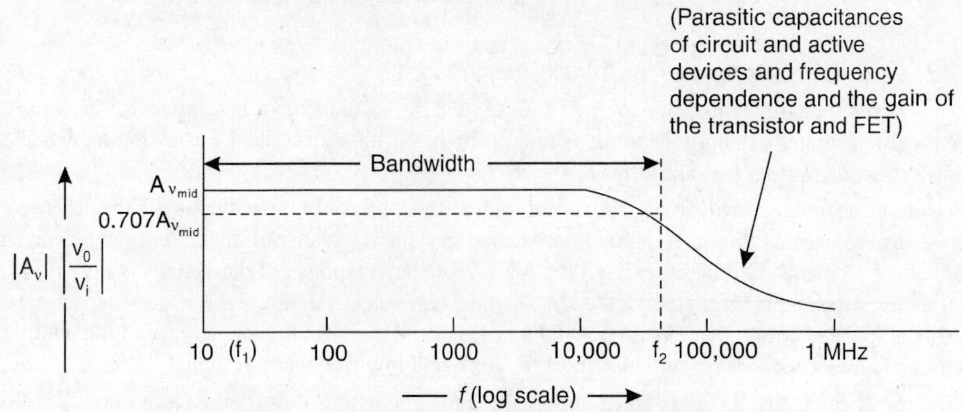

(*c*) Gain-frequency plot of a transformer-coupled emplifier

Fig. 21-3.

The drop in gain for the transformer-coupled amplifier, you require a basic understanding of transformer action and transformer equivalent circuit. Thus at low frequencies, the drop in gain is due to the shortening effect (across the input terminals of the transformer) produced by the magnetising *inductive reactance. The gain must obviously be zero at $f = 0$, since at this point there is no longer a changing flux estabilished through the core to induce a secondary or output voltage.

The high-frequency response of a transformer-coupled amplifier is controlled primarily by the stray capacitance between the turns of primary and secondary windings.

For the direct-coupled amplifier, there are no coupling or bypass capacitors to cause a drop in gain at low frequencies.

The high frequencies response is determined by either the parasitic capacitances of the circuit or the frequency dependance of the gain of the active device.

Notice from Fig. 21-5, that there is a band of frequencies in which the magnitude of the gain is either equal or relatively close to the midband value. To fix the frequency boundaries of relatively high gain, we chose 0.707 $A_{v\,mid}$ to be the gain at cut-off levels. The corresponding frequencies f_1 and f_2 are generally called *corner*, *cut-off, band, break* or half-power frequencies. The multiplying factor 0.707 was chosen because at this level, the output power is one-half the mid-band power output. Thus, at mid-frequencies

$$P_{o_{mid}} = \frac{|v_o^2|}{R} = \frac{|A_{v_{mid}} \; v_i|^2}{R_o}$$

and at the half power frequencies,

$$P_{o_{HPF}} = \frac{\left|0.707\, A_{v_{mid}}\; v_i\right|^2}{R_o} = \frac{0.5\left|A_{v_{mid}}\; v_i\right|^2}{R_o}$$

Thus, $$P_{o_{HPF}} = 0.5\, P_{o_{mid}}$$

The bandwidth (also called passband) of each amplifier is determined by f_1 and f_2. Thus,

$$\text{bandwidth (BW)} = f_2 - f_1$$

21-4. Normalized Gain-versus-Frequency Plot

For applications in the field of communication engineering, a decibel plot of the voltage gain versus frequency is more useful than that shown in Fig. 21-3. However, it may be noted that before obtaining the logarithmic plot, the curve is generally normalized as shown in Fig. 21-4. in this firgure, the gain at each frequency is divided by the mid-band value. Obviously, the mid-band value is then 1 as indicated. At all the half-power frequencies the resulting level is $0.707 = 1/\sqrt{2}$. A decibel plot can now be obtained by applying the equation,

$$\left.\frac{A_v}{A_{v_{mid}}}\right|_{dB} = 20 \log_{10} \frac{A_v}{A_{v_{mid}}}$$

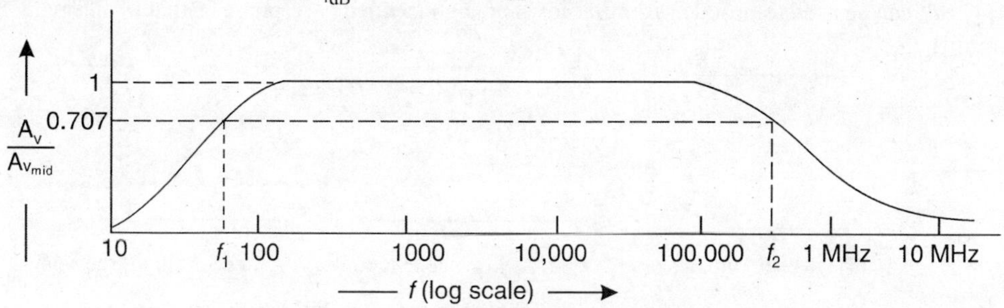

Fig. 21-4.

At mid-band frequences, $20 \log_{10} 1 = 0$ and at the cut-off frequencies, $20 \log_{10} 1/\sqrt{2} = -3\text{dB}$. Both values are clearly indicated in the resulting decibel plot of Fig. 21-5. It may be carefully noted that the smaller the fraction ratio, the more negative the decibel level.

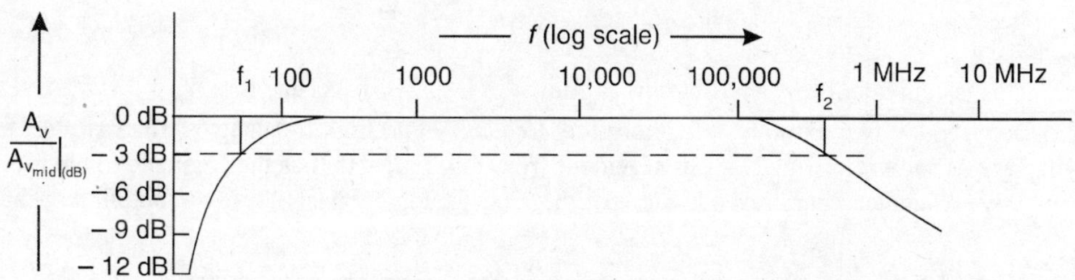

Fig. 21-5.

It will be interesting to know that most amplifiers introduce a 180° phase-shift between input and output signals. But this is only the case in the midband region. At low frequencies there is a phase-shift such that V_o lays V_i by an increased angle. At high frequencies, the phase-shift will drop below 180°. Fig. 21-6 shows a standard phase plot for an RC coupled amplifier.

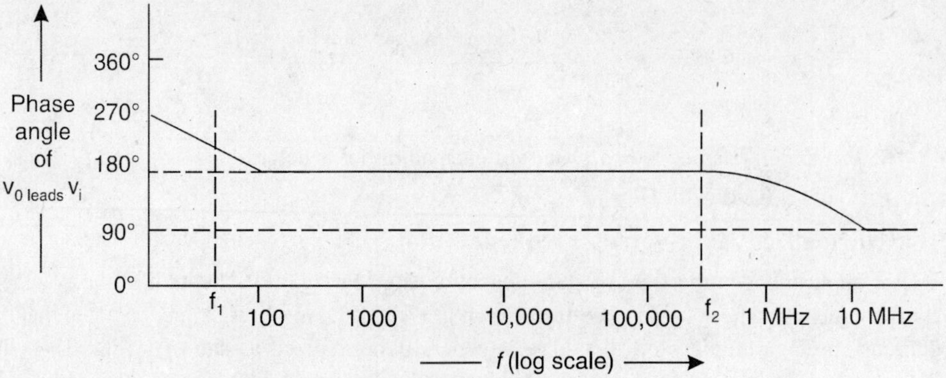

Fig. 21-6.

21-5. Low-Frequency Analysis

We have already mentioned that in the low-frequency region of the single-stage BJT or FET amplifier, it is the *R-C* combinations formed by the circuit capacitors C_C, C_E and C_S and the resistive parameters of the circuit that determine the cut-off frequencies. A series *R-C* circuit shown in Fig. 21-7 (*a*), can be used to produce a similar low and high frequency response as that of an single-stage amplifier.

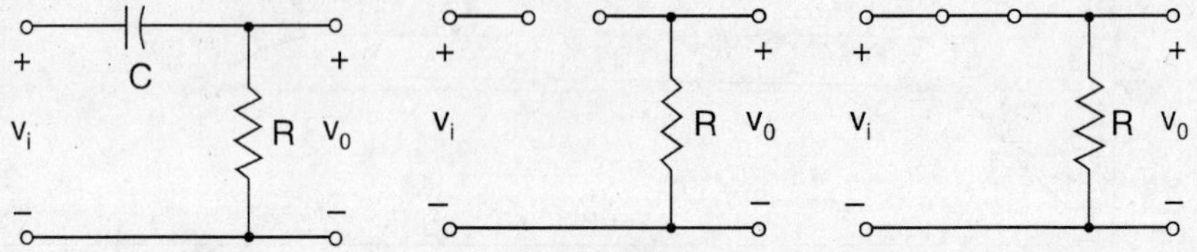

(*a*) R-C circuit (*b*) At low frequency (*c*) At high frequency

Fig. 21-7.

We know that at very low frequencies ($f \approx 0$), the capacitive reactance,

$$X_C = \frac{1}{2\pi f\, C} = \frac{1}{2\pi(0)C} = \infty$$

This is equivalent to an open circuit as shown in Fig. 21-7 (*b*). Therefore, $v_o = 0$ V. At high frequencies, the capacitive reactance,

$$X_C = \frac{1}{2\pi f\, C} \approx 0$$

This is equivalent to a short-circuit as shown in Fig. 21-7(*c*). Therefore, $v_o = v_i$.

Between the two extremes (*i.e.*, very low frequency and high frequencies), the ratio $A_v = v_o/v_i$ will vary as shown in Fig. 21-8. It is evident from this figure that as the frequency increases, the capacitive reactance decreases and more of the input voltage appear across the output terminals.

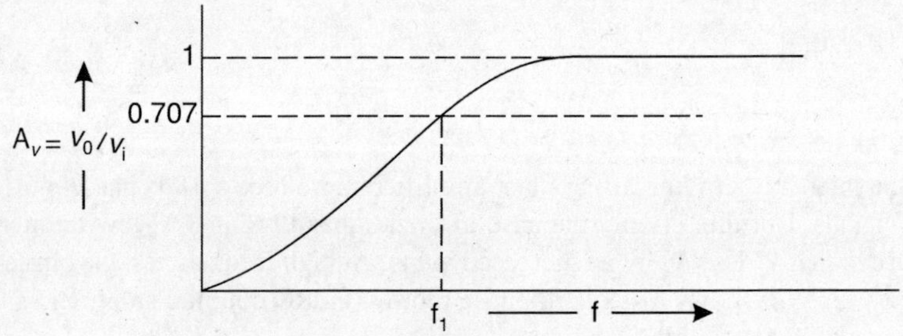

Fig. 21-8.

If we apply a voltage-division rule to the R-C circuit of Fig. 21-7 (*a*), we get,

$$\boldsymbol{v_o} = \frac{\boldsymbol{R v_i}}{\boldsymbol{R + X_C}}$$

Here the magnitude of v_o is given by the equation,

$$v_o = \frac{R\, v_i}{\sqrt{R^2 + X_C^2}}$$

If $X_C = R$ (a special case), the output voltage,

$$v_o = \frac{Rv_i}{\sqrt{R^2 + X_C^2}} = \frac{Rv_i}{\sqrt{R^2 + R^2}} = \frac{Rv_i}{\sqrt{2R^2}} = \frac{Rv_i}{\sqrt{2}R}$$

$$= \frac{1}{\sqrt{2}} v_i$$

∴ The magnitude of the voltage gain,

$$|A_V| = \frac{v_o}{v_i} = \frac{1}{\sqrt{2}} = 0.707$$

This equation implies that a frequency corresponding to which $X_C = R$, the output will be 0.707 (or 70.7%) of the input for the circuit shown in Fig. 21-7 (*a*).

The frequency at which the condition $X_C = R$ can be obtained as follows,

Let f_1 = the frequency at which $X_C = R$. Then,

$$X_C = \frac{1}{2\pi f_1 C} = R$$

or

$$f_1 = \frac{1}{2\pi RC} \qquad ...(ii)$$

In terms of logarithms,

$$G_v = 20 \log_{10} A_v = 20 \log_{10} 0.707 = 20 \log_{10} \frac{1}{\sqrt{2}} \text{ dB}$$

$$= -3\ d\text{B}$$

while at $A_v = v_o/v_i = 1$,

$$G_v = 20 \log_{10} A_v = 20 \log_{10} 1 = 0 d\text{B}$$

This helps us to recognise in Fig. 21-8 that there is a – 3 dB drop in gain from the mid-band level when $f = f_1$,

Notice that the gain equation (*i*) can also be written as,

$$A_v = \frac{v_o}{v_i} = \frac{R}{R - jX_C} = \frac{1}{1 - j(X_C/R)}$$

$$= \frac{1}{1 - j\left(\frac{1}{\omega CR}\right)} \qquad (\because X_C = 1/\omega C)$$

$$= \frac{1}{1 - j\left(\frac{1}{2\pi f CR}\right)} \qquad ...(iii)\ (\because \omega = 2\pi f)$$

Since, $f_1 = 1/2\pi RC$, therefore the equation (*iii*) can be rewritten as,

$$A_v = \frac{1}{1 - j\left(\frac{f_1}{f}\right)} \qquad ...(iv)$$

In the magnitude and phase form, the above equation can be written as,

$$A_v = \frac{v_o}{v_i} = \frac{1}{\sqrt{1+\left(\frac{f_1}{f}\right)^2}} \angle \tan^{-1}\left(\frac{f_1}{f}\right)$$

For the magnitude when $f = f_1$,

$$|A_v| = \frac{v_o}{v_i} = \frac{1}{\sqrt{1+(1)^2}} = \frac{1}{\sqrt{2}} = 0.707 \rightarrow -3dB$$

In the logarithmic form, the gain in dB is,

$$A_{v\,(dB)} = 20 \log_{10} \frac{1}{\sqrt{1+\left(\frac{f_1}{f}\right)^2}}$$

$$= 20 \log_{10} \left[1+\left(\frac{f_1}{f}\right)^2\right]^{-1/2}$$

$$= 20\left(-\frac{1}{2}\right) \log_{10} \left[1+\left(\frac{f_1}{f}\right)^2\right]$$

$$= -10 \log_{10} \left[1+\left(\frac{f_1}{f}\right)^2\right] \qquad ...(v)$$

For frequencies where $f << f_1$, the ratio $f_1/f >> 1$,
therefore, the equation (v) can be written as,

$$A_{v(dB)} = -10 \log_{10} \left(\frac{f_1}{f}\right)^2$$

$$= -20 \log_{10} \left(\frac{f_1}{f}\right) \qquad ...(iv)$$

Using equation (vi) we can easily obtain the followings :

$$\text{At } f = f_1 \; ; \; \frac{f_1}{f} = 1 \text{ and } -20 \log_{10} 1 = 0 \text{ dB}$$

$$\text{At } f = \frac{1}{2} f_1 \; ; \; \frac{f_1}{f} = 2 \text{ and } -20 \log_{10} 2 = -6 \text{ dB}$$

$$\text{At } f = \frac{1}{4} f_1 \; ; \; \frac{f_1}{f} = 4 \text{ and } -20 \log_{10} 4 = -12 \text{ dB}$$

$$\text{At } f = \frac{1}{10} f_1 \; ; \; \frac{f_1}{f} = 10 \text{ and } -20 \log_{10} 10 = -20 \text{ dB}$$

If we plot a graph for the frequency range from $0.1\,f$ to f_1, we will get a straight line indicated by AB as shown in Fig. 21-9.

Notice that the graph has been plotted using a log scale. Fig. 21-9 also shown another straight line (BC) for the condition 0dB.

As mentioned earlier, the straight line segments (called a symptotes) are only accurate for 0dB when $f_1 >> f$. However, we know that when $f = f_1$, there is a 3dB drop from the mid-band level. This information along with the straight line segments can be used to sketch a fairly accurate plot of the frequency response as indicated by a dotted curve in the same figure.

It will be interesting to know that the piecewise linear plot of the asymptotes and associated break points is called a *Bode plot* of the magnitude versus frequency.

Based on our calculations and the plot, we may summerize the discussion through the following points :

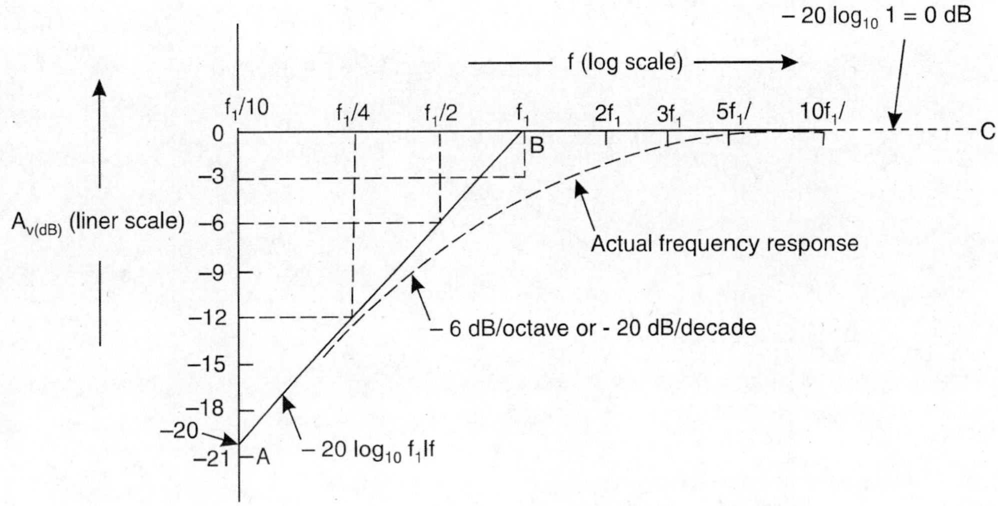

Fig. 21-9.

(*i*) A change in frequency be a factor of 2 (equivalent to 1 octave), results in a 6 dB change in the ratio. This is seen clearly by the change in gain from $f_1/2$ to f_1.

(*ii*) For a 10 : 1 change in frequency (equivalent to 1 decade), there is a 20 dB change in the ratio. This is demonstrated clearly by the change in gain from $f_1/10$ to f_1.

Further it is also evident from the above discussion that we can easily obtain a decibel plot for any given function of the form $[A_{v(dB)} = -20 \log_{10} (f_1/f)]$ by the folowing procedure.

(*i*) Find f_1 from the circuit parameters

(*ii*) Sketch two asymptotes–one along the OdB line and the other drawn through f_1 sloped at 6dB/octave or 20dB/decade.

(*iii*) Find the 3dB point corresponding to f_1 and sketch a smooth curve.

Example 21-8. *For the network shown in Fig. 21-10,* ***(a)*** *Determine the break frequency,* ***(b)*** *sketch the asymptotes, and* ***(c)*** *locate the – 3dB point and sketch the frequency response curve.*

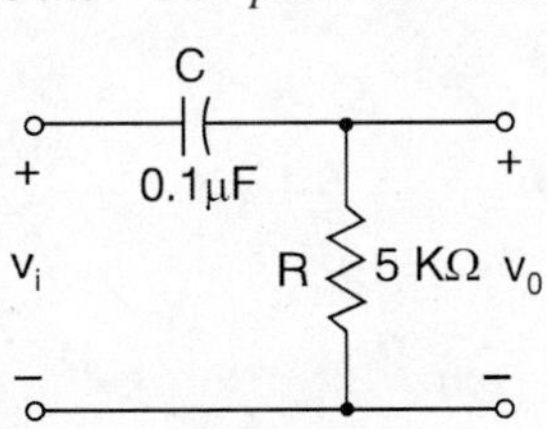

Fig. 21-10.

Solution. Given : $R = 5\ K\Omega = 5 \times 10^3\ \Omega$, $C = 0.1\ \mu F = 0.1 \times 10^{-6}$ F.

(*a*) *Break frequency*

We know that the break frequency,

$$f_1 = \frac{1}{2\pi RC} = \frac{1}{2\pi \times (5 \times 10^3) \times (0.1 \times 10^{-6})} \approx 318.5 \text{ Hz}$$

(*b*) *Sketch of asymptotes*

Following the procedure given in Art. 21-5, we can obtain the asymptotes by sketching two straight lines AB and BC as shown in Fig. 21-11.

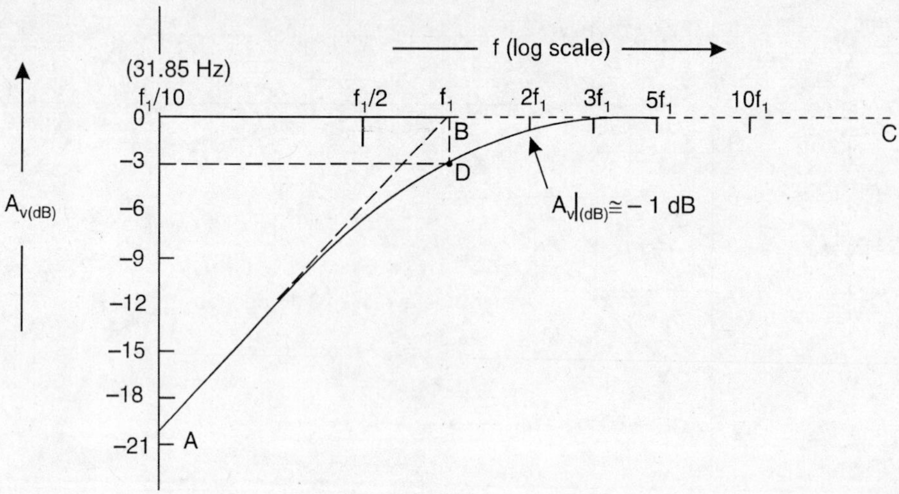

Fig. 21-11.

(*c*) *Frequency response curve*

This can be obtained by finding the – 3dB point (point D) and then joining the points AD and C as shown in Fig. 21-11.

21-6. Determination of Gain from the Frequency Plot

We have already discussed in the last article, that how we can obtain a low frequency response for a given RC circuit. Now let us see how we can use this plot to obtain a gain in terms of its magnitude and phase plots.

The gain at any frequency can be obtained from the frequency plot, by using the equation,

$$A_{v(dB)} = 20 \log_{10} \frac{v_o}{v_i}$$

Rearranging this eqation, we get,

$$\frac{A_{v(dB)}}{20} = \log_{10} \frac{v_o}{v_i}$$

or

$$A_v = \frac{v_o}{v_i} = 10^{\left(\frac{A_{v(dB)}}{20}\right)}$$

For example, if $A_{v(dB)} = -$ 3dB, then,

$$A_v = \frac{v_o}{v_i} = 10^{\left(-\frac{3}{20}\right)} = 10^{(-0.15)} \approx 0.707 \text{ as expected.}$$

The phase angle (θ) can be determined by using the equation,

$$\theta = \tan^{-1}\left(\frac{f_1}{f}\right) \quad ...(i)$$

For frequencies, $f << f_1$, from equation (*i*), we get,

$$\theta = \tan^{-1}\left(\frac{f_1}{f}\right) \rightarrow 90°$$

For example if $f_1 = 100\,f$, then

$$\theta = \tan^{-1}(100) = 89.4°$$

On the other hand, if $f_1 = f$, then again using equation (*i*), we get,

$$\theta = \tan^{-1}\left(\frac{f_1}{f}\right) = \tan^{-1} 1 = 45°$$

Finally for $f >> f_1$,

$$\theta = \tan^{-1}\left(\frac{f_1}{f}\right) \rightarrow 0°$$

For example if $f = 100\,f_1$, then

$$\theta = \tan^{-1}\left(\frac{f_1}{f}\right) = \tan^{-1}\left(\frac{1}{100}\right) = \tan^{-1}(0.01) = 0.573°$$

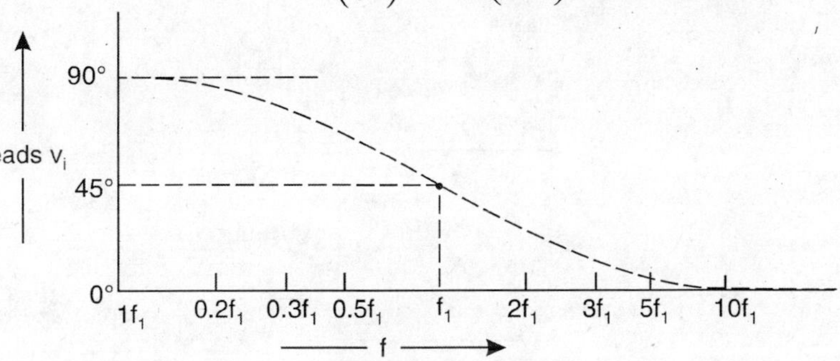

Fig. 21-12.

Fig. 21-12, shows a plot of $\theta = \tan^{-1}(f_1/f)$. If we add additional 180° phase shift introduced the amplifier, the phase plot shown in Fig. 21-13 will be obtained.

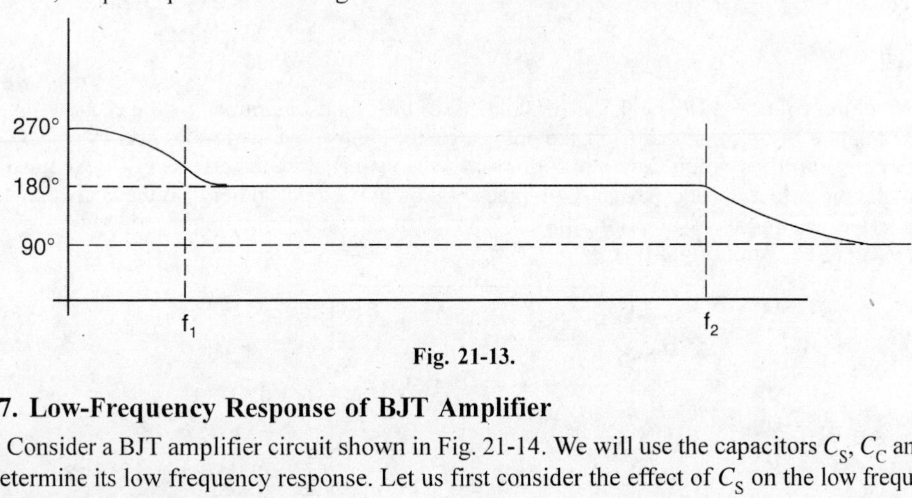

Fig. 21-13.

21-7. Low-Frequency Response of BJT Amplifier

Consider a BJT amplifier circuit shown in Fig. 21-14. We will use the capacitors C_S, C_C and C_E to determine its low frequency response. Let us first consider the effect of C_S on the low frequency response, followed by C_C and C_E.

(i) Determination of low-frequency cut-off f_{LS} : Since, C_S is normally connected between the applied source (V_S) and the BJT, we can obtain a general form of RC configuration as shown in Fig. 21-15.

Notice that the total resistance of the circuit is $R_s + R_i$. The lower cut-off frequency due to the effect of C_S can be obtained form the equation of the form $f = 1/2\pi RC$. Thus,

$$f_{L_S} = \frac{1}{2\pi(R_S + R_i)C_S} \quad ...(i)$$

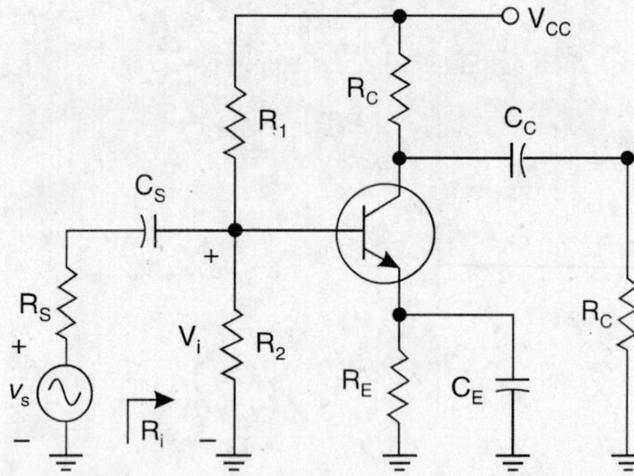

Fig. 21-14.

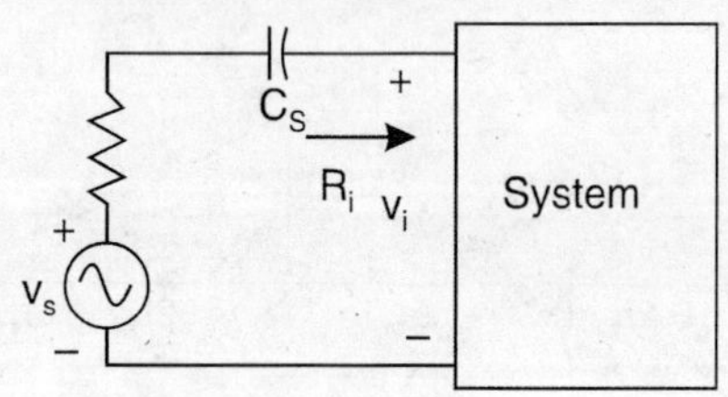

Fig. 21-15.

We know that at mid- or high- frequencies, the reactance of the capacitor (X_C) will be sufficiently small to permit a short circuit approximation for the element. Then the voltage v_i is given by the equation,

$$v_i \Big|_{mid} = \frac{R_i V_S}{R_i + R_S} \quad ...(iii)$$

At f_{LS} the voltage v_i will be 0.707 (or 0.707%) of the value determined by the eqation (*ii*). This is based on the assumption that C_S is the only capacitor controlling the low frequency response.

For the amplifier circuit shown in Fig. 21-14, when we analyse the effects of C_S, we must take the assumption that C_E and C_C can be replaced by a short circuit, otherwise the analysis will be complicated. Using this assumption, the a.c. equivalent circuit for the input section of Fig. 21-14, will appear as shown in Fig. 21-16.

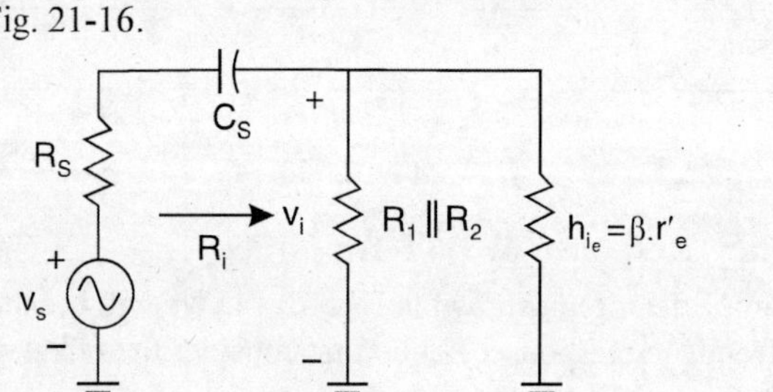

Fig. 21-16.

From the circuit shown in Fig. 21-16, we find that the value of R_i can be obtained by using the equation.

$$R_i = R_1 \parallel R_2 \parallel \beta r'_e$$

The voltage v_i applied to the input of BJT,

$$v_i = \frac{R_i\, v_S}{R_i + R_S - jX_{C_S}} \qquad \text{...(iii)}$$

(ii) Determination of low-frequency cut-off f_{LC} : Since, the coupling capacitor is normally connected between the output of BJT and the applied load, the R-C configuration that determines the low cut-off frequency due to C_C appears as shown in Fig. 21-17.

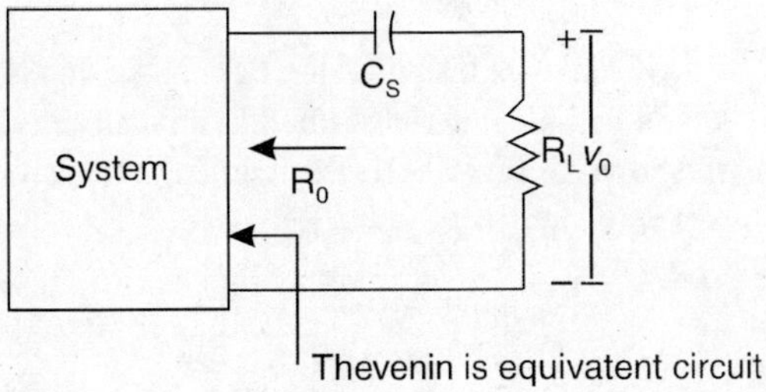

Fig. 21-17.

Notice that total resistance of the circuit is $(R_0 + R_L)$. The lower cut-off frequency due to the effect of C_C can be obtained from the equation,

$$f_{L_C} = \frac{1}{2\pi (R_0 + R_L) C_C}$$

Ignoring the effects of C_S and C_E, the output voltage V_0 will be 70.7% of its midband value at f_{L_C}. For the amplifier circuit of Fig. 21-14, the a.c. equivalent circuit for the output section with $v_i = ov$ appears as shown in Fig. 21-18.

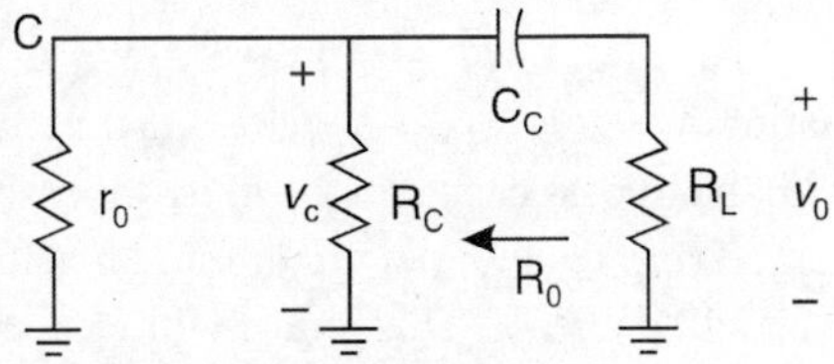

Fig. 21-18.

Using the ac equivalent circuit of Fig, 21-18, we find that

$$R_0 = R_C\, r_0$$

(iii) Determination of low-frequency cut-off f_{L_E} :

The R-C configuration that determines the lower cut-off frequency due to C_E appears as shown in Fig. 21-19.

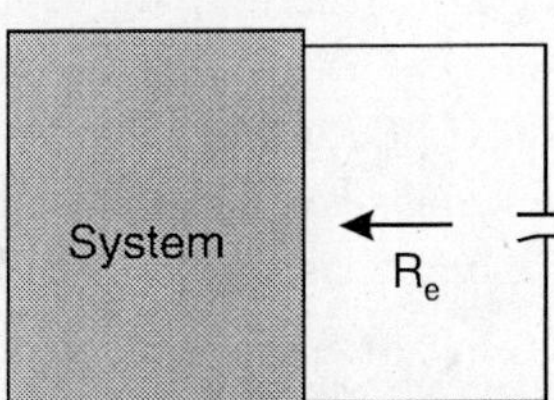

Fig. 21-19.

Notice that the equivalent resistance is indicated by R_e. The lower cut-off frequency due to the effect of C_E can be determined from the equation,

$$f_{L_E} = \frac{1}{2\pi R_e C_E}$$

For the amplifier circuit of Fig. 21-14, the a.c. equivalent circuit as 'seen' by C_E appears in Fig. 21-20.

Using the a.c. equivalent circuit of Fig. 21-20, we find that

$$R_e = R_E \parallel \left(\frac{R'_S}{\beta} + r_e' \right)$$

where, $$R'_S = R_S \parallel R_1 \parallel R_2$$

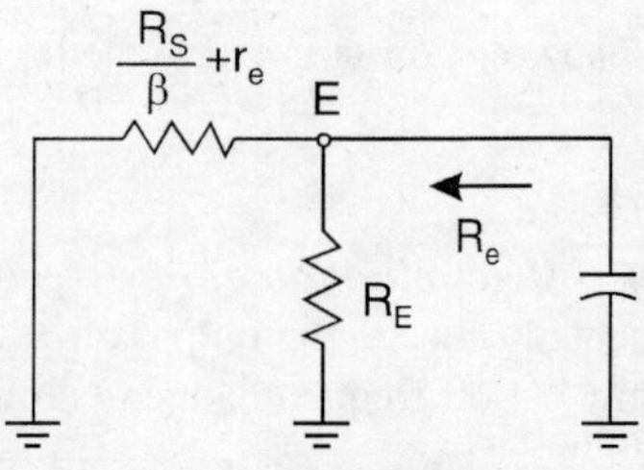

Fig. 21-20.

The effect of C_E on the gain can be better described in a quantitative manner by recalling that the gain for the configuration shown in Fig. 21-21, is given by the equation,

$$A_v = -\frac{R_C}{r_e' + R_E} \quad \text{...(iv)}$$

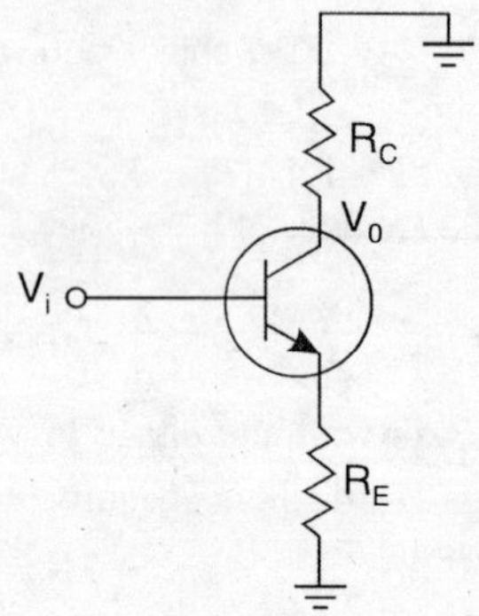

Fig. 21-21.

It is evident from the equation $A_v = -R_C/(r_e' + R_E)$ that maximum gain is availably only where $R_E = 0$. At low frequencies, with the bypass capacitor C_E in its "open-circuit" equivalent state, all of the R_E appears in the gain equation (*iv*), resulting in the minimum gain. As the frequency increases, the reactance of the capacitor C_E will decrease. This inturn will reduce the parallel impedance of R_E and C_E. Corresponding to a certain frequency, a situation is reached where R_E is effectively "shorted out" by C_E. The result is a maximum or msidband gain given by the equation $A_v = -R_C/r_e'$. At f_{L_E}, the gain will be 3 dB below the midband value.

Before proceeding further, we must take note of the following important points regarding the role of C_S, C_E and C_C capacitors in the low-frequency response of the BJT amplifier.

(i) C_S, C_C and C_E will affect only the low-frequency response.

(ii) At the midband frequency level, the short-circuit equivalents for the capacitors can be inserted.

(iii) Although each capacitor value (C_S, C_C or C_E) will affect the gain Av = v_0/v_i in a similar frequency range, the *highest low frequency cut-off* determined by C_S, C_C and C_E will have the greatest impact since it will be the last encountered before the mid-band level.

(iv) If the cut-off frequency estabilished by, each capacitor are relatively, for apart, the highest cut-off freaquency will essentially determine the lower cut-off frequency for the entire system.

(*v*) If there are two or more 'high' cut-off frequencies, the effect will be to raise the lower cut-off frequency and reduce the resulting bandwidth of the system.

(*vi*) If the cut-off frequencies are sufficiently separated, the effect of one on the other can be ignored.

Example 21-9. *For the BJT amplifier circuit shown in Fig. 21-22, determine the lower cut-off frequencies for C_S, C_C and C_E. Also sketch the dB plot for the low-frequency response.*

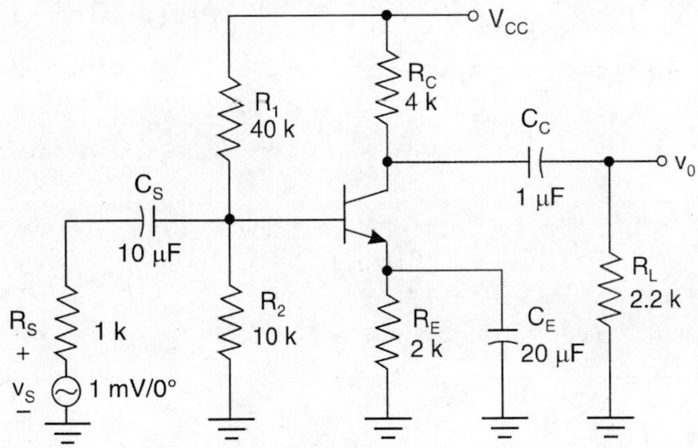

Fig. 21-22.

Solution. Given : $R_C = 4$ kΩ, $R_1 = 40$ kΩ, $R_2 = 10$ kΩ, $R_E = 2$ kΩ, $R_S = 1$ kΩ, $R_L = 2.2$ kΩ, $C_S = 10\ \mu F$, $C_E = 20\ \mu F$, $C_C = 1\ \mu F$, $\beta = 100$ and $r_0 = \infty$.

We know that $\beta R_E = 100 \times 2\text{ k}\Omega = 200\text{ k}\Omega$

Since, βR_E (= 200 kΩ) >> 10 R_2 (= 100 kΩ). Therefore,

$$V_B \approx \left(\frac{R_2}{R_2 + R_1}\right) V_{CC} = \frac{10\text{ k}\Omega \times 20\text{ V}}{10\text{ k}\Omega + 40\text{ k}\Omega} = \frac{200}{50} = 4\text{V}$$

We also know that,

$$I_E = \frac{V_E}{R_E} = \frac{V_B - 0.7\text{ V}}{R_E} = \frac{4 - 0.7\text{ V}}{2\text{ k}\Omega} = 1.65\text{ mA}$$

and

$$r_e' = \frac{26\text{ mV}}{I_E} = \frac{26\text{ mV}}{1.6\text{ mA}} = 15.76\ \Omega$$

∴

$$\beta r_e' = 100 \times 15.76 = 1576\ \Omega \text{ or } 1.576\text{ k}\Omega$$

Further we know that midband gain,

$$A_v = \frac{v_O}{v_i} = \frac{-R_C \| R_L}{r_e'} = -\frac{(4\text{ k}\Omega) \| (2.2\text{ k}\Omega)}{15.76\ \Omega} = -90$$

The input resistance,

$$R_i = R_1 \| R_2 \| \beta r_e'$$
$$= (40\text{ k}\Omega) \| (10\text{ k}\Omega) \| (1.576\text{ k}\Omega)$$
$$\approx 1.32\text{ k}\Omega$$

The voltage applied to the input of BJT,

$$v_i = \frac{R_i\, v_S}{R_i + R_S}$$

or $$\frac{v_i}{v_S} = \frac{R_i}{R_i + R_S} = \frac{1.32\text{ k}\Omega}{1.32\text{ k}\Omega + 1\text{ k}\Omega} = 0.569$$

Therefore, the stage gain,

$$\therefore \quad A_{v_S} = \frac{v_0}{v_i} = \frac{v_0}{v_i}\cdot\frac{v_i}{v_s} = (-90)\times(0.569) = -15.21$$

Determination of f_{L_S}

We know that input resistance,

$$R_i = R_1 \parallel R_2 \parallel \beta r'_e = (40\text{ k}\Omega) \parallel (10\text{ k}\Omega) \parallel (1.576\text{ k}\Omega)$$
$$\approx 1.32\text{ k}\Omega$$

and the lower cut-off frequency,

$$f_{L_S} = \frac{1}{2\pi(R_S + R_i)C_S}$$
$$= \frac{1}{2\pi(1\text{ k}\Omega + 1.32\text{ k}\Omega)\times(10\mu F)}$$
$$\approx 6.86\text{ Hz } \textbf{Ans.}$$

Determination of f_{L_C}

We know that lower cut-off frequency due to the effect of C_C,

$$f_{L_C} = \frac{1}{2\pi(R_C + R_L)C_C}$$
$$= \frac{1}{2\pi(4\text{ k}\Omega + 2.2\text{ k}\Omega)\times 1\,\mu F}$$
$$\approx 25.68\text{ Hz } \textbf{Ans.}$$

Determination of f_{L_E}

$$R'_S = R_S \parallel R_1 \parallel R_2 = (1\text{ k}\Omega) \parallel (40\text{ k}\Omega) \parallel (10\text{ k}\Omega)$$
$$\approx 0.889\text{ k}\Omega$$

and $$R_e = R_E \parallel \left(\frac{R'_S}{\beta} + r'_e\right)$$
$$= (2\text{k}\Omega) \parallel \left(\frac{0.889\text{ k}\Omega}{100} + 15.76\ \Omega\right)$$
$$= (2\text{ k}\Omega) \parallel (8.89\ \Omega + 15.76\ \Omega)$$
$$= (2\text{ k}\Omega) \parallel (24.65\ \Omega)$$
$$\approx 24.35\ \Omega$$

We also know that the lower cut-off frequency due to the effect of C_E,

$$f_{L_E} = \frac{1}{2\pi R_e C_E}$$
$$= \frac{1}{2\pi\times(24.35\ \Omega)\times(20\ \mu F)}$$
$$= 327\text{ Hz } \textbf{Ans.}$$

dB plot for the low-frequency response

We know that the dB plots are usually normalized by dividing the voltage gain (A_v) by the magnitude of the midband gain. As determined in part (*a*), the midband gain (or the overall stage gain) is 51.21. So the ratio ($|A_v/A_{v.(mid)}|$) will be 1 in the midband region.

Let f_{L_E} = The lower cut off frequency f_1

Then an asymptote at-6dB/octave can be drawn for the frequency (f_{L_E}) as shown in Fig. 21-23 to form the Bode plot. Also notice that this line is also the envelop for the actual low frequency response. At f_1, the actual curve is –3dB down from the midband level as defined by 0.707 $A_{v_{mid}}$ level. This allows us to sketch the actual frequency response curve as shown in Fig. 21-23.

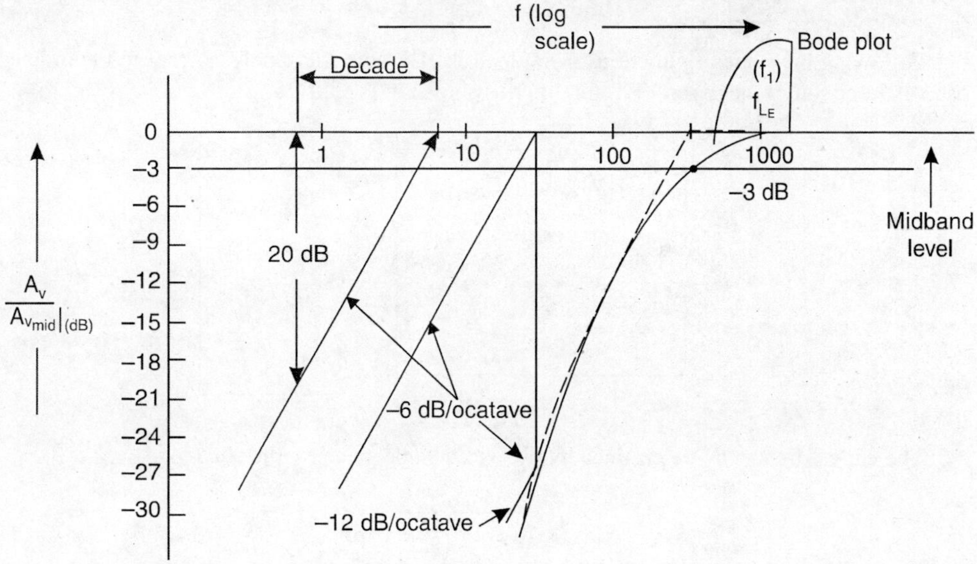

Fig. 21-23.

Notice from Fig. 21-23., that a –6 dB/octave asymptote has been drawn at each frequency determined in part (*a*). This demostrates clearly that it is f_{L_E} for the given amplifier circuit that determines the – 3 dB point. The f_{L_C} begins to affect the shape of the envelop around – 24 dB. The magnitude plot shows clearly that the slope of the resultant asymptote is the sum of the asymptotes having the same sloping direction in the same frequency interval. Notice that the slope has dropped to – 12 dB/octave for frequencies less than f_{L_C} . As a matter of fact, it could drop to – 18 dB/octave if the three cutoff frequencies f_{L_S}, f_{L_C} and f_{L_E} were closer together.

21-8. Low-frequency Response of FET Amplifier

We have already discussed the low-frequency response of a BJT amplifier in the last article. The analysis of FET amplifier in the low-frequency region is quite similar to that of the BJT amplifier.

Fig. 21-24, shows the circuit of a typical FET amplifier. Notice there are again three capacitors of primary concern : C_G, C_C and C_S. Although, Fig. 21-24, is used to estabilish fundamental

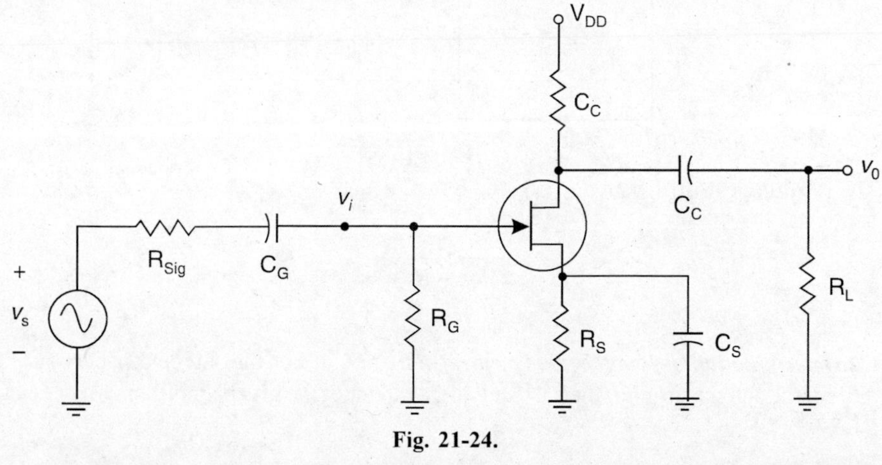

Fig. 21-24.

equations for low-frequency analysis, but the procedure and conclusions can be applied to other FET amplifier configurations as well.

(1) *Determination of low-frequency cut-off due to C_G* : We know that C_G is the coupling capacitor between the source and FET.

Following the same argument as mentioned in BJT amplifier analysis, we can obtain the a.c. equivalent circuit of the given FET amplifier as shown in Fig. 21-25.

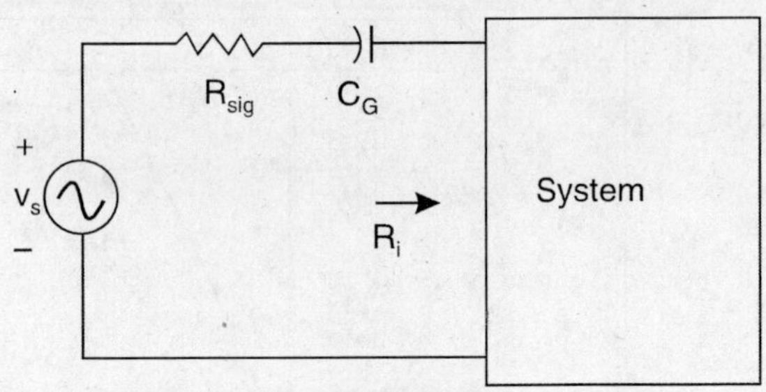

Fig. 21-25.

The cut-off frequency determined by C_G will then be given by the equation,

$$f_{L_G} = \frac{1}{2\pi(R_{sig} + R_i)C_G}$$

We know that for the FET amplifier circuit shown in Fig. 21-24,

$$R_i = R_G$$

As a matter of fact, $R_G >> R_{sig}$. In such a situation, the lower cut-off frequency is determined primarily by R_G and C_G. Thus,

$$f_{L_G} \approx \frac{1}{2\pi R_G C_G}$$

(2) *Determination of low-frequency cut off due to C_C* : For the coupling capacitor (C_C) between the FET and the load, the a.c. equivalent circuit of the given amplifier circuit is as shown in Fig. 21-26.

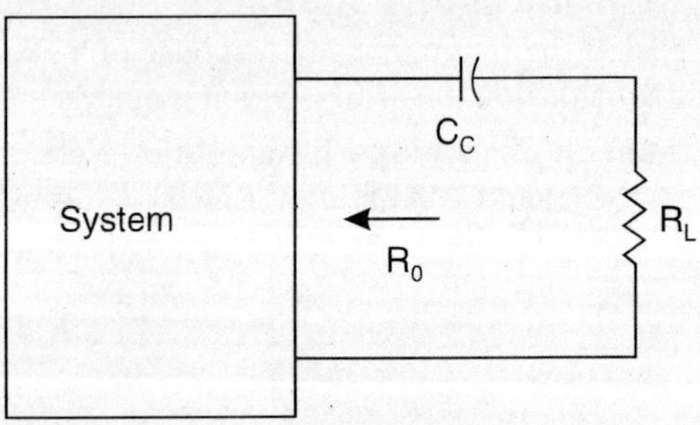

Fig. 21-26.

The resulting cut-off frequency is given by the equation,

$$f_{L_C} = \frac{1}{2\pi(R_0 + R_L)C_C}$$

where, $R_0 = R_D \parallel r_d$

(3) *Determination of low frequency cut-off due to C_S* : For the source capacitor (C_S) between the FET source and the ground, the a.c. equivalent circuit of the given amplifier circuit is a shown in Fig. 21-27.

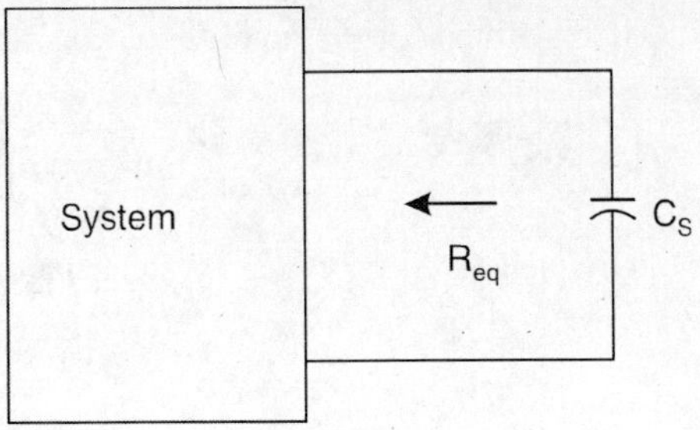

Fig. 21-27.

The cut-off frequency determined by C_S will then be given by the equation,

$$f_{L_S} = \frac{1}{2\pi R_{eq} C_S} \quad ...(i)$$

where,

$$R_{eq} = \frac{R_S}{1 + \dfrac{(1 + g_m r_d)\cdot R_S}{(r_d + R_D \| R_L)}} \quad ...(ii)$$

If $r_d \approx \infty$, then the equation (*ii*) reduces to,

$$R_{eq} = R_S \,\|\, \frac{1}{g_m}$$

Example 21-10. *For the FET amplifier circuit shown in Fig. 21-28, determine the lower cut-off frequencies for C_S, C_C and C_G. Also sketch the dB plot for the low-frequency response. Given for the FET, I_{DSS} = 8 mA, V_P = – 4V and $r_d = \infty$.*

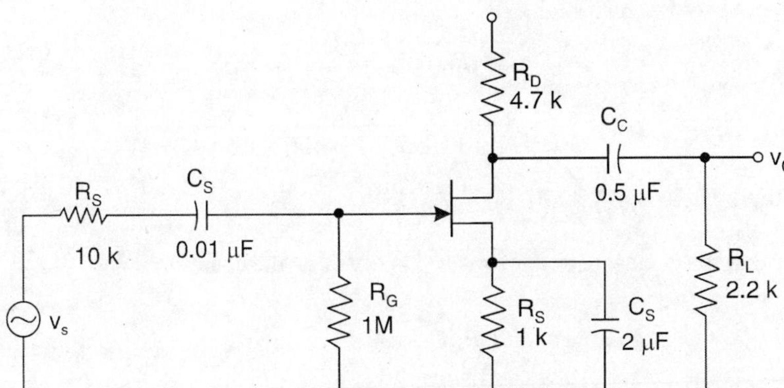

Fig. 21-28.

Solution. Given, $R_D = 4.7\ k\Omega$, $R_G = 1\ k\Omega$, $R_S = 1\ k\Omega$, $C_G = 0.01\ \mu F$, $C_C = 0.5\ \mu F$, $C_S = 2\ \mu F$, $R_{sig} = 10\ k\Omega$, $R_L = 2.2\ k\Omega$, I_{DSS} = 8 mA, V_P = – 4V and $r_d = \infty$.

Plotting the transfer curve of $I_D = I_{DSS}(1 - V_{GS}/V_P)^2$ and superimposing the curve defined by $V_{GS} = -I_D R_S$ will result in an intersection at V_{GSQ} = – 2 V and I_{DQ} = 2 mA. Besides this, we know that,

$$g_{mo} = \frac{2 I_{DSS}}{|V_P|} = \frac{2 \times 8\,\text{mA}}{4\ \text{V}} = 4\ \text{mS}$$

and

$$g_m = g_{mo}\left(1 - \frac{V_{GSQ}}{V_P}\right)$$

$$= 4\text{mS}\left(1 - \frac{-2\text{V}}{-4\text{V}}\right)$$

$$= 2\text{mS}$$

Determination of f_{L_G}

We know that lower cut-off frequency, f_{L_G} is given by the equation,

$$f_{L_G} = \frac{1}{2\pi(R_{sig} + R_i)C_G}$$

$$= \frac{1}{2\pi(R_{sig} + R_G)C_G} \qquad ...(\because R_i \approx R_G)$$

$$= \frac{1}{2\pi(10\text{ k}\Omega + 1\text{ M}\Omega) \times (0.01\mu F)}$$

$$\approx 15.8\text{ Hz } \textbf{Ans.}$$

Determination of f_{L_C}

We know that lower cut-off frequency, f_{L_C} is given by the equation,

$$f_{L_C} = \frac{1}{2\pi(R_0 + R_L)C_C}$$

$$= \frac{1}{2\pi[(R_D \| r_d) + R_L]C_C} \qquad ...(\because R_0 = R_D \| r_d)$$

$$= \frac{1}{2\pi(R_D + R_L)C_C} \qquad ...(r_d = \infty)$$

$$= \frac{1}{2\pi(4.7\text{ k}\Omega + 2.2\text{ k}\Omega) \times 0.5\ \mu F}$$

$$\approx 46.13\text{ Hz } \textbf{Ans.}$$

Determination of f_{L_S}

We know that lower cut-off frequency, f_{L_S} is given by the equation,

$$f_{L_S} = \frac{1}{2\pi R_{eq} C_S} \qquad ...(i)$$

Where, $$R_{eq} = R_S \| \frac{1}{g_m} = (1\text{ k}\Omega) \| \frac{1}{2\text{ mS}}$$

$$= (1\text{ k}\Omega) \| (0.5\text{ k}\Omega)$$

$$= 333.33\ \Omega$$

Substituting this value of R_{eq} in equation (*i*), we get,

$$f_{L_S} = \frac{1}{2\pi \times (333.3\ \Omega) \times (2\ \mu F)}$$

$$= 238.73\text{ Hz.}$$

A comparison of all the three lower cut-off frequencies (f_{L_G} = 15.8 Hz, f_{L_C} = 46.13 Hz and f_{L_S} = 238.73 Hz) shows that f_{L_S} is the largest cut-off frequency. Therefore, f_{L_S} defines the low cut-off frequency for the FET amplifier circuit shown in Fig. 21-28.

dB plot for the low frequency response

We know that the midband gain of the FET amplifier is given by the equation,

$$A_{v\,mid} = \frac{v_0}{v_i} = -\,g_m\,(R_D \,\|\, R_L)$$

Substituting the values of g_m, R_D and R_L in the above equation, we get,

$$A_{v\,mid} = -\,(2\text{ mS}) \times (4.7\text{ k}\Omega \,\|\, 2.2\text{ k}\Omega)$$
$$= (-\,2\text{ mS}) \times (1.499\text{ k}\Omega)$$
$$\approx -\,3$$

Using the midband gain to normalize the response of the circuit, the dB plot for the FET amplifier circuit can be sketched as shown in Fig. 21-29.

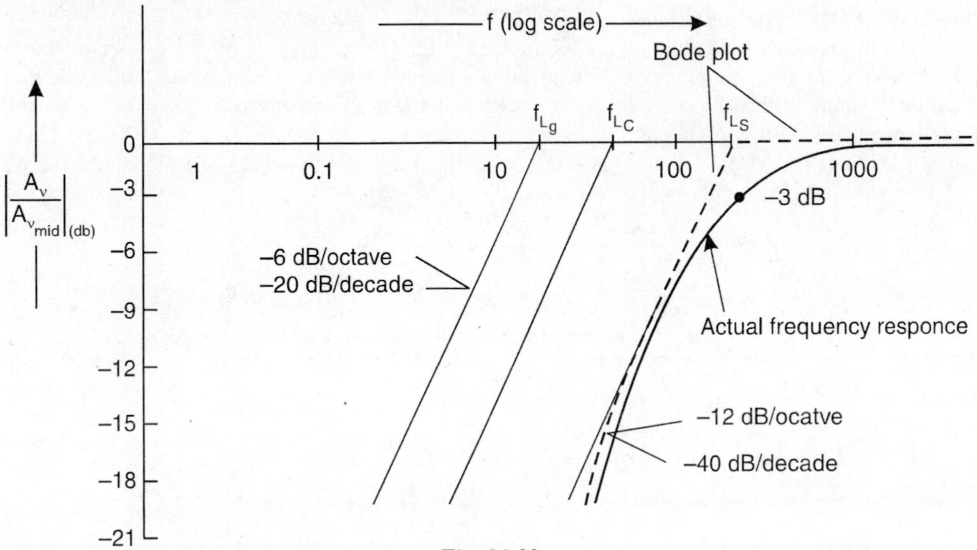

Fig. 21-29.

21-9. Miller Effect Capacitance

Strictly speaking, in the high frequency region, the capacitive elements of importance are the interelectrode capacitances and the wiring capacitances between the leads of the circuit. The interelectrode capacitances are the internal capacitances between the terminals of the active device (BJT or FET). Refer to Fig. 21-30(*a*) and (*b*).

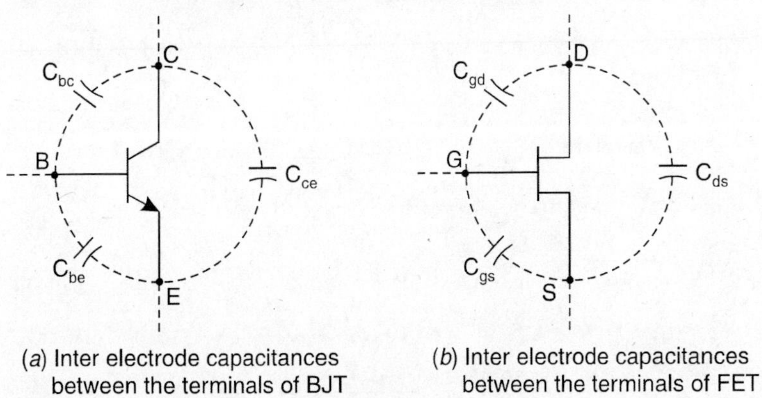

(*a*) Inter electrode capacitances between the terminals of BJT

(*b*) Inter electrode capacitances between the terminals of FET

Fig. 21-30.

The wiring capacitances are the capacitances between the leads of the circuit on the input and output sides of the circuit. Refer to Fig. 21-31.

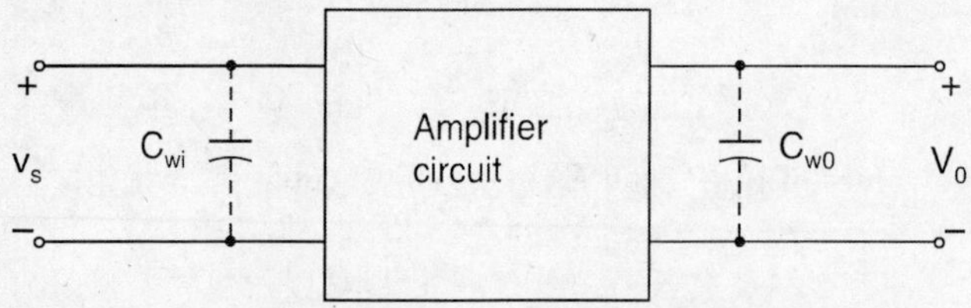

Fig. 21-31. *Illustrating wiring capacitances between the input and output leads of the amplifier circuit.*

The large capacitors of the amplifiers circuit that controlled the low frequency response have all been replaced by their short-circuit equivalent due to their very low reactance at high frequences.

It may be carefully noted that inverting amplifiers introduce a phase shift of 180° between input and output. This results in a negative value of voltage gain (A_v). Because of this phase shift, the input and output capacitance is increased by a capacitance level sensitive to (1) the interelectrode capacitance between the input and output terminals of the active device (BJT or FET) and (2) the gain of the amplifier. This feedback capacitance as illustrated in Fig. 21-23 is designated by C_f.

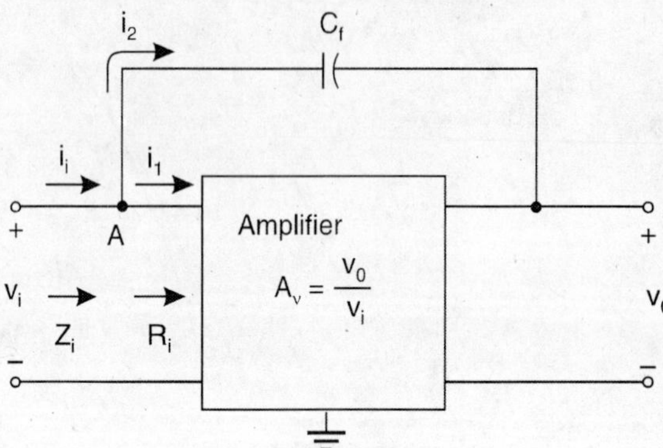

Fig. 21-32.

Applying Kirchoff's current Law at point A, we get,

$$i_i = i_1 + i_2 \qquad ...(i)$$

Using Ohm's law, it is possible to write the three currents, i_i, i_1 and i_2 as follows,

$$i_i = \frac{v_i}{R_i} \text{ and } i_1 = \frac{v_i}{r_i}$$

Also,

$$i_2 = \frac{v_i - v_O}{X_{C_f}}$$

Since, $v_O = A_v\, v_i$, therefore,

$$i_2 = \frac{v_i - A_v \cdot v_i}{X_{C_f}} = \frac{(1 - A_v)v_i}{X_{C_f}}$$

Substituting the values of i_i (= v_i/Z_i), i_1 (= V_i/R_i) and i_2 [= $(1 - A_v)v_i/X_{C_f}$] in equation (*i*), we get,

$$\frac{v_i}{R_i} = \frac{v_i}{R_i} + \frac{(1 - A_v)\, v_i}{X_{C_f}}$$

Taking out v_i common from R.H.S. and cancelling it out from two sides we get,

$$\frac{1}{Z_i} = \frac{1}{R_i} + \frac{1 - A_v}{X_{C_f}}$$

$$= \frac{1}{R_i} + \frac{1}{\dfrac{X_{C_f}}{1 - A_v}} \qquad ...(ii)$$

But $$\frac{X_{C_f}}{1 - A_v} = \frac{1}{\omega C_f (1 - A_v)} = \frac{1}{\omega (1 - A_v) C_f} \qquad ...(iii)$$

Let $(1 - A_v) = C_M$ (*i.e.*, Miller effect capacitance),
Then equation (*iii*) can be rewritten as,

$$\frac{X_{C_f}}{1 - A_v} = \frac{1}{\omega C_M} = X_{C_M}$$

where, X_{C_M} = capacitive reactance due to the Miller effect capacitance C_M.

Substituting the value of $X_{C_f}/(1 - A_v)$ $(= X_{C_M})$ in equation (*iii*) we get,

$$\frac{1}{Z_i} = \frac{1}{R_i} + \frac{1}{X_{C_M}} \qquad ...(iii)$$

The above equation helps us to establish an equivalent circuit of Fig. 21-32. The resulting equivalent circuit is as shown in Fig. 21-33.

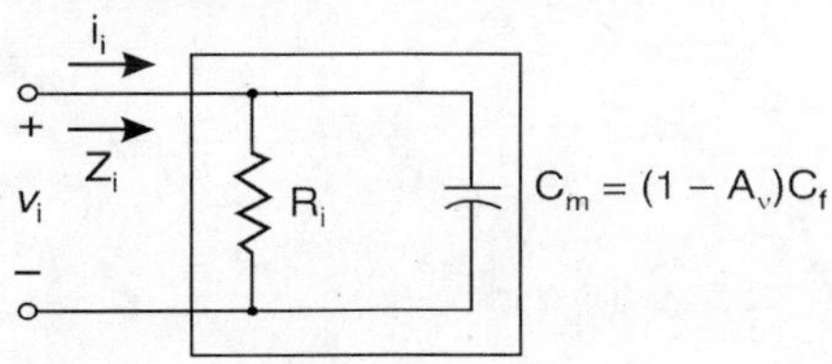

Fig. 21-33.

Notice that the equivalent circuit of Fig. 21-32, includes the some R_i that we have dealt with in previous chapters (especially chapter-16), with the addition of feedback capacitor magnified by the gain of the amplifier. Any interelectrode capacitance at the input terminals to the amplifier will simply be added in parallel with the elements of Fig. 21-33.

In general, the Miller effect input capacitance is given by the equation,

$$C_{M_i} = (1 - A_v)\, C_f \qquad ...(iv)$$

It is evident from the above equation that for any inverting amplifier, the input capacitance will be increased by a Miller effect capacitance sensitive to the gain of the amplifier and the interelectrode capacitance connected between the input and output terminals of the active device (BJT or FET).

At high frequencies, the gain A_v will be a function of the level of C_{M_i}. However, since the maximum gain is the midband value, using the midband value will result in the highest level of C_M. This represents the worst-case scenario. Hence, remember the value of A_v to be used in equation (*iv*) is midband gain value.

If you examine equation (*iv*) carefully, you will find that if we use a positive value for A_v, it would result in a negative capacitance. It is because of this reason that we use an inverting amplifier configuration in out discussion.

Let us now focus on the output capacitance of the amplifier. We will use the circuit shown in Fig. 21-34 to determine the output of Miller effect capacitance.

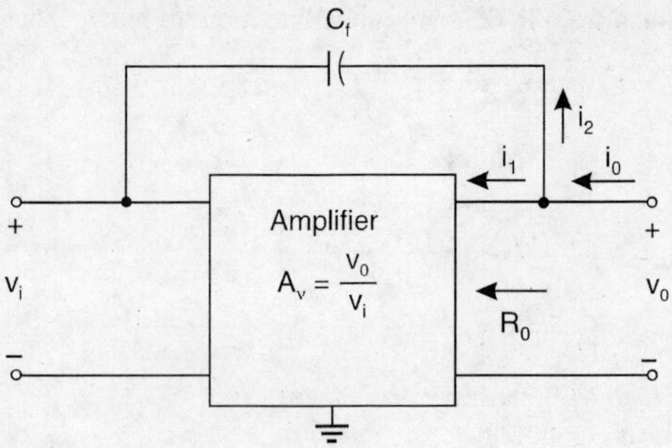

Fig. 21-34.

Applying Kirchoff's current Law at point B (refer to Fig. 21-34), we get,

$$i_o = i_1 + i_2$$

With
$$i_1 = \frac{v_o}{r_o} \text{ and } i_2 = \frac{v_o - v_i}{X_{C_f}}$$

$\therefore$
$$i_o = \frac{v_o}{R_o} + \frac{v_o - v_i}{X_{C_f}}$$

Since, $R_o >> X_{C_f}$, therefore the term v_o/R_o is negligible as compared to the term $(v_o - v_i)/X_{C_f}$. Hence,

$$i_o \approx \frac{v_o - v_i}{X_{C_f}}$$

Since,
$$\frac{v_o}{v_i} = A_v$$

$\therefore$
$$v_i = v_o/A_v$$

Substituting this vlaue of V_i in equation (v), we get,

$$i_o = \frac{v_o - \frac{v_o}{A_v}}{X_{C_f}} = \frac{v_o\left(1 - \frac{1}{A_v}\right)}{X_{C_f}}$$

or
$$\frac{i_o}{v_o} = \frac{1 - \frac{1}{A_v}}{X_{C_f}}$$

or
$$\frac{v_o}{i_o} = \frac{X_{C_f}}{1 - \frac{1}{A_v}}$$

Substituting X_{C_f} $(= 1/\omega c_f)$ in the above equation, we get,

$$\frac{v_o}{i_o} = \frac{1}{\omega c_f\left(1 - \frac{1}{A_v}\right)} = \frac{1}{\underbrace{\omega\left(1 - \frac{1}{A_v}\right)C_f}_{C_{M_o}}} = \frac{1}{WC_{M_o}} \quad ...(vi)$$

where, $$C_{M_o} = \left(1 - \frac{1}{A_v}\right) C_f \quad ...(vii)$$

Usually $A_v >> 1$, therefore equation (*vii*) reduces, to,

$$C_{M_o} = C_f$$

So to conclude our discussion, we can say that the Miller effect capacitance,

$$C_{M_i} = (1 - A_v)\, C_f$$

and $$C_{M_o} = C_f$$

21-10. High-Frequency Response of BJT Amplifier

Strictly speaking, at the high frequency end, there are two factors that will define the – 3 dB point :

(1) The circuit capacitance (parasitic and introduced), and

(2) Frequency dependence of h_{fe} (or β).

Now we shall study the effect of both these factors.

1. *Circuit parameters:* In the high-frequency region, the RC circuit of concern is the one shown in Fig. 21-35. At increasing frequencies, the reactance X_C will decrease in magnitude. Because of this, it results in a shortening effect across the output. This inturn results in a decrease in gain.

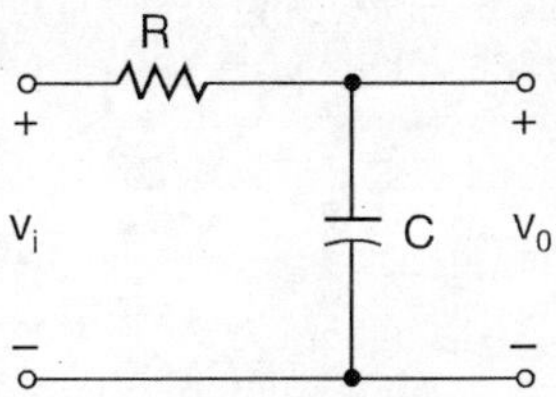

Fig. 21-35.

The deviation of an equation for the corner frequency for the RC circuit shown in Fig. 21-35 follows along similar line to that applied for low-frequency region. The most-significant difference is in the general form of A_v as shown below,

$$A_v = \frac{v_o}{v_i} = \frac{1}{1 + j\left(\frac{f}{f_2}\right)} \quad ...(i)$$

This equation results in a magnitude plot as shown in Fig. 21-36. Notice that the plot drops off at 6 dB/octave with increasing frequency. Also notice that f_2 is in the denominator of the frequency ratio in equation (*i*) rather than the numerator as occurred for f_1 in equation (*iv*) [page 616].

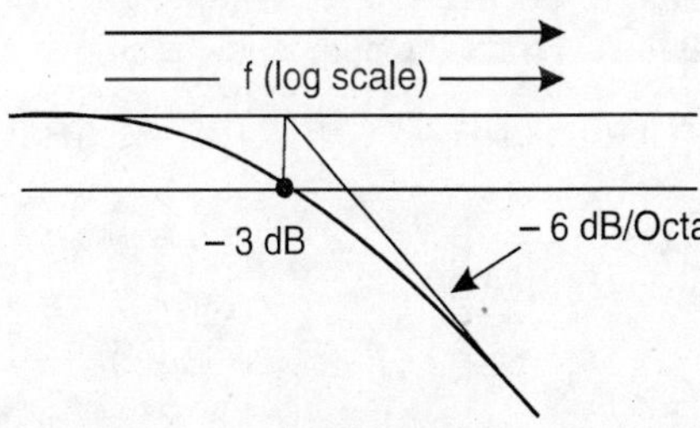

Fig. 21-36.

Fig. 21-37 shows a BJT amplifier circuit indicating the various parasitic capacitances (C_{be}, C_{bc}, C_{ce}) of the BJT along with the wiring capacitances (C_{W_i} and C_{W_o}) introduced during construction.

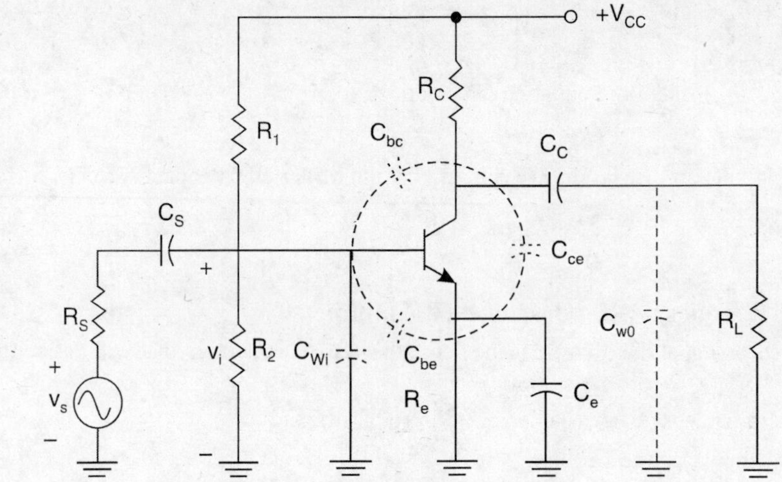

Fig. 21-37.

Fig. 21-38 shows the high-frequency equivalent circuit of Fig. 21-37. Notice the absence of C_E, C_C and C_E. These are assumed to be in the short-circuit state at high frequencies. The capacitance C_i includes the input wiring capacitance C_{w_i}, the transition capacitance C_{be} and the Miller capacitance C_{M_i}. Similarly, C_0 includes the output wiring capacitance C_{W_o}, the parasitic capacitance C_{ce} and the output Miller capacitance C_{M_o}.

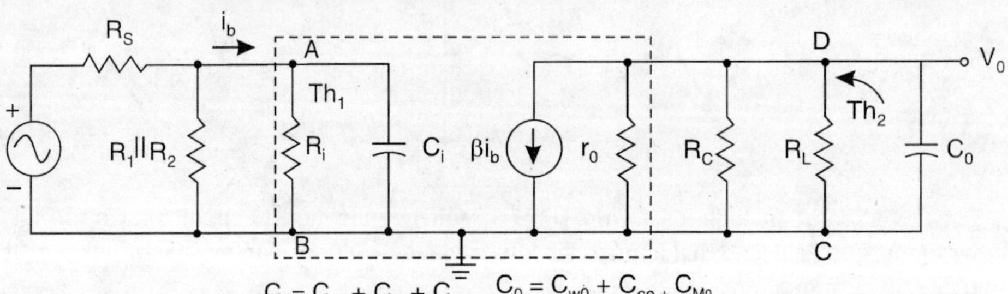

Fig. 21-38.

It may be noted that the capacitance C_{be} is the largest of the parasititc capacitances while C_{ce} is the smallest. As a matter of fact, the manufacturer simply provide the levels of C_{be} and C_{bc} in the specifications data sheet. The value of C_{ce} is not included unless it will affect the response of a particular BJT in a specific area of application.

In order to analyse the ac equivalent circuit of Fig. 21-38, let us Thevenise the input and output circuits at the terminals AB and CD respectively. The resulting, Thevenin's equivalent circuit for the input side is shown in Fig. 21-39 (*a*) and for the output side is shown in Fig. 21-39 (*b*).

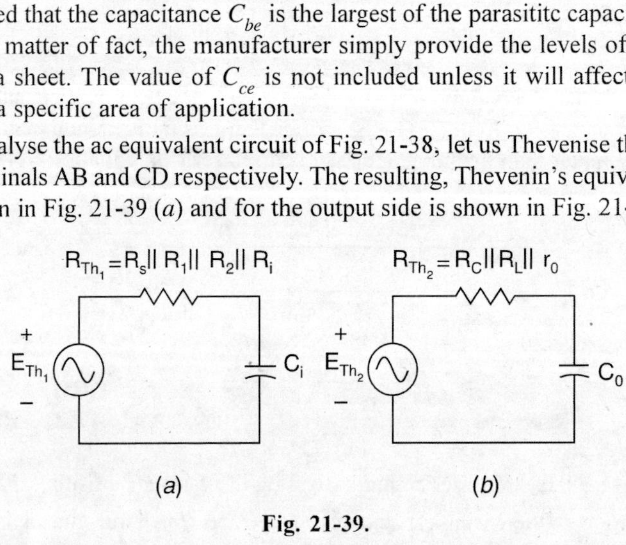

Fig. 21-39.

For the input circuit shown in Fig. 21-39 (*a*), the higher cut-off frequency is given by the equation,

$$f_{H_i} = \frac{1}{2\pi R_{Th_1} C_i} \quad ...(ii)$$

where, $R_{Th_1} = R_S \| R_1 \| R_2 \| R_i$

and $C_i = C_{W_i} + C_{be} + C_{M_i}$

$$= C_{W_i} + C_{be} + (1 - A_v) C_{bc}$$

At very high frequencies the effect of C_i is to reduce the total impedance of the parallel combination of R_1, R_2, R_i and C_i. As a result of this, voltage across C_i reduces. This reduces the value of I_b because of which the gain for the complete system is reduced.

Let us now focus on the Thevinin's equivalent circuit on the output side (Fig. 21-39 (*b*)). For the output circuit, the higher cutoff frequency is given by the equation,

$$f_{H_o} = \frac{1}{2\pi R_{Th_2} C_o}$$

Where, $R_{Th_2} = R_C \| R_L \| r_O$

and $C_o = C_{W_o} + C_{ce} + C_{M_o}$

Notice that at very high frequencies, the capacitive reactance of C_o will decrease and consequently reduce the total impedance of the output parallel branches of Fig. 21-38. The net result is that v_0 will also decline towards *O* as the reactance X_C becomes smaller. Both the frequencies f_{H_i} and f_{H_o} will define a – 6 dB/octave asymptote each such as the one shown in Fig. 21-36. If the parasitic capacitors were the only elements to determine the *high cut off frequency, the lowest frequency would be the determining factor.* However, the decrease in h_{fe} (or β) with frequency must also have to be considered whether its break frequency is lower than f_{H_i} or f_{Ho}.

2. h_{fe} (or β) *variation.* The variation of h_{fe} (or β) with frequency is given by the equation,

$$h_{fe} = \frac{h_{fe}\ mid}{1 + j\left(\frac{f}{f_\beta}\right)} \quad ...(iii)$$

The use of h_{fe} rather than β is due to the fact that manufacturers typically use the hybrid parameters when covering the issue in their specification data sheets.

The only undefined quantity, f_β is determined by a set of parameters employed in the *hybrid* π or *Giacoletto model.* The hybrid π (or Giacoletto model) is frequently applied to represent a BJT in the high-frequency region. Such a circuit is as shown in Fig. 21-40.

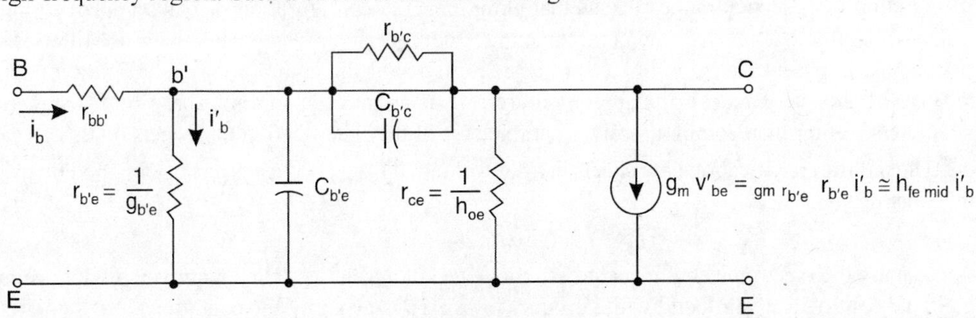

Fig. 21-40.

The various parameters shown in the hybrid-π model.

$r_{bb'}$ = The resistance that includes (*i*) the base contact (*ii*) base bulk and (*iii*) base spreading resistance. The base contact resistance is due to the actual connection to the base. The bulk resistance is the resistance from the external terminal to the active region of the BJT while the base spreading resistance is the actual resistance with in the active base region.

$r_{b'c}$, r_{ce}, and $r_{b'c}$ = The resistance between the indicated terminals when the BJT is in active region.

$C_{b'c}$ = Transition capacitive

$C_{b'e}$ = diffusion capacitive

The equation for high-frequency cut-off,

$$f_\beta \,(\text{or } f_{h_{fe}}) = \frac{g_{b'c}}{2\pi(C_{b'e} + C_{b'c})} \qquad ...(iv)$$

Since, $g_m = h_{fe_{mid}} \cdot g_{b'e}$, therefore $g_{b'e} = g_m / h_{fe_{mid}}$. Substituting $g_{b'e}$ ($= g_m / h_{fe_{mid}}$) in equation (*iv*), we get,

$$\therefore \qquad f_\beta = \frac{1}{h_{fe_{mid}}} \cdot \frac{g_m}{2\pi/(C_{b'e} + C_{b'c})} \qquad ...(v)$$

Further since,

$$g_m = h_{fe_{mid}} \, g_{b'e} = h_{fe_{mid}} \frac{1}{r_{b'e}}$$

$$\approx \frac{h_{fe_{mid}}}{h_{ie}} = \frac{\beta_{mid}}{\beta_{mid}\, r_e'} = \frac{1}{r_e'}$$

Also using the approximations,

$$C_{b'e} \approx \text{Cbc} \text{ and } C_{b'c} \approx C_{bc}$$

Using the subsitutions for g_m ($= 1/r'_e$), $C_{b'e}$ ($= C_{be}$), $C_{b'c}$ ($= C_{bc}$) in equation (*v*), we get,

$$f_\beta = \frac{1}{2\pi\beta_{mid}\, r_e'(C_{be} + C_{bc})} \qquad ...(iv)$$

It is evident from equation (*iii*) that f_β is a function of r'_e which in turn is a function of the bias conditions.

Now suppose we extract the multiplying factor $h_{fe_{mid}}$ from equation (*iii*), then we find the basic format of equations (*i*) and (*iii*) is the same. This reveals that h_{fe} will drop off from its mid band value with a 6 dB/octave slope as shown in Fig. 21-41.

Fig. 21-41 also shows a plot of h_{fb} (or α) versus frequency. Notice that a small change in h_{fb} for the chosen frequency range reveals that common-base configuration displays improved high-frequency characteristics over the common emitter configuration. Also recall that the Miller effect capacitance is absent due to noninverting characteristics of the common-base configuration. It is because of this very reason that manufacturers usually specify common-base high-frequency parameters (rather than common-emitter parameters) in the specification datasheets of BJT.

Thus, if the specification data sheets lists the valued of f_α and α, then f_β can be obtained by using the equation,

$$f_\beta = f_\alpha (1 - \alpha)$$

Usually the electronic engineers have to make use of a quantity called gain-band width product in circuit design. The gain-band width product for a BJT is given by the condition,

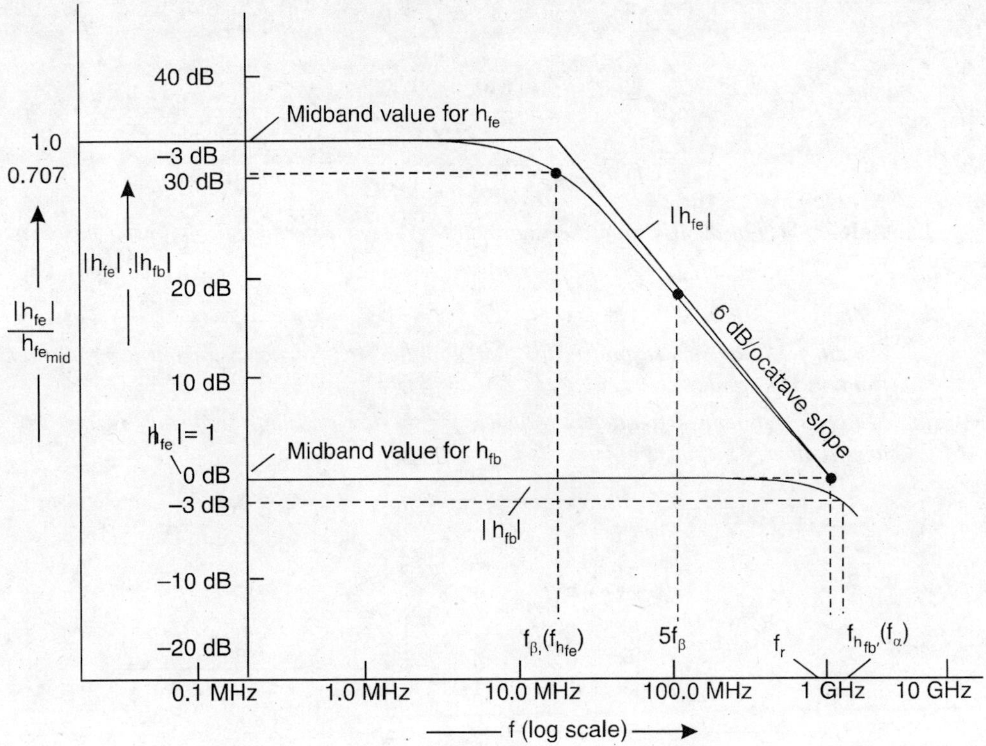

Fig. 21-41.

$$\left| \frac{h_{fe_{mid}}}{1 + j\left(\dfrac{f}{f_\beta}\right)} \right| = 1$$

So that $$\left| h_{fe} \right|_{dB} = 20 \log_{10} \left| \frac{h_{fe_{mid}}}{1 + j\left(\dfrac{f}{f_\beta}\right)} \right| = 20 \log_{10} 1 = 0 \text{ dB}$$

The frequency at which $\left| h_{fe} \right|_{dB} = 0$ dB is designated by f_T and is shown in Fig. 21-41. The magnitude of h_{fe} at the defined condition point ($f_T >> f_\beta$) is given by the equation,

$$\frac{h_{fe_{mid}}}{\sqrt{1 + \left(\dfrac{f_T}{f_\beta}\right)^2}} \approx \frac{h_{fe_{mid}}}{\dfrac{f_T}{f_\beta}} = 1$$

So that $$f_T \approx h_{fe_{mid}} \cdot f_\beta \qquad ...(vii)$$

Notice R.H.S. of the above equation. It consists of two factors $h_{fe_{mid}}$ (*i.e.*, the gain) and f_β (which is the bandwidth). It is because of this reason that f_T is considered as gain-bandwidth product.

Now equation (*vii*), can be rewritten as,

$$f_T \approx \beta_{mid}\, f_\beta \qquad ...(viii)$$

or $$f_\beta = \frac{f_T}{\beta_{mid}}$$

Substituting the value of f_β from equation (*vi*) into an equation (*viii*) we get,

$$f_T = \beta_{mid} \cdot \frac{1}{2\pi\beta_{mid}\, r_e' (C_{be} + C_{bc})}$$

or

$$f_T \approx \frac{1}{2\pi r_e' (C_{be} + C_{bc})}$$

Example 21-11. *For the BJT amplifier shown in Fig. 21-42, determine the following parameters :*

(a) f_{H_i} *and* f_{H_o}

(b) f_β *and* f_T

(c) Sketch the frequency response (dB plot) for the high-frequency region using the results from part (a) and (b).

(d) Sketch the combined frequency response for the low and high frequency region. For the low frequency response use results from Example 24.9.

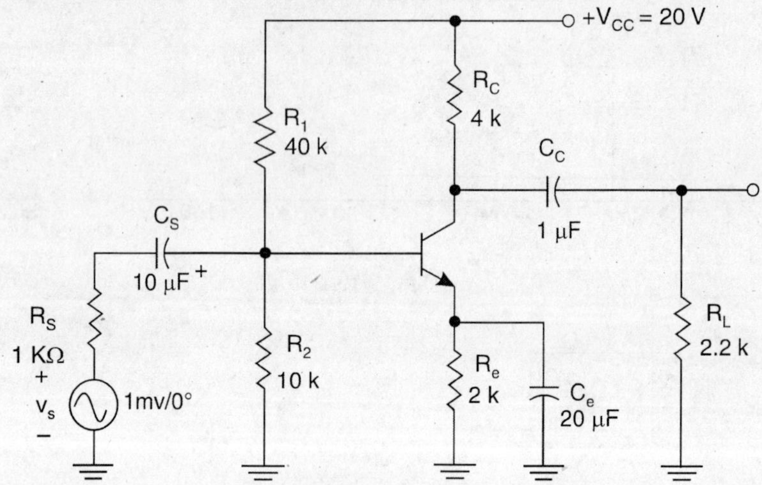

Fig. 21-42.

Solution. Given: *Transistor parameters:* $\beta = 100$, $r_o = \infty$, $C_{be} = 36$ pF, $C_{bc} = 4$ pF, $C_{ce} = 1$ pF, $C_{W_i} = 6$ pF and $C_{W_o} = 8$ pF.

(a) *Values of* f_{H_i} *and* f_{H_o}

As shown earlier in Example 24.9,

$$R_i = \beta\, r'_e = 1.576 \text{ k}\Omega.$$

$$A_{v_{mid}} = -90$$

For the input circuit of the a.c. equivalent circuit (refer to Fig. 21-37),

$$R_{Th_1} = R_S \,\|\, R_1 \,\|\, R_2 \,\|\, R_3 \,\|\, R_i$$

$$= (1\text{ k}\Omega) \,\|\, (40\text{ k}\Omega) \,\|\, (10\text{ k}\Omega) \,\|\, (1.576\text{ k}\Omega)$$

$$\approx 0.568\text{ k}\Omega$$

We know that the input capacitance,

$$C_i = C_{W_i} + C_{bc} + (1 - A_v) C_{be}$$

$$= 6\text{ pF} + 36\text{ pF} + (1 - (-90))\, 4\text{ pF}$$

$$= 406\text{ pF}$$

We also know that higher cutoff frequency,

$$f_{H_i} = \frac{1}{2\pi R_{Th_1} C_i}$$

$$= \frac{1}{2\pi \times (0.568 \text{ k}\Omega) \times (406 \text{ pF})}$$

$$= 690.15 \text{ kHz } \textbf{Ans.}$$

For the output circuit of a.c. equivalent circuit (refer to Fig. 21-38),

We know that,

$$R_{Th_2} = R_C \parallel R_L = (4 \text{ k}\Omega) \parallel (2.2 \text{ k}\Omega) = 1.419 \text{ k}\Omega$$

and the output capacitance,

$$C_o = C_{W_o} + C_{ce} + C_{M_o}$$

$$= 8 \text{pF} + 1 \text{pF} + \left(1 - \frac{1}{A_v}\right) C_{bc}$$

$$= 8 \text{pF} + 1 \text{ pF} + \left(1 - \frac{1}{-90}\right) \times 4 \text{ pF}$$

$$= 13.04 \text{ pF}$$

$$\therefore \quad f_{H_o} = \frac{1}{2\pi R_{Th_2} C_o}$$

$$= \frac{1}{2\pi \times (1.419 \text{ k}\Omega)(13.04 \text{ pF})}$$

$$= 8.6 \text{ MHz. } \textbf{Ans.}$$

(*b*) *Values of f_β and f_T*

We know that,

$$f_\beta = \frac{1}{2\pi \, \beta_{mid} \, r'_e (C_{be} + C_{bc})}$$

$$= \frac{1}{2\pi \times 100 \times (15.76 \ \Omega) \times (36 \text{ pF} + 4 \text{ pF})}$$

$$= \frac{1}{2\pi \times 100 \times (15.76 \ \Omega) \times (40 \text{ pF})}$$

$$= 2.5 \text{ MHz } \textbf{Ans.}$$

and

$$f_T = \beta_{mid} \, f_\beta$$

$$= 100 \times (2.52 \text{ MHz})$$

$$= 252 \text{ MHz } \textbf{Ans.}$$

(*c*) *Frequency plot (or dB plot)*

Using the same approach as mentioned in earlier examples, we can obtain the high-frequency response as indicated by the curve DEF shown in Fig. 21-43.

(*d*) *Combined Frequency plot*

Using the results obtained in Example 21-9 for the low frequency region, the complete graph low appears as ABCDEF shown in Fig. 21-43.

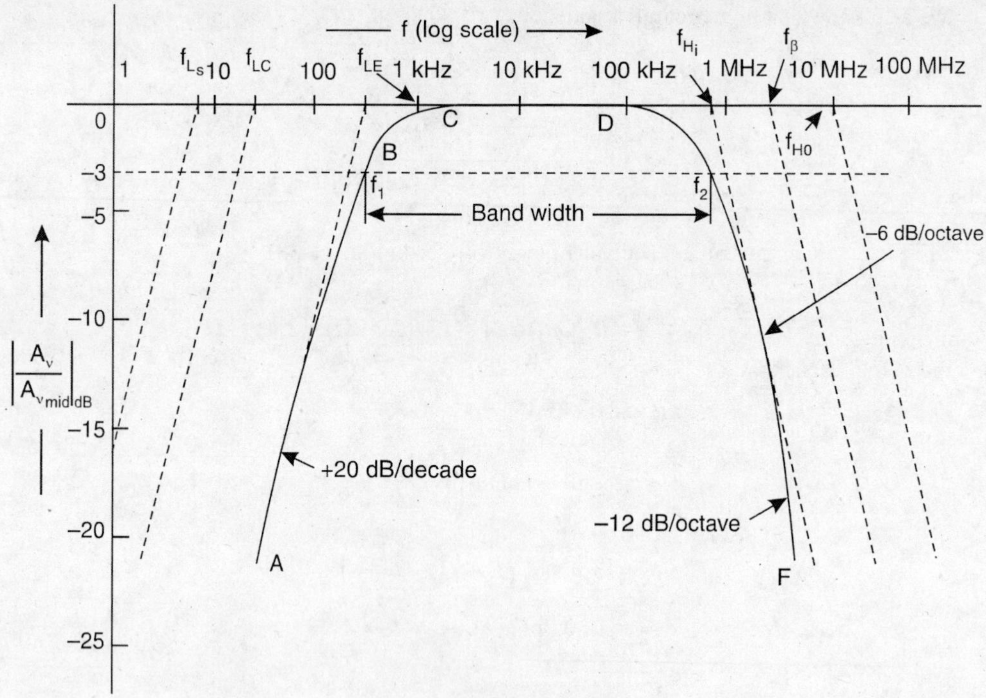

Fig. 21-43.

21-11. High-Frequency Response of FET Amplifier

As a matter of fact, the high-frequency response of the FET amplifier will proceed in the same manner as that for the BJT amplifier. Fig. 21-44 shows a JFET amplifier. As shown there are interelectrode and wiring capacitances that will determine the high-frequency characteristics of the amplifier. The values of capacitors C_{gs} and C_{gd} vary from typically 1 to 10 pF while the capacitance C_{ds} ranges from 0.1 to 1 pF.

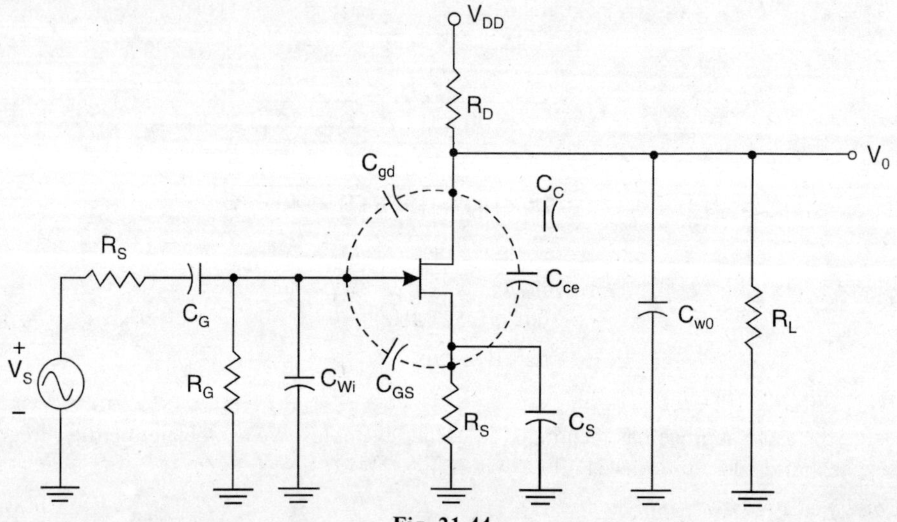

Fig. 21-44.

Notice that the circuit shown in Fig. 21-43, is an inverting amplifier. Therefore, the Miller effect capacitance will appear in the high-frequency a.c. equivalent circuit as shown in Fig. 21-45. At high-frequences, C_i will approach a short-circuit equivalent and V_{gs} will drop in value and reduce the overall gain. At frequencies where C_o approaches its short-circuit equivalent, the parallel output voltage V_o will drop in magnitude.

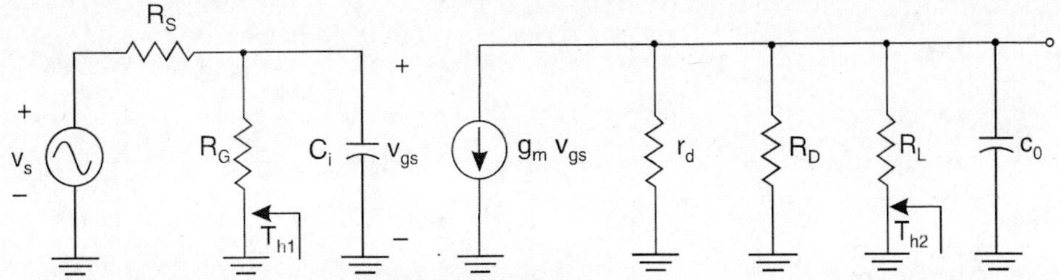

Fig. 21-45.

The cut-off frequency defined by the input circuit can be obtained by the first finding the Thevenin's equivalent circuit of the input section as shown in Fig. 21-46. For this circuit,

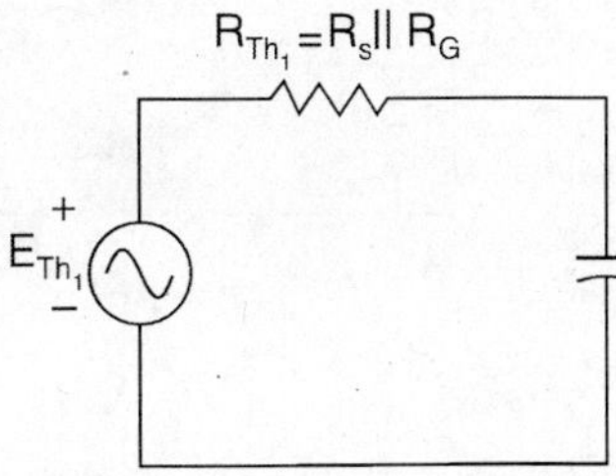

Fig. 21-46.

$$f_{H_i} = \frac{1}{2\pi R_{Th_2} C_i}$$

where, $R_{Th_2} = R_s \parallel R_G$

$$C_i = C_{W_i} + C_{gs} + C_{M_i}$$

and $C_{M_i} = (1 - A_v)\, C_{gd}$

Similarly, for the output circuit, the Thevenin's equivalent circuit is as shown in Fig. 21-47. For this circuit,

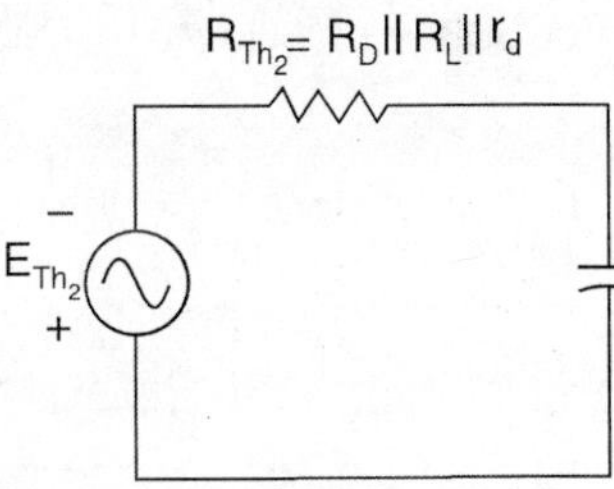

Fig. 21-47.

$$f_{H_o} = \frac{1}{2\pi R_{Th_2} C_o}$$

where, $R_{Th_2} = R_D \parallel R_L \parallel r_d$

$$C_o = C_{W_o} + C_{ds} + C_{M_o}$$

and $C_{M_o} = \left(1 - \frac{1}{A_v}\right) C_{gd}$

21-12. Multistage Frequency Effects

It has been experimentally observed that for a second transistor stage connected directly to the output of the first stage, there will be a significant change in the overall frequency response. In the high frequency region the output capacitance C_o must now include the wiring capacitance (C_{W_i}), parasitic capacitance (C_{be}) and Miller capacitance (C_{M_i}) of the following stage. Further there will be additional low-frequency cut-off levels due to the second stage that will further reduce the overall gain of the system in this region.

It may be carefully noted that for each additional stage the upper cut-off frequency will be determined primarily by that stage having the lowest cut-off frequency. The low-frequency is primarily determined by that stage having the highest low-frequency cut-off frequency. This indicates how one poorly designed stage can offset an otherwise well designed cascaded system.

Let us now study the effect of increasing the number of identical stages. Consider the situation shown in Fig. 21-48. This shows the frequency response of a single stage ($n = 1$), two identical stages ($n = 2$) and three identical stages ($n = 3$). As seen for a single stage, the cut-off frequencies are f_1 and f_2. For two identical stages in cascade, the drop-off rate in the high- and low-frequency regions has increased to –12 dB/octave or – 40 dB/decade. Notice that at f_1 and f_2, the decibel drop is now – 6 dB rather than –3 dB. The –3 dB points have shifted to f'_1 and f'_2 as indicated. Also notice the resulting drop in bandwidth $(f'_2 - f'_1)$.

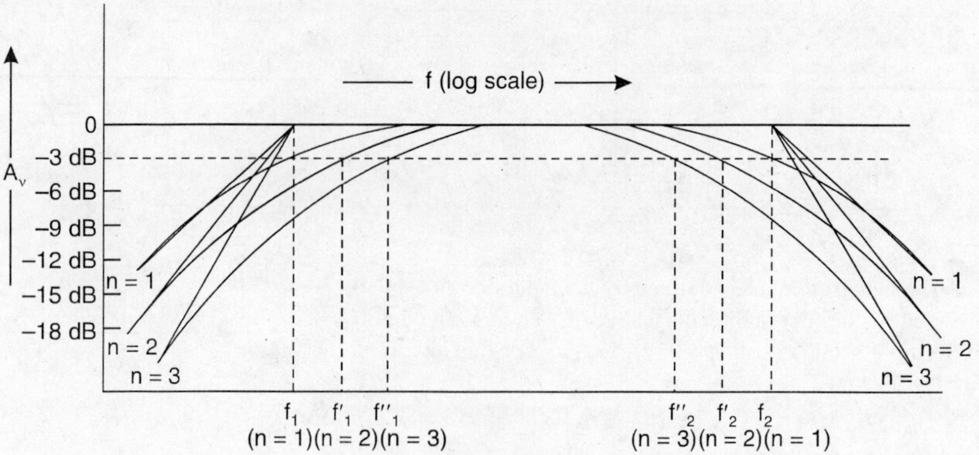

Fig. 21-48.

In the three-stage system of identical stages, f_1 and f_2 will have –18 dB/octave or –60 dB/decade drop-off rate. The decibel drop at f_1 and f_2 is now –9 dB. The –3 dB points have shifted to f_1'' and f_2''. As a result there is a further reduction in bandwidth $(f_2'' - f_1'')$.

Let us now develop and equation for each band frequency as a function of the number of identical stages (n).

For the low-frequency region,

$$A_{v_{low\,(overall)}} = A_{v_{1\,low}}\, A_{v_{2\,low}}\, A_{v_{3\,low}} \,.........\, A_{v_{n\,low}}$$

Since, each stage is identical, therefore,

$$A_{v_{1\,low}} = A_{v_{2\,low}} = A_{v_{3\,low}} = ... = A_{v_{n\,low}} = A_{v_{low}}$$

$$\therefore \quad A_{v_{low\,(overall)}} = (A_{v_{low}})^n$$

or
$$\frac{A_{v_{low}}}{A_{v_{mid}}}(overall) = \left(\frac{A_{v_{low}}}{A_{v_{mid}}}\right)^n$$

$$= \frac{1}{\left(1 - j\left(\frac{f_1}{f}\right)\right)^n} \qquad ...(i)$$

Setting the magnitude of this result is equal to (–3 dB level), *i.e.*, $\frac{1}{\sqrt{2}}$, the equation (i) can be written as,

$$\frac{1}{\left[\sqrt{1 + \left(\frac{f_1}{f_1'}\right)^2}\right]^n} = \frac{1}{\sqrt{2}}$$

or
$$\left\{\left[1 + \left(\frac{f_1}{f_1'}\right)^2\right]^{1/2}\right\}^n = \left\{\left[1 + \left(\frac{f_1}{f_1'}\right)^2\right]^n\right\}^{1/2} = (2)^{1/2}$$

or $$\left[1+\left(\frac{f_1}{f_1'}\right)^2\right]^n = 2$$

or $$1+\left(\frac{f_1}{f_1'}\right)^2 = (2)^{\frac{1}{n}}$$

Rearranging the above equation, we get,

$$f'_1 = \frac{f_1}{\sqrt{2^{1/n}-1}}$$

Similarly, for a high-frequency region, we can show that,

$$f'_2 = \left(\sqrt{2^{1/n}-1}\right)f_2$$

Table 21-2.

No.	$\sqrt{2^{1/n}-1}$
1	1
2	0.64
3	0.51
4	0.43
5	0.39

Table 21-2 lists the magnitude of $\sqrt{2^{1/n}-1}$ for various values of n. As seen from this table, for $n = 2$, the value of $\sqrt{2^{1/n}-1}$ factor is 0.64. This means the upper cut-off frequency,

$$f'_2 = 0.64\,f_2$$

That is 64% of the value obtained for a single-stage. Also the lower cut-off frequency,

$$f'_1 = \frac{1}{0.64}\,f_1 = 1.56\,f_1$$

For $n = 3$ (*i.e.*, 3-stage amplifier), the upper cut-off frequency,

$$f'_2 = 0.51\,f_2$$

and the lower cut-off frequency,

$$f'_1 = \frac{1}{0.51}\,f_1 = 1.96\,f_1$$

SUMMARY

In this chapter, you have learned that :

1. It is more convenient to express the voltage and power gain of electronic amplifiers in decibels (dB).
2. It is more convenient to study the preformance of *RC* coupled, transformer coupled and direct coupled amplifiers from the Bode plot-gain versus frequnecy plot.
3. The lower cut-off frequency due to C_S of a common emitter BJT amplifier is given by,

$$f_{L_S} = \frac{1}{2\pi(R_S + R_i)C_S}$$

and due to C_C,

$$f_{L_C} = \frac{1}{2\pi(R_0 + R_L)C_C}$$

and due to C_E,

$$f_{L_E} = \frac{1}{2\pi R_e C_E}$$

4. The high frequency response of a BJT amplifier is affected by the interelectrode capacitances and the wiring capacitance between the leads of the network. It is also affected by the variation of h_{fe} (or β) with frequency.
5. Increasing the number of stages leads to raising the lower cut-off frequency and lower the upper cut-off frequency. In other words, it leads to reduction in the overall bandwidth.

GLOSSARY

Bel : A term derived from the surname of Alexander Graham Bell.

Decibel : A term defined as 10 decibels = 1 bel.

Interelectrode capacitance : A capacitance that exists between the terminals of an active device (BJT or FET).

NUMERICAL PROBLEMS

1. For the BJT amplifier circuit shown in Fig. 21-49, determine r'_e, $A_{v_{mid}}$, R_i, $A_{v_{s\,mid}}$, f_{L_s}, f_{L_C} and f_{L_E}. Also determine the lower cut-off frequency. Sketch the low frequency response on the Bode plot.

   Given: $C_{W_i} = 5\ pF;\ C_{W_o} = 8\ pF;\ C_{bc} = 12\ pF;\ C_{be} = 40\ pF;\ C_{ce} = 8\ pF.$

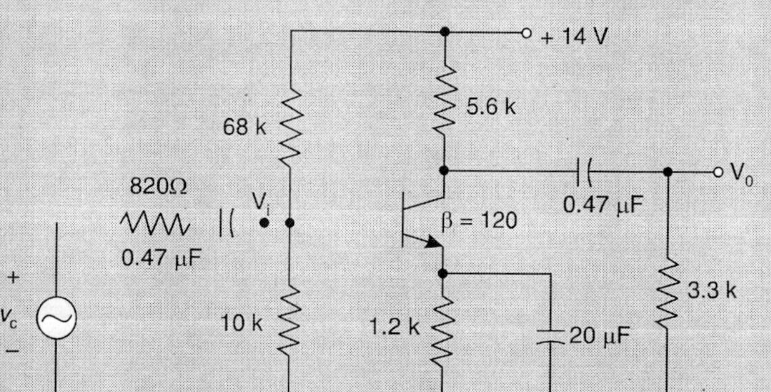

Fig. 21-49.

(**Ans.** $r'_e = 28.5\Omega$, $A_{v_{mid}} = -72.9$, $R_i = 2.5\ k\Omega$, $A_{v_s} = -54.7$, $f_{L_s} = 103.4\ Hz$, $f_{L_C} = 38.1 Hz$ $f_{L_E} = 235.8\ Hz$.)

2. For the circuit shown in Fig. 21-49, determine the values of f_{H_i} and f_{H_o}. Assuming that $C_{b'e} = C_{be}$ and $C_{b'c} = C_{bc}$. Find f_β and f_T. Sketch the frequency response for the high frequency region using a Bode plot and determine the cut-off frequency.

   (**Ans.** $f_{H_i} = 278.6\ kHz$; $f_{H_o} = 1.42\ MHz$.)

DESCRIPTIVE QUESTIONS

1. What are the factors that affect the low frequency response of **(*a*)** BJT amplifier **(*b*)** FET amplifier. Justify with reasons.
2. Derive an expression for the lower cut-off frequency of a BJT amplifier.
3. What is Miller effect capacitance ? Explain with example.
4. Outline the approach to obtain the high frequency response of a BJT amplifier.
5. State Miller's theorem.
   (Electronic Circuits - I, Anna University, April/May 2008, Nov./Dec 2007)
   (Electronic Circuits - VTU June/July 2008)
6. Draw general frequency response curve of an amplifier.
   (Electronic Circuits - I, Anna University, Nov/Dec 2007)
7. Draw the high frequency equivalent of CE amplifier.
   (Electronic Circuits - I, Anna University, April/May 2008)
8. Draw the frequency response of video amplifier.
   (Electronic Circuits - I, Anna University, April/May 2008)

MULTIPLE CHOICE QUESTIONS

1. In a single-stage RC coupled common-emitter amplifier, the phase-shift at the lower 3 dB frequency is,
   (*a*) zero (*b*) 135° (*c*) 180° (*d*) 225°
   (*UPSC Engg. Services* 2001)
2. If the applied a.c. power to a system is 5 μW at 100 mV and the output power is 48 W, the power gain in decibels is ....
   (*a*) 69.8 dB (*b*) 96.8 dB (*c*) 86.9 dB (*d*) 89.6 dB
3. The frequency that is 1 octave above 5 kHz is
   (*a*) 10 kHz (*b*) 1 kHz (*c*) 20 kHz (*d*) 100 kHz
4. The frequency that is 2 decade above 1 kHz is
   (*a*) 10 kHz (*b*) 100 kHz (*c*) 1 MHz (*d*) none of these
5. The normalized gain of an amplifier is defined as
   (*a*) A_v (*b*) $A_{v_{mid}}$ (*c*) $A_v/A_{v_{mid}}$ (*d*) $10 \log A_v/A_{v_{mid}}$

ANSWERS

1. (*b*) **2.** (*a*) **3.** (*a*) **4.** (*b*) **5.** (*c*)

❑ ❑ ❑

22

CHAPTER

Large Signal (or Power) Amplifiers

OUTLINE

Objectives

After completing this chapter, you should be able to :

- Compare voltage amplifier and power amplifier.
- Understand the parameters of a transistor, which determines the performance of power amplifiers.
- Draw the a.c. load line of a power amplifier circuit.
- Describe class-A and class -B amplifiers.
- Calculate the collector efficiency and power rating of a transistor in class A configuration.
- Understand the operation of a transformer coupled class-A amplifier.
- Explain the operation of a class-B push-pull amplifier.
- Understand the characteristics of a class-C amplifier.

22-1. Introduction

A ***power amplifier is an amplifier, which is capable to providing a large amount of power to the load such as loudspeaker, or motor,*** *etc*. It is essential in almost all electronic systems, where a large amount of power is required to be supplied to the load. The power amplifier, is used as a last stage in as electronic system. For example, a *public address system* (PAS) consists of a microphone, a multistage amplifier, a power amplifier and a loudspeaker. The microphone converts the sound waves into electrical signal, which is of very low voltage (usually of few millivolts). This signal is *insufficient* to drive the loudspeaker. Therefore, this signal is first raised to a sufficiently high value (a few volts) by passing it through a multistage small-signal (or voltage) amplifier. This signal is then used to drive the power amplifier, because it is incapable of delivering a large amount of power to the loudspeakers.

A power amplifier is more commonly known *as audio amplifier*. The audio amplifiers are used in public address systems, tape recorders, stereo systems, television receivers, radio receivers, broadcast transmitters, *etc*. It will be interesting to know that *a power amplifier does not actually amplify the power*. As a matter of fact, *it takes power from the d.c. power supply connected to the output circuit and converts it into useful a.c. signal power*. The power is fed to the load. The type of a.c. power developed, at the output of a power amplifier, is controlled by the input signal. Thus we can say that actually a power amplifier is a *d.c. to a.c. power converter,* whose action is controlled by the input signal.

The power amplifiers, are also known as *large signal amplifiers*. The term 'large signal' for the power amplifiers arises because these amplifiers use a large part of their a.c. load line for operation. It is in contrast to the small signal amplifiers, which use only 10% of their a.c. load line for operation. The small signal amplifiers are commonly known as voltage amplifiers.

22-2. Difference Between Voltage Amplifier and Power Amplifier

We have already discussed in the last article that a small signal amplifier is called voltage amplifier and a large signal amplifier as a power amplifier. It will be interesting to know that a voltage amplifier produces an increased voltage into a high resistance load. On the other hand, power amplifier produces a large increase in output power or an appreciable amount of current flow into a relatively low resistance load. Following are some of the important differences in the construction of a voltage and power amplifier circuit :

(*i*) The transistor used in voltage amplifier has a large value of current gain (β) – typically 100 as compared to that of a power amplifier. The current gain (β) of a transistor used in power amplifier is, usually, between 20 and 50.

(*ii*) In a voltage amplifier, the input resistance of a transistor is required to be quite low as compared to its output resistance. On the other hand, in a power amplifier, the input resistance of a transistor is generally very large as compared to its output resistance.

(*iii*) The *physical size* of the transistor used in voltage amplifiers is usually small and is known as low or medium-power transistor. On the other hand, power amplifiers employ larger size transistors called *power transistors*. Such transistors are needed because a considerable amount of heat is dissipated within the transistor and hence, a large surface area is required for heat dissipation.

(*iv*) The voltage amplifiers, usually employ *RC coupling* for interstage connection, whereas power amplifiers use *transformer coupling*.

22-3. Performance Parameters

As a matter of fact, there are three parameters namely *circuit efficiency*, *distortion and power dissipation* capability of a transistor, which determines the performance of power amplifiers. These are discussed as below :

(1) *Circuit efficiency* : It is also called the overall efficiency or conversion efficiency of the power amplifier. The major criterion for a power amplifier is not the power gain, but the maximum a.c. power output. We have already, discussed that an amplifier is basically an energy convertion device, *i.e.*, it converts d.c. power supply into a.c. power output. Therefore, the ability of a power amplifier, to convert d.c. power supply from the source into a.c. power is a measure of its effectiveness and is called its ***circuit efficiency***.

Thus, circuit efficiency may be defined as the maximum a.c. output power to the d.c. input power. Mathematically, the circuit efficiency,

$$\eta = \frac{\textbf{Maximum a.c. output power}}{\textbf{d.c. input power}}$$

The maximum theoretical values for circuit efficiency depend upon the way in which the load is coupled to the transistor and the class of operation of the amplifier. Its value may lie, anywhere, from 25% to 90%.

(2) *Distortion* : A good amplifier should produce an output, which does not differ from the input in any respect, except amplitude. In other words, the amplifier output is expected to be enlarged but faithful reproduction of the input. However, in actual practice, it is not possible to construct such an ideal amplifier, whose output is exact reproduction of the input. The output is always found to be different from the input either in its waveform or frequency content. This difference between the output and input of an amplifier is called ***distortion***.

Two types of distortions namely *amplitude* (or harmonic) *distortion* and *crossover distortion* are considered to be important in power amplifiers. The amplitude distortion results from the *non-linearity* of the transistor. It occurs because of the fact that transistor output may not increase, equally, for all portions of the input signal during positive and negative half cycles. On the other hand, crossover distortion (which usually exists in class B amplifiers) occurs when transistors *do not operate in the correct phase* with each other.

(3) *Power dissipation capability* : The ability of power transistor to dissipate heat is known as *power dissipation capability* or *power rating*. The heat is generated within the transistor due to excessive current passing through it during amplification. If the heat generated, within the transistor, is greater than its power dissipation capability, it is likely to burn out. The power dissipation capability of a transistor may be increased by connecting a mass of metal (called *heat sink*) to the transistor case. The increased surface area at the transistor case allows the heat to escape more easily into the surrounding air. It keeps the case temperature within permissible limits.

22-4. A.C. Load Line

We know that every amplifier experiences two types of loads namely a d.c. load and an a.c. load. Consequently, the amplifier has two types of load lines, *i.e.*, d.c. load line and a.c. load line. We have already discussed about d.c. load line in chapter on transistor biasing circuits. In this chapter, we shall discuss about a.c. load line.

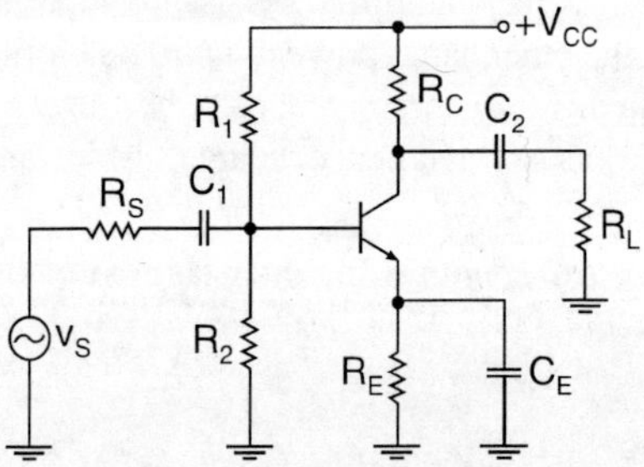

(*a*) Common emitter amplifier circuit.

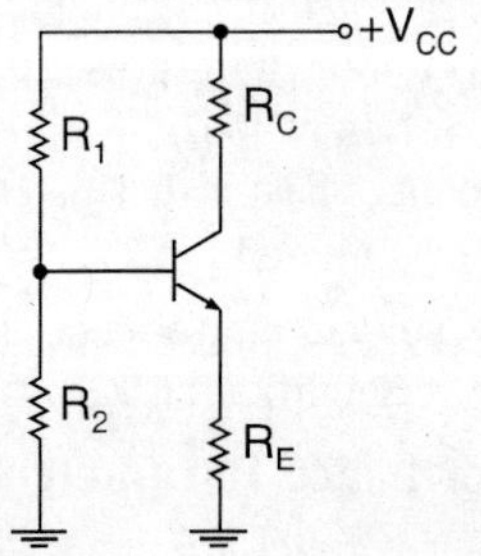

(*b*) d.c. equivalent circuit.

Fig. 22-1.

Consider a common emitter amplifier circuit as shown in Fig. 22-1 (*a*). Its d.c. equivalent circuit is shown in Fig. 22-1 (*b*). With the help of this circuit, we can determine the upper and lower ends of the d.c. load line. The upper end is given by the relations,

$$V_{CE} = 0 \text{ and } I_C = \frac{V_{CC}}{R_C + R_E}$$

It is shown in Fig. 22-2 (*a*) as point *A*. Similarly, the lower end is given by the relations,

$$I_C = 0 \text{ and } V_{CE} = V_{CC}$$

It is shown in Fig. 22-2 (*a*) as point *B*. Joining the points *A* and *B*, by a straight line, we get the d.c. load line.

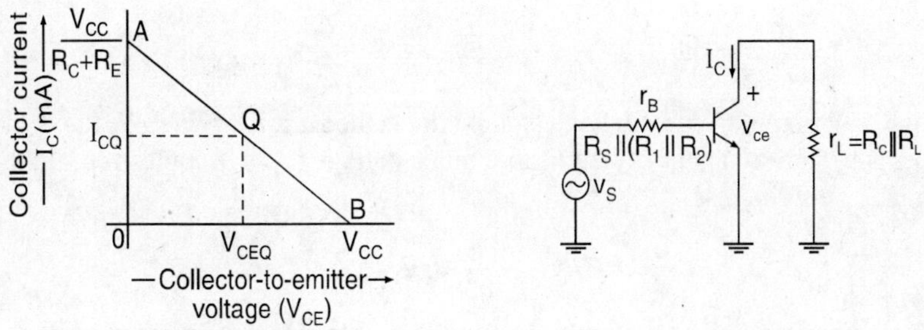

(*a*) D.C. load line. (*b*) A.C. eqvivalent.

Fig. 22-2.

The d.c. equivalent circuit may also be used to determine the *Q*-point values of collector current (I_{CQ}) and collector-to-emitter voltage (V_{CEQ}). We know that voltage drop across the resistor R_2,

$$V_{R2} = V_{CC}\left(\frac{R_2}{R_1 + R_2}\right)$$

∴ *Q*-point emitter current,

$$I_{EQ} = \frac{V_{R2} - V_{BE}}{R_E}$$

and the *Q*-point collector current,

$$I_{CQ} = I_{EQ}$$

The *Q*-point collector-to-emitter voltage,

$$V_{CEQ} = V_{CC} - I_{CC}(R_C + R_E)$$

Fig. 22-2 (*b*) shows the a.c. equivalent circuit of the common emitter amplifier. In this Fig. the resistance driving the base,

$$r_B = R_S \parallel (R_1) \parallel R_2)$$

and a.c. load resistance seen by the collector,

$$r_L = R_C \parallel R_L$$

Now summing up the voltages around the collector loop of the a.c. equivalent circuit,

$$v_{ce} + i_c \cdot r_L = 0$$

$$\therefore \quad i_c = -\frac{v_{ce}}{r_L} \quad \text{...}(i)$$

where, i_c = A.C. collector current, and

v_{ce} = A.C. collector-to-emitter voltage,

Now let I_{CQ} = The d.c. collector current at *Q*-point

V_{CEQ} = The d.c. collector-emitter voltage at *Q*-point.

$$I_{C\,(sat)} = \text{The d.c. saturation current, and}$$

$$V_{CE\,(cut\text{-}off)} = \text{The d.c. cut-off voltage.}$$

Then the a.c. collector current,

$$i_c = \Delta I_C = I_C - I_{CQ}$$

and the a.c. collector-to-emitter voltage,

$$v_{ce} = \Delta V_{CE} = V_{CE} - V_{CEQ}$$

Substituting the values of i_c and v_{ce} in equation (*i*),

$$I_C - I_{CQ} = -\frac{V_{CE} - V_{CEQ}}{r_L} = -\frac{V_{CE}}{r_L} + \frac{V_{CEQ}}{r_L}$$

or

$$I_C = I_{CQ} + \frac{V_{CEQ}}{r_L} - \frac{V_{CE}}{r_L} \qquad ...(ii)$$

The above equation is an equation of a straight line called ***a.c. load line***. We know that when the transistor goes into saturation, the collector-to-emitter voltage (V_{CE}) becomes zero. In that case, the equation (*ii*) may be written as

$$I_C = I_{CQ} + \frac{V_{CEQ}}{r_L} = I_{C(sat)}$$

The above expression gives the upper end of the a.c. load line. Similarly, when the transistor goes into cut-off, the collector current (I_C) becomes zero. In that case, the equation (*ii*) may be written as

$$0 = I_{CQ} + \frac{V_{CEQ}}{r_L} - \frac{V_{CE}}{r_L}$$

$$= I_{CQ} + \frac{V_{CEQ}}{r_L} - \frac{V_{CE(cut-off)}}{r_L} \qquad ...(\because V_{CE} + V_{CE\,(cut\text{-}off)} \text{ at } I_C = 0)$$

or

$$V_{CE\,(cut\text{-}off)} = V_{CEQ} + I_{CQ} \cdot r_L$$

The above equation gives us the lower end of a.c. load line. By joining the upper and lower ends, we get a complete a.c. load line as shown in Fig. 22-3 (*a*). The locus of a.c. load line gives all the possible a.c. operating points. At any instant, during a.c. cycle, the operating point of the transistor is, somewhere, along the a.c. load line. The exact location may be determined by the amount of change from the *Q*-point.

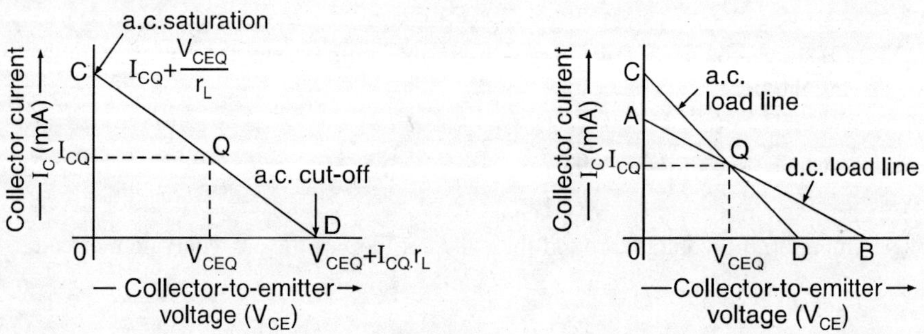

(*a*) A.C. load line.

(b) A.C. and D.C. load lines.

Fig. 22-3.

Fig. 22-3 (*b*) shows the a.c. load line superimposed on the d.c. load line. It may be noted that the slope of a.c. load line (equal to $1/r_L$) is greater than that of d.c. load line. Both the load lines intersect at *Q*-point.

Notes : 1. The a.c. load line for emitter follower (*i.e.*, common collector amplifier), common base amplifier and swamped amplifier may be obtained by a derivation almost identical to that given for the common emitter amplifier.

2. For emitter follower, the saturation value of current,

$$I_{C\,(sat)} = I_{CQ} + \frac{V_{CEQ}}{r_E}$$

where, r_E is the a.c. equivalent load resistance at the emitter (*i.e.*, equal to $R_E \parallel R_L$). The a.c. collector-to-emitter cut-off voltage,

$$V_{CE\,(cut\text{-}off)} = V_{CEQ} + I_{CQ} \cdot r_E$$

3. For a common base amplifier, the values of $I_{C(sat)}$ and $V_{CE(cut\text{-}off)}$ are approximately the same as those given for a common emitter amplifier.

4. For swamped amplifier, the value of saturation current is given by the relation.

$$I_{C(sat)} = I_{CQ} + \frac{V_{CEQ}}{r_E + r_L}$$

where, r_L = A. C. equivalent load resistance and

r_E = Unbypassed emitter resistance.

The a.c. collector-to-emitter cut-off voltage is given by the relation,

$$V_{CE\,(cut\text{-}off)} = V_{CEQ} + I_{CQ}\,(r_E + r_L)$$

Example 22-1. *Fig. 22-4 shows the circuit of a common emitter amplifier.*

Determine the values of collector current and collector-to-emitter voltage at the points of saturation and cut-off under a.c. signal operation. Also draw the a.c.load line. Assume V_{BE} = 0.7 V.

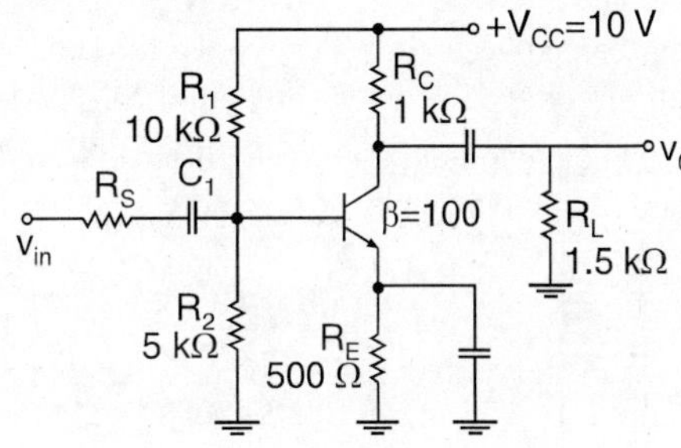

Fig. 22-4.

Solution. V_{CC} = 10 volts; R_1 = 10 kΩ; R_2 = 5 kΩ; R_C = 1 kΩ = 1000 Ω; R_E = 500 Ω; R_L = 1.5 kΩ = 1500 Ω; β = 100 and V_{BE} = 0.7 V.

Collector current and collector-to-emitter voltage

First of all, we shall determine the *Q*-point values of collector current and collector-to-emitter voltage for the amplifier current. We know that voltage drop across resistor R_2,

$$V_{R_2} = V_{CC}\left(\frac{R_2}{R_1 + R_2}\right) = 10\left(\frac{5}{10+5}\right) = 3.33 \text{ V}$$

and the value of emitter current,

$$I_{EQ} = \frac{V_{R_2} - V_{BE}}{R_E} = \frac{3.33 - 0.7}{500} \text{ A} = 5.26 \times 10^{-3} \text{ A} = 5.26 \text{ mA}$$

∴ Value of collector current,

$$I_{CQ} = I_{EQ} = 5.26 \times 10^{-3} \text{ A} = 5.26 \text{ mA}$$

We also know that the collector-to-emitter voltage,

$$V_{CEQ} = V_{CC} - I_{CQ}\,(R_C + R_E)$$
$$= 10 - (5.26 \times 10^{-3}) \times (1000 + 500) = 2.11 \text{ V}$$

Thus, the *Q*-point is located at I_C = 5.26 mA and V_{CE} = 2.11 V as shown in Fig. 22-5.

Now we shall find the values of collector current and collector-to-emitter voltage at saturation and cut-off points. We know that at saturation point, the collector current,

$$I_{C\,(sat)} = I_{CQ} + \frac{V_{CEQ}}{r_L}$$

$$= 5.26 \times 10^{-3} + \frac{2.11}{1000 \parallel 1500} \qquad ...(\because r_L = R_C \parallel R_L)$$

$$= 8.78 \times 10^{-3}\ \text{A} = 8.78\ \text{mA}\ \textbf{Ans.}$$

and the collector-to-emitter voltage,

$$V_{CE\,(sat)} = 0\ \textbf{Ans.}$$

Similarly, the collector current at cut-off point

$$I_{C\,(cut\text{-}off)} = 0\ \textbf{Ans.}$$

and the collector-to-emitter voltage,

$$V_{CE\,(cut\text{-}off)} = V_{CEQ} + I_{CQ} \cdot r_L$$

$$= 2.11 + (5.26 \times 10^{-3}) \times 600\ \text{V}$$

$$= 5.27\ \text{V}\ \textbf{Ans.}$$

A.C. load line

The a.c. load line may be obtained by plotting the points of a.c. saturation and a.c. cut-off and then joining them by a straight line as shown in Fig. 22-5.

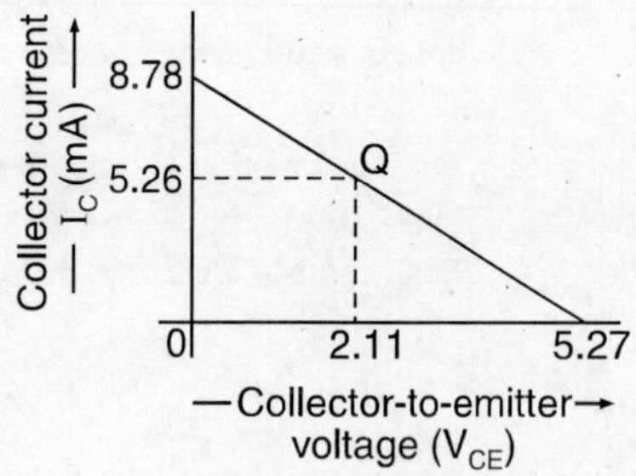

Fig. 22-5. *A.C. load line.*

22-5. What does the a.c. load line tell us ?

The a.c. load line is used to tell us the maximum output voltage swing for a given common-emitter amplifier. In other words, the a.c. load line will tell us the maximum possible peak (V_{PP}) output voltage from a given amplifier. The maximum V_{PP} is know as *compliance* of the amplifier.

The compliance of an amplifier is found by obtaining the maximum values of I_C and V_{CE} from there respective values of I_{CQ} and V_{CEQ}. The maximum possible variations are illustrated in Fig. 22-6.

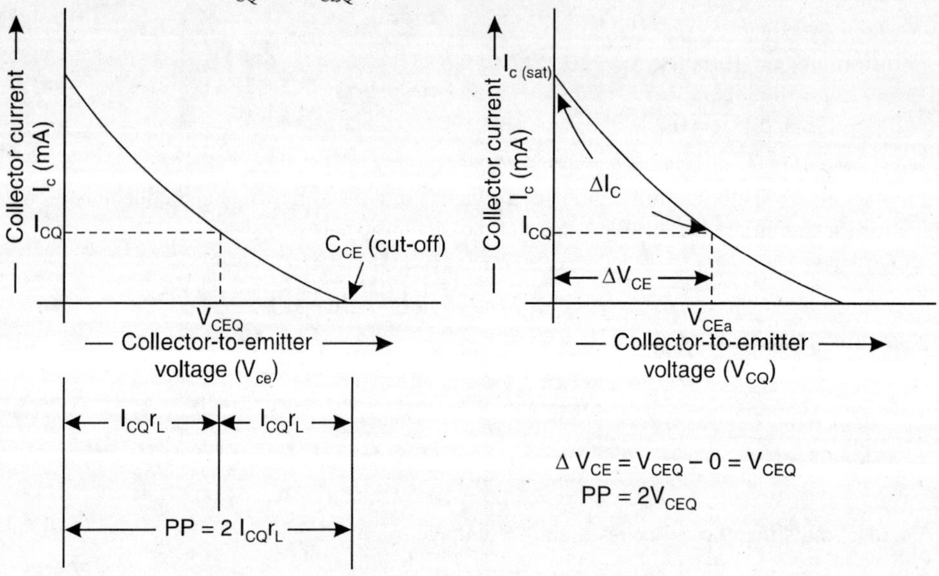

Fig. 22-6.

As seen from this figure, the maximum possible variation for V_{CE} is equal to the difference between $V_{CE\,(\text{cut-off})}$ and V_{CEQ}. Since, this variation is equal to $I_{CQ}\ r_L$, the maximum peak output voltage from the amplifier is equal to $I_{CQ}\ r_L$. Two times this value will gives us the maximum peak-to-peak variation of the output voltage. Thus,

$$PP = 2\ I_{CQ}\ r_L \qquad ....(i)$$

$$= 2\ V_{CEQ} \qquad ....(ii)$$

It may be noted that having two different equations (*i*) and (*ii*), may seam a bit confusing at first, but it really is not so strange. The equation (*i*) sets the limit in terms of $V_{CE(cut\text{-}off)}$. In other words, if we expected the value obtained by this equation, the output voltage will try to exceed $V_{CE(cut\text{-}off)}$ which it cannot do. The result of exceeding the value obtained by equation (*i*) is shown in Fig. 22-7 (*a*), this is called cut-off clipping, because the output voltage is clipped off at the value of $V_{CE(cut\text{-}off)}$.

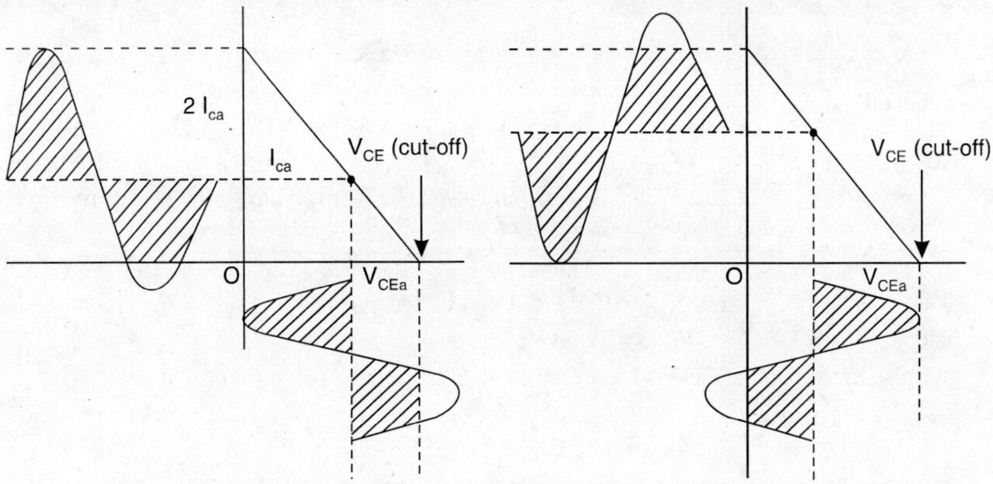

(*a*) Cut - off clipping (*a*) Saturation clipping

Fig. 22-7.

Similarly, equation (*ii*) sets the limit in terms of $I_{C(sat)}$. If we exceed the value obtained by this equation, the output will experience saturation clipping Refer to Fig. 22-7 (*b*).

So the point is that when determining the output compliance for a given amplifier, solve both the equations (i) and (*ii*). The lower of the two results is the compliance of the amplifier.

Example 22-2. *A common emitter amplifier shown in Fig. 22-8 has V_{CC} = 20 V, R_1 = 10 kΩ, R_2 = 1.8 kΩ, R_C = 620 Ω, R_E = 200 Ω, R_L = 1.2kΩ and h_{FE} = 180. Determine the compliance (PP) of the amplifier.*

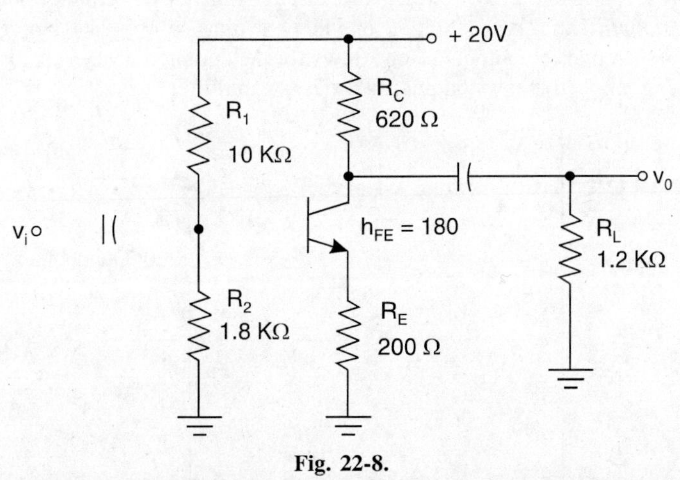

Fig. 22-8.

Solution. In order to find the compliance (*i.e.*, maximum peak-to-peak voltage), we need to calculate I_{CQ} and V_{CEQ}. We can follow the same approach as given in Example 22-4 Alternatively follow as below :

We know that base voltage,

$$V_B = V_{CC} \times \frac{R_2}{R_1 + R_2} = 20 \times \frac{1.8\ k}{10\ k + 1.8\ k} = 3.05\ V$$

and voltage at the emitter,

$$V_E = V_B - V_{BE} = 3.05 - 0.7 = 2.35\ \text{V}$$

∴ Value of emitter current,

$$I_E = \frac{V_E}{R_E} = \frac{2.35}{200} = 0.01175\ \text{A or } 11.75\ \text{mA}$$

and collector current,

$$I_C \approx I_E = 11.75\ \text{mA}$$

We also known that the voltage from collector-to-emitter,

$$V_{CE} = V_{CC} - I_E (R_C + R_E) = 20 - (0.01175) \times (620 + 200)\ \text{V}$$
$$= 20 - 9.635\ \text{V7}$$
$$= 10.365\ \text{V}$$

Thus, $I_{CQ} = 0.01175$ A or 11.75 mA

and $V_{CEQ} = 10.36$ V

We also know that the a.c. load resistance,

$$r_L = R_C \,||\, R_L = 620 \,||\, 1.2\ \text{k}$$
$$= 408.8\ \text{W}\Omega$$

Now let us solve for both the compliance values,

$$PP = 2\, I_{CQ}\, r_L = 2 \times 0.01175 \times 408.8 = 9.61\ \text{V}$$
$$PP = 2\, V_{CEQ}\, r_L = 2 \times 10.365 = 20.73\ \text{V}$$

Since, the overall compliance is always equal to smaller of the two values obtained, therefore the maximum *PP* for the amplifier show in Fig 22-8 is 9.61 V. If the value is exceeded, cut-off clipping will occur.

22-6. Classification of Power Amplifiers

The classification of power amplifiers is based on transistor biasing condition and amplitude of the input signal. It also takes into account the portion of the cycle for which the transistor conducts. On the above basis, following are the three main classifications of power amplifiers :

(1) ***Class-A amplifier :*** It is an amplifier in which the transistor bias and amplitude of the input signal is such that the output current flows for the complete cycle (*i.e.*, 360°) of the input signal. Fig. 22-9 (*a*) shows output for a class-A amplifier.

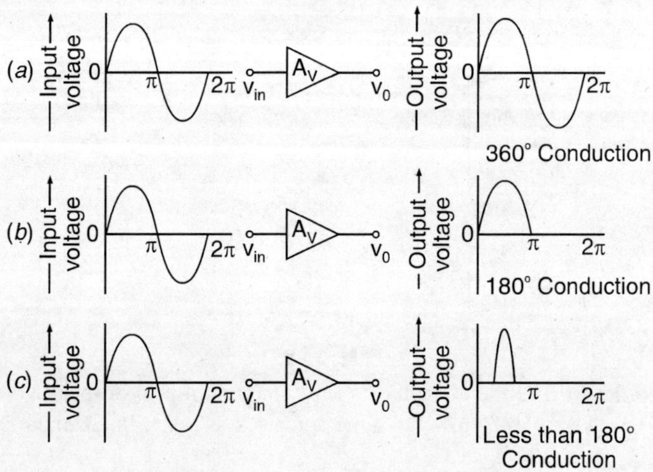

Fig. 22-9. *Classifications of power amplifiers.*

(2) ***Class-B amplifier:*** It is an amplifier in which the transistor bias and amplitude of the input signal are such that the output current flows for only one half cycle (*i.e.*, 180°) of the input signal. Fig. 22-9 (*b*) shows the output for a class-B amplifier.

(3) ***Class-C amplifier:*** It is an amplifier in which the transistor bias and amplitude of the input signal are such that output current flows for less than half cycle (*i.e.*, less than 180°) of the input signal. Fig. 22-9 (*c*) shows the output for a class-C amplifier.

In addition to the above three main types, there is another amplifier called class-AB amplifier. The characteristics of such an amplifier lies in-between those of class-A and class-B amplifiers. Thus in a class-AB amplifier, the output current flows for more than half cycle but less than the complete cycle (*i.e.*, current flows for more than 180° but less than 360°). The class-A and class-B amplifiers are called *audio power amplifiers* because they are capable of amplifying the signals with frequencies ranging from 20 Hz to 20 kHz.

22-7. Class-A Amplifier

The class-A amplifier is, basically, a common emitter amplifier as shown in Fig. 22-10. This circuit is called direct coupled class-A power amplifier, in order to distinguish it from a transformer coupled power amplifiers. The only difference between this circuit and the small signal version, considered earlier, is that the signals handled by the power amplifier circuit are in the range of volts. And the transistor used is a power transistor, capable of operating in the range of few watts. In the class-A amplifier, the transistor bias and amplitude of the input signal are such that the output current flows for the complete cycle (*i.e.*, 360°) of the input signal. This condition is achieved by locating the *Q*-point somewhere near the centre of the load line. However, in order to obtain the maximum output signal, the *Q*-point is set at the centre of the load line.

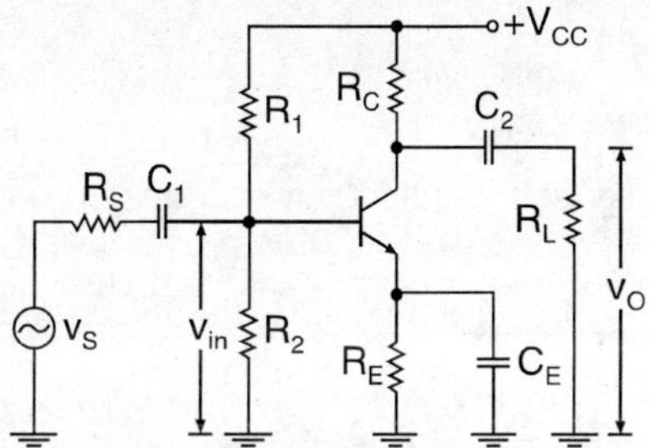

Fig. 22-10. *Circuit for class-A amplifier.*

Fig. 22-11 shows the a.c. load line and the variation of collector current and the collector-to-emitter voltage above and below their *Q*-point values for maximum output. It may be noted from this Fig. that ideally the collector current can vary from its *Q*-point value (*i.e.*, I_{CQ}) to its saturation value (*i.e.*, $I_{C\,(sat)}$) = 2 I_{CQ} and down to its cut-off value (*i.e.*, zero). The maximum or peak value of the collector current is I_{CQ} and that of the collector-to-emitter voltage is V_{CEQ}. It is the largest signal possible from a class-A amplifier. If the signal is too large, the amplifier is driven further. In that case, the amplifier will clip at cut-off and saturation.

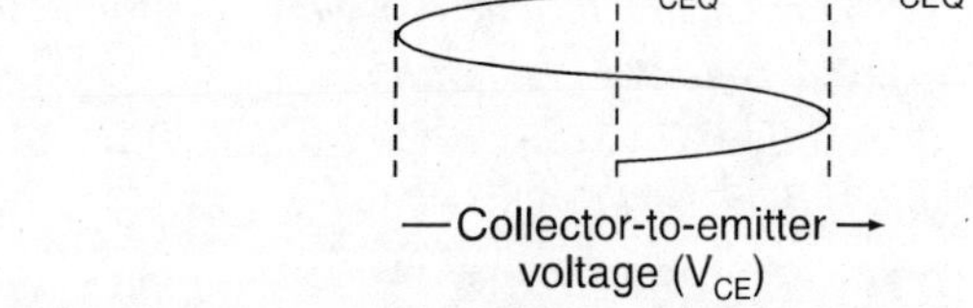

Fig. 22-11. *Maximum output for class-A amplifier.*

Now we shall obtain the expressions for amplifier parameters such as voltage gain, current gain and power gain for class-A amplifier.

Voltage gain : The voltage gain for a class-A power amplifier may be obtained in the same way as discussed for a small-signal amplifier in chapter 26. It is given by the relation,

$$A_v = \frac{r_L}{r'_e}$$

where, r_L = A.C. load resistance whose value is equal to the parallel combination of collector resistance (R_C) and load resistance (R_L), *i.e.*, $r_L = R_C \parallel R_L$. If R_L is not connected then $r_L = R_C$.

r'_e = A.C. emitter diode resistance.

It will be interesting to know that the formula for a.c. emitter diode resistance, $r'_e = 25/I_E$, where I_E is in milliamperes is not void for large signal or power amplifier.

The large signal a.c. emitter diode resistance is determined graphically from the transconductance curve as shown in Fig. 22-12. It is done by using the relation,

$$r'_e = \frac{\Delta V_{BE}}{\Delta I_C}$$

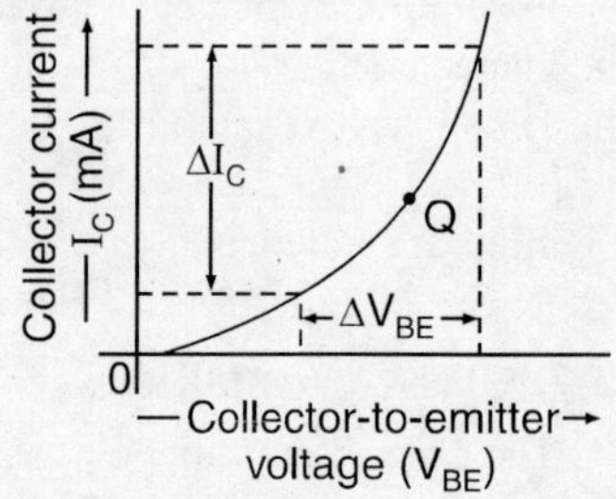

Fig. 22-12. *Transconductance curve.*

Current gain : The current gain of a transistor is the ratio of a.c. collector current (i_c) to the a.c base current (i_b). Mathematically, the current gain,

$$A_i = \frac{i_c}{i_b} = \beta$$

Power gain : We know that the a.c. input power to the base of transistor,

$$p_{in} = v_{in} \cdot i_b$$

and the a.c. output power from the collector,.

$$p_o = -v_o \cdot i_c$$

The negative sign in the above equation indicates that the phase of input signal is reversed at the output.

∴ Power gain,

$$A_p = \frac{p_o}{p_{in}} = \frac{-v_o \cdot i_c}{v_{in} \cdot i_b} = -\frac{v_o}{v_{in}} \times \frac{i_c}{i_b}$$

$$= -A_v \cdot A_i = -\left(\frac{r_L}{r'_e}\right) \times \beta$$

where, A_v = Voltage gain, and

A_i = Current gain.

Example 22-3. *Find the large-signal voltage gain and power gain of the amplifier shown in Fig. 22-13.*

Assume that the value of r'_e has been found to be 8Ω from the graphical data.

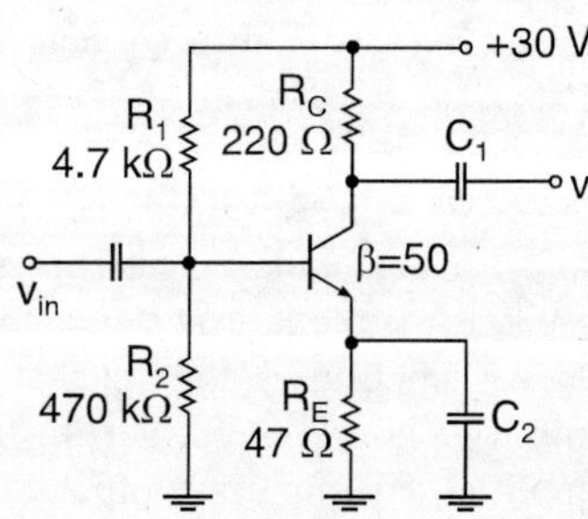

Fig. 22-13.

Solution. Given : $r'_e = 8\Omega$; $R_C = 220\ \Omega$; $R_E = 47\ \Omega$; R_1 = 4.7 kΩ; $R_2 = 470\ \Omega$ and $\beta = 50$.

We know that a.c. load resistance,

$$r_L = R_C \parallel R_L = R_C = 220\ \Omega$$

and voltage gain,

$$A_v = \frac{r_L}{r'_e} = \frac{200}{8} = 27.5$$

We also know that current gain,

$$A_i = \beta = 50$$

∴ Power gain,

$$A_p = A_v \cdot A_i = 22\text{-}5 \times 50 = 1375 \textbf{ Ans.}$$

22-8. Characteristics of Class-A Amplifier

Though the class-A amplifiers have a number of characteristics, yet the following are important from the subject point of view :

- The output current flows during the entire cycle of the a.c. input signal.
- The operation of the amplifier is restricted to smaller central region of the load line, so that it can operate in the linear region of the load line. The large-signals may shift the *Q*-point into non-linear regions near saturation or cut-off and hence produce amplitude distortion.
- Since the transistor operates over the linear region of the load line, therefore the output waveform is almost similar to the input waveform.
- The a.c. power output per active device (or transistor) is smaller than that of class-B or class-C amplifier.
- The overall efficiency or circuit efficiency of the amplifier circuit is an important parameter. It is defined as the ratio of a.c. power delivered to the load to the total power supplied by the d.c. source. Mathematically, the overall efficiency,

$$\eta_o = \frac{\text{A.C. power delivered to the load}}{\text{Total power supplied by the d.c. source}}$$

$$= \frac{\text{Average a.c. output power}}{\text{Average d.c. input power}}$$

  The maximum possible overall efficiency of a class-A amplifier with series fed resistive load is 25%.

- The collector efficiency of the amplifier circuit is another important parameter. It is defined as the ratio of a.c. power delivered to the load, to the power supplied by the d.c. source to the transistor. Mathematically, collector circuit efficiency,

$$\eta_C = \frac{\text{A.c. power delivered the load}}{\text{Power supplied by the d.c. source to the transistor}}$$

$$= \frac{\text{Average a.c. output power}}{\text{Average d.c. input power to the transistor}}$$

  The maximum possible value of collector efficiency for a series fed resistive load is 50%.

- If we use a transformer coupled load instead of a direct coupled resistive load, the maximum possible overall efficiency increases to 50%.

22-9. Power Relations for Class-A Amplifier

Fig. 22-14 (*a*) shows the output circuit of a single stage Class-A amplifier. In this circuit, the load itself is shown connected as a collector resistor. When the load is connected to the amplifier circuit in this configuration, it is referred to as direct coupled load.

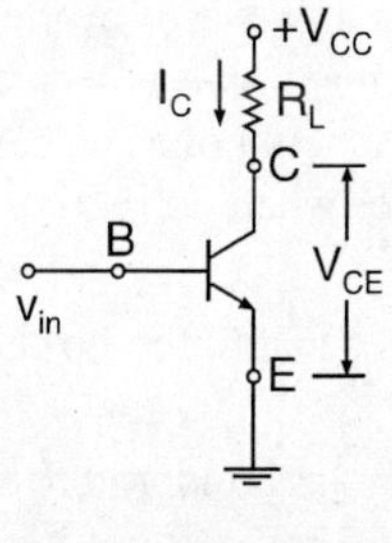

(*a*) Output-circuit of a class-A amplifier.

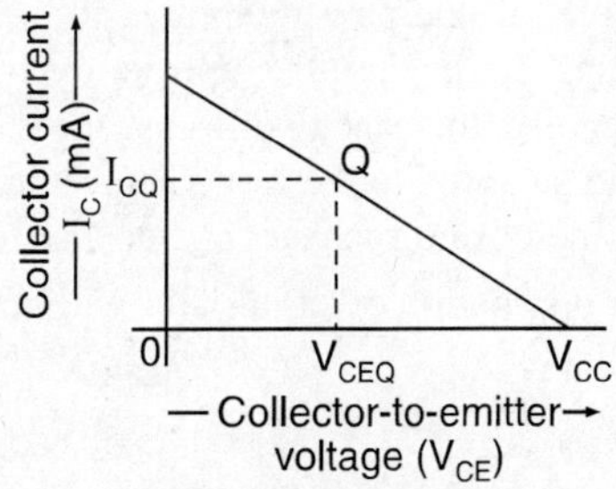

(*b*) A.C. load line.

Fig. 22-14.

Fig. 22-14 (*b*) shows its a.c. load line with *Q*-point located at its centre. We know that when an a.c. input signal is applied, the *Q*-point shifts up and down from its centre position. Accordingly, the output current will increase or decrease from its quiescent value (*i.e.*, I_{CQ}). Similarly, the collector-to-emitter voltage (V_{CE}) will also increase or decrease from its quiescent value (*i.e.*, V_{CEQ}). It may be noted that a long as the signal variations are limited to the linear region of the load line, average value of the collector current is I_{CQ}. It is true because the positive and negative input signal swings produce equal changes in I_{CQ}.

The total average power supplied to the output circuit by the collector (or V_{CC}) supply is given by the relation,

$$P_{in\,(dc)} = V_{CC} \cdot I_{CQ} = 2\,V_{CEQ} \cdot I_{CQ} \qquad ...(\because V_{CC} = 2\,V_{CEQ})$$

This power is utilized within the load resistor and transistor in the following ways :

(*i*) The power consumed in the load resistor (R_L) whose value is given by the relation,

$$P_{RL\,(dc)} = (I_{CQ})^2 \cdot R_L$$

It is dissipated in the form of heat.

(*ii*) The power consumed within the transistor, whose value is given by the relation,

$$P_{tr\,(dc)} = P_{in\,(dc)} - P_{RL\,(dc)}$$

The power consumed by the transistor is further utilized in the following two parts :

(*a*) A part of power is utilized to develop a.c. power across the load resistor. It constitutes the useful a.c. power output and its value is given by the relation,

$$P_{O\,(ac)} = I^2 \cdot R_L = \frac{V^2}{R_L}$$

$$= \frac{V_p^2}{2R_L} \qquad ...(\because V_p = \sqrt{2}\,V)$$

where, V = The r.m.s. value of a.c. output voltage,

V_p = The peak or maximum value of a.c. output voltage, and

I = The r.m.s. value of a.c. output current.

(*b*) The remaining part is dissipated in the form of heat within the transistor collector region. Its value is designated as $P_{C\,(dc)}$ and it goes as a waste.

It is evident from above, that the complete equation may be written as follows :

$$P_{in\,(dc)} = P_{RL\,(dc)} + \underbrace{P_{O(ac)} + P_{C(dc)}}_{Pt_{r\,(dc)}}$$

or

$$V_{CC} \cdot I_{CQ} = I_{CQ}^2 \cdot R_L + I^2 \cdot R_L + P_{C\,(dc)}$$

It will be interesting to know that under zero-signal or no signal condition, the a.c. output power is zero. Therefore, all the power supplied to the transistor goes as a waste in the form of heat. It means that a transistor in class-A amplifier operation dissipates maximum power under zero-signal condition. Otherwise, the heating of transistor is reduced by the amount of the a.c. power converted by the stage and supplied to the load. Thus, a transistor is cooler when delivering power to a load than that when there is no such a.c. power transfer.

The power dissipation of a transistor with no input signal is equal to the product of its *Q*-point current and voltage, *i.e.*,

$$P_{tr\,(dc)} = V_{CEQ} \cdot I_{CQ} \qquad \text{... (Under zero signal condition)}$$

Note : When the peak-to-peak value of the output voltage (V_{PP}) is known, the a.c. power output (or a.c. load power),

$$P_{O(ac)} = \frac{V_{PP}^2}{8\,R_L} \qquad ...(\because V_{PP} = 2\,V_P)$$

22-10. Overall and Collector Efficiency of Class-A Amplifier

We have already discussed in Art. 22-8 that overall efficiency is defined as the ratio of a.c. power delivered to the load to the total power supplied by the d.c. source. Mathematically, the overall efficiency,

$$\eta_o = \frac{\text{A.C. Power delivered to the load}}{\text{Total power supplied by the d.c. source}}$$

$$= \frac{\text{Average a.c. output power}}{\text{Average d.c. input power}} = \frac{P_{o\,(ac)}}{P_{in\,(dc)}}$$

We know that average a.c. output power,

$$P_{o\,(ac)} = \frac{V_p^2}{2R_L} = \frac{V_p \cdot I_p}{2} \qquad ...(\because V_p = I_p \cdot R_L)$$

and the average d.c. input power,

$$P_{in\,(dc)} = V_{CC} \cdot I_{CQ}$$

$$\therefore \quad \eta_o = \frac{V_p \cdot I_p / 2}{V_{CC} \cdot I_{CQ}} = \frac{V_p \cdot I_p}{2V_{CC} \cdot I_{CQ}} \qquad ...(i)$$

Now for the largest possible output signal, from a class-A amplifier, the *Q*-point must be located at the centre of a.c. load line. As a result of this, the maximum or peak value of the output voltage (V_p) is equal to V_{CEQ} and the peak value of output current (I_p) is equal to I_{CQ} , *i.e.*,

$$V_p = V_{CEQ} \quad \text{and} \quad I_p = I_{CQ}$$

Substituting these values of V_p and I_p in equation (*i*)

$$\eta_o = \frac{V_{CEQ} \cdot I_{CQ}}{2V_{CC} \cdot I_{CQ}} = \frac{V_{CEQ} \cdot I_{CQ}}{2\,(2V_{CEQ}) \cdot I_{CQ}} = 0.25 = 25\%$$

The above value is the maximum possible value of overall efficiency and is designated as $\eta_{o\,(max)}$. It means that the maximum value of overall efficiency for class-A amplifier is 25%. However, in actual practice, it is always found to be less than this value. The collector-efficiency for a class-A amplifier is given by the relation,

$$\eta_o = \frac{\text{Average a.c. output power}}{\text{Average d.c. input power to the transistor}}$$

$$= \frac{P_{o\,(ac)}}{P_{tr\,(dc)}}$$

The maximum value of collector efficiency is obtained when the output signal is the largest possible. Under this condition, the average a.c. output power,

$$P_{o\,(ac)} = \frac{V_p \cdot I_p}{2} = \frac{V_{CEQ} \cdot I_{CQ}}{2}$$

The d.c. input power to the transistor is maximum under zero signal condition and its value is,

$$P_{tr\,(dc)} = V_{CEQ} \cdot I_{CQ}$$

$$\therefore \quad \eta_{C\,(max)} = \frac{V_{CEQ} \cdot I_{CQ}}{2\,(V_{CEQ} \cdot I_{CQ})} = 0.5 = 50\%$$

It means that maximum collector efficiency is 50%. In other words, the maximum a.c. power, which is transistor can deliver is $P_{tr\,(dc)}/2$. Incidentally, the power dissipation of a transistor under zero signal condition, is maximum and is equal to $P_{in\,(dc)}/2$. It is evident from the above discussion that a class-A amplifier, with a direct coupled load, can supply only one-half the amount of a.c.

power to the load, which its transistor can dissipate. The a.c. power input is four times the amount of a.c. signal power in the load.

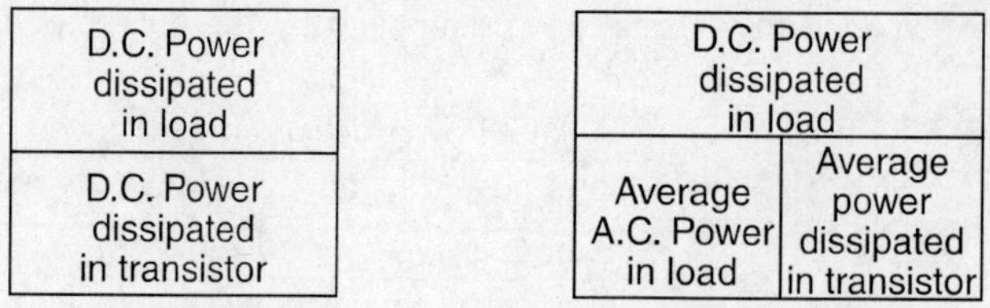

(*a*) No input signal. (*b*) Full input signal.

Fig. 22-15.

These facts are shown in Fig. 22-15 (*a*) and (*b*). Fig. 22-15 (a) shows that power relationships in a class-A amplifier, when there is no input signal. It is evident from the Fig. that the d.c. input power is utilized is two ways. One-half is dissipated in the load resistor, while the other half is dissipated in the transistor. Similarly, Fig. 22-15 (*b*) shows the power relationships, when there is a full input signal. Here it indicates that one-fourth (or 25%) of the total d.c. input power is converted into a.c. output power, while the remaining power is dissipated as a heat in load resistor and transistor.

Example 22-4. *A certain power transistor meant for class-A operation has a zero signal power dissipation of 20 watts. If the a.c. output power is 5 W, find* **(a)** *collector efficiency, and* **(b)** *power rating of the transistor.*

Solution. Given : $P_{tr\,(dc)} = 20$ W and $P_{o\,(ac)} = 5$ W

(*a*) *Collector efficiency*

We know that collector efficiency of class-A amplifier,

$$\eta_e = \frac{P_{o\,(ac)}}{P_{tr\,(dc)}} = \frac{5}{20} = 0.25 = 25\% \textbf{ Ans.}$$

(*b*) *Power rating of the transistor*

We know that zero signal condition represents the condition under which the transistor dissipates maximum power. Therefore, power rating of the transistor

$$= 20 \text{ W } \textbf{Ans.}$$

Example 22-5. *The maximum collector dissipation of a transistor used in a class A amplifier is 10 W. When a signal is applied, the collector efficiency of the circuit is 32%. Calculate the a.c. power output.*

Solution. Given : $P_{c\,(dc)} = 10$ W and $\eta_c = 32\% = 0.32$

Let $P_{o\,(ac)}$ = The a.c. power output.

We know that collector efficiency, (η_c),

$$0.32 = \frac{P_{o\,(ac)}}{P_{tr\,(dc)}} = \frac{P_{o\,(ac)}}{P_c + P_{o\,(ac)}} = \frac{P_{o\,(ac)}}{10 + P_{o\,(ac)}}$$

or

$$0.32 \times (10 + P_{o\,(ac)}) = P_{o\,(ac)}$$

$$3.2 + 0.32 \times P_{o\,(ac)} = P_{o\,(ac)}$$

$$\therefore \quad P_{o\,(ac)} = \frac{3.2}{(1 - 0.32)} = 4.7 \text{ W } \textbf{Ans.}$$

Example 22-6. *An amplifier has a collector efficiency of 50% and operates from a 24 V supply. If the output power is 3.5 W, what is the total power dissipated within the circuit? What is most likely source for most of this power loss ?*

Solution. Given : $\eta_c = 50\% = 0.5$; $V_{CC} = 24$ V; $P_{o\,(ac)} = 3.5$ W.

We know that collector efficiency (η_c),

$$0.5 = \frac{P_{o\,(ac)}}{P_{tr\,(dc)}} = \frac{3.5}{P_{tr\,(dc)}}$$

$$\therefore \qquad P_{tr\,(dc)} = \frac{3.5}{0.5} = 7 \text{ W}$$

We also know that,

$$P_{tr\,(dc)} = P_{c\,(dc)} + P_{o\,(ac)}$$

or $$7 = P_{c\,(dc)} + 3.5$$

or $$P_{c\,(dc)} = 7 - 3.5 = 3.5 \text{ W}$$

The power $P_{c\,(dc)}$ = 3.5 W is dissipated in the form of heat within the transistor collector region.

Example 22-7. *A single-stage, class-A amplifier has V_{CC} = 20 volt, V_{CEQ} = 10 volt, I_{CQ} = 600 mA and collector load resistor, R_L = 16 Ω. The a.c. output current varies by ± 300 mA with the a.c. input signal.*

*Determine **(a)** the power supplied by the d.c. source to the amplifier circuit, **(b)** d.c. power consumed by the load resistor, **(c)** a.c power developed across the load resistor, **(d)** d.c. power delivered to the transistor, **(e)** d.c. power wasted in transistor collector, **(f)** overall efficiency, and **(g)** collector efficiency.*

Solution. Given : V_{CC} = 20 volt; V_{CEQ} = 10 volt; I_{CQ} = 600 mA = 0.6 A; R_L = 16 Ω and I_p = 300 mA = 0.3 A.

(***a***) *Power supplied by the d.c. source to the amplifier circuit*

We know that power supplied by the d.c. source to the amplifier circuit,

$$P_{in\,(dc)} = V_{CC} \cdot I_{CQ} = 20 \times 0.6 = 12 \text{ W } \textbf{Ans.}$$

(***b***) *D.C. power consumed by the load resistor*

We know that d.c. power consumed by the load resistor,

$$P_{RL\,(dc)} = (I_{CQ})^2 \times R_L = (0.6)^2 \times 16 \text{ W} = 5.76 \text{ W } \textbf{Ans.}$$

(***c***) *A.C. power developed across the load resistor*

We know that r.m.s. value of the collector current,

$$I = \frac{I_p}{\sqrt{2}} = \frac{0.3}{\sqrt{2}} = 0.212 \text{ A}$$

∴ A.C. power developed across the load resistor,

$$P_{o\,(ac)} = I^2 \cdot R_L = (0.212)^2 \times 16 = 0.72 \text{ W } \textbf{Ans.}$$

(***d***) *D.C. power delivered to the transistor*

We know that d.c. power delivered to the transistor,

$$P_{tr\,(dc)} = P_{in\,(dc)} - P_{RL\,(dc)} = 12 - 5.76 \text{ W} = 6.24 \text{ W } \textbf{Ans.}$$

(***e***) *D.C. power wasted in the transistor collector*

We know that d.c. power wasted in the transistor collector,

$$P_{c\,(dc)} = P_{tr\,(dc)} - P_{o\,(ac)} = 6.24 - 0.72 \text{ W} = 5.52 \text{ W } \textbf{Ans.}$$

(***f***) *Overall efficiency*

We know that overall efficiency,

$$\eta_o = \frac{P_{o\,(ac)}}{P_{in\,(ac)}} = \frac{0.72}{12} = 0.06 = 6\% \textbf{ Ans.}$$

(*g*) *Collector efficiency*

We know that collector efficiency,

$$\eta_o = \frac{P_{o\,(ac)}}{P_{tr\,(ac)}} = \frac{0.72}{6.24} = 0.115 = 11.5\% \textbf{ Ans.}$$

22-11. Transformer Coupled Class-A Amplifier

We have already discussed that the overall efficiency of a direct coupled class-A amplifier does not exceed 25%. The main reason for this is that load resistance is directly connected in the output circuit of the amplifier. Due to this, the quiescent current (I_{CQ}) passes through the resistance.

This current represents a considerable waste of power, whose value is equal to $I_{CQ}^2 \cdot R_L$, because it does not contribute to a.c. output power. This problem is solved by using a suitable transformer for coupling the load to the amplifier stage as shown in Fig. 22-16. Since, the primary winding of a transformer has a low resistance, therefore the power absorbed in the winding is negligible as compared to the resistive load.

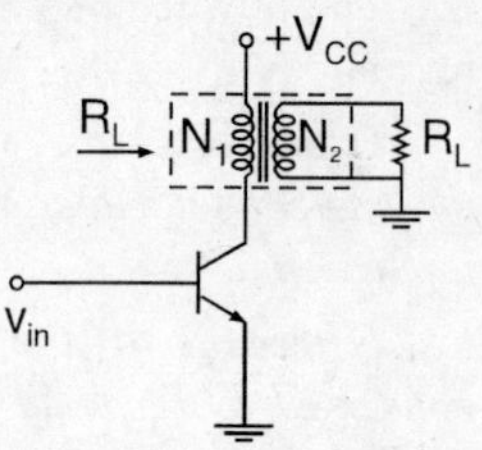

Fig. 22-16. *Transformer coupled class-A amplifier.*

The transformer coupled class-A amplifier is also known as single-ended class-A amplifier. The term 'single-ended' is used to differentiate this type from the push-pull amplifiers. The function of a transformer is to match the low impedance load (such as loudspeaker) to that of the output impedance of the amplifier. The impedance matching property follows from the relations :

$$V_1 = \left(\frac{N_1}{N_2}\right) V_2 \text{ and } I_1 = \left(\frac{N_2}{N_1}\right) I_2$$

where, V_1 = Primary voltage; V_2 = Secondary voltage

I_1 = Primary current; I_2 = Secondary current

N_1 = Primary turns and N_2 = Secondary turns

It may be noted that if the value of secondary turns (N_2) is less than of primary turns (N_1), then the above equations show that the transformer reduces the voltage in proportion to the turns ratio N_2/N_1 (equal to *n*) and steps up the current in the same ratio. The ratio of these two equations gives,

$$\frac{V_1}{I_1} = \left(\frac{1}{n^2}\right)\frac{V_2}{I_2}$$

Now, V_1/I_1 represents the effective load resistance or the resistance seen looking into the primary of the transformer and is designated as R'_L. On the other hand, V_2/I_2 represents the value of load resistance (*i.e.*, R_L). Therefore, the equation (*i*) may be rewritten as,

$$R'_L = \left(\frac{1}{n^2}\right) R_L$$

or $$R'_L = a^2 \cdot R_L$$

where, $$a = \frac{1}{n}\left(\text{or } \frac{N_1}{N_2}\right)$$

and is called *step down turns ratio*. Usually, the value of '*a*' is made much larger than unity. It is necessary to make the load resistance appear as a large effective resistance seen from the transformer primary.

It may be noted that in an ideal transformer, there is no primary drop. As a result of this, all the supply voltage appears as the collector-to-emitter voltage of the transistor (*i.e.*, $V_{CC} = V_{CEQ}$). Therefore, total power drawn from the d.c. supply is equal to the d.c. power developed within the transistor. Thus, the overall efficiency and collector efficiency becomes equal, *i.e.*,

$$\eta_o = \frac{P_{o\,(ac)}}{V_{CC} \cdot I_{CQ}} = \frac{P_{o\,(ac)}}{V_{CEQ} \cdot I_{CQ}} = \eta_c$$

The maximum value of overall or collector efficiency of a transformer coupled class-A amplifier is 50%. This can be seen from the proof given below :

Proof

We know that average power delivered by the d.c. supply,

$$P_{in\,(dc)} = V_{CC} \cdot I_{CQ}$$

and power developed in the transistor under zero-signal condition,

$$P_{tr\,(dc)} = V_{CEQ} \cdot I_{CQ} = V_{CC} \cdot I_{CQ}$$

For the largest possible output signal, the peak (or maximum) value of the output voltage developed across the load resistor,

$$V_p = V_{CEQ}$$

and the peak value of output current,

$$I_p = I_{CQ}$$

∴ Overall efficiency,

$$\eta_o = \frac{P_{o\,(ac)}}{P_{in\,(dc)}} = \frac{V_p \cdot I_p / 2}{V_{CC} \cdot I_{CQ}} = 0.5 = 50\%$$

and collector efficiency,

$$\eta_c = \frac{P_{o\,(ac)}}{P_{tr\,(dc)}} = \frac{V_p \cdot I_p / 2}{V_{CC} \cdot I_{CQ}} = 0.5 = 50\%$$

Example 22-8. *Calculate the effective resistance seen looking into the primary of a 15 : 1 transformer connected to an output load of 8 Ω.*

Solution. Given : $a = N_1/N_2 = 15$ and $R_L = 8\ \Omega$.

We know that the value of effective resistance,

$$R'_L = a^2 \cdot R_L = (15)^2 \times 8 = 1800\ \Omega = 1.8\ \text{k}\Omega \ \textbf{Ans.}$$

Example 22-9. *What turns ratio is required to match a 16 Ω speaker load to an amplifier, so that the effective load resistance is 10 kΩ ?*

Solution. Given : $R_L = 16\ \Omega$ and $R'_L = 10\ \text{k}\Omega = 10 \times 10^3\ \Omega$

Let a = Required turns ratio

We know that the value of effective resistance (R'_L)

$$10 \times 10^3 = a^2 \cdot R_L = a^2 \cdot 16$$

or

$$a = \sqrt{\frac{(10 \times 10^3)}{16}} = \sqrt{625} = 25$$

Thus, we need a transformer with turns ratio of 25 : 1. **Ans.**

Example 22-10. *A single-ended class-A amplifier has a transformer coupled load of 8 Ω. If the transformer turns ratio (N_1/N_2) is 10, find the maximum power output delivered to the load. Take the zero signal collector current of 500 mA.*

Solution. Given : $R_L = 8\ \Omega$; $a = N_1/N_2 = 10$ and $I_{CQ} = 500$ mA = 0.5 A.

We know that the effective load, as seen by the primary of the transformer,

$$R'_L = a^2 \cdot R_L = (10)^2 \times 8 = 800\ \Omega$$

and the maximum power delivered to the load,

$$P_{o\,(ac)} = \frac{1}{2} I_{CQ}^2 \times R'_L = \frac{1}{2} \times (0.5)^2 \times 800 \text{ W} = 100 \text{ W } \textbf{Ans.}$$

Example 25-11. *A class - A CE amplifier operates from V_{CC} = 20 V draws a current I_{CQ} = 5 A and feeds a load of 40 W through a step up transformer of n_1/n_2, = 3.16. Find the efficiency of the amplifier when it is properly matched for maximum power supply.*

(Electronic Circuits - I, Anna University, April/May 2008)

Solution. Given: V_{CC} = 20 V, I_{CQ} = 5 A, n_1/n_2, = 3.16, R_L = 40 Ω

We know that,
$$n = \frac{n_2}{n_1}$$

We know that effective load resistance (R'_L),

$$R'_L = \frac{R_L}{(n)^2} = \frac{40}{(3.16)^2} = 4\Omega$$

and the power dissipated,

$$P_{DC} = V_{CC}\, I_{CQ} = 20 \times 5 = 100 \text{ W}$$

For maximum power,

$$P_{ac} = \frac{1}{2} \cdot \frac{V^2{}_{CC}}{R_L^1} = \frac{1}{2} \frac{(20)^2}{4} = 50 \text{ W}$$

We know that efficiency,

$$\%\, y = \frac{P_{ac}}{P_{DC}} \times 100 = \frac{50}{100} \times 100 = 50\% \qquad \textbf{Ans.}$$

Example 22-12. *A transistor rated for a maximum collector dissipation of 100 mW operates a single-ended class-A output stage from a 10 volt supply. Calculate the approximate values of **(a)** maximum undistorted a.c. output power, **(b)** the quiescent current, and **(c)** turns-ratio of the output transformer, if the load resistance is 16 Ω. Assume overall and collector efficiency = 0.5.*

Solution. Given : $P_{tr\,(dc)}$ = 100 mW = 0.1 W; V_{CC} = 10 volt; R_L = 16 Ω and $\eta_o = \eta_c = 0.5$.

(*a*) *Maximum undistorted a.c. output power*

We know that maximum collector efficiency (η_c)

$$0.5 = \frac{P_{o\,(ac)}}{P_{tr\,(dc)}} = \frac{P_{o\,(ac)}}{0.1}$$

$$\therefore \quad P_{o\,(ac)} = 0.5 \times 0.1 = 0.05 \text{ W} = 50 \text{ mW } \textbf{Ans.}$$

(*b*) *Quiescent collector current*

We know that a.c. output power, ($P_{o\,(ac)}$),

$$0.05 = \frac{1}{2} V_{CC} \cdot I_{CQ} = \frac{1}{2} \times 10 \times I_{CQ} = 5\, I_{CQ}$$

$$\therefore \quad I_{CQ} = 0.05/5 = 0.01 \text{ A } \textbf{Ans.}$$

(*c*) *Transformer turns ratio*

Let a = Transformer turns ratio

We know that effective load resistance i.e., the load resistance seen looking into the primary of the transformer,

$$R'_L = \frac{V_{CC}}{I_{CQ}} = \frac{10}{0.01} = 1000\ \Omega$$

and effective load resistance (R'_L),

$$1000 = a^2 \cdot R_L = a^2 \times 16$$

$$a = \sqrt{1000/16} = 8 \textbf{ Ans.}$$

Example 22-13. *A power transistor working in class-A amplifier is supplied from a.c. supply of 10 V. If the maximum change in collector current is 50 mA, find the power transferred to a 4 Ω speaker when (**a**) speaker is directly connected; and (**b**) speaker is connected through the transformer to the transistor circuit.* (*U.P.S.C., Civil Services*, 1986)

Solution. Given : $V_{CC} = 10$ volt; $I_P = 50$ mA = 0.05 A and $R_L = 4\ \Omega$.

(***a***) *Speaker directly connected*

Fig. 22-17 (*a*) shows the speaker directly connected to the transistor circuit. Now we know that the peak (or maximum) value of the change in collector-to-emitter voltage is equal to V_{CC} (*i.e.*, 10 V). Moreover the r.m.s. value of output current,

$$I = \frac{I_P}{\sqrt{2}} = \frac{0.05}{\sqrt{2}} = 0.035 \text{ A}$$

∴ Average power delivered to the speaker,

$$P_{o\,(ac)} = I^2 \cdot R_L = (0.035)^2 \times 4 = 4.9 \times 10^{-3} \text{ W} = 4.9 \text{ mW} \textbf{ Ans.}$$

(***b***) *Speaker connected through the transformer*

Fig. 22-17 (*b*) shows the speaker connected through the transformer to the transistor circuit. In this case, also the maximum or peak change in collector-to-emitter voltage is equal to V_{CC} (*i.e.*, 10V). We know that the effective load resistance seen looking directly into the primary of a transformer.

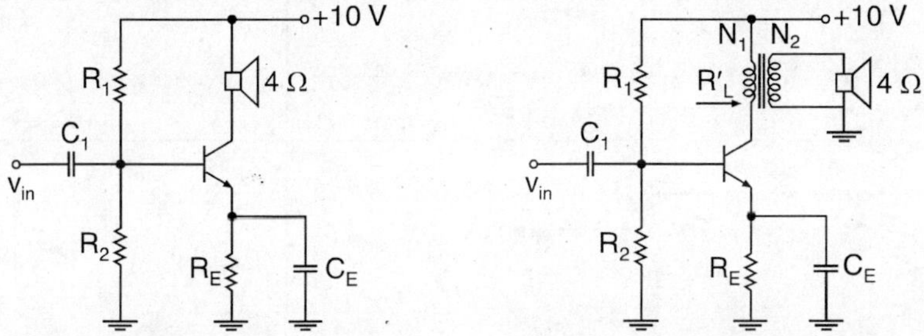

(*a*) Speaker directly connected. (*b*) Speaker connected through a transformer.

Fig. 22-17.

$$R'_L = \frac{V_1}{I_1} = \frac{V_{CC}}{I_{CQ}} = \frac{10}{0.05} = 200\ \Omega$$

We also know that the effective load resistance (R'_L),

$$200 = a^2 \cdot R_L = a^2 \times 4 \text{ or } a = \sqrt{200/4} = \sqrt{50} = 7$$

∴ Output transformer has a turns ratio (N_1/N_2) equal to 7 : 1. The transformer secondary voltage,

$$V_2 = \frac{V_1}{a} = \frac{10}{7} = 1.43 \text{ V}$$

and peak value of secondary current,

$$I_{2P} = \frac{V_2}{R_L} = \frac{1.43}{4} = 0.36 \text{ A}$$

∴ r.m.s. value of secondary current,

$$I_2 = 0.36/\sqrt{2} = 0.255 \text{ A}$$

and the average power transferred to the speaker,

$$I_2^2 \times R_L = (0.255)^2 \times 4 = 0.26 \text{ W} = 260 \text{ mW} \textbf{ Ans.}$$

22-12. Class-B Amplifier

We have already discussed in Art. 22-6 that in a class-B, amplifier, the transistor bias and amplitude of the input signal is such that the output current flows only for one-half cycle (*i.e.*, 180°) of the input signal. This condition is achieved by setting the *Q*-point at cut-off (or lower end) of the load line.

Fig. 22-18 (*a*) illustrates the biasing condition on the transconductance curve, while Fig. 22-18 (*b*) shows the biasing condition on the a.c. load line. The variation of collector current with the input signal is also shown in both these figures. It may be noted that with no input signal, the transistor is biased at cut-off. At this point, there is no current flow through the transistor and hence no power is dissipated by the transistor. However, when the signal is applied, the transistor handles an average current, which increases it for large input signals.

Thus, contrary to class-A amplifier operation in which the power dissipation of a transistor is maximum with no input signal and minimum with largest input signal, the operation of class-B amplifier circuit is to increase transistor dissipation for increased input signal. Since, the average current in class-B amplifier is less than that in class-A, therefore the amount of power dissipated by the transistor is less in class-B amplifier.

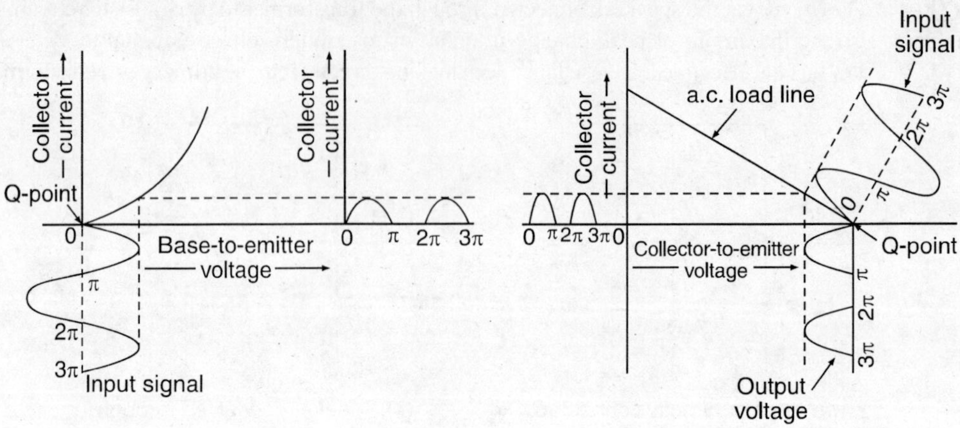

(*a*) Q-point on transconductance curve. (*b*) Q-point on a.c. load line.

Fig. 22-18.

22-13. Characteristics of Class-B Amplifier

Following are some of the important characteristics of class-B amplifier :

(*i*) The output current flows only for one-half of the cycle (*i.e.*, 180°) of the input signal.

(*ii*) The transistor dissipates no power with zero input signal. However, it increases with the increase in the amplitude of input signal. It is contrary to class-A amplifier operation, where the transistor dissipation is maximum with no input signal and minimum with the largest input signal.

(*iii*) The average current drawn by the circuit in class-B operation is smaller than that in classA. As a result of this, the amount of power dissipated by the transistor is less in class-B. Thus, the overall efficiency of the circuit is higher than that of class-A. Its maximum value has been found to be equal to 78.5%.

22-14. Power Relations for Class-B Amplifier

Following are some of the power relations for class-B amplifier which are important from the subject point of view :

(1) ***D.C input power :*** The input power comes from the d.c. source (*i.e.*, the V_{CC} supply) and is given by the relation,

$$P_{in\,(dc)} = V_{CC} \cdot I_{dc}$$

where, I_{dc} is the average value of current drawn from the V_{CC} supply. If I_{CP} is the peak (or maximum) value of average value of current, then

$$I_{dc} = I_{CP}/\pi$$

$$\therefore \quad P_{in\,(dc)} = V_{CC} \cdot I_{CP}/\pi$$

(2) ***D.C. power loss in load resistor:*** Its values is given by the relation,

$$P_{RL\,(dc)} = I_{dc}^2 \cdot R_L = \left(\frac{I_{CP}}{\pi}\right)^2 \cdot R_L \qquad ...(\because I_{dc} = I_{CP}/\pi)$$

(3) ***A.C. power output in load resistor :*** Its value is given by the relation,

$$P_{o\,(ac)} = I^2 \cdot R_L = V^2/R_L = V_p^2/2\,R_L$$

where, I = The r.m.s. value of a.c. output current,

V = The r.m.s. value of a.c. output voltage, and

V_p = The peak (or maximum) value of a.c. output voltage.

(4) ***Power dissipated within the transistor :*** Its value is given by the relation,

$$P_{c\,(dc)} = P_{in\,(dc)} - P_{RL\,(dc)} - P_{o\,(ac)}$$

(5) ***Overall efficiency :*** Its value is given by the relation,

$$\eta_o = \frac{P_{o\,(ac)}}{P_{in\,(dc)}} = \frac{I^2 \cdot R_L}{V_{CC} \cdot I_{dc}}$$

22-15. Maximum Value of A.C. Output Power in Class-B Amplifier

Fig. 22-19 shows the location of a Q-point for class-B amplifier alongwith the maximum variation of collector-to-emitter voltage (V_{CE}) and collector current (I_C) with the largest possible signal.

It may be noted from this figure that the peak or maximum value of the output current is I_{CP} and that of the output voltage is V_{CEQ} or V_{CC}.

We know that the a.c. output power in a class-B amplifier is developed only during one-half cycle of the input signal. Therefore its value is given by the relation,

$$P_{o\,(ac)} = \frac{1}{2}\left(\frac{V_{CC}}{\sqrt{2}}\right) \cdot \left(\frac{I_{CP}}{\sqrt{2}}\right) = \frac{1}{4} \times V_{CP} \cdot I_{CP}$$

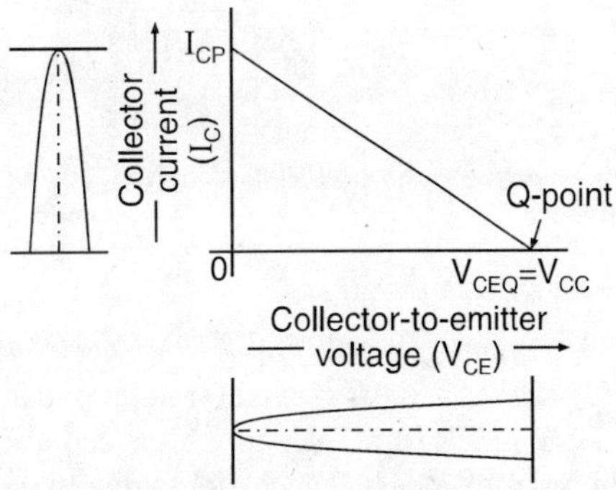

Fig. 22-19. *Illustrating maximum a.c. output power in class-B amplifier.*

22-16. Maximum Overall Efficiency for Class-B Amplifier

We have already discussed in Art. 22-14 that the overall efficiency for a class-B amplifier,

$$\eta_o = \frac{P_{o\,(ac)}}{P_{in\,(dc)}}$$

We also know that maximum power developed, $P_{o\,(ac)} = \frac{1}{4} V_{CP} \cdot I_{CP}$

$\therefore$ Maximum value of overall efficiency,

$$\eta_{o\,(max)} = \frac{\frac{1}{4} \cdot V_{CP} \cdot I_{CP}}{V_{CC} \cdot I_{dc}} = \frac{\frac{1}{4} \cdot V_{CP} \cdot I_{CP}}{V_{CC} \cdot I_{CP}/\pi} \qquad ...\left(\because I_{dc} = \frac{I_{CP}}{\pi}\right)$$

$$= \frac{\frac{1}{4} V_{CC} \cdot I_{CP}}{V_{CC} \cdot I_{CP}/\pi} = \frac{\pi}{4} \qquad \text{... } (\because V_{CP} = V_{CC})$$

$$= 0.785 = 78.5\%$$

22-17. Class-B Push-Pull Amplifier

We have already discussed in previous articles that in a class-B amplifier, the transistor conducts only for one-half-cycle and remains off during the other half cycle. It means that we obtain an output signal, whose either positive or negative half cycles are missing. This type of output signal gives a large distortion. In order to avoid this, we use two transistors connected in a push-pull arrangement. In this arrangement, one of the transistors conducts during one half cycle and the other conducts during the second half cycle. With push-pull arrangement, it is possible to build class-B amplifier, which have low distortion, large a.c. output power and high efficiency.

Fig. 22-20 shows a transformer coupled class-B push-pull amplifier. The circuit consists of two centre-tapped transformers T_1 and T_2 and two identical transistors Q_1 and Q_2. The transformer (T_1) is an input transformer and is called ***phase splitter***. It is required to produce two signal voltages, which are 180° out of phase with each other. These two signal voltages, with opposite polarity, drive the input of transistors Q_1 and Q_2. The transformer (T_2) is an output transformer and is required to couple the a.c. output signal from the collector to the load.

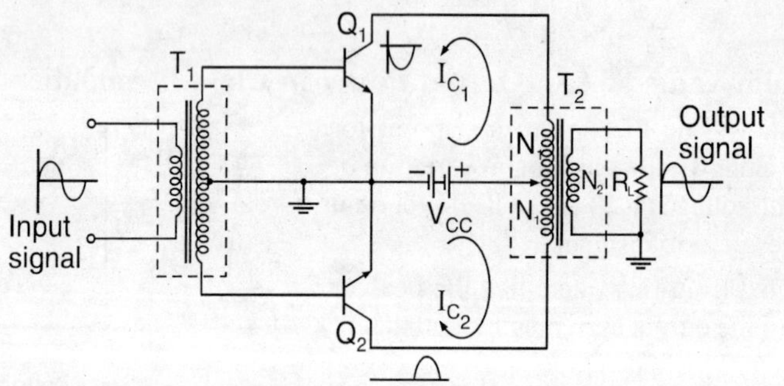

Fig. 22-20. *Transformer coupled class-b push-pull amplifier.*

The transistors Q_1 and Q_2 are biased at cut-off, which is done by providing zero bias on the base of each transistor. The zero bias is achieved by connecting the emitter and base terminals together. In order to get a balanced circuit, the two emitters are connected to the centre tap of transformer T_1 secondary and the V_{CC} supply, to the centre tap of transformer T_2 secondary.

Now we shall discuss the working of push-pull amplifier circuit. When there is no input signal, both the transistors Q_1 and Q_2 are cut-off. Hence, no current is drawn from the $V_{(CC)}$ supply. Thus, there is no power wasted on standby ant the power dissipation in both the transistors is, practically, zero. It will be interesting to know that even if both the transistors Q_1 and Q_2 are biased, little above their cut-off points, the transformers still operate near zero flux. It is because of the fact that d.c. bias current, for each transistor, flows in opposite directions through the secondary of transformer T_2. The magnetic flux set up by each of these currents results in opposite flux through the magnetic core, so that the net flux in the perfectly matched case is zero. Thus, the transformer is not required to handle a large flux due to d.c. bias currents. As a result of this, the transformer can have a small size core, biased to operate near zero flux.

Now consider only the positive half cycle of the input signal. During this, the base of transistor Q_1 is positive and that of Q_2 is negative. As a result of this, the Q_1 conducts, while the transistor Q_2 is OFF. When the input signal goes negative, Q_1 turns off and Q_2 conducts. Thus at any instant, only one transistor in the circuit is conducting. In other words, each transistor handles one-half of the input signal. Then the output transformer joins these two halves and produces a full-sine wave in the load resistor. However, it has been found that joining of outputs of two transistors (*i.e.*, Q_1 and Q_2) is never perfect and hence the output is distorted.

The class-B push-pull amplifier is known as double ended class-B amplifier. This amplifier is frequently employed in the output (or final) stages of the amplifier circuits. These are used in public address systems (PAS's) tape recorders and stereo amplifiers.

22-18. Advantages of Class-B Push-Pull Amplifier

Though the class-B push-pull amplifier has a number of advantages over class-A amplifier, yet the following are important from thc subject point of view :

(1) The circuit efficiency of a class-B push-pull amplifier is 78.5%, which is much higher than that of class-A whose value is 25%. It is mainly due to the reason that no power is drawn from the d.c. supply under no signal condition.

(2) The use of push-pull system in the class-B amplifier eliminates even order harmonics in the a.c. output signal.

(3) Because of the absence of even harmonics, the circuit gives more output, per device, for a given amount of distortion.

(4) There is no d.c. component in the output signal. It is because of the fact, that d.c. components of two collector currents, through the two-halves of the primary of the T_2-transformer flow in opposite directions. As a result of this, there is no possibility of the core saturation of the T_2-transformer, even at the peak value of the signals. Thus we can use smaller sized cores in the transformers, without affecting the circuit performance.

22-19. Crossover Distortion

We know that transistors in a class-B push-pull amplifier are biased at cut-off. It means that when the d.c. bias voltage is zero, the input signal voltage must exceed the barrier voltage before a transistor conducts. In other words, a transistor does not conduct, until the input signal voltage exceeds 0.7 V for silicon and 0.3 V for germanium transistors. Because of this, there is a time interval between the positive and negative alternations of the input signal when neither transistor is conducting as shown in Fig. 22-21. The resulting distortion in the output signal is quite common and is called ***crossover distortion***.

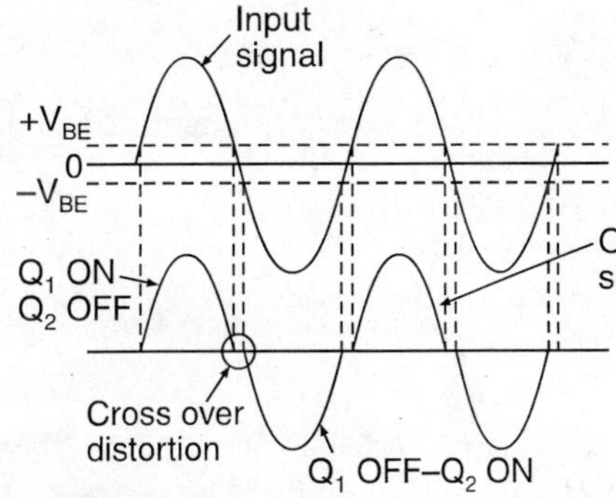

Fig. 22-21. *Illustrating crossover distortion.*

The crossover distortion may be avoided by applying a slight forward bias (equal to 0.7 V for silicon and 0.3 V for germanium transistors) to the base-emitter junction of both the transistors of the amplifier circuit. It causes the transistor to conduct immediately, when the a.c. input signal is applied. The application of slight forward bias shifts the Q-point slightly above the cut-off. In that case, each transistor operates for more than one half cycle. The resulting operation of the transistor is called class-AB operation.

22-20. Efficiency of Class-B Push-Pull Amplifier

Consider a class-B push-pull amplifier circuit as shown in Fig. 22-20. The a.c. load line, the operating point and the output current and voltage for the push-pull circuit are as shown in Fig. 22-22. In this figure, the a.c. load line for Q_2-transistor has been plotted upside down with respect to that of Q_1-transistor in order to get the combined output.

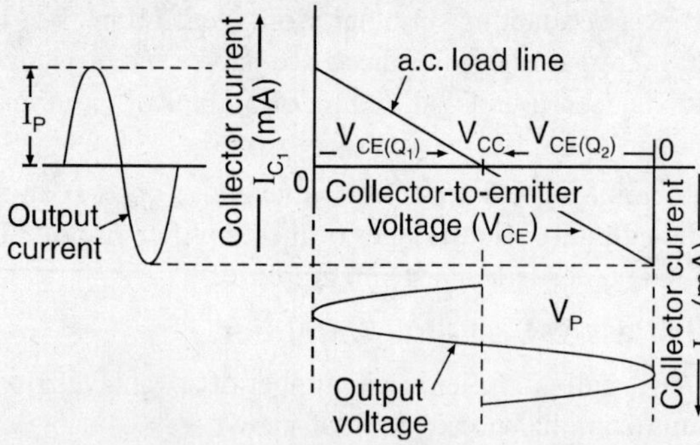

Fig. 22-22. *Illustrating a.c. load line output current and voltage waveforms for the class-B push-pull amplifier.*

Now let I_C = Average (or d.c.) value of output current of Q_1 or Q_2 transistor.
I_P = Peak or maximum value of output current,
V_C = Average (or d.c.) value of *output voltage, and
V_P = Peak (or maximum) value of the output voltage.

Then a.c. power delivered to the load resistor,

$$P_{o\,(ac)} = V_C \cdot I_C = \left(\frac{V_P}{\sqrt{2}}\right) \cdot \left(\frac{V_P}{\sqrt{2}}\right) = \frac{V_P \cdot I_P}{2}$$

and total power dissipation of the two transistors,

$$2P_{C\,(dc)} = P_{in\,(ac)} = P_{o\,(ac)} = V_C \cdot I_C - V_P \cdot I_P/2$$

$$= V_{CC} \cdot \frac{2I_P}{\pi} - \frac{V_P \cdot I_P}{2} \qquad \ldots\left(\because I_C = \frac{2I_P}{\pi}\right)$$

$$= 2\,I_P\left(\frac{V_{CC}}{\pi} - \frac{V_P}{4}\right)$$

∴ Overall efficiency,

$$\eta_o = \frac{P_{o\,(ac)}}{P_{in\,(dc)}} = \frac{\dfrac{V_P \cdot I_P}{2}}{V_{CC}\left(\dfrac{2I_P}{\pi}\right)} = \frac{\pi}{4} \cdot \frac{V_P}{V_{CC}} = 0.785\,\frac{V_P}{V_{CC}}$$

For the largest possible output signal, the peak value of the output voltage is equal to the V_{CC} supply (*i.e.*, $V_P = V_{CC}$). In that case, the overall efficiency is maximum, and its value,

$$\eta_{o\,(max)} = 0.785 = 78.5\%$$

The value of collector efficiency is equal to the overall efficiency, whose maximum value is also 78.5%.

Note : The class B amplifier has the same output power characteristic as the class A amplifier, By formula, a.c. power output,

$$P_{O(ac)} = \frac{V_P^2}{2\,R_L} \; or \; \frac{V_{PP}^2}{8\,R_L}$$

To calculate the maximum possible output power, we need to determine its compliance. The compliance of a class *B* amplifier is found as

$$PP = 2\,V_{CEQ}$$

$$= V_{CC} \qquad \ldots(\because V_{CEQ} = V_{CC}/2)$$

* Here we have assumed that the transformer turns ratio is $(N_1 + N_1) : N_2$. It means that turns ratio for each transistor (Q_1 or Q_1) is $N_1 = N_2$. In other words, each transistor uses half the secondary of the output transformer.

Example 22-14. Fig. 22-23 shows the circuit of a class-*B* push-pull amplifier. What is the power drawn from the d.c. source it the peak output voltage ,(V_P) is 16 V.

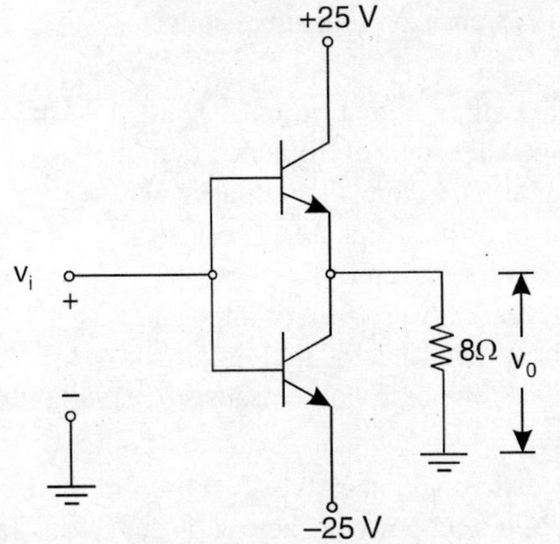

Fig. 22-23.

Solution. Given $R_L = 8\ \Omega$; $(V_P) = 16$ V.

We know that power drawn from the source

$$\approx \text{a.c. power output}$$

$$= \frac{V_P^2}{2R_L} = \frac{16^2}{2 \times 8} = 16 \text{ W } \textbf{Ans.}$$

Example 22-15. *A class-B output stage has an efficiency of 60%. If the maximum collector dissipation of each transistor is 2.5 W, calculate the d.c. input power and the a.c. output power.*

Solution. Given: $\eta_o = 60\% = 0.6$; Maximum collector dissipation of each transistor $P_{c\,(dc)} =$ 2.5W.

Let $P_{in\,(dc)}$ = The d.c. input power, and

$P_{o\,(ac)}$ = The a.c. output power.

We know that total collector (or power) dissipation in two transistors

$$2P_{c\,(dc)} = 2 \times 2.5 = 5 \text{ W}$$

We also know that overall efficiency, (η),

$$0.60 = \frac{P_{o\,(ac)}}{P_{in\,(dc)}}; \qquad \therefore\ P_{o\,(ac)} = 0.6 \times P_{in\,(dc)} \qquad ...(i)$$

and the d.c. input power,

$$P_{in\,(dc)} = 2P_{c\,(dc)} + P_{o\,(ac)} = 5 + 0.6 \times P_{in\,(dc)}$$

or $$(1 - 0.6)\, P_{in\,(dc)} = 5; \quad \therefore\ P_{in\,(dc)} = \frac{5}{1 - 0.6} = 12.5 \text{ W } \textbf{Ans.}$$

From equation (*i*), $$P_{o\,(ac)} = 0.6 \times P_{in\,(dc)} = 0.6 \times 12.5 = 7.5 \text{ W } \textbf{Ans.}$$

22-21. Transformerless Class-B Push-Pull Amplifier

The push-pull amplifier discussed in Art. 22-17 has two major drawbacks namely it requires a bulky and expensive transformer and also input signal voltage 180° out of phase with each other. The second requirement necessitates the use of an input centre-tapped transformer called phase splitter. It makes the driver circuitry of the amplifier complicated. The transformerless amplifier removes both the above mentioned drawbacks while retaining all the advantages of push-pull configuration.

Fig. 22-24 shows a transformerless class-B push-pull amplifier. In this figure, the input centre tapped transformer is removed by an input driver called phase splitter or phase inverter. It consists of an NPN transistor with equal collector and emitter resistances.

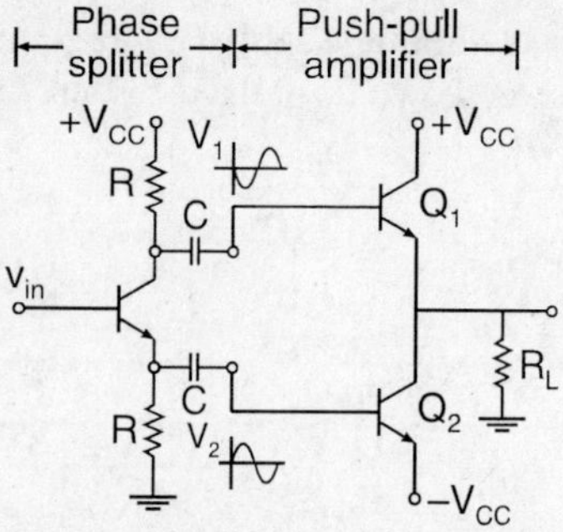

Fig. 22-24. *Transformerless class-B push-pull amplifier.*

It may be noted that the emitter output of phase splitter produces an inphase signal, while that of collector output produces an out of phase signal. Therefore, the output signals are equal in amplitude, but opposite in phase. The load is directly connected to the amplifier circuit. The removal of output transformer does not create any problem of impedance mismatch between the output impedance of the circuit and the load resistance. These days, the high-power transistors available have an output impedance of the order of 4 to 8 ohms.

The split supply (*i.e.*, positive and negative V_{CC}) used in the circuit gives us an advantage of making the d.c. component of output power as zero. It is due to this fact that we have connected the load directly with the amplifier circuit.

22-22. Complementary Symmetry Class-B Push-Pull Amplifier

Fig. 22-25 shows another class-B push-pull amplifier, which requires neither an input nor an output transformer. This arrangement uses transistors having complementary symmetry in the emitter follower configuration. The term 'complementary' means that the circuit uses two identical transistors – one NPN and the other PNP. Both these transistors have identical input and output characteristics. The term 'symmetry' means that the biasing resistors are equal. As a result of this, the emitter-base junction of each transistor is biased with the same voltage. It may be noted that the input signal is capacitively coupled, whereas the output signal is direct coupled.

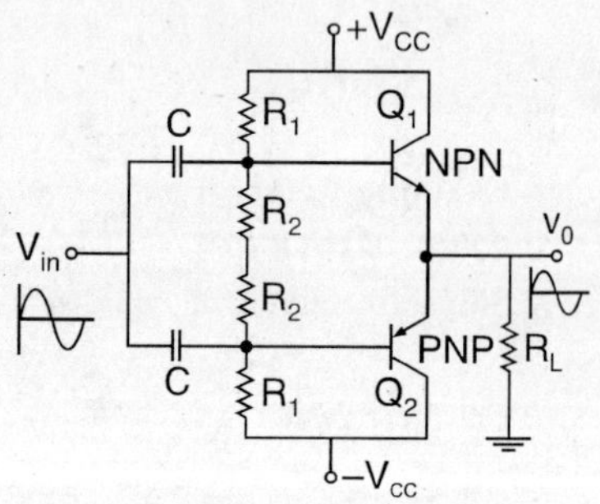

Fig. 22-25. *Complementary symmetry class-B push-pull amplifier.*

The circuit operation may be understood that with zero input signal, none of the transistor conducts. Thus, the output voltage is zero. During the positive input half-cycle, Q_1 conducts and Q_2 is off. Similarly, during the negative input half-cycle, Q_1 is driven into cut-off and Q_2 conducts.

It may be noted that amplifier circuit has a unity gain because of the emitter follower configuration. Moreover, there is no phase inversion of the output signal. The split supply (*i.e.*, equal positive and negative V_{CC}) used in the circuit gives us an advantage that the d.c. component of output voltage can be made zero. Thus, only a.c. component of the power is available across the load resistor (R_L).

22-23. Class-C Amplifier

The class-C amplifier is the most efficient power amplifier, which can produce more load power than that of either class-A or class-B amplifier. To amplify a sinusoidal signal, a class-C amplifier has to be tuned to the sinusoidal frequency. Because of this fact, the class-C amplifier is called a tuned amplifier or a narrow band circuit. It means that it can amplify only the resonant frequency and those frequencies which are closer to it. In order to avoid the need for large inductors and capacitors, in the resonant circuit, the class-C amplifiers are used to amplify the signals at radio frequencies (*i.e.*, frequencies above 20 kHz). Thus, class-C amplifier is basically a radio-frequency (RF) power amplifier and not an audio power amplifier like class-A and class-B amplifier.

22-24. Characteristics of Class-C Amplifier

Following are some of the important characteristics of class-C amplifier :

- The output current flows only during a part of the positive (or negative) half cycle of the input signal. This condition is achieved by biasing the transistor below cut-off.
- The output signal does not resemble the input signal because it consists of narrow pulses.
- The class-C amplifier is the most efficient power amplifier and its overall efficiency, under certain conditions, may approach even 100%.

SUMMARY

In this chapter you have learned that :

1. A power amplifiers is an amplifier, which is capable of providing a large amount power to the load.
2. The power amplifiers are classified as class-A, class -B and class -C amplifiers.
3. For a class -A amplifier, the output current flows during the entire cycle of the a.c. input signal.
4. The efficiency of a class-A amplifier is increased by using transformer coupled load instead of direct coupled.
5. For a class-B amplifier, the output current flows during only one-half cycle of the input signal.
6. Class-B push-pull amplifier uses two or more transistor to achieve the reproduction of complete input cycle at the output.
7. Class-C amplifier is the most efficient power amplifier and it reproduces only less than a half cycle of the input a.c. signal.

GLOSSARY

Amplitude distortion : Distortion in the amplifier of the output compared to the input of an amplifier due to the non-linearity of the transistor.

Circuit efficiency : The ratio of the maximum a.c. output to the d.c. input power.

Class-A amplifier : An amplifier in which the transistor bias and amplitude of the input signal is such that the output current flows for the complete cycle of the input signal.

Class-B amplifier : An amplifier in which the transistor bias and amplitude of the input signal is such that the output current flows for only one half cycle of the input signal.

Class-B push-pull amplifier : A class-B amplifier in which two transistor are connected in a push-pull arrangement so that the output current flows for the complete cycle of input signal.

Class-C amplifier : An amplifier in which the transistor bias and amplitude of the input signal is such that the output current flows for less than half cycle of the input signal.

Harmonic distortion : Refer amplitude distortion.

Power amplifier : An amplifier, which is capable of providing a large amount of power to the load.

Power dissipation capabilities : The ability of power transistor to dissipate heat.

Single-ended class-A amplifier : Transformer coupled class-A amplifier.

NUMERICAL PROBLEMS

1. Determine the collector current and the collector-to-emitter voltage at the points of saturation and cut-off in Fig. 22-26. under a.c. signal operation and hence draw the a.c. load line. Assume $V_{BE} = 0.7$ V. **(Ans.** 1.4 mA; 4 V)

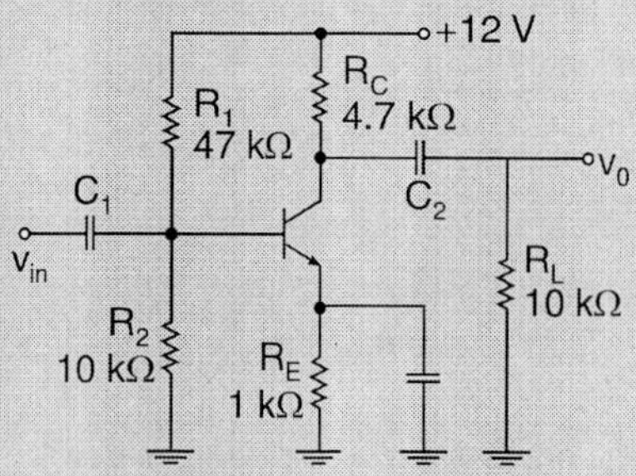

Fig. 22-26.

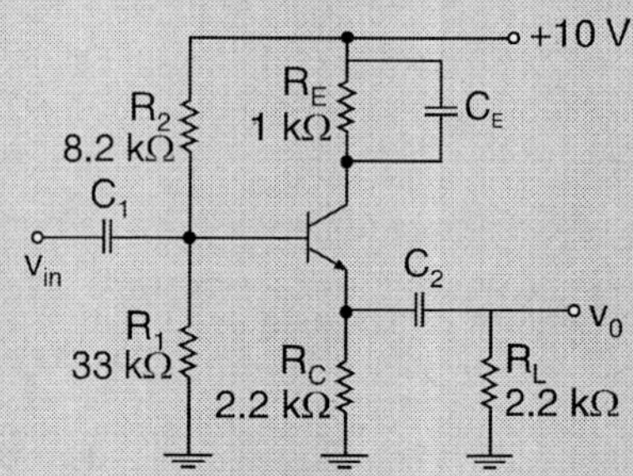

Fig. 22-27.

2. Fig. 22-27 shows a single-stage common emitter amplifier. Find the collector current and collector-to-emitter voltage at saturation and cut-off points. Also draw the a.c. load line. Assume $V_{BE} = 0.7$ V. **(Ans.** 1.29 mA, 5.8 V)
3. Find the large-signal voltage gain and power gain for the common emitter amplifier shown in Fig. 22-28. Assume that r'_e has been found to be 10 Ω from the graphical data. **(Ans.** 10; 250)
4. Fig. 22-29 shows a single-stage common emitter amplifier. Find the large signal voltage gain and power gain for this circuit. Assume that r'_e has been found to be 50 Ω from the graphical data. **(Ans.** 5.28; 2.64)

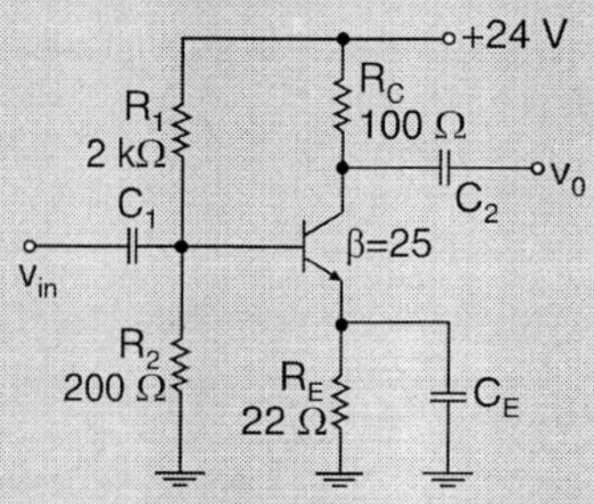

Fig. 22-28.

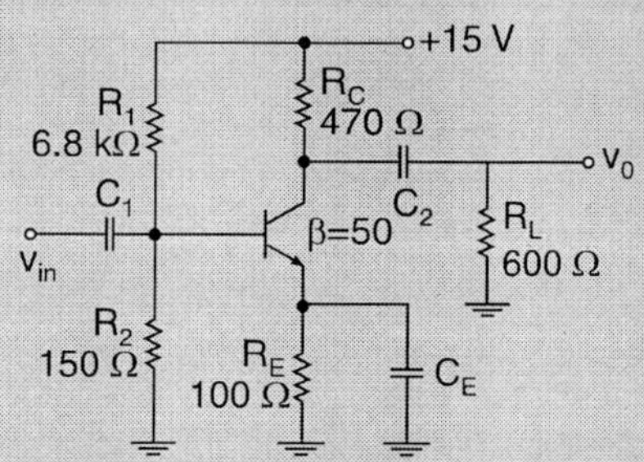

Fig. 22-29.

5. A certain power transistor meant for Class-A operation has a zero signal power dissipation of 10 watts. If the a.c. output power is 4 W , find **(*a*)** collector efficiency, and **(*b*)** power rating of the transistor. **(Ans.** 25%; 10 W)
6. An amplifier has an efficiency of 31% and power output of 12 W. Is this likely to be a class-A or class-B amplifier and what is the d.c. input power to the circuit ? **(Ans.** Class A, 38.7 W)
7. An amplifier has an efficiency of 32% and a collector dissipation of 0.8 W. Calculate the d.c. power input and a.c. power output of the circuit. **(Ans.** 1.176 W, 0.376 W)
8. A class-A transformer coupled amplifier uses a 25 : 1 transformer to drive 4 Ω load. Calculate the effective a.c load (seen by the transistor connected to the larger turns side of the transformer). **(Ans.** 2500 Ω)
9. What turns ratio transformer is needed to couple to an 8 Ω load so that it appears as a 10 kΩ effective load ? **(Ans.** 35 : 1)

10. A power transistor of a transformer coupled class-A amplifier has to deliver a maximum of 4 W of low frequency power to a 4 Ω load with a power supply 24 V. Find the values of (***a***) d.c. input power, (***b***) quiescent voltage and current, (***c***) peak value of collector current, and (***d***) transformer turns ratio. Take $\mu = 0.4$.

(**Ans.** (*a*) 8 W; (*b*) 24 V, 0.33 A; (*c*) 0.4 A; and (*d*) 4.2)

11. A power transistor of a class-A amplifier is supplied from a d.c. supply of 10 V. If the maximum collector current is 100 mA, find the power transferred to a 4 Ω load, if load is (***a***) directly coupled, and (***b***) transformer coupled. (**Ans.** (*a*) 20 mW; (*b*) 500 mW)

12. A power transistor used in class -A amplifier is supplied from a 10 V battery. If the maximum collector current change is 100 mA, calculate the power transferred to 4 Ω loud speaker when it is

(***a***) directly connected in the collector

(***b***) transformer coupled for maximum power transfer. Find the turns ratio of the transformer in the second case. (*Bangalore Uni.B.Sc. April/May* 2001)

(**Ans.** (*a*) 20 mW, transformer turns ratio = 5 : 1)

[**Hint :** Refer to Example 22-13.]

DESCRIPTIVE QUESTIONS

1. Why a power amplifier is called a large signal amplifier ? Explain it briefly.

2. What is the difference between the voltage amplifier and power amplifier ?

3. How do you classify the power amplifiers ? With the help of suitable sketches or otherwise compare class-A, class-B and class-C amplifiers.

4. Explain the meaning of class-A, class-B and class-C as applied to power amplifiers.

5. Draw a neat circuit diagram of a class-A direct coupled amplifier using NPN transistor and explain clearly its operation.

6. Draw the circuit diagram of a single ended class-A power amplifier. Explain its operation.

7. Draw neatly the configuration of push-pull amplifier and explain its working. Derive an expression for its efficiency.

8. Show that the maximum collector efficiency of class-B amplifier is 78.6 per cent.

9. Give the advantages and disadvantages of push-pull configuration in power amplifiers.

10. (***a***) With a neat circuit diagram, explain the principle of operation of a push-pull amplifier circuit using transistors.

(***b***) Derive the equation for conversion efficiency of class-B power amplifier.

(***c***) Clearly explain the reason why most of the power amplifier used in practice are designed to operate in class AB stage.

11. Derive and expression for the maximum conversion efficiency of a class-B push-pull amplifier. (*Electronic Circuits VTU June/July 2008*)

12. Discuss the complementary -symmetry class-B amplifier and obtain its efficiency.

(*Electronic Circuits - I, Anna University, April/May 2008*)

13. Describe the operation of class-AB amplifier to avoid cross-over distortion.

(*Electronic Circuits - I, Anna University, April/May 2008*)

14. What is cross-over distortion? How it can be eliminated?

(*Electronic Circuits - I, Anna University, April/May 2008*)

15. With a neat circuit diagram, explain the operation of transformer coupled class-A power amplifier. *(Electronic Circuits - I, Anna University, April/May 2008)*

16. What are the factors affecting thermal stability of power amplifiers? Explain its significance. *(Electronic Circuits - I, Anna University, April/May 2008)*

17. Draw the class-B push pull emitter follower circuit and explain its operation. *(Electronic Circuits - I, Anna University, April/May 2008)*

18. What is called as crossover distortion and how to minimize this distortion? *(Electronic Circuits - I, Anna University, Nov/Dec 2007)*

19. Explain class-*A RC*-coupled power amplifier. *(Electronic Circuits - I, Anna University, Nov/Dec 2007)*

20. Class-C tuned amplifier. *(Electronic Circuits - II, Anna University, Nov/Dec 2007*

21. With neat circuit diagram, explain amplifier and derive its efficiency *(Electronic Circuits - II, Anna University, May/June 2007)*

MULTIPLE CHOICE QUESTIONS

1. In a class-A amplifier, the current in the output circuit flows for :

(*a*) 360° (*b*) 180° (*c*) 90° (*d*) less than 90°

2. The maximum collector circuit efficiency of class-A amplifier with a transformer coupled load is

(*a*) 25% (*b*) 50% (*c*) 78.6% (*d*) 95.4%

3. Which of the following is a main function of the transformer used in the output of a power amplifiers :

(*a*) To increase the output power

(*b*) To increase the voltage gain

(*c*) To match the load resistance with the output resistance of a transistor

(*d*) None of the above

4. Class-A power amplifiers, as compared to class-B amplifiers :

(*a*) Have more efficiency

(*b*) Have less distortion

(*c*) Are not susceptible to power supply hum

(*d*) None of the above

5. In which of the following amplifiers, does the amplifying device becomes cooler when the input device is changed from no signal to a rated value.

(*a*) Class-A amplifier (*b*) Class-B amplifier

(*c*) Class-AB amplifier (*d*) Class-C amplifier

6. High-power efficiency of a push-pull amplifier is due to the fact that

(*a*) each transistor conducts on different cycles of input

(*b*) transistors are placed in CE configuration

(*c*) there is no quiescient collector current

(*d*) low forward biasing voltage is required.

7. The circuit efficiency of a class-A amplifier can be increased by using
(*a*) low d.c. power input (*b*) direct-coupled load
(*c*) low-rating transistor (*d*) transformer-coupled load.

8. In a class-A amplifier, worst-case condition occurs with
(*a*) zero signal input (*b*) maximum signal input
(*c*) high load resistance (*d*) transformer coupling.

9. The output of a class-B amplifier
(*a*) is distortion-free
(*b*) consists of positive half-cycle only
(*c*) is like the output of a full-wave rectifier
(*d*) comprises short-duration current pulses.

10. The maximum overall efficiency of a transformer-coupled class-A amplifier is—per cent.
(*a*) 78.5 (*b*) 25 (*c*) 50 (*d*) 85

11. A transistor audio amplifier is found to have an overall efficiency of 70 per cent. Most probably, it is a .....amplifier.
(*a*) class-B push-pull (*b*) single-stage class-C
(*c*) transformer-coupled class-A (*d*) direct-coupled class-A

12. The main purpose of using transformer coupling in a class-A amplifier is to make it more
(*a*) distortion-free (*b*) bulky (*c*) costly (*d*) efficient.

13. A class-B push-pull amplifier has the main advantage of being free from
(*a*) any circuit imbalances (*b*) unwanted noise
(*c*) even-order harmonic distortion (*d*) d.c. magnetic saturation effects.

14. Crossover distortion occurs in........amplifiers.
(*a*) push-pull (*b*) class-A (*c*) class-B (*d*) class AB

15. The maximum overall efficiency of a class-B push-pull amplifier cannot exceed —per cent.
(*a*) 100 (*b*) 78.5 (*c*) 50 (*d*) 85

16. The dissipation at the collector is zero in the quiescent. State and increases with excitation in the case of a
(*a*) class A series-fed amplifier (*b*) class A transistor coupled amplifier
(*c*) class AB amplifier (*d*) class B amplifiers

17. Class AB operation is often used in power (large signal) amplifiers in order to,
(*a*) get maximum efficiency (*b*) remove even harmonics
(*c*) overcome cross-over distortion (*d*) reduce collector dissipation

ANSWERS

1. (*a*)	**2.** (*b*)	**3.** (*c*)	**4.** (*b*)	**5.** (*a*)
6. (*c*)	**7.** (*d*)	**8.** (*a*)	**9.** (*b*)	**10.** (*c*)
11. (*a*)	**12.** (*d*)	**13.** (*c*)	**14.** (*a*)	**15.** (*b*)
16. (*d*)	**17.** (*a*)			

❑❑❑

23

CHAPTER

Tuned Amplifiers

OUTLINE

Objectives

After completing this chapter, you should be able to :

- Calculate the resonant frequency of a parallel resonant circuit.
- Understand the bandwidth of a resonant circuit.
- Find the Q-factor of a given resonant circuit.
- Describe a single-tuned voltage amplifier and its frequency response.
- Explain the frequency response of a double-tuned voltage amplifier.
- Understand stagger-tuned voltage amplifier.

23-1. Introduction

In our day to day work, sometimes it is desired that an amplifier should be a selective one, *i.e.*, the amplifier should select a desired frequency (or a narrow band of frequencies) and amplify it to the desired levels. For example, radio and television stations transmit at a particular radio frequency (*i.e.*, a frequency in the range of 30 kHz to 300 MHz) assigned to them. The radio or a television receiver is required to pick up and amplify the desired radio frequency signal, while rejecting all others. This job cannot be done by the audio amplifiers discussed in the last chapter. It is because of the fact that such amplifiers have resistive loads and consequently their gains is independent of the signal frequency over a large bandwidth. In other words, since the audio amplifier amplifies a wide range of frequencies equally well, therefore it cannot be used for the selection of a specific desired frequency, while rejecting all other frequencies.

In order to pick up and amplify the desired radio frequency signal, the resistive load in the audio amplifier is replaced by a tuned circuit (also called a parallel resonant circuit) as shown in Fig. 23-1. The tuned circuit is capable of selecting as particular frequency and rejecting the others.

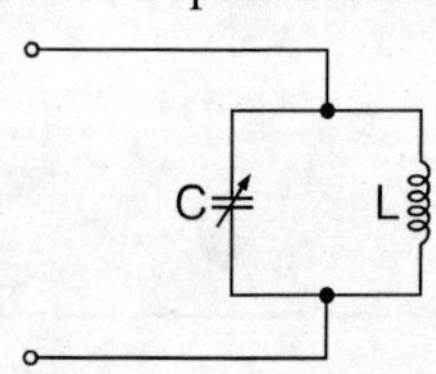

Fig. 23-1. *A tuned circuit.*

Thus, the use of tuned circuit in the transistor amplifier circuit, makes possible the selection and amplification of a particular desired radio frequency. Such an amplifier is called a ***tuned voltage amplifier***.

Thus an amplifier, which amplifies a specific frequency (or a narrow band of frequencies), is called a *tuned voltage amplifier.* It serves the following two purposes :

(*i*) Selection of a desired radio frequency signal.

(*ii*) Amplification of the selected signal to a suitable voltage level.

23-2. Parallel Resonant Circuit

A parallel resonant circuit consists of an inductor (L) and a capacitor (C) connected in parallel to each other with respect to a supply source. Fig. 23-2 shows a parallel resonant circuit connected across an a.c. supply source of variable frequency. Here, the resistance R represents the coil resistance. Its value is, usually, of the order of few ohms and therefore may be neglected as compared to the impedance of the resonant circuit.

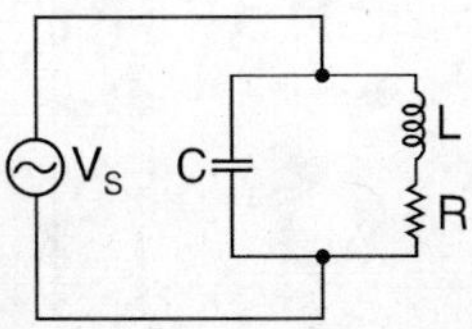

Fig. 23-2. *A parallel resonant circuit connected to an a.c. supply source.*

Now consider the frequency of the a.c. supply source to be varied suitably. As a result of this, the circuit will encounter different impedance at different frequencies. As the frequency is increased, the inductive reactance (X_L) is also increased and the capacitive reactance (X_C) is decreased. There is a certain frequency of the applied a.c. voltage at which the inductive reactance is equal to the capacitive reactance. This frequency is called resonance frequency and is designated by the symbol f_o. At resonance frequency, the circuit is said to be in electrical resonance. Under resonance condition, the ***impedance of the resonance circuit becomes maximum (and the line current i.e., the current drawn from the source) is minimum.***

The expression for resonant frequency may be obtained from the condition,

$$X_L = X_C$$

Substituting the values of X_L equal to $2\pi f_o \cdot L$ and X_C equal to $1/2\pi f_o \cdot C$ in this equation,

$$2\pi f_o \cdot L = 1/\pi f_o \cdot C$$

$$\therefore \quad f_0 = \frac{1}{2\pi\sqrt{L \cdot C}} = \frac{0.159}{\sqrt{L \cdot C}}$$

If the value of inductance is in henrys and the capacitance is by farads, then the resonant frequency is expressed in hertz.

Example 23-1. *A parallel resonant circuit has an inductance of 150 μH and a capacitor of 100 pF. Find the value of resonant frequency.* *(Electronic Circuits, Anna Univ., May/Jun 2007)*

Solution. Given: $L = 150\ \mu H = 150 \times 10^{-6}$ H and $C = 100$ pF $= 100 \times 10^{-12}$ F.

We know that the resonant frequency,

$$f_o = \frac{0.159}{\sqrt{L \cdot C}} = \frac{0.159}{\sqrt{(150 \times 10^{-6}) \times (100 \times 10^{-12})}} \text{Hz}$$

$$= 1.3 \times 10^6 \text{ Hz} = 1.3 \text{ MHz} \textbf{ Ans.}$$

Example 23-2. *Calculate the resonant frequency of a class-C tuned amplifier whose capacitor value is 10 pF and inductor value is 1mH.*

(Electronic Circuits - II, Anna University, May/June 2007)

Solution. Given: $C = 10\ pF = 10 \times 10^{-12}$, $L = 1mH = 1 \times 10^{3}\ H$

We know that resonant frequency,

$$f_0 = \frac{1}{2\pi\sqrt{LC}} = \frac{1}{2\pi\sqrt{1 \times 10^3\ (10 \times 10^{-12})}}$$

$$= 1.59 \times 10^3 \text{ Hz} = 1.59 \text{ kHz} \qquad \textbf{Ans.}$$

23-3. Resonance Curve

It is a curve, which shows the variation of circuit impedance (or circuit current) with the change in frequency of the applied voltage.

Fig. 23-3 (*a*) shows the variation of circuit impedance (Z_p) with the change in frequency of the applied voltage. It is evident from this curve, that the impedance is maximum at the resonance and its value is equal to *L/ CR, i.e.*,

$$Z_p = L/CR$$

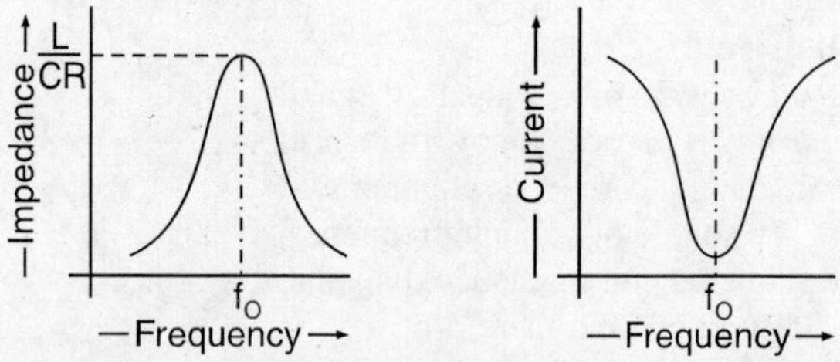

(*a*) Impedance versus frequency. (*b*) Current versus frequency.

Fig. 23-3. *Resonance curves of a parallel tuned circuit.*

As the frequency is changed, above or below the resonance, the value of impedance decreases rapidly. Fig. 23-3 (*b*) shows the resonance curve, in terms of the variation of circuit current with the change in frequency of the applied voltage. It is evident from the curve that the circuit current is minimum at resonance frequency and its value is equal to *V*/(*L*/*CR*). As the frequency is changed, above or below the resonance, the value of circuit current increases rapidly.

Example 23-2. *A parallel resonant circuit consists of a capacitor of 100 pF and an inductor of 100 μH. The inductor has a resistance of 5 Ω. Find the value of frequency at which the circuit will resonate and the circuit impedance at resonance.*

Solution. Given : $C = 100$ pF $= 100 \times 10^{-12}$ F; $L = 100$ μH $= 100 \times 10^{-6}$ H and R = 5 Ω.

Resonant frequency

We know that the resonant frequency,

$$f_o = \frac{0.159}{\sqrt{L \cdot C}} = \frac{0.159}{\sqrt{(100 \times 10^{-6}) \times (100 \times 10^{-12})}}$$

$$= 1.59 \times 10^6 = 1.59 \text{ MHz } \textbf{Ans.}$$

Circuit impedance at resonance

We also know that the circuit impedance at resonance,

$$Z_p = \frac{L}{CR} \frac{100 \times 10^{-6}}{(100 \times 10^{-12}) \times 5} = 2 \times 10^5 \ \Omega = 200 \text{ k}\Omega \ \textbf{Ans.}$$

23-4. Bandwidth of a Parallel Resonant Circuit

We have already discussed in the last article about the variation of circuit impedance (or current) of a parallel resonant circuit. This variation gives the resonance circuit an ability to discriminate between the frequency at resonance and those at off resonance. This discriminating ability of a resonance circuit is expressed in terms of its ***bandwidth.***

The bandwidth of a resonant circuit is given by the band of frequencies, which lie between two points (*A* and *B*) on either side of the resonance curve where the impedance drops to $1/\sqrt{2}$ or 0.707 of its maximum value at resonance as shown in Fig. 23-4.

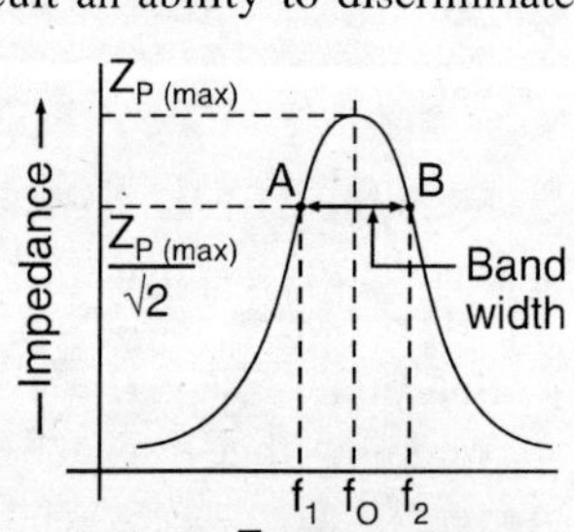

Fig. 23-4. *Bandwidth of a parallel resonance circuit.*

The resonant circuit gives practically, the same response as at resonance, for the band of frequencies, which lie within the bandwidth. This band is also called the pass-band of the resonant circuit. Thus if f_1 is the frequency corresponding to point *A* and f_2, corresponding to the point *B*, then the bandwidth,

$$\text{B.W.} = f_2 - f_1 = \Delta f$$

The points marked *A* and *B* on the resonance curve shown in Fig. 23-4 are known as half power points, because the power developed at each of these points in half the power developed at resonance. Consequently, the frequencies f_1 and f_2 are known as lower half-power and upper half-power frequencies respectively.

23-5. Sharpness of Resonance

The resonance curve, of a resonant circuit, is required to be as sharp as possible in order to provide a high selectivity. The sharp resonance curve means that the impedance falls off rapidly as the frequency is varied above and below the resonant frequency. Mathematically, the sharpness of a resonance curve is defined as ratio of the bandwidth of the circuit to its resonant frequency, *i.e.*,

Sharpness of resonance

$$= \frac{\text{Bandwidth}}{\text{Resonant frequency}} = \frac{BW}{f_o} = \frac{f_2 - f_1}{f_o} = \frac{1}{Q_o} \quad ...(i)$$

where, Q_o is called the **quality factor** or simply ***Q-factor*** of the circuit at resonance. Its value is equal to the ratio of inductive reactance (or capacitive reactance) at resonance to the circuit resistance. Mathematically, the *Q-factor*,

$$Q_o = \frac{X_L}{R} = \frac{\omega_o \cdot L}{R} = \frac{2\pi f_o \cdot L}{R}$$

where, L = Value of circuit inductance, and

R = Value of circuit resistance or coil resistance.

It is evident from equation (*i*) that the sharpness of resonance (*i.e.*, selectivity) depends upon the value of *Q*-factor. Higher the value of Q_o, more selective is the resonant circuit. The value of Q_o, in turn, depends upon the value of coil resistance (*R*). Smaller the value of coil resistance, higher is the value Q_o.

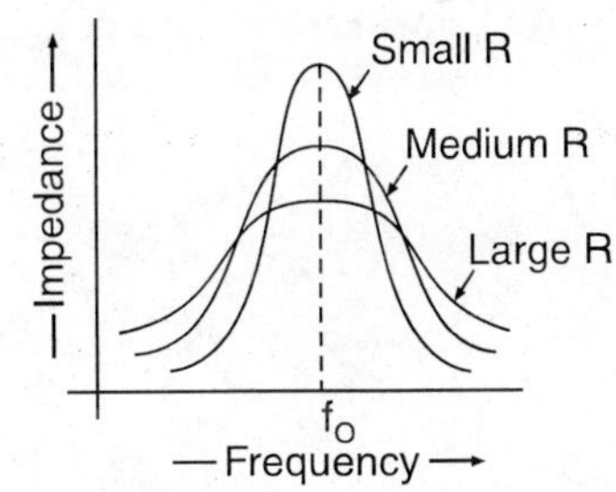

Fig. 23-5. *Effect of coil resistance (R) on the sharpness of resonance curves.*

Fig. 23-5 shows the effect of coil resistance (*R*) on the sharpness of the resonance curve. It is evident from this curve that when the coil resistance is small, the resonance curve is very sharp. On the other hand, if the coil resistance is large, the resonance curve is less sharp. It may be noted from equation (*i*), that the bandwidth,

$$\text{BW} = \frac{f_o}{Q_o} \quad ...(ii)$$

or

$$f_o = \text{BW} \times Q_o \quad ...(iii)$$

It is evident from equation (*ii*) that higher the value of quality factor (Q_o), the lower will be the value of bandwidth. Thus we conclude that a *higher value of quality factor provides a higher selectivity, but a smaller bandwidth, whereas a lower value of quality factor provides a poor selectivity but a larger bandwidth.*

Example 23-3. *Calculate the bandwidth between the half-power points of circuit which resonates at 1MHz and has a Q of 100.* (*UPSC Engg. Services* 2002)

Solution. Given f_o = 1MHz and Q_o = 100

We know that bandwidth of a circuit,

$$BW = \frac{f_o}{Q_o} = \frac{1\ MHz}{100} = 0.01\ MHz = 10\ kHz \ \textbf{Ans.}$$

Example 23-4. *A tuned circuit has resonant frequency of 1600 kHz and a bandwidth of 10 kHz. What is the value of its Q-factor?*

Solution. Given: f_o = 1600 kHz and BW = 10 kHz.

Let Q_o = Value of quality factor.

We know that the bandwidth (BW),

$$10 = \frac{1600}{Q_o}$$

or $Q_o = 1600/10 = 160$ **Ans.**

Example 23-5. *A tuned amplifier has its maximum gain at a frequency of 2MHz and has a bandwidth of 50 kHz. Calculate the Q-factor.*

Solution. Given: $f_o = 2$ MHz $= 2 \times 10^6$ Hz and BW = 50 kHz $= 50 \times 10^3$ Hz.

Let Q_o = The value of Q-factor,

We know that the bandwidth (BW_o),

$$50 \times 10^3 = \frac{f_o}{Q_o} = \frac{2 \times 10^6}{Q_o}$$

or $$Q_o = \frac{2 \times 10^6}{50 \times 10^3} = 40 \text{ **Ans.**}$$

Example 23-6. *A circuit is resonant at 455 kHz and has a 10 kHz bandwidth. The inductive reactance is 1255 Ω. What is the parallel impedance of the circuit at resonance ?*

Solution. Given: $f_o = 455$ kHz $= 455 \times 10^3$ Hz; BW = 10 kHz $= 10 \times 10^3$ Hz and $X_L = 1255\ \Omega$.

Let Z_P = Value of impedance at resonance

We know that the value of bandwidth (BW),

$$10 \times 10^3 = \frac{f_o}{Q_o} = \frac{455 \times 10^3}{Q_o}$$

$\therefore$ $$Q_o = \frac{455 \times 10^3}{10 \times 10^3} = 45.5$$

and value of Q-factor (Q_o),

$$45.5 = \frac{X_L}{R} = \frac{1255}{R}$$

$\therefore$ $$R = 1255/45.5 = 27.6\ \Omega$$

We also know that the inductive reactance (X_L),

$$1255 = 2\pi f_o \cdot L = 2\pi \times (455 \times 10^3) \times L = 2859 \times 10^3\, L$$

$$L = 1255/\,(2859 \times 10^3) = 0.439 \times 10^{-3}$$

and value of capacitive reactance at resonance,

$$X_C = X_L\ 1255\ \Omega$$

or $$\frac{1}{2\pi f_o \cdot C} = 1255$$

$$\frac{1}{2\pi \times (455 \times 10^3) \times C} = 1255$$

or $$\frac{1}{(2859 \times 10^3)\, C} = 1255$$

$\therefore$ $$C = \frac{1}{(1255 \times 2859 \times 10^3)} = 278.7 \times 10^{-12} \text{ F}$$

and value of circuit impedance at resonance,

$$Z_P = \frac{L}{CR} = \frac{0.439 \times 10^{-3}}{(278.7 \times 10^{-12}) \times 27.6} = 57 \times 10^3\ \Omega = 57\ k\Omega\ \textbf{Ans.}$$

Example 26-7. *Design a tuned amplifier using FET to have f_0 = 1 MHz, 3 dB bandwidth is 10 kHz and maximum gain is – 10, FET has gm = 5 mA/V, r_d = 10 kΩ.*

We know that the maximum gain of the amplifier (A_i(max))

$$A_{i\ (\max)} = -g_m R$$

Therefore,
$$R = -\frac{(-10)}{5\ mA/V} = 2 \times 10^3 \Omega$$

We also know that the 3 *dB* bandwidth of the amplifier, (*BW*),

$$BW = \frac{1}{2\pi.R.C}$$

$$C = \frac{1}{2\pi.R.BW} = \frac{1}{2\pi \times (2 \times 10^3) \times (10 \times 10^3)} = 7.957n$$

We also know that the resonant frequency (f_0),

$$f_0 = \frac{1}{2\pi\sqrt{LC}}$$

$$1 \times 10^6 = \frac{1}{2\pi\sqrt{L \times 7.957 \times 10^{-9}}}$$

or
$$L = 3.183\ MH$$

We know that,
$$R = R_p \parallel r_d \quad \text{or} \quad \frac{1}{R} = \frac{1}{R_p} + \frac{1}{rd}$$

$$\frac{1}{R} = \frac{1}{R} - \frac{1}{R_p} = \frac{1}{1} - \frac{1}{rd}$$

$\therefore$
$$R_p = 2.5\ \text{kW}$$

$$Q_0 = \frac{R_p}{\omega_0.L} = \frac{2.5\,k\Omega}{2 \times \pi \times (1 \times 10^6) \times (3.183 \times 10^{-6})}$$

We also know that the series resistor is given by,

$$R_s = \frac{\omega_0 L}{Q_0} = \frac{2 \times \pi \times (1 \times 10^6) \times (3.183 \times 10^{-6})}{125} = 0.16\Omega$$

Fig. 23-6 shows the *FET* tuned amplifier with the desired values of the frequency, bandwidth and maximum gain.

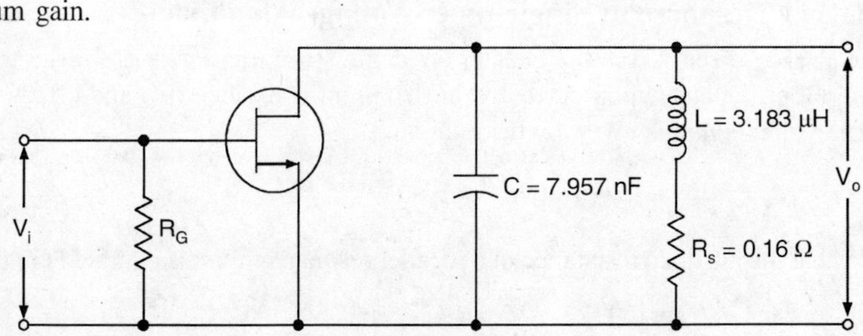

Fig. 23-6

23-6. Single-tuned Voltage Amplifier

Fig. 23-7 shows the circuit of a single-tuned voltage amplifier using a bipolar junction transistor (BJT). The circuit shown in Fig. 23-7 (*a*) is called capacitively coupled tuned amplifier because the output is taken through a coupling capacitor (C_c). Similarly, the circuit shown in Fig. 23-7 (*b*) is called inductively coupled tuned amplifier because the output is taken across an inductor. Both these circuits consist of a transistor amplifier and tuned circuit as the load. The values of capacitance (C) and inductance (L) of the tuned circuit are selected in such a way that the resonant frequency of the tuned circuit is equal to the frequency to be selected and amplified. The resistors R_1, R_2 and R_E are called biasing resistors. These resistors provide the d.c. operating currents and voltages for the transistor.

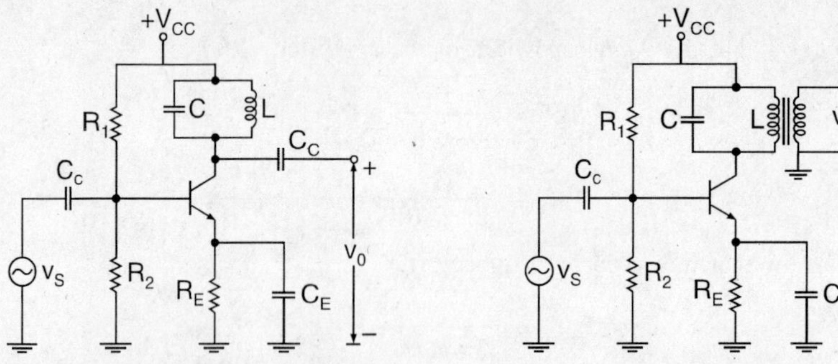

(*a*) Capacitively coupled. (*b*) Inductively coupled.

Fig. 23-7. *Single-tuned voltage amplifier using field-effect transistor.*

It will be interesting to know that the other active devices like field-effect transistor (FET's) and metal-oxide semiconductor field-effect transistors (MOSFET's) may also be used in tuned voltage amplifier circuit instead of bipolar junction transistor. Fig. 23-8 shows the circuit of single-tuned voltage amplifier using a field effect transistor.

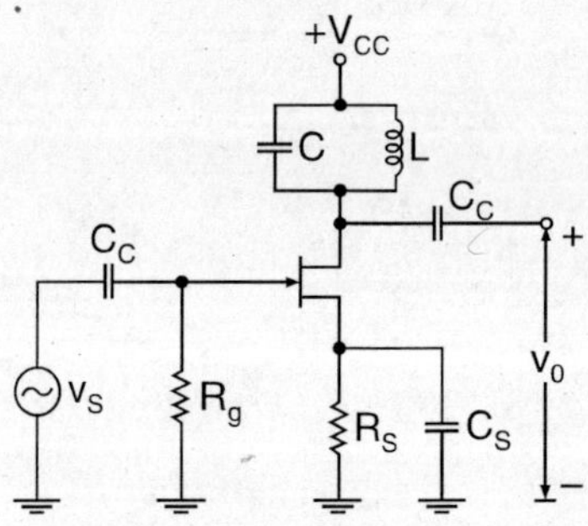

Fig. 23-8. *Single-tuned voltage amplifier using field-effect transistor.*

The working of tuned voltage amplifier may be understood by considering a radio frequency signal, to be amplified, applied at the input of the amplifier. The resonant frequency, of the tuned circuit, is made equal to the frequency of the input signal by changing the value of capacitance (C) or inductance (L). When the frequency of the tuned circuit becomes equal to that of the input signal, a large signal appears across the output terminals. It may be noted that if the input signal is a complex wave (*i.e.*, it contains many frequency components), in that case the signal with frequency equal to the resonant frequency will be amplified. And all the other frequencies will be rejected by the tuned circuit.

23-7. Frequency Response of Single-tuned Voltage Amplifier

We have already discussed in the chapter on single stage transistors amplifier that the voltage gain of an amplifier depends upon current gain (β), input resistance (R_i) and effective or a.c. load resistance. The voltage gain is given by the relation,

$$A_v = \beta \times \frac{r_L}{R_i}$$

We know that the a.c. load resistance of a parallel resonant circuit (*i.e.*, tuned circuit) is given by the relation,

$$r_L = Z_p = \frac{L}{CR}$$

where, L = Value of inductance,
C = Value of capacitance, and
R = Value of effective resistance of the inductor.

∴ Voltage gain of a tuned voltage amplifier is given by the relation,

$$A_v = \beta \times \frac{\frac{L}{CR}}{R_i}$$

We also know that the value of the quantity L/CR (called impedance of the tuned circuit) is very high at the resonant frequency and it decreases as the frequency changes above or below the resonant frequency. Therefore, voltage gain of a tuned amplifier is very high at the resonant frequency and decreases as the frequency changes above or below the resonant frequency.

The above facts are shown in the form of a voltage gain versus frequency plot shown in Fig. 23-9. Such a plot is called frequency response curve of a tuned voltage amplifier.

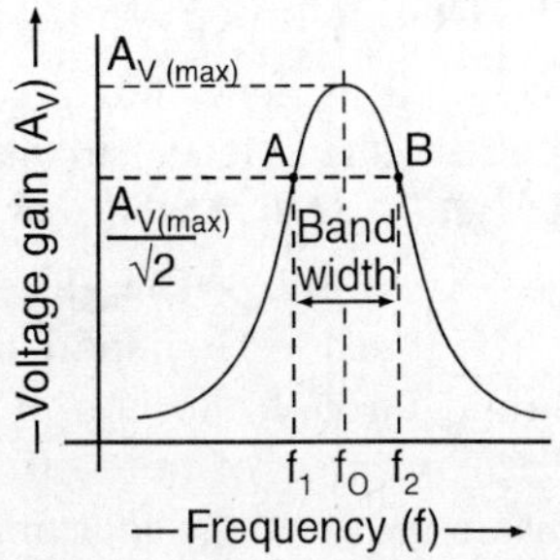

Fig. 23-9. Frequency response curve.

The bandwidth (BW) of an amplifier is equal to the frequency difference between the points A and B on either side of the resonant frequency, where the value of voltage gain drops to $1/\sqrt{2}$ of its maximum value at resonance. Thus, bandwidth

$$\text{B.W.} = \Delta f = f_2 - f_1 = \frac{f_o}{Q_o}$$

where, Q_o is the quality factor (or Q-factor) of the tuned circuit.

23-8. Limitations of Single-tuned Voltage Amplifier

The tuned voltage amplifier are used in the wireless communication systems. Generally, the communication transmitters do not transmit the audio (*i.e.*, voice or music) signal directly. Instead the audio signal is raised to some high frequency level. This high frequency is in the radio frequency (RF) range and it serves as a carrier of the audio signal. The process of raising the audio signal to the radio frequency level is called **modulation**. The communication transmitters transmit the modulated wave.

The modulated wave has a relatively narrow band of frequencies centred around the carrier frequency (f_c) as shown in Fig. 23-10 (*a*). The frequencies (f_1 and f_2) in the narrow band actually represent the information and hence is an important part of the transmitted signal.

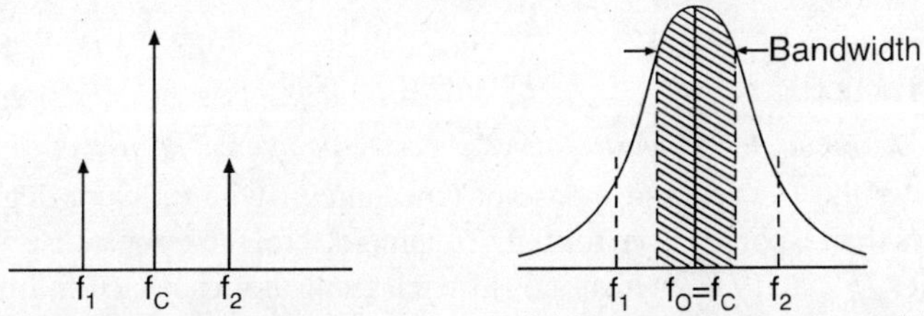

(*a*) Frequency spectrum of the transmitted signal.

(*b*) Frequency response of the tuned voltage amplifier.

Fig. 23-10.

The tuned voltage amplifier, in communication receivers, is used to select the desired carrier frequency and amplifying the complete band of frequencies around the selected carrier frequency. In other words, tuned voltage amplifiers are required to be highly selective. But the high selectivity requires a tuned circuit with a high *Q*-factor. We also know that a high-*Q* circuit will give a high voltage gain. But, at the same time, it will give reduced bandwidth (because bandwidth is reciprocal to the *Q*-factor). It means that a tuned voltage amplifier with reduced bandwidth may not be able to amplify equally the complete band of the transmitted signal as shown in Fig. 23-10 (*b*). In other words, narrow bandwidth or smaller pass-band of the amplifier will result in a poor reproduction of the audio signal. It is the major limitation of a single tuned, voltage amplifier and is overcome by using a double tuned circuit. In double-tuned voltage amplifiers, we use two tuned circuits which are inductively coupled to each other.

23-9. Double-tuned Voltage Amplifier

Fig. 23-11 shows the circuit of a double tuned voltage amplifier. It consists of a transistor amplifier with two tuned circuits. One of the tuned circuits (*i.e.*, L_1 C_1) is shown as the collector load and other (*i.e.*, L_2C_2) as the output. The resistor R_1, R_2 and R_E are used to provided d.c. currents and voltages for the transistor operation.

The signal to be amplified is applied at the input terminal through the coupling capacitor (C_C).The resonant frequency of the tuned circuit L_1C_1 is made equal to that of the signal (by adjusting L_1or C_1). Under these conditions, the tuned circuit offers a very high impedance to the input signal. As a result of this, a large output appears across the tuned circuit L_1C_1. The output from this tuned circuit is inductively coupled to the L_2C_2 tuned circuit.

The double tuned voltage amplifiers are extensively used in intermediate frequency (*IF*) amplifiers in radio and television receivers.

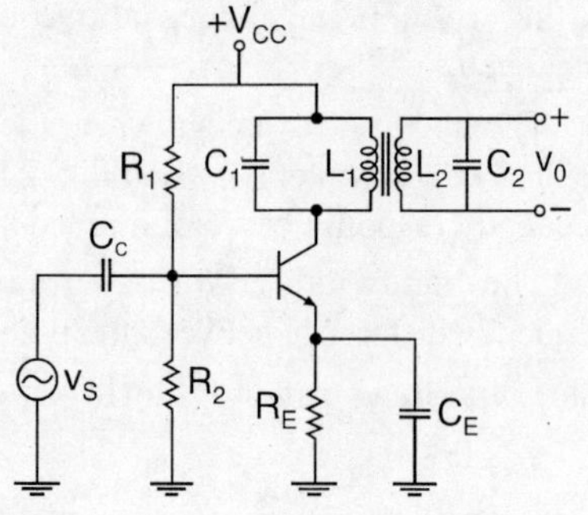

Fig. 23-11. *Double tuned voltage amplifier.*

23-10. Frequency Response of Double-tuned Voltage Amplifier

The frequency response of a double tuned voltage amplifier depends upon its degree of coupling (also called coefficient of coupling) between the two tuned circuits. The degree of coupling gives us an idea of the amount of energy transferred between the two tuned circuits.

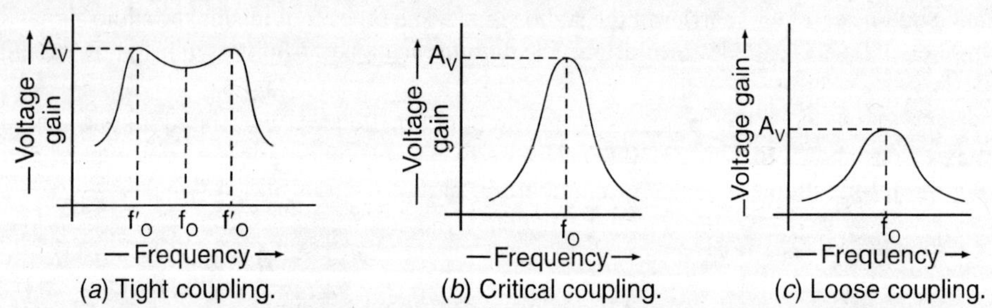

Fig. 23-12. *Frequency response of tuned circuits with different degrees of coupling.*

Fig. 23-12 shows the frequency response of tuned circuit with different degrees of coupling. Fig. 23-12 (*a*) shows the response curve for tight coupling. Here, the resonance is shown to occur at two new frequencies (f'_o and f''_o) which are different from the resonance frequency (f_o). As the degree of coupling is decreased, the two frequencies (f'_o and f''_o) come closer and merge into one frequency (f_o) at critical coupling. The frequency response curve for such type of coupling is shown in Fig. 23-12 (*b*). When the degree of coupling is decreased below the critical coupling, a single peak of reduced height is obtained as shown in Fig. 23-12(*c*).

The tuned circuits of a double-tuned voltage amplifier are tightly coupled to each other. Therefore, the frequency response of a double-tuned voltage amplifier is shown in Fig. 23-13. This type of response curve provides a high selectivity, high gain and relatively large bandwidth to the tuned voltage amplifier.

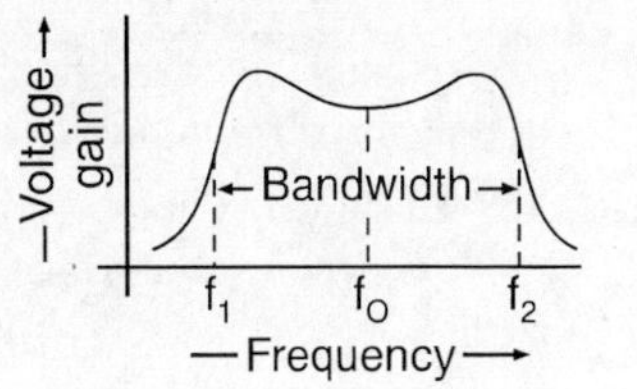

Fig. 23-13. *Frequency response of a double-tuned voltage amplifier.*

23-11. Stagger-tuned Voltage Amplifier

It has been observed that if two or more tuned circuits, which are synchronously tuned are cascaded, the overall bandwidth decreases. However, if the different tuned circuits which are cascaded, are tuned to slightly different frequencies, it is possible to obtain an increased bandwidth with a flat pass-band with steep sides. This technique is known as ***stagger-tuning.***

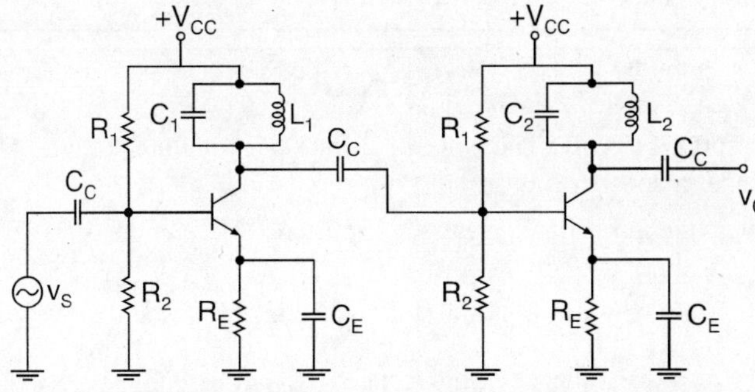

Fig. 23-14. *A two-stage stagger tuned voltage amplifier.*

Fig. 23-14 shows a two-stage tuned voltage amplifier. The stagger tuning in this circuit, may be achieved by resonating the tuned circuits L_1C_1 and L_2C_2 to slightly different frequencies.

Fig. 23-15 shows the frequency response of a stagger-tuned amplifier. In this figure, curve '*a*' shows the gain versus frequency response of the L_1C_1 tuned circuit. Similarly, the curve '*b*' shows the response of the L_2C_2 tuned circuit. The curve '*c*' indicates the combined (or the overall) response of the circuit. It is evident from this curve that the amplifier has a greater bandwidth and flatter pass-band. It has been found that more the number of stages used, the flatter will be the pass-band and steeper will be the gain fall-off outside the pass-band. It may be noted that because of stagger-tuning, there is a loss of voltage gain (the gain reduces from A_v to A'_v). If an optimum stagger-tuning is employed, the response curve of the amplifier is very close to a rectangular response curve. Such a response curve is known as *Butterworth response*.

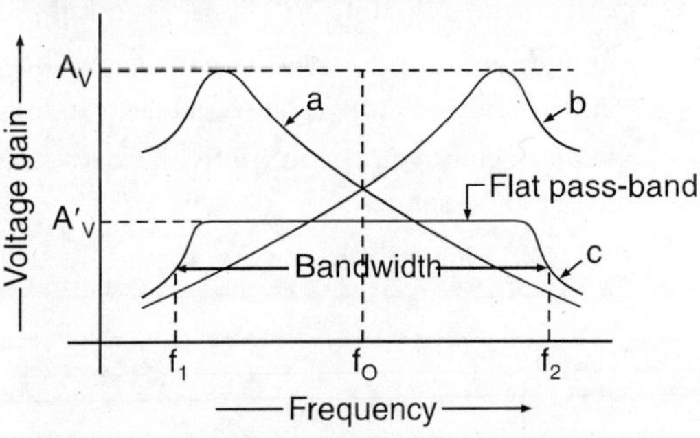

Fig. 23-15. *Frequency response of a stagger-tuned voltage amplifier.*

23-12. Neutralization and its need in tuned amplifiers

As a matter of fact, in tuned *RF* amplifiers, the interjunction capacitance, C_{bc} becomes dominant, *i.e.*, the reactance becomes low. Due to this, a feedback from collector to base is created. So the circuit becomes unstable if the feedback signal is in phase with the input signal. In order to prevent the circuit to oscillate, the gain of the amplifier is to be reduced to make amplifier stable.

Instead of loosing circuit performance by the reduction of gain, L.A. Hazeltive introduced a circuit in which troublesome effert of collector to-base capacitance of the transistor is neutralized by introducing a signal which cancels the signal coupled through to collertor -to- base capacitance. This is called *neutralization*.

SUMMARY

In this chapter you have learned that :

1. The tuned voltage amplifiers are used to amplify a specific frequency or a narrow band of frequencies, for which it is tuned.
2. Under resonance condition, the impedance of the resonance circuit becomes maximum and the line current becomes minimum.
3. A double-tuned voltage amplifier has a wider bandwidth compared to single-tuned voltage amplifier.
4. A stagger-tuned voltage amplifier gives a flat pass-band with an increased bandwidth than a double-tuned voltage amplifier.

GLOSSARY

Bandwidth of resonant circuit : The range of frequencies for which the circuit gives practically same response as at resonance.

Double-tuned voltage amplifier : A tuned amplifier, which has two resonance circuits (*i.e.,* LC circuit) and can show wider bandwidth.

Pass-band : The frequency range of a tuned amplifier, for which the response is higher than or equal to $\frac{1}{\sqrt{2}}$ times the response at resonance frequency.

Quality factor (Q-factor) : The ratio of inductive reactance (or capacitive reactance) at resonance frequency, to the circuit resistance.

Resonance curve : A curve, which shows the variation of circuit impedance (or circuit current) with the change in frequency of the applied voltage.

Resonance frequency : The frequency at which the impedance of the resonance circuit becomes maximum and the line current is minimum.

Single-tuned voltage amplifier : Tuned voltage amplifier which is tuned for only one resonance frequency. It uses only one resonance circuit.

Stagger-tuned voltage amplifier : Cascaded tuned amplifiers, which gives wider bandwidth and flat pass-band.

Tuned voltage amplifier : An amplifier, which amplifiers a specific frequency.

NUMERICAL PROBLEMS

1. A parallel resonant circuit has an inductance of 100 μH and a capacitor of 100 pF. Find the value of resonant frequency. (**Ans.** 1.59 MHz)
2. A parallel *LC* circuit consists of a capacitor of 3000 pF and an inductor of 100 μH. The inductor has a resistance of 5 Ω . Find the value of resonant frequency and the circuit impedance at resonance. (**Ans.** 291 kHz and 6.66 kΩ)
3. A tuned circuit has resonant frequency of 800 kHz and a bandwidth of 10 kHz. What is the value of its *Q*-factor? (**Ans.** 80)
4. A parallel resonant circuit has a bandwidth of 15 kHz and *Q*-factor of 120. What is the resonant frequency of the circuit? (**Ans.** 1800 kHz)

DESCRIPTIVE QUESTIONS

1. What is a tuned voltage amplifier? Explain it briefly.
2. Derive an expression for the resonance frequency of a parallel resonant circuit.
3. Explain the meaning of a resonance curve. Plot a resonance curve for a parallel tuned circuit.
4. Discuss the frequency response of a single-tuned amplifier.
5. Draw the circuit diagram of a double-tuned voltage amplifier and explain how its frequency response is better than that of a single-tuned voltage amplifier.
6. What is stagger-tuning? How it is employed in the tuned voltage amplifier?
7. Draw the circuit of a single tuned amplifier. Draw its frequency response. Using small signal equivalent circuit derives expressions for gain at responance and 3 *dB* bandwidth.
   (*Electronic Circuits - II, Anna University, Nov/Dec 2007*)
8. Draw the equivalent circuit of capacitance coupled single tuned amplifier and derive the expression fir the bandwidth.(*Electronic Circuits - II, Anna University, May/June 2007*)
9. Write shorts notes on: Stagger tuned amplifier
   (*Electronic Circuits - II, Anna University, May/June 2007*)
10. Discuss the working of single tuned and synchronously tuned amplifiers.
    (*Electronic Circuits - II, Anna University, May/June 2007*)
11. Explain class 'C' tuned amplifier derives its efficiency.
    (*Electronic Circuits - II, Anna University, May/June 2007*)
12. Brief the relation between bandwidth and *Q*-factor
    (*Electronic Circuits - II, Anna University, May/June 2007*)

MULTIPLE CHOICE QUESTIONS

1. Which of the following amplifiers should be used to amplify a radio frequency signal?
   (*a*) Audio-frequency amplifier (*b*) Push-pull amplifier
   (*c*) Video amplifier (*d*) Tuned voltage amplifier
2. An amplifier has a pass-band from 745 kHz to 755 kHz, and a resonant frequency of 750 kHz. Its bandwidth is
   (*a*) 5 kHz (*b*) 10 kHz (*c*) 20 kHz (*d*) None of the above
3. If the *Q*-factor of a resonant circuit is doubled, its bandwidth is
   (*a*) Halved (*b*) Doubled (*c*) Quadrupled (*d*) Remains the same
4. A tuned circuit has resonant frequency of 1 MHz. To get a bandwidth of 20 kHz, its *Q*-factor must be equal to
   (*a*) 20 (*b*) 50 (*c*) 100 (*d*) 500
5. Stagger-tuning in tuned amplifiers is achieved by using
   (*a*) Double tuned circuit
   (*b*) Tuned circuits which are tuned to same frequency
   (*c*) Tuned circuits which are tuned to slightly different frequencies
   (*d*) None of the above
6. If a Q of a single-stage tuned amplifier is doubled, than it bandwidth will
   (*a*) remain same (*b*) become half (*c*) become double (*d*) become four times
   (*UPSC Engg-Services* 2000)

ANSWERS

1. (*d*) **2.** (*b*) **3.** (*a*) **4.** (*b*) **5.** (*c*) **6.** (*b*)

❑❑❑

24

CHAPTER

Feedback Amplifiers

OUTLINE

Objectives

After completing this chapter, you should be able to :

- Know the difference between positive feedback and negative feedback.
- Calculate the values of amplifier gain (A'_v) and feedback ratio (β) of the feedback amplifier.
- List the adavantages and disadvantages of negative feedback.
- Know the affect of negative feedback on the B.W. and distortion in amplifiers.
- List the different types of feedback connections.
- Compare and contract the characteristics of different types of feedback connections.

24-1. Introduction

An ideal linear amplifier, in the mid-frequency region, provides an output signal, which is an *exact replica* of the input signal. However, the practical amplifiers depart from the ideal for many reasons, *e.g.*, non-linearity of transistor characteristics, parameter variation, temperature effects, *etc*. Some of these factors can be minimized by improving the involved basic devices. But there is a limit, beyond which this approach does not work. The best approach is to use the *principle of feedback* to achieve a desired degree of improvement. The *feedback* is a process of injecting some energy from the output and then return it back to the input. The amplifiers, which use the feedback principle, are called *feedback amplifiers*.

A feedback amplifier, basically, consists of two parts namely amplifier circuit and feedback circuit (or network). Depending upon whether the feedback signal increases or decreases the input signal, there are two basic types of feedback in amplifiers:

(1) *Positive feedback* : If the feedback signal (voltage or current) is applied in such a way that it is *in phase with the input signal* and thus increases it, then it is called a *positive feedback*. Sometimes, it is also known as *regenerative feedback* or *direct feedback*.

The positive feedback increases the gain of amplifier. However, if produces excessive distortion due to which it is seldom used in amplifiers. The positive feedback is used in oscillators.

(2) *Negative Feedback* : If the feedback signal (*i.e.*, voltage or current) is applied in such a way that it is *out of phase with the input signal* and thus decreases it, then it is called a *negative feedback*. Sometimes, it is also called as *degenerative feedback* or *inverse feedback*.

The negative feedback reduces gain of the amplifier. However, it improves the amplifier performance in many other respects. Thus a negative feedback is frequently used in small-signal as well as with large-signal amplifier circuits.

24-2. Principle of Feedback Amplifiers

Fig. 24-1 shows an ordinary amplifier, *i.e.*, an amplifier which does not employ any feedback.

We know that the voltage gain of such an amplifier is given by the ratio of the output voltage (V_o) to the input voltage (V_{in}). Mathematically, the voltage gain,

$$\boxed{A_v = \frac{V_o}{V_{in}}}$$

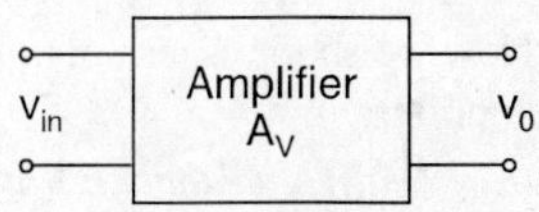

Fig. 24-1. *An ordinary amplifier with voltage gain.*

The voltage gain (A_v) is also known as *open loop gain*. Now let us add a feedback circuit to the amplifier as shown in Fig. 24-2. The addition of feedback circuit changes the values of amplifier gain, its input as well as the output voltages.

Let V'_o = Output voltage of feedback amplifier.

β = Fraction of the output voltage fed to the input. It is called feedback ratio and is different from common emitter transistor current gain, and

A'_v = Voltage gain of feedback amplifier.

The feedback circuit injects a fraction (β) of the output voltage (V'_o) and return it to the input voltage, whose value changes to $V_{in} \pm \beta \cdot V'_o$. It will be interesting to know that the value at input valtage is $V_{in} + \beta V'_0$ if the feedback voltage is in phase with the input voltage. And it is equal to $V_{in} - \beta \cdot V'_0$, if the feedback voltage is out of phase with the input voltage. This input voltage is amplified by the circuit and its value at the output,

$$A_v (V_{in} + \beta \cdot V'_o) = V'_o$$

Rearranging the above equation,

$$V'_o (1 - \beta \cdot A_v) = A_v \cdot V_{in}$$

or

$$\frac{V'_o}{V_{in}} = \frac{A_v}{1 - \beta \cdot A_v} = A'_v$$

where, A'_v is equal to V'_o/V_{in} and is the voltage gain of the feedback amplifier.

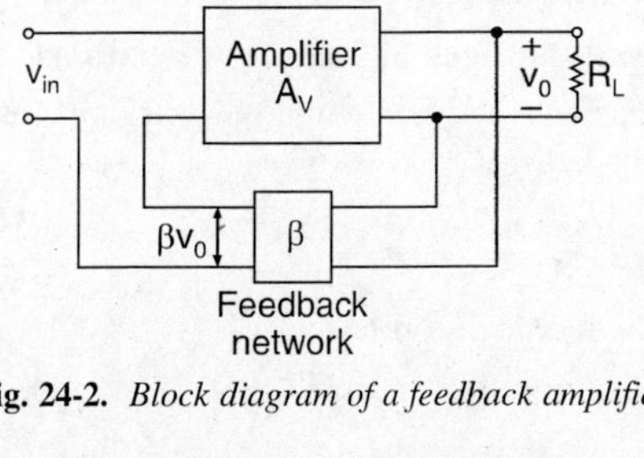

Fig. 24-2. *Block diagram of a feedback amplifier.*

For a positive feedback, the feedback ratio (β) is positive and for negative feedback, it is negative. Therefore, voltage gain of the feedback amplifier

$$A'_v = \frac{A_v}{1 - \beta \cdot A_v} \qquad \text{...(For positive feedback)}$$

$$= \frac{A_v}{1 - (-\beta \cdot A_v)} = \frac{A_v}{1 + \beta \cdot A_v} \qquad \text{...(For negative feedback)}$$

The term $(\beta \cdot A_v)$ is called *loop gain*. The difference between unity and the loop gain (*i.e.*, $1 - \beta \cdot A_v$) is called return difference (or *desentivity*). The amount of feedback introduced into an amplifier may also be expressed in decibels by the expression,

Feedback in dB,

$$= 20 \log_{10}\left(\frac{A'_v}{A_v}\right) = 20 \log_{10}\left(\frac{1}{1 + \beta \cdot A_v}\right)$$

Now we shall discuss in more details the effect of negative feedback and positive feedback on the amplifiers.

(a) Negative feedback : It may be noted from the voltage gain expression of negative feedback that as the term $(1 + \beta \cdot V_v) > 1$, the value of A'_v is smaller that A_v. In other words, gain of a negative feedback amplifier (A'_v) is smaller than the open loop gain (A_v). Thus, the negative feedback reduces the gain of an amplifier. It is due to this fact that a negative feedback is called *degenerative* feedback. The value of closed loop gain (A'_v) may also be looked from the following angle:

We know that the term $\beta \cdot A_v >> 1$. Therefore, $1 + \beta \cdot A_v \approx \beta \cdot A_v$.

Thus,
$$A'_v = \frac{A_v}{\beta \cdot A_v} = \frac{1}{\beta}$$

The above relation indicates that the voltage gain of a negative feedback amplifier (A'_v) depends only upon the feedback ratio (β). Though the value of voltage gain of a negative feedback amplifier (A'_v) is quite low as compared to (A_v), yet it is found to be very stable. It means that the voltage gain (A'_v) is not affected by any change in device parameters, supply voltage and from aging of circuit components *etc.*

It may be noted that stability of voltage gain (A'_v) comes only if the feedback network is made up of stable elements such as resistors, inductors and capacitors. If the feedback network has an active device, the gain will not remain stable.

(b) Positive feedback : It may be noted from the voltage gain expression of positive feedback that as the term $(1 - \beta \cdot A_v) < 1$, therefore $A'_v > A_v$. It shows that voltage gain of a positive feedback amplifier (A'_v) is greater than open loop gain (A_v). It is due to this fact that a positive feedback is called *regenerative* feedback.

A very interesting situation occurs, when the product $\beta \cdot A_v$ becomes equal to unity. In that case $1 - \beta \cdot A_v$ is equal to zero. Therefore, the voltage gain of a positive feedback amplifier becomes-*infinity*. It means that the circuit should give an output without any input. But in actual practice, it is not possible. As a matter of fact, the amplifier becomes an oscillator, which supplies its own input.

24-3. Advantages and Disadvantages of Negative Feedback

Though there are numerous advantages of employing negative feedback in the amplifier, yet some of the important are given below :

(*i*) Increased stability.

(*ii*) Increased bandwidth.

(*iii*) Less amplitude and harmonic distortion.

(*iv*) Decreased noise.

(*v*) Less frequency distortion.

(*vi*) Less phase distortion.

(*vii*) Input and output resistances can be modified as desired.

The only disadvantage of negative feedback is that it reduces the amplifier gain. However, this disadvantage is outweighed due to numerous advantages obtained through negative feedback.

Example 24-1. *The voltage gain of certain amplifier without feedback is 400. If the feedback ratio (β) is equal to 0.1, find the voltage gain of the amplifier with negative feedback.*

Solution. Given: $A_v = 400$ and $\beta = 0.1$

We know that the voltage gain of an amplifier with negative feedback,

$$A'_v = \frac{A_v}{1 + \beta \cdot A_v} = \frac{400}{1 + (0.1 \times 400)} = \frac{400}{41} = 9.76 \textbf{ Ans.}$$

Example 24-2. *An amplifier has a voltage gain of 1000. With negative feedback, the voltage gain reduces to 10. Calculate the fraction of the output that is feedback to the input.*

(Electronic Circuits-II, Anna Univ., May/June 2007)

Solution. Given: $A_v = 1000$ and $A'_v = 10$

Let β = Feedback ratio.

We know that the voltage gain with feedback (A'_v)

$$10 = \frac{A_v}{1 + \beta \cdot A_v} = \frac{1000}{1 + (\beta \times 1000)}$$

or

$$1 + 1000\,\beta = \frac{1000}{10} = 100$$

$$\therefore \quad \beta = \frac{99}{1000} = 0.099 \textbf{ Ans.}$$

Example 24-3. *The voltage gain of an amplifier without feedback and with negative feedback respectively are 100 and 20. Determine the percentage of negative feedback (β).*

(UPSC Engg. Services 2000)

Solution. Given : $A_v = 100$ and $A'_v = 20$

We know that the voltage gain with feedback (A'_v),

$$20 = \frac{A_v}{1 + \beta \cdot A_v} = \frac{100}{1 + (\beta \times 100)}$$

or

$$1 + 100\,\beta = \frac{100}{20} = 5$$

$$\therefore \quad \beta = \frac{4}{100} = 0.04 \textbf{ Ans.}$$

Example 24-4. *An amplifier with negative feedback gives an output of 12.5 V with an input of 1.5 V. When feedback is removed, it requires 0.25 V input for the same output. Find **(a)** value of voltage gain without feedback; and **(b)** value of β, if the input and output are in phase and β is real.*

(Electronic Circuits-II, Anna Univ., May/June 2007)

Solution. Given: $V'_o = V_o = 12.5$ volt; $V'_{in} = 1.5$ volt and $V_{in} = 0.25$ volt.

(*a*) *Voltage gain without feedback*

We know that voltage gain of an amplifier without negative feedback

$$A_v = \frac{V_o}{V_{in}} = \frac{12.5}{0.25} = 50 \textbf{ Ans.}$$

(***b***) *Value of β when input and output are in phase*

We know that the voltage gain with negative feedback,

$$A'_v = \frac{V'_o}{V'_{in}} = \frac{12.5}{1.5} = 8.33 \text{ V}$$

and the voltage gain with negative feedback, (A'_v)

$$8.33 = \frac{A_v}{1 + \beta \cdot A_v} = \frac{50}{1 + (\beta \times 50)}$$

or $$8.33\,(1 + 50\,\beta) = 50$$

$$1 + 50\,\beta = \frac{50}{8.33} = 6$$

∴ $$\beta = \frac{(6-1)}{50} = 0.1 \textbf{ Ans.}$$

Example 24-5. *With a negative feedback, an amplifier gives an output of 10 V with an input of 0.5V. When the feedback is removed, it requires 0.25V input fo the same output. Calculate* (***a***) *gain without feedback, and* (***b***) *feedback fraction β.*

(*Electronic Circuits - ll, Anna University, May/June 2007*)

Solution. Given: $V'_0 = V_0 = 10$ volt; $V'_{in} = 0.5$ volt and $V_{in} = 0.25$ volt.

(***a***) *Calculation of Voltage gain without feedback*

We know that voltage gain of an amplifier without negative feedback

$$A_V = \frac{V_0}{Vin} = \frac{10}{0.25} = 40$$

(***b***) *Calculation of voltage gain and feedback fraction β with feedback*

We know that the voltage gain with negative feedback.

$$A'_V = \frac{V'_0}{V'_{in}} = \frac{10}{0.5} = 20 \text{ V}$$ **Ans.**

and the voltage gain with negative feedback, (A'_v)

$$20 = \frac{Av}{1 + \beta.Av} = \frac{40}{1 + \beta \times 40}$$

$$20\,(1 + \beta \times 40) = 40$$

$$1 + 40\beta = \frac{40}{20} = 20$$

Therefore, $$\beta = \frac{20-1}{40} = 0.475$$ **Ans.**

Example 24-6. *An engineer designs an amplifier to have a voltage gain of 80, but when constructed it gives only a voltage gain of 60. A technician suggests that a positive feedback may be introduced to increase the gain to the required level,* (***a***) *How much feedback should be given to increase the voltage gain to the required level?* (***b***) *What disadvantage may arise due to this type of operation?* (***c***) *What percentage of feedback will cause oscillation?*

(*Bangalore University, B.Sc. Elect., May* 2001)

Solution. Given: $A_v = 60$ and $A'_v = 80$

(***a***) *Value of feedback ratio*

Let β = Value of feedback ratio required to increase the voltage gain from 60 to 80.

We know that voltage gain with positive feedback (A'_v)

$$80 = \frac{A_v}{1 - \beta \cdot A_v} = \frac{60}{1 - \beta \cdot 60} = \frac{60}{1 - 60\,\beta}$$

$$80\,(1 - 60\,\beta) = 60$$

$$1 - 60\,\beta = \frac{60}{80} = 0.75$$

$$\therefore \quad 60\,\beta = 1 - 0.75 = 0.25$$

or $$\beta = \frac{0.25}{60} = 0.004 \textbf{ Ans.}$$

(*b*) *Disadvantage of employing positive feedback*

We know that the amplifier with positive feedback cannot have a stable voltage gain, because its gain increases with feedback. The increased voltage gain will further increase the feedback. This process will continue till the feedback ($\beta \cdot A_v$) becomes equal to unity, Under this condition, the amplifier will start oscillating. In other words, the amplifier becomes an oscillator. **Ans.**

(*c*) *Percentage of feedback which causes oscillation*

We know that the condition for oscillation,

$$\beta \cdot A_v = 1$$

or $$\beta \times 60 = 1$$

$$\therefore \quad \beta = 1/60 = 0.017 = 1.7\% \textbf{ Ans.}$$

Example 24-7. *An amplifier with negative feedback has a voltage gain of 100. It is found that without feedback an input signal of 50 mV is required to produce a given output; whereas with feedback, the input signal must be 0.6 V for the same output. Calcualte the value of voltage gain without feedback and feedback ratio.* (*Bangalore Univ. B.Sc. April/May* 2001)

Solution. Given : $A'_v = 100$; $v_{in} = 50$ mV

Let A_v = voltage gain without feedback,

v_o = output voltage without feedback,

v'_o = output voltage with feedback,

v'_{in} = inpt voltage with feedback (= 0.06 V given)

We know that the voltage gain with feedback, (A'_v),

$$100 = \frac{v'_o}{v'_{IN}} = \frac{v'_o}{0.6}; \quad \therefore\ v'_o = 60 \text{ V}$$

$$A_v = \frac{v'_o}{v'_{IN}} = \frac{v'_o}{50 \text{ mV}} = \frac{60}{50 \text{ mV}} = 1200 \textbf{ Ans.}$$

We also know that voltage gain with feedback (A'_v)

$$100 = \frac{A_v}{1 + \beta \cdot A_v} = \frac{1200}{1 + \beta \times 1200}$$

$$\therefore \quad \beta = \frac{\left(\dfrac{1200}{100} - 1\right)}{1200} = 9.1 \times 10^{-3} \text{ or } 0.9\% \textbf{ Ans.}$$

24-4. Gain Stability

We have already discussed in the last article that the use of negative feedback reduces that voltage gain of an amplifier. Moreover, the voltage gain of a feedback amplifier (A'_v) depends on the feedback network (*i.e.*, $A'_v = 1/\beta$). It means that the voltage gain of a feedback amplifier does not change with the variation in transistor parameters due to temperature, aging, *etc*. Thus, the use of negative feedback provides a greater stability to the voltage gain of the amplifier.

We know that the voltage gain of an amplifier with negative feedback,

$$A'_v = \frac{A_v}{1 + \beta \cdot A_v}$$

Taking natural logarithms on both sides of the above expression,

$$\log_e A'_v = \log_e \left(\frac{A_v}{1 + \beta \cdot A_v} \right)$$

$$= \log_e A_v - \log_e (1 + \beta \cdot A_v)$$

Differentiating both side,

$$\frac{dA'_v}{A'_v} = \frac{dA_v}{A_v} - \left(\frac{\beta \cdot dA_v}{1 + \beta \cdot A_v} + \frac{A_v \cdot d\beta}{1 + \beta \cdot A_v} \right)$$

The derivative of β (*i.e.*, $d\beta$) is zero, because the feedback network, usually, consists of stable elements such as resistor, capacitors and inductors. Therefore,

$$\frac{dA'_v}{A'_v} = \frac{dA_v}{A_v} - \frac{\beta \cdot dA_v}{1 + \beta \cdot A_v}$$

$$= dA_v \left(\frac{1}{A_v} - \frac{\beta}{1 + \beta \cdot A_v} \right)$$

$$= \frac{1}{1 + \beta \cdot A_v} \times \frac{dA_v}{A_v}$$

The termd dA'_v/A'_v represents the fractional change in amplifier voltage gain with feedback and dA_v/A_v the fractional change in voltage gain without feedback. The term $1/(1 + \beta \cdot A_v)$ is called *sensitivity. It indicates the ratio of percentage change in voltage gain with feedback to the percentage change in voltage gain without feedback.*

Mathematically, the sensitivity

$$= \frac{\dfrac{dA'_v}{A'_v}}{\dfrac{dA_v}{A_v}} = \frac{1}{1 + \beta \cdot A_v}$$

The reciprocal of the term sensitivity is called desensitivity. Thus, desensitivity = $1 + \beta \cdot A_v$

The term 'desensitivity' indicates the factor by which the voltage gain has been reduced due to feedback. It may be proved that

Desensitivity $\approx \dfrac{A_v}{A'_v}$ and

Sensitivity $\approx \dfrac{A'_v}{A_v}$

Example 24-8. *An amplifier has an open loop gain of 800 and a feedback ratio of 0.05. If the open loop gain changes by 20% due to temperature, find the percentage change in closed loop gain.*

Solution. Given: $A_v = 800$; $\beta = 0.05$ and $dA_v / A_v = 20$

We know that the percentage change in closed loop gain,

$$\frac{dA'_v}{A'_v} = \frac{1}{1 + \beta \cdot A_v} \times \frac{dA_v}{A_v} = \frac{1}{1 + 0.05 \times 800} \times 20 = 0.5\ \%\ \textbf{Ans.}$$

Example 24-9. *An amplifier with a closed loop gain of 100 is required and this gain should not vary more than 1%, when the inherent gain of the amplifier without feedback varies by 20%. Find the values of A_v and β.*

Solution. Given: $A'_v = 100$; $dA'_v/A'_v = 1\% = 0.01$ and $dA_v/A_v = 20\% = 0.20$.

We know that,

$$\frac{dA'_v}{A'_v} = \frac{1}{1+\beta \cdot A_v} \times \frac{dA_v}{A_v}$$

$$0.01 = \frac{1}{1+\beta \cdot A_v} \times 0.2$$

or $$1+\beta \cdot A_v = \frac{0.2}{0.01} = 20$$

We also know that the closed loop gain (A'_v),

$$100 = \frac{A_v}{1+\beta \cdot A_v} = \frac{A_v}{20}$$

∴ $$A_v = 100 \times 20 = 2000 \textbf{ Ans.}$$

and $$\beta = (1+\beta \cdot A_v)/A_v = 20/2000 = 0.01 \textbf{ Ans.}$$

24-5. Increased Bandwidth

Consider an amplifier, whose bandwidth is required to be found out when it is subjected to negative feedback.

Let BW = Bandwidth of an amplifier without feedback,

and BW′ = Bandwidth of an amplifier with feedback,

A_v = Voltage gain of the amplifier without feedback, and

β = Feedback ratio.

We have already discussed that the negative feedback increases the bandwidth of an amplifier by a factor equal to $(1 + \beta \cdot A_v)$.

∴ $$BW' = (1+\beta \cdot A_v)\, B.W. \qquad ...(i)$$

It may be noted that voltage gain of a feedback amplifier reduces by the same factor by which the bandwidth is increased (*i.e.*, $1 + \beta \cdot A_v$). Thus, the product of voltage gain and bandwidth of an amplifier without feedback and with feedback remains the same. We know that bandwidth of an amplifier is given by the separation between the upper and lower 3 dB frequencies.

Now let f_1 = the lower 3 dB frequency, and

f_2 = the upper 3 dB frequency.

∴ Bandwidth of the amplifier,

$$B.W. = f_2 - f_1$$

It can be proved easily that with negative feedback the upper (f_2') and lower (f_1') 3 dB frequencies of an amplifier are given by the relation,

$$f_2' = (1+\beta \cdot A_v) f_2$$

and $$f_1' = \frac{f_1}{(1+\beta \cdot A_v)}$$

It is evident from the above relations that the upper 3 dB frequency (f_2') is greater than f_2 by a factor of $(1 + \beta \cdot A_v)$. Similarly, f_1' is smaller than f_1 by a factor of $(1 + \beta \cdot A_v)$. Since, the product of voltage gain and bandwidth is the same without feedback and with feedback, therefore

$$A'_v \times B.W.' = A_v \times B.W.$$

or $$A'_v (f_2' - f_1') = A_v (f_2 - f_1)$$

The above facts are illustrated in the frequency response graph (*i.e.*, gain versus frequency graph) as shown in Fig. 24-3.

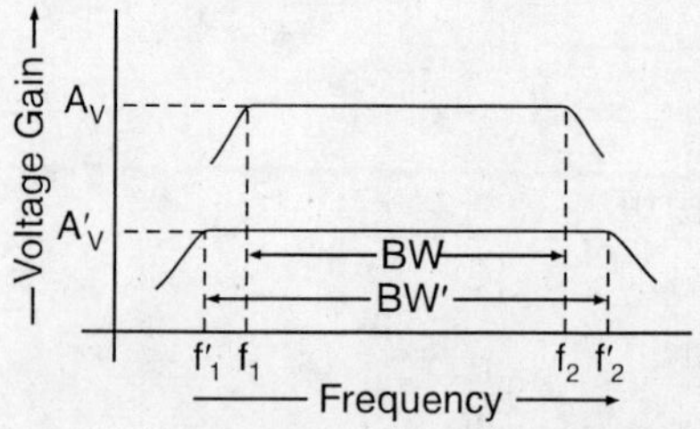

Fig. 24-3. *Gain versus frequency graph of an amplifier with and without feedback.*

In this figure, the voltage gain is plotted along the vertical axis and the frequency along the horizontal axis. It is evident from the plot that the negative feedback has increased the bandwidth from BW to BW′.

Example 24-10. *An amplifier has a mid-frequency gain of 100 and a bandwidth of 200 kHz.* **(a)** *What will be the new bandwidth and gain, if 5% negative feedback is introduced.* **(b)** *What should be the amount of feedback, if the bandwidth is to be restricted to 1 MHz.*

Solution. Given: $A_v = 100$; B.W. = 200 kHz = 200×10^3 Hz and $\beta = 5\% = 0.05$.

(*a*) *New bandwidth and gain with 5% negative feedback*

We know that the new bandwidth (*i.e.*, the bandwidth with 5% or 0.05 negative feedback),

$$\text{B.W.}' = (1 + \beta \cdot A_v)\,\text{B.W.}$$
$$= [1 + (0.05 \times 100)] \times 200 \times 10^3 \text{ Hz}$$
$$= 1200 \times 10^3 \text{ Hz} = 1200 \text{ kHz} \textbf{ Ans.}$$

and gain with feedback,

$$A'_v = \frac{A_v}{1 + \beta \cdot A_v} = \frac{100}{1 + (0.05 \times 100)} = 16.7 \textbf{ Ans.}$$

(*b*) *Amount of feedback with BW′ equal to 1 MHz* (1×10^6 *Hz*)

Let β = Amount of feedback to restrict the bandwidth to 1 MHz.

We know that bandwidth with feedback (B.W.′),

$$1 \times 10^6 = (1 + \beta \cdot A_v)\,\text{B.W.}$$
$$= (1 + \beta \times 100) \times (200 \times 10^3)$$

$\therefore$ $$1 + 100\,\beta = (1 \times 10^6)/(200 \times 10^3) = 5$$

or $$\beta = 4/100 = 0.04 = 4\% \textbf{ Ans.}$$

Example 24-11. *An amplifier has a midband gain of 1500 and a bandwidth of 4 MHz. The midband gain reduces to 150 when a negative feedback is applied. Determine the value of feedback factor and the bandwidth.* (*UPSC Engg. Services* 1998)

Solution. Given : $A_v = 1500$; B.W. = 4 MHz and $A'_v = 150$.

Let, b = feedback factor and

B.W.′ = bandwidth with feedback

We know that the voltage gain with feedback (A'_v),

$$150 = \frac{A_v}{1 + \beta \cdot A_v} = \frac{1500}{1 + (\beta \times 1500)}$$

$$\therefore \qquad \beta = \frac{\left(\dfrac{1500}{150} - 1\right)}{1500} = 6 \times 10^{-3} \text{ or } 0.6\% \textbf{ Ans.}$$

We also know that bandwidth with feedback,

$$\text{B.W.}' = (1 + \beta \cdot A_v) \text{ B.W.} = (1 + (6 \times 10^{-3}) \times 1500)\ (4 \times 10^6) \text{ Hz}$$
$$= 10 \times 4 \times 10^6 \text{ Hz} = 40 \times 10^6 \text{ Hz}$$
$$= 40 \text{ MHz } \textbf{Ans.}$$

Example 24-12. *(a) An amplifier has a value of R_{in} = 4.2 kΩ, A_v = 220, and β = 0.01. Determine the value of input resistance of the feedback amplifier.*

(b) The amplifier in part (a) had cut-off frequencies of f_1 = 1.5 kHz and f_2= 501.5 kHz before the feedback path was added. What are the new cut-off frequencies for the circuit?

Solution. Given: R_{in} = 4.2 kΩ, A_v = 220 and β = 0.01

(*a*) *Input resistance of a feedback amplifier*

We know that input resistance of a feedback amplifier

$$R'_i = (1 + \beta \cdot A_v)\, R_i = (1 + 0.01 \times 220) \times 4.2 \text{ k}\Omega$$
$$= 13.44 \text{ k}\Omega \textbf{ Ans.}$$

(*b*) *New cut-off frequencies (f_1' and f_2')*

We know that lower 3 dB frequency,

$$f_1' = \frac{f_1}{1 + \beta \cdot A_v} = \frac{1.5 \text{ kHz}}{1 + 0.01 \times 220} = 0.469 \text{ kHz} = 469 \text{ Hz } \textbf{Ans.}$$

and the upper 3 dB frequency,

$$f_2' = (1 + \beta A_v) f_2 = (1 + 0.01 \times 220)\ 501.5 \text{ kHz} = 1604.8 \text{ kHz } \textbf{Ans.}$$

24-6. Decreased Distortion

Consider a large amplitude signal applied to a stage of an amplifier, so that the operation of an active device (*i.e.*, transistor) extend slightly beyond its range of linear operation. As a result of this, the output signal is slightly distorted. Now if a negative feedback is introduced to the amplifier stage, the voltage gain reduces. But if the input signal is increased, by the same amount by which the gain is reduced, the output signal amplitude remains the same (*i.e.*, as it was without feedback). Now if we measure the distortion in both the cases, it will be found that distortion has reduced due to feedback by a factor (1 + β · A_v). Mathematically, amplifier distortion with feedback,

$$D' = \frac{D}{1 + \beta \cdot A_v}$$

where, D is the amplifier distortion without feedback.

It may be noted that the input signal to the feedback amplifier may be the actual signal available from an external source or it may be an output of an amplifier preceding the feedback stage. To increase the input of the feedback amplifier by a factor (1 + β · A_v), we can either increase the nominal gain of the preamplifying stages or add a new stage. It will be interesting to know that the full benefit of the feedback amplifier, in reducing distortion, can be obtained. This can be done if the preamplifying stages do not introduce any additional distortion because of the increased output.

Example 24-13. *An RC coupled amplifier has a voltage gain of 1000, f_1 = 50 Hz, f_2 = 200 kHz and a distortion of 5%, without feedback. Find the amplifier voltage gain, f_1', f_2' and distortion when a negative feedback is applied with feedback ratio of 0.01.*

Solution. Given: $A_v = 1000$; $f_1 = 50$ Hz; $f_2 = 200$ kHz; $D = 5\%$ and $\beta = 0.01$.

Voltage gain with feedback

We know that voltage gain with feedback,

$$A'_v = \frac{A_v}{1 + \beta \cdot A_v} = \frac{1000}{1 + (0.01 \times 1000)} = 90.9 \textbf{ Ans.}$$

f_1' and f_2'

We know that the lower 3 dB frequency,

$$f_1' = \frac{f_1}{1 + \beta \cdot A_v} = \frac{50}{1 + (0.01 \times 1000)} = 4.5 \text{ Hz } \textbf{Ans.}$$

and the upper 3 dB frequency,

$$f_2' = (1 + \beta \cdot A_v) f_2 = [1 + (0.01 \times 1000)] \times 200 \text{ kHz}$$
$$= 2200 \text{ kHz} = 2.2 \text{ MHz } \textbf{Ans.}$$

Distortion with feedback

We know that distortion with feedback,

$$D' = \frac{D}{1 + \beta \cdot A_v} = \frac{5}{1 + (0.01 \times 1000)} = 0.45\% \textbf{ Ans.}$$

24-7. Decreased Noise

It has been observed that there are many sources of noise in an amplifier depending upon the active device used in the circuit. It will be interesting to know that with the use of feedback, the magnitude of noise (N) is reduced by a factor of $(1 + \beta \cdot A_v)$ in a similar manner as distortion. Thus, noise with feedback in given by the relation,

$$N' = \frac{N}{1 + \beta \cdot A_v}$$

and decrease in noise,

$$= \frac{N - N'}{N} = 1 - \frac{N'}{N}$$

$$= 1 - \frac{\frac{N}{1 + \beta \cdot A_v}}{N} = 1 - \frac{1}{1 + \beta \cdot A_v}$$

Example 24-14. *A negative feedback is used to reduce the noise from an amplifier by 80%. (**a**) What must be the percentage of negative feedback to accomplish this, if the input voltage gain is 100? (**b**) What will be the voltage gain with feedback?*

Solution. Given: $A_v = 100$; Reduction of noise = 80% = 0.8.

(*a*) *Percentage of negative feedback*

Let β = Value of negative feedback to reduce the noise of amplifier by 80%.

We know that the reduction of noise of an amplifier with feedback,

$$0.8 = 1 - \frac{1}{1 + \beta \cdot A_v} = 1 - \frac{1}{1 + \beta \times 100}$$

$$\frac{1}{1 + 100\beta} = 1 - 0.8 = 0.2$$

$$1 + 100\beta = 1/0.2 = 5$$

$$\therefore \quad \beta = 4/100 = 0.04 \textbf{ Ans.}$$

(*b*) *Voltage gain with feedback*

We also know that voltage gain with feedback,

$$A'_v = \frac{A_v}{1 + \beta \cdot A_v} = \frac{100}{1 + (0.04 \times 100)} = 20 \textbf{ Ans.}$$

24-8. Types of Feedback Connections

There are four basic types (or ways) of connecting the feedback signal from an amplifier output to its input. These types are given below :

(*i*) Voltage-series feedback connection.

(*ii*) Voltage-shunt feedback connection.

(*iii*) Current-series feedback connection.

(*iv*) Current-shunt feedback connection.

It means that both voltage and current can be feedback to the input either in series or parallel. In the feedback connection types, the term 'voltage' refers to connecting the output voltage as input to the feedback network. The term 'current' refers to tapping off some output current through the feedback network. The term 'series' refers to connecting the feedback signal in series with the input signal voltage, and the term 'shunt' refers to connecting the feedback signal in shunt (parallel) with an input current source.

It has been observed that the *series feedback connections tend to increase the input resistance, while the shunt feedback connections tend to decrease the input resistance*. Moreover, the *voltage feedback will tend to decrease the output resistance, while the current feedback tends to increase the output resistance*. As a matter of fact, higher input resistance and lower output resistance is desired for most cascade amplifiers. Both of these characteristics are obtained by using the voltage-series feedback connection.

24-9. Voltage-series Feedback Connection

It is also called a *shunt-derived series-fed feedback* connection. A block diagram of a voltage-series feedback connection is as shown in Fig. 24-4. In this connection, a fraction of the output voltage is applied in series with the input voltage through the feedback network. However, the input to the feedback network is in parallel with output of the amplifier.

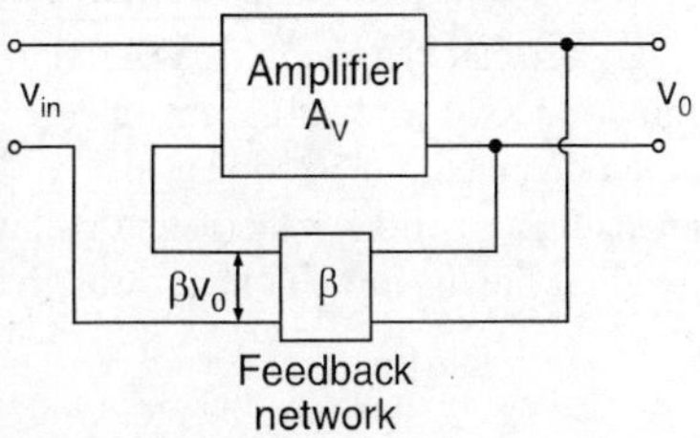

Fig. 24-4. *Voltage-series feedback connections.*

It can be shown easily that the voltage-series feedback connection increase the input resistance and decreases the output resistance of the feedback amplifier.

The input resistance of a feedback amplifier is given by the relation,

$$R'_i = (1 + \beta \cdot A_v)\, R_i$$

where, β = The feedback fraction,

A_v = The voltage gain of a amplifier without feedback, and

R_i = The input resistance of an amplifier without feedback.

Similarly, the output resistance of a feedback amplifier is given by the relation,

$$R'_o = \frac{R_o}{1 + \beta \cdot A_v}$$

where, R_o is the output resistance of an amplifier without feedback.

As usual, the voltage-series feedback connection reduces the voltages gain of a feedback amplifier and its value is given by the relation,

$$A'_v = \frac{A_v}{1 + \beta \cdot A_v}$$

Example 24-15. *Calculate the voltage gain, input and output resistances of a voltage-series feedback amplifier having $A_v = 300$, $R_i = 1.5$ kΩ, $R_o = 50$ kΩ and $\beta = 1/15$.*

Solution. Given: $A_v = 300$; $R_i = 1.5$ kΩ; $R_o = 50$ kΩ and $\beta = 1/15$.

Voltage gain

We know that voltage gain of a voltage-series feedback amplifier,

$$A'_v = \frac{A_v}{1 + \beta \cdot A_v} = \frac{300}{1 + \left(\frac{1}{15}\right) \times 300} = 14.3 \textbf{ Ans.}$$

Input resistance

We know that input resistance,

$$R'_i = (1 + \beta \cdot A_v)R_i = \left[1 + \left(\frac{1}{15}\right) \times 300\right] 1.5 = 31.5 \text{ k}\Omega \textbf{ Ans.}$$

Output resistance

We also know that output resistance,

$$R'_i = \frac{R_o}{1 + \beta \cdot A_v} = \frac{50}{1 + \left(\frac{1}{15}\right) \times 300} = 2.4 \text{ k}\Omega \textbf{ Ans.}$$

24-10. Voltage-shunt Feedback Connection

It is also called a *shunt-derived shunt-fed feedback* connection. A block diagram of a voltage-shunt feedback connection is shown in Fig. 24-5. In this connection, a fraction of the output voltage is applied in parallel with the input voltage through the feedback network.

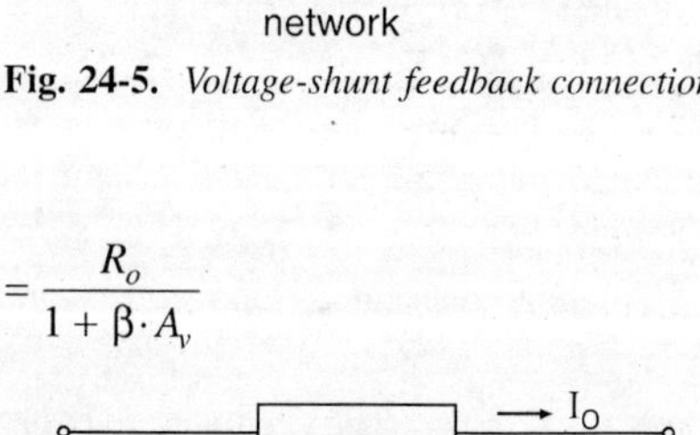

Fig. 24-5. *Voltage-shunt feedback connection.*

It can be shown easily that the voltage-shunt feedback connection decreases both input and output resistances of the feedback amplifier by a factor equal to $(1 + \beta \cdot A_v)$. Thus,

$$R'_i = \frac{R_i}{1 + \beta \cdot A_v} \text{ and } R'_o = \frac{R_o}{1 + \beta \cdot A_v}$$

24-11. Current-series Feedback Connection

It is called a *series-derived series-fed feedback* connection. A block diagram of a current-series feedback connection is shown in Fig. 24-6. In this connection a fraction of the output current is converted into a proportional voltage by the feedback network and then applied in series with the input.

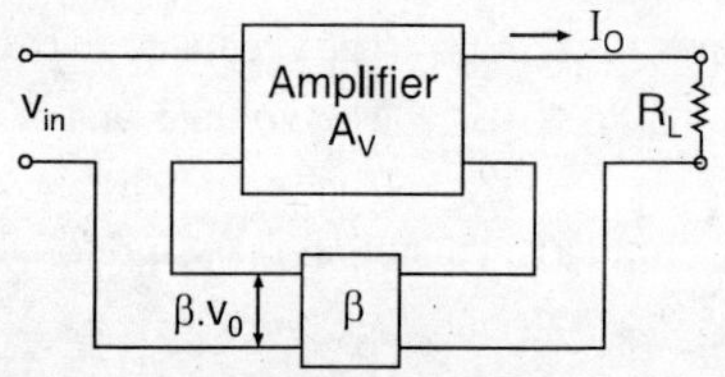

Fig. 24-6. *Current-series feedback connection.*

It can be shown easily that the current-series feedback connection increases both the input resistance and the output resistance of a feedback amplifier by a factor equal to $(1 + \beta \cdot A_v)$. Thus,

$$R'_i = (1 + \beta \cdot A_v)\, R_i \text{ and } R'_o = (1 + \beta \cdot A_v)\, R_o$$

24-12. Current-shunt Feedback Connection

It is also called a *series-derived shunt-fed feedback* connection. A block diagram of a current-shunt feedback connection is shown in Fig. 24-7. In this connection, a fraction of the output current is converted into a proportional voltage by the feedback network and then applied in parallel with the input voltage.

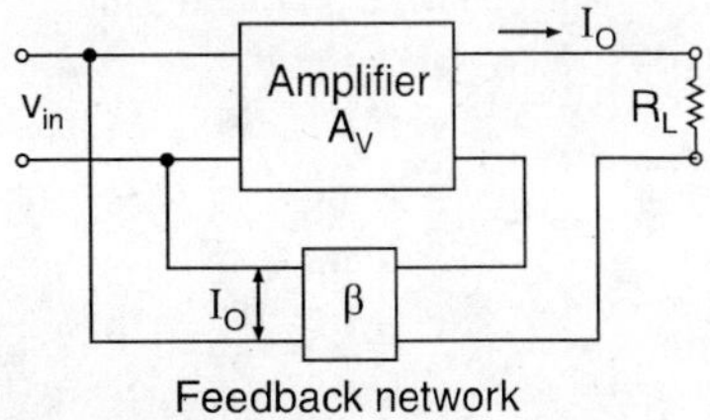

Fig. 24-7. *Current-shunt feedback connection.*

It can be shown easily that the current-series feedback connection decreases the input resistance but increase the output resistance by a factor equal to $(1 + \beta \cdot A_v)$. Thus

$$R'_i = \frac{R_i}{1 + \beta \cdot A_v} \text{ and } R'_o = (1 + \beta \cdot A_v) \cdot R_o$$

24-13. Comparison of Feedback Connections

We have already discussed in the last articles that a series feedback always increase the input resistance, while the shunt feedback decreases its value. Similarly, a voltage feedback (either series or shunt type) always decreases the output resistance, whereas a current feedback (either series or shunt type) always increases its value. All these facts are summarized in Table 24-1. It may be noted that the four types of feedback connectional decrease the voltage gain, distortion and noise. However, the bandwidth increases in all the types of connections.

Table 24-1. *Effect of feedback on input and output resistances*

S. No.	*Types of feedback connection*	*Input resistance*	*Output resistance*
1.	Voltage-series	increases	decreases
2.	Voltage-shunt	decreases	decreases
3.	Current-series	increases	increases
4.	Current-shunt	decreases	increases

24-14. Negative Feedback in Transistor Amplifiers

We have already discussed in the last articles that negative feedback is a method for feeding a portion of amplified output back to the input so as to *decrease* the value of input signal. We have also discussed that we can feedback either voltage or current (series or shunt) back to the input. The negative feedback is widely used in electronic amplifiers to achieve desired performance. It is used in all modern high gain transistor (and integrated circuit) amplifier used in radio and television receivers, stereo systems, *etc*. The negative feedback is also used in regulated power supply circuits and in those circuits where impedance matching is required. We shall now discuss some important transistor amplifier circuits employing negative feedback.

24-15. Common Emitter Amplifier Without Emitter Bypass Capacitor

Fig. 24-8 shows a circuit of a common emitter amplifier without emitter bypass capacitor. This circuit has a current-series feedback. The current through resistor R_E results in a voltage, which is in opposite phase to that of the input signal. As a result of this, the output voltage (v_o) is reduced.

In order to remove current-series feedback, the emitter resistor must be removed or bypassed by a capacitor (as is usually done).

It can be shown easily that voltage gain of a common emitter amplifier without feedback,

$$A_v = \frac{R_C}{r'_e}$$

and voltage gain of a common emitter feedback amplifier,

$$A'_v = \frac{R_C}{(r'_e + R_E)}$$

Similarly, the feedback ratio,

$$\beta = \frac{R_E}{R_C}$$

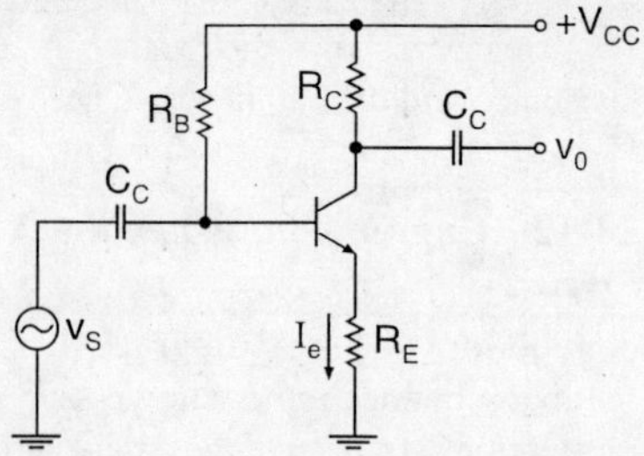

Fig. 24-8. *Common-emitter amplifier without emitter bypass capacitor.*

Example 24-16. *An amplifier with current series feedback has the form shown in Fig. 24-9. The transistor has h_{fe} = 100, h_{ie} = 2 kΩ, R_c = 470 Ω and R_e consists of two 100 Ω resistors, one is bypassed by capacitor C_E.*

Calculate the values of voltage gain and input resistance of the amplifier with and without feedback.

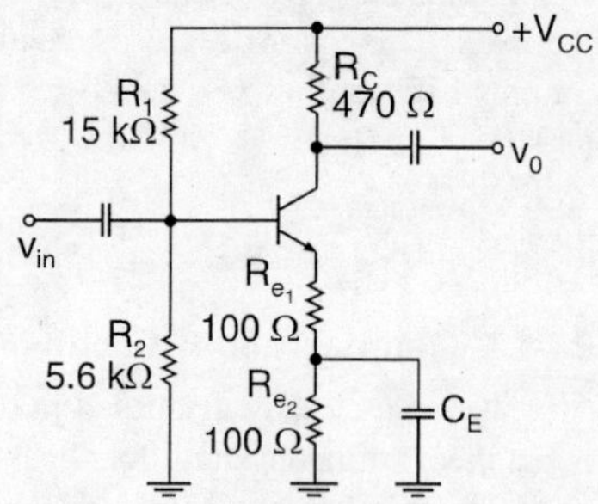

Fig. 24-9.

Solution. Given: h_{fe} = 100 Ω; h_{ie} = 2 kΩ = 2000 Ω; R_{e_1} = 100 Ω; R_{e_2} = 100 Ω; R_1 = 15 kΩ = 15000Ω; R_2 = 5.6 kΩ = 5600 Ω and R_c = 470 Ω.

Voltage gain and input resistance without feedback

We know that the given amplifier circuit is called a swamped amplifier and its voltage gain,

$$A_v = \frac{A_i \cdot R_c}{R_i} = \frac{h_{fe} \cdot R_c}{h_{ie}} \qquad ...(\because A_i = h_{fe};\ R_i = h_{ie})$$

$$= \frac{100 \times 470}{2000} = 23.5 \textbf{ Ans.}$$

and input resistance,

$$R_i = (R_1 \parallel R_2) \parallel h_{ie} = (15000 \parallel 5600) \parallel 2000\ \Omega$$

$$= 1342\ \Omega \textbf{ Ans.}$$

Voltage gain and input resistance with feedback

We also know that the feedback ratio,

$$\beta = \frac{R_e}{R_c} = \frac{R_{e_1}}{R_c} = \frac{100}{470} = \frac{1}{4.7}$$

∴ Voltage gain of the feedback amplifier,

$$A'_v = \frac{A_v}{1 + \beta \cdot A_v} = \frac{23.5}{1 + \left(\frac{1}{4.7} \times 23.5\right)}$$

$$= 3.92 \textbf{ Ans.}$$

and the input resistance,

$$R'_i = R_i\,(1 + \beta \cdot A_v)$$

$$= 1342\left[1 + \left(\frac{1}{4.7} \times 23.5\right)\right] = 8052\ \Omega \textbf{ Ans.}$$

24-16. Emitter Follower

Fig. 24-10 shows the circuit of an emitter follower or a common collector amplifier. This circuit is also a case of current-series feedback.

The negative feedback is produced in the same way as discussed for the common emitter amplifier without emitter bypass capacitor. In an emitter follower, the feedback ratio is equal to unity, (*i.e.*, $\beta = 1$). Therefore, the overall voltage gain of an emitter follower is approximately unity (*i.e.*, $A_v = A'_v = 1$).

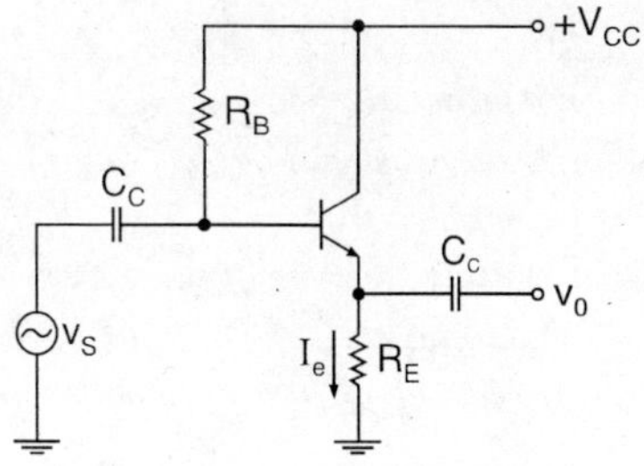

Fig. 24-10. *Emitter follower.*

24-17. Collector Feedback Biased Common Emitter Amplifier

Fig. 24-11 shows the circuit, of a collector feedback biased common emitter amplifier. This circuit is also a case of voltage-shunt feedback.

The feedback is obtained by coupling a portion of the output voltage to the base through the resistor R_B. The voltage-shunt feedback connection provides a stabilized amplifier overall gain. But it reduces the input and output resistances. It can be shown that the feedback ratio,

$$\beta = \frac{R_C}{R_B}$$

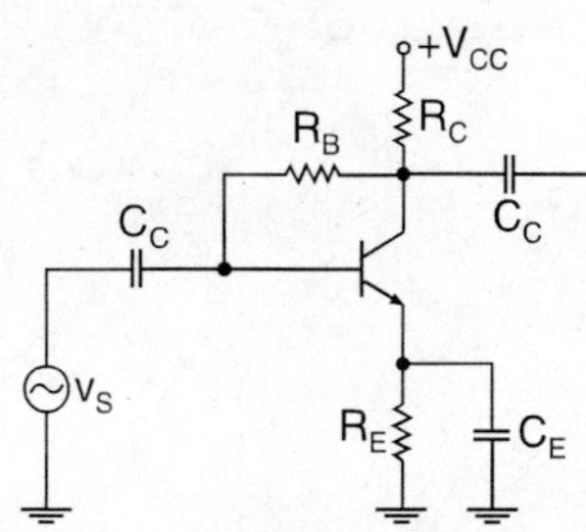

Fig. 24-11. *Collector feedback biased common emitter amplifier.*

24-18. Negative Feedback in Multistage Amplifiers

It will be interesting to know that the negative feedback is very useful in multistage transistor amplifiers. Fig. 24-12 shows the two cascaded stages of a transistor amplifier with voltage-series feedback. The cascaded amplifier stages, with transistor Q_1 and Q_2, provides an overall voltage gain. A feedback network consists of resistors R_3 and R_4 coupled by a capacitor C_6 (to block d.c.) between output and input.

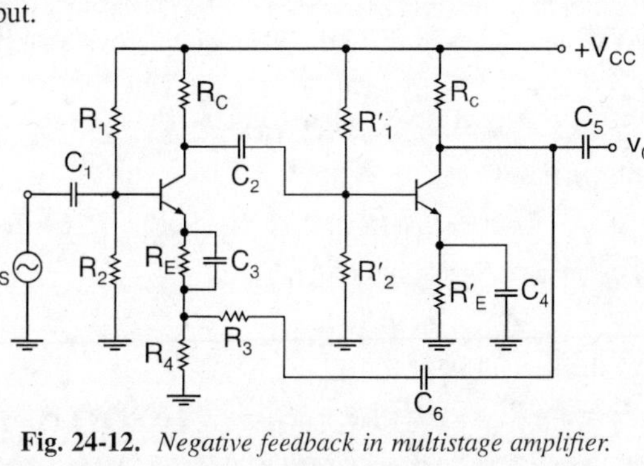

Fig. 24-12. *Negative feedback in multistage amplifier.*

The feedback signal is taken from the collector of the second stage and is connected to the emitter (neglecting the bypassed R_E resistor) of the first amplifier stage. The negative feedback results from the connection, since the inphase output signal of the second stage connected through the feedback network to the emitter opposes the input signal between the base-emitter junction of the transistor of the first stage.

It is also possible to obtain the current-shunt feedback in a multistage amplifier. In order to obtain current-shunt feedback, the feedback signal is taken from the emitter of the second stage (through a feedback network consisting of a resistor and capacitor) and is connected to the base of the first stage.

Example 24-17. *Find the voltage gain, input resistance and output resistance of a series-shunt pair type two stage feedback amplifier using transistors with* h_{fe} *= 99 and* h_{ie} *= 2 kΩ. The circuit components are as follows:*

Load resistance of first stage $(R_c) = 22\ k\Omega$

Emitter resistance of first stage $(R_4) = 100\ \Omega$

Biasing resistors of second stage $(R_1$ *and* $R_2) = 220\ k\Omega$ *and* $22\ k\Omega$

Load resistance of second stage $(R'_c) = 4.7\ k\Omega$

Feedback resistor from collector of second stage to emitter of first stage

$(R_3) = 7.8\ k\Omega$

(Electronic Circuits-II, Anna Univ., May/June 2007)

Solution. Given: $h_{fe}\ (= h_{fe_1} = h_{fe_2}) = 99$; $h_{ie}\ (= h_{ie_1} = h_{ie_2}) = 2\ \text{k}\Omega$; $R_c = 22\ \text{k}\Omega$; $R_4 = 100\ \Omega$; $R_1 = 220\ \text{k}\Omega$; $R_2 = 22\ \text{k}\Omega$; $R'_c = 4.7\ \text{k}\Omega$; $R_3 = 7.8\ \text{k}\Omega$.

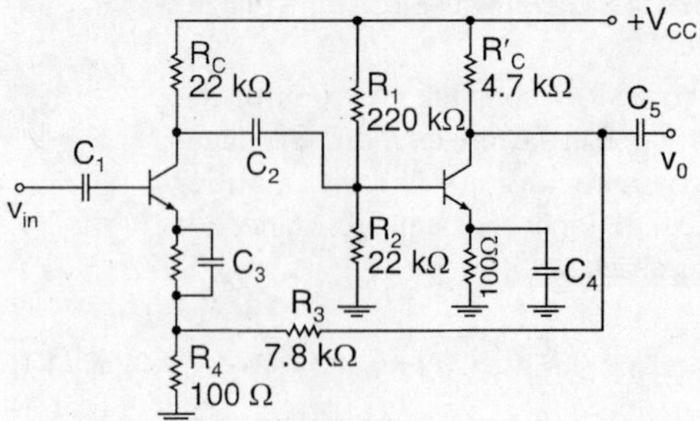

Fig. 24-13.

Fig. 24-13 shows the circuit arrangement of the series-shunt pair type two stage feedback amplifier.

Voltage gain, input and output resistance without feedback

These values may be obtained by disconnecting the feedback network *i.e.*, resistors R_3 and R_4 from emitter of the first stage. In that case, it becomes a simple two stage amplifier. We know that input resistance of the first stage,

$$R_i = h_{ie_1} = 2\ \text{k}\Omega\ \textbf{Ans.}$$

and output resistance of the first stage,

$$R_{o_1} = R_c \parallel (R_1 \parallel R_2) \parallel h_{ie_2} = 22 \parallel (220 \parallel 20) \parallel 2 = 1.67\ \text{k}\Omega$$

Similarly, input resistance of the second stage,

$$R_{i_2} = h_{ie} = 2\ \text{k}\Omega$$

and output resistance of the second stage,

$$R_{o_2} = R'_c \parallel (R_3 + R_4) = 4.7 \parallel (7.8 + 0.1) = 2.95\ \text{k}\Omega\ \textbf{Ans.}$$

Voltage gain, input and output resistance with feedback

We know that the voltage gain of the first stage without feedback,

$$A_{v_1} = \frac{h_{fe} \cdot R_{o_1}}{h_{ie}} = \frac{99 \times 1.67}{2} = 82.7$$

and voltage gain of second stage without feedback,

$$A_{v_2} = \frac{h_{fe} \cdot R_{o_2}}{h_{ie}} = \frac{99 \times 2.95}{2} = 146$$

∴ Overall gain of the cascaded amplifier, without feedback,

$$A_v = A_{v_1} \cdot A_{v_2} = 82.7 \times 146 = 13155$$

Now let us connect the feedback network back at its place. In that case, the value of feedback ratio,

$$\beta = \frac{R_4}{R_3 + R_4} = \frac{0.1}{7.8 + 0.1} = \frac{1}{79}$$

and the input resistance of the amplifier with feedback,

$$R_{i_1} = R_i(1 + \beta \cdot A_v) = \left[1 + \left(\frac{1}{79} \times 12074\right)\right] = 308 \text{ k}\Omega \textbf{ Ans.}$$

Similarly, output resistance of the amplifier with feedback,

$$R'_{o_2} = \frac{R_{o_2}}{1 + \beta \cdot A_v} = \frac{2.95}{1 + \left(\frac{1}{79} \times 12074\right)} = 0.0192 \text{ k}\Omega = 19.2\ \Omega \textbf{ Ans.}$$

and the overall voltage gain of the amplifier with feedback,

$$A'_v = \frac{A_v}{1 + \beta \cdot A_v} = \frac{12074}{1 + \left(\frac{1}{79} \times 12074\right)} = 78.5 \textbf{ Ans.}$$

SUMMARY

In this chapter, you have learned that :

1. A feedback amplifier consists of two parts : amplifier circuit and feedback circuit.
2. If the feedback signed is applied in such a way that it is in-phase with the input signal, it is called positive feedback.
3. If the feedback signal is applied in such a way that it is out-of-phase with the input signal, it is called negative feedback.
4. The negative feedback in amplifiers (*i*) increases the stability, (*ii*) increases the bandwidth, and (*iii*) reduces the frequency and phase distortion.

GLOSSARY

Feedback : A process of injecting some energy from the output of the amplifier and then return it back to the input.

Negative feedback : A connection scheme where the feedback signal is out-of-phase with the input signal.

Positive feedback : A connection scheme where the feedback signal is in-phase with the input signal.

NUMERICAL PROBLEMS

1. An amplifier has a gain of 2000. If the feedback ratio is 0.04, then find the voltage gain of the amplifier with negative feedback. (**Ans.** 24.7)
2. An amplifier has a value of $A_v = 1000$ and $\beta = 0.22$. Determine the value of A'_v for the circuit. (**Ans.** 4.52)
3. An amplifier has a value of $A_v = 588$ and $A'_v = 10.6$. Determine the value of feedback fraction β. (**Ans.** 0.093)
4. An amplifier has a gain of 4000. With negative feedback, the gain reduces to 25. Calculate the fraction of the output that is feedback to the input. (**Ans.** 0.04)
5. An amplifier with negative feedback has a gain of 50. It is found that without feedback, an input signal of 0.1 V is required to produce a given output, whereas with feedback, the input signal must be 0.8 V for the same output. Calculate the amount of voltage gain and feedback ratio. (**Ans.** $A_v = 400$; $\beta = 0.018$)
6. An amplifier has a voltage gain of 500. A technician decides that 10% negative feedback should be employed to reduce distortion. (*a*) What will be the voltage gain with feedback? (*b*) What should be the feedback in order to double the gain that existed in (*a*)? (**Ans.** (*a*) 9.8; (*b*) 4.9%)
7. An amplifier has an open loop gain of 200 and feedback ratio of 0.1. If the open loop gain changes by 20% due to variation of transistor parameters, find the percentage change in closed loop gain. (**Ans.** 0.95%)
8. The midband voltage gain of a certain amplifier is 500. Its upper half-power frequency (f_2) is 20 kHz and the lower half-power frequency (f_1) is 20 Hz. What will be the upper and lower half power frequencies and voltage gain, if 2% negative feedback in introduced? (**Ans.** $f_2 = 220$ kHz; $f_1 = 1.8$ Hz, $A_v = 45.5$)
9. An amplifier with 2.5 kΩ input resistance and 50 kΩ output resistance has a voltage gain of 100. The amplifier is now modified to provide 5% negative voltage feedback in series with the input. Calculate (*a*) the voltage gain, (*b*) the input resistance, and (*c*) the output resistance with feedback. (**Ans.** (*a*) 16.67, (*b*) 15 kΩ, and (*c*) 8.3 kΩ)
10. Fig. 24-14 shows a transistor feedback amplifier. If the transistor has $h_{fe} = 100$ and $h_{ie} = 2$ kΩ, find the voltage gain, and input impedance of the amplifier with feedback and without feedback. (**Ans.** $A_v = 83.3$; $R_i = 12$ kΩ; $A'_v = 500$ and $R_i = 2$ kΩ)

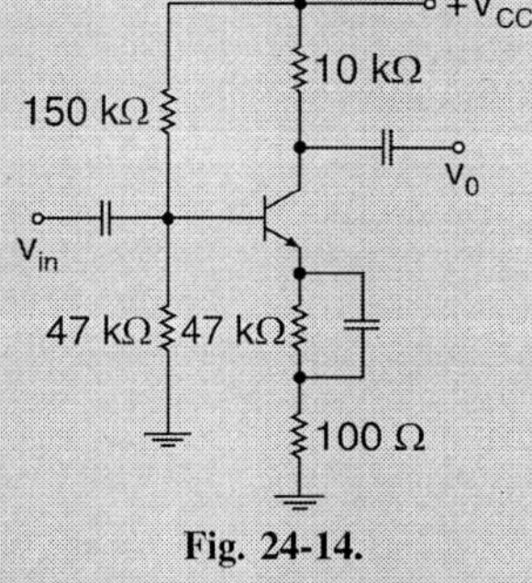

Fig. 24-14.

11. Find the voltage gain, input resistance and output resistance of a series-shunt pair type two stage feedback amplifier shown in Fig. 24-15. The amplifier uses transistors with $h_{fe} = 99$ and $h_{ie} = 2$ kΩ. (**Ans.** $A'_v = 96.5$; $R'_{i1} = 45.4$ kΩ and $R'_{02} = 130$ kΩ).

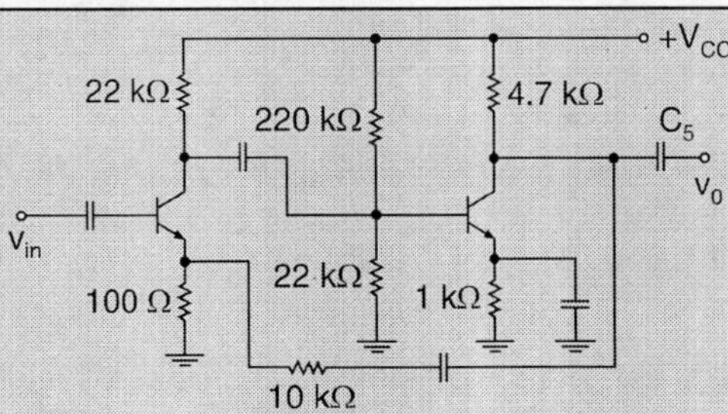

Fig. 24-15.

DESCRIPTIVE QUESTIONS

1. What do you understand by the term 'negative feedback' in amplifiers?
2. Compare the negative feedback and positive feedback.
3. Enumerate the effects of negative feedback on the various characteristics of the amplifier.
4. Describe with necessary derivations, the effect of negative feedback on the bandwidth and distortion in an amplifier.
5. What is the effect of a voltage series negative feedback in the following performance measures of a BJT amplifier :
   (*i*) Input resistance (*ii*) Output resistance
   (*iii*) Bandwidth (*iv*) Distortion and noise (*v*) Gain stability
5. Derive an expression for input and output resistance of voltage series feedback amplifier.
   (*Electronic Circuits, VTU June/July 2008*)
6. Mention the advantage of negative feedback.
   (*Electronic Circuits, VTU June/July 2008*)
7. Define negative and positive feedback.
   (*Electronic Circuits - II, Anna University, Nov/Dec 2007*)
8. Draw the circuit of a current series feedback amplifier and explain. Derive expressions for input and output impedance. How does it improve the stability of the amplifier?
   (*Electronic Circuits - II, Anna University, Nov/Dec 2007*)
9. Draw the block diagrams of the four possible feedback topologies and explain.
   (*Electronic Circuits - II, Anna University, Nov/Dec 2007*)
10. Prove that the bandwidth of the amplifier increases with negative feedback.
    (*Electronic Circuits - II, Anna University, Nov/Dec 2007*)
11. Compare negative and positive feedback.
    (*Electronic Circuits - II, Anna University, May/June 2007*)
12. With neat circuit diagram, explain the current series feedback amplifier and also derive the expression for feedback ratio, input resistance, output resistance and voltage gain.
    (*Electronic Circuits - II, Anna University, May/June 2007*)
13. Explain in detail, the effect of negative feedback on gain, bandwidth, distortion, noise, input resistance and output resistance.
    (*Electronic Circuits - II, Anna University, May/June 2007*)
14. Draw a block diagram of voltage shunt Feedback amplifier and give its input and output resistance. (*Electronic Circuits - II, Anna University, May/June 2007*)

15. Discuss the effects of negative feedback on gain, bandwidth, distortion and noise.
(Electronic Circuits - II, Anna University, May/June 2007)

16. Explain Nyquist stability criterion to analyse the stability of feedback amplifiers.
(Electronic Circuits - II, Anna University, May/June 2007)

17. Derive the expressions for current gain, voltage gain, input impedance and output impedance for an emitter follower circuit.
(Electronic Circuits - II, Anna University, April/May 2008)

18. Derive the expression for A_V, A_I, R_{in} and R_0 of the emitter follower.
(Electronic Circuits, VTU June/July 2008)

19. Discuss the effect of negative feedback on the frequency of an amplifier.
(Electronic Circuits - II, Anna University, June 2009)

20. Using the block diagram, derive the closed loop form transfer ratio of a feedback system in terms of the open-loop gain.
(Electronic Circuits - II, Anna University, June 2009)

MULTIPLE CHOICE QUESTIONS

1. For amplifiers using negative feedback, the bandwidth
(*a*) lowers its input and output resistances
(*b*) increases its input and output resistances
(*c*) lowers its input resistance and increases the output resistance

2. In a current-series feedback amplifier, the input resistance
(*a*) increases (*b*) decreases (*c*) remains unchanged

3. In an amplifier, if you employ voltage-shunt negative feedback
(*a*) the gain increases (*b*) the input resistance decreases
(*c*) the output resistance decreases (*d*) the bandwidth decreases

4. A common-emitter (CE) transistor amplifier employing an *n*-bypassed series emitter resistance is a case of
(*a*) current-series feedback (*b*) voltage-shunt feedback
(*c*) voltage-series feedback (*d*) current-shunt feedback

5. The advantage of using negative feedback in an amplifier is that its gain can be made practically independent of
(*a*) temperature changes (*b*) age of components
(*c*) frequency (*d*) all of the above.

6. Feedback in an amplifier always helps to
(*a*) control its output (*b*) increase its gain
(*c*) decrease its input impedance (*d*) stabilize its gain.

7. The only drawback of using negative feedback in amplifiers is that it involves
(*a*) gain sacrifice (*b*) gain stability
(*c*) temperature sensitivity (*d*) frequency dependence.

8. Closed-loop gain of a feedback amplifier is the gain obtained when
(*a*) its output terminals are closed (*b*) negative feedback is applied
(*c*) feedback loop is closed (*d*) feedback factor exceeds unity.

9. A large sacrifice factor in a negative feedback amplifiers leads to
(*a*) inferior performance
(*b*) increased output impedance
(*c*) characteristics impossible to achieve without feedback
(*d*) precise control over output.

10. Negative feedback in an amplifier
(*a*) lowers its lower 3-dB frequency (*b*) raises its upper 3-dB frequency
(*c*) increases its bandwidth (*d*) all of the above.

11. Regarding negative feedback in amplifiers which statement is WRONG ?
(*a*) it widens the separation between 3-dB frequencies
(*b*) it incerases the gain-bandwidth product
(*c*) it improves gain stability
(*d*) it reduces distortion.

12. Negative feedback reduces distortion in an amplifier only when it
(*a*) comes as part of input signal (*b*) is part of its output.
(*c*) is generated within the amplifier (*d*) exceed a certain safe level.

13. An amplifier with no feedback has a gain-bandwidth product of 4 MHz. Its closed-loop gain is 40. The new band-width is
(*a*) 100 kHz (*b*) 160 MHz
(*c*) 10 MHz (*d*) 20 kHz.

14. The shunt-derived series-fed feedback in an amplifier
(*a*) increases its output impedance (*b*) decreases its output impedance
(*c*) increases its input impedance (*d*) both (*b*) and (*c*).

15. A feedback amplifier has a closed gain of –200. It should not vary more than 50% despite 25% variation in amplifier gain A without feedback. The value of A is
(*a*) 800 (*b*) – 800
(*c*) 1000 (*d*) – 1000

16. The gain of a negative feedback amplifier is 40 dB. If the attenuation of the feedback path is 50 dB, then the gain of the amplifier without feedback is
(*a*) 78.92 (*b*) 146.32
(*c*) 215.51 (*d*) 317.23

17. In a common emitter amplifier, the unbypassed emitter resistor provides,
(*a*) voltage-shunt feedback (*b*) current-series feedback
(*c*) negative-voltage feedback (*d*) positive-current feedback.

ANSWERS

1. (*b*)	**2.** (*a*)	**3.** (*a*)	**4.** (*b*)	**5.** (*d*)
6. (*a*)	**7.** (*a*)	**8.** (*c*)	**9.** (*c*)	**10.** (*d*)
11. (*b*)	**12.** (*c*)	**13.** (*a*)	**14.** (*d*)	**15.** (*d*)
16. (*b*)	**17.** (*b*)			

❑❑❑

25

CHAPTER

Sinusoidal Oscillators

OUTLINE

Objectives

After completing this chapter, you should be able to :

- Know the difference types of oscillators.
- List different types of oscillators.
- Understand how the osicllations are produced in a LC circuit.
- Calculate the frequency of oscillatory circuit.
- List the factors that affect the frequency stability of an oscillator.
- Define the Barkhausen criterion for sustained oscillations.
- List the various types of LC oscillators (or radio-frequency oscillators).
- Sketch and describe the circuit operation of tuned-collector oscillator.
- Sketch and explain the circuit operation of tuned-drain (FET) oscillator.
- Sketch and describe the Hartley Oscillator circuit operation.

- Calculate the frequency of oscillation for the Colpitts oscillator.
- Know the difference between radio frequency and audio-frequency oscillators.
- Describe the operation of phase-shift oscillator.
- Calculate the frequency of oscillation of phase-shift oscillator.
- Calculate the frequency of oscillation of Wien-bridge oscillator.
- Explain the circuit operation of tunnel diode oscillator.

25-1. Introduction

We have already discussed in the previous chapter on Feedback Amplifiers that if the feedback signal is applied in such a way that it is in phase with the input signal and thus increases it, then it is called a *positive feedback, degenerative feedback* or *direct feedback.*

The use of positive feedback results in a feedback amplifier having closed-loop gain (A'_v) greater than the open-loop gain (A_v). It results in *instability* and operates as an *oscillatory circuit*. An oscillator circuit provides a, constantly, varying amplified output signal of any desired frequency.

25-2. Comparison Between an Amplifier and an Oscillator

Fig. 25-1(a) show a block diagram of an amplifier, while (b) shows that of an oscillator. We know that an amplifier is a device, which produces an output signal with a similar waveform as that of the input. But its power level is generally high. This additional power is supplied by an external d.c. source. Thus an amplifier is essentially an energy convention device, *i.e.*, a device, which gets energy from the d.c. source and converts it into an a.c. energy at the same frequency as that of the input signal. The d.c. to a.c. conversion is controlled by the input signal. It means that *if there is no input signal then no energy convertion take place. Thus, there is not output signal.*

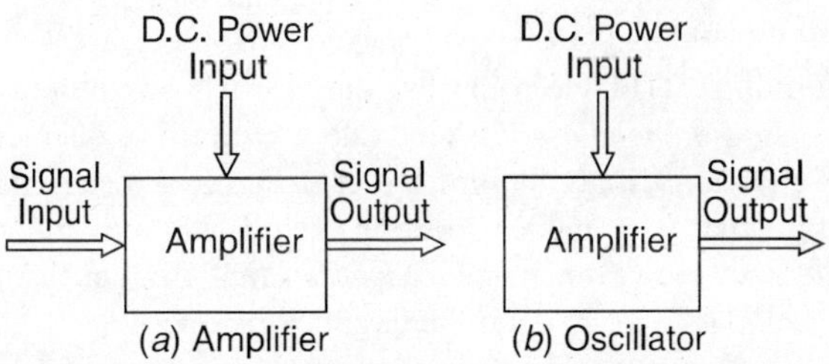

Fig. 25-1

On the other hand, *an oscillator is a device, which produces an output signal, without any input signal of any desired frequency*. It keeps producing an output signal, so long as the d.c. power is supplied. It is evident from the above discussion that *an oscillator does not require any external signal to start or maintain energy convertion process.*

25-3. Classification of Oscillators

The electronic oscillators may be broadly classified into the following two categories :

(*i*) ***Sinusoidal or harmonic oscillators :*** The oscillators, which provide an output having *a sine waveform*, are called sinusoidal or harmonic oscillators. Such oscillators can provide output at frequencies ranging from 20 Hz to 1 GHz (1 GHz = 10^{12} Hz).

(*ii*) ***Non-sinusoidal or relaxation oscillators :*** The oscillators, which provide an output having a square, rectangular or sawtooth wavform, are called non-sinusoidal or relaxation oscillators. Such oscillators can provide output at frequencies ranging from zero to 20 MHz. (1 MHz = 10^6 Hz).

In this chapter, we shall discuss only the sinusoidal oscillators. The non-sinusoidal oscillators will be discussed separately in the next chapter. The sinusoidal oscillators may be further sub-divided into the following types :

(*a*) ***Tuned circuit oscillators :*** These oscillators use a tuned-circuit consisting of inductors (L) and capacitors (C) and are used to generate high-frequency signals. Thus, they are also known as *radio frequency* (R.F.) *oscillators*. Such oscillators are *Hartley, Colpitts, Clapp-oscillators* etc.

(*b*) ***RC oscillators :*** There oscillators use resistors and capacitors and are used to generate low-or audio-frequency signals. Thus, they are also known as *audio-frequency* (A.F.) *oscillators*. Such oscillators are *phase-shift* and *Wien-bridge* oscillators.

(*c*) ***Crystal oscillators :*** These oscillators use *quartz crystals* and are used to generate highly stabilized output signal with frequencies up to 10 MHz. The *pierce oscillator* is an example of a cystal oscillator.

(*d*) ***Negative-resistance oscillators :*** These oscillators use *negative-resistance characteristic* of the devices such as tunnel diodes. A tuned diode oscillator is an example of negative resistance oscillator.

25-4. Applications of Sinusoidal Oscillators

The sinusoidal oscillators have a variety of applications in telecommunications and electronics. The most important application in telecommunications is the use of sine wave as a *carrier signal* (used in modulating music and speech signals) in both wire and wireless transmissions. The sine wave signals are also used in testing frequency response of various electronic systems and equipments such as amplifiers, filters, analogue communication channels, closed-loop control systems *etc*.

25-5. Nature of Sinusoidal Oscillations

The nature of oscillations, produced by sinusoidal oscillators, may be of two types, namely damped oscillations and undamped oscillations.

(1) ***Damped oscillations :*** The electrical oscillations, whose amplitude goes on decreasing with time are called *damped oscillations*. The waveform of such oscillations is shown in Fig. 25-2 (*a*). The damped oscillations are produced by those oscillator circuits in which power losses take place continuously during each oscillation without any means for compensating the same. However, it may be noted that frequency of oscillations remains unchanged because it depends upon the circuit parameters.

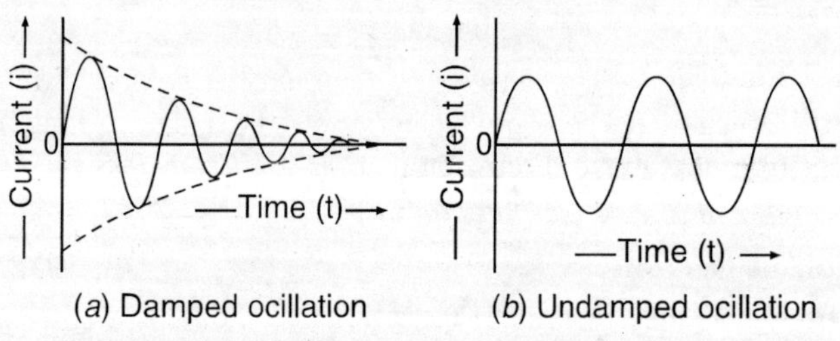

Fig. 25-2. *Nature of sinusoidal oscillations.*

(2) ***Undamped oscillations :*** The electrical oscillations, whose amplitude remains constant with time, are called *undamped* oscillations. The waveform of such oscillations is shown in Fig. 25-2(*b*). The undamped oscillations are produced by those oscillator circuits in which either there are no power losses or if they have any power loss then they have provision for compensating those losses. It may be noted that an oscillator is required to produce undamped oscillations for utilizing in various telecommunication and electronics applications.

25-6. The Oscillatory Circuit

A circuit, which produces electrical oscillations of any desired frequency, is known as an *oscillatory circuit*. This circuit consists of two reactive elements namely inductor (L) and a capacitor (C) connected in parallel with each other as shown in Fig. 25-3. Thus such a circuit is also known as LC circuit or *tank circuit*. Both the reactive devices are capable of storing electrical energy. It is a well known fact that an inductor stores energy in its magnetic field, whenever a current flows through it. Similarly, a capacitor stores energy in its electric field, whenever a voltage (or potential difference) is applied across its plates. Both the reactive elements are supposed to be an ideal one (*i.e.*, they do not have any power losses). Now we shall discuss as to how an LC circuit generates oscillations.

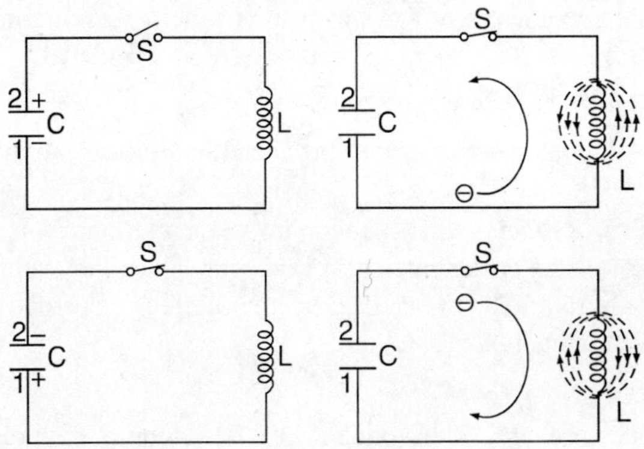

Fig. 25-3. *LC circuit acting as an oscillatory circuit.*

Suppose a capacitor has been charged from a d.c. voltage source with a polarity as shown in Fig. 25-3 (*a*). As the switch (S) is open, it cannot discharge through the coil (L).

When the switch (S) is closed as shown in Fig. 25-3 (*b*), the capacitor discharge through the coil (L). The discharging the due to the flow of electrons from plate 1 to plate 2 through the coil (L) as indicated by the arrow (note that conventional current direction is from plate 2 to plate 1). This current flow sets up *magnetic field* around the coil. This magnetic field stores the energy released by the electric field. Because of the inductive effect, the current grows up slowly towards the maximum value. This situation occurs, when the capacitor is fully discharged. At this instant, the *electrical energy stored in the capacitor becomes zero and the energy stored in magnetic field is maximum.* In other words, we say that electrical energy across the capacitor is completely converted into magnetic field energy around the coil.

Once the capacitor is discharged completely, the magnetic field around the coil begins to collapse and produces a counter *e.m.f.* According to Lenz's Law, the counter *e.m.f.*, keeps the electrons moving in the same direction as shown in Fig. 25-3 (*b*). This again charges the capacitor, though in the opposite direction, as shown in Fig. 25-3 (*c*). It may be noted that plate 1 has become positive whereas the plate 2, is still negative. When the capacitor is charged completely, in the opposite direction, the magnetic field around the coils is also collapsed completely. *At this instant, the energy previously stored in the magnetic field, has now converted into the energy stored in the electric field of the capacitor.*

After this, the capacitor starts discharging in the opposite direction, so that the electrons now move from plate 2 to plate 1 as shown in Fig. 25-3 (*d*). The electric field starts collapsing whereas magnetic field starts building up again, but in the reverse direction. Fig. 25-3 (*d*) shows the condition when the capacitor has fully discharged once again.

The above mentioned sequence of charging and discharging of a capacitor results in an alternating motion of electrons or an alternating current. As a result of this, the energy is alternatively stored in the electric field of the capacitor and the magnetic field of the coil. *This interchange of energy between the capacitor and coil continues to be repeated, and results in the production of electrical oscillations.*

If there are no power losses in the tank circuit, the waveform of the electrical oscillations will be undamped (*i.e.*, of constant amplitude). However, in a practical tank circuit, the waveform of the oscillations is damped (*i.e.*, of decreasing amplitude). It is because of the energy lost by the L and C element, which may take place because of the following two reasons :

(1) Some energy is lost in the form of heat produced in the resistance of the coil, connecting wires and within the dielectric material of the capacitor.

(2) Some energy is lost from the circuit due to radiation in the form of electromagnetic waves.

As a result of these losses, the amplitude of oscillating current in the tank circuit decreases gradually and eventually it becomes zero. The waveform of such an oscillating current will be of the form as shown in Fig. 25-2 (*a*).

25-7. Frequency of Oscillatory Circuit

The frequency (*f*) of the oscillatory current, produced in the tank circuit, is determined by the following two factors:

(1) ***Capacitance of the capacitor :*** The frequency of oscillation (f_o) is inversely propórtional to the square root of the capacitance of a capacitor. Thus, larger the value of capacitance, greater is the time required for removal of the discharge current and hence lower is its frequency. Mathematically, the frequency,

$$f_o \propto 1/\sqrt{C} \qquad ...(i)$$

(2) ***Self-inductance of the coil :*** The frequency of oscillation (f_o) is proportional to the square root of the self-inductance of the coil. Thus larger the inductance, greater is the inductance effect and hence longer the time required by the current to stop flowing during the discharge of the capacitor. Mathematically, the frequency,

$$f_o \propto 1/\sqrt{L} \qquad ...(ii)$$

Combining the equations (*i*) and (*ii*) we get

$$f_o \propto \frac{1}{\sqrt{L \cdot C}} = \frac{1}{2\pi\sqrt{L \cdot C}}$$

$$= \frac{0.159}{\sqrt{L \cdot C}} \text{ in Hertz}$$

If *L* is in μH and *C* in μF, then the frequency,

$$f_o = \frac{\mathbf{159}}{\sqrt{L \cdot C}}$$

It may be noted that the damped oscillations, so produced, are not useful in actual practice. We need continuous undamped oscillations for practical applications such as radio and cable transmission. The *undamped oscillations can be produced if some additional energy is supplied in correct phase and correct direction to the tank circuit for making up the power losses continuously occurring in the circuit.*

25-8. Frequency Stability of an Oscillator

The frequency stability of an oscillator is a measure of its ability to maintain a constant frequency, over a long time interval. However, it has been found that if an oscillator is set at some particular frequency, it does not maintain it for a longer period. In other words, the frequency of an oscillator changes slowly (or drifts away) from the initially set value. Sometimes, the change in frequency is uniform in one direction (*i.e.*, either increasing or decreasing). But at some other times, it may be changing quite erratically. The change in oscillation frequency may arise due of the following factors :

(*i*) ***Operating point of the active device :*** The operating point of the active device (*i.e.*, bipolar transistor or FET) is selected in such a way so that its operation in non-linear region, changes the values of device parameters which, in turn, affects the frequency stability of the oscillator.

(2) ***Circuit components :*** The values of circuit components (*i.e.*, resistor, inductors and capacitors) change with the variation in temperature. Since, such changes take place slowly, they also cause a drift in oscillator frequency.

(3) ***Supply voltage :*** The changes in d.c. supply voltage applied to the active device, shift the oscillator frequency. This problem can be avoided by using a highly regulated power supply.

(4) ***Output load :*** A change in the output load may cause a change in the Q-factor of the tank circuit, thereby causing a change in oscillator output frequency.

(5) ***Interelement capacitances and stray capacitances :*** Any change in the interelement capacitances of a transistor (particularly collector-to-emitter capacitance), cause changes in the oscillator output frequency and thus affect the frequency stability. Similarly, the stray capacitances also affect the frequency stability of an oscillator. The effect of change in interelement capacitances can be neutralized by putting an additional capacitor across the corresponding elements. However, it is difficult to avoid the effect of stray capacitances.

25-9. The Barkhausen Criterion

We have already discussed in the chapter of Feedback Amplifiers, that the overall voltage gain of a positive feedback amplifier is given by the expression,

$$A'_v = \frac{A_v}{1 - \beta \cdot A_v}$$

A_v = Gain of an amplifier without feedback also called open loop-gain, and

* $\beta \cdot A_v$ = Product of feedback fraction and open loop-gain. It is called loop-gain

It will be interesting to know that if the product ($\beta \cdot A_v$) is made equal to unity, then the denominator of equation (*i*) is zero. It means that the gain (A'_v) will increase to infinity. Of course, the output of the feedback amplifier, in actual practice, cannot be infinite. Therefore, the condition (*i.e.*, $1 - \beta \cdot A_v = 0$) represents that there will be an output voltage, whose frequency is completely different from the input signal. In other words, the circuit has stopped amplifying and started oscillating. It may be noted that the oscillations will not be maintained, if the value of $\beta \cdot A_v$ is less than unity.

We know that an amplifier reverses the phase of an input signal at its output. It means that an amplifier causes a phase shift of 180° between the input and output signals. In order to provide positive feedback, the feedback network must provide a phase shift of 180°, so as to provide a signal with a phase shift of 360° or 0° at the amplifier input.

The above two conditions for positive feedback (*i.e.*, $\beta \cdot A_v = 1$ *and the net phase shift around the loop equal to 360° or 0°) are called Barkhausen criterion for oscillation.* Mathematically, the Barkhausen criterion may be stated as follows:

$$\beta \cdot A_v = 1 \qquad \text{... (i)}$$

and

$$\angle \beta \cdot A_v = 0°$$

Thus, if $\beta \cdot A_v$ is a complex quantity (*i.e.*, a quantity which contains both real and imaginary parts) then its real part must be equal to unity and imaginary part equal to zero, *i.e.*,

$$\boxed{\beta \cdot A_v = 1 + j_0}$$

However, an oscillator in which the quantity $\beta \cdot A_v$ is exactly unity, is not realizable in practice. In every practical oscillator, the quantity $\beta \cdot A_v$ is slightly larger than unity. Thus, an oscillator circuit must satisfy the two conditions mentioned above to produce sustained oscillations. The conditions are (*a*) loop gain ($\beta \cdot A_v$) ≥ 1 and (*b*) phase-shift between the input and output signals must be equal to 360° or 0°. It may be noted that if a circuit does not satisfy any one of these conditions, the oscillations will not be produced.

* $\beta \cdot A_v$ is actually a complex quantity in positive feedback amplifiers, because of the use of inductive elements.

25-10. Tuned Circuit Oscillators

These are also known as *LC oscillators, resonant circuit oscillators or tank circuit oscillators*. These oscillators are used to produce an output with frequencies ranging from 1 MHz to 500 MHz. Hence, they are also known as R.F. oscillators. A *bipolar transistor or an FET is used as an amplifier with tuned circuit oscillators. With an amplifier and an LC tank circuit, we can feedback a signal with right amplitude and phase to maintain oscillations.

It will be interesting to know that most of the oscillators used in radio transmitters and receivers are of LC oscillators type. Depending upon the way, the feedback is used in the circuit, the LC oscillators are of the following five types :

(*i*) ***Tuned-collector*** **or** ***Armstrong oscillator :*** It uses inductive feedback from the collector of a transistor to the base. The LC circuit is in the collector circuit of the transistor.

(*ii*) ***Tuned base oscillator :*** It also uses inductive feedback. But the LC circuit is in the base circuit.

(*iii*) ***Harley oscillator :*** It also uses inductive feedback.

(*iv*) ***Colpitts oscillator :*** It uses capacitive feedback.

(*v*) ***Clapp oscillator :*** It also uses capacitive feedback.

We shall now, discuss all the above mentioned tuned circuit oscillators in detail in the following pages.

25-11. Tuned-Collector Oscillator

It is also called *Armstrong Oscillator*. Fig. 25-4 shows the circuit of such an oscillator. In this circuit, the collector drives an LC tank circuit. The feedback signal is taken from a small secondary winding (L_1) and feedback to the base. There is a phase shift of 180° in the transformer and another 180° phase shift is provided by the transistor amplifier. It means that the total phase shift around the loop is zero and hence the feedback is positive. Ignoring the loading effect of base, the feedback fraction,

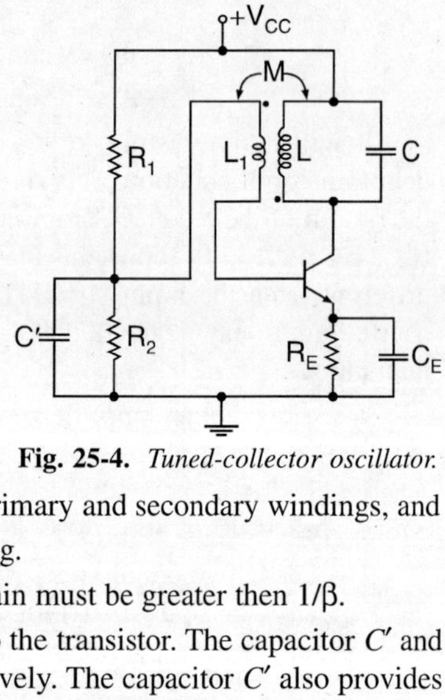

Fig. 25-4. *Tuned-collector oscillator.*

$$\beta \approx M/L$$

where, M = The mutual inductance between the primary and secondary windings, and

L = Self-inductance of the primary winding.

It may be noted that for oscillations to start, the voltage gain must be greater then $1/\beta$.

The resistors R_1, R_2 and R_E are used to provide d.c. bias to the transistor. The capacitor C' and C_E act as bypass capacitors for the resistors R_2 and R_E respectively. The capacitor C' also provides a.c. ground for the secondary transformer (L_1). The operation of the circuit may be explained as follows :

When the circuit is energized by switching on the supply, the collector current starts increasing to the quiescent value. This current charges the capacitor (C). When this capacitor is fully charged, it discharges through the primary winding (L) producing magnetic field around the primary coil. When the capacitor is fully discharged, the magnetic field collapses and charges the capacitor in the reverse direction. Then again the capacitor discharges in the reverse direction. This process continues again and again. As a result of this, the oscillations are produced.

The frequency of oscillations produced by the tank circuit depends upon the values of primary inductance (L) and capacitance (C). Its value given by the expression,

* The operational amplifiers (Op Amps) cannot be used as an amplifier with LC oscillators, because its unity gain frequency (f_{unity}) is much smaller than the frequency range of LC oscillators.

$$f_o = \frac{1}{2\,\pi\sqrt{L \cdot C}} = \frac{0.159}{\sqrt{L \cdot C}}\ \text{Hz}$$

These oscillations induce some voltage in secondary winding (L_1) by mutual induction. The magnitude of the voltage induced depends upon the number of turns of secondary winding (L_1) and the coefficient of coupling (k) between the coils L and L_1. However, the frequency of the induced voltage is the same as that of the oscillations produced in the tank circuit. The voltage across L_1 winding appears between the base and emitter terminals of the transistor. This voltage is amplified at the collector of a transistor and is partially used for overcoming the losses, which take place in the tank circuit. The remaining voltage may be utilized for any other purpose.

Example 25-1. *A tuned collector oscillator is tuned to operate at 22 kHz by a variable capacitor set to 2 nF. What is the value of the tuned circuit inductance.*

Solution. Given f_o = 22 kHz = 22 × 10^3 Hz and C = 2 nF = 2 × 10^{-9} F

We know that for a tuned collector oscillator the frequency of oscillations (f_o),

$$22 \times 10^3 = \frac{0.159}{\sqrt{L \cdot C}} = \frac{0.159}{\sqrt{L \times (2 \times 10^{-9})}}$$

$$\therefore \quad L = \frac{\left(\dfrac{0.159}{22 \times 10^3}\right)^2}{(2 \times 10^{-9})} = 0.026 \text{ H } \textbf{Ans.}$$

Example 25-2. *A tuned collector oscillator operates at 2.2 MHz. At what frequency will it work if its tuned circuit capacitance is reduced by 50%.*

Solution. Given: f_o = 2.2 MHz = 2.2 × 10^6 Hz.

Let f_o' = The frequency at which the tuned circuit will work when the capacitance is reduced by 50%.

We know that frequency of oscillations, for a tuned collector oscillator,

$$f_o = \frac{0.159}{\sqrt{L \cdot C}} \quad \text{or} \quad f_o \propto \frac{1}{\sqrt{C}} \qquad ...(i)$$

and when the capacitance is reduced by 50%, the frequency of oscillations,

$$f_o' \propto \frac{1}{\sqrt{C/2}} = \frac{\sqrt{2}}{\sqrt{C}} = \frac{1.414}{\sqrt{C}} \qquad ...(ii)$$

Dividing equation (*ii*) by (*i*), we get

$$\frac{f_o'}{f_o} = \frac{1.414/\sqrt{C}}{1/\sqrt{C}} = 1.414;$$

$$\therefore \quad f_o' = 1.414 f_o = 1.414 \times 2.2 \times 10^6 = 3.11 \times 10^6 \text{ Hz} = 3.11 \text{ MHz } \textbf{Ans.}$$

25-12. Tuned-Drain (FET) Oscillator

Fig. 25-5 shows the tuned drain (FET) oscillator. It is similar to the tuned collector transistor oscillator shown in Fig. 25-4. This FET has a high input resistance and amplification factor than that of a bipolar transistor. Therefore, FET oscillators are more commonly used in actual practice. The frequency of this oscillator is given by the relation.

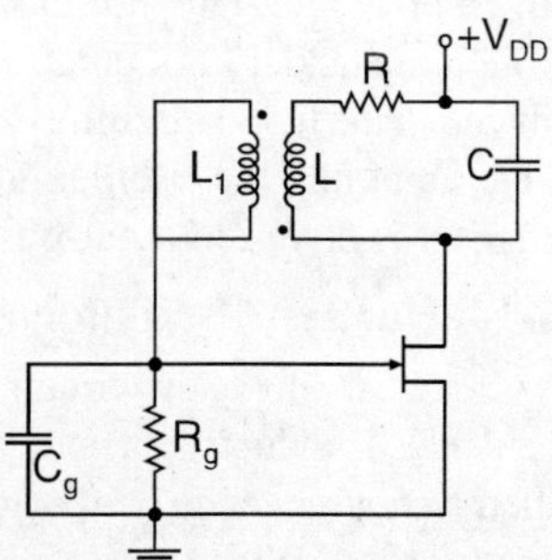

Fig. 25-5. *Tuned-drain (FET) oscillator*

$$f_o = \frac{1}{2\,\pi\sqrt{L \cdot C}} \qquad ...(i)$$

and the condition (*i.e.*, $\beta \cdot A_v = 1$) is equivalent to

$$\mu = L / M \qquad ...(ii)$$

where, μ = Amplification factor for the FET, and

M = Multual inductance between the coils L_1 and L.

If we take into account, the coil resistance (R), then the above expression for frequency modifies to

$$f_o' = \frac{1}{2\,\pi\sqrt{L \cdot C}}\left(1 + \frac{R}{r_D}\right) \qquad ...(iii)$$

where, r_D = The a.c. drain resistance.

and the condition ($\beta \cdot A_v = 1$) is equivalent to

$$g_m = \frac{\mu \cdot R \cdot C}{\mu \cdot M - L} \qquad ...(iv)$$

where, g_m is the value of transconductance for an FET.

The equation (*iv*) gives the minimum value of transconductance required for sustained oscillations. The circuit operation may be understood from the following discussion.

When the circuit is first energized by switching on the supply, the gate bias (provided by the parallel combination of the $R_g \cdot C_g$) is zero and the FET operates with a large value of transconductance (g_m). The loop gain is therefore greater than unity and the amplitude of oscillations starts growing. As the oscillations grow, the bias becomes negative and the value of transconductance (g_m) decreases. Finally, the amplitude stabilizes itself at that value for which the loop gain for oscillation frequency is reduced to unity.

25-13. Tuned-Base Oscillator

It is also known as *tickler oscillator*. Fig. 25-6 shows the circuit of such an oscillator. In this circuit, the tuned circuit is put in the base circuit. The feedback signal is supplied to the base through the coil L_1 (called tickler coil). There is a phase shift of 180° by the transistor amplifier and another 180° phase shift is provided by the transformer. It means that the total phase shift, around the loop, is zero and hence, the feedback is positive.

The resistors R_1, R_3 and R_E are used to provide d.c. bias to the transistor. The capacitors C' and C_E act as coupling and bypass capacitors. The operation of the circuit may be understood from the following discussion.

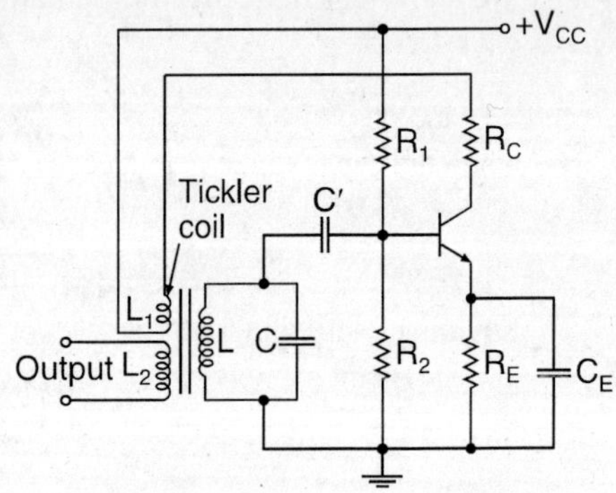

Fig. 25-6. *Tuned-base oscillator.*

When the circuit is energized, by switching on the supply, the collector current starts increasing to a quiscient value. The increasing collector current, through the tickler coil, creates a changing magnetic field around it. As a result of this, a voltage is induced in the tuned circuit. Because of the correct positive feedback of the coils, and sufficient gain of the amplifier, the oscillations start growing. The sustained oscillations are produced, when the transistor is driven into cut-off due to the voltage developed across the capacitor (C).

25-14. Hartley Oscillator

Fig. 25-7 shows the circuit of a Hartley oscillator. Its tank circuit consists of two coils L_1 and L_2. The coil L_1 is inductively coupled to coil L_2 and the combination works as an autotransformer. A coil called *radio frequency choke* (RFC) is connected between the collector and the V_{CC} supply. It acts as a load for the collector and also permits an easy flow of d.c. current. But blocks a.c. current. The feedback between the output and input circuits is accomplished through autotransformer action,

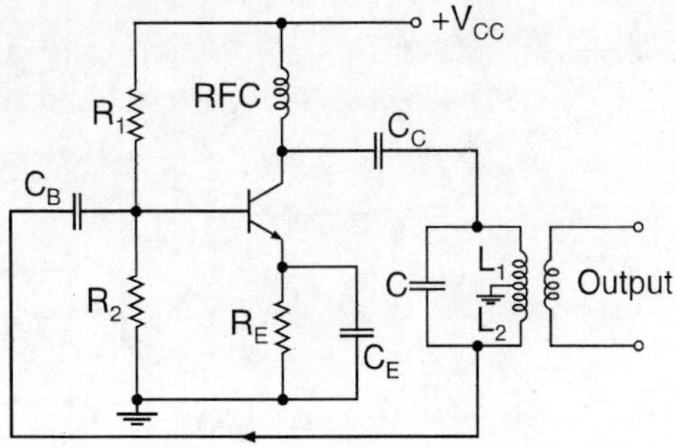

Fig. 25-7. *Hartley oscillator*

which also introduces a phase shift of 180°. The phase reversal between the output and input voltage occurs because they are taken from the opposite ends of the coils (L_1 and L_2) with respect to the tap, which is grounded. It may be noted that the tap on the combination of L_1 and L_2 coils is actually connected to the transistor emitter terminal *via* ground and through the capacitor (C_E). Since, the transistor also introduces a phase shift of 180°, therefore, the total phase shift is 360° and hence, the feedback is positive. Ignoring the loading effects of the base, the feedback fraction is given by the relation,

$$\beta = L_2 / L_1$$

For oscillations to start, the voltage gain must be greater than $1/\beta$ which is equal to L_1/L_2.

The capacitor C_C, connected between the collector and the tuned circuit, is called *coupling capacitor.* It permits only the a.c. currents to pass to the tank circuit. In other words, the capacitor C_C blocks the d.c. currents. The capacitor C_B, called *blocking capacitor*, further blocks the d.c. currents reaching at the base. The resistors R_1, R_2 and R_E are used to provide d.c. bias to the transistor.

When the circuit is energised, by switching on the supply, the collector current flows. The oscillations are produced because of positive feedback from the tank circuit. The frequency of oscillations is given by the relation,

$$f_o = \frac{1}{2\pi\sqrt{L \cdot C}}$$

where, $L = L_1 + L_2 + 2M$

$= L_1 + L_2$

(If the mutual inductance M is neglected)

The output from the tank may be taken out by means of another coil coupled either to or coil as shown in the figure 28.8. Sometimes, a field-effect transistor (FET) may also be used as an amplifying device instead of a bipolar transistor. Fig. 25-8 shows the circuit of a Hartley oscillator using the FET.

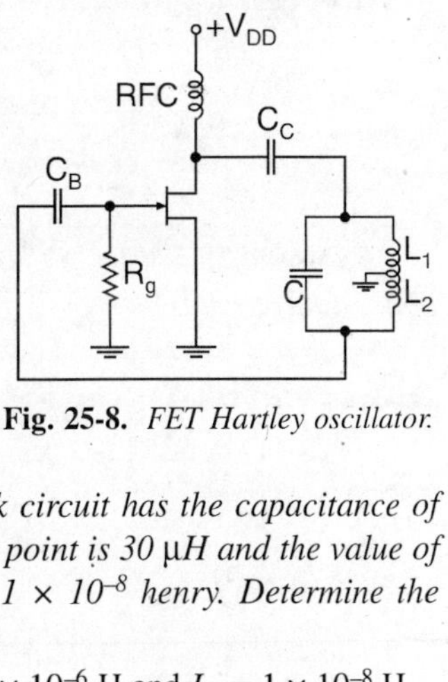

Fig. 25-8. *FET Hartley oscillator.*

Example 25-3. *In a transistorized Hartley oscillator, the tank circuit has the capacitance of 100 pF. The value of inductance between the collector and tapping point is 30 μH and the value of inductance between the tapping point and the transistor base is 1 × 10^{–8} henry. Determine the frequency of oscillations. Neglect the mutual inductance.*

Solution. Given: $C = 100$ pF $= 100 \times 10^{-12}$ F; $L_1 = 30$ μH $= 30 \times 10^{-6}$ H and $L_2 = 1 \times 10^{-8}$ H.

We know that the frequency of oscillations,

$$f_o = \frac{1}{2\pi\sqrt{L \cdot C}} = \frac{1}{2\pi\sqrt{(L_1 + L_2)\, C}}$$

$$= \frac{1}{2\pi\sqrt{[(30 \times 10^{-6}) + (1 \times 10^{-8})] \times (100 \times 10^{-12})}}$$

$= 2.9 \times 10^6$ Hz $= 2.9$ MHz **Ans.**

Example 25-4. *Calculate the* **(i)** *operating frequency, and* **(ii)** *feedback fraction for the hartley oscillator shown in Fig. 25-9 the mutual inductance between the coils, M = 20 μH, and L_1 = 1000 μH and L_2 = 100 μH.* *(Dip. IETE, June* 2001)

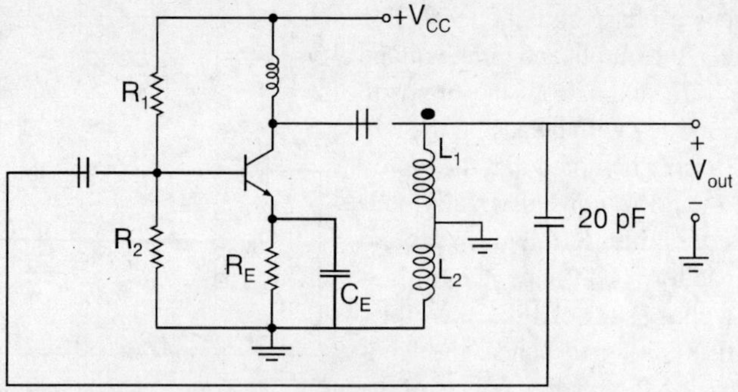

Fig. 25-9.

Solution. Given: $L_1 = 1000\ \mu H$; $L_2 = 100\ \mu H$; $M = 20\ \mu H$, and . C - 20 pF

We know that frequency of oscillations of a given circuit (a Hartley oscillator),

(i) Operating frequency

$$f_o = \frac{1}{2\pi\sqrt{LC}} = \frac{1}{2\pi\sqrt{(L_1 + L_2 + 2M)C}}$$

$$= \frac{1}{2\pi\sqrt{[(1000 \times 10^{-6}) + (100 \times 10^{-6}) + (20 \times 10^{-6})] \times (20 \times 10^{-12})]}}$$

$= 1.06 \times 10^6$ or 1.06 MHz **Ans.**

(ii) Feedback fraction

$$\beta = L_2 / L_1 = \frac{100\ \mu H}{1000 \mu H}$$

$$\Rightarrow \qquad \beta = \frac{1}{10}$$

Example 25-5. *For the Hartley oscillator shown in Fig. 25-10, find the frequency of oscillations of the circuit.*

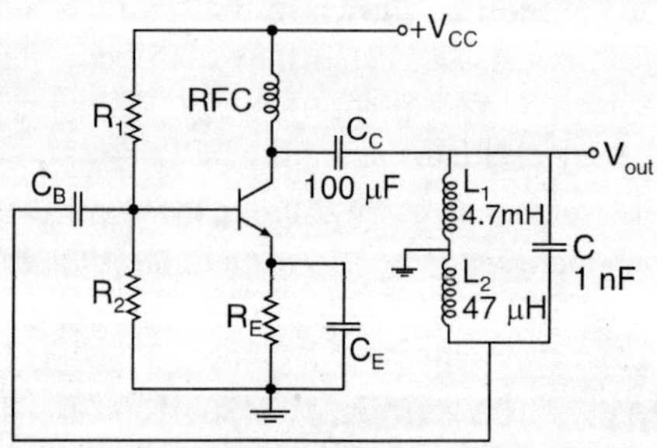

Fig. 25-10.

Solution. Given: $C = 1\text{nF} = 1 \times 10^{-9}$ F; $L_1 = 4.7\text{mH} = 4.7 \times 10^{-3}$ H; and $L_2 = 47\mu\text{H} = 47 \times 10^{-6}$

We know that frequency of oscillations,

$$f_o = \frac{1}{2\pi\sqrt{L \cdot C}} = \frac{1}{\sqrt{2\pi (L_1 + L_2) \cdot C}}$$

$$= \frac{1}{2\pi\sqrt{[(4.7 \times 10^{-3})] + (47 \times 10^{-6}) \times (1 \times 10^{-9})\ C}}$$

$$= \frac{1}{2\pi \times 2.179 \times 10^{-6}} = 0.0731 \times 10^6 \text{ Hz} = 73.1 \text{ kHz } \textbf{Ans.}$$

Example 25-6. *A Hartley oscillator is designed with L_1 = 2 mH, L_2 = 20 μH and a variable capacitance. Determine the range of capacitance values, if the frequency of oscillation is varied between 950 and 2050 kHz.*

Solution. Given: $L_1 = 2 \text{ mH} = 2 \times 10^{-3}$ H; $L_2 = 20\ \mu\text{H} = 20 \times 10^{-6}$ H; f_o = 950 to 2050 kHz = 950×10^3 Hz to 2050×10^3 Hz.

Let C_1 = Value of capacitance required for a frequency of 950 kHz, and

C_2 = Value of capacitance required for a frequency of 2050 kHz.

We know that the resonant frequency (f_o),

$$950 \times 10^3 = \frac{1}{2\,\pi\sqrt{L_1 \cdot L_2) \cdot C_1}}$$

$$= \frac{1}{2\,\pi\,\sqrt{[(2 \times 10^{-3}) + (20 \times 10^{6})] \cdot C_1}}$$

$$= \frac{1}{2\,\pi\,\sqrt{(2.02 \times 10^{-3}) \cdot C_1}}$$

or $$C_1 = \frac{1}{4\,\pi^2\,\sqrt{(2.02 \times 10^{-3}) \times (950 \times 10^3)^2}}$$

$$= 13.9 \times 10^{-12} \text{ F} = 13.9 \ pF$$

and $$2050 \times 10^3 = \frac{1}{2\,\pi\sqrt{(L_1 + L_2) \cdot C_2}}$$

$$= \frac{1}{2\,\pi\sqrt{[(2 \times 10^{-3}) + (20 \times 10^{-6}) \cdot C_2}}$$

$$= \frac{1}{2\,\pi\sqrt{(2.02 \times 10^{-3}) \cdot C}}$$

or $$C_2 = \frac{1}{4\,\pi^2 \times (2.02 \times 10^{-3}) \times (2050 \times 10^3)^2}$$

$$= 2.98 \times 10^{-12} \text{ F} = 2.98 \ pF$$

Hence the range of capacitance values required is 2.98 pF to 13.9 pF **Ans.**

Example 25-7. *In a Hartley oscillator if L_1 = 0.1 mH, L_2 = 10 μH and mutual inductance between the coils equal to 20 μH. calculate the value of capacitor C of the oscillatory circuit to obtain frequency of 4110 kHz and also find the condition for sustained oscillations.*

(*Bangalore Univ. B.Sc Electronics May* 2001)

Solution. Given : L_1 = 0.1 mH; L_2 = 10 μH; M = 20 μH, f_o = 4110 kHz.

Let C = value of the capacitor required.

We know that the frequency of oscillation of a Hartley oscillator circuit (f_o),

$$4110 \times 10^3 = \frac{1}{2\pi\,\sqrt{LC}} = \frac{1}{2\pi\,\sqrt{(L_1 + L_2 + 2M)\,C}}$$

or $$4110 \times 10^3 = \frac{1}{2\pi\sqrt{[(0.1 \times 10^{-3}) + (10 \times 10^{-6}) + (2 \times 10^{-6})] \times C}}$$

∴ $$C = 1.154 \times 10^{-11} \text{ or } 11.54 \text{ pF}$$

The condition for sustained oscillation is that voltage gain

$$A_V = \frac{1}{\beta} = \frac{1}{L_2 / L_1} = \frac{L_1}{L_2} = \frac{0.1 \times 10^{-3}}{10 \times 10^{-6}} = 10 \textbf{ Ans.}$$

25-15. Colpitts Oscillator

It is similar to the Hartley oscillator, except for one difference. The colpitts oscillator uses tapped capacitance instead of tapped inductance used in Hartley oscillator. Fig. 25-11 (*a*) shows the Colpitts oscillator circuit, whereas Fig. 25-11 (*b*) shows its a.c. equivalent circuit. The tank circuit is made up of two capacitors C_1 and C_2 connected in series with each other across a fixed inductance (L). The resistors (R_1, R_2, and R_E), capacitors (C_C and C_B) and RFC have the function as mentioned for Hartley oscillator. The feedback between the output and input circuit is accomplished by the voltage developed across the capacitor C_2. Ignoring the loading effect of the base, the feedback fraction,

$$\beta = C_1/C_2$$

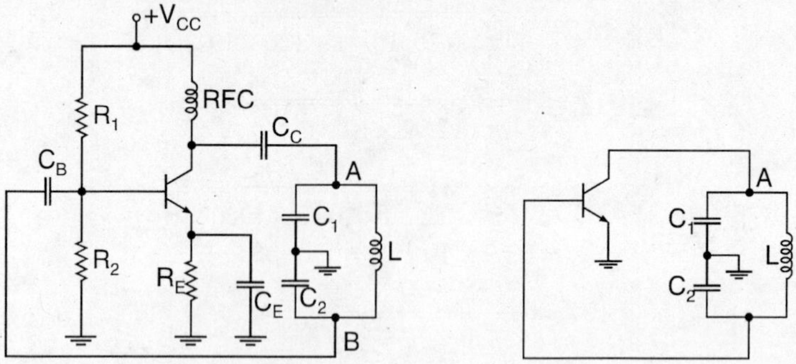

(*a*) Colpitts oscillator. (*b*) A.C. equivalent circuit.

Fig. 25-11.

For oscillation to start, the voltage gain (A_v) must be greater than $1/\beta$ (C_2 or C_1). Or in other words, $A_v > C_2 / C_1$. The frequency of oscillations (neglecting mutual inductance), is given by the relation,

$$f_o = \frac{1}{2\pi\sqrt{L \cdot C}}$$

(Where, C is the effective value of capacitance and equal to $\dfrac{C_1 \cdot C_2}{C_1 + C_2}$)

$$\therefore \qquad f_o = \frac{1}{2\pi\sqrt{L \times \dfrac{C_1 \cdot C_2}{C_1 + C_2}}}$$

It may be noted that in a colpitts oscillator, the capacitors C_1 and C_2 act as a simple alternating voltage divider. Therefore points A and B are 180° out of phase with each other. Another phase shift of 180° is provided by the transistor itself. Thus there is a total phase shift of 360° between the emitter-base and collector-base circuits.

When the circuit is energized by switching on the supply, the capacitors C_1 and C_2 are charged. These capacitors discharge through the coil (L), which sets up the oscillations of frequency $f_o = 1/2\pi\sqrt{L \cdot C}$ where $C = C_1 \cdot C_2 / (C_2 + C_1)$.The oscillations across the capacitor C_2 are feedback to the base-emitter junction and appear in an amplified form at the collector. Because of the positive feedback, the oscillations of constant amplitude are produced. The colpitts oscillators are widely used as a signal generator for frequencies between 1 MHz and 500 MHz.

Example 25-8. *It is required to generate an output frequency equal to 100 kHz using a Colpitts oscillator. Determine the value of C_1 and C_2, given L = 0.5 mH.* (*Bangalore Univ. Jan.* 2004)

Solution. Given f_o = 100 kHz = 100×10^3 Hz, L = 0.5 mH = 0.5×10^{-3} H.

Let $C' = C_1 = C_2$.The value of capacitance required. We know that the frequency (f_o),

$$100 \times 10^3 = \frac{1}{2\pi\sqrt{L.C}} = \frac{1}{2\pi\sqrt{(0.5 \times 10^{-3}) \times C'/2}} \qquad \left(\because C = \frac{C'.C'}{C' + C'} = \frac{C'}{2}\right)$$

or
$$\left(100 \times 10^3\right)^2 = \frac{1}{4\pi^2 \times \left(0.5 \times 10^{-3}\right) \times (C'/2)}$$

$$\therefore \quad C' = \frac{1}{4\pi^2 \times \left(0.5 \times 10^{-3}\right) \times \left(100 \times 10^3\right)^2} = 0.01 \times 10^{-6} \text{ F}$$

$$= 0.01 \ \mu\text{F} \ \textbf{Ans.}$$

Example 25-9. *In a transistor Colpitts oscillator, $C_1 = 0.001$ μF; $C_2 = 0.01$ μF and L = 5 μH. Find the required gain for oscillation and the frequency of oscillation.*

Solution. Given: $C_1 = 0.0001$ μF = 0.001×10^{-6} F; $C_2 = 0.01$ μF = 0.01×10^{-6} F and $L = 5$ μH = 5×10^{-6} H.

We know that voltage gain required to produce oscillations,

$$A_v > \frac{C_2}{C_1} = \frac{0.01 \times 10^{-6}}{0.001 \times 10^{-6}} = 10 \ \textbf{Ans.}$$

and the capacitance,

$$C = \frac{C_1 \cdot C_2}{C_1 + C_2} = \frac{(0.001 \times 10^{-6}) \times (0.01 \times 10^{-6})}{(0.001 \times 10^{-6}) + (0.01 \times 10^{-6})}$$

$$= 9.091 \times 10^{-10} \text{ F}$$

∴ Frequency of oscillations,

$$f_o = \frac{1}{2\pi\sqrt{L \cdot C}}$$

$$= \frac{1}{2\pi \times \sqrt{(5 \times 10^{-6}) \times (9.091 \times 10^{-10})}} \text{ Hz}$$

$$= 2.37 \times 10^6 \text{ Hz} = 2.37 \text{ MHz} \ \textbf{Ans.}$$

Example 25-10. *For the transistor Colpitts oscillator shown in Fig. 25-12, find the frequency of oscillations.*

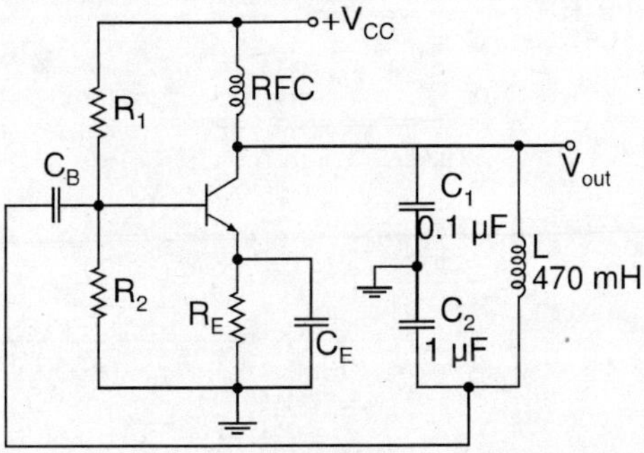

Fig. 25-12.

Solution. Given: $C_1 = 0.1$ μF = 0.1×10^{-6} F; $C_2 = 1$ μF = 1×10^{-6} F and $L = 470$ μH = 470×10^{-6} H.

We know that frequency of oscillations,

$$f_o = \frac{1}{2\pi\sqrt{L \cdot C}} = \frac{1}{2\pi\sqrt{L \times \dfrac{C_1 \cdot C_2}{C_1 + C_2}}}$$

$$= \frac{1}{2\pi\sqrt{(470 \times 10^{-6}) \times \frac{(0.1 \times 10^{-6}) \times (1 \times 10^{-6})}{[(0.1 \times 10^{-6}) + (1 \times 10^{-6})]}}}$$

$$= \frac{1}{2\,\pi\sqrt{(470 \times 10^{-6}) \times \frac{0.1 \times 10^{-6}}{1.1}}} = \frac{1}{2\,\pi \times 6.537 \times 10^{-6}}$$

$= 0.0244 \times 10^6$ Hz = 24.4 kHz **Ans.**

Example 25-11. *A Colpitts oscillator is designed with C_1 = 100 pF and C_2 = 7500 pF. The inductance is variable. Determine the range of inductance values, if the frequency of oscillation is to vary between 950 kHz and 2050 kHz.* *(Electronic Circuits-II, Univ., May/June 2007*

Solution. Given: $C_1 = 100$ pF $= 100 \times 10^{-12}$ F; $C_2 = 7500$ pF $= 7500 \times 10^{-12}$ F and $f_o = 950$ kHz to 2050 kHz $=950 \times 10^3$ Hz to 2050×10^3 Hz.

Let L_1 = Value of inductance required for a frequency of 950×10^3 Hz, and

L_2 = Value of inductance required for a frequency of 2050×10^3 Hz.

We know that the effective value of capacitance,

$$C = \frac{C_1 \cdot C_2}{C_1 + C_2} = \frac{(100 \times 10^{-12}) \times (7500 \times 10^{-12})}{(100 \times 10^{-12}) + (7500 \times 10^{-12})} \text{ F}$$

$$= 98.68 \times 10^{-12} \text{ F}$$

and the resonant frequency (f_o),

$$950 \times 10^3 = \frac{1}{2\,\pi\sqrt{L_1 \cdot C}} = \frac{1}{2\,\pi \times \sqrt{L_1 \times (98.68 \times 10^{-12})}}$$

or
$$(950 \times 10^3)^2 = \frac{1}{4\,\pi^2 \cdot L_1 \times (98.68 \times 10^{-12})}$$

∴
$$L_1 = \frac{1}{4\,\pi^2 \times (98.68 \times 10^{-12}) \times (950 \times 10^3)^2}$$

$$= 284 \times 10^{-6} \text{ H} = 284\ \mu\text{H}$$

and
$$2050 \times 10^3 = \frac{1}{2\,\pi\sqrt{L_2 \cdot C}}$$

$$= \frac{1}{2\,\pi\sqrt{L_2 \times (98.68 \times 10^{-12})}}$$

or
$$(2050 \times 10^3)^2 = \frac{1}{4\,\pi^2 \times L_2 \times (98.68 \times 10^{-12})}$$

∴
$$L_2 = \frac{1}{4\,\pi^2 \times (98.68 \times 10^{-12}) \times (2050 \times 10^3)^2}$$

$$= 61 \times 10^{-6} \text{H} = 61\ \mu\text{H}$$

Thus the range of inductance required for the variation of the desired frequency is from 61 μH to 284 μH. **Ans.**

Example 25-12. *The frequency of oscillation of a Colpitts oscillator is given by,*

$$f_o = \frac{1}{2\,\pi\sqrt{\left[L\left(\frac{C_1 C_2}{C_1 + C_2}\right)\right]}}$$

where L, C_1 and C_2 are the frequency - determining components. Such a circuit operates at 450 kHz with $C_1 = C_2$. What will be the oscillation frequency if the value of C_2 is doubled?

Solution. Given: $f_o = 450$ kHz $= 450 \times 10^3$ Hz; and $C_1 = C_2 = C$ (say)

Let f_o = The frequency when the value of C_2 is doubled.

We know that the frequency of oscillations of a Colpitts oscillator,

$$f_o = \frac{1}{2\pi\sqrt{\left[L\left(\dfrac{C_1C_2}{C_1+C_2}\right)\right]}} = \frac{1}{2\pi\sqrt{L\cdot\dfrac{C\cdot C}{C+C}}}$$

$$= \frac{1}{2\pi\sqrt{\dfrac{L.C}{2}}} = \frac{1}{2\pi}\cdot\sqrt{\frac{2}{L\cdot C}} \qquad ...(i)$$

when the value of C_2 is doubled, the frequency of oscillation,

$$f_o' = \frac{1}{2\pi\sqrt{\left[L\left(\dfrac{C_1\cdot 2C_2}{C_1+2C_2}\right)\right]}} = \frac{1}{2\pi\sqrt{L\cdot\dfrac{2\,C\cdot C}{3\,C}}}$$

$$= \frac{1}{2\pi\sqrt{\dfrac{2\,L.C}{3}}} = \frac{1}{2\pi}\cdot\sqrt{\frac{2}{2\,LC}} \qquad ...(ii)$$

Dividing equation (*ii*) by (*i*),

$$\frac{f_o'}{f_o} = \frac{\dfrac{1}{2\pi}\cdot\sqrt{\dfrac{3}{2\,LC}}}{\dfrac{1}{2\pi}\cdot\sqrt{\dfrac{2}{LC}}} = \sqrt{\frac{3}{2LC}\times\frac{LC}{2}} = \sqrt{\frac{3}{4}} = 0.866$$

$\therefore \qquad f_o' = 0.866 \times f_o = 0.866 \times 450 \times 10^3 = 389 \times 10^3$ Hz $= 389.7$ kHz **Ans.**

25-16. FET Colpitts Oscillator

Fig. 25-13 shows the circuit of a field-effect transistor (FET) oscillator. In this circuit, the feedback signal is applied to the gate terminal of FET. Since, the gate has a high input resistance, therefore the loading effect on the tank circuit is much less than that with a bipolar transistor.

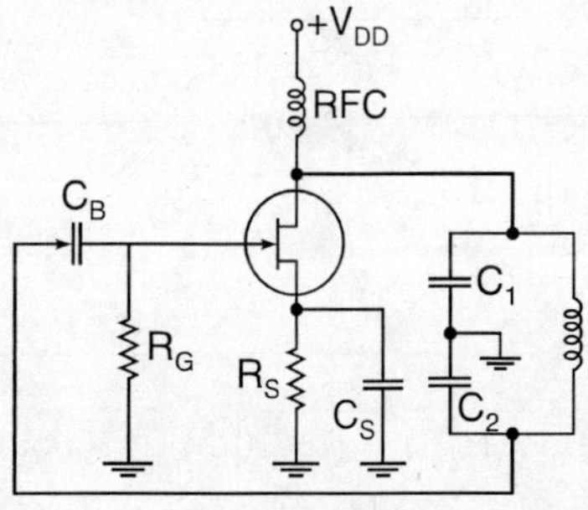

Fig. 25-13.

The operation of the FET Colpitt's oscillator circuit may be explained in the same way as discussed for transistor Colpitt's oscillator.

25-17. Clapp Oscillator

It is an improved version of the Colpitts oscillator. Fig. 25-14 shows the circuit of a Clapp oscillator. The circuit differs from the Colpitts oscillator only in one respect, that it contains one additional capacitor (C_3) connected in series with the inductor. The addition of capacitor (C_3) improves the frequency stability and eliminates the effect of transistor parameters and stray capacitances.

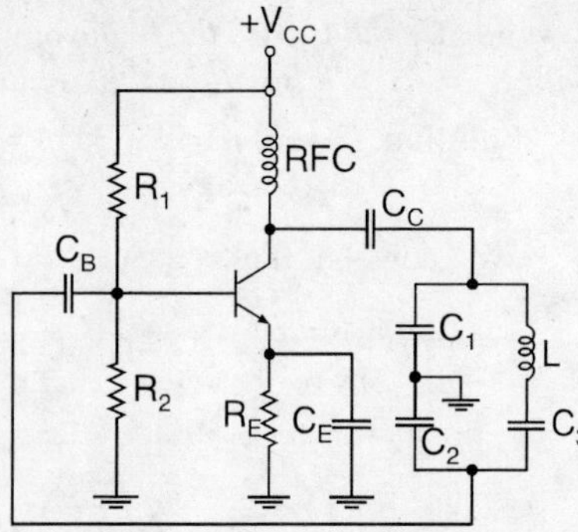

Fig. 25-14. *Transistor clapp oscillator.*

The operation of Clapp oscillator circuit is in the same way as that of Colpitts oscillator. The frequency of oscillator is given by the relation,

$$f_o = \frac{1}{2\,\pi\sqrt{L \cdot C}}$$

where,

$$C = \frac{1}{\dfrac{1}{C_1} + \dfrac{1}{C_2} + \dfrac{1}{C_3}}$$

Usually, the value of C_3 is much smaller than C_1 and C_2. As a result of this, C is approximately equal to C_3. Therefore, the frequency of oscillation,

$$f_o = \frac{1}{2\,\pi\sqrt{L \cdot C_3}}$$

Example 25-13. *Calculate the frequency of oscillations for the Clapp Oscillator shown in Fig. 25-15.* *(Electronic Circuits-II, Anna Univ., May/June 2007)*

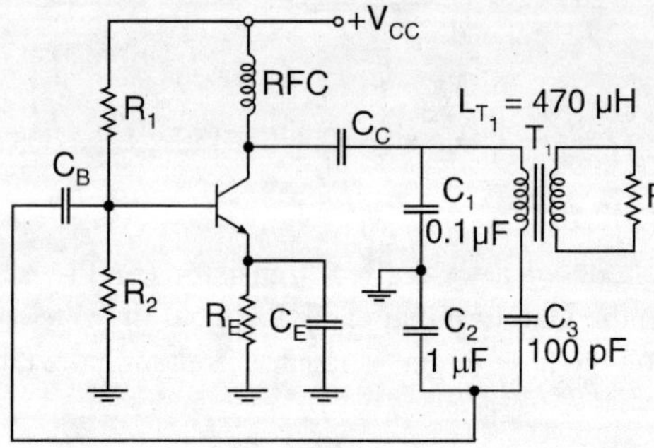

Fig. 25-15.

Solution. Given: $C_1 = 0.1\ \mu\text{F} = 0.1 \times 10^{-6}$ F; $C_2 = 1\ \mu\ \text{F} = 1 \times 10^{-6}$ F; $C_3 = 100\ \text{pF} = 100 \times 10^{12}$ F and $L_{T_1} = L = 470\ \mu\text{H} = 470 \times 10^{-6}$ H

We know that the capacitance,

$$C = \frac{1}{\dfrac{1}{C_1} + \dfrac{1}{C_2} + \dfrac{1}{C_3}} = \frac{1}{\dfrac{1}{(0.1 \times 10^{-6})} + \dfrac{1}{(1 \times 10^{-6})} + \dfrac{1}{(100 \times 10^{-12})}}$$

$$= \frac{1}{(10 \times 10^{6}) + (1 \times 10^{6}) + (10{,}000 \times 10^{6})} = \frac{1}{10011 \times 10^{6}} = 99.9 \times 10^{-12}\ \text{F} = 99.9\ \text{pF}$$

$\therefore$ Frequency of oscillation

$$f_o = \frac{1}{2\,\pi\sqrt{LC}} = \frac{1}{2\,\pi \times \sqrt{(470 \times 10^{-6}) \times (99.9 \times 10^{-12})}} = 734.5\ \text{kHz}\ \textbf{Ans.}$$

25-18. Crystal Oscillator

It is basically a tuned oscillator. The crystal oscillator uses a piezoelectric crystal as a resonant tank circuit. The crystal is, usually, made of quartz material and provides a high degree of frequency stability and accuracy. Therefore, the crystal oscillators are very useful in those applications, where frequency stability is very essential. The crystal oscillators are widely used in communication transmitters, digital watches and clock, *etc*.

25-19. Quartz Crystal

A quartz crystal has a very peculiar property known as *piezoelectric effect*. According to this effect, when an a.c. voltage is applied across a quartz crystal, it vibrates at the frequency of the applied voltage. Conversely, if a mechanical force is applied to vibrate a quartz crystal, it generates an a.c. voltage. The other materials, which exhibit the piezoelectric effect are Rochelle salt and tourmaline. The quartz material is preferred because it is inexpensive and readily available in nature. Moreover, its properties lie in between those of Rochelle salt and tourmaline.

25-20. Characteristics of a Quartz Crystal

The natural shape of a quartz crystal is a hexagonal prism with pyramids at the ends as shown in Fig. 25-16 (*a*). In order to get a usable crystal out of it, we have to cut a rectangular slab out of a natural crystal as shows in Fig. 25-6 (*b*). There are a number of ways to cut the natural crystal namely *X*-cut, *Y*-cut, *XY*-cut, AT-cut *etc*. Different cuts have different piezoelectric properties (*i.e.*, resonant frequency, temperature coefficient, *etc*.).

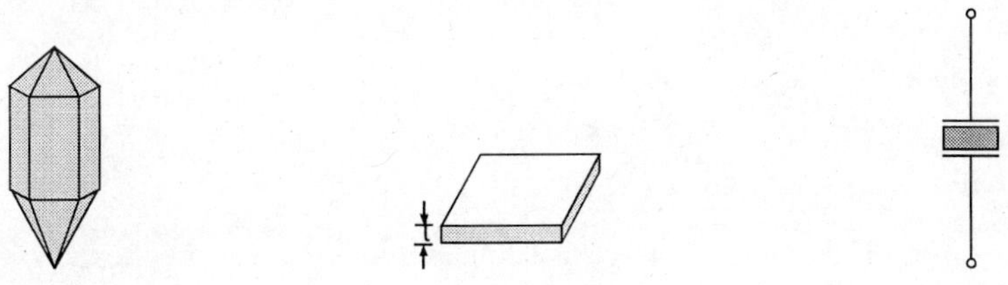

(*a*) Natural shape of a Quartz crystal. (*b*) Rectangular slab. (*c*) Crystal symbol.

Fig. 25-16.

For use in electronic circuits, the slab is mounted between two metal plates and housed in a package, which is equal to the size of postage stamp. In actual practice, the package as a whole is known as crystal and its symbol is shown in Fig. 25-16 (*c*). When a crystal is placed across an a.c. source, it starts vibrating. The amount of vibration depends upon the frequency of the applied voltage. By changing the frequency, we can find a frequency at which the crystal vibrations reach its maximum value. The frequency, at which it happens, is called resonant frequency of the crystal.

The quartz crystals are available with resonant frequencies ranging from 15 kHz to 10 MHz. For higher frequencies (*i.e.*, frequencies up to 100 MHz), the crystal can be made to vibrate on overtones (*i.e.*, multiples of fundamental frequency). However, for frequencies greater than 10 MHz, the crystals are not useful.

25-21. Electrical Equivalent Circuit of a Crystal

Fig. 25-17 (*a*) and (*b*) shows the symbol and electrical equivalent circuit of a crystal respectively. It consists of a series R-L-C_1 circuit in parallel with a capacitance (C_2). When the crystal mounted across the a.c. source is not vibrating, it is equivalent to the capacitance C_2. However, when the crystal is vibrating, it acts like a tuned circuit R-L-C_1.

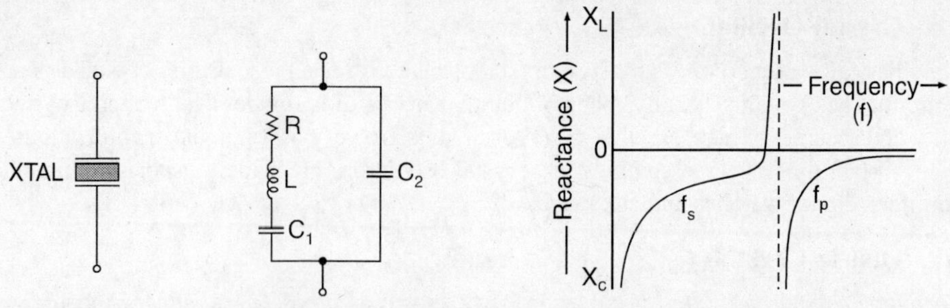

(*a*) Symbol. (*b*) Equivalent circuit. (*c*) Reactance versus frequency.

Fig. 25-17.

Fig. 25-17 (*c*) shows the graph of a reactance (X_L or X_C) versus frequency (f). It is evident from this graph that the crystal has two closely spaced resonant frequencies. The first one is the series resonant frequency (f_s), which occurs when reactance of the inductance (L) is equal to the reactance of the capacitance C_1. In that case, the impedance of the equivalent circuit is equal to the resistance R and the frequency of oscillation is given by the relation,

$$f_s = \frac{1}{2\,\pi\sqrt{L \cdot C_1}}$$

The second one is the parallel resonant frequency (f_p), which occurs when the reactance of R-L-C_1 branch is equal to the reactance of capacitor C_2. At this frequency, the crystal offers a very high impedance to the external circuit and the frequency of oscillation is given by the relation.

$$f_p = \frac{1}{2\,\pi\sqrt{L \cdot C_1}}$$

where
$$C = \frac{C_1 \cdot C_2}{(C_1 + C_2)}$$

The value of C_2 is usually very large as compared to C_1. Therefore the value of C is approximately equal to C_1 and hence the series resonant frequency is approximately equal to the parallel resonant frequency (*i.e.*, $f_s = f_p$).

Example 25-14. *A crystal has following parameters: L = 0.5 H, C_s = 0.06 pF, C_p = 1 pF and R = 5 KΩ. Find the series and resonant frequencies and corresponding Q – factor of the crystal.* (*Electronic Circuits - II, Anna University, May/June 2007*)

Solution. Given: L = 0.5 *H*, C_s = 0.06 *pF*, C_p = 1 *pF*, *R* = 5 kΩ

We know that the value of series resonant frequencies,

$$f_s = \frac{1}{2\pi\sqrt{LC_s}}$$

$$= \frac{1}{2\pi\sqrt{0.5 \times 0.06 \times 10^{-12}}}$$

$$= 918.9 \; kHz$$ **Ans.**

We know that the effective value of C,

$$C_{eq} = \frac{C_p \,.\, C_s}{C_p + C_S}$$

$$= \frac{1 \times 10^{-12} \times 0.06 \times 10^{-12}}{(1 \times 10^{-12}) + 0.06 \times 10^{-12}}$$

$$= 5.66 \times 10^{-14\mathrm{F}}$$

and the value of parallel resonant frequencies,

$$f_p = \frac{1}{2\pi\sqrt{L\,C_{eq}}}$$

$$= \frac{1}{2\pi\sqrt{0.5 \times 5.663 \times 10^{-14}}}$$

$$= 946.05 \ kHz$$ **Ans.**

We know that the value of Q – factor of the crystal,

$$Q = \frac{\omega_s L}{R} = \frac{2\pi f_s L}{R}$$

$$= \frac{2\pi \times 918.9 \times 10^3 \times 0.5}{5 \times 10^3}$$

$$= 577.4$$ **Ans.**

Example 25-15. *A crystal has the following parameters: L = 0.33 H, C = 0.065 pF, C_2 = 1.0 pF and R = 5.5 kΩ. Find the series resonant frequency and Q-factor of the crystal.*

Solution. Given: L = 0.33 H; C_1 = 0.065 pF = 0.065 × 10^{-12} F; C_2 = 1.0 pF and R = 5.5 kΩ = 5.5 × 10^3Ω

Series resonant frequency

We know that the series resonant frequency,

$$f_s = \frac{1}{2\,\pi\sqrt{L \cdot C_1}} = \frac{1}{2\,\pi \times \sqrt{0.33 \times (0.065 \times 10^{-12})}}$$

$$= 1.09 \times 10^6 \text{ Hz}$$ **Ans.**

Q-factor of the crystal

We know that the Q-factor

$$= \frac{2\,\pi\,f_s \cdot L}{R} = \frac{2\,\pi \times (1.09 \times 10^6) \times 0.33}{5.5 \times 10^3} = 411$$ **Ans.**

Example 25-16. *The parameters of a crystal oscillator equivalent circuit are L_S = 0.8 H; C_S = 0.08 pF; R_S = 5 kΩ and C = 1.0 pF, Determine the resonance frequencies, F_s and f_P.*

(*UPSC Engg. Services* 1999)

Solution. Given : $L = L_S$ = 0.8 H; $C_1 = C_S$ = 0.08 pF; $R = R_S$ = 5 kΩ and $C_2 = C$ = 1.0 pF.

We know that the series resonance frequency,

$$f_S = \frac{1}{2\pi\sqrt{LC_1}} = \frac{1}{2\pi\sqrt{(0.8 \times 0.08 \times 10^{-12})}} = 0.629 \times 10^6 \ Hz$$

$$= 0.629 \text{ MHz}$$ **Ans.**

and the parallel resonance frequency,

$$f_P = \frac{1}{2\pi\sqrt{LC}} = \frac{1}{2\pi\sqrt{0.8 \times \left(\frac{(0.08 \times 10^{-12}) \times (1.0 \times 10^{-12})}{(0.08 \times 10^{-12}) + (1.0 \times 10^{-12})}\right)}}$$

$$= \frac{1}{2\pi\sqrt{0.8 \times 0.0741 \times 10^{-12}}} \text{ Hz}$$

$$= 0.654 \times 10^6 \text{ Hz or } 0.654 \text{ MHz}$$ **Ans.**

Note. The frequencies f_s and f_p are very close to each other. For practical purpose $f_s \approx f_p$.

25-22. *Q*-Factor and Frequency Stability of a Crystal

The crystal has a very high value of *Q*-factor as compared to the discrete *LC* tank circuits. For example, the *Q*-factor of a typical crystal is 20,000 as compared to a maximum of about 100 for high quality *LC* tank circuits. The extremely high values of the *Q*-factor of a crystal produces oscillators with very stable frequency values.

The frequency of an oscillator tends to change slightly with time. This change in frequency with time is called *frequency drift*. The frequency drift is produced by temperature aging etc. In a crystal oscillator, the frequency drift is very small *i.e.*, less than 0.0001% (*i.e.*, 1 part in 10^6) per day. It is because of this fact that the *crystals are used as a basic timing device in digital watches.*

25-23. Crystal Oscillator Circuit

A variety of crystal oscillator circuits is possible such as crystal-controlled tuned collector-oscillator, Colpitts crystal oscillator Clapp crystal oscillator *etc*. But the pierce crystal oscillator is the most commonly used circuit in actual practice. Fig. 25-18 shows the transistor pierce crystal oscillator. In this circuit, the crystal in connected as a series element in the feed-back path from collector to the base.

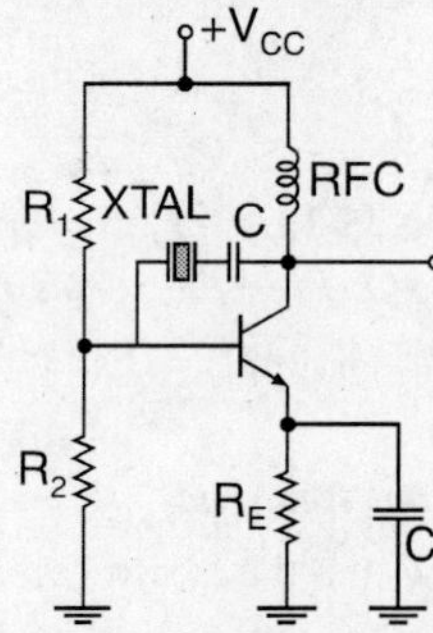

Fig. 25-18. *Transistor pierce crystal oscillator.*

The resistors R_1, R_2 and R_3 provide a voltage-divider stabilized d.c. bias circuit. The capacitor C_E provides a.c. bypass of the emitter resistor and RFC (radio frequency choke) coil provides for d.c. bias while decoupling any a.c. signal on the power lines from affecting the output signal. The coupling capacitor *C* has negligible impedance at the circuit operating frequency. But it blocks any d.c. between collector and base.

The circuit frequency of oscillation is set by the series-resonant frequency of the crystal and its value is given by the relation,

$$f_o = \frac{1}{2\,\pi\sqrt{L \cdot C_1}} \qquad ...(i)$$

It may be noted that the changes in supply voltage, transistor device parameters *etc*. have no effect on the circuit operating frequency, which is held stabilized by the crystal.

Fig. 25-19 shows the FET version of the pierce crystal oscillator. The advantage of using a FET is that it provides a higher input resistance to the crystal. However, the circuit frequency of oscillation is the same as given by equation (*i*).

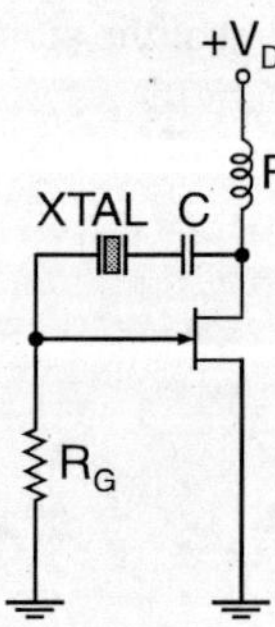

Fig. 25-19. *FET pierce crystal oscillator.*

25-24. RC Oscillators

These oscillators employ resistors and capacitors and are used to generate low-or audio-frequency signals. Hence, they are also known as *audio-frequency (A.F.) oscillators*. The tuned (or *LC*) oscillators are not suitable at low-frequencies because the size of inductors (*L*) and capacitors (*C*) becomes very large and therefore are not practicable to manufacture in actual practice.

25-25. Basic Principle of RC Oscillators

We know that a single stage of common-emitter (CE) amplifier not only amplifies the input signal, but also introduces a phase shift of 180°. If we take a fraction of the output and feed it back to the input, without considering the phase shift, the negative feedback takes place. Due to the negative feedback, the output voltage of the amplifiers decreases. However, for producing oscillations, we must have positive feedback of sufficient magnitude. This occurs only when the fraction of output voltage is feedback with the same phase as that of the input signal. This condition may be achieved in any one of the following two ways :

(1) A fraction of the output of a single-stage amplifiers is passed through a phase-shift network, before feeding back to the input. The phase-shift network gives another phase-shift of 180° in addition to the phase-shift of 180° introduced by the amplifier. Thus, there is a total phase-shift of 360°, which is also equal to 0°. The RC oscillator, utilizing this principle, is known as **phase-shift oscillator**.

(2) A second-stage of the amplifier is used for producing another 180° phase shift in addition to the phase-shift of 180° produced by the first stage. Thus, there is a total phaseshift of 360°, which is also equal to 0°. A fraction of the output from the second stage is feedback to the input of the first stage without producing any further phase-shift. The RC oscillator, utilizing this principle, is known as **Wien-bridge oscillator**.

25-26. Phase-Shift Oscillator

Fig. 25-20 shows the circuit of a phase shift oscillator using field-effect transistor (FET). The circuit is drawn to show clearly the amplifier and feedback network. The amplifier stage is self-biased with a capacitor by-passed source resistor (R_S) and drain bias resistor (R_D). The expression for voltage gain of the amplifier is given by their relation,

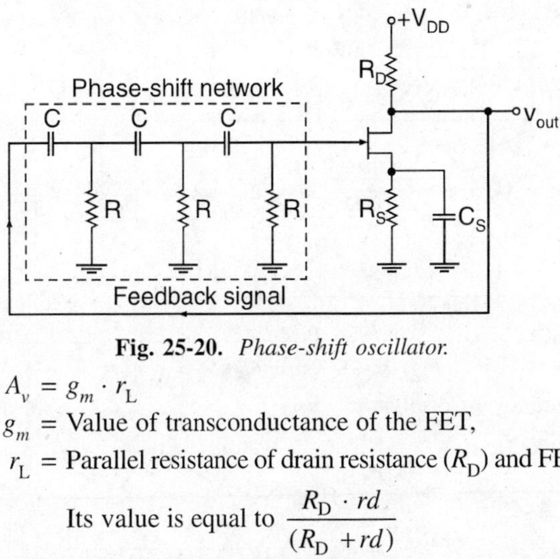

Fig. 25-20. *Phase-shift oscillator.*

$$A_v = g_m \cdot r_L$$

where, g_m = Value of transconductance of the FET,

and r_L = Parallel resistance of drain resistance (R_D) and FET output resistance (rd).

Its value is equal to $\dfrac{R_D \cdot rd}{(R_D + rd)}$

The feedback network consists of three *identical RC sections. Each section produces a phase-shift of 60°. Therefore, the net phase shift of the feedback network is 180°. Since, the amplifier stage also introduces a phase-shift of 180°, therefore the total phase shift between the input and output circuit is 360° or 0°.

When the circuit is energised, by switching on the supply, the circuit starts oscillating. The oscillations may start due to the minor variation in d.c. supply or the inherent noise in the FET. The variation in the gate current is amplified in the drain circuit. Then it is feedback through the phase

* In actual practice, it is difficult to get a feedback network of three identical RC sections, which produces a total phase shift of 180°. It is because of the fact, that each RC section in the feedback network loads down the previous one.

shift network and finally applied to the gate. The oscillations will be maintained if the loop gain (*i.e.*, $A_v \cdot \beta$) is at least equal to unity. However, to start the oscillations, the loop gain must be greater than unity. Using classical network analysis, it can be shown that circuit frequency of oscillation is given by the relation,

$$f_o = \frac{1}{2\pi CR\sqrt{6}}$$

and the feedback fraction of the RC phase-shift network,

$$\beta = 1/29$$

Using the value of β (equal to 1/29) we find that for the loop gain ($A_v \cdot \beta$) to be greater than unity, the gain of the amplifier stage must be greater than 1/β or 29.

The phase shift oscillator is particularly useful for generating signals in the audio frequency range. Usually, it is used for generating fixed-spot frequencies, which can be selected by switching in preset RC networks. For the variable frequency oscillator, the three capacitors are ganged and varied simultaneously.

Example 25-17. *Draw the circuit of RC phase shift oscillator. With R = 6 kΩ, C = 1 500 pF and R_C = 18 kΩ, obtain its operating frequency.*

(*Electonic circuits - II, Anna university, Nov/Dec 2007*)

Solution. Given: $R = 6\ k\Omega = 6 \times 10^3\ \Omega$, $C = 1500\ pF = 1500 \times 10^{-12}\ F$, $R_D = 18\ k\Omega = 18 \times 10^3\Omega$,

Fig Shows the *RC* phase shift oscillator circuit.

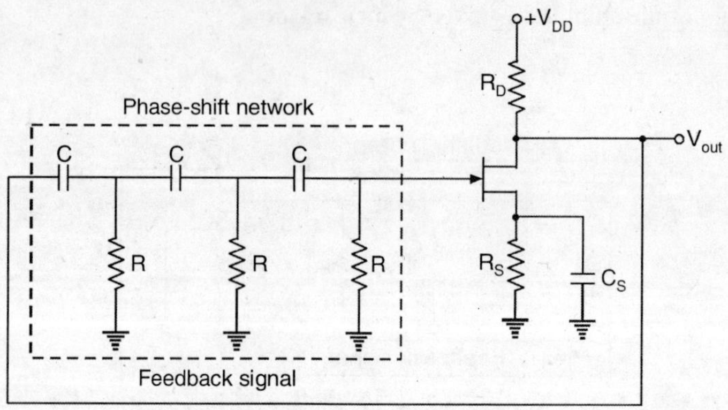

Fig. 25-21 *Phase-Shift OScillator*

We know the frequency of oscillators

$$f_0 = \frac{1}{2\pi CR\sqrt{6}}$$

$$f_0 = \frac{1}{2\pi\, x\,(1500 \times 10^{-12}) \times 6 \times 10^3 \times 2.44}$$

$$= 7250\ Hz \text{ or } 7.25\ kHz.$$ **Ans.**

Example 25-18. *It is desired to design a phase shift oscillator using an FET having g_m = 5000 μmhos, r_d = 40 kΩ and feedback circuit resistor (R) = 10 kΩ. Select the value of capacitor for oscillator operation at 1 kHz and the drain resistance for A_v > 29 to ensure oscillator action.*

Solution. Given: $g_m = 5000\ \mu$mhos $= 5000 \times 10^{-6}$ mhos; $r_d = 40$ kΩ; $R = 10$ kΩ $= 10 \times 10^3$and $f_o = 1$ kHz $= 1 \times 10^3$ Hz.

Value of capacitor

Let C = Value of the capacitor.

We know that the frequency of oscillator (f_o),

$$1 \times 10^3 = \frac{1}{2\,\pi\, C \cdot R\sqrt{6}} = \frac{1}{2\,\pi \cdot C\,(10 \times 10^3)\sqrt{6}} = \frac{1}{1.539 \times 10^5\, C}$$

or $$C = \frac{1}{(15.39 \times 10^5) \times (1 \times 10^3)} = 6.5 \times 10^{-9}\ \text{F}$$

$$= 0.0065\ \mu\text{F}\ \textbf{Ans.}$$

Value of drain resistance

Let R_D = Value of drain resistance.

We know that to start oscillations, the amplifier voltage gain (A_v) must be greater than 29. Therefore, let us assume that the value of amplifier voltage gain is 40. We also know that the amplifier voltage gain (A_v),

$$40 = g_m \cdot r_L = (5000 \times 10^{-6})\ r_L$$

$$\therefore \quad r_L = \frac{40}{(5000 \times 10^{-6})} = 8 \times 10^3\ \Omega = 8\ \text{k}\Omega$$

We also know that the parallel resistance (r_L),

$$\therefore \quad 8 = \frac{R_D \cdot r_d}{R_D + r_d} = \frac{R_D \times 40}{R_D + 40}$$

$$8\,(R_D + 40) = 40\,R_D$$

or $$32\,R_D = 320$$

$$\therefore \quad R_D = 320/32 = 10\ \text{k}\Omega\ \textbf{Ans.}$$

Example 25-19. *Design a phase shift oscillator to operate at a frequency of 2 kHz using a JFET. Assume that* $\mu = 50$ *and* $r_d = 5\ \text{k}\Omega$ *and the phase shift network does not load the amplifier.*

(*a*) Find the minimum value of R_D to be used in the drain circuit.

(*b*) *Find the value of the RC product.*

(*c*) *Choose reasonable values of R and C to make this oscillator work satisfactorily.*

Solution. Given: $f_o = 2\ \text{kHz} = 2 \times 10^3$ Hz; $\mu = 50$ and $r_d = 5\ \text{k}\Omega$.

(*a*) *Maximum value of* R_D

We know that the FET amplification factor (μ),

$$50 = g_m \cdot r_d = g_m \times 5$$

$$\therefore \quad g_m = 50/5 = 10\ \text{m ohm.}$$

We also know that in order to start oscillations, the amplifier voltage gain (μ) is 50, therefore the circuit will oscillate. Let us assume that the amplifier voltage gain (A_v) is 40.

Now the amplifier voltage gain (A_v),

$$40 = g_m \cdot r_L = 10 \times r_L$$

$$\therefore \quad r_L = 40/10 = 4\ \text{k}\Omega$$

We also know that a.c. load resistance (r_L)

$$4 = R_D \parallel r_d = \frac{R_D \times r_d}{R_D + r_d} = \frac{R_D \times 5}{R_D + 5}$$

or $$4\,R_D + 20 = 5\,R_D$$

$$\therefore \quad R_D = 20\ \text{k}\Omega\ \textbf{Ans.}$$

(*b*) *Value of the RC product*

We know that the frequency of oscillation (f_o),

$$2 \times 10^3 = \frac{1}{2\pi\, C \cdot R\sqrt{6}} = \frac{0.065}{RC}$$

$$RC = \frac{0.065}{(2 \times 10^3)} = 3.25 \times 10^{-5} \textbf{ Ans.}$$

(*c*) *Value of R and C*

Let us choose R equal to 10 kΩ or 10×10^3 Ω. Then the value of capacitor,

$$C = \frac{(3.25 \times 10^{-5})}{(10 \times 10^3)} = 3.25 \times 10^{-9} \text{ F} = 3.25 \text{ nF}$$

∴ The values of R and C, which make the oscillator to work satisfactorily are

$$R = 10 \text{ k}\Omega \text{ and } C = 3.25 \text{ nF} \textbf{ Ans.}$$

Example 25-20. *Prove that the feedback 'β' of the phase-shift network is given by:*

$$-\beta = \frac{1}{1 - 5\,\alpha^2 - j\,(6\alpha - \alpha^3)}$$

where,

$$\alpha = \frac{1}{\omega CR}$$

Assume that the network does not load the amplifier.

Solution. Fig. 25-22 shows the phase-shift network of an RC phase-shift oscillator. In this circuit, V_o represents the output voltage and V_f represents the negative of feedback voltage (V_f); *i.e.*,$V'_f = V_f$. The negative sign with V_f indicates that it is 180° out of phase with respect to V_o or V'_f

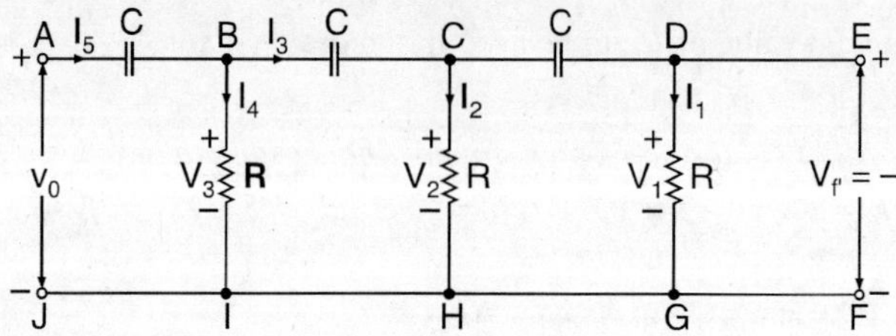

Fig. 25-22.

In order to find the value of feedback 'β', consider the distribution of voltages and currents within the circuit as shown in the Fig. 28-22. Then apply Kirchoff's Voltage Law (*i.e.*, sum up the voltages around the various meshes) and determine the ratio $V_f / V_o = -V_f / V_o = \beta$

Summing up the voltages around the mesh *DEFG*,

$$V'_f = I_1 \cdot R$$

or

$$I_1 = \frac{V'_f}{R} \qquad \ldots (i)$$

and around the mesh *CDGH*,

$$V_2 = \frac{I_1}{j\,\omega\,C} + I_1 \cdot R = I_1\left(\frac{1}{j\,\omega\,C} + R\right).$$

Now let

$$\alpha = \frac{1}{\omega\,C\,R} \text{ or } \frac{1}{\omega\,C} + R \cdot \alpha$$

$$\therefore \qquad V_2 = I_1\left(\frac{R \cdot \alpha}{j} + R\right) = I_1 \cdot R\left(\frac{\alpha}{j} + 1\right)$$

We also know that

$$I_2 = \frac{V_2}{R} = I_1\left(\frac{\alpha}{j} + 1\right) \qquad ...(ii)$$

It may be noted that the current,

$$I_3 = I_2 + I_1 = I_1\left(\frac{\alpha}{j} + 1\right) + I_1$$

$$= I_1\left(\frac{\alpha}{j} + 2\right) \qquad ...(iii)$$

Similarly, summing up the voltages around the mesh *BCHI,*

$$V_3 = \frac{I_3}{j\,\omega\,C} + I_2 \cdot R = I_3 \cdot \frac{R \cdot \alpha}{j} + I_2 \cdot R \qquad ...(\because 1/\omega C = R \cdot \alpha)$$

Substituting the values I_2 and I_3 in the above equation from equations (*ii*) and (*iii*),

$$V_3 = \left[I_1\left(\frac{\alpha}{j} + 2\right)\right]\left(\frac{R \cdot \alpha}{j}\right) + I_1\left(\frac{\alpha}{j} + 1\right)R$$

$$= I_1\left(\frac{R \cdot \alpha^2}{j^2} + \frac{2R \cdot \alpha}{j} + \frac{R \cdot \alpha}{j} + R\right)$$

$$= I_1 \cdot R\left(\frac{\alpha^2}{j^2} + \frac{3\alpha}{j} + 1\right)$$

and

$$I_4 = \frac{V_3}{R} = I_1\left(\frac{\alpha^2}{j^2} + \frac{3\alpha}{j} + 1\right) \qquad ...(iv)$$

It may also be noted that the current,

$$I_5 = I_3 + I_4$$

$$= I_1\left(\frac{\alpha}{j} + 2\right) + I_1\left(\frac{\alpha^2}{j^2} + \frac{3\alpha}{j} + 1\right)$$

$$= I_1\left(\frac{\alpha^2}{j^2} + \frac{4\alpha}{j} + 3\right) \qquad ...(v)$$

Now summing up the voltages around, the mesh *ABIJ*,

$$V_o = \frac{I_5}{j\,\omega\,C} + I_4 \cdot R = I_5 \cdot \frac{R \cdot \alpha}{j} + I_4 \cdot R$$

Substituting the values of I_4 and I_5 in the above equation from equations (*iv*) and (*v*)

$$V_o = I_1\left(\frac{\alpha^2}{j^2} + \frac{4\alpha}{j} + 3\right) \cdot \left(\frac{R \cdot \alpha}{j}\right) + I_1 \cdot \left(\frac{a^2}{j^2} + \frac{3\alpha}{j} + 1\right)R$$

$$= I_1 \cdot R\left(\frac{\alpha^3}{j^3} + \frac{4\alpha^2}{j^2} + \frac{3\alpha}{j} + \frac{\alpha^2}{j^2} + \frac{3\alpha}{j} + 1\right)$$

$$= V'_f\left(\frac{\alpha^3}{j^3} + \frac{5\alpha^2}{j^2} + \frac{6\alpha}{j} + 1\right) \qquad ...(\because I_1 \cdot R = V'_f)$$

$$= V'_f\,(j\alpha^3 - 5\alpha^2 - j6\alpha + 1) \qquad ...(\because j^2 = -1)$$

$$= V'_f\,[1 - 5\alpha^2 - j\,(6\alpha - \alpha^3)]$$

$$\therefore \qquad \frac{V'_f}{V_o} = \frac{1}{1 - 5\alpha^2 - j(6\alpha - \alpha^3)}$$

Now, since the actual feedback factor is the ratio of V_f/V_o, therefore

$$\frac{V_f}{V_o} = \beta = \frac{-V'_f}{V_o}$$

$$\text{and} \qquad -\beta = \frac{V'_f}{V_o} = \frac{1}{1 - 5\alpha^2 - j(6\alpha - \alpha^3)}$$

Example 25-21. *Obtain the expression for the frequency of oscillations for a phase-shift oscillator and show that gain of an amplifier must exceed 29 in this oscillator. Given the feedback factor,*

$$-\beta = \frac{1}{1 - 5\alpha^2 - j(6\alpha - \alpha^3)}$$

$$\text{where} \qquad \alpha = \frac{1}{\omega R C}$$

Solution. Given: The feedback factor,

$$-\beta = \frac{1}{1 - 5\alpha^2 - j(6\alpha - \alpha^3)} \qquad ...(i)$$

The expression for the frequency of oscillations may be obtained from the above expression by equating the reactive part equal to zero *i.e.*,

$$6\alpha - \alpha^3 = 0$$

$$\therefore \qquad \alpha = \sqrt{6}$$

$$\text{or} \qquad \frac{1}{\omega C R} = \sqrt{6}$$

$$\omega = \frac{1}{C R \sqrt{6}}$$

$$\text{or} \qquad 2\pi f = \frac{1}{C R \sqrt{6}} \qquad ...(\because \omega = 2\pi f)$$

$$\therefore \qquad f = \frac{1}{2\pi C R \sqrt{6}} \textbf{ Ans.}$$

Substituting the value of α (equal to $\sqrt{6}$) in equation (*i*),

$$-\beta = \frac{1}{1 - 5(\sqrt{6})^2 - j(6\sqrt{6} - (\sqrt{6})^3)}$$

$$= \frac{1}{1 - 30} = -\frac{1}{29}$$

$$\text{or} \qquad \beta = \frac{1}{29}$$

Now, we know that the Barkhausen condition for positive feedback is that the loop gain $(A_v \cdot \beta)$ must be equal to or greater than unity *i.e.*,

$$A_v \cdot \beta \geq 1$$

$$\text{or} \qquad A_v \geq 1/\beta = 29 \textbf{ Ans.}$$

25-27. Transistor Phase Shift Oscillator

It has been found that by using a transistor as the active element of the amplifier stage, the output of the feedback network is loaded appreciably by the relatively low input resistance (*i.e.*, h_{ie}) of the transistor. This problem can be removed by using an emitter follower input stage followed by a common-emitter amplifier stage. But if the single transistor stage is required, then the use of voltage-shunt feedback is more suitable.

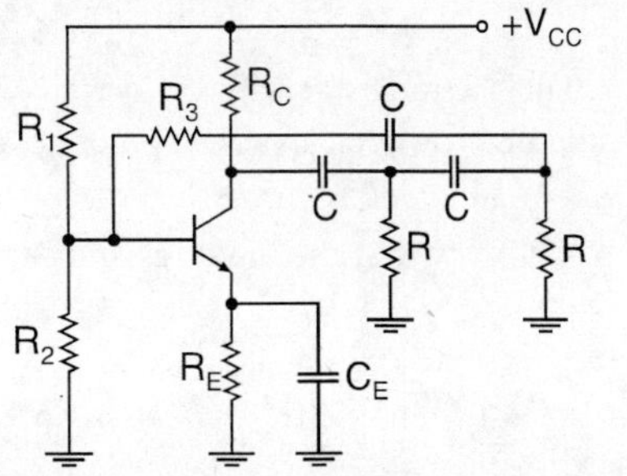

Fig. 25-23. *Transistor phase shift oscillator.*

Fig. 25-23 shows the circuit of a transistor phase shift oscillator using voltage-shunt feedback connection. In this connection, the feedback signal is coupled through the feedback resistor (R_s) in series with the amplifier stage input resistance R_i (whose value is equal to the parallel combination of resistors R_1, R_2 and h_{ie}). In order to make the calculations simpler, the value of resistor R_3 is made equal to the difference of resistor R_2 and R_i, *i.e.*,

$$R_3 = R - R_i$$

where, R is the value of resistor in the phase shift network. The frequency of oscillation for the transistor phase shift oscillators is given by relation,

$$f = \left(\frac{1}{2\pi\, RC}\right) \times \frac{1}{\sqrt{6 + 4\,(R_C / R)}}$$

For the loop gain to be greater than unity, the current gain of the transistor,

$$h_{fe} > 23 + 29\frac{R_c}{R} + 4\frac{R}{R_c}.$$

Example 25-22. *Select the value of capacitor C and transistor gain h_{fe} to provide an oscillator frequency of $f_o = 2$ kHz. The circuit values are $h_{ie} = 2$ kΩ; $R_1 = 20$ kΩ; $R_2 = 80$ kΩ; $R_C = 10$ kΩ and $R = 8$ kΩ. Also find the value of feedback resistorR_3.*

Solution. Given: $f_o = 2$ kHz $= 2 \times 10^3$ Hz; $h_{ie} = 2$ kΩ; $R_1 = 20$ kΩ; $R_2 = 80$ kΩ; $R_C = 10$ kΩ $= 10 \times 10^3$ and R = 8 kΩ = 8×10^3Ω.

Value of capacitor C

We know that frequency of oscillation (f_o)

$$2 \times 10^3 = \frac{1}{2\pi\, RC} \times \frac{1}{\sqrt{6 + 4\,R_C / R}}$$

$$= \left(\frac{1}{2\pi \times (8 \times 10^3) \times C}\right) \times \frac{1}{\sqrt{6 + 4 \times (10 \times 10^3)/(8 \times 10^3)}}$$

$$= \left(\frac{1}{16\pi \times 10^3 \times C}\right) \times \frac{1}{\sqrt{6 + 5}}$$

$$= \left(\frac{1}{16\pi \times 10^3 \times C}\right) \times \frac{1}{3.32}$$

$$\therefore \quad C = \frac{1}{(16\pi \times 10^3) \times 3.32 \times (2 \times 10^3)}$$

$$= 3.0 \times 10^{-9}\ \text{F} = 0.003\ \mu\text{F}\ \textbf{Ans.}$$

Value of transistor gain h_{fe}

We know that current gain of a transistor,

$$h_{fe} \geq 23 + 29\frac{R}{R_C} + 4\frac{R_C}{R}$$

$$\geq 23 + 29 \times \left(\frac{8 \times 10^3}{10 \times 10^3}\right) + 4\left(\frac{10 \times 10^3}{8 \times 10^3}\right)$$

$$\geq 23 + 23.2 + 5 = 51.2$$

Thus, a transistor with $h_{fe} \geq 51.2$ will provide sufficient loop gain for the circuit to operate as an oscillator. However, in actual practice, a transistor with h_{fe} greater than 60 should be selected. **Ans.**

Value of feedback resistor R_3

We know that the input resistance of the amplifier stage,

$$R_i = (R_1 \parallel R_2) \parallel h_{ie} = (20 \parallel 80) \parallel 2 = 1.8 \text{ k}\Omega = 1.8 \times 10^3 \ \Omega$$

and the value of feedback resistor,

$$R_3 = R - R_i = (8 \times 10^3) - (1.8 \times 10^3) \ \Omega$$

$$= 6.2 \times 10^3 \ \Omega = 6.2 \text{ k}\Omega \ \textbf{Ans.}$$

25-28. Wien-bridge Oscillator

In the previous articles, we have discussed, phase shift oscillators. They have one drawback, *i.e.*, when they are used to have a variable frequency circuit, it is difficult to adjust equally the capacitor values of the phase-shift network simultaneously. An oscillator circuit, which is more useful for variable frequency operation, is the Wein-bridge oscillator.

Fig. 25-24 shows the circuit of a Wein-bridge oscillator using transistors. However, we may also use FET's instead of bipolar transistors. The circuit consists of a two-stage RC coupled amplifier, which provides an approximately 360° or 0° phase-shift. So the feedback network has no need to introduce any additional phase-shift. The feedback network consists of $C_1 - R_1$, $C_2 - R_2$ (called a lead-lag network) and $R_3 - R_4$ (called a voltage-divider). The lead-lag network provides a positive feedback to the input of the first stage and the voltage-divider, the negative feedback to the emitter of Q_1 transistor.

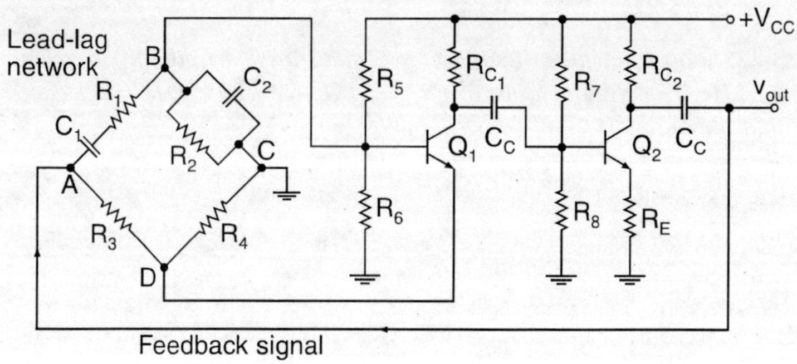

Fig. 25-24. *Wein-bridge oscillator.*

It can be shown by a simple analysis that the frequency of oscillations for the Wein-bridge oscillator is given by the relation,

$$f_o = \frac{1}{2\pi\sqrt{R_1 R_2 C_1 C_2}}$$

$$= \frac{1}{2\pi RC} \quad \text{...(If } R_1 = R_2 = R \text{ and } C_1 = C_2 = C)$$

and the ratio of resistor R_3 to R_4,

$$R_3/R_4 = 2$$

Thus, a ratio of R_3 to R_4 greater than 2 will provide a sufficient gain for the circuit to oscillate at the desired frequency.

It has been observed that in a FET Wein-bridge oscillator, the oscillator frequency can be varied by varying two capacitors C_1 and C_2 simultaneously. These are variable air gang capacitors. The frequency range of the oscillator can be changed by switching the different preset values of resistors R_1 and R_2 into the circuit. On the other hand, in a transistor Wein-bridge oscillator, because of the low input resistance of the transistor, the range changes are made by switching in the preset values of capacities C_1 and C_2. For continuous frequency variation, the values of resistors R_1 and R_2 are changed by using a dual potentiometer. The operation of a Wein-bridge oscillator may be explained as follows:

When the circuit is energised by switching on the supply, a small random oscillations appearing at the base of Q_1 transistor are amplified, at its collector. These oscillations are further amplified at the collector of Q_2 transistor. Since, the oscillations at the collector of Q_2 transistor have been inverted twice, therefore these oscillations are in phase with the input signal. A part of the output signal from the collector of Q_2 transistor is feedback to the Wein-bridge, which is further amplified. The process continues, till sustained oscillations are produced.

The Wein-bridge oscillator is a standard oscillator circuit for generating low frequencies in the range of 20 Hz to about 1 MHz. It is used in all commercial audio signal generators. The Wein-bridge oscillator has an advantage that it gives an extremely pure sine wave output, good frequency stability and a highly stabilized amplitude. A greater amplitude stability may be obtained by using a thermistor (a device with negative temperature coefficient) in place of the resistor R_3.

Example 25-23. *A Wein-bridge oscillator is used for operation at f_o = 10 kHz. If the value of R is 100 kΩ, find the value of the capacitor C.*

Solution. Given: $f_o = 10 \text{ kHz} = 10 \times 10^3 \text{ Hz}$; $R = 100 \text{ k}\Omega = 100 \times 10^3 \ \Omega$

We know that frequency of oscillation (f_o)

$$10 \times 10^3 = \frac{1}{2\pi RC} = \frac{1}{2\pi \times (100 \times 10^3) \times C}$$

$$= \frac{1}{628.3 \times 10^3 C}$$

$$\therefore \quad C = \frac{1}{(628.3 \times 10^3) \times (10 \times 10^3)} = 159 \times 10^{-12} \text{ F} = 159 \text{ pF } \textbf{Ans.}$$

25-29. Tunnel Diode Oscillator

A tunnel diode may be used to generate very high frequency signal (*i.e.*, frequencies of above 10 G/Hz). Fig. 25-25 shows the circuit of a tunnel diode oscillator. The circuit consists of a tunnel diode in series with a resistor (*R*) and a switch (*S*) connected across a supply source (*V*). The value of resistor (*R*) is selected in such a way that it biases the tunnel diode in the midway of the negative resistance region.

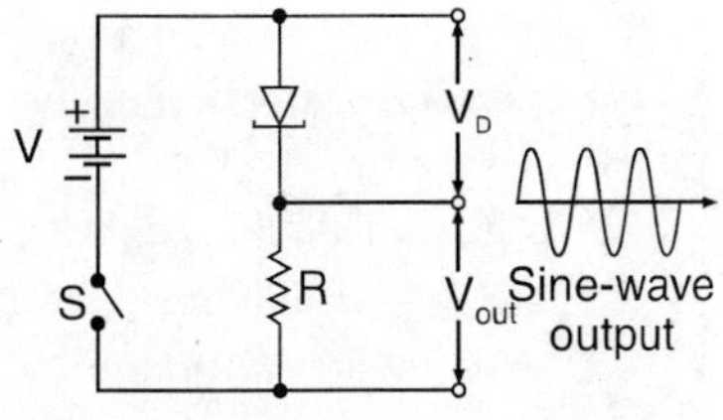

Fig. 25-25. *Tunnel diode oscillator.*

Fig. 25-26 shows the V-I characteristic of a tunnel diode. The curve *AB* is the negative-resistance region. It is evident that the *Q*-point is set at the middle of the curve *AB*. The *Q*-point can move between the points *A* and *B* during the circuit operation. The point *A* is called the peak-point and the point *B* as the valley point. The values of currents and voltages corresponding to the points *A* and *B* is indicated in the Fig. 25-25.

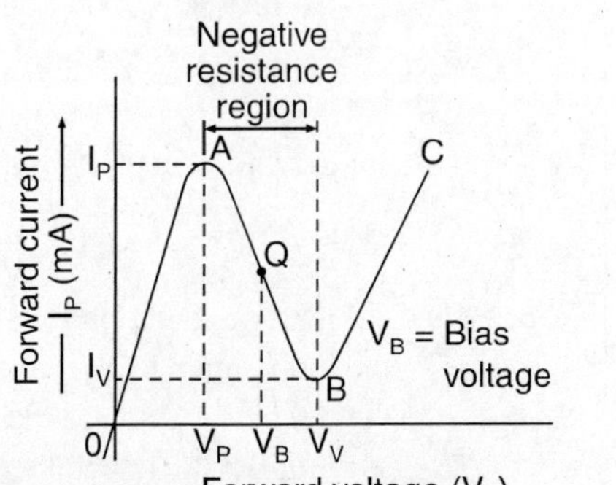

Fig. 25-26. *V-I characteristic of tunnel diode.*

When the switch (*S*) is closed, the circuit current rises immediately towards the constant value, whose value is determined by the value of resistor (*R*) and the diode resistance. However, as the voltage drop across the tunnel diode (V_D) exceeds the peak-point voltage (V_P), the tunnel diode is driven into negative resistance region. In this region, the current starts decreasing, till the voltage V_D becomes equal to the valley-point voltage (V_V). At this point, a further increase in voltage V_D drives the diode into the positive resistance region (*i.e.*, *BC* curve). As a result of this, the circuit current tends to increase. This increase in circuit current will increase the voltage drop across resistor (*R*) which will reduce the voltage (V_D).

This brings the diode back into the negative resistance region. The decrease in voltage V_D increases the value of current in the circuit, till point *A* is reached where the voltage V_D is equal to the voltage V_P. This completes one cycle of operation. In this way, the circuit will continue to oscillate back and fourth in the negative-resistance region and will produce a sine wave output as shown in the Fig. 25-26.

The tunnel diode oscillator discussed above is not used in practice. In fact a circuit shown in Fig. 25-27 is more commonly used. In this circuit, the resistor, R_1 sets the proper biasing for the diode, whereas resistor R_2 sets the proper current level for the tank circuit. The parallel combination of resistor, (R_P), inductor, (*L*) and a capacitor (*C*) forms a tank circuit, which resonates at the selected frequency.

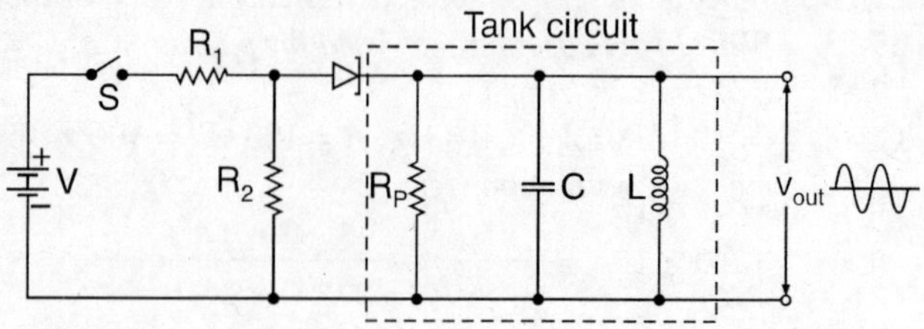

Fig. 25-27. *Practical tunnel diode oscillator.*

SUMMARY

In this chapter, you have learned that :

1. Oscillator is a device (or a circuit) which produces an output signal of any desired frequency without any input signal.

2. The oscillators are of two types : (1) sinusoidal, and (2) non-sinusoidal (or relaxation) oscillators

3. The sinusoidal oscillators can also be broadly classified into two types :
(1) radio-frequency (*rf*) oscillators, and (2) audio-frequency (*af*) oscillators.

4. The frequency of an *RF* oscillator is given by,

$$f_o = \frac{1}{2\pi\sqrt{LC}}$$

where, L = inductance in henrys and
C = value of capacitance in farads.

If *L* is in μ*H* and *C* in μ*F*, then frequency of oscillator,

$$f_o = \frac{159}{\sqrt{LC}}$$

5. The Barkhausen condition for sustained oscillations are
(*i*) the loop gain $\beta.\ A_v = 1$ and
(*ii*) the net phase-shift around the loop must be 360° or 0°.
To start with, $\beta\ A_v > 1$ and other that its value must come down to unity.

6. The condition for Hartley oscillator to oscillate is that the feedback fraction,

$$\beta = \frac{L_2}{L_1}$$

For oscillations to start, the voltage gain (A_v), must be greater than ($1/\beta$), *i.e.* equal to L_1/L_2.

7. The frequency of oscillations of a Hartley oscillator circuit is given by,

$$f_o = \frac{1}{2\pi\sqrt{LC}} = \frac{1}{2\pi\sqrt{(L_1 + L_2)C}}$$

8. The condition for Colpitts oscillator to oscillate is that the feedbacks fraction

$$\beta = \frac{C_1}{C_2}$$

For oscillations to start, the voltage gain (A_v) must be greater than $1/\beta$, *i.e.*, equal to (C_2/C_1).

9. The frequency of oscillations of a Colpitts oscillator circuit is given by,

$$f_o = \frac{1}{2\pi\sqrt{LC}} = \frac{1}{2\pi\sqrt{L \times \left(\frac{C_1 . C_2}{C_1 + C_2}\right)}}$$

10. The frequency of oscillations of *RC* (or phase-shift) oscillator is given by,

$$f_o = \frac{1}{2\pi\, CR\sqrt{6}}$$

and the feedback fraction,

$$\beta = 1/29$$

11. The frequency of oscillations of Wien-bridge oscillator is given by,

$$f_o = \frac{1}{2\pi\, RC}$$

and the feedback fraction

$$R_3/R_4 = 2.$$

GLOSSARY

Audio-frequency (AF) oscillator: An oscillator that produces frequency output from 1 Hz to 20 kHz.

Barkhausen criterion : A condition for sustained oscillations, *i.e.*, $\beta.\, A_v = 1$ net phase shift around the loop is zero.

Colpitts Oscillator : A circuit in which the tank circuit consists of two capacitors (connected in series) and an inductor in parallel.

Clapp Oscillator : A circuit in which the tank circuit consists of two capacitors (connected in series) in parallel to a combination of inductors in series with a capacitor.

Hartley Oscillator : A circuit in which the tank circuit consist of two inductors (connected in series) and a capacitor in parallel.

Oscillator : A device (or a circuit) which produces an output signal of any desired frequency without any input signal.

Quartz Crystal : A crystal of Rochelle salt or tourmaline which can produce oscillations when subjected to a mechanical force.

Radio-frequency (RF) oscillators: An oscillator that produces frequency output from 1 MHz and higher.

NUMERICAL PROBLEMS

1. A tuned-collector oscillator circuit used in a local oscillator of radio receiver has a fixed inductance of 100 μH and a fixed capacitor of 1000 pF. Calculate the frequency of oscillations. (**Ans.** 503.3 kHz)

2. The resonant circuit of a tuned-collector transistor oscillator has a resonance frequency of 2 MHz. If the value of capacitance is 820 pF, determine the value of inductance required. (**Ans.** 7.7 μH)

3. Determine the operating frequency of a transistor Hartley Oscillator, if L_1 = 100 μH; L_2 = 1 mH; M = 20 μH and C = 20 pF. (**Ans.** 1.05 MHz)

4. The frequency of oscillation of a Hartley Oscillator is given by:

$$f_o = \frac{1}{2\pi\sqrt{(L_1 + L_2)\,C}}$$

Calculate the oscillation frequency if C = 0.01 μF and $L_1 = L_2$ = 25 mH. (**Ans.** 7.117 kHz)

5. Calculate the frequency of oscillations for the Clapp Oscillator shown in Fig. 25-28. (**Ans.** 87.7 kHz)

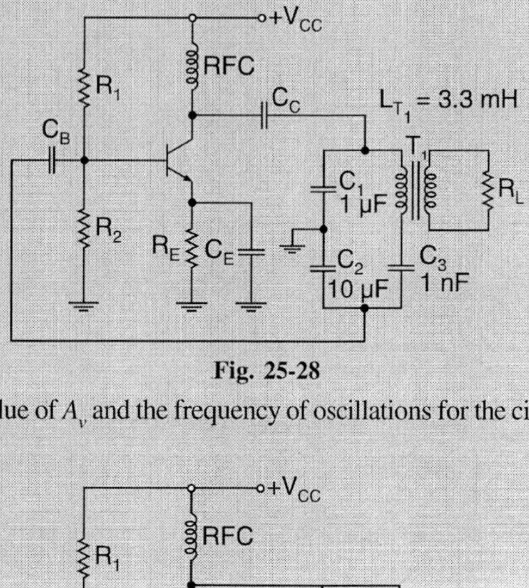

Fig. 25-28

6. Calculate the value of A_v and the frequency of oscillations for the circuit shown in Fig. 25-29. (**Ans.** 5.53 MHz)

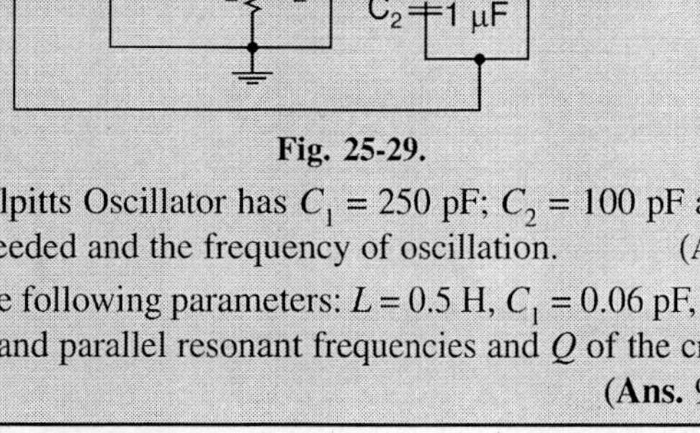

Fig. 25-29.

7. A transistor Colpitts Oscillator has C_1 = 250 pF; C_2 = 100 pF and L = 60 μH. Find the value of gain needed and the frequency of oscillation. (**Ans.** 2.5 and 2.43 MHz)

8. A crystal has the following parameters: L = 0.5 H, C_1 = 0.06 pF, C_2 = 1 pF and R = 5 kΩ. Find the series and parallel resonant frequencies and Q of the crystal. (**Ans.** 918.9 kHz and 946 kHz)

9. An FET phase-shift oscillator uses three identical RC sections in the feedback network. The values of the components are $R = 100$ kΩ and $C = 0.01$ μF. Determine the frequency of oscillations. (**Ans.** 64.97 Hz)

10. A transistor RC phase-shift oscillator has a frequency of 100 kHz. If the value of R is 10 kHz, find the value of C and gain necessary for sustained oscillations. (**Ans.** 41.67 pF, 29)

11. A Wein-bridge oscillator has a frequency of 500 Hz. If the value of C is 1000 pF, determine the value of R. (**Ans.** 318.3 kΩ)

DESCRIPTIVE QUESTIONS

1. Explain briefly, what do you mean by an electronic oscillator? Give two applications of an oscillator.

2. What are the different types of oscillators?

3. Explain the meaning of the terms
(*a*) damped oscillations, and
(*b*) undamped oscillations.

4. Why an LC tank circuit, once excited, does not produce sustained oscillations? Explain it briefly.

5. What are the factors which affect the frequency stability of an oscillator?

6. Give the two Barkhausen conditions required for sinusoidal oscillations to be sustained.

7. Draw the circuit diagram of the tuned-oscillator. Explain the circuit operation.

8. Explain briefly, how the oscillations are maintained in Hartley Oscillator.

9. Discuss the criteria for oscillation for a Colpitts Oscillator.

10. Draw a neat circuit diagram of a transistorized Colpitts Oscillator with a suitable equivalent circuit. Give the expressions for the following:
(*i*) Frequency of oscillations, and
(*ii*) Minimum gain for sustained oscillations.

11. Explain with the neat circuit diagram, the working of a phase-shift oscillator using three sections of *RC* network. State the expression for the frequency of oscillations and the minimum gain of the amplifier for sustained oscillations.

12. Why do we need three RC networks for a phase-shift oscillator? Can it be two or four?

13. With a neat circuit diagram, explain the principle of operation of a RC phase-shift oscillator using FET.

14. Explain the operation of a transistorized Wien-bridge oscillator with the help of neat circuit diagram. How is amplitude stability achieved in this circuit?

15. Define Loop gain. (*Electronic Circuits - II, Anna University, Nov/Dec 2007*)

16. State Barkhausen's criterion for oscillation. (*Electronic circuits - II, Anna University, Nov/Dec 2007*)

17. Crystal Oscillators possess high degree of frequency stability-Justify. (*Electronic Circuits - II, Anna University, Nov/Dec 2007*)

18. Define 'Q' of tank Circuit. (*Electronic Circuits - II, Anna University, Nov/Dec 2007*)

19. What type of feedback is applied for oscillators? (*Electronic Circuits - II, Anna University, Nov/Dec 2007*)

20. Draw the circuit diagram of Wein-bridge oscillator. Explain the need for providing both positive and negative feedback. (*Electronic Circuits - II, Anna University, Nov/Dec 2007*)

21. Draw the circuit of a Colpitts Oscillator. Derive expressions of frequency of oscillation and condition for oscillation.

(*Electronic Circuits - II, Anna University, Nov/Dec 2007*)

22. What are the advantages of LC oscillator and RC oscillator?

(*Electronic Circuits - II, Anna University, May/June 2007*)

23. Draw the circuit of RC phase shift oscillator and explain its working. Also derive the expression for frequency of oscillator and condition for maintenance of oscillator.

(*Electronic Circuits - II, Anna University, May/June 2007*)

24. What is piezoelectric effect? Draw the equivalent circuit of quartz. Describe the crystal oscillator with neat circuit diagram.

(*Electronic Circuits - II, Anna University, May/june 2007*)

25. Draw the electrical equivalent circuit of the quartz crystal oscillator and write the expression for series and parallel resonant frequencies.

(*Electronic Circuits - II, Anna University, May/june 2007*)

26. Explain clapp oscillator and derive its frequency of oscillation.

(*Electronic Circuits - II, Anna University, Maj/June 2007*)

27. Exaplain Wein bridge oscillator circuit and derives its frequency of oscillator.

(*Electronic Circuits - II, Anna University, May/June 2007*)

MULTIPLE CHOICE QUESTIONS

1. An electronic oscillator is
(*a*) just like an alternator (*b*) nothing but an amplifier
(*c*) an amplifier with amplifier (*d*) a converter of a.c. to d.c. energy

2. The frequency of oscillation of an elementary *LC* oscillatory circuit depends on
(*a*) coil resistance (*b*) coil inductance
(*c*) capacitance (*d*) both (*a*) and (*c*)

3. For sustaining oscillations in an oscillator
(*a*) feedback factor should be unity (*b*) phase shift should be 0°
(*c*) feedback should be negative (*d*) both (*a*) and (*b*).

4. If Barkhausen criterion is not fulfilled by an oscillator circuit, it will
(*a*) stop oscillating (*b*) produce damped waves continuously
(*c*) becomes an amplifier (*d*) produce high-frequency whistles.

5. For sustained oscillations in an oscillator, the loop gain should be
(*a*) zero (*b*) less than unity
(*c*) greater than unity

6. Which of the following factors do not affect the frequency stability of an oscillator?
(*a*) output load
(*b*) inter-element capacitances and stray capacitances
(*c*) temperature variation
(*d*) coil size

7. A transistor Hartley Oscillator uses
(*a*) resistive feedback (*b*) inductive feedback
(*c*) capacitive feedback (*d*) none of the above

8. Crystal oscillators are superior to tuned *LC* oscillators mainly because of their

(*a*) exceptionally high degree of frequency stability

(*b*) high value of *Q*-factor

(*c*) size of the crystal

(*d*) both (*a*) and (*b*) above

9. In a RC phase-shift oscillator, the expression for frequency of oscillation is :

(*a*) $f = 1/2\pi\sqrt{LC}$ (*b*) $f = 1/2\pi\, RC\sqrt{6}$

(*c*) $f = 1/\pi\, CR$ (*d*) None of the above

10. In a transistor Hartley oscillator

(*a*) inductive feedback is used (*b*) untapped coil is used

(*c*) entire coil is in the output circuit (*d*) no capacitor is used

11. A Hartley oscillator is used for generating

(*a*) very low frequency oscillation (*b*) radio-frequency oscillation

(*c*) microwave oscillation (*d*) audio-frequency oscillation

12. A Colpitts oscillator uses

(*a*) tapped coil (*b*) inductive feedback

(*c*) tapped capacitance (*d*) no tuned *LC* circuit

13. In *RC* phase-shift oscillator circuits.

(*a*) there is no need for feedback

(*b*) feedback factor is less than unity

(*c*) pure sine wave output is possible

(*d*) transistor parameters determine oscillation frequency

14. Wien bridge oscillator is most often used whenever

(*a*) wide range of high purity sine waves is to be generated

(*b*) high feedback ratio is needed

(*c*) square output waves are required

(*d*) extremely high resonant frequencies are required.

15. The primary advantage of a crystal oscillator is that

(*a*) it can oscillate at any frequency

(*b*) it gives a high output voltage

(*c*) its frequency of oscillation remains almost constant

(*d*) it operates on a very low d.c. supply voltage

16. The RC network shown in Fig. 25-30 can provide a maximum theoretical phase-shift of

(*a*) 90° (*b*) 180°

(*c*) 270° (*d*) 360° (*UPSC Engg. Services* 2001)

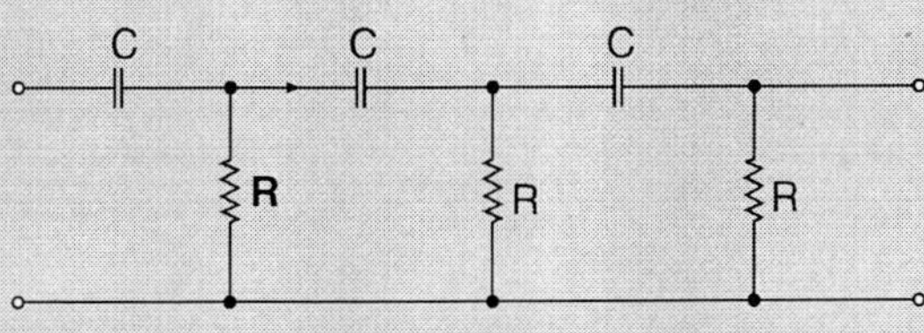

Fig. 25-30.

17. Which one of the following circuit is most suitable as an oscillator at a frequency of 100 Hz.

(*a*) Hartley oscillator (*b*) colpitts oscillator

(*c*) Crystal oscillator (*d*) Wein bridge oscillator

(*UPSC Engg. Services* 2001)

18. The Barkhausen criterion for sustained oscillation is given by

(*a*) $A\beta = 1$ (*b*) $|A\beta| \geq 1$

(*c*) $|A\beta| < 1$ (*d*) $< A\beta = 180°$

19. An oscillator always needs an amplifier with

(*a*) positive feedback (*b*) negative feedback

(*c*) both types of feedback (*d*) an *LC* tank circuit.

20. Crystal oscillators are often used for providing

(*a*) square wave output

(*b*) triangular wave output

(*c*) stable carrier wave in radio transmitters

(*d*) frequency multiplication.

21. A *CE* amplifier can be converted into an oscillator by

(*a*) providing adequate positive feedback

(*b*) phase shifting the output by 180° and feeding this phase shifted output to the input.

(*c*) using only a series tuned circuit as a load on the amplifier

(*d*) using a negative resistance device as a load on the amplifier.

(*e*) all of above (*UPSC Engg Services.* 2002)

ANSWERS

1. (*c*)	**2.** (*d*)	**3.** (*d*)	**4.** (*a*)	**5.** (*c*)
6. (*d*)	**7.** (*b*)	**8.** (*d*)	**9.** (*b*)	**10.** (*a*)
11. (*b*)	**12.** (*c*)	**13.** (*c*)	**14.** (*a*)	**15.** (*c*)
16. (*b*)	**17.** (*d*)	**18.** (*a*)	**19.** (*a*)	**20.** (*c*)
21. (*e*)				

❑❑❑

26

CHAPTER

Non-Sinusoidal Oscillators

OUTLINE

Objectives

After completing this chapter, you should be able to :

- Appreciate the difference between sinusoidal and non-sinusoidal oscillators.
- List the various types of non-sinusoidal oscillators.
- Understand the various switching times of a transistor.
- Describe the various factors contributing to transistor switching times delays.
- Sketch and describe the operation of transistor astable multi-vibrator (*i.e.*, a multi-vibrator using discrete components).
- Sketch the waveforms at the base and collector of a transistor astable multivibrator.
- Sketch and describe the operation of transistor monostable multivibrator.
- Sketch the waveforms at the base and collector of a transistor monostable multivibrator.
- Sketch and explain the circuit operation by sketching the waveforms at the inputs and outputs of a transistor bistable multivibrator.
- Explain the circuit operation of a transistor schmitt trigger.
- Describe the circuit operation of a 555-timer in monostable and astable mode.
- Explain the operation of Blocking oscillator.

26-1. Introduction

The oscillators, which generate waveforms other than sine waveform, are called non-sinusoidal oscillators. The waveforms generated by nonsinusiodal oscillators include *square, triangular rectangular and sawtooth waveforms.* These oscillators use one or two bipolar junction transistors (or field-effect transistors). The transistor is used as a switch, which operates in two states namely saturation and cut-off state. When the switch operates, there is a sharp transition from one state to another. Usually, a transistor remains in the cut-off state for a certain period of time, which is followed by another interval of time during which it remains in saturation. Since, the transistor alternately supplies power to the load, and relaxes when it is in cut-off state, therefore the non-sinusoidal oscillators are also known as ***relaxation oscillators***.

26-2. Classification of Non-sinusoidal Oscillators

The nonsinusoidal oscillators may be classified into the following three types :

(*i*) Multivibrators

(*ii*) Blocking oscillators and

(*iii*) Sawtooth or sweep generators.

The multivibrators and blocking oscillators are discussed in this chapter, while the sawtooth or sweep generators will be discussed in the chapter on Time-Base Circuits.

26-3. Transistor as a Switch

We have already discussed in Art 26-2 that nonsinusoidal oscillators employ one or two transistors, which are used as a switch. The switch operates between two states namely *saturation* and cut-off state. The saturation state occurs when both the junctions (*i.e.*, emitter-base junction and collector base junction) of a transistor are forward biased. On the other hand, the cut-off state occurs, when both the junctions are reverse biased.

Consider a common emitter transistor switch as shown in Fig.26-1. A pulse input (v_{in}) as shown in the Fig. 26-1 is applied at the based of the transistor. It is used to control the voltage (or state of the switch) between collector and emitter of the transistor. It may be noted that at time $t < T_1$, , the input voltage (v_{in}) is equal to $-V_1$ and the emitter-base diode is reverse biased. Since, the collector-base diode is also reverse biased, therefore the transistor is cut-off, and practically no current exists anywhere in the circuit. As a result of this, the collector-to-emitter or the output voltage (v_o) is approximately equal to the supply voltage (V_{CC}). Thus, when the collector current (i_C) is zero and $v_o = V_{CC}$, the transistor acts as an open switch.

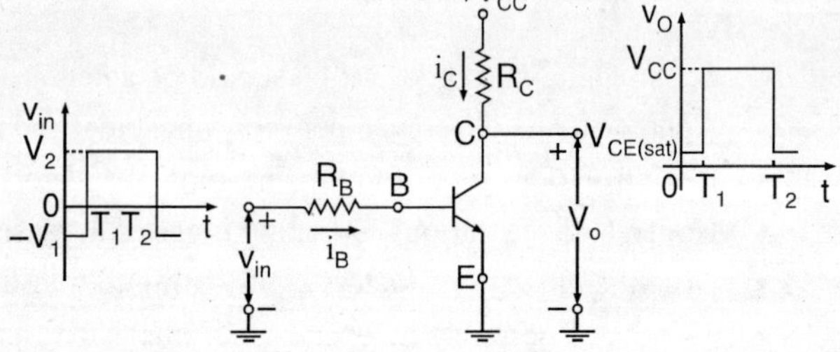

Fig. 26-1. *A Transistor switch with input and output waveforms.*

It will be interesting to know that when the input voltage becomes equal to *V_2 for the time interval $T_1 < t < T_2$, both the emitter-base and collector-base diodes are forward biased and the transistor is saturated. In this interval (*i.e.*, $T_1 < t < T_2$) the output voltage, $v_o = V_{CE(sat)}$. The value of $V_{CE(sat)}$ for silicon transistor is 0.2 V and the collector current is maximum. Its value is approximately equal to V_{CC}/R_C. These values indicate that the transistor acts as a closed switch and the current in the closed switch is determined by the supply voltage (V_{CC}) and the load resistor (R_C).

At time $T = T_2$, the input waveform returns back to $-V_1$.

This causes the transistor to switch back to cut-off state. We may summarize the above two results as follows :

$$i_C = 0 \quad \text{and} \quad v_{CE} = V_{CC} \qquad \text{...(Open switch)}$$

$$i_C = V_{CC}/R_C \quad \text{and} \quad v_{CE} = V_{CE(sat)} \qquad \text{...(Closed switch)}$$

* The value of V_2 is selected to ensure that transistor is in saturation. It is achieved by setting the value of base current (I_B) greater than $I_{B(sat)}$. The value of $I_{B(sat)}$ is equal to $V_{CC}/\beta \cdot R_C$ where β is the current gain of the transistor.

26-4. Transistor Switching Times

It has been observed that the application of a pulse input at the input (*i.e.*, base-emitter junction of a transistor is not followed directly by a change in output voltage. In other words, there is always some delay before the output changes as compared to the input. This phenomenon is illustrated in Fig. 26-2.

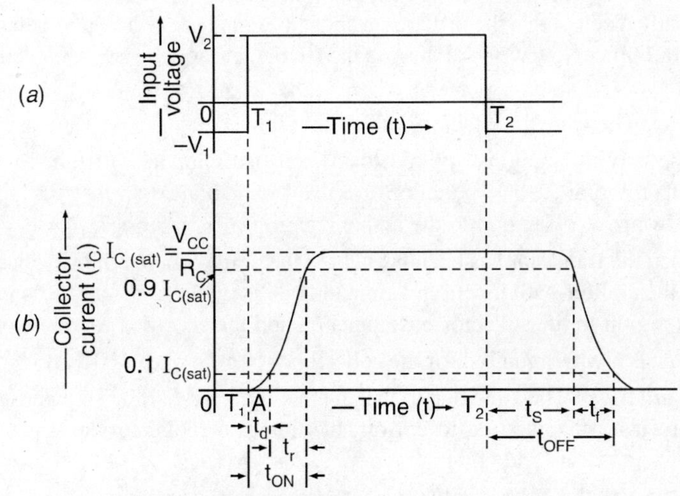

(*a*) Input pulse waveform

(*b*) Collector current waveform with various time delays

Fig. 26-2. *Transistor switching times.*

Fig. 26-2(*a*) shows the waveform of the input pulse applied to the transistor and Fig. 26-2(*b*) indicates the resulting waveform of collector current (i_C) alongwith the various time delays involved. Following are the important time delays in the collector current waveform, which may be noted from the Fig.26-2.

(*i*) ***Delay time*** **(t_d) :** It may be noted that the collector current does not quickly respond to the input signal at time $t = T_1$. Instead, there is a delay indicated by T_1A. The time that elapses during this delay, together with the time required for the collector current to rise to 10% of its maximum (*i.e.*, saturation) value $I_{C(sat)}$ equal to V_{CC}/R_C is called the delay time (t_d). The value of delay time for a transistor 2N 2222 A is about 10 nanoseconds.

(*ii*) ***Rise time*** **(t_r) :** The time required for the collector current to rise from 10% to 90% of the maximum value is called rise time (t_r). The value of rise time for 2 *N* 2222 A transistor is about 25 nanoseconds.

(*iii*) ***Time-ON time*** **(t_{ON}) :** The sum of the delay time (t_d) and the rise time (t_r) is called the turn-ON time. Mathematically, the turn-ON time,

$$t_{ON} = t_d + t_r$$

For 2N2222 A transistor, the total turn-ON time is equal to 35 nanoseconds.

(*iv*) ***Storage time*** **(t_s) :** It may be noted from the Fig. 26-2 that when the input signal returns back to its initial state (*i.e.*, $-V_1$) at time $t = T_2$, the collector current again fails to respond immediately. The interval, which elapses between the transition of the input voltage waveform and the time when collector current has dropped to 90% of its maximum value is called the storage time. The value of storage time for 2 N 2222. A transistor is about 225 nanoseconds.

(*v*) ***Fall-time*** **(t_f) :** The storage interval is followed by fall time (t_p). It is the time required for the collector current to fall from 90% to 10% of its maximum value. Its value for 2N2222 A transistor is about 60 nanoseconds.

(*vi*) ***Turn-OFF time*** (t_{OFF}) **:** It is the sum of storage time (t_s) and the fall time (t_f). Mathematically, the turn-OFF time,

$$t_{\text{OFF}} = t_s + t_f$$

For 2N2222 A transistor the total turn-OFF time is equal to 285 nanoseconds.

26-5. Factors Contributing to Various Time Delays in a Transistor

We have already discussed in the last article about various time delays like delay time, rise time, storage time and fall time. Now we shall discuss the factors, in more detail, which contribute to these time delays.

(*a*) ***Delay time :*** There are three factors, which contribute delay time (t_d) and are as given below :

(1) When the driving signal is applied to the transistor input, a certain time is required to charge-up the emitter-base junction transistor capacitance so that the transistor may be brought from cut-off region to the active regain.

(2) Even when the transistor has been brought to the point, where minority carriers have began to cross the emitter-base junction, a time interval is required before these carriers can cross the base region to the collector-base junction and the recorded as a collector current.

(3) Some time is always required for the collector current to rise to 10% of its maximum value.

(*b*) ***Rise and fall time :*** These are due to the fact that if a base-current step is used to saturate the transistor or return it from saturation to cut-off, the transistor collector current must traverse the active region.

It has been observed that the collector current increases or decreases along an exponential curve as shown in Fig. 26-3.

(*c*) ***Storage time :*** It results from the fact that a transistor, in saturation, has a saturation charge of excess minority carriers stored in the base-region. A transistor cannot respond, until this saturation charge has been removed. The storage time can be reduced by adding a capacitor (*C*) across the base resistor (R_B). It can also be eliminated by not allowing the transistor to enter into saturation by adding a Schottky diode across the collector and base terminals of the transistor.

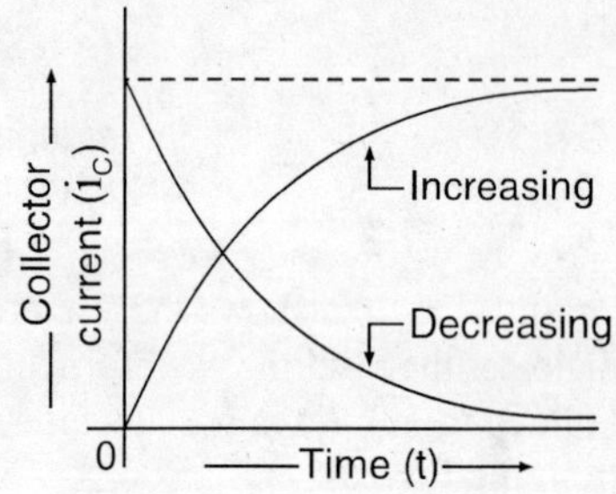

Fig. 26-3. *Increase or decrease of collector current.*

26-6. Multi-vibrators

These are two stage amplifiers, which are coupled in such a way that the output of one amplifier provides an input for the other. The amplifiers use positive feedback (also called regenerative feedback) which is supplied in such a manner that the transistor of one stage is driven into saturation (*i.e.*, ON state) and the other to cut-off (*i.e.*, OFF state). The action is reversed after a certain time, depending upon the circuit conditions, *i.e.*, the saturated transistor is driven into cut-off and the cut-off transistor is driven into saturation.

26-7. Types of a Multi-vibrators

There are three basic types of multivibrators depending upon the type of coupling network used.

(*i*) Astable multivibrator.

(*ii*) Monostable multivibrator.

(*iii*) Bistable multivibrator.

An astable multivibrator uses capacitive coupling and monostable multivibrator uses resistive-capacitive coupling, while the bistable multivibrator uses resistive coupling. Now we shall discuss all the above three types of multivibrators one by one in the following pages.

26-8. Astable Multivibrator

It is also called a *free-running relaxation oscillator* and is commonly used to generate square waveform. Fig. 26-4 shown the circuit of a collector-coupled astable multivibrator using two identical NPN transistors Q_1 and Q_2. It is possible to have $R_{C_1} = R_{C_2}$ R_C, $R_1 = R_1 = R_2 = R$ and $C_1 = C_2 = C$. In that case, the circuit is known as symmetrical astable multivibrator. The transistor Q_1 is forward biased by the V_{CC} supply through resistor R_1. Similarly, the transistor Q_2 is forward biased by the V_{CC} supply through resistor R_2. The output of transistor Q_1 is coupled to the input of transistor Q_2 through the capacitor C_1. Similarly, the output of transistor Q_2 is coupled to the input of transistor Q_1 through the capacitor C_2.

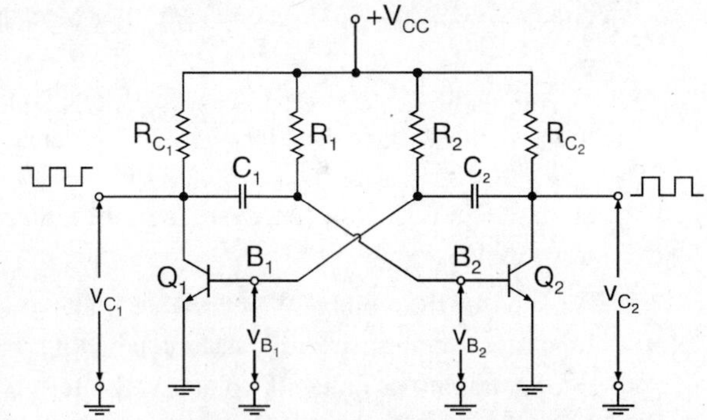

Fig. 26-4. *Astable multivibrator.*

As a matter of fact, the astable multivibrator may be thought of as two common-emitter amplifying stages. Each stage provides a feedback through a capacitor at the input of the other. Since, the amplifying stage introduces a 180° phase-shift and another 180° phase-shift is introduced by a capacitor, therefore the feedback signal has a total phase-shift of 360° or 0°. Thus, the feedback is positive and the circuit works as an oscillator. In other words, because of capacitive coupling, none of the transistor can remain permanently cut-off or saturated. Instead the circuit has two quasi-stable states (ON and OFF) and it makes periodic transition between these two states.

The output of an astable multivibrator is available at the collector terminal of either transistor (*i.e.*, Q_1 and Q_2) as shown in the Fig. 26-5. However, the two outputs are 180° out of phase with each other. Therefore, one of the output is said to be the complement of the other.

26-9. Circuit Operation of Astable Multivibrator

The circuit operation of an astable multivibrator is easy to understand with the waveforms shown in Fig. 26-5. These waveforms are at the base and collector of transistors Q_1 and Q_2. Fig. 26-5(*a*) shows the waveform for the base voltage of transistor Q_1 (*i.e.*, v_{B_1}) and Fig. 26-5(*b*) for the collector voltage of transistor Q_1 (*i.e.*, v_{C_1}). Similarly. Fig. 26-5 (*c*) shows the waveform for the base voltage of transistor Q_2 (*i.e.*, v_{B_2}) and Fig. 26-5(*d*) for the collector voltage of transistor Q_2 (*i.e.*, v_{C_2}). The circuit operation may be explained as follows :

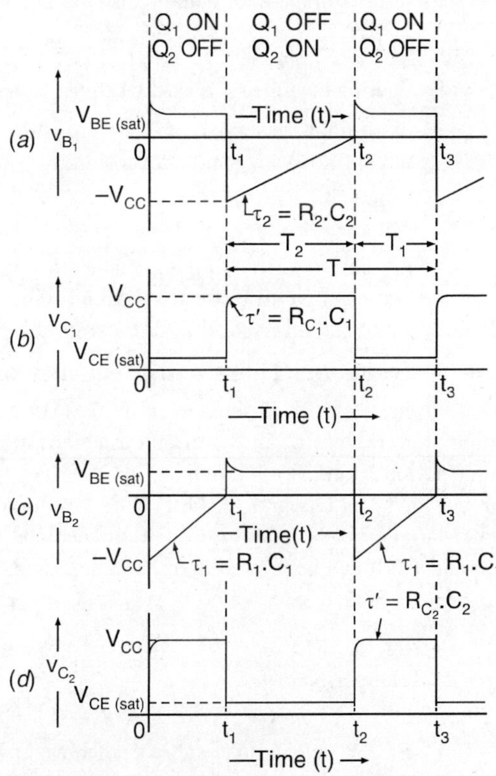

Fig. 26-5. *Wavefroms at the base and collector of transistors Q_1 and Q_2 in astable multivibrator.*

(1) When the d.c. power supply (V_{CC}) is switched ON, (say at $t = 0$) one of the transistor will start conducting more than the other due to some imbalance in the circuit. Let us suppose that transistor Q_1 start conducting more than that of transistor Q_2. Then because of positive feedback, the transistor Q_1 will be driven into saturation and transistor Q_2 to cut-off. Thus at $t > 0$, the transistor Q_1 is ON and Q_2 is OFF. Thus at $t > 0$, $v_{B_1} = V_{BE(sat)}$ (*i.e.*, 0.7 V for silicon transistor), $v_{C_1} = V_{CE\ (Sup)}$ (*i.e.*, 0.3 V for silicon transistor), v_{B_1} is negative and $V_{C_2} = V_{CC}$.

(2) During the time $t > 0$ (*i.e.*, when Q_1 is ON and Q_2 is OFF), the capacitor C_1 is charging towards the voltage V_{CC} through R_1. The charging takes place exponentially with time constant $\tau_1 = R_1 . C_1$. Since, the base of transistor Q_2 is directly connected to capacitor , C_1, as shown in Fig. 26-4 (page 4). Therefore, the voltage V_{B_2} also increases exponentially towards V_{CC}.

(3) As soon as the voltage V_{B_2} increases, above the cut-in voltage (*i.e.*, 0.5 V for silicon transistor), the transistor Q_2 starts conducting. It occurs at $t = t_1$. As the transistor Q_2 goes into saturation, its collector voltage (v_{C_1}) falls to $v_{CE(sat)}$. The fall in voltage v_{C_2} causes an equal fall, (*i.e.*, $V_{CC} - V_{CE(sat)} = V_{CC}$) in voltage v_{B_1} because the two are capacitively coupled. The fall in voltage v_{B_2} cuts-off the transistor Q_1 and its collector voltage (v_{C_1}) starts rising towards V_{CC} with a time constant $\tau' = R_C . C_1$. The rise in voltage V_{C_1} is coupled through capacitor C_1 to the base of the transistor Q_2, causing a small overshoot in voltage v_{B_2} . Soon the voltage V_{B_2} settles at $V_{BE(sat)}$, *i.e.*, 0.7 V level. Thus at $t \geq t_1$, the transistor Q_1 is OFF and Q_2 is ON. The voltage levels at this instant are v_{B_1} is negative, $v_{C_1} = V_{CC}$, $v_{B_2} = V_{BE} = V_{BE(sat)}$ and $v_{C_2} = V_{CE(sat)}$.

(4) During the time $t > t_1$ (*i.e.*, when Q_1 is OFF and Q_2 is ON) the voltage v_{B_1} rises exponentially with time constant $\tau_2 = R_2 \cdot C_2$ towards V_{CC}. At $t = T_2$, the voltage v_{B_1} reaches the cut-in level (*i.e.*, 0.5 V) and a reverse transition takes place (*i.e.*, Q_1 turns ON and Q_2 turns OFF). The voltage levels for $t > t_2$ are $v_{B_1} = v_{BE(sat)}$, $v_{C_1} = v_{CE(sat)}$; v_{B_2} is negative and $v_{C_2} = V_{CC}$. Thus, the voltage levels for $t > t_2$ are the same as for $t > 0$.

It may be noted that for an astable multivibrator, using PNP transistors, the voltages and currents are negative to those for a circuit with NPN transistors. Thus all the waveform shown in Fig. 26-5 must be inverted, if PNP transistors are used.

26-10. Switching Times and Frequency of Oscillation

We have already discussed in the last article the waveforms at the base and collector of the transistors Q_1 and Q_2 in the circuit astable multivibrator. Each cycle of the waveform consists of time period T_1 and T_2 as shown in Fig. 26-5. The time period T_1 represents the duration for which the transistor Q_2 is ON and Q_1 OFF. Similarly, time period T_2 represents the duration for which the transistor Q_1 is ON and Q_2 is OFF. Both the time periods T_1 and T_2 depend upon charge of capacitors C_1 and C_2. It can be proved that the time periods,

$$T_1 = 0.69\ R_1 \cdot C_1 \qquad \text{...(}i.e.\text{, ON time for } Q_1\text{)}$$

and

$$T_2 = 0.69\ R_2 \cdot C_2 \qquad \text{...(}i.e.\text{, ON time for } Q_2\text{)}$$

∴ Total period of the wave,

$$T_1 = T_1 + T_2 = 0.69\ (R_1 \cdot C_1 + R_2 \cdot C_2)$$

If $R_1 = R_2 = R$ and $C_1 = C_2 = C$, then we have a symmetrical astable multivibrator, whose time periods $T_1 = T_2$ and the total time period,

$$T = 0.69\ R \cdot C\ + 0.69\ R \cdot C = 1.38\ R \cdot C$$

and the frequency of oscillation is given by the reciprocal of the period, *i.e.*,

$$f = \frac{1}{T} = \frac{1}{1.38\,R \cdot C}$$

It is evident from the above expression that the frequency of oscillation may be varied by adjusting the values of R and C. However, it is more practicable to adjust R than C. It may be noted that if the values of resistors R_1 and R_2 have not been selected with d.c. current gain of the transistor (h_{FE} or β) in mind, the oscillations may not take place. To ensure oscillations, the value of resistor

$$R_1 \le h_{FE(min)} \cdot R_{C_1}$$

and $$R_2 \le h_{FE(mn)} \cdot R_{C_2}$$

where, $h_{FE\,(min)}$ is the minimum value of d.c. current gain of transistors Q_1 and Q_2. If the values of $R_{C_1} = R_{C_2} = R_C$ and $R_1 = R_2 = R$ then

$$R \le h_{FE(min)} \cdot R_C$$

and $$f = \frac{1}{1.38\,R \cdot C}$$

Example 26-1. *An astable multivibrator has component values, $R_1 = R_2 = 20\ k\Omega$ and $C_1 = C_2 = 100\ pF$. Calculate the frequency of oscillator.*

Solution. Given: $R_1 = R_2 = R = 20\ k\Omega = 20 \times 10^3\Omega$ and $C_1 = C_2 = C = 100\ pF = 100 \times 10^{-12}$ F.

We know that the frequency of oscillator,

$$f = \frac{1}{1.38\,R\,C}$$

$$= \frac{1}{1.38\,(20 \times 10^3) \times (100 \times 10^{-12})}\ \text{Hz}$$

$$= 0.362 \times 10^6\ \text{Hz} = 362\ \text{kHz}$$ **Ans.**

Example 26-2. *Determine the period and frequency of oscillation for an astable multivibrator with component values $R_1 = 2\ k\Omega$; $R_2 = 20\ k\Omega$ and $C_1 = 0.01\ \mu F$, $C_2 = 0.05\ \mu F$.*

Solution. Given: $R_1 = 2\ k\Omega = 2 \times 10^3\Omega$; $R_2 = 20\ k\Omega = 20 \times 10^3\ \Omega$; $C_1 = 0.01\ \mu F = 0.01 \times 10^{-6}$F and $C_2 = 0.05\ \mu F = 0.05 \times 10^{-6}$ F.

Period of oscillation

We know that period of oscillation,

$$T = 0.69\,(R_1 \cdot C_1 + R_2 \cdot C_2)$$

$$= 0.69\,[(2 \times 10^3) \times (0.01 \times 10^{-6}) + (20 \times 10^3) \times (0.05 \times 10^{-6})]$$

$$= 0.69\,[(0.02 \times 10^{-3}) + (1 \times 10^{-3})]$$

$$= 0.7 \times 10^{-3} = 0.7\ \text{ms}$$ **Ans.**

Frequency of oscillation

We know that frequency of oscillation,

$$f = \frac{1}{T} = \frac{1}{0.7 \times 10^{-3}}$$

$$= 1.43 \times 10^3\ \text{Hz} = 1.43\ \text{kHz}$$ **Ans.**

Example 26-3 *Determine the value of capacitors to be used in an astable multivibrator to provide a train of pulse 2 μs wide at a repetition rate of 100 kHz, if $R_1 = R_2 = 20k\Omega$.*

(*Electronic Circuits -II, Anna University, June 2009, Nov/Dec 2007*)

Solution. Given: Pulse width (T_1) = 2 μ sec = 2×10^{-6} sec; f = 100 kHz = 100×10^3 and $R_1 = R_2 = 20$ kΩ = 20×10^3 Ω.

Fig. 26-6 shows the waveform of the required pulse train.

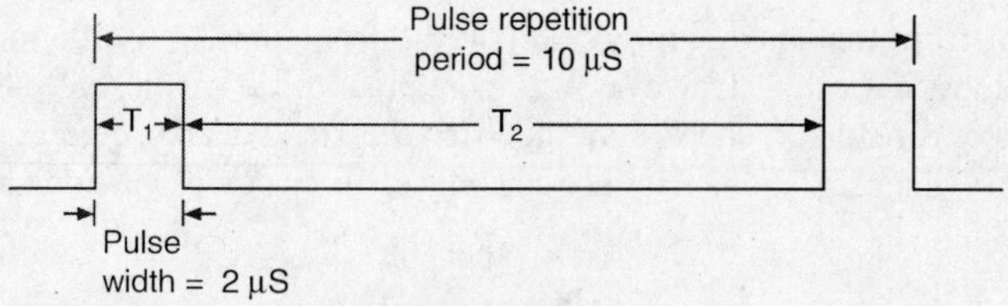

Fig. 26-6.

Let C_1 = Value of the capacitor required to provide the desired pulse width, and

C_2 = Value of the capacitor required to provide the desired repetition rate.

We know that frequency of oscillation (f),

$$100 \times 10^3 = \frac{1}{T}$$

or
$$T = \frac{1}{(100 \times 10^3)} 10 \times 10^{-6}$$

We also know that pulse width (T_1).

$$\therefore \quad 2 \times 10^{-6} = 0.69 \; R_1.C_1$$

$$= 0.69 \times (20 \times 10^3) \times C_1$$

$$\therefore \quad C_1 = \frac{2 \times 10}{0.69 \times (20 \times 10^3)} = 145 \; pF \qquad \textbf{Ans.}$$

and the time period (T)

$$10 \times 10^{-6} = T_1 + T_2 = (2 \times 10^{-6}) + T_2$$

$$\therefore \quad T_2 = (10 \times 10^{-6}) - (2 \times 10^{-6}) = 8 \times 10^{-6} \text{ sec}$$

We also know that the time period (T_2)

$$8 \times 10^{-6} = 0.69 \; R_2. \; C_2 = 0.69 \times (20 \times 10^3) \; C_2$$

$$C_2 = \frac{8 \times 10^{-6}}{0.69 \times (20 \times 10^{-3})} = 579.5 \; pF \textbf{ Ans.}$$

Example 26-4. *Determine the value of capacitors to be used in an astable multivibrator to provide a train of pulse 1 μ sec wide at a repetition rate of 100 kHz. Given $R_1 = R_2 = 10$ kΩ.*

Solution. Given: Pulse width (T_1) = 1 μsec = 1×10^{-6} sec; f = 100 kHz = 100×10^3 H_z and $R_1 = R_2$= 10 kΩ = 10×10^3 Ω.

Fig. 26-7 shows the waveform of the required pulse train.

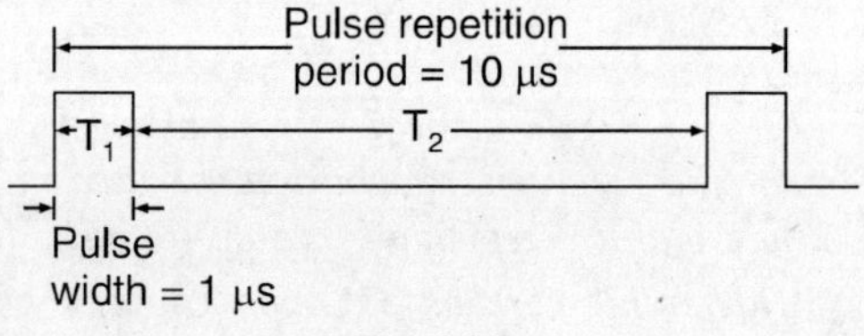

Fig. 26-7.

Let C_1 = Value of the capacitor required to provide the desired pulse width, and
C_2 = Value of the capacitor required to provide the desired repetition rate.

We know that frequency of oscillation (f),

$$100 \times 10^3 = \frac{1}{T}$$

or
$$T = \frac{1}{(100 \times 10^3)} = 10 \times 10^{-6} \text{ sec}$$

We also know that pulse width (T_1),

$$1 \times 10^{-6} = 0.69\, R_1 \cdot C_1 = 0.69 \times (10 \times 10^3) \times C_1$$

$$\therefore \quad C_1 = \frac{1 \times 10^{-6}}{0.69 \times (10 \times 10^3)} = 145 \times 10^{-12} \text{ F}$$

$$= 145 \text{ pF } \textbf{Ans.}$$

and the time period (T),

$$10 \times 10^{-6} = 10^{-6} = T_1 + T_2 = 1 \times 10^{-6} + T_2$$

$$\therefore \quad T_2 = (10 \times 10^{-6}) - (1 \times 10^{-6}) = 9 \times 10^{-6} \text{ sec}$$

We also know that the time period (T_2),

$$9 \times 10^{-6} = 0.69\, R_2 \cdot C_2 = 0.69 \times (10 \times 10^3)\, C_2$$

$$\therefore \quad C_3 = \frac{9 \times 10^{-6}}{0.69 \times (10 \times 10^3)} = 1304 \times 10^{-12} \text{ F}$$

$$= 1304 \text{ pF } \textbf{Ans.}$$

Example 26-5. *A collector coupled astable multivibrator is to be used to produce an unsymmetrical square wave of durations* τ_A = 310 μsec and τ_B = 250 μsec, V_{CC} = 15 V; $I_{C(sat)}$ = 5 mA, $h_{E(min)}$ = 20.

Find the values of circuits components and output waveform time constant.

Solution. Given: T_2 (τ_A)310 μsec = 310 × 10^{-6} sec; T_2 (τ_B) = 250 μsec = 250 × 10^{-6} sec; V_{CC} = 15 volt; $I_{C(sat)}$ = 5 mA = 5 × 10^{-3} A and $h_{FC(min)}$ = 20.

Value of circuit components

We know that the value of collector resistance,

$$R_C = \frac{V_{CC}}{I_{C(sat)}} = \frac{15}{5 \times 10^{-3}} = 3000\ \Omega$$

∴ Collector resistance of transistors Q_1 and Q_2,

$$R_{C_1} = R_{C_2} = R_C = 3000\ \Omega \ \textbf{Ans.}$$

We also know that the value of base current required to saturate the transistor,

$$I_{B(sat)} = \frac{I_{C(sat)}}{h_{FE(min)}} = \frac{5 \times 10^{-3}}{20} \text{ A} = 0.25 \times 10^{-3} \text{ A}$$

Assuming a base over drive factor of 2, the value of base current required,

$$I_B = 2\,I_{B(sat)} = 2 \times 0.25 \times 10^{-3}\text{ A} = 0.5 \times 10^{-3}\text{ A}$$

$$\therefore \quad R_1 = R_2 = R = \frac{V_{CC}}{I_B} = \frac{15}{0.5 \times 10^{-3}}$$

$$= 30 \times 10^3\Omega = 30\text{ k}\Omega\ \textbf{Ans.}$$

We know that the time period (T_1)

$$310 \times 10^{-6} = 0.69\,R_1 \cdot C_1 = 0.69 \times (60 \times 10^3) \times C_1$$

$$\therefore \quad C_1 = \frac{310 \times 10^{-6}}{0.69 \times (30 \times 10^3)} = 14.98 \times 10^{-9}\text{ F}$$

$$= 14980\text{ pF}\ \textbf{Ans.}$$

Similarly, the time period (T_2),

$$250 \times 10^{-6} = 0.69\,R_2 \cdot C_2 = 0.69 \times (30 \times 10^3)\,C_2$$

$$\therefore \quad C_2 = \frac{250 \times 10^{-6}}{0.69 \times (60 \times 10^3)} = 1208 \times 10^{-9}\text{ F} = 12080\text{ pF}\ \textbf{Ans.}$$

Output waveform time constants

We know that the time constant,

$$\tau_1 = R_1 \cdot C_1 = (30 \times 10^3) \times (14980 \times 10^{-12})\text{ sec}$$

$$= 449.4 \times 10^{-6}\text{ sec} = 449.4\ \mu\text{sec}\ \textbf{Ans.}$$

and the time constant,

$$\tau_2 = R_2 \cdot C_2 = (30 \times 10^3) \times (12080 \times 10^{-12})\text{ sec}$$

$$= 362.4 \times 10^{-6}\text{ sec} = 362.4\ \mu\text{sec}\ \textbf{Ans.}$$

Similarly, the time constant,

$$\tau' = R_{C_1} \cdot C_1 = 3000 \times (7.49 \times 10^{-9})\text{ sec}$$

$$= 22.5 \times 10^{-6}\text{ sec} = 22.5\ \mu\text{sec}\ \textbf{Ans.}$$

and the time constant,

$$\tau'' = R_{C_2} \cdot C_2 = 3000 \times (6.04 \times 10^{-9})\text{ sec}$$

$$= 18.1 \times 10^{-6}\text{ sec} = 18.1\ \mu\text{sec}\ \textbf{Ans.}$$

26-11. Monostable Multivibrator

It is also called *one shot* or *univibrator* and can be used to generate a gating pulse, whose width can be controlled. The monostable moltivibrator provides a single pulse of desired duration in response to an *external trigger. The external trigger causes the circuit to go to the quasi-stable state. After a certain interval of time, the circuit returns to its original stable state.

Fig. 26-8 shows the circuit of a monostable multivibrator using NPN transistors. Here, the output of transistor Q_2 is coupled to the base of transistor Q_1 through the resistance R_1. On the other hand, the output of transistor Q_1 is coupled to the base of transistor Q_2 through the capacitor C_2. The capacitor C_1 is known as commutating capacitor or speed up capacitor. Its function is to speed up the circuit in making abrupt transitions between the ON and OFF states. The base of transistor Q_2 is returned to the V_{CC} supply through a resistor R_3, while the base of transistor Q_1 is connected to the negative supply through a resistor R_2. The advantage of this biasing is that it keeps the transistor Q_1 OFF and Q_2 ON. This state is known as a *stable state of the monostable multivibrator.*

* It is short duration pulse of suitable amplitude to initiate operation of an electronic circuit. Such a pulse may be generated by a differeneiator (R–C circuit) or by a monostable multivibrator.

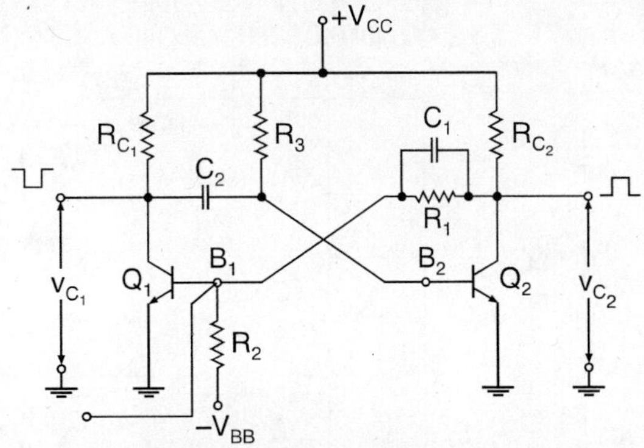

Fig. 26-8. *Monostable multivibrator.*

The output of a monostable multivibrator is available at the collector terminal of either transistor (*i.e.*, Q_1 or Q_2) as shown in the Fig. 26-8. However, the two outputs are complement of each other, *i.e.*, when one of the output is at V_{CC} level, the other is at $V_{CC(sat)}$ level.

The circuit operation may be explained by keeping in mind that initially the circuit is in its stable state, *i.e.*, Q_1 is OFF and Q_2 is ON. Now Let us see as to what happens when the trigger is applied :

(*i*) When a positive trigger pulse of sufficient amplitude is applied to the base of transistor Q_1, it overrides the reverse bias provided by the V_{BB} supply and gives it a forward bias. Because of this, the transistor Q_1 starts conducting.

(*ii*) As the transistor Q_1 conducts, its collector voltage falls due to the voltage drop across resistor R_{C_1}. This fall in voltage is coupled through capacitor C_2, which decreases the forward bias of transistor Q_2.

(*iii*) Because of the reduced forward bias, the collector current of transistor Q_2 starts decreasing and its collector voltage rises exponentially towards $V_{CC} \cdot R_1/(R_1 + R_{C_2})$ with a time constant $t_2 = C_1 \cdot (R_1 \parallel R_{C_2})$.

(*iv*) The rising collector voltage of transistor Q_2 is coupled to the base of transistor Q_1 through resistor R_1, where it further increases its forward bias. Because of the increased forward bias, the transistor Q_1 conducts more. This action is cummulative because of the positive feedback, and the collector voltage of harristor Q_1 falls to $V_{CC\,(sat)}$.

(*v*) Immediately, the capacitor C_2 starts charging exponentially towards V_{CC} with a time constant $\tau_1 = R_3 \cdot C_2$. As C_1 charges, the voltage at the base of the transistor Q_2 decreases. As C_1charges further, the transistor Q_2 is pulled out from the cut-off and the reverse transition takes place, *i.e.*, Q_2 turns ON and Q_1 turns OFF.

(*vi*) When the transistor Q_2 starts conducting, its collector voltage falls because of the drop across resistor R_{C_2}. This drop is coupled to the base of the transistor Q_1, whose collector voltage rises towards V_{CC} with time constant $\tau_3 = R_{C_1} \cdot C_2$. Finally, the transistor Q_1 turns fully ON and the transistor Q_1 goes OFF. The circuit remains in this stable state till another pulse is applied.

Fig. 26-9 shows the waveforms at the base and collector of the transistor Q_1 and Q_2 of a monostable multivibrator. The width or duration of the pulse obtained at the collector or output of either transistor (Q_1 or Q_2) of the monostable multivibrator is given by the expression,

$$\tau_p = 0.69\, R_3 \cdot C_1$$

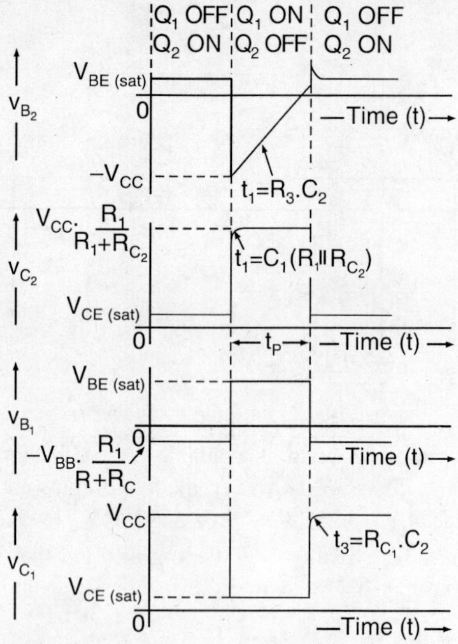

Fig. 26-9. *Waveforms at the base and collector of transistor Q_1 and Q_2 of a monostable multivibrator.*

Example 26-6. *A 20 kHz, 75% duty cycle square wave is used to trigger continuously, a monostable multivibrator with a triggered pulse duration (t_p) of 5 μs. What will be the duty cycle of the waveform at the output (Q) of the monostable?*

Solution. Given: Frequency of square wave, $f = 20$ kHz $= 20 \times 10^3$ Hz.

We know that time period of the square wave,

$$T = \frac{1}{f} = \frac{1}{20 \times 10^3} = 0.05 \times 10^{-3} \text{ s} = 50 \text{ μs}.$$

Since, the duty cycle of the square wave is 75%, therefore the waveform for this is as shown in Fig. 26-10(*a*). Now the monostable multivibrator is triggered once each time a new pulse arrives. The monostable multivibrator remains triggered only for a duration, $t_p = 5$ μs. The waveform of the output Q of the monostable multivibrator is as shown in Fig. 26-10(*b*).

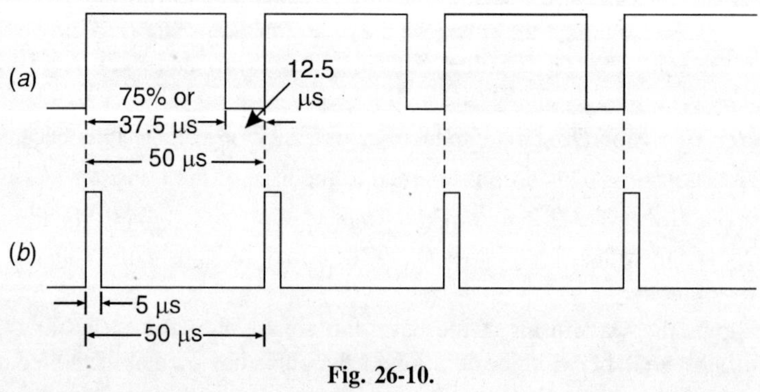

Fig. 26-10.

It is evident from the Fig. 26-10(*b*) that the duty cycle of the output (Q) of the monostable multivibrator is 5 μs/50 μs = 10%. **Ans.**

Example 26-7. *A monostable multivibrator is required to convert a 100 kHz, 30% duty cycle square wave to a 100 kHz, 50% duty cycle square wave. Find the values of R_3 and C_1.*

Solution. We know that a 100 kHz, 30 % duty cycle square wave is applied at the trigger input of the monostable multivibrator. This waveform is shown in Fig. 26-11(*a*). In order to convert it to a square waveform with 50% duty cycle, we want that the monostable multivibrator should have,

$$t_p = 50\% \text{ of the time period of 100 kHz waveform.}$$

Now time period of 100 kHz square wave,

$$T = \frac{1}{f} = \frac{1}{100 \text{ kHz}} = \frac{1}{100 \times 10^{-3}} = 10 \times 10^{-6} \text{ s} = 10 \text{ μs}.$$

$$t_p = 50\% \text{ of } 10 \text{ μs}, \textit{ i.e.}, \frac{1}{2} \times 10 \text{ μs} = 5 \text{ μs}$$

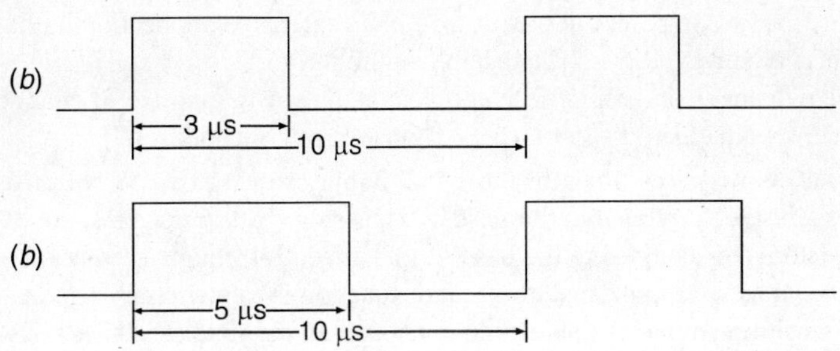

Fig. 26-11.

Fig. 26-11(*b*) shows the waveform produced at the output of monostable multivibrator.

We know that pulse width (t_p),

$$5 \times 10^{-6} = 0.69\, R_3 \cdot C_1 \qquad ...(i)$$

Let us select $C_1 = 0.001$ μF (*i.e.*, 1 nF), then, from equation (*i*)

$$5 \times 10^{-6} = 0.69 \times R_3 \times (0.001 \times 10^{-6})$$

$$\therefore \quad R_3 = \frac{5 \times 10^{-6}}{0.69 \times (0.001 \times 10^{-6})} = 7246.4\ \Omega$$

$$= 7.246 \text{ k}\Omega \textbf{ Ans.}$$

Note. In the above example, the choice of `C' should be such that the value of R_3 turns out to be in kΩ. The reason for this is that we want the current in the circuit to be in milliamperes.

Example 26-8. *A collector-coupled monostable multi using NPN silicon transistor has the following parameters : V_{CC} = 12 V, V_{BB} = 3V, R_C = 2 k, R_1 R_3 = R = 20 k, h_{FE} = 30, r_{bb}' 200 Ω, C = 1000 pF, neglect I_{CBO}. **(a)** Calculate and plot to scale the waveshapes at each base and collector, **(b)** find the width of output pulse.* (*Nagpur Univ. Winter* 2003)

Solution. Given: $R_C = 2\ k\ \Omega$; $R_1 = R_3 = R = 20\ k\Omega$; $h_{FE} = 30$, $r_{bb}' = 200\ \Omega$; $C = 1000$ pF. Refer to Fig. 26-12.

The width of the duration of the pulse,

$$\begin{aligned} T &= 0.69\ C_1\ R_3 \\ &= 0.69 \times 1000\ \text{pF} \times 20\ \text{k} \\ &= 13.8 \times 10^{-6}\ \text{s} \\ &= 13.8\ \mu\text{s} \end{aligned}$$

Note: For sketching waveforms refer to Fig. 26-9.

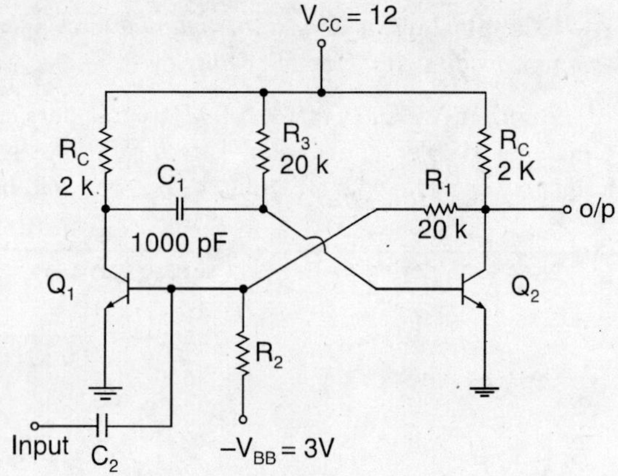

Fig. 26-12.

26-12. Bistable Multivibrator

It is also called *flip-flop, Eccles-Jordan circuit, trigger-circuit or binary.* The bistable multivibrator is used for counting and storing of binary information in computer circuits. It also finds extensive application in the generation and processing of pulse type waveforms. The bistable multivibrator has *two stable states* and can stay in any one of these two states, indefinitely, as long as the power is supplied. It changes to other state only when it receives a trigger (a short direction pulse) from outside.

Fig. 26-13 shows the circuit of a bistable multivibrator using two NPN transistors. Here, the output of a transistor Q_2 is coupled to the base of transistor Q_1 through a resistor R_1. Similarly, the output of a transistor Q_1 is coupled to the base of transistor Q_2 through a resistor R_2. The capacitors C_1 and C_2 are known as *speed up* capacitors. Their function is to increase the speed of the circuit in making abrupt transition from one stable state to another stable state. The base resistors (R_3 and R_4) of both the transistors are connected to a common source ($-V_{BB}$). The output of a bistable multivibrator is available at the collector terminal of both the transistors Q_1 and Q_2. However, the two outputs are the complements of each other.

The operation of the circuit may be explained as given below :

(*i*) When V_{CC} supply is switched on one of the transistor will start conducting more than the other. Then because of the feedback action, this transistor will be driven into saturation and the other to cut-off. Let us assume that the transistor Q_1 is in saturation (*i.e.*, ON) and Q_2 is cut-off (*i.e.*, OFF). It is a stable state of the circuit and will remain in this state, till a trigger pulse is applied from outside. A negative pulse applied to the set input will turn OFF the transistor Q_1 and Q_2 to ON. A similar action can also be achieved by applying a positive pulse at reset input.

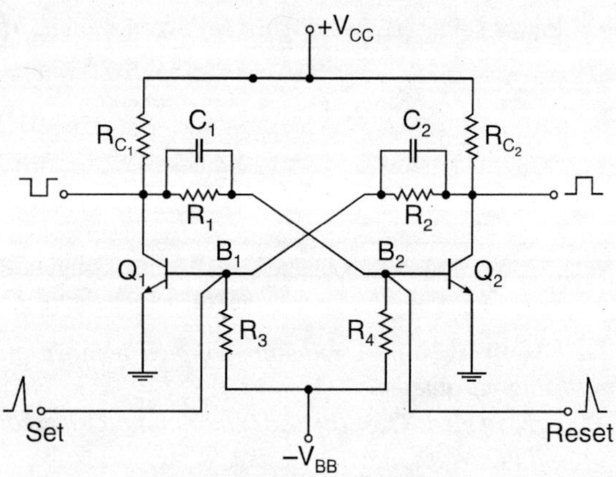

Fig. 26-13. *Bistable multivibrator.*

(*ii*) Suppose a positive pulse is applied at the reset input. It will cause the transistor Q_2 to conduct. As the collector voltage of Q_2 falls, it cuts-off the transistor Q_1. Thus the circuit switches to the other stable state, *i.e.*, a state where Q_1 is OFF $Q_2$and is ON.

(*iii*) Now if a positive pulse is applied at the set input, it will switch the circuit back to its original stable state, *i.e.*, Q_1 ON and Q_2 OFF.

Fig. 26-14 shows the waveforms at the collector of transistor Q_1 and Q_2 of the bistable multivibrator in response to the trigger pulses applied at the set and reset input. It is evident from these waveforms, that the output waveforms are the complement of each other.

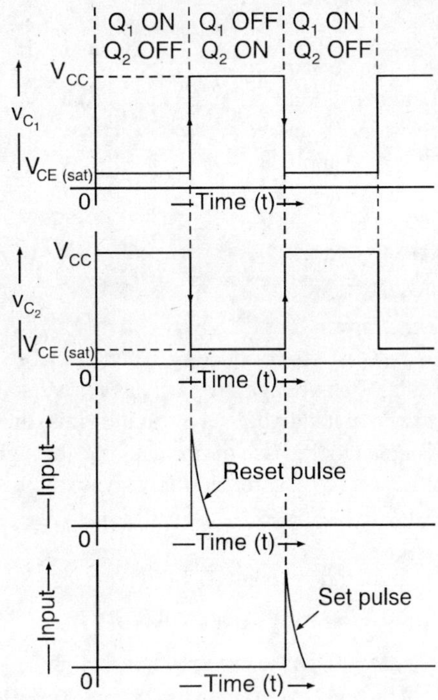

Fig. 26-14. *Waveforms at the collector of transistor Q_1 and Q_2.*

Notes: 1. There are two methods of triggering bistable multivibrators namely unsymmetrical triggering and symmetrical triggering. The unsymmetrical triggering requires two trigger inputs called set and reset. These are two independent trigger sources. This method of triggering is used in computer logic circuits. The symmetrical triggering requires a common trigger source and is used in counter circuits. It may be noted that we have explained the operation of a bistable multivibrator using a method of unsymmetrical triggering.

2. The trigger pulses may be applied either at the base or collector of the transistors, as it does not make any difference.

26-13. Applications of Multivibrators

The multivibrators are used in a variety of electronic circuits such as in computers, control systems and communication systems. The *astable multivibrators are used as square-wave generators, voltage-to-frequency converters and in pulse synchronization, as clock for binary logic signals, etc.* The monostable multivibrators are used for the generation of *well defined pulses,* the logic design of pulse delay, variable pulse width *etc*. The bistable multivibrators are used as memory elements in shift registers, counters, *etc*.

26-14. Schmitt Trigger

Fig. 26-15 shows the circuit of a Schmitt trigger (after the name of its inventor). It may be noted that the circuit, is, somewhat, like multivibrator circuits. The Schmitt trigger is used for *waveshaping circuits.* It can be used for generation of square wave from a sine-wave input. Basically, the circuit has two opposite operating states as in all other multivibrator circuits. However, the trigger signal is not, typically, a pulse waveform but a slowly varying a.c. voltage. The Schmitt trigger is level sensitive and switches the output state at two distinct trigger levels. One of the triggering levels is called a *lower-trigger level* (abbreviated as L.T.L) and the other as *upper-trigger level* (abbreviated as U.T.L.).

The circuit of a Schmitt trigger consist of two identical transistors Q_1 and Q_2 coupled through an emitter R_E. The resistors R_1 and R_2 form a voltage divider across the V_{CC} supply and ground. These resistors provide a small forward bias on the base of transistor Q_2. The operation of the circuit may be explained as follows :

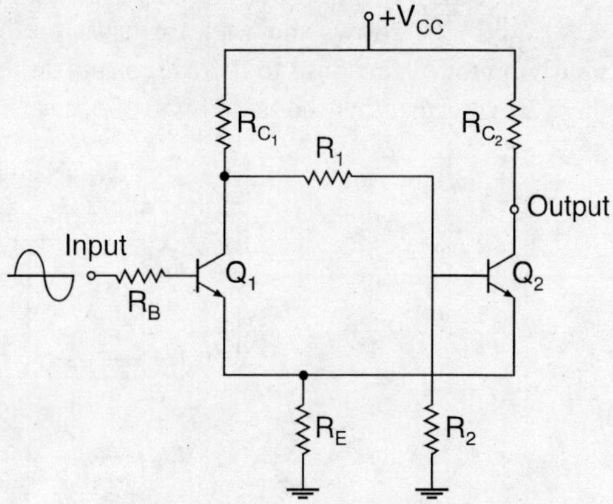

Fig. 26-15. *Schmitt Trigger.*

Let us suppose that initially there is no signal at the input. Then as soon as the power supply V_{CC} is switched on, the transistor Q_2 starts conducting. The flow of its current through resistor R_3 produces a voltage drop across it. This voltage drop acts as a reverse bias across the emitter-base junction of transistor Q_1 due to which it cuts-off. As a result of this, the voltage at its collector rises to V_{CC}. This rising voltage is coupled to the base of transistor Q_2 through the resistor R_1. It increases the forward bias at the base of transistor Q_2 and therefore drives it into saturation and holds it there. At this instant, the collector voltage, levels are $v_{C_1} = V_{CC}$ and $v_{C_2} = v_{CE(sat)}$ as shown in Fig. 26-16.

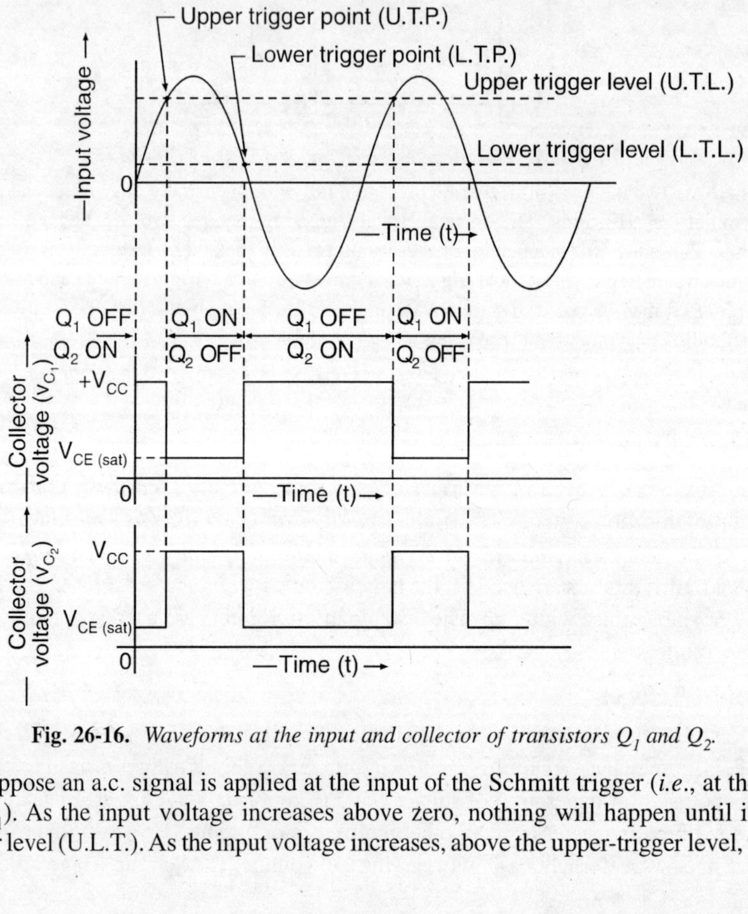

Fig. 26-16. *Waveforms at the input and collector of transistors Q_1 and Q_2.*

Now suppose an a.c. signal is applied at the input of the Schmitt trigger (*i.e.*, at the base of the transistor Q_1). As the input voltage increases above zero, nothing will happen until it crosses the upper trigger level (U.L.T.). As the input voltage increases, above the upper-trigger level, the transistor

Q_1 conducts. *The point, at which it starts conducting, is known as upper-trigger point* (U.T.P.). As the transistor Q_1 conducts, its collector voltage falls below V_{CC}. This fall is coupled through resistor R_1 to the base of transistor Q_2 which reduces its forward bias. This in turn reduces the current of transistors Q_2 and hence, the voltage drop across the resistor R_E. As a result of this, the reverse bias of transistor Q_1 is reduced, and it conducts more. As the transistor Q_1 conducts more heavily, its collector current further reduces due to which the transistor Q_2 conducts near cut-off. This process continues till the transistor Q_1 is driven into saturation and Q_2 into cut-off. At this instant, the collector voltage levels are $V_{C_1} = V_{CE\,(sat)}$ and $V_{C_2} = V_{CC}$ as shown in the Fig. 26-16.

The transistor Q_1 continue to conduct *till the input voltage falls below the lower-trigger level* (L.T.L.). It will be interesting to know that when the input voltage becomes equal to the lower-trigger level, the emitter-base junction of transistor Q_1 becomes reverse biased. As a result of this, its collector voltages starts rising towards V_{CC}. This rising voltage increases the forward bias across transistor Q_2, due to which it conducts. *The points, at which transistor Q_2 starts conducting is called lower-trigger point* (L.T.P.). Soon the transistor Q_2 is driven into saturation and Q_1 to cut-off. This completes one cycle. The collector voltage levels at this instant are $V_{C_1} = V_{CC}$ and $V_{C_2} = V_{CE(sat)}$. No change in state will occur during the negative half cycle of the input voltage.

It is evident from the above discussion that the output of a Schmitt trigger is a positive going pulse, whose width depends upon the time during which transistor Q_1 is conducting. The conduction time is set by the upper- and lower-trigger levels.

These days Schmitt trigger is also available in the form of integrated circuit (IC) Fig. 26-17 shows a schmitt trigger inverter connected as an oscillate (*i.e.*, as an astable multivibrator) The signal at V_{out} is an approximate square wave with a frequency that depends upon the values of *R* and *C*. The relationship between the frequency and RC values is shown in Fig. 26-18 for these different Schmitt trigger integral circuits.

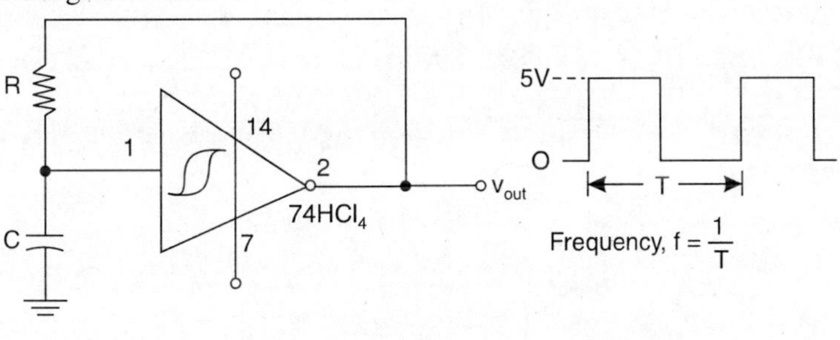

Fig. 26-17.

IC	Frequency	Remark
7414	≈ 0.8/RC	R ≤ 500 Ω
74LS14	≈ 0.8/RC	R ≤ 2 K Ω
74HCI4	≈ 0.8/RC	R ≤ 10 M Ω

Fig. 26-18.

As seen from Fig. 26-18 for 7414, Schmitt trigger inverter, the output frequency is given by $f \approx 0.8$ / RC. Notice that for this IC, the value of R ≤ 500 Ω. For 74 LS 14 IC, the output frequency is again 0.8 / RC but the value of R ≤ 2 kW. On the other hand for 74 HC 14, the square were output frequency is given by 1.2/ RC where the value of R ≤ 10 MΩ.

Example 26-9. *Design a Schmitt trigger circuit to have V_{cc} = 12 V, UTP = 5 V, LTP = 3V and Ic = 2 mA using two silicon NPN transistors with $h_{FF(min)}$ = 100 and $I_2 = 0.1/Il_{c2}$*

(*Electronic Circuits - II, Anna University, Nov/Dec 2007*)

Solution. Given: UTP = 5V, LTP = 3 V, V_{CC} = 12 V

$$V_i = V_{B2} = \text{UTP} = 5V \qquad \text{when } Q_2 \text{ is "ON"}$$

$$V_E = V_i - V_{BE1} = V_{B2} - V_{BE2} = 5 - 0.7 = 4.3\ V$$

Let $I_{C2} = I_{E2} = 1$ mA ...in "ON" state

Then, $$R_E = \frac{V_E}{I_{E_2}} = \frac{4.3}{1\times10^{-3}} = 4.3\,k\Omega$$

Now $$I_{C2}\,R_{C2} = V_{CC} - V_E - V_{CE2(sat)}$$

Let $V_{CE2(sat)}$ = 0.2 V, then

$$(1 \times 10^{-3})\,R_{C2} = 12 - 4.3 - 0.2$$

or $$R_{C2} = 7.5 \text{ kW}$$

Now $$I_2 = 0.1 \times I_{C2} = (1 \times 10^{-3}) / 10 = 1 \times 10^{-4} \text{ A}$$

$$R_2 = \frac{V_{B2}}{I_1} = \frac{5}{1\times10^{-4}} = 50\,k\Omega$$

$$I_{B2} = \frac{I_{C2}}{h_{fe(min)}} = \frac{1\times10^{-3}}{100} = 10\mu\ A$$

Therefore, $$I_2 + I_{B2} = \frac{V_{CC} - V_{B2}}{R_{C1} + R_1}$$

or $$R_{C1} + R_1 = \frac{12-5}{1\times10^{-4} + 10\times10^{-6}} = 63.6363\times10^3 = 63.6\,k\Omega \qquad ...(1)$$

Now $V_{B2} = V_{B1}$ = LTP = 3 V and Q_1 is **"ON"**

$$I_1 = \frac{V_{B2}}{R_2} = \frac{3}{50\times10^3} = 60\,\mu A$$

and $$I_{C1} \approx I_{E1} = \frac{V_{B1} - V_{BE1}}{R_E} = \frac{3-0.7}{4.3\times10^3} = 5.348\times10^{-4}\ A$$

$$V_{CC} = R_{C1}\,(I_{C1} + I_1) + I_1\,(R_1 + R_2) \qquad ...(2)$$

using equation (1) in equation (2)

$$V_{CC} = I_{C1}\,R_{C1} + I_1\,(R_C + R_1) + I_1\,R_2$$

$$12 = 5.348 \times 10^{-4}\,Rc1 + 60 \times 10^{-6} \times 63.6363 \times 10^3 + 10^{-6} \times 50 \times 10^3$$

$$R_{C1} = 9.7 \text{ k}\Omega$$

$$R_1 = 63.6 - 9.7 = 53.9 \text{ k}\Omega.$$

Thus, when Q_2 is "ON" $V_0 = V_{CC} - I_{C2(ON)}\,R_{C2} = 12 - 1 \times 10^{-3} \times 7.5 \times 10^3 = 4.5$ V

And when Q_2 is OFF, $V_0 = V_{CC} = 12\ V$

The Schmitt trigger circuit with the desired values of components is shown in Fig. 26-19 below :

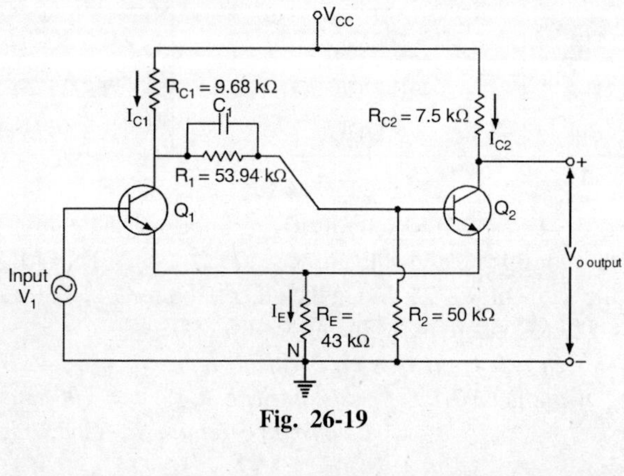

Fig. 26-19

26-15. The 555-Timer

The 555-Timer is a very popular IC used in electronic industry. This IC can be used in the monostable multivibrator made as well as astable multivibrator made. Let us study this device in detail.

Fig. 26-20 shows the functional block diagram of a 555-timer. As seen from this diagram, it has two comparators labelled A and B, an R - S flip-flop labelled as latch, an inverted output buffer and these resistors of 5 kΩ each. The name 555-timer has come from the three resistor of 5 kΩ each within the timer circuits. The comparators are the devices whose output are + 5 V when the voltage on the positive (+) input is greater than the voltage on the negative (–) input and (OV) when the negative (–) input is greater than the voltage on the positive (+) input.

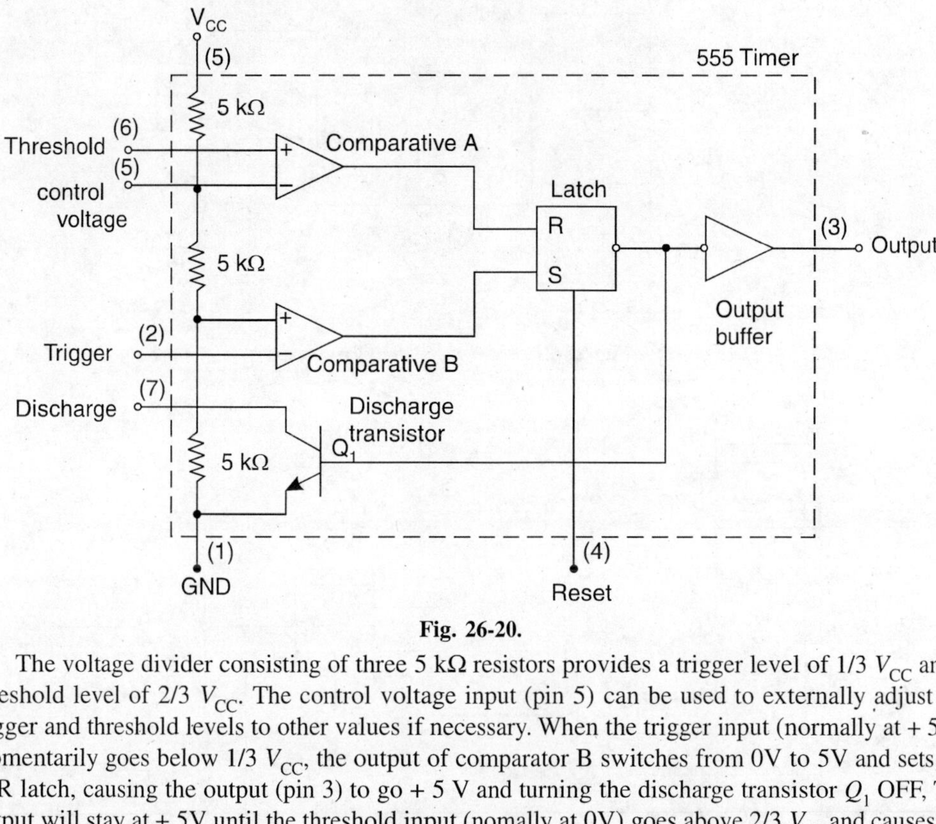

Fig. 26-20.

The voltage divider consisting of three 5 kΩ resistors provides a trigger level of 1/3 V_{CC} and a threshold level of 2/3 V_{CC}. The control voltage input (pin 5) can be used to externally adjust the trigger and threshold levels to other values if necessary. When the trigger input (normally at + 5 V) momentarily goes below 1/3 V_{CC}, the output of comparator B switches from 0V to 5V and sets the S–R latch, causing the output (pin 3) to go + 5 V and turning the discharge transistor Q_1 OFF, The output will stay at + 5V until the threshold input (nomally at 0V) goes above 2/3 $V_{CC}$and causes the output of comparator A to switch from 0C to + 5 V. This resets the latch causing the output to go back to 0V and turning the discharge transistor 'ON' the external reset input can be used to reset the latch independent of the threshold input the trigger and throshold inputs (pins 2 and 6) are controlled by external components connected to produce either monostable or astable action.

26-16. Monostable Operation of 555-Timer

The 555-timer can be set up as a monostable multivibrator as shown in Fig 26-21. Notice that it makes use of external resistor (R) and a capacitor (C). The pulse width of the output is given by the equation,

$$t_p = 1.1\ R_1\ C_1$$

As seen from Fig. 26-20, the control voltage input is not used and is connected to a decoupling capacitor, C_2 to prevent noise from affecting the trigger and throshold levels.

The operation of this circuit may be explained as follows. Before a trigger pulse is applied, the output is 0V and the discharge transistor Q_1 is ON, keeping C_1 discharged as shown in Fig. 26-21. When a negative going trigger pulse is applied at it, the output goes to + 5 V and the discharge.

Transistor turns OFF, allowing capacitor C_1 to begin charging through R_1 as shown in Fig. 26-22.

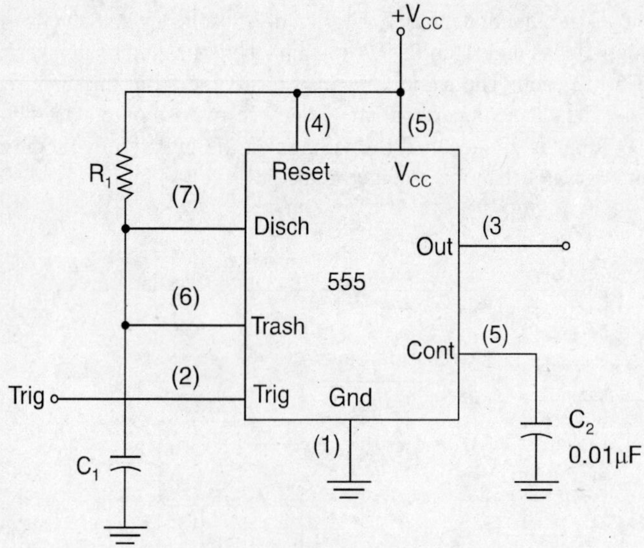

Fig. 26-21. *555-Timer when Triggered*

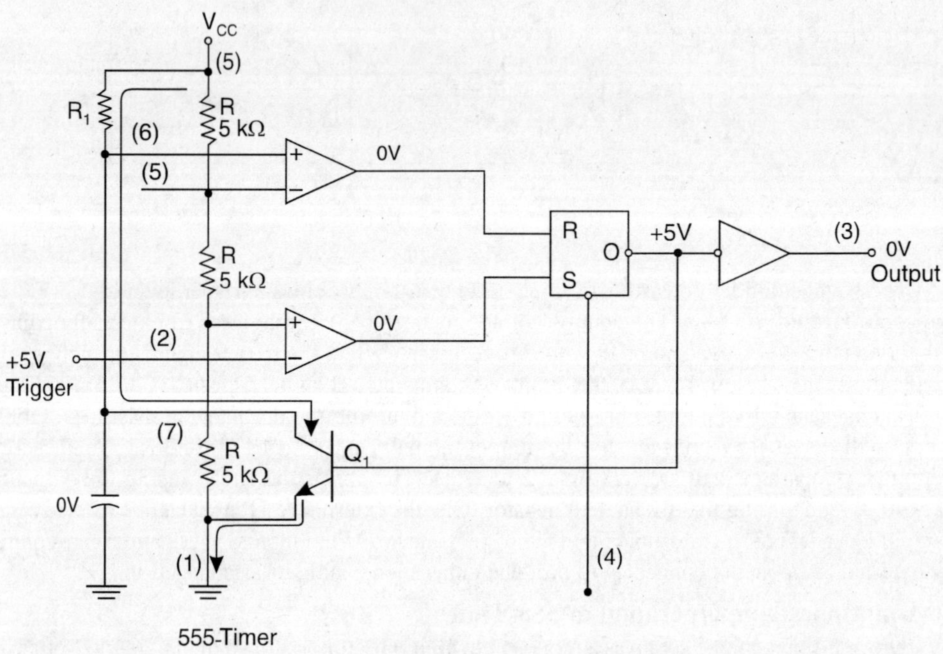

Fig. 26-22.

When C_1 changes to 1/3 V_{CC}, the output goes back to 0V at t_1 and Q_1 turns ON immediately discharging C_1 as shown in Fig. 26-23. It is obvious that the charging rate of C_1 determines how long the output is at + 5 V level.

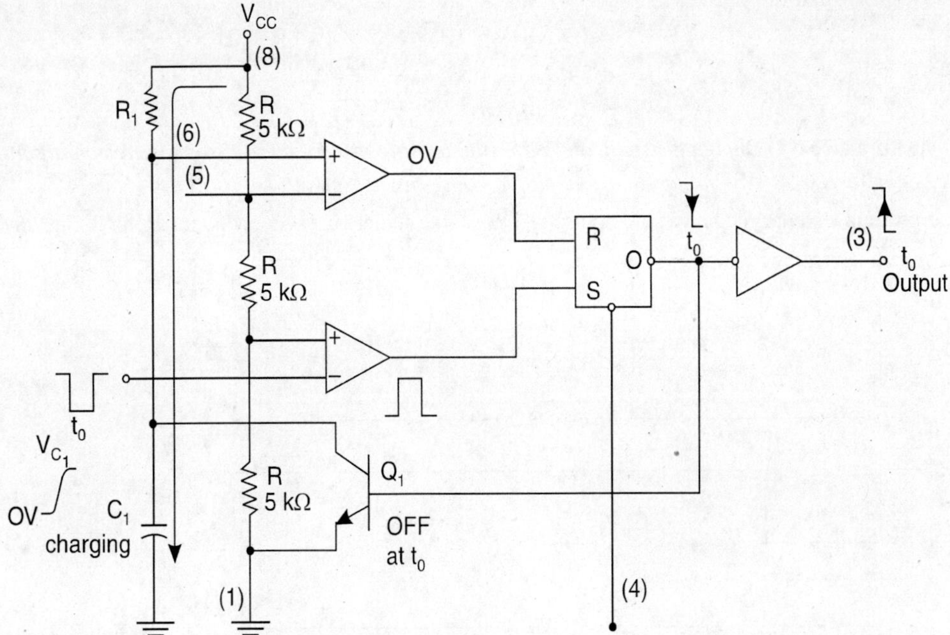

Fig. 26-23. *555-timer when Triggered.*

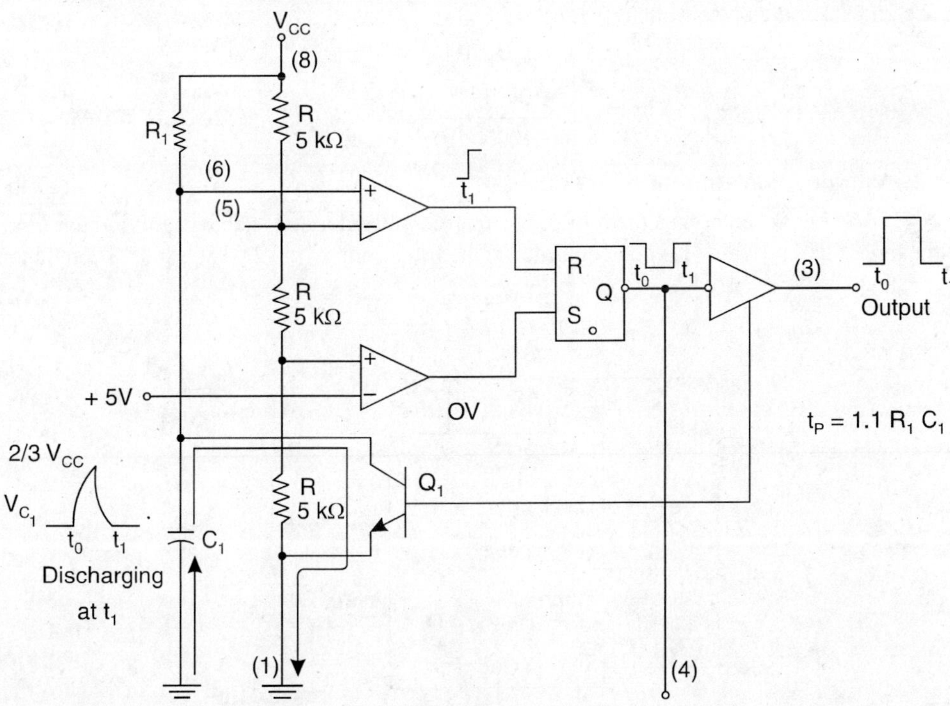

Fig. 26-24.

Example 26-10. *Determine the value of pulse width for a 555-timer operating in a monostable mode with $R_1 = 2.2\ k\Omega$ and $C_1 = 0.01\mu F$.*

Solution Given: $R_1 = 2.2\ k\Omega$ and $C_1 = 0.01\ \mu F$

We know that the pulse width

$$t_p = 1.1\ R_1\ C_1 = 1.1 \times (2.2 \times 10^3) \times (0.01 \times 10^{-6})\text{s}$$
$$= 24.2 \times 10^{-6}\ \text{s}$$
$$= 24.2\ \mu\text{s}\ \textbf{Ans.}$$

Example 26-11. *Design a circuit using 555-timer that will stretch a 1 μS negative going pulse that occurs every 60 μS into a 10 μS negative going pulse. Select C = 1000 pF.*

Solution. Given: A negative going pulse of 1 μs duration that occurs every 60 μs. Refer to Fig. 26-23.

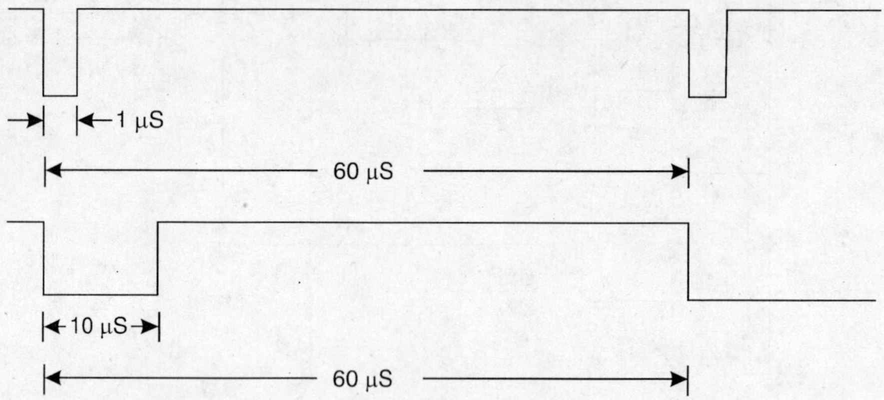

Fig. 26-25.

We know that the pulse width (t_p)

$$10 \times 10^{-6} = 1.1 \times R_1 \times (1000 \times 10^{-12})$$

$$\therefore \quad R_1 = \frac{10 \times 10^{-6}}{1.1 \times (1000 \times 10^{-12})} = 9.09 \times 10^3\ \Omega \text{ or } 9.09\ \text{k}\Omega\ \textbf{Ans.}$$

26-17. Astable Operation of 555-Timer

Fig. 26-26 shows an arrangement of 555-Timer connected as an astable multivibrator. Notice that the threshold input (pin 6) is non connected to the trigger input (pin 2). The external components

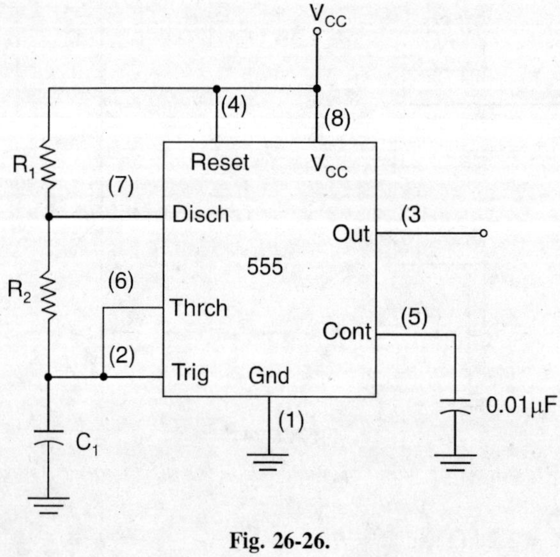

Fig. 26-26.

R_1, R_2 and C_1 form the tuning network that sets the frequency of oscillation. The 0.01 μ F capacitor, C_2 connected to the control (pin 5) input is strictly for decoupling and has no effect on the circuit. operation. In same cases, it can be left off even.

The circuit operation may be explained as follows. Initially when the power is turned ON, the capacitor (C_1) is uncharged and thus the trigger voltage (Pin 2) is at 0V. This causes the output of comparator B to be at + 5 V and the output of comparator A to be at 0V forcing the output of the latch and the base of Q_1 to be at 0V. As a result, the transistor Q_1 is OFF. Now C_1 begins charging through R_1 and R_2.This situation is shown in Fig. 26-27.

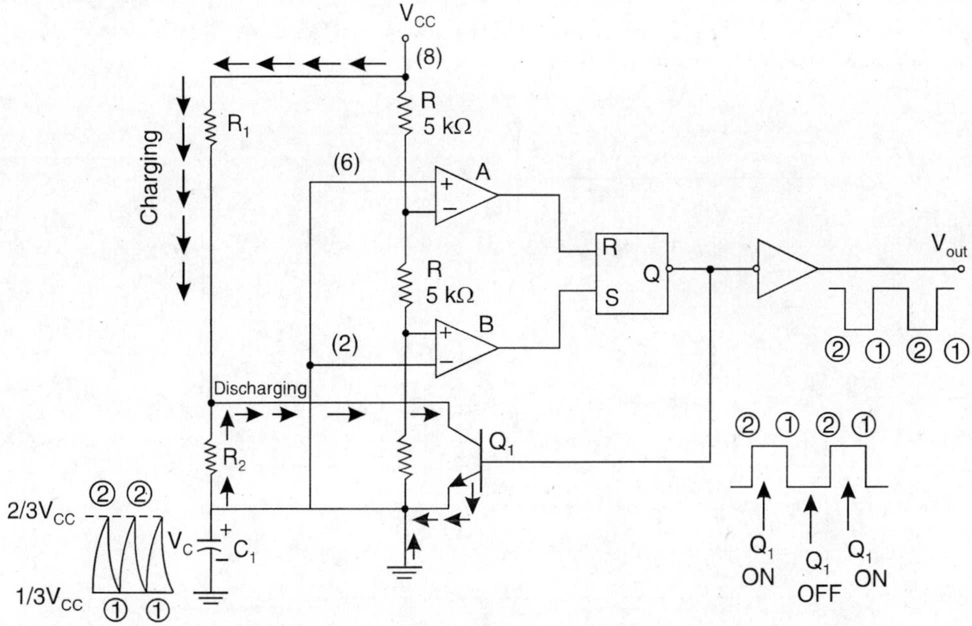

Fig. 26-27. *555-Timer operation in astable mode.*

When the capacitor voltage reaches 1/3 V_{CC}, comparator B switches to its O output state and when the capacitor reaches 2/3 V_{CC}, comparator A switches to its + 5 V state. This resets the latch causing the base of Q_1 to go + 5 V. As a result Q_1 turns ON. This sequence creates a discharge path for the capacitor through R_2 and the transistor as shown in Fig. 26-27. The capacitor now begins to discharge causing comparator A to go 0V. At the point where the capacitor discharges down to 1/3 V_{CC}, comparator B switches to + 0V level. This sets the latch, making the base of Q_1 to go down to 0V. As a result Q_1 turns OFF. Another charging cycle begins and the entire process repeats. This produces a rectangular wave output whose duty cycle depends on the values of R_1 and R_2.

The frequency of oscillator of a 555-timer in astable mode is given by the expression,

$$f = \frac{1.44}{(R_1 + 2R_2)\, C_1} \quad ...(i)$$

By selecting R_1 and R_2, the duty cycle of the output can be adjusted. Since, C_1 charges through $R_1 + R_2$ and discharges only through R_2, a duty cycle approacshing 50% can be achieved of $R_2 >> R_1$ so that charging and discharging times are approximately equal.

From equation (*i*), the time period,

$$T = \frac{1}{f} = \frac{(R_1 + 2\,R_2)\,C_1}{1.44} \frac{(R_1 + R_2)\,C_1}{1.44} + \frac{R_2\,C_1}{1.44}$$

$$= 0.7\,(R_1 + R_2)\,C_1 + 0.7\,R_2\,C_1$$

$$= t_1 + t_2$$

Notice that t_1 is the charging time and t_2 is the discharging time. These timings are shown in Fig. 26-28.

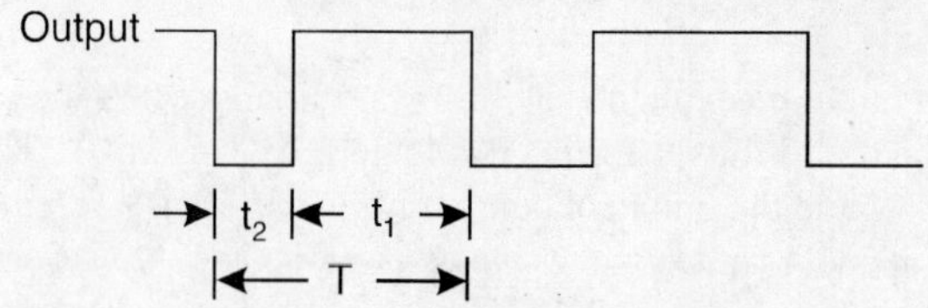

Fig. 26-28.

Example 26-12. *Fig. 26-29 shows the 555-timer in an astable mode. Determine the values of* t_1, t_2 *and duty cycle of the output waveform.*

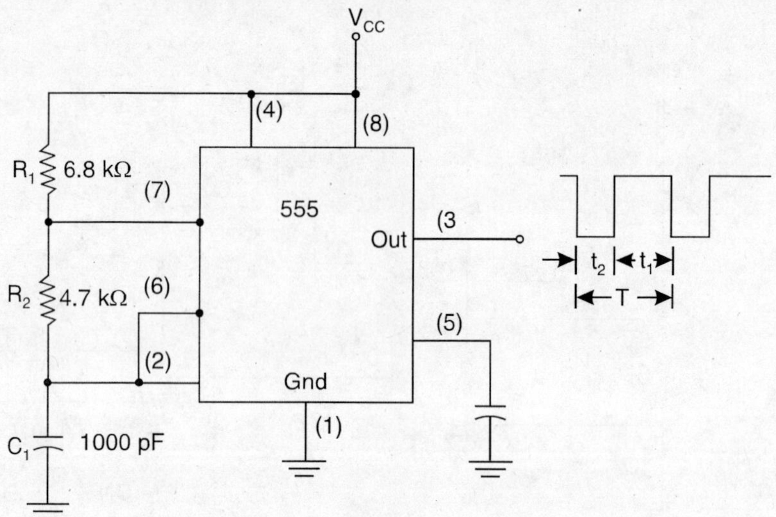

Fig. 26-29.

Solution. Given: $R_1 = 6.8$ kΩ; $R_2 = 4.7$ kΩ and $C_1 = 1000$ pF.

We know that the time interval.

$$t_2 = 0.7\,R_2\,C_1 = 0.7 \times (4.7 \times 10^{+3}) \times (1000 \times 10^{-12})\text{ s}$$
$$= 3.29 \times 10^{-6}\text{ s or } 3.29\ \mu\text{s}.$$

and the time interval

$$t_1 = 0.7\,(R_1 + R_2)\,C_1 = 0.7\,[(6.8 \times 10^3) + (4.7 \times 10^3)] \times (1000 \times 10^{-12})$$
$$= 8.05 \times 10^{-6}\text{ s or } 8.05\ \mu\text{s}.$$

We also known that duty cycle of the waveform,

$$\text{duty cycle} = \frac{t_1}{T} \times 100 = \frac{8.05\ \mu s}{(t_1 + t_2)} \times 100$$

$$= \frac{8.05\ \mu s}{8.05 + 3.29\ \mu s} \times 100$$

$$= 71\%\ \textbf{Ans.}$$

Example 26-13. *Calculate the frequency and duty cycle of the output of Astable multivibrator shown in Fig. 26-30 and draw the waveform obtained across the capacitor,* C_1.

(*Bangalore University* 2000)

Solution. Given: $R_1 = 27\ k\Omega$; $R_2 = 56\ k\Omega$ and $C_1 = 0.01\ \mu F$.

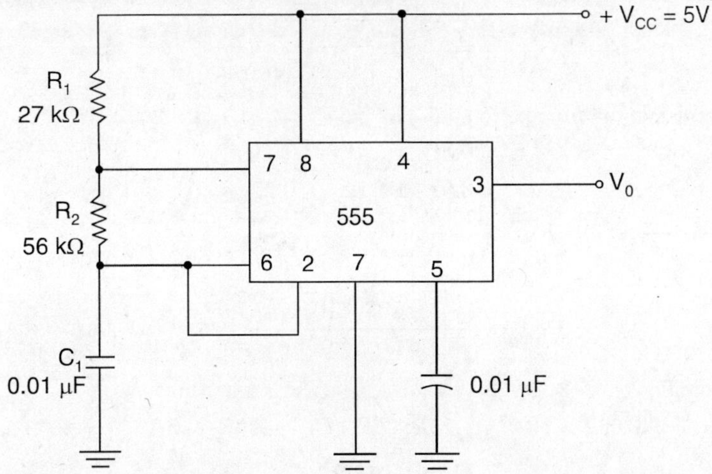

Fig. 26-30.

We know that the time interval, $t_2 = 0.7\ R_2\ C_1 = 0.7 \times (56 \times 10^3) \times (0.01 \times 10^{-6}) = 0.392$ ms

and the time interval, $t_1 = 0.7\ (R_1 + R_2)\ C_1 = 0.7 \times (27 \times 10^3 + 56 \times 10^3) \times (0.01 \times 10^{-6})$

$= 0.581$ ms

$\therefore$ Timer period, $T = t_1 + t_2 = 0.581 + 0.392 = 0.973$ ms

and the frequency, $f = \dfrac{1}{T} = \dfrac{1}{0.973\ ms} = 1.03$ kHz **Ans.**

Fig. 26-31 shows the waveform across the capacitor along with the output waveform.

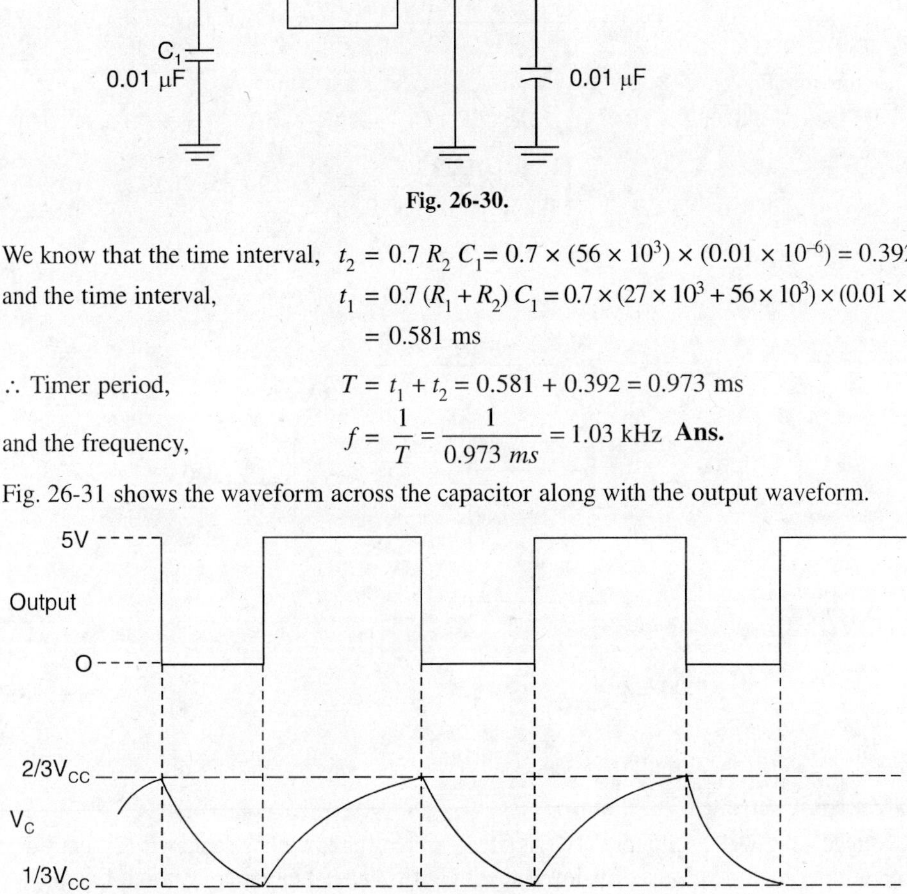

Fig. 26-31.

Example 26-14. *Design an astable multivibrator that will oscillate at 50 kHz, 60% duty cycle. Select C = 0.0022 µ F.*

Solution. Given: $f = 50$ kHz $= 50 \times 10^3$ Hz; duty cycle = 60%, C = 0.0022 µ F.

We know that the time period,

$$T = \frac{1}{f} = \frac{1}{50 \times 10^3} = 0.02 \times 10^{-3}\ s \text{ or } 0.02 \text{ ms}$$

and the duty cycle,

$$0.6 = \frac{t_1}{T} = \frac{t_1}{0.02 \times 10^{-3}}$$

$$\therefore \quad t_1 = 0.012 \times 10^{-3} \text{ s or } 0.012 \text{ ms}$$

We also know that the time period, (T)

$$0.02 \text{ ms} = t_1 + t_2 = 0.012 + t_2$$

$$\therefore \quad t_2 = 0.02 - 0.012 = 0.008 \text{ ms}$$

Now the time duration (t_2)

$$0.008 \times 10^{-3} = 0.7\, R_2\, C_1 = 0.7 \times R_2 \times (0.0022 \times 10^{-6})$$

$$\therefore \quad R_2 = \frac{0.008 \times 10^{-3}}{(0.7 \times 0.0022 \times 10^{-6})} = 5.19 \text{ k}\Omega \textbf{ Ans.}$$

The time duration (t_1),

$$0.012 \times 10^{-3} = 0.7\,(R_1 + R_2)\, C_1 = 0.7\,(R_1 + 5.19 \text{ k}) \times (0.0022 \times 10^{-6})$$

$$\therefore \quad R_1 = \left(\frac{0.012 \times 10^{-3}}{0.7 \times (0.0022 \times 10^{-6})} - 5.19\, k\right) = (7.79 \text{ k} - 5.19\, k)$$

$$= 2.6 \text{ k}\Omega \textbf{ Ans.}$$

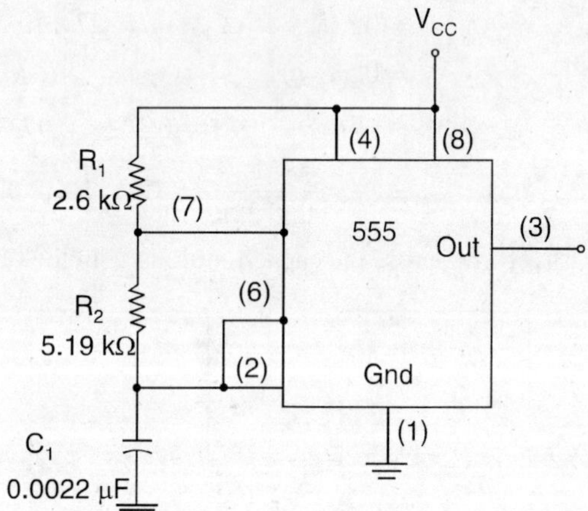

Fig. 26-32.

Fig. 26-32 shows the circuit to produce 50 kHz square wave with 60% duty cycle

26-18. Voltage-Controlled Oscillator

A voltage-controlled oscillator (VCO) is a free-running circuit whose output freequency depands on a given input control voltage. The lower the inpout voltage, the higher the input freequency and vice-versa. The 555-timer discussed in Article 26-15 can be used to operate as VCO as show in Fig. 26-33.

Notice that this circuit is the same as shown in Fig. 26-24 with the exception of the potentiometer R_3, the trigger level is at 1/3 V_{CC} and a threshold level is at 2/3 V_{CC}. With the potentiometer R_3 in place, we can externally adjust the trigger and threshold level to other values. It may be carefully noted that when the control voltage (V_{CON}) input is reduced, the difference between the threshold voltage and the trigger level is also reduced. For example, when the control voltage (V_{CON}) is +10 V, the threshold voltage is +10V and trigger voltage is +5 V, the difference being +5 V. When the control voltage (V_{CON}) is +5V, the threshold voltage is +5 V, and trigger voltage is +2.5 V, the difference being +2.5 V. This means by reducing the control voltage (V_{CON}), we are reducing the difference between the threshold voltage and trigger voltage.

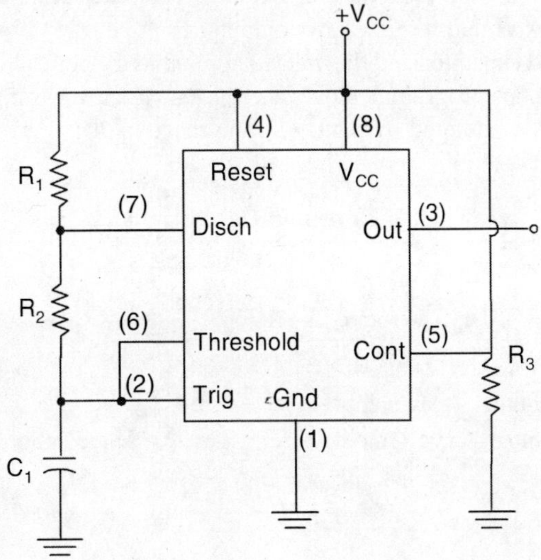

Fig. 26-33 *555-timer connected as VCO*

When the difference between the threshold voltage and trigger voltage is recued, it takes less time for the capacitor C, to charge or discharge between the two values. Therefore, the cycle time is reduced and the operating frequency is increased. Thus, we are able to control the operating frequency by varying the control voltage. Fig. 26-34 shows the trigger (or capacitor) voltage and output waveforms.

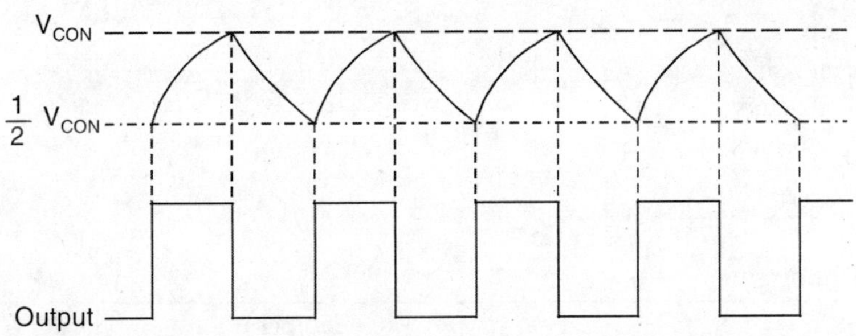

Fig. 26-34 *Trigger input output voltage waveforms of a 555-timer as a VCD*

Fig. 26-35 shows the specification datasheet of NE/SE566, a general purpose voltage controller oscillator manufactured by philips semi conductors. This device simultaneously produces a square-wave output as well as triangular-wave output upto a frequency of IMHz. The frequency of oscillation of this device is determined by external resistor and capacitor and the voltage applied to the control terminal (V_c). It may be noted carefully that the peak-to-peak value of voltage for the triangular-wave out put for NE566/SE566 is 2.4 (max). On the other hand, for the sequence wave output, the peak-to-peak value of voltage ($V_{P.P}$) is 5.4V. The maximum operating voltage for the device is 26 V. For further details please refer to the specification datasheet of the device.

Types of VCOs

There are two types of VCOs : **(1)** RC relaxation oscillator, and **(2)** LC oscillator. The VCO discussed in the last article is an RC relaxation oscillator. This type of oscillator is limited the frequencies up to I MHz. For higher frequencies, LC oscillator type of VCOs are more suitable.

FUNCTION GENERATOR

The NE/SE566 Function Generator is a voltage-controlled oscillator of exceptional linearity with buffered square wave and triangle wave outputs. The frequency of oscillation is determined by an external resistor and capacitor and the voltage applied to the control terminal.

The oscillator can be programmed over a ten-to-one frequency range by proper selection of an external resistance and modulated over a ten-to-one range by the control voltage, with exceptional linearity.

Pin Configurations

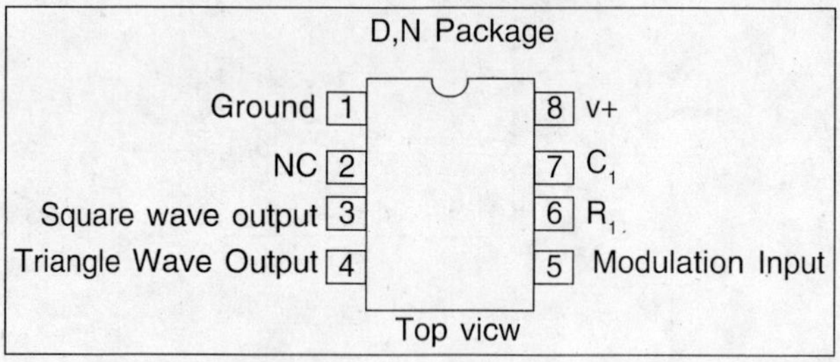

Features

- Wide range of operating voltage (up to 24V; single or dual)
- High linearity of modulation
- Highly stable centre frequency (200ppm/°C typical)
- Highly linear triangle wave output
- Frequency programming by means of a resistor or capacitor, voltage or current
- Frequency adjustable over 10-to-1 range with same capacitor

Applications

- Tone generators
- Frequency shift keying
- FM modulators
- Clock generators
- Signal generators
- Function generators.

Ordering Information

Description	Temperature range	Order code	DWG #
8-Pin Plastic Small Outlin (SO) Package	0 to +70°C	NE566D	0174C
14-Pin Ceramic Dual In-Line Package (CERDIP)	0 to +70°C	NE566F	0581B
8-Pin Plastic Dual In-Line Package (DIP)	0 to +70°C	NE566N	0404B
8-Pin Plastic Dual In-Line Package (DIP)	–55°C to +125°C	SE566N	0404B

Block Diagram

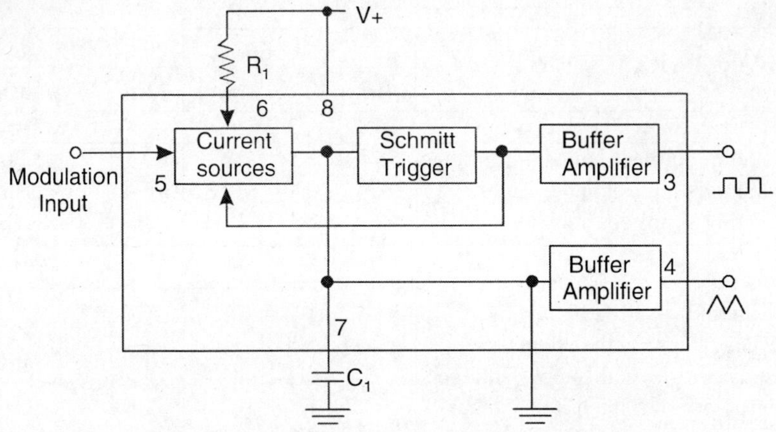

Fig. 26-35 *Specifications datasheet - NE/SE 566*

Equivalent Schematic

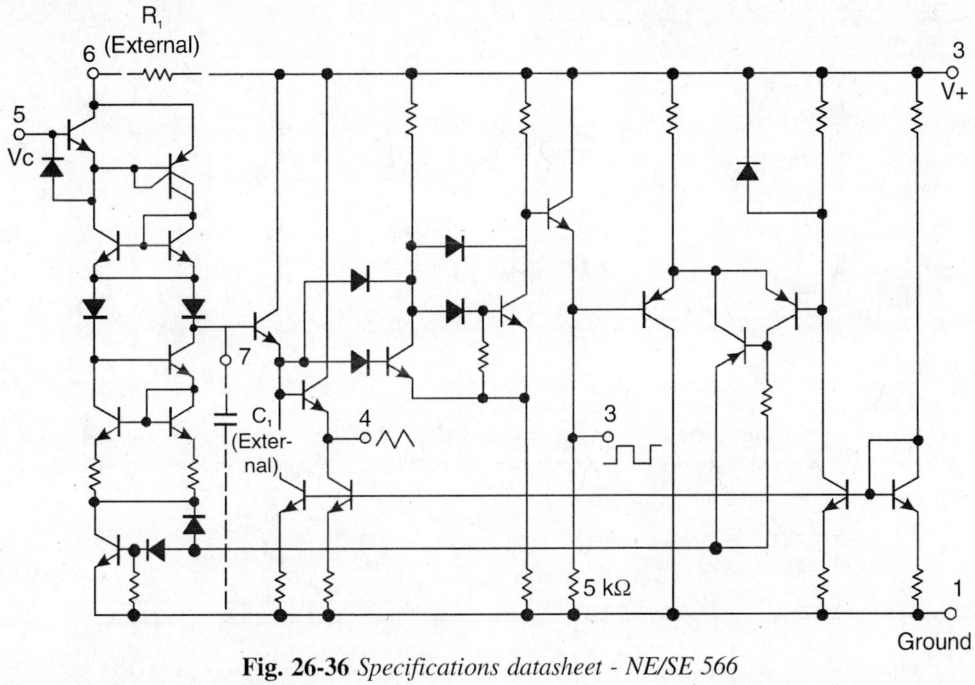

Fig. 26-36 *Specifications datasheet - NE/SE 566*

Absolute Maximum Ratings

SYMBOL	PARAMETER	RATING	UNIT
V+	Maximum operating voltage	26	V
V_{IN}, V_C	Input voltage	3	V_{P-P}
T_{STG}	Storage temperature range	–65 to +150	°C
T_A	Operating ambient temperature range		
	NE566	0 to +70	°C
	SE566	–55 to +125	°C
P_D	Power dissipation	300	mW

DC Electrical Characteristics

TA = 25°C, VCC = ±6V, unless otherwise specified.

SYMBOL	PARAMETER	SE566			NE566			UNIT
		Min	Typ	Max	Min	Typ	Max	
General								
T_A	Operating ambient temperature range	-55		125	0		70	°C
V_{CC}	Operating supply voltage	±6		±12	±6		±12	V
I_{CC}	Operating supply current		7	12.5		7	12.5	mA
VCO[1]								
f_{MAX}	Maximum operating frequency		1			1		MHz
	Frequency drift with temperature		500			600		ppm/°C
	Frequency drift with supply voltage		0.1	1		0.2	2	%/V
	Control terminal input impedance[2]		1			1		MΩ
	FM distortion (±10% deviation)		0.2	0.75		0.4	1.5	%
	Maximum sweep rate		1			1		MHz
	Sweep range		10:1			10:1		
Output								
	Triangle wave output							
	impedance		50			50		Ω
	voltage	1.9	2.4		1.9	2.4		V_{P-P}
	linearity		0.2			0.5		%
	Square wave input							
	impedance		50			50		Ω
	voltage	5	5.4		5	5.4		V_{P-P}
	duty Cycle	45	50	55	40	50	60	%
t_R	Rise time		20			20		ns
t_F	Fall Time		50			50		ns

NOTES:

1. The external resistance for frequency adjustment (R_1) must have a value between 2kΩ and 20kΩ.
2. The bias voltage (V_C) applied to the control terminal (Pin 5) should be in the range $V_+ \le V_C \le V_+$.

Typical Performance Characteristics

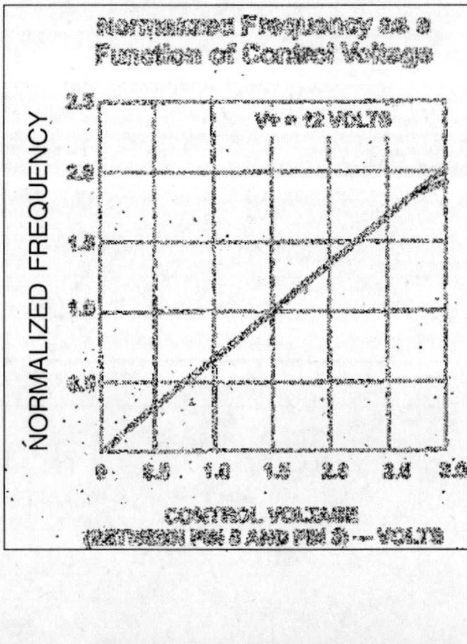

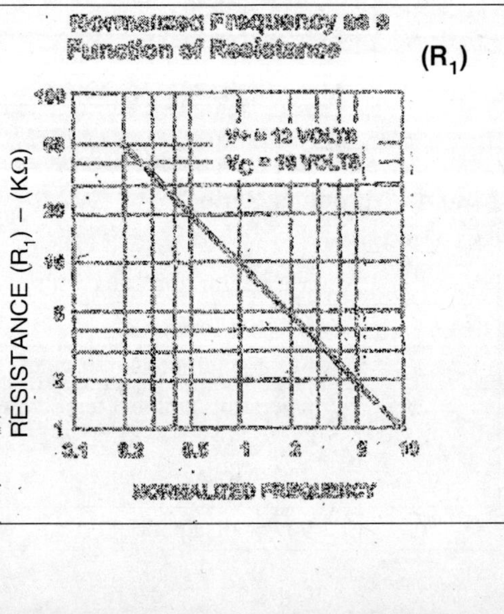

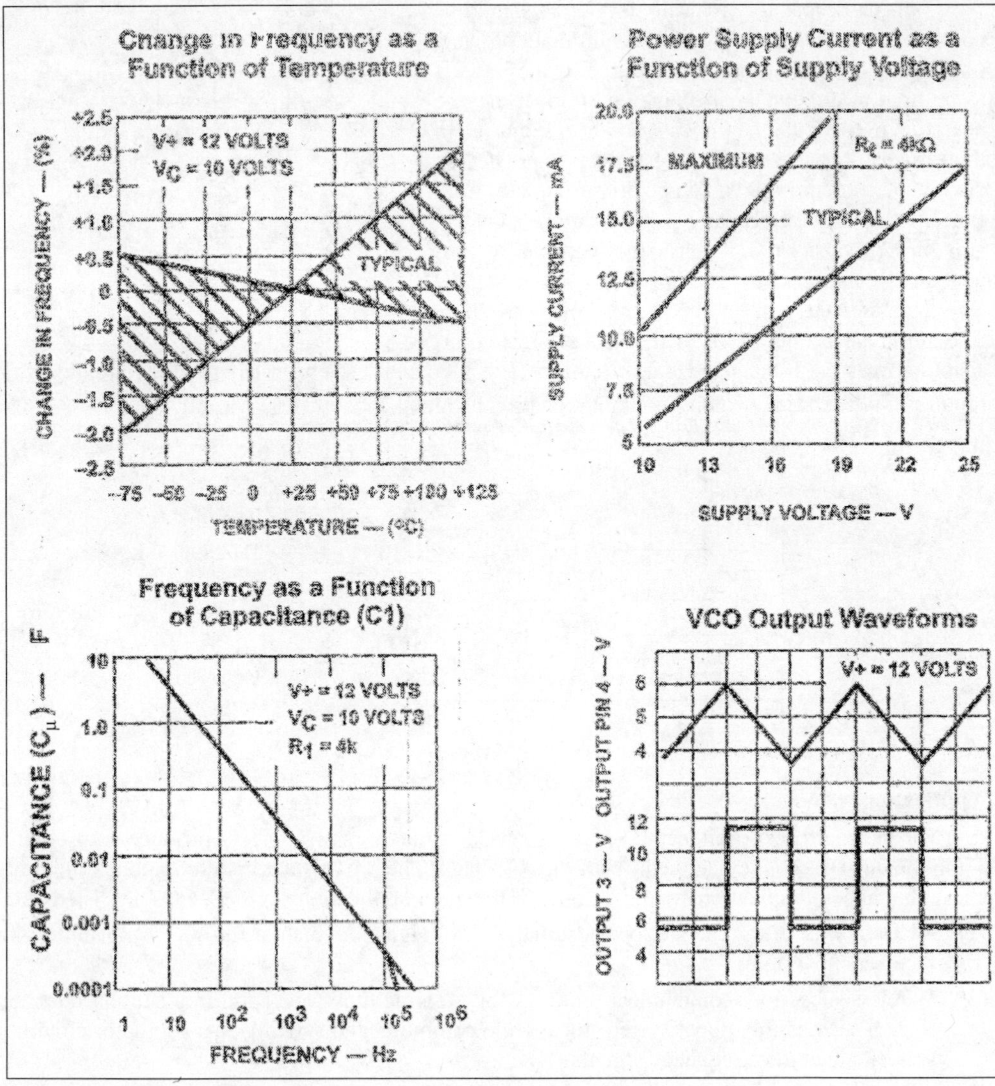

Operating Instructions

The NE/SE566 Function Generator is a general purpose voltage-controlled oscillator designed for highly linear frequency modulation. The circuit provides simultaneous square wave and triangle wave outputs at frequencies up to 1MHz. A typical connection diagram is shown in Figure 1. The control terminal (Pin 5) must be biased externally with a voltage (V_C) in the range

$$V_+ \leq V_C \leq V_+$$

where, V_{CC} is the total supply voltage. In Figure 1, the control voltage is set by the voltage divider formed with R_2 and R_3. The modulating signal is then AC coupled with the capacitor C_2. The modulating signal can be direct coupled as well, if the appropriate DC bias voltage is applied to the control terminal. The frequency is given approximately by

$$f_O = \frac{2[(V+) - (V_C)]}{R_1 C_1 V+}$$

and R_1 should be in the range $2\ k\Omega < R_1 < 20\ k\Omega$.

A small capacitor (typically 0.001μF) should be connected between Pins 5 and 6 to eliminate possible oscillation in the control current source. If the VCO is to be used to drive standard logic circuitry, it may be desirable to use a dual supply as shown in Figure 2. In this case the square wave output has the proper DC levels for logic circuitry. RTL can be driven directly from Pin 3. For DTL or TTL gates, which require a current sink of more than 1mA, it is usually necessary to connect a 5kW resistor between Pin 3 and negative supply. This increases the current sinking capability to 2mA. The third type of interface shown uses a saturated transistor between the 566 and the logic circuitary. This scheme is used primarily for TTL circuitry which requires a fast fall time (<50ns) and a large current sinking capability.

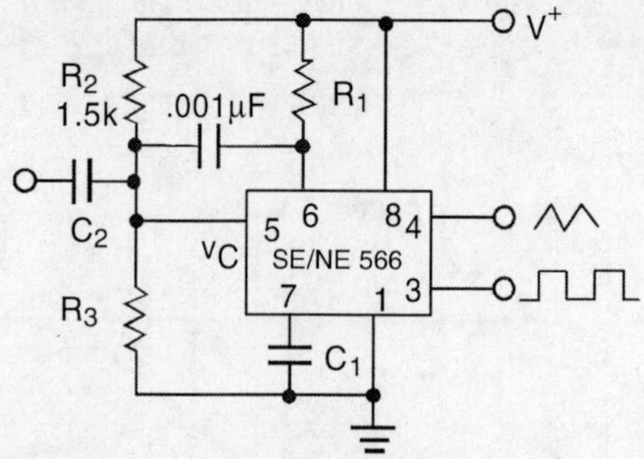

Fig. 26-37.

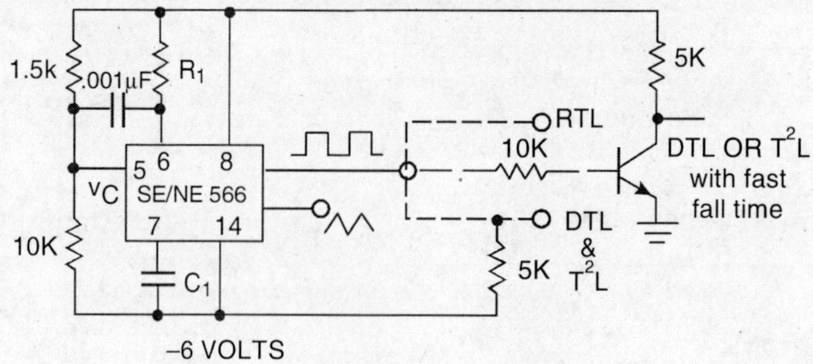

Fig. 26-38.

Applications of VCOs

Voltage controlled oscillators (VCOs) are used in many applications to produce an oscillating signal having a frequency defined by an input voltage. The VCO is a critical component in almost every digital communication systems. Some of the main applications of the VCOs are listed here :

(1) VCOs are used to generate local oscillator (LC) signals, which are used by transmitter and receiver subsystems.

(2) Wireless subscriber communication units, for example those operating in a cellular telephone system such as the global system for mobile communication (GSM) use oscillator circuits to generate the radio frequency signals.

(3) VCOs are used within many synthesiser and tuner circuits such as those found in TVs and in wireless communication sets.

(4) A high frequency VCO is extremely important for applications such as processor clock generation, distribution, and frequency synthesis.

(5) VCO is a main part in phase-locked used loop (PLL) circuits. These PLL circuits are used in many types of measurements electromechanical control, microprocessor and communication applications.

Phase-locked loop

Phase-locked loop (abbreviated as PLL) is a control system that generates a periodic output signal that has a constant phase relationship with a periodic input signal. A phase-locked loop circuit responds to both the frequency and the phase of the input signals automatically raising or lowering the frequency of a controlled oscillator until it is matched to the reference in both frequency and phase.

Phase-locked loop circuit is widely used in radio, telecommunications, computers and other electronic applications where it is desired to stabilize a generated signal or to detect signals in the presence of noise. Since, an integrated circuit can hold a complete phase-locked loop building block, the technique is widely used in modern electronic devices, with signal frequencies from a fraction of a cycle per second upto many gigahertz.

The circuit for a phase-locked loop comprises a voltage-controlled oscillator (VCO), a phase comparator, an amplifier and a low pass filter as shown in Fig. 26-39.

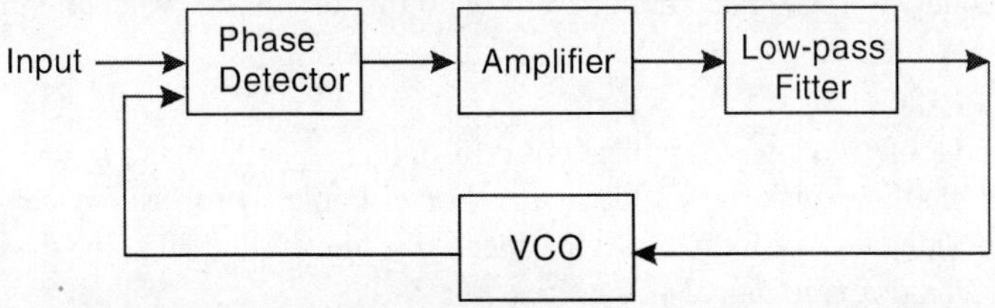

Fig. 26-39 *Block diagram of a typical phase-locked loop.*

The VCO is already discussed in Art. 26-18 the amplifier is discussed in chapter-16. Now we shall discuss about the phase detector circuit.

Phase detector : The phase detector is core element of a phase locked loop. It gives the phase difference in the loop to be detected and the resultant error voltage to be produced. Detecting the phase difference is very important in many applications such as motor control, servo mechanisms and demodulations.

There is a variety of different circuit, that can be used as phase detectors, some use analogue techniques, while other use digital techniques. The phase detectors may be classified into two categories namely

(*a*) Phase only sensitive detector

(*b*) Phase - frequency detector

Now we will discuss about these two types in the following pages.

(*a*) ***Phase only sensitive detector :*** This simply produces an output that is proportional to the phase difference between the two signals. When the phase difference between the two incoming signals is steady it produces a constant voltage. When there is a frequency difference between the two signals it produces a varying voltage. A familiar example of phase detector is the exclusive OR (X OR) gate. Fig. 26-40 shows the operation of the phase detector. One of the inputs of an Exclusive - OR gate corries the reference frequency signal (first) while the other input is a feedback signal. Since, the output of an Exclusive OR gate produces logic '1' when the input are difference and a logic '0' When its inputs are the same, the wave form at the output of Exclusive - OR gate is as shown in fig.

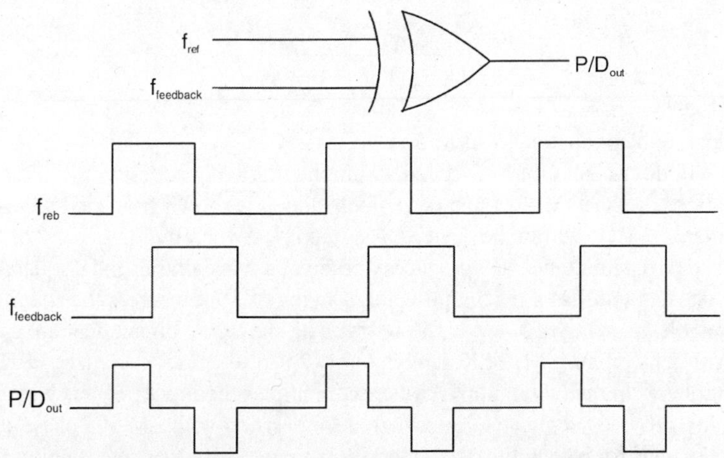

Fig. 26-40 *Illustrating the operation of a phase detector.*

Phase frequency sensitive detector

A phase frequency sensitive detector is a digital logic circuit that acts as our extended version of the phase detector. It generates an output which contains a frequency error between the two input signals. These detectors are normally used where the signal to noise ratios are very high and signal level transitions are well defined and predictable. Normally these phase frequency detectors are made up of D latches, NAND gate and a tristate gate that drives the voltage-controlled oscillator (VCO).

Applications of PLL

Phase-locked loops have wide applications in different areas such as communications, wireless systems, digital circuit and disk drive electronics. For example, wireless transceivers must still be able to generate a wide range of frequencies in order to up convert the out going data for transmission and down convert the received signals for processing.

Although there are numerous applications of PLL in the electronics engineering but following are important from the subject point of view.

- Data and Tape Synchronization
- FSK Modulation
- Frequency Synthesizer
- Frequency Multiplication and division
- Telemetry receivers
- Coherent demodulators
- Robotics and radio control.
- Modems
- FM Demodulation
- Tone decoding
- SCA Demodulation
- Signal regeneration
- Satellite

Voltage controlled oscillator

Within the phase locked loop circuit, the performance of the voltage controlled oscillator (VCO) is of paramount importance. In order that the PLL can meet its full specification, a well designed oscillator is essential. Designing a new VCO requires careful designing and the following important criteria.

(*i*) Tuning range
(*ii*) Tuning gain-tuning shift for a given tuning voltage range
(*iii*) Low phase noise.

To know more about the VCO and its operation please refer to article.

Commercially available PLL-LM565

There are different PLL devices available in the market. Here are some of manufactures and their model.

❑ Signetics	NE 560 to NE 567
❑ Micro electronics	MC 4046 COS/MOS
❑ NET Electronics	NTE 989
❑ National Semiconductors	LM 565

Now we will discuss about LM 565 in detail in the following pages.

Specifications datasheet of LM 565 phase-locked loop is shown from next pages. The voltage controlled oscillator (VCO) within the device has a maximum operating frequency of 500 kHz. The VCO within the device simultaneously produces a triangular wave output and a square-wave output. The triangular-wave output has a maximum voltage swing of $3V_{pp}$ whereas the square-wave output has a maximum voltage swing of $5.4\ V_{pp}$. The device data shows the circuit diagrams for (*i*) 2.4 kHz synchronous AM demodulator, (*ii*) FSK demodulator (2.025 to 2.225 kHz), (*iii*) FSK demodulator with dc restoration (*iv*) frequency multiplier where it multiples the input frequency by a factor of 10. The datasheet also provides application information for desising with LM 565 phase-locked loop. It also provides equations for calculating the important parameters such as free-running frequency, the loop-the natural frequency and damping factor of a loop-filter.

LM565/LM565C
Phase Locked Loop

General Description

The LM565 and LM565C are general purpose phase locked loops containing a stable, highly linear voltage controlled oscillator for low distortion FM demodulation, and a double balanced phase detector with good carrier suppression. The VCO frequency is set with an external resistor and capacitor, and a tuning range of 10:1 can be obtained with the same capacitor. The characteristics of the closed loop system—bandwidth, response speed, capture and pull in range—may be adjusted over a wide range with an external resistor and capacitor. The loop may be broken between the VCO and the phase detector for insertion of a digital frequency divider to obtain frequency multiplication. The LM565H is specified for operation over the "55°C to +125°C military temperature range. The LM565CN is specified for operation over the 0°C to +70°C temperature range.

Features

- 200 ppm/°C frequency stability of the VCO
- Power supply range of ±5 to ±12 volts with 100 ppm/% typical
- 0.2% linearity of demodulated output
- Linear triangle wave with in phase zero crossings available
- TTL and DTL compatible phase detector input and square wave output
- Adjustable hold in range from ±1% to > ±60%

Applications

- Data and tape synchronization
- Modems
- FSK demodulation
- FM demodulation
- Frequency synthesizer
- Tone decoding
- Frequency multiplication and division
- SCA demodulators
- Telemetry receivers
- Signal regeneration
- Coherent demodulators

Connection Diagrams

Metal Can Package

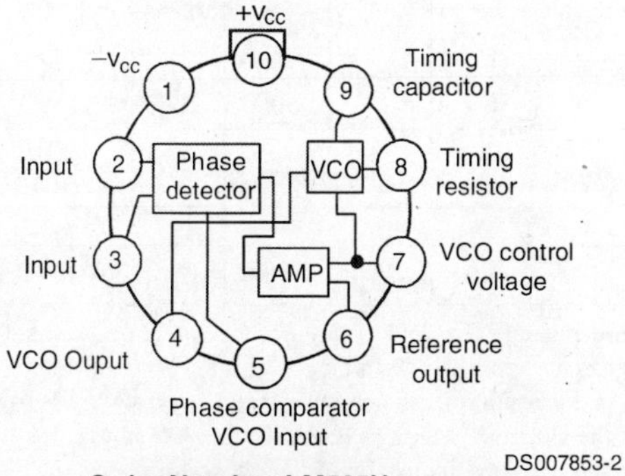

Order Number LM565H
See NS Package Number H10C

Fig. 26-41

Dual-in-Line Package

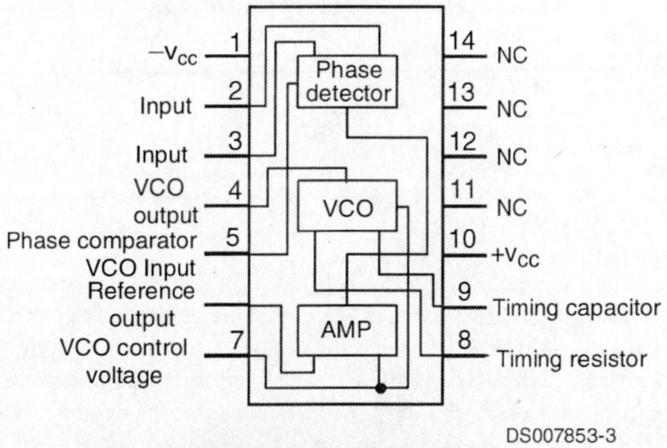

Order Number LM565CN
See NS Package Number N14A

Fig. 26-42

Absolute Maximum Ratings (Note 1)

If Military/Aerospace specified devices are required, please contact the National Semiconductor Sales Office/ Distributors for availability and specifications.

Supply Voltage ±12V
Power Dissipation (Note 2) 1400 mW
Differential Input Voltage ±1V

Operating Temperature Range
LM565H –55°C to +125°C
LM565CN 0°C to +70°C
Storage Temperature Range –65°C to +150°C
Lead Temperature
(Soldering, 10 sec.) 260°C

Electrical Characteristics

AC Test Circuit, TA = 25°C, VCC = ±6V

Parameter	Conditions	ML565			LM565C			
		Min	Typ	Max	Min	Typ	Max	UNIT
Power Supply Current			8.0	12.5		8.0	12.5	mA
Input Impedance (Pins 2, 3)	$-4V < V_2, V_3 < 0V$	7	10			5		kΩ
VCO Maximum Operating Frequency	$C_o = 2.7$ pF	300	500		250	500		kHz
VCO Free-Running Frequency	$C_o = 1.5$ nF $R_o = 20$ kW $f_o = 10$ kHz	–10	0	+10	–30	0	+30	%
Operating Frequency Temperature Coefficient			–100			–200		ppm/°C
Frequency Drift with Supply Voltage			0.1	1.0		0.2	1.5	%/V
Triangle Wave Output Voltage		2	2.4	3	2	2.4	3	Vp-p
Triangle Wave Output Linearity			0.2			0.5		%
Square Wave Output Level		4.7	5.4		4.7	5.4		Vp-p
Output Impedance (Pin 4)			5			5		kΩ
Square Wave Duty Cycle		45	50	55	40	50	60	%
Square Wave Rise Time			20			20		ns
Square Wave Fall Time			50			50		ns
Output Current Sink (Pin 4)		0.6	1		0.6	1		mA
VCO Sensitivity	fo = 10 kHz		6600			6600		Hz/V
Demodulated Output Voltage (Pin 7)	±10% Frequency Deviation	250	300	400	200	300	450	mVp-p
Total Harmonic Distortion	±10% Frequency Deviation		0.2	0.75		0.2	1.5	%
Output Impedance (Pin 7)			3.5			3.5		kΩ
DC Level (Pin 7)		4.25	4.5	4.75	4.0	4.5	5.0	V
Output Offset Voltage $\|V_7 - V_6\|$			30	100		50	200	mV
Temperature Drift of |V7 – V6 |			500			500		µV/°C
AM Rejection		30	40			40		dB
Phase Detector Sensitivity K_D			0.68			0.68		V/radian

Note 1: Absolute Maximum Ratings indicate limits beyond which damage to the device may occur. Operating Ratings indicate conditions for which the device is functional, but do not guarantee specific performance limits. Electrical Characteristics state DC and AC electrical specifications under particular test conditions which guarantee specific performance limits. This assumes that the device is within the Operating Ratings. Specifications are not guaranteed for parameters where no limit is given, however, the typical value is a good indication of device performance.

Note 2: The maximum junction temperature of the LM565 and LM565C is +150°C. For operation at elevated temperatures, devices in the TO-5 package must be derated based on a thermal resistance of +150°C/W junction to ambient or +45°C/W junction to case. Thermal resistance of the dual-in-line package is +85°C/W.

Typical Performance Characteristics

Power Supply Current as a Function of Supply Voltage

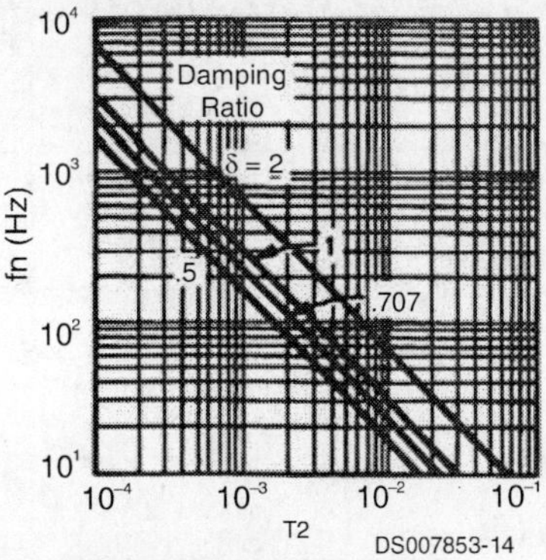

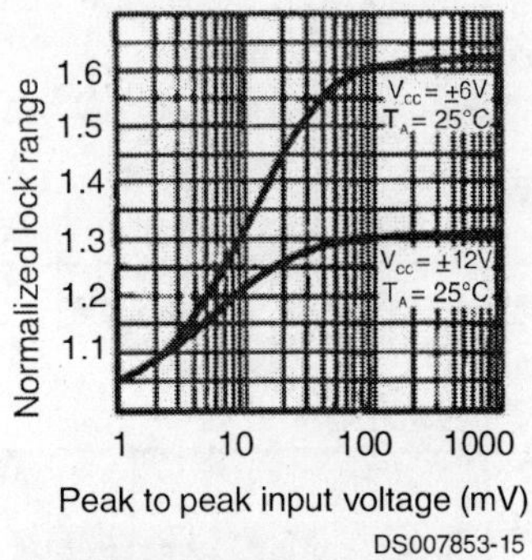

VCO Frequency Function of Temperature

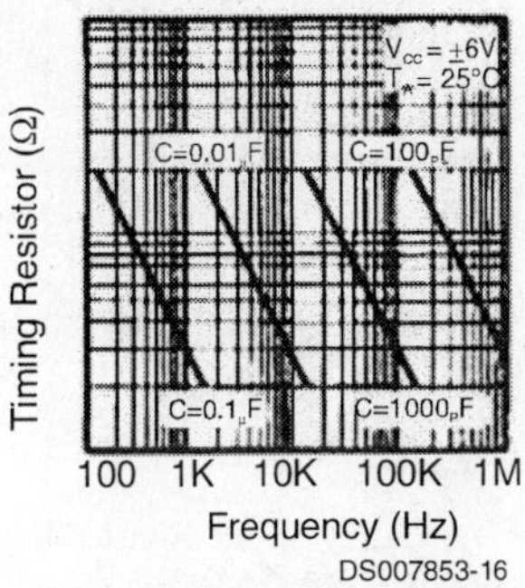

Oscillator Output Waveforms

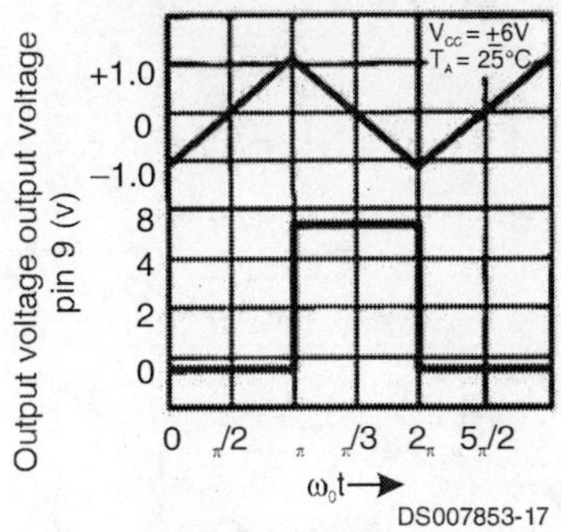

Phase Shift vs Frequency

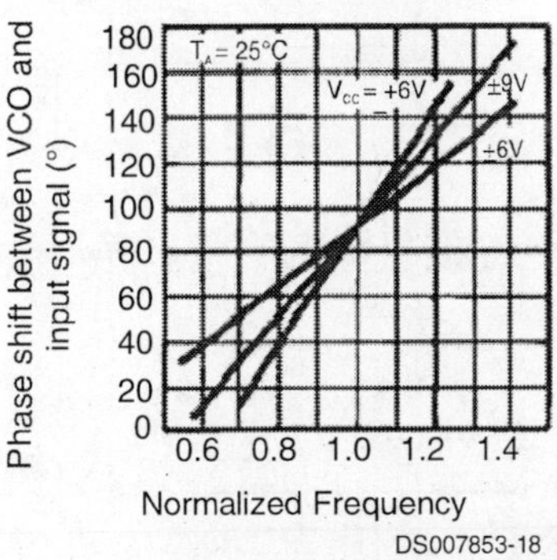

VCO Frequency as a Function of Temperature

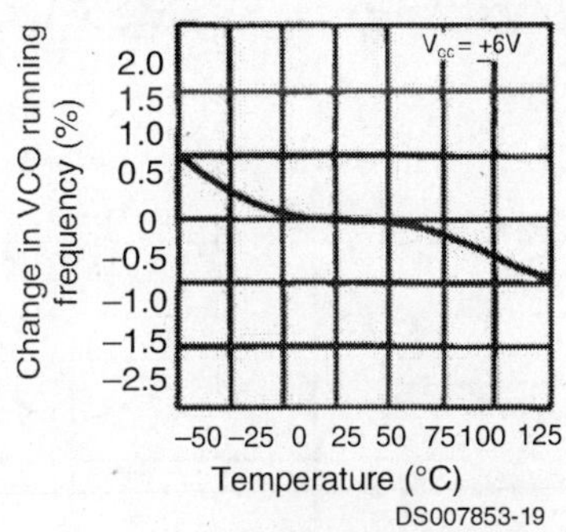

Loop Grain vc Load Resistance

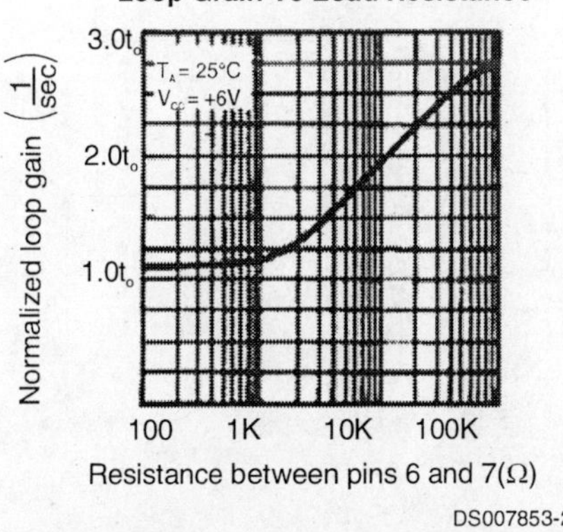

Hold in Range as a Function of R_{6-7}

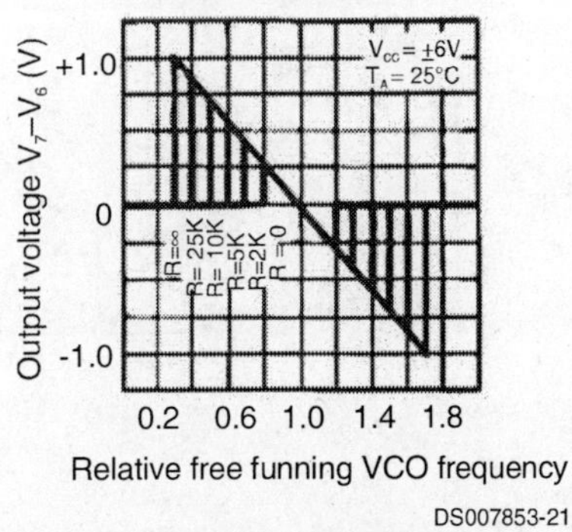

Schemmatic Diagram

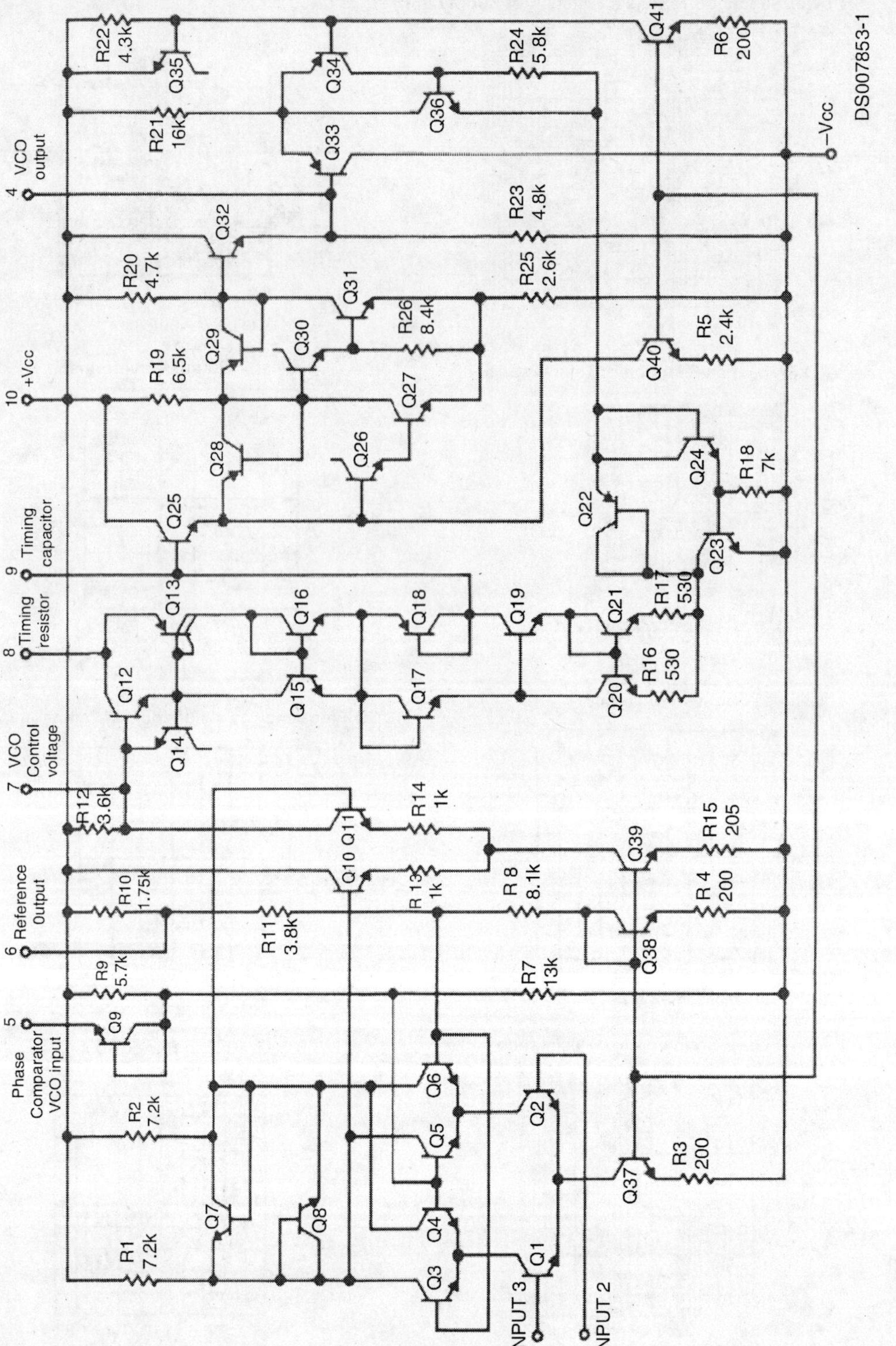

Fig. 29.43

AC Test Circuit

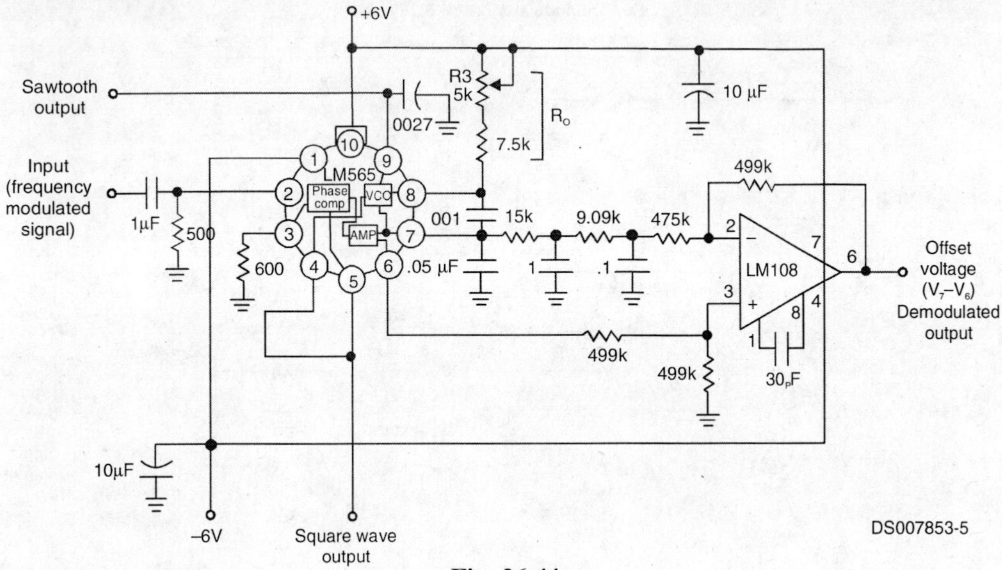

Fig. 26-44

Note : S_1 open for output offset voltage (V_7–V_6) measurement.

Typical Applications

2400 Hz Synchronous AM Demodulator

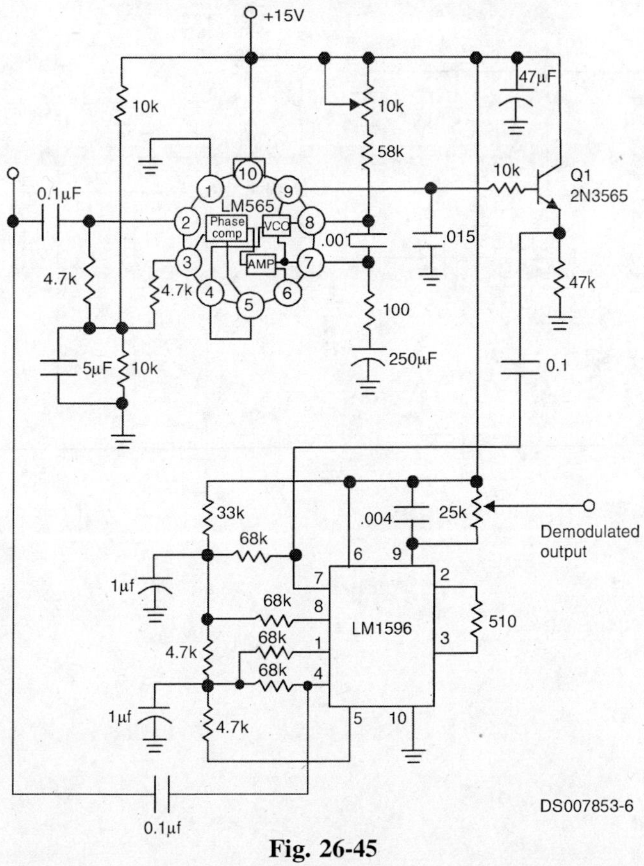

Fig. 26-45

Typical Applications (continued)

FSK Demodulator (2025-2225 cps)

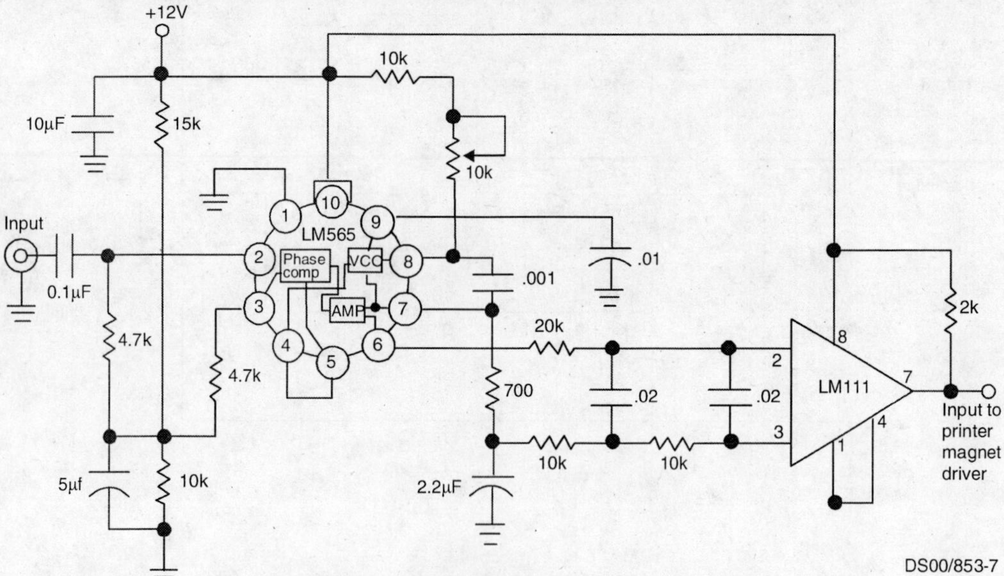

Fig. 26-46

FSK Demodulator with DC Restoration

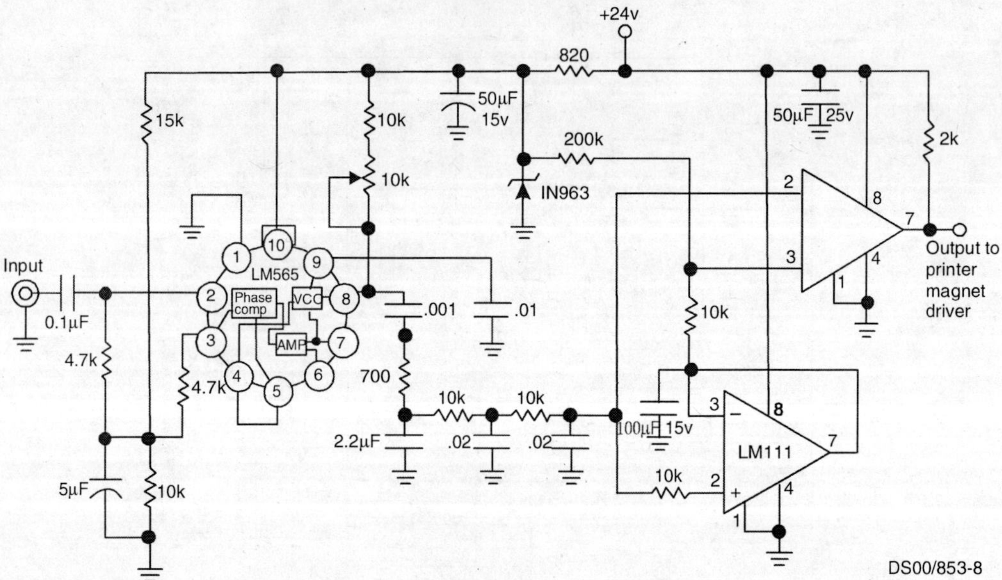

Fig. 26-47

Typical Applications (continued)

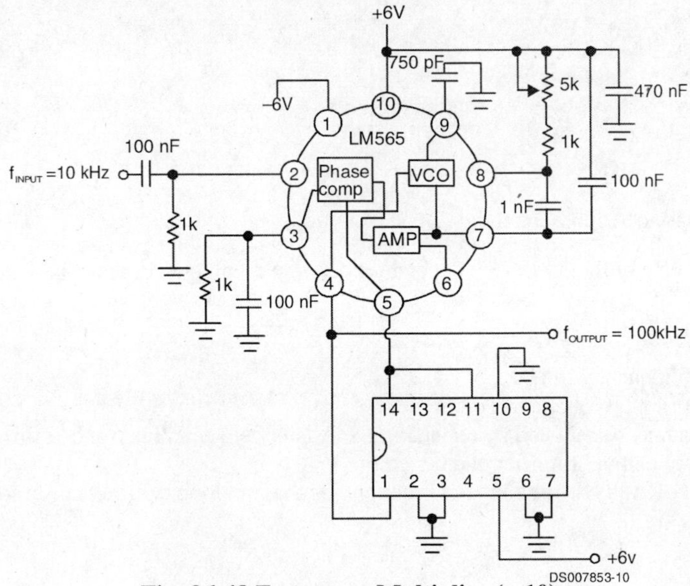

Fig. 26-48 Frequency Multiplier (× 10)

IRIG Channel 13 Demodulator

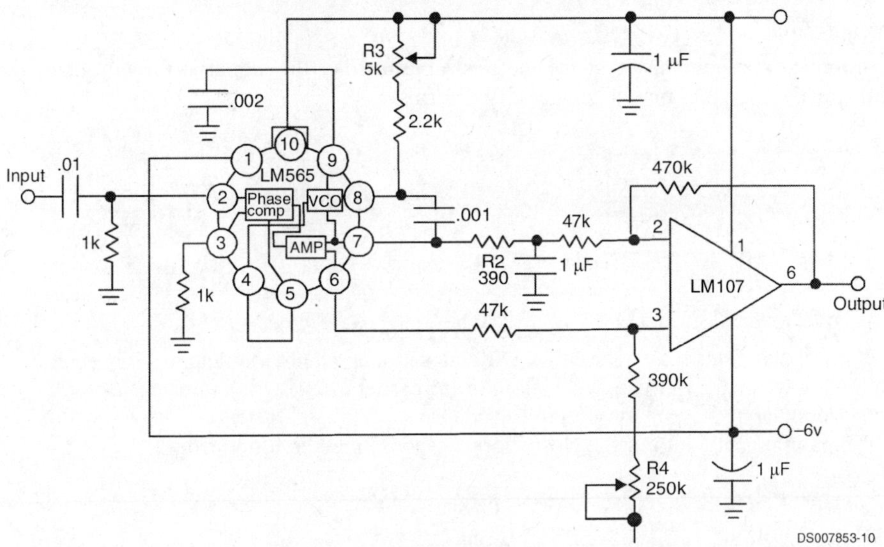

Fig. 26-49

Applications Information

In designing with phase locked loops such as the LM565, the important parameters of interest are :

Free running Frequency

$$f_0 \cong \frac{0.3}{R_o C_o}$$

LOOP GAIN: relates the amount of phase change between the input signal and the VCO signal for a shift in input signal frequency (assuming the loop remains in lock). In servo theory, this is called the "velocity error coefficient."

$$\text{Loop gain} = K_0K_D = \frac{1}{\text{sec}}$$

$$K_0 = \text{oscillator sensitivity} \quad \frac{\text{radians / sec}}{\text{volt}}$$

$$K_D = \text{phase detector sensitivity} \quad \frac{\text{volts}}{\text{radian}}$$

The loop gain of the LM565 is dependent on supply voltage, and may be found from :

$$K_0K_D = \frac{33.6\, f_o}{V_C}$$

f_o = VCO frequency in Hz
V_c = total supply voltage to circuit

Loop gain may be reduced by connecting a resistor between pins 6 and 7; this reduces the load impedance on the output amplifier and hence the loop gain.

HOLD IN RANGE: the range of frequencies that the loop will remain in lock after initially being locked.

$$f_H = \pm \frac{8\, f_0}{V_C}$$

f_o = free running frequency of VCO
V_c = total supply voltage to the circuit

The Loop Filter

In almost all applications, it will be desirable to filter the signal at the output of the phase detector (pin 7); this filter may take one of two forms:

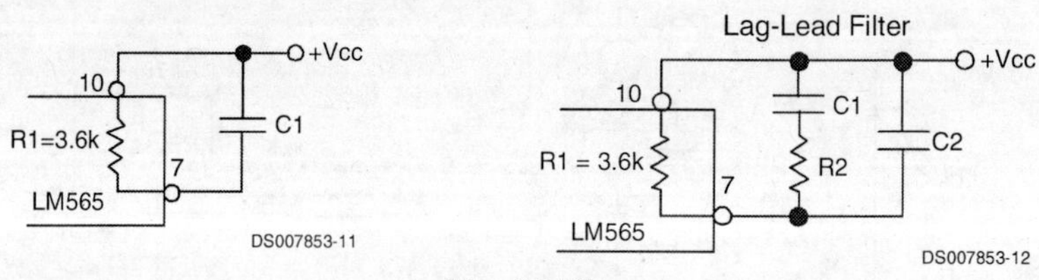

Fig. 26-50

A simple lag filter may be used for wide closed loop bandwidth applications such as modulation following where the frequency deviation of the carrier is fairly high (greater than 10%), or where wideband modulating signals must be followed.

The natural bandwidth of the closed loop response may be found from :

$$f_n = \frac{1}{2\pi}\sqrt{\frac{K_0K_D}{K_1C_1}}$$

Associated with this is a damping factor :

$$\delta = \frac{1}{2}\sqrt{\frac{1}{R_1C_1K_0D_D}}$$

For narrow band applications where a narrow noise bandwidth is desired, such as applications involving tracking a slowly varying carrier, a lead lag filter should be used. In general, if $1/R_1C_1 < K_o K_D$, the damping factor for the loop becomes quite small resulting in large overshoot and possible instability in the transient response of the loop. In this case, the natural frequency of the loop may be found from

$$f_n = \frac{1}{2\pi}\sqrt{\frac{K_0K_D}{\tau_1 + \tau_2}}$$

$$\tau_1 + \tau_2 = (R_1 + R_2)C_1$$

R_2 is selected to produce a desired damping factor d, usually between 0.5 and 1.0. The damping factor is found from the approximation:

$$\delta = \pi\tau_2 f_n$$

These two equations are plotted for convenience.

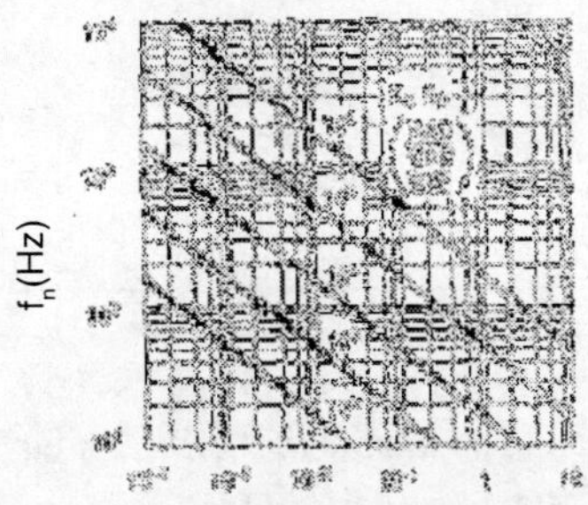

Filter Time Constant vs Natural Frequency

Capacitor C_2 should be much smaller than C1 since its function is to provide fitering of carrier. In general $C_2 \le 0.1C_1$.

SUMMARY

In this chapter you have learned that :

1. Transistor can be operated as a switch when the switch is closed, $i_C = V_{CC}/R_C$ and $V_{CE(sat)}$ and when the switch is open, $i_C = 0$ and $V_{CE} = V_{CC}$.
2. When transistor is operated as switch has certain time delays : The time delays are (*i*) rise time (*ii*) fall time and (*iii*) storage time delay
3. There are three basic types of multivibrators : (*i*) astable (*ii*) monostable and (*iii*) bistable multivibrators
4. The astable multivibrator is also called free running relaxation oscillator. It is commonly used to generate a square waveform which is used as a clock in electronic circuits. The frequency of oscillation for a transistor astable multivibrator is given by,

$$F = \frac{1}{0.69\,(R_1\,C_1 + R_2\,C_2)}$$

$$= \frac{1}{1.38\,RC} \qquad (\text{if } R_1 = R_2\,R,\ C_1 = C_2 = C)$$

To ensure oscillations,

$$R_1 \le h_{FE(\min)} \cdot R_{C_1}$$

$$R_2 \le h_{FE(\min)} \cdot R_{C_2}$$

The astable multivibrator has no stable states.

5. A monostable multivibrator is also called one-shot. It is used to generate a gating pulse whose width can be controlled. The monostable multivibrator has one stable state.
6. A bistable multivibrator is also called a flip-flop or a latch. It has two stable states and can stay in any one of these states as long as the power is supplied. It changes to other state only when it receives a trigger from outside
7. The 555-timer IC can be used in monostable as well as in astable mode. In monostable mode, the pulse width duration is obtained by,

$$t_p = 1.1\,R_1\,C_1$$

When used in astable mode, the frequency of oscillation is given by,

$$f = \frac{1.44}{(R_1 + 2R_2)\,C_1}$$

8. Blocking oscillator is an oscillator which is used to produce high current pulses of short duration.

GLOSSARY

Blocking Oscillator : a circuit used to produce high current pulse of short duration.
Astable multivibrator : a device or a circuit used to produce a square waveform for clock generators.
Monostable multivibrator : a device or a circuit used to produce a gating pulse of desired width. It needs a trigger input.
Bistable multivibrator : a device or a circuit used as a storage device in digital circuits. It is also called flip-flop or latch.
Schmitt trigger : a device or a circuit for producing square wave output. It is designed specifically for responding to slowly varying inputs.

NUMERICAL PROBLEMS

1. What is the frequency of an astable multivibrator having the component values of $R_1 = R_2 = R = 10$ kΩ and $C_1 = C_2 = C = 1000$ pF? (**Ans.** 72.5 kHz)
2. Determine the value of resistors to be used to provide a symmetrical oscillation of 100 kHz, if $C_1 = C_2 = C = 330$ pF. (**Ans.** 22 kΩ)
3. Determine the period and frequency of oscillation for an astable multivibrator with component values of $R_1 = R_2 = 4.7$ kΩ and $C_1 = C_2 = 270$ pF. (**Ans.** 1.75 μs and 57.1 kHz)
4. Select the value of R and C for the monostable multivibrator to give an output pulse width of 400 ns. (**Ans.** Choose $C = 100$ pF, then $R = 5.78$ kΩ)
5. Refer to Fig. 26-51. For the self-biased binary NPN transistor, worst case values of

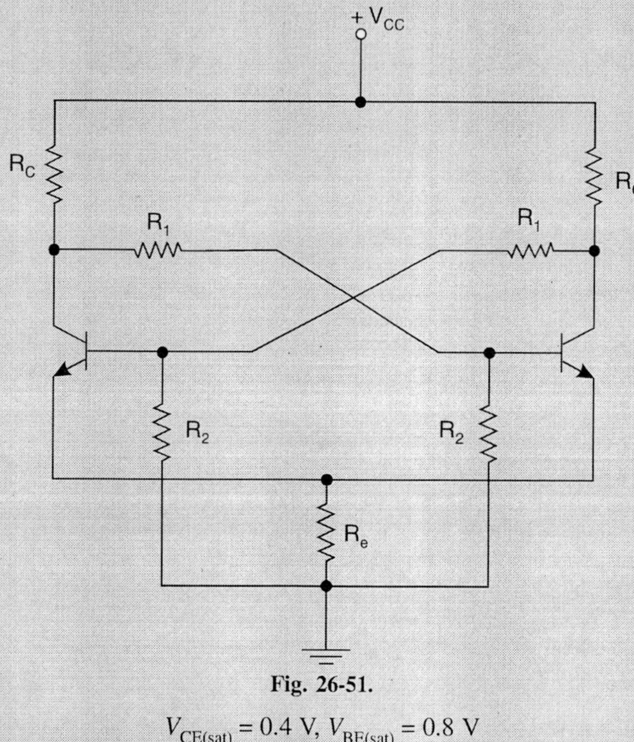

Fig. 26-51.

$$V_{CE(sat)} = 0.4 \text{ V}, V_{BE(sat)} = 0.8 \text{ V}$$

and zero base -to-emitter cut off. The circuit parameters are $V_{CC} = 20$ V, $R_1 = 30$ k, $R_e = 390$ Ω, $R_C = 4.7$ k, $R_2 = 15$ k.

(*a*) Find stable state current and voltages.

(*b*) Find min. value of h_{FE} required to give the values in part (*a*).

(*c*) As the temperature is increased, what is the max. value to which I_{CBO} can increase before the condition is reduced where neither transistor is off ?

(*Nagpur Univ. Winter* 2003)

6. The fixed bias binary uses PNP Si transistors with worst case (maximum) values of $V_{CE(sat)} = -0.5$ V, $V_{BE(sat)} = -1.0$ V, $I_{CBO} = -10$ nA at 25°C and zero base to emitter voltage at cut-off . The circuit parameters are $V_{CC} = V_{BB} = 6$ V, $R_C = 1.2$ k, $R_1 = 4.7$ k and $R_2 = 27$ k.

(*a*) Find $h_{FE(min)}$ and verify that one transistor is OFF and the other is ON. Find the stable-state currents and voltages.

(*b*) If the reverse saturation current doubles every 10° C, what is the maximum temperature at which one transistor will remain OFF ? (*Nagpur University. Winter* 2003)

7. The emitter coupled monostable multivibrator uses PNP transistor with $h_{FE} = 40$. Neglect the voltages of a forward biased junction and assume $r_{bb}' = 0$. The circuit parameters are: $V_{CC} = 12$ V, $R_{C_1} = 3.6$ k, $R_{C_2} = 2.2$ k, $R_e = 1$ k and $R = 68$ k.

The bias V is adjusted to be midway between V_{min} and V_{max} A trigger is applied at $t = 0$ Calculate all the voltage levels and plot the waveforms. (*Nagpur Univ. Winter* 2003)

8. The self-biased binary used NPN silicon transistors having worst case (maximum) values of $V_{CE(sat)} = 0.4$ V and $V_{BE(sat)} = 0.8$V and zero base-to-emitter voltage for cut-off. The usual circuit parameters of $V_{CC} = 20$ V, $R_C = 4.7$ kΩ; $R_1 = 30$ kΩ; $R_2 = 15$ k Ω and $R_C = 390$ Ω.

(*a*) Draw a neat circuit and insert the parameters indicated and find the stable state currents and voltages.

(*b*) Find the minimum values of h_{FE} required to give the values in part (*a*).

(*c*) As the temperature is increased, what is the maximum values to which I_{CBO} can increases before the condition is reached where neither transistor is OFF ?

(*AMIE, Winter* 1993)

9. Two monostables A and B triggered simultaneously produces pulse widths T_A and T_B respectively. Give a suitable circuit which will give the output pulse width $| T_A - T_B |$. Also give a circuit which will indicate through LEDle which monostable has a longer pulse width. (*AMIE., Summer* 2000)

10. A schmitt trigger circuit is to be designed with a VTP of 5V and LTP of 3 V. The NPN silicon transistors employed have $h_{FE(min)} = 100$ and I_C to be 2 mA. The available supply voltage is 12 V. Design a suitable circuit. (*AMIE., Summer* 1999)

11. Calculate the frequency and duty cycle of the output of Astable multivibrator and drawn the waveform obtained across the capacitor, C_1 Refer to fig. 26-52.

(*Bangalore Univ. B.Sc April/May* 2001)

Ans. $f = 85$ kHZ, duty cycle = 59.5%.

12. Explain the working of the astable circuit show in Fig. 26-53. Assume that the saturation voltages for the Op-Amp are $\pm V_{CC}$ Derive an expression for the output pulse width of the monostable

Drawn the waveform when triggers are applied at times t_1 and $t_2 < t$.

Suggest a suitable trigger circuit if an NPN transistor is used for the, switch S.

(*AMIE., Summer* 2000)

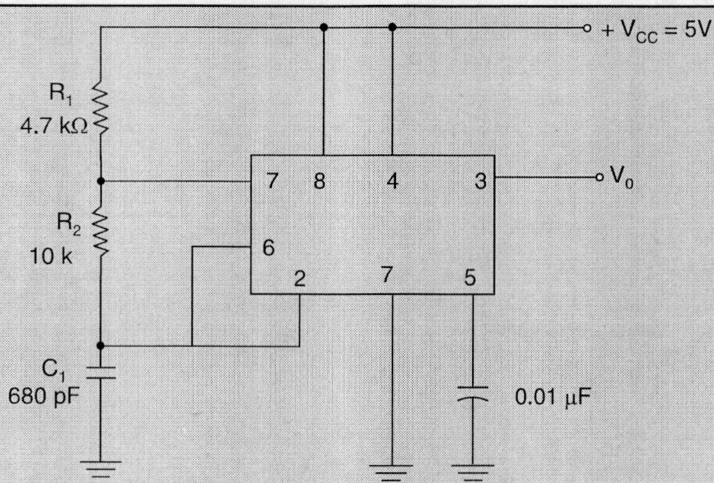

Fig. 26-52.

(*AMIE., Summer 2000*)

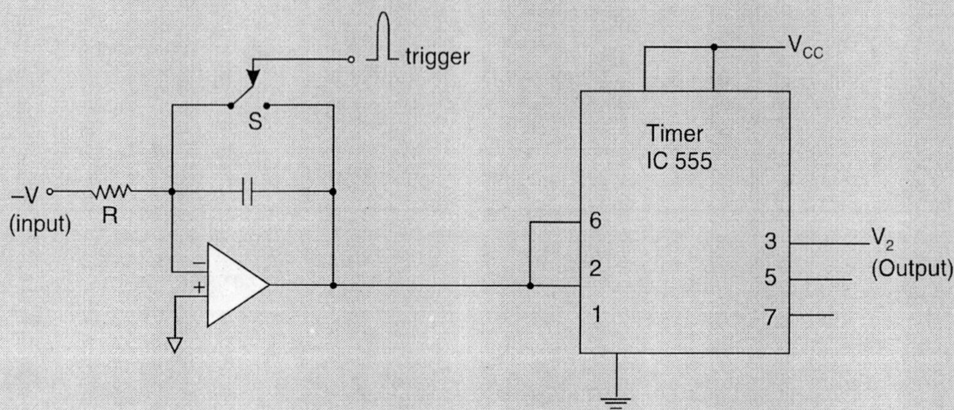

Fig. 26-53.

DESCRIPTIVE QUESTIONS

1. What is a nonsinusoidal oscillator? Explain it briefly.

2. Explain the switching action of a transistor.

3. What is a multivibrator? Explain the difference between the three types of multivibrators.

4. With a neat sketch, explain the working of an astable multivibrator.

5. Draw the switching waveforms for the astable multivibrator.

6. What is monostable multivibrator? Explain its working with the help of waveforms.

7. With a neat sketch, explain the operation of a bistable multivibrator.

8. How a Schmitt trigger is different from a multivibrator?

9. What is a blocking oscillator? Explain how it can be used to generate high-current pulses of short duration.

10. Draw any one type of a Schmitt trigger circuit and explain with waveforms.

11. Drawn the circuit of an emitter coupled transistor monostable multivibrator and explain its operation. Show how this multivibrator can be modified to control the width of the output pulse. (*AMIE., Summer* 1999)

12. Draw the circuit of an astable multivibrator and explain its working. (*AMIE.*, 1996)

13. Draw the circuit of a Schmitt trigger and explain its working. Show how the hysteresis loop in the circuit stabiliges the operation of Schmitt trigger. (*AMIE., Winter* 1996)

14. Explain the working of a self-biased transistor bistable multivibrator circuit. What is the purpose of the speed-up capacitors ? (*AMIE, Winter* 98)

15. Explain the working of a collector-coupled monostable multivibrator. Give expression for its gate width. (*AMIE., Winter* 1998)

16. Draw the circuit of a Schmitt trigger and explain its principle of operation. Discuss how the hysteresis loop in the circuit stabilizes the operation of the Schmitt trigger. (*AMIE., Winter* 1997)

17. (*a*) Distinguish between a bistable and a monostable multivibrator.
(*b*) Discuss the principle of operation of a transistor bistable multivibrator.
(*c*) Discuss the principle of operation of transistor monostable multivibrator. (*AMIE., Winter* 1997)

18. Explain the operation of an astable multivibrator. (AMIE., Summer 1995)

19. Explain the hysteresis loop and hysteresis range associated with Schmitt trigger. How can the hysteresis voltage be reduced in the emitter coupled Schmitt trigger? (*AMIE., Summer* 1995)

20. Explain with the circuit diagram, the collector coupled monostable multivibrator. (*AMIE., Summer* 1995)

21. Sketch an emitter-coupled transistor Schmitt trigger circuit using NPN transistor and briefly explain its operation. Define the terms : upper trigger point (UTP), lower-trigger point (LTP) and hysteresis related to the Schmitt trigger circuit. (*AMIE., Summer* 1997)

22. Draw the circuit diagram of a collector-coupled mono and explain its working. (*AMITE., Summer* 1996)

23. Explain the working of self-biased transistor bistable multivibrator. What is a retriggerable monoshot ? (*AMIE., Summer* 1996)

24. (*a*) Explain the operation of a bistable multivibrator
(*b*) Give a typical triggering circuit by bistable multivibrator. (*AMIE., Summer* 1994)
(*c*) Draw the circuit diagram of a collector coupled astable multivibrator using transistors and explain its principle of operation. Draw the waveforms at the base and collector of one transistor. (*AMIE., Winter* 2001)

25. (*a*) Draw the schematic of DC based monostable circuit. Explain how the output pulse width is controlled Also examine the concepts of positive and negative triggering in respect of input-output relations in a monostable multivibrator.

26. (*a*) Compare a Schmitt trigger with a binary circuit.
(*b*) Draw a neat circuit diagram of a Schmitt trigger and explain the various methods of minimizing the hysteresis in the circuit.
(*c*) What are delay lines ? Describe their application areas. (*AMIE., Winter* 1994)

27. (*a*) Draw the internal operational circuit diagram of an IC timer 555.
(*b*) Explain the operation of an IC 555 as an astable multivibrator.
(*c*) Derive an expression for frequency of the output pulse waveform of the circuit.
(*d*) What are the two comparator reference voltage of IC 555 if $V_{CC} = +9$ volts.
(*AMIE., Winter* 1997)

28. (*a*) Sketch a transistor Schmitt trigger circuit and briefly explain its operation giving input output waveforms.
(*b*) Draw the circuit diagram of a 555 based astable multivibrator. Explain its operation. Also discuss about the changes in the circuit to generate a 50% duty cycle square waveform. (*AMIE., Winter* 2000)

29. (*a*) Draw an operational internal circuits of a timer IC 555.
(*b*) Show that the timer works as a Schmitt trigger when pins 2 and 5 are tied together to an input voltage (v_i) and the output (v_o) is taken at pin 3.

30. (*a*) Use a suitable integrator circuit in the feedback path of the above Schmitt trigger to make a pulse generator.
(*b*) Derive an expression for the frequency of the output pulse waveform.

31. Explain with the aid of a neat circuit diagram, the working of a triggered blocking oscillator with base timing. Derive an expression for the pulse duration, t_p. Enlist the drawbacks of the circuit. (*AMIE., Winter* 1997, 1998)

31. (*a*) Explain with a neat circuit diagram, how a blocking oscillator can be used to produce high current pulses of short duration.
(*b*) Enumerate the various methods of triggering a bistable multivibrator and describe one such method fully.
(*c*) Explain how a BJT can be used as a switch. (*AMIE., Winter* 1994)

32. Give the circuit schematic off 555-timer connected as an astable multivibrator. Describe its operation. (*Electronic Circuits, VTU June/July 2008*)

33. What is the other name of Astable Multivibrator.
(*Electronic Circuits–II, Anna University, Nov/Dec 2007*)

34. State the use of Schmitt trigger. (*Electronic Circuits–II, Anna University, Nov/Dec 2007*)

35. What is Blocking Oscillator? (*Electronic Circuits–II, Anna University, Nov/Dec 2007*)

36. Draw and explain the circuit of astable multivibrator and hence draw the waveforms at salient points. Derive expressions for frequency of oscillation.
(*Electronic Circuits - II, Anna University, Nov/Dec 207*)

37. Draw and explain the circuit of a Monostable multivibrator with suitable triggering arrangements. Give the waveforms at different modes.
(*Electronic Circuits–II, Anna University, Nov/Dec 2007*)

38. Derive an expression of the pulse width of Monostable multivibrator.
(*Electronic Circuits–II, Anna University, Nov/Dec 2007*)

39. State any two applications of Monostable multivibrator.
(*Electronic Circuits–II, Anna University, Nov/Dec 2007*)

40. With a neat circuit, explain the operation of Schmitt trigger.
(*Linear integrated Circuits, Anna University, April/May 2008*)

41. Discuss the operation of IC 555 as a monostable multivibrator. Draw the waveform and explain. (*Linear Integrated Circuits, Anna University, April/May 2008*)

42. Give the application of bistable multivibrator.

(*Electronic Circuits–II, Anna University, May/June 2007*)

43. What is the necessity for damping in a monostable-blocking oscillator?

(*Electronic Circuits–II, Anna University, May/June 2007*)

44. Explain the operation of collectro coupled astable multivibrator and also derive the expression of T for a collector astable multivibrator.

(*Electronic Circuits–II, Anna University, May/June 2007*)

45. What is Schmitt trigger? Explain the operation of Schmitt using neat circuit diagram. Given the mathematical analysis of the Schmitt trigger.

(*Electronic Circuits–II, Anna University, May/June 2007*)

46. With neat circuit diagram, explain the working of a monosatable blocking oscillator using emitter timing. Draw the equivalent circuit. Also derive the expression for pulse width.

(*Electronic Circuits–II, Anna University, May/June 2007*)

47. With neat diagram, explain the working of a Schmitt trigger circuit. Draw the corresponding wave forms. (*Electronic Circuit-II Anna University, May/June 2007*)

48. With neat circuit diagram, explain monostable blocking oscillators with base and emitter timing. Draw necessary wave forms.

(*Electronics Circuit–II Anna University, May/June 2007*)

49. Define rise time. (*Electronic Circuits–I, Anna University, Nov/Dec 2007*)

50. Define duty cycle (*Electronic Circuits–II, Anna University, May/June 2007*)

51. Give the relation between rise time and bandwidth.

(*Electronic Circuits–II, Anna University, May/June 2007*)

MULTIPLE CHOICE QUESTIONS

1. When a transistor is driven into saturation, its collector voltage is

(*a*) Zero (*b*) $V_{CE(sat)}$

(*c*) V_{CC} (*d*) None of the above

2. An astable multivibrator is also called free-running relaxation oscillator because it

(*a*) oscillates continuously between two unstable states

(*b*) requires no d.c. power for its operation

(*c*) runs and relaxes alternately

(*d*) none of the above

3. The frequency of oscillation of an astable multivibrator depends on

(*a*) transistor current gain (h_{FE}) (*b*) value of collector load resistors

(*c*) duration of the input pulse (*d*) values of *R* and *C*

4. A monostable multivibrator

(*a*) gives one output pulse for one input trigger pulse

(*b*) gives two output pulses for one input trigger pulse

(*c*) has one stable state

(*d*) both (*a*) and (*c*) above

5. A bistable multivibrator

(*a*) has two stable states (*b*) has two unstable states

(*c*) oscillates between two stable states without any trigger pulse
(*d*) is used for generating square wave

6. Non-sinusoidal wave forms
(*a*) are departments from sine waveform (*b*) have low mark-to-space ratio
(*c*) are much easier to generate (*d*) are unfit for digital operation.

7. A relaxation oscillator is one which
(*a*) has two stable states (*b*) relaxes indefinitely
(*c*) produces non-sinusoidal output (*d*) oscillates continuously.

8. A square pulse has a mark-to-space ratio of
(*a*) 1 : 1 (*b*) 1 : 2 (*c*) 2 : 1 (*d*) 1 : 4

9. Which of the following statement is WRONG ? In a multivibrator
(*a*) output is available continuously (*b*) feedback between two stages is 100%
(*c*) positive feedback is employed (*d*) when one transistor is ON, the other is OFF.

10. An monostable multivibrator circuit
(*a*) has no stable state
(*b*) gives two output pulses for one input trigger pulse
(*c*) returns to its stand-by states automatically
(*d*) has no energy-storage element.

11. A bistable multivibrator circuit
(*a*) has two unstable states
(*b*) has one energy-storage element
(*c*) switches between its two states automatically
(*d*) is not an oscillator.

12. The frequency of oscillation of an astable multivibrator depends mainly on the
(*a*) value of collector load resistors (*b*) *RC* values of the circuit
(*c*) value of transistor β (*d*) width of the input pulse.

13. An monostable multivibrator is frequently used
(*a*) in memory and timing circuits (*b*) for producing triangular waves
(*c*) in counting circuits (*d*) for regeneration of distorted waves.

14. A bistable multivibrator is a oscillator.
(*a*) triggered (*b*) free-running (*c*) sine-wave (*d*) sawtooth

15. A blocking oscillator
(*a*) is a triggered oscillator (*b*) is an amplifier with negative feedback
(*c*) generates sinusoidal waves (*d*) produces sharp and narrow pulse.

16. In an astable multivibractor
(*a*) $\beta = 1$ (*b*) $\beta A = 1$ (*c*) $\beta > 1$ (*c*) $\beta < 1$.

17. In order to generate a square wave from a sinusoidal input signal one can use
(*a*) Schmitt trigger circuit (*b*) Clipper and amplifiers
(*c*) Monostable multivibrators (*d*) both (*a*) and (*b*).
(*UPSC Engg Services* 2002)

18. In every practical oscillator, the loop-gain is slightly larger than unity and the amplitude of the oscillations is limited by the
(*a*) magnitude of the loop gain
(*b*) onset of non-linearity
(*c*) magnitude of the gain of the amplifier
(*d*) feedback transmission factor. (*UPSC Engg. Services* 2000)

19. In an astable multivibrator, the base resistor of 100 kΩ and capacitors of 100 pF are used. The frequency of output will be

(*a*) 1.2 kHz (*b*) 5.5 kHz (*c*) 6.9 kHz (*d*) 12.8kHz

(*AMIE., Winter* 1999)

20. For a Schmitt trigger invertor the upper and lower trip voltage are 3 V and 1 V and higher and lower states are 15 V and 2 V. The output for a sinusoidal input of 10 V peak will be between

(*a*) 1 V and 3 V (*b*) 2V and 15 V (*c*) 3 V and 15 V (*d*) 10 V and 15 V

(*AMIE., Summer* 1999)

21. A transistor used as a switches is operated in

(*a*) active region (*b*) cut-off region
(*c*) saturation region (*d*) cut-off and saturation region.

(*AMIE., Summer* 2001)

22. A 1 ms pulse can be stretched to 1 s pulse by using.

(*a*) an astable multivibrator (*b*) a monostable multivibrator
(*c*) a monostable multivibrator (*d*) a Schmitt trigger circuit

(*UPSC Engg Services* 2000)

23. A monostable multivibrator circuit show in Fig.26-54. The value of C would be nearly.

(*a*) 0.001 μF (*b*) 0.01 μF (*c*) 0.1μF (*d*) 1 μF

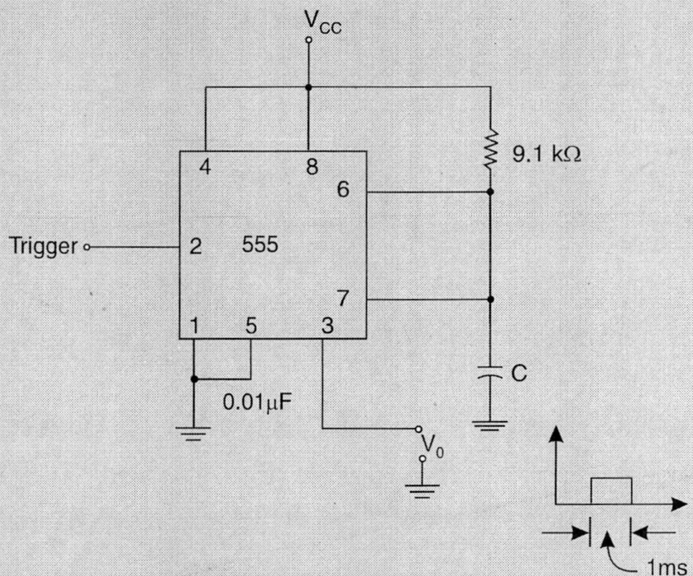

Fig. 26-54.

24. The circuit shown in Fig. 26-55 represent

(*a*) an astable multivibrator (*b*) a monostable multivibrator
(*c*) voltage-controlled oscillator (*d*) ramp generator.

(*UPSC Engg Services* 1999)

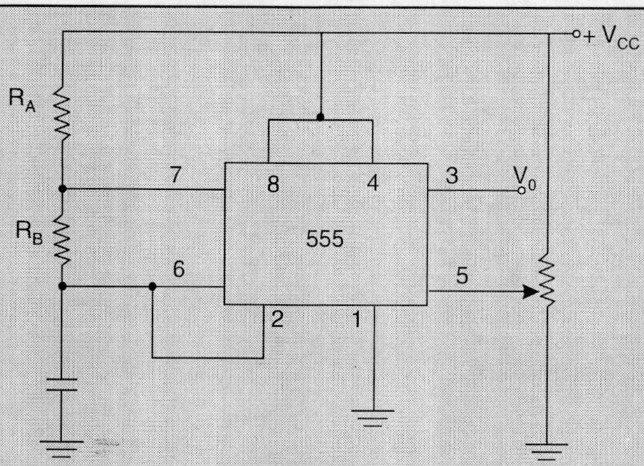

Fig. 26-55.

25. A Schottky diode is connected between the base and collector of the transistor to

(*a*) prevent cut-off (*b*) prevent saturation

(*c*) decrease power dissipation (*d*) none of these.

ANSWERS

1. (*b*)	**2.** (*c*)	**3.** (*d*)	**4.** (*d*)	**5.** (*a*)
6. (*a*)	**7.** (*c*)	**8.** (*a*)	**9.** (*a*)	**10.** (*c*)
11. (*d*)	**12.** (*b*)	**13.** (*d*)	**14.** (*a*)	**15.** (*d*)
16. (*a*)	**17.** (*a*)	**18.** (*c*)	**19.** (*c*)	**20.** (*b*)
21. (*d*)	**22.** (*b*)	**23.** (*c*)	**24.** (*c*)	**25.** (*b*)

❑❑❑

27

CHAPTER

Linear Wave Shaping Circuits

OUTLINE

Objectives

After completing this chapter, you should be able to :

- Know what is wave shaping ?
- Sketch the different types of non-sinusoidal waveforms.
- Describe the operation of differentiating circuit.
- Explain the generation of narrow pulses from square wave using differentiator.
- Describe the operation of integrating circuit.
- Explain the generation of a triangular wave from a square wave using integrator.

27-1. Introduction

A process by which nonsinusoidal waveforms are altered in passing through the circuit elements (such as diodes, resistors, inductors and capacitors) is called ***waveshaping***. The waveshaping is used to perform any one of the following functions :

- To generate one wave from the other wave.
- To limit the voltage level of the waveform to some preset value and suppressing all other voltage levels in excess of the preset level.
- To cut-off the positive and negative portions of the input waveform.
- To hold the waveform to a particular d.c. level.

The waveshaping is important in most of the signal process systems and is performed by the circuits known as **differentiators, integrators, limiters, clippers and clampers.**

The differentiator circuits are used to generate sharp narrow pulses either from distorted pulse waveforms or from rectangular waveforms. The integrator circuits are required to generate a voltage, which increases or decreases linearly with time (called a *ramp*). The limiters (also called amplitude limiters) are used to limit the voltage level of the waveform to some preset value and suppressing all other voltage levels in excess of this preset level. The limiters can also be used to protect a circuit against any possible damage due to excess voltage. These are an important part of signal processing circuit in television and radio receivers. The clipper circuit is required to cut-off the positive and negative portions of an input waveform. On the other hand, a clamper circuit is used to hold (or restore) the waveform to a particular d.c. level. The clamper circuits are also very useful in television receivers.

27-2. Different Non-sinusoidal Wave Shapes

We know that any waveform, whose shape is different from that of a standard sinusoidal wave, is called a nonsinusoidal waveform. Some of the important nonsinusoidal waveforms are *pulse, square, symmetrical square, triangular and sawtooth, etc*. All these waveforms are a part of standard voltage or current signals such as step, ramp, and exponential and are discussed as below :

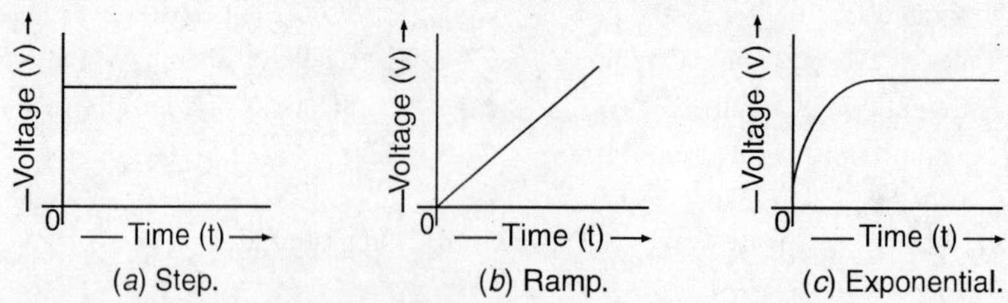

Fig. 27-1. *Standard voltage signals.*

(*i*) ***Step voltage :*** A voltage, which maintains the zero value for all times $t < 0$ and maintains the value V for all times $t > 0$ is called a step voltage. The transition between the two voltage levels takes place at $t = 0$ and is accomplished in a very short interval. Thus as shown in Fig. 27-1 (*a*), $v = 0$ immediately before $t = 0$ and $v = V$ immediately after $t = 0$.

(*ii*) ***Ramp voltage :*** A voltage, which is zero for $t < 0$ and increases linearly with time for $t > 0$, is called a ramp voltage. It is also called a sweep voltage. Thus as shown in Fig. 27-1 (*b*), $v = 0$ immediately before $t = 0$ and $v = \alpha \cdot t$ after $t = 0$. Here, α is a constant of proportionality.

(*iii*) ***Exponential voltage :*** A voltage, which is zero for $t < 0$ and increases nonlinearily with time as shown in Fig. 27-1 (*c*), is called an exponential voltage.

Now we shall discuss the pulse, square, symmetrical square, triangular and sawtooth waveforms in more details.

(*a*) ***Pulse waveform :*** Fig. 27-2 (*a*) shows the waveform of an ideal pulse. The pulse amplitudes is V and the pulse duration is t_p. It is evident from Fig. 27-2 (*b*) and (*c*) that the pulse may be considered as the sum of the step voltage + V, whose discontinuity occurs at $t = 0$ and a step voltage – V, whose discontinuity occurs at $t = t_p$. The pulse waveforms find extensive use in almost every field of electronics such as communication, computers, defence equipments *etc*.

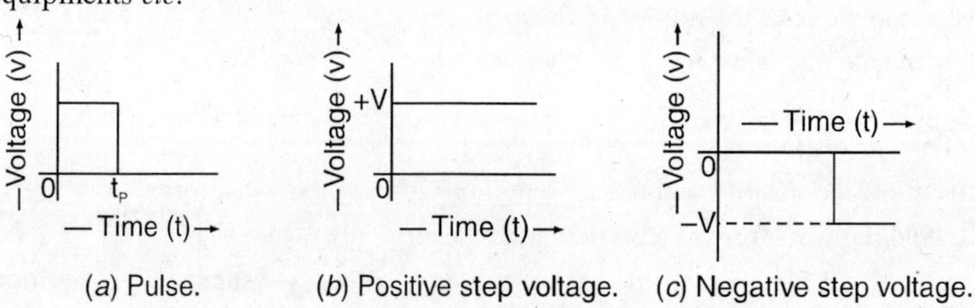

Fig. 27-2.

(*b*) ***Square waveform :*** A waveform, which maintains itself at one constant voltage level V_1 for time T_1 and at another constant level V_1 for time T_2 and is repetitive with a period $T = T_1 + T_2$ as shown in Fig. 27-3 (*a*) is called a *square waveform*. The square waveforms are used in digital electronic circuits, radars and as synchronising pulses in television.

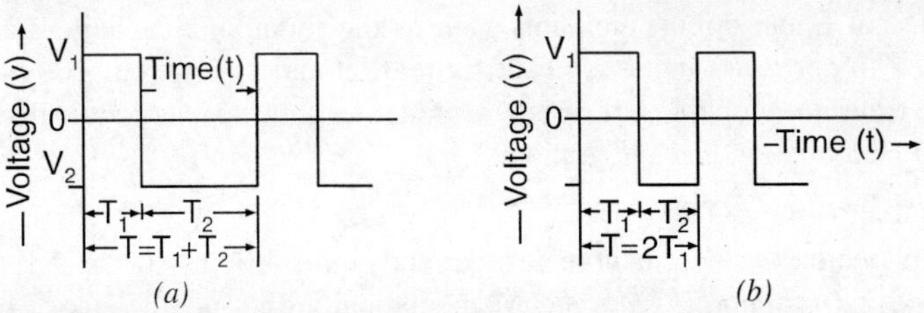

Fig. 27-3.

(*c*) ***Symmetrical square waveform :*** A square waveform for which $T_1 = T_2 = T/2$ as shown in Fig. 27-3. (*b*) is called a symmetrical square waveform. It may be noted that because of the symmetry, the voltage levels V_1 and V_2 are equal and opposite, *i.e.*, $V_1 = - V_2$. The symmetrical square waveforms are very useful in digital electronic circuits.

(*d*) ***Triangular waveform :*** A waveform, which increases linearly with time to a voltage level V for a time $T/2$ and then decreases linearly to its original level for a time $T/2$ and is repetitive with a period T as shown in Fig. 27-4 (*a*) is called *triangular waveform*. It may be noted from the figure, that a triangular wave may be considered as the sum of ramp voltage, which increases at a rate of $2V/T$ for a time $T/2$ and the ramp voltage which decreases at a rate of $- 2V/T$ for the remaining time $T/2$. The *triangular waveforms* are used in scanning circuits, where a uniform left-to-right scan is required as in computer displays. These are also used in timing circuits for electronic applications.

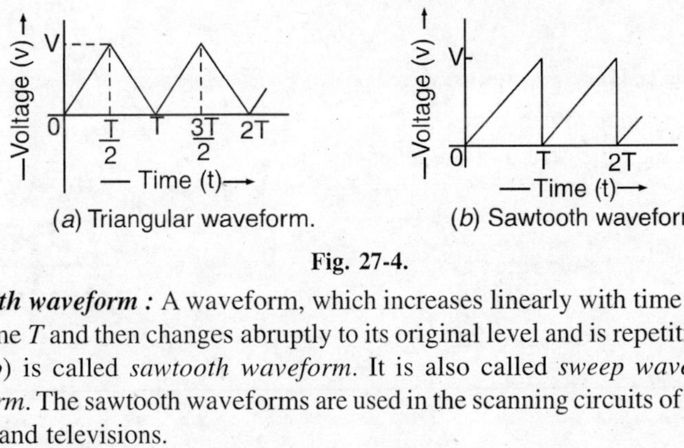

(*a*) Triangular waveform. (*b*) Sawtooth waveform.

Fig. 27-4.

(*e*) ***Sawtooth waveform :*** A waveform, which increases linearly with time to a voltage level V for a time T and then changes abruptly to its original level and is repetitive as shown in Fig. 27-4 (*b*) is called *sawtooth waveform*. It is also called *sweep waveform* or *time-base waveform*. The sawtooth waveforms are used in the scanning circuits of cathode ray oscilloscopes and televisions.

27-3. Types of Wave Shaping Circuits

Following two types of wave shaping circuits are important from the subject point of view :

(1) ***Linear wave shaping circuits :*** The circuits, which make use of only linear circuit elements such as the inductors, capacitors and resistors are known as *linear wave shaping circuits*. Such circuits are used to perform functions of differentiation and integration.

(2) ***Nonlinear wave shaping circuits :*** The circuits, which (in addition to linear circuit elements) make use of nonlinear circuit elements such as diodes and transistors are known as *nonlinear wave shaping circuits*. Such circuits are used to perform functions of amplitude limiting, clipping and clamping.

Now we shall discuss both the above mentioned types of wave shaping circuits one by one in the following pages.

27-4. Linear Wave Shaping Circuits

We have already discussed in the last article that the linear wave shaping circuits are those which make use of linear circuit elements such as the inductors, capacitors and resistors. Such circuits are used to perform functions of differentiation and integration. The linear wave shaping circuits use the transient properties of resistors, inductors and capacitors to realise the mathematical operations of differentiation and integration.

27-5. Differentiating Circuit

A circuit in which the output voltage is directly proportional to the derivative of the input voltage is called a *differentiating circuit*. Mathematically, the output voltage of a differentiating circuit,

$$v_o \propto \frac{dv_i}{dt}$$

$$= k \cdot \frac{dv_i}{dt}$$

where, k is a constant of proportionality.

A differentiating circuit may be realised by a simple RC series circuit as shown in Fig. 27-5. In this circuit, the output is taken across the resistor (R). The circuit is designed in such a way that reactance of the capacitor ($X_C = 1/2\pi f_C$) is very large as compared to the value of resistor (R), *i.e.*,

$$X_C >> R$$

$$\frac{1}{2\pi f_C} >> R$$

or $$\frac{1}{2\pi RC} >> f \quad ...(i)$$

where, f = Fequency of the input signal, and

C = Value of capacitance of the capacitor.

We know that the time period of the input signal,

$$T = \frac{1}{f}$$

$\therefore$ Equation (i) may be rewritten as

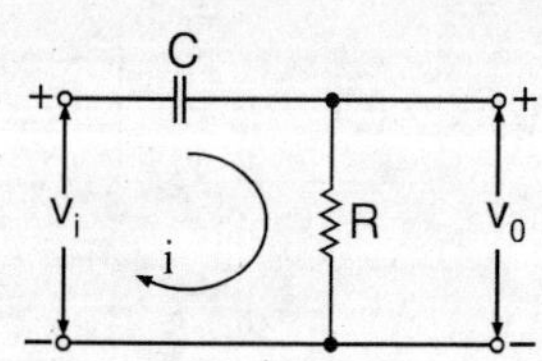

Fig. 27.5. *Differentiating circuit.*

$$\frac{1}{2\pi RC} << \frac{1}{T}$$

$$2\pi RC << T$$

or $$\tau << T \quad ...(ii)$$

where, $\tau = RC$ is called time constant of the circuit.

It is evident from equation (ii) that if time constant of the RC circuit (*i.e.*, $\tau = R.C$) is made much smaller than the time period of the input signal waveform, then the output of the circuit.

$$v_o = R.\,C.\,\frac{dv_i}{dt}$$

where, $C \cdot dv_i/dt$ is the value of current through the circuit.

In actual practice, it has been found that a time constant of about 1 to 10 per cent of the time period of the input signal waveform is usually sufficient to provide good differentiation, *i.e.*,

$$\tau \leq 0.1\,T \quad \text{...(For good differentiation)}$$

It will be interesting to know that in a differentiating circuit, the average (or d.c.) level of the output signal is always zero and it does not depend on the average level of the input signal. As a result of this, the output must extend in positive and negative directions, with respect to the zero-voltage axis, and must be equal to the area which is below the zero axis.

Sometimes, the differentiating circuit is also known as a 'high-pass' filter. The name 'high-pass' filter is designated because the circuit blocks the low-frequencies and allows high-frequencies to pass through it. It is due to the reason that reactance of the capacitor decreases with the increasing frequency. As a result of this, at very high frequencies, the capacitor acts as a short circuit and virtually all the input appears at the output.

Example 27-1. *Prove that the average (or d.c.) value of the output of a differentiating circuit is zero.* (*A.M.I.E., Summer*, 1987)

Solution. Consider a differentiating circuit as shown in Fig. 27-6.

Let v_i = Input voltage,

v_o = Output voltage,

i = Circuit current,

C = Value of capacitance of the capacitor,

q = Charge on the capacitor, and

R = Value of resistor.

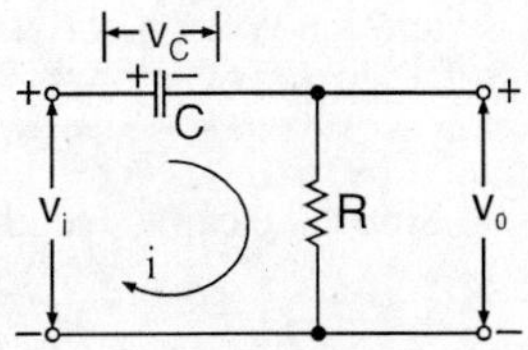

Fig. 27.6

We know that voltage across the capacitor,

$$v_C = q/C$$

Applying Kirchoff's Voltage Law to the differentiating circuit,

$$v_i = v_C + v_o$$

$$v_i = \frac{q}{C} + v_o \qquad \ldots\left(\because v_c = \frac{q}{C}\right)$$

Differentiating the above equation with respect to t,

$$\frac{dv_i}{dt} = \frac{1}{C} \times \frac{dq}{dt} + \frac{dv_o}{dt}$$

The term dq/dt in the above equation represents the rate of flow of charge with time, *i.e.*, the current i. Therefore, replacing dq/dt by i in the above equation,

$$\frac{dv_i}{dt} = \frac{i}{C} + \frac{dv_o}{dt}$$

$$= \frac{v_o}{RC} + \frac{dv_o}{dt} \qquad \ldots(\because i = v_o/R)$$

Multiplying the above equation on both sides by dt and integrating it over one period T,

$$\int_{t=0}^{T} dv_i = \int_{t=0}^{T} \left(\frac{v_o}{RC}\right) dt + \int_{t=0}^{T} dv_o$$

$$v_i(T) - v_i(0) = \frac{1}{RC} \int_{t=0}^{T} v_o \cdot dt + v_o(T) - v_i(0)$$

Under steady state conditions, the output waveform as well as the input signal is repetitive with a period T, so that $v_o(T) = v_o(0)$ and $v_i(T) = v_i(0)$. Therefore,

$$\frac{1}{RC} \int_0^T v_o \cdot dt = 0$$

or

$$\int_0^T v_o \cdot dt = 0$$

The above integral represents the area under the output waveform over one cycle. Since, the value of the integral is zero, therefore the average level of the output signal is always zero.

Example 27-2. *A 1 μF capacitor is charged from a 6V source through a 10 W resistor. If the capacitor has an initial voltage of –3 V, calculate its voltage after 8 ms.* *(AMIE., Winter* 1998)

Solution. Given : $C = 1\ \mu F$; $V_i = 6$ V; $R = 10\ \text{k}\Omega$. Initial voltage on the capacity (v_o) = – 3 V; $t = 8$ ms.

Let v_f = voltage at the capacitor after 8 ms

We know that time constant of a circuit,

$$\tau = R.C = 10\ \text{k}\Omega \times 1\ \mu F$$
$$= (10 \times 10^{+3}) \times (1 \times 10^{-6}) = 10 \text{ ms}$$

and the voltage across the capacitor after 8 ms,

$$v_f = 6(1 - e^{-8/10}) = 6(1 - 0.449) - 3.3 \text{ V}$$

Since, the capacitor already holds –3 V ; therefore the actual output voltage = 3.3 – 3.0 = 0.3 V **Ans.**

27-6. Applications of a Differentiating Circuit

Following are some of the important applications of a differentiating circuit, which are important from the subject point of view :

(1) To generate a series of narrow pulses (called pips) from the rectangular or square waveforms. The pipes are used as trigger pulses or synchronization pulses in circuits used in television and cathode ray oscilloscopes.

(2) To generate a step from a ramp input.

(3) To generate a square wave from a triangular wave input.

All the applications are discussed in more details one by one in the following pages.

27-7. Generation of Narrow Pulses from Square Wave

Fig. 27-7(*a*) shows the square waveform, which acts as input to the differentiating circuit. It is assumed, that time constant ($\tau = R \cdot C$) of the differentiating circuit is much smaller than the time period of the input waveform (*i.e.*, $\tau << T$). We know that a square waveform may be considered to be the sum of positive steps and negative voltage steps. Now, as a matter of fact a differentiating circuit for a step voltage input, should produce a sharp spike of infinite amplitude for an abrupt change and zero for a constant part of the step voltage at its output. It is true because, the exact derivative of a discontinuity is infinite and the derivative of a constant is zero. Thus under ideal conditions, the output of a differentiating circuit should be as shown in Fig. 27-7(*b*). But in actual practice, we get an output as shown in Fig. 27-7(*c*). Here, the output is shown to rise instantaneously to an amplitude of 2V and then decays down to zero. The difference in the ideal and the real outputs may be explained as below :

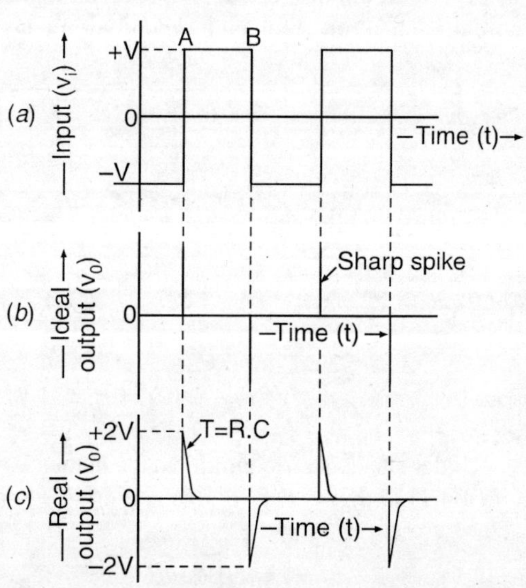

Fig. 27-7. *Generation of narrow pulses from a square wave.*

It has been observed that whenever, there is an abrupt change (*i.e.*, discontinuity) in voltage at the input of the capacitor, it acts as a short-circuit instantaneously and a full change in the input voltage appears across the output. It is because of the fact, that a capacitor, requires an infinite charging current for the voltage across it to change instantaneously. Since the charging current remains finite, therefore it act as short-circuit and passes the step change (or discontinuity) in the input voltage as such to its output. Secondly, the capacitor blocks any constant (or d.c.) input voltage and hence does not allow it to reach at the output.

Thus with square wave input, when the input changes abruptly from a voltage level – V to + V (the total change being 2 V), the differentiating circuit produces it at its output as such. During the constant part AB, the output decays exponentially with a time constant τ (equal to $R \cdot C$). At a time t = 5 τ, the output decays to 0.6% of its initial value. Thus, we can assume that the output has reached its steady state in a time interval equal to 5τ.

Note. If a square wave is applied to a differentiating circuit whose time constant (τ) is not much smaller as compared to the time period of the input waveform, the output is not a train of narrow pulses. But it consists of a square wave whose upper and lower portions are tilted towards the zero-voltage axis. And if a differentiating circuit has $\tau \geq T$, the square wave will be passed through the circuit without any change.

Example 27-3. *A square-wave whose peak-to-peak value is 1V extends ± 0.5 V w.r.t. ground. The half period is 0.1 sec. This voltage is impressed upon an RC differentiating circuit whose time constant is 0.2 sec. What are the steady state max. and min. values of the output voltage.*

(*Nagpur Univ. Winter* 2003)

Solution. Given: Square-wave input with peak-to-peak value = 1V. Half-period ($T/2$) = 0.1 s. Time constant of RC circuit (τ) = 0.25. Refer to Fig. 27-8 for better view.

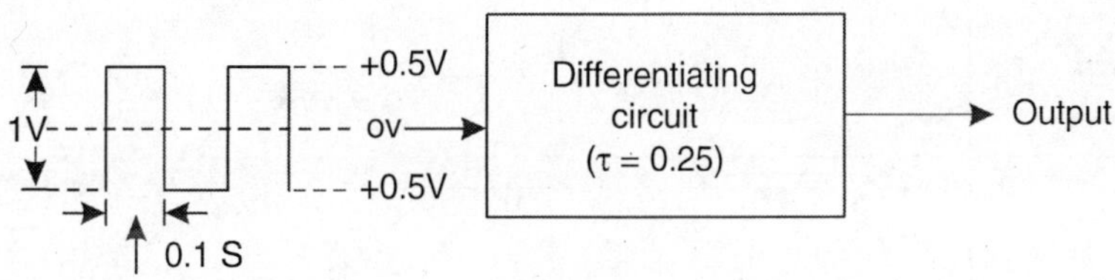

Fig. 27-8.

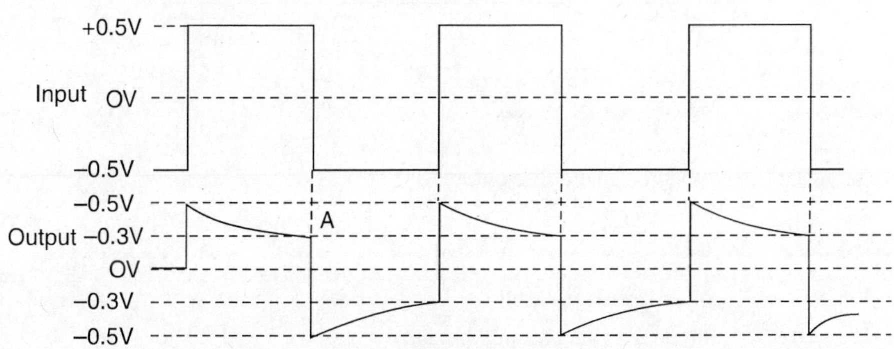

Fig. 27-9.

We know that corresponding to point A the capacitor will charge to a voltage,

$$v_c = 0.5 \cdot e^{-t/\tau} = 0.5 \times e^{-0.1/0.2} = 0.5 \times 0.606 = 0.3\,\text{V}$$

Similarly, when the input goes to – 0.5, the capacitor will charge to a voltage of – 0.3. Therefore, the output waveform can be obtained as shown in Fig. 27-9.

27-8. Generation of Voltage-step from a Ramp

Fig. 27-10(a) shows a ramp waveform, which acts as an input to the differentiating circuit. As a matter of fact, that circuit should produce an output waveform as shown in Fig. 27-10(b). It indicates that under ideal conditions, the output should be a step with amplitude equal to $\alpha \cdot \tau$, where α is a constant which indicates the slope of ramp voltage and τ is a time constant of the circuit (equal to the product $R \cdot C$).

In actual practice, the output of the circuit increases exponentially in accordance with the relation,

$$v_o = \alpha \cdot \tau (1 - e^{-t/\tau})$$

If $t \geq 5t$ (*i.e.*, 5 times the time constant of the circuit), the term is negligible as compared to unity. Therefore, the circuit output,

$$v_o = \alpha \cdot \tau$$

It means that output reaches its steady state value (equal to $\alpha \cdot \tau$) after a time $t = 5\tau$. The waveform of the actual or real output is as shown in Fig. 27-10 (c).

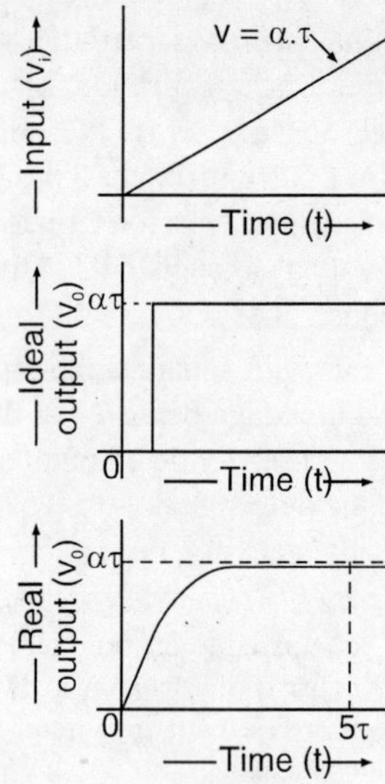

Fig. 27-10. *Generation of voltage-step from a ramp.*

Example 27-4. *The pulse from a high-voltage generator is as shown in Fig. 27-11. This is applied to the RC differentiator with RC = 250 ps. If the maximum value observed at the output of the differentiator is 50 V, what is the peak value of the input voltage ?* (*AMIE., Summer* 1997)

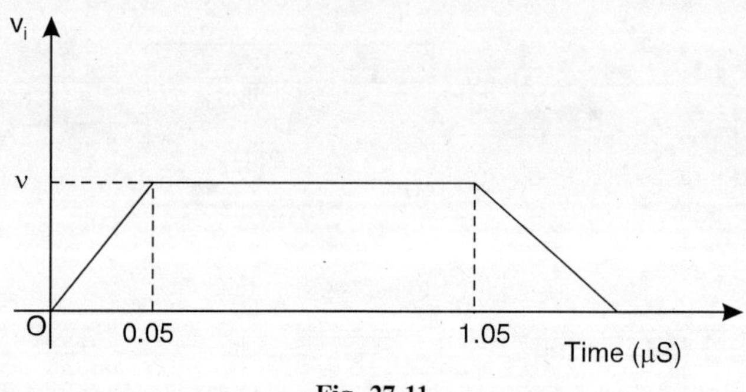

Fig. 27-11.

Solution. Given time constant of the differentiator = 250 ps and v_o = 5 V.

Let V_P = peak value of the input voltage.

We know that steady state value of the output, (v_o)

$$50 \text{ V} = \alpha \cdot \tau = \alpha \cdot (250 \times 10^{-12})$$

$$\therefore \quad \alpha = \frac{50}{250 \times 10^{-12}} = 0.2 \times 10^{12} \text{ V/s}$$

$\therefore$ At $t = 0.05$ μs, the peak value of input voltage

$$V_P = 0.2 \times 10^{12} \times (0.05 \times 10^{-6}) = 0.01 \times 10^6 \text{ V}$$

$$= 10 \text{ kV } \textbf{Ans.}$$

27-9. Generation of Square Wave from a Triangular Wave

Fig. 27-12(a) shows the triangular waveform, which acts as an input to the differentiating circuit. It is assumed that time constant of the circuit ($\tau = R \cdot C$) is much smaller than the time period of the input waveform (*i.e.*, $\tau << T$). As a matter of fact, the circuit should produce an output waveform, which changes abruptly to a new value, each time, the slope of the input waveform changes to a new value. For example, consider points *A*, *B*, *C* and *D* of the input waveform. Then ideally the circuit should have discontinuities corresponding to these points as shown in Fig. 27-12(*b*).

In actual precice, whenever the slope of the input voltage abruptly changes, the circuit output takes certain finite time (equal to 5τ) to reach to its steady-state level. It may be noted that smaller the value of time constant of the differentiating circuit, lesser will be the time taken by its output to reach to its steady-state level. However, a smaller time constant reduces the amplitude of the output voltage.

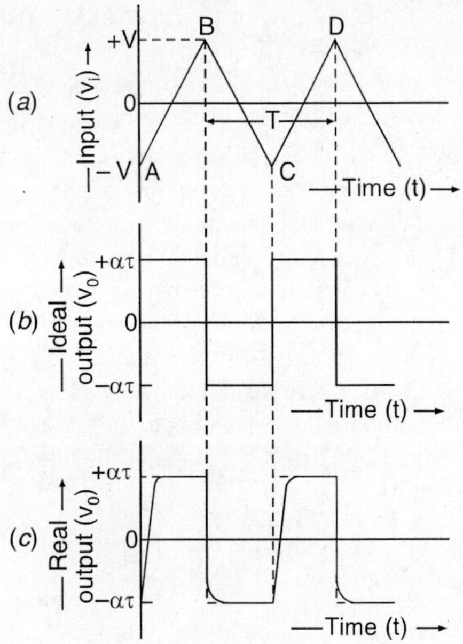

Fig. 27-12. *Generation of a square wave from a tringular wave.*

27-10. Integrating Circuit

A circuit, in which the output voltage is directly proportional to the integral of the input voltage, is called an integrating circuit. Mathematically, the output voltage of an integrating circuit,

$$= v_o \,\alpha \int v_i \cdot dt$$

$$= k \int v_i \cdot dt$$

where, *k* is a constant of proportionality.

The integrating circuit may be realised by a simple RC series circuit as shown in Fig. 27-13 in which the output is taken across a capacitor (*C*). The circuit is designed in such a way that the time constant of the circuit ($\tau = R \cdot C$) is greater than or equal to 10 times the period of the input signal waveform, *i.e.*,

$$\tau \geq 10\,T \qquad \text{...(For good integration)}$$

The output of the integrated circuit is given by the relation,

$$v_o = \frac{1}{RC} \int v_i \cdot dt$$

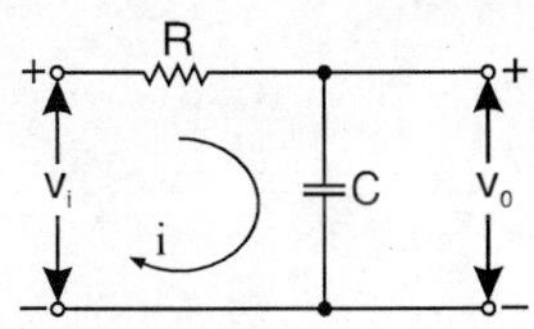

Fig. 27-13 *Integrating circuit.*

It will be interesting to know that in an integrating circuit, the average (or d.c.) level of the output signal is the same as that of the input signal (*i.e.*, it is not zero as in a differentiating circuit). Sometimes, the integrating circuit is also called a low-pass filter. *The name 'low-pass' filter is designated because of the fact that the circuit passes low-frequencies readily but attenuates high-frequencies*. The attenuation of high frequencies is due to the reactance of capacitor, which decreases with the increasing frequency. At very high frequencies, the capacitor acts as a short circuit and therefore, the output drops to zero.

27-11. Applications of an Integrating Circuit

Following are some of the important applications of an integrating circuit, which are important from the subject point of view :

(*i*) To perform mathematical integration in analog computers.

(*ii*) To generate a triangular wave from a square wave.

(*iii*) To generate a sawtooth wave from a rectangular wave.

Now we shall discuss the use of integrating circuit to generate a triangular and a sawtooth in the following pages.

27-12. Generation of Triangular Wave from a Square Wave

We know that a square wave as shown in Fig. 27-14(*a*) may be considered as a sum of two voltage steps, of opposite polarities, which are displaced in time by $T/2$, where T is the time period of a square wave. It has been found that the response of an integrating circuit to a voltage step is exponential with a time constant $\tau = RC$. However, if the value of time constant is very large, as compared to time period of the input waveform (*i.e.*, $\tau >> T$), the response is almost a straight line as shown in Fig. 27-14(*b*). Here the line LM is the response to a voltage step AB. Similarly, the line MN represents the response corresponding to the voltage step CD, and so on. In that case, the slope of the line is given by the relation,

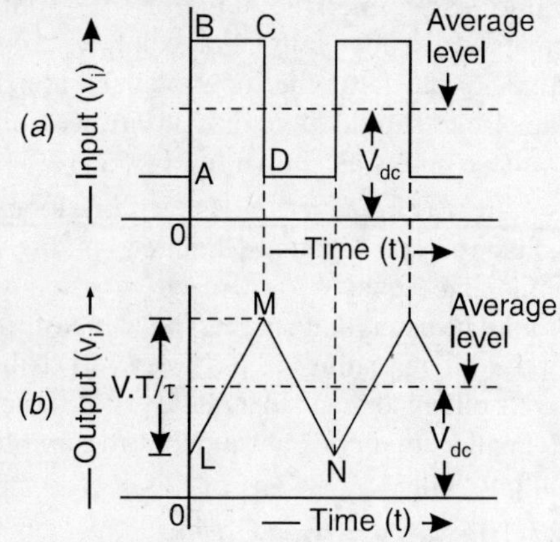

Fig. 27-14. *Generation of a triangular wave from a square wave.*

$$\text{slope} = \frac{V_f - V_i}{\tau}$$

where, V_i = Initial value, and

V_f = Final value.

Now suppose, the input initial value is $-V$ and the final value is $+V$, then the slope of the response curve is equal to,

$$\text{slope} = \frac{V - (-V)}{\tau} = \frac{2V}{\tau}$$

Then in a time equal to $T/2$, the total change of the output voltage is given by the relation,

$$v_o = \frac{2V}{\tau} \times \frac{T}{2} = \frac{V \cdot T}{\tau}$$

However, in actual practice, it has been found that the amplitude of the output voltage is always less than VT/τ.

Example 27-5. *Obtain the response from a cascaded network as shown in Fig.* 27-15.

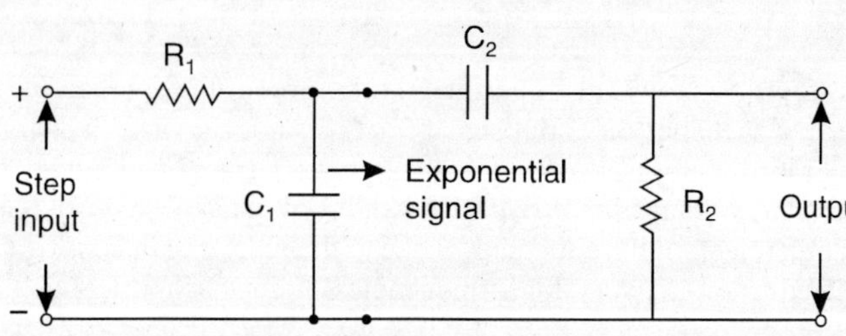

Fig. 27-15.

What is the effect of time constant ? (*AMIE., Summer* 1994)

Solution. Given : Circuit shown in Fig. 27-15. Notice that the circut consists of an integrator followed by a differentiator circuit. The cascaded response is as shown in Fig. 27-16.

27-13. Generation of Sawtooth Wave from a Rectangular Wave

Fig. 27-17(*a*) shows the waveform of a rectangular wave. It is a modified version of a square wave. The response of an integrating circuit to a rectangular wave is similar to that discussed for a square wave in the last article, except the output waveform, which is a sawtooth wave (instead of a triangular wave) as shown in Fig. 27-17(*b*),

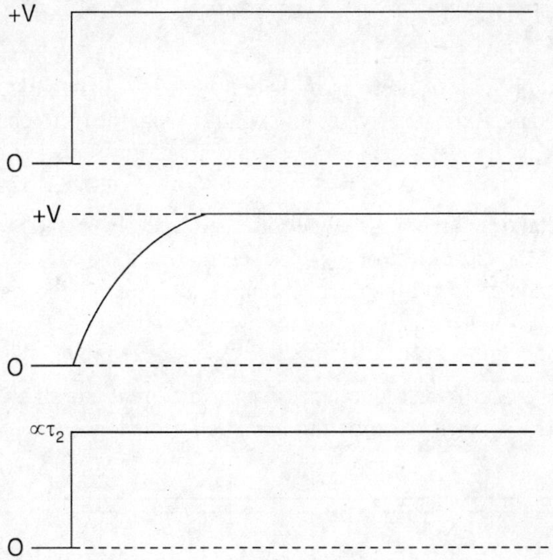

Fig. 27-16.

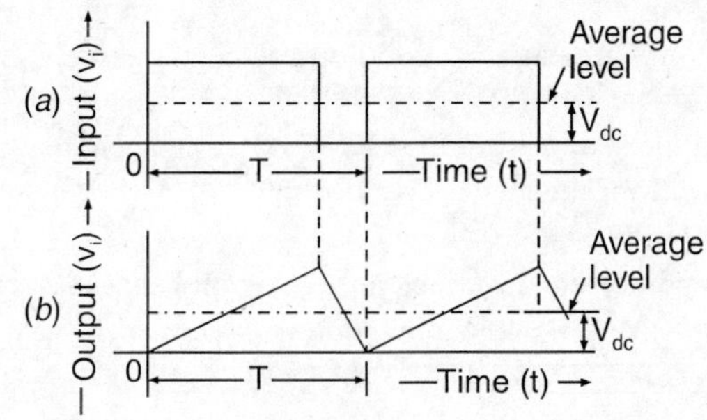

Fig. 27-17.

Note: The average levels in the waveforms in Fig. 27-14 and 27-17 are indicated to remind the reader that the integrating circuit does not disturb these at the output.

SUMMARY

In this chapter you have learned that :

1. There are several different types of waveforms in the field of electronics: pulse, square, symmetrical square, triangular and sawtooth.
2. A differentiating circuit can be used to generate narrow spikes from a square wave input. It can also be used to generate a square wave from a triangular wave.
3. An integrating circuit can be used to generate a triangular wave from a square wave. It can also be used to generate a sawtooth wave from rectangular wave.

GLOSSARY

Differentiator : An *RC* circuit with an input applied to the capacitor and output taken across the resistor.

Integrator : An *RC* circuit with an input applied to the resistor and output taken across the capacitor.

Time constant : The product of *R* and *C*.

NUMERICAL PROBLEMS

1. The pulse from a high voltage generator rises linearly for 0.04 μs and than remains constant for 2 μs. The rate of rise of pulse is measured with RC differentiating circuit whose time constant is 400 ps. If the positive output voltage from the differentiator has a maximum value of 100 V, what is the peak value of the generator ? Derive clearly the expressions used. (*AMIE., Winter* 1996)

2. The pulse from a high voltage generator 'a magnetron' rises linearly for 0.05 μs and then remians constant for 1 μs. The rate of rise of pulse is measured with an *RC* differentiating circuit whose time constant is 250 ps. If the positive output voltage from the differentiator has a maximum value of 100 V, what is the peak voltage of generator ? Develop clearly, any expression employed ? (*AMIE., Winter* 1993)

3. A sinusoidal input, $v_i(t) = V_m \sin Wt$ is fed to a passive *RC* differentiator. If $WRC = 0.01$, determine (*i*) the phase error and (*ii*) the magnitude reduction factor. (*AMIE., Summer* 1997)

DESCRIPTIVE QUESTIONS

1. What do you mean by the term 'wave shaping'? Also discuss, why it is needed in practical applications ?

2. Discuss the necessary criteria for a good differentiating circuit. Give some applications of this circuit.

3. Show by drawing the neat waveforms as to how we can generate
(*i*) narrow pulses from square wave, and
(*ii*) square wave from triangular wave

4. Enumerate the conditions under which an RC circuit behaves as an integrator.

5. Sketch the output waveforms for an integrating circuit when the input signal is
(*i*) a square wave and
(*ii*) a rectangular wave

6. Obtain the conditions under which a low-pass *RC* circuit behaves as an integrator. Sketch the output voltage waveform, if the input to this circuit is a square wave. (*AMIE., Summer* 1993)

7. If a step voltage of amplitude *V* is impressed on a high-pass *RC* circuits, show that the output voltage is given by :

$$v_o = Ve^{-t/RC}$$

Hence or otherwise, obtain the expression for the percent tilt in the output. (*AMIE., Summer* 1993)

8. Show how an *RC* circuit can be used as (*a*) an integrator and (*b*) a differentiator. (*AMIE., Winter* 1993)

9. Obtain the response of a highpass *R-C* circuit for the following waveforms:
(*a*) Single step pulse
(*b*) Ramp waveform.
What is the effect of time constant. (*AMIE., Summer* 1994)

10. (*a*) Obtain the response of a low-pass R-C circuit to the following inputs (*i*) single step pulse (*ii*) ramp waveform
(*b*) Obtain the conditions under which a low-pass *R-C* circuit behaves as an integrator. Sketch the output voltage waveform if the input to this circuit is a symmetrical square wave. (*AMIE., Winter* 1994)

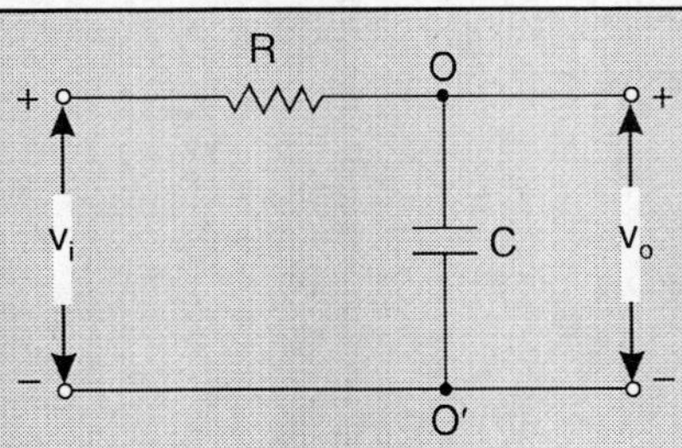

Fig. 27-18.

11. Plot the capacitor voltage and current versus time for a series *RC* circuit with a switch and a d.c. voltage source. Explain the shape of the graph. Derive equations for the capacitor voltage when $t = RC$ and when $t = 5\ RC$. State any assumptions made by you. (*AMIE., Winter* 1998)

12. (*a*) Obtain the response of a low pass *RC* circuit for the following waveforms: (*i*) Ramp input, (*ii*) Square wave input.

(*b*) Explain how a low pass filter can be used as an integrator.(*AMIE., Summer* 1995)

13. A pulse is applied to a low-pass *RC* circuit. Prove by direct integration that the area under the pulse is the same as the area under the output waveform across capacitor. Explain this result physically. (*Nagpur Univ., Winter* 2003)

14. Show how the response of a lowpass *RC* circuit for ramp input for different time constant. How can it be modified for square wave input ? (*AMIE., Winter* 1996)

MULTIPLE CHOICE QUESTIONS

1. The criteria for a good differentiating circuit is that its time constant ($\tau = R \cdot C$) should be

(*a*) much larger than the period of the input waveform
(*b*) less than or equal to the period of the input waveform
(*c*) much smaller than the period of the input waveform

2. If the input to a differentiating circuit is a triangular wave, then the output wave will be

(*a*) square wave (*b*) rectangular wave
(*c*) sawtooth wave (*d*) none of the above

3. The criteria for good integrating circuit is that its time constant ($\tau = R \cdot C$) should be

(*a*) equal to the period of the input waveform
(*b*) much larger than the period of the input waveform
(*c*) much smaller than the period of the input waveform

4. The main function of a clipping circuit is to

(*a*) remove a certain portion of the input signal above or below a certain level
(*b*) restore d.c. level to the signal
(*c*) suppress amplitude variations in the input signal voltage
(*d*) both (*a*) and (*c*) above

5. The clamper circuit is used to

(*a*) introduce a d.c. level to a.c. signal
(*b*) suppress variations in amplitude of the input signal
(*c*) obtain an output which is integral of the input signal

6. A capacitor of 100 μF is charged to 10 V through a resistance of 10 kΩ. It will be fully charged in

(*a*) 5 s (*b*) 0.1 s (*c*) 1 s (*d*) 0.5 s

7. An integrator has a transfer function whose
(*a*) amplitude increases linearly with the frequency
(*b*) amplitude varies inversely with the frequency
(*c*) amplitude remains constant
(*d*) amplitude decreases linearly with the frequency *(AMIE., Summer 2000)*

8. A triangular waveform can be generated by
(*a*) differentiating a square waveform
(*b*) integrating a square waveform
(*c*) differentiating a sinusoidal waveform
(*d*) integrating a sinusoidal waveform. *(AMIE., Winter 2001)*

ANSWERS

1. (*c*)	**2.** (*a*)	**3.** (*b*)	**4.** (*d*)	**5.** (*a*)
6. (*a*)	**7.** (*a*)	**8.** (*b*)		

❑❑❑

28

CHAPTER

Time Base Circuits

OUTLINE

Objectives

After completing this chapter, you should be able to :

- Know what is a time base signal.
- Explain the difference between the free running and triggered time base generators.
- Sketch the waveforms and calculate the frequency of a sweep circuit unijunction transistor switch.
- Describe the cirucit operation of transistor constant current sweep.
- Explain the circuit operation of a Miller sweep circuit.
- Describe the circuit opearation of a Bootstrap sweep circuit.

28-1. Introduction

The circuits, which provide a linear time scale, are called time base circuits or time base generators. In such circuits, either the output voltage (or the output current) is a linear function of time, over a specified time interval. These circuits are extremely important in most of the electronic systems. For example, a linear time base voltage is required on the deflection plates of cathode ray oscilloscope, to sweep the electron beam, from left to right across the screen. Similarly, a linear time base current waveform is required in the deflection coils of a television receiver. Because of the sweep application, the circuits are also sometimes known as ***sweep circuits*** (or sweep generators).

28-2. General Features of a Time Base Signal

Fig. 28-1 (*a*) shows a typical time base signal. Here the voltage starts from some initial value as indicated by point *A* in the figure. Then it increases linearly with time to a maximum value indicated by point *B* in the figure. After this, the voltage returns back to its initial value as indicated by point *C* in the figure.

The time, required by the voltage from its initial value to reach its maximum value, is known as sweep time and is designated by the symbol T_s. Similarly, the time taken by the voltage to its initial

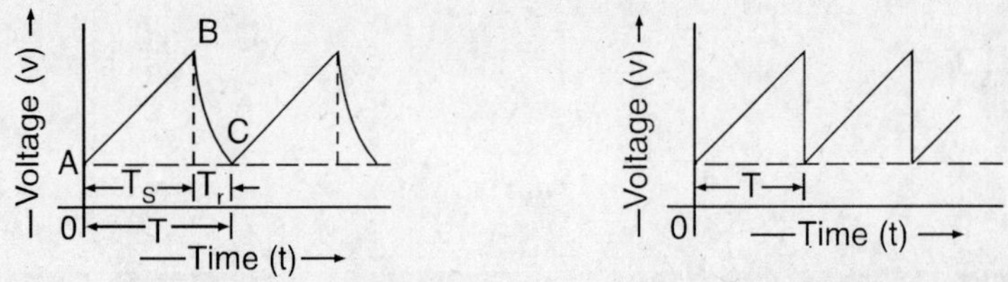

(*a*) General sweep voltage. (*b*) Sawtooth voltage waveform.

Fig. 28-1. *Sweep waveforms.*

value is called a *restoration time* and is designated by the symbol . T_r The restoration time is also known as *return time or the flyback time.*

In most of the applications, the shape of the waveform during the restoration time, as well as the restoration time itself, is of no importance. However, in certain applications, it is desired that the restoration time is very short in comparison with the time occupied by the linear portion of the waveform (*i.e.*, $T_r << T_s$). In that case, the restoration time is extremely short and a new voltage is initiated at the instance the previous one is terminated. The waveform, in such a case, is as shown in Fig. 28-1 (*b*) and is known as sawtooth waveform. It is customary to refer to waveforms shown in Fig. 28-1 (*a*) and (*b*) as sweep waveforms even in applications not involving the deflection of an electron beam.

It will be interesting to know that the circuits, which generate time base signals, do not ordinarily provide sweep voltages that are precisely linear. A most useful way of expressing the deviation from linearity is the slope error or sweep-speed error which is given by the relation,

$$e_s = \frac{\text{Difference in slope at the beginning and end of sweep}}{\text{Initial value of slope}}$$

There is one another term called ***'sweep speed'***, which is used to compare the performance of different sweep generating circuits. In all these circuits, a capacitor is an essential element. The charging and discharging of a capacitor generates a sweep voltage. Suppose a capacitor (*C*) is charged by a constant current (*I*). Then the voltage across the capacitor at any time (v_C) is given by the relation,

$$v_c = \left(\frac{I}{C}\right) t$$

The rate of change of capacitor voltage with time is known as sweep speed. Thus,

$$\textbf{Sweep speed} = \frac{dv_c}{dt} = \frac{d}{dt}\left(\frac{I}{C}\right) t = \frac{I}{C}$$

The sweep speed is required to be constant in most of the applications.

28-3. Types of Time Base Circuits

Following two types of time base circuits are important from the subject point of view:

(*i*) ***Free running time base generator :*** A circuit in which the periodic sawtooth waveform is generated, without the application of any signal, is called a free running time base generator. Such a circuit is required to display a periodic waveform. It may be noted that in a free running time base generator, the sweep time (T_s) must be larger than the period of the waveform to be displayed.

(*ii*) ***Triggered time base generator :*** A circuit in which the linear waveform with a prescribed duration of time is generated by the application of a trigger signal, is called a triggered time base generator. Such a circuit is required to display widely separated narrow-width pulser or a waveform which may not be periodic but may occur at irregular intervals.

The operation of the circuit may be understood from the condition, that the switch (S) is initially assumed to be closed. At this instant, the capacitor voltage and hence the output voltage (v_o) is zero, when the switch (S) is opened, the capacitor voltage starts towards the supply voltage (V) in accordance with the equation,

$$V_s = V(1 - e^{-t/RC}) \qquad \text{...}(i)$$

However, the capacitor is never allowed to charge to a voltage equal to the supply voltage. It is because of the fact that after a time interval (T_s), when the sweep voltage has attained the value (V_s), the switch (S), again closes. The resulting sweep voltage waveform is as shown in Fig. 28-2 (*b*). The sweep-speed error for the exponential sweep circuit is given by the relation,

$$e_s = \frac{V_s}{V}$$

It is evident from the above relation that smaller the value of sweep voltage (V_s), lower will be the value of sweep-speed error and hence, higher is the value of linearity. Due to this reason, the exponential charging sweep circuit is used only in those applications, which require lower sweep voltages (usually of the order of volts or tens of volts). If $t/RC << 1$, then we can expand equation (*i*) as follows :

$$v_s = \frac{V \cdot t}{R \cdot C}\left(1 - \frac{t}{2\,RC} + \frac{t^2}{6\,R^2C^2} \ldots\right)$$

$$= \frac{V \cdot t}{R \cdot C} \qquad \text{... (Retain first term and neglect others)}$$

Since, $v_s = V_s$ when $t = T_s$, therefore

$$V_s = \frac{V \cdot T_s}{R \cdot C}$$

or

$$\frac{V_s}{V} = \frac{T_s}{R \cdot C}$$

∴ Sweep-speed error,

$$e_s = \frac{V_s}{V} = \frac{T_s}{R \cdot C} = \frac{T_s}{\tau}$$

where, τ is the time constant and is equal to ($R \cdot C$). It is obvious from this relation that for good linearity (*i.e.*, smaller sweep-speed error), the time constant (τ) must be very large as compared to sweep duration (T_s).

28-4. Methods of Generating a Time Base Waveform

Although there are a number of methods of generating a time base waveform, yet the following are important from the subject point of view :

(*i*) ***Exponential charging :*** In this method, a capacitor is charged through a resistor to a voltage, which is small in comparison with the supply voltage.

(*ii*) ***Constant current charging :*** In this method, a capacitor is charged linearly from a constant-current source.

(*iii*) ***Miller integration :*** In this method, an operational amplifier (Op Amp) based integrator is used to convert a step into a ramp waveform.

(*iv*) ***Boot strap :*** In this method, a constant current is approximated by maintaining nearly constant voltage across a fixed resistor in series with a capacitor.

28-5. Exponential Sweep Circuit

Fig. 28-2 shows a simplest circuit to generate a time base signal. It is known as exponential sweep circuit. This circuit consists of a series RC circuit connected to a supply voltage V. A switch S

(which is usually is transistor switch) is connected across the capacitor *C*. The output voltage is taken across the capacitor (*C*).

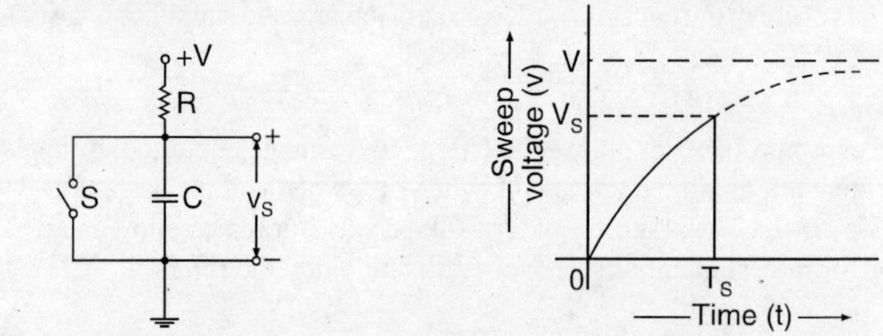

(*a*) Sweep circuit. (*b*) Seep voltage waveform.

Fig. 28-2.

28-6. Sweep Circuit Using Transistor Switch

We have already discussed in the last article about the exponential sweep circuit. We know that the circuit requires an electronic switch *S*. Many devices are available to serve as a switch *S* such as bipolar junction transistor, field-effect transistor, unijunction transistor, thyratron etc.

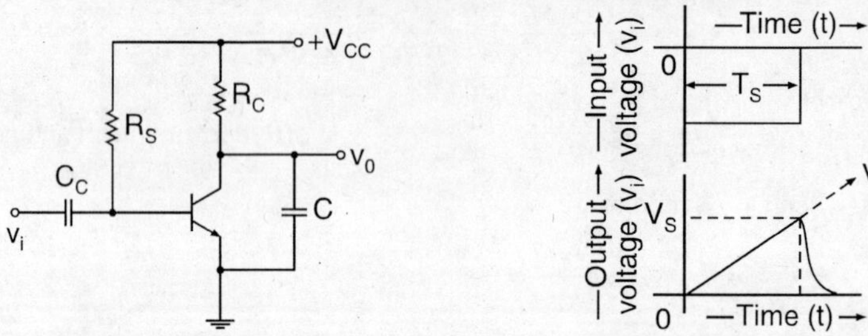

(*a*) Basic transistor switch sweep circuit. (*b*) Associated input and output waveforms.

Fig. 28-3.

Fig. 28-3 (*a*) shows the basic circuit of a transistor switch sweep circuit. This circuit requires a gating waveform (v_i) as shown in Fig. 28-3 (*b*). It may be obtained from a monostable multivibrator (*i.e.*, one shot) or an astable multivibrator. If the 'gating' waveform is obtained from a monostable multivibrator, the sweep generated by the circuit is called *triggered sweep* and the circuit is known as *triggered voltage time base generator*. However, if the 'gating' waveform is obtained from an astable multivibrator, the sweep generated is called *free running sweep* and the circuit is known as *free running voltage time base generator.*

Initially, the transistor is biased ON and operates in the saturation region. Thus, when there is no input (*i.e.*, $v_i = 0$), the output voltage is zero. (Actually its value is equal to $V_{CE\ (sat)}$). When the 'gating' pulse (*i.e.*, a negative pulse) is applied, the transistor turns OFF. As a result of this, the capacitor voltage rises to a target value V_{CC} with a time constant $R_C \cdot C$. The charging curve ignoring $V_{CE(sat)}$ is governed by relation,

$$v_o(t) = V_{CC}\,(1 - e^{-t/Rc \cdot C})$$

If $t/R_c \cdot C$ <<1, the above equation may be expanded into a power series in $t/R_c \cdot C$. Then retaining only the first term of the power series, the output voltage,

$$v_o(t) = V_{CC} \frac{t}{R_c \cdot C} \qquad ...(i)$$

At $t = T_s$, the output voltage,

$$v_o(T_s) = V_{CC} \cdot \frac{T_s}{R_c \cdot C} = V_s \qquad ...(ii)$$

The equation (i) represents an approximately linear waveform. The slope error or sweep-speed error or such a waveform is given by the relation,

$$e_s = \frac{V_s}{V_{CC}} = \frac{T_s}{R_c \cdot C}$$

It is evident from the above relation that the output or sweep voltage level (V_s) must be a small fraction of the supply voltage (V_{CC}) and this is not a desirable feature, if a certain sweep voltage level is desired. This shortcoming can be remedied by the Miller sweep and the Bootstrap sweep circuits.

It may be noted that the transistor switch if OFF only for the gating time (T_s). At the end of time T_s, the capacitor discharges, and the voltage is again zero. It is also possible to generate a negative-going sweep using transistor switch. But in that case, we have to use a PNP transistor instead of a NPN transistor in the sweep circuit.

28-7. A Transistor Constant Current Sweep

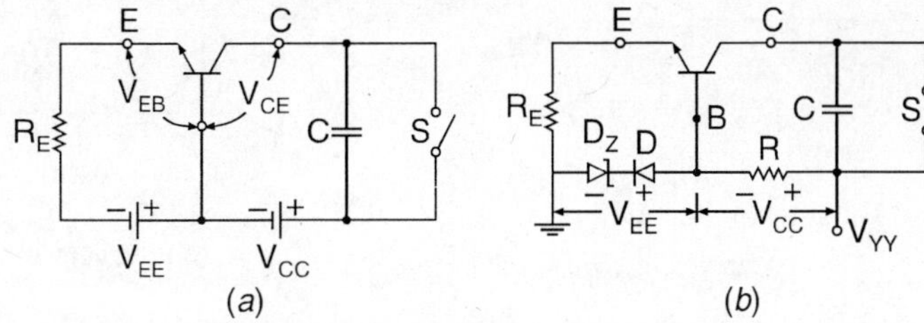

Fig. 28-4. *A constant current sweep circuit.*

Fig. 28-4(a) shows a circuit to generate a sweep using constant current obtained from a common base transistor. We know that except for very small values of collector-to-base voltage, the collector current of a transistor in the common base configuration is very nearly constant, when the emitter current is held fixed. This characteristic may be used to generate a quite linear sweep by causing a constant current to flow into a capacitor. The value of emitter current in Fig. 28-4(a) is given by the expression,

$$I_E = \frac{V_{EE} - V_{EB}}{R_E}$$

If the emitter-to-base voltage (V_{BE}) remains constant with times after the switch (S) opened, then the collector current will be a constant whose nominal value,

$$I_C = h_{FB} \cdot I_B = \alpha \cdot I_E$$

where, h_{FB} (or α) is the common base current gain of the transistor.

The constant current sweep circuit discussed above has one drawback, that it makes the sweep rate as a function of temperature. Since the emitter-base junction voltage (V_{BE}) for a fixed current decreases by about 2 mV/°C, therefore the *sweep speed increases with the temperature.

Fig. 28-4(b) shows a constant current sweep circuit, which employs a single supply V_{YY} and provides a considerable temperature compensation. The emitter supply voltage is now equal to the voltage (V_Z) across the zener diode (D_Z) plus the drop (V_D) across the diode D. The collector supply voltage V_{CC} is now equal to the drop across resistance R, *i.e.*, $V_{CC} = V_{YY} - V_{EE}$. The diode D is made of the same material as that of the transistor (*i.e.*, both silicon or germanium). Hence, the diode (D) serve to compensate for the temperature dependent emitter-top-base voltage (V_{BE}). If the drops across

* We know that expression for sweep speed is I/C. In the present case, $I = I_C = \alpha \cdot I_E$. Therefore, sweep speed is $\alpha \cdot I_E/C$. Since, $I_E = (V_{EE} - V_{EB})/R_E$, therefore sweep speed $= \alpha\,(V_{EE} - V_{EB})/R_E$. It is evident from this relation that as the voltage V_{EB} decreases, the sweep speed increases.

diode and emitter-base junction are always the same, then the voltage across resistor (R_E) remains equal to (V_Z) and the emitter current is equal to V_Z/R_E, which is constant. Thus the sweep speed is essentially independent of temperature.

28-8. Miller Sweep Circuit

It is also called *Miller integrator.* Fig. 28-5 shows the circuit of a Miller integrator or sweep circuit. Here, the transistor Q_1 acts as a switch and transistor Q_2 is a common emitter amplifier, *i.e.*, a high gain amplifier. The operation of the circuit may be understood from the condition that initially, the transistor Q_1 is ON and Q_2 is OFF. At this instant, the voltage across the capacitor and the output voltage is equal to V_{CC}.

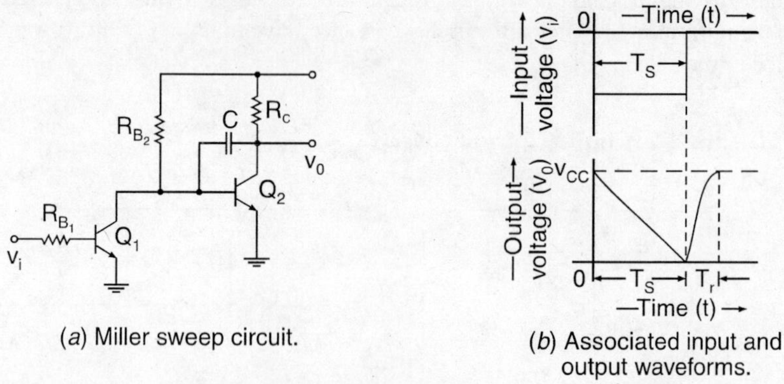

Fig. 28-5.

Now suppose a pulse of negative polarity as shown in Fig. 28-5(*b*) is applied at the base of the transistor Q_1. As a result of this, the emitter-base junction of the transistor Q_1 is reverse biased and it turns OFF. This causes the transistor Q_2 to turn ON. As the transistor Q_2 conducts, the output voltage begins to decrease towards zero. Since, the capacitor (C) is coupled to the base of transistor Q_2, therefore the rate of decrease of the output voltage is controlled by the rate of discharge of capacitor (C). The time constant of the discharge is given by the relation,

$$\tau = R_B \cdot C$$

Since, the value of time constant is very large, therefore the discharge current practically remains constant. As a result of this, the run down of the collector voltage is linear. When the input pulse is removed, the transistor Q_1 turns ON and Q_2 turns OFF. It will be interesting to know that as the transistor Q_1 turns OFF, the capacitor (C) charges quickly, through resistor R_C to V_{CC} with a time constant (τ) equal to $R_C.C$. The waveform of the generated sweep or the output voltage(v_0). is shown in Fig. 28-5(*b*).

It may be noted from the circuit that we have shown only one stage of amplification. But in actual practice more than one stage of amplification is used. Usually, the amplifier stage uses emitter followers at the input and output to increase the low-frequency input and output resistances (*i.e.*, R_i and R_o). Sometimes, the resistor R_C in the Miller sweep circuit is replaced by a diode. The diode forward resistance (R_f) helps the capacitor to charge quickly from V_{CC}. This reduces the flyback time (T_r) of the generated sweep. The Miller sweep circuit provides an excellent sweep linearity as compared to other sweep circuits.

28-9. Bootstrap Sweep Circuit

Fig 28-6(*a*) shows a practical form of a bootstrap sweep circuit. Here the transistor Q_1 acts as a switch and Q_2 as an emitter follower (*i.e.*, a ***unity gain amplifier***). The operation of the circuit may be explained as follows.

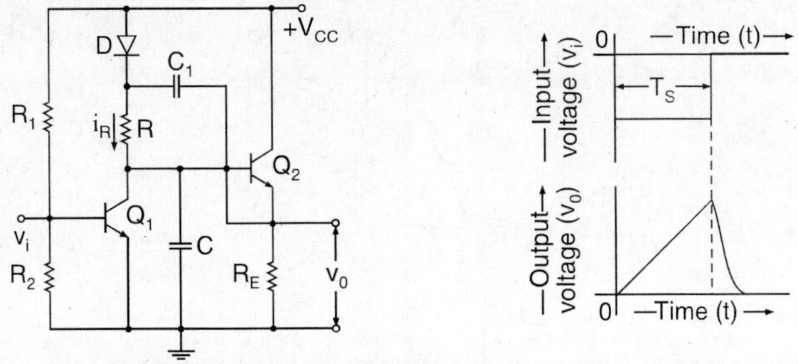

(*a*) Bootstrap sweep circuit. (*b*) Associated waveforms.

Fig. 28-6.

Initially, the transistor Q_1 is ON and Q_2 is OFF. Therefore, the capacitor C_1 is charged to V_{CC} through the diode forward resistance (R_F). At this instant, the output voltage (v_o) is zero. When a negative pulse as shown in Fig. 28-6 (*b*) is applied to the base of transistor Q_1, it turns OFF. Since, transistor Q_2 is an emitter follower, therefore the output voltage (v_o) is the same as the base voltage of transistor Q_2. Thus, as the transistor Q_1 turns OFF, the capacitor (C_1) starts charging this capacitor (C) through resistor (R). As a result of this, both the base voltage of Q_2 and the output voltage begins to increase from zero. As the output voltage increases, the diode D becomes reverse biased. It is because of the fact that the output voltage is coupled through the capacitor (C_1) to the diode. Since, the value of capacitor (C_1) is much larger than that of capacitor (C), therefore the voltage across capacitor (C_1) practically remains constant. Thus, the voltage drop across the resistor (R) also remains constant. Because of this, the current i_R through the resistor R also remains constant. It means that capacitor C is charged with a constant current. This causes the voltage across the capacitor C (and hence the output voltage) to increase linearly with time.

It is evident from the above discussion that the circuit pulls itself up by its own bootstrap and hence it is known as *bootstrap sweep circuit*. When the negative pulses removed from the input of transistor Q_1, the capacitor C discharges rapidly through Q_1 and the output voltage returns to zero. Then capacitor C_1 again charges to voltage V_{CC} through diode D. The waveform of the output voltage is as shown in Fig. 28-7 (*b*).

28-10. Current Time Base Generator

We have already discussed the methods and circuits by means of which we can generate a voltage that varies linearly with the time. Such a voltage may be applied to a resistor, thereby generating a current that increases linearly with time. However, the more accurate method is to cause a linearly time varying current to flow through an inductor. Such circuits are used in radar and television displays.

Fig. 28-7(*a*) shows a simple circuit of a current time base generator. Here an inductor (L) in series with a transistor is connected across the V_{CC} supply. The transistor operates as a switch in this circuit. The gating waveform at the base operates between two levels V_1 and V_2 as shown on Fig. 28-7(*b*). The lower level (V_1) keeps the transistor in cut-off, while the upper level drives the transistor into saturation.

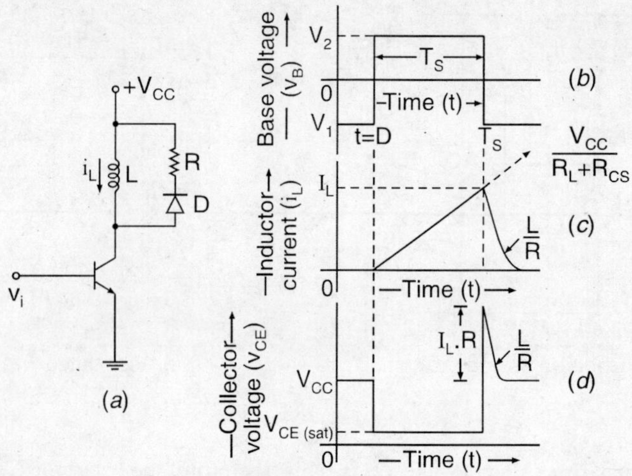

Fig. 28-7. *A current time base circuit.*

The operation of the circuit may be understood from the condition that when the transistor switch is turned ON, then neglecting the effect of small saturation resistance (R_{CS}), the current through an inductor (i_L) increases linearly with the time. The diode D does not conduct during the sweep, because it is reverse biased. The growth of current through an inductor is given by the relation,

$$i_L = \frac{V_{CC}}{R_L + R_{CS}}\left[1 - e^{\left(\frac{R_L + R_{CS}}{L}\right)t}\right]$$

where, R_L is the coil resistance.

Thus, if $t < T_S << L/(R_L + R_{CS})$, the above expression may be approximated as,

$$i = \frac{V_{CC}}{L} \cdot t$$

It may be noted that the sweep terminates at $t = T_S$, then the gating signal drives the transistor to cut-off. The inductor current then continues to flow through the diode (D) and resistor (R), until it reduces to zero as shown in Fig. 28-7 (*c*).

Fig. 28- 7 (*d*) shows the waveform of the collector-to-emitter voltage (v_{CE}) of a transistor. It may be noted that before the transistor is turned ON and some time after it has been turned OFF, the voltage $v_{CE} = V_{CC}$. But during the time interval, in which the transistor is ON, the voltage v_{CE} is quite low, *i.e.*, equal to $V_{CE\,(sat)}$. It may also be noted that at the moment the transistor is turned OFF, a large spike of amplitude $I_L \cdot R$ appears across the inductor. This voltage must be kept below the breakdown voltage, so that it does not damage the transistor.

The sweep-speed error for the current time base circuit is given by the relation,

$$e_S = \frac{T_S\,(R_L + R_{CS})}{L}$$

It is evident from the above expression that to maintain linearity the time constant $L/(R_L + R_{CS})$ must be kept large enough as compared to the sweep time (T_S).

SUMMARY

In this chapter you have learned that :

1. The circuits in which output voltage (or circuit) is a linear function of time over a specified time interval are called time base circuits.
2. There are two types of time base circuits : (*i*) free running and (*ii*) triggered time base generator.
3. The time base signal can be generated by (*i*) charging a capacitor through a resistor to a voltage which is small in comparison with supply voltage, (*ii*) charging a capacitor linearly from a constant current source, (*iii*) using Miller integrator to convert a step into ramp waveform.

GLOSSARY

Bootstrap sweep circuit : A Time base circuit that pulls itself by its own bootstrap.

Current time base generator : An inductor placed in series with a transistor across the V_{CC} supply.

Miller Integrator : A circuit used to generate a time base signal.

Sweep Circuit : A circuit in which an electronic switch (say transistor) is connected across a capacitor. The capacitor is charged/discharged by opening/closing a switch electronically.

Time base signal : A signal required in deflection coils of television or radar receivers.

NUMERICAL PROBLEMS

1. In a simple UJT sweep circuit, the values of R and C are 47 kΩ and 0.2 μF respectively. Determine the period and frequency of oscillation. Assume $\eta = 0.56$.

   (**Ans.** 7.7 ms and 129.6 Hz)

2. Determine the value of R for a UJT relaxation oscillator having $C = 0.01$ μF, $\eta = 0.60$ and $f = 10$ kHz. (**Ans.** 10.9 kΩ)

3. The square-wave shown in Fig. 28-8 (*a*) is used as excitation for the R-L circuits of Fig. 28-8 (*b*). Sketch the waveform of the output voltage, v_o when L/R is much smaller than either T_1 or T_2. (*AMIE., Summer* 1993)

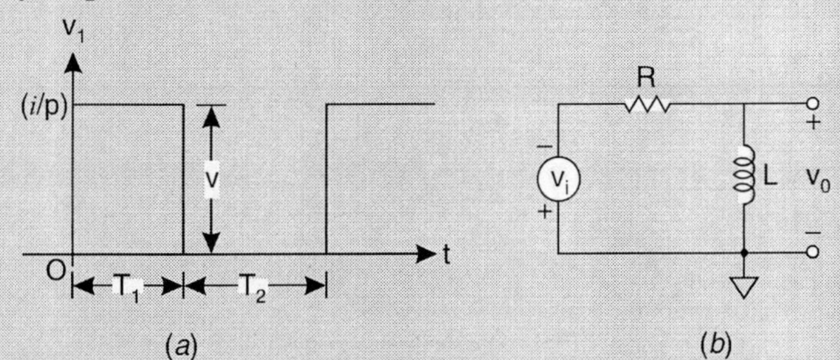

Fig. 28-8.

4. It is desired to drive the coil in the following head of a magnetic storage drum with the coil current waveform indicated in Fig. 28-9. The coil inductance is 0.4 mH and the coil resistance is 1.5 Ω and the self-resonant frequency of the coil is 5 MHz.

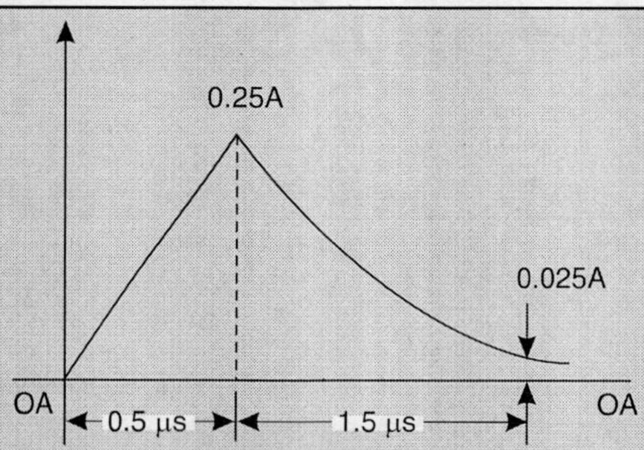

Fig. 28-9.

Assume that the external stray capacitance across the coil is 25 pF. Calculate (*a*) the damping resistance and (*b*) Sketch the current waveform of the driver. (*AMIE., Winter* 1993)

DESCRIPTIVE QUESTIONS

1. What is the need of a time base signal?
2. Describe the following terms with reference to a time base signal
   (*a*) Sweep time (*b*) Flyback time (*c*) Slope error (*d*) Sweep speed
3. Explain the difference between a free running time base generator and a triggered time base generator.
4. What are different methods of generating a time base signal?
5. Explain how a sweep can be generated by charging a capacitor *C* through resistor *R* directly from the source *V*.
6. Describe the working of a sweep circuit using transistor switch. Also draw the associated waveforms.
7. Draw a circuit for generating sweep using unijunction transistor switch. Give expressions for sweep period and frequency.
8. Explain the working of a Miller integrator type of sawtooth voltage generator.
9. Give the simple Miller sweep circuit using NPN transistor.
10. Explain the working of a Bootstrap sweep circuit. Also give the expression for sweep-speed error.
11. Draw the circuit of a current time base generator. Explain the working of the circuit with the associated waveforms.
12. Define and explain the following terms related to sweep generators : (*i*) Sweep speed error (*ii*) transmission error (*iii*) displacement error. (*AMIE., Summer* 1996)
13. Explain the principle of Miller sweep generator. (*AMIE., Summer* 1996)
14. Explain following terms associated with a time-base signal : slope error, displacement error and transmission error. (*W* 98)
    Explain the principle of a Miller time-base generator. (*W* 98)
15. Explain with the aid of neat circuit diagram, the principle of working of a current sweep generator. Show that a trapezoidal driving voltage is required for linear current sweep and explain clearly, its working. (*AMIE., Winter* 1993, 1996)

16. Sketch a circuit to generate a ramp wave.

Explain its operation. Also deduce the expressions for output frequency.

(*AMIE., Summer* 2001)

17. Give the block diagram of a general purpose sweep system and explain the function of each block. Explain how a unijunction transistor can be used to generator sweep waveforms. (*AMIE., Winter* 1996)

18. Explain the working of any bootstrap type time-base generator giving neat diagrams.

(*AMIE., Summer* 1997)

19. Sketch the circuit of a simple RC ramp generator. Briefly, explain its operation and its limitatitons. Also sketch the typical input and output waveforms. (*AMIE., Summer* 2002)

20. (*a*) How will you generate the ramp waveform using (*i*) The Miller circuit (*ii*) The Bootstrap circuit.

(*b*) Compare the waveforms generated using Miller and Bootstrap circuits.

(*c*) Draw the circuit diagram of a free running blocking oscillator and explain its operation.

List its applications. (*AMIE., Summer* 1994)

21. Sketch a Miller integrator circuit connected to operate as a ramp generator. Show the input and output waveforms and explain the circuit operation. (*AMIE., Winter* 2000)

22. Explain the working of any bootstrap type time-base generator giving neat diagrams.

(*AMIE., Summer* 2000)

23. (*a*) Draw the practical circuits of a bootstrap time-base generator. Discuss the factors which determine sweep speed and linearity of this sweep.

(*b*) Distinguish clearly between Miller and Bootstrap time-base generators.

(*AMIE., Winter* 1994)

24. Explain with suitable diagrams and waveforms the working of a UJT relaxation oscillator. Also derive its frequency of oscillation.

(*Electronic Circuits–II, Anna University, Nov/Dec 2007*)

25. Discuss briefly the operation of a current time base generator with necessary diagram.

(*Electronic Circuits–II, Anna University, May/June 2007*)

26. Define time base generator. (*Electronics Circuit–II, Anna University, May/June 2007*)

27. Explain with circuit diagram and waveforms of the transistor voltage time base generator that uses Bootstrapping. (*Electronic Circuits–II, Anna University, Nov/Dec 2007*)

28. What does bootstrapping mean ?

(*Electronic Circuits–I, Anna University, April/May 2008*)

MULTIPLE CHOICE QUESTIONS

1. Which the following time base circuit is suitable to display widely separated narrow-width pulses?

(*a*) Voltage time base generator (*b*) Current time base generator

(*c*) Free running time base generator (*d*) Triggered time base generator

2. The sweep speed of a simple RC circuit is given by

(*a*) I/C (*b*) $I \cdot t/C$

(*c*) T_S/RC (*d*) $V \cdot T_S/RC$

3. The sweep period in a sweep circuit using unijunction transistor is

(*a*) directly proportional to *R* (*b*) inversely proportional to *R*

(*c*) directly proportional to *C* (*d*) directly proportional to the product

4. If you require good linearity, which type of sweep generator you will use?

(*a*) An RC circuit with SCR (*b*) A UJT sweep circuit

(*c*) A Miller integrating type circuit (*d*) A 555-times circuit

5. Which of the following sweep circuit is used in television receivers for display?

(*a*) Voltage time base generator (*b*) Current time base generator

(*c*) Bootstrap sweep circuit (*d*) None of the above

6. In the circuit shown in Fig. 28-10 if $R_1 >> R_P$ and the pulses can saturate the transistor *Q*, then the output voltage, V_o will be :

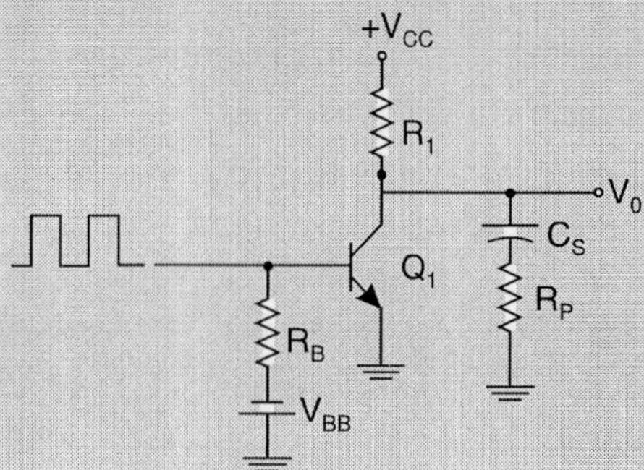

Fig. 28-10.

(*a*)  (*b*) 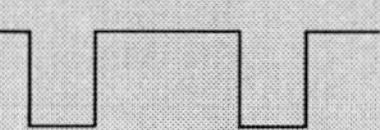

(*c*)  (*d*) 

ANSWERS

1. (*d*) **2.** (*a*) **3.** (*d*) **4.** (*c*) **5.** (*b*)

6. (*a*)

❑❑❑

29

CHAPTER

Operational Amplifiers (OP-AMPs)

OUTLINE

Objectives

After completing this chapter, you should be able to :

- Sketch the circuit symbol of an Op-amp.
- Identify the different types of Op-amp packages.
- List the important parameters of an Op-amp.
- Know the typical values of gain, input resistance, output resistance and common-mode rejection ratio (CMRR) of an Op-amp.
- Understand the relationship between slow rate and maximum frequency which an Op-amp can handle.
- Explain the operation of inverting amplifier and non-inverting amplifier.
- Design a summing amplifier to add different input voltages.

29-1. Introduction

An operational amplifier (or Op-amp in short) is a linear integrated circuit (IC) that has a very high voltage gain, a high input impedance and a low output impedance. The internal circuit of a very popular Op-amp (μA 741) is shown in Fig. 29-1(*a*). Its schematic symbol is shown in Fig. 29-1(*b*) and its pin diagram in Fig. 29-1(*c*).

As seen from Fig. 29-1(*b*), Op-amp is provided with two input terminals. One of the terminals, labelled '-' is known as ***inverting input terminal***. The word "inverting" implies that if a signal is applied to the '-' terminal of the op-amp, it appears with the opposite polarity at the output, *i.e.*, a sinusoidal signal will experience a phase shift of 180°. The other input terminal, labelled "+" is called the ***non-inverting input terminal***. In other words, a signal applied to this terminal of the op-amp is amplified without inversion. The op-amp actually amplifies the difference between the voltages applied to its two input terminals.

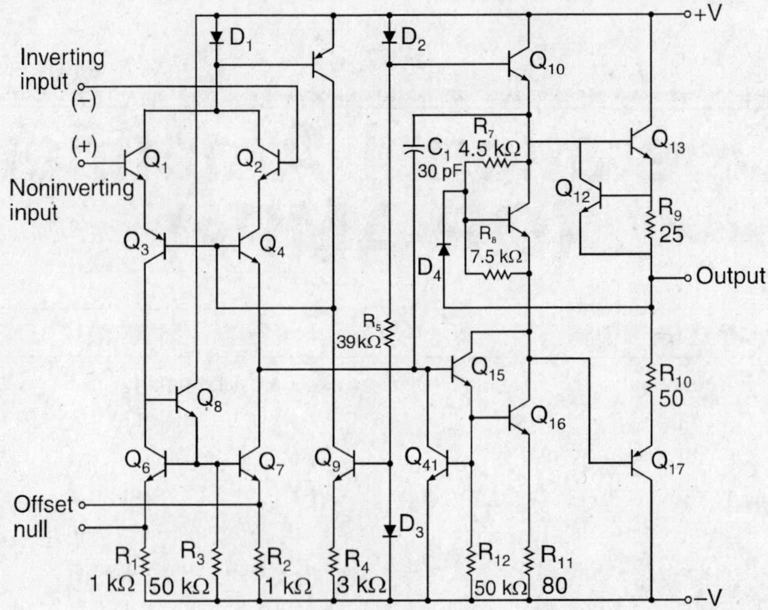

(a) Schematic diagram of a 741 Ic chip.

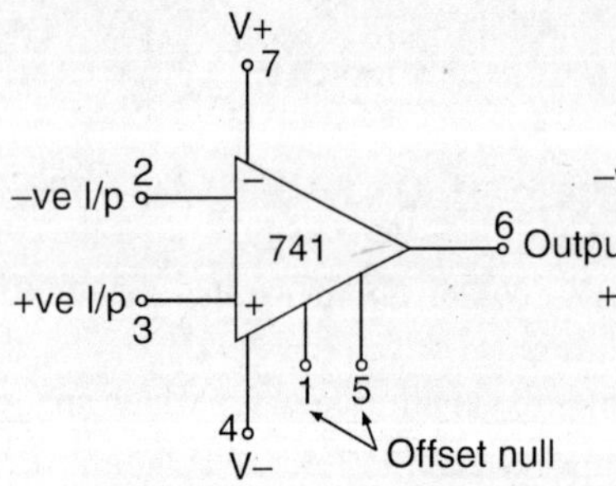

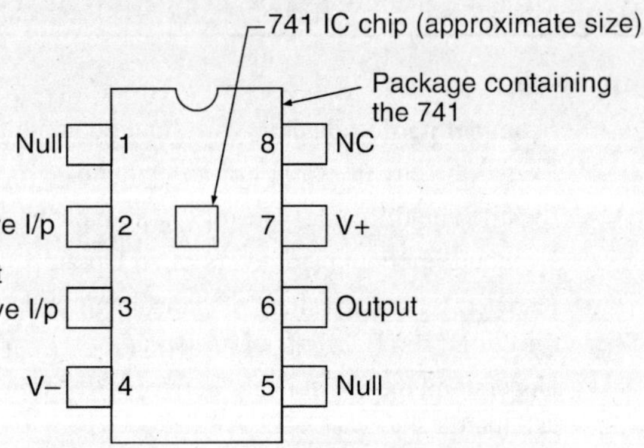

(b) Circuit symbol for a 741. (b) The 741 chip is shown packaged in an 8-pin dual-in-line package (top view).

Fig. 29-1. *OP-AMP.*

Two further terminals are provided for the connection of positive and negative power supply voltages. Both polarities are necessary so that the output voltage can vary either side of zero volts. Most op-amps operate satisfactory from a wide range of supply voltages and a few designed to operate from a single-polarity supply.

It may be carefully noted that the schematic diagram of the µA 741 op-amp shown in Fig. 29-1(*a*), that it makes use of all bipolar transistors. Such op-amps are called bipolar type op-amps. Some of the op-amps have a FET input circuit with the remainder of the circuit using bipolar transistors. Such op-amps are called FET type op-amps.

29-2. Operational Overview

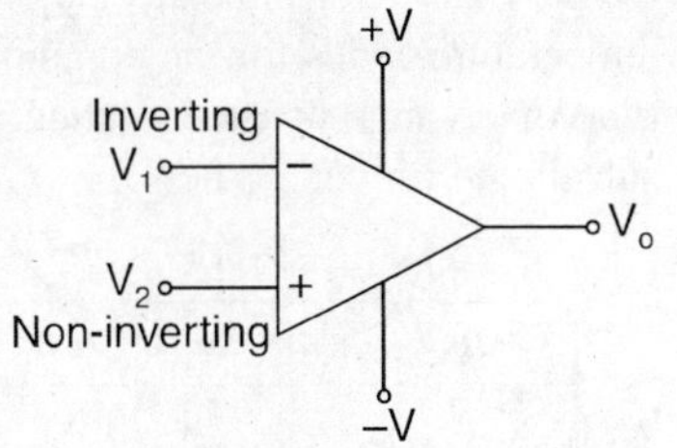

Fig. 29-2. *Illustrating the overview of an OP-AMP.*

The input circuit of an op-amp, which is driven by inverting and non-inverting inputs, is called a ***differential amplifier***. This means that the op-amp amplifies the difference between the voltages applied at the non-inverting input and inverting input. This point is illustrated in Fig. 29-2. In this Fig., v_1 is the voltage applied at the inverting input of the op-amp and v_2 is the voltage applied at the non-inverting input. The difference between the two voltages *i.e.*, $v_2 - v_1$ (or $v_2 - v_1$) *acts* as an input to the op-amp. Mathematically, the difference voltage that will be applied is,

$$v_{in} = v_2 - v_1$$

The output (v_o) from the op-amp for a given pair of input voltages depends upon the factors such as (1) the gain of the amplifier, (2) the polarity relationship between v_1 and v_2 and values of supply voltage + *V* and – *V*.

29-3. OP-AMP Supply Voltages

The op-amp has two supply voltage inputs, labelled as + *V* and – *V*. These two supply pins are normally connected in any one of the following three ways as shown in Fig. 29-3(*a*), (*b*), and (*c*).

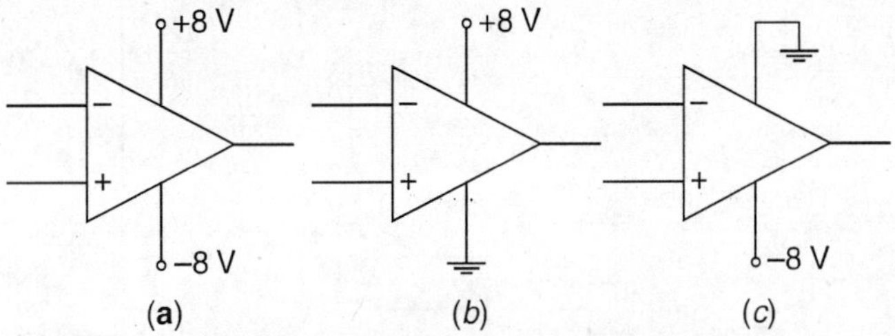

Fig. 29-3. *OP-AMP supply voltages.*

As seen from the Fig. 29.3, one way is to have + *V* and – *V* set to equal voltages (say 8 volts) that are of opposite polarities as shown in Fig. 29-3(*a*). The other way is to provide a single supply voltage to one of the supply pins and grounding the other as shown in Fig. 29-3(*b*) and (*c*). In this Figure, the supply voltage is connected to + *V* pin while the – *V* pin is grounded. Whereas in Fig. 29-3(*c*), the supply voltage is connected to –*V* pin and the + *V* is grounded. The choice among the three methods of connecting supply voltage to an op-amp depends on the particular application.

29-4. IC Identification

There are several types of op-amps produced in the form of integrated circuit (IC) by different manufacturers. These op-amps are identified using a seven character identification (ID) code. This code is shown in Fig. 29-4. The code has three parts namely *prefix*, *designator* and *suffix*. The prefix is used to identify the particular manufacturer. A listing of the most common prefixes is given in Table 29-1.

Prefix	*Designator*	*Suffix*
MC	741 C	N

Fig. 29-4. *An OP-AMP ID code.*

Table 29-1. *Manufacturer Prefixes*

Prefix	*Manufacturer*
AD	Analog Devices
CA	RCA
LM	National Semiconductor
MC	Motorola
NE/SE	Signetics
OP	Precision Monolithics
RC/RM	Ratheon
SG	Silicon General
TI	Texas Instruments
UA	Fairchild

The designator code indicates two things: (1) the three-digit number indicates the specific type of op-amp,

(2) the final letter indicates the operating temperature range. Table 29-2 gives the commonly used temperature codes for the op-amps. The designator code is used not only to determine the specific type of op-amp you are dealing with, but also to help you to determine which op-amps can be substituted for each other.

Table 29-2. *Temperature codes*

Code	*Temperature range (°C)*	*Application*
C	0 to 70	Commercial
I	– 25 to 85	Industrial
M	– 55 to 125	Military

Table 29-3. *Suffix codes*

Code	*Package type*
D	Plastic Dual-in-line (DIP)
J	Ceramic DIP
N, P	Plastic DIP with longer lead

29-5. OP-AMP Packages

The op-amps are available in two packages: (1) dual-in-line (DIP) and metal can as shown in Fig. 29-5. Metal cans (type TO-5) are available with 8, 10 or 12 leads. The DIP packages of op-amps have 8 or 14 pins. The DIP package is more common than metal can. The type of package is an important factor when considering whether or not the op-amp may be substituted for another.

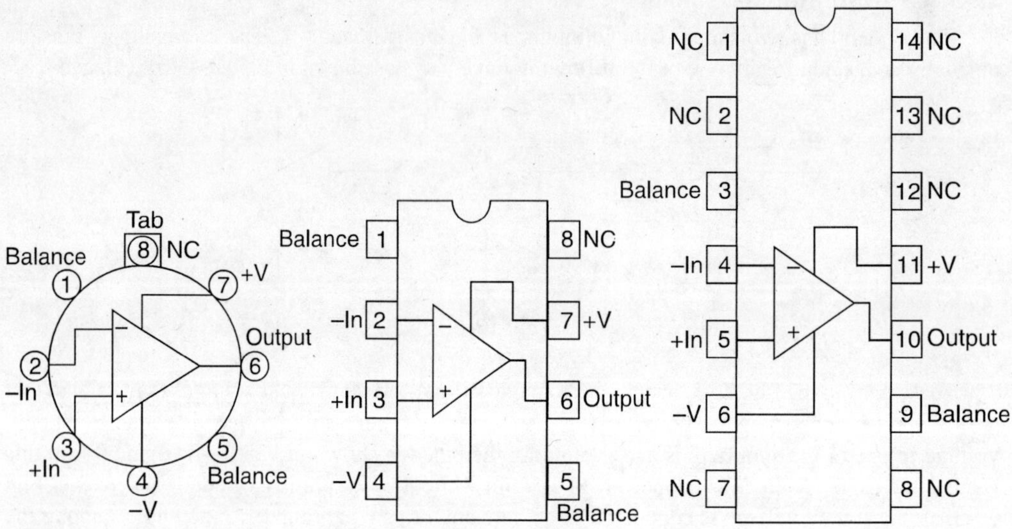

(*a*) Metal can TO-5. (*b*) 8 pin DIP. (*c*) 14 pin DIP.

Fig. 29-5. *OP-AMP packages.*

29-6. OP-AMP Parameters

Following are some of the important op-amp parameters which are necessary to understand the fundamental concepts of op-amp operation.

(1) *Gain :* The ideal op-amp would have an infinite open-loop differential gain. But in all practical circuits using Op-amp, the gain falls short of the infinite value. Practical op-amps have open-loop gains which vary considerably from one type to another but may be somewhere between 25,000 to 3,00,000. The gain is often expressed in decibels (dB). If v_1 is the input and is the output-voltage of the op-amp, then gain in decibels.

$$\text{dB} = 20 \log_{10} \frac{v_2}{v_1}$$

(2) *Input Resistance :* Ideally the input resistance of an op-amp is infinitely high. But in practice, the input resistance may be any value between 250 kΩ and 40 MΩ for the op-amps with bipolar transistor input and 10^{12} Ω for the op-amps with field-effect transistor (FET) input.

(3) *Output Resistance* : Basically an op-amp is a voltage amplifier. Therefore, it should have its input resistance as low as possible. Practical op-amps have the output resistance of the order of 100 Ω.

(4) *Common-mode Rejection Ratio (CMRR)* : The output voltage of an op-amp is proportional to the difference between the voltages applied to the inverting and non-inverting terminals. Ideally when the two voltages are equal, the output voltage should be zero. A signal applied to both the input terminals is called a common-mode signal. It is nearly always an unwanted noise voltage. The ability of an op-amp to suppress common-mode signals is expressed in terms of its common-mode rejection ratio (CMRR). Mathematically, the common-mode rejection ratio is given by the relation,

$$CMRR = \frac{\textbf{Differential voltage gain}}{\textbf{common mode voltage gain}} = \frac{A_{dm}}{A_{cm}}$$

where, A_{dm} = Differential voltage gain, and

A_{cm} = Common-mode voltage gain.

The common-mode rejection ratio can also be expressed in decibels. Thus,

$$CMRR = 20 \log_{10} \frac{A_{dm}}{A_{cm}} \text{ dB}$$

Typically an op-amp might have a common-mode rejection ratio of 90 dB.

(5) *Slew rate* : The slew rate of an op-amp is the maximum rate at which its output voltage is capable of changing. It is expressed in volts per second (V/μs). For example, μA 741 op-amp has a slew rate of 0.5 V/μs. This means that the output from μA 741 op-amp can change by 0.5 V every micro-second. When a signal at a given frequency is applied to the op-amp, the maximum permissible output voltage is determined by the slew rate. And if we try to develop the output voltage higher than the maximum permissible value, the signal waveform will be distorted. This fact is illustrated in Fig. 29-6. In this Fig. we have shown the input (a square waveform) and output waveforms of an op-amp with inadequate slew rate.

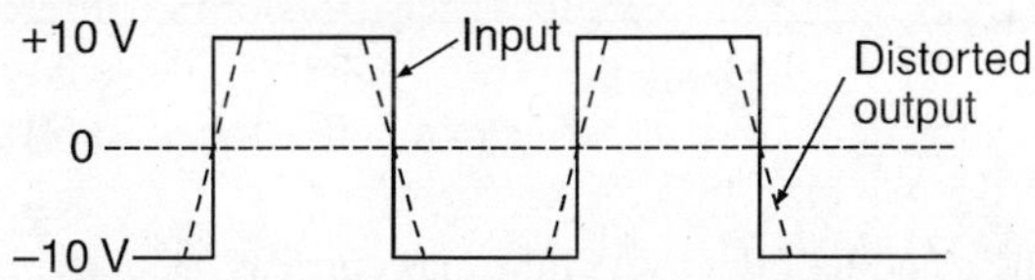

Fig. 29-6. *Illustrating the effect of inadequate slew rate.*

Typically the slew rate for op-amps is in the range of 0.3 – 12 V/μs. Since, frequency is a function of time, the slew rate can be used to determine the maximum operating frequency of the op-amp as follows :

$$f_{max} = \frac{\textbf{slew rate}}{2\pi V_{pk}}$$

where, V_{pk} is the maximum or peak value of the applied input signal.

(6) *Bandwidth* : The open-loop voltage gain of an op-amp is not constant at all frequencies but falls off at high frequencies because of the capacitive effects. The gain frequency characteristics of op-amps are specified by the manufacturers in two ways: ***(i)*** the bandwidth for large signals over which less than 5% distortion is obtained and ***(ii)*** the frequency at which the gain falls to unity.

Fig. 29-7 shows the gain-frequency characteristic of a μA 741 op-amp. As seen from this Fig., we find that at frequencies below 10 Hz, the voltage gain is constant (approximately 106 dB). But at higher frequencies, the gain falls off at the constant rate of 6 dB/octave. Its value becomes 0 dB (*i.e.*, = 1) at a frequency (f_t) of 1 MHz. Other op-amps have similar gain-frequency characteristics but with different values of low frequency gain and frequency (f_t) at which the gain reduces to 0 dB.

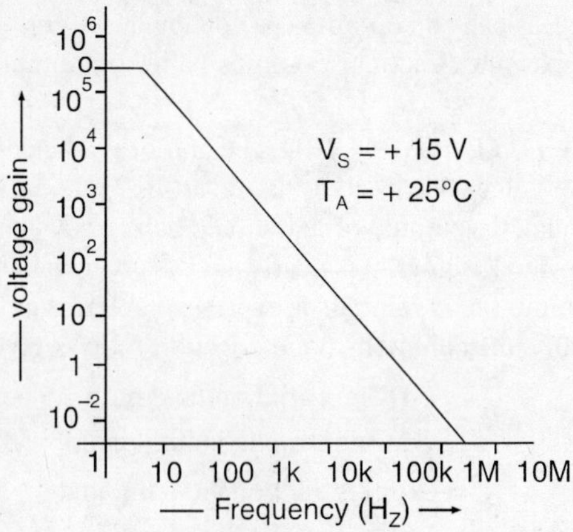

Fig. 29-7.

(7) ***Input Offset Voltage :*** Ideally speaking, the output voltage of an op-amp should be zero when the value of an applied voltage at both the input terminals is zero. However, for any practical op-amp, it is found that the output voltage does exist for zero input voltage. This voltage arises because of small imbalances within the op-amp. But for the convenience of discussion, we assume that the output is processed by an input offset voltage (V_{OS}).

The input offset voltage of an op-amp is equal to the output voltage for zero input voltage divided by the open-loop voltage gain of the amplifier. The input offset voltage is in the range of 10 μV to 7 mV for op-amps with bipolar transistor input type and 0.5 to 15 mV for op-amps with field-effect transistor (FET) input type. Although the offset voltage is small, yet it is amplified by the circuit and may not be of negligible amplitude at the output. Most op-amps are provided with some means of nulling out the effects of input offset voltage.

(8) ***Input Bias Current :*** The input circuit of an op-amp is always a differential amplifier and in many cases bipolar transistors are employed. Both transistors of the differential amplifier must be provided with a base bias current. To keep the circuit balanced, these base currents should be of equal value. However, in all practical op-amps due to manufacturing imperfections, there is always some difference between the two base bias currents.

The input bias current is one-half the sum of the bias currents taken by each input of the op-amp. These are the bias currents for the input transistors. Typically, the input bias current is 10 to 50 nA for a bipolar transistor input op-amp and as low as 10 to 100 pA for the FET input op-amp.

(9) ***Input Offset Current :*** The difference between the bias currents is known as the input offset current. Typical values are in the range of 3-20 nA for bipolar transistor input op-amps and a few picoamperes (pA) for FET input op-amps.

Example 29-1. *Calculate the common-mode rejection ratio (CMRR) of an op-amp, that has a differential gain of 2 00 000 and a common-mode gain of 6.33.*

Solution. Given : A_{dm} = 200 000 and A_{cm} = 6.33

We know that common-mode rejection ratio,

$$CMRR = 20 \log_{10}\left(\frac{A_{dm}}{A_{cm}}\right) \text{ in dB}$$

$$= 20 \log_{10}\left(\frac{200\ 000}{6.33}\right) \text{ in dB}$$

$$= 90 \text{ dB } \textbf{Ans.}$$

Example 29-2. *An op-amp has a CMRR of 90 dB. If its differential voltage gain is 30,000, calculate its common-mode gain.*

Solution. Given : *CMRR* = 90 dB and A_{dm} = 30 000

Let A_{cm} = The common-mode voltage gain of the op-amp.

We know that common-mode rejection ratio (CMRR)

$$90 = 20 \log_{10}\left(\frac{A_{dm}}{A_{cm}}\right)$$

$$= 20 \log_{10}\left(\frac{30\,000}{A_{cm}}\right)$$

or

$$4.5 = \log_{10}\left(\frac{30\,000}{A_{cm}}\right)$$

or

$$10^{4.5} = \frac{30\,000}{A_{cm}}; \quad \therefore A_{cm} = \frac{30\,000}{10^{4.5}} = 0.95 \textbf{ Ans.}$$

Example 29-3. *The amplifier shown in Fig. 29-8 is being used to amplify an input signal to a peak output voltage of 100 mV. What is the maximum operating frequency of the amplifier? The slew rate for µA 741 op-amp is 0.5 V/µs.*

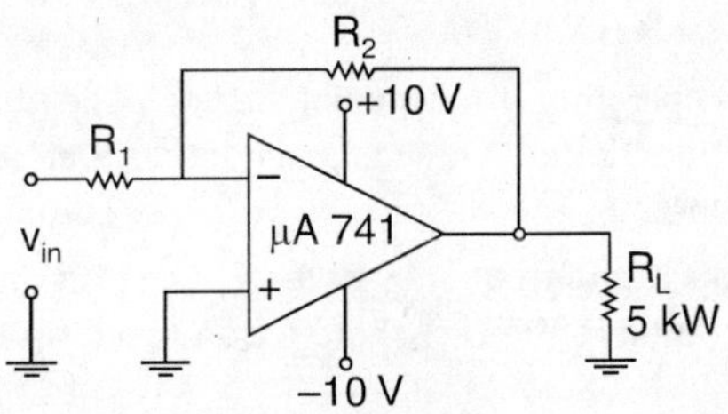

Fig. 29-8.

Solution. Given : Slew rate = 0.5 V/µs = 0.5 × 10^6 V/s; V_{pk} = 100 mV = 0.1 V.

We know that maximum operating frequency for the amplifier,

$$f_{max} = \frac{\text{slew rate}}{2\pi V_{pk}} = \frac{0.5 \times 10^6}{2\pi \times 0.01}$$

$$= 796 \times 10^3 \text{ Hz} = 796 \text{ Hz } \textbf{Ans.}$$

Example 29-4. *A 50 Hz square wave is to be amplified by an op-amp to have an output voltage swing ± 10 V. Two op-amps are available:* **(a)** *the µA 741 which has a slew rate of 0.5 V/µs, and* **(b)** *the TLO 81 with a slew rate of 13 V/µs. Determine if either or both of the op-amps are suitable for the purpose.*

Solution. Given : The frequency of the square wave (*f*) = 50 kHz = 50 × 10^3 Hz; Output voltage swing = ± 10 V.

We know that time-period of the square waveform,

$$T = \frac{1}{f} = \frac{1}{50 \times 10^3} = 20 \times 10^{-6} = 20 \text{ µs}$$

Now, in order to amplify the input signal the op-amp must change its output voltage from 10 V to – 10 V (from – 10 V to + 10 V), *i.e.*, a total voltage change of 20 V in one half-period, *i.e.*, 20/2 = 10 µs.

(*a*) *μA 741*

For this op-amp, the slew rate is 0.5 V/μs. Therefore in 10 μs, the output voltage of μA 741 can only change by 10 × 0.5 = 5 V. But the requirement is of 20 V. Therefore, μA 741 op-amp is not suitable.

(*b*) *TLO 81*

For this op-amp, the slew rate is 13 V/μs. Therefore in 10 μs, the output voltage of TLO 81 can change 10 × 13 = 130 V. But the requirement is only of 20 V. Therefore TLO 81 op-amp is suitable.

Alternative solution. Given $V_{pk} = 10$ V.

The maximum operating frequency for the μA 741 op-amp,

$$f_{max} = \frac{\text{slew rate}}{2\pi V_{pk}} = \frac{0.5 \text{ V/μs}}{2\pi \times 10}$$

$$= \frac{0.5 \times 10^6}{20\pi} = 7957.7 \text{ Hz} = 7.957 \text{ Hz}$$

Similarly, the maximum operating frequency for the TLO 81 op-amp,

$$f_{max} = \frac{\text{slew rate}}{2\pi V_{pk}} = \frac{13 \text{ V/μs}}{2\pi \times 10}$$

$$= \frac{13 \times 10^6}{20\pi} = 206.9 \text{ KHz}$$

Comparing the maximum operating frequency of the μA 741 (7.957 kHz) with the frequency of the input signal (50 kHz) we find that the value is much lower than that of the input signal. Hence, the μA 741 is not a suitable op-amp.

However, the maximum operating frequency of the TLO 81 op-amp is much higher (206.9 kHz) than the frequency of the input signal (50 kHz), therefore TLO 81, op-amp can be used to amplify 50 kHz square wave signal.

29-7. The OP-AMP as a Voltage Amplifier

We have already mentioned in Art 29-1, that op-amp has a very high voltage gain. But it will be interesting to know that an op-amp is never used without the application of negative or positive feedback. When an op-amp is used as a voltage amplifier, a large amount of negative feedback must be applied to specify accurately the voltage gain. There are two ways in which an op-amp can be connected to act as an amplifier–one method gives an inverted gain and the other gives a non-inverting gain. The method of connecting the op-amp which produces inverting gain is called ***inverting amplifier***. Similarly, the method of connecting the op-amp which produces non-inverting gain is called ***non-inverting amplifier***. Now we will discuss both the above types of voltage amplifiers one by one in the following pages.

29-8. Inverting Amplifier

Fig. 29-9 shows the connection method for producing the inverted gain using op-amp. Such a configuration is called *inverting amplifier*. The op-amp circuit makes use of a single resistor (R_1) and a single feedback resistor (R_2). Notice that resistor R_1 is connected between the input terminal of the amplifier circuit and inverting input terminal (–) of the op-amp. While the resistor R_2 is connected between the inverting input terminal (–) and the output terminal of the op-amp. The non-inverting terminal (+) is connected to the ground. The

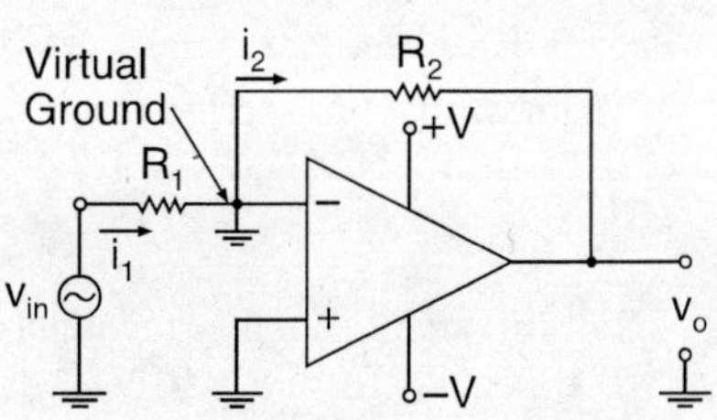

Fig. 29-9. *Inverting Amplifier.*

inverting amplifier produces a 180° phase shift in voltage from input to output. Thus, the input and output signals of the inverting amplifier are not in-phase with each other.

Now we know that the op-amp gain without any feedback (called open-loop gain) is very high. This means that the voltage at the inverting (*i.e.*, '–') terminal must be very small. As a matter of fact, the input voltage at the inverting terminal will be very nearly at the same potential as the non-inverting terminal. Now since, the non-inverting input is grounded, the inverting input of an op-amp is also at the ground potential and is referred to as *virtual ground.*

Since, the inverting terminal is virtual ground, therefore the entire input voltage appears across R_1 and the input current, $i_1 = v_{in}/R_1$. Moreover, the input resistance (R_{in}) of an op-amp is very high (ideally infinite), so very little current flows into the op-amp. Practically, the current drawn by the op-amp is negligible or nearly zero. Thus, all the input current flows through R_2 (*i.e.*, $i_1 = i_2$). And the voltage developed across R_2 is equal to the output voltage (v_o) of the circuit. Using this knowledge, let us develop expressions for the voltage gain, input resistance, output resistance and common mode rejection ratio of the inverting amplifier.

We have already mentioned above that the input voltage of an op-amp,

$$v_{in} = i_1 \cdot R_1, \quad \text{or} \quad i_1 = v_{in}/R_1$$

And output voltage,

$$v_o = -\,i_2 \cdot R_2, \quad \text{or} \quad i_2 = -\,v_o/R_2$$

The negative sign indicates that output is at a lower potential than that of a virtual ground. It is obvious because current through resistor R_2 is flowing from the side connected to the inverting input to the side connected to the output (see Fig. 29-7). Since, the input- and output-currents are equal (*i.e.*, $i_1 = i_2$), because op-amp input current is negligible.

$$\therefore \qquad \frac{v_{in}}{R_1} = \frac{v_o}{R_2}$$

$$\text{or} \qquad v_o = \frac{R_2}{R_1} v_{in}, \quad \text{or} \quad \frac{v_o}{v_{in}} = -\frac{R_1}{R_2} \qquad \text{...}(i)$$

Now recall that voltage gain (A_v) of an amplifier is defined as the ratio of output voltage to the input voltage. Mathematically,

$$A_v = \frac{\text{Output voltage}}{\text{Input voltage}} = \frac{v_o}{v_{in}}$$

Substituting the value of (v_o/v_{in}) from equation (*i*) into the above equation, we get,

$$A_v = -\frac{R_2}{R_1}$$

The above equation indicates that voltage gain of an inverting amplifier can be determined by dividing the value of R_2 by R_1. The negative sign indicates that the inverting amplifier output is 180° out of phase with respect to the input.

Now that we have discussed a basic op-amp circuit, let us give practical meaning to few terms.

(*a*) ***The open-loop and closed-loop gain :*** The open-loop voltage gain (A_{OL}) of an op-amp is the gain that is measured when there is no feedback path (*i.e.*, no physical connection) between the output and input circuit. When a feedback path is present, such as the R_2 connection in the inverting amplifier, the resulting circuit gain is referred to as the closed-loop gain (A_{CL}).

(*b*) ***Input Resistance (R_{in}) :*** Although, the op-amp has an extremely high input resistance (ideally infinite) yet the inverting amplifier does not have that high value. Referring to Fig. 29-10, we find that the voltage source "sees" the overall resistance to be as the parallel combination of input resistance (R_1) that is going to virtual ground and the op-amp input resistance. But since, the input resistance of an op-amp is very high, therefore input resistance of an inverting amplifier is,

$$R_{in} \approx R_1$$

It may be noted carefully that the value of R_1 will always be much less than the input resistance of the op-amp, therefore, the overall input resistance of an inverting amplifier will be much lower than the op-amp input resistance.

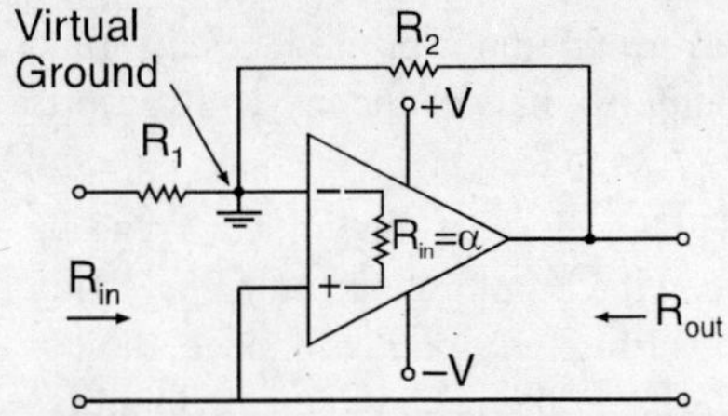

Fig. 29-10. *Illustrating input and output resistance of an inverting amplifier.*

(*c*) ***Output Resistance :*** Referring to Fig. 29-10, again, we find that output resistance of an inverting amplifier is the parallel combination of R_2 and the output resistance of op-amp itself. Since, the output resistance of an op-amp is usually very low (of the order of 100 Ω), the presence of R_2 reduces the overall output resistance of the inverting amplifier below 100 Ω.

(*d*) ***Common-mode Rejection ratio :*** We have already defined common-mode rejection ratio (CMRR) of the op-amp as the ratio of differential gain to common-mode gain. Since, the differential gain of the inverting amplifier (A_{CL}) is much less than the differential gain of the op-amp (A_{OL}), therefore CMRR of the inverting amplifier will be much lower than that of the op-amp. Mathematically, CMRR of the inverting amplifier is,

$$CMRR = \frac{A_{CL}}{A_{CM}}$$

where, = The closed loop voltage gain of the inverting amplifier, and

= The common-mode gain of the op-amp.

Example 29-5. *Determine the maximum allowable value of v_{in} for the circuit shown in Fig. 29-11. Assume that the gain of the amplifier is 200.*

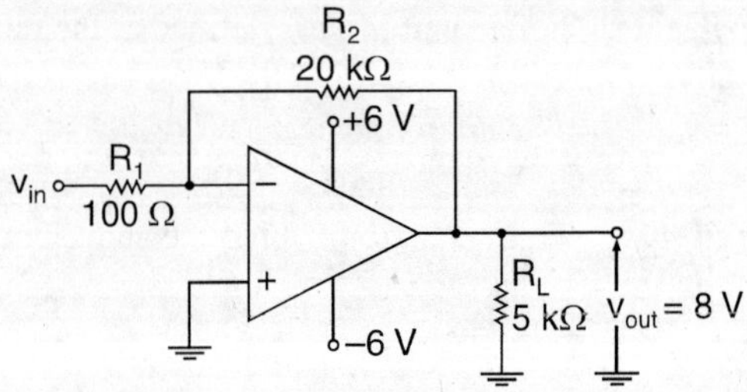

Fig. 29-11.

Solution. Given: $A_{CL} = 200$ and $v_{out} = 8$ V

We know that voltage gain of an inverting amplifier, (A_{CL}),

$$200 = -\frac{v_o}{v_{in}} = -\frac{8}{v_{in}}$$

$$v_{in} = -\frac{8}{200} = 0.04 = 40 \text{ mV} \textbf{ Ans.}$$

Example 29-6. *Determine the peak-to-peak output voltage for the circuit shown in Fig. 29-12. Also determine the maximum possible output values for the amplifier. Assume the gain of the amplifier is 150.*

Solution. Given: $A_{CL} = 150$

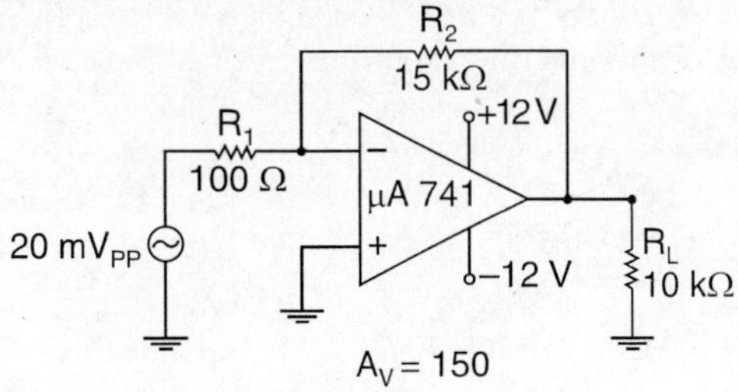

Fig. 29-12.

We know that voltage gain of an inverting amplifier, (A_{CL}),

$$150 = \frac{v_{out}}{v_{in}} = \frac{v_{out}}{200 \text{ m } V_{pk}}$$

$$v_{out} = 150 \times 20 \text{ m } V_{pk} = 3000 \text{ m } V_{pk} = 3 \ V_{pk} \ \textbf{Ans.}$$

According to the guidelines when dealing with 741, if the load resistance is between 2 k Ω to 10 kΩ, the maximum possible output values are

$$V_{pk\,(+)} = (+\text{ V}) - 2 \text{ V} = 12 - 2 = 10 \text{ V (maximum)}$$

and $$V_{pk\,(-)} = (-\text{ V}) + 2 \text{ V} = -12 + 2 = -10 \text{ V (maximum)}$$

Thus, maximum possible output value could be between + 10 V and – 10 V. **Ans.**

Example 29-7. *Fig. 29-13 shows an inverting amplifier with an input voltage v_i.*

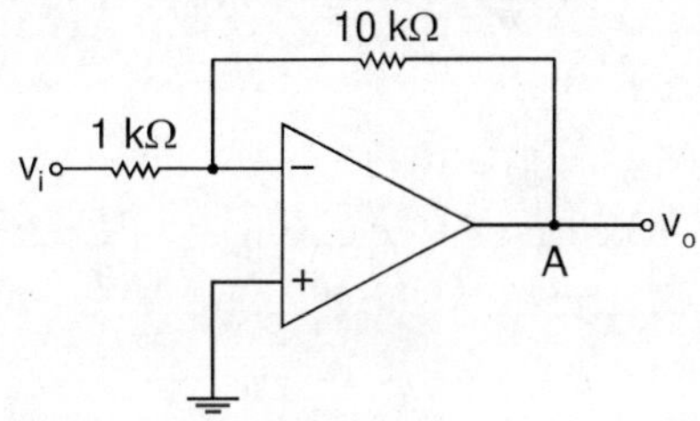

Fig. 29-13.

The input voltage is increased from 0.1 V to 0.4 V. what will be the voltage at node A in each case & why ? (*Bangalore Univ. B.Sc Electronics.* May 2001)

Solution. Given : The input voltage increased from 0.1 V to 0.4V.

$$R_1 = \text{k}\Omega; R_2 = 10 \text{ k}\Omega;$$

We know that voltage gain of the inverting amplifier,

$$\text{A}_{CL} = \frac{R_2}{R_1} = \frac{10 \text{ k}}{1 \text{ k}} = 10$$

Thus, when the input voltage is increased from 0.1 V to 0.4 V, the voltage at node A increases from 1 V to 4 V **Ans.**

Example. 29-8. *Fig. 29-14 shows an inverting amplifier with an input voltage of 1 V.*

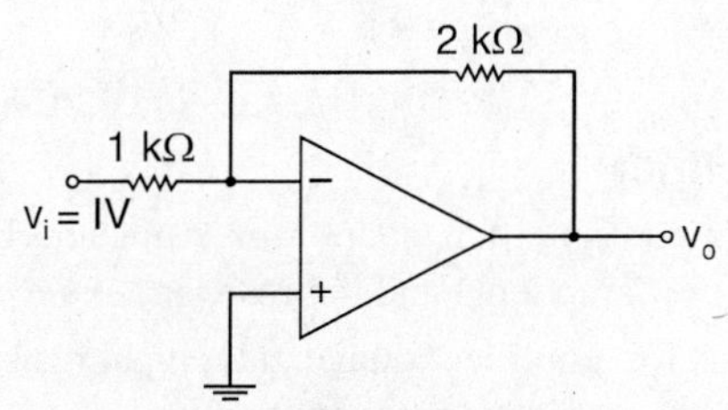

Fig. 29-14.

What will be value of output voltage? Justify. (*Bangalore Univ. B.Sc Electronics May* 2001)

Solution. Given : $R_1 = 1$ kΩ; $R_2 = 2$ kΩ and $V_i = 1$V.

We know that the gain of the inverting amplifier,

$$A_{CL} = \frac{R_2}{R_1} = \frac{2\text{ k}}{1\text{ k}} = 2$$

∴ Output voltage of the inverting amplifier,

$$v_o = A_{CC} \times V_i = 2 \times 1 = 2 \text{ V } \textbf{Ans.}$$

Example 29-9. *For the circuit shown in Fig. 29-15, find the closed-loop gain, input impedance, common-mode rejection ratio (CMRR) and maximum operating frequency.*

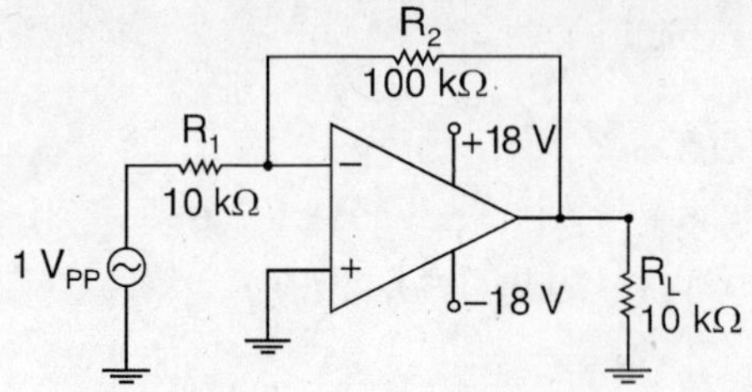

Op-Amp Paramaters.

$$A_{CM} = 0.001$$
$$A_{OL} = 180{,}000$$
$$Z_{in} = 1 \text{ M}\Omega$$
$$Z_{out} = 80\ \Omega \text{ (max)}$$
$$\text{Slew rate} = 0.5 \text{ V}/\mu\text{s}$$

Fig. 29-15.

Solution. We know that closed-loop gain,

$$A_{CL} = \frac{R_2}{R_1} = \frac{100\text{k}\Omega}{10\text{k}\Omega} = 10 \textbf{ Ans.}$$

and input impedance of the circuit,

$$Z_{in} \approx R_1 = 10 \text{ k}\Omega \textbf{ Ans.}$$

We also know that output impedance of the circuit,

$$Z_{out} \approx Z_{out} \text{ of the op-amp} = 80\ \Omega \textbf{ Ans.} \qquad (\because Z_{in} \text{ of op-amp} >> R_1)$$

and the common-mode rejection ratio,

$$CMRR = \frac{A_{CL}}{A_{CM}} = \frac{10}{0.001} = 10\ 000 \textbf{ Ans.}$$

To calculate the maximum operating frequency for the inverting amplifier circuit, we need to determine its peak output voltage.

With values of $v_{in} = 1\ V_{PP}$ and $A_{CL} = 10$, the peak-to-peak output voltage,

$$v_{out} = 1 \times 10 = 10\ V_{PP}$$

Thus, the peak output voltage is 5 and the maximum operating frequency

$$f_{max} = \frac{\text{slew rate}}{2\pi V_{pk}} = \frac{0.5 \text{ V}/\mu\text{s}}{2\pi \times 5}$$

$$= \frac{0.5/10^{-6} \text{ (V/s)}}{2\pi \times 5} = \frac{0.5 \times 10^6}{10\pi}$$

$$= 15.9 \times 10^3 \text{ Hz} = 15.9 \text{ kHz } \textbf{Ans.}$$

29-9. Non-Inverting Amplifier

Fig. 29-16 shows the basic configuration of a non-inverting amplifier. Note that the input signal is applied to the non-inverting op-amp input and the resistor R_1 is returned to ground.

Let us develop expressions for amplifier voltage gain, input resistance, and output resistance in the same way as we did in the case of inverting amplifier.

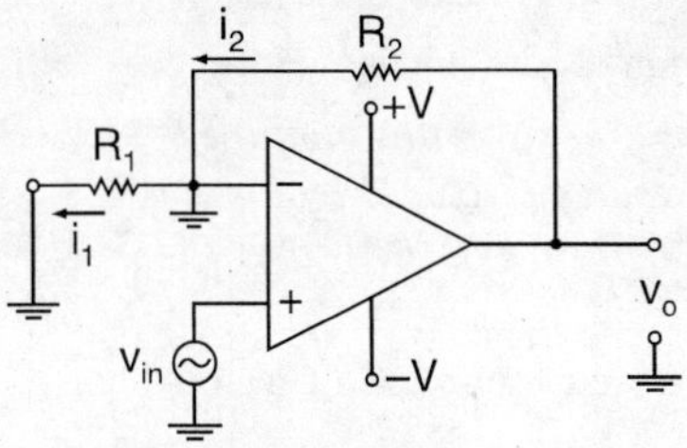

Fig. 29-16.

(1) ***Voltage gain :*** We know that the op-amp gain without feedback (called open-loop gain) is very high. This means that voltage at the inverting terminal should nearly be the same as that at the non-inverting terminal. Since, the non-inverting input is at a potential of v_{in}, therefore inverting input is also at the same potential. If i_1 is the current through resistor R_1, then its value,

$$i_1 = \frac{v_{in}}{R_1} \quad \text{or} \quad v_{in} = i_1 \cdot R_1 \qquad ...(i)$$

Since, the voltage drop across R_2 is equal to the difference between V_{in} and v_0, therefore current through resistor R_2,

$$i_2 = \frac{v_o - v_{in}}{R_2} \quad \text{or} \quad i_2 \cdot R_2 = v_o - v_{in}$$

$$\therefore \qquad v_o = v_{in} + i_2 \cdot R_2 \qquad ...(ii)$$

Now recall that op-amp draws a negligible input current for its operation. For all practical purposes, its value is considered to be zero, therefore, $i_1 = i_2$.

Replacing i_2 by i_1 in equation (*ii*), we get

$$v_o = v_{in} + i_1 \cdot R_2$$

The amplifier voltage gain,

$$A_v = \frac{\text{Output voltage}}{\text{Input voltage}} = \frac{v_o}{v_{in}}$$

Substituting the values of v_o and v_{in} from equation (*ii*) and (*i*) into the above equation,

$$A_v = \frac{v_{in} + i_1 \cdot R_2}{i_1 \cdot R_1} = \frac{i_1 \cdot R_1 + i_1 \cdot R_2}{i_1 \cdot R_1} = \frac{R_1 + R_2}{R_1}$$

or

$$A_v = 1 + \frac{R_2}{R_1}$$

The above equation shows that voltage gain of a non-inverting amplifier will always be greater than the equivalent gain of the inverting amplifier by a value of 1. For example, if an inverting amplifier has a gain of 100, the equivalent non-inverting amplifier will have a gain of 101.

(2) ***Input resistance :*** The non-inverting amplifier has extremely high input resistance because the input signal is applied directly to the op-amp (*i.e.*, not through a resistor R_1 as it happens in the case of inverting amplifier). As a matter of fact, the presence of feedback network causes the amplifier input resistance to be greater than the input resistance of the op-amp itself.

(3) ***Output Resistance :*** The output resistance of a non-inverting amplifier is lower than the output resistance of the op-amp itself. This characteristic is similar to that of the inverting amplifier.

(4) ***Common-mode Rejection Ratio (CMRR) :*** The common-mode rejection ratio (CMRR) for a non-inverting amplifier is determined in the same way as we do it for the inverting amplifier. Thus,

$$CMRR = \frac{A_{CL}}{A_{CM}}$$

where, A_{CL} = The closed-loop voltage gain, and

A_{CM} = The common-mode voltage gain.

Example 29-10. *For the non-inverting amplifier shown in Fig. 29-17, find the values of closed-loop gain (A_{CL}), common mode rejection ratio (CMRR), and the maximum operating frequency.*

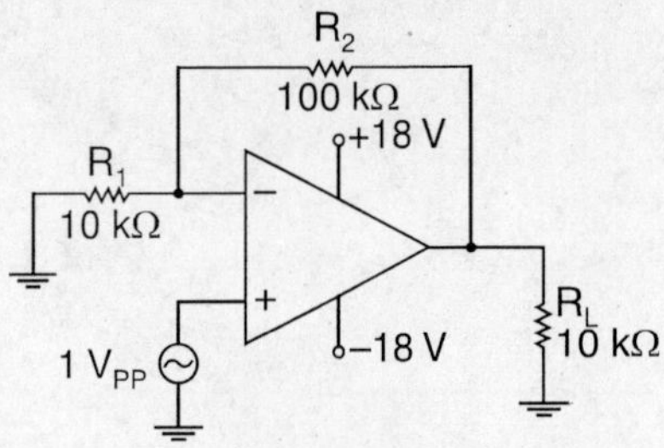

Fig. 29-17.

Solution. We know that closed-loop gain of a non-inverting amplifier,

$$A_{CL} = \left(1 + \frac{R_2}{R_1}\right) = \left(1 + \frac{100\text{ k}\Omega}{10\text{ k}\Omega}\right) = (1 + 10) = 11 \textbf{ Ans.}$$

and the common-mode rejection ratio,

$$CMRR = \frac{A_{CL}}{A_{CM}} = \frac{11}{0.001} = 11{,}000 \textbf{ Ans.}$$

To calculate the maximum operating frequency for the non-inverting amplifier circuit, we need to determine its peak output voltage.

With values of $v_{in} = 1\ V_{PP}$ and $A_{CL} = 11$, the peak-to-peak output voltage,

$v_{out} = 1 \times 11\ V_{PP}$

Thus, the peak output voltage is 5.5 V_{pk} and the maximum operating frequency,

$$f_{max} = \frac{\text{slew rate}}{2\pi V_{pk}} = \frac{0.5\text{ V}/\mu\text{s}}{2\pi \times 5.5}$$

$$= \frac{0.5 \times 10^6}{2\pi \times 5.5} = 14.47 \times 10^3 \text{ Hz}$$

$$= 14.47 \text{ kHz } \textbf{Ans.}$$

29-10. Non-Inverting Amplifier as a Buffer

The extremely high input resistance and extremely low output resistance of the non-inverting amplifier makes it very useful as a buffer. You may recall that the emitter follower and source follower are circuits that can be used to match a source with high input resistance to a load with low resistance. The non-inverting amplifier can be used for the same purpose. However, the major difference is that the basic non-inverting amplifier is capable of producing high voltage gain values while the emitter follower and source follower have voltage gain values which are less than '1'.

29-11. The Voltage Follower

The voltage follower circuit can be obtained from the non-inverting amplifier by removing resistors R_1 and R_2. The resulting circuit is as shown in Fig. 29-18. Notice that the input voltage is applied at the non-inverting input of an op-amp, and the inverting input is connected to the op-amp output. Because of the direct connection between the inverting input and output terminals, a 100% voltage-series feedback is applied. The voltage gain of a voltage follower may be obtained by using the equation of a non-inverting amplifier, *i.e.*,

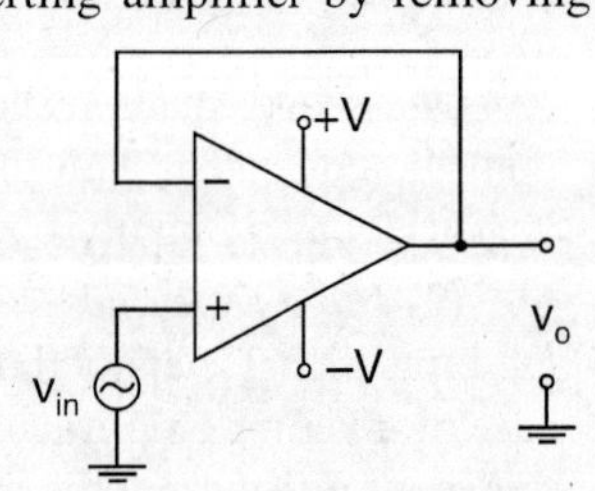

Fig. 29-18.

$$A_v = 1 + \frac{R_2}{R_1} = 1, \quad \because \text{ for a voltage follower, } R_1 = R_2 = 0$$

The voltage follower has a very high input resistance and a very low output resistance. The main application of a voltage follower is as a means of connecting a source with high input resistance to a load with low output resistance, *i.e.*, to act as a buffer amplifier.

The common-mode rejection ratio (CMRR) of a voltage follower circuit is given by the relation,

$$CMRR = \frac{A_{CL}}{A_{CM}}$$

Since, the closed loop gain (A_{CL}) is equal to '1', therefore the common-mode rejection ratio,

$$CMRR = \frac{1}{A_{CM}}$$

Example 29-11. *For the voltage follower shown in Fig. 29-19, find the values of A_{CL}, CMRR and f_{max}.*

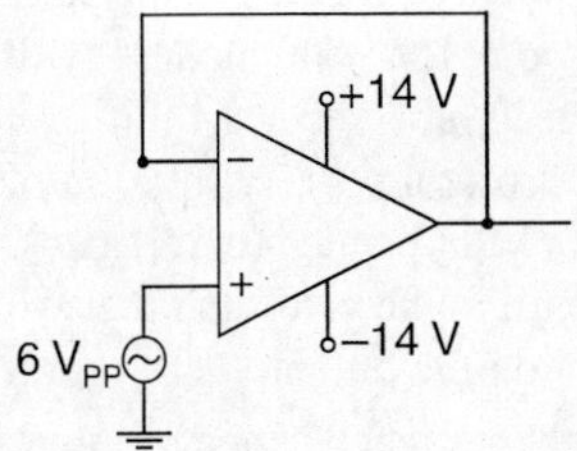

Op-Amp parameters.

$A_{CM} = 0.001$
$A_{OL} = 180{,}000$
$Z_{in} = 1\ M\Omega$
$Z_{out} = 80\ \Omega$ (max)
Slew rate $= 0.5\ V/\mu s$.

Fig. 29-19.

Solution. We know that for a voltage follower,

$$A_{CL} = 1 \textbf{ Ans.}$$

and the common-mode rejection ratio,

$$CMRR = \frac{A_{CL}}{A_{CM}} = \frac{1}{0.001} = 1000 \textbf{ Ans.}$$

To calculate the maximum operating frequency for the voltage follower, we need to determine its peak output voltage.

With values of $v_{in} = 6\ V_{PP}$ and $A_{CL} = 1$, the peak-to-peak output voltage,

$$v_{out} = 6 \times 1 = 6\ V_{PP}$$

Thus the peak output voltage is $3\ V_{pk}$ and the maximum operating frequency,

$$f_{max} = \frac{\text{slew rate}}{2\pi V_{pk}} = \frac{0.5\ V/\mu s}{2\pi \times 5} = \frac{0.5 \times 10^6}{2\pi \times 3} = 26.5 \text{ kHz } \textbf{Ans.}$$

29-12. Summing Amplifier

The summing amplifier is an op-amp circuit that accepts several inputs and provides an output that is proportional to the sum of inputs. The basic summing amplifier is shown in Fig. 29-20.

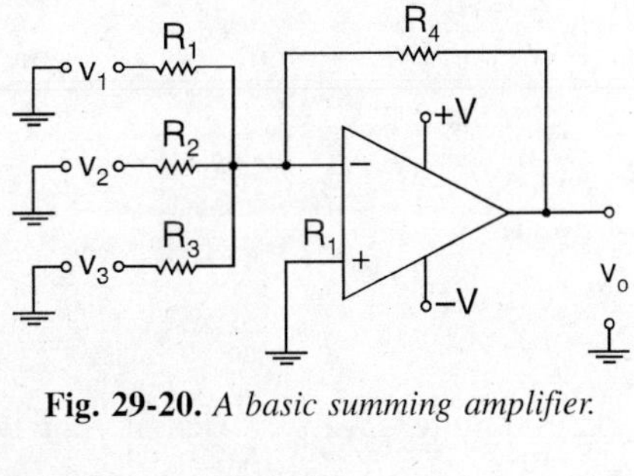

Fig. 29-20. *A basic summing amplifier.*

The operation of the circuit may be understood as follows. Let us suppose we have an input voltage only at V_1, then output voltage,

$$v_o = -\frac{R_4}{R_1} \cdot v_1$$

Similarly, if v_2, is the only input, the output voltage is,

$$v_o = -\frac{R_4}{R_2} \cdot v_2$$

And if v_3 is the only input, the output voltage is,

$$v_o = -\frac{R_4}{R_3} \cdot v_3$$

However, if all the three voltages are present, the output voltage would be,

$$v_o = \left(-\frac{R_4}{R_1} \cdot v_1\right) + \left(-\frac{R_4}{R_2} \cdot v_2\right) + \left(-\frac{R_4}{R_3} \cdot v_3\right)$$

And if $R_1 = R_3 = R_4 = R$, then output voltage,

$$v_o = -(v_1 + v_2 + v_3)$$

It is evident from the above equation that the output voltage is equal to the sum of three input voltages. The negative sign indicates that the output voltage is 180° out of phase with respect to the three input voltages. It may be noted carefully, that there is no practical limit on the number of summing inputs. As long as the amplifier is capable of providing the correct output voltage, the circuit can have any number of inputs. The equation (*i*) can be expanded (or reduced). Moreover the resistor used at the amplifier input do not necessarily have the same value. In other words, the inputs can be weighted so that certain inputs have more of effect on the output.

A summing amplifier can be used to produce the mathematical average of input voltages. This is obtained by setting the ratio of the feedback resistor to the input resistor equal to the reciprocal of the number of inputs. This is because of the reason that the average of several numbers is obtained by adding the numbers and then dividing by the quantity of numbers we have. Such a summing amplifier is known as an ***averaging amplifier***.

Example. 29-12. *Fig. 29-21 shows the circuit of a summing amplifier.*

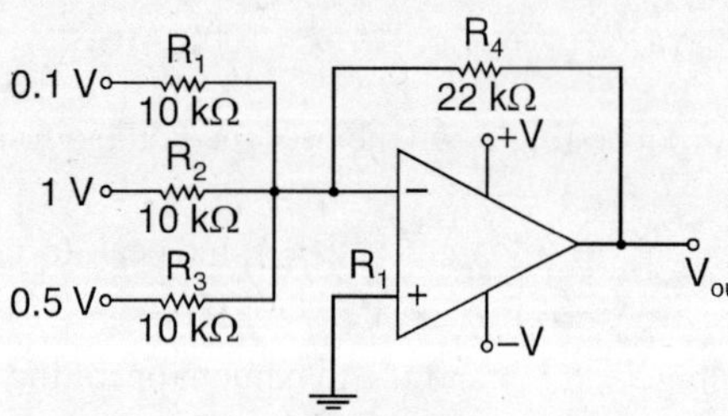

Fig. 29-21.

Determine the value of output voltage for this circuit.

Solution. Given : $v_1 = 0.1$ V; $v_2 = 1$ V; $v_3 = 0.5$ V; $R_1 = 10$ kΩ; $R_2 = 10$ kΩ; $R_3 = 10$ kΩ and $R_4 = 22$ kΩ.

We know that output voltage for the summing amplifier,

$$v_{out} = \left(-\frac{R_4}{R_1} \cdot v_1\right) + \left(-\frac{R_4}{R_2} \cdot v_2\right) + \left(-\frac{R_4}{R_3} \cdot v_3\right)$$

$$= \left(-\frac{22\text{ k}\Omega}{10\text{ k}\Omega} \times 0.1\right) + \left(-\frac{22\text{ k}\Omega}{10\text{ k}\Omega} \times 1\right) + \left(-\frac{22\text{ k}\Omega}{10\text{ k}\Omega} \times 0.5\right)$$

$$= -(0.22 + 2.2 + 1.1) = 3.52 \text{ V}$$ **Ans.**

Example. 29-13. *Design an Amp circuit that will produce an output equal to* $-(4\,V_1 + V_2 + 0.1\,V_3)$.

Write an expression for the output and sketch its output waveform when $V_1 = 2\,\omega$, $V_2 = +5$ *V dc and* $V_3 = -100$ *V dc.* (*Banglore Univ. B.Sc. April/May* 2001)

Solution. Given : $V_1 = 2 \text{ Sin } \omega t$, $V_2 = +5$ dc and $V_3 = -100$ Vdc.

The Op-Amp circuit that will produce an output equal to $-(4\ V_1 + V_2 + 0.1\ V_3)$ is shown in Fig. 29-22. Notice that V_1 bias a given of 4, V_2 has a unity gain and V_3 a gain of 0.1.

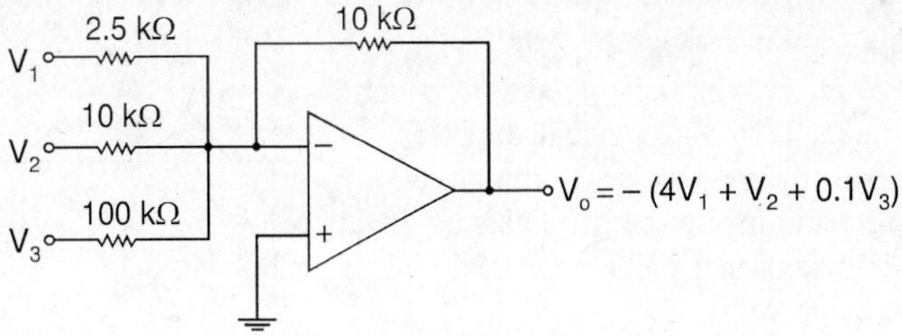

Fig. 29-22.

Now if $V_1 = 2 \sin \omega t$

$V_2 = +5$ Vdc

and $V_3 = -100$ Vdc

Then output voltage,

$$V_o = -(4\ V_1 + V_2 + 0.1\ V_3)$$
$$= -(4 \times 2 \sin \omega t + 5 + 0.1 \times (-100\ \text{V})$$
$$= -(8 \sin \omega t + 5 + 10\ V)$$
$$= -(15 + 8 \sin \omega t)$$

Fig. 29-23. shows the sketch of the output waveform.

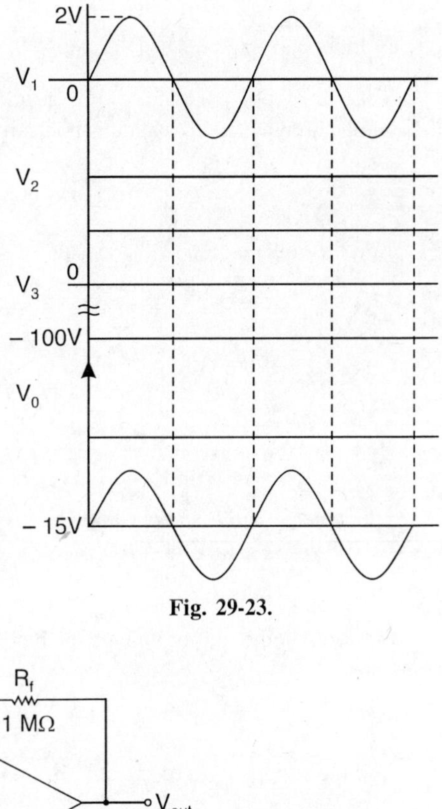

Fig. 29-23.

Example 29-14. *Calculate the output voltage of a these input summing amplifier. Given : $R_1 = 200$ kΩ, $R_2 = 250$ kΩ; $R_3 = 500$ kΩ; $R_f = 1$ MΩ, $V_1 = -2$ V, $V_2 = +2$V, $V_3 = +1$ V.*

(Bangalore Univ. July 2003)

Solution. Given : $V_1 = -2$ V, $V_2 = +2$ V, $V_3 = +1$ V; $R_1 = 200$ k Ω, $R_2 = 250$ kΩ, $R_3 = 500$ kΩ, $R_f = 1$ MΩ.

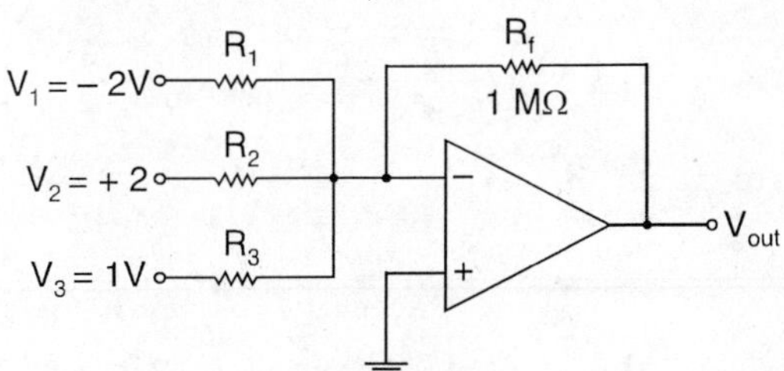

Fig. 29-24.

We know that output voltage for the summing amplifier (refer to Fig. 29-24),

$$v_{out} = \left(-\frac{R_f}{R_1}\right) v_1 + \left(-\frac{R_f}{R_2}\right) v_2 + \left(-\frac{R_f}{R_3}\right) v_3$$

$$= \left(-\frac{1\ M\Omega}{200\ k\Omega}\right)(-2V) + \left(-\frac{1\ M\Omega}{250\ k\Omega}\right)(+2V) + \left(\frac{-1\ M\Omega}{500\ \text{k}\Omega}\right)(-1\ V)$$

$$= (-5) \times (-2\ \text{V}) + (-4)(+2\ \text{V}) + (-2)(-1\ \text{V})$$
$$= 10\ \text{V} - 8\ \text{V} + 2\ \text{V}$$
$$= 4\ \text{V}\ \textbf{Ans.}$$

29-13. Differential Amplifier

Fig. 29-25 shows the circuit of a differential amplifier. Notice that the voltages are applied at both the op-amp inputs (*i.e.*, inverting and non-inverting inputs) through the resistors. And the difference between these two voltages is amplified.

Suppose that voltage 'v' volts is applied to terminal 1 and zero volts to terminal 2. The difference in the potentials at the inverting and non-inverting op-amp terminals is very nearly zero. Therefore the inverting terminal must be at zero potential. This means that the input voltage v_1 is developed across resistor R_1.

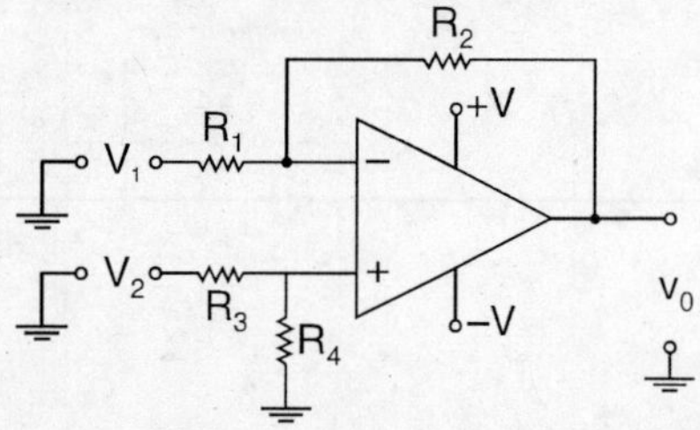

Fig. 29-25. *A differential amplifier.*

$\therefore$ Input current,
$$i_1 = \frac{v_1}{R_1}$$

Since, the input resistance of the op-amp is high, the current, (*i.e.*, i_1) flows through resistor R_2. the voltage drop across R_2 which is the output voltage V_{out} of the circuit,

$$= -v_1 \cdot \left(\frac{R_2}{R_1}\right)$$

And the voltage gain of the circuit,

$$A_v = \frac{v_{out}}{v_1} = \frac{-v_1 \cdot (R_2 / R_1)}{v_1} = -\frac{R_2}{R_1}$$

Conversely if the voltage applied to input terminals 1 and 2 are respectively zero and v_2 volts, the voltage appearing at the non-inverting terminal

$$= v_2 \cdot \left(\frac{R_4}{R_3 + R_4}\right)$$

This voltage will also appear at the inverting terminal. So, the voltage across resistor R_1,

$$= -v_2 \cdot \left(\frac{R_4}{R_3 + R_4}\right)$$

In this case, the output voltage of the circuit,

$$v_{out} = v_2 \cdot \left(\frac{R_4}{R_3 + R_4}\right) + \left\{\left(-v_2 \cdot \frac{R_4}{R_3 + R_4}\right) \times \left(-\frac{R_2}{R_1}\right)\right\}$$

And the voltage gain of the circuit,

$$A_v = \frac{v_{out}}{v_2} = \left(\frac{R_4}{R_3 + R_4}\right)\left(1 + \frac{R_2}{R_1}\right)$$

Lastly, if the voltages applied to the terminals 1 and 2 are v_1 and v_2 volts respectively, the difference between the two voltages is amplifier. If $v_1 > v_2$, a voltage $(v_1 - v_2)$ is amplified $(-R_2/R_1)$ times. On the other hand, if $v_2 > v_1$, a voltage $(v_2 - v_1)$ is amplified by $[R_4/(R_3 + R_4)] \cdot [1 + (R_2/R_1)]$ times

If $R_1 = R_3$ and $R_2 = R_4$, then the voltage gain,

$$A_v = \frac{v_{out}}{v_1 - v_2} = \frac{R_2}{R_1}; \quad \therefore \quad v_{out} = \frac{R_2}{R_1}(v_2 - v_1)$$

29-14. OP-AMP Frequency Response

We have already discussed in Art. 29-6 that slew rate of an op-amp is a measure of how fast the output voltage can change. It is expressed in volts/microseconds (V/μs). The maximum operating frequency of an op-amp is,

$$f_{max} = \frac{\text{slew rate}}{2 \cdot \pi \cdot v_{pk}}$$

Thus, the peak output voltage (v_{pk}) limits the maximum operating frequency. When the maximum output frequency of an op-amp is exceeded, the result is distorted output waveform. It will be interesting to know that if the operating frequency of an op-amp is increased beyond a certain point, it will,

(*i*) Decrease the maximum output voltage swing.

(*ii*) Decrease the open-loop voltage gain.

(*iii*) Decrease the input resistance and.

(*iv*) Increase the output resistance.

29-15. Frequency versus Gain Characteristic of an Op-amp

Fig. 29-26 shows the frequency versus gain characteristic of a typical op-amp (such as 741). As seen from this Fig., the op-amp gain remains stable from 0 Hz to a certain cut-off frequency (which is less than 10 Hz). Then the gain reduces by a factor of 100 when the frequency is increased by a factor of 10 (*i.e.*, the gain drops at the rate of 20 dB per decade).

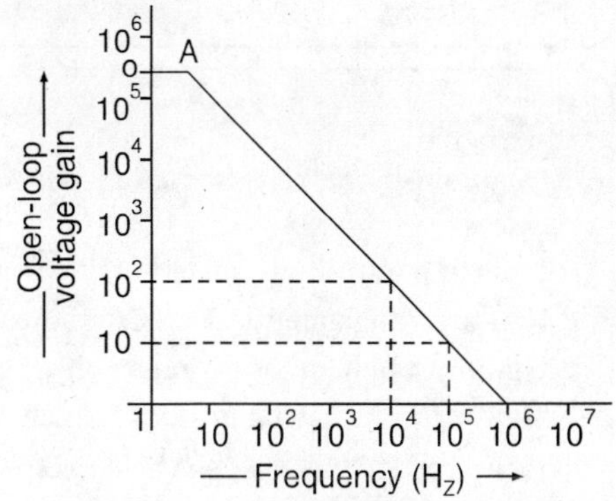

Fig. 29-26. *Op-amp frequency vs. gain characteristic.*

Since the op-amp is a d.c. amplifier, it will exhibit mid-band voltage gain at 0 Hz. As the frequency of operation is increased above 0 Hz, a point (such as 'A' shown in Fig. 29-26) is reached where the gain starts dropping. This drop in gain is due to internal values of input and output capacitance. It may be noted that as the frequency is increased to 10^6 Hz (*i.e.*, 1 MHz), the op-amp gain is reduced to '1' (*i.e.*, unity). This means that the 741 op-amp cannot be operated above 10^6 Hz. The frequency range from 0 to 10^6 Hz is referred to as the bandwidth of an op-amp. If we wish to operate the op-amp at a gain of 100, the bandwidth is 10^4 Hz (or 10 kHz). And if the gain required is 10, then the bandwidth is 10^5 Hz (or 100 kHz). This indicates that the gain and bandwidth of an op-amp are inversely proportional to each other. Based on this fact, we can make the following two statements :

(1) *The higher the gain of an op-amp, the narrower is its bandwidth, and*

(2) *The lower the gain of an op-amp, the wider is its bandwidth.*

It is evident from the above discussion, that if we want a wider bandwidth, we have to settle for less gain. On the other hand, if we want a high gain, we have to settle for a narrower bandwidth.

SUMMARY

In this chapter you have learned that :

1. An operational amplifier has a very high voltage gain, a high input impedance and a low output impedance.

2. The Op-amp makes use of differential amplifiers as an input stage.

3. The common-mode rejection ratio (CMRF) is the ratio of differential gain to the common-mode gain. Typical, its value is 90 dB.

4. The maximum operating frequency of an Op-amp can be obtained by using the equation,

$$f_{max} = \frac{\textit{slew rate}}{2\pi V_{pk}}$$

where V_{pk} is the maximum or peak value of the applied input signal.

5. The gain of an inverting amplifier is given by,

$$A_V = 1 - \frac{R_2}{R_1}$$

6. The gain of non-inverting amplifier is given by,

$$A_V = 1 + \frac{R_2}{R_1}$$

7. The voltage gain of an Op-amp configured as a differential amplifier is given by,

$$A_V = \left(\frac{R_4}{R_3 + R_4}\right) \cdot \left(1 + \frac{R_2}{R_1}\right)$$

8. If $R_1 = R_3$ and $R_2 = R_4$, then,

$$A_v = \frac{R_2}{R_1}$$

GLOSSARY

Common-mode rejection ratio (CMRR) : a ratio of differential gain to common-mode gain.

Inverting amplifier : an Op-amp circuit with input applied at the negative (–) terminal.

Non-inverting amplifier : an Op-amp circuit with input applied at the positive (+) terminal.

Op-amp : an IC with very high voltage gain, high input impedance and low output impedance.

Slew rate : the maximum rate at which the output voltage of an Op-amp is capable of changing Typically its value for IC 741 is 0.5 V/μs.

Voltage Follower : An Op-amp configured as a unity gain amplifier.

NUMERICAL PROBLEMS

1. A certain Op-amp has a open-loop gain of 100,000 and a common-mode gain of 0.25. Determine the CMRR and express it in decibels. (**Ans.** 400,000 and 112 dB)

2. Fig. 29-27 shows the circuit of an inverting amplifier. Determine the value of maximum operating frequency for this circuit. (**Ans.** 79.6 kHz)

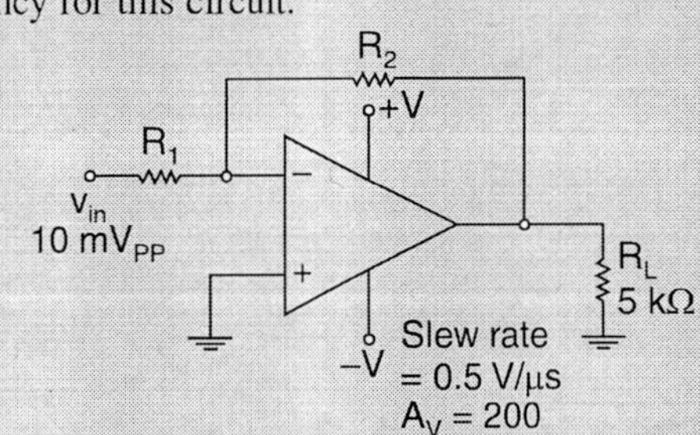

Fig. 29-27.

3. Determine the peak-to-peak output voltage value for the circuit shown in Fig. 29-28. Also determine the maximum possible output values for the amplifier. Assume that the gain of the amplifier is 150. (**Ans.** 15 V_{PP}, + 9 V and – V)

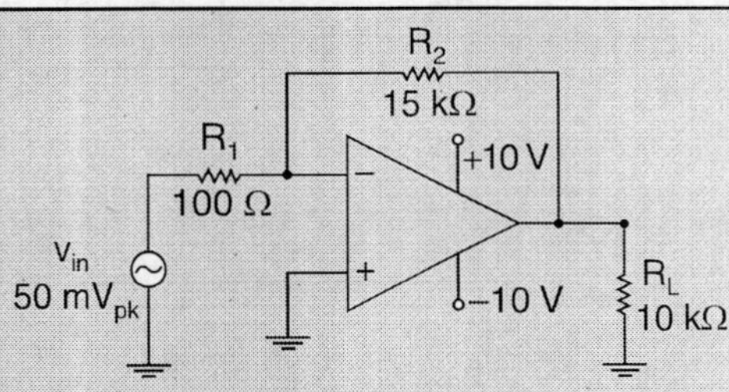

Fig. 29-28.

4. Determine the maximum operating frequency for the amplifier shown in Fig. 29-29.

(**Ans.** 19.8 kHz)

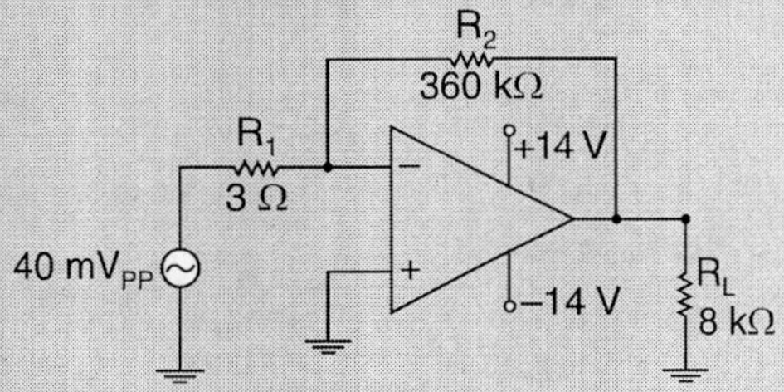

Fig. 29-29.

5. Fig. 29-30 shows the Op-amp configuration. Determine the value of R_2 required to produce a gain of 100. (**Ans.** 220 kΩ)

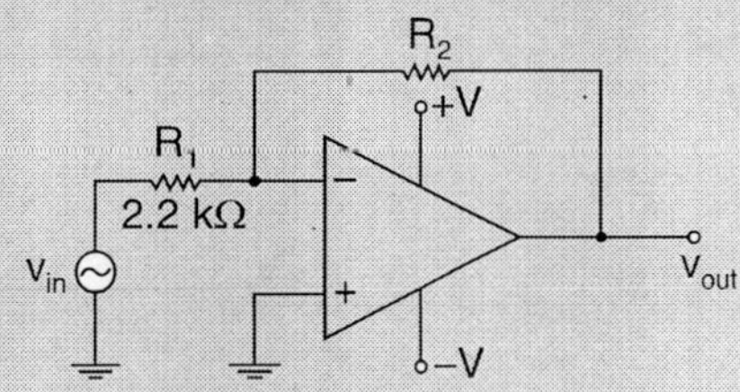

Fig. 29-30.

6. Determine the gain of the amplifier in Fig. 29-31. The open-loop voltage gain is 100,000.

(**Ans.** 22.3)

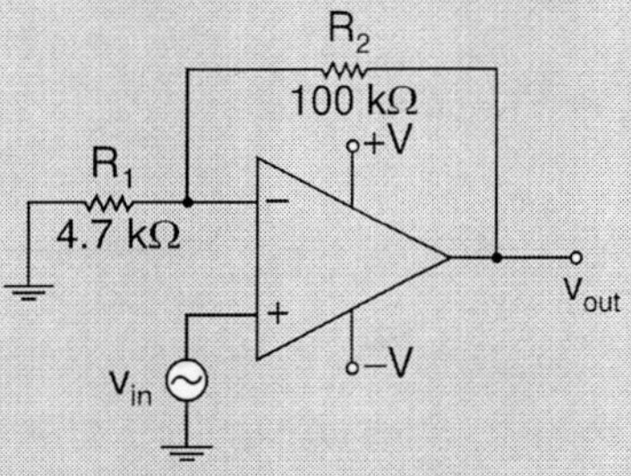

Fig. 29-31

7. The Op-amp shown in Fig. 29-32. has the following parameters : A_{OL} = 50,000, Z_{in} = 4 MΩ, and Z_{out} = 50 Ω. (*a*) Find the values of input and output impedance, (*b*) Determine the closed-loop voltage gain. (**Ans.** (*a*) 1 kΩ and 50 Ω (*b*) – 100)

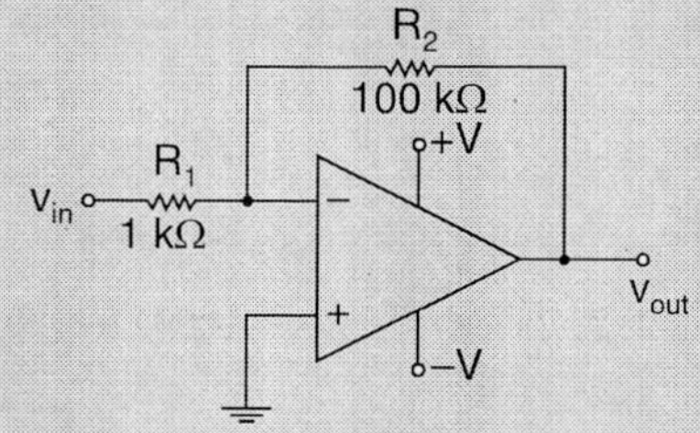

Fig. 29-32.

DESCRIPTIVE QUESTIONS

1. Describe some of the characteristics of a practical Op-amp.
2. Define the term "common-mode rejection ratio".
3. For a given value of differential gain, does a higher CMRR results in a higher or lower common-mode gain? Explain briefly.
4. How does the input impedance of a non-inverting amplifier compare to the input impedance of the op-amp itself.
5. What is a voltage follower ? Describe its main characteristics.
6. Draw the circuit of a summing amplifier using inverting amplifier configuration. Write an equation for the output voltage for this circuit.
7. What is a differential amplifier ? Can it be used in a single-ended input configuration ?
8. Define unity gain bandwidth of an OP-Amp.
   (*Linear Integrated Circuits, Anna University, April/May 2008*)
9. Define slew rate. What causes it?
   (*Linear Integrated Circuits, Anna University, April/May 2008*)
10. Draw the circuit diagram of a non-inverting amplifier.
   (*Linear Integrated Circuits, Anna University, April/May 2008*)
11. Define CMRR. (*Electronic Circuits–I, Anna University, Nov/Dec 2007*)
12. Write the improving methods of CMRR.
   (*Electronic Circuits–I, Anna University, Nov/Dec 2007*)
13. What is meant be Gain-Bandwidth product?
   (*Electronic Circuits–I, Anna University, April/May 2008*)

MULTIPLE CHOICE QUESTIONS

1. Which of the following characteristics do not necessarily apply to the Op-amp :
   (*a*) high gain (*b*) low power
   (*c*) high input impedance (*d*) low output impedance
2. The output voltage of a particular Op-amp increases 8 V in 12 μs in response to a step voltage on the input. The slew rate is
   (*a*) 0.667 V/μs (*b*) 0.75 V/μs (*c*) 1.5 V/μs (*d*) 96 V/μs
3. The voltage gain of a voltage follower is :
   (*a*) unity (*b*) less than unity (*c*) greater than unity (*d*) variable

4. The highest possible impedance is achieved with the :
(*a*) inverting amplifier (*b*) non-inverting amplifier
(*c*) differential amplifier (*d*) voltage follower

5. The voltage gain of a non-inverting Op-amp amplifier :
(*a*) greater than unity (*b*) less than unity
(*c*) always greater than unity (*d*) both (*a*) & (*c*)

6. An Op-amp can be classified as........ amplifier.
(*a*) linear (*b*) low-r_m (*c*) positive feedback (*d*) *RC*-coupled

7. An ideal Op-amp has
(*a*) infinite A_v (*b*) infinite R_i (*c*) zero R_0 (*d*) all the above.

8. Op-amp have become very popular in industry mainly because
(*a*) they are dirt cheap
(*b*) their external characteristics can be changed to suit any application
(*c*) of their extremely small size
(*d*) they are available in different packages.

9. Since input resistance of an ideal Op-amp is infinite
(*a*) its output resistance is zero
(*b*) its output becomes independent of load resistance
(*c*) its input current is zero
(*d*) it becomes a current -controlled device.

11. When an input voltage of 1 V is applied to an Op-amp having $A_v = 10^6$ and bias supply of + 15 V, the output voltage available is
(*a*) 15×10^6 V (*b*) 10^6V (*c*) 15 μ V (*d*) 15 V.

12. An inverting amplifier has $R_f = 2$ M and $R_1 = 2$ k, Its scale factor is
(*a*) 1000 (*b*) – 1000 (*c*) 10^{-3} (*d*) -10^{-3}

13. In an inverting amplifier, the two input terminals of an ideal Op-amp are at the same potential because
(*a*) the two input terminals are directly shorted internally
(*b*) the input impedance of the Op-amp is in finity
(*c*) common-mode rejection ratio is infinity
(*d*) the open-loop gain of the Op-amp is infinity

14. The open -loop gain of an operational amplifier is 10^5. An input signal of 1 mV is applied to the inverting input with the non-inverting input connected to the ground. The supply voltages are ± 10 V. The output of the amplifier will be
(*a*) + 100 V (*b*) – 100 V
(*c*) + 10 V (approximately) (*d*) – 10 V (approximately)

15. An ideal Op-amp has
(*a*) infinite input and output impedance
(*b*) very low input and output impedance
(*c*) low input and infinite output impedance
(*d*) infinite input and zero output impedance.

16. For an ideal difference amplifier, the common mode Rejection Ratio (CMRR) should be:
(*a*) As high as possible (*b*) As low as possible
(*c*) Constant (*d*) unity

17. The input stage of an Op-Amp is usually a
(*a*) differential amplifier (*b*) class-B push-pull amplifier
(*c*) CE amplifier (*d*) Level shifter

18. The unity gain frequency of an op-Amp equal

(*a*) 10^3 Hz (*b*) 10^6 Hz (*c*) 10^2 Hz (*d*) 10 Hz

19. An Op-Amp passesses

(*a*) very large input resistance and very large output resistance
(*b*) very large input resistance and very small output resistance.
(*c*) very small input resistance and very small output resistance.
(*d*) very small input resistance and very large output resistance

(*UPSC Engg. Service* 2002)

20. In a 741 Op-amp, There is 20 dB/decade fall-off starting at a relatively low frequency. This is due to :

(*a*) applied load (*b*) internal compensation
(*c*) impedance of the source (*d*) power dissipation in the chip.

(*UPSE Engg. Service* 2001)

21. The input differential stage of Op-amp 741 is biased at about 10 μA current. Such a low current of the input stage gives

(*a*) high CMRR (*b*) high differential gain
(*c*) low CMRR (*d*) high input impedance.

(*UPSC Engg Services* 2002)

22. A differential amplifier is invariably used in the input stage of all Op-amp. This is done basically to provide the Op-amp with a very high

(*a*) CMRR (*b*) bandwidth (*c*) slew-rate (*d*) open-loop gain

(*UPSC Engg. Services* 2001)

23. The slew rate of an Op-amp is 0.5 $V/_{\mu s}$ The maximum frequency of a sinusoidal input of 2 V_{rms} that can be handled without excessive distortion is

(*a*) 3 kHz (*b*) 30 kHz (*c*) 200 kHz (*d*) 2 MHz

24. In a circuit if the open loop gain is 10^6 and output voltage is 10 V, the differential voltage should be

(*a*) 10 μV (*b*) 0.1 V (*c*) 100 μV (*d*) 1 μV.

(*UPSC Engg. Services* 2002)

25. An Op-amp has a common-mode gain of 0.01 and a differential mode gain of 10^5. Its common mode rejection ratio would be :

(*a*) 10^{-7} (*b*) 10^{-3} (*c*) 10^3 (*d*) 10^7

26. The voltage gain versus frequency curve of an Op-amp is shown in Fig. 29-33.

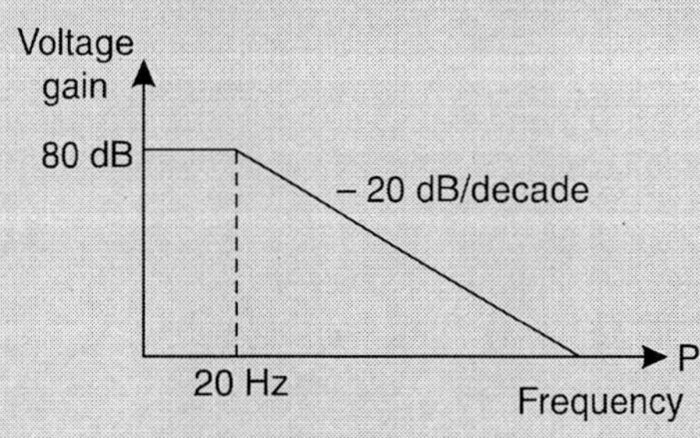

Fig. 29-33.

The gain-based width product of the Op-amp is

(*a*) 200 Hz (*b*) 200 MHz (*c*) 200 kHz (*d*) 2 MHz.

(*UPSC Engg. Services* 1999)

27. Active load is used the collector of the differential amplifier of an Op-amp to

(*a*) increase the output resistance

(*b*) increase the differential gain, A_d

(*c*) increase maximum peak-to-peak output voltage

(*d*) eliminate load resistance from the circuit.

28. The output voltage of the circuit shown in Fig. 29-34 is

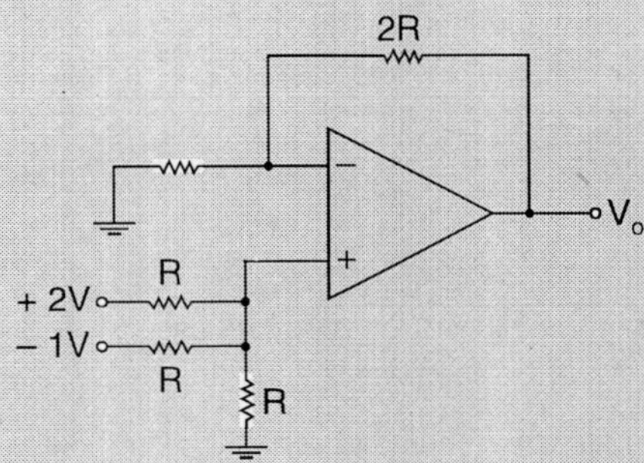

Fig. 29-34.

(*a*) 1.0 V (*b*) 1.5 V (*c*) 2.0 V (*d*) 2.5 V

29. In the circuit shown in Fig 33-35, the value of output, v_0 is

(*a*) + 3 V (*b*) – 3 V (*c*) – 7 V (*d*) + 7 V

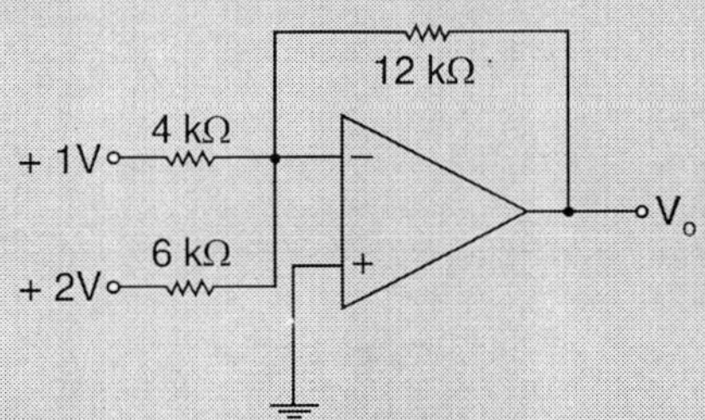

Fig. 29-35.

30. When in a negative scaler, both R_1 and R_f are reduced to zero, the circuit functions as

(*a*) integrator (*b*) subtractor (*c*) comparator (*d*) unity follower.

31. The two input terminals of an Op-amp are known as

(*a*) positive and negative (*b*) differential and non-differential

(*c*) inverting and non-inverting (*d*) high and low.

32. In the inverting Op-amp circuit shown in Fig. 29-36 the resistance

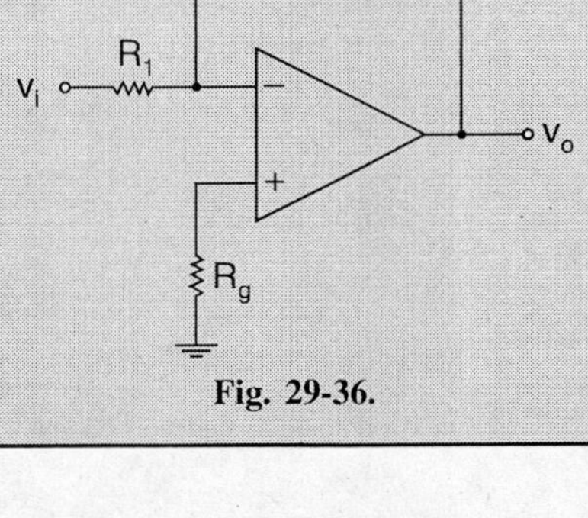

Fig. 29-36.

R_g is chosen as $R_1 \parallel R_2$ in order to

(*a*) increase the gain (*b*) reduce offset voltage

(*c*) reduce offset current (*d*) increase CMRR.

33. The v_0 of the Op-amp circuit shown in Fig. 29-37. is :

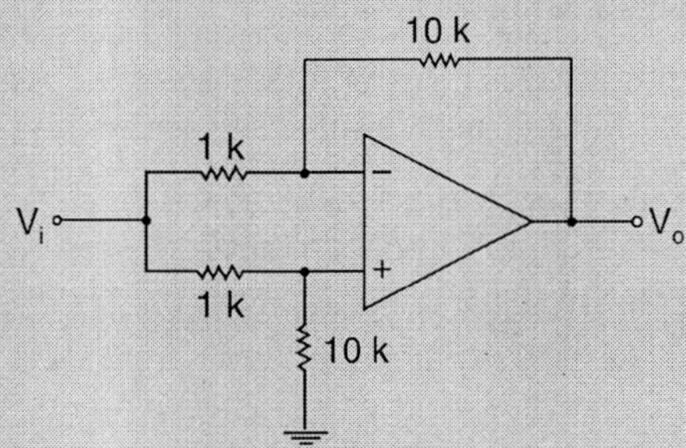

Fig. 29-37.

(*a*) 11 v_i (*b*) 10 v_i (*c*) v_i (*d*) zero

(*UPSC Engg Services*, 1999)

34. A circuit using an Op-amp is shown in Fig. 29-38. It has :

(*a*) voltage-series feedback (*b*) voltage-shunt feedback

(*c*) current-shunt feedback (*d*) current-series feedback.

(*UPSC Engg. Services*, 2001)

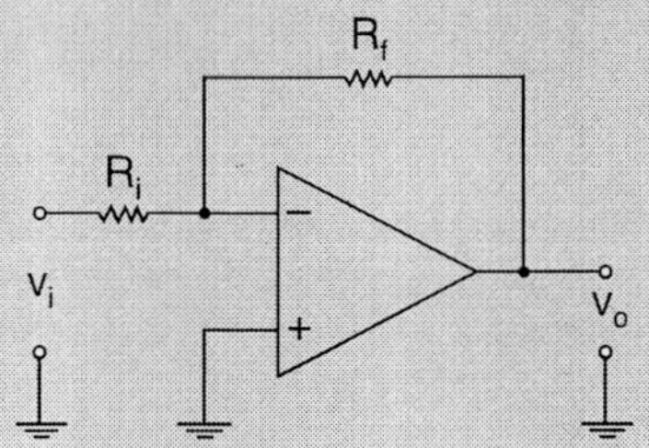

Fig. 29-38.

35. A non-inverting Op-amp is shown in Fig. 29-39.

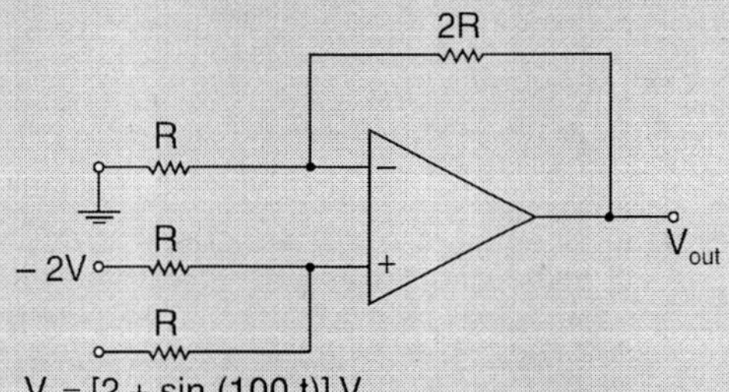

Fig. 29-39.

The output voltage v_0, for an input, $v_i = [2 + \sin(100\,t)]$ V is

(*a*) $\frac{3}{2}\sin(100t)$ (*b*) $3\sin(100\,t)$

(*c*) $2\sin(100\,t)$ (*d*) $3\sin(100\,t) + \frac{1}{2}$

(*UPSC Engg Services* 2002)

36. The Op-amp circuit shown in Fig 29-40 has an input impedance of 1 MΩ and an open-loop gain of 10^5. The output impedance seen by source V_S is

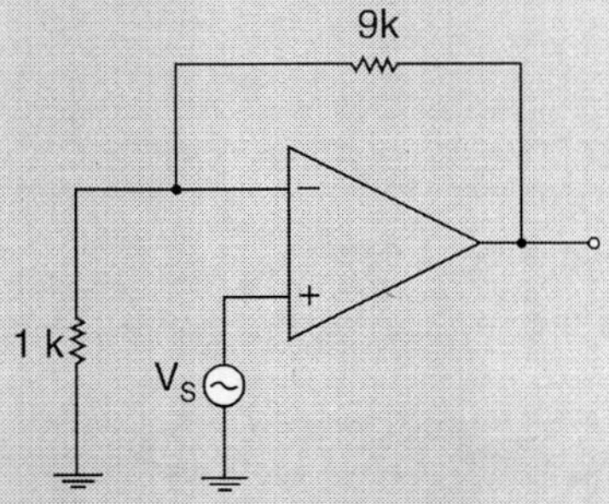

Fig. 29-40.

(*a*) 10^{11} Ω (*b*) 10^{11} Ω

(*c*) 10 k Ω (*d*) – 1000

37. An Op-amp with an open-loop gain of 10,000, R_{in} = 2 k Ω and R_0 = 500 Ω is used in the non-inverting configuration shown in Fig. 29-41. The output resistance R_0 is

(*a*) 250.5 Ω (*b*) 21 Ω

(*c*) 2 Ω (*d*) 0.998 Ω

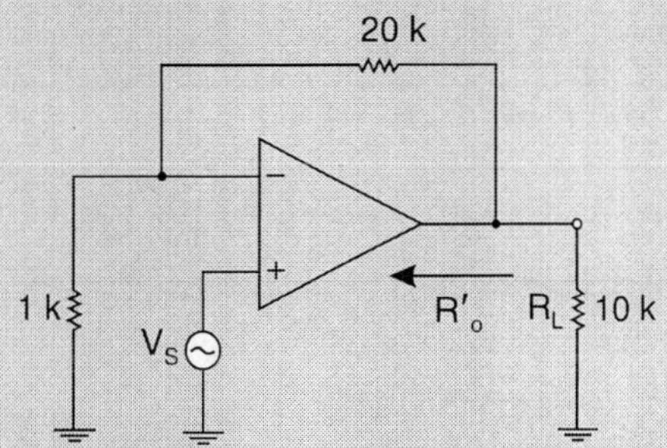

Fig. 29-41.

ANSWERS

1. (*b*)	**2.** (*a*)	**3.** (*a*)	**4.** (*d*)	**5.** (*d*)
6. (*a*)	**7.** (*d*)	**8.** (*b*)	**9.** (*c*)	**10.** (*a*)
11. (*d*)	**12.** (*b*)	**13.** (*b*)	**14.** (*b*)	**15.** (*d*)
16. (*a*)	**17.** (*a*)	**18.** (*b*)	**19.** (*b*)	**20.** (*b*)
21. (*d*)	**22.** (*d*)	**23.** (*b*)	**24.** (*a*)	**25.** (*d*)
26. (*c*)	**28.** (*c*)	**29.** (*d*)	**30.** (*d*)	**31.** (*c*)
32. (*a*)	**33.** (*b*)	**34.** (*a*)	**35.** (*b*)	**36.** (*a*)
37. (*d*)				

❑❑❑

30

CHAPTER

OP-AMP Applications

OUTLINE

Objectives

After completing this chapter, you should be able to :

- State the purpose of a comparator.
- Explain the operation of comparator circuit.
- Describe the operation of integrator.
- Calculate the cut-off frequency of the integrator and the frequency at which it starts to lose its linear characteristics.
- Describe the operation of differentiator.
- Discuss the use of Op-amp in an audio amplifier.
- Describe the operation of the Op-amp based Wein-Bridge oscillator.
- Explain the operation of voltage controlled oscillator.
- Analyze the operation of low-pass filter.
- Describe and analyze the operation of high-pass filter.
- Describe how a low-pass filter and high pass filter can be connected together to form a band-pass filter and a band reject filter.

30-1. Introduction

We have already discussed in the last chapter about an Op-amp that it is an integrated circuit with very high gain high input impedance and low output impedance. In this chapter we shall study about the applications of an Op-amp. An Op-amp can be used as a constant gain multiplier, voltage summing amplifier voltage buffer amplifier, basic amplifier in instrumentation circuits, comparator, in oscillator circuits and active filters. We shall discuss some of these applications in the following pages.

30-2. Comparators

The comparator is a circuit that is used to compare two voltages and provide an output indicating the relationship between those two voltages. Generally speaking, comparators are used to compare either,

(*i*) Two changing voltages to each other, as in comparing two sine waves, or

(*ii*) A changing voltage to a set d.c. reference voltage.

Fig. 30-1(*a*) shows the circuit of an Op-amp comparator. It may be noted that there is no feed-back path in the circuit. In this circuit, the input voltage is applied to the non-inverting input terminal and a set d.c. reference voltage (V_{ref}) if applied to the inverting terminal of the Op-amp. The operation of the circuit is as follows.

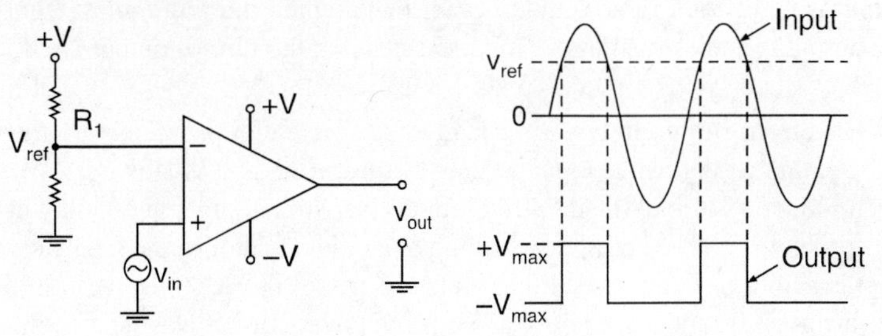

(*a*) Circuit. (*b*) Input and output waveforms.

Fig. 30-1. *OP-AMP comparator.*

As long as the input voltage is below V_{ref}, the comparator output is approximately " $-V_{max}$" volts. But if the input voltage equals V_{ref} or exceeds it, the comparator output changes to " $+V_{max}$" volts. Thus, depending upon the value of input voltage, the comparator produces a d.c. voltage that indicates the polarity (or magnitude) relationship between the two input voltages as shown in Fig. 30-1 (*b*). A special case arises if the inverting input of the Op-amp is grounded and the input signal voltage is applied to the non-inverting input. In such a situation, $V_{ref} = 0$, and the output is equal to $-V_{max}$, so long the input signal is below 0. As soon as, the input signal crosses zero the output switches to $+V_{max}$. Such a circuit is called zero-level detector. This can be used as a squaring circuit to produce a square wave from a sine wave.

30-3. Integrator

Technically speaking, *integrator is a circuit whose output is proportional to the area of its input waveform.* Fig. 30-2 shows the circuit of an Op-amp integrator. The RC circuit itself also acts as a simple integrator. But the problem with such a simple circuit is that the output voltage is not a linear triangular output as it should be. The function of the Op-amp is to linearize the output. It may be noted from the diagram that the inverting input to the Op-amp is held at a virtual ground by the differential amplifier in the Op-amp input circuit. Therefore, the input current, $I_1 = V_{in}/R_1$. Because of the high input impedance of the Op-amp, virtually all of I_1 will flow to the capacitor.

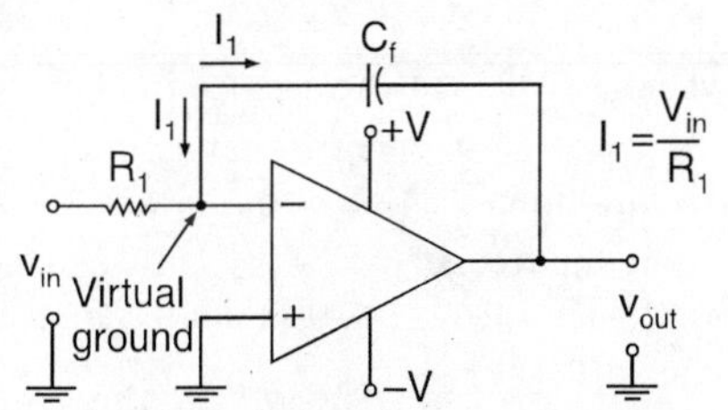

Fig. 30-2. *OP-AMP Integrator.*

Assuming that V_{in} is constant for a given period of time R_1 and is a fixed value, I_1 can also be assumed to be a constant value. Since, virtually all of I_1 flows to the capacitor, C is being charged by a constant current source. Thus, as long as V_{in} is constant, the capacitor will charge/discharge at a linear rate. This produces a linear ramp output as shown in Fig. 30-3.

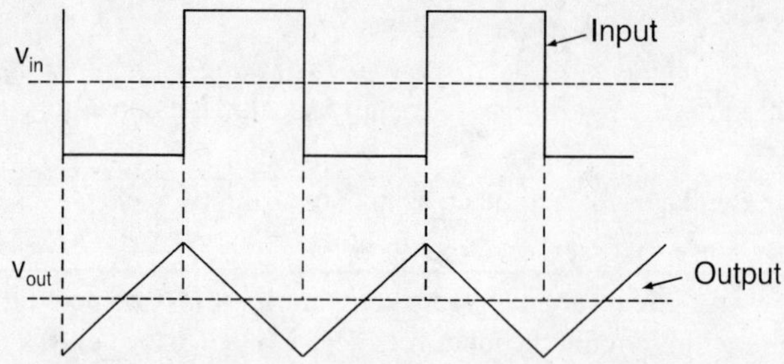

Fig. 30-3. *Input versus output of an OP-AMP integrator.*

It may be noted from the Fig. 33.3 shown above that when the input goes positive (*i.e.*, from – *V* to + *V*), the output is a negative ramp. On the other hand, when the input goes negative (*i.e.*, from + *V* to – *V*), the output is a positive ramp. This indicates that the circuit output is 180° out of phase with the input.

The output wavefrom shown in Fig. 30-3 lead us to a potential problem with the Op-amp integrator As seen from this waveform the mean value of the output is 0 V. However, in actual practice this may not been properly compensated for the input offset voltage, the output may not centre it self around 0 V. This problem can be rectified by adding a parallel resistor in the feedback path as shown in Fig 30-4. But the addition of this resistor has a disadvantage. The RC feedback circuit like any RC circuit will have a cut-off frequency. In this case it is the lower cut-off frequency with which we are concerned with. As the frequency decreases, the reactance of the capacitor will increase. At some point, the reactance of the capacitor will be greater than the value of R_f. As a result, the integration action of the circuit will stop.

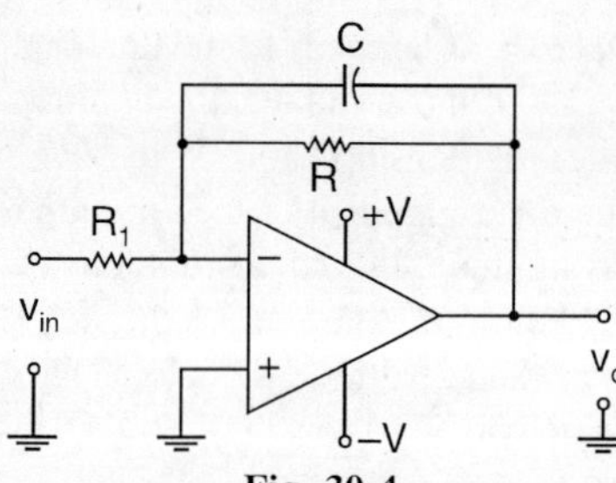

Fig. 30-4.

The cut-off frequency of an integrator is given by the equation,

$$f_1 = \frac{1}{2\pi R_f C_f}$$

where, R_f = resistance in ohms and
C_f = capacitance in farads

It will be interesting to know that the integrator will start to lose its linear output characteristics before the frequency (f_1) given by equation (*i*) is reached. Thus for an optimum output, the value of X_C should always be less than 0.1 R_f, *i.e.*,

$$X_C < 0.1 R_f$$

This leads to the point that the integrator circuit should not be operated below the frequency,

$$f_{min} = 10 f_1 = \frac{10}{2\pi R_f C_f}$$

Example 30-1. *Fig. 30-5 shows an integrator with a lower cut-off frequency of 159 Hz. Determine the value of the capacitance required in the circuit.*

Solution. Given : $R_1 = 1\,\text{k}\Omega$; $R_2 = 100\ \text{k}\Omega$ and $f_1 = 159$ Hz

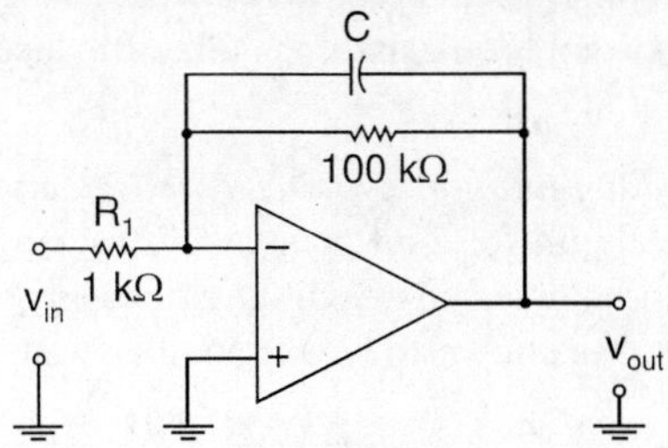

Fig. 30-5.

Let C = value of capacitance (F).

We know that the cut-off frequency of an integrator circuit (f_1),

$$159 = \frac{1}{2\,\pi\, R_f\ C_f} = \frac{1}{2\,\pi \times (100 \times 10^3) \times C}$$

$$\therefore \quad C = \frac{1}{2\,\pi \times (100 \times 10^3) \times 159} = 1.0 \times 10^{-8} \text{ or } 0.01\ \mu\text{F} \ \textbf{Ans.}$$

Example 30-2. *Determine the lower cut-off frequency of an integrator circuit shown in Fig. 30-6.*

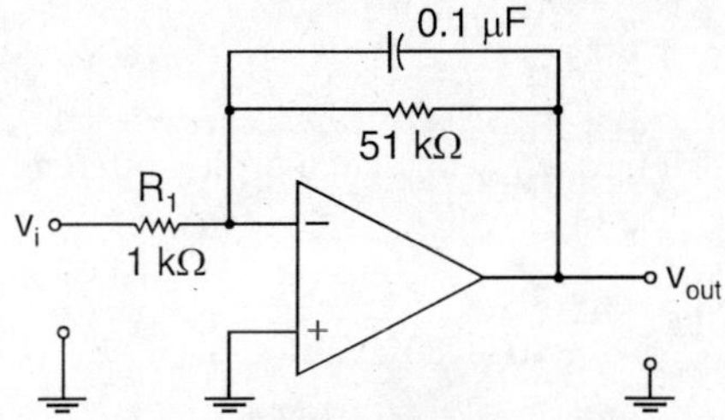

Fig. 30-6.

Also determine the frequency at which the integrator will start to lose its linear characteristics.

Solution. Given : $R_1 = 1\ \text{k}\Omega$; $R_f = 51\text{k}\Omega$ and $C_f = 0.1\ \mu\text{F}$.

We know that the cut-off frequency of an integrator circuit,

$$f = \frac{1}{2\,\pi\, R_f\ C_f} = \frac{1}{2\,\pi \times (51 \times 10^3) \times (0.1 \times 10^{-6})} = 312 \text{ Hz} \ \textbf{Ans.}$$

The minimum frequency at which the integrator will start to lose its linearly,

$$f_{min} = 10\, f_1 = 10 \times 312 = 3120 \text{ Hz } \textbf{Ans.}$$

30-4. The Miller Integrator

The integrator circuit shown is Fig. 30-7 can also be represented as shown in Fig. 30-8. Notice that the feedback capacitor may be broken into two capacitors, one at the input and the other at the output. The circuit shown in Fig. 30-8 is more commonly known as Miller integrator.

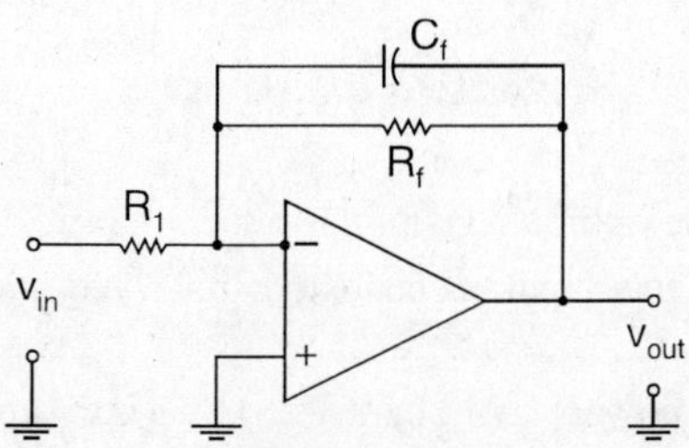

Fig. 30-7.

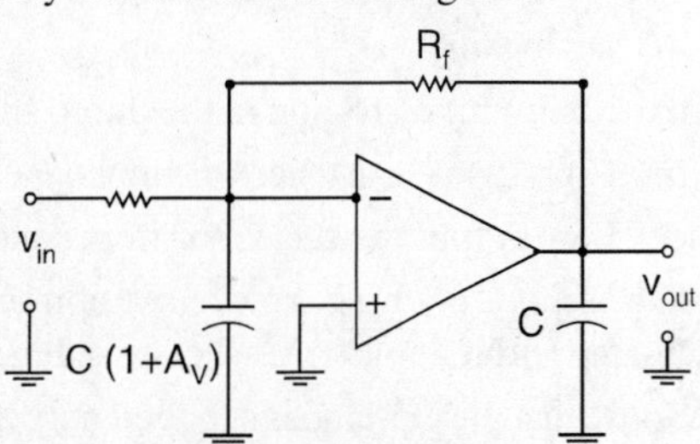

Fig. 30-8.

As the operating frequency increases, the feedback capacitor will act more and more like a short circuit. Eventually a stage will come when the capacitor would short out the feedback resistor completely. In such a situation the closed loop gain of the circuit reduces to zero.

30-5. Differentiator

Technically speaking, *the differentiator is a circuit whose output is proportional to the rate of change of its input signal.* The circuit for a differentiator is as shown in Fig. 30-9. Notice that the position of capacitor and resistor is reversed in a differentiator in comparison to the integrator. The input signal for the differentiator is shown to be a triangular waveform. The resulting output is a square waveform.

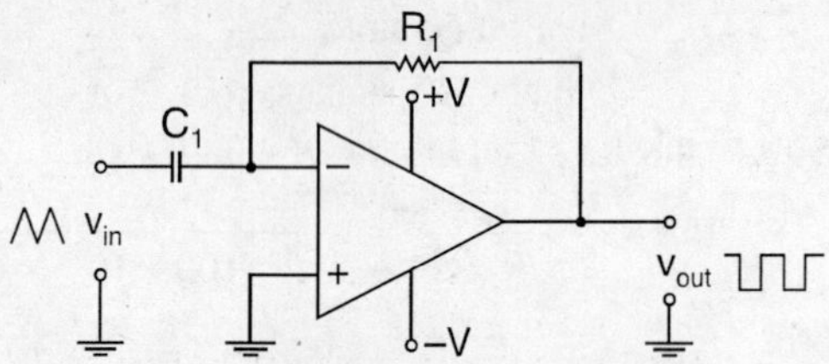

Fig. 30-9.

The cut-off frequency of a differentiate is given by the equation,

$$f_2 = \frac{1}{2\pi R_1 C_1}$$

There is a limit on the high frequency operation of the differentiation. The circuit will start to lose its operating characteristics at one-tenth the value of f_2, *i.e.,*

$$f_{max} = \frac{f_2}{10} = \frac{1}{20\pi R_1 C_1}$$

Example. 30-3. *A differentiator has R_1 = 10 kΩ and C_1 = 0.01 μF. Determine its cut-off frequency and its maximum linear operating frequency.*

Solution. Given: R_1 = 10 kΩ and C_1 = 0.01 μF.

We know that cut-off frequency of a differentiator,

$$f_2 = \frac{1}{2\pi R_2 C_2} = \frac{1}{2\pi \times (10 \times 10^3) \times (0.01 \times 10^{-6})} = 159 \text{ Hz}$$

and the maximum linear operating frequency,

$$f_{max} = \frac{f_2}{10} = \frac{1.59 \; kHz}{10} = 159 \text{ Hz} \quad \textbf{Ans.}$$

30-6. Audio Amplifier

As a matter of fact, in most communication receivers, the final output stage is the audio amplifier. The ideal audio amplifier will have the following characteristics :

(*i*) High gain

(*ii*) Minimum distortion in the audio frequency range (*i.e.*, 20 Hz to 20 kHz range).

(*iii*) High input resistance (or impedance).

(*iv*) Low output resistance (or impedance) to provide optimum coupling to the speaker.

The use of Op-amp in an audio amplifier will fulfill the requirements listed above very nicely. An Op-amp audio amplifier is shown in Fig. 30-10.

Notice that the Op-amp is supplied only from + *V* volt power supply, the – *V* terminal is grounded. Because of this, the output will be between the limits of (+ *V* – 1) volt and + 1 volt approximately.

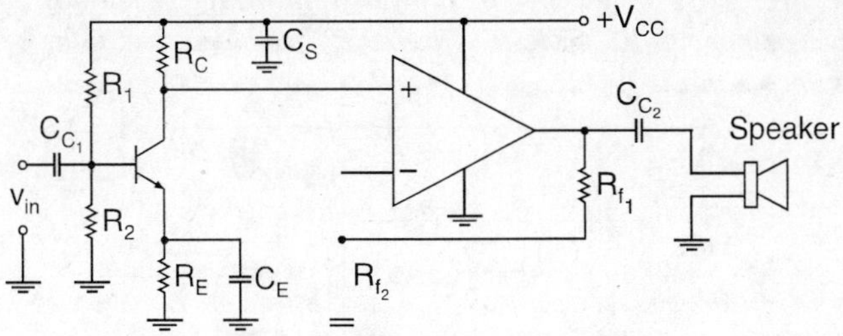

Fig. 30-10. *An OP-AMP audio amplifier.*

Also notice the use of a coupling capacitor (C_{C_2}) between the Op-amp and speaker. This capacitor is necessary to reference the speaker signal around ground. The capacitor C_S, is included in the V_{CC} line to prevent any transient current caused by the operation of Op-amp from being coupled back to Q_1 through the power supply. The high gain requirement is accomplished by the combination of two amplifier stages. The high R_{in}/Low R_{out} of the audio amplifier is accomplished by the Op-amp itself, as the low distortion characteristic.

30-7. OP-AMP Based Oscillator Circuits

We have already discussed about sinusoidal oscillators in Chapter-25. There we defined the oscillator as a circuit that produces an output waveform without any external signal source. The only input to an oscillator is the d.c. power supply. As such, the oscillator can be viewed as being a signal generator. We have also discussed in the same chapter, about the different types of oscillator circuits (like Wein-bridge Oscillator, Colpitts Oscillator and Crystal Oscillator) using bipolar junction transistor and field-effect transistor. Now we will study these oscillator circuits using Op-amp.

30-8. OP-AMP Based Wein-bridge Oscillator

Fig. 30-11 shows the basic version of a Wein-bridge Oscillator. The circuit uses an Op-amp and RC bridge circuit. Note the basic bridge connection carefully. Resistors R_1, R_2, and capicitors C_1 and C_2 form the frequency adjustment elements while resistors R_3 and R_4 form part of the feedback path. The Op-amp output is connected as the bridge input as points '*a*' and '*c*'. The bridge circuit output at points '*b*' and '*d*' is the input to the Op-amp. Fig. 30-12 shows an alternative way of connecting the RC bridge circuit to the Op-amp. In a typical Wein-bridge Oscillator, $R_1 = R_2 = R$, and $C_1 = C_2 = C$. This means that the two RC circuits will have the same cut-off frequency. The cut-off frequency is given by (1/2πRC).

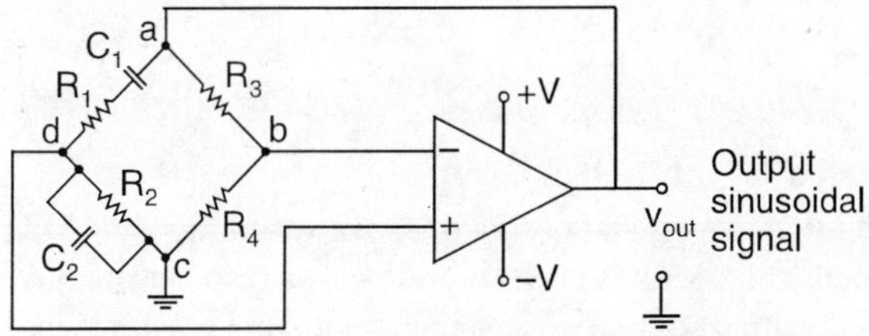

Fig. 30-11. *An alternative Wein-Bridge Oscillator.*

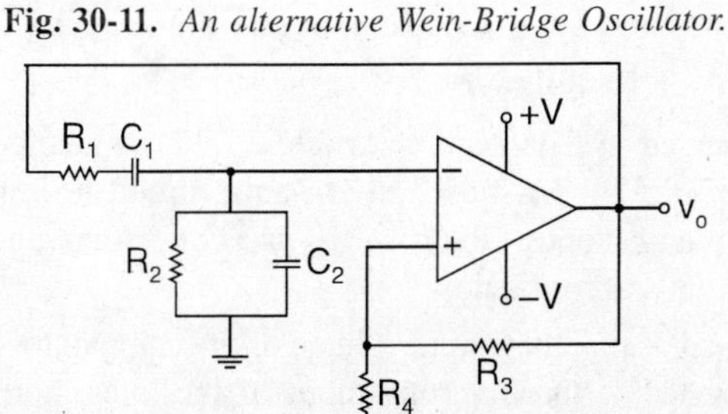

Fig. 30-12. *An alternative Wein-Bridge Oscillator.*

Example 30-4. *Fig. 30-13 shows the circuit of a Wein-bridge Oscillator using Op-amp as an amplifier. Notice the components R_1, C_1, R_2, C_2, R_3 and R_4 are connected in the bridge configuration in the same way as shown in Fig. 30-12.*

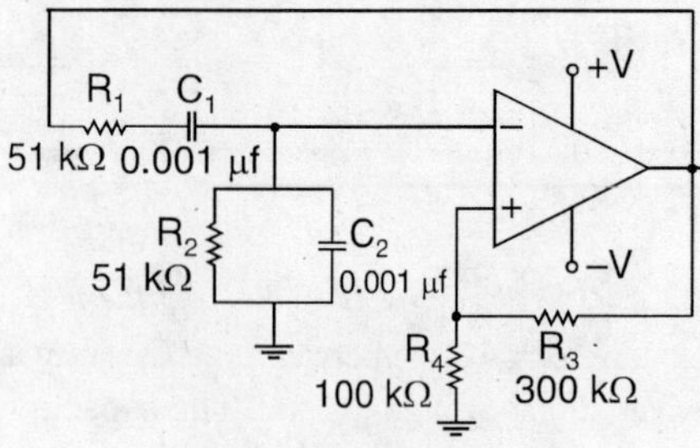

Fig. 30-13.

Calculate the frequency of the Wein-bridge oscillator.

Solution. Given: $R_1 = R_2 = 51\ k\Omega = 51 \times 10^3 \Omega$ and $C_1 = C_2 = C = 0.001\ \mu F = 0.001 \times 10^{-6}$ F.

We know that the frequency of oscillations,

$$f_0 = \frac{1}{2\pi\ RC} = \frac{1}{2\pi \times (51 \times 10^3) \times (0.001 \times 10^{-6})} = 3120.7 \text{ Hz } \textbf{Ans.}$$

30-9. OP-AMP Based Colpitts Oscillator

An Op-amp based Colpitts Oscillator is as shown in Fig. 30-14. Here the Op-amp provides the basic amplification needed while a LC feedback network of Colpitts configuration sets the oscillator frequency.

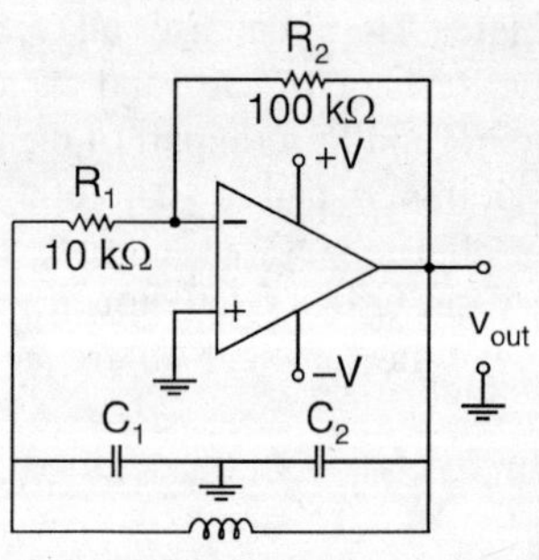

Fig. 30-14. *An OP-AMP based Colpitts Oscillator.*

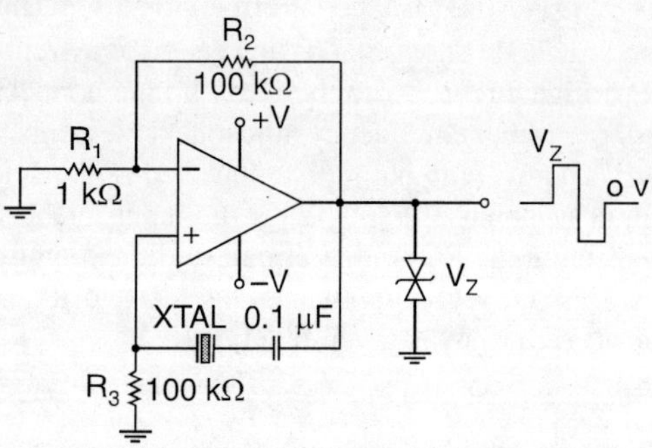

Fig. 30-15. *An OP-AMP based Crystal Oscillator.*

30-10. OP-AMP Based Crystal Oscillator

An Op-amp can be used in a crystal oscillator as shown in Fig. 30-15. The crystal is connected in the series resonant path and operates at the crystal series resonant frequency. The present circuit has a high gain so that an output square-wave signal results are as shown in the Fig.30-15. A pair of Zener diodes is shown at the output to provide output amplitude at exactly the Zener voltage (V_Z).

30-11. A Triangular-Wave Oscillator

We have already discussed in Chapter-27 (Article 27-12), a method to generate a triangular-wave from a square-wave. Fig. 30-16 shows an Op-amp circuit to implement that method. The circuit makes use of two Op-amps: one of them is used as a comparator and the other as an integrator. The operation of the circuit is as given below :

To begin with, let us assume that the output voltage of the comparator is at its maximum negative level. This output is connected to the inverting input of the integrator through resistor, R_1. This produces a positive-going ramp on the output of the integrator. When the ramp voltage reaches the

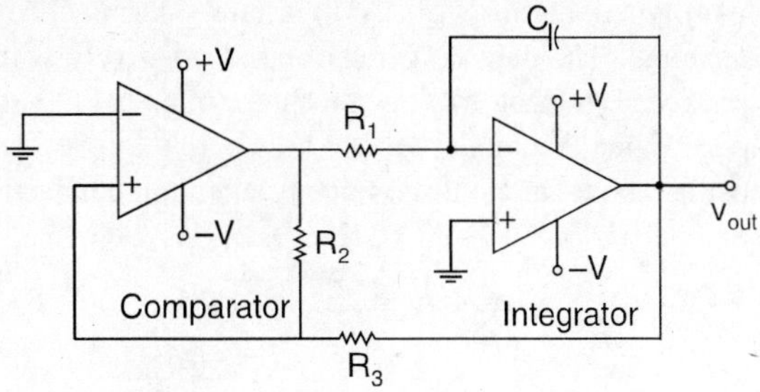

Fig. 30-16. *A triangular-wave oscillator.*

upper trigger point (UTP), the comparator switches to its maximum positive level. This positive level causes the integrator ramp to change to a negative going direction. The ramp continues in this direction until the lower trigger point (LTP) of the comparator is reached. At this point, the comparator output switches back to its maximum negative level and the cycle repeats. This action is shown in Fig. 30-17.

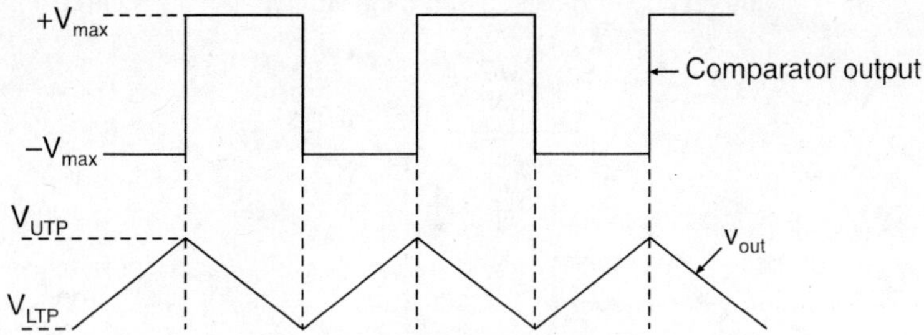

Fig. 30-17. *Square and triangular waveforms.*

Since, the comparator produces square-wave output, therefore, circuit shown in Fig. 30-16 can be used as both a triangular-wave oscillator and a square-wave oscillator. Devices of this type are commonly known as function generators because they produce more than one output function. The output amplitude of the square-wave is set by the output swing of the comparator. While the output amplitude of the triangular-wave is set by the resistors R_2 and R_3 by establishing the UTP and LTP voltages according to the following formulae :

$$V_{\text{UTP}} = +V_{max}\left(\frac{R_3}{R_2}\right)$$

$$V_{\text{LTP}} = -V_{max}\left(\frac{R_3}{R_2}\right)$$

It may be noted that the comparator output levels, $+V_{max}$ are equal. The frequency of both waveforms depends on the R_1C time constant as well as the amplitude-setting resistor R_2 and R_3. By varying R_1, the frequency of oscillation can be adjusted without changing the output amplitude.

$$f = \frac{1}{4\,R_1C}\left(\frac{R_2}{R_3}\right)$$

30-12. A Voltage-Controlled Sawtooth Oscillator (VCO)

The voltage-controlled oscillator (VCO) is an oscillator whose frequency can be changed by a variable d.c. control voltage. The VCOs can be either sinusoidal or nonsinusoidal. One way to build a voltage-controlled sawtooth oscillator is as shown in Fig. 30-18 (*a*). This circuit makes use of an Op-amp integrator that uses a switching device called programmable unijunction transistor (abbreviated as PUT) in parallel with the feedback capacitor to terminate each ramp at a prescribed level and effectively "reset" the circuit.

The programmable unijunction transistor (PUT) is a three terminal device, *i.e.*, it has an anode, a cathode and a gate terminal. The gate is always biased positively with respect to the cathode. When the anode voltage exceeds the gate voltage by approximately 0.7 V, the PUT turns on and acts as a forward biased diode. When the anode voltage falls below this level, the PUT turns off. Also, the value of current must be above the holding value to maintain conduction.

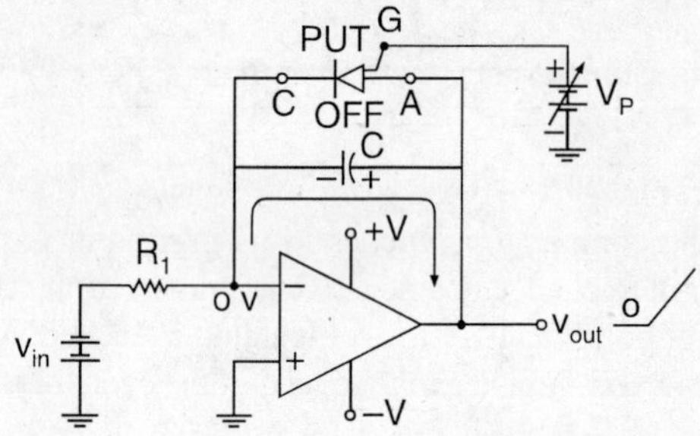

(*a*) Illustrating the charging of a capacitor when PUT is OFF.

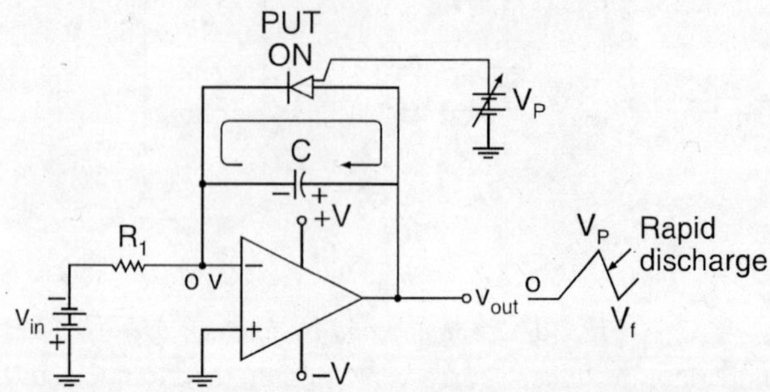

(*b*) Illustrating the discharging of a capacitor when PUT is ON.

Fig. 30-18. *A voltage controlled sawtooth oscillator.*

The operation of the circuit may be explained as below. The negative d.c. input voltage, $-V_{IN}$, produces a positive-going ramp on the output. During the time that ramp is increasing the circuit acts as a regular integrator. When the ramp voltage (*i.e.*, voltage at PUT anode) exceeds the gate voltage by 0.7 V, the PUT turns on. This forces the capacitor to discharge rapidly as shown in Fig. 30-18 (*b*). However, the capacitor does not discharge completely to zero because of the PUT's forward voltage, V_F. Discharging of the capacitor continues until the current through PUT drops below the holding value. At this point, PUT turn off and capacitor begins to charge again, thus generating a new output ramp. The cycle repeats and the resulting output is a repetitive sawtooth waveform as shown in the Fig. 30-18. It may be noted that the sawtooth amplitude and period can be adjusted by varying the PUT gate voltage.

The frequency of smooth curve is given by the relation :

$$f = \frac{|V_{IN}|}{R_1 C}\left(\frac{1}{V_P - V_F}\right)$$

It is evident from the above equation that the frequency depends upon the time constant "$R_1 C$" of the integrator and the peak voltage set by the PUT. The time period of the sawtooth wave is the reciprocal of the frequency, *i.e.*,

$$T = \frac{V_P - V_F}{|V_{IN}|/R_1 C}$$

30-13. A Square-Wave Relaxation Oscillator

Fig. 30-19 shows the circuit of a basic relaxation oscillator. Its operation depends upon charging and discharging of a capacitor. Notice that the Op-amp's inverting input is the capacitor voltage and the noninverting input is a portion of the output feed-back through resistors R_2 and R_3.

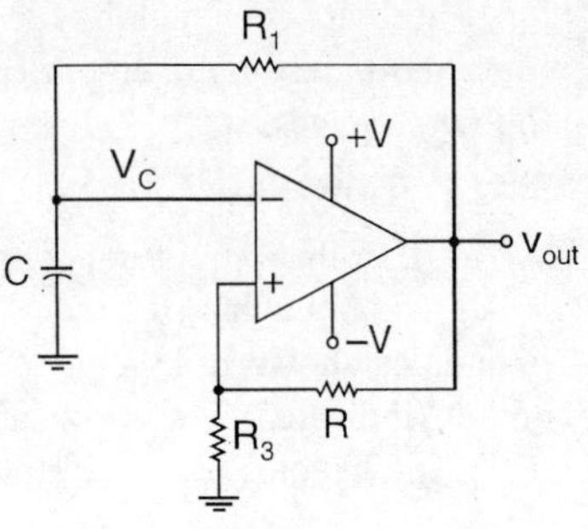

Fig. 30-19. *A square-wave relaxation oscillator.*

When the circuit is first turned on, the capacitor is uncharged. Because of this, the inverting input is at 0 V. This makes the output a positive maximum, and capacitor begins to charge towards V_{out} through R_1. When the capacitor voltage reaches a value equal to the feedback voltage on the non-inverting input, the op-amp switches to the maximum negative stage. At this point, the capacitor begins to discharge from + V_f toward – V_f. When the capacitor voltage reaches – V_f, the Op-amp switches back to the maximum positive state. This action continues to repeat and a square wave output is obtained as shown in Fig. 30-20.

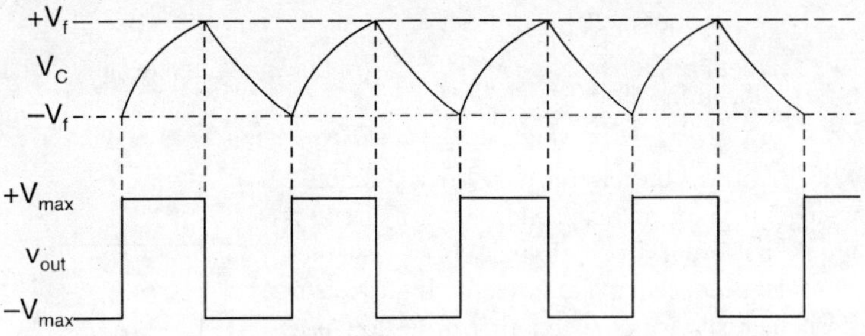

Fig. 30-20. *Waveforms for a square-wave relaxation oscillator.*

30-14. High-Impedance Voltmeter

Fig. 30-21 shows the circuit of a high impedance voltmeter. In such a circuit, the closed loop-gain depends on the internal resistance of the meter, R_M. The input voltage will be amplified and the output voltage will cause a proportional current to flow through the meter. By adding a small series potentiometer in the feedback loop, the meter can be calibrated to provide a more accurate reading.

The high input impedance of the Op-amp reduces the circuit loading that is caused by the use of the meter. Although this type of circuit would cause some circuit loading, it would be much more accurate than a VOM (Volt-Ohm-Meter) with an input impedance of 20 kΩ/V.

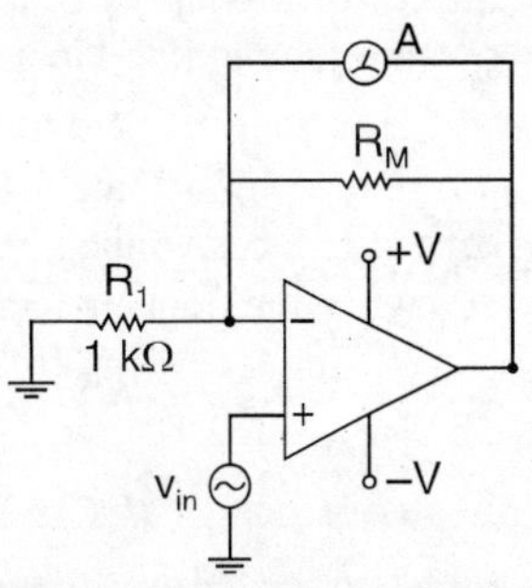

Fig. 30-21. *A high impedance voltmeter.*

30-15. Active Filters

We have already discussed in Chapter-27, about tuned voltage amplifiers (or simply tuned amplifiers). Such amplifiers are designed to amplify only those frequencies that are within certain range. As long as the input signal is within the specified range, it will be amplified. If it goes outside of this frequency range, amplification will be drastically reduced. The tuned amplifier circuits using Op-amp are generally referred to as active filters. Such circuits do not require the use of inductors. The frequency response of the circuit is determined by resistor and capacitor values.

We have already mentioned above that Op-amps can be used to build active filter circuits. A filter circuit can be constructed using passive components like resistors and capacitors. But an active filter, in addition to the passive components makes use of an Op-amp as an amplifier. The amplifier in the active filter circuit may provide voltage amplification and signal isolation or buffering.

There are four major types of filters namely, low-pass filter, high-pass filter, and band-pass filter and band-stop or notch filter. All these four types of filters are discussed one by one in the following pages.

30-16. Low-Pass Filter

A filter that provides a constant output from d.c. up to a *cut-off frequency* (f_{OH}) and then passes no signal above that frequency is called an ideal low-pass filter. The ideal response of a low-pass filter is as shown in Fig. 30-22 (*a*). Notice that the response shows that the filter has a constant output (indicated by a horizontal line '*ab*') from d.c. or zero frequency up to a cut-off frequency (f_{OH}).And beyond f_{OH}, the output is zero as indicated by the vertical line '*bc*' in the Fig..

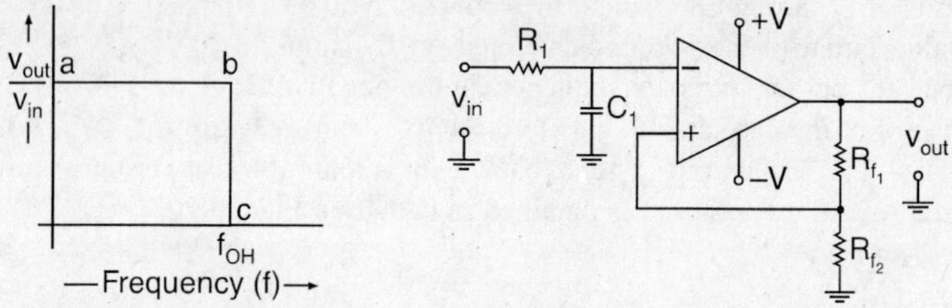

(*a*) Ideal response of a low-pass filter.

(*b*) Low-pass active filter circuit.

Fig. 30-22. *Low-pass filter.*

Fig. 30-22 (*b*) shows the circuit of a low-pass active filter using a single resistor and capacitor. Such a circuit is also referred to as first-order (or single-pole) low-pass filter. It is called first-order because it makes use of a single resistor and a capacitor. The response of such a first-order low pass filter is as shown in Fig. 30-23. Notice that the response below the cut-off frequency (f_{OH}) shows a constant gain (indicated by a horizontal line '*ab*').

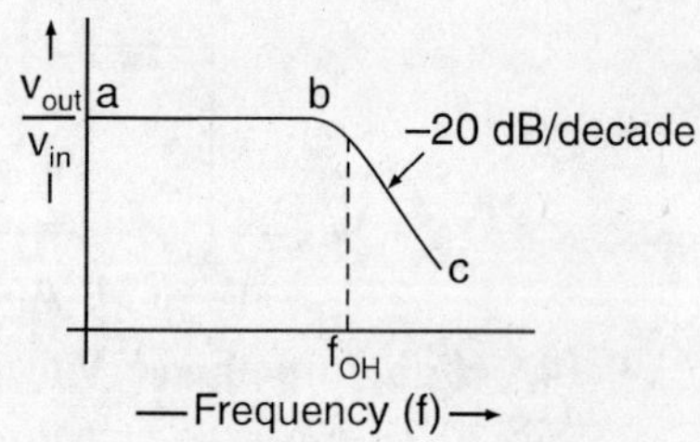

Fig. 30-23. *Response of a first-order low-pass active filter.*

However, beyond the cut-off frequency, the gain does not reduce immediately to zero as expected in Fig. 30-22 (*a*) but reduces with a slope of 20 dB/decade (means that the output voltage reduces by a factor of 100 when the frequency increases by a factor of 10). The voltage gain for a low-pass filter below the cut-off frequency (f_{OH}) is given by the relation,

$$A_v = 1 + \frac{R_3}{R_1}$$

And the cut-off frequency is determined by the relation,

$$f_{OH} = \frac{1}{2\pi R_1 C_1}$$

It is possible to connect two sections of the filter together as shown in Fig. 30-24 (*a*). Such a circuit is called second-order (or two-pole) low pass filter. Fig. 30-24 (*b*) shows another configuration of the second-order low-pass filter.

Each circuit shown in Fig. 30-24 has two RC circuits, $R_1 - C_1$ and $R_2 - C_2$. As the operating frequency increases beyond f_2, each circuit will be dropping the closed-loop gain by 20 dB, giving a total roll-off rate of 40 dB/decade when operated above f_2. The cut-off frequency for each of the circuit is given by,

$$f_2 = \frac{1}{2\pi\sqrt{R_1 R_2 C_1 C_2}}$$

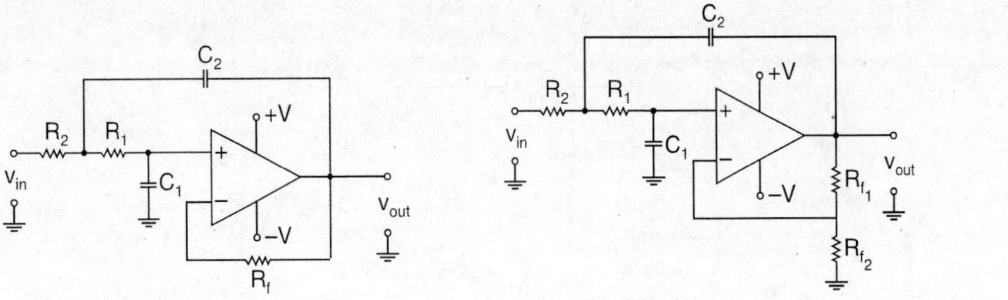

(*a*) Second order low-pass filter. (*b*) Another configuration of second-order low-pass filter.

Fig. 30-24.

30-17. High-Pass Filter

As a matter of fact, there is very little difference between the high-pass filter and the low-pass filter. Fig. 30-25 (*a*) shows the circuit of a first order (or single-pole) high-pass filter and Fig. 30-25 (*b*), the circuit of a second-order (or two-pole) high-pass filter. Notice that the only thing that has changed is the positions of the capacitors and resistors. The value of cut-off frequencies f_1 and f_2 is obtained by using the same equations we used for low-pass filter.

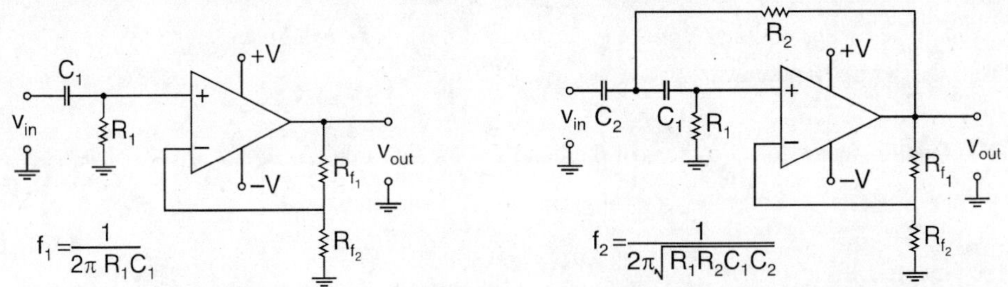

(*a*) A first-order high pass filter. (*b*) A second-order high pass filter.

Fig. 30-25.

Fig. 30-26 shows the gain versus frequency response of a high-pass filter. Notice that the solid line indicates the ideal response while the dashed line, corresponds to the actual response of the filter circuit. The ideal curve indicates that the filter has a zero output for the frequencies below f_{OL} (indicated by the line '*ab*'). And beyond f_{OL}, it has a constant output. The actual response curve may correspond to the roll-off gain by 20 dB/decade for first order or 40 dB/decade for second-order low-pass filters.

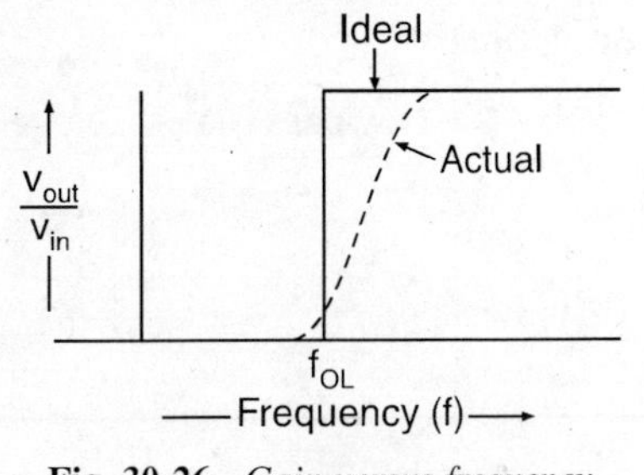

Fig. 30-26. *Gain versus frequency response of a high-pass filter.*

30-18. Band-Pass Filter

A band-pass filter is the one that is designed to pass all frequencies within its bandwidth. A simple way to construct a band-pass filter is to cascade a high-pass filter as shown in Fig. 30-27.

The first stage of the band-pass filter will pass all frequencies that are below its cut-off value, f_2. All the frequencies passed by the first stage will head into the second stage. This stage will pass all frequencies above its value of f_1. The result of this circuit action is as shown in Fig. 30-28. Note that the only frequencies that will pass through the amplifier are those that fall within the pass band of both amplifiers. The values of f_1 and f_2 can be obtained by using the relations, $1/2\pi\, R_1C_1$ and, $1/2\pi\, R_2C_2$. Then bandwidth,

$$BW = f_2 - f_1$$

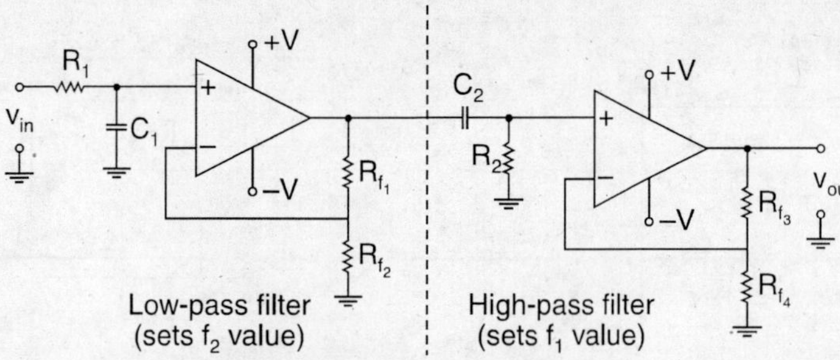

Fig. 30-27. *A band-pass filter circuit.*

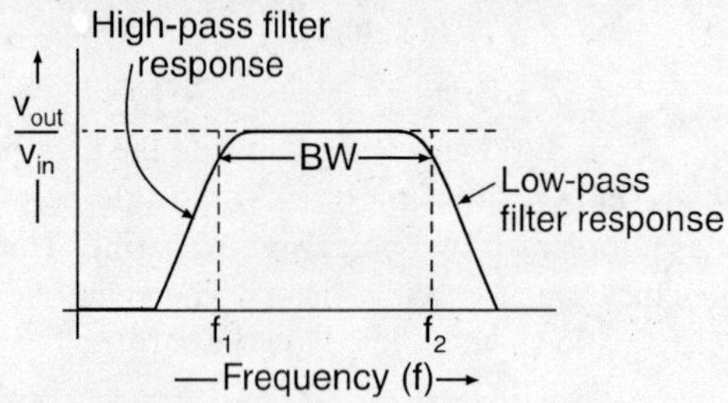

Fig. 30-28. *Gain versus frequency response of a band-pass filter.*

And the centre frequency,

$$f_o = f_1 \cdot f_2$$

The Quality-factor (or *Q*-factor) of the band-pass filter circuit,

$$Q = \frac{f_o}{BW}$$

30-19. Notch Filter

The notch filter is designed to block all frequencies that fall within its bandwidth. Fig. 30-29 (*a*) shows a block diagram and 30-29 (*b*), the gain versus frequency response curve of a multistage notch filter.

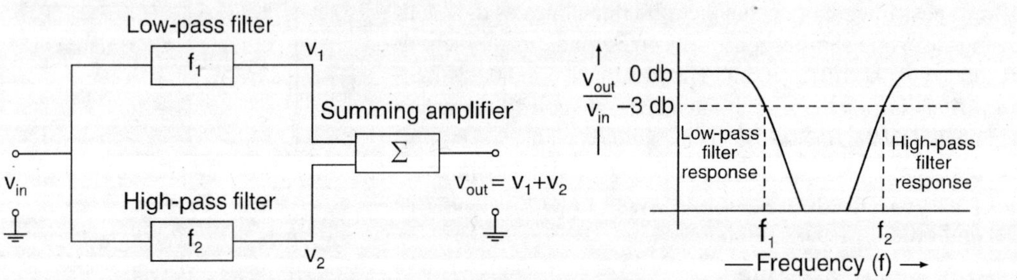

(*a*) Block diagram. (*b*) Gain versus frequecny response curve.

Fig. 30-29. *Notch filter.*

The block diagram shows that the circuit is made up of a high-pass filter, a low-pass filter and a summing amplifier. The summing amplifier produces an output that is equal to a sum of the filter output voltages. The circuit is designed in such a way so that the cut-off frequency, f_1 (which is set by a low-pass filter) is lower in value than the cut-off frequency, f_2 (which is set by high-pass filter). The gap between the values of f_1 and f_2 is the bandwidth of the filter.

When the circuit input frequency is lower than f_1, the input signal will pass through low-pass filter to the summing amplifier. Since, the input frequency is below the cut-off frequency of the high-pass filter, v_2 will be zero. Thus, the output from the summing amplifier will equal the output from the low-pass filter. When the circuit input frequency is higher than f_2, the input signal will pass through the high-pass filter to the summing amplifier. Since, the input frequency is above the cut-off frequency of the low-pass filter, v_1 will be zero. Now the summing amplifier output will equal the output from the high-pass filter.

It is evident from the above discussion that frequencies below f_1 and those above f_2, have been passed by the notch filter. But when the circuit frequency between f_1 and f_2, neither of the filters will produce an output. Thus, v_1 and v_2 will be both zero and the output from the summing amplifier will also be zero.

The frequency analysis of the notch filter is identical to the band-pass filter. First, determine the cut-off frequencies of the low-pass and the high-pass filters. Then using these calculated values, determine the bandwidth, centre frequency and Q values of the circuit.

SUMMARY

In this chapter you have learned that :

1. Comparator is a circuit that is used to compare two voltage and provide an output indicating the relationship between the two voltages.
2. The integrator is a circuit whose output is proportional to the area of its input waveform.
3. The differentiator is a circuit whose output is proportional to the rate of change of its input signal.
4. Op-amp can be used in amplifiers and oscillator circuits to provide gain as well as the 180° phase-shift.
5. The cut-off frequency of a single-pole low-pass filter is given by,

$$f_2 = \frac{1}{2\pi\, RC}$$

the maximum overall gain is,

$$A_{CL} = \frac{R_{f_1}}{R_{f_2}} + 1 = 1.25 \text{ or } 2\ dB$$

6. The cut-off frequency of a single-pole high-pass filter and the maximum overall gain is given by the same set of equations as for the low-pass filter.
7. The cut-off frequency of a double-pole low-pass filter is given by,

$$f_2 = \frac{1}{2\pi\sqrt{R_1 R_2 C_1 C_2}}$$

and the maximum overall gain,

$$A_{CL} = \frac{R_{f_1}}{R_{f_2}} + 1 = 1.56 \text{ (or } 4\ dB)$$

8. For double pole high -pass filter the cut-off frequency and maximum gain are given by the same equation does that of a double-pole low-pass filter.

GLOSSARY

Active filter : a tuned amplifier circuit using Op-amp.
Audio amplifier : the final audio stage in electronic communication receives
Band-pass filter : a filter designed to pass all frequencies that fall between lower cut-off (f_1) and upper cut-off (f_2) frequencies.
Band-stop filter : a filter designed to block all frequencies that fall between f_1 and f_2
Comparator : a circuit whose d.c. output is determined by the polarity relationship of its inputs.
High-pass filter : a filter designed to pass all frequencies above it cut-off frequency.
Low-pass filter : a filter to pass all frequencies below its cut-off frequency
Notch filter : same as band-stop filter.
Integrator : a circuit whose output is proportional to the area of its input waveform.
Differentiator : a circuit whose output is proportional to the rate of change of its input signal.

NUMERICAL PROBLEMS

1. Determine the lower cut-off frequency and the frequency at which the circuit shown in Fig. 30-30 will start to lose its linear output characteristics. (**Ans.** 79.6 kHz, 796 kHz.)

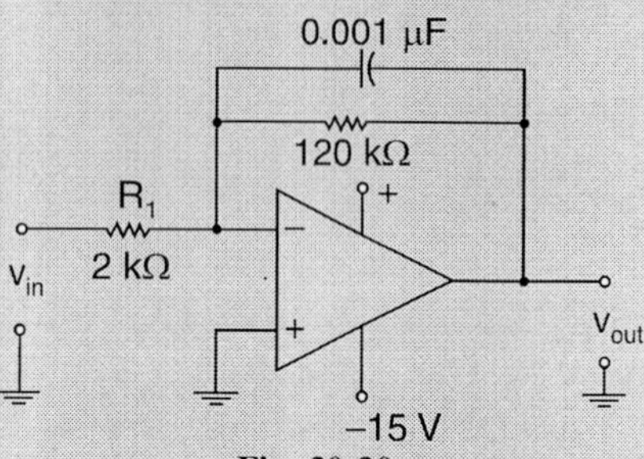

Fig. 30-30.

2. A differentiator has values of $R_1 = 2.2$ kΩ; $C_1 = 0.15$ μF. Determine its maximum linear operating frequency and its cut-off frequency. (**Ans.** 48.2 Hz and 482 Hz)
3. Using 741 Op-Amp, design an astable multivibrator to produce ± 9 V, square wave output. (*AMIE., Winter* 1998)

DESCRIPTIVE QUESTIONS

1. What is a comparator? How it can be used to produce a square-wave at the output from a sine-wave?
2. Draw the circuit diagram of an integrator using Op-amp. Label it neatly.
3. What is a differentiator? Explain briefly.
4. Draw the circuit diagram of an Op-amp based Wein-bridge Oscillator.
5. Draw the circuit diagram of an Op-amp based Colpitts Oscillator and Crystal Oscillator.
6. Explain the function of a square-wave relaxation oscillator with the help of a suitable circuit diagram.
7. What is the difference between a low-pass and a high-pass filter? Explain it with the help of gain-versus-frequency response graph.
8. Sketch the gain-versus-frequency curves for a band-pass and notch-filter.
9. Give any four applications of a comparator. (*Linear Integrated Circuits, Anna University, April/May 2008*)
10. Draw the input and output waveform of miller integrator circuit. (*Electronics Circuit-II, Anna University, May/June 2007*)
11. Methods of improving slew rate. (*Linear integrated Circuits, Anna University, April/May 2008*)

MULTIPLE CHOICE QUESTIONS

1. The purpose of a comparator is to
   (*a*) amplify an input voltage
   (*b*) detect the occurrence of a changing input voltage
   (*c*) maintain a constant output when the d.c. input voltage changes
   (*d*) produce a change in output when an input voltage equals the reference voltage
2. The Op-amp comparator circuit uses
   (*a*) positive feedback (*b*) negative feedback
   (*c*) regenerative feedback (*d*) no feedback
3. The feedback path in an Op-amp differentiator consists of
   (*a*) a resistor (*b*) a capacitor
   (*c*) a resistor and a capacitor in series (*d*) a resistor and a capacitor in parallel
4. The feedback path in an Op-amp integrator consists of
   (*a*) a resistor (*b*) a capacitor
   (*c*) a resistor and a capacitor in series (*d*) a resistor and a capacitor in parallel
5. For the high-pass circuit to act as a differentiator, the time constant must be
   (*a*) small
   (*b*) very small in comparison to the time period of the input signal
   (*c*) very high in comparison to the time period of the input signal
   (*d*) of moderate value
6. In the circuit shown in Fig. 30-31.

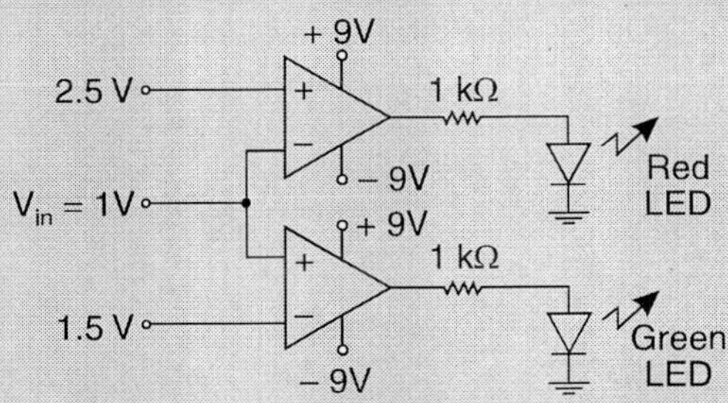

Fig. 30-31.

   (*a*) only red will glow (*b*) only green will glow
   (*c*) both red and green will glow (*d*) neither red nor green will glow
7. The low frequency gain of the low pass filter shown in Fig. 30-32. at an input frequency one decade above the cut-off is

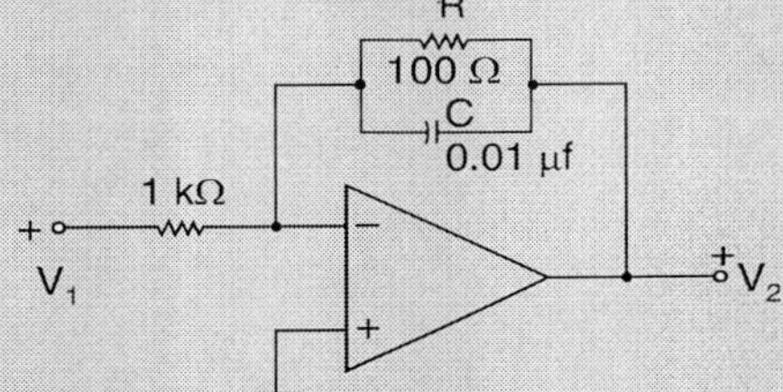

Fig. 30-32.

   (*a*) 10 dB (*b*) 20 dB
   (*c*) 30 dB (*d*) 40 dB

8. A unit positive step is applied at the input of the circuit shown in Fig. 30-33. After 20 seconds, the output v_0 will be

(*a*) + 20 V (*b*) + 10 V

(*c*) – 10 V (*d*) – 20 V

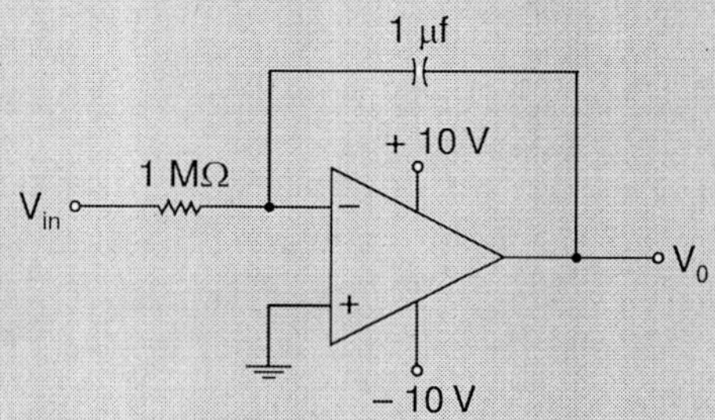

Fig. 30-33.

9. Which one of the following correctly depicts the output of a low-pass filter to a pulse input.

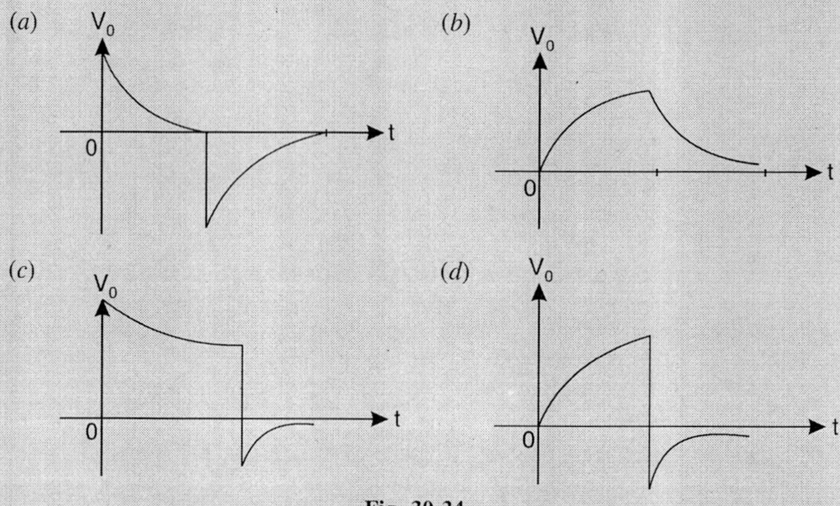

Fig. 30-34.

10. A circuit with Op-amp is shown in Fig. 30-35. The voltage v_0 is,

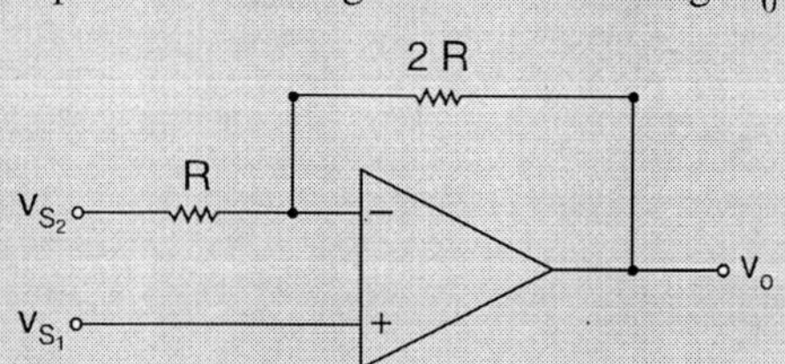

Fig. 30-35.

(*a*) $3V_{S_1} - 6V_{S_2}$ (*b*) $2V_{S_1} - 3V_{S_2}$

(*c*) $2V_{S_1} - 2V_{S_2}$ (*d*) $3V_{S_1} - 2V_{S_2}$

11. The output voltage v_0 of the circuit shown in Fig. 30-36, is

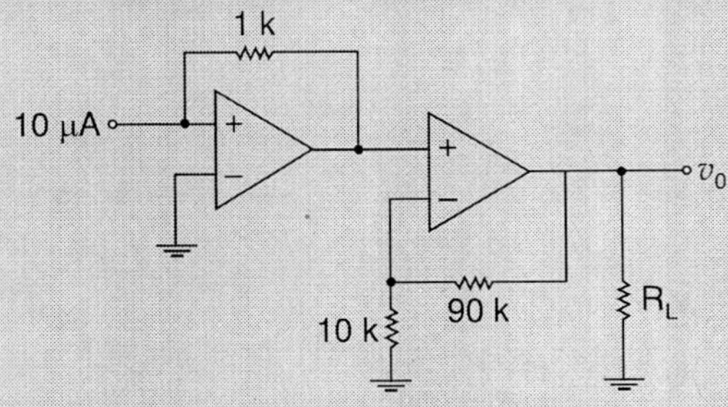

Fig. 30-36.

(*a*) – 100 V (*b*) – 100mV (*c*) 10 V (*d*) 10 mV

(*UPSC Engg. Services*, 1999)

12. An Op-amp circuit is shown in Fig. 30-37. The current 'I' is :

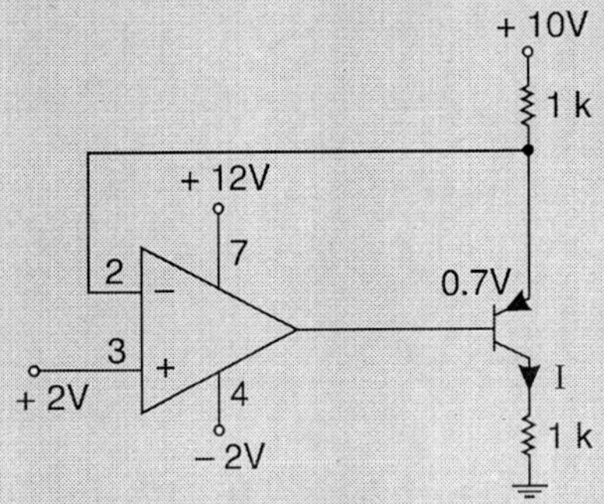

Fig. 30-37.

(*a*) zero (*b*) 6 mA (*c*) 10 mA (*d*) 12 mA

13. A circuit is shown in Fig 30-38. The largest value of R_L that can be used in this circuit is

(*a*) 100 Ω (*b*) 400 Ω (*c*) 2 kΩ (*d*) 20 kΩ

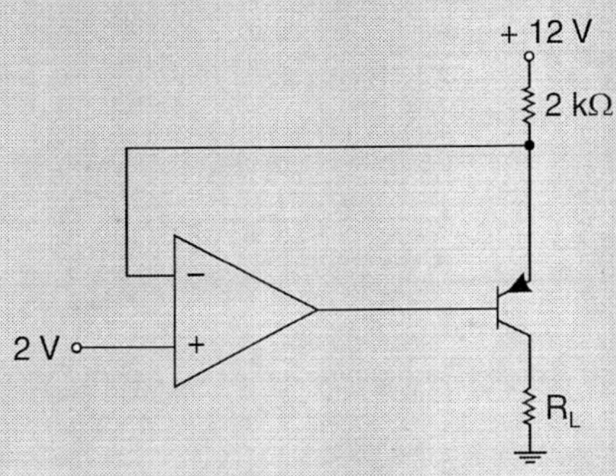

Fig. 30-38.

(*UPSC Engg. Services* 2001)

14. An Op-amp circuit shown in Fig. 30-39, the output v_o, will be (Assume ideal Op-amp)

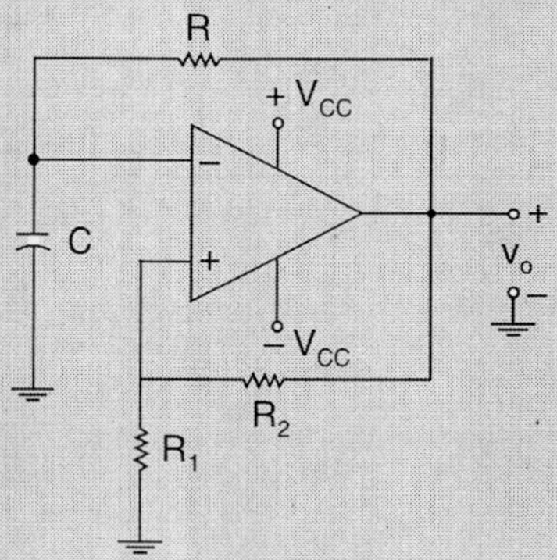

Fig. 30-39.

(*a*) equal to zero because the input is zero

(*b*) dependent on element values hence nothing can be predicted without a knowledge of element values

(*c*) a square wave varying between + V_{CC} and – V_{CC}.

(*d*) a sinusodial wave of amplifier V_{CC}. (*UPSC Engg. Services*, 2002)

15. The effect of a finite gain of an operational amplifier used in an integrator is that

(*a*) it would not integrate

(*b*) the slope of output will very with time

(*c*) the final values of the output voltage will reduce

(*d*) there will be instability in the circuit. (*UPSC Engg. Services* 2001)

16. In an PLL, lock occurs when the

(*a*) input frequency and the VCO frequency are the same

(*b*) phase error is 180°

(*c*) VCO frequency is double the input frequency.

(*d*) phase error is 90°. (*UPSC Engg Services*, 2001)

17. The Op-amp circuit shown in Fig 30-40. is,

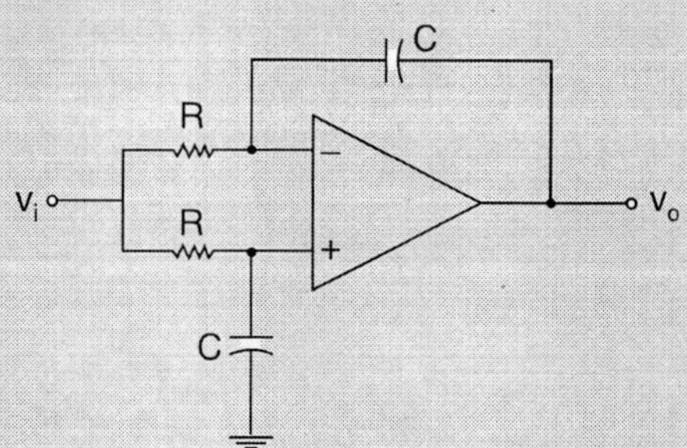

Fig. 30-40.

(*a*) a high-pass circuit (*b*) a low-pass circuit

(*c*) a band-pass circuit (*d*) an all-pass circuit

(*UPSC Engg. Services* 2000)

18. The essential blocks of a phase-locked loop (PLL) are phase detector, amplifier,
(*a*) high-pass filter and crystal controlled oscillator
(*b*) low-pass filter and crystal controlled oscillator
(*c*) high-pass filter and voltage controlled oscillator
(*d*) low-pass filter and voltage controlled oscillator (*UPSC Engg Services* 2000)

19. A second-order band-pass active filter can be obtained by cascading a low-pass second-order section having cut-off frequency, f_{OH} with a high-pass second-order section having cut-off frequency f_{OL}, provided

(*a*) $f_{OH} > f_{OL}$ (*b*) $f_{OH} < f_{OL}$ (*c*) $f_{OH} = f_{OL}$ (*d*) $f_{OH} \le \frac{1}{2} f_{OL}$

(*UPSC Engg Services* 1999)

20. Higher order active filters are used for
(*a*) bandwidth (*b*) gain in the passband
(*c*) impedance (*d*) higher roll-off rate

21. In the circuit shown in Fig. 30-41, the output voltage will be
(*a*) 9 V (*b*) 10 V (*c*) 11 V (*d*) 12 V

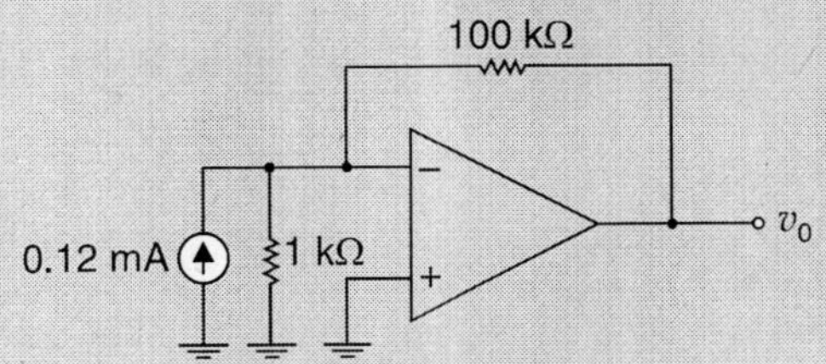

Fig. 30-41.

22. An Op-amp is used in the circuit as shown in Fig. 30-42. The current I_O is,

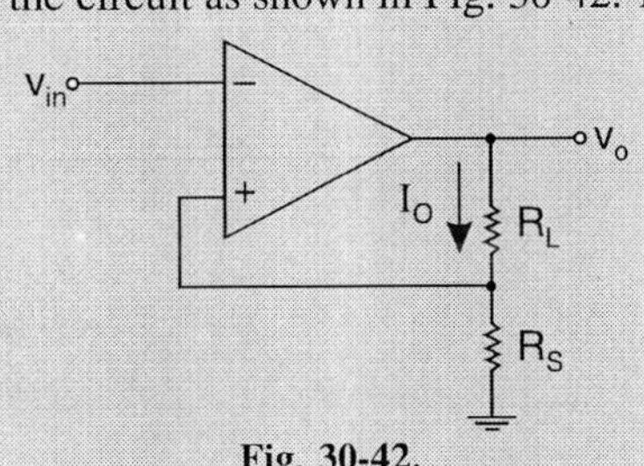

Fig. 30-42.

(*a*) $V_{in} \times \frac{R_L}{R_S\,(R_L + R_S)}$ (*b*) $\frac{V_{in}}{R_S}$

(*c*) $\frac{V_{in}}{R_L}$ (*d*) $V_{in}\left(\frac{1}{R_S} + \frac{1}{R_L}\right)$

Hint. $V_o = V_{in} + I_o \,.\, R_L$

$\therefore$ $I_o = \frac{V_o - V_{in}}{R_L}$,

But $A_V = 1 + \frac{R_L}{R_S} = \frac{V_o}{V_{in}}$

$\therefore$ $V_o = A_V \,.\, V_{in}$

and $I_o = \frac{(A_V - 1)\,V_{in}}{R_L} \approx \frac{A_v \cdot V_{in}}{R_L} = \frac{V_{in}}{R_S \cdot R_L}(R_S + R_L)$

23. In the Op-amp circuit shown in Fig. 30-43. $V_i > O$ and $i\ I_o.\ e^{-\alpha v}$ the output, v_o will be proportional to,

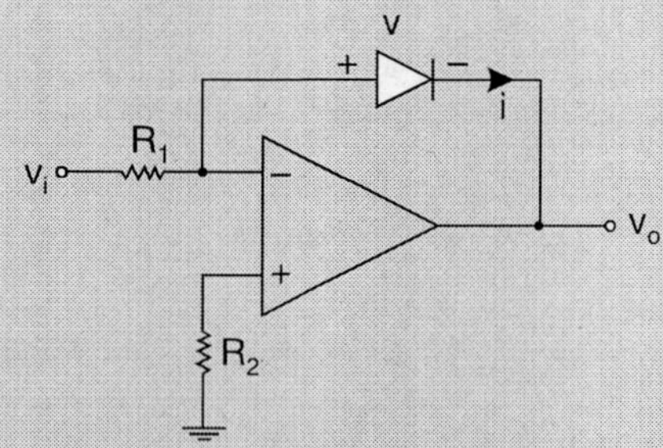

Fig. 30-43.

(*a*) $\sqrt{v_i}$ (*b*) v_i (*c*) e^{kv_i} (*d*) $l_n\ (kv_i)$

24. In the circuit shown in Fig 30-44. The current through the resistance R is :

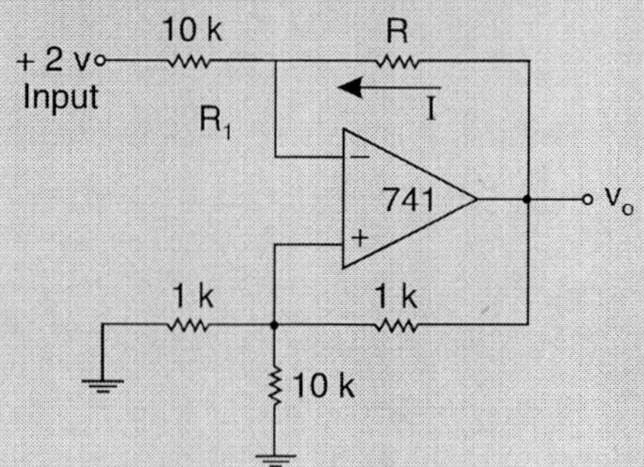

Fig. 30-44.

(*a*) 100 μA (*b*) – 100 μA (*c*) 1 μA (*d*) – 1 μA.

(*UPSC Engg. Services* 2001)

25. An amplifiers of gain A is pridged by a capacitance 'C' as shown in Fig. 30-45. The effective input capacitance is :

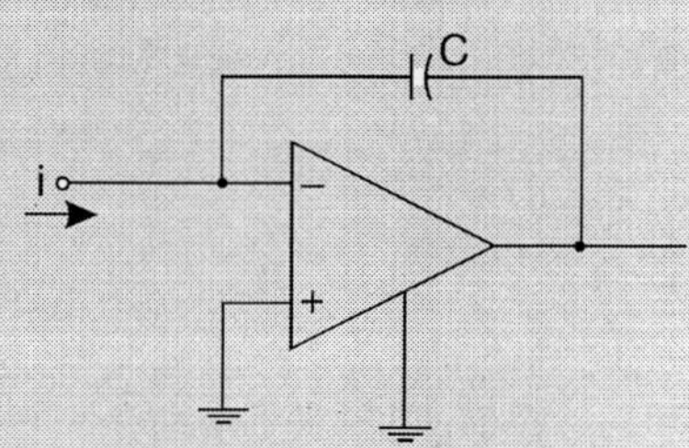

Fig. 30-45.

(*a*) C/A (*b*) C (1– A) (*c*) C (1 + A) (*d*) C A

ANSWERS

1. (*c*)	**2.** (*d*)	**3.** (*a*)	**4.** (*b*)	**5.** (*b*)
6. (*a*)	**7.** (*b*)	**8.** (*c*)	**9.** (*b*)	**10.** (*d*)
11. (*d*)	**12.** (*a*)	**13.** (*b*)	**14.** (*c*)	**15.** (*a*)
16. (*a*)	**17.** (*d*)	**18.** (*d*)	**19.** (*a*)	**20.** (*d*)
21. (*d*)	**22.** (*d*)	**23.** (*d*)	**24.** (*c*)	**25.** (*c*)

❑❑❑

31

CHAPTER

Regulated Power Supplies

OUTLINE

Objectives

At the end of this chapter, you should be able to :

- Know what is voltage regulator.
- Understand the concept of voltage regulation.
- Describe the difference between line regulation and load regulation.
- List the various types of voltage regulators.
- Explain the operation of zener diode shunt regulator.
- Calculate the optimum value of current limiting resistor in a zener diode voltage regulator.
- Describe the circuit operation of transistor shunt and transistor series regulators.
- Sketch the circuit of controlled transistor series regulator.

- Explain the operation of transistor current regulator.
- Describe the operation of variable feedback regulator.
- Sketch and explain circuit operation of basic Op-amp series and Op-amp shunt regulator circuits.
- Lost the various types of switching regulators.
- Describe the operation of step-down, step-up and inverting type switching regulators.
- Know how the fixed positive and negative voltage IC regulators are selected and connected to produce the desired output voltage.
- Design the circuit using adjustable positive or negative output linear voltage regulators.
- Sketch the basic d.c. power supply circuit.

31-1. Introduction

It is a well known fact, that all the electronic devices (*i.e.*, diodes, transistors, integrated circuits etc.) and circuits require a constant d.c. voltage for their operation. The d.c. voltage can be supplied from dry cells or batteries. But these are expensive as compared to conventional regulated d.c. power supplies. The regulated d.c. power supply is made by converting the domestic a.c. supply to d.c. supply. The essential elements, which constitute a regulated d.c. power supply, are given below :

(*i*) Rectifier,

(*ii*) Filter, and

(*iii*) Voltage regulator.

The rectifier and filter has already been discussed in the last chapter. Now we shall study the voltage regulator in the following pages.

31-2. Voltage Regulator

We have already discussed in the last article that a regulated d.c. power supply consists of three elements namely rectifier, filter and voltage regulator. The rectifier is the front or the first element of a d.c. power supply, the filter is the second (or intermediate element) and the voltage regulator is the last element. (Refer to Fig. 31-1.) A rectifier is used to convert the a.c. output of the transformer to pulsating d.c. (Fig. 31-2(*a*)). This pulsating d.c. is then applied to a filter which reduces the variations in the rectifier ouput [Fig. 31-2 (*b*)]. The concept of a voltage regulator may be understood from the following discussion.

Fig. 31-1.

Consider a combination of full-wave rectifier and a filter circuit connected to an a.c. input voltage source. When the a.c. input voltage, at the rectifier input, changes above or below its normal value, it will cause a change in d.c. voltage produced at the filter output. A similar change in d.c. voltage may also occur, when the load resistance connected at the filter output changes above or below its normal value. It means that d.c. output voltage fluctuates (or changes) whenever the a.c. input voltage or the load resistance varies above or below the normal values.

A fluctuating d.c. voltage may result in an erratic operation of electronic devices and circuits. For example, it may cause a frequency shift in an oscillator, change of calibration in instruments, distortion in the output of power amplifiers, *etc*. In order to avoid fluctuations, a voltage regulator circuit is used. The voltage regulator is connected between the filter and the load. Its function is to maintain a constant output d.c. voltage inspite of the a.c. input voltage fluctuations or changes in load resistance values.

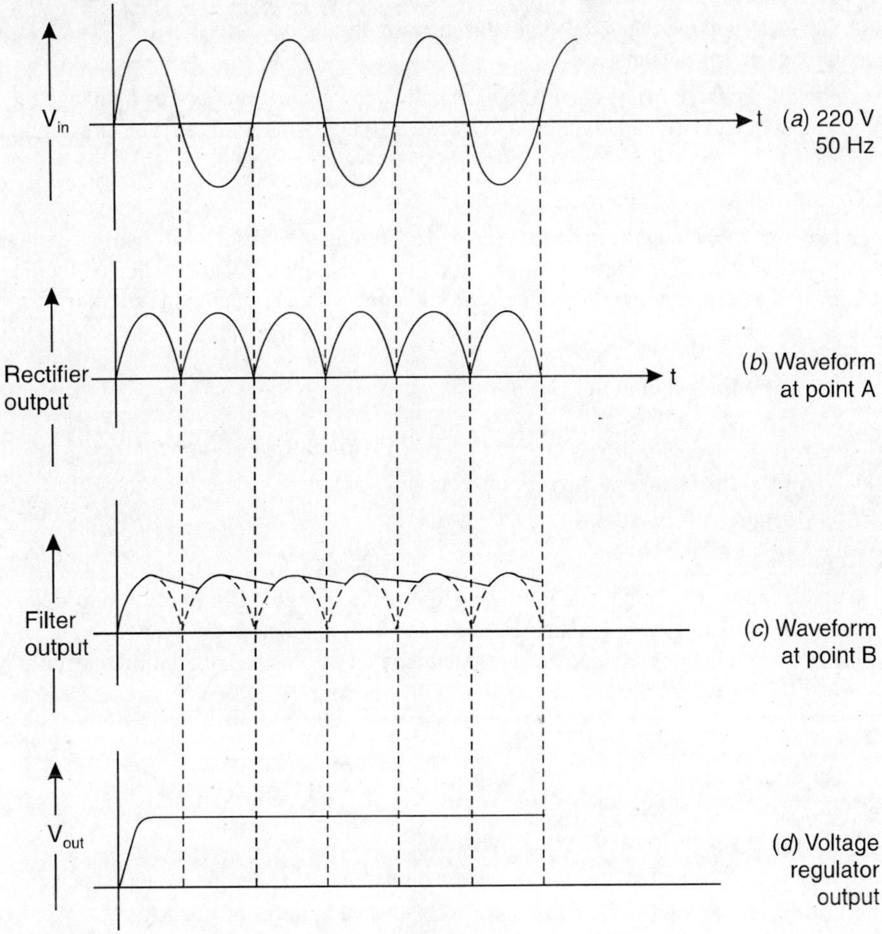

Fig. 31-2.

31-3. Voltage Regulation

An ideal voltage reulator will maintain a constant d.c. output voltage regardless of changes in either its input voltage or its load current demand. For example consider a +5 *Vdc* regulator shown in Fig 31-3. A change in the regulator input voltage shown as ΔV_S in Fig. 31-3 (*a*) would not appear at the regular output. It may be carefully noted that ΔV_S could be either a change in the steady-state (d.c.) value of V_S or could be some ripple voltage. In either case, the change in V_L for the ideal regulator would be *OV*. However during this discussion, we have assumed that the value of input voltage to the voltage regulator (V_S) does not decrease below the value required to maintain the operation of the voltage regulator. As a thumb rule, a regulator cannot have 5 Vdc output if the input is less then 5 Vdc.

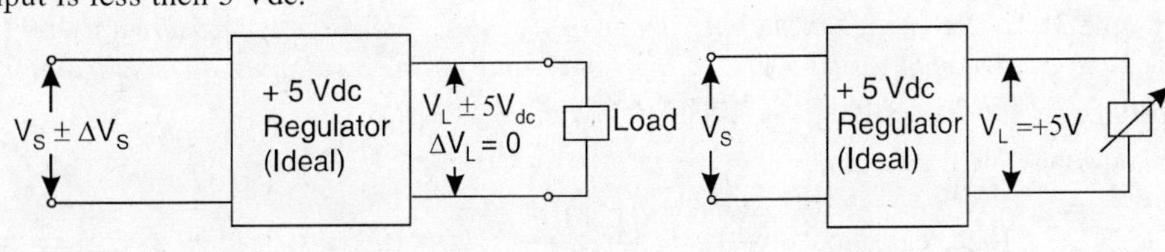

Fig. 31-3.

Fig. 31-2 shows how an ideal voltage regulator would respond to a change in load current demand. The ideal voltage regulator maintains a constant output voltage (*i.e.*, $\Delta V_L = 0$) despite the changes in load current demand.

Accordingly we have two ways of specifying the quality of regulation for a voltage regulator : (1) line regulation, and (2) load regulation. Now we will explain both these measures of regulation in more detail.

Line Regulation

In actual practice, a change in input voltage to a voltage regulator will cause a change in its output or load voltage. The *line regulation* rating of a voltage regulator indicates the change in output voltage that will occur per unit change in the input voltage. The line regulation of a voltage regulator is given by,

$$\textbf{Line regulation} = \frac{\Delta V_L}{\Delta V_S}$$

where, ΔV_L = the change in output voltage usually in millivolts or microvolts

and ΔV_S = the change in input voltage, usually in volts.

The line regulation is expressed as (V/V), or $\mu V/V$.

Load Regulation

In practice the voltage regulator will also experience a slight change in output voltage when there is a change in load current demand. The load regulation indicates the change in output voltage that will occur per unit change in load current. Mathematically, the load regulation may be expressed as,

$$\textbf{Load regulation} = \frac{V_{NL} - V_{FL}}{\Delta I_L}$$

where, V_{NL} = no load output voltage.

V_{FL} = the full-load output voltage and

ΔI_L = the change in load current demand

The load regulation of a voltage regulator is expressed in terms of μV/mA.

Ideally speaking, whether it is line regulation or load regulation, the lower the two values are, the better is the quality of regulation.

Some people prefer another way of expressing voltage regulation as given below. Here the voltage regualtion is expressed as the ability of the regulator to maintain a constant output voltage inspite of a.c. input voltage fluctuations and changes in load resistance.

Mathematically,

$$\textbf{\% regulation} = \frac{V_{NL} - V_{FL}}{V_{FL}} \times 100$$

where, V_{NL} = no load voltage and

V_{FL} = full-load voltage

Example 31-1. *The change in output voltage of a voltage regulator is measured at 100 μV when the input voltage changes by 5 V. Determine the value of line regulation for this regulator.*

Solution. Given : $\Delta V_L = 100\ \mu V = 100 \times 10^{-6}$ V and $\Delta V_{in} = 5\ V$

We know that line regulation,

$$\text{Line regulation} = \frac{\Delta V_L}{\Delta V_S} = \frac{100 \times 10^{-6}}{5} = 20 \times 10^{-6}\ V/V$$

$$= 20\ \mu\text{V/V}\ \textbf{Ans.}$$

Example 31-2. *A voltage regulator has a line regulator of 1.4 μV/V. Determine the change in its output voltage of the input voltage changes by 10 V.*

Solution. Given : Line regulation = 1.4 mV/V = 1.4×10^{-6} V/V and ΔV_S = 10 V.

Let ΔV_L = change in output voltage of the regulator,

We know that line regulation,

$$1.4 \times 10^{-6} = \frac{\Delta V_L}{\Delta V_S} = \frac{\Delta V_L}{10}$$

$$\therefore \quad \Delta V_L = (1.4 \times 10^{-6}) \times 10 = 14 \times 10^{-6} \text{ V or } 14 \text{ μV } \textbf{Ans.}$$

Example 31-3. *A voltage regulator is rated for an output (or load) current of I_L = 0 to 40 mA. Under no load conditions, the output voltage from the circuit is 8V. Under full-load conditions, the output voltage drops to 7.995 V. What is the value of load regulation for this voltage regulator.*

Solution. Given : ΔI_L = 40 mA; V_{NL} = 8 V and V_{FL} = 7.995 V

We know that load regulation of a voltage regulator,

$$\text{load regulation} = \frac{V_{NL} - V_{FL}}{\Delta I_L} = \frac{8 - 7.995}{40 \text{ mA}} = \frac{0.005}{40 \text{ mA}}$$

$$= \frac{5 \text{ mV}}{40 \text{ mA}} = 0.125 \text{ mV/mA}$$

$$= 125 \text{ μV/mA. } \textbf{Ans.}$$

Example 31-4. *A voltage regulator has a load regulation of 10 μV/mA. Under no-load conditions, the output voltage is 5 V. Calculate the voltage under full-load condition, if the load current is 20 mA.*

Solution. Given : Load regulation = 10 μV/mA: V_{NL} = 5 V, ΔI_L = 20 mA.

Let V_{FL} = output voltage at full-load.

We know that the load regulation of a voltage regulation,

$$10 \text{ μV/mA} = \frac{V_{NL} - V_{FL}}{\Delta I_L} = \frac{5 - V_{FL}}{20 \text{ mA}}$$

$$\therefore \quad V_{FL} = 5 \text{ V} - (10 \text{ μV/mA}) \times 20 \text{ mA}$$

$$= 5 \text{ V} - 200 \text{ μV} = 4.9998 \text{ V } \textbf{Ans.}$$

Example. 31-5. *A 10 V regulated d.c. power supply has a regulation of 0.002 per-cent. Find the magnitude of variation in output voltage.*

Solution. Given V_0 = 10 volts; Regulation = 0.002% = 0.00 002.

We know that a regulation of 0.00002 means, that the output voltage will vary within the limits of 0.002% of 10 V. Therefore, magnitude of variation

$$= 0.00002 \times 10 = 0.00\,002 \text{ V} = 0.2 \text{ mV}$$

This indicates that the variation in output voltage is 0.2 mV. However, since the variation is equal on the increasing or decreasing side, therefore the magnitude of variation may also be expressed as ± 0.1 mV. **Ans.**

31-4. Types of Voltage Regulators

Although there are many types of voltage regulators, yet the following ones are important from the subject point of view :

- Zener diode shunt regulator.
- Transistor series regulator.
- Transistor current regulator
- Transistor shunt regulator.
- Controlled transistor series regulator.
- Op-amp series regulator

- Op-amp shunt regulator
- Switching regulators
- Monolithic regulator.

31-5. Zener Diode Shunt Regulator

Fig. 31-4 shows a circuit of zener diode shunt regulator. Since, the zener is connected in parallel (or shunt) with the load, therefore the circuit is known as shunt regulator. A resistance, (R_S) is connected in series with the zener to limit current in the circuit. Therefore, the resistance R_S is also known as series current limiting resistor.

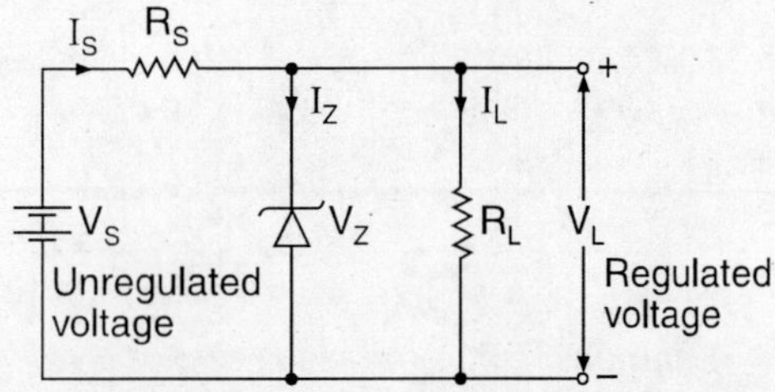

Fig. 31-4. *Zener diode shunt regulator.*

The output voltage (V_L) is taken across the load resistance (R_L). For proper operation, the input voltage (V_S) must be greater than the zener voltage (V_Z). This ensures that zener operates in the reverse breakdown region.

The input current (*i.e.*, current through the limiting resistor) is given by the relation,

$$I_S = \frac{V_S - V_Z}{R_S}$$

where, V_S = The d.c. input voltage to the regulator circuit, and
V_Z = The zener voltage.

The ideal zener diode acts as a constant voltage source of voltage (V_Z). However, a practical zener diode has a finite value of resistance called zener resistance (r_Z). Because of the zener resistance, there is a voltage drop across it which is equal to $I_Z \cdot r_Z$. Therefore, the voltage across the terminals of the zener diode (which is equal to load voltage),

$$V_L = V_Z + I_Z \cdot r_Z \qquad ...(i)$$

If the zener resistance is negligible, then the load voltage,

$$V_L = V_Z \qquad ...(ii)$$

and the current through the load resistance, is given by the relation,

$$I_L = \frac{V_L}{R_L} \qquad ...(iii)$$

We know that input current is the sum of zener current and load current *i.e.*,

$$I_S = I_Z + I_L \quad \text{or} \quad I_Z = I_S - I_L$$

Example 31-6. *Fig. 31-5 shows the circuit of a zener diode shunt regulator.*

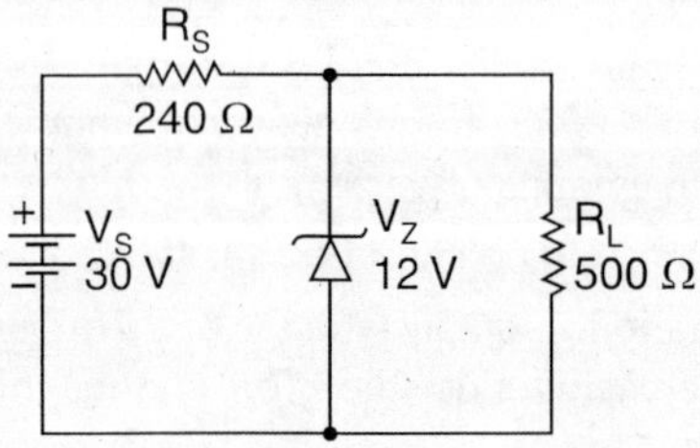

Fig. 31-5.

Find **(a)** *the load voltage,* **(b)** *voltage drop across series resistance, and* **(c)** *current through the zener diode.*

Solution. Given : V_S = 30 volts; R_S = 240 Ω; V_Z = 12 volts and R_L = 500 Ω.

(a) *Load Voltage*

We know that the load voltage

$$V_L = V_Z = 12 \text{ V} \textbf{ Ans.}$$

(*b*) *Voltage drop across series resistance*

We know that current through series resistance R_S.

$$I_S = \frac{V_S - V_Z}{R_S} = \frac{30 - 12}{240} = 0.075 \text{ A}$$

and voltage drop across series resistance R_S.

$$= I_S \cdot R_S = 0.075 \times 240 = 18 \text{ V } \textbf{Ans.}$$

(*c*) *Current through zener diode*

We also know that current through R_L,

$$= I_L = \frac{V_L}{R_L} = \frac{12}{500} = 0.024 \text{ A}$$

and current through zener diode,

$$I_Z = I_S - I_L = 0.075 - 0.024 = 0.051 \text{ A } \textbf{Ans.}$$

31-6. Working of Zener Diode Shunt Regulator

It may be discussed under the two heads, namely regulation with varying input voltage and regulation with varying load resistance.

(i) Regulation with varying input voltage : Consider the regulator circuit as shown in Fig. 31-6 (*a*). Here the load resistance (R_L) is kept fixed and the input voltage (V_S) varies within the limits. As the input voltage increases, the input current (I_S) also increases. This increases the current through zener diode, without affecting the load current (I_L). The increase in input current will also increase the voltage drop across series resistance (R_S), thereby keeping the load voltage (V_L) as constant. On the other hand, if the input voltage is decreased, the input current also decreases. As a result of this, the current through zener will also decrease. Consequently, the voltage drop across series resistance will be reduced. Thus, the load voltage (V_L) and load current (I_L) remains constant.

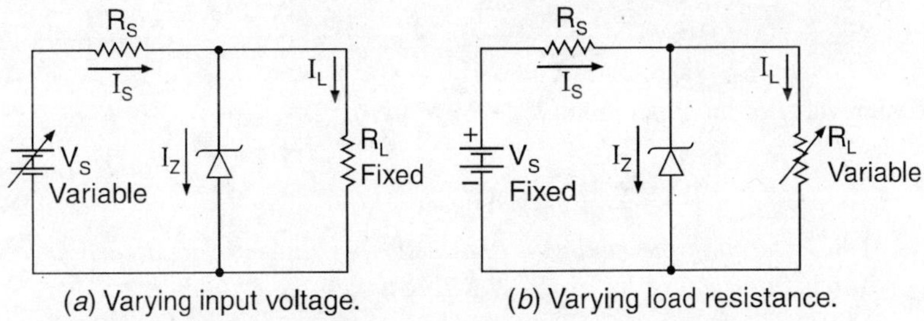

Fig. 31-6. *Working of zener diode shunt regulator.*

(ii) Regulation with varying load resistance : Consider the regulator circuit as shown in Fig. 31-6 (*b*). Here the input voltage (V_S) is kept fixed and the load resistance (R_L) varies. The variation of load resistance changes the current (I_L) through it, thereby changing voltage (V_L) across it. When the load resistance decreases, the load current increases. This causes the zener current to decrease. As a result of this, the input current and the voltage drop across series resistance remains constant. Thus, the load voltage (V_L) is also kept constant. On the other hand, if the load resistance increases, the load current decreases. As a result of this, the zener current increases. This again keeps the values of input current and voltage drop across series resistance as constant. Thus the load voltage remains constant.

Example 31-7. *A zener diode shown in Fig. 31-7 is used in the circuit has a breakdown voltage of 5.1 V, the zener resistance (r_Z) of 10 Ω and the minimum and maximum values of current through it are 1 mA and 15 mA respectively.*

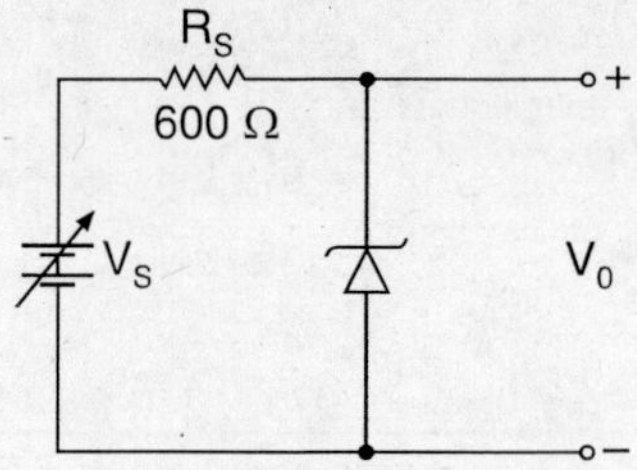

Fig. 31-7.

Determine the minimum and maximum values of the input voltage, which can be regulated by the zener shunt regulator.

Solution. Given : $V_Z = 5.1$ volts; $r_Z = 10\ \Omega$; $I_{Z(min)} = 1\text{ mA} = 1 \times 10^{-3}$ A; $I_{Z(max)} = 15\text{ mA} = 15 \times 10^{-3}$ A and $R_S = 600\ \Omega$.

Fig. 31-8 shows the equivalent circuit of the given zener diode shunt regulator.

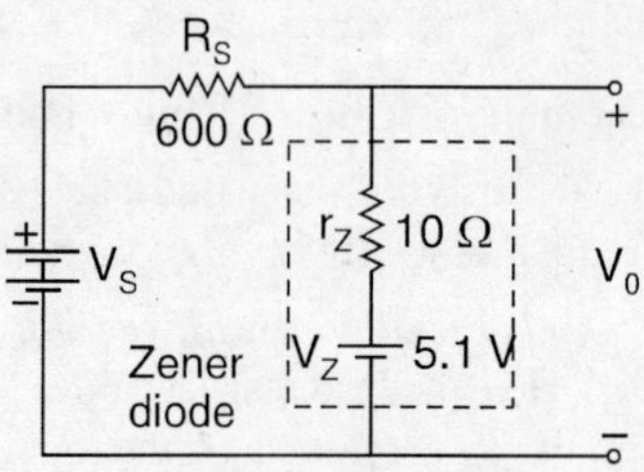

Fig. 31-8.

It may be noted that the zener diode is replaced by a battery of 5.1 V in series with a resistance of 10 Ω.

Minimum value of input voltage

We known that minimum value of output voltage,

$$V_{O\,(min)} = V_Z + I_{Z(min)} \cdot r_Z = 5.1 + (1 \times 10^{-3})\ 10 = 5.11\text{ V}$$

and the minimum value of input voltage,

$$V_{S\,(min)} = I_{Z\,(min)} \cdot R_S + V_{O\,(min)} = (1 \times 10^{-3}) \times 600 + 5.11 = 5.71\text{ V } \textbf{Ans.}$$

Maximum value of input voltage

We know that maximum value of output voltage,

$$V_{O\,(max)} = V_Z + I_{Z\,(max)} \cdot r_Z = 5.1 + (15 \times 10^{-3}) \times 10\text{ V}$$
$$= 5.25\text{ V}$$

and the maximum value of the input voltage,

$$V_{S\,(max)} = I_{Z\,(max)} \cdot R_S + V_{O\,(max)} = (15 \times 10^{-3}) \times 600 + 5.25\text{ V}$$
$$= 14.25\text{ V } \textbf{Ans.}$$

Example 31-8. *Determine the minimum and maximum values of load current for which the zener diode shunt regulator shown in Fig. 31-9 will maintain regulation.*

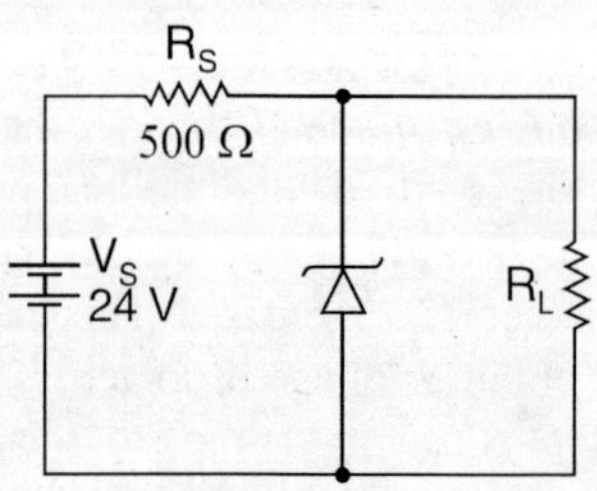

Fig. 31-9.

What is the minimum value of load resistance (R_L) that can be used in the circuit. The zener diode has a V_Z equal to 12 V, $I_{Z\,(min)}$ of 3 mA, $I_{Z\,(max)}$ of 90 mA and r_Z is zero.

Solution. Given : $V_S = 24$ volts; $R_S = 500\ \Omega$; $V_Z = 12$ volts; $I_{Z\,(min)} = 3$ mA; $I_{Z\,(max)} = 90$ mA and $r_Z = 0$.

Minimum value of load current

We know that current through series resistance,

$$I_S = \frac{V_S - V_Z}{R_S} = \frac{24 - 12}{500} = 0.024 \text{ A} = 24 \text{ mA}$$

Thus, we see that the input current (equal to 24 mA) is much less than the maximum value of zener current (equal to 90 mA). Therefore, we can assume that all the input current flows through zener diode. In that case, the minimum value of load current is zero. **Ans.**

Maximum value of load current

We know that the value of load current will be maximum, when the zener current is minimum. Therefore, maximum value of load current,

$$I_{L\,(max)} = I_S - I_{Z\,(min)} = 24 - 3 = 21 \text{ mA } \textbf{Ans.}$$

Minimum value of load resistance

We also know that minimum value of load resistance is obtained by using the maximum value of load current. Therefore, minimum value of load resistance,

$$R_{L\,(min)} = \frac{V_Z}{I_{L\,(max)}} = \frac{12}{21} = 0.571 \text{ k}\Omega = 571\ \Omega \textbf{ Ans.}$$

31-7. Optimum Value of Current Limiting Resistor

As a matter of fact, in a zener diode shunt regulator circuit, the value of current limiting resistor must be properly selected to fulfil the following two requirements :

(*i*) When the input voltage is minimum and the load current is maximum, sufficient current must be supplied to keep the zener diode within its breakdown region (or regulating region).

(*ii*) When the input voltage is maximum and the load current is minimum, the zener current must not increase the maximum rated value.

The optimum value of current limiting resistor can be determined by using the following two equations for maximum and minimum values of zener current :

$$I_{Z\,(max)} = \frac{V_{S\,(max)} - V_Z}{R_S} - I_L\,(min) \qquad ...(i)$$

$$I_{Z\,(min)} = \frac{V_{S\,(min)} - V_Z}{R_S} - I_L\,(max) \qquad ...(ii)$$

From equation (*i*), we find that the minimum value of current limiting resistor,

$$R_{S\,(min)} = \frac{V_{S\,(max)} - V_Z}{I_{Z\,(max)} + I_{L\,(min)}}$$

and from equation (*ii*), the maximum value of current limiting resistor,

$$R_{S\,(max)} = \frac{V_{S\,(min)} - V_Z}{V_{Z\,(min)} + I_{L\,(max)}}$$

The value of current limiting resistor (R_S) should be chosen in such a way that its value is between $R_{S\,(max)}$ and $R_{S\,(min)}$, *i.e.*,

$$R_{S\,(min)} < R_S < R_{S\,(max)}$$

Example 31-9. *For a zener shunt regulator, if $V_Z = 10$ V, $R_S = 1$ kΩ, $R_L = 2$ kΩ and the input voltage varies from 22 to 40 V, find the maximum and minimum values of zener current.*

Solution. Given : $V_Z = 10$ volts; $R_S = 1 \text{ k}\Omega = 1 \times 10^3\ \Omega$; $R_L = 2 \text{ k}\Omega = 2 \times 10^3\ \Omega$; $V_{S\,(min)} = 22$ volts and $V_{S\,(max)} = 40$ volts.

Maximum value of zener current

We know that load current,

$$I_L = \frac{V_Z}{R_L} = \frac{100}{2 \times 10^3} = 5 \times 10^{-3} \text{ A} = 5 \text{ mA}$$

and maximum value of zener current,

$$I_{Z\,(max)} = \frac{V_{S\,(max)} - V_Z}{R_S} - I_{L\,(min)} = \frac{40 - 10}{1 \times 10^3} - 5 \times 10^{-3} \qquad ...(\because I_{L\,(min)} = I_L)$$

$$= 25 \times 10^{-3} \text{ A} = 25 \text{ mA } \textbf{Ans.}$$

Minimum value of zener current

We also know that minimum value of zener current,

$$I_{Z\,(min)} = \frac{V_{S(min)} - V_Z}{R_S} - I_{L(max)} = \frac{22 - 10}{1 \times 10^3} - (5 \times 10^{-3}) \qquad ...(\because I_{L\,(max)} = I_L)$$

$$= 7 \times 10^{-3} \text{ A} = 7 \text{ mA } \textbf{Ans.}$$

Example 31-10. *A 10 V zener diode is used to regulate the voltage across a variable load resistor. The input voltage varies between 13 and 16 V. The load current (I_L) varies between 10 and 85 mA. The minimum zener current is 15 mA. Find **(i)** the maximum value of R_S, and **(ii)** the maximum power dissipated by the zener diode, using this value of R_S. (A.M.I.E., Summer, 1987)*

Solution. Given : V_Z = 10 volts; $V_{S\,(min)}$ = 13 volts; $V_{S\,(max)}$ = 16 volts; $I_{L\,(min)}$ = 10 mA = 10×10^{-3} A; $I_{L\,(max)}$ = 85 mA = 85×10^{-3} A; and $I_{Z\,(min)}$ = 15 mA = 15×10^{-3} A.

(*i*) *Maximum value of R_S*

We know that maximum value of R_S,

$$R_{S\,(max)} = \frac{V_{S\,(min)} - V_Z}{I_{Z\,(min)} + I_{L\,(max)}} = \frac{13 - 10}{(15 \times 10^{-3} + 85 \times 10^{-3})} = 30\ \Omega \textbf{ Ans.}$$

(*ii*) *Maximum power dissipated*

We also know that maximum value of zener current,

$$I_{Z\,(max)} = \frac{V_{S\,(max)} - V_Z}{R_{S\,(max)}} - I_{L\,(min)}$$

$$= \frac{16 - 10}{30} - (10 \times 10^{-3})$$

$$= 0.2 - 0.01 = 0.19 \text{ A}$$

and maximum power dissipated by the zener,

$$P_{Z\,(max)} = I_{Z\,(max)} \times V_Z = 0.19 \times 10 = 1.9 \text{ W } \textbf{Ans.}$$

Example 31-11. *A zener regulator has to handle supply voltage variation from 19.5 volts to 22.5 volts. Find the magnitude of regulating resistance, if the load resistance is 6 kΩ. The zener diode has the following specifications :*

Breakdown voltage = 18 volts

Minimum zener diode current = 2 μA

Maximum power dissipation = 60 mW

Zener resistance = 20 Ω. *(A.M.I.E., Winter, 1986)*

Solution. Given : $V_{S\,(min)}$ = 19.5 volts; $V_{S\,(max)}$ = 22.5 volts; R_L = 6 kΩ; V_Z = 18 volts; $I_{Z\,(min)}$ = 2 μA = 2×10^{-6} A; $P_{Z\,(max)}$ = 60 mW = 60×10^{-3} W and r_Z = 20 Ω.

Let $I_{Z\,(max)}$ = Maximum value of zener current.

We know that maximum value of power dissipation $P_{Z(max)}$,

$$60 \times 10^{-3} = I_{Z\,(max)}^2 \times r_Z = I_{Z\,(max)}^2 \times 20$$

$$\therefore \quad I_{Z\,(max)}^2 = \frac{60 \times 10^{-3}}{20} = 3 \times 10^{-3}$$

and $$I_{Z\,(max)} = \sqrt{(3 \times 10^{-3})} = 0.0548 \text{ A} = 54.8 \times 10^{-3} \text{ mA}$$

We also know that load current,

$$I_L = \frac{V_L}{R_L} = \frac{V_Z}{R_L} = \frac{18}{6 \times 10^3} = 3 \times 10^{-3} \text{ A} = 3 \text{ mA}$$

$\therefore$ Maximum value of regulating (or series resistance),

$$R_{S\,(max)} = \frac{V_{S\,(min)} - V_Z}{I_{Z\,(min)} + I_{L\,(max)}} = \frac{19.5 - 18}{(2 \times 10^{-6}) + (3 \times 10^{-3})} \quad ...(\because I_{L\,(max)} = I_L)$$

$$= 500 \text{ ohm}$$

and minimum value of regulating resistance,

$$R_{S\,(min)} = \frac{V_{S(max)} - V_Z}{I_{Z(max)} + I_{L(min)}} = \frac{22.5 - 18}{(54.8 \times 10^{-3}) + 3 \times 10^{-3}} \quad ...(\because I_{L\,(min)} = I_L)$$

$$= 77.8 \text{ ohm} \quad ...(i)$$

$\therefore$ Magnitude of regulating resistance should be between 77.8 ohms and 500 ohms, *i.e.*, $77.8 < R_S < 500$ ohms. **Ans.**

Example 31-12. *Fig. 31-10 shows the circuit of a zener diode voltage regulation.*

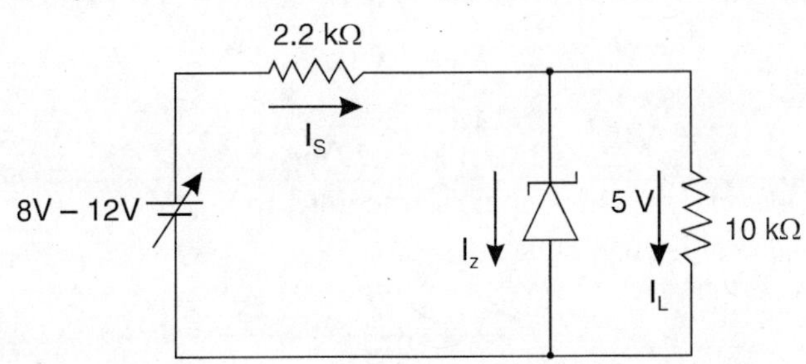

Fig. 31-10.

Find the maximum and minimum value of zener current in the circuit.

(*Bangalore Univ. B.Sc. Electronics, May* 2001)

Solution. Given : $V_{S\,(min)} = 8$ V; $V_{S\,(max)} = 12$ V; $R_S = 2.2$ kW; $V_Z = 5$ V and $R_L = 10$ kΩ.

We know that minimum value of input current,

$$I_{S\,(min)} = \frac{V_{S\,(min)} - V_Z}{R_S} = \frac{8 - 5}{2.2 \text{ k}} = 1.364 \text{ mA}$$

and maximum value of input current,

$$I_{S\,(max)} = \frac{12 - 5}{2.2 \text{ k}} = 3.182 \text{ mA}$$

We also know that the value of load current,

$$I_L = \frac{V_L}{R_L} = \frac{5}{10 \text{ k}} = 0.5 \text{ mA}$$

$\therefore$ The minimum value of zener current,

$$I_{Z\,(min)} = I_{S\,(min)} - I_L = 1.364 - 0.5 = 0.864 \text{ mA} \textbf{ Ans.}$$

and maximum value of zener current,

$$I_{Z\,(max)} = I_{S\,(max)} - I_L = 3.182 - 0.5 = 2.682 \text{ mA } \textbf{Ans.}$$

Example 31-13. *Design the zener regulator for the following specifications :*

Output voltage = 5 V

Load current = 20 mA

Zener power dissipation = 500 mW

Input voltage = 12 ± 3 V (*Bangalore Univ. Jan.* 2004)

Solution: Given: $V_O = V_L = 5$ V; $I_L = 20$ mA; $P_{Z\,(max)} = 500$ mW and $V_S = 12 \text{ V} \pm 3$ V.

Let $I_{Z\,(max)}$ = Max. value of current through the zener.

We know that maximum power dissipation of a zener, $P_{Z\,(max)}$;

$$500 \times 10^{-3} = V_Z \cdot I_{Z\,(max)} = 5\, I_{Z\,(max)}$$

$$\therefore \quad I_{Z\,(max)} = \frac{500 \times 10^{-3}}{5} = 100 \times 10^{-3} = 100 \text{ mA}$$

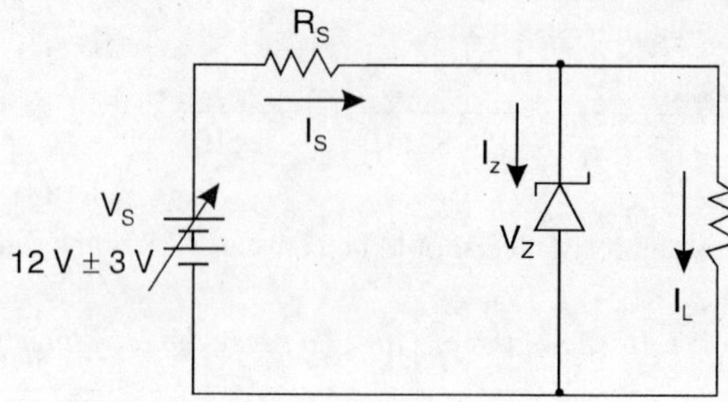

Fig. 31-11.

$$I_{S\,(max)} = I_L + I_Z$$
$$= 20 + 100 = 120 \text{ mA}$$

We also know that the maximum input current, (refer to Fig. 31-11), and the minimum value of regulating resistance,

$$R_{S\,(min)} = \frac{V_{S(max)} - V_Z}{I_{Z(max)} + I_{L(min)}} = \frac{15 - 5}{120 \text{ mA}} = 83.3\ \Omega \text{ Ans.}$$

Notice that if we fix the value of $R_S = 83.3\ \Omega$, the minimum value of zener current,

$$I_Z = \frac{V_{S(min)} - V_Z}{R_S} - I_{L(max)} = \frac{9 - 5}{83.3} - (20 \times 10^{-3}) = 28.02 \text{ mA}$$

So when the input voltage varies from the normal 12 V within the range of ± 3 V, the zener current will vary from 28.02 mA to maximum 100 mA if the current limiting resistor is at least 83.3 Ω. But to be more on the safeside, you can select $R_S = 220\ \Omega$. **Ans.**

Note :- In the present case, the statement does not give any value for $I_{Z\,(min)}$. That value will put a maximum limit on the value of R_S. So the chosen valve of R_S should not be very high as well.

Example 31-14. *Design and draw a Zener regulator circuit to meet the following specifications:*

Load voltage = 8 V

Input voltage = 30 V

Load current = 0–50 mA

$I_{z(min)}$ = *5 mA*

P_z = *1 watts*

(*Electronic Circuits - I, Anna University, Nov/Dec 2007*)

Solution. Given: $V_L = 8$ V, $V_{in} = 30$ V, $I_L = 0 - 50$ mA, $I_{z(min)} = 5$ mA

Fig. 31-12 shows a zener diode regulator circuit.

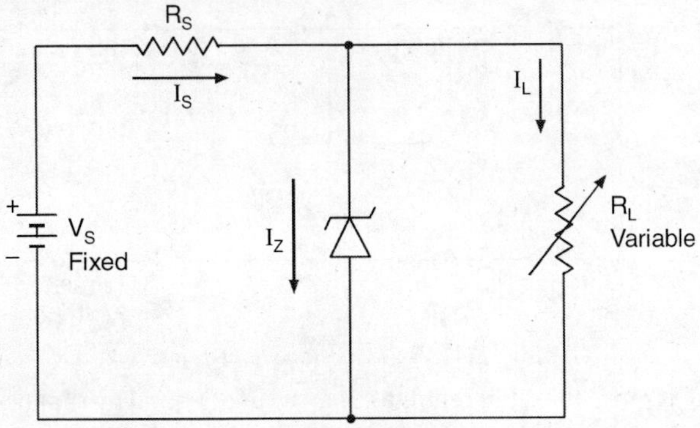

Fig. 31-12.

Input voltage is constant,

$$V_{inmin} = V_{inmax} = 30 \text{ V}$$

$$I_{Lmin} = 0 \text{ A}, I_{Lmax} = 50 \text{ mA}$$

Maximum value of Zener current,

$$I_{z(max)} = \frac{P_z}{V_0} = \frac{1}{8} = 0.125 \text{ mA}$$

Minimum value of load resistor,

$$R_{max} = \frac{V_{min} - V_0}{I_{Lmax} + I_{zmin}} = \frac{30 - 8}{(50 \times 10^{-3}) + (5 \times 50^{-3})} = 400$$

Maximum value of current limiting resistor,

$$R_{min} = \frac{V_{in\,max} - V_0}{I_{Lmin} + I_{zmax}} = \frac{30 - 8}{0 + (125 \times 10^{-3})} = 176\ \Omega$$

This for any value of R between 176 Ω to 400 Ω, the circuit shown in Fig. 31-12 will act as a regulator circuit.

31-8. Disadvantages of Zener-diode Shunt Regulator

Following are the disadvantages of a zener-diode shunt regulator :

(*i*) The maximum load current, which can be supplied to load resistor (R_L) is limited to $I_{Z\,(max)} - I_{Z\,(min)}$, which is usually of few milliamperes.

(*ii*) A large amount of power is wasted in the zener-diode and the series resistance (R_S) in comparison with the load power.

(*iii*) The regulation factor and the output resistance are not very low.

These drawbacks are overcome in a transistorized voltage regulators, which will be discussed in the following pages.

31-9. Transistor Shunt Regulator

Fig. 31-13 shows a circuit of a transistor shunt regulator. Since, the transistor is connected in shunt with the load resistance, therefore the circuit is known as shunt regulator. It may be noted from the regulator circuit, that the load voltage (V_L) is equal to the sum of zener voltage (V_Z) and the base-to-emitter voltage (V_{BE}) of a transistor, *i.e.*,

$$V_L = V_Z + V_{BE} \text{ or } V_{BE} = V_L - V_Z \qquad ...(i)$$

Since, the voltage for a given zener diode is fixed, therefore any increase or decrease in load voltage will have a corresponding effect on the base-to-emitter voltage.

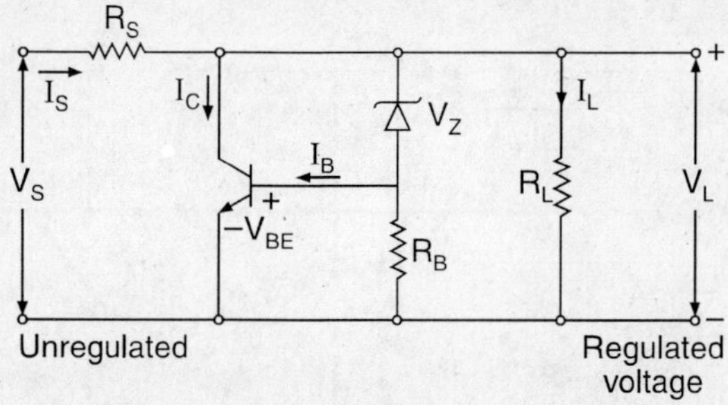

Fig. 31-13. *Transistor shunt regulator.*

The working of a transistor shunt regulator may be understood by supposing that the unregulated input voltage (V_S) increases. Because of this, the load voltage also increases. As a result of this and from equation (*i*) above, we find that base-to-emitter voltage also increases. And the base current of the transistor (I_B) increases. Due to this, the collector current of a transistor (I_C) also increases. This causes the input current (I_S) to increase, which in turn increases the voltage drop across series resistance (V_{RS}). Consequently, the load voltage decreases. It is true because sum of the voltage drop across series resistance (V_{RS}) and the load voltage (V_L) is equal to input voltage (V_S) at all times, *i.e.*,

$$V_S = V_{R_S} + V_L \quad \text{or} \quad V_L = V_S - V_{R_S} \qquad ...(ii)$$

The above discussion may be summarized in the form of an equation as given below :

$$V_S \uparrow \rightarrow V_L \uparrow \rightarrow \; V_{BE} \uparrow \rightarrow I_C \; \rightarrow V_{R_S} \uparrow \rightarrow V_L \downarrow$$

Here arrows in the upward direction show the increase in the value, while the arrow in the downward direction indicates the decrease in value. The transistor shunt regulator is suitable for the loads in which the load current varies from a finite minimum value to a finite maximum value.

Example 31-15. *Determine the regulated voltage and circuit currents for the shunt regulator of Fig. 31-14.*

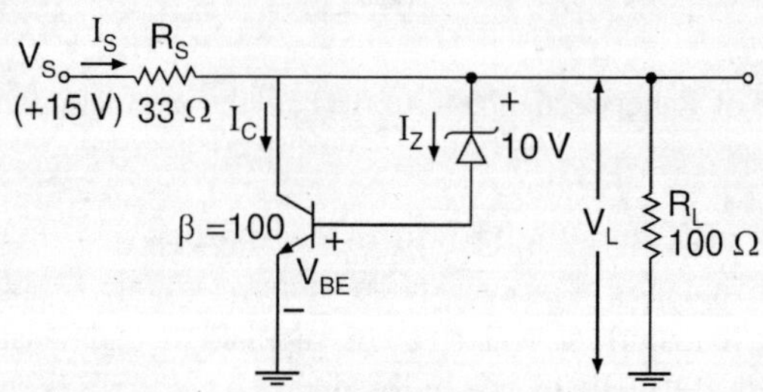

Fig. 31-14.

Solution. We know that load voltage,

$$V_L = V_Z + V_{BE} = 10 + 0.7 = 10.7 \text{ V}$$

and the value of load current,

$$I_L = \frac{V_L}{R_L} = \frac{10.7}{100} = 0.107 \text{ A} = 107 \text{ mA}$$

The current through, R_S,

$$I_S = \frac{V_S - V_L}{R_S} = \frac{15 - 10.7}{33} = 0.1303 \text{ A} = 130.3 \text{ mA}$$

∴ The collector current,

$$I_C = I_S - I_L = 130.3 - 107 = 23.3 \text{ mA}$$

The base current for the transistor (or current through the zener),

$$I_B = I_Z = \frac{I_C}{\beta} = \frac{23.3 \text{ mA}}{100} = 0.233 \text{ mA} = 233 \ \mu\text{A} \ \textbf{Ans.}$$

31-10. Transistor Series Regulator

Fig. 31-15 shows a circuit of a transistor series regulator. Since, *the transistor is connected in series with the load*, therefore the circuit is known as a *series regulator*. This circuit is also called as *emitter follower*, because its transistor behaves as an emitter follower. In other words, the transistor behaves as a variable resistance whose value is determined by the amount of base current. It may be noted from the circuit that the load voltage (V_L) is equal to the difference of zener voltage (V_Z) and the base-to-emitter voltage (V_{BE}), *i.e.*,

$$V_L = V_Z - V_{BE} \quad \text{or} \quad V_{BE} = V_Z - V_L \qquad ...(i)$$

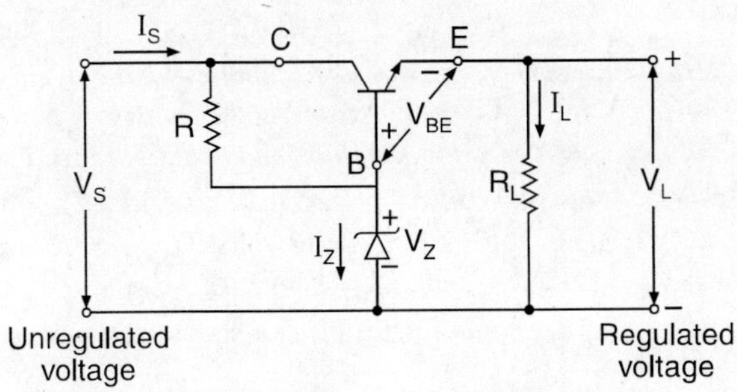

Fig. 31-15. *Transistor series regulator.*

The working of a transistor series regulator may be understood by supposing that value of load resistance is increased. Because of this, the load current decreases and the load voltage tends to increase. However, as seen from the equation (*i*), that any increase in the load voltage will decrease the base-to-emitter voltage (V_{BE}) because the zener voltage is fixed. As a result of this, the forward bias of the transistor is reduced, which reduces its level of conduction. This increases the collector-to-emitter voltage of the transistor. This in turn will slightly decrease the input current to compensate for the increase in the value of load resistance, so that load voltage may remain at a constant value. The circuit action may be summarized in the form of the following equation :

$$V_L \uparrow \rightarrow V_{BE} \downarrow \rightarrow V_{CE} \uparrow \rightarrow V_Z \downarrow$$

The output of a transistor series regulator is approximately equal to zener voltage (V_Z). This regulator can also be used for larger load currents.

Example 31-16. *Calculate the output voltage and Zener current in the regulator circuit shown in Fig. 31-16.*

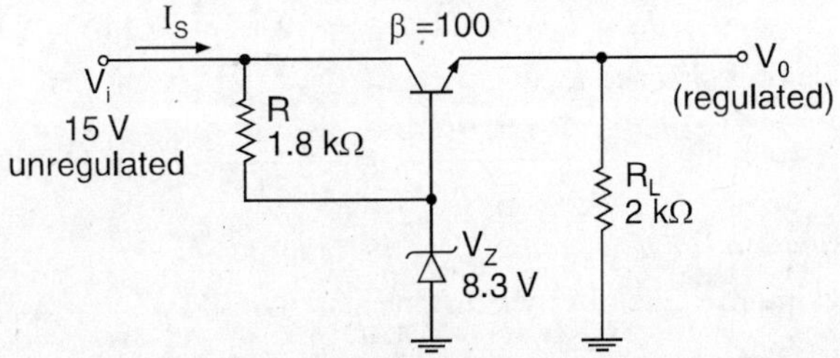

Fig. 31-16.

Solution. Given : $V_i = V_S = 15$ V; $V_Z = 8.3$ V; $\beta = 100$, $R = 1.8$ kΩ and $R_L = 2$ k.Ω

We know that the load voltage,

$$V_L = V_Z - V_{BE} = 8.3 - 0.7 = 7.6 \text{ V}$$

and collector-to-emitter voltage,

$$V_{CE} = V_S - V_L = 15 - 7.6 = 7.4 \text{ V}$$

∴ Current through resistor, R

$$I_R = \frac{V_S - V_Z}{R} = \frac{15 - 8.3}{1.8 \times 1000} = 3.72 \times 10^{-3} \text{ A} = 3.72 \text{ mA}$$

Also the load current,

$$I_L = \frac{V_L}{R_L} = \frac{7.6}{2000} = 3.8 \times 10^{-3} = 3.8 \text{ mA}$$

and base current,

$$I_B = \frac{I_L}{\beta} = \frac{3.8 \text{ mA}}{50} = 0.076 \text{ mA} = 76 \text{ μA} \qquad (\because I_C \approx I_E = I_L)$$

∴ Current through the zener,

$$I_Z = I_R - I_B = 3.72 - 0.076 = 3.644 \text{ mA } \textbf{Ans.}$$

Example 31-17. *A transistor series regulator is to supply a load current of 0 to 2 A at 12 V. The unregulated d.c. input varies from 15 V to 20 V. Assuming the transistor β = 100, V_{BE} = 0.5 V, zener voltage, V_Z = 12.5 V and $I_{Z\,(min)}$ = 1 mA, find the maximum value of the resistance (R) and power rating of resistance R, zener diode and transistor.*

Solution. Given : $I_{L(min)} = 0$; $I_{L\,(max)} = 2$ A; $V_L = 12$ volts; $V_{S\,(min)} = 15$ volts; $V_{S\,(max)} = 20$ volts; $\beta = 100$; $V_{BE} = 0.5$ volts; $V_Z = 12.5$ volts; and $I_{Z\,(min)} = 1$ mA.

Fig. 31-17 shows the circuit diagram of the transistor series regulator.

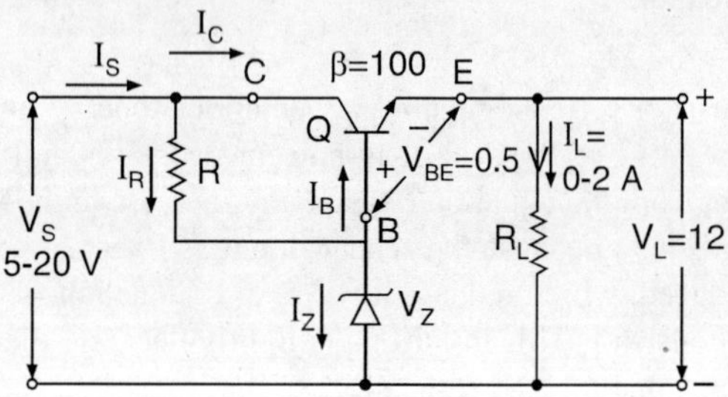

Fig. 31-17.

Maximum value of resistance R

We know that the maximum value of base current of a series transistor,

$$I_{B\,(max)} = \frac{I_{C\,(max)}}{\beta} = \frac{I_{L\,(max)}}{\beta} = \frac{2}{100} = 0.02 \text{ A} = 20 \text{ mA}$$

and current through the resistance R,

$$I_R = I_{B\,(max)} + I_{Z\,(min)} = 20 + 1 = 21 \text{ mA} = 21 \times 10^{-3} \text{ A}$$

∴ Maximum value of resistance R,

$$R_{max} = \frac{V_{S\,(min)} - V_Z}{I_R} = \frac{15 - 12.5}{21 \times 10^{-3}} = 119 \ \Omega \ \textbf{Ans.}$$

Maximum power dissipation (or power rating) of the zener diode

We know that maximum value of zener current,

$$I_{Z(max)} = \frac{V_{S\,(max)} - V_Z}{I_R} = \frac{20 - 12.5}{119} = 63 \text{ mA} = 63 \times 10^{-3} \ \textbf{Ans.}$$

and the maximum power dissipation (or power rating) of a zener diode,

$$P_{Z\,(max)} = V_Z \times I_{Z\,(max)} = 12.5 \times 63 \times 10^{-3} = 0.79 \text{ W } \textbf{Ans.}$$

Maximum power dissipation (or power rating) of resistance R

We know that maximum power dissipation of resistance R,

$$P_{R\,(max)} = (V_{S\,(max)} - V_Z) \cdot I_{Z\,(max)} = (20 - 12.5) \times 63 \times 10^{-3} \text{ W} = 0.47 \text{ W } \textbf{Ans.}$$

Maximum power dissipation (or power rating) of the transistor

We also know that maximum value of collector-to-emitter voltage of a transistor.

$$V_{CE\,(max)} = V_{S\,(max)} - V_L = 20 - 12 = 8 \text{ V}$$

and maximum power dissipation of the transistor,

$$P_{D\,(max)} = V_{CE\,(max)} \times I_L\,(max) = 8 \times 2 = 16 \text{ W } \textbf{Ans.}$$

31-11. Controlled Transistor Series Regulator

Fig. 31-18 shows a controlled transistor series regulator circuit. It is also known as feedback voltage regulator circuit. This circuit employs two transistors namely Q_1 and Q_2. The Q_1 transistor is a series regulator transistor. It is also called control transistor or pass transistor. And the Q_2 transistor functions both as a voltage comparator and d.c. amplifier. Therefore, it is known as an amplifier transistor.

The action of a controlled transistor series regulator may be studied by supposing that there is an increase in load voltage either due to input voltage variation or change of load (V_{BE}) of Q_2 transistor. Because of this, the collector current of Q_2 transistor (I_{C_2}) increases. As a result of this, total current (I_{ST}) flowing through the resistance (R_S) increases. This, in turn decreases the collector current of Q_2 transistor. Since, the base of Q_1 transistor is tied to the collector of Q_2 transistor,

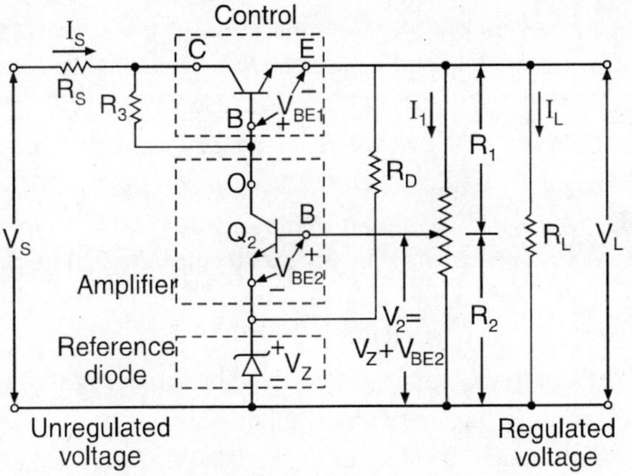

Fig. 31-18. *Controlled transistor series regulator.*

therefore base voltage of Q_1 transistor, with respect to ground, decreases. This in turn decreases the base-to-emitter voltage (V_{BE1}) of the transistor. Therefore, the collector-to-emitter voltage (V_{CE1}) of Q_1 transistor has to increase in order to maintain the same emitter current. If the change in collector-to-emitter voltage of a transistor can be made equal to the variation in input voltage (*i.e.*, ΔV_S), the load voltage will remain constant. Since, the collector-to-base voltage of Q_1 transistor is equal to its collector-to-emitter voltage, we can assume that if the value of ΔV_S is dropped across the resistance R_3 , the load voltage will remain constant. It may be noted that by changing the tapping on the potentiometer ($R_1 - R_2$), the d.c. load voltage of the regulator circuit can be varied. The circuit action may be summarized in the form of following equation :

$$V_L \uparrow \rightarrow V_{BE_2} \uparrow \rightarrow I_{C_2} \uparrow \rightarrow V_{BE1} \downarrow \rightarrow V_{CE1} \uparrow \rightarrow V_L \downarrow$$

The regulated load voltage for a controlled transistor series regulator is given by the relation,

$$V_L = (V_Z + V_{BE_2})\left(1 + \frac{R_1}{R_2}\right) = V_2 \times \left(1 + \frac{R_1}{R_2}\right)$$

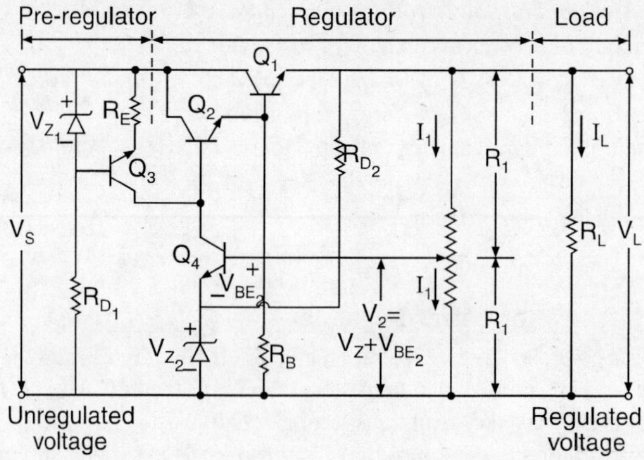

Fig. 31-19. *A series regulator using pre-regulator and Darlington Pair.*

The current gain of the series transistor can be increased by using a Darlington Pair. This can be increased further by increasing the value of the resistance R_3, which is achieved by using a pre-regulator. A pre-regulator is a constant-current source and therefore its equivalent resistance (R_3) is very large. Such a circuit, which incorporates both the above features is shown in Fig. 31-19.

Example 31-18. *Design a series regulated power supply to provide a nominal output voltage of 12 volts and supply load current of 200 milliamperes. The nominal unregulated input voltage is 30 volts and source resistance is 10 ohms. Ratings of the active devices used are as follow:*

Control transistor (silicon) : $h_{fe} = 150;\ h_{ie} = 1.5\ k\Omega\ ;\ V_{BE} = 0.7$ *volt;* $I_C = 10$ *mA*

Amplifier transistor (silicon) : $h_{fe} = 100;\ h_{ie} = 250\ \Omega\ ;\ V_{BE} = 0.7$ *volt*

Reference zener diode : $V_R = 6$ *volts;* $R_Z = 10$ *ohms at* $I_Z = 20$ *mA*

Sketch the complete circuit and obtain reasonable values of the components you use.

Solution. Given : $V_L = 12$ volts; $I_L = 200$ mA; $V_S = 30$ volts; $R_S = 10\ \Omega$; $\beta_1 (= h_{fe1}) = 150$; $I_{C1} = 10$ mA; $V_{BEI} = 0.7$ volts; $\beta_2 (= h_{fe2}) = 100$; $V_{BE2s} = 0.7$ volts; $V_Z (= V_R) = 6$ volts; $R_Z = 10\ \Omega$ and $I_Z = 20$ mA; $I_D = 10$ mA $= 10 \times 10^{-3}$ A; $I_1 = 10$ mA $= 10 \times 10^{-3}$ A.

Fig. 31-20 shows the complete circuit of a transistor series regulator power supply. Let us choose the collector current of Q_2 transistor (I_{C_2}) and the current I_D equal to 10 mA each. Now we shall use these values to determine the values of the resistances R_D, R_1 and R_2.

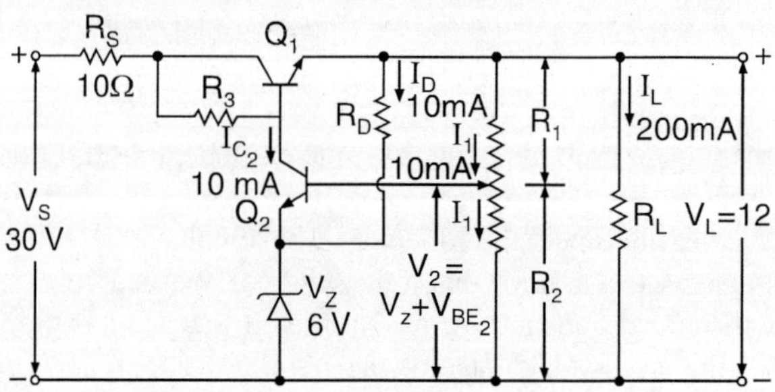

Fig. 31-20.

Value of resistance R_D

We know that the value of resistance

$$R_D = \frac{V_L - V_Z}{I_D} = \frac{12 - 6}{10 \times 10^{-3}}\ \Omega = 600\ \Omega \textbf{ Ans.}$$

Value of resistance R_1 and R_2

We know that d.c. load voltage (V_L),

$$12 = (V_Z + V_{BE_2})\left(1+\frac{R_1}{R_2}\right) = (6+0.7)\left(1+\frac{R_1}{R_2}\right) = 6.7\left(1+\frac{R_1}{R_2}\right)$$

$$\therefore \quad 1+\frac{R_1}{R_2} = \frac{12}{6.7} = 1.79$$

$$\text{or} \quad \frac{R_1}{R_2} = 1.79 - 1 = 0.79 \quad \text{... } (i)$$

We also know that base current of transistor,

$$I_{B2} = \frac{I_{C_2}}{\beta_2} = \frac{10}{100} = 0.1 \text{ mA}$$

Since, we require the value of current through potentiometer ($R_1 - R_2$) to be very large than that of base current (*i.e.*, I_1 I_{B2}), therefore we select the current I_1 equal to 10 mA. Then the voltage drop across resistance R_2 ,

$$V_2 = V_Z + V_{BE_2} = 6 + 0.7 = 6.7 \text{ V}$$

$$\therefore \text{ Value of resistance} \quad R_1 = \frac{V_Z - V_2}{I_1} = \frac{12-6.7}{10\times10^{-3}}\Omega = 530\ \Omega \textbf{ Ans.}$$

Now substituting the value of R_1 in equation (*i*),

$$\frac{530}{R_2} = 0.79$$

$$\text{or} \quad R_2 = \frac{530}{0.79} = 670\ \Omega \textbf{ Ans.}$$

Value of resistance R_3

We know that base current of Q_1 transistor,

$$I_{B_1} = \frac{I_L + I_1 + I_D}{h_{fe_1}} = \frac{200+10+10}{150} = 1.5 \text{ mA} = 1.5\times10^{-3} \text{ A}$$

and current through resistance R_3,

$$I = I_{B_1} + I_{C_2} = 1.5 + 10. = 11.5 \text{ mA} = 11.5\times10^{-3} \text{ A}$$

$$\therefore \text{ Value of resistance} \quad R_3 = \frac{V_S - (V_{BE1} + V_L)}{I} = \frac{30-(0.7+12)}{11.5\times10^{-3}}\Omega$$

$$= 1500\ \Omega \text{ or } 1.5 \text{ k}\Omega \textbf{ Ans.}$$

31-12. Short-Circuit Protection against Overload

As a matter of fact, a practical regulator circuit has to be always protected against short-circuit overload. A typical short-circuit overload protection circuit is shown in Fig. 31-21. It employs two diodes D_1 and D_2 , and a sensing resistance R connected as shown in the figure. The diodes D_1 and D_2 normally remain open.

When the output terminals of a regulator circuit are shorted, then current through Q_1-transistor increases. This increases voltage drop across sensing resistance R. When the drop across resistance R is such that two diodes start conducting, then current flowing through resistance R is given by the relation,

$$I = \frac{2V_D - V_{BE_1}}{R} \quad \text{...}(i)$$

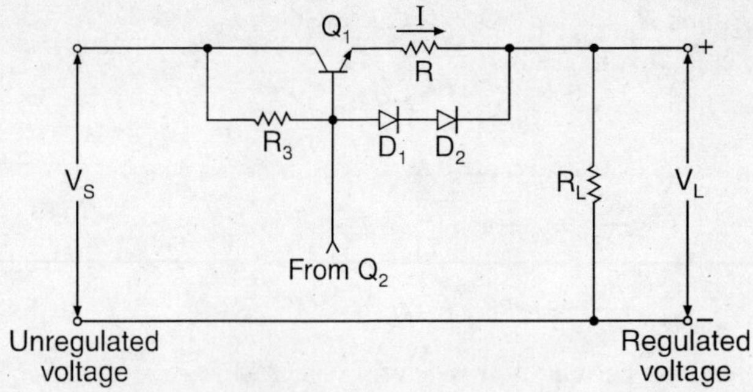

Fig. 31-21 *Short-circuit protection against overload.*

where, V_D = Voltage drop across each diode, and

V_{BE_1} = Base-to-emitter voltage of Q_1-transistor.

There can be no further increase in current through resistance R or the Q_1-transistor. Thus, the current I given by the equation (i) is the maximum current, which will flow through Q_1-transistor under short-circuit condition. Thus, the series transistor (Q_1) is protected from being damaged.

31-13. Transistor Current Regulator

The main function of a current regulator is to maintain a fixed current through the load despite variations in the terminal voltage. Such a circuit employing a zener diode and a PNP transistor is shown in Fig. 31-22.

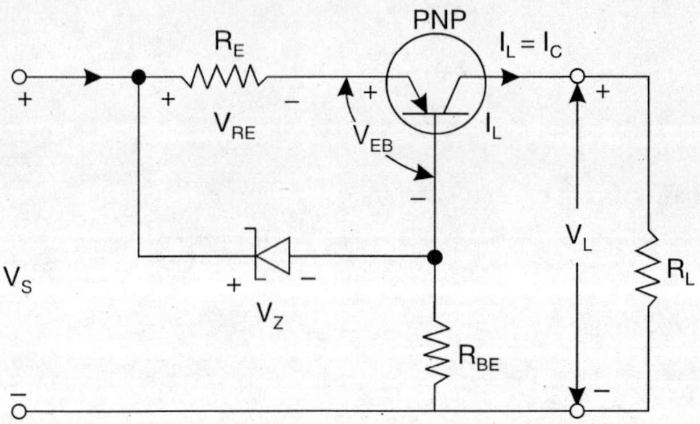

Fig. 31-22.

Suppose due to drop in V_L, current I_L (= I_C) is decreased, then I_E (≈ I_C) will also decrease. Hence, drop across R_E, *i.e.*, V_{BE} will decrease, as per Kirchoff's Voltage law,

$$-V_{RE} - V_{BE} + V_Z = 0 \text{ or } V_{BE} = V_Z - V_{RE}$$

(Fixed)

Hence, a decrease in V_{RE} will decrease V_{BE} and therefore, the conductivity of the transistor will also decrease. This process keeps I_L at a fixed level.

31-14. Variable Feedback Regulator

The voltage regulators considered so far provide a non-adjustable output voltage. This would be fine if only single value of regulated voltage is required.

Fig. 31-23 shows a feedback regulator which provides different values of regulated dc voltage. In Fig. 31-23, Q_1 is the pass transistor and Q_2 is the feedback transistor whose purpose is to provide and sample output or load voltage. It offsets any change in the output voltage. Since, potentiometer R_3 is connected in parallel with zener diode D, it has zener voltage, V_Z aplied across it. Voltage across the variable resistor varies from O to V_Z. Capacitor C ensures that voltage across D and R_3 does not change suddenly.

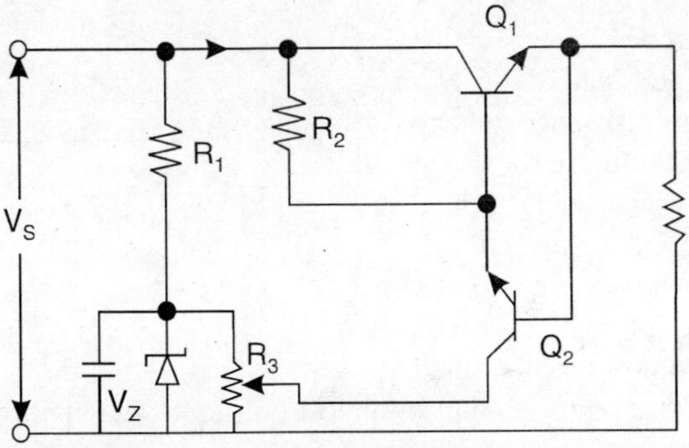

Fig. 31-23.

Voltage at the base of Q_2 is 0.7 V more positive than the voltage at the emitter. Its emitter voltage and hence, the base voltage can be changed with the help of R_3. Since, the base of Q_2 is tied to the output, it is responsible for providing output or load voltage. The voltage V_{CE_1} across the pass transistor is given by the difference of input voltage and output voltage. The current through Q_1 is equal to the load current. R_2 prevents saturation of the transistors where as R_1, limits the current flowing through D.

The working of feedback regulator may be explained as follows :

Since, base voltage of Q_2 is directly related to V_{out} it will change if V_{out} changes. The base and collector of Q_2 are 180° out of phase with each other. If base voltage increases due to increase in V_{out} collector voltage would decrease. Now, collector of Q_2 controls base of Q_1. As the base voltage of Q_1 decreases, its collector-emitter resistance increases which lowers the load current. This in turn, lowers the output voltage thereby offsetting the attempted increases in V_{out}. The opposite of these steps provides the action of an attempted decrease in output voltage.

Example 31-19. *In the variable feedback regulator circuit of Fig. 31-23. $V_S = 25$ V, $V_Z = 15$ V and $R_L = 1$ kΩ. If the wiper of R_3 is adjusted half-way and assuming silicon transistor, compute (i) V_{out} (ii) I_L (iii) I_{E1} (iv) P_1.*

Solution. Given $V_S = 25$ V; $V_Z = 15$ V and $R_L = 1$ kΩ

(*i*) V_{out}

We know that output voltage,

$$V_{out} = \text{Voltage at the wiper} + V_{BE_2} = (15/2) + 0.7 = 8.2 \text{ V} \textbf{ Ans.}$$

(*ii*) I_L

We know that load current,

$$I_L = V_{out}/R_L = 8.2/1 \text{ k}\Omega = 8.2 \text{ mA}$$

(*iii*) I_{E_1}

We know that, $I_{E_1} = I_L = 8.2$ mA **Ans.**

(*iii*) P_1

We know that $V_{CE1} = V_S - V_{out} = 25 - 8.2 = 16.8$ V; $\therefore P_1 = 16.8 \text{ V} \times 8.2 \text{ mA} = 140$ mW **Ans.**

31-15. Basic OP-AMP Series Regulator

It circuit is shown in Fig. 31-24 and its operation may be explained as follows :

The potentiometer R_2-R_3 senses any change in output voltage V_{out}. When V_{out} attempts to decrease because of decrease in V_S or because of the increase in I_L, a proportional voltage decrease is applied to the inverting output of the Op-amp by the potentiometer. Since, the other Op-amp input is held by the Zener voltage at a fixed reference voltage V_{REF}, a small difference voltage (called error voltage) is developed across the two inputs of the Op-amp. This difference voltage is amplified and Op-amp's output voltage increases. This increases in voltage is applied to the base of Q_1 causing the emitter voltage (= V_{out})to increase till the voltage to the inverting input again equals the reference (Zener) voltage. This action offsets attempted decrease in the output voltage thus keeping it almost constant. The opposite action occurs if the output voltage tries to increase.

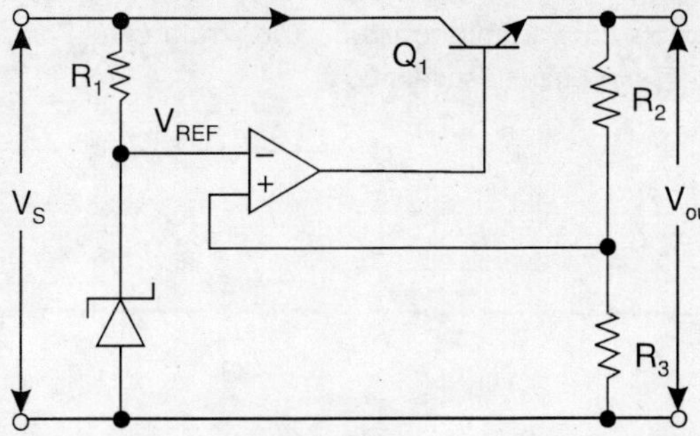

Fig. 31-24.

It will be seen that the Op-amp of Fig. 31-24 is actually connected as a non-inverting amplifier where V_{REF} is the input at the noninverting terminal and the R_2/R_3 voltage divider forms the negative feedback network. The closed-loop voltage gain is given by $A = 1 + (R_2/R_3)$. Neglecting base-emitter voltage of Q_1, we get

$$V_{out} = V_{REF}\,(1 + R_2/R_3)$$

It is seen that V_{out} depends on Zener voltage and potential divider resistors R_2 and R_3 but is independent of input voltage V_S.

31-16. Basic Op-amp Shunt Regulator

Such a shunt type linear regulator is shown in Fig. 31-25. Here, the control element is a series resistor R_2 and a transistor Q_1 in parallel with the load. In such a regulator, regulation is achieved by controlling the current through Q_1.

The circuit opearation may be explained as follows. When output voltage tries to decrease due to change in either the input voltage or load current or temperature, the attempted decrease is sensed by R_3 and R_4 and applied to the non-inverting input of the Op-amp. The resulting difference in voltage reduces the Op-amp's output, driving Q_1 less thus reducing its collector current (shunt current), and increasing its collector-to-emitter resistance. Since, collector-to-emitter resistance acts as a voltage divider with R_1, this action offsets the attempted decrease in output voltage and hence, maintains it at a constant value. The opposite action occurs when output voltage tries to increase. The shunt regulator is less efficient than the series type but offers inherent short-circuit protection.

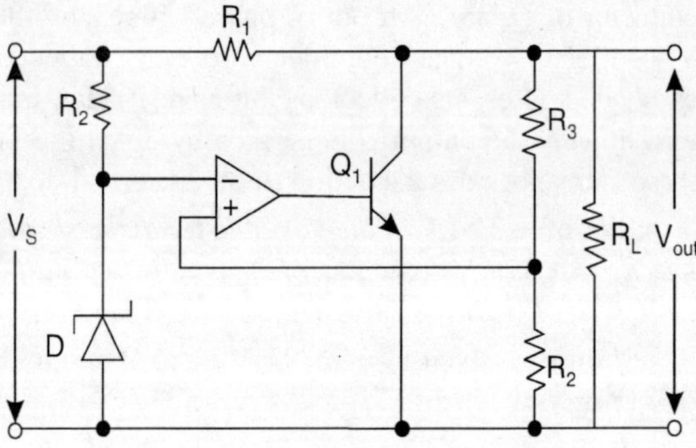

Fig. 31-25.

31-17. Switching Regulators

In the linear regulators considered so far, the control element, *i.e.*, the transistor conducts all the time, the amount of conduction varying with changes in output voltage or current. Due to continuous power loss, the efficiency of such a regulator is reduced to 50 per cent or less.

A switching regulator is different because its control element operates like a switch, *i.e.*, either it is saturated (closed) or cut-off (open). Hence, there is no unnecessary wastage of power which results in higher efficiency of 90% or more.

Switching regulators are of three basic types : (*i*) step-down regulator, (*ii*) step-up regulator, and (*iii*) inverting regulator.

31-18. Step-down Switching Regulator

In this regulator (Fig. 31-26), V_{out} is always less than V_S. An unregulated positive d.c. voltage is applied to the collector of the *NPN* transistor. A series of pulses from oscillator is sent to the base of transistor *T* which gets saturated (closed) on each of the positive pulses. It is so because *NPN* transistor needs a positive voltage pulse on its base in order to turn ON. A saturated transistor acts as a closed switch, hence it allows V_S to send current through *L* and charge *C* to the value of output voltage during the on-time (T_{ON}) of the pulse. The diode D_1 is reverse-biased at this point and hence, does not conduct.

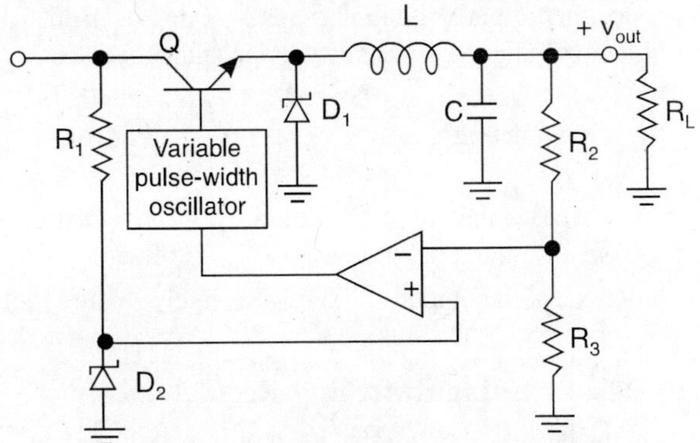

Fig. 31-26.

Eventually when positive pulse turns to zero, *Q* is cut-off and acts like an open switch during the off period (T_{OFF}) of the pulse. The collapsing magnetic field of the coil produces self-induced voltage and keeps the current flowing by returning energy to the circuit.

The value of output voltage depends on input voltage and pulse width, *i.e.*, on-time of the transistor. When on-time is increased relative to off-time, *C* charges more thus increasing V_{out}. When T_{ON} is decreased, *C* discharges more thus decreasing V_{out}. By adjusting the duty cycle (T_{ON}/T) of the transistor, V_{out} can be varied.

$$\therefore \qquad V_{out} = V_S\,(T_{ON}/T)$$

where, *T* is the period of the ON-OFF cycle of the transistor and is related to frequency by $T = 1/f$. Also, $T = T_{ON} + T_{OFF}$ and the ratiro (T_{ON}/T) is called the ***duty cycle***.

The regulating action of the circuit may be explained as follows :

When V_{out} tries to decrease, on-time of the transistor is increased causing an additional charge on the capacitor *C* to offset the attempted decrease. When V_{out} tries to increase, T_{ON} of the transistor is decreased causing *C* to discharge enough to offset the attempted increase.

31-19. Step-up Switching Regulator

The circuit is shown in Fig. 31-27. When transistor *Q* turns ON on the arrival of the positive pulse at its base, voltage across *L* increases quickly to $V_S - V_{CE\,(sat)}$ and magnetic field of *L* expands quickly. During on-time of the transistor, V_L keeps decreasing from its initial maximum value. The longer transistor is ON, the smaller V_L becomes.

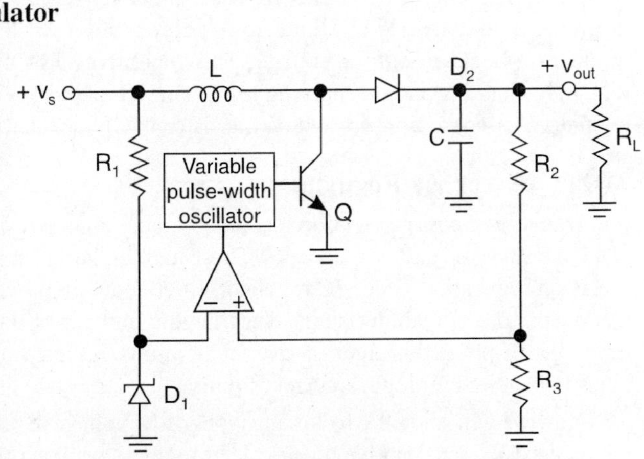

Fig. 31-27.

When transistor turns OFF, magnetic field of *L* collapses and its polarity reverses so that its voltage adds to the input voltage thus producing an output voltage greater than the input voltage. During off-time of the transistor, D_2 is forward-biased and allows C to charge. The variations in V_{out} due to charging and discharging action are sufficiently smoothed by filtering action of L and C.

It may be noted that shorter the on-time of the transistor, greater the inductor voltage and hence, greater the output voltage (because greater V_L)adds to V_S. On the other hand, the longer the on-time, the smaller the inductor voltage and hence, lesser the output voltage (because smaller V_L adds to V_S).

The regulating action may be explained as follows :

When V_{out} tries to decrease (because of either increasing load or decreasing V_S transistor on time decreases thereby offsetting attempted decrease in V_{out}. When V_{out} tries to increase, on-time increase and attempted increase in V_{out} is offset.

As seen, the output voltage is inversely related to the duty cycle.

$$\therefore \quad V_{out} = V_S \, (T/T_{ON})$$

31-20. Inverting Switching Regulator

The basic diagram of such a regulator is shown in Fig. 31-28. This regulator provides an output voltages that is opposite in polarity to the input voltages.

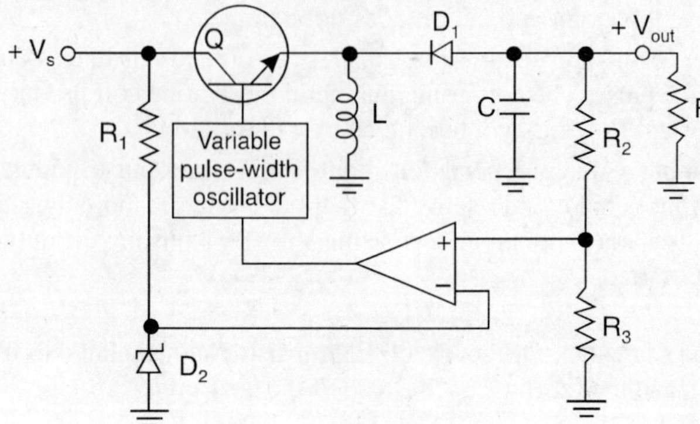

Fig. 31-28.

When transistor Q turns ON by the positive pulse, the inductor voltage V_L jumps to $V_S - V_{CE\,(sat)}$ and the magnetic field of the inductor expands rapidly. When transistor is ON, the diode D_2 is reverse-biased and V_L decreases from its initial maximum value. When transistor turns OFF, the magnetic field collapses and inductor's polarity reverses. This forward-biases D_2, charges C and produces a negative output voltage. This repetitive ON-OFF action of the transistor produces a repetitive charging and discharging that is smoothed by LC filter action. As in the case of a step-up regulator, lesser the time for which transistor is ON, greater the output voltage and *vice versa*.

31-21. IC Voltage Regulators

Due to low-cost fabrication technique, many commercial integrated-circuit (*IC*) regulators are available since the past two decades. These include fairly simple, fixed-voltage types of high quality precision regulators. These *IC* regulators have much improved preformances as compared to those made from discrete components. They have a number of unique build-in features such as current limiting self-protection against overtemperature, remote control operation over a wide range of input voltages and foldback current limiting.

Now we will study the following types of *IC* voltage regulators : **(1)** fixed positive linear voltage regulators, **(2)** fixed negative linear voltage regulators, **(3)** adjustable positive linear voltage regulators, and **(4)** adjustable negative linear voltage regulators.

31-22. Fixed Positive Linear Voltage Regulators

There are many *IC* regulators available in the market that produces a fixed positive output voltage. But 7800 series of *IC* regulators is representative of three terminal devices that are available with several fixed positive output voltages making them useful in a wide range of applications. Fig. 31-29 (*a*) shows a standard configuration of a fixed positive voltage *IC* regulator of 7800 series. Notice that it has three terminals labelled as input, output and ground. The last two digits (marked xx) in the part number designate the input voltage. For example, *IC* 7805 is a + 5V regulator. Similarly *IC* 7812 is a +12 V regulator and *IC* 7815 is a +15 V regulator. The capacitor C_1 (typically 0.33 μF) is required only if the power supply filter is located more than 3 inches from the *IC* regulator. The capacitor C_2 (typically 0.01 μF) acts basically as a line filter to improve transient response.

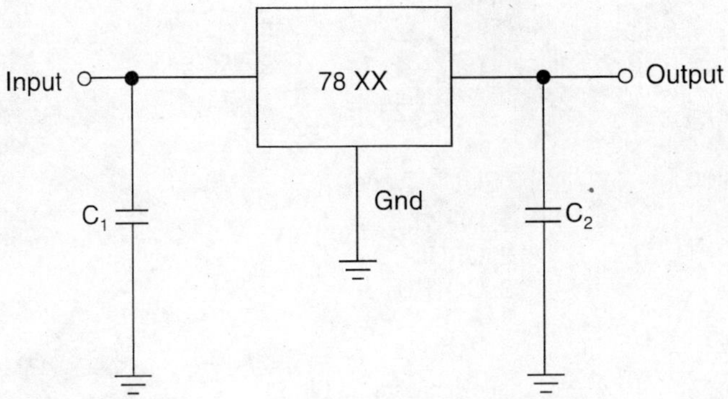

(*a*)

Part Number	Output Voltage
7805	+5.0 V
7806	+6.0 V
7808	+8.0 V
7809	+9.0 V
7812	+12.0 V
7815	+15.0 V
7818	+18.0 V
7824	+24.0 V

(*b*)

Fig. 31-29.

Fig. 31-29(*b*) shows the part number and the output voltage of 7800 series *IC* voltage regulators. As seen from this figure, the 7800 series has *IC* regulators that can produce output voltages ranging from +5.0 to +24.0 volt. It may be carefully noted that although these regulators are designed primarily to produce fixed output voltage but they can be used with external components to obtain adjustable output voltage and current.

Fig. 31-30 shows the circuit indicating the use of 78 XX as an adjustable voltage regulator. The output voltage is given by the equation,

$$V_{out} = V_{fixed} + \left(\frac{V_{fixed}}{R_1} + I_Q\right)R_2$$

For example, for a 7805 *IC* regulator, V_{fixed} = 5V. Let $R_1 = R_2 = 1\ \text{k}\Omega$ and I_Q = 2 mA, then its output voltage is,

$$V_{out} = 5 + \left(\frac{5}{1\ \text{k}\Omega} + 5\ \text{mA}\right) \times 1\ \text{k}\Omega$$

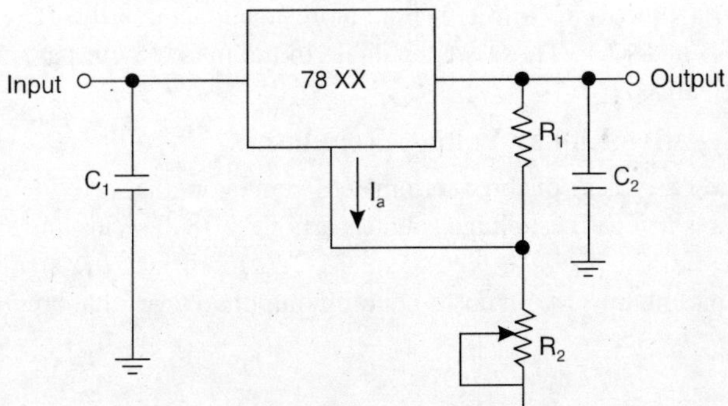

Fig. 31-30.

Thus output voltage of *IC* 7805 regulator can be adjusted anywhere between 5 V to 15 V.

This example indicates that the output of *IC* 7805 regulator is adjusted to 15 V using external resistances R_1 and R_2.

The standard 7800 series can produce output current in excess of A when used with adequate heat sink. It is available in aluminium can package TO-3 (indicated by *K* in the part number) and plastic package TO-220 (indicated by *T* in the part number), The 78L00 series can provide up to 100 mA and is available in TO-92 and metal TO-39 low profile packages. The 78M00 series can provide upto 0.5 A and is available in plastic TO-202 package. Fig. 31-31 shows the typical metal and plastic packages for the IC voltages regulators.

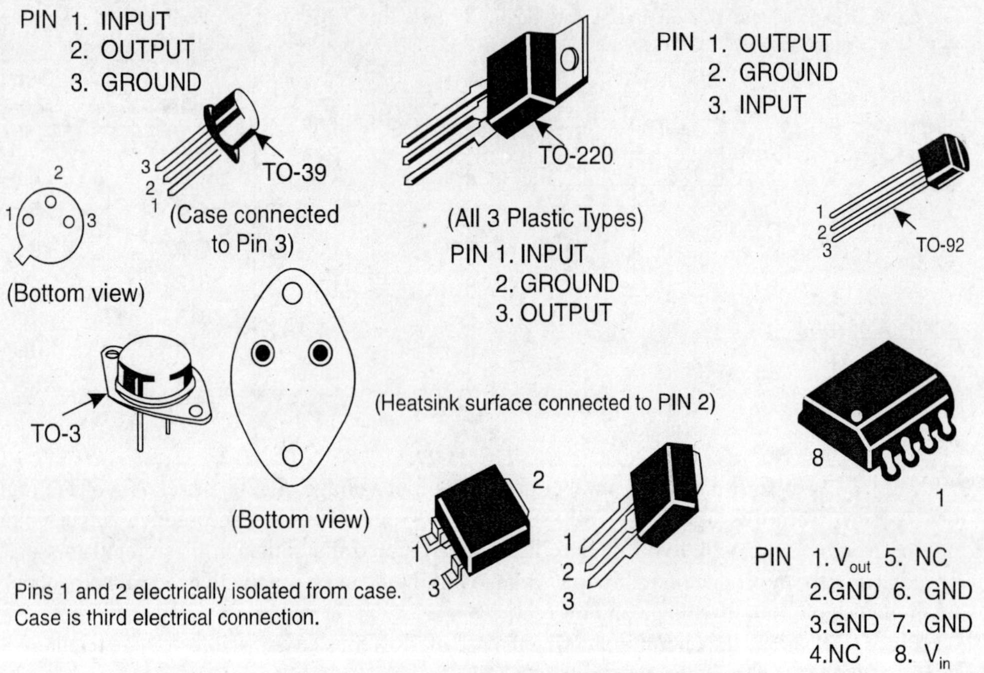

Fig. 31-31. Typical metal and plastic packages

It may be noted that the input voltage for the *IC* regulator must be at least 2V above the ouptut voltage. This is required in order to maintain regulation. The input voltage should not be more than 35 or 40 volts depending upon the part number. The circuit inside all the *IC* regulators have internal thermal overload protection and short-circuit current-limiting features. Thermal overload in a *IC* regulator occurs whenever the internal power dissipation becomes excessive and the temperature of the device exceeds a certain value.

In India, Bharat Electronic Limited, Bangalore manufactures the *IC* voltage regulators with output voltages of 5 and 12 V. These are available in the market with part numbers BEL 7805 and BEL 7812 respectively.

31-23. Fixed Negative Linear Voltage Regulators

The 7900 series is typical of three-terminal *IC* regulators that provide a fixed negative output voltage. This series is a negative-voltage counterpart of the 7800 series and shares most of the same features, charateristics and package types. Fig. 31-32(*a*) indicates the standard configuration and Fig. 31-32(*b*), the part numbers with corresponding output voltages that are available in 7900 series.

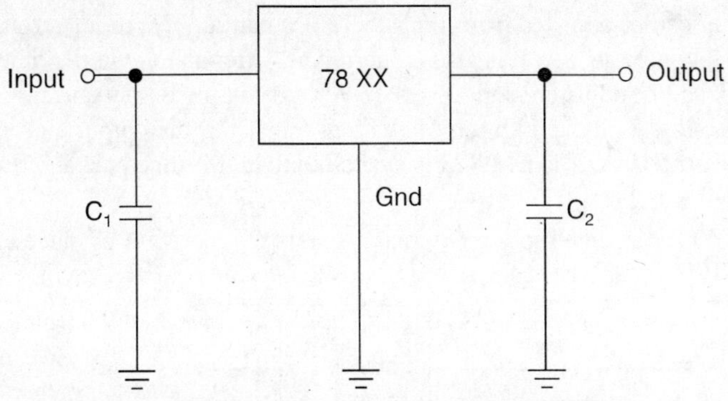

(a)

Part Number	Output Voltage
7905	– 5.0 V
7905.2	– 5.2 V
7906	– 6.0 V
7908	– 8.0 V
7912	– 12.0 V
7915	– 15.0 V
7918	– 18.0 V
7924	– 24.0 V

(b)

Fig. 31-32.

The capacitor C_1 (typically 0.22 μF) is required only if the power supply filter is located more than 3 inches away from the *IC* regulator. The capacitor C_2 (typically 1 μF) is required for stability of the output voltage. Both capacitors C_1 and C_2 must be solid tantalum capacitors.

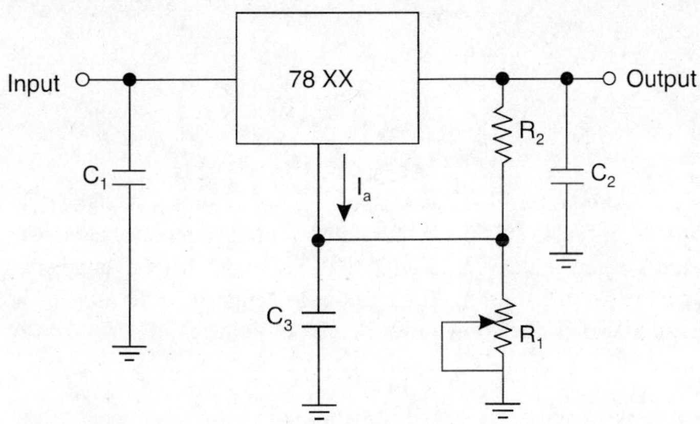

Fig. 31-33.

Fig. 31-33 shows the use of 79XX to produce an adjustable output voltage. The capacitor C_3 (typically 25 μF) improves the transient response of the output voltage. The output voltage is given by the equation,

$$V_{out} = V_{fixed}\left(\frac{R_1 + R_2}{R_2}\right)$$

The recommended value of R_2, for 7905 is 300 Ω, for 7915, its value is 750 Ω and for 7915 is 1 kΩ.

In india, BEL manufactures IC regulators with output voltage of –5 V and –12 V. These are available in the market with part numbers BEL 7905 and BEL 7915 respectively.

31-24. Adjustable Positive Output Linear Voltage Regulators

We can adjust the output voltage of 7800 series *IC* regulators higher than their fixed (or set) voltages. For example, the output voltage of 7805 can be adjusted higher than 5 V. But the performance and reliability of 7800 series to produce voltage higher than its fixed value is not considered to be good

The LM 317 and LM 723 are IC regulators whose output voltage can be adjusted over a wide range. The output voltage of LM 317 can be adjusted from 1.2 V to 37 V, it can supply output current of 100 mA and is available in TO-92 package, *i.e.*, it is also a 3 terminal IC regulator. On the other hand, the output voltage of LM 723 can be adjusted from 2 V to 37 V, it can supply output current of 150 mA without external transistor. But with the addition of external transistor, the output current capability can be increased in excess of 10 A. The LM 723 is available in dual-in-line package and in a metal can package.

Fig. 31-34 (*a*) shows the LM 317 connected to the external resistors R_1 and R_2 to produce an adjustable output voltatge.

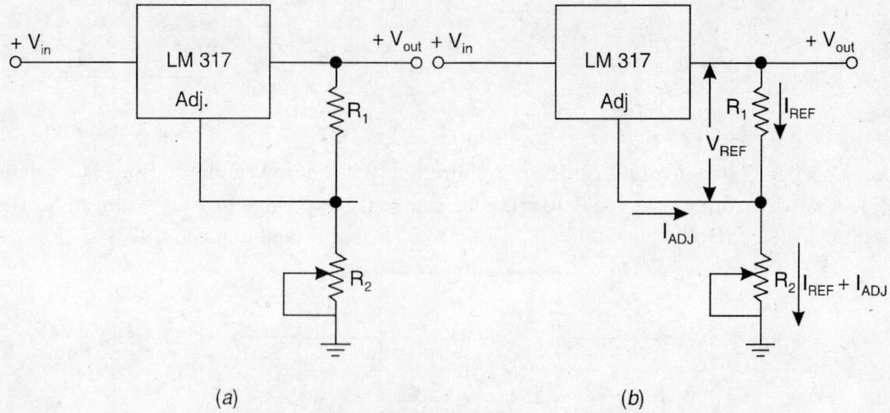

Fig. 31-34.

In opeation, the LM 317 develops a constant 1.25 V references voltage (V_{REF}) between the output and adjustment terminal. This constant reference voltage produces a constant current, (I_{REF}) through R_1, regardless of the value of R_2, Fig. 31-34 (*b*). Notice that the value of current through R_2 is the sum of I_{REF} and I_{ADJ} is a very small current at the adjustment terminal. The value of I_{ADJ} is typically around 100 μA. It can be shown that the output voltage.

$$V_{out} = V_{REF}\left(1+\frac{R_2}{R_1}\right) + I_{ADJ}\cdot R_2$$

It is evident from the above equation that the output voltage is a function of R_1 and R_2. Usually the value of R_1 is recommended to be around 220 Ω. Once the value of R_1 is set, the output voltage is adjusted by varying R_2.

Example 31-20. *Calculate the minimum and maximum output voltages for the IC voltage regulator shown in Fig.31-35. Assume* I_{ADJ} *= 100 μA,* V_{in} *= + 35 V.*

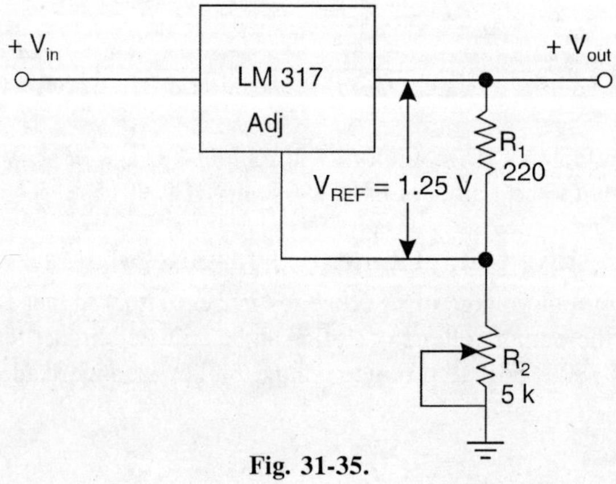

Fig. 31-35.

Solution. Given : $I_{ADJ} = 100\ \mu A$, $V_{in} = +35\ V$

The equation for output voltage of the *IC* voltage regulator is given by,

$$V_{out} = V_{REF}\left(1 + \frac{R_2}{R_1}\right) + I_{ADJ} \cdot R_2$$

When R_2 is set at its minimum value of (*i.e.,* 0 Ω), the output voltage,

$$V_{out\,(min)} = 1.25\left(1 + \frac{0}{220}\right) + (100 \times 10^{-6}) \times 0$$

$$= \mathbf{1.25\ V}$$

When R_2 is set at its maximum values (*i.e.,* 5 kΩ), the output voltage,

$$V_{out\,(max)} = 1.25\left(1 + \frac{5000}{220}\right) + (100 \times 10^{-6}) \times 5000$$

$$= 29.66 + 0.5 = 30.16\ \text{V.}\quad \textbf{Ans.}$$

Example 31-21. *Determine the regulated voltage in the circuit of Fig. 31-36 with $R_1 = 220\ \Omega$ and $R_2 = 1.5\ k\Omega$.*

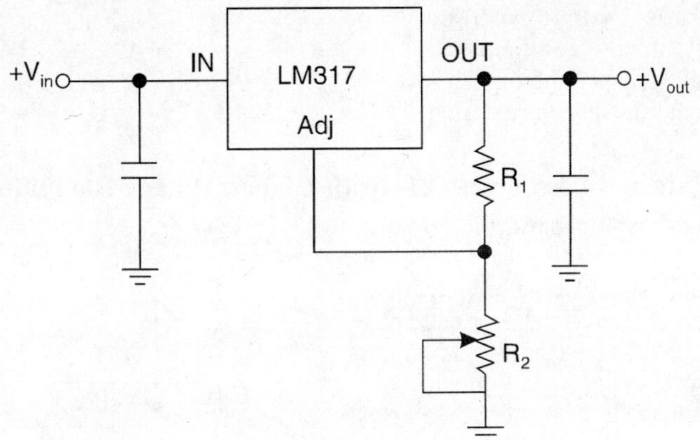

Fig. 31-36.

Solution. Given : $R_1 = 220\ \Omega$ and $R_2 = 1.5\ k\Omega = 1500\ \Omega$.

We know that output voltage of an adjustable voltage regulator,

$$V_o = 1.25\left(\frac{R_2}{R_1} + 1\right) = 1.25\left(\frac{1500}{220} + 1\right) = 7.82\ \text{V}\ \textbf{Ans.}$$

Example 31-22. *Fig. 31-37 shows the circuit of an adjustable voltage regulator using LM 317 integrated circuit. If the value of R_2 is adjusted to 2.4 kΩ.*

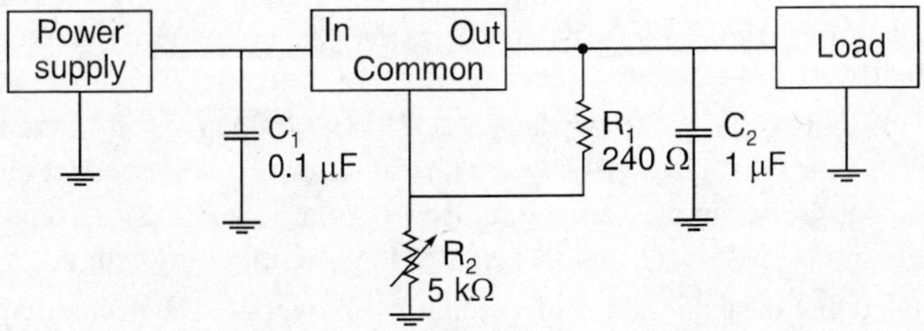

Fig. 31-37.

Determine the regulated d.c. output voltage for the circuit.

Solution. Given : $R_1 = 240\ \Omega$ and $R_2 = 2.4\ k\Omega = 2400\ \Omega$

We know that the regulated d.c. output voltage,

$$V_o = 1.25\left(\frac{R_2}{R_1}+1\right) = 1.25\left(\frac{2400}{240}+1\right) = 13.75 \text{ V } \textbf{Ans.}$$

31-25. Adjustable Negative Output Linear Voltage Regulators

A good example of this type of regulators is LM 337. The regulator is a negative output counter part of LM 317. The LM 337 (like LM 317) requires two external resistors for adjustment of output voltage as shown in Fig. 31-38. The output voltage can be adjusted anywhere from – 1.2 V to – 37 V depending upon the external resistor values.

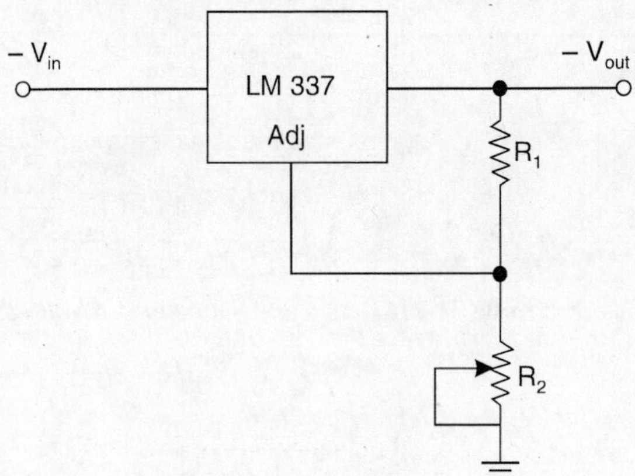

Fig. 31-38.

The LM 723 can also be used as an adjustable negative output voltage regulator. The output voltage of this *IC* regulator can be adjusted anywhere from – 2.0 V to – 37 V depending upon the external resistor values.

31-26. Use of External Pass Transistor with Linear Voltage Regulators

We have already mentioned in the last two articles that a linear voltage regulator (7800 and 7900 series) is capable of delivering only a certain amount of output current to a load. For example, the 7800 series regulators can handle a maximum output current of at least 1.3 A and typical 2.5 A. If the load current exceeds the maximum allowable value, there will be a **thermal overload** and the regulator will **shut down.** *A thermal overload condition means that there is excessive power dissipation inside the regulator.*

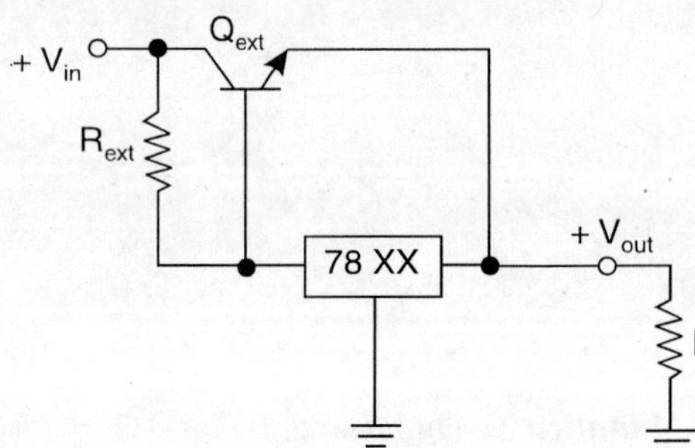

Fig. 31-39.

If an application requires a larger value of load current than the maximum current that the regulator can deliver, we will have to use an external pass transistor as shown in Fig. 31-39. The value of R_{ext} (current-sensing resistor) determines the value of current as which the external pass transistor Q_{ext} begins to conduct because it sets the base to emitter voltage of the transistor. As long as the current is less than the value set by R_{ext} the transistor is off and the regulator operates normally. This is because the voltage drop across R_{ext} is less than 0.7 V (*i.e.*, the base to emitter voltage required to turn T_{ext} on). The value of R_{ext} is determined by the equation $R_{ext} = 0.7\ V/I_{max}$ where I_{max} is the value of current that the voltage regulator is to handle **internally.**

When the current is sufficient to produce at least a 0.7 drop across R_{ext}, the transistor Q_{ext} turns on and conducts any current in excess of I_{max}. The transistor Q_{ext} will conduct current depending on the load requirement. For example, if the total load current is 5 A and I_{max} was selected to be 1 A, then the external pass transistor (Q_{ext}) will conduct 4 A of current through it.

It may be noted that the external pass transistor is a power with heat sink that must be capable of handling a maximum power given by the equation,

$$P_{ext} = I_{ext}(V_{in} - V_{out})$$

One major drawback of the circuit shown in Fig. 31-39 is that the external pass transistor is not protected from excessive current, such as would result from a shorted output. This drawback can be overcome by using an additional current limiting circuit as shown in Fig. 31-40.

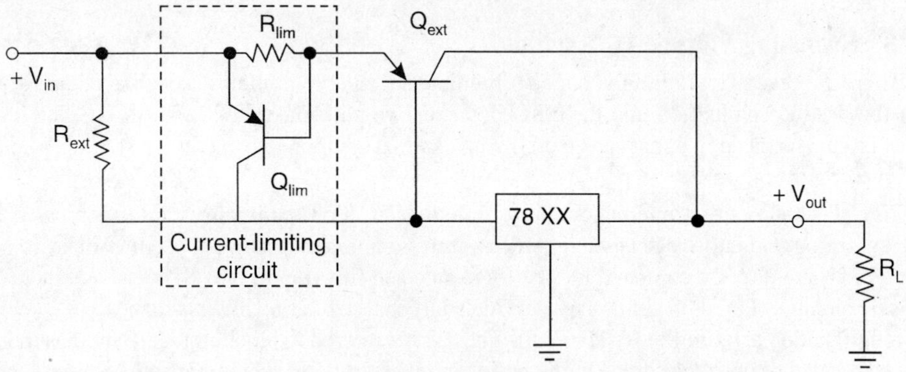

Fig. 31-40.

The operation of this circuit may be explained as follows. The current sensing resistor, R_{lim} sets the base-to-emitter voltage of Q_{lim}. The base-to-emitter voltage of Q_{ext} is now determined by $\left(V_{R_{ext}} - V_{R_{lim}}\right)$ because they have opposite polarities. So for normal operation, the drop across R_{ext} must be sufficient to overcome the opposing drop across R_{lim}.

If the current through Q_{lim} exceeds a certain maximum value, ($I_{ext\,(max)}$) because of a shorted output or a faulty load, the voltage across R_{lim}, reaches 0.7 V and turns Q_{lim} on. As a result, Q_{lim} now conducts current away from Q_{ext} and through the regulator. This forces a thermal overload to occur and shut down the regulator. Remember, the *IC* voltage regulator circuitry is internally protected from thermal overload as part of its design. This way the external pass transistor is protected from excessive current.

31-27. Use of Linear Voltage Regulators as a Current Regulator

The 3-terminal linear voltage regulator can be used as a current source when an application requires that a constant current be supplied to a variable load. The basic circuit is shown in Fig. 31-41. Here R_1 is the current-setting resistor. The regulator provides a fixed output voltage, V_{out} between the ground terminal and the output terminal. However, it may be noted that the ground pin of the regulator is not connected to the circuit ground. The constant current suplied to the load, is given by the equation,

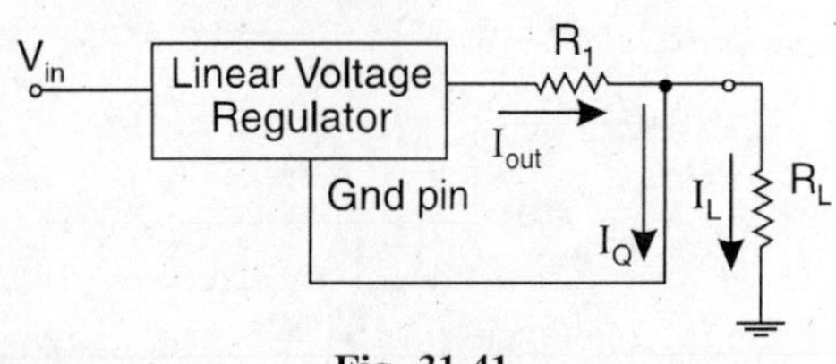

Fig. 31-41.

$$I_L = \frac{V_{out}}{R_L} + I_Q$$

Usually the current, I_Q is very small as compared to the output current and hence can be neglected, therefore

$$I_L \approx \frac{V_{out}}{R_L}$$

For example, if we use 7805 regulator to provide a constant current of 1 A to a variable load, then

$$R_1 = \frac{5}{1} = 5\ \Omega$$

Please note that input voltage must be at least 2 V greater than the output voltage. Thus, for 5 V regulator, V_{in} must be greater than 7 V.

31-28. Switching Voltage IC Regulators

There are several switching voltage *IC* regulators available in the market. The choice depends upon the desired application and the cost. However, we will illustrate it with the *IC* 78S40. This device is a universal device that can be used with external components to provide step-up, step down and inverting operation.

Fig. 31-42 shows that internal circuitry of the *IC* 78C40. The circuitry can be compared to the basic switching regulators discussed in Art. As seen from this diagram, the oscillator and comparator functions are directly comparable. The logic gate and flip-flop in the 78S40 we not included in the basic circuit of Fig. 31-42, but they provide additional regulation. Transistor Q_1 and Q_2 perform the same function as Q in the basic circuit. The 1.25 V reference block in the 78S40 has the same purposes as that of the zener diode in the basic circuit and diode D_1 in the 78S40 corresponds to D_1 in the basic circuit.

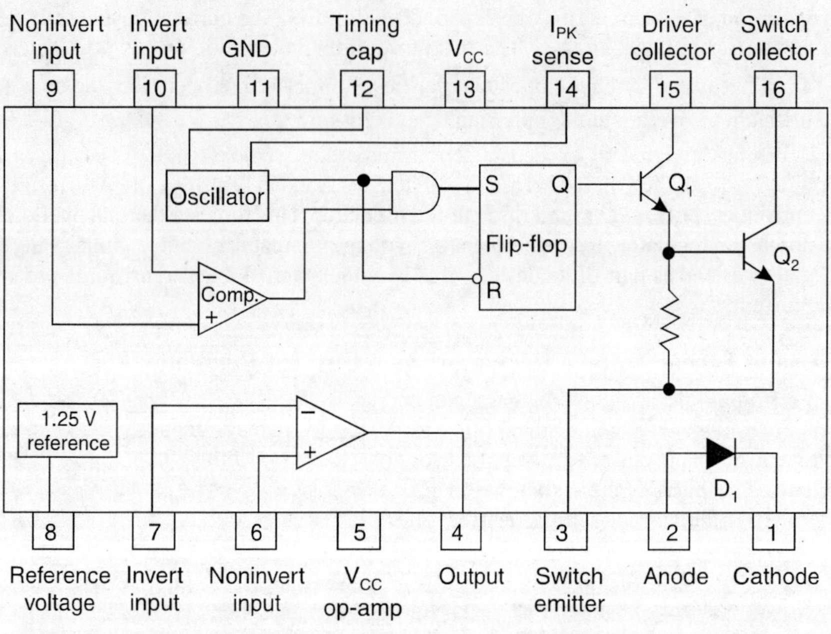

Fig. 31-42.

The circuitry of 78S40 has also an uncommitted or unused Op-amp. We require external circuitry to operate this device as a regulator.

Fig. 31-43 shows the external connections of the IC 78S40 for a step-down switching regulator configuration. Note that in this configuration the circuit produces an output voltage which is less than input voltage.

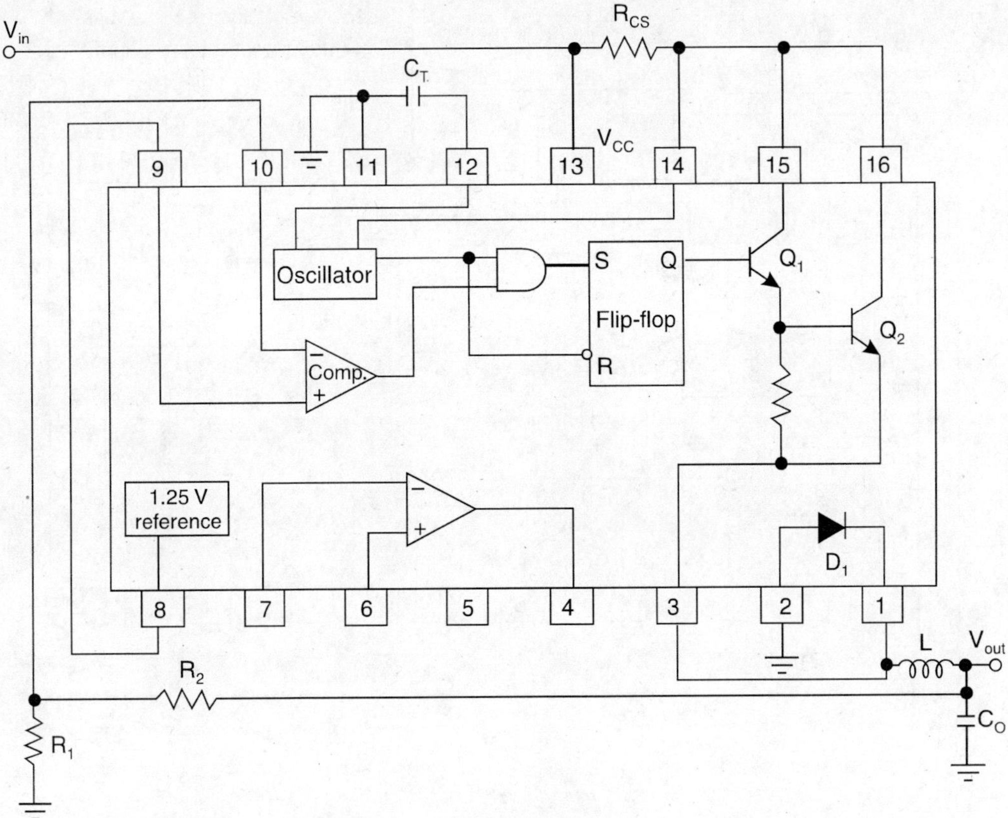

Fig. 31-43.

Fig. 31-44 shows the IC 78S40 connected to the external components for a step-up switching regulator configuration. In this case the output voltage is greater than the input voltage. An inverting configuration is also possible but is not shown here.

In both the circuits of Fig. 31-43 and Fig. 31-44, the capacitor C_T (called timing capacitor) controls the pulse width and frequency of the oscillator and thus establishes the on-time of transistor Q_1 The voltage across the resistor R_{CS} (called current-sensing resistor) is used internally by the oscillator to vary the duty cycle based on the desired peak current. The voltage divider made up of R_1 and R_2 reduces the output voltage to a value equal to the reference voltage. If the output voltage (V_{out}) exceeds its set value, the output of the comparator switches to the low state, disabling the gate to turn off until the output decreases. This regulating action is in addition to that produced by the duty cycle variation of the oscillator.

31-29. Complete d.c. Power Supply Circuit

Fig. 31-45 shows a complete d.c. power supply circuit. It consists of a step down transformer, full-wave bridge rectifier, a π-filter, a transistor series voltage regulator. The unregulated a.c. voltage voltage is fed from the transformer through a bridge rectifier. It is then filtered by a π-filter (CLC filter) and finally regulated by a transistor series regulator.

It is possible to take either smaller or greater d.c. output voltage than that produced by the d.c. power supply circuit. The smaller ouptut can be taken by using a voltage divider at the output terminals of the d.c. supply. A d.c. output voltage, greater than the normal value, may be obtained by using voltage multiplier. A voltage multiplier produces a d.c. voltage equal to a multiple of the maxium input voltage (*i.e.*, 2 V_m, 3 V_m and so on).

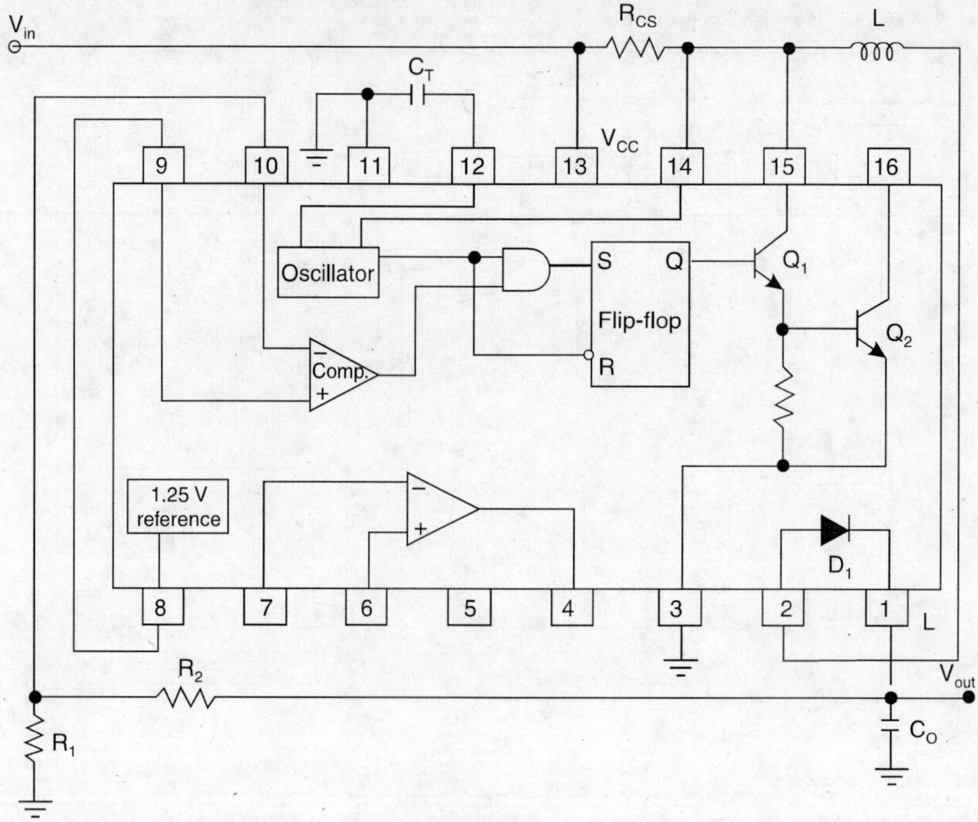

Fig. 31-44.

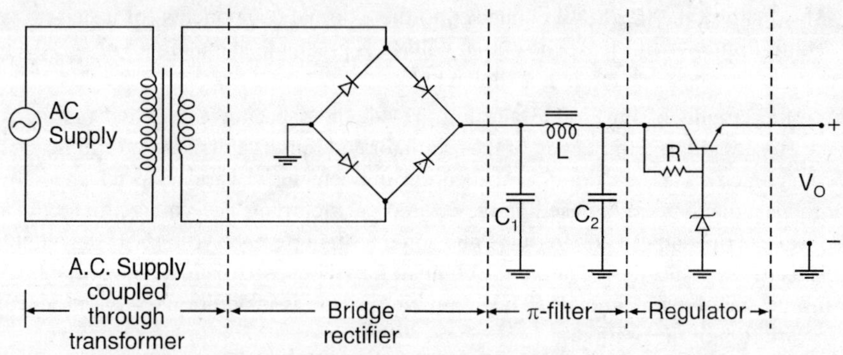

Fig. 31-45.

Such power supplies are used for high-voltage/low current devices like cathode-ray tubes (*i.e.*, the picture tubes in TV receivers, oscilloscopes and computer displays).

31-30. Switched-Mode Power Supply

(Electronic Circuits - I, Anna University, April/May 2008)

A switched-mode power supply (or simply SMPS) is an electronic power supply unit (PSU) that incorporates a switching regulator in order to provide the required output voltage. An SMPS is actually a power converter that transmits power from a source (*e.g.*, a battery or the electrical power grid) to a load (*e.g.*, a personal computer) with ideally no loss. The function of the converter is to provide a reliable output voltage often at a different level than the input voltage. In such a device, power handling electronics components are continuously switching on and off with high frequency in order to provide the transfer of electric energy via energy storage components (inductors and capacitors). By varying duty cycle, frequency or a relative phase of these transitions the average value of output voltage or current is controlled.

Classification of Switch Mode Power Supplies

The switch mode power supplies (SMPS) can broadly be classified into four types according to the input and output waveforms :

(*i*) AC in, DC out: rectifier, off-line converter input stage

(*ii*) DC in DC out: voltage converter, or current converter, or DC to DC converter

(*iii*) AC in, AC out: frequency change, cycloconverter, transformer

(*iv*) DC in AC out: inverter

Fig. 31-46 shows a block diagram of SMPS. As seen from this diagram, the SMPS has several sections namely input rectifier and filter, inverter chopper, output transformer, output rectifier and filter and chopper controller. All these sections are explained below.

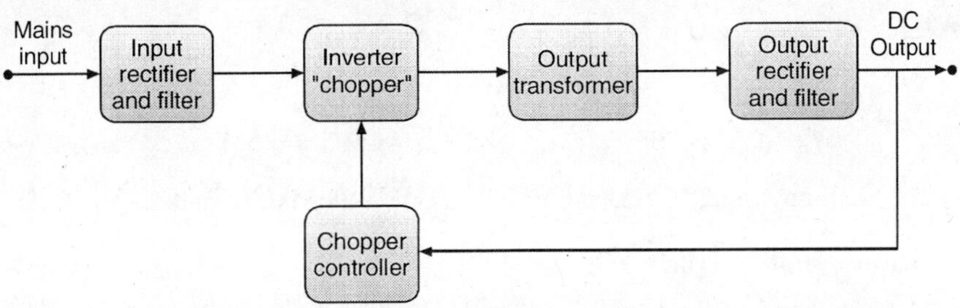

Fig. 31-46. *AC-DC SMPS with output voltage regulation*

(1) ***Input rectifier stage*** **:** If the SMPS has an AC input, then the first stage is to convert the input to DC. This is called **rectification**. The rectifier circuit can be configured as a voltage doubler by the addition of a switch operated either manually or automatically.

(2) ***Chopper*** **:** The inverter stage converts DC, whether directly from the input or from the rectifier stage, to AC by running it through a power oscillator, whose output transformer is very small with few windings at a frequency of tens or hundreds of kilohertz (kHz). The frequency is usually chosen to be above 20 kHz, to make it in audible to humans. The output voltage is optically coupled to the input and thus very tightly controlled.

(3) ***Voltage converter and output rectifier*** **:** If the output is required to be isolated from the input, as is usually the case in mains power supplies, the inverted AC is used to drive the primary winding of a high-frequency transformer. This converts the voltage up or down to the required output level on its secondary winding. The output transformer in the block diagram servers this purpose.

If a DC output is required, the AC output from the transformer is rectified. The rectified output is then smoothed by a filter consisting of inductors and capacitors. For higher switching frequencies, components with lower capacitance and inductance are needed.

(4) ***Regulation*** **:** A feedback circuit shown in the block diagram, monitors the output voltage and compares it with a reference voltage, Depending on design/safety requirements, the controller may or may not contain an isolation mechanism (such as opto-couplers) to isolate it from the DC output. Switching supplies in computers, TVs and VCRs have these opto-couplers to tightly control the output voltage.

31-31. Uninterrupted Power Supply (UPS)

An uninterrupted power supply (UPS) is an electrical equipment that provides emergency power to a load when the main supply voltage fails. A UPS is typically used to protect computers, data centres, telecommunication equipment or other electrical equipment where an unexpected power disruption could cause injuries, fatalities, serious business disruption and/or data loss.

UPS units range in size from units designed to protect a single computer without a monitor to large units powering entire data centres.

Primary role of UPS: The primary role of UPS is to provide short term power when the inpower source fails. However, most UPS units are also capable in correcting common mains power supply problems such as surge, spikes, noise, frequency instability and harmonic distortion.

Types of UPS

The UPS can be categorized into many different types on the basis of the problems they address:

(*i*) Off-line/Standby

(*ii*) Line-interactive

(*iii*) Double conversion/On-line

(*iv*) Hybrid technology/Double Conversion on demand

(*v*) Ferro-resonant

(*vi*) DC Power

(*vii*) Rotary

Now we shall discuss all these type of UPS in the following pages:

31-32. Offline/standby UPS

Fig. 31-47 shows an offline/standby UPS. It is absersated as SPS A typical protection time for this type of UPS is 0-20 minutes. With this type of UPS, a user's equipment is normally connected directly to incoming utility power with the some voltage transient clamping devices used in a common surge protected power strip connected across the power line.

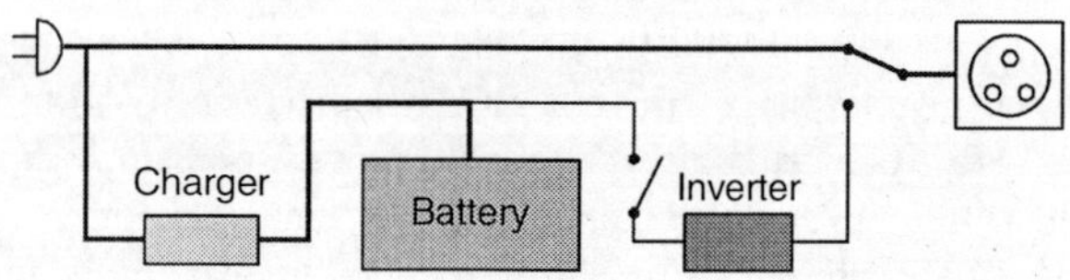

Fig. 31-47. *Offline/stand by UPS running on normal AC power*

When the incoming power supply voltage falls below predetermined level, the SPS turns on its internal DC-AC inverter circuitry which is powered from an internal storage battery as shown in fig. The SPS then mechanically switches the connected equipment on to its DC-AC inverter output.

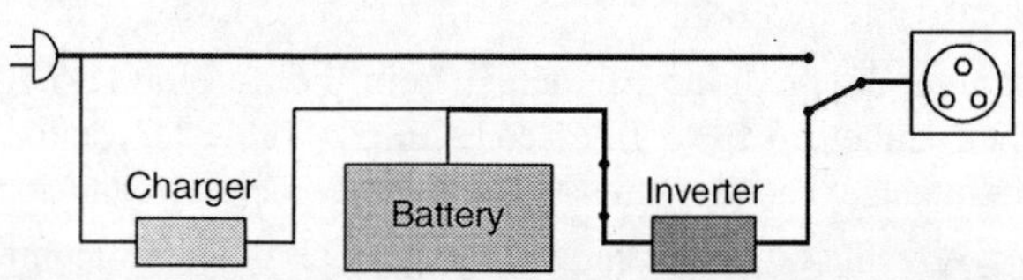

Fig. 31-48. *Offline/standby UPS runing in over/under-voltage or power loss*

The switch over time can be as long as 25 ms depending upon the amount of time it takes the standby UPS to detect the lost utility voltage.

31-33. Line Interactive UPS

This UPS is similar in operation to a standby UPS but with the addition of a *multi-tap variable-voltage autotransformer* as shown in Fig. 31-49. The typical protection time for line interactive UPS is 5 – 30 minutes. The capacity can be expanded for several hours. The line interactive UPS is able to tolerate continuous under voltage brown outs and over voltage surges without consuming the limited reserve battery power. It campers sates by auto-selecting different power taps on the auto transformer as shown in Fig. 31-49 (*b*), (*c*) and (*d*) as changing the auto transformer tap can course a very brief output power disruption. This is because of chirp of a UPS for a moment as it switches to a battery before changing the selected power tap.

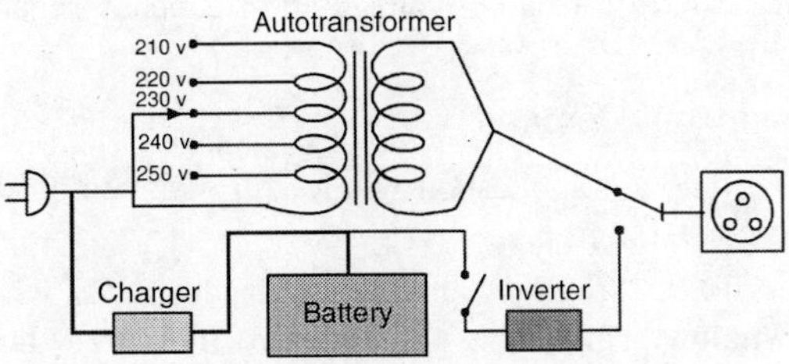

(*a*) Line--interactive UPS. Illustrating Normal AC power

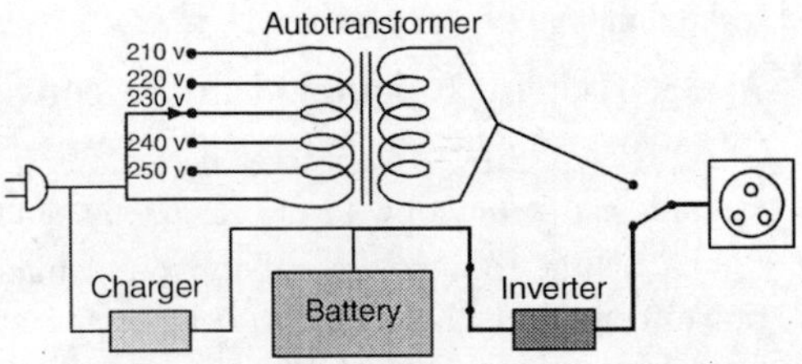

(*b*) Line-Interactive UPS : Illustrating large over/under-voltage or loss of AC power

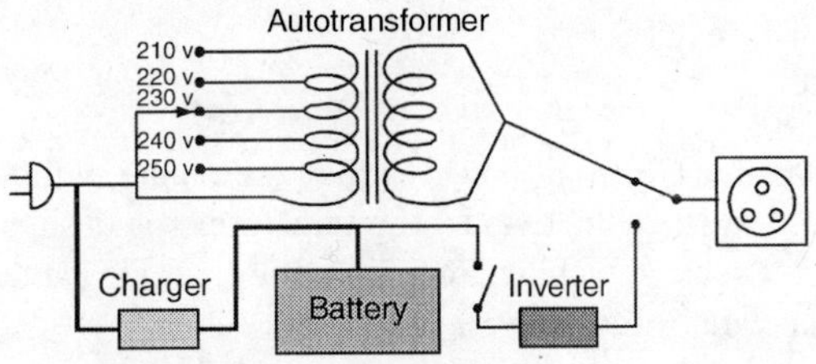

(*c*) Line-Interactive UPS : Illustrating small over voltage condition.

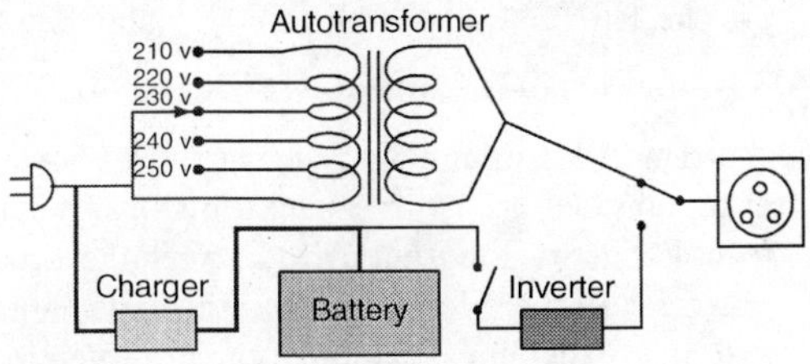

(*d*) Line-Interactive UPS : Illustrating small under voltage condition

Fig. 31-49.

It may be carefully noted that in low-voltage condition, the UPS will use more current than normal so that it may need a higher current circuit than a normal device. For example, to power a 1000 watt device at 230 volts the UPS will draw about 4.35A.

If a burn out occurs, the voltage drops to 200 V, the UPS will draw 5 A. This also works in reverse, *i.e.*, in an over voltage condition, the UPS will need fewer amperes of current.

31-34. Double Conversion/On-line UPS

This UPS is ideal for environments where electrical isolation is necessary or for an equipment that is very sensitive to power fluctuations. The on-line UPS is generally more expensive but may be necessary when the power environment is noisy such as in industrial settings, for large equipment loads like data centres. Typical protection time for online UPS is 5 – 30 minutes but can be expanded to several hours.

The basic technology of online UPS is the same as in standby or Line-interactive UPS. However, it costs much more due to its having much greater current AC-to-DC battery charger/rectifier and with the rectifier and inverter designed to run continuously with improved cooling systems. It is called a double conversion UPS due to the rectifier directly driving the inverter, even when powered from normal AC current.

It may be carefully noted that in an online UPS, the batteries are always connected to the inverter so that no power transfer switches are necessary. When power loss occurs, the rectifier simply drops out of the circuit and the batteries keep the power steady and unchanged. When power is restored, the rectifier resumes carrying most of the load and begins charging the batteries. However, the charging current may be limited to prevent the high power rectifier from over heating the batteries and boiling off the electrolyte.

Advantage of Online UPS

The main advantage of online UPS is its ability to provide an electrical firewall between the incoming mains power supply and sensitive electronic equipment. While the standby and line-inter active UPs merely fitter the input power supply, the double-conversion UPS provided a layer of insulation from power quality problems. It allows control of output voltage and frequency regardless of input voltage and frequency.

31-35. Hybrid Topology / Double conversion on Demand

This type of UPS is targeted towards high efficiency application whole still maintaining the features and protection level offered by double conversion.

A hybrid (double conversion on demand) UPS operates as an off-line/standby UPS when power conditions are with in a certain present window. This allows the UPS to achieve very high efficiency ratings. When the power conditions fluctuate out side the predefined windows, the UPS switches to online/double conversion operation. In the double conversion mode, the UPS can adjust for voltage variations without having to use battery power, can filter out line noise and control frequency. Some of the examples of the hybrid / double conversion on demand UPS design are the HP R 8000, HP R 12000/3 and Eaton Blade UPS.

31-36. Ferro-Resonant UPS

This UPS operates in the same way as a standby UPS unit however they are online with the exception that a ferro-resonant transformer is used to filter the output. The transformer is designed to hold energy long enough to cover the time between switching from line power to the battery power and effectively eliminate the transfer time. Many ferro-resonant UPS are 82-88% efficient and after excellent isolation. Typical protection time for ferro-resonant UPS is 5 – 15 minutes but capacity can be expanded to several hours.

It will be interesting to know that the ferro-resonant transformer has three windings—one for ordinary mains power, the second one for rectified battery power and the third for output ac power to the load.

The ferro-resonant UPS used to be the dominant type of UPS. The power rating is limited to 150 kVA range. This UPS is still mainly used in oil and gas. Petrochemical, chemical, utility and heavy industry markets due to robust nature of UPS.

31-37. DC Power UPS

A UPS designed for powering DC equipment is very similar to online UPS except that it does not need an output inverter and often the power device does not need a power supply. Rather than converting AC to DC to change batteries, there DC to AC to power the external device, and then back to DC inside the powered device, some equipment accepts DC power directly and allows one or more conversion steps to be eliminated. The equipment is more commonly known as rectifier.

31-38. Rotary UPS

This UPS uses the inertia of a high-mass spinning fly wheel to provide short-term ride through in the event of power loss. The flywheel also acts as a buffer against power spikes and sags Typical protection time for rotary UPS is 20 – 60 seconds but the capacity can be expanded to several seconds. The rotary UPS is used for applications requiring 10000 watts of protection to justify the expense.

SUMMARY

In this chapter you have learned that :

1. The function of a voltage regulator is to maintain a constant d.c. output voltage inspite of the a.c. input voltage flutuations or changes in load resistance.
2. The line regulation rating of a voltage regulator indicates the change in input voltage that will cause a change in the output voltage. It is expressed as the ratio $\Delta V_L / \Delta V_S$.
3. The load regulation indicates the change in output voltage that will occur per unit change in load current, *i.e.*, $(V_{NL} - V_{FL})/\Delta I_L$
4. Zener diode shunt regulator/can be used to regulate the output voltage for the varying input voltage and varying load resistance.
5. The zener diode shunt regulator has two main drawbacks :
   (*i*) the maximum load current which can be supplied to load resistor (R_L) is limited to few milliamperes,
   (*ii*) a large amount of power is wasted in the zener diode and the series resistance (R_S) in comparison with load power.
6. The transistor series regulator can be used for larger load currents.
7. Variable feedback regulator circuit is used to provide adjustable output voltage.
8. Switching regulators are of three basic types : (*i*) step-down, (*ii*) step-up and (*iii*) inverting regulator.
9. IC linear voltage regulator are available that can provide fixed positive, fixed negative, adjustable positive and adjustable negative output.
10. External pass transistor can be used with the IC voltage regulators to provide larger values of load current.

GLOSSARY

Current limiting resistor : A resistance placed in series with a zener diode to limit the amount of current that can be drawn through the zener diode.

Fixed negative voltage regulator : A regulator with a predetermined negative d.c. output voltage.

Fixed positive voltage regulator : A regulator with a predetermined positive d.c. output voltage.

IC Voltage regulator : A device that is used to hold the output voltage from a d.c. power supply relatively constant over a range of line and load currents.

Line regulation : A rating that indicates the change in regulator output voltage that will occur per unit change in input voltage.

Load regulation : A rating that indicates the change in regulator output voltage that will occur per unit change in load current.

Regulated d.c. power supply : A d.c. power supply with extremely low internal resistance.

Regulation : A rating that indicates the maximum change in regulator output voltage that can occur when input voltage and load current are varied over their entire rated ranges.

Shunt Feedback regulator : A voltage regulator that uses an error detection circuit to control the conduction of a shunt regulating transistor.

Shunt regulator : A voltage regulator in which the primary active device is in parallel with the load.

Voltage regulator : A circuit designed to maintain a constant d.c. load voltage despite changes in line voltage or load current demand.

NUMERICAL PROBLEMS

1. A zener diode shunt regulator circuit is to be designated to maintain a constant load current of 400 mA and voltage of 40 volts. The input voltage is 90 ± 5 V. The zener diode voltage is 40 V and its dynamic resistance is 2.5 Ω. Find the following quantities for the regulator : (*a*) the series dropping resistance (R_S), (*b*) zener power dissipation and (*c*) load resistance (R_L). Assume the zener current to be 10% of load current.

(*A.M.I.E., Winter* 1985) (**Ans.** 112.5 Ω, 3.6 W and 100 Ω)

2. Over what range of input voltage will the zener shunt regulator shown in Fig. 31-50 maintains 30 V across load resistor. Assume R_S = 200 Ω and $I_{Z(max)}$ = 25 mA.

(**Ans.** 33 to 38 V)

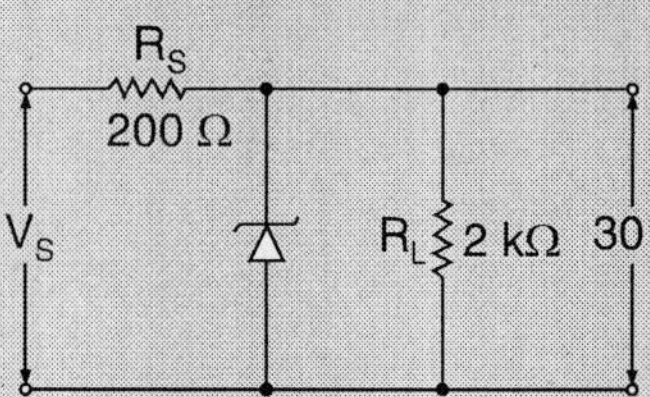

Fig. 31-50.

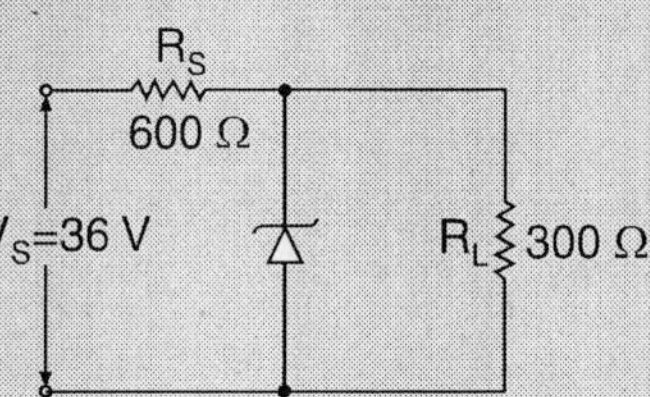

Fig. 31-51.

3. A 36 V d.c voltage is applied through a series resistor of 600 Ω to a load 300 of Ω shunted by a zener diode as shown in Fig. 31-51. If V_Z = 8 V and r_Z = 10 Ω , find (*a*) d.c. load voltage; (*b*) power dissipated in the zener and (*c*) maximum current, that a regulator can deliver and still regulate. (**Ans.** 8.19 V, 156 mV and 46.5 mA)

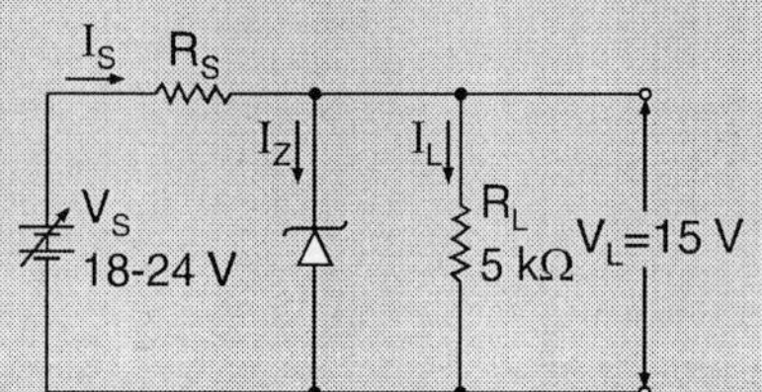

Fig. 31-52.

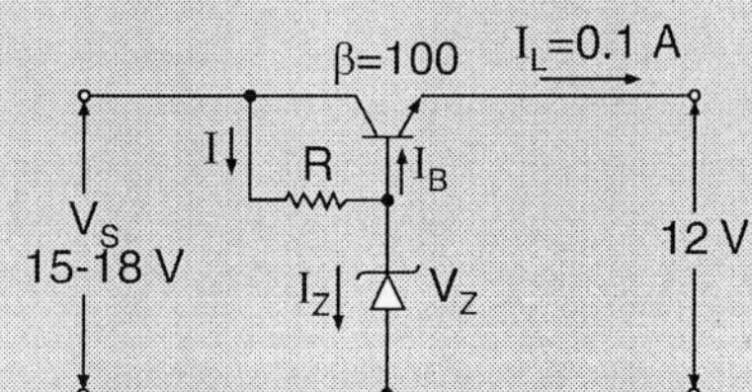

Fig. 31-53.

4. Draw the output waveform for the given input waveform. Also draw the output waveform when the polarity of the bias voltage is reversed. (*Banglore University*, 2001)

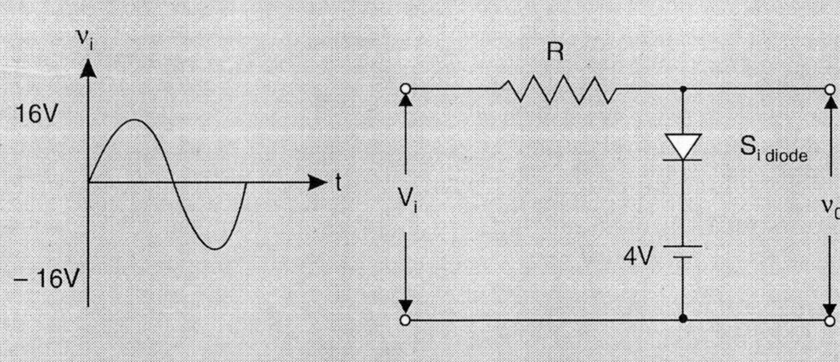

Fig. 6-54.

5. A zener diode voltage regulating circuit is as shown in Fig. 31-53. The zener diode used has a zener voltage (V_Z) of 15 V and minimum current $I_{Z\,(min)}$ of 2 μA, a power dissipation of 120 mW and a zener resistance of 40 Ω. If the load resistance is 5 kΩ and the input voltage varies from 18 to 24 V, find the value of R_S. (**Ans.** 158 Ω < R_S < 1000 Ω)

6. Assuming the base current (I_B) to be 20% of for V_S = 15 volts and V_{EB} = 0.6 volts, estimate the values of R, power in the transistor, and zener diode for the circuit shown in Fig. 31-54. (**Ans.** 400 Ω, 0.17 W and 0.6 W)

7. Determine the regulated voltage and circuit currents for the shunt regulator of Fig. 31-55.

(**Ans.** V_L = 8.9 V; I_L = 89 mA; I_S = 109 mA; I_C = 20 mA and I_Z = 0.2 mA)

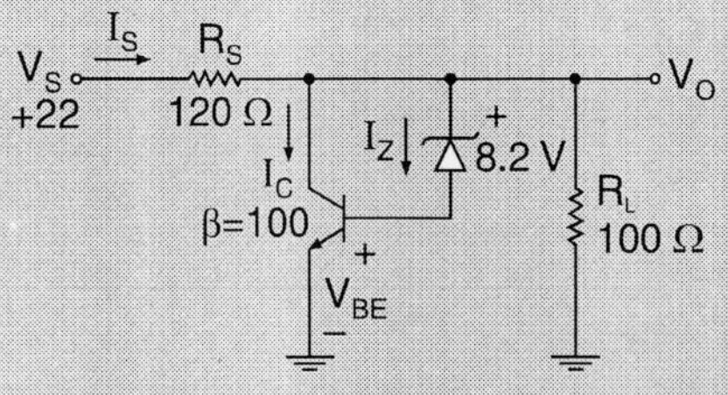

Fig. 31-55.

8. Calculate the output voltage and zener current in the regulator circuit of Fig. 31-56.

(**Ans.** V_0 = 11.3 V; I_Z = 36 mA)

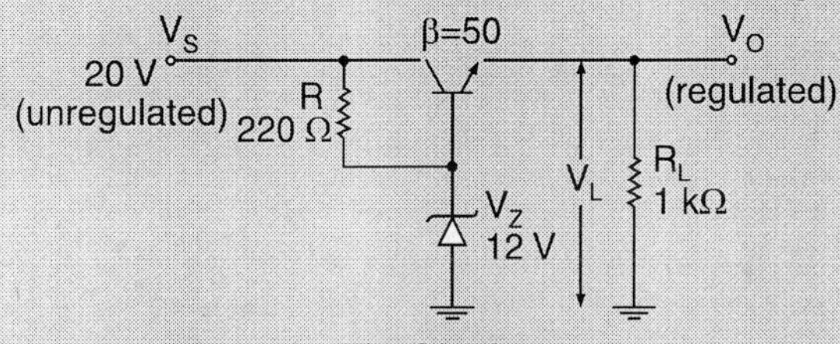

Fig. 31-56.

9. For the circuit shown in Fig. 31-57, find the ratings of R and zener, assuming the zener current to be 5 mA under normal full load condition.

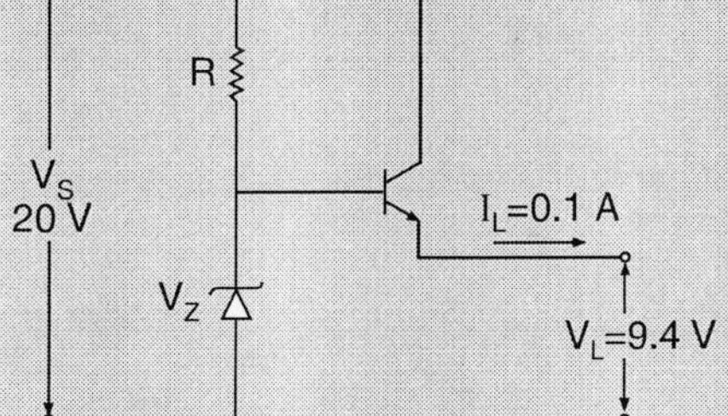

Fig. 31-57.

The available transistors are of β = 50; V_{BE} = 0.6 V and having the following specifications :

Type 1: V_{CEO} = 30 V, $I_{C\,(max)}$ = 2.5 A, P = 30 W

Type 2: V_{CEO} = 25 V $I_{C\,(max)}$ = 3A, P = 60 W

Type 3: V_{CEO} = 15 V, $I_{C\,(max)}$ = 5 A, P = 50 W

Choose the type of transistor suitable for the above circuit. (*A.M.I.E., Summer*, 1984)

(**Ans.** 400 Ω; 0.25 W and transistor type 1)

10. The LM 317 is being used in a circuit with adjustment resistance values of $R_1 = 510\ \Omega$ and $R_2 = 6.834\ \text{k}\Omega$ (adjusted potentiometer value). Determine the output voltage for the circuit. (**Ans.** 14.4 V)

11. The Zener diode of Fig. 31-58 has the following rating :

$V_z = 15$ V at $I_z = 17$ mA

$r_z = 14\ \Omega$ at $I_z = 17$ mA

$I_{z\,(min)} = 0.25$ mA $I_{z\,(max)} = 66$ mA

What would be the load voltage when load current IL varies from 1 mA to 60 mA. Also calculate the voltage regulation of the regulator. (**Ans.** 14.846 V, 15.672 V, 5.27%)

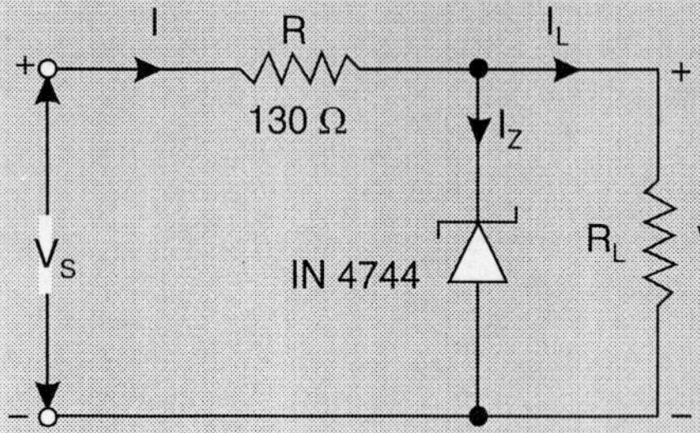

Fig. 31-58.

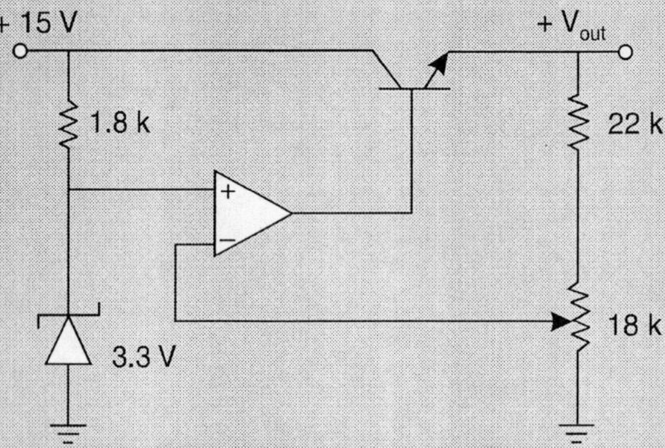

Fig. 31-59.

12. Determine the output voltage, V_{out} for the op-amp series regulator shown in Fig. 31-59. (**Ans.** 7.3 V)

13. Calculate the minimum and maximum output voltages for the IC voltage regulator shown in Fig. 31-60. Assume $I_{ADJ} = 50$ mA, $V_{in} = +35$ V, $V_{REF} = 1.2$ V (**Ans.** 1.25 V, 12.71 V)

14. Fig. 31-61 shows the circuit of a current regulator. What value of R_1 is necessary to provide a constant current of 1 A. (**Ans.** 12 Ω)

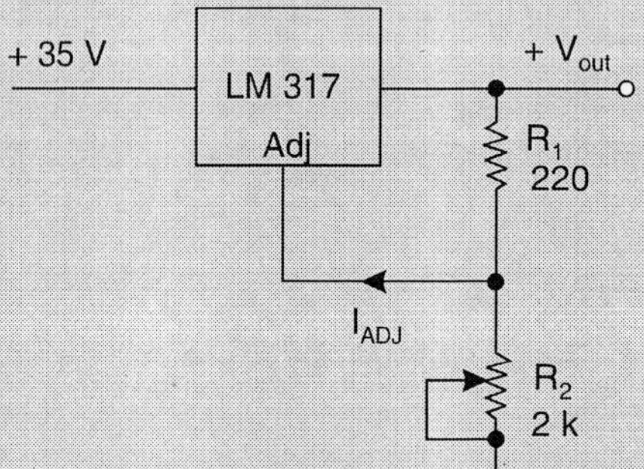

Fig. 31-60.

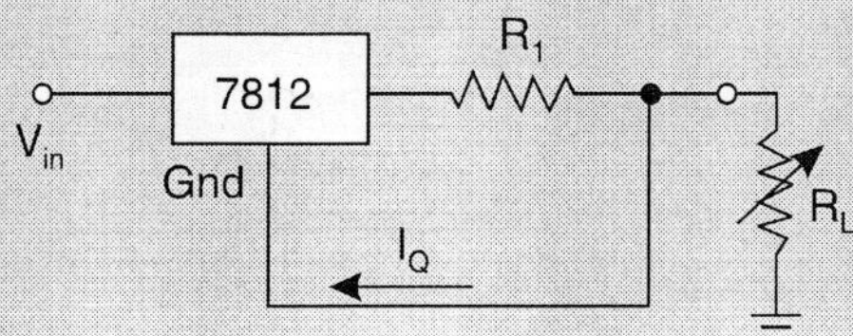

Fig. 31-61.

DESCRIPTIVE QUESTIONS

1. Describe a zener shunt regulator circuit.

2. What are the disadvantages of zener shunt regulator ?

3. Draw circuit diagram of transistor shunt regulator. Explain it briefly.

4. Draw circuit diagram of a transistorized series voltage regulator.

5. Describe the monolithic IC voltage regulators.

6. Explain how zener diode acts as a regulator.

(*Electronic Circuits - I, Anna University, April/May 2008*)

7. Draw and explain the working principles of a SMPS circuit with its output waveforms.

(*Electronic Circuits - I, Anna University, April/May 2008*)

8. Define voltage regulation.

(*Electronic Circuits - I, Anna University, April/May 2008*)

9. Explain the procedure for designing a shunt type voltage regulator.

(*Electronic Circuits - I, Anna University, Apri/May 2008*)

10. What is a switching regulator? What are the types of switching regulator? Explain them in detail with suitable figures.

(*Electronic Circuits - I, Anna University, Apri/May 2008*)

11. Write short notes on switching mode power supply.

(*Electronic Circuits - I, Anna University, Apri/May 2008*)

12. With neat sketch explain the Switched Mode Power supplies.

(*Electronic Circuits - I, Anna University, Apri/May 2008*)

13. What are the various block that form a Basic Voltage Regulator? Explain the series shunt voltage regulator. List advantages of Voltage Regulator?

(*Linear Integrated Circuits, Anna University, April/May 2008*)

14. Draw the functional block diagram of switching regulator and explain.

(*Linear Integrated Circuits, Anna University, April/May 2008*)

15. Zener diode can be used as voltage regulator-Justify it.

(*Electronic Devices, Anna University, Nov/Dec 2007*)

MULTIPLE CHOICE QUESTIONS

1. Which of the following is not an essential element of a d.c. power supply.

(*a*) Rectifier (*b*) Filter

(*c*) Voltage regulator (*d*) Voltage amplifier

2. A voltage regulator is a circuit which

(*a*) Converts the a.c. voltage to d.c. voltage

(*b*) Smoothens the a.c. variation in d.c. output voltage

(*c*) Maintains a constant d.c. output voltage inspite of the fluctuations in a.c. input voltage or load current

(*d*) None of the above

3. The percentage voltage regulation of voltage supply providing 100 V unloaded and 95 V at full load is

(*a*) 5.3% (*b*) 5.0%

(*c*) 0.53% (*d*) None of the above

4. Which of the following voltage regulator is preferred for providing large values of load current.

(*a*) Zener diode shunt regulator (*b*) Transistor series regulator

(*c*) Transistor shunt regulator (*d*) None of the above

5. A pre-regulator is used with a transistor series regulator in order to

(*a*) Increase the input resistance of the series transistor

(*b*) Decrease the input resistance of the series transistor

(*c*) Increase the output resistance of the series transistor
(*d*) Increase the current gain of the series transistor

6. The main job of a voltage regulator is to provide a nearly—output voltage.
(*a*) sinusoidal (*b*) constant (*c*) smooth (*d*) fluctuating

7. A 10-V d.c. regulator power supply has a regulation of 0.005 per cent. Its output voltage will vary within an envelope of .... millivolt.
(*a*) ± 2.5 (*b*) ± 0.5 (*c*) ± 5 (*d*) ± 0.05

8. An ideal voltage regulator has a voltage regulation of
(*a*) 1 (*b*) 100 (*c*) 50 (*d*) 0.

9. In a Zener diode shunt voltage regulator, the diode regulates so long as it is kept in .... condition.
(*a*) forward (*b*) reverse (*c*) loaded (*d*) unloaded

10. The power efficiency of a switching voltage regulator is much higher than that of a linear regulator because it operates.
(*a*) in saturation (*b*) in cut-off
(*c*) like a switch (*d*) on high duty cycle.

11. A transistor series voltage regulator is called emitter-follower regulator because the emitter of the pass transistor follows the ....voltage.
(*a*) output (*b*) input (*c*) base (*d*) collector

12. In an Op-amp series voltage regulator, output voltage depends on
(*a*) Zener voltage (*b*) voltage divider resistors
(*c*) output voltage (*d*) both (*a*) and (*b*)

13. In a feedback series regulator circuit, the output voltage is regulated by controlling the
(*a*) magnitude of input voltage
(*b*) gain of the feedback transistor
(*c*) reference voltage
(*d*) voltage drop across the series pass transistor

14. An Op-amp shunt regulator differs from the series regulator in the sense that its control element is connected in
(*a*) series with line resistor (*b*) parallel with the resistor
(*c*) parallel with load resistor (*d*) parallel with input voltage.

15. A switching voltage regulator can be of the following type :
(*a*) step-down (*b*) step-up (*c*) inverting (*d*) all of the above

16. In an inverting type switching regulator, output voltage is .... input voltage.
(*a*) lesser than (*b*) greater than (*c*) equal to (*d*) opposite to.

17. The output voltage of a step-down type switching voltage regulator depends on
(*a*) input voltage (*b*) duty cycle
(*c*) transistor on time (*d*) all of the above.

18. As compared to voltage regulators made up of discrete components, IC regulators have the inherent advantage/s of
(*a*) self protection against over-temperature
(*b*) remote control
(*c*) current limiting
(*d*) all of the above.

19. A 12 V monolithic regulator is adjusted to obtain a higher output voltages as shown in Fig. 31-62. The v_o will be

(*a*) 12 V (*b*) 17 V (*c*) 24 V (*d*) 29 V

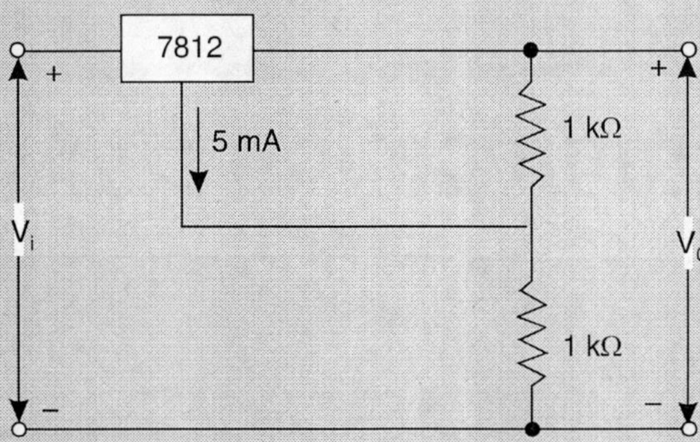

Fig. 31-62.

20. A three terminal monolithic IC regulator can be used as

(*a*) an adjustable output voltage regulator alone
(*b*) an adjustable output voltage regulator and a current regulator
(*c*) a current regulator and a power switch
(*d*) a current regulator alone

21. In a switched-mode power supply (SMPS), after conversion of a.c. supply to a highly filtered d.c. voltage, a switching transistor is switched ON and OFF at a very high speed by a pulse width modulator (PWM) which generates very high frequency square pulses. The frequency of the pulses is typically in the range of

(*a*) 100 Hz-200 Hz (*b*) 500 Hz-1 kHz
(*c*) 2 kHz-5 kHz (*d*) 20 kHz-50kHz

(*UPSC Engg. Services*. 2002)

ANSWERS

1. (*d*)	**2.** (*c*)	**3.** (*a*)	**4.** (*b*)	**5.** (*d*)
6. (*b*)	**7.** (*a*)	**8.** (*d*)	**9.** (*b*)	**10.** (*c*)
11. (*c*)	**12.** (*d*)	**13.** (*d*)	**14.** (*a*)	**15.** (*d*)
16. (*d*)	**17.** (*d*)	**18.** (*d*)	**19.** (*d*)	**20.** (*d*)
21. (*d*)				

❑ ❑ ❑

32

CHAPTER

Communication Systems

OUTLINE

Objectives

At the end of this chapter, you should be able to :

- Describe the need for modulaton.
- Define the modulation, amplitude modulation, frequency modulation and phase modulation.
- Understand the difference between amplitude modulation (AM), frequency modulation (FM) and phase modulation (PM).
- Sketch the waveforms of the amplitude modulated, frequency modulated and phase modulated waveforms in time-domain and frequency-domain.

- Calculate the value of percentage modulation or modulation index.
- Know the upper and lower side frequencies in the AM wave.
- Analyze the modulated carrier wave.
- Sketch the block diagram of a AM transmitter and explain its operation.
- Explain the operation of modulating amplifier circuit.
- Calculate the modulation index for FM signal.
- Explain what demodulation is ?
- Sketch the circuit diagram and explain the circuit operation for demodulation of AM and FM signals.
- Sketch the block diagram of superheterodyne receiver and explain its operation.
- Compare the characteristics of AM and FM signals.

32-1. Introduction

For successful transmission and reception of intelligence (code, voice, music, *etc.*) by the use of radio waves, two processes are essential :

(*i*) Modulation, and (*ii*) Demodulation.

Speech and music, *etc.* are sent thousands of kilometres away by a radio transmitter. The scene in front of a television camera is also sent many kilometres away to viewers. Similarly, a Moon probe or Venus probe checking its environments, sends the information it gathers millions of kilometres through space to receivers on earth. In all these cases, the carrier is the high-frequency radio wave. The intelligence, *i.e.,* sight, sound or other data collected by the probe is impressed on the radio wave and is carried along with it to the destination.

Modulation is the process of **combining the low-frequency signal with a very high-frequency radio wave called carrier wave (CW)**. The resultant wave is called **modulated carrier wave**. This job is done at the transmitting station.

Demodulation is the process of **separating** or **recovering the signal from the modulated carrier wave**. It is just the opposite of modulation and is performed at the receiving end.

32-2. What is a Carrier Wave ?

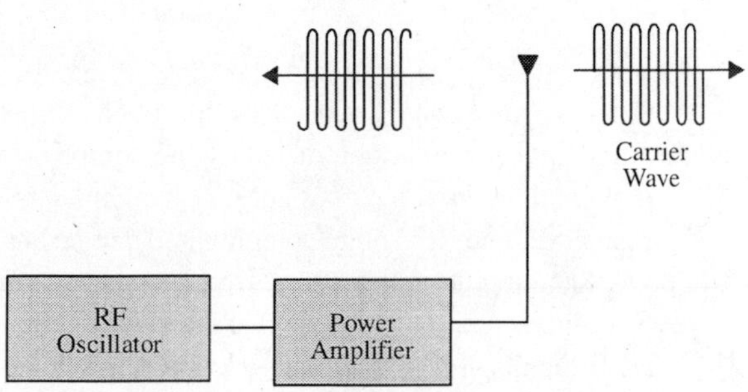

Fig. 32-1.

It is a high-frequency undamped radio wave produced by radio-frequency oscillators. As seen from Fig. 32-1, the output of these oscillators is first amplified and then passed on to an antenna. This antenna radiates out these high-frequency (electromagnetic) waves into space. These waves have constant amplitude and travel with the velocity of light. They are inaudible, *i.e.*, by themselves they cannot produce any sound in the loudspeaker of a receiver. As their name shows, their job is to **carry the signal** (audio or video) from transmitting station to the receiving station. The resultant wave is called **modulated** carrier wave.

32-3. Radio Frequency Spectrum

Radio frequencies used by different communication systems extend from very low frequencies to extra high frequencies as tabulated below along with their acronym abbreviations.

Table No. 32-1

Frequency	*Designation*	*Abbreviation*	*Uses*
3-30 kHz	very low frequency	VLF	long distance telegraphy broadcasting
30-300 kHz	low frequency	LF	long distance point-to-point service, navigational aids, sound broadcasting and line carrier systems.
300 kHz-3MHz	medium frequency	MF	sound broadcasting, ship-shore services and line carrier systems
3-30 MHz	high frequency	HF	medium and long-distance point-to-point services, sound broadcasting, linear carrier systems.
30-300 MHz	very high frequency	VHF	short-distance communication, TV and
300 MHz-3GHz	ultra high frequency	UHF	sound broadcasting, radar
3-30 GHz	super high frequency	SHF	outer-space radio communication,
30-300 GHz	extra high frequency systems and radar.	EHF	point-to-point microwave communication

Generally, none of the frequencies above 300 GHz is classified as radio waves.

32-4. Sound

It is a sort of disturbance which requires some physical medium for its propagation. Human voice consists of a series of compressions and rarefactions which travel through air with a velocity of about 345 m/s. The frequency range of human voice is from 20-4000 Hz which lies within the audible range of 20 to 20,000 Hz. Variations in human voice can be converted into corresponding variations in electric current with the help of a microphone as shown in Fig.32-2.

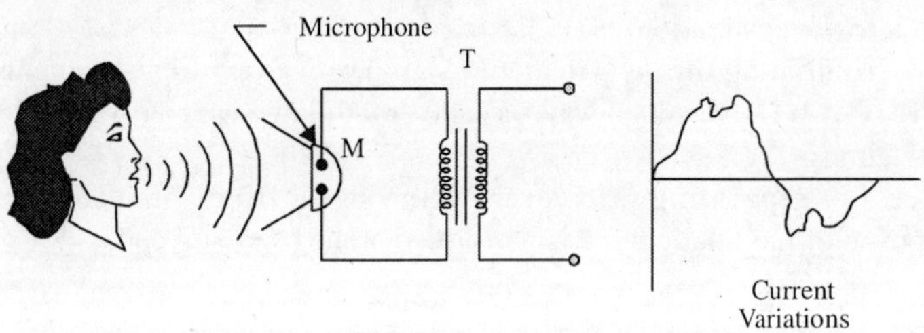

Fig. 32-2.

When a sound wave strikes the microphone, it produces AF sound current. The positive half-cycles of sound current are produced by the compressions and negative half cycles by rarefactions.

As seen, human voice does not produce pure sinusoidal current because it does not consist of one frequency alone. It is quite complex and can be analysed to consist of a fundamental (or lowest frequency) and its integral multiple frequencies called **overtones** or **harmonics**.

We are interested in two main characteristics of sound :

(*i*) **Intensity**—It is the energy content of the wave. It depends on its amplitude. In fact, intensity of a wave is directly proportional to the square of its amplitude, *i.e.,* $I \propto a^2$. Sensation of loudness felt by a listener depends directly on the intensity of the wave falling on his ears.

(*ii*) **Frequency**—It produces the sensation called pitch. Audible sounds have a frequency range from 20 Hz to 20,000 Hz.

Though every sound has complex frequency structure, we will consider only single-frequency sound whose current wave is a pure sine wave as shown in Fig. 32-3. It will be used as the modulating signal when discussing the process of modulation.

32-5. Need for Modulation

Sometimes, beginners question the necessity of modulation, *i.e.*, using a carrier wave to carry the low-frequency signal from one place to another. Why not transmit the signals directly and save lot of botheration? Unfortunately, there are three main hurdles in the process of such direct transmission of audio-frequency signals :

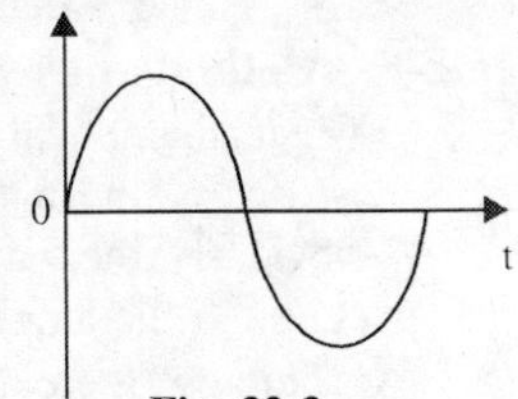

Fig. 32-3

(*i*) They have relatively short range,

(*ii*) If everybody started transmitting these low-frequency signals directly, mutual interference will render all of them ineffective

(*iii*) Size of antennas required for their efficient radiation would be large, *i.e.*, about 75 km as explained below.

For efficient radiation of a signal, the minimum length of an antenna is one quarter wavelength ($\lambda/4$). The antenna length L is connected with the frequency of the signal wave by the relation $L = 75 \times 10^6/f$ metres. For transmitting an audio signal of $f = 1000$ Hz, $L = 75 \times 10^6/10^3 = 75{,}000$ m = 75 km ! In view of this immense size of antenna length, it is impractical to radiate audio-frequency signals directly into space.

Hence, the solution lies in modulation which enables a low-frequency signal to travel very large distances through space with the help of a high-frequency carrier wave. These carrier waves need reasonably-sized antennas and produce no interference with other transmitters operating in the same area.

32-6. Radio Broadcasting

Let us see how radio broadcasting stations broadcast speech or music, *etc.*, from their broadcasting studios.

First, the speech or music which consists of a series of compressions and rarefactions is translated into a tiny varying electric current with the help of a crystal microphone. The frequency of variations of this current lies in the audio-range, hence it is known as **audio frequency signal**. The audio-frequency signal cannot be radiated out from the antenna directly because transmission at audio-frequencies is not practical. For this purpose, oscillations of very high frequency or radio-frequency are produced with the help of any one of the oscillators. The electromagnetic waves so produced are of constant amplitude but of extremely high frequency. These waves, which are neither seen nor heard, travel through space with the velocity of light, *i.e.*, 3×10^8 m/s (approx). The audio-frequency signal which is to be broadcast, is then superimposed on the *RF* waves, which are known as carrier waves (because they carry *A.F.* signal through space to distant places). In a way, the carrier waves can be likened to a horse and the audio-frequency signal to a rider. The process by which *AF* signal or information is impressed on the carrier wave is known as modulation. The horse and rider travel through space. At the receiving end, they strike the receiving aerial and enter the receiver which separates the horse from the rider. The horse, *i.e.*, carrier wave is returned and the rider, *i.e.*, audio-frequency signal is converted back into sound. This process by which the *R.F.* waves and *A.F.* waves are separated is known as ***detection*** or ***demodulation*** (because it is the reverse of modulation).

32-7. Modulation

It is the process of ***combining an audio-frequency (AF) signal with a radio frequency (RF) carrier wave.*** The AF signal is also called a ***modulating wave*** and the resultant wave produced is called ***modulated wave***.

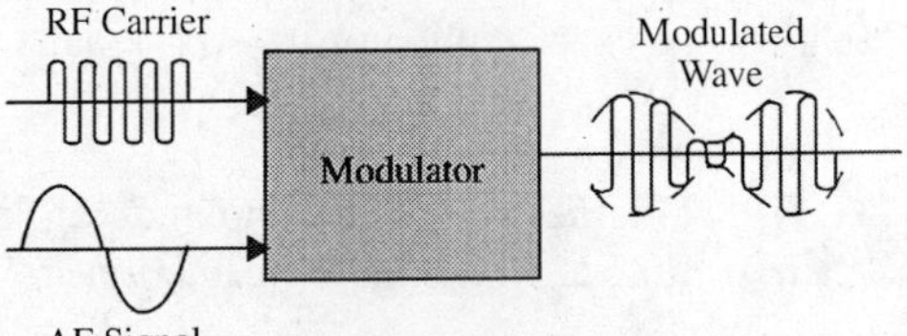

Fig. 32-4.

During modulation, some characteristic of the carrier wave is varied in time with the modulating signal and is accomplished by combining the two.

32-8. Methods of Modulation

The mathematical expression for a sinusoidal carrier wave is

$$e = E_C \sin(\omega_c t + \phi) = E_C \sin(2\pi f_c t + \phi)$$

Obviously, the waveform can be varied by any of its following three factors or parameters :

(1) E_C — the amplitude, **(2)** f_c — the frequency, **(3)** ϕ — the phase.

Accordingly, there are three types of sine-wave modulations known as :

(1) ***Amplitude Modulation (AM)***

Here, the information or *AF* signal changes the amplitude of the carrier wave without changing its frequency or phase.

(2) ***Frequency Modulation (FM)***

In this case, the information signal changes the frequency of the carrier wave without changing its amplitude or phase.

(3) ***Phase Modulation (PM)***

Here, The information signal changes the phase of the carrier wave without changing its other Parameters.

32-9. Amplitude Modulation

In this case, the amplitude of the carrier wave is varied in proportion to the instantaneous amplitude of the information signal or AF signal. Obviously, the amplitude (and hence the intensity) of the carrier wave is changed ***but not its frequency.*** Greater the amplitude of the AF signal, greater the fluctuations in the amplitude of the carrier wave.

The process of amplitude modulation is shown graphically in Fig. 32-5. For the sake of simplicity, the *AF* signal has been assumed sinusoidal [Fig. 32-5 (*a*)]. The carrier wave by which it is desired to transmit the *AF* signal is shown in Fig. 32-5 (*b*). The resultant wave called modulated wave is shown in Fig. 32-5 (*c*).

The function of the modulator is to mix these two waves in such a way that (*a*) is transmitted along with (*b*). All stations broadcasting on the standard broadcast band (550-1550 kHz) use *AM* modulation.

If you observe the envelope of the modulated carrier wave, you will realize that it is an exact replica of the *AF* signal wave.

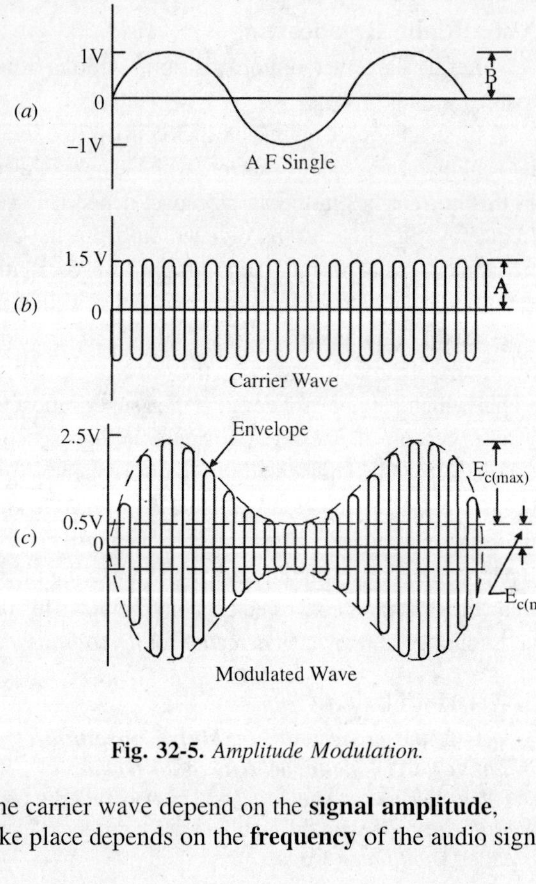

Fig. 32-5. *Amplitude Modulation.*

In summary

(*i*) **Fluctuations** in the amplitude of the carrier wave depend on the **signal amplitude**,

(*ii*) **Rate** at which these fluctuations take place depends on the **frequency** of the audio signal.

32-10. Percent Modulation

It indicates the degree to which the AF signal modulates the carrier wave

$$m = \frac{\text{maximum value of signal wave}}{\text{maximum value of carrier wave}} \times 100 = \frac{\text{signal amplitude}}{\text{carrier amplitude}} \times 100$$

$$= \frac{B}{A} \times 100 \quad \text{—Fig. 32-5}$$

The ratio B/A expressed as a fraction is called **modulation index (MI)**

$$m = M.I. \times 100$$

From Fig. 32-6, it is seen that $B = 1\ V$ and $A = 1.5$ V

$$\therefore \quad m = \frac{1}{1.5} \times 100 = 32\text{-}7\%$$

Modulation may also be defined in terms of the values referred to the modulated carrier wave.

$$m = \frac{E_{c(max)} - E_{c(min)}}{E_{c(max)} + E_{c(min)}} \times 100$$

where $E_{c(max)}$ and $E_{c(min)}$ are the maximum and minimum values of the amplitude of the ***modulated carrier wave***.

Again, from Fig. 32-5 we see that

$$m = \frac{2.5-0.5}{2.5+0.5} \times 100$$

$$= \frac{2}{3} \times 100 = 32\text{-}7\%$$

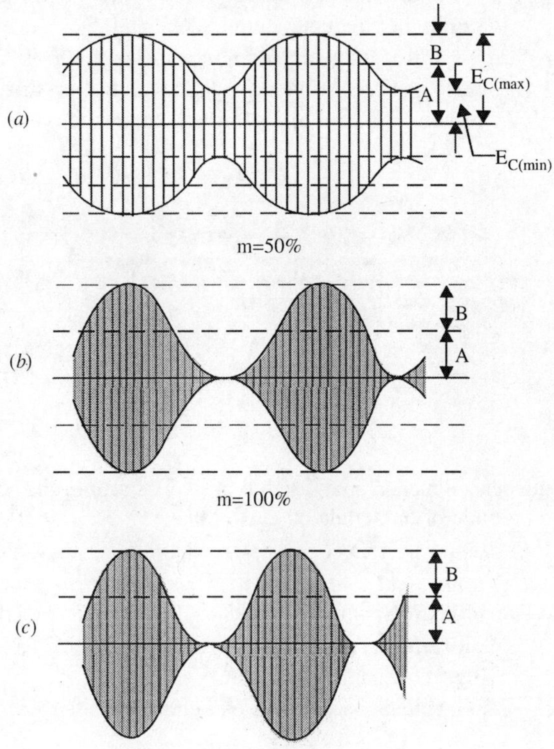

Fig. 32-6. *Percent Modulated wave*

Fig. 32-6 shows a modulated wave with different degrees of modulation. As before, both the signal and carrier waves are assumed to be sine waves. Smallest value of $m = 0$, *i.e.*, when amplitude of the modulating signal is zero. It means that $m = 0$ for an unmodulated carrier wave. Maximum value of $m = 1$ when $B = A$. Value of m can vary anywhere from 0 to 100% without introducing distortion. Maximum undistorted power of a radio transmitter is obtained when $m = 100\%$. If m is less than 100 per cent, power output is reduced though the power content of the carrier is not. Modulation in excess of 100 per cent produces severe distortion and interference (called splatter) in the transmitter output.

Example 32-1 *A modulated carrier wave has maximum and minimum amplitudes of 750 mV and 250 mV. Calculate the value of percentage modulation.*

Solution. The modulated wave is shown in Fig. 32-7.

Here, $E_{c(max)} = 750$ mV and

$E_{c(min)} = 250$ mV

$$\therefore \quad m = \frac{E_{c(max)} - E_{c(min)}}{E_{c(max)} + E_{c(min)}} \times 100$$

$$= \frac{750 - 250}{750 + 250} \times 100 = \mathbf{50\%}$$

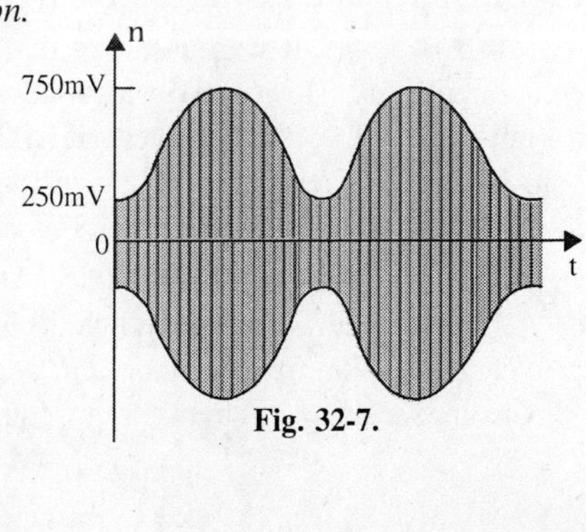

Fig. 32-7.

32-11. Upper and Lower Side Frequencies

An unmodulated carrier wave consists of only one single-frequency component of frequency f_c. When it is combined with a modulating signal of frequency FM, heterodyning action takes place.

As a result, two additional frequencies called **side frequencies** are produced. The *AM* wave is found to consist of three frequency components :

(*i*) The original carrier frequency component, f_c .

(*ii*) A higher frequency component $(f_c + f_m)$. It is called the sum component.

(*iii*) A lower frequency component $(f_c - f_m)$. It is called the difference component.

The two new frequencies are called the **upper-side frequency** (*USF*) and **lower-side frequency** (*LSF*) respectively and are symmetrically located around the carrier frequency. The modulating frequency remains unchanged but does not appear in the amplifier output because the amplifier's load presents practically zero impedance to this low frequency.

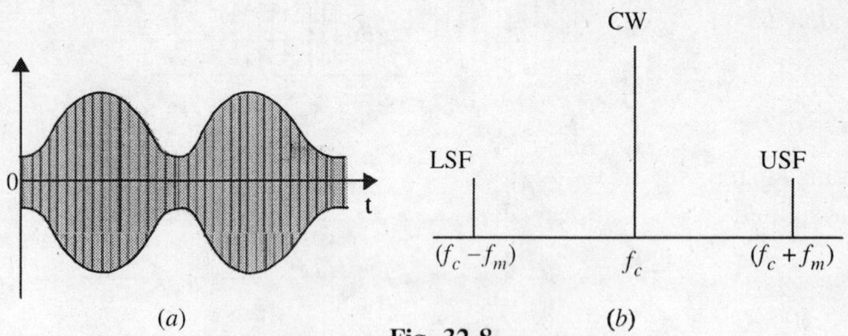

Fig. 32-8.

These are shown in time domain in Fig. 32-8 (*a*) and in frequency domain in Fig. 32-8 (*b*).

The amplitude of the side frequencies depends on the value of *m*. The amplitude of each side frequency = *mA*/2 where *A* is the amplitude of unmodulated carrier wave.

Example 32-2. *A 10-MHz sinusoidal carrier wave of amplitude 10 mV is modulated by a 5 kHz sinusoidal audio signal wave of amplitude 6 mV. Find the frequency components of the resultant modulated wave and their amplitudes. (Electronics & Comm. Engg.; Madras Univ. 1991)*

Solution. Here, $f_c = 10$ MHz and $f_m = 5$ kHz = 0.005 MHz. The modulated carrier contains the following frequencies :

(*i*) original carrier wave of frequency

f_c = **10 MHz**

(*ii*) USF of frequency

= 10 + 0.005 = **10.005 MHz**

(*iii*) LSF of frequency

= 10 – 0.005 = **9.995 MHz**

The frequency spectrum is shown in Fig. 32-9

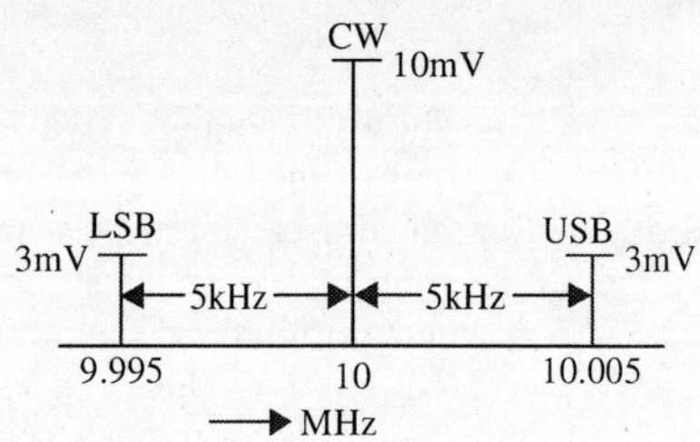

Fig. 32-9.

$$\text{Here, } m = \frac{B}{A} = \frac{6}{10} = 0.6 = 0.6$$

Amplitude of *LSF* = *USF* = *mA*/2 = 0.6 × 10/2 = **3 mV** as shown.

32-12. Upper and Lower Sidebands

In Art 32-11, it was assumed that the modulating signal was composed of one frequency component only. However, in a broadcasting station, the modulating signal is the human voice (or music) which contains waves with a frequency range of 20-4000 Hz. Each of these waves has its own *LSF* and *USF*. When combined together, they give rise to an upper-side **band** (*USB*) and a lower-side **band** (LSB) as shown in Fig. 32-10. The *USB*, in fact, contains all sum components of the signal and carrier frequency whereas *LSB* contains their difference components.

The channel width (or bandwidth) is given by the difference between extreme frequencies, i.e., between maximum frequency of *USB* and minimum frequency of *LSB*. As seen,

Channel width = 2 × maximum frequency of modulating signal = $2 \times f_{m\,(max)}$

Example 32-3. *An audio signal given by 15 sin 2π (2000 t) amplitude-modulates a sinusoidal carrier wave 60 sin 2π (100,000) t.*

Determine :

(a) modulation index, (b) percent modulation,

(c) frequencies of signal and carrier, and

(d) frequency spectrum of the modulated wave.

(Electronics & Telecom Engg., Jadavpur Univ. 1991)

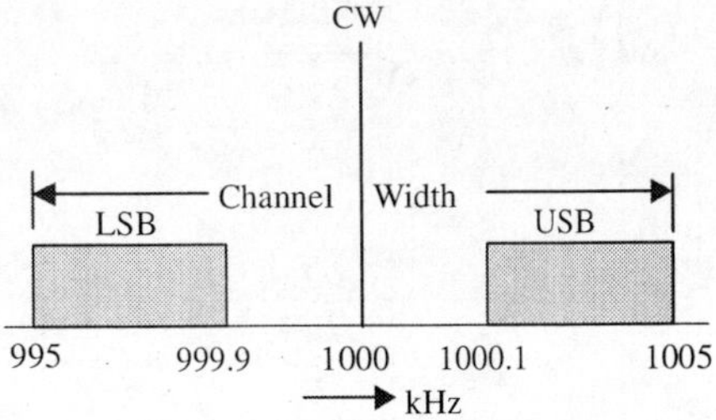

Fig. 32-10.

Solution. Here, $B = 15$ and $A = 60$

(*a*) $M.I. = \frac{B}{A} = \frac{15}{60} = 0.25$

(*b*) $m = M.I. \times 100 = 0.25 \times 100 = 25\%$

(*c*) $f_m = 2000$ Hz — by inspection of the given equation

$f_c = 100{,}000$ Hz — by inspection of the given equation

(*d*) The three frequencies present in the modulated *CW* are

(*i*) 100,000 Hz = 100 kHz

(*ii*) 100,000 + 2000 = 102,000 Hz = 102 kHz

(*iii*) 100,000 – 2000 = 98,000 Hz = 98 kHz

Example 32-4. *A bandwidth of 15 MHz is available for AM transmission. If the maximum audio signal frequency used for modulating the carrier is not to exceed 15 kHz, how many stations can broadcast within this band simultaneously without interfering with each other?*

(Electronics & Telecom Engg.; Pune Univ. 1991)

Solution. *BW* required by each station

$$= 2 f_{m(max)} = 2 \times 15 = 30 \text{ kHz}$$

Hence, the number of station which can broadcast within this frequency band without interfering with one another is

$$= \frac{15 \text{ MHz}}{30 \text{ kHz}} = 500$$

Example 32-5. *In a broadcasting studio, a 1000 kHz carrier is modulated by an audio signal of frequency range, 100-5000 Hz. Find* **(i)** *width or frequency range of sidebands* **(ii)** *maximum and minimum frequencies of USB* **(iii)** *maximum and minimum frequencies of LSB, and* **(iv)** *width of the channel.* *(Electronics & Comm. Engg. IERE, London)*

Solution. (*i*) Width of sideband = 5000 – 100 = 4900 Hz

(*ii*) Max. frequency of USB

= 1000 + 5 = 1005 kHz

Min. frequency of *USB*

= 1000 + 0.1 = 1000.1 kHz

(*iii*) Max. frequency of *LSB*

= 1000 – 0.1 = 999.9 kHz

Min. frequency of *LSB*

= 1000 – 5 = 995 kHz

(*iv*) Width of channel = 1005 – 995 = 10 kHz

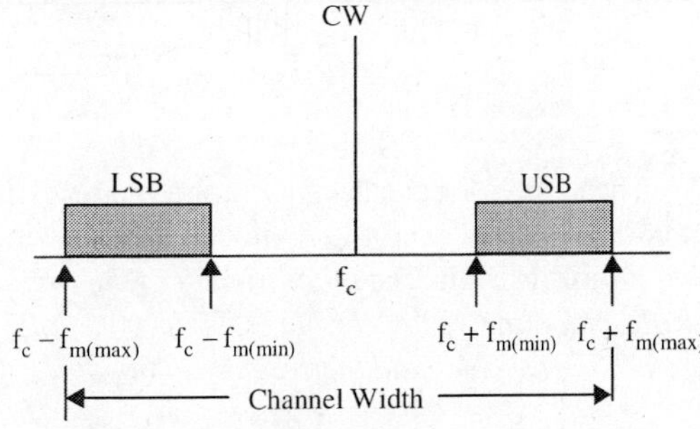

Fig. 32-11.

32-13. Mathematical Analysis of a Modulated Carrier Wave

The equation of an unmodulated carrier wave is

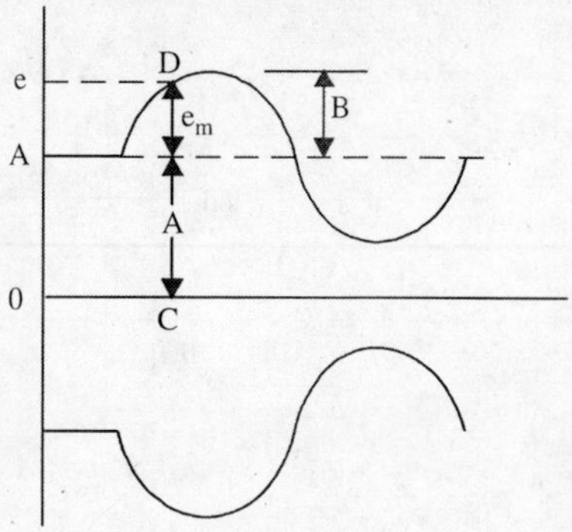

Fig. 32-12.

$$e_c = E_c \sin 2\pi f_c t = \text{A} \sin 2\pi f_c t = \text{A} \sin \omega t$$

where, A is the constant amplitude of the carrier wave and $\omega = 2\pi f_c$.

Let the equation of the single-frequency sinusoidal modulating signal be

$$e_m = E_m \sin 2\pi f_m t = B \sin 2\pi f_m t$$

$$= B \sin Pt \qquad \text{— where, } p = 2\pi f_m$$

As seen from Fig. 32-12, the amplitude of the modulated carrier wave at any instant is

$$= \text{A} + e_m \qquad (\because \text{ A is constant})$$

$$= (A + B \sin Pt)$$

Hence, its instantaneous value is given by

$$e = (A + B \sin pt)\sin \omega t = A \sin \omega t + B \sin \omega t . \sin pt$$

$$= A \sin \omega t + \frac{B}{2} . 2 \sin \omega t . \sin pt = A \sin \omega t + \frac{B}{2} \cos(\quad p)t \quad \cos(\quad p)t) \; *$$

$$= A \sin \omega t + \frac{B}{2} \cos (\omega - p)\, t - \cos (\omega + p) t$$

$$= A \sin 2 \pi f_c t + \frac{B}{2} \cos 2\pi (f_c - f_m) t - \frac{B}{2} \cos 2\pi (f_c + f_m) t$$

As seen from Art. 32-10, $m = B/A$ or $B = mA$

$$\therefore \; e = \text{A} \sin 2\pi f_c t + \frac{mA}{2} \cos 2\pi (f_c - f_m) t - \frac{mA}{2} \cos \pi 2 (f_c + f_m) t$$

It is seen that the modulated wave contains three components :

(*i*) $A \sin 2\pi f_c t$ — the original carrier wave

(*ii*) $\frac{mA}{2} \cos 2\pi (f_c + f_m) t$ — upper side frequency

(*iii*) $\frac{mA}{2} \cos 2\pi (f_c - f_m) t$ — lower side frequency

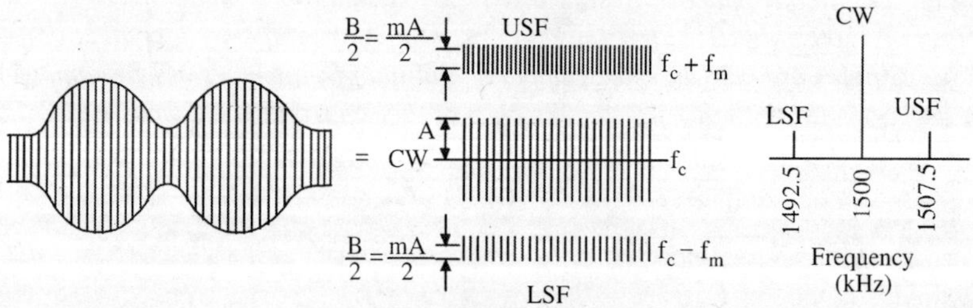

Fig. 32-13.

These three frequencies are not a mathematical fiction but they actually exist. In fact, with the help of a narrow band filter, we can separate side frequencies from the carrier wave.

Example 32-6. *The tuned circuit of the oscillator in an AM transmitter uses a 40 mH coil and a 1 nF capacitor. If the carrier wave produced by the oscillator is modulated by audio frequencies upto 10 kHz, calculate the frequency band occupied by the side bands and channel width.*

(Electronics & Telecom. Engg., Calcutta Univ. 1990)

* $2 \sin A . \sin B = \cos (A - B) - \cos (A + B)$

Solution. The resonant frequency is given by

$$f_c = \frac{1}{2\pi\sqrt{LC}} = \frac{1}{2\pi\sqrt{40\times10^{-6}\times1\times10^{-9}}} \cong 796\,\text{kHz}$$

The frequency range occupied by the sidebands is from **796 kHz to 806 kHz**. The channel width = 2 ×10 = **20 kHz**

32-14. Power Relations in an AM Wave

As discussed in Art. 32-13, a modulated carrier wave consists of the following three components :

(i) Original carrier wave of amplitude A

(ii) USF wave of amplitude ($mA/2$)

(iii) LSF wave of amplitude ($mA/2$)

Now, power radiated out by a wave through an antenna is proportional to (amplitude)^2.

Carrier power, $P_C \propto A^2 = KA^2$

USB power, $P_{USB} \propto \left(\frac{B}{2}\right)^2 = \frac{KB^2}{4}$; LSB power, $P_{LSB} \propto \left(\frac{B}{2}\right)^2 = \frac{KB^2}{4}$

Total sideband power $P_{SB} = 2\times\frac{KB^2}{4} = \frac{KB^2}{2}$

Total power radiated out from the antenna is $P_T = P_c + P_{SB} = KA^2 + \frac{KB^2}{2}$

Substituting, $B = mA$, we get $P_T = KA^2 + \frac{K}{2}(mA)^2 = KA^2\left(1+\frac{m^2}{2}\right)$

Now $P_C = KA^2$

(i) $\therefore\ P_T = P_C\left(1+\frac{m^2}{2}\right) = P_C\left(\frac{2+m^2}{2}\right)$ *(ii)* $P_C = P_T\left(\frac{2}{2+m^2}\right)$

(iii) $P_{SB} = P_T - P_C = P_C\left(1+\frac{m^2}{2}\right) - P_C = \frac{m^2}{2}P_C = \left(\frac{m^2}{2+m^2}\right)P_T$

(iv) $P_{USB} = P_{LSB} = \frac{1}{2}P_{SB} = \frac{m^2}{4}P_C = \frac{1}{2}\left(\frac{m^2}{2+m^2}\right)P_T$

Let us consider the case when m = 1 or 100%

(i) $P_T = 1.5\,P_C = 1.5\times$ carrier power *(ii)* $P_C = \frac{2}{3}P_T = \frac{2}{3}\times$ total power radiated

(iii) $P_{USB} = P_{LSB} = \frac{1}{4}P_C$ = 25% of carrier power

(iv) $P_{SB} = \frac{1}{2}P_C$ = 50% of carrier power *(v)* $\frac{P_{USB}}{P_T} = \frac{0.25\,P_C}{1.5\,P_C} = \frac{1}{6}$

It means that **single side-band contains 1/6th of the total power radiated out by the transmitter.** That is why single-side band (SSB) transmission is more power efficient.

Example 32-7. *The total power content of an AM wave is 1500 W. For a 100 percent modulation, determine*

(i) power transmitted by carrier, *(ii) power transmitted by each side band.*

(Electronics & Comm. Engg., Kerala Univ. 1991)

Solution. P_T = 1500 W, P_C = ? P_{USB} = ? P_{LSB} = ?

(i) $P_C = P_T\left(\frac{2}{2+m^2}\right) = P_T\left(\frac{2}{2+1^2}\right) = \frac{2}{3}P_T = \frac{2}{3}\times1500 = 1000$ W

(*ii*) $P_{USB} = P_{LSB} = \frac{m^2}{4} P_C = \frac{1^2}{4} \times 1000 = 250 \text{ W}$

As seen, $P_{USB} = \frac{1}{6} P_T$

Example 32-8. *The total power content of an AM wave is 2.64 kW at a modulation factor of 80%. Determine the power content of*

(*i*) *carrier,* (*ii*) *each sideband.* (*Electronics & Comm., Roorkee Univ. 1991*)

Solution. (*i*) $P_C = P_T \left(\frac{2}{2+m^2}\right) = 2.64 \times 10^3 \left(\frac{2}{2+0.8^2}\right) = 2000 \text{ W}$

(*ii*) $P_{USB} = P_{LSB} = \frac{m^2}{4} P_c = \frac{0.8^2}{4} \times 2000 = 320 \text{ W}$

Example 32-9. *A transmitter used for radio telephone has an unmodulated carrier power of 10 kW and can be modulated to a maximum of 80 percent by a single-frequency signal before overloading. Find the value to which carrier power can be increased if a 50 percent modulation limit is imposed.* (*Electronics & Comm. Engg. Andhra Univ 1988*)

Solution. $P_T = P_C \left(\frac{2+m^2}{2}\right) = 10\left(\frac{2+0.8^2}{2}\right) = 13.2 \text{ kW}$

Now, when $m = 0.5$, P_T is still 13.2 kW. Hence, new value of carrier power is given by

$$13.2 = P_C \left(\frac{2+0.5^2}{2}\right); \quad P_C = 11.73 \text{ kW}$$

It is seen that P_C can be increased from 10 kW to 11.73 kW with a total power limit of 13.2 kW and $m = 0.5$

Example 32-10. *A certain transmitter radiates 10 kW of power with the carrier unmodulated and 11.8 kW with the carrier sinusoidally modulated.*

(*a*) *Find the modulation factor,*

(*b*) *If another wave modulated to 40% is also transmitted, calculate the radiated power.*

Solution. (*a*) $11.8 = 10\,(1 + m^2/2)$; $m = 0.6$ or 60%

(*b*) $P_T = 10\left(1 + \frac{0.6^2}{2} + \frac{0.4^2}{2}\right) = 12.6 \text{ kW}$

32-15. Forms of Amplitude Modulation

As shown in Art. 32-11, **one carrier** and **two sidebands** are produced in AM generation. It is found that it is not necessary to transmit all these signals to enable the receiver to reconstruct the original signal. Accordingly, we may attenuate or altogether remove the carrier or any one of the sidebands without affecting the communication process. The advantages would be

(*i*) less transmitted power, and (*ii*) less bandwidth required

The different suppressed component systems are :

(*a*) DSB-SC

It stands for ***double-sideband suppressed carrier*** system [Fig. 32-14 (*a*)]. Here, carrier component is suppressed thereby saving enormous amount of power. As seen from Art. 32-14, carrier signal contains 32-7 per cent of the total transmitted power for $m = 1$, Hence, power saving amounts to 32-7% at 100% modulation.

(*b*) SSB-TC

As shown in Fig. 32-14 (*b*), in this case, one sideband is suppressed but the other sideband and carrier are transmitted. It is called ***single sideband transmitted carrier*** system. For $m = 1$, power saved is 1/6 of the total transmitted power.

(c) SSB-SC

This is the **most dramatic suppression of all** because it suppresses one sideband and the carrier and transmits only the remaining sideband as shown in Fig. 32-14 (*c*). In the standard or double-sideband full-carrier (*DSB.FC*) *AM*, carrier conveys **no information but contains maximum power**. Since, the two sidebands are exact images of each other, they carry the same audio information. Hence, **all information is available in one sideband only**. Obviously carrier is ***superfluous*** and one side band is ***redundant.*** Hence, one sideband and the carrier can be discarded with no loss of information. The result is *SSB* signal. The advantage of *SSB-SC* system are as follows ;

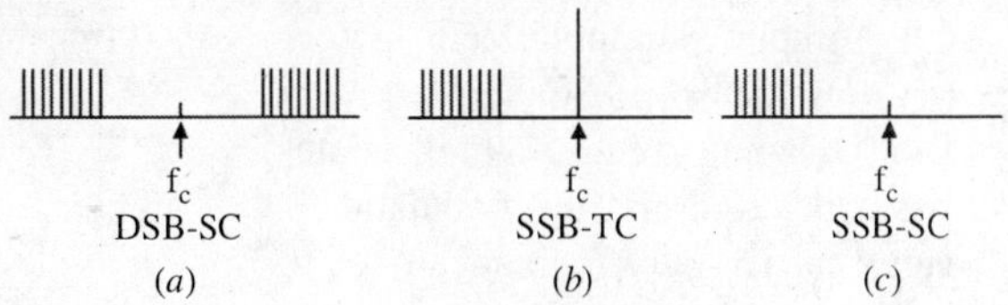

Fig. 32-14.

(*i*) Total saving of 83.3% in transmitted power (32-7% due to suppression of carrier wave and 16.6% due to suppression of one sideband). Hence, power is conserved in an SSB transmitter.

(*ii*) Bandwidth required is reduced by half, *i.e.*, 50%. Hence, twice as many channels can be multiplexed in a given frequency range.

(*iii*) The size of power supply required is very small. This fact assumes vital importance particularly in a spacecraft.

(*iv*) Since, the SSB signal has narrower bandwidth, a narrower passband is permissible within the receiver, thereby limiting the noise pick up.

However, the main reason for wide spread use of *DSB-FC* (rather than *SSB-SC*) transmission in broadcasting is the relative simplicity of its modulating equipment.

32-16. Generation of SSB

Three main systems are employed for the generation of (*iii*) *SSB*.

(*i*) Filter method, (*ii*) Phase cancellation method and (*iii*) The 'third' method.

All these methods use some form of balanced modulator to suppress the carrier.

Example 32-11. *In an AM wave, calculate the power saving when the carrier and one sideband are suppressed corresponding to*

(*i*) $m = 1$ (*ii*) $m = 0.5$ *(Electronics and Comm. Engg., Osmania Univ 1986)*

Solution. (*i*) When $m = 1$

$$P_T = P_C\left(1+\frac{m^2}{2}\right) = 1.5\,P_C\,; \qquad P_{LSB} = P_{USB} = \frac{m^2}{4}\,P_C = 0.25\,P_C$$

$$\therefore \quad \text{saving} = P_T - P_{USB} = 1.5\,P_c - 0.25\,P_c = 1.25\,P_c$$

$$\text{saving} = \frac{1.25}{1.5 P_C} \times 100 = 83.3\%$$

(*ii*) When $m = 0.5$

$$P_T = P_C\left(1+\frac{m^2}{2}\right) = P_C\left(\frac{2+0.5^2}{2}\right) = 1.125 P_C$$

$$P_{USB} = P_{LSB} = \frac{m^2}{4}\,P_C = \frac{0.5^2}{4}\,P_C = 0.0625\,P_C$$

$$\therefore \quad \text{saving} = \frac{1.125\,P_c - 0.0625\,P_C}{1.125\,P_C} \times 100 = 94.4\%$$

32-17. Methods of Amplitude Modulation

There are two methods of achieving amplitude modulation :

(*i*) Amplifier modulation,

(*ii*) Oscillator modulation.

Block diagram of Fig. 32-15 illustrates the basic idea of amplifier modulation.

Here, carrier and *AF* signal are fed to an amplifier and the result is an *AM* output . The modulation process takes place in the active device used in the amplifier.

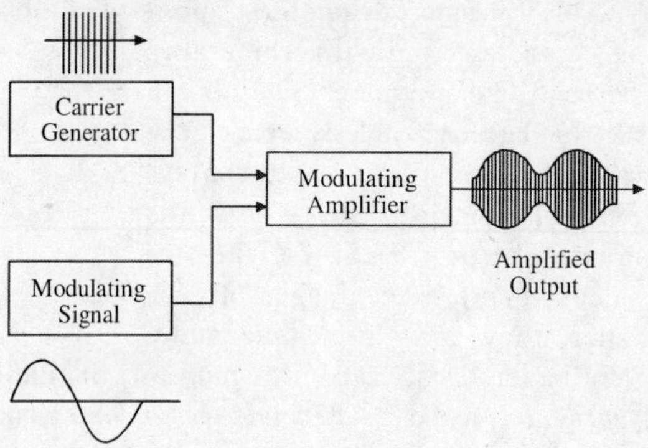

Fig. 32-15.

32-18. Block Diagram of an AM Transmitter

Fig. 32-16 shows the block diagram of a typical transmitter. The carrier wave is supplied by a crystal-controlled oscillator at the carrier frequency. It is followed by a tuned buffer amplifier and an *RF* output amplifier.

The source of *AF* signal is a microphone. The audio signal is amplified by a low level audio amplifier and, finally, by a power amplifier. It is then combined with the carrier to produce a modulated carrier wave which is ultimately radiated out in the free space by the transmitter antenna as shown.

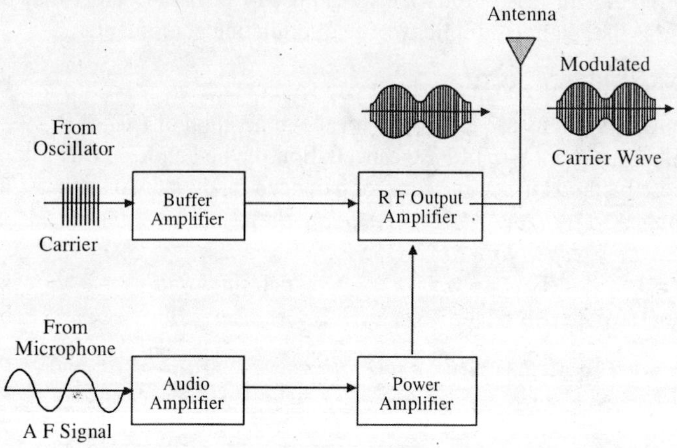

Fig. 32-16.

32-19. Modulating Amplifier Circuit

It is shown in Fig. 32-17. The carrier signal from a crystal oscillator is coupled to the base of Q through transformer T_1. The low-frequency modulating signal is also coupled to the base of Q by capacitor C_1. The voltage divider $R_1 - R_2$, as usual, provides proper forward bias for the transistor whereas C_2 and C_3 are bypass capacitors.

The modulating signal applied across R_2 causes variations in the base bias in step with its own variations. The amplitude of the transistor *RF* output changes according to the changes in *AF* modulating signal. The $L_1 C_4$ circuit is kept tuned to the carrier frequency. The amplitude-modulated carrier or *RF* current in L_1 induces similar current in L_2 which is connected to an antenna. Finally, the *AM* carrier wave is radiated out in space by the transmitting antenna.

32-20. Frequency Modulation OR CAD Simulation Diagram.

As the name shows, in this modulation, it is only the **frequency of the carrier which is changed and not its amplitude.**

The amount of change in frequency is determined by the amplitude of the modulating signal whereas rate of change is determined by the frequency of the modulating signal. As shown in Fig. 32-17, in an *FM* carrier, information (or intelligence) is carried as variations in its frequency. As seen, frequency of the modulated carrier increases as the signal amplitude increases but decreases as the signal amplitude decreases. It is at its highest frequency (point *H*) when the signal amplitude is at its maximum positive value and is at its lowest frequency (point *L*) when signal amplitude has maximum negative value. When signal amplitude is zero, the carrier frequency is at its normal frequency f_0 (also called ***resting*** or ***centre*** frequency.)

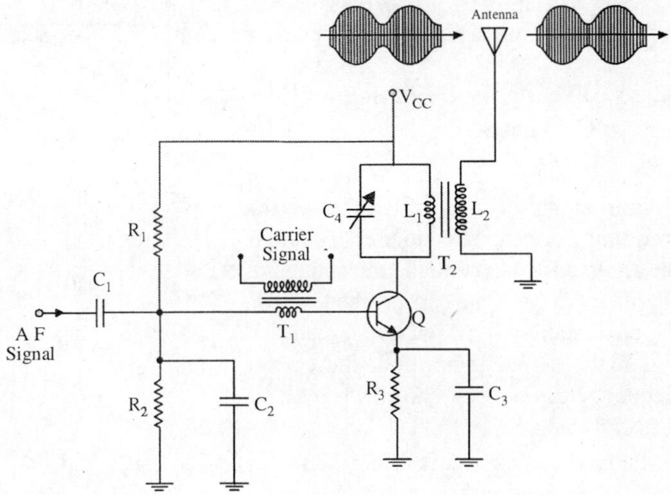

Fig. 32-17.

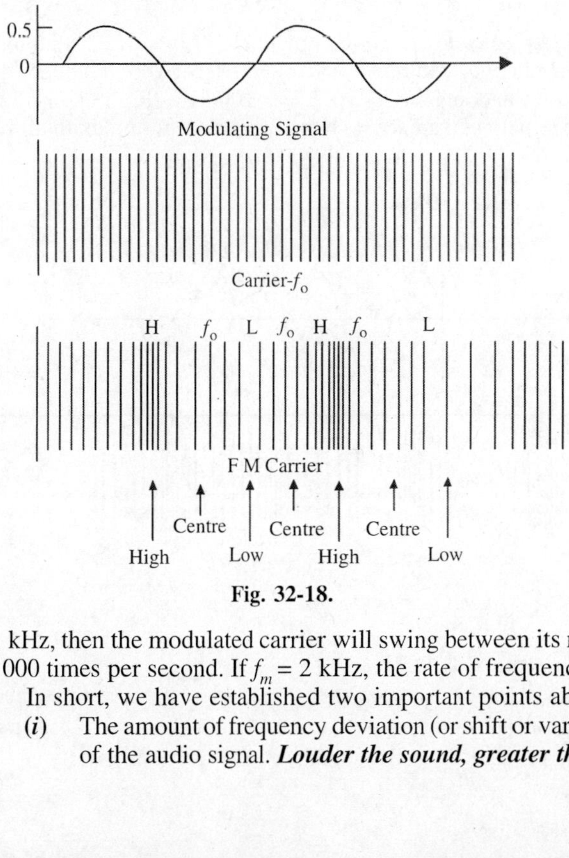

Fig. 32-18.

The fact that amount of change in frequency depends on signal amplitude is illustrated in Fig. 32-19, where *R* stands for resting frequency. Here, signal amplitude is almost double of that in Fig. 32-18 ***though its frequency is the same.*** This louder signal causes greater frequency change in modulated carrier as indicated by increased bunching and spreading of the waves as compared with relatively weaker signal of Fig. 32-18.

The rate at which frequency shift takes place depends on the signal frequency as shown in Fig. 32-20. For example, if the modulating signal is 1 kHz, then the modulated carrier will swing between its maximum frequency and lowest frequency 1000 times per second. If f_m = 2 kHz, the rate of frequency swing would be twice as fast :

In short, we have established two important points about the nature of frequency modulation:

(*i*) The amount of frequency deviation (or shift or variation) depends on the amplitude (loudness) of the audio signal. ***Louder the sound, greater the frequency deviation*** and ***vice-versa.***

However, for the purposes of *FM* broadcasts, it has been internationally agreed to restrict maximum deviation to **75 kHz** on each side of the centre frequency for sounds of maximum loudness. Sounds of lesser loudness are permitted ***proportionately less frequency deviation.***

(*ii*) The rate of frequency deviation depends on the **signal frequency**.

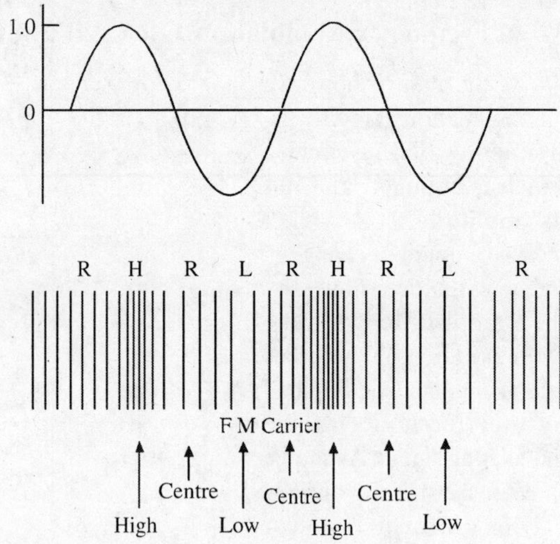

Fig. 32-19.

32-21. Frequency Deviation and Carrier Swing

The frequency of an *FM* transmitter without signal input is called the **resting frequency** or **centre frequency** (f_0) and is the allotted frequency of the transmitter. In simple words, it is the carrier frequency on which a station is allowed to broadcast. When the signal is applied, the carrier frequency deviates up and down from its resting value f_0.

This change or shift either above or below the resting frequency is called frequency deviation (Δ_f).

The total variation in frequency from the lowest to the highest is called carrier swing (CS). Obviously,

carrier swing = 2 × frequency deviation of $CS = 2 \times \Delta_f$

A maximum frequency deviation of 75 kHz is allowed for commercial *FM* broadcast stations in the 88 to 168 MHz *VHF* band. Hence, *FM* channel width is 2 × 75 = 150 kHz. Allowing a 25 kHz guard band on either side, the channel width becomes = 2(75 + 25) = 200 kHz (Fig. 32-21). This guard band is meant to prevent interference between adjacent channels. However, a maximum frequency deviation of 25 kHz is allowed in the sound portion of the TV broadcast.

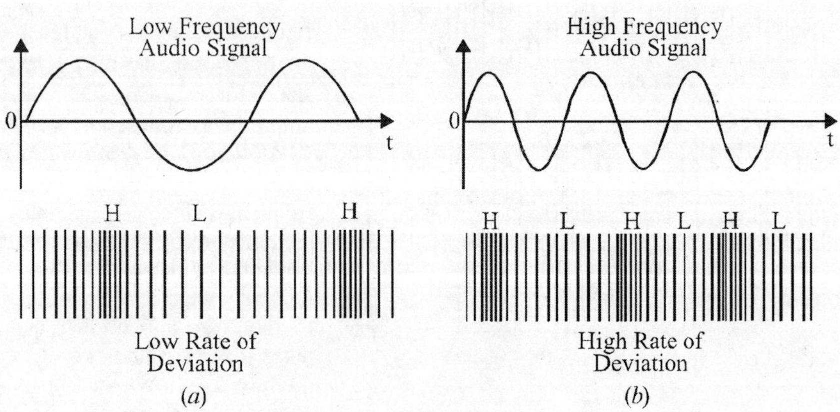

Fig. 32-20.

In *FM*, the highest audio frequency transmitted is 15 kHz.

Consider an FM carrier of resting frequency 100 MHz. Since, $(\Delta f)_{max} = 75$ kHz, the carrier frequency can swing from the lowest value of 99.925 MHz to the highest value of 100.075 MHz. Of course, deviations lesser than 75 kHz corresponding to relatively softer sounds are always permissible.

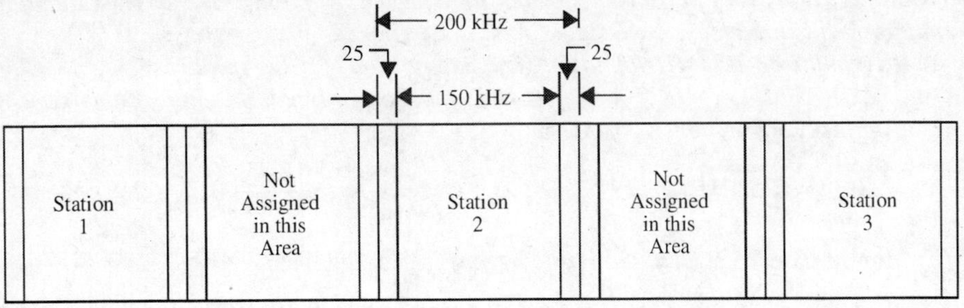

Fig. 32-21.

32-22. Modulation Index

It is given by the ratio $m_f = \dfrac{\text{frequency deviation}}{\text{modulation frequency}} = \dfrac{\Delta f}{f_m}$

Unlike amplitude modulation, this **modulation index can be greater than unity**. By knowing the value of m_f, we can calculate the number of significant sidebands and the bandwidth of the *FM* signal.

32-23. Deviation Ratio

It is the worst-case modulation index in which maximum permitted frequency deviation and maximum permitted audio frequency are used.

$$\therefore \qquad \text{Deviation ratio} = \frac{(\Delta f)}{f_{m(max)}}$$

Now, for *FM* broadcast stations, $(\Delta f)_{max} = 75$ kHz and maximum permitted frequency of modulating audio signal is 15 kHz.

$$\therefore \qquad \text{deviation ratio} = \frac{75\,\text{kHz}}{15\,\text{kHz}} = 5$$

For sound portion of commercial TV deviation ratio $= \dfrac{25\,\text{kHz}}{15\,\text{kHz}} = 1.67$

32-24. Percent Modulation (FM)

When applied to *FM*, this term has slightly different meaning than when applied to *AM*. In *FM*, it is given by the ratio of actual frequency deviation to the maximum allowed frequency deviation.

$$m = \frac{(\Delta f)_{actual}}{(\Delta f)_{max}}$$

Obviously, 100% modulation corresponds to the case when actual deviation equals the maximum allowable frequency deviation. If, in some case, actual deviation is 50 kHz, then

$$m = \frac{50}{75} = \frac{2}{3} = 0.667 = 66.7\%$$

Value of $m = 0$ corresponds to zero deviation, *i.e.*, unmodulated carrier wave. It is seen from the above equation that $m \propto (\Delta f)_{actual}$. It means that when **frequency deviation** (*i.e.*, signal loudness) **is doubled, modulation is doubled**.

Example 32-12. *What is the modulation index of an FM carrier having a carrier swing of 100 kHz and a modulating signal of 5 kHz ?*

Solution. $CS = 2 \times \Delta f$

$$\therefore \qquad \Delta f = \frac{CS}{2} = \frac{100}{2} = 50\text{ kHz} \quad \therefore \quad m_f = \frac{\Delta f}{f_m} = \frac{50}{5} = 10$$

Example. 32-13. *An FM transmission has a frequency deviation of 18.75 kHz. Calculate percent modulation if it is broadcast*

(*i*) *in the 88-108 MHz band* (*ii*) *as a portion of a TV broadcast.*

(Elect. and Comm. Engg., Madurai Kamaraj Univ. 1990)

Solution. (*i*) For this transmission band,

$(\Delta f)_{max} = 75$ kHz $\qquad \therefore \; m = \dfrac{18.75}{75} \times 100 = 25\%$

(*ii*) In this case, $(\Delta f)_{max} = 25$ kHz $\qquad \therefore \; m = \dfrac{18.75}{25} \times 100 = 75\%$

Example 32-14. *An FM signal has a resting frequency of 105 MHz and highest frequency of 105.03 MHz when modulated by a signal of frequency 5 kHz. Determine*

(*i*) *frequency deviation,* (*ii*) *carrier swing,* (*iii*) *modulation index,*
(*iv*) *percent modulation,* (*v*) *lowest frequency reached by the FM wave.*

(Electronics and Comm. Engg., Osmania Univ. 1992)

Solution. (*i*) $\Delta f = 105.03 - 105 = 0.03$ MHz = 30 kHz

(*ii*) $CS = 2 \times \Delta f = 2 \times 30 = 60$ kHz (*iii*) $m_f = \dfrac{30}{5} = 6$

(*iv*) $m = \dfrac{30}{75} \times 100 = 40\%$ (*v*) lowest frequency = 105 – 0.03 = 104.97 MHz

32-25. FM Sidebands

In *FM*, when a carrier is modulated, a number of sidebands are formed.* Though theoretically their number is infinite, their strength becomes negligible after a few sidebands. They lie on both sides of the centre frequency spaced f_m apart as shown in Fig.32-22. Sidebands at equal distances from f_0 have equal amplitudes. If f_0 is the centre frequency and f_m the frequency of the modulating signal, then *FM* carrier contains the following frequencies :

(*i*) f_0 (*ii*) $f_0 \pm f_m$ (*iii*) $f_0 \pm 2f_m$ (*iv*) $f_0 \pm 3f_m$ and so on

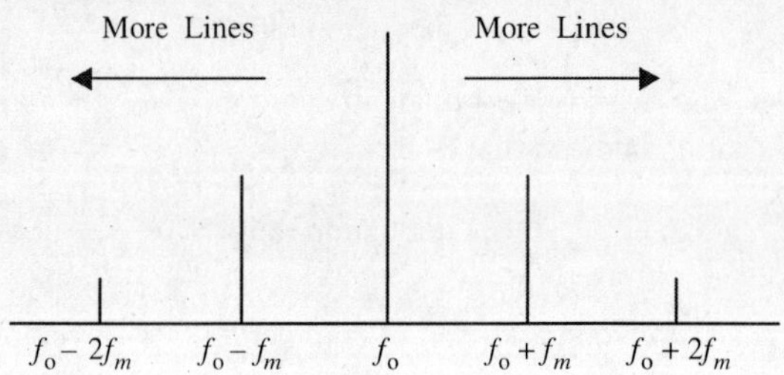

Fig. 32-22.

The bandwidth occupied by the spectrum is $BW = 2nf_m$ where n is the highest order of the significant sideband.

Another approximate expression for spectrum bandwidth is $BW = 2\,(1 + m_f)f_m$

Now, $m_f = \dfrac{\Delta f}{f_m}$, hence $\qquad BW = 2\,(\Delta_f + f_m)$

This expression is based on the assumption that sidebands having amplitudes less than 5% of the unmodulated carrier wave are negligble or when m_f is at least 6.

32-26. Modulation Index and Number of Sidebands

It is found that the number of sidebands

(*i*) depends ***directly*** on the amplitude of the modulating signal,

(*ii*) depends ***inversely*** on the frequency of the modulating signal.

Since, frequency deviation is directly related to the amplitude of the modulating signal, the above two factors can be combined in one factor called modulation index.

Hence, number of sidebands depends on $m_f = \Delta f \,/\, f_m$

* An AM signal has only two side frequencies for each modulating frequency (Art. 32-11)

Obviously, the number of pairs of sidebands

(*i*) *increases* as frequency deviation (or amplitude of modulating signal) *increases.*

(*ii*) *increases* as the modulating signal frequency *decreases.*

Table No. 32-2

m_f	0.5	1	2	3	4	5	6	7
n	2	3	4	6	7	8	9	11
$BW = 2nf_m$	$4f_m$	$6f_m$	$8f_m$	$12f_m$	$14f_m$	$16f_m$	$18f_m$	$22f_m$
m_f	8	9	10	11	12	13	14	15
n	12	13	14	15	16	17	18	19
$BW = 2nf_m$	$24f_m$	$26f_m$	$28f_m$	$30f_m$	$32f_m$	$34f_m$	$36f_m$	$38f_m$

Table 32-2 summarizes the relation between the number (n) of sidebands on either side of the carrier and the corresponding bandwidth of the spectrum.

Example 32-15. *A 5 kHz audio signal is used to frequency-modulate a 100 MHz carrier causing a frequency deviation of 20 kHz. Determine*

(*i*) *modulation index* (*ii*) *bandwidth of the FM signal.*

Solution. (*i*) $m_f = \dfrac{\Delta f}{f_m} = \dfrac{20}{5} = 4$

As seen from Table 32-2 $BW = 14f_m = 14 \times 5 = 70$ kHz

Note. We cannot use the alternate expression for *BW* given in Art. 32-25 above because $m_f < 6$

Example 32-16. *In an FM circuit, the modulation index is 10 and the higest modulation frequency is 20 kHz. What is the approximate bandwidth of the resultant FM signal ?*

(Applied Electronics, Mumbai Univ. 1990)

Solution. Since, the value of m_f is more than 6, we will use the expression

$$BW = 2\,(\Delta_f + f_m)$$

Now, $m_f = \dfrac{\Delta f}{f_m}$ or $10 = \dfrac{\Delta f}{20}$ $\therefore$ $\Delta_f = 200$ kHz

$\therefore$ $BW = 2\,(200 + 20) = 440$ kHz

32-27. Mathematical Expression for FM Wave

The unmodulated carrier is given by $e_c = A \sin 2\pi f_0 t$

The modulating signal is given by $e_m = B \sin 2\pi\, f_m t$

The modulated carrier frequency f swings around the resting frequency f_0, thus

$$f = f_0 + \Delta f .\ \sin 2\pi f_m t$$

Hence, equation for the frequency-modulated wave becomes

$$e = A \sin 2\pi f t = \text{A} \sin [2\pi(f_0 + \Delta f.\ \sin 2\pi f_m t)t]$$

$$= A \sin\left(2\pi f_0\, t + \frac{\Delta f}{f_m} + \cos 2\pi f_m t\right) = A \sin\,(2\pi f_0 t + m_f \cos 2\pi f_m t)$$

32-28. Demodulation or Detection

When the *RF* modulated waves, radiated out from the transmitter antenna, after travelling through space, strike the receiving aerials, they induce very weak *RF* currents and voltages in them. If these high-frequency currents are passed through headphones or loudspeakers, they produce no effect on them because all such sound-producing devices are unable to respond to such high frequencies due to large inertia of their vibrating discs, *etc*. Neither will such *RF* currents produce any effect on human ear because their frequencies are much beyond the audible frequencies (20 to 20,000 Hz approximately). Hence, it is necessary to demodulate them first in order that the sound-producing devices may be actuated by audio-frequency current similar to that used for modulating the carrier wave at the broadcasting station.

This process of ***recovering AF signal from the modulated carrier wave is known as demodulation or detection.***

The demodulation of an *AM* wave involves two operations :

(*i*) rectification of the modulated wave and

(*ii*) elimination of the *RF* component of the modulated wave.

However, demodulation of an *FM* wave involves three operations **(*i*)** conversion of frequency changes produced by modulating signal into corresponding amplitude changes, **(*ii*)** rectification of the modulating signal, and **(*iii*)** elimination of *RF* component of the modulated wave.

32-29. Essentials of AM Detection

For recovering the *AF* waveform from modulated wave (a mixture of *AF* wave and *RF* carrier), it is essential to find some way of reducing (or better, eliminating) one half of the modulated wave. The result of this elimination (or rectification) would be that the average value of the wave would not be zero because, now, the impulse would be all in one direction as shown in Fig. 32-23. If this new wave is now passed through a headphone shunted by a suitable capacitor, then *AF* wave will pass through the headphone whereas the *RF* wave will be by-passed by the capacitor (because the high inductance of magnet coils of the headphones will offer tremendous impendance to *RF* currents). Hence, two will become separated.

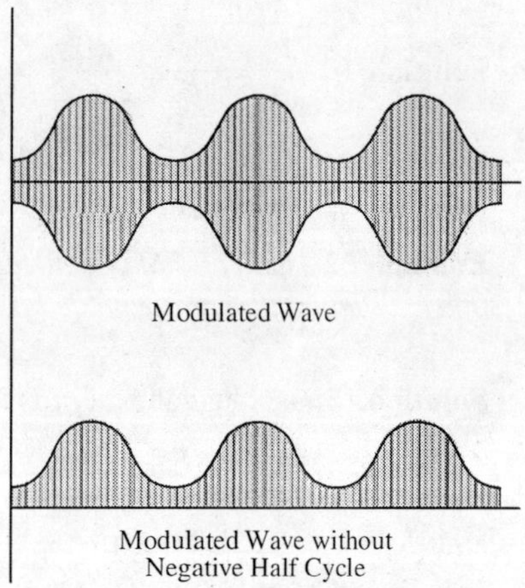

Fig. 32-23.

32-30. Diode Detector for AM Signals

Diode detection is also known as ***envelope-detection*** or ***linear detection***. In appearance, it looks like an ordinary half-wave rectifier circuit with capacitor input as shown in Fig. 32-24. It is called **envelope detection** because it recovers the *AF* signal envelope from the composite signal. Similarly, diode detector is called **linear detector** because its output **is *proportional to the voltage of the input* signal.***

Circuit Action

Of the various *RF* voltages induced in the receiver aerial, only those having the ***same*** frequency as the resonant frequency of *LC* circuit are tuned in due to electromagnetic induction between coils L_1 and L. By varying C, the resonant frequency of the *LC* circuit can be varied and hence *RF* signal of any desired frequency can be tuned in. This input signal is rectified by the diode and passed on to the low-pass filter RC_1.

---

* A detector circuit in which the rectified output is proportional to the square of the input signal voltage is called square-law or non-linear detector.

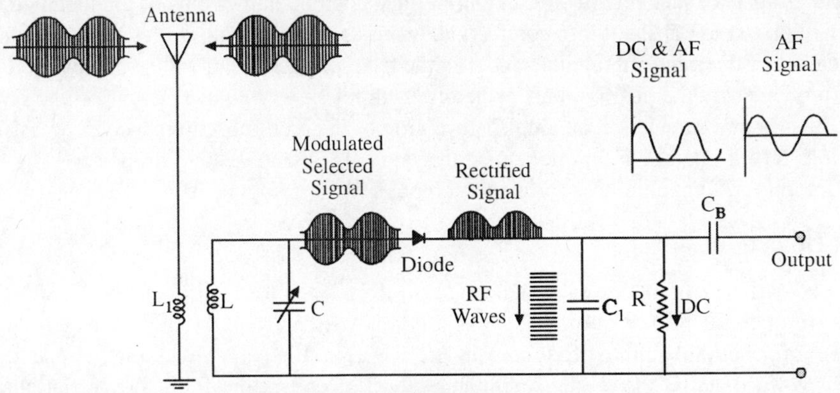

Fig. 32-24.

The *RF* carrier wave is filtered out by low-reactance capacitor C_1 which is called *RF* filter capacitor or *RF* by-pass capacitor. The dc component of the remaining signal is shunted out through *R* because it cannot pass through blocking capacitor C_B. But the low, frequency *AF* signal can easily get through C_B and becomes available across the output. When passed through a suitable device, say, a headphone, the original sound can be heard.

Advantages

Diode detectors are extensively used in AM broadcast receivers because they have the following advantages :

- They can handle comparatively large input signals;
- They can be operated as linear or power detectors;
- They rectify with negligible distortion and hence, have good linearity;
- They are well-adopted for use in simple automatic-gain control circuits.

Disadvantages

However, the disadvantages are that

- They do not have the ability to amplify the rectified signal by themselves as is done by a transistor detector (Art. 32-31). However, it is not a very serious drawback since signal amplification can be affected both before and after rectification;
- While conducting, the diode consumes some power which reduces the *Q* of its tuned circuit as well as its gain and selectivity.

32-31. Transistor Detectors for AM Signals

Transistors can be used as detector amplifiers, *i.e.*, both for rectification and amplification. As shown in Fig. 32-25, the *RF* signal is applied at the base-emitter junction where rectification takes place. The amplification of the recovered signal takes place in the emitter-collector circuit.

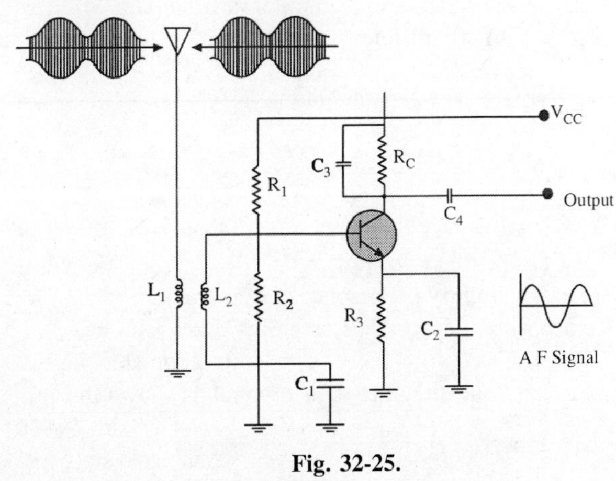

Fig. 32-25.

It should be noted that when used as a detector, the transistor has to be biased at cut-off or operated as class-B. As is usual, C_2 ac grounds R_3 whereas voltage divider R_1 — R_2 establishes proper bias which holds the transistor just at cut-off. Obviously, only positive half-cycles of the modulated *RF* input signal will drive the transistor into conduction, whereas negative half-cycles would be cut-off. Thus, rectification of the input signal takes place in the base-emitter circuit. The small variations in base current produced by the positive half-cycle of the input signal produce large (β time) variations in the collector current. Hence, an amplified version of the rectified signal appears across R_C – C_3 combination. C_3 eliminates *RF* component so that only *AF* signal voltage drops across R_C. The capacitor C_4 permits *AF* signal to pass through but blocks its dc component.

32-32. FM Detection

As discussed earlier, an *FM* carrier signal contains information (or intelligence we wish to convey) in the form of frequency variations above and below the centre frequency of the carrier. For recovering the information, we must first convert the *FM* signal in such a way that it appears as a modulated *RF* voltage across the diode. A simple method of converting frequency variations into voltage variations is to make use of the principle that reactance (of coil or capacitor) varies with frequency. When an *FM* signal is applied to an inductor, the current flowing through it varies in amplitude according to the changes in frequency of the applied signal. Now, changes in frequency of the *FM* signal depend on the amplitude of the modulating *AF* signal. Hence, the current in the inductor varies as per the amplitude of the original modulating signal. In this way, frequency changes in *FM* signal are converted into amplitude changes in current. These changes in current when passed through a resistor produce corresponding changes in voltage.

Hence, we find that ultimately, frequency variations in *FM* signal are converted into voltage changes. Also, there exists a linear relation between the two – something essential for distortion-less demodulation.

FM demodulation may be carried out with the help of (*i*) ratio detector, and (*ii*) quadrature detector.

32-33. Quadrature Detector

This detector depends on the frequency/phase relationship of a tuned circuit. It uses only one tuned circuit and is becoming increasingly popular in the integrated *FM* strips.

Theory

Let us first consider the general principle. A sinusoidal current is given by the equation

$$i = I_m \sin\theta = I_m \sin \omega t$$

Suppose, it flows through a circuit shown in Fig. 32-27 (*a*). The voltage V_L across the inductor (assumed pure) leads the current *I* by 90°

$$\therefore \quad V_L = V_L \cos \omega t$$

The voltage V_z across the parallel tuned circuit will be in phase with *I* at resonance. However, at frequencies slightly different (±1%) from the resonant frequency, the phase angle *f* will be given by

$$\tan\phi = \frac{yQR}{R} = y\,Q \quad \text{(assuming } f_0/f \cong 1\text{).} \qquad \text{— Fig. 32-26}$$

Here, $$\frac{\omega}{\omega_0} - \frac{\omega_0}{\omega} = \frac{f}{f_0} - \frac{f_0}{f} \cong \frac{2\Delta_f}{f_0}$$

where, $\Delta f = f - f_0$

f_0 = resonant frequency

f = slightly off-resonance frequency

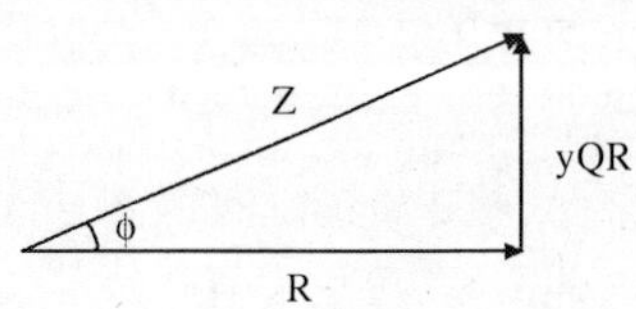

Fig. 32-26.

Hence, equation for V_z is given by $V_z = V_z \sin(\omega t - \phi)$.

When the two voltages V_L and V_z are applied as inputs to a multiplier, the output voltage V_0 is found to be proportional to their product as shown in Fig. 32-27(*b*).

$$V_0 \propto V_L . V_z \propto \cos \omega t . \sin(\omega t - \phi)$$

$$\propto \sin(2\omega t - \phi) - \sin\phi$$

A low-pass filter is used to reject the double frequency component sin $(2\omega t - \phi)$ and select only the low-frequency sin ϕ component.

$\therefore \quad V_0 \propto \sin\phi \propto \tan\phi$ — since, ϕ is very small

$\propto yQ \propto y \propto 2.\dfrac{\Delta f}{f_0}$ — where, V_m is the modulating AF voltage.

Now, $\Delta_f = f - f_0 \propto k\, V_m$

$\therefore \quad V_0 \propto V_m$

It shows that output voltage V_0 is proportional to the original modulating signal voltage.

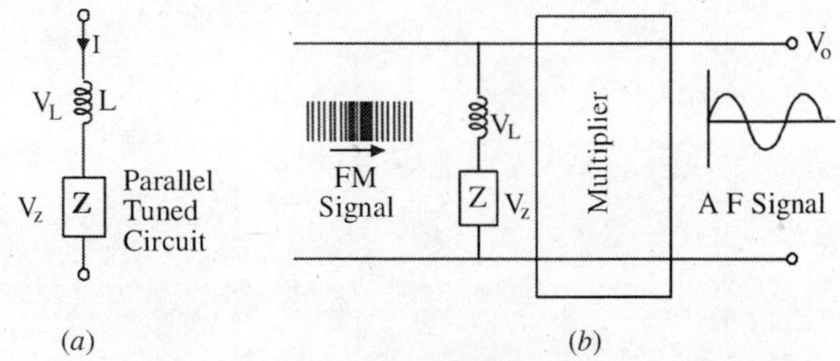

Fig. 32-27.

32-34. Frequency Conversion

Let us first consider its necessity for radio broadcasting purposes and the basic principle of heterodyning action.

(*i*) Need

It is very difficult to design amplifiers which give uniformly high gain over a wide range of radio frequencies used in commercial broadcast stations. However, it is possible to design amplifiers which can provide high-gain uniform amplification over a narrow band of comparatively lower frequencies called intermediate frequencies (*IF*). Hence, it is necessary to convert the modulated *RF* carrier into modulated *IF* carrier by using a frequency converter.

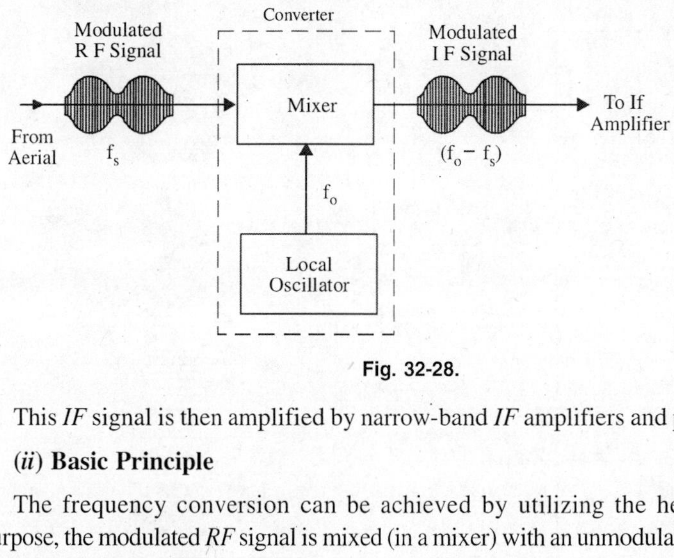

Fig. 32-28.

This *IF* signal is then amplified by narrow-band *IF* amplifiers and passed on to the *AM* detectors.

(*ii*) Basic Principle

The frequency conversion can be achieved by utilizing the heterodyne principle. For this purpose, the modulated *RF* signal is mixed (in a mixer) with an unmodulated *RF* signal produced by local oscillator as shown in Fig. 32-28.

The oscillator and the mixer may be either two separate devices or may be combined into one device called converter. The process of combining two ac signals of different frequencies in order to obtain a signal of new frequency is called heterodyning action.

(*iii*) Heterodyning Action

Suppose the carrier signal of frequency f_s is heterodyned with another signal of frequency f_0, then two additional signals are produced whose frequencies are :

(*i*) $f_0 + f_s$—the sum component (*ii*) $f_0 - f_s$— the difference component

Usually, the sum frequency is removed by bandpass filtering. The difference frequency (also called ***beat frequency***) is retained and forms the *IF* frequency in *AM* receivers.

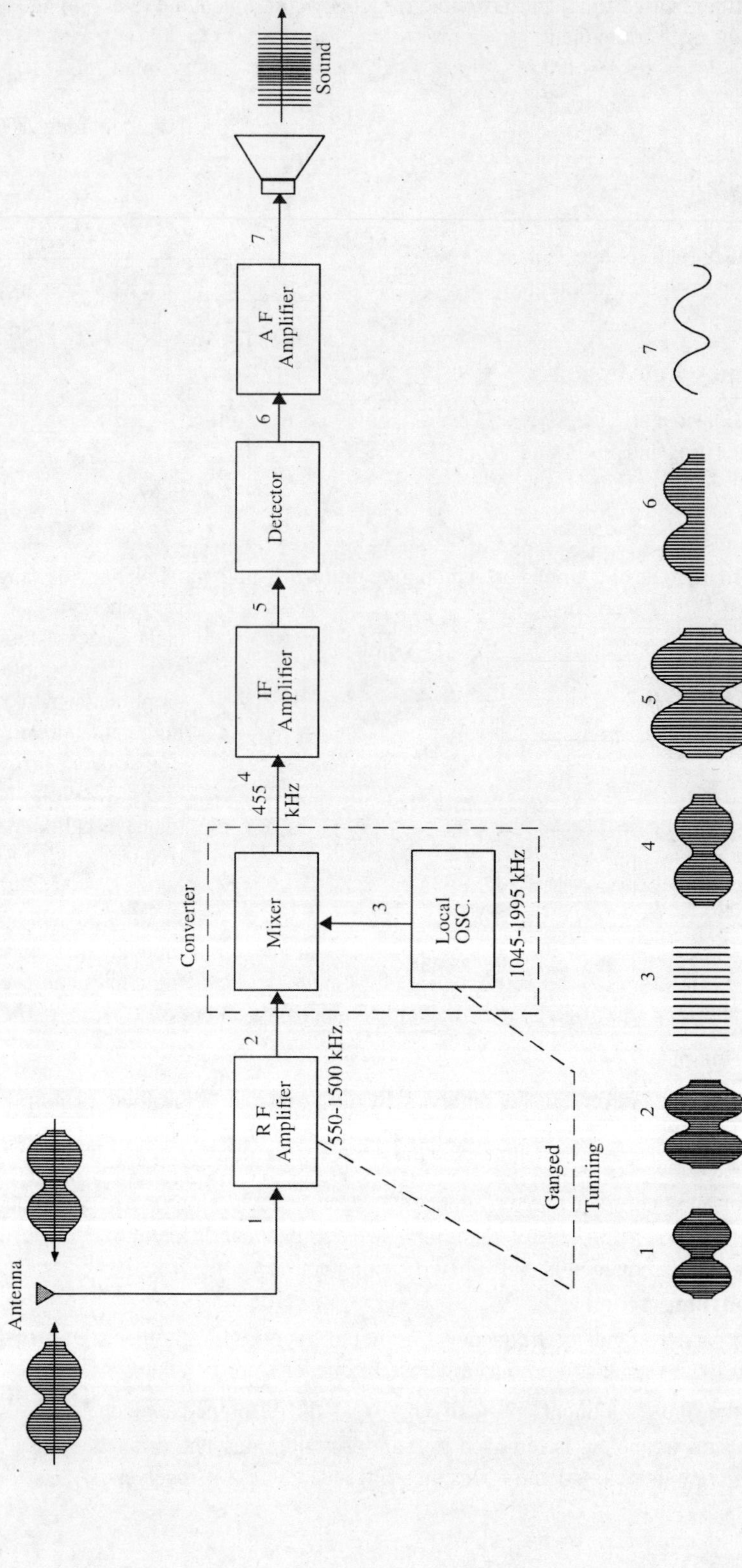
Antenna
1
R F Amplifier
550-1500 kHz
2
Converter
Mixer
3
Local OSC.
1045-1995 kHz
Ganged Tunning
455 kHz
4
IF Amplifier
5
Detector
6
A F Amplifier
7
Sound
1
2
3
4
5
6
7

Fig. 32-29.

Suppose, an *AM* carrier signal of frequency 1500 kHz is mixed with an unmodulated carrier (produced by local oscillator) of frequency 1955 kHz. Then, following two frequencies are produced

(*i*) 3455 kHz, and (*ii*) 455 kHz

The **higher frequency** is generally filtered out leaving behind the difference frequency of 455 kHz which forms the *IF* frequency.

32-35. Superhetrodyne AM Receiver

It is also referred to as superhet and is extensively used in modern AM receivers. Its block diagram with signals is shown in Fig. 32-29. The operation of this receiver is as under :

(*i*) Let us assume that the incoming signal frequency is 1500 kHz. It is first amplified by the *R.F.* amplifier.

(*ii*) Next, it enters a mixer circuit which is so designed that it can conveniently combine two radio frequencies—one fed into it by the *R.F.* amplifier and the other by a local oscillator.

(*iii*) The local oscillator is an *RF* oscillator whose frequency of oscillation can be controlled by varying the capacitance of its capacitor. In fact, the tuning capacitor of the oscillator is ganged with the capacitor of the input circuit so that the difference in the frequency of the selected signal and oscillator frequency is always constant. Usually, the difference is maintained at 455 kHz. If signal frequency is 1,500 kHz, then oscillator frequency can be either 1,955 or 1,045 kHz. Let us suppose that it is 1,955 kHz. In fact, local oscillator frequency is always higher than the frequency of the incoming signal.

(*iv*) When two alternating currents of these two different frequencies are combined in the mixer transistors, then phenomenon of beats is produced. In the present case, the beat frequency is 1955-1500 = 455 kHz. Since, this frequency is lower than the signal frequency but still above the range of audio frequencies, it is called ***intermediate frequency*** **(*IF*).**

(*v*) The 455 kHz output of the mixer is then passed on to the *IF* amplifier which is ***fixed-tuned*** to 455 kHz frequency. In practice, one or more stages of *IF* amplification may be used.

(*vi*) The output of *IF* amplifier is demodulated by a detector which provides the audio signal.

(*vii*) This audio signal is amplified by the audio-frequency (*AF*) amplifier whose output is fed to a loud-speaker which reproduces the original sound.

32-36. Standard Superhet AM Receiver

Fig. 32-30 shows the block diagram of a standard superhet *AM* receiver alongwith its seven-transistor circuit diagram. Usually, transistor radios are all-wave radios having two or more bands. With the help of a band switch, the coils and trimmers of any one band can be connected to the circuit at will. For the sake of simplicity, the coils and trimmers of one band only have been shown in Fig. 32-30. The operation of various stages and the functions of various radio components used in the radio circuit of Fig. 32-30 are described below.

(*i*) Frequency Changer or Converter

It is the first stage* in a transistor radio and has two sections—oscillator and mixer. A single silicon transistor Q_1 performs both the above functions. L_4 is the oscillator coil whereas L_2 is the signal input coil. Resistors R_1 and R_2 form a voltage divider for giving base bias to Q_1

The signals intercepted by the aerial are picked up by coil L_1 which is tuned to the transmission frequency of the desired station with the help of ganged capacitor C_1 and its fine-tuning trimmer C_2. This signal is then induced into the secondary winding L_2 which passes it on to the base of Q_1 *via* C_3. The local oscillations produced by L_4 and ganged capacitor C_7 are given to the emitter of Q_1 *via* C_4.

As seen, Q_1 has been used in *CE* configuration for producing oscillations. The positive feedback required for producing oscillations is given from the collector of Q_1 to its emitter through L_3, L_4 and C_4.

* Sometimes, there is an RF stage before it (Fig. 32-29)

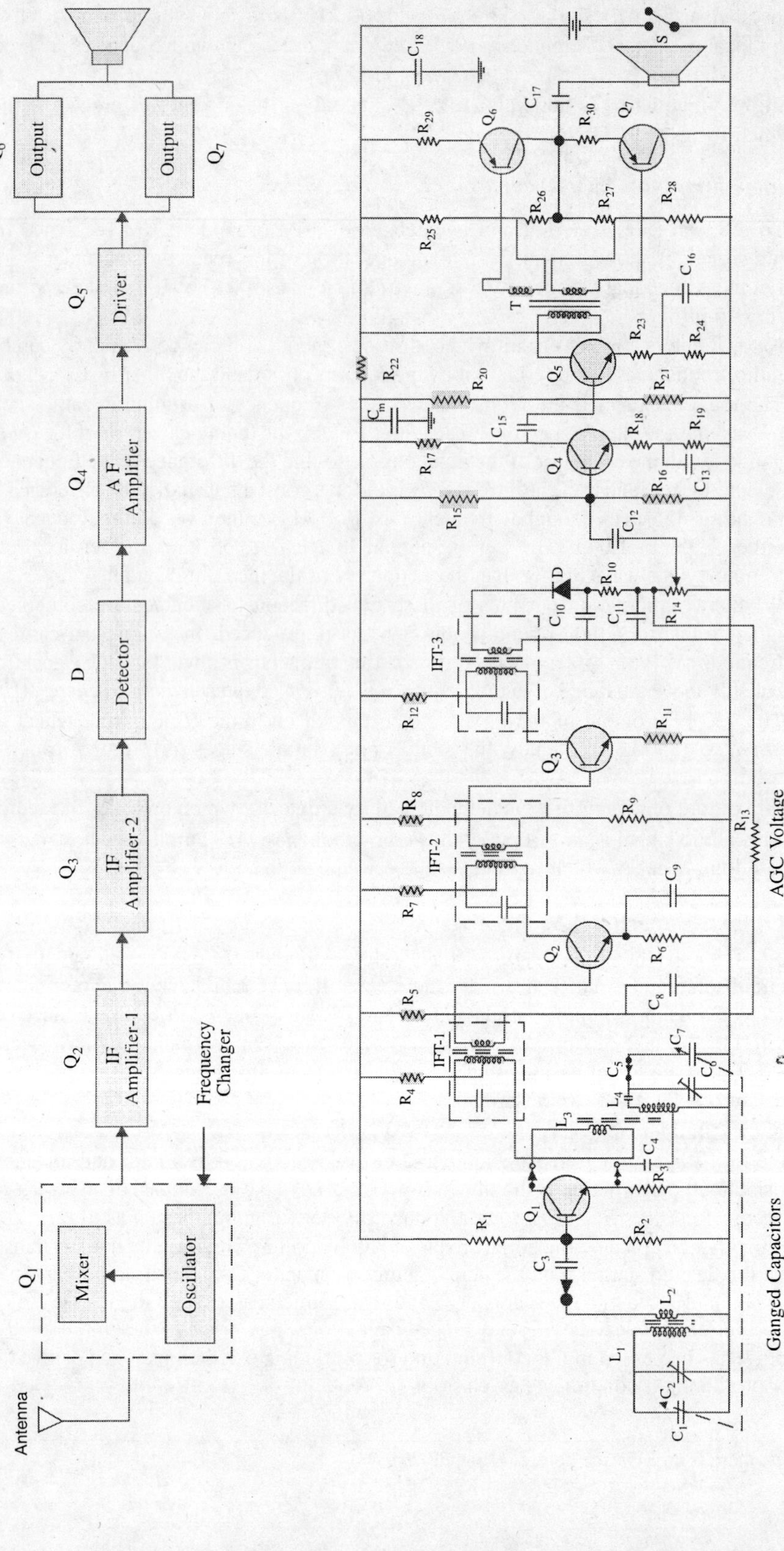
Antenna
Mixer
Oscillator
Frequency Changer
IF Amplifier-1
IF Amplifier-2
Detector
A F Amplifier
Driver
Output
Output
IFT-1
IFT-2
IFT-3
AGC Voltage
Ganged Capacitors

Fig. 32-30.

The transistor Q_1 mixes the incoming signal frequency f_s and oscillation frequency f_0 thereby producing the sum and difference frequencies of $(f_0 + f_s)$ and $(f_0 - f_s)$ respectively. The difference frequency called *IF* frequency is then passed on to the *IF* transformer IFT-1 connected in the collector of Q_1.

(*ii*) IF Amplifier

As the name indicates, its function is to amplify the *IF* signal obtained from the frequency changer. Usually, two stages are used to obtain sufficient gain. The gain of the first stage is controlled by the automatic gain control circuit (*AGC*) with the help of dc voltage drop across R_{10}. This dc voltage is obtained from the detector diode. Three fixed-tuned *IF* transformers are used—one at the collector of the frequency changer and the other two for the two *IF* amplifier stages. These transformers are also used as collector load as well as for inter-stage-coupling.

Resistor R_5 gives base bias to Q_2 whereas capacitor C_8 decouples the *AGC* voltage. As usual, R_8 and R_9 form the voltage divider circuit for giving base bias to Q_3.

The amplified *IF* signal obtained from *IFT-2* is further amplified by *IFT-3* of the second *IF* amplifier stage before it is passed on to the detector diode.

(*iii*) Detector Circuit

It consists of a diode* which rectifies the *IF* signal. Resistor R_{13}, capacitors C_{10} and C_{11} together filter out *RF*. The audio signal is given to the following audio section through potentiometer R_{14} which is also used as a volume control. Capacitor C_{12} blocks the dc component in the rectified signal. This dc voltage is used for *AGC via* R_{10}. The main purpose of using *AGC* is to minimize the variations in sound due to changes in signal strength (because of fading) and to prevent overloading of second IF stage because of a very strong signal from a close-by transmitter.

(*iv*) Audio Frequency (AF) Amplifier

The audio section consists of three stages, *i.e.,* audio amplifier, driver and output which feeds a loud speaker. The audio-amplifiers amplify the audio signal given to Q_4 through volume control and C_{12}, R_{15} and R_{16} are biasing resistors whereas R_{19} is the emitter resistor with a by-pass capacitor C_{13}. The amplified signal is developed over collector load resistor R_{17}. This amplified signal is capacitively-coupled to the driver stage through C_{15}.

(*v*) Driver

The driver stage further amplifies the audio signal obtained from the previous *AF* amplifier till it is large enough to drive the output stage. Transformer *T* (having two secondary windings) is used for coupling to the output stage. The emitter resistor of Q_5 is split into two parts (R_{23} and R_{24}) with only R_{24} being by-passed by C_{16}. This provides some negative feedback and reduces distortion.

(*vi*) Output Stage

The function of this stage is to amplify the audio signal received from the driver stage and hence, provide sufficient power to drive the loudspeaker. Class-B push-pull amplifier employing transistors Q_6 and Q_7 is used for economizing battery consumption. Resistors R_{25} and R_{26} provide base bias to Q_6 and R_{27} and R_{28} to Q_7 whereas R_{29} and R_{30} are the emitter resistors of the two respective transistors. The audio signal is given *via* C_{17} to the loudspeaker which converts it into sound similar to the one transmitted by the broadcasting transmitter.

---

* Sometimes, transistors are used as detectors.

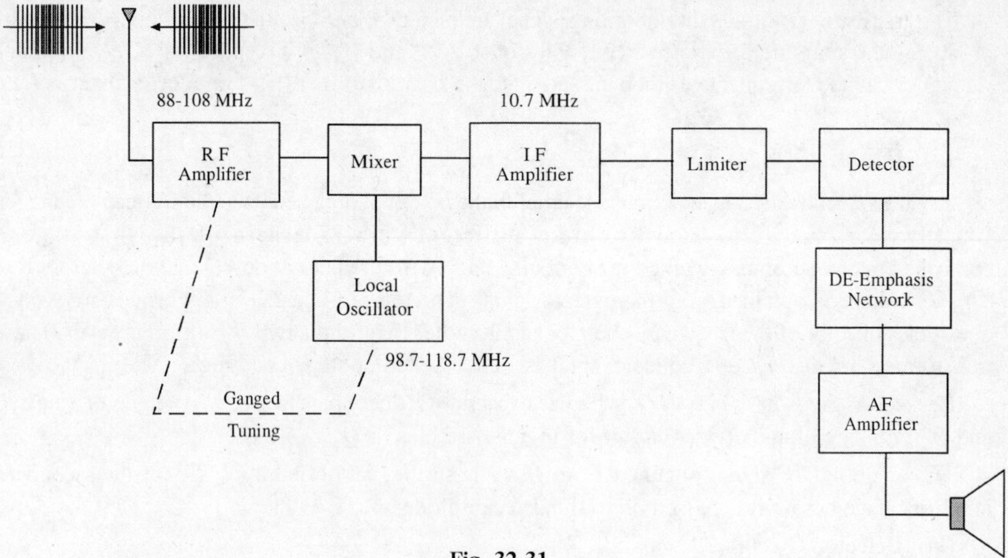

Fig. 32-31.

32-37. FM Receiver

The block diagram for such a receiver is shown in Fig. 32-31. These receivers operate in the *VHF* band of 88-108 MHz and have an *IF* of 10.7 MHz with a bandpass of 2 ×75 kHz = 150 kHz.

Like *AM* receivers, such receivers also have a frequency converter and *IF* amplifiers.

The function of the **limiter** is to remove all ***amplitude variations*** (caused by noise) from *IF* signal which might have crept into the *FM* signal. This removal of amplitude variations is necessary for distortionless demodulation. **Limiter is a sort of clipping circuit**.

The de-emphasis network ***reduces the amplitude of high frequencies in the audio signal*** which was earlier increased by the pre-emphasis network at the transmitting station. It serves to re-establish the **tonal balance** of the speech or music, *etc*., lost in the pre-emphasis. Without it, the sound signal would have a heavy treble effect. It is, in fact, a low-pass filter having a time constant of 75 μs.

32-38. Comparison Between AM and FM

Frequency modulation ***(FM)*** has the following advantages as compared to amplitude modulation (*AM*) :

(*i*) All transmitted power in ***FM*** is useful whereas in ***AM*** most of it is in carrier which ***serves no useful purpose.***

(*ii*) It has high signal-to-noise (S/N) ratio. It is due to two reasons : firstly, there happens to be less noise at ***VHF*** band and secondly, ***FM*** receivers are fitted with amplitude limiters which remove amplitude variations caused by noise.

(*iii*) Due to 'guard-band' there is hardly any adjacent-channel interference.

(*iv*) Since only transmitter ***frequency*** is modulated in ***FM***, only fraction of a watt of audio power is required to produce 100% modulation as compared to high power required in *AM*.

However, FM has the following disadvantages :

(*a*) It requires **much wider channel** — almost 7 to 15 times as large as needed by AM.

(*b*) It requires complex and expensive transmitting and receiving equipment.

(*c*) Since FM reception is limited to only line of sight, area of reception for FM is much smaller than for *AM*.

32-39. The Four Fields of FM

There are four major areas of application for *FM* transmission :

(*i*) First use is in **FM** broadcast band 88-108 MHz with 200 kHz channels in which commercial *FM* stations broadcast programmes to their listeners.

(*ii*) Second use is in TV. Though video signal is amplitude-modulated, sound is transmitted by a separate transmitter which is frequency-modulated.

(*iii*) Third use is in the mobile or emergency services which transmit voice frequencies (20-4000 Hz) only.

(*iv*) Fourth use is in the amateur bands where again only voice frequencies are transmitted.

SUMMARY

In this chapter you have learned that :

1. Modulation is a process of combining a low frequency signal with a very high frequency signal called Carrier Wave.
2. Demodulation is the process of recovering the signal from the modulated carrier wave.
3. The modulation is required to transmit the signals over a longer distance and also transmitting information over different channels.
4. The modulation can be analog or digital. There are three types of analog (sine-wave) modulation- amplitude modulation(AM), frequency modulation(FM) and phase modulation(PM).
5. In AM, the information signal changes the amplitude of the carrier wave without changing its frequency or phase.
6. In frequency modulation, the information signal changes the frequency of the carrier wave without changing the amplitude or the phase
7. In phase modulation, the information signal changes the phase of the carrier wave without changing its amplitude or frequency
8. The modulation index (or percentage of modulation) is calculated by using the equation

$$m = \frac{\max imum\ value\ of\ signal\ wave}{\max imum\ vale\ of\ caries\ wave} \times 100$$

9. An un-modulated carrier wave consists of only one single frequency component of frequency, f_c.
10. The AM wave consists of three components: the original carrier frequency component (f_c), the upper side frequency component(f_c+f_m) and the lower frequency component (f_c- f_m).
11. The channel width (or bandwidth) is given by the difference between the upper side band and lower side band frequencies. It is equal to the 2 times the value of maximum frequency of the modulating signal.
12. The amplitude modulated wave can be transmitted without transmitting the carrier frequency. So we can have double side band, suppressed carrier(DSB-SC), and single side band, suppressed carrier (SSB-SC).
13. In frequency modulation, change or shift above or below the resting frequency is called frequency deviation (*Äf*). The rate of frequency deviation depends on signal frequency. As per the International agreement, the maximum deviation is restricted to 75kHz.
14. The modulation index in frequency deviation is given by,

$$m_f = \frac{frequency\ deviation}{\mathrm{mod}\,ulation\ frequency} = \frac{\Delta f}{f_m}$$

15. In frequency modulation, the modulation index and the number of sidebands depends (*i*) directly on the amplitude of the modulating signal, and (*ii*) inversely on the frequency of the modulating signal.

16. The process of recovering the audio frequency signal from the modulated carrier wave is known as demodulation.

17. The superhetrodyne receiver is used to demodulate the signals in modern AM receivers. The basic blocks are RF amplifier, Local oscillator, mixer, IF amplifier, detector, AF amplifier and speaker.

GLOSSARY

Amplitude modulation is a process in which the audio frequency signal changes the amplitude of the carrier wave without changing the frequency or phase.

Carrier wave is a high frequency un-damped radio wave produced by RC oscillators.

Channel width difference between the maximum frequency of upper side band(USB) and minimum frequency of lower side band (LSB).

Demodulation is the process of recovering audio frequency signal from the modulated carrier wave.

Frequency modulation is the process in which the signal changes the frequency of the carrier wave without changing its amplitude or phase.

Modulation the process of combining the low frequency signal with a very high-frequency radio wave.

Modulation index is the ration of frequency deviation to the modulation frequency (*i.e.*, f/f_m).

Phase Modulation is the the process in which the signal changes the phase of the carrier wave without changing its other two parameters.

Superhetrodyne receiver is the receiver used in amplitude modulated receivers to detect audio frequency signal.

NUMERICAL PROBLEMS

1. A carrier wave has an amplitude of 500 mV. A modulating signal causes its amplitude to vary from 200 mV to 800 mV. What is the percentage modulation? **[60%]**

2. A 100.MHz carrier having an amplitude of 50 V is amplitude-modulated by a 5 kHz audio signal having an amplitude of 20 V. Find
(*i*) modulation index,
(*ii*) percent modulation,
(*iii*) components of modulated wave,
(*iv*) the amplitudes of sidebands.
[(*i*) 0.4 (*ii*) 40 (*iii*) 100 MHz, 100.005 MHz, 99.995 MHz (*iv*) 10 V]

3. An audio signal given by 30 sin $(2\pi \times 2500t)$ is used for modulating a carrier wave given by the equation 60 sin $(2\pi \times 200{,}000t)$. Find
(*i*) percent modulation,
(*ii*) frequencies of the signal and the carrier,
(*iii*) frequency spectrum of the modulated wave.
[(*i*) 50 (*ii*) 2500 Hz , 200,000 Hz (*iii*) 200; 202.5; 197.5 kHz)]

4. How many AM broadcast stations can be accommodated in a 5 MHz bandwidth if each station transmits a signal modulated by an audio signal having a maximum frequency of 5 kHz. **[500]**

5. A 100 kHz bandwidth is to accomodate 5 AM broadcasts simultaneously. What is the maximum modulating frequency permissible for each station? **[10** kHz]

6. Total power content of an AM signal is 3000 W. For 100 percent modulation, calculate
(*i*) carrier power (*ii*) power in each sideband [(*i*) **2 kW** (*ii*) **0.5 kW]**

7. The total power content of an AM wave is 1320 W. What is the percent modulation if each sideband contains 160 W ? **[0.8]**

8. An SSB-SC signal contains 2 kW. Find power contained in
(*i*) carrier (*ii*) sideband. [(*i*) **0** (*ii*) **2 kW**]

9. As compared to DSB-FC, Calculate the power saving in SSB-SC when percent modulation is
(*i*) 1 (*ii*) 0.8 (*iii*) 0.5 **[**(*i*) **83.3%** (*ii*) **87.87%** (*iii*) **94.4%]**

10. What is the modulation index of an FM carrier having a carrier swing of 120 kHz and a modulating signal of 10 kHz. **[6]**

11. An FM signal has a resting frequency of 100 MHz and the highest frequency of 100.05 MHz when modulated by an audio signal of 5 kHz. Determine
(*i*) frequency deviation , (*ii*) carrier swing,
(*iii*) modulation index, (*iv*) percent modulation. **[**(*i*) **50 kHz** (*ii*) **100 kHz** (*iii*) **10** (*iv*) **32-7%]**

12. A 10 kHz audio signal is used to frequency-modulate a 100 MHz carrier causing a frequency deviation of 50 kHz. Determine
(*i*) modulation index, (*ii*) bandwidth of FM signal. [(*i*) **5** (*ii*) **160 kHz)]**

13. A carrier of 1MHz with 400 Watt of its power is amplitude modulated with a sinusoidal signal of 2500 Hz. The depth of modulation is 75%. Calculate the side band frequencies, the band width, the power in the side bands and the total power in the modulated wave.
[(*a*) 1002.5 kHz, 997.5 kHz (*b*) 5000 kHz] [*Electronics Engg; Bangalore Univ. 2004*]

14. The total power content of an AM wave is 2.64 kW at a modulation factor of 80%. Determine the power content of **(*i*)** carrier **(*ii*)** each side band. [(*i*) **2000 W** (*ii*) **320 W**]
[*Electronics Engg. Bangalore Univ. 2002*]

15. A 100MHz carrier wave is frequency modulated by a 10 kHz sinusoidal modulating signal. If the maximum frequency deviation is 50 kHz. Find the modulation index. **[5]**
[*Electronics Engg. Bangalore Univ., 2002*]

16. A 500 W, 1MHz carrier is amplitude modulated with a sinusoidal signal of 1 kHz. The depth of modulation is 60%. Calculate the band width, power in the side bands and the total power transmitted.
[(*i*) 2 kHz **(*ii*) 38.14** **(*iii*) 576.27** W] [*Electronics Engg., Bangalore Univ., 2003*]

17. A carrier signal has a peak amplitude of 100V. Modulation index is 40%. Power is developed across a load of 100 Ω. Determine the value of transmitted power. [**108** W]
(*Electronics Engg. Bangalore Univ., 2003*)

DESCRIPTIVE QUESTIONS

1. What is the difference between the three types of modulation-amplitude modulation, frequency modulation and phase modulation.

2. Sketch the waveform of a amplitude modulated wave fr m - 0, 50% and 100%

3. Sketch the frequency spectrum of the AM modulated wave showing the posito of the carrier wave and the two side bands.

4. What do you mean by double side band-suppressed carrier and single band-suppressed carrier ? Explain in detail.

5. Sketch the block diagram of AM transmitter and explain its operation.

6. Sketch the frequency spectrum of frequency modulated wave.

7. Sketch the block diagram of a superheterodyne receiver and explain it operation.

MULTIPLE CHOICE QUESTIONS

1. The main purpose of modulation is to
(*a*) combine two waves of different frequencies
(*b*) achieve wave-shaping of the carrier wave
(*c*) transmit low-frequency information over long distances efficiently
(*d*) produce sidebands.

2. Demodulation
(*a*) is performed at the transmitting station
(*b*) removes side-bands
(*c*) rectifies modulated signal
(*d*) is opposite of modulation.

3. In amplitude modulation
(*a*) carrier frequency is changed
(*b*) carrier amplitude is changed
(c) three sidebands are produced
(d) fidelity is improved.

4. 100% modulation is produced in AM when carrier
(*a*) frequency equals signal frequency
(*b*) frequency exceeds signal frequency
(*c*) amplitude equals signal amplitude
(*d*) amplitude exceeds signal amplitude.

5. For a given carrier wave, maximum undistorted power is transmitted when value of modulation is
(*a*) 1 (*b*) 0.8
(*c*) 0.5 (*d*) 0.

6. In an AM wave with 100 percent modulation, each sideband carries — of the total transmitted power.
(*a*) one-half (*b*) one-sixth
(*c*) one-third (*d*) two-third

7. Given a carrier frequency of 100 kHz and a modulating frequency of 5 kHz, the bandwidth of AM transmission is____ kHz.
(*a*) 5 (*b*) 200
(*c*) 10 (*d*) 20

8. When modulation of an AM wave is decreased
(*a*) percentage carrier power is decreased
(*b*) percentage carrier power is increased
(*c*) total transmitted power is increased
(*d*) percentage sideband power is unaffected.

9. In AM transmission, power content of the carrier is maximum when m equals
(*a*) 0 (*b*) 1
(*c*) 0.8 (*d*) 0.5.

10. In AM transmission with m = 1, suppression of carrier cuts power dissipation by a factor of
(*a*) 6 (*b*) 2
(*c*) 3 (*d*) 4

11. As compared to DSB-FC 100% modulated transmission, power saving in SSB-SC system is ________ per cent.
(*a*) 94.4 (*b*) 50
(*c*) 100 (*d*) 83.3

12. In an AM transmission with 100% modulation, 66.7% of power is saved when ___________ is/are suppressed.
(*a*) carrier (*b*) carrier and USB
(*c*) carrier and LSB (*d*) USB and LSB

13. In an AM system, full information can be conveyed by transmitting only
(*a*) the carrier (*b*) the upper sideband
(*c*) the lower sideband (*d*) any one sideband.

14. A 10 MHz carrier of peak value 10 V is amplifide modulated by a 10 kHz signal of amplitude 6 V. The amplitude of each side band frequency is
(*a*) 3 V (*b*) 4 V
(*c*) 5 V (*d*) 6 V

15. The block diagram of an AM receiver is shown in Fig. 32-32. The missing block and its location are respectively

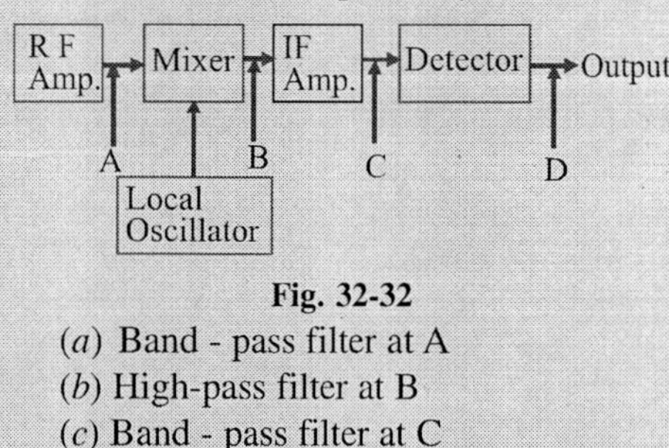

Fig. 32-32

(*a*) Band - pass filter at A
(*b*) High-pass filter at B
(*c*) Band - pass filter at C
(*d*) Low-pass filter (audio amplifier) at D.

16. In FM transmission, amplitude of the modulating signal determines

(*a*) rate of frequency variations

(*b*) amount of frequency shift

(*c*) total balance of transmission

(*d*) distance of broadcast.

17. In FM, when frequency deviation is doubled

(*a*) modulation is doubled

(*b*) modulation is halved

(*c*) carrier swing is halved

(*d*) modulation index is decreased.

18. In an FM broadcast in VHF band, channel width is ___________

(*a*) 75 (*b*) 25

(*c*) 88 (*d*) 200

19. Modern FM receivers use de-emphasis circuit for

(*a*) reducing high frequency noise

(*b*) reducing the amplitude of high frequencies in the audio signal

(*c*) increasing the amplitude of higher modulating frequencies

(*d*) making demodulation easy

20. One of the serious disadvantages of FM transmission is its

(*a*) high static noise

(*b*) limited line-of-sight range

(*c*) expensive equipment

(*d*) adjacent channel interference.

21. In India, the maximum frequency deviation is commercial FM broadcasting signals is about,

(*a*) 5 kHz (*b*) 15 kHz

(*c*) 7.5 kHz (*d*) 200 kHz

22. The correct sequence of subsystems in an FM receiver is

(*a*) mixer, RF amplifier, limiter, IF amplifier, discriminator, audio amplifier

(*b*) RF amplifier, mixer, IF amplifier, limiter, discriminator, audio amplifier

(*c*) RF amplifier, mixer, limiter, discriminator, IF amplifier, audio amplifier

(*d*) mixer, IF amplifier, limiter, audio amplifier, discriminator.

ANSWERS

1. (*c*) **2.** (*d*) **3.** (*b*) **4.** (*c*) **5.** (*a*) **6.** (*b*) **7.** (*c*) **8.** (*b*) **9.** (*a*) **10.** (*c*) **11.** (*d*)
12. (*a*) **13.** (*d*) **14.** (*a*) **15.** (*d*) **16.** (*b*) **17.** (*a*) **18.** (*d*) **19.** (*b*) **20.** (*b*) **21.** (*d*) **22.** (*b*)

❑❑❑

33

CHAPTER

Number Systems and Codes

OUTLINE

Objectives

After Completing this chapter, you should be able to :

- Know the decimal, binary, octal and hexadecimal-number systems.
- Convert numbers from decimal-to-binary and from binary-to-decimal.
- Convert numbers from binary-to-octal and from octal-to-binary.

- Convert numbers from binary-to-hexadecimal and from hexadecimal-to-binary.
- Express decimal numbers using the BCD code.
- Understand the difference between the BCD code and the straight binary code.
- Describe the parity method for error detection.
- Determine parity (odd or even) of digital data.

33-1. Introduction

The number systems are used quite frequently in the field of digital electronics and computers. However, the type of number system used in computers could be different at different stages of the usage. For example, when a user key-in some data into the computer, he/she will do it using decimal number system *i.e.*, the system we all have used for several years of doing arithmetic problems. But when the information goes inside the computer, it needs to be converted to a form suitable for processing data by the digital circuitry. Similarly, when the data has to be displayed on the monitor for the user, it has to be again in the decimal number system. Hence, the conversion from one number system to another one is an important issue to be understood.

Now we shall start our studies in this chapter with the different types of number systems. We will understand how the counting is done in different number systems. We will also discuss, how to convert one number system to another one.

33-2. Types of Number Systems

There are several number systems but the following are the important ones in the field of digital electronics :

(*i*) Decimal number systems

(*ii*) Binary number system

(*iii*) Octal number system

(*iv*) Hexadecimal number system.

Now we shall study all these number systems one by one in the following pages.

33-3. Decimal Number System

The decimal number system is composed of 10 numerals or symbols. These symbols are 0, 1, 2, 3, 4, 5, 6, 7, 8 and 9. Using these 10 symbols as digits of a number, we can express any quantity how big or small it could be. The decimal system is also called the base- 10 system because it has 10 digits. The decimal number system has evolved naturally as a result of the fact that people have 10 fingers on their hands. The word "digit" is derived from the Latin word for "finger".

The decimal number system is a positional value system in which the value of a digit depends on its position. For example, consider the decimal number 462. We know that the digit 4 actually represents 4 hundreds, the 6 represents 6 tens and the 2 represents 2 units. Now since, the digit 4 carries the most weight of the three digits, therefore, it is referred to as the *most-significant digit* (*MSD*). On the other hand, the digit 2 carries the least weight and is called the *least-significant weight* (*LSD*).

Let us consider another example, say the decimal number 26.48. This number is actually equal to 2 tens plus 6 units plus 4 tenths plus 8 hundredths. It can be thought of to be equal to (2 * 10 + 6 * 1 + 4 * 0.1 + 8 * 0.01). The decimal point is used to separate the integer and fractional parts of the number.

As a matter of fact, the various positions relative to the decimal point carry weights that can be expressed as powers of 10. This is illustrated in Fig. 33-1 where the number 3547.216 is represented. The decimal point separates the positive powers of 10 from the negative powers.

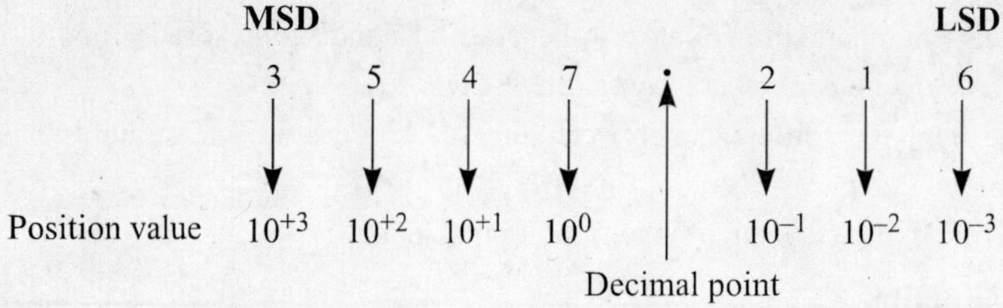

Fig. 33-1. *Illustrating position values as powers of 10 in a decimal number system.*

It is evident from the above diagram that the number 3547.216 is equal to :

$$(3\times10^{+3})+(5\times10^{+3})+(4\times10^{+1})+(7\times10^{0})+(2\times10^{-1})+(1\times10^{-2})+(6\times10^{-3})$$

Thus, in general any decimal number is equal to the sum-of-products of each digit value and its position value.

33-4. Counting in Decimal Number System

We have already discussed in the last article that the decimal system is composed of 10 digits, *i.e.*, 1, 2, 3, 4, 5, 6, 7, 8 and 9. Thus, when counting in the decimal number system, we start with 0 in the units position and take each digit in progression until we reach 9. After reaching 9 we have finished with the decimal digits. To count beyond 9, we form two digit combinations, with the second digit, *i.e.*, "1" in the tens position followed by 0, 1, 2, 3, 4, 5, 6, 7, 8 and 9 in the units position. Thus we count beyond 9 as 10, 11, 12, 13 and so on. This process continues until we reach 19.

To count beyond 19, we take the digit 2 at the tens position and digits 0, 1, 2, ... at the units position. Thus we count as 20, 21, 22, and so on until the count of 99 is reached. After reaching at 99, we have again finished with the decimal digits. To count beyond 99, we form three digit combinations with the second digit, *i.e.*, 1 at the hundreds position, first digit, *i.e.*, 0 at the tens position and another 0 at the units position. So we count as 100, 101, 102 and so on. This process is shown in Fig. 33-2.

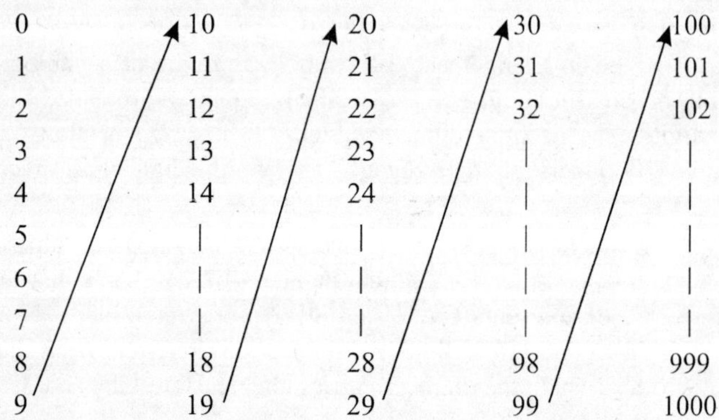

Fig. 33-2. *Counting in a decimal number system.*

33-5. The Binary Number System

We have already discussed in the last article about the most commonly used number system (*i.e.*, the decimal number system) in day-to-day life. Unfortunately, the decimal system cannot be conveniently implemented in digital systems. The major reason for this is that it is very difficult to design digital electronic equipment which can work with 10 different voltage levels (each voltage level representing one decimal character, 0 through 9). On the other hand, it is very easy to design simple but accurate digital electronic circuits which operate with only two voltage levels. It is because of this reason, that almost every digital electronic system uses the binary (or base-2) number system as the basic number system of its operations. However, it may be noted that other number systems are also used often with the binary number system.

As discussed in the last chapter (Art. 33-10), the binary number system has two digits, 0 and 1. These digits are also called *bits*. The binary number system can be used to represent any quantity that can be represented in decimal or other number systems. However, it may be noted carefully, that it will take a greater number of bits to express a given quantity.

Like decimal system, the binary number system is also a position-value system where each bit has its own value or weight expressed as a power of 2. This is illustrated in Fig. 33-3 by considering a binary number 1101.101. As shown in the diagram, places to the left of the binary point are positive powers of 2. Whereas the places to the right of the binary point are negative powers of 2.

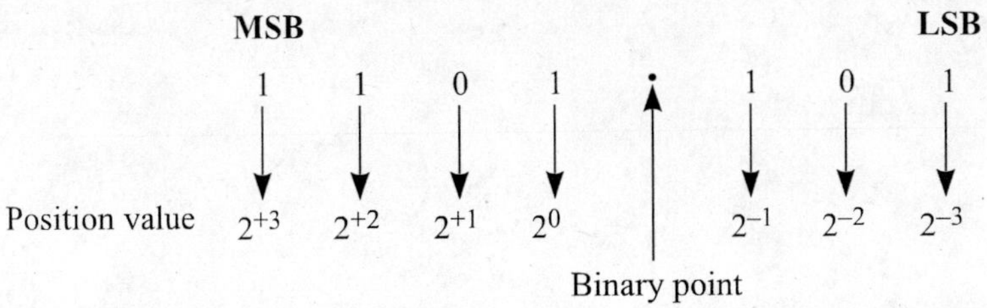

Fig. 33-3. *Illustrating a position values as powers of 2 in a binary number system.*

It may be noted that in the binary number indicated in Fig. 33-3 there are four bits to the left of the binary point. These four bits represent the integer (or whole number) part of the number. There are three bits to the right of the binary point. These three bits represent the fractional part of the binary number. The bit at the left-most position (*i.e.*, the bit with the largest position value) is called the most-significant bit (MSB). Similarly, the bit at the right-most position (*i.e.*, the bit with the smallest position value) is called the least significant bit (LSB).

33-6. Counting in the Binary Number System

In order to count in binary number system, first we look at how we count in decimal system. We start at 0 and count up to 9 before we run out of digits. We then start another digit position (to the left) and continue counting 10 through 99. At this point we have exhausted all two-digit combinations, so a third digit is needed to count from 100 through 999.

A comparable situation occurs when we count in binary, except that we have only two digits called bits. Thus we begin counting as 0, 1. At this point we have used both the digits. To count beyond this, we form two digit combinations: second digit (*i.e.*, 1) followed by the first digit (*i.e.*, 0) and next second digit followed by second digit again. So we continue to count as 10, 11. Now we have exhausted all two-digit combinations, so we need three digit combinations. The three-digit combinations again start with the second binary digit followed by the first binary digit repeated two times (*i.e.*, 100). Thus, we continue to count as 100, 101, 110 and 111. Now again we have exhausted all the three-digit combinations and hence to continue counting, we need to form four-digit combinations and this process continues.

Table 33-1 shows a binary count of zero through fifteen. Notice the pattern with which the 1s and 0s alternate (or toggle) in each column. Thus in the column with position value of 2^0 (*i.e.*, the right most column), the 1s and 0s toggle each time we go to the next count. However, in the column with position value of 2^{+1}, the 1s and 0s toggle after two counts. Similarly, in the columns with position value of 2^{+2} and 2^{+3} , the 1s and 0s toggle after four and eight counts respectively.

Table 33-1. *Counting in Binary*

Decimal Number	*Binary number*			
	2^{+3}	2^{+2}	2^{+1}	2^{0}
0	0	0	0	0
1	0	0	0	1
2	0	0	1	0
3	0	0	1	1
4	0	1	0	0
5	0	1	0	1
6	0	1	1	0
7	0	1	1	1
8	1	0	0	0
9	1	0	0	1
10	1	0	1	0
11	1	0	1	1
12	1	1	0	0
13	1	1	0	1
14	1	1	1	0
15	1	1	1	1

It may be noted from the Table 33-1 that, we need four bits or digits to count from zero to 15 (*i.e.*, $2^4 - 1$). In general with n bits, we can count up to a number equal to $2^n - 1$. Thus,

Largest decimal number= $2^n - 1$ For example, if $n = 5$, then we can count from zero to $(2^5 - 1) = 32 - 1 = 31$. Similarly, if $n = 6$, we can count from zero to $(2^6 - 1) = 64 - 1 = 63$.

Note. In order to avoid confusion among the different number systems, we will add subscript 2 to the number represented in the binary system and subscript 10 to the numbers represented in the decimal system. For example, the number 1101.101_2 is a binary number and 16.538_{10} is a decimal number. Another way is to write B for binary and D for decimal at the end of the numbers. For example, 1010B and 526D are the binary and decimal numbers respectively.

Example 33-1. *What is the largest decimal value that can be represented by* **(*a*)** *a 8-bit binary,* **(*b*)** *a 33-bit binary number?*

Solution.

(*a*) *Largest decimal number for a 8-bit number*

We know that largest decimal value for a 8-bit binary number,

$$= 2^8 - 1 = 256 - 1 = 255 \textbf{ Ans.}$$

(*b*) *Largest decimal number for a 33-bit number*

We know that the largest decimal value for a 33-bit binary number,

$$= 2^{16} - 1 = 65536 - 1 = 65535 \textbf{ Ans.}$$

33-7. Binary-to-Decimal Conversion

We have already discussed the representation of numbers in decimal and binary number system. In order to understand the binary-to-decimal conversion, we will study the conversion of integer (or whole) numbers and the fractional numbers separately in the following pages.

33-8. Integer Binary-to-Decimal Conversion

Following is the procedure for converting an integer (or whole) binary number to its equivalent decimal number.

Step 1 Write the binary number.

Step 2 Directly under the binary number, write the position values or weights of each bit working from right to left.

Step 3 If a zero appears in a digit position, cross-out the weight for that position.

Step 4 Add the remaining weights to obtain the decimal equivalent.

For example, let us consider the conversion of binary 101 to its decimal equivalent.

Step 1	1	0	1
	↓	↓	↓
Step 2	2^2	2^1	2^0
Step 3	2^2	$\cancel{2^1}$	2^0

Step 4 $2^2 + 2^0 = 4 + 1 = 5_{10}$ **Ans.**

Consider another example of converting binary 10101 to its decimal equivalent.

Step 1	1	0	1	0	1
	↓	↓	↓	↓	↓
Step 2	2^4	2^3	2^2	2^1	2^0
Step 3	2^4	$\cancel{2^3}$	2^2	$\cancel{2^1}$	2^0

Step 4 $2^4 + 2^2 + 2^0 = 16 + 4 = 1 = 21_{10}$ **Ans.**

Example 33-2. *Convert the binary number 01010110_2 to its equivalent decimal number.*

Solution. Given the binary number = 01010110_2

Step 1	0	1	0	1	0	1	1	0
	↓	↓	↓	↓	↓	↓	↓	↓
Step 2	2^7	2^6	2^5	2^4	2^3	2^2	2^1	2^0
Step 3	2^7	2^6	2^5	2^4	2^3	2^2	2^1	2^0

Step 4 $2^6 + 2^4 + 2^2 + 2^1 = 64 + 16 + 4 + 2 = 86_{10}$ **Ans.**

33-9. Fractional Binary-to-Decimal Conversion

We have already discussed the integer binary-to-decimal conversion. Let us now see how a binary fraction can be converted into corresponding decimal equivalents. Consider for example the conversion of fractional binary number 0.1010 to decimal.

Step 1	0	1	0	1	0
		↓	↓	↓	↓
Step 2		2^{-1}	2^{-2}	2^{-3}	2^{-4}
Step 3		2^{-1}	~~2^{-2}~~	2^{-3}	~~2^{-4}~~

Step 4 $\quad 2^{-1}+2^{-3}=0.5+0.125=0.625_{10}$ **Ans.**

Example 33-3. *Convert the binary number 0.1100 to its equivalent decimal number.*

Solution. Given: the binary number = 0.1100

Step 1	0	1	1	0	0
		↓	↓	↓	↓
Step 2		2^{-1}	2^{-2}	2^{-3}	2^{-4}
Step 3		2^{-1}	2^{-2}	~~2^{-3}~~	~~2^{-4}~~

Step 4 $\quad 2^{-1}+2^{-2}=0.5+0.25=0.75$

$\therefore \quad 0.1100_2=0.75_{10}$ **Ans.**

Example 33-4. *Convert the binary number 1101.0110 to its equivalent decimal number.*

Solution. Given: the binary number = 110.0110

Step 1	1	1	0	1	0	1	1	0
	↓	↓	↓	↓	↓	↓	↓	↓
Step 2	2^3	2^2	2^1	2^0	2^{-1}	2^{-2}	2^{-3}	2^{-4}
Step 3	2^3	2^2	~~2^1~~	2^0	~~2^{-1}~~	2^{-2}	2^{-3}	~~2^{-4}~~

Step 4 $\quad 2^3+2^2+2^0+2^{-2}+2^{-3}=8+4+1+0.25+0.125=9.375$

$\therefore \quad 1101.0110_2=9.375_{10}$ **Ans.**

33-10. Decimal-to Binary Conversion

The conversion from decimal-to-binary is usually performed by a digital computer for ease of interpretation by the person reading the number. On the other hand, when a person enters a decimal number into a digital computer, that number must be converted to binary before it can be operated on. There are two methods of decimal-to-binary conversion :

(*i*) Sum-of-weights method, and

(*ii*) Repeated division by-2 method

Both these method can be used for converting the integer and fractional decimal numbers. The methods are discussed below one by one in the following pages.

33-11. Sum-of-weights method for Integer Decimal-to-Binary Conversion

This method makes use of binary weights. To find the binary number that is equivalent to a given decimal number is to determine the set of binary weights whose sum is equal to the decimal number. An easy way to remember binary weights is that the lowest weight is 1 (*i.e.*, 2^0). By doubling this weight, we get the next higher weight. Thus, a list of eight binary weights would be (read from right to left) :

128, 64, 32, 16, 8, 4, 2, 1

Consider, for example, the conversion of decimal number to its binary equivalent using sum-of-weights method. Then we know that, the number 9 can be expressed as the sum of binary weights as follows :

$$9 = 8 + 1 \quad \text{or} \quad 9 = 2^3 + 2^1$$

Placing 1s in the appropriate weight positions, 2^3 and 2^0 and 0s in the 2^2 and 2^1 positions, we can write the binary number for decimal 9 as:

2^3	2^2	2^1	2^0
↓	↓	↓	↓
1	0	0	1

$\therefore \quad 9_{10} = 1001_2$

Example 33-5. *Convert each of the following decimal numbers to their binary equivalents using sum-of-weights methods:*

(*a*) 17 (*b*) 24 (*c*) 61 (*d*) 93

Solution.

(*a*) Given: The decimal number = 17

We know that the decimal number can be expressed by the sum-of-weights as follows:

$$17 = 16 + 1 = 2^4 + 2^0$$

Placing 1s in the appropriate weight positions, 2^4 and 2^0 and 0s in the 2^3, 2^2 and 2^1 positions, determines the binary number for decimal 17. Thus,

2^4	2^3	2^2	2^1	2^0
↓	↓	↓	↓	↓
1	0	0	0	1

$\therefore \quad 17_{10} = 10001_2$ **Ans.**

(*b*) Given: The decimal number = 24

We know that the decimal number,

$$24 = 16 + 8 = 2^4 + 2^3$$

Placing 1s in the appropriate weight positions, 2^4 and 2^3 and 0s in the 2^2, 2^1 and 2^0 positions, determines the binary number for decimal 24. Thus,

2^4	2^3	2^2	2^1	2^0
↓	↓	↓	↓	↓
1	1	0	0	0

$\therefore \quad 24_{10} = 11000_2$ **Ans.**

(*c*) Given: The decimal number = 61

We know that the decimal number,

$$61 = 32 + 16 + 8 + 4 + 1$$

Placing 1s in the appropriate weight positions, 2^5, 2^4, 2^3 and 2^2 and 2^0 and 0s in the 2^1 position determines the binary number for decimal 61. Thus

2^5	2^4	2^3	2^2	2^1	2^0
↓	↓	↓	↓	↓	↓
1	1	1	1	0	1

$\therefore \quad 61_{10} = 111101_2$ **Ans.**

(*d*) Given: The decimal number = 93

We know that the decimal number,

$$93 = 64 + 16 + 8 + 4 + 1$$

Placing 1s in the appropriate weight positions, 2^6, 2^4, 2^3, 2^2 and 2^0 and 0s in the 2^5 and 2^1 position determines the binary number for decimal number 93. Thus

2^6	2^5	2^4	2^3	2^2	2^1	2^0
↓	↓	↓	↓	↓	↓	↓
1	0	1	1	1	0	1

$\therefore \quad 93_{10} = 1011101_2$ **Ans.**

33-12. Sum-of-weights method for Conversion of Fractional Decimal-to-Binary

This method makes use of position weights. In order to find a binary fractional number that is equivalent to a given decimal fractional number, we have to determine the set of binary weights whose sum is equal to the decimal number. An easy way to remember fractional binary weights is that the most significant weight is 0.5 (2^{-1}) and that by dividing by 2, we get the next lower weight. Thus, a list of four fractional binary weights would be: 0.5, 0.25, 0.125 and 0.0625.

Consider for example, the conversion of the decimal fraction 0.625 to its equivalent binary. We know that,

$$\begin{aligned} 0.625 &= 0.5 + 0.125 \\ &= 2^{-1} + 2^{-2} \\ &= 0.101 \end{aligned}$$

There is a 1 in the 2^{-1} position, a 0 in the 2^{-1} position and a 1 in the 2^{-3} position.

Example 33-6. *Convert the decimal fraction 0.375 by using sum-of-weights method to its equivalent binary fraction.*

Solution. Given: The decimal fraction = 0.375

We know that the decimal fraction,

$$\begin{aligned} 0.375 &= 0.25 + 0.125 \\ &= 2^{-2} + 2^{-3} = 0.011_2 \end{aligned}$$

There is a 0 in the 2^{-1} position, a 1 in the 2^{-2}, 2^{-3}, positions.

Thus, $\quad 0.375_{10} = 0.011_2$ **Ans.**

33-13. Repeated Division-by-2 Method for Integer Decimal-to-Binary Conversion

It is a systematic way of converting integer numbers from decimal-to-binary. The procedure for conversion is as given below :

Step 1 Begin by dividing the given decimal number by 2.

Step 2 Devide each resulting quotient by 2 until there is a 0 whole number quotient.

Step 3 The remainders generated by each division form the binary number. The first remainder to be produced is the least significant bit (LSB) in the binary number, and the last remainder to be produced is the most significant bit (MSB). In other words, reading the remainders from bottom-to-top constitutes the required binary number.

In order to illustrate the procedure, let us consider the conversion of decimal 10 to its equivalent binary number.

			Top ↑
$10 \div 2 = 5$	with a remainder	0	(LSB)
$5 \div 2 = 2$	with a remainder	1	
$2 \div 2 = 1$	with a remainder	0	
$1 \div 2 = 0$	with a remainder	1	(MSB)
			Bottom

Reading the remainders from bottom to top, we find that the binry number for the given decimal number 10 is 1010.

$$\therefore \quad 10_0 = 1010_2$$

Example 33-7. *Convert each of the following decimal numbers using repeated-division by-2 method.*

(*a*) 19 and (*b*) 45

Solution.

(*a*) Given the decimal number = 19.

We know that the conversion from decimal-to-binary can be done by using can be done by using repeated-division-by 2 method.

			Top ↑
$19 \div 2 = 9$	with a remainder	1	(LSB)
$9 \div 2 = 4$	with a remainder	1	
$4 \div 2 = 2$	with a remainder	0	
$2 \div 2 = 1$	with a remainder	0	(MSB)
$1 \div 2 = 0$	with a remainder	1	
			Bottom

Reading the remainders from bottom to top, the binary number for the decimal number 19 is 10011.

$$\therefore \quad 19_{10} = 10011_2 \quad \textbf{Ans.}$$

(*b*) *Given the decimal number = 45.*

We know that the conversion from deciaml-to-binary can be done by using repeated division-by-2 method.

			Top ↑
$45 \div 2 = 22$	with a remainder	1	(LSB)
$22 \div 2 = 11$	with a remainder	0	
$11 \div 2 = 5$	with a remainder	1	
$5 \div 2 = 2$	with a remainder	1	
$2 \div 2 = 1$	with a remainder	0	(MSB)
$1 \div 2 = 0$	with a remander	1	
			Bottom

Reading the remainders from bottom to top, the binary number for the decimal number 45 is 101101_2.

Note that most computers or digital systems deal with groups of 4, 8, 16 or 32 bits, therefore we should keep all our answers in that form. Thus, by adding leading zeros to the binary number 101101 will not change its numeric value. Therefore, the 8-bit answer is:

$$101101_2 = 00101101_2$$

33-14. Repeated Multiplication-by-2 for Fractional Decimal to Binary Conversion

We have already seen in the last article that the repeated division-by-2 method can be used to convert integer (or whole number) decimal to its equivalent binary. Decimal fractions can be converted to binary by repeated multiplication-by-2. Following is the procedure that is used for conversion:

Step 1 Begin by multiplying the given decimal fraction by 2 and then multiplying each resulting fractional part of the product by 2.

Step 2 Repeat step 1 until the fractional product is zero or until the desired number of decimal places is reached.

Step 3 The carried bits or carries generated by the multiplication produce the binary number. The first carry produced is the most-significant bit (MSB) and the last carry is the least significant bit (LSB). In other words, reading from top to bottom constitutes the required fractional binary.

In order to illustrate the procedure, consider the conversion of 0.3125 to its equivalent binary fraction.

			Top ↓
$0.3125 \times 2 = 0.625$	with a carry	0	MSB
$0.625 \times 2 = 1.25 = 0.25$	with a carry	1	
$0.25 \times 2 = 0.50 = 0.50$	with a carry	0	
$0.50 \times 2 = 1.00 = 0$	with a carry	1	LSB
			Bottom

Since, the fractional part is zero, so we will stop the repeated multiplication. Reading from top to bottom, the required binary number is 0101.

$\therefore\ 0.3125_{10} = 0.101_2$

Example 33-8. *Convert the decimal fraction 0.9028 to its equivalent binary fraction (up to 4 binary places) using repeated multiplication-by-2 method.*

Solution. Given: The decimal fraction = 0.9028

			Top ↓
$9028 \times 2 = 1.8056 = 0.8056$	with a carry of	1	MSB
$0.8056 \times 2 = 1.6112 = 0.6112$	with a carry of	1	
$0.6112 \times 2 = 1.2224 = 0.2224$	with a carry of	1	
$0.2224 \times 2 = 0.4448 = 0.4448$	with a carry of	0	LSB
			Bottom

and so on.........

We can continue this process further if we wish because the binary fractional part has not reduced to zero. However, we stop here because the desired number of binary places is reached. Reading from top to bottom, we find that the binary fraction is 0.1110.

$\therefore\ 0.9028_{10} = 0.1110_2$ **Ans.**

Example 33-9. *Convert the number 47.25 to its equivalent binary.*

(UP.Tech. Univ., 2004-05)

Solution. Given: the number = 47.25_{10}

We will convert the given decimal number into its equivalent binary by converting first the integer number 47_{10} and then the fraction number 0.25_{10}.

Conversion of integer 47_{10}

We know that the conversion of an integer from decimal-to-binary can be done by using repeated division-by-2 method.

			Top
$47 \div 2 = 23$	with a remainder	1	↑ (LSB)
$23 \div 2 = 11$	with a remainder	1	
$11 \div 2 = 5$	with a remainder	1	
$5 \div 2 = 2$	with a remainder	1	
$2 \div 2 = 1$	with a remainder	0	
$1 \div 2 = 0$	with a remainder	1	(MSB)
			Bottom

Reading the remainders from bottom to top, the binary number for the decimal number 47 is 101111_2.

Conversion of fraction 0.25_{10}

We know that the conversion of a fraction from decimal-to-binary can be done by using repeated multiplication.

			Top
$0.25 \times 2 = 0.50 = 0.50$	with a carry of	0	(MSB)
$0.50 \times 2 = 1.00 = 0$	with a carry of	1	↓ (LSB)
			Bottom

We will stop here because the binary fraction part has reduced to 0. Reading from top to bottom, we find that the binary fraction of 0.25_{10} is 0.01_2.

Combining the integer and fractional parts of the number, 47.25_{10} we get,

$$(47.25)_{10} = (101111.01)_2 \quad \textbf{Ans.}$$

33-15. Octal Number System

The octal number system provides a convenient way to express binary numbers and codes. This system is composed of eight digits 0, 1, 2, 3, 4, 5, 6 and 7. After reaching 7 we have finished with the digits. To count beyond 7, we form two digit combinations in the same way as discussed for decimal and binary number system. Thus beyond 7, we count as 10, 11, 12, 13, 14, 15, 16, 17, 20, 21...and so on.

It may be noted that counting in octal is similar to counting in decimal, except that the digits 8 and 9 are not used. To distinguish octal numbers from decimal numbers, we will use the subscript 8 to indicate the octal number. For example, 148 is in octal is equivalent to 12 in decimal.

With N octal digits, we can count from 0 up to $(8^N - 1)$, for a total of 8^N different counts. For example with 2 octal digit positions we can count from 00_8 to 77_8 which is 0_{10} to 63_{10} for a total of $8^2 = 64_{10}$ different octal numbers. Similarly with three octal digit positions, we can count from 000_8 to 777_8 which is 0_{10} to 511_{10} for a total of $8^3 = 512_{10}$ different octal numbers. Fig. 33-4 shows the position weight of each digit in an octal number.

8^{+5} 8^{+4} 8^{+3} 8^{+2} 8^{+1} • 8^{-1} 8^{-4} 8^{-3} 8^{-2}

↑

Octal pont

Fig. 33-4. *Position weights in an octal number.*

Table 33-2 shows the binary and octal numbers corresponding to first ten decimal numbers.

Table 33-2. ***Binary and Octal Numbers corresponding to first ten decimal numbers***

Decimal	*Binary*	*Octal*
0	000	0
1	001	1
2	010	2
3	011	3
4	100	4
5	101	5
6	110	6
7	111	7
8	1000	10
9	1001	11
10	1010	12

Please note that to signify an octal number, a subscript 8 or the letter O is used (*i.e.*, 17_8 or 17_0). However the use of subscript 8 is more common in actual practice.

33-16. Octal-to-Decimal Conversion

The octal-to-decimal conversion is a useful tool in the field of digital electronics (or computers). Let us study how the conversion from octal to decimal is done. Since the octal number system has a base of eight, each successive digit has a position weight of 8_0, the next higher digit has a position weight of 8^1, the next of 8^2 and so on. For a fractional octal, the first digit to the right of octal point, has a position weight of 8^{-1}, the next has a position weight of 8^{-2} the next 8^{-3} and so on.

Following is the procedure to convert an octal number to its decimal equivalent.

Step 1 Write the octal number.

Step 2 Directly under the octal number write the position weight of each digit working from right to left.

Step 3 Multiply each octal digit by its position weight and take sum of the products.

In order to illustrate the method, consider an example, of converting 23748.

The evaluation of an octal number in terms of its decimal equivalent is accomplished by multiplying each digit by its weight and summing the products, consider for example, the conversion of octal number 2374 to its equivalent decimal. The conversion is shown below:

Step 1 2 3 7 4

↓ ↓ ↓ ↓

Step 2 8^{+3} 8^{+2} 8^{+1} 8^0

Step 3 $= (2 \times 8^3) + (3 \times 8^2) + (7 \times 8^1) + (4 \times 8^0)$

$$= (2\times512)+(3\times64)+(7\times8)+(4\times1)$$

$$=1024+192+56+4 \;=1276_{10}$$

$\therefore\ 2374_8 = 1276_{10}$

Consider another example of converting 372.6 to its equivalent decimal. Then,

Step 1	3	7	2	6
	↓	↓	↓	↓
Step 2	8^{+3}	8^{+2}	8^{+1}	8^{0}

Step 3

$$= (3\times8^{+3})+(7\times8^{+2})+(2\times8^{+1})+(6\times8^{0})$$

$$= (3\times64)+(7\times8)+(2\times1)+(6\times0.125)$$

$$= 192 + 56 + 2 + 0.750 = 250.75$$

$\therefore\ 3726_8 = 250.75_{10}$

Example 33-10. *Convert the number 326_6 to its equivalent decimal number.*

Solution. Given the octal number = 326.

We know that, in order to convert an octal number to its decimal equivalent, we multiply each digit by its position weight and then sum the products, thus,

Step 1	3	2	6
	↓	↓	↓
Step 2	8^{+2}	8^{+1}	8^{0}

Step 3

$$= (3\times8^{+2}) + (2\times8^{+1}) + (6\times8^{0})$$

$$= (3 \times 64) + (2 \times 8) + (6 \times 1)$$

$$= 192 + 16 + 6 = 214_{10}$$

$\therefore\ 326_8 = 214_{10}$ **Ans.**

33-17. Decimal-to-Octal Conversion

We have already discussed in the last article, the method of converting the octal number to its decimal equivalent. Now we shall study the conversion of decimal-to-octal conversion. In order to make the study more effective we will consider the conversion of integer and fractional numbers separately.

33-18. Integer Decimal-to-Octal Conversion

In order to convert an integer decimal number to its equivalent octal number, we use a repeated division by-8 method.

Step 1 Begin by dividing the given decimal number by 8.

Step 2 Divide each resulting quotient by 8 until there is a zero whole number quotient.

Step 3 The remainders are generated by each division from the octal number. The first remainder to be produced is the least significant digit (LSD) in the octal number and the last remainder to be produced is the most significant digit (MSD). In other words, reading the remainders from bottom-to-top constitutes the octal number.

If a calculator is being used to perform the divisions in the conversion process, the results will include a decimal fraction of the remainder. The remainder can be obtained by multiplying the fraction by 8. Let us illustrate the procedure by considering the conversion of 266_8 to its decimal equivalent.

266 ÷ 8 = 33.25	with a remainder	0.25 → 0.25 × 8 = 2	Top ↑ LSB
33 ÷ 8 = 4.125	with a remainder	0.125 → 8.125 × 8 = 1	
4 ÷ 8 = 0.5 = 0	with a remainder	0.5 → 0.5 × 4 = 1	MSB Bottom

Since, the quotient is zero, therefore we will stop the division by 8. Reading the remainders from bottom to top, we find that the octal equivalent of the decimal number 266 is 412.

33-19. Fractional Decimal to Octal Conversion

In order to convert a fractional decimal number to an equivalent octal number, we use a repeated multiplication-by-8-method.

Step 1 Begin by multiplying the given decimal fraction by 8 and thus multiplying each resulting fractional part of the product by 2.

Step 2 Repeat step 1 until the fractional product is zero with the desired number of decimal places is reached.

Step 3 The carried digits or carries generated by the multiplication produce the octal number.

For example consider the conversion of fraction 0.25 to its octal equivalent. The first carry produced is the most-significant digit (MSD). In other words reading from top to bottom constitutes the required fractional octal.

In order to illustrate the procedure, consider the conversion of 0.3125 to its equivalent octal.

03125 × 8 = 2.5 = 0.5	with a carry	2	Top ↓ MSD
0.5 × 8 = 4.0 = 0	with a carry	4	LDS Bottom

Since, the fractional part is zero, so we will stop the repeated multiplication, Reading from top to bottom, the required octal number is 24.

$\therefore\ 0.3125_{10} = 0.24_8$

33-20. Octal-to Binary Conversion

The primary advantage of the octal number system is the case with which conversion can be made between binary and octal numbers. The conversion from octal-to-binary is performed by converting each octal digit to its 3-bit binary equivalent. Table 33-3 shows the 3-bit binary equivalent for the octal digits 0, 1, 2,...7.

Table 33-3. *Octal-to-binary conversion*

Octal	*Binary*
0	000
1	001
2	010
3	011
4	100
5	101
6	110
7	111

Using Table 33-3, we can convert any octal number to binary by individually converting each digit. Consider, for example, the conversion of 13 to its equivalent binary. From the Table 33-3, we find that octal digit 1 is equivalent to 001 and octal digit 3 is equivalent to 011. Thus

$$\begin{array}{cc} 1 & 3 \\ \downarrow & \downarrow \\ \overbrace{001} & \overbrace{011} \end{array}$$

$\therefore\ 13_8 = 001011_2$ or simply 1011_2

Similarly, suppose we need to convert the octal 526_8 to its binary equivalent. Again from the table 33-3, we find that the octal digit 5 it equivalent to binary 101, the octal digit 2 is binary 010 and octal digit 6 is 110. Thus,

$$\begin{array}{ccc} 5 & 2 & 6 \\ \downarrow & \downarrow & \downarrow \\ \overbrace{101} & \overbrace{010} & \overbrace{110} \end{array}$$

Hence, the binary equivalent for the octal 526_8 is 101010110_2

$\therefore\ 526_8 = 101010110_2$

Example 33-11. *Convert the following octal numbers to their binary equivalent.*

(*a*) 321_8 (*b*) 4653_8 (*c*) 13274_8

Solution.

(*a*) Given: The octal number = 321

We know that to convert the given octal number to its binary equivalent, we have to convert each octal digit to its 3-bit binary equivalent. Thus,

$$\begin{array}{ccc} 3 & 2 & 1 \\ \downarrow & \downarrow & \downarrow \\ \overbrace{011} & \overbrace{010} & \overbrace{001} \end{array}$$

Thus, the octal 321_8 is equivalent to the binary 011010001_2.

Dropping off the zero at the left most position, the binary number can be written as 11010001.

$\therefore\ 321_8 = 1101000_2$ **Ans.**

(*b*) Given: The octal number = 4653_8

We know that to convert the given octal number to its binary equivalent, we have to convert each octal digit to its 3-bit binary equivalent. Thus,

$$\begin{array}{cccc} 4 & 6 & 5 & 3 \\ \downarrow & \downarrow & \downarrow & \downarrow \\ \overbrace{100} & \overbrace{110} & \overbrace{101} & \overbrace{011} \end{array}$$

Thus, the octal 4653_8 is equivalent to the binary 100110101011_2.

$\therefore\ 4653_8 = 100110101011_2$ **Ans.**

(*c*) Given: The octal number = 13274_8

We know that to convert the given octal number to its binary equivalent, we have to convert each octal digit by its 3-bit binary equivalent. Thus,

1	3	2	7	4
↓	↓	↓	↓	↓
001	011	010	111	100

Thus, the octal number 13274 is equivalent to binary 001011010111100.

Dropping off the zero at the left-most position, the resulting binary number can be written as 1011010111100.

$\therefore$ $13274_8 = 1011010111100_2$ **Ans.**

Example 33-12. *Convert each of the following decimal numbers to octal equivalent by repeated diversion-by-8 method.*

(*a*) 142 (*b*) 221 (*c*) 435

Solution.

(*a*) Given: The decimal number = 142

We know that in order to convert the given number 142_{10} to its octal equivalent, we have to use the repeated diversion-by-8 method. Thus,

			Top ↑ LSD
$142 \div 8 = 17.75 = 17$	with a remainder	$0.75 \rightarrow 0.75 \times 8 = 6$	
$17 \div 8 = 2.125 = 2$	with a remainder	$0.105 \rightarrow 0.125 \times 8 = 1$	
$2 \div 8 = 0.25 = 0$	with a remainder	$0.25 \rightarrow 0.25 \times 8 = 2$	
			MSD Bottom

Since, the quotient is zero, therefore we will stop the diversion-by-8. Reading the remainders from bottom to top, we find that the octal equivalent of the decimal number 142 is 216.

$\therefore$ $142_{10} = 216_8$ **Ans.**

(*b*) Given: The decimal number = 221

We know that in order to correct the given number 221_{10} to its octal equivalent, we have to use the repeated diversion-8 method. Thus,

			Top ↑ LSD
$221 \div 8 = 27.625 = 27$	with a remainder	$0.625 \rightarrow 0.625 \times 8 = 5$	
$27 \div 8 = 3.375 = 3$	with a remainder	$0.375 \rightarrow 0.375 \times 8 = 3$	
$3 \div 8 = 0.375 = 0$	with a remainder	$0.375 \rightarrow 0.375 \times 8 = 3$	
			MSD Bottom

Since, the quotient is zero, therefore we will stop the diversion by 8. Reading the remainders from bottom to top, we find that the octal equivalent of 221 is 325.

$\therefore$ $13274_8 = 1011010111100_2$ **Ans.**

(*c*) Given: The decimal number = 435

			Top ↑ LSD
$435 \div 8 = 54.375 = 54$	with a remainder	$0.375 \rightarrow 0.375 \times 8 = 3$	
$54 \div 8 = 6.75 = 6$	with a remainder	$0.75 \rightarrow 0.75 \times 8 = 6$	
$6 \div 8 = 0.75 = 0$	with a remainder	$0.75 \rightarrow 0.75 \times 8 = 6$	
			MSD Bottom

Since, the quotient is zero, therefore we will stop the diversion by 8. Reading the remainders from bottom to top, we find the octal equivalent of 435 is 663.

$\therefore\ 435_{10} = 663_8$ **Ans.**

Example 33-13. Convert each of the following octal numbers to their equivalent binary numbers

(*a*) 624 (*b*) 326 (*c*) 476

Solution.

(*a*) Given: The octal number = 624

We know that to convert the given octal number to its binary equivalent, we have to convert each octal digit to its binary equivalent. Thus,

6	2	4
↓	↓	↓
110	010	100

$\therefore\ 624_8 = 110010100_{10}$ **Ans.**

(*b*) Given: The octal number = 326

We know that to convert the given octal number to its binary equivalent we have to convert each octal digit to its binary equivalent. Thus,

3	2	6
↓	↓	↓
011	010	110

$\therefore\ 326_8 = 011010110_2$ **Ans.**

(*c*) Given: The octal number = 476

4	7	6
↓	↓	↓
100	111	110

$476_8 = 100111110_2$ **Ans.**

Example 33-14. *Convert the decimal number 2497.50_{10} to its equivalent octal.*

Solution. Given: The decimal number 2497.50.

We know that we can convert the integer part of the given number to its equivalent octal by the repeated division by 2 method. Whereas the fractional part is to be converted to its equivalent octal by the repeated multiplication-by-8 method.

Conversion of integer part

				Top ↑
$2497 \div 8 = 312.13 = 312$	with a remainder	$0.13 \rightarrow 0.13 \times 8 = 1$		LSD
$312 \div 8 = 39.00 = 39$	with a remainder	$0 \rightarrow 0 \times 8 = 0$		
$39 \div 8 = 4.88 = 4$	with a remainder	$0.88 \rightarrow 0.88 \times 8 = 7$		
$4 \div 8 = 0.5 = 0$	with a remainder	$0.5 \rightarrow 0.5 \times 8 = 4$		MSD
				Bottom

$\therefore\ 2497_{10} = 4701_8$

Conversion of fractional part

$$0.50 \times 8 = 4.00 = 0 \text{ with a carry } 4$$

Since, the fractional part is zero, so we stop the repeated multiplication. Thus,

$$0.5_{10} = 0.4_8$$

Combining the integer and the fractional part of the numbers, we get,

$$2497.50_{10} = 4701.4_8 \quad \textbf{Ans.}$$

33-21. Binary-to-Octal Conversion

Conversion of a binary integer number to an octal integer number is the reverse of the octal-to-binary conversion. Following is the procedure for converting a given binary number to its octal conversion.

In order to illustrate the procedure, consider the conversion of 10111010 to octal.

∴ $101111010_2 = 472_8$ **Ans.**

Let us consider another example, the conversion of 011010110 of octal.

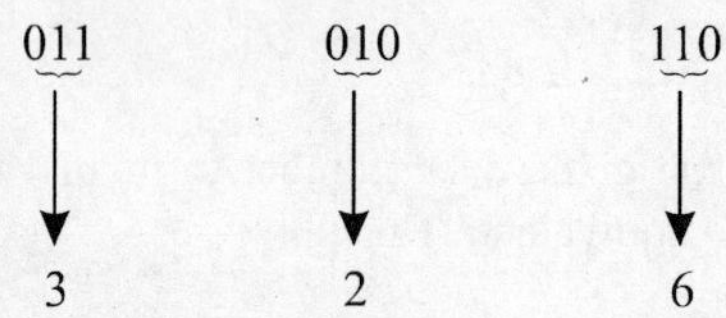

Note that we formed 3-bit combination while moving right to left by starting from the right most position. For the left-most group, since it does not have the 3-bit combination, so we added one zero from out side to complete the combination.

∴ $011010101102_2 = 326_8$ **Ans.**

Example 33-15. *Convert the number* 10111001_2 *to octal*

Solution. Given: The binary number = 10111001

We know that in order to convert the given binary number to its equivalent octal, we have to form 3-bit combinations while moving right to left by starting from the right most position.

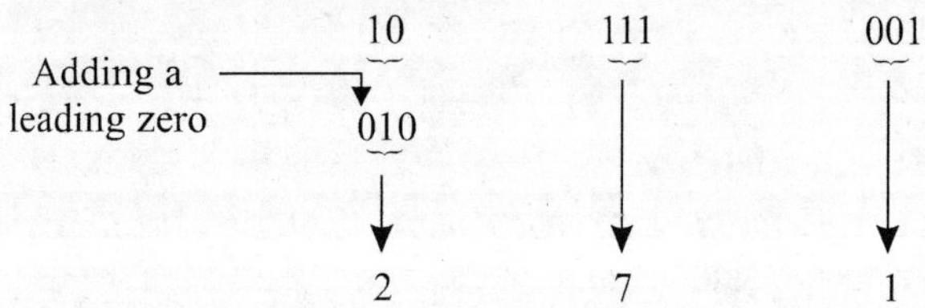

Note that for the left most group, since it does not have the 3-bit combination, so we have added one zero from out side to complete the combination.

∴ $10111001_2 = 271_8$ **Ans.**

Example 33-16. *Convert the following binary numbers to their octal equivalents.*

(*a*) *110101111* (*b*) *100110010* (*c*) *10111111001*

Solution.

(*a*) Given: The binary number = 110101111

We know that in order to convert the given binary number to its octal equivalent, we have to form 3-bit combinations while moving right to left by starting from the right most position. Thus,

110 → 6   101 → 5   111 → 7

$\therefore\ 110101111_2 = 657_8$ **Ans.**

(*b*) Given: The binary number = 100110010

We know that in order to convert the given binary number to its octal equivalent, we have to form 3-bit combinations while moving right to left by starting from the right most position. Thus,

100 → 4   110 → 6   010 → 2

$\therefore\ 100110010_2 = 462_8$ **Ans.**

(*c*) Given: The binary number, 10111111001

We know that in order to convert the given binary number to its octal equivalent, we have to form 3-bit combinations while moving right to left by starting from the right most position. Thus,

10 (Adding a leading zero) → 010 → 2   111 → 7   111 → 7   001 → 1

Note that for the left most group, since it does not have the 3-bit combination, so we have added one zero from ourside to complete the combination.

$\therefore\ 10111111001_2 = 2771_8$ **Ans.**

33-22. Usefulness of Octal System

We have already discussed the octal number system and conversion from the binary and decimal numbers to octal and vice versa. The ease with which conversions can be made between octal and binary makes the octal system attractive as "shorthand" means of expressing large binary numbers. In computer work, binary number with up to 64 bits are not uncommon. These binary numbers, as we shall see, do not always represent a numerical quantity but are often some type of code that conveys non numerical information. In computers, binary numbers might represent :

(*i*) actual numerical data
(*ii*) numbers corresponding to a location called (address) in memory
(*iii*) an instruction code
(*iv*) a code representing alphabetic and other non numerical characters
(*v*) group of bits representing the status of devices internal or external to the computer.

When dealing with a large quantity of binary numbers of many bits, it is convenient and more efficient for us to write the numbers in octal rather than binary. However keep in mind that the digital circuits and systems work strictly in binary. We use octal numbers only as a convenience for the operators of the system.

Example 33-17. *Convert the decimal 486 to its eight binary equivalent by first converting it to octal.*

Solution. Given: The decimal number = 486

			Top
$486 \div 8 = 60$	with a remainder	6	↑ LSB
$60 \div 8 = 7$	with a remainder	4	
$4 \div 8 = 0$	with a remainder	7	MSB
			Bottom

Now in order to convert the octal 746_8, to its binary equivalent, we replace each octal digit by its 3-bit equivalent binary number, *i.e,*

$$\begin{array}{ccc} 7 & 4 & 6 \\ \downarrow & \downarrow & \downarrow \\ \overbrace{111} & \overbrace{100} & \overbrace{110} \end{array}$$

$\therefore \; 486_{10} = 111100110_2$ **Ans.**

33-23. Hexadecimal Number System

The hexadecimal number system like octal number system is used primarily as a "shorthand" way of displaying binary numbers because it is very easy to convert between binary and hexadecimal. As you are probably aware, long binary numbers are difficult to read and write because it is easy to drop or transpose a bit. Since computers and microprocessors understand only 1s and 0s, it is necessary to use these digits when you program in "machine language". Imagine writing a bit instruction for a microprocessor system in 1s and 0s. It is much more efficient to use the hexadecimal or octal. Hexadecimal is frequently used in computer and microprocessor applications.

The hexadecimal system is composed of 16 digits which includes numerals and alphabetic characters. The 16 digits are 0, 1, 2, 3,...7, 8, 9, A, B, C, D, E and F. Most digital systems process binary data in groups that are multiples of four bits, making the hexadecimal number very convenient because each hexadecimal digit represent a 4-bit binary number as shown in Table 33-4.

Table 33-4. ***Binary and hexadecimal numbers for the first 16 decimal numbers***

Decimal	*Binary*	*Hexadecimal*
0	0000	0
1	0001	1
2	0010	2
3	0011	3
4	0100	4
5	0101	5
6	0110	6
7	0111	7
8	1000	8
9	1001	9
10	1010	A
11	1011	B
12	1100	C
13	1101	D
14	1110	E
15	1111	F

Note that to signify a hexadecimal number, a subscript 16 or the letter H is used (*i.e.*, $D7_{16}$ or D7H). The use of subscript is more common while solving tutorial problems whereas the use of letter H is more common while writing computers programs.

In order to count in hexadecimal beyond F, We form two digit combinations by starting with the second digit followed by the first digit, *i.e.*,

10, 11, 12, 13, 14, 15, 16, 17, 18, 19, 1A, 1B, 1C, 1D, 1E, 1F, 20, 21, 22, 23, 24, 25, 26, 27, 28, 29, 2A, 2B, 2C, 2D, 2E, 2F, 30, 31....and so on.

With two hexadecimal digits, you can count up to FF_{16} which is equivalent to decimal 255. To count beyond this, we form three digit combinations. For instance 100_{16} is decimal 256, 101_{16} is decimal 257, and so forth. The maximum 4-digit hexadecimal number is $FFFF_{16}$ which is equivalent to decimal 65,535.

33-24. Hexadecimal-to-Decimal Conversion

We have already discussed the representation of numbers in hexadecimal number system. Now we shall study how a hexadecimal integer (or a whole) number can be converted to its equivalent hexadecimal number.

Following is the procedure to convert a hexadecimal number to its equivalent decimal number:

Step 1 Write the hexadecimal number

Step 2 Directly under the hexadecimal number, write the position weight of each digit working from right to left.

Step 3 Multiply the decimal value of each hexadecimal digit by its position weight and take sum of the products.

In order to illustrate the method, let us take an example, we will convert E5 to its equivalent decimal number,

Step 1 E 5

↓ ↓

Step 2 16^{+1} 16^{0}

Step 3 $(E\times16^{+1})+(5\times16^{0})$

$=(E\times16)+(5\times1)$

$=(14\times16)+(5\times1)$

$= 224 + 5 = 229$

$\therefore$ $E5_{16} = 229_{10}$ **Ans.**

Let us take another example of a hexadecimal number 0.12_{16} and convert to its equivalent decimal fraction.

Step 1 0 . 1 2

↓ ↓

Step 2 16^{-1} 16^{-2}

Step 3 $1\times16^{-1}+2\times16^{-2}$

$= 0.0625 + 0.0078$

$= 0.0703$ **Ans.**

Example 33-18. *Convert the hexadecimal number $2A6_{16}$ to its equivalent decimal number.*

Solution. Given: The hexadecimal number = 2A6

We know that to convert 2A6 to its equivalent decimal number, we will multiply the decimal value of each hexadecimal digit by its position weight and then table the sum of these products.

Step 1	2	A	6
	↓	↓	↓
Step 2	16^{+2}	16^{+1}	16^{+0}

Step 3 $(2\times16^{+2})+(A\times16^{+1})+(6\times16^{0})$

$= (2\times256)+(10\times16)+(6\times1)$

$= 678$

$\therefore\ 2A6_{16} = 678_{10}$ **Ans.**

Example 33-19. *Convert the hexadecimal number B2F8 to its equivalent decimal number.*

Solution. Given: The hexadecimal number = B2F8

We know that to convert B2F8 to its equivalent decimal number, we will multiply the decimal value of each hexadecimal digit by its position weight and then take the sum of these products.

Step 1	B	2	F	8
	↓	↓	↓	↓
Step 2	16^{+3}	16^{+2}	16^{+1}	16^{+0}

Step 3 $(B \times 16^{+3}) + (2 \times 16^{+2}) + (F \times 16^{+1}) + (8 \times 16^{0})$

$= (B \times 4096) + (2 \times 256) + (F \times 16) + (8 \times 1)$

$= (11 \times 4096) + (2 \times 256) + (15 \times 16) + (8 \times 1)$

$= 45056 + 512 + 240 + 8$

$= 45816$

$\therefore\ B2F8_{16} = 45816_{10}$ **Ans.**

33-25. Decimal-to-Hexadecimal Conversion

It is a systematic way of converting numbers from decimal to hexadecimal. The procedure for conversion is as given below:

Step 1 Begin by dividing the given decimal number by 16.

Step 2 Divide each resulting quotient by 16 until there is a zero whole number quotient.

Step 3 The remainders generated by each division from the hexadecimal number. The first remainder to be produced is the least-significant digit (LSD) in the hexadecimal number and the last remainder to be produced is the most-significant digit (MSD). In other words, reading the remainders from bottom-to-top constitutes the hexadecimal number.

Note that any remainders that are greater than 9 are represented by the letters A through F.

If a calculator is being used to perform the divisions in the conversion process, the results will include a decimal fraction of a remainder. The remainder can be obtained by multiplying the fraction by 16.

Let us illustrate the procedure for conversion of the decimal number 650 to its equivalent hexadecimal number.

$$650 \div 16 = 40.625$$

So in this case 40 is a quotient and 0.625 is a fraction remainder. The remainder becomes

$$0.625 \times 16 = 10 = \text{A (hexadecimal number)}$$

Thus the complete conversion is as shown below:

$650 \div 16 = 40.625 = 40$	with a remainder	$0.625 \rightarrow 0.625 \times 16 = 10\ (= \text{A})$	Top ↑ LSD
$40 \div 16 = 2.5 = 2$	with a remainder	$0.5 \rightarrow 0.5 \times 16 = 8$	
$2 \div 16 = 0.125 = 0$	with a remainder	$0.125 \rightarrow 0.125 \times 16 = 2$	MSD Bottom

Since, the quotient is zero, therefore we will stop the division by 16. Reading the remainders from bottom to top, we find that the hexadecimal equivalent of the decimal number 650 is 28A.

$\therefore\ 650_{10} = 28\text{A}_{16}$ **Ans.**

Example 33-20. *Convert the number 151_{10} to its equivalent hexadecimal number.*

Solution. Given: The number $= 151_{10}$

We know that we can use the repeated division by 16 for converting a decimal number to its hexadecimal equivalent. Using division by hand, we get,

$151 \div 16 = 9$	with a remainder	7	Top ↑ LSD
$9 \div 16 = 0$	with a remainder	9	MSD Bottom

Since, the quotient is zero, so we will stop the further division. Reading the remainders from bottom to top, the hexadecimal number is 97.

$\therefore\ 151_{10} = 97_{16}$ **Ans.**

Example 33-21. *Convert the number 498_{10} to its equivalent hexadecimal number*

We know that we can use the repeated division by 16 for converting a decimal number to its hexadecimal equivalent. Using division by hand, we get,

$498 \div 16 = 31$	with a remainder	2	Top ↑ LSD
$31 \div 16 = 1$	with a remainder	15 (= F)	MSD
$1 \div 16 = 0$	with a remainder of	1	Bottom

Since, the quotient is zero, so we will stop the further division. Reading the remainders from bottom to top, the hexadecimal number is 1F2.

$\therefore\ 498_{10} = 1\text{F}2_{16}$ **Ans.**

Example 33-22. *Convert the following decimal numbers to their hexadecimal equivalent:*
(*a*) 2890 (*b*) 4019

Solution. (*a*) Given: The decimal number = 2890

$2890 \div 16 = 180.625 = 180$	with a remainder	$0.625 = 0.625 \times 16 = 10\ (= A)$		Top ↑ LSD
$180 \div 16 = 11.25 = 11$	with a remainder	$0.25 = 0.25 \times 16 = 4$		
$11 \div 16 = 0.6875 = 0$	with a remainder	$0.6875 = 0.6875 \times 16 = 11\ (= B)$		MSD Bottom

Since, the quotient is zero, so we will stop the further division. Reading from bottom to top, the equivalent hexadecimal number of a decimal number 2890 is B4A.

$\therefore\ 2890_{10} = B4A_{16}$ **Ans.**

(*b*) Given: The decimal number = 4019

$4019 \div 16 = 251.875 = 251$	with a remainder	$0.875 = 0.875 \times 16 = 14\ (= E)$		Top ↑ LSD
$251 \div 16 = 15.688 = 15$	with a remainder	$0.688 = 0.688 \times 16 = 11\ (= B)$		
$15 \div 16 = 0.938 = 0$	with a remainder	$0.738 = 0.738 \times 16 = 15\ (= F)$		MSD Bottom

Since, the quotient is zero, so we will stop the further division. Reading from bottom to top, the equivalent hexadecimal under of a decimal number 4019 is FBE.

$\therefore\ 4019_{10} = FBE_{16}$ **Ans.**

33-26. Hexadecimal-to-Binary Conversion

Like the octal number system, the hexadecimal number system is used primarily as a "shorthand" method for representing binary numbers. For conversion, each hexadecimal digit is covered to its four-bit binary equivalent.

The procedure for converting hexadecimal number to its binary equivalent is as follows:

Step 1 Write the hexadecimal number

Step 2 Write the 4-bit binary equivalent for each hexadecimal digit

Step 3 Check if there are any zeros on the left most position of the binary number obtained. Drop off the zeros and write the answer as a binary number.

In order to illustrate the above procedure, Consider the conversion of $2D6_{16}$ to its equivalent binary.

Step 1	2	D	6
	↓	↓	↓
Step 2	0010	1101	0110
Step 3	001011010110		
	↓ Drop off the leading zeros		

Therefore, the required binary number is, 1011010110

$\therefore\ 2D6_{16} = 1011010110_2$ **Ans.**

Example 33-23. *Convert the following hexadecimal numbers of their binary equivalent:*
(*a*) 9F2 (*b*) 2A6

Solution. (*a*) Given the hexadecimal number = 9F2

	9	F	2
Step 1	↓	↓	↓
Step 2	1001	1111	0010
Step 3	100111110010		

$\therefore$ $9F2_{16} = 1001\ 1111\ 0010_2$ **Ans.**

(*b*) Given: The hexadecimal number = 2A6.

	2	A	6
Step 1	↓	↓	↓
Step 2	0010	1010	0110
Step 3	00101010010		
	↓		
	Drop off the leading zeros		

$\therefore$ $9A6_{16} = 1010100110_2$ **Ans.**

33-27. Binary-to-Hexadecimal Conversion

Conversion from binary to-hexadecimal is just the reverse of the process discussed in the last article (*i.e.*, reverse of the process for hexadecimal to-binary conversion.

The procedure for converting hexadecimal number to its binary equivalent is an follows:

Step 1 Write the binary number.

Step 2 Starting from the right most position, group the binary number into groups of four-bits. If necessary, we can add zeros at the left-most position.

Step 3 Convert each 4-bit binary group to its equivalent hexadecimal digit.

In order to illustrate the procedure, consider the conversion of binary 10111 to its equivalent hexadecimal.

Right-most position

Step 1	1 0 1	1 1
Step 2	0001	0111
Three zeros are added to form a Four-bit Group	↓	↓
Step 3	1	7

$\therefore$ $10111_2 = 17_6$ **Ans.**

Example 33-24. *Convert the following binary numbers to their hexadecimal equivalents.*

(*a*) 10100110 (*b*) 1111 110000 (*c*) 100110000010.

Solution.

(*a*) Given: The binary number = 10100110.

Step 1 1 0 1 0 0 1 1 0

Step 2 1010 0110

Step 3 A 6

$\therefore$ $10100110_2 = A6_{16}$ **Ans.**

(*b*) Given: The binary number = 1111110000

Step 1 1 1 1 1 1 1 0 0

Step 2 0011 1111 0000

Two zeros added to form 4-bit group

Step 3 3 F 0

$\therefore$ $1111110000_2 = 3F0_{16}$ **Ans.**

Example 33-25. *Convert the following bases :*

(a) *$(11011.011)_2$ = ($)_{16}$*

(b) *$(2AC9)_{16}$ = ($)_7$* *(UP Tech. Univ. Elect.Engg, 1st Sem. 2006-07)*

Solution.

(*a*) *Conversion of $(11011.011)_2$ its equivalent hexadecimal*

Step 1 1 1011 • 011

Add 3 zeros to complete 4-unit group: 000

Add one zero to complete 4-unit group: 0110

1 B 6

Thus, the hexadecimal equivalent of the given binary number 11011.011 is 1B6. *i.e.*, $(11011.011)_2 = (1B6)_{16}$ **Ans.**

(*b*) *Conversion of $(2AC9)_{16}$ to its equivalent number with base 7.*

In order to convert the given number $(2AC9)_{16}$ to its equivalent number with base 7, let us convert the number $(2AC9)_{16}$ to its decimal equivalent and then the decimal equivalent to a number with base 7.

Step 1	2	A	C	C
	↓	↓	↓	↓
Step 2	16^3	16^2	$16^{\prime}$	16°

$$\text{Step 3} \quad = (2\times16^3)+(A\times16^2)+(C\times16^1)+(9\times16^0)$$
$$= (2\times4096)+(10\times256)+(12\times16)+(9\times1)$$
$$= 8192+2560+192+9$$
$$= (10953)_{10}$$

Let us now convert the decimal number 10953 to its equivalent number with base 7. This can be done by dividing 10953 repeatly-by-7. Thus,

			Top
$10953 \div 7 = 1560$	with a remainder of	1	↑ LSB
$1560 \div 7 = 222$	with a remainder of	6	
$222 \div 7 = 31$	with a remainder of	5	
$31 \div 7 = 4$	with a remainder of	3	MSD
$4 \div 7 = 0$	with a remainder of	4	
			Bottom

Reading the remainders from bottom to top, we get 43561.

$\therefore$ $(2AC9)_{16} = (10953)_{10} = (43561)_7$ **Ans.**

33-28. Hexadecimal-to-Octal Conversion

There are two methods for converting hexadecimal number to its octal equivalent. The first method is to convert the given hexadecimal number to binary equivalent and then from binary to octal. The second method is to convert the given hexadecimal number to its decimal equivalent and then from decimal to octal. However, the first method is much simpler and convenient.

The procedure for conversion of hexadecimal-to-octal is as given below :

Step 1 Write the hexadecimal number

Step 2 Replace each hexadecimal digit by its 4-bit binary equivalent. This will give us the binary equivalent of the given hexadecimal number.

Step 3 Form 3-bit combinations by starting from the right-most position. Replace each 3-bit combination by its octal equivalent. This will give us the octal equivalent for the given hexadecimal number.

In order to illustrate the procedure, consider the conversion of 5C2 to its octal equivalent.

Step 1	5	C	2
	↓	↓	↓
Step 2	0101	1100	0010

Step 3 010 111 000 010

↓ ↓ ↓ ↓

2 7 0 2

Thus, the octal equivalent for the given hexadecimal number is 2702_8

$\therefore$ $5C2_{16} = 2702_8$ **Ans.**

Example 33-26. *Convert the hexadecimal number $8AD9_{16}$*

Step 1	8	A	D	9
	↓	↓	↓	↓
Step 2	1000	1010	1101	1001

Step 3 100 010 011 011 001

↓ ↓ ↓ ↓ ↓

4 2 3 3 1

Thus, the octal equivalent for the given hexadecimal is 42331_8.

$\therefore$ $8AD9_{16} = 42331_8$ **Ans.**

Example 33-27. *Convert $(A7.3B)_{16}$ into its equivalent.*

Solution. Given: The hexadecimal number = A7.3B

First of all we will convert $A7.3B_{16}$ to its equivalent binary and then to its octal equivalent.

A 7	3 B	— Hexadecimal number
10100111	00111011	— Binary equivalent
2 4 7	1 6 6	— Octal equivalent

Thus, we find that, octal equivalent of hexadecimal (A7.3B) is 247.166.

$\therefore$ $A7.3B_{16} = 247.166_8$ **Ans.**

Example 33-28. *Convert the number $(1E7B)_{16}$ to its equivalent octal.*

(*UP Tech. Univ., Elect Engg: 2nd Sem. 2004-05*)

Solution. Given: the hexadecimal number = 1E7B

Step 1	1	E	7	B
	↓	↓	↓	↓
Step 2	0001	1110	0111	1011

Step 3 0 0 01 111 001 111 011

↓ ↓ ↓ ↓ ↓

Drop of leading zero 1 7 1 7 3

Thus, we find that octal equivalent of $(1E7B)_{16}$ is $(17173)_8$

$\therefore$ $(1E7B)_{16} = (17173)_8$ **Ans.**

33-29. Octal-to-Hexadecimal Conversion

Like hexadecimal-to-octal there are two methods for converting octal-to-hexadecimal. The first method is to convert the given octal number to its binary equivalent and then from binary to hexadecimal. The second method is to convert octal-to-hexadecimal first and then from decimal-to-hexadecimal. However, the first method is much simpler and convenient.

The procedure for conversion from octal-to-hexadecimal is as given below:

Step 1 Write the octal number.

Step 2 Replace each octal digit by its 3-bit binary equivalent. This will give us the binary equivalent of the given octal number.

Step 3 Form the 4-bit combinations by starting from the right-most position. Replace each 4-bit combination by its hexadecimal equivalent. This will give us the hexadecimal equivalent for the given octal number.

In order to illustrate the procedure, consider the conversion of 321_8 to its hexadecimal equivalent.

Step 1 3 2 1

Step 2 011 010 001

Step 3 01101 0001

D 1

Thus, the hexadecimal equivalent for the given octal number is $D1_{16}$

$\therefore$ $321_8 = D1_{16}$ **Ans.**

Example 33-29. *Convert the octal number 1024_8 to its hexadecimal equivalent.*

Solution. Given: The octal number = 1024.

Step 1 1 0 2 4

Step 2 001 000 010 100

Step 3 0010 0001 0100

2 1 4

Thus, the hexadecimal equivalent for the given octal number is 214_{16}.

$\therefore$ $1024_8 = 214_{16}$ **Ans.**

Example 33-30. *Convert the number $(2715)_8$ to its equivalent hexadecimal number:*

(UP Tech. Univ., 2004-05)

Solution. Given the number $(2715)_8$

Step 1	2	7	1	5
	↓	↓	↓	↓
Step 2	010	111	001	101

Step 3   0 1 0 1   1 1 0 0   1 1 0 1

              ↓   ↓   ↓

              5   C   D

Thus the hexadecimal equivalent for the given octal number is $5CD_{16}$.

$\therefore$ $(2715)_8 = 5CD_{16}$ **Ans.**

Example 33-31. *Convert the following number to the given base:*

(*i*) $(62)_{10} = (?)_2 = (?)_8$

(*ii*) $(AFB)_{16} = (?)_2 = (?)_8$

Solution.

(*i*) $(62)_{16} = (?)_2$

		Top	
$62 \div 2 = 31$	with a remainder	0	
$31 \div 2 = 15$	with a remainder	1 ↑	LSB
$15 \div 2 = 7$	with a remainder	1	
$7 \div 2 = 3$	with a remainder	1	
$3 \div 2 = 1$	with a remainder	1	
$1 \div 2 = 0$	with a remainder	1	MSB
		Bottom	

$\therefore$ $(62)_{10} = 111110_2$ **Ans.**

In order to convert the binary 111110 to its equivalent octal, start from the right-most position and form a groups of 3-bits. Thus,

1 1 1	1 1 0
↓	↓
7	6

$\therefore$ $(62)_{10} = 111110_2 = (76)_8$ **Ans.**

(*ii*) $(AFB)_{16} = (?)_2 = (?)_8$

In order to convert the given hexadecimal number AFB to its equivalent binary, replace each hexadecimal digit by its-4-bit binary equivalent. Thus

A	F	B
↓	↓	↓
1 0 1 0	1 1 1 1	1 0 1 1

$\therefore$ $(AFB)_{10} = 101011111011_2$ **Ans.**

In order to convert the binary number 101011111011 to its octal equivalent, start from the right-most position and form groups of 3-bits and replace each 3-bits group by its equivalent octal digit. Thus,

$$\underbrace{101}_{\downarrow\ 5}\quad \underbrace{011}_{\downarrow\ 3}\quad \underbrace{111}_{\downarrow\ 7}\quad \underbrace{011}_{\downarrow\ 3}$$

$\therefore$ $(AFB)_{16} = 101011111011_2 = (5373)_8$ **Ans.**

Example 33-32. *Convert the following numbers:*

(***a***) $(6089.25)_{10} = (\quad)_8$

(***b***) $(A6B.F5)_{16} = (\quad)_2$

(***c***) $(375.37)_8 = (\quad)_2$ (*UP Tech. Univ., 2006-07*)

Solution.

(***a***) *Conversion of* $(6089.25)_{10}$ *to its octal equivalent*

Given; the decimal number 6089.25

We will first convert the integer number 6089_{10} to its octal equaivalent and then the fractional part 0.25_{10}. We know that the conversion of a decimal integer to its octal equivalent can be done by using division-by-8 method.

		Top
$6089 \div 8 = 761$	with a remainder of	1 ↑ (LSB)
$761 \div 8 = 95$	with a remainder of	1
$95 \div 8 = 11$	with a remainder of	7
$11 \div 8 = 1$	with a remainder of	3
$1 \div 8 = 0$	with a remainder of	1 (MSB)
		Bottom

Reading the remainders from bottom-to-top, the octal equivalent of decimal integer 6089 is 13711.

We also know that the conversion of a decimal fraction to its octal equivalent can be done by using repeated multiplication-by-8 method. Thus,

$$0.25 \times 8 = 2 = 0 \text{ with a carry of } 1 \text{ (M8B)}$$

Since the fraction part is reduced to 0 already, so we will stop. Therefore, octal equivalent of $(0.25)_{10}$ is $(0.11)_8$.

Combining the integer and fractional parts together, we get

$$(6089.25)_{10} = (13711.1)_8 \quad \textbf{Ans.}$$

(***b***) *Conversion of* $(A6B.F5)_{16}$ *to its binary equivalent.*

Step 1	A	6	B	•	F	5
	↓	↓	↓		↓	↓
Step 2	1010	0110	1011	•	1111	0101

Step 3 101001101011. 11110101

Therefore, the required binary number is 101001101011.111110101

$$(A6B.F5)_{16} = (101001101011.1111010101)_2 \quad \textbf{Ans.}$$

(*c*) *Conversion of* $(375.37)_8$ *to its binary equivalent*

Step 1	3	7	5	•	3	7
	↓	↓	↓		↓	↓
Step 2	011	111	101		011	111

Step 3 011 111 101.011 111

↓

Drop off the leading zero

$$\therefore \quad (375.37)_8 = (11111101.011111)_2 \quad \textbf{Ans.}$$

Example 33-33. Convert the following numbers into desired bases:

(*a*) $(4320)_{10} = (?)_2$

(*b*) $(1201102)_5 = (?)_{10}$ (*UP Tech. Univ. Basic Elect., 1st Sem (CO), 2003-04*)

Solution.

(*a*) *Conversion of* $(4320)_{10}$ *to its binary equivalent*

We know that conversion from decimal to binary can be done by using repeated-division-by 2 method.

			Top
4320 ÷ 2 = 2160	with a remainder	0	↑ (LSB)
2160 ÷ 2 = 1080	with a remainder	0	
1080 ÷ 2 = 540	with a remainder	0	
540 ÷ 2 = 270	with a remainder	0	
270 ÷ 2 = 135	with a remainder	0	
135 ÷ 2 = 67	with a remainder	1	
67 ÷ 2 = 33	with a remainder	1	
33 ÷ 2 = 16	with a remainder	1	
16 ÷ 2 = 8	with a remainder	0	
8 ÷ 2 = 4	with a remainder	0	
4 ÷ 2 = 2	with a remainder	0	
2 ÷ 2 = 1	with a remainder	0	
1 ÷ 2 = 0	with a remainder	1	(MSB)
			Bottom

Reading the remainders from bottom to top, the binary number for the given decimal number 4320 is 1000011100000.

$$\therefore \quad (4320)_{10} = (1000011100000)_2 \quad \textbf{Ans.}$$

(**b**) *Conversion of* $(1201102)_3$ *to its equivalent decimal.*

We know that any number with any base value can be expressed as the sum of digits multiplied by their corresponding position weights. Thus the least significant digit of the given number $(1201102)_3$ is 3^0, the next digit 0 has a position value of 3^1 and so on. The conversion can be carried out as follows:

Step 1	1	2	0	1	1	0	2
	↓	↓	↓	↓	↓	↓	↓
Step 2	3^6	3^5	3^4	3^3	3^2	3^1	3^0

Step 3 $(1\times3^6)+(2\times3^5)+(0\times3^4)+(1\times3^3)+(1\times3^2)$

$= 729 + 486 + 2719$

$= (1251)_{10}$ **Ans.**

Example 33-34. *Obtain the following conversions: (a)* $(10111)_2 = (?)_8$, *(b)* $(9BC8)_{16} = (?)_{10} = (?)_8$ *(UP Tech Univ. Basic Elect., 2nd Sem. 2003-04)*

Solution.

(**a**) *Conversion of* $(10111)_2$ *to its equivalent octal.*

Step 1 10 111

adding a leading zero → 010

010 → 2, 111 → 7

Note that for the left most 2-bit group, since it does not have the 3-bit combination, so we have added one zero from our side to complete the combination.

$\therefore$ $(10111)_2 = (27)_8$ **Ans.**

(**b**) *Conversion of* $(9BC8)_{16}$ *to its decimal and octal equivalents*

Let us convert the given hexadecimal number to its decimal equivalent.

Step 1	9	B	C	8
	↓	↓	↓	↓
Step 2	16^3	16^2	16^1	16^0

Step 3 $(9\times16^3)+(B\times16^2)+(C\times16^1)+(8\times16^0)$

$= 9 \times 4096 + (11 \times 256) + (12 \times 16) + (8 \times 1)$

$= (39980)_{10}$

$\therefore$ $(9BC8)_{16} = (39980)_{10}$ **Ans.**

Let us now convert the given hexadecimal number 9BC8 to its octal equivalent. We will do this by first converting $(9BC8)_{16}$ to its binary equivalent and then the binary-to-octal.

In order to convert the given hexadecimal number 9BC8 to its equivalent binary, replace each hexadecimal digit by its 4-bit binary equivalent. Thus,

9	B	C	8
↓	↓	↓	↓
1001	1011	1100	1000

$\therefore$ $(9BC8)_{16} = (1001101111001000)_2$

In order to convert the binary 1001101111001000 to its octal equivalent, start from right-most position and form groups of 3-bits each and replace each 3-bit group by its octal equivalent digit. Thus,

001	001	101	111	001	000
↓	↓	↓	↓	↓	↓
1	1	5	7	1	0

add two zeros to form 3-bit group

$\therefore$ $(9BC8)_{10} = (115710)_8$ **Ans.**

Example 33-35. *Convert the following numbers into desired bases:*

(***a***) $(3451)_{16} = (?)_2$, (***b***) $1745.246_8 = (?)_{16}$ (*UP Tech., Besic Elect, 1st Sem. 2003-04*)

Solution. (***a***) *Conversion of* $(3451)_{10}$ *to its binary equivalent.*

We know that the conversion from decimal-to-binary can be done by using repeated division-by-2 method.

Division		Remainder	
$3451 \div 2 = 1725$	with a remainder of	1	Top ↑ (LSB)
$1725 \div 2 = 862$	with a remainder of	1	
$862 \div 2 = 431$	with a remainder of	0	
$431 \div 2 = 215$	with a remainder of	1	
$215 \div 2 = 107$	with a remainder of	1	
$107 \div 2 = 53$	with a remainder of	1	
$53 \div 2 = 26$	with a remainder of	1	
$26 \div 2 = 13$	with a remainder of	0	
$13 \div 2 = 6$	with a remainder of	1	
$6 \div 2 = 3$	with a remainder of	0	
$3 \div 2 = 1$	with a remainder of	1	
$1 \div 2 = 1$	with a remainder of	1	(MSB) Bottom

Reading the remainders from bottom to top, we find that the binary number for the given decimal number 3457 is 110101111011.

$\therefore$ $(3451)_{10}$ $(110101111011)_2$ **Ans.**

(***b***) *Conversion of* $(1745.246)_8$ *to its hexadecimal equivalent.*

In order to convert the given octal number 1745.246 to its hexadecimal equivalent, we will convert it to its binary equivalent first and then to hexadecimal.

In order to convert the octal-to-binary, convert each octal digit to its 3-bit binary equivalent. Thus,

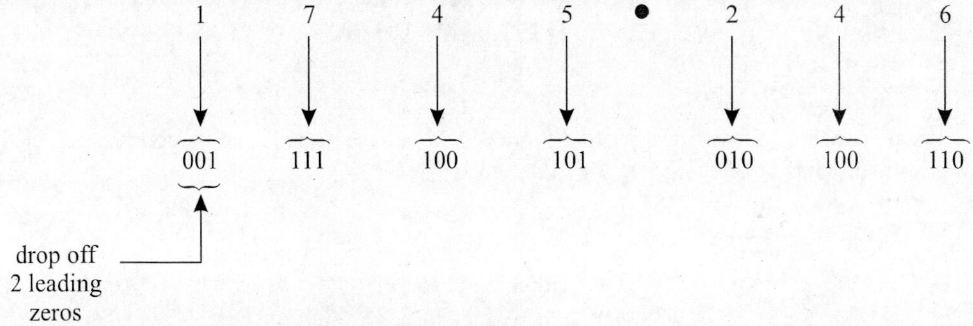

$(1745.246)_8 = (1111100101.010100110)_2$

Next starting from the binary point convert the binary-to-hexadecimal by forming 4-bit groups.

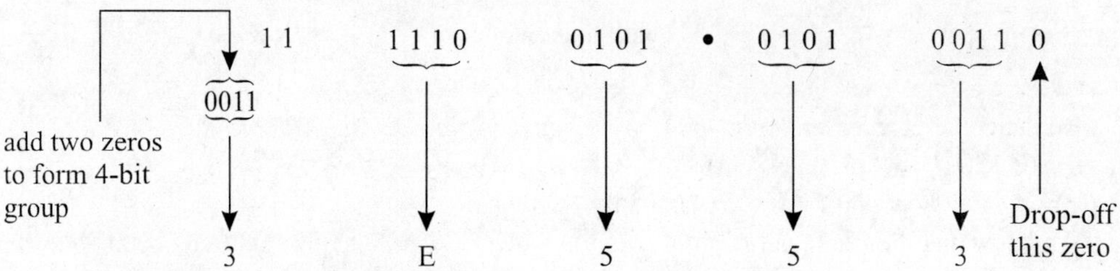

$\therefore\ (1111100101.010100110)_2 = (3E5.53)_{16}$

or $(1745.246)8 = (3E5.53)_{16}$ **Ans.**

Example 33-36. *Convert the following numbers as indicated:*

(*a*) $(1001)_8 = (\quad)_2$ (*b*) $(2CCD)_{16} = (\quad)_5$

(*c*) $(0.45)_{10} = (\quad)_8$ (*d*) $(345)_8 = (\quad)_{10}$

(*e*) $(7841)_9 = (\quad)_{10}$(*UP. Tech. Univ., Elect. Engg I Sem. 2005-06*)

Solution.

(***a***) *Conversion of* $(1001)_8$ *to its equivalent binary.*

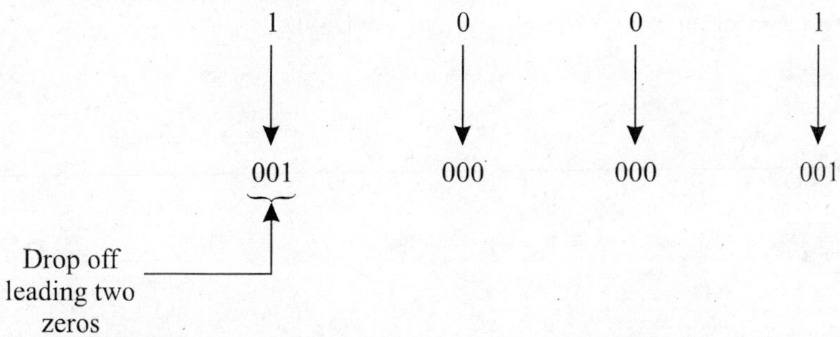

$\therefore\ (1001)_8 = (1000000001)_2$ **Ans.**

(***b***) *Convert of* $(2CCD)_{16} = (\quad)_5$

Let us convert the given hexadecimal number to its decimal equivalent.

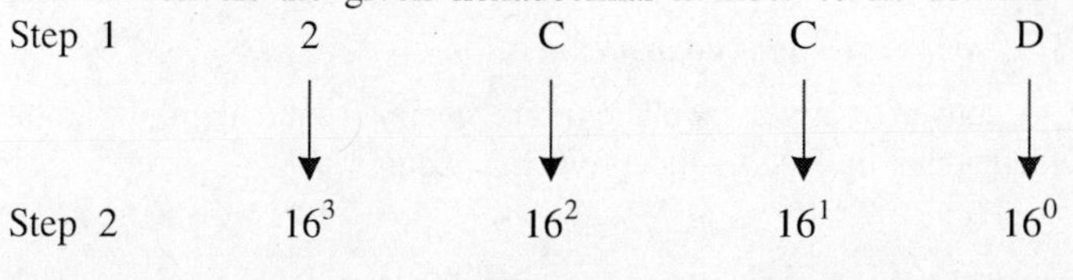

Step 3 $(2\times16^3)+(C\times16^2)+(C\times16^1)+(D\times16^0)$

$= (2 \times 4096) + (12 \times 256) + (12 \times 16) + (13 \times 1)$

$= 8192 + 3072 + 192 + 13$

$= 11469$

$\therefore\ (2CCD)_{16} = (11469)_{10}$

Let us now convert this number to a number with a base 5. This can be done by repeatedly division-by-5. Thus,

		Top ↑	
$11469 \div 5 = 2293$	with a remainder of	4	(LSB)
$2293 \div 5 = 458$	with a remainder of	3	
$458 \div 5 = 91$	with a remainder of	3	
$91 \div 5 = 81$	with a remainder of	1	
$18 \div 5 = 3$	with a remainder of	3	
$3 \div 5 = 0$	with a remainder of	3	(MSB)
		Bottom	

Reading the remainders from bottom to top, the number is 331334.

$\therefore\ (2CCD)_{16} = (331334)_5$ **Ans.**

(*c*) *Conversion of* $(0.45)_{10}$ *to its octal equivalent.*

We know that the fractional point of a decimal number can be converted to its octal equivalent by the repeated multiplication-by-8 method. Thus,

		Top ↓
$0.45 \times 8 = 3.50 = 0.5$	with a carry	3
$0.5 \times 8 = 4.00 = 0$	with a carry	4
		Bottom

Reading the caries from top to bottom, we get 0.34.

$\therefore\ (4.45)_{10} = (0.34)_8$ **Ans.**

(*d*) *Conversion of* $(345)_8$ *to its decimal equivalent.*

We know that in order to convert an octal number to its decimal equivalent, we multiply each digit by its position weight and then sum the products. Thus,

Step 1	3	4	5
	↓	↓	↓
Step 2	8^2	8^1	8^0

Step 3 $= (3 \times 8^2) + (4 \times 8^1) + (5 \times 8^0)$

$= (3 \times 64) + 32 + 5$

$= 192 + 32 + 5$

$= 229$

$\therefore\ (345)_8 = (229)_{10}$ **Ans.**

(*e*) *Conversion of* $(7841)_9$ *to its decimal equivalent.*

We know that in order to convert a given number to its decimal equivalent, we multiply each digit by its position weight and then sum the products. Thus,

Step 1 7 8 4 1

↓ ↓ ↓ ↓

Step 2 9^3 9^2 9^1 9^0

Step 3 $= (7 \times 9^3) + (8 \times 9^2) + (4 \times 9^1) + (1 \times 9^0)$

$= 7 \times 729 + 8 \times 81 + 4 \times 9 + 1 \times 1$

$= 5103 + 648 + 36 + 1$

$= 5788$

$\therefore \quad (7841)_9 = (5788)_{10}$ **Ans.**

Example 33-37. *Determine the value of x for the given number,*

$(25)_x = (85)_{10}$ *(UP Tech., Univ. Elect. Engg 1st Sem. 2004-05)*

Solution. Given: $(25)_x = (85)_{10}$

We know that,

2 5

↓ ↓

$2 \times x^1$ $5 \times x^0$

$\therefore \ (25)_x = 2x + 5 \times 1 = 2x + 5$

We also know that,

$$(25)_x = 2x + 5 = 85$$

$$\therefore \quad 2x = 85 - 5 = 80$$

$$\text{or} \quad x = \frac{80}{2} = 40 \quad \textbf{Ans.}$$

Example 33-38. *Convert the following members to their equivalents as shown below:*

(*a*) $(4021.25)_{10} = (\ \ )_2$

(*b*) $(\text{A6F . CD})_{16} = (\ \ \ )_8$ *(UP. Tech. Univ., Elect Engg. 2nd Sem. 2005-06)*

Solution. (*a*) *Conversion of $(4021.25)_{10}$ to its equivalent binary.*

We will convert the given number $(4021.25)_{10}$ to its equivalent binary by first converting the integer part 4021 followed by the fraction part 0.25.

		Top
$4021 \div 2 = 2010$	with a remainder	1 ↑ (LSB)
$2010 \div 2 = 1005$	with a remainder	0
$1005 \div 2 = 502$	with a remainder	1
$502 \div 2 = 251$	with a remainder	0
$251 \div 2 = 125$	with a remainder	1
$125 \div 2 = 62$	with a remainder	1
$62 \div 2 = 31$	with a remainder	0
$31 \div 2 = 15$	with a remainder	1
$15 \div 2 = 7$	with a remainder	1
$7 \div 2 = 3$	with a remainder	1
$3 \div 2 = 1$	with a remainder	1
$1 \div 2 = 0$	with a remainder	1 (MSB)
		Bottom

Reading remainders from bottom to bottom to top, number is 110010011101.

$\therefore\ (4021)_{10} = (111110110101)_2$

Let us now convert the fractional part 0.25_{10} to its binary equivalent. This can be done by repeatedly multyplying-by-2.

		Top
$0.25 \times 2 = 0.5 = 0.5$	with a carry	1 ↓
$0.5 \times 2 = 1.0 = 0$	with a carry	1 ↓
		Bottom

Reading carry from top to bottom, the fractional binary number is 0.11

$\therefore\ (0.25)_{10} = (0.11)_2$

Thus $(4021.25)_{10} = (11111011010101.11)_2$ **Ans.**

(***b***) *Conversion of (A6F.CD)$_{16}$ to its octal equivalent*

Step 1	A	6	F	•	C	D		
Step 2	1010	0110	1111		1100	1101		
Step 3	1010	001	101	111	•	110	011	01 → 010 (Add one zero to form 3-bit group)
Step 4	5	1	5	7	•	6	3	2

$\therefore\ (A6F.CD)_{16} = (5157.632)_8$ **Ans.**

33-30. Binary-Coded-Decimal Number System

Binary coded decimal (BCD) is a way to express each of the decimal digits with a ***binary code***. Since there are only ten code groups in the BCD system, it is very easy to convert between the decimal and BCD. Because we like to read and write in decimal, the BCD system provides an excellent interface to binary systems. Examples of such interfaces are keypad inputs and digital readouts.

The BCD system is also called the 8421 code. Binary coded decimal means that each decimal digit, 0 through 9, is represented by a binary code of four bits. The designation 8421 indicates the binary weights of four bits (2^3, 2^2, 2^1, 2^0). The ease of conversion between BCD (or 8421) code numbers and familiar decimal numbers is the main advantage of this code. There are ten binary combinations that represent the ten decimal digits as shown in Table 33-5.

Table 33-5. ***BCD code for ten decimal digits***

Decimal Digit	0	1	2	3	4	5	6	7	8	9
BCD	0000	0001	0010	0011	0100	0101	0110	0111	1000	1001

It should be noted that with four bits, sixteen numbers (0000 through 1111) can be represented but in the BCD (or 8421) code, only ten of these are used.

There are six code combinations that are not used. These are 1010, 1011, 1100, 1101, 1110 and 1111. These six code combinations are referred to as invalid in the BCD (or 8421) code.

The procedure for converting a given decimal number to its BCD equivalent is as follows :

Step 1 Write the decimal number.

Step 2 Convert each decimal digit to its 4-bit binary equivalent.

Step 3 Write the binary number as an answer.

In order to illustrate the procedure, consider the conversion of decimal number 35.

Step 1 3 5

Step 2 0011 0101

Step 3 00110101

Drop off these zeros

$\therefore$ 35_{10} = 110101 BCD

Example 33-39. *Convert the following decimal numbers to binary-coded-decimal (BCD) code:*

(a) 843 and *(b) 2469.*

Solution.

(*a*) Given: The decimal number = 843

We know that to convert the decimal number, we need to change each digit to its equivalent 4-bit binary, *i.e.*,

Step 1 8 4 3

Step 2 1000 0100 0011

Step 3 100001000011 BCD

$\therefore$ 843_{10} = 100001000011 BCD **Ans.**

Step 1 2 4 6 9

Step 2 0010 0100 0110 1001

Step 3 0010 0100 0110 1001

$\therefore$ 2469_{10} = 0010010001101001 BCD **Ans.**

Example 33-40. *Encode* $(39, 584)_{10}$ *into 8421 BCD number.*

Solution. Given: A decimal number = 39, 584

We know that in order to convert a given decimal number to its equivalent BCD number, we have to replace each decimal digit by its 4-bit binary equivalent. Thus

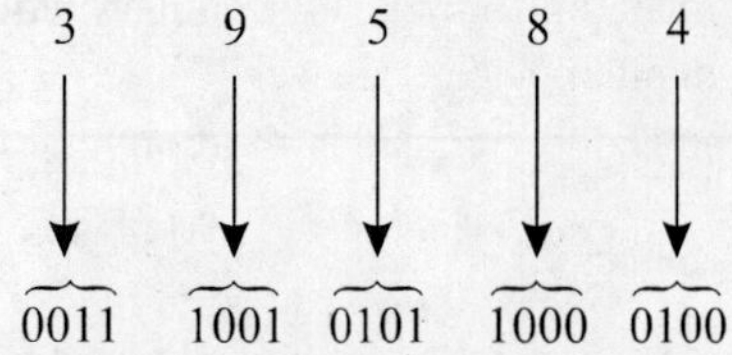

$\therefore\ 39584_{10} = 111001010110000100_{BCD}$ **Ans.**

33-31. Determination of a Decimal Number from a BCD number

Following is the procedure for the determination of a decimal number from a given BCD numbers.

Step 1 Write the BCD number.

Step 2 Start at the right-most bit and break the code into groups of four bit each. We can add zeros at the left-most position to complete the four-bit group.

Step 3 Convert each four-bit group to its equivalent decimal digit.

Note that if the four-bit group is in the range of 1010 to 1111, it is a forbidden code group which indicates an error in the BCD number.

Let us illustrate the procedure by considering the BCD number 1101010001

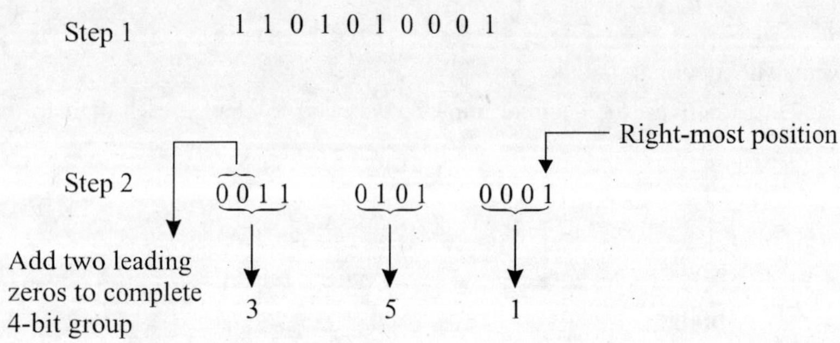

$\therefore$ 1101010001 BCD = 351_{10} **Ans.**

Example 33-41. *Convert each of the following BCD numbers to their equivalent BCD numbers* (*a*) *1000011000111001* (*b*) *1001010011111000*

Solution.

(*a*) Given: The BCD number 1000011000111001

Step 1	1 0 0 0 0 1 1 0 0 0 1 1 1 0 0 1
Step 2	1 0 0 0 0 1 1 0 0 0 1 1 1 0 0 1
Step 3	8 6 3 9

∴ 1000011000111001 BCD = 8639_{10} **Ans.**

(*b*) Given: The BCD number 1001010011111000

Step 1	1 0 0 1 0 1 0 0 1 1 1 1 1 0 0 0
Step 2	1 0 0 1 0 1 0 0 1 1 1 1 1 0 0 0
Step 3	9 4 (Fifteen indicates error) 8

Fifteen indicates error

Therefore, we find that 100101001111000 BCD is not a valid code.

33-32. Comparison of BCD and Binary

It is important to realize that BCD is not another number system like binary, octal, decimal and hexadecimal. It is, in fact, the decimal system with each digit encoded in its binary equivalent. It is also important to understand that a BCD number is not the same as a straight binary number. a straight binary code takes the complete decimal number and represents it in binary; the BCD code converts each decimal digit to binary individually. In order to illustrate this point consider the number 173 and compare its straight binary and BCD codes.

173 = 0001 0111 0011

The BCD code requires 12 bits while the straight binary code requires only eight bits to represent 173. BCD requires more bits than straight binary to represent decimal numbers of more than one digit. This is because BCD does not use all possible four-bit groups, as pointed out earlier, and is therefore somewhat efficient.

The main advantage of the BCD code is the relative ease of converting to and from decimal. Only the four-bit code groups for the decimal digits 0 through 9 need to be remembered. This case of conversion is especially important from a hardware standpoint because in a digital system it is the logic circuits that perform the conversions to and from decimal.

33-33. Comparison of Number systems

Table 33-6 gives the representation of the decimal number 1 through 15 in the binary, octal, hex number systems and in BCD code for comparison purpose. Study it carefully and make sure you understand how it was obtained. Especially note how the BCD representation always uses four for each decimal digit.

Table 33-6. ***Comparison of different number systems***

Decimal	*Binary*	*Octal*	*Hexadecimal*	*BCD*
0	0	0	0	0000
1	1	1	1	0001
2	01	2	2	0010
3	10	3	3	0011
4	100	4	4	0100
5	101	5	5	0101
6	110	6	6	0110
7	111	7	7	0111
8	1000	10	8	1000
9	1001	11	9	1001
10	1010	12	A	0001 0000
11	1011	13	B	0001 0001
12	1100	14	C	0001 0010
13	1101	15	D	0001 0011
14	1110	16	E	0001 0100
15	1111	17	F	0001 0101

Example 33-42. *How many bits are required to represent the decimal numbers in the range from 0 to 999 using **(a)** Straight binary code, and **(b)** BCD code?*

Solution.

Given: The decimal numbers in the range from 0 to 999.

(***a***) *Number of bits required using straight binary code*

We know that the maximum number of decimal numbers which we can represent with 9 bits is 2^9 (*i.e.*, 512) and with 10 bits is 2^{10} (*i.e.*, 1024). Since the given range is from 0 to 999, therefore we require at least 10 bits to represent the decimal numbers in the range from 0 to 999. **Ans.**

(***b***) *Number of bits required using BCD code*

We know that the number of bits using BCD code depends upon the maximum value of the given decimal number (*i.e.*, 999). First each digit is to be converted to its equivalent 4-bit BCD code, *i.e.*,

$$\begin{array}{ccc} 9 & 9 & 9 \\ \downarrow & \downarrow & \downarrow \\ \overbrace{1001} & \overbrace{1001} & \overbrace{1001} \end{array}$$

Note that we need 12-bits to represent the decimal number 999 in BCD code.

∴ Number ozzf bits required to represent the decimal numbers in the range from 0 to 999 = 12 **Ans.**

33-34. The Byte

The byte is a group (or string) of 8-bits. The byte is very commonly used in the field of microprocessors and microcomputers. The reason for this is that the microprocessors and microcomputers handle and store data and information in groups of eight bits. Thus instead of specifying the data as 8-bit, 33-bit, 32-bit etc., we say it as 1-byte, 2-byte, 3-byte respectively.

It will be interesting to know that a group of four-bits is called a ***nibble***. Hence a byte contains two nibbles. Another point worth noting is that we need two hexadecimal digits to represent a byte.

Example 33-43. *What is the largest decimal value that can be represented in binary using (**a**) one byte, and (**b**) two bytes?*

Solution.

(*a*) Given, a largest binary number of one byte long.

We know that one byte is 8 bits. So the largest binary value will be equivalent to decimal.

$$2^8 - 1 \quad 256 - 1 = 255 \quad \textbf{Ans.}$$

(*b*) Given a largest number of two byte long.

We know that two bytes is 16 bites. So the largest binary value will be equivalent to decimal,

$$2^{16} - 1 = 65{,}536 - 1 = 65{,}535 \quad \textbf{Ans.}$$

Example 33-44. *How many bytes are needed to represent the decimal value 856469 in BCD?*

Solution. We know that each decimal digit can be converted to its 4-bit BCD code. Since the given number 856469 is a 6-digit number, therefore this number requires 6 × 4 = 24 bits. Further since a byte is equivalent to 8-bits, therefore 24-bits are equal to 3-bites. this is shown below.

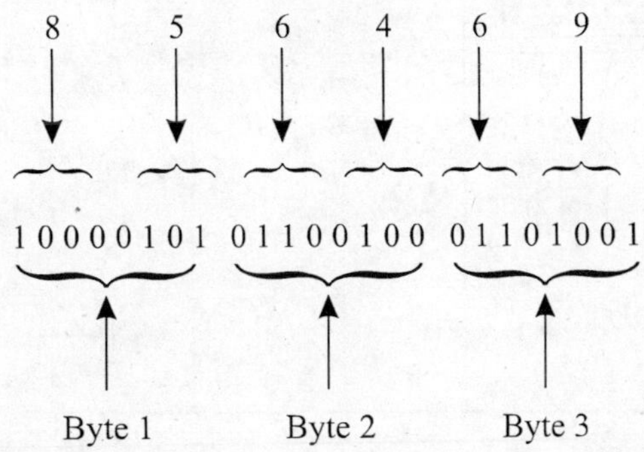

33-35. Alphanumeric codes

In addition to numerical data, a computer must be able to handle non-numerical information used in day-to-day processing. In other words, a computer should recognize codes that represent numbers, letters of the alphabet, punctuation marks, and other special characters. These codes are called alphanumeric codes. A complete alphanumeric code would include the 26 lowercase letters. 26 uppercase letters, 10 numeric digits, 7 punctuation marks, and anywhere from 20 to 40 other characters, such as +, /, #, %, * and so on. We can say that an alphanumeric code represents all of the various characters and functions that are found on a computer keyboard. The most widely used alphanumeric code is the *American standard code for Information Interchange (ASCII)*. We will now discuss the ASCII code in the following pages.

33-36. ASCII Code

We have already discussed in the last article that ASCII stands for American Standard Code for Information Interchange.

The ASCII code is used for the transfer of a alphanumeric information between a computer and input/output devices such as monitors or printers. A computer also uses it internally to store the information that an operator types in at the computer's keyboard.

The ASCII code pronounced "asked is a seven-bit code, and so it has 2^7 (=128) possible code groups. This is more than enough to represent all of the standard keyboard characters as well as control functions such as the <RETURN> and <LINEFEED> functions. Table 33-7 shows a complete list of the ASCII code. Table 33-7 shows a partial list of ASCII code. You may find it more convenient to use.

Table 33-7. American standard Code for Information Interchange

					MSB			
LSB	*000*	*001*	*010*	*011*	*100*	*101*	*110*	*111*
0000	NUL	DLE	SP	0	@	P		p
0001	SOH	DC_1	!	1	A	Q	a	q
0010	STX	DC_2	”	2	B	R	b	r
0011	ETX	DC_3	#	3	C	S	c	s
0100	EOT	DC_4	$	4	D	T	d	t
0101	ENQ	NAK	%	5	E	U	e	u
0110	ACK	SYN	&	6	F	V	f	v
0111	BEL	ETB	‘	7	G	W	g	w
1000	BS	CAN	(	8	H	X	h	x
1001	HT	EM	)	9	I	Y	i	y
1010	LF	SUB	*	:	J	Z	j	z
1011	VT	ESC	+	;	K	[	k	{
1100	FF	FS	’	<	L	\	I	|
1101	CR	GS	-	=	M	]	m	}
1110	SO	RS	.	>	N	↑	n	~
1111	SI	US	/	?	O	—	o	DEL

Definitions of control abbreviations:

ACK Acknowledge
BEL Bell
BS Backspace
CAN Cancel
CR Carriage return
DC_1-DC_4 Direct control
DEL Delete idle
DLE Data link escape
EM End of medium
ENQ Enquiry
EOT Ens of transmission block
ESC Escape
ETB End of transmission block
ETX End text
FF Form feed
FS Form separator
GS Group separator
HT Horizontal tab
LF Line feed
NAK Negative acknowledge
NUL Null
RS Record separator
SI Shift out
SO Shift out
SOH Start of heading
SP Space
STX Start text
SUB Substitute
SYN Synchronous idle
US Unit separator
VT Vertical tab

This table is arranged in such a way that the least significant bits (LSB) appear along a column and most-significant bits (MSB) along the horizontal row right at the top.

The least-significant bits (LSB) are a 4-bit group while most-significant bits (MSB) are a 3-bit group.

Let us see how to use this table to final the ASCII code for letter "A". First locate the position of "A" in the ASCII table, identify the corresponding MSB-group and LSB-group bits. For the letter "A" we find that MSB are 100 and LSB are 0001. Therefore,

ASCII code for letter "A" = 100 0001

Similarly, the ASCII code for '**P**' is 1110000 and for '**ESC**' is 0011011.

Table 33-8. *Partial listing of ACII code*

Character	*Seven-Bit ASCII*	*Hex*	*Character*	*Seven-Bit ASCII*	*Hex*
A	1000001	41	Y	1011001	59
B	1000010	42	Z	1011010	5A
C	1000011	43	0	0110000	30
D	1000100	44	1	0110010	32
E	1000101	45	2	0110010	33
F	1000110	46	3	0110011	33
G	1000111	47	4	0110100	34
H	1000111	48	5	0110101	35
I	1001001	49	6	0110110	36
J	1001010	4A	7	0110111	37
K	1001011	4B	8	0111000	38
L	1001100	4C	9	0111001	39
M	1001101	4D	blank	0100000	20
N	1001110	4E		0101110	2E
O	1001111	4F	(	0101000	28
P	1010000	50	+	0101011	2B
Q	1010001	51	$	0100100	24
R	1010010	52	*	0101010	2A
S	1010010	53	)	0101001	29
T	1010100	54	—	0101101	2D
U	1010101	55	/	0101111	2F
V	1010110	56	,	0101100	3D
W	1010111	57	=	0111101	0D
X	1011000	58	<RETURN>	0001101	0D
			<LINEFEED>	0001010	0A

Example 33-45. *The following is a message encoded in ASCII code. What is the message?*
1001000 1000101 1001100 1001100 1001111

Solution. Given: The message encoded in ASCII code.

1001000 1000101 1001100 1001100 1001111

First of all, let us convert seven-bit code to its equivalent hexadecimal number. The resulting number can be obtained as follows.

100 1000 100 0101 100 1100 100 1100 100 1111

↓ ↓ ↓ ↓ ↓ ↓ ↓ ↓ ↓ ↓

4 8 4 5 4 C 4 C 4 F

Thus, the resulting values are 48, 45, 4C, 4C and 4F.

Now locate these hexadecimal values in Table 33-8 and determine the character represented by each. The results are

"HELLO"

The ASCII code is used extensively to exchange information between the computer and its peripheral devices.

33-37. Parity Method for Error Detection

As a matter of fact, the movement of binary data and codes from one location to another is the most frequent operation performed in digital electronic systems. For example:

(*i*) The transmission of digitized voice over a mobile phone

(*ii*) The storage of data in and retrieval of data from external memory devices such as magnetic tape and disk.

(*iii*) The transmission of information from a computer to a remote use of terminal or another computer over telephone lines.

Whenever information is transmitted from one electronic system (say transmitter) to another system (say receiver), there is a possibility that errors can occur such that the receiver does not receive the identical information that was sent by the transmitter. The major reason for any transmission errors is *electrical noise.* The noise consists of spurious fluctuations in voltage or current that are present in all electronic systems to varying degrees. Fig. 33-4 is a simple illustration of a type of transmission error.

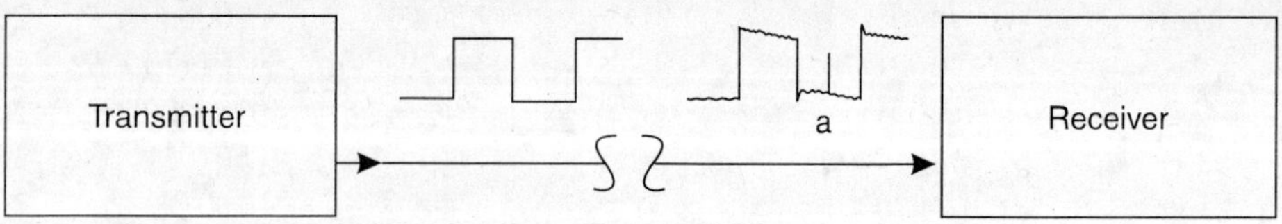

Fig. 33-4. *Illustration of a producing noise an error in the transmission of digital data.*

The transmitter sends a relatively noise-free serial digital signal over a signal line to a receiver. However, by the time the signal reaches the receiver, it contains a certain degree of noise superimposed on the original signal. Occasionally, the noise is large enough in amplitude that it will alter the logic level of the signal as it does at point '*a*'. When this occurs, the receiver may incorrectly interpret that bit as a logic 1, which is not what the transmitter had sent.

Most modern digital electronic equipment is designed to be relatively error-free, and the probability of errors such as shown in Fig. 33-5 is very low. However, we must realize that digital electronic systems often transmit mullions, of bits per second, so that even a very low rate of occurrence of errors can produce an occasional error that might prove to be bothersome, if not disastrous. Due to this reason, many digital electronic systems employ some method for detection (and correction) of errors. One of the simplest and mostwidely used methods for error detection is the ***Parity Method***.

Before we understand the parity method, it is essential to study about the parity bit.

33-38. Parity Bit

Strictly speaking, a *Parity bit* is an extra bit that is attached to a code group that is being transmitted from one location to another. The parity bit is made either 0 or 1, depending on the number of 1s that are contained in the code group. Two different methods are used for attaching an extra bit. These are discussed below.

The Even Parity Method

In this method, the value of the parity bit is chosen so that the total number of 1s in the code group (including the parity bit) is an *even* number. For example, suppose that the code group is 1000101. This is the ASCII character "E". The code group has *three* 1s. Therefore, we will add a parity bit of 1 to make the total number of 1s an even number. The *new* code group. Including the parity bit, thus becomes 11000101.

The Odd Parity Method

In this method, the value of the parity bit is chosen so that the total number of 1s in the code group (including the parity bit) is an *odd number*. For example, suppose the code group is 1000101.

The code group has three 1s. Therefore we will add a parity bit of 0 to keep the total number of 1s an odd number. The new code group including the party bit becomes 01000101.

Example 33-46. *Assign the proper even parity bit to the following code groups:*

(*a*) 1010

(*b*) 111000

(*c*) 101101

Solutions :

(*a*) Given: the code group 1010.

From the given code group we find that it has even number of 1s. Therefore we attach a 0 as the parity bit at most-significant bit position.

0 1010

Code group

Parity bit

(*b*) Given: The code group 111000

For a given code group we find that it has odd number of 1s. Therefore we attach a 1 as the parity bit at the most-significant bit position, *i.e.*,

1 111000

Code group

Parity bit

(*c*) Given the code group 101101

For a given code group we find that it has even number of 1s. Therefore we attach a 0 as the parity bit at the most-significant bit position.

0 101101

Code group

Parity bit

Example 33-47. *Computers often communicate with other remote computers over the telephone lines-this is how the communication takes place over the internet. When one computer is transmitting a message to another, the information is usually encoded in ASCII. What actual bit strings would a computer transmit to send the message TECHNOLOGY using ASCII with even parity.*

Solution. Given, the message TECHNOLOGY.

In order to find the bit strings, we will first look up the ASCII codes for each character in the message by using the Table 33-8. Then for each code, we count the number of 1s. If it is an even number, attach a 0 as a parity bit at the most-significant bit position. If it is an odd number, attach a 1. Thus the resulting 8-bit codes will all having an even number of 1s (including parity) are:

Table 33-8

Character	*Parity bit*	*Code group*
T–	1	1010100
E–	1	1000101
C–	1	1000011
H–	0	1001000
N–	0	1001110
O–	1	1001111
L–	1	1001100
O–	0	1001110
G–	0	1000111
Y–	0	1011001

Example 33-48. *The following is a message encoded in ASCII code.*

1001000 1000101 101100 1001100 1001111

What is the message?

Solution. Given the message encoded in the ASCII code:

1001000 1000101 101100 1001100 1001111

We know that to read the message we have to convert the given data string into the corresponding characters by using the ASCII code Table 33-9. From this table we find that,

1001000	Corresponds to	'H'
1000101	Corresponds to	'E'
1001100	Corresponds to	'L'
1001100	Corresponds to	'L'
1001111	Corresponds to	'O'

Therefore, the given data string represents the message "HELLO"

33-39. Applications of the Number Systems

We have already discussed a variety of number systems in this chapter. Which number system is used depends on how the data were developed and how they are to be used. Now we will go through several applications that depend on the translation and interpretation of these digital representations.

Example 33-49. *Typically digital thermometers use BCD to drive their digital displays. How many BCD bits are required to drive a 3-digit thermometer display? (a) What 12-bits are sent to display for a temperature of 157°C?*

Solution.

(*a*) *Number of BCD bits required to drive a 3-digit thermometer display*

Given, a digital thermometer using BCD for display

We know that each BCD bit requires a 4-bit code, therefore 3-digit display will have $3 \times 4 = 12$-bits. **Ans.**

(*b*) *12-bit BCD code sent for display*

Given: The temperature = 157°C

We know that to find the 12-bit BCD code representing a temperature of 157°C, we need to convert each digit to its 4-bit BCD code. Thus,

1	5	7
↓	↓	↓
0001	0101	0111

$\therefore$ 157°C = $0001\ 0101\ 0111_{BCD}$ **Ans.**

Example 33-50. *A particular brand of compact disk (CD) player has the capability of converting 12-bit signals from a CD into their equivalent analog values.* ***(a)*** *What are the largest and smallest hex values that can be used in the CD system.* ***(b)*** *How many different analog values can be represented by the system.*

Solution. (*a*) *Number of analog values*

We know that the CD player has a capability of converting 12-bit signals from a CD into their equivalent analog values, and

Number of analog values in decimal = 2^{12} = 4096 **Ans.**

(***b***) *Largest and smallest hex values*

We know that the largest decimal value is, $2^{12}-1$ = 4095 and the smallest decimal value is 0.

The largest value in hexadecimal number system can be obtained by using the repeated division by 16 as given below:

			Top ↑
4095 ÷ 16 = 255.9375 = 225	with a remainder	$0.9375 \rightarrow 0.9375 \times 16 = 15 \rightarrow$ F	LSD
225 ÷ 16 = 15.9375 = 15	with a remainder	$0.9375 \rightarrow 0.9375 \times 16 = 15 \rightarrow$ F	
15 ÷ 16 = 0.9375 = 0	with a remainder	$0.9375 \rightarrow 0.9375 \times 16 = 15 \rightarrow$ F	MSD
			Bottom

Since, the quotient is zero, therefore we stop the repeated-division by 16. Reading the remainders from bottom-to-top, we find that the required hexadecimal number is FFF.

$\therefore$ Largest value in hex = FFF_{16}
Smallest value in hex = 0_{16} **Ans.**

Example 33-51. *Most PC-compatible computer systems use a 33-bit address code to identify each of over 1 million memory locations.* ***(a)*** *How many hex characters are required to identify the address of each memory location?* ***(b)*** *What is the 5-digit hex address of the 500th memory location?* ***(c)*** *If 60 memory locations are used for data storage starting at location 000FFH, what is the location of the last data item?*

Solution. (*a*) *Number of hexadecimal characters*

We know that each hexadecimal digit represents 4 bits. Therefore 33-bit requires five hexadecimal digits. **Ans.**

(*b*) *5-digit hexadecimal address of 500th memory location.*

We know that the decimal number 500 is to be converted to its equivalent hexadecimal number by using repeated division-by-16 method, *i.e.*,

			Top
$500 \div 16 = 31$	with a remainder	4	↑ LSD
$31 \div 16 = 1$	with a remainder	15 (= F)	
$15 \div 16 = 0$	with a remainder	15 (= F)	MSD
			Bottom

$\therefore\ 500_{10} = FF4_{16}$.

Now we know that the first memory location is 00000 and the 500th location is 00FF3 (*i.e.*, 00FF4 minus 1).

$\therefore$ The 5-digit hexadecimal address of the 500th memory location is 00FF3 **Ans.**

(*c*) *Location of the last data item*

We know that the data is stored from the memory location 00FF4 onwards. If 60 locations are to be filled up, then, 00FF4 + 60 = 01054H. But we have to subtract 1 from the value 01054H because the memory location 00FF4H received the first data item, so we needed 59 more memory spaces

$\therefore$ The location of the last data item = (01054 – 1) H = 01053H **Ans.**

Example 33-52. *A chemical processing plant uses a computer to monitor the temperature and pressure of four chemical tanks as shown in Fig. 33-5. Whenever a temperature or a pressure exceeds the danger limit, an internal tank sensor applied a "1" to its corresponding output to the computer. If all conditions are OK, then all outputs are zero. (a) If the computer reads the binary string 00101000, what problems exist. (b) What problems exist if the computer is reading 55 H? (c) What hexadecimal number is read by the computer if the temperature and pressure in both tanks B and D are high? (d) Tanks A and B are taken out of their use and their sensor outputs are connected to 1's. A computer programmer must write a program to ignore these new circuit conditions. The computer program must check that the value read is always less than what decimal equivalent when no problem exists.*

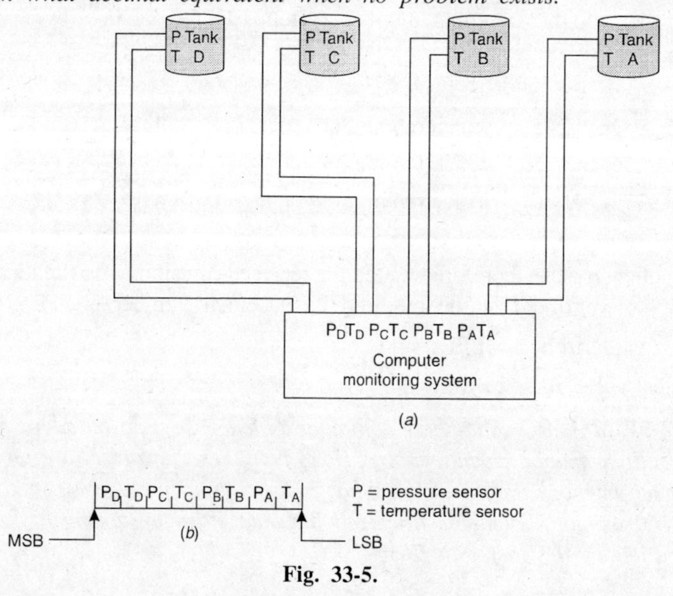

Fig. 33-5.

Solution. Given: A chemical processing plant using a computer to monitor the temperature and pressure of four chemical tanks.

(*a*) *Problems that exist if computer reads the binary string 00101000.*

We know that the computer reads the data in the format shown below

P_D	T_D	P_C	T_C	P_B	T_B	P_A	T_A

MSB LSB

By comparing the corresponding values of pressure and temperature with data input, we find P_B and P_C are 1. This implies that the pressure in tanks B and C are dangerously high. **Ans.**

0	0	1	0	1	0	0	0
P_D	T_D	P_C	T_C	P_B	T_B	P_A	T_A

(*b*) *Problems that exist if the computer is reading 55H.*

We know that the computer reads the data 55 H which is read in the format shown below:

0	1	0	1	0	1	0	1
P_D	T_D	P_C	T_C	P_B	T_B	P_A	T_A

By comparing the corresponding values of pressure and temperature with the data input, we find that the temperature of all the four tanks A, B, C and D are dangerously high. **Ans.**

(*c*) *Hexadecimal number read by the computer if the temperature and pressure in both tanks B and D are high.*

We know that the computer must read 1 for the temperature and pressure bits for both tanks B and D. The data read by the computer should be as shown below:

1	1	0	0	1	1	0	0
P_D	T_D	P_C	T_C	P_B	T_B	P_A	T_A

From the above diagram we find that the binary value is 11001100. The hexadecimal equivalent of this binary number is CC_{16} or CCH **Ans.**

(*d*) *Decimal number read by the computer with A and B out of use.*

We know that the tanks A and B are taken out of their use and their outputs are connected to 1's. This means P_A, T_A, P_B and T_B all are '1'. Now if T_C goes high, then the number read by the computer is as indicated below:

0	0	0	1	1	1	1	1
P_D	T_D	P_C	T_C	P_B	T_B	P_A	T_A

From the above diagram, we find that the binary number read by the computer is 00011111. This is equivalent to decimal number 31_{10}. Thus so long the computer reads below 31_{10}, there is no problem with the pressure and temperature of the tanks C and D. **Ans.**

SUMMARY

In this chapter, we have learn that :

1. Using an n-bit number, we can represent decimal values ranging from 0 to $2^n - 1$.
2. The octal number system consists of eight digits, 0 through 7.
3. A decimal number can be converted to octal by using the repeated division-by-8 method.
4. Octal-to-binary conversion is accomplished by simply replacing each octal digit with its 3-bit binary equivalent. The process is reversed for binary-to-octal conversion.
5. The hexadecimal number system consists of 16 digits and characters, 0 through 9 and A through F.
6. One hexadecimal digit represents a four-bit binary number and its primary usefulness is in simplifying bit patterns and making them easier to read.
7. A decimal number can be converted to hexadecimal by the repeated division-by-16 method.
8. A decimal number is converted to Binary-Coded-Decimal (BCD) by replacing each decimal digit with the appropriate 4-bit binary code.
9. The ASCII code is a 7-bit alphanumeric code that is widely used in computer systems for input and ouput information.
10. The parity method for error detection attaches a special parity bit to each transmitted group of bits.

GLOSSARY

Octal number system : A number system that has a base of 8, and uses digits from 0 to 7 to express and octal number.

Hexadecimal number syste : A number system that has a base of 16, and uses digits from 0 through 9 plus letters A through F to express a hexadecimal number.

Straight binary coding : Representation of a decimal number by its equivalent binary number.

Binary-coded-decimal (BCD) code : Four bit code used to represent each digit of a decimal number by its four-bit binary equivalent.

Byte : Group of eight bits.

Alphanumeric codes : Codes that represent numbers, letters, punctuation marks and special characters.

ASCII (American Standard Code for Information Interchange) : Seven-bit alphanumeric code used by most computer manufacturers.

Parity method : Scheme used for error detection during the transmission of data.

Parity bit : Additional bit that is attached to each code group so that the total number of 1s being transmitted is always odd (or always even).

DESCRIPTIVE QUESTIONS

1. Why is the binary number system commonly used in digital electronics?
2. How are the position values (or weighting factors) determined for each binary position in a base-2 (*i.e.*, binary) number system.

3. Why is hexadecimal number system used instead of the octal number system when working with 8-and 33-bit digital computers?

4. Why is Digital electronics considered important? State the advantages and disadvantages of using binary numbers. (*UP Tech. Univ., Sem I., 2003-04*)

5. What is BCD? Where it is used. (*UP Tech. Univ., Sem I., 2003-04*)

6. How does BCD differ from the straight binary number system?

7. Why ASCII code required by digital computer systems?

8. Why can't the parity method detect a double error in transmitted data?

TUTORIAL PROBLEMS

1. What is the largest decimal value that can be represented by (*a*) a 10-bit (*b*) a 33-bit number. (**Ans.** (*a*) 1048, (*b*) 1,098,304)

2. Convert the following binary numbers to their decimal equivalents.

(*a*) 11011 (*b*) 10110101

(*c*) 1011111 (*d*) 10001101 (**Ans.** 27,181,191,141)

3. Convert the following decimal numbers to their binary equivalent.

(*a*) 45 (*b*) 76

(*c*) 177 (*d*) 975

(**Ans.** (*a*) 101101, (*b*) 1001100, (*c*) 10110001, (*d*) 1111001111)

4. Convert the following octal numbers to their decimal equivalent.

(*a*) 24.6 (*b*) 372

(*c*) 547 (*d*) 2374

(**Ans.** (*a*) 20.75, (*b*) 250, (*c*) 359, (*d*) 1276)

5. Convert each of the following decimal numbers to their octal

(*a*) 266 (*b*) 255

(*c*) 372 (*d*) 919

(**Ans.** (*a*) 412, (*b*) 377, (*c*) 564, (*d*) 1627)

6. Convert each of the following octal values to binary:

(*a*) 13 (*b*) 25

(*c*) 140 (*d*) 7526

(**Ans.** (*a*) 001011, (*b*) 010101, (*c*) 001100000, (*d*) 111101010110)

7. Convert each of the following binary numbers to their octal equivalent.

(*a*) 110101 (*b*) 101111001

(*c*) 100110011010 (*d*) 11010000100

(**Ans.** (*a*) 65, (*b*) 571, (*c*) 4632, (*d*) 3204)

8. When a large decimal number is to be converted to a binary, it is sometimes easier to convert it first to octal and then from octal to binary. Try this method for 2048_{10}.

(**Ans.** (*a*) 4000_8 100 000 000 000)

9. Convert each of the following hexadecimal values to their decimal equivalent.

(*a*) E5 (*b*) 2AF

(*c*) 1BC2 (*d*) B2F8

(**Ans.** (*a*) 229, (*b*) 687, (*c*) 7106, (*d*) 45,816)

10. Convert each of the following decimal values to their hexadecimal equivalent values.

(*a*) 214 (*b*) 423

(*c*) 650 (*d*) 2591

(**Ans.** (*a*) D6, (*b*) 1A7, (*c*) 28A, (*d*) A1F)

11. List the hexadecimal numbers in sequence from 380 to 3A0.

12. How many hexadecimal digits are required to represent decimal numbers upto, 1,098,304.

(**Ans.** 4 hex digits)

13. Encode each of the following decimal numbers to their BCD equivalent.

(*a*) 479 (*b*) 1204

(**Ans.** (*a*) 010001111001, (*b*) 0001001000001000)

MULTIPLE CHOICE QUESTIONS

1. What range of decimal values can be represented by a eight-bit binary number?

(*a*) 0 to 63 (*b*) 0 to 127

(*c*) 0 to 255 (*d*) 0 to 511

2. What is the next number in the octal counting sequence: 824, 825, 826, 827,

(*a*) 828 (*b*) 82A

(*c*) 820 (*d*) 830

3. What range of decimal values can be represented by a four-digit hexadecimal number?

(*a*) 0 to 1023 (*b*) 0 to 4095

(*c*) 0 to 8191 (*d*) 0 to 65535.

4. What is next four numbers in the hexadecimal counting sequence D9A, D9B, D9C, D9D,....

(*a*) D9E, D9F, D90, D91 (*b*) D9E, D9F, DA0, DA1

(*c*) D9E, E10, E11, E12 (*d*) D9E, D9F, D10, D11.

5. The number of bits required to represent an eight-digit decimal number in BCD is

(*a*) 8 (*b*) 16

(*c*) 24 (*d*) 32

ANSWERS

1. (*c*) **2.** (*d*) **3.** (*d*) **4.** (*b*) **5.** (*d*)

❑❑❑

34

CHAPTER

Logic Gates

OUTLINE

Objectives

After Compeleting this chapter, you should be able to :

- Describe the operation and construct truth tables of an INVERTER, the AND gate and the OR gate.
- Describe the operation and construct truth tables of the NAND gate, the NOR gate, the exclusive –OR gate and the exclusive –NOR gate.
- Express the logical operation of INVERTER, AND, OR, NAND, NOR, XOR and XNOR gates in terms of a Boolean equation.
- Construct timing diagram showing the proper time relationships of inputs and outputs for the various logic gates.
- Use the logic gates to implement a circuit represented by a Boolean equation.
- List specific devices that contain the various logic gates.
- Use each logic gate in simple applications.

34-1. Introduction

Logic gates are the basic building blocks for forming digital electronic circuitry. A logic gate has one output terminal and one or more input terminals. Its output would be HIGH (1) or Low (0) depending on the digital level (s) at the input terminal (s). The logic gates can be used to design digital systems that will evaluate digital input levels and produce a specific output response based on that particular logic circuit design.

We will study the following logic gates in this chapter :

(1) INVERTER (or NOT gate) **(2)** AND gate **(3)** OR gate **(4)** NAND gate
(5) NOR gate **(6)** Exclusive-OR gate **(7)** Exclusive-NOR gate

The INVERTER, AND and OR gates are considered to be the **basic** logic gates in the field of digital electronics. The NAND, NOR, Ex-OR and Ex-NOR gates are constructed using the basic logic gates.

Now before we start out studies on logic gates, let us know about Boolean constant, variables, truth tables and Boolean expressions. These terms are very important to understand the operation of Logic gates.

34-2. Boolean Constants and Variables

Boolean algebra differs in a major way from an ordinary algebra in that Boolean constants and variables are allowed to have only two possible values, 0 or 1. A Boolean variable is a quantity that may, at different times, be equal to either 0 or 1. Boolean variables are often used to represent the voltage level present on a wire or at the input/output terminals of a circuit. For example, in a certain digital system the Boolean value of 0 might be assigned to any voltage in the range from 0 to 0.8 V while the Boolean value of 1 might be assigned to any voltage in the range 2 to 5 V.

Thus, Boolean 0 and 1 do not present actual numbers but instead represent the state of voltage, or what is called its LOGIC LEVEL. A voltage in a digital circuit is said to be at the logic level 0 or the logic level 1. Several other items are used synonymously with 0 and 1. Some of the more common ones are shown in Table 34-1. We will use the 0/1 and LOW/HIGH designations most of the time.

Boolean algebra is a means for expressing the relationship between the logic circuit's inputs and outputs. The inputs are considered logic variables whose logic levels at any time determine the output levels. In all our discussion to follow, we shall use letter symbols to represent logic variables. For example "A" might represent a certain digital circuit input and at any time we must have either "A" = 0 or "A" = 1.

Table 34-1. *Boolean constants and Variables*

Logic 0	*Logic 1*
False	True
Off	On
Low	High
No	Yes
Open-switch	Closed-switch

Boolean algebra is relatively easy to work with as compared to an ordinary algebra because of only two possible values. It may be noted that in Boolean algebra there are no fractions, decimals, negative numbers, square roots, cube roots, logarithms, imaginary numbers and so on. As a matter of fact, in Boolean algebra there are only three basic operations, NOT, AND and OR.

These basic operations are called logic operations. Digital circuits called logic gates can be constructed from diodes, transistors, and resistors connected in such a way that the circuit output is the result of a basic logic operation (OR, AND and NOT) performed on the inputs. We will use Boolean algebra in this chapter to describe the logical operation of basic logic gates. In Chapter 35 we will use Boolean algebra to analyse and design logic circuits using combinations of basic logic gates.

34-3. Truth Tables

A truth table is a means for describing how a logic circuit's output depends on the logic levels present at the circuit's inputs. Table 34-2 illustrates a truth table for one type of two-input logic circuit. The table lists all possible combinations of logic levels present at inputs (*A* and *B*) along with the corresponding output level (X). The first row in the table shows that when A and B are both at the 0 level, the output X is at the level 1 or, equivalently, in the 1 state. The second row shows that when input *B* is changed to the 1 state, so that $A = 0$ and $B = 1$, the output *X* becomes a 0. Similarly, the third row shows that when input *A* is changed to the 1 state and *B*, to 0 state, the output *X* becomes a 1. And the last row shows when both the inputs *A* and *B* are changed to "1" state, the output *X* becomes a 0.

Table 34-2. *Two-inputs truth table*

Inputs		*Output*
A	***B***	***X***
0	0	0
0	1	1
1	0	1
1	1	0

Tables 34-3 and 34-4 show the truth tables for a typical three-and four-input logic circuits. Again, each table lists all possible combinations of input logic levels on the left, with the resultant logic level for output *X* on the right. It may be noted carefully that the actual values for *X* will depend on the type of logic circuit.

Table 34-3

Inputs			*Output*
A	*B*	*C*	*X*
0	0	0	0
0	0	1	1
0	1	0	0
0	1	1	1
1	0	0	0
1	0	1	1
1	1	0	0
1	1	1	1

Table 34-4

Inputs				*Output*
A	*B*	*C*	*D*	*X*
0	0	0	0	0
0	0	0	1	0
0	0	1	0	1
0	0	1	1	0
0	1	0	0	0
0	1	0	1	0
0	1	1	0	1
0	1	1	1	0
1	0	0	0	0
1	0	0	1	0
1	0	1	0	1
1	0	1	1	0
1	1	0	0	0
1	1	0	1	0
1	1	1	0	1
1	1	1	1	0

As seen from the Table 34-3, 34-4 and 34-5, there are 4 rows for the two-input truth table, 8 rows for a three-input truth table, and 16 rows for the four-input truth table. In general, the list of all possible input combinations is equal to 2^N for N-input truth table. Moreover the list of all possible input combinations follows the binary counting sequence, and so it is an easy matter to write down all of the combinations without missing any. For example, in a two-input truth table, the sequence of combinations is 00, 01, 10 and 11.

34-4. Boolean Expression

We have already discussed in Chapter 33 that digital or logic circuits operate in the binary mode where each input and output voltage is either a 0 or 1. The 0 and 1 designations represent predefined voltage ranges. The characteristic of logic circuits allows us to use Boolean algebra as a tool for the analysis and design of digital circuits. Boolean algebra is a relatively simple mathematical tool that allows us to describe the relationship between logic circuit output (s) and its inputs as an algebraic equation. This algebraic equation is known as Boolean expression or simply Boolean equation.

34-5. The INVERTER

The INVERTER is also known as a NOT gate. It is a logic gate that performs inversion (also called complementation) operation. It changes one logic level to the opposite logic level, *i.e.*, from HIGH to LOW level or from LOW to HIGH level. In terms of bits, it changes a 1 to 0 and a 0 to 1.

Fig. 34-1(*a*) shows a logic symbol of an INVERTER. Note that input is shown on the left (labeled as *A*) and the output is on the right (labeled as *X*).

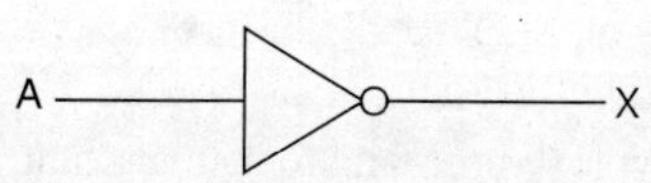

(a) logic symbol

Input	*Output*
A	*X*
0	1
1	0

(b) truth table

Fig. 34-1. *An INVERTER.*

Logical Operation **:** When the INVERTER input is LOW (0), the output, X is HIGH (1). On the other hand, if the input is HIGH (1), the output is LOW (0). This operation is summarized in the form of a truth table as shown in Fig. 34-1 (b).

34-6. Pulsed Operation of an INVERTER

Consider a pulse input to an INVERTER as shown in Fig. 34-2. In order to study the pulse operation, we will consider the time instances t_1 and t_2 along the time axis. Thus at t_1, the input goes from LOW(0) to HIGH(1) therefore, the output goes from HIGH(1) to LOW(0). At t_2, the input goes from HIGH(1) to LOW(0) therefore, the output goes from LOW(0) to HIGH(1).

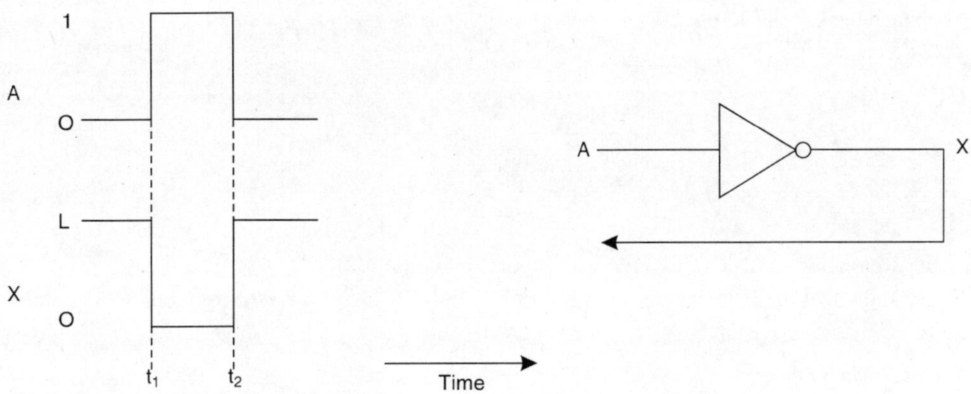

Fig. 34-2. *Illustrating the pulsed operation of an INVERTER.*

It is evident from the figure shown above that an INVERTER produces an output that is opposite or complement of its input.

Example 34-1. *Fig. 34-3 shows a waveform applied to the input of an INVERTER, Sketch the output waveform corresponding to the input.*

Fig. 34-3.

Solution. Given: An INVERTER with a waveform *A* at its input.

We know that for a given waveform, an INVERTER produces an inverted output waveform. Thus the output waveform it exactly opposite to the input as shown in Fig. 34-4.

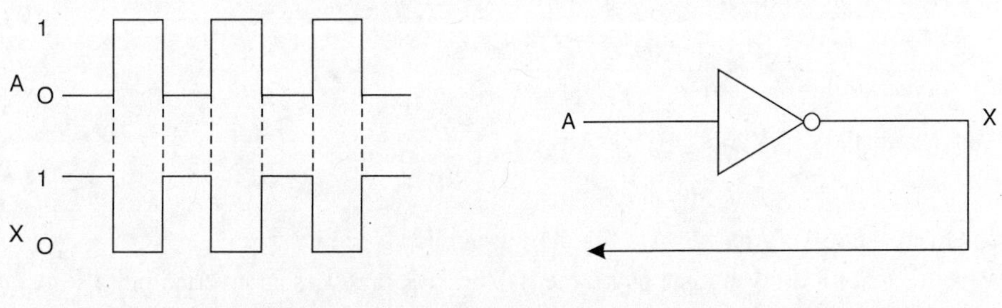

Fig. 34-4.

34-7. Boolean Expression for an INVERTER

We have already discussed in Article 34-5 that the INVERTER output is complement of its input, i.e., output is 0 if the input is 1. The output is 1 if the input is 0. This operation of the INVERTER (or NOT gate) can be expressed in the form of Boolean expression as follows:

Let A = the input variable and

X = the output variable,

Then $X = \overline{A}$ ...(*i*)

Equation (*i*) can be read as *X equals A bar*. It states that the output *X* is complement or opposite of the input *A*, *i.e.*, $X = 0$ if $A = 1$, and $X = 1$ if $A = 0$.

Example 34-2. *Write the Boolean equation at X for the circuit shown in Fig. 34-5. Evaluate the expression if A = 1.*

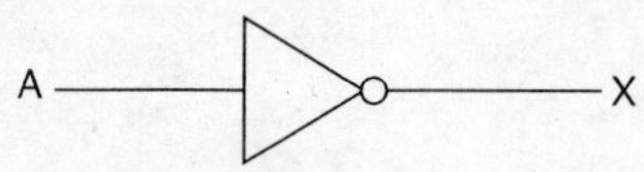

Fig. 34-5.

Solution. Given: An INVERTER with input *A* and output *X*.

We know that the Boolean equation for an INVERTER is given by the relation:

$$X = \overline{A} \quad \text{...(i)}$$

Value of X when A = 1

Substituting $A = 1$ in equation (*i*), we get

$$X = \overline{1} = 0 \textbf{ Ans.}$$

Example 34-3. ***(a)*** *Write the Boolean equation at X and Z for an INVERTER circuit shown in Fig. 34-6. Evaluate the value of X and Z when A = 0.*

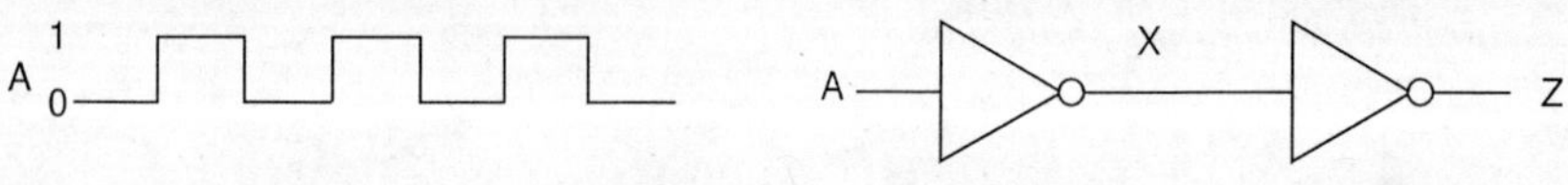

Fig. 34-6.

(b) *Sketch the waveforms at X and Z if the timing waveform shown in Fig. 34-6 is input at A.*

Solution. Given: ***(a)*** An INVERTER circuit with input labeled as *A*.

We know that the Boolean expression for the INVERTER '1' is given by the relation,

$$X = \overline{A} \textbf{ Ans.}$$

and the Boolean expression for the INVERTER '2' is,

$$Z = \overline{X} \textbf{ Ans.}$$

Value of X and Z if A = 0

We know that, $X = \overline{A} = \overline{0} = 1$ **Ans.**

and $Z = \overline{X} = \overline{1} = 0$ **Ans.**

(b) Given, the waveform shown in Fig. 34-6 is applied at the input *A*.

We know that for a given input pulse, the INVERTER produces an inverted pulse at its output. Therefore the waveforms obtained at *X* and *Z* are as shown in Fig. 34-7.

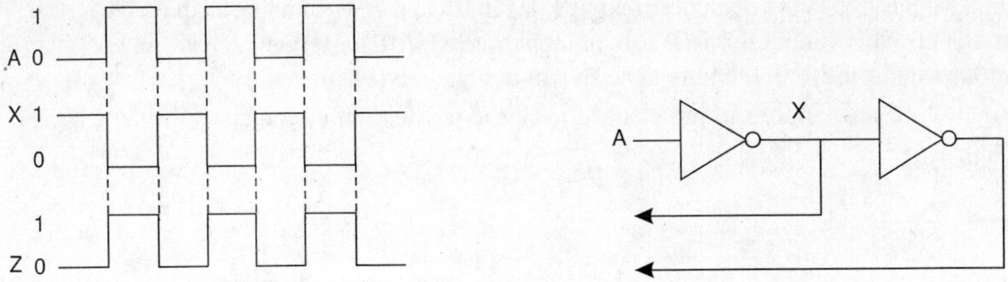

Fig. 34-7.

34-8. Application of an INVERTER

Fig. 34-8 shows a very simple application of an INVERTER. Here the INVERTER is used for producing 1's complement of an 8-bit number. The bits of the given binary number are applied to the INVERTER inputs and the 1's complement of the number appears on the outputs.

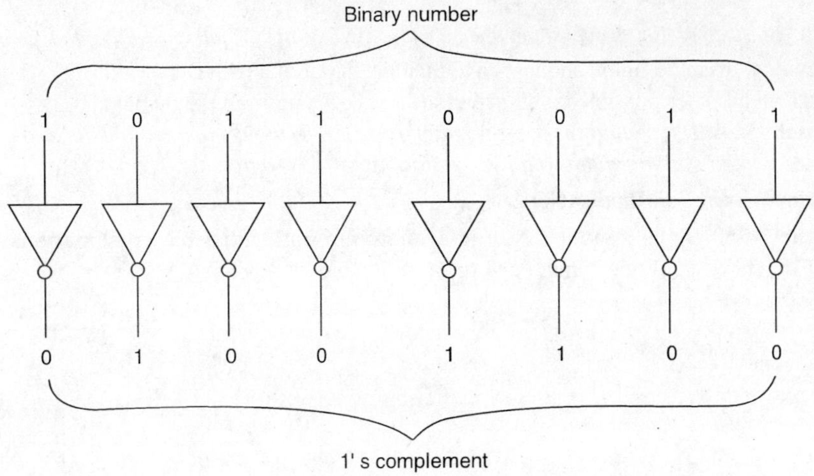

Fig. 34-8. *INVERTER for producing 1's complement.*

34-9. The AND Gate

The AND gate is one of the basic gates from which all logic functions are constructed. It is composed of 2 or more inputs and a single output. Fig. 34-9 (a) shows a logic symbol for a 2-input AND gate. Note that inputs are shown on the left (labeled as *A* and *B*) and the output is on the right labeled as *X*).

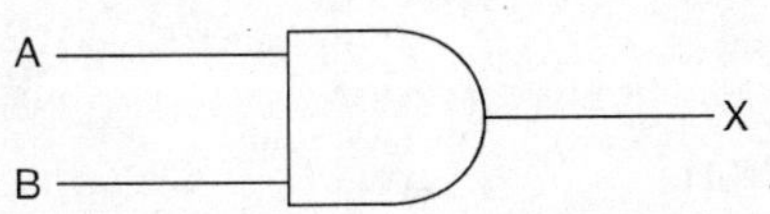

(*a*) Logic symbol

Inputs		*Output*
A	*B*	*X*
0	0	0
0	1	0
1	0	0
1	1	1

(*b*) truth table

Fig. 34-9. *A 2-input AND gate.*

Logical operation. An AND gate produces a HIGH(1) output *only* when *all* of the inputs are HIGH(1). When any of the inputs is LOW(0), the output is LOW(0).

In a 2-input AND gate operation, output X is HIGH(1) if input A and B both are HIGH(1). The output X is LOW(0) if input A is LOW(0), or input B is LOW(0), or if A and B both are LOW(0). This operation is summarized in the truth table shown in Fig. 34-9 (b).

Fig. 34-10 shows the resulting output for all the possible input combinations (*i.e.*, 00, 01, 10 and 11).

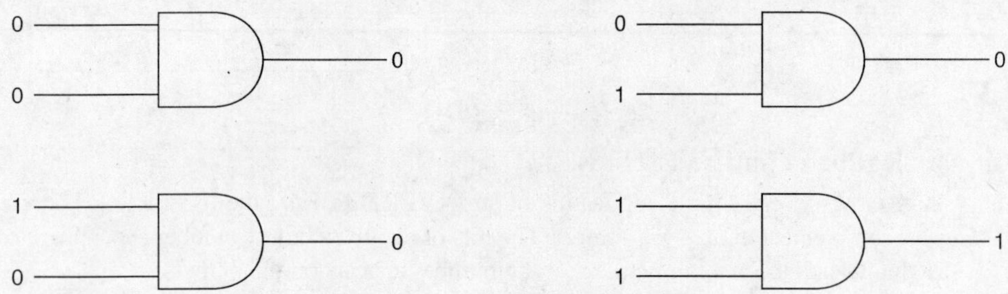

Fig. 34-10. *Illustrating logic operation of a 2-input AND gate for all possible input combinations.*

It is evident from the figure shown above that the output of an AND gate is 1 only when both the inputs are 1. In all other possible combinations (*i.e.*, 00, 01, and 10) the output is 0.

Example 34-4. *Determine the total number of possible input combinations of a 3-input AND gate. Sketch a logic symbol and develop a truth table for this logic gate.*

Solution. Given: A 3-input AND gate.

We know that for a 3-input AND gate, there are eight possible combinations ($2^3 = 8$). Fig. 34-11 (a) shows the logic symbol and (b) a truth table for a 3-input AND gate.

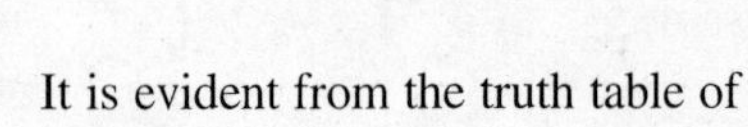

(a) Logic symbol

Inputs			Output
A	*B*	*C*	*X*
0	0	0	0
0	0	1	0
0	1	0	0
0	1	1	0
1	0	0	0
1	0	1	0
1	1	0	0
1	1	1	1

(b) Truth table

Fig. 34-11.

It is evident from the truth table of a 3-input AND gate that output is 1 only when all the three inputs are 1. The output is 0 if any of the three inputs is 0 or all the three inputs are 0.

Example 34.5. *Determine the total number of possible input combinations of a 4-input AND gate. Sketch a logic symbol and develop a truth table for this logic gate.*

Solution. Given: A 4-input AND gate.

We know that for a 4-input AND gate, there are sixteen possible combination ($2^4 = 16$). Fig. 34-12 (a) shows the logic symbol and (b) a truth table for a 4-input AND gate.

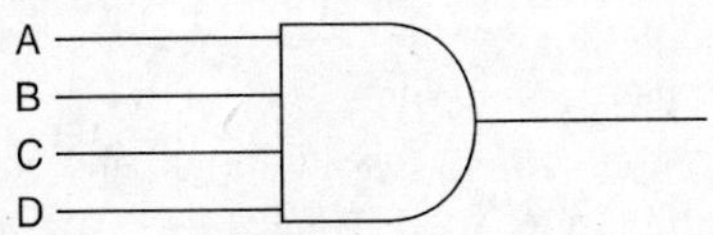

(a) Logic symbol

A	*B*	*C*	*D*	*X*
Inputs				*Output*
0	0	0	0	0
0	0	0	1	0
0	0	1	0	0
0	0	1	1	0
0	1	0	0	0
0	1	0	1	0
0	1	1	0	0
0	1	1	1	0
1	0	0	0	0
1	0	0	1	0
1	0	1	0	0
1	0	1	1	0
1	1	0	0	0
1	1	0	1	0
1	1	1	0	0
1	1	1	1	1

(b) Truth table

Fig. 34-12.

It is evident from the truth table of a 4-input AND gate that output is 1 only when all the four inputs are 1. The output is 0 if any of the four inputs is 0 or all the four inputs are 0.

34-10. Pulsed Operation of an AND Gate

It will be interesting to know that in a majority of applications, the inputs to a logic gate (of any type) are not stationery voltage levels, but voltage waveforms that change frequently between 1 and 0 logic levels. Now let us study the operation of an AND gate with pulse waveform inputs *A* and *B* as shown in Fig. 34-13. Keep it in mind that an AND gate obeys the truth table operation (refer to Fig. 34-9 (b) regardless of whether its inputs are constant levels or levels that change back and forth.

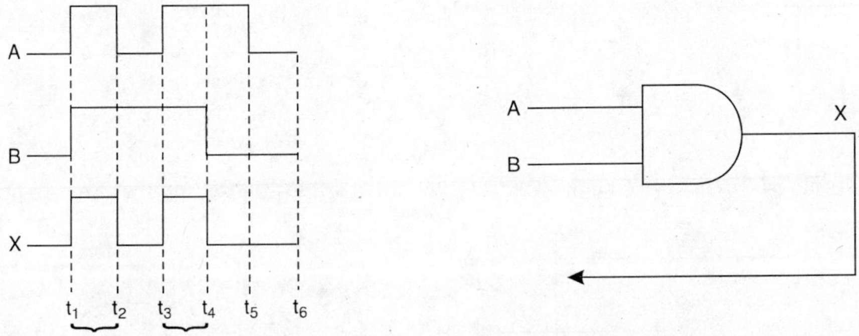

A and *B* both are 1 during these two intervals, therefore, output *X* is 1.

Fig. 34-13. *Illustrating pulsed operation of a 2-input AND gate.*

At $t = t_1$, both inputs *A* and *B* goes 1, therefore output, *X* changes from 0 to 1. At $t = t_2$, input *A* goes 0, $B = 1$, therefore *X* goes 0. At $t = t_3$, *A* goes 1, $B = 1$, therefore *X* goes 1. At $t = t_4$, $A = 1$, *B* goes 0, therefore *X* goes 0. At $t = t_5$, *A* goes 0, $B = 0$, therefore $X = 0$, at $t = t_6$, $A = 0$, $B = 0$, therefore $X = 0$. The resulting output waveform at *x* is shown in Fig. 34-13.

Example 34-6. *Fig. 34-14 shows the two input waveforms, A and B applied to an AND gate.*

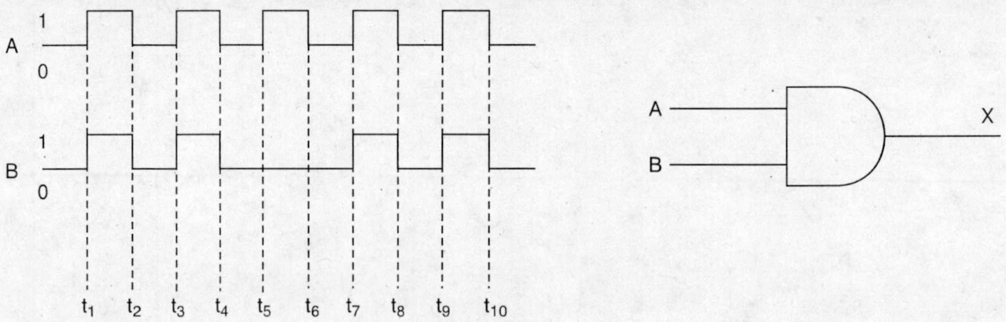

Fig. 34-14.

Sketch the resulting output waveform at X.

Solution. Given: A two-input AND gate with the waveforms at *A* and *B*.

We know that an AND gate output is 1 only when both the inputs *A* and *B* are 1. When any one of the inputs is 0, the output is 0. The output is also 0 when both the inputs are 0. Thus at $t = t_0$, $A = 0$, $B = 0$, therefore $X = 0$. At $t = t_1$, *A* goes 1, *B* goes 1, therefore *X* also goes 1. At $t = t_2$, *A* goes 0, *B* goes 0, therefore *X* also goes 0, At $t = t_3$, *A* and *B* again goes 1, therefore *X* goes 1. At $t = t_4$, both *A* and *B* goes 0, therefore *X* goes 0. At $t = t_5$, *A* goes 1, $B = 0$, but *X* remains 0. At $t = t_6$, *A* goes 0, $B = 0$, therefore *X* still remains at 0. At $t = t_7$ *A* goes 1, *B* goes 1, therefore *X* goes 1. At $t = t_8$, both *A* and *B* goes 0, therefore *X* also goes 0. At $t = t_9$, both *A* and *B* goes 1, therefore *X* goes 1. At $t = t_{10}$, both *A* and *B* goes 0, therefore *X* goes 0. The resulting input waveform at *X* is shown in Fig. 34-15.

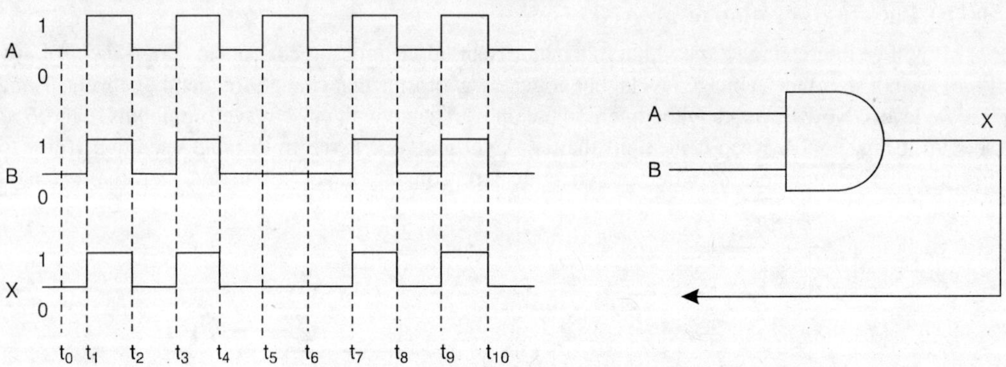

Fig. 34-15.

Example 34-7. *Fig. 34-16 shows the two input waveforms A and B applied to an AND gate.*

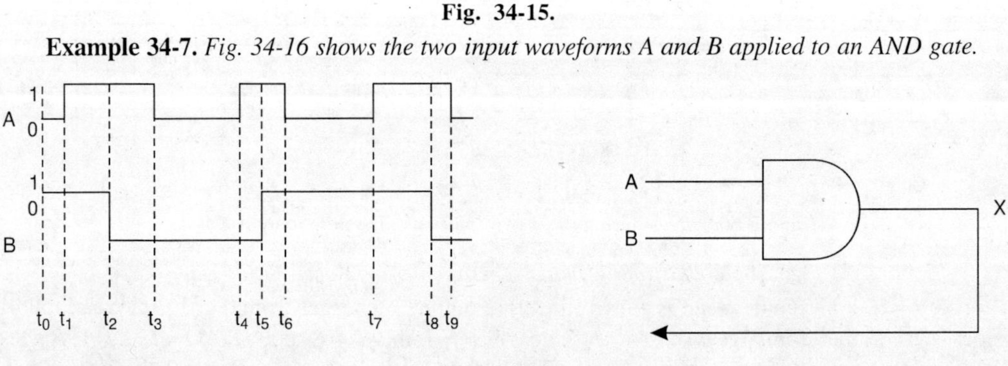

Fig. 34-16.

Sketch the resulting output waveform showing its proper relation to the inputs.

Solution. Given: A two-input AND gate with waveforms at A and B.

We know that an AND gate output is 1 only when both the inputs A and B are 1. When anyone of the inputs is 0, the output is 0. Thus at $t = t_0$, $A = 0$, $B = 1$, therefore $X = 0$. At $t = t_1$, A goes 1, $B = 1$, therefore X goes 1. At $t = t_2$, $A = 1$, B goes 0, there X goes 0. At $t = t_3$, A goes 0, $B = 0$, therefore $X = 0$. At $t = t_4$, A goes 1, $B = 0$, therefore $X = 0$. At $t = t_5$, $A = 1$, B goes 1, therefore X goes 1. At $t = t_6$, A goes 0, $B = 1$, therefore, X goes 0. At $t = t_7$, A goes 1, $B = 1$, therefore X goes 1. At $t = t_8$, $A = 1$, B goes 0, therefore X goes 0. At $t = t_9$, A goes 0, $B = 0$, therefore $X = 0$. The resulting output waveform at X is shown in Fig. 34-17.

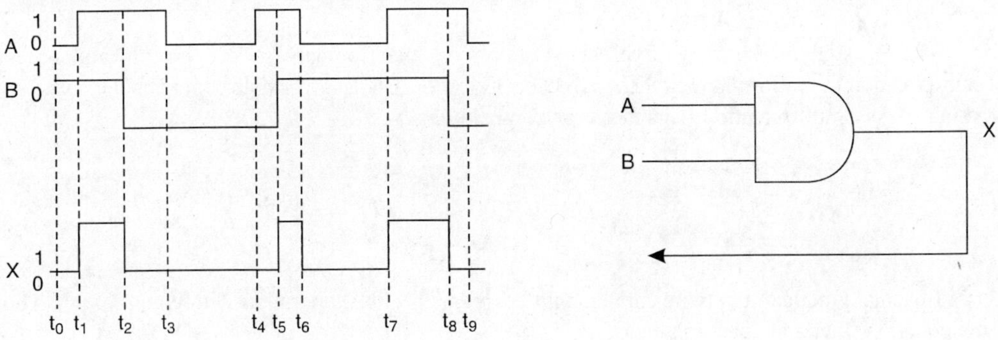

Fig. 34-17.

Example 34-8. *Fig. 34-18 shows the input waveforms A, B and C applied to an AND gate.*

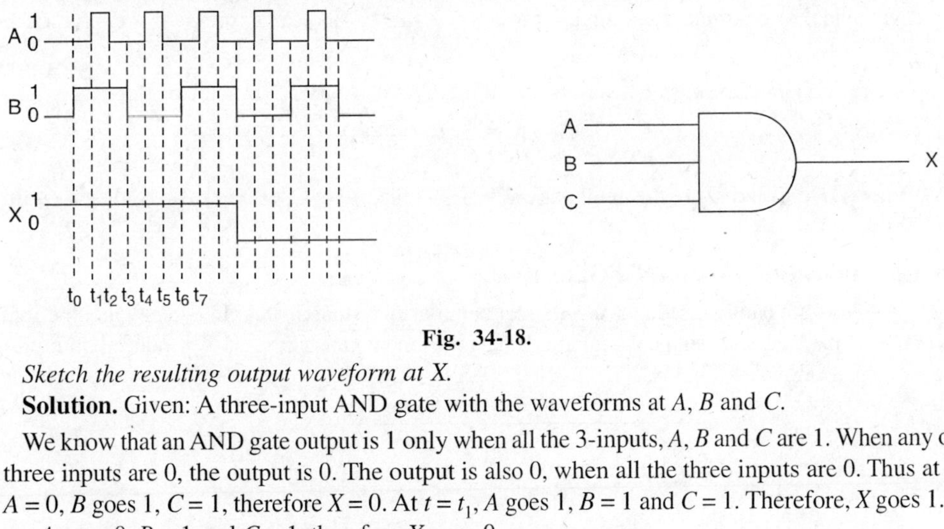

Fig. 34-18.

Sketch the resulting output waveform at X.

Solution. Given: A three-input AND gate with the waveforms at A, B and C.

We know that an AND gate output is 1 only when all the 3-inputs, A, B and C are 1. When any one of three inputs are 0, the output is 0. The output is also 0, when all the three inputs are 0. Thus at $t = t_1$, $A = 0$, B goes 1, $C = 1$, therefore $X = 0$. At $t = t_1$, A goes 1, $B = 1$ and $C = 1$. Therefore, X goes 1. At $t = t_2$, A goes 0, $B = 1$ and $C = 1$, therefore X goes 0.

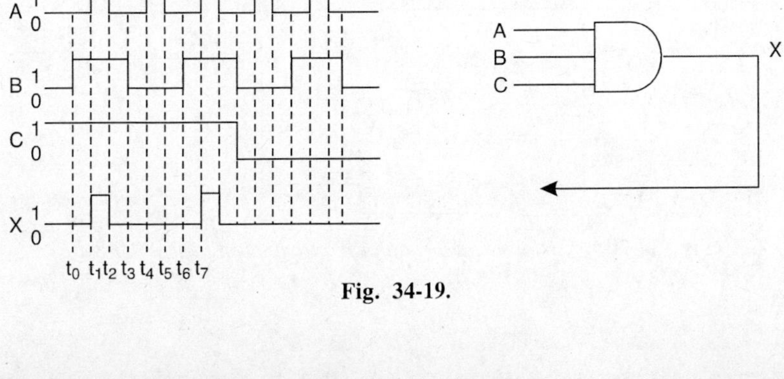

Fig. 34-19.

At $t = t_3$, $A = 0$, B goes 0, $C = 1$, therefore $X = 0$. At $t = t_4$, A goes 1, $B = 0$, $C = 1$, therefore $X = 0$. At $t = t_5$, A goes 0, $B = 0$, $C = 1$, therefore $X = 0$. At $t = t_6$, $A = 0$, B goes 1, $C = 1$, therefore $X = 0$. At $t = t_7$, A goes 1, $B = 1$, $C = 1$, therefore X goes 1 and so on... The resulting output waveform at X is shown in Fig. 34-19.

34-11. Boolean Expression for an AND Gate

We have already discussed in the Article 34-9 that for a 2-input AND gate, output is 1 only if both the inputs *A and B* are 1. If either of the inputs *A and B* is 0, the output is 0. This operation of an AND gate can be expressed in the form of a Boolean expression as follows:

$$X = A \, . \, B \qquad \text{...}(i)$$

The "." sign in the above expression stands for the Boolean AND operation and not the multiplication operation. However the AND operation on Boolean variables operates the same way as in ordinary multiplication. This fact is shown below:

$$0.0 = 0$$
$$0.1 = 0$$
$$1.0 = 0$$
$$1.1 = 0$$

In actual practice, it is more convenient to express the AND operation without a "." sign. Thus, the equation (i) can also be written as,

$$X = AB \qquad \text{...}(ii)$$

To extend the AND operation to more than two input variables. Simply use a new letter for each input variable. For example, for a three input AND gate, the Boolean expression is expressed as

$$X = ABC \qquad \text{...}(iii)$$

Where A, B and C are the input variables. Similarly for a 4-input AND gate, the Boolean expression is

$$X = ABCD \qquad \text{...}(iv)$$

Where A, B, C and D are the input variables. Thus this can be extended to an AND gate with any number of inputs.

34-12. Application of an AND Gate

An AND gate can be used in a simple seat belt alarm system in a car to detect when the ignition switch is on *and* the seat belt is unbuckled. If the ignition switch is ON, a 1 is produced on input A of the AND gate shown in Fig. 34-20.

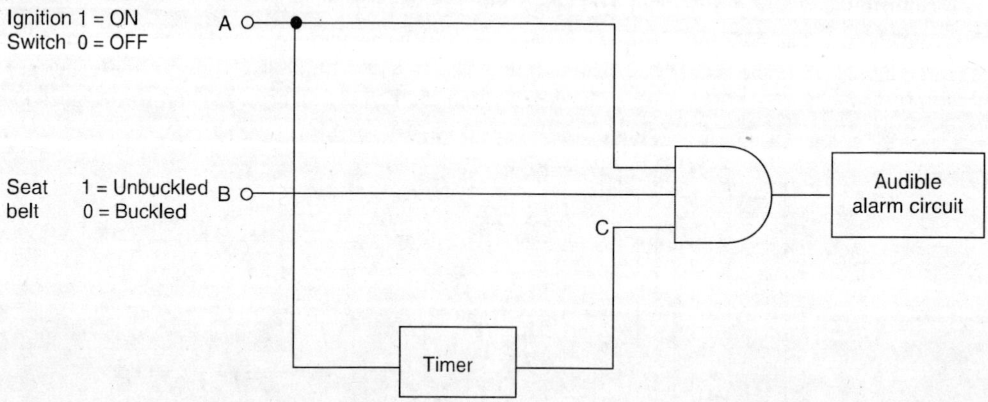

Fig. 34-20. *A simple seat belt alarm circuit using an AND gate.*

If the seat belt is not properly buckled, 1 is produced on input *B* of the AND gate. Also when the ignition switch is turned ON, a *timer* is started that produces a 1 on input *C* for 30 seconds.

If all the three conditions exist, *i.e.*, if the ignition switch is ON, *and* the seat belt is unbuckled *and* the time is running, the output of the AND gate is 1. This will activate the audible alarm to remind the driver (say to buckle the seat belt).

34-13. The OR Gate

Like AND gate, the OR gate, is another basic logic gate from which all logic functions are constructed. It is composed of two or more inputs and a single output. Fig. 34-21 (a) shows a logic symbol for a 2-input OR gate. Notice that the inputs are shown on the left labelled as *A* and *B* and the output is on the right (labelled as *X*).

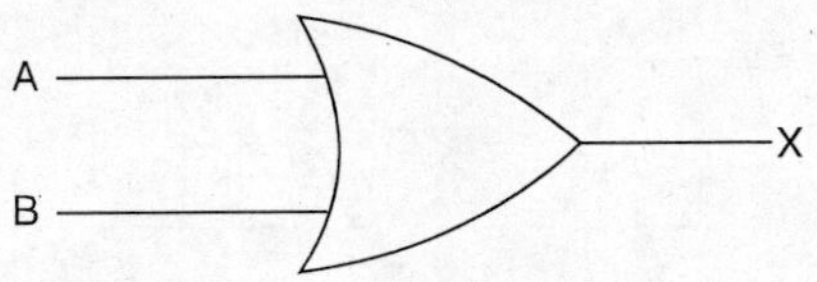

(a) Logic symbol

Inputs		Output
A	*B*	*X*
0	0	0
0	1	1
1	0	1
1	1	1

(b) Truth table

Fig. 34-21. *A 2-input OR gate.*

Logical operation of the OR gate. The logical operation of the OR gate may be described as below: the OR gate produces a HIGH(1) on the output when *any* of the inputs is HIGH(1). The output is LOW(0) when all the inputs are LOW(0).

In a 2-input OR gate operation, output *X* is HIGH(1) if either input *A* or input *B* is HIGH(1), or if both *A* and *B* are HIGH(1). The output, *X* is LOW(0) if both *A* and *B* are LOW(0), this operation is summarized in the truth table shown in Fig. 34-21 (b).

Notice that the table shows the resulting output for all the possible input combinations, *i.e.*, 00, 01, 10 and 11.

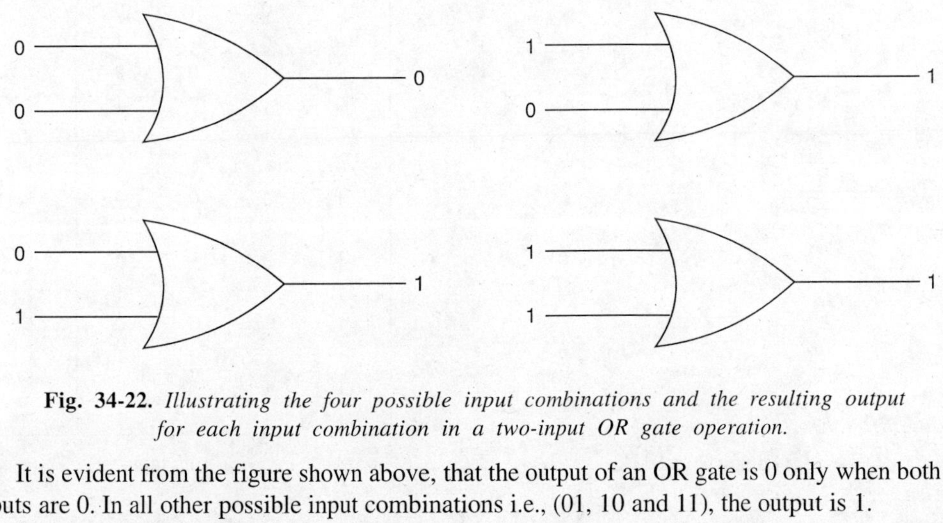

Fig. 34-22. *Illustrating the four possible input combinations and the resulting output for each input combination in a two-input OR gate operation.*

It is evident from the figure shown above, that the output of an OR gate is 0 only when both the inputs are 0. In all other possible input combinations i.e., (01, 10 and 11), the output is 1.

Example 34-9. *Determine the total number of possible input combinations of a 3-input OR gate. Also draw a logic symbol and develop the truth table for this logic gate.*

Solution. Given: A 3-input OR gate.

We know that for a 3-input OR gate, there are eight possible input combinations ($2^3 = 8$). Fig. 34-23 (a) shows the logic symbol and 34-23 (b) a truth table for a 3-input OR gate.

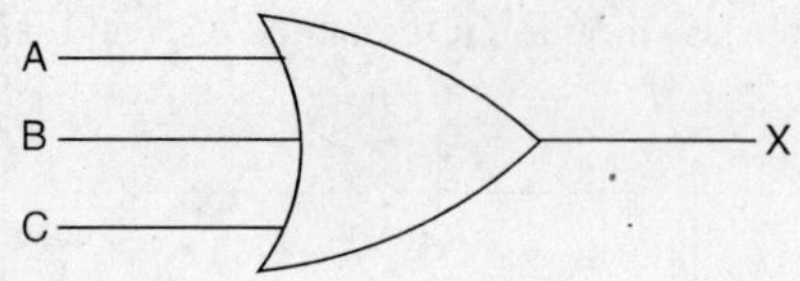

(a) Logic symbol

Inputs			Output
A	*B*	*C*	*X*
0	0	0	0
0	0	1	1
0	1	0	1
0	1	1	1
1	0	0	1
1	0	1	1
1	1	0	1
1	1	1	1

(b) Truth table

Fig. 34-23. *A 3-input OR gate.*

Note that a 3-input OR gate output is 1 when any of the three inputs, *A*, *B* and *C* is 1. The output is 0 only when all the three inputs are 0.

Example 34-10. *Determine the total number of possible input combinations of a 4-input OR gate. Also draw a logic symbol and develop the truth table for this logic gate.*

Solution. Given: A 4-input OR gate.

We know that for a 4-input OR gate, there are sixteen possible input combinations ($2^4 = 16$). Fig. 34-24 (a) shows the logic symbol and 34-24 (b) a truth table for a 4-input OR gate.

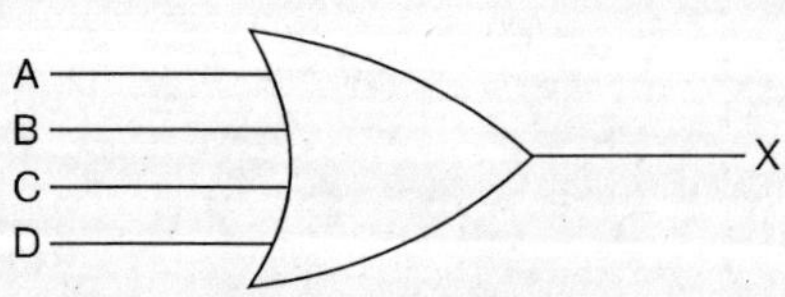

(a) Logic symbol

Inputs				Output
A	*B*	*C*	*D*	*X*
0	0	0	0	0
0	0	0	1	1
0	0	1	0	1
0	0	1	1	1
0	1	0	0	1
0	1	0	1	1
0	1	1	0	1
0	1	1	1	1
1	0	0	0	1
1	0	0	1	1
1	0	1	0	1
1	0	1	1	1
1	1	0	0	1
1	1	0	1	1
1	1	1	0	1
1	1	1	1	1

(b) Truth table

Fig. 34-24. *A 4-input OR.*

Note that a 4-input OR gate output is 1 when any of the four inputs *A*, *B*, *C* and *D* is 1. The output is 0 only when all the for inputs are 0.

34-14. Pulsed Operation of an OR Gate

We know that in a majority of applications, the inputs to a logic gate (of any type) are not stationery voltage levels but are voltage waveforms, that change frequently between 1 and 0 logic levels. Now let us study the operation of an OR gate with pulse waveform inputs. Keep it in mind that an OR gate obeys the truth table operation regardless of whether its inputs are constants levels or levels that change back and forth.

Consider an OR gate with two-inputs *A* and *B* as shown in Fig. 34-25. In order to determine the output level, we will look at the inputs with respect to each other with reference to the truth table shown in Fig. 34-21.

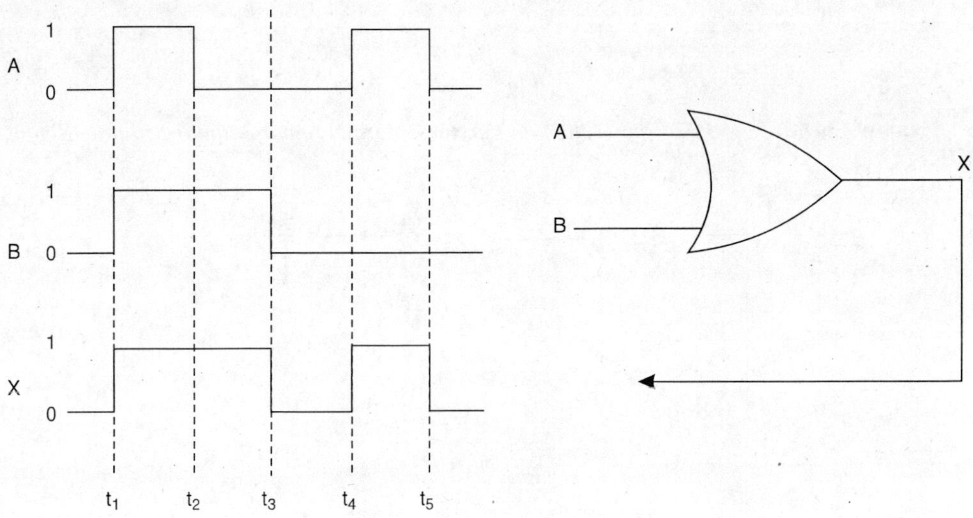

Fig. 34-25. *Illustrating pulsed operation of a two-input OR gate.*

At $t = t_1$, both the inputs *A* and *B* goes 1, therefore the output, *X* changes from 0 to 1. At $t = t_2$, input *A* goes *O*, *B* is 1, therefore *X* is 1, at $t = t_3$, *A* is 0, *B* goes 0, therefore *X* goes 0. At $t = t_4$, *A* goes 1, *B* is 0, therefore *X* is 1. The resulting output waveform is as shown in Fig. 34-11.

Example 34-11. *Fig. 34-26 shows the two input waveforms A and B applied to an OR gate.*

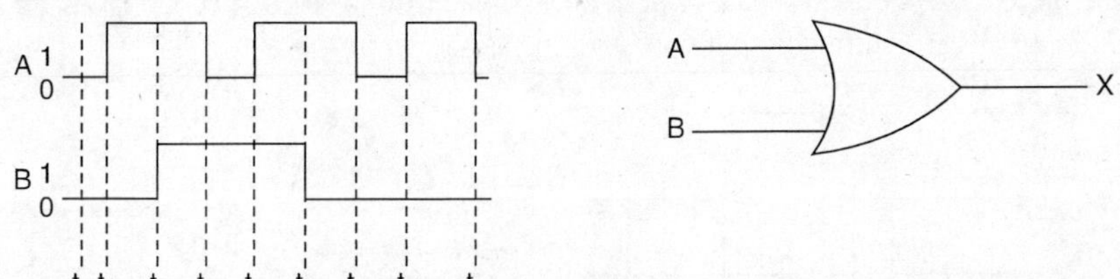

Fig. 34-26.

Sketch the resulting output waveform at X.

Solution. Given: The OR gate with the waveforms *A* and *B* at its two inputs.

We know that an OR gate output is 1 whenever *A* is 1 or *B* is 1 or both *A* and *B* are 1. The OR gate output is 0 when both *A and B* are 0. Thus at $t = t_0$, $A = 0$ and $B = 0$, therefore $X = 0$. At $t = t_1$, *A* goes 1, $B = 0$, therefore *X* goes 1. At $t = t_2$, $A = 1$, *B* goes 1, therefore *X* remains 1. At $t = t_3$, *A* goes 0, $B = 1$, therefore X still remains at 1. At $t = t_4$, A goes 1, $B = 1$, therefore X again remains 1. At $t = t_5$, $A = 1$, B goes 0. However the output X still remains at 1. At $t = t_6$, A goes 0, $B = 0$, therefore X goes 0. At $t = t_7$, A goes 1, $B = 0$, therefore X goes 1, at $t = t_8$, A goes 0, $B = 0$, therefore X goes 0. The resulting output waveform is as shown in Fig. 34-27.

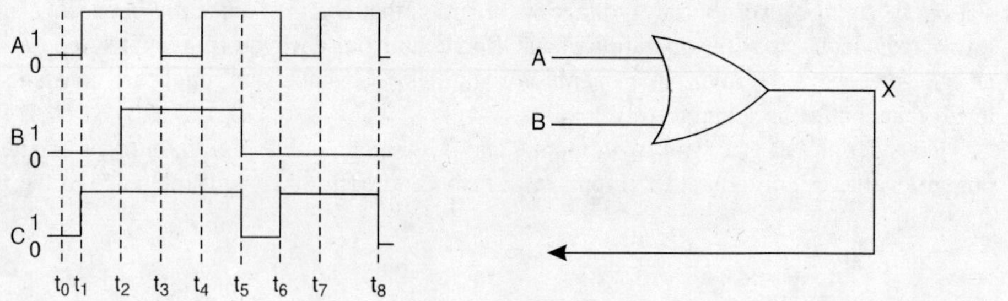

Fig. 34-27.

Example 34-12. *Fig. 34-28 shows the two input waveforms, A and B applied to an OR gate.*

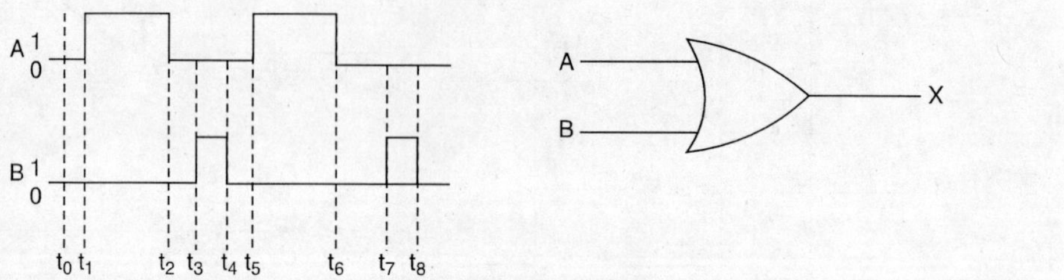

Fig. 34-28.

Sketch the resulting output waveform at X.

Solution. Given: Two-input OR gate with waveforms applied at inputs A and B. We know that the OR gate output is 1 whenever A is 1 or B is 1 or both A and B are 1. The OR gate output is 0 when both A and B are 0. Thus at $t = t_0$, $A = 0$ and $B = 0$, therefore $X = 0$, at $t = t_1$, A goes 1, $B = 0$, therefore X goes 1. At $t = t_2$, A goes 0, $B = 0$, therefore X goes 0. At $t = t_3$, $A = 0$ and B goes 1, therefore X goes 1. At $t = t_4$, $A = 0$, B goes 0, therefore X goes 0. At $t = t_5$, A goes 1, $B = 0$, therefore X goes 1. At $t = t_6$, A goes 0, $B = 0$, therefore X goes 0. At $t = t_7$, $A = 0$, B goes 1, therefore X goes 1. At $t = t_8$, $A = 0$, B goes 0, therefore X goes 0. The resulting output waveform is as shown in Fig. 34-29.

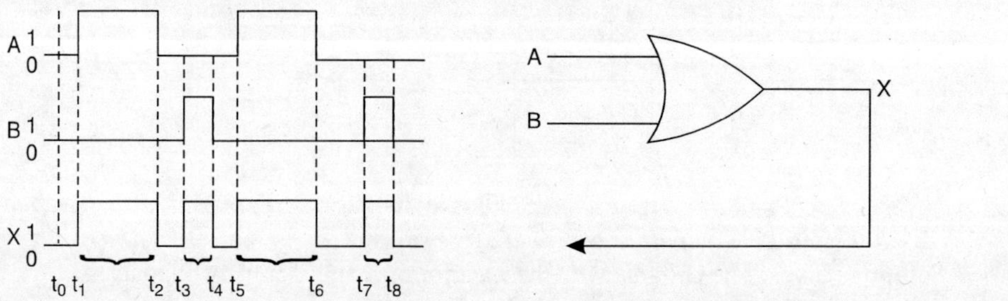

Fig. 34-29.

Example 34-13. *Fig. 34-30 shows the three input waveforms, A, B and C applied to an OR gate.*

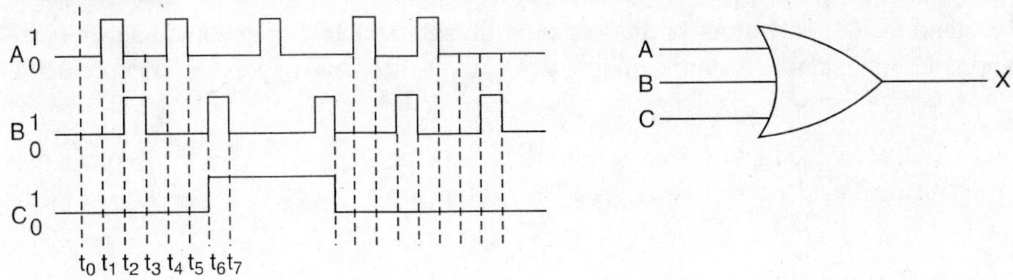

Fig. 34-30.

Sketch the resulting output waveform at X.

Solution. Given: a 3-input OR gate with waveforms at A, B and C inputs.

We know that an OR gate output is 1, whenever A is 1 or B is 1 or C is 1 or all A, B and C are 1. The OR gate output is 0 when all A, B and C are 0. Thus at $t = t_0$, $A = 0$, $B = 0$ and $C = 0$, therefore $x = 0$. At $t = t_1$, A goes 1, $B = 0$ and $C = 0$, therefore X goes 1. At $t = t_2$, A goes 0, $B = 0$ and $C = 0$, therefore X goes 0. At $t = t_3$, $A = 0$, B goes 1, $C = 0$, therefore X goes 1. At $t = t_4$, $A = 0$, B goes 0, $C = 0$, therefore X goes 0. At $t = t_5$, A goes 1, $B = 0$, $C = 0$, therefore X goes 1. At $t = t_6$, A goes 0, $B = 0$, $C = 0$, therefore X goes 0. At $t = t_7$, $A = 0$, B goes 1, C goes 1, therefore X goes 1 and so on... The resulting output waveform is as shown in Fig. 34-31.

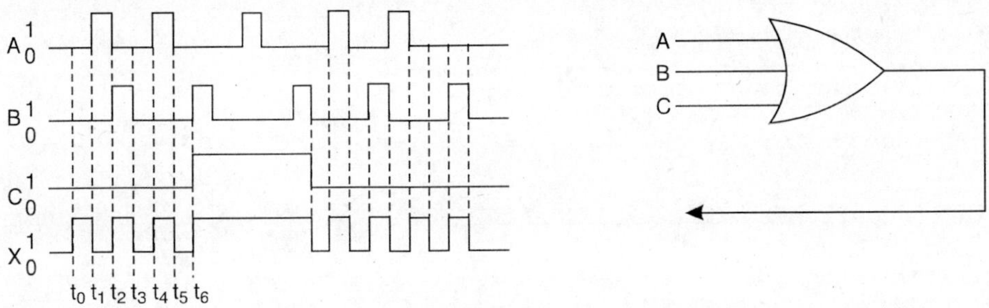

Fig. 34-31.

34-15. Boolean Expression for an OR Gate

We have already discussed in the last article that for a two-input OR gate, the output is 1, whenever either of the inputs A and B is 1 or both the inputs A and B are 1. If both the inputs are 0, the output is 0. This operation of an OR gate can be expressed in the form of a Boolean expression as follows,

$$X = A + B$$

The "+" sign in the above expression stands for the Boolean OR operation or the Boolean addition operation. The OR operation on Boolean variables operates the way as shown below:

$$0 + 0 = 0$$

$$0 + 1 = 1$$

$$1 + 0 = 1$$

$$1 + 1 = 1$$

Notice that the OR operation produces $1 + 1 = 1$, not $1 + 1 = 2$. In Boolean algebra, we can never have a result greater than 1.

The expression, $X = A + B$ is read as $X = A$ OR B. The key theory to remember is that the "+" sign stands for the OR operation and not for ordinary addition.

To extend the OR operation for more than two-input variables, simply use a new letter for each input variable. For example for a three-input OR gate, the Boolean expression is expressed as:

$$X = A + B + C \qquad ...(5)$$

where A, B, C are the input variables. Similarly for a four-input OR gate, the Boolean expression is

$$X = A + B + C + D \qquad ...(6)$$

where A, B, C and D are the input variables. Thus this approach can be extended to an OR gate with any number of inputs.

34-16. Application of an OR Gate

An OR gate can be used in a simplified portion of an intrusion detection and alarm system as shown in Fig. 34-32. The circuit shown in Fig. 34-32 consists of three sensors which can be placed on the doors and windows within a room (say with one door and two windows). The sensors are magnetic switches that produces a 1 output when open and a 0 output when closed. As long as the windows and the door are secured, the switches are closed. Therefore all the three inputs of the OR gate are 0 (Due to this the output is 0 and the alarm is de-activated). When one of the windows or the door is opened, a 1 is produced on the input to the OR gate. As a result of this, the OR gate output becomes 1. This activates an alarm circuit to warn of the intrusion.

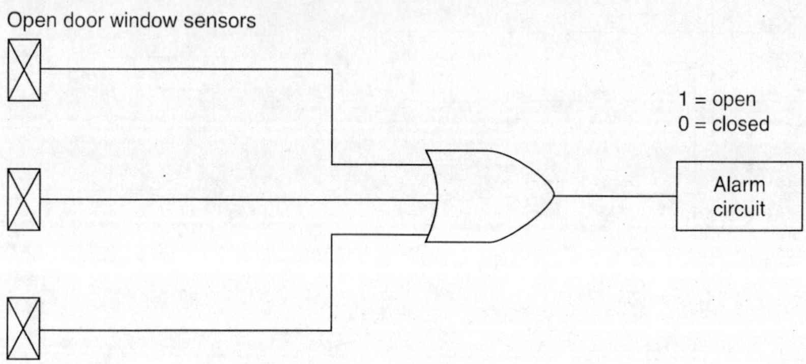

Fig. 34-32. *A simplified instruction detection system using an OR gate.*

34-17. The NAND and NOR Gate

We have already discussed the NOT, AND and OR gates in the previous articles. Now we will study two other types of logic gates that are used extensively in digital circuitry. These logic gates are the NAND and NOR gates. The NAND and NOR gates actually combine the basic operations AND, OR, and NOT, which make it relatively easy to describe them using Boolean algebra operations. Now we will discuss both these logic gates one by one in the following pages.

34-18. The NAND Gate

The term NAND is a construction of NOT-AND. It implies an AND function with a complemented (or inverted) output. The standard logic symbol for a 2-input NAND gate is as shown in Fig. 34-33 (a). Notice that this symbol is the same as AND gate symbol except for a small circle (or a bubble) on its output. This small circle denotes the invention operation. Thus the NAND gate operates like an AND gate followed by INVERTER as shown in Fig. 34-33 (b).

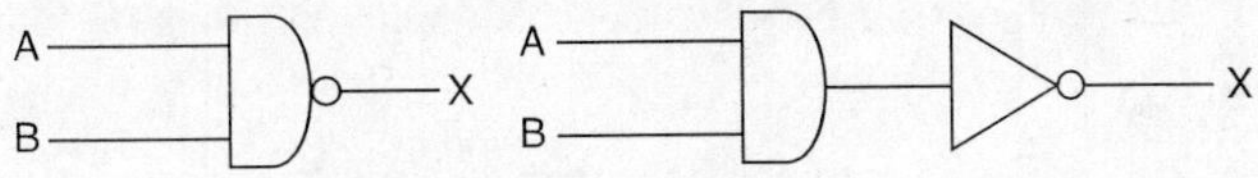

Input		Output
A	*B*	*X*
0	0	1
0	1	1
1	0	1
1	1	0

(a) Logic symbol (b) NOT-AND equivalent (c) Truth table

Fig. 34-33. 2-input NAND gate A.

Logical Operation of the NAND gate. The logical operation of the NAND gate is described as below. The NAND gate produces a LOW(0) output only when all the inputs are HIGH(1). When any of the inputs is LOW(0), the output will be HIGH(1).

In a 2-input NAND gate operation, output *X* is LOW(0) if inputs *A* and *B* are HIGH(1). The output *X* is HIGH(1) if either *A* or *B* is LOW(0) or if both *A* and *B* are LOW(0). This operation can be summarized in the truth table shown in Fig. 34-33 (c).

Fig. 34-34 illustrates the logical operation of a 2-input NAND gate for all four input combinations, *i.e.*, 00, 01, 10 and 11.

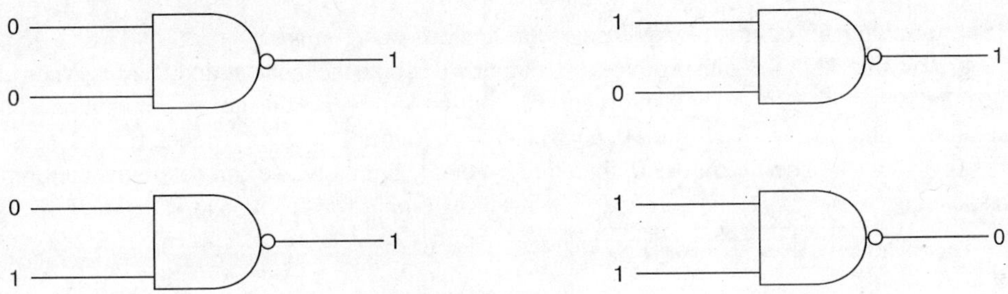

Fig. 34-34. Illustrating the four possible input combinations and the resulting output for each in a two-input NAND gate operation.

It is evident from the figure shown above that the output of a NAND gate is 0 only when both the inputs are 1. In all other possible input combinations (*i.e.*, 00, 01 and 10), the output is 1.

34-19. Pulsed Operation of a NAND Gate

As discussed earlier in the case of AND and OR gates, the inputs to the logic gates (of any type) are not stationery voltage levels but are voltage waveforms that change frequently between 1 and 0 logic levels. Now we will study the operation of NAND gate with pulse waveform inputs. Keep it in mind that a NAND gate obeys the truth table operation regardless of whether its inputs are constant levels or levels that change back and forth.

Consider a two waveforms *A* and *B* applied at the inputs of a 2-input NAND gate as shown in Fig. 34-35. In order to determine the output level, we will look at the inputs with respect to each other with reference to the truth table shown in Fig. 34-33 (b).

At $t = t_1$, input A goes 1, $B = 1$, therefore X goes 0. At $t = t_2$, $A = 1$, B goes 0, therefore X goes 1. At $t = t_3$, A goes 0, $B = 0$, therefore $X = 1$. At $t = t_4$, A goes 1, $B = 0$, therefore $X = 1$. At $t = t_4$, A goes 0, $B = 0$, therefore $X = 1$. At $t = t_5$, $A = 0$, B goes 1, therefore $X = 1$. At $t = t_6$, A goes 1, $B = 1$, therefore $X = 0$. At $t = t_7$, A goes 0, $B = 0$, therefore X goes 1. At $t = t_8$, $A = 0$, B goes 0, therefore $x = 0$. At $t = t_9$, A goes 1, $B = 0$, therefore $X = 1$. The resulting output waveform at X is as shown in Fig. 34-35.

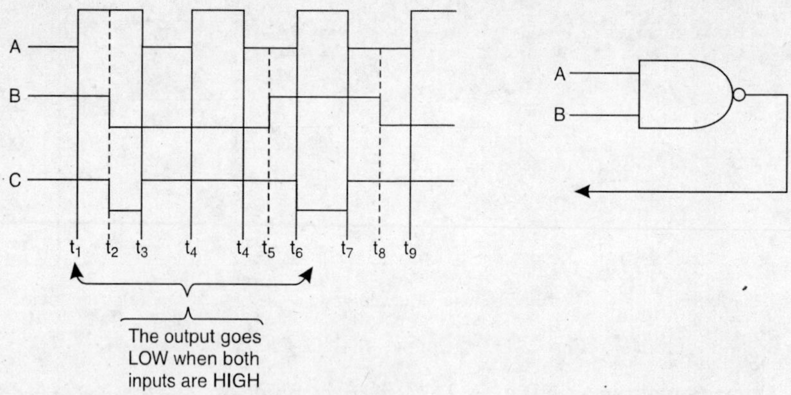

Fig. 34-35. *Illustrating pulsed operation of a 2-input NAND gate.*

Example 34-14. *Fig. 34-36 shows the two waveforms applied to the NAND gate inputs. Determine the resulting output waveform.*

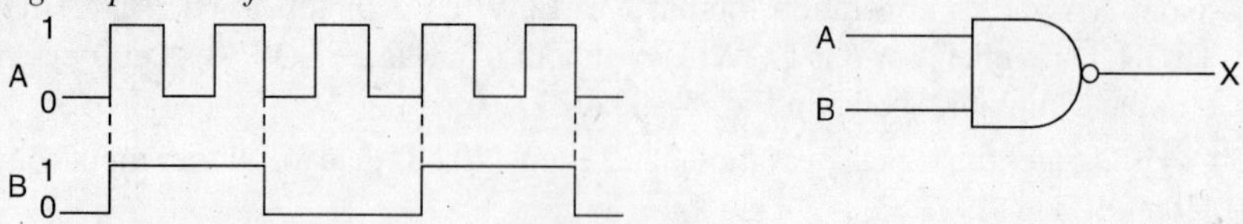

Fig. 34-36.

Solution. Given: A 2-input NAND gate with *A* and *B* waveforms.

We know that a NAND gate produces a 0 output when both the inputs *A* and *B* are 1. When either of the inputs *A* or *B* is 0, or both *A* and *B* are 0 the output will be 1. Thus at $t = t_1$, both *A* and *B* goes 1, therefore *X* goes 0. At $t = t_2$, *A* goes 0, $B = 1$, therefore *X* goes 1. At $t = t_3$, *A* goes 1, $B = 1$, therefore *X* goes 0. At $t = t_4$, *A* goes 0, *B* goes 0, therefore *X* goes 1. Similarly, we can determine the output *X* corresponding to $t = t_5, t_6, t_7, t_8, t_9$ and t_{10}. The resulting output waveform is as shown in Fig. 34-37.

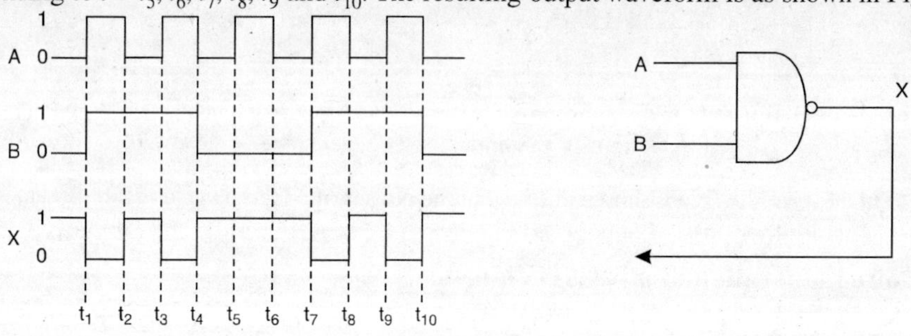

Fig. 34-37.

Example 34-15. *Sketch the output waveform for the 3-input NAND gate shown in Fig. 34-38 showing its proper time relationships to the inputs.*

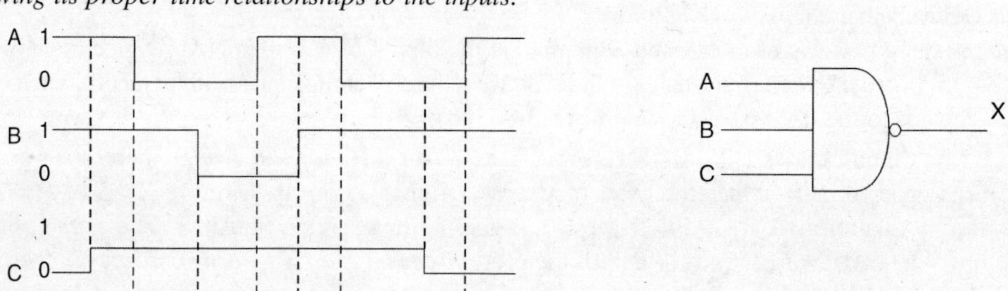

Fig. 34-38.

Solution. Given *A* 3-input NAND gate with *A, B* and *C* waveforms.

We know that a NAND gate produces 0 output only when all the three inputs, *A, B* and *C* are 1. The output is 1, if any of the three inputs *A, B, C* is 0, or all the three inputs are 0.

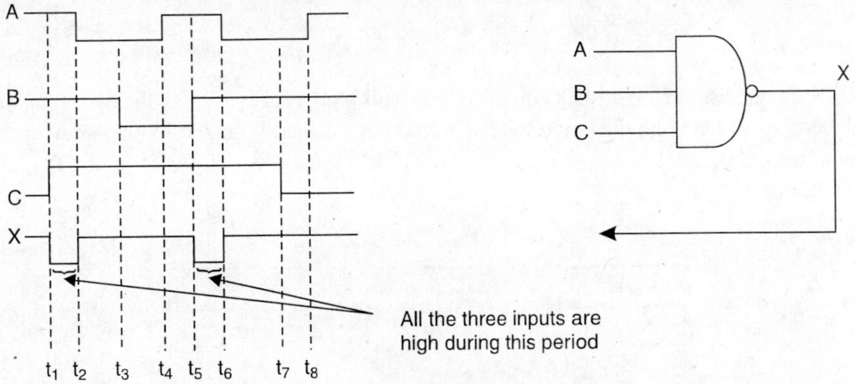

Fig. 34-39.

Thus at $t = t_1$, $A = 1$, $B = 1$, C goes 1, therefore X goes 0. At $t = t_2$, A goes 0, $B = 1$, $C = 1$, therefore X *goes 1.* At $t = t_3$, $A = 0$, B goes 0, $C = 1$, therefore $X = 1$. At $t = t_4$, A goes 1, $B = 0$, $C = 1$, therefore $X = 1$, At $t = t_3$, $A = 1$, B goes 1, $C = 1$, therefore X goes 0. At $t = t_6$, A goes 0, $B = 1$, $C = 1$, therefore X goes 1. At $t = t_7$, $A = 0$, $B = 1$, C goes 0, therefore $X = 1$. At $t = t_8$, A goes 1, $B = 1$, $C = 0$, therefore $X = 1$. The resulting output waveform is shown in Fig. 34-39.

A faster way of sketching the waveform is to find the time period during which all the three inputs are 1. During this we know that X is 0. For all other possibilities, X is 1.

34-20. Application of a NAND Gate

Fig. 34-40 shows the application of a NAND gate in a drug manufacturing plant that uses two tanks to store a liquid chemical. Each tank has a sensor that detects when the chemical level drops to 25% of full. The sensors produce a 5V level when the tanks are more than one-quarter full. When the level in the tank drops to one-quarter full, the sensor produces a 0 V level.

As seen from the figure, the NAND gate with its two inputs is connected to the tank level sensors. The output of the NAND gate is connected to the supply voltage (+V) through the green LED (a light emitting diode) and a resistor (R). The operation of the system can be explained as follows: As long as both sensor outputs are HIGH(I), indicating that both tanks are more than one-quarter full, the NAND gate output is LOW(0). This turns on the LED. However if either of the two sensor outputs goes LOW(0), the NAND gate output goes HIGH(1), which turns the LED off. Thus, a single LED is used to indicate when both tanks are more than one-quarter full.

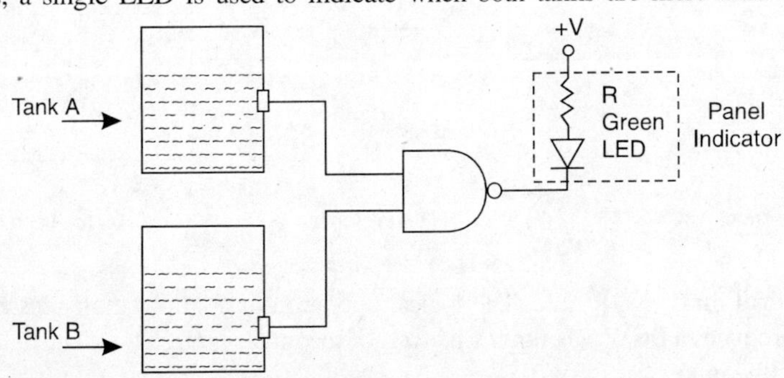

Fig. 34-40. *An electronic system to detect liquid level in a drug manufacturing plant.*

34-21. Boolean Expression for the NAND Gate

We have already discussed in Art 34-18 that in a 2-input NAND gate, output, X is 0 only if inputs A and B are 1. The output, X is 1 is either A or B or if both A and B are 0. This logical operation of the NAND gate can be expression in the form of a Boolean expression as follows:

$$X = \overline{A \cdot B}$$
$$= \overline{AB}$$

The above expression is read as X equals *AB bar.* This expression says that the two input variables, A and B are first ANDed and then inverted (or complemented) as indicated by the bar over the AND expression. If we evaluate this expression for all possible values of two input variables, the results are as shown below:

Table 34-5

A	B	$\overline{AB} = X$
0	0	$\overline{0 \cdot 0} = \overline{0} = 1$
0	0	$\overline{0 \cdot 1} = \overline{0} = 1$
1	0	$\overline{1 \cdot 0} = \overline{0} = 1$
1	1	$\overline{1 \cdot 1} = \overline{1} = 0$

To extend the Boolean expression for the NAND gate, for more than two-input variables, simply use a new letter for each input variable. For example, for 3-input NAND gate, the Boolean expression is given by the question:

$$X = \overline{ABC}$$

where A, B, C, and D are the input variables. Similarly, for a 4-input NAND gate, the Boolean expression is given by the equiation,

$$X = \overline{ABCD}$$

where A, B, C, and D are the input variables.

34-22. The NOR Gate

The term NOR is a contraction of NOT-OR. It implies an OR function with a complemented (or inverted) output. The standard logic symbol for a 2-input NOR gate is as shown in Fig. 34-41(*a*). Notice that this symbol is the same as OR gate symbol except for a small circle (or a bubble) on its output. This small circle denotes the inversion operation. Thus the NOR gate operates like an OR gate followed by INVERTER as shown in Fig. 34-41 (*b*).

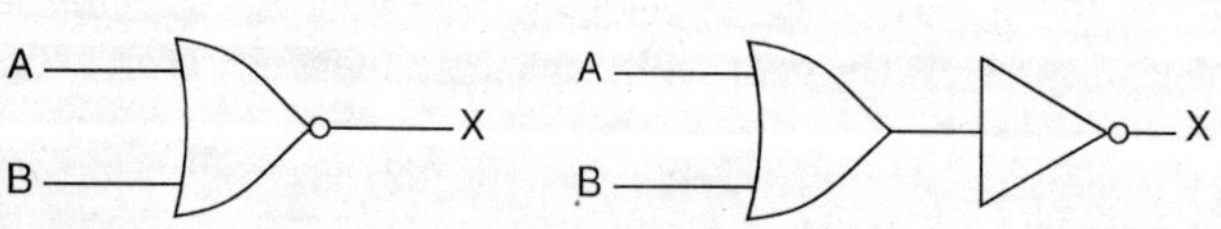

Inputs		Output
A	B	X
0	0	1
0	1	0
1	0	0
1	1	0

(a) Logic symbol (b) NOT-OR equivalent (c) Truth table

Fig. 34-41. *A 2-input NOR gate.*

Logical operation of NOR gate. The logical operation of a NOR gate is described as below. The NOR gate produces a LOW (0) output when *any* of the inputs is HIGH(1). The output is HIGH(1) only when all of the inputs are LOW(0).

In a 2-input NOR gate operation, output *X* is LOW(0) if either input *A* or input *B* is HIGH(1) or if both the inputs *A* and *B* are HIGH(I). The output *X* is HIGH(1) if both *A* and *B* LOW(0). This operation can be summarized in the truth table shown in Fig. 17.41 (*c*).

Fig. 17.42 illustrates the logical operation of a 2-input NOR gate for all four input combinations, *i.e.*, 00, 01, 10 and 11.

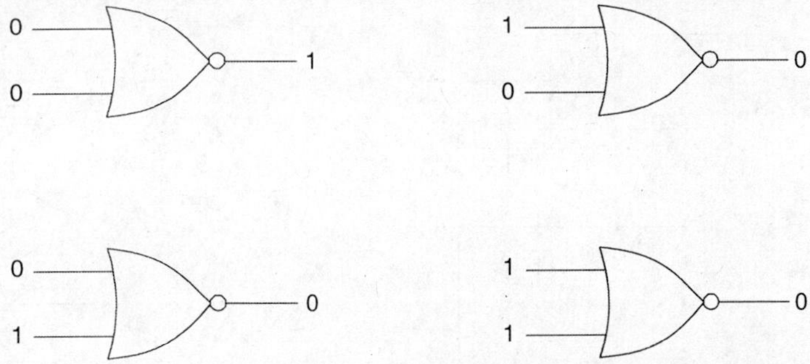

Fig. 34-42. *Illustratin the four possible input combinations and the resulting output for each in a two-input NOR gate operation.*

It is evident from the figure shown above that the output of a NOR gate is 1 only when both the inputs are 0. In all other possible input combinations (*i.e.*, 11, 01 and 10), the output is 0.

34.23 Pulsed Operation of a NOR Gate

Consider two waveforms *A* and *B* applied at the inputs of a 2-input NOR gate as shown in Fig. 34.23. In order to determine the output level, will look at the inputs with respect to each other with reference to the truth table shown in Fig. 17.41(*c*).

At $t = t$, *A* and *B* both are 0, therefore *X* is 1, At $t = t_1$, *A* goes 1, *B* is 0, therefore *X* goes 0, At $t = t_2$, *A* goes 0, *B* is 0, therefore *X* goes 1. At $t = t_3$, *A* is 0, *B* goes 1, therefore *X* goes 0. At $t = t_4$, *A* is 0, *B* goes 0, therefore *X* goes 1, At $t = t_5$, *A* goes 1, *B* is 0, therefore *X* goes 0. At $t = t_6$, *A* goes 0, *B* is 0, therefore output *X* goes 1. The resulting output waveform is as shown in Fig. 34-43.

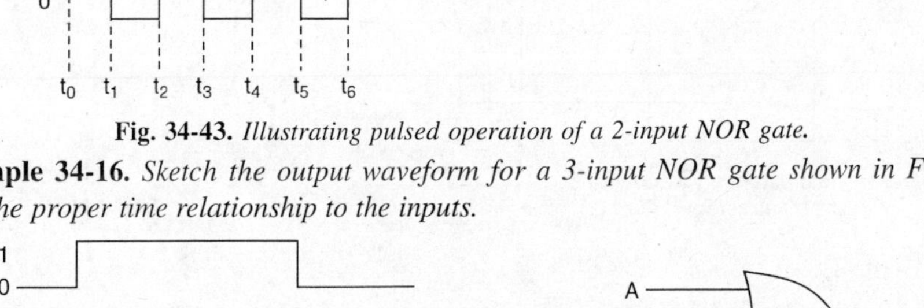

Fig. 34-43. *Illustrating pulsed operation of a 2-input NOR gate.*

Example 34-16. *Sketch the output waveform for a 3-input NOR gate shown in Fig. 34-44. Showing the proper time relationship to the inputs.*

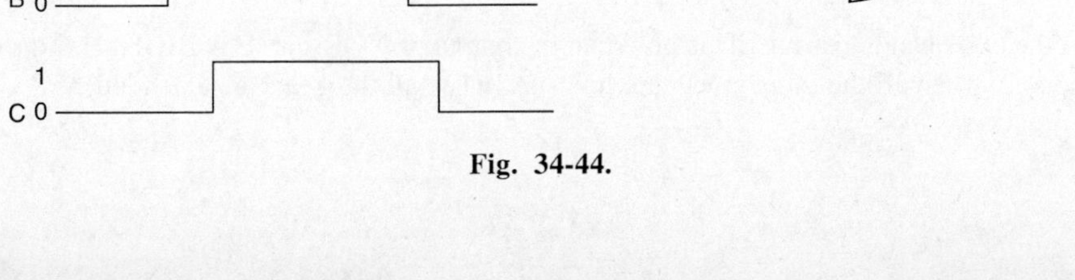

Fig. 34-44.

Solution. Given a 3-input NOR gate with waveforms *A*, *B* and *C*.

We know that in a NOR gate, the output is 0 when any of the inputs is 1. The output is 1, only when all the inputs are 0. With this point in mind, we note that before $t = t_1$, all the three inputs *A*, *B* and *C* are 0. Therefore, output is 1. At $t = t_1$, *A* goes 1, $B = 0$, $C = 0$ therefore *X* goes 0. Between t_1 and t_6, one of the inputs *A, B* and *C* is 1, therefore output is 0. At $t = t_6$, $A = 0$, $B = 0$, *C* goes 0, therefore *X* goes 1.

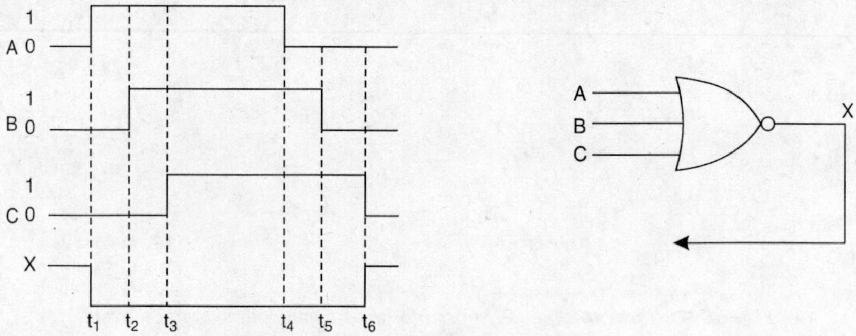

Fig. 34-45.

The resulting output waveform at *X* is as shown in Fig. 34-45.

34-24. Application of a NOR Gate

Fig. 34-46 shows a NOR-gate circuit required to indicate the status of the landing gear prior landing in an aircraft, as a part of its functional monitoring system. In this circuit, when a landing gear is extended, its sensor produces a LOW(0) voltage. When a landing gear is retracted, its sensor produces a HIGH(1) voltage.

A green LED display turns on if all the three gears are properly extended when the "gear down" switch is activated in preparation for landing. A red LED display turns on if any of the gears fail to extend properly prior to landing.

It is evident from the above diagram that LED is on when the output of NOR-gate 1 is LOW(0). This could be possible only if any of the inputs to NOR-gate 1 is HIGH(1), (*i.e.*, when any of the gears is retracted).

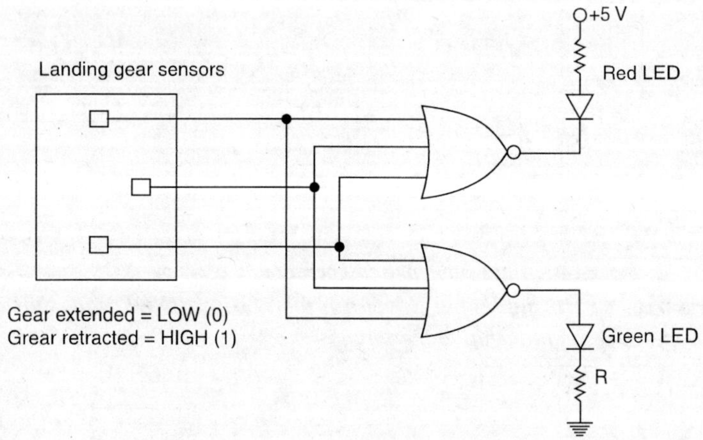

Fig. 34-46. *A NOR-gate circuit to indicate the status of the landing gear prior to landing in an aircraft.*

On the other hand, green LED is on when the output of NOR-gate 2 is HIGH(1). This could be possible only when all the three inputs are low (*i.e.*, when all the gears are extended).

34-25. Boolean Expression for a NOR Gate

We have already discussed in Art 34-22 that in a 2-input NOR gate, the output X is LOW(0) if either input A or input B is HIGH(1) or if both A and B are HIGH(1). The output, X is HIGH(1) if both A and B are LOW(0). This logical operation of the NOR gate can be expressed in the form of a Boolean expression as follows:

$$X = \overline{A + B}$$

This equation says that the two input variables are first ORed and then complemented. It is read as complement of A OR B (or A OR B bar). If we evaluate the expression $\overline{A+B}$ for all possible values of two input variables, the results are as shown below.

Table 34-6

A	B	$\overline{A+B} = X$
0	0	$\overline{0+0} = \overline{0} = 1$
0	1	$\overline{0+1} = \overline{1} = 0$
1	0	$\overline{1+0} = \overline{1} = 0$
1	1	$\overline{1+1} = \overline{1} = 0$

To extend the Boolean expression for the NOR gate for more than 2-input variables, simply use a new letter for each input variables. For example, for a 3-input NOR gate, the Boolean expression is given by the equation,

$$X = \overline{A + B + C}$$

where A, B and C are the input variables, Similarly. For a 4-input NOR gate, the Boolean expression is,

$$X = \overline{A + B + C + D}$$

where A, B, C and D are the input variables.

34-26. The Exclusive-OR and Exclusive-NOR Gate

The Exclusive-OR and Exclusive-NOR gates are actually formed by a combination of other logic gates (*i.e.*, INVERTER, AND and OR gates). However, because of their fundamental importance in many applications, the Exclusive -OR and Exclusive -NOR gates are treated as basic logic gates with their unique symbols. Both these logic gates are discussed one by one in the following pages.

34-27. The Exclusive-OR Gate

The Exclusive -OR (abbreviated as XOR) gate has only two-inputs. Unlike the other logic gates we have discussed earlier, the Exclusive-OR gate has never more than two inputs. Fig. 34-47 (*a*) shows the logic symbol of an Exclusive-OR gate.

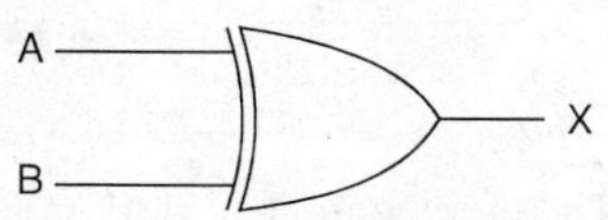

(a) Logic symbol

Inputs		*Output*
A	*B*	*X*
0	0	0
0	1	1
1	0	1
1	1	0

(b) Truth table

Fig. 34-47. *An Exclusive-OR gate.*

Logic operation. The logical operation of an exclusive-OR gate may be described as below: The output of an Exclusive-OR gate is HIGH(1) only when the two inputs are at opposite logic levels. The output is LOW(0) when the two inputs are at same logic levels. This operation can be stated as follows with reference to inputs A and B and output X. In an exclusive-OR gate operation, output X is HIGH(1) if input A is LOW(0) and input B is HIGH(1), or if input A is HIGH(1) and input B is

LOW(0). The output X is LOW(0) if A and B are both HIGH(1) or both LOW(0). This operation can be summarized in the truth table shown in Fig. 34-47(b)

Fig. 34-48 shows the four possible input combinations (*i.e.*, 00, 01, 10 and 11) and the resulting output for the exclusive-OR gate.

It is evident from the figure shown above that the output of the Exclusive-OR gate is 1 only when the inputs are at opposite logic levels (*i.e.*, 01 or 10). The output is 0 when both the inputs are at same, logic levels (*i.e.*, 00 or 11).

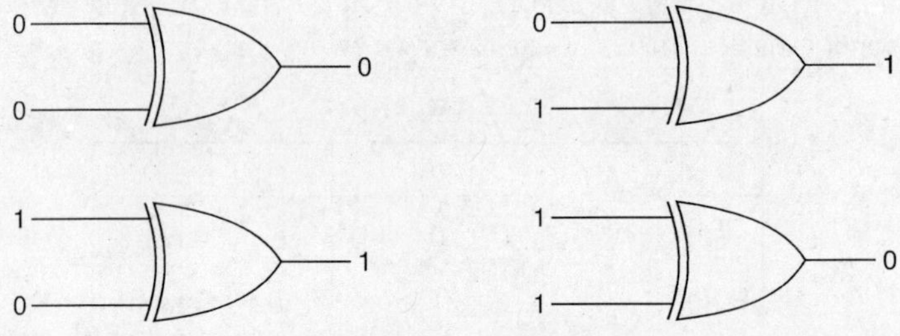

Fig. 34-48. *Illustrating the four possible input combinations and the resulting output for each in a Exclusive-OR gate.*

34-28. Pulse Operation of an Exclusive-OR Gate

Consider waveform A and B applied at the inputs of an Exclusive-OR gate as shown in Fig. 34-49.

We know that in Exclusive-OR gate the output is HIGH(1) if the input A is LOW(0) and input B is HIGH(1), or if input A is HIGH(1) and input B is LOW(0). The output X is LOW(0) if A and B are both HIGH(1) or both LOW(0).

In other words, the output of an Exclusive-OR gate is HIGH(1) when the input A and input B are at opposite logic levels. The output is LOW(0) if the input A and input B are at the same logic level. Let us apply these points to sketch the output waveform.

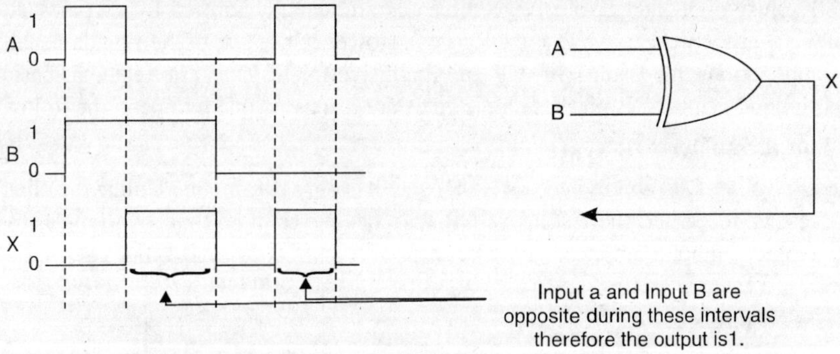

Fig. 34-49. *Illustrating the pulsed operation of an exclusive-OR gate.*

At t_1, input A goes 1, input B also goes 1, Therefore output X is 0, At t_2, A goes 0, B is 1, therefore output is 1. At t_3, A is 0, B goes 0, therefore X goes 0. At t_4, A goes 1, B is 0, therefore X gears 1. At t_5, A goes 0, B is 0, therefore X goes 0. The resulting output waveform is as shown in Fig. 34-50.

Another way of sketching the output waveform is to identify the time intervals of the waveforms A and B, during which the inputs are at different logic levels. The output X is 1 during these time intervals. For the rest, the output X is 0. Refer to the Fig. 34-49.

Example 34-17 *Sketch the output waveform for an Exclusive-OR gate with the input waveforms shown in Fig. 34-50.*

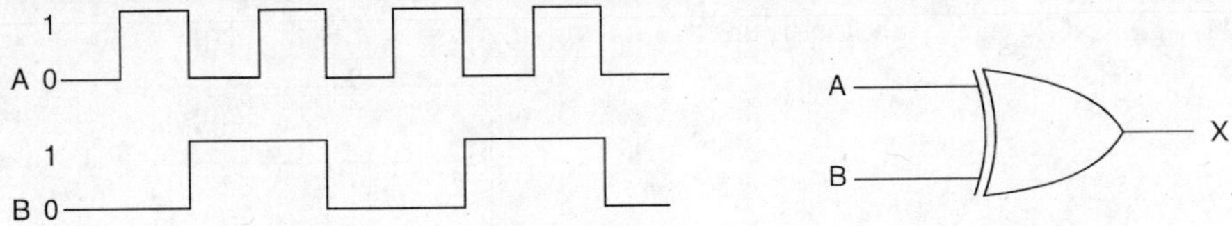

Fig. 34-50.

Solution. Given an Exclusive-OR gate with two input waveforms.

We know that an Exclusive-OR gate produces a HIGH(1) output whenever its two inputs are at opposite logic levels. On the other hand, the Exclusive-OR gate produces a LOW(0) output whenever the two inputs are at the same logic levels. Using this we can identify the time intervals from the given input waveforms, during which the two inputs are at opposite logic levels or at same logic levels. We can sketch an output 1 and 0 respectively for these time intervals as shown in Fig. 34-51.

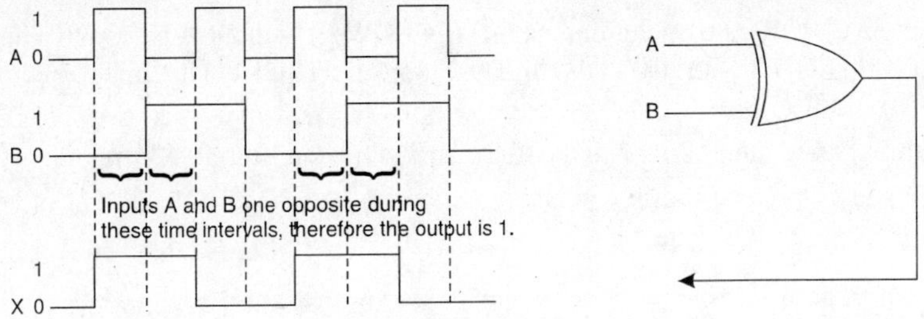

Fig. 34-51.

34-29. Application of an Exclusive-OR Gate

Fig. 34-52 shows an exclusive-Or gate being used to detect the failure in either one of the identical circuits operating in parallel. As long as both the circuits are operating properly, the outputs of both the circuits are always the same. Therefore, the output of the exclusive-OR gate is LOW(0). However, if one of the circuits fails, the outputs of the two circuits will be at opposite levels, for some time. Therefore, the exclusive-OR gate will produce a HIGH(1) output indicating failure in one of the circuits.

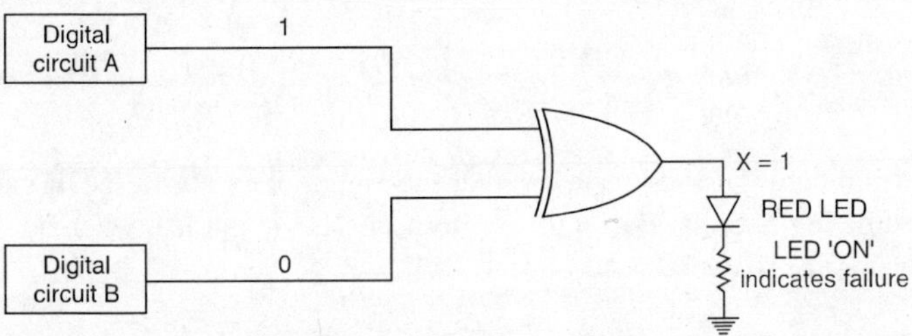

Fig. 34-52. *An-Exclusive-OR gate to detect failure is one of the identical circuits operating in parallel.*

34-30. Boolean Expression for an Exclusive-OR Gate

The Boolean expression for an Exclusive-OR gate is given by the equation,

$$X = \overline{A}B + A\overline{B}$$

Note that unlike the INVERTER, AND and OR gates we have two terms in the Boolean Expresses of the Exclusive-OR gate.

34-31. The Exclusive-NOR Gate

The Exclusive-NOR (abbreviated as XNOR) gate has also only two inputs like Exclusive-OR gate. Fig. 34-53 (*a*) shows the logic symbol of an Exclusive-NOR(XNOR) gate. The bubble on the output of the XNOR gate symbol indicates that its output is opposite to the output of the XOR gate.

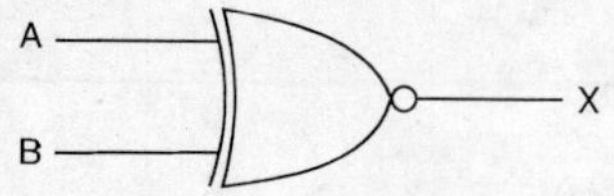

(a) Logic symbol

Inputs		*Output*
A	*B*	*X*
0	0	1
0	1	0
1	0	0
1	1	1

(b) Truth table

Fig. 34-53. *The Exclusive-NOR gate.*

Logical Operation. The logical opration of an Exclusive-NOR(X NOR) gate may be described as below: The output of an XNOR gate is HIGH(1) only when the two inputs are at the same logic level. The output is LOW(0) when the two inputs are at opposite logic levels. This opration can be stated as follows with reference to inputs *A* and *B* and output *X*.

In an Exclusive-NOR gate operation, output *X* is LOW(0) if input *A* is LOW(0) and input *B* is HIGH(1), or if *A* is HIGH(1) and *B* is LOW(0), The output *X* is HIGH(1) if *A* and *B* are both HIGH(1) or both LOW(0).

This opration can be summarized in the truth table shown in Fig. 34-53 (b).

Fig. 34-54 shows the four possible input combinations (*i.e.*, 00, 01, 10 and 11) and the resulting output for the Exclusive-NOR gate.

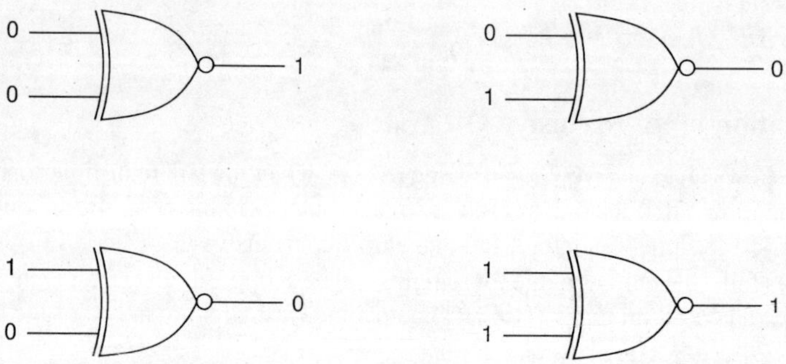

Fig. 34-54. *Illustrating the four possible input combinations and the resulting output for each in an Exclusive-NOR gate.*

It is evident from the figure shown above that the output of exclusive-NOR gate is 1 only when the inputs are at same logic levels (*i.e.*, both 0 or both 1). The output is 0 when both the inputs are at opposite logic levels (*i.e.*, 01 or 10).

34-32. Pulsed Operation of an Exclusive-NOR Gate

Consider waveforms *A* and *B* applied at the inputs of an Exclusive NOR gate as shown in Fig. 34-55. We know that in Exclusive-NOR gate, the output *X* is LOW(0) if input *A* is LOW(0) and input *B* is HIGH(1), or if *A* is HIGH(1) and *B* is LOW(0). The output *X* is HIGH(1) if *A* and *B* are both HIGH(1) or both LOW(0).

In other words, the output of an Exclusive-NOR gate, is HIGH(1), when the input *A* and input *B* are both at the same logic levels. The output is LOW(0), when the input *A* and input *B* are at different logic levels. Let us apply these points to sketch the output waveform.

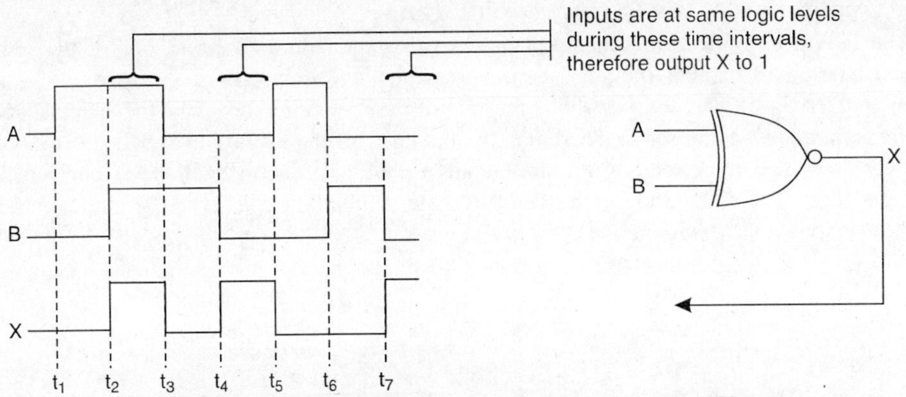

Fig. 34-55. *Illustrating Pulsed operation of an Exclusive-NOR gate.*

At t_1, input *A* goes 1, *B* is 0, therefore *X* is 0, At t_2, *A* is 1, *B* goes 1, therefore *X* goes 1. at t_3, *A* goes 0, *B* is 1, therefore *X* goes 0. At t_4 *A* is 0, *B* goes 0, therefore *X* goes 1. At t_5, *A* goes 1, *B* is 0. Therefore *X* goes 0. At t_6, *A* goes 0, *B* goes 1, therefore *X* is 0. At t_7, *A* is 0, *B* goes 0, therefore *X* goes 1. The resulting output waveform is as shown in figure.

Another way of sketching the output waveform is to identify the time intervals of the waveforms *A* and *B* during which the inputs are at same logic levels. The output *X* is 1 during these intervals. For the rest, the output *X* is 0. The resulting output waveform is shown in Fig. 34-54.

Example 34-18 *Sketch an output waveform for an Exclusive-NOR gate with the input waveforms as shown in Fig. 34-56.*

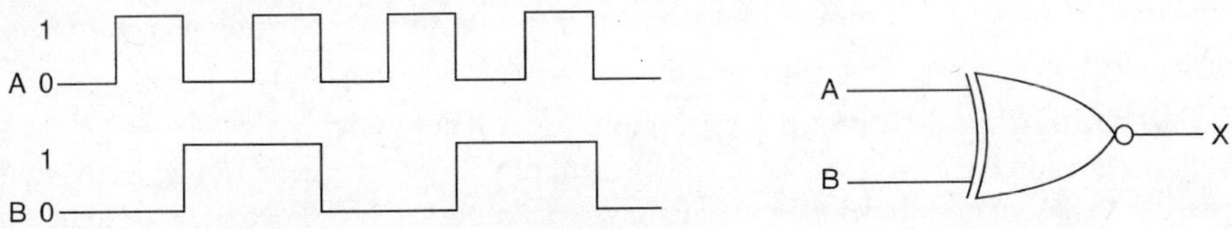

Fig. 34-56.

Solution. Given the Exclusive-NOR gate with two input waveforms.

We know that an Exclusive-NOR gate produces a HIGH(1) output whenever the two inputs are at same logic levels. On the other hand, the Exclusive-NOR gate produces a LOW(0) output whenever the two inputs are at opposite logic levels. Using this, we can identify the time intervals from the given input waveforms during which the two inputs are at the same and opposite logic levels. Sketch an output 1 and 0 respectively for these time intervals as shown in Fig. 34-57.

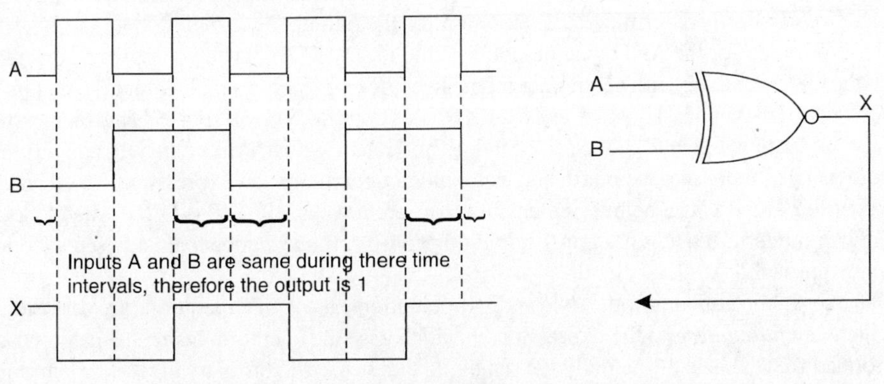

Fig. 34-57.

34-33. Application of an Exclusive-NOR Gate

The Exclusive-NOR gate can also be used to detect failure in either one of the identical circuits oprating in parallel (*i.e.*, the same application as discussed for exclusive-OR gate) as shown in Fig. 34-58. As long as both the circuits are oprating properly the outputs of both the circuits are always the same. Therefore, the output *X* of the exclusive NOR gate is HIGH(1). However, if one of the circuits fails, the outputs of the two circuits will be at opposite levels for some time. Therefore, the Exclusive-NOR gate will produce a LOW(0) output indicating a failure in one of the circuits.

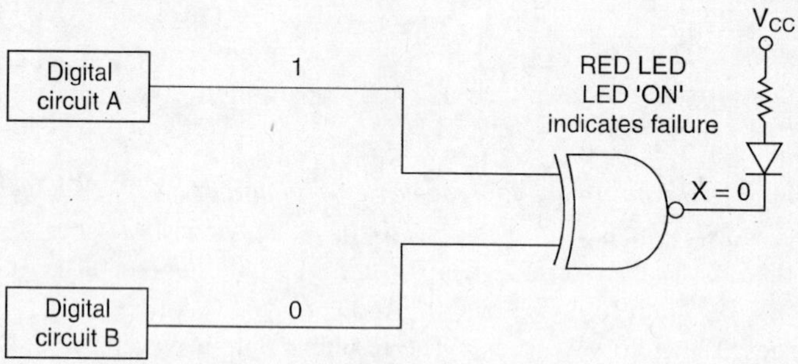

Fig. 34-58. *An Exclusive-NOR gate to detect failure in one of the identical circuits 1operating in parallel.*

34-34. Boolean Expression for an Exclusive-NOR Gate

The Boolean expression for an Exclusive-NOR gate is given by the equation,

$$X = \overline{AB} + AB$$

Note that like Exclusive-OR gate, we have two terms in the Boolean expression of the Exclusive NOR gate.

34-35. Determination of Boolean Expression for a Logic Circuit

We have already discussed in last articles that the INVERTER (or NOT gate), AND and OR gates are the basic building blocks of a digital system. A digital system consists of logic circuits. Further any logic circuit, no matter how complex it may be, is completely described using the Boolean oprations. For example, consider the circuit shown in Fig. 34-59.

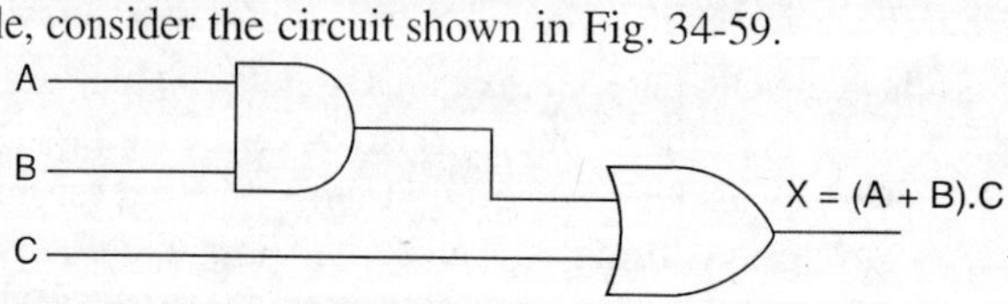

Fig. 34-59. *Logic circuit with its Boolean expression.*

This circuit has three inputs, *A*, *B*, *C* and a single output, *X*. Utilizing the Boolean expression for each gate, we can easily determine the expression for the output.

The expression for the AND gate output in written as *A, B*. This AND output is connected as an input to the OR gate along with *C*, another input. The OR gate operates on its inputs so that its output is the OR sum of the inputs. Thus, we can express the OR output as $X = A.\,B + C$. This final expression could also be written as $X = C + A.\,B$, since it does not matter which term of the OR sum written first.

Occasionally, there may be confusion as to which operation is an expression is performed first. The expression $A.\,B + C$ can be interpreted in two different ways: **(1)** *A. B* is ORed with *C*, or **(2)** *A* is ANDed with the term $B + C$. In order to avoid this confusion, the students are advised to follow the following approach.

If an expression contains both AND and OR operations, the AND operations are performed first, unless there are *parentheses* n the expression, in which case the operation inside the parentheses is to be performed first. This is the actually the same rule that is used-in ordinary algebra to determine the order of operations.

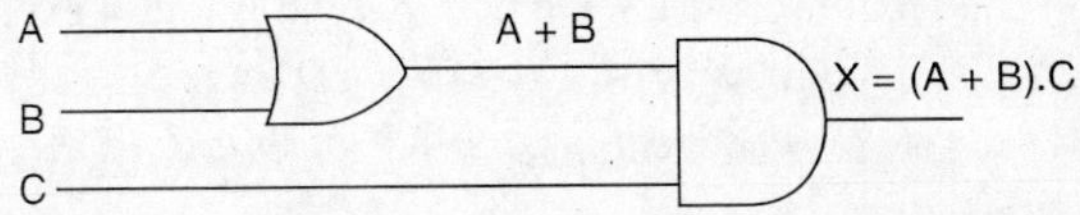

Fig. 34-60. *Logic circuit diagram whose output expression requires parentheses.*

In order to illustrate this point further, let us consider the logic circuit shown in Fig. 34-60. The expression for the OR gate output is simply $A + B$. This output serves as an input to the AND gate along with another input, C. Thus, we express the output of the AND gate as $X = (A + B).\ C$. Note the use of parentheses here to indicate that A and B are ORed first, before their OR sum is ANDed with C. Without the parentheses it would be interpreted incorrectly, since $A + B.\ C$ means that A is ORed with the product $B.\ C$.

34-36. Boolean Expression for a Logic Circuit Containing INVERTERs

Whenever an INVERTER is present in a logic circuit diagram, its output expression is simply equal to the input expression with a bar over it. Fig. 34-61 shows two examples using INVERTER. In Fig. 34-61(*a*), input A is field through an INVERTER, whose output is fed to an OR gate together with B, so that the OR Output is equal to $\overline{A} + B$. Note the bar is over the A alone, indicating that A is first inverted and then ORed with B.

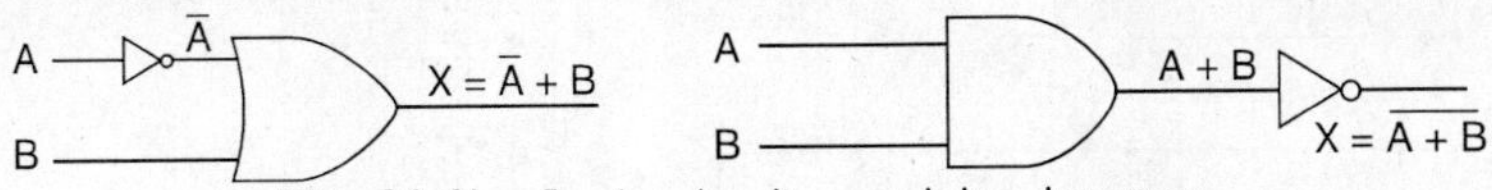

Fig. 34-61. *Logic circuit containing inverters.*

Now consider the circuit shown in Fig. 34-61(*b*). In this circuit, the output of the OR gate is equal to $A + B$ and is fed through the INVERTER. The INVERTER output is therefore equal to $(A + B)$, since it inverts the complete input expression. Note that the bar covers the entire expression $\overline{(A + B)}$. This is important because, as will be shown later, the expressions $\overline{(A + B)}$, and $(\overline{A} + \overline{B})$ are not equivalent. The expression $\overline{(A + B)}$ means that A is ORed with B and then their OR sum is inverted, whereas the expression $(\overline{A} + \overline{B})$ indicates that A is inverted and B is inverted and the results are then ORed together.

Example 34-19. *Write the Boolean expression for output X for the logic circuit shown in Fig. 34-62.*

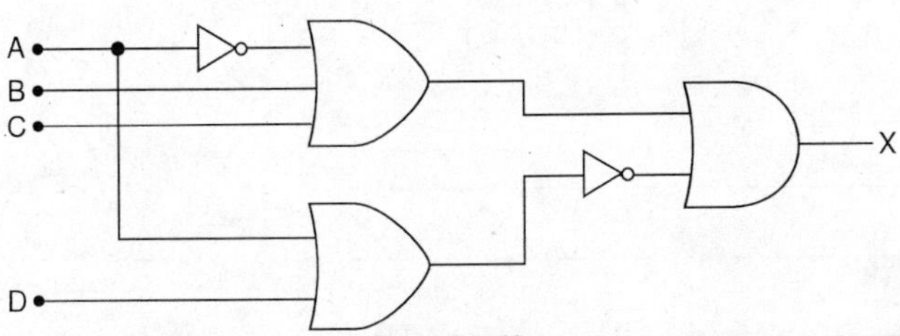

Fig. 34-62.

Solution. Given: The logic circuit containing INVERTER, AND and OR gates of Fig. 34-62.

In order to determine the Boolean expression for output X, determine the output of each logic gate while starting from the left-most inputs and work towards the final output writing the expression for each gate.

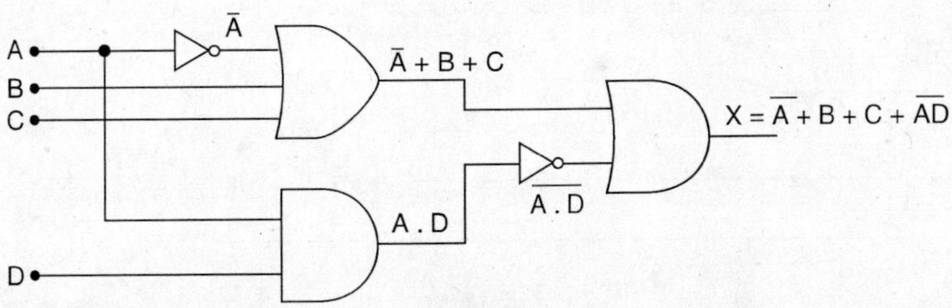

Fig. 34-63.

As seen from circuit shown in Fig. 34-63, we find that the Boolean expression for the output,

$$X = \overline{A} + B + C + \overline{AD} \quad \textbf{Ans.}$$

Example 34-20. *Write the Boolean expression for output X of the logic circuit shown in Fig. 34-64.*

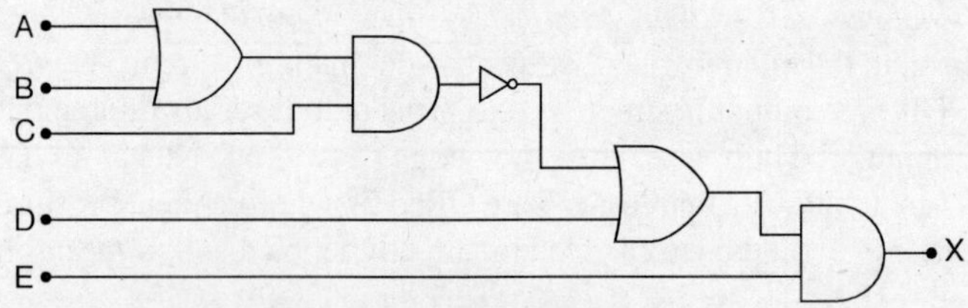

Fig. 34-64.

Solution. Given, the logic circuit containing INVERTER, AND OR and gates of Fig. 34-61.

We know that to determine the Boolean expression for the final output *X*, we start from left-most inputs and work towards the right as shown in Fig. 34-65.

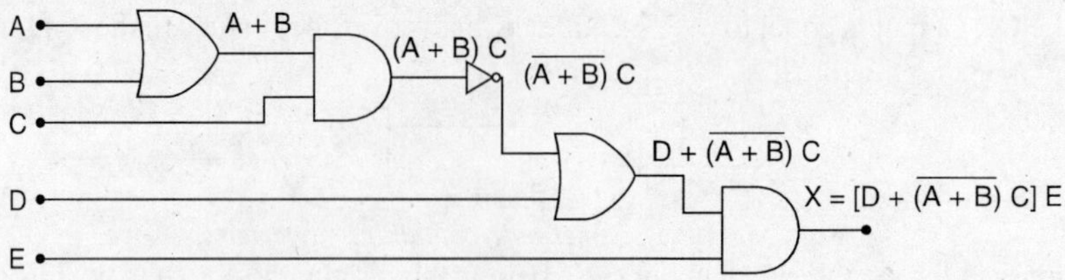

Fig. 34-65.

Writing the Boolean expression for each gate, we find that the Boolean expression for the final output.

$$X = [D + \overline{(A + B)C}]E \quad \textbf{Ans.}$$

Example 34-21 *Determine the Boolean expression for output X for the logic circuit shown in Fig. 34-66.*

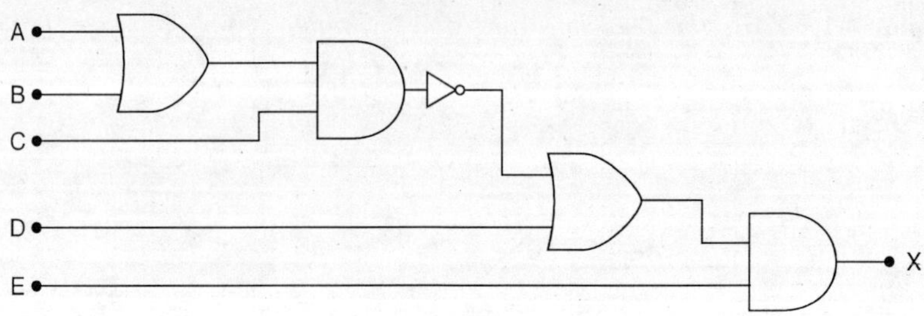

Fig. 34-66.

Solution. Given, the logic circuit containing INVERTER, AND and OR gates of Fig. 34-66.

We know that to determine the Boolean expression for the final output *X*, we start from left-most inputs and work towards the light as shown in Fig. 34-67.

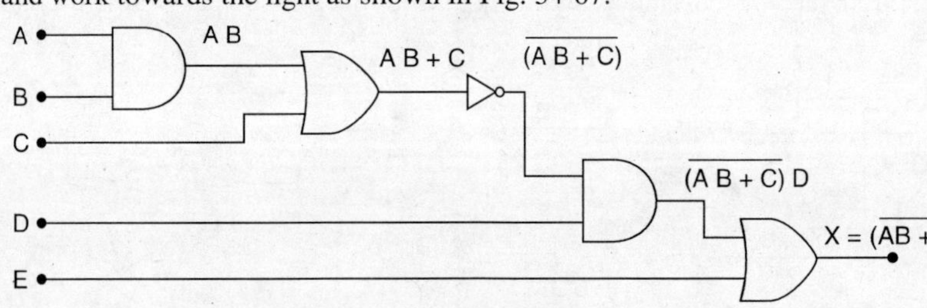

Fig. 34-67.

Writing the Boolean expression for each gate, we find that the Boolean expression for the final output.

$$X = \overline{(AB + C)}\, D + E \quad \textbf{Ans.}$$

34-37. Evaluating the Boolean Expression for the Logic Circuit Output

Once the Boolean expression for a circuit output has been obtained, the output logic level can be determined for any set of input levels. Fog example, suppose that we want to know the logic level of the output X for the circuit in Fig. 34-68(a) for the case where $A = 0$, $B = 1$, $C = 1$, and $D = 1$. As ordinary algebra, the value of X can be found by substituting the values of the variables into the expression and performing the indicted operations as follows:

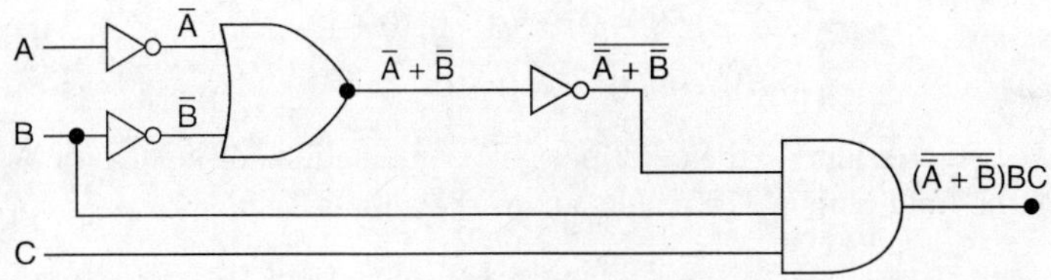

Fig. 34-68.

We know that the Boolean expression for the final output X can be written as,

$$X = (\overline{\bar{A} + \bar{B}})\, BC \qquad ...(1)$$

Let us suppose we want to know the logic level of the output X for the case where $A = 0$, $B = 1$ and $C = 1$. Substituting the values of the variables A, B and C into the expression.

$$X = (\overline{\bar{0} + \bar{1}})\,1 \cdot 1$$

$$= (\overline{1 + 0})\,1 \cdot 1$$

$$= (\bar{1})\,1 \cdot 1$$

$$= 0.1.1$$

$$= 0$$

In general, we must follow the following rules when evaluating the Boolean expression of a logic circuit.

(*i*) First, perform all inversions of single terms; *i.e.*, is, $\bar{0} = 1$ or $\bar{1} = 0$.

(*ii*) Then perform all operations within parentheses.

(*iii*) Perform an AND operating before an OR operation unless parentheses indicate otherwise.

(*iv*) If an expression has a bar over it, perform the operations of the expression first and then invert the result.

34-38. Determination of a Truth Table for a Logic Circuit

We have already discussed in the last article the determination and evaluation of a Boolean expression for a given logic circuit. Once we have evaluated a Boolean expression for all the possible combinations of the input variables, we can summarize the results in the form of a truth table.

In order to illustrate the concept, consider a circuit shown in Fig. 34-69. We know that the Boolean expression for this circuit can be written as,

$$X = (A + B)(B + \bar{C})$$

Let us evaluate Boolean expression for all the possible values of input variables, A, B and C.

(*i*) When $A = 0$, $B = 0$, $C = 0$, the final output,

$$X = (A + B)(B + \bar{C}) = (0 + 0)(0 + \bar{0}) = (0 + 0)(0 + 1) = 0.1 = 0$$

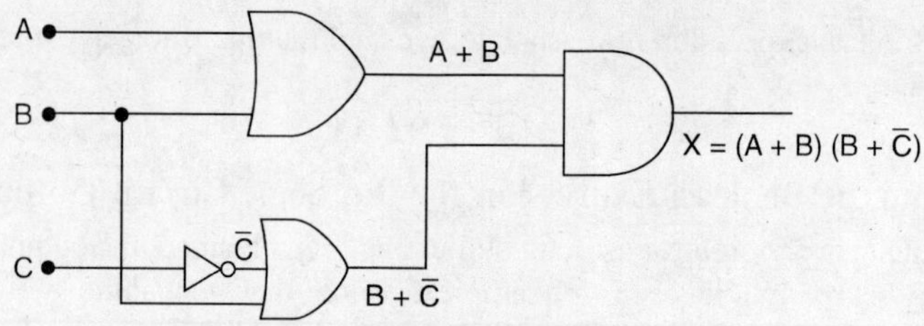

Fig. 34-69.

(*ii*) When $A = 0$, $B = 0$, $C = 1$, the final output,

$$X = (A + B)(B + \overline{C}) = (0 + 0)(0 + \bar{1})$$
$$= (0 + 0)(0 + 0) = 0.0 = 0$$

Similarly, we can consider all the other possible combinations of input variables and determine the logic level of the final output. The results are as shown in a Table 34-7.

Table 34-7

Inputs			*Outputs*
A	*B*	*C*	*X*
0	0	0	0
0	0	1	0
0	1	0	1
0	1	1	1
1	0	0	1
1	1	0	1
1	1	1	1

34-39. Determining Output Logic Level from a Circuit Diagram

As a matter of fact, the output logic level for given input levels can also be determined directly from the circuit diagram *without* using the Boolean expression. This technique is useful for technicians during the troubleshooting or testing of a logic system. It also tells them what each gate output is supposed to be as well as the final output. In order to illustrate this point, let us consider the circuit shown in Fig. 34-70. The input levels for this circuit are $A = 1$, $B = 0$, $C = 1$, and $D = 0$.

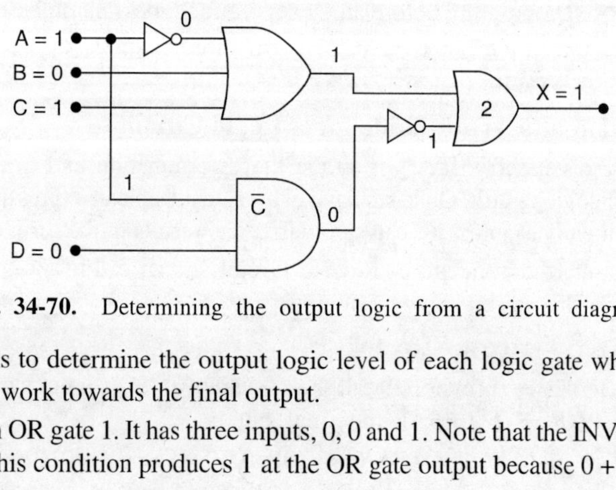

Fig. 34-70. Determining the output logic from a circuit diagram.

The procedure is to determine the output logic level of each logic gate while starting from the left-most inputs and work towards the final output.

Let us start from OR gate 1. It has three inputs, 0, 0 and 1. Note that the INVERTER changes this input $A = 1$ to a 0. This condition produces 1 at the OR gate output because $0 + 0 + 1 = 1$. The AND gate has inputs of 1 and 0 which produces a 0 output because 1.0 = 0. This 0 is inverted to 1 and applied to an OR gate 2 along with the 1 from the OR gage 1 output. The 1 and 1 inputs to the OR gate 2 produces an output of 1 because 1 + 1 = 1.

Example 34-22. *Determine the output logic level for a digital circuit shown in Fig. 34-71, having A = 0, B = 1 and C = 1.*

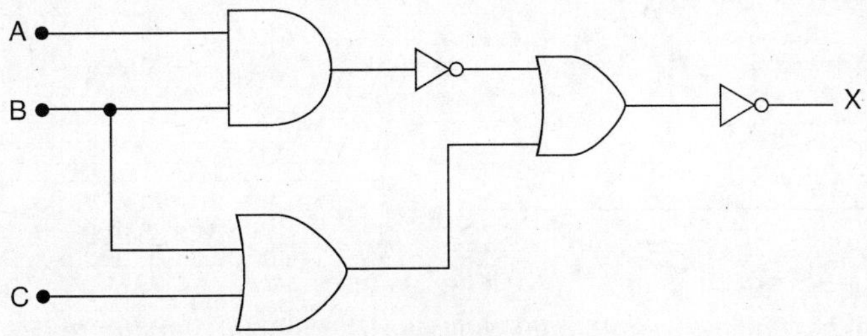

Fig. 34-71.

Solution. Given: The digital circuit shown in Fig. 34-71 with $A = 0$, $B = 1$ and $C = 1$.

Let us determine the output logic level of each logic gate while starting from the left-most inputs and work towards the output. The resulting outputs are shown as in Fig. As seen, the final output, $X = 0$. **Ans.**

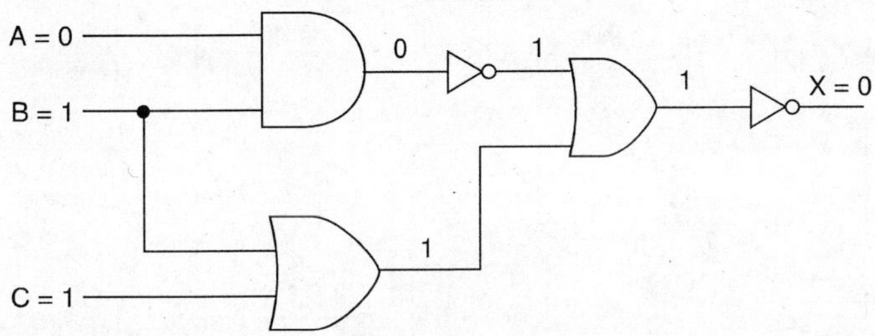

Fig. 34-72.

Example 34-23. *Determine the output logic level for a logic circuit shown in Fig. 34-73 when A = 1, B = 0, C = 1 and D = 1.*

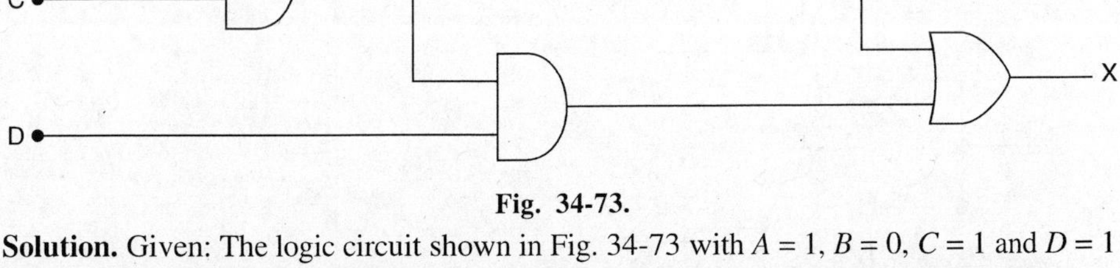

Fig. 34-73.

Solution. Given: The logic circuit shown in Fig. 34-73 with $A = 1$, $B = 0$, $C = 1$ and $D = 1$

Let us determine the output logic level of each logic gate while starting from the left-most inputs and work towards the output. The resulting outputs are as shown in Fig. 34-74. As seen the final output, $X = 1$ **Ans.**

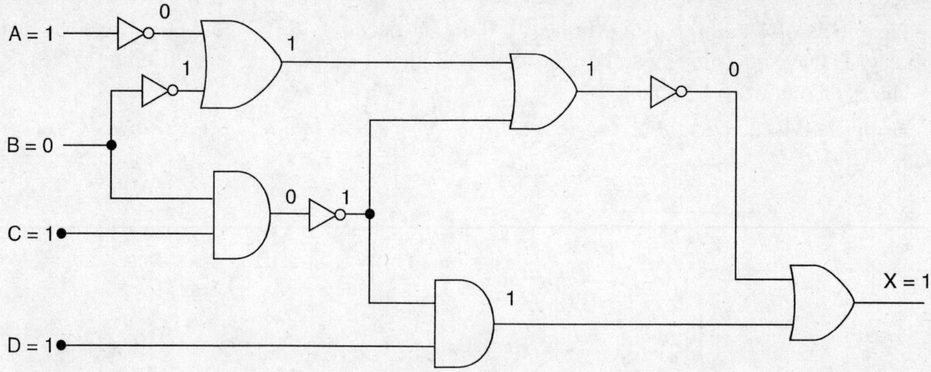

Fig. 34-74.

Example 34-24. *For the logic circuit shown in Fig. 34-75. Simplify the expression for X and determine what conditions (states) for A, B, C and D make X = 0.* (*UP Tech. Univ., Sem. II, 2003-04*)

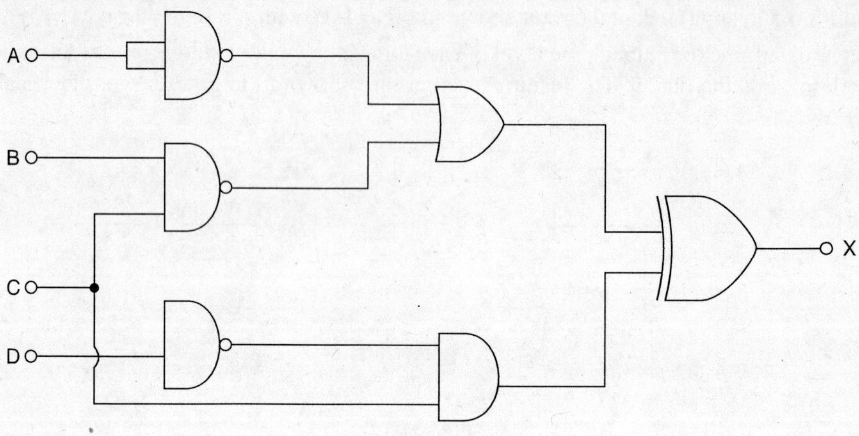

Fig. 34-75.

Solution. Given: The logic circuit containing AND, OR NAND, NOR, and EXOR gates of Fig. 34-75.

We know that to determine the boolean expression on for the output *X*, we start from the left-most inputs and work towards right as shown in Fig. 34-76.

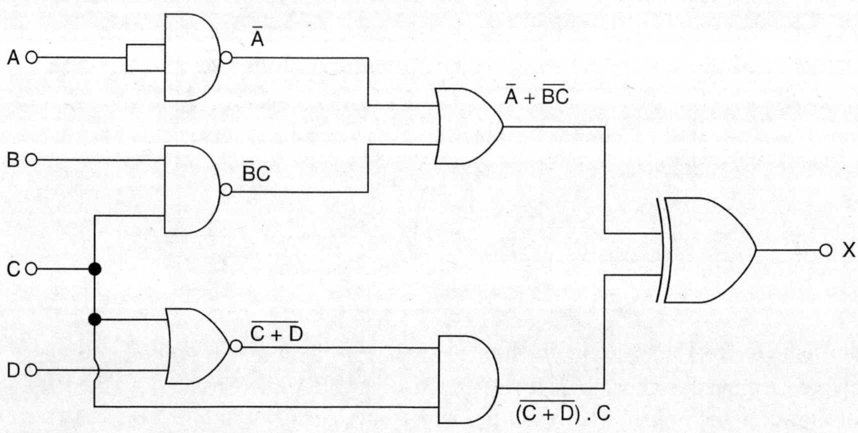

Fig. 34-76.

From the above diagram, we find that the output,

$$X = (\overline{\overline{A} + \overline{BC}}) \cdot \{\overline{(C + D)} \cdot C\} + (\overline{A} + \overline{BC})\{\overline{\overline{C + D} \cdot C}\}$$

From the above expression, we find that $X = 0$ if both the inputs of exclusive OR gate are 0 or 1,

i.e., $\overline{A} + \overline{BC} = 0$ and $\overline{(C + D)} \cdot C = 0$ ...(*i*)

or $\overline{A} + \overline{BC} = 1$ and $\overline{(C + D)} \cdot C = 1$ ...(*ii*)

Using equation (*i*), we find that $\overline{A} + \overline{BC} = 0$ is possible only if $\overline{A}$ and $\overline{BC}$ both are zero. This means $A = 1$; $\overline{BC} = 0$ means $B = C = 1$.

Similarly, $\overline{(C + D)} \cdot C = 0$ is possible if either $\overline{C + D} = 0$ or $C = 0$ or $\overline{C + D} = 0$ and $C = 1$ or $\overline{C + D} = 1$ and $C = 0$.

$\overline{C + D} = 0$ $\quad C = 1, D = 0$ or $C = 0, D = 1$

$\overline{C + D} = 1$ $\quad C = D = 0$ -

Thus, we find that $X = 0$ if $A = 1, B = 1, C = 1, D = 0$ or if $A = 1, B = 1, C = 0$ and $D = 0$.

Similarly, we can find another set of inputs from equation (*ii*). Thus,

$\overline{A} + \overline{BC} = 0$ means $A = 1; B = 1, C = 0$ or $A = 1, B = 0$ and $C = 1$ and $\overline{(C + D)} \cdot C = 1$ means $C = 1$; and $D = 0$

Thus, X is equal to 0 if $A = 1, B = 1, C = 1$ and $D = 0$ or when $A = 1, B = 1, C = 0$ and $D = 0$. Notice that it is the same condition as determined from equation (*i*).

Thus, $X = 0$ $\quad$ if $A = B = 1, \; C = 1, \; D = 0$ $\quad$ **Ans.**

$\qquad\qquad$ if $A = B = 1, \; C = D = 0$

34-40. Implementing Logic Circuits From Boolean Expressions

It will be interesting to know that if the operation of a circuit is defined by a Boolean expression, a logic circuit can be implemented directly from that expression. For example, suppose e need a circuit that is defined by $X = A \,.\, B \,.\, C$. Then we immediately know that all that is needed is a three input AND gate. If we need a circuit that is defined by $X = A + B$, we will use a two-input OR gate with an INVERTER on one of the inputs. The same reasoning used for these simple cases can be extended to more complex logic circucits.

Now suppose that we want to construct a circuit whose output is $Y = AC + \overline{B}C + AB\overline{C}$. This Boolean expression contains three terms $(AC, \overline{B}C, AB\overline{C})$, which are ORed together. This tells us that a three-input OR gate is required with inputs that are equal to AC, $\overline{B}C$, and $AB\overline{C}$, respectively. This is illustrated in Fig. 34-27, where a three input OR gate is drawn with inputs labeled as $AC, \overline{B}C$, and $AB\overline{C}$.

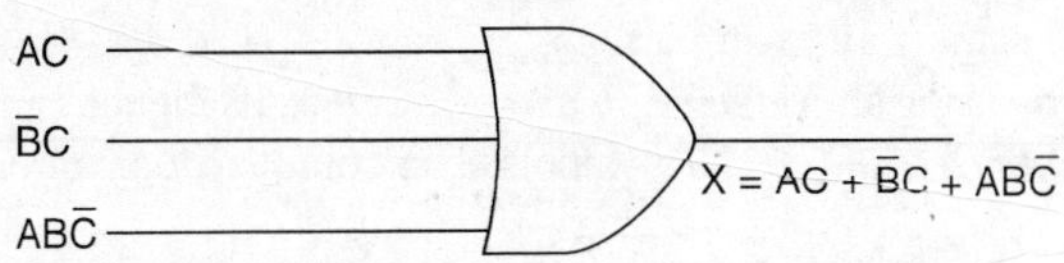

(a)

Fig. 34-77.

Notice that, that each OR gate input is an AND product term, which means that an AND with appropriate inputs can be used to generate each of these terms. This is shown in Fig. 34-77 which is the final circuit diagram. Notice the use of INVERTERS to produce the $\overline{B}$ and $\overline{C}$ terms required in the expression.

The general approach can always be followed, although we shall find that there are some other more efficient techniques that can be employed. From now onwards, however, this straight forward method will be used to minimize the number of new logic circuits that are to be studied.

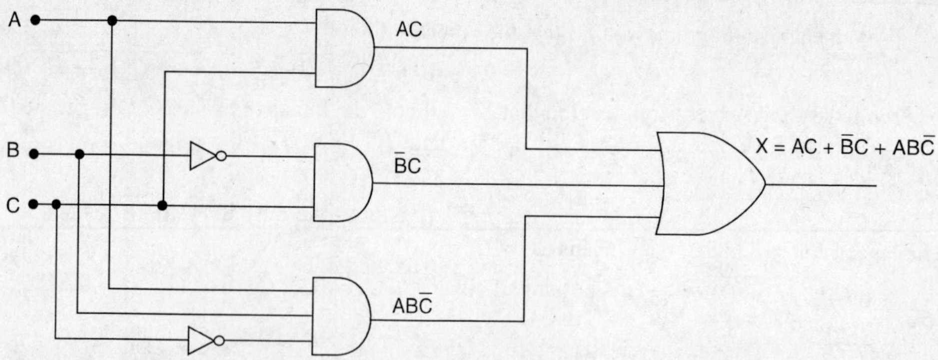

(b)

Fig. 34-77. *Illustrating the implementation of logic circuits from Boolean expressions.*

Example 34-24. Sketch the logic circuit for the Boolean expressions:

(*a*) $(\overline{A}+\overline{B})\cdot C$ (*b*) $\overline{A}B+\overline{C}$

Solution.

(*a*) Given: The Boolean expression,

$$= (\overline{A}+\overline{B})\,C$$

In order to implement this expression as a logic circuit, we find that it requires three input variables (*A*, *B*, and *C*). The term $\overline{A}$ is obtained by using an INVERTER with *A* as an input as shown in Fig. 34-78. Similarly, the term $\overline{B}$ is obtained by using another INVERTER with *B* as its input. Next the term $(\overline{A}+\overline{B})$ is obtained by ORing the outputs of two INVERTERs. Finally an AND gate is required to produce $(\overline{A}+\overline{B})\ C$.

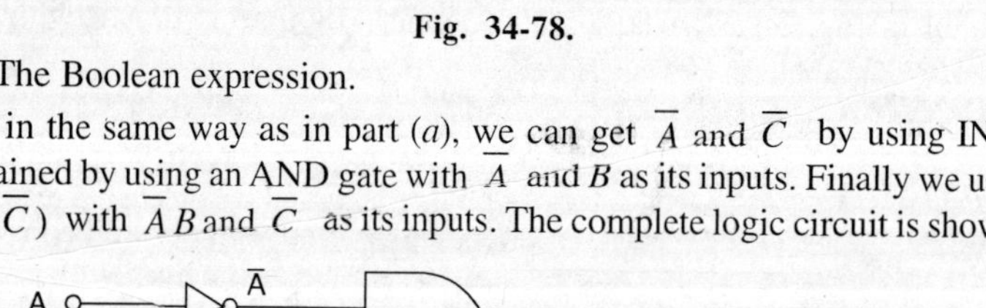

Fig. 34-78.

(*b*) Given: The Boolean expression.

Proceeding in the same way as in part (*a*), we can get $\overline{A}$ and $\overline{C}$ by using INVERTERs. The term $\overline{A}B$ is obtained by using an AND gate with $\overline{A}$ and *B* as its inputs. Finally we use on OR gate to produce $(\overline{A}B+\overline{C})$ with $\overline{A}B$ and $\overline{C}$ as its inputs. The complete logic circuit is shown in Fig. 34-79.

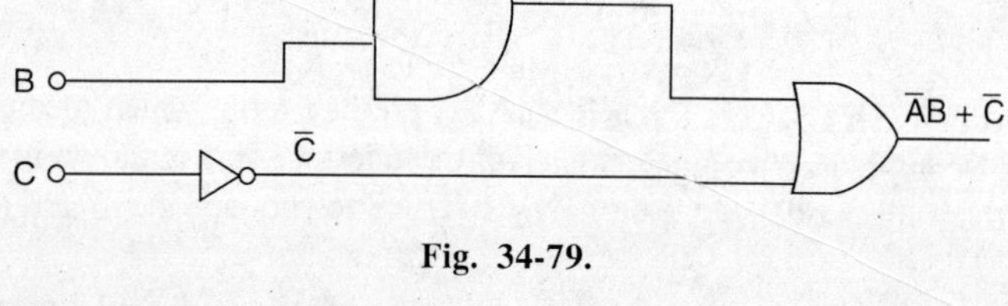

Fig. 34-79.

Example 34-25. *Construct a logic circuit that can implement the Boolean expression:*

$$X = A\overline{B} + \overline{A}B$$

Solution. Given: The Boolean expression is,

$$X = A\overline{B} + \overline{A}B$$

In order to implement this expression as a logic circuit, we find that it requires two input variables, A and B. The $\overline{A}$ and $\overline{B}$ are obtained by using INVERTERs and the terms $A\overline{B}$ and $\overline{A}B$ by using two AND gates as shown in Fig. 34-80. Finally the output, X, is obtained by using an OR gate with $(\overline{A}B)$ and $(A\overline{B})$ as its inputs.

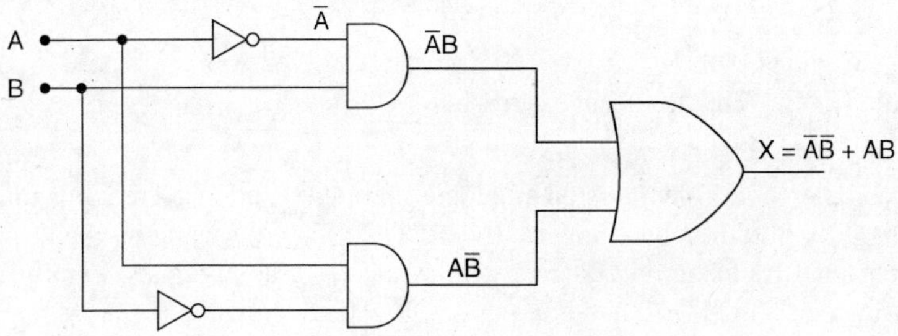

Fig. 34-80.

Example 34-26. *Construct a logic circuit that can implement the Boolean expression:*

$$X = \overline{A}\,\overline{B} + AB$$

Solution. Given: The Boolean expression is,

$$X = \overline{A}\,\overline{B} + AB$$

In order to implement this expression as a logic circuit, we find that it requires two input variables, A and B. The $\overline{A}$ and $\overline{B}$ are obtained by using INVERTERs and the terms $A\overline{B}$ and AB are obtained by using AND gates as shown in Fig. 34-81. Finally the output, X is obtained by using an OR gate with $(\overline{A}\,\overline{B})$ and (AB) as its inputs.

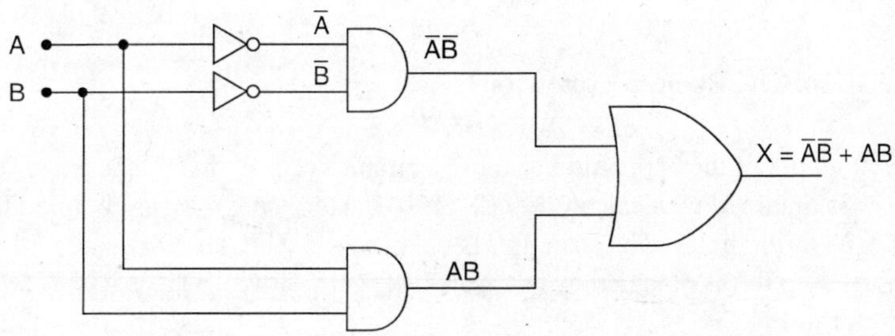

Fig. 34-81.

Example 34-27. *Construct a logic circuit that can implement the Boolean expression.*

$$X = AB + \overline{B}C$$

Solution. Given: The Boolean expression.

$$X = AB + \overline{B}C$$

In order to implement this expression as a logic circuit, we find that it requires three variables A, B and C. The $\overline{B}$ is obtained by using an INVERTER and the terms AB and $\overline{B}C$ are obtained by using AND gates as shown in Fig. 34-82. Finally, the output X is obtained by using an OR gate with (AB) and $(\overline{B}C)$ as its inputs,

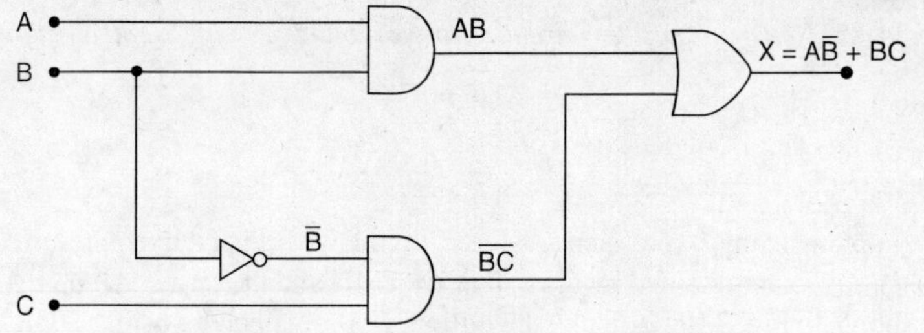

Fig. 34-82.

Example 34-28. *Construct a logic circuit using INVERTER, AND and OR gates for the Boolean expression* $X = \overline{W + P\overline{Q}}$

Solution. Given: The Boolean expression,

$$X = \overline{W + P\overline{Q}}$$

In order to implement this expression as a logic circuit, we find that it requires three variables W, P and Q. The $\overline{Q}$ is obtained by using an INVERTER, $P\overline{Q}$ by using an AND gates. The term $W + P\overline{Q}$ is obtained by using on OR gate with W and $P\overline{Q}$ as its inputs. Finally the output X is obtained by using an INVERTER at the output of OR gate as shown in Fig. 34-83.

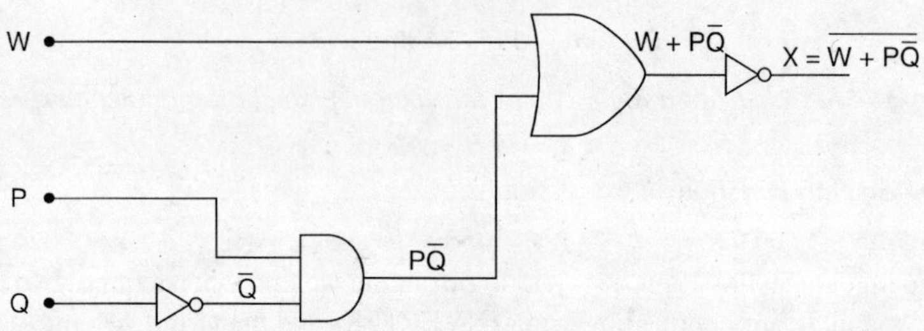

Fig. 34-83.

Example 34-29. *Construct a logic circuit using INVERTER, AND and OR gates for the Boolean expression,*

$$X = MN(P + \overline{N})$$

Solution. Given: The Boolean expression

$$X = MN(P + \overline{N})$$

In order to implement the expression as a logic circuit, we find that it requires three variables M, N, and P. The $\overline{N}$ is obtained by using an INVERTER. The term $(P + \overline{N})$ is obtained by using an OR gate with P and $\overline{N}$ as its inputs. The term MN is obtained by using an AND gate with M and N as its inputs. Finally the output X is obtained by using an AND gate with (MN) gate with (MN) and $(P + \overline{N})$ as its inputs. Refer to Fig. 34-84.

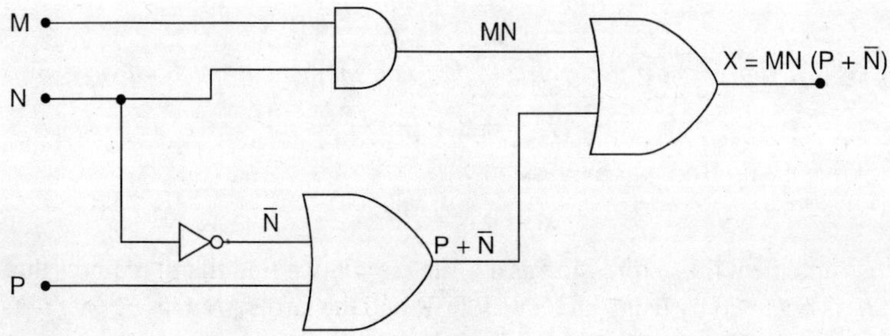

Fig. 34-84.

Example 34-30. *Construct a logic circuit using INVERTER, AND and OR gates for the Boolean expression.*

$$X = \overline{AB(C+D)}$$

Solution. Given: The Boolean expression,

$$X = \overline{AB\,(C+D)}$$

In order to implement this expression as a logic circuit, we find that it requires for variables, *A*, *B*, *C* and *D*. The term (*AB*) is obtained by using an AND gate with *A* and *B* as its inputs. The term $(C + D)$ is obtained by using an OR gate with *C* and *D* as its inputs. Next, the term *AB* (*C* + *D*) is obtained by using an AND gate with (*AB*) and (*C* + *D*) with its inputs. Finally, the output *X* is obtained by using an INVERTER as shown in Fig. 34-85.

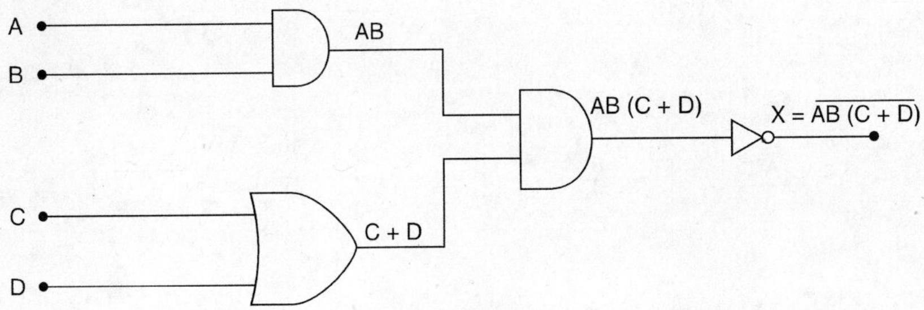

Fig. 34-85.

Example 34-31. *Construct a logic circuit using INVERTER, AND and OR gates for the Boolean expression,*

$$X = \overline{(A + B + \overline{C}\overline{D}\overline{E})} + \overline{B}C\overline{D}$$

Solution. Given: The Boolean expression,

$$X = \overline{(A + B + \overline{C}\overline{D}\overline{E})} + \overline{B}C\overline{D}$$

In order to implement this expression as a logic circuit, we find that it requires five variables. *A*, *B*, *C*, *D* and *E*. $\overline{B}, \overline{C}, \overline{D}$ and $\overline{E}$ are obtained by using INVERTERS. The term *A* + *B* is obtained by using an OR gate with *A* and *B* as its inputs. The term $\overline{C}\,\overline{D}\,\overline{E}$ is obtained by using an AND gate with $\overline{C}, \overline{D}$ and $\overline{E}$ as its inputs. The term $\overline{B}\,C\,\overline{D}$ is obtained by using another 3-input AND gate $\overline{B}$, *C* and $\overline{D}$ as its inputs. Next the term $\overline{A + B + \overline{C}\overline{D}\overline{E}}$ is obtained by using an INVERTER, Finally, the output, *X* is obtained by using an OR gate as shown in Fig. 34-86.

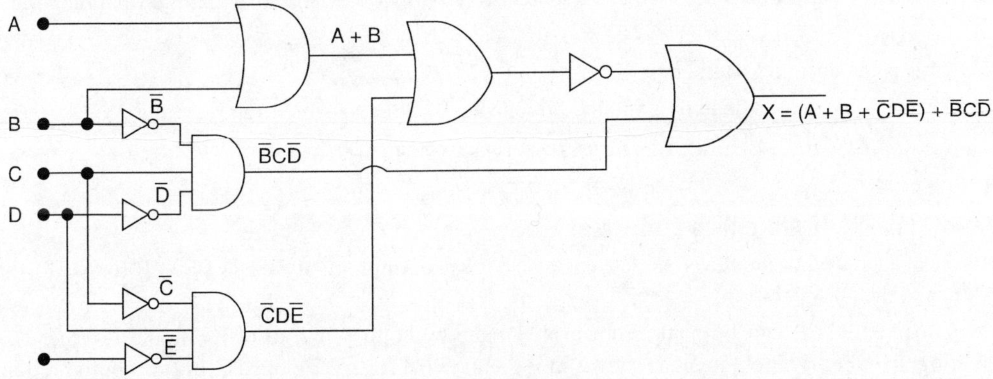

Fig. 34-86.

34-41. The Universality of NAND and NOR Gates

The universality of NAND gates means that it can be used as an INVERTER and that combinations of NAND gates can be used to implement the AND, OR and NOR operations. Similarly, the NOR gate can be used to implement the AND, OR and NOR operations. Similarly, the NOR gate can be used to implement the INVERTER, AND, OR and NAND operations. Now we will study one by one in the following pages as how the NAND and NOR gates can be used as universal gates.

34-42. The NAND Gate a Universal Logic Gate

The NAND gate is a universal logic gate because it can be used to implement the INVERTER (or NOT), the AND and the OR operation.

An INVERTER can be made from a NAND gate by connecting all of its inputs together and creating, in effect, a single input as shown in Fig. 34-87 (*a*) for a 2-input gate. In this configuration, the NAND simply acts as INVERTER because its output is,

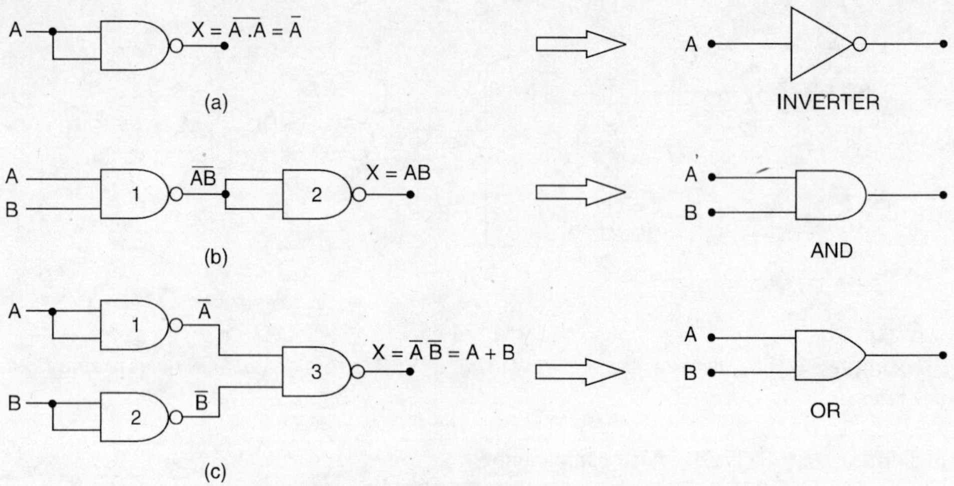

Fig. 34-87.

An AND operation can be performed by connecting 2 NAND gates as shown in Fig 34-87(*b*). Here, the NAND gate 2 is used as an INVERTER so that the final output,

$$X = \overline{\overline{AB}} = AB$$

This is the desired AND function.

An OR operation can be performed by connecting 3 NAND gates as shown in Fig. 34-87(*c*), Here, NAND gates 1 and 2 are used as INVERTERs to invert the inputs, so that the final output is.

$$X = \overline{\bar{A} \cdot \bar{B}}$$

This is the desired OR operation.

Using DeMorgan's theorem (explained in chapter 4), it is possible to show that

$$\overline{\bar{A} \cdot \bar{B}} = \bar{\bar{A}} + \bar{\bar{B}} = A + B$$

$$\therefore \quad X = A + B$$

34-43. The NOR Gate as a Universal Gate

Like the NAND gate, the NOR gate can also be used to implement the INVERTER (or NOT), the AND and the OR operation.

An INVERTER can be made from a NOR gate by connecting all of its inputs together and creating, in effect, a single input as shown in Fig. 34-88 (*a*) for a 2-input gate. In this configuration, the NOR gate simply acts as an INVERTER because its output is,

$$X = \overline{A + A} = \bar{A}$$

An OR operation can be performed by connecting a 2 NOR gates as shown in Fig. 34-88 (*b*), Here, the NOR gate 2 is used as an INVERTER, so that the final output,

$$X = \overline{\overline{A+B}} = A + B$$

An AND operation can be performed by connecting 3 NOR gates as shown in Fig. 34-88 (*c*). Here the NOR gates 1 and 2 are used as INVERTERs to invert the inputs, so that the final output is,

$$X = \overline{\overline{A} + \overline{B}}$$

Using DeMorgan's theorem, it is possible to show that,

$$\overline{\overline{A} + \overline{B}} = \overline{\overline{A}} \cdot \overline{\overline{B}} = AB$$

$$\therefore \quad X = AB$$

This is the desired AND function.

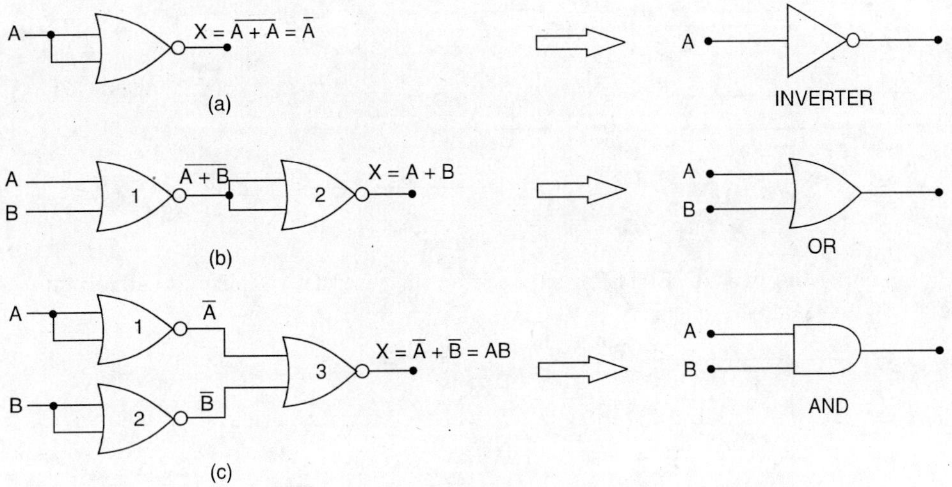

Fig. 34-88.

Example 34-32. *Sketch a NAND-NAND logic circuit for the Boolean equation,*

$$y = A\overline{B} + AC + BD$$

Solution. Given: The Boolean equation,

$$y = A\overline{B} + AC + BD$$

Let us first of all, implement this circuit using the basic logic gates, *i.e.*, using AND, OR and INVERTER gates. The $\overline{B}$ is obtained by INVERTERTing B and $A\overline{B}$ by ANDing the variables A and $\overline{B}$. Similarly, AC is obtained by ANDing variables A and C. Next the term BD is obtained by ANDing the variables B and D Finally, the terms $A\overline{B}$, AC and BD are ORed to get Y output. Fig. 34-89 shows the logic circuit for the given Boolean expression using basic logic gates.

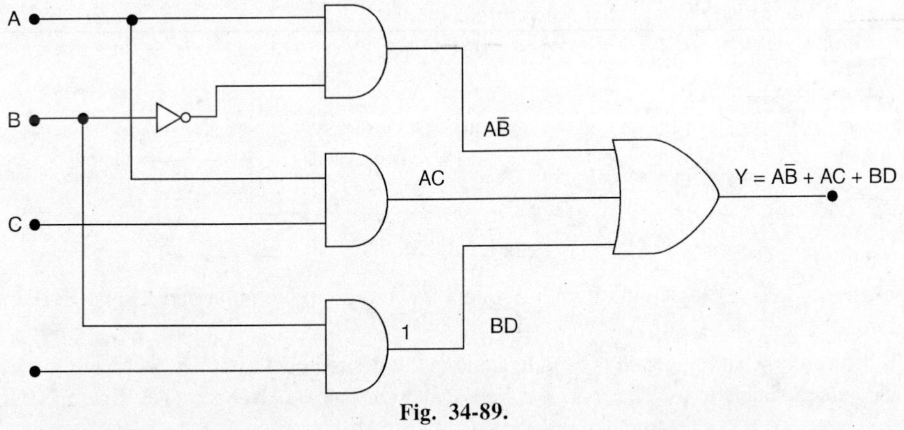

Fig. 34-89.

Now we can replace each logic gate by its NAND gate equivalent. Notice that the INVERTER is replaced by a single NAND gate each AND gate is replaced by two NAND gates while an OR gate is replaced by three NAND gates. The resulting circuit is as shown in Fig. 34-90.

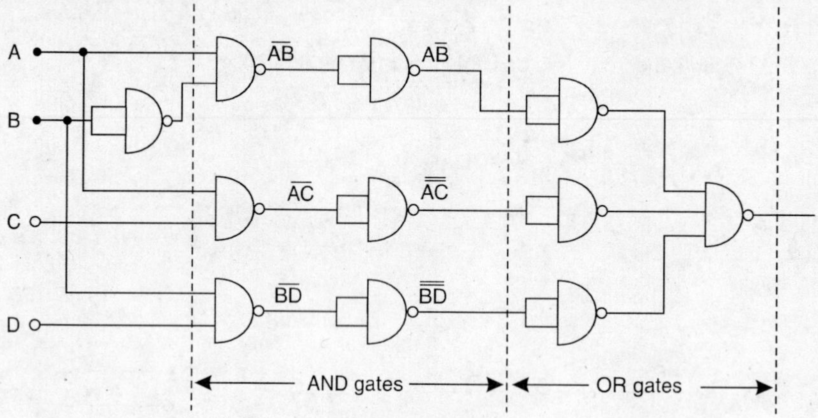

Fig. 34-90.

Removing the two INVERTERs, on the same line we can simplify the circuit shown in Fig. 34-90 to the one shown in Fig. 34-91.

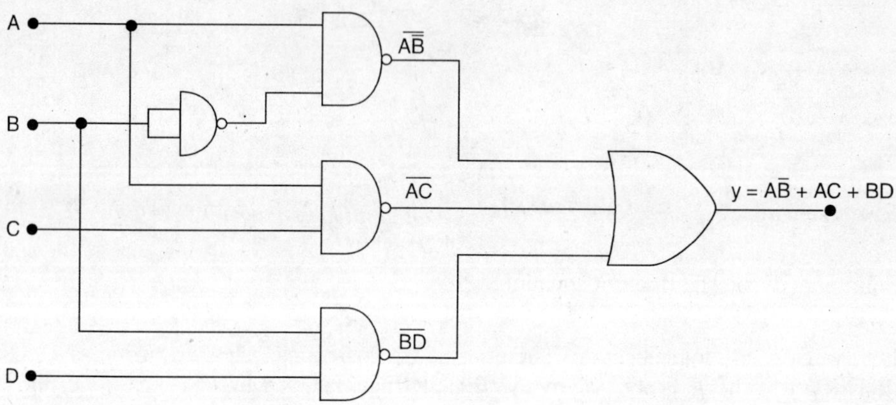

Fig. 34-91.

Example 34-33. *In the combinational logic circuit shown in Fig. 34-92, the input square wave is phase-inverted at the output when the control input is 1. It is passed on as it is when the control input is zero. Determine the logic gate used in the given figure to achieve the desired operation.*

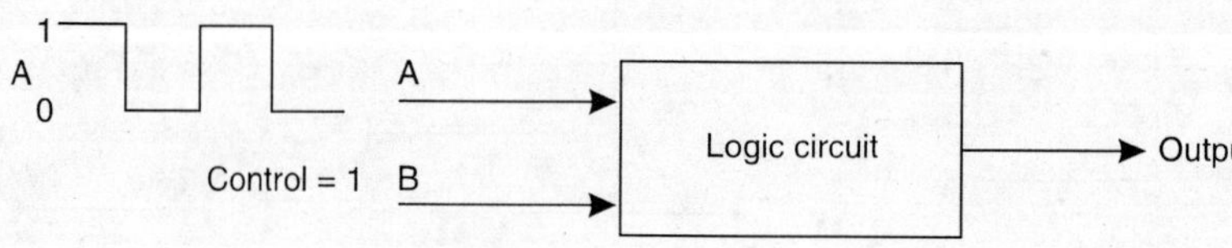

Fig. 34-92.

Solution. Given: A logic circuit with a square waveform at one input and a logic HIGH at the other input.

To determine what logic gate is used to achieve the desired operation, we will have to examine the logic gate available to us. These are the AND gate, OR gate, NAND gate, NOR gate, XOR gate and XNOR gate. Since, the question says the squares waveform is inverted at the output when the control input is 1, and it is passed as it is when the control input is zero, we will have to rule out the possibility of an AND gate and an OR gate because these logic gates don't invert the waveforms. Let us continue with the remaining four logic gates, *i.e.*, NAND gate, NOR gate, XOR gate and XNOR gate.

We know that the Boolean expression for a NAND gate is given by the expression,

$$X = \overline{AB}$$

In the present situation B is the control input. Now if $B = 1$, then $X = \overline{A \cdot 1} = \overline{A}$, *i.e.*, the square wave input is inverted at the NAND gate output. But if $B = 0$, then $X = \overline{A \cdot 0} = \overline{0} = 1$. This means the circuit will produce HIGH output, if control input B is 0, which is not a desired function. Therefore, the logic circuit with in the box of a given circuit is **not** a NAND gate.

Let us not examine the Boolean expression for a NOR gate. We know that the Boolean expression for a NOR gate is,

$$X = \overline{A+B}$$

If $B = 1$, then $X = \overline{A+1} = \overline{1} = 0$

If $B = 0$, then $X = \overline{A+0} = \overline{A}$

So we find that if control input (B) is 1, the circuit output is 0 (LOW) and if control input (B) is 0, the circuit output is the opposite of the input, *i.e.*, the inverted square waveform. But this is again not a desired operation. Therefore, we *rule out* the possibility of a NOR gate also.

Let us now examine the Boollean expression for an XOR gate. We know that Boolean expression for an XOR gate is,

$$X = A\overline{B} + \overline{A}B$$

If $B = 1$, then $X = A \cdot \overline{1} + \overline{A} \cdot 0 + \overline{A} = \overline{A}$

If $B = 0$, then $X = A \cdot \overline{0} + \overline{A} \cdot 0 = A \cdot 1 + 0 = A$

So we find that if the logic circuit makes use of an XOR gate, then if the control input B is 1, the logic gate output is opposite to the input A *i.e.* the output will be an inverted square wave. On the other hand if B is 0, the logic gate output is the same as input A *i.e.*, the square waveform is passed as it is. This is the desired operation of the given logic circuit. Therefore the logic gate which we need is an XOR gate. **Ans.**

34-44. Using Integrated Circuit Logic Gates

As a matter of fact, all the logic gates (which have been discussed in this chapter) are available in various configurations in TTL (Transistor-Transistor Logic) and CMOS (Complementary Metal-Oxide Semiconductor) Logic families. Thus, TTL IC 7404 and CMOSIC 74C04 are hex (six) INVERTERs. The TTLIC 7400 and CMOSIC 74 C00 are quad (four) two-input NAND gates. The TTL IC 7402 and CMOS IC 74 02 are qual two-input NOR gates. Other popular NAND and NOR gates are available in three; four-and eight-input configurations. The TTLIC 7486 and CMOSIC 74C86 all quad Exclusive-OR (XOR) gates. Similarly, the TTLIC 74266 and CMOSIC 74C266 are quad Exclusive-NOR (XNOR) gates.

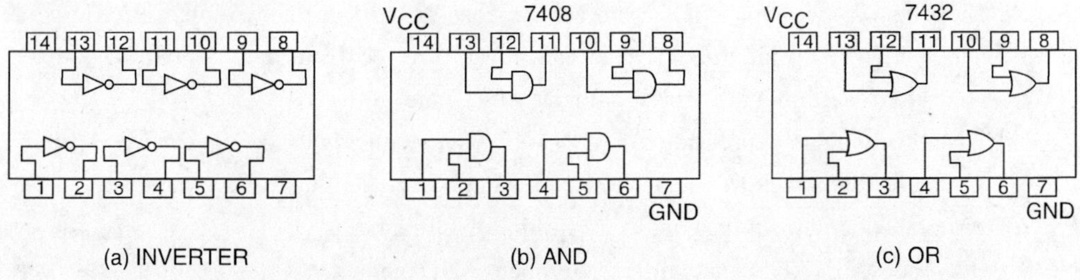

(a) INVERTER (b) AND (c) OR

Fig. 34-93.

You can consult a TTL or CMOS data manual for the availability and pin configuration of these ICs. The pin configurations for the TTL hex INVERTER, the quad AND and the quad OR gates are given in Fig. 34-93. Similarly the pin configurations for the TTL quad NAND, quad NOR, gates are shown in Fig. 34-94.

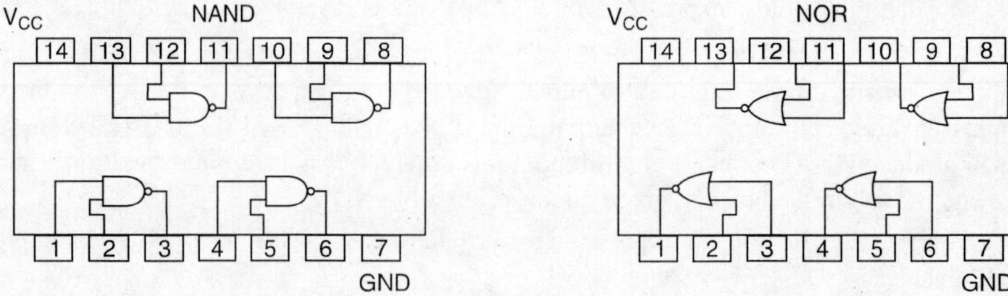

Fig. 34-94.

SUMMARY

In this chapter you have learned that:

1. Boolean algebra is a powerful mathematics tool used in design and analysis of digital/logic circuits.
2. The INVERTER, AND and OR are the basic Boolean operations.
3. An INVERTER (or a NOT) gate produces an output that is the opposite logic level as the input. An AND gate produces a HIGH output when all inputs are HIGH. An OR gate produces a HIGH output when any input is HIGH.
4. A NAND gate is the same as an AND gate with its output connected to an INVERTER. A NOR gate is the same as an OR gate with its output connected to an INVERTER.
5. NAND gates can be used to implement any of the basic Boolean expression. Similarly NOR gates can also be used for this purpose.
6. An Exclusive-OR gate has the Boolean expression, $X = \overline{A}B + A\overline{B}$. Its output X will be HIGH only when its inputs A and B are at opposite logic levels.
7. An Exclusive-NOR gate has the Boolean expression, $X = \overline{AB} + AB$. Its output X will be HIGH only when its inputs A and B are at same logic levels.

DESCRIPTIVE QUESTIONS

1. If you were to build a truth table for an eight-input AND gate, how many different combinations of inputs would you have? Sketch the table with thee input combinations. Indicte the output level for each input combination.
2. Show how a two-input NOR gate can be constructed from a two-input NAND gate.
3. Show how a two-input NOR gate can be constructed from a two-input NOR gate.
4. Describe the truth table for a 5-input AND gate.
5. Write the output expression for a NAND gate with inputs A, B, C and D.
6. Write the output expression for a NOR gate with inputs A, B, C and D.
7. Write the output expression for a 3-input NOR gate.
8. How does a NAND gate differs from an AND gate?
9. How does an exclusive-OR gate differ from on OR gate in its logical operation?

10. Suppose that you have an unknown 2-input gate. That is either OR gate or an AND gate. What combinations of input levels should you apply to the gate's inputs to determine which type of gate it is?

11. What are universal gates and why? *(UP Tech. Univ., Sem I., 2006-07)*

12. Prove that the NAND and NOR gate is a universal gate. *(UP Tech. Univ., Sem II., 2005-06)*

13. (*a*) Realize AND, OR, NOT using only NAND gates.

(*b*) Using NOR gates only, realize Ex-OR gate.*(UP Tech. Univ., Sem I., 2005-06)*

14. Realize a 2-input NOR gate using 2-input NAND gate. *(UP Tech. Univ., Sem II, 2004-05)*

15. Explain what do you mean by universal logic gate. *(UP Tech. Univ., Sem I, 2004-05)*

TUTORIAL PROBLEMS

1. Fig. 34-95 shows a waveform applied to the input of an INVERTER. Sketch the output waveform corresponding to the input.

Fig. 34-95.

2. A network of cascaded INVERTERS is shown in Fig. 34-96. If a logic 1 is applied to a point A determine the logic levels at a points *B* through *F*.

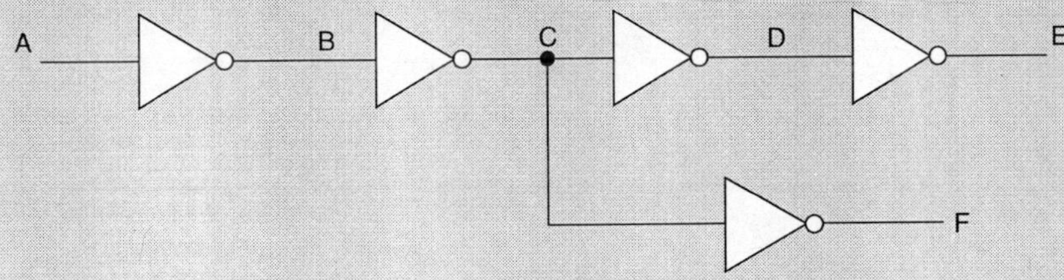

Fig. 34-96.

(**Ans.** $B = 0, C = 1, D = 0, E = 1, F = 0$)

3. Determine the output, *X* for a 2-input AND gate with the input waveforms shown in Fig. 34-97.

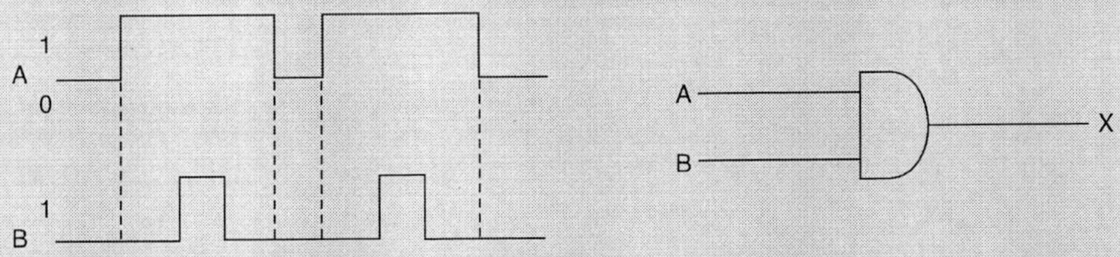

Fig. 34-97.

4. Determine the output, *X* for a 2-input AND gate with the input waveforms shown in Fig. 34-98.

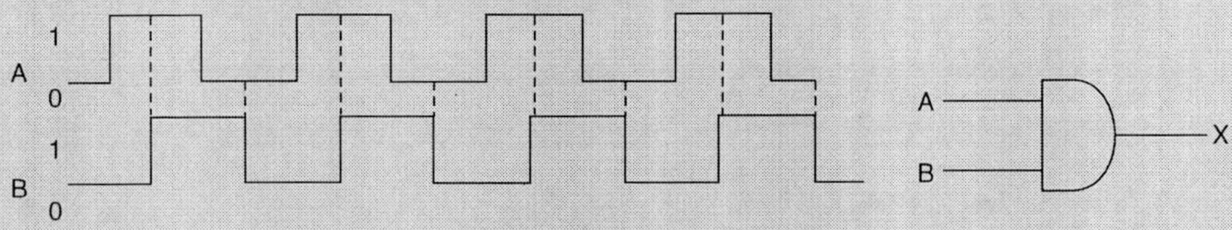

Fig. 34-98.

5. Fig. 34-99 shows the input waveforms applied to a 3-input AND gate. Show the output waveform in proper relation to the inputs with a timing diagram.

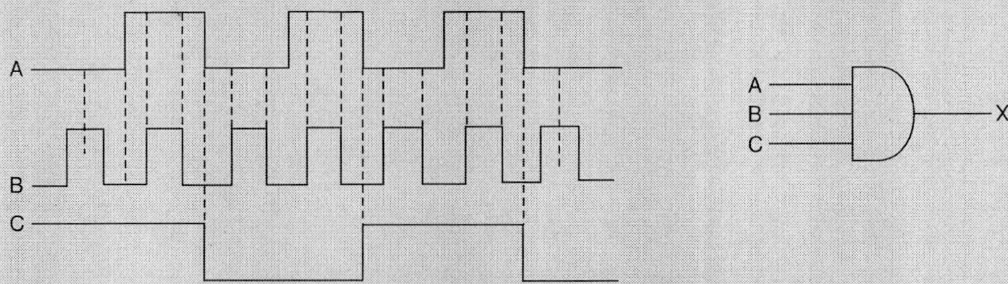

Fig. 34-99.

6. Fig. 34-100 shows the input waveform applied to a 4-input AND gate. Show the output waveform in proper relation to the inputs with a timing diagram.

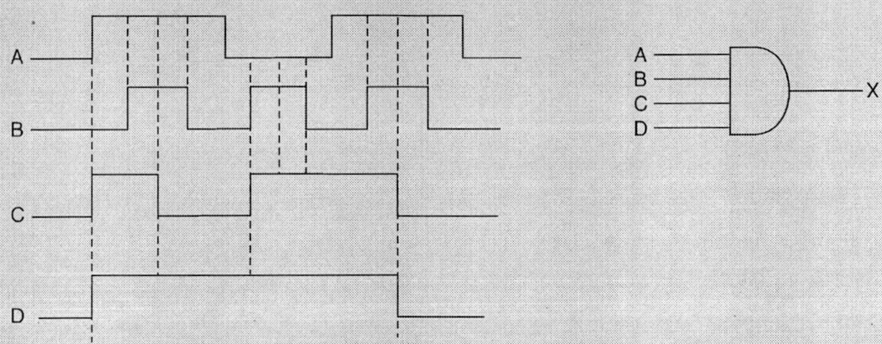

Fig. 34-100.

7. Sketch the output waveform *X* for the 2-input OR gate shown in Fig. 34-101

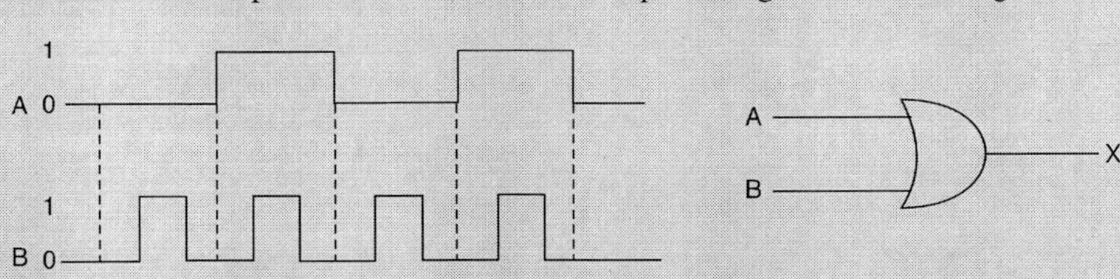

Fig. 34-101.

8. Determine the output, X for a 2-input OR gate with the input waveforms shown in Fig. 34-102

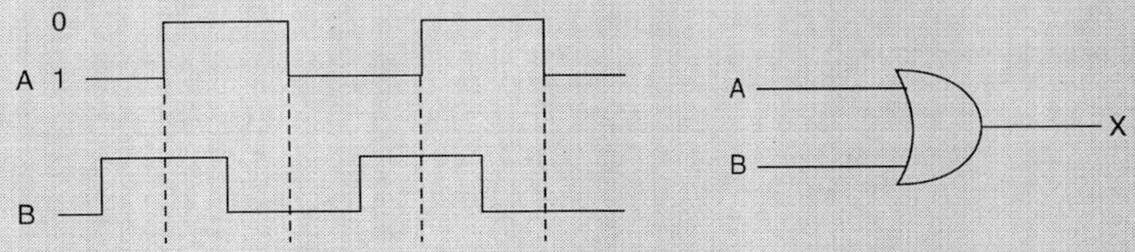

Fig. 34-102.

9. Fig. 34-103 shows the input waveform at A for a 2-inut AND gate. Sketch the input waveform at B that will produce the output X as shown in the figure.

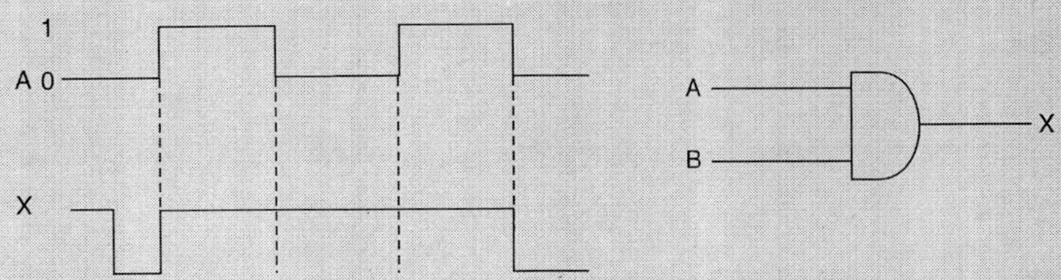

Fig. 34-103.

10. Fig. 34-104 shows the input waveform at A for a 2-input OR gate. Sketch the input waveform at B that will produce the output at X as shown in the figure.

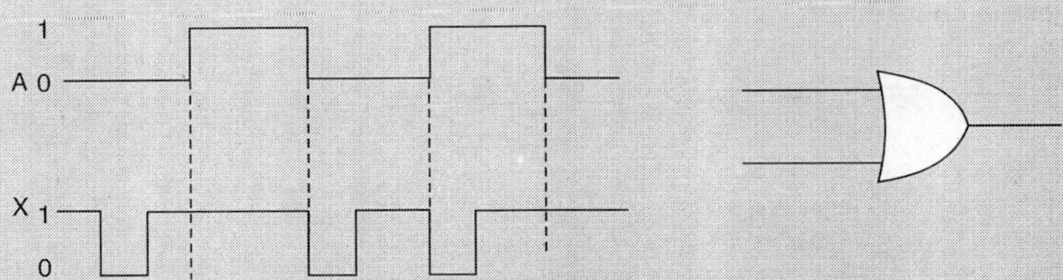

Fig. 34-104.

11. Determine the output for a 2-input OR gate when the input waveforms are as shown in Fig. 34-105 and draw a timing diagram.

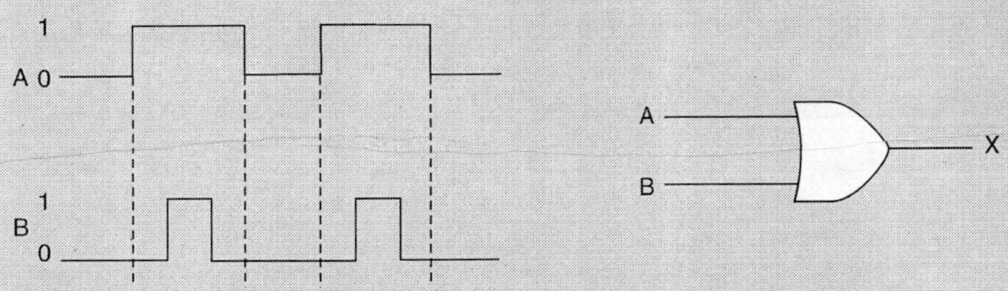

Fig. 34-105.

12. Fig. 34-99 shows the input waveforms applied to a 3-input OR gate. Show the output waveform in proper relation to the inputs with a timing diagram.

13. Fig. 34-100 (Problem 6) shows the input waveforms applied to a 4-input OR gate. Show the output waveform in proper relation to the inputs with a timing diagram.

14. For the set of input waveforms shown in Fig. 34-106 determine the output waveform *X*, for the 2-input NAND gate and draw the timing diagram.

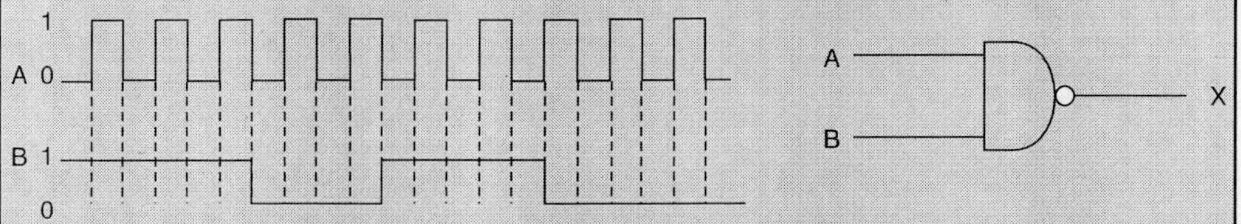

Fig. 34-106.

15. Determine the 3-input NAND gate output for the input waveforms shown in Fig. 34-107 draw the timing diagram.

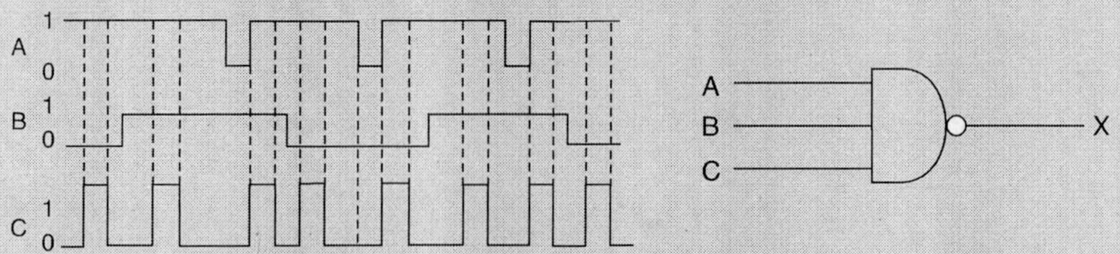

Fig. 34-107.

16. Fig. 34-108 shows the waveforms at a 4-input NAND gate. Show the output waveform in proper relation to the inputs with a timing diagram.

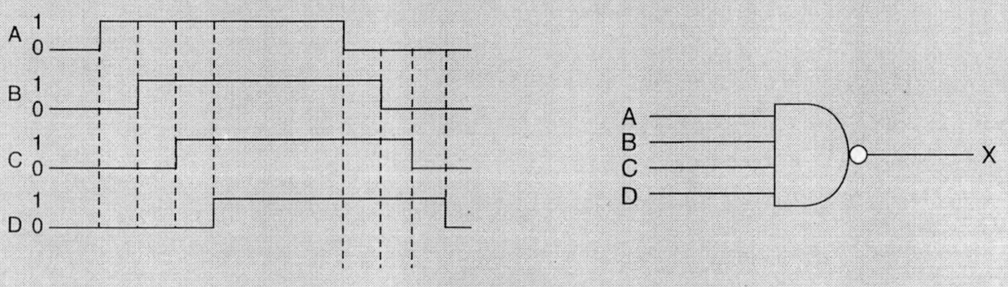

Fig. 34-108.

17. Repeat Problem 12 for a 2-input NOR gate.

18. Repeat Problem 13 for a 3-input NOR gate.

19. Determine the 2-input NOR gate output for the input waveforms shown in Fig. 34-109 and draw the timing diagram.

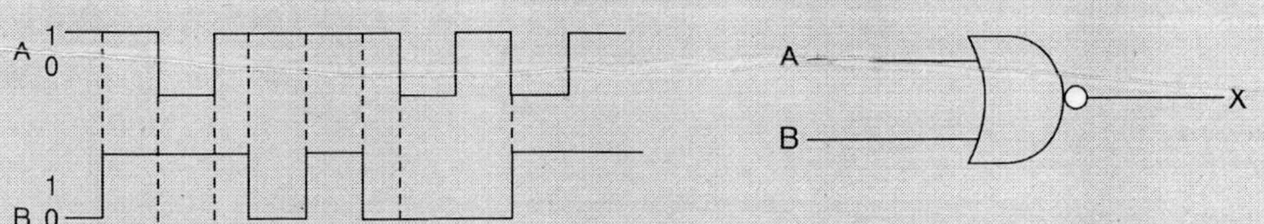

Fig. 34-109.

20. Determine the output of an exclusive-OR gate for the input waveforms shown in Fig. 34-110 and draw the timing diagram.

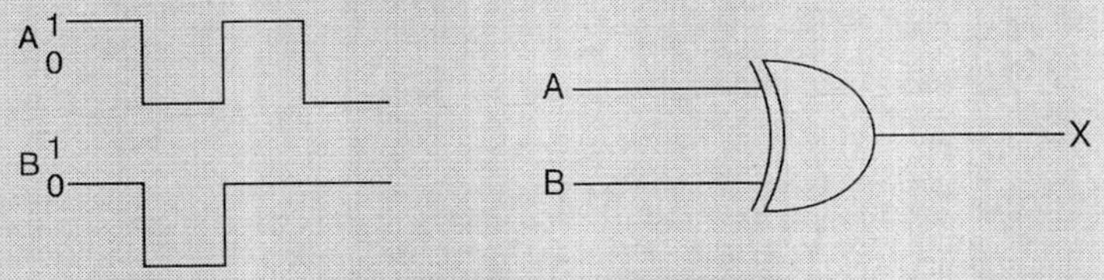

Fig. 34-110.

21. Determine the output of an exclusive - NOR gate for the input waveforms shown in Fig. 34-111 and draw the timing diagram.

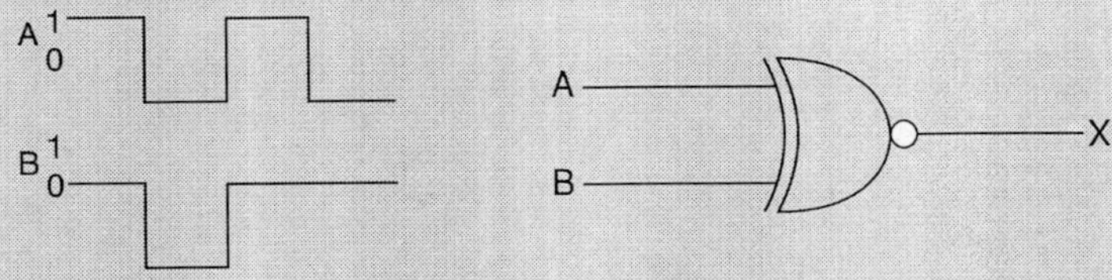

Fig. 34-111.

MULTIPLE CHOICE QUESTIONS

1. The number of table entries needed for a five-input logic circuit is:
(*a*) 4 (*b*) 8 (*c*) 16 (*d*) 32

2. Which of the following input combination will produce a HIGH at the output of a two-input AND gate
(*a*) both inputs are LOW
(*b*) one of the inputs is HIGH and the other one is LOW
(*c*) both inputs are HIGH
(*d*) none of these.

3. The output of a two-input NAND gate is HIGH if
(*a*) both input are LOW
(*b*) one of the input is HIGH and the other one is LOW
(*c*) both inputs are HIGH
(*d*) none of these.

4. An exclusive-OR function is expressed as
(*a*) $\overline{A}\overline{B} + AB$ (*b*) $\overline{A}B + A\overline{B}$
(*c*) $(\overline{A} + B)(A + \overline{B})$ (*d*) $(\overline{A} + \overline{B})(A + B)$

5. The AND operation can be produced with
(*a*) two NAND gates (*b*) three NAND gates
(*c*) one NOR gates (*d*) three NOR gates.

6. The OR operation can be produced with
(*a*) two NOR gates
(*b*) three NAND gates
(*c*) four NAND gates
(*d*) both (*a*) & (*b*)

7. All Boolean expression can be implemented with

(*a*) NAND gates only

(*b*) NOR gates only

(*c*) combinations of NAND and NOR gates

(*d*) combinations of AND gates, OR gates and INVERTERS

8. To get an output 1 from circuit of Fig. 34-112. The inputs ABC must be

(*a*) 010 (*b*) 100

(*c*) 101 (*d*) 110.

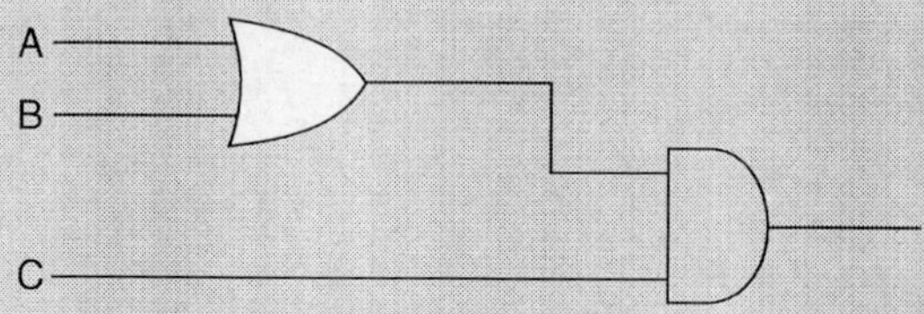

Fig. 34-112.

9. Which of the following gates will have an output of 1?

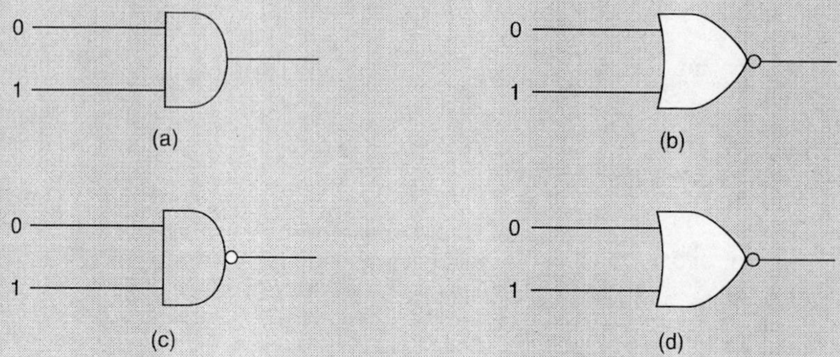

Fig. 34-113.

10. The logic gate which produces a 0 or low-level output when one or both of the inputs are 1 is called......gate.

(*a*) AND (*b*) OR

(*c*) NOR (*d*) NAND.

ANSWERS

1. (*d*)	**2.** (*c*)	**3.** (*a*)	**4.** (*b*)	**5.** (*a*)
6. (*a*)	**7.** (*e*)	**8.** (*b*)	**9.** (*c*)	**10.** (*c*)

❑ ❑ ❑

35

CHAPTER

Boolean Algebra and Logic Simplification

OUTLINE

Objectives

After copulating this chapter, you should be able to:

- Write Boolean equations for combinational logic applications.
- Utilize Boolean Algebra laws and rules for simplifying combinational logic circuits.
- Apply DeMorgan's theorem to complex Boolean equations to arrive at simplified equivalent equations.
- Design single-gate logic circuits by utilizing the universal capability of NAND and NOR gates.
- Troubleshoot combinational logic circuits.
- Implement sum-of-products expressions utilizing AND-OR-INVERT gates.
- Utilize the Karnaugh mapping procedure to systematically reduce complex Boolean equations to their simplest form.
- Describe the steps involved in solving a complete system design application.

35-1. Introduction

Boolean algebra is a convenient and systematic way of expressing and analysing the operation of logic circuits. In the last chapter, we have already introduced the Boolean operations and expressions in terms of their relationships to INVERTER, AND, OR, NAND, NOR, exclusive-OR (XOR) and exclusive-NOR (XNOR) gates.

This chapter covers the laws, rules and theorems of Boolean algebra and their application to digital circuits. We will learn how to simplify the logic circuits using the methods of Boolean algebra and Karnaugh maps.

35-2. Laws and Rules of Boolean Algebra

Boolean algebra is an important mathematical tool for designing and analysing digital systems. A basic knowledge of Boolean Algebra is indispensable to the study and analysis of logic circuits. As in other areas of mathematics, there are certain well developed rules and laws that must be followed in order to apply Boolean Algebra properly. Let us now study some of the important laws and rules of Boolean algebra. First we will study the laws of Boolean algebra and then the rules.

35-3. Laws of Boolean Algebra

Broadly speaking there are 3 basic categories laws of Boolean algebra: **(1)** the **Commutative laws**, **(2)** the **Associative laws,** and **(3)** the **Distributive laws**. Each of these categories is illustrated with two or three Boolean variables, but the number of Boolean variables is not limited to this.

(1) **Cummutative laws.** There are 2 laws under this category—***(i)*** Cummutative law of addition and ***(ii)*** Cummutative law of multiplication.

(i) Cummutative law of addition. This law for 2 variables is written algebraically as

$$A + B = B + A \qquad ...(35\text{-}1)$$

This law states that the order in which the variables are ORed makes no difference. Remember the addition and the OR operation are the same when applied to logic circuits. Fig. 35-1 illustrates the Cummutative law as applied to the OR gate. It shows that it doesn't matter to which input each variable is applied.

Fig. 35-1. *Illustrating cummutative law of addition.*

(*ii*) ***Cummutative law of multiplication*** **:** This law for 2 variables is written algebraically as,

$$AB = BA \qquad ...(35\text{-}2)$$

This law states that the order in which the variables are ANDed makes no difference. Fig. 35-2 illustrates the law as applied to the AND gate.

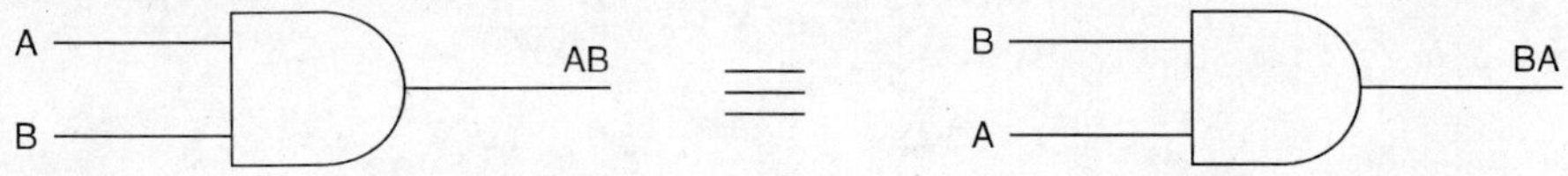

Fig. 35-2. *Illustrating cummutative law of multiplication.*

(**2**) ***Associative laws*****.** Like cummutative laws, there are 2 laws under this category: (*i*) associative law of addition and (*ii*) associative law of multiplication.

(*i*) ***Associative law of addition*** **:** This law for 3 variables is written algebraically as,

$$A + (B + C) = (A + B) + C \qquad ...(35\text{-}3)$$

This law states that in ORing of more than 2 variables, the result is the same regardless of the grouping of the variables. Fig. 35-3 illustrates this law as applied to OR gates.

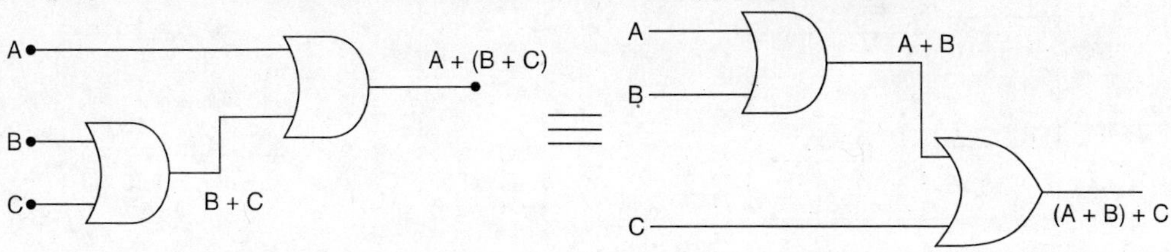

Fig. 35-3. *Illustrating associative law of addition.*

(*ii*) ***Associative law of multiplication*** **:** This law for 3 variables is written algebraically as :

$$A\,(BC) = (AB)\,C \qquad ...(35\text{-}4)$$

This law states that it makes no difference in what order the variables are grouped when ANDing more than 2 variables. Fig. 35-4 illustrates this law when applied to AND gates.

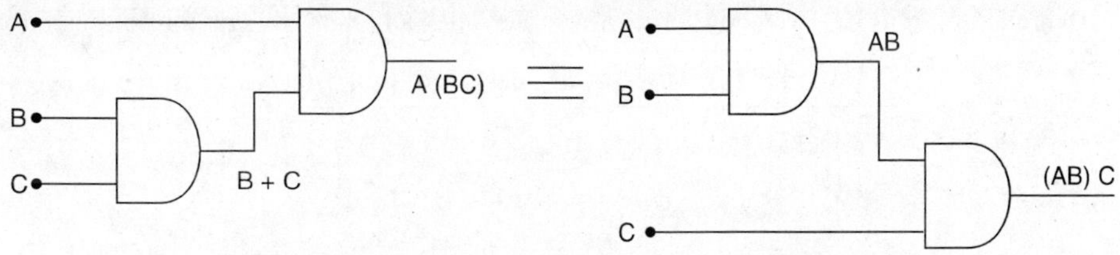

Fig. 35-4. *Illustrating associative law of multiplication.*

3. ***Distributive Law*** **:** This law is the same as in ordinary algebra. The distributive law for 3 variables is written as follows :

$$A\,(B + C) = AB + AC \qquad ...(35\text{-}5)$$

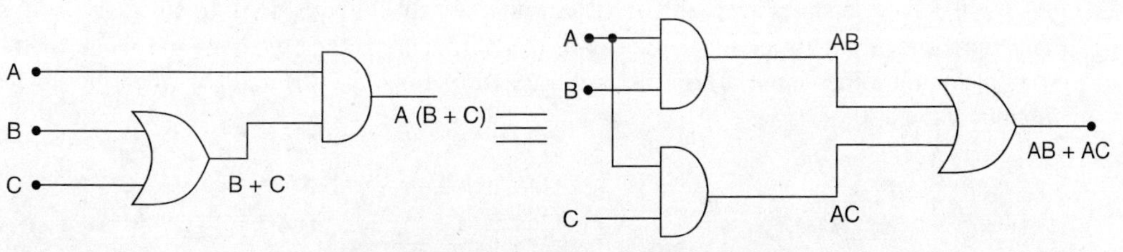

Fig. 35-5. *Illustrating a distributive law.*

This law states that ORing 2 or more variables and ANDing the result with a single variable is equivalent to ANDing the single variable with each of the two or more variables and then ORing the products. The distributive law also expresses the process of **factoring** in which a common variable *A* is factored out of the product terms. Fig. 35-5 illustrates the distributive law in terms of gate implementation.

Example 35-1. *Realize the following expression using Ex-OR/Ex-NOR gates and basic gates if required.*

$$f = \bar{A}B\bar{C} + \bar{A}\bar{B}C + A\bar{C}D + AC\bar{D}$$ *(U.P. Tech. Univ., Sem, II 2003-04)*

Solution. Given the Boolean expression,

$$f = \bar{A}B\bar{C} + \bar{A}\bar{B}C + A\bar{C}D + AC\bar{D}$$

Applying the distributive law,

$$f = \bar{A}\,(B\bar{C} + \bar{B}C) + A\,(\bar{C}D + C\bar{D})$$
$$= \bar{A}\,(B \oplus C) + A\,(C \oplus D)$$

The above equation can be implemented using Ex-OR and other gates as shown in Fig. 35-6.

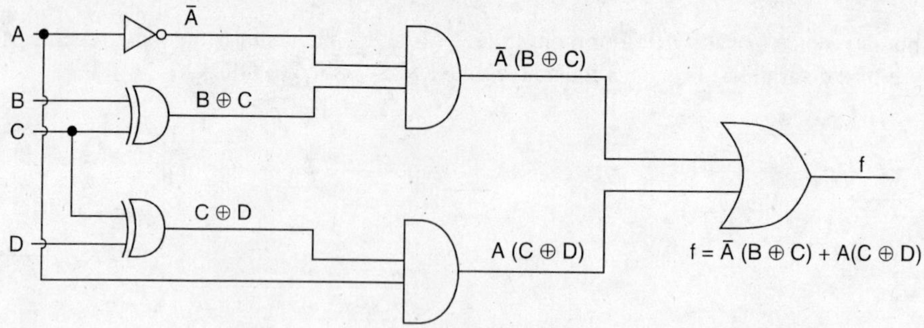

Fig. 35-6.

35-4. Rules of Boolean Algebra

Table 35-1 lists the 12 basic rules that are useful in manipulating and simplifying Boolean expressions. We will view the rules 1 through 9 in terms of their application to logic gates. Whereas rules 10 through 12 will be derived in terms of simpler rules and the laws previous discussed.

Table 35-1. *Twelve Basic rules of Boolean algebra*

1. $A\,.\,0 = 0$	7. $A\,.\,A = A$
2. $A\,.\,1 = A$	8. $A\,.\,\bar{A} = 0$
3. $A + 0 = A$	9. $\bar{\bar{A}} = A$
4. $A + 1 = 1$	10. $A + AB = A$
5. $A + A = A$	11. $A + \bar{A}B = A + B$
6. $A + \bar{A} = 1$	12. $(A + B)\,(A + C) = A + BC$

(1) A . 0 = 0 This rule states a variable ANDed with ϕ is always equal to 0. In other words, if at any time one input to an AND gate is 0, the output is 0, regardless of the value of the variable on the other input. This rule is illustrated in Fig. 35-7 (*a*) and (*b*) where the lower input is fixed at 0.

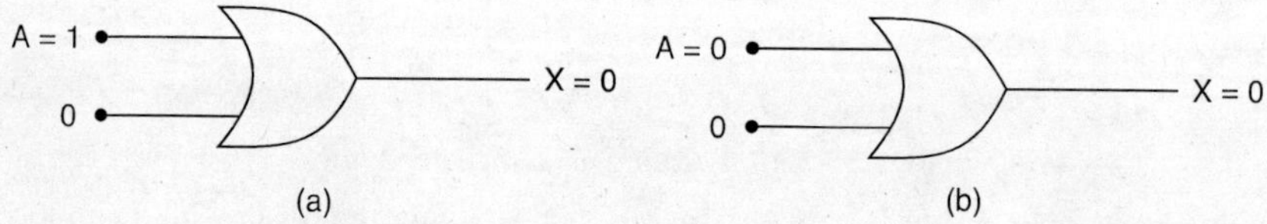

Fig. 35-7. *Illustrating Rule 1: A . 0 = 0.*

(2) $A \cdot 1 = A$. This rule states that a variable ANDed with 1 is always equal to the variable. If the variable A is 0, the output of the AND gate is 0. If the variable A is 1, the output of the AND gate is 1, because both input are now 1s. This rule is illustrated in Fig. 35-8 (a) and (b) where the lower input is fixed at 1.

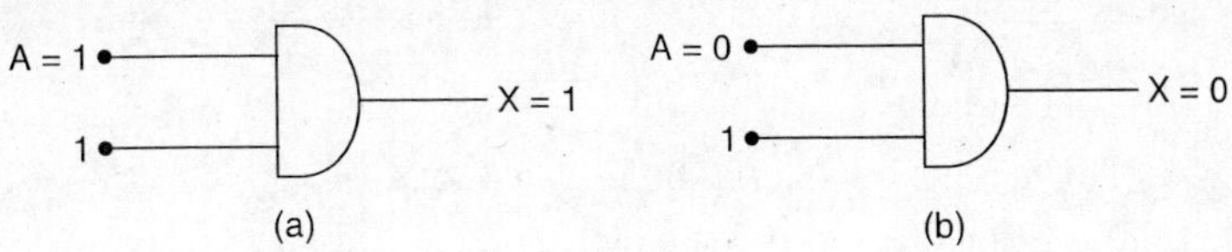

Fig. 35-8. *Illustrating Rule 2: $A \cdot 1 = A$.*

(3) $A + 0 = 0$. This rule states that a variable ORed with 0 is always equal to the variable. If A is 1, the output is 1, which is equal to A. If A is 0, the output is 0, which is also equal to A. This rule is illustrated in Fig. 35-9 (a) and (b) where the lower input is fixed at 0.

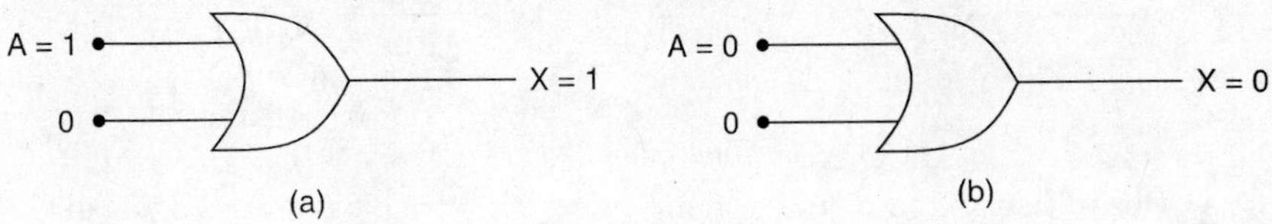

Fig. 35-9. *Illustrating Rule 3: $A + 0 = 0$.*

(4) $A + 1 = 1$. This rule states that a variable ORed with 1 is always equal to 1. A 1 on an input to an OR gate produces a 1 on the output, regardless of the value of the variable on the other input. This rule is illustrated in Fig. 35-10 (a) and (b) where the lower input is fixed at 1.

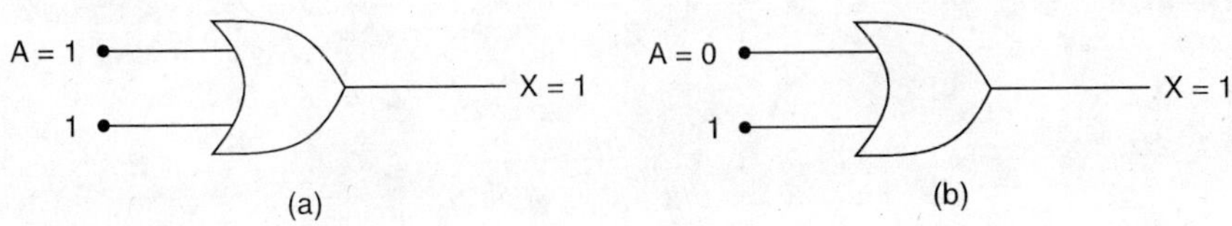

Fig. 35-10. *Illustrating Rule 4: $A + 1 = 1$.*

(5) $A + A = A$. This rule states that a variable ORed with itself is always equal to the variable. If A is 0, then $0 + 0 = 0$, and if A is 1, then $1 + 1 = 1$. This is illustrated in Fig. 35-11 (a) and (b) where both inputs are the same variable.

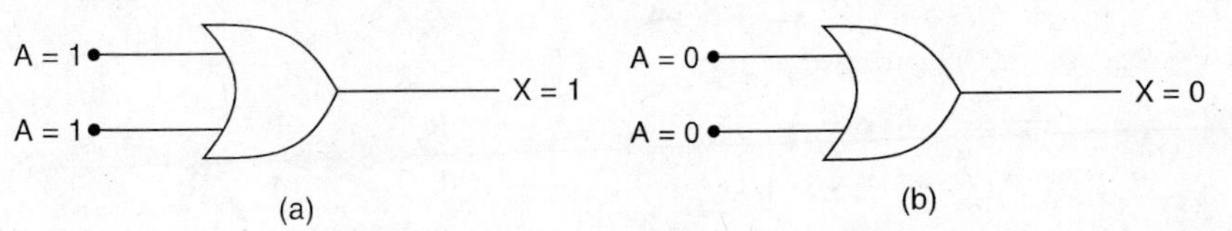

Fig. 35-11. *Illustrating Rule 5: $A + A = A$.*

(6) $A + \overline{A} = 1$. This rule states that a variable ORed with its complement is always equal to 1. If A is 0, then $0 + \overline{0} = 0 + 1 = 1$. If A is 1, then $1 + \overline{1} = 1 + 0 = 1$. This is illustrated in Fig. 35-12 (a) and (b) where one input is the complement of the other.

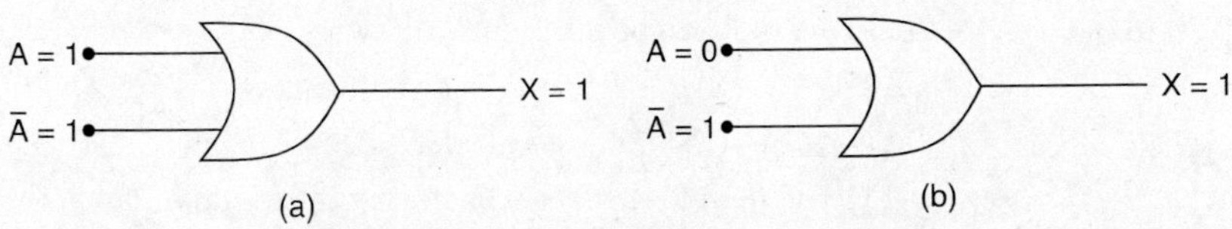

Fig. 35-12. *Illustrating Rule 6: $A + \overline{A} = 1$.*

(7) $A \,.\, A = A$. This rule states that a variable ANDed with itself is always equal to the variable. If A is 0, then $0 \,.\, 0 = 0$, and if A is 1, then $1 \,.\, 1 = 1$. This is illustrated in Fig. 35-13 (a) and (b).

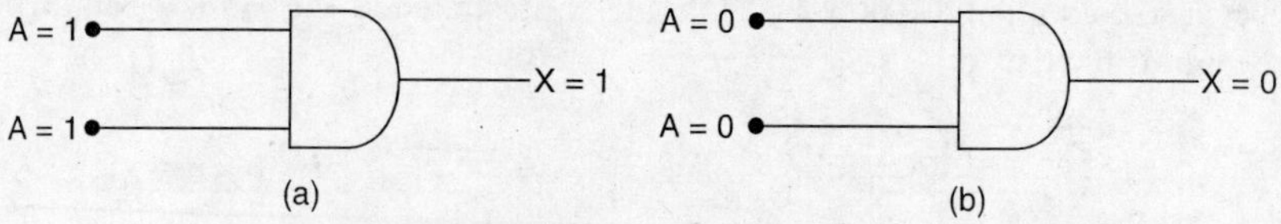

Fig. 35-13. *Illustrating Rule 7: $A \,.\, A = A$.*

(8) $A \,.\, \overline{A} = 0$. This rule states that a variable ANDed with its complement is always equal to 0. Either A or $\overline{A}$ will always be 0, and when a 0 is applied at the input of an AND gate, the output will be 0. This rule is illustrated in Fig. 35-14 (a) and (b).

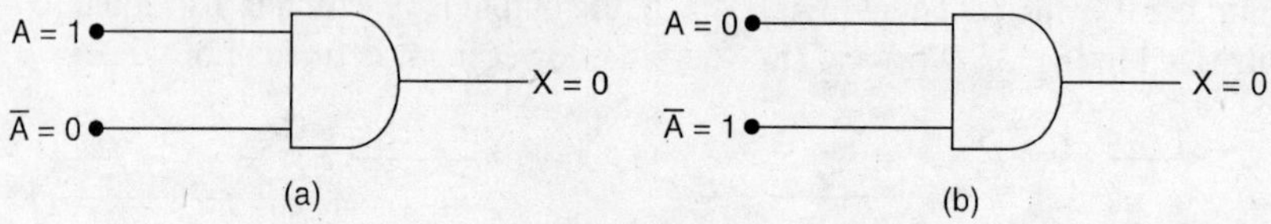

Fig. 35-14. *Illustrating Rule 8: $A.\overline{A} = 0$.*

(9) $\overline{\overline{A}} = A$. This rule states that a double complement of a variable is always equal to the variable. If we start with the variable A and complement (or invert) it once, we get $\overline{A}$. If we then take $\overline{A}$ and complement it again, we get A, which is the original variable, this rule is illustrated in Fig. 35-15 (a) and (b).

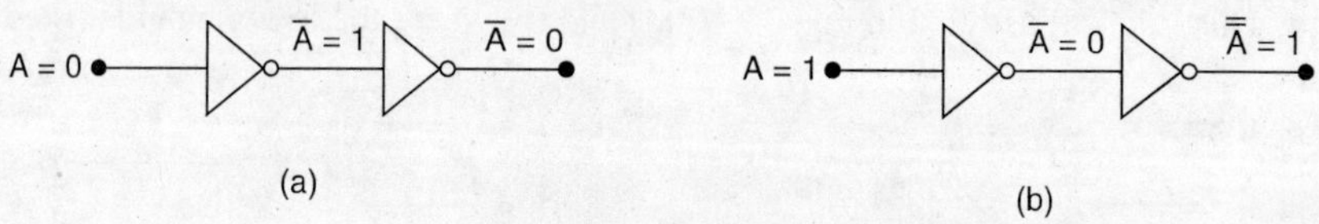

Fig. 35-15. *Illustrating Rule 9: $\overline{\overline{A}}.A$.*

(10) $A + AB = A$. This rule can be proved by applying the distributive law, rule 4 and rule 2 as follows:

$$\begin{aligned} A + AB &= A\,(1 + B) && \text{— Factoring (distributive law)} \\ &= A \,.\, 1 && \text{— Rule 4: } (1 + B) = 1 \\ &= A && \text{— Rule 2: } A \,.\, 1 = A \end{aligned}$$

(11) $A + \overline{A}B = A + B$. This rule can be proved as follows:

$$\begin{aligned} A + \overline{A}B &= (A + AB) + \overline{A}B && \text{— Rule 10: } A = A + AB \\ &= (AA + AB) + \overline{A}B && \text{— Rule 7: } A = AA \\ &= AA + AB + A\overline{A} + \overline{A}B && \text{— Rule 8: Adding } A\overline{A} = 0 \\ &= (A + \overline{A})\,(A + B) && \text{— Factoring (distributive law)} \\ &= 1.(A + B) && \text{— Rule 6: } A + \overline{A} = 1 \\ &= A + B && \text{— Rule 2: drop the 1} \end{aligned}$$

(12) $(A + B)\,(A + C) = A + BC$. This rule can be proved as follows:

$$\begin{aligned} (A + B)\,(A + C) &= AA + AC + AB + BC && \text{— Distributive law} \\ &= A + AC + AB + BC && \text{— Rule 7: } AA = A \\ &= A\,(1 + C) + AB + BC && \text{— Factoring (distributive law)} \\ &= A.1 + AB + BC && \text{— Rule 4: } 1 + C = 1 \end{aligned}$$

$= A\,(1 + B) + BC$ — Factoring (distributive law)

$= A.1 + BC$ — Rule 4: $1 + B = 1$

$= A + BC$ — Rule 2: $A\,.\,1 = A$

Example 35-2. *Using the Truth table, show that the Boolean expression $A + AB = A$.*

Solution. Given the Boolean expression $A + AB = A$. From the given expression, we find that we have 2 variables *i.e.*, A and B. These can have 4 possible combinations *i.e.*, 00, 01, 10 and 11. This is shown in the first two columns of the truth table shown in Table 35-2. The third column for AB is obtained by ANDing the first two columns. The fourth column of the truth table is obtained by ORing the first column and the third column. An inspection of the first and the fourth column indicates that there are the same. This verifies that $A + AB = A$.

Table 35-2

A	B	AB	$A + AB$
0	0	0	0
0	1	0	0
1	0	0	1
1	1	1	1

These two columns are equal

Example 35-3. *Using the truth table, show that the Boolean expression,* $A + \overline{A}B = A + B$.

Solution. Given: The Boolean expression $A + \overline{A}B = A + B$.

From the given expression, we find that we have two variables, *i.e.*, A and B. These two variables can have 4 possible combinations, *i.e.*, 00, 01, 10 and 11. This is shown in the first two columns of the truth table shown in Table 35-3. The third column $(\overline{A})$ is obtained by complementing the first column. The fourth column $(\overline{A}B)$ is obtained by ANDing the second and the third column. The fifth column is obtained by adding the first and fourth column. The sixth column is obtained by adding the first two columns.

An inspection of the fifth and sixth columns indicate that these are the same. This verifies that $A + \overline{A}B = A + B$.

Table 35-3

A	B	$\overline{A}$	$\overline{A}B$	$A + \overline{A}B$	$A + A$
0	0	1	0	0	0
0	1	1	1	1	1
1	0	0	0	1	1
1	1	0	0	1	1

Check these columns are equal

Example 35-4. *Using the truth table, show that the Boolean expression, $(A + B)(A + C) = A + BC$.*

Solution. Given the Boolean expression

$$(A + B)\,(A + C) = A + BC$$

From the given expression, notice that there are 3 variables, A, B and C. These three variables can have 8 possible combinations listed in Table 35-4 in the first three columns. The fourth column is obtained by ORing the first two. The fifth column is obtained by ORing the first and third column. The sixth column is obtained by

Table 35-4

A	B	C	$A + B$	$A + C$	$(A + B)(A + C)$	BC	$A + BC$
0	0	0	0	0	0	0	0
0	0	1	0	1	0	0	0
0	1	0	1	0	0	0	0
0	1	1	1	1	1	1	1
1	0	0	1	1	1	0	0
1	0	1	1	1	1	0	0
1	1	0	1	1	1	0	0
1	1	1	1	1	1	1	1

Check these two colums are equal

ANDing the fourth and fifth column. The seventh column is obtained by ANDing the second and third columns. Finally the eighth column is obtained by ORing the first and seventh column. Notice that the sixth and the eighth columns are the same. This verifies that $(A + B)(A + C) = A + BC$.

Example 35-5. *Using the Boolean laws and rules simplify the expression,*

$$y = A\bar{B}D + A\bar{B}\bar{D}$$

Solution. Given: The expression

$$y = A\bar{B}D + \overline{ABD}$$

To simplify this expression, factor out the common variables $A\bar{B}$ by applying distributive law (3 – 5),

$$\therefore \quad y = A\bar{B}\,(D + \bar{D})$$

Applying Rule 5 $(A + \bar{A} = 1)$ to the term $D + \bar{D}$ we get,

$$y = A\bar{B}\,(D + \bar{D}) = A\bar{B}\,.1$$

Applying Rule 2 (A . 1 = A) to the right term, we get,

$$y = A\bar{B} \text{ **Ans.**}$$

Example 35-6. *Using the Boolean laws and rules, simplify the expression,*

$$x = ACD + \bar{A}BCD$$

Solution. Given: The Boolean, expression,

$$x = ACD + \bar{A}BCD$$

We know that the given expression can be simplified by applying the Boolean laws and rules repeatedly. Thus, applying the distributive law (3 – 5) to the terms on the right, we get,

$$x = ACD + \bar{A}BCD$$

$$= (A + \bar{A}B)\,CD$$

Now applying Rule 11: $(A+\overline{A}B = A+B)$, we get,

$$x = (A+\overline{A}B)\ CD$$
$$= (A+B)\ CD \textbf{ Ans.}$$

Example 35-7. *Using the Boolean laws and rules, simplify the logic expression*

$$Z = (\overline{A}+B)\ (A+B).$$

Solution. Given: The logic expression

$$Z = (\overline{A}+B)\ (A+B) \qquad ...(i)$$

To simplify this expression, let us apply the distributive law to the two terms in parenthesis,

$$Z = \overline{A}A + \overline{A}B + AB + BB \qquad ...(ii)$$

Applying Rule 8: $(\overline{A}A = 0)$ to the first term and Rule 7: $(BB = B)$ to the last term, equation (*ii*) can be written as,

$$Z = \overline{A}B + AB + B$$

Factoring out the variable B from all the three terms of the above equation, we get,

$$Z = (\overline{A}+A+1)\ B$$

Applying Rule 4 and 6 $(\overline{A}+A+1=1)$ to the term in parenthesis, we get,

$$Z = 1\ .\ B$$

Applying Rule 2: $(1\ .\ B = B)$ to the right side, we get,

$$Z = B \textbf{ Ans.}$$

Example 35-8. *Using Boolean algebra techniques, simplify the expression*

$$f = AB + A\ (B+C) + B\ (B+C).$$ *(UP Tech. Univ., Sem. II, 2006-07)*

Solution. Given: The expression,

$$f = AB + A\ (B+C) + B\ (B+C)$$

Applying the distributive law to the second and third terms in the expression on the right:

$$f = AB + AB + AC + BB + BC$$

Applying Rule 7: $(BB = B)$ to the fourth term on the right,

$$f = AB + AB + AC + B + BC$$

Applying Rule 5: $(AB + AB = AB)$ to the first two terms,

$$f = AB + AC + B + BC$$

Applying Rule 10: $(B + BC = B)$ to the last two terms

$$f = AB + AC + B$$

Applying Rule 10: $(AB + B = B)$ to the first and third terms

$$f = B + AC \textbf{ Ans.}$$

Note: This approach is not necessarily the only approach to reach at the result.

Example 35-9. *Using laws and rules of Boolean algebra, simplify the Boolean expression,*

$$Z = [A\overline{B}\ (C+BD) + \overline{AB}]\ C.$$

Solution. Given: The Boolean expression,

$$Z = [A\overline{B}\ (C+BD) + \overline{AB}]\ C$$

Applying the distributive law to the terms on the right within brackets:

$$Z = (A\overline{B}C + A\overline{B}BD + \overline{AB})\ C$$

Applying Rule 8: $(\overline{B}B = 0)$ to the second term on the right within parenthesis:

$$Z = (A\overline{B}C + A.0.D + \overline{AB})\ C$$

Applying Rule 1: ($A \,.\, 0 \,.\, D = 0$) to the second term on the right within parenthesis:

$$Z = (A\overline{B}C + 0 + \overline{AB})\ C$$

Applying Rule 3: (drop the zero) within parenthesis,

$$Z = (A\overline{B}C + \overline{AB})\ C$$

Applying the distributive law,

$$Z = A\overline{B}CC + \overline{AB}C$$

Applying Rule 7: ($CC = C$) to the first term on the right,

$$Z = A\overline{B}C + \overline{AB}C$$

Factoring out $\overline{B}C$ from the first and second terms on the right,

$$Z = (A + \overline{A})\ \overline{B}C$$

Applying Rule 6: ($A + \overline{A} = 1$),

$$Z = 1.\overline{B}C$$

Applying Rule 2 (drop the 1),

$$Z = \overline{B}C \textbf{ Ans.}$$

Example 35-10. *Using Boolean laws and rules simplify the expression,*

$$Z = \overline{A}BC + A\overline{BC} + \overline{ABC} + A\overline{B}C + ABC.$$

Solution. Given the Boolean expression,

$$Z = \overline{A}BC + A\overline{BC} + \overline{ABC} + A\overline{B}C + ABC$$

Factoring BC out of the first and last terms on the right,

$$Z = (\overline{A} + A)\ BC + A\overline{BC} + \overline{ABC} + A\overline{B}C$$

Applying Rule 6: ($\overline{A} + A = 1$) to the firm term within parenthesis and factoring $A\overline{B}$ out from the second and last terms on the right,

$$Z = 1.BC + A\overline{B}\ (\overline{C} + C) + \overline{ABC}$$

Applying Rule 2: (drop the 1) to the first term and Rule 6 ($\overline{C} + C = 1$) to the term within parenthesis,

$$Z = BC + A\overline{B}.1 + \overline{ABC}$$

Applying Rule 2: (drop the 1) to the second term on the right,

$$Z = BC + A\overline{B} + \overline{ABC}$$

Factoring $\overline{B}$ out from the second and third terms on the right,

$$Z = BC + (A + \overline{AC})\ \overline{B}$$

Applying Rule 11: ($A + \overline{AC} = A + \overline{C}$) to the term in parenthesis,

$$Z = BC + (A + \overline{C})\ \overline{B}$$

Using the distributive and commutative laws, we get,

$$Z = BC + A\overline{B} + \overline{BC} \textbf{ Ans.}$$

Example 35-11. *Simplify the following logic expression using Boolean algebra:*

(*a*) $f = AB + A\ (B + C) + B\ (B + C)$

(*b*) $f + A\overline{B}\overline{C}D + \overline{A}\overline{B}D + BC\overline{D} + \overline{A}B + B\overline{C}$. *(U.P. Tech. Univ., Sem II 2006-07)*

Solution. (*a*) Given, $f = AB + A\ (B + C) + B\ (B + C)$

$$= AB + AB + AC + B.B + BC$$

$$= AB + AC + B + BC \qquad (\because AB + AB = AB \text{ and } B.B = 1)$$

$$= AB + AC + B\,(1 + C)$$

$$= AB + AC + B \qquad (\because 1 + C = 1)$$

$$= B + AB + AC$$

$$= B\,(1 + A) + AC$$

$$= B + AC \text{ **Ans.**} \qquad (\because 1 + A = 1)$$

(*b*) Given,

$$f = A\bar{B}\bar{C}D + \bar{A}\bar{B}D + BC\bar{D} + \bar{A}B + B\bar{C}$$

$$= \bar{B}D\,(A\bar{C} + \bar{A}) + B\,(\bar{A} + \bar{C}) + BC\bar{D}$$

$$= \bar{B}D\,(\bar{A} + \bar{C}) + B\,(\bar{A} + \bar{C}) + BC\bar{D}$$

$$= (\bar{A} + \bar{C})\,(\bar{B}D + B) + BC\bar{D}$$

$$= (\bar{A} + \bar{C})\,(B + D) + BC\bar{D}$$

$$= \bar{A}B + \bar{A}D + B\bar{C} + \bar{C}D + BC\bar{D} \text{ **Ans.**}$$

35-5. De-Morgan's Theorems

De-Morgan, a mathematician proposed the following two theorems that are an important part of Boolean algebra:

Theorem 1. ***The complement of a product of variables is equal to the sum of the complements of the variables, for two variables, the theorem 1 is given by the expression,***

$$\overline{XY} = \bar{X} + \bar{Y} \qquad ...(i)$$

Theorem 2. ***The complement of a sum of variables is equal to the product of the complements of the variables. For two variables, the theorem 2 is given by the expression,***

$$\overline{X + Y} = \overline{XY} \qquad ...(ii)$$

Let us now examine the two theorems in terms of logic circuits. First consider theorem 1, i.e.,

$$\overline{XY} = \bar{X} + \bar{Y}$$

The left hand side of the above equation can be viewed as the output of a NAND gate whose inputs are X and Y. On the other hand, the right hand side the equation is the result of first inverting X and Y and then putting them through an OR gate. These two representations are equivalent and are illustrated in Fig. 35-16 (*a*). This implies that an OR gate with INVERTERs on each of its inputs is equivalent to a NAND gate. When the OR gate with inverted inputs is used to represent the NAND function, it is usually drawn as shown in Fig. 35-16 (b), where the small circles (or bubbles) represent inversion.

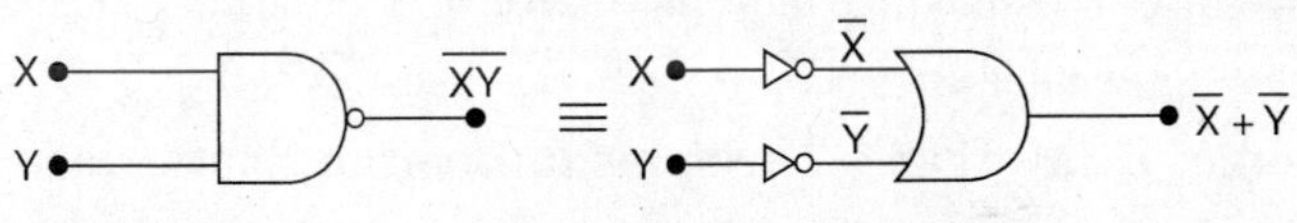

(a)

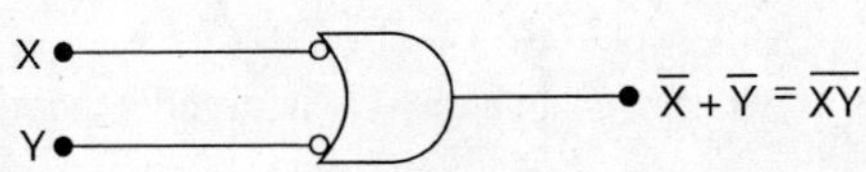

(b)

Fig. 35-16.

Let us now consider the theorem 2; *i.e.*,

$$\overline{X+Y} = \overline{X}\,.\,\overline{Y}$$

The left hand side of the above equation can be viewed as the output of a NOR gate whose inputs are *X* and *Y*. On the other hand, the right hand side of the equation is the result of first inverting both *X* and *Y* and then putting them through an AND gate. These two representations are equivalent and are illustrated in Fig. 35-17 (a), this implies that an AND gate with INVERTERs on each of its inputs is equivalent to a NOR gate. When the AND gate with inverted inputs is used to represent the NOR function, it is usually drawn as shown in Fig. 35-17 (b), where the small circles (or bubbles) on the inputs represent the inversion operation.

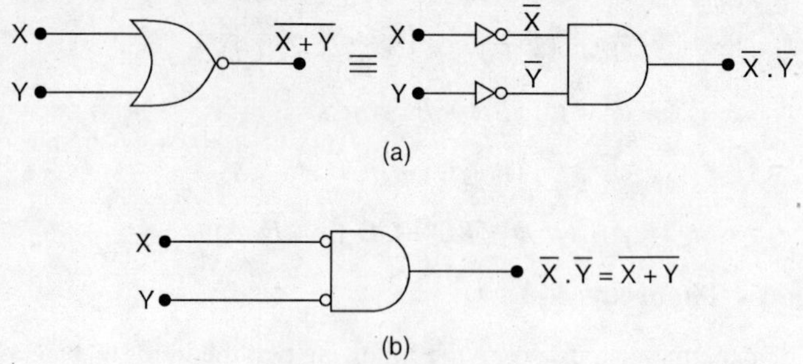

Fig. 35-17.

It may be carefully noted that each variable in De Morgan's theorem as stated above in Equations (*ii*) and (*iii*) can also represent a combination of other variables. For example, *X* can be equal to the term *AB* + *C*, and *Y* can be equal to the term *A* + *BC*. So if you apply De Morgan's theorem for two variables as stated by :

to the expression, $\overline{XY} = \overline{X} + \overline{Y}$

$$\overline{(AB+C)\,(A+BC)}$$

Then we will get the result,

$$\overline{(AB+C)\,(A+BC)} = \overline{(AB+C)} + \overline{(A+BC)}$$

Notice that the right hand side of the above expression have two terms: $\overline{AB+C}$ and $\overline{(A+BC)}$. Applying De-Morgan's theorem $\overline{X+Y} = \overline{XY}$ individually to each term again, we get,

$$\overline{(AB+C)} + \overline{(A+BC)} = (\overline{AB})\,\overline{C} + \overline{A}\,(\overline{BC})$$

Notice that we still have two terms on the right hand side of the above expression. This means De-Morgan's theorem can again be applied. These terms are $\overline{AB}$ and $\overline{BC}$. A final application of De-Morgan's theorem gives,

$$(\overline{AB})\,\overline{C} + \overline{A}\,(\overline{BC}) = (\overline{A}+\overline{B})\,\overline{C} + \overline{A}\,(\overline{B}+\overline{C})$$

Although this result can be simplified further by the use of Boolean rules and laws yet De Morgan's theorems cannot be used any more.

35-6. Application of De-Morgan's Theorems to Boolean Expressions

In order to study the procedure for the application of De-Morgan's theorems and Boolean algebra, let us consider a specific Boolean expression :

$$\overline{(A+\overline{BC})+D\,(\overline{E+\overline{\overline{F}}})}$$

***Step* 1.** Identify the terms to which we can apply De Morgan's theorems and think of each term as a single variable, let $\overline{A+B\overline{\overline{C}}} = X$ and $\overline{D\,(E+\overline{\overline{F}})} = Y$.

***Step* 2.** Since, $\overline{X+Y} = \overline{XY}$, therefore,

$$\overline{\overline{(A+BC)}+[D\,\overline{(E+\overline{\overline{F}})}]} = [\overline{\overline{A+B\overline{\overline{C}}}}]\;\overline{D\,\overline{(E+\overline{\overline{F}})}}$$

***Step* 3.** Use Rule 9: $\overline{\overline{A}} = A$ to cancel the double bars over the left term of the right-hand side of the above expression (Note that this is not a part of De-Morgan's theorem):

$$[\overline{\overline{A+B\overline{C}}}]\,[\overline{D\,\overline{(E+\overline{\overline{F}})}}] = [A+B\overline{C}]\,[\overline{D\,\overline{(E+\overline{\overline{F}})}}]$$

***Step* 4.** In the right term of the right-hand side of the above expression, let $Z = \overline{E+\overline{\overline{F}}}$. Then,

$$[A+B\overline{C}]\,[\overline{D\,\overline{(E+\overline{\overline{F}})}}] = [A+B\overline{C}]\,[\overline{DZ}]$$

***Step* 5.** Since, $\overline{DZ} = \overline{DZ}$, therefore,

$$[A+B\overline{C}]\,[\overline{D\,\overline{(E+\overline{\overline{F}})}}] = [A+B\overline{C}]\,[\overline{D}+\overline{\overline{E+\overline{F}}}]$$

***Step* 6.** Use Rule 9: $\overline{\overline{A}}+A$ to cancel the double bar over the $E+\overline{F}$ part of the term:

$$(A+B\overline{C})\,(\overline{D}+\overline{\overline{E+\overline{F}}}) = (A+B\overline{C})\,(\overline{D}+E+\overline{F})$$

Example 35-12. *Apply De-Morgan's theorems to the following Boolean expressions with 3 and 4 variables:*

(*a*) $\overline{XYZ}$ (*b*) $\overline{X+Y+Z}$

(*c*) $\overline{WXYZ}$ (*d*) $\overline{W+X+Y+Z}$

Solution. (*a*) Given: The Boolean expression $\overline{XYZ}$.

We know that the given expression by applying De-Morgan's theorem 1 can be written as

$$\overline{XYZ} = \overline{X}+\overline{Y}+\overline{Z} \textbf{ Ans.}$$

(*b*) Given, the Boolean expression $\overline{X+Y+Z}$

We know that the given expression by applying De-Morgan's theorem 1 can be written as,

$$\overline{X+Y+Z} = \overline{XYZ} \textbf{ Ans.}$$

(*c*) Given the Boolean expression $\overline{WXYZ}$

We know that the given expression, by applying De-Morgan's theorem 1 can be written as,

$$\overline{WXYZ} = \overline{W}+\overline{X}+\overline{Y}+\overline{Z} \textbf{ Ans.}$$

(*d*) Given the Boolean expression, $W+X+Y+Z$

We know that the given expression, by applying De-Morgan's theorem 2 can be written as,

$$\overline{W+X+Y+Z} = \overline{WXYZ} \textbf{ Ans.}$$

Example 35-13. *Using Boolean algebra simplify the following Boolean expression.*

$$Z = \overline{AB+AC}+\overline{AB}C$$

Solution. Given: The Boolean expression,

$$Z = \overline{AB+AC}+\overline{AB}C$$

Applying De-Morgan's theorem 2 to the first term on the right

$$Z = \overline{(AB)}.\overline{(AC)}+\overline{AB}C$$

Applying De-Morgan's theorem 1 to each term on the right in the parenthesis,

$$Z = (\overline{A}+\overline{B}).(\overline{A}+\overline{C})+\overline{AB}C$$

Applying the distributive law to the two terms on the right in the parenthesis

$$Z = \overline{AA}+\overline{AC}+\overline{AB}+\overline{BC}+\overline{AB}C$$

Applying Rule 7 ($\overline{AA} = \overline{A}$) to the first term, and applying Rule 10,

$$[\bar{A} + \overline{ABC} = \bar{A}B\,(1+C) = \overline{AB}]$$

to the third and last term on the right,

$$Z = \bar{A} + \overline{AC} + \overline{AB} + \overline{BC}$$

Applying Rule 10 $[\bar{A} + \overline{AC} = \bar{A}\,(1+\bar{C}) = \bar{A}]$ to the first and second terms on the right,

$$Z = \bar{A} + \overline{AB} + \overline{BC}$$

Applying Rule 10, $[\bar{A} + \overline{AB} = \bar{A}\,(1+\bar{B}) = \bar{A}]$ to the first and second terms on the right,

$$Z = \bar{A} + \overline{BC}$$

Example 35-14. *Simplify the Boolean expression,*

$$Z = \overline{(\bar{A} + C).(B + \bar{\bar{D}})}$$

to one having only single variables inverted.

Solution. Given: The expression,

$$Z = \overline{(\bar{A} + C).(B + \bar{D})}$$

Applying De-Morgan's theorem 1 to the two terms in parenthesis:

$$Z = \overline{(\bar{A} + C)} + \overline{(B + \bar{D})}$$

Applying De-Morgan's theorem 2 to the two terms in parenthesis:

$$\therefore \qquad Z\ (\bar{\bar{A}}.\bar{C}) + (\bar{B}.\bar{\bar{D}})$$

Applying Rule 9: $[\bar{\bar{A}}.\bar{C} = A\bar{C}$ and $\bar{B}\bar{\bar{D}} = \bar{B}D]$ to the first and second terms in parenthesis:

$$Z = (A\bar{C}) + (\bar{B}.D)$$

$$= A\bar{C} + \bar{B}D \ \textbf{Ans.}$$

Example 35-15. *Simplify the following expression using De-Morgan's theorems and Boolean algebra laws and rules:*

(a) $W = \overline{RST}\ \overline{(R + + S + T)}$

(b) $X = \overline{ABC} + \bar{A}BC + ABC + A\overline{BC} + A\bar{B}C$

(c) $Y = (B + \bar{C})\ (\bar{B} + C) + \overline{\bar{A} + B + \bar{C}}$

(d) $Z = \overline{(C + D)} + \bar{A}C\bar{D} + A\overline{BC} + \overline{AB}CD + AC\bar{D}.$

Solution. (*a*) Given: The expression

$$q = \overline{RST}\ \overline{(R + S + T)}$$

$$q = \overline{RST}\ \overline{(R + S + T)}$$

Applying De Morgan's Theorems and to the terms on the right, we get,

$$q = (\bar{R} + \bar{S} + \bar{T})\ (\bar{R}.\bar{S}.\bar{T})$$

Applying distributive law to the terms on the right, we get,

$$q = \overline{RRST} + \overline{SRST} + \overline{TRST}$$

Applying Rule 7: $(A\,.\,A = A)$ to all the three terms on the right, we get,

$$q = \overline{RST} + \overline{RST} + \overline{RST}$$

Applying Rule 5: $(A + A = A)$, we get,

$$q = \overline{RST}\ \textbf{Ans.}$$

(***b***) Given: The expression

$$x = \overline{ABC} + \bar{A}BC + ABC + A\overline{BC} + A\bar{B}C$$

Adding another term ABC from our own side, into the given expression, we get,

$$= \overline{ABC} + \bar{A}BC + ABC + A\overline{BC} + A\bar{B}C + ABC$$

$$(\because ABC + ABC = ABC)$$

Combining the 1st term with 4th term, 2nd with the 3rd term, and 5th term with the last term,

$$x = (\overline{A}\,\overline{B}\,\overline{C} + A\overline{B}\,\overline{C}) + (\overline{A}BC + ABC) + (A\overline{B}C + ABC)$$

$$= (\overline{A} + A)\ \overline{B}\,\overline{C} + (\overline{A} + A)\ BC + A\ (\overline{B} + B)\ C$$

$$= 1.\overline{BC} + 1.BC + A.1.C$$

$$= \overline{BC} + BC + AC \ \textbf{Ans.}$$

(*c*) Given: The expression

$$Z = (B + \overline{C})\ (\overline{B} + C) + \overline{\overline{A} + B + \overline{C}}$$

$$Z = (B + \overline{C})\ (\overline{B} + C) + \overline{\overline{A} + B + \overline{C}}$$

$$= (B\overline{B} + \overline{B}\,\overline{C} + BC + C\overline{C}) + \overline{\overline{A} + B + \overline{C}}$$

$$= 0 + \overline{B}\,\overline{C} + BC + 0 + \overline{\overline{A} + B + \overline{C}}$$

$$= \overline{B}\,\overline{C} + BC + \overline{\overline{A}}\,\overline{B}\,\overline{\overline{C}}$$

$$= \overline{B}\,\overline{C} + BC + A\overline{B}C$$

$$= \overline{B}\,\overline{C} + (B + A\overline{B})\ C$$

$$= \overline{BC} + (B + A)\ C.$$

Example 35-16. *Simplify each of the following expressions using De-Morgan's theorems :*

(*a*) $\overline{\overline{A}B\overline{C}}$ (*b*) $\overline{\overline{A} + \overline{B}C}$ (*c*) $\overline{AB\overline{CD}}$.

Solution. (*a*) Given: The expression $= \overline{\overline{A}B\overline{C}}$

Applying De-Morgan's theorem 1, we get,

$$\overline{\overline{A}B\overline{C}} = \overline{\overline{A}} + \overline{B} + \overline{\overline{C}}$$

Using Rule 9: $\overline{\overline{A}} = A$ to cancel the double bars over the first and third term of the right-side of the above expression,

$$\overline{\overline{A}B\overline{C}} = A + \overline{B} + C \ \textbf{Ans.}$$

(*b*) Given: The expression $= \overline{\overline{A} + \overline{B}C}$

Applying De-Morgan's Theorem 2, we get,

$$\overline{\overline{A} + \overline{B}C} = \overline{\overline{A}}.\overline{\overline{B}C}$$

Using Rule 9: $= \overline{\overline{A}} = A$ to cancel the double bars over A, we get,

$$\overline{\overline{A} + \overline{B}C} = A.\overline{\overline{B}C}$$

Applying De-Morgan's Theorem 1, we get,

$$\overline{\overline{A} + \overline{B}C} = A.(\overline{\overline{B}} + \overline{C})$$

Again using Rule 9: $\overline{\overline{A}} = A$ to cancel the double bars over B, we get,

$$\overline{\overline{A} + \overline{B}C} = A.(B + \overline{C}) \ \textbf{Ans.}$$

(*c*) Given: The expression $= \overline{AB\overline{CD}}$

Applying De-Morgan's Theorem 1, we get,

$$\overline{AB\overline{CD}} = \overline{A} + \overline{B} + \overline{\overline{CD}}$$

Again using Rule 9: $\overline{\overline{A}} = A$ to cancel the double bars on the third term of the right-hand side, we get,

$$\overline{AB\overline{CD}} = \overline{A} + \overline{B} + CD \ \textbf{Ans.}$$

Example 35-17. *Simplify each of the following expressions using De-Morgan's Theorems.*

(*a*) $\overline{A\,\overline{(B+\overline{C})}\,D}$ (*b*) $\overline{(M+\overline{N})\,(\overline{M}+N)}$ (*c*) $\overline{\overline{\overline{AB}C}D}$

Solution. (*a*) Given: The expression,

$$= \overline{A\,\overline{(B+\overline{C})}\,D}$$

Applying De-Morgan's theorem 1, we get,

$$\overline{A\,\overline{(B+\overline{C})}\,D} = \overline{A} + \overline{\overline{(B+\overline{C})}} + \overline{D}$$

Using Rule 9: $\overline{\overline{A}} = A$ to cancel the double bars over the second term on the right-hand side, we get,

$$\overline{A\,\overline{(B+\overline{C})}\,D} = \overline{A} + (B+\overline{C}) + \overline{D}$$

$$= \overline{A} + B + \overline{C} + \overline{D} \textbf{ Ans.}$$

(*b*) Given: The expression,

$$= \overline{(M+\overline{N})\,(\overline{M}+N)}$$

Applying De-Morgan's Theorem 1, we get

$$\overline{(M+\overline{N})\,(\overline{M}+N)} = \overline{(M+\overline{N})} + \overline{(\overline{M}+N)}$$

Applying De-Morgan's theorem 2 on the two terms on the right-hand side, we get,

$$\overline{(M+\overline{N})\,(\overline{M}+N)} = \overline{M}\,\overline{\overline{N}} + \overline{\overline{M}}\,\overline{N}$$

Using Rule 9: $\overline{\overline{A}} = A$ to cancel the double bars over M and N, we get

$$\overline{(M+\overline{N})\,(\overline{M}+N)} = \overline{M}N + M\overline{N} \textbf{ Ans.}$$

(*c*) Given: The expression,

$$\overline{\overline{\overline{AB}C}D} = \overline{\overline{\overline{AB}C}} + \overline{D}$$

Using Rule 9: $\overline{\overline{A}} = A$ to cancel the double bar over the first term on the right-hand side, we get,

$$\overline{\overline{\overline{AB}C}D} = (\overline{AB})\,C + \overline{D}$$

Applying again De-Morgan's Theorem 1 to the first term on the right-hand side, we get,

$$\overline{\overline{\overline{AB}C}D} = (\overline{A}+\overline{B})\,C + \overline{D}$$

$$= \overline{A}C + \overline{B}C + \overline{D} \textbf{ Ans.}$$

Example 35-18. *Apply De-Morgan's theorem to the each of the following Boolean expressions:*

(*a*) $\overline{\overline{(A+B)}+\overline{C}}$ (*b*) $\overline{(\overline{A}+B)+CD}$ (*c*) $\overline{(A+B)\,\overline{C}\,\overline{D}+(E+\overline{F})}$.

Solution. (*a*) Given the Boolean expression: $\overline{\overline{(A+B)}+\overline{C}}$

Let $\overline{A+B} = X$ and $\overline{C} = Y$. The expression $\overline{\overline{A+B}+\overline{C}}$ is of the form $\overline{X+Y} = \overline{X}\,\overline{Y}$ and can be rewritten as,

$$\overline{\overline{(A+B)}+\overline{C}} = \overline{\overline{(A+B)}}\;\overline{\overline{C}}$$

Using Rule 9 $(\overline{\overline{A}} = A)$ to cancel the double bars over the terms $(A+B)$ and C on the right, we get,

$$\overline{\overline{(A+B)}+\overline{C}} = (A+B)\,C \textbf{ Ans.}$$

(*b*) Given the Boolean expression: $\overline{(\overline{A}+B)+CD}$

Let $\overline{A}+B = X$ and $CD = Y$. The expression $\overline{(\overline{A}+B)+CD}$ is of the form $\overline{X+Y} = \overline{X}\,\overline{Y}$ and can be rewritten as,

$$\overline{(\overline{A}+B)+CD} = \overline{(\overline{A}+B)}\;(\overline{CD})$$

Let $\overline{A} = Z$, then the term $\overline{\overline{A}+B}$ on the right is of the form $\overline{X+Y} = \overline{X}\,\overline{Y}$. Similarly the term

$\overline{CD}$ on the right is of the form $\overline{XY} = \overline{X} + \overline{Y}$. Therefore, the expression $(\overline{\overline{A}+B})\;\overline{CD}$ can be re- written as,

$$(\overline{\overline{A}+B})\;\overline{CD} = (\overline{\overline{A}}\,\overline{B})\;(\overline{C}+\overline{D})$$

Using Rule 9: $(\overline{\overline{A}} = A)$ to cancel the double bars on the right term $\overline{\overline{A}}\,\overline{B}$, we get,

$$(\overline{\overline{A}+B})\;\overline{CD} = A\overline{B}\;(\overline{C}+\overline{D})$$

$$\therefore \quad \overline{\overline{A}+B+CD} = A\overline{B}\;(\overline{C}+\overline{D}) \textbf{ Ans.}$$

(*c*) Given: The expression: $\overline{(A+B)\,\overline{C}\,\overline{D}+E+\overline{F}}$

Let $(A+B)\,\overline{C}\,\overline{D} = X$ and $E+\overline{F} = Y$. The expression $\overline{(A+B)\,\overline{C}\,\overline{D}+(E+\overline{F})}$ is of the form $\overline{X+Y} = \overline{X}\,\overline{Y}$ and can be rewritten as,

$$\overline{(A+B)\,\overline{C}\,\overline{D}+(E+\overline{F})} = [\overline{(A+B)\,\overline{C}\,\overline{D}}]\,[\overline{E+\overline{F}}]$$

Next, apply De-Morgan's theorem 1 to the term $\overline{(A+B)\,\overline{C}\,\overline{D}}$ and $\overline{E+\overline{F}}$. Let $A+B = P$, $\overline{C} = Q$ and $\overline{D} = R$. Then the right term $\overline{(A+B)\,\overline{C}\,\overline{D}}$ is of the form $\overline{PQR} = \overline{P}+\overline{Q}+\overline{R}$. Similarly, let $E = M$ and $\overline{F} = N$. Then right the term $\overline{E+\overline{F}}$ is of the form $\overline{M+N} = \overline{M}\,\overline{N}$. Both the terms $\overline{(A+B)\,\overline{C}\,\overline{D}}$ and $\overline{E+\overline{F}}$ can be rewritten as,

$$[\overline{(A+B)\,\overline{C}\,\overline{D}}]\,[\overline{E+\overline{F}}] = [\overline{(A+B)}+\overline{\overline{C}}+\overline{\overline{D}}]\,[\overline{E}\,\overline{\overline{F}}]$$

Using Rule 9 $(\overline{\overline{A}} = A)$ to cancel the double bars on *C*, *D* and *F* in the right term, we get,

$$[\overline{(A+B)\,\overline{C}\,\overline{D}}]\,[\overline{E+\overline{F}}] = [\overline{(A+B)}+C+D]\,[\overline{E}F]$$

Next applying De Morgan's theorem again to the term $\overline{A+B}$ on the right we get,

$$[\overline{(A+B)}+C+D]\,[\overline{E}F] = (\overline{A}\,\overline{B}+C+D)\;\overline{E}F$$

$$\therefore \quad \overline{(A+B)\,\overline{C}\,\overline{D}+(E+\overline{F})} = (\overline{A}\,\overline{B}+C+D)\;\overline{E}F \textbf{ Ans.}$$

Example 35-19. *Apply De Morgan's theorem to the following expressions:*

(*a*) $\overline{(A+B+C)\;D}$ (*b*) $\overline{ABC+DEF}$ (*c*) $\overline{A\overline{B}+\overline{C}D+EF}$.

Solution. (*a*) Given: The expression $= \overline{(A+B+C)\;D}$

Let $A+B+C = X$ and $D = Y$

Then the given expression is of the form $\overline{XY} = \overline{X}+\overline{Y}$ and can be rewritten as,

$$\overline{(A+B+C)\;D} = \overline{A+B+C}+\overline{D}$$

Next applying De-Morgan's theorem 2 to the term $\overline{A+B+C}$,

$$\overline{(A+B+C)}+\overline{D} = \overline{A}\,\overline{B}\,\overline{C}+\overline{D}$$

$$\therefore \quad \overline{(A+B+C)\;D} = \overline{A}\,\overline{B}\,\overline{C}+\overline{D} \textbf{ Ans.}$$

(*b*) Given: The expression $\overline{ABC+DEF}$

Let $ABC = X$ and $DEF = Y$

Then the given expression $\overline{ABC+DEF}$ is of the form $\overline{X+Y} = \overline{X}\,\overline{Y}$ and, can be rewritten as,

$$\overline{ABC+DEF} = \overline{ABC}\,.\,\overline{DEF}$$

Applying De-Morgan's theorem 1 to each of the terms $\overline{ABC}$ and $\overline{DEF}$,

$$(\overline{ABC})\;(\overline{DEF}) = (\overline{A}+\overline{B}+\overline{C})\;(\overline{D}+\overline{E}+\overline{F})$$

$$\therefore \quad \overline{ABC+DEF} = (\overline{A}+\overline{B}+\overline{C})\;(\overline{D}+\overline{E}+\overline{F})$$

(*c*) Given: The expression, $\overline{A\overline{B}+\overline{C}D+EF}$

Let $A\overline{B} = X$, $\overline{C}D = Y$ and $EF = Z$.

Then the expression, $\overline{A\overline{B} + \overline{C}D + EF}$ is of the form $\overline{X + Y + Z} = \overline{X}\,\overline{Y}\,\overline{Z}$ and can be rewritten as

$$\overline{A\overline{B} + \overline{C}D + EF} = (\overline{A\overline{B}}).(\overline{\overline{C}D}).(\overline{EF})$$

Next applying De-Morgan's theorem 1 to each of the terms $\overline{A\overline{B}}, \overline{\overline{C}D}$ and $\overline{EF}$, we get,

$$(\overline{A\overline{B}})\ (\overline{\overline{C}D})\ (\overline{EF}) = (\overline{A} + B)\ (C + \overline{D})\ (\overline{E} + \overline{F})$$

$$\therefore \quad \overline{A\overline{B} + \overline{C}D + EF} = (\overline{A} + B)\ (C + \overline{D})\ (\overline{E} + \overline{F}) \textbf{ Ans.} \qquad (\because \overline{\overline{B}} = B \text{ and } \overline{\overline{C}} = C)$$

Example 35-20. *Simplify the Boolean expression,*

$$T(x, y, z) = (x + y)\ [\overline{\overline{x}\ (\overline{y} + \overline{z})}] + \overline{x}\,\overline{y} + \overline{x}\,\overline{z} \quad \textit{(Nagpur University, Mar/Apr. 1998)}$$

Solution. Given: The Boolean expression:

$$\begin{aligned}
T(x, y, z) &= (x + y)\ [\overline{\overline{x}\ (\overline{y} + \overline{z})}] + \overline{x}\,\overline{y} + \overline{x}\,\overline{z} \\
&= (x + y)\ [(\overline{\overline{x}} + \overline{(\overline{y} + \overline{z})}] + \overline{x}\,\overline{y} + \overline{x}\,\overline{z} \\
&= (x + y)\ (x + (\overline{\overline{y}}.\overline{\overline{z}})) + \overline{x}\,\overline{y} + \overline{x}\,\overline{z} \\
&= (x + y)\ (x + yz) + \overline{x}\,\overline{y} + \overline{x}\,\overline{z} \\
&= x.x + xyz + xy + yyz + \overline{x}\,\overline{y} + \overline{x}\,\overline{z} \\
&= x + xyz + xy + yz + \overline{x}\,\overline{y} + \overline{x}\,\overline{z} \\
&= x\ (1 + y + yz) + yz + \overline{x}\,\overline{y} + \overline{x}\,\overline{z} \\
&= x + yz + \overline{x}\,\overline{y} + \overline{x}\,\overline{z} \\
&= (x + \overline{x}\,\overline{y}) + yz + \overline{x}\,\overline{z} \\
&= x + \overline{y} + yz + \overline{x}\,\overline{z} \\
&= (x + \overline{x}\,\overline{z}) + (\overline{y} + yz) \\
&= (x + \overline{z}) + (\overline{y} + z) \\
&= x + \overline{y} + (\overline{z} + z) \\
&= x + \overline{y} + 1 \\
&= x + \overline{y}
\end{aligned}$$

$$\therefore \quad T(x, y, z) = x + \overline{y} \textbf{ Ans.}$$

35-7 Boolean Analysis of Logic Circuits

Boolean algebra provides a concise way to express the operation of a logic circuit formed in the form of a (Boolean expression). Using the Boolean expression, the output of a logic circuit can be readily determined for various combinations of intput values. To derive the Boolean expression of a given logic circuit, we begin at the left-most inputs and work towards the final output, writing the expression for each logic gate.

Once the Boolean expression for a given logic circuit has been determined, a truth table (*i.e.*, a table that shows the output of the logic circuit for all possible values of the input variables) is developed. The procedure requires that we evaluate the Boolean expression for all possible combinations of values for the input variables.

This method of analysis is convenient and useful for logic circuits with fewer number of logic gates.

Example 35-21. *Determine the Boolean expression for the logic circuit shown in Fig. 35-18.*

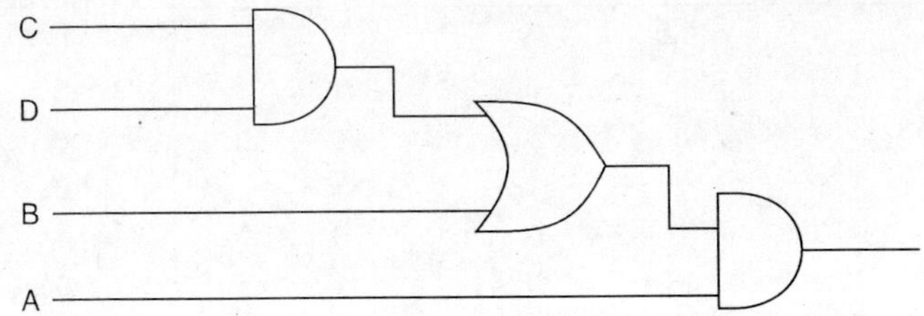

Fig. 35-18.

Analysis this circuit by evaluating the Boolean expression for all possible input combinations of values for the input variables.

Solution. Given: The logic circuit shown in Fig. 35-18.

Let us redraw the given logic circuit as shown in Fig. 35-19 and then determine the Boolean expression by beginning from the left most inputs and work towards the final output.

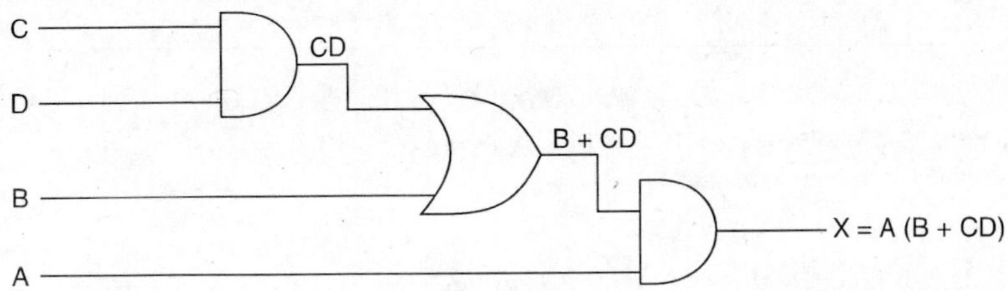

Fig. 35-19.

Thus, we find that the expression for the left-most AND gate with inputs C and D is CD. Next the output of the left-most AND gate is one of the inputs to the OR gate and B is the other input. Therefore the expression of the OR gate is $B + CD$. Finally, the output of the OR gate is one of the inputs to the right-most AND gate and A is the other input. Therefore, the expression for this AND gate is $A\ (B + CD)$. This is the final output expression for the entire circuit. Thus the output expression for the given circuit,

$$X = A\ (B + CD) \textbf{ Ans.}$$

Let us now evaluate the expression $A\ (B + CD)$. First, find the values of the variables that makes the expression equal to 1, using the laws and rules of Boolean algebra. In the present case, we find that the expression $A\ (B + CD) = 1$ only if $A = 1$ and $B + CD = 1$ because

$$A\ (B + CD) = 1.1 = 1$$

Inputs				Output
A	*B*	*C*	*D*	*A* (*B* + *CD*)
0	0	0	0	0
0	0	0	1	0
0	0	1	0	0
0	0	1	1	0
0	1	0	0	0
0	1	0	1	0
0	1	1	0	0
0	1	1	1	0
1	0	0	0	0
1	0	0	1	0
1	0	1	0	0
1	0	1	1	1
1	1	0	0	1
1	1	0	1	1
1	1	1	0	1
1	1	1	1	1

Fig. 35-20.

Now determine when the $B + CD$ term equals 1. The term $B + CD = 1$ if either $B = 1$ or $CD = 1$ or if both B and CD equal 1 because

$$B + CD = 1 + 0 = 1$$

$$B + CD = 0 + 1 = 1$$

$$B + CD = 1 + 1 = 1$$

Next the term $CD = 1$ only if $C = 1$ and $D = 1$.

To summarize, the expression $A\,(B + CD) = 1$ when $A = 1$ and $B = 1$ regardless of the values of C and D or when $A = 1$ and $C = 1$ and $D = 1$, regardless of the value of B. For all other possible combinations of the variables, the expression $A\,(B + CD) = 0$. The result is shown in the form of a truth table as shown in Fig. 35-20.

35-8. Simplification of Logic Circuits

We have already discussed in the last article that Boolean algebra is a powerful tool to analyse logic circuits, this is achieved by evaluating the Boolean expression for all possible combinations of values for the input variables.

Many times, in the application of Boolean algebra, we have to reduce a particular expression to its simplest form or change its form to a more convenient one to implement the expression most efficiently using logic gates. Broadly speaking there are two methods to achieve this:

(1) Simplification using Boolean algebra, and **(2)** Simplification using Karnaugh maps, both these methods are discussed one by one in the following pages.

35-9. Simplification of Logic Circuits using Boolean Algebra

We know that Boolean algebra can be used to simplify the Boolean expression for a given logic circuit. Such a simplification leads to the simplification of a logic circuit. This means that the logic circuit with a simplified Boolean expression contains fewer gates and therefore will be smaller and cheaper to implement than the original logic circuit.

Now we will illustrate the simplification method through examples (that follow in the subsequent pages).

Notice that the examples contain two essential steps :

(*i*) The original Boolean expression is put into sum-of-products form by repeated application of Boolean laws, rules and theorems.

(*ii*) Once the original Boolean expression is in this form, the product terms are checked for common factors. The factory is performed whenever possible and usually it leads to elimination of one or more terms from the original Boolean expression. It may be noted that this simplification method depends on a thorough knowledge of Boolean algebra and considerable practice in its application.

Example 35-22. *Determine the Boolean expression for the output of the logic circuit shown in Fig. 35-21.*

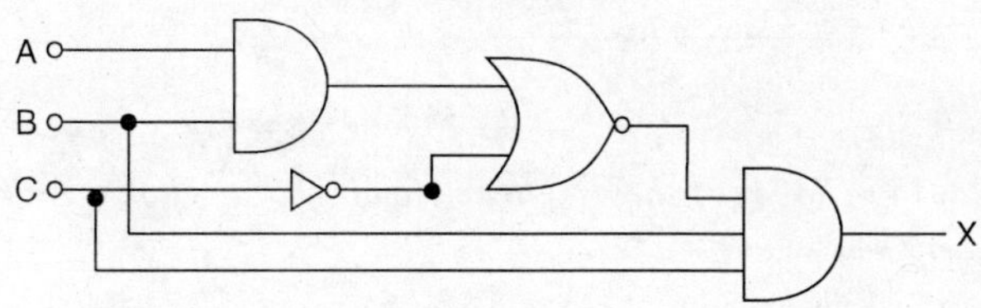

Fig. 35-21.

Solution. Given: The logic circuit shown in Fig. 35-21.

First of all, let us determine the Boolean expression for the output of the logic circuit, we know that the output of the left-most AND gate with inputs A and B is AB. Next the output of inverter is $\bar{C}$. Next the output of an OR gate with AB and $\bar{C}$ as its inputs is $\overline{(AB + \bar{C})}$. Finally, the output of right-most AND gate with $\overline{(AB + \bar{C})}$, B and C as its inputs is,

$$X = BC\ \overline{(AB + \bar{C})} \qquad ...(i)$$

Fig. 35-22 shows the resulting outputs at all the logic gate outputs.

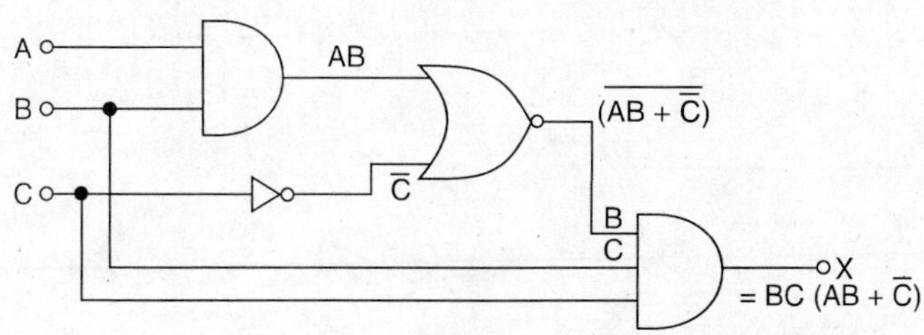

Fig. 35-22.

The Boolean expression in equation (*i*) can be simplified as follows by applying De-Morgan's theorems. Boolean laws and rules

$$X = BC\ \overline{(AB + \bar{C})}$$

$$= BC\ (\overline{AB}.\bar{\bar{C}})$$

$$= BC\ (\overline{AB}.C) \qquad (\because \bar{\bar{C}} = C)$$

$$= BC\ [(\bar{A} + \bar{B})]\ C$$

$$= BC\,(\bar{A}C + \bar{B}C)$$

$$= \bar{A}BC + BC\bar{B}C$$

$$= \bar{A}BC + 0 \qquad (\because B\bar{B} = 0)$$

$$= \bar{A}BC \qquad ...(ii)$$

The Boolean expression in equation (*ii*) can be implemented by an INVERTER and a three-input AND gate as shown in Fig. 35-23. Notice that the given logic circuit used four logic gates. But after simplification the logic circuit makes use of only two logic gates.

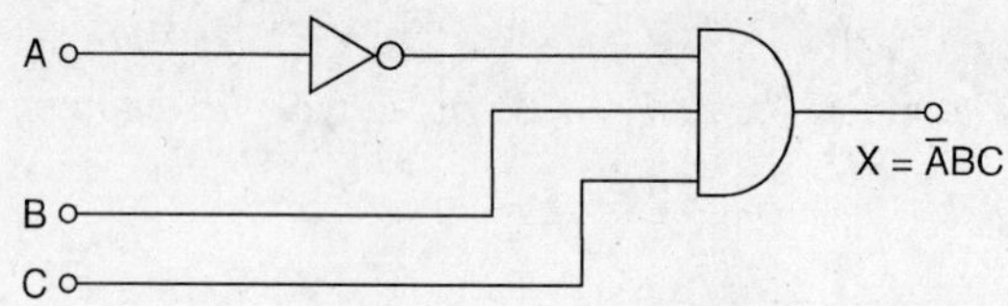

Fig. 35-23.

Example 35-23. *Fig. 35-24 (a) shows a logic circuit with the inputs A and B. The waveforms at the two inputs are also shown in Fig. 35-24 (b).*

Determine the minimized circuit and then sketch the waveform at the output.

Solution. Given: The logic circuit shown in Fig. 35-24.

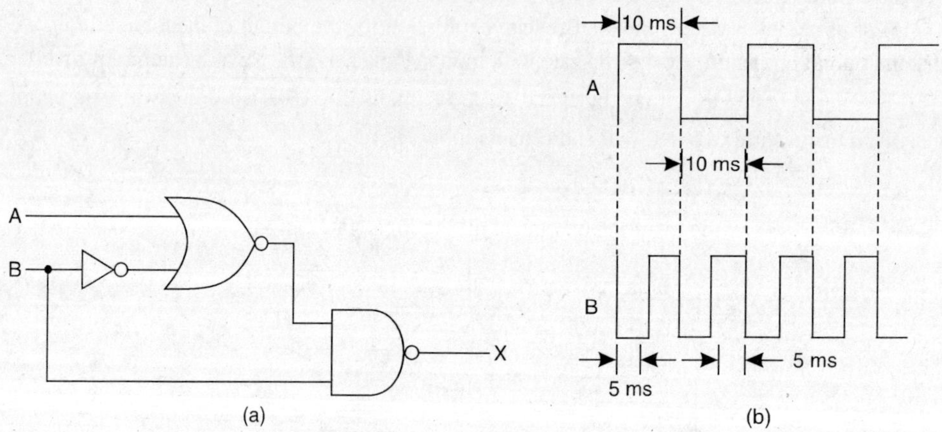

Fig. 35-24.

First of all, let us redraw the given logic circuit as shown in Fig. 35-25. Notice the input to the INVERTER gate is B and therefore its output is $\bar{B}$. Next input to NOR gate are A and $\bar{B}$. Therefore, the output of this logic gate is $\overline{A + \bar{B}}$. Finally the inputs to the NAND gate are $\overline{A + \bar{B}}$ and B. Therefore, the output of this logic gate is,

$$X = \overline{(\overline{A + \bar{B}})\,.\,B}$$

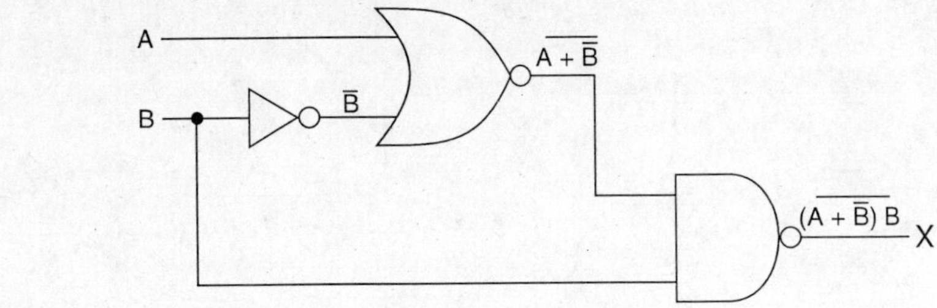

Fig. 35-25.

Applying the De-Morgan's theorem '1' to the term on the right, we get,

$$X = (\overline{\overline{A+\overline{B}}} + \overline{B})$$

Applying Rule 9: $(\overline{\overline{A}} = A)$ on the first term within parenthesis, we get,

$$X = A + \overline{B} + \overline{B}$$

Applying Rule 5: $(A + A = A)$ to the last two terms on the right, we get,

$$X = A + \overline{B}$$

Thus, the minimized circuit for the given logic circuit is as shown in Fig. 35-26.

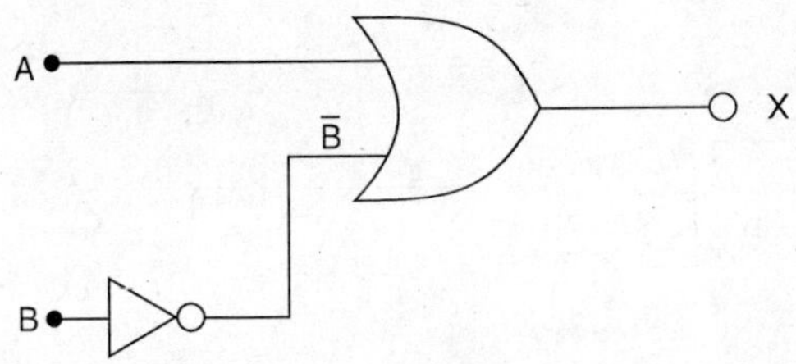

Fig. 35-26.

The resulting waveform at X is obtained by adding the waveforms at A and $\overline{B}$ and is as shown in Fig. 35-27.

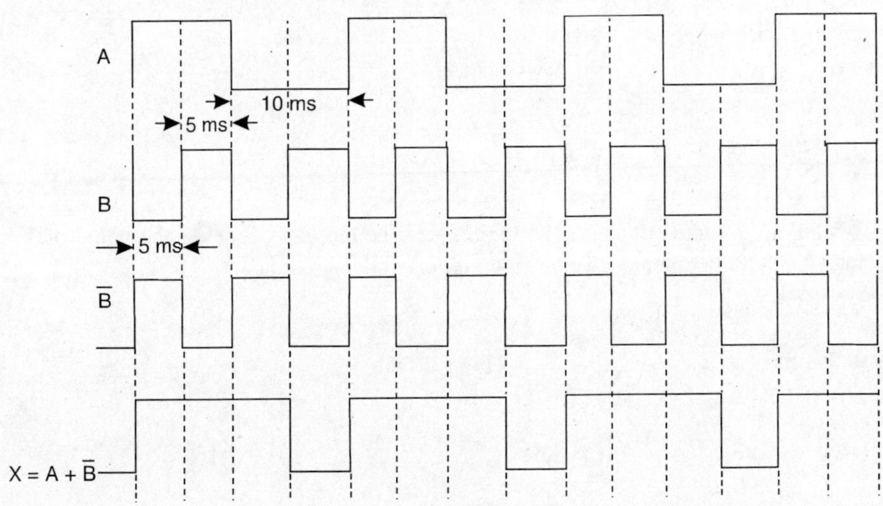

Fig. 35-27.

Example 35-24. *Using Boolean algebra minimize the logic circuit shown in Fig. 35-28.*

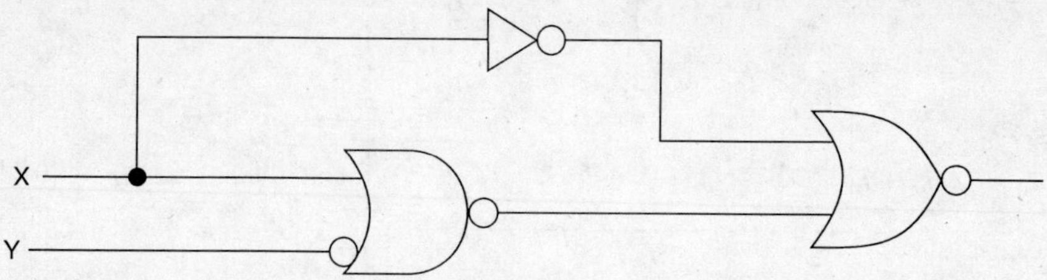

Fig. 35-28.

Solution. Given: The logic circuit shown in Fig. 35-28. First of all, let us redraw the given, circuit as shown in Fig. 35-29. Notice that all the logic gates have been numbered for convenience. The output of INVERTER gate (1) is $\overline{X}$ and of logic gate (2) is $\overline{X + \overline{Y}}$. Finally the output of logic gate (3) is,

$$f = \overline{(\overline{X} + \overline{X + \overline{Y}})}$$

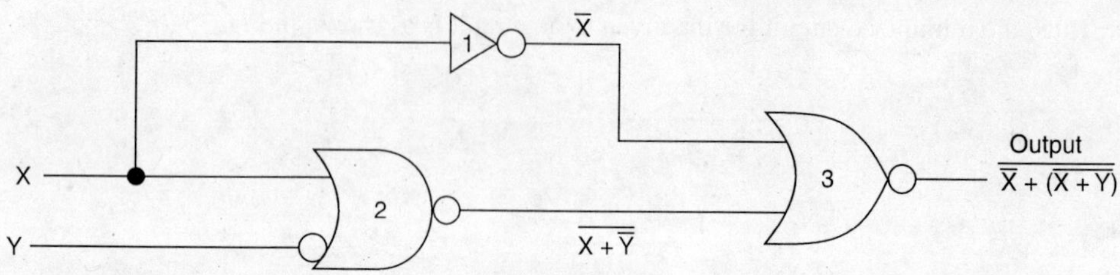

Fig. 35-29.

Applying De-Morgan's theorem '2' to the right side, we get,

$$f = \overline{\overline{X}} . \overline{\overline{(X + \overline{Y})}}$$

Applying Rule 9: $(\overline{\overline{A}} = A)$ to the terms on the right we get,

$$f = X\ (X + \overline{Y})$$

Applying Distributive law, we get

$$f = X . X + X\overline{Y}$$

$$= X + X\overline{Y}$$

$$= X\ (1 + \overline{Y})$$

$$= X \textbf{ Ans.}$$

∴ A minimized logic circuit has no logic gate. The output is equal to the value of *X* itself as shown in Fig. 35-30. In other words the given logic circuit is redundant.

Fig. 35-30.

Example 35-25. *Determine the Boolean expression for the output, X of a logic circuit shown in Fig. 35-31.*

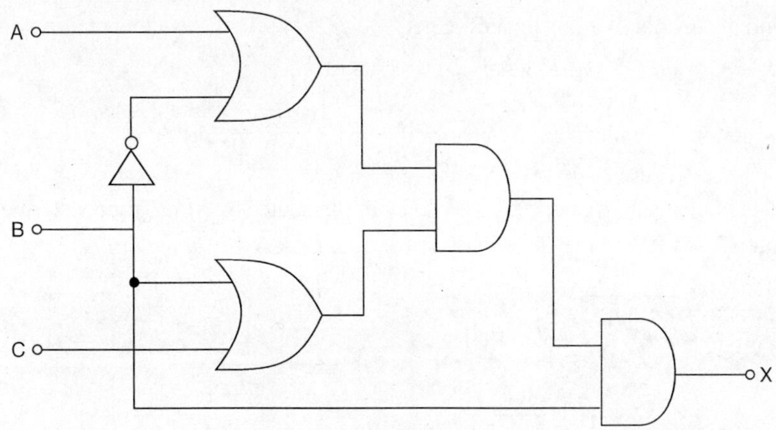

Fig. 35-31.

Minimize the Boolean expression for the output X using Boolean algebra.

Solution. Given: The logic circuit shown in Fig. 35-31.

First of all, let us determine the Boolean expression for the output of the given logic circuit. The circuit has been redrawn as shown in Fig. 35-32. Notice that each logic gate has been assigned a number for convenience.

We know that the input to INVERTER gate (1), is B and therefore, its output is $\overline{B}$. The inputs to the OR gate (2) are A and $\overline{B}$. Therefore its output is $A + \overline{B}$. The inputs to OR gate (3) are B and C. Therefore, its output is $B + C$. Next the inputs to AND gate (4) are $(A + \overline{B})$ and $(B + C)$. Therefore its output is $(A + \overline{B})\ (B + C)$. Finally, the inputs to AND gate (5) are $(A + \overline{B})\ (B + C)$ and B.

Therefore, its output,

$$X = (A + \overline{B})\ (B + C)\ B \qquad ...(i)$$

Let us now minimize this function by using laws and rules of Boolean algebra.

Applying distributing law of Boolean algebra to equation (*i*), we get,

$$X = (A + \overline{B})\ (BB + BC)$$

$$= (A + \overline{B})\ (B + BC) \qquad (\because BB = B)$$

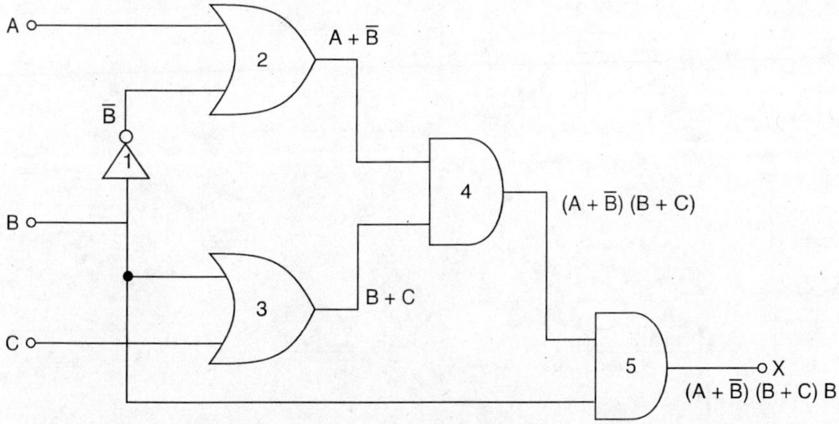

Fig. 35-32.

Applying Rule 10: $(A + AB = A)$ to the term $(B + BC)$, we get,

$$X = (A + \overline{B})\ B$$

Applying the distributive law again, we get,

$$X = AB + B\overline{B}$$

$$= AB \qquad (\because B\overline{B} = 0)$$

$\therefore$ The minimized Boolean expression for the given logic circuit,

$$X = AB \textbf{ Ans.}$$

Note. The minimized function, $X = AB$ can be implemented by using just one 2-input AND gate.

Example 35-26. *Determine the output, Y for the logic circuit shown in Fig. 35-33.*

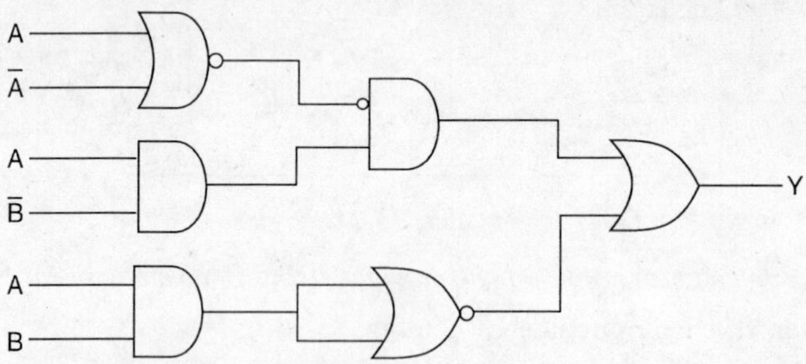

Fig. 35-33.

Simplify the Boolean expression using Boolean algebra laws, rules and theorems.

Solution. Given: The logic circuit shown in Fig. 35-33. First of all let us determine the Boolean expression for the output of the given logic circuit. The circuit has been redrawn as shown in Fig. 35-34 with each logic gate labelled with a number for convenience. We know that the output of NOR logic gate (1), is $\overline{A + \overline{A}}$.

Next, the output of AND gate (2) is $A\overline{B}$ and the output of AND gate 3 is AB. Next it is the logic gate (4) with one of its input (from logic gate 1) is inverted and the other input is $A\overline{B}$. The output of this logic gate is

$$\overline{\overline{(A + \overline{A})}.A\overline{B}}$$

Next the NOR gate (5) is acting as an INVERTER because two of its inputs are shorted together. Therefore the output of this logic gate is $\overline{AB}$.

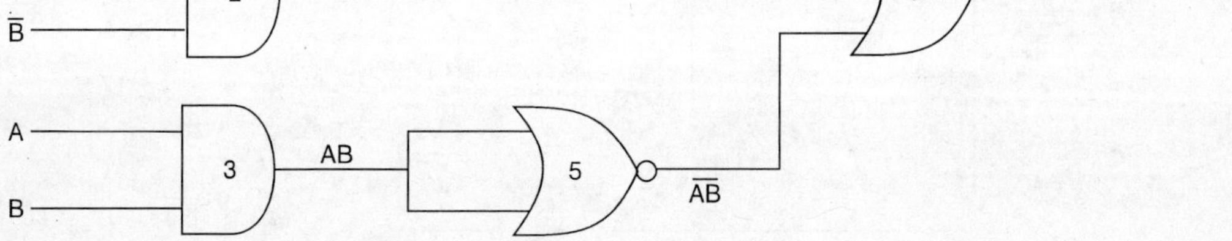

Fig. 35-34.

Finally, the output of OR gate (6) is

$$Y = [(A + \overline{\overline{A}}).(A\overline{B})] + \overline{AB}$$

Le us now simplify this Boolean expression using Boolean algebra.

Applying Rule 9: $(\overline{\overline{A}} = A)$ to the term $(A + \overline{\overline{A}})$, we get,

$$Y = [(A + \overline{A}).(A\overline{B})] + \overline{AB}$$

Applying Rule 6: $(A + \overline{A} = 1)$, to the first term on right within parenthesis, we get,

$$Y = [1\ (A\overline{B})] + \overline{AB}$$

Applying Rule 2: $(A \,.\, 1 = A)$ to the first term on the right, we get,

$$Y = A\overline{B} + \overline{AB}$$

Applying De-Morgan's theorem 2 to the second term on the right, we get,

$$Y = A\overline{B} + \overline{A} + \overline{B}$$

Factoring out the variable $\overline{B}$,

$$Y = (A + 1)\ \overline{B} + \overline{A}$$

Applying Rule 4: $(A + 1 = 1)$ to the first term on the right, we get,

$$Y = \overline{B} + \overline{A}$$

Example 35-27. *Determine the output for the logic circuit shown in Fig. 35-35.*

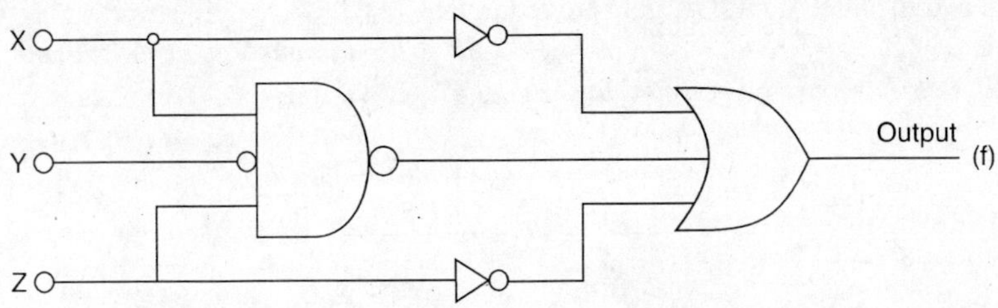

Fig. 35-35.

Minimize the output function using Boolean algebra and sketch the minimized logic circuit.

Solution. Given: The logic circuit shown in Fig. 35-35.

First of all, let us determine the Boolean expression for the given logic circuit. The circuit has been redrawn as shown in Fig. 35-36, with each logic gate labelled with a number for convenience. Notice the inputs to the logic gate (1), one of its inputs is X, the other is inverted Y (*i.e.*, $\overline{Y}$) and the third input is Z. The output of this logic gate is $\overline{X\overline{Y}Z}$, next the output of INVERTER (2) is $\overline{X}$ and that of INVERTER (3) is $\overline{Z}$. Finally the output of OR gate (4).

$$f = \overline{X} + \overline{X\overline{Y}Z} + \overline{Z}\ \text{i[}$$

Now we will minimize this logic function using Boolean algebra.

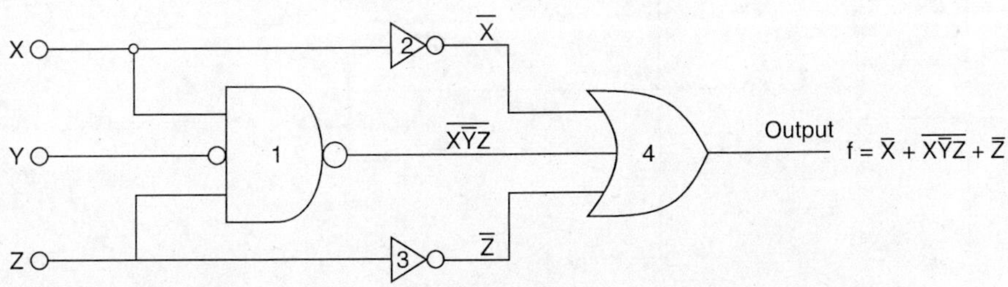

Fig. 35-36.

We know that the output of the logic circuit,

$$f = \overline{X} + \overline{X\overline{Y}Z} + \overline{Z}$$

Applying De-Morgan's theorem 1 to the second term on the right,

$$f = \overline{X} + (\overline{X} + \overline{\overline{Y}} + \overline{Z}) + \overline{Z}$$

Applying rule 9 : $(\overline{\overline{A}} = A)$ to the second term on the right within parenthesis, we get,

$$f = \overline{X} + (\overline{X} + Y + \overline{Z}) + \overline{Z}$$

Applying Rule 5: $(A + A = A)$ to all the terms on the right,

$$f = (\overline{X} + \overline{X}) + Y + (\overline{Z} + \overline{Z})$$
$$= \overline{X} + Y + \overline{Z}$$

Using De-Morgan's theorem 1 in the reverse order, this can be rewritten as

$$f = \overline{X\overline{Y}Z}$$

This function can be implemented by a three-input NAND gate as shown in Fig. 35-37.

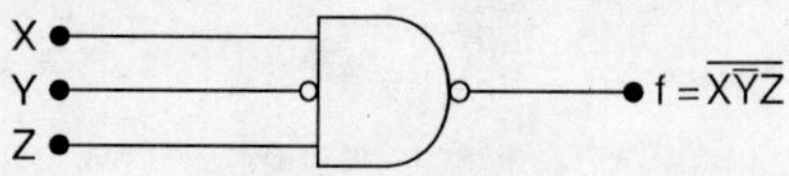

Fig. 35-37.

Note. If we compare the logic circuit of Fig. 35-36 with that of Fig. 35-37, we find that the two INVERTERs and the three input OR gate are redundant.

Example 35-28. *The circuit shown in Fig. 35-38 is used to implement the function* $Z = f(A, B) = A + B$. *What values should be selected for I and J ?*

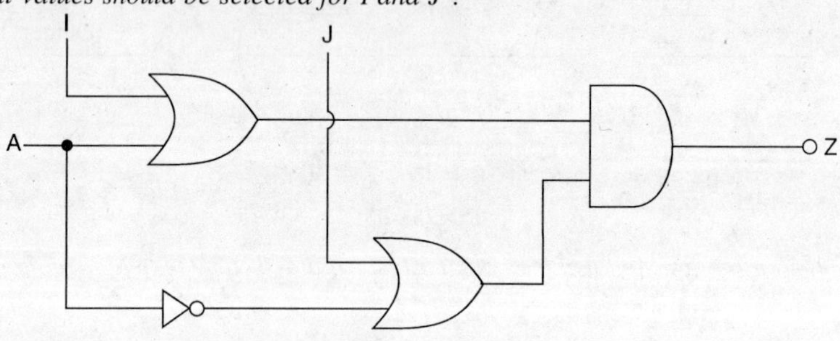

Fig. 35-38.

Solution. Given: The logic circuit of Fig. 35-38.

First of all let us, redraw the circuit as shown in Fig. 35-39. Notice that each logic gate has been assigned a number for convenience.

We know that output of gate 1 (OR gate),

$$= A + I$$

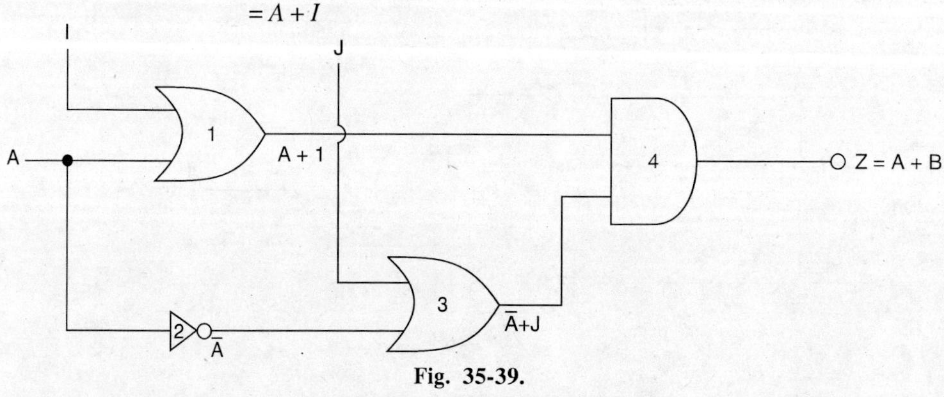

Fig. 35-39.

and output of gate 2 (NOT gate),

$$= \overline{A}$$

We also know that output of gate 3 (OR gate),

$$= \overline{A} + J$$

and output of gate 4 (AND gate)

$$Z = (\overline{A} + J).(A + I)$$

$$= \overline{A}A + JA + \overline{A}I + JI$$

$$= 0 + JA + \overline{A}I + JI \qquad (\because A\,\overline{A} = B)$$

Since, the output Z is given to be equal to $A + B$, therefore

$$A + B = JA + \overline{A}I + JI$$

Comparing the terms on left hand and right hand side in the above equation, we find that,

$$JA = A$$

and $$B = \overline{A}I + JI$$

From equation (i), we find that if $J = 1$ then $1 \,.\, A = A$. Therefore, $J = 1$

From equation (ii), we find that,

$$B = (\overline{A} + J)\ I = (\overline{A} + 1)\ I = 1.I = I$$

Thus, $J = 1$ and $I = B$ **Ans.**

Example 35-29. *Determine the logic function realized by the circuit shown in Fig. 35-40.*

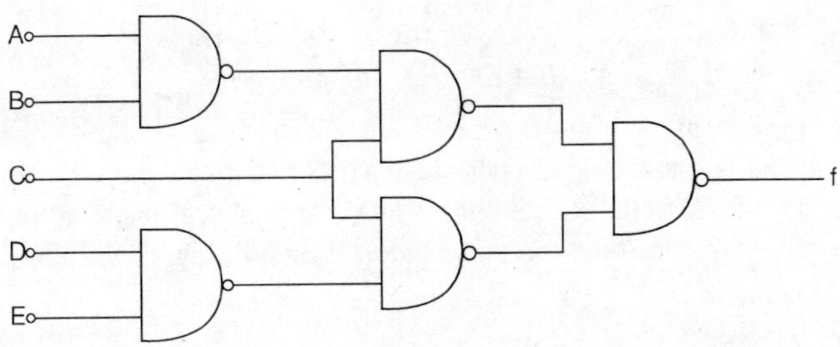

Fig. 35-40.

Minimize the output function, f using Boolean algebra.

Solution. Given: The logic circuit shown in Fig. 35-40.

First of all, let us redraw the given logic circuit as shown in Fig. 35-41. Notice that all the logic gates have been assigned numbers for the convenience.

We know that NAND gate (1) has its inputs A and B. The output of this logic gate is $\overline{AB}$. Similarly, the NAND gate 2 has D and E as its inputs. Therefore the output of this logic gate is $\overline{DE}$. Next, the NAND gate (3) has $(\overline{AB})$ and C as its inputs The output of this logic gate is $\overline{C\overline{AB}}$. Similarly, the NAND gate (4), has $\overline{DE}$ and C as its inputs. Therefore, the output of this logic gate is $\overline{C\overline{DE}}$. Finally, the NAND gate (5) has $\overline{(C\overline{AB})}$ and $\overline{(C\overline{DE})}$ as its inputs. The output of this logic gate is,

$$f = \overline{\overline{(C\overline{AB})} \,.\, \overline{(C\overline{DE})}}$$

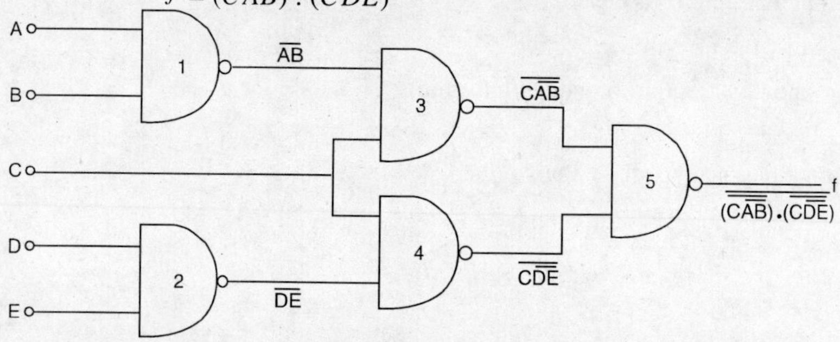

Fig. 35-41.

Let us minimize this function using Boolean algebra. Applying De-Morgan's theorem '1' to the terms within parenthesis, we get,

$$f = \overline{\overline{C\overline{AB}}} + \overline{\overline{C\overline{DE}}}$$

Applying Rule 9: $(\overline{\overline{A}} = A)$ to both the terms on the right, we get,

$$f = C\overline{AB} + C\overline{DE} \qquad ...(ii)$$

Applying De-Morgans theorem '1' again to the terms $\overline{AB}$ and we get,

$$f = C\,(\overline{A} + \overline{B}) + C\,(\overline{D} + \overline{E})$$

$$= C\,\{\overline{A} + \overline{B} + \overline{D} + \overline{E}\} \qquad ...(iii)$$

Note. If we implement equation (*iii*) using NAND gates we can do so by using only 3 logic gates, as shown below:

$$f = C\,(\overline{A} + \overline{B} + \overline{D} + \overline{E}) = C\,(\overline{ABDE})$$

The circuit to implement the logic function $f = C\,(\overline{ABDE})$ is as shown in Fig. 35-42. But the limitation of this design is that we need one four-input NAND gate and two 2-input NAND gates. This means we need two ICs-as the four input NAND gate is not available in the same IC as that of 2-input NAND gate. So we conclude, the logic circuit shown in Fig. 35-42 could be a better choice for hardware implementation.

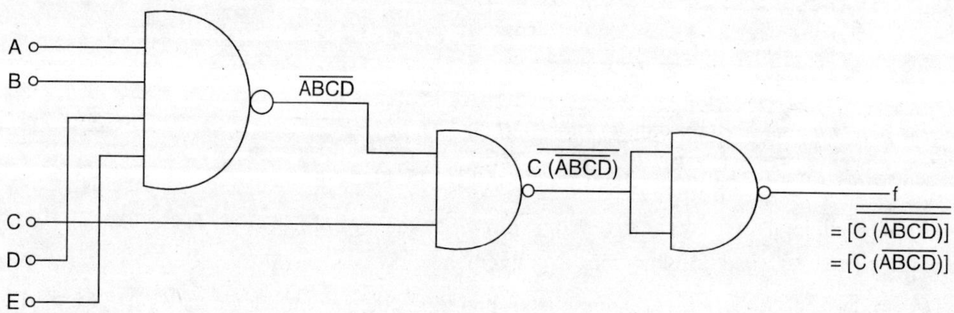

Fig. 35-42.

Example 35-30. *Fig. 35-43 shows a combinational logic circuit. Obtain a simplified Boolean expression for this circuits at V, W, X, Y and Z in terms of input variables A, B and C.*

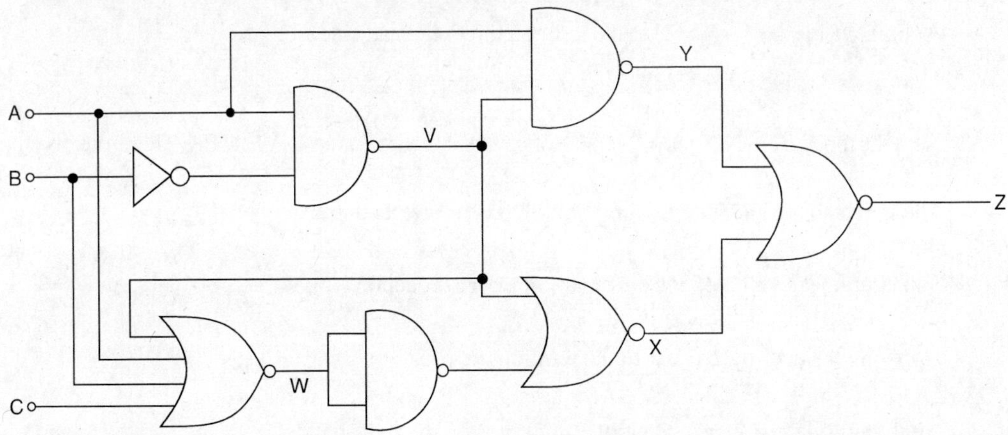

Fig. 35-43.

Solution. Given: The logic circuit shown in Fig. 35-43.

First of all redraw the given logic circuit as shown in Fig. 35-44. Notice that all the logic gates have been assigned numbers for convenience.

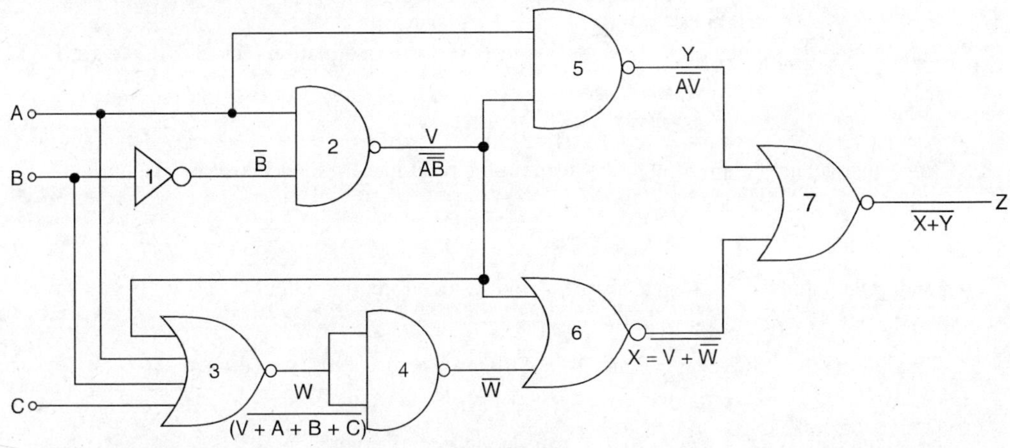

Fig. 35-44.

We know that the input to INVERTER gate (1) is B, therefore its output is $\overline{B}$. Next the inputs to NAND gate (2) are A and $\overline{B}$. Therefore its output.

$$V = \overline{A\overline{B}}$$

Applying De-Morgan's theorem '1' to the term on the right, we get,

$$V = \overline{A} + \overline{\overline{B}}$$

$$= \overline{A} + B \quad \textbf{Ans.} \qquad \ldots(i)$$

Next the inputs to 4-input NOR gate (3) are V, A, B and C. Therefore, its output,

$$W = \overline{V + A + B + C} \qquad \ldots(ii)$$

Substituting $V\ (= \overline{A} + B)$ from equation (*i*) into equation (*ii*), we get,

$$W = \overline{(\overline{A} + B) + A + B + C} = \overline{(\overline{A} + A) + (B + B) + C}$$

Applying Rule 5: $(A + A = A)$ and Rule 6: $(A + \overline{A} = 1)$ to be term on the right-hand side of the above equation, we get,

$$W = \overline{1 + B + C}$$

Applying Rule 4: $(A + 1 = 1)$ to the term on the right hand side, we get,

$$W = \overline{1}$$

$$= 0 \textbf{ Ans.} \qquad ...(iii)$$

Next the input to NAND gate (4) is $\overline{W}$. Notice that it is wired as an INVERTER. Therefore, its output is $\overline{W}$.

The inputs to NAND gate (5) are V and A. Therefore, its output,

$$Y = \overline{AV} \qquad ...(iv)$$

Substituting the value of V from equation (i) into equation (iv), we get,

$$Y = \overline{AV} = \overline{A\,(\overline{A} + B)}$$

Applying the distributive law to the term on the right, we get,

$$Y = \overline{A\overline{A} + AB}$$

Applying Rule 8: $(A.\overline{A} = 0)$ to the first term on the right-hand side of the above equation, we get,

$$y = \overline{O + AB}$$

$$= \overline{AB} \textbf{ Ans.} \qquad ...(v)$$

The inputs to the NOR gate (6) V are $\overline{W}$, therefore, the output of this logic gate,

$$X = \overline{V + \overline{W}}$$

Substituting the values of $V\ (= \overline{A} + B)$ and W (= 0) from equation (i) and (iii), we get,

$$X = \overline{\overline{A} + B + \overline{O}}$$

$$= \overline{\overline{A} + B + 1}$$

Applying Rule: 4 $(A + 1 = 1)$ to the term on the right-hand side of the above equation,

$$X = \overline{1}$$

$$= 0 \textbf{ Ans} \qquad ...(vi)$$

Finally the inputs to NOR gate (7) are Y and X, therefore, the output of this logic gate is

$$Z = \overline{X + Y} \qquad ...(vii)$$

Substituting the value of $X = 0$ and $Y = \overline{AB}$ into equation (vii), we get

$$Z = \overline{0 + \overline{AB}}$$

Applying Rule 3: $(A + 0 = A)$ to the term on the right, we get,

$$Z = \overline{\overline{AB}}$$

Applying Rule 9: $(\overline{\overline{A}} = A)$ to the term on the right, we get,

$$Z = AB \textbf{ Ans.}$$

35-10. Standard Forms of Boolean Expressions

Strictly speaking, all Boolean expressions, regardless of their form, can be converted into either of two standard forms: **sum-of-products (SOP) form** or the **product-of-sums (POS) form**. Both these standards forms of Booleans expression are discussed one by one in the following pages.

35-11. The Sum-of-Products (SOP) Form

In Boolean algebra, **a product term** (also called min-term) is the product of literals (or Boolean variables). In logic circuits, a product term is produced by AND operation with no OR operations. Some examples of product terms are:

$$AB,\ A\overline{B},\ AB\overline{C},\ A\overline{B}C\overline{D}$$

A product term is equal to 1 if only if each of the literals in the term is 1 A product term is equal to 0 when one or more of the literals are 0.

When two or more product terms are ORed (*i.e.*, summed by Boolean addition), the resulting expression is called Sum-of-Products (SOP). Some examples of sum-of-products are:

$$AB + BCD$$

$$ABC + CDE + \overline{B}C\overline{D}$$

$$A\overline{B} + \overline{A}B\overline{C} + AC$$

It may be carefully noted that sum-of-products expression can contain a single variable term However, in a sum-of-products expression, a single overbar cannot extend over more than one variable, although more than one variable in a term can have an overbar. For example, $\overline{A}B\overline{C}$ can be a term in sum-of-products expression but not $\overline{ABC}$.

35-12. Domain of a Boolean Expression

The **domain** of a Boolean expression is the set of variables contained in the expression in either complemented or uncomplemented form. For example, the domain of the expression $A\overline{B} + \overline{A}B\overline{C}$ is the set of variables, *A*, *B* and *C*. Similarly the domain of $AB + BCD + C\overline{D} + D$ is the set of variables. *A*, *B*, *C* and *D*.

35-13. Implementation of Sum-of-Products (SOP) Expression

In order to implement a sum-of-products expression, we require ORing the outputs of two of more AND gates. A product term is produced by an AND operation and the addition of two or more product terms is produced by an OR operation. Therefore, sum-of-products expression can be implemented by AND-OR logic.

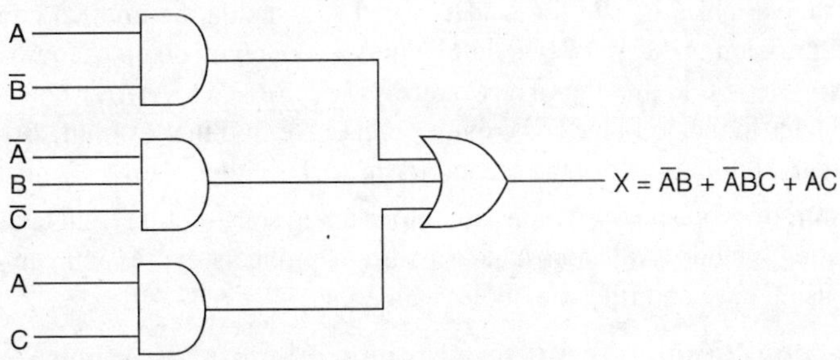

Fig. 35-45. *Implementation of sum-of-products expression* $A\overline{B} + \overline{A}BC + AC$ *using AND-OR logic.*

Fig. 35-45 shows the implementation of sum-of-products expression $A\overline{B} + \overline{A}BC + AC$. Notice that the number of AND gates is equal to the number of terms in the sum-of-products expression. The outputs of AND gates feed into the inputs of an OR gate. The output *X* of the OR gate equal the sum-of-products expression.

35-14. Conversion of a General Expression to Sum-of-Products (SOP) Form

It will be interesting to know that any logic expression can be changed into sum-of-products form by applying Boolean algebra techniques. For example, the expression $A\,(BC + D)$ can be converted to sum-of-products form by applying the distributive law:

$$A\,(BC + D) = ABC + AD.$$

Example 35-31. *Convert each of the following Boolean expressions to sum-of-products form:*

(*a*) $(A + B)\,(C + \overline{B})$ **(*b*)** $(A + \overline{B}C)\,C$ **(*c*)** $(A + C)\,(B + AC)$.

Solution. (*a*) Given, the Boolean expression $(A + B)\,(C + \overline{B})$

We know that the given Boolean expression can be converted to sum-of-products form by applying the distributive law,

$$(A + B)\,(C + \overline{B}) = AC + A\overline{B} + BC + B\overline{B}$$
$$= AC + A\overline{B} + BC + 0 \qquad (\because B\overline{B} = 0)$$
$$= AC + A\overline{B} + BC \textbf{ Ans.}$$

(*b*) Given, the Boolean expression $(A + \overline{B}C)\,C$

We know that the given Boolean expression can be converted to sum-of-products form by applying the distributive law,

$$(A + \overline{B}C)\,C = AC + \overline{B}CC$$
$$= AC + \overline{B}C \qquad (\because C.C = C)$$

(*c*) Given, the Boolean expression, $(A + C)\,(B + AC)$

We know that the given Boolean expression can be converted to sum-of-products form by applying the distributive law,

$$(A + C)\,(B + AC) = AB + AAC + BC + ACC$$
$$= AB + AC + BC + AC \qquad (A.A = A;\ C.C = C)$$
$$= AB + AC + BC \textbf{ Ans.} \qquad (AC + AC + AC)$$

35-15. The Standard Sum-of-Products (SOP) Form

A standard sum-of-products (SOP) expression is one in which all the variables in the domain appear in each product term of the expression. For example, $AB\overline{C}D + \overline{A}\,\overline{B}C\overline{D} + \overline{A}B\overline{C}D$ is a standard sum-of-products form because all the four variables *A*, *B*, *C* and *D* appear in each product term in the expression.

On the other hand, a sum-of-products expression in which some of the product terms do not contain all of the variables in the domain, is called non-standard sum-of-products expression. For example, the expression $\overline{A}B\overline{C} + A\overline{B}D + AB\overline{C}D$ has a domain made up of four variables *A*, *B*, *C* and *D*. From the expression find that the first two terms (*i.e.*, $\overline{A}B\overline{C}$ and $A\overline{B}D$) does not have a complete set of four variables in the domain. To be more precise, we find that from the first term $\overline{A}B\overline{C}$, either *D* or $\overline{D}$ is missing. Similarly from the second term $A\overline{B}D$, either *C* or $\overline{C}$ is missing.

Standard sum-of-product expressions are important in constructing truth tables and in Karnaugh-map simplification method. Any non-standard sum-of-products expression can be converted to the standard form using laws and rules of Boolean algebra.

35-16. Converting Product Terms to a Standard Sum-of-Products (SOP) Form

We have already discussed in the last article that any non-standard sum-of-products expression can be converted to the standard form using laws and rules of Boolean algebra. Following is the procedure for converting a non-standard sum-of-products expression to a standard form.

(*i*) Identify the non-standard product terms in the given expression. Multiply each such product term by a term made up of the sum of a missing variable and its complement. This results in two product terms. As we know that we can multiply anything by 1 without changing its value.

(*ii*) Repeat step 1 until all resulting product terms contain all variables in the domain in either complemented or uncomplemented form. In converting a product term to a standard form, the number of product terms is doubled for each missing variable.

Let us illustrate the procedure through the following example. Consider the conversion of Boolean expression $A\overline{B}C + \overline{A}B + AB\overline{C}D$ to a standard sum-of-products form.

We know that the domain of the given sum-of-product expression is *A*, *B*, *C* and *D*. From the expression. We find that the first two product terms *i.e.*, $A\overline{B}C$ and $\overline{A}B$ are not in the standard form, while the third term $AB\overline{C}D$ is already a standard product term.

For conversion of a non-standard product terms to a standard form, let us take one term at a time.

The first term $A\overline{B}C$ has a missing variable D or $\overline{D}$. So multiply the first term by $D+\overline{D}$ as follows:

$$A\overline{B}C = A\overline{B}C\,(D+\overline{D}) = A\overline{B}CD + A\overline{B}C\overline{D}$$

From the above expression we find that a single non-standard product term $A\overline{B}C$ has produced two standard product terms, $A\overline{B}CD$ and $A\overline{B}C\overline{D}$.

It may be noted that a variable added to its complement equals 1 (*i.e.*, $D+\overline{D}=1$) *i.e.*, Rule 6 of Boolean algebra (refer to Table 35-1). Therefore there is no violation in the way we have converted a non-standard product term to a standard form.

The second term $\overline{A}B$ has a missing variable C or $\overline{C}$ and D or $\overline{D}$. In order to convert this non-standard product term to a standard form, we will multiply $\overline{A}B$ first by $C+\overline{C}$, and then by $D+\overline{D}$. Thus first multiplying $\overline{A}B$ by $C+\overline{C}$, we get,

$$\overline{A}B = \overline{A}B\,(C+\overline{C}) = \overline{A}BC + \overline{A}B\overline{C}$$

Multiplying the two resulting terms by $D+\overline{D}$, we get,

$$\overline{A}B\overline{C} + \overline{A}B\overline{C}\,. = \overline{A}BC\,(D+\overline{D}) + \overline{A}B\overline{C}\,(D+\overline{D})$$

$$= \overline{A}BCD + \overline{A}BC\overline{D} + \overline{A}B\overline{C}D + \overline{A}B\overline{C}\,\overline{D}$$

or $$\overline{A}B = \overline{A}BCD + \overline{A}BC\overline{D} + \overline{A}B\overline{C}D + \overline{A}B\overline{C}\,\overline{D}$$

Thus we find that a non-standard term $\overline{A}B$ has produced four product terms in the standard form.

The standard sum-of-products (SOP) form of the given expression,

$$A\overline{B}C + \overline{A}B + AB\overline{C}D = A\overline{B}CD + A\overline{B}C\overline{D} + \overline{A}BCD + \overline{A}BC\overline{D} + \overline{A}B\overline{C}D + \overline{A}B\overline{C}D + AB\overline{C}D.$$

Example 35-32. *Convert the Boolean expression* $A\overline{B}C + \overline{B}C\overline{D} + A\overline{C}D$ *to a standard sum-of-products (SOP) form.*

Solution. Given, the Boolean expression $A\overline{B}C + \overline{B}C\overline{D} + A\overline{C}D$.

We know that the domain of the given expression is A, B, C and D, from the expression we find that all the three product terms are in the non-standard form. Let us convert these product terms to a standard form taking one term at a time.

The first term $A\overline{B}C$ has a missing variable D or $\overline{D}$. So multiply the first term by $D+\overline{D}$ as follows,

$$A\overline{B}C = A\overline{B}C\,(D+\overline{D}) = A\overline{B}CD + A\overline{B}C\overline{D}$$

The second term $\overline{B}C\overline{D}$ has a missing variable A or $\overline{A}$. So multiply the second term by $A+\overline{A}$ as follows,

$$\overline{B}C\overline{D} = (A+\overline{A})\,\overline{B}C\overline{D} = A\overline{B}C\overline{D} + \overline{A}\,\overline{B}C\overline{D}$$

The third term $A\overline{C}D$ has a missing variable B or $\overline{B}$. So multiply the third term by $B+\overline{B}$ as follows,

$$A\overline{C}D = A\overline{C}D\,(B+\overline{B}) = AB\overline{C}D + A\overline{B}\,\overline{C}D$$

Thus the standard SOP (i.e., sum-of-products) form of the given expression,

$$A\overline{B}C + \overline{B}C\overline{D} + A\overline{C}D = A\overline{B}CD + A\overline{B}C\overline{D} + A\overline{B}C\overline{D} + \overline{A}\,\overline{B}C\overline{D} + AB\overline{C}D + A\overline{B}\,\overline{C}D \text{ **Ans.**}$$

35-17. Binary Representation of a Standard Product Term

Strictly speaking, a standard product term is equal to 1 for only one combination of variable values. For example the product term $A\overline{B}C\overline{D}$ is equal to 1 when $A=1$; $B=0$, $C=1$ and $D=0$ as shown below. The product term $A\overline{B}C\overline{D}$ is 0 for all other combinations of values for the variables.

$$A\overline{B}C\overline{D} = 1.\overline{0}.1.\overline{0} = 1.1.1.1 = 1$$

It may be noted that the product term $A\overline{B}C\overline{D}$ has a binary value 1010 (i.e., decimal 10).

The student must remember that a product term is implemented with an AND gate whose output is 1 if and only if each of its inputs is 1. INVERTERs are used to produce the complements of the variables wherever required.

A sum-of-products (SOP) expression is equal to 1 if and only if one or more of the sum terms is equal to 1.

Example 35-33. *Determine the binary values for which the following sum-of-products (SOP) expression is equal to 1,*

$$A\overline{B}CD + AB\overline{C}D + \overline{A}\overline{B}\overline{C}\overline{D}$$

Solution. Given, the SOP expression,

$$A\overline{B}CD + AB\overline{C}D + \overline{A}\overline{B}\overline{C}\overline{D}$$

We know that the given SOP expression is equal to 1 when any or all of the three product terms are equal to 1.

The product term $A\overline{B}CD$ is equal to 1 when $A = 1$, $B = 0$, $C = 1$ and $D = 0$, *i.e.*,

$$A\overline{B}CD = 1.\overline{0}.1\overline{0} = 1.1.1.1 = 1$$

The term $AB\overline{C}D$ is equal to 1 when $A = 1$, $B = 1$, $C = 0$ and $D = 1$, *i.e.*,

$$AB\overline{C}D = 1.1.\overline{0}.1 = 1.1.1.1 = 1$$

The term $\overline{A}\overline{B}\overline{C}\overline{D}$ is equal to 1 when $A = 0$, $B = 0$, $C = 0$ and $D = 0$, i.e.,

$$\overline{A}\overline{B}\overline{C}\overline{D} = \overline{0}.\overline{0}.\overline{0}.\overline{0} = 1.1.1.1 = 1.$$

35-18. The Product-of-Sums (POS) Form

In Boolean algebra, a **sum term** (also called a max-term) is the sum of literals (or Boolean variables). In logic circuits, a sum term is produced by an OR operation with no AND operations involved. Some examples of such terms are:

$$A + B,\ A + \overline{B},\ \overline{A} + B + C \text{ and } \overline{A} + B + \overline{C} + D.$$

A sum term is equal to 1 when one or more of the literals in the term are 1. A sum term is equal to 0 if and only if each of the literals is 0.

When two or more sum terms are multiplied, the resulting expression is a **product-of-sums (POS)**. Some examples of product-of-sums form are:

$$(A + \overline{B})\ (\overline{A} + B + C)$$

$$(A + B)\ (A + \overline{B} + C)\ (A + \overline{C})$$

$$(\overline{A} + B + \overline{C})\ (C + \overline{D} + E)\ (B + C + \overline{D})$$

It may be carefully noted that product-of-sums (POS) expression can contain a single variable term. However in product-of-sums (POS) expression, a single overbar cannot extend over more than one variable, although more than one variable in a term can have an overbar. For example, a product-of-sums (POS) expression can have the term $\overline{A} + \overline{B} + \overline{C}$ but not $\overline{A + B + C}$.

35-19. Implementation of a Product-of-Sums (POS) Expression

In order to implement a product-of-sums (POS) expression, we require ANDing the outputs of two or more OR gates. A sum term is produced by an OR operation and the product of two or more sum terms is produced by an AND operation. Therefore, a product-of-sums expression can be implemented by AND-OR logic.

Fig. 35-46 shows the implementation of product-of-sums expression, $(A + \overline{B})\ (B + \overline{C} + D)\ (\overline{A} + C)$. Notice that the number of OR gates is equal to the number of terms in the product-of-sums expression. The output of OR gates feed into the inputs of an AND gate. The output X of the AND gate equals the product-of-sums expression.

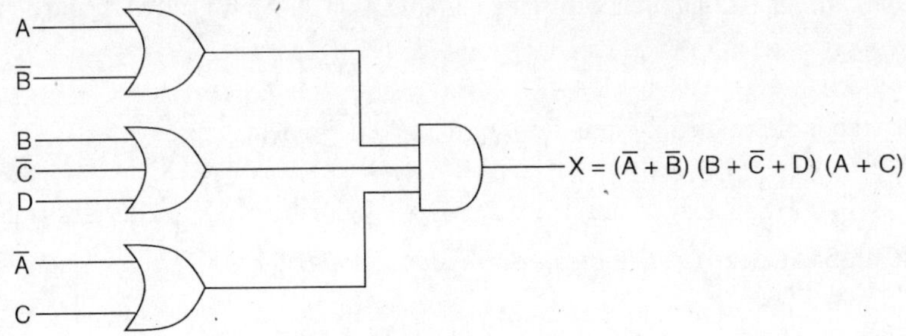

Fig. 35-46. *Implementation of a product-of-sums expression,* $(A + \overline{B})\ (B + \overline{C} + D)\ (\overline{A} + C)$ *using AND-OR logic.*

35-20. The Standard Product-of-Sums (POS) Form

A standard product-of-sums (POS) expression is one in which all the variables in the domain appear in each sum term of the expression. For example, the expression,

$$(A + \overline{B} + C + \overline{D})\ (\overline{A} + B + \overline{C} + D)\ (A + B + \overline{C} + D)$$

is a standard product-of-sums expression.

On the other hand, a product-of-sums expression, in which some of the sum terms do not contain all of the variables in the domain, is called non-standard product-of-sums expression, for example the expression,

$$(\overline{A} + B + C)\ (B + \overline{C} + \overline{D})\ (A + \overline{C} + D)$$

is a non-standard product-of-sums expression, because all the three sum terms do not have all the four variables.

Any non-standard product-of-sums expression can be converted to the standard form using laws and rules of Boolean algebra.

35-21. Converting a Sum Term to a Standard Product-of-Sums (POS) Form

We have already discussed in the last article that any non-standard product-of-sums expression can be converted to the standard form using laws and rules of Boolean algebra. Following is the procedure for converting a non-standard product-of-sums expression to a standard form.

(*i*) Identify the non-standard sum terms in the given expression. To each non-standard sum term, add a product term consisting of a missing variable and its complement. This results in two sum terms. As we know we can add 0 to anything without changing its value.

(*ii*) Apply rule 12 of the Boolean algebra from Table 35-1: $A + BC = (A + B)\ (A + C)$.

(*iii*) Repeat step 1 until all resulting sum terms contain all variables in the domain in either complemented or uncomplemented form.

Let us illustrate the procedure by the following example, consider the conversion of the Boolean expression,

$$(\overline{A} + B + \overline{C})\ (B + \overline{C} + D)\ (A + \overline{B} + \overline{C} + D)$$

into a standard product-of-sums (POS) form.

We know that the domain of the given product-of-sums expression is A, B, C and D. From the expression, we find that the first two sum terms, *i.e.*, $(\overline{A} + B + \overline{C})$ and $(B + \overline{C} + D)$ are not a standard POS form while the third term $(A + \overline{B} + \overline{C} + D)$ is already a standard term. For conversion of the non-standard sum terms to a standard form, let us take one term at a time.

The first term $\overline{A} + B + \overline{C}$ has a missing variable D or $\overline{D}$. So we add $D\overline{D}$ in this term and apply rule 12 of the Boolean algebra (From Table 35-1) as follows:

$$\overline{A} + B + \overline{C} = \overline{A} + B + \overline{C} + D\overline{D} = (\overline{A} + B + \overline{C} + D)\ (\overline{A} + B + \overline{C} + \overline{D})$$

The second term $B+\overline{C}+D$ has a missing variable A or $\overline{A}$. So we add $A\overline{A}$ and apply rule 12 as follows:

$$B+\overline{C}+D = A\overline{A}+B+\overline{C}+D = (A+B+\overline{C}+D)\ (\overline{A}+B+\overline{C}+D)$$

Thus, the standard product-of-sums form of the given Boolean expression,

$(\overline{A}+B+\overline{C})\ (B+\overline{C}+D)\ (A+\overline{B}+\overline{C}+D)$

$= (\overline{A}+B+\overline{C}+D)\ (\overline{A}+B+\overline{C}+\overline{D})\ (A+B+\overline{C}+D)\ (\overline{A}+B+\overline{C}+D)\ (A+\overline{B}+\overline{C}+D).$

Example 35-34. *Convert the Boolean expression* $(A+\overline{B})\ (B+C)$ *to a standard product-of-sums form.*

Solution. Given the Boolean expression, $(A+\overline{B})\ (B+C)$

We know that the domain of the given expression is A, B and C. From the expression, we find that both the sum terms i.e., $(A+\overline{B})$ and $(B+C)$ are in the non-standard form. Let us convert these sum terms to a standard form taking one term at a time.

The first term $(A+\overline{B})$ has a missing variable C or $\overline{C}$. So we add $C\overline{C}$ and apply rule 12 as follows.

$$A+\overline{B} = A+\overline{B}+C\overline{C} = (A+\overline{B}+C)\ (A+\overline{B}+\overline{C})$$

The second term $(B+C)$ has a missing variable A or $\overline{A}$. So we add $A\overline{A}$ and apply rule 12 as follows.

$$B+C = A\overline{A}+B+C = (A+B+C)\ (\overline{A}+B+C)$$

Thus, the standard product-of-sums form of the given Boolean expression,

$$(A+\overline{B})\ (B+C) = (A+\overline{B}+C)\ (A+\overline{B}+\overline{C})\ (A+B+C)\ (\overline{A}+B+C) \textbf{ Ans.}$$

35-22. Binary Representation of a Standard Sum Term

Strictly speaking, a standard sum term is equal to 0 for only one combination of variable values. For example, the sum term $A+\overline{B}+C+\overline{D}$ is 0 when $A=0$, $B=1$, $C=0$ and $D=1$ as shown below. The sum term $A+\overline{B}+C+\overline{D}$ is 1 for all other combinations of values for the variables,

$$A+\overline{B}+C+\overline{D} = 0+\overline{1}+0+\overline{1} = 0+0+0+0 = 0$$

It may be noted that the sum term has a binary value 0101 (*i.e.*, decimal 5).

The student must remember that a sum term is implemented with an OR gate whose output is 0 if and only if each of its inputs is 0. INVERTERs are used to produce the complements of the variables wherever required.

A product-of-sums (POS) expression is equal to 0 if and only if one or more of the sum terms in the expression is equal to 0.

Example 35-35. *Obtain the standard sum-of-products form of the given function,*

$$f = a + a'bc + bc' \qquad \textit{(U.P. Tech. Univ. Sem. I, 2004-05)}$$

Solution. Given, the function,

$$\begin{aligned} f &= a + a'bc + bc' \\ &= a + \overline{a}bc + b\overline{c} \\ &= a\,(b+\overline{b})\,(c+\overline{c}) + \overline{a}bc + (a+\overline{a})\,b\overline{c} \\ &= (ab + a\overline{b})\,(c+\overline{c}) + \overline{a}bc + ab\overline{c} + \overline{a}b\overline{c} \\ &= abc + ab\overline{c} + a\overline{b}c + a\overline{b}\overline{c} + \overline{a}bc + ab\overline{c} + \overline{a}b\overline{c} \\ &= abc + ab\overline{c} + a\overline{b}c + a\overline{b}\overline{c} + \overline{a}bc + \overline{a}b\overline{c} \ \textbf{Ans.} \end{aligned}$$

Example 35-36. *Convert the following Boolean function into canonical form:*

(*a*) $AB + B\overline{C}D + \overline{A}D$ (*b*) $(A+\overline{B})\ (\overline{C}+\overline{D})\ (\overline{B}+\overline{C})$

(*c*) $(\overline{B}+\overline{C})$. *(U.P. Tech. Univ. Sem II, 2005-06)*

Solution. We know that a canonical form is a form in which the sum term (or product term) contains all the variables. It is also called standard sum-of-products (or product-of-sums) form. Following the steps given in Article 35-16, let us convert the given Boolean function into its canonical form.

(*a*) *Conversion of* $AB + B\bar{C}D + \bar{A}D$ *into canonical form*

$$f = AB + B\bar{C}D + \bar{A}D$$
$$= AB\,(C + \bar{C})\,(D + \bar{D}) + (A + \bar{A})\,(B\bar{C}D) + \bar{A}\,(B + \bar{B})\,(C + \bar{C})\,D$$
$$= (ABC + AB\bar{C})\,(D + \bar{D}) + (AB\bar{C}D + \bar{A}B\bar{C}D)$$
$$+ (\bar{A}BC + \bar{A}\bar{B}C + \bar{A}B\bar{C} + \bar{A}\bar{B}\bar{C})\,D$$
$$= ABCD + AB\bar{C}D + ABC\bar{D} + AB\bar{C}\bar{D} + AB\bar{C}D + \bar{A}B\bar{C}D$$
$$+ \bar{A}BCD + \bar{A}\bar{B}CD + \bar{A}B\bar{C}D + \bar{A}\bar{B}\bar{C}D \quad \textbf{Ans.}$$

(*b*) *Conversion of* $(A + \bar{B})\,(\bar{C} + \bar{D})\,(\bar{B} + \bar{C})$ *into canonical form*

The given expression is in non-standard product-of-sums form. Its canonical form can be obtained as follows:

$$f = (A + \bar{B})\,(\bar{C} + \bar{D})\,(\bar{B} + \bar{C})$$
$$= (A + \bar{B} + C\bar{C} + D\bar{D})\,(A\bar{A} + B\bar{B} + \bar{C} + \bar{D})\,(A\bar{A} + \bar{B} + \bar{C} + D\bar{D})$$
$$= (A + \bar{B} + C\bar{C} + D\bar{D})\,(A\bar{A} + B\bar{B} + \bar{C} + \bar{D})\,(A\bar{A} + \bar{B} + \bar{C} + D\bar{D})$$
$$= (A + \bar{B} + C + D\bar{D})\,(A + \bar{B} + \bar{C} + D\bar{D}) \cdot (A\bar{A} + B + \bar{C} + \bar{D})\,(A\bar{A} + \bar{B} + \bar{C} + \bar{D})$$
$$\cdot\,(A\bar{A} + \bar{B} + \bar{C} + D)\,(A\bar{A} + \bar{B} + \bar{C} + \bar{D})$$
$$= (A + \bar{B} + C + D)\,(A + \bar{B} + C + \bar{D})\,(A + \bar{B} + \bar{C} + D)\,(A + \bar{B} + \bar{C} + \bar{D})$$
$$\cdot\,(A + B + \bar{C} + \bar{D})\,(\bar{A} + B + \bar{C} + \bar{D})\,(A + \bar{B} + \bar{C} + \bar{D})\,(\bar{A} + \bar{B} + \bar{C} + \bar{D})$$
$$\cdot\,(A + \bar{B} + \bar{C} + D)\,(\bar{A} + \bar{B} + \bar{C} + D)\,(A + \bar{B} + \bar{C} + \bar{D})\,(\bar{A} + \bar{B} + \bar{C} + \bar{D}) \quad \textbf{Ans.}$$

(*c*) *Conversion of* $(\bar{B} + \bar{C})$ *into canonical form*

$$f = (\bar{B} + \bar{C})$$
$$= (A\bar{A} + \bar{B} + \bar{C} + D\bar{D})$$
$$= (A + \bar{B} + \bar{C} + D\bar{D})\,(\bar{A} + \bar{B} + \bar{C} + D\bar{D})$$
$$= (A + \bar{B} + \bar{C} + D)\,(\bar{A} + \bar{B} + \bar{C} + \bar{D}) \cdot (\bar{A} + \bar{B} + \bar{C} + D)\,(\bar{A} + \bar{B} + \bar{C} + \bar{D}) \quad \textbf{Ans.}$$

Example 35-37. *Realize the following function using NOR gate only,*

$$Z = \overline{ABC}D + \bar{A}B\bar{C}\bar{D} + \bar{A}B\bar{C}D + \bar{A}BCD + A\bar{B}C\bar{D} + ABCD$$

(*U.P. Tech. Univ., Sem. I, 2004-05*)

Solution. Given, the Boolean function,

$$Z = \overline{ABC}D + \bar{A}B\bar{C}\bar{D} + \bar{A}B\bar{C}D + \bar{A}BCD + A\bar{B}C\bar{D} + ABCD$$

Notice that the expression is a standard sum-of-products expression. In order to implement this function using NOR gates we need to simplify this function. This can be done using Boolean algebra or Karnaugh map. Let us simplify using Boolean algebra.

$$Z = \overline{ABC}D + \bar{A}B\bar{C}\bar{D} + \bar{A}B\bar{C}D + \bar{A}BCD + A\bar{B}C\bar{D} + ABCD$$
$$= (\bar{A} + \bar{B} + \bar{C})\,D + \bar{A}B\bar{C}\,(\bar{D} + D) + (\bar{A} + A)\,BCD + A\bar{B}C\bar{D}$$
$$= \bar{A}D + \bar{B}D + \bar{C}D + \bar{A}B\bar{C} + BCD + A\bar{B}C\bar{D}$$
$$= D\,(\bar{B} + BC) + \bar{A}D + \bar{C}D + \bar{A}B\bar{C} + A\bar{B}C\bar{D}$$
$$= \bar{B}D + CD + \bar{A}D + \bar{C}D + \bar{A}B\bar{C} + A\bar{B}C\bar{D}$$
$$= \bar{B}\,(D + AC\bar{D}) + CD + \bar{A}D + \bar{C}D + \bar{A}B\bar{C}$$
$$= \bar{B}D + A\bar{B}C + CD + \bar{A}D + \bar{C}D + \bar{A}B\bar{C}$$
$$= \bar{B}D + A\bar{B}C + (C + \bar{C})\,D + \bar{A}D + \bar{A}B\bar{C}$$

$$= \overline{B}D + A\overline{B}C + D + \overline{A}D + \overline{A}B\overline{C}$$

$$= (\overline{B} + 1 + \overline{A})\ D + A\overline{B}C + \overline{A}B\overline{C}$$

$$= D + A\overline{B}C + \overline{A}B\overline{C}$$

$$= \overline{\overline{D + A\overline{B}C + \overline{A}B\overline{C}}}$$

$$= \overline{\overline{D}\,.\,\overline{A\overline{B}C}\,.\,\overline{\overline{A}B\overline{C}}}$$

$$= \overline{\overline{D}\,.(\overline{A} + \overline{\overline{B}} + C)\,.(\overline{\overline{A}} + \overline{B} + \overline{\overline{C}})}$$

$$= \overline{\overline{D}\,.(\overline{A} + B + C)\,.(A + \overline{B} + C)}$$

$$= \overline{\overline{D}} + \overline{(\overline{A} + B + C)} + \overline{(A + \overline{B} + C)}$$

$$= D + \overline{\overline{A} + B + C} + \overline{A + \overline{B} + C}$$

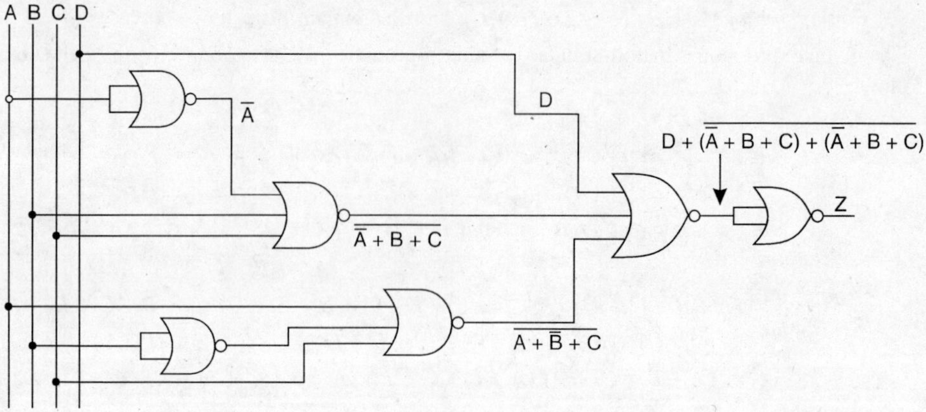

Fig. 35-47.

Fig. 35-47 shows the implementation of the simplified Boolean function. It is possible to replace the three-input-NOR gates by those with only two-inputs.

Example 35-38. *Determine the binary values of the variables for which the following standard product-of-sums (POS) expression is equal to 0:*

$$(A + \overline{B} + C + D)\ (A + B + \overline{C} + D)\ (\overline{A} + \overline{B} + \overline{C} + D)$$

Solution. Given, the POS expression,

$$(A + \overline{B} + C + D)\ (A + B + \overline{C} + D)\ (\overline{A} + \overline{B} + \overline{C} + D)$$

We know that the given POS expression equals 0 when any of the three sum terms equals 0. The term $A + \overline{B} + C + D$ is equal to 0, when $A = 0$, $B = 1$, $C = 0$ and $D = 0$, *i.e.*,

$$A + \overline{B} + C + D\ = 0 + \overline{1} + 0 + 0 = 0 + 0 + 0 + 0 = 0 \text{ **Ans.**}$$

The term $A + B + \overline{C} + D$ is equal to 0 when $A = 0$, $B = 0$, $C = 1$ and $D = 0$ *i.e.*,

$$A + B + \overline{C} + D\ = 0 + 0 + \overline{1} + 0 = 0 + 0 + 0 + 0 = 0 \text{ **Ans.**}$$

The term $\overline{A} + \overline{B} + \overline{C} + \overline{D}$ is equal to 0 when $A = 1$, $B = 1$, $C = 1$ and $D = 1$, *i.e.*,

$$\overline{A} + \overline{B} + \overline{C} + \overline{D}\ = \overline{1} + \overline{1} + \overline{1} + \overline{1} = 0 + 0 + 0 + 0 = 0 \text{ **Ans.**}$$

35-23. Converting Standard Sum-of-Products (SOP) to Standard Product-of-Sums (POS)

It will be interesting to know that the binary values of the product terms in a given standard sum-of-products (SOP) expression are not present in the equivalent standard product-of-sums (POS) expression. Also, the binary values that are not represented in the SOP expression are present in the equivalent POS expression. Therefore, to convert from standard SOP to standard POS, we take the following steps:

(*i*) Evaluate each product term in the SOP expression, *i.e.*, determine the binary numbers which represent the product terms.

(*ii*) Determine all the binary numbers not included in the evaluation in step 1.

(*iii*) Write the equivalent sum term for each binary number from step 2 and express in POS form

A similar procedure can be used to convert a standard POS expression to a standard SOP expression.

Example 35-39. *Convert the following sum-of-products (SOP) expression to an equivalent product-of-sums (POS) expression:*

$$\overline{A}\overline{B}\overline{C} + \overline{A}\overline{B}C + \overline{A}BC + A\overline{B}C + ABC$$

Solution. Given, the SOP expression,

$$\overline{A}\overline{B}\overline{C} + \overline{A}\overline{B}C + \overline{A}BC + A\overline{B}C + ABC$$

The evaluation of each product term in the given expression is

$$000 + 001 + 011 + 101 + 111$$

Since, there are three variables in the domain of this expression (i.e., A, B and C), there are total of 2^3 (= 8) possible combinations. The given SOP expression contains five of these combinations, so the POS must contain the other three which are 010, 100 and 110. Since, these are binary values that make the sum term 0, therefore the equivalent POS expression is

$$(\overline{A} + B + \overline{C})\ (A + \overline{B} + \overline{C})\ (A + B + \overline{C})$$ **Ans.**

35-24. Boolean Expressions and Truth Tables

As a matter a fact, all standard Boolean expressions (SOP or POS) can be easily converted into truth table format. This is done by using binary values for each term in the expression. The conversion into a truth table format is required because of the fact that truth table is a common way of presenting the logical operation of a circuit. The truth table can also be used to determine a standard SOP or POS expression.

35-25. Converting Sum-of-Products (SOP) Expressions to Truth Table Format

We have already discussed in Article 35-17 that on SOP (sum-of-products) expression is equal to 1 only if at least one of the product terms is equal to 1. We know that a truth table is simply a list of the possible combinations of input variable values and the corresponding output values (i.e., 1 or 0). For a Boolean expression with a domain of 2 variables, there are four different combinations of those variables (because $2^2 = 4$). Similarly, for a Boolean expression with a domain of 3 variables, there are eight different combinations of those variables (because $2^3 = 8$). For a Boolean expression, with a domain of 4 variables, there are sixteen different combinations of those variables (because $2^4 = 16$) and so on.

Following is the procedure to construct a truth table for the given SOP expression.

(*i*) List all the possible combinations of binary values of the variables in the given expression.

(*ii*) Convert the SOP expression to standard form if it is not already so.

(*iii*) Place a 1 in the output column (X) for each binary value that makes the standard SOP expression a 1 and place a 0 for all the remaining values.

In order to illustrate the procedure, consider the construction of a truth table for the standard SOP expression,

$$\overline{A}\overline{B}C + A\overline{B}\overline{C} + ABC$$

We know that there are 3 variables in the domain, so there are eight possible combinations of binary values as shown in the left three columns of the Table 35-5. The binary values that make the product terms in the expression equal to 1 are 001 for $\overline{A}\overline{B}C$, 100 for $A\overline{B}\overline{C}$, and 111 for ABC. For each of these binary values, a 1 is placed in the output column as shown in the table above. For each of the remaining binary combinations, a 0 is placed in the output column.

Table 35-5.

Inputs			*Output*
A	***B***	***C***	***X***
0	0	0	0
0	0	1	1
0	1	0	0
0	1	1	0
1	0	0	1
1	0	1	0
1	1	0	0
1	1	1	1

Example 35-40. *Develope a truth table for the following Boolean expression:*

$$\overline{A}B + AB\overline{C} + \overline{AC} + A\overline{B}C$$

Solution. Given, the Boolean expression,

$$\overline{A}B + AB\overline{C} + \overline{AC} + A\overline{B}C$$

We know that the given expression has 3 variables in its domain. A quick look at the expression indicates that it is not a standard SOP (sum-of-products) form. Let us simplify the Boolean expression first and then convert it into a standard SOP form.

$$\overline{A}B + AB\overline{C} + \overline{AC} + A\overline{B}C = \overline{A}B + AB\overline{C} + \overline{A} + \overline{C} + A\overline{B}C$$

The first term on the right side has a missing variable C or $\overline{C}$. So we multiply this term by $C + \overline{C}$. The third term has two missing variables B or $\overline{B}$ and C or $\overline{C}$. So we multiply this term by $B + \overline{B}$ and $C + \overline{C}$. The fourth term again has two missing variables, A or $\overline{A}$ and B or $\overline{B}$. So we multiply this term by $A + \overline{A}$ and $C + \overline{C}$. Thus,

$$\overline{A}B + AB\overline{C} + \overline{A} + \overline{C} + A\overline{B}C$$

$$= \overline{A}B\,(C + \overline{C}) + AB\overline{C} + \overline{A}\,(B + \overline{B})\,(C + \overline{C}) + (A + \overline{A})\,(B + \overline{B})\,\overline{C} + A\overline{B}C$$

$$= \underline{\overline{A}BC} + \underset{\times}{\overline{A}B\overline{C}} + \underset{?}{AB\overline{C}} + \underline{\overline{A}BC} + \underset{\times}{\overline{A}B\overline{C}} + \underset{\times\times}{\overline{A}\overline{B}C} + \underset{o}{\overline{A}\overline{B}\overline{C}} + \underset{?}{AB\overline{C}}$$

$$+ A\overline{B}\overline{C} + \underset{\times}{\overline{A}B\overline{C}} + \underset{o}{\overline{A}\overline{B}\overline{C}} + A\overline{B}C$$

Combining the similar terms, we get,

$$\overline{A}B + AB\overline{C} + \overline{A} + \overline{C} + A\overline{B}C = \overline{A}BC + \overline{A}B\overline{C} + AB\overline{C} + \overline{A}\overline{B}C + \overline{A}\overline{B}\overline{C} + A\overline{B}\overline{C} + A\overline{B}C$$

The equation on the right above is a standard SOP expression. Since there are 3 variables in the domain, therefore there are eight possible combinations of binary values as shown in the left three columns of the Table 35-6. The binary values that make the product terms in the expression equal to 1 are 011 (for $\overline{A}BC$), 010 (for $\overline{A}B\overline{C}$), 110 (for $AB\overline{C}$), 001 (for $\overline{A}\overline{B}C$), 000 (for $\overline{A}\overline{B}\overline{C}$), 100 (for ($A\overline{B}\overline{C}$), and 101 (for $A\overline{B}C$). For each of these seven binary values, we place a 1 in the output column as shown in the table above. The only remaining binary combination is 111 and we place a 0 for this combination in the output column.

Table 35-6

Inputs			*Output*
A	*B*	*C*	*X*
0	0	0	1
0	0	1	1
0	1	0	1
0	1	1	1
1	0	0	1
1	0	1	1
1	1	0	1
1	1	1	0

35-26. Converting Product-of-Sums (POS) Expressions to a Truth Table Format

We have already discussed in Article 35-21 that a POS (product-of-sums) expression is equal to 0 only if at least one of the sum terms is equal to 0. Following is the procedure to construct a truth table for the given POS expression,

(*i*) List all the possible combinations of binary values of the variables.

(*ii*) Convert the POS expression to standard form if it is not already so.

(*iii*) Place a 0 in the output column (X) for each binary value that makes the expression a 0 and place a 1 for all the remaining binary values.

In order to illustrate the procedure, consider the construction of a truth table for a standard POS expression,

$$(\overline{A} + \overline{B} + C)\ (A + \overline{B} + \overline{C})\ (A + B + C)$$

We know that there are 3 variables in the domain, so there are eight possible combinations of binary values as shown in the left three columns of the Table 35-7. The binary values that make the sum terms in the expression equal to 0 are 110 (for $\overline{A} + \overline{B} + C$), 011 (for $A + \overline{B} + \overline{C}$) and 000 (for $A + B + C$). For each of these binary values, a 0 is placed in the output column as shown in the table. For each of the remaining binary combinations, a 1 is placed in the output column.

Table 35-7.

Inputs			*Output*
A	*B*	*C*	*X*
0	0	0	0
0	0	1	1
0	1	0	1
0	1	1	0
1	0	0	1
1	0	1	1
1	1	0	0
1	1	1	1

Example 35-41. *Develop a truth table for the following standard POS (product-of-sums) expression:*

$$(\overline{A} + \overline{B} + \overline{C})\ (\overline{A} + B + \overline{C})\ (A + \overline{B} + \overline{C})$$

Solution. Given, the standard POS expression,

$$(\bar{A}+\bar{B}+\bar{C})\,(\bar{A}+B+\bar{C})\,(A+\bar{B}+\bar{C})$$

We know that there are 3 variables in the domain of the given POS expression. So there are eight possible combinations of binary values as shown in the left three columns of the Table 35-8.

Table 35-8.

Inputs			*Output*
A	*B*	*C*	*X*
0	0	0	1
0	0	1	1
0	1	0	1
0	1	1	0
1	0	0	1
1	0	1	0
1	1	0	1
1	1	1	0

The binary values that make the sum terms in the expression equal to 0 are 111 (for $\bar{A}+\bar{B}+\bar{C}$), 101 (for $\bar{A}+B+\bar{C}$), 011 (for $A+\bar{B}+\bar{C}$). For each of these binary values, a 0 is placed in the output column as shown in the table. For each of the remaining binary combinations, a 1 is placed in the output column. **Ans.**

35-27. Determining Standard Expressions from a Truth Table

We have already discussed in the previous articles the procedure for converting a standard SOP or POS expression. Now we will study the method of determining standard SOP and POS expressions from a truth table.

To determine the standard SOP expression represented by a truth table, list the binary values of the input variables for which the output is 1. Convert each binary value to the corresponding product term by replacing each 1 with the corresponding variable and each 0 with the corresponding variable complement. For example, 1001 is converted to a product term as follows.

$$1001 \rightarrow A\bar{B}\bar{C}D$$

To determine a standard POS expression represented by a truth table, list the binary values for which the output is 0. Convert each binary value to the corresponding sum term by replacing each 1 with the corresponding variable complement and each 0 with the corresponding variable. For example, the binary value 1010 is converted to a sum term as follows:

$$1010 \rightarrow \bar{A}+B+\bar{C}+D$$

Example 35-42. *Fig. 35-48 shows a truth table for a typical digital circuit.*

Inputs			*Output*
A	*B*	*C*	*X*
0	0	0	0
0	0	1	0
0	1	0	1
0	1	1	1
1	0	0	0
1	0	1	0
1	1	0	1
1	1	1	1

Fig. 35-48.

Determine the standard SOP (sums-of-products) expression and the equivalent standard POS (product-of-sums) expression for the given truth table.

Solution. Given, the truth table of Fig. 35-48.

Standard SOP expression

From the given truth table we find, that there are four 1s in the output column and the corresponding binary values are, 010, 011, 110 and 111. These binary values are converted to product terms as follows :

$$010 \rightarrow \overline{A}B\overline{C}$$

$$011 \rightarrow \overline{A}BC$$

$$110 \rightarrow AB\overline{C}$$

$$111 \rightarrow ABC$$

Thus, the resulting standard SOP expression for the output X is,

$$X = \overline{A}B\overline{C} + \overline{A}BC + AB\overline{C} + ABC \text{ **Ans.**}$$

Standard POS expression

From the given truth table, we find that the output is 0 for binary values 000, 001, 100 and 101. These binary values are converted to sum terms as follows:

$$110 \rightarrow A + B + C$$

$$001 \rightarrow A + B + \overline{C}$$

$$100 \rightarrow \overline{A} + B + C$$

$$101 \rightarrow \overline{A} + B + \overline{C}$$

The resulting standard POS expression for the output,

$$X = (A + B + C)\,(A + B + \overline{C})\,(\overline{A} + B + C)\,(\overline{A} + B + \overline{C}) \text{ **Ans.**}$$

35-28. The Karnaugh Map (K-Map)

The Karnaugh map (or simply a *K*-map) is similar to a truth table because it presents all the possible values of input variables and the resulting output for each value. However, instead of being organised into columns and rows like a truth table, the Karnaugh map is an array of squares (or cells) in which each square represents a binary value of the input variables. The squares are arranged in a way so that simplification of given expression with two-three, and four variables Karnaugh maps to illustrate the principles. Karnaugh map-with five-variables is beyond the scope of this book. For higher number of variables, a Quine-McClusky method can be used. This method is also beyond the scope of this book.

The number of squares in a Karnaugh map is equal to the total number of possible input variable combination (as is the number of rows in a truth table). For two variables, the number of square is $2^2 = 4$, for three variables, the number of squares is $2^3 = 8$ and for four variables, the number of squares is $2^4 = 16$.

35-29. The Two-variable Karnaugh Map

Fig. 35-49 (*a*) shows a two-variable Karnaugh map. As seen, it is an array of four squares. In this cases, *A* and *B* are used for two variables although any other two letters could be used. The binary values of *A* (*i.e.*, 0 and 1) are indicated along the left side as $\overline{A}$ and *A* (notice the sequence) and the binary values of *B* are indicated across the top as $\overline{B}$ and *B*. The value of a given square is the value of *A* at the left in the same row combined with the value of *B* at the top in the same column. For example, a square in the upper left corner has a value of $\overline{A}\overline{B}$ and a square in the lower right corner has a value of *AB*. Fig. 35-49 (*b*) shows the standard product terms represented by each square in the Karnaugh map.

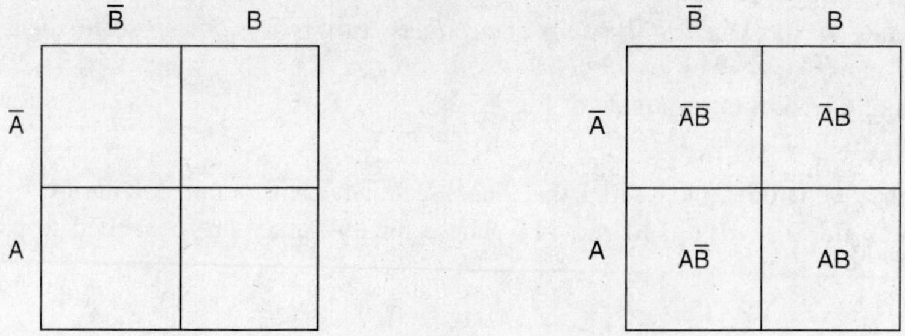

(a) (b)

Fig. 35-49.

35-30. The Three-variable Karnaugh Map

Fig. 35-50 (*a*) shows a three-variable Karnaugh map. As seen it is an array of eight squares. In this case, *A*, *B* and *C* are used for the variables although any other three letters could be used. The value of *A* and *B* are along the left side (notice the sequence carefully) and the values of *C* are across the top.

The value of a given square is the value of *A* and *B* at the left in the same row combined with the value of *C* at the top in the same column. For example, a square in the upper left corner has a value of $\overline{A}\overline{B}\overline{C}$ and a square in the bottom right corner has a value of $A\overline{B}C$. Fig. 35-50 (*b*) shows the product terms that are represented by each square in the Karnaugh map.

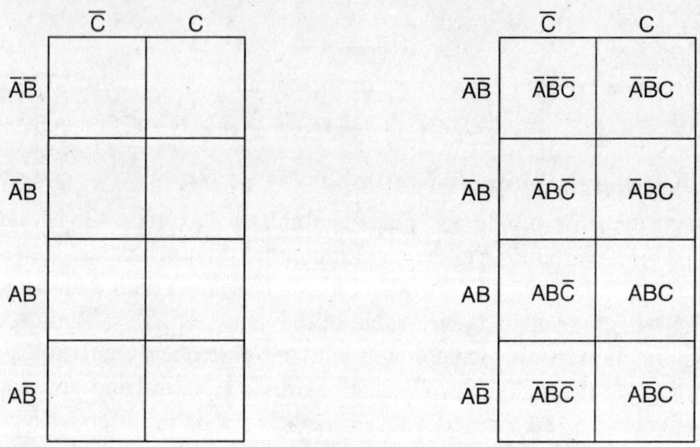

(a) (b)

Fig. 35-50.

35-31. The Four-variable Karnaugh Map

Fig. 35-51 (*a*) shows a four-variable Karnaugh map. As seen, it is an array of sixteen squares. In this case *A*, *B*, *C* and *D* are used for the variables. The values of *A* and *B* are along the left side, and the values of *C* and *C* are across the top. The sequence of the variable values may be noted carefully.

The value of a given square is the value of *A* and *B* at the left in the same row combined with the values of *C* and *D* at the top in the same column. For example, a square in the upper right corner has a value $\overline{A}\overline{B}\overline{C}\overline{D}$ and a square in the lower left corner has a value $\overline{A}\overline{B}\overline{C}\overline{D}$. Fig. 35-51 (*b*) shows the standard product terms that are represented by each square in the four-variable Karnaugh map.

	$\bar{C}\bar{D}$	$\bar{C}D$	CD	$C\bar{D}$
$\bar{A}\bar{B}$				
$\bar{A}B$				
AB				
$A\bar{B}$				

(a)

	$\bar{C}\bar{D}$	$\bar{C}D$	CD	$C\bar{D}$
$\bar{A}\bar{B}$	$\bar{A}\bar{B}\bar{C}\bar{D}$	$\bar{A}\bar{B}\bar{C}D$	$\bar{A}\bar{B}CD$	$\bar{A}\bar{B}C\bar{D}$
$\bar{A}B$	$\bar{A}B\bar{C}\bar{D}$	$\bar{A}B\bar{C}D$	$\bar{A}BCD$	$\bar{A}BC\bar{D}$
AB	$AB\bar{C}\bar{D}$	$AB\bar{C}D$	$ABCD$	$ABC\bar{D}$
$A\bar{B}$	$A\bar{B}\bar{C}\bar{D}$	$A\bar{B}\bar{C}D$	$A\bar{B}CD$	$A\bar{B}C\bar{D}$

(b)

Fig. 35-51.

35-32. Square Adjacency in Karnaugh Map

We have already discussed a two-variable Karnaugh map, a three-variable Karnaugh map and a four-variable Karnaugh map. Now we shall discuss the concept of square adjacency in a Karnaugh map.

It will be interesting to know that the squares in a Karnaugh map are arranged in such a way that there is only a single-variable change between adjacent squares. Adjacency is defined as a single-variable change. It means the squares that differ by only one variable are adjacent in a Karnaugh map shown in Fig. 35-50 (*b*), the $\bar{A}\bar{B}\bar{C}$ square is adjacent to $\bar{A}B\bar{C}$ square, the $\bar{A}BC$ square and the $AB\bar{C}$ square. It may be carefully noted that square with values that differ by more than one variable are not adjacent. For example, the $\bar{A}B\bar{C}$ square is not adjacent to the $\bar{A}\bar{B}C$ square, the $A\bar{B}\bar{C}$ square or the $A\bar{B}C$ square. In other words, each square is adjacent to the squares that are immediately next to it on any of its four sides. However, a square is not adjacent to the squares that diagonally touch any of its corners.

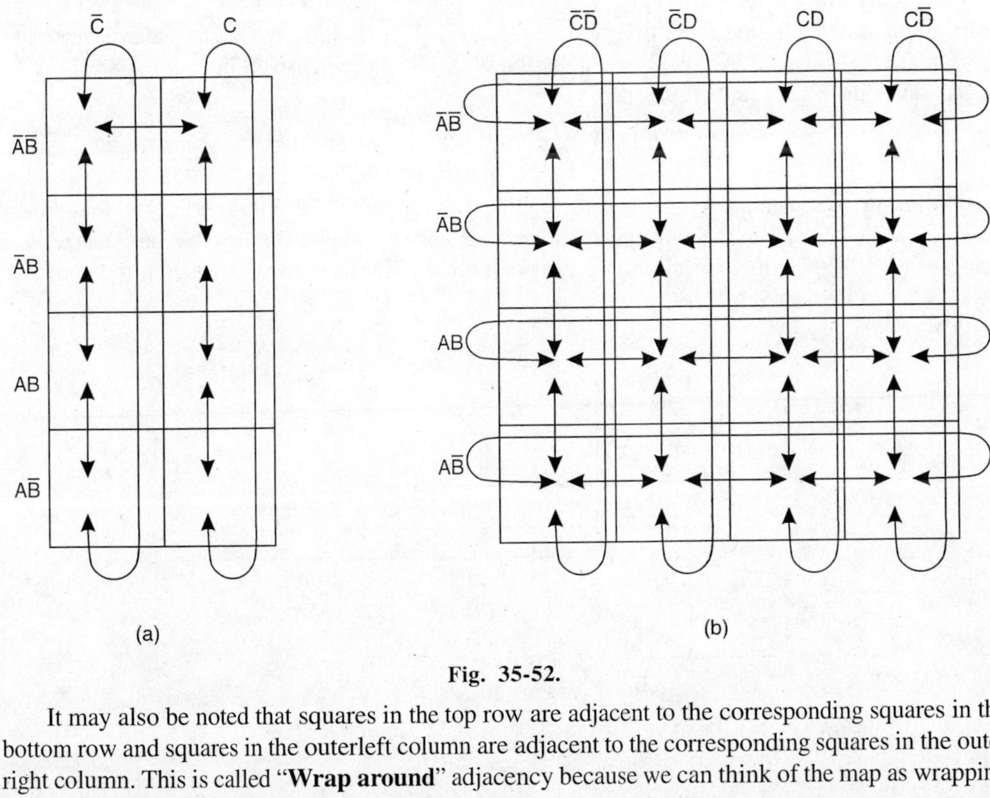

(a) (b)

Fig. 35-52.

It may also be noted that squares in the top row are adjacent to the corresponding squares in the bottom row and squares in the outerleft column are adjacent to the corresponding squares in the outer right column. This is called "**Wrap around**" adjacency because we can think of the map as wrapping around from top to bottom to form a cylinder or from left to right to form a cylinder. Fig. 35-51 (*a*) and (*b*) shows the square adjacencies with a three-variable and a four-variable Karnaugh maps respectively.

Notice the square adjacencies in a four variable Karnaugh map: Here for example, the square $\overline{A}\overline{B}\overline{C}\overline{D}$ is adjacent to $\overline{A}\overline{B}\overline{C}D$ square, $AB\overline{C}\overline{D}$ square, $A\overline{B}C\overline{D}$ square and $A\overline{B}C\overline{D}$ square. Similarly $\overline{A}B\overline{C}D$ square is adjacent to $\overline{A}\overline{B}\overline{C}D$ square, $\overline{A}B\overline{C}\overline{D}$ square, $\overline{A}BCD$ square and $AB\overline{C}D$ square.

35-33. Mapping a Standard SOP Expression on the Karnaugh Map

Consider a *SOP* (sum-of-products) expression. $\overline{A}\overline{B}\overline{C} + \overline{A}B\overline{C} + A\overline{B}C + A\overline{B}\overline{C}$. In order to map this expression on the Karnaugh map, we need a three variable Karnaugh-map because the given expression has three variables *A*, *B* and *C*. Then select the first product term $\overline{A}\overline{B}\overline{C}$ and enter 1 in the corresponding square (*i.e.*, the first row and the first column) as shown in Fig. 35-53.

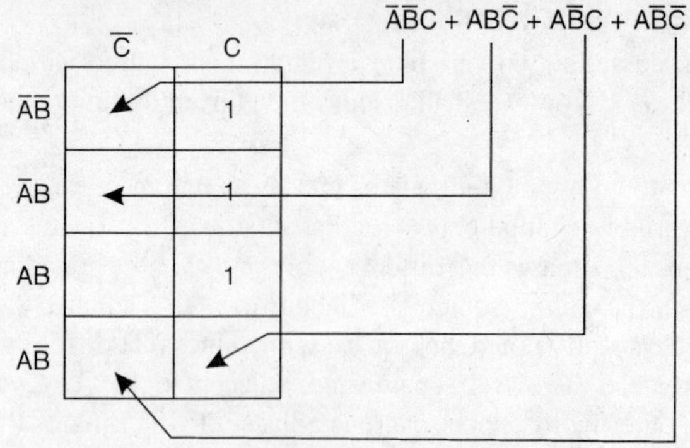

Fig. 35-53.

Similarly, for the second product term, $\overline{A}B\overline{C}$ place a 1 in the second row and first column. Repeat this process for the other two product terms, *i.e.*, $A\overline{B}C$ and $A\overline{B}\overline{C}$. The squares that do not have 1 are the squares for which the expression is 0. Usually when working with sum-of-product expressions, the 0's are left off the map.

Example 35-43. *Map the following SOP expression on the Karnaugh map:*

$$\overline{A}\overline{B}C + \overline{A}BC + AB\overline{C} + ABC.$$

Solution. Sketch a three variable Karnaugh map as shown in Fig. 35-54. Select the first product term $\overline{A}\overline{B}C$ and enter 1 in the corresponding square. Similarly enter 1 for the other products terms in the given *SOP* expression. Check that number of 1s in the Karnaugh map is equal to the number of product terms in the given *SOP* expression.

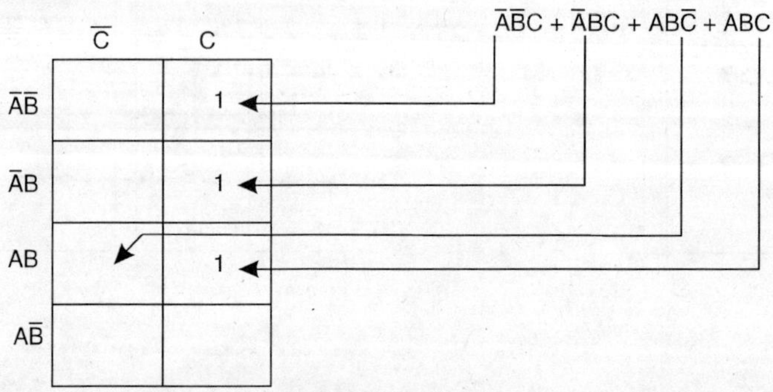

Fig. 35-54.

Example 35-44. *Draw the K-map for the given function,*

$$f = xy + xz + x'yz'.$$ (*U.P. Tech. Univ., Sem. I, 2004-05*)

Solution. Given, the logic function,

$$f = xy + xz + x'yz'$$
$$= xy + xz + \overline{x}y\overline{z}$$

Let us convert this function into a standard sum-of-products (*SOP*) form,

$$f = xy\ (z + \overline{z}) + x\ (y + \overline{y})\ z + \overline{x}y\overline{z}$$
$$= xyz + xy\overline{z} + xyz + x\overline{y}z + \overline{x}y\overline{z}$$
$$= xyz + xy\overline{z} + x\overline{y}z + \overline{x}y\overline{z} \qquad ...(i)$$

Now sketch a three variable Karnaugh map as shown in Fig. 35-55. Select the first product term *xyz* and enter 1 in the corresponding square. Similarly enter 1 for the other product terms in the *SOP*, equation (*i*). Check that the number of 1s in the Karnaugh map is equal to the number of product terms in the *SOP* expression.

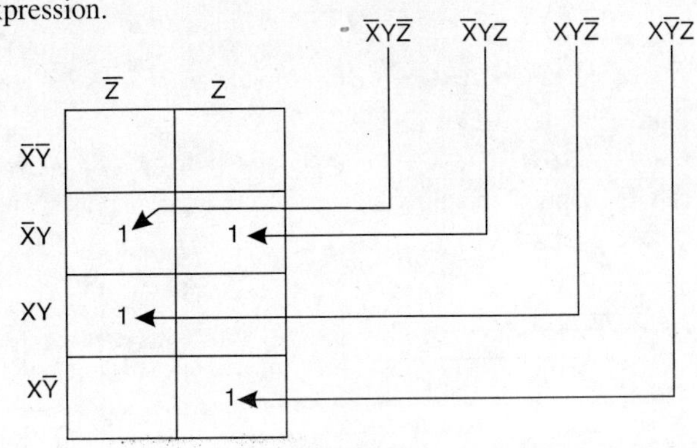

Fig. 35-55.

Example 35-45. *Map the following standard sum-of-products (SOP) expression on a Karnaugh map:* $\overline{A}BC\overline{D} + ABC\overline{D} + AB\overline{C}\overline{D} + \overline{A}\overline{B}\overline{C}\overline{D} + AB\overline{C}D + \overline{A}\overline{B}\overline{C}D + A\overline{B}C\overline{D} + ABCD$

Solution. Sketch a four variable Karnaugh map. Select the first product term $\overline{A}BC\overline{D}$ from the given *SOP* expression. Check that number of 1s in the Karnaugh map is equal to the number of product terms in the given *SOP* expression.

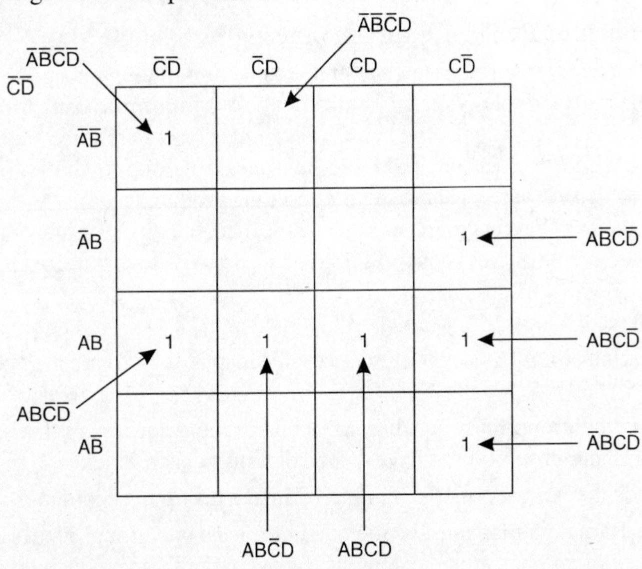

Fig. 35-56.

35-34. Mapping a Non-standard *SOP* Expression on the Karnaugh Map

A non-standard sum-of-products (*SOP*) expression is a one that has product terms with one or more missing variables. In such a case, Boolean expression must first be converted to a standard form by a procedure explained in Article 35-16.

Let us consider an example to illustrate the procedure for mapping a non-standard *SOP* expression on the Karnaugh map. Suppose we have the *SOP* expression:

$$\overline{A} + \overline{AB} + AB\overline{C}$$

As seen, this expression is obviously not in standard form because each product term does not have three variables. The first term is missing two variables, the second term is missing one variable and the third term is standard. In order to convert the given non-standard *SOP* expression to a standard form, we multiply the first product term by $B + \overline{B}$ and $C + \overline{C}$, and the second term by $C + \overline{C}$. Expanding the resulting expression, we get,

$$\overline{A}\,(B + \overline{B})\,(C + \overline{C}) + \overline{AB}\,(C + \overline{C}) + AB\overline{C}$$

$$= \overline{A}BC + \overline{A}\overline{B}C + \overline{A}B\overline{C} + \overline{A}\overline{B}\overline{C} + A\overline{B}C + \overline{A}\overline{B}\overline{C} + A\overline{B}C$$

Rearranging the expression for our convenience, we get

$$\overline{A}\overline{B}\overline{C} + \overline{A}\overline{B}C + \overline{A}B\overline{C} + \overline{A}BC + AB\overline{C} + A\overline{B}\overline{C} + A\overline{B}C$$

This expression can be mapped on a three-variable Karnaugh map as shown in Fig. 35-57.

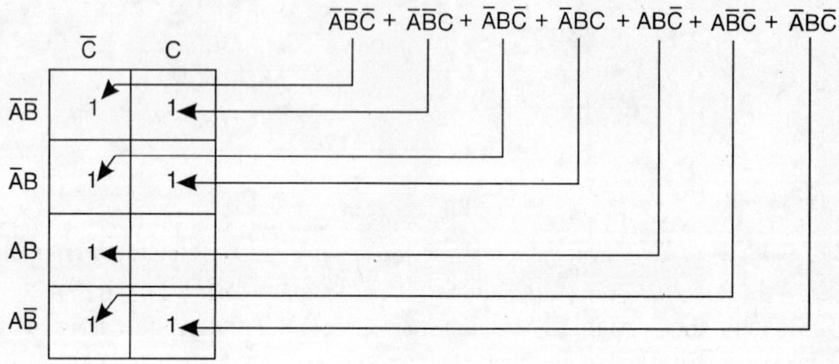

Fig. 35-57.

35-35. Simplification of Boolean Expression using Karnaugh Map

The process that result in an expression containing the minimum number of possible terms with the minimum number of variables is called **simplification** or **minimization**. After the *SOP* (or the Boolean) expression has been mapped on the Karnaugh map, there are three steps in the process of obtaining a minimum *SOP* expression. The three steps are: (*a*) grouping the 1s, (*b*) determining the product term for each group and (*c*) summing the resulting product terms.

(a) Grouping the 1s : We can group the 1s on the Karnaugh map according to the following rules by enclosing those adjacent squares containing 1s. The objective is to maximize the size of the groups and to minimize the number of groups.

(*i*) A group must contain either 1, 2, 4, 8 or 16 squares. In the case of two-variable Karnaugh map, 4 squares is the maximum group, for three-variable map, 8 squares are the maximum group and so on.

(*ii*) Each square in the group must be adjacent to one or more squares in that same group but all squares in the same group do not have to be adjacent to each other.

(*iii*) Always include the largest possible number of 1s in a group in accordance with rule 1.

(*iv*) Each 1 on the Karnaugh map must be included in at least one group. The 1s already in a group can be included in another group as long as the overlapping group include non-common 1s.

(b) Determining the Product Term for each group: Following are the rules that are applied to find the minimum product terms and the minimum sum-of-products expression:

(*i*) *Group the squares that have 1s:* Each group of squares containing 1s creates one product term composed of all variables that occur in only one form (either uncomplemented or complemented) within the group are eliminated. These are known as contradictory variables.

(*ii*) In order to determine the minimum product term for each group, we need to look at the standard methodology for three-variable and four-variable Karnaugh map respectively.

- *For a three-variable K-map:* (1) for 1-square group we get a three-variable product term, (2) for a 2-square group, we get a two-variable product term, (3) for a 4-square product term, we get a one-variable product term.
- *For a four-variable K-map:* (1) for a 1-square group, we get a four-variable product term, (2) for a 2-square group, we get a three-variable product term, (3) for a 4-square group, we get a two-variable product term, and (4) for a 8-square group, we get a one-variable product term.

(c) Summing the resulting product terms: When all the minimum product terms are derived from the Karnaugh map, these are summed to form the minimum sum-of-products expression.

Note. In some cases, there may be more than one way to group the 1s to form the product terms. Whatever be the way, the minimal expression must have the same number of product terms and each product term, the same number of Boolean variables.

The examples given below will help you to understand and apply the simplification of the *SOP* expression using Karnaugh map.

Example 35-46. *Simplify the following Boolean expression using the Karnaugh mapping technique:*

$$X = \overline{A}B + \overline{A}\overline{B}C + AB\overline{C} + A\overline{B}\overline{C}$$

Solution. The first step is to map the given Boolean expression on the Karnaugh map. Notice that there are three variables *A*, *B* and *C* in the Boolean expression, therefore we need a three-variable Karnaugh map.

The Boolean expression to be mapped is

$$X = \overline{A}B + \overline{A}\overline{B}C + AB\overline{C} + A\overline{B}\overline{C}$$

Note that the given Boolean expression is a non-standard *SOP* expression because the first product term $\overline{A}B$ has the variable *C* missing in it. This can be converted into a standard *SOP* form by modifying the expression as below:

$$X = \overline{A}B.1 + \overline{A}\overline{B}C + AB\overline{C} + A\overline{B}\overline{C}$$
$$= \overline{A}B\,(C + \overline{C}) + \overline{A}\overline{B}C + AB\overline{C} + A\overline{B}\overline{C} \qquad (C + \overline{C} = 1)$$
$$= \overline{A}BC + \overline{A}B\overline{C} + \overline{A}\overline{B}C + AB\overline{C} + A\overline{B}\overline{C} \qquad \text{...(1)}$$

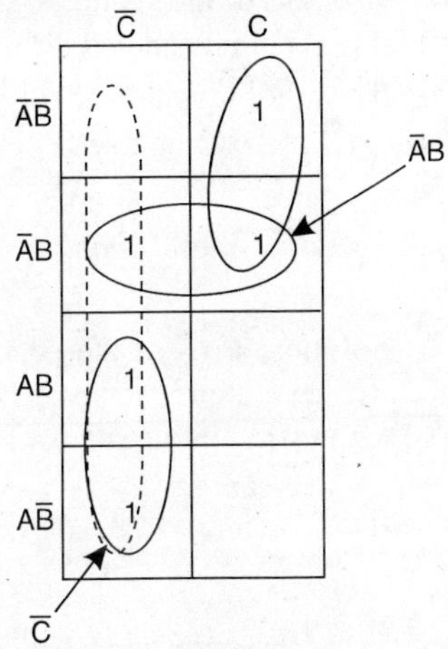

Fig. 35-58.

Equation (1) can be mapped on the Karnaugh map shown in Fig. 35-58. In order to simplify the given expression, the 1s can be grouped together as shown by the loop around the 1s. The four 1s in the first column are grouped together and the term we get is $\overline{C}$. This is because of the fact that the squares within this group contain both *A* and $\overline{A}$ and *B* and $\overline{B}$, so these variables are eliminated. Similarly, the two 1s in the second row are grouped together and the term we get is $\overline{A}B$. This is because of the fact that squares in this group contain both *C* and $\overline{C}$ which is eliminated. Summing up the two-product terms, the simplified expression is,

$$X = \overline{A}C + \overline{A}B + A\overline{C}$$

Example 35-47. *Simplify the following SOP expression using the Karnaugh mapping procedure:*

$$X = \bar{A}B\bar{C}D + A\bar{B}\bar{C}D + \bar{A}\bar{B}\bar{C}D + AB\bar{C}D + AB\bar{C}\bar{D} + ABCD$$

Solution. First of all, notice that the given *SOP* expression is already in the standard from *i.e.*, all the product terms in the given expression have all four variables *A*, *B*, *C* and *D*.

Next sketch a four-variable Karnaugh map. Select the first product term $(\bar{A}B\bar{C}D)$ from the given expression an enter 1 in Fig. 35-59. Similarly, enter 1 for the other product terms in the given *SOP* expression to complete the mapping. In order to simplify the given *SOP* expression, the 1s can be grouped together as shown be the loop around the 1s. The four 1s in the second column are grouped together and the product term get $\bar{C}D$. This is because of the fact that squares within this group contain both *A* and $\bar{A}$ and *B* and $\bar{B}$, so these variables are eliminated.

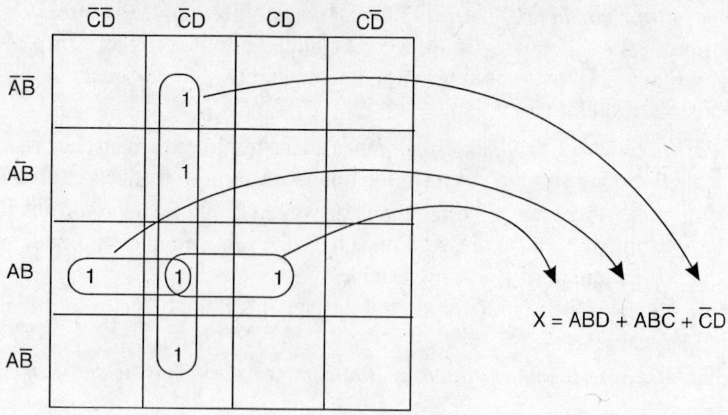

Fig. 35-59.

The two 1s in the first and second columns can be grouped together. This group contains both *D* and $\bar{D}$, so this variable is eliminated and the resulting product term is $AB\bar{C}$. Similarly, the two 1s in the second and third columns can be grouped together. This group contains both *C* and $\bar{C}$, so this variable is eliminated and the resulting product term is *ABD*.

The resulting minimal or simplified *SOP* expression is obtained by summing up the three product terms $\bar{C}D$, $AB\bar{C}$ and *ABD* as shown below:

$$X = ABD + AB\bar{C} + \bar{C}D$$

Example 35-48. *Simplify the following SOP expression using the Karnaugh mapping technique.*

$$X = B\bar{C}\bar{D} + \bar{A}B\bar{C}D + AB\bar{C}D + \bar{A}BCD + ABCD$$

Solution. First of all, notice that the given *SOP* expression is in the non-standard *SOP* form because the first product term $(B\bar{C}\bar{D})$ has a variable *A* and $\bar{A}$ missing in it. Let us convert the given *SOP* expression into a standard *SOP* expression into a standard *SOP* form as shown below:

$$X = 1.B\bar{C}\bar{D} + \bar{A}B\bar{C}D + AB\bar{C}D + \bar{A}BCD\ ABCD$$

$$= (\bar{A} + A)\ BCD + ABCD + ABCD + ABCD + ABCD \quad (\because \bar{A} + A = 1)$$

$$= \bar{A}B\bar{C}\bar{D} + AB\bar{C}\bar{D} + \bar{A}B\bar{C}D + AB\bar{C}D + \bar{A}BCD + ABCD$$

This expression can be mapped on to the four-variable Karnaugh by entering 1 for each product term in the corresponding square as shown in Fig. 35-60.

In order to simplify the *SOP* expression, the 1s can be grouped together as shown by the loop around the 1s. *The four* 1s looped together form the first and second columns, contain both *A* and $\bar{A}$ and *D* and $\bar{D}$, so these variables are eliminated and the resulting product term is $B\bar{C}$. Similarly, the four 1s looped together form the second and the third column contain both *A* and $\bar{A}$ and *C* and $\bar{C}$ so these variables are eliminated and the resulting product term is *BD*. The resulting simplified *SOP* expression is the sum of the product term $B\bar{C}$ and *BD*, *i.e.*,

$$X = B\overline{C} + BD$$

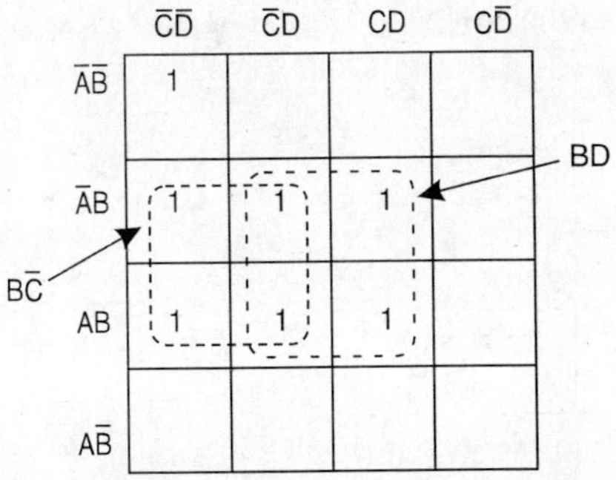

Fig. 35-60.

Example 35-49. *Fig. 35-61 shows a Karnaugh map of a sum-of-products (SOP) function. Determine the simplified SOP function.* (*UPSC Civil Services, 2000*)

	$\overline{C}\overline{D}$	$\overline{C}D$	CD	$C\overline{D}$
$\overline{A}\overline{B}$	1	1		1
$\overline{A}B$		1		
AB			1	
$A\overline{B}$	1	1		1

Fig. 18-61.

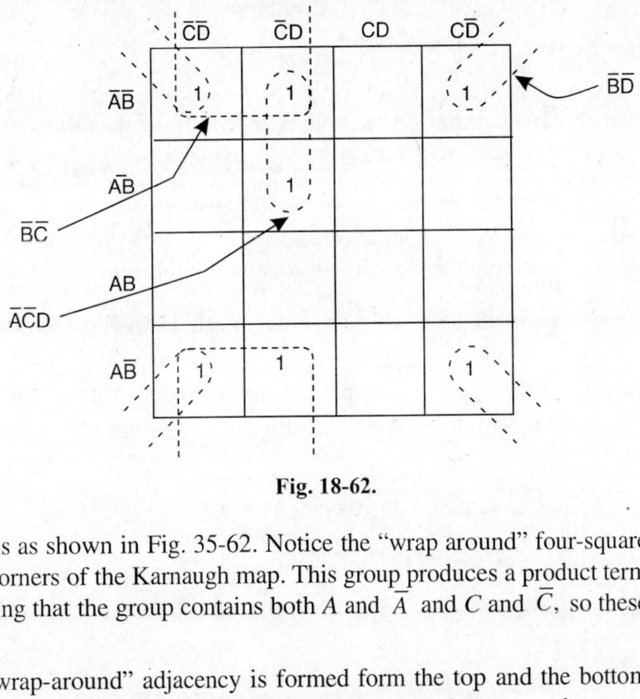

Fig. 18-62.

Solution. The grouping of 1s is as shown in Fig. 35-62. Notice the "wrap around" four-square group that includes the 1s on four corners of the Karnaugh map. This group produces a product term $\overline{B}\overline{D}$. This is determined by observing that the group contains both A and $\overline{A}$ and C and $\overline{C}$, so these variables are eliminated.

Another group of four with "wrap-around" adjacency is formed form the top and the bottom rows of the Karnaugh map. This group overlaps with the previous group and produces a product term $\overline{B}\overline{C}$. This is determined by observing that this group contains both A and $\overline{A}$ and D and $\overline{D}$, so these variables are eliminated.

The remaining 1 is absorbed in a overlapping group of two squares. This group produces a three-variable term $\overline{A}\overline{C}D$. This is determined by observing that this group contains both B and $\overline{B}$, so this variable is eliminated. This resulting simplified *SOP* function is the sum of the product terms $\overline{B}\overline{D} + \overline{B}\overline{C}$ and $\overline{A}\overline{C}D$, *i.e.*,

$$X = \overline{B}\overline{D} + \overline{B}\overline{C} + \overline{A}\overline{C}D$$

Example 35-50. *Minimize the K-map shown in Fig. 35-63.* (*U.P. Tech. Univ., Sem. II, 2004-05*)

Solution. Given, the *K*-map as shown in Fig. 35-63. Let us redraw the given *K*-map as shown in Fig. 35-64. We know that larger the group of 1s, the simpler the resulting term will be:

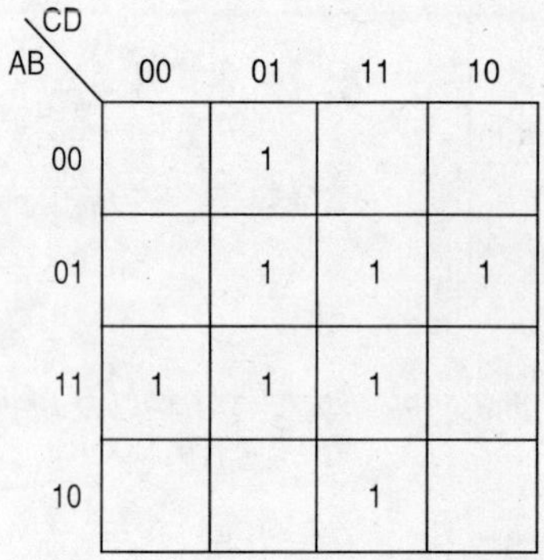

Fig. 18-63.

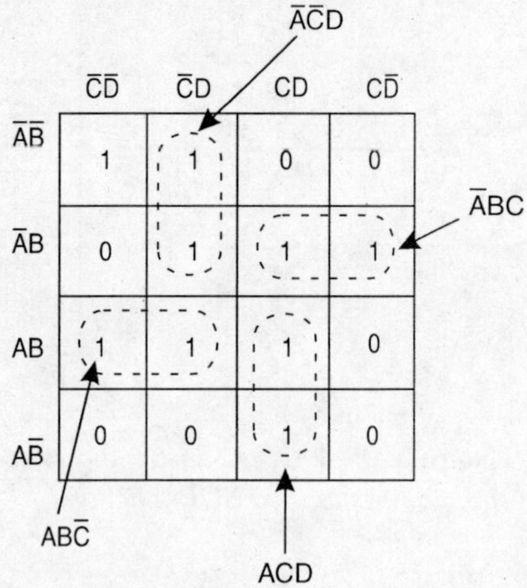

Fig. 18-64.

The grouping of 1s for the given *K*-map is as shown in Fig. 35-64. This gives 4 product terms namely $\overline{A}\overline{C}D$, $\overline{A}BC$, $AB\overline{C}$ and ACD. Therefore, the minimized function is,

$$f = \overline{A}\overline{C}D + \overline{A}BC + AB\overline{C} + ACD \textbf{ Ans.}$$

Note. Although we can combine four 1s together in the centre of the *K*-map. This produces a product term *BD*. But since all the 1s within this group have already been taken care of in other groups, therefore the grouping of four 1s is redundant.

35-36. Mapping Directly on Karnaugh Map from a Truth Table

It is possible to map directly on Karnaugh map from a truth table. Recall that a truth table. Recall that a truth table gives the output of a Boolean expression for all possible input variable combinations. Let us illustrate direct mapping through an example of a Boolean expression and its truth table representation.

Inputs			Output
A	*B*	*C*	*X*
0	0	0	0
0	0	1	1
0	1	0	0
0	1	1	1
1	0	0	0
1	0	1	1
1	1	0	1
1	1	1	1

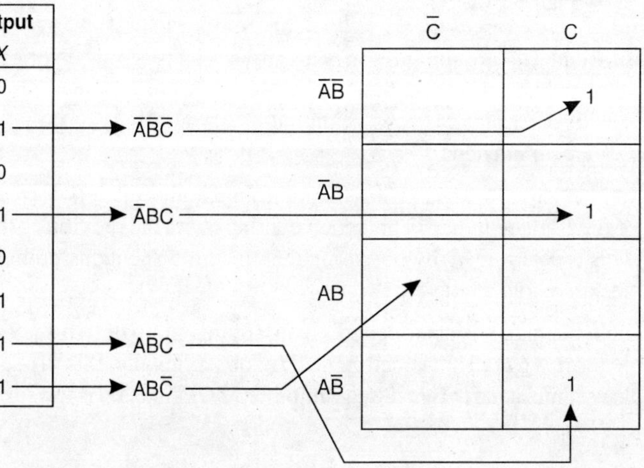

Fig. 35-65.

Let $X = \overline{A}\,\overline{B}C + \overline{A}BC + A\overline{B}C + AB\overline{C}$. Then its truth table can be indicated as shown in Fig. 35-65 (*a*) and its Karnaugh mapping is shown in Fig. 35-65 (*b*). Notice in the truth table that the output X is 1 for four different input variable combinations.

It is evident from the Fig. 35-65 (*a*) and (*b*) that truth table and Karnaugh map are simply different ways to represent a logic function.

Example 35-51. *Implement the following Boolean expression using minimum number of 3-input NAND gates.*

$$f(A, B, C, D) = (1, 2, 3, 4, 7, 9, 10, 12)$$

Solution. The given Boolean function indicates that its output is 1 corresponding to the term indicated within the expression i.e., 1, 2, 3, 4, 7, 9, 10 and 12. This is shown in Fig. 35-66 (*a*). We can map these values directly on to the four-variable Karnaugh map as shown in Fig. 35-66 (*b*). In order to simplify the Boolean expression represented on the Karnaugh map, group the 1s as shown in the Fig. 35-66 (*b*). The group of two squares in the first coloum produces a product term $B\overline{C}\,\overline{D}$. This is determined by observing that the group contains both A and $\overline{A}$, so this variable is eliminated. Another group of two squares in the second coloum produces term $\overline{B}\,CD$. The variable A is eliminated because the group contains both A and $\overline{A}$.

Decimal Number	Inputs				Outputs	
	A	B	C	D	F	
0	0	0	0	0	0	
1	0	0	0	1	1	→ $\overline{A}\,\overline{B}\,\overline{C}D$
2	0	0	1	0	1	→ $\overline{A}\,\overline{B}C\overline{D}$
3	0	0	1	1	1	→ $\overline{A}\,\overline{B}CD$
4	0	1	0	0	1	→ $\overline{A}B\overline{C}\,\overline{D}$
5	0	1	0	1	0	
6	0	1	1	0	0	
7	0	1	1	1	1	→ $\overline{A}BCD$
8	1	0	0	0	0	
9	1	0	0	1	1	→ $A\overline{B}\,\overline{C}D$
10	1	0	1	0	1	→ $A\overline{B}C\overline{D}$
11	1	0	1	1	0	
12	1	1	0	0	1	→ $AB\overline{C}\,\overline{D}$
13	1	1	0	1	0	
14	1	1	1	0	0	
15	1	1	1	1	0	

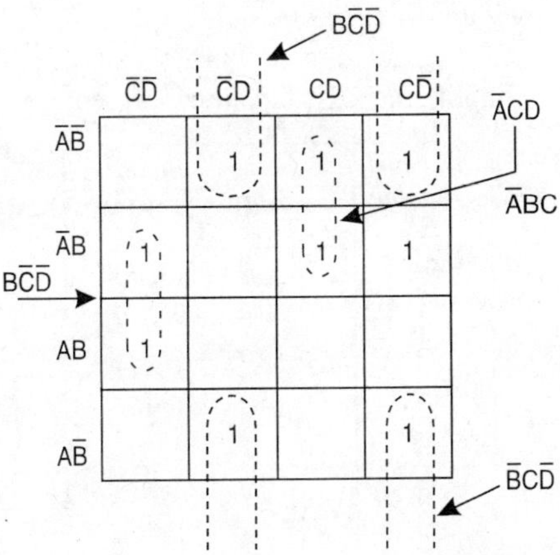

Fig. 35-66.

Another group of two squares in the third coloumn produces the term $\overline{A}\,CD$. The variable B is eliminated because the group contains both B and $\overline{B}$. Still another group of two squares in the fourth coloumn produces the term $\overline{B}\,C\overline{D}$. The variable A is eliminated because the group contains both A and A. Thus the resulting simplified expression,

$$f(A, B, C, D) = B\overline{C}\,\overline{D} + \overline{B}\,\overline{C}\,D + \overline{A}\,C\,D + \overline{B}\,C\,\overline{D}$$

This can be implemented using 3-input NAND gate as shown in Fig. 35-67.

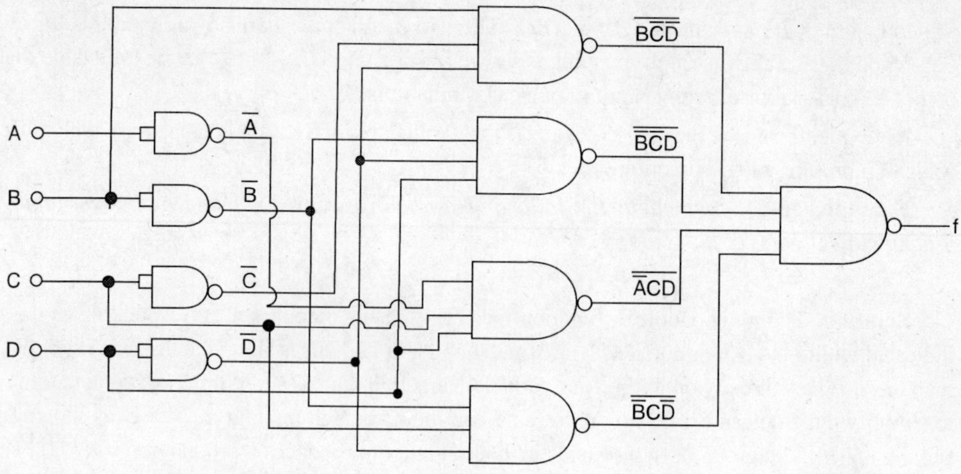

Fig. 35-67.

Example 35-52. *Minimize using K-map, the function,*

$$f\ (A,\ B,\ C,\ D) = \quad (1, 3, 5, 7, 9, 11, 13, 15)$$

Then convert the minimized function to POS form. (*UP Tech. Univ. Sem I, 2005-06*)

Solution. Given: The logic function,

$$f\ (A,\ B,\ C,\ D) = \quad (1, 3, 5, 7, 9, 11, 13, 15)$$

The given Boolean function indicates that its output is 1 corresponding to the term indicated within the expression, *i.e.*, 1, 3, 5, 7, 9, 11, 13 and 15. This is shown in Fig. 35-68 (*a*). We can map these values directly on the four variable Karnaugh map as shown in Fig. 35-68 (*b*). In order to simplify the logic function represented on the Karnaugh map, group the 1s as shown in Fig. 35-68 (*b*).

Decimal Number	Inputs A	B	C	D	Outputs f	
0	0	0	0	0	0	
1	0	0	0	1	1	→ $\overline{A}\overline{B}\overline{C}D$
2	0	0	1	0	0	
3	0	0	1	1	1	→ $\overline{A}\overline{B}CD$
4	0	1	0	0	0	
5	0	1	0	1	1	→ $\overline{A}B\overline{C}D$
6	0	1	1	0	0	
7	0	1	1	1	1	→ $\overline{A}BCD$
8	1	0	0	0	0	
9	1	0	0	1	1	→ $A\overline{B}\overline{C}D$
10	1	0	1	0	0	
11	1	0	1	1	1	→ $A\overline{B}CD$
12	1	1	0	0	0	
13	1	1	0	1	1	→ $AB\overline{C}D$
14	1	1	1	0	0	
15	1	1	1.	1	1	→ $ABCD$

(a)

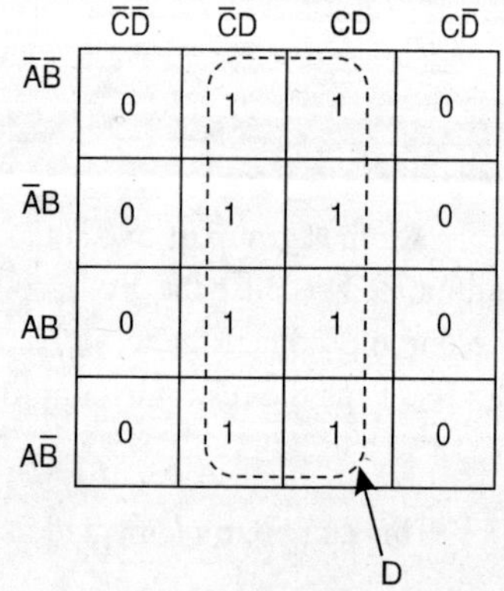

(b)

Fig. 35-68.

As seen from Fig. 35-68 (*b*), the eight squares containing 1s can be grouped together. This produces the term *D*. Therefore, the minimized function $f = D$.

Note that it is already in POS form. So no further action is required.

Example 35-53. *Convert SOP function :*

$$f(A, B, C, D) = A\bar{B}C + BCD + ACD + \bar{A}B$$

into POS form using K-map. (*UP Tech. Univ. Sem II, 2003-04*)

Solution. Given, the SOP expression is,

$$f(ABCD) = A\bar{B}C + BCD + ACD + \bar{A}B$$

Let us convert it into a standard SOP form.

$$\begin{aligned} f(A,B,C,D) &= A\bar{B}C(D+\bar{D}) + (A+\bar{A})BCD + A(B+\bar{B})CD + \bar{A}B(C+\bar{C})(D+\bar{D}) \\ &= A\bar{B}CD + A\bar{B}C\bar{D} + ABCD + \bar{A}BCD + ABCD + A\bar{B}CD \\ &\quad + \bar{A}BCD + \bar{A}B\bar{C}D + \bar{A}BC\bar{D} + \bar{A}B\bar{C}\bar{D} \end{aligned}$$

Sketch a four-variable Karnaugh (or K) map as shown in Fig. 35-69 select the product terms one by one and enter 1 for all the terms as shown in Fig. 35-69.

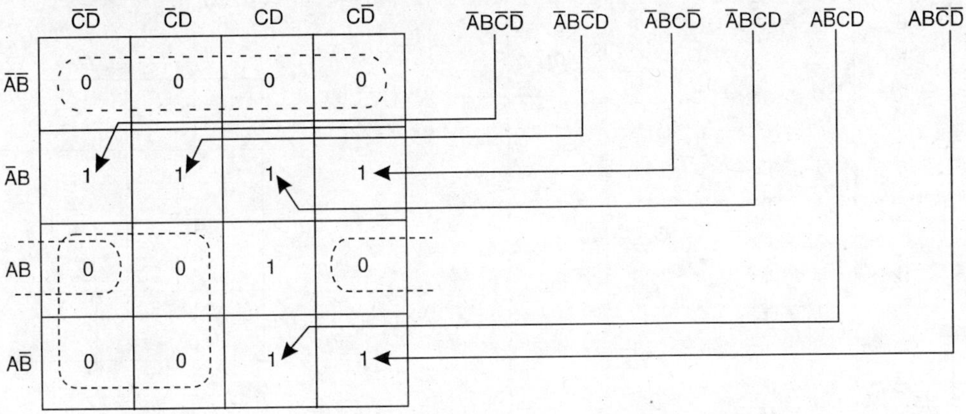

Fig. 35-69.

Since, there are 4 variables in the domain of SOP expression. (*i.e.*, *A*, *B*, *C* and *D*), there are total of 2^4 (= 16) possible conbinations. The given SOP expression contains 7 of these combinations. So the POS must contain the other 9 terms. These are $\bar{A}\bar{B}\bar{C}\bar{D}$, $\bar{A}\bar{B}\bar{C}D$, $\bar{A}\bar{B}C\bar{D}$, $\bar{A}\bar{B}CD$, $AB\bar{C}\bar{D}$, $AB\bar{C}D$, $ABC\bar{D}$ $A\bar{B}\overline{CD}$ and $A\bar{B}\bar{C}D$. Notice that we have entered 0's for all these 9 terms.

Grouping the four 0's in the top row, we get $\bar{A}\bar{B}$. Similarly, grouping the four 0's in the lower left corner of K-map, we get $A\bar{C}$. The 0 in the left-most column third row and a zero on the right most coloumn, third row can be combined together to producing a term $AB\bar{D}$.

Therefore, the SOP expression for zero entries is,

$$f = \bar{A}\bar{B} + A\bar{C} + AB\bar{D}$$

In product-of-sum (POS) form, this expression is,

$$f = (A+B)\cdot(\bar{A}+C)\cdot(\bar{A}+\bar{B}+D)$$ **Ans.**

Example 35-54. *Minimize the following boolean function using K-map.*

$$f(w, x, y, z) = \Sigma_m (0, 2, 5, 7, 8, 10, 13, 15)$$

(*UP Tech. Sem I (CO) 2003-04*)

Solution. Given, the Boolen function,

$$f(w, x, y, z) = \Sigma_m (0, 2, 5, 7, 8, 10, 13, 15)$$

The given Boolean function indicates that its output is 1 corresponding to the term indicated within the expression, *i.e.*, 0, 2, 5, 7 8, 10, 13 and 15. This is shown in Fig. 35-70 (*a*). We can map these values directly on the four-variable Karnaugh map as shown in Fig. 35-70 (*b*). In order to simplify the Boolean expression represented on the Karnaugh map, group the 1's as shown in Fig. 35-70 (*b*). The group of 1's or the four corners of the Karnaugh map produced a product term $\overline{x}\,\overline{z}$. This is determined by observing that the group contains both ($\overline{w}$ and w) and (y and $\overline{y}$), so these variable are eliminated. Another group of four squares in the centre of karnaugh map produced a term *xz* the variables *w* and *y* are eliminated because the group contains both ($\overline{w}$ and w) and (y and $\overline{y}$).

Decimal Number	W	X	Y	Z	Outputs f	
0	0	0	0	0	1	→ $\overline{W}\overline{X}\overline{Y}\overline{Z}$
1	0	0	0	1	0	
2	0	0	1	0	1	→ $\overline{W}\overline{X}Y\overline{Z}$
3	0	0	1	1	0	
4	0	1	0	0	0	
5	0	1	0	1	1	→ $\overline{W}X\overline{Y}Z$
6	0	1	1	0	0	
7	0	1	1	1	1	→ $\overline{W}XYZ$
8	1	0	0	0	1	→ $W\overline{X}\overline{Y}\overline{Z}$
9	1	0	0	1	0	
10	1	0	1	0	1	→ $W\overline{X}Y\overline{Z}$
11	1	0	1	1	0	
12	1	1	0	0	0	
13	1	1	0	1	1	→ $WX\overline{Y}Z$
14	1	1	1	0	0	
15	1	1	1	1	1	→ $WXYZ$

(a)

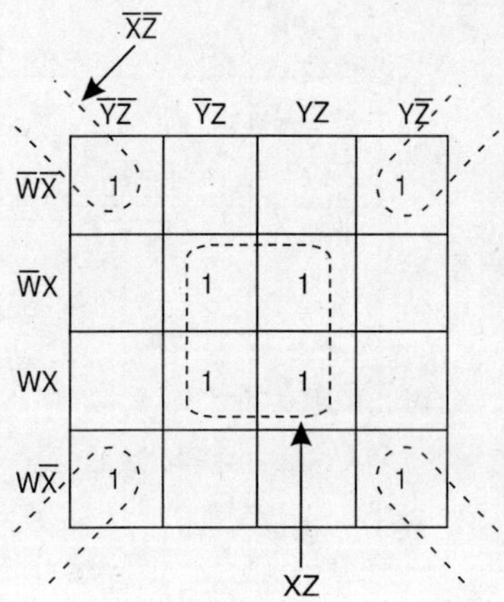

(b)

Fig. 35-70.

Thus the minimized Boolean function is,

$$f = \overline{x}\,\overline{z} + xz \quad \textbf{Ans.}$$

Example 35-55. *Simplify the following expression using K-map.*

$$f(A, B, C, D) = \pi_m\,(0,1,3,6,7,8,9,11,13,14,15)$$

(*UP Tech. Univ. Sem II, 2006-07*)

Solution. Given the logic expression,

$$f(A, B, C, D) = \pi_m\,(0,1,3,6,7,8,9,11,13,14,15)$$

Notice the expression is given in terms of product-of sums (and not sum-of products). This is evident from the symbol π_m (instead of *m*). The given expression indicates that its output is 0 corresponding to the term indicated within the expression. This is shown in Fig. 35-71 (*a*). We can map these values directly on to the four-variable Karnaugh map as shown in Fig. 35-71 (*b*). In order to simplify the Boolean expression represented on the Karnaugh map group the zero's as shown in the Fig. 35-71 (*b*).

Decimal Number	A	B	C	D	Outputs f	
0	0	0	0	0	0	→ $\overline{A}\overline{B}\overline{C}\overline{D}$
1	0	0	0	1	0	→ $\overline{A}\overline{B}\overline{C}D$
2	0	0	1	0	1	
3	0	0	1	1	0	→ $\overline{A}\overline{B}CD$
4	0	1	0	0	1	
5	0	1	0	1	1	
6	0	1	1	0	0	→ $\overline{A}BC\overline{D}$
7	0	1	1	1	0	→ $\overline{A}BCD$
8	1	0	0	0	0	→ $A\overline{B}\overline{C}\overline{D}$
9	1	0	0	1	0	→ $A\overline{B}\overline{C}D$
10	1	0	1	0	1	
11	1	0	1	1	0	→ $A\overline{B}CD$
12	1	1	0	0	1	
13	1	1	0	1	0	→ $AB\overline{C}D$
14	1	1	1	0	0	→ $ABC\overline{D}$
15	1	1	1	1	0	→ $ABCD$

(a)

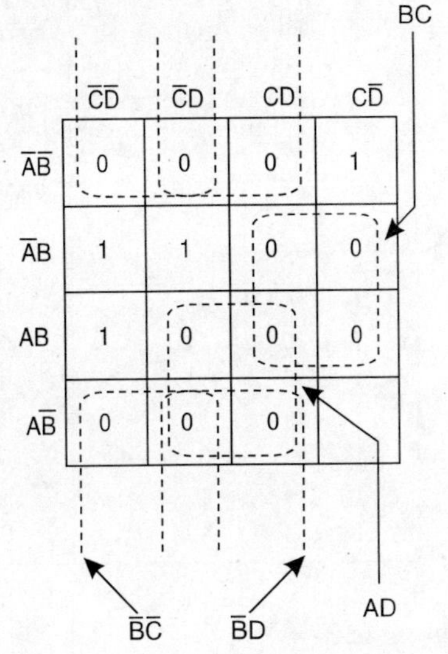

(b)

Fig. 35-71.

Grouping the 0's as shown, the product terms are,

$$f(A, B, C, D) = \overline{B}\,\overline{C} + \overline{B}\,D + BC + AD$$

and the sum terms are,

$$f(A, B, C, D) = (\overline{B} + \overline{C}) \cdot (B + \overline{D}) \cdot (B + C) \cdot (A + D) \quad \textbf{Ans.}$$

35-37. "Don't Care" Conditions

In digital systems design sometimes a situation arises in which some input variable conditions are not allowed. For example in a *BCD* (binary coded decimal) code, there are six invalid combinations: 1010, 1011, 1100, 1101, 1110 and 1111. Since, these unallowed states will never occur in an application application involving the *BCD* code, they can be treated as "terms either 1 or 0 may by assigned to the output.

Now, we shall discuss as how the "don't care" terms can be used to advantage on the Karnaugh map for simplifying the Logic equations.

Consider for example, a combinational circuit which products a '1' output corresponding to a *BCD* input equal and greater than 6. The output is 0 corresponding to a *BCD* input less than 6. The truth table for this situation is as shown in Fig. 35-72 (*a*).

We know that we can place 1*s* directly from the trough table on the Karnaugh map. Similarly we can place *X* for the "don't care" entries directly on the Karnaugh map as shown in Fig. 35-72 (*b*). When grouping the 1*s*, *Xs* can be treated as 1s make a larger grouping or as 0*s*, if they cannot be used to advantage Recall the larger the group of 1*s* the simpler the resulting term will be. Taking advantage of the "don't care" and using them as 1*s*, the resulting expression for the output is $A + BC$. However, if the "don't cares" are not used as 1*s*, the resulting expression is $A\overline{B}\,\overline{C} + \overline{A}BC$. Thus we can see the advantage of using "don't care" terms to get simplest logic expression.

Inputs A	B	C	D	Outputs X
0	0	0	0	0
0	0	0	1	0
0	0	1	0	0
0	0	1	1	0
0	1	0	0	0
0	1	0	1	0
0	1	1	0	1
0	1	1	1	1
1	0	0	0	1
1	0	0	1	1
1	0	1	0	x
1	0	1	1	x
1	1	0	0	x
1	1	0	1	x
1	1	1	0	x
1	1	1	1	x

(*a*)

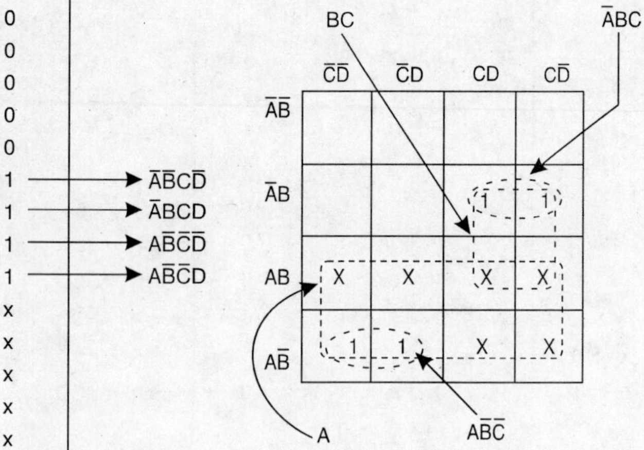

(*b*)

Fig. 35-72.

Example 35-56. *Consider the Karnaugh map shown in Fig. 35-73 (a). Determine the logic function represented by the map and simplify it in the minimal form.*

Solution. We know that when grouping the 1*s*, X*s* can be treated as 1*s* to make a larger grouping or as 0*s* if they cannot be used to advantage. Recall, the larger the group of 1*s*, the simpler the resulting term will be. Taking advantage of the X*s* and using them as 1*s*, the grouping of 1*s* and X*s* is as shown in Fig. 35-73.

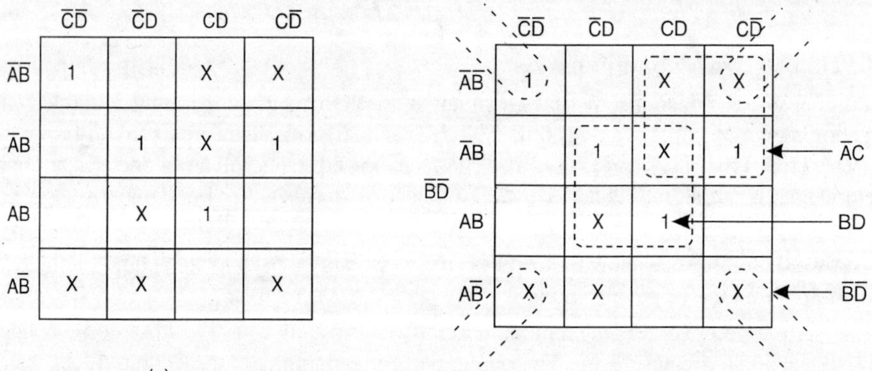

Fig. 35-73.

Notice the "wrap-around" four-square group that includes the 1*s* and X*s* on four corners of the Karnaugh map. This group produces a term $\bar{B}\bar{D}$. This is determined by observing that the group contains both A and $\bar{A}$ and C and $\bar{C}$, so these variables are eliminated.

Another group of four squares containing 1*s* and X*s* around the centre of Karnaugh map is formed. This group produces a product term BD. This is determined by observing that the group contains both A and $\bar{A}$ and C and $\bar{C}$, so these variables are eliminated.

Another group of four containing 1*s* and X*s* if formed near the top right corner of Karnaugh map. This group overlaps with the perevius group and produces a produces a product term $\overline{A}\,C$. This is determined by observing that the group contains both B and $\overline{B}$ and D and $\overline{D}$, so these variables are eliminated.

The resulting simplified logic function is the sum of the three product terms: $\overline{B}\,\overline{D}, BD$ and $\overline{A}\,C$, *i.e.*,

$$X = \overline{B}\,\overline{D} + BD + \overline{A}\,C$$

Example 35-57. *For the K-map shown in Fig. 35-74, find the minimized expression.*

	$\overline{C}\overline{D}$	$\overline{C}D$	CD	$C\overline{D}$
$\overline{A}\overline{B}$	0	0	X	1
$\overline{A}B$	0	X	X	0
AB	1	X	1	1
$A\overline{B}$	1	1	X	X

X→Don't care

Fig. 35-74.

(*UPSC Engg. Service Electronic & Telecom Engg. 2002*)

Solution. Given: The K-map shown in Fig. 35-74.

The grouping of 1s is as shown in Fig. 35-74. Notice the "wrap around" eight square groups that includes the 1s on the lower two rows of the Karnaugh map. The group produces as product term *A*. This is determined by observing that the group contains B and $\overline{B}$, C and $\overline{C}$, D and $\overline{D}$, so all these variables are eliminated.

Another group of four with "wrap-around" adjacency is formed from the top and the bottom rows of the Karnaugh map. This group overlaps with the previous group and produces a product term $\overline{B}C$. This is determined by obseving that this group contains both A and $\overline{A}$ and D and $\overline{D}$, so these variables are eliminated.

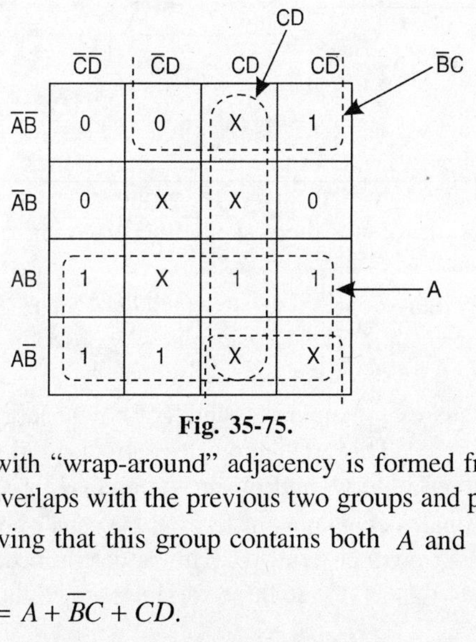

Fig. 35-75.

Another group of four with "wrap-around" adjacency is formed from the third column of the Karnaugh map. This group overlaps with the previous two groups and produces a product term *CD*. This is determined by observing that this group contains both A and $\overline{A}$ and B and $\overline{B}$, so these variables are eliminated.

Minimized expression $= A + \overline{B}C + CD$.

Example 35-58. Simplify the following expression using K-map and implement the output using fundamental gates.

$$f\,(A,\ B,\ C,\ D) = {}_{m}(1,3,4,6,8,9,11,13,15) + {}_{d}(0,2,14)$$

(UP Tech. Univ. sem II, 2006-07)

Solution. The given Boolean function indicates that its output is 1 corresponding to the term indicated with in the expression ${}_{m}(1,3,4,6,8,9,11,13,15)$, *i.e.*, 1, 3, 4, 6, 8, 9, 11, 13 and 15. The Boolean function also indicates that the output is X (*i.e.*, could be 1 or 0) corresponding to the terms 0, 2 and 14. This is shown in Fig. 35-76 (*a*). We can map these values directly on the four-variable Karnaugh map as shown in Fig. 35-76 (*b*). In order to simplify the Boolean expression represented on the Karnaugh map, group the 1's as shown in Fig. 35-76 (*b*).

Decimal Number	Inputs *A*	*B*	*C*	*D*	Output *f*	
0	0	0	0	0	X	→ $\bar{A}\bar{B}\bar{C}\bar{D}$
1	0	0	0	1	1	→ $\bar{A}\bar{B}\bar{C}D$
2	0	0	1	0	X	→ $\bar{A}\bar{B}C\bar{D}$
3	0	0	1	1	1	→ $\bar{A}\bar{B}CD$
4	0	1	0	0	1	→ $\bar{A}B\bar{C}\bar{D}$
5	0	1	0	1	0	
6	0	1	1	0	1	→ $\bar{A}BC\bar{D}$
7	0	1	1	1	0	
8	1	0	0	0	1	→ $A\bar{B}\bar{C}\bar{D}$
9	1	0	0	1	1	→ $A\bar{B}\bar{C}D$
10	1	0	1	0	0	
11	1	0	1	1	1	→ $A\bar{B}CD$
12	1	1	0	0	0	
13	1	1	0	1	1	→ $AB\bar{C}D$
14	1	1	1	0	X	→ $ABC\bar{D}$
15	1	1	1	1	1	→ $ABCD$

(a)

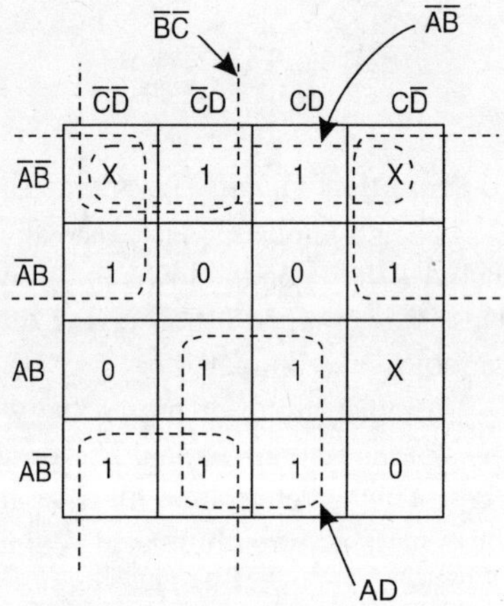

(b)

Fig. 35-76.

We know that when grouping 1*s*, *Xs* can be treated as 1*s* to make a larger grouping or as 0'*s*, if they cannot be used to advantage. Recall, the larger the group of 1*s*; the simpler the resulting term will be. Taking advantage of *Xs* and using them as 1*s*, the grouping of 1*s* and *Xs* is as shown in Fig. 35-76 (*b*).

Notice the "wrap around" four-square groups that includes 1'*s* and *X's* in the first row of Karnaugh map. This group produces a term $\bar{A}\,\bar{B}$. This is determined by observing that the group contains both (C and $\bar{C}$), and (D and $\bar{D}$), so these variables are eliminated.

Another group of four squares containing 1'*s* and *Xs* from the left-most and right-most columns (from the top two rows) is formed. This group produces a product term $\bar{A}\,\bar{D}$. This is determined by observing that the group contains both ($\bar{B}$ and B) and ($\bar{C}$ and C), so these variables are eliminated.

Another group of four squares containing in 1s from the center-bottom of the Karnaugh map is formed. This group produces a product term *AD*. This is determined by observing that the group contains both (B and $\bar{B}$) and ($\bar{C}$ and C), so these variables are eliminated.

Another group of four squares containing 1*s* from the top left and bottom-left rows of the Karnaugh map is formed. This group produces a product term $\overline{B}\,\overline{C}$. This is determined by observing that the group contains both (*A* and $\overline{A}$) and (*D* and $\overline{D}$), So these variables are eliminated.

The resulting simplified logic function is the sum of four product terms : $\overline{A}\,\overline{B}$, $\overline{A}\,\overline{D}$, *AD* and $\overline{B}\,\overline{C}$, *i.e.*,

$$f = \overline{A}\,\overline{B} + \overline{A}\,\overline{D} + AD + \overline{B}\,\overline{C} \textbf{ Ans.}$$

Example 35-59 *Minimize the function given below using K-map:*

$$f(A,B,C,D) \;=\; {}_m(0,2,3,6,8,9,12,15) + \quad d(1,4,10,11).$$

(*UP Tech. Univ. Sem II, 2003-04*)

Solution. Given the logic function.

The given logic function indicates that its output is 1 corresponding to terms indicated within the expression, *i.e.*, 0, 2, 3, 6, 8, 9, 12 and 15. The output is *X* (*i.e.*, 0 or 1) corresponding to the terms 1, 4, 10 and 11, as shown in Fig. 35-77 (*a*).

Decimal Number	Inputs A	B	C	D	Outputs f	
0	0	0	0	0	1	→ $\overline{A}\overline{B}\overline{C}\overline{D}$
1	0	0	0	1	X	→ $\overline{A}\overline{B}\overline{C}D$
2	0	0	1	0	1	→ $\overline{A}\overline{B}C\overline{D}$
3	0	0	1	1	1	→ $\overline{A}\overline{B}CD$
4	0	1	0	0	X	→ $\overline{A}B\overline{C}\overline{D}$
5	0	1	0	1	0	
6	0	1	1	0	1	→ $\overline{A}BC\overline{D}$
7	0	1	1	1	0	
8	1	0	0	0	1	→ $A\overline{B}\overline{C}\overline{D}$
9	1	0	0	1	1	→ $A\overline{B}\overline{C}D$
10	1	0	1	0	X	→ $A\overline{B}C\overline{D}$
11	1	0	1	1	X	→ $A\overline{B}CD$
12	1	1	0	0	1	→ $AB\overline{C}\overline{D}$
13	1	1	0	1	0	
14	1	1	1	0	0	
15	1	1	1	1	1	→ ABCD

(a)

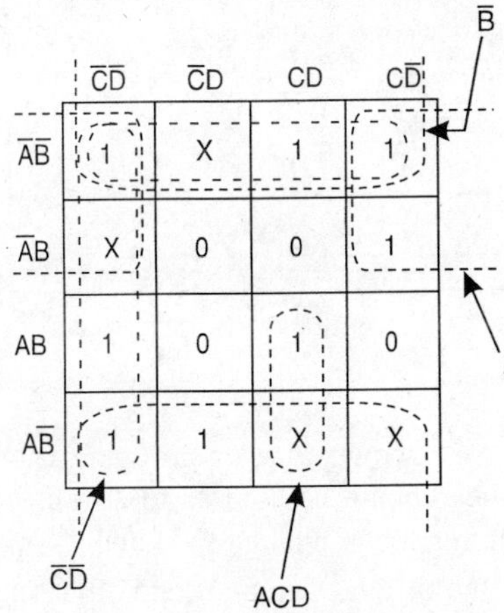

(b)

Fig. 35-77.

We can map these values directly on the Karnaugh map shown in Fig. 35-77 (*b*).

Grouping the 1*s* in the top and bottom rows of Karnaugh map, we get the term $\overline{B}$. Grouping 1*s* in the left-most column of Karnaugh map, we get the term $\overline{C}\,\overline{D}$. Similarly grouping the 1*s* in the top-two rows from the left-most and right-most columns, we get the term $\overline{A}\,\overline{D}$. The grouping of 1 with *X* in the inner right column produces a product term *ACD*. Therefore the minimized expression is,

$$f = \overline{B} + \overline{A}\,\overline{D} + \overline{C}\,\overline{D} + ACD$$

Example 35-60 *Minimize the given Boolean function using*

$$f(A,B,C,D) \;=\; {}_m(3,4,5,7,9,13,14,15) + \quad d(0,2,8).$$

Implement the minimized function only using NAND gates. (*UP Tech. Univ. Sem II, 2005-06*)

Solution. Givn, the Boolean function,

$$f(A, B, C, D) = \sum_m (3,4,5,7,9,13,14,15) + \sum d(0,2,8).$$

The given Boolean function indicates that its output is 1 corresponding to the terms 3, 4, 5, 7, 9, 13, 14 and 15 and is *X* (*i.e.*, 0 or 1) corresponding to the terms 0, 2 and 8. This is shown in Fig. 35-78 (*a*) We can map these values directly on the four-variable Karnaugh map as shown in Fig. 35-78 (*b*).

Decimal Number	Inputs *A*	*B*	*C*	*D*	Outputs *f*	
0	0	0	0	0	X	
1	0	0	0	1	0	
2	0	0	1	0	X	
3	0	0	1	1	1	→ $\bar{A}\bar{B}CD$
4	0	1	0	0	1	→ $\bar{A}B\bar{C}\bar{D}$
5	0	1	0	1	1	→ $\bar{A}B\bar{C}D$
6	0	1	1	0	0	
7	0	1	1	1	1	→ $\bar{A}BCD$
8	1	0	0	0	X	
9	1	0	0	1	1	→ $A\bar{B}\bar{C}D$
10	1	0	1	0	0	
11	1	0	1	1	0	
12	1	1	0	0	0	
13	1	1	0	1	1	→ $AB\bar{C}D$
14	1	1	1	0	1	→ $ABC\bar{D}$
15	1	1	1	1	1	→ $ABCD$

(a)

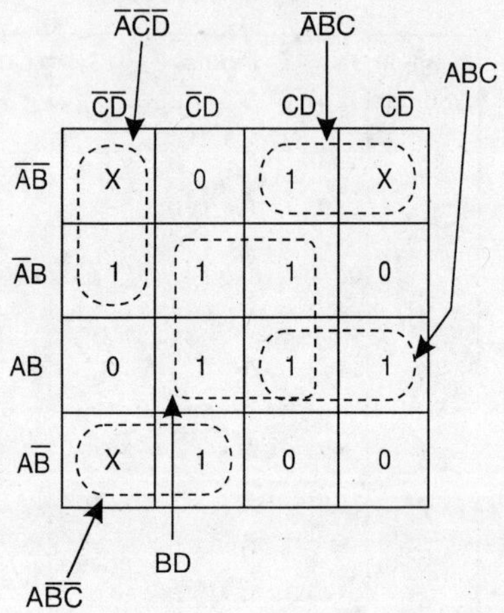

(b)

Fig. 35-78.

We know that when grouping 1*s*, *Xs* can be treated as 1*s* to make a larger grouping of 1*s* or as 0*s* if they cannot be used as to advantage. Taking advantages of *Xs*, and using then as 1*s* whenever required the grouping of 1'*s* and *Xs* is a shown in Fig. 35-78 (*b*).

Notice the "wrap-around" four square group that include all 1*s* in the centre of Karnaugh map. This group produces a term *BD*. This is determined by observing that the group contains (*A* and $\bar{A}$) and ($\bar{C}$ and *C*), so these variables are eliminated.

Similarly notice the "wrap-around" two square group that contain *X* and 1 in the left-most column of the Karnaugh map. This group produces a product term $\bar{A}\bar{C}\bar{D}$. This is determined by observing that the group contains $\bar{B}$ and *B*, so this variable is eliminated.

Following a similar approach, the other "wrap-arounds" produces $\bar{A}BC$, ABC and $A\bar{B}\bar{C}$ terms. Thus the minimized Boolean expression is

$$f = BD + \bar{A}\bar{C}\bar{D} + \bar{A}\bar{B}C + ABC + A\bar{B}\bar{C}$$

In order to implement this function using only NAND gates, let us represent the function as

$$f = \overline{\overline{BD + \bar{A}\bar{C}\bar{D} + \bar{A}\bar{B}C + ABC + A\bar{B}\bar{C}}}$$

$$= \overline{\overline{BD} \cdot \overline{\bar{A}\bar{C}\bar{D}} + \overline{\bar{A}\bar{B}C} + \overline{ABC} + \overline{A\bar{B}\bar{C}}}$$

the above expression can be implemented using NAND gates circuit as shown in Fig. 35-79.

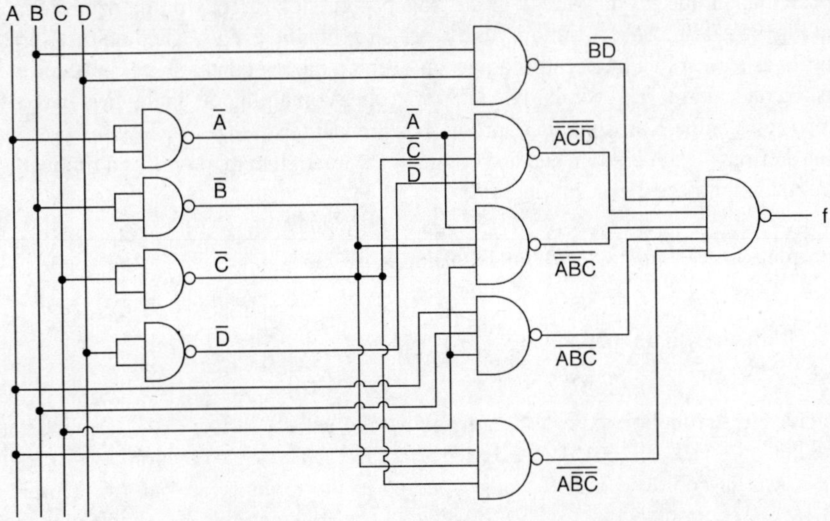

Fig. 35-79.

Example 35-61 *Minimize the given Boolean function with the help of Boolean Algebra rules:*

$$f\ (A,\ B,\ C,\ D) = \bar{A}\,\bar{B}\,\bar{C}\,D + \bar{A}\,\bar{B}\,C\,D + \bar{A}\,B\,\bar{C}\,D + \bar{A}\,BCD + A\,\bar{B}\,\bar{C}\,\bar{D}$$
$$+A\,\bar{B}\,\bar{C}\,D + A\,\bar{B}\,C\,\bar{D} + AB\bar{C}\,\bar{D} + AB\,\bar{C}\,D + ABC\bar{D}$$

(UP Tech. Univ., Sem II 2005-06)

Solution. Given the Boolean function,

$$f\,(A, B, C, D) = \bar{A}\,\bar{B}\,\bar{C}\,D + \bar{A}\,\bar{B}\,C\,D + \bar{A}\,B\,\bar{C}\,D + \bar{A}\,BCD + A\,\bar{B}\,\bar{C}\,\bar{D}$$
$$+A\,\bar{B}\,\bar{C}\,D + A\,\bar{B}\,C\,\bar{D} + AB\bar{C}\,\bar{D} + AB\,\bar{C}\,D + ABC\bar{D}$$

The can be minimized by using Boolean algebra rules as follows.

$$f\ (A,\ B,\ C,\ D) = \bar{A}\,\bar{B}\,(\bar{C} + C)\,D + \bar{A}\cdot B\,(\bar{C} + C)\,D + A\,\bar{B}\,\bar{C}\,(\bar{D} + D)$$
$$+A\,(\bar{B} + B)\,C\bar{D} + AB\bar{C}\,(\bar{D} + D)$$
$$= \bar{A}\,\bar{B}D + \bar{A}\,BD + A\,\bar{B}\,\bar{C} + AC\bar{D} + AB\bar{C}$$
$$= \bar{A}\,(\bar{B} + B)\,D + A\,(\bar{B} + B)\,\bar{C} + AC\,\bar{D}$$
$$= \bar{A}D + A\bar{C} + AC\bar{D}$$
$$= \bar{A}D + A\,(\bar{C} + C\bar{D})$$
$$= \bar{A}D + A\,(\bar{C} + \bar{D}) = \bar{A}D + A\bar{C} + A\bar{D}\quad \textbf{Ans.}$$

Note: This question can be solved alternatively by using Karnaugh map as shown in Fig. 35-80.

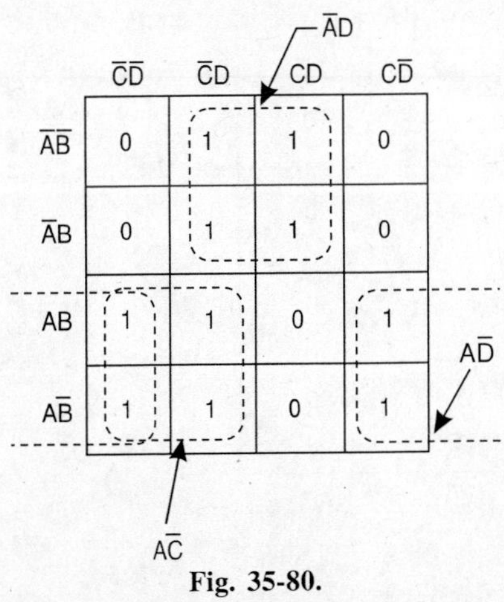

Fig. 35-80.

We can map all the product terms in the given expression directly on the four-variable map as shown in Fig. 35-80. In order to simplify the Boolean expression represented on the Karnaugh map, group the 1*s* as shown in Fig. 35-80. The group of four squares containing 1*s* in the top two rows, produces a product term $\overline{A}D$. Similarly, the four squares containing 1'*s* in the first two coloums and bottom two-rows of the Karnaugh map, produces a product term, $A\overline{C}$. Finally the group of squares containing 1*s* from the 1st column and last column of Karnaugh map, produces a product term, $A\overline{D}$. Therefore, the minimized Boolean function,

$$f\ (A,\ B,\ C,\ D) = \overline{A}D + A\overline{C} + A\overline{D} \quad \textbf{Ans.}$$

Example 35-62 *Minimize the following using K-map.*

$$f = \ _m(1, 2, 5, 7, 9, 15) + \phi \ \ _m(0, 3, 4, 6)$$ (*UP Tech. Univ., Sem I, 2006-07*)

Solution. Given, the logic function,

$$f = \ _m(1, 2, 5, 7, 9, 15) + \phi \ \ _m(0, 3, 4, 6)$$

The given logic function indicates that its output is 1 corresponding to the term indicated within the expression $_m(1, 2, 5, 7, 9, 15)$, *i.e.*, 1, 2, 5, 7, 9 and 15. The logic function also indicates that the output is *X* (*i.e.*, it could be 1 or 0) corresponding to the terms 0, 3, 4 and 6. This is shown in Fig. 35-81(*a*). We can map these values directly on the four variable Karnaugh map as shown in Fig. 35-81(*b*). In order to simplify the logic function represented on the Karnaugh map, group the 1*s* as shown in Fig. 35-81 (*b*).

We know that when grouping 1*s*, *Xs* can be treated as 1*s* to make a larger grouping or as 0*s*, if they cannot be used to advantage. Recall, the larger the group of 1*s*, the simpler the resulting term will be. Taking advantage of *Xs* and using then as 1*s*, the grouping of 1*s* and *Xs* is as shown in Fig. 35-81 (*b*).

Notice the "wrap-around" eight square groups in the top-two rows of Karnaugh map. This group produces a product term $\overline{A}$.

Decimal Number	Inputs A	B	C	D	Outputs f	
0	0	0	0	0	X	
1	0	0	0	1	1	→ $\overline{A}\overline{B}\overline{C}D$
2	0	0	1	0	1	→ $\overline{A}\overline{B}C\overline{D}$
3	0	0	1	1	X	
4	0	1	0	0	X	
5	0	1	0	1	1	→ $\overline{A}B\overline{C}D$
6	0	1	1	0	X	
7	0	1	1	1	1	→ $\overline{A}BCD$
8	1	0	0	0	0	
9	1	0	0	1	1	→ $A\overline{B}\overline{C}D$
10	1	0	1	0	0	
11	1	0	1	1	0	
12	1	1	0	0	0	
13	1	1	0	1	0	
14	1	1	1	0	0	
15	1	1	1	1	1	→ ABCD

(a)

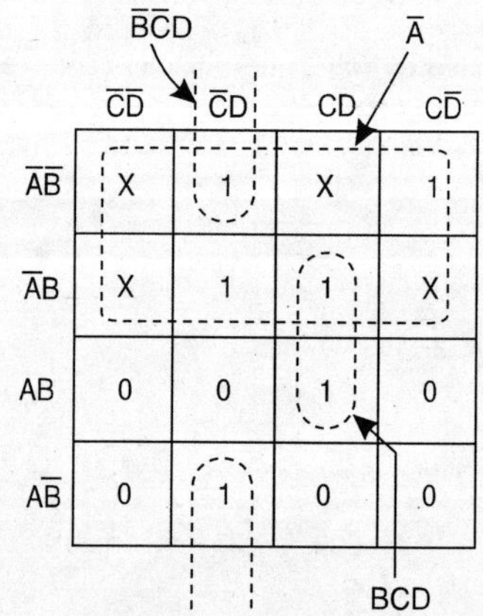

(b)

Fig. 35-81.

A "wrap-around" two squares in the third column, produces a product term of *BCD*. Similarly, a "wrap-around" two squares in the second column (1st and last row) of Karnaugh map, produces a product term of $\overline{B}\,\overline{C}\,D$. Therefore the minimized logic function,

$$f = \overline{A} + BCD + \overline{B}\,\overline{C}\,D \quad \textbf{Ans.}$$

SUMMARY

In this chapter, you have learnt that :

1. There are several Boolean laws and rules that provide the means to form equivalent logic circuits and/or simplification of logic circuits.

2. De-Morgan's theorems are used in logic simplification whenever the inversion bars cover more than one variable in the original Boolean expression. The two theorems are:

$$\overline{X + Y} = \overline{X}\,\overline{Y}$$

$$\text{and} \quad \overline{XY} = \overline{X} + \overline{Y}$$

3. The Sum-of-Products (SOP) and Product-of-Sums (POS) are two general forms for the logic expressions.

4. The Karnaugh map is a graphical method for representing a logic circuit's truth table and generating a simplified expression for the circuit output.

GLOSSARY

Adjacency. Characteristic of cells in the Karnaugh map in which there is a single variable change from one cell to another cell next to it any of its four sides.

Associative law. In addition (ORing) and multiplication (ANDing) of three or more variables, the order in which the variables are grouped makes no difference.

Cell. An area on the Karnaugh that represents a unique combination of variables in a product term.

Commutative law. In addition (ORing) and multiplication (ANDing) of two variables, the order in which the variables are ORed and ANDed makes no difference.

Distributive law. ORing several variables and then ANDing the result with the single variable is equivalent to ANDing the single variable with each of the several variables and then ORing the products.

Domain. All the Boolean variables in any given Boolean equation or an expression.

Don't care. A combination of input literals (or variables) that cannot occur and can be used as a 1 or a 0 on a Karnaugh map.

Karnaugh map (or K-map). An arrangement of cells representing the combination of literals (or variables) in a Boolean expression and used for a systematic simplification of the expression.

Literal. A variable or a complement of a variable.

Product-of-sums (POS). A Boolean expression that is basically the ANDing of ORed terms.

Sum-of-products (SOP). A Boolean expression that is basically the O-Ring of ANDed terms.

Sum term. The Boolean sum of two or more literals or product terms) equivalent to an OR operation.

Variable. A symbol used to represent a logical quantity that can have a value of 1 or 0. It is usually designated by an italic letter e.g. *A*, *B*, *C* and *D*.

DESCRIPTIVE QUESTIONS

1. Define each of the following terms:

   (*a*) Karnaugh map (*b*) Sum-of-Products form

   (*c*) Product-of-Sums form (*d*) Don't care state.

2. Use Boolean algebra to prove than Exclusive NOR (XNOR) output expression is the exact inverse of the XOR output expression.
3. Why is De-Morgan's theorem important in the simplification of Boolean expressions?
4. Use De-Morgan's theorem to prove that a NOR gate with inverted inputs is equivalent to an AND gate.
5. Explain briefly how De-Morgan's theorem is applied in the simplification of Boolean expression.
6. Sketch a 3-variable Karnaugh map and label each cell according to its binary value.
7. Sketch a 4-variable Karnaugh map and label each cell according to its binary value.
8. What do you understand by min terms and max terms? Explain briefly with suitable truth table.

   [**Hint:** minterm is the same as product term and max term is the same as sum term]

9. What do you understand by the term "Canonical form"?

   [**Hint:** The Boolean function expressed as a sum of min terms or product of max. terms are said to be in canonical form]

10. Write note on the following:

    (*a*) De-Morgan's theorem (*b*) Caronical form of Boolean function.

    (*UP Tech. Univ., Sem I., 2005-06*)

TUTORIAL PROBLEMS

1. Using the trugh table, show that the Boolean expression,

   (*a*) $AB + A\overline{B} = A$ (*b*) $\overline{A} + AB = \overline{A} + B$

   (*c*) $\overline{A} + B = \overline{A}B$ (*d*) $A(\overline{A} + \overline{B}) = AB$

2. Simplify the following Boolean expressions to a minimum number of literals:

   (*a*) $\overline{X}\,\overline{Y} + XY + \overline{X}Y$ (*b*) $\overline{X}Y + X\overline{Y} + XY + \overline{X}\,\overline{Y}$

   (**Ans.** (*a*) $\overline{X} + \overline{Y}$; (*b*) 1)

3. Using the Boolean laws and rules, simplify the following expressions:

   (a) $Z = (\overline{A} + B)(A + \overline{B})$ (b) $Y = \overline{A}C + AB\overline{C}$

   (c) $Y = \overline{A}\,\overline{B}C\overline{D} + \overline{A}\,\overline{B}\,\overline{C}$ (d) $Y = \overline{A}D + ABD$

   (**Ans.** (*a*) AB, (*b*) $A\overline{C}$, (*c*) $\overline{A}\,\overline{B}\,\overline{D}$ (*d*) $(\overline{A} + B)D$)

4. Simplify the following Boolean function using Boolean algebra:

   $$X = A + \overline{A}B + \overline{A}\,\overline{B} \quad \textbf{Ans.}$$

5. (*a*) Write a logic expression for the output *y* in terms of logic variables *A*, *B* and *C* from the truth table shown in Fig. 35-82

   (*b*) Give a suitable circuit for realizing *Y* with a minimum number of logic gates.

A	B	Y
0	0	$\overline{C}$
0	1	C
1	0	1
1	1	0

Fig. 35-82.

6. Simplify the Boolean expressions:

(*a*) $X = ABC + AB\overline{C} + A\overline{B}C$

(*b*) $X = (A + B + AB)(A + C)$ (**Ans.** (*a*) $A(B + C)$, (*b*) $A + BC$)

7. Prove the following identity in the truth table:

$$XY + YZ + Y\overline{Z} = X\overline{Z} + Y\overline{Z}$$

8. Simplify the following Boolean equations using rules of Boolean algebra:

$$x = (A + \overline{B}\,\overline{C})(\overline{B + C})$$ (**Ans.** $\overline{B}\,\overline{C}$)

9. Design a logic circuit to give an output,

$$x = (\overline{AB} + \overline{A}C)(\overline{AC + C})$$

10. Simplify the following Boolean expressions:

(*a*) $y = AB\overline{C} + A\overline{B}\,\overline{C} + \overline{A}\,\overline{C}\,\overline{B} + A\overline{B}$

(*b*) $y = AB\overline{C}\,\overline{D} + AB\overline{C}D + ABC + ABD + CD$

11. Determine by means of truth table, the validity of De-Morgan's theorem for three variables:

$$\overline{ABC} = \overline{A} + \overline{B} + \overline{C}$$

12. Simplify the following Boolean expression using De-Morgan's theorem:

(*a*) $z = \overline{A + \overline{B}C}$

(*b*) $y = \overline{(A + BC)(D + EF)}$

(**Ans.** (*a*) $\overline{A}(B + \overline{C})$; (*b*) $\overline{A}\,\overline{B} + \overline{A}\,\overline{C} + \overline{D}\,\overline{E} + \overline{D}\,\overline{F}$)

13. Use De Morgan's theorems to simplify the following Boolean expressions:

(*a*) $y = \overline{R\overline{S}T + \overline{Q}}$

(*b*) $y = \overline{A + \overline{B} + \overline{C}D}$

(**Ans.** (*a*) $(\overline{R} + S + \overline{T})Q$; (*b*) $y = \overline{A}B(C + \overline{D})$)

14. Determine the Boolean expression for the circuit shown in Fig. 35-83. Use De-Morgan's theorem and then Boolean algebra rules to simplify the equation. Draw the Simplified circuit.

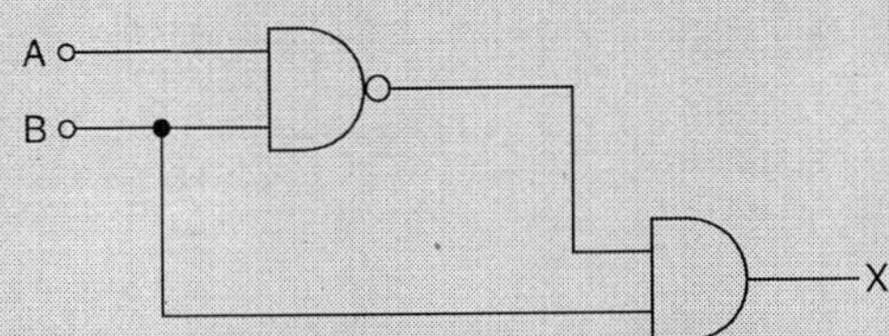

Fig. 35-83.

15. Repeat Problem 6 for the circuit shown in Fig. 35-84 (**Ans.** $\overline{B} + \overline{C}$)

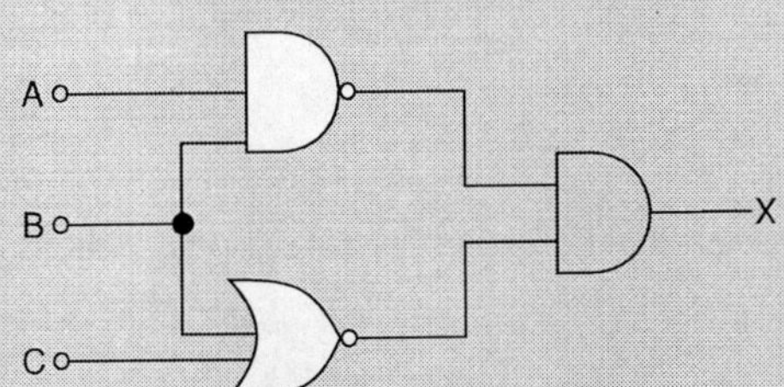

Fig. 35-84.

16. Repeat Problem 6 for the circuit shown in Fig. 35-85. (**Ans.** $\overline{A} + \overline{B} + C + D$)

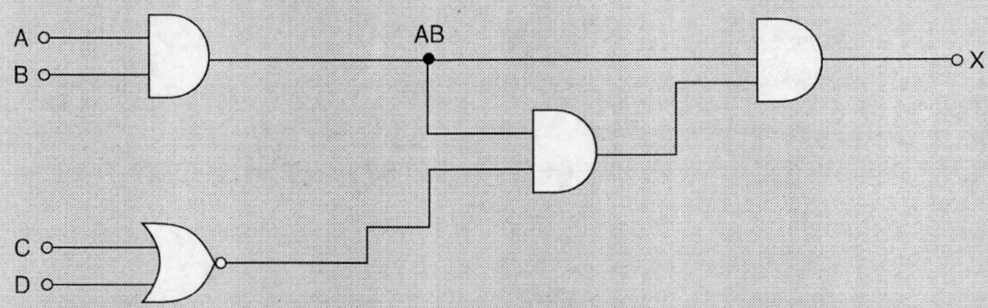

Fig. 35-85.

17. Determine the Boolean expression for the logic circuit shown in Fig. 35-86. Simplify the Boolean expression using Boolean Laws and De-Morgan's theorem. Redraw the logic circuit using the simplified Boolean expression.

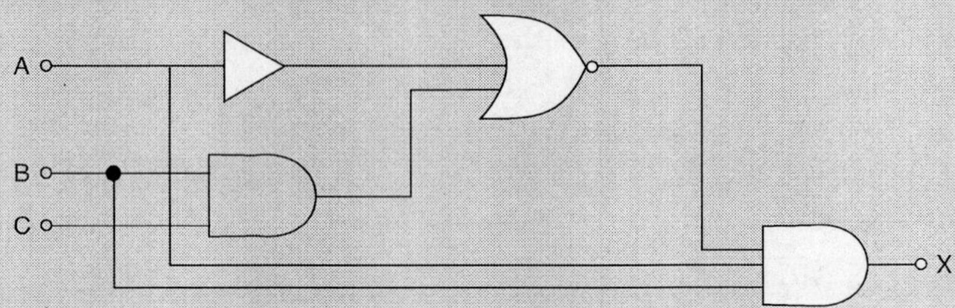

Fig. 35-86.

18. Determine the output, *X* of a logic circuit shown in Fig. 35-87. Simplify the output expression using Boolean Laws and theorems. Redraw the logic circuit with the simplified expression. (**Ans.** $\overline{B}\,\overline{C}\,D + AC$)

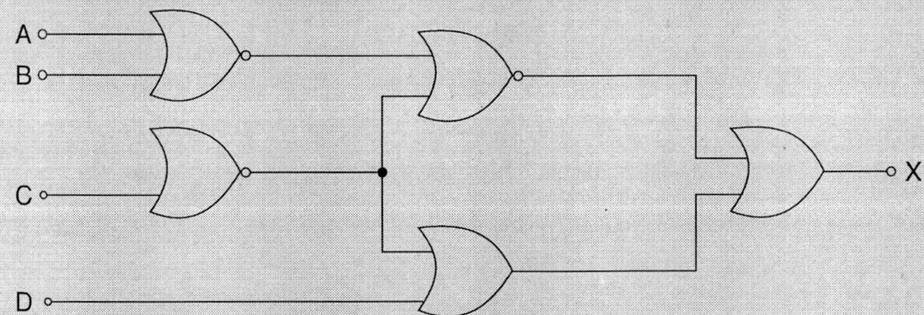

Fig. 35-87.

19. Consider the logic circuit shown in Fig. 35-88. Determine the Boolean expression at the circuit output *X*, simplify it. Using the simplified expression, redraw the logic circuit.

(**Ans.** $A(\overline{B} + C)$)

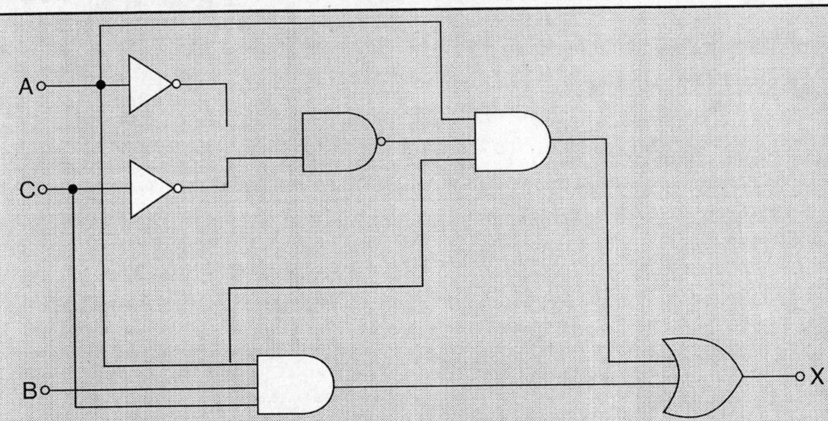

Fig. 35-88.

20. Given the following Boolean expression: $f = x\bar{y}z + \bar{x}\bar{y}z + \bar{w}xy + w\bar{x}y + wxy$

(*a*) Obtain the truth table of the function.

(*b*) Draw the logic diagram using original Boolean expression.

21. Express the Boolean function, $f = A + \bar{B}C$ in sum-of min-terms.

22. Express the Boolean function, $f = xy + \bar{x}z$ in product-of max-terms.

23. Express each function in sum of-product terms and product of-sum terms:

(*a*) $(AB + C)(B + AC)$ (*b*) $(\bar{A} + B)(\bar{B} + C$

24. Express the Boolean function:

$f(A, B, C) = \quad (0,1,2,5,8,9,10)$ in product-of-sums form.

25. Express the following in sum-of min terms and product of max terms:

(*a*) $f(A, B, C, D) = \quad (0,2,6,11,13,14)$ (*b*) $f(x, y, z) = \Pi(0,3,6,7)$

(**Ans.** (*a*) sum-of min terms $\bar{A}\bar{B}\bar{C}\bar{D} + \bar{A}\bar{B}C\bar{D} + \bar{A}BC\bar{D} + A\bar{B}CD + ABC\bar{D}$,

Product-of max-terms: $(\bar{A} + B + C + D)(A + \bar{B} + \bar{C} + D)(A + \bar{B} + \bar{C} + D)$

$(A + \bar{B} + C + \bar{D})(A + B + \bar{C} + \bar{D})(A + B + C + D)$;

(*b*) Sum-of min-terms $= \bar{X}\bar{Y}Z + \bar{X}Y\bar{Z} + X\bar{Y}\bar{Z} + X\bar{Y}Z$

Product of max-terms: $(\bar{X} + \bar{Y} + \bar{Z})(\bar{X} + \bar{Y} + Z)(\bar{X} + Y + Z)(X + Y + Z))$

26. Fig. 35-89 shows a three-variable Karnaugh map. Group the 1*s* and hence obtain the minimized Boolean expression.

	$\bar{C}$	C
$\bar{A}\bar{B}$	1	
$\bar{A}B$		1
AB	1	1
$A\bar{B}$		

Fig. 35-89.

	$\bar{C}\bar{D}$	$\bar{C}D$	CD	$C\bar{D}$
$\bar{A}\bar{B}$	1	1		
$\bar{A}B$	1	1	1	1
AB				
$A\bar{B}$		1	1	

Fig. 35-90.

27. Fig. 35-90 shows a four-variable Karnaugh-map. Group the 1*s* and hence obtain the minimized Boolean expression. (**Ans.** $\overline{A}B + \overline{A}\,\overline{C} + A\overline{B}D$)

28. A truth table has output 1*s* for these inputs : $ABCD = 0011$, $ABCD = 0110$, $ABCD = 1000$ and $ABCD = 1100$, and 0*s* for the other inputs. Draw the Karnaugh map and find the simplified Boolean equation for the truth table. (**Ans.** $A\overline{C}\,\overline{D} + \overline{A}\,\overline{B}CD + \overline{A}BC\overline{D}$)

29. Determine the minimized expression for the Karnaugh map shown in Fig. 35-91. (**Ans.** $B\overline{C} + BD + \overline{B}C$)

	$\overline{C}\,\overline{D}$	$\overline{C}D$	CD	$C\overline{D}$
$\overline{A}\,\overline{B}$	0	0	1	1
$\overline{A}B$	1	1	1	0
AB	1	X	X	X
$A\overline{B}$	0	0	X	X

Fig. 35-91.

	$\overline{C}\,\overline{D}$	$\overline{C}D$	CD	$C\overline{D}$
$\overline{A}\,\overline{B}$	1	1		1
$\overline{A}B$	1	1		1
AB				1
$A\overline{B}$				

Fig. 35-92.

30. Determine the simplified Boolean expression for the Karnaugh map shown in Fig. 35-92. *(UPSC Engg. Services 1995)* (**Ans.** $\overline{A}\,\overline{C} + \overline{A}\,\overline{D} + ABC$)

31. Simplify the following Boolean functions using Karnaugh map:

(*a*) $f(A, B, C) = (0, 2, 3, 4, 6)$ (**Ans.** $\overline{C} + \overline{A}B$)

(*b*) $f(A, B, C, D) = 0, 1, 2, 4, 5, 7, 11, 15$ (**Ans.** $\overline{A}\,\overline{C} + \overline{A}\,\overline{B}\,\overline{D} + \overline{A}BD + ACD$)

32. Simplify the Boolean function:

$$f(w, x, y, z) = (0, 1, 2, 4, 5, 6, 8, 9, 12, 13, 14)$$

and draw the logic diagram for simplified Boolean expression. (**Ans.** $\overline{C} + \overline{A}\,\overline{D} + B\overline{D}$)

33. Minimize the following function given below using the Karnaugh map:

$$f(A, B, C, D) = (0, 1, 2, 3, 5, 6, 8, 10, 15)$$

and realize the function using NAND gates. (**Ans.** $\overline{AB} + \overline{B}\,\overline{D} + \overline{A}C\overline{D} + ABCD$)

34. Simplify the following Boolean function in product-of-sums form by means of four variable map. Draw the logic diagram using gates:

$$f(A,\ B,\ C,\ D) = (2, 3, 4, 5, 6, 7, 11, 14, 15)$$

(**Ans.** $\overline{A}B + CD + \overline{A}C + BC$)

35. Given the logical function of the three variables:

$$f(A, B, C) = (A + \overline{B} + C)(\overline{A} + B + C)$$

Express *f* in the standard sum-of-products form.

(**Ans.** $\overline{A}\,\overline{B}C + \overline{A}\,\overline{B}C + \overline{A}B\overline{C} + A\overline{B}\,\overline{C} + AB\overline{C} + ABC$)

36. Simplify the following using the Karnaugh map:

$$f = \overline{A}\,\overline{B}\,\overline{C} + \overline{B}C + \overline{A}B$$ (**Ans.** $\overline{A} + \overline{B}C$)

37. Design a combinational circuit with three inputs *A*, *B*, *C* and three outputs *x*, *y*, *z*. When the binary input is 0, 1, 2 or 3, the binary output is one greater than the input. When the binary input is 4, 5, 6 or 7, the binary output is one less than the input.

38. Seven switches operate the lamp in the following way :- if switches 1, 3, 5 and 7 are closed and switch 2 is open or if switches 2, 4 and 5 are closed and switch 3 is open, or if all seven switches are closed, the lamp will light. Use NOT, AND and OR gates to show how the switches must be connected.

39. Design a logic circuit that will allow a signal to pass the output only when control inputs B and C are both HIGH, otherwise the output will stay LOW.

40. Design a logic circuit with inputs P, Q and R so that the output S is high whenever P is 0 or whenever $Q = R = 1$.

MULTIPLE CHOICE QUESTIONS

1. Boolean algebra is essentially based on

(*a*) symbols (*b*) logic (*c*) truth (*d*) numbers

2. Different variables used in Boolean can have values of

(*a*) 0 or 1 (*b*) LOW or HIGH (*c*) true or false (*d*) ON or OFF

3. According to Boolean algebra, $(A + \overline{A})$ equals

(*a*) A (*b*) 1 (*c*) 0 (*d*) $A\overline{A}$

4. According to Absorptive law of Boolean algebra, $(A + AB)$ equals,

(*a*) A (*b*) B (*c*) AB (*d*) A

5. When we De-Morganise $\overline{A}\,\overline{B}$, we get,

(*a*) $\overline{A}\,\overline{B}$ (*b*) $\overline{A} + \overline{B}$ (*c*) $\overline{A + B}$ (*d*) $A + B$

6. The expression $\overline{ABC}$ can be simplified to

(*a*) $\overline{A}\,B\,\overline{C}$ (*b*) $AB + BC + CA$ (*c*) $AB + \overline{C}$ (*d*) $\overline{A} + \overline{B} + \overline{C}$

7. According to Boolean Algebra

8. While obtaining minimal sum-of-products expression,

(*a*) all don't cares are ignored

(*b*) all don't cares are treated as logic 1*s*

(*c*) all don't cares are treated as logic 0*s*

(*d*) only such don't cares that help minimization are treated as logic 1*s*.

9. Which of the following Boolean algebra rules is correct

(*a*) $A \cdot \overline{A} = 1$ (*b*) $A + A \cdot B = A + B$

(*c*) $A + \overline{A}B = A + B$ (*d*) $A(A + B) = B$

10. Simplified form of Boolean expression $(A + \overline{B} + \overline{A}B)C$ is

(*a*) 1 (*b*) 0 (*c*) C (*d*) $\overline{C}$

11. Which of the following is not true?

(*a*) $A \cdot A = A$ (*b*) $A \cdot \overline{A} = 0$ (*c*) $A + 1 = 1$ (*d*) $A + \overline{A} = 0$

12. The minimum number of NAND gates required to implement the Boolean function $A + \overline{A}B + \overline{A}\overline{B}C$ is

(*a*) three (*b*) two (*c*) one (*d*) zero

13. The Boolean expression $(\overline{A} + B)(A + B)$ when simplified yields to

(*a*) A (*b*) B (*c*) $\overline{A}$ (*d*) $\overline{B}$

14. The Boolean expression for the shaded area in the Venn diagram is

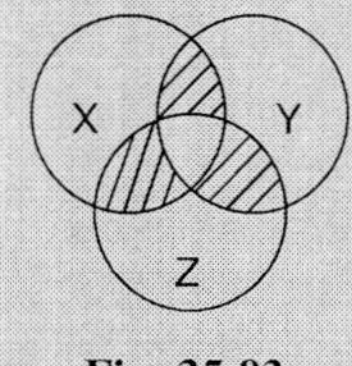

Fig. 35-93.

(*a*) $XY + YZ + XZ$ (*b*) $XY\overline{Z} + \overline{X}YZ + X\overline{Y}Z$

(*c*) $XYZ + \overline{X}\,\overline{Y}\,\overline{Z}$ (*d*) $\overline{X}\,\overline{Y}\,\overline{Z} + X\overline{Y}\,\overline{Z}$

15. $X + \overline{X}Y$ is reduced to

(*a*) X (*b*) $X + \overline{Y}$ (*c*) $X + Y$ (*d*) $\overline{X} + Y$

16. A four-variable Karnaugh map is

(*a*) 8 min-terms (*b*) 16 min-terms (*c*) 32 min-terms (*d*) 24 min-terms

17. Minimum number of literals in the expression, $ABC + AB + BC$ is

(*a*) 1 (*b*) 2 (*c*) 3 (*d*) 4

18. $\overline{ABCD} + B\overline{C}D + \overline{AC} + A$ is equivalent to

(*a*) 1 (*b*) $\overline{C}$ (*c*) $\overline{A} + C$ (*d*) $A + \overline{C}$

19. In Karnaugh map, the Gray reflected code is used to number

(*a*) only rows (*b*) only column

(*c*) both rows and columns (*d*) the diagonal elements.

20. Boolean expression $\overline{xy}z + yz + xz$ can be reduced to

(*a*) $\overline{x}$ (*b*) z (*c*) y (*d*) $x + y$

21. The Karnaugh map for a Boolean function is shown in Fig. 35-94. The number of essential prime implicates for the function is

	$\overline{C}\overline{D}$	$\overline{C}D$	CD	$C\overline{D}$
$\overline{A}\overline{B}$	1	0	1	1
$\overline{A}B$	1	0	0	0
AB	1	0	0	0
$A\overline{B}$	0	1	0	1

Fig. 35-94.

(*a*) 4 (*b*) 5 (*c*) 6 (*d*) 8

22. In Boolean algebra, if $f = (A + B)(\overline{A} + C)$ then,

(*a*) $f = AB + \overline{A}C$ (*b*) $f = AB + \overline{AB}$

(*c*) $f = AC + \overline{A}B + BC$ (*d*) $f = AA + \overline{A}B$

23. Which one of the following is equivalent to the Boolean expression: $y = \overline{A}\,\overline{B} + \overline{BC} + \overline{CA}$

(*a*) $\overline{AB + BC + CA}$ (*b*) $(\overline{A} + \overline{B})(\overline{B} + \overline{C})(\overline{A} + \overline{C})$

(*c*) $\overline{(A + B)(B + C)(C + A)}$ (*d*) $(\overline{A + B})(\overline{B + C})(\overline{C + A})$

ANSWERS

1. (*b*)	**2.** (*a*)	**3.** (*b*)	**4.** (*a*)	**5.** (*d*)
6. (*d*)	**7.**	**8.** (*d*)	**9.** (*c*)	**10.** (*c*)
11. (*d*)	**12.** (*a*)	**13.** (*b*)	**14.** (*b*)	**15.** (*c*)
16. (*b*)	**17.** (*c*)	**18.** (*a*)	**19.** (*c*)	**20.** (*b*)
21. (*a*)	**22.** (*c*)	**23.** (*d*)		

❑❑❑

36

CHAPTER

Arithmetic Operations and Circuits

OUTLINE

Objectives

After Completing of this chapter, you should be able to :

- Do binary addition, subtraction, multiplication and division on the given two binary numbers.
- Add and subtract the given two hexadecimal numbers.
- Manipulate the signed binary numbers using the 2's complement system.
- Understand the usage of full-adders in the design of parallel binary adders.

36-1. Introduction

Many digital systems such as computers, video game machines, PDAs, recent mobile phones have a capability to do a variety of arithmetic operations.

In order to understand the functioning of such a digital system, it is important to know the various methods of performing arithmetic operations and the circuits that can do that task. We will begin our study by learning first, how the various arithmetic operations are performed on binary numbers using "pencil and paper". Then we will study the actual integrated circuits that perform these operations in digital systems.

36-2. Binary Addition

The addition of two binary numbers is performed in exactly the same manner as the addition of decimal numbers. Let us review the decimal addition. Consider the addition of two decimal numbers 264_{10} and 173_{10}.

		MSD		LSD
Carry	→	1		
		2	6	4
	+	1	7	3
		4	3	7

Recall that for decimal addition we start from the least-significant digit side. Thus, the sum of digits 4 and 3 produce 7. Next the digits in the second position are added. This produces13. So we write 3 in the second position and a carry of 1 in the third position. Finally the digits in the third position are added along with the carry. This produces a digit 4 in the third position.

We can follow the same general steps for binary addition. Unlike decimal addition, there are only four cases that can occur in binary addition. These four cases are:

$$0 + 0 = 0$$
$$1 + 0 = 1$$
$$1 + 1 = 10 = 0 + \text{carry of } 1 \text{ into next position.}$$
$$1 + 1 + 1 = 11 = 1 + \text{carry of } 1 \text{ into next position.}$$

The last case occurs when the two bits in a certain position are 1 and there is a carry from the previous position.

Consider for example the addition of the two binary numbers at a time (decimal equivalents are shown in parenthesis)

(*a*)	(*b*)	(*c*)
Carry → 1		
1 1 (3)	1 0 0 (4)	1 1 0 (6)
+ 1 1 (3)	+ 1 0 (2)	1 0 0 (4)
1 1 0 (6)	1 1 0 (+ 6)	1 0 1 0 (10)

As a matter of fact, it is not necessary to consider the addition of more than two binary numbers at a time. It is because of the fact that in all digital systems the circuitry that actually performs the addition can handle only two numbers at a time. Whenever we need to add more than two numbers, first the two numbers are added together and then their sum is added to the third number and so on. It is not a serious drawback since digital computers can typically perform an addition operation in few microseconds or less.

It will be interesting to know that addition is the most important arithmetic operation in digital systems. The operations of subtraction, multiplication and division are carried out using addition as their basic operation.

36-3. 1's and 2's Complements of Binary Numbers

The 1's complement and 2's complement of a binary number are important because they allow us to represent the negative numbers. The method of 2's complement system (discussed later in Art. 36-7) is commonly used to handle negative numbers. Let us now study how the 1's complement and 2's complement of a binary number is obtained.

Obtaining 1's complement of a binary number. The 1's complement of a binary number is obtained by simply changing (or complementing) all 1s to 0s and all 0s to 1s. For example, the 1's complement of 10110010 is obtained as follows:

MSB							LSB	
1	0	1	1	0	0	1	0	— Binary number
↓	↓	↓	↓	↓	↓	↓	↓	
0	1	0	0	1	1	0	1	—1's complement

∴ The 1's complement of 10110010 = 01001101

Obtaining 2's Complement of a binary number. The 2's complement of a binary number is obtained by adding 1 to the LSB (least-significant bit) of the 1's complement. For example, 2's complement of 10110010 is obtained as follows:

MSB							LSB	
1	0	1	1	0	0	1	0	— Binary number
↓	↓	↓	↓	↓	↓	↓	↓	
0	1	0	0	1	1	0	1	— 1's complement
+							1	
0	1	0	0	1	1	1	0	— 2's complement

∴ The 2's complement of 10110010 is 01001110

An alternative method of obtaining 2's complement of a binary number is as follows:

(*i*) Start at the right with the LSB (least-significant bit) and write the bits as they are up to and including the first 1.

(*ii*) Take 1's complement of the remaining bits.

Let us illustrate the procedure by finding the 2's complement of 10111000.

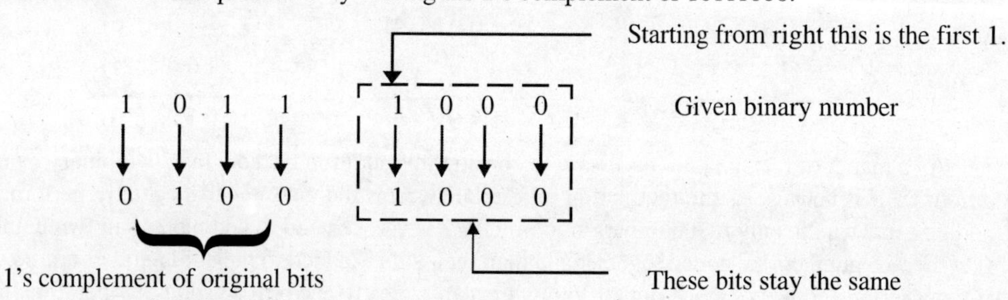

∴ 2's complement of 10111000 is 01001000.

Example 36-1. *Determine the 2's complement of the following 8-bit binary numbers*:

(*a*) 00010110 (*b*) 11111100 (*c*) 10010001

Solution.

(*a*) **2's complement of 00010110**

	0	0	0	1	0	1	1	0	— Given binary number
	↓	↓	↓	↓	↓	↓	↓	↓	
	1	1	1	0	1	0	0	1	— 1's complement
+								1	
	1	1	1	0	1	0	1	0	— 2's complement.

∴ The 2's complement of 00010110 is 11101010 **Ans.**

(*b*) **2's complement of 11111100**

	1	1	1	1	1	1	0	0	— Given binary number
	↓	↓	↓	↓	↓	↓	↓	↓	
	0	0	0	0	0	0	1	1	— 1's complement
+								1	
	0	0	0	0	0	1	0	0	— 2's complement

∴ The 2's complement of 11111100 is 00000100 **Ans.**

(*c*) **2's complement of 10010001**

	1	0	0	1	0	0	0	1	— Given binary number
	↓	↓	↓	↓	↓	↓	↓	↓	
	0	1	1	0	1	1	1	0	— 1's complement
+								1	
	0	1	1	0	1	1	1	1	— 2's complement

∴ The 2's complement of 10010001 is 01101111 **Ans.**

36-4. Representing Signed Numbers

A signed number is a number which consists of both sign and magnitude information. The sign indicates whether the number is positive or negative and the magnitude is the value of the number.

There are three methods of representing signed numbers. These are:

(*i*) The sign magnitude system

(*ii*) The 1's complement system

(*iii*) The 2's complement system.

All these methods are discussed one by one in the following pages.

36-5. Sign-Magnitude System

In this system, the left-most bit of a binary number is the sign-bit. The sign-bit is 0 for positive and a 1 for negative binary numbers. The magnitude bits are in true (or uncomplemented) binary for both positive and negative numbers.

For example, the decimal number +13 is expressed as a binary number in a sign- magnitude system as

$$+13 = 0\ \underbrace{1101}$$

Sign bit (0) — Magnitude bits (1101)

The decimal number –13 is expressed as

$$-13 = 1\ \underbrace{1101}$$

Sign bit (1) — Magnitude bits (1101)

Notice the only difference between + 13 and – 13 is the sign bit because the magnitude bits are in true binary for both positive and negative numbers.

Example 36-2. *Express the decimal number – 25 as a binary number using sign-magnitude system.*

Solution. Given the decimal number = – 25

$$-25 = 1\ \underbrace{11001}$$

Sign bit (1) — Magnitude bits (11001)

Notice that the sign bit is 1 because the given decimal number is negative.

$\therefore \quad -25_{10} = 111001_2$ **Ans.**

36-6. 1's Complement System

The 1's complement system for representing the signed numbers works as explained below :

If a number is positive— the sign bit is 0 and the magnitude of a binary number is represented in its true binary form.

If a number is negative— the sign bit is 1 and the magnitude of a binary number is represented as 1's complement of a corresponding positive number.

For example consider the decimal number +9. Its magnitude is represented as 1001, *i.e.*, +9 =1001. When we attach a sign bit of 0, the complete signed number becomes 01001. However, if we consider a decimal number -9_{10}, it must be represented in 1's complement form.

$$\begin{array}{ccccc} +9 = & 1 & 0 & 0 & 1 \\ & \downarrow & \downarrow & \downarrow & \downarrow \\ & 0 & 1 & 1 & 0 \end{array} \quad \text{— 1's complement}$$

When we attach a sign bit of 1, the complete signed number becomes,

$$-9 = 1\ 0110$$

Sign bit (1)

36-7. 2's Complement System

The 2's complement system for representing the signed numbers works as explained below:

If a number is positive— the sign bit is 0 and the magnitude of a binary number is represented in its true binary form.

If a number is negative— the sign bit is 1 and the magnitude of a binary number is represented as 2's complement of a corresponding positive number.

For example consider the decimal number +19. Its magnitude is represented as 10011. When we attach a sign bit of 0, the complete signed number becomes 010011. However, if we consider -19_{10}, it must be represented in 2's complement form.

19	= 1	0	0	1	1		— Binary equivalent of 19_{10}
	0	1	1	0	0		— 1's complement
	+				1		
	0	1	1	0	1		— 2's complement

When we attach a sign bit of 1, the complete signed number becomes,

$$-19_{10} = 1\ 01101_2$$

Sign bit (arrow pointing to the leading 1)

36-8. Special Case in 2's Complement System

We have already discussed the 2's complement system in the last article. Now we will study some special cases in this system of representing signed binary numbers.

Whenever a signed binary number has a 1 in the sign bit and all 0s for the magnitude bits, its decimal equivalent is -2^N, where N is the number of bits in the magnitude. For example,

$$1000 = -2^3 = -8$$

$$10000 = -2^4 = -16$$

$$100000 = -2^5 = -32$$

and so on.

Thus, we can state that the complete range of the values that can be represented in the 2's complement system having N magnitude bits is.

$$-2^N \text{ to } +(2^N-1)$$

In other words, there are 2^{N+1} different values including zero.

Fig. 36-1 lists all the signed numbers that can be represented in four bits using the 2's complement system (note that there are three magnitude bits and one sign bit). The sequence starts at $-2^N = -2^3 = -8 = 1000$ and proceed upwards to $+(2^N-1) = +2^3-1 = +7 = 0111$.

The 2's complement system is the most widely used method of representing signed binary numbers and performing arithmetic operations in computer systems. Most computer systems are based on 8-bit or 136-bit numbers. In a 8-bit system, the total number of different combinations of bits is 2^8 (decimals 256). On the other hand in a 136-bit system, the number is (2^{16} equals 65, 536).

In a 8-bit system, we have seven bits for magnitude and one bit for representing the sign. So the range of positive numbers in an 8-bit system is 0000 0000 to 01111111 (*i.e.*, 0 to 127). On the other hand, the range of negative numbers is $1111\ 1111_2$ to 1000 0000 (*i.e.*, – 1 to – 128).

Decimal Value	*Signed binary in 2's complement*
$+7 = 2^3 - 1$	0111
+6	0110
+5	0101
+4	0100
+3	0011
+2	0010
+1	0001
0	0000
–1	1111
–2	1110
–3	1101
–4	1100
–5	1011
–6	1010
–7	1001
$-8 = -2^3$	1000

Fig. 36-1.

Example 36-3. *Determine the range of signed decimal values that can be represented in* 12 *bits* (*including the sign bit*).

Solution. We know that a 12-bit signed number will have one sign-bit and eleven bits to represent magnitude.

The largest negative value is,

$$1000\ 0000\ 0000 = -2^{11} = -2048_{10}$$

and the largest positive value is,

$$0111\ 1111\ 1111 = +2^{11} - 1 = +2047$$

Thus, the complete range is – 2048 to + 2047; this is a total of 4096 different values, including zero. Alternatively, since there are eleven magnitude bits (N = 11), then there are $2^{N+1} = 2^{12} = 4096$ different values. **Ans.**

Example 36-4. *Determine the largest negative decimal value that can be represented by a two-byte number* ?

Solution. We know that a byte is a group of eight bits. So a two-byte number is a sixteen – bit number. Now since MSB (*i.e.*, most significant bit) is used to represent a sign, then there are fifteen bits for magnitude.

∴ The largest negative value in a sixteen-bit number is,

$$= -2^{15} = -32,768 \quad \textbf{Ans.}$$

Example 36-5. *List an* 8-*bit representation of numbers by starting with positive number* + 7 to – 8, *using* 2'*s complement*.

Solution. Given the decimal numbers from + 7 to – 8.

Decimal Number	*Representation using 2's complement*
+ 7	0000 0111
+ 6	0000 0110
+ 5	0000 0101
+ 4	0000 0100
+ 3	0000 0011
+ 2	0000 0010
+ 1	0000 0001
0	0000 0000
– 1	1111 1111
– 2	1111 1110
– 3	1111 1101
– 4	1111 1100
– 5	1111 1011
– 6	1111 1010
– 7	1111 1001
– 8	1111 1000

Fig. 36-2.

We know that for all positive decimal numbers starting from +7 to 0 will have a sign bit as zero and the magnitude part as in the true binary form. Therefore the representation of these numbers is as shown in Fig. 36-2.

We also know that for all negative decimal numbers starting from – 1 to – 8 will have have a sign bit of '1' and the magnitude put in the 2's complement form.

Following the steps explained earlier we can obtain 2's complement of the magnitude bit of all the negative number. Attaching a sign bit-1 with the magnitude part, the number will become as shown in Fig. 36-2.

Example 36-6. *Express the decimal number –39 as an 8-bit number in sign- magnitude, 1's complement and 2's complement systems.*

Solution. Given the decimal number = – 39

Representation of –39 in sign- magnitude system.

First of all, let us write the 8-bit binary number for + 39. This can be written as:

$$+ 39 = 00100111$$

In the sign magnitude system, – 39 is produced by changing the sign-bit (i.e. MSB) to a 1 and leaving the magnitude bits unchanged. The number is.

$$\underset{\text{Sign-bit}}{1}\ \underbrace{0100111}_{\text{Magnitude}} = -39$$

$\therefore$ – 39 = 10100111 **Ans.**

Representation of –39 in 1's complement system.

In 1's complement system, – 39 is produced by taking 1's complement of + 39. Thus,

+ 39 =	0	0	1	0	0	1	1	1	— Binary number
	↓	↓	↓	↓	↓	↓	↓	↓	
	1	1	0	1	1	0	0	0	—1's complement

∴ – 39 = 11011000 **Ans.**

Representation of – 39 in 2's complement system.

In 2's complement system, – 39 is produced by taking 2's complement of +39. Thus.

+ 39 =	0	0	1	0	0	1	1	1	— Binary number
	1	1	0	1	1	0	0	0	— 1's complement
+								1	
	1	1	0	1	1	0	0	1	— 2's complement

∴ – 39 = 11011001 **Ans.**

Example 36-7. *Using 2's complement system show the binary representation of* – 43.875_{10} *and* – 0.712_{10} (*Nagpur University Oct/ Nov. 1997*)

Solution. Given : the numbers – 43.875_{10} and – 0.712_{10}

Binary representation of – 43.875

Let us use 2's complement system to represent the given number into its equivalent binary.

0	1	0	1	0	1	1	.	1	1	1	— Binary equivalent of + 43.875_{10}
↓	↓	↓	↓	↓	↓	↓		↓	↓	↓	
1	0	1	0	1	0	0	.	0	0	0	— Complement each bit to form 1's complement
										+ 1	— add 1 to form 2's complement
1	0	1	0	1	0	0	.	0	0	1	— 2's complement of 43.875

Thus the binary representation of – 43.875 in 2's complement system is 1010100.001 **Ans.**

Binary representation of – 0.712.

Again use 2's complement system to represent the given number into its equivalent binary.

0	.	1	0	1	1	0	1	1	0	— Binary equivalent of + 0.712_{10}
↓		↓	↓	↓	↓	↓	↓	↓	↓	
1		0	1	0	0	1	0	0	1	— Complement each bit to form 1's complement
									+ 1	— add 1 to form 2's complement
1	.	0	1	0	0	1	0	1	0	— 2's complement of + 0.712_{10}

Thus the binary representation of – 0.712_{10} in 2's complement system is 1.01001010_{10} **Ans.**

36-9. Evaluation of Signed Numbers

We have already discussed in Art 36-5, 36-6 and 36-7 the three different systems for representing signed numbers. Now we will study the evaluation of signed numbers, i.e. the reverse process. This means given the signed number expressed in any one of three systems: sign-magnitude, 1's complement or 2's complement system, how to obtain the decimal value of the signed number. This is discussed below separately for each signed number separately.

36-10. Evaluation of Numbers Represented in Sign-magnitude System

Positive and negative numbers in the sign-magnitude system are evaluated by summing the position values (or weights) in all the *magnitude* bit positions where there are 1s and ignoring those position values where there are 0s. The sign is determined by examination of the sign bit.

Consider for example the signed number 10010101 represented in sign magnitude system. The seven magnitude bits and their position values are as follows:

0 0 1 0 1 0 1	— Magnitude bits of signed number
↓ ↓ ↓ ↓ ↓ ↓ ↓	
2^6 2^5 2^4 2^3 2^2 2^1 2^0	— Position values of the bits
$\cancel{2^6}$ $\cancel{2^5}$ 2^4 $\cancel{2^3}$ 2^2 $\cancel{2^1}$ 2^0	— Ignore position, values which correspond to 0's
$2^4 + 2^2 + 2^1$	— Sum the position, values which correspond to 1's.

$= 16 + 4 + 1$

$= 21$

Since the sign bit is 1, therefore the decimal value is -21_{10}.

Example 36-8. *Determine the decimal value of the sign-magnitude number* 01110111.

Solution. Given the number represented in sign magnitude system = 01110111. The decimal value of this number can be obtained as shown below:

1 1 1 0 1 1 1	— Magnitude bits of the signed number
↓ ↓ ↓ ↓ ↓ ↓ ↓	
2^6 2^5 2^4 2^3 2^2 2^1 2^0	— Position value of the bits
2^6 2^5 2^4 $\cancel{2^3}$ 2^2 2^1 2^0	— Ignore position values, which correspond to 0's
$2^6 + 2^5 + 2^4 + 2^2 + 2^1 + 2^0$	— Sum the position values which correspond to 1s.

$= 64 + 32 + 16 + 4 + 2 + 1$

$= 119$

Since, the sign bit is 0, therefore the decimal value is +119 **Ans.**

36-11. Evaluation of a Number Represented in 1's Complement System

Positive numbers in the 1's complement system are evaluated by summing the position values (or weights) in all bit positions where there are 1s and ignoring those positions where there are 0s. Negative numbers are evaluated by assigning a *negative value* to the position value of the sign bit, summing all the position values where there are 1s and adding 1 to the result.

Consider for example, the evaluation of the signed number 00010111_2. The decimal value of this number is determined as shown below:

0 0 0 1 0 1 1 1	— 8-bits of the 1's complement number
↓ ↓ ↓ ↓ ↓ ↓ ↓ ↓	
-2^7 2^6 2^5 2^4 2^3 2^2 2^1 2^0	— Position value of the bits
↑	

Note that the sign bit is assigned the –ve value.

$-\cancel{2^7}\ \cancel{2^6}\ \cancel{2^5}\ 2^4\ \cancel{2^3}\ 2^2\ 2^1\ 2^0$ — Ignore position values which correspond to 0s

$2^4 + 2^2 + 2^1 + 2^0$ — Sum the position values which correspond to 1s

$= 16 + 4 + 2 + 1$

$= 23$

∴ The decimal value of the signed number 00010111(represented in 1's complement system) is + 23.

Example 36-9. *Determine the decimal value of the* 1*'s complement number* 11101011.

Solution. Given, the 1's complement number = 11101011

In order to determine its decimal value, we write the number and the position value under each bit. Recall that the sign bit is to be assigned a negative position value. Sum all the position values where there are 1s and ignore those positions where there are 0s. If the sum is negative add 1 to the result.

1	1	1	0	1	0	1	1
↓	↓	↓	↓	↓	↓	↓	↓
-2^7	2^6	2^5	2^4	2^3	2^2	2^1	2^0
-2^7	2^6	2^5	$\cancel{2^4}$	2^3	$\cancel{2^2}$	2^1	2^0

$= -\ 2^7 + 2^6 + 2^5 + 2^3 + 2^1 + 2^0$

$= -\ 128 + 64 + 32 + 8 + 2 + 1$

$= -\ 21$

Since, the number is negative, we add 1 to the result, therefore the final number,

$$= -21 + 1 = -20_{10} \quad \textbf{Ans.}$$

36-12. Evaluation of a Number Represented in 2's Complement System

Positive and negative numbers in 2's complement system are evaluated by summing the position values (or weights) in all bit positions where there are 1s and ignoring those positions where there are 0s. The position value of the sign bit in the negative number is given a negative value.

Consider for example, the evaluation of 2's complement number 01010110_2. The decimal value of this number is determined as shown below:

0	1	0	1	0	1	1	0	— 8-bits of 2's complement number.
↓	↓	↓	↓	↓	↓	↓	↓	
-2^7	2^6	2^5	2^4	2^3	2^2	2^1	2^0	— Position value of the bits
↑								

Note that the sign bit is assigned the –ve value.

$-\cancel{2^7}\ 2^6\ \cancel{2^5}\ 2^4\ \cancel{2^3}\ 2^2\ 2^1\ \cancel{2^0}$ — Ignore position values which correspond to 0s.

$2^6 + 2^4 + 2^2 + 2^1$ — Sum the position values which correspond to 1s

$= 64 + 16 + 4 + 2$

$= +\ 86$

∴ The decimal value of the 2's complement number is + 86.

Now let us consider another signed number 11101000_2 represented in 1's complement system. The decimal value of this number is determined as shown below :

$$\begin{array}{cccccccc} 1 & 1 & 1 & 0 & 1 & 0 & 0 & 0 \\ \downarrow & \downarrow & \downarrow & \downarrow & \downarrow & \downarrow & \downarrow & \downarrow \\ -2^7 & 2^6 & 2^5 & 2^4 & 2^3 & 2^2 & 2^1 & 2^0 \\ \uparrow \end{array}$$

— 8-bits of the signed number.

— Position value of the bits.

Note that the sign bit is assigned the –ve value.

$-2^7 \quad 2^6 \quad 2^5 \quad \not{2^4} \quad 2^3 \quad \not{2^2} \quad \not{2^1} \quad \not{2^0}$ — Ignore position values which correspond to 0s.

$-2^7 + 2^6 + 2^5 + 2^3$ — Sum the position values which correspond to 1s.

$= -128 + 64 + 32 + 8$

$= -24$

Since, the sum is negative, so we add 1 to the result. Therefore the final number is,

$$-24 + 1 = 23$$

Thus, the decimal value of the signed number 11101000 (represented in 1s complement system) is – 23.

To summarize if the sign bit is positive sum the position values in all the magnitude bit positions where there are 1s and ignore those positions where there are 0s. However, if the sign bit is negative, then first complement all the magnitude bits, then sum all the magnitude bit positions where there are 1s and ignore those positions where there are 0s.

Now let us consider another 2's complement number 10101010. The decimal value of this number is determined as shown below.

$$\begin{array}{cccccccc} 1 & 0 & 1 & 0 & 1 & 0 & 1 & 0 \\ \downarrow & \downarrow & \downarrow & \downarrow & \downarrow & \downarrow & \downarrow & \downarrow \\ -2^7 & 2^6 & 2^5 & 2^4 & 2^3 & 2^2 & 2^1 & 2^0 \\ \uparrow \end{array}$$

— 2's complement number.

— Position value of the bits

Note that the sign-bit is assigned a –ve value.

$-2^7 \quad \not{2^6} \quad 2^5 \quad \not{2^4} \quad 2^3 \quad \not{2^2} \quad 2^1 \quad \not{2^0}$ — Ignore position values which correspond to 0's

$= -128 + 32 + 8 + 2$ — Sum the position values which correspond to 1's

$= -86$

$\therefore$ The decimal value of the 2's complement number 10101010 is – 86.

To summarize, if the sign-bit is positive, sum the position values in all the magnitude bit positions where there are 1s and ignore those positions where there are 0s. However if the sign-bit is negative, then first complement all the magnitude bits, then add one to the least-significant bit. Then sum all the magnitude bit positions where there are 1s and ignore those positions where there are 0s.

Example 36-10. *Determine the decimal value of the 2's complement number* 111010111.

Solution. Given the 2's complement number = 111010111.

$$\begin{array}{ccccccccc} 1 & 1 & 1 & 0 & 1 & 0 & 1 & 1 & 1 \\ \downarrow & \downarrow & \downarrow & \downarrow & \downarrow & \downarrow & \downarrow & \downarrow & \downarrow \\ -2^8 & 2^7 & 2^6 & 2^5 & 2^4 & 2^3 & 2^2 & 2^1 & 2^0 \\ \uparrow \end{array}$$

— 2's complement number

— Position value of the bit.

Note that the sign bit is assigned -ve value

$-2^8 \quad 2^7 \quad 2^6 \quad \not{2}^5 \quad 2^4 \quad \not{2}^3 \quad 2^2 \quad 2^1 \quad 2^0$ — Ignore position values which correspond to 0s.

$-2^8 + 2^7 + 2^6 + 2^4 + 2^2 + 2^1 + 2^0$ — Sum the position values which correspond to 1s.

$= -256 + 128 + 64 + 16 + 4 + 2 + 1$

$= -41$

∴ The decimal value of the 2's complement number 111010111_2 is -41_{10} **Ans.**

Example 36-11. *Determine the range of unsigned decimal numbers that can be represented in a byte.*

Solution. We know that a byte is a group of eight bits. Since, we are interested in unsigned decimal numbers here, therefore there is no sign bit. This means we can use all the eight bits to represent a magnitude of a number. Therefore the values of unsigned decimal numbers range from.

$$00000000_2 = 0_{10} \text{ to } 11111111_2 = 255_{10}$$

This is a total of 256 different numbers which we could have predicted, since $2^8 = 256$ **Ans.**

36-13. Addition of Two Binary Numbers

We have already discussed the three different methods of representing binary numbers namely (1) signed system (2) 1's complement system and (3) 2's complement system. Now let us study the addition of two binary numbers. We will use 2's complement system for representing binary numbers as this is the system used in all digital systems to do arithmetic operations.

In order to illustrate the addition operation in 2's complement system, we will consider the following five different types of cases :

(*i*) Both the binary numbers are positive

(*ii*) One of the numbers is positive and the other one is negative. It will be assumed that negative number is smaller than the positive number.

(*iii*) Again one of the numbers is positive and the other one is negative. But in this case the negative number is larger than the positive number.

(*iv*) Both the numbers are negative.

(*v*) Both the numbers are equal and opposite.

All these five different cases for addition are discussed one by one in the following pages.

36-14. Addition of Two Positive Numbers

Consider the addition of two positive numbers say +7 and +5. The addition operation for these two numbers is as shown below:

$$\begin{array}{rcl} +7 \rightarrow & \boxed{0} & 0111 \\ +5 \rightarrow & \boxed{0} & 0101 \\ \hline & \boxed{0} & 1100 \quad \text{(Sum} = +12) \end{array}$$

Sign bits ↑

Notice that the sign bits of both the numbers are 0 and the sum of sign bit is 0. This indicates that sum is positive. Also notice that both numbers have same number of bits. This must always be done while performing arithmetic operations in 2's complement system.

36-15. Addition of a Positive Number and a Small Negative Number

Consider the addition of two binary numbers where one of the numbers is positive and the other one is a negative number. It is also under consideration that negative number is a smaller number. Let us suppose we have the two numbers as +7 and –5. Remember that the number – 5 will be in its 2's complement form. Thus, +5 (00101) must be converted to – 5(11011). The addition operation for this situation is as shown below.

```
                 Sign bits
+ 7 → 0  0111
– 5 → 1  1011
      ---------
    1 0  0010
```

This carry is disregarded; the result is 0010 (= +2).

Notice that the sign bit of negative number is 1. Further the sign bits also participate in the addition process. As a matter of fact, a carry is generated in the last position of addition. This carry is always disregarded, so that the final sum is 0010 which is equal to +2.

36-16. Addition of a Positive Number and a Large Negative Number

Consider the addition of two numbers where one of the numbers is positive and the other one is negative-the negative number is larger than the positive number. Let us suppose we have two numbers –7 and +5. Remember that number –7 will be in its 2's complement form. Thus +7 (00111) must be converted to – 7 (*i.e.*, 11001). The addition operation of this situation is as shown below:

```
                 Sign bits
– 7 → 1  1001
+ 5   0  0101
      ---------
      1  1110    (Sum = – 2)
```

Negative sign bit

Notice that the sum here has a sign bit of 1, indicating a negative number. Since the sum is negative, it is in the 2's complement form.

To find the true magnitude of the sum, we must take 2's complement of the result, 11110

```
  1  1  1  1  0    — Result
  0  0  0  0  1    — 1's complement
+             1
  -------------
  0  0  0  1  0    — 2's complement
```

Thus, 2's complement of the result (11110) is 00010 = +2. Hence the result 11110 represents –2.

36-17. Addition of Two Negative Numbers

Let us now study the addition of two negative numbers. In order to illustrate the procedure, consider the two numbers as –7 and – 5. Remember that both numbers –7 and –5 will be in their 2's complement form. Thus +7 (00111) must be converted to –7 (*i.e.*, 11001). Similarly, +5 (00101) must be converted to – 5 (*i.e.*, 11011). The addition operation of –7 and –5 is as shown below :

```
                          Sign bits
  - 7 →  1   1001
  - 5 →  1   1011
        1 1  0100
        ↑
        This carry is disregarded; the result is 10100 (sum = –12).
```

Notice that the final result is negative and in 2's complement form with a sign bit of 1. Taking 2's complement of result (10100), we get

	1	0	1	0	0	— result
	0	1	0	1	1	— 1's complement of result
+					1	
	0	1	1	0	0	— 2's complement of result

Thus, we find that 2's complement of result is 01100 = +12. Therefore, the sum of two negative numbers – 7 and – 5 is – 12.

36-18. Addition of Two Equal and Opposite Numbers

Consider the addition of two equal and opposite numbers say + 7 and – 7. The addition of these two numbers is as shown below.

```
  + 7 →  0   0111
  - 7 →  1   1001
        1 0  0000
        ↑ ↑
        | Sign bits
        disregard this carry; the result is 00000 (Sum = + 0)
```

Notice that -7 is in 2's complement form. We find that the result is +0 as expected.

Example 36-12. *Perform the following addition operation in the 2's complement system. Use eight (including the sign bit) for each number. Check your results by converting binary result back to decimal.*

(*a*) 32_{10} *and* 18_{10} (*b*) $+21_{10}$ *and* -13_{10}

(*c*) -28_{10} *and* $+38_{10}$ (*d*) -36_{10} *and* -15_{10}

(*e*) $+17_{10}$ *and* -17_{10}

Solution.

(*a*) Addition of 32_{10} and 18_{10}. We know that in this case both the numbers are positive. Therefore, the addition is carried out as follows:

```
                          Sign bits
  + 32₁₀ →  0  0100000
  + 18₁₀ →  0  0010010
    50₁₀    0  0110010      (Sum = + 50)
```

Notice that the sign bits of both numbers are 0 and the sum of sign bit is 0. Therefore, sum is positive. Thus, using 2's complement system we find that,

$$32_{10} + 18_{10} = 50_{10} \textbf{ Ans.}$$

***(b)* Addition of $+21_{10}$ and -13_{10}.** We know that in order to add -13_{10} to $+21_{10}$, we will have to find 2's complement of -13_{10}. Thus

		Sign bits	
+ 21 →		0	0010101
– 13 →		1	1110011
+ 8	1 (carry)	0	0001000

This carry is disregarded; the result is 00001000 (= 8_{10}).

Thus, using 2's complement we find that $21_{10} - 13_{10}$ is $+8_{10}$ **Ans.**

***(c)* Addition of -28_{10} and $+38_{10}$.** In order to add -28_{10} to $+38_{10}$, we must represent -28_{10} in its 2's complement form. Thus

		Sign bits	
-28_{10} →		1	11100100
$+38_{10}$ →		0	00100110
$+10_{10}$	1 (carry)	0	0001010

This carry is disregarded; the result is 00001010 (= 10_{10}).

Thus, using 2's complement, we find that $-28_{10} + 38_{10}$ is $+10_{10}$ **Ans.**

***(d)* Addition of -36_{10} and -15_{10}.** In order to add -36_{10} and -15_{10}, we must represent these numbers to their 2's complement form. Thus

		Sign bits		
– 36 →		1	1100100	
– 15 →		1	1110001	
– 51	1 (carry)	1	1001101	(Sum = – 51)

This carry is disregarded; the result is 11001101 (= -51_{10}).

Notice that the find result is negative and in 2's complement form with a sign bit of 1. Taking 2's complement of result (11001101), we get

	1	1	0	0	1	1	0	1	— result
	0	0	1	1	0	0	1	0	— 1's complement of result
+								1	
	0	0	1	1	0	0	1	1	— 2's complement of result

Thus, we find 2's complement of result is 00110011 = -51_{10} **Ans.**

***(e)* Addition of $+17_{10}$ and -17_{10}.** We know that this is a case of two equal and opposite numbers. The addition of these two numbers is as shown below:

		Sign bits	
+ 17 →		0	0010001
– 17 →		1	1101111
	1 (carry)	0	0000000

disregard this carry; the result is 00000000 (= 0_{10}).

Thus, using 2's complement system we find that the result is + 0 as expected.

36-19. Arithmetic Subtraction

Subtraction is a special case of addition. When subtracting one binary number (called subtrahend), from another binary number (called the minuend), we can use the following two step procedure:

(*i*) Negate (*i.e.*, find 2's complement of) the subtrahend- This will change the subtrahend to its equivalent value of opposite sign.

(*ii*) Add this to the minuend- The result of this addition will represent the difference between the subtrahend and the minuend.

It may be carefully noted that both numbers have the same number of bits in their representations. This is valid in all 2's complement arithmetic operations.

In order to illustrate the subtraction, let us consider the situation where +5 is to be subtracted from +7. The subtraction is shown as below.

+7 → 00111 (minuend)

+5 → 00101 (subtrahend)

If we take 2's complement of the subtrahend, we get 11011 which represents –5. Now add this to the minuend.

Sign bits

+ 7 → 0 0111 (minuend)

– 5 → 1 1011 (2's complement of subtrahend)

1 0 0010

disregard, so the result is 00010 = + 2

It is evident from the above discussion that when the subtrahend is changed to its 2's complement, it actually becomes –5, so we are adding –5 and +7. As a matter of fact, this is the same as subtracting +5 from +7. This is actually the same situation as discussed in Art 36-15. Thus, we find that any subtraction operation actually becomes one of addition when the 2's complement system is used. This feature of 2's complement system has made it the most widely used methods available. This is possible because it allows addition and subtraction to be performed by the same circuitry.

Let us consider another example of subtraction. This time we will illustrate the procedure of subtracting + 7 from – 5.

Sign bits

– 5 → 1 1011 (minuend)

– 7 → 1 1001 (2's complement of + 7)

1 1 0100

disregard; so the result is 10100 = (Equal to – 12)

Notice that – 5 is in 2's complement form. Also note that we have negated +7 to perform subtraction. Since, the result (10100) is negative, it has to converted to its 2's complement, which is 01100 (i.e.+12). Therefore, the subtraction of + 7 from – 5 is obviously – 12.

The student is strongly recommended to verify the results of the above procedure for the following subtractions:

(*a*) +7 – (– 5) (*b*) – 7 – (+5)

(*c*) – 7 – (– 5) (*d*) + 5 – (– 5)

You must remember always that when the result has a sign bit of 1, it is negative and in 2's complement form.

Example 36-13. *Subtract* 78_{10} *from* 97_{10} *by using 2's complement arithmetic*

(Nagpur Univ. Oct/Nov. 1997)

Solution. Given the two decimal numbers 78_{10} and 97_{10}.

In order to subtract 78 from 97, we will have to find its 2's complement. Then add this number to the binary equivalent of 97. Thus

Sign bit (↓ first bit)

+ 78 →	0	1	0	0	1	1	1	0	— binary equivalent of 78
	1	0	1	1	0	0	0	1	— 1's complement of 78
+								1	
	1	0	1	1	0	0	1	0	— 2's complement of 78

Now,

Sign bits (↓ first column)

97 →		0	1	1	0	0	0	0	1	— binary equivalent of 97
– 78		1	0	1	1	0	0	1	0	— 2's complement of 78
19	+									
	~~1~~	0	0	0	1	0	0	1	1	

↑ disregard; so the result is 10011.
This is equivalent to 19_{10}.

$\therefore \quad 97_{10} - 78_{10} = 19_{10}$ **Ans.**

Example 36-14 *Using 2's complement method, perform the subtraction* $101_2 - 111_2$.

(Nagpur University, Oct/Nov. 1997)

Solution. Given the two binary numbers 101_2 and 111_2. In order to subtract 111_2 from 101_2, we will have to find the 2's complement of 101_2 and then add this to the binary 101_2. Thus, 2's complement of 111_2 is 001_2.

Sign bits (↓ first column)

∴	0	101
+	1	001
	1	110

Since, the sign bit of the result is 1, the result is in 2's complement . Thus, taking 2's complement of 1110, we get 0010. Therefore, the result is -0010_2 or -2_{10} **Ans.**

36-20. Arithmetic Overflow

We have already discussed in previous articles, the addition and subtraction of two binary numbers using 2's complement system. In each of the examples used to illustrate the addition or subtraction, the numbers consisted of one sign bit and four magnitude bits. The results also consisted of a sign bit and four magnitude bits. Any carry into the sixth bit was disregarded. In all of the cases we have considered, the magnitude of the answer was small enough to fit into four bits. Let us look at the addition of +9 and +8.

	Sign bits		
+ 9 →	0	1001	
+ 8 →	0	1000	
	1	0001	
Incorrect sign bit	↑	↑	Incorrect magnitude

It is evident from the above that the answer has a negative sign bit, which is obviously incorrect because we are adding two positive numbers. Notice that the result of +9 and +8 should be +17, and the magnitude 17 requires more than four bits. Because of this reason, the magnitude overflows into the sign bit position. The overflow condition can occur only when two positive or two negative numbers are being added. Further, the overflow condition produces an incorrect result. The overflow can be detected by checking the sign bit of the result is the same an the sign bits of the number being added.

It may be noted that overflow can occur while performing subtraction if minuend and subtrahend are of different signs. For example, if we are subtracting –8 from +9, the –8 is negated to become +8 and is added to + 9 just as shown above. This causes an overflow and produces an erroneous negative result since the magnitude is too large

It will be interesting to know that all computers have a special circuit to detect any overflow condition whenever two numbers are added or subtracted.

36-21. Arithmetic Multiplication

As a matter of fact, the multiplication of binary numbers is done in the same manner as the multiplication of decimal numbers. But the process is much simpler because we are multiplying by 0 or 1 and no other digits. Let us illustrate the multiplication by a following example.

				1	0	1	1	← multiplicand = 11_{10}
				1	1	0	1	← multiplier = 13_{10}
				1	0	1	1	Partial products
			0	0	0	0		
		1	0	1	1			
	1	0	1	1				
1	0	0	0	1	1	1	1	← Final product = 143_{10}

As seen, the multiplicand and the multiplier are in true binary form and no sign bit is used. Notice that the steps used in multiplication process are exactly the same as used in decimal multiplication.

(*i*) First, we examine the least-significant bit (LSB) of the multiplier. In our case it is 1. This 1 when multiplied by the multiplicand produces 1011. This is written as the first partial product.

(*ii*) Next the second bit of the multiplier is examined. Since, it is 0, when multiplied by the multiplicand, it produces 0000. This is written as the second partial product. Notice that it is shifted one place to the left relative to the first partial product.

(*iii*) Next the third bit of the multiplier is examined. Since, it is 1 and so 1011 is written as the third partial product. Again note that this third partial product is shifted one place to the left relative to the second one.

(*iv*) Next, the fourth bit is 1 and so the last partial product is 1011 shifted again one position to the left.

(*v*) Finally the four partial products are then summed to produce the final product.

It may be carefully noted that most digital circuits can add only two binary numbers at a time. Because of this reason, the partial products formed during multiplication cannot be all added together at the same time. Instead, they are added together two at a time. That is the first partial product is added to the second, their sum is added to the third and so on. The process of addition is shown below.

Add	1 0 1 1	← first partial product	
	0 0 0 0	←	second partial product shifted left
Add	0 1 0 1 1	← sum of first two partial products	
	1 0 1 1	←	third partial product shifted left
Add	1 1 0 1 1 1	← sum of first three partial products	
	1 0 1 1	←	fourth partial product shifted left
	1 0 0 0 1 1 1 1	←	sum of four partial products. This equals final total product

36-22. Multiplication in 2's Complement System

We have already discussed in last article that the multiplication of binary numbers is carried out in the same manner as for decimal numbers. This method is used by the computer (that use the 2's complement representation), provided that both the multiplicand and the multiplier are put in true binary form. If the two numbers to be multiplied are positive, they are already in true binary form and are multiplied as they are. The result product is also positive and is given a sign bit of 0.

On the other hand, when the two numbers are negative, they will be in 2's complement form. The 2's complement of each number is taken to convert it into a positive number and then the two numbers are multiplied . The product is kept as a positive number and is given a sign bit of 0. It may be noted that when one of the numbers is positive and the other is negative, the negative number is first converted to a positive magnitude by taking its 2's complement. The product will be in true –magnitude form. However, the product must be negative, since the original numbers are of opposite sign. Thus, the product is then changed to 2's complement form and is given a sign bit of 1

36-23. Binary Division

As a matter of fact the process for dividing one binary number (called dividend) by another (called divisor) is the same as that which is followed by decimal numbers. The actual process(refered to as long division) is simpler in binary because when we are checking to see how many times the divisor goes into the dividend, there are only two possibilities, 0 and 1. Let us illustrate the division process by the following simple division examples,

	0011	
	11 $\overline{)\,1001}$	1001 → dividend
	011	11 → divisor
First partial remainder ⟶	0011	0011 → quotient
	11	$\dfrac{\text{dividend}}{\text{divisor}}$ = quotient
Second partial remainder ⟶	0	

The above example shows that we have 1001_2 divided by 11_2. This is equivalent to 9 ÷ 3 in decimal. Notice that the resulting quotient is $0011_2 = 3_{10}$.

In digital computers, the division is carried out by using repeated subtraction using 2's complement system. Following is the procedure used for division in most of the digital circuits.

(i) Determine if the signs of the dividend and divisor are the same or different. This determines what the sign of the quotient will be. Intialize the quotient to 0.

(ii) Subtract the divisior from the dividend using 2's complement addition to get the first partial remainder and add 1 to the quotient. If this partial remainder is positive go to step 3. However if the partial remainder is zero or negative, the division is complete.

(iii) Subtract the divisor from the partial remainder and add 1 to the quotient. If the result is positive, repeat for the next partial remainder. If the result is zero or negative, the division is complete.

It is evident from the above procedure that we continue to subtract the divisor from the dividend and the partial remainders until there is a zero or negative result.

Example 36-15. *Divide the number* 221_{10} *by* 17_{10} *by converting the original decimal number to its 8-bit binary equivalent.*

Solution. Given the two decimal numbers 221 and 17. We know that the 8-bit binary equivalent of 221_{10} is,

$$221_{10} = 11011101_2$$

and that of 17_{10} is

$$17_{10} = 0001\ 0001_2$$

Let us now divide 11011101_2 by $0001\ 0001_2$ as shown below.

```
                     1101          ←    quotient
                 ____________
0001 0001       / 1101 1101        ←    dividend
    ↑            – 1000 1
    |           ____________
divisor ─┘         101 01
                 – 100 01
                ____________
                     1 0001
                   – 1 0001
                ____________
                          0
```

From the division process shown above we find that, $11011101_2 \div 00010001_2 = 1101_2 = 13_{10}$. **Ans.**

Note. Check that $221_{10} \div 17_{10} = 13_{10}$ which is found to be the same as through the binary division.

36-24. BCD Addition

We have already discussed in chapter 2 that many computers and calculators use the BCD code to represent decimal numbers. Recall that this code takes each decimal digit and represents it by a four-bit code ranging from 0000 to 1001. The addition of decimal numbers that are in BCD form can be best understood by considering the two cases that can occur when two decimal digits are added.

***(i)* Sum equals 9 or less :** Let us consider the addition of 5 and 4 using BCD to represent each digit.

5	0101	← BCD code for 5
+ 4	0100	← BCD code for 4
9	1001	← BCD code for 9

As seen the addition is carried out in normal binary addition and the sum is 1001 which is the BCD code for 9. Consider another example,

54	0101 0100	← BCD code for 54
+ 33	0011 0011	← BCD code for 33
87	1000 0111	← BCD code for 87

It may be noted that the four-bit codes for 4 and 3 are added in binary to produce 0111 which is a BCD code for 7. Similarly, if we add the second decimal digit positions, it produces 1000 which is BCD code for 8. The total is 1000 0111 which is BCD code for 87.

It may be carefully noted that in none of the examples discussed above, the sums of the pairs of decimal digits exceeded 9. Therefore no decimal digits were produced. For these cases, the BCD addition process is straight forward. In fact the addition process is the same as binary addition.

(ii) Sum Greater than 9 : Let us consider the addition of 7 and 5 using BCD to represent each digit.

7	0111	← BCD code for 7
+ 5	0101	← BCD code for 5
+ 12	1100	← Invalid BCD code

It may be carefully noted that the sum 1100 does not exist in the BCD code. As a matter of fact, it is one of the six forbidden or invalid four-bit code groups. This has occurred because of the sum of two digits exceed 9. Whenever this occurs, the sum must be corrected by the addition of six (0110) to take into account the skipping of the six invalid code groups. Thus,

	0111	← BCD code for 7
	+ 0101	← BCD code for 5
	1100	← Invalid BCD code
	+ 0110	← Add 6 for correction
0001	0010	← BCD code for 12.
1	2	

It is evident from the illustration above that if 0110 is added to the invalid BCD code, it produces a correct BCD result. It may be noted that with the addition of 0110, a carry is produced into the second decimal position. The addition must be performed whenever the sum of the two decimal digits is greater than 9.

Example 36-16. *Add the following decimal numbers after converting each to its BCD code.*

(*a*) 74 + 23

(*b*) 147 + 380

Solution. (*a*) Given the decimal numbers 74 and 23.

74	0111	0100	← BCD code for 74
+ 23	0010	0011	← BCD code for 23
97	1001	0111	← BCD code for 97

(*b*) Given the decimal numbers 147 and 380.

147	0001	0100	0111	← BCD code for 147
380	0011	1000	0000	← BCD code for 380
+				
527	0100	1100	0111	← Invalid code in 2nd digit
	1 ←	+0110		← Add 6 to convert
	0101	0010	0111	← Convert BCD sum
	↑ 5	↑ 2	↑ 7	

Notice that the addition of the four-bit codes for 4 and 8 digits results in invalid sum. This is converted by adding 0110. This generates a carry of 1, which is carried over to be added to the BCD sum of the third-position digits.

36-25. Hexadecimal Arithmetic

We have already discussed hexadecimal numbers in chapter 2. Recall that it is a method of representing groups of four-bits as a single digit. Hexadecimal numbers are used extensively in computer programming using machine-language and expressing addresses of computer memories. When working in these areas, you will encounter situations where hexadecimal numbers must be added or subtracted.

36-26. Hexadecimal Addition

Addition can be done directly with hexadecimal numbers by remembering that the hexadecimal digits 0 through 9 are equivalent to decimal digits 0 through 9 and that hexadecimal digits A through F are equivalent to decimal numbers 10 through 15. When adding two hexadecimal numbers use the following procedure.

(*i*) Add the two hexadecimal digits in decimal, mentally inserting the decimal equivalent for those digits larger than 9.

(*ii*) If the sum is 15 or less, it can be directly expressed as a hexadecimal digit.

(*iii*) If the sum is greater than or equal to 16, subtract 16 and carry a 1 to the next digit position.

Example 36-17. *Find the sum of the following points of hexadecimal numbers*:

(*a*) 4F *and* 2D (*b*) 91 B *and* 6F2

(*c*) FFF + 0FF (*d*) A7C5 and 2DA8

Solution.

(*a*)

```
    4  F
  + 2  D
  ------
    7  C
```

Notice that F + D = 15 + 13 = 28, which is 12 with a carry of 1, i.e (28 – 16 =12), 12 is written down as C. Next, 4 + 2 + Carry = 7.

∴ 4F + 2D = 7C **Ans.**

(*b*)

```
     9  1  B
   + 6  F  2
  ----------
  1  0  0  D
```

Notice that B + 2 = 11 + 2 = 13 which is equivalent to hexadecimal number D. Further 1 + F = 1 + 15 = 16, which is 0 with a carry of 1, i.e (16 – 16 = 0). Next 9 + 6 + carry = 16, which is equal to 0 with a carry of 1.

∴ 91B + 6F2 = 100D **Ans.**

(*c*)

$$\begin{array}{cccc} & F & F & F \\ + & 0 & F & F \\ \hline 1 & 0 & E & E \end{array}$$

Notice that F + F = 15 +15 =30 ,which is 14 with a carry of 1, i.e (30 – 16 = 14), 14 is written as E. Next F + F = 30 again E with a carry of 1. Finally F+ carry = 15+1=16 which in hexadecimal is equal to 0 with a carry of 1.

∴ FFF + 0FF = 10EE **Ans.**

(*d*)

$$\begin{array}{ccccc} & A & 7 & C & 5 \\ + & 2 & D & A & 8 \\ \hline & D & 5 & 6 & D \end{array}$$

Notice that 5 + 8 = 13 which is equivalent to hexadecimal D. Next the second position digits C + A = 12 + 10 = 22 which is equal to 6 with a carry of 1, i.e (22 – 16 = 6). Similarly, the third position digits 7 + D + carry = 7 + 13 + 1 = 21 which is equal to 5 with a carry of 1, i.e (21 – 16 = 5). Finally the fourth position digits, A + 2 + carry = 10 + 2 + 1 = 13 which is equivalent to hexadecimal D.

∴ A7C5 + 2DA8= D56D **Ans.**

Example 36-18. *Add the following hexadecimal numbers* $2A3.06_{16}$ *and* $001B.20_{16}$.

(*Nagpur University, Oct/Nov. 1997*)

Solution. Given : The two hexadecimal numbers 2A3.06 and 001B.20.

Since both the numbers are positive, we can add them straightaway. Thus

$$\begin{array}{ccccccc} & & 2 & A & 3 & . & 0 & 6 \\ + & 0 & 0 & 1 & B & . & 2 & 0 \\ \hline & 0 & 2 & B & E & . & 2 & 6 \end{array}$$

∴ $2A3_{16} + 001B.20_{16} = 02BE.26_{16}$ **Ans.**

36-27. Hexadecimal Subtraction

Since, a hexadecimal number can be used to represent a binary number, it can also be used to represent the 2's complement of a binary number. For example the hexadecimal representation of $D6_{16}$ is 11010110. The 2's complement of this number is 00101010 which in decimal is equivalent to $2A_{16}$.

We know that 2's complement allows us to subtract by adding binary numbers. We can also use this method for hexadecimal subtraction.

Consider for example, the subtraction of 21 from 75 as shown below.

	21_{16}		← Hexadecimal number
0010		0001	← Binary number
1101		1111	← Take 2's complement
	DF		← Convert back to Hex

Now add DF_{16} to 75, Thus,

$$\begin{array}{r} 7\;5 \\ +\;D\;F \\ \hline \not{1}\;5\;4 \end{array}$$ → Drop carry as in 2's complement addition.

∴ The difference 75 – 21 = 54_{16}

Example 36-19. *Perform the following subtractions on the pairs of hexadecimal numbers.*

(*a*) D7 – A8 (*b*) A05C – 24CA

Solution.

(*a*)

A	8	← Hexadecimal number
1010	1000	← Binary number
0101	1000	← Take 2's complement
5	8	← Convert back to Hexadecimal

Now, add 58_{16} to D7. Thus,

$$\begin{array}{r} D\;7 \\ +\;5\;8 \\ \hline \not{1}\;2\;F \end{array}$$ ← Drop carry as in 2's complement addition.

Note that 7 + 8 = 15 which is hexadecimal F. Similarly D + 5 = 12.Dropping carry we get 2.

∴ The difference D7 – A8 = $2F_{16}$ **Ans.**

(*b*)

2	4	C	A	← Hexadecimal number
0010	0100	1100	1010	← Binary number
1101	1011	0011	0110	← Take 2's complement
D	B	3	6	← Convert back to Hexadecimal

Now add DB 36_{16} to $A05C_{16}$. Thus,

$$\begin{array}{r} A\;0\;5\;C \\ +\;D\;B\;3\;6 \\ \hline \not{1}\;7\;B\;9\;2 \end{array}$$ Drop carry as in 2's complement addition.

Note that adding digits at the first position C + 6 = 2 with a carry of 1.Next adding digits at the second position along with a carry, 5 + 3 + carry = 9. Next adding digits at the third-position, 0 + B = B. Finally adding digits at the fourth position, A + D = 17 = 7 + carry. Dropping carry as in 2's complement addition we get 7B92.

∴ The difference, $A05C_{16}$ – 24 CA_{16} = $7B92_{16}$ **Ans.**

36-28. Arithmetic Circuits

As a matter of fact, all the arithmetic operations and procedures discussed in the previous articles can be implemented using adders formed from basic logic gates. For a large number of digits, we can use medium-scale integration (MSI) circuits. The MSI circuits have several adders within a single integrated package.

36-29. The Half Adder

Figure 36-3 (a). shows the logic symbol of a half-adder. As seen from this figure, we find that the half-adder accepts two binary digits on its inputs and produces two digits on its outputs: a sum bit (S) and a carry bit (C_{out}). Fig 36-3 (b) shows the truth table for the half-adder.

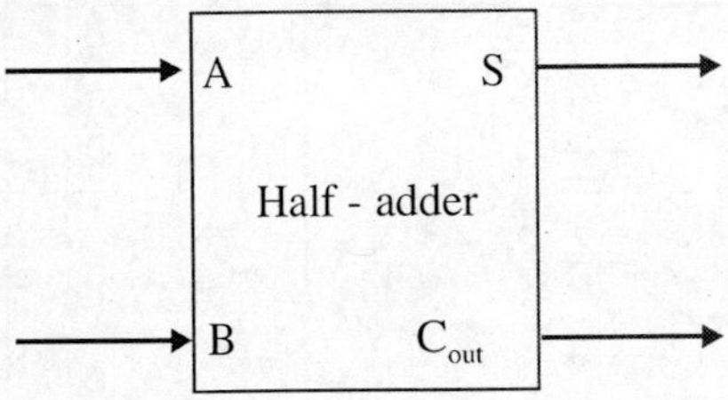

(*a*) Logic symbol

Inputs		*Outputs*	
A	***B***	***S***	C_{out}
0	0	0	0
0	1	1	0
1	0	1	0
1	1	0	1

(*b*) Truth table

Fig. 36-3. *A half-adder*

The half- adder follows the basic rules for binary addition:

$$0 + 1 = 0$$
$$0 + 1 = 1$$
$$1 + 0 = 1$$
$$1 + 1 = 10 \text{ or } 0 \text{ with a carry of } 1$$

The Boolean expression for the sum output (S) can be expressed by the equation.

$$S = A\overline{B} + \overline{A}B$$
$$= A \oplus B \qquad ...(1)$$

and the Boolean expression for the carry output,

$$C_{ont} = AB \qquad ...(2)$$

The equations (1) and (2) can be implemented by a logic circuit shown in Fig 36-4 . As seen from this figure, the sum output (s) is obtained from an excluse-OR gate while the carry output (C_{out}) is the output of a two-input AND gate.

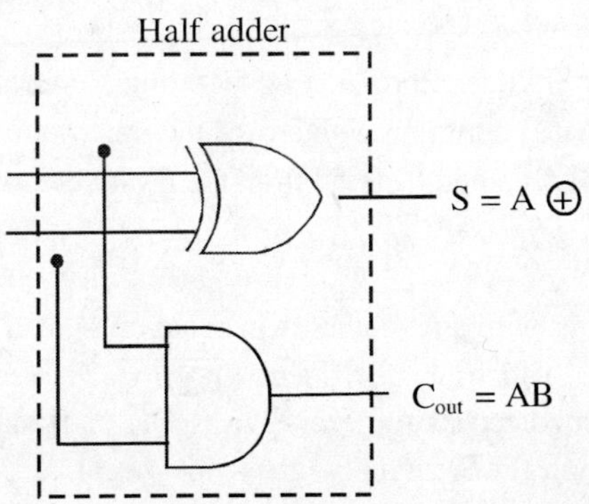

Fig. 36-4. *Logic implementation of a half-adder*

36-30. The Full – Adder

Fig 36-5 (a) shows the logic symbol of a full-adder. As seen from this figure, we find that the full-adder accepts three inputs-two input bits A and B and a carry input (C_{in}). It has two outputs: (1) a sum output (s) and a carry- output (C_{out}). Fig 36-5 (b) shows the truth table for the full-adder.

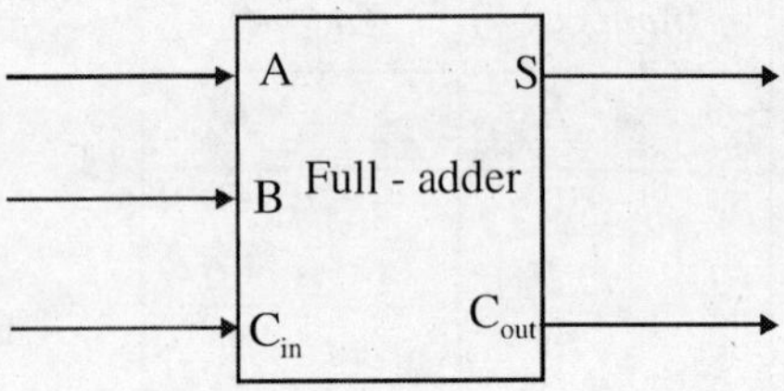

(*a*) Logic symbol

Inputs			*Outputs*	
A	*B*	C_{in}	*S*	C_{out}
0	0	0	0	0
0	0	1	1	0
0	1	0	1	0
0	1	1	0	1
1	0	0	1	0
1	0	1	0	1
1	1	0	0	1
1	1	1	1	1

(*b*) Truth table

Fig. 36-5. *Full-adder*

The full-adder also follows the same basic rules for binary addition as the half adder, Thus,

0 + 0 + 0	=	0	with carry	0
0 + 0 + 1	=	1	with carry	0
0 + 1 + 0	=	1	with carry	0
0 + 1 + 1	=	0	with carry	1
1 + 0 + 0	=	1	with carry	0
1 + 0 + 1	=	0	with carry	1
1 + 1 + 0	=	0	with carry	1
1 + 1 + 1	=	1	with carry	1

The Boolean expression for the sum output (S) can be obtained from the truth table (shown in Fig. 36-5 (b)) by summing the terms for which *S*=1.Thus, the sum,

$$S = \bar{A}\bar{B}C_{in} + \bar{A}B\bar{C}_{in} + A\bar{B}\bar{C}_{in} + ABC_{in} \qquad \text{----(3)}$$

Let us now try to simplify this expression by factoring. Unfortunately none of the terms in the expression has two variables in common with any of the other terms. However, $\bar{A}$ can be factored from the first two terms and A can be factored from the last two terms. Thus,

$$S = \bar{A}\,(\bar{B}C_{in} + B\bar{C}_{in}) + A\,(\bar{B}\bar{C}_{in} + BC_{in}) \qquad ...(4)$$

Notice that the first terms within parentheses in equation (4) can be recognized as the exclusive-OR combination of B and C_{in}, which can be written as $B \oplus C_{in}$. The second term within the parentheses in equation (4) can be recognized as the exclusive -NOR of B and C_{in}, which can be written as $\overline{B \oplus C_{in}}$. Thus the expression for S can be written as,

$$S = \bar{A}\,(B \oplus C_{in}) + A\,(\overline{B \oplus C_{in}}) \qquad ...(5)$$

Let $X = B \oplus C_{in}$, then equation (5) can be rewritten as,

$$= A \oplus X \qquad \text{---(6)}$$

Replacing X with $B + C_{in}$, equation (6) can be written as,

$$S = A \oplus (B \oplus C_{in}) \qquad \text{---(7)}$$

$$= A \oplus B \oplus C_{in}$$

Next consider the carry output, (C_{out}) in the truth table of Fig 36-5 (b). The boolean expression for the carry output can be obtained from the truth table by summing the terms for which $C_{out} = 1$. Thus the carry output,

$$C_{out} = \overline{A}BC_{in} + A\overline{B}C_{in} + AB\overline{C}_{in} + ABC_{in}$$

This expression can be simplified by factoring. In order to do so let us add the ABC_{in} term 3 times since it has common factors with each of the other terms. Hence the expression for carry output,

$$C_{out} = \left(\overline{A}BC_{in} + ABC_{in}\right) + \left(A\overline{B}C_{in} + ABC_{in}\right) + \left(AB\overline{C}_{in} + ABC_{in}\right)$$

$$= BC_{in}\ (\overline{A} + A) + AC_{in}\ (\overline{B} + B) + AB\ (\overline{C}_{in} + C_{in})$$

Since $\overline{A} + A = \overline{B} + B = \overline{C}_{in} + C_{in} = 1$, therefore the above expression can be simplified to,

$$C_{ont} = BC_{in} + AC_{in} + AB \qquad \text{...(8)}$$

The equations (7) and (8) can be implemented by using logic gates. From equation (7) we find that to implement to full-adder sum output function, two 2-input exclusive –OR gates can be used. The first exclusive-OR gate generates the term $A \oplus B$, and the second has its inputs the output of the first exclusive –OR gate and the input carry as shown in the Fig 36-6. Similarly from equation (8) we find that to implement the full- adder carry output function, three 2-input AND gates followed by a 3-input OR gate can be used. The complete circuit of a full adder is shown in Fig 36-6

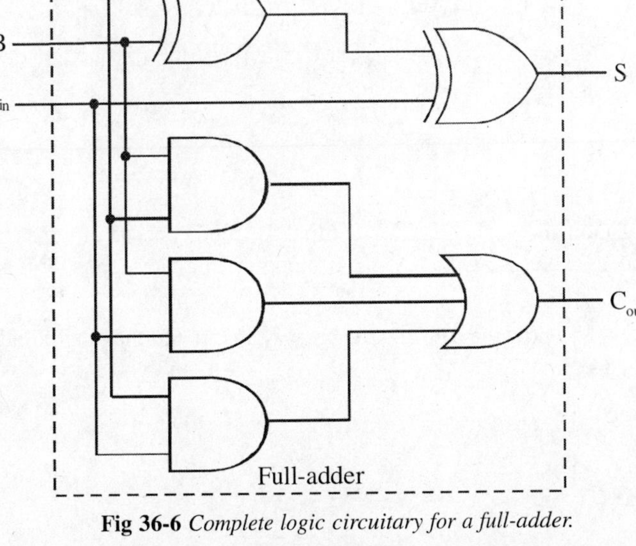

Fig 36-6 *Complete logic circuitary for a full-adder.*

Notes.

1. We simplified the expressions for S and C_{out} using algebraic methods. The K-map can also be used for this purpose. We leave it as an exercise for the students to try it out.
2. Several other implementations can be used to produce the same expressions for S and C_{out}.

36-31. Implementation of Full-adder using Two Half-adders

We have already discussed in the previous articles about the implementation of half-adder and a full-adder circuit using logic gates. Let us now study how a full adder can be implemented using half-adder circuit as a building block.

We know that the sum output of a half-adder with two-inputs (A and B) as shown in Fig 36-7 is given by the expression.

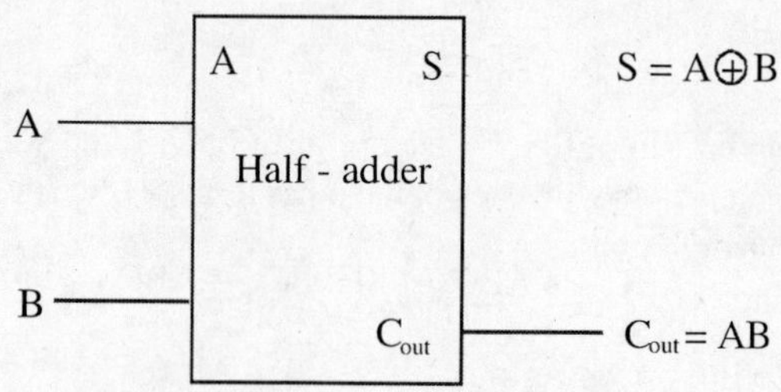

Fig. 36-7. *Logic symbol of a Half-adder*

$$S = A \oplus B$$

and the carry output,

$$C_{out} = AB$$

For a full-adder, the sum output,

$$S = A \oplus B \oplus C_{in}$$

and the carry output,

$$\begin{aligned} C_{out} &= \bar{A}BC_{in} + A\bar{B}C_{in} + AB\bar{C}_{in} + ABC_{in} \\ &= (\bar{A}B + A\bar{B})\, C_{in} + AB\,(\bar{C}_{in} + C_{in}) \\ &= (A \oplus B)\, C_{in} + AB\,(1) \\ &= (A \oplus B)\, C_{in} + AB \end{aligned}$$

The above equation shows that the carry output function consists of two terms. The first term is ANDing of C_{in} and an exclusive-OR of A and B inputs. While the second term is an ANDing of two inputs A and B.

Now Fig 36-8 shows how the sum and the carry output functions of a full-adder can be obtained by using two half-adders.

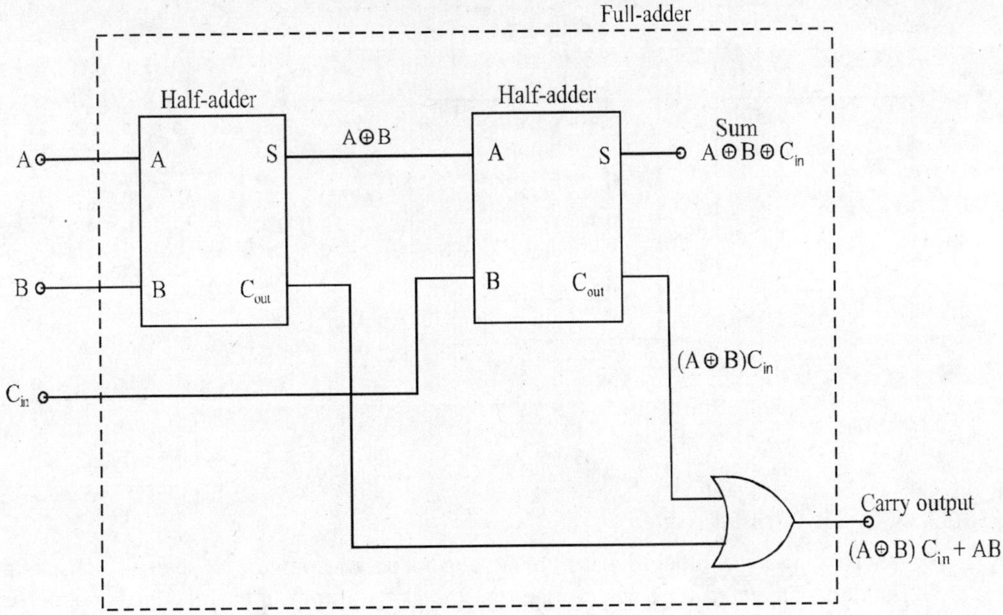

Fig 36-8 *Implementation of a full-adder using two half-adders.*

36-32 Parallel Binary Adders

As a matter of fact, a single full adder is capable of adding two 1-bit numbers and an input carry. If we need to add binary numbers with more than one bit, we require additional full-adders. It may be carefully noted that when one binary number is added to another, each column generates a sum bit and a 1 or 0 carry bit to the next column to the left. This process is illustrated below with a 2-bit numbers 11 and 01.

```
        ┌──── Carry bit from the right column
        ▼
        1
        1  1
    +   0  1
    ─────────
     1  0  0
     ▲
```

In this case, the carry bit from second column becomes a sum bit.

In order to add two binary numbers, a full adder is required for each bit in the number. Thus, for two-bit numbers we need two full-adders. Similarly, for the addition of four-bit numbers, we need four full-addres and so on.

Fig 36-9 shows a block diagram of a basic 2-bit parallel adder using two full- adders. Notice that the carry output of each full-adder is connected to the carry input of the next higher-order full adder. Further notice that either a half-adder can be used for the least significant position or the carry input of a full-adder can be connected to ground (0). This is because of the reason that there is no carry input to the least significant position.

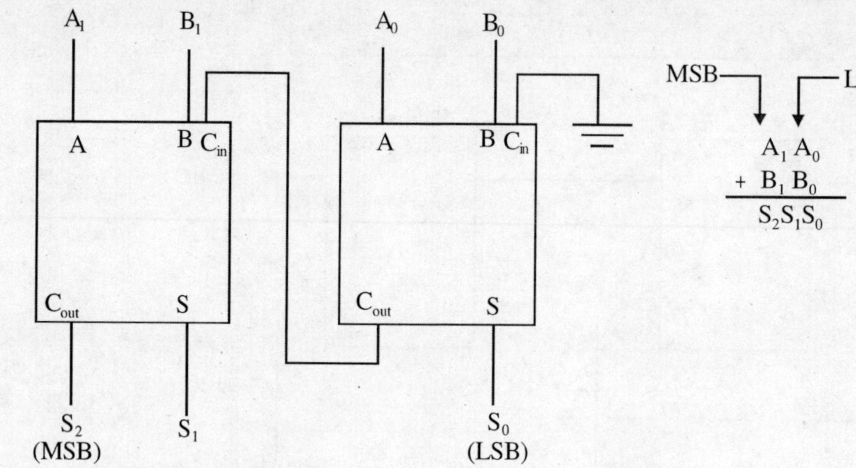

Fig. 36-9. *Block diagram of a basic 2-bit parallel adder.*

36-33. Four-bit Parallel Adder

Fig 36-10 shows the block diagram of a four-bit parallel adder. As seen, it consists of four full-adders connected together. Notice that the carry output of each adder is connected to the carry input of next higher stage. These are called internal carries. Also notice that the least-significant bits (LSBs) of the two binary numbers (being added) go into the right-most full-adder whereas the higher order bits are applied as shown to the successive higher-order adders. The most-significant bits (MSBs) of the two numbers (being added) are applied to the left-most full-adder.

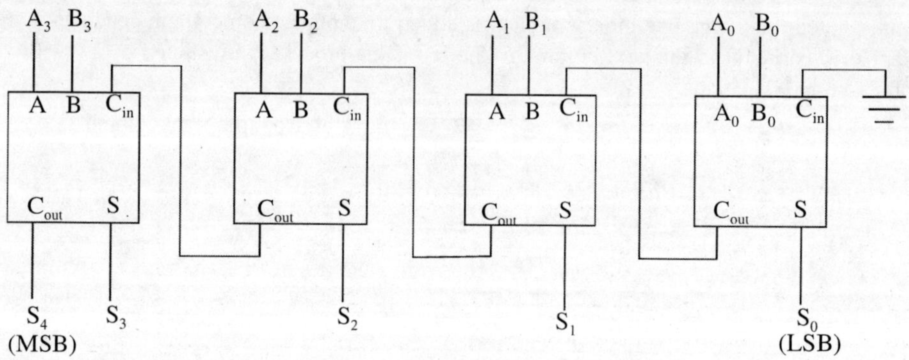

Fig. 36-10. *Block diagram of a four-bit parallel adder.*

Fig – 36-11 shows the logic symbol for a 4-bit parallel adder. Notice that the input carry is labelled as C_0. and the output carry at C_4. This is done to keep in line with most manufacturer's datasheets.

In terms of the method used to handle carries in a parallel adder, there are two types :

(*i*) The ripple carry adder and

(*ii*) The carry look-ahead adder

A ripple carry adder is one in which the carry output of each full-adder is connected to the carry input of the next higher order stage. The sum and the output carry of any stage cannot be produced until the input carry occurs. This process causes a time delay in the addition.

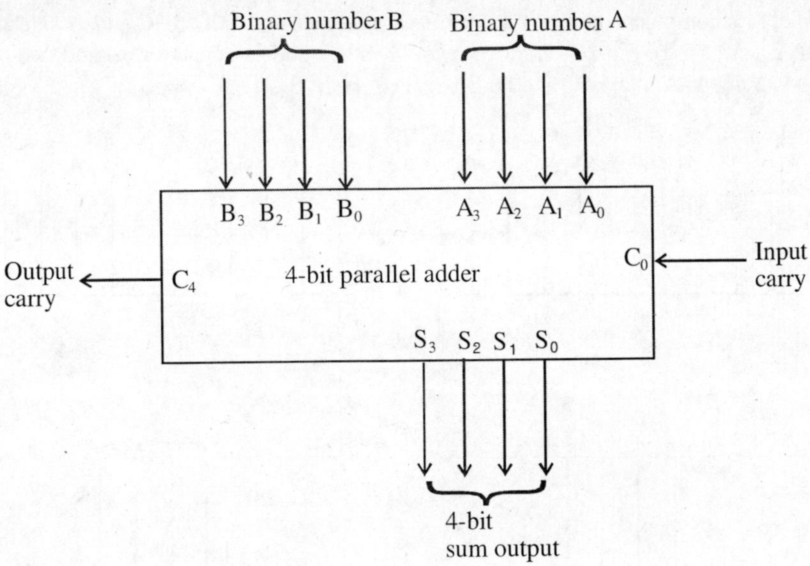

Fig. 36-11. *Logic symbol of a 4-bit parallel adder.*

The ripple carry delay can be eliminated by making use of a method called look-ahead carry addition. The look-ahead carry adder anticipates the output carry of each stage and based on the input bits of each stage, produces the output carry by either carry generation or carry propagation. Carry generated occurs when an output carry is produced (or generated) internally by the full- adder. A carry is generated only when both input bits are 1s. On the other hand, carry propagation occurs when the input carry is rippled to become the output carry. An input carry may be propagated by the full-adder when either or both of the input bits are 1s.

36-34. Integrated Circuits Parallel Adders

There are several parallel adders that are available as ICs (integrated-circuits) in the market. Table 36-1 shows the most commonly used parallel adders. All these adders contain four interconnected full-adders and the look-ahead carry circuitry needed for high speed operation.

Table 36-1. *IC parallel Adders*

Device	*Family*	*Description*
7483A	TTL	4-bit binary full-adder
74LS83A	LSTTL	4-bit binary full-adder
74283	TTL	4-bit binary full-adder
74LS283	LSTTL	4-bit binary full-adder
74HC283	CMOS	4-bit binary full-adder

The 74283s are identical to 7483s except that they have V_{CC} and ground pins 16 and 8 respectively. In all new integrated-circuit devices, it has become a common standard to have the power and ground pins at the corners of the integrated circuit parallel adders. For pin configuration of 7483/74283, refer to Appendix C.

36-35. Cascading Parallel Adders

In order to accomplish addition of large binary numbers, two or more IC adders can be connected together (*i.e.*, cascaded). Fig 36-12 shows two 74LS83A adders connected to add two 8-bit binary numbers, $A_7 A_6 A_5 A_4 A_3 A_2 A_1 A_0$ and $B_7 B_6 B_5 B_4 B_3 B_2 B_1 B_0$.

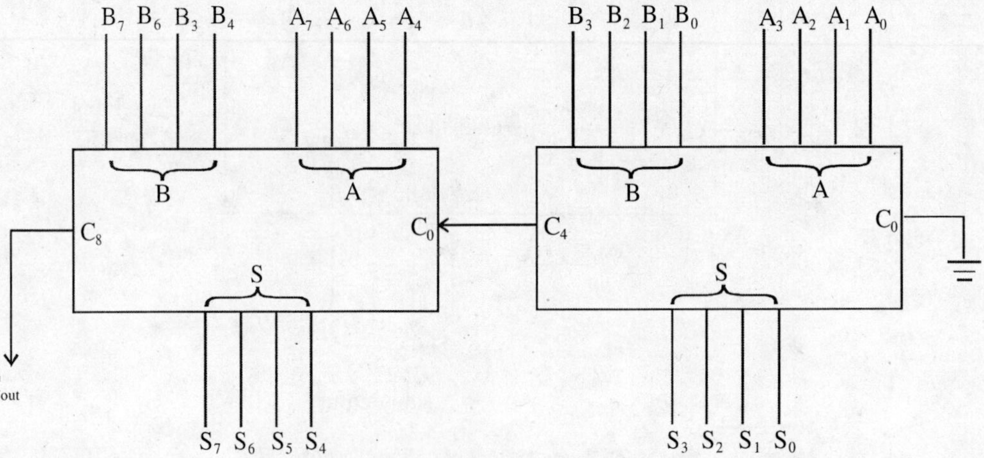

Fig. 36-12. *Cascading of two 74LS283As to form an 8-bit parallel adder.*

It may be noted from the figure shown above that the adder on the right adds the lower-order bits of the numbers. On the other hand, the adder on the left adds the higher-order bits plus the C_4 carry out of the lower-order adder. The eight sum outputs ($S_7 S_6 S_5 S_4 S_3 S_2 S_1 S_0$) are the resultant sum of the two 8-bit numbers. C_8 is the carry output (C_{out}) of the MSB position. The C_8 can be used as a carry input to the third adder stage if larger binary numbers are to be added.

36-36. 2's Complement Adder/Subractor Circuit

We have already mentioned earlier that arithmetic operations like subtraction, multiplication and division can be performed by adders and logic gates. Let us see how the parallel adder can be used to do addition and subtraction using 2's complement.

Recall that positive 2's complement numbers are exactly the same as regular true binary numbers. These can be added using regular binary addition. However, subtraction in 2's complement arithmetic is performed by converting the number to be subtracted to a negative number in the 2's complement and then using regular binary addition. Therefore, once we have the numbers in 2's complement form, we can use a parallel adder to get the answer.

For example, to subtract 9 from 18, we would first convert 9 to a 2's complement number, *i.e.*,

$+9 \rightarrow$	0 0 0 0 1 0 0 1	
	1 1 1 1 0 1 1 0	— 1's complement of 9
	+ 1	
	1 1 1 1 0 1 1 1	— 2's complement of 9

The we odd 18 + (– 9), *i.e.*,

18 →	0 0 0 1 0 0 1 0	
– 9	1 1 1 1 0 1 1 1	— 2's complement of 9
+		
	1 0 0 0 0 1 0 0 1	

disregard the carry. The result is $00001001_2 = 9_{10}$

Fig. 36-13 shows a circuit to implement a 2's complement adder/subtractor using 74283 adders. As seen from the circuit, 8-bit number on A inputs ($A_0 – A_7$) is brought directly into parallel adders. The other 8-bit binary number comes in on $B_0 – B_7$ lines. If the B number is to be subtracted, the complementing switch will be in position 'S'. this causes each bit in the B-number to be complemented (i.e it produces 1's complement of B number). At the same time, the lower-order parallel added receives $C_0 = 1$. This has the effect of adding a 1 to the already complemented B number, making it a 2's complement number. Notice that if the complementry switch is in position 'A', no complementation is done and the circuit performs normal addition.

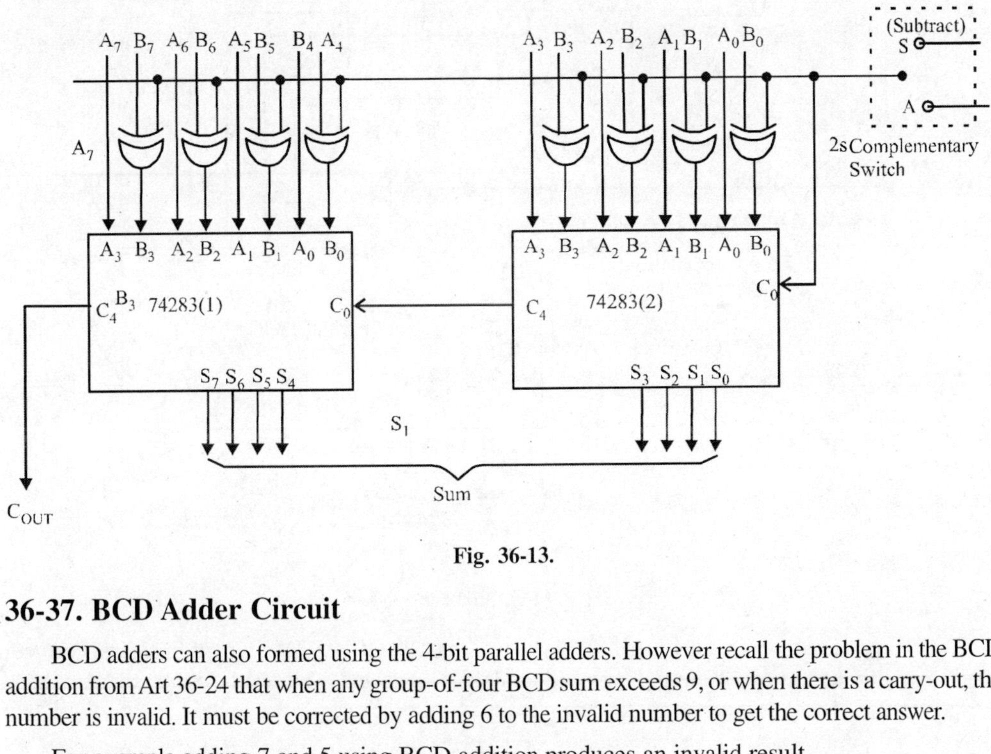

Fig. 36-13.

36-37. BCD Adder Circuit

BCD adders can also formed using the 4-bit parallel adders. However recall the problem in the BCD addition from Art 36-24 that when any group-of-four BCD sum exceeds 9, or when there is a carry-out, the number is invalid. It must be corrected by adding 6 to the invalid number to get the correct answer.

For example adding 7 and 5 using BCD addition produces an invalid result.

	0111	← BCD code for 7
	0101	← BCD code for 5
	1100	← Invalid BCD code
	+ 110	← Add 6 for correction
0001	0010	← BCD code for 12
1	2	

Checking for a sum greater than 9 (or a carry out) can be done easily using logic gates. Then whenever an invalid sum occurs, it can be corrected by adding 6 (i.e 0110) via the connections as shown in Fig. 36-14.

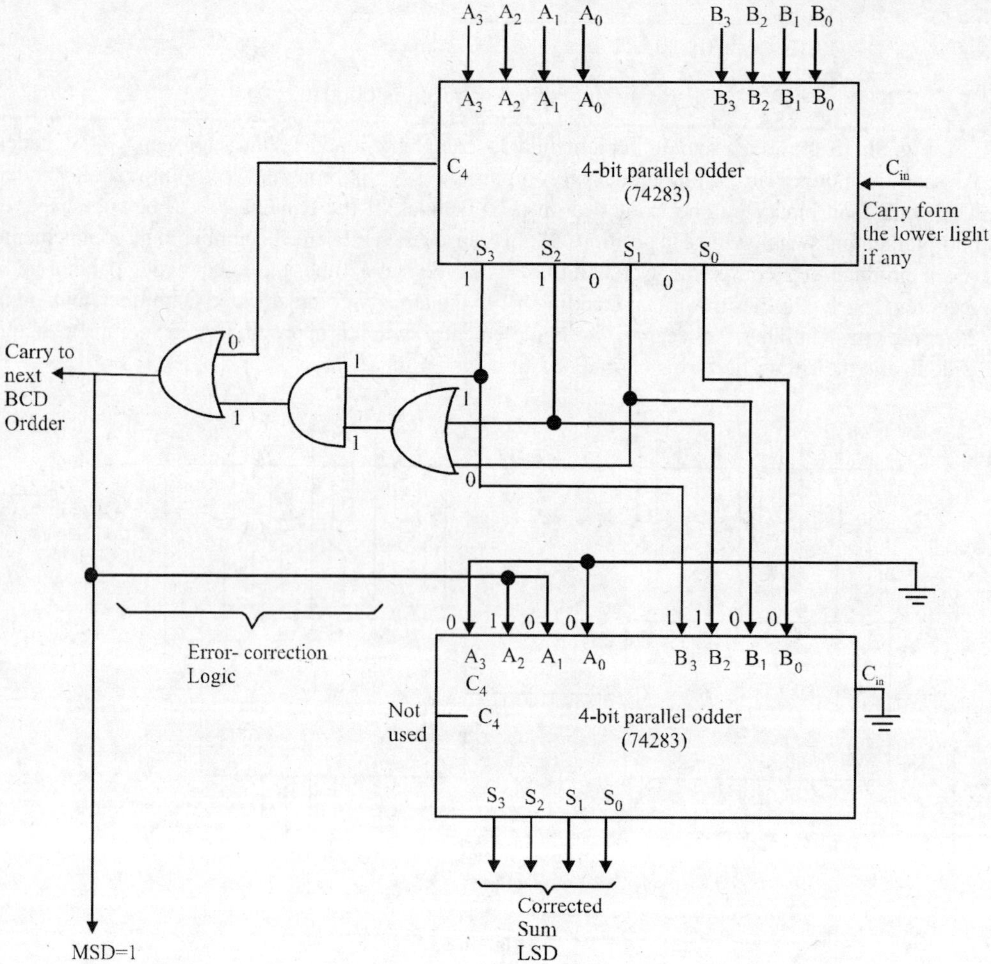

Fig. 36-14.

A 4-bit BCD adder is available in a single IC package. The IC74583 has an internal error correction logic circuitry to add two 4-bit numbers and produce a corrected 4-bit result with carry output.

SUMMARY

In this chapter, you have learned that :

1. To represent signed numbers in binary, a sign bit is attached as the most-significant bit. A positive sign is represented by a 0 bit and a negative sign by a 1 bit.
2. The 1's complement of a binary number is obtained by complementing all the bits of a given number.
3. The 2's complement of a binary number is obtained by adding 1 to 1's complement.
4. In an addition operation, overflow is possible when both numbers are positive or when both numbers are negative. An incorect sign bit in the sum indicates the occurrence of an overflow.

5. Subtraction can be performed on signed binary numbers by negating the subtrahend and adding it to the minuend.
6. A full-adder performs the additions on two bits plus a carry input. A parallel binary adder is made up of cascaded full-adders.

GLOSSARY

Augend : It is a number to which an addend is added.

Addend : It is a number to be added to another number.

Carry : A digit or a bit that is generated when two numbers are added and the result is greater than the base for the number system being used.

Full adder : A digital circuit that adds two bits and an input carry to produce a sum and an output carry.

Half adder : A digital circuit that adds two bits and produces a sum and an output carry. It cannot handle input carries.

Minuend : A number from which the subtrahend is to be subtracted.

Look-ahead carry : It is an ability of some parallel adders to predict, without having to wait for the carry to propagate through the full adders.

Over flow : During the process of adding signed binary numbers a carry of 1 is generated from the MSB position of the number into the sign bit position.

Parallel adder : A digital circuit (made from full adders) that is used to add all the bits of the given two numbers simultaneously.

Sign-magnitude system : It is a system for representing signed binary numbers where MSB represents sign of a number and the remaining bits represent the true binary value.

Subtrahend : A number that is used to be subtracted from the minuhend.

2's Coomplement system : It is a system for representing negative binary numbers. In this system the MSB (equals 1) represents the sign of a number and the remaining bits represent 2's complement of a number.

DESCRIPTIVE QUESTIONS

1. Under what circumstances would you use a half-adder instead of a full- adder circuit.
2. Reconstruct the half-adder circuit of Fig 36-4 using (*a*) only NOR gates (*b*) only NAND gates.
3. The circuit shown in Fig 36-15 is an attempt to build a half-adder. Will the sum and carry output function properly? (Hint: Write the Boolean expression for the sum (S) and carryout (C_{out}) terminals.

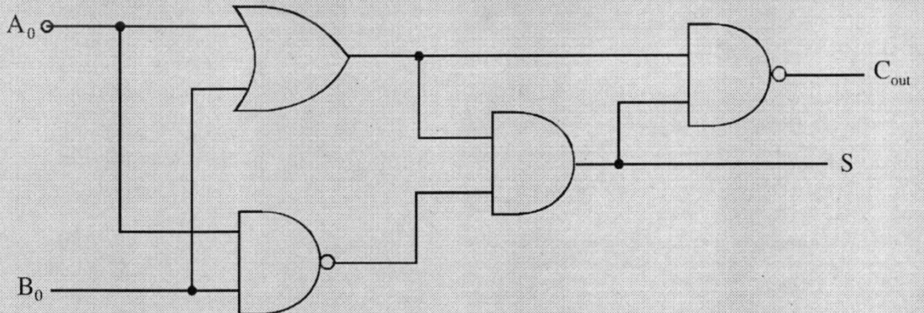

Fig. 36-15.

4. Use the K-map to prove that the carry output (C_{out}) function of the full-adder whose truth table is given in Fig 36-5 (*b*) can be implemented using Fig 36-16.

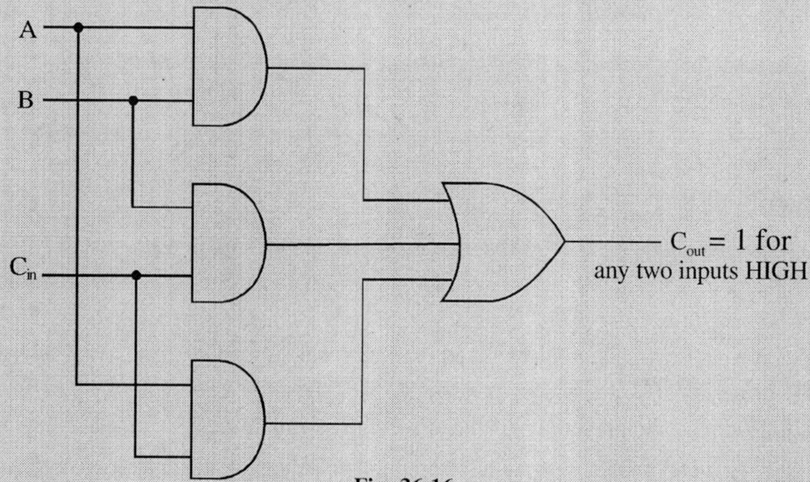

Fig. 36-16.

5. Draw the truth table for a three-input adder. Explain clerly the meaning of input and output symbols of the table. Write the Boolean expression for sum and carry.
6. Draw the schematic of a BCD adder using 4-bit binary adders. Explain the operation with suitable examples.
7. Explain the working of a half-adder circuit with the help of truth tables. Realize a full-adder using half-adders and/or gates.
8. Using 4-bit parallel binary adder, design a circuit to perform BCD addition. Show different conditions with different examples.
9. Sketch a block diagram of a 4-bit full-adder using four full-adders.

TUTORIAL PROBLEMS

1. Determine the 2's complement of the following 8-bit numbers:
   (*a*) 01011100 (*b*) 00100110 (*c*) 00001111
   (**Ans.** (*a*) 10100100, (*b*) 11011010, (*c*) 11110001).
2. Express the decimal number – 37 as a binary number using sign-magnitude system.
   (**Ans.** 1100101).
3. Determine the range of signed decimal values that can be represented in 16 bits (including the sign bit). (**Ans.** -2^{15} to $(2^{15}-1)$).
4. Determine the largest negative decimal value that can be represented by a 4-byte number.
   (**Ans.** -2^{-31}).
5. List in order, all the signed numbers that can be represented in five bits using the 2's complement system.
6. Perform the following operations in the 2's complement system. Use eight bits (including the sign bit) for each number.
   (*a*) Add 22 to 6 (*b*) Add 15 to – 4 (*c*) Add 36 to – 84 (*d*) Add – 48 to – 80
   (**Ans.** (*a*) 00011100_2, (*b*) 00001011_2, (*c*) 1010000_2, (*d*) 10000000_2).

7. Memory locations in personal computers are usually given in hexadecimal. If a computer programmer writes a program that requires 100 memory locations, determine the last memory location that is used if the program starts at the location 3C9FH where H = base 16 hexadecimal. (**Ans.** 3D02 H).

8. A particular model of personal computer indicates in its owner's manual that the following memory locations are used for storage of operating system subroutines: 07A2BH to 0AD68H inclusive and 02D8H to 03E00H inclusive. Determine the total number of memory locations used for this purpose in hexadecimal number. (**Ans.** 62BD H).

MULTIPLE CHOICE QUESTIONS

1. The 1's complement of 10111001 is

(*a*) 01000111 (*b*) 01000110 (*c*) 11000110 (d) 10101010

2. The 2's complement of 11001000 is

(*a*) 00110111 (*b*) 00110001 (*c*) 01001000 (*d*) 00111000

3. In the 2's complement system, the number 10010011_2 is equal to

(*a*) -19_{10} (*b*) $+109_{10}$ (*c*) $+91_{10}$ (*d*) -109_{10}

4. A binary number B can be subtracted from another binary number A by

(*a*) adding A and B and takign 2's complement of sum

(*b*) adding A with 2's complement of B

(*c*) adding 2's complement of A with B

(*d*) adding 2's complement of both A and B (*AMIE., Winter 2000*)

5. A certain memory location holds the hexadecimal data 77. If this represents an unsigned number, the decimal value is:

(*a*) + 77 (*b*) + 119 (*c*) – 119 (*d*) None of these

6. A certain memory location holds the hexadecimal data E5. If this represents a signed number, the decimal value is

(*a*) + 229 (*b*) – 229 (*c*) – 27 (*d*) None of these

7. The owner's manual for a small microcomputer system states that the computer has usable memory locations at the following addresses: 0200 through 03FF and through 4000 through 7FD0. The total number of available memory locations are:

(*a*) 01FF (*b*) 3FD0 (*c*) 41CF (*d*) None of these

8. In a half-adder having two inputs (A and B) and two outputs (Sum (S) and carry (C_{out})), the Boolean expression for S and C_{out} in terms of A and B. are:

(*a*) $S = \overline{A}B + A\overline{B}; C = A.B$ (*b*) $S = AB + \overline{A}B; C = A + B$

(*c*) $S = \overline{AB} + AB; C = A + \overline{B}$ (*d*) $S = \overline{A}B + A\overline{B}; C + \overline{A} + B$

9. A full-adder can be made out of:

(*a*) two half-adders (*b*) two half-adders and a NOT gate

(*c*) two half-adders and an OR gate (*d*) two half-adders and an AND gate

(*UPSC Engg. Services 1997*)

10. The m-bit parallel adder consists of

(*a*) m + 1 full adders (*b*) m/2 full adders

(*c*) (m – 1) full adders (*d*) m full adders (*AMIE., Winter 2000*)

11. A full adder can be constructed by using

(*a*) 2 AND gates, 3 XOR gates and OR gate

(*b*) 3 AND gates, 2 XOR gates and an OR gate

(*c*) 3 AND gates, 2 XOR gates and 2 or gates

(*d*) 2 AND gates, 2 XOR gates and 3 OR gates (*AMIE., Winter 2000*)

12. A carry look ahead adder is frequently used for addition because it.

(*a*) is faster (*b*) is more accurate (*c*) uses fewer gates (*d*) costs less

ANSWERS

1. (*b*)	**2.** (*d*)	**3.** (*b*)	**4.** (*b*)	**5.** (*b*)	**6.** (*c*)
7. (*c*)	**8.** (*a*)	**9.** (*c*)	**10.** (*b*)	**11.** (*b*)	**12.** (*a*)

APPENDIX A

Resistor and Capacitor Values

TYPICAL STANDARD RESISTOR VALUES

Ω	Ω	Ω	kΩ	kΩ	kΩ	MΩ	MΩ
—	10	100	1	10	100	1	10
—	12	120	1.2	12	120	1.2	—
—	15	150	1.5	15	150	1.5	15
—	18	180	1.8	18	180	1.8	—
—	22	220	2.2	22	220	2.2	22
2.7	27	270	2.7	27	270	2.7	—
3.3	33	330	3.3	33	330	3.3	—
3.9	39	390	3.9	39	390	3.9	—
4.7	47	470	4.7	47	470	4.7	—
5.6	56	560	5.6	56	560	5.6	—
6.8	68	680	6.8	68	680	6.8	—
—	82	820	8.2	82	820	—	—

TYPICAL STANDARD CAPACITOR VALUES

pF	*pF*	*pF*	*pF*	μ*F*	μ*F*	μ*F*	μ*F*	μ*F*	μ*F*	μ*F*
5	50	500	5000		0.05	0.5	5	50	500	5000
—	51	510	5100		—	—	—	—	—	—
—	56	560	5600		0.056	0.56	5.6	56	—	5600
—	—	—	6000		0.06	—	6	—	—	6000
—	62	620	6200		—	—	—	—	—	—
—	68	680	6800		0.068	0.68	6.8	—	—	—
—	75	750	7500		—	—	—	75	—	—
—	—	—	8000		—	—	8	80	—	—
—	82	820	8200		0.082	0.82	8.2	82	—	—
—	91	910	9100		—	—	—	—	—	—
10	100	1000		0.01	0.1	1	10	100	1000	10000
—	110	1100		—	—	—	—	—	—	—
12	120	1200		0.012	0.12	1.2	—	—	—	
—	130	1300		—	—	—	—	—	—	—
15	150	1500		0.015	0.15	1.5	15	150	1500	
—	160	1600		—	—	—	—	—	—	
18	180	1800		0.018	0.18	1.8	18	180	—	
20	200	2000		0.02	0.2	2	20	200	2000	
22	220	2200		—	0.22	2.2	22	—	—	
24	240	2400		—	—	—	—	240	—	
—	250	2500		—	0.25	—	25	250	2500	
27	270	2700		0.027	0.27	2.7	27	270	—	
30	300	3000		0.03	0.3	3	20	300	3000	
33	330	3300		0.033	0.33	3.3	33	330	3300	
36	360	3600		—	—	—	—	—	—	
39	390	3900		0.039	0.39	3.9	39	—	—	
—	—	4000		—	—	4	—	400	—	
43	430	4300		—	—	—	—	—	—	
47	470	4700		0.047	0.47	4.7	47	—	—	

APPENDIX B

Resistor and Capacitor Values

Device Listing

The following device specification sheets are located on the pages indicated.

(Courtesy : Fairchild Semiconductor, TSC, National Semiconductor)

1N4001 - 1N4007

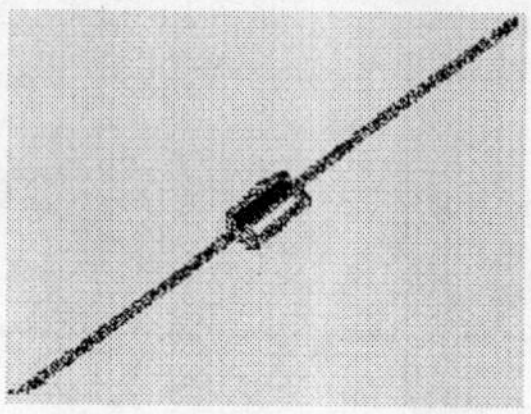

Features

- Low forward voltage drop.
- High surge current capability.

1.0 Ampere General Purpose Rectifiers

Absolute Maximum Ratings*

T_A = 25°C unless otherwise noted

Symbol	Parameter	Value	Units
I_O	Average Rectified Current 0.375 " lead length @ T_A = 75° C	1.0	A
$i_{f\ (surge)}$	Peak Forward Surge Current 8.3 ms single half-sine-wave superimposed on rated load (JEDEC method)	30	A
PD	Total Device Dssipation Derate above 25°C	2.5 20	W mW/°C
$R_{\theta JA}$	Thermal Resistance, Junction to Ambient	50	°C/W
T_{stg}	Storage Temperature Range	–55 to +175	°C
T_J	Operating Junction Temperature	–55 to + 150	°C

*These ratings are limiting values above which the serviceability of any semiconductor device may be impaired.

Electrical Characteristics

T_A = 25°C unless otherwise noted

Parameter	Device							Units
	4001	4002	4003	4004	4005	4006	4007	
Peak Repetitive Reverse Voltage	50	100	200	400	600	800	1000	V
Maximum RMS Voltage	35	70	140	280	420	560	700	V
DC Reverse Voltage (Rated V_R)	50	100	200	400	600	800	1000	V
Maxmimum Reverse Current								
@ rated VR T_A = 25°C				5.0				μA
T_A = 100°C				500				μA
Maximum Forward voltage @1.0 A				1.1				V
Maximum Full Load Reverse Current,				30				μA
Full Cycle T_A = 75° A								
Typical Junction Capacitance				15				pF
V_R = 4.0 V, f = 1.0 MHz								

ZENERS
1N746A - 1N759A

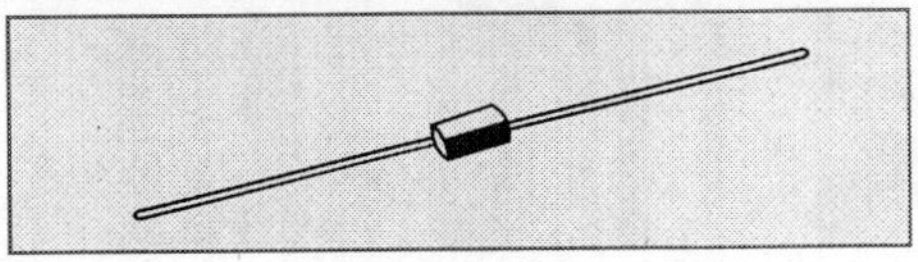

Absolute Maximum Ratings*

T_A = 25°C unless otherwise noted

Symbol	Parameter	Value	Units
P_D	Power Dissipation	500	mW
T_{STG}	Storage Temperature Range	–65 to +200	°C
T_J	Operating Junction Temperature	+ 175	°C
	Lead Temperature (1/16" from case for 10 seconds)	+ 230	°C

* These ratings are limiting values above which thge servicebility of the diode may be impaired.

NOTES:

1) These ratings are based on a maximum junction temperature of 200 degrees C.

2) These are steady state limits. The factory should be consulted on applications involving pulsed or low duty cycle operations.

Electrical Characteristics

T_A = 25°C unless otherwise noted

Device	V_Z(V)	$Z_Z(\Omega)$	@ I_Z(mA)	$I_{R1}(\mu A)$	@ V_R(V)	I_{R2}(mA)	@ V_R(V) TA = 150°C	T_C (%/°C)	I_{ZRM}*(mA)
1N746A	3.3	28	20	10	1.0	30	1.0	−0.070	110
1N747A	3.6	24	20	10	1.0	30	1.0	−0.065	100
1N748A	3.9	23	20	10	1.0	30	1.0	−0.060	95
1N749A	4.3	22	20	2.0	1.0	30	1.0	+/− 0.055	85
1N750A	4.7	19	20	2.0	1.0	30	1.0	+/− 0.030	75
1N751A	5.1	17	20	1.0	1.0	20	1.0	+/− 0.030	70
1N752A	5.6	11	20	1.0	1.0	20	1.0	+ 0.038	65
1N753A	6.2	7.0	20	0.1	1.0	20	1.0	+ 0.045	60
1N754A	6.8	5.0	20	0.1	1.0	20	1.0	+ 0.050	55
1N755A	7.5	6.0	20	0.1	1.0	20	1.0	+ 0.058	50
1N756A	8.2	8.0	20	0.1	1.0	20	1.0	+ 0.062	45
1N757A	9.1	10	20	0.1	1.0	20	1.0	+ 0.068	40
1N758A	10	17	20	0.1	1.0	20	1.0	+ 0.075	35
1N759A	12	30	20	0.1	1.0	20	1.0	+ 0.077	38

*I_{ZRM}(Maximum Zener Current Rating) Values shown are based on the JEDEC rating of 400 milliwatts. Where the actual zener voltage (VZ) is known at the operating point, the maximum zener current may be increased and is limited by the derating curve.

PN2222A MMBT2222A PZT2222A

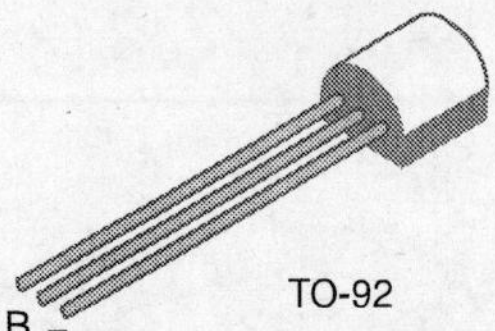

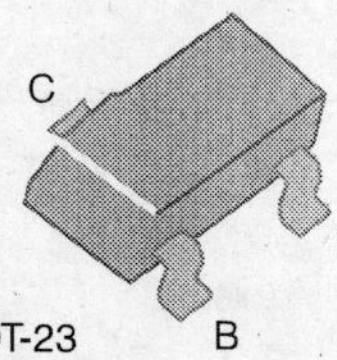

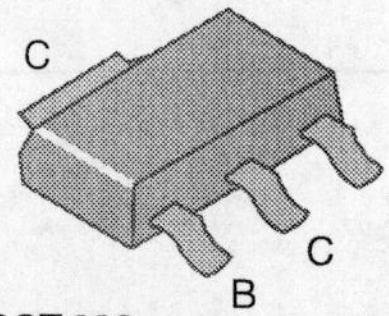

NPN General Purpose Amplifier

This device is for use as a medium power amplifier and switch requiring collector currents up to 500 mA. Sourced from Process 19.

Absolute Maximum Ratings*

T_A = 25°C unless otherwise noted

Symbol	Parameter	Value	Units
V_{CEO}	Collector-Emitter Voltage	40	V
V_{CBO}	Collector-Base Voltage	75	V
V_{EBO}	Emitter-Base Voltage	6.0	V
I_C	Collector Current - Continuous	1.0	A
T_J, T_{stg}	Operating and Storage Junction Temperature Range	–55 to +150	°C

*These ratings are limiting values above which the serviceability of any semiconductor device may be impaired.

NOTES:

1) These ratings are based on a maximum junction temperature of 150 degrees C

2) These are steady state limits. The factory should be consulted on applications involving pulsed or low duty cycle operations.

Thermal Characteristics

T_A = 25°C unless otherwise noted

Symbol	Characteristics	Max			Units
		PN2222A	*MMBT2222A	**PZT2222A	
P_D	Total Device Dssipation	625	350	1,000	mW
	Derate above 25°C	5.0	2.8	8.0	mW/°C
$R_{\theta JC}$	Thermal Resistance, Junction to Case	83.3			°C/W
$R_{\theta JA}$	Thermal Resistance, Junction to Ambient	200	357	125	°C/W

*Device mounted on FR-4 PCB 1.6" X 1.6" X 0.06."

**Device mounted on FR-4 PCB 36 mm X 18 mm X 1.5 mm; mounting paid for the collector lead min. 6 cm^2.

PN2222A / MMBT2222A / PZT2222A

NPN General Purpose Amplifier

(continued)

Electrical Characteristics T_A = 25°C unless otherwise noted

Symbol	Parameter	Test Conditions	Min	Max	Units
OFF CHARACTERISTICS					
$V_{(BR)CEO}$	Collector-Emitter Breakdown Voltage*	I_C = 10 mA, I_B = 0	40		V
$V_{(BR)CBO}$	Collector-Base Breakdown Voltage	I_C = 10 μA, I_E = 0	75		V
$V_{(BR)EBO}$	Emitter-Base Breakdown Voltage	I_E = 10 μA, I_C = 0	6.0		V
I_{CEX}	Collector Cutoff Current	V_{CE} = 60 V, $V_{EB(OFF)}$ = 3.0 V		10	nA
I_{CBO}	Collector Cutoff Current	V_{CB} = 60 V, I_E = 0,		0.01	μA
		V_{CB} = 60 V, I_E = 0, T_A = 150°A		10	μA
I_{EBO}	Emitter Cutoff Current	V_{EB} = 3.0 V, I_C = 0		10	nA
I_{BL}	Base Cutoff Current	V_{CE} = 60 V, $V_{EB(OFF)}$ = 3.0 V		20	nA
ON CHARACTERISTICS					
h_{FE}	DC Current Gain	I_C = 0.1 mA, V_{CE} = 10 V	35		
		I_C = 1.0 mA, V_{CE} = 10 V	50		
		I_C = 10 mA, V_{CE} = 10 V	75		
		I_C = 10 mA, V_{CE} = 10 V, T_A = -55°C	35		
		I_C = 150 mA, V_{CE} = 10 V*	100	300	
		I_C = 150 mA, V_{CE} = 1.0 V*	50		
		I_C = 500 mA, V_{CE} = 10 V*	40		
$V_{CE(sat)}$	Collector-Emitter Saturation	I_C = 150 mA, I_B = 15 mA		0.3	V
	Voltage*	I_C = 500 mA, I_B = 50 mA		1.0	V
$V_{BE(sat)}$	Base-Emitter Saturation Voltage*	I_C = 150 mA, I_B = 15 mA	0.6	1.2	V
		I_C = 500 mA, I_B = 50 mA		2.0	V
SMALL SIGNAL CHARACTERISTICS					
f_T	Current Gain - Bandwidth Product	I_C = 20 mA, V_{CE} = 20 V, f = 100 MHz	300		MHz
C_{obo}	Output Capacitance	V_{CB} = 10 V, I_E = 0, f = 100 kHz		8.0	pF
C_{ibo}	Input Capacitance	V_{EB} = 0.5 V, I_C = 0, f = 100 kHz		25	pF
$rb'C_C$	Collector Base Time Constant	I_C = 20 mA, V_{CB} = 20 V, f = 31.8 MHz		150	pS
NF	Noise Figure	I_C = 100 μA, V_{CE} = 10 V		4.0	dB
		R_S = 100 kΩ, f = 1.0 kHz			
$Re_{(hie)}$	Real part of Common-Emitter	I_C = 20 mA, V_{CB} = 20 V		60	Ω
	High Frequency Input Impedance	f = 300 MHz			
SWITCHING CHARACTERISTICS					
t_d	Delay Time	V_{CC} = 30 V, $V_{BE(OFF)}$ = 0.5 V		10	ns
t_r	Rise Time	I_C = 150 mA, I_{B1} = 15 mA		25	ns
t_s	Storage Time	V_{CC} = 30 V, I_C = 150 mA		225	ns
t_f	Fall Time	I_{B1} = I_{B2} = 15 mA		60	ns

*Pulse Test: Pulse Width ≤ 300 μs, Duty Cycle ≤ 2.0%

Spice Model

NPN (Is=14.34f Xti=3 Eg=1.11 Vaf=74.03 Bf= 255.9 Ne=1.307 Ise=14.34f Ikf=0.2847 Xtb=1.5 Br=6.092 Nc=2 Isc=0 Ikr=0 Rc=1 Cjc=7.306p Mjc=0.3416 Vjc=.75 Fc=.5 Cje=22.01p Mje=.377 Vje=.75 Tr=46.91n Tf=411.1p Itf=.6 Vtf=1.7 Xtf=3 Rb=10)

2N3904 MMBT3904 PZT3904

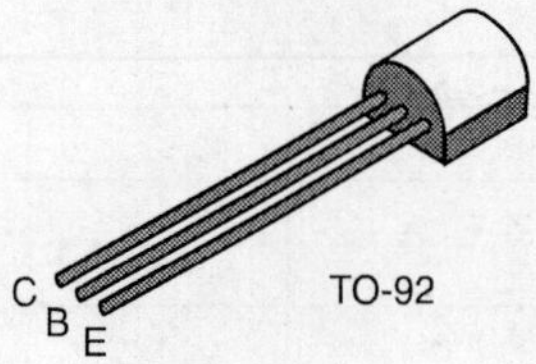

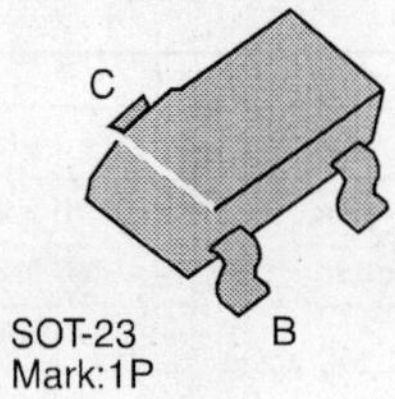

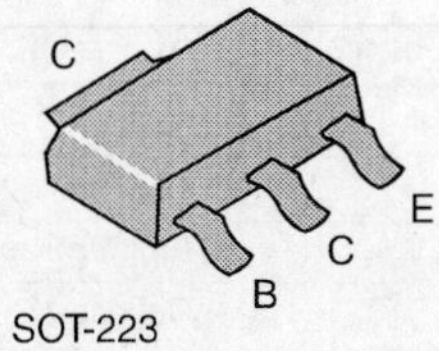

NPN General Purpose Amplifier

This device is designed as a general purpose amplifier and switch. The useful dynamic range extends to 100 mA as a switch and to 100 MHz as an amplifier.

Absolute Maximum Ratings*

T_A = 25°C unless otherwise noted

Symbol	Parameter	Value	Units
V_{CEO}	Collector-Emitter Voltage	40	V
V_{CBO}	Collector-Base Voltage	60	V
V_{EBO}	Emitter-Base Voltage	6.0	V
I_C	Collector Current - Continuous	200	mA
T_J, T_{stg}	Operating and Storage Junction Temperature Range	–55 to +150	°C

*These ratings are limiting values above which the serviceability of any semiconductor device may be impaired.

NOTES:

1) These ratings are based on a maximum junction temperature of 150 degrees C

2) These are steady state limits. The factory should be consulted on applications involving pulsed or low duty cycle operations.

Thermal Characteristics

T_A = 25°C unless otherwise noted

Symbol	Characteristics	Max			Units
		PN3904	*MMBT3904	**PZT3904	
P_D	Total Device Dssipation	625	350	1,000	mW
	Derate above 25°C	5.0	2.8	8.0	mW/°C
$R_{\theta JC}$	Thermal Resistance, Junction to Case	83.3			°C/W
$R_{\theta JA}$	Thermal Resistance, Junction to Ambient	200	357	125	°C/W

*Device mounted on FR-4 PCB 1.6" × 1.6" × 0.06."

**Device mounted on FR-4 PCB 36 mm × 18 mm × 1.5 mm; mounting paid for the collector lead min. 6 cm^2.

NPN General Purpose Amplifier

(continued)

Electrical Characteristics $T_A = 25°C$ unless otherwise noted

Symbol	Parameter	Test Conditions	Min	Max	Units
OFF CHARACTERISTICS					
$V_{(BR)CEO}$	Collector-Emitter Breakdown Voltage*	$I_C = 1.0$ mA, $I_B = 0$	40		V
$V_{(BR)CBO}$	Collector-Base Breakdown Voltage	$I_C = 10$ μA, $I_E = 0$	60		V
$V_{(BR)EBO}$	Emitter-Base Breakdown Voltage	$I_E = 10$ μA, $I_C = 0$	6.0		V
I_{BL}	Base Cutoff Current	$V_{CE} = 30$ V, $V_{EB} = 3$ V		50	nA
I_{CEX}	Collector Cutoff Current	$V_{CE} = 30$ V, $V_{EB} = 3$ V		50	nA
ON CHARACTERISTICS					
h_{FE}	DC Current Gain	$I_C = 0.1$ mA, $V_{CE} = 1.0$ V	40		
		$I_C = 1.0$ mA, $V_{CE} = 1.0$ V	70		
		$I_C = 10$ mA, $V_{CE} = 1.0$ V	100	300	
		$I_C = 50$ mA, $V_{CE} = 1.0$ V	60		
		$I_C = 100$ mA, $V_{CE} = 1.0$ V	30		
$V_{CE(sat)}$	Collector-Emitter Saturation Voltage	$I_C = 10$ mA, $I_B = 1.0$ mA		0.2	V
		$I_C = 50$ mA, $I_B = 5.0$ mA		0.3	V
$V_{BE(sat)}$	Base-Emitter Saturation Voltage	$I_C = 10$ mA, $I_B = 1.0$ mA	0.65	0.85	V
		$I_C = 50$ mA, $I_B = 5.0$ mA		0.95	V
SMALL SIGNAL CHARACTERISTICS					
f_T	Current Gain - Bandwidth Product	$I_C = 10$ mA, $V_{CE} = 20$ V, f = 100 MHz	300		MHz
C_{obo}	Output Capacitance	$V_{CB} = 5.0$ V, $I_E = 0$, f = 1.0 MHz		4.0	pF
C_{ibo}	Input Capacitance	$V_{EB} = 0.5$ V, $I_C = 0$, f = 1.0 MHz		8.0	pF
NF	Noise Figure	$I_C = 100$ μA, $V_{CE} = 5.0$ V, $R_S = 1.0$ kΩ, f = 10 Hz to 15.7 kHz		5.0	dB
SWITCHING CHARACTERISTICS					
t_d	Delay Time	$V_{CC} = 3.0$ V, $V_{BE} = 0.5$ V		35	ns
t_r	Rise Time	$I_C = 10$ mA, $I_{B1} = 1.0$ mA		35	ns
t_s	Storage Time	$V_{CC} = 3.0$ V, $I_C = 10$ mA		200	ns
t_f	Fall Time	$I_{B1} = I_{B2} = 1.0$ mA		50	ns

*Pulse Test: Pulse Width ≤ 300 μs, Duty Cycle ≤ 2.0%

Spice Model

NPN (Is=6.734f Xti=3 Eg=1.11 Vaf=74.03 Bf= 416.4 Ne=1.259 Ise=6.734f Ikf=66.78m Xtb=1.5 Br=.7371 Nc=2 Isc=0 Ikr=0 Rc=1 Cjc=3.638p Mjc=.3085 Vjc=.75 Fc=.5 Cje=4.493p Mje=.2593 Vje=.75 Tr=239.5n Tf=301.2p Itf=.4 Vtf=4 Xtf=2 Rb=10)

2N5457 MMBF5457
2N5458 MMBF5458
2N5459 MMBF5459

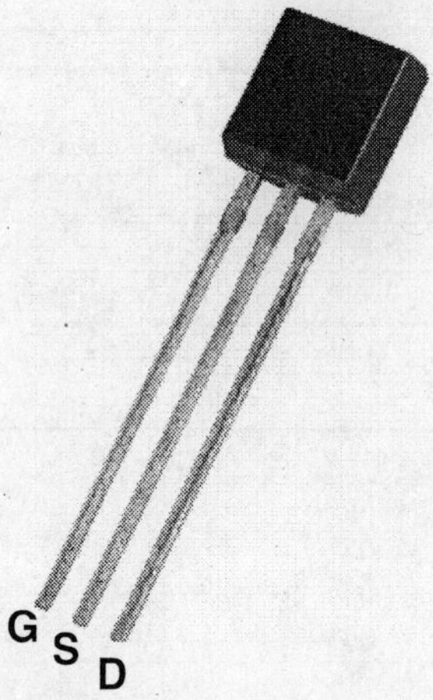

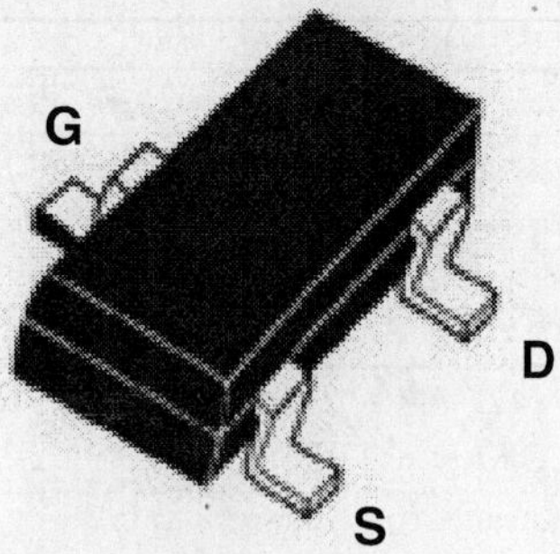

SOT-23

Mark: 6D/61S/6L

N-Channel General Purpose Amplifier

This device is a low level audio amplifier and switching transistor and can be used for analog switching aplications. Sourced from Process 55.

Absolute Maximum Ratings*

TA = 25°C unless otherwise noted

Symbol	Parameter	Value	Units
V_{DG}	Drain-Gate Voltage	25	V
V_{GS}	Gate-Source Voltage	– 25	V
I_{GF}	Forward Gate Current	10	mA
T_J, T_{stg}	Operating and Storage Junction Temperature Range	–55 to +150	°C

*These ratings are limiting values above which the serviceability of any semiconductor device may be impaired.

NOTES:

1) These ratings are based on a maximum junction temperature of 150 degrees C

2) These are steady state limits. The factory should be consulted on applications involving pulsed or low duty cycle operations.

Thermal Characteristics

TA = 25°C unless otherwise noted

Symbol	Characteristics	Max		Units
		2N5457-5459	*MMBF5457-5459	
P_D	Total Device Dssipation	625	350	mW
	Derate above 25°C	5.0	2.8	mW/°C
$R_{\theta JC}$	Thermal Resistance, Junction to Case	125		°C/W
$R_{\theta JA}$	Thermal Resistance, Junction to Ambient	357	556	°C/W

*Device mounted on FR-4 PCB 1.6" × 1.6" × 0.06."

N-Channel General Purpose Amplifier

(continued)

Electrical Characteristics TA = 25°C unless otherwise noted

Symbol	Parameter	Test Conditions		Min	Typ	Max	Units
OFF CHARACTERISTICS							
$V_{(BR)GSS}$	Gate-Source Breakdown Voltage	$I_G = 10\ \mu A, V_{DS} = 0$		– 25			V
I_{GSS}	Gate Reverse Current	$V_{GS} = -15\ V, V_{DS} = 0$				– 1.0	nA
		$V_{GS} = -15\ V, V_{DS} = 0, T_A = 100°C$				– 200	nA
$V_{GS(off)}$	Gate Source Cut-off Voltage	$V_{DS} = 15\ V, I_D = 10\ nA$	**5457**	– 0.5		– 6.0	V
			5458	– 1.0		– 7.0	V
			5459	– 2.0		– 8.0	V
V_{GS}	Gate-Source Voltage	$V_{DS} = 15\ V, I_D = 100\ \mu A$	**5457**		– 2.5		V
		$V_{DS} = 15\ V, I_D = 200\ \mu A$	**5458**		– 3.5		V
		$V_{DS} = 15\ V, I_D = 400\ \mu A$	**5459**		– 4.5		V
ON CHARACTERISTICS							
I_{DSS}	Zero-Gate Voltage Drain Current*	$V_{DS} = 15\ V, V_{GS} = 0$	**5457**	1.0	3.0	5.0	mA
			5458	2.0	6.0	9.0	mA
			5459	4.0	9.0	16	mA
SMALL SIGNAL CHARACTERISTICS							
g_{fs}	Forward Transfer Conductance*	$V_{DS} = 15\ V, V_{GS} = 0, f = 1.0\ kHz$					
			5457	1000		5000	μmhos
			5458	1500		5500	μmhos
			5459	2000		6000	μmhos
g_{os}	Output Conductance	$V_{DS} = 15\ V, V_{GS} = 0, f = 1.0\ kHz,$			10	50	μmhos
C_{iss}	Input capacitance	$V_{DS} = 15\ V, V_{GS} = 0, f = 1.0\ kHz$			4.0	pF	
C_{rss}	Reverse Transfer Capacitance	$V_{DS} = 15\ V, V_{GS} = 0, f = 1.0\ kHz$			8.0	pF	
NF	Noise Figure	$V_{DS} = 15\ V, V_{GS} = 0, f = 1.0\ kHz$ $R_G = 1.0$ megohm, BW = 1.0 Hz			5.0	dB	

*Pulse Test: Pulse Width ≤ 300 μs, Duty Cycle ≤ 2.0%

1N5391 THRU **1N5399**

1.5 AMPS. Silicon Rectifiers

Voltage Rangde
50 to 1000 Volts
Current
1.5 Amperes

DO-15

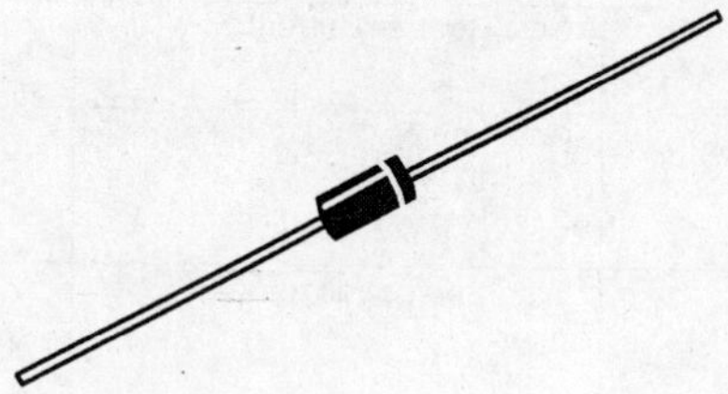

Features

- ✧ Low forward voltage drop
- ✧ High current capability
- ✧ High reliability
- ✧ High surge current capability

Mechanical Data

- ✧ Cases: Molded plastic
- ✧ Epoxy: UL94V-O rate flame retardant
- ✧ Lead: Axial leads, solderable per MIL-STD-202, Method 208 guaranteed
- ✧ Polarity: Color band denotes cathode end
- ✧ High temperature soldering guaranteed: 260°C/10 seconds/3.75",(9.5mm) lead lengths at 5 lbs., (2.3kg) tension
- ✧ Weight: 0.40 gram

Maximum Ratings and Electrical Characteristics

Rating at 25°C ambient temperature unless otherwise specified.
Single phase, half wave, 60 Hz, resistive or inductive load.
For capacitive load, derate current by 20%

Type Number	Symbol	1N 5391	1N 5392	1N 5393	1N 5395	1N 5397	1N 5398	1N 5399	Units
Maximum Recurrent Peak Reverse Voltage	V_{RRM}	50	100	200	400	600	800	1000	V
Maximum RMS Voltage	V_{RMS}	35	70	140	280	420	560	700	V
Maximum DC Blocking Voltage	V_{DC}	50	100	200	400	600	800	1000	V
Maximum Average Forward Rectified Current 3.75 (9.5mm) Lead length @ T_A = 75°C	$I_{(AV)}$	1.5							A
Peak Forward Surge Current, 8.3 ms Single Half Sine-wave Superimposed on Rated Load (JEDEC method)	I_{FSM}	50							A
Maximum Instantaneous Forward Voltage @ 1.5A	V_F	1.1	1.0						V
Maximum DC Reverse Current @ T_A = 25°C at RAted DC Blocking Voltage @ T_A = 100°C Maximum Full Load Reverse Current, Full	I_R	5.0 50							uA uA
Cycle Average 3.75"(9.5mm) Lead length @ TA = 75°C	HT_{IR}	30							uA
Typical Junction Capacitance (Note 1)	Cj	50							pF
Typical Thermal Resistance (Note 2)	$R_{\theta JA}$	60							°C/W
Operating Temperature Range	T_J	–65 to +150							°C
Storage Temperature Range	T_{STG}	–65 to +150							°C

Notes: 1. Measured at 1 MHz and Applied Reverse Voltage of 4.0 V D.C.
2. Mount on Cu-Pad Size 10mm × 10mm on P.C.B.

LM741
Operational Amplifier

General Description

The LM741 series are general purpose operational amplifiers which feature improved performance over industry standards like the LM709. They are direct, plug-in replacements for the 709C, LM201, MC1439 and 748 in most applications.

The amplifiers offer many features which make their application nearly foolproof: overload protection on the input and output, no latch-up when the common mode range is exceeded, as well as freedom from oscillatons.

The LM741C is identical to the LM741/LM741A except that the LM741C has their performance guaranteed over a 0°C to +70°C temperature range, instead of –55°C to +125°C.

mA741M . . . J PACKAGE
(TOP VIEW)

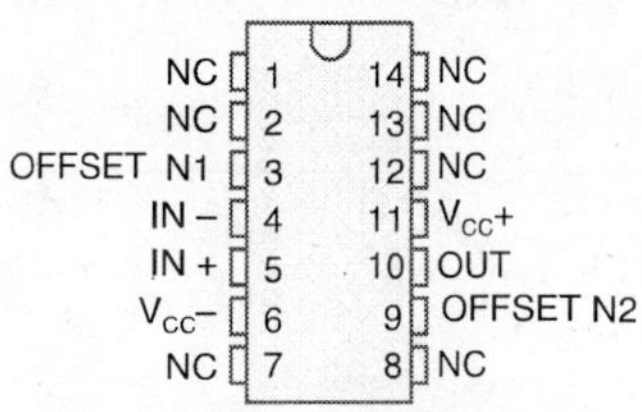

μA741M . . . JG PACKAGE
μA741C, μA741I . . . D,P, OR PW PACKAGE
(TOP VIEW)

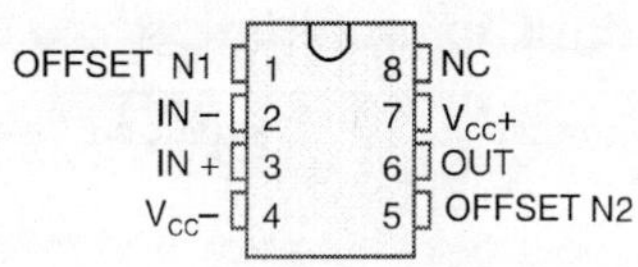

μA741M . . . U PACKAGE
(TOP VIEW)

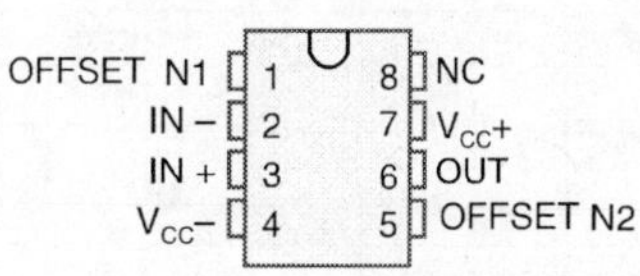

μA741M . . . U PACKAGE
(TOP VIEW)

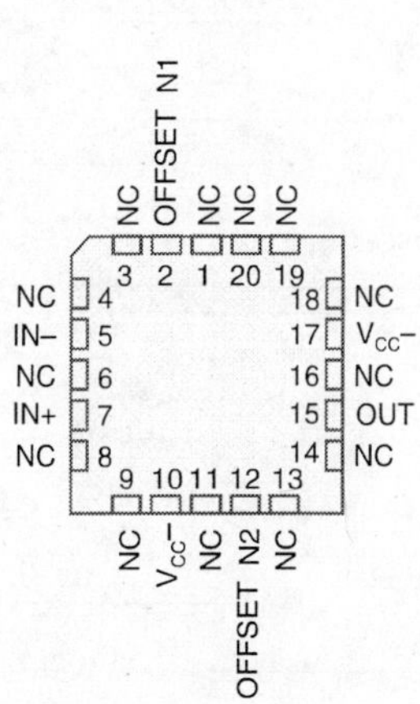

NC – No internal connection

Absolute Maximum Ratings (Note 2)

If Military/Aerospace specified devices are required, please contact the National Semiconductor Sales Office/ Distributions for availability and specifications.

(Note 7)

	LM741A	LM741	LM741C
Supply Voltage	±22V	±22V	±18V
Power Dissipation (Note 3)	500 mW	500 mW	500 mW
Differential Input Voltage	±30V	±30V	±30V
Input Voltage (Note 4)	±15V	±15V	±15V
Output Short Circuit Duration	Continuous	Continuous	Continuous
Operating Temperature Range	–55°C to +125°C	–55°C to +125°C	0°C to +70°C
Storage Temperature Range	–65°C to +150°C	–65°C to +150°C	–65°C to +150°C
Junction Temperature	150°C	150°C	100°C
Soldering Information			
N-Package (10 seconds)	260°C	260°C	260°C
J-or H-Package (10 seconds)	300°C	300°C	300°C
M-Package			
Vapor Phase (60 seconds)	215°C	215°C	215°C
Infrared (15 seconds)	215°C	215°C	215°C

See AN-450 "Surface Mounnting Methods and Their Effect on Product Realiablity" for other methods of soldering surface mount devices.

	LM741A	LM741	LM741C
ESD Tolerance (Note 8)	400V	400V	400V

Electrical Characteristics (Note 5)

Parameter	Conditions	LM741A			LM741			LM741C			Units
		Min	Typ	Max	Min	Typ	Max	Min	Typ	Max	
Input Offset Voltage	$T_A = 25°C$										
	$R_S \leq 10\ k\Omega$					1.0	5.0		2.0	6.0	mV
	$R_S \leq 50\ \Omega$	0.8	3.0								mV
	$T_{AMIN} \leq T_A \leq T_{AMAX}$										
	$R_S \leq 50\Omega$										
	$R_S \leq 10\ k\Omega$						6.0			7.5	mV
Average Input Offset Voltage Drift				15							mV/°C
Input Offset Voltage Adjustment Range	$T_A = 25°C$, $V_S = \pm 20V$	±10				±15		15±			mV
Input Offset Current	$T_A = 25°C$		3.0	30		20	200		20	200	nA
	$T_{AMIN} \leq T_A \leq T_{AMAX}$			70		85	500			300	nA
Average Input Offset Current Drift			0.5								nA/°C
Input Bias Current	$T_A = 25°C$		30	80		80	500		80	500	nA
	$T_{AMIN} \leq T_A \leq T_{AMAX}$			0.210			1.5			0.8	μA
Input Resistance	$T_A = 25°C$, $V_S = \pm 20V$	1.0	6.0		0.3	2.0		0.3	3.0		MΩ
	$T_{AMIN} \leq T_A \leq T_{AMAX}$, $V_S = \pm 20V$	0.5									MΩ
Input Voltage Range	$T_A = 25°C$							±12	±13		V
	$T_{AMIN} \leq T_A \leq T_{AMAX}$				±12	±13					V

Electrical Characteristics (Note 5) (Continued)

Parameter	Conditions	LM741A			LM741			LM741C			Units
		Min	Typ	Max	Min	Typ	Max	Min	Typ	Max	
Large signal Voltage Gain	$T_A = 25°C$, $R_L \le 2$ kΩ										
	$V_S = \pm 20V$, $V_O = \pm 15V$	50									V/mV
	$V_S = \pm 15V$, $V_O = \pm 10V$				50	200		20	200		V/mV
	$T_{AMIN} \le T_A \le T_{AMAX}$										
	$R_L \ge 2$ kΩ,										
	$V_S = \pm 20V$, $V_O = \pm 15V$	32									V/mV
	$V_S = \pm 15V$, $V_O = \pm 10V$				25			15			V/mV
	$V_S = \pm 5V$, $V_O = \pm 2V$	10									V/mV
Output Voltage Swing	$V_S = \pm 20V$			15							mV/°C
	$R_L \ge 10$ kΩ	±16									V
	$R_L \ge 2$ kΩ	±15									V
	$V_S = \pm 15V$										
	$R_L \ge 10$ kΩ				±12	±14		±12	±14		V
	$R_L \ge 2$ kΩ				±10	±13		±10	±13		V
Output Short Circuit	$T_A = 25°C$	10	25	35		25			25		mA
Current	$T_{AMIN} \le T_A \le T_{AMAX}$	10		40							mA
Common-Mode	$T_{AMIN} \le T_A \le T_{AMAX}$										
Rejection Ratio	$R_S \le 10$ kΩ, $V_{CM} = \pm 12V$				70	90		70	90		dB
	$R_S \le 50$ Ω, $V_{CM} = \pm 12V$	80	95								dB
Supply Voltage Rejection	$T_{AMIN} \le T_A \le T_{AMAX}$		0.5								nA/°C
Ratio	$V_S = \pm 20V$ to $V_S = \pm 5V$										
	$R_S \le 50$Ω	86	96								dB
	$R_S \le 10$ kΩ				77	96		77	96		dB
Transient Response	$T_A = 25°C$, Unity Gain										
Rise Time			0.25	0.8		0.3			0.3		µs
Overshoot			6.0	20		5			5		%
Bandwidth (Note 6)	$T_A = 25°C$	0.437	1.5								MHz
Slew Rate	$T_A = 25°C$, Unity Gain	0.3	0.7			0.5			0.5		V/µs
Supply Current	$T_A = 25°C$					1.7	2.8		1.7	2.8	mA
Power Consumption	$T_A = 25°C$										
	$V_S = \pm 20V$		80	150							mW
	$V_S = \pm 15V$					50	85		50	85	mW
LM741A	$V_S = \pm 20V$										
	$T_A = T_{AMIN}$			165							mW
	$T_A = T_{AMAX}$			135							mW
LM741	$V_S = \pm 15V$										
	$T_A = T_{AMIN}$					60	100				mW
	$T_A = T_{AMAX}$					45	75				mW

Note 2: "Absolute Maximum Ratings" indicate limits beyond which damage to the device may occur. Operating Ratings indicate conditions for which the device is functional, but do not guarantee specific performance limits.

Electronic Characteristics (Note 5) (Continued)

Note 3: For operation at elevated temperatures, these devices must be derated based on thermal resistance, and T_j max. (listed under "Absolute Maximum Ratings"). $T_j = T_A + (\theta_{jA} P_D)$.

Thermal Resistance	Cerdip (J)	DIP (N)	HO8 (H)	SO-8 (M)
θ_{jA} (Junction to Ambient)	100°C/W	100°C/W	170°C/W	195°C/W
θ_{jA} (Junction to Case)	N/A	N/A	25°C/W	N/A

Note 4: For supply voltages less than ±15V, the absolute maximum input voltage is equal to the supply voltage.

Note 5: Unless otherwise specified, these specifications apply for $V_S = \pm 15V$, $-55°C \le T_A \le +125°C$ (LM741/LM741A). For the LM741C/LM741E, these specifications are limited to $0°C \le T_A \le +70°C$.

Note 6: Calculated value from: BW (MHz) = 0.35/Rise Time(μs).

Note 7: For military specifications see RETS741X for LM741 and RETS741AX for LM741A.

Note 8: Human body model, 1.5 kΩ in series with 100 pF.

Schematic Diagram

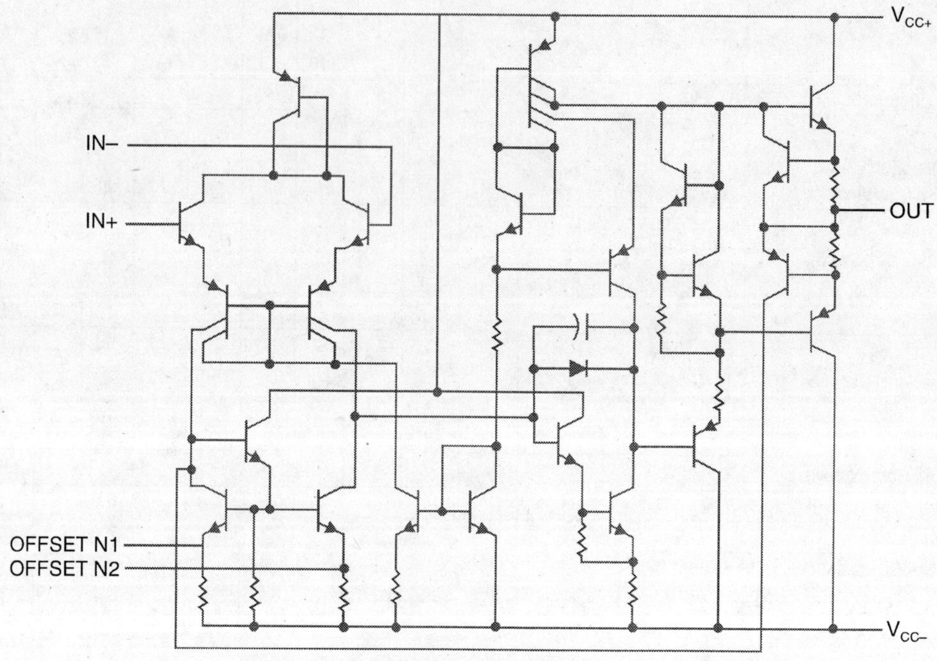

Component count	
Transistors	22
Resistors	11
Diode	1
Capacitor	1

7800 SERIES
(SERIES μA7800)

Positive-Voltage Regulators

October 1993 - revised Octorber 1999

- **3-Terminal Regulators**
- **Output Current Up to 1.5 A**
- **No External Components**
- **Internal Thermal Overload Protection**
- **High power Dissipation Capability**
- **Internal Short-Circuit Current Limiting**
- **Output Transistor Safe-Area Compensation**

NOMINAL OUTPUT VOLTAGE	REGULATOR
5V	μA7805
6V	μA7806
8V	μA7808
8.5V	μA7885
9V	μA7809
10V	μA7810
12V	μA7812
15V	μA7815
18V	μA7818
20V	μA7820
24V	μA7824
27V	μA7827

DESCRIPTION

This series of fixed-voltage monolithic integrated-circuit voltage regulators designed for a wide range of applications. These applications include on-card regulation fof elemination of noise and distribution problems associated with single-point regulation. Each of these regulation can deliver up to 1.5 amperes of output current. The internal current limiting and termal shutdown features of these regulators make them essentially immune to overload.

absolute maximum ratings over operating temperature range (unless otherwise noted)

		μA78--	UNIT
Input voltage	μA7824, μA7827	40	V
	All others	35	
Continuous total dissipation at 25°C freepair temperature		2	W
Continuous total dissipation at (or below) 25°C case temperature		15	
Operating free-air, case, or virtual junctions temperature range		0 to 150	°C
Storage temperature range		–65 to 150	
Lead temperature 1.6 mm (1/16 inch) from case for 10 seconds		260	

Recommended operating conditions

Parameter		MIN	MAX	UNIT
Input voltage V_1	μA7805	7	25	V
	μA7806	8	25	
	μA7808	10.5	25	
	μA7885	10.5	25	
	μA7809	11.5	27	
	μA7810	12.5	28	
	μA7812	14.5	30	
	μA7815	17.5	30	
	μA7818	21	33	
	μA7820	23	36	
	μA7824	27	38	
	μA7827	30	40	
Output current, I_O			1.5	A
Operating virtual junction temperature, T_J		0	125	°C

Index

D

M

N

O

P

Q

R

S

T

U

V

W

Z